POLYMER HANDBOOK

POLYMER HANDBOOK

SECOND EDITION

J. BRANDRUP • E. H. IMMERGUT, Editors

with the collaboration of

W. McDOWELL

A WILEY-INTERSCIENCE PUBLICATION

JOHN WILEY & SONS, New York • Chichester • Brisbane • Toronto

Library of Congress Cataloging in Publication Data

Brandrup, J   ed.
  Polymer handbook.

  "A Wiley Interscience publication."
  Includes bibliographies.
  1. Polymers and polymerization—Tables, etc.
I. Immergut, E. H., joint ed. II. McDowell, W.,
joint ed. III. Title.

QD388.B7   1974   547′.84′0202   74-11381
ISBN 0-471-09804-3

Printed in the United States of America

10 9

CONTRIBUTORS

Aggarwal, S. L. — The General Tire & Rubber Company
Research and Development Division
Akron, Ohio

Allegra, G. — Istituto di Chimica dell' Universita
Trieste, Italy

Andreeva, L. N. — Institute of High Molecular Weight Compounds
at the Academy of Sciences of the USSR
Leningrad, USSR

Bareiss, R. E. — Editorial Office:
"Die Makromolekulare Chemie"
Mainz, Germany

Barrales-Rienda, J. M. — Instituto de Plásticos y Caucho
Patronato "Juan de la Cierva" C. S. I. C.
Madrid, Spain

Bassi, I. W. — Montecatini Edison S. p. A.
Centro Ricerche di Milano
Milano, Italy

Bello, A. — Instituto de Plásticos y Caucho
Patronato "Juan de la Cierva" C. S. I. C.
Madrid, Spain

Bohn, L. — Hoechst AG
Frankfurt/Main, Germany

Booth, C. — University of Manchester
Department of Chemistry
Manchester, Great Britain

Bührer, H. G. — Midland Macromolecular Institute
Midland, Mich.

Burg, K. H. — Hoechst AG
Frankfurt/Main, Germany

Burrell, H. — Inmont Corporation
Clifton, N. J.

Chapiro, A. — Laboratoire de Chimie des Radiations du C. N. R. S.
92-Bellevue, France

Chiang, R. † — Monsanto Company
St. Louis, Missouri

Collins, E. A. — B. F. Goodrich Chemical Company
Technical Center
Avon Lake, Ohio

Cooper, D. R. — University of Manchester
Department of Chemistry
Manchester, Great Britain

Daniels, C. A. — B. F. Goodrich Chemical Company
Technical Center
Avon Lake, Ohio

Dawkins, J. V. — Duke University
Department of Chemistry
Durham, N. C.

Elias, H. -G. — Midland Macromolecular Institute
Midland, Mich.

Ende, H. A. — BASF AG
Ludwigshafen, Germany

Fester, W. — Hoechst AG
Frankfurt/Main, Germany

Fleischer, D. — Hoechst AG
Frankfurt/Main, Germany

Fuchs, O. — Hoechst AG
Frankfurt/Main, Germany

Gerrens, H. — BASF AG
Ludwigshafen, Germany

Goodman, M. — University of California, San Diego
La Jolla, Calif.

Grassie, N. — The University
Chemistry Department
Glasgow, Scotland

Guzmán, G. M. — Universidad Politécnica de Barcelona
Barcelona, Spain

Hinkamp, P. E. — Dow Chemical Company
Designed Polymers Research
Midland, Mich.

Hirsch, G. — BASF AG
Ludwigshafen, Germany

Huglin, M. B. — University of Salford
Department of Chemistry
Salford, England

Ivin, K. J. — The Queen's University of Belfast
Department of Chemistry
Belfast, N. Ireland

Iwama, M. — Kyoto University
Institute for Chemical Research
Kyoto, Japan

Kamada, K. — Kyoto University
Institute for Chemical Research
Kyoto, Japan

Klärner, P. E. O. — BASF AG
Kunststoff-Laboratorium
Ludwigshafen, Germany

Korus, R. — University of Waterloo
Department of Chemical Engineering
Waterloo, Ontario, Canada

Krigbaum, W. R. — Duke University
Department of Chemistry
Durham, N. C.

Kurata, M. — Kyoto University
Institute for Chemical Research
Kyoto, Japan

Lawton, E. L. — Monsanto
Triangle Park Development Center, Inc.
Research Triangle Park, N. C.

CONTRIBUTORS

Lee, W.A. Royal Aircraft Establishment
Farnborough, Hants, Great Britain

Lindemann, Martin K. C.S. Tanner Co.
Greenville, S.C.

Luft, G. Technische Hochschule Darmstadt
Institut für Chemische Technologie
Darmstadt, Germany

Magill, J.H. University of Pittsburgh
Department of Metallurgical and Materials
Engineering
Pittsburgh, Pa

Masson, J.C. Monsanto
Triangle Park Development Center, Inc.
Research Triangle Park, N.C.

Matonis, V.A. Monsanto Company
Indian Orchard, Mass.

Miller, Robert L. Midland Macromolecular Institute
Midland, Mich.

O'Driscoll, K.F. University of Waterloo
Department of Chemical Engineering
Waterloo, Ontario, Canada

Peebles, Jr., L.H. Office of Naval Research
Boston, Mass.

Massachusetts Institute of Technology
Department of Chemical Engineering
Cambridge, Mass.

Pflüger, R. BASF AG
Anwendungstechnische Abteilung Kunststoffe
Ludwigshafen, Germany

Ringwald, E.L. Monsanto
Triangle Park Development Center, Inc.
Research Triangle Park, N.C.

Rothe, M. Universität Mainz
Organisch-Chemisches Institut
Mainz, Germany

Rudd, J.F. The Dow Chemical Company
DIG Physical Research
Midland, Mich.

Rutherford, R.A. Rubber and Plastics Research
Association of Great Britain
Shawbury, Shrewsbury, Great Britain

Scotney, A. The University
Chemistry Department
Glasgow, Scotland

Sextro, G. Hoechst AG
Frankfurt/Main, Germany

Shafrin, E.G. U.S. Naval Research Laboratory
Laboratory for Chemical Physics
Washington, D.C.

Sperati, C.A. E.I. DuPont de Nemours and Co., Inc.
Plastics Department
Parkersburg, West Virginia

Stannett, V. North Carolina State University
Department of Chemical Engineering
Raleigh, N.C.

Stempel, G.H. Oak Ridge Drive
Cuyahoga Falls, Ohio

Sütterlin, N. Röhm GmbH
Darmstadt, Germany

Suhr, H.-H. Hoechst AG
Frankfurt/Main, Germany

Treiber, E. Svenska Träforskningsinstitutet
Stockholm, Sweden

Tsunashima, Y. Kyoto University
Institute for Chemical Research
Kyoto, Japan

Tsvetkov, V.N. Institute of High Molecular Weight Compounds
at the Academy of Sciences of the USSR
Leningrad, USSR

Ueyama, N. University of California, San Diego
La Jolla, Calif.

Ulbricht, J. Technische Hochschule für Chemie
Leuna-Merseburg, DDR

Wilkes, C.E. B.F. Goodrich Company
Research and Development
Brecksville, Ohio

Wilski, H. Hoechst AG
Frankfurt/Main, Germany

Wolf, B.A. Universität Mainz
Institut für Physikalische Chemie
Mainz, Germany

Wood, Lawrence A. National Bureau of Standards
Washington, D.C.

Wunderlich, W. Röhm GmbH
Darmstadt, Germany

Yasuda, H. Camille Dreyfus Laboratory
Research Triangle Institute
Research Triangle Park, N.C.

Young, Lewis, J. Koppers Company, Inc.
Law/Patent Department
Pittsburgh, Pa

PREFACE

Eight years have passed since the publication of the first edistion of this POLYMER HANDBOOK. This according to many publications is exactly the time it takes for our knowledge in the natural science to double. We can confirm this statement with nearly every table in this Handbook, which indicates that progress is still occurring in all areas of Polymer Science.

This increase in the volume of literature has created a considerable number of practical problems since it was desirable to maintain a one volume Handbook. Therefore, a number of restrictions has to be imposed on the content in addition to enlarging the book format. Firstly, this edition is limited to synthetic polymers plus poly(saccharides) and derivatives. Secondly, spectroscopic data were eliminated since a number of other publications make this information available in the meantime. Within this framework, a number of new tables were added in order to complete the Handbook. Thus, tables on activation volumes, activation enthalpies and entropies of stereocontrol in free radical polymerization, isomorphous polymer pairs, compatible polymers, heat capacities, refractive indices, polymolecularity correction factors, Schulz-Blaschke and Huggins coefficients, Fikentscher K-values relative viscosity conversion tables and tables on several important polymers like poly(vinyl chloride), poly(tetrafluoroethylene), poly(oxymethylene), poly(methyl methacrylate), and poly(amide 6) appear for the first time.

It is the purpose of this Handbook to help in the search for data and constants needed in theoretical and experimental polymer work. This objective sets the framework for the contents of the second edition, as it did for the first one. Therefore, we repeat the following statements from the preface to the first edition:

"First of all, only fundamental constants and parameters were compiled rather than data of interest to the polymer engineer of fabricator. Data of fundamental interest are interpreted as data which are either physical or chemical constants of the polymer molecule within reasonable or predictable limits, or are constants of existing physical laws describing the properties and the behavior of polymers. Constants which depend to a major extent on the particular processing conditions, or sample history, were not compiled as they can be found in existing plastics handbooks and encyclopedias. Within these limits the selection of tables to be compiled was governed by two principles: sufficient data should exist to make a compilation worthwhile, and their scientific basis should be commonly accepted.

No critical evaluation of published values was attempted since this would have been an impossible undertaking within a reasonable time or with the manpower available for the task. Therefore, the authors of the individual tables were urged to list all data found in the literature except values which in their judgement, were obviously erroneous. The reader is requested to use the data with this restriction in mind and to consult the original literature references for details. Whereas a complete compilation of existing data was attempted, there will, no doubt, be some omissions. The users are asked to send to the editors any data they might be aware of but cannot find in the Handbook. This would help to make the Handbook a more reliable information source, and, therefore, an increasingly useful tool for polymer scientists.

The inclusion of the table on practical data of commercial polymers may seem to be inconsistent with the selection principles expressed earlier, but, we felt that a selected listing of such data would remind all of us that polymer "science is not a religion to be worshipped for its own sake. Science is not an ornament on society's chest. It must be woven into the fabric of life to help with the world's needs". (W.J. Sparks - Presidential Address, American Chemical Society Meeting, Detroit, 1965)".

The editors would like to thank everyone who contributed to the second edition either by submitting tables or giving advice, encouragment or support. J.B. thanks Farbwerke Hoechst AG for generous help and for the permission to undertake this task.
Thanks are also due Dr. W. McDowell who helped bear the editorial burden of the second edition by conscienciously proofreading and checking the various tables.

Spring 1974   The Editors

TABLE OF CONTENTS

TABLE OF CONTENTS

We asked for comments about the first edition and suggestions for improvements. Here they are:

General Comments

The task of reviewing this massive work is similar to setting out to review the ABC rail guide. However, just as the possession of an ABC is vital to those who travel a lot on trains, so is the Polymer Handbook an essential piece of equipment for polymer users.
W. A. Holmes-Walker -

Publication of the Polymer Handbook marks another significant milestone in the growth of macromolecular chemistry since its inception early in this century. . . . . It is a much needed addition to the polymer literature and merits a place besides the standard chemical reference works.
G. M. Kline, Science, September 1966. -

Dieses Handbuch der Polymerchemie kommt einem ausgesprochenen Bedürfnis entgegen, indem es dem Polymerchemiker erstmals die Möglichkeit gibt, die von ihm benötigten Daten nachzuschlagen, statt sie aus der rasch anwachsenden Literatur in mühsamer Arbeit zusammensuchen zu müssen. Es gehört in die Hand jedes Polymerchemikers und wird sich bald als unersetzlich erweisen.
H. Luessi, Chmica, October 1966 -

Das Handbuch ist für die Forschung und das Labor ein hervorragendes zeitsparendes Hilfsmittel.
C. Mierisch, Textil-Praxis, October 1967 -

What Perry's "Chemical Engineers' Handbook" is to the chemical engineer, and the "Handbook of Chemistry and Physics" is to the chemist, the "Polymer Handbook" will be to the polymer scientist. This book will be a gold mine of polymer chemical data for those who know how to use it.
P. F. Bruins, Polytechnic Institute of Booklyn, Modern Plastics, February 1967 -

Dieses Werk ist sowohl für den Synthetiker, den Applikations- und Physiko-Chemiker wie auch für den Analytiker eine sehr wertvolle Zusammenstellung . . . .
H. Batzer, Kunststoffe-Plastics, 1966 -

For persons who work with polymers, this handbook will prove to be a valuable addition to the working literature.
W. W. West, Chevron Research Company -

The compilation has been a tremendous effort by a large number of distinguished workers, and the reward for their careful and painstaking work will be the appreciation of their colleagues, whose life will now be made much easier.
C. E. H. Bawn, Nature, 1966 -

Specific Comments and Suggestions

Topic: Nomenclature

Eine weitere Schwierigkeit war die Nomenklatur, die bei weitem nicht einheitlich ist.
Mierisch, Textilpraxis 1967 -

A brief introductory chapter outlines some well thought out rules for polymer nomenclature.
W. W. West, The Vartex -

Es bleibt zu hoffen, daß die Herausgeber in einer folgenden Auflage zu einer einheitlichen Nomenklatur finden.
K. F. Elgert, Kunststoffe, 1968 -

The first problem facing the editors is nomenclature. They have adopted a sensible compromise, adopting several systems, rather than forcing a single system into areas where it was difficult to use.
J. W. S. H. Hearle, Skinners Record, September 1966 -

The book is well arranged, the system of nomenclature is clearly set out, the index is adequate.
Res. Ass. Brit. Paint Colour Varnish Manuf. (England), October 1966 -

An introductory chapter on "Nomenclature rules" provides a useful guide to the system used in classifying and naming the polymers . . . . . This is particularly helpful in the case of the confusing situation with regard to the naming of polyurethanes, polycarbonates, and copolymers.
G. M. Kline, Science, Vol. 153 -

Topic: Selection of Tables

. . . . such topics as the physical properties of monomers and solvents should be omitted, this information is readily available in organic chemistry texts.
D. G. H. Ballard, Polymer, 1966 -

Besonders interessante Kapitel für den Synthetiker und auch den Applikationschemiker dürften sein: . . . , sowie die physikalischen Eigenschaften von einigen wichtigen Polymeren, Oligomeren, Monomeren und Lösungsmitteln.
H. Batzer, Kunststoffe - Plastics, May 1966 -

A remarkable collection of data on polymerization catalysts and inhibitors, properties of monomers and solvents ; . . . The above listing does not begin to illustrate the depth or utility of the information available. For example, not only are physical properties of monomers included, but also copolymerization reactivity ratios...
P. F. Bruins, Modern Plastics, February 1967 -

Sehr zu begrüßen ist schließlich eine Sammlung von physikalischen Daten von Oligomeren.
Br., Gummi, Asbest, Kunststoffe, March 1967 -

Physical constants of some important polymers. Here are data on eleven materials, but - rather unfortunately - each contribution is by a different author and the treatment correspondingly varies.
W. J. Roff, Textile Inst. Ind., October 1966 -

Den Herausgebern wäre allerdings zu empfehlen, in das Kapitel VI auch Polyamid 6 aufzunehmen.
K. Gehrke, Plaste Kautschuk, December 1966 -

There is also lack of information on polyvinyl chloride, Nylon 6, the higher Nylons and some of the newer materials such as polypyromellitimides
D. G. H. Ballard, Polymer, 1966 -

Ask for more remote data in the field and it still may be there, easily found and attractively tabulated.

.... a veritable treasure house of information.... The editors aimed to compile only fundamental constants and parameters and so in some ways can be forgiven for paying only sparse attention to data relating to deformation and flow.
D. R. J. Hill, British Plastics, 1967 -

Topic: Form

The authors and editors are to be congratulated on packing so much into a single volume. Yet it would seem .... that the same information might have been compressed without loss of anything but blank space, into a still more compact presentation.
W. J. Roff, Textile Inst. Ind., October 1966 -

.... thanks to a clear lay-out (and) by following the simple nomenclature rules, information can be extracted with very little difficulty.
W. A. Holmes-Walker -

Die Vielseitigkeit und Fülle des Zahlenmaterials, das dieses Handbuch in wohlgeordneter, konzentrierter Form enthält ....
H. Reichert, Faserforsch. Textiltechn., 1966 -

Die Tabellen sind übersichtlich dargestellt und mit kurzen prägnanten Einführungen versehen.
H. Lüssi, Chimia, 1966 -

Das Buch ist ohne Frage von Bedeutung als Nachschlagewerk; aber wenn man sich orientieren möchte, stört die kleine Schrift sehr. Ob man daran nichts ändern könnte?
W. Scheele, Kautschuk Gummi, 1967 -

The camera-copy typed manuscript is small but legible.
W. W. West, The Vortex, May 1967 -

They can be forgiven for directly reprinting a large part of the copolymerization section from G. E. Ham "Copolymerization". The type script is so different from the rest of the book that there must be a certain aesthetic displeasure associated with it: it is as if the pages were colored green.
But if the editors saved time and trouble by this method and intend to amend this in their next edition, they surely cannot be blamed.
D. R. J. Hill, British Plastics, February 1967 -

The system of numbering the pages afresh in each section seems to offer no advantages to the reader and probably most users would prefer to see consecutive pagnination throughout the book.
W. J. Roff, Textile Inst. Ind., October 1966 -

Die Seitenzahlen sind nicht fortlaufend; dies verbessert die Übersichtlichkeit und erleichtert den Gebrauch.
H. Batzer, Kunststoffe-Plastics, May 1966 -

Topic: Subject Index

An extensive subject index adds greatly to the value of the book.
C. W. H. Bawn, Nature, 1966 -

.... ein allerdings sicher noch ausbaufähiges Register ....
Br., Gummi, Asbest, Kunststoffe, March 1967 -

Ein ausführliches Sachregister am Schluß des Buches erleichtert dem Leser das Auffinden der Konstanten und Daten .....
Melliand Textilberichte, September 1966 -

POLYMER HANDBOOK

I. NOMENCLATURE RULES – UNITS

I. NOMENCLATURE RULES - UNITS

1. Nomenclature Rules

During the time of preparation of the first edition no universally accepted set of rules for naming all classes of polymers existed. We had to devise our own scheme in order to tabulate polymers in an easy way. Since then, the IUPAC Macromolecular Nomenclature Commission has published tentative rules to name regular single-strand organic polymers. These rules have been applied in this Handbook. Since the Commission stated that a number of common names are well established by usage and that they do not intend to supplant them immediately by structure-based names, we have adhered to these common names as much as possible for simple molecules. Thus, for instance, poly(methyl methacrylate), poly(styrene), and their derivatives are named with their common names. In general, structure-based nomenclature was applied less strictly in smaller tables in order to facilitate ease of reading of these tables. The big advantage of the structure-based nomenclature lies in naming complicated macromolecules. These are mainly condensation type polymers and here we have adhered strictly to the rules given by the Commission.

According to these tentative rules (reprinted below with the permission of the publisher) polymers can be named as follows:

(1) Find the underlined constitutional repeating unit (CRU) which is independent of the way it was prepared (rule 1.21, page I-4).

(2) Wherever possible, the CRU should be a bivalent group. This principle of minimizing the number of free valences supersedes all orders of seniority, discussed below (rule 2.12, page I-5).

(3) The CRU is written from left to right beginning with the subunit of highest seniority and proceeding in a direction defined by the shortest path to the subunit equal or next in seniority (rule 2.11, page I-5). Ring and chain atoms are counted individually (rule 2.13, page I-6). For subunits of equal path length see rule 2.14, page I-6 and 2.32, page I-9).

Example:

poly(oxyethyleneoxyterephthaloyl) not poly(oxyterephthaloyloxyethylene)

but:

poly(oxyterephthaloyloxydecamethylene)

(4) Start to name the CRU with the subunit of highest seniority. The descending order of seniority among the types of bivalent groups is
- heterocyclic rings (names according to rule 2.2, page I-7).
- chains containing heteroatoms (names according to rule 2.3, page I-8).
- carbocyclic rings (names according to rule 2.4, page I-9).
- chains containing only carbon

This order is unaffected by the presence of rings, atoms, or groups that are not part of the main chain (rule 2.12, page I-5).

Example:

poly(oxyethylene) not poly(ethyleneoxy)

poly(1,4-phenyleneethylene) not poly(ethylenephenylene)
or poly(methylenephenylenemethylene).

(5) The order of seniority among heterocyclic rings is given in rule 2.23, page I-7.

(6) The descending order of seniority of heteroatoms in the main chain is O, S, Se, Te, N, P, As, Sb, Bi, Si, Ge, Sn, Pb, B, Hg (rule 2.31, page I-8).

(7) The name of the subunit should include in one single name as many as possible of
- the main chain atoms or rings and
- substituents (rule 1.21, page I-4).

Example:

poly(ethylidene) not poly(methylmethylene)

(8) The subunits are named wherever possible according to the definitive rules for nomenclature of organic chemistry (International Union of Pure and Applied Chemistry "Nomenclature of Organic Chemistry". Sections A, B and C combined, Butterworths, London 1971. See also "Tentative Rules for the Nomenclature of Organic Chemistry", Section E, Fundamental Stereochemistry, J. Org. Chem. 35, 2849 (1970)).

Problems still exist in naming three-dimensional polymers, statistical, block- and graft-copolymers. These polymers were named according to their source. Rules of the first edition were applied. Thus, the different components of statistical copolymers and co-condensates are separated by the infix -co-. Components of block-copolymers are separated by the symbol: and the components of graft-copolymers by the symbol +. If monomers yield several polymeric structures a source name was also preferred. Examples are: poly(acrolein), poly(glutardialdehyde), poly(2-formyl-Δ^5-dihydropyran).

NOMENCLATURE OF REGULAR SINGLE-STRAND ORGANIC POLYMERS
TENTATIVE RECOMMENDATION OF IUPAC MACROMOLECULAR DIVISION
IUPAC-INFORMATION BULLETIN NOVEMBER 1972*

Contents

A single strand polymer is composed of molecules whose constitutional units can be chosen such that all of them have no more than two terminal atoms.
See ref. 5.

INTRODUCTION

In 1952, the Subcommission on Nomenclature of the IUPAC Commission on Macromolecules published (1) a report on the nomenclature of macromolecules that included a method for the systematic naming of linear organic polymers on the basis of structure. A later report (2) dealing with steric regularity utilized this system of nomenclature. When the first report was issued, the skeletal rules were adequate for most needs; indeed, most polymers could at that time be reasonably named on the basis of the substance used in producing the polymer. In the intervening years, however, the rapid growth of the polymer field has dictated a need for modification and expansion of the earlier rules. This report presents an updating of those rules. Necessarily, a great many changes in detail were required, since it is desirable that organic polymer nomenclature adhere as much as possible to the Definitive Rules for the Nomenclature of Organic Chemistry (3, 4).

These rules are designed to name, uniquely and unambiguously, the structures of linear regular organic polymers whose repeating structures can be written within the framework of ordinary chemical principles; stereochemistry is not considered in this report. As with organic nomenclature, this nomenclature describes chemical structures rather than substances. It is realized that polymeric substances ordinarily include many structures, and that a complete description of even a single polymer molecule would include an itemization of terminal groups, branching, random impurities, degree of steric regularity, chain imperfections, etc. Nonetheless, it is useful to think of a substance as being represented by a single structure that may itself be hypothetical. To the extent that the polymer structure can be portrayed as a chain of regularly repeating structural or constitutional repeating units (the terms are synonymous), the structure can be named by these rules; in addition, provision has been made for including end groups in the name.

In this report, the fundamental principles and the basic rules of the structure-based nomenclature are given first, accompanied by detailed extensions and applications. An appendix is included containing a limiting list of acceptable source-based names, along with the corresponding structure-based names, of common polymers. The Commission sees no objection to the continued use of such source-based names where they are clear and unambiguous, but prefers the use of the structure-based nomenclature detailed in these rules.

*Reprinted with permission of the publisher.

FUNDAMENTAL PRINCIPLES

This nomenclature system rests upon the selection of a preferred constitutional repeating unit (5)   (abbreviation: CRU) of which the polymer is a multiple; the name of the polymer is simply the name of this repeating unit prefixed by poly. The unit itself is named wherever possible according to the Definitive Rules for the Nomenclature of Organic Chemistry(3).   For single-strand polymers, this unit is a bivalent group.

In using this nomenclature, the steps to be followed in sequence are (1) identify the CRU, (2) orient the CRU, and (3) name the CRU. Identification and orientation must always precede the selection of the name of the polymer.

(1)   Identification of the Constitutional Repeating Unit

There are many ways to write the CRU for most polymer structures. In simple cases, these units are readily identified:

$$\text{---}\!\!\left[\text{CHCH}_2\right]_n \qquad \text{The CRU's are:} \qquad \text{---CHCH}_2\text{---} \quad \text{and} \quad \text{---CH}_2\text{CH---}$$
$$\overset{|}{\text{CH}_3} \overset{|}{\text{CH}_3}\overset{|}{\text{CH}_3}$$

In more complex cases, it is often necessary to draw a large segment of the chain and from it choose all of the possible CRU's. For example, in the polymer

$$\text{-OCHCH}_2\text{ OCHCH}_2\text{ OCHCH}_2\text{ OCHCH}_2\text{ OCHCH}_2\text{ OCHCH}_2\text{-}$$
$$\overset{|}{\text{F}}\overset{|}{\text{F}}\overset{|}{\text{F}}\overset{|}{\text{F}}\overset{|}{\text{F}}\overset{|}{\text{F}}$$

the CRU's are

$$\text{-OCHCH}_2\text{-} \quad \text{-CH}_2\text{OCH-} \quad \text{-OCH}_2\text{CH-} \quad \text{-CH}_2\text{CHO-} \quad \text{-CHOCH}_2\text{-} \quad \text{-CHCH}_2\text{O-}$$
$$\overset{|}{\text{F}}\overset{|}{\text{F}}\overset{|}{\text{F}}\overset{|}{\text{F}}\overset{|}{\text{F}}\overset{|}{\text{F}}$$

To allow construction of a unique name, a single CRU must be selected. The rules following have been designed to specify both seniority among subunits, i.e. the point at which to begin writing the CRU, and the direction along the chain in which to continue to the end of the CRU. The preferred constitutional repeating unit will be one beginning with the subunit of highest seniority (see Rule 2). From this subunit, one proceeds toward the subunit next in seniority. In the preceding example, the subunit of highest seniority is an oxygen atom and the subunit next in seniority is a substituted $\text{-CH}_2\text{CH}_2\text{-}$unit. The parent CRU will therefore be either $\text{-OCH}_2\text{CH}_2\text{-}$ or $\text{-CH}_2\text{CH}_2\text{O-}$. Further choice in this case is based on the lowest locant for substitution, so that the CRU is

$$\text{-OCHCH}_2\text{-} \quad \text{rather than} \quad \text{-OCH}_2\text{CH-} \quad \text{or} \quad \text{-CH}_2\text{CHO-} \quad \text{rather than} \quad \text{-CHCH}_2\text{O-}$$
$$\overset{|}{\text{F}}\overset{|}{\text{F}}\overset{|}{\text{F}}\overset{|}{\text{F}}$$

(2)   Orientation of the Constitutional Repeating Unit

The CRU is written to read from left to right. In the above example, the preferred CRU is therefore $\text{-OCHCH}_2\text{-}$.
$$\overset{|}{\text{F}}$$

(3)   Naming the Constitutional Repeating Unit

The name of the CRU is formed by citing, in order, the name of the largest subunits within the CRU (Rule 1.21). In the example, the oxygen atom is called oxy and the $\text{-CH}_2\text{CH}_2\text{-}$ (preferred to $\text{-CH}_2\text{-}$ because it is larger and can be named as a unit) is called ethylene; the latter unit substituted with one fluorine atom is called 1-fluoroethylene. The CRU in question is therefore named oxy(1-fluoroethylene), and the corresponding polymer is

$$\text{---}\!\!\left[\text{OCHCH}_2\right]_{\underline{n}} \qquad \text{Poly[oxy(1-fluoroethylene)]}$$
$$\overset{|}{\text{F}}$$

The rules that follow are essentially directions for the selection of the CRU in a given polymer.

Rule 1. THE BIVALENT CONSTITUTIONAL REPEATING UNIT

Regular single-strand polymer chains can usually be represented as multiples of a bivalent repeating unit which can itself be named. The name of the polymer is poly(bivalent constitutional repeating unit). In those cases in which a choice is possible between a bivalent and a higher-valent CRU, the bivalent unit is always selected. The principle of minimizing the number of free valences in the CRU supersedes all orders of seniority. (See also Rule 2.12).

$$\text{-C=CH-} \qquad \text{is preferred} \qquad \text{=CH-CH=}$$

1.1   The Generic Name

Linear polymers of unspecified chain length will be named by prefixing poly to the name, placed in parentheses or brackets, of the structural repeating unit of the polymer, i.e., the smallest unit of which the polymer is a multiple. If the name of the repeating unit is "ABC", the corresponding polymer name is

$$\text{---}\!\!\left[\text{ABC}\right]_{\underline{n}} \qquad \text{Poly(ABC)}$$

Where it is desired to specify chain length, the appropriate Greek prefix (deca,, docosa, etc.) may be used in place of poly. For a single-strand polymer, the CRU is a bivalent group and is named within the restriction of directional citation by the IUPAC organic nomenclature rules (3, 4).

1.2   Simple Constitutional Repeating Units

1.21 -   The CRU may contain one or more subunits. Among the possible subunits of combinations of adjacent subunits, the largest possible bivalent group, based on main chain atoms and rings only, is to be named (see also Rule 2.11). When the largest bivalent group includes the entire CRU its name, prefixed by <u>poly</u>, is the name of the polymer.

$+CH_2 +_n$ Poly(methylene)

$+OCH_2CH_2 +_n$ Poly(oxyethylene)

The name of a CRU or any subunit has no relationship to the manner in which the unit was prepared; the name is simply that of the largest identifiable unit and any locants for unsaturation, substituents, etc., are dictated by the <u>structure</u> of the unit.

$+CH=CHCH_2CH_2 +_n$ Poly(1-butenylene) (not poly(2-butenylene), which gives a higher locant to the double bond, nor poly(vinyleneethylene), which identifies less than the largest unit in the CRU)

1.22 -   Identification of the preferred CRU rests on (a) the kinds of atoms or rings in the main chain or (b) on the location of substituents when there is only one kind of main chain atom or ring. Orientation of the CRU in case (a) is determined by the rules of seniority given in Rule 2; in case (b), lowest locants (except when fixed numbering applies; see Rule 1.24) are given to substituents in alphabetical order (Rule 2.42).

$+CHCH_2CH_2 +_n$ Poly(trimethylene − $\underline{d}_1$)

Poly(3-bromo-2'-chloro-<u>p</u>−terphenyl-4, 4''−ylene)

<u>After</u> the CRU and its orientation, reading left to right, have been established, the CRU or its constituent subunits are named to include as many as possible, in order, of (a) the main chain atoms or rings and (b) the substituents within a single name (see also Rule 3.1).

$+CH +_n$ Poly(ethylidene) (not poly(methylmethylene))

$+CHCH_2 +_n$ Poly(1-phenylethylene) (not poly(benzylidene-methylene) or poly(1-phenyldimethylene))

$+C-C-CH_2-CH_2 +_n$ Poly(1, 2-dioxotetramethylene) (not poly(succinyl), since substituent positions 1, 2 are preferred to 1, 4, and identification and orientation of the CRU precede formation of the name)

$+OCCH_2CH_2C +_n$ Poly(oxysuccinyl) (not poly[oxy(1, 4-dioxotetramethylene)], since succinyl is an approved name (3))

$+CCH_2C(CH_2)_3 +_n$ Poly(1, 3-dioxohexamethylene) (not poly(malonyltrimethylene) because the six-carbon chain is the largest unit that can itself be named)

Unsaturation in an acyclic repeating unit is indicated wherever possible by the use of an unsaturated bivalent group name rather than a saturated multivalent group name. This procedure will lead to a name for the group having the minimum number of free valences (see Rule 2.12).

$+CH=CH +_n$ Poly(vinylene) (not $+CH-CH +_n$, poly(ethanediylidene))

1.23 -   If, after identification and orientation, the CRU is found to contain one or more acyclic bivalent groups having more than two hetero atoms in the main chain, these groups may often be advantageously named by replacement nomenclature (3). The main chain of the group is named and numbered as though the entire chain was an acyclic hydrocarbon and the hetero atoms named by means of prefixes "aza", "oxa", etc., with locants to fix their positions.

$+O-CH_2-CH_2-NH-CH_2-S-CH_2-CH_2-NH- \bigcirc +_n$

Replacement name: Poly(1-oxa-6-thia-4, 9-diaza-1, 9-nonanediyl-1, 3-cyclohexylene)

Systematic name: Poly(oxyethyleneiminomethylenethioethyleneimino-1, 3-cyclohexylene)

See Rules 2.14 and 2.32 for other examples of the use of replacement nomenclature.

1.24 -   Bivalent groups having fixed numbering retain that numbering in naming the CRU (see also Rules 2.22 and 2.41).

Poly(2,4-pyridinediyl)

For most acyclic and monocarbocyclic bivalent groups, preference in lowest numbers is given to the carbon atoms having the free valences. In other families of compounds, notably the polycyclic hydrocarbons, bridged hydrocarbons, spiro hydrocarbons, and heterocyclic ring systems, numbering is fixed for the ring system. Free valences in groups are numbered as low as possible, consistent with the fixed numbering. Since direction through the bivalent group is a requisite parameter in naming polymers, the same fixed numbering is retained for either direction of progress through the group in generating the polymer name.

Poly(2,7-naphthylene) (not poly(7,2-naphthyl-ene) or poly(3,6-naphthylene))

Poly(tricyclo[2.2.1.0^{2,6}]hept-3,5-ylene)

Poly(2H-furo[3,2-b]pyran-2,6-diyl)

Poly(5-oxaspiro[3.5]non-2,7-ylene)

Rule 2.   BIVALENT CONSTITUTIONAL REPEATING UNITS HAVING TWO OR MORE SUBUNITS

Many regular single-strand polymers can be represented as multiples of bivalent repeating units, such as –ABC–, that consist of a series of smaller sub-units, –A–, –B–, and –C–. The prototype name of the polymer is poly-(ABC), where (ABC) stands for the names of A, B, and C, taken in that order. This rule is concerned with the seniority fo subunits in identifying the preferred CRU for a given polymer structure.

2.1   Seniority of Subunits and Direction of Citation

2.11 -   Polymers having CRU's containing two or more subunits are named with the prefix poly followed in parentheses or brackets by the names of the largest possible subunits cited in order from left to right as they appear in the CRU. The CRU is written from left to right beginning with the subunit of highest seniority and proceeding in a direction defined by the shorter path to the subunit next in seniority.

Poly(oxyterephthaloylhydrazoterephthaloyl) (not poly(oxycarbonyl-1,4-phenylenebicarbamoyl-1,4-phenylenecarbonyl))

2.12-   The principle of minimizing the number of free valences in the CRU supersedes all orders of seniority. Wherever possible, the CRU in a linear polymer should be a bivalent group. The starting point for the unit is at a single free valence adjacent or nearest to the subunit of highest seniority and citation will be in the direction of the shorter path toward that subunit or subunit combinations of highest seniority.

Poly(methylidyne-4,2-piperidinediylidene-1,4-cyclohexanediylidene-2-ethanyl-1-ylidene)

For citation of the first subunit, the order of seniority among the types of bivalent groups is (1) heterocyclic rings (see Rule 2.2), followed by (2) chains containing hetero atoms (see Rule 2.3), (3) carbocyclic rings (see Rule 2.4), and (4) chains containing only carbon, in that order. This order is unaffected by the presence of rings, atoms, or groups that are not part of the main chain, even though such substituents could be expressed as part of a trivial name for a bivalent group.

Poly(4,2-pyridinediyl-imino-1,4-cyclohexylene-benzylidene)

2.13 - Choice of direction along the main chain of the CRU is determined by the shorter path, counting ring and chain atoms individually, from the subunit of highest seniority to the subunit next in seniority.

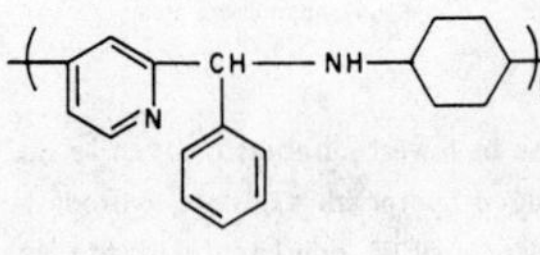

Poly(4,2-pyridine-
diylbenzylidene-
imino-1,4-cyclo-
hexylene)

The possible paths between subunits of first and second seniority necessarily involve subunits of lesser seniority. Except in cases where two paths are of equal length (see Rule 2.14), the number rather than the nature of the atoms involved is the determining factor.

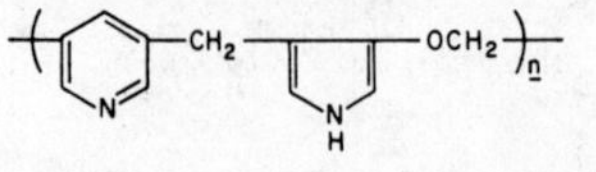

Poly(3,5-pyridinediylmethylene-3,4-pyrrolediyloxymethylene) (not poly-
(3,5-pyridinediylmethyleneoxy-3,4-pyrrolediylmethylene), in which the
longer —CH₂O— path between rings is followed)

Where a ring constitutes all or part of a path, the shortest continuous chain of atoms in the ring is selected

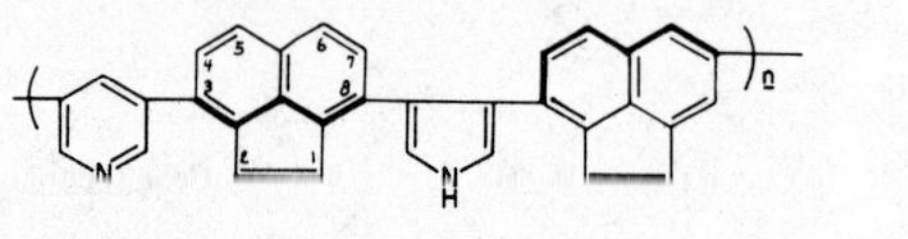

Poly(3,5-pyridinediyl-3,8-acenaphthylenylene-3,4-pyrrolediyl-3,7-acenaph-
thylenylene) (heavy line denotes path followed)

2.14 - When the choice of path determining direction of citation involves paths of equal length to subunits of equal seniority in the normal order of precedence, the choice of path depends upon the kind of subunits in the paths themselves. This condition applies to chains typified by the following generalized structures, where A, B, and C are subunits in that order of decreasing seniority, separated by paths of differing lengths x and y that contain units of lower seniority than C:

-C-y-B-x-A-x-B-y-C-

-B-y-A-x-A-y-B-

-B-y-A-x-A-x-A-y-B-

The choice of direction is from a subunit A to the nearest part of a path x having highest seniority, or if two paths x are identical in every respect, to the nearest part of path y having highest seniority, etc., until some point of difference is encountered. (See also Rules 2.24 and 2.32).
Examples of direction of citation based on the constituent parts of paths of equal length:

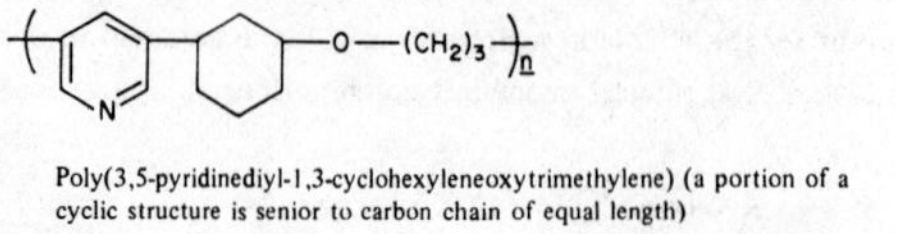

Poly(3,5-pyridinediyl-1,4-phenylenemethyleneoxymethyleneiminomethyl-
eneoxy-1,4-phenylenemethylene) (choice of path from heterocyclic ring
to O determined by position of phenylene)

—OCH₂OCH₂NHCH₂CH₂SCH₂NHCH₂CH₂—ₙ

Poly(oxymethyleneoxymethyleneiminoethylenethiomethyleneiminoethyl-
ene) or Poly(1,3-dioxa-8-thia-5,10-diazo dodecamethylene (choice of path
from O to S determined by position of NH)

Poly(3,5-pyridinediyl-1,3-cyclohexyleneoxytrimethylene) (a portion of a
cyclic structure is senior to carbon chain of equal length)

Where substituents controle the choice of CRU, the order of seniority is that given in Rule 2.42.

—SCH₂CHCH₂SCH₂CH₂CH₂—ₙ
       Cl

Poly[thio(2-chlorotrimethylene)thiotrimethylene]

—SCH₂CH₂SCH₂CHCH₂CHCH₂—ₙ
              NH₂   COOH

Poly[thioethylenethio(2-amino-4-carboxypentamethylene)]
(direction determined by alphabetical order)

—SCH₂CH₂SCHCH₂CH₂CHCH₂—ₙ
          COOH   NH₂

Poly[thioethylenethio(4-amino-1-carboxypentamethylene)]
(direction determined by lowest locants takes precedence over alphabetical order)

$$\leftarrow SCHCH_2SCH_2CH_2CHCH_2CH \rightarrow_n$$
$$\underset{I}{|} \qquad \underset{Cl}{|} \quad \underset{Br}{|}$$

Poly[thio(1-iodoethylene)thio(5-bromo-3-chloropentamethylene)]
(direction determined by the lower locant in the cited subunit after
beginning the CRU)

2.2   Heterocyclic Rings

2.21 -   Bivalent CRU's having two or more subunits that include a heterocyclic ring system in the main chain are named by citing first the heterocyclic ring bivalent group of highest seniority and proceeding by the shorter path in descending order of preference to (a) another of the same heterocyclic ring
(see Rule 2.24)

Poly [3,5-pyridinediylmethylene-3,5-pyridinediyl(tetrahydro-2H-pyran-
3,5-diyl)]

(b) the heterocyclic ring next in seniority (see Rule 2.23)

Poly(2,6-morpholinediyl-3,5-pyridinediyl-2,8-thianthrenediyl)

(c) the senior acyclic bivalent group containing a hetero atom in the main chain (see Rule 2.31)

Poly(3,5-pyridinediylmethyleneoxy-1,4-phenylene)

(d) the senior carbocyclic ring system (see Rule 2.41)

Poly(3,5-pyridinediyl-1,4-phenylene-1,2-cyclopentylene)

and (e) the senior acyclic bivalent group containing only carbon in the main chain (see Rule 2.42)

Poly(3,5-pyridinediylcarbonyloxymethylene)

2.22 -   Consistent with the fixed numbering of heterocyclic rings, the points of attachment to the main chain of the CRU should have the lowest permissible
locants.

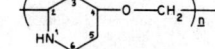

Poly(2,4-piperidinediyloxy-
methylene)

Poly(4,2-piperidinediyloxy-
methylene)

Where there is a choice, the point of attachment at the left side of the ring should have the lowest permissible number.

Poly(4H-1,2,4-triazole-3,5-
diylmethylene)

2.23 -   Among heterocyclic ring systems, the descending order of seniority is (a) a ring system with nitrogen in the ring; (b) a ring system containing nitrogen
and a hetero atom other than nitrogen as high as possible in the order given in Rule 2.31; (c) a ring system containing the greatest number of rings; (d) a
ring system having the largest individual ring; (e) a ring system having the largest number of hetero atoms; (f) a ring system containing the greatest
variety of hetero atoms; and (g) the ring system having the greatest number of hetero atoms highest in the order given in Rule 2.31. This order is that
followed in Rule B-2 of the IUPAC Rules (3).

Poly(4,2-pyridinediyl-4H-1,2,4-triazole-3,5-diylmethylene)

Further examples of the application of seniority in heterocyclic ring systems are

Phenoxazine   >   Furazan   >   Thiazole   >

Phenazine   >   Carbazole   >   Cinnoline   >

Quinazoline   >   Phthalazine   >   Purine   >

Pyramidine   >   Pyridine   >   Pyrrole   >

Phenoxathiin   >   Furan   >   Thiophene

When two heterocyclic subunits differ only in degree of unsaturation, the senior subunit is that having the least hydrogenated ring system.

Poly(3,5-pyridinediyl-2,4-piperidinediyl)

Among assemblies of identical heterocyclic rings, the ring of highest seniority is that having lowest numbers for the points of attachment between the rings within the assembly consistent with the fixed numbering of the parent ring.

Poly[(3,3'-biquinoline)-6,6'-diyl] (not poly[(6,6'-biquinoline)-3,3'-diyl])

Poly[(2,3'-bipyridine)-4,5'-diyl] (not poly[(3,2'-bi-pyridine)-5,4'-diyl] or poly-[(4,3'-bipyridine)-2,5'-diyl])

Further choice is based on the number and kind of substituting groups (see Rule 2.42).

Poly[(4-chloro-3,3'-bi-pyridine-5,5'-diyl)meth-ylene]

2.24 -   When the CRU contains two identical rings of highest seniority or more than two such rings of highest seniority separated by identical paths, the direction of citation is determined by the shorter path to the subunit of second seniority. Further choice is based on the shorter path from that subunit to the subunit of third seniority, etc., as indicated in the order of seniority in Rule 2.21.

Poly(4,2-piperidinediylmethylene-4,2-piperidinediyl-1,2-cyclopentylene-ethylene-1,2-cyclopentylenemethylene) (numbers in rings show order of preference in formation of name)

2.3   Hetero Atoms in Chains

2.31 -   Complex bivalent CRU's in which the senior subunit is a hetero atom or an acyclic chain with a hetero atom in the main chain are named by citing first the hetero atom of highest seniority and proceeding by the shorter path in descending order of seniority to (a) another hetero atom of the same kind; (b) the hetero atom next in seniority; (c) the senior carbocyclic ring system (see Rule 2.41); and (d) the senior acyclic bivalent group containing only carbon in the main chain (see Rule 2.42). For the most common hetero atoms the descending order of seniority is O, S, Se, Te, N, P, As, Sb, Bi, Si, Ge, Sn, Pb, B, and Hg; other hetero atoms may be placed within this order as indicated by their positions in the periodic table.

$$\left[NHCCH_2 \underset{\overset{\|}{O}}{} SiH_2 (CH_2)_3 \right]_n$$

Poly[imino(1-oxoethylene)silylenetrimethylene]

Poly(oxymethylene-
iminocarbonylthio-
1,3-phenyleneethyl-
ene)

Poly(oxyiminomethylenehydrazomethylene)

Parentheses must be used in some cases to prevent ambiguity.

Poly[thio(carbonyl)] ("carbonyl" is enclosed in parentheses to dif-
ferentiate the structure from $\underset{\underset{n}{\parallel}}{\overset{S}{C}}$, poly(thiocarbonyl))

The direction of bonding in unsymmetrical single-atom radicals (e.g. =N- or -N= for nitrilo) is indicated by the endings of the names of the adjacent subunits in the CRU

Poly(nitrilo-2-propen-3-yl-1-ylidene-1,4-phenylene-1-propen-1-yl-3-
ylidenenitrilo-1,4-phenylene)

Poly(oxycarbonylnitrilo-1,3-propanediylidenenitrilocarbonyl)

Direction in a bivalent group such as azoxy ($-\overset{\underset{}{\overset{O}{\uparrow}}}{N}=N-$ or $-N=\overset{\underset{}{\overset{O}{\uparrow}}}{N}-$) is indicated by the prefixes ONN or NNO, respectively, in that order of seniority.

Poly[oxymethylene-ONN-azoxy(chloromethylene)]

The unsymmetrical bivalent group -N=N-NH-, designated "diazoamino" under the IUPAC Organic Rules (3), in the present directional nomenclature for polymers is called "azoimino".

Among hetero atoms of the same kind, the hetero atom of highest seniority is the one most highly substituted, with the order of substituent seniority be-ing that given in Rule 2.42.

Poly(sulfonyl-1,4-phenylenesulfinylmethylenethiotrimethylene)

Poly[(methylimino)methyleneimino-1,3-phenylene)]

2.32 -   If the CRU contains two or more hetero atoms of highest seniority or more than two such hetero atoms separated by identical paths, the direction of cita-tion is determined by the shorter path to the subunit of second seniority. Further choice is based on the shorter path from that subunit to the subunit of third seniority, etc., as indicated in the order of seniority in Rule 2.31.

Poly(oxymethyleneoxymethyleneoxymethyleneimino-1,3-phenylenemeth-
yleneiminomethylene) or poly(1,3,5-trioxa-7-azaheptamethylene-1,3-
phenylene-2-azatrimethylene) (the shorter path x between the NH group
and the ring has been taken)

2.4   Carbocyclic Rings and Carbon Chains

2.41 -   Constitutional repeating units in which the senior subunit is a carbocyclic ring system are named by citing first the carbocyclic ring of highest seniority and proceeding by the shorter path, in descending order of seniority, to (a) another of the same carbocycle; (b) the carbocyclic system next in seniority; and (c) the acyclic bivalent group appearing earliest in the alphabet. Carbocyclic ring system seniority is based on complexity, with the ring system of highest seniority being that containing the largest number of rings. Further order of seniority is based on (a) the largest individual ring at the first point of difference; (b) the largest number of atoms common to the rings; (c) the lowest locant numbers at the first point of difference for ring junc-tions, and (d) the least hydrogenated ring. The basis for further choice is found in Rule C-14.1 of the IUPAC Rules (3). The direction of citation in CRU's having two or more carbocycles of highest seniority is determined in a manner analogous to that of Rule 2.32.

Poly(2,7-naphthyl-
ene-1,4-phenylene-
1,3-cyclohexylene)

Examples of the application of seniority rules in carbocyclic ring systems:
(a) Largest number of rings:

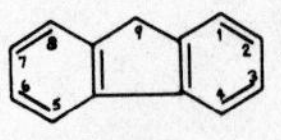

Fluorene   is senior to   Benzocyclooctene

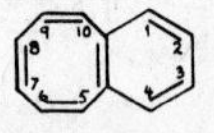

(b) Largest individual ring at the first point of difference

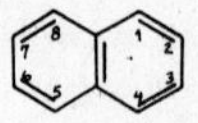

Naphthalene   is senior to   Indene

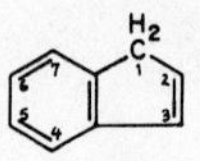

(c) Largest number of atoms common to the rings:

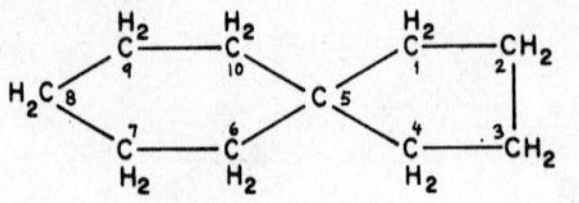

is senior to

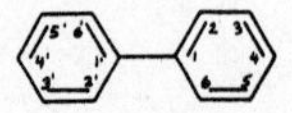

Spiro[4,5]decane   Biphenyl

(d) Lowest locant numbers at the first point of difference for ring junctions:

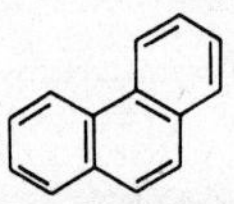

Phenanthrene   is senior to   Anthracene

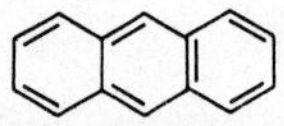

(e) Lowest state of hydrogenation:

is senior to    is senior to

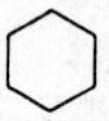

More than one numbering method may be in use in certain ring systems, such as the spiro hydrocarbons. Generally, in a specific ring system, a ring with unprimed locants is senior to one with primed locants. Points of attachment to the main chain of the CRU receive lowest permissible numbers.

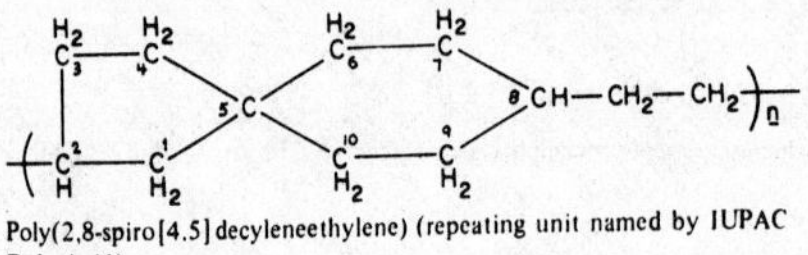

Poly(2,8-spiro[4,5]decyleneethylene) (repeating unit named by IUPAC Rule A-41)

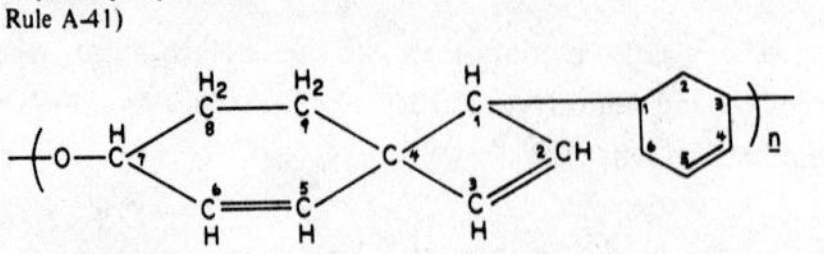

Poly(oxyspiro[3,5]nona-2,5-dien-7,1-ylenecyclohex-4-en-1,3-ylene)

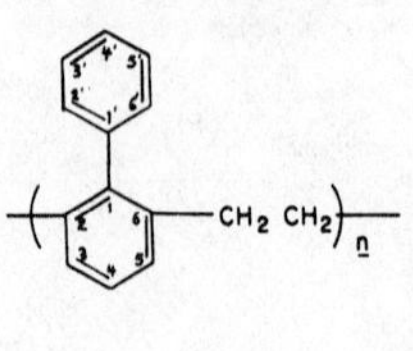

Poly(2,6-biphenylyleneethylene)

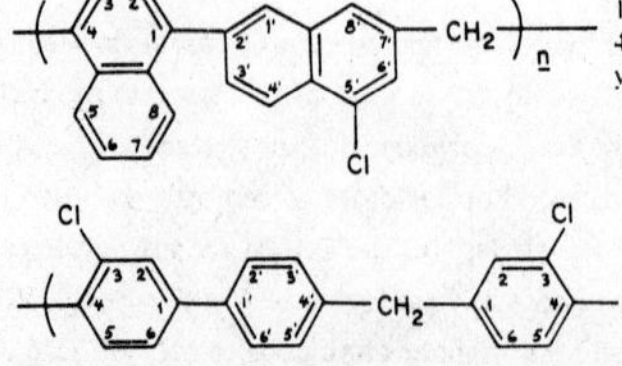

Poly[(5'-chloro-1,2'-binaphthyl-4,7'-ylene)methylene]

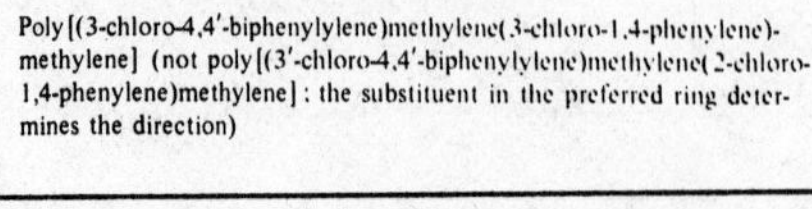

Poly[(3-chloro-4,4'-biphenylylene)methylene(3-chloro-1,4-phenylene)-methylene] (not poly[(3'-chloro-4,4'-biphenylylene)methylene(2-chloro-1,4-phenylene)methylene]: the substituent in the preferred ring determines the direction)

2.42 -   When equal paths lead through two of the same acyclic subunits, choice of direction is determined, in descending order, by (a) the acyclic chain with
          the largest number of substituents

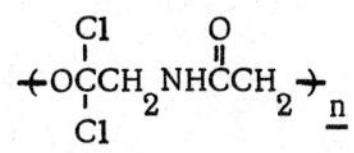

Poly[oxy(1, 1-dichloroethylene)imino(1-oxoethylene)]

(b) the chain having substituents with lowest locants

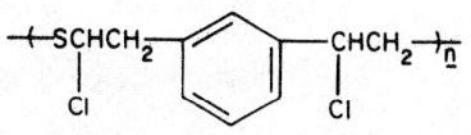

Poly[thio(1-chlo-
roethylene)-1,3-
phenylene(1-chlo-
roethylene)]

and (c) the alphabetical order of substituents

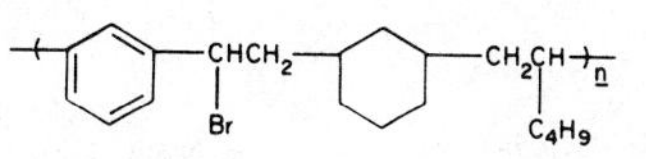

Poly[1,3-phenyl-
ene(1-bromoethyl-
ene)-1,3-cyclohex-
ylene(2-butyleth-
ylene)]

Rule 3. SUBSTITUENTS

3.1   Substituents to acyclic or cyclic subunits in the main chain of the CRU are included within the trivial name of the subunit wherever such name is ap-
          proved by the IUPAC Organic Rules (3) (see also Rule 1.22).

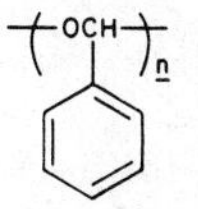

Poly(oxybenzyl-
idene)

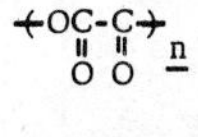

Poly(oxyoxalyl)

3.2   Substituents along the main chain other than those included in the name of a subunit are denoted by means of prefixes appended to the name of the sub-
          unit to which they are bound. In bivalent groups not having numbering fixed by other criteria, lowest locants appear at the left side of the bivalent
          group as written in the CRU (See also Rule 2.14).

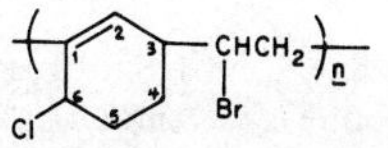

Poly[(6-chloro-1-cyclohexen-1,3-
ylene)(1-bromoethylene)] (not Poly-
[(6-chloro-1-cyclohexen-3,1-ylene)-
(2-bromoethylene)])

Functional derivatives clearly a part of the CRU are named as substituents to the appropriate subunit by the use of prefixes.

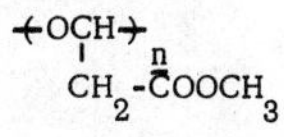

Poly[oxy[2-(methoxycarbonyl)ethylidene]]

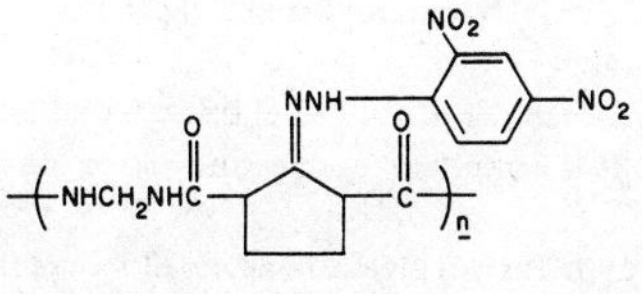

Poly[iminomethyleneiminocarbonyl[2-[(2,4-dinitrophenyl)hydrazono]-
1,3-cyclopentylene]carbonyl]

3.3   Salts and onium compounds of polymers are named by placing the appropriate prefix or suffix together with the name of the CRU in the enclosed part of
          the polymer name.

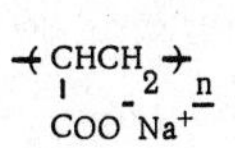

Poly(sodium 1-carboxylatoethylene)

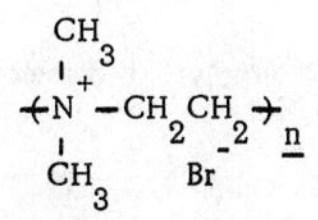

Poly[(dimethyliminio)ethylene bromide]

Certain substituents are frequently expressed as part of a trivial name. The subunit thus named can itself be further substituted without altering the original trivial name.

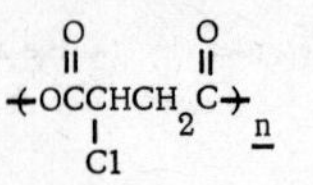

Poly[oxy(2-chlorosuccinyl)]

The same (in this case, double-bonded oxygen) substituents not expressed in a trivial name have no special seniority.

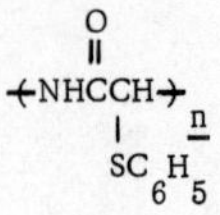

Poly[imino[1-oxo-2-(phenylthio)-ethylene]]

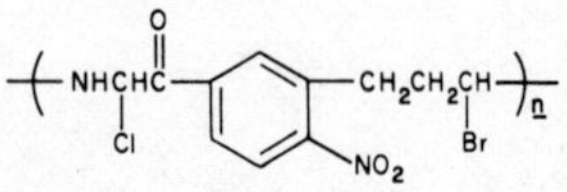

Poly[imino(1-chloro-2-oxoethylene)(4-nitro-1,3-phenylene)(3-bromotrimethyl-ene)]

Substituents in unknown positions in specific subunits are named in the usual way but either without locants or with an x locant.

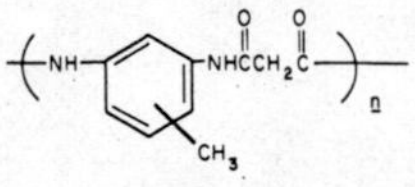

Poly[imino(methyl(or x-methyl)-1,3-phenylene)imino-malonyl]

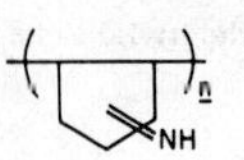

Poly(x-imino-1,2-cyclopentylene) (the x is required to differentiate this structure from 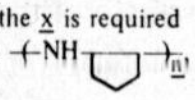 poly(imino-1,2-cyclopentylene))

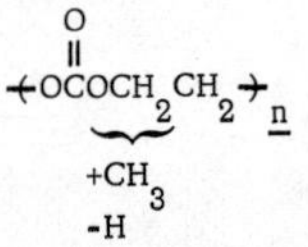

Poly[oxycarbonyloxy(methylethylene)]   (position of the methyl not stated)

3.4   End groups may be specified by prefixes placed ahead of the name of the polymer. The end group designated by α is that attached to the left side of the CRU written as described in the preceding rules; the other end group is designated by ω.

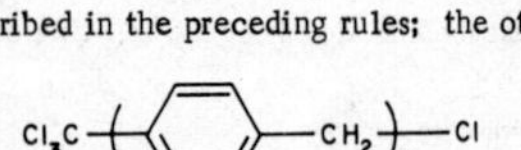

α-(Trichloromethyl)-ω-chloro-poly(1,4-phenylenemethyl-ene)

The Commission acknowledges its debt to the Committee on Nomenclature of the Division of Polymer Chemistry of the American Chemical Society, whose efforts resulted in a general updating and extension of the 1952 IUPAC Polymer Rules (1). The updated rules appeared in Macromolecules, 1, 193 (1968).

APPENDIX A

Recommended Trivial Names for Common Polymers

The Commission recognized that a number of common polymers have semisystematic or trivial names that are well established by usage; it is not intended that they be immediately supplanted by the structure-based names. Nonetheless, it is hoped that for scientific communication the use of semisystematic or trivial names for polymers will be kept to a minimum.

For the following idealized structural representations, the semisystematic or trivial names given are approved for use in scientific work; The corresponding structure-based names are given as alternative names. Equivalent names for close analogs of these polymers (e.g., other alkyl ester analogs of poly(methyl acrylate) are also acceptable. Where the semisystematic name is an obvious source-based name, the polymer referred to is that derived from the indicated source.

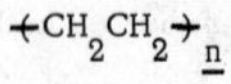

polyethylene   poly(methylene)

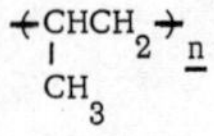

polypropylene   poly(propylene)

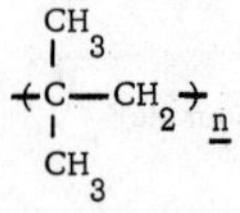

polyisobutylene   poly(1,1-dimethylethylene)

polybutadiene   poly(1-butenylene)

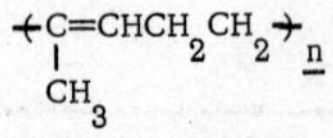

polyisoprene   poly(1-methyl-1-butenylene)

Structure		
$-(CHCH_2)_n-$ (with phenyl)	polystyrene	poly(1-phenylethylene)
$-(CHCH_2)_n-$	polyacrylonitrile	poly(1-cyanoethylene)
$\quad$ CN		
$-(CHCH_2)_n-$	poly(vinyl alcohol)	poly(1-hydroxyethylene)
$\quad$ OH		
$-(CHCH_2)_n-$	poly(vinyl acetate)	poly(1-acetoxyethylene)
$\quad$ OOCCH$_3$		
$-(CHCH_2)_n-$	poly(vinyl chloride)	poly(1-chloroethylene)
$\quad$ Cl		
F $-(CCH_2)_n-$ F	poly(vinylidene fluoride)	poly(1,1-difluoroethylene)
$-(CF_2CF_2)_n-$	poly(tetrafluoroethylene)	poly(difluoromethylene)
(dioxane ring with CH$_2$, O, C$_3$H$_7$)	poly(vinyl butyral)	poly[(2-propyl-1,3-dioxane-4,6-diyl)-methylene]
$-(CHCH_2)_n-$ COOCH$_3$	poly(methyl acrylate)	poly[1-(methoxycarbonyl)ethylene]
CH$_3$ $-(C-CH_2)_n-$ COOCH$_3$	poly(methyl methacrylate)	poly[1-(methoxycarbonyl)-1-methyl-ethylene]
$-(OCH_2)_n-$	polyformaldehyde	poly(oxymethylene)
$-(OCH_2CH_2)_n-$	poly(ethylene oxide)	poly(oxyethylene)
$-(O-\phi)_n-$	poly(phenylene oxide)	poly(oxy-1,4-phenylene)
$-(OCH_2CH_2OOC-\phi-CO)_n-$	poly(ethylene terephthalate)	poly(oxyethyleneoxyterephthaloyl)
$-(NH(CH_2)_6NHCO(CH_2)_4CO)_n-$	poly(hexamethylene adipamide)	poly(iminohexamethyleneiminoadipoyl)
$-(NHCO(CH_2)_5)_n-$	poly(ϵ-caprolactam)	poly[imino(1-oxohexamethylene)]

REFERENCES

1. International Union of Pure and Applied Chemistry, Physical Chemistry Division, Commission on Macromolecules, Subcommission on Nomenclature, "Report on Nomenclature in the Field of Macromolecules", J. Polymer Sci., 8 (3), 257-77 (1952).
2. International Union of Pure and Applied Chemistry, Physical Chemistry Division, Commission on Macromolecules, Subcommission on Nomenclature, "Report on Nomenclature Dealing with Steric Regularity in High Polymers", Pure Appl. Chem. 12, 645-56 (1966).
3. International Union of Pure and Applied Chemistry, "Nomenclature of Organic Chemistry", Sections A, B, and C combined. Butterworths, London, 1971.
4. International Union of Pure and Applied Chemistry, "Tentative Rules for the Nomenclature of Organic Chemistry. Section E. Fundamental Stereochemistry" IUPAC Inform. Bull. No. 35, 36 (1969). Also published in J. Org. Chem. 35, 2849 (1970) and elsewhere.
5. International Union of Pure and Applied Chemistry, Macromolecular Division, Nomenclature Commission, "Basic Definitions of Terms Relating to Polymers" IUPAC Inform. Bull. Apps. No. 13, 1-12 (1971).

2. Units

The International System of Units (1) is used in this Handbook as far as possible, since this system will become obligatory in many European countries within the next few years and since it is supported by the National Bureau of Standards (2) and the American Society for Testing and Materials (3).

Only cursory information is given here for units needed in this Handbook. Detailed information may be found in the following References:

1. "SI-units and recommendations for the use of their multiples and of certain other units", ISO 1000-1973.
2. "The International System of Units (SI)", Editors C.H. Page, P. Vigoreux, Natl. Bur. Std. Spec. Publ. 330, 1972.
3. "Standard Metric Practice Guide", ASTM E-380-72.

2.1 International Units and Conversion Factors

Quantity	SI-unit	Selection of multiples of the SI-unit	Accepted units used with SI	Units accepted temporarily	Units outside the SI which should not be used
length	m (metre)	km, cm, mm, μm, nm		$1\ \text{Å} = 10^{-10}\ \text{m}$	$1\ \mu = 10^{-6}\ \text{m}$
area	m^2	km^2, dm^2, cm^2, mm^2			
volume	m^3	dm^3, cm^3, mm^3	1		
time	s (second)	ks, ms, μs, ns	d, h, min		
frequency	Hz (hertz)	THz, GHz, MHz, kHz			
mass	kg (kilogram)	Mg, g, mg, μg	t		$1\ \gamma = 10^{-9}\ \text{kg}$
density	kg/m^3	mg/m^3; kg/dm^3; g/cm^3	t/m^3, kg/l		
force	N (newton)	MN, kN, mN, μN			$1\ \text{dyn} = 10^{-5}\ \text{N}$; pond, kilopond
pressure	Pa (pascal)	GPa, MPa, kPa, mPa, μPa		$1\ \text{bar} = 10^5\ \text{Pa}$, $1\ \text{atm} = 101325\ \text{Pa}$	$1\ \text{torr} = \dfrac{101325}{760}\ \text{Pa}$
stress	Pa or N/m^2	GPa; MPa or N/mm^2; kPa			
viscosity (dynamic)	Pa · s	mPa · s			poise; $1\ \text{P} = 0.1\ \text{Pa} \cdot \text{s}$
viscosity (kinematic)	m^2/s	mm^2/s			stokes; $1\ \text{St} = 10^{-4}\ m^2/s$
surface tension	N/m	mN/m			dyn/cm
energy, work, heat	J (joule)	TJ, GJ, MJ, kJ, mJ	eV		$1\ \text{erg} = 10^{-7}\ \text{J}$, $1\ \text{cal} = 4.1868\ \text{J}$
power	W (watt)	GW, MW, kW, mW, μW			
temperature	K (kelvin), °C				$1\ ^\circ\text{K} = 1\ \text{K}$
thermal conductivity	W/mK				
heat capacity	J/K	kJ/K			
specific heat	J/(kg · K)	kJ/(kg · K)			
entropy	J/K	kJ/K			
amount of matter	mol (mole)				Mole

2.2 SI-Prefixes

Factor	Prefix	Symbol	Factor	Prefix	Symbol
10^{12}	tera	T	10^{-1}	deci	d
10^{9}	giga	G	10^{-2}	centi	c
10^{6}	mega	M	10^{-3}	milli	m
10^{3}	kilo	k	10^{-6}	micro	μ
10^{2}	hecto	h	10^{-9}	nano	n
10^{1}	deka	da	10^{-12}	pico	p
			10^{-15}	femto	f
			10^{-18}	atto	a

2.3 Conversion Factors

LENGTH		m	in	ft	yd	thou or mil
1 m	=	1	$3.937 \cdot 10^{1}$	3.281	1.094	$3.937 \cdot 10^{4}$
in	=	$2.540 \cdot 10^{-2}$	1	$8.333 \cdot 10^{-2}$	$2.778 \cdot 10^{-2}$	$1.0 \cdot 10^{3}$
ft	=	$3.048 \cdot 10^{-1}$	12	1	$3.334 \cdot 10^{-1}$	$1.2 \cdot 10^{4}$
yd	=	$9.144 \cdot 10^{-1}$	36	3	1	$3.6 \cdot 10^{4}$
thou or mil	=	$2.540 \cdot 10^{-5}$	$1.0 \cdot 10^{-3}$	$8.334 \cdot 10^{-5}$	$2.779 \cdot 10^{-5}$	1

AREA

		m^2	sq. in	sq. ft	sq. yd	ar
1 m^2	=	1	$1.550 \cdot 10^3$	$1.076 \cdot 10^1$	1.196	$1.0 \cdot 10^{-2}$
sq. in	=	$6.452 \cdot 10^{-4}$	1	$6.944 \cdot 10^{-3}$	$7.716 \cdot 10^{-4}$	$6.452 \cdot 10^{-6}$
sq. ft	=	$9.290 \cdot 10^{-2}$	$1.440 \cdot 10^2$	1	$1.111 \cdot 10^{-1}$	$9.290 \cdot 10^{-4}$
sq. yd	=	$8.361 \cdot 10^{-1}$	$1.296 \cdot 10^3$	9	1	$8.361 \cdot 10^{-3}$
ar	=	$1.0 \cdot 10^2$	$1.550 \cdot 10^5$	$1.076 \cdot 10^3$	$1.196 \cdot 10^2$	1

VOLUME

		m^3	l (liter)	cu. in	cu. ft	cu. yd	gal (US)	gal (UK)
1 m^3	=	1	10^3	$6.102 \cdot 10^4$	$3.531 \cdot 10^1$	1.308	$2.642 \cdot 10^2$	$2.20 \cdot 10^2$
1 (liter)	=	10^{-3}	1	$6.102 \cdot 10^1$	$3.531 \cdot 10^{-2}$	$1.308 \cdot 10^{-3}$	$2.642 \cdot 10^{-1}$	$2.20 \cdot 10^{-1}$
cu. in	=	$1.639 \cdot 10^{-5}$	$1.639 \cdot 10^{-2}$	1	$5.787 \cdot 10^{-4}$	$2.143 \cdot 10^{-5}$	$4.329 \cdot 10^{-3}$	$3.605 \cdot 10^{-3}$
cu. ft	=	$2.832 \cdot 10^{-2}$	$2.832 \cdot 10^1$	$1.728 \cdot 10^3$	1	$3.703 \cdot 10^{-2}$	7.481	6.229
cu. yd	=	$7.646 \cdot 10^{-1}$	$7.645 \cdot 10^2$	$4.666 \cdot 10^4$	$2.7 \cdot 10^1$	1	$2.020 \cdot 10^2$	$1.682 \cdot 10^2$
gal (US)	=	$3.785 \cdot 10^{-3}$	3.785	$2.310 \cdot 10^2$	$1.337 \cdot 10^{-1}$	$4.951 \cdot 10^{-3}$	1	$8.327 \cdot 10^{-1}$
gal (UK)	=	$4.546 \cdot 10^{-3}$	4.546	$2.774 \cdot 10^2$	$1.605 \cdot 10^{-1}$	$5.946 \cdot 10^{-3}$	1.201	1

MASS

		kg	lbm	ton (UK) (long ton)	cwt (UK) (long cwt)	ton (US) (short ton)	ounce	grain
1 kg	=	1	2.205	$9.842 \cdot 10^{-4}$	$1.968 \cdot 10^{-2}$	$1.102 \cdot 10^{-3}$	$3.527 \cdot 10^1$	$1.543 \cdot 10^4$
lbm	=	$4.536 \cdot 10^{-1}$	1	$4.464 \cdot 10^{-4}$	$8.929 \cdot 10^{-3}$	$5.0 \cdot 10^{-4}$	$1.6 \cdot 10^1$	$7.0 \cdot 10^3$
ton (UK)	=	$1.016 \cdot 10^3$	$2.240 \cdot 10^3$	1	$2.0 \cdot 10^1$	1.120	$3.584 \cdot 10^4$	$1.568 \cdot 10^7$
cwt (UK)	=	$5.080 \cdot 10^1$	$1,120 \cdot 10^2$	$5.0 \cdot 10^{-2}$	1	$5.600 \cdot 10^{-2}$	$1.792 \cdot 10^3$	$7.840 \cdot 10^5$
ton (US)	=	$9.072 \cdot 10^2$	$2.0 \cdot 10^3$	$8.929 \cdot 10^{-1}$	$1.786 \cdot 10^1$	1	$3.2 \cdot 10^4$	$1.40 \cdot 10^7$
ounce	=	$2.835 \cdot 10^{-2}$	$6,250 \cdot 10^{-2}$	$2.790 \cdot 10^{-5}$	$5.580 \cdot 10^{-4}$	$3.125 \cdot 10^{-5}$	1	$4.375 \cdot 10^2$
grain	=	$6.480 \cdot 10^{-5}$	$1.429 \cdot 10^{-8}$	$6.378 \cdot 10^{-8}$	$1.276 \cdot 10^{-6}$	$7.143 \cdot 10^{-8}$	$2.286 \cdot 10^{-3}$	1

DENSITY

		kg/m^3	$Mg/m^3 = g/cm^3$	lbm/cu. ft	lbm/cu. in	lbm/gal (UK)	lbm/gal (US)
1 kg/m^3	=	1	$1.0 \cdot 10^{-3}$	$6.243 \cdot 10^{-2}$	$3.613 \cdot 10^{-5}$	$1.002 \cdot 10^{-2}$	$8.345 \cdot 10^{-3}$
$Mg/m^3 = g/cm^3$	=	$1.0 \cdot 10^3$	1	$6.243 \cdot 10^1$	$3.613 \cdot 10^{-2}$	$1.002 \cdot 10^1$	8.345
lbm/cu. ft	=	$1.602 \cdot 10^1$	$1.602 \cdot 10^{-2}$	1	$5.789 \cdot 10^{-4}$	$1.605 \cdot 10^{-1}$	$1.337 \cdot 10^{-1}$
lbm/cu. in	=	$2.768 \cdot 10^4$	$2.768 \cdot 10^1$	$1.728 \cdot 10^3$	1	$2,774 \cdot 10^2$	$2.310 \cdot 10^2$
lbm/gal (UK)	=	$9.978 \cdot 10^1$	$9.978 \cdot 10^{-2}$	6.229	$3.605 \cdot 10^{-3}$	1	$8.327 \cdot 10^{-1}$
lbm/gal (US)	=	$1.198 \cdot 10^2$	$1.198 \cdot 10^{-1}$	7.480	$4.329 \cdot 10^{-3}$	1.201	1

FORCE

		N ($kg\ m/s^2$)	kgf, kp	lbf	dyn	tonf (UK) (long ton)	tonf (US) (short ton)
1 N	=	1	$1.020 \cdot 10^{-1}$	$2.248 \cdot 10^{-1}$	$1.0 \cdot 10^5$	$1.004 \cdot 10^{-4}$	$1.124 \cdot 10^{-4}$
kp	=	9.807	1	2.205	$9.807 \cdot 10^5$	$9.842 \cdot 10^{-4}$	$1.102 \cdot 10^{-3}$
lbf	=	4.448	$4.536 \cdot 10^{-1}$	1	$4.448 \cdot 10^5$	$4.464 \cdot 10^{-4}$	$5.0 \cdot 10^{-4}$
dyn	=	$1.0 \cdot 10^{-5}$	$1.020 \cdot 10^{-6}$	$2.248 \cdot 10^{-6}$	1	$1.004 \cdot 10^{-9}$	$1.124 \cdot 10^{-9}$
tonf (UK)	=	$9.964 \cdot 10^3$	$1.016 \cdot 10^3$	$2.240 \cdot 10^3$	$9.964 \cdot 10^8$	1	1.120
tonf (US)	=	$8.896 \cdot 10^3$	$9.072 \cdot 10^2$	$2.0 \cdot 10^3$	$8.896 \cdot 10^8$	$8.929 \cdot 10^{-1}$	1

PRESSURE

		Pa (N/m^2) ($kg/m\ s^2$)	bar	kgf/cm^2 kp/cm^2 at	atm	Torr	psi $lbf/sq.\ in$	mbar
1 Pa	=	1	$1.0 \cdot 10^{-5}$	$1.020 \cdot 10^{-5}$	$9.869 \cdot 10^{-6}$	$7.501 \cdot 10^{-3}$	$1.450 \cdot 10^{-4}$	$1.0 \cdot 10^{-2}$
bar	=	$1.0 \cdot 10^5$	1	1.020	$9.869 \cdot 10^{-1}$	$7.501 \cdot 10^2$	$1.450 \cdot 10^1$	$1.0 \cdot 10^3$
kp/cm^2	=	$9.807 \cdot 10^4$	$9.807 \cdot 10^{-1}$	1	$9.678 \cdot 10^{-1}$	$7.356 \cdot 10^2$	$1.422 \cdot 10^1$	$9.807 \cdot 10^2$
atm	=	$1.013 \cdot 10^5$	1.01325	1.033	1	$7.60 \cdot 10^2$	$1.470 \cdot 10^1$	$1.01325 \cdot 10^3$
Torr	=	$1.333 \cdot 10^2$	$1.333 \cdot 10^{-3}$	$1.360 \cdot 10^{-3}$	$1.316 \cdot 10^{-3}$	1	$1.934 \cdot 10^{-2}$	1.333 2
psi	=	$6.895 \cdot 10^3$	$6.895 \cdot 10^{-2}$	$7.031 \cdot 10^{-2}$	$6.805 \cdot 10^{-2}$	$5.171 \cdot 10^1$	1	$6.895 \cdot 10^1$
mbar	=	$1.0 \cdot 10^2$	$1 \cdot 10^{-3}$	$1.020 \cdot 10^{-3}$	$9.869 \cdot 10^{-4}$	0.7501	$1.450 \cdot 10^{-2}$	1

STRESS		$\dfrac{Pa}{(N/m^2)}$ $(kg/m\ s^2)$	$\dfrac{N/mm^2}{(MPa)}$	daN/cm^2	daN/mm^2	kgf/cm^2 kp/cm^2	kgf/mm^2 kg/cm^2	psi
1 Pa	=	1	$1.0 \cdot 10^{-6}$	$1.0 \cdot 10^{-5}$	$1.0 \cdot 10^{-7}$	$1.020 \cdot 10^{-5}$	$1.020 \cdot 10^{-7}$	$1.450 \cdot 10^{-4}$
N/mm^2	=	$1.0 \cdot 10^{6}$	1	$1.0 \cdot 10^{1}$	$1.0 \cdot 10^{-1}$	$1.020 \cdot 10^{1}$	$1.020 \cdot 10^{-1}$	$1.450 \cdot 10^{2}$
daN/cm^2	=	$1.0 \cdot 10^{5}$	$1.0 \cdot 10^{-1}$	1	$1.0 \cdot 10^{-2}$	1.020	$1.020 \cdot 10^{-2}$	$1.450 \cdot 10^{1}$
daN/mm^2	=	$1.0 \cdot 10^{7}$	$1.0 \cdot 10^{1}$	$1.0 \cdot 10^{2}$	1	$1.020 \cdot 10^{2}$	1.020	$1.450 \cdot 10^{3}$
kp/cm^2	=	$9.807 \cdot 10^{4}$	$9.807 \cdot 10^{-2}$	$9.807 \cdot 10^{-1}$	$9.807 \cdot 10^{-3}$	1	$1.0 \cdot 10^{-2}$	$1.422 \cdot 10^{1}$
kp/mm^2	=	$9.807 \cdot 10^{6}$	9.807	$9.807 \cdot 10^{1}$	$9.807 \cdot 10^{-1}$	$1.0 \cdot 10^{2}$	1	$1.422 \cdot 10^{3}$
psi	=	$6.895 \cdot 10^{3}$	$6.895 \cdot 10^{-3}$	$6.895 \cdot 10^{-2}$	$6.895 \cdot 10^{-4}$	$7.031 \cdot 10^{-2}$	$7.031 \cdot 10^{-4}$	1

Multiples: GPa, MPa, kPa, mPa, μPa

VISCOSITY (dynamic)		$\dfrac{Pa\ s}{(N\ s/m^2)}$ $(kg/s\ m)$	$\dfrac{mPa\ s}{(mN\ s/m^2)}$	cP	$kp\ s/m^2$
1 Pa s	=	1	$1.0 \cdot 10^{3}$	$1.0 \cdot 10^{3}$	$1.020 \cdot 10^{-1}$
mPa s	=	$1.0 \cdot 10^{-3}$	1	1.0	$1.020 \cdot 10^{-4}$
cP	=	$1.0 \cdot 10^{-3}$	1.0	1	$1.020 \cdot 10^{-4}$
$kp\ s/m^2$	=	9.807	$9.807 \cdot 10^{3}$	$9.807 \cdot 10^{3}$	1
$kp\ h/m^2$	=	$3.530 \cdot 10^{4}$	$3.530 \cdot 10^{7}$	$3.530 \cdot 10^{7}$	$3.60 \cdot 10^{3}$
lbm/ft s	=	1.488	$1.488 \cdot 10^{3}$	$1.488 \cdot 10^{3}$	$1.518 \cdot 10^{-1}$
lbm/ft h	=	$4.134 \cdot 10^{-4}$	$4.134 \cdot 10^{-1}$	$4.134 \cdot 10^{-1}$	$4.215 \cdot 10^{-5}$
lbf s/sq. ft	=	$4.788 \cdot 10^{1}$	$4.788 \cdot 10^{4}$	$4.788 \cdot 10^{4}$	4.882

		$kp\ h/m^2$	lbm/ft s	lbm/ft h	lbf s/sq. ft
1 Pa s	=	$2.833 \cdot 10^{-5}$	$6.720 \cdot 10^{-1}$	$2.419 \cdot 10^{3}$	$2.089 \cdot 10^{-2}$
mPa s	=	$2.833 \cdot 10^{-8}$	$6.720 \cdot 10^{-4}$	2.419	$2.089 \cdot 10^{-5}$
cP	=	$2.833 \cdot 10^{-8}$	$6.720 \cdot 10^{-4}$	2.419	$2.089 \cdot 10^{-5}$
$kp\ s/m^2$	=	$2.778 \cdot 10^{-4}$	6.590	$2.372 \cdot 10^{4}$	$2.048 \cdot 10^{-1}$
$kp\ h/m^2$	=	1	$2.372 \cdot 10^{4}$	$8.540 \cdot 10^{7}$	$7.373 \cdot 10^{2}$
lbm/ft s	=	$4.215 \cdot 10^{-5}$	1	$3.60 \cdot 10^{3}$	$3.103 \cdot 10^{-2}$
lbm/ft h	=	$1.171 \cdot 10^{-8}$	$2.778 \cdot 10^{-4}$	1	$8.634 \cdot 10^{-6}$
lbf s/sq. ft	=	$1.356 \cdot 10^{-3}$	$3.217 \cdot 10^{1}$	$1.158 \cdot 10^{5}$	1

VISCOSITY (kinematic)		m^2/s	mm^2/s	m^2/h	cSt	sq. ft/s	sq. ft/h
1 m^2/s	=	1	$1.0 \cdot 10^{6}$	$3.60 \cdot 10^{3}$	$1.0 \cdot 10^{6}$	$1.076 \cdot 10^{1}$	$3.875 \cdot 10^{4}$
mm^2/s	=	$1.0 \cdot 10^{-6}$	1	$3.60 \cdot 10^{-3}$	1.0	$1.076 \cdot 10^{-5}$	$3.875 \cdot 10^{-2}$
m^2/h	=	$2.778 \cdot 10^{-4}$	$2.778 \cdot 10^{2}$	1	$2.778 \cdot 10^{2}$	$2.990 \cdot 10^{-3}$	$1.076 \cdot 10^{1}$
cSt	=	$1.0 \cdot 10^{-6}$	1.0	$3.60 \cdot 10^{-3}$	1	$1.076 \cdot 10^{-5}$	$3.875 \cdot 10^{-2}$
sq. ft/s	=	$9.290 \cdot 10^{-2}$	$9.290 \cdot 10^{4}$	$3.345 \cdot 10^{2}$	$9.290 \cdot 10^{4}$	1	$3.60 \cdot 10^{3}$
sq. ft/h	=	$2.581 \cdot 10^{-5}$	$2.581 \cdot 10^{1}$	$9.290 \cdot 10^{-2}$	$2.581 \cdot 10^{1}$	$2.778 \cdot 10^{-4}$	1

SURFACE TENSION		$\dfrac{N/m}{(kg/s^2)}$	mN/m	kgf/m kp/m	dyn/cm
1 N/m	=	1	$1.0 \cdot 10^{3}$	$1.020 \cdot 10^{-1}$	$1.0 \cdot 10^{3}$
mN/m	=	$1.0 \cdot 10^{-3}$	1	$1.020 \cdot 10^{-4}$	1.0
kp/m	=	9.807	$9.807 \cdot 10^{3}$	1	$9.807 \cdot 10^{3}$
dyn/cm	=	$1.0 \cdot 10^{-3}$	1.0	$1.020 \cdot 10^{-4}$	1

ENERGY		J $(N\,m)$ $(kg\,m^2/s^2)$	kW h	kgf m / kp m	Ps h	lbf ft
1 J	=	1	$2.778 \cdot 10^{-7}$	$1.020 \cdot 10^{-1}$	$3.777 \cdot 10^{-7}$	$7.376 \cdot 10^{-1}$
kW h	=	$3.60 \cdot 10^{6}$	1	$3.671 \cdot 10^{5}$	1.360	$2.655 \cdot 10^{6}$
kp m	=	9.807	$2.724 \cdot 10^{-6}$	1	$3.704 \cdot 10^{-6}$	7.233
PS h	=	$2.648 \cdot 10^{6}$	$7.355 \cdot 10^{-1}$	$2.70 \cdot 10^{5}$	1	$1.953 \cdot 10^{6}$
lbf ft	=	1.356	$3.766 \cdot 10^{-7}$	$1.363 \cdot 10^{-1}$	$5.120 \cdot 10^{-7}$	1
erg	=	$1.0 \cdot 10^{-7}$	$2.778 \cdot 10^{-14}$	$1.020 \cdot 10^{-8}$	$3.777 \cdot 10^{-14}$	$7.375 \cdot 10^{-8}$
HP h	=	$2.685 \cdot 10^{6}$	$7.457 \cdot 10^{-1}$	$2.737 \cdot 10^{5}$	1.014	$1.980 \cdot 10^{6}$
kcal	=	$4.187 \cdot 10^{3}$	$1.163 \cdot 10^{-3}$	$4.269 \cdot 10^{2}$	$1.581 \cdot 10^{-3}$	$3.088 \cdot 10^{3}$
BTU	=	$1.055 \cdot 10^{3}$	$2.931 \cdot 10^{-4}$	$1.076 \cdot 10^{2}$	$3.985 \cdot 10^{-4}$	$7.782 \cdot 10^{2}$

		erg	HP h	kcal	BTU
1 J	=	$1.0 \cdot 10^{7}$	$3.725 \cdot 10^{-7}$	$2.388 \cdot 10^{-4}$	$9.478 \cdot 10^{-4}$
kW h	=	$3.60 \cdot 10^{13}$	1.341	$8.598 \cdot 10^{2}$	$3.412 \cdot 10^{3}$
kp m	=	$9.807 \cdot 10^{7}$	$3.653 \cdot 10^{-6}$	$2.342 \cdot 10^{-3}$	$9.295 \cdot 10^{-3}$
PS h	=	$2.648 \cdot 10^{13}$	$9.863 \cdot 10^{-1}$	$6.324 \cdot 10^{2}$	$2.510 \cdot 10^{3}$
lbf ft	=	$1.356 \cdot 10^{7}$	$5.051 \cdot 10^{-7}$	$3.238 \cdot 10^{-4}$	$1.285 \cdot 10^{-3}$
erg	=	1	$3.725 \cdot 10^{-14}$	$2.388 \cdot 10^{-11}$	$9.478 \cdot 10^{-11}$
HP h	=	$2.685 \cdot 10^{13}$	1	$6.412 \cdot 10^{2}$	$2.544 \cdot 10^{3}$
kcal	=	$4.187 \cdot 10^{10}$	$1.560 \cdot 10^{-3}$	1	3.968
BTU	=	$1.055 \cdot 10^{10}$	$3.930 \cdot 10^{-4}$	$2.520 \cdot 10^{-1}$	1

HEAT		J $(N\,m)$ $(kg\,m^2/s^2)$	kcal	BTU	CHU centigrade heat unit
1 J	=	1	$2.388 \cdot 10^{-4}$	$9.478 \cdot 10^{-4}$	$5.262 \cdot 10^{-4}$
kcal	=	$4.187 \cdot 10^{3}$	1	3.968	2.203
BTU	=	$1.055 \cdot 10^{3}$	$2.520 \cdot 10^{-1}$	1	$5.552 \cdot 10^{-1}$
CHU	=	$1.900 \cdot 10^{3}$	$4.539 \cdot 10^{-1}$	1.80	1
W h	=	$3.60 \cdot 10^{3}$	$8.598 \cdot 10^{-1}$	3.412	1.894

POWER		W (J/s) $(kg\,m^2/s^3)$	kgf m/s / kp m/s	PS HP (metr.)	HP
1 W	=	1	$1.020 \cdot 10^{-1}$	$1.360 \cdot 10^{-3}$	$1.341 \cdot 10^{-3}$
kp m/s	=	9.807	1	$1.333 \cdot 10^{-2}$	$1.315 \cdot 10^{-2}$
PS	=	$7.355 \cdot 10^{2}$	$7.5 \cdot 10^{1}$	1	$9.863 \cdot 10^{-1}$
HP	=	$7.457 \cdot 10^{2}$	$7.604 \cdot 10^{1}$	1.014	1
erg/s	=	$1.0 \cdot 10^{-7}$	$1.020 \cdot 10^{-8}$	$1.360 \cdot 10^{-10}$	$1.341 \cdot 10^{-10}$
ft lbf/s	=	1.356	$1.383 \cdot 10^{-1}$	$1.843 \cdot 10^{-3}$	$1.818 \cdot 10^{-3}$
kcal/h	=	1.163	$1.186 \cdot 10^{-1}$	$1.581 \cdot 10^{-3}$	$1.560 \cdot 10^{-3}$
BTU/h	=	$2.931 \cdot 10^{-1}$	$2.988 \cdot 10^{-2}$	$3.985 \cdot 10^{-4}$	$3.930 \cdot 10^{-4}$

multiples: GW, MK, kW, mW, µW.

		erg/s	ft lbf/s	kcal/h	BTU/h
1 W	=	$1.0 \cdot 10^{7}$	$7.376 \cdot 10^{-1}$	$8.598 \cdot 10^{-1}$	3.412
kp m/s	=	$9.807 \cdot 10^{7}$	7.233	8.432	$3.346 \cdot 10^{1}$
PS	=	$7.355 \cdot 10^{9}$	$5.425 \cdot 10^{2}$	$6.324 \cdot 10^{2}$	$2.510 \cdot 10^{3}$
HP	=	$7.457 \cdot 10^{9}$	$5.50 \cdot 10^{2}$	$6.412 \cdot 10^{2}$	$2.544 \cdot 10^{3}$
erg/s	=	1	$7.375 \cdot 10^{-8}$	$8.598 \cdot 10^{-8}$	$3.412 \cdot 10^{-7}$
ft lbf/s	=	$1.356 \cdot 10^{7}$	1	1.166	4.626
kcal/h	=	$1.163 \cdot 10^{7}$	$8.578 \cdot 10^{-1}$	1	3.968
BTU/h	=	$2.931 \cdot 10^{6}$	$2.162 \cdot 10^{-1}$	$2.520 \cdot 10^{-1}$	1

THERMAL CONDUCTIVITY		$\dfrac{\text{W/m K}}{(\text{kg m/s}^3\text{ K})}$	kcal/m h $^{\circ}$C	BTU/ft h $^{\circ}$F	BTU/in h $^{\circ}$F	BTU in/ sq. ft h $^{\circ}$F
1 W/m K	=	1	$8.598 \cdot 10^{-1}$	$5.778 \cdot 10^{-1}$	$4.815 \cdot 10^{-2}$	6.933
kcal/m h $^{\circ}$C	=	1.163	1	$6.720 \cdot 10^{-1}$	$5.60 \cdot 10^{-2}$	8.064
BTU/ft h $^{\circ}$F	=	1.731	1.488	1	$8.333 \cdot 10^{-2}$	$1.2 \cdot 10^{1}$
BTU/in h $^{\circ}$F	=	$2.077 \cdot 10^{1}$	$1.786 \cdot 10^{1}$	$1.2 \cdot 10^{1}$	1	$1.44 \cdot 10^{2}$
BTU in/sq. ft h$^{\circ}$F	=	$1.442 \cdot 10^{-1}$	$1.24 \cdot 10^{-1}$	$8.333 \cdot 10^{-2}$	$6.944 \cdot 10^{-3}$	1

SPECIFIC HEAT, HEAT CAPACITY		$\dfrac{\text{J/kg K}}{(\text{m}^2/\text{s}^2\text{ K})}$	kJ/kg K	kcal/kg $^{\circ}$C	BTU/lbm $^{\circ}$F
1 J/kg K	=	1	$1.0 \cdot 10^{-3}$	$2.389 \cdot 10^{-4}$	$2.389 \cdot 10^{-4}$
kJ/kg K	=	$1.0 \cdot 10^{3}$	1	$2.389 \cdot 10^{-1}$	$2.389 \cdot 10^{-1}$
kcal/kg $^{\circ}$C	=	$4.187 \cdot 10^{3}$	4.187	1	1.0
BTU/lbm $^{\circ}$F	=	$4.187 \cdot 10^{3}$	4.187	1.0	1
Wh/kg K	=	$3.60 \cdot 10^{3}$	3.60	$8.598 \cdot 10^{-1}$	0.8598

II. POLYMERIZATION AND DEPOLYMERIZATION

DECOMPOSITION RATES OF ORGANIC FREE RADICAL INITIATORS

J.C. Masson

Monsanto Triangle Park Development Center, Inc.
Research Triangle Park, North Carolina

<u>Contents</u>   <u>Page</u>

A. INTRODUCTION

The decomposition of most organic free radical initiators follows first order kinetics. With some of the peroxide compounds, however, higher order decompositions are observed. Generally, the higher order reaction is caused by reaction of radicals with the initiator (induced decomposition) and may be eliminated by 1.) extrapolation of rate data to zero concentration, 2.) addition of a monomer or other "radical trap".

Decomposition rate (k_d) data in these tables are reported for first order kinetics

$$-dI/dt = k_d I$$

      where I = initiator concentration
      and t = time

The decomposition rate constant k_d is related to half life $(t_{1/2})$ by the following equation:

$$t_{1/2} = 0.693/k_d$$

Figure 1 relates k_d in $[sec^{-1}]$ to half-life for the range of k_d found in the tables.

For some of the initiators listed, the enthalpy $(\Delta H^{\neq})$ is given (see notes) rather than the Arrhenius activation energy (E_a). The two quantities are related by the equation $E_a = \Delta H^{\neq} + RT$ where R is the gas constant (in kJ/mol-deg.) and T the absolute temperature (94). Assuming that k_d is linear with respect to $1/T$ and that the activation energy E_a and the decomposition rate constant k_d for one temperature are known, k_d for any temperature can be calculated from the following expression:

$$\log k_2 = \log k_1 - \frac{E_a (T_2 - T_1)}{2.303 R (T_2 T_1)}$$

Where given by the author, the overall equation for k_d in terms of the frequency factor (A) and activation energy (E_a) has been included. Thus, for any temperature (converted to K) the k_d may be calculated:

$$k_d = A \exp [-E_a/RT]$$

The data have been arranged into seven tables. Within each table individual initiators are listed according to the following criteria:

    I.   Initiators

        a. according to increasing number of carbon atoms.
        b. for compounds containing equal number of carbons, alphabetically, neglecting trivial prefixes.
        c. Table 7 - Miscellaneous initiators are listed alphabetically.

    II. For each initiator, solvents are listed alphabetically. Additives present in the vapor phase are in parenthesis.

    III. For a given solvent, all measurements reported by one investigator are listed in a series, with the activation energy listed opposite the lowest temperature.

As <u>Chemical Abstracts</u> does not systematically list initiators under a general subject index, it was necessary to peruse specific journals for articles of interest. Other references were found from literature citations in the journals checked.

<u>Abbreviations:</u>   DMSO dimethyl sulfoxide; DMAC dimethylacetamide; THF tetrahydrofuran; DMF dimethylformamide

RELATIONSHIP OF HALF LIFE $(t_{1/2})$ TO RATE CONSTANT (k)

(Half Lives are to the Left of Each Vertical Line)

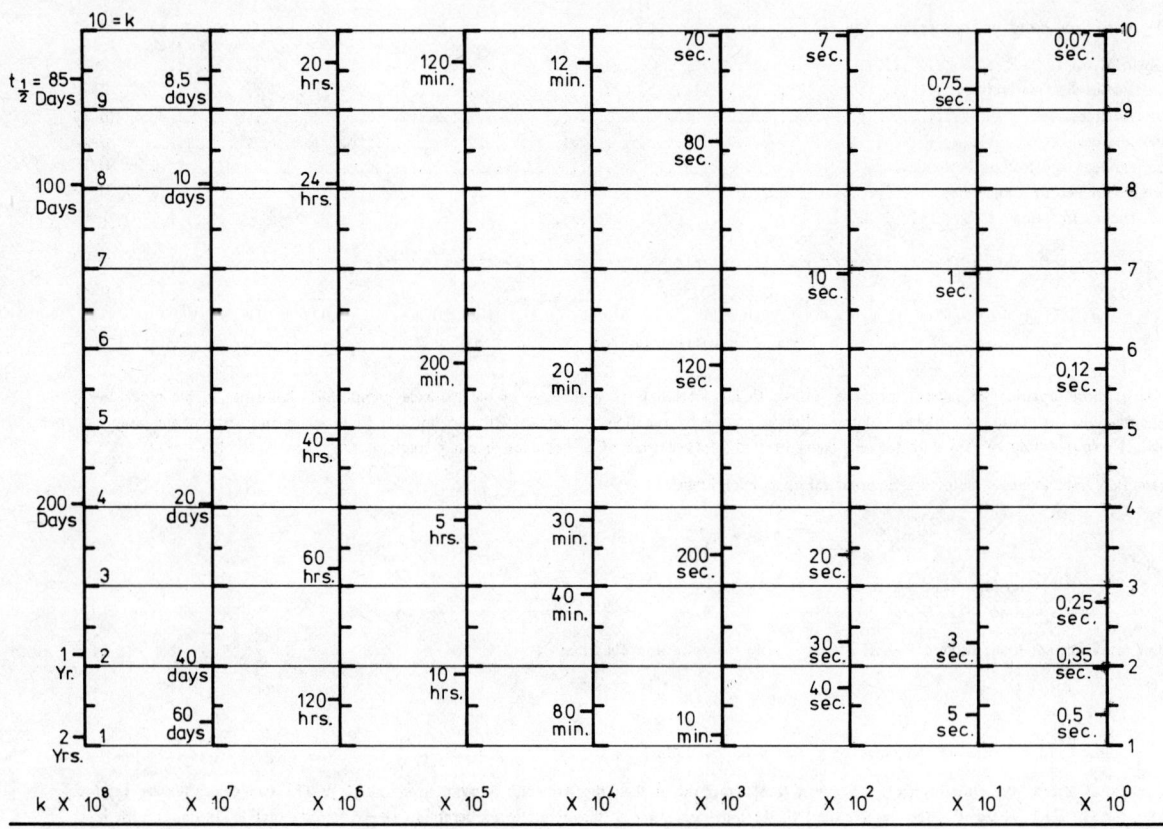

Number of C atoms	Initiator	Solvent	T $^\circ$C	k_d [s^{-1}]	E_a [kJ mol^{-1}]	Notes	References
		1 - AZO-NITRILES					
5	2-cyano-2-propyl-azo-formamide	chlorobenzene	100	1.5×10^{-5}			93
		toluene	100	2.1×10^{-5}	144.3		93
			110	6.8×10^{-5}			93
		xylene	100	2.1×10^{-5}	144.3		93
			120	2.4×10^{-4}			93
8	2,2'-azo-bis-isobutyronitrile (2,2'-azo-bis-2-methylpropionitrile)	acetic acid	79.9	1.43×10^{-4}			146
			79.9	1.48×10^{-4}		v2	146
			79.9	1.62×10^{-4}		v3	146
			80	1.52×10^{-4}		a	62
			82	1.50×10^{-4}		a	2
			82	1.49×10^{-4}		a	175
		acetonitrile	79.9	1.25×10^{-4}			146
			79.9	1.24×10^{-4}		v2	146
		tert-amyl alcohol	80.2	1.40×10^{-4}			61
		aniline	80.2	1.68×10^{-4}			61
		benzene	40.0	5.44×10^{-7}	128.4	a, t_2	69
			45.2	1.12×10^{-6}		a, t_2	69
			50.0	2.64×10^{-6}		a, t_2	69
			55.0	5.19×10^{-6}		a, t_2	69
			60.5	1.16×10^{-5}		a, t_2	69
			69.5	3.78×10^{-5}		a, t_2	69
			40	4.83×10^{-7}	123.4		66
			50	2.085×10^{-6}			66
			60	8.45×10^{-6}			66
			70	3.166×10^{-5}			66
			78	8.023×10^{-5}			66
		benzene or toluene	37	2.83×10^{-7}	128.9		39
			43	7.35×10^{-7}			39
			50	2.16×10^{-6}			39
			60	9.15×10^{-6}			39
			100	1.52×10^{-3}			39
			T [K]	$1.58 \times 10^{15} \exp[-128.9\ \mathrm{kJ}/RT]$			39
		n-butanol	82	1.54×10^{-4}		a	175
			82	1.55×10^{-4}		a	2
		isobutanol	82	1.66×10^{-4}		a	2, 175
			80.2	$1.67\text{-}1.76 \times 10^{-4}$			61
		di-n-butyl phthalate	80	2.64×10^{-4}	122.2	y4	236
			90	6.47×10^{-4}		y4	236
			100	1.78×10^{-3}		y4	236
			110	4.88×10^{-3}		y4	236
			120	1.43×10^{-2}		y4	236
			127	2.48×10^{-2}		y16	236
			137	5.43×10^{-2}		y16	236
			145	1.24×10^{-1}		y32	236
		carbon tetrachloride	40	2.15×10^{-7}	128.4	a, t_2	69
			60	4.00×10^{-6}		a, t_2	69
			77	1.21×10^{-4}		a	62
		cyclohexanone	82	1.43×10^{-4}		a	2, 175
		N,N-dimethylaniline	66.82	3.483×10^{-5}			26
			72.27	6.914×10^{-5}			26
			80	1.83×10^{-4}		a	62
		dioxan/water (80/20) pH 7.0	65.3	2.53×10^{-5}	141.0		186
			70.0	3.20×10^{-5}			186
			75.0	8.5×10^{-5}			186
			80.0	1.62×10^{-4}			186
		dioxan/water (80/20) pH 10.7	75.0	1.72×10^{-4}			186
		dodecanethiol	80	1.46×10^{-4}		a	62
		ethyl acetate	40	4.7×10^{-7}	128.5	a, t_2	69
			60	9.36×10^{-6}		a, t_2	69

Number of C atoms	Initiator	Solvent	T °C	k_d [s^{-1}]	E_a [kJ mol^{-1}]	Notes	References
		AZO-NITRILES (Cont'd.)					
8	2,2'-azo-bis-isobutyronitrile (Cont'd.)	diethylene glycol monobutyl ether	66.82	2.442×10^{-5}			26
		maleimide (solid state)	65	4.97×10^{-5}		a	185
			72	2.07×10^{-4}		a	185
		methyl methacrylate	70	3.1×10^{-5}			216
			70	1.27×10^{-4}		v4	216
		nitrobenzene	80	1.98×10^{-4}	132.2	a	226
			100	2.24×10^{-3}		a	226
		1-nitrobutane	82	1.45×10^{-4}		a	2,175
		propylene carbonate	72.27	5.821×10^{-5}			26
		styrene	50	2.97×10^{-6}	127.6	a	132
			70	4.72×10^{-5}		a	132
			T[K]	1.29×10^{15} exp[-127.6 kJ/RT]			132
			68.4	3.8×10^{-5}			234
		toluene	70	4.0×10^{-5}	121.3		47
			80.4	1.55×10^{-4}			47
			90.0	4.86×10^{-4}			47
			100.0	1.60×10^{-3}			47
			105.0	2.61×10^{-3}			47
			69.8	3.8×10^{-5}	142.3		61
			80.2	$1.72-1.60 \times 10^{-4}$			61
			80.3	1.30×10^{-4}		m$_2$	234
		xylene	50	2.0×10^{-6}		a	175
			77	9.46×10^{-5}		a	175
			82	1.44×10^{-4}		a	175
			80	1.53×10^{-4}	131.0	a	62
			80	1.50×10^{-4}		a, t$_4$	62
			82	1.45×10^{-5}		a	2
10	1,1'azo-bis-1-cyclobutanenitrile	mesitylene	130.4	$5.20 \pm .16 \times 10^{-5}$	134.3		101
			141.6	$1.60 \pm .03 \times 10^{-4}$			101
		xylene	120.4	$2.05 \pm .05 \times 10^{-5}$			101
	2,2'-azo-bis-2-methylbutyronitrile	toluene	69.8	2.3×10^{-5}	129.7		61
			80.2	$8.4-9.0 \times 10^{-5}$			61
		xylene	80	9.97×10^{-5}	123.0	a	62
	2,2'-azo-bis-2-ethylpropionitrile	nitrobenzene	80	8.3×10^{-5}	143.1	a	226
			100	1.08×10^{-3}		a	226
12	4,4'-azo-bis-4-cyanopentanoic acid	water	80	8.97×10^{-5}	142.3	a	62
	1,1'-azo-bis-1-cyclopentane nitrile	toluene	80.3	$7.45 \pm .39 \times 10^{-5}$	141.4		101
			89.2	$2.43 \pm .06 \times 10^{-4}$			101
			95.1	$5.18 \pm .11 \times 10^{-4}$			101
	2,2'-azo-bis-2-cyclopropylpropionitrile (mp 64-65)	toluene	44.2	$3.50 \pm .011 \times 10^{-5}$	117.2	w	57
			49.5	$7.53 \pm .005 \times 10^{-5}$		w	57
			59.2	$2.68 \pm .06 \times 10^{-4}$		w	57
	2,2'-azo-bis-2-cyclopropylpropionitrile (mp 76-77)	toluene	44.2	$3.90 \pm .05 \times 10^{-5}$	108.8	w	57
			49.5	8.17×10^{-5}		w	57
			59.3	$2.46 \pm .02 \times 10^{-4}$		w	57
	2,2'-azo-bis-2,3-dimethylbutyronitrile	toluene	69.8	2.6×10^{-5}	133.9		61
			80.2	1.02×10^{-4}			61
	2,2'-azo-bis-2-methylvaleronitrile	toluene	69.8	4.2×10^{-5}	138.1		61
			80.2	$1.65-1.74 \times 10^{-4}$			61
14	2,2'-azo-bis-2-cyclobutylpropionitrile (mp 38-42)	toluene	80.5	1.51×10^{-4}		w	104
	2,2'-azo-bis-2-cyclobutylpropionitrile (mp 81.5-82.5)	toluene	80.5	1.51×10^{-4}		w	104
	1,1'-azo-bis-1-cyclohexane nitrile	chlorobenzene	79.97	8.42×10^{-6}	140.2		179
			100.12	1.107×10^{-4}			179
		DMSO	80	1.01×10^{-5}	135.6	h	135
			85	2.01×10^{-5}			135
			90	3.89×10^{-5}			135
			95	6.83×10^{-5}			135

Notes page II-40; References page II-41

Number of C atoms	Initiator	Solvent	T °C	k_d [s^{-1}]	E_a [kJ mol^{-1}]	Notes	References
		AZO-NITRILES (Cont'd.)					
14	1,1'-azo-bis-1-cylohexane nitrile (Cont'd.)	nitrobenzene	100	1.14×10^{-4}			226
		toluene	80.3	6.5×10^{-6}	121.3		101
			95.2	5.44×10^{-5}			101
			102.4	1.26×10^{-4}			101
		xylene	77	5.31×10^{-6}		a	175
	2,2'-azo-bis-2,4-dimethylvaleronitrile	xylene	77	5.77×10^{-4}		a	175
	dimethyl-4,4'-azo-bis-cyanopentanoate (meso)	DMAC	77.9	1.43×10^{-4}	133.9	a	205
			85.0	3.76×10^{-4}		a	205
			90.2	6.80×10^{-4}		a	205
			99.7	2.05×10^{-3}		a	205
	(n_D+)		77.6	1.46×10^{-4}	133.5	a	205
			85.9	4.03×10^{-4}		a	205
			90.0	6.90×10^{-4}		a	205
			99.2	2.04×10^{-3}		a	205
	(n_D-)		77.9	1.49×10^{-4}	133.5	a	205
			85.4	3.90×10^{-4}		a	205
			90.2	6.95×10^{-4}		a	205
			99.0	1.97×10^{-3}		a	205
	4,4'-azo-bis-4-cyano-1-methyl-piperidine	DMSO	80	9.7×10^{-6}	136.4	h	135
			85	1.75×10^{-5}			135
			90	3.64×10^{-5}			135
			95	6.47×10^{-5}			135
	2,2'-azo-bis-2-propyl-butyronitrile	nitrobenzene	80	2.55×10^{-4}	128.9	a	226
			100	2.72×10^{-3}		a	226
	2,2'-azo-bis-2,3,3-trimethylbutyro-nitrile (mp 114-16)	toluene	79.9	7.42×10^{-5}	146.4	w	57
			89.0	2.59×10^{-4}		w	57
	2,2'-azo-bis-2,3,3-trimethylbutyro-nitrile (mp 116-18)	toluene	79.9	1.05×10^{-4}	125.5	w	57
			88.9	3.09×10^{-4}		w	57
	2,2'-azo-bis-2-methylhexylonitrile	toluene	80.2	1.58×10^{-4}			61
	2,2'-azo-bis-2,4-dimethylvaleronitrile	toluene	69.8	1.98×10^{-4}	121.3		61
			80.2	7.1×10^{-4}			61
	2,2'-azo-bis-2,4-dimethylvaleronitrile (mp 56-57)	toluene	59.7	8.05×10^{-5}	121.3	w	57
			69.9	2.89×10^{-4}		w	57
	2,2'-azo-bis-2,4-dimethylvaleronitrile (mp 74-76)	toluene	69.8	1.98×10^{-4}	121.3	w	57
			80.2	7.1×10^{-4}		w	57
	2,2'-azo-bis-2-isopropylbutyronitrile	toluene	80.5	$1.01 \pm .05 \times 10^{-4}$			50
15	4-cyano-1-methylpiperidine-4,4'-azo-4'-cyano-1,1'-dimethylpiperidinium nitrate	DMSO	80	1.76×10^{-4}	132.6	h	135
			85	3.31×10^{-5}			135
			90	6.23×10^{-5}			135
			95	1.156×10^{-4}			135
16	4,4'-azo-bis-4-cyano-1,1'-dimethyl-piperidinium nitrate	DMSO	80	2.84×10^{-5}	124.7	h	135
			85	5.11×10^{-5}			135
			90	9.87×10^{-5}			135
			95	1.626×10^{-4}			135
	1,1'-azo-bis-1-cycloheptane-nitrile	acetic acid	69.5	3.28×10^{-4}			101
		toluene	48.9	$2.69 \pm .05 \times 10^{-5}$	115.1		101
			58.9	$9.72 \pm .12 \times 10^{-5}$			101
			67.3	$2.69 \pm .12 \times 10^{-4}$			101
	2,2'-azo-bis-2-methylheptylonitrile	toluene	80.0	$1.63 \pm .1 \times 10^{-4}$			50
		xylene	80.0	1.78×10^{-4}	126.4	a	62
	1,1'-azo-bis-1-(2-methylcyclohexane)-nitrile	toluene	80.2	$7.43 \pm .18 \times 10^{-6}$			101
	1,1'-azo-bis-1-cyclohexanecarbonitrile	xylene	80	4.7×10^{-6}	166.9	a	62
	2,2'-azo-bis-2-cyclopentylpropionitrile (mp 72.2-74.5)	toluene	80.5	1.31×10^{-4}		w	104
	2,2'-azo-bis-2-cyclopentylpropionitrile (mp 96.3-97.6)	toluene	80.5	1.30×10^{-4}		w	104
	2,2'-azo-bis-2,4,4-trimethylvaleronitrile (mp 67.5-69)	toluene	40	$1.175 \pm .01 \times 10^{-4}$	113.5	w	50
			50	$4.45 \pm .14 \times 10^{-4}$		w	50
	2,2'-azo-bis-2,4,4-trimethylvaleronitrile (mp 94.5-95.5)	toluene	40	$6.95 \pm .1 \times 10^{-5}$	121.3	w	50
			50	$2.89 \pm .14 \times 10^{-4}$		w	50

Number of C atoms	Initiator	Solvent	T °C	k_d [s^{-1}]	E_a [kJ mol^{-1}]	Notes	References
	AZO-NITRILES (Cont'd.)						
16	2,2'-azo-bis-2-isopropyl-3-methylbutyronitrile	toluene	80.5	$1.325 \pm .035 \times 10^{-4}$			50
18	1,1'-azo-bis-1-cyano-4,4'-dimethylcyclohexane	DMSO	80	1.51×10^{-5}	132.2	h	135
			85	2.79×10^{-5}			135
			90	5.40×10^{-5}			135
			95	9.77×10^{-5}			135
	2,2'-azo-bis-2-cyclohexylpropionitrile	toluene	80.3	8.3×10^{-6}			61
			80.5	2.27×10^{-4}			104
	1,1'-azo-bis-1-cyclooctanenitrile	toluene	36.6	$5.35 \pm .08 \times 10^{-5}$	108.4		101
			45.4	$1.45 \pm .06 \times 10^{-4}$			101
			49.7	$2.60 \pm .08 \times 10^{-4}$			101
20	2,2'-azo-bis-2-benzylpropionitrile	toluene	80	1.16×10^{-4}			54
	2,2-azo-bis-2-isobutyl-4-methylvaleronitrile	toluene	60.1	$3.78 \pm .18 \times 10^{-4}$			50
			80.5	5.28×10^{-3}			50
	2,2'-azo-bis-2-(4-chlorobenzyl)propionitrile	toluene	80	8.8×10^{-5}			54
	2,2'-azo-bis-2-(4-nitrobenzyl)propionitrile	toluene	80	1.00×10^{-4}			54
22	1,1'-azo-bis-1-cyclodecanenitrile	toluene	50.8	$5.40 \pm .11 \times 10^{-5}$			101
			60.2	$1.70 \pm .03 \times 10^{-4}$			101
			69.5	$5.69 \pm .02 \times 10^{-4}$			101
	2 - MISCELLANEOUS AZO-DERIVATIVES						
5	2,3-diazobicyclo[2,2,1]hept-2-ene	isooctane	131.5	5.44×10^{-6}	157.7		180
			142.3	1.82×10^{-5}			180
			164.1	1.66×10^{-4}			180
			180.8	8.35×10^{-4}			180
		toluene	164.1	1.64×10^{-4}			180
6	2,2'-azo-bis-propane	vapor	250	7.67×10^{-3}	171.1	a	110
			260	1.67×10^{-2}		a	110
			270	3.35×10^{-2}		a	110
			280	6.52×10^{-2}		a	110
			290	1.28×10^{-1}		a	110
	2,2'-dichloro-2,2'-azo-bis-propane	silicone oil	185	4.8×10^{-4}			240
			199	2.1×10^{-3}			240
			221	1.6×10^{-2}			240
	triazobenzene	nitrobenzene	105	3.5×10^{-7}	140.2		45
			115	1.20×10^{-6}			45
			130	6.25×10^{-6}			45
			145	2.50×10^{-5}			45
		tetralin	105	4.0×10^{-7}	136.0		45
			115	1.34×10^{-6}			45
			130	6.01×10^{-6}			45
			145	2.47×10^{-5}			45
8	azo-bis-isobutyramidine	chloroform	60	7.1×10^{-6}			184
		DMSO	70	3.68×10^{-6}			21
		methanol	60	1.45×10^{-6}			184
	azo-bis-isobutyramidine · 2HCl (AIBA)	water	60.0	3.70×10^{-5}	128.9	h	233
			70.0	1.33×10^{-4}			233
			75.0	2.58×10^{-4}			233
			80.0	5.13×10^{-4}			233
8	AIBA	water	40.15	2.49×10^{-6}	122.6	h	237
			50.05	9.79×10^{-6}			237
			60.20	4.03×10^{-5}			237
			70.10	1.52×10^{-4}			237
			100.10	4.96×10^{-3}			237
			70	1.52×10^{-4}			21
	AIBA - kaolin adduct	water	50.0	1.37×10^{-5}	133.9	h	233
			60.0	6.16×10^{-5}			233
			70.0	2.28×10^{-4}			233
			80.0	9.53×10^{-4}			233

Number of C atoms	Initiator	Solvent	T $^{\circ}$C	k_d [s^{-1}]	E_a [kJ mol^{-1}]	Notes	References

MISCELLANEOUS AZO-DERIVATIVES (Cont'd.)

Number of C atoms	Initiator	Solvent	T $^{\circ}$C	k_d [s^{-1}]	E_a [kJ mol^{-1}]	Notes	References
8	AIBA - bentonite adduct	water	50.0	1.80×10^{-5}	126.4	h	233
			55.0	3.97×10^{-5}			233
			65.0	1.53×10^{-4}			233
			70.0	3.21×10^{-4}			233
			80.0	1.04×10^{-3}			233
	azo-bis-isobutyramidine · 2HNO$_3$	DMSO-cumene	60	4.86×10^{-5}			21
			70	1.53×10^{-4}			21
	2,2'-azo-bis-isobutane	vapor	180	5.01×10^{-5}		m$_4$	221
			190	1.53×10^{-4}		m$_4$	221
			210	1.05×10^{-3}		m$_4$	221
		diphenyl ether/iso-quinoline 90/10	165	2.782×10^{-5}	176.6	h	171
			175	8.74×10^{-5}			171
			185	2.513×10^{-4}			171
			190	4.143×10^{-4}			171
			195	6.731×10^{-4}			171
			200	1.093×10^{-3}			171
9	α-phenylethyl-azo-methane	diphenyl ether	151	4.35×10^{-5}	161.5		149
			161	1.16×10^{-4}			149
			171	3.48×10^{-4}			149
		hexadecane	161	1.20×10^{-4}			149
10	2,2'-azo-bis-methyl-2-methylpropionate	collidine	80	1.67×10^{-4}		a	226
		1,2-dichlorobenzene	80	1.44×10^{-4}		a	226
		diethyl oxalate	80	1.52×10^{-4}		a	226
		ethylene glycol	80	1.24×10^{-4}		a	226
		nitrobenzene	80	1.61×10^{-4}	129.3	a	226
			100	1.73×10^{-3}		a	226
		undecane	80	1.04×10^{-4}		a	226
		xylene	80	1.09×10^{-4}	149.8	a	62
	N,N'-azo-piperidine	silicone oil	181	1.7×10^{-3}			240
			228	4.3×10^{-2}			240
	2,2'-diacetoxy-2,2'-azo-propane	silicone oil	258	1.46×10^{-2}			240
	dimethyl-2,2'-azo-isobutyrate	paraffin	99.2	1.02×10^{-3}			240
	2,3,7,8-tetraazahexacyclo-[7.4.1.0^{4,12}.0^{6,11}.0^{10,13}]tetradeca-2,7-diene	acetonitrile-d$_3$	215	1.18×10^{-5}	188.7	x$_1$	174
			230	6.65×10^{-5}		x$_1$	174
			250	3.40×10^{-4}		x$_1$	174
11	α-phenylethyl-azo-2-propane	diphenyl ether/benzo-quinone	143.2	1.50×10^{-4}			220
12	azo-bis-(N,N'-dimethyleneisobutyr-amidine) · 2HNO$_3$	DMSO	60	2.06×10^{-4}			21
			70	6.64×10^{-4}			21
			75	1.08×10^{-3}			21
		DMSO-cumene	70	6.97×10^{-4}			21
		DMSO-tetralin	60	2.13×10^{-4}			21
	azo-bis-(N,N'-dimethyleneisobutyr-amidine)	DMSO	80	1.79×10^{-4}			21
		DMSO-cumene	75	8.04×10^{-5}			21
			80	1.39×10^{-4}			21
			85	2.6×10^{-4}			21
		DMSO-tetralin	60	1.09×10^{-5}			21
	azo-bis-(1-carbomethoxy-3-methylpropane)	benzene	36	3.05×10^{-7}			68
			45	1.31×10^{-6}			68
			55	4.54×10^{-6}			68
			65	1.82×10^{-5}			68
		carbon tetrachloride	45	7.18×10^{-7}			68
			55	3.79×10^{-6}			68
			65	1.02×10^{-5}			68
	4,4'-azo-bis-(4-cyanopentanoic acid) (meso)	DMAC	77.6	1.33×10^{-4}	133.5	a	205
			85.3	3.70×10^{-4}		a	205
			90.9	7.17×10^{-4}		a	205
			99.8	2.02×10^{-3}		a	205

Number of C atoms	Initiator	Solvent	T $^{\circ}$C	k_d [s^{-1}]	E_a [kJ mol^{-1}]	Notes	References
			MISCELLANEOUS AZO-DERIVATIVES (Cont'd.)				
12	4,4'-azo-bis-(4-cyanopentanoic acid) (Cont'd.) $(n_D \underline{+})$	DMAC	77.9	1.53×10^{-4}	134.3	a	205
			85.8	4.11×10^{-4}		a	205
			90.1	7.00×10^{-4}		a	205
			99.1	2.09×10^{-3}		a	205
	$(n_D \,+)$		78.0	1.55×10^{-4}	134.3	a	205
			85.4	4.05×10^{-4}		a	205
			90.0	6.98×10^{-4}		a	205
			99.0	2.00×10^{-3}		a	205
	$(n_D \,-)$		77.7	1.51×10^{-4}	134.3	a	205
			86.0	4.34×10^{-4}		a	205
			90.5	7.33×10^{-4}		a	205
			99.7	2.24×10^{-3}		a	205
	2,2'-dicyclopropyl-2,2'-azopropane	diphenyl ether/isoquinoline 90/10	145.0	7.56×10^{-5}	158.2	h	171
			150.0	1.273×10^{-4}			171
			155.0	2.147×10^{-4}			171
			160.0	3.771×10^{-4}			171
			165.0	6.07×10^{-4}			171
			170.0	1.033×10^{-3}			171
	2,2'-azo-bis-(ethyl-2-methylpropionate)	chlorobenzene	70	4.13×10^{-5}	123.0	a	139
			92.6	5.93×10^{-4}		a	139
		nitrobenzene	80	1.56×10^{-4}		a	226
	azo-bis-isobutanoldiacetate	cyclohexane	170.0	4.86×10^{-5}			189
			180.0	1.27×10^{-4}			189
			189.0	3.47×10^{-4}			189
			199.0	1.01×10^{-3}			189
16	1,1'-azo-bis-1-phenylethane	dodecane	97.3	3.175×10^{-5}			26
		ethylbenzene, toluene	100.4	$5.45 \underline{+} .05 \times 10^{-5}$	136.4		107,98
			110.3	$1.69 \underline{+} .01 \times 10^{-4}$			107,98
		ethylbenzene	105.02	8.473×10^{-5}			152
			105.28	9.02×10^{-5}			183
	$(1,1'-d_2)$		105.28	7.62×10^{-5}			183
	$1,1,1,1',1',1'-d_6)$		105.02	7.623×10^{-5}			152
		N-methyl-N-benzylaniline	97.3	4.135×10^{-5}			26
		N-methylpropionamide	97.3	3.688×10^{-5}			26
		diphenylmethane	97.3	3.995×10^{-5}			26
		propylene carbonate	97.3	3.294×10^{-5}			26
	1,1'-azo-bis-1-chloro-1-phenylethane	acetophenone	74.8	8.6×10^{-4}			240
			83	1.95×10^{-3}			240
		benzophenone	67.8	2.8×10^{-4}			240
			86.2	3.3×10^{-3}			240
			98.2	1.40×10^{-2}			240
		paraffin	94.5	2.4×10^{-3}			240
			112.3	5.2×10^{-2}			240
			120.5	1.03×10^{-1}			240
			129.3	1.6×10^{-1}			240
			137	5.0×10^{-1}			240
		toluene	64	1.8×10^{-4}			240
			74	7.2×10^{-4}			240
			64	1.79×10^{-4}	135.1	a	239
			74	7.20×10^{-4}		a	239
	1,1'-azo-bis-1-chloro-1-(3-bromophenyl)-ethane	toluene	59	1.64×10^{-4}	131.8	a	239
			69	6.64×10^{-4}		a	239
	1,1'-azo-bis-1-chloro-1-(4-bromophenyl)-ethane	toluene	59	1.77×10^{-4}	127.2	a	239
			69	6.75×10^{-4}		a	239
	1,1'-azo-bis-1-chloro-1-(4-chlorophenyl)-ethane	toluene	59	1.97×10^{-4}	118.8	a	239
			69	6.93×10^{-4}		a	239
	1,1'-azo-bis-1,1-dicyclopropylethane	diphenyl ether/isoquinoline 90/10	120	6.33×10^{-5}	149.0	h	171
			130	2.03×10^{-4}			171
			140	6.07×10^{-4}			171
			150	1.695×10^{-3}			171

Number of C atoms	Initiator	Solvent	T °C	k_d [s^{-1}]	E_a [kJ mol^{-1}]	Notes	References
	MISCELLANEOUS AZO-DERIVATIVES (Cont'd.)						
17	3,7-diphenyl-1,2-diaza-1-cycloheptene	xylene	61.0	3.80×10^{-5}			138
			70.0	1.34×10^{-4}			138
			80.0	4.27×10^{-4}			138
			89.3	1.59×10^{-3}			138
			100.2	4.04×10^{-3}			138
			80	4.26×10^{-4}		a	44
			100.2	3.94×10^{-3}		a	44
18	1,1'-azo-bis-cumene	benzene	58.91	1.94×10^{-4}		v5	217
		dodecane	60.23	1.47×10^{-4}			217
		toluene	40.04	9.46×10^{-6}	121.3	h	217
			49.52	4.33×10^{-5}			217
			59.42	1.62×10^{-4}			217
			69.23	5.80×10^{-4}			217
	1,1'-azo-bis-3-chlorocumene	toluene	36.0	8.8×10^{-6}	115.9		200
			42.8	2.46×10^{-5}			200
			48.2	5.55×10^{-5}			200
			61.1	2.796×10^{-4}			200
	1,1'-azo-bis-4-chlorocumene	toluene	36.0	1.11×10^{-5}	112.5		200
			42.8	2.90×10^{-5}			200
			48.2	6.52×10^{-5}			200
	1,1'-azo-bis-4-fluorocumene	toluene	36.0	5.7×10^{-6}	115.9		200
			42.8	1.31×10^{-5}			200
			48.2	2.94×10^{-5}			200
			61.1	1.880×10^{-4}			200
	1,1'-azo-bis-1-chloro-1-(4-tolyl)ethane	acetophenone	74.8	2.1×10^{-3}			240
			79.5	3.0×10^{-3}			240
			86	7.4×10^{-3}			240
		paraffin	105	1.66×10^{-3}			240
			109.5	2.5×10^{-2}			240
			126	6.4×10^{-2}			240
			135.5	1.7×10^{-1}			240
		toluene	59	1.03×10^{-4}			240
			69	4.0×10^{-4}			240
			59	1.03×10^{-4}	126.8	a	239
			69	3.93×10^{-4}			239
	1,1'-azo-bis-1-(4-methoxyphenyl)ethane	ethylbenzene	100.4	$7.15 \pm .1 \times 10^{-5}$	149.8		98
			110.3	$2.48 \pm .02 \times 10^{-4}$			98
	1,1'-azo-bis-1-phenylpropane	ethylbenzene	100.4	$2.35 \pm .1 \times 10^{-5}$	135.1		98
			110.3	$7.2 \pm .2 \times 10^{-5}$			98
	2,2'-azo-bis-2-phenyl-hexafluoropropane	toluene	45.0	8.36×10^{-6}	137.2	h	173
			50.0	2.16×10^{-5}			173
			55.0	4.92×10^{-5}			173
			60.0	8.81×10^{-5}			173
			67.0	2.46×10^{-4}			173
			43.8	6.3×10^{-5}	115.5		42
			53.8	2.4×10^{-4}			42
	3,8-diphenyl-1,2-diaza-1-cyclooctene	tetralin	143	9.52×10^{-5}	153.6		137
			151	2.09×10^{-4}			137
			165	8.77×10^{-4}			137
			173	1.82×10^{-3}			137
19	phenyl-azo-diphenylmethane	decalin	124.5	3.44×10^{-5}	142.3		46
			144.5	2.69×10^{-4}			46
20	1,1'-azo-bis-1-acetoxy-1-phenylethane	paraffin	131	3.2×10^{-2}			240
			142	4.3×10^{-2}			240
			149.5	8.0×10^{-2}			240
	1,1'-azo-bis-1,1-dicyclopropylbutane	diphenyl ether/isoquinoline 90/10	120.0	4.26×10^{-5}	159.0	h	171
			125.0	7.94×10^{-5}			171
			130.0	1.49×10^{-4}			171
			135.0	2.697×10^{-4}			171
			140.0	4.774×10^{-4}			171
			147.0	1.055×10^{-3}			171

Number of C atoms	Initiator	Solvent	T °C	k_d [s^{-1}]	E_a [kJ mol^{-1}]	Notes	References
			MISCELLANEOUS AZO-DERIVATIVES (Cont'd.)				
20	2,2'-azo-bis-2-(4-tolyl)-propane	toluene	36.0	8.4×10^{-6}	110.0		200
			42.8	1.62×10^{-5}			200
			61.1	2.08×10^{-4}			200
	1,1'-azo-bis-1,1,1-tricyclopropylmethane	diphenyl ether/isoquinoline 90/10	105.0	7.89×10^{-5}	143.5	h	171
			110.0	1.374×10^{-4}			171
			115.0	2.36×10^{-4}			171
			120.0	4.58×10^{-4}			171
			125.0	7.76×10^{-4}			171
			130.0	1.38×10^{-3}			171
			135.0	2.39×10^{-3}			171
		decalin/isoquinoline 90/10	118.5	3.60×10^{-4}			171
		isoquinoline	118.5	3.30×10^{-4}			171
		cumene/isoquinoline 90/10	118.5	4.45×10^{-4}			171
22	1,1'-azo-bis-1-phenyl-3-methylbutane	ethylbenzene	100.4	7.6×10^{-5}	139.3		98
			110.3	2.42×10^{-4}			98
25	phenyl-azo-triphenylmethane	acetic acid	43.30	5.7×10^{-5}	117.2		52
			64.00	8.4×10^{-4}			52
		anisole	25.0	2.58×10^{-6}	120.5	b, h	103
			50.2	1.31×10^{-4}		b	103
			74.7	3.03×10^{-3}		b	103
			25.0	3.0×10^{-6}	118.3		114
			74.5	2.9×10^{-3}			114
			80.1	6.8×10^{-3}			114
			85.9	1.1×10^{-2}			114
		benzene	25.0	4.29×10^{-6}	112.1	b, h	103
			49.6	1.24×10^{-4}			103
			74.7	3.12×10^{-3}		b	103
		benzonitrile	25.0	2.62×10^{-6}	121.3	b, h	103
			50.3	1.56×10^{-4}		b	103
			74.7	3.14×10^{-3}			103
		chlorobenzene	25.0	3.77×10^{-6}	118.4	b, h	103
			49.6	1.67×10^{-4}		b	103
			74.7	3.93×10^{-3}		b	103
		cyclohexane	25.0	4.22×10^{-6}	102.5	h	103
			49.6	9.90×10^{-5}		b	103
			74.7	1.75×10^{-3}		b	103
		diethyl malonate	25.1	3.1×10^{-6}	116.6		114
			74.5	2.8×10^{-3}			114
			80.1	5.9×10^{-3}			114
			85.9	1.0×10^{-2}			114
		nitrobenzene	25.0	2.6×10^{-6}	118.7		114
			74.5	3.0×10^{-3}			114
			80.1	5.8×10^{-3}			114
			85.9	9.8×10^{-3}			114
		pyridine	53.35	1.74×10^{-4}	129.7		52
			64.00	8.0×10^{-4}			52
		toluene	43.8	6.3×10^{-5}	115.5		42
			53.8	2.4×10^{-4}			42
			45.45	$8.48 \pm .11 \times 10^{-5}$	122.6		75
			55.55	3.51×10^{-4}			75
			43.30	6.0×10^{-5}	113.0		52
			53.35	2.25×10^{-4}			52
			53.3	2.25×10^{-4}	113.0		46
	3-bromophenyl-azo-triphenylmethane	toluene	53.8	1.14×10^{-4}	125.1		42
			64.0	4.58×10^{-4}			42
	4-bromophenyl-azo-triphenylmethane	toluene	53.35	1.05×10^{-4}	117.2		52
			64.30	4.28×10^{-4}			52
	4-hydroxyphenyl-azo-triphenylmethane	acetic acid	54.00	1.42×10^{-4}	133.9		52
			64.00	6.2×10^{-4}			52
		pyridine	54.00	1.52×10^{-4}	133.9		52

Number of C atoms	Initiator	Solvent	T °C	k_d [s^{-1}]	E_a [kJ mol^{-1}]	Notes	References
		MISCELLANEOUS AZO-DERIVATIVES (Cont'd.)					
25	4-hydroxyphenyl-azo-triphenylmethane (Cont'd.)	pyridine	64.00	6.7×10^{-4}			52
		toluene	54.00	1.70×10^{-4}	121.3		52
			64.00	6.4×10^{-4}			52
	2-nitrophenyl-azo-9-phenylfluorene	toluene	45.45	1.01×10^{-4}	119.7		75
			55.55	$3.71 \pm .04 \times 10^{-4}$			75
	4-nitrophenyl-azo-9-phenylfluorene	toluene	45.45	$2.06 \pm .02 \times 10^{-4}$	111.7		75
			55.55	$7.53 \pm .02 \times 10^{-4}$			75
	2,4-dinitrophenyl-azo-9-phenylfluorene	toluene	55.55	$2.06 \pm .01 \times 10^{-4}$	119.2		75
			64.94	$6.92 \pm .04 \times 10^{-4}$			75
	2-nitrophenyl-azo-triphenylmethane	toluene	64.94	$1.46 \pm .01 \times 10^{-4}$	123.8		75
			75.06	$5.26 \pm .08 \times 10^{-4}$			75
	3-nitrophenyl-azo-triphenylmethane	toluene	53.8	5.8×10^{-5}	110.9		42
			64.0	1.99×10^{-4}			42
	4-nitrophenyl-azo-triphenylmethane	decane	50	2.017×10^{-5}			168
			60	9.988×10^{-5}			168
			70	3.350×10^{-4}			168
			77.5	1.16×10^{-3}			168
		1-decene	60	1.038×10^{-4}			168
		dodecane	50	1.933×10^{-5}			168
			60	9.166×10^{-5}			168
			70	3.250×10^{-4}			168
		1-eicosene	60	7.60×10^{-5}			168
		heptane	50	2.716×10^{-5}			168
			60	1.043×10^{-4}			168
			70	3.900×10^{-4}			168
		hexadecane	50	1.716×10^{-5}			168
			60	8.133×10^{-5}			168
			70	2.966×10^{-4}			168
			77.5	1.09×10^{-3}			168
		1-hexadecene	60	8.25×10^{-5}			168
		hexane	50	2.750×10^{-5}			168
			60	1.195×10^{-4}			168
		1-hexene	60	1.143×10^{-4}			168
		nonane	50	2.200×10^{-5}			168
			60	9.900×10^{-5}			168
			70	3.710×10^{-4}			168
		octadecane	60	7.780×10^{-5}			168
			70	2.783×10^{-4}			168
		1-octadecene	60	7.58×10^{-5}			168
		octane	60	1.015×10^{-4}			168
			70	3.650×10^{-4}			168
			77.5	1.28×10^{-3}			168
		1-octene	60	1.123×10^{-4}			168
		pentane	50	3.600×10^{-5}			168
			60	1.280×10^{-4}			168
		tetradecane	50	1.866×10^{-5}			168
			60	8.800×10^{-5}			168
			70	3.116×10^{-4}			168
		1-tetradecene	60	9.02×10^{-5}			168
		toluene	64.94	$2.58 \pm .04 \times 10^{-4}$	123.0		75
			75.06	$9.19 \pm .04 \times 10^{-4}$			75
			53.35	5.7×10^{-5}	113.0		52
			64.30	2.25×10^{-4}			52
	2,4-dinitrophenyl-azo-triphenylmethane	toluene	75.06	$1.90 \pm .02 \times 10^{-4}$	122.6		75
			84.98	$6.13 \pm .03 \times 10^{-4}$			75
26	1,1'-azo-bis-1,2-diphenylethane (meso)	ethylbenzene	96.56	2.75×10^{-5}	138.5	a, h	162
			106.47	9.03×10^{-5}		a	162
			115.28	2.42×10^{-4}		a	162
	(D, L)		96.56	3.22×10^{-5}	138.9	a, h	162
			106.47	1.04×10^{-4}		a	162
			115.28	2.84×10^{-4}		a	162

DECOMPOSITION RATES OF INITIATORS

Number of C atoms	Initiator	Solvent	T °C	k_d [s^{-1}]	E_a [kJ mol^{-1}]	Notes	References
	MISCELLANEOUS AZO-DERIVATIVES (Cont'd.)						
26	azo-bis-diphenylmethane	toluene	64.0	3.40×10^{-4}	111.3		46,107
			54.0	1.01×10^{-4}			107
	1,1'-azo-bis-1-(4-tolyl)-cyclohexane	toluene	36.0	8.7×10^{-6}	107.1		200
			42.8	2.01×10^{-5}			200
			61.1	2.009×10^{-4}			200
	4-methoxyphenyl-azo-triphenylmethane	toluene	54.00	2.13×10^{-4}	117.2		52
			64.00	7.6×10^{-4}			52
	3-tolyl-azo-triphenylmethane	toluene	43.8	$7.4 \pm .2 \times 10^{-5}$	113.8		42
			53.8	$2.77 \pm .1 \times 10^{-4}$			42
	4-tolyl-azo-triphenylmethane	toluene	43.30	$6.9 \pm .1 \times 10^{-5}$	100.4		52
			53.35	$2.25 \pm .05 \times 10^{-4}$			52
27	4-acetaminophenyl-azo-triphenylmethane	toluene	54.00	1.46×10^{-4}	125.5		52
			64.00	5.9×10^{-4}			52
40	3,10,13,20-tetraphenyl-1,2,11,12-tetra-aza-1,11-cycloeicosadiene	xylene	110	$7.20 \pm .35 \times 10^{-5}$	145.6		38
			120	$2.30 \pm .09 \times 10^{-4}$			38
			130	$6.90 \pm .15 \times 10^{-4}$			38
44	3,12,15,24-tetraphenyl-1,2,13,14-tetra-aza-1,13-cyclotetracosadiene	ethylbenzene	112.95	8.761×10^{-5}			158
		xylene	110.8	8.9×10^{-5}	126.4		48
			119.8	2.2×10^{-4}			48
			110	$7.63 \pm .4 \times 10^{-5}$	143.9		38
			120	$2.46 \pm .12 \times 10^{-4}$			38
			130	$7.12 \pm .22 \times 10^{-4}$			38
	3 - ALKYL PEROXIDES						
2	methyl peroxide	vapor	T[K]	$1.6 \times 10^{15} \exp[-147.7\ \text{kJ/RT}]$			72
		methanol	T[K]	$4.1 \times 10^{15} \exp[-153.9\ \text{kJ/RT}]$			225
4	ethyl peroxide	styrene	60	1.2×10^{-9}	147.3		92
		vapor	140.2	1.75×10^{-4}	131.8	a	127
			147.8	3.60×10^{-4}		a	127
			160.0	1.08×10^{-3}		a	127
			176.5	3.78×10^{-3}		a	127
			184.5	7.16×10^{-3}		a	127
			145.9	6.69×10^{-4}			187
			145.9	5.02×10^{-4}		t_8	187
		vapor (toluene)	200	3.58×10^{-2}	132.6		87
			210	6.76×10^{-2}			87
			218	1.47×10^{-1}			87
			226	2.23×10^{-1}			87
			234	3.86×10^{-1}			87
			245	6.43×10^{-1}			87
6	propyl peroxide	vapor	146.5	2.50×10^{-4}	132.2	a	128
			155.3	6.00×10^{-4}		a	128
			166.8	1.95×10^{-3}		a	128
			175.4	4.10×10^{-3}		a	128
	isopropyl peroxide	styrene	60	$6. \times 10^{-10}$	154.8		92
8	tert-butyl peroxide	acetic acid	115	1.2×10^{-5}	139.7	h	192
			120	2.19×10^{-5}			192
			125	2.98×10^{-5}			192
			130	6.29×10^{-5}			192
		acetonitrile	95	9.53×10^{-7}			208
			125	3.89×10^{-5}		d	208
			115	1.19×10^{-5}	129.7	h	192
			120	2.21×10^{-5}			192
			125	3.47×10^{-5}			192
			130	5.63×10^{-5}			192
		benzene	80	7.81×10^{-8}	142.3	m_2, u_2	129
			130	$2.48\text{-}3.04 \times 10^{-5}$		m_2, u_2	129
			100	8.8×10^{-7}	146.9	a	126
			100	8.75×10^{-7}			197
			115	5.66×10^{-6}		a	126

Notes page II-40; References page II-41

Number of C atoms	Initiator	Solvent	T °C	k_d [s^{-1}]	E_a [kJ mol^{-1}]	Notes	References
		ALKYL PEROXIDES (Cont'd.)					
8	tert-butyl peroxide (Cont'd.)	benzene	130	3.00×10^{-5}		a	126
			120	1.39×10^{-5}		z(0.98)	33
			120	7.6×10^{-6}		z(864)	33
			120	7.5×10^{-6}		z(1620)	33
			120	3.7×10^{-6}		z(3480)	33
			120	1.10×10^{-5}	147.7	h	192
			125	1.99×10^{-5}			192
			130	3.22×10^{-5}			192
			135	6.19×10^{-5}			192
		benzhydrol	125	8.7×10^{-5}			140
		tert-butanol	120	1.41×10^{-5}	143.5	h	192
			125	2.49×10^{-5}			192
			130	4.30×10^{-5}			192
			135	7.32×10^{-5}			192
		2-butanol	125	4.8×10^{-5}			140
		n-butyl mercaptan	125	1.5×10^{-5}			140
		tri-n-butylamine	125	$1.7 \pm .3 \times 10^{-5}$			64
			135	$4.2 \pm .4 \times 10^{-5}$			64
			145	$1.60 \pm .21 \times 10^{-4}$			64
			T[K]	$2.8 \times 10^{14} \exp[-146.4 \text{ kJ/RT}]$			123
		tert-butylbenzene	125	$1.5 \pm .2 \times 10^{-5}$			64
			135	$5.0 \pm .3 \times 10^{-5}$			64
			145	$1.51 \pm .22 \times 10^{-4}$			64
			T[K]	$2.8 \times 10^{14} \exp[-146.4 \text{ kJ/RT}]$			123
		carbon tetrachloride	120	9×10^{-6}		z(0.98)	33
			120	2.4×10^{-6}		z(1930)	33
			120	2.3×10^{-6}		z(2890)	33
			120	8.6×10^{-7}		z(5525)	33
		cumene	125	$1.6 \pm .1 \times 10^{-5}$			64
			135	$5.2 \pm .3 \times 10^{-5}$			64
			145	$1.56 \pm .13 \times 10^{-4}$			64
			T[K]	$2.8 \times 10^{14} \exp[-146.4 \text{ kJ/RT}]$			123
		cyclohexane	95	2.48×10^{-7}	170.7	h	208
			120	6.3×10^{-6}			192
			125	1.52×10^{-5}			192
			130	2.59×10^{-5}			192
			135	4.64×10^{-5}			192
		cyclohexanol	125	2.4×10^{-5}			140
		cyclohexene	120	7.6×10^{-6}	156.1	h	192
			125	1.38×10^{-5}			192
			130	2.81×10^{-5}			192
			135	4.41×10^{-5}			192
			120	8.3×10^{-6}		z(0.98)	33
			120	6.2×10^{-6}		z(1275)	33
			120	3.77×10^{-6}		z(2890)	33
			120	2.65×10^{-6}		z(5725)	33
		cyclohexylamine	125	5.50×10^{-5}	157.3	h	143
		dimethylaniline	120	9.6×10^{-6}			192
			125	1.89×10^{-5}			192
			130	3.41×10^{-5}			192
			135	5.84×10^{-5}			192
		ethyl benzoate	120	1.07×10^{-5}	148.5	h	192
			125	1.92×10^{-5}			192
			130	3.39×10^{-5}			192
			135	5.90×10^{-5}			192
		N-ethylcyclohexylamine	125	4.01×10^{-5}			143
		2-methyl-2-butanol	120	1.26×10^{-5}	149.4	h	192
			125	2.34×10^{-5}			192
			130	4.47×10^{-5}			192
			135	6.80×10^{-5}			192

Number of C atoms	Initiator	Solvent	T °C	k_d [s^{-1}]	E_a [kJ mol^{-1}]	Notes	References
		ALKYL PEROXIDES (Cont' d.)					
8	tert-butyl peroxide (Cont' d.)	methyl methacrylate	T[K]	$2.8 \times 10^{14} \exp[-146.4 \text{ kJ/RT}]$			123
		N-methylpiperidine	125	1.54×10^{-5}			143
		norbornanol	125	4.4×10^{-5}			140
		piperidine	125	3.49×10^{-5}			143
		2-octanol	125	5.5×10^{-5}			140
		1-propanol	125	2.8×10^{-5}			140
		styrene	T[K]	$2.8 \times 10^{14} \exp[-146.4 \text{ kJ/RT}]$			123
		tetrahydrofuran	120	9.7×10^{-6}	155.2	h	192
			125	1.84×10^{-5}			192
			130	3.39×10^{-5}			192
			135	5.76×10^{-5}			192
			125	1.5×10^{-5}			140
		toluene	100	6.82×10^{-7}			197
			120	1.34×10^{-5}		z(0.98)	33
			120	9.5×10^{-6}		z(2000)	33
			120	8.0×10^{-6}		z(2850)	33
			120	5.7×10^{-6}		z(5170)	33
			125	1.6×10^{5}			140
			125	1.62×10^{-5}			143
		triethylamine	120	7.9×10^{-6}	169.9	h	192
			125	1.69×10^{-5}			192
			130	3.15×10^{-5}			192
			135	5.55×10^{-5}			192
		vapor (acetone)	127.5	7.4×10^{-6}			238
			131	1.13×10^{-5}			238
			146	6.0×10^{-5}			238
			151.5	1.03×10^{-4}			238
			162	3.6×10^{-4}			238
			167	5.2×10^{-4}			238
			145	1.3×10^{-4}	165.3		125
			T[K]	$5.9 \times 10^{16} \exp[-165.3 \text{ kJ/RT}]$			125
		vapor (carbon tetrachloride)	150	8.58×10^{-4}		l	130
		vapor (chloroform)	150	1.167×10^{-3}		l	130
		vapor (dichlorodifluoro-methane)	150	$>2.00 \times 10^{-4}$			130
		vapor (trichloroethylene)	150	2.35×10^{-3}		l	130
		vapor (methylene chloride)	150	1.017×10^{-3}		l	130
		vapor (3-pentanone)	145	1.5×10^{-4}	165.3		125
			T[K]	$6.8 \times 10^{16} \exp[-165.3 \text{ kJ/RT}]$			125
		vapor (silicon tetrafluoride)	160	1.05×10^{-3}	113.0	a, i(2.6)	117
			160	1.25×10^{-3}		a, i(33)	117
			160	1.27×10^{-3}		a, i(100)	117
		vapor (toluene)	148	9.0×10^{-5}			88
			158	2.5×10^{-4}			88
		vapor	103.2	5.6×10^{-7}			227
			111.9	2.03×10^{-6}			227
			120.2	6.39×10^{-6}			227
			129.5	1.98×10^{-5}			227
			138.5	6.00×10^{-5}			227
			145.4	1.24×10^{-4}			227
			125	1.1×10^{-5}			64
			135	3.6×10^{-5}			64
			145	1.15×10^{-4}			64
			129.6	1.64×10^{-5}			86
			141.0	6.28×10^{-5}			86
			152.5	2.25×10^{-4}			86
			166.8	8.92×10^{-4}			86
			130	1.82×10^{-5}	159.0	a	96
			140	5.75×10^{-5}		a	96
			150	1.75×10^{-4}		a	96

Number of C atoms	Initiator	Solvent	T °C	k_d [s^{-1}]	E_a [kJ mol^{-1}]	Notes	References

ALKYL PEROXIDES (Cont'd.)

Number of C atoms	Initiator	Solvent	T °C	k_d [s^{-1}]	E_a [kJ mol^{-1}]	Notes	References
8	tert-butyl peroxide (Cont'd.)	vapor	160	4.88×10^{-4}		a	96
			170	1.35×10^{-3}		a	96
			129	1.97×10^{-5}			88
			138	4.3×10^{-5}			88
			149	1.30×10^{-4}			88
			152	1.62×10^{-4}			88
			139.7	6.0×10^{-5}	163.6		63
			147.2	1.43×10^{-4}			63
			154.6	3.22×10^{-4}			63
			159.8	5.53×10^{-4}			63
			T[K]	$3.2 \times 10^{16} \exp[-163.6 \text{ kJ/RT}]$			63
			145	1.3×10^{-4}	161.5		125
			T[K]	$1.9 \times 10^{16} \exp[-161.5 \text{ kJ/RT}]$			175,125
			149.5	$1.79 \pm .06 \times 10^{-4}$	156.5	i(37-132)	83
			160	4.00×10^{-4}	154.8	a, i(2.6)	117
			160	4.53×10^{-4}		a, i(33)	117
			160	4.83×10^{-4}		a, i(100)	117
		vapor (He)	280	7.7	154.8	i(10)	111
			290	1.51×10^{1}		i(10)	111
			300	2.77×10^{1}		i(10)	111
			310	4.87×10^{1}		i(10)	111
			320	8.34×10^{1}		i(10)	111
			330	1.38×10^{2}		i(10)	111
			340	2.13×10^{2}		i(10)	111
			350	3.22×10^{2}		i(10)	111
		KBr pellets	109	3.8×10^{-7}	159.0	h	241
			127	2.96×10^{-5}			241
			149	4.4×10^{-4}			241
	sec-butyl peroxide	toluene	100	2.7×10^{-6}			197
		vapor	100	1.5×10^{-6}			197
	butyl peroxide	styrene	60	3.3×10^{-9}	142.3		92
	1-hydroxybutyl-n-butyl peroxide	α-methylstyrene	79.4	1.7×10^{-5}	102.9		32
			99.4	1.06×10^{-4}			32
			109.9	2.9×10^{-4}			32
	1-hydroxyisobutyl-isobutyl peroxide	α-methylstyrene	79.4	3.7×10^{-5}	83.7	c	32
			99.4	2.0×10^{-4}		c	32
			109.9	4.8×10^{-4}		c	32
	1-hydroxyisobutyl-1-d-isobutyl-1,1-d$_2$ peroxide	α-methylstyrene	99.5	6.9×10^{-5}	89.5		25
			109.8	1.8×10^{-4}			25
			122.0	4.2×10^{-4}			25
10	tert-amyl peroxide	bulk	125	5.7×10^{-5}			202
			132.2	1.15×10^{-4}			202
		decalin	125	2.8×10^{-5}			202
		octane	125	3.0×10^{-5}			202
		triethylamine	125	3.5×10^{-5}			202
		vapor	132.2	7.2×10^{-5}	154.8-171.5	i(200-225)	63
			136.7	1.15×10^{-4}		i(200-225)	63
			142.2	2.16×10^{-4}		i(200-225)	63
			149.2	4.8×10^{-4}		i(100-200)	63
			136.7	1.34×10^{-4}		i(440-610)	63
			142.2	2.41×10^{-4}		i(440-610)	63
			149.7	5.61×10^{-4}		i(440-610)	63
13	tert-butyl-α-cumyl peroxide	tert-butylbenzene	138	1.48×10^{-4}		a	80
			158	9.62×10^{-4}		a	80
		cumene	138	1.44×10^{-4}		a	80
			158	8.88×10^{-4}		a	80
		dodecane	128	4.44×10^{-5}	146.4	a	80
			138	1.39×10^{-4}		a	80
			148	3.21×10^{-4}		a	80
			158	8.88×10^{-4}		a	80

DECOMPOSITION RATES OF INITIATORS

Number of C atoms	Initiator	Solvent	T °C	k_d [s^{-1}]	E_a [kJ mol^{-1}]	Notes	References

ALKYL PEROXIDES (Cont'd.)

Number of C atoms	Initiator	Solvent	T °C	k_d [s^{-1}]	E_a [kJ mol^{-1}]	Notes	References
16	2,5-dimethyl-2,5-di(tert-butylperoxy)-hexane	benzene	115	1.15×10^{-5}	166.9	a	126
			130	6.86×10^{-5}		a	126
			145	4.75×10^{-4}		a	126
	2,5-dimethyl-2,5-di(tert-butylperoxy)-3-hexyne	benzene	115	3.91×10^{-6}	156.9	a	126
			130	2.35×10^{-5}		a	126
			145	1.14×10^{-4}		a	126
			160	6.17×10^{-4}		a	126
17	n-butyl-4,4-bis(tert-butylperoxy)-valerate	dodecane	100	5.83×10^{-6}		a	8
			115	3.53×10^{-5}		a	8
			130	2.91×10^{-4}		a	8
18	cumyl peroxide	benzene	115	1.56×10^{-5}	170.3	a	126
			130	1.05×10^{-4}		a	126
			145	6.86×10^{-4}		a	126
		tert-butylbenzene	158	1.72×10^{-3}		a	80
		cumene	T[K]	$4.31 \times 10^{14} \exp[-144.3\,\text{kJ}/RT]$			67
			138	2.57×10^{-4}		a	80
			158	1.52×10^{-3}		a	80
		dodecane	100	0.75×10^{-5}		a	80
			138	2.31×10^{-4}		a	80
			148	5.37×10^{-4}		a	80
			158	1.83×10^{-3}		a	80
		diisopropylcarbinol	138	3.16×10^{-4}		a	80

4 - ACYL PEROXIDES

Number of C atoms	Initiator	Solvent	T °C	k_d [s^{-1}]	E_a [kJ mol^{-1}]	Notes	References
4	acetyl peroxide	acetic acid	55.2	2.8×10^{-6}	126.4	d, e	106,176
			64.9	9.9×10^{-6}		d, e	106,176
			75.2	3.75×10^{-5}		d, e	106,176
			85.2	1.30×10^{-4}		d, e	106,176
			73.2	2.62×10^{-5}		a	100
		benzene	35	9.5×10^{-7}			68
			55	3.14×10^{-6}			68
			65	1.27×10^{-5}			68
			50	1.22×10^{-6}	136.0	a	126
			70	2.39×10^{-5}		a	126
			85	1.73×10^{-4}		a	126
			70	2.38×10^{-5}		a	16
			60.3	5.0×10^{-6}			4
			80	8.7×10^{-5}			20
			55.2	2.6×10^{-6}	135.1	c, e	106,176
			64.9	1.07×10^{-5}		c, e	106,176
			75.2	4.65×10^{-5}		c, e	106,176
			85.2	1.62×10^{-4}		c, e	106,176
		n-butanol	60.3	3.4×10^{-5}			4
		sec-butanol	60.3	3×10^{-5}			4
		tert-butanol	60.3	3.1×10^{-6}	133.9		4
			80.3	4.9×10^{-5}			4
		carbon tetrachloride	26	1.08×10^{-7}			68
			46	4.84×10^{-7}			68
			65	2.11×10^{-6}			68
			80	5.5×10^{-5}			20
		chloroform	80.3	~ 5			4
		cyclohexane	55.2	2.1×10^{-6}	131.4	c, e	106,176
			64.9	8.3×10^{-6}		c, e	106,176
			75.2	3.60×10^{-5}		c, e	106,176
			85.2	1.27×10^{-4}		c, e	106,176
		cyclohexene	60	4.5×10^{-6}	133.5		20
			70	1.77×10^{-5}			20
			80	7.0×10^{-5}			20
			90	2.28×10^{-4}			20
			100	7.61×10^{-4}			20

ACYL PEROXIDES (Cont'd.)

Number of C atoms	Initiator	Solvent	T °C	k_d [s^{-1}]	E_a [kJ mol^{-1}]	Notes	References
4	acetyl peroxide (Cont'd.)	cyclopentene	70	1.60×10^{-5}	137.2		20
			80	7.0×10^{-5}			20
			90	2.55×10^{-4}			20
			100	7.25×10^{-4}			20
		cumene	80	7.6×10^{-5}			20
		decane	80	6.85×10^{-5}			168
		n-dodecane	60	2.3×10^{-6}			70
			80	6.15×10^{-5}			168
		ethanol	60.3	1.01×10^{-4}	129.7		4
			80.3	1.40×10^{-3}			4
		heptane	80	7.72×10^{-5}			168
		hexadecane	80	5.39×10^{-5}			168
		n-hexane	60	3.4×10^{-6}			70
		1-hexene	70	2.35×10^{-5}	132.6		20
			80	8.7×10^{-5}			20
			90	3.05×10^{-4}			20
			100	9.83×10^{-4}			20
		2-methyl-1-pentene	80	9.0×10^{-5}	126.8		20
			90	3.12×10^{-4}			20
			100	9.81×10^{-4}			20
		n-octadecane	60	1.9×10^{-6}			70
		n-octane	60	2.9×10^{-6}			70
			80	7.34×10^{-5}			168
		isooctane	60	2.9×10^{-6}			70
			55.2	2.35×10^{-6}	134.7	c, e	106,176
			64.9	9.4×10^{-6}		c, e	106,176
			75.2	4.03×10^{-5}		c, e	106,176
			85.2	1.49×10^{-4}		c, e	106,176
		1-pentene	70	2.45×10^{-5}			20
			80	9.4×10^{-5}			20
			90	3.22×10^{-4}			20
		propionic acid	64.9	1.4×10^{-5}			106,176
			85.2	1.66×10^{-4}		d, e	106,176
		n-tetradecane	60	2.0×10^{-6}			70
			80	5.90×10^{-5}			168
		toluene	60.3	5×10^{-6}	129.7		4
			55.2	2.7×10^{-6}	133.9	c, e	106,176
			64.9	1.14×10^{-5}		c, e	106,176
			75.2	4.70×10^{-5}		c, e	106,176
			85.2	1.59×10^{-4}		c, e	106,176
			73.2	3.06×10^{-5}	138.1	a	100
			85.5	1.72×10^{-4}		a	100
			80	7.33×10^{-5}	129.7	a	113
		vapor (toluene)	88.0	3.12×10^{-4}	123.4		105
			134.7	3.1×10^{-2}			105
			150.7	1.18×10^{-1}			105
			161.7	2.77×10^{-1}			105
			170.7	6.10×10^{-1}			105
			184.2	1.76			105
6	propionyl peroxide	acetic acid	65.0	3.8×10^{-5}	123.0	c, e	51
			85.0	4.3×10^{-4}		c, e	51
		acetic anhydride	65.0	3.5×10^{-5}	128.9	d, e	51
			85.0	4.5×10^{-4}		c, e	51
		benzene	65.0	1.88×10^{-5}	129.3	c, e	51
			85.0	2.40×10^{-4}		c, e	51
			50	2.72×10^{-6}	127.6	a	126
			70	4.30×10^{-5}		a	126
			85	2.89×10^{-4}		a	126
		benzonitrile	65.0	3.9×10^{-5}	130.5	d, e	51
			85.0	5.1×10^{-4}		d, e	51

Number of C atoms	Initiator	Solvent	$T\ ^{\circ}C$	$k_d\ [s^{-1}]$	$E_a\ [kJ\ mol^{-1}]$	Notes	References
		ACYL PEROXIDES (Cont' d.)					
6	propionyl peroxide (Cont' d.)	dioxane	65.0	4.5×10^{-5}	116.7	c, e	51
			85.0	4.5×10^{-4}		c, e	51
		n-hexane	65.0	1.50×10^{-5}	123.8		51
			85.0	1.72×10^{-4}			51
		isooctane	65.0	9.8×10^{-6}	130.5	c, e	51
			86.5	1.44×10^{-4}		c, e	51
		nitrobenzene	65.0	3.7×10^{-5}	120.9	c, e	51
			85.0	4.1×10^{-4}			51
		toluene	65.0	1.87×10^{-5}	130.1		51
			85.0	2.54×10^{-4}			51
		vapor	65.0	1.0×10^{-5}	125.5		51
			85.0	1.6×10^{-4}			51
			99.4	8×10^{-4}	125.5		85
			134.4	2.6×10^{-2}			85
			152.2	1.22×10^{-1}			85
			176.4	8.0×10^{-1}			85
			190,9	2,33			85
			T[K]	$2.5 \times 10^{14}\ exp[-125.5kJ/RT]$			85
	2-iodopropionyl peroxide	acetone	56	2.19×10^{-4}		s	77
		benzene	62.5	$2.40\text{-}2.81 \times 10^{-4}$		s	77
			62.5	7.12×10^{-4}		n, s	77
			62.5	$2.36 \pm .07 \times 10^{-4}$	108.8	p	77
		n-butyl vinyl ether	62.5	2.47×10^{-4}		s	77
		cyclohexene	62.5	2.7×10^{-4}		s	77
		95% ethanol	62.5	4.0×10^{-4}		s	77
8	butyryl peroxide	acetic acid	65.0	4.7×10^{-5}	125.1	c, e	51
			85.0	5.6×10^{-4}		c, e	51
		acetic anhydride	65.0	4.3×10^{-5}			51
			85.0	5.5×10^{-4}		c, e	51
		benzene	65.0	2.24×10^{-5}	131.4	d, e	51
			85.0	3.02×10^{-4}		c, e	51
		benzonitrile	65.0	4.3×10^{-5}	131.4	d, e	51
			85.0	5.8×10^{-4}			51
		dioxane	65.0	4.6×10^{-5}	116.3	c, e	51
			85.0	4.6×10^{-4}		c, e	51
		hexane	65.0	1.14×10^{-5}	131.4	c, e	51
			85.0	1.53×10^{-4}		c, e	51
		isooctane	65.0	1.11×10^{-5}	133.5	c, e	51
			85.0	1.56×10^{-4}		c, e	51
		toluene	65.0	2.14×10^{-5}	130.5	d, e	51
			85.0	2.87×10^{-4}		c, e	51
		vapor	65.0	1.6×10^{-5}	123.8		51
			85.0	2.0×10^{-4}			51
			96.7	8.6×10^{-4}	123.8		85
			127.4	1.5×10^{-2}			85
			158.9	3.0×10^{-1}			85
			178.9	1.27			85
			T[K]	$1.9 \times 10^{14}\ exp[-123.8\ kJ/RT]$			85
	isobutyryl peroxide	acetonitrile	40	6.81×10^{-4}		m_3	167
		benzene	40	2.38×10^{-4}		m_3	167
			40	2.40×10^{-4}			148
		benzonitrile	40	4.2×10^{-4}			148
		tert-butanol	40	2.51×10^{-4}			148
		carbon tetrachloride	40	7.8×10^{-5}			148
			45	1.58×10^{-4}			148
			50	3.05×10^{-4}			148
			55	5.61×10^{-4}			148
			60	7.67×10^{-4}			148
		chlorobenzene	40	1.73×10^{-4}			148
		chloroform	40	7.5×10^{-5}			148

Notes page II-40; References page II-41

Number of C atoms	Initiator	Solvent	T °C	k_d [s^{-1}]	E_a [kJ mol^{-1}]	Notes	References
				ACYL PEROXIDES (Cont'd.)			
8	isobutyryl peroxide (Cont'd.)	cyclohexane	40	4.5×10^{-5}			148
			40	4.70×10^{-5}			167
		fluorobenzene	40	1.23×10^{-4}			148
		isooctane	25	$3.35 \pm .03 \times 10^{-6}$	114.2		84
			35	$1.54 \pm .02 \times 10^{-5}$			84
			45	$6.14 \pm .02 \times 10^{-5}$			84
			55	2.26×10^{-4}			84
			T[K]	2.8×10^{14} exp [-114.2 kJ/RT]			84
			40	3.2×10^{-5}			148
		isopropanol	40	3.05×10^{-5}			148
		nitrobenzene	40	5.80×10^{-4}			148
		nujol	40	4.63×10^{-5}			167
		tetralin	40	1.75×10^{-4}			148
		toluene	40	1.43×10^{-4}			148
		vapor	40	$\sim 1 \times 10^{-5}$			167
		p-xylene	40	1.40×10^{-4}			148
	cyclopropane formyl peroxide	carbon tetrachloride	64.5	4.4×10^{-6}			34
			70.4	9.3×10^{-6}			34
			77.8	2.31×10^{-5}			34
	diacetyl succinoyl diperoxide	styrene	60	5.2×10^{-6}	125.5		215
			73.5	2.3×10^{-5}			215
			85	9.3×10^{-5}			215
	succinoyl peroxide	acetone	70	2.80×10^{-5}	99.6	a	126
			85	1.21×10^{-4}		a	126
			100	4.36×10^{-4}		a	126
9	acetyl benzoyl peroxide	chlorobenzene	70	2×10^{-5}			155
10	5-bromo-2-thenoyl peroxide	carbon tetrachloride	75	1.53×10^{-5}		a, m_1	232
	4-bromo-2-thenoyl peroxide	carbon tetrachloride	75	1.14×10^{-5}		a, m_1	232
	5-chloro-2-thenoyl peroxide	carbon tetrachloride	75	1.58×10^{-5}		a, m_1	232
	α-chloropropionyl m-chlorobenzoyl peroxide	acetonitrile	41	3.05×10^{-5}		m_3	167
		cyclohexane	41	1.51×10^{-5}		m_3	167
	cyclobutane formyl peroxide	carbon tetrachloride	65	5.15×10^{-5}			34
			70	$8.95, 6.63 \times 10^{-5}$			34
			75	1.41×10^{-4}			34
	cyclopropane acetyl peroxide	carbon tetrachloride	14	9.45×10^{-5}			34
			25	$9.75, 10.57 \times 10^{-5}$			34
			44.5	$5.01 \pm .23 \times 10^{-4}$	101.7		24
			56.5	$2.64 \pm .01 \times 10^{-4}$			24
			44.5	$6.5-8.0 \times 10^{-4}$		n	24
	diacetyladipoyl diperoxide	styrene	60	6.6×10^{-6}			215
			73.5	4.73×10^{-5}			215
			85	1.84×10^{-4}			215
	2-thenoyl peroxide	carbon tetrachloride	75	2.21×10^{-5}		a, m_1	232
	3-thenoyl peroxide	carbon tetrachloride	75	2.14×10^{-5}		a, m_1	232
11	benzoyl isobutyryl peroxide	acetonitrile	41	4.06×10^{-4}			167
		cyclohexane	41	1.63×10^{-5}			167
			70	3.05×10^{-4}			167
			40	1.45×10^{-5}	112.5	h, j	164
			50	5.398×10^{-5}		j	164
			60	1.924×10^{-4}		j	164
			70	6.872×10^{-4}		j	164
	m-chlorobenzoyl isobutyryl peroxide	acetonitrile	41	1.03×10^{-3}			167
		cyclohexane	40	3.486×10^{-5}	111.3	h, j	164
			50	1.362×10^{-4}		j	164
			60	4.378×10^{-4}		j	164
			41	4.40×10^{-5}			167
	p-chlorobenzoyl isobutyryl peroxide	cyclohexane	50	1.029×10^{-4}		j	164
			55	1.651×10^{-4}			164
			60	3.21×10^{-4}			164
			65	5.638×10^{-4}			164

DECOMPOSITION RATES OF INITIATORS

Number of C atoms	Initiator	Solvent	T °C	k_d [s^{-1}]	E_a [kJ mol^{-1}]	Notes	References
			ACYL PEROXIDES (Cont'd.)				
11	p-fluorobenzoyl isobutyryl peroxide	cyclohexane	55	1.117×10^{-4}			164
			60	2.038×10^{-4}			164
			65	4.893×10^{-4}			164
			70	9.603×10^{-4}			164
	5-methyl-bis-2-thenoyl peroxide	carbon tetrachloride	75	2.92×10^{-5}		a	232
	p-nitrobenzoyl isobutyryl peroxide	cyclohexane	40	8.1×10^{-5}			164
			45	1.336×10^{-4}			164
			50	2.889×10^{-4}			164
			55	4.725×10^{-4}			164
			60	8.921×10^{-4}			164
12	β-allyloxypropionyl peroxide	toluene	70	2.01×10^{-5}		a	191
			80	8.62×10^{-5}		a	191
			90	2.53×10^{-4}		a	191
		p-xylene	70	2.32×10^{-5}		a	191
			80	8.88×10^{-5}		a	191
			90	2.95×10^{-4}		a	191
	cyclobutane acetyl peroxide	carbon tetrachloride	65	1.37×10^{-5}			34
			70	$2.13, 3.08 \times 10^{-5}$			34
			75	3.83×10^{-5}			34
	cyclopentane formyl peroxide	carbon tetrachloride	40	1.50×10^{-5}			34
			45	2.55×10^{-5}			34
			50	4.96×10^{-5}			34
			55	$8.17, 7.85 \times 10^{-5}$			34
	hexanoyl peroxide	toluene	77	1.186×10^{-4}			209
	5-hexenoyl peroxide	toluene	60.1	1.06×10^{-5}	129.3	c	209
			70.4	4.15×10^{-5}		c	209
			76.4	8.59×10^{-5}		c	209
			85.0	2.668×10^{-4}		c	209
	4-methoxybenzoyl isobutyryl peroxide	cyclohexane	55	5.465×10^{-5}			164
			60	1.024×10^{-4}			164
			65	1.876×10^{-4}			164
			70	3.208×10^{-4}			164
	4-methylbenzoyl isobutyryl peroxide	cyclohexane	40	9.61×10^{-6}		j	164
			50	3.619×10^{-5}		j	164
			60	1.305×10^{-4}		j	164
			70	4.772×10^{-4}		j	164
	4-methyl-2-thenoyl peroxide	carbon tetrachloride	75	2.92×10^{-5}		a, m$_1$	232
	5-methyl-2-thenoyl peroxide	carbon tetrachloride	75	4.21×10^{-5}		a, m$_1$	232
14	2-azidobenzoyl peroxide	benzene	50	2.5×10^{-5}		d$_2$	161
			80	9.4×10^{-4}		d$_2$	161
	benzoyl peroxide	acetic acid	75	7.53×10^{-5}		a, r	74
		acetone	50	2.25×10^{-6}	111.3	a	126
			70	2.63×10^{-5}		a	126
			85	1.34×10^{-4}		a	126
			100	5.83×10^{-4}		a	126
		acetonitrile	70	1.76×10^{-5}		t$_1$	124
		acetophenone	70	1.15×10^{-5}	126.4	a	5
			80	4.32×10^{-5}		a	5
			94.5	2.30×10^{-4}		a	5
		allyl alcohol	80	3.80×10^{-4}		a, r	74
		anisole	30	1.42×10^{-7}		a	109
		benzaldehyde	80	5.50×10^{-5}		a, r	74
			90	1.71×10^{-4}		a, r	74
		benzene	30	4.80×10^{-8}	116.3	a	109
			55	1.14×10^{-6}		a, r	74
			60	2.76×10^{-6}		a, r	74
			60	2.0×10^{-6}	124.3	m$_2$	14
			80	2.5×10^{-5}		m$_2$	14
			70	1.38×10^{-5}		a	213
			78	2.30×10^{-5}		c	230

Notes page II-40; References page II-41

Number of C atoms	Initiator	Solvent	T °C	k_d [s^{-1}]	E_a [kJ mol^{-1}]	Notes	References
				ACYL PEROXIDES (Cont'd.)			
14	benzoyl peroxide (Cont'd.)	benzene	78	1.67×10^{-5}		t_9	230
			79.8	3.48×10^{-5}			170
			80	4.8×10^{-5}			20
			66	7.72×10^{-6}	129.7	a	131
			72.5	1.87×10^{-5}		a	131
			78	3.77×10^{-5}		a	131
			70	1.17×10^{-5}	133.9	a	73
			75	2.62×10^{-5}		a	73
			80	4.39×10^{-5}		a	73
			50.8	4.28×10^{-7}	123.8	a, t_2	69
			54.9	8.53×10^{-7}		a, t_2	69
			60.9	1.66×10^{-6}		a, t_2	69
			65.6	3.22×10^{-6}		a, t_2	69
			71.0	5.94×10^{-6}		a, t_2	69
			75.8	1.19×10^{-5}		a, t_2	69
			70	1.48×10^{-5}		a	126
			85	8.94×10^{-5}		a	126
			100	4.96×10^{-4}		a	126
			70	1.03×10^{-5}		a, m_2	16
			70	1.18×10^{-5}		t_1	124
			75	1.48×10^{-5}	128.0		12
			75	1.66×10^{-5}	124.3	m_3	12
			85	4.7×10^{-5}			12
			85	5.5×10^{-5}		m_3	12
			100	2.28×10^{-4}			12
			100	2.56×10^{-4}		m_3	12
			79	2.58×10^{-5}		a, t_6	120
			80	3.35×10^{-5}		a	7
		benzyl alcohol	80	4.44×10^{-4}		a, r	74
		bromobenzene	80.2	8.15×10^{-5}			231
				2.19×10^{-5}		t_9	231
				3.84×10^{-5}		m_1	231
				3.55×10^{-5}		m_3	231
				4.34×10^{-5}		m_5	231
		butanol	80	6.06×10^{-4}		a, r	74
		butanone	80	4.64×10^{-5}		a, r	74
		di-n-butyl phthalate	117	2.78×10^{-3}	120.1	y_{16}	236
			127	7.44×10^{-3}			236
			137	1.72×10^{-2}			236
			147	3.89×10^{-2}			236
		carbon tetrachloride	75	1.07×10^{-5}		a, r	74
			79	1.69×10^{-5}		t_6, a	121
		chlorobenzene	70	1.35×10^{-5}		t_1	124
			80	4.64×10^{-5}		a, r	74
			80.2	2.85×10^{-5}			231
			80.2	2.36×10^{-5}		t_9	231
			80.2	3.52×10^{-5}		c, m_1	231
			80.2	2.62×10^{-5}		c, m_3	231
		chloroform	30	5.47×10^{-8}		a	109
		cumene	80	3.69×10^{-5}		a, r	74
			85	6.39×10^{-5}		a, r	74
			90	1.19×10^{-4}		a, r	74
			30	7.30×10^{-8}		a	109
			45	1.85×10^{-7}	120.5		67
			60	1.45×10^{-6}			67
			80	1.70×10^{-5}			67
			T[K]	1.20×10^{13} exp[-120.5 kJ/RT]			67
			100	2.5×10^{-4}		a, t_7	122
		cyclohexane	80	7.72×10^{-5}		a, r	74
		decalin	80	2.26×10^{-4}		a, r	74

Number of C atoms	Initiator	Solvent	T °C	k_d [s^{-1}]	E_a [kJ mol^{-1}]	Notes	References
		ACYL PEROXIDES (Cont'd.)					
14	benzoyl peroxide (Cont'd.)	dioxane	70	1.30×10^{-5}		t_i	74
			80	6.72×10^{-4}		a, r	74
			80	4.20×10^{-4}		a, r	6
			80	4.18×10^{-5}		a, m_1	232
		ethylbenzene	30	3.61×10^{-8}		a	109
			75	1.81×10^{-5}		a, r	74
			80	3.33×10^{-5}		a, r	74
			85	5.56×10^{-5}		a, r	74
			90	1.01×10^{-4}		a, r	74
			80	3.15×10^{-5}		c	229
		90% formic acid	80	6.94×10^{-4}		a, r	74
		n-heptane	80	3.11×10^{-5}		a, r	74
		isopropylbenzene	80	3.34×10^{-5}		c	229
		methyl acetate	49.2	6.28×10^{-7}	123.8	a, t_2	69
			53.9	1.0×10^{-6}		a, t_2	69
		methylcyclohexane	80	5.25×10^{-5}		a, r	74
		4-methyl-2-pentanone	80	4.28×10^{-5}		a, r	74
		α-methylstyrene	70	3.02×10^{-5}		a	213
		nitrobenzene	80	4.58×10^{-5}		a, r	74
			30	6.61×10^{-8}	117.6	a	109
		n-pentanol	80	1.48×10^{-4}		a, r	74
		phenol	80	6.25×10^{-4}		a, r	74
		propionic acid	80	3.19×10^{-5}		a, r	74
		styrene	34.8	3.89×10^{-8}		a	10
			49.4	5.28×10^{-7}		a	10
			61.0	2.58×10^{-6}		a	10
			74.8	1.83×10^{-5}		a	10
			100.0	4.58×10^{-4}		a	10
		poly(styrene)	56.4	3.8×10^{-7}		a	17
			64.6	1.47×10^{-6}		a	17
			76.7	9.27×10^{-6}		a	17
			83.4	2.50×10^{-5}		a	17
			98.5	1.41×10^{-4}		a	17
			70.9	2.86×10^{-6}		a	15
			80.1	1.11×10^{-5}		a	15
			89.5	3.33×10^{-5}		a	15
		tetralin	80	3.72×10^{-5}		a, r	74
		toluene	30	4.94×10^{-8}	120.5	a	109
			49.0	6.0×10^{-7}	123.8	a, t_2	69
			55.1	1.31×10^{-6}		a, t_2	69
			60.2	2.83×10^{-6}		a, t_2	69
			65.1	5.69×10^{-6}		a, t_2	69
			70.3	1.10×10^{-5}		a, t_2	69
		poly(vinyl chloride)	64.6	6.3×10^{-7}		a	17
			76.7	5.11×10^{-6}		a	17
			83.4	1.44×10^{-5}		a	17
			98.5	9.33×10^{-5}		a	17
		p-xylene	80	3.10×10^{-5}		c	229
	3-bromobenzoyl peroxide	benzene	60	1.1×10^{-6}			116
			80	1.22×10^{-5}			116
			80	2.60×10^{-5}		a	7
		dioxane	80	2.57×10^{-5}		a, m_1	6,232
	4-bromobenzoyl peroxide	dioxane	80	3.23×10^{-5}		a, m_1	6,232
	4-tert-butylbenzoyl peroxide	dioxane	80	6.06×10^{-5}		a, m_1	232
	2-chlorobenzoyl peroxide	acetophenone	80	3.88×10^{-4}	123.0	a	5
		benzene	80	3.12×10^{-4}		a	7
	3-chlorobenzoyl peroxide	acetophenone	80	2.85×10^{-5}	128.4	a	5
		dioxane	80	2.63×10^{-5}		a, m_1	6
	4-chlorobenzoyl peroxide	acetophenone	80	3.83×10^{-5}	127.2	a	5
		benzene	80	2.17×10^{-5}		a	7

Number of C atoms	Initiator	Solvent	T $^{\circ}$C	k_d [s^{-1}]	E_a [kJ mol^{-1}]	Notes	References

ACYL PEROXIDES (Cont'd.)

Number of C atoms	Initiator	Solvent	T $^{\circ}$C	k_d [s^{-1}]	E_a [kJ mol^{-1}]	Notes	References
14	4-chlorobenzoyl peroxide (Cont'd.)	benzene	50	6.2×10^{-7}	128.9	a	126
			85	6.64×10^{-5}		a	126
			100	3.86×10^{-4}		a	126
		dioxane	80	3.62×10^{-5}		a, m_1	6
		styrene	34.8	8.3×10^{-8}		a	10
			49.4	8.3×10^{-7}		a	10
			61.0	3.33×10^{-6}		a	10
			74.8	2.22×10^{-5}		a	10
			100.0	4.17×10^{-4}		a	10
	cyclohexane formyl peroxide	benzene	30	9.64×10^{-5}	84.5	a, h	219
			35	1.46×10^{-4}			219
			40	3.10×10^{-4}			219
			45	5.11×10^{-4}			219
			50	7.77×10^{-4}			219
		carbon tetrachloride	35	6.6×10^{-5}			148
			45	2.11×10^{-4}			148
			50	4.45×10^{-4}			148
			60	1.30×10^{-3}			148
			35	2.87×10^{-5}			34
			40	$5.22, 5.29 \times 10^{-5}$			34
			45	9.67×10^{-5}			34
	diacetylsebacoyl diperoxide	carbon tetrachloride	60	1.04×10^{-5}		x_1	215
			73.5	5.20×10^{-5}		x_1	215
			85	2.30×10^{-4}		x_1	215
	2,4-dichlorobenzoyl peroxide	benzene	70	9.70×10^{-5}		a, j	16
			70	1.24×10^{-4}		a, k	16
			50	1.08×10^{-5}	117.6	a	126
			70	1.37×10^{-4}		a	126
			85	7.69×10^{-4}		a	126
		styrene	34.8	3.88×10^{-6}		a	10
			49.4	2.39×10^{-5}		a	10
			61.0	7.78×10^{-5}		a	10
			74.8	2.78×10^{-4}		a	10
			100.0	4.17×10^{-3}		a	10
	cyclopentane acetyl peroxide	carbon tetrachloride	65	1.48×10^{-5}			34
			70	3.20×10^{-5}			34
			75	4.97×10^{-5}			34
	heptanoyl peroxide	toluene	77	1.24×10^{-4}		a	19
	6-heptenoyl peroxide	toluene	70	5.33×10^{-5}		a, k	19
			70	5.01×10^{-5}		a, j	19
			77	1.07×10^{-4}		a, j	19
			85	2.88×10^{-4}		a, k	19
	2-iodobenzoyl peroxide	chloroform	22	1.86×10^{-3}			82
	2-iodobenzoyl 4-nitrobenzoyl peroxide	acetone	25	3.0×10^{-4}			78
		acetonitrile	25	2.1×10^{-4}			78
		benzene	25	5.7×10^{-5}			78
		carbon tetrachloride	25	3.4×10^{-5}			78
		chloroform	25	2.8×10^{-4}			78
		nitrobenzene	25	6.2×10^{-4}			78
	3-methylbenzoyl peroxide	dioxane	80	4.38×10^{-5}		a, m_1	232
	4-methylbenzoyl peroxide	dioxane	80	6.11×10^{-5}		a, m_1	232
	2-nitrobenzoyl peroxide	acetophenone	59.3	5.80×10^{-5}	119.7	a	5
			80	1.34×10^{-3}		a	5
		methyl iodide	24.95	1.78×10^{-5}	81.2	h	82
			45.05	1.50×10^{-4}			82
	3-nitrobenzoyl peroxide	acetophenone	80	3.80×10^{-5}	126.4	a	5
	4-nitrobenzoyl peroxide	acetophenone	80	4.33×10^{-5}	126.8	a	5
	3,5-dinitrobenzoyl peroxide	acetophenone	80	1.87×10^{-5}	130.5	a	5
15	benzoyl phenylacetyl peroxide	benzene	20	1.10×10^{-4}	90.8		222
			25	2.05×10^{-4}			222

Number of C atoms	Initiator	Solvent	$T \,^{\circ}C$	$k_d \, [s^{-1}]$	$E_a \, [kJ \, mol^{-1}]$	Notes	References
	ACYL PEROXIDES (Cont'd)						
15	benzoyl phenylacetyl peroxide (Cont'd.)	benzene	25	3.15×10^{-4}		v_1	222
			25	6.38×10^{-4}		n	222
			30	3.88×10^{-4}			222
			35	6.67×10^{-4}			222
	4-tert-butylbenzoyl isobutyryl peroxide	cyclohexane	55	6.543×10^{-5}		t_{10}	164
			60	1.171×10^{-4}		t_{10}	164
			65	2.663×10^{-4}		t_{10}	164
			70	5.016×10^{-4}		t_{10}	164
	3-cyanobenzoyl benzoyl peroxide	dioxane	80	2.73×10^{-2}		a, m_1	6
	3-methoxybenzoyl benzoyl peroxide	dioxane	80	4.82×10^{-2}		a, m_1	6
	4-methoxybenzoyl benzoyl peroxide	dioxane	80	7.57×10^{-5}		a, m_1	6
	4-methoxybenzoyl 3-bromobenzoyl peroxide	dioxane	80	4.43×10^{-2}		a, m_1	6
	4-methoxybenzoyl 3,5-dinitrobenzoyl peroxide	benzene	51	1.02×10^{-5}			108
		nitrobenzene	51	9.61×10^{-4}			108
			51	1.67×10^{-3}		t_5	108
	4-methoxybenzoyl 4-nitrobenzoyl peroxide	benzene	70	2.08×10^{-5}		a	89
			70	0.00×10^{-5}		a, v_1	89
16	3,5-dibromo-4-methoxybenzoyl peroxide	benzene	60	9.5×10^{-7}			95
			60	6.1×10^{-7}			116
			80	9.4×10^{-6}			116
	caprylyl peroxide	benzene	50	3.44×10^{-6}	128.9	a	126
			70	5.78×10^{-5}		a	126
			85	3.78×10^{-4}		a	126
		mineral oil	$T[K]$	$9.8 \times 10^{15} \, exp[-140.1 \, kJ/RT]$			1
	3-cyanobenzoyl peroxide	dioxane	80	1.70×10^{-2}		a, m_1	6
	4-cyanobenzoyl peroxide	acetophenone	80	2.43×10^{-5}		a	5
		dioxane	80	2.03×10^{-5}		a, m_1	6
	cycloheptane formyl peroxide	carbon tetrachloride	35	7.85×10^{-5}			34
			40	$1.63, 1.34 \times 10^{-5}$			34
			45	2.02×10^{-4}			34
	cyclohexane acetyl peroxide	carbon tetrachloride	65	1.27×10^{-5}			34
			70	2.76×10^{-5}			34
			75	3.61×10^{-5}			34
			54.4	$3.1 \pm .1 \times 10^{-6}$			24
			64.3	$1.19 \pm .05 \times 10^{-5}$			24
			71.8	$2.95 \pm .11 \times 10^{-5}$			24
	2-ethyl-4-methyl-2-pentenoyl peroxide	mineral oil	$T[K]$	$7.1 \times 10^{16} \, exp[-138.4 \, kJ/RT]$			1
	2-ethylhexanoyl peroxide	mineral oil	$T[K]$	$1.2 \times 10^{14} \, exp[-106.4 \, kJ/RT]$			1
	2-ethyl-2-hexenoyl peroxide	mineral oil	$T[K]$	$1.6 \times 10^{16} \, exp[-136.3 \, kJ/RT]$			1
	2-iodophenylacetyl peroxide	acetone	0	2.60×10^{-5}			79
		chloroform	0	3.98×10^{-5}		c, e, q	79
		toluene	0	1.3×10^{-5}		q	79
	2-methoxybenzoyl peroxide	acetophenone	50	6.0×10^{-5}	113.8	a	5
			80	2.15×10^{-3}		a	5
	3-methoxybenzoyl peroxide	acetophenone	80	6.42×10^{-5}	120.9	a	5
		dioxane	80	5.75×10^{-5}		a, m_1	6
	4-methoxybenzoyl peroxide	acetophenone	80	1.56×10^{-4}	120.1	a	5
		dioxane	80	1.18×10^{-4}		a, m_1	6
	2-methylbenzoyl peroxide	acetophenone	70	9.02×10^{-5}	126.4	a	5
	3-methylbenzoyl peroxide	acetophenone	80	4.70×10^{-5}	126.4	a	5
		dioxane	80	4.40×10^{-5}		a, m_1	6
	4-methylbenzoyl peroxide	acetophenone	80	5.92×10^{-5}	125.1	a	5
		dioxane	80	6.13×10^{-5}		a, m_1	6
	endo-norbornane-2-carbonyl peroxide	carbon tetrachloride	44.5	$6.1 \pm .2 \times 10^{-6}$		j	23
			53.9	$2.83 \pm .17 \times 10^{-5}$		j	23
			65.9	$1.25 \pm .07 \times 10^{-4}$		j	23
			44.5	$9.1 \pm 1.3 \times 10^{-6}$		j, m_2	23
			53.9	$4.33 \pm .13 \times 10^{-5}$		j, m_2	23
			65.9	$1.28 \pm .18 \times 10^{-4}$		j, m_2	23

ACYL PEROXIDES (Cont'd.)

Number of C atoms	Initiator	Solvent	T °C	k_d [s^{-1}]	E_a [kJ mol^{-1}]	Notes	References
16	exo-norbornane-2-carbonyl peroxide	carbon tetrachloride	44.5	$4.68\pm1 \times 10^{-5}$		j,	23
			53.9	$2.05\pm .13 \times 10^{-4}$		j	23
			65.9	$8.18\pm .77 \times 10^{-4}$		j	23
			44.5	$7.2 \pm .6 \times 10^{-5}$		j, m$_2$	23
			53.9	$1.60\pm .3 \times 10^{-4}$		j, m$_2$	23
			65.9	$8.48\pm .17 \times 10^{-4}$		j, m$_2$	23
	endo-norbornene-5-carbonyl peroxide	carbon tetrachloride	44.5	$6.30\pm .27 \times 10^{-5}$		j	23
			53.9	$1.21\pm .07 \times 10^{-4}$		j	23
			65.9	$7.18\pm .55 \times 10^{-4}$		j	23
			44.5	$2.22\pm .10 \times 10^{-5}$		j, m$_2$	23
			53.9	$4.52\pm .28 \times 10^{-5}$		j, m$_2$	23
			65.9	$2.37\pm .02 \times 10^{-4}$		j, m$_2$	23
	exo-norbornene-5-carbonyl peroxide	carbon tetrachloride	44.5	$6.58\pm .52 \times 10^{-5}$		j	23
			53.9	$1.21\pm .12 \times 10^{-4}$		j	23
			65.9	$8.42\pm .17 \times 10^{-4}$		j	23
			44.5	$2.58\pm .15 \times 10^{-5}$		j, m$_2$	23
			53.9	$1.20\pm .06 \times 10^{-4}$		j, m$_2$	23
			65.9	$7.10\pm .32 \times 10^{-4}$		j, m$_2$	23
	phenylacetyl peroxide	acetonitrile	20	1.02×10^{-3}			167
		benzene	20	4.76×10^{-4}			167
		carbon tetrachloride	20	1.86×10^{-4}			167
		cyclohexane	20	1.60×10^{-4}			167
		toluene	0	2.50×10^{-5}	96.2	f, u$_1$	59
			18	3.34×10^{-4}		f, u$_1$	59
	triptoyl peroxide	benzene	80	1.42×10^{-4}			102
18	apocamphoyl peroxide	benzene	80	2.3×10^{-4}			102
	5-tert-butylthenoyl peroxide	carbon tetrachloride	75	4.03×10^{-5}		a, m$_1$	232
	dibenzoyl succinoyl diperoxide	unknown	70	7.7×10^{-6}	125.5	x$_1$	214
			75	1.42×10^{-5}		x$_1$	214
			85	5.4×10^{-5}		x$_1$	214
	nonanoyl peroxide	mineral oil	T[K]	8.4×10^{14} exp[-127.1 kJ/RT]			1
	2-nonenoyl peroxide	mineral oil	T[K]	1.6×10^{15} exp[-128.8 kJ/RT]			1
	3-nonenoyl peroxide	mineral oil	T[K]	3.7×10^{14} exp[-108.0 kJ/RT]			1
19	dibenzoyl itaconyl diperoxide	unknown	70	5.63×10^{-4},	83.7	x$_2$	214
				1.87×10^{-5}	115.1	x$_2$	214
			75	8.63×10^{-4}	83.7	x$_2$	214
				3.62×10^{-5},	115.1	x$_2$	214
			85	1.69×10^{-3}		x$_2$	214
				1.00×10^{-4},		x$_2$	214
	dibenzoyl α-methylsuccinoyl diperoxide	unknown	70	2.74×10^{-4}	82.8	x$_2$	214
				1.26×10^{-5}	116.3	x$_2$	214
			75	4.26×10^{-4}		x$_2$	214
				2.40×10^{-5}		x$_2$	214
			85	9.60×10^{-4}		x$_2$	214
				8.7×10^{-5}		x$_2$	214
20	decanoyl peroxide	benzene	60	1.53×10^{-5}	127.2	a	126
			70	5.67×10^{-5}		a	126
			85	3.80×10^{-4}		a	126
		mineral oil	T[K]	2.7×10^{15} exp[-131.9 kJ/RT]			1
	dioctanoyl α-bromosuccinoyl diperoxide	unknown	70	1.52×10^{-4},	96.2	x$_2$	214
				1.82×10^{-5}	126.4	x$_2$	214
			75	2.74×10^{-4}		x$_2$	214
				3.83×10^{-5}		x$_2$	214
			85	6.97×10^{-4}		x$_2$	214
				1.15×10^{-4},		x$_2$	214
	dioctanoyl α-chlorosuccinoyl diperoxide	unknown	70	2.02×10^{-4},	95.4	x$_2$	214
				2.42×10^{-5}	126.4	x$_2$	214
			75	3.23×10^{-4},		x$_2$	214
				4.83×10^{-5}		x$_2$	214
			85	8.50×10^{-4}.		x$_2$	214

Number of C atoms	Initiator	Solvent	T °C	k_d [s^{-1}]	E_a [kJ mol^{-1}]	Notes	References
		ACYL PEROXIDES (Cont'd.)					
20	dioctanoyl α-chlorosuccinoyl diperoxide	unknown	85	1.58×10^{-4}		x_2	214
	4-ethyl-2-octenoyl peroxide	mineral oil	T[K]	$8.2 \times 10^{14} \exp[-127.4 \text{ kJ/RT}]$			1
21	dioctanoyl itaconoyl diperoxide	unknown	70	3.23×10^{-4}	92.9	x_2	214
				2.80×10^{-5}	127.6	x_2	214
			75	5.47×10^{-4}		x_2	214
				5.60×10^{-5}		x_2	214
			85	1.28×10^{-3}		x_2	214
				1.92×10^{-4}		x_2	214
	dioctanoyl α-methylsuccinoyl diperoxide	unknown	70	5.48×10^{-4}	96.2	x_2	214
				5.88×10^{-5}	131.0	x_2	214
			75	9.58×10^{-4}		x_2	214
				1.06×10^{-4}		x_2	214
			85	2.61×10^{-3}		x_2	214
				3.84×10^{-4}		x_2	214
22	benzoyl 2-[trans-2-(3-nitrophenyl)vinyl] benzoyl peroxide	chlorobenzene	70	1.87×10^{-4}		t_9	175
	benzoyl 2-[trans-2-(4-nitrophenyl)vinyl] benzoyl peroxide	chlorobenzene	70	1.18×10^{-4}		t_9	175
		THF	70	6.2×10^{-5}		t_9	175
	benzoyl 2-[trans-2-(4-nitrophenyl)vinyl]-4-nitrobenzoyl peroxide	chlorobenzene	70	1.54×10^{-5}		t_9	175
	benzoyl 2-[trans-2-(phenyl)vinyl]benzoyl peroxide	chlorobenzene	35	9.42×10^{-5}	84.5	h, t_9	175
			70	3.02×10^{-3}		t_9	175
		methanol	35	9.77×10^{-4}		t_9	175
	4-benzylidenebutyryl peroxide	acetophenone	50	7.92×10^{-5}	99.6		133
			55	1.45×10^{-4}			133
			65	4.13×10^{-4}			133
			T[K]	$1.07 \times 10^{12} \exp[-99.6 \text{ kJ/RT}]$			133
			55	1.38×10^{-4}	98.7	m_2	133
			65	4.02×10^{-4}		m_2	133
			T[K]	$7.14 \times 10^{11} \exp[-98.7 \text{ kJ/RT}]$		m_2	133
		benzene	50	2.40×10^{-5}	91.2	h	194
			60	6.80×10^{-5}			194
			70	1.845×10^{-4}			194
		carbon tetrachloride	60	3.47×10^{-5}	112.5	m_2	133
			70	1.06×10^{-4}		m_2	133
			T[K]	$1.42 \times 10^{13} \exp[-112.5 \text{ kJ/RT}]$		m_2	133
		nitrobenzene	60	3.38×10^{-4}			133
		propylene carbonate	40	7.36×10^{-5}	89.5		133
			50	2.13×10^{-4}			133
			T[K]	$5.93 \times 10^{10} \exp[-89.5 \text{ kJ/RT}]$			133
			50	2.08×10^{-4}		m_2	133
			40	7.23×10^{-5}	82.0	h	194
			50	2.093×10^{-4}			194
			60	5.117×10^{-4}			194
		toluene	70	1.64×10^{-4}			133
	4-tert-butylbenzoyl peroxide	dioxane	80	6.08×10^{-2}		a, m_1	6
	cis-4-tert-butylcyclohexane formyl peroxide	butane	40	1.25×10^{-5}		z(0.98)	145
			40	1.69×10^{-5}		z(1010)	145
		carbon tetrachloride	40.0	$8.65 \pm .35 \times 10^{-5}$	76.1	h	35
			45.45	$1.32 \pm .04 \times 10^{-4}$			35
			50.7	$2.35 \pm .05 \times 10^{-4}$			35
	trans-4-tert-butylcyclohexane formyl peroxide	carbon tetrachloride	40.0	$4.25 \pm .23 \times 10^{-5}$	81.6	h	35
			44.7	$7.10 \pm .38 \times 10^{-5}$			35
			48.9	$1.14 \pm .04 \times 10^{-4}$			35
	trans-4-(4-chlorobenzylidene)butyryl peroxide	benzene	50	2.10×10^{-5}	95.0	h	194
			60	6.56×10^{-5}			194
			70	1.75×10^{-4}			194
		propylene carbonate	40	5.33×10^{-5}			194
			50	1.448×10^{-4}			194
			60	2.791×10^{-4}			194

Number of C atoms	Initiator	Solvent	T °C	k_d [s^{-1}]	E_a [kJ mol^{-1}]	Notes	References
				ACYL PEROXIDES (Cont'd.)			
22	trans-4-(4-fluorobenzylidene)butyryl peroxide	benzene	50	2.67×10^{-5}	94.1	h	194
			60	8.44×10^{-5}			194
			70	2.119×10^{-4}			194
		propylene carbonate	40	6.53×10^{-5}			194
			50	1.685×10^{-4}			194
			60	4.540×10^{-4}			194
	1-naphthoyl peroxide	benzene	54.6	1.01×10^{-4}			170
			59.9	1.86×10^{-4}			170
			64.5	3.0×10^{-4}			170
	4-nitrobenzoyl-2-[trans-2-(4-nitrophenyl)vinyl]benzoyl peroxide	chlorobenzene	70	6.06×10^{-5}		t_9	175
	5-phenylpentanoyl peroxide	acetophenone	77	2.37×10^{-4}			133
		benzene	77	1.054×10^{-5}	127.6	h, m_2	194
		carbon tetrachloride	70	2.76×10^{-5}	130.1	m_2	133
			77	7.19×10^{-5}		m_2	133
			85	1.87×10^{-4}		m_2	133
			T[K]	1.76×10^{15} exp[-130.1 kJ/RT]		m_2	133
		propylene carbonate	60	4.41×10^{-5}			133
			60	2.80×10^{-5}		m_2	133
24	dibenzoyl 2-bromosebacoyl diperoxide	unknown	70	$4.80 \times 10^{-5}{}'$	87.9	x_2	214
				3.87×10^{-5}	110.8	x_2	214
			75	$7.64 \times 10^{-4}{}'$		x_2	214
				7.22×10^{-5}		x_2	214
			85	$2.00 \times 10^{-3}{}'$		x_2	214
				2.02×10^{-4}		x_2	214
	dioctanoyl 2-bromosebacoyl diperoxide	unknown	70	6.46×10^{-4}	99.6	x_2	214
				$5.23 \times 10^{-5}{}'$	129.7	x_2	214
			75	$1.171 \times 10^{-3}{}'$		x_2	214
				1.08×10^{-4}		x_2	214
			85	$3.00 \times 10^{-3}{}'$		x_2	214
				3.72×10^{-4}		x_2	214
	lauroyl peroxide	benzene	30	2.56×10^{-7}		a	99
			40	4.91×10^{-7}			68
			50	2.19×10^{-6}			68
			60	9.17×10^{-6}			68
			70	2.86×10^{-5}			68
			60	1.51×10^{-5}	127.2	a	126
			70	5.58×10^{-5}		a	126
			85	3.75×10^{-4}		a	126
			70	4.33×10^{-5}		a, j	16
		carbon tetrachloride	40	2.91×10^{-7}			68
			50	1.15×10^{-6}			68
			60	4.75×10^{-6}			68
			70	1.87×10^{-5}			68
		ethyl acetate	40	6.03×10^{-7}			68
			50	2.70×10^{-6}			68
			61	1.05×10^{-5}			68
			70	3.99×10^{-5}			68
		ethyl ether	30	1.97×10^{-6}		a	99
		styrene	34.8	2.06×10^{-7}		a	10
			49.4	2.25×10^{-6}		a	10
			61.0	1.42×10^{-5}		a	10
			74.8	1.00×10^{-4}		a	10
			100.0	2.39×10^{-3}		a	10
		mineral oil	T[K]	2.2×10^{16} exp[-137.9 kJ/RT]			1
	trans-4-(4-methoxybenzylidene)butyryl peroxide	benzene	50	2.03×10^{-4}			194
	trans-4-(4-methylbenzylidene)butyryl peroxide	benzene	50	6.58×10^{-5}	90.4	c, h	194
			60	1.815×10^{-4}		c	194
			70	4.729×10^{-4}		c	194

DECOMPOSITION RATES OF INITIATORS

Number of C atoms	Initiator	Solvent	T °C	k_d [s^{-1}]	E_a [kJ mol^{-1}]	Notes	References
	ACYL PEROXIDES (Cont'd.)						
26	2-phenoxybenzoyl peroxide	acetophenone	65	8.18×10^{-5}	121.3	a	5
28	myristoyl peroxide	benzene	70	3.38×10^{-5}		a	16
36	menthylphthaloyl peroxide	dioxane	55	1.15×10^{-4}			65
	5 - HYDROPEROXIDES AND KETONE PEROXIDES						
4	sec-butyl hydroperoxide	toluene	172.0	2.65×10^{-5}			199
			182.3	4.9×10^{-5}			199
	tert-butyl hydroperoxide	benzene	154.5	4.29×10^{-6}	170.7	h	76
			161.7	9.27×10^{-6}			76
			169.3	2.0×10^{-5}			76
			174.6	4.0×10^{-5}			76
			172.3	1.09×10^{-5}			199
			182.6	3.1×10^{-5}			199
		cumene	182.6	8.1×10^{-5}		c	199
		cyclohexane	100	1.2×10^{-7}			199
			172	1.08×10^{-4}		c	199
		dodecane	86.1	1.32×10^{-6}	128.4	b	41
			98.5	5.55×10^{-6}		b	41
		heptane	172	1.41×10^{-4}			199
		n-octane	149.8	8×10^{-6}	163.2	f	118
			159.9	2.5×10^{-5}		f	118
			169.6	6.9×10^{-5}		f	118
			179.6	1.82×10^{-5}		f	118
		toluene	100	5.7×10^{-8}			199
			172.5	9.2×10^{-6}			199
			181.5	2.69×10^{-5}			199
			192.6	8.3×10^{-5}			199
			204.5	1.52×10^{-4}		c	199
			214.9	3.24×10^{-4}			199
		vapor	570	3.4×10^{-1}			188
			670	4.95			188
			773	8.4×10^{1}			188
			873	5.66×10^{2}			188
			973	2.58×10^{3}			188
	tert-butyl hydroperoxide/cobalt 2-ethyl-hexanoate	chlorobenzene	0	1.01×10^{-3}			190
			0	8.29×10^{-5}		v	190
			0	5.40×10^{-4}		v_7^{10}	190
			25	2.3×10^{-3}			198
			25	1.1×10^{-3}		v_1	198
	tert-butyl hydroperoxide/cobalt stearate		45	1.4×10^{-3}			198
			45	1.0×10^{-3}		v_1	198
	methyl ethyl ketone peroxides	ethyl acetate	70	1.28×10^{-6}			13
6	cyclohexyl hydroperoxide	benzene	70	0			91
		benzene/styrene (50/50)	70	1.27×10^{-3}		a	91
		cyclohexane	130	2.38×10^{-3}		a	91
			140	1.16×10^{-2}		a	91
			150	3.20×10^{-2}		a	91
		benzene	80				119
		cyclohexane	80	reaction order			119
		cyclohexene	80	varies from			119
		dimethylheptadiene	80	1.5 to 2.0			119
		1-octene	80				119
8	1,4-dimethylcyclohexane hydroperoxide	1,4-dimethylcyclohexane	120	1.4×10^{-5}	137.2	c	224
	n-octyl hydroperoxide	white oil	150	9.29×10^{-5}	112.5	a	177
	2,4,4-trimethylpentyl 2-hydroperoxide	white oil	150	9.29×10^{-5}	112.5	a	177
(8)x)	poly(phenyleneethyl hydroperoxide)/manganese resinate	chlorobenzene	26	2.00×10^{-4}	46.0	t_3	212
			30	2.75×10^{-4}		t_3	212
			35	3.67×10^{-4}		t_3	212
9	cumene hydroperoxide	mesitylene			98.7		58
		styrene			101.3		58

Number of C atoms	Initiator	Solvent	T °C	k_d [s^{-1}]	E_a [kJ mol^{-1}]	Notes	References
				HYDROPEROXIDES AND KETONE PEROXIDES (Cont'd.)			
9	cumene hydroperoxide (Cont'd.)	toluene	125	9×10^{-6}			199
			139	3×10^{-5}			199
			182.3	6.45×10^{-5}			199
		white oil	150	1.34×10^{-4}	121.3	a	177
10	decalin hydroperoxide	acetic acid	130	5.5×10^{-4}			228
		chlorobenzene	130	10% in 3 hours			228
		decalin	130	1.41×10^{-5}	124.3		228
		1,2-dichlorobenzene	130	4.48×10^{-5}	126.4		228
		ethylene glycol	130	1.65×10^{-4}	117.2		228
		nitrobenzene	130	4.73×10^{-5}			228
		pyridine	130	$2.31\text{-}2.89 \times 10^{-4}$			228
	pinane hydroperoxide	benzene	130	7.08×10^{-6}	123.8	a	18
			145	2.72×10^{-5}		a	18
			160	9.17×10^{-5}		a	18
	1-phenyl-2-methylpropyl-1-hydroperoxide	benzene	133.8	3.18×10^{-6}	122.2	h	76
			143.9	8.95×10^{-6}			76
			153.9	2.0×10^{-5}			76
			163.7	4.03×10^{-5}			76
			174	9.77×10^{-5}			76
	1-phenyl-2-methylpropyl-2-hydroperoxide	benzene	144.2	5.04×10^{-6}	125.5	h	76
			154.5	1.21×10^{-5}			76
			165.6	2.92×10^{-5}			76
			176.0	6.97×10^{-5}			76
	tetralin hydroperoxide	poly(butene)	170	2.17×10^{-3}	78.2	t_3	71
		n-butyl stearate	170	1.47×10^{-4}	125.5	t_3	71
		2-ethyl-1-hexene	130	1.08×10^{-4}	82.8	e, t_3	71
			170	1.26×10^{-3}		t_3	71
		1-hexadecene	170	7.92×10^{-4}	117.2	t_3	71
		mineral oil	135.6	4.2×10^{-5}	131.4	t_3	71
			150.6	1.00×10^{-4}		e, t_3	71
			170	4.82×10^{-4}		t_3	71
		n-octadecane	170	2.54×10^{-4}	119.2	t_3	71
		isooctane	170	1.31×10^{-4}		t_3	71
		octyl ether	170	1.45×10^{-3}	121.3	t_3	71
		poly(propylene)	170	2.50×10^{-3}		t_3	71
		n-tetradecane	170	2.32×10^{-4}		t_3	71
		tetralin	T[K]	$2.27 \times 10^9 \exp[-102.1 \text{ kJ/RT}]$			112
		2,2,4-trimethyl-1-pentene	170	1.67×10^{-3}		t_3	71
		white oil	150	1.34×10^{-4}	121.3		177
$(10)_n$	poly(cumyleneethyl hydroperoxide) (m.w. 1100)	toluene	130	2.79×10^{-5}			210
			140	4.0×10^{-5}			210
			150	5.78×10^{-5}			210
				6 - PERESTERS AND PEROXY CARBONATES			
4	dimethyl peroxalate	pentane	25	1.7×10^{-5}			203
5	tert-butyl percarbamate	chlorobenzene	90	6.6×10^{-6}			141
	tert-butyl performate	chlorobenzene	130.8	5.43×10^{-5}	159.0	h	218
			140.8	1.70×10^{-4}			218
			140.6	1.80×10^{-4}		m_2	218
			140	2.12×10^{-3}	64.0	v_8	218
			140	5.06×10^{-4}		v_1	218
		4-chlorotoluene	140.6	1.61×10^{-4}			218
			140	1.62×10^{-3}		v_8	218
		cumene	140.6	1.77×10^{-3}		b	218
			140	1.02×10^{-3}		b	218
6	tert-butyl peracetate	tert-amyl alcohol	75	2.8×10^{-6}			193
		benzene	85	1.2×10^{-6}			193
			85	2.18×10^{-6}	151.9	a	126
			100	1.54×10^{-5}		a	126
			115	1.02×10^{-4}		a	126

Number of C atoms	Initiator	Solvent	T °C	k_d [s^{-1}]	E_a [kJ mol^{-1}]	Notes	References

PERESTERS AND PEROXY CARBONATES (Cont'd.)

Number of C atoms	Initiator	Solvent	T °C	k_d [s^{-1}]	E_a [kJ mol^{-1}]	Notes	References
6	tert-butyl peracetate (Cont'd.)	benzene	130	5.69×10^{-4}		a	126
		n-butanol	75	2.65×10^{-5}			193
		2-butanol	75	1.13×10^{-4}		c	193
			75	5.3×10^{-5}		t_7	193
			75	4×10^{-7}		m_3	193
		chlorobenzene	60	2.31×10^{-8}	159.0	a, h	81
		hexane	130.1	5.08×10^{-4}			169
		paraffin	130.1	3.13×10^{-4}			169
	tert-butyl trichloroperacetate	chlorobenzene	60	1.19×10^{-5}	125.9	a, h	81
			66.8	2.75×10^{-5}	126.8	h	28
			77.0	1.00×10^{-4}			28
	diethyl peroxydicarbonate	tert-butanol	45	1.25×10^{-5}	133.9-138.1	c	18
			55	5.7×10^{-5}		c	18
		2,2'-oxydiethylene-bis	40	6.94×10^{-6}	127.2	a	90
		(allyl carbonate)	50	2.86×10^{-5}		a	90
			60	1.28×10^{-4}		a	90
	diethyl peroxalate	pentane	25	2.6×10^{-5}			203
7	tert-butyl peracrylate	benzene	90	1.12×10^{-5}	86.2	a	211
			100	1.61×10^{-5}		a	211
			110	4.85×10^{-5}		a	211
	tert-butyl perpropionate	benzene	80	1.53×10^{-6}		a	213
			90	6.04×10^{-6}		a	213
		α-methylstyrene	70	3.02×10^{-7}		a	213
			90	4.85×10^{-6}		a	213
8	tert-butyl perisobutyrate	benzene	78	3.77×10^{-5}		a	37
			70	6.69×10^{-5}	140.6	a	126
			85	5.33×10^{-5}		a	126
			100	3.50×10^{-4}		a	126
		bulk	70	4.12×10^{-5}		a	37
		chlorobenzene	90.6	8.13×10^{-5}	140.6	h	136
			100.7	2.75×10^{-4}			136
			110.0	8.92×10^{-4}			136
		cumene	90.6	6.9×10^{-5}	133.1	h	136
			100.7	2.35×10^{-4}			136
			110.0	6.57×10^{-4}			136
	tert-butylperoxyisopropyl carbonate	benzene	90	6.64×10^{-6}		a	9
			100	2.21×10^{-5}		a	9
			110	6.87×10^{-5}		a	9
	tert-butyl permethacrylate	benzene	90	1.92×10^{-5}	137.0	a	211
			100	7.66×10^{-5}		a	211
			110	1.92×10^{-4}		a	211
	diisopropyl peroxalate	pentane	25	6.0×10^{-5}			203
	diisopropyl peroxydicarbonate	benzene	54.0	5.0×10^{-5}		m_2	60
		di-n-butyl phthalate	77	1.39×10^{-3}		y_{16}	236
			87	4.09×10^{-3}		y_{16}	236
			97	1.25×10^{-2}		y_{16}	236
			107	3.54×10^{-2}		y_{16}	236
			117	7.98×10^{-2}		y_{16}	236
		ethylbenzene	54.3	4.5×10^{-5}			60
			54.3	5.2×10^{-5}		m_2	60
		2,2'-oxydiethylene-bis	40	6.39×10^{-6}	117.6	a	90
		(allyl carbonate)	50	2.28×10^{-5}		a	90
			60	9.44×10^{-5}		a	90
		toluene	50	3.03×10^{-5}		a	90
	ethyl tert-butyl peroxalate	benzene	45	4.48×10^{-5}	112.5	c, h	31
			55	1.63×10^{-4}		c	31
			65	5.93×10^{-4}		c	31
9	tert-butyl 5-bromo-2-perthenoate	carbon tetrachloride	99.2	2.24×10^{-6}	143.5	a	176
			112.0	1.18×10^{-5}		a	176
			124.5	4.30×10^{-5}		a	176

Number of C atoms	Initiator	Solvent	T °C	k_d [s^{-1}]	E_a [kJ mol^{-1}]	Notes	References
				PERESTERS AND PEROXY CARBONATES (Cont'd.)			
9	tert-butyl 5-chloro-2-perthenoate	carbon tetrachloride	99.2	2.29×10^{-6}	143.5	a	176
			112.0	9.56×10^{-6}		a	176
			124.5	4.42×10^{-5}		a	176
	di-(tert-butylperoxy)-carbonate	chlorobenzene	99.95	6.72×10^{-5}	133.1		27
			110.1	2.13×10^{-4}			27
			120.1	6.05×10^{-4}			27
		1,2-dichlorobenzene	120.1	5.98×10^{-4}			27
		cumene	99.95	6.72×10^{-5}			27
		isopropyl ether	99.95	2.76×10^{-4}			27
	tert-butyl perpivalate	benzene	50	9.77×10^{-6}	119.7	a	126
			70	1.24×10^{-4}		a	126
			85	7.64×10^{-4}		a	126
		chlorobenzene	58.6	3.35×10^{-5}	125.5	h	28
			64.3	7.01×10^{-5}			28
			74.8	2.79×10^{-4}			28
			60	3.85×10^{-5}	128.0	a, h	81
			60.6	4.00×10^{-5}		t_9	156
			74	1.93×10^{-4}		t_9	156
		cumene	45	3.53×10^{-6}	106.7	h	160
			55	1.73×10^{-5}			160
			65	4.53×10^{-5}			160
			64.6	5.81×10^{-5}	115.5		159
			75.6	2.10×10^{-4}			159
			84.6	5.94×10^{-4}			159
		dioxane/water (90/10)	60.6	6.26×10^{-5}		t_9	156
		isooctane	60.6	1.97×10^{-5}		t_9	165
			73.9	8.82×10^{-5}		t_9	165
	tert-butyl perpivalate-d$_9$	chlorobenzene	60.6	3.24×10^{-5}		t_9	156
			74.0	1.57×10^{-4}		t_9	156
		dioxane/water (90/10)	60.6	5.05×10^{-5}		t_9	156
		isooctane	60.6	1.62×10^{-5}		t_9	165
			73.9	8.82×10^{-5}		t_9	165
	tert-butyl 2-perthenoate	carbon tetrachloride	99.2	3.3×10^{-6}	147.3	m_2	176
			112.0	1.66×10^{-5}		m_2	176
			124.5	6.87×10^{-5}		m_2	176
	tert-butyl 1-pyrrolidine-percarboxylate	chlorobenzene	90	7.59×10^{-5}			141
	tert-butyl N-succinimido-percarboxylate	acetonitrile	100	9.9×10^{-5}			142
		benzene	100	3.79×10^{-5}	113.4	h, m_2	142
		chlorobenzene	90	1.32×10^{-5}			141
			100	4.91×10^{-5}		m_2	142
		cumene	90	$\sim 3.3 \times 10^{-6}$			141
		cyclohexane	100	1.10×10^{-5}		t_9	142
		cyclohexene	100	9.0×10^{-6}			142
		methanol	100	6.00×10^{-3}			142
		methylene chloride	90	9.24×10^{-5}			141
		nitrobenzene	100	3.41×10^{-4}		m_2	142
10	di(tert-butylperoxy)-oxalate	benzene	35.0	6.77×10^{-5}	106.7	c, h	30
			45.0	2.61×10^{-4}		c	30
			55.0	9.3×10^{-4}		c	30
	bis(2-nitro-2-methylpropyl)-peroxy-dicarbonate	toluene	50	2.22×10^{-5}		a	90
	tert-butyl 5-methyl-2-perthenoate	carbon tetrachloride	99.2	4.6×10^{-6}	137.7	a, m_2	176
			112.0	1.94×10^{-5}		a, m_2	176
			124.5	7.95×10^{-5}		a, m_2	176
	di-tert-butyl peroxalate	tert-butanol	25.0	1.83×10^{-5}			197
		n-pentane	37.8	1.01×10^{-4}			197
		styrene	35	5.5×10^{-5}	75.3		215
			45	1.71×10^{-4}			215
			55	6.01×10^{-4}			215

DECOMPOSITION RATES OF INITIATORS

Number of C atoms	Initiator	Solvent	T $^{\circ}$C	k_d [s^{-1}]	E_a [kJ mol^{-1}]	Notes	References
			PERESTERS AND PEROXY CARBONATES (Cont'd)				
10	β-methyl-β-phenyl-β-peroxy-propiolactone	carbon tetrachloride	106.8	5.65×10^{-6}	131.8	h, v_9	195
			126	4.59×10^{-5}			195
			134	9.71×10^{-5}		c, v_9	195
11	tert-butyl-4-chloroperbenzoate	phenyl ether	100.0	3.89×10^{-6}	164.4	a	53
			110.1	1.85×10^{-5}		a	53
			120.2	6.39×10^{-5}		a	53
			130.9	2.42×10^{-4}		a	53
	tert-butyl-N-(2-chlorophenylperoxy) carbamate	toluene	87.0	3.5×10^{-5}	157.3	h	36
			95.5	1.48×10^{-4}			36
			103.3	3.3×10^{-4}			36
	tert-butyl-N-(3-chlorophenylperoxy) carbamate	toluene	78.0	2.6×10^{-5}	115.5	h	36
			87.0	7.8×10^{-5}			36
			96.6	2.03×10^{-5}			36
			102.7	4.62×10^{-4}			36
	tert-butyl-N-(4-chlorophenylperoxy) carbamate	toluene	73.0	5.57×10^{-5}	119.7	h	36
			78.5	1.42×10^{-4}			36
			87.0	2.75×10^{-4}			36
			92.5	5.37×10^{-4}			36
	tert-butyl-N-(2,5-dichlorophenylperoxy) carbamate	toluene	88.0	2.75×10^{-5}	128.9	h	36
			95.0	5.58×10^{-5}			36
			103.3	1.48×10^{-4}			36
			114.5	4.82×10^{-4}			36
	tert-butyl 2,2-dimethyl-pentanoate	cumene	45	6.2×10^{-6}	105.4	h	160
			55	2.15×10^{-5}			160
	tert-butyl 5-ethyl-2-perthenoate	carbon tetrachloride	99.2	5.39×10^{-6}	130.5		176
			112.0	2.12×10^{-5}			176
			124.5	7.91×10^{-5}			176
	tert-butyl 2-iodoperbenzoate	chlorobenzene	85.0	$4.02 \pm .04 \times 10^{-5}$			134
			102.4	$2.58 \pm .02 \times 10^{-4}$			134
			118.8	$1.32 \pm .02 \times 10^{-3}$			134
	tert-butyl 4-nitroperbenzoate	phenyl ether	110.1	7.56×10^{-6}	172.8		53
			120.2	3.19×10^{-5}			53
			130.9	1.11×10^{-4}			53
			141.5	3.92×10^{-4}			53
	tert-butyl-N-(3-nitrophenylperoxy) carbamate	toluene	78.0	8.4×10^{-6}	133.1	h	36
			88.7	2.43×10^{-5}			36
			98.0	1.01×10^{-4}			36
			106.0	2.36×10^{-4}			36
	tert-butyl-N-(4-nitrophenylperoxy) carbamate	toluene	73.0	6.4×10^{-6}	113.8	h	36
			87.0	3.11×10^{-5}			36
			98.0	8.75×10^{-5}			36
			106.0	2.38×10^{-4}			36
	tert-butyl perbenzoate	acetic acid	100.0	3.83×10^{-5}	130.1	a	56
			110.0	1.14×10^{-4}		a	56
		benzene	100	1.07×10^{-5}	145.2	a	126
			115	6.22×10^{-5}		a	126
			130	3.50×10^{-4}		a	126
			110.0	3.50×10^{-5}	144.3	a	56
			119.4	1.04×10^{-4}		a	56
			130.0	3.30×10^{-4}		a	56
		bromobenzene	119.4	1.32×10^{-4}		a	56
		n-butanol	90.0	9.27×10^{-5}	120.5	a	56
			100.0	2.70×10^{-4}		a	56
		n-butyl acetate	110.0	1.06×10^{-4}	123.4	a	56
			119.4	2.67×10^{-4}		a	56
			110.0	3.61×10^{-5}	148.5	a, m_2	56
			119.4	1.10×10^{-4}		a, m_2	56
		tert-butylbenzene	119.4	1.03×10^{-4}		a	56
		n-butyl ether	100.0	7.80×10^{-5}	99.2	a	56
			110.0	1.80×10^{-4}		a	56

Notes page II-40; References page II-41

Number of C atoms	Initiator	Solvent	T °C	k_d [s^{-1}]	E_a [kJ mol^{-1}]	Notes	References
				PERESTERS AND PEROXY CARBONATES (Cont'd.)			
11	tert-butyl perbenzoate (Cont'd.)	chlorobenzene	110.0	3.83×10^{-5}	141.8	a	56
			119.4	1.11×10^{-4}		a	56
			120	1.31×10^{-4}		m_2	22
			135	6.74×10^{-4}		m_2	22
			150	3.12×10^{-3}		m_2	22
		4-chlorotoluene	110.0	3.25×10^{-5}	144.3	c	55
			119.4	9.80×10^{-5}		c	55
			130.0	3.06×10^{-4}		c	55
			60	3.85×10^{-7}	140.2	a, h	81
		ethylbenzene	119.4	1.07×10^{-4}		a	56
		methyl benzoate	119.4	7.80×10^{-5}		a	56
		phenyl ether	100.0	6.94×10^{-6}	156.9	a	53
			110.1	2.28×10^{-5}		a	53
			120.2	9.00×10^{-5}		a	53
			130.9	2.92×10^{-4}		a	53
		xylene	119.4	1.09×10^{-4}	141.4	a	56
			130.0	3.42×10^{-4}		a	56
	tert-butyl-N-(4-bromophenylperoxy) carbamate	toluene	70.8	3.32×10^{-5}	127.6	h	36
			79.8	1.04×10^{-4}			36
			84.0	1.75×10^{-4}			36
			96.0	7.70×10^{-4}			36
	tert-butyl percarboxycyclohexane	chlorobenzene	100.1	2.75×10^{-5}	131.0	h	181
			111.4	9.65×10^{-5}			181
			120.0	2.39×10^{-4}			181
		cumene	79.6	1.86×10^{-5}		z(1)	163
			79.6	1.42×10^{-5}		z(2030)	163
			79.6	1.15×10^{-5}		z(4050)	163
	tert-butyl per-2-methylphenylacetate	isooctane	60.6	6.60×10^{-5}		m_2	165
			73.9	3.233×10^{-4}		m_2	165
	tert-butyl-N-(phenylperoxy) carbamate	toluene	51.2	3.4×10^{-6}	139.7	h	36
			67.7	2.31×10^{-5}			36
			77.7	1.15×10^{-4}			36
			90.7	6.41×10^{-4}			36
		chlorobenzene	T[K]	1.51×10^{16} exp[-136.0 kJ/RT]			40
12	4-bromocumyl perpropionate	benzene	70	1.48×10^{-6}	129.7	a	213
			80	5.73×10^{-6}		a	213
			90	1.81×10^{-5}		a	213
		α-methylstyrene	70	1.99×10^{-6}	121.3	a	213
			80	4.93×10^{-6}		a	213
			90	1.55×10^{-5}		a	213
	tert-butyl bicyclo[2,2,1]heptane-2-percarboxylate	cumene	85	1.16×10^{-6}	155.2	h	160
			100	1.02×10^{-5}			160
			110	3.76×10^{-5}			160
			110	4.6×10^{-5}	150.2	h	223
			120	1.56×10^{-4}			223
			130	4.99×10^{-4}			223
	tert-butyl endo-bicyclo[2,2,1]-heptane percarboxylate	chlorobenzene	94.0	1.75×10^{-5}	137.2	h	147
			101.9	6.20×10^{-5}			147
			109.6	1.7×10^{-4}			147
			120.5	5.72×10^{-4}			147
			94.5	9.69×10^{-5}			182
			101.9	1.96×10^{-5}			182
			111.9	6.53×10^{-5}			182
			100.1	4.61×10^{-5}	149.0	h	181
			111.4	1.96×10^{-4}			181
			120.0	5.40×10^{-4}			181
		cumene	94.0	1.87×10^{-5}	154.0	h	147
			101.9	5.30×10^{-5}			147
			109.6	1.10×10^{-4}			147
			120.3	4.3×10^{-4}			147

Number of C atoms	Initiator	Solvent	T °C	k_d [s^{-1}]	E_a [kJ mol^{-1}]	Notes	References

PERESTERS AND PEROXY CARBONATES (Cont'd.)

Number of C atoms	Initiator	Solvent	T °C	k_d [s^{-1}]	E_a [kJ mol^{-1}]	Notes	References
12	tert-butyl endo-bicyclo[2,2,1]-heptane percarboxylate (Cont'd.)	cumene	94.5	5.27×10^{-5}	130.1	h	182
			101.9	1.12×10^{-4}			182
			111.9	3.57×10^{-4}			182
	tert-butyl exo-bicyclo[2,2,1]-heptane-2-percarboxylate	chlorobenzene	94.5	8.86×10^{-5}			182
			101.9	2.28×10^{-4}			182
			112.1	8.85×10^{-4}			182
			96.0	1.33×10^{-4}			147
			100.1	1.90×10^{-4}	129.3	h	181
			111.4	6.79×10^{-4}			181
			120.0	1.64×10^{-3}			181
		cumene	84.4	2.19×10^{-5}	129.3	h	147
			94.7	7.29×10^{-5}			147
			100.1	1.37×10^{-4}			147
			108.6	3.6×10^{-4}			147
			113.5	6.19×10^{-4}			147
			94.5	6.05×10^{-5}	131.4	h	182
			101.9	1.53×10^{-4}			182
			112.1	4.72×10^{-4}			182
	tert-butyl endo-bicyclo[2,2,1]-hept-5-ene-2-percarboxylate	cumene	94.5	3.11×10^{-5}	136.4	h	182
			101.9	7.35×10^{-5}			182
			112.0	2.48×10^{-4}			182
	tert-butyl exo-bicyclo[2,2,1]-hept-5-ene-2-percarboxylate	chlorobenzene	94.5	5.59×10^{-5}			182
			101.9	1.25×10^{-4}			182
			112.1	5.00×10^{-4}			182
		cumene	94.5	4.48×10^{-5}	138.1	h	182
			101.9	1.18×10^{-4}			182
			112.1	3.97×10^{-4}			182
	tert-butyl 3-chlorophenyl-peracetate	chlorobenzene	79.6	4.05×10^{-5}	123.0	h	178
			90.7	1.44×10^{-4}			178
			100.5	4.38×10^{-4}			178
		cumene	79.6	2.98×10^{-5}		z(0.98)	207
			79.6	1.99×10^{-5}		z(6090)	207
	tert-butyl 4-chlorophenyl peracetate	chlorobenzene	79.6	8.44×10^{-5}	117.2	h	178
			90.7	2.95×10^{-4}			178
			100.5	8.19×10^{-4}			178
		cumene	79.6	6.54×10^{-5}		z(0.98)	207
			79.6	4.45×10^{-5}		z(6090)	207
	tert-butyl 2,2-diethylbutyrate	cumene	45	1.23×10^{-5}	103.3	h	160
			55	4.92×10^{-5}			160
			65	1.46×10^{-4}			160
	tert-butyl 4-methoxyperbenzoate	phenyl ether	100.0	1.07×10^{-5}	149.8	a	53
			110.1	4.17×10^{-5}		a	53
			120.2	1.28×10^{-4}		a	53
			130.9	4.28×10^{-4}		a	53
	tert-butyl 4-methylperbenzoate	phenyl ether	100.0	9.42×10^{-6}	151.0	a	53
			110.1	3.19×10^{-5}		a	53
			120.2	1.06×10^{-4}		a	53
			130.9	3.25×10^{-4}		a	53
	tert-butyl 2-methylsulfonylperbenzoate	chlorobenzene	105	6.68×10^{-6}			22
			120	5.57×10^{-5}			22
			135	2.76×10^{-4}			22
			150.6	2.05×10^{-3}			22
	tert-butyl 2-(methylthio)perbenzoate	chlorobenzene	60	8.08×10^{-4}	95.0	a, h	97
			39.4	2.59×10^{-4}		b	134
			50.1	$2.42 \pm .04 \times 10^{-4}$		c	134
			50.2	$1.88 \pm .01 \times 10^{-4}$		m_2	134
			69.8	$1.96 \pm .03 \times 10^{-3}$			134
	tert-butyl 4-(methylthio)perbenzoate	chlorobenzene	120.4	$1.75 \pm .05 \times 10^{-4}$			134
	tert-butyl 4-nitrophenylperacetate	chlorobenzene	79.6	2.5×10^{-5}	124.7	h	178
			90.7	8.9×10^{-5}			178
			100.5	2.83×10^{-4}			178

Number of C atoms	Initiator	Solvent	T °C	k_d [s^{-1}]	E_a [kJ mol^{-1}]	Notes	References
				PERESTERS AND PEROXY CARBONATES (Cont'd.)			
12	tert-butyl 4-nitrophenylperacetate (Cont'd.)	chlorobenzene	80	3.77×10^{-5}			169
		decane	77.5	6.30×10^{-6}			168
		dodecane	77.5	5.81×10^{-6}			168
			100	1.31×10^{-4}			168
		hexadecane	77.5	5.11×10^{-6}			168
		octane	77.5	6.42×10^{-6}			168
			100	1.60×10^{-4}			168
		tetradecane	77.5	5.56×10^{-6}			168
	di-tert-butyl per-2-chlorosuccinoate	unknown	105	2.74×10^{-4}	102.9	x_2	214
				1.22×10^{-5}'	143.5	x_2	214
			115	5.90×10^{-4}		x_2	214
				3.83×10^{-5}'		x_2	214
			125	1.52×10^{-3}		x_2	214
				1.00×10^{-4}'		x_2	214
	di-tert-butyl persuccinoate	styrene	105	1.93×10^{-5}	154.8	x_1	215
			115	6.7×10^{-5}		x_1	215
			125	2.53×10^{-4}		x_1	215
	tert-butyl phenylperacetate	chlorobenzene	60	6.79×10^{-6}	120.1	a, h	81
			77.0	6.85×10^{-5}	117.6	h	28
			88.6	2.45×10^{-4}		m_2	28
			79.6	1.05×10^{-4}	116.7	h	178
			90.7	3.53×10^{-4}			178
			100.5	1.003×10^{-3}			178
			79.6	1.07×10^{-4}		z(0.98)	157
			79.6	9.1×10^{-5}		z(4050)	157
			79.6	1.02×10^{-4}		z(0.98)	163
			79.6	9.0×10^{-4}		z(4050)	163
			85.0	1.945×10^{-4}		m_2	165
		cumene	79.6	6.6×10^{-5}		z(0.98)	157
			79.6	5.6×10^{-5}		z(4050)	157
			79.6	6.78×10^{-5}		z(0.98)	163,207
			79.6	4.73×10^{-5}		z(6090)	163,207
		decane	77.5	3.00×10^{-5}			168
		dodecane	77.5	2.75×10^{-5}			168
		hexadecane	77.5	2.60×10^{-5}			168
		isooctane	85.0	1.08×10^{-4}			165
			95.2	3.527×10^{-4}			165
		octane	77.5	3.11×10^{-5}			168
		paraffin oil	85.0	1.150×10^{-4}			165
		tetradecane	77.5	2.66×10^{-5}			168
	tert-butyl 2-propylperpenten-2-oate (cis)	cumene	94.9	2.78×10^{-5}	137.2	h	172
			100.1	5.4×10^{-5}		z(0.98)	172
			100.1	4.26×10^{-5}		z(1100)	172
			100.1	2.31×10^{-5}		z(4000)	172
			110.1	1.72×10^{-4}			172
	(trans)	cumene	94.9	2.47×10^{-5}	143.9	h	172
			100.1	4.7×10^{-5}		z(0.98)	172
			100.1	3.13×10^{-5}		z(1100)	172
			100.1	1.42×10^{-5}		z(3830)	172
			110.1	1.64×10^{-4}			172
	tert-butyl-N-(3-tolylperoxy) carbamate	toluene	64.0	4.58×10^{-5}	102.9	h	36
			70.7	9.17×10^{-5}			36
			78.0	2.03×10^{-4}			36
			88.5	5.78×10^{-4}			36
	4-chlorocumyl perpropionate	benzene	70	1.46×10^{-6}	129.3	a	213
			80	5.68×10^{-6}		a	213
			90	1.79×10^{-5}		a	213
		α-methylstyrene	70	1.50×10^{-6}	118.8	a	213
			80	4.54×10^{-6}		a	213
			90	1.47×10^{-5}		a	213

Number of C atoms	Initiator	Solvent	T °C	k_d [s^{-1}]	E_a [kJ mol^{-1}]	Notes	References
				PERESTERS AND PEROXY CARBONATES (Cont'd.)			
12	cumyl perpropionate	benzene	70	2.60×10^{-6}	101.7	a	213
			80	6.15×10^{-6}		a	213
			90	1.90×10^{-5}		a	213
		α-methylstyrene	70	6.31×10^{-6}	106.7	a	213
			80	1.71×10^{-5}		a	213
			90	4.94×10^{-5}		a	213
	4-iodocumyl perpropionate	benzene	70	1.37×10^{-6}	133.1	a	213
			80	5.76×10^{-6}		a	213
			90	1.79×10^{-5}		a	213
		α-methylstyrene	70	1.54×10^{-6}	125.5	a	213
			80	4.96×10^{-6}		a	213
			90	1.62×10^{-5}		a	213
	4-nitrocumyl perpropionate	benzene	70	1.36×10^{-6}	133.5	a	213
			80	5.32×10^{-6}		a	213
			90	1.78×10^{-5}		a	213
		α-methylstyrene	70	1.35×10^{-6}	124.7	a	213
			80	4.07×10^{-6}		a	213
			90	1.43×10^{-5}		a	213
13	benzyl (tert-butylperoxy) oxalate	benzene	45	3.65×10^{-5}	111.3	c, h	31
			55	1.33×10^{-4}		c	31
			65	4.69×10^{-4}		c	31
	tert-butyl bicyclo[2,2,2]octane-1-percarboxylate	cumene	65	7.5×10^{-6}	119.2	h	160
			75	2.82×10^{-5}			160
			85	8.47×10^{-5}			160
			80	5.1×10^{-5}	120.1	h	223
			90	1.67×10^{-4}			223
			100	4.86×10^{-4}			223
	tert-butyl endo-bicyclo[2,2,1]-2-methyl-heptane-2-percarboxylate	cumene	75	7.35×10^{-5}	126.8	h	206
			80	1.39×10^{-4}			206
			85	2.64×10^{-4}			206
			90	4.81×10^{-4}			206
			95	8.41×10^{-4}			206
	tert-butyl exo-bicyclo[2,2,1]-2-methyl-heptane-2-percarboxylate	cumene	60	7.78×10^{-5}	116.7	h	206
			65	1.44×10^{-4}			206
			70	2.75×10^{-4}			206
			75	4.89×10^{-4}			206
			80	8.93×10^{-4}		c	206
	tert-butyl 1,4-dimethylcyclohexane-1-percarboxylate (cis)	cumene	60.0	7.52×10^{-5}	115.2	h	206
			70.0	2.66×10^{-4}			206
			80.0	8.49×10^{-4}		c	206
	(trans)	cumene	60.0	6.93×10^{-5}	116.3	c, h	206
			70.0	2.52×10^{-4}		c	206
			80.0	7.79×10^{-4}			206
	tert-butyl 3-methoxyphenyl peracetate	chlorobenzene	79.6	9.9×10^{-5}	122.2	h	178
			90.7	3.45×10^{-4}			178
			100.5	1.051×10^{-3}			178
	tert-butyl 4-methoxyphenyl peracetate	chlorobenzene	56.0	4.57×10^{-5}	105.4	h	178
			60.3	9.9×10^{-5}			178
			70.2	3.06×10^{-4}			178
			79.3	7.99×10^{-4}			178
		cumene	60	6.85×10^{-5}		z(0.98)	207
			60	5.31×10^{-5}		z(4050)	207
			79.6	5.942×10^{-4}			207
		decane	77.5	2.52×10^{-4}			168
		dodecane	77.5	2.60×10^{-4}			168
		isooctane	60.5	4.72×10^{-5}			169
		octane	77.5	2.53×10^{-4}			168
		paraffin oil	60.5	4.20×10^{-5}			169
		tetradecane	77.5	2.68×10^{-4}			168

Notes page II-40; References page II-41

PERESTERS AND PEROXY CARBONATES (Cont'd.)

Number of C atoms	Initiator	Solvent	T °C	k_d [s^{-1}]	E_a [kJ mol^{-1}]	Notes	References
13	tert-butyl 4-methylphenyl peracetate	chlorobenzene	70.4	8.67×10^{-5}	110.9	h	178
			79.6	2.37×10^{-4}			178
			90.7	7.95×10^{-4}			178
		cumene	79.6	1.649×10^{-4}		z(0.98)	207
			79.6	1.180×10^{-4}		z(6080)	207
		decane	77.5	9.08×10^{-5}			168
		dodecane	77.5	8.57×10^{-5}			168
		hexadecane	77.5	1.032×10^{-4}			168
		octane	77.5	8.83×10^{-5}			168
		tetradecane	77.5	7.75×10^{-5}			168
	di-tert-butyl perglutarate	styrene	105	2.65×10^{-5}		x_1	215
			115	1.06×10^{-4}		x_1	215
			125	3.00×10^{-4}		x_1	215
	4-nitrobenzyl (tert-butylperoxy) oxalate	benzene	45	1.30×10^{-5}	116.7	c, h	31
			55	4.89×10^{-5}		c	31
			65	1.89×10^{-4}		c	31
14	tert-butyl α-methylpercinnamate (cis)	cumene	99.6	1.23×10^{-4}	128.0	f, h	150
			109.8	3.70×10^{-4}		f	150
	(trans)	cumene	99.6	3.5×10^{-5}	125.5	f, h	150
			109.8	1.03×10^{-4}		f	150
	tert-butyl 2-phenyl-3-perbutenoate	chlorobenzene	60	2.9×10^{-3}	96.2	c, h	81
	tert-butyl 4-phenyl-3-perbutenoate	chlorobenzene	60	1.15×10^{-4}	98.3	c. h, g	81
	tert-butyl phenyldimethylperacetate	chlorobenzene	60	9.6×10^{-4}	109.2	c, h	81
		isooctane	40.6	2.95×10^{-5}		t_9	165
			60.6	3.059×10^{-4}		t_9	165
			60.6	4.064×10^{-4}		m_2	165
	dibenzyl peroxolate	pentane	25	6.7×10^{-5}			203
	di-tert-butyl adipate	styrene	105	3.50×10^{-5}	151.0	x_1	215
			115	1.23×10^{-4}		x_1	215
			125	3.05×10^{-4}		x_1	215
	dicyclohexyl peroxydicarbonate	benzene	50	5.4×10^{-5}		c	201
		α-methylstyrene	50	5.9×10^{-5}			201
	4-methoxybenzyl (tert-butylperoxy) oxalate	benzene	45	6.69×10^{-5}	109.6	c, h	31
			55	2.48×10^{-4}		c	31
			65	8.27×10^{-4}		c	31
15	tert-butyl 1-adamantyl percarboxylate	cumene	45	5.15×10^{-6}	115.5	h	160
			55	2.05×10^{-5}			160
			65	7.40×10^{-5}			160
			64.6	6.87×10^{-5}	116.7	h	159
			74.6	2.44×10^{-4}			159
			84.6	7.78×10^{-4}			159
			60	5.4×10^{-5}	124.7	h	223
			70	2.12×10^{-4}			223
			80	7.20×10^{-4}			223
	tert-butyl α,β-dimethyl percinnamate (cis)		95.0	1.02×10^{-4}	118.8	a, h	151
			103.2	2.29×10^{-4}		a	151
			110.1	5.15×10^{-4}		a	151
	(trans)		85.1	3.9×10^{-5}	122.6	a, h	151
			95.0	1.17×10^{-4}		a	151
			103.2	2.84×10^{-4}		a	151
			110.0	6.05×10^{-4}		a	151
			110.0	7.90×10^{-4}		a, v_6	151
	tert-butyl 1-pernaphthoate	chlorobenzene	110.1	8.69×10^{-4}			154
	tert-butyl 4-tert-butylperbenzoate	chlorobenzene	100.1	$3.81 \pm .02 \times 10^{-5}$			134
			119.8	$4.50 \pm .09 \times 10^{-4}$			134
			135.9	$2.38 \pm .03 \times 10^{-3}$			134
	di-tert-butyl perpimelate	styrene	105	4.45×10^{-5}	150.6	x_1	215
			115	1.05×10^{-4}		x_1	215
			125	5.06×10^{-4}		x_1	215

Number of C atoms	Initiator	Solvent	T °C	k_d [s^{-1}]	E_a [kJ mol^{-1}]	Notes	References
				PERESTERS AND PEROXY CARBONATES (Cont'd.)			
16	cumyl N-phenylperoxy carbamate	xylene	T[K]	1.26×10^{14} exp[-117.6 kJ /RT]			40
	dibenzyl peroxydicarbonate	toluene	50	2.92×10^{-5}		a	90
	di-tert-butyl perphthalate	benzene	100	1.08×10^{-5}	157.	a	126
			115	7.83×10^{-5}		a	126
			130	4.80×10^{-4}		a	126
	di-tert-butyl persuberate	styrene	115	7.86×10^{-5}	159.0	x_1	215
			125	2.81×10^{-4}		x_1	215
17	tert-butyl 2-(phenylthio)perbenzoate	acetone	25	1.89×10^{-5}		m_2	22
			40	1.22×10^{-4}		m_2	22
		acetonitrile	25	1.01×10^{-4}		m_2	22
			40	5.44×10^{-4}		m_2	22
		tert-butanol	25	5.26×10^{-5}		m_2	22
			40	3.24×10^{-4}		m_2	22
		chlorobenzene	39.3	4.98×10^{-5}		b	134
			53.8	$5.23 \pm .06 \times 10^{-4}$			134
			53.8	$4.58 \pm .07 \times 10^{-4}$		m_2	134
			70.0	$2.62 \pm .04 \times 10^{-3}$			134
			25	1.76×10^{-5}		m_2	22
			40	1.03×10^{-4}		m_2	22
		cyclohexane	25	9.8×10^{-7}		m_2	22
			40	6.9×10^{-6}		m_2	22
		DMSO	25	1.11×10^{-4}		m_2	22
			40	6.02×10^{-4}		m_2	22
		ethanol	25	2.31×10^{-4}		m_2	22
			40	1.65×10^{-3}		m_2	22
		isopropanol	25	1.33×10^{-4}		m_2	22
			40	7.25×10^{-4}		m_2	22
		methanol	25	8.21×10^{-4}		m_2	22
			40	4.75×10^{-3}		m_2	22
	tert-butyl dibenzothiophene-4-per-carboxylate	chlorobenzene	105	2.79×10^{-5}	128.0	h	22
			119.4	1.29×10^{-4}			22
			135.1	6.07×10^{-4}			22
	di-tert-butyl perazelate	styrene	105	5.13×10^{-5}	154.8		215
			115	1.20×10^{-4}			215
			125	5.10×10^{-4}			215
18	tert-butyl diphenylperacetate	chlorobenzene	60	4.44×10^{-4}	101.7	a, h	81
		cumene	40.3	2.75×10^{-5}	104.6	h	136
			49.6	9.81×10^{-5}			136
			59.9	3.10×10^{-4}			136
			70.4	1.04×10^{-3}			136
		decane	77.5	1.66×10^{-3}			168
		dodecane	77.5	1.688×10^{-3}			168
		hexadecane	77.5	1.632×10^{-3}			168
		octane	77.5	1.651×10^{-3}			168
		tetradecane	77.5	1.627×10^{-3}			168
	tert-butyl 2-(phenylthiomethyl)-perbenzoate	chlorobenzene	98.7	4.64×10^{-5}	134.7	h	154
			120.1	4.20×10^{-4}			154
	tert-butyl thioxanthone-4-percarboxylate	chlorobenzene	120	1.89×10^{-4}	156.5	h, m_2	22
			135	1.15×10^{-3}		m_2	22
	di-tert-butyl persebacate	styrene	105	5.10×10^{-5}	154.8	x_1	215
			115	1.20×10^{-4}		x_1	215
			125	5.00×10^{-4}		x_1	215
	di-n-heptyl persuccinate	styrene	60	6.1×10^{-6}	131.0	x_1	215
			73.5	3.85×10^{-5}		x_1	215
			85	1.70×10^{-4}		x_1	215
19	tert-butyl diphenylmethylperacetate	chlorobenzene	60	1.9×10^{-3}	103.3	a, h	81
	tert-butyl diphenylperglycidate (cis)	cumene	60.0	2.8×10^{-5}	112.1	h	204
			70.0	5.5×10^{-5}			204
			80.0	1.68×10^{-4}			204

Number of C atoms	Initiator	Solvent	T °C	k_d [s^{-1}]	E_a [kJ mol^{-1}]	Notes	References
		PERESTERS AND PEROXY CARBONATES (Cont'd.)					
19	tert-butyl diphenylperglycidate (trans)	cumene	60.0	1.6×10^{-5}	117.2	h	204
			70.0	3.9×10^{-5}			204
			80.0	1.20×10^{-4}			204
	tert-butyl α-phenylpercinnamate (cis)	cumene	99.6	1.52×10^{-4}	136.8	h	150
			109.8	4.41×10^{-4}			150
	(trans)	cumene	99.6	8.2×10^{-5}	133.9	h	150
			109.8	2.59×10^{-4}			150
20	di-n-heptyl peradipate	styrene	60	1.02×10^{-5}	133.1	x_1	215
			73.5	6.57×10^{-5}		x_1	215
			85	2.98×10^{-4}		x_1	215
21	tert-butyl 8-(phenylthio)-1-pernaphthoate	chlorobenzene	50	3.83×10^{-5}	99.2	h, m_2	154
			70	3.44×10^{-4}		m_2	154
			70	3.12×10^{-4}			154
			80	9.82×10^{-4}		m_2	154
22	2,5-dimethylhexyl 2,5-di(peroxybenzoate)	benzene	100	1.87×10^{-5}	154.0	a	126
			115	1.25×10^{-4}		a	126
			130	7.14×10^{-4}		a	126
23	tert-butyl 2,6-di(phenylthio)-perbenzoate	chlorobenzene	40.1	7.07×10^{-5}			154
24	tert-butyl triphenylperacetate	chlorobenzene	34.9	7.8×10^{-4}			153
		cumene	25.7	1.7×10^{-4}	100.8	h	153
			34.9	5.8×10^{-4}			153
			45.5	2.3×10^{-3}			153
	di-tert-butyl 2,3-diphenylpersuccinate	cumene	70.1	9.83×10^{-5}	125.5	c, h	196
			79.9	3.29×10^{-4}		c	196
			90.0	1.20×10^{-3}		c	196
	di-n-heptyl persebacate	styrene	60	1.17×10^{-5}	131.0	x_1	215
			73.5	7.13×10^{-5}		x_1	215
			85	3.03×10^{-5}		x_1	215
25	tert-butyl 2-(2,2-diphenylvinyl)perbenzoate	chlorobenzene	90.0	1.32×10^{-4}			175
			90.3	7.45×10^{-5}		t_9	175
			100.0	3.27×10^{-4}			175
			105.5	4.64×10^{-4}		t_9	175
			119.3	1.54×10^{-3}		t_9	175
		cyclohexane	90.0	3.5×10^{-5}			175
		methanol	90.4	2.3×10^{-3}		t_9	175
	tert-butyl 2-percarboxybenzalfluorene	chlorobenzene	90	3.75×10^{-4}			175
		methanol	90	2.47×10^{-3}			175
		7 - MISCELLANEOUS INITIATORS					
	N-(1-cyanocyclohexyl)pentamethylene-keteneimine	chlorobenzene	80.0	3.25×10^{-6}			179
			89.2	1.007×10^{-5}			179
			100.1	4.025×10^{-5}			179
	dibenzyl hyponitrite	paraffin	61.5	6.5×10^{-4}			240
			68.5	3.7×10^{-3}			240
			75.5	8.3×10^{-3}			240
			80.5	1.45×10^{-2}			240
			132	8.7×10^{-1}			240
	1,4-dimethyl-1,4-diphenyltetrazene-2	benzophenone	121	3.3×10^{-4}			240
			149	3.45×10^{-3}			240
		cumene	120	2.3×10^{-4}			240
			130	6.5×10^{-4}			240
			140	1.67×10^{-3}			240
		paraffin	164.2	2.6×10^{-2}			240
			174.3	5.6×10^{-2}			240
			186	1.43×10^{-1}			240
			194	2.5×10^{-1}			240
		silicone oil	126	3.7×10^{-4}			240
			139	1.92×10^{-3}			240
	potassium persulfate	0.1 m NaOH	50	9.5×10^{-7}	140.2	a	242
			60	3.16×10^{-6}		a	242

DECOMPOSITION RATES OF INITIATORS

Number of C atoms	Initiator	Solvent	$T\ ^{\circ}C$	$k_d\ [s^{-1}]$	$E_a\ [kJ\ mol^{-1}]$	Notes	References
		MISCELLANEOUS INITIATORS (Cont'd.)					
	potassium persulfate (Cont'd.)	0.1 m NaOH	70	2.33×10^{-5}		a	242
			80	9.16×10^{-5}		a	242
			90	3.5×10^{-4}		a	242
		water (pH 3)	50	1.66×10^{-6}		a	242
		water	80	6.89×10^{-5}		a, v_{11}	115
			80	5.78×10^{-5}		a, v_{11}, v_{12}	115
	1-pentanesulfonyl azide	diphenyl ether	166	4.46×10^{-4}			166
	1,4-butanedisulfonyl azide	diphenyl ether	163	5.02×10^{-4}			166
	1,6-hexanedisulfonyl azide	diphenyl ether	163	5.02×10^{-4}			166
	1,9-nonanedisulfonyl azide	diphenyl ether	150	8.84×10^{-5}			166
			160	2.25×10^{-4}			166
			170	4.45×10^{-4}			166
	1,10-decanedisulfonyl azide	diphenyl ether	163	4.45×10^{-4}			166
	1,4-dimethylcyclohexane-α,α'-disulfonyl azide	diphenyl ether	163	4.82×10^{-4}			166
	m-xylene-α,α'-disulfonyl azide	diphenyl ether	163	6.09×10^{-4}			166
	p-xylene-α,α'-disulfonyl azide	diphenyl ether	163	5.78×10^{-4}			166
	benzenesulfonyl azide	naphthalene	110	3.6×10^{-6}	152.3		235
			120	1.07×10^{-5}			235
			125	1.97×10^{-5}			235
			130	3.41×10^{-5}			235
			135	6.08×10^{-5}			235
	p-bromobenzenesulfonyl azide	naphthalene	120	1.36×10^{-5}			235
	p-chlorobenzenesulfonyl azide	naphthalene	120	1.15×10^{-5}			235
	p-methoxybenzenesulfonyl azide	naphthalene	120	1.31×10^{-5}			235
	p-nitrobenzenesulfonyl azide	naphthalene	120	1.60×10^{-5}			235
	p-toluenesulfonyl azide	1,4-dichlorobutane	145	1.70×10^{-4}			166
		dimethyl terephthalate	155	3.23×10^{-4}			166
		diphenyl ether	130	3.30×10^{-5}			166
			145	1.44×10^{-4}			166
			155	3.43×10^{-4}			166
		hexanoic acid	155	2.97×10^{-4}			166
		naphthalene	120	1.12×10^{-5}			235
		nitrobenzene	155	3.97×10^{-4}			166
		1-octanol	155	3.63×10^{-4}			166
		tetradecane	155	3.80×10^{-4}			166
	p-toluenesulfonyl-p-tolylsulfone	acetonitrile	29.3	3.9×10^{-5}	103.8	h	144
			39.3	1.45×10^{-4}			144
			49.3	5.4×10^{-4}			144
		dioxane	29.3	2.1×10^{-5}	115.5	h	144
			39.5	1.01×10^{-4}			144
			49.3	3.9×10^{-4}			144

C. NOTES

a	k_d converted to sec^{-1} from author's units	m_3	methyl methacrylate added to minimize induced decomposition
b	k_d values for several concentrations averaged	m_4	isobutene added to minimize induced decomposition
c	k_d increases with increasing initiator concentration	m_5	acenaphthalene added to minimize induced decomposition
d	k_d decreases with increasing initiator concentration	n	trichloroacetic acid added
d_2	after 1st half life; rate slower initially	o	addition of trichloroacetic acid did not affect k_d
e	k_d listed is for lowest initiator concentration	p	degassed; the rate is independent of concentration
f	k_d is extrapolated value for zero initiator concentration	q	addition of trichloroacetic acid increased k_d several fold
g	k_d has been corrected for induced decomposition	r	not inhibited, but initiator concentration low enough (.01-.09 molar)
h	$\Delta H^{\neq}$ not E_a		so that higher order decomposition is unimportant
i	pressure (number gives mbar)	s	solvent not degassed
j	iodometric analysis	t_1	2,6-di-tert-butylphenol added to inhibit induced decomposition
k	infrared analysis	t_2	α,α-diphenyl-β-picrylhydrazyl added to inhibit induced decomposition
l	k_d is limiting value with respect to additive concentration	t_3	phenyl-α-naphthylamine added to inhibit induced decomposition
m_1	3,4-dichlorostyrene added to minimize induced decomposition	t_4	tetrachloroquinone added to inhibit induced decomposition
m_2	styrene added to minimize induced decomposition	t_5	1,3,5-trinitrobenzene added to inhibit induced decomposition

NOTES (Cont'd.)

t_6 I_2 added to inhibit induced decomposition
t_7 O_2 added to inhibit induced decomposition
t_8 5-20% NO_2 added to inhibit induced decomposition
t_9 galvanoxyl added to inhibit induced decomposition
t_{10} α,γ-bisdiphenylene-β-phenylallyl added to inhibit induced decomposition
u_1 in absence of oxygen
u_2 from initiation rate data
v_1 acetic acid added
v_2 $CuCl_2$ added
v_3 CuCl added
v_4 0.1 M $AgClO_4$ / M AIBN added
v_5 3.9 M thiophenol added
v_6 tert-butyl mercaptan added

v_7 2.5 M cyclohexane added
v_8 ~4 x 10^{-3} M pyridine added
v_9 2 x 10^{-2} M pyridine added
v_{10} 2.5 M cumene added
v_{11} buffered with sodium pyrophosphate
v_{12} saturated with ethyl acetate
w stereoisomers
x_1 actual rate divided by two because of 2 identical peroxide groups
x_2 each peroxide group has different k_d
y measured in differential scanning calorimeter; subscript is heating rate in deg./min.
z pressure (number gives bar)

D. REFERENCES

1. J. E. Guillet, T. R. Walker, M. F. Meyer, J. P. Hawk, E. B. Towne, Ind. Eng. Chem., Prod. Res. Develop., 3, 257 (1964).
2. L. M. Arnett, J. Am. Chem. Soc., 74, 2027 (1952).
3. L. M. Arnett, J. H. Peterson, J. Am. Chem. Soc., 74, 2031 (1952).
4. W. M. Thomas, M. T. O'Shaughnessy, J. Polymer Sci., 11, 455 (1953).
5. A. T. Blomquist, A. J. Buselli, J. Am. Chem. Soc., 73, 3883 (1951).
6. C. G. Swain, J. T. Clarke, W. H. Stockmeyer, J. Am. Chem. Soc., 72, 5426 (1950).
7. D. J. Brown, J. Am. Chem. Soc., 70, 1208 (1948).
8. S. W. Butaka, L. L. Zabrocki, M. F. McLaughlin, J. R. Kolcznski, O. L. Mageli, Ind. Eng. Chem. Prod. Res. Develop., 3, 261 (1964).
9. W. A. Strong, Ind. Eng. Chem. Prod. Res. Develop., 3, 264 (1964).
10. L. E. Redington, J. Polymer Sci., 3, 503 (1948).
11. B. Baysal, A. V. Tobolsky, J. Polymer Sci., 8, 529 (1952).
12. A. Conix, G. Smets, J. Polymer Sci., 10, 525 (1953).
13. M. R. Gopalan, M. Santhappa, J. Polymer Sci., 25, 333 (1957).
14. J. C. Bevington, J. Toole, J. Polymer Sci., 28, 413 (1958).
15. H. C. Haas, J. Polymer Sci., 39, 493 (1959).
16. A. I. Lowell, J. R. Price, J. Polymer Sci., 43, 1 (1960).
17. H. C. Haas, J. Polymer Sci., 55, 33 (1961).
18. D. F. Doehnert, O. L. Mageli, Mod. Plastics, 36 No. 6, 142 (1959).
19. R. C. Lamb, P. W. Ayers, M. K. Toney, J. Am. Chem. Soc., 85, 3483 (1963).
20. H. J. Shine, J. A. Waters, D. M. Hoffman, J. Am. Chem. Soc., 85, 3613 (1963).
21. G. S. Hammond, R. C. Neuman, Jr., J. Am. Chem. Soc., 85, 1501 (1963).
22. D. L. Tuleen, W. G. Bentrude, J. C. Martin, J. Am. Chem. Soc., 85, 1938 (1963).
23. H. Hart, F. J. Chloupek, J. Am Chem. Soc., 85, 1155 (1963).
24. H. Hart, R. A. Cipriani, J. Am. Chem. Soc., 84, 3697 (1962).
25. L. J. Durham, H. S. Mosher, J. Am Chem. Soc., 84, 2811 (1962).
26. R. C. Petersen, J. H. Markgraf, S. D. Ross, J. Am. Chem. Soc., 83, 3819 (1961).
27. M. M. Martin, J. Am. Chem. Soc., 83, 2869 (1961).
28. P. D. Bartlett, D. M. Simons, J. Am. Chem. Soc., 82, 1753 (1960).
29. P. D. Bartlett, E. P. Benzing, R. E. Pincock, J. Am. Chem. Soc., 82, 1762 (1960).
30. P. D. Bartlett, R. E. Pincock, J. Am. Chem. Soc., 82, 1769 (1960).
31. L. J. Durham, H. S. Mosher, J. Am. Chem. Soc., 82, 4537 (1960).
33. C. Walling, G. Metzger, J. Am Chem. Soc., 81, 5365 (1959).
34. H. Hart, D. P. Wyman, J. Am. Chem. Soc., 81, 4891 (1959).
35. H. H. Lau, H. Hart, J. Am. Chem. Soc., 81, 4897 (1959).
36. E. L. O'Brien, F. M. Beringer, R. B. Mesrobian, J. Am. Chem. Soc., 81, 1506 (1959).
37. N. A. Milas, A. Golubovic, J. Am. Chem. Soc., 80, 5994 (1958).
38. C. G. Overberger, I. Tashlick, M. Vernstein, R. G. Hiskey, J. Am. Chem. Soc., 80, 6556 (1958).
39. J. P. Van Hook, A. V. Tobolsky, J. Am. Chem. Soc., 80, 779 (1958).
40. E. L. O'Brien, F. M. Berringer, R. B. Mesrobian, J. Am. Chem. Soc., 79, 6238 (1957).
41. B. K. Morse, J. Am. Chem. Soc., 79, 3375 (1957).
42. S. Solomon, C. H. Wang, S. G. Cohen, J. Am. Chem. Soc., 79 4104 (1957).
43. C. Walling, J. Pellon, J. Am. Chem. Soc., 79, 4786 (1957).
44. C. G. Overberger, J. G. Lombardino, I. Tashlich, R. G. Hiskey, J. Am. Chem. Soc., 79, 2662 (1957).
45. K. E. Russel, J. Am. Chem. Soc., 77, 3487 (1955).
46. S. G. Cohen, C. H. Wang, J. Am. Chem. Soc., 77, 3628 (1955).
47. M. Talât-Erben, S. Bywater, J. Am. Chem. Soc., 77, 3712 (1955).
48. C. G. Overberger, M. Lapkin, J. Am. Chem. Soc., 77, 4651 (1955).
49. G. S. Hammond, J. N. Sen, C. E. Boozer, J. Am. Chem. Soc., 77, 3244 (1955).
50. C. G. Overberger, W. F. Hale, M. B. Berenbaum, A. B. Finestone, J. Am. Chem. Soc., 76, 6185 (1954).
51. J. Smid, A. Rembaum, M. Szwarc, J. Am. Chem. Soc., 78, 3315 (1956).
52. S. G. Cohen, C. H. Wang, J. Am. Chem. Soc., 75, 5504 (1953).
53. A. T. Blomquist, I. A. Berstein, J. Am. Chem. Soc., 73, 5546 (1951).
54. C. G. Overberger, H. Biletch, J. Am. Chem. Soc., 73, 4880 (1951).
55. A. T. Blomquist, A. F. Ferris, J. Am. Chem. Soc., 73, 3408 (1951).
56. A. T. Blomquist, A. F. Ferris, J. Am. Chem. Soc., 73, 3412 (1951).
57. C. G. Overberger, M. B. Berenbaum, J. Am. Chem. Soc., 73, 2618 (1951).
58. V. Stannett, R. B. Mesrobian, J. Am. Chem. Soc., 72, 4125 (1950).
59. P. D. Bartlett, J. E. Leffler, J. Am. Chem. Soc., 72, 3030 (1950).
60. S. G. Cohen, D. B. Sparrow, J. Am. Chem. Soc., 72, 611 (1950).
61. C. G. Overberger, M. T. O'Shaughnessy, H. Shalit, J. Am. Chem. Soc., 71, 2661 (1949).
62. F. M. Lewis, M. S. Matheson, J. Am. Chem. Soc., 71, 747 (1949).
63. J. H. Raley, F. F. Rust, W. E. Vaughan, J. Am. Chem. Soc., 70, 88 (1948).
64. J. H. Raley, F. F. Rust, W. E. Vaughan, J. Am. Chem. Soc., 70, 1336 (1948).
65. C. S. Marvel, R. L. Frank, E. Prill, J. Am. Chem. Soc., 65, 1647 (1943).
66. C. E. H. Bawn, D. Verdin, Trans. Faraday Soc., 56, 815 (1960).
67. H. C. Bailey, G. W. Godin, Trans. Faraday Soc., 52, 68 (1956).
68. C. E. H. Bawn, R. G. Halford, Trans. Faraday Soc., 51, 780 (1955).
69. C. E. H. Bawn, S. F. Mellish, Trans. Faraday Soc., 47, 1216 (1951).
70. W. Braun, L. Rajbenbach, F. R. Eirich, J. Phys. Chem., 66, 1591 (1962).

REFERENCES (Cont'd.)

71. J. R. Thomas, O. L. Harle, J. Phys. Chem., 63, 1027 (1959).
72. P. L. Hanst, J. G. Calvert, J. Phys. Chem., 63, 104 (1959).
73. B. Barnett, W. E. Vaughan, J. Phys. Chem., 51, 926 (1947).
74. B. Barnett, W. E. Vaughan, J. Phys. Chem., 51, 942 (1947).
75. S. G. Cohen, F. Cohen, C. H. Wang, J. Org. Chem. 28, 1479 (1963).
76. R. R. Hiatt, W. M. J. Strachan, J. Org. Chem., 28, 1893 (1963).
77. J. E. Leffler, J. S. West, J. Org. Chem., 27, 4191 (1962).
78. W. Honsberg, J. E. Leffler, J. Org. Chem., 26, 733 (1961).
79. J. E. Leffler, A. F. Wilson, J. Org. Chem., 25, 424 (1960).
80. M. S. Kharasch, A. Fono, W. Nudenberg, J. Org. Chem., 16, 105 (1951).
81. P. D. Bartlett, R. R. Hiatt, J. Am. Chem. Soc., 80, 1398 (1958).
82. J. E. Leffler, R. D. Faulkner, C. C. Petropoulos, J. Am. Chem. Soc., 80, 5435 (1958).
83. L. Batt, S. W. Benson, J. Chem. Phys., 36, 895 (1962).
84. J. Smid, M. Szwarc, J. Chem. Phys., 29, 432 (1958).
85. A. Rembaum, M. Szwarc, J. Chem. Phys., 23, 909 (1955).
86. R. K. Brinton, D. H. Volman, J. Chem. Phys., 20, 25 (1952).
87. R. E. Rebbert, K. J. Laidler, J. Chem. Phys., 20, 574 (1952).
88. J. Murawski, J. S. Roberts, M. Szwarc, J. Chem. Phys., 19, 698 (1951).
89. J. E. Leffler, J. Am. Chem. Soc., 72, 67 (1950).
90. F. Strain. W. E. Bissinger, W. R. Dial, H. Rudoff, B. J. DeWitt, H. C. Stevens, J. H. Langston, J. Am. Chem. Soc., 72, 1254 (1950).
91. A. Farkas, E. Passaglia, J. Am. Chem. Soc., 72, 3333 (1950).
92. W. A. Pryor, D. M. Huston, T. R. Fiske, T. L. Pickering, E. Ciuffarin, J. Am. Chem. Soc., 86, 4237 (1964).
93. J. C. Bevington, A. Wahid, Polymer, 4, 129 (1963).
94. A. A. Frost, R. G. Pearson, "Kinetics and Mechanism," Wiley, New York (1953) p. 97.
95. J. C. Bevington, T. D. Lewis, Polymer, 1, 1 (1960).
96. F. W. Birss, C. J. Danby, C. Hinshelwood, Proc. Roy. Soc. (London), A, 239, 154 (1957).
97. J. C. Martin, W. G. Bentrude, Chem. Ind. (London), 1959, 192.
98. S. G. Cohen, S. J. Groszos, D. B. Sparrow, J. Am. Chem. Soc., 72, 3947 (1950).
99. W. E. Cass, J. Am. Chem. Soc., 72, 4915 (1950).
100. S. D. Ross, M. A. Fineman, J. Am. Chem. Soc., 73, 2176 (1951).
101. C. G. Overberger, H. Biletch, A. B. Finestone, J. Lilker, J. Herbert, J. Am. Chem. Soc., 75, 2078 (1953).
102. P. D. Bartlett, F. D. Greene, J. Am. Chem. Soc., 76, 1088 (1954).
103. M. G. Alder, J. E. Leffler, J. Am. Chem. Soc., 76, 1425 (1954).
104. C. G. Overberger, A. Lebovits, J. Am. Chem. Soc., 76, 2722 (1954).
105. A. Rembaum, M. Szwarc, J. Am. Chem. Soc., 76, 5975 (1954).
106. M. Levy, M. Steinberg, M. Szwarc, J. Am. Chem. Soc., 76, 5978 (1954).
107. S. G. Cohen, C. H. Wang, J. Am. Chem. Soc., 77, 2457 (1955).
108. J. E. Leffler, C. C. Petropoulos, J. Am. Chem. Soc., 79, 3068 (1957).
109. W. E. Cass, J. Am. Chem. Soc., 68, 1976 (1946)..
110. H. C. Ramsperger, J. Am. Chem. Soc., 50, 714 (1928).
111. F. P. Lossing, A. W. Tickner, J. Chem. Phys., 20, 907 (1952).
112. A. Robertson, W. A. Waters, J. Chem. Soc., 1948, 1578.
113. O. J. Walker, G. L. E. Wild, J. Chem. Soc., 1937, 1132.
114. J. E. Leffler, R. A. Hubbard, II, J. Org. Chem., 19, 1089 (1954).
115. P. D. Bartlett, K. Nozaki, J. Polymer Sci., 3, 216 (1948).
116. J. C. Bevington, J. Toole, L. Trossarelli, Makromol. Chem., 32, 57 (1959).
117. A. N. Bose, C. Hinshelwood, Proc. Roy. Soc. (London), A, 249, 173 (1959).

118. E. R. Bell, J. H. Raley, F. F. Rust, F. H. Seubold, W. E. Vaughan, Discussions Faraday Soc., 10, 242 (1951).
119. L. Bateman, H. Hughes, A. L. Morris, Discussions Faraday Soc., 14, 190 (1953).
120. G. S. Hammond, J. Am. Chem. Soc., 72, 3737 (1950).
121. G. S. Hammond, L. M. Soffer, J. Am. Chem. Soc., 72, 4711 (1950).
122. G. A. Russell, J. Am. Chem. Soc., 78, 1044 (1956).
123. J. A. Offenbach, A. V. Tobolsky, J. Am. Chem. Soc., 79, 278 (1957).
124. G. S. Hammond, U. S. Nandi, J. Am. Chem. Soc., 83, 1213 (1961).
125. G. O. Pritchard, H. O. Pritchard, A. F. Trotman-Dickenson, J. Chem. Soc., 1954, 1425.
126. O. L. Mageli, S. D. Butaka, D. J. Bolton, Wallace & Tiernan, Lucidol Division, Bulletin 30.30, "Evaluation of Organic Peroxides from Half-Life Data".
127. E. J. Harris, A. C. Egerton, Proc. Roy. Soc., (London) A 168, 1 (1938).
128. E. J. Harris, Proc. Roy. Soc., (London) A 173, 126 (1939).
129. J. K. Allen, J. C. Bevington, Proc. Roy. Soc., (London) A 262, 271 (1961).
130. G. Archer, C. Hinshelwood, Proc. Roy. Soc., (London) A 261, 293 (1961).
131. J. H. McClure, R. E. Robertson, A. C. Cuthbertson, Can. J. Res., B20, 103 (1942).
132. J. W. Breitenbach, A. Schindler, Monatsh. Chem., 83, 724 (1952).
133. R. C. Lamb, F. F. Rogers, G. D. Dean, F. W. Voight, J. Am. Chem. Soc., 84, 2635 (1962).
134. W. G. Bentrude, J. C. Martin, J. Am. Chem. Soc., 84, 1561 (1962).
135. R. C. Neuman, R. P. Pankratz, J. Org. Chem., 36, 4046 (1971).
136. P. D. Bartlett, L. B. Gortler, J. Am. Chem. Soc., 85, 1864 (1963).
137. C. G. Overberger, I. Tashlick, J. Am. Chem. Soc., 81, 217 (1959).
138. C. G. Overberger, J. G. Lambardino, J. Am. Chem. Soc., 80, 2317 (1958).
139. G. S. Hammond, J. R. Fox, J. Am. Chem. Soc., 86, 1918 (1964).
140. E. S. Huyser, C. J. Bredeweg, J. Am. Chem. Soc., 84, 2401 (1964).
141. E. Hedaya, R. L. Hinman, L. M. Kibler, S. Theodoropulos, J. Am. Chem. Soc., 86, 2727 (1964).
142. T. Koenig, W. Brewer, J. Am. Chem. Soc., 86, 2728 (1964).
143. E. S. Huyser, C. J. Bredeweg, R. M. Van Scoy, J. Am. Chem. Soc., 86, 4148 (1964).
144. J. L. Kice, N. E. Pawlowski, J. Am. Chem. Soc., 86, 4898 (1964).
145. C. Walling, H. N. Moulden, J. H. Waters, R. C. Newman, J. Am. Chem. Soc., 87, 518 (1965).
146. J. K. Kochi, D. M. Mog, J. Am. Chem. Soc., 87, 522 (1965).
147. P. D. Bartlett, M. McBride, J. Am. Chem. Soc., 87, 1727 (1965).
148. R. C. Lamb, J. G. Pacifici, P. W. Ayers, J. Am. Chem. Soc., 87, 3928 (1965).
149. S. Seltzer, F. T. Dunne, J. Am. Chem. Soc., 87, 2628 (1965).
150. L. A. Singer, N. P. Kong, J. Am. Chem. Soc., 88, 5213 (1966).
151. R. M. Fantazier, J. A. Kampmeier, J. Am. Chem. Soc., 88, 5219 (1966).
152. S. Seltzer, E. J. Hamilton, Jr., J. Am. Chem. Soc., 88, 3775 (1966).
153. J. P. Lorand, P. D. Bartlett, J. Am. Chem. Soc., 88, 3294 (1966).
154. T. H. Fischer, J. C. Martin, J. Am. Chem. Soc., 88, 3382 (1966).
155. C. Walling, Z. Cekovic, J. Am. Chem. Soc., 89, 6681 (1967).
156. T. Koenig, R. Wolf, J. Am. Chem. Soc., 89, 2948 (1967).
157. R. C. Neuman, J. V. Behar, J. Am. Chem. Soc., 89, 4549 (1967).
158. S. G. Mylonakis, S. Seltzer, J. Am. Chem. Soc., 90, 5487 (1968).
159. J. P. Lorand, S. D. Chodroff, R. W. Wallace, J. Am. Chem. Soc., 90, 5266 (1968).

REFERENCES (Cont'd.)

160. R. C. Fort Jr., R. E. Franklin, J. Am. Chem. Soc., 90, 5267 (1968).

161. J. E. Leffler, H. H. Gibson Jr., J. Am. Chem. Soc., 90, 4117 (1968).

162. S. E. Scheppelle, S. Seltzer, J. Am. Chem. Soc., 90, 358 (1968).

163. R. C. Neuman, J. V. Behar, J. Am. Chem. Soc., 91, 6024 (1969).

164. R. C. Lamb, J. R. Sanderson, J. Am. Chem. Soc., 91, 5034 (1969).

165. T. Koenig, R. Wolf, J. Am. Chem. Soc., 91, 2574 (1969).

166. D. S. Breslow, M. F. Sloan, N. R. Newburg, W. B. Renfrow, J. Am. Chem. Soc., 91, 2273 (1969).

167. C. Walling, H. P. Warts, J. Milovanovic, C. G. Pappiaonnou, J. Am. Chem. Soc., 92, 4927 (1970).

168. W. A. Pryor, K. Smith, J. Am. Chem. Soc., 92, 5403 (1970).

169. T. Koenig, J. Huntington, R. Cruthoff, J. Am. Chem. Soc., 92, 5413 (1970).

170. J. E. Leffler, R. G. Zepp, J. Am. Chem. Soc., 92, 3713 (1970).

171. J. C. Martin, J. W. Timberlake, J. Am. Chem. Soc., 92, 978 (1970).

172. R. C. Newman Jr., G. D. Holmes, J. Am. Chem. Soc., 93, 4242 (1971).

173. J. B. Levy, E. J. Lehmann, J. Am. Chem. Soc., 93, 5790 (1971).

174. K. Shen, J. Am. Chem. Soc., 93, 3064 (1971).

175. T. W. Koenig, J. C. Martin, J. Org. Chem., 29, 1520 (1964).

176. R. D. Schuetz, J. L. Shea, J. Org. Chem., 30, 844 (1965).

177. J. R. Thomas, J. Am. Chem. Soc., 77, 246 (1955).

178. P. D. Bartlett, C. Ruchardt, J. Am. Chem. Soc., 82, 1756 (1960).

179. C. S. Wu, G. S. Hammond, J. M. Wright, J. Am. Chem. Soc., 82, 5386 (1960).

180. S. G. Cohen, R. Zand, C. Steel, J. Am. Chem. Soc., 83, 2895 (1961).

181. P. D. Bartlett, R. E. Pincock, J. Am. Chem. Soc., 84, 2445 (1962).

182. M. M. Martin, D. C. De Jongh, J. Am. Chem. Soc., 84, 3526 (1962).

183. S. Selzer, J. Am. Chem. Soc., 83, 2625 (1961).

184. S. N. Gupta, U. S. Nandi, J. Polymer Sci., A-1 8, 3019 (1970).

185. T. Kagiya, M. Izu, S. Kawai, K. Fukui, J. Polymer Sci., A-1 6, 1719 (1968).

186. K-P. S. Kwei. J. Polymer Sci., A3, 2387 (1965).

187. C. Leggett, J. C. J. Thynne, Trans. Faraday Soc., 63, 2504 (1967).

188. S. W. Benson, G. N. Spokes, J. Phys. Chem. 72, 1182 (1968).

189. G. A. Mortimer, J. Org. Chem., 30, 1632 (1965).

190. W. H. Richardson, J. Org. Chem., 30, 2804 (1965).

191. R. C. Lamb, J. G. Pacifici, P. W. Ayers, J. Org. Chem., 30, 3099 (1965).

192. E. S. Huyser, R. M. Van Scoy, J. Org. Chem., 33, 3524 (1968).

193. C. Walling, J. C. Azar, J. Org. Chem., 33, 3888 (1968).

194. R. C. Lamb, L. P. Spadafino, R. G. Webb, E. B. Smith, W. E. McNew, J. G. Pacifici, J. Org. Chem., 31, 147 (1966).

195. F. D. Greene, W. Adam, G. A. Knudsen Jr., J. Org. Chem., 31, 2087 (1966).

196. L. M. Bobroff, L. B. Gortler, D. J. Sahn, H. Wiland, J. Org. Chem., 31, 2678 (1966).

197. R. Hiatt, T. Mill, K. C. Irwin, J. K. Castleman, J. Org. Chem., 33, 1421 (1968).

198. R. Hiatt, K. C. Irwin, C. W. Gould, J. Org. Chem., 33, 1430 (1968).

199. R. Hiatt, K. C. Irwin, J. Org. Chem., 33, 1436 (1968).

200. P. Kovacic, R. R. Flynn, J. F. Gormish, A. H. Kappelman, J. R. Shelton, J. Org. Chem., 34, 3312 (1969).

201. D. E. Van Sickle, J. Org. Chem., 34, 3446 (1969).

202. E. S. Huyser, K. J. Jankauskas, J. Org. Chem., 35, 3196 (1970).

203. R. A. Sheldon, J. K. Kochi, J. Org. Chem., 35, 1223 (1970).

204. A. Padwa, N. C. Das, J. Org. Chem., 34, 816 (1969).

205. C. G. Overberger, D. A. Labianca, J. Org. Chem., 35, 1762 (1970).

206. W. G. Schindel, R. E. Pincock, J. Org. Chem., 35, 1789 (1970).

207. R. C. Newman Jr., J. V. Behar, J. Org. Chem., 36, 654 (1971).

208. C. Walling, D. Bristol, J. Org. Chem., 36, 733 (1971).

209. R. C. Lamb, W. E. McNew Jr., J. R. Sanderson, D. C. Lunney, J. Org. Chem., 36, 174 (1971).

210. G. S. Kolesnikov, A. Y. Chuchin., Polymer Sci, U.S.S.R., 7, 1931 (1965).

211. G. A. Nosayev, O. N. Romanlsova, Polymer Sci. U.S.S.R., 8, 14 (1966).

212. Y. A. Chuchin, V. A. Lazarev, M. B. Fromberg, Polymer Sci, U.S.S.R., 10, 2968 (1968).

213. L. M. Aparovich, T. I. Yurzhenko, Polymer Sci. U.S.S.R., 10, 1313 (1968).

214. A. I. Prisyazhnyuk, S. S. Ivanchev, Polymer Sci. U.S.S.R., 12, 514 (1970).

215. S. G. Yerigova, S. S. Ivanchev, Polymer Sci. U.S.S.R., 11, 2377 (1969).

216. C. H. Bamford, R. Denyer, J. Hobbs, Polymer 8, 493 (1967).

217. S. F. Nelsen, P. D. Bartlett, J. Am. Chem. Soc., 88, 137 (1966).

218. R. E. Pincock, J. Am. Chem. Soc., 86, 1820 (1964).

219. R. C. Lamb, J. G. Pacifici, J. Am. Chem. Soc., 86, 914 (1964).

220. S. Seltzer, J. Am. Chem. Soc., 85, 14 (1963).

221. J. B. Levy, B. K. W. Copeland, J. Am. Chem. Soc., 82, 5314 (1960).

222. T. Suehiro, H. Tsuruta, S. Hibino, Bull. Chem. Soc. Japan, 40, 674 (1967).

223. L. B. Humphrey, B. Hodgson, R. E. Pincock, Can. J. Chem. 46, 3099 (1968).

224. V. Stannett, R. B. Mesrobian, Discussions Faraday Soc., 14, 242 (1953).

225. Y. Takezaki, C. Takeuchi, J. Chem. Phys. 22, 1527 (1954).

226. K. Ziegler, W. Deparade, W. Meye, Annalen 567, 141 (1950).

227. A. R. Blake, K. O. Kutschke, Can. J. Chem., 37, 1462 (1959).

228. C. F. H. Tipper, J. Chem. Soc., 1953, 1675.

229. W. R. Foster, G. H. Williams, J. Chem. Soc., 1962, 2862.

230. G. B. Gill, G. H. Williams, J. Chem. Soc., 1965, 995.

231. G. B. Gill, G. H. Williams, J. Chem. Soc., 1965, 7127.

232. R. D. Schuetz, D. M. Teller, J. Org. Chem., 27, 410 (1962).

233. H. G. G. Dekking, J. App. Polymer Sci., 9, 1641 (1965).

234. C. G. Overberger, P. Fram, T. Alfrey Jr., J. Polymer Sci., 6, 539 (1951).

235. K. Takemoto, R. Fujita, M. Imoto, Makromol. Chem., 112, 116 (1968).

236. K. E. J. Barrett, J. App. Polymer Sci., 11, 1617 (1967).

237. T. J. Dougherty, J. Am. Chem. Soc., 83, 4849 (1961).

238. M. T. Jaquiss, J. S. Roberts, M. Szwarc, J. Am. Chem. Soc., 74, 6005 (1952).

239. S. Goldschmidt, B. Acksteiner, Annalen, 618, 173 (1958).

240. J. C. McGowen, T. Powell, Rec. Trav. Chim., 81, 1061 (1962).

241. H. A. Bent, B. Crawford Jr., J. Am. Chem. Soc., 79, 1793 (1957).

242. I. M. Kolthoff, I. K. Miller, J. Am. Chem. Soc., 73, 3055 (1951).

PROPAGATION AND TERMINATION CONSTANTS
IN FREE RADICAL POLYMERIZATION

R. Korus and K. F. O'Driscoll
Department of Chemical Engineering,
University of Waterloo,
Waterloo, Ontario, Canada

In a free radical addition polymerization,

$$P_n^{\cdot} + M \xrightarrow{k_p} P_{n+1}^{\cdot} \qquad (1)$$

$$2\, P_n^{\cdot} \xrightarrow{k_t} \text{dead polymer} \qquad (2)$$

where $P_n^{\cdot}$ is a propagating chain of any length n and M is the monomer. The rate constants are defined by the equations:

$$R_p = -\frac{d[M]}{dt} = k_p\,[P\cdot]\,[M] \qquad (3)$$

$$R_t = -\frac{d[P\cdot]}{dt} = g\, k_t\,[P\cdot]^2 \qquad (4)$$

$$\text{where } [P\cdot] = \sum_{n=1}^{\infty} [P_n^{\cdot}] \qquad (5)$$

It should be noted that the rate constants k_p and k_t are usually assumed to be independent of chain length. In the American literature, the value of g in Equation (4) usually is 2 while European researchers commonly make g unity. In this chapter, g = 1.

Simultaneous determination of absolute values of both k_p and k_t from a single experiment has not been reported. In practice, the ratio k_p^2/k_t is determined from measurements of molecular weight as a function of rate of polymerization for a low conversion polymerization. The ratio k_p/k_t is determined from non-steady state measurements of the average lifetime τ of the growing polymer chain in a photochemically initiated polymerization. This lifetime may be defined by noting that the concentration of chains present must be related to their average lifetime and rate of disappearance by

$$[P\cdot] / \tau = R_t \qquad (6)$$

which yields from equations (3) and (4)

$$\frac{k_p}{k_t} = \frac{\tau R_p}{[M]} \qquad (7)$$

By combining the separately determined ratios, k_p^2/k_t and k_p/k_t, the individual propagation and termination rate constants may be calculated. Alternatively, the rate of initiation, R_i may be measured as the rate of initiator disappearance and equated to R_t. This gives from equations (6) and (3)

$$k_p = \frac{R_p}{R_i\, \tau\, [M]} \qquad (8)$$

There is a large degree of imprecision inherent in measuring τ and in combining data from different experiments which helps to explain the scatter in the data tabulated here.

Details of theory and experiments are given in the general references (1): The tabulated data refer to two basically different methods of measuring τ : method A uses a rotating sector with intermittent photoinitiation over many periods of rising and falling free radical concentration; method B effectively measures only a single rise (or fall) in the radical concentration. A variety of experimental techniques have been used to sense changes occurring during polymerization:

B1 - dilatometry	B3 - interferometry	B5 - temperature change (thermistor)	B7 - light scattering
B2 - dielectric constant	B4 - temperature change (thermocouple)	B6 - viscosity	B8 - monomer pressure

Method C in the tabulation refers to recalculated values and method D to values obtained in emulsion polymerization by application of the Smith-Ewart theory (1f). It should be noted that methods B-7 and D measure only k_p.

The two monomers, styrene and methyl methacrylate, have been so extensively studied that the data for them are presented as Arrhenius plots. The scatter visible in Figures 1 and 2 should serve as a warning against casual acceptance of any single number.

In copolymerization where four propagation rate constants are commonly used, two treatments have been used to desribe the variation of polymerization rate with monomer feed composition: one assumes the existence of three simple termination reactions (two-terminations and one cross-termination) related by a "φ-factor", and the other regards the termination rate constant as a diffusion controlled parameter which is composition dependent. Given the uncertainty in this area, a tabulation of rate constants is not yet warranted. A single paper has appeared recently applying the rotating sector technique to polymerization (1e).

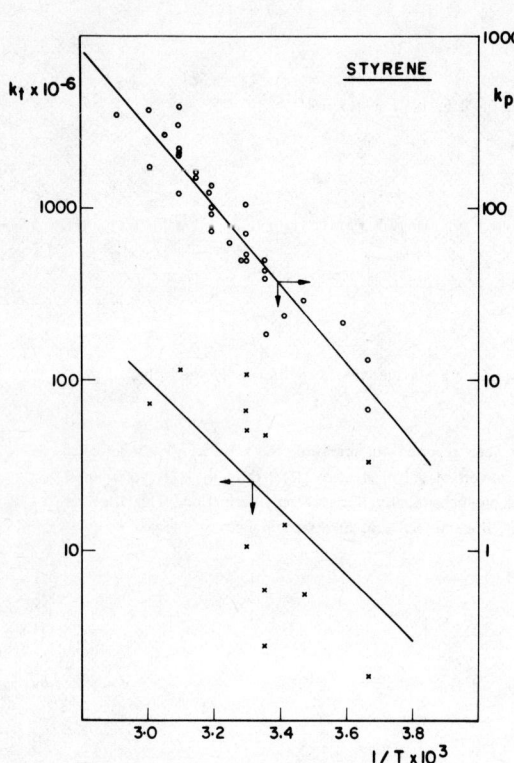

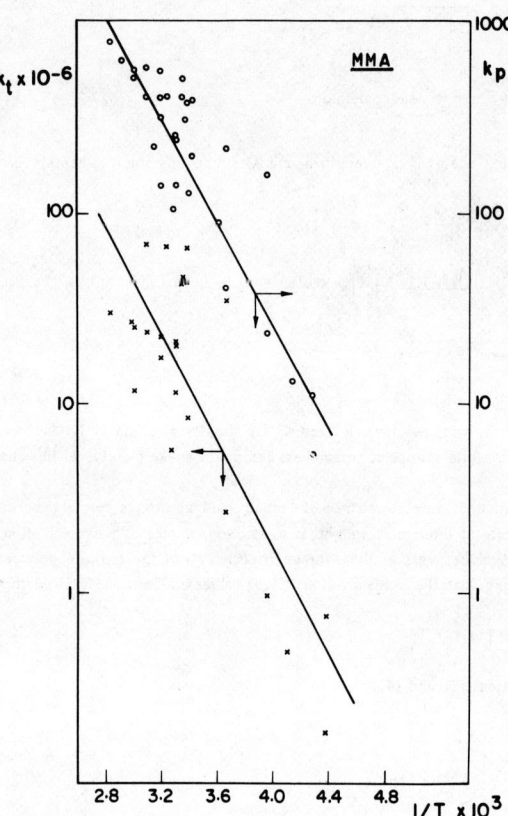

Arrhenius plots of all data for styrene for k_p (o) and k_t (x).
Solid line is least-squares calculated assuming all points of equal value.

Arrhenius plots of all data for methyl methacrylate for k_p (o) and k_t (x).
Solid line is least-squares calculated assuming all points of equal value.

Monomer	k_p [1/mol·s]	$k_t \cdot 10^{-6}$ [1/mol·s]	$k_p/k_t \cdot 10^6$	Temp. [°C]	Method	Remarks	References
1.1 DIENES							
Butadiene	8.4	-	-	10	D		25
Chloroprene	220	-	-	40	D	quoted in Ref. 1d	68
	105	-	-	10	D		7
	228	-	-	25	D		7
	423	-	-	35	D		7
Dimethylbutadiene	120	-	-	60	D		77
Isoprene	2.8	-	-	5	D		26
1.2 OLEFINS							
Ethylene	470±30	1050 ±50	-	83	A	solvent: benzene	58
	18.6±2	455 ±50	0.041	-20.01	A		72
	242	540	-	83	C	using results of Ref. 58	72
1.3.1 ACRYLIC DERIVATIVES							
Acrylamide	17200±3000	16.3±0.7	-	-	A	aqueous solvent, pH=1	86
	6000±1000	3.3±0.6	-	-	A	aqueous solvent, pH=5.5	86
	4000±1000	1.0±0.2	-	-	A	aqueous solvent, pH=13	86
	18000±1500	14.5±2.0	-	25	A	solvent: water	50
Acrylic acid,							
--, butyl ester	13	0.018	-	25	A		14
	-	-	820(680)	15	B2		23
	-	-	59	25	B4		34
	2100	330	-	25	B4		62
	14.5	0.018	-	35	A		14
	-	-	1250	30	B1		57
	-	-	1410	30	B1	solvent: benzene	57
	-	-	833	30	B1	solvent: ethyl acrylate	57
	-	-	840	30	B1	solvent: n-butyl propionate	57
--, methyl ester	1300	75	17	15	A	from unpublished results of Ross and Melville	23
	880	260	3.4	15	A	from unpublished results of Matheson	23
	-	-	12(10)	15	B2		23
	1580	55	28	25	A		20
	580	6.5	-	25	B1		55
	720	4.3	-	30	A		22
	1000	3.55	283	50	A		56
	2090	9.5	-	60	A		22
Acrylonitrile	-	-	1.9	20	B4	solvent: DMF	92
	3000-5000	-	-	0	D		48
	127	12.2	10.4	25	B4		53
	-	-	3.4	25	A	solvent: DMF	61
	52	5	-	25	B4		69
	51	1.8	29	25	B4,B5	solvent: DMF	80
	20000	-	-	40	D		48
	24	23	-	50	-	primary radical termination in a precipitating medium	38
	1960	782	-	60	A	solvent: DMF	49
	23000	2800	-	15	A	in water	60
	14500	2000	-	25	A	in water	59
	28000	3700	-	25	A	in water	60
	32500	4400	-	30	A	in water	60
	1910	290	-	25	A	solvent: dimethyl sulfoxide	74
	11600	19000	-	60	A	solvent: dimethyl sulfoxide	74
	178	24.9	-	0	A	solvent: DMF	75
	382±230	47.6±22	-	25	A	solvent: DMF	75
	660	86.5	-	66	A	solvent: DMF	75
	3300±300	1200 ±170	2.7	50	A	solvent: DMF, [M] = 3.8 m/1	96
	3200±400	300 ±50	10.8	50	A	solvent: DMS, [M] = 3.8 m/1	96
	3000±600	240 ±70	13.0	50	A	ethylene carbonate, [M] = 3.8 m/1	96

Monomer	k_p [1/mol·s]	$k_t \cdot 10^{-6}$ [1/mol·s]	$k_p/k_t \cdot 10^6$	Temp. [°C]	Method	Remarks	References
N,N-Dimethylacrylamide	11000	38	296	50	A		82

1.3.2 METHACRYLIC DERIVATIVES

Monomer	k_p [1/mol·s]	$k_t \cdot 10^{-6}$ [1/mol·s]	$k_p/k_t \cdot 10^6$	Temp. [°C]	Method	Remarks	References
Methacrylamide	800	16.5	-	25	A	solvent: water	76
Methacrylonitrile	~21	~27	-	30	-		30
	-	-	1.04	20	A		44
	26	21	-	25	A		44
	-	-	1.51	30	A		44
Methacrylic acid,							
--, benzyl ester	1250	41.9	29.9	30	A		94
--, n-butyl ester	369	10.2	-	30	A		28
	-	-	91	30	B1	solvent: ethyl acetate	78
--, 2-chloroethyl ester	254	6.71	38.5	30	A		94
--, 2-cyclohexylethyl ester	1190	32.8	36.3	30	A		94
--, ethyl ester	126	7.35	17.2	30	A		94
--, isobutyl ester	-	-	91	30	B1	solvent: ethyl acetate	78
--, isopropyl ester	121	4.52	26.7	30	A		94
--, 2-methoxyethyl ester	249	9.30	26.8	30	A		94
--, methyl ester	-	-	7.78	5	A		8
	143	12.2	-	30	A		8
	-	-	16.06	50	A		8
	367	18.7	-	60	A		8
	310±20	66 ±4	-	23.6	A		13
	410±50	68 ±12	-	35.9	A		13
	580±60	69 ±10	-	50.5	A		13
	-	-	40	30	B1		57
	-	-	14	30	B1	solvent: ethyl acetate	57
	-	-	46	30	B1	solvent: diethyl ether	57
	404	17.6	23	40	B1	solvent: ethyl acetate	57
	251	21	-	30	C	using results of Ref. 8	67
	322	22.5	-	40	C	using results of Ref. 8	67
	410	24	-	50	C	using results of Ref. 8	67
	515	25.5	-	60	C	using results of Ref. 8	67
	640	27.5	-	70	C	using results of Ref. 8	67
	800	30.5	-	80	C	using results of Ref. 8	67
	11.1	0.760	14.6	-45	B1	solvent: ethyl acetate	81
	5.43	0.184	29.4	-45	B1	solvent: ethyl acetate	81
	13.2	0.488	27	-30	B1	solvent: ethyl acetate	81
	24	0.979	24.5	-20	B1	solvent: ethyl ecetate	81
	160	-	4.3	-20	A	also rate constants given as a function of viscosity for various solvents	101
	390	-	10	20	A	also rate constants given as a function of viscosity for various solvents	101
	560	-	14	40	A		101
	260	21	12.3	25	A	solvent: benzene [M]=4.69 m/l	90
	270	21	12.9	25	A	solvent: fluorobenzene [M]=4.69 m/l	90
	280	19.5	14.45	25	A	solvent: chlorobenzene [M]=4.69 m/l	90
	285	17.5	16.2	25	A	solvent: anisole [M]=4.69 m/l	90
	310	17	18.2	25	A	solvent: bromobenzene [M]=4.69 m/l	90
	330	17	19.5	25	A	solvent: benzonitrile [M]=4.69 m/l	90
	260	21	12.3	25	A	solvent: benzene [M] = 4.69 m/l	98
	340	17	20.2	25	A	solvent: methyl benzoate [M] = 4.69 m/l	98
	240	11.5	20.9	25	A	solvent: methyl phenylacetate [M] = 4.69 m/l	98
	335	9	38.6	25	A	solvent: dimethyl phthalate [M] = 4.69 m/l	98
	270	22	11.9	25	A	solvent: dimethyl carbonate [M] = 4.69 m/l	98
	330	16	20.0	25	A	solvent: diethyl oxalate [M] = 4.69 m/l	98
	180±50	20 ±6	-	20	B7	$\underline{DP}=(2-8) \times 10^3$	103
	500±250	63 ±32	-	20	B7	$\underline{DP}=(3-5) \times 10^4$	103
	517	37	14	10	B1	solvent: methanol	102
	527	23	23	10	B1	solvent: pyridine	102
	28	4.4	6	10	B1	solvent: DMF	102
	41.6	2.69	-	0	B6	assuming biradical initiation	10

Monomer	k_p [1/mol·s]	$k_t \cdot 10^{-6}$ [1/mol·s]	$k_p/k_t \cdot 10^{6}$	Temp. [°C]	Method	Remarks	References
Methacrylic acid (Cont'd.)							
--, methyl ester (Cont'd.)	220	35	6.2	0	A	also rate constants given as a function of viscosity for various solvents	98
	92	-	-	5	D		71
	-	-	6.75	15	B3		24
	-	-	9.8	15	B4		34
	200	-	-	20	B7	k_p found as a linear function of mol. wt.	97
	128	8.44	15.2	22	B1		36
	384	44	8.7	22.5	B4		65
	512.6	46.6	-	25	A		35
	410	42.7	9.6	25	A	also rate constants given as a function of viscosity for various solvents	100
	248	22.7	10.9	30	A		15
	-	-	13.6	30	B4		34
	-	-	7.41	30	B5		42
	-	-	14	30	A		43
	-	-	14	30	B1	solvent: ethyl acetate	78
	141	11.6	12.1	30	A		94
	106	5.7	-	32	B4		63
	-	-	8.08	35	B5		42
	-	-	8.92	40	B5		42
	140	-	-	40	D		71
	224	-	-	45	D		73
	-	-	19	50	A		43
	573	2.0	-	60	C	termination by combinat. using results of Ref. 8 and 32	37
	573	11.9	-	60	C	termination by disprop. using results of Ref. 8 and 32	37
--, nonyl ester	-	-	180	30	B1	solvent: ethyl acetate	78
--, phenyl ester	411	11.9	34.6	30	A		94
--, 2-phenylethyl ester	228	1.88	123	30	A		94
--, 3-phenylpropyl ester	149	0.813	181	30	A		94
--, n-propyl ester	467	45.1	-	30	A		29
--, tert-butyl ester	350	14	-	25	A		64

1.3.4 VINYL HALIDES

Monomer	k_p [1/mol·s]	$k_t \cdot 10^{-6}$ [1/mol·s]	$k_p/k_t \cdot 10^{6}$	Temp. [°C]	Method	Remarks	References
Tetrafluoroethylene	7400	7.4×10^{-5}	-	40	-	large active chain end concentration measured by addition of inhibitor in aqueous solution polymerization	89
	9100	8.7×10^{-5}	-	50	-	large active chain end concentration measured by addition of inhibitor in aqueous solution polymerization	89
	-	-	16.6	30	B8	solvent: water	93
	-	-	16.4	40	B8	solvent: water	93
	-	-	17.3	50	B8	solvent: water	93
	-	-	19.6	60	B8	solvent: water	93
	-	-	20.5	70	B8	solvent: water	93
Vinyl bromide	5.8	42	0.138	-50	A		88
	22.7	92	0.246	-30	A		88
	570	385	-	20	-	quoted in Ref. 88	52
	995	705	-	50	-	quoted in Ref. 88	52
Vinyl chloride	6200	1100	-	25	A	solvent: tetrahydrofuran	33
	11000	2100	-	50	A	solvent: tetrahydrofuran	33
	11000	2100	71	25	A		61
	3130	2300	-	25	A		85
Vinylidene chloride	2.3	0.023	-	15	A		19
	8.6	0.175	-	25	A		19
	36.8	1.80	-	35	A		19

Monomer	k_p [1/mol·s]	$k_t \cdot 10^{-6}$ [1/mol·s]	$k_p/k_t \cdot 10^{6}$	Temp. [°C]	Method	Remarks	References

1.3.5 VINYL ESTERS

Monomer	k_p	$k_t \cdot 10^{-6}$	$k_p/k_t \cdot 10^{6}$	Temp.	Method	Remarks	References
Vinyl acetate	2000	220	-	-15	B6		12
	2800	220	-	0	B6		12
	-	-	3.1(3.5)	15	B2		23
	-	-	20	15	B3		24
	910(680)	57(36)	16(19)	15	B1		54
	586(556)	3040(2860)	-	15.9	A		2
	670	2500	-	15.9	A		4
	700	2600	-	15.9	A	solvent: n-hexane	5
	-	-	26	16	B4		34
	559	51.8	10.8	20	B5		47
	1100	80	-	25	A		3
	1012	58.8	17.2	25	A		9
	1000	59	-	25	A		16
	895	24	37.2	25	B4	at 4% conversion	39
	2640	116.8	22.6	50	A		9
	9500-19000	380-760	-	60	C	using results of Ref. 9 and 16	70
	7730	-	-	60	-		41
	-	-	12.1	30	B5		42
	-	-	13.4	35	B5		42
	-	-	14.4	40	B5		42

1.4 STYRENE DERIVATIVES

Monomer	k_p	$k_t \cdot 10^{-6}$	$k_p/k_t \cdot 10^{6}$	Temp.	Method	Remarks	References
Styrene	-	-	0.65(0.68)	15	B2		23
	-	-	2.0	15	B3		24
	40±20	80 ±40	0.5	15	A	dimension of k_p and k_t: [kg/mol·s]	51
	-	-	0.55	15	B1		95
	24	14	-	20	A	from copolymerization data with sulfur dioxide	27
	-	-	0.6	20	B4		92
	51.9	10.5	4.93	30	A		17
	72.5	66.5	-	30	A		45
	106	108	0.983	30	A		83
	102	-	-	40	D	quoted in Ref. 1d	68
	390	-	-	50	D		6
	209	115	-	50	C	using results of Ref. 21	66
	6.91	1.83	-	0	B6		7
	18.7	2.79	-	25	B6		7
	29.2	5.55	-	15	B1		18
	39.5	5.96	-	25	B1		18
	13.2	33.2	-	0	A		21
	-	-	0.538	10	A		21
	44	47.5	-	25	A		21
	55	50.5	1.089	30	A		21
	123	-	1.895	50	A		21
	176	72	-	60	A		21
	22	-	-	5	D		26
	139	-	-	40	D		26
	-	-	0.55	20	A		31
	-	-	2.36	30	A		31
	-	-	2.06;2.28	30	B5		42
	-	-	2.50	35	B5		42
	-	-	2.77	40	B5		42
	63	-	-	35	D		46
	96	-	-	40	D		46
	151	-	-	45	D		46
	223	-	-	50	D		46
	270	-	-	55	D		46
	376	-	-	60	D		46
	164	-	-	45	D		73
	206	-	-	50	D		73

Monomer	k_p [1/mol·s]	$k_t \cdot 10^{-6}$ [1/mol·s]	$k_p/k_t \cdot 10^{6}$	Temp. [°C]	Method	Remarks	References
Styrene (Cont'd.)	51.0	-	-	30.5	D		79
	120.5	-	-	40.5	D		79
	300.3(291.0)	-	-	50.2	D		79
	357.9	-	-	70.9	D		79
	50	-	-	25	D		71
	75	-	-	40	D		71
--, p-bromo-	186	46	4.03	30	A		83
--, p-chloro-	150	77	1.95	30	A		83
--, p-cyano-	219	35	6.25	30	A	solvent: dimethylacetamide	83
--, p-fluoro-	112	127	0.927	30	A		83
--, p-methoxy-	2.92	1.06	-	0	B6		11
	71	33	2.18	30	A		83
--, o-methyl-	60	-	-	50	D		79
--, p-methyl-	84	66	1.28	30	A		83
	140	-	-	50	D		79

1.5 VINYL HETEROAROMATICS

Monomer	k_p	$k_t \cdot 10^{-6}$	$k_p/k_t \cdot 10^{6}$	Temp.	Method	Remarks	References
N-Vinylcarbazole	0.84	0.092	91	-30	A		88
	2.2	0.110	20	-10	A		88
	-	-	17.5	0	A		88
	6.0	0.306	19.6	10	A		88
	-	-	13.7	30	A		88
2-Vinylpyridine	186	33	5.6	25	A,B4		84
	96.5	8.9	10.9	25	B4		84
4-Vinylpyridine	12	3	-	25	B6		40
	-	-	19	25	B5		80
5-Vinylpyridine, 2-methyl-	47	3.5	13.4	25	B5		80
	87.3	53	1.65	20	B1	in bulk	102
	122	66	1.85	20	B1	solvent: methanol, [M]=1.95 m/l	102
	138	59	2.35	20	B1	solvent: methanol, [M]=4.83 m/l	102
	209	43	4.90	20	B1	solvent: 50% (molar) aqueous methanol	102
	17.3	1.2	14.8	20	B1	solvent: acetic acid	102
N-Vinylpyyrolidone	953	65	14.5	20	B4		87
	-	-	35	25	B1	solvent: methanol	95

REFERENCES

1a. C. H. Bamford, W. G. Barb, A. D. Jenkins, P. F. Onyon, "The Kinetics of Vinyl Polymerization by Radical Mechanism" Butterworths Scientific Publications, London, 1958.

1b. A. Weissberger, "Technique of Organic Chemistry", Vol. VIII, Part II, Chapter XX, "Investigation of Rates and Mechanisms of Reactions", Interscience Publishers, New York, 1963.

1c. A. M. North, "Kinetics of Free Radical Polymerization", Pergamon Press, Oxford, (1966).

1d. Kh. S. Bagdasaryan, "Theory of Free Radical Polymerization", translated from the Russian Second Edition (1966) by J. Schmorak, Israel Program for Scientific Translations (1968).

1e. J. P. Marano, L. H. Shendalman, C. A. Walker, J. Polymer Sci. A-1, 8, 3461 (1970).

1f. W. V. Smith, R. H. Ewart, J. Chem. Phys. 16, 592 (1948).

2. G. M. Burnett, H. W. Melville, Nature 156, 661 (1945).

3. C. G. Swain, P. D. Bartlett, J. Am. Chem. Soc. 68, 2381 (1946).

4. G. M. Burnett, H. W. Melville, Proc. Roy. Soc. (London) A 189, 456 (1947).

5. G. M. Burnett, H. W. Melville, Proc. Roy. Soc. (London) A 189, 494 (1947).

6. W. V. Smith, J. Am. Chem. Soc. 70, 3695 (1948).

7. C. H. Bamford, M. J. S. Dewar, Proc. Roy. Soc. (London) A 192, 308 (1948).

8. M. S. Matheson, E. E. Auer, E. B. Bevilacqua, E. J. Hart, J. Am. Chem. Soc. 71, 497 (1949).

9. M. S. Matheson, E. E. Auer, E. B. Bevilacqua, E. J. Hart, J. Am. Chem. Soc. 71, 2610 (1949).

10. C. H. Bamford, M. J. S. Dewar, Proc. Roy. Soc. (London) A 197, 356 (1949).

11. D. W. E. Axford, Proc. Roy. Soc. (London) A 197, 374 (1949).

12. G. Dixon-Lewis, Proc. Roy. Soc. (London) A 198, 510 (1949).

13. M. H. Mackay, H. W. Melville, Trans. Faraday Soc. 45, 323 (1949).

14. H. W. Melville, A. F. Bickel, Trans. Faraday Soc. 45, 1049 (1949).

15. L. Valentine, Thesis, Aberdeen 1949, through Ref. 36.

16. H. Kwart, H. S. Broadbent, P. D. Bartlett, J. Am. Chem. Soc. 72, 1060 (1950).

17. H. W. Melville, L. Valentine, Trans. Faraday Soc. 46, 210 (1950).

18. G. M. Burnett, Trans. Faraday Soc. 46, 772 (1950).

19. J. D. Burnett, H. W. Melville, Trans. Faraday Soc. 46, 976 (1950).

20. Ross, Thesis, Aberdeen 1950, through Ref. 28.

21. M. S. Matheson, E. E. Auer, E. B. Bevilacqua, E. J. Hart, J. Am. Chem. Soc. 73, 1700 (1951).

22. M. S. Matheson, E. E. Auer, E. B. Bevilacqua, E. J. Hart, J. Am. Chem. Soc. 73, 5395 (1951).

23. T. G. Majury, H. W. Melville, Proc. Roy. Soc. (London) A 205, 496 (1951).

REFERENCES (Cont' d.)

24. N. Grassie, H. W. Melville, Proc. Roy. Soc. (London) A 207, 285 (1951).

25. M. Morton, P. P. Salatiello, H. Landfield, J. Polymer Sci. 8, 215 (1952).

26. M. Morton, M. P. Salatiello, H. Landfield, J. Polymer Sci. 8, 279 (1952).

27. W. G. Barb, Proc. Roy. Soc. (London) A 212, 177 (1952).

28. G. M. Burnett, P. Evans, H. W. Melville, Trans. Faraday Soc. 49, 1096 (1953).

29. G. M. Burnett, P. Evans, H. W. Melville, Trans. Faraday Soc. 49, 1105 (1953).

30. P. J. Flory, "Principles of Polymer Chemistry", Cornell University Press, New York 1953, p. 158.

31. S. Fujii, Bull. Chem. Soc. Japan 27, 216 (1954).

32. J. C. Bevington, H. W. Melville, R. P. Taylor, J. Polymer Sci. 14, 463 (1954).

33. G. M. Burnett, W. W. Wright, Proc. Roy. Soc. (London) A 211, 41 (1954).

34. W. J. Bengough, H. W. Melville, Proc. Roy. Soc. (London) A 225, 330 (1954).

35. B. R. Chinmayanandam, H. W. Melville, Trans. Faraday Soc. 50, 73 (1954).

36. G. M. Burnett, "Mechanism of Polymer Reactions", Interscience Publishers, New York, 1954, p. 230,233.

37. J. L. O' Brien, F. Gormick, J. Am. Chem. Soc. 77, 4757 (1955).

38. J. Durup, M. Magat, J. Polymer Sci. 18, 586 (1955).

39. W. J. Bengough, H. W. Melville, Proc. Roy. Soc. (London) A 230, 429 (1955).

40. P. F. Onyon, Trans. Faraday Soc. 51, 400 (1955).

41. M. Matsumoto, M. Maeda, Kobunshi Kagaku 12, 428 (1955).

42. H. Miyama, Bull. Chem. Soc. Japan 29, 711,715 (1956).

43. S. Fujii, S. Tanaka, S. Sutani, J. Polymer Sci. 20, 586 (1956).

44. N. Grassie, E. Vance, Trans. Faraday Soc. 52, 727 (1956).

45. E. A. Nicholson, R. G. W. Norrish, Discussions Faraday Soc. 22, 104 (1956).

46. E. Bartholome, H. Gerrens, R. Herbeck, H. Weitz, Z. Elektrochem. 60, 334 (1956).

47. H. Miyama, Bull. Chem. Soc. Japan 29, 720 (1956), 30, 10 (1957).

48. W. M. Thomas, R. L. Webb, J. Polymer Sci. 25, 124 (1957).

49. C. H. Bamford, A. D. Jenkins, R. Johnston, Proc. Roy. Soc. (London) A 241, 364 (1957).

50. F. S. Dainton, M. Tordoff, Trans. Faraday Soc. 53, 499 (1957).

51. K. Ueberreiter, G. Sorge, Z. Phys. Chem. (Frankfurt) 13, 158 (1957).

52. Kryszewski, Roczniki Chemii 31, 893 (1957).

53. W. J. Bengough, J. Polymer Sci. 28, 475 (1958).

54. W. J. Bengough, Trans. Faraday Soc. 54, 868 (1958).

55. W. J. Bengough, A. C. K. Smith, Trans. Faraday Soc. 54, 1553 (1958).

56. Z. A. Sinitsyna, Kh. S. Bagdasaryan, Zh. Fiz. Khim. 32, 1319 (1958).

57. S. W. Benson, A. M. North, J. Am. Chem. Soc. 81, 1339 (1959).

58. Z. Laita, Z. Machacek, J. Polymer Sci. 38, 459 (1959).

59. F. S. Dainton, D. G. L. James, J. Polymer Sci. 39, 299 (1959).

60. F. S. Dainton, R. S. Eaton, J. Polymer Sci. 39, 313 (1959).

61. W. J. Bengough, S. A. McIntosh, R. A. M. Thomson, Nature 184, 266 (1959).

62. W. J. Bengough, H. W. Melville, Proc. Roy. Soc. (London) A 249, 445 (1959).

63. W. J. Bengough, H. W. Melville, Proc. Roy. Soc. (London) A 249, 455 (1959).

64. D. H. Grant, N. Grassie, Trans. Faraday Soc. 55, 1042 (1959).

65. P. Hyden, H. W. Melville, J. Polymer Sci. 43, 201 (1960).

66. G. Henrici-Olive, S. Olive, Makromol. Chem. 37, 71 (1960).

67. G. V. Schulz, G. Henrici-Olive, S. Olive, Z. Phys. Chem. (Frankfurt) 27, 1 (1960).

68. Z. Maňásek, A. Řežábek, International Symposium on Macromolecular Chemistry, Section 2 (Russian version, 1960), through Ref. 1d, p. 117.

69. W. J. Bengough, Proc. Roy. Soc. (London) A 260, 205 (1961).

70. G. V. Schluz, D. J. Stein, Makromol. Chem. 52, 1 (1962).

71. P. E. M. Allen, G. M. Burnett, J. M. Downer, J. R. Majer, Makromol. Chem. 58, 169 (1962).

72. W. Rabel, K. Ueberreiter, Ber. Bunsenges. 67, 710 (1963).

73. H. Gerrens, Ber. Bunsenges. 67, 741 (1963).

74. E. F. T. White, M. J. Zissell, J. Polymer Sci. A 1, 2189 (1963).

75. N. Colebourne, E. Collinson, D. J. Currie, F. S. Dainton, Trans. Faraday Soc. 59, 1357 (1963).

76. F. S. Dainton, W. D. Sisley, Trans. Faraday Soc. 59, 1369 (1963).

77. M. Morton, W. E. Gibbs, J. Polymer Sci. A 1, 2679 (1963).

78. A. M. North, G. A. Reed, J. Polymer Sci. A 1, 1311 (1963).

79. K. P. Paoletti, F. W. Billmeyer, jr., J. Polymer Sci. A 2, 2049 (1964).

80. A. F. Revzin, Kh. S. Bagdasaryan, Zh. Fiz. Khim. 38, 1020 (1964).

81. J. Hughes, A. M. North, Trans. Faraday Soc. 60, 960 (1964).

82. A. M. North, A. M. Scallan, Polymer 5, 447 (1964).

83. M. Imoto, M. Kinoshita, M. Nishigaki, Makromol. Chem. 86, 217 (1965).

84. W. I. Bengough, W. Henderson, Trans. Faraday Soc. 61, 141 (1965).

85. W. I. Bengough, R. A. M. Thomson, Trans. Faraday Soc. 61, 1735 (1965).

86. D. J. Currie, F. S. Dainton, W. S. Watt, Polymer 6, 451 (1965).

87. V. A. Agasandyan, E. A. Trosman, Kh. S. Bagdasaryan, A. D. Litmonovich, V. Ya. Shtern, Vysokomoekul. Soedin. 8, 1580 (1966).

88. J. Hughes, A. M. North, Trans. Faraday Soc. 62, 1866 (1966).

89. A. N. Plyusnin, N. M. Chirkor, Teoret. i Eksp. Khimiya 2, 777 (1966).

90. C. H. Bamford, S. Brumby, Makromol. Chem. 105, 122 (1967).

91. F. Hrabák, M. Bezděk, V. Hynková, Z. Pelzbauer, J. Polymer Sci. C 16, 1345 (1967).

92. Kh. S. Bagdasaryan, É. A. Trosman, Kinetika i Kataliz 8, 43 (1967).

93. E. F. Nosov, Kinetika i Kataliz 8, 680 (1967).

94. K. Yokota, M. Kani, Y. Ishii, J. Polymer Sci. A-1, 6, 1325 (1968).

95. Yu. L. Spirin, T. S. Yatsimirskaya, Teoret. i Eksp. Khimiya 4, 849 (1968).

96. J. Ulbricht, J. Polymer Sci. C 16, 3747 (1968).

97. G. N. Kornienko, A. Chervenka, I. M. Bel' govskii, N. S. Yenikolopyan, Vysokomolekul. Soedin. A 11, 2703 (1969).

98. J. P. Fischer, G. Muecke, G. V. Schulz, Ber. Bunsenges. Physik. Chem. 73, 154 (1969).

99. C. H. Bamford, S. Brumby, Chem. & Ind. 1969, 1020 (1969).

100. G. M. Burnett, G. G. Cameron, M. M. Zafar, European Polymer J. 6, 823 (1970).

101. J. P. Fischer, G. V. Schulz, Ber. Bunsenges. Physik. Chem. 74, 1077 (1970).

102. A. V. Angelova, Yu. L. Spirin, R. Ye. Koval' chuk, Vysokomolekul. Soedin. A 12, 2708 (1970).

103. S. V. Kozlov, I. M. Bel' govskii, N. S. Enikolopyan, Vysokomolekul. Soedin. B 13, 46 (1971).

INHIBITORS AND INHIBITION CONSTANTS IN FREE RADICAL POLYMERIZATION [*]

J. Ulbricht

Technische Hochschule für Chemie
Leuna-Merseburg, DDR

The inhibition constant k_z is the rate constant of the reaction of a polymer radical P· with an inhibitor molecule Z:

$$P \cdot \; + \; Z \; \xrightarrow{\;k_z\;} \; X$$

X may represent unreactive or reactive products which can terminate or partially regenerate polymer chains as well as cause copolymerization of Z with the monomer M. The reaction of the inhibitor reduces more or less the overall rate of the polymerization.

The determination of the inhibition constant requires an exact kinetic analysis of each separate process, which has not been carried out in most cases. Generally only the ratio k_z/k_p is obtained, where k_p is the rate constant of propagation reaction:

$$P \cdot \; + \; M \; \xrightarrow{\;k_p\;} \; P \cdot$$

Inhibitor	Monomer	T [°C]	k_z/k_p	k_z [1/mol · s]	References
Acetophenone	methyl acrylate	50	-	< 0.25	10
	vinyl acetate	50	0.0091	16	11,13
Aniline	methyl acrylate	50	0.0001	< 0.1	10,13
	vinyl acetate	50	0.015	26	11,13
--, o-nitro-	methyl acrylate	50	0.00341	-	10
--, m-nitro-	methyl acrylate	50	0.00421	4.21	10,13
--, p-nitro-	methyl acrylate	50	0.00267	2.67	10,13
	vinyl acetate	50	4.97	8500	11,13
Anisol					
--, m-nitro	methyl acrylate	50	0.00727	7.2	10,13
--, p-nitro-	styrene	50	0.035	-	15
--, 1,3,5-trinitro-	styrene	50	20.3	-	14
Anthracene	acrylonitrile	50	2.67	2670	12
	methyl acrylate	50	0.098	100	10
	vinyl acetate	50	20.9	36000	11,13
Benzene					
--, bromo-	vinyl acetate	50	0.0019	3.3	11,13
--, diazoamino-	methyl acrylate	50	0.00419	-	10
--, o-dinitro-	methyl acrylate	50	0.018	-	10
	styrene	50	2.82	-	15
	vinyl acetate	45	96	-	1,2
--, m-dinitro-	methyl acrylate	44.1	-	58	6
		50	0.0309	30.9	10,13
	methyl methacrylate	44.1	0.0048	2.2	4
	styrene	50	5.17	-	15
	vinyl acetate	45	105	-	1,2
		50	66.0	113000	11,13
--, p-dinitro-	styrene	50	13.52	-	15
	vinyl acetate	45	267	-	1,2
		50	68.5	116000	11,13
--, nitro-	methyl acrylate	50	0.00464	4.64	10,13
	styrene	50	0.326	-	15
	vinyl acetate	45	19	-	1
		50	11.2	19300	11,13
--, nitro-d_5-	methyl acrylate	50	0.00468	-	10
--, p-nitro-chloro-	styrene	50	0.364	-	15
--, 1,3,5-trinitro-	methyl acrylate	50	0.204	204	10,13
	styrene	50	64.2	-	14
	vinyl acetate	45	890	-	2
		50	404	760000	11,13

[*] reprinted from the first edition

Inhibitor	Monomer	T [$^\circ$C]	k_z/k_p	k_z [1/mol $\cdot$ s]	References
Benzoic acid					
--, p-nitro-	methyl acrylate	50	0.0107	10.7	10,13
	vinyl acetate	50	24.9	43000	11,13
--, -, ethyl ester	styrene	50	1.68	-	15
--, 1,3,5-trinitroethyl ester	styrene	50	57.2	-	14
Benzonitrile	methyl acrylate	50	-	< 0.1	13
	vinyl acetate	50	0.0041	7.1	11,13
Benzophenone	methyl acrylate	50	-	< 0.15	13
	vinyl acetate	50	0.030	50	11,13
p-Benzoquinone	allyl acetate	80	50	-	3
	acrylonitrile	50	0.91	910	12
	methyl acrylate	44.1	-	1200	6
	methyl methacrylate	44.1	5.5	2400	4
		60	4.5	-	5
	styrene	50	518 ± 25	-	16
		60	227	-	5
		90	560	-	3
--, chloro-	styrene	50	720 ± 70	-	16
--, 2,5-dichloro-	methyl acrylate	44.1	-	10200	6
	methyl methacrylate	44.1	-	5500	6
	styrene	50	> 0.0002	-	18
--, 2,6-dichloro-	methyl acrylate	44.1	-	16700	6
	methyl methacrylate	44.1	-	16500	6
	styrene	50	$> 10^{-4}$	-	18
--, 2,3-dimethyl-	styrene	50	120 ± 20	-	16
--, 2,5-dimethyl-	styrene	40	106	-	17
(Phlorone)	styrene	50	82 ± 10	-	16
		60	61	-	17
		90	43	-	3
--, methoxy-	styrene	50	193 ± 10	-	16
--, methyl-	styrene	50	266 ± 15	-	16
--, tetrabromo-	styrene	50	618 ± 10	-	18
(Bromanil)					
--, tetrachloro-	methyl acrylate	44.1	-	2000	6
(Chloranil)	methyl methacrylate	44.1	0.26	120	4
	styrene	50	2040	-	18
--, tetraiodo-	styrene	50	2740 ± 30	-	18
(Iodanil)					
--, tetramethyl-	allyl acetate	80	4.1	-	3
(Duroquinone)	styrene	90	0.68	-	3
--, trichloro-	styrene	50	$> 10^4$	-	18
--, trimethyl-	styrene	50	25 ± 5	-	16
Benzoyl chloride	methyl acrylate	50	-	< 0.15	13
	vinyl acetate	50	0.037	64	11,13
Diphenyl	methyl acrylate	50	-	< 0.2	13
	vinyl acetate	50	0.027	46	11,13
--, nitro-	methyl acrylate	50	0.00604	-	10
Diphenylamine	vinyl acetate	50	0.014	24	11,13
Diphenylpicrylhydrazyl	methyl methacrylate	44.1	2000	-	4
Durene					
--, dinitro-	vinyl acetate	45	1.3	-	1
Ferric chloride in N,N-dimethylformamide	acrylonitrile	60	3.33	6500	9
	methacrylonitrile	60	30.8	620	9
	methyl acrylate	60	-	6800	9
	methyl methacrylate	60	-	5000	9
	styrene	60	536	94000	9
	vinyl acetate	60	-	235000	9
Fluorene	methyl acrylate	50	0.00602	-	10
Furfurylidene malononitrile	methyl acrylate	44.1	-	2900	6
	methyl methacrylate	44.1	1.2	550	4
Methane					
--, triphenyl-	methyl acrylate	50	0.00283	-	10

Inhibitor	Monomer	T [°C]	k_z/k_p	k_z [1/mol·s]	References
$(C_6H_5)CH_{0,3}D_{0,7}$ Naphthalene	methyl acrylate	50	0.00144	-	10
	methyl acrylate	50	-	< 0.2	13
	vinyl acetate	50	0.144	25.5	11,13
--, 1,5-dinitro-	methyl acrylate	50	0.0151	-	10
α-Naphthol	methyl acrylate	50	0.00496	-	10
--, 2-nitro-	methyl acrylate	50	0.0111	-	10
β-Naphthol					
--, 1-nitro-	methyl acrylate	50	0.0154	-	10
Oxygen	methyl methacrylate	50	33000	10^7	7
	styrene	50	14600	10^6-10^7	8
Phenanthrene	methyl acrylate	50	0.0026	2.6	10,13
	vinyl acetate	50	0.57	98.0	11,13
Phenol	methyl acrylate	50	0.0002	< 0.2	10,13
	vinyl acetate	50	0.012	21	11,13
--, 2,4-dinitro-	methyl acrylate	50	0.0649	-	10
--, o-nitro-	methyl acrylate	50	0.0108	-	10
--, m-nitro-	methyl acrylate	50	0.00562	5.62	10,13
--, p-nitro-	methyl acrylate	50	0.00426	4.26	10,13
	vinyl acetate	50	9.07	15500	11,13
Picramide	styrene	50	11.8	-	14
Picric acid	methyl acrylate	50	0.319	-	10
	styrene	50	211	-	14
Picryl chloride	styrene	50	58.5	-	14
Stilbene	methyl acrylate	50	0.00196	-	10
Sulfur	methyl acrylate	44.1	-	1100	6
	methyl methacrylate	44.1	0.075	40	4
	vinyl acetate	45	470	-	2
Toluene	vinyl acetate	50	0.0012	2.1	11,13
--, p-bromo-	vinyl acetate	50	0.0021	3.6	11,13
--, 2,4-dinitro-	methyl acrylate	50	0.0188	-	10
	styrene	50	1.543	-	15
--, o-nitro-	methyl acrylate	50	0.00323	-	10
	styrene	50	0.055	-	15
--, m-nitro-	methyl acrylate	50	0.00412	4.12	10,13
--, p-nitro-	methyl acrylate	50	0.00486	4.86	10,13
	styrene	50	0.203	-	15
	vinyl acetate	45	20	-	1
		50	10.8	18600	10,13
--, 1,3,5-trinitro-	methyl acrylate	44.1	-	105	6
		50	0.0596	-	10
	methyl methacrylate	44.1	0.05	23	4
	styrene	50	14.6	-	14
	vinyl acetate	45	890	-	2
Tolunitrile	vinyl acetate	50	0.0039	6.7	11,13
p-Xylene	vinyl acetate	50	0.0015	2.6	11,13

REFERENCES

1. P. D. Bartlett, H. Kwart, J. Am. Chem. Soc. 72, 1051 (1950).
2. P. D. Bartlett, H. Kwart, J. Am. Chem. Soc. 74, 3969 (1952).
3. P. J. Flory, "Principles of Polymer Chemistry", Cornell University Press, New York (1953), p. 172.
4. J. L. Kice, J. Am. Chem. Soc. 76, 6274 (1954).
5. J. C. Bevington, N. A. Ghanem, H. W. Melville, J. Chem Soc. (London) 1955, 2822.
6. J. L. Kice, J. Polymer Sci. 19, 123 (1956).
7. G. V. Schulz, G. Henrici, Makromol. Chem. 18/19, 437 (1956).
8. G. Henrici-Olive, S. Olive, Makromol. Chem. 24, 64 (1957).
9. C. H. Bamford, A. D. Jenkins, R. Johnston, Proc. Roy. Soc. (London) A 239, 214 (1957); J. Polymer Sci. 29, 355 (1958).

10. Z. A. Sinytsina, Kh. S. Bagdasaryan, Zh. Fiz. Khim. 32, 2663 (1958).
11. Z. A. Sinytsina, Kh. S. Bagdasaryan, Zh. Fiz. Khim. 34, 1110 (1960).
12. Z. A. Sinytsina, Kh. S. Bagdasaryan, Zh. Fiz. Khim. 34, 2736 (1960).
13. Kh. S. Bagdasaryan, Z. A. Sinytsina, J. Polymer Sci. 52, 31 (1961).
14. F. Tüdös, J. Kende, M. Azori, J. Polymer Sci. 53, 17 (1961), A1 1369 (1963).
15. F. Tüdös, J. Kende, M. Azori, Vysokomolekul. Soedin. 4, 1262 (1962).
16. F. Tüdös, L. Simandi, Vysokomolekul. Soedin. 4, 1271 (1962).
17. F. Tüdös, L. Simandi, Vysokomolekul. Soedin. 4, 1425 (1962).
18. F. Tüdös, L. Simandi. M. Azori, Vysokomolekul. Soedin. 4, 1431 (1962).

TRANSFER CONSTANTS TO MONOMER, POLYMER, CATALYST AND
SOLVENT IN FREE RADICAL POLYMERIZATION

Lewis J. Young
Koppers Company, Inc.
Law/Patent Department
Pittsburgh, Pennsylvania

Contents

A. INTRODUCTION

The transfer reaction in radical polymerization describes a process in which further growth of the individual polymer molecule is prevented but which does not interfere with the kinetic chain. The polymer radical reacts with an atom from another molecule (monomer, solvent, catalyst, etc.) forming dead polymer and a new radical.

$$P\cdot + RS \longrightarrow PR + S\cdot$$

This new radical can continue the kinetic chain. If the activity of the latter radical is similar to that of the former, the transfer process will have no influence upon the overall polymerization kinetics, but the molecular weight of the polymer produced will be less than it would be without this transfer reaction. In other words, a single kinetic chain produces several polymer molecules with the aid of the transfer reaction.

The quantitative treatment of this transfer process defines a dimensionless transfer constant C as the ratio of the rate of the transfer reaction to the rate of propagation and is determined by the following expression (Ref. 127).

$$\frac{1}{P_n} = \frac{1}{P_{n,o}} + C_X \frac{X}{M}$$

P_n = number average molecular weight obtained in the presence of transfer agent X.

$P_{n,o}$ = number average molecular weight obtained in the absence of transfer agent X.

C_X = transfer constant to transfer agent X.

M = monomer concentration

X = transfer agent concentration.

This expression is limited to cases, where $\frac{1}{P_{n,o}}$ is a constant.

This holds only under certain conditions, for instance, if the rate of propagation is dependent on the square root of the monomer concentration (145). A more general equation is given by the following (Ref. 145):

$$\frac{1}{P_n} = A \frac{R_p}{M^2} + \Sigma C_X \frac{X}{M}$$

Where $A = \dfrac{k_{t,k} + 2k_{t,d}}{2k_p^2}$

and $\Sigma C_X \dfrac{X}{M} = C_M + C_I \dfrac{I}{M} + C_S \dfrac{S}{M} + C_P \dfrac{P}{M} + \ldots$

R_p = overall rate of polymerization

$k_{t,d}$ = constant of termination by disproportionation

$k_{t,k}$ = constant of termination by combination

M = concentration of monomer

I = concentration of catalyst

k_p = propagation constant

S = concentration of solvent

P = concentration of polymer

C_M = transfer constant to monomer $(k_{tr,M}/k_p)$

C_I = transfer constant to catalyst $(k_{tr,I}/k_p)$

C_P = transfer constant to polymer $(k_{tr,P}/k_p)$

C_S = transfer constant to solvent $(k_{tr,S}/k_p)$

The transfer constant is obtained by measuring the number average molecular weight of the polymer at different concentrations of the transfer agent X, all other variables being kept constant. The first part of the right side of this equation is calculated first and used with its numerical value. The slope of the plot $\Sigma C_X X / M$ against X / M yields the transfer constant. Details of this evaluation can be found in the article of G. Henrici-Olivé (145).

The transfer constant of a very reactive molecule can also be determined from the rate of disappearance of transfer agent and monomer.

$$\frac{d\ln X}{d\ln M} = C_X$$

The conversion dependence of the transfer agent concentration (J.P. Fischer, Farbwerke Hoechst AG, Frankfurt (Main))

The decrease of the transfer agent concentration with increasing monomer conversion is important in practice because very reactive transfer agents are used up preferentially giving a broad molecular weight distribution if they cannot be replenished.

In analogy to quickly decomposing initiators with their criterion of "half-life" one can define in addition to the transfer constant C_X a "half-conversion" $U_{1/2}$ as that monomer conversion where the transfer agent is half consumed:

$$U_{1/2} = 100\,(1 - 0.5^{1/C_X}).$$

The following table demonstrates (with some examples for a calibration curve) that the "half-conversion" $U_{1/2}$ decreases with increasing transfer constant C_X:

Table: The "half-conversion" of transfer agents

C_X	$U_{1/2}$ (%)	C_X	$U_{1/2}$ (%)
0.1	99.9	5	13.0
0.2	96.8	10	6.7
0.5	75.0	20	3.4
1	50.0	50	1.4
2	29.3	100	0.7

The transfer agent must be replenished at this monomer conversion at the latest, if polymerization is to proceed still yielding a narrow molecular weight distribution.

The tables contain all data found in the literature. Some data (indicated by the letters J, R, or S) have to be used with caution especially if the correct molecular weight average has not been used, or if possible retardation of the polymerization rate has not been taken into account. Whereever possible, these errors have been mentioned. Unless otherwise noted, constants are assumed to have been determined in bulk, using azo-type initiators.

B. TRANSFER CONSTANTS TO MONOMERS

Monomer	T °C	$C_M \times 10^4$	Remarks	References
Acetic acid, allyl ester	80	176		29
		700		27
Acrylamide	25	0.12	D	92
		0.16		74
		0.2		75
	60	0.2	E	103
		0.6	E	103
Acrylamide, N,N-dimethyl-	50	1.5		272
Acrylic acid, benzyl ester	55	0.883		291
	60	0.905		291
	65	0.955		291
Acrylic acid, ethyl ester	50	0.193		306
	65	0.332		306
	70	0.351		306
Acrylic acid, methyl ester	55	0.275	C,AA	221
	60	0.036	B,AA	221
		0.325	C,AA	221
	65	0.10	C,AA	221
		0.11	C,F10	123
		0.11	C,AA	221
		0.37	C,AA	221
	70	0.01	C,F10	322
		0.072	B,AA	221
		0.16	C,AA	221
		0.18	C,AA	221
	75	0.405	C,AA	221
		0.224	C,AA	221
		0.25	C,F10	123
		0.25	C,AA	221
Acrylic acid, tetrahydrofurfuryl ester	40	4.0	C,F2	319
Acrylic acid, 2-chloro-, ethyl ester	60	3.0		342
Acrylic acid, thio-, methyl ester	60	560		140
Acrylonitrile	20	0.18	E	299
	25	0.105		284
	30	1.5	F8	196
	40	0.17		284
	50	0.050	F18	410
		0.27	F14	376
		8.2	F16	362
	60	0.26		95
Acrylonitrile (Cont'd.)		0.3		23
Anisole, p-vinyl	0	0.198		13
	60	0.74		53
Anthranilic acid, vinyl ester	70	80		139
Benzoic acid, vinyl ester	60	6.0		321,393
	80	7.0		26
	95	2.0		321
	50	4.0		213
1,3-Butadiene, 2-chloro-	50	2.32	F2	153
1,3-Butadiene, 1,1,2-trichloro-	25	16	E	230
1-Butene	40	3.1		334
	50	5.1		334
	60	7.3		334
2-Butene, cis-	40	3.2		334
	50	4.9		334
	60	11.2		334
2-Butene, trans-	40	3.0		334
	50	5.2		334
	60	10.8		334
3-Buten-2-one, 3-methyl-	80	4.00	C	84
Butyric acid, vinyl ester	50	26.7		71
	80	22.3		200
Carbamic acid, vinyl-, ethyl ester	60	0.25		117
Decanoic acid, vinyl ester	50	45.5		71
Ethylene	60	0.4	W5	102
		1.4	W4-W5	102
		4.2	W4	102
	83	5.0	D,w	204
		5.32	D,w	205
	110	1.1	W4-W5	102
		3.5	W4-W5	102
		9.0	W4	259
	130	0.0	C,F21,W5	102
		1.6	W4-W5	102
		4.7	W4-W5	102
		11.2	W4	102
	50-70	5	W3	218

Monomer	T °C	$C_M \times 10^4$	Remarks	References
Ethylene, chloro-	20	3.2		94
	25	3.2		315
		7.0		425
	30	6.25	Y	58
	40	19.4	F5	463
	50	6.4		246
		7.8	F3	379
		8.5	F4	379
		11		379
		13.5	C,F26	392
		10.8	Y	58
	60	12.3	F3	379
		12.8	F4	379
	70	23.8	Y	58
Glutaramic acid, N,N-diisobutyl-, vinyl ester	60	42	C,F10	106
Glutaramic acid, N,N-diisopentyl-, vinyl ester	60	51	C,F10	107
	70	51	C,F10	107
	80	51	C,F10	107
Hexanoic acid, vinyl ester	80	36		200
5-Hexen-3-yn-2-ol, 2-methyl-	60	5.0		226
Isobutyric acid, vinyl ester	80	46		200
Lauric acid, vinyl ester	50	45.5		71
Maleic anhydride	75	750	C	175
Maleic anhydride/methyl methacrylate	60	100		367
Methacrylic acid, bornyl ester	60	2.85		162
-", butyl ester	50	0.54		233
	60	0.14		264
-", tert-butyl ester	70	0.35		9
-", 2,3-epoxypropyl ester	60	0.59		198
-", ethyl ester	45	0.248	CC	193
	60	0.259	CC	193
	80	0.456	CC	193
	90	0.442	CC	193
-", hexadecyl ester	70	0.14		138
-", isobornyl ester	60	1.85		162
-", isobutyl ester	35	0.189	DD	193
	50	0.179	DD	193
	60	0.14	F2	264
		0.165	DD	264
		0.224	DD	193
	80	0.301	DD	193
	100	0.382	DD	193
Methacrylonitrile	25	2.08	Z	125
	60	5.81	Z	125
	70	8.00	Z	125
	80	10.05	Z	125
Methyl methacrylate	0	0.128		69
	30	0.148		18
		0.117		69
		0.260	D	224
	50	0.10	F2	147, 327
		0.15		145
		0.477	D	224
		0.85		69
	60	0.07		274
		0.10	C	34, 290, 317
		0.103		316
	65	0.18		145
	70	0.20	C,F10	123
		0.17		267
		0.20		145
		0.23		440
		0.265		123
		0.29	C,F17	123
		0.30		104
		0.45		224
		0.807	D	459
	75	1.37		123
		0.27	C,F10	123
		0.33	C	123
		0.60	C,F11	123
		0.70	C,F2	145
	80	0.25		123
		0.40		147
	90	0.10		145
	100	0.38		145
	120	0.58		
Naphthalene, 1-vinyl-	50	290		216
	60	310		216
	70	300		216
1,4-Pentadiene, 1,1,2,3,3,4,5,5-octafluoro-	110	500	W10	68
Phenol, o-vinyl-	70	130		186
2-Picoline, 5-vinyl-	70	6.7		11
1-Propene, 3-chloro-	80	1600		27
-", 2-methyl-	40	2.5		334
	50	4.4		334

Monomer	T °C	$C_M \times 10^4$	Remarks	References
1-Propene, 2-methyl- (Cont'd.)	60	6.9		334
Propionic acid, vinyl ester	--	3.6		158
	50	48.9		71
Pyridine, 2-vinyl-	15-35	0	D	38
--, 4-vinyl-	25	6.7		283
2-Pyrrolidinone, 1-vinyl-	20	4.0		58
Stearic acid, vinyl ester	50	69.8		71
Styrene	0	0	BB	360
		0.108		17,69
	25	0.279		69,245
		0.358		17
	27	0.31	A	145
	30	0.2	BB	318,360
		0.32		245
	45	0.3	BB	360
	50	0.35	H	354,355
		0.40	F2	318,327
		0.50	C,H	388
		0.6	C	145
		0.62	A,K	145
		0.65	B,C,H	388
		0.78	A	146
	60	0.07	C	467
		0.6	C	34,171,223
		0.6		145,240
		0.6	BB	318,360
		0.75		467
		0.79	A	145
		0.85	A,F	145
		0.92		447
		1.1		53
		1.37	C	234
	67.8	1.0		34
	70	0	C,F10	123
		0.6		104
		0.6	BB	360
		0.8	B,H	388
		0.96		67
		1.16	A	145
		1.35	A	145
		2.0	C,F2	123
	75	0	C,F10	123
		1.6	C	466
		5.0	C,F2	123
		5.00	C,F11	123

Monomer	T °C	$C_M \times 10^4$	Remarks	References
Styrene (Cont'd.)	80	0.7		98
		0.75	BB	360
		1.00		318
	80.3	4.0		34
	90	0.85	BB	360
		1.25	C,F	145
		1.47	A	145
		1.79	A	145
	99	1.5		98
	100	1.72	A	145
		1.8	A,F	145
		1.83	A	145
		2.80	A	145
	110	1.40		318
	117	2.45	A	145
	132	3.0	A,F	145
		3.4	A	145
		5.33	A	145
Styrene, p-bromo-	50	23		192
--, o-chloro-	30	0.25		62
	50	0.25	C	60
		0.28	C,H	60
--, p-iodo-	50	1.9	F1	56
Succinimide, N-vinyl-	55	0.55	C	93
Valeric acid, 4-methyl-, vinyl ester	80	24.8		200
Vinyl acetate	--	1.9		158
	-60	0.40	D	464
	-40	0.60	D	464
	-20	0.70	D	464
	-20	0.30	D	241
	0	0.50	D	241
		0.90	I4	86
	20	0.96	D	464
		0.94		328
	25	0.90	D	241
		1.3	I7	86
		1.45	I7	86
		2.4	D	202
		10.7		69
	40	1.29		12
		1.32		328
	45	2.0		51
	50	0.25		466
		0.41		265
		0.54	W14	265

Monomer (Cont'd.)	T °C	$C_M \times 10^4$	Remarks	References
Vinyl acetate (Cont'd.)		0.61	W15	265
		1.29		12
		4.55	C	71
		20		225
	60	1.75		12
		1.8		466
		1.9		124
		1.91		328
		1.93		348
		2.0		231
Vinyl acetate (Cont'd.)	60	2.1	I12	145
		2.4		294
		2.5	C	86
		2.5		167
		2.6		167
	65	2.8	C	80
		2.1	C	466
	70	2.4		403
		2.9	I13	86

C. TRANSFER CONSTANTS TO POLYMERS

Polymer	T °C	$C_P \times 10^4$	Remarks	References
Acrylamide, N,N-dimethyl-				
Poly(N,N-dimethylacrylamide)	50	0.61		272
Acrylic acid, ethyl ester				
Poly(methyl methacrylate)	60	12800	M6	120
Acrylic acid, methyl ester				
Poly(isoprene)	75	12.6	C,F2	308
--, chlorinated	60	0.5	L	212
Poly(methyl acrylate)		1.0	L	105
Poly(methyl methacrylate)	60	18000	M6	120
Acrylonitrile				
Cellulose	60	1.0	L,N	361
		11	L,M4	361
		20	L,M5	361
Poly(acrylonitrile)	50	4.7	F14	376
	60	3.5	L	135
Poly(methyl methacrylate)	60	0.2	N	25
		240	M2	25
		900	M3	25
		1270	M	25
Poly(sarcosine)	60	400		23
1,3-Butadiene				
Poly(1,3-butadiene)	50	11		142
Ethylene, chloro-				
Poly(vinyl chloride)	50	5	L	211
Hexanoic acid, vinyl ester				
Poly(oxyethylene), dodecyl ether	60	780		278
Methacrylic acid, butyl ester				
Poly(methyl methacrylate)	60	7700	M6	120
Methacrylic acid, dodecyl ester				
Poly(methyl methacrylate)	60	12800	M6	120
Methacrylic acid, ethyl ester				
Poly(isoprene), chlorinated	70	18.3	C,F2	308
Poly(vinyl chloride)	70	21.0	C,F28	308
Methyl methacrylate				
Poly(ethylene)	50	0.6		209
Poly(isoprene), chlorinated	80	23.4	C,F17	307
Poly(methyl methacrylate)	40	1.5	N	326
	50	360	M	326
		0.22	N	147
		1.5	N	326
		1.5	M	327
		360	M	326
		360	N	327
		1000	M	147
	60	0.1		105
		1.5	N	326
		2.1		262
	80	360	M	326
	80	2.48		262
	90	0.22	N	147
		1000	M	147
Poly(propylene)	50	0.04	C	292
		1.0		209
	130	0.42		292

Polymer	T°C	$C_p \times 10^4$	Remarks	References
Methyl methacrylate (Cont'd.)				
Poly(styrene)	50	0.75		148
	60	2.20		262
	80	2.95		262
Poly(vinyl acetate)	60	2.0		263
	80	2.8		263
Poly(vinyl chloride)	--	11		250
	70	10.0	C,F28	308
Poly(vinyl urethane)	50	17	C,F8	157
Rubber, natural	50	10.9		258
		11.0	C	258
2-Pyrrolidinone, 1-vinyl-				
Dextran	50	5		333
		5.87		203
Styrene				
Poly(oxadipoyloxy-2,2-dibromo-methyl trimethylene)	60	4.05	L	325
Poly(oxyethylene), dodecyl ether	60	20		278
Poly(oxyhexamethyleneoxy sebacoyl)	80	13	C,N	394
Poly(methyl methacrylate)	50	0.4	N	327
		<0.3	M	326
		1110	N	326
		1140	M	327
	60	16.4	M1	64
		17.5	M1	64
		57000	M7	120
		320000	M6	120
	80	3.74		262
	100	6.04		262
Poly(propylene)	50	0.025	C	292
	66	26		486
	130	0.3		292
		0.30		207
Poly(styrene)	50	1.9		144,148
		4.5		327
		14.0		327
		16.6		176
	55	15		177
	60	0.8	L	105
		1.9		73
		3.1		105
		15.4	M1	64
		15.8	M1	64
		16.6		176
	73.5	1.0	C	8
	85	1.4	C	8

Polymer	T°C	$C_p \times 10^4$	Remarks	References
Styrene (Cont'd.)				
Poly(styrene) (Cont'd.)	90	5.8		327
	100	2.0	N	276
	110	9.2		177
Poly(vinyl acetate)	130	10.8		176
	154	1.8		43
	100	1.5		43
Poly(vinyl acetate)	100	6.6		263
Poly(vinyl chloride)	130	9.2		263
Poly(2-vinylpyridine)	--	160		250
	50	8-10		279
Vinyl acetate				
Poly(oxyethylene)	60	17	N	277
		40	M	277
--, dodecyl ether	60	10	M8	277
		40	M9	277
Poly(methyl methacrylate)	60	750		271
Poly(styrene)	60	21		263
	75	26		263
	40	12		263
Poly(vinyl acetate)	60	15		263
	75	19		263
	0	0.5		97
		1.7		350
	11	2	O	145
	21	4	O	145
	31	16	O	145
	40	11.2		12
		30.9		43
		32.0		263
	50	0.06	W15	265
		0.11	W14	265
		0.15		265
		3		43
		10.2		12
	60	1.2		124
		1.4		168,169
		1.8		348
		1.9		329
		2.5		165
		3.0		105
		4.0		12
		6.8		105
		8.0		105
	70	47.0	O	263
		2		403

Remarks and References page II-98

Polymer	$T\,^\circ C$	$C_P \times 10^4$	Remarks	References
Vinyl acetate (Cont'd.)				
Poly(vinyl acetate) (Cont'd.)				
Bisulfite ion	70	4		403
Hydrogen peroxide	-15	0.36		97
	60-70	3.5	P	350
Poly(vinyl acetate-co-vinyl chloride)	60	0.21	VCL-Part	166
	60	3.0	VOAC-Part	166

D. TRANSFER CONSTANTS TO CATALYSTS

Catalyst	$T\,^\circ C$	$C_I \times 10^4$	Remarks	References
Acrylamide				
Bisulfite ion	75	0.17	F13	352
Hydrogen peroxide	25	0.0005	D	92
Acrylic acid, benzyl ester				
Isobutyronitrile, 2,2'-azobis-	55	0		291
	60	0		291
	65	0		291
Acrylic acid, ethyl ester				
Isobutyronitrile, 2,2'-azobis-	65	0		306
Acrylic acid, methyl ester				
Benzoyl peroxide	55	0.0143		221
	60	0.0246		221
	65	0.0375		221
	70	0.01	C,F10	322
		0.05		221
2-Butanone, peroxide	65	0.05	F10	123
		0.05		221
	70	0.077		221
	75	0.113	F10	123
		0.113		221
tert-Butyl hydroperoxide	60	0.01		221
	70	0.0266		221
tert-Butyl peroxide	65	0.00047		221
	70	0.00082		221
	75	0.00111		221
Acrylonitrile				
Isobutyronitrile, 2,2'-azobis-	50	0	F16	362
	60	0		95
Benzoic acid, vinyl ester				
Benzoyl peroxide	80	0.0527		26
Isobutyronitrile, 2,2'-azobis-	80	0		26
3-Buten-2-one, 3-methyl-				
Benzoyl peroxide	80	0.0509		84
Ethylene				
Azoethane, 1,1'-dimethyl-	83	0.5	D,W	204
		0.51	D,W	205
Ethylene, chloro-				
Valeronitrile, 2,2'-azobis[2,4,4-trimethyl-	25	0.85		425
Maleic anhydride				
Benzoyl peroxide	75	2.63		175
Methacrylonitrile				
Isobutyronitrile, 2,2'-azobis-	60	0		125
Methyl methacrylate				
Acetophenone, 2-diazo-2-phenyl-	70	0.0		459
p-Anisoyl peroxide	60	0.037		316
Benzoyl peroxide	50	0.01		145
	60	0.02	I2	34
	65	0.0025	F2	145
		0.00698		123
2-Butanone peroxide	70	0.0033	F17	123
		0.0033	F10	123
		0.0040	F11	123
		0.0092	F11	123
	75	0.00553	F11	123
		0.00667	F2	123
		0.0071	F10	123
		0.0089		123
tert-Butyl hydroperoxide	80	0.0111	F17	123
		0.0128		123
tert-Butyl peroxide	60	1.27 x [cat]		34
	20	< 0.0001	C,D	35
Butyronitrile, 2-ethyl-, 2,2'-azobis-	60	0		317
-, 2-methyl-, 2,2'-azobis-	60	0		317
-, 2,3,3-trimethyl-, 2,2'-azobis-	60	0		317
Cinnamoyl peroxide	60	0.009		316

Methyl methacrylate (Cont'd.)

Catalyst	T °C	C_I	Remarks	References
Methyl methacrylate (Cont'd.)				
Cyclohexanecarbonitrile, 1,1'-azodi-	60	0		317
Cyclopropanecarboxylic acid, 1-methyl-2-(9-anthryl)-, methyl ester	70	0.002		440
Hydrogen peroxide	60	0.046	F11	290
Hydroperoxide, α,α-dimethylbenzyl	60	0.33		34
Isobutyronitrile, 2,2'-azobis-	50	0	F2	327
	60	0.0	F10	34,290,317
Methane, diazodiphenyl-	70	0.0		267
Palmitoyl peroxide	60	0		316
		0.16	I11	145
Peroxide, bis(m-chlorobenzoyl)	60	0.003		316
--, bis(o-chlorobenzoyl)	60	0.019		316
		0.35	F10	290
		0.8	I10	145
--, bis(p-chlorobenzoyl)	60	0.009		316
--, bis(m-nitrobenzoyl)	60	0.012		316
--, bis(p-nitrobenzoyl)	60	0.144		316
--, bis(m-phenylazobenzoyl)	50	0.00001	C	282
2-Tetrazene, 1,1,4,4-tetramethyl-	30	0.038	D	483
o-Toluoyl peroxide	60	0.046		316
Valeronitrile, 2-methyl-, 2,2'-azobis-	60	0.06	I11	145
		0		317
Styrene				
Acetyl peroxide	70	0		87,217
p-Anisyl peroxide	70	0.074		87
Benzoyl peroxide	22	0.1	D	65
	50	0.13		58
	60	0.048		171
		0.055		240
		0.101	C	234
	70	0		217
		0.075		87
		0.12		67
		0.18		58,65
	80	0.13		58,65
2-Butanone, peroxide	50	0.46	H	388
	70	0.0667	F2	123
		0.1250	F10	123
		0.1670	F11	123
	75	0.1250	F2	123
		0.1670	F10	123
		0.2000	F11	123
		0.243		123
tert-Butyl hydroperoxide	60	0.035		171
	70	0.051	H	388

Catalyst	T °C	C_I	Remarks	References
Styrene (Cont'd.)				
tert-Butyl hydroperoxide (Cont'd.)	70	0.060		399
		0.063	F6	400
		0.064		400
		0.066	F2,F5	400
Butyl peroxide	80	0.003		385
	60	0.00076	F9	302
		0.00092	F2	302
sec-Butyl peroxide	80	0.0029	F9	302
Palmitoyl peroxide	60	0.0004	F2	301
tert-Butyl peroxide	80	0.0021	F2	301
	60	0.00023	F9	303
		0.0003		303
		0.0006	F2,F12	303
		0.00086		300
		0.0013		304
	70	0.039	F2	400
	80	0.0022	F2	303
	50	0.0033		385
Butyryl peroxide	70	0.018		87
Cinnamoyl peroxide	70	1.10		87
Crotonyl peroxide	70	0.146		87
Cyclohexanone peroxide	60	0.062		33
Ethyl peroxide	60	0.00066	F2	301
Formamide, 2-cyano-2-propylazo-	80	0.0024	F2	301
Furoyl peroxide	100	0.17		47
Hexanoyl peroxide	70	0.23		87
Hydroperoxide, α,α-dimethylbenzyl	70	0.166		87
	40	0.052	H	388
	50	0.069	H	388
	60	0.063		171
--, α,α-dimethylbenzyl, p-iso-propyl-	70	0.082	H	388
		0.10		399
Isobutyronitrile, 2,2'-azobis-	70	0.033		388
	50	0	F2	327
		0	H	388
Isopropyl peroxide	60	0		171
		0		318
		0.012		468
		0.16		447
	60	0.0003	F2	301
Lauroyl peroxide	80	0.0015	F2	301
	70	0		217
p-Menth-8-yl hydroperoxide	84	0.024		87
		0		217
	50	0.048	H	388

Remarks and References page II-98

Styrene (Cont'd.)

Catalyst	$T\ ^{\circ}C$	C_I	Remarks	References
Myristoyl peroxide	70	0		217
	70	0.116		87
2-Naphthoyl peroxide	70	0.178		87
Nickel peroxide	60	0.00265	C,F2	266
Octanoyl peroxide	70	0		217
		0.098		87
Oleoyl peroxide	70	0.154		87
Palmitoyl peroxide	70	0.142		87
Peroxide, bis(p-acetoxybenzoyl)	70	0.187		87
-", bis(m-bromobenzoyl)	70	0.465		87
-", bis(o-bromobenzoyl)	50	1.0		58,65
	70	2.17		87
-", bis(p-bromobenzoyl)	70	0.193		87
-", bis(p-tert-butylbenzoyl)	70	0		87
-", bis(m-chlorobenzoyl)	70	0.346		87
-", bis(o-chlorobenzoyl)	22	2.0	D	65
	70	1.91		87
-", bis(p-chlorobenzoyl)	70	0.216		217
-", bis(p-cyanobenzoyl)	70	0.804		87
-", bis(2,4-dichlorobenzoyl)	60	2.9		217
	70	2.6		217
-", bis(α,α-dimethylbenzyl)	50	0.01	H	388
-", bis(m-fluorobenzoyl)	70	0.246		87
-", bis(o-fluorobenzoyl)	70	0.40		87
-", bis(p-fluorobenzoyl)	70	0.219		87
-", bis(hydroxyheptyl)	50	<0.005	H	388
-", bis(m-iodobenzoyl)	70	0.262		87
-", bis(p-methoxycarbonyloxy-benzoyl)	70	0.208		87
-", bis(m-nitrobenzoyl)	70	6.2		87
-", bis(o-nitrobenzoyl)	70	7.4		87
-", bis(m-phenylazobenzoyl)	70	5.5	C	281
	90	4.8	C	281
-", bis(5-phenyl-2,4-pentadienoyl)	70	5.24		87
-", bis(2-thiophenecarbonyl)	50	0.23		65
	70	0.38		65
-", tert-butoxymaleoyl	70	1.52		87
-", tert-butoxyphthaloyl	70	0.018		87
-", butylidenebis[tert-butyl	80	0.00077	C	411
	90	0.00140	C	411

Styrene (Cont'd.)

Peroxide (Cont'd.)

Catalyst	$T\ ^{\circ}C$	C_I	Remarks	References
-", sec-butylidenebis[tert-butyl	80	0.00072	C	411
	90	0.00115	C	411
-", isobutylidenebis[tert-butyl	80	0.0083	C	411
	90	0.00155	C	411
-", isopropylidenebis[tert-butyl	80	0.00040	C	411
	90	0.00105	C	411
Pinanyl hydroperoxide	70	0.026	H	388
Pivalonitrile	60	0.000038		468
Propylene, oxidized	60	1.01		396
	70	1.14		396
Propyl peroxide	60	0.00084		305
Sorboyl peroxide	70	1.19		87
Stearoyl peroxide	70	0.154		87
Succinonitrile, tetramethyl-	60	0.000037		468
2-Tetrazene, 1,1,4,4-tetramethyl-	30	0.038	D	483
o-Toluoyl peroxide	70	0.175		87
p-Toluoyl peroxide	50	0.17		65
	70	0.003		87
9-Undecenoyl peroxide	70	0.19		65
	70	0.065		87
Valeronitrile, 2,2'-azobis[2,4,4-trimethyl-	25	0.59		426

Vinyl acetate

Catalyst	$T\ ^{\circ}C$	C_I	Remarks	References
Benzoyl peroxide	60	0.032		465
		0.09		231
		0.15		80
Isobutyronitrile, 2,2'-azobis-	65	0.040		466
	50	0.025		466
	60	0.055		466
Lauroyl peroxide	60	0.10		80
Palmitoyl peroxide	60	0.10		80
Peroxide, bis(m-bromobenzoyl)	60	0.17	I3	145
	60	0.24		80
-", bis(o-bromobenzoyl)	60	0.6	I3	145
	60	0.25		80
-", bis(p-bromobenzoyl)	60	3.5	I3	145
	60	0.17		80
-", bis(o-chlorobenzoyl)	60	0.17	I3	145

E. TRANSFER CONSTANTS TO SOLVENTS AND ADDITIVES

Solvent	T°C	$C_S \times 10^4$	Remarks	References
Acetic acid, allyl ester				
Benzene	80	21.0		273
p-Benzoquinone	80	520000	J	27
--, 2,3,5,6-tetrachloro-	80	1600000	J	27
--, 2,3,5,6-tetramethyl-	80	41400	J	27
--, 2,3,5-trichloro-	80	550000	J	27
Carbon tetrachloride	100	20000	C	210
Phosphorus trichloride	40	10000	C,D	214
Acrylamide				
Iron(II)chloride	25	42600		75
Isopropyl alcohol	50	19	F13	353
	80	7.2	F13,S	353
Methanol	30	0.13	C,F13	130
Propionamide	25	220	E	103
	60	64	E	103
Water	25	0.204	C,F13	437
	40	5.8	F13	444
Acrylic acid, butyl ester				
Aniline, N,N-dimethyl-	50	380		161
Acrylic acid, ethyl ester				
Acetic acid	50	0.176		472
	80	0.537	A	471
	100	1.05	A	472
Acetone	40	0.207		472
	60	0.27	C	134
	80	1.10	A	471
	100	2.30	A	472
Acetonitrile	50	0.158		472
	60	0.245		472
	80	0.55	A	471
	100	1.43	A	472
Aniline, N,N-dimethyl-	50	2300	F16	228
Benzene	50	0.016		472
	60	0.22		472
		0.27	C	133
		0.45	C	134
	80	0.525	A	471
	100	2.2	A	472
--, bromo-	60	0.163	A	471
	80	0.685	A	472
	100	3.34	A	472
--, chloro-	40	0.054	A	472

Solvent	T°C	$C_S \times 10^4$	Remarks	References
Acrylic acid, ethyl ester (Cont'd.)				
Benzene, chloro-	80	0.37	A	471
	100	1.68	A	472
--, ethyl-	40	0.688		472
	60	11.6		472
	80	16.80	A	471
Carbon tetrachloride	100	28.7	A	472
Phosphorus trichloride	50	1.44		472
2-Butanone	60	0.151	C	134
	80	1.92		472
Butyl alcohol	90	3.29	A	471
	40	4.45	A	472
	60	1.28		472
	80	2.91		472
sec-Butyl alcohol	100	5.85	A	471
	45	12.6	A	472
	63	10.6		472
tert-Butyl alcohol	80	18.5		472
	100	22,20	A	471
	40	31.5	A	472
Butyric acid	60	0.068		472
Carbon tetrachloride	80	0.17	A	472
	100	0.712	A	471
	80	1.64	A	472
	40	0.855	A	471
Chloroform	60	0.332		472
	100	0.90		472
	70	3.2		416
Cumene	80	1.13	C	134
	100	1.55	A	471
	40	2.80	A	472
Cyclohexane	60	0.195		472
	70	0.89	C	134
	80	1.57	C	134
	100	1.49	A	471
	50	4.74	A	472
	60	11.7		472
	80	13.8		472
	90	22.2	A	471
	50	26.0	A	472
	60	28.9		472
	80	0.48		472
	80	0.61	C	134
	80	1.22	A	471

Acrylic acid, ethyl ester (Cont'd.)

Solvent	T°C	$C_S \times 10^4$	Remarks	References
Cyclohexane (Cont'd.)	100	3.08	A	472
Ethyl acetate	50	0.298		472
	60	0.448		472
		0.69	C	134
	80	0.89	A	471
Formic acid	100	1.82	A	472
	80	0.046	A	471
Hexane	50	0.524		472
	60	0.593		472
	80	0.97	A	471
Isobutyl alcohol	100	1.46	A	472
	45	2.1		472
	63	3.31		471
	80	4.65	A	472
Isopropyl alcohol	100	8.06	A	445
	-	21	C	134
Methanol	60	0.32	C	472
		1.70	4	134
Toluene	40	0.88		471
	70	1.84	C	471
	80	2.60	A	470
	100	6.80	A	332
Acrylic acid, methyl ester				
Acetone	60	0.23	C	470
	80	0.622		332
Acetophenone		1.1	C	111
	50	<2.5		15
Aluminum, triethyl	60	480	F2,S	156
Aniline	50	<1.0		15
--, N,N-dimethyl-	50	60	F16	228
--, m-nitro-	50	42.1		15
--, p-nitro-	50	26.7		15
Anisole, m-nitro-	50	72.7		15
Anthracene	50	1000		332
Benzene	80	0.326		111
		0.45	C	111
--, chloro-	80	0.52	C	332
		0.986		111
--, o-dichloro-	80	0.71	C	332
--, m-dinitro-	50	309		15
--, ethyl-	80	6.056		332
--, nitro-	50	46.4		15
--, 1,3,5-trinitro-	50	2040		15
Benzoic acid, p-nitro-	50	107		15
Benzonitrile	50	<1.0		15

Acrylic acid, methyl ester (Cont'd.)

Solvent	T°C	$C_S \times 10^4$	Remarks	References
Benzophenone	50	<1.5		15
p-Benzoquinone	50	8100		15
Benzoyl chloride	50	<1.5		15
Biphenyl	50	<2.0		15
2-Butanone	80	3.238	C	332
Butyl alcohol	80	3.61		111
sec-Butyl alcohol	80	2.747		332
tert-Butyl alcohol	80	14.14		332
Carbon tetrabromide	80	0.389		109
	40	3500	F2	109
	50	3500	F2	109
	60	4100	F2	109
Carbon tetrachloride	80	1.25	C	111
		1.266	C	183
		1.323		332
		1.55		183
Chloroform	80	2.100		183
		2.144		332
		2.333	C	183
Cumene		2.5	C	111
	80	6.966		332
Cyclohexane		16.2	C	111
	80	0.027	C	332
Cyclohexanone	80	1.2	C	111
Diphenylamine-T	80	5.5	C	111
Ethane, 1,2-dichloro-	60	300	F2	46
	80	0.82	C	183
--, 1,1,2,2-tetrachloro-	80	1.00		183
		0.932		332
--, 1,1,1-trichloro-		1.55		183
Glutaric acid, 2,4-dimethyl-		1.561		183
--,--, dimethyl ester	80	0.574	C	332
2,4,6-Heptanetricarboxylic acid trimethyl ester	60	0.45		212
Isobutyl alcohol	60	0.54		212
Isobutyric acid, methyl ester	80	2.496		332
Naphthalene	60	1.4		212
2,6-Octadiene, 2,6-dimethyl-	50	<2.0		15
Oleic acid, methyl ester	60	42		324
Phenanthrene	80	3.66		441
Phenol	50	26		15
	50	<2.0		15
--, m-nitro-	50	56.2		15
--, p-nitro-	50	42.6		15

Solvent	T°C	$C_S \times 10^4$	Remarks	References
Acrylonitrile (Cont'd.)				
Aniline, N,N-dimethyl- (Cont'd.)	60	870	J	24
		964	F2,G	229
		21800	F16,G,J	229
Anthracene	50	18000		15
Arabinose	60	13.0	F8	361
Benzene	60	2.46	G	95
--, bromo-	60	1.36	G,S	95
--, tert-butyl-	60	1.93	G	95
--, chloro-	60	0.79	G,S	95
--, ethyl-	60	35.73	G,S	95
--, iodo-	60	5.19	G,S	95
Benzoic acid, vinyl ester	65	1400		136
p-Benzoquinone	50	13000		15
Borane, tributyl	60	6470	F7	156
2-Butanone	60	6.43	G	95
--, 3-methyl-	60	21.08	G	95
1-Buten-3-yne	50	3800	F7	377
Butyl alcohol	60	15.42	G	95
sec-Butyl alcohol	60	97.55	G	95
tert-Butyl alcohol	60	0.44	G	95
Butyric acid, 4-hydroxy-γ-lactone	50	0.658		375
		0.74		374
Cadmium, dibutyl	60	55000	F7	156
Carbonic acid, cyclic ethylene ester	50	0.073		293
		0.33		197
		0.39		374
		0.474		375
		0.5		404
		0.511	E	458
		1.0		356
	60	0.128		293
Carbon tetrabromide	50	400		493
	60	500		20
Carbon tetrachloride	40	1900	F7,J	24
	60	0.85	G	96
	80	1.13	C	336
Chloroform	60	5.64	G	95
	80	5.90	C	336
Copper(II)chloride	35	180000	U(0.01)	402
		190000	U(0.1)	402
		320000	U(1.6)	402
	60	1000000	F7	37
Copper(II)sulfate	35	1900	F13,J	402
		2800	J,V(0.0001)	402
		3000	J,U(0.01)	402

Remarks and References page II-98

Solvent	T°C	$C_S \times 10^4$	Remarks	References
Acrylic acid, methyl ester (Cont'd.)				
Phosphine, tributyl-	60	1890	F2	156
Phosphorus trichloride	40	1000	C,D	214
Silane, tetraethyl-	60	33.1	F2	156
Stearic acid, methyl ester	60	0.751		441
Toluene	60	2.7		21,24
	60	1.775		332
	80	2.7	C	111
--, m-nitro-	50	41.2		15
--, p-nitro-	50	48.6		15
Triethylamine	60	400		24
Tripropylamine	60	470	F2	156
Acrylic acid, tetrahydrofurfuryl ester				
Carbon tetrachloride	40	1.0	C,F2	319
Acrylonitrile				
Acetaldehyde	--	14	F	436
	50	47	F7	377
Acetamide, N,N-dimethyl-	50	4.945		375
		5.05		197
Acetic acid	50	0.81	F14	377
Acetone	50	1.7	F14	377
	60	1.13	F	95
Acetonitrile	50	0.7	F14	377
	60	2.0	F15	86
Acrolein, bis(2-ethoxyethyl)acetal	60	120	F7	499
--, bis(2-butoxyethyl)acetal	60	90.5	F7	499
Aluminum, hydrodiisobutyl	60	3940	F7	156
--, triethyl	60	590	F7	156
	100	170000	F7	247
--, triisobutyl	100	280000	F7	247
Aniline	40	32.0	F2,G	229
	50	44.0	F2,G	229
	60	9600	F16,G,J	227,229
	60	12200	F16,G,J	229
--, N,N-diethyl-	40	215	F2,G	229
	50	58100	F16,G,J	229
		359	F2,G	229
		93800	F16,G,J	227,229
	60	547	F2,G	229
--, N,N-dimethyl-	40	143200	F16,G,J	229
		605	F2,G	229
	50	11900	F16,G,J	229
		708	F2,G	229
		1040	F16,J	161
		15000	F16,J	228
		15400	F16,G,J	227,229

Acrylonitrile (Cont'd.)

Solvent	T °C	$C_S \times 10^4$	Remarks	References
Copper(II)sulfate (Cont'd.)				
-,-	35	10700	J,V(0.001)	402
		13500	J,U(0.1)	402
		39300	J,V(0.01)	402
		136000	J,U(1.0)	402
		210000	J,V(0.1)	402
Crotonaldehyde	50	47	F7	377
Crotononitrile, trans-	50	19	F7	377
Cumene	60	41.41	G	95
Cyclohexane	60	2.06	G	95
-, methyl-	60	2.31	G	95
Dimethylamine	50	175	F7	377
m-Dioxane, 5,5-dimethyl-2-vinyl-	60	2.20	F7	498
-, 4-methyl-2-vinyl-	60	4.40	F7	498
-, 2-vinyl-	60	2.71	F7	498
1,3-Dioxolane, 2-vinyl-	60	16.60	F7	498
1,3-Dioxolane-4-methanol, 2-vinyl-	60	2.40	F7	498
Diphenylamine-T	60	700	F7	61
Erythritol	60	12.8	F8	361
Ethane, 1,2-dichloro-	60	1.47	G	95
-, 1,1,2,2-tetrachloro-	60	3.11	G	95
-, 1,1,1-trichloro-	60	1.25	G	95
-, 1,1,2-trichloro-	60	1.68	G	95
Ethanol, 2,2'-iminodi-	30	10.1	F13	476
-, 2,2',2''-nitrilotri-	30	76.0	F13	476
Ether, dodecyl vinyl	50	4.95	C	4
Ethyl acetate	60	2.54	G	95
Formamide, N,N-dimethyl-	20	1.4	C,J,F7	248
	25	4.97	D,J	91
	40	3.24	E	458
	50	1.0		347
		2.70		374
		2.78		197
		2.8	F7	377
		2.83		375
		10		356
	60	2.412		347
		4.494		22
		5.0	F15	86
Formic acid	50	1.8	F14	377
-, N-methyl-	50	0.5	F14	377
Glucose	60	6.9	F8	361
α,D-Glucoside, methyl-	60	20	F7	220
-,-, 6-deoxy-6-iodo-	60	50	F7	220
-,-, 6-deoxy-6-mercapto-	60	1300	F7	220
-,-, 6-deoxy-6-phthalimido-	60	50	F7	220
-,-, 2,3-di-O-benzyl-	60	90	F7	220

Acrylonitrile (Cont'd.)

Solvent	T °C	$C_S \times 10^4$	Remarks	References
α,D-Glucoside, methyl- (Cont'd.)				
-,-, 2,3,4,6-tetra-O-acetyl-	60	30	F7	220
-,-, 6,-O-(p-toluenesulfonyl)-	60	10	F7	220
-,-, 6-O-triphenylmethyl-	60	80	F7	220
β,D-Glucoside, methyl-	60	20	F7	220
-,-, 6-deoxy-6-dipropylamino-	60	1100	F7	220
Glutaronitrile, 2,4-dimethyl-	50	0.6	F14	376
Glyceraldehyde	60	32.1	F8	361
Glycerol	60	23.5	F8	361
2,4,6-Heptanetricarbonitrile	50	1.0	F14	376
1-Heptanol, 2,2,3,3,4,4,5,5,6,6,7,7-dodecafluoro-	60	1.9	F2	442
1,5-Hexadien-3-yne	50	11700	F7	377
Hydrocyanic acid	50	0.81	F14	377
	--	6.2	F	436
Indium, triethyl-	60	2220	F7	156
Iron(III)chloride	60	33300	F7	19
Isobutyl alcohol	60	24.06	G	95
Isobutyronitrile	50	1.3	F14	376
	60	1.8	I6	376
		3.5		135
Lactonitrile	50	4.0	F15	377
Lead, tetraethyl-	60	1	F14	156
Magnesium perchlorate	50	243	F7	376
		< 0.05		6
Mercury, diethyl-	60	72.2	F7	156
Methane, dichloro-	60	3.06	G	95
-, nitro-	60	6.0	F15	86
Methanol	50	0.5	F14	377
Methylamine	50	175	F7	377
2,6-Octadiene, 2,6-dimethyl-	60	450		6
1-Pentanol, 2,2,3,3,4,4,5,5-octafluoro-	60	1.9	F2	442
1-Pentanol, 4-methyl-	60	11.79	G	95
Piperidine, 1-ethyl-	60	3300	J	24
-, 1-methyl-	60	2300	J	24
1-Propanol, 2,2,3-tetrafluoro-	60	1.5	F2	442
Silane, tetraethyl-	60	21.0	F7	156
Sorbitol	60	6.1	F8	361
Stibine, tributyl-	60	111000	F7	156
Succinonitrile	60	0.6		86
Sulfur dioxide	50	0		362
Tin, tetrabutyl-	60	80.8	F7	156
Toluene	50	1.153	G	347
	60	2.632	G	347
		3.2		21

Solvent	T °C	$C_S \times 10^4$	Remarks	References
Acrylonitrile (Cont'd.)				
Toluene (Cont'd.)	60	5.83	G	24,95
o-Toluidine, N,N-dimethyl-	40	272	F2,G	229
	50	334	F2,G	229
	60	30200	F16,G,J	227,229
	60	463	F2,G	229
Tributylamine	60	6700	J	24
Triethylamine	60	1700	T	20
		1900		20
		3800		22
		5900		24
		6600		24
Trimethylamine	50	175	F7,J	377
	60	790	J	24
Tripropylamine	60	4280	F7	156
		10500	J	24
Zinc, diethyl	60	16000	F7	156
Zinc chloride	50	0.006	F13	410
Acrylonitrile/styrene (38.5-61.5 mol-%)				
Methanol	65	1.4		116
Anisole, p-vinyl-				
Anisole, p-isopropyl-	60	3.40		407
Benzonitrile, p-isopropyl-	60	40.9		407
Cumene	60	4.28		407
-,- p-bromo-	60	11.8		407
-,- p-tert-butyl-	60	3.64		407
-,- p-chloro-	60	8.52		407
Benzoic acid, vinyl ester				
Benzene	60	1.5		393
	80	0.4		26
Benzoic acid, isopropyl ester	60	1.0		393
2-Butanone	80	29.0		26
Carbon tetrachloride	80	730		26
Chloroform	80	105		26
Cyclohexane	80	2.3		26
1,3-Butadiene, 2-chloro-				
1-Butene, 3-chloro-3-methyl-	50	5.47	F1	152
2-Butene, 1-chloro-3-methyl-	50	4.57	F2	152
2,6-Octadiene, 2,7-dichloro-	50	5.24	F2	152
3-Buten-2-one, 3-methyl-				
Benzene	80	2.489	C	84
-,- ethyl-	80	6.934	C	84
Cyclohexane, methyl-	80	0.500	C	84
Toluene	80	3.282	C	84

Solvent	T °C	$C_S \times 10^4$	Remarks	References
Butyraldehyde, divinyl acetal				
Aniline, N,N-dimethyl-	50	1060		430
Benzene	50	27.7		430
Butyl alcohol	50	3500		430
tert-Butyl alcohol	50	37.2		430
1,3-Dioxolane, 2-propyl-	50	708		430
Butyric acid, vinyl ester				
Benzene	80	3.28		201
Carbamic acid, vinyl-, ethyl ester				
Acetone	60	8.5		117
Benzene	60	1.25		117
p-Dioxin, 2,3-dihydro-/maleic anhydride				
Benzene	60	11500	R	169
Carbon tetrabromide	60	10000000	R	169
Carbon tetrachloride	60	170000	R	169
Chloroform	60	36000	R	169
p-Dioxane	60	10700	R	169
Toluene	60	60000	R	169
m-Xylene	60	99000	R	169
Ethylene				
Acetamide, N,N-diethyl-	130	125	C, W5	454
-,-, N,N-diisopropyl-	130	57	C, W5	454
-,-, N,N-dimethyl-	130	182	C, W5	454
-,-, N-ethyl-	130	115	C, W5	454
-,-, N-methyl-	130	61	C, W5	454
Acetic acid, butyl ester	130	89	C, W5	454
-,-, tert-butyl ester	130	40	C, W5	454
-,-, methyl ester	130	40	C, W5	454
Acetic acid, chloro-				
-,-, methyl ester	90	610	C, Q1	434
		1210	C, Q2	434
		2170	C, Q3	434
		2150	C, Q4	434
		2580	C, Q5	434
	120	630	C, Q1	434
		1160	C, Q2	434
		1880	C, Q3	434
		1990	C, Q4	434
		2090	C, Q5	434
	130	1120	C, W5	454
Acetic acid, cyano-				
-,-, methyl ester	130	6700	C, W5	454
Acetic acid, dichloro-				
-,-, methyl ester	55	1800	C, Q1	435
		7900	C, Q2	435

Solvent	T °C	$C_S \times 10^4$	Remarks	References
Ethylene (Cont'd.)				
Acetic acid, dichloro-				
--,--, methyl ester (Cont'd.)	55	11100	C,Q3	435
		13000	C	435
	90	1900	C,Q1	435
		6900	C,Q2	435
		9400	C,Q3	435
		10500	C	435
	120	2000	C,Q1	435
		5900	C,Q2	435
		8100	C,Q3	435
		11100	C	435
Acetic acid, trichloro-				
--,-- methyl ester	55	700	C,Q1	484
		16100	C,Q2	484
		27500	C,Q3	484
		38000	C,Q4	484
		65700	C	484
	90	1000	C,Q1	484
		17000	C,Q2	484
		27700	C,Q3	484
		39000	C,Q4	484
		65200	C	484
	120	1200	C,Q1	484
		15400	C,Q2	484
		23600	C,Q3	484
		33000	C,Q4	484
		54100	C	484
Acetic anhydride	130	130	C,W5	454
Acetone	130	160	C,W17	456
		165	C,F21,W5	259
		168	C,W5	456
	200	282	C,W5	455
Acetonitrile	130	110	C,Q1	159
Benzaldehyde	130	1970	C,F21,W5	457
Benzene	20	0.629	E,w	143
	83	20	D,w	204
		21	D,w	205
	130	9.4	C,W5	260
	50-70	18	w3	218
--, ethyl-	130	430	C,W17	456
		480	C,W5	52
		520	C,W5	454
		560	C,W5	52
Benzene-D6	200	500	C,W5	455
Benzoic acid, butyl ester	130	5.6	C,W5	260
	130	140	C,W5	454

Solvent	T °C	$C_S \times 10^4$	Remarks	References
Ethylene (Cont'd.)				
Benzoic acid, ethyl ester	130	55	C,W5	454
-- methyl ester	130	53	C,W5	454
Butane	130	40	C,F21,W5	259
		45	C,W17	456
--, 1-iodo-	100	94500	Q1,W14	187
		113000	Q2,W14	187
		126000	Q3,W14	187
2-Butanone	130	550	C,W17	456
		600	C,F21,W5	259
	200	750	C,W5	455
--, 3-methyl-	130	840	C,F21,W5	259
1-Butene	130	330	C,W5	52
		470	C,W5	52
		560	C,F21,W5	259
--, 2-methyl-	189	900	B,W5	443
	200	570	C,W5	455
--, 3-methyl-	130	210	C,F21,W5	259
		530	C,F21,W5	259
2-Butene	130	1200	C,F21,W5	259
--, 1,4-dichloro-	130	250	C,F21,W5	259
		380	C,W5	454
--, 2-methyl-	130	4100	C,W5	454
	130	470	C,F21,W5	259
tert-Butyl alcohol	30	0	E,W4	482
Butylamine	130	2	C,F21,W5	259
Butyraldehyde	130	220	C-W5	454
Butyric acid, methyl ester	130	3250	C,W5	457
Butyronitrile	130	220	C,W5	454
	130	520	C,F21,W5	259
Carbon tetrachloride	20	215	E,w	143
	50	772	C,Q1	391
	70	974	C,Q1	391
		7000	C,W4	215
		32000	C	210
	90	740	Q1,W12	453
		1210	C,Q1	391
		20200	Q2,W12	453
		32500	Q3,W12	453
		34000	W12	453
	130	9800	C,F21,W5	259
	140	1600	C,Q1	170
		1700	C,Q1,F19	170
		1800	C,Q1,W13	170
		2200	C,Q1,F20	170
		22000	C,Q2	170
		23000	C,Q2,F19	170

Ethylene (Cont'd.)

Solvent	$T^\circ C$	$C_S \times 10^4$	Remarks	References
Carbon tetrachloride (Cont'd.)	140	30000	C, Q1, W1	170
		36000	C, Q3	170
		38000	C, Q2, F20	170
		39000	C, Q3, F19	170
		60000	C, Q2, W13	170
		60000	C, Q3	170
		61000	C, Q3, F20	170
		70000	C, Q4, F19	170
		90000	C, Q4, F20	170
		100000	C, Q2, W1	170
		110000	C, Q3, W1	170
		130000	C, Q3, W13	170
		140000	C, Q4, W1	170
		180000	C	170
Chloroform	28	2100	E, Q1, W	244
		13000	E, Q2, W	244
		15000	E, Q3, W	244
	70	8000	C, W4	215
	80	30000	C	210
		2470	C, Q1, W12	101
		15500	C, Q2, W12	101
		24900	C, Q3, W12	101
		31200	C, Q4, W12	101
		43900	C, W12	101
	101	1500	E, Q1, W	244
		4500	E, Q3, W	244
		5400	E, Q2, W	244
	103	2890	C, Q1, W12	101
		15400	C, Q2, W12	101
		23800	C, Q3, W12	101
		29800	C, Q4, W12	101
		41100	C, W12	101
	130	2900	C, F21, W5	259
	140	3210	C, Q1, W12	101
		15200	C, Q2, W12	101
		22200	C, Q3, W12	101
		28000	C, Q4, W12	101
		37600	C, W12	101
Cumene	130	500	C, F21, W5	259
Cyclohexane	130	80	C, F21, W5	259
		90	C, W17	456
		91	C, F21, W5	259
--, methyl-	200	190	C, W5	455
Cyclopentane	130	110	C, F21, W5	259
	130	109	C, W17	456
	189	127	C, F21, W5	259

Ethylene (Cont'd.)

Solvent	$T^\circ C$	$C_S \times 10^4$	Remarks	References
Cyclopentane (Cont'd.)	200	228	C, W5	455
Cyclopropane	130	0	C, F21, W5	259
Decane	130	120	C, F21, W5	259
	189	425	B, W5	443
1-Decene	189	1090	B, W5	443
Dibutylamine	130	1070	C, W5	454
Dimethylamine	130	1900	C, F21, W5	259
p-Dioxane	130	320	C, F21, W5	259
Ethane	130	6	C, F21, W5	259
--, 1,1-bis(dimethylamino)-	130	1070	C, W5	457
--, 1-bromo-2-chloro-	130	390	C, W5	454
--, chloro-	70	120	C, W4	215
--, 1,2-dibromo-	130	1250	C, W5	454
--, 1,1-dichloro-	70	1500	C, W4	215
--, 1,2-dichloro-	130	110	C, F21, W5	259
--, iodo-	65	151000	Q1	189
		159000	Q2	189
		165000	Q3	189
	100	110000	Q1	452
		136000	Q1, W14	188
		143000	Q2, W14	188
		148000	Q3, W14	188
		149000	Q3	452
--, 1,1,1-trichloro-	70	500	C, W4	215
Ethyl acetate	130	45	C, W5	454
	200	121	C, W17	455
Ethyl alcohol	20	5.71	E, W	143
	130	68	C, W17	456
		69	C, F21, W5	259
		75	C, W5	454
	125-135	190	C, W2	381
	200	135	C, W5	455
Ethylene oxide	130	7	C, F21, W5	259
Formaldehyde	130	560	C, W5	457
Formamide, N,N-dimethyl-	130	260	C, F21, W5	259
Formic acid, methyl ester	130	42	C, W5	454
Furan, tetrahydro-	130	288	C, W5	454
		289	C, F21, W5	259
Heptaldehyde	200	401	C, W5	455
	130	2600	C, W17	457
		3900	C, W5	457
Heptane	200	4800	C, W5	457
	130	80	C, F21, W5	259
Hexane	50-70	90	W3	218
	130	68	C, F21, W5	259
	189	225	B, W5	443

Remarks and References page II-98

Ethylene (Cont'd.)

Solvent	T °C	$C_S \times 10^4$	Remarks	References
1-Hexene	189	900	B,W5	443
—, 2-ethyl-	130	3300	C,F21,W5	259
Hydrogen	130	159	C,W5	454
	200	160	C,F21,W5	259
	130	400	C,W5	455
Isobutyronitrile	130	1070	C,F21,W5	259
Isocyanic acid, butyl ester	130	212	C,W5	454
Isopropyl alcohol	130	130	C,W17	456
	130	140	C,F21,W5	259
	100	144	C,W5	454
	125–135	570	C,W1	381
Isothiocyanic acid, butyl ester	200	234	C,W5	455
Methane	130	750	C,W5	454
	130	0	C,W5	259
—, bromochloro-	100	1600	C,Q1,W11	2
		4500	C,Q2,W11	2
		7500	C,Q3,W11	2
		8000	C,Q4,W11	2
	130	10000	C,W5	454
—, chloro-	140	1600000	Q1,W11,II	1
		4700000	Q2,W11,II	1
—, dichloro-	70	4	C,W4	215
	70	700	C,W4	215
—, dimethoxy-	130	360	C,F21,W5	259
—, iodo-	65	73	C,W5	454
		41000	Q1	189
		45000	Q3	189
		45000	Q2	189
	100	41000	Q1	189
		45000	Q2	189
		45000	Q3	189
Methanol	130	21	C,F21,W5	259
Methylamine	130	53	C,W5	454
1-Octene	130	360	C,W5	52
Pentane, 2,2,4-trimethyl-	130	64	C,F21,W5	259
Pentene	189	900	B,W5	443
—, 4,4-dimethyl-	130	175	C,W5	454
—, 4-methyl-	130	310	C,W5	454
Phosphine	130	207000	C,W5	457
—, dibutyl-	130	36000	C,W5	454
—, tributyl-	130	4500	C,W5	454
—, triphenyl-	130	50	C,W5	454
Phosphorous acid, dimethyl ester	130	5100	C,F21,W5	259
Propane	130	27	C,F21,W5	259
	200	27.6	C,W17	456
Propane (Cont'd.)	130	31	C,F21,W5	259
—, 2-chloro-	200	65.2	C,W5	455
—, 2-chloro-2-methyl-	70	250	C,W4	215
—, 2,2-dimethyl-	70	40	C,W4	215
	130	8	C,F21,W5	259
—, 2-iodo-	65	570000	Q1	189
		590000	Q2	189
		607000	Q3	189
	100	455000	Q1	189
		470000	Q2	189
		483000	Q3	189
—, 2-methyl-	130	50	C,F21,W5	259
—, 1,1,1,2,2,3,3,3-octafluoro-	200	72	C,W5	454
1-Propene	130	136	C,W5	455
	130	4	C,F21,W5	259
	130	110	C,W5	52
	130	122	C,W5	454
		150	C,W5	52
	200	200	C,W5	455
Propionaldehyde	130	2300	C,W17	457
	200	3300	C,W5	457
	200	2880	C,W5	457
Propionic acid, methyl ester	90	63000	C,Q1	485
		92000	C,Q3	485
		108000	C,Q5	485
	120	78000	C,Q1	485
		112000	C,Q3	485
		139000	C,Q5	485
—, 3-cyano-, methyl ester	130	790	C,W5	454
Silane, tetramethyl-	130	0	C,F21,W5	259
Tetradecane	189	580	B,W5	443
1-Tetradecene	189	1760	B,W5	443
Toluene	130	130	C,W5	52
	130	154	C,W5	454
Tributylamine	200	180	C,W17	456
	130	220	C,W5	455
Tridecane	130	820	C,W5	454
Trimethylamine	130	140	C,F21,W5	259
	130	180	C,F21,W5	259
	130	330	C,W5	454
Water	20	1.71	E,W	143
p-Xylene	130	300	C,F21,W5	259
	130	317	C,W5	454
		400	C,W17	456
	200	434	C,W5	455

Solvent	T°C	$C_S \times 10^4$	Remarks	References
Ethylene, bromo-				
Carbon tetrachloride	60	50		417
Ethylene, chloro-				
Acetaldehyde	50	110		164
Aniline, N,N-dimethyl-	50	2700		161
Benzene	35	1500		219
Butyraldehyde	50	350	W8	500
		420	W7	500
		500	W15	500
		580		500
Carbon tetrabromide	50	47000	F26	492
		500000		66
	60	3300	Q1	418
		18500	Q2	418
		74500	Q3	418
		121500	Q4	418
Cyclohexane, 1,2-epoxy-4-vinyl-	--	264		428
Ethane, 1,2-dichloro-	25	4.0		425
	50	4.5	C,F26	392
Ether, dodecyl vinyl	50	156		4
Furan, tetrahydro-	25	16		425
	40	30		246
	50	24		164
Heptane, 2,4,6-trichloro-	50	5		211
Oxalic acid, diethyl ester	25	1.4		315
Pentane, 2,4-dichloro-	25	9.0	G	315
	50	5		211
Ethylene, 1,1-dichloro-				
Phosphorus trichloride	60	14	D	214
Ethylene, tetrafluoro-				
Ethanol	100	800	C,Q5,W2	3
		820	C,Q6,W2	3
		860	C,Q4,W2	3
		880	C,Q3,W2	3
Isopropyl alcohol	100	1540	C,Q3,W2	3
		1660	C,Q4,W2	3
		1700	C,Q5,W2	3
Methanol	100	350	C,Q3,W2	3
		390	C,Q4,W2	3
Hexanoic acid, vinyl ester				
Benzene	80	4.9		201
1-Hexene				
Carbon tetrachloride	140	41000	C,Q1,F19	170
Cyclohexanol	125-135	390	C	381
Ethyl alcohol	30-35	170	D	381

Solvent	T°C	$C_S \times 10^4$	Remarks	References
Isobutyric acid, vinyl ester				
Carbon tetrachloride	80	4.49		201
Maleic acid, diethyl ester				
Stearamide, N-allyl-	90	22.4	J	173
Maleic anhydride/methyl methacrylate				
Acetone	60	1.6	HH	367
Maleic anhydride/styrene				
Aniline, N,N-dimethyl-	60	930		489
Carbon tetrachloride	70	1.04	HH	365
Methacrylic acid, butyl ester				
Benzene	60	0.158		264
Methacrylic acid, 2-(diethylamino)ethyl ester/styrene				
Carbon tetrachloride	80	23.6		335
Toluene	80	13.3		335
Methacrylic acid, ethyl ester				
Acetic acid	80	0.095		83
Acetone	80	0.102		83
Acetophenone	80	0.281		83
Benzene	80	0.081		83
--, chloro-	80	0.436		83
--, ethyl-	80	1.428		83
2-Butanone	80	0.252		83
Butyl alcohol	80	0.454		83
sec-Butyl alcohol	80	1.604		83
tert-Butyl alcohol	80	0.417		83
Carbon tetrachloride	80	0.901		83
		5.640	C	183
Chloroform	80	0.703		83
Cumene		2.360	C	183
Cyclohexane	80	2.067		83
Ethane, 1,2-dichloro-	80	0.928		83
--, 1,1,2,2-tetrachloro-		1.821	C	183
	80	0.311		83
--, 1,1,1-trichloro-		1.820	C	183
Ethyl acetate	80	0.536		83
Ethyl alcohol	80	0.919		83
Heptane	80	0.429		83
2-Heptanone	80	0.865		83
Isobutyl alcohol	80	0.702		83
2,4-Pentanedione	80	0.445		83
Toluene	80	0.236		83
	80	0.436		83

Remarks and References page II-98

Solvent	T °C	$C_S \times 10^4$	Remarks	References
Methacrylic acid, hexadecyl ester				
Carbon tetrachloride	70	0.983	C	137
Cumene	70	2.05	C	137
Methacrylic acid, isobutyl ester				
Benzene	60	0.165		264
Carbon tetrachloride	80	1.971	C	183
Chloroform	80	1.110	C	183
Ethane, 1,2-dichloro-	80	0.510	C	183
--, 1,1,2,2-tetrachloro-	80	0.510	C	183
Methacrylonitrile				
Carbon tetrabromide	100	900	A,F7	109
Carbon tetrachloride	120	1000	A,F7	109
Iron(II)chloride	60	30800	F7	19
Isobutyraldehyde	60	0,001		70
Toluene	60	0.996		21
Methyl methacrylate				
Acetaldehyde	60	6.5	A	86
Acetic acid	80	0.24		31
--, 1,1-dimethyl-2,2,2-trinitro- ethylester	45	520		110
Acetone	60	0.195		81
	80	0.275	A	81
Acetonitrile, (m-bromophenyl)-	60	3.89		462
--, (p-bromophenyl)-	60	2.78		461
--, (m-chlorophenyl)-	60	4.28		462
--, (p-chlorophenyl)-	60	3.21		461
--, (p-methoxyphenyl)-	60	19.0		461
--, phenyl-	60	5.18		461
--, m-tolyl-	60	5.75		462
--, p-tolyl-	60	7.83		461
Acetylene, p-bromophenyl-	60	41.3	K	151
--, p-chlorophenyl-	60	38.9	K	151
--, p-nitrophenyl-	60	127.5	K	151
Aluminum, hydrodiisobutyl	60	3600	S	156
triethyl	60	1240	S	156
Aniline	60	4.2	F2	424
	80	6.3	F2	424
	100	9.0	F2	424
	50	18	F16	228
		30.4	H	409
		430		161
--, N,N-dimethyl-	60	11.3	A,F2	424
	70	10.8		270
	80	17.0	A,F2	424
Methyl methacrylate (Cont'd.)				
Aniline, N,N-dimethyl- (Cont'd.)	100	20,0	A,F2	424
--, N,N-divinyl-	60	340		82
--, N-methyl-	80	7.0	F2	424
	100	10.0	F2	424
	60	13.3	F2	424
p-Anisaldehyde	60	1.11		495
Anisole	25	0	D	16
--, p-ethynyl-	60	40.3	K	151
--, p-isopropyl-	60	3.46		406
--, p-methyl-	60	0.57		496
Anthracene	50	0		15
Azobenzene	50	100	C	282
Benzaldehyde	60	2.5		86
--, p-bromo-		0.86		495
--, m-chloro-	60	1.43		495
--, p-chloro-	60	0.96		495
--, p-cyano-	60	1.03		495
Benzene	25	2.06		16
	50	0	D	147,327
	52	0.036		81
	60	0.027		81
		0.040		181
	75	0.83	C	123
	80	0.33	C	31
--, bromo-		0.075	A	81
--, tert-butyl-	90	0.080		145
--, chloro-		0.24	I1	147
	25	0.036		16
	60	0	D	31
--, p-diisopropyl-	80	0.260	A	16
--, m-dinitro-	60	0	D	81
--, ethyl-	60	0.074		31
	60	0.200	A	81
	50	0.207		406
	52	5.72	D	15
	60	52		81
--, ethynyl-	80	0.501		81
		0.766		81
	60	1.311		31
Benzonitrile		1.350	A	145
	60	2.1		150
	60	21.9	K	151
		22.3	J	150
--, fluoro-	25	0	D	16
	25	0	D	16
	60	0.162		150

Methyl methacrylate (Cont'd.)

Solvent	T °C	$C_s \times 10^4$	Remarks	References
Benzonitrile, p-hydroxy-	50	6,0		479
--, p-isopropyl-		4,99	D	406
p-Benzoquinone	44.1	55000		194
	50	57000		15
	60	45000		42
--, 2,3,5,6-tetrachloro-	44.1	2600		194
Benzyl ether	25	10,4	D	368
	60	8,0		368
Bibenzyl	50	0,0	F2	354
Borane, tributyl-	60	7,45		156
Butane, 1-chloro-	80	1,20	A	31
--, 1,1,1-trinitro-	45	8300		110
1,4-Butanediol	60	0,61	F24	421
	80	1,07	F24	421
2-Butanone	60	0,45	C	290
	70	0,56	C	123
	75	0,83	C	123
	80	0,70	A	31
1-Butene	40	3,1		287
	50	5,1		287
--, 3,4-epoxy-2-methyl-	80	29,6		269
2-Butene, cis-	40	3,2		287
	50	4,9		287
--, trans-	40	3,0		287
	50	5,2		287
3-Buten-2-ol, 1-chloro-3-methyl-	80	18,8		269
Butyl alcohol	60	0,394		479
	80	0,25	A	31
sec-Butyl alcohol	60	0,259		479
	80	0,85	A	31
tert-Butyl alcohol	60	0,085		81
	80	0,100	A	31
Butylamine, N-nitro-	45	0,152		81
	60	0		110
Butyl ether	60	0,8		86
1-Butyne, 1-phenyl-	60	10,8		150
Butyraldehyde	50	1,47	W8	500
		2,25	W7	500
		3,40		500
Carbon tetrabromide	60	2700	F2	109
	80	3300	F2	109
	100	4600	F2	109
Carbon tetrachloride	20	0,2	C,D	35
	30	20	D,F2	451
	60	0,925		81
	60	2,40	C	181

Methyl methacrylate (Cont'd.)

Solvent	T °C	$C_s \times 10^4$	Remarks	References
Carbon tetrachloride (Cont'd.)	60	5		86
	80	2,393	A	31
		2,421	I1	81
	--	3,3		145
		24,4	D	427
Chloroform	60	0,454		81
	80	1,77	C	181
		1,129		81
		1,400	A	31
	--	1,9	I1	145
		1,9	D	427
Copper(II)chloride	60	10500000	F7	36
Cumene	60	1,9		7
		2,56	A	406
	80	1,9	I1	31
		2,4		145
--, p-bromo-	60	3,71	A	406
--, p-tert-butyl-	60	2,74		406
--, p-chloro-	60	3,07	A	406
Cyclohexane	60	12		86
	80	0,10	A	31
--, methyl-	80	0,195	A	31
Cyclotetrasiloxane, octamethyl-	50	2,5	C	450
p-Dioxane	79.5	0,080		474
	80	0,222	A	31
Diphenylamine	44.1	0		194
Diphenylamine-T	60	0,3	F2	46
Disiloxane, 1,1-dimethoxy-3,3,3-tri-methyl-1-phenyl-	79.5	0,032		474
--, hexamethyl-	79.5	0,104		474
Ethane, 1,2-dichloro-	60	0,35		81
	80	0,756		110
--, nitro-	45	2,0		81
--, 1,1,2,2-tetrachloro-	60	0,155	A	31
	80	0,200		81
--, 1,1,1-trichloro-	80	0,235		31
--, 1,1,1-trinitro-	45	0,600	A	110
		1400		
Ethyl acetate	60	0,100	C	316
		0,13	A	290,316
		0,132	C	316
		0,134		290
		0,155		290
		0,156	C	290
		0,46	I10	145
	70	0,55	C	123

Remarks and References page II-98

Methyl methacrylate (Cont'd.)

Solvent	T °C	$C_s \times 10^4$	Remarks	References
Ethyl acetate (Cont'd.)	75	0.83	C	123
Ethyl alcohol	80	0.240	A	31
	60	0.40	A,F2	423
	80	0.625	A,F2	423
	100	0.80	A	49
Ethylene glycol	60	0.28	F24	421
	80	0.60	F24	421
Glycerol	60	0.152		479
Heptane	50	1.8		206
1-Heptanol, 2,2,3,3,4,4,5,5,6,6,7,7-dodecafluoro-	60	2.8	F2	442
Hydrazyl, 2,2-diphenyl-1-picryl	44.1	20000000		194
Hydroquinone	45	7.0		30
	131	100		398
Indium, triethyl-	60	332	A	156
Isobutyl alcohol	60	0.10		81
	80	0.505		479
	80	0.229	A	81
Isobutyric acid	80	0.250	A	31
--, methyl ester	80	0.900		31
Isopropyl alcohol	60	0.26		275
	60	0.583		81
	80	1.907		81
Lead, tetraethyl	100	3.00	A	49
	130	8	C	337
	130	43	C	337
	130	45	C	480
	60	3.14	S	156
Malononitrile, furfurylidene-	44.1	12000		194
Mercury, diethyl	60	0.898		156
Methane, bromotrichloro-	30	830	D,Q3	311
	30	12000	D	311
	30	45000	D,Q4	311
--, bromotrinitro-	45	10000		110
--, dichloro-	60	0.100		81
	80	0.217		81
--, nitro-	60	2.0		86
--, trinitro-	45	5400		110
--, triphenyl-	60	4.0		44
Methanol	60	0.2	A,F2	423
	80	0.33	A,F2	423
	100	0.45	A	49
1-Naphthol	45	<5.0		30
2,6-Octadiene, 2,6-dimethyl-	60	6.7		324
Oleic acid, methyl ester	60	1.68		441
		8.0		7

Methyl methacrylate (Cont'd.)

Solvent	T °C	$C_s \times 10^4$	Remarks	References
Pentane, 2,2,4-trimethyl-	50	1.2		206
1-Pentanol, 2,2,3,3,4,4,5,5-octa-fluoro-	60	2.36	F2	442
2-Pentanone, 4,4-dimethyl-5,5,5-trinitro-	45	400		110
--, 4-methyl-	80	0.700	A	31
3-Pentanone	60	0.833		81
	80	1.729		81
		1.775	I1	145
	79.5	2.7		206
Pentasiloxane, dodecamethyl-	50	0.145		474
Phenol	50	2.5		479
--, o-bromo-	50	5.0		479
--, p-bromo-	50	5.0		479
--, o-chloro-	50	3.5		479
--, o-ethyl-	50	7.2		479
--, p-ethyl-	50	7.5		479
--, o-isopropyl-	50	13.3		479
--, p-isopropyl-	50	13.3		479
--, p-methoxy-	45	<5.0		30
--, 2,3,4,6-tetramethyl-	45	11.0		30
Phenyl ether	25	6.42	D	368
		9.13		368
Phosphine, octyl-	60	23000		295
--, phenyl-	60	161000	S	295
--, tributyl-	60	30.6		156
Piperidine, 1-nitroso-	45	8.2	A	110
Propane, 1,2-dichloro-	80	0.675	A	31
--, 1,1-dinitro-	45	68		110
--, 2,2-dinitro-	45	15		110
--, 1-nitro-	45	5		110
1,2-Propanediol	60	0.48	F24	421
	80	0.86	F24	421
1-Propanol, 2,2,3,3-tetrafluoro-	60	1.86	F2	442
1-Propene, 2-methyl-	40	2.5		287
	50	4.4		287
Propionic acid, 2,2,2-trinitroethyl ester	45	3000		110
Propionitrile, 3-phosphino-	60	14000		295
--, 3,3'-phosphinylidenedi-	60	13000		295
Propyl alcohol	60	0.69	A,F2	423
	80	0.84	A,F2	423
	100	1.25	A	49
Pyrocatechol	45	<5		30
--, p-tert-butyl-	45	9		30
Pyrogallol	45	26		30

Solvent	T °C	$C_s \times 10^4$	Remarks	References
Methyl methacrylate (Cont'd.)				
Silane, (α-bromotolyl)trimethoxy-	79.5	29.8		474
-,-, chlorotrimethyl-	30	144	D,F2	451
-,-,	50	2.2		249
-,-, dichlorodimethyl-	30	175	D,F2	451
-,-,	50	2.45		249
-,-, dichloromethyl-	79.5	15.2		474
-,-, dimethoxymethylphenyl-	79.5	0.20		474
-,-, tetraethyl-	60	5.75		156
-,-, tetramethyl-	30	5.0	D,F2	451
-,-,	50	1.3		249
-,-, trichloromethyl-	30	185	D,F2	451
-,-,	50	2.7		249
-,-, triisopropoxymethyl-	79.5	0.697		474
-,-, trimethoxymethyl-	79.5	0.331		474
-,-, trimethoxyphenyl-	79.5	0.053		474
Silicon chloride	30	256	D,F2	451
	50	3.0		249
Stearamide, N-allyl-	90	3.01	C,J	173
Stearic acid, methyl ester	60	0.282		441
Stibine, tributyl	60	<1.0		156
Tin, tetrabutyl	60	1.32		156
Toluene	20	0.04	C,D	35
	52	0.084		81
	60	0.170		21, 81
		0.190		316
		0.202	C	32, 316, 317
		0.250	C	316
		0.26	C	275
		0.400	C	316
		0.45	C	496
	70	0.567	C	123
	80	0.292		32
		0.303		81
		0.52	I1	145
		0.525	A	31
		0.91	C	123
-, p-bromo-	60	0.58		496
-, m-chloro-	60	0.48		496
-, p-chloro-	60	0.49		496
-, α-chloro-	60	4.17	C	181
-, p-ethynyl-	60	64.4	K	151
-, p-nitro-	44.1	0		194
-, 2,4,6-trinitro-	44.1	500		194
m-Tolunitrile, α-cyano-	60	4.55		462
p-Tolunitrile	60	0.73		496
-, α-cyano-	60	1.74		461

Solvent	T °C	$C_s \times 10^4$	Remarks	References
Methyl methacrylate (Cont'd.)				
Tributylamine	70	25.7		270
Triethylamine	20	1.5	C,D	35
	60	8.3	J	24
Tripropylamine	60	1900		25
	60	14.6		156
Trisiloxane, 1,1,1,3,5,5,5-hepta-methyl-	79.5	0.284		474
-,-, 3-(β-methylphenethyl)-	79.5	2.08		474
-,-, 3-phenyl-	79.5	0.232		474
-,-, 3-propyl-	79.5	0.189		474
-,-, 3-(3,3,3-trifluoropropyl)-	79.5	0.205		474
-,-, 1,1,1,5,5,5-hexamethyl-3-phenyl-	79.5	1.84		474
-,-, octamethyl-	79.5	0.032		474
Water	60	0		268
		0.002		479
m-Xylene	80.5	0	F7,F24	48
p-Xylene	60	0.43		496
2,4-Xylenol	60	0.50		496
	50	6.5		479
Naphthalene, 1-vinyl-				
Naphthalene	60	30-50		216
Nonanoic acid, vinyl ester				
Toluene	50	11.1	R	357
	60	13.9	R	357
	70	15.2	R	357
1-Octene				
Butyl alcohol	117-18	270	C	381
sec-Butyl alcohol	117-18	520	C	381
Ethyl alcohol	115-18	230	C	381
Isopropyl alcohol	30-5	630	D	381
	120-5	520	C	381
Methanol	116-20	110	C	381
Phthalimide, N-vinyl-				
Chloroform	55	3.4		313
Ethane, 1,2-dichloro-	55	0.35		313
Methane, dichloro-	55	0.97		313
2-Picoline, 5-vinyl-				
Allyl ether	70	21.4		11
1-Propene				
Carbon tetrachloride	100	50000-100000	C	210
Chloroform	100	10300	C	210
1-Propene, 3-chloro-				
Carbon tetrachloride	100	4800	C	210

Solvent	T °C	$C_S \times 10^4$	Remarks	References
1-Propene, 3-chloro- (Cont'd.)				
Hydrochloric acid	80	18000	Q1	238
		54000	Q2	238
1-Propene, 2-methyl-				
Carbon tetrachloride	100	170000	C	210
Propionic acid, vinyl ester				
Toluene	--	1.4		158
Stearic acid, vinyl ester				
Toluene	50	20.7	R	357
	70	21.6	R	357
Styrene				
Acetaldehyde	60	8.5		86
Acetamide, N,N-dimethyl-	50	0.743		375
	60	4.6		113
Acetic acid	40	2.22	C	385
--, methyl ester	100	0.33		72
--, bromo-	60	430		112
	68	300		149
--, chloro-	68	200		149
--,-, methyl ester	60	0.3	A	128
	100	0.75	A	128
--, dibromo-, ethyl ester	90	2700	C	64
--, dichloro-, ethyl ester	60	1.3	A	128
--, iodo-	68	8000		149
--, phenyl-	60	6.0	A	128
--,-, ethyl ester	100	8.8		72
--, tribromo-	--	30000		178
	90	24000	C	64
--,-, ethyl ester	90	105000	C	64
--, trichloro-	--	270		178
	60	66.0	A	128
--,-, ethyl ester	60	65.0	A	128
	90	90.0	C	64
	100	145.0	A	128
Acetic anhydride	60	0.7	A	128
Acetoacetic acid, ethyl ester	100	1.3		72
--, 2-acetyl, ethyl ester	100	3.0		72
Acetone	60	4.1	C	182
		< 0.5	A	128
--, oxime	60	2.2	A	128
Acetonitrile	60	0.44	A	128
--, (m-bromophenyl)-	60	66.7		462
--, (p-bromophenyl)-	60	68.5		461
--, (m-chlorophenyl)-	60	65.2		462
--, (p-chlorophenyl)-	60	66.0		461

Solvent	T °C	$C_S \times 10^4$	Remarks	References
Styrene (Cont'd.)				
Acetonitrile (Cont'd.)				
--, (p-methoxyphenyl)-	60	51.0		461
--, phenyl-	60	45.1		461
--, m-tolyl-	60	48.5		462
--, p-tolyl-	60	49.2		461
Acetyl bromide	60	8600	S	112
Acetyl chloride, chloro-	60	3300		112
Acetylene, p-bromophenyl-	60	188.5	K	151
--, p-chlorophenyl-	60	161	K	151
--, p-nitrophenyl-	60	3130	K	151
Adipic acid, diallyl ester	60	6.0		446
--, di-(3,3,5,5-tetramethyl-4-nitrosocyclohexyl) ester	60	11000	F9,J	314
Allyl alcohol	60	1.5	A	128
Aluminum, hydrodiisobutyl	100	269000		156
		280000	A	155
--, triethoxy-	100	< 0.1	A	155,156
--, triethyl	100	80500		156
		170000	A	155
--, triisobutoxy-	100	< 1.0	A	155
--, triisobutyl	110	285000	A	155
Aniline	50	20	F2	154
	60	2.0	A	128
--, N,N-dimethyl-	50	12	F2	154
		53		161
--, N,N-divinyl-	60	130		82
--, N-methyl-	50	13	F2	154
--, 2,4,6-trinitro-	50	118000		370,372
p-Anisaldehyde	60	2.86		495
Anisole, p-ethynyl-	60	60.0	K	151
--, m-isopropyl-	60	5.23		405
--, p-isopropyl-	60	3.23		406
--, p-methyl-	60	0.78		496
--, 2,4,6-trinitro-	50	203000		370,372
Anthracene	44.1	20000	J	159
--, dihydro-	50	610	D	271
2H-Azepin-2-one, hexahydro-	130	0.61	C,I	108
Azobenzene	70	750	C	281
Benzaldehyde	60	4.54		495
		5.5		86
	100	2.6	A	298
--, p-bromo-	60	12.2		495
--, m-chloro-	60	13.7		495
--, p-chloro-	60	8.63		495
	100	5.6	A	298
--, p-cyano-	60	76.7		495

Styrene (Cont'd.)

Benzene

Solvent	$T^\circ C$	$C_S \times 10^4$	Remarks	References
Benzene	35	3.9	F2	119
	40	5.8	F2	119
	50	0.01		327
	60	0.018	A	127
		0.023		79
		0.028	C	255
		0.04	A,15	145
		1.92	H	415
	70	5.50	C	123
	75	6.67	C	123
	80	0.061	15	185
		0.121	C	255
		0.156		273
	100	0.184	A	127
		0.23	A,15	145
		0.31	A	235
		0.42	A	235
	132	0.81	A	235
		0.89	A	127
--, bromo-	60	1.5	A,15	145
		1.78	C	182
	155	3	A	237
--, sec-butyl-	60	6.22	H	415
--, tert-butyl-	60	0.04	A,15	145
		0.06	A	127
	80	0.193	15	185
	100	0.55	A	127
--, chloro-	60	0.133	C	255
		1.50	C	182
	80	0.235	C	255
		0.874	15	185
	100	0.54	A	235
	140	0.6		370,372
--, 2,4,6-trinitro-	50	585000		58
--, 1,2-dibromoethyl-	60	1950		112
--, p-dibutyl-	60	7.02	H	415
--, p-di-sec-butyl-	60	10.70	H	415
--, p-di-tert-butyl-	60	0.87	H	415
--, m-dichloro-	140	0.2		58
		1.4		67
--, o-dichloro-	60	3.4	C	182
--, p-dichloro-	60	2.6	C	182
--, diethyl-, (mixture)	100	3.35	A	235
		6.33	A	235
	132	5.13	A	235

Styrene (Cont'd.)

Benzene (Cont'd.)

Solvent	$T^\circ C$	$C_S \times 10^4$	Remarks	References
--, p-diisopropyl-	60	3.30		497
		6.60		406
--, ethyl-	60	0.67	A	127
		0.70		79
		0.710	C	255
		0.83	A,15	145
		2.7	C	363
	80	1.07	15	185
		1.113	C	255
	100	1.38	A	235
		1.62	A	127
		2.2	A,15	145
	132	2.33	A	235
		2.31	A	235
		2.9	A	127
		4.9	A,15	145
--, ethynyl-	60	96.3	K	151
		98.3	J	150
--, sec-hexyl-	60	12.76	H	415
--, sec-pentyl-	60	9.43	H	415
--, tri-sec-butyl-	60	13.30	H	415
--, 1,3,5-trinitro-	40	948000		372
	50	643000		370,372,373
	60	351000		372
Benzo[B] chrysene	44.1	130000	J	159
Benzoic acid				
--, m-(phenylazo)-, methyl ester	70	700	C	281
--, 2,4,6-trinitro-, ethyl ester	50	572000		370,372
Benzoin	60	40	A	128
Benzonitrile	60	5.3		150
--, p-isopropyl-	60	18.6		406
Benzo[A]pyrene	44.1	140000	J	159
p-Benzoquinone	60	2270000	J	42
	80	5660000	J	27
--, 2-anilino-	45	5320000	J	199
--, 2,5-dimethyl-	80	430000	J	27
--, 2-methyl-	80	2100000	J	27
--, 2,3,5,6-tetrachloro-	80	9500000	J	27
--, 2,3,5,6-tetramethyl-	80	6700	J	27
--, 2,3,5-trimethyl-	80	260000	J	27
Benzyl ether	25	71.6	D	368
	60	62.4		368
Bibenzyl, α,α'-dibromo-	60	3020		112

Solvent	T °C	$C_S \times 10^4$	Remarks	References
Styrene (Cont'd.)				
Bicyclohexyl, 3,3,3',3',5,5,5',5'- octamethyl-4,4'-dinitroso-	60	11200	F9,J	314
	80	12000	F9,J	314
Borane, tributoxy-	100	< 0.1		156
--, tributyl-	100	34.8		156
Butane, 1-bromo-	60	0.06	A	128
	100	0.35	A	128
--, 1-chloro-	60	0.04	A	128
	100	0.37	A	128
--, 2-chloro-	60	1.2	A,S	128
	100	0.3	A	128
--, 2,2-dimethyl-	60	0.43	A	304
--, 1-iodo-	60	1.85	A	128
	100	5.5	A	128
1,4-Butanediol	60	3.2	F24	422
	80	5.6	F24	422
	100	9.3	F24	422
2-Butanone	60	4.98	C	182
	70	8.60	C	123
	75	12.00	C	123
1-Butene	100	2.6	A	358
--, 3,3-dimethyl-2-phenyl-	110	10	A	433
--, 2,4-diphenyl-	110	70	A	433
--, 3,4-epoxy-2-methyl-	80	17.4		269
--, 2-methyl-	100	3.1	A	358
--, 3-methyl-	100	6.9	A	358
2-Butene	100	2.0	A	358
--, 1,4-dichloro-	80	51	C	112
--, 2,3-dimethyl-	100	5.4	A	358
--, 2-methyl-	100	2.9	A	358
3-Buten-2-ol, 1-chloro-3-methyl-	80	8.2	A	269
Butyl alcohol	40	11.2	F24	119
	50	6.5		430
	60	0.06	A,C	128
		1.6	A,C	323,385
sec-Butyl alcohol	60	0.562		479
tert-Butyl alcohol	50	6.6		430
	60	0.22		422
		6.65	A	385
	80	0.345		422
	100	0.55	A	261
	130	1.0	A	261
Butylamine	60	0.5	A	128
tert-Butyl ether	60	2.6	C,F2	301
	80	1.0	C,F2	301
tert-Butyl isocyanide	100	33.0		475

Solvent	T °C	$C_S \times 10^4$	Remarks	References
Styrene (Cont'd.)				
1-Butyne, 1-phenyl-	60	34.3		150
Butyraldehyde	--	2.7	W8	412
	--	14.3		412
	50	2.7	F29,W8	414
		3.7	F29,W16	414
		4.0	F2,W7	413
		4.7	F29,W7	414
		8.0	F2,W15	413
		8.0	F29,W15	414
		11.7	F2	413
		14.3	F29	414
--, diallyl acetal	60	5.7	A	128
Butyric acid, 4-hydroxy-, γ-lactone	100	11.0	A	128
Cadmium, dibutyl	60	20.2		446
Carbonic acid, cyclic ethylene ester	50	0.409		375
--, diallyl ester	100	1170		156
--, di-(3,3,5,5-tetramethyl-4- nitrosocyclohexyl)ester	50	0.235	E	458
	60	6.2		446
Carbon tetrabromide	60	13000	F27,J	314
	80	14000	F9,J	314
	40	18000	F2	109
	60	17800	C	64
		22000	F2	109
		25000	F1	109
		136000	A	128
		4200000		487
Carbon tetrachloride	70	18000		61
	80	23000	F2	109
	90	25100	C	64
	100	23500	A	128
	--	80	W8	412
	--	100		412
	60	69	Q36	63
		84	C,W8	401
		87	C	113,183
		90	C,W7	401
		92	A,C	128,129,210
		98	C	401
		100		366
		109	Q142	63
		110		223
		120		416
		122	A,X	114
		148	X	115
	76	18	C	236

Styrene (Cont'd.)

Solvent	T°C	$C_S \times 10^4$	Remarks	References
Carbon tetrachloride (Cont'd.)	76	117	A	236
	80	133	I5	185
	85	34.1		163
	95	148	Q38	63
	100	186	Q142	63
Chloroform	100	185	A	128,129
	132	304	A	129
	140	300		58
	60	0.5	A	128,210
		0.566	C	255
		3.40	C	182
	68	4.0		149
	80	0.50	I5	185
		0.916	C	255
m-Cresol	50	11		247,248
o-Cresol	50	33		247,248
	60	43		121
p-Cresol	50	22.1		247,248
	60	39		121
--, α-phenyl-	60	<5		121
Cumene	60	0.8		7
		0.82	A	127
		1.04	A,I5	145
		3.88		406
	80	1.31	I5	185
		2.00	A	127
	100	2.90	A,I5	145
--, m-bromo-	60	8.29		405
--, p-bromo-	60	7.57		406
--, p-tert-butyl-	60	3.46		406
--, p-chloro-	60	6.90		406
Cyclohexane	60	0.024	A	127
		0.031		79
		0.04	A,I5	145
		0.063	C	255
		8.5	F23	119
	80	0.066	I5	185
		0.083	C	255
		0.156		273
	100	0.16	A	127
		0.23	A,I5	145
		0.31	A	235
	132	0.81	A	235
		0.87	A	127
		1.5	A,I5	145
Cyclohexanone	60	7.90	C	182

Styrene (Cont'd.)

Solvent	T°C	$C_S \times 10^4$	Remarks	References
Cyclopentanone	60	3.30	C	182
1-Cyclopentene-1-carboxylic acid, 3,3,5,5-tetramethyl-4-nitroso-, diester with pyrocatechol	60	13000	F27,J	314
	80	10000	F9,J	314
Cyclotetrasiloxane, octamethyl-	50	4.0	C	450
Dibenzo[def,p]chrysene	44.4	60000	J	159
Dibenzo[a,j]naphthacene	44.4	130000	J	159
Dibenzo[a,l]naphthacene	44.4	130000	J	159
p-Dioxane	60	0	A,S	128
		2.28	C	4
		2.75	C	182
	100	0.8	C	220
Diphenylamine-T	60	0.9	F2	46
Disiloxane, hexamethyl-	79.5	0.387		474
Ethane, 1,2-dibromo-	60	0.988	C	255
--, 1,2-dichloro-	80	1.914	C	255
	60	0.333	C	255
		4.12	C	182
--, pentaphenyl-	70	1.1		67
	80	1.137	C	255
		9.8	C	112
--, 1,1,2,2-tetrachloro-	100	3.84	A	235
	60	20000	A	127
--, 1,1,2,2-tetrachloro-1,2-difluoro-	100	10.8	A	235
	90	1.13	A	14
--, 1,1,2-trichloro-1,2,2-trifluoro-	90	0.84	A	14
Ethanehexacarboxylic acid, hexaethyl ester	50	<50		89
1,1,2-Ethanetricarboxylic acid	100	0.94	A	72
Ether, benzyl methyl	68	6.0		149
--, p-bromobenzyl methyl	68	6.0		149
--, p-chlorobenzyl methyl	68	4.0		149
--, p-cyanobenzyl methyl	68	20.0		149
--, dodecyl vinyl	60	3.32	C	4
		4.11	C	4
Ethyl acetate	60	15.5	F10	119
	70	5.5	C	123
	75	6.67	C	123
	100	0.39		72
Ethyl alcohol	60	1.32	F2	423
		1.611		479
	80	2.60	F2	423

Remarks and References page II-98

Styrene (Cont'd.)

Solvent	T °C	$C_S \times 10^4$	Remarks	References
Ethylene, 1,1-diphenyl-	110	450	A	432
Ethylene glycol	60	1.36	F24	422
	80	2.70	F24	422
	100	4.70	F24	422
--, bis(m-phenylazobenzoate)	70	900	C	281
Ethyl ether	60	5.64	C	4
Fluorene	60	75.0	A	127
	100	124.0	A	127
Formamide, N,N-dimethyl-	40	1.09	E	458
	50	0.869		375
	60	4.0		113
	100	1.08		256
Furan, tetrahydro-	50	0.50	C	449
α,D-Glucoside, methyl-				
--, 6-deoxy-6-mercapto-	100	55000		220
--, 2,3-di-O-benzyl-	100	62		220
--, 2,3,4,6-tetra-O-acetyl-	100	2.0		220
--, 6-O-(p-toluenesulfonyl)-	100	2.0		220
--, 2,3,4-tri-O-acetyl-6-deoxy-6-iodo-	100	50		220
--, 6-O-triphenylmethyl-	100	21		220
β,D-Glucoside, methyl-				
--, 6-deoxy-6-dipropylamino-	100	22		220
Heptane	60	0.42	A	127
	100	0.95	A	127
1-Heptanol, 2,2,3,3,4,4,5,5,6,6,7,7-dodecafluoro-	60	13.33	F2	442
1-Heptene	100	2.7	A	358
2-Heptene	100	3.2	A	358
Hexane	100	0.9	A	358
1-Hexene	100	2.5	A	358
2-Hexene	100	3.6	A	358
Hydrochloric acid	100	0		238
Hydroquinone	60	3.6	A	128
Indium, triethyl-	100	17600		156
Iron(III) chloride	60	5360000		19
Isobutyl alcohol	60	0.17	A	385
		0.497		479
Isobutyl alcohol-D	100	2.9	A	261
Isobutyraldehyde	130	7.8	A	261
		8.2	W8	412
		21.0		412
Isobutyric acid	60	2.5	C	385
		4.6	A	385
Isobutyronitrile	100	2.7	A	261
	130	3.5	A	261

Styrene (Cont'd.)

Solvent	T °C	$C_S \times 10^4$	Remarks	References
Isophthalic acid, diallyl ester	60	3.5		446
Isopropyl alcohol	60	3.05	F2	423
	80	4.00	F2	423
	100	1.7	A	261
		6.0	F2	423
Isopropyl alcohol-D	130	2.7	A	261
Isopropyl-1-D alcohol-D	100	1.6	A	261
	100	0.78	A	261
Lead, tetraethyl	100	1.24		156
Malonic acid, diallyl ester	60	5.2		446
--, diethyl ester	60	0.47		128
	100	0.46	A	72
--, dimethyl ester	100	0.42		72
--, acetyl-, diethyl ester	100	1.2		72
--, bromo-, diethyl ester	60	700	A	128
	100	1200	A	128
--, butyl-, diethyl ester	100	0.82		72
--, dibromo-, diethyl ester	600	12000	A	128
--, dichloro-, diethyl ester	60	30.0	A	128
	100	62.0	A	128
--, diethyl-, diethyl ester	100	0.88	A	72
--, ethyl-, diethyl ester	100	0.72	A	72
--, phenyl-, diethyl ester	100	3.5	A	72
Mercury, diethyl	100	0.335		156
Methane, bis(2-chloroethoxy)-	80	6.0	C	112
--, bromotrichloro-	80	76000	D,Q2	312
		77000	D,Q2	195
		2400000	D,Q3	312
		2780000	D,Q3	195
	60-80	650000		439
	80	9450	I5	185
--, dibromo-	60	110		112
--, dichloro-	60	0.15	A	128
	80	9.5	C	112
	100	11.8	A	128
--, diiodo-	60	710		112
--, diphenyl-	60	2.3	A	127
	100	4.2	A	127
--, nitro-	60	10		86
--, triphenyl-	60	3.5	A	127
Methanetricarboxylic acid	80	6.0		412
	100	8.0	A	412
Methanol	60	0.91	C	385
		0.296	A	385
	100	0.74	F2	479
	80	1.10	F2	423

Styrene (Cont'd.)
Methanol (Cont'd.)

Solvent	T °C	$C_S \times 10^4$	Remarks	References
-, (4-biphenylyl)phenyl(p-vinyl-phenyl)-	100	1.22	F2	423
-, bis(4-biphenylyl)(p-vinyl-phenyl)-	50	3.5		57
Naphthalene	50	3.5		57
-, decahydro-	60	11		216
-, 2-isopropenyl-	60	0.4	A	127
	80	56		127
	99	70		98
		69		98
		86		98
-, 2-methoxy-	60	<5.0		121
-, 1,2,3,9-tetrahydro-1-phenyl-	60	10000	C	467
1-Naphthol	60	480	J	50,121
-, 2,4-dichloro-	60	490	J	121
1-Naphthol-D	60	75		50,121
2-Naphthol	60	77		122
Naphtho[2,3-a]pyrene	44.4	2400000	J	159
2,6-Octadiene, 2,6-dimethyl-	60	2.0	C	324
2-Octene	100	2.8	A	358
Oleic acid, methyl ester	60	3.15	C	441
		3.52		441
		455		239
Oxalic acid, diallyl ester	70	420	C	141
-, diethyl ester	90	6.64	A	441
	60	4.2		446
Oxetane, 3,3-bis(chloromethyl)-	60	13.5	F22	119
Pentane, 1-chloro-	50	8.0	C	449
-, 2,2,4-trimethyl-	60	0.49	A	128
2,4-Pentanedione	100	<10	A	155
	100	2.0		72
1-Pentanol, 2,2,3,3,4,4,5,5-octa-fluoro-	60	11.36	F2	442
3-Pentanone	60	2.6	A	385
Pentasiloxane, dodecamethyl-	79.5	0.285		474
1-Pentene	100	2.3	A	358
-, 4,4-dimethyl-2-phenyl-	110	10	A	433
-, 2,4-diphenyl-	110	170	A	433
-, 4-methyl-2,4-diphenyl-	110	2900	A	433
-, 4-methyl-2-phenyl-	110	10	A	433
-, 2,4,4-triphenyl-	110	2600	A	433
2-Pentene		4.2	A	358
-, 2-methyl-	100	6.3	A	358
-, 4-methyl-	100	6.9	A	358
Phenol	50	8.1		247,248
	60	14		122

Styrene (Cont'd.)
Phenol (Cont'd.)

Solvent	T °C	$C_S \times 10^4$	Remarks	References
-, p-benzyloxy-	60	290	J	50,121
-, p-tert-butyl-	60	26		121
-, o-chloro-	60	6.0		122
-, p-chloro-	60	-11		121
-, 2,6-di-tert-butyl-	60	49		121
-, 2,6-diisopropyl-	60	310	J	121
-, p-fluoro-	60	54		121
-, m-methoxy-	60	<5		121
-, o-methoxy-	60	43		122
-, p-methoxy-	60	260	J	121
-, o-phenyl-	60	<5		121
-, 2,3,4,6-tetramethyl-	60	580	J	50,121
-, 2,4,6-trinitro-	50	2110000		371
Phenol-D, p-benzyloxy-	60	10		50
-, 2,3,4,6-tetramethyl-	60	20		50
Phenyl ether	25	7.94	D	368
	60	7.86		368
Phosphine, dibutyl-	100	20800		296
-, diethyl-	100	13500		296
-, octyl-	60	36000	A	295
-, phenyl-	60	439000	A,S	295
-, tributyl-	100	24.4		156
Phosphoric acid, tributyl ester	100	<0.1		156
Phosphorus, white (P4)	25	400	E	99
Phosphorus trichloride	57	250	C,D	214
	75	800		494
o-Phthalic acid				449
-, bis(2-methylallyl)ester	60	6.3		446
-, diallyl ester	60	6.3		446
-, di-(3,3,5,5-tetramethyl-4-nitrosocyclohexyl) ester	80	12000	F9,J	314
Piperidine	60	13200	F27,J	314
Propane, 1-chloro-2,3-epoxy-	60	1.0	A	128
-, 1-chloro-2-methyl-	50	7.5	C	449
	60	1.4	A	128
	100	3.0	A	128
-, 1,1,1,3-tetrabromo-3-phenyl-	90	36500	C	64
-, 1,1,1-tribromo-	90	24100	C	64
1,2-Propanediol	60	2.08	F24	422
	80	3.90	F24	422
1,3-Propanediol, 2,2-bis(bromo-methyl)-, diacetate	100	6.80	F24	422
1,3-Propanedione, 1,3-diphenyl-	60	4.05		325
	60	7.0	A	128

Styrene (Cont'd.)

Solvent	T °C	$C_S \times 10^4$	Remarks	References
1-Propene, 3-chloro-2-methyl-	60	24.0	A	128
--, 2-methyl-	100	1.7	A	358
Propionaldehyde, diallyl acetal	60	12.3		446
Propionic acid	60	0.05	A	128
		4.3		385
		4.5	C	323
		4.65	A	385
Propionitrile, 3-phosphino-	60	50000	A	295
--, 3,3'-phophinylidenedi-	60	50000	A	295
Propyl alcohol	60	2.00	F2	423
	80	3.14	F2	423
	100	3.60	F2	423
Propylene oxide	50	1.6	C	449
2-Propyn-1-ol	60	7.0	A	128
Pyridine	60	0.6	A	128
Pyrocatechol	60	1340	J	50,121
--, p-tert-butyl-	60	3600	J	50,121
Pyrocatechol-D	60	260		50
--, p-tert-butyl-	60	370		50
Pyrogallol	60	10400	J	50,121
Pyrogallol-D	60	1600		50
Sebacic acid, diallyl ester	60	4.8		446
Silane, chlorotrimethyl-	50	12.5		249
--, dichlorodimethyl-	50	17.8		249
--, tetraethyl-	100	8.12		156
--, tetramethyl-	50	3.1		249
--, trichloromethyl-	50	19.2		249
--, triethyl-	70	33.4		90
	80	36.8		90
--, trimethoxymethyl-	79.5	0.23		474
--, triphenyl-	70	2.44		90
	80	1.3		90
Silicon chloride	50	20.0		249
Stearamide, N-allyl-	90	5.82	A,J	173
		8.30	C,J	173
Stearic acid, methyl ester	60	1.06		441
		526		239
--, p-bromo-	70	15.6	C	141
--, o-chloro-	90	0.676	A	441
Stibine, tributyl	100	58.0		156
Stilbene, dibromo-	60	3020	A	112
Styrene, α-bromo-	70	10000		172
--, β-bromo-	70	2000		172
--, dibromo-	60	1950		112
--, α-ethyl-	110	10		439

Styrene (Cont'd.)

Solvent	T °C	$C_S \times 10^4$	Remarks	References
Styrene (Cont'd.)				
--, α-methyl-	60	0.86		431
	74	95		343
	80	3.2		98
		4.9		98
	99	5.6		98
		8.5		98
Succinic acid, diallyl ester	110	1.67		431
Succinonitrile, tetraphenyl-	60	5.4		446
Terephthalamide, N,N'-dimethyl-N,N'-dinitroso-	50	28000		89
	40	1400		160
	60	2000		160
Terephthalic acid, diallyl ester	60	4.5		446
Tin, tetrabutyl	100	3.71		156
Toluene	60	0.105	C	255
		0.121		21,24,79
		0.125	A	127
		0.134		78
		0.16	A,I5	145
		0.21		318
		0.82		496
		1.10	C	363
		2.05		363
	80	0.15	C	385
		0.298	I5	185
		0.3	C	385
		0.308	C	255
		0.310		76,77
		0.313		78,273
	100	0.53	A	235
		0.55	C	385
		0.645	A	127
		0.72	A	235
		0.8	A,I5	145
--, p-bromo-	132	1.12	A	235
--, m-chloro-	60	1.30		496
--, o-chloro-	60	1.25		496
--, p-chloro-	70	0.62		67
--, α-chloro-	140	1.8		67
--, α,α-dichloro-	60	1.07		496
--, p-ethynyl-	60	1.56	A	128
--, α,α,α-trichloro-	60	50.0	A	128
--, 2,4,6-trinitro-	60	72.0	K	151
	50	57.5	A	128
		146000	A	370,372

Styrene (Cont'd.)

Solvent	T°C	$C_S \times 10^4$	Remarks	References
p-Toluidine	50	78	F2	154
--, N,N-dimethyl-	50	16	F2	154
--, N-methyl-	50	11	F2	154
m-Tolunitrile, α-cyano-	60	91.4		462
p-Tolunitrile	60	2.07		496
--, α-cyano-	60	112		461
s-Triazine, trimethyl-	60	0.468		88
Triethylamine	60	1.4	W9	364
		3.0	W6	364
		7.1		24
		7.5		363,364
Tripropylamine	100	24.2		156
Trisiloxane, octamethyl-	79.5	0.069		474
Urea, 1,3-bis(3,3,5,5-tetramethyl-4-nitrosocyclohexyl)-				
Water	60	14500	F9,J	314
	80	15000	F9,J	314
	60	0.006		479
m-Xylene	60	0.78		496
p-Xylene	60	0.84		496
--, α,α'-dibromo-	60	150		112
2,6-Xylenol	60	110		121
Zinc, diethyl	100	3660		156

Styrene, p-chloro-

Solvent	T°C	$C_S \times 10^4$	Remarks	References
Anisole, p-isopropyl-	60	3.86		406
Benzaldehyde	100	1.9	A	298
--, p-chloro-	100	9.7	A	298
Benzene, p-diisopropyl-	60	3.62		497
		7.24		406
Benzonitrile, p-isopropyl-	60	8.84		406
Carbon tetrabromide	60	52000	F2	109
Carbon tetrachloride	60	45		366
Cumene	60	3.44		406
--, p-bromo-	60	5.71		406
--, p-tert-butyl-	60	3.52		406
--, p-chloro-	60	4.97		406

Styrene, p-iodo-

Solvent	T°C	$C_S \times 10^4$	Remarks	References
Benzene	50	0.2		56

Styrene, p-methyl-

Solvent	T°C	$C_S \times 10^4$	Remarks	References
Anisole, p-isopropyl-	60	3.27		408
Benzene, p-diisopropyl-	60	3.67		408
		7.34		406
Benzonitrile, p-isopropyl-	60	26.0		408
Cumene	60	4.12		408
--, p-bromo-	60	9.23		408

Styrene, p-methyl- (Cont'd.)

Solvent	T°C	$C_S \times 10^4$	Remarks	References
Cumene (Cont'd.)				
--, p-tert-butyl-	60	3.59		408
--, p-chloro-	60	7.67		408

Styrene, pentafluoro-

Solvent	T°C	$C_S \times 10^4$	Remarks	References
Benzene, bromo-	60	0		469
--, chloro-	60	0		469
--, ethyl-	60	5.35		469
--, fluoro-	60	0.117		469
2-Butanone, 3-methyl-	60	1.09		469
Cumene	60	6.76		469
Furan, tetrahydro-	60	1.53	A	469
2-Pentanone, 4-methyl-	60	1.61		469
Toluene	60	0.033		469
		1.20		469

Styrene/Styrene, p-chloro-

Solvent	T°C	$C_S \times 10^4$	Remarks	References
Carbon tetrachloride	60	15	EF	366
		115	EE	366

Succinimide, N-vinyl-

Solvent	T°C	$C_S \times 10^4$	Remarks	References
Acetic acid	55	0.077	C,F25	93
Ethane, 1,2-dichloro-	55	0.127	C,F26	93

Valeric acid, 4-methyl-, vinyl ester

Solvent	T°C	$C_S \times 10^4$	Remarks	References
Benzene	80	6.2		201

Vinyl acetate

Solvent	T°C	$C_S \times 10^4$	Remarks	References
Acetaldehyde	30	400		346
	45	530	J	86
	50	390		460
	60	0.72		257
		200	I8	145
		220		232
		500		460
		570	F2	86
		660	J	86
	70	610		460
	75	700	J	86
	60	40		466

Acetamide, N-butyl-

Solvent	T°C	$C_S \times 10^4$	Remarks	References
Acetic acid	50	0.180		465
	60	0.166	C	466
		0.200		466
		1.0	I9	145
		1.13		289
		10	F2	86
--, allyl ester	65	0.170	C	466
	60	85	F2	86
		94	J	86

Remarks and References page II-98

Vinyl acetate (Cont'd.)

Acetic acid (Cont'd.)

Solvent	T °C	$C_S \times 10^4$	Remarks	References
--, benzyl ester	60	80	F2,J	86
--, butyl ester	50	13.2		208
--, sec-butyl ester	50	4.4		208
--, tert-butyl ester	60	8.0		86
	50	1.5		208
--, isobutyl ester	60	6.2		232
	50	9.1		208
--, isopropyl ester	50	3.5		208
	60	3.1		145,232
		8.0		86
--, methyl ester	67.5	9.0	J	86
	75	10	J	145,231
	60	1.56		145,232
		1.6		145,232
		2.5		86
--, propyl ester	50	6.2		208
	70	3.4	J	145,232
--, 1,3,3,3-tetrachloropropyl ester	70	423.2	C	389
--, bromo-	60	489	C,J	386
	70	4450	C	389
--, chloro-	70	2550	C	389
--, cyano-, methyl ester	60	5000	J	86
--, dichloro-, ethyl ester	60	210		86
--, phenyl-	60	400	J	86
--, trichloro-	70	1445	C	389
--, -, ethyl ester	60	4400		86
--, -, trifluoro-, ethyl ester	50	30	C	71
Acetic anhydride	60	8.0		86
Acetoacetic acid, ethyl ester	70	80.4	C	390
Acetone	60	1.5		86
		11.70	C	289
		12.0		145
	70	25.6	C	330
	75	42	C	184
--, bromo-	60	10		86
Acetonitrile	60	2100	J	86
--, phenyl-	50	91.5		15
Acetophenone	60	100	J	86
	70	62.0	C	390
--, 3'-hydroxy-	45	1405	J	51
--, 3'-hydroxy-D-	45	1290	J	51
Aniline	50	149		15
--, N,N-dimethyl-	50	210	F2	154
--, N-methyl-	50	260	F2	154
		360	F2	154

Vinyl acetate (Cont'd.)

Solvent	T °C	$C_S \times 10^4$	Remarks	References
Aniline, p-nitro-	50	48600		15
m-Anisaldehyde	60	2500	J	86
o-Anisaldehyde	60	420	J	86
p-Anisaldehyde	60	370	F2,J	86
Anisole	60	10	J	86
Anthracene	20	668000	J	41
	30	603000	J	41
	40	455000	J	41
	50	205700		15
	60	364000	J	41
	60	278000	J	41
--, 9-phenyl-	50	273000	J	41
Benzaldehyde	60	230	F2	145,355
	70	460		86
	75	540		86
		421	C,J	390
		600	J	86
--; m-chloro-	60	860	J	86
--; o-chloro-	60	390	J	86
--; p-chloro-	60	340	J	86
--; m-cyano-	60	1070	J	86
--; p-cyano-	60	610	J	86
--; p-isopropyl-	50	540	J	41
Benz[A]anthracene	60	117000	J	263
Benzene	60	1.07		86
		1.2	J	294
		1.2		80
		2.4	C	289
		2.96		294
		20	C	390
--, bromo-	70	5.27	C	86
	75	1.4	J	263
		1.40		184
		3.6	C	15
--, tert-butyl-	50	18.9		265
	60	8.0	W7	265
--; chloro-		9.2	W15	265
		10.0		389
		134.2	C	184
		25.2	C	289
		3.61	W7	265
		5.6	W15	265
		6.8		265
		8.0		289
		8.35	J	86
		80		

Vinyl acetate (Cont'd.)

Solvent	T°C	$C_S \times 10^4$	Remarks	References
Benzene (Cont'd.)				
--, chloro- (Cont'd.)	70	2.61	C	389
	75	12.7	C	184
--, o-dichloro-	75	42	C	184
--, p-dichloro-	70	11.8	C	389
--, m-dinitro-	45	1050000	C,J	28
	50	645700		15
--, o-dinitro-	45	960000	C,J	28
--, p-dinitro-	45	2670000	C,J	28
	50	662800		15
--, ethyl-	60	55.15		289
	60	100	J	86
--, nitro-	50	110300		15
--, 1,3,5-trinitro-	45	8900000	C,J	28
	50	4342800		15
Benzo[b]chrysene	50	283000	J	41
Benzoic acid	60	50	J	86
--, ethyl ester	60	26	J	86
--, p-nitro-	50	245700	J	15
Benzoic anhydride	60	130	J	86
Benzoin	60	800	J	86
Benzonitrile	50	40.6	J	15
--, m-hydroxy-	45	820	J	51
--, m-hydroxy-D-	45	805	J	51
Benzo[GHI]perylene	50	18200	J	370
Benzophenone	50	286		15
Benzo[a]pyrene	50	306000	J	41
Benzo[e]pyrene	50	8400	J	370
p-Benzoquinone, 2,3,5,6-tetramethyl-	45	950000	J	27
Benzoyl chloride	50	366		15
	70	300	C	389
Benzyl alcohol	70	556	C	390
Biphenyl	50	263		15
	60	6.4		132
Borane, tributyl-	10	9000		191
Butane, 1-bromo-	60	50		86
	70	1100	C	389
--, 1-chloro-	60	10		86
--, 1-iodo-	60	800		86
2,3-Butanedione	60	670	F2,J	86
2-Butanone	60	73.80		289
	70	63.6	C	390
	75	165	C	184
--, 3-methyl-	60	118.16		289
1-Butene, 3-chloro-3-methyl-	50	28900	F2	152
2-Butene, 1-chloro-3-methyl-	50	4300	F2	152
1-Buten-3-yne	60	0	J	232
Butyl alcohol	60	20.0	I9	145
		20.39		289
sec-Butyl alcohol	70	29.1	C	390
	60	31.74	C	289
tert-Butyl alcohol	70	6.21	C	390
	75	95.0	C	184
	60	0.46	I9	289
		0.5		145
		1.3		86
Butyl ether	70	12.1	C,J	390
	60	76		86
3-Butyn-2-ol, 2-methyl-	60	400	J	86
Butyraldehyde	40	440	W8	500
		510	W7	500
		590	W15	500
		670		500
	60	650		232,355
		1000		86
Butyric acid, ethyl ester	70	388	C	390
--, methyl ester	50	45	C	71
	60	18		145,232
		19		86
Carbon tetrabromide	60	7390000	F2	109,145
Carbon tetrachloride	70	28740	C	389
	0	1500		346
	20	4700	E	309
	45	7600	J	86
	60	800	Q1	10
		1300	Q2	10
		4700	Q3	10
		6700	Q4	10
		7300	F2	86
		8000	Q5	10
		9000	Q6	86
		9600	J	210
		10000	Q7	10
		10500	Q8	10
		10700		389
Chloral	70	2023	C	86
	75	10500	J	86
Chloral hydrate	60	5000	C	389
	70	4927	C	389
Chloroform	70	4312	C	345
	--	100		346
	30	140		

Remarks and References page II-98

Vinyl acetate (Cont'd.) — **Chloroform (Cont'd.)**

Solvent	T °C	$C_S \times 10^4$	Remarks	References
	60	125.18	I9	289
	130		I9	145
	150		J	86
	160			210
	170		F2	86
Chrysene	70	554	C	389
m-Cresol	50	3360	J	370
m-Cresol-D	45	375	J	51
p-Cresol	45	85	J	61
p-Cresol-D	45	710	J	51
Crotonaldehyde	45	130	J	51
Cumene	60	1800	J	355
	60	89.9		289
	70	100		7,86
	70	139	C	390
Cyclohexane	75	356	C	184
	60	6.59	I9	289
		7.0		145
-, methyl-	60	100	F2	86
	60	11.75		289
		24		86
1,3-Cyclohexanedione, 5,5-dimethyl-	70	5580	C	390
Cyclohexanol	70	127	C	390
Cyclohexanone	60	180	F2	86
	75	670	C	184
Cyclohexene	60	620	J	86
		1600	F2	86
Dibenz[a,h]anthracene	75	770	J	86
Dibenzo[def,mno]chrysene	50	87000	J	370
	40	1565000	J	41
	50	1292000	J	41
	60	1054000	J	41
Dibenzo[def,p]chrysene	50	736000		41
Diethylene glycol	70	85.3	C	390
p-Dioxane	60	20	F2	86
	70	49.1	C	390
Diphenylamine	50	138		15
	60	240		46
Diphenylamine-D	60	170		46
Diphenylamine-T	60	230		46
Ethane, 1,1-dibromo-	60	1100		86
-, 1,2-dibromo-	70	134	C	389
-, 1,1-dichloro-	60	65		86
-, 1,2-dichloro-	60	5		86
		7	I9	145
		7.18		289

Vinyl acetate (Cont'd.) — **Ethane (Cont'd.)**

Solvent	T °C	$C_S \times 10^4$	Remarks	References
-, 1,2-dichloro-	70	10.2	C	389
-, hexachloro-	70	1210	C	389
-, pentachloro-	70	1348	C	389
-, 1,1,2,2-tetrabromo-	60	6000		86
-, 1,1,2,2-tetrachloro-	60	107.03		289
	60	160		86
-, 1,1,1-trichloro-	70	67.72	C	389
-, 1,1,2-trichloro-	60	71.11		289
1,1-Ethanediol, diacetate	60	35.98		289
	60	40		86
Ether, benzyl methyl	60	280	F2,J	86
-, bis(2-chloroethyl)	70	245	C	389
-, dodecyl vinyl	60	57.2	C	4
		73.5	C	4
-, ethyl	60	45.3	C	4
Ethyl acetate	20	1.52		328
	40	2.11		328
	50	2.9		208
		12		71
	60	1.07	C	289
		1.25	C	80
		2.6		145,232
		2.96		328
		3.3		86
		3.4		420
		7.8		390
Ethyl alcohol	70	25	F2	86
	60	26.3	C	390
Ethylene, tetrabromo-	70	2800	C	389
-, tetrachloro-	70	465	C	389
-, tribromo-	70	34720	C	389
-, trichloro-	70	3810	C	389
Ethylene glycol	70	83.0	C,J	390
Fluorene	60	4700	J	86
-, 9-phenyl-	70	3610	C,J	390
Formamide, N,N-dimethyl-	70	16240	C,J	390
Formic acid, ethyl ester	60	50		86
-, methyl ester	50	22	C	71
2-Furaldehyde	60	3		86
Furfuryl alcohol	60	15000	J	86
Glycolic acid, methyl ester	70	2520	C,J	390
Heptane	60	300		86
1-Heptanol, 2,2,3,3,4,4,5,5,6,6,7,7-dodecafluoro-	50	17.0		208
	60	33.3	F2	442

Vinyl acetate (Cont'd.)

Solvent	T°C	$C_S \times 10^4$	Remarks	References
1,5-Hexadien-3-yne	60	0		232
Hexanoic acid, 2-ethyl-, ethyl ester	50	65	C	71
Hydroquinone	50	7000	J	338
-, 2,5-di-tert-butyl-	50	38000	J	338
-, 2,6-dimethyl-	50	29000	J	338
-, tetramethyl-	50	74000	J	338
Isobutyl alcohol	60	21.75		289
	70	32.4	C	390
Isobutyric acid	60	5.02		289
-, ethyl ester	50	160	C	71
-, methyl ester	60	58		145,232
		71		420
		86		86
Isobutyronitrile	60	100		86
Isopropyl alcohol	70	44.6	C	390
Lactic acid, ethyl ester	60	700		86
-, methyl ester	60	640		86
Lauric acid, ethyl ester	50	105	C	71
Malonic acid, dimethyl ester	60	17		86
p-Mentha-1,8-diene	50	1900	J	86
Mesitol	50	5000	J	338
Methane, bromotrichloro-	25	≫10000	D,S	39
-, dichloro-	60	6000	C	39
-, iodo-		>400000	C	39
-, nitro-	70	6303	F2,J	389
-, tribromo-	60	4	C	86
-, triphenyl-	75	1230	C	389
Methanol	70	2300		86
	60	2600		389
		34760	C	45
		700		5
	--	9.0		5
	10	10.0		5
		10.5		190
-, m-chloro-		3.4		145,231
-, p-chloro-	10	2.26		320
	60	4.3		86
		6.0		320
		5.5		320
Naphthacene	50	8620000	J	370
Naphthalene	50	1150	J	370
		1457		15
-, decahydro-	70	1715	C,J	390
		48		86
Naphtho[1,2,3,4-def]chrysene	50	135000	J	41
Nonanoic acid, ethyl ester	50	80	C	71
2,6-Octadiene, 2,7-dichloro-	50	3200	F2	152
-, 2,6-dimethyl-	60	430		324
		700		7
Octanoic acid, ethyl ester	50	70	C	71
Oleic acid, methyl ester	60	217		441
		1000	J	86
Orthoformic acid, trimethyl ester	90	358		441
	--	7.6		345
Oxalic acid, diethyl ester	30	8.0		346
-, dimethyl ester	60	4.0		86
	60	1.0		86
		2.0		145,232
Paraldehyde	70	136	C	390
Pentane, 2,2,4-trimethyl-	50	8.0		208
2,4-Pentanedione	60	10		86
1-Pentanol, 2,2,3,3,4,4,5,5-octafluoro-	60	26.1	F2	442
2,Pentanone, 4-methyl-	60	34.52		289
3-Pentanone	60	10.0	I9	145
Pentyl acetate	60	114.39		289
Pentyl alcohol	70	7.2	C	390
	75	87.0	C	184
	75	56	C	184
Perylene	50	288000	J	41
Phenanthrene	50	870	J	370
	50	5600	C	15
Phenol	70	3380	C,J	390
	45	215	J	51
	50	120		15
	60	600	J	86
-, m-chloro-	45	205	J	51
-, p-chloro-	45	400	J	51
-, p-nitro-	50	88600		15
-, 2,3,4,6-tetramethyl-	45	11300	J	51
Phenol-D	45	35	J	51
-, m-chloro-	45	70	J	51
-, p-chloro-	45	80	J	51
-, 2,3,4,6-tetramethyl-	45	800	J	51
Phosphorus trichloride	60	15000	C,D	214
Propane, 2-bromo-2-methyl-	60	150		86
-, 2-chloro-2-methyl-	60	26		86
1-Propanol, 2,2,3,3-tetrafluoro-	60	7,11	F2	442
1-Propene, 3-chloro-	60	3100		86
-, 3-chloro-2-methyl-	60	400	J	86
Propionaldehyde	60	950	F2	86
	70	457	C	390

Remarks and References page II-98

Solvent	$T\ ^\circ C$	$C_S \times 10^4$	Remarks	References
Vinyl acetate (Cont'd.)				
Propionic acid, ethyl ester	50	40	C	71
--, methyl ester	60	23		232
		26		420
Pyrene	50	11500	J	370
Pyrocatechol, 4-(1,1,3,3-tetramethyl-butyl)-	50	11000	J	338
Pyrogallol	50	50000	J	338
Salicylic acid	70	296	C	390
Stearamide, N-allyl-	90	620.1	C,J	173
Stearic acid, ethyl ester	50	140	C	71
--, methyl ester	60	38.2		441
Succinic acid, diethyl ester	60	41		420
--, dihexyl ester	60	40		420
--, dimethyl ester	60	31		420
m-Tolualdehyde	60	570	J	86
p-Tolualdehyde	60	440	J	86
Toluene	50	3.3		158
		12.0		15
		14.9	R	357
		100	C	71
		123		71
	60	17.8	R	357
		20.75	C	80
		20.89		21,59,289
		21.6		357
		34	J	86

Solvent	$T\ ^\circ C$	$C_S \times 10^4$	Remarks	References
Vinyl acetate (Cont'd.)				
Toluene (Cont'd.)				
	60	35	F2	86
		49.0	W15	265
		69.0		265
	70	21.1	R	357
		21.8	C	390
		29.2		357
	75	66	C	184
--, bromo-	80	91.6		273
--, o-chloro-	50	20.6		15
--, p-chloro-	70	92,86	C	389
--, α-chloro-	70	195	C	389
	60	450	J	86
	70	584	C	389
--, p-nitro-	75	118	C	184
p-Toluidine	50	106300		15
--, N,N-dimethyl-	50	750	F2	154
--, N-methyl-	50	380	F2	154
Toluonitrile	50	830	F2	154
Triethylamine	50	38.3		15
	50	360	F2	154
Triphenylene	60	370	J	24,145
Veratraldehyde	50	160	J	370
Xylene	60	550	J	86
	50	14.9	J	15
	70	278	C	390
	75	166	C	184

F. TRANSFER CONSTANTS TO SULFUR COMPOUNDS

Modifier	$T\ ^\circ C$	C_X	Remarks	References
Acrylamide, N-octadecyl-/acrylonitrile (1:1 mol ratio)				
1-Dodecanethiol	60	0.539	C,F24	174
Acrylic acid, methyl ester				
1-Butanethiol	30	1.53	A	397
	60	1.69	A	69,397
Ethanethiol	50	0.78	Q1	330
		0.94	Q1	478
		1.57		330
		1.65	Q2	478
		1.79	Q2	330
		5.00	Q3	330

Modifier	$T\ ^\circ C$	C_X	Remarks	References
Acrylic acid, methyl ester (Cont'd.)				
2-Propanethiol	60	0.544	Q1	477
		0.668	Q2	477
		0.700	Q3	477
		0.656	Q4	477
Acrylonitrile				
1-Dodecanethiol	50	0.73		378
Hydrogen sulfide	50	0.13		378
Methyl sulfide	20	0.1	C,F8	248
Methyl sulfoxide	30	0.47		196
	40	0.0000812	E	458
	50	0.11		404

Table (cont'd. — right portion)

Modifier	T °C	C_X	Remarks	References
1,3-Butadiene/styrene (Cont'd.)				
Formic acid, thio-, dithiobis- (Cont'd.)				
--,-, O,O'-diheptyl ester	5	0.41	B,H,J	383
--,-, O,O'-dihexyl ester	50	1.12	C,H,J	382
	5	0.74	B,H,J	383
--,-, O,O'-diisobutyl ester	50	1.78	C,H,J	382
	5	1.87	B,H,J	383
	-5	1.573	H	180
--,-, O,O'-diisopentyl ester	50	6.43	C,H,J	382
	5	1.21	B,H,J	383
	5	7.0	B,H,J	341
	-5	1.85	H	180
--,-, O,O'-diisopropyl ester	50	4.38	C,H,J	382
	5	2.83	B,H,J	383
	5	3.4	B,H,J	341
		12.5	B,H	100
		0.40-1.31	B,H,J,GG4	100
		0.80-3.01	B,H,J,GG3	100
		1.40-3.70	B,H,J,GG2	100
		1.80-3.50	B,H,J,GG1	100
		1.80-6.0	B,H,J,GG5	100
	--	9.78		179
--,-, O,O'-dimethyl ester	-5	1.185	H	180
	50	9.78	C,H,J	382
	5	16.43	B,H,J	383
	-5	15.86	H	180
--,-, O,O'-dineopentyl ester	50	27.04	C,H,J	382
	5	1.01	B,H,J	383
--,-, O,O'-dioctyl ester	50	3.40	C,H,J	382
	5	0.23	B,H,J	383
--,-, O,O'-dipentyl ester	50	0.64	C,H,J	382
	5	1.45	B,H,J	383
	-5	1.70	H	180
--,-, O,O'-dipropyl ester	50	3.20	C,H,J	382
	5	4.42	B,H,J	383
	-5	1.815	H	180
1-Heptanethiol	50	9.20	C,H,J	382
--, 1,1,3,3,5,5,6-heptamethyl-	5	0.64	B,H,J	384
1-Hexanethiol	5	1.22	B,H,J	384
--, 1,1,3,3,5,5-hexamethyl-				
1-Octanethiol	5	4.1	H	54, 55
	-5	2.7-4.7	H	54
--, 1,1,3,5,5,7-octamethyl-	5	0.34	B,H,J	384
p-Dioxin, 2,3-dihydro-/maleic anhydride				
1-Butanethiol	60	6300	R	169

Remarks and References page II-98

Table (cont'd. — left portion)

Modifier	T °C	C_X	Remarks	References
Acrylonitrile (Cont'd.)				
Methyl sulfoxide (Cont'd.)	50	0.29		197
Methyl tetrasulfide	80	0.795	J	375
	50	0.69		85
p-Toluenethiol	50	0.97		378
Acrylonitrile/1,3-butadiene (10-90%)				
1-Hexanethiol	5	1.5	C,H	380
--, 1,1,3,3,5,5-hexamethyl-				
1-Pentanethiol	5	3.3	C,H	380
--, 1,1,3,3-tetramethyl-				
Acrylonitrile/1,3-butadiene (20-80%)				
1-Hexanethiol	5	1.2	C,H	380
--, 1,1,3,3,5,5-hexamethyl-				
1-Pentanethiol	5	2.1	C,H	380
--, 1,1,3,3-tetramethyl-				
Acrylonitrile/1,3-butadiene (30-70%)				
1-Hexanethiol	5	1.1	C,H	380
--, 1,1,3,3,5,5-hexamethyl-				
1-Pentanethiol	5	1.8	C,H	380
--, 1,1,3,3-tetramethyl-				
Acrylonitrile/styrene (38.5-61.5 mol-%)				
1-Pentanethiol	50	1.02	C,H	380
--, 1,1,3,3,4-pentamethyl-				
1,3-Butadiene				
1-Butanethiol, 1,1,3,3-tetramethyl-	5	5.3	H	242
	50	3.7	H	242
1-Octanethiol	5	21.8	H	242
	50	16.0	H	242
	50	19.0	H	243
--, 1,1,3,3,5,5,7,7-octamethyl-	50	3.0	H	242
1-Tetradecanethiol	50	19.5	H	242
1,3-Butadiene/styrene				
1-Dodecanethiol	5	0.66	B,H,J	384
Formic acid, thio-, dithiobis-				
--,-, O,O'-dibutyl ester	5	2.42	B,H,J	383
	-5	1.75	H	180
--,-, O,O'-di-sec-butyl ester	50	5.34	C,H,J	382
	5	1.65	B,H,J	383
	-5	1.565	H	180
--,-, O,O'-dicyclohexyl ester	5	1.78	B,H,J	383
	50	3.71	C,H,J	382
--,-, O,O'-diethyl ester	5	8.43	B,H,J	383
	-5	4.23	H	180
	50	16.04	C,H,J	382

TRANSFER CONSTANTS

Modifier	T°C	C_X	Remarks	References
Ethylene				
1-Butanethiol	130	5.8	C,W5	457
Ethyl sulfide	130	0.027	C,W5	454
Methyl sulfoxide	130	12	C,F21,W5	259
2-Propanethiol, 2-methyl-	130	15.0	C,W5	457
Sulfur hexafluoride	130	0.0000	C,F21,W5	259
Methyl methacrylate				
Acetic acid, dithiodi-, diethyl ester	60	0.00065		86
-, mercapto-, methyl ester	60	0.63		274
Acetophenone, 3'-mercapto-	45	4.2		30
-, 4'-mercapto-	45	2.6		30
Anisole, 4,4'-dithiodi-	50	0.0044		285
	60	0.0052		285
p-Anisoyl disulfide	60	14.6		369
Benzenesulfonic acid				
-, compound with pyridine	60	0.00365		490
-, thio-, S-phenyl ester	60	0.0196	A	491
Benzenethiol	45	4.7		30
	60	2.7		274
-, m-bromo-	45	3.8		30
-, p-bromo-	45	5.6		30
-, m-chloro-	45	3.5		30
-, p-chloro-	45	4.6		30
Benzoyl disulfide	60	10.0		369
Benzyl disulfide	25	0.134		368
	50	0.016	F2	354
	60	0.00627		368
Benzyl sulfide	25	0.0183	D	368
	50	0.0098		252
	50	0.0098	F2	354
	60	0.00154		368
Benzyl sulfone	50	0.0040		252
Benzyl sulfoxide	50	0.0039		252
Benzyl tetrasulfide	50	0.084	F2,J	354
Benzyl trisulfide	50	0.021	F2,J	354
1-Butanethiol	60	0.66		274
		0.67	A	69,397
Carbanilide, thio-	50	0.008	F2	251
-,-, 4,4'-bis(dimethylamino)-	50	0.22	C,F2	251
-,-, 4,4'-dichloro-	50	0.09	C,F2	251
-,-, 3,3'-dimethoxy-	50	0.35	C,F2	251
-,-, 3,3'-dimethyl-	50	0.33	C,F2	251
-,-, 4,4'-dimethyl-	50	0.21	C,F2	251
-,-, 3,3'-dinitro-	50	0.13	C,F2	251
Carbon disulfide	50	0.55	A	251
	60	0.00007		288

Modifier	T°C	C_X	Remarks	References
Methyl methacrylate (Cont'd.)				
2,13-Dioxa-7,8-dithia-3,12-disila-tetradecane, 3,3,12,12-tetramethoxy-	79.5	0.00258		474
Disulfide, bis(p-bromobenzoyl)-	60	16.7		369
-, bis(p-bromophenyl)-	50	0.0046		285
	60	0.0098		285
-, bis(p-chlorobenzoyl)-	60	10.0		369
-, bis(p-chlorophenyl)-	50	0.0072		285
	60	0.0117		285
-, bis(p-cyanobenzoyl)-	60	290		369
-, bis(dimethylthiocarbamoyl)-	70	0.0115		104
-, bis(p-nitrobenzoyl)-	60	694		369
-, bis(o-nitrophenyl)-	50	0.0176		285
	60	0.0508		285
-, bis(p-nitrophenyl)-	50	0.0127		285
	60	0.0193		285
Ethanethiol, 2-amino-, hydrochloride	50	0.17	F7	473
	70	0.15	F7	473
Ethanol, 2-mercapto-	50	0.43	F2	473
	60	0.62		274
	70	0.40	F2	473
Ethyl disulfide	60	0.00013		310
Formamidine, N,N'-diphenyl- -,-, 1,1'-dithiobis-	50	0.22	C,F2	251
Methyl sulfoxide	60	0.0000071	A	438
2-Naphthalenethiol	45	1.3		30
	60	3.1		274
1-Pentanethiol	50	0.8	H	344
Phenyl disulfide	25	0.0174	D	368
	50	0.0038		285
	60	0.0007		310
Phenyl sulfide	50	0.00176		368
	50	0.0085		285
Phenyl sulfone	50	0.0110	A	491
Phenyl sulfoxide	60	0.0003	A	491
	25	0.00554	D	368
	50	0.0064		252
	60	0.00132		368
	50	0.0006		252
	50	0.0013		252
1-Propanethiol, 3-[1,3,3,3-tetramethyl-1-(trimethylsiloxy)disiloxanyl]-	79.5	0.128		474
-, 3-(trimethoxysilyl)-	79.5	0.693		474
2-Propanethiol	60	0.38		274
-, 2-methyl-	60	0.18		274

Modifier	T°C	C_X	Remarks	References
Methyl methacrylate (Cont'd.)				
Propionic acid, 3-mercapto-	50	0.39	F2	473
	70	0.38	F2	473
Sulfide, benzoyl dimethylthiocarbamoyl	20	5.5	D	280
	60	0.08	A	280
Sulfur	--	1.3	C	118
Sulfur	44.1	0.075		194
Thiosulfuric acid, S-butyl ester	50	0.106	F2,J	354
-,-, sodium salt	60	0.00226		490
-,-, S-isopropyl ester				
-,-, sodium salt	60	0.00167		490
-,-, S-phenyl ester				
-,-, compound with pyridine	60	0.374		490
-,-, potassium salt	60	0.00537		490
-,-, S-propyl ester				
-,-, sodium salt	60	0.00190		490
m-Toluenethiol	45	4.7		30
p-Toluenethiol	45	7.4		30
α-Toluenethiol	50	0.027	F2,J	354
p-Toluoyl disulfide	60	11.0		369
p-Tolyl disulfide	50	0.0031		285
	60	0.0044		285
Styrene				
Acetic acid, dithiodi-	50	<0.005		297
	99	0.2		96
-,-, diethyl ester	60	0.015		86
-,-, dimethyl ester	99	0.1		96
-,-, mercapto-	99	>14.0	J,R	96
-,-, butyl ester	50	26.0	A	146
-,-, ethyl ester	60	58.0	A	126
-,-, methyl ester	60	0.63		323
	99	27.6		96
-,-, phenyl-				
-,-, P,P'-dithiodi-	50	0.24		89
-,-, thio-	99	>14.7	J	96
Aniline, 2,2'-dithiodi-	25	3.8		297
	50	3.0		297
		3.4		297
	75	3.0		297
-, 4,4'-dithiodi-	50	0.24		89
-, N-dodecyl-, 4,4'-dithiodi-	50	0.21		89
Anisole, 2,2'-dithiodi-	50	3.0		89
-, 4,4'-dithiodi-	50	0.18		89
-, p-methyl-, α,α'-dithiobis-	99	0.021		96
p-Anisoyl disulfide	60	96.0		369

Modifier	T°C	C_X	Remarks	References
Styrene (Cont'd.)				
Anthranilic acid, 4,4'-dithiobis-	50	3.0		89
Benzenesulfonic acid				
-,- compound with pyridine	60	0.00617		490
-,- thio-, S-phenyl ester	50	1.60	A	491
	60	1.67	A	491
Benzenethiol	99	0.08		96
	99	25.1		96
-,- o-ethoxy-	99	0.13		96
-,- p-ethoxy-	99	0.21		96
2-Benzimidazolethiol	50	0.01		89
Benzoic acid, 2,2'-dithiodi-	50	<0.005		89
-,-, diethyl ester	99	0.2		96
-,-, dimethyl ester	50	0.11		297
-,-, 4,4'-dithiodi-	50	0.17		89
-,-, diethyl ester	50	0.11		89,297
-,-, o-mercapto-	99	>14.7		96
-,-, methyl ester	99	17.0	J	96
-,-, thio-	99	6.23		96
Benzothiazole, 2,2'-dithiobis-	25	2.3		297
	50	2.1		297
		2.3		297
	75	2.4		297
	99	2.73		96
-,- 2,2'-thiobis-	50	<0.005		297
2-Benzothiazolethiol	50	0.03		297
	99	0.26		96
Benzo[b]thiophene, 4,5,6,7-tetrahydro-4-(2-thienyl)-	60	0.011		419
Benzoyl disulfide	50	<0.005		89
	60	0.0107		304
		36.0		369
	99	0.11		96
Benzyl alcohol, o,o'-dithiobis-	50	0.58		297
-,- P,P'-dithiobis-	50	0.09		89
Benzyl diselenide	60	2.0	A	128
Benzyl disulfide	50	0.02		89
		0.03		297
	60	0.00878	A	368
		0.01	A	128
Benzyl sulfide	99	0.011		96
	25	0.0548	D	368
	50	0.069		368
	60	0.00335		368
Benzyl sulfone	50	0.052		252
Benzyl sulfoxide	50	0.051		252

Remarks and References page II-98

Styrene (Cont'd.)

Modifier	T °C	C_X	Remarks	References
1-Butanethiol	25	5.4	D, Q1	340
	60	13.2	D	340
		21.0	A	395
		22.0	A	69, 397
		25		131
	70	15.0	A	395
	80	17.0	A	395
	99	15.4		96
--, 1,1,3-tetramethyl-	5	5.5	H	242
		6.4	H	222
1-Butanethiol-D	50	4.3	H	242
	60	5.2	A	395
	70	4.0	A	395
	80	7.0	A	395
Butyl disulfide	25	0.00079		448
	55	0.00154		448
	60	0.0024		304
	99	0.0068		96
	150	0.022	A	359
sec-Butyl disulfide	50	< 0.005		89
tert-Butyl disulfide	50	< 0.005		89
Butyl sulfide	60	0.00014		304
tert-Butyl sulfide	60	0.0022		304
Carbanilic acid, p,p'-dithiodi-, diethyl ester	60	0.025		304
Carbon disulfide	50	0.24		89
Carbonic acid, dithio-	60	0.00066	A	288
--,--, S,S'-bis(carboxymethyl) ester trithio-	50	0.36		297
--,--, S,S'-bis(carboxymethyl) ester	50	0.21		297
2,13-Dioxa-7,8-dithia-3,12-disilatetra-decane, 3,3,12,12-tetramethoxy-	79.5	0.0118		474
Disulfide, bis(p-bromobenzoyl)	60	745		369
--, bis(α-bromo-o-tolyl)	50	1.0		297
--, bis(p-chlorobenzoyl)	60	196		369
--, bis(chlorobenzyl)	50	< 0.005		297
--, bis(2-chloroethyl)	50	0.01		297
--, bis(α-chloro-o-tolyl)	50	1.3		297
--, bis(p-cyanobenzoyl)	60	3190		369
--, bis(diethylthiocarbamoyl)	60	0.724		286
--, bis(dimethylthiocarbamoyl)	60	1.11		286
	70	0.0136		104
	80	0.568	J	40
		0.620		481

Styrene (Cont'd.)

Disulfide (Cont'd.)

Modifier	T °C	C_X	Remarks	References
--; bis(dimethylthiocarbamoyl) (Cont'd.)	95	0.780		481
		0.860	J	40
	115	0.939	J	40
		1.035		481
	130	0.984	J	40
		1.150		481
--; bis(2-ethylhexyl)	50	< 0.005		297
--; bis(N-ethyl-N-phenylthio-carbamoyl)	60	1.75		286
--; bis(1-methylheptyl)	99	0.0104		96
--; bis(morpholinothiocarbonyl)	60	6.1		286
--; bis(1-naphthylmethyl)	99	0.033		96
--; bis(p-nitrobenzoyl)	60	6650		369
--; bis(o-nitrophenyl)	50	---		89
--; bis(1-phenylethyl)	50	< 0.005		89
--; bis(β-(2-pyridyl)ethyl)	50	0.03		297
--; bis(2,3,5,6-tetramethylphenyl)	50	0.73		297
--; bis(2,4,6-triisopropylphenyl)	50	0.12		89
1-Dodecanethiol	60	14.8	A	69,126
		19.0		59
	100	13.0		156
	110	26.0		156
Ethanethiol	50	17.1		331
Ethanol, 2,2'-dithiodi-, dithiobis-	50	< 0.005		297
--,--, di(chloroacetate)	50	< 0.005		297
Ether, ethyl 3-mercaptopropyl	60	14.1	A	126
	100	21.0		69
Ethyl disulfide	99	13.7		69
Formamidine, N,N'-diphenyl-, 1,1'-dithiobis-		0.0045		96
Formic acid, thio-, dithiobis-, O,O'-diisopropyl ester	50	6.72	C,F2	251
1-Heptanethiol	50	5.3		297
1-Hexanethiol	99	7.5		96
--, 1,1,3,3,5,5-hexamethyl-	99	15.1		96
	99	15.3		96
Hexyl disulfide	25	3.2		297
	50	2.9		297
Hydrogen sulfide	75	2.4		297
Isobutyl disulfide	99	0.0104		96
Isopropyl disulfide	70	5.0		145
Lauryl disulfide	60	0.0020		304
	60	0.00066		304
	60	0.00023	A	128

Styrene (Cont'd.)

Modifier	T°C	C_x	Remarks	References
Lepidine, 2,2'-dithiodi-	50	0.04		89
Mesityl disulfide	50	0.69		89
Methyl disulfide	60	0.0094		304
Methyl sulfoxide	40	0.0000693	E	458
	50	0.548		375
	60	0.0000242		429
		0.000048	A	438
Morpholine, 4,4'-dithiodi-	50	<0.005		297
1-Naphthalenemethanethiol	25	12.7		297
	50	18.3		297
	75	15.7		297
	99	24.6		96
1-Naphthalenethiol	99	0.15		96
2-Naphthalenethiol	99	0.18		96
1-Naphthoyl disulfide	50	0.34		89
1-Naphthyl disulfide	99	1.57		96
2-Naphthyl disulfide	25	0.17		297
	75	0.19		297
	99	0.29		297
	99	0.36		96
1-Octadecanethiol	99	14.7		96
Octadecyl disulfide	99	0.024		96
1-Octanethiol	5	19.3	H	339
	50	23.0	H	242
-,- 1,1,3,5,5,7,7-octamethyl-	50	19.0	H	242
2-Octanethiol	99	4.7	H	96
1,4,5-Oxadithiepane	50	3.2	A	359
1-Pentanethiol	150	0.057	H	344
Phenetole, 2,2'-dithiodi-	50	20.0		96
-,- 4,4'-dithiodi-	99	0.075		96
Phenyl disulfide	99	0.33		297
	50	0.06	A	491
Phenyl disulfone	60	0.102	A	368
		0.0103		491
		0.111	A	285
	50	0.147		491
Phenyl sulfide	60	0.022	A	491
	25	0.025	A	368
	50	0.0325	D	252
Phenyl sulfone	60	0.056		368
	60	0.00548		252
Phenyl sulfoxide	50	0.021		252
	50	0.024		252
1-Propanethiol, 3-(trimethoxysilyl)-	79.5	5.90	H	474
2-Propanethiol, 2-methyl-	50	4.0	A	344
	50	3.1	A	261
	60	3.7	A	69,126

Styrene (Cont'd.)

Modifier	T°C	C_x	Remarks	References
2-Propanethiol, 2-methyl- (Cont'd.)	60	4.6		59
	100	1.8	A	261
		2.3	A	69,126
Propionic acid, 3,3'-dithiodi-, dipropyl ester	50	<0.005		297
-,-, 2-mercapto-	--	7.7		178
-,-, 3-mercapto-	--	9.4		178
Propyl disulfide	50	6.0		297
Pyridine, 2,2'-dithiodi-	60	0.00234		304
Quinoline, 2,2'-dithiodi-	50	0.01		89
Sulfide, ethyl 2,4-diphenylbutyl	50	0.05		331
-,- ethyl phenethyl	50	30.0		331
1-Tetradecanethiol	50	7.15		242
Thiosulfuric acid, S-butyl ester	50	19.0	H	490
-,- sodium salt	60	0.173		490
-,- S-isopropyl ester, sodium salt	60	0.407		490
-,- S-phenyl ester	60	6.28		490
-,- compound with pyridine	60	0.763		490
potassium salt	60	0.150		96
-,- S-propyl ester, sodium salt	60	0.07		96
p-Toluenethiol	99	25.5		96
α-Toluenethiol	99	26.0		369
-,- p-methoxy-	99	46.3		297
p-Toluoyl disulfide	60	0.22		297
o-Tolyl disulfide	25	0.23		297
p-Tolyl disulfide	50	0.28		297
	75	0.32		297
	50	0.11		96
	99	0.15		297
2,6-Xylyl disulfide	50	0.69		297
Vinyl acetate				
Acetic acid, dithiodi-, diethyl ester	60	1.41		86
-,- thio-, diethyl ester	60	1.5	J	351
-,-, S-(2-hydroxyethyl)ester, acetate	60	0.0132		140
Acetyl disulfide	60	0.29		86
1-Butanethiol	60	48.0		140,397
Butyl disulfide	60	1.0	J	86,351
Butyl sulfide	60	0.026	J	351
1,4,5-Oxadithiepane	60	0.25-2.5	J	351
Sulfur	45	470.0	J	28
α-Toluenethiol	70	0.885	C	390

Remarks and References page II-98

G.   REMARKS

A. Thermal initiation
B. Hydroperoxide initiation
C. Peroxide initiation
D. Photoinitiation
E. γ-Ray initiation
F. Solution polymerization in:

F 1.	Acetonitrile	F16.	Sulfur dioxide
F 2.	Benzene	F17.	Toluene
F 3.	Butyl acetate	F18.	Zinc chloride
F 4.	Butyl chloride	F19.	Octane
F 5.	Chlorobenzene	F20.	Methanol
F 6.	o-Dichlorobenzene	F21.	Propane
F 7.	Dimethylformamide	F22.	Diethyl oxalate
F 8.	Dimethyl sulfoxide	F23.	Cyclohexane
F 9.	p-Dioxane	F24.	Butyl alcohol
F10.	Ethyl acetate	F25.	Acetic acid
F11.	Ethyl methyl ketone	F26.	1, 2-Dichloroethane
F12.	Heptane	F27.	Ethylbenzene
F13.	Water	F28.	Cyclohexanone
F14.	Magnesium perchlorate	F29.	Parabutyraldehyde
F15.	Succinonitrile		

G. Heterogeneous polymerization
H. Emulsion polymerization
I. Recalculated from data of:

I 1. Basu, Sen, and Palit, 1950
I 2. Baysal and Tobolsky, 1952
I 3. Chadha and Misra, 1958
I 4. Dixon-Lewis, 1949
I 5. Gregg and Mayo, 1953
I 6. Ham, 1956
I 7. Kwart, Broadbent, and Bartlett, 1950
I 8. Matsumoto, et. al., 1959
I 9. Palit and Das, 1954
I10. Palit, Nandi, and Saha, 1954
I11. Saha, Nandi, and Palit, 1956
I12. Stein and Schulz, 1960
I13. Wheeler, Lavin, and Crozier, 1952

J. Apparent transfer constant; retardation occurred
K. Average value
L. Estimated from model compounds
M. For end groups:

M 1. -Tribromomethyl   M 2. -Butylamino

M. For end groups   (Cont'd.)

M 3.	-Diethylamino	M 6.	-Thioglycolate
M 4.	-Hydroxymethyl	M 7.	-CH(OH)CH$_2$-SH
M 5.	-Aldehyde	M 8.	-Dodecyl
		M 9.	-2-Hydroxyethyl

N. For middle groups
O. For side chain
P. For main chain
Q (). Telomerization. (number of monomer units in transferring chain)
R. Calculated from viscosity average molecular weight
S. Value uncertain
T. In presence of 0.4 mol/liter lithium nitrate
U. In presence of ( ) mol/liter sodium chloride
V. In presence of ( ) mol/liter sodium bromide
W. Under pressure of:

W 1.	200 psi	(13.8 bar)	W10.	168000 psi	(11583.6 bar)
W 2.	250 psi	(17.2 bar)	W11.	600 psi	(41.4 bar)
W 3.	1470 psi	(101.4 bar)	W12.	2500 psi	(172.4 bar)
W 4.	5000 psi	(344.8 bar)	W13.	18 psi	(1.2 bar)
W 5.	20000 psi	(1379.0 bar)	W14.	735 psi	(50.7 bar)
W 6.	26600 psi	(1834.1 bar)	W15.	14200 psi	(979.1 bar)
W 7.	28400 psi	(1958.2 bar)	W16.	42600 psi	(2937.3 bar)
W 8.	56000 psi	(3861.2 bar)	W17.	34500 psi	(2378.8 bar)
W 9.	64600 psi	(4454.2 bar)			

X. Corrected for loss of low molecular weight polymer
Y. CM = 125   x exp (-7300/RT)
Z. CM = 3.0 x exp (-5673/RT)
AA. CM = 0.4 x exp (-6219/RT)
BB. CM = 0.2 x exp (-5400/RT)
CC. CM = 6.4 x exp (-4120/RT)
DD. CM = 4.77 x exp (-3540/RT)
EE. Radical is styrene
FF. Radical is p-chlorostyrene
GG. 1/C varies with polym. rate and emulsifier of:

GG1.	K caprate	GG4.	K palmitate
GG2.	K laurate	GG5.	K rosinate
GG3.	K myristate		

HH. C is combined constant for copolymerization, =

$$(C_{S1} xR_1 xM_1 + C_{S2} xR_2 xM_2)/(R_1 xM_1 xM_1 + 2xM_1 xM_2 + R_2 xM_2 xM_2)$$

II. In presence of FeBr$_2$.

H.   REFERENCES

1. I. B. Afanas'ev, T. N. Eremina, Zh. Org. Khim. 2, 1832 (1966).

2. I. B. Afanas'ev, T. N. Eremina, E. D. Safronenko, Zh. Org. Khim. 1, 844 (1965); J. Org. Chem. USSR 1, 849 (1965).

3. I. B. Afanas'ev, E. D. Safronenko, A. A. Beer, Vysokomolekul. Soedin., Ser. B, 9, 802 (1967).

4. G. Akazome, S. Sakai, K. Maurai, Kogyo Kagaku Zasshi 63, 592 (1960); from CA 56, 4924F (1962).

5. L. Alexandru, M. Oprish, Polymer Sci. USSR 3, 99 (1962).

6. P.W. Allen, G. P. Mcsweeney, Trans. Faraday Soc. 54, 715 (1958).

7. P.W. Allen, F. M. Merrett, J. Scanlan, Trans. Faraday Soc. 51, 95 (1955).

8. U. N. Anistmov, S. S. Ivanchev, A. I. Yurzhenko, Vysokomolekul. Soedin., Ser. A, 9, 692 (1967).

9. I. A. Arbusova, L. I. Medvedeva, Izv. Akad. Nauk SSSR, Ser. Khim. 1957, 1349;

10. T. Asahara, T. Makishima, Kogyo Kagaku Zasshi 69, 2173 (1966).

11. C. Aso, M. Sugabe, Kogyo Kagaku Zasshi 68, 1970 (1965).

12. R. Autrata, J. Muller, Collect. Czech. Chem. Commun. 24, 3442 (1959).

13. D. W. E. Axford, Proc. Roy. Soc., Ser. A 197, 374 (1949).

14. T. M. Babchinitser, K.K. Mozgova, V. V. Korshak, Dokl. Akad. Nauk SSSR 173, 575 (1967); from CA 67, 117314E (1967).

15. Kh. S. Bagdasar'ian, Z. A. Sinitsina, J. Polymer Sci. 52, 31 (1961).

16. C. H. Bamford, S. Brumby, Makromol. Chem. 105, 122 (1967).

17. C. H. Bamford, M. J. S. Dewar, Proc. Roy. Soc., Ser. A 192, 309 (1948).

18. C. H. Bamford, M. J. S. Dewar, Proc. Roy. Soc., Ser. A 197, 356 (1949).

19. C. H. Bamford, A. D. Jenkins, R. Johnston, Proc. Roy. Soc., Ser. A 239, 214 (1957).

20. C. H. Bamford, A. D. Jenkins, R. Johnston, Proc. Roy. Soc., Ser. A 241, 364 (1957).

21. C. H. Bamford, A. D. Jenkins, R. Johnston, Trans. Faraday Soc. 55, 418 (1959).

22. C. H. Bamford, A. D. Jenkins, R. Johnston, E. F. T. White, Trans. Faraday Soc. 55, 168 (1959).

23. C. H. Bamford, A. D. Jenkins, E. F. T. White, J. Polymer Sci. 34, 271 (1959).

24. C. H. Bamford, E. F. T. White, Trans. Faraday Soc. 52, 716 (1956).

25. C. H. Bamford, E. F. T. White, Trans. Faraday Soc. 54, 268 (1958).

26. S. Banerjee, M. S. Muthana, J. Polymer Scid. 37, 469 (1959).

27. P. D. Bartlett, G. S. Hammond, H. Kwart, Discussions Faraday Soc. 2, 342 (1947).

28. P. D. Bartlett, H. Kwart, J. Am. Chem. Soc. 74, 3969 (1952).

29. P. D. Bartlett, F. A. Tate, J. Am. Chem. Soc. 75, 91 (1953).

30. S. C. Barton, R. A. Bird, K. E. Russell, Can. J. Chem. 41, 2737 (1963).

31. S. Basu, J. N. Sen, S. R. Palit, Proc. Roy. Soc., Ser. A 202, 485 (1950).

32. S. Basu, J. N. Sen, S. R. Palit, Proc. Roy. Soc., Ser. A 214, 247 (1952).

33. B. Baysal, J. Polymer Sci. 33, 381 (1958).

34. B. Baysal, A. V. Tobolsky, J. Polymer Sci. 8, 529 (1952).

35. I. M. Bel'govskii, L. S. Sakhonenko, N. S. Yenikolopyan, Vysoko-molekul. Soedin. 8, 369 (1966).

36. W. I. Bengough, W. H. Fairservice, Trans. Faraday Soc. 61, 1206 (1965).

37. W. I. Bengough, W. H. Fairservice, Trans. Faraday Soc. 63, 382 (1967).

38. W. I. Bengough, W. Henderson, Trans. Faraday Soc. 61, 141 (1965).

39. W. I. Bengough, R. A. M. Thomson, Trans. Faraday Soc. 56, 407 (1960).

40. J. Beniska, E. Staudner, J. Polymer Sci., C, 16, 1301 (1967).

41. T. Berezhnykh-Földes, F. Tüdös, European Polymer J. 2, 219 (1966).

42. J. C. Bevington, N. A. Ghanem, H. W. Melville, J. Chem. Soc. 1955, 2822.

43. J. C. Bevington, G. M. Guzman, H. W. Melville, Proc. Roy. Soc., Ser. A 221, 437 (1954).

44. J. C. Bevington, H. G. Troth, Trans. Faraday Soc. 58, 2005 (1962).

45. J. C. Bevington, H. G. Troth, Trans. Faraday Soc. 59, 127 (1963).

46. J. C. Bevington, H. G. Troth, Trans. Faraday Soc. 59, 1348 (1963).

47. J. C. Bevington, A. Wahid, Polymer 3, 585 (1962).

48. B. R. Bhattacharyya, U. S. Nandi, J. Polymer Sci., A-1, 4, 2675 (1966).

49. B. R. Bhattacharyya, U. S. Nandi, Makromol. Chem. 116, 8 (1968).

50. R. A. Bird, G. A. Harpell, K. E. Russell, Can. J. Chem. 40, 701 (1962).

51. R. A. Bird, K. E. Russell, Can. J. Chem. 43, 2123 (1965).

52. L. Bogetich, G. A. Mortimer, G. W. Daues, J. Polymer Sci. 61, 3 (1962).

53. E. P. Bonsall, L. Valentine, H. W. Melville, J. Polymer Sci. 7, 39 (1951).

54. C. Booth, L. R. Beason, J. Polymer Sci. 42, 93 (1960).

55. C. Booth, L. R. Beason, J. T. Bailey, J. Appl. Polymer Sci. 5, 116 (1961).

56. D. Braun, T.-O. Ahn, W. Kern, Makromol. Chem. 53, 154 (1962).

57. D. Braun, G. Arcache, R. J. Faust, W. Neumann, Makromol. Chem. 114, 51 (1968).

58. J. W. Breitenbach, Makromol. Chem. 8, 147 (1952).

59. J. W. Breitenbach, Z. Elektrochem. 60, 286 (1956).

60. J. W. Breitenbach, H. Edelhauser, R. Hochrainer, Monatsh. Chem. 97, 217 (1966).

61. J. W. Breitenbach, H. Karlinger, Monatsh. Chem. 82, 245 (1951).

62. J. W. Breitenbach, O. F. Olaj, Makromol. Chem. 96, 83 (1966).

63. J. W. Breitenbach, O. F. Olaj, K. Kuchner, H. Horacek, Makromol. Chem. 87, 295 (1965).

64. J. W. Breitenbach, O. F. Olaj, A. Schindler, Monatsh. Chem. 91, 205 (1960).

65. J. W. Breitenbach, A. Schindler, Monatsh. Chem. 83, 724 (1952).

66. J. W. Breitenbach, A. Schindler, Monatsh. Chem. 86, 437 (1955).

67. J. W. Breitenbach, A. Schindler, Monatsh. Chem. 88, 53 (1957).

68. D. W. Brown, J. E. Fearn, R. E. Lowry, J. Polymer Sci. A3, 16 (1965).

69. G. M. Burnett, Quart. Revs. (London) 4, 292 (1950).

70. G. M. Burnett, F. L. Ross, J. N. Hay, Makromol. Chem. 105, 1 (1967).

71. A. J. Buselli, M. K. Lindemann, C. E. Blades, J. Polymer Sci. 28, 485 (1958).

72. J. I. G. Cadogan, D. H. Hey, J. T. Sharp, J. Chem. Soc. B 1966, 933.

73. M. Cantow, G. Meyerhoff, G. V. Schulz, Makromol. Chem. 49, 1 (1961).

74. E. A. S. Cavell, Makromol. Chem. 54, 70 (1962).

75. E. A. S. Cavell, I. T. Gilson, J. Polymer Sci., A-1, 4, 541 (1966).

76. R. N. Chadha, G. S. Misra, Indian J. Phys. 28, 37 (1954).

77. R. N. Chadha, G. S. Misra, Current Sci. (India) 23, 186 (1954).

78. R. N. Chadha, G. S. Misra, Makromol. Chem. 14, 97 (1954).

79. R. N. Chadha, G. S. Misra, Current Sci. (India), 24, 26 (1955).

80. R. N. Chadha, G. S. Misra, Trans. Faraday Soc. 54, 1227 (1958).

81. R. N. Chadha, J. S. Shukla, G. S. Misra, Trans. Faraday Soc. 53, 240 (1957).

82. E. Y. C. Chang, C. C. Price, J. Am. Chem. Soc. 83, 4650 (1961).

83. S. R. Chatterjee, S. N. Khanna, S. R. Palit, J. Indian Chem. Soc. 41, 622 (1964).

84. A. K. Chaudhuri, Makromol. Chem. 31, 214 (1959).

85. S. Chubachi, P. K. Chatterjee, A. V. Tobolsky, J. Org. Chem. 32, 1511 (1967).

86. J. T. Clarke, R. O. Howard, W. H. Stockmayer, Makromol. Chem. 44/46, 427 (1961).

87. W. Cooper, J. Chem. Soc. 1952, 2408.

88. A. T. Coscia, R. L. Kugel, J. Pellon, J. Polymer Sci. 55, 303 (1961).

89. A. J. Costanza, R. J. Coleman, R. M. Pierson, C. S. Marvel, C. King, J. Polymer Sci. 17, 319 (1955).

90. J. Curtice, H. Gilman, G. S. Hammond, J. Am. Chem. Soc. 79, 4754 (1957).

91. F. S. Dainton, R. G. Jones, Trans. Faraday Soc. 63, 1512 (1967).

92. F. S. Dainton, M. Tordoff, Trans. Faraday Soc. 53, 499 (1957).

93. N. V. Daniel, A. F. Nikolaev, Vysokomolekul. Soedin 8, 465 (1966).

94. F. Danusso, D. Sianesi, Chim. Ind. (Milan) 37, 695 (1955).

95. S. K. Das, S. R. Chatterjee, S. R. Palit, Proc. Roy. Soc., Ser. A 227, 252 (1955).

96. V. A. Dinaburg, A. A. Vansheidt, Zh. Obshch. Khim. 24, 840 (1954).

97. G. Dixon-Lewis, Proc. Roy. Soc., Ser. A 198, 510 (1949).

98. K. W. Doak, M. A. Deahl, I. H. Christmas, 137th ACS Meeting, Cleveland, Ohio, Abstr. Papers, Vol. 1, No. 1, 151 (April 1960).

99. H. Drawe, A. Henglein, Makromol. Chem. 84, 203 (1965).

100. E. Dvorak, F. Hrabak, J. Polymer Sci., C 16, 1051 (1967).

101. B. A. Englin, R. Kh. Freidlina, Izv. Akad. Nauk SSSR, Ser. Khim. 1965 (13), 425; from CA 63, 4122E (1965).

102. B. Erussalimsky, N. Tumarkin, F. Duntoff, S. Lyubetzky, A. Goldenberg, Makromol. Chem. 104, 288 (1967).

103. T. A. Fadner, H. Morawetz, J. Polymer Sci. 45, 475 (1960).

104. T. E. Ferington, A. V. Tobolsky, J. Am. Chem. Soc. 77, 4510 (1955).

105. T. G. Fox, S. Gratch, Ann. N. Y. Acad. Sci. 57, 367 (1953).

106. G. N. Freidlin, K. A. Solop, Vysokomolekul. Soedin. 7, 1060 (1965).

107. G. N. Freidlin, K. A. Solop, Vysokomolekul. Soedin. 8, 1151 (1966).

108. T. M. Frunze, V. V. Korshak, E. L. Baranov, B. V. Lokshin, Vysokomolekul. Soedin. 8, 455 (1966).

109. N. Fuhrman, R. B. Mesrobian, J. Am. Chem. Soc. 76, 3281 (1954).

110. T. R. Fukuto, J. P. Kispersky, U. S. Dept. Com., Office Tech. Serv., Pb Rept. 147, 271 (1953).

111. S. Gadkary, S. Kapur, Makromol. Chem. 17, 29 (1955).

112. J. A. Gannon, E. M. Fettes, A. V. Tobolsky, J. Am. Chem. Soc. 74, 1854 (1952).

113. M. H. George, J. Polymer Sci. A2, 3169 (1964).
114. M. H. George, P. F. Onyon, Trans. Faraday Soc. 59, 134 (1963).
115. M. H. George, P. F. Onyon, Trans. Faraday Soc. 59, 1390 (1963).
116. H. Gerrens, H. Ohlinger, R. Fricker, Makromol. Chem. 87, 209 (1965).
117. L. Ghosez, G. Smets, J. Polymer Sci. 37, 445 (1959).
118. G. P. Gladyshev, G. V. Leplyanin, Vysokomolekul. Soedin., Ser. A, 9, 2438 (1967).
119. J. E. Glass, N. L. Zutty, J. Polymer Sci., A-1, 4, 1223 (1966).
120. M. S. Gluckman, M. J. Kampf, J. L. O'Brien, T. G. Fox, R. K. Graham, J. Polymer Sci. 37, 411 (1959).
121. M. P. Godsay, G. A. Harpell, K. E. Russell, J. Polymer Sci. 57, 641 (1962).
122. M. P. Godsay, D. H. Lohmann, K. E. Russell, Chem. Ind. (London) 1959, 1603.
123. M. R. Gopalan, M. Santhappa, J. Polymer Sci. 25, 333 (1957).
124. W. W. Graessley, H. Mittelhauser, R. Maramba, Makromol. Chem. 86, 129 (1965).
125. N. Grassie, E. Vance, Trans. Faraday Soc. 52, 727 (1956).
126. R. A. Gregg, D. M. Alderman, F. R. Mayo, J. Am. Chem. Soc. 70, 3740 (1948).
127. R. A. Gregg, F. R. Mayo, Discussions Faraday Soc. 2, 328 (1947).
128. R. A. Gregg, F. R. Mayo, J. Am. Chem. Soc. 70, 2373 (1948).
129. R. A. Gregg, F. R. Mayo, J. Am. Chem. Soc. 75, 3530 (1953).
130. V. F. Gromov, A. V. Matveeva, P. M. Khomikovskii, A. D. Abkin, Vysokomolekul. Soedin., Ser. A, 9, 1444 (1967).
131. J. Guillot, A. Guyot, Compt. Rend., Ser. C, 266, 1209 (1968).
132. H. C. Haas, H. Husek, J. Polymer Sci. A2, 2297 (1964).
133. Y. Hachihama, H. Sumitomo, Technol. Repts. Osaka Univ. 5, 491 (1955); from CA 51, 8474B (1957).
134. Y. Hachihama, H. Sumitomo, Technol. Repts. Osaka Univ. 5, 497 (1955); from CA 51, 8474C (1957).
135. G. E. Ham, J. Polymer Sci. 21, 337 (1956).
136. G. E. Ham, E. L. Ringwald, J. Polymer Sci. 8, 91 (1952).
137. D. Hardy, K. Nytrai, N. Fedorova, G. Kovacs, Polymer Sci. USSR 4, 584 (1963).
138. D. Hardy, K. Nytrai, N. Fedorova, G. Kovacs, Vysokomolekul. Soedin. 4, 1872 (1962).
139. D. Hardy, V. Spiegel, K. Nytrai, Polymer Sci. USSR 2, 528 (1961).
140. G. Hardy, J. Varga, K. Nytrai, I. Tsajlik, L. Zubonyai, Vysokomolekul. Soedin. 6, 758 (1964).
141. S. A. Harrison, W. E. Tolberg, J. Am. Oil Chem. Soc. 30, 114 (1953); calc. by F. R. Mayo and C. W. Gould (1964).
142. R. A. Hayes, J. Polymer Sci. 13, 583 (1954).
143. E. J. Henley, C. Chong, J. Polymer Sci. 36, 511 (1959).
144. G. Henrici-Olive, S. Olive, Makromol. Chem. 37, 71 (1960).
145. G. Henrici-Olive, S. Olive, Fortschr. Hochpolymer. Forsch. 2, 496 (1961).
146. G. Henrici-Olive, S. Olive, Makromol. Chem. 53, 122 (1962).
147. G. Henrici-Olive, S. Olive, G. V. Schulz, Makromol. Chem. 23, 207 (1957).
148. G. Henrici-Olive, S. Olive, G. V. Schulz, Z. Phys. Chem. (Frankfurt) 20, 176 (1959).
149. R. Hiddema, Proefschrift Rijks Universitat Groningen, 1953, Ref. 16 in Breitenbach (1956).
150. K. Higashiura, Kogyo Kagaku Zasshi 69, 349 (1966).
151. K. Higashiura, M. Oiwa, J. Polymer Sci., A-1, 6, 1857 (1968).
152. F. Hrabak, M. Bezdek, Makromol. Chem. 115, 43 (1968).
153. F. Hrabak, M. Bezdek, Collect. Czech. Chem. Commun. 33, 278 (1968).
154. F. Hrabak, L. Jiresova, Collect. Czech. Chem. Commu. 26, 1283 (1961).
155. T. Huff, E. Perry, J. Am. Chem. Soc. 82, 4277 (1960).
156. T. Huff, E. Perry, J. Polymer Sci. A-1, 1553 (1963).
157. F. Ide, K. Nakatsuka, H. Tamura, Kobunshi Kagaku 23, 45 (1965); from CA 64, 17716G (1966).
158. F. Ide, K. Okano, S. Nakano, K. Nakstsuka, Shikizai Kyokaishi 40, 571 (1967); from CA 69, 3162Q (1968).
159. J. L. Ihrig, S. P. Sood, J. Polymer Sci. A3, 1573 (1965).
160. M. Imoto, K. Higashiura, Kobunshi Kagaku 17, 468 (1960); from CA 55, 22900G (1961).
161. M. Imoto, T. Otsu, T. Oda, H. Takatsugi, M. Matsuda, J. Polymer Sci. 22, 137 (1956).
162. M. Imoto, T. Otsu, K. Tsuda, T. Ito, J. Polymer Sci. A2, 1407 (1964).
163. M. Imoto, K. Takemoto, H. Azuma, Makromol. Chem. 114, 210 (1968).
164. M. Imoto, K. Takemoto, Y. Nakai, Makromol. Chem. 48, 80 (1961).
165. S. Imoto, T. Kominami, Kobunshi Kagaku 15, 60 (1958); from CA 53, 8690G (1959).
166. S. Imoto, T. Kominami, Kobunshi Kagaku 15, 279 (1958); from CA 54, 2803A (1960).
167. S. Imoto, J. Ukida, T. Kominami, Kobunshi Kagaku 14, 127 (1957); from CA 52, 1670 A (1958).
168. S. Imoto, J. Ukida, T. Kominami, Kobunshi Kagaku 14, 384 (1957); from CA 52, 5024D (1958).
169. S. Iwatsuki, K. Nishio, Y. Yamashita, Kogyo Kagaku Zasshi 70, 384 (1967).
170. V. Jaacks, F. R. Mayo, J. Am. Chem. Soc. 87, 3371 (1965).
171. D. H. Johnson, A. V. Tobolsky, J. Am. Chem. Soc. 74, 938 (1952).
172. M. H. Jones, Can. J. Chem. 34, 108 (1956).
173. E. F. Jordan, jr., B. Artymyshyn, A. N. Wrigley, J. Polymer Sci., A-1, 6, 575 (1968).
174. E. F. Jordan, jr., G. R. Riser, W. E. Parker, A. N. Wrigley, J. Polymer Sci., A-2, 4, 975 (1966).
175. R. M. Joshi, Makromol. Chem. 53, 33 (1962).
176. H. Kaemmerer, F. Rocaboy, Compt. Rend., Ser. AB 256, 4440 (1963).
177. H. Kaemmerer, F. Rocaboy, Makromol. Chem. 72, 76 (1964).
178. K. Kaeriyama, Nippon Kagaku Zasshi 88, 783 (1967); from CA 69, 19607Z (1968).
179. M. Kalfus, J. Kopytowski, S. Lesniak, Z. Skupinska, Polimery 9 (2), 54 (1964); from CA 62, 9325E (1965).
180. S. Kamenar, Chem. Zvesti 14, 525 (1960); from Chem. Zentr. 133, 493 (1961).
181. S. L. Kapur, J. Sci. Ind. Res. 108, 180 (1951).
182. S. L. Kapur, J. Polymer Sci. 11, 399 (1953).
183. S. L. Kapur, S. D. Gadkary, J. Sci. Ind. Res. 17B, 152 (1958)
184. S. L. Kapur, R. M. Joshi, J. Polymer Sci. 14, 489 (1954).
185. K. Katagiri, K. Uno, S. Okamura, J. Polymer Sci. 17, 142 (1955).
186. M. Kato, H. Kamogawa, J. Polymer Sci. A-1, 4, 2771 (1966).
187. V. Ya. Katsobashvili, R. Ya. Chernaya, I. B. Afanas'ev, Vysokomolekul. Soedin., Ser. B, 9, 342 (1967).
188. V. Ya. Katsobashvili, E. D. Safronenko, I. B. Afanas'ev, Vysokomolekul. Soedin. 7, 823 (1965).
189. V. Ya. Katsobashvili, E. D. Safronenko, I. B. Afanas'ev, Vysokomolekul. Soedin. 8, 282 (1966).
190. H. Kawakami, N. Mori, K. Kawashima, M. Sumi, Kogyo Kagaku Zasshi 66, 88 (1963); from CA 59, 4042B (1963).
191. H. Kawakami, N. Mori, M. Sumi, Kobunshi Kagaku 20, 408 (1963); from CA 61, 13422D (1964).
192. W. Kern, D. Braun, Makromol. Chem. 27, 23 (1958).
193. S. N. Khanna, S. R. Chatterjee, U. S. Nandi, S. R. Palit, Trans. Faraday Soc. 58, 1827 (1962).
194. J. L. Kice, J. Am. Chem. Soc. 76, 6274 (1954).
195. W. J. Kirkham, J. C. Robb, Trans. Faraday Soc. 57, 1757 (1961).
196. H. Kitagama, Kobunshi Kagaku 20, 5 (1963); from Makromol. Chem. 64, 229 (1963).
197. H. Kiuchi, M. Watanabe, Kobunshi Kagaku 21, 37 (1964); from CA 61, 7107F (1964).
198. I. M. Kochnov, M. F. Sorokin, Polymer Sci. USSR 6, 869 (1964).
199. M. Kubin, L. Zikmund, Collect. Czech. Chem. Commun. 32, 535 (1967).

200. C. J. Kurian, M. S. Muthana, Makromol. Chem. 29, 1 (1959).
201. C. J. Kurian, M. S. Muthana, Makromol. Chem. 29, 19 (1959).
202. H. Kwart, H. S. Broadbent, P. D. Bartlett, J. Am. Chem. Soc. 72, 1060 (1950).
2o3. K. S. Kwei, F. R. Eirich, J. Phys. Chem. 66, 828 (1962).
204. Z. Laita, J. Polymer Sci. 38, 247 (1959).
205. Z. Laita, Z. Machacek, J. Polymer Sci. 38, 459 (1959).
206. M. Lazar, J. Pavlinec, Chem. Zvesti 15, 428 (1961); from CA 55, 22896C (1961).
207. M. Lazar, J. Pavlinec, J. Polymer Sci. A2, 3197 (1964).
208. M. Lazar, J. Pavlinec, Z. Manasek, Collect. Czech. Chem. Commun. 26, 1380 (1961).
209. M. Lazar, R. Rado, J. Pavlinec, J. Polymer Sci. 53, 163 (1961).
210. F. M. Lewis, F. R. Mayo, J. Am. Chem. Soc. 76, 457 (1954).
211. D. Lim, M. Kolinsky, J. Polymer Sci. 53, 173 (1961).
212. D. Lim, O. Wichterle, J. Polymer Sci. 29, 579 (1958).
213. M. Litt, V. Stannett, Makromol. Chem. 37, 19 (1960).
214. J. R. Little, P. F. Hartman, J. Am. Chem. Soc. 88, 96 (1966).
215. J. R. Little, L. W. Hartzel, F. O. Guenther, F. R. Mayo, private communication to C. Walling, "Free radicals in solution", John Wiley and Sons, N. Y., 1957, p. 257.
216. S. Loshaek, E. Broderick, P. Bernstein, J. Polymer Sci. 39, 223 (1959).
217. A. I. Lowell, J. R. Price, J. Polymer Sci. 43, 1 (1960).
218. S. G. Lyubetskii, B. A. Dolgoplosk, B. L. Erusalimskii, Polymer Sci. USSR 3, 164 (1962).
219. Z. Machacek, F. Cermak, Chem. Prum. 16, 604 (1966); from CA 66, 18907Y (1967).
220. G. Machell, G. N. Richards, J. Chem. Soc. 1961, 3308.
221. V. Mahadevan, M. Santhappa, Makromol. Chem. 16, 119 (1955).
222. K. L. Mallik, Naturwissenschaften 45, 385 (1958).
223. T. Manabe, T. Utsumi, S. Okamura, J. Polymer Sci. 58, 121 (1962).
224. M. S. Matheson, E. E. Auer, E. B. Bevilacqua, E. J. Hart, J. Am. Chem. Soc. 71, 497 (1949).
225. M. S. Matheson, E. E. Auer, E. B. Bevilacqua, E. J. Hart, J. Am. Chem. Soc. 71, 2610 (1949).
226. S. G. Matsoyan, N. N. Morlyan, F. S. Kinoyan, Vysokomolekul. Soedin. 7, 1159 (1965).
227. M. Matsuda, S. Abe, N. Tokura, J. Polymer Sci. A2, 3877 (1964).
228. M. Matsuda, K. Matsumoto, N. Tokura, Kogyo Kagaku Zasshi 68, 1269 (1965).
229. M. Matsuda, N. Tokura, J. Polymer Sci. A2, 4281 (1964).
230. M. Matsuda, S. Fujii, J. Polymer Sci., A-1, 5, 2617 (1967).
231. M. Matsumoto, M. Maeda, J. Polymer Sci. 17, 438 (1955).
232. M. Matsumoto, J. Ukida, G. Takayama, T. Eguchi, K. Mukumoto, K. Imai, Y. Kazusa, M. Maeda, Makromol. Chem. 32, 13 (1959).
233. B. Matuska, J. Kossler, V. Srajer, Collect. Czech. Chem. Commun. 23, 1456 (1958).
234. J. A. May, jr., W. B. Smith, J. Phys. Chem. 72, 216 (1968).
235. F. R. Mayo, J. Am. Chem. Soc. 65, 2324 (1943).
236. F. R. Mayo, J. Am. Chem. Soc. 70, 3689 (1948).
237. F. R. Mayo, J. Am. Chem. Soc. 75, 6133 (1953).
238. F. R. Mayo, J. Am. Chem. Soc. 76, 5392 (1954).
239. F. R. Mayo, C. W. Gould, J. Am. Oil Chem. Soc. 41, 25 (1964).
240. F. R. Mayo, R. A. Gregg, M. S. Matheson, J. Am. Chem. Soc. 73, 1691 (1951).
241. V. V. Mazurek, V. G. Gasan-Zade, G. T. Nesterchuk, Polymer Sci. USSR 6, 1587 (1964).
242. E. J. Meehan, I. M. Kolthoff, H. R. Sinha, J. Polymer Sci. 16, 471 (1955).
243. E. J. Meehan, I. M. Kolthoff, P. R. Sinha, J. Polymer Sci. A2, 4911 (1964).
244. F. W. Mellows, M. Burton, J. Phys. Chem. 66, 2164 (1962).
245. H. W. Melville, L. Valentine, Trans. Faraday Soc. 46, 210 (1950).
246. H. S. Mickley, A. S. Michaels, A. L. Moore, J. Polymer Sci. 60, 121 (1962).
247. E. B. Milovskaya, T. G. Zhuravleva, Vysokomolekul. Soedin., Ser. A, 9, 1128 (1967).
248. E. B. Milovskaya, T. G. Zhuravleva, L. V. Zamoyskaya, J. Polymer Sci., C, 16, 899 (1967).
249. Y. Minoura, Y. Enomoto, Kogyo Kagaku Zasshi 70, 1021 (1967).
250. Y. Minoura, Y. Hayashi, M. Imoto, Kobunshi Kagaku 15, 260 (1958); from CA 54, 2803F (1960).
251. Y. Minoura, T. Sugimura, J. Polymer Sci., A-1, 4, 2721 (1966).
252. Y. Minoura, T. Sugimura, T. Hirahara, Kogyo Kagaku Zasshi 70, 357 (1967).
253. Y. Minoura, N. Yasumoto, T. Ishii, Kogyo Kagaku Zasshi 65, 1299 (1962); from CA 58, 1538D (1963).
254. Y. Minoura, N. Yasumoto, T. Ishii, Makromol. Chem. 71, 159 (1964).
255. G. S. Misra, R. N. Chadha, Makromol. Chem. 23, 134 (1957).
256. G. S. Misra, R. C. Rastogi, V. P. Gupta, Makromol. Chem. 50, 72 (1961).
257. T. Miyake, M. Matsumoto, Kogyo Kagaku Zasshi 62, 1101 (1959); from CA 57, 15342A (1962).
258. Y. Mori, K. Sato, Y. Minoura, Kogyo Kagaku Zasshi 61, 462 (1958); from CA 55, 4021F (1961).
259. G. A. Mortimer, J. Polymer Sci., A-1, 4, 881 and 1895 (1966).
260. G. A. Mortimer, L. C. Arnold, J. Am. Chem. Soc. 84, 4986 (1962).
261. M. Morton, J. A. Cala, I. Piirma, J. Am. Chem. Soc. 78, 5394 (1956).
262. M. Morton, I. Piirma, J. Am. Chem. Soc. 80, 5596 (1958).
263. M. Morton, I. Piirma, J. Polymer Sci. A1, 3043 (1963).
264. A. S. Nair, M. S. Muthana, Makromol. Chem. 47, 114, 128 (1961).
265. H. Nakamoto, Y. Ogo, T. Imoto, Makromol. Chem. 111, 93 (1968).
266. T. Nakata, T. Otsu, M. Imoto, J. Polymer Sci. A3, 3383 (1965).
267. T. Nakaya, K. Ohashi, M. Imoto, Makromol. Chem. 114, 201 (1968).
268. U. S. Nandi, P. Ghosh, S. R. Palit, Nature 195, 1197 (1962).
269. A. F. Nikolaev, N. V. Meiya, G. A. Balaev, Vysokomolekul. Soedin. 7, 2122 (1965).
270. K. Noma, Y. Tajima, M. Niwa, Sci. Eng. Rev. Doshisha Univ. 3, 91 (1962); from CA 59, 2955C (1963).
271. R. G. W. Norrish, J. P. Simons, Proc. Roy. Soc., Ser. A 251, 4 (1959).
272. A. M. North, A. M. Scallan, Polymer 5, 447 (1964).
273. K. Nozaki, Discussions Faraday Soc. 2, 337 (1947).
274. J. L. O'Brien, F. Gornick, J. Am. Chem. Soc. 77, 4757 (1955).
275. J. L. O'Brien, J. R. Panchak, T. G. Fox, Abstracts of 124th ACS Meeting, Chicago, 1953.
276. S. Okamura, K. Katagiri, Makromol. Chem. 28, 177 (1958).
277. S. Okamura, K. Katagiri, T. Motoyama, J. Polymer Sci. 43, 509 (1960).
278. S. Okamura, T. Motoyama, J. Polymer Sci. 58, 221 (1962).
279. S. Okamura, K. Takeya, Kobunshi Kagaku 15, 353 (1958); from CA 54, 8143 (1960).
280. M. Okawara, T. Nakai, E. Imoto, Kogyo Kagaku Zasshi 69, 973 (1966).
281. O. F. Olaj, J. W. Breitenbach, I. Hofreiter, Makromol. Chem. 91, 264 (1966).
282. O. F. Olaj, J. W. Breitenbach, I. Hofreiter, Makromol. Chem. 110, 72 (1967).
283. P. F. Onyon, Trans. Faraday Soc. 51, 400 (1955).
284. P. F. Onyon, J. Polymer Sci. 22, 19 (1956).
285. T. Otsu, Y. Kinoshita, M. Imoto, Makromol. Chem. 73, 225 (1964).
286. T. Otsu, K. Nayatani, Makromol. Chem. 27, 149 (1955).
287. T. Otsu, A. Shimizu, M. Imoto, J. Polymer Sci., B2, 973 (1964).
288. T. Otsu, K. Tsuda, N. Kita, Mem. Fac. Eng., Osaka City Univ. 7, 95 (1965); from CA 66, 95434U (1967).
289. S. R. Palit, S. K. Das, Proc. Roy. Soc., Ser. A 226, 82 (1954).
290. S. R. Palit, U. S. Nandi, N. G. Saha, J. Polymer Sci. 14, 295 (1954).
291. S. K. Patra, D. Mangaraj, Makromol. Chem. 111, 168 (1968).
292. J. Pavlinec, M. Lazar, J. Polymer Sci., C, 22, 297 (1968).
293. L. H. Peebles, J. Polymer Sci. A3, 341 (1965).

294. L. H. Peebles, J. T. Clarke, W. H. Stockmayer, J. Am. Chem. Soc. 82, 4780 (1960).

295. J. J. Pellon, J. Polymer Sci. 43, 537 (1960).

296. E. Perry, J. Polymer Sci. 54, S-46 (1961).

297. R. Pierson, A. Costanza, A. Weinstein, J. Polymer Sci. 17, 221 (1955).

298. G. Platau, F. R. Eirich, R. B. Mesrobian, A. E. Woodward, J. Polymer Sci. 39, 357 (1959).

299. A. Prevot-Bernas, J. Sebban-Danon, J. Chim. Phys. 53, 418 (1956).

300. W. A. Pryor, J. Phys. Chem. 67, 519 (1963).

301. W. A. Pryor, D. M. Huston, T. R. Fiske, T. L. Pickering, E. Ciuffarin, J. Am. Chem. Soc. 86, 4237 (1964).

302. W. A. Pryor, G. L. Kaplan, J. Am. Chem. Soc. 86, 4234 (1964).

303. W. A. Pryor, A. Lee, C. E. Witt, J. Am. Chem. Soc. 86, 4229 (1964).

304. W. A. Pryor, T. L. Pickering, J. Am. Chem. Soc. 84, 2705 (1962).

305. W. A. Pryor, E. P. Pultinas, jr., J. Am. Chem. Soc. 85, 133 (1963).

306. P. V. T. Raghuram, U. S. Nandi, J. Polymer Sci., A-1, 5, 2005 (1967).

307. S. P. Rao, M. Santhappa, Current Sci. (India) 34(6), 174 (1965);

308. S. P. Rao, M. Santhappa, J. Polymer Sci., A-1, 5, 2681 (1967).

309. A. I. Restaino, W. N. Reed, J. Polymer Sci. 36, 499 (1959).

310. E. H. Riddle, "Monomeric Acrylic Ester", Reinhold Publishing Corp., N.Y., 1954, ref. 14, p. 64.

311. J. C. Robb, E. Senogles, Trans. Faraday Soc. 58, 708 (1962).

312. J. C. Robb, D. Vofsi, Trans, Faraday Soc. 55, 558 (1959).

313. M. E. Rozenberg, A. F. Nikolaev, A. V. Pustovalova, Vysokomolekul. Soedin. 8, 1155 (1966).

314. L. V. Ruban, A. L. Buchachenko, M. B. Neiman, Yu. V. Koknanov, Vysokomolekul. Soedin. 8, 1642 (1966).

315. M. Ryska, M. Kolinsky, D. Lim, J. Polymer Sci., C, 16, 621 (1967).

316. N. G. Saha, U. S. Nandi, S. R. Palit, J. Chem. Soc. 1956, 427.

317. N. G. Saha, U. S. Nandi, S. R. Palit, J. Chem. Soc. 1958, 7.

318. N. G. Saha, U. S. Nandi, S. R. Palit, J. Chem. Soc. 1958, 12.

319. I. Sakurada, K. Noma, Y. Ofuji, Kobunshi Kagaku 20, 481 (1963); from CA 63, 8487 C (1965).

320. I. Sakurada, Y. Sakaguchi, K. Hashimoto, Kobunshi Kagaku 19, 593 (1962); from CA 61, 16159D (1964).

321. G. F. Santee, R. H. Marchessault, H. G. Clark, J. J. Kearny, V. Stannett, Makromol. Chem. 73, 177 (1964).

322. M. Santhappa, V. M. Iyer, Current Sci. (India) 24, 173 (1955).

323. M. Santhappa, V. S. Vaidhyanathan, Current Sci. (India) 23, 259 (1954).

324. J. Scanian, Trans. Faraday Soc. 50, 756 (1954).

325. E. Schonfeld, I. Waltcher, J. Polymer Sci. 35, 536 (1959).

326. G. V. Schulz, G. Henrici, S. Olive, J. Polymer Sci. 17, 45 (1955).

327. G. V. Schulz, G. Henrici, S. Olive, Z. Elektrochem. 60, 296 (1956).

328. G. V. Schulz, L. Roberts-Nowakowska, Makromol. Chem. 80, 36 (1964).

329. G. V. Schulz, D. J. Stein, Makromol. Chem. 52, 1 (1962).

330. G. P. Scott, C. C. Soong, W.-S. Huang, J. L. Reynolds, J. Org. Chem. 29, 83 (1964).

331. G. P. Scott, J. C. Wang, J. Org. Chem. 28, 1314 (1963).

332. J. N. Sen, U. S. Nandi, S. R. Palit, J. Indian Chem. Soc. 40, 729 (1963).

333. K. P. Shen, F. R. Eirich, J. Polymer Sci. 53, 81 (1961).

334. A. Shimizu, T. Otsu, M. Imoto, Bull. Chem. Soc. Japan 38, 1535 (1965).

335. T. Shimomura, Y. Kuwabara, E. Tsuchida, I. Shinohara, Kogyo Kagaku Zasshi 71, 283 (1968).

336. T. Shimomura, E. Tsuchida, I. Shinohara, Mem. School Sci. Eng., Waseda Univ. 8, 37 (1964); from CA 63, 14985A (1965).

337. T. Shimomura, E. Tsuchida, I. Shinohara, Mem. School Sci. Eng., Waseda Univ. 29, 1 (1965); from CA 65, 10675A (1966).

338. M. Simonyi, F. Tudös, J. Pospisil, European Polymer J. 3, 101 (1967).

339. P. R. Sinha, K. L. Mallik, J. Indian Chem. Soc. 34, 424 (1957).

340. C. Sivertz, J. Phys. Chem. 63, 34 (1959).

341. B. Skrabal, L. Rosik, Chem. Prumysl 8, 46 (1958).

342. G. Smets, L. Convent, X. van der Borght, Makromol. Chem. 23, 162 (1957).

343. G. Smets, L. Dehaes, Bull. Soc. Chim. Belges 59, 13 (1950).

344. W. V. Smith, J. Am. Chem. Soc. 68, 2059 (1964).

345. R. D. Spencer, M. B. Fulton, U.S. Dept. Com., Office Tech. Serv., PB Rept. 144,900 (1961).

346. R. D. Spencer, M. B. Fulton, B. H. Beggs, Astracts of 137th ACS Meeting, Cleveland, 1960.

347. N. T. Srinivasan, M. Santhappa, Makromol. Chem. 26, 80 (1958).

348. D. J. Stein, Makromol. Chem. 76, 170 (1964).

349. D. J. Stein, G. V. Schulz, Makromol. Chem. 38, 248 (1960).

350. D. J. Stein, G. V. Schulz, Makromol. Chem. 52, 249 (1962).

351. W. H. Stockmayer, R. O. Howard, J. T. Clarke, J. Am. Chem. Soc. 75, 1756 (1953).

352. T. J. Suen, Y. Jen, J. V. Lockwood, J. Polymer Sci. 31, 481 (1958).

353. T. J. Suen, A. M. Schiller, W. N. Russell, Advances in Chem. Series No. 34, "Polymerization and polycondensation processes", ACS, Wash., D.C., 1962, pp. 217-24.

354. T. Sugimura, Y. Ogata, Y. Minoura, J. Polymer Sci., A-1, 4, 2747 (1966).

355. G. Takayama, Kobunshi Kagaku 15, 89 (1958); from CA 53, 868D (1959).

356. W. M. Thomas, E. H. Gleason, J. J. Pellon, J. Polymer Sci. 17, 275 (1955).

357. C. F. Thompson, W. S. Port, L. P. Witnauer, J. Am. Chem. Soc. 81, 2552 (1959).

358. A. P. Titov, I. A. Livshits, Zh. Obshch. Khim. 29, 1605 (1959).

359. A. V. Tobolsky, B. Baysal, J. Am. Chem. Soc. 75, 1757 (1953).

360. A. V. Tobolsky, J. Offenbach, J. Polymer Sci. 16, 311 (1955).

361. T. Toda, J. Polymer Sci. 58, 411 (1962).

362. N. Tokura, M. Matsuda, F. Yazaki, Makromol. Chem. 42, 108 (1960).

363. A. C. Toohey, K. E. Weale, Trans. Faraday Soc. 58, 2439 (1962).

364. A. C. Toohey, K. E. Weale, Trans. Faraday Soc. 58, 2446 (1962).

365. E. Tsuchida, Y. Ohtani, H. Nakadai, I. Shinohara, Kogyo Kagaku Zasshi 70, 573 (1967).

366. E. Tsuchida, Y. Ohtani, H. Nakadai, I. Shinohara, Kogyo Kagaku Zasshi 69, 1230 (1966).

367. E. Tsuchida, T. Shimomura, K. Fujimori, Y. Ohtani, I. Shinohara, Kogyo Kagaku Zasshi 70, 566 (1967).

368. K. Tsuda, S. Kobayashi, T. Otsu, Bull. Chem. Soc. Japan 38, 1517 (1965).

369. K. Tsuda, T. Otsu, Bull. Chem. Soc. Japan 39, 2206 (1966).

370. F. Tudös, T. Berezhnykh-Földes, European Polymer J. 2, 229 (1966).

371. F. Tudös, I. Kende, M. Azori, J. Polymer Sci. 53, 17 (1961).

372. F. Tudös, I. Kende, M. Azori, J. Polymer Sci. A1, 1353 (1963).

373. F. Tudös, I. Kende, M. Azori, J. Polymer Sci. A1, 1369 (1963).

374. J. Ulbricht, Faserforsch. Textiltech. 10, 166 (1959).

375. J. Ulbricht, Faserforsch. Textiltech. 11, 62 (1960).

376. J. Ulbricht, Z. Phys. Chem. (Frankfurt) 221, 346 (1962).

377. J. Ulbricht, B. Sandner, Faserforsch. Textiltech. 17, 208 (1966).

378. J. Ulbricht, R. Sourisseau, Faserforsch. Textiltech. 16, 213 (1965).

379. T. Uno, K. Yoshida, Kobunshi Kagaku 15, 819 (1958); from CA 54, 20298D (1960).

380. C. A. Uraneck, J. E. Bulreigh, J. Appl. Polymer Sci. 12, 1075 (1968).

381. W. H. Urry, F. W. Stacey, E. S. Huyser, O. O. Juveland, J. Am. Chem. Soc. 76, 450 (1954).

382. V. Vaclavek, Chem. Prumysl 10, 103 (1960).

383. V. Vaclavek, J. Appl. Polymer Sci. 11, 1881 (1967).

384. V. Vaclavek, J. Appl. Polymer Sci. 11, 1903 (1967).

385. V. S. Vaidhyanathan, M. Santhappa, Makromol. Chem. 16, 140 (1955).

386. R. L. Vale, W. G. P. Robertson, J. Polymer Sci. 33, 518 (1958).

387. B. M. E. van der Hoff, J. Polymer Sci. 44, 241 (1960).

388. B. M. E. van der Hoff, J. Polymer Sci. 48, 175 (1960).

389. A. A. Vansheidt, G. Khardi, Acta Chim. Acad. Sci. Hung. 20, 261 (1959); from CA 54, 6180B (1960).

390. A. A. Vansheidt, G. Khardi, Acta Chim. Acad. Sci. Hung. 20, 381 (1959); from CA 54, 11552F (1960).

391. F. K. Velichko, I. P. Lavrent'ev, Yu. P. Chizhov, Izv. Akad. Nauk SSSR, Ser. Khim. 1966, 172; from CA 64, 12485C (1966).

392. G. Vidotto, A. Crosato-Arnaldi, G. Talmini, Makromol. Chem. 114, 217 (1968).

393. M. Vrancken, G. Smets, Makromol. Chem. 30, 197 (1959).

394. J. Vuillemenot, B. Barbier, G. Riess, A. Banderet, J. Polymer Sci. A3, 1969 (1965).

395. L. A. Wall, D. W. Brown, J. Polymer Sci. 14, 513 (1954).

396. R. A. Wallace, K. L. Hadley, J. Polymer Sci., A-1, 4, 71 (1966).

397. C. Walling, J. Am. Chem. Soc. 70, 2561 (1948).

398. C. Walling, E. R. Briggs, J. Am. Chem. Soc. 68, 1141 (1946).

399. C. Walling, Y. Chang, J. Am. Chem. Soc. 76, 4878 (1954).

400. C. Walling, L. Heaton, J. Am. Chem. Soc. 87, 38 (1965).

401. C. Walling, J. J. Pellon, J. Am. Chem. Soc. 79, 4776 (1957).

402. M. Watanabe, H. Kiuchi, J. Polymer Sci. 58, 103 (1962).

403. O. L. Wheeler, E. Lavin, R. N. Crozier, J. Polymer Sci. 9, 157 (1952).

404. E. F. T. White, M. J. Zissell, J. Polymer Sci. A1, 2189 (1963).

405. T. Yamamoto, Bull. Chem. Soc. Japan 40, 642 (1967).

406. T. Yamamoto, T. Otsu, J. Polymer Sci. B4, 1039 (1966); see also Bull. Chem. Soc. Japan 40, 2449 (1967).

407. T. Yamamoto, T. Otsu, Koguo Kagaku Zasshi 70, 2403 (1967).

408. T. Yamamoto, T. Otsu, M. Imoto, Kogyo Kagaku Zasshi 69, 990 (1966).

409. S. D. Yevstratova, M. F. Margaritova, S. S. Medvedev, Polymer Sci. USSR 5, 681 (1964).

410. M. Yoshida, K. Tanouchi, Kobunshi Kagaku 20, 545 (1963); from Makromol. Chem. 71, 216 (1964).

411. V. V. Zaitseva, V. D. Enal'ev, A. I. Yurzhenko, Vysokomolekul. Soedin., Ser. A, 9, 1958 (1967).

412. V. M. Zhulin, M. G. Gonikberg, V. N. Zagorbinina, Dokl. Akad. Nauk SSSR 163, 106 (1965); from CA 63, 11706G (1965).

413. V. M. Zhulin, M. G. Gonikberg, V. N. Zagorbinina, Izv. Akad. Nauk SSSR, Ser. Khim. 1966, 827.

414. V. M. Zhulin, M. G. Bonikberg, V. N. Zagorbinina, Izv. Akad. Nauk SSSR, Ser. Khim. 1966, 997.

415. M. G. Zimina, N. P. Apukhtina, Kolloid Zh. 21, 181 (1959).

416. J. Aoyagi, K. Kitamura, I. Shinohara, Kogyo Kagaku Zasshi 73, 2045 (1970).

417. J. Aoyagi, I. Shinohara, Kogyo Kagaku Zasshi 74, 1191 (1971).

418. T. Asahara, M. C. Chou, Bull. Chem. Soc. Japan 42, 1373 (1969).

419. C. Aso, T. Kunitake, M. Shinsenji, H. Miyazaki, J. Polymer Sci., A-1, 7, 1497 (1969).

420. J. C. Bevington, M. Johnson, J. P. Sheen, European Polymer J. 8, 209 (1972).

421. G. C. Bhaduri, U. S. Nandi, Makromol. Chem. 128, 176 (1969).

422. G. C. Bhaduri, U. S. Nandi, Makromol. Chem. 128, 183 (1969).

423. B. R. Bhattacharyya, U. S. Nandi, Makromol. Chem. 149, 231 (1971).

424. B. R. Bhattacharyya, U. S. Nandi, Makromol. Chem. 149, 243 (1971).

425. D. Braun, F. Weiss, Makromol. Chem. 138, 83 (1970).

426. D. Braun, F. Weiss, Angew. Makromol. Chem. 15, 127 (1971).

427. C-T. Chen, W-D. Huang, J. Chin. Chem. Soc. 16, 46 (1969); from CA 71, 102247U (1969).

428. Y. Choshi, A. Tsuji, G. Akazome, K. Murai, Kogyo Kagaku Zasshi 71, 908 (1968); from CA 69, 67745Q (1968).

429. N. N. Dass, M. H. George, European Polymer J. 6, 897 (1970).

430. H. J. Dietrich, M. A. Raymond, J. Macromol. Sci. Chem., A6, 191 (1972).

431. J. P. Fischer, Makromol. Chem. 155, 211 (1972).

432. J. P. Fischer, Makromol. Chem. 155, 227 (1972).

433. J. P. Fischer, W. Lüders, Makromol. Chem. 155, 239 (1972).

434. R. Kh. Freidlina, A. B. Terent'ev, N. S. Ikonnikov, Dokl. Akad. Nauk SSR 193, 605 (1970).

435. R. Kh. Freidlina, A. V. Terent'ev, N. S. Ikonnikov, Izv. Akad. Nauk SSSR, Ser. Khim. 1970, 554.

436. S. S. Frolov, T. M. Slivchenko, Izv. Vyssh. Ucheb. Zaved., Khim. Khim. Tekhnol. 14, 1264 (1971); from CA 76, 25660X (1972).

437. I. Geczy, H. I. Nasr, Acta Chim. (Budapest) 70, 319 (1971); from CA 76, 100150M (1972).

438. S. N. Gupta, U. S. Nandi, J. Polymer Sci., A-1, 8, 1493 (1970).

439. D. A. J. Harker, R. A. M. Thomson, I. R. Walters, Trans. Faraday Soc. 67, 3057 (1971).

440. M. Imoto, K. Ohashi, Makromol. Chem. 117, 117 (1968).

441. E. F. Jordan, jr., B. Artymyshyn, A. N. Wrigley, J. Polymer Sci., A-1, 7, 2605 (1969).

442. I. Kar, B. M. Mandal, S. R. Palit, J. Polymer Sci., A-1, 7, 2829 (1969).

443. S. Kobayashi, Kogyo Kagaku Zasshi 72, 2511 (1969).

444. C. Kwang-Fu, Kobunshi Kagaku 29, 233 (1972).

445. I. M. Likhterova, E. M. Lukina, Zh. Obshch. Khim. 42, 194 (1972); from CA 76, 127504S (1972).

446. A. Matsumoto, M. Oiwa, J. Polymer Sci., A-1, 10, 103 (1972).

447. J. A. May, jr., W. B. Smith, J. Phys. Chem. 72, 2993 (1968).

448. L. A. Miller, V. Stannett, J. Polymer Sci., A-1, 7, 3159 (1969).

449. Y. Minoura, M. Mitoh, Kogyo Kagaku Zasshi 74, 747 (1971).

450. Y. Minoura, A. Tabuse, Kogyo Kagaku Zasshi 74, 990 (1971).

451. Y. Minoura, H. Toshima, J. Polymer Sci., A-1, 7, 2837 (1969).

452. M. Modena, P. Piccardi, European Polymer J. 7, 1 (1971).

453. N. Mogi, M. Shindo, Kogyo Kagaku Zasshi 73, 786 (1970).

454. G. A. Mortimer, J. Polymer Sci., A-1, 8, 1513 (1970).

455. G. A. Mortimer, J. Polymer Sci., A-1, 8, 1535 (1970).

456. G. A. Mortimer, J. Polymer Sci., A-1, 8, 1543 (1970).

457. G. A. Mortimer, J. Polymer Sci., A-1, 10, 163 (1972).

458. I. G. Murgulescu, I. Vlagiu, Rev. Roum. Chim. 14, 411 (1969); from CA 71, 71773F (1969).

459. T. Nakaya, M. Tanaka, M. Imoto, Makromol. Chem. 149, 221 (1971).

460. J. H. Ok, S. B. Pak, Hwahak Kwa Hwahak Kongop 1971, 239 (Korean); from CA 77, 20147U (1972).

461. T. Ota, S. Masuda, Kogyo Kagaku Zasshi 73, 2020 (1970).

462. T. Ota, S. Masuda, C. Aoyama, M. Ebisudani, Kogyo Kagaku Zasshi 74, 994 (1971).

463. G. S. Park, D. G. Smith, Trans. Faraday Soc. 65, 1854 (1969).

464. F. Patat, P. Mehnert, Monatsh. Chem. 98, 538 (1967).

465. S. P. Potnis, A. M. Deshpande, Makromol. Chem. 125, 48 (1969).

466. S. P. Potnis, A. M. Deshpande, Makromol. Chem. 153, 139 (1972).

467. W. A. Pryor, J. H. Coco, Macromolecules, 3, 500 (1970).

468. W. A. Pryor, T. R. Fiske, Macromolecules 2, 62 (1969).

469. W. A. Pryor, T. -L. Huang, Macromolecules 2, 70 (1969).

470. E. Pulat, C. B. Senvar, Commun. Fac. Sci. Univ. Ankara, Ser. B, 15(3), 25 (1968); from CA 72, 55934Z (1970).

471. P. V. T. Raghuram, U.S. Nandi, J. Polymer Sci., A-1, 7, 2379 (1969).

472. P. V. T. Raghuram, U. S. Nandi, J. Polymer Sci., A-1, 8, 3079 (1970).

473. K. K. Roy, D. Pramanick, S. R. Palit, Makromol. Chem. 153, 71 (1972).

474. J. C. Saam, D. J. Gordon, J. Polymer Sci., A-1, 8, 2509 (1970).

475. T. Saegusa, Y. Ito, N. Yasuda, Polymer J. 1, 591 (1970).

476. S. K. Saha, A. K. Chaudhuri, J. Polymer Sci., A-1, 10, 797 (1972).

477. G. P. Scott, A. M. R. Elghoul, J. Polymer Sci., A-1, 8, 2255 (1970).

478. G. P. Scott, F. J. Foster, Macromolecules 2, 428 (1969).

479. R. B. Seymour, J. M. Sosa, V. J. Patel, J. Paint Technol. 43, (563), 45 (1971).

480. T. Shimomura, E. Tsuchida, I. Shinohara, Kogyo Kagaku Zasshi 71, 1074 (1968); from CA 69, 97249Y (1968).

481. E. Staudner, J. Beniska, European Polymer J. Suppl. 1969, 537.

482. F. Suganuma, H. Mitsui, S. Machi, M. Hagiwara, T. Kagiya, J. Polymer Sci., A-1, 6, 3127 (1968).

483. K. Sugiyama, T. Nakaya, M. Imoto, J. Polymer Sci., A-1, 10, 205 (1972).

484. A. B. Terent'ev, N. S. Ikonnikov, R. Kh. Freidlina, Izv. Akad. Nauk SSSR, Ser. Khim 1971, 73.

485. A. B. Terent'ev, N. S. Ikonnikov, R. Kh. Freidlina, Dokl. Akad. Nauk SSSR 196, 1373 (1971).

486. A. S. Tevlina, H. S. Kolesnikov, S. N. Sividova, V. V. Ryltsev, Vysokomolekul. Soedin., Ser. A, 9, 2473 (1967); Polymer Sci. USSR 9, 2797 (1967).

487. R. A. M. Thomson, I. R. Walters, Trans. Faraday Soc. 67, 3046 (1971).

488. R. A. M. Thomson, I. R. Walters, J. R. King, J. Polymer Sci., B, 10, 63 (1972).

489. E. Tsuchida, T. Tomono, Kogyo Kagaku Zasshi 73, 2040 (1970).

490. M. Tsunooka, M. Fujii, N. Ando, M. Tanaka, N. Murata, Kogyo Kagaku Zasshi 73, 805 (1970).

491. M. Tsunooka, T. Higuchi, M. Fujii, M. Tanaka, N. Murata, Kogyo Kagaku Zasshi 73, 596 (1970).

492. G. Vidotto, S. Brugnaro, G. Talamini, Makromol. Chem. 140, 249 (1970).

493. G. Vidotto, S. Brugnaro, G. Talamini, Makromol. Chem. 140, 263 (1970).

494. F. Yamada, M. Kanbe, I. Shinohara, Memo. Sch. Sci. Eng., Waseda Univ. 1969, No. 33, 67; from CA 73, 88255C (1970).

495. T. Yamamoto, M. Hasegawa, T. Otsu, Bull. Chem. Soc. Jap. 42, 1364 (1969).

496. T. Yamamoto, S. Nakamura, M. Hasegawa, T. Otsu, Kogyo Kagaku Zasshi 72, 727 (1969).

497. T. Yamamoto, T. Otsu, J. Polymer Sci., A-1, 7, 1279 (1969).

498. N. Yamashita, T. Seita, M. Yoshihara, T. Maeshima, Kogyo Kagaku Zasshi 74, 2157 (1971).

499. N. Yamashita, T. Seita, M. Yoshihara, T. Maeshima, J. Polymer Sci., B, 9, 641 (1971).

500. V. M. Zhulin, M. G. Gonikberg, A. L. Goff, V. N. Zagorbinina, Vysokomolekul. Soedin., Ser. A, 11, 777 (1969).

COPOLYMERIZATION REACTIVITY RATIOS

Lewis J. Young

Koppers Company, Inc.

Law/Patent Department

Pittsburgh, Pennsylvania

When a monomer 1 is copolymerized with a monomer 2 the relationship between the composition of the copolymer and the composition of the monomer mixture is given by eq. (1)

$$\frac{dm_1}{dm_2} = \frac{M_1(r_1M_1 + M_2)}{M_2(r_2M_2 + M_1)} \qquad (1)$$

where m_1 = the moles of monomer 1 entering the copolymer, m_2 = the moles of monomer 2 entering the copolymer, M_1 = the moles of monomer 1 in the monomer mixture, M_2 = the moles of monomer 2 in the monomer mixture, and r_1 and r_2 are the monomer reactivity ratios.

The monomer reactivity ratios r_1 and r_2, for any monomer pair are the ratios of the rate constants of different propagation reactions as defined by

$$\sim M_1 \cdot + M_1 \longrightarrow \sim M_1M_1 \cdot \qquad \text{Rate constant } k_{11} \qquad (2)$$

$$\sim M_1 \cdot + M_2 \longrightarrow \sim M_1M_2 \cdot \qquad \text{Rate constant } k_{12} \qquad (3)$$

$$\sim M_2 \cdot + M_2 \longrightarrow \sim M_2M_2 \cdot \qquad \text{Rate constant } k_{22} \qquad (4)$$

$$\sim M_2 \cdot + M_1 \longrightarrow \sim M_2M_1 \cdot \qquad \text{Rate constant } k_{21} \qquad (5)$$

$$r_1 = k_{11}/k_{12} \qquad (6)$$

$$r_2 = k_{22}/k_{21} \qquad (7)$$

$\sim M\cdot$ represents a polymer chain ending in a radical derived from monomer M.

The values of r_1 and r_2 have been collected from the literature and are presented here together with references. Some of the original papers have not been examined and in these cases the reference to the original papers includes the reference in Chemical Abstracts. The values which have been obtained from Chemical Abstracts should be used with caution for it is sometimes difficult to determine to which of the two monomers the values of r_1 and r_2 refer.

Each monomer pair is listed under both monomers.

In cases (indicated by the symbol P) where the abstract did not give the reactivity ratios and the original journal was not readily available, we have decided that it would at least be of value to list these systems in this table.

Negative reactivity ratios are reported in the literature, but have never been satisfactorily explained, except as an indication of faulty data.

The column "Remarks" is intended to indicate the conditions of copolymerization. Unless otherwise shown, a free-radical system was used.

COPOLYMERIZATION PARAMETER

M1	M2	R1	+/-	R2	+/-	TEMP.	REMARKS	REF.
ACENAPHTHYLENE	ACRYLIC ACID, METHYL ESTER	4		0.1		60	I7	1326
	ACRYLONITRILE	2.56		0.02		60	I7	1326
	1,3-BUTADIENE	1.5		0.48		65	I7	1385
	ETHER, BUTYL VINYL	0.04	0.02	20		-78	H10	390
		0.14	0.03	6.0	1.0	-20	H10	390
		0.24	0.04	4.2	0.8	0	H10	390
		0.38	0.04	1.3	0.3	30	H10	390
	ETHYLENE, CHLORO-	64		0.001			Q	391
	METHACRYLONITRILE	2.38		0.15		60	I7	1326
	METHYL METHACRYLATE	2.25		0.44		60	I7	1326
	STYRENE	0.33		3.81		90	A	987
	VINYL ACETATE	4.4	0.3	0.3	0.1	30	H10,I7	841
		0.99	0.08	1.005	0.08	60		608
ACETALDEHYDE	ACETALDEHYDE, CHLORO-	0.6	0.1	0.28	0.05	-78	H14	432
	BUTYRALDEHYDE	0.92		1.07		-55	G8	944
		1.33		0.88		-60	G8	944
		1.97		0.59		-78	G8	944
	CHLORAL	0.18	0.05	0.00		-78	H14	431
	ETHYLENE OXIDE	0.40	0.12	1.05	0.40	-20	H10	728
	OXETANE, 3,3-BIS(CHLOROMETHYL)-	0.10	0.09	1.16	0.46	-5	H10	728
ACETALDEHYDE, DIVINYL ACETAL	VINYL ACETATE	0.99	0.08	1.005	0.08	60		608
ACETALDEHYDE, CHLORO-	ACETALDEHYDE	0.28	0.05	0.6	0.1	-78	H14	432
	CHLORAL	85	5	0.00		-50	H14	431
		280	10	0.00		-78	H14	431
ACETALDEHYDE, DICHLORO-	CHLORAL	0.65		1.50	0.08	-78	H40,I2	627
ACETAMIDE, 2-ACRYLAMIDO-	ACRYLIC ACID	0.40		0.55		60		313
	ETHER, 2-AMINOETHYL VINYL	3.3		0		60		313
ACETAMIDE, N-ALLYL-	ACRYLIC ACID	0.07		1.3		80	I12	857
	METHACRYLIC ACID	0.04		16.9		80	I12	857
ACETAMIDE, N-ALLYL-2-CHLORO-	ACRYLIC ACID	0.04		3.6		80	I12	857
	METHACRYLIC ACID	0.05		12.2		80	I12	857

M1	M2	R1	+/-	R2	+/-	TEMP.	REMARKS	REF.
ACETANILIDE, N-VINYL-								
	METHYL METHACRYLATE	0.005		8.90			Q	666
	STYRENE	0.009		22.3			Q	666
	VINYL ACETATE	0.054	0.017	13.0	0.66	75		367
		1.60	0.13	0.15	0.015	70		367
ACETANILIDE, 4'-VINYL-								
	METHYL METHACRYLATE	0.50	0.05	0.30	0.05	65		317
ACETIC ACID, ALLYL ESTER								
	ACRYLIC ACID, 2-CHLOROETHYL ESTER	0		5.5	1.0	60		560
	ACRYLIC ACID, METHYL ESTER	0		5		60		620
	ACRYLONITRILE	0.0		1.9		0	D,L3	389
		0.05		8.2			E,Q	983
				8.0			D	389
	ETHYLENE, CHLORO-			1.2		40		620
		0		1.16		40		654
		0		6.6		60		620
	ETHYLENE, 1,1-DICHLORO-	<0.0075		<0.13		30	C	87
	MALEIC ANHYDRIDE			85	2	60		855
	METHACRYLIC ACID, ETHYL ESTER			98	8	60		855
	METHYL METHACRYLATE			23		60		620
	1-PROPENE, 3-CHLORO-2-METHYL-	0.058	0.014	11.1	2.6	133	E	281
		0.15	0.02	4.5	0.81	153	E	281
	STYRENE	0.00		90	10	60		620
	VINYL ACETATE	0.45	0.15	0.60	0.15	60	C	560
		0.7		1.0		60		834
ACETIC ACID, 2,3-EPOXYPROPYL ESTER								
	FURAN, TETRAHYDRO-	0.25		2.5		0	H10	738
ACETIC ACID, CHLORO-, ALLYL ESTER								
	METHACRYLIC ANHYDRIDE	0		42		60		372
	METHYL METHACRYLATE	0		50	2	60		372
		0		50		75	C	178
ACETIC ACID, CHLORO-, VINYL ESTER								
	ACRYLIC ACID	0.6		0.1		20	E	326
	ACRYLONITRILE	0.09		0.34		80	C	1070
	ETHER, BUTYL VINYL	1.0		0.5			Q	494
	ETHER, ETHYL VINYL	1.01	0.02	0.46	0.02	60		870
	ETHER, PHENYL VINYL	0.98	0.01	0.68	0.015	60		870
	ETHYLENE, CHLORO-	1.0	0.005	0.8	0.01	60		1213
	LAURIC ACID, VINYL ESTER	0.4		1.6		60		325
	MALONONITRILE, METHYLENE-	0.87		0.65	0.02	0	E	289
	STYRENE	<0.004		0.13	0.05	40		1070
		0.03		45		80	C	870
	VINYL ACETATE	0.05	0.02	1.0	0.005	60		1070
		1.20		0.73		60		256

COPOLYMERIZATION PARAMETER

M1	M2	R1	+/-	R2	+/-	TEMP.	REMARKS	REF.
ACETIC ACID, DICHLORO-, VINYL ESTER								
	ACRYLONITRILE	0.18		0.25		80		1070
	ETHYLENE, CHLORO-	0.7		1.25		60	C	1213
	STYRENE	0.28		20		80	C	1070
ACETIC ACID, DIETHYLPHOSPHONYL-, ALLYL ESTER								
	FUMARIC ACID, 1,3-BUTYLENEGLYCOL POLYESTER	0.075		10.0		80		906
	FUMARIC ACID, ETHYLENEGLYCOLPHOSPHINATE POLYESTER	0.15	0.06	1.73	0.03	80	Q	604
ACETIC ACID, METHACRYLAMINO-, ETHYL ESTER								
	METHYL METHACRYLATE	0.90		1.09		70	C	772
ACETIC ACID, METHACRYLIMINODI-, DIMETHYL ESTER								
	METHYL METHACRYLATE	0.013		16.67		70	C	772
ACETIC ACID, PHENYL-, VINYL ESTER								
	METHYL METHACRYLATE	0.03	0.01	26.4	2.5	60	C,17	1135
	VINYL ACETATE	0.92	0.07	0.96	0.07	60	C,17	1135
ACETIC ACID, THIO-, VINYL ESTER								
	ACRYLIC ACID, METHYL ESTER	0.23		0.8		50		1249
	CARBAZOLE, 9-VINYL-	0.5		0.622		60		329
	CARBONIC ACID, CYCLIC VINYLENE ESTER	12.9		0.04		60		755
	PHTHALIMIDE, N-VINYL-	3.2		0.23		60		329
	STYRENE	0.25		4		50		1249
	SUCCINIMIDE, N-VINYL-	3.1		0.095		60		329
	VINYL ACETATE	5.5		0.05		50		1249
ACETIC ACID, TRIFLUORO-, VINYL ESTER								
	VINYL ACETATE	0.2		0.25		70		1203
		0.32		0.6		60	C	312
ACETONITRILE								
	ETHYLENE OXIDE	0		7.8		70	H20,R	230
	PROPANE, 1-CHLORO-2,3-EPOXY-	0		6.1		70	H11,R	232
		0		0.2		70	H49,R	232
		0		2.5		70	H11,R	232
		0		2.5		70	H11,17,R	232
				4.0		70	H11,17,R	232
		0	0.02	4.2	0.4	70	H20	231
ACETOPHENONE, 4'-HYDROXY-, METHACRYLATE								
	STYRENE	0.49		0.15		60		742
		0.49		0.15		60		739

M1	M2	R1	+/-	R2	+/-	TEMP.	REMARKS	REF.
ACETOPHENONE, 4'-ISOPROPENYL-	STYRENE	0.27		0.95		60		1078
ACETOPHENONE, 4'-ISOPROPENYL-, OXIME	STYRENE	0.64		0.91		60		1078
ACETOPHENONE, 4'-VINYL-	METHYL METHACRYLATE	1.37		0.32		60		1078
	STYRENE	1.15	0.10	0.25	0.03	60		1078
ACETOPHENONE, 4'-VINYL-, OXIME	METHYL METHACRYLATE	0.50		0.98		60		1078
	STYRENE	0.54		1.04		60		1078
ACETYLENE, DIPHENYL-	ACRYLIC ACID, METHYL ESTER	0		55	5	60	O	212
	ACRYLONITRILE	0		13.6	1.0	60	O	212
ACETYLENEDICARBONITRILE	STYRENE	0		1.40		55	I5	158
ACROLEIN	ACRYLAMIDE	1.65	0.1	0.19	0.02	50		488
		1.69	0.1	0.21	0.02	50	I6	845
	ACRYLIC ACID	2.0	0.05	0.76	0.02	20	C	843
		0.50	0.05	0.76	0.20	50	L	845
		2.40	0.30	1.15	0.20	54	I11,HH	198
		6.70	0.50	0.05	0.05	75	I11,HH	198
			3.00	0.00		80	I11,HH	198
	ACRYLIC ACID, BUTYL ESTER	1.6		0.6		50	C,H	1252
		1.2		0.6		60	I10	1251
	ACRYLIC ACID, ETHYL ESTER	1.6		0.6		50	I10	1251
		1.6		0.6		60	C,H	1252
	ACRYLIC ACID, 2-ETHYLHEXYL ESTER	0.0	0.05	7.7	0.2	50	I10	1251
	ACRYLIC ACID, METHYL ESTER	1.2		0.6		60	L	845
		1.6		0.6		50	I10	1251
		10.0		0.2	0.2	20	C,H	1252
	ACRYLONITRILE	1.09	0.05	0.77	0.1	20	C	843
		1.09	0.05	0.77	0.1	50	L	845
		1.28	0.036	0.60	0.18	50	I6	840
		1.60	0.04	0.52	0.02	50		488
		1.60	0.04	0.52	0.02	50	I6	845
	BENZENESULFONIC ACID, P-VINYL-, SODIUM SALT	0.23	0.12	0.05	0.03	54	I11,HH	198
		0.26	0.03	0.025	0.025	54	I11,HH	198
		0.33	0.15	0.32	0.05	52	I11,HH	198
	ETHER, PHENYL VINYL	0.92	0.18	0.86	0.10	20	H2O	875

COPOLYMERIZATION PARAMETER

M1	M2	R1	+/-	R2	+/-	TEMP.	REMARKS	REF.
ACROLEIN	ETHER, THYMYL VINYL	1.82	0.03	0.18	0.07	20	H2O	875
	ETHER, M-TOLYL VINYL	1.11		0.71		20	H2O	875
	ETHER, O-TOLYL VINYL	0.53	0.04	0.32	0.11	20	H2O	875
	ETHER, P-TOLYL VINYL	0.99	0.05	0.96	0.05	20	H2O	875
	ETHYLENE, CHLORO-	5.4		0.04		60	I5	1115
	MALEIMIDE	3.2		0.12		60	I17	1286
	METHACRYLONITRILE	0.72		1.20		50		488
		0.72	0.06	1.20	0.08	50	I10	845
	METHYL METHACRYLATE	0.2	0.06	10.0	0.08	20	C	843
		0.5	0.05	1	0.2	50	C,H	1252
		0.8		1.2		60	I10	1251
	3,6-PYRIDAZINEDIONE, 1,2-DIHYDRO-	16		0		60	I17	1286
	3(2H)-PYRIDAZINONE, 6-HYDROXY-	16		0		60	I17	1289
	3(2H)-PYRIDAZINONE, 6-HYDROXY-, ACETATE	4		0		60	I17	1289
	3(2H)-PYRIDAZINONE, 6-METHYL-	8		0		60	I17	1289
	PYRIDINE, 2-VINYL-	4.0		0.32		50	I6	845
	STYRENE	0.034		0.25		50	C,H	1252
		0.25		0.22		60	I10	1251
		0.33				50	I10	1346
	VINYL ACETATE	3.3	0.1	0.1	0.05	50	L	845
		3.33	0.1	0.1	0.05	20	C	843
ACROLEIN, BIS(2-BUTOXYETHYL)ACETAL	ACRYLONITRILE	0.01		7.2		60	I6	1437
ACROLEIN, BIS(2-ETHOXYETHYL)ACETAL	ACRYLONITRILE	0.01		7.56		60	I6	1437
ACROLEIN, DIETHYL ACETAL	ACRYLONITRILE	0.01		12			Q	1112
ACROLEIN, 2-CHLORO-	PROPIONALDEHYDE, 3-(ALLYLOXY)-	10		0.01		60	I17	1287
	3(2H)-PYRIDAZINONE, 6-HYDROXY-	30		0		60	I17	1289
	3(2H)-PYRIDAZINONE, 6-HYDROXY-, ACETATE	20		0		60	I17	1289
	3(2H)-PYRIDAZINONE, 6-METHYL-	48		0		60	I17	1289
	STYRENE	0.18		0.06		60	I17	1288
		0.18	0.026	0.078	0.022	60		241
ACROLEIN, 3-CYANO-	STYRENE	0		0.20		60	I5,R	913
	STYRENE, ALPHA-METHYL-	0		0.033		80	C,I10	1405
ACRYLAMIDE	ACROLEIN	0.19	0.02	1.65	0.1	50	I6	488
		0.21	0.02	1.69	0.1	50	I6	845
		0.76	0.02	2.0	0.05	20	C	843

REMARKS PAGE II-366, REFERENCES PAGE II-368

M1	M2	R1	+/-	R2	+/-	TEMP.	REMARKS	REF.
ACRYLAMIDE	ACROLEIN	0.76	0.02	2.0	0.05	50	L	845
	ACRYLAMIDE, N-(HYDROXYMETHYL)-	2.9	0.4	0.9	0.2		I11,O	469
	ACRYLIC ACID	0.05		0.72		-15	E,IL,W	327
		0.25		0.77		-15	E	327
		0.48	0.06	1.73	0.21	5	E	1163
		0.60	0.02	1.43	0.03	60	C,HH	119
		0.62				25	C	1309
		1.38		0.36		25	B,I11	893
		1.4				60	I7	1309
	ACRYLIC ACID, METHYL ESTER	1.30	0.05	0.05	0.05	60	C,I15	41
	ACRYLIC ACID, POTASSIUM SALT	0.78		1.35		90	C,I11,L	473
	ACRYLIC ACID, SODIUM SALT	1.4		0.84		90	C,I11	473
		1.10	0.05	0.35	0.03	60	C,I15	1309
		0		12		60	C	119
	ACRYLIC ACID, 3-BROMO-, CIS-	1.04	0.27	6.5	0.5	20	F,I73	1297
	ACRYLIC ACID, 3-BROMO-, TRANS-			0.94	0.6	20	F,I73	1297
	ACRYLONITRILE	1.3		0.8	0.16	40	F	929
		1.3		0.8		60	I46	561
		1.3		0.8		60	I47	561
		1.3		0.8		60	I36	561
		1.357		0.875		60	I48	561
		0.0		1.0		30		369
						60		770
	BENZENEBORONIC ACID, P-VINYL-	13.8	0.5	0.05	0.02		C,I11,Q	89
	CARBONIC ACID, CYCLIC VINYLENE ESTER	0		6.7	0.6	20	F,I73	1297
	CINNAMIC ACID, CIS-	54	2	3	0.2	20	F,I73	1297
	CINNAMIC ACID, TRANS-	0				30	B,I11	1309
	CROTONIC ACID	4.72		0.11	0.3	20	F,I73	1297
	CROTONIC ACID, CIS-	0		6.5				1273
	CROTONIC ACID, TRANS-	5.32		0.12	0.4	20	Q	1297
		14.9		0			Q	1273
	ETHENESULFONIC ACID, SODIUM SALT	70	16	0	0.1	50	C,I12,L	138
	ETHYLENE, BROMO-	19.6		0.15		50		991
	ETHYLENE, CHLORO-	4.89	0.08	0.09	0.08	40		616
	ETHYLENE, 1,1-DICHLORO-	7.0		0.22		60	I6,O	41
	FUMARIC ACID	2.50		0.47		25	C,I26,Q	577
	IMIDAZOLE, 2-METHYL-1-VINYL-	0.68		0				578
	ISOTHIOCYANIC ACID, VINYL ESTER	2.2		0.01				1070
	MALEIC ACID	12.8		0.3		25	C,I11	674
		0.25					I6,O	577
	MALEIC ANHYDRIDE	0.75				70	E	1125
		0.74				50		1125
	METHACRYLAMIDE	0.12	0.11	1.1	0.20	25	D,I11	194
	METHACRYLIC ACID	0.18	0.02	0.25	0.03	60	I11	1416
		0.5	0.01	1.36	0.02	60	I49	1416
		1	0.01			25	B,I11	1309
		0.04		2.13		60	C,I15	1309
	METHACRYLIC ACID, CADMIUM SALT	0.07		2.28		30	C,I15	1426
	METHACRYLIC ACID, CALCIUM SALT	0.24		0.71		30	F,I11	1052
	METHACRYLIC ACID, LITHIUM SALT	0.5		0.73		30	C,I11	1425
	METHACRYLIC ACID, MAGNESIUM SALT	0.2		0.92		30	C,I11	1425
							C,I11	1426

M1	M2	R1	+/-	R2	+/-	TEMP.	REMARKS	REF.
ACRYLAMIDE								
	METHACRYLIC ACID, POTASSIUM SALT	0.41		0.46		30	C,I11	1425
	METHACRYLIC ACID, SODIUM SALT	0.42		0.59		30	C,I11	1425
	METHACRYLIC ACID, ZINC SALT	0.04		1.2		30	C,I11	1426
	METHYL METHACRYLATE	0.44	0.06	2.6	0.21	70	I12	1382
		0.82	0.07	2.53	0.22	70	I72	1382
		2.45	0.35	2.55	0.4	70	I10	1382
	2-PENTENOIC ACID, CIS-	0		7.1	0.5	20	F,I73	1297
	2-PENTENOIC ACID, TRANS-	0		11	0.6	20	F,I73	1297
	2-PICOLINE, 5-VINYL-	0.56	0.09	0.01	0.09	60		41
	2-PICOLINIUM METHYL SULFATE, 1-METHYL-5-VINYL-	0.19		2.7		48		878
	STYRENE	0.2		1.05		50	I15	1170
		0.3	0.09	1.44	0.22	70	I12	1382
		0.59	0.1	1.13	0.05	70	I72	1270
		0.59	0.09	1.13	0.17	70	I72	1382
		1.38		1.27		70	I10	1382
ACRYLAMIDE, N-(1-ACETYL-3-METHYLBUTYL)-	STYRENE	0.25	0.1	1.87	0.1	60	I7	425
ACRYLAMIDE, N-(1-ACETYL-3-METHYLBUTYL)-2-METHYL-	STYRENE	0.25	0.05	1.07	0.05	60	I7	425
ACRYLAMIDE, N-BENZYL-2-METHYL-	ACRYLIC ACID, METHYL ESTER	0.46	0.16	0.65	0.12	60		1445
ACRYLAMIDE, N-(1-BENZYL-2-OXOPROPYL)-	STYRENE	0.48	0.05	2.06	0.05	60	I6	425
ACRYLAMIDE, N-(1-BENZYL-2-OXOPROPYL)-2-METHYL-	STYRENE	0.57	0.1	1.30	0.1	60	I7	425
ACRYLAMIDE, N-(CARBAMOYLMETHYL)-2-METHYL-	ACRYLIC ACID	1.5		0.4		60	I12	1204
ACRYLAMIDE, N-CYCLOHEXYL-2-METHYL-	ACRYLONITRILE					70	I7,P	16
ACRYLAMIDE, N,N-DIBUTYL-	METHYL METHACRYLATE	0.42	0.04	1.85	0.05	50	I7	117
	STYRENE	0.32	0.02	1.65	0.03	50	I7	117
ACRYLAMIDE, N,N-DIMETHYL-	ACRYLIC ACID	0.5	0.1	0.4	0.05	75		826
	METHYL METHACRYLATE	0.42	0.1	2.3	0.24	70	I12	1383

M1	M2	R1	+/-	R2	+/-	TEMP.	REMARKS	REF.
ACRYLAMIDE, N,N-DIMETHYL-								
	METHYL METHACRYLATE	0.45	0.08	1.8	0.18	50		710
		0.51	0.07	2.04	0.11	70	I10	1383
	STYRENE	0.23	0.13	1.23	0.43	50		710
		0.42	0.05	1.33	0.08	70	I12	1383
		0.44	0.08	1.28	0.15	70	I10	1383
ACRYLAMIDE, N,2-DIMETHYL-								
	ACRYLIC ACID, METHYL ESTER	0.22	0.1	0.57	0.15	60	C	1445
	METHYL METHACRYLATE	0.24		1.54		70		772
		0.24	0.02	1.54	0.05	70	I27	660
ACRYLAMIDE, N-(1,1-DIMETHYL-3-OXOBUTYL)-								
	METHYL METHACRYLATE	0.57	0.03	1.68	0.06	60		179
	STYRENE	0.49	0.06	1.77	0.08	60		179
ACRYLAMIDE, N-ETHYL-N-(1,1-DIHYDROPERFLUOROBUTYL)-								
	METHYL METHACRYLATE	0.89		0.77		66	C	513
ACRYLAMIDE, N-ETHYL-2-METHYL-								
	ACRYLIC ACID, METHYL ESTER	0.24	0.05	0.53	0.04	60	C	1445
	METHYL METHACRYLATE	0.11		1.75		70		772
ACRYLAMIDE, N-(2-HYDROXY-1,1-BIS(HYDROXYMETHYL)ETHYL)-2-METHYL- *								
	STYRENE	1.29		0.47		60		1104
ACRYLAMIDE, N-(HYDROXYMETHYL)-								
	ACRYLAMIDE	0.9	0.2	2.9	0.4		I11,Q	469
	ACRYLIC ACID, BUTYL ESTER	0.61	0.17	0.87	0.05		K,Q	467
	ACRYLIC ACID, ETHYL ESTER	1.4	0.2	1.4	0.2		K,Q	468
	ACRYLIC ACID, METHYL ESTER	1.9	0.7	1.3	0.2		K,Q	467
	ACRYLONITRILE	1.2	0.1	0.7	0.1		K,Q	468
		2.33	0.1	0.98	0.05		Q	1361
	ETHYLENE, CHLORO-	23.5		0		40		617
ACRYLAMIDE, N-(HYDROXYMETHYL)-2-METHYL-								
	ACRYLONITRILE	1.22	0.3	0.15	0.1	20	F	1200
ACRYLAMIDE, N-HYDROXYMETHYL-2-METHYL-, BUTYRATE								
	METHYL METHACRYLATE	0.50	0.07	1.65	0.13	70	C,I7	51
ACRYLAMIDE, N-ISOPROPYL-								
	CARBONIC ACID, THIO-, O-TERT-BUTYL S-VINYL ESTER	1.17	0.05	0.3	0.05	70		1176

COPOLYMERIZATION PARAMETER

M1 / M2	R1	+/-	R2	+/-	TEMP.	REMARKS	REF.
ACRYLAMIDE, N,N'-ISOPROPYLIDENEBIS-							
ACRYLONITRILE	2.44		0.57		60	I15	1313
STYRENE	0.565		1.57		60	I15	1313
ACRYLAMIDE, N-METHOXYMETHYL-2-METHYL-							
METHYL METHACRYLATE	0.27	0.03	1.66	0.02	70	I27	660
ACRYLAMIDE, N,N'-METHYLENEBIS-							
ACRYLONITRILE	1.50	0.12	0.64	0.12	30	F,I11	370
SEMICARBAZIDE, 1-(METHACRYLCYLAMIDINO)-, HYDROCHLORIDE	2.9	0.25	0.091	0.016	60		1420
ACRYLAMIDE, 2-METHYL-N-(1-METHYL-2-OXOPROPYL)-							
STYRENE	0.01	0.02	0.91	0.02	60	I7	425
ACRYLAMIDE, 2-METHYL-N-(1-NAPHTHYL)-							
ACRYLONITRILE					70	I7,P	16
ACRYLAMIDE, N-(1-METHYL-2-OXOPROPYL)-							
STYRENE	0.05	0.02	1.83	0.02	60	I7	425
ACRYLAMIDE, 2-METHYL-N-(2-OXOPROPYL)-							
STYRENE	0.13	0.1	1.15	0.1	60	I7	425
ACRYLAMIDE, N-OCTADECYL-							
ACRYLONITRILE	1.44	0.019	1.10	0.035	60	C,I14	454
ETHYLENE, 1,1-DICHLORO-	1.37	0.008	0.438	0.008	60	C,I14	454
METHYL METHACRYLATE	0.44	0.04	3.85	0.17	60	I7	1239
STYRENE	0.2	0.05	1.41	0.1	80	I7	1239
VINYL ACETATE	6.11	0.045	0.027	0.009	60	C,I14	454
ACRYLAMIDE, N-OCTYL-							
METHYL METHACRYLATE	0.24	0.06	3.50	0.06	50	I7	117
STYRENE	0.20	0.05	2.70	0.10	50	I7	117
ACRYLAMIDE, N-TERT-OCTYL-							
METHYL METHACRYLATE	0.24	0.04	3.6	0.04	50	I7	117
STYRENE	0.25	0.06	2.8	0.10	50	I7	117
ACRYLANILIDE, 4'-((2-CHLOROETHYL)SULFAMOYL)-							
ACRYLONITRILE	1.00	0.05	0.20	0.03	70	I6	680
STYRENE	0.38	0.06	0.69	0.06	100	I5	680

REMARKS PAGE II-366, REFERENCES PAGE II-368

M1	M2	R1	+/-	R2	+/-	TEMP.	REMARKS	REF.
ACRYLANILIDE, 4'-((2-CHLOROETHYL)SULFAMOYL)-2-METHYL-								
	ACRYLONITRILE	1.03	0.03	0.05	0.03	70	I6	680
	STYRENE	0.33	0.07	1.02	0.08	70	I6	680
ACRYLANILIDE, 4'-CHLORO-2-METHYL-								
	METHYL METHACRYLATE	0.24		0.61		70	C	772
ACRYLANILIDE, 2,4'-DIMETHYL-								
	ACRYLONITRILE	0.67		0.39		70	I7,P	16
	METHYL METHACRYLATE					70	C	772
ACRYLANILIDE, 4'-METHOXY-2-METHYL-								
	METHYL METHACRYLATE	0.57		0.14		70	C	772
ACRYLANILIDE, 2-METHYL-4'-NITRO-								
	ACRYLONITRILE					70	I7,P	16
ACRYLIC ACID								
	ACETAMIDE, 2-ACRYLAMIDO-	0.55		0.40		60	I12	313
	ACETAMIDE, N-ALLYL-	1.3		0.07		80	I12	857
	ACETAMIDE, N-ALLYL-2-CHLORO-	3.6		0.04		80		857
	ACETIC ACID, CHLORO-, VINYL ESTER	0.1		0.6		20	E	326
	ACROLEIN	0.00		6.70	3.00	80	I11,HH	198
		0.05	0.05	2.40	0.50	75	I11,HH	198
		1.15	0.20	0.50	0.30	54	I11,HH	198
	ACRYLAMIDE	0.36		0.62		-15	E,I1,W	327
		0.72		1.4		25	B,I11	1309
		0.77		1.38		60	C,I15	1309
		1.43	0.03	0.05		60	I7	893
		1.73	0.21	0.25		-15	E	327
		0.4		0.60		5	E	327
		0.4		0.48		25	C	119
	ACRYLAMIDE, N-(CARBAMOYLMETHYL)-2-METHYL-	0.58		1.5		60	C,HH	1163
	ACRYLAMIDE, N,N-DIMETHYL-	1.15		0.5		60	I12	1204
	ACRYLIC ACID, BUTYL ESTER	0.42	0.05	1.07	0.02	75	C,I12	826
	ACRYLIC ACID, METHYL ESTER	1.1	0.02	0.78	0.06	50	C,K	766
				0.98	0.1	45	C,I6	1018
				0.95		45	C	1377
						45		1377
	ACRYLONITRILE	1.15	0.06	0.35	0.02	50	C,P	12
		6.0	2.0	0.13	0.02	50	C,I11	402
	BENZAMIDE, N-ALLYL-	4.7		0.05		80	C	1070
	BENZENESULFONIC ACID, P-VINYL-, SODIUM SALT	0.10	0.02	1.0	0.2	80	I12	857
	BUTYRAMIDE, N-ALLYL-3-METHYL-	4.3		0.05		70	C,I11	301
	BUTYRIC ACID, 2-METHYLENE-, ETHYL ESTER	0.9	0.1	0.45	0.02	80	I12	857
	CARBAMIC ACID, DIETHYL-, VINYL ESTER	5.55		0.09	0.05	80	I7	896
	ETHYLENE, CHLORO-	6.4		0.025		80		316
		6.8		0.107		50	I10	1236
						60	I57	541

M1	M2	R1	+/-	R2	+/-	TEMP.	REMARKS	REF.
ACRYLIC ACID								
	ETHYLENE, CHLORO-	8.2		0.027		50	I7	935
	ETHYLENE, 1,1-DICHLORO-	1.26		0.46		50	I10	1236
	FORMAMIDE, N-ALLYL-	2.6		0.06		80	I12	857
	1-HEPTANOL, 2,2,3,3,4,4,5,5,6,6,7,7-DODECAFLUORO-, ACRYLATE	3.8	0.6	1.8	0.3	30	E,I28	1105
	HYDROQUINONE, VINYL-, DIBENZOATE	0.44	0.13	0.95	0.002	60	I2	482
	MALONONITRILE, METHYLENE-	0.26	0.06	0.29	0.08	50	R	289
	METHACRYLIC ACID, BUTYL ESTER	0.29		3.67		50	C,I12	766
	METHYL METHACRYLATE	0.11		9.40	1.0	30	D	855
		0.12		2.02		60	LL	1378
		0.225		2.64		50	I17	1247
		0.24	0.03	1.25	0.85	70	I6,L	84
		0.25	0.04	2.89	0.06	50		486
		0.30	0.02	1.86	0.04	50	I7	486
		0.31	0.03	1.5	0.08	50	I5,L	486
		0.31	0.04	2.46	0.04	50	I10,L	486
		0.33	0.05	2.32	0.04	50		486
				1.5	0.03	60	I29	1378
				2.3		50	I63	1247
				2.17		50	C	1247
	PHOSPHONIC ACID, ALPHA-PHENYLVINYL-	0.98	0.08	0.44		70	I7	524
	1-PROPENE	9.7		0		50		749
	2-PROPEN-1-OL, 2-CHLORO-, ACETATE	1.0	0.2	0.0	0.1	100		484
	2-PYRROLIDINONE, 1-VINYL-	1.3	0.01	0.15	0.02	75	C	1012
	STYRENE	0.05	0.03	0.25	0.04	45	C	1377
		0.07	0.01	0.25	0.02	50	C,I6	485
		0.08	0.02	1.1	0.01	45	I10	1377
		0.13		0.75		50	I5	485
		0.14		0.90		50	Q	485
		0.15		0.25		50	I29	775
		0.15		0.70		50	I6	485
		0.15		1.03		60	C	485
		0.25		0.25		60		166
		0.25		0.15		70		168
		0.35		0.15		70		1212
		0.45		0.22		80		299
	VINYL ACETATE	2	0.1	0.25	0.05	80	C	1070
		10	1	0.1		70		32
				0.01	0.003	70	C	119
ACRYLIC ACID, ALLYL ESTER								
	ACRYLIC ACID, METHYL ESTER	0.33	0.03	0.52	0.04	60	C	1367
	ACRYLONITRILE	1.12	0.21	0.73	0.06	-40	G17,15	197
	1-PROPENE, 3-CHLORO-	10.4	0.3	0.08	0.01	60	C	1367
	STYRENE	0.14	0.02	0.61	0.03	60	C	1367
	VINYL ACETATE	17.35	0.5	0.111	0.01	60	C	1367
ACRYLIC ACID, BENZYL ESTER								
	ACRYLONITRILE	0.631	0.07	1.490	0.08	25	E	83
		0.636		0.294		60		63
	METHYL METHACRYLATE	0.34	0.03	1.7	0.17	60		1157
	1-PROPENE, 3-CHLORO-	9.90	0.14	0.058	0.003	60	C	457

REMARKS PAGE II-366, REFERENCES PAGE II-368

M1	M2	R1	+/-	R2	+/-	TEMP.	REMARKS	REF.
ACRYLIC ACID, BENZYL ESTER	STYRENE	0.20		0.55		60		742
	STYRENE	0.20		0.55		60		740
	STYRENE	0.25	0.02	0.5	0.05	60		1157
ACRYLIC ACID, 2-BUTOXYETHYL ESTER	STYRENE	0.17	0.06	0.76	0.13	60		570
ACRYLIC ACID, BUTYL ESTER	ACROLEIN	0.6		1.6		50	C,H	1252
	ACROLEIN	0.6		1.6		60	I10	1251
	ACRYLAMIDE, N-(HYDROXYMETHYL)-	0.87	0.05	0.61	0.17	45	K,Q	467
	ACRYLIC ACID	0.78		1.15		50	C,K	1018
	ACRYLIC ACID	1.07		0.58		60	C,I12	766
	ACRYLONITRILE	0.75	0.18	1.52	0.03	60	C,I14	453
	ACRYLONITRILE	0.89	0.08	1.2	0.1	60		941
	ACRYLONITRILE	0.9	0.1	1.0	0.1		I,Q	948
	ACRYLONITRILE	1.005	0.005	1.003	0.012	60	C	662
	ACRYLONITRILE	1.005	0.005	1.003	0.012	60	C	661
	ACRYLONITRILE	1.2	0.1	1.0	0.1		K,Q	948
	ACRYLONITRILE	1.2	0.1	1.0	0.1		K,Q	948
	ANTHRACENE 9-VINYL-	3.76		0.163		78	C	478
	1,3-BUTADIENE	0.08	0.02	0.99	0.07	5	K	1025
	2-BUTANOL, 1-(P-VINYLPHENYL)-	0.17		0.4		60	I5	1093
	3-BUTEN-2-ONE	0.65	0.07	1.6	0.1	50		184
	1,3-DIOXOLANE-4-METHANOL, 2,2-DIMETHYL-, ACRYLATE	1.32	0.15	0.45	0.15	50	C,I2	195
	1,3-DIOXOLANE-4-METHANOL, 2,2-DIMETHYL-, 4-(METHYLENESUCCINA* TE)	1.32	0.15	0.45	0.15	50	C,I2	195
	1,3-DIOXOLANE-4-METHANOL, 2,2-DIMETHYL-, METHYL FUMARATE	2.69	0.55	0	0.01	50	C,I2	195
	ETHYLENE	11.9	2.5	0.03		70	M3	148
	ETHYLENE, BROMO-	3.7		0.19		60		1098
	ETHYLENE, BROMO-	13.8		0.018		0	F,K	1098
	ETHYLENE, CHLORO-	4.4		0.07		45		963
	ETHYLENE, 1,1-DICHLORO-	0.46	0.13	0.84	0.2	70	K	575
	ETHYLENE, 1,1-DICHLORO-	0.58	0.03	0.87	0.01	0		575
	ETHYLENE, 1,1-DICHLORO-	0.83	0.02	0.88	0.10	50	C	453
	FUMARIC ACID, DIESTER WITH HYDRACRYLONITRILE	2.65	0.35	0.03		90	C	1137
	METHACRYLIC ACID	0.35		1.31		50	C,I12	766
	METHACRYLIC ACID, BUTYL ESTER	0.3		2.2		50	C,I12	766
	METHYL METHACRYLATE	0.20		1.74		60		95
	METHYL METHACRYLATE	0.37	0.1	1.8		60		305
	1-PROPENE, 3-CHLORO-	5.83	1.32	0.10	0.17	60	C	457
	PYRIDINE, 2-VINYL-	2.51	0.05	0.097	0.04	60	C	955
	PYRIDINE, 4-VINYL-	0.46	0.09	5.15	0.09	60	C	274
	STYRENE	0.15		0.76		60		120
	STYRENE	0.15		0.8		60		1096
	STYRENE	0.19	0.04	0.48	0.04	25	C,D	55
	STYRENE	0.21	0.01	0.66	0.02	60	I7	120
	STYRENE	0.34		0.82	0.01	60		570
	STYRENE, O-CHLORO-	0.2	0.05	2.25	0.2	60	C,K	1186
	SUCCINIC ACID, METHYLENE-, DIMETHYL ESTER	0.40		0.94		40	Q	169

COPOLYMERIZATION PARAMETER

M1	M2	R1	+/-	R2	+/-	TEMP.	REMARKS	REF.
ACRYLIC ACID, BUTYL ESTER								
	VINYL ACETATE	3.07	0.3	0.06	0.01	70		1324
ACRYLIC ACID, SEC-BUTYL ESTER								
	STYRENE	0.28	0.02	0.97	0.05	60		570
ACRYLIC ACID, 2-BUTYNYL ESTER								
	ACRYLIC ACID, METHYL ESTER	0.60	0.14	0.09	0.015	-40	G2,I5	196
	ACRYLONITRILE	0.30	0.07	1.14	0.25	-40	G2,I5	196
	METHYL METHACRYLATE	2.78	0.45	0.26	0.05	-40	G2,I5	196
ACRYLIC ACID, CALCIUM SALT								
	S-TRIAZINE, HEXAHYDRO-1,3,5-TRIACRYLOYL-	0.56	0.02	1.71	0.03	20	F	295
ACRYLIC ACID, 2-CHLOROETHYL ESTER								
	ACETIC ACID, ALLYL ESTER	5.5	1.0	0.9	0.1	60		560
	ACRYLIC ACID, METHYL ESTER	0.9	0.1	0.92	0.05	60		560
	P-DIOXIN, 2,3-DIHYDRO-	0.95	1.0	0.0	0.1	60	C	212
	ETHER, DODECYL VINYL	12.9		0.0		60		438
	FUMARONITRILE	1.97	0.08	0.013	0.010	60		436
	MALEIC ANHYDRIDE	10.2	0.3	0.11	0.03	60	I7	438
	2-PROPEN-1-OL, 2-METHYL-, ACETATE	6.2	0.5	0.03	0.05	60		560
		4	1	0		60		
	STYRENE	0.08	0.01	0.59	0.03	60	C	211
		0.10	0.01	0.54	0.01	60		560
		0.12		0.43		60		740
		0.12		0.43		60		742
	STYRENE, O-CHLORO-	0.533	0.015	1.71	0.08	40	C,K	1186
	SUCCINIC ANHYDRIDE, METHYLENE-	0.065		2.7			I7,Q	2
		0.34		2.25			I5,Q	2
ACRYLIC ACID, DODECYL ESTER								
	ACRYLONITRILE	1.3	0.1	3.2	0.5	60		941
ACRYLIC ACID, 2,3-EPOXYPROPYL ESTER								
	ACRYLIC ACID, 2-ETHYLHEXYL ESTER	1.08		0.98		138		492
	ACRYLONITRILE	1.0		1.01		60		413
	STYRENE	0.17		0.60		60		413
		0.25		0.73		138		492
	VINYL ACETATE	7.6		0.0029		60	I5	1166
ACRYLIC ACID, 2-(2-(ETHOXYETHOXY)ETHOXY)ETHYL ESTER								
	VINYL ACETATE	5		0.065		60	I5	1166
ACRYLIC ACID, 2-(2-ETHOXYETHOXY)ETHYL ESTER								
	VINYL ACETATE	1.25		0.032		60	I5	1166

REMARKS PAGE II-366, REFERENCES PAGE II-368

M1	M2	R1	+/-	R2	+/-	TEMP.	REMARKS	REF.
ACRYLIC ACID, 2-ETHOXYETHYL ESTER								
	VINYL ACETATE	3.95		0.027		60	I5	1166
ACRYLIC ACID, ETHYL ESTER								
	ACROLEIN	0.6		1.6		50	C,H	1252
	ACROLEIN	0.6		1.2		60	I10	1251
	ACRYLAMIDE, N-(HYDROXYMETHYL)-	1.4	0.2	1.4	0.2		K,Q	468
	ACRYLIC ACID, SODIUM SALT	5.7		1.5		50	C,I11/12	403
	ACRYLIC ACID, 2-HYDROXY-, ETHYL ESTER, ACETATE	1.0	0.1	1.0	0.05	60	C	993
	ACRYLONITRILE	0.67	0.02	1.17	0.01	50	I6	122
	ACRYLONITRILE	0.93		1.12		70	C,J	803
	ACRYLONITRILE	0.95		0.44		80	C	1070
	ACRYLONITRILE	1.0		10		25	I36	122
	ANTHRACENE 9-VINYL-	4.0		0.66		50	I11,L	122
	ANTHRACENE 9-VINYL-	3.43		0.274		78	C	478
	P-DIOXIN, 2,3-DIHYDRO-	37.9		0.002	0.001		C	433
	ETHANESULFONIC ACID, 2-HYDROXY-, METHACRYLATE	0.3	1.4	3.2	0.6	60	I55	1241
	ETHER, 2-CHLOROETHYL VINYL	5.0	0.05	0.15		50	I7	464
	ETHYLENE, 1,1-DICHLORO-	0.8		0.5		40	F,K	1053
	ETHYLENE, 1,1-DIPHENYL-	8.3	0.5	0.03	0.01	36	F,K	240
	FUMARIC ACID, DIESTER WITH HYDRACRYLONITRILE	14.7	0.9	0.026	0.015	90	C	1137
	LINOLEIC ACID, METHYL ESTER	7.4	0.4	0.16	0.02	60	C	619
	LINOLENIC ACID, METHYL ESTER	0.24	0.12	2.03	0.12	60		619
	METHACRYLIC ACID, 2-(DIETHYLAMINO)ETHYL ESTER	0.28	0.02	2.0	0.88	60	P,Q	943
	METHYL METHACRYLATE	0.47		1.83		60		305
	2,5-NORBORNADIENE	3.05		0.01		60	Q	590
	2,5-NORBORNADIENE	3.05		0.01		50		95
	10,12-OCTADECADIENDIC ACID, TRANS,TRANS-, METHYL ESTER	1.9	0.2	0.064	0.003	50		1081
	10,12-OCTADECADIENDIC ACID, TRANS,TRANS-, METHYL ESTER	2.3		0.48		60		1082
	1-PROPENE, 3-CHLORO-	7.73	0.35	0.084	0.05	70		619
	1-PROPEN-2-OL, ISOCYANATE	0.79	0.11	0.15	0.100	60	C	30
	1-PROPEN-2-OL, ISOCYANATE	24		0		60	Q	457
	3,6-PYRIDAZINEDIONE, 1,2-DIHYDRO-	0.19	0.06	0.23	0.07	75	I117	419
	PYRIDINE, 2-VINYL-	0.16	0.04	1.01	0.14	60		1127
	PYRIDINE, 2-VINYL-	0.17		0.77	0.05	60		1012
	STYRENE	0.17		0.77		60		570
	STYRENE	0.19		0.79		50		740
	STYRENE	0.20		0.80		70		742
	STYRENE					60		1011
	STYRENE					80		1011
	STYRENE, O-CHLORO-	0.048	0.05	4.01	0.43	80	C	1070
	STYRENE, METHYL-, 1:2 ORTHO:PARA MIXTURE	0.15	0.2	0.48	0.06	80	C,K	1186
	SULFIDE, ISOBUTYL VINYL	0.35		0.05		60	A	41
	THIANTHRENE, 2-VINYL-	0.07		0.314		60		1225
	VERSATIC 10 ACID, VINYL ESTER	6.46	0.50	0.10	0.10	60	I7	363
								954
ACRYLIC ACID, ETHYL ESTER, STANNIC CHLORIDE COMPLEX								
	ETHYLENE, CHLORO-	0.99		0.01		50	D	1428

COPOLYMERIZATION PARAMETER

M1	M2	R1	+/-	R2	+/-	TEMP.	REMARKS	REF.
ACRYLIC ACID, 2-ETHYLHEXYL ESTER	ACROLEIN	0.6		1.6		60	I10	1251
	ACRYLIC ACID, 2,3-EPOXYPROPYL ESTER	0.98		1.08		138		492
	ETHYLENE, CHLORO-	4.15	0.6	0.16	0.03	63	C	938
		8.25	0.75	0.002	0.002	-50	H38,I20	938
	STYRENE	0.26	0.02	0.94	0.07	60		570
		0.29		0.91		138		492
	STYRENE, O-CHLORO-	0.17	0.1	2.65	0.23	40	C,K	1186
	VINYL ACETATE	2.1		0.28			K,Q	1143
		7.5	0.15	0.04	0.1		C,I7,Q	1308
ACRYLIC ACID, 3-HYDROXY-2,2-BIS(HYDROXYMETHYL)PROPYL ESTER, * TRINITRATE	METHYL METHACRYLATE	0.39		1.5		60		1439
	STYRENE	0.26		0.38		60		1439
ACRYLIC ACID, 4-HYDROXYBUTYL ESTER	STYRENE	0.73		0.72		45	C,K	1018
ACRYLIC ACID, 2-HYDROXYETHYL ESTER	1-PROPENE, 3-CHLORO-	8.85	1.32	0.016	0.030	60	C	457
ACRYLIC ACID, 2-HYDROXYETHYL ESTER, NITRATE	STYRENE	0.08		0.48		60		1307
ACRYLIC ACID, 2-HYDROXYPROPYL ESTER	VINYL ACETATE	4.55		0.01		60	I5	1166
ACRYLIC ACID, X-HYDROXYBUTYL ESTER, NITRATE	STYRENE	0.2		0.41		60		1307
ACRYLIC ACID, ISOBUTYL ESTER	METHYL METHACRYLATE	0.20		1.71		60		95
ACRYLIC ACID, ISOPROPYL ESTER	STYRENE	0.26		0.76		60		740
		0.26		0.76		60		742
ACRYLIC ACID, 2-METHOXYETHYL ESTER	STYRENE	0.21	0.02	0.72	0.03	60		570
ACRYLIC ACID, METHYL ESTER	ACENAPHTHYLENE	0.1		4		60	I7	1326
	ACETIC ACID, ALLYL ESTER	5		0		60		620

M1	M2	R1	+/-	R2	+/-	TEMP.	REMARKS	REF.
ACRYLIC ACID, METHYL ESTER								
	ACETIC ACID, THIO-, VINYL ESTER	0.8		0.23		50	O	1249
	ACETYLENE, DIPHENYL-	55	5	0		60	C	212
	ACROLEIN	0.2		10.0		20		843
	ACROLEIN	0.6		1.6		50	C,H	1252
	ACROLEIN	0.6	0.2	1.2		60	I10	1251
	ACROLEIN	7.7	0.2	0.0	0.05	20		488
	ACROLEIN	7.7		0		50	L	845
	ACRYLAMIDE	0.05		1.30		60		41
	ACRYLAMIDE, N-BENZYL-2-METHYL-	0.65	0.05	0.46	0.05	60		1445
	ACRYLAMIDE, N,2-DIMETHYL-	0.57	0.12	0.22	0.16	60		1445
	ACRYLAMIDE, N-ETHYL-2-METHYL-	0.53	0.15	0.24	0.1	60		1445
	ACRYLAMIDE, N-(HYDROXYMETHYL)-	1.3	0.04	1.9	0.7	45		467
	ACRYLIC ACID	0.95	0.2	1.1	0.06	45	K,Q	1377
	ACRYLIC ACID	0.98	0.02	0.42	0.02	60	C	1377
	ACRYLIC ACID	0.52	0.02	0.33	0.03	-40	C,16	1367
	ACRYLIC ACID, ALLYL ESTER	0.09	0.04	0.60	0.14	60	C	196
	ACRYLIC ACID, 2-BUTYNYL ESTER	0.9	0.015	0.9	0.1	60	G2,I5	560
	ACRYLIC ACID, 2-CHLOROETHYL ESTER	0.92	0.1	0.95	0.03	60		212
	ACRYLIC ACID, 2-CHLORO-, 2-ETHOXYETHYL ESTER	0.1	0.05	1.2	0.1	0	P,Q	900
	ACRYLIC ACID, 2-CYANO-, METHYL ESTER	0.4	0.1	1.1	0.02	60	I7	507
	ACRYLONITRILE	0.67		1.26		-78	E,I8	816
	ACRYLONITRILE	0.7		3.6		50	E,I8	40
	ACRYLONITRILE	0.70		1.02		80	I17	816
	ACRYLONITRILE	0.71	0.02	0.50	0.47	50		439
	ACRYLONITRILE	0.76	0.012	1.17	0.1	65		455
	ACRYLONITRILE	0.83		0.84		50	C	1436
	ACRYLONITRILE	0.84	0.05	1.5	0.1	-78	E,I18	724
	ACRYLONITRILE	0.9	0.05	1.1		60	K	600
	ACRYLONITRILE	0.95		1.40		50		815
	ACRYLONITRILE	1.04	0.02	0.78	0.02	20	F,I11	941
	ACRYLONITRILE	1.22	0.20	0.70	0.20	30	F,I11	439
	ACRYLONITRILE	1.25	0.15	0.86	0.05	50	J	1072
	ACRYLONITRILE	1.54	0.05	0.75	0.05			1074
	ACRYLONITRILE							439
	ACRYLOYL CHLORIDE	5.8		0.04		25	F,I11,L	914
	ADIPIC ACID, DIVINYL ESTER	0.345		2.3		45		894
	ANISOLE, O-VINYL-	4	1	0.1	0.1	60	C,I10	877
	ANISOLE, P-VINYL-	0.04		2.6		40		1186
	ANTHRACENE 9-VINYL-	0.07		2		40		1186
	ATROPIC ACID, METHYL ESTER	2.97		0.082		76	C	478
	BENZENE, ETHYNYL-	0.055		1.0		60		571
	BENZENESULFONIC FLUORIDE, M-VINYL-	0.62	0.02	0.27	0.04	75		212
	BENZENESULFONIC FLUORIDE, P-VINYL-	0.14		1.50		75	I29	332
	BENZOTHIAZOLE, 2-VINYLTHIO-	0.20		4.0		60	I29	332
	1,3-BUTADIENE	0.1		1.0		5		487
	1,3-BUTADIENE, 2-CHLORO-	0.05	0.02	0.76	0.04	60	K	1025
	2-BUTANOL, 1-(P-VINYLPHENYL)-	0.078	0.010	11.1	1.5	60	C	214
	CARBAMIC ACID, DIETHYL-, VINYL ESTER	0.18		0.48		80	I5	1093
	CARBAZOLE, 9-VINYL-	4.45		0.01		30		316
	CARBONIC ACID, CYCLIC (HYDROXYMETHYL)ETHYLENE ESTER, ACRYLATE	0.43	0.02	0.11	0.02	75	I5	1129
		0.50		0.050		75	I7	331
				0.050			I7	331

COPOLYMERIZATION PARAMETER

M1: ACRYLIC ACID, METHYL ESTER

M2	R1	+/-	R2	+/-	TEMP.	REMARKS	REF.
CARBONIC ACID, CYCLIC (HYDROXYMETHYL)ETHYLENE ESTER, MALEATE-	9.10	0.40	0.10	0.02	50	C,I2	199
CARBONIC ACID, CYCLIC (HYDROXYMETHYL)ETHYLENE ESTER, 4-(ME * THYLENESUCCINATE)-	9.10	0.40	0.10	0.02	50	C,I2	199
CHROMIUM, TRICARBONYL-P1-(BENZYL ACRYLATE)-	0.63	0.3	0.56	0.11	70	I38	1357
CHROMIUM, TRICARBONYL-P1-(STYRENE)-	0.75		0.001		70	I7	1355
P-DIOXIN, 2,3-DIHYDRO-	23.	2.	0.07	0.05	60	Q	433
ETHENESULFONIC ACID, ALLYL ESTER	10.7	0.2	0.11	0.02	70	I7	209
ETHENESULFONIC ACID, BUTYL ESTER	5.0	1.5	0	0.03	80	C	754
ETHER, 2-BROMOVINYL ETHYL	31.5	2.5	0		50	C	786
ETHER, BUTYL VINYL	3.6		0		60	I7	1325
ETHER, ETHYL VINYL	3.3		0		50	N,O	620
ETHER, METHYL 1-PHENYLVINYL	0.17	0.01	0		60		572
ETHERS, C8-C18 ALKYL VINYL					50	P	11
ETHYLENE	11		0.2		150	M7,R	142
ETHYLENE, 1,1-BIS(P-CHLOROPHENYL)-	0.092	0.006	0		60	O	212
ETHYLENE, 1,1-BIS(P-METHOXYPHENYL)-	0.049	0.005	0		60	O	212
ETHYLENE, CHLORO-	4		0.06		50	P	543
	4		0.06		45		964
	4.4	0.5	0.12	0.01	45		963
	5		0.083		50	C	600
	9.0				50		620
							620
							165
ETHYLENE, 1,1-DICHLORO-	0.6		0.85		60	C	373
	0.8		0.85		60	I7,FF	1220
	0.84	0.06	0.7	0.10	60		373
	1		0.99		60	I7,FF	214
	1		1		60	C	620
	1.1		0.3		70		32
	3.67		0.1		60	I7,FF	373
	38		0.005		60	KK	1220
ETHYLENE, 1,2-DIMETHOXY-	4.8	0.4	0.02	0.02	60	G17,I21	526
ETHYLENE, 1,1-DIPHENYL-	0.092	0.010	0.02		60		434
	0.102	0.006	0		50	Q	409
ETHYLENE, TETRACHLORO-			0		60		212
ETHYLENE, TRICHLORO-	200		0		60	N,O	620
	830		0		60	N,O	211
	33		0		60	N,O	620
ETHYLENE GLYCOL, BIS(2-CHLOROACRYLATE)					60	P,Q	900
FERROCENE, VINYL-	0.63	0.03	0.82	0.05	70	I7	1268
FERROCENEETHANOL, ACRYLATE	0.687	0.043	0.76	0.027	65	I7	1358
FERROCENEMETHANOL, ACRYLATE	4.46	0.2	0.14	0.02	70	I7	1267
FERROCENEMETHANOL, METHACRYLATE	0.82	0.07	0.08	0.03	70	I7	1267
3,5-HEPTADIEN-2-ONE	0.35	0.06	2.3	0.2	60		272
1,3,5-HEXATRIENE, 2,3,4,5-TETRACHLORO-, TRANS-	0.27	0.04	3.17	0.27	60	Q	18
1-HEXENE	8.5	2	0		60		212
1-HEXYNE	11.2	2	0		60		212
HYDROQUINONE, VINYL-, DIBENZOATE	0.46		0.75		78	I7	465
IMIDAZOLE, 2-METHYL-1-VINYL-	1.28		0.05		70	C,I10	579
ISOCYANIC ACID, VINYL ESTER	1.38		0.14		60		420
ISOPRENE	0.12	0.27	0.75	1.0	50		377
MALEIC ANHYDRIDE	2.5		0.02		75	N,O'	620
	2.8	0.05	0.02		75	C	1033
METHACRYLAMIDE	2.0		0.22	0.0	65		187

ACRYLIC ACID, METHYL ESTER

M2	R1	+/-	R2	+/-	TEMP.	REMARKS	REF.
METHACRYLANILIDE	0.66	0.22	1.05	0.19	60		1445
METHACRYLIC ACID, 2-BUTYNYL ESTER	0.56	0.08	0.35	0.05	-40	G2,15	196
METHACRYLIC ACID, 2-(DIETHYLAMINO)ETHYL ESTER						P,Q	943
METHACRYLIC ANHYDRIDE	0.16	0.02	4.75	0.15	60	C,I7	1145
METHYL METHACRYLATE	0.34	0.1	1.69	0.4	60	C,I7	96
	0.35	0.1	1.8	0.13	65		305
	0.36	0.15	2.23	0.15	50	D	1458
	0.45	0.1	2.3	0.4	48		1458
	0.5	0.1	2.3	0.05	130	A	532
	0.504	0.01	1.91	0.05	130	MM	863
	0.93		0.99		50	D,MM	1458
	0.96		0.96		48	H2O,I7	1458
	1		1		20		551
METHYL METHACRYLATE-METHYL-D	1.5	0.5	0.3	0.005	65	G3	187
METHYL METHACRYLATE-METHYL-14C	4.5	0.08	0.1	0.1	-30	A	551
2,4-PENTADIENOIC ACID, METHYL ESTER	0.50	0.02	2.3	0.5	130	C	863
PHENANTHRENE, 2-VINYL-	0.43	0.07	2.00	0.10	60		78
PHENANTHRENE, 3-VINYL-	0.50	0.03	2.0	0.2	60	C	272
PHOSPHONIC ACID, HYDROXYMETHYL-, DIETHYL ESTER, ACRYLATE	0.1	0.05	2.0	0.2	60	C	783
PHTHALIC ACID, DIALLYL ESTER	0.8	0.09	1.75	0.25	60		783
2-PICOLINE, 5-VINYL-	0.88		0.29	0.05	75	C,CC	760
	11.5		0.05	0.10	80	C,CC	1292
	12.9		0.058	0.10	60	Q	1292
1-PROPENE, 3-CHLORO-	0.172	0.007	0.88		60	C	940
	0.35	0.05	0.70		120	C	105
	5.450	0.070	0.071	0.011	90	C	456
	6.460	0.570	0.404	0.073	60	C	456
	8.450	0.370	0.054	0.039	30	C	456
	9.100	0.930	-0.020	0.100	60		456
2-PROPENE-1-SULFONIC ACID, ALLYL ESTER	22	4	0.37	0.09	60	I7	824
1-PROPEN-2-OL, ISOCYANATE	5.3	0.7	0.11	0.04	60	Q	1201
2-PROPEN-1-OL, 2-CHLORO-, ACETATE	0.60	0.03	0.08	0.08	100	N,O	419
PYRIDINE, 5-ETHYL-2-VINYL-	0.8		0	0.05	60		333
PYRIDINE, 2-VINYL-	0.7		1.16		50		484
	0.179	0.006	1.58		60	C	939
	0.168	0.003	2.03		60		939
	0.20	0.09	2.14	0.49	60		7
PYRIDINE, 4-VINYL-	0.15		1.7	0.2	60		377
	0.22	0.01	0.2		60		939
PYRIDINIUM FLUOROBORATE, 1-VINYL-	1.5	0.16	0.041	0.024	60	I31	223
2-PYRROLIDINONE, 1-VINYL-	0.27	0.07	0.133	0.006	60		1070
SILANE, TETRAVINYL-	8.87	0.06	2.5	0.3	70		884
SORBONITRILE	0.30	0.01	0.03		-78	C	272
STEARIC ACID, VINYL ESTER	5.8	0.05	0.9	0.60	0	Q	1050
STYRENE	0.07	0.02	0.90	0.04	60		815
	0.13		0.68		60	E,I18	610
	0.14		0.7		-78	H10,I16	570
	0.15		0.4		60	C	600
	0.16	0.02	1.0	0.1	60		816
	0.18	0.02	0.75	0.1	60	E,I18,AA	407
	0.18	0.03	0.75	0.03	25	C	168
	0.18		0.75	0.11			559
						E	149

COPOLYMERIZATION PARAMETER

M1	M2	R1	+/-	R2	+/-	TEMP.	REMARKS	REF.
ACRYLIC ACID, METHYL ESTER	STYRENE	0.20	0.05	0.75	0.1	70		34
		0.238	0.02	0.825	0.0	131	E,18	559
		0.3		0.65		0	H,17,Q	816
		0.4	0.2	2.2	0.2		H20,17	358
		0.4	0.2	2.2	0.2	20	Q,H,124	551
		1		1				358
		5.0		0.13		-78	E,18,AA	816
	STYRENE, O-BROMO-	0.01		6.9		40		1186
	STYRENE, P-BROMO-	0.002		6		40		1186
	STYRENE, M-CHLORO-			4.8		40		1186
	STYRENE, O-CHLORO-	0.01	0.03	6	0.4	40	C,K	1186
	STYRENE, P-CHLORO-	0.15	0.03	3.9	1.4	40	C	1186
	STYRENE, 2,5-DICHLORO-	0.25	0.04	3.4	0.28	70	C	555
				4.27		60		7
	STYRENE, O-FLUORO-	0.01		3.8		40		1186
	STYRENE, P-FLUORO-	0.04		2.6		40		1186
	STYRENE, METHYL-, 1:2 ORTHO:PARA MIXTURE	0.15	0.20	0.59	0.06	80		41
	STYRENE, M-METHYL-	0.14		1.65		40		1186
	STYRENE, O-METHYL-	0.27		0.98		40		1186
	STYRENE, P-METHYL-	0.17		1.54		40		1186
	SUCCINIMIDE, N-VINYL-	1.2		0.4		60	C	279
	SULFIDE, ISOBUTYL VINYL	0.35		0.05		60	A	1225
	SULFIDE, METHYL VINYL	0.35		0.05		60		788
	SULFIDE, PENTACHLOROPHENYL VINYL	0.89	0.04	0.25	0.03	80		360
	SULFIDE, PHENYL VINYL	0.40		0.05	0.02	60		785
	2,4,8,10-TETRAOXOSPIRO(5.5)UNDECANE, 3,9-DIVINYL-	3.14		0.23		60	C,17	1123
	9-UNDECENOIC ACID, VINYL ESTER	3.69	0.12	0.031	0.12	60	C	591
	VINYL ACETATE	3	0.3	0.5	0.06	60	P,Q	1069
		6.3	0.4	0.031	0.006	60	C,K	77
		6.7	2.2	0.029	0.011	50		1263
		9	2.5	0.1	0.1	60	C	763
ACRYLIC ACID, METHYL-D ESTER	METHYL METHACRYLATE	0.42	0.06	2.2	0.4	130	A	863
ACRYLIC ACID, METHYL-14C ESTER	METHYL METHACRYLATE	0.43	0.02	2.00	0.10	60	C	78
	STYRENE	0.21	0.01	0.74	0.04	60	C	78
ACRYLIC ACID, 1-METHYL-2-NITROPROPYL ESTER	STYRENE	0.13		0.58		80	126	930
ACRYLIC ACID, 2-METHYL-2-NITROPROPYL ESTER	STYRENE	0.08		0.71		80	126	930
ACRYLIC ACID, 2-NITROBUTYL ESTER	ACRYLONITRILE	1.58		0.90		70	C	945

REMARKS PAGE II-366, REFERENCES PAGE II-368

M1 / M2	R1	+/-	R2	+/-	TEMP.	REMARKS	REF.
ACRYLIC ACID, 2-NITROBUTYL ESTER							
METHYL METHACRYLATE	0.22		1.60		70	C	945
STYRENE	0.115		0.58		70	C	945
ACRYLIC ACID, 9-OCTADECEN-1-YL ESTER							
METHYL METHACRYLATE	0.3	0.03	2.23	0.01	60	I7	1239
ACRYLIC ACID, OCTADECYL ESTER							
ACRYLONITRILE	0.4	0.12	1.61	0.34	50	C,P	12
	0.68	0.18	1.74	0.04	60	I7	1239
	1.2	0.1	4.1	0.8	60	C,I14	453
	1.01	0.01	0.91	0.05	60		941
ETHYLENE, 1,1-DICHLORO-	1.01	0.01	0.91	0.05	60	C	453
METHYL METHACRYLATE	0.48	0.12	2.36	0.04	60		1239
STYRENE	0.18	0.45	0.44	0.07	60	I7	1239
	0.31	0.31	0.79	0.09	60	I45	1239
	0.34	0.06	0.75	0.12	60	I7	1239
ACRYLIC ACID, OCTYL ESTER							
ACRYLONITRILE	0.83	0.23	1.93	0.08	50	C,P	12
	4.8		0.12		60	C,I14	453
ETHYLENE, CHLORO-					45		963
ETHYLENE, 1,1-DICHLORO-	0.70	0.01	0.87	0.02	60	C	453
STYRENE, METHYL-, 1:2 ORTHO:PARA MIXTURE	0.16	0.13	0.49	0.06	80		41
ACRYLIC ACID, POTASSIUM SALT							
ACRYLAMIDE	0.84		1.4		90	C,I11	473
	1.35		0.78		90	C,I11,L	473
ACRYLIC ACID, PROPYL ESTER							
METHYL METHACRYLATE	0.29	0.1	1.61	0.1	60		305
1-PROPENE, 3-CHLORO-	7.56	0.10	0.091	0.013	60	C	457
ACRYLIC ACID, 2-PROPYNYL ESTER							
ACRYLONITRILE	1.49		0.95		60		1306
METHYL METHACRYLATE	0.05		1.12		60		1306
STYRENE	0.24		0.45		60		1306
ACRYLIC ACID, SODIUM SALT							
ACRYLAMIDE	0.35	0.03	1		60	C,I15	1309
	1.5		1.10	0.05	60	C	119
ACRYLIC ACID, ETHYL ESTER	0.77		5.7		50	C,I11/12	403
ACRYLONITRILE	0.34	0.23	0.21		50	C,I12	402
BENZENESULFONIC ACID, P-VINYL-, SODIUM SALT	5.8		2.3	1.2	40	C,I11	301
ETHENESULFONIC ACID, SODIUM SALT	2		0		60		138
VINYL ACETATE			0.01		60	C	119

COPOLYMERIZATION PARAMETER

M1 / M2	R1	+/-	R2	+/-	TEMP.	REMARKS	REF.
ACRYLIC ACID, TETRAHYDROFURFURYL ESTER							
ACRYLONITRILE	0.665		0.689		70	C	701
PROPANE, 1-CHLORO-2,3-EPOXY-	0.56	0.05	0.12	0.01	0	H1o	706
STYRENE	0.485		0.501		70	C	701
ACRYLIC ACID, TETRAHYDROFURFURYL ESTER							
ACRYLONITRILE	0.773	0.07	0.602	0.10	60		63
	0.91		1.577		25	E	83
ACRYLIC ACID, 2-BROMO-, ETHYL ESTER							
STYRENE	0.5		0.06		60		1429
	0.5		0.06		60		1430
ACRYLIC ACID, 2-BROMO-, METHYL ESTER							
METHYL METHACRYLATE	2.08		0.23		60	17	96
ACRYLIC ACID, 3-BROMO-, CIS-							
ACRYLAMIDE	12	0.5	0		20	F,173	1297
ACRYLIC ACID, 3-BROMO-, TRANS-							
ACRYLAMIDE	6.5	0.6	0		20	F,173	1297
ACRYLIC ACID, 3-BUTOXY-, BUTYL ESTER							
STYRENE	0.01		78		60		829
ACRYLIC ACID, 2-CHLORO-, 2-ETHOXYETHYL ESTER							
ACRYLIC ACID, METHYL ESTER						P,Q	900
METHYL METHACRYLATE						P,Q	900
STYRENE						P,Q	900
ACRYLIC ACID, 2-CHLORO-, ETHYL ESTER							
STYRENE	0.3		0.08		60		1429
	0.3		0.08		60		1430
ACRYLIC ACID, 2-CHLORO-, METHYL ESTER							
ACRYLONITRILE	2.0	0.3	0.15	0.02	70		23
MALONONITRILE, METHYLENE-	0.41	0.14	0.091	0.05	50		289
METHACRYLIC ACID, BUTYL ESTER	1.2		0.3		60/80	P	916
METHACRYLIC ACID, ETHYL ESTER	2.00		0.15		60/80	P	916
METHYL METHACRYLATE	2.00		0.15		70	P,Q	916
	0.30	0.2	0.25	0.04	60		23
STYRENE					60	C,17	97
					60	C,17	96
					70	C	259

REMARKS PAGE II-366, REFERENCES PAGE II-368

M1	M2	R1	+/-	R2	+/-	TEMP.	REMARKS	REF.
ACRYLIC ACID, 2-CYANO-, METHYL ESTER	ACRYLIC ACID, METHYL ESTER	1.2		0.1		60	I7	507
	METHYL METHACRYLATE	0.13		0.10		60		747
		0.25		0.04		60		507
	1-OCTENE	0.63	0.06	-0.17	0.04	60	I7	507
	STYRENE	0.03		0.01		60	I7	507
	STYRENE, ALPHA-METHYL-	0.001		0.05		60	I7	507
	VINYL ACETATE	0.5		0.005		60	I7	507
ACRYLIC ACID, 3-ETHOXY-, ETHYL ESTER	ACRYLONITRILE	0.02	0.02	10.5	1.5	80	C	786
	STYRENE	0		23.5	1	80	C	786
		0.01		24		60		829
ACRYLIC ACID, 2-FLUOROMETHYL-, ETHYL ESTER	METHYL METHACRYLATE	0.513	0.046	1.05	0.04	60	I2	779
	STYRENE	0.091	0.017	0.341	0.013	60	I2	779
ACRYLIC ACID, 2-FORMAMIDO-	ACRYLONITRILE	2.5		0.01		60	I15	545
ACRYLIC ACID, 2-HYDROXY-, ETHYL ESTER, ACETATE	ACRYLIC ACID, ETHYL ESTER	1.0	8.05	1.0	0.1	60	C	993
	METHYL METHACRYLATE	0.65	0.05	1.65	0.07	60	C	993
	STYRENE	0.20	0.05	0.57	0.05	60	C	993
	VINYL ACETATE	5.4	0.5	0.08	0.03	60	C	993
ACRYLIC ACID, 2-METHOXY-, METHYL ESTER	ACRYLONITRILE	0.3		0.15		60		1341
	STYRENE	0.51		1.1		60		1430
		0.51		1.1		60		1341
ACRYLIC ACID, 3-PROPOXY-, PROPYL ESTER	STYRENE	0.01		42		60		829
ACRYLIC ACID, THIO-, BENZYL ESTER	STYRENE	0.2		0.15		60		1133
ACRYLIC ACID, THIO-, BUTYL ESTER	STYRENE	0.34		0.19		60		1133
ACRYLIC ACID, THIO-, TERT-BUTYL ESTER	STYRENE	0.45		0.21		60		1133

M1	M2	R1	+/-	R2	+/-	TEMP.	REMARKS	REF.
ACRYLIC ACID, THIO-, ETHYL ESTER								
	STYRENE	0.28		0.19		60		1133
ACRYLIC ACID, THIO-, METHYL ESTER								
	1,3-BUTADIENE	0.20	0.05	0.35	0.01	70		594
ACRYLIC ACID, THIO-, PROPYL ESTER								
	STYRENE	0.32		0.19		60		1133
ACRYLIC ANHYDRIDE								
	METHACRYLONITRILE	0.9		0.4		35	I13	891
	STYRENE	0.1		0.17		35	I13	891
ACRYLONITRILE								
	ACENAPHTHYLENE	0.02		2.56		60	I7	1326
	ACETIC ACID, ALLYL ESTER	1.9		0.0		0	D,L3	389
		8.0		0.05		0	D	389
		8.2		0.0			E,Q	983
	ACETIC ACID, CHLORO-, VINYL ESTER	0.34		0.09	0.04	80	C	1070
	ACETIC ACID, DICHLORO-, VINYL ESTER	0.25		0.18	0.04	80	C	1070
	ACETYLENE, DIPHENYL-	13.6	1.0	1.60		60	D	212
	ACROLEIN	0.52		1.60		50	I6	845
		0.52	0.02	1.60		50	I6	488
		0.60	0.18	1.28	0.036	50	I6	840
		0.77	0.1	1.09	0.05	20	C	843
		0.77	0.1	1.09	0.05	50	L	845
		7.2		0.01		60	I6	1437
		7.56		0.01		60	I6	1437
		12		0.01		60	Q	1112
	ACROLEIN, BIS(2-BUTOXYETHYL)ACETAL	0.8		1.3		60	I46	561
	ACROLEIN, BIS(2-ETHOXYETHYL)ACETAL	0.8		1.3		60	I48	561
	ACROLEIN, DIETHYL ACETAL	0.8		1.3		60	I47	561
	ACRYLAMIDE	0.8		1.3		60	I36	561
		0.875		1.357		30	F	369
		0.94	0.16	1.04	0.27	40	I7,P	929
	ACRYLAMIDE, N-CYCLOHEXYL-2-METHYL-	0.7		1.2		70	K,Q	16
	ACRYLAMIDE, N-(HYDROXYMETHYL)-	0.98	0.1	2.33	0.1	20	F	468
	ACRYLAMIDE, N-(HYDROXYMETHYL)-2-METHYL-	0.15	0.05	1.22	0.1	60		1361
	ACRYLAMIDE, N,N'-ISOPROPYLIDENEBIS-	0.57	0.1	2.44	0.3	30	F	1200
	ACRYLAMIDE, N,N'-METHYLENEBIS-	0.64	0.12	1.50	0.12	70	I15	1313
	ACRYLAMIDE, 2-METHYL-N-(1-NAPHTHYL)-	1.10	0.035	1.44	0.019	60	F,I11	370
	ACRYLAMIDE, N-OCTADECYL-	0.20	0.03	1.00	0.05	70	I7,P	16
	ACRYLANILIDE, 4'-((2-CHLOROETHYL)SULFAMOYL)-	0.05	0.03	1.03	0.03	70	C,I14	454
	ACRYLANILIDE, 4'-((2-CHLOROETHYL)SULFAMOYL)-2-METHYL-					70	I6	680
	ACRYLANILIDE, 2,4'-DIMETHYL-					70	I6	680
	ACRYLANILIDE, 2-METHYL-4'-NITRO-					70	I7,P	16
						70	I7,P	16
	ACRYLIC ACID	0.13	0.02	6.0	2.0	50	C,P	12
						80	C	1070
		0.35		1.15		50	C,I11	402

M1	M2	R1	+/-	R2	+/-	TEMP.	REMARKS	REF.
ACRYLONITRILE								
	ACRYLIC ACID, ALLYL ESTER	0.73	0.06	1.12	0.21	-40	G17,15	197
	ACRYLIC ACID, BENZYL ESTER	0.294	0.08	0.636	0.07	60		63
		1.490		0.631		25	E	83
	ACRYLIC ACID, BUTYL ESTER	1.0	0.1	1.2	0.1		K,Q	948
		1.0	0.1	0.9	0.1		I,Q	948
		1.0	0.1	1.2	0.1		K,Q	948
		1.003	0.012	1.005	0.005	60	C	661
		1.003	0.012	1.005	0.005	60	C	662
		1.2	0.1	0.89	0.08	60		941
	ACRYLIC ACID, 2-BUTYNYL ESTER	1.52	0.03	0.75	0.18	60	C,I14	453
	ACRYLIC ACID, DODECYL ESTER	1.14	0.25	0.30	0.07	-40	G2,15	196
	ACRYLIC ACID, 2,3-EPOXYPROPYL ESTER	3.2	0.5	1.3	0.1	60		941
	ACRYLIC ACID, ETHYL ESTER	1.01		1.0		60		413
		0.44		0.95		80	C	1070
		0.66		4.0		50	I11,L	122
		1.12		0.93		70	C,J	803
		1.17	0.01	0.67	0.02	50	I6	122
		10		1.0		25	I36	122
	ACRYLIC ACID, METHYL ESTER	0.04		5.8		25	F,I11,L	914
		0.50	0.47	0.71	0.012	80	F,I11	455
		0.70	0.20	1.22	0.20	20	J	1072
		0.75	0.05	1.54	0.05	50	K	439
		0.78	0.02	1.04	0.02	50		439
		0.84		0.83		65		724
		0.86	0.05	1.25	0.15	30	F,I11	1074
		1.02	0.02	0.70	0.02	50	I17	439
		1.1		0.9		-78	E,I18	815
		1.1		0.4		0	E,I18	816
		1.17		0.76		50		1436
		1.26	0.1	0.67	0.1	60		40
		1.40	0.1	0.95	0.05	60		941
		1.5	0.1	0.84	0.05	50	C	600
		3.6		0.7		-78	E,I8	816
		0.90		1.58		70	C	945
	ACRYLIC ACID, 2-NITROBUTYL ESTER	1.61	0.34	0.4	0.12	50	C,P	12
	ACRYLIC ACID, OCTADECYL ESTER	1.74	0.04	0.68	0.18	60	I7	1239
		4.1	0.8	1.2	0.1	60	C,I14	453
	ACRYLIC ACID, OCTYL ESTER	1.93	0.08	0.83	0.23	50	C,P	12
	ACRYLIC ACID, 2-PROPYNYL ESTER	0.95		1.49		60	C,I14	453
	ACRYLIC ACID, SODIUM SALT	0.21		0.77		50	C,I12	1306
	ACRYLIC ACID, TETRAHYDROFURFURYL ESTER	0.689	0.10	0.665	0.07	70	C	402
	ACRYLIC ACID, TETRAHYDROFURYL ESTER	0.602		0.773		25		701
	ACRYLIC ACID, 2-CHLORO-, METHYL ESTER	1.577		0.91		70	E	63
	ACRYLIC ACID, 3-ETHOXY-, ETHYL ESTER	0.15	0.02	2.0	0.3	80		23
		10.5	1.5	0.02	0.02	70	C	786
	ACRYLIC ACID, 2-FORMAMIDO-	0.01		2.5		60	I15	545
	ACRYLIC ACID, 2-METHOXY-, METHYL ESTER	0.15		0.3		60		1341
	ACRYLONITRILE, 2-HYDROXY-, 2-METHOXY-, ACETATE	0.09	0.17	7.4	0.40	60		1070
	ACRYLOYL CHLORIDE	0.37	0.25	1.93	0.25	60		1341
		1.2		1		50	I7	1375
	BETA-ALANINE, N-ALLYL-	4.25	0.15	0.09	0.03	50	C,I36	1372

M1	M2	R1	+/-	R2	+/-	TEMP.	REMARKS	REF.
ACRYLONITRILE	ALLYL ALCOHOL	0.03		1.99		RT		190
		0.6		0		20	D,L3	389
		1.6		0.01		0	F,I36	1303
		2.6	0.53	0.05	0.10	25	D	389
		3.96	0.3	0.11	0.01	80	D	736
		5.5		0.1		60	E,Q	983
	ALLYL ETHER	4.9		0.01		60		72
	AMMONIUM CHLORIDE, ALLYLDIMETHYLDODECYL	13.1		0.4		60	I17	1246
		14.5		0.5		60	C,J	1246
	AMMONIUM CHLORIDE, ALLYLTRIETHYL-	8.3		0.8		60	I17	1246
		18.8		0.03		60	C,J	1246
	AMMONIUM CHLORIDE, DIETHYL(2-HYDROXYETHYL)-, METHACRYLATE	0.13	0.1	0.85	0.15	70	I12	1109
		0.24	0.05	0.9	0.2	70	T12	1109
	ANILINE, N,N-DIVINYL-	0.42		0.003		70		162
	ANISOLE, P-PROPENYL-/MALEIC ANHYDRIDE	0.007	0.0024	45	2.9	60	17	1259
	ANISOLE, P-3-BUTENYL-	3.0		0		70		67
	ATROPIC ACID, METHYL ESTER	0.075		6.7		60		571
	BENZENE, ALLYL-	3.5		0		70		67
	BENZENE, ETHYNYL-	0.26	0.03	0.33	0.05	60		212
	BENZENE, PROPENYL-, CIS-	1.61	0.13	-0.20	0.06	60		1065
	BENZENE, PROPENYL-, TRANS-	0.57	0.002	0.001	0.021	60		1065
	BENZENESULFONIC ACID, P-VINYL-	0.10	0.02	1.20	0.10	45	C,I11	440
		0.07	0.02	1.5	0.04	70	C,I11	301
	BENZENESULFONIC ACID, P-VINYL-, POTASSIUM SALT	0.05	0.01	1.40	0.04	45	C,I11	440
	BENZENESULFONIC ACID, P-VINYL-, SODIUM SALT	0.05	0.02	1.5	0.02	40	C,I11	301
		0.15	0.02	0.55	0.03	45	I17	442
	BENZOIC ACID, VINYL ESTER	5.0	0.05	0.05	0.05	75	C	166
	BENZOIC ACID, P-VINYL-	0.076	0.001	1.8	0.05	50	I6	735
	BENZOIC ACID, P-VINYL-, SODIUM SALT	0.09	0.08	0.26	0.06	50	F,I11	735
	BENZYL ALCOHOL, ALPHA,ALPHA-DIETHYL-P-VINYL-	0.04		0.38		60	I66	1092
	BENZYL ALCOHOL, ALPHA,ALPHA-DIMETHYL-P-VINYL-	0.05		0.41		60	I66	1092
	BENZYL ALCOHOL, ALPHA-ETHYL-ALPHA-METHYL-P-VINYL-	0.06		0.39		60	I66	1092
	BENZYL ALCOHOL, ALPHA-METHYLENE-, ACETATE	0.08	0.01	0.4	0.05	75	C	166
	1,3-BUTADIENE	0.0	0.04	0.46	0.03	50	K	1022
		0.02		0.35	0.08	50		352
		0.02		0.3		40	E,I15	833
		0.03		0.28		5	K	243
		0.04		0.18		5	K	243
		0.05	0.01	0.40	0.02	50	H38	60
		0.06	0.01	0.35	0.01	50		353
		0.075		0.1		40	E,I15	833
		0.25		0.35		60	I2	311
	1,3-BUTADIENE, 2-CHLORO-	0.005	0.011	0.33	0.53	60		621
		0.01	0.01	6.07	0.53	50		353
		0.034	0.0	6.07	0.23	50		352
		0.045	0.004	6.93	0.03	50		353
	1,3-BUTADIENE, 2-FLUORO-	0.05	0.01	5.35	0.10	60		211
	1,3-BUTADIEN-1-OL, ACETATE	0.07	0.03	4.80		50	I6	840
	2-BUTANOL, 1-(P-VINYLPHENYL)-	0.0		0.50		50	K	733
	1-BUTENE	0.04		0.7		70	C	815
	1-BUTENE, 4-(P-CHLOROPHENYL)-	8.0		0.31		60	I5	1093
		4.0		0.10		60		865
				0		70		67

M1	M2	R1	+/-	R2	+/-	TEMP.	REMARKS	REF.
ACRYLONITRILE								
	1-BUTENE, 4-CYCLOHEXYL-	4.5		0		70		67
	1-BUTENE, 3,3-DIMETHYL-1-PHENYL-, TRANS-	18.8		0		60		1065
	1-BUTENE, 3-METHYL-1-PHENYL-, CIS-	18.0	0.5	-0.38	0.32	60		1065
	1-BUTENE, 3-METHYL-1-PHENYL-, TRANS-	3.37	0.20	-0.16	0.50	60		1065
	1-BUTENE, 1-PHENYL-, CIS-	2.58	0.09	-0.02	0.34	60		1065
	1-BUTENE, 1-PHENYL-, TRANS-	0.98	0.02	0.10	0.06	60		1065
	1-BUTENE, 4-PHENYL-	3.5		0		70		67
	2-BUTENE, CIS-	14.0		0.00		60		865
	2-BUTENE, TRANS-	14.0		0.00		60		865
	2-BUTENE, 2,3-DIMETHYL-	4.0	0.03	-0.26	0.04	60		1066
	2-BUTENE, 2-METHYL-	2.7	0.04	-0.27	0.15	60		1066
	1-BUTENE:SULFUR DIOXIDE (1:1 COMPLEX)	1.0		0.8		-78	B,I2	342
	3-BUTENOIC ACID, 3-HYDROXY-, BETA-LACTONE	7.52	0.08	1.78	0.22	60	Q	421
	3-BUTEN-2-ONE	0.61	0.04	1.39	0.06	50		560
	3-BUTEN-2-ONE, 3-METHYL-	0.63	0.06	1.2		80	I6	775
	1-BUTEN-3-YNE	0.3	0.08	0.70	0.14	60	Q	1070
	1-BUTEN-3-YNE	0.36	0.01	0.60	0.02	60	C	784
	1-BUTEN-3-YNE	0.13	0.10	4.65	1	60	I6	840
	1-BUTEN-3-YNE, 2-METHYL-	0.23	0.01	0.47		30		784
	BUTYRALDEHYDE, 2-METHYLENE-	0.33		1.5	0.5	30	C,Q	140
	CARBAMIC ACID, VINYL-, 2,3-EPOXYPROPYL ESTER	0.237	0.004	0.04		60	I5	418
	CARBAMIC ACID, VINYL-, 2,3-EPOXYPROPYL ESTER	0.066		0.09		70	I5	1129
	CARBAZOLE, 9-VINYL-	0.28	0.02	0.09	0.02	60	D	1102
	CARBAZOLE, 9-VINYL-	0.35		0.09		70		1102
	CARBONIC ACID, CYCLIC VINYLENE ESTER	14.9	0.1	0.085	0.001	60	P,Q	511
	CINNAMIC ACID, METHYL ESTER	6	2	0		60	C	1370
	CINNAMIC ACID, ALPHA-CYANO-, ETHYL ESTER	18	2	-0.15	0.1	60	C	211
	CINNAMONITRILE	6.4	1	0	0.5	60	C	1374
	CROTONALDEHYDE	8.0		0.0105	0.0105	60	M3	1374
	CROTONIC ACID	25	4.2	0.09		60	I6	394
	CROTONONITRILE, CIS-	21	10	0		60	C	840
	CYCLOBUTANE, METHYLENE-	10	3	0.02	0.03	50	I6	211
	CYCLOBUTANE, 1-METHYLENE-3-PHENYL-	1.01		0.02		60	I7	840
	CYCLOBUTANECARBONITRILE, 3-METHYLENE-	1.01		0		60	I7	1134
	CYCLOBUTANECARBOXYLIC ACID, 3-ACETYL-2,2-DIMETHYL-, VINYL ESTER *	0.85		0		60	I7	934
	CYCLOBUTANECARBOXYLIC ACID, 3-METHYLENE-	0.65		0		60	I7	934
	CYCLOBUTANECARBOXYLIC ACID, 1-METHYL-3-METHYLENE-, METHYL ESTER *	0.65		0.02		60	I7	934
		30	12	0.02		60	I6	840
		0.75		0		60	I7	934
		0.75		0		60	I7	934
	CYCLOHEPTENE	6.9	0.35	-0.24	0.12	60		1066
	1,3-CYCLOHEXADIENE	0.02		0.18		60	C	682
	1,3-CYCLOHEXADIENE	0.026	0.014	0.20	0.01	60	C	248
	1,3-CYCLOHEXADIENE	0.05	0.2	0.11	0.02	65		1058
	1,3-CYCLOHEXADIENE, 1-METHOXY-	1.00		0.50		200	C	682
	1,3-CYCLOHEXADIENE, 1-METHYL-	0.48	0.1	0.13	0.1	60		1434
	CYCLOHEXANE, ALLYL-	0.18	0.05	0.18	0.05	65		1058
	CYCLOHEXANE, METHYLENE-	6.0		0		70		67
	CYCLOHEXANE, METHYLENE-	1.13		0		60	I7	1134

M1	M2	R1	+/-	R2	+/-	TEMP.	REMARKS	REF.
ACRYLONITRILE	CYCLOHEXENE	8.0	0.74	-0.42	0.37	60		1066
	3-CYCLOHEXENE-1-CARBOXYLIC ACID, 1-METHYL-, 2,3-EPOXYPROPYL * ESTER	8.0	0.74	-0.42	0.37	60		1066
	CYCLOOCTENE	2.8	0.15	-0.28	0.21	60	16	1066
	1,3-CYCLOPENTADIENE, 2-(4-PENTENYL)-	0.1		0.9		60	17	1315
	CYCLOPENTANE, METHYLENE-	0.39		0		60		1134
	CYCLOPENTENE	4.8	0.36	-0.16	0.34	60		1066
	4-CYCLOPENTENE-1,3-DIONE	3.67		0.21		60	17,L	324
	DECANOIC ACID, VINYL ESTER	4.0	0.3	0.04	0.03	60	C,16	58
	M-DIOXANE, 5,5-DIMETHYL-2-VINYL-	2.25		0.08		60	16	1438
	M-DIOXANE, 4-METHYL-2-VINYL-	4.65		0.04		60	16	1438
	M-DIOXANE, 2-VINYL-	4		0.04		60	16	1438
	P-DIOXIN, 2,3-DIHYDRO-	5.9	0.1	0.001	0.002	60	Q	433
	1,3-DIOXOLANE, 2-VINYL-	4.62		0.03		60	16	1351
	1,3-DIOXOLANE-4-METHANOL, 2,2-DIMETHYL-, ACRYLATE	4.8		0.02		60	16	1438
	1,3-DIOXOLANE-4-METHANOL, 2-VINYL-	0.9	0.09	0.84	0.34	50	C,16	195
		2.45		0.04		60	16	1351
		4.25		0.05		60	16	1438
	1,3-DIOXOLAN-4-OL, 2-VINYL-, ACRYLATE	0.95		0.89		60	16	1352
	DISILANE, PENTAMETHYLVINYL-	3	1	0	0.01	60		837
	DISILOXANE, 1,1,3,3-PENTAMETHYL-3-VINYL-	8.0	1.2	0.1	0.0004	50		773
	ETHANOL, 2-(2-(VINYLOXY)ETHOXY)-	0.59	0.02	0.0021		60		1070
	ETHENESULFONIC ACID	1.5		0.15		60	Q	546
		4.52		0.22		60	I36	37
	ETHENESULFONYL FLUORIDE	5.78	0.11	0.15	0.03	50	17	811
		7.7	0.51	0.13	0.05	70	17	811
	ETHER, TERT-BUTYL VINYL	0.14	0.04	0.0032	0.0002	60	C	1070
	ETHER, 2-CHLOROETHYL VINYL	1.0	0.2	-0.07	0.1	60		435
	ETHER, DODECYL VINYL	0.82	0.05	0.0	0.2	50	C	13
	ETHER, ETHYL VINYL	0.7	0.2	0.03	0.02	80	C	786
		5.0	0.2	0.03	0.02	80	C	848
	ETHER, ISOBUTYL VINYL	1.05	0.1	0	0.05	60	N,O	622
	ETHER, METHYL 1-PHENYLVINYL	0.06	0.005	0		50	D	1409
	ETHER, OCTADECYL VINYL	0.85	0.05	0.0		60		572
	ETHER, OCTYL VINYL	0.81	0.05	0.0	0.2	50	C	13
	ETHER, PHENYL VINYL	2.5		0.23	0.2	50	C	13
	ETHERS, C8-C16 ALKYL VINYL	7.0				60	C,P	1180
	ETHYLENE	0.024	0.003	0		50	E,I2	955
		0.014	0.002	0		20		309
	ETHYLENE, 1,1-BIS(P-CHLOROPHENYL)-	0.94	0.02	0.007	0.003	60	C	214
	ETHYLENE, 1,1-BIS(P-METHOXYPHENYL)-	2.25		0.055		60	C	214
	ETHYLENE, BROMO-	2.7		0.04		0	F,K	1098
		2.8		0.04		60		1098
	ETHYLENE, CHLORO-	3.28	0.7	0.02	0.03	40	H38	62
		3.6	0.5	0.052	0.02	60		817
		3.7	0.06	0.074	0.02	50		560
		4.0	0.2	0.040	0.009	38		958
							I26	165
							E,I7	1008
	ETHYLENE, 1,1-DICHLORO-	0.44	0.05	0.40	0.05	40	I31/I60	215
		0.48	0.02	0.95	0.02	40	I14/I31	215
		0.50	0.03	0.76	0.03	40	I31/I45	215
		0.57	0.01	0.69	0.02	40	I11/I58	215

REMARKS PAGE II-366, REFERENCES PAGE II-368

M1	M2	R1	+/-	R2	+/-	TEMP.	REMARKS	REF.
ACRYLONITRILE	ETHYLENE, 1,1-DICHLORO-	0.58	0.02	0.42	0.03	40	I11/I14	215
		0.585		0.390		32,8	K	589
		0.63	0.02	1.70	0.03	40	I11/I15	215
		0.63	0.03	1.43	0.02	40	I31/I58	215
		0.65	0.03	0.95	0.03	40	I31/I61	215
		0.65	0.03	0.49	0.03	40	I11/I59	215
		0.65	0.03	1.30	0.02	40	I31/I59	215
		0.66	0.03	1.74	0.02	40	I12/I31	215
		0.70	0.03	1.80	0.03	40	I15/I31	215
		0.73	0.03	1.39	0.03	40	I11/I12	215
		0.91	0.10	0.37	0.10	60	C	558
		1.20		0.49		45	C,I14,L	764
	ETHYLENE, 1,2-DIMETHOXY-	0.65	0.05	0.01	0.01	60		434
	ETHYLENE, 1,1-DIPHENYL-	0.028	0.003	0		60	C	211
		0.028	0.003	0		60	C	214
	ETHYLENE, FLUORO-	24	2	0.001		30	H38	1165
		24	2	0.001		30	H38,I38	879
	ETHYLENE, NITRO-	44.0	0.2	0.005	0.2	25	E,I7	1007
	ETHYLENE, TETRACHLORO-	0.01	0.01	65	15	78	E,I5	1060
	ETHYLENE, TRICHLORO-	456		0		60	C	1310
		470	70	0		75	N,O	210
	ETHYLENE GLYCOL, DIMETHACRYLATE	62.1	3.5	0		60	C	1311
	ETHYLENE OXIDE	67		0		60	N,Q	620
		0.22	0.01	1.24	0.10	60		71
		0		0.2		70	HII,R	230
		0		5.1		70	H2O,R	230
		12		0		70	C,R	230
	FATTY ACIDS, C6-C18, VINYL ESTERS	4.0	0.3	0.04	0.03	60	C,I6,5	58
	FERROCENE, VINYL-	0.16	0.05	0.15	0.05	70	I7	1268
	FORMIC ACID, VINYL ESTER	0.04	0.05	3.0		70	Q	775
	FUMARIC ACID, 1,3-BUTYLENEGLYCOL POLYESTER	3.0	0.2	0.04	0.005	60	C	166
	FUMARIC ACID, DIETHYL ESTER	1.03		1.12	0.40	60	N,O	908
	FURAN, 2-METHYL-	8		0		60	E,Q	620
	FURAN, 2-VINYL-	1.893		0.044		60	C	1005
	D-GLUCITOL, 1:2:3,4:5,6-TRI-O-ALLYLIDENE-	0.037	0.004	0.82	0.06	70	C	74
	D-GLUCITOL, 1:2:3,4:5,6-TRI-O-(2-BUTENYLIDENE)-	0.7		0.26		80	C	1350
	D-GLUCOFURANOSE, 1:2:5,6-DI-O-ISOPROPYLIDENE-, 3-METHACRYLATE *	0.3		0.36		80	C	1350
		0.3		0.36		80	C	1350
	D-GLUCOSE, METHACRYLOYL-	0.25	0.15	0.07	0.03	50	I6	498
	GLYCINE, N-ALLYL-	3.9	0.025	4.1	0.125	50	C,I36	1372
	1,5-HEXADIEN-3-YNE	0.03	0.3	0.04	0.03	60	I6	840
	HEXANOIC ACID, VINYL ESTER	4.0	2	0.01	0.01	30	C,I6	58
	HEXANOIC ACID, 2-ETHYL-, VINYL ESTER	12		4.05		70		166
	1,3,5-HEXATRIENE, 2,3,4,5-TETRACHLORO-, TRANS-	12.2	2.4	0		60		17
	1-HEXENE	5.09	0.96	-0.67	2.17	60	C	212
	1-HEXENE, 1-PHENYL-, CIS-	1.09	0.03	-0.09	0.29	60	C	1065
	1-HEXENE, 1-PHENYL-, TRANS-	0.17	0.01	0.63	0.4	70	C	1065
	1-HEXEN-3-YNE	5.4	0.3	0		70		784
	5-HEXEN-3-YN-2-OL, 2-METHYL-	0		0		60	C,P	607
	1-HEXYNE	0		3.64		60		212
	HYDRACRYLIC ACID, BETA-LACTONE					25	G10,I2	1243
						30	G10,I6	1243

M1 = ACRYLONITRILE

M2	R1	+/-	R2	+/-	TEMP.	REMARKS	REF.
HYDRACRYLIC ACID, BETA-LACTONE	0.84		5		40	G10,I6	1243
	1.2		0		50		1244
	0.33	0.08	0.23	0.08	-60	G24,I6	1243
HYDRAZINE, 1-ACRYLOYL-2-BENZOYL-	0.33	0.08	0.23	0.08	60	I6	950
HYDRAZINIUM HYDROXIDE, 2-METHACRYLOYL-1,1,1-TRIMETHYL-, INNER SALT *	0.03		0.16		60	I6	950
INDENE	0.19	0.03	0.45	0.06	60	Q	1070
ISOCYANIC ACID, VINYL ESTER	0.03	0.03	0.29	0.05	50		420
ISOPRENE	0.05	0.02	1.4	0.02	50		352
	0.36		0.04		60		353
ISOTHIOCYANIC ACID, VINYL ESTER	4.0	0.3	0		60	C,I6	1070
LAURIC ACID, VINYL ESTER	5		0		60		58
LINOLEIC ACID, METHYL ESTER	12		0		60		618
LINOLENIC ACID, METHYL ESTER	6		-0.604		60		618
MALEIC ACID, DIETHYL ESTER	0.527	1.8	-0.5	0.4	70	N,O	620
MALEIC ANHYDRIDE	6.2		2.0		70		620
MALEIMIDE, N,2-DIMETHYL-	0.06	0.04	1.7	0.3	30	C	553
MALONONITRILE, BENZYLIDENE-	0.15		2.5		70	C	1374
METHACRYLALDEHYDE	0.093	0.03	0.515	0.2	70	C	315
METHACRYLANILIDE	9.55	0.47	0.810	0.10	-40	C,I11	525
METHACRYLIC ACID	0.206	0.02	1.15	0.02	60	I7,P	16
	1.055		1.08		25	C,I62	886
METHACRYLIC ACID, ALLYL ESTER	0.31		0.04		60	G17,I5	197
METHACRYLIC ACID, BENZYL ESTER	1.10	0.20	1.3	0.01	-40	E	63
METHACRYLIC ACID, BUTYL ESTER	0.14	0.02	1.6	0.2	60	C	83
METHACRYLIC ACID, 2-BUTYNYL ESTER	0.15	0.1	1.85	0.1	70		679
METHACRYLIC ACID, 2-CHLOROETHYL ESTER	0.19	0.08	1.32	0.1	70	G2,I15	196
METHACRYLIC ACID, 2-(DIETHYLAMINO)ETHYL ESTER	0.14	0.001	1.04	0.03	60	C	64
METHACRYLIC ACID, 2,3-EPOXYPROPYL ESTER	0.21		0		70	P,Q	943
METHACRYLIC ACID, ISOBUTYL ESTER	4.1	0.04	1.44	0.15	50	I12	1109
METHACRYLIC ACID, NOVOLAK RESIN ESTER	0.19		1.75		60		1109
METHACRYLIC ACID, PENTAMETHYLDISILOXANYL METHYL ESTER	0.15	0.03	0.710	0.01	60	C	414
METHACRYLIC ACID, 2-PROPYNYL ESTER	0.154		1.01		25	I6	679
METHACRYLIC ACID, TETRAHYDROFURYL ESTER	0.477	0.05	1.01	0.20	25	E	984
METHACRYLIC ACID, 2,4,8,10-TETRAOXASPIRO(5.5)UNDEC-3,9-YLENE* DIETHYLENE ESTER	0.477		0.21		60	E	632
METHACRYLIC ANHYDRIDE	1.94	0.05	2.68	0.20	60		1306
METHACRYLONITRILE	0.32		1.77		60		63
	0.33		1.69		70	16	83
	0.43	0.05	0.50	0.05	-50	C,J	83
	0.9	0.1	0.05	0.02	-50	H21,I12	66
	1.0		2.8	0.4	50	G	724
METHACRYLOYL CHLORIDE	0.35		0.13		-78	I7	354
METHYL METHACRYLATE	0.01		1.20		26	E	803
	0.03		0.85		26	H36	204
	0.09	0.05	1.01			H40	204
	0.10		1.30		-20/-52	I16,Q	609
	0.100	0.070	1.351	0.133	40	E	918
						C	456

M1 — ACRYLONITRILE

M2	R1	+/-	R2	+/-	TEMP.	REMARKS	REF.
METHYL METHACRYLATE	0.12		1.34		-12	H38,124	1084
	0.12		1.34		12	H38	1084
	0.12		1.34		26	H39	1084
	0.12		1.34		26	H38	1084
	0.12		1.34		70		147
	0.12		1.34		90		1083
	0.13	0.05	1.16	0.22	60	H38	62
	0.14		1.25			Q	871
	0.15		1.45		-80	H38,12	1084
	0.15		1.05		26	H14	1084
	0.15	0.02	1.09	0.21	70	C,I62	886
	0.15	0.03	1.65	0.40	20	F,I11	1072
	0.15	0.07	1.20	0.14	60	C	558
	0.150	0.080	1.224	0.100	80	C	456
	0.160	0.100	1.186	0.120	100	C	456
	0.37		1.46		70	M4	147
	0.45		1.46		90	M4	1084
	0.45		2.01		70	M3	147
	0.46		2.01		90	M3	1083
	0.9	0.08	1.14	0.02	100	C,I17	448
	1.3		0.32		-60	G	65
	1.5	0.1	0.05	0.02	-40	H22,12	204
	2.0	0.1	0.10	0.03	-15	H21,12	204
	3.0	0.2	0.06	0.01	20	H21,12	204
	4.0	0.1	0.03	0.01	-30	H21,12	204
	4.5	0.5	0.02	0.02	-78	H22,12	204
	5	0.5	0.07	0.01	-50	H21,12	204
	5		0.14		-12	H42	1084
	5.0		0.14		-12	G2	1084
	6.5	0.5	0.03	0.01	-78	H21,12	204
	7	0.5	0.02		-78	H21,12	204
	7		0.40		-12	H41	1084
	7.0		0.39		-8	G17	1084
	7.9	1.6	0.25	0.25	-78	E,I8	636
	<1		0.05		-30	G3	551
	>1		>1		20	H	551
	>1		>1		-55	G3,P	265
	10				-80	H42,I5	1084
MYRISTIC ACID, VINYL ESTER	4.0	0.3	0.30	0.03	60	C,I6	58
NAPHTHALENE, 1-(3-BUTENYL)-	3.6		0.04		70		67
NAPHTHALENE, 2-ISOPROPENYL-	0.05	0.01	0.23	0.02	99		213
NONANOIC ACID, VINYL ESTER	3.57	0.16	0.059	0.095	60	C	591
2,5-NORBORNADIENE	0.65	0.02	0.47	0.08	50	C	248
	0.67		0.08		50		1082
	0.67		0.08		100		1081
5-NORBORNENE-2-CARBOXYLIC ACID, 2,3-EPOXYPROPYL ESTER	1.4		0		60		416
5-NORBORNENE-2-CARBOXYLIC ACID, METHYL ESTER	1.5	0.5	0.2	0.1	60		458
5-NORBORNENE-2-CARBOXYLIC ACID, 2-METHYL-, 2,3-EPOXYPROPYL ESTER *	1.5	0.5	0.2	0.1	60		458
2-NORPINANE-2-ETHANOL, ACRYLATE	1.1	0.1	0.9	0.1	60	C	600
10,12-OCTADECADIENOIC ACID, TRANS,TRANS-, METHYL ESTER	0.4		0.0		60		618
OCTADECANOIC ACID, 12-OXO-, VINYL ESTER	3.11	0.05	0.0	0.02	60	C	593
OCTADECANOIC ACID, 9(10)-OXO-, VINYL ESTER	2.96	0.12	0.0	0.091	60	C	593

M1	M2	R1	+/-	R2	+/-	TEMP.	REMARKS	REF.
ACRYLONITRILE	OCTANOIC ACID, VINYL ESTER	4.00	.3	0.04	0.03	60	C,I6	58
	OXALIC ACID, ETHYL VINYL ESTER	2.0		0.2		60		479
	PALMITIC ACID, VINYL ESTER	4.0	0.3	0.04	0.03	60	C,I6	58
	1,4-PENTADIENE	1.116		<0.01		60		88
	1,4-PENTADIENE, 3,3-DIMETHYL-	3.31		0		50	CC	155
	2,4-PENTADIENENITRILE	0.10	0.03	40	3	50	I6	840
	2,4-PENTANEDIONE, 3-ETHYLIDENE-	15		0		50	I7	1339
	1-PENTENE, 5-CYCLOHEXYL-	4.5		0		70		67
	1-PENTENE, 2-METHYL-	1.29		0		60	C,J	497
	1-PENTENE, 3-METHYL-1-PHENYL-, CIS-	22.6	1.0	-0.34	0.43	60		1065
	1-PENTENE, 3-METHYL-1-PHENYL-, TRANS-	6.15	0.50	-0.47	0.33	60		1065
	1-PENTENE, 1-PHENYL-, CIS-	3.04	0.24	-0.05	0.51	60		1065
	1-PENTENE, 1-PHENYL-, TRANS-	1.04	0.03	-0.002	0.22	60		1065
	1-PENTENE, 5-PHENYL-	3.5		0		70		67
	1-PENTENE, 2,4,4-TRIMETHYL-	1.22		0		60	C,J	497
	2-PENTENOIC ACID, 4-HYDROXY-, GAMMA-LACTONE	113	14	0		60	Q	421
	3-PENTENOIC ACID, 4-HYDROXY-, GAMMA-LACTONE	14.5	0.5	0		60	Q	421
	PHENETHYL ALCOHOL, ALPHA-METHYL-P-VINYL-	0.10		0.53		60		42
	PHOSPHONIC ACID, ALLYL-, DIETHYL ESTER	4.3		0.32		40		889
	PHOSPHONIC ACID, ALPHA-PHENYLVINYL-	0.69	0.18	0.32	0.07	70	C	524
	PHOSPHONIC DICHLORIDE, VINYL-	6		0.05		25	E	1192
	PHTHALIC ACID, DIALLYL ESTER	2.54	0.16	0.049	0.004	80	C	612
		3.72	0.15	0.057	0.004	60	C	612
	2-PICOLINE, 5-VINYL-	0.10	0.05	1.10	0.2	60	C	1059
		0.16	0.003	2.3	0.04	60	Q	940
	2-PICOLINIUM CHLORIDE, 5-VINYL-	0.15		2.7		60	F,I11,Q	235
	1-PROPANESULFONIC ACID, 3-HYDROXY-, ACRYLATE, SODIUM SALT	1.8	0.02	0.55	0.01	50	I17	1373
	1-PROPANESULFONIC ACID, 3-HYDROXY-, METHACRYLATE, SODIUM SALT *	1.8	0.02	0.55	0.01	50	I17	1373
	1-PROPENE	0.8		0.1		60	H10/I4,Q	480
	1-PROPENE, 1-CHLORO-, CIS-	29.5		0.005		60		745
	1-PROPENE, 1-CHLORO-, TRANS-	18.0		0.01		60		745
	1-PROPENE, 2-CHLORO-	1.4		0.01		60		745
	1-PROPENE, 3-CHLORO-	1.0		0		0	D,L3	389
		3.0	0.2	0.05	0.01	60	C	166
		4.5		0.1		0	D	389
	1-PROPENE, 3-CHLORO-2-METHYL-	5.5		0.0		60		620
	1-PROPENE, 2-METHYL-	1.14	0.04	0.0		80	I7,L	6
		1.26	0.04	0.0		60	I7,L	6
		1.28	0.04	0.0		60	I6	6
	1-PROPENE, 3,3,3-TRICHLORO-	0.80		0.00		60		865
		1.8	0.2	0.02	0.02	50	H38	724
	2-PROPENE-1-SULFONAMIDE, N-(2-CHLOROETHYL)-	12.2	1.2	0.100	0.015	60		60
	2-PROPENE-1-SULFONIC ACID	2.8		0.43		60		211
		1.85	0.01	0.18	0.01	45	I17	417
	2-PROPENE-1-SULFONIC ACID, 2,3-EPOXYPROPYL ESTER	3.1		0.38		60		441
	2-PROPENE-1-SULFONIC ACID, SODIUM SALT	0.69	8.05	0.28	0.05	60	I17	417
		1.00	0.01	-0.07	0.02	60	I17	648
		1.25	0.01	-0.10	0.02	45	I40	441
		4.94	0.06		0.06	45	F,J	441
	1-PROPENE-1,2,3-TRICARBOXYLIC ACID, TRANS-, TRIMETHYL ESTER	5.50	0.50	-0.10	0.10	30	C,I6	595
	1-PROPEN-2-OL, ISOCYANATE	0.24	0.02	-0.10	0.11	60	Q	419

REMARKS PAGE II-366, REFERENCES PAGE II-368

M1	M2	R1	+/-	R2	+/-	TEMP.	REMARKS	REF.
ACRYLONITRILE								
	PROPIONALDEHYDE, 3-(ALLYLOXY)-	3.9		0.01		-78	H21	1287
		5		0.015		60		1287
	2-PROPYN-1-OL, ACETATE	6.15		0.05		60		1305
	2H-PYRAN-3-CARBOXYLIC ACID, 5,6-DIHYDRO-2,6-DIMETHYL-, VINYL* ESTER	6.15		0.05		60		1305
	3(2H)-PYRIDAZINONE, 2(P-CHLOROPHENYL)-6-HYDROXY-	15		0		60	I17	1290
	3(2H)-PYRIDAZINONE, 6-HYDROXY-2-PHENYL-	12		0		60	I17	1290
	3(2H)-PYRIDAZINONE, 6-HYDROXY-2-(P-TOLYL)-	12		0		60	I17	1290
	PYRIDINE, 5-ETHYL-2-VINYL-	0.02	0.02	0.43	0.05	60		422
	PYRIDINE, 2-VINYL-	0.05	0.01	21.9	5.52	60	P,Q	1080
		0.05	0.01	21.9	5.52	60		534
		0.086	0.002	2.13	0.03	60		533
	PYRIDINE, 4-VINYL-	0.113	0.005	0.47	0.09	60	C,I11,Q	1073
		0.77		0.83		60		422
	PYRIDINE, 4,4'-VINYLENEDI-	0.113		0.41		60	Q	871
	PYRIDINIUM FLUOROBORATE, 1-VINYL-	0.95	0.05	0.02	0.05	60		422
	SILANE, ALLYLDIMETHYLPHENYL-	1.06		0.20		50	C,K	603
	SILANE, ALLYLTRIETHOXY-	2.24		0.2		60	I31	223
	SILANE, ALLYLTRIMETHYL-	1.7		0.1		70	C	1317
	SILANE, 2-BUTENYLTRIETHOXY-	3.98		0		50	C	957
	SILANE, (1-CHLOROVINYL)TRIETHOXY-	10.0		0		70	C	1317
	SILANE, (1-CYCLOHEXEN-4-YL)TRIETHOXY-	0.7		0		50	C	957
	SILANE, DIETHOXYETHYLVINYL-	12.0		0		50	C	957
	SILANE, DIETHOXYMETHYLVINYL-	9.0		0		50	C	957
	SILANE, DIETHOXYPHENYLVINYL-	6.0		0		50	C	957
	SILANE, DIMETHYLPHENYLVINYL-	8.3		0.18		50	C	957
	SILANE, TRIETHOXY(5-NORBORNEN-2-YL)-	3.84		0		50	C	957
	SILANE, TRIETHOXYPROPENYL-	0.7		0		70	C	957
	SILANE, TRIETHOXYVINYL-	20.0		0		50	C	957
		4.5		0		50	C	1316
		5	1	0		50	C	957
	SILANE, TRIISOPROPOXYVINYL-	6.5		0		60	C	957
	SILANE, TRIMETHOXYVINYL-	6.0		0.08		50	C	957
	SILANE, TRIMETHYLVINYL-	3.85		0.07		50	C	849
		3.9		0.04		70	C	957
	STEARIC ACID, VINYL ESTER	4.0	3	0.064	0.005	60	C,I6	58
		4.20	0.3	0.04	0.03	60	C	591
	STERAMIDE, N-ALLYL-	4.3	0.02	0.03		70		1050
	2,2'-STILBENEDISULFONIC ACID, 4,4'-BIS(3-PHENYLUREIDO)-, DI * SODIUM SALT	3.61	0.087	0.118	0.084	80	C,I14	454
		3.61	0.087	0.118	0.084	80	C,I14	454
	STYRENE	0.00		0.33	0.03	75	R	319
		0.02	0.02	0.3	0.08	-78	E	919
		0.02	0.02	0.45	0.08	50		560
		0.02	0.02	0.47	0.03	40	H38	62
		0.03	0.03	0.33		65		297
		0.03	0.03	0.36	0.04	86,5		297
		0.03		0.52	0.08	0/15	E	919
		0.03		0.41	0.08	20	E	167
		0.04	0.04	0.40	0.05	67/80	K	258
						75	C	259
						60	C	212

COPOLYMERIZATION PARAMETER

M1	M2	R1	+/-	R2	+/-	TEMP.	REMARKS	REF.
ACRYLONITRILE	STYRENE	0.04	0.04	0.41	0.08	60	C	558
		0.05	0.01	0.4	0.1	30	D	1102
		0.05	0.02	0.38	0.03	41,5		297
		0.058	0.02	0.398	0.02	50		352
		0.06	0.01	0.39		70	C,J	803
		0.07		0.37	0.02	99		213
		0.07	0.006	0.37		90		1083
		0.13		0.43	0.03	90		958
		0.14		0.55		50	M4	1083
		0.17		0.09		90	M3	1083
		0.21	0.01	0.62	0.02	90	E,I15	833
		0.28		0		40	G22/G21	1120
		0.35	0.04	0.05	0.02	110	E	919
		0.385	0.015	0.45	0.01	-20	I15,L	285
		0.41	0.03	0.03	0.02	65	G22/G21	1120
		1.5		0.2		130	I15,L	285
		2		0.5		65	H12,I21	847
		5		2		0	H12,I21	847
		12		0.7		80	H7,I21	847
		15	1	0.05	0.02	80	H7,I21	847
		20	2	0.02	0.01	0	H21,I2	204
		33		0.005		-45	H21,I2	204
		33.0		0.005		-70/0	E,I6	2
						-78	E,I6	635
	STYRENE, 2,5-DICHLORO-	0.22	0.05	0.07	0.05	67,5		297
		0.25	0.11	0.09	0.06	86,5		297
	STYRENE, ALPHA-METHYL-	0.26	0.02	0.11	0.02	38,5	A	1051
		0.03		0.17		20		1051
		0.04		0.20		50		713
		0.04		0.2		60		1051
		0.045	0.02	0.1	0.02	75	C,R	260
		0.06	0.03	0.25	0.06	70	I16,Q	609
	STYRENE, M-METHYL-	0.07		0.43		75	Q	40
	STYRENE, O-METHYL-	0.06		0.33			Q	40
	STYRENE, P-METHYL-	0.05	0.04	0.33	0.1		Q	40
	STYRENE, 2,4,6-TRIMETHYL-	0.98	0.2	0.16	0.02	138		207
	STYRENES, CHLOROMETHYL-	0.06		0.67		60		41
	SUCCINIC ACID, METHYLENE-	0.25		1.57		50	C,I11	676
	SUCCINIC ACID, METHYLENE-, DIPOTASSIUM SALT	0.43		0.10		50	C,I11	676
	SUCCINIC ANHYDRIDE, METHYLENE-	0.13		4.8			I7,Q	2
	SUCCINIMIDE, N-VINYL-	0.54		0.16			C	279
	SULFIDE, BUTYL VINYL	0.097		0.046		60		1338
	SULFIDE, TERT-BUTYL VINYL	0.13		0.01		60		1338
	SULFIDE, ETHYL VINYL	0.065		0.055		60	A	1225
	SULFIDE, ISOBUTYL VINYL	0.068		0.05		60		1338
	SULFONE, BUTYL VINYL	1.1	0.2	0.2	0.1	-78	G3	265
	SULFOXIDE, ETHYL VINYL	3		0.2		60		1226
	TAURINE, N-METHACRYLOYL-N-METHYL-, SODIUM SALT	0.75		1.25		60	C,I17	1435
	2,4,8,10-TETRAOXASPIRO(5.5)UNDECANE, 3,9-BIS(1-CHLOROVINYL)-	0.58		0.8		80	C,I6	1353
	2,4,8,10-TETRAOXASPIRO(5.5)UNDECANE, 3,9-DIISOPROPENYL-	0.89		0.04		80	C	1349
	2,4,8,10-TETRAOXASPIRO(5.5)UNDECANE, 3,9-DIPROPENYL-	4		0.04		80	C	1349

M1	M2	R1	+/-	R2	+/-	TEMP.	REMARKS	REF.
ACRYLONITRILE	2,4,8,10-TETRAOXASPIRO(5.5)UNDECANE, 3,9-DISTYRYL-	2.32		0.03		80	C,16	1353
	2,4,8,10-TETRAOXASPIRO(5.5)UNDECANE, 3,9-DISTYRYL-	2.3		0.056		60	C,16	1123
	2,4,8,10-TETRAOXOSPIRO(5.5)UNDECANE, 3,9-DIVINYL-	7.11		0.02		60	C,16	1345
	P-TOLUENESULFONAMIDE, N-METHYL-N-VINYL-	0.42		0		60	C	279
	P-TOLUENESULFONAMIDE, N-METHYL-N-VINYL-	0.42		0		60	C	280
	TRISILANE, 1,1,1,2,3,3,3-HEPTAMETHYL-2-VINYL-	2	1	0		60	I6	837
	TRISILOXANE, 1,1,3,5,5,5-HEPTAMETHYL-3-VINYL-	8.0	1.2	0.1	0.01	50		773
	TRISILOXANE, 1,1,5,5,5-HEXAMETHYL-3-TRIMETHYLSILOXY-3-VINYL- *	8.0	1.2	0.1	0.01	50		773
	9-UNDECENOIC ACID, VINYL ESTER	1.82	0.04	0.0	0.10	60	C	591
	VERSATIC 10 ACID, VINYL ESTER	5.22	0.20	0	0.10	60	I7	954
	VINYL ACETATE	3.88		0.009	0.013	25	P,Q	1071
	VINYL ACETATE	4.05	0.3	0.061		60	D	947
	VINYL ACETATE	4.2		0.05		50		622
	VINYL ACETATE	4.2		0.05		50	C,16	220
	VINYL ACETATE	5.6	0.5	0.03	0.03	40	C,16	219
	VINYL ACETATE	6	2	0.02	0.02	-196	H38	62
	VINYL ACETATE	6.0		0.2		60	C	260
	VINYL ACETATE	6.0		0.07		70	E	400
	VINYL ACETATE	6.0		0.07		50	C,J	497
	VINYL ACETATE	6.0		0.024		50		23
	VINYL ETHER	0.938	0.9		0.01	50		88
	VINYL SULFONE	1.94		0.364		50	CC	155
ACRYLONITRILE, TIN COMPLEX	ETHYLENE, CHLORO-	0.5		0.01		35	D	1428
ACRYLONITRILE, 2-CHLORO-	METHYL METHACRYLATE	0.43	0.03	0.17	0.03	55		302
	STYRENE	0.075	0.01	0.055	0.03	60		302
	STYRENE	0.08	0.02	0.1	0.03	80	C	134
	STYRENE	0.13		0.06		60		1430
	STYRENE	0.13		0.06		60		1429
ACRYLONITRILE, 2-HYDROXY-, ACETATE	ACRYLONITRILE	7.4	0.40	0.09	0.17	60		1070
	STYRENE	0.16	0.04	0.2	0.04	60	Q	727
	STYRENE	0.2		0.16		60		1430
	STYRENE	0.20	0.052	0.19	0.056	60		1070
ACRYLONITRILE, 2-METHOXY-	ACRYLONITRILE	1.93		0.37		60		1341
	STYRENE	0.35		0.53		60		1341
	STYRENE	0.35		0.35		60		1430
ACRYLOPHENONE	STYRENE	1.10	0.10	0.107		60		746

M1 / M2	R1	+/-	R2	+/-	TEMP.	REMARKS	REF.
ACRYLOYL CHLORIDE							
ACRYLIC ACID, METHYL ESTER	2.3		0.345		45		894
ACRYLONITRILE	1	0.25	1.2	0.25	50	I7	1375
ETHYLENE, CHLORO-	2.65		0.017		50	I24	1236
ETHYLENE, 1,1-DICHLORO-	1.12		0.5		50	I24	1236
METHYL METHACRYLATE	1.51		0.48		45		894
STYRENE	0.03	0.01	0.10	0.01	67	I7	892
ADIPIC ACID, DIALLYL ESTER							
STYRENE	0.087	0.007	35.5	0.9	60	C	1293
ADIPIC ACID, DIVINYL ESTER							
ACRYLIC ACID, METHYL ESTER	0.1	0.1	4	1	60	C,I10	877
ADIPIC ACID, METHYL VINYL ESTER							
STYRENE	0.02	0.01	49.5	0.7	80	C	268
VINYL ACETATE	0.66	0.04	0.57	0.04	80	C	268
BETA-ALANINE, N-ALLYL-							
ACRYLONITRILE	0.09	0.03	4.25	0.15	50	C,I36	1372
ALLYL ALCOHOL							
ACRYLONITRILE	0		0.6		0	D,L3	389
	0.01		1.6		20	F,I36	1303
	0.05		2.6		20	D	389
	0.1		5.5			E,Q	983
	0.11	0.10	3.96	0.53	25	D	736
	1.99		0.03		RT		190
METHACRYLIC ACID, ETHYL ESTER			108	5	60		855
METHYL METHACRYLATE			91	5	60		855
ALLYL ESTERS							
METHYL METHACRYLATE						P,Q	854
ALLYL ETHER							
ACRYLONITRILE	0.01	0.01	4.9	0.3	80		72
2-PICOLINE, 5-VINYL-	0		80		70		72
ALLYL ETHYL PHOSPHITE							
FUMARIC ACID, 1,3-BUTYLENEGLYCOL POLYESTER	0.035		5.5		80		906
	0.035	0.035	5.50	2.50		Q	604
AMMONIUM CHLORIDE, ALLYLDIMETHYLDODECYL							
ACRYLONITRILE	0.4		13.1		60	I17	1246
	0.5		14.5		60	C,J	1246

REMARKS PAGE II-366, REFERENCES PAGE II-368

M1	M2	R1	+/-	R2	+/-	TEMP.	REMARKS	REF.
AMMONIUM CHLORIDE, ALLYLTRIETHYL-								
	ACRYLONITRILE	0.03		18.8		60	C,J	1246
		0.8		8.3		60	I17	1246
AMMONIUM CHLORIDE, DIETHYL(2-HYDROXYETHYL)-, METHACRYLATE								
	ACRYLONITRILE	0.85	0.15	0.13	0.1	70	I12	1109
		0.9	0.2	0.24	0.05	70	I12	1109
	STYRENE	0.25	0.05	0.3	0.02	70	I12	1109
		0.55	0.1	0.43	0.1	70	I12	1109
ANILINE, N,N-DIMETHYL-P-VINYL-								
	METHYL METHACRYLATE	0.11	0.02	0.205	0.02	60		1024
	STYRENE	0.62		0.89				1040
		0.62		0.89		60	Q	1168
		0.84	0.05	1.015	0.06	60		1024
	STYRENE, ALPHA-METHYL-	31	19	0.035	0.015	5	H20,I32	752
ANILINE, N,N-DIMETHYL-P-(P-VINYLPHENYLAZO)-								
	STYRENE	3.8		0.4		60		125
ANILINE, N,N-DIVINYL-								
	ACRYLONITRILE	0.003		0.42		60		162
	METHYL METHACRYLATE	0.01		2.0		60		162
	STYRENE	0.45		13.0		60		162
	STYRENE, P-METHYL-	0.05		11.8		60		162
	VINYL ACETATE	2.0		0.1		60		162
ANILINE, N-(ALPHA-PHENYL-P-VINYLBENZYLIDENE)-								
	STYRENE	1.88	0.08	0.36	0.02	60		125
ANILINE, N-VINYLOXY-								
	METHYL METHACRYLATE	0.75	0.05	0.07	0.02	60		874
O-ANISALDEHYDE								
	STYRENE	0		0.1		0	H10	1402
P-ANISALDEHYDE								
	STYRENE	0	0.01	0.1	0.02	0	H10	1402
P-ANISIC ACID, VINYL ESTER								
	BENZOIC ACID, VINYL ESTER	0.92		1.28		60		397
	BENZOIC ACID, P-BROMO-, VINYL ESTER	0.82		1.19		60		397
	BENZOIC ACID, P-CHLORO-, VINYL ESTER	0.73		0.92		60		397
	BENZOIC ACID, P-CYANO-, VINYL ESTER	0.75		0.93		60		397
	VINYL ACETATE	1.45		0.78		30		397

COPOLYMERIZATION PARAMETER

M1	M2	R1	+/-	R2	+/-	TEMP.	REMARKS	REF.
ANISOLE, P-PROPENYL-/MALEIC ANHYDRIDE								
	ACRYLONITRILE	45	2.9	0.007	0.0024	60	I7	1259
	ETHYLENE, 1,1-DICHLORO-	53.5	6.1	0.006	0.0055	60	I7	1259
	METHACRYLIC ACID, 2-CHLOROETHYL ESTER	11.5	1.2	0.095	0.014	60	I7	1259
	METHACRYLONITRILE	29.8	2.6	0.028	0.008	60	I7	1259
ANISOLE, P-3-BUTENYL-								
	ACRYLONITRILE	0		3.0		70		67
ANISOLE, P-(DIAZOMETHYL)-								
	TOLUENE, ALPHA-DIAZO-	1.2	0.1	0.62	0.1	40	174	1396
ANISOLE, P-ISOPROPENYL-								
	ETHER, 2-CHLOROETHYL VINYL	1.1		0.42		RT	H20,I7	592
		1.3		0.73		RT	H20,I22	592
ANISOLE, P-PROPENYL-								
	ANISOLE, P-VINYL-	0.04	0.02	1.2	0.2	30	H10,I24	652
	INDENE	0.3		1.6		0	H23	1110
	STYRENE	2.8	0.2	0.45	0.2	30	H10,I24	652
	STYRENE, O-CHLORO-	0	0.01	22	8	70	C	21
		18	3	0.03	0.005	0	H20,I32	21
ANISOLE, P-PROPENYL-, CIS-								
	ANISOLE, P-PROPENYL-, TRANS-	1.02	0.08	0.29	0.04	0	H10,I23	1215
		2.96	0.05	0.72	0.02	0	H10,I2	1215
	ANISOLE, P-VINYL-	0	0.02	9.03	0.65	-78	H10,I2	1215
		0	0.02	4.42	0.27	-78	H10,I123	1215
		0	0.02	3.6	0.23	-78	H19,I23	1215
		0	0.03	7.73	0.57	0	H10,I2	1215
	STYRENE	8.24	0.86	0.42	0.12	0	H10,I23	1215
ANISOLE, P-PROPENYL-, TRANS-								
	ANISOLE, P-PROPENYL-, CIS-	0.29	0.04	1.02	0.08	0	H10,I123	1215
		0.72	0.02	2.96	0.05	0	H10,I2	1215
	ANISOLE, P-VINYL-	0	0.01	3.17	0.05	-78	H10,I2	1215
		0	0.07	1.4	0.1	-78	H10,I123	1215
		0.01	0.02	2.11	0.24	0	H10,I2	1215
		1.56	0.11	1.71	0.09	-78	H19,I2	1215
	STYRENE			0.13	0.04	0	H10,I123	1215
ANISOLE, M-VINYL-								
	STYRENE	1.1	0.15	0.90	0.15	0	H20,I33	752
	STYRENE, P-CHLORO-	2.6	0.4	0.38	0.05	0	H20,I33	752
	STYRENE, ALPHA-METHYL-	0.3	0.1	5	1	5	H20,I33	752

REMARKS PAGE II-366, REFERENCES PAGE II-368

M1	M2	R1	+/-	R2	+/-	TEMP.	REMARKS	REF.
ANISOLE, O-VINYL-								
	ACRYLIC ACID, METHYL ESTER	2.6		0.04		40	H10,I28	1186
	ANISOLE, P-VINYL-	0.35	0.03	2.9	0.7	30	H10,I28	382
		0.35	0.09	3.9	0.1	-20	H10,I28	382
		0.45	0.05	6.4	0.4	-50	H10,I28	382
	ETHYLENE, 1,1-DIPHENYL-	0		0		0	G17,I5	1449
	STYRENE	20				40	G17,I7	1449
		3.6	0.8	0.11	0.04	-20	H10,I28	382
		3.9	0.7	0.20	0.02	30	H19,I28	382
ANISOLE, P-VINYL-								
	ACRYLIC ACID, METHYL ESTER	2	0.2	0.07	0.02	40	H10,I24	1186
	ANISOLE, P-PROPENYL-	1.2		0.04	0.02	30	H19,I23	652
	ANISOLE, P-PROPENYL-, CIS-	3.6	0.23	0	0.02	-78	H10,I23	1215
		4.42	0.27	0	0.03	-78	H10,I23	1215
		7.73	0.57	0	0.02	-78	H10,I2	1215
		9.03	0.65	0	0.03	-78	H10,I23	1215
	ANISOLE, P-PROPENYL-, TRANS-	1.4	0.1	0	0.02	-78	H19,I2	1215
		1.71	0.09	0.01	0.07	-78	H10,I2	1215
		2.11	0.24	0	0.01	0	H10,I2	1215
		3.17	0.05	0		-78	H10,I2	1215
	ANISOLE, O-VINYL-	2.9	0.07	0.35	0.03	30	H10,I28	382
		3.9	0.1	0.35	0.09	-20	H10,I28	382
		6.4	0.4	0.45	0.05	-50	H10,I28	382
	BENZONITRILE, P-VINYL-	0.093	0.01	0.85		30		505
	CARBAZOLE, 9-VINYL-	0.09	0.005	23.2	1.9	0	H10	1147
		0.13		21.4	0.81	25	H10	1147
	ETHER, 2-CHLOROETHYL VINYL	1.55		9.32		0	H10,I2	1284
		2.36		5.06		-36	H10,I2	1284
		4.37	0.3	2.81	0.3	-78	H10,I2	1285
		4.55	0.35	3.08	0.26	-78	H10,I3	1285
		6.93	0.5	1.73	0.19	-78	H19,I2	1285
		7.8	0.2	1.56	0.04	-78	H19,I3	1285
		9.25		1.63		-78	H10,I3	1285
		11.		2		0	H31,I24	1284
		11.		2		30	H31,I24	470
		12.1		1.24		30	H19,I3	471
	ETHYLENE, 1,1-DIPHENYL-	0.3		0		0	G17,I5	1284
		0.32		0.29	0.03	40	G17,I7	1449
	METHYL METHACRYLATE	0.32	0.05	0.29		60		115
	STYRENE	0.045		19		60	G18,I5	1024
		0.05	0.02	10.9	0.8	25	G17,I2	862
		0.13	0.02	4.1	0.5	0	G18,I5	965
		0.23	0.02	2.9	0.2	1	G4,I5	965
		0.79		1.05		0	Q	965
		0.82		1.05		60		1040
		0.85	0.07	1.16	0.09	60		1168
		0.93		1.0		30		1024
		1.50		1.13		60	E	505
		5.6	0.22	0.70	0.10	25	631	
		5.6		0.48		0	H23,I22	149
		7.6	0.08	0.22	0.08	0	H23,I22	965

COPOLYMERIZATION PARAMETER

M1	M2	R1	+/-	R2	+/-	TEMP.	REMARKS	REF.
ANISOLE, P-VINYL-	STYRENE	11	1	0.34	0.05	0	H46,I33	965
		11.5	0.7	0.38	0.04	0	H23,I33	965
		12	2	-0.33	0.03	0	H23,I2	965
		14	3	-0.12	0.07	0	H23,I33	965
		19	3	-0.04	0.04	0	H20,I33	965
		29	5	-0.02	0.07	0	H20,I33	965
		31	6	-0.00	0.03	0	H23,I32	965
		35	5	-0.12	0.012	0	H23,I33	965
		38		0.5		30	H10,I28	382
		46	3	0.05	0.04	0	H23,I32	965
		72		0.01		0	H23,I2	965
		100		0.01		0	H20,I32	752
	STYRENE, M-BROMO-	0.25		1.4		30		699
	STYRENE, P-BROMO-	0.43		1.1		30		505
	STYRENE, M-CHLORO-	0.20		1.9		30		699
	STYRENE, P-CHLORO-	0.48		0.70		30		505
	STYRENE, ALPHA-METHYL-	0.58	0.03	0.86	0.08	60		1024
	STYRENE, P-METHYL-	15	5	0.30	0.1	5	H20,I32	752
		0.68	0.06	2.2	0.1	0	G18,I5	965
		0.72	0.02	1.93	0.03	0	G4,I5	965
		1.54	0.10	0.52	0.06	0	H23,I22	965
		1.9	0.3	0.14	0.14	0	H20,I33	965
		3.7	0.40	-0.03	0.07	0	H23,I29	965
		4.3		0.3		30	H31,I24	470
		4.3		0.3		30	H31,I24	471
		10.		0.1		30	H31,I32	471
ANTHRACENE, 1-VINYL-	STYRENE	0.57	0.04	0.81	0.03	65		477
ANTHRACENE 9-VINYL-	ACRYLIC ACID, BUTYL ESTER	0.163		3.76		78	C	478
	ACRYLIC ACID, ETHYL ESTER	0.274		3.43		78	C	478
	ACRYLIC ACID, METHYL ESTER	0.082		2.97		76	C	478
	METHYL METHACRYLATE	0.071		3.81		78	C	477
	STYRENE	0.25		2.12		100		477
	STYRENE	0.3		2.2		70		172
ARSINE, DIPHENYL(P-VINYLPHENYL)-	STYRENE	1.18	0.04	0.46	0.03	65	I2	126
ATROPIC ACID, BUTYL ESTER	STYRENE	0.23	0.03	0.09	0.03	60		748
ATROPIC ACID, 2-CHLOROETHYL ESTER	STYRENE	0.30	0.03	0.06	0.03	60		748

M1 / M2	R1	+/-	R2	+/-	TEMP.	REMARKS	REF.
ATROPIC ACID, ETHYL ESTER							
METHYL METHACRYLATE	0.03	0.02	0.31	0.03	65		175
STYRENE	0.19	0.03	0.04	0.03	60		748
	0.19	0.04	0.04	0.03	65		175
ATROPIC ACID, METHYL ESTER							
ACRYLIC ACID, METHYL ESTER	1.0		0.055		60		571
ACRYLONITRILE	6.7		0.075		60		571
METHACRYLONITRILE	0.25		0.19		60	R	573
METHYL METHACRYLATE	0		0.21		60		571
STYRENE	0	0.03	0.3	0.1	65		175
	0.4	0.2	0.03	0.02	65		174
	0.45	0.03	0.06	0.03	60		748
	1.0		0.055		60		571
ATROPIC ACID, PROPYL ESTER							
STYRENE	0.19	0.03	0.08	0.03	60		746
ATROPONITRILE							
STYRENE	0.7		0.02		80	C	538
2H-AZEPIN-2-ONE, HEXAHYDRO-							
2-PIPERIDONE	0.64		0.16		90	17,Q	762
2-PYRROLIDINONE	0.75		5.0			G18	514
2H-AZEPIN-2-ONE, HEXAHYDRO-7-(1-CYCLOHEXENYL)-							
STYRENE	0.36		2		60	17	1332
2H-AZEPIN-2-ONE, HEXAHYDRO-1-METHACRYLOYL-							
STYRENE	0		1.86	0.10	130	C,I	270
	0		1.0		70/80		1031
2H-AZEPIN-2-ONE, HEXAHYDRO-1-VINYL-							
ETHER, BUTYL VINYL	1.39	0.01	0		60		882
ETHER, ISOPROPYL VINYL	1.88	0.02	0		60		882
ETHER, PHENYL VINYL	2.53	0.03	0.39	0.3	60	C	881
MALEIC ANHYDRIDE/STYRENE	0.21	0.01	1.73	0.06	65		1221
VINYL ACETATE	0.31		0.63		65		1183
AZIRIDINE, 1-ACRYLOYL-							
STYRENE	0.25		0.56		70	I10	1421
AZIRIDINE, 1-CROTONOYL-, TRANS-							
STYRENE	0		20		70	I10	1421

M1	M2	R1	+/-	R2	+/-	TEMP.	REMARKS	REF.
AZIRIDINE, 1-METHACRYLOYL-								
	STYRENE	0.41		0.47		70	I10	1421
AZIRIDINE, 1-PHENYL-								
	HYDRACRYLIC ACID, BETA-LACTONE	7.5		0.15		0	I63	459
1-AZIRIDINECARBOXAMIDE, N-ISOPROPENYL-								
	STYRENE	0.01	0.03	15.7	0.8	60	I5	418
1-AZIRIDINECARBOXAMIDE, N-VINYL-								
	METHYL METHACRYLATE	0.08	0.04	1.70	0.10	60	I5	418
	STYRENE	0.05	0.04	14.5	0.9	60	I5	418
1-AZIRIDINECARBOXYLIC ACID, ALLYL ESTER								
	STYRENE	0		50		70	I10	1421
BENZALDEHYDE								
	1,3-CYCLOHEXADIENE, 1-METHYL-	0.01	0.01	0.84	0.02	-78	H10	73
	INDENE	0.007	0.007	0.38	0.02	-78	H10	73
	ISOPRENE	0.04	0.05	0.65	0.23	0	H10	73
	STYRENE	0.0	0.02	0.3	0.02	0	H10	1402
				0.35		50	H11	228
		0.01	0.01	0.27	0.06	0	H10	73
		0.5		1.8		50	H10	228
BENZALDEHYDE, O-CHLORO-								
	STYRENE	0		0.53		0	H10	1402
BENZALDEHYDE, P-CHLORO-								
	STYRENE	0	0.02	0.45	0.08	0	H10	1402
BENZALDEHYDE, M-NITRO-								
	STYRENE	0		0.8		0	H10	1402
BENZALDEHYDE, O-NITRO-								
	STYRENE	0		1.52		0	H10	1402
BENZALDEHYDE, P-NITRO-								
	STYRENE	0		2.13		0	H10	1402
BENZAMIDE, N-ALLYL-								
	ACRYLIC ACID	0.05		4.7		80	I12	857
	METHACRYLIC ACID	0.03		13.3		80	I12	857

M1	M2	R1	+/-	R2	+/-	TEMP.	REMARKS	REF.
BENZAMIDE, P-VINYL-								
	STYRENE	1.0		0.47		100	I5	57
BENZAMIDE, P-VINYL-, LITHIUM CHLORIDE COMPLEX								
	STYRENE	1.6		0.10		100	G17,I6	57
BENZENE								
	VINYL ACETATE	0		500	100	60	I7	767
BENZENE, ALLYL-								
	ACRYLONITRILE	0		3.5		70		67
BENZENE, 1-CHLORO-4-(EPOXYETHYL)-								
	BENZENE, (EPOXYETHYL)-	0.40	0.15	1.1	0.2	-20	H10,I20	717
BENZENE, M-DIVINYL-								
	METHYL METHACRYLATE	0.61		0.41		70	C	1423
	STYRENE	0.60		0.65		60	C	1044
		0.60		0.65		61	C	1042
		0.88		0.60		80	C,EE	1041
		1.08		1.27		80	C,EE	1041
BENZENE, O-DIVINYL-								
	2-PICOLINE, 5-VINYL-	0.45	0.15	0.90	0.10	60	C,I7	69
	STYRENE	1.0		0.92		80	C	1039
BENZENE, P-DIVINYL-								
	METHYL METHACRYLATE	1.3		0.62		70	C	1423
	STYRENE	0.5		0.14		80	C	1039
		1.46		0.77		70	C	1034
		2.08		0.77		80	C	1041
BENZENE, (EPOXYETHYL)-								
	BENZENE, 1-CHLORO-4-(EPOXYETHYL)-	1.1	0.2	0.40	0.15	-20	H10,I20	717
	HYDRACRYLIC ACID, BETA-LACTONE	0.1	0.1	15.0		50	H14	171
	OXETANE, 3,3-BIS(CHLOROMETHYL)-	0.8		0.8		0	H10,I13	47
	SUCCINIC ANHYDRIDE	0.33		0.0		150	H80,I22	828
	TOLUENE, P-(EPOXYETHYL)-	0.65	0.1	1.2	0.2	-20	H10,I20	717
	1,3,5-TRIOXANE	0		2.7	0.3	35		488
BENZENE, 1-(2,3-EPOXYPROPOXY)-3-METHOXY-								
	PROPANE, 1,2-EPOXY-3-PHENOXY-	0.6	0.2	0.9	0.2	85	H44	1363
		0.88	0.1	1.24	0.2	85	H40	1363
		0.98	0.1	1	0.2	85	H13	1363
		1.05	0.15	0.9	0.05	30	G13,I17	1362

M1	M2	R1	+/-	R2	+/-	TEMP.	REMARKS	REF.
BENZENE, 1-(2,3-EPOXYPROPOXY)-4-METHOXY-								
	PROPANE, 1,2-EPOXY-3-PHENOXY-	0.85	0.05	1.19	0.02	30	G13,I17	1362
		1.37	0.2	0.71	0.1	85	H13	1363
		1.49	0.2	0.7	0.1	85	H44	1363
		1.75	0.2	0.91	0.1	85	H40	1363
BENZENE, (1,3-EPOXYPROPYL)-								
	OXETANE, 3,3-BIS(CHLOROMETHYL)-	2.7		0.05		0	H10,I3	49
BENZENE, (2,3-EPOXYPROPYL)-								
	OXETANE, 3,3-BIS(CHLOROMETHYL)-	1.1		0.45		0	H10,I3	49
BENZENE, ETHYNYL-								
	ACRYLIC ACID, METHYL ESTER	0.27	0.04	0.62	0.02	60		212
	ACRYLONITRILE	0.33	0.05	0.26	0.03	60		212
	ETHYLENE, 1,1-DICHLORO-	1.4		0.1		60		211
	METHACRYLIC ACID	0.07	0.01	2.3	0.05	60		291
	METHACRYLIC ACID, BENZYL ESTER	0.24	0.01	2.5	0.2	60	C	291
	METHACRYLIC ACID, BUTYL ESTER	1.7		0.21		60		1242
	METHACRYLIC ACID, ETHYL ESTER	2.1		0.23		60		1242
	METHACRYLIC ACID, ISOBUTYL ESTER	1.9		0.27		60		1242
	METHACRYLIC ACID, PROPYL ESTER	1.4		0.22		60		1242
	METHYL METHACRYLATE	0.20	0.05	1.5		60		475
	PYRIDINE, 2-VINYL-	0.2		4.0	0.07	60	C	782
BENZENE, 1-METHOXY-3-(VINYLOXY)-								
	ETHER, PHENYL VINYL	1.1	0.07	0.71	0.058	-78	H20,I2	1191
BENZENE, 1-METHOXY-4-(VINYLOXY)-								
	ETHER, PHENYL VINYL	2.1	0.14	0.42	0.06	-78	H20,I2	1191
BENZENE, PROPENYL-, CIS-								
	ACRYLONITRILE	-0.20	0.06	1.61	0.13	60	H20,I21	1065
	STYRENE	0.03	0.05	1.95	0.1	0		650
	STYRENE, P-CHLORO-	0.32	0.02	1.0	0.1	0	H20,I33	761
BENZENE, PROPENYL-, TRANS-								
	ACRYLONITRILE	0.001	0.021	0.57	0.002	60	H10,I24	1065
	STYRENE	0.07	0.02	1.8	0.2	30	H20,I21	652
	STYRENE, P-CHLORO-	0.2	0.1	1.4	0.15	0		650
		0.32	0.04	0.74	0.06	0	H20,I33	761
BENZENE, 2,3,5,6-TETRACHLORO-P-DIVINYL-								
	STYRENE			2.62		49,2		810

M1 / M2	R1	+/-	R2	+/-	TEMP.	REMARKS	REF.
BENZENE, 1,2,4-TRIVINYL-							
STYRENE	1.12		1.80		70	C	1034
BENZENEBORONIC ACID, P-VINYL-							
ACRYLAMIDE	1.0		0.0		60		770
BENZENESULFONAMIDE, N-(2-CHLOROETHYL)-P-VINYL-							
STYRENE	1.25	0.25	0.25	0.1	85	I10	428
BENZENESULFONAMIDE, P-VINYL-							
STYRENE	1.07		0.24		90	B,I6	1045
BENZENESULFONIC ACID, MALEIMIDO ESTER							
STYRENE	0.03		0.02		35	I10	1319
BENZENESULFONIC ACID, P-VINYL-							
ACRYLONITRILE	1.20	0.10	0.10	0.02	45	C,I11	440
BENZENESULFONIC ACID, P-VINYL-, POTASSIUM SALT							
ACRYLONITRILE	1.5	0.4	0.07	0.02	70	C,I11	301
STYRENE	0.54		-0.06		90	B	1045
	0.56		0		90	B	1045
	0.93		0.02		110	B,I6	1045
BENZENESULFONIC ACID, P-VINYL-, SODIUM SALT							
ACROLEIN	0.025	0.025	0.26	0.03	54	I11,HH	198
	0.05	0.03	0.23	0.12	54	I11,HH	198
	0.32	0.05	0.33	0.15	52	I11,HH	198
ACRYLIC ACID	1.0	0.2	0.10	0.02	70	C,I11	301
ACRYLIC ACID, SODIUM SALT	2.3	1.2	0.34	0.23	40	C,I11	301
ACRYLONITRILE	0.55	0.03	0.15	0.02	45	I17	442
	1.40	0.04	0.05	0.01	45	C,I11	440
	1.5	0.2	0.05	0.02	40	C,I11	301
BENZENESULFONIC FLUORIDE, M-VINYL-							
ACRYLIC ACID, METHYL ESTER	1.50		0.14		75	I29	332
STYRENE	1.25		0.8		75	I2	332
BENZENESULFONIC FLUORIDE, P-VINYL-							
ACRYLIC ACID, METHYL ESTER	4.0		0.20		75	I29	332
STYRENE	1.30		0.25		75	I29	332
BENZOFURAN							
BENZOTHIOPHENE	3.75	0.25	0.5	0.15	-78	H23,I3	1455

COPOLYMERIZATION PARAMETER

M1	M2	R1	+/-	R2	+/-	TEMP.	REMARKS	REF.
BENZOFURAN								
	INDENE	0.15	0.01	1.85	0.25	-78	H23,I3	1455
	STYRENE	0.95	0.1	0.80	0.1	30	H10,I24	653
BENZOFURAN, 4,7-DIMETHYL-								
	INDENE	1.3	0.1	0.7	0.1	-78	H23,I3	1455
BENZOIC ACID, ALLYL ESTER								
	ISOPHTHALIC ACID, DIALLYL ESTER	1.02		0.69		60	C	1296
	PHOSPHONOUS DICHLORIDE, PHENYL-	0.003		0.05		60		1368
	1-PROPENE, 3-CHLORO-	2.5		1.25		60	C	834
	TEREPHTHALIC ACID, DIALLYL ESTER	0.89		0.77		60	C	1296
BENZOIC ACID, MALEIMIDO ESTER								
	STYRENE	0.01		0.03		35	I10	1319
BENZOIC ACID, VINYL ESTER								
	ACRYLONITRILE	0.05	0.05	5.0	0.05	75	C	166
	P-ANISIC ACID, VINYL ESTER	1.28		0.92		60		397
	BENZOIC ACID, P-BROMO-, VINYL ESTER	0.71		1.18		60		397
	BENZOIC ACID, P-CHLORO-, VINYL ESTER	0.85		0.84		60		397
	BENZOIC ACID, P-CYANO-, VINYL ESTER	0.67		0.89		60		397
	ETHYLENE, CHLORO-	0.28		0.72		45		964
		0.3		1.8		60		1213
		0.5		1.7		40	C,I11,J	503
	ETHYLENE, 1,1-DICHLORO-	0.1	0.02	7.0	1	50	C	166
		0.1	0.02	7.0	1	50	C	775
	MALONONITRILE, METHYLENE-	0.008		0.10		43		289
	METHYL METHACRYLATE	0.07	0.01	20.3	2.5	60	C,I7	1135
	1-PROPENE, 3-CHLORO-	0.46	0.16	0.88	0.47	60	C	457
	2-PYRROLIDINONE, 1-VINYL-	0.44	0.09	2.45	0.1	60	C	1019
	STYRENE	0.05		38		80		82
		0.99		0.35		60		1019
	VINYL ACETATE	1.13	0.13	0.7	0.09	60	C,I7	1135
		1.5		0.7		80		82
		1.74	0.07	0.66	0.07	30	C	397
BENZOIC ACID, P-BROMO-, VINYL ESTER								
	P-ANISIC ACID, VINYL ESTER	1.19		0.82		60		397
	BENZOIC ACID, VINYL ESTER	1.18		0.71		60		397
	BENZOIC ACID, P-CHLORO-, VINYL ESTER	0.96		1.11		60		397
	P-TOLUIC ACID, VINYL ESTER	1.04		0.64		60		397
	VINYL ACETATE	1.70		0.60		30		397
BENZOIC ACID, P-CHLORO-, VINYL ESTER								
	P-ANISIC ACID, VINYL ESTER	0.92		0.73		60		397
	1-AZIRIDINECARBOXAMIDE, N-ISOPROPENYL-	0.84		0.85		60		397
	BENZOIC ACID, P-CYANO-, VINYL ESTER	0.78		1.02		60		397

M1	M2	R1	+/-	R2	+/-	TEMP.	REMARKS	REF.
BENZOIC ACID, P-CHLORO-, VINYL ESTER								
	P-TOLUIC ACID, VINYL ESTER	0.89		0.74		60		397
	VINYL ACETATE	1.85		0.62		30		397
BENZOIC ACID, P-CYANO-, VINYL ESTER								
	P-ANISIC ACID, VINYL ESTER	0.93		0.75		60		397
	BENZOIC ACID, VINYL ESTER	0.89		0.67		60		397
	BENZOIC ACID, P-BROMO-, VINYL ESTER	1.11		0.96		60		397
	BENZOIC ACID, P-CHLORO-, VINYL ESTER	1.02		0.78		60		397
	P-TOLUIC ACID, VINYL ESTER	0.86		0.65		60		397
	VINYL ACETATE	1.63		0.61		30		397
BENZOIC ACID, P-HYDROXY-, METHACRYLATE								
	STYRENE	0.84		0.13		105	C,I10	1150
		1.24		0.32		105	C	1150
BENZOIC ACID, P-(1-HYDROXYETHOXY)-, VINYL ESTER, ACETATE								
	ETHYLENE, CHLORO-	1.18		1.06		50	I5	1250
BENZOIC ACID, P-ISOPROPENYL-, METHYL ESTER								
	1,3-BUTADIENE	0.266	0.047	0.551	0.033	60		1256
BENZOIC ACID, P-MALEIMIDO-, ETHYL ESTER								
	ETHYLENE, CHLORO-	0.05		3.3		60		1343
BENZOIC ACID, P-VINYL-								
	ACRYLONITRILE	1.8		0.076	0.001	50	I6	735
	METHYL METHACRYLATE	0.40		1.20		60		1078
	STYRENE	0.28	0.05	1.04		60		1078
BENZOIC ACID, P-VINYL-, 2,3-EPOXYPROPYL ESTER								
	STYRENE	0.95	0.10	0.40	0.02	70	C	429
		0.95	0.10	0.40	0.02	70		430
BENZOIC ACID, P-VINYL-, SODIUM SALT								
	ACRYLONITRILE	0.26	0.06	0.09	0.08	50	F,I11	735
BENZOIC ACID, P-(VINYLOXY)-, VINYL ESTER								
	ETHER, P-CHLOROPHENYL VINYL	0.56		0.43		0	H20	1250
	ETHYLENE, CHLORO-	0.89		0.61		50	I5	1250
BENZONITRILE								
	PROPANE, 1-CHLORO-2,3-EPOXY-	0	0.02	2.8	0.2	70	H20	231

COPOLYMERIZATION PARAMETER

M1	M2	R1	+/-	R2	+/-	TEMP.	REMARKS	REF.
BENZONITRILE, P-((4-AMINO-6-VINYL-S-TRIAZIN-2-YL)AMINO)-								
	STYRENE	0.31		0.64		60	117	1453
BENZONITRILE, P-VINYL-								
	ANISOLE, P-VINYL-	0.85		0.093		30		505
	METHYL METHACRYLATE	1.41	0.13	0.22	0.02	60		1024
	STYRENE	1.16	0.13	0.28	0.025	60		1168
	STYRENE	1.16		0.28		60		1024
	STYRENE, M-BROMO-	1.2		0.19		30		505
	STYRENE, M-CHLORO-	1.1		0.56		30		699
	STYRENE, M-CHLORO-	1.20		0.60		30		699
	STYRENE, P-CHLORO-	1.4		0.34		30		505
BENZOPHENONE, 4-VINYL-								
	STYRENE	3.0	0.3	0.18	0.08	60		129
BENZOTHIAZOLE, 2-(ALLYLTHIO)-								
	METHYL METHACRYLATE	0		3.15		60		1381
BENZOTHIAZOLE, 2-VINYLTHIO-								
	ACRYLIC ACID, METHYL ESTER	1.0		0.1		60		487
BENZOTHIOPHENE								
	BENZOFURAN	0.5	0.15	3.75	0.25	-78	H23,I3	1455
	INDENE	0.08	0.002	2.4	0.025	-78	H23,I3	1455
BENZOYL CHLORIDE, P-VINYL-								
	STYRENE	1.80	0.20	0.17	0.03	60		430
BENZYL ALCOHOL, ALPHA,ALPHA-DIETHYL-P-VINYL-								
	ACRYLONITRILE	0.38		0.04		60	166	1092
BENZYL ALCOHOL, ALPHA, ALPHA-DIMETHYL-P-VINYL-								
	ACRYLONITRILE	0.41		0.05		60		1092
	STYRENE	0.56	0.05	0.50	0.05	50	166	128
	STYRENE	1.25		0.79		60	166	1092
	STYRENE, P-CHLORO-	0.53		1.24		60	166	1092
BENZYL ALCOHOL, ALPHA-ETHYL-ALPHA-METHYL-P-VINYL-								
	ACRYLONITRILE	0.39		0.06		60	166	1092
	STYRENE, P-CHLORO-	0.33		1.43		60	166	1092
BENZYL ALCOHOL, ALPHA-METHYLENE-, ACETATE								
	ACRYLONITRILE	0.4	0.05	0.08	0.01	75	C	166

M1	M2	R1	+/-	R2	+/~	TEMP.	REMARKS	REF.
BENZYL ALCOHOL, ALPHA-METHYLENE-, ACETATE								
	1,3-BUTADIENE	0.29		0.13		50	C,K	91
1,1'-BIINDENE								
	INDENE	0.045	0.008	2.07	0.04	-72	H23	586
	STYRENE	0.17	0.05	1.95	0.1	-72	H23,I3	1279
	STYRENE, ALPHA-METHYL-	0.15	0.15	3.6	0.2	-72	H23,I3	1279
BIPHENYL, 4-FLUORO-4'-VINYL-								
	STYRENE	0.88	0.07	0.50	0.15	70		106
BIPHENYL, 4-VINYL-								
	STYRENE	0.92	0.08	0.98	0.04	70		106
		1.4	0.1	0.89	0.1	60	I7	1248
BISMUTHINE, DIPHENYL(P-VINYLPHENYL)-								
	STYRENE	1.35	0.10	0.63	0.05	65	I2	126
BORAZINE, 2,4,6-TRIALLYL-1,3,5-TRIPHENYL-								
	METHYL METHACRYLATE	0		8.6		80		768
	STYRENE	0		42.9		80		768
BORAZINE, 1,3,5-TRIPHENYL-2,4,6-TRIVINYL-								
	METHYL METHACRYLATE	0		0.70		80		768
	STYRENE	0		0.26		80		768
	VINYL ACETATE	0		0		80		768
1,3-BUTADIENE								
	ACENAPHTHYLENE	0.48		1.5		65	I7	1385
	ACRYLIC ACID, BUTYL ESTER	0.99	0.07	0.08	0.02	5	K	1025
	ACRYLIC ACID, METHYL ESTER	0.76	0.04	0.05	0.02	5	K	1025
	ACRYLIC ACID, THIO-, METHYL ESTER	0.35	0.01	0.20	0.05	70		594
	ACRYLONITRILE	0.1		0.06		40	E,I15	833
		0.18		0.03		5	K	243
		0.28		0.02		5	K	243
		0.3		0.02		40	E,I15	833
		0.33		0.25		60		621
		0.35		0.075		60	I2	311
		0.35	0.01	0.05	0.01	50		353
		0.35	0.08	0.0	0.04	50		352
		0.40	0.02	0.04	0.01	50	H38	60
		0.46	0.03	0.0		50	K	1022
		0.551	0.033	0.266	0.047	60		1256
	BENZOIC ACID, P-ISOPROPENYL-, METHYL ESTER	0.13		0.29		50	C,K	91
	BENZYL ALCOHOL, ALPHA-METHYLENE-, ACETATE	0.0		2.86		50	K	353
	1,3-BUTADIENE, 2-CHLORO-	0.059	0.014	3.41		50		353
	1,3-BUTADIENE, 2,3-DIMETHYL-	0.85		0.63	0.07	5	K	290

COPOLYMERIZATION PARAMETER

1,3-BUTADIENE

M1	M2	R1	+/-	R2	+/-	TEMP.	REMARKS	REF.
1,3-BUTADIENE	1,3-BUTADIENE, 2,3-DIMETHYL-	1.26		0.78		-18	K	732
	1,3-BUTADIENE, 2-PHENYL-	2.58	0.36	1.90	0.09	0	H62,I2	583
	1-BUTANOL, 2,2,3,3,4,4,4-HEPTAFLUORO-, ACRYLATE	0.35		0.07		50	C,K	438
	1-BUTENE	0.15		1.81		-40	H46,I18	113
	1-BUTENE	0.19		0.26		-40	H4,I2	114
	1-BUTENE	1.81		0.15		-78	H56,I2	114
	1-BUTENE, 2-METHYL-	6.7		0.6		-40	H46,I18	113
	1-BUTENE, 3-METHYL-	0.05		2.0		-40	H46,I18	113
	2-BUTENE, CIS-	0.15		0.45		-40	H46,I18	113
	2-BUTENE, CIS-	0.17		0.51		-78	H56,I2	114
	2-BUTENE, 2,3-DIMETHYL-	0.45		0.15		-40	H56,I18	114
	2-BUTENE, 2,3-DIMETHYL-	0.9		0.38		-40	H46,I18	113
	BUTYRALDEHYDE, 2-METHYLENE-	0.6	0.06	1.34	0.08		Q	1162
	CHALCONE	0.78	0.12	0.03		60		851
	CHALCONE, 2-CHLORO-	0.78	0.12	-0.02		59		851
	CINNAMIC ACID, O-CHLORO-, METHYL ESTER	0.78	0.12	-0.01	0.05	60	K	851
	CINNAMIC ACID, O-CHLORO-, METHYL ESTER	1.07		-0.02	0.05	60		851
	CINNAMIC ACID, P-CHLORO-, METHYL ESTER	1.20	0.3	-0.03	0.05	80		851
	CINNAMIC ACID, ALPHA-CYANO-, ETHYL ESTER	2.73		0.00		80	K,Q	658
	CYCLOBUTANECARBOXYLIC ACID, 3-ACETYL-2,2-DIMETHYL-, VINYL ESTER *	0.25		0		35	K,Q	658
	1,3-CYCLOHEXADIENE	0.25		1.1		35	G30	1320
	ETHYLENE, CHLORO-	0.8		0.035		50	C,K	962
	ETHYLENE, 1,1-DICHLORO-	8.8	0.2	0.05		50	K	1025
	ETHYLENE, 1,1-DIPHENYL-	0.09		0		5	G5,I5	1450
	ETHYLENE, 1,1-DIPHENYL-	0.09		0		5	G19,I5	1450
	ETHYLENE, 1,1-DIPHENYL-	0.1		0		0	G5,I7	1450
	ETHYLENE, 1,1-DIPHENYL-	0.13		0		40	G17,I5	1450
	ETHYLENE, 1,1-DIPHENYL-	0.71		0		0	G18,I7	1450
	ETHYLENE, 1,1-DIPHENYL-	54		0		40	G17,I7	1450
	ETHYLENE, TRICHLORO-	12.32	0.31	-0.007	0.022	80	C	310
	FORMIC ACID, VINYL ESTER	5.0	0.01	0.2	0.05	120	F	994
	FUMARIC ACID, DIETHYL ESTER	2.13		0.25			Q,T	587
	FUMARIC ACID, DINONYL ESTER	2.02		0.32			Q,T	587
	FUMARIC ACID, DITHIO-, DIMETHYL ESTER	0.0106	0.0175	-0.0014	0.027	50	C,I7	598
	GLUTARONITRILE, 2-METHYLENE-	0.00		0.00		25	F,K,DD	789
	ISOPRENE	0.24		0.02		25	H4,Q	789
	ISOPRENE	0.12		3.4			K	137
	ISOPRENE	0.75		0.85		5	H25,I7,Q	290
	ISOPRENE	0.92		1.25			K	276
	ISOPRENE	0.94		1.06		-18	H26,I7,Q	732
	ISOPRENE	0.99		1.37			H73,Q	276
	ISOPRENE	1		1			H43	898
	ISOPRENE	1		1		-15	H43	765
	ISOPRENE	1		1		13	H43	765
	ISOPRENE	1		1		20	H43	765
	ISOPRENE	1.0		0.9		48	H25,Q	765
	ISOPRENE	1.0		1.0			H24,L	898
	ISOPRENE	1.15		0.59		30	H27,I7,Q	136
	ISOPRENE	1.88	0.05	0.55	0.05		H74,Q	276

REMARKS PAGE II-366, REFERENCES PAGE II-368

M1	M2	R1	+/-	R2	+/-	TEMP.	REMARKS	REF.
1,3-BUTADIENE	ISOPRENE	2.0		0.6			H73,Q	898
		2.3	0.1	1.15	0.05	30	H25	136
		2.80		0.60			H74,Q	898
		3.38	0.14	0.47	0.03	50	G17,19	796
		3.6		0.11			G17,18,Q	135
		4.5		0.13			G17,15,Q	135
		4.50		0.80			H75,Q	898
	MALEIC ACID, DIETHYL ESTER	8.08		0.11			Q,T	587
	MALEIC ACID, DINONYL ESTER	5.36		0.12			Q,T	587
	METHACRYLIC ACID	0.201		0.526			C,K	267
	METHACRYLIC ACID, NONYL ESTER	0.76		0.32		50	Q	588
	METHACRYLONITRILE	0.36	0.07	0.04	0.04	5	K	211
	METHYL METHACRYLATE	0.53	0.05	0.06	0.03	5	K	1025
		0.70		0.32			Q	588
		0.75	0.05	0.25	0.03	90		560
	NONANOIC ACID, VINYL ESTER	26.3	10.0	0.02	0.02	60	C	591
	2-NORPINANE-2-ETHANOL, ACRYLATE	1.1		0.2			C,Q	601
	2,4-PENTADIENENITRILE	5		1.70		50	K	1022
	2,4-PENTADIENOIC ACID, METHYL ESTER	5.8	0.05	0.05	0.05	50	H76,I7	101
	PHOSPHONIC ACID, (DIMETHYLENEETHYLENE)DI-, TETRAETHYL ESTER	0.05		0.05		70		1395
	PHOSPHONOUS DICHLORIDE, PHENYL-	0.003	0.006	0.474	0.031	60	139	793
	2-PICOLINE, 5-VINYL-	0.606	0.07	0.412	0.026		K,Q	562
		1.30	0.01	0.72	0.03	30		189
		1.32	0.01				Q	971
	1-PROPENE, 2-METHYL-	0		43	15	-100	H4,I3	483
		0.01	0.01	115		-103		956
		1.0		8.0		-78	E	286
		1.2		14.0		-78	E,L1	286
		2.0		30.0		-78	E,L1	734
	1-PROPENE-1,2,3-TRICARBOXYLIC ACID	0.37	0.03	0.00	0.01	50	K,P	595
	1-PROPENE-1,2,3-TRICARBOXYLIC ACID, TRANS-, TRIETHYL ESTER	0.65	0.05	0.02		60	C	734
	1-PROPENE-1,2,3-TRICARBOXYLIC ACID, TRANS-, TRIMETHYL ESTER	0.40	0.03	0.00	0.03	50	K,C	595
	STEARIC ACID, VINYL ESTER	34.5	6.0	0.034	0.034	60	C	591
	STILBENE, TRANS-	0		0		0	G17,15	1452
	STYRENE	50		4.6	0.5	40	G17,I7	1452
		0.13	0.1	8	1	-35	H76,Q	1122
		0.2	0.12	9.8		-15	H17,15,N	909
		0.44		0.68		25	G32,15	1397
		0.50		1.68		20	U	776
		0.6		0.5		50	G32,15	1397
		1.12		0.01	0.01	43	G32,15	1397
		1.30		0.58	0.15	50	H38	60
		1.35		0.38		-18		353
		1.37		0.38		-18	K	731
		1.37		0.64		5	K	290
		1.38	0.03	0.78	0.01	60	K	290
		1.39	0.15	0.825	0.086	30		560
		1.39	0.2	0.5	0.1	50		189
		1.4		0.44		5	C,K	628
		1.40	0.08	0.38	0.08	5		730
		1.40	0.08	0.23	0.07	50	K	730
		1.48						352

COPOLYMERIZATION PARAMETER

M1	M2	R1	+/-	R2	+/-	TEMP.	REMARKS	REF.
1,3-BUTADIENE								
	STYRENE	1.59	0.05	0.44	0.03	50	K	393
		1.8	0.4	0.6	0.1	45	K	644
		1.83		0.65		45	K	290
		1.83	0.32	0.65	0.16	45	K	644
		3.0		0.06		29	G17,I7	657
		5.5		0.3		25	H17,I8,N	909
		7.0		-0.1		30	G17,I20	547
		10		-0.04		50	G32,I7	1397
		12.5		0.1		25	H17,I2,N	909
		20		-0.05		50	G17,I7	529
		27.5		<0.04		40	G4,I21	449
	STYRENE, P-CHLORO-	1.07		0.42		50	K	1022
	STYRENE, 2,5-DICHLORO-	0.46	0.01	0.46	0.01	50	K	1022
	STYRENE, 2,4-DIMETHYL-	0.65		0.2		70	C	26
	STYRENE, ALPHA-METHYL-	16.65	0.1	0.06	0.04	20	G17,I9,O	1017
		0.12		1.45			K	790
	9-UNDECENOIC ACID, VINYL ESTER	1.6	0.5	0.010	0.01	12,8		257
		37.9	4.0	0.015	0.015	60	C	591
1,3-BUTADIENE, 2-CHLORO-								
	ACRYLIC ACID, METHYL ESTER	11.1	1.5	0.078	0.010	60	C	214
	ACRYLONITRILE	4.80	0.03	0.05	0.01	50	I6	840
		5.35	0.20	0.045	0.004	60		211
		6.07	0.53	0.005	0.011	50		353
		6.07	0.53	0.01	0.01	50		352
		6.93	0.23	0.034	0.0	50		353
	1,3-BUTADIENE	2.86		0.0		50	K	353
		3.41	0.07	0.059	0.014	50	C,K	472
	1,3-BUTADIENE, 2,3-DICHLORO-	0.355	0.055	2.15	0.25	40	Q	795
	1,3-BUTADIENE, 2-FLUORO-	3.70		0.22			I7	508
	1,3-BUTADIENE, HEXACHLORO-	3.70		0.10		40/60		508
	1,3-BUTADIENE, HEXAFLUORO-	5.47		0.10		50		493
		5.47		0.10		50/70		508
	3-BUTEN-2-ONE, 3-METHYL-	5.52		0.1		50	K	1030
	BUTYRIC ACID, VINYL ESTER	3.6	0.2	0.1	0.05	40	C,K	1002
	ETHER, ISOPROPYL VINYL	90		0.02		65	C	645
	ETHYLENE, 1,1-DIPHENYL-	0.164	0.043	11.45	0.54	65	C	214
	FORMIC ACID, VINYL ESTER	3.17	0.16	-0.00	0.05	65	C	493
		15.0		-0.035		60	C	1002
	FUMARIC ACID, DIETHYL ESTER	30		0.01		65	C	214
	1,3,5-HEXATRIENE, 2,3,4,5-TETRACHLORO-, TRANS-	6.65	0.37	0.027	0.010	60	C	542
	ISOPRENE	3.6	0.17	0.2	0.07	70	K	353
		2.82	0.22	0.063	0.051	50		352
	MALONONITRILE, METHYLENE-	3.65	0.11	0.133	0.025	50		289
	METHACRYLIC ACID	0		0.20	0.06	40		289
	METHYL METHACRYLATE	0.010		0.0017		40	C,K	1030
		0.016		0.0048		40	K	242
	PHOSPHONIC ACID, ALPHA-PHENYLVINYL-	2.7	0.2	0.15	0.05	40	C	214
	PROPIONIC ACID, VINYL ESTER	3.9	0.25	0.18	0.06	60	I10	1091
		6.12	0.2	0.080	0.007	60	C	1002
		7.5		0.1		60		
		70		0.05		65		

REMARKS PAGE II-366, REFERENCES PAGE II-368

M1	M2	R1	+/-	R2	+/-	TEMP.	REMARKS	REF.
1,3-BUTADIENE, 2-CHLORO-								
	PYRIDINE, 2-VINYL-	5.19	0.03	0.06	0.01	60		534
		5.19	0.03	0.06	0.01	60		533
		5.195	0.003	0.064	0.001	50		536
	QUINOLINE, 2-VINYL-	2.10	0.13	0.38	0.03	50		536
		2.10	0.13	0.38	0.03	60		534
		2.10	0.13	0.38	0.03	60		533
	STYRENE	0.04	0.01	6.9	0.5	0	H30,I7	758
		0.06	0.01	12.8	0.3	0	H10,I9	758
		0.065	0.01	16.0	0.05	0	H30,I22	758
		0.15	0.05	33.0	0.5	0	H10,I23	758
		0.24		15.6		-18	G3	263
		2.4		15.6			H,Q	266
		5.22	0.64	0.00		50	K	353
		6.3		0.005		50	Q	266
		6.3	0.1	0.00		70		353
		7	2	0.05		60	C	26
		8.11	0.34	0.052	0.02	60	C	211
	VINYL ACETATE	18		0.2	0.01	30	H71	1063
		50		0.01		65	C	1002
1,3-BUTADIENE, 2,3-DICHLORO-								
	1,3-BUTADIENE, 2-CHLORO-	2.15	0.25	0.355	0.055	40	C,K	472
	ETHYLENE, 1,1-DIPHENYL-	4.5	0.45	0		60	C	211
		4.5	0.45	17		60	C	214
	METHYL METHACRYLATE	10.3	1.5	0.073	0.015	60	C	214
	STYRENE	10.8	1.2	0.041	0.012	60	C	211
1,3-BUTADIENE, 2,3-DIMETHYL-								
	1,3-BUTADIENE	0.63		0.85		5	K	290
		0.78		1.26		-18	K	732
	ETHYLENE, 1,1-DIPHENYL-	0.23	0.04	0		40	G17,I7	1451
	ISOPRENE			17	5	50	H	797
	STILBENE, TRANS-	0.84		1.18		-18	K	732
		0		0		0	G17,I5	1452
	STYRENE	8.5		0		40	G17,I7	1452
		0.92	0.02	0.42	0.02	-18	K	732
1,3-BUTADIENE, 2-FLUORO-								
	ACRYLONITRILE	0.50	0.10	0.07	0.03	50	K,Q	733
	1,3-BUTADIENE, 2-CHLORO-	0.22		3.70			Q	795
	1,3-BUTADIENE, HEXAFLUORO-	0.22		3.70			Q	508
		2.93		0.24		40		508
	ISOPRENE	2.05	0.19	0.19	0.10	50	K	733
	METHYL METHACRYLATE	1.54	0.08	0.64	0.08	50	K	733
	STYRENE	1.55	0.10	0.50	0.10	50	K	733
		1.61	0.24	0.16	0.08	5	K	733
	STYRENE, ALPHA-METHYL-	1.71	0.19	0.38	0.11	50	K	733

COPOLYMERIZATION PARAMETER

M1	M2	R1	+/-	R2	+/-	TEMP.	REMARKS	REF.
1,3-BUTADIENE, HEXACHLORO-								
	1,3-BUTADIENE, 2-CHLORO-	0.10		5.47		40/60	I7	508
	STYRENE	0		>1		70		23
1,3-BUTADIENE, HEXAFLUORO-								
	1,3-BUTADIENE, 2-CHLORO-	0.10		5.47		50		493
		0.10		5.52		50/70	K	508
	1,3-BUTADIENE, 2-FLUORO-	0.24		2.93		40		508
	ISOPRENE	0.78	0.05	1.19	0.12	40		508
1,3-BUTADIENE, 2-PHENYL-								
	1,3-BUTADIENE	1.90	0.09	2.58	0.36	0	H62,I2	583
1,3-BUTADIENE, 1,1,2-TRICHLORO-								
	STYRENE	1.18	0.08	0.07	0.03	70	K	269
		1.18		0.07		60		520
1,3-BUTADIEN-1-OL, ACETATE								
	ACRYLONITRILE	0.7		0.0		70	C	315
	VINYL ACETATE	>9		0.0		70	C	315
1,3-BUTADIENOLS, ACETATE, MIXED ISOMERS								
	METHYL METHACRYLATE	1.2		0.35		70	I2	1275
	STYRENE	2.4		0.3		70	I2	1275
BUTANE, 1,4-BIS(P-VINYLPHENYL)-								
	STYRENE	0.78		1.13		61	C	1042
BUTANE, 1,4-DIINDEN-1-YL-								
	INDENE	0.5	0.1	0.5	0.1	-78	H23,I3	1280
BUTANE, 1,2-EPOXY-								
	FURAN, TETRAHYDRO-	0.58	0.02	0.58	0.02	0	H10,I4	1207
		0.8		0.19		0	H83,I24	1154
		0.96		0.25		0	H82,I24	1154
	OXETANE, 3,3-BIS(CHLOROMETHYL)-	2.0		0.3		0	H10,I3	49
	PROPYLENE OXIDE	0.39		0.71		0	H83,I24	1154
		0.5		0.76		0	H82,I24	1154
BUTANE, 1,3-EPOXY-								
	OXETANE, 3,3-BIS(CHLOROMETHYL)-	3.4	0.65	0.04	0.17	0	H10,I3	49
	OXETANE, 3-(CHLOROMETHYL)-3-METHYL-	3.75		0.25		0	H13	284

M1	M2	R1	+/-	R2	+/-	TEMP.	REMARKS	REF.
BUTANE, 2,3-EPOXY-2-METHYL-								
	OXETANE, 3-(CHLOROMETHYL)-3-METHYL-	0.31	0.04	3.6	0.1	-50	H10	283
		0.4	0.14	2.2	0.2	0	H13	283
		0.75	0.1	1.77	0.2	80	H11	1118
1-BUTANOL, 2,2,3,3,4,4,4-HEPTAFLUORO-, ACRYLATE								
	1,3-BUTADIENE	0.07		0.35		50	C,K	838
	METHYL METHACRYLATE	0.25		1.4		50	C,K	838
	STYRENE	0.07		0.33		50	C,K	838
2-BUTANOL, 1-(P-VINYLPHENYL)-								
	ACRYLIC ACID, BUTYL ESTER	0.4		0.17		60	I5	1093
	ACRYLIC ACID, METHYL ESTER	0.48		0.18		60	I5	1093
	ACRYLONITRILE	0.31		0.04		60	I5	1093
1-BUTENE								
	ACRYLONITRILE	0.10		8.0		60		865
	1,3-BUTADIENE	0.15		1.81		-78	H56,I18	114
		0.26		0.19		-40	H56,I2	114
		0.26		0.19		-40	H4,I2	114
		1.81		0.15		-40	H46,I18	113
	1-BUTENE, 3-METHYL-	8.5		0.013		60	H12	491
	2-BUTENE, TRANS-	0.15	0.05	3.1	0.7	-20/-3	H	629
	1-DECENE	1.5		0.70		23	H54,I21	565
	ETHYLENE	0.006		3.25	0.38	130-220	M5	112
		0.015		35.5		-78	H61	1079
		0.019		85.0		23	H54,I21	565
		0.019		29.5			G,Q	688
		0.019		32.5		-20	H61	1124
		0.043		29.60		0/75	H28	690
		0.016		26.96		0/75	H29	690
	ETHYLENE, CHLORO-	<0.64		3.6			M6,Q	236
	1-PROPENE	0.21		3.2		130-220	H1	656
				3.4		60		936
		0.20		4.5		23	G,P,Q	625
		0.227		4.39		0-75	H54,I21	565
		0.252		4.04		0-75	H28	690
		0.45		3.3		30	H29	690
		0.5	0.2	2.4	0.4	60	H57	185
		0.50		1.62		60	H24	1269
		0.7		0.7		-78	H12,I20	346
		0.8		4.3		30	H61	1079
	1-PROPENE, 3-CHLORO-			0.1	0.05	-78	H57,M12	185
	STYRENE	5.0	1.0			30	C	404
	VINYL ACETATE	5.51		1.98		50	H12	827
		0.34		2.0		60		865
1-BUTENE, 4-(P-CHLOROPHENYL)-								
	ACRYLONITRILE	0		4.0		70		67

M1	M2	R1	+/-	R2	+/-	TEMP.	REMARKS	REF.
1-BUTENE, 4-CYCLOHEXYL-	ACRYLONITRILE	0		4.5		70		67
1-BUTENE, 3,3-DIMETHYL-	ETHYLENE, CHLORO-	0.0		5.0	0.7	70		23
1-BUTENE, 3,3-DIMETHYL-1-PHENYL-, TRANS-	ACRYLONITRILE	0		18.8		60		1065
1-BUTENE, 3,3-DIMETHYL-2-PHENYL-	STYRENE			30		110	A,O	1189
1-BUTENE, 2,4-DIPHENYL-	STYRENE			1.6		110	A,O	1189
1-BUTENE, 3,4-EPOXY-2-METHYL-	METHYL METHACRYLATE	0		56		60-140	C	630
	STYRENE	0		77		60-140	C	630
1-BUTENE, 2-METHYL-	1,3-BUTADIENE	0.6		6.7		-40	H46,I18	113
1-BUTENE, 3-METHYL-	1,3-BUTADIENE	2.0		0.05		-40	H46,I18	113
	1-BUTENE	0.013		8.5		60	H12	491
	ETHYLENE	0.001		243		10	H45	1087
1-BUTENE, 3-METHYL-1-PHENYL-, CIS-	ACRYLONITRILE	-0.38	0.32	18.0	0.5	60		1065
1-BUTENE, 3-METHYL-1-PHENYL-, TRANS-	ACRYLONITRILE	-0.16	0.50	3.37	0.20	60		1065
1-BUTENE, 2-NITRO-	SORBIC ACID, 2-CYANO-, BUTYL ESTER	0.42	0.1	0.75	0.1	-30	G13,I6	1216
		0.44		1.12	0.2	0	G13,I5	1216
		0.5	0.3	1.2	0.3	0	G13,I6	1216
		0.55	0.2	0.85	0.2	0	G13,I45	1216
		0.57	0.2	1	0.2	24	G25,I45	1216
		0.75	0.25	0.75	0.25	0	G13,I6	1216
		0.8	0.25	0.65	0.25	0	G27,I45	1216
		0.8	0.25	0.65	0.25	0	G28,I6	1216
		0.8	0.4	1.5	0.4		G13,I6	1216
		0.84	0.1	0.6	0.1	90	G13,I6	1216

M1	M2	R1	+/-	R2	+/-	TEMP.	REMARKS	REF.
1-BUTENE, 2-NITRO-								
	SORBIC ACID, 2-CYANO-, BUTYL ESTER	0.85	0.2	0.6	0.2	0	G26,I45	1216
		0.9	0.1	0.98	0.1	0	G24,I6	1216
	SORBIC ACID, 2-CYANO-, METHYL ESTER	1.04		2.14		0	G13,I7	1216
		0.4	0.15	1.5	0.15	0	G13,I6	1216
1-BUTENE, 1-PHENYL-								
	STYRENE, P-CHLORO-	0		0.88	0.30	0	H20,I33	761
1-BUTENE, 1-PHENYL-, CIS-								
	ACRYLONITRILE	-0.02	0.34	2.58	0.09	60		1065
1-BUTENE, 1-PHENYL-, TRANS-								
	ACRYLONITRILE	0.10	0.06	0.98	0.02	60		1065
1-BUTENE, 4-PHENYL-								
	ACRYLONITRILE	0		3.5		70		67
2-BUTENE, CIS-								
	ACRYLONITRILE	0.00		14.0		60	H56,I18	865
	1,3-BUTADIENE	0.15		0.45		-78	H46,I18	114
		0.45		0.15		-40	H56,I2	113
		0.51		0.17		-40		114
	ETHYLENE, CHLORO-	0.001		8.8		60		936
	MALEIC ANHYDRIDE	0		0.016	0.005	60	I7	669
		0		0.016	0.005	60	I7	668
	VINYL ACETATE	0.07		8.0		60		865
2-BUTENE, TRANS-								
	ACRYLONITRILE	0.00		14.0		60		865
	1-BUTENE	3.1	0.7	0.15	0.05	-20/-3	H	629
	ETHYLENE			7.5		130-220	M1	656
	ETHYLENE, CHLORO-	0.02		7.3		60		936
	MALEIC ANHYDRIDE	0		0.030	0.009	60	I7	669
		0		0.030	0.009	60	I7	668
	VINYL ACETATE	0.07		7.0		60		865
2-BUTENE, 1,3-DICHLORO-								
	STYRENE	0	0.03	15	3.7	70	C	229
	VINYL ACETATE	4.8	0.9	0.0		50	C	233
2-BUTENE, 1,4-DIINDEN-1-YL-								
	INDENE	0.75	0.05	0.2	0.05	-78	H23,I3	1280

M1	M2	R1	+/-	R2	+/-	TEMP.	REMARKS	REF.
2-BUTENE, 2,3-DIMETHYL-	ACRYLONITRILE	-0.26	0.04	4.0		60		1066
	1,3-BUTADIENE	0.38		0.9	0.03	-40	H46,I18	113
2-BUTENE, 2-METHYL-	ACRYLONITRILE	-0.27	0.15	2.7	0.04	60		1066
	MALEIC ANHYDRIDE	0		0.005	0.002	60	I7	669
	STYRENE	0.65	0.19	1.55	0.31	-78	H23,I9	1223
	STYRENE	0.8	0.11	1.85	0.25	-78	H23,I42	1223
	STYRENE	1.7	0.2	1.65	0.21	-78	H23,I13	1223
1-BUTENE:SULFUR DIOXIDE (1:1 COMPLEX)	ACRYLONITRILE	0.8		1.0		-78	B,I2	342
3-BUTENOIC ACID, 2-ETHYL-2-METHYL-, ETHYL ESTER	VINYL ACETATE	2.2	0.4	0.1	0.1	60	C	933
		3.2	0.50	0.3	0.2	60	C	933
3-BUTENOIC ACID, 3-HYDROXY-, BETA-LACTONE	ACRYLONITRILE	0		7.52	0.08	60	Q	421
	ETHYLENE, CHLORO	0.2		5.74		60	I7	614
3-BUTEN-2-OL, 1-CHLORO-3-METHYL-	METHYL METHACRYLATE	0		90		60-140	C	630
	STYRENE	0		110		60-140	C	630
3-BUTEN-2-ONE	ACRYLIC ACID, BUTYL ESTER	1.6	0.1	0.65	0.07	50		184
	ACRYLONITRILE	1.39	0.06	0.63	0.06	50	I6	840
	ACRYLONITRILE	1.78	0.22	0.61	0.04	60		560
	ETHER, PHENYL VINYL	4.4		0.01		70		376
	ETHYLENE, CHLORO-	8.3		0.10		70	C,Q	22
	ETHYLENE, 1,1-DICHLORO-	1.8		0.55		70		22
	GLUTARONITRILE, 2-METHYLENE-	5.05	0.95	1.24	0.76	60	Q	946
	STYRENE	0.27	0.04	0.35	0.06	60		1114
	STYRENE	0.35	0.02	0.29	0.04	60		560
	STYRENE, 2,5-DICHLORO-	0.5		2.0		70		22
	S-TRIAZINE, 4,6-DIAMINO-2-VINYL-	0.26		1.2		60		759
	VINYL ACETATE	7.00	0.04	0.05	0.15	70	C	315
3-BUTEN-2-ONE, 3-METHYL-	ACRYLONITRILE	0.70	0.14	0.36	0.08	80	C	1070
	1,3-BUTADIENE, 2-CHLORO-	1.2		0.3			Q	775
	ETHYLENE, 1,1-DICHLORO-	0.1	0.05	3.6	0.2	40	C,K	1030
		4.5		0.15			Q	775
		4.5	0.1	0.15		60	C	166
	STYRENE	0.29	0.06	0.44	0.10	80	C	1070

M1 / M2	R1	+/-	R2	+/-	TEMP.	REMARKS	REF.
3-BUTEN-2-ONE, 3-METHYL-							
STYRENE	0.66		0.32	0.01	80	C	1070
STYRENE, ALPHA-METHYL-	1.7		0.03	0.10		C,Q	1070
1-BUTEN-3-YNE							
ACRYLONITRILE	0.60	0.02	0.13	0.01	60		784
	4.65	1	0.23	0.10	50		840
1-PROPENE, 2-METHYL-	0.13		8.0		-100	H10,N	206
1-BUTEN-3-YNE, 2-METHYL-							
ACRYLONITRILE	0.47	0.01	0.33	0.01	60		784
PYRIDINE, 2-VINYL-	0.55	0.1	1.65	0.05	60		784
BUTYRALDEHYDE							
ACETALDEHYDE	0.59		1.97		-78	G8	944
	0.88		1.33		-60	G8	944
	1.07		0.92		-55	G8	944
ISOCYANIC ACID, BUTYL ESTER	10		0		-75	G17,19	1331
	70		5		-75	G19,154	1331
	500		10		-75	G19,15	1331
	0		10		-75	G17,154	1331
STYRENE	0.5		0.3			H11,17,Q	1108
	0.6		1.8			H10,17,Q	1108
			0.4		-10	H10	1182
BUTYRALDEHYDE, DIVINYL ACETAL							
VINYL ACETATE	1.06	0.01	1.005	0.015	60		608
BUTYRALDEHYDE, 2-METHYLENE-							
ACRYLONITRILE	0	0.08	0.237			C,Q	140
1,3-BUTADIENE	1.34		0.6	0.06		Q	1162
BUTYRAMIDE, N-ALLYL-3-METHYL-							
ACRYLIC ACID	0.05		4.3		80	I12	857
METHACRYLIC ACID	0.05		8.1		80	I12	857
BUTYRIC ACID, VINYL ESTER							
1,3-BUTADIENE, 2-CHLORO-	0.02		90		65	C	1002
CARBAZOLE, 9-VINYL-	0.059	0.020	1.28	0.06	100	C	1001
ETHYLENE, CHLORO-	0.28		2		60		1213
	0.55	0.05	1.75	0.03	50	Q	176
	0.65	0.04	1.35	0.05	60		799
METHYL METHACRYLATE	0.03	0.01	25	2.5	60	C,17	1135
1-PROPENE, 3-CHLORO-	0.31	0.35	1.15	0.59	60	C	457
	0.97	0.07	1	0.07	60	C,17	1135
VINYL ACETATE	1.25	0.15	1.35	0.15	60	17	954

M1	M2	R1	+/-	R2	+/-	TEMP.	REMARKS	REF.
BUTYRIC ACID, 2,3-EPOXY-, ALLYL ESTER								
	BUTYRIC ACID, 2,3-EPOXY-, METHYL ESTER	0.35	0.27	0.11	0.01		H10,Q	864
BUTYRIC ACID, 2,3-EPOXY-, METHYL ESTER								
	BUTYRIC ACID, 2,3-EPOXY-, ALLYL ESTER	0.11	0.01	0.35	0.27		H10,Q	864
BUTYRIC ACID, HEPTAFLUORO-, VINYL ESTER								
	STYRENE	0.017		8.73			C,Q	1070
BUTYRIC ACID, 3-HYDROXY-, BETA-LACTONE (COPOLYMERIC WITH * ACRYLONITRILE)								
BUTYRIC ACID, 4-HYDROXY-, GAMMA-LACTONE	HYDRACRYLIC ACID, BETA-LACTONE	0.2	0.1	6.3	2.8	60	G18,I15	867
	HYDRACRYLIC ACID, BETA-LACTONE	0.36	0.10	18	2	0	H13	924
				4.5		0	H10	980
	OXETANE, 3,3-BIS(CHLOROMETHYL)-	0.04		1.4		0	H10,I20	979
BUTYRIC ACID, 2-METHYLENE-	STYRENE	0.31	0.01	0.68	0.01	70		1212
BUTYRIC ACID, 2-METHYLENE-, ETHYL ESTER	ACRYLIC ACID	0.45	0.05	0.9	0.1	80	I7	896
BUTYRIC ACID, 2-METHYLENE-, METHYL ESTER	METHYL METHACRYLATE	0.1		2.03		30,60		98
		0.96		1.16		30		98
	STYRENE	0.20	0.02	0.90	0.07	70	G18	175
		0.21	0.02	0.82	0.08	65		174
BUTYRIC ACID, 3-METHYL-2-METHYLENE-, METHYL ESTER	STYRENE	0		1.85	0.07	65		174
		0		1.94	0.03	70		175
CADMIUM, DIACETATOBIS(5-VINYL-2-PICOLINE)-	STYRENE	0.47	0.04	0.56	0.05	60	I6	952
CARBAMIC ACID, DIETHYL-, VINYL ESTER	ACRYLIC ACID	0.09		5.55		80		316
	ACRYLIC ACID, METHYL ESTER	0.01		4.45		80		316
	MALEIC ANHYDRIDE	0.035		0		80		316
	STYRENE	0.03	0.01	32	5	66		802
	VINYL ACETATE	0.25	0.08	1.8	0.4	66		802

M1	M2	R1	+/-	R2	+/-	TEMP.	REMARKS	REF.
CARBAMIC ACID, DIETHYLDITHIO-, S-VINYL ESTER								
	STYRENE	0.143		4.05		60		681
CARBAMIC ACID, DIETHYLTHIO-, S-VINYL ESTER								
	STYRENE	0.14	0.03	4.4	0.6	66		802
	VINYL ACETATE	1.5	0.3	0.16	0.08	66		802
CARBAMIC ACID, ISOPROPENYL-, 2,3-EPOXYPROPYL ESTER								
	METHYL METHACRYLATE	0.12		1.70		60	I5	418
CARBAMIC ACID, N-METHYL-N-VINYL THIO-, S-ETHYL ESTER								
	STYRENE	0.025	0.01	13.0	3.0	66		802
	VINYL ACETATE	1.3	0.2	0.6	0.1	66		802
CARBAMIC ACID, VINYL-, TERT-BUTYL ESTER								
	CARBONIC ACID, THIO-, O-TERT-BUTYL S-VINYL ESTER	0.12	0.05	5.1	0.1	70		1176
CARBAMIC ACID, VINYL-, 2,3-EPOXYPROPYL ESTER								
	ACRYLONITRILE	1.5	0.5	0.066	0.004	60	I5	418
	METHYL METHACRYLATE	0.10		2.75		60	I5	418
	STYRENE	0.02	0.02	10.4	0.02	60	I5	418
	VINYL ACETATE	3.2	0.3	0.26	0.04	60	I5	418
CARBAMIC ACID, VINYL-, ETHYL ESTER								
	2-PYRROLIDINONE, 1-VINYL-	0.42		2		65	I15	287
	VINYL ACETATE	0.33		0.33		65	I15	287
CARBAZOLE, 9-VINYL-								
	ACETIC ACID, THIO-, VINYL ESTER	0.622		0.5		60	I7	329
	ACRYLIC ACID, METHYL ESTER	0.050		0.50		75	I5	331
	ACRYLONITRILE	0.11	0.02	0.43	0.02	30	I5	1129
		0.04	0.02	0.28	0.02	30	D	1129
		0.09		0.35		30		1102
		0.09		0.35		60		1102
	ANISOLE, P-VINYL-	21.4	0.81	0.13	0.005	25	H10	1147
		23.2	1.9	0.09	0.01	0	H10	1147
	BUTYRIC ACID, VINYL ESTER	1.28	0.06	0.059	0.020	100	C	1001
	ETHYLENE, CHLORO-	4.77		0.17		0	Q	391
	ETHYLENE, 1,1-DICHLORO-	3.7		0.020		75	I7	331
	FORMIC ACID, VINYL ESTER	4.22	0.16	0.196	0.004	100	C	1001
	METHYL METHACRYLATE	0.04	0.01	2.0	0.1	75	I7	331
		0.07	0.03	2.7	0.3	30	I5	1129
		0.20		2.0		70	C	29
	1-PROPENE, 3-CHLORO-	>1		0		70		30
	1-PROPENE, 2,3-DICHLORO-	>1		0		70		30
		>1		0		70		24
	PROPIOLONITRILE	0.075	0.005	0.03	0.005	30	I5	1129

CARBAZOLE, 9-VINYL-

M2	R1	+/-	R2	+/-	TEMP.	REMARKS	REF.
PROPIONIC ACID, VINYL ESTER	1.68	0.14	0.076	0.018	100	C	1001
STYRENE	0.012	0.002	5.5	0.8	70	C	29
	0.035		5.7		75	I7	331
	>7		0.01	0.01	30	H52,I5/6	949
STYRENE, P-CHLORO-	0.023	0.003	7	0.2	30	I5	1129
STYRENE, 2,5-DICHLORO-	0.016		8.0		70	Q	33
	0.016	0.002	8	0.5	60		32
SUCCINIMIDE, N-VINYL-	0.3		1.05			I31,Q	1217
VINYL ACETATE	0.15	0.03	0.03	0.01	65	C	1389
	2.68	0.10	0.126	0.032	100	C	1001
	3.02	0.24	0.152	0.018		C	1001
	3.9	0.2	0.13	0.03	30	I5	1129

CARBONIC ACID, CYCLIC (HYDROXYMETHYL)ETHYLENE ESTER, ACRYLATE

M2	R1	+/-	R2	+/-	TEMP.	REMARKS	REF.
ACRYLIC ACID, METHYL ESTER	1.04	0.10	0.965	0.15	50	C,I2	199
METHACRYLIC ACID, HEXYL ESTER	0.66	0.08	1.50	0.13	50	C,I2	199
METHYL METHACRYLATE	0.521	0.03	1.83	0.20	50	C,I2	199
STYRENE	0.20	0.06	0.80	0.06	50	C,I2	199
VINYL ACETATE	10.20	0.40	0.049	0.012	50	C,I2	199

CARBONIC ACID, CYCLIC (HYDROXYMETHYL)ETHYLENE ESTER, MALEATE

M2	R1	+/-	R2	+/-	TEMP.	REMARKS	REF.
ACRYLIC ACID, METHYL ESTER	0.10	0.02	9.10	0.40	50	C,I2	199
METHYL METHACRYLATE	0		7.5	0.5	50	C,I2	199
STYRENE	0.10	0.05	0.74	0.03	50	C,I2	199
VINYL ACETATE	0.09	0.02	0.05	0.01	50	C,I2	199

CARBONIC ACID, CYCLIC (HYDROXYMETHYL)ETHYLENE ESTER, 4-(METHYLENESUCCINATE)

M2	R1	+/-	R2	+/-	TEMP.	REMARKS	REF.
ACRYLIC ACID, METHYL ESTER	1.44	0.40	0.45	0.04	50	C,I2	199
METHYL METHACRYLATE	0.24	0.15	1.38	0.25	50	C,I2	199
STYRENE	0.25	0.10	0.18	0.05	50	C,I2	199
VINYL ACETATE	3.35	0.05	0.02	0.01	50	C,I2	199

CARBONIC ACID, CYCLIC VINYLENE ESTER

M2	R1	+/-	R2	+/-	TEMP.	REMARKS	REF.
ACETIC ACID, THIO-, VINYL ESTER	0.04	0.01	12.9	1.1	60	C,I11,Q	755
ACRYLAMIDE	0.05	0.02	13.8	0.5		P,Q	89
ACRYLONITRILE	0.085	0.001	14.9	0.1	70		511
	0.160	0.001	0.185	0.001	50	I26	1370
ETHER, ISOBUTYL VINYL	0		13		60		1
ETHYLENE, CHLORO-	0.09		5.2		80		1403
METHYL METHACRYLATE	0.005		70		70		349
	0.01		67		20	I11	581
2-PYRROLIDINONE, 1-VINYL-	0.4		0.7		60		349
STYRENE	0		<20.0		60		458
SULFIDE, METHYL VINYL	0.05	0.04	10.6	1.2	60		458
	0.0579		3.71		55		584
VINYL ACETATE	0.13	0.1	7.3	0.7	60		458

M1 / M2	R1	+/-	R2	+/-	TEMP.	REMARKS	REF.
CARBONIC ACID, CYCLIC VINYLENE ESTER							
VINYL ACETATE	0.15		4.00		70	C	314
	0.27		3.0		70		349
	3.075		0.036		25	F,Ill	512
CARBONIC ACID, DIVINYL ESTER							
STYRENE, P-CHLORO-	0.036		27.5		60	I7	0496
VINYL ACETATE	1.33	0.08	0.77	0.05	60	I7	0496
CARBONIC ACID, ETHYL VINYL ESTER							
STYRENE, P-CHLORO-	0.025	0.017	39.8	0.7	60	I7	0496
VINYL ACETATE	0.79	0.07	0.87	0.04	60	I7	0496
CARBONIC ACID, THIO-, O-TERT-BUTYL S-VINYL ESTER							
ACRYLAMIDE, N-ISOPROPYL-	0.3	0.05	1.17	0.05	70		1176
CARBAMIC ACID, VINYL-, TERT-BUTYL ESTER	5.1	0.1	0.12	0.05	70		1176
METHACRYLIC ACID, TERT-BUTYL ESTER	0.15	0.05	1.7	0.05	70		1176
METHYL METHACRYLATE	0.17	0.05	1.4	0.05	70		1176
2-PYRROLIDINONE, 1-VINYL-	3.94	0.05	0.12	0.01	70		1176
STYRENE	0.2	0.05	3	0.1	70		1176
VINYL ACETATE	11	0.1	0.04	0.01	70		1176
CARBON MONOXIDE							
ETHYLENE	0		0.147		120-130	C,M3	183
	0		0.500		135	C,I21,M8	183
	0		0.042		20	E,M2	182
	0		0.045		20	E,M7	183
	0.15		0.16		70	E,L	121
ETHYLENE, CHLORO-	0.15		13.47	0.02	60	I7	750
VINYL ACETATE	0.33	0.05	0.24	0.05	60		646
3-CARENE							
METHYL METHACRYLATE	0.22		3.58			Q	813
CHALCONE							
1,3-BUTADIENE	0.03		0.78	0.12	60		851
CHALCONE, 2-CHLORO-							
1,3-BUTADIENE	-0.02		0.78		59	K	851
	-0.04	0.05	0.78	0.12	60		851
CHLORAL							
ACETALDEHYDE	0.00		0.18	0.05	-78	H14	431
ACETALDEHYDE, CHLORO-	0.00		85	5	-50	H14	431
	0.00		280	10	-78	H14	431
ACETALDEHYDE, DICHLORO-	1.50		0.65		-78	H40,I2	627

COPOLYMERIZATION PARAMETER

M1	M2	R1	+/-	R2	+/-	TEMP.	REMARKS	REF.
CHLORAL								
	ISOCYANIC ACID, BUTYL ESTER	26		0		-75	G19,I9	1331
				0		-75	G19,I54	1331
	ISOCYANIC ACID, PHENYL ESTER	0.5		0		-75	G19,I9	1331
				0		-75	G19,I54	1331
CHROMIUM, TRICARBONYL-PI-(BENZYL ACRYLATE)-								
	ACRYLIC ACID, METHYL ESTER	0.56	0.11	0.63	0.3	70	I38	1357
	STYRENE	0.1	0.02	0.34	0.13	60	I38	1357
CHROMIUM, TRICARBONYL-PI-(STYRENE)-								
	ACRYLIC ACID, METHYL ESTER	0		0.75		70	I7	1355
	STYRENE	0		1.39		70	I7	1355
CINNAMIC ACID								
	METHYL METHACRYLATE	0		3.8		70	I7	84
	STYRENE	0		1.85	0.03	60		86
CINNAMIC ACID, CIS-								
	ACRYLAMIDE	6.7	0.6	0		20	F,I73	1297
CINNAMIC ACID, TRANS-								
	ACRYLAMIDE	3	0.2	0		20	F,I73	1297
CINNAMIC ACID, BENZYL ESTER								
	STYRENE	0.16	0.03	1.45	0.03	60		748
CINNAMIC ACID, BUTYL ESTER								
	STYRENE	0.08	0.03	1.95	0.03	60		748
		0.10		0.87		80	C	1070
CINNAMIC ACID, TERT-BUTYL ESTER								
	STYRENE	0.10	0.03	1.83	0.03	60		748
CINNAMIC ACID, 2-CHLOROETHYL ESTER								
	STYRENE	0.16	0.03	1.48	0.03	60		748
CINNAMIC ACID, ETHYL ESTER								
	STYRENE	0.17	0.09	1.49	0.16	70	C	1121
		0.05		2.7	0.3	75	I7	806
2-PICOLINE, 5-VINYL-								
	STYRENE	0.10	0.03	1.50	0.03	60		748

M1	M2	R1	+/-	R2	+/-	TEMP.	REMARKS	REF.
CINNAMIC ACID, METHYL ESTER								
	ACRYLONITRILE	0		6	2	60		211
		0		1.9	0.2	60		211
	STYRENE	0.10	0.03	2.03	0.03	60	O	748
CINNAMIC ACID, PHENYL ESTER								
	STYRENE	0.08	0.03	1.53	0.03	60		748
CINNAMIC ACID, VINYL ESTER								
	METHACRYLONITRILE	0.15	0.06	4.6	0.4	70	17	806
	2-PYRROLIDINONE, 1-VINYL-	1.20		0.01		70	17	806
	STYRENE	0.25	0.1	1.25	0.1	70	17	806
	VINYL ACETATE	1.20	0.10	0.04		70	17	806
CINNAMIC ACID, O-CHLORO-, METHYL ESTER								
	1,3-BUTADIENE	-0.02	0.05	1.07	0.12	60		851
		-0.03	0.05	1.20	0.12	80		851
CINNAMIC ACID, P-CHLORO-, METHYL ESTER								
	1,3-BUTADIENE	0.00	0.05	2.73	0.3	80		851
CINNAMIC ACID, ALPHA-CYANO-								
	METHYL METHACRYLATE	0		17.0		70		84
CINNAMIC ACID, ALPHA-CYANO-, BENZYL ESTER								
	STYRENE	0		0.43		80	C,R	1199
CINNAMIC ACID, ALPHA-CYANO-, BUTYL ESTER								
	STYRENE	0		0.53		80	C,R	1199
CINNAMIC ACID, ALPHA-CYANO-, CYCLOHEXYL ESTER								
	STYRENE	0		0.44		80	C,R	1199
CINNAMIC ACID, ALPHA-CYANO-, ETHYL ESTER								
	ACRYLONITRILE	-0.15	0.1	18	2	70	C	1374
	1,3-BUTADIENE	0		0.25		35	K,O	658
	STYRENE	0		0.3		80	Q	538
		0		0.51		80	C,R	1199
CINNAMIC ACID, ALPHA-CYANO-, 2-ETHYLHEXYL ESTER								
	STYRENE	0		0.59		80	C,R	1199

COPOLYMERIZATION PARAMETER

M1	M2	R1	+/-	R2	+/-	TEMP.	REMARKS	REF.
CINNAMIC ACID, ALPHA-CYANO-, HEXYL ESTER								
	STYRENE	0		0.42		80	C,R	1199
CINNAMIC ACID, ALPHA-CYANO-, METHYL ESTER								
	STYRENE	0		0.45		80	C,R	1199
CINNAMONITRILE								
	ACRYLONITRILE	0.56	0.5	6.4	1	70	C	1374
	CINNAMONITRILE, P-CHLORO	0.62	0.05	0.87	0.01	25	H47,I2	1255
	CINNAMONITRILE, M-CHLORO-	0.78	0.08	1.03	0.01	25	H47,I2	1255
	CINNAMONITRILE, M-METHOXY-	1.56	0.14	1.09	0.03	25	H47,I2	1255
	CINNAMONITRILE, P-METHOXY-	1.25	0.09	0.71	0.04	25	H47,I2	1255
	CINNAMONITRILE, M-METHYL-	1.19	0.15	0.78	0.1	25	H47,I2	1255
	CINNAMONITRILE, P-METHYL-		0.04	0.85	0.02	25	H47,I2	1255
	STYRENE	0		2.2		80	C	538
	STYRENE	0.1	0.1	0.9	0.2	90	C	118
CINNAMONITRILE, P-CHLORO								
	CINNAMONITRILE	0.87	0.01	0.56	0.05	25	H47,I2	1255
CINNAMONITRILE, M-CHLORO-								
	CINNAMONITRILE	1.03	0.01	0.62	0.08	25	H47,I2	1255
CINNAMONITRILE, M-METHOXY-								
	CINNAMONITRILE	1.09	0.03	0.78	0.14	25	H47,I2	1255
CINNAMONITRILE, P-METHOXY-								
	CINNAMONITRILE	0.71	0.04	1.56	0.09	25	H47,I2	1255
CINNAMONITRILE, M-METHYL-								
	CINNAMONITRILE	0.78	0.1	1.25	0.15	25	H47,I2	1255
CINNAMONITRILE, P-METHYL-								
	CINNAMONITRILE	0.85	0.02	1.19	0.04	25	H47,I2	1255
CITRACONIC ANHYDRIDE								
	METHYL METHACRYLATE	-0.06	0.08	31.8	2.5	30	D	709
	STYRENE	0.01	0.01	0.15	0.02	60	C	861
	STYRENE	0.01		0.15		60	C	166
COUMARIN								
	ETHYLENE, 1,1-DICHLORO-	0		1		70		23

REMARKS PAGE II-366, REFERENCES PAGE II-368

M1 / M2	R1	+/-	R2	+/-	TEMP.	REMARKS	REF.
CROTONALDEHYDE							
ACRYLONITRILE	0		8.0		60	M3	394
	0.0105	0.0105	25	4.2	50	I6	840
ETHYLENE, CHLORO-	0.02		1.93		60	I5	1115
ETHYLENE, 1,1-DICHLORO-	0		17		70		217
2-PYRROLIDINONE, 1-VINYL-	0.03		0.5		80		1004
STYRENE	0.03		14.7		100		393
	0.07		14.7		100	M2	393
	0.12		14.7		100	M3	393
CROTONAMIDE							
VINYL ACETATE	2.0	0.05	0.01	0.01	110		1003
CROTONAMIDE, N-(HYDROXYMETHYL)-							
VINYL ACETATE	0.045	0.1	0.01	0.01	110		1003
CROTONIC ACID							
ACRYLAMIDE	0		54	2	30	B,I11	1309
ACRYLONITRILE	0.065	0.005	21	10	60	D	211
ETHYLENE, 1,1-DICHLORO-	0.02	0.02	35	5	60	C	166
METHYL METHACRYLATE	0		4.3		70		84
2-PYRROLIDINONE, 1-VINYL-	0		0.85	0.05	65		996
STYRENE	0		20		60		211
VINYL ACETATE			0.33			Q	319
	0.01		0.3		70	C	997
	0.01		0.3		70	C	955
	0.01	0.01	0.33	0.05	68	C	166
CROTONIC ACID, CIS-							
ACRYLAMIDE	0.11		4.72		20	Q	1273
	4	0.3	0			F,I73	1297
IMIDAZOLE, 2-METHYL-1-VINYL-	0.6		0.09			Q	1273
CROTONIC ACID, TRANS-							
ACRYLAMIDE	0.12		5.32		20	Q	1273
	6.5	0.4	0			F,I73	1297
IMIDAZOLE, 2-METHYL-1-VINYL-	0.49		0.19			Q	1273
CROTONIC ACID, BUTYL ESTER							
ETHYLENE, 1,1-DICHLORO-	0		33		70		217
CROTONIC ACID, METHYL ESTER							
STYRENE	0.01		26.0		60	C,I7	928
	0.01		26.0		60	C,I7	641

M1	M2	R1	+/-	R2	+/-	TEMP.	REMARKS	REF.
CROTONIC ACID, PROPYL ESTER								
	ETHYLENE, 1,1-DICHLORO-	0		20		70		217
CROTONIC ACID, 2-PROPYNYL ESTER								
	STYRENE	0		3.42		60		1306
CROTONIC ACID, 4-HYDROXY-, GAMMA-LACTONE								
	STYRENE	0		8.5	5.6	60		458
CROTONONITRILE, CIS-								
	ACRYLONITRILE	0.09	0.03	10	3	50	I6	840
	CROTONONITRILE, TRANS-	0.79	0.31	1.55	0.11	0	H21,I9	515
		0.92	0.21	2.64	0.06	-78	G16,I54	515
		0.94	0.04	1.05	0.02	-78	G6,I5	515
		0.97	0.04	1.13	0.02	-78	G19,I5	515
		0.98	0.10	0.96	0.04	-78	G15,I5	515
		1.48	0.54	4.09	0.18	-78	G16,I5	515
	MALEIC ANHYDRIDE	0		0		60	I7	669
CROTONONITRILE, TRANS-								
	CROTONONITRILE, CIS-	0.96	0.04	0.98	0.10	-78	G15,I5	515
		1.05	0.02	0.94	0.04	-78	G6,I5	515
		1.13	0.02	0.97	0.04	-78	G19,I5	515
		1.55	0.11	0.79	0.31	0	H21,I9	515
		2.64	0.06	0.92	0.21	-78	G16,I54	515
		4.09	0.18	1.48	0.54	-78	G16,I5	515
	STYRENE	0		20.0	0.5	60	C,R	444
		0		24.0		60	C,I7	641
		0		24.0		60	C,I7	928
CYANAMIDE, DIALLYL-								
	VINYL ACETATE	1.62	0.01	0.6.	0.02		K,Q	1386
CYCLOBUTANE, 3-(EPOXYETHYL)-1,1,2,2-TETRAFLUORO-								
	PROPANE, 1-CHLORO-2,3-EPOXY-	0.408	0.007	1.45	0.15	20	H13	1360
CYCLOBUTANE, METHYLENE-								
	ACRYLONITRILE	0		1.01		60	I7	934
		0		1.01		60	I7	1134
CYCLOBUTANE, 1-METHYLENE-3-PHENYL-								
	ACRYLONITRILE	0		0.85		60	I7	934
CYCLOBUTANECARBONITRILE, 3-METHYLENE-								
	ACRYLONITRILE	0.02		0.65		60	I7	934

M1	M2	R1	+/-	R2	+/-	TEMP.	REMARKS	REF.
CYCLOBUTANECARBOXYLIC ACID, 3-ACETYL-2,2-DIMETHYL-, VINYL ESTER *	ACRYLONITRILE	0.143	0.1	30	12	50	16	840
	1,3-BUTADIENE	0.015	0.046	3.40	0.04	60	C	591
	ETHYLENE, CHLORO-	0.446	0.015	37.8	6.5	60	C	591
	ETHYLENE, 1,1-DICHLORO-	0.03	0.028	1.458	0.04	60	C	591
			0.028	3.00	0.18	60	C	591
	STYRENE	0.01	0.01	65	17	60	C	591
CYCLOBUTANECARBOXYLIC ACID, 3-METHYLENE-	ACRYLONITRILE	0		0.75		60	I7	934
CYCLOBUTANECARBOXYLIC ACID, 1-METHYL-3-METHYLENE-, METHYL ESTER *	ACRYLONITRILE	0		0.78		60	I7	934
CYCLOBUTENE, HEXAFLUORO-	ETHYLENE, FLUORO-	0		3	0.6	30	H38	1165
		0		3	0.6	30	H38,I38	879
1,5,9-CYCLODODECATRIENE, TRANS-,TRANS-,CIS-	METHYL METHACRYLATE	0.2		4.84		80	Q	1422
	STYRENE	0.2		4.84		90	C,I2	1415
	STYRENE	0.22		2.54		90	C,I7	1414
CYCLOHEPTASILOXANE, TETRADECAMETHYL-	CYCLOTETRASILOXANE, HEPTAMETHYLVINYL-	17.21	0.09	0.40	0.01	90	G10	445
	CYCLOTETRASILOXANE, 2,4,6,8-TETRAMETHYL-2,4,6,8-TETRAVINYL-	1.13	0.04	0.68	0.01	90	G10	445
		1.33	0.04	0.58	0.01	150	G11	445
CYCLOHEPTENE	ACRYLONITRILE	-0.24	0.12	6.9	0.35	60		1066
	MALEIC ANHYDRIDE	0		0.068	0.002	60	I7	668
1,3-CYCLOHEXADIENE	ACRYLONITRILE	0.11	0.02	0.05	0.2	65		1058
		0.18		0.02		60	C	682
		0.20	0.01	0.026	0.014	200	C	248
		0.50		1.00		50	C	682
	1,3-BUTADIENE	1.1		0.8		50	G30	1320
	CYCLOHEXENE, 1-VINYL-	0.12	0.09	2.52	0.72	0	H19,I3	1208
	ISOPRENE	0.2	0.05	1.5	0.1	30	G17,I20	203
		0.6	0.05	0.4	0.05	0	H23,I2	203
		0.6	0.05	0.4	0.05	30	H17,I7	203
		1.64		0.5		50	G30,I7	1419
	STYRENE	0.36	0.05	0.53	0.05	0	G17	1058
		0.77	0.10	0.53	0.10	0	H10	1058

M1 / M2	R1	+/-	R2	+/-	TEMP.	REMARKS	REF.
1,3-CYCLOHEXADIENE, 1-METHOXY-							
ACRYLONITRILE	0.13	0.1	0.48	0.1	60		1434
STYRENE	0	0.2	4.68	0.2	60		1434
1,3-CYCLOHEXADIENE, 1-METHYL-							
ACRYLONITRILE	0.18	0.05	0.18	0.05	65		1058
BENZALDEHYDE	0.84	0.02	0.01	0.01	-78	H10	73
CYCLOHEXANE, ALLYL-							
ACRYLONITRILE	0		6.0		70		67
CYCLOHEXANE, 1,2-DIVINYL-							
STYRENE	0.01		30		0	H4,I2	1138
CYCLOHEXANE, 1,2-EPOXY-4-VINYL-							
ETHYLENE, CHLORO-	0.2		3.14		60	Q	1107
VINYL ACETATE	0		2.26				1169
CYCLOHEXANE, METHYLENE-							
ACRYLONITRILE	0		1.13		60	I7	1134
CYCLOHEXANE, 1,1,2,3,3,4,4-OCTAFLUORO-5,6-DIMETHYLENE-							
STYRENE	0.15		0.34		50		1155
CYCLOHEXANE, VINYL-							
MALEIC ANHYDRIDE	0.008		0.22		65	I7	1148
1-PROPENE	0.049		80			H54,Q	1427
CYCLOHEXANEACETIC ACID, ALPHA-METHYLENE-, METHYL ESTER							
STYRENE	0		1.65	0.05	65		174
CYCLOHEXASILOXANE, DODECAMETHYL-							
CYCLOTETRASILOXANE, HEPTAMETHYLVINYL-	1.54	0.02	0.39	0.01	90	G10	445
CYCLOTETRASILOXANE, 2,4,6,8-TETRAMETHYL-2,4,6,8-TETRAVINYL-	0.23	0.02	5.09	0.17	90	G10	445
	0.26	0.01	1.93	0.10	130	G11	445
	0.33	0.02	6.78	0.17	130	G10	445
	0.37	0.03	2.85	0.42	165	G11	445
	0.61	0.05	2.27	0.12	230	G12	445
CYCLOHEXENE							
ACRYLONITRILE	-0.42	0.37	8.0	0.74	60		1066
MALEIC ANHYDRIDE	0		0.083	0.005	60	I7	668

M1 / M2	R1	+/-	R2	+/-	TEMP.	REMARKS	REF.
CYCLOHEXENE, 1-VINYL-							
1,3-CYCLOHEXADIENE	2.52	0.72	0.12	0.09	0	H19,I3	1208
CYCLOHEXENE, 4-VINYL-							
ETHYLENE, 1,1-DICHLORO-	0		1.8		70		217
3-CYCLOHEXENE-1-CARBOXYLIC ACID, 1-METHYL-, 2,3-EPOXYPROPYL * ESTER							
ACRYLONITRILE	0.02		18.4		100		416
STYRENE	0		21.0		100		416
3-CYCLOHEXENE-1,5-DICARBOXYLIC ACID, 1,2-DIMETHYL-, BIS(2,3 * -EPOXYPROPYL) ESTER							
STYRENE	0		17.2		100		416
1,3-CYCLOOCTADIENE, CIS-,CIS-							
STYRENE	0.32	0.05	1.9	0.05	0	H23	1304
	0.48	0.05	1.63	0.05	0	H20	1304
	0.5	0.05	1.64	0.05	-20	H20	1304
	0.56	0.05	1.66	0.05	-20	H23	1304
	0.42	0.05	0.56	0.05	0	H23	1304
	0.52	0.05	0.52	0.05	-20	H23	1304
STYRENE, P-CHLORO-	0.7	0.05	0.45	0.05	0	H20	1304
	0.84	0.05	0.32	0.05	-20	H20	1304
CYCLOOCTENE							
ACRYLONITRILE	-0.28	0.21	2.8	0.15	60	I7	1066
MALEIC ANHYDRIDE	0		0.067	0.006	60		668
1,3-CYCLOPENTADIENE							
1,3-CYCLOPENTADIENE, METHYL- (53/47 2/1-MIXT.)	0.36	0.26	8.5	3.5	-78	H69,I2	516
	0.42	0.23	14.9	5.6	-78	H19,I3	516
1,3-CYCLOPENTADIENE, 1,3-DIMETHYL-	0.3	0.1	6.85	1.1	-78	H10,I2	1141
PHENOL, O-VINYL-	1.79	0.27	0.36	0.23	-25	H10,I3	1224
	2.99	0.21	0.78	0.08	-25	H10,I2	1224
1,3-CYCLOPENTADIENE, METHYL- (53/47 2/1-MIXT.)							
1,3-CYCLOPENTADIENE	8.5	3.5	0.36	0.26	-78	H69,I2	516
	14.9	5.6	0.42	0.23	-78	H19,I3	516
1,3-CYCLOPENTADIENE, 1,3-DIMETHYL-							
1,3-CYCLOPENTADIENE	6.85	1.1	0.3	0.1	-78	H10,I2	1141
1,3-CYCLOPENTADIENE, 1-METHYL-							
1,3-CYCLOPENTADIENE, 2-METHYL-	0.01	0.05	1.1	0.2	-78	H10,I2	70

M1 / M2	R1	+/-	R2	+/-	TEMP.	REMARKS	REF.
1,3-CYCLOPENTADIENE, 2-METHYL-							
1,3-CYCLOPENTADIENE, 1-METHYL-	1.1	0.2	0.01	0.05	-78	H10,I2	70
1,3-CYCLOPENTADIENE, 2-(4-PENTENYL)-							
ACRYLONITRILE	0.9		0.1		60	I6	1315
CYCLOPENTANE, METHYLENE-							
ACRYLONITRILE	0		0.39		60	I7	1134
CYCLOPENTENE							
ACRYLONITRILE	-0.16	0.34	4.8	0.36	60	I7	1066
MALEIC ANHYDRIDE	0		0.010	0.003	60		668
4-CYCLOPENTENE-1,3-DIONE							
ACRYLONITRILE	0.21		3.67		60	I7,L	324
ETHYLENE, 1,1-DICHLORO-	0.15		2.4	0.6	65	I7	1049
METHYL METHACRYLATE	0.083		7.4		65	I7	324
STYRENE	0.415		0.024		50		1048
STYRENE, P-CHLORO-	0.02	0.01	0.32	0.06	50	I7	1049
CYCLOPROPANECARBOXYLIC ACID, 2-ISOPROPENYL-, ETHYL ESTER							
CYCLOPROPANECARBOXYLIC ACID, 2-METHYL-2-VINYL-, ETHYL ESTER	2.5	0.25	0.42	0.1	60	I7	1272
CYCLOPROPANECARBOXYLIC ACID, 2-VINYL-, ETHYL ESTER	1.97	0.1	0.51	0.06	60	I7	1272
CYCLOPROPANECARBOXYLIC ACID, 2-METHYL-2-VINYL-, ETHYL ESTER							
CYCLOPROPANECARBOXYLIC ACID, 2-ISOPROPENYL-, ETHYL ESTER	0.42	0.1	2.5	0.25	60	I7	1272
CYCLOPROPANECARBOXYLIC ACID, 2-VINYL-, ETHYL ESTER	0.84	0.1	1.27	0.25	60	I7	1272
CYCLOPROPANECARBOXYLIC ACID, 2-VINYL-, ETHYL ESTER							
CYCLOPROPANECARBOXYLIC ACID, 2-ISOPROPENYL-, ETHYL ESTER	0.51	0.06	1.97	0.1	60	I7	1272
CYCLOPROPANECARBOXYLIC ACID, 2-METHYL-2-VINYL-, ETHYL ESTER	1.27	0.25	0.84	0.1	60	I7	1272
CYCLOPROPENE, TETRACHLORO-							
STYRENE	0		5.7	0.4	60		1214
VINYL ACETATE	0		0.72	0.02	80		1214
CYCLOTETRASILOXANE, HEPTAMETHYLVINYL-							
CYCLOHEPTASILOXANE, TETRADECAMETHYL-	0.40	0.01	17.21	0.09	90	G10	445
CYCLOHEXASILOXANE, DODECAMETHYL-	0.39	0.01	1.54	0.02	90	G10	445
CYCLOTETRASILOXANE, 2,4,6,8-TETRAMETHYL-2,4,6,8-TETRAVINYL-							
CYCLOHEPTASILOXANE, TETRADECAMETHYL-	0.58	0.01	1.33	0.04	150	G11	445
CYCLOHEPTASILOXANE, TETRADECAMETHYL-	0.68	0.01	1.13	0.04	90	G10	445
CYCLOHEXASILOXANE, DODECAMETHYL-	1.93	0.10	0.26	0.01	130	G11	445

M1	M2	R1	+/-	R2	+/-	TEMP.	REMARKS	REF.
CYCLOTETRASILOXANE, 2,4,6,8-TETRAMETHYL-2,4,6,8-TETRAVINYL-								
	CYCLOHEXASILOXANE, DODECAMETHYL-	2.27	0.12	0.61	0.05	230	G12	445
		2.85	0.42	0.37	0.03	165	G11	445
		5.09	0.17	0.23	0.02	90	G10	445
		6.78	0.17	0.33	0.02	130	G10	445
CYCLOTRISILOXANE, HEXAMETHYL-								
	1-OXA-2,5-DISILACYCLOPENTANE, 2,2,5,5-TETRAMETHYL-	0.146	0.014	8.11	0.13	40	G	633
		0.31	0.07	4.10	0.77	100	G	633
CYCLOTRISILOXANE, HEXAPHENYL-								
	1-OXA-2,5-DISILACYCLOPENTANE, 2,2,5-TRIMETHYL-5-PHENYL-	1.47	0.25	0.58	0.16	25	G	633
DECANOIC ACID, VINYL ESTER								
	ACRYLONITRILE	0.04	0.03	4.0	0.3	60	C,I16	58
	ETHYLENE, CHLORO-	0.2		4.7		40	C,I11,J	503
1-DECENE								
	1-BUTENE	0.70		1.5		23	H54,I21	565
	1-HEXENE	0.90		1.3		23	H54,I21	565
	STYRENE	0.16		5.0		50	H12	827
M-DIOXANE								
	OXETANE, 3,3-BIS(CHLOROMETHYL)-	0		13		0	H10,I3	50
M-DIOXANE, 4-(CHLOROMETHYL)-								
	OXETANE, 3,3-BIS(CHLOROMETHYL)-	0		13		0	H10,I3	50
M-DIOXANE, 5,5-DIMETHYL-2-VINYL-								
	ACRYLONITRILE	0.08		2.25		60	I6	1438
M-DIOXANE, 4-METHYL-								
	OXETANE, 3,3-BIS(CHLOROMETHYL)-	0		13		0	H10,I3	50
M-DIOXANE, 4-METHYL-2-VINYL-								
	ACRYLONITRILE	0.04		4.65		60	I6	1438
M-DIOXANE, 4-PHENYL-								
	OXETANE, 3,3-BIS(CHLOROMETHYL)-	0		12.5		0	H10	977
	1,3,5-TRIOXANE	0		46		0	H10,I3	50
				9	2	35		488

M1	M2	R1	+/-	R2	+/-	TEMP.	REMARKS	REF.
M-DIOXANE, 2-VINYL-								
	ACRYLONITRILE	0.04		4		60	I6	1438
P-DIOXANE								
	OXETANE, 3,3-BIS(CHLOROMETHYL)-	0		1.85		0	H10	977
				7		0	H10,I3	50
1,3-DIOXEPANE								
	1,3-DIOXOLANE, 4-(CHLOROMETHYL)-	14.2	1.5	0.12	0.05	0	H10	716
	STYRENE	24.5	2.7	0.37	0.07	0	H10,I2	719
M-DIOXIN, 4-METHYL-								
	MALEIC ANHYDRIDE	0.18		0		50	R	481
	STYRENE	0.05		2.5		50	R	481
P-DIOXIN, 2,3-DIHYDRO-								
	ACRYLIC ACID, 2-CHLOROETHYL ESTER	0.002	0.1	12.9	1.0	60		438
	ACRYLIC ACID, ETHYL ESTER	0.002	0.001	37.9	1.4			433
	ACRYLIC ACID, METHYL ESTER	0.001	0.05	23.	2.			433
	ACRYLONITRILE	0.001	0.002	5.9	0.1		Q	433
	MALEIMIDE	1.28		0			Q	665
	METHYL METHACRYLATE	0.02	0.02	15.3	0.6		Q	433
1,3-DIOXOLANE								
	1,3-DIOXOLANE, 4-(CHLOROMETHYL)-	1.8		0.15	0.05	0	H10	437
		1.8	0.2	0.15	0.04	0	H62	1061
		4.5	1.0	0.08	0.05	0	H16,I10	1061
		9.3	1.0	0.07	0.10	0	H20,I3	1061
		10.7	0.3	0.18	0.03	0	H10	718
		10.7	0.3	0.18		0	H10	716
		12.0	1.0	0.12		0	H10,I2	1061
		13.3	1.3	0.13	0.04	0	H16,I2	1061
		14.0	2.0	0.09	0.04	0	H16,I22	1061
		20	5	0.14	0.08	0	H20	1061
		21	2	0.37	0.10	0	H10,I22	1061
		23.0	6.0	0.23	0.15	0	H20,I22	1061
		28		0.30	0.20	0	H16,I3	1061
	FURAN, TETRAHYDRO-	0.24	0.03	195	10	0	H20,I2	1061
		0.245		40		0	H20,I3	1061
		0.25	0.05	28	4	0	H10	718
		0.71	0.10	25	3	0	H10,I2	1061
	OXETANE, 3,3-BIS(CHLOROMETHYL)-	0.65		1.5		0	H10	437
		0.65	0.05	1.5	0.1	0	H10	718
		1.5	0.1	0.94	0.05	0	H20,I2	1442
	2-OXETANONE, 4,4-DIMETHYL-	0.08		1.2		0	H10	1061
	PROPANE, 1-CHLORO-2,3-EPOXY-	1.9	0.2	0.35	0.05	25	H10,I2	720
	STYRENE	6.5	0.85	0.65	0.07	0	H10,I3	719
	TRIMETHYLENE OXIDE	0.01		90		0	H20,I3	1061
		0.013		165		0	H10,I2	1061

M1	M2	R1	+/-	R2	+/-	TEMP.	REMARKS	REF.
1,3-DIOXOLANE								
	TRIMETHYLENE OXIDE	0.03		100		0	H20,I2	1061
	1,3,5-TRIOXANE	0.47	0.15	1.36	0.03	35		488
		1.36	0.03	0.47	0.15	35	C,I22	489
		1.75		0.57		70	H63	544
1,3-DIOXOLANE, 4-(CHLOROMETHYL)-								
	1,3-DIOXEPANE	0.12	0.05	14.2	1.5	0	H10	716
	1,3-DIOXOLANE	0.07	0.05	9.3	1.0	0	H20,I3	1061
		0.08	0.04	4.5	1.0	0	H16,I10	1061
		0.09	0.04	14.0	2.0	0	H16,I22	1061
		0.12	0.03	12.0	1.0	0	H10,I2	1061
		0.13	0.04	13.3	1.3	0	H16,I2	1061
		0.14	0.08	20	5	0	H20	1061
		0.15		1.8		0	H10	437
		0.15	0.05	1.8	0.2	0	H62	1061
		0.18	0.10	10.7	0.3	0	H10	718
		0.18	0.10	10.7	0.3	0	H10	716
		0.23	0.15	23.0	6.0	0	H20,I22	1061
		0.30	0.20	28	10	0	H16,I3	1061
		0.37	0.10	21	2	0	H10,I22	1061
	1,3-DIOXOLANE, 4-METHYL-	0.45	0.08	3.3	0.4	0	H10	716
	STYRENE	0.38	0.05	10.5	1.5	0	H10,I2	719
	1,3,6-TRIOXOCANE	0.08	0.02	75	15	0	H10	716
1,3-DIOXOLANE, 4-METHYL-								
	1,3-DIOXOLANE, 4-(CHLOROMETHYL)-	3.3	0.4	0.45	0.08	0	H10	716
	OXETANE, 3,3-BIS(CHLOROMETHYL)-			2.5		0	H10	977
1,3-DIOXOLANE, 2-VINYL-								
	ACRYLONITRILE	0.02		4.8		60	I6	1438
		0.03		4.62		60	I6	1351
	MALEIC ANHYDRIDE	0.12		0.17		60	C	1348
	METHYL METHACRYLATE	0.03		28.9		60	I7	1351
	STYRENE	0.02		44.9		60	I7	1351
1,3-DIOXOLANE-4-METHANOL, 2,2-DIMETHYL-, ACRYLATE								
	ACRYLIC ACID, BUTYL ESTER	0.45	0.15	1.32	0.15	50	C,I2	195
	ACRYLONITRILE	0.84	0.34	0.9	0.09	50	C,I6	195
	METHACRYLIC ACID, HEXYL ESTER	0.47	0.06	2.37	0.01	50	C,I2	195
	PYRIDINE, 2-VINYL-	0.12	0.1	0.836	0.18	50	C,I2	195
	STYRENE	0.17	0.02	0.389	0.04	50	C,I2	195
1,3-DIOXOLANE-4-METHANOL, 2,2-DIMETHYL-, 4-(METHYLENESUCCINA* TE)								
	ACRYLIC ACID, BUTYL ESTER	2.4	1.1	0.17	0.07	50	C,I2	195
	METHACRYLIC ACID, HEXYL ESTER	0.85	0.20	1.06	0.12	50	C,I2	195
	PYRIDINE, 2-VINYL-	0.16	0.11	1.06	0.10	50	C,I2	195
	STYRENE	0.18	0.04	0.14	0.01	50	C,I2	195

M1 / M2	R1	+/-	R2	+/-	TEMP.	REMARKS	REF.
1,3-DIOXOLANE-4-METHANOL, 2,2-DIMETHYL-, METHYL FUMARATE							
ACRYLIC ACID, BUTYL ESTER	0		2.69	0.55	50	C,12	195
METHACRYLIC ACID, HEXYL ESTER	0		17.5	2.5	50	C,12	195
PYRIDINE, 2-VINYL-	0		0.76	0.06	50	C,12	195
STYRENE	0		0.17	0.05	50	C,12	195
1,3-DIOXOLANE-4-METHANOL, 2-VINYL-							
ACRYLONITRILE	0.04		2.45		60	16	1351
	0.05		4.25		60	16	1438
STYRENE	0.03		20.4		60	17	1351
VINYL ACETATE	1.8		0.51		60	17	1351
1,3-DIOXOLAN-4-OL, 2-VINYL-, ACRYLATE							
ACRYLONITRILE	0.89		0.95		60		1352
STYRENE	0.22		0.81		60		1352
1,3-DIOXOLAN-4-OL, 2-VINYL-, METHACRYLATE							
STYRENE	0.51		0.49		60		1352
DISILANE, PENTAMETHYLVINYL-							
ACRYLONITRILE	0		3	1	60	16	837
STYRENE	0		50	10	60	17	837
DISILOXANE, 1-(CHLOROMETHYL)-1,1,3,3-TETRAMETHYL-3-(P-VINYL PHENYL)- *							
STYRENE	1.08	0.16	1.01	0.08	80	12	308
DISILOXANE, 1,1,3,3-PENTAMETHYL-3-VINYL-							
ACRYLONITRILE	0.1	0.01	8.0	1.2	50		773
STYRENE	0.10	0.01	60	9	50		773
VINYL ACETATE	0.010	0.001	0.99	0.15	70		773
DISILOXANE, 1,1,3,3-PENTAMETHYL-3-(P-VINYLPHENYL)-							
STYRENE	1.2	0.4	1.04	0.06	80		306
DISILOXANE, 1-(P-VINYLPHENYL)-							
STYRENE	1.2	0.4	1.04	0.06	80		371
DODECANAMIDE, N-(ACRYLAMIDOMETHYL)-							
METHYL METHACRYLATE	1.39	0.03	0.61	0.04	80	17	462
STYRENE	0.88	0.02	0.24	0.07	80	17	462
VINYL ACETATE	10.0	0.8	0.05	0.03	80	17	462

M1	M2	R1	+/-	R2	+/-	TEMP.	REMARKS	REF.
ETHANE, 1,2-BIS(P-VINYLPHENYL)-								
	STYRENE	0.87		1.05		61	C	1042
ETHANE, 1,2-DIINDEN-1-YL-								
	INDENE	0.4	0.05	0.6	0.05	-78	H23,I3	1280
ETHANESULFONIC ACID, 2-HYDROXY-, METHACRYLATE								
	ACRYLIC ACID, ETHYL ESTER	3.2	0.6	0.3	0.05	60	I55	1241
	ETHYLENE, 1,1-DICHLORO-	3.6	0.5	0.22	0.03	60	I55	1241
	METHACRYLIC ACID, ETHYL ESTER	2	0.4	1	0.1	60	I55	1241
	STYRENE	0.6	0.2	0.37	0.03	60	I55	1241
ETHANESULFONIC ACID, 2-HYDROXY-, SODIUM SALT, METHACRYLATE								
	METHACRYLIC ACID, 2-HYDROXYETHYL ESTER	0.7	0.1	1.6	0.1	6	I11	1241
ETHANOL, 2-(VINYLOXY)-								
	ETHANOL, 2-(VINYLOXY)-, ACETATE	1.23	0.06	0.93	0.01	60		1299
	3-PENTANONE, 1-HYDROXY-5-(VINYLOXY)-, ACETATE	0.93	0.05	0.83	0.2	60		1299
ETHANOL, 2-(VINYLOXY)-, ACETATE								
	ETHANOL, 2-(VINYLOXY)-	0.93	0.01	1.23	0.06	60		1299
	2-PYRROLIDINONE, 1-VINYL-	1.02		0.08		60		1394
ETHANOL, 2-(2-(VINYLOXY)ETHOXY)-								
	ACRYLONITRILE	0.0021	0.0004	0.59	0.02	60		1070
ETHENESULFONAMIDE, N-(2-CHLOROETHYL)-								
	STYRENE	0.075		3		62		1234
ETHENESULFONIC ACID								
	ACRYLONITRILE	0.15	0.02	1.5			Q	546
	METHYL METHACRYLATE	0.22	0.03	4.52		60	I36	37
		0		1		70	C	315
ETHENESULFONIC ACID, ALLYL ESTER								
	ACRYLIC ACID, METHYL ESTER	0.07	0.02	10.7	0.2	60	I7	209
	STYRENE	0.23	0.03	1.6	0.1	60	I7	209
	VINYL ACETATE	3.60	0.20	0.38	0.08	60	I7	209
ETHENESULFONIC ACID, BUTYL ESTER								
	ACRYLIC ACID, METHYL ESTER	0.11	0.03	5.0	1.5	70	C	754
	ETHYLENE, CHLORO-	0.30	0.05	0.35	0.05	70	C	754
	ETHYLENE, 1,1-DICHLORO-	0.065	0.007	7.5	0.6	80	C	754
	STYRENE	0.13	0.03	2.5	1.0	90	C	754

COPOLYMERIZATION PARAMETER

M1 / M2	R1	+/-	R2	+/-	TEMP.	REMARKS	REF.
ETHENESULFONIC ACID, BUTYL ESTER							
VINYL ACETATE	0.20	0.05	0.04	0.01	170	P,Y	960
					70	C	754
ETHENESULFONIC ACID, 2,3-EPOXYPROPYL ESTER							
METHYL METHACRYLATE	0.01	0.13	14.4	2.4	70	I5	415
STYRENE	0.03	0.08	1.19	0.11	70	I5	415
ETHENESULFONIC ACID, PROPYL ESTER							
ETHYLENE, CHLORO-	0.40		0.22		70	I2	937
ETHENESULFONIC ACID, SODIUM SALT							
ACRYLAMIDE	0		14.9		50		138
ACRYLIC ACID, SODIUM SALT	0		5.8		60		138
ETHENESULFONYL FLUORIDE							
ACRYLONITRILE	0.13	0.05	7.7	0.51	70	I7	811
	0.15	0.03	5.78	0.11	50	I7	811
ETHER, ALLYL PHENYL							
ETHYLENE, CHLORO-	0.30		2.53		60		714
METHYL METHACRYLATE	0.06		66.		60		714
ETHER, 2-AMINOETHYL VINYL							
ACETAMIDE, 2-ACRYLAMIDO-	0		3.3		60		313
ETHER, BENZYL VINYL							
ETHER, BUTYL VINYL	1.61	0.08	0.72	0.05	-78	H4,I2	1447
ETHER, BIS 2-(VINYLOXY ETHYL)							
STYRENE	0		16.8	0.5	60		1456
ETHER, BIS(P-VINYLPHENYL)							
STYRENE	1.06		0.94		61	C	1042
ETHER, 2-BROMOVINYL ETHYL							
ACRYLIC ACID, METHYL ESTER	0.02	0.02	31.5	2.5	80	C	786
STYRENE			37.5	2.0	80	C	786
ETHER, 1-BUTENYL ETHYL, CIS-							
ETHER, 1-BUTENYL ETHYL, TRANS-	1.23	0.06	0.92	0.01	-78	H10,I3	726
ETHER, ISOBUTYL VINYL	0.81	0.03	0.26	0.04	-78	H10,I3	726

REMARKS PAGE II-366, REFERENCES PAGE II-368

M1	M2	R1	+/-	R2	+/-	TEMP.	REMARKS	REF.
ETHER, 1-BUTENYL ETHYL, TRANS-	ETHER, 1-BUTENYL ETHYL, CIS-	0.92	0.01	1.23	0.06	-78	H10,I3	726
	ETHER, ISOBUTYL VINYL	1.54	0.14	0.68	0.03	-78	H10,I3	726
ETHER, 1-BUTENYL ISOPROPYL, CIS-	ETHER, 1-BUTENYL ISOPROPYL, TRANS-	1.18	0.05	0.49	0.02	-78	H10,I3	726
	ETHER, ISOBUTYL VINYL	0.93	0.06	0.32	0.05	-78	H10,I3	726
ETHER, 1-BUTENYL ISOPROPYL, TRANS-	ETHER, 1-BUTENYL ISOPROPYL, CIS-	0.49	0.02	1.18	0.05	-78	H10,I3	726
	ETHER, ISOBUTYL VINYL	0.90	0.05	0.50	0.06	-78	H10,I3	726
ETHER, P-TERT-BUTYLPHENYL VINYL	ETHYLENE, CHLORO-	0.138		1.08		60		1178
	ETHYLENE, 1,1-DICHLORO-	0.048		2.01		60		1178
ETHER, BUTYL PROPENYL, CIS-	ETHER, BUTYL VINYL	4.0	1.0	0.5	0.2	-78	H10,I3	649
		4.0	1.0	0.5	0.2	-78	H10,I3	651
ETHER, TERT-BUTYL PROPENYL, CIS-	ETHER, TERT-BUTYL VINYL	0.28	0.08	2.2	0.4	-76	H10,I2	356
ETHER, BUTYL PROPENYL, TRANS-	ETHER, BUTYL VINYL	2.3	1.0	0.8	0.3	-78	H10,I3	651
		2.3	1.0	0.8	0.3	-78	H10,I3	651
		2.3	1.0	0.8	0.3	-78	H10,I3	649
ETHER, BUTYL STYRYL	ETHER, BUTYL VINYL	0.45	0.1	0.25	0.1	-78	H10,I2	1116
ETHER, BUTYL VINYL	ACENAPHTHYLENE	1.3	0.3	0.38	0.04	30	H10	390
		4.2	0.8	0.24	0.04	0	H10	390
		6.0	1.0	0.14	0.03	-20	H10	390
		20		0.04	0.02	-78	H10	390
	ACETIC ACID, CHLORO-, VINYL ESTER	0.46	0.015	1.01	0.01	60		870
		0.5	0.02	1.0	0.02	60	0	494
	ACRYLIC ACID, METHYL ESTER	0		3.6	0.01	50	I7	1325
	2H-AZEPIN-2-ONE, HEXAHYDRO-1-VINYL-	0		1.39	0.08	60		882
	ETHER, BENZYL VINYL	0.72	0.05	1.61	1.0	-78	H4,I2	1447
	ETHER, BUTYL PROPENYL, CIS-	0.5	0.2	4.0	1.0	-78	H10,I3	651
		0.5	0.2	4.0	1.0	-78	H10,I3	649
	ETHER, BUTYL PROPENYL, TRANS-	0.8	0.3	2.3	1.0	-78	H10,I3	551
		0.8	0.3	2.3	1.0	-78	H10,I3	651
		0.8	0.3	2.3	1.0	-78	H10,I3	649

COPOLYMERIZATION PARAMETER

M1	M2	R1	+/-	R2	+/-	TEMP.	REMARKS	REF.
ETHER, BUTYL VINYL								
	ETHER, BUTYL STYRYL	0.25	0.1	0.45	0.1	-78	H10,I2	1116
	ETHER, TERT-BUTYL VINYL	0.19	0.02	9.67	0.05	-78	H4,I2	1447
	ETHER, CYCLOHEXYL VINYL	0.29	0.02	3.8	0.02	-78	H4,I2	1447
	ETHER, ETHYL STYRYL	0.45	0.1	0.25	0.1	-78	H10,I2	1116
	ETHER, ETHYL VINYL	0.45	0.1	1.02	0.1	20	H21	495
	ETHER, HEXYL VINYL	2.05	0.05	0.75	0.02	-78	H4,I2	1447
	ETHER, ISOBUTYL VINYL	0.95	0.02	1.38	0.03	-78	H4,I2	1447
	ETHER, ISOPROPYL VINYL	0.73	0.02	1.48	0.01	-78	H4,I2	1447
		0.38	0.07	2.77	0.1	-78	H4,I2	1447
		0.55	0.05	0.14	0.05	20	H21	495
	ETHER, ALPHA-METHYLBENZYL VINYL	0.38	0.1	1.4	0.1	-78	H4,I2	1447
	ETHER, METHYL STYRYL	0.55	0.01	0.15	0.05	-78	H10,I2	1116
	ETHER, METHYL VINYL	5.67	0.02	0.47	0.02	-78	H4,I2	1447
	ETHER, PHENYL VINYL	2.17	0.15	0.24	0.17	-78	H4,I2	1447
	ETHER, PROPYL STYRYL	0.45	0.1	0.25	0.1	-78	H10,I2	1116
	ETHER, PROPYL VINYL	1.35	0.15	0.99	0.74	-78	H4,I2	1447
	ETHYLENE, CHLORO-	0.024		2.6		60		1178
	ETHYLENE, 1,1-DICHLORO-	0.012		1.73		60		1178
	FUMARIC ACID, 1,3-BUTYLENEGLYCOL POLYESTER	0		1.8		60		908
	MALEIC ANHYDRIDE	0		0.045	0.5	50	C,I26	1328
	METHACRYLIC ACID, PIPERIDIDE	0		0.075			E,Q	1301
	METHACRYLIC ACID, 2-(VINYLOXY)ETHYL ESTER	0.82		0.004		26	H37,I30	550
	METHYL METHACRYLATE	0.2	0.1	1.6	0.2	60		493
	PHTHALIMIDE, N-VINYL-	0.03		3.27		60	I24	694
	2-PYRROLIDINONE, 1-VINYL-	0.18	0.03	2.97	0.01	60		882
	SILANE, TRIMETHYLVINYLOXY-	0.32	0.09	1.4	0.1	-78	H19,I2	1330
			0.15	1.04	0.09	-78	H19,I23	1330
	STYRENE	0.0		15.0	6.0	100		988
		14.4	0.4	0.34	0.03	-78	H84	1257
		18	0.5	0.44	0.07	20	H84	1257
	SUCCINIMIDE, N-VINYL-	0		15		60	C	279
	VINYL ACETATE	0		3.7		50	I7	1325
		0.16		2.5			I5	1166
		0.2	0.05	0.71	0.1	60	Q	494
ETHER, TERT-BUTYL VINYL								
	ACRYLONITRILE	0.0032	0.0002	0.14	0.04	60	C	1070
	ETHER, TERT-BUTYL PROPENYL, CIS-	2.2	0.4	0.28	0.08	-76	H10,I2	356
	ETHER, BUTYL VINYL	9.67	0.05	0.19	0.02	-78	H4,I2	1447
	SILANE, TRIMETHYLVINYLOXY-	1.45	0.09	0.24	0.06	-78	H19,I23	1330
		1.6	0.04	0.21	0.04	-78	H19,I2	1330
ETHER, 2-CHLOROETHYL VINYL								
	ACRYLIC ACID, ETHYL ESTER	0.15		5.0		50	I7	464
	ACRYLONITRILE	-0.07	0.1	1.0	0.2	60		435
	ANISOLE, P-ISOPROPENYL-	0.42		1.1		RT	H20,I7	592
		0.73		1.3		RT	H20,I22	592
	ANISOLE, P-VINYL-	1.24		12.1		0	H19,I3	1284
		1.56	0.04	7.8	0.2	-78	H19,I3	1285
		1.63		9.25		0	H10,I3	1284
		1.73	0.19	6.93	0.5	-78	H19,I2	1285

REMARKS PAGE II-366, REFERENCES PAGE II-368

M1	M2	R1	+/-	R2	+/-	TEMP.	REMARKS	REF.
ETHER, 2-CHLOROETHYL VINYL								
	ANISOLE, P-VINYL-	2		11		30	H31,I24	†70
		2		11.		30	H31,I24	471
		2.81	0.3	4.37	0.3	-78	H10,I2	1285
		3.08	0.26	4.55	0.35	-78	H10,I3	1285
		5.06		2.36		-36	H10,I2	1284
		9.32		1.55		0	H10,I2	1284
	ETHER, ISOBUTYL VINYL	0.34	0.14	2.03	0.55	30	H20,I7	224
		0.5		2.0		30	H31,I24	470
		0.58		2.17		-78	H19,I2	1284
		0.58		2.17		0	H19,I2	1284
		0.70	0.05	1.90	0.05	25	H3?	239
	ETHER, METHYL VINYL	0.21	0.06	2.67	0.06	-10	H37,I2,V	357
	ETHER, OCTADECYL VINYL	0	0.001	2.59	0.2	30	H20,I7	224
	ETHYLENE, CHLORO-	0		6.1	0.3	60	I7	1340
	METHACRYLIC ANHYDRIDE	8.0	1.0	0.06	0.02	-30		372
	STYRENE	12.0	1.0	0.06	0.02	25	H32,I24	143
		16.0	1.5	0.04	0.01	-30	H34,I24	143
		24.0	2.0	0.08	0.03	25	H10,I24	143
		30.5	3.0	0.11	0.04	25	H33,I24	143
		36	8.0	0.12	0.05	-30	H10,I22	143
		51.0	15	3		30	H31,I24	470
				0	0.01	25	H33,I7	143
	STYRENE, P,ALPHA-DIMETHYL-	1.7		0.64		RT	H20,I22	592
		1.7		0.51		RT	H20,I7	592
		1.7		0.54		RT	H20,I7	592
	STYRENE, ALPHA-METHYL-	1.02	0.1	1	0.1	-78	H19,I3	1285
		2.0	0.2	0.75	0.07	-30	H33,I24	143
		2.05	0.33	0.68	0.14	-78	H10,I3	1285
		2.5		0.76		-23	H19,I3	1284
		2.6		0.34		RT	H20,I22	40
		3.3		0.42		RT	H20,I22	40
		3.3		0.42		RT	H20,I22	592
		3.46	0.25	0.46	0.07	-78	H19,I2	1285
		3.9	0.15	0.48	0.04	0	H33,I24	143
		4.5	0.5	0.07	0.01	25	H33,I24	143
		5		0.42		-23	H19,I2	1284
		5.0		0.33	0.05	RT	H20,I7	40
		5.03	0.66	0.31	0.05	30	H20,I7	224
		5.72	0.7	0.42		-78	H10,I2	1285
		6.02		0.42		-23	H19,I3	1284
	STYRENE, P-METHYL-	2.31	0.17	1.12	0.06	-78	H19,I3	1285
		3.31		1.12		0	H19,I3	1284
		8		0.88		-78	H19,I3	1284
		8.8	0.45	0.4	0.03	-78	H10,I3	1285
		10.1		0.5		-78	H19,I2	1285
		18.2		0.4		0	H19,I2	1284
		45	1.8	5	0.07	30	H31,I24	470
ETHER, M-CHLOROPHENYL VINYL	ETHER, PHENYL VINYL	0.19	0.021	4.6	0.5	-7b	H20,I2	1191

COPOLYMERIZATION PARAMETER

M1	M2	R1	+/-	R2	+/-	TEMP.	REMARKS	REF.
ETHER, P-CHLOROPHENYL VINYL	BENZOIC ACID, P-(VINYLOXY)-, VINYL ESTER	0.43		0.56		0	H2O	1250
	ETHER, PHENYL VINYL	0.27	0.023	3.6	0.34	-78	H2O,I2	1191
	ETHYLENE, CHLORO-	0.74		1.97		20	H10,I24	1262
	ETHYLENE, 1,1-DICHLORO-	0.9		1.53		20	H10,I7	1262
		0.63		1		60	H10,I7	1178
		0.144		2.43		60		1178
ETHER, 2-CHLOROVINYL ETHYL, CIS-	ETHER, 2-CHLOROVINYL ETHYL, TRANS-	1.39	0.04	0.28	0.04	-78	H10,I3	1335
	ETHER, ISOBUTYL VINYL	26		0		-78	H10,I3	1335
ETHER, 2-CHLOROVINYL ETHYL, TRANS-	ETHER, 2-CHLOROVINYL ETHYL, CIS-	0.28	0.04	1.39	0.04	-78	H10,I3	1335
ETHER, CYCLOHEXYL VINYL	ETHER, BUTYL VINYL	3.8	0.02	0.29	0.02	-78	H4,I2	1447
	2-PYRROLIDINONE, 1-VINYL-	0		3.84		60		881
ETHER, DODECYL VINYL	ACRYLIC ACID, 2-CHLOROETHYL ESTER	0.013	0.010	1.97	0.08	60		436
	ACRYLONITRILE	0.0	0.2	0.82	0.05	50	C	13
	ETHYLENE, CHLORO-	0.15	0.2	1.93	0.15	50	C	9
	ETHYLENE, 1,1-DICHLORO-	0.00	0.2	1.30	0.015	50	C	10
	FUMARONITRILE	0.004	0.003	0.019	0.003	60	I7	436
	MALEIC ANHYDRIDE	-0.046	0.054	0.046	0.052	50	A,I7	15
	METHYL METHACRYLATE	-0.0		1.0		100	Q	11
	STYRENE	0.0	0.05	27	2	70		14
		0.0	0.05	56	5	50		14
	VINYL ACETATE	0	0.23	3.67	0.45	50	C	12
ETHER, ETHYL 3,3-DIMETHYL-1-BUTENYL, TRANS-	ETHER, ISOBUTYL VINYL	0.01		100		-78	H10,I3	726
ETHER, ETHYL 1-HEPTENYL, CIS-	ETHER, ETHYL 1-HEPTENYL, TRANS-	1.16	0.05	0.88	0.03	-78	H10,I3	726
	ETHER, ISOBUTYL VINYL	0.46	0.05	0.17	0.05	-78	H10,I3	726
ETHER, ETHYL 1-HEPTENYL, TRANS-	ETHER, ETHYL 1-HEPTENYL, CIS-	0.88	0.03	1.16	0.05	-78	H10,I3	726
	ETHER, ISOBUTYL VINYL	0.59	0.09	2.10	0.21	-78	H10,I3	726
ETHER, ETHYL 3-METHYL-1-BUTENYL, CIS-	ETHER, ETHYL 3-METHYL-1-BUTENYL, TRANS-	1.76	0.08	0.23	0.01	-78	H10,I3	726
	ETHER, ISOBUTYL VINYL	0.09	0.04	1.71	0.20	-78	H10,I3	726

REMARKS PAGE II-366, REFERENCES PAGE II-368

M1	M2	R1	+/-	R2	+/-	TEMP.	REMARKS	REF.
ETHER, ETHYL 3-METHYL-1-BUTENYL, TRANS-	ETHER, ETHYL 3-METHYL-1-BUTENYL, CIS-	0.23	0.01	1.76	0.08	-78	H10,13	726
	ETHER, ISOBUTYL VINYL	0.09	0.01	14.44	0.21	-78	H10,13	726
ETHER, ETHYL 4-METHYL-1-PENTENYL, CIS-	ETHER, ETHYL 4-METHYL-1-PENTENYL, TRANS-	0.46	0.15	0.74	0.07	-78	H10,13	726
	ETHER, ISOBUTYL VINYL	0.032	0.038	0.20	0.10	-78	H10,13	726
ETHER, ETHYL 4-METHYL-1-PENTENYL, TRANS-	ETHER, ETHYL 4-METHYL-1-PENTENYL, CIS-	0.74	0.07	0.46	0.15	-78	H10,13	726
	ETHER, ISOBUTYL VINYL	0.33	0.03	2.48	0.10	-78	H10,13	726
ETHER, ETHYL 2-METHYLPROPENYL	ETHER, ETHYL PROPENYL, CIS-	0.02	0.02	24	2.4	-78	H10,13	1336
	ETHER, ETHYL PROPENYL, TRANS-	0.04	0.02	19.1	1.8	-78	H10,13	1336
	ETHER, ISOBUTYL VINYL	0.00	0.01	1.48	0.03	-78	H10,13	726
ETHER, ETHYL PROPENYL, CIS-	ETHER, ETHYL 2-METHYLPROPENYL	24	2.4	0.02	0.02	-78	H10,13	1336
	ETHER, ETHYL PROPENYL, TRANS-	1.37	0.29	0.72	0.16	-78	H10,13	726
	ETHER, ETHYL VINYL	4.0	0.5	0.35	0.10	-76	H10,12	356
	ETHER, ISOBUTYL VINYL	2.25	0.10	0.23	0.07	-78	H10,13	726
ETHER, ETHYL PROPENYL, TRANS-	ETHER, ETHYL 2-METHYLPROPENYL	19.1	1.8	0.04	0.02	-78	H10,13	1336
	ETHER, ETHYL PROPENYL, CIS-	0.72	0.16	1.37	0.29	-78	H10,13	726
	ETHER, ETHYL VINYL	0.94	0.10	0.94	0.10	-76	H10,12	356
	ETHER, ISOBUTYL VINYL	1.44	0.02	0.56	0.03	-78	H10,13	726
ETHER, ETHYL STYRYL	ETHER, BUTYL VINYL	0.25	0.1	0.45	0.1	-78	H10,12	1116
ETHER, ETHYL VINYL	ACETIC ACID, CHLORO-, VINYL ESTER	0.68	0.01	0.98	0.02	60	N,O	870
	ACRYLIC ACID, METHYL ESTER	0		3.3		60	N,O	620
	ACRYLONITRILE	0		5.0		60	C	622
		0.03	0.02	0.7	0.2	80	C	786
		0.03	0.02	0.7	0.2	80		848
	ETHER, BUTYL VINYL	0.75	0.02	2.05	0.05	-78	H4,12	1447
		1.02	0.1	0.45	0.5	20	H21	495
	ETHER, ETHYL PROPENYL, CIS-	0.35	0.10	4.0	0.5	-76	H10,12	356
	ETHER, ETHYL PROPENYL, TRANS-	0.94	0.10	0.94	0.10	-76	H10,12	726
	ETHER, ISOBUTYL VINYL	1.30	0.02	0.92	0.02	-78	H10,13	726
	ETHYLENE	0		5.4		75	M3	1413
		0		2.7		75	M2	1413
	ETHYLENE, CHLOROTRIFLUORO-	0		0.008		20	E	1400
	ETHYLENE, 1,1-DICHLORO-	0		3.2		60	N,O	620

M1	M2	R1	+/-	R2	+/-	TEMP.	REMARKS	REF.
ETHER, ETHYL VINYL								
	METHYL METHACRYLATE	0.01		37	20	60		372
	STYRENE	0		90	40	60		560
	VINYL ACETATE	0		80		80	c	786
		0		3.0	0.1	60	c	622
ETHER, HEXYL VINYL								
	ETHER, BUTYL VINYL	1.38	0.03	0.95	0.02	-78	H4,I2	1447
ETHER, ISOBUTYL PROPENYL, CIS-								
	ETHER, ISOBUTYL PROPENYL, TRANS-	1.16	0.03	0.59	0.02	0	H20,I23	725
		1.85	0.09	0.33	0.03		H10,I3	725
		2.21	0.20	0.46	0.01	-78	H10,I3	725
		2.88	0.33	0.27	0.08	-78	H77,I2	725
		4.45	0.53	0.25	0.09	-78	H10,I2	725
		5.94	0.50	0.17	0.06	0	H37,I2	725
	ETHER, ISOBUTYL VINYL	0.29	0.05	1.56	0.16	0	H37,I2	725
		0.81	0.03	0.44	0.03	0	H20,I23	725
		0.85	0.12	0.29	0.08	-78	H10,I3	725
		1.78	0.08	0.13	0.04	-78	H77,I2	725
		2.17	0.09	0.18	0.04	-78	H10,I3	725
		2.20	0.14	0.29	0.05	-78	H10,I3	725
ETHER, ISOBUTYL PROPENYL, TRANS-								
	ETHER, ISOBUTYL PROPENYL, CIS-	0.17	0.06	5.94	0.50	0	H37,I2	725
		0.25	0.09	4.45	0.53	-78	H10,I3	725
		0.27	0.08	2.88	0.33	-78	H77,I2	725
		0.33	0.03	1.85	0.09	-78	H10,I3	725
		0.46	0.01	2.21	0.03	0	H10,I23	725
		0.59	0.02	1.16	0.03	0	H37,I2	725
	ETHER, ISOBUTYL VINYL	0.10	0.04	31.4	2.3	0	H10,I2	725
		0.35	0.04	2.64	0.16	-78	H10,I3	725
		0.65	0.01	1.67	0.02	-78	H77,I2	725
		0.71	0.02	1.05	0.03	0	H10,I3	725
		0.79	0.06	0.96	0.14	0	H20,I23	725
		0.90	0.03	1.04	0.05	-78	H10,I3	725
ETHER, ISOBUTYL VINYL								
	ACRYLONITRILE	0.185	0.05	1.05	0.1	50	D	1409
	CARBONIC ACID, CYCLIC VINYLENE ESTER	0.26	0.001	0.160	0.001	50	126	1
	ETHER, 1-BUTENYL ETHYL, CIS-	0.68	0.04	0.81	0.14	-78	H10,I3	726
	ETHER, 1-BUTENYL ETHYL, TRANS-	0.32	0.03	1.54	0.06	-78	H10,I3	726
	ETHER, 1-BUTENYL ISOPROPYL, CIS-	0.50	0.05	0.93	0.05	-78	H10,I3	726
	ETHER, 1-BUTENYL ISOPROPYL, TRANS-	1.48	0.06	0.90	0.15	-78	H10,I3	726
	ETHER, BUTYL VINYL	1.90	0.1	0.73	0.05	-78	H4,I2	1447
		2.0	0.05	0.70		25	H31	239
		2.03		0.5		30	H31,I24	470
		2.17		0.34		30	H20,I7	224
	ETHER, 2-CHLOROETHYL VINYL	2.17	0.55	0.58	0.14	-78	H19,I2	1284
				0.58		0	H19,I2	1284

M1	M2	R1	+/-	R2	+/-	TEMP.	REMARKS	REF.
ETHER, ISOBUTYL VINYL	ETHER, 2-CHLOROVINYL ETHYL, CIS-	0		26		-78	H10,13	1335
	ETHER, ETHYL 3,3-DIMETHYL-1-BUTENYL, TRANS-	100		0.01		-78	H10,13	726
	ETHER, ETHYL 1-HEPTENYL, CIS-	0.17	0.05	0.46	0.05	-78	H10,13	726
	ETHER, ETHYL 1-HEPTENYL, TRANS-	2.10	0.21	0.59	0.09	-78	H10,13	726
	ETHER, ETHYL 3-METHYL-1-BUTENYL, CIS-	1.71	0.20	0.09	0.04	-78	H10,13	726
	ETHER, ETHYL 3-METHYL-1-BUTENYL, TRANS-	14.44	0.21	0.09	0.01	-78	H10,13	726
	ETHER, ETHYL 4-METHYL-1-PENTENYL, CIS-	0.20	0.10	0.032	0.038	-78	H10,13	726
	ETHER, ETHYL 4-METHYL-1-PENTENYL, TRANS-	2.48	0.03	0.33	0.03	-78	H10,13	726
	ETHER, ETHYL 2-METHYLPROPENYL	1.48	0.07	0.00	0.01	-78	H10,13	726
	ETHER, ETHYL PROPENYL, CIS-	0.23	0.03	2.25	0.10	-78	H10,13	726
	ETHER, ETHYL PROPENYL, TRANS-	0.56	0.02	1.44	0.02	-78	H10,13	726
	ETHER, ETHYL VINYL	0.92	0.04	1.30	0.02	-78	H10,13	726
	ETHER, ISOBUTYL PROPENYL, CIS-	0.13	0.04	1.78	0.08	0	H77,12	725
		0.18	0.05	2.17	0.09	-78	H10,12	725
		0.29	0.08	2.20	0.14	-78	H10,13	725
		0.44	0.03	0.85	0.12	0	H20,123	725
		1.56	0.16	0.81	0.03	0	H20,123	725
		0.96	0.14	0.29	0.05	-78	H10,13	725
	ETHER, ISOBUTYL PROPENYL, TRANS-	1.04	0.05	0.79	0.06	-78	H10,13	725
		1.05	0.03	0.90	0.03	0	H77,12	725
		1.67	0.02	0.71	0.02	-78	H10,12	725
		2.64	0.16	0.65	0.01	0	H37,12	725
		31.4	2.3	0.35	0.04	-78	H10,13	725
	ETHER, ISOPROPYL PROPENYL, TRANS-	0.18	0.08	2.37	0.19	-78	H10,13	726
	ETHER, ISOPROPYL VINYL	0.22	0.03	2.70	0.11	-78	H10,13	726
	ETHYLENE, CHLORO-	0.02	0.01	2.0	0.2	50	Q	775
		0.02	0.01	0.10	0.2	60	C	166
	FERROCENE, VINYL-	9.7	1	0.1	0.1	50	H10,12	1139
	FUMARIC ACID, 1,3-BUTYLENEGLYCOL POLYESTER	0		2.0	0.7	60		908
	PROPANE, 1,2-EPOXY-3-METHOXY-	7.3		1.03	0.06	50	H10,I10	443
	SILANE, TRIMETHYLVINYLOXY-	0.34	0.02	1.37	0.05	-78	H19,12	1330
		0.35	0.03	1.08		-78	H19,123	1330
	STYRENE	0.01		50		0	Q	775
		0.1	0.1	17.0		0	E	989
		10.0	4.0	0.3	0.3	0	E,X	989
ETHER, ISOPENTYL VINYL	FUMARIC ACID, 1,3-BUTYLENEGLYCOL POLYESTER	0		3.8	0.7	60		908
ETHER, ISOPROPYL PROPENYL, CIS-	ETHER, ISOPROPYL PROPENYL, TRANS-	2.04	0.05	0.48	0.02	-78	H10,13	726
	ETHER, ISOPROPYL VINYL	0.80	0.40	1.1	0.2	-76	H10,12	356
ETHER, ISOPROPYL PROPENYL, TRANS-	ETHER, ISOBUTYL VINYL	2.37	0.19	0.18	0.08	-78	H10,13	726
	ETHER, ISOPROPYL PROPENYL, CIS-	0.48	0.02	2.04	0.05	-78	H10,13	726
	ETHER, ISOPROPYL VINYL	0.19	0.05	4.9	0.4	-78	H10,12	356

COPOLYMERIZATION PARAMETER

M1	M2	R1	+/-	R2	+/-	TEMP.	REMARKS	REF.
ETHER, ISOPROPYL VINYL								
	2H-AZEPIN-2-ONE, HEXAHYDRO-1-VINYL-	11.45	0.54	1.88	0.02	60		882
	1,3-BUTADIENE, 2-CHLORO-	0.14	0.05	0.164	0.043	65		645
	ETHER, BUTYL VINYL	2.77	0.1	0.55	0.05	20	H21	495
	ETHER, ISOBUTYL VINYL	2.70	0.11	0.38	0.07	-78	H4,I2	1447
	ETHER, ISOPROPYL PROPENYL, CIS-	1.1	0.2	0.80	0.03	-76	H10,I3	726
	ETHER, ISOPROPYL PROPENYL, TRANS-	4.9	0.4	0.19	0.40	-76	H10,I2	356
	2-PYRROLIDINONE, 1-VINYL-	0		1.68	0.05	60	H10,I2	882
ETHER, ALPHA-METHYLBENZYL VINYL								
	ETHER, BUTYL VINYL	1.4	0.1	0.38	0.1	-78	H4,I2	1447
ETHER, METHYL 1-PHENYLVINYL								
	ACRYLIC ACID, METHYL ESTER	0		0.17	0.01	60		572
	ACRYLONITRILE	0		0.06	0.005	60		573
	METHACRYLONITRILE	0		0.55		60	R	573
	METHYL METHACRYLATE	0		2.5	0.3	60		572
	STYRENE	0		2.7	0.6	60		572
ETHER, METHYL STYRYL								
	ETHER, BUTYL VINYL	0.15	0.05	0.55	0.1	-78	H10,I2	1116
ETHER, METHYL VINYL								
	ETHER, BUTYL VINYL	0.47	0.02	5.67	0.02	-78	H4,I2	1447
	ETHER, 2-CHLOROETHYL VINYL						H37,I2,V	357
	STYRENE	0.01		100		-10	Q	775
ETHER, OCTADECYL VINYL								
	ACRYLONITRILE	0.0	0.2	0.85	0.05	50	C	13
	ETHER, 2-CHLOROETHYL VINYL	2.67	0.06	0.21	0.06	30	H20,I7	224
	ETHYLENE, CHLORO-	-0.1	0.2	2.10	0.20	50	C	9
	ETHYLENE, 1,1-DICHLORO-	0.0	0.3	1.50	0.15	50	C	10
	METHYL METHACRYLATE	0		1		50		11
	VINYL ACETATE	0	0.35	4.50	0.58	50	C	12
ETHER, OCTYL VINYL								
	ACRYLONITRILE	0.0	0.2	0.81	0.05	50	C	13
	ETHYLENE, CHLORO-	0.1	0.1	1.90	0.20	50	C	9
	ETHYLENE, 1,1-DICHLORO-	0.0	0.2	1.35	0.15	50	C	10
	METHYL METHACRYLATE	0.0		1		50		11
	STYRENE	0	0.04	65	5	70	C	14
	VINYL ACETATE	0	0.31	3.47	0.51	50	C	12
ETHER, PENTYL VINYL								
	FUMARIC ACID, 1,3-BUTYLENEGLYCOL POLYESTER	0		2.7	0.7	60		908

REMARKS PAGE II-366, REFERENCES PAGE II-368

M1	M2	R1	+/-	R2	+/-	TEMP.	REMARKS	REF.
ETHER, PHENYL VINYL	ACETIC ACID, CHLORO-, VINYL ESTER	0.8	0.02	1.0	0.005	60		870
	ACROLEIN	0.86	0.10	0.92	0.18	20	H2O	875
	ACRYLONITRILE	0.23		2.5		60		1180
	2H-AZEPIN-2-ONE, HEXAHYDRO-1-VINYL-	0.39	0.3	2.53	0.03	60	C	881
	BENZENE, 1-METHOXY-3-(VINYLOXY)-	0.71	0.058	1.1	0.07	-78	H2O,12	1191
	BENZENE, 1-METHOXY-4-(VINYLOXY)-	0.42	0.06	2.1	0.14	-78	H2O,12	1191
	3-BUTEN-2-ONE	0.01		4.4				376
	ETHER, BUTYL VINYL	0.24	0.17	2.17	0.15	-78	C,Q	1447
	ETHER, M-CHLOROPHENYL VINYL	4.6	0.5	0.19	0.021	-78	H2O,12	1191
	ETHER, P-CHLOROPHENYL VINYL	1.53		0.9		20	H10,17	1262
		1.97		0.74		20	H10,I24	1262
	ETHER, M-TOLYL VINYL	3.6	0.34	0.27	0.023	-78	H2O,12	1191
	ETHER, P-TOLYL VINYL	0.79	0.067	1.5	0.056	-78	H2O,12	1191
	ETHYLENE, CHLORO-	0.55	0.014	1.76	0.032	-78	H2O,12	1191
		0.16		1.06		60		1179
		0.161		1.059		60		714
	ETHYLENE, 1,1-DICHLORO-	0.43		1.93		60		1179
		0.041		1.634		60		1178
		0.041		1.63		60		1178
	METHYL METHACRYLATE	0.38	0.12	2.37	0.42	60	Q	986
		0.01		3.5		60	C,Q	376
	2-PYRROLIDINONE, 1-VINYL-	0.13		140		60		714
	STYRENE	0.22	0.001	4.43	0.001	60	C,Q	881
		0.01		1.7		60	C,Q	376
ETHER, PROPYL STYRYL	ETHER, BUTYL VINYL	0.25	0.1	0.45	0.1	-78	H10,12	1116
ETHER, PROPYL VINYL	ETHER, BUTYL VINYL	0.99	0.74	1.35	0.15	-78	H4,12	1447
	FUMARIC ACID, 1,3-BUTYLENEGLYCOL POLYESTER	0		1.6	0.5	60		908
ETHER, THYMYL VINYL	ACROLEIN	0.18	0.07	1.82	0.03	20	H2O	875
ETHER, M-TOLYL VINYL	ACROLEIN	0.71		1.11		20	H2O	875
	ETHER, PHENYL VINYL	1.5	0.056	0.79	0.067	-78	H2O,12	1191
	ETHYLENE, CHLORO-	0.131		1.23		60		1178
	ETHYLENE, 1,1-DICHLORO-	0.036		1.91		60		1178
ETHER, O-TOLYL VINYL	ACROLEIN	0.32	0.11	0.53	0.04	20	H2O	875
	ETHYLENE, CHLORO-	0.138		1.33		60		1178
	ETHYLENE, 1,1-DICHLORO-	0.041		2.08		60		1178

COPOLYMERIZATION PARAMETER

M1 / M2	R1	+/-	R2	+/-	TEMP.	REMARKS	REF.
ETHER, P-TOLYL VINYL							
ACROLEIN	0.96	0.05	0.99	0.05	20	H2O	875
ETHER, PHENYL VINYL	1.76	0.032	0.55	0.014	-78	H2O,I2	1191
ETHYLENE, CHLORO-	0.128		1.17		60		1178
ETHYLENE, 1,1-DICHLORO-	0.034		1.96		60		1178
ETHERS, C8-C16 ALKYL VINYL							
ACRYLONITRILE					50	C,P	955
MALEIC ANHYDRIDE					50	C,P	12
ETHERS, C8-C18 ALKYL VINYL							
ACRYLIC ACID, METHYL ESTER					50	P	11
ETHYLENE, CHLORO-					50	C,P	12
ETHYL, 1,1-DICHLORO-					50	C,P	12
METHYL METHACRYLATE					50	C,P	12
STYRENE					50	C,P	12
VINYL ACETATE					50	C,P	12
ETHYLAMINE, N,N-DIMETHYL-2-(VINYLOXY)-							
METHYL METHACRYLATE						P,Q	170
ETHYLENE							
ACRYLIC ACID, BUTYL ESTER	0.03	0.01	11.9	2.5	70	M3	148
ACRYLIC ACID, METHYL ESTER	0.2		11		150	M7,R	142
ACRYLONITRILE	0		7.0		20	E,I2	309
1-BUTENE	3.2	0.38	0.64		130-220	M1	656
	3.25				130-220	M5	112
	3.6		0.16			M6,Q	236
	26.96		0.043		0/75	H29	690
	29.5		0.019			G,Q	688
	29.60		0.019		0/75	H28	690
	32.5		0.019		-20	H61	1124
	35.5		0.006		-78	H61	1079
	85.0		0.015		23	H54,I21	565
	243		0.001		10	H45	1087
1-BUTENE, 3-METHYL-	7.5				130-220	M1	656
2-BUTENE, TRANS-	0.042		0		20	E,M2	182
CARBON MONOXIDE	0.045		0		20	E,M7	183
	0.147		0		120-130	C,M3	183
	0.16	0.02	0		70	E,L	121
	0.500		0		135	C,I21,M8	183
ETHER, ETHYL VINYL	2.7		0		75	M2	1413
	5.4		0		75	M3	1413
ETHYLENE, CHLORO-	0.05		4.16		0	H38,I15	642
	0.08		4.70		0	C	642
	0.16		1.85			Q	296
	0.16		1.85		60	M10	246
	0.16		1.85		70	M10	226
	0.2		2.0		50	H71	643
	0.2		1.85		70	M5	226

M1	M2	R1	+/-	R2	+/-	TEMP.	REMARKS	REF.
ETHYLENE	ETHYLENE, CHLORO-	0.20	0.02	1.85	0.2	70	M2	245
		0.20	0.02	1.85	0.2	70	L,M10	225
		0.21	0.02	3.21	0.22	50	C,M14	1046
		0.24	0.07	3.60	0.30	90	M3	1083
		0.24	0.07	3.60	0.30	90	M3	148
		0.285	0.075	1.13	0.32	70	I13,M10	225
		0.3		2.0		60	C,I21	744
		0.3		2		60	C	1113
		1.2		0		60	F,JJ	744
	ETHYLENE, CHLOROTRIFLUORO-	0.070	0.0015	0.001	0.002	-78	H38	794
		0.116	0.001	0.004	0.001	-40	H38	794
		0.173	0.001	0.009	0.001	0	H38	794
		0.250	0.0025	0.0285	0.001	60	E,M2	1302
	ETHYLENE, 1,1-DICHLORO-	0.03		15.2		30	H38,I38	879
	ETHYLENE, FLUORO-	1.7	0.1	0.3	0.03	30	H38	1165
	ETHYLENE, TETRAFLUORO-	1.7	0.1	0.3	0.03	30	M3	148
	FUMARIC ACID, DIETHYL ESTER	4.39	0.77	0.16	0.05	160	K	396
	MALEIC ACID, DIETHYL ESTER	0.15		0.85		80	D	1240
	MALEIC ANHYDRIDE	0.38		0.1		25	D,I71	1240
		0.61		0.024		25	M7,R	142
	METHYL METHACRYLATE	0.25		10		150	M7,R	142
		0.25		10		150	E,D	576
		0		17		40	M7,R	142
	1-OCTENE	0.2				150	M1	656
		3.2				130/220	M5	112
	1-PENTENE	3.66	0.15			130/220	H45	1388
		33.2		0.0145		-10	H45	853
		33.2		0.0145		-10	H61	852
		41.0		0.015		-20	H5	852
		48.2		0.018		-20	I7,M5	1412
	PHOSPHONIC ACID, METHYL-, DIALLYL ESTER	0.29	0.02			80	I7,M3	1412
		0.41	0.04			80	I7,M2	1412
		0.49	0.04	0		80	I7,M2	1411
	PHOSPHONIC ACID, VINYL-, DIETHYL ESTER	0.17	0.03	0		80	I7,M3	1411
	PHOSPHONIC ACID, VINYL-, DIPHENYL ESTER	0.04	0.004	0		80	I7,M3	1411
		0.055	0.007	0		140	I7,M2	1411
	1-PROPENE	0.11	0.02			80	H35,120	273
		0.9	0.1	1.9	0.2	120	H61,120	250
		3		<0.62		21	M1	656
		3.2	0.15			130/220	M5	112
		3.43				130/220	H2/3	912
		4.7	0.6	0.21	0.02	20/60	H29	690
		5.61		0.145		0/75	H29	689
		5.61		0.145		75	H53,144	45
		6.4	0.2	0.7	0.2	30	H28	690
		7.08		0.088		0/75	H28	623
		7.08	0.20	0.08		25	H24,144	45
		7.2	0.2	0.28	0.003	30	H7,144	45
		8		0.18	0.04	30	H28,120	1172
		10.3	2.3	0.025	0.005	26	H15	582
		10.7		0.022		30	H5	154
		14.8		0.037		30	H45	691
		15.0		0.04		-20		

M1	M2	R1	+/-	R2	+/-	TEMP.	REMARKS	REF.
ETHYLENE	1-PROPENE	15.0		0.070		30	H4	154
		15.7		0.110		75	H18	692
		15.7		0.110		75	H12	692
		15.72		0.11		0/75	H18	690
		15.72		0.11		0/75	H12	690
		16		0.04			H45,Q	1387
		16		0.062		70	H28,121	474
		16		0.062		70	H67,121	474
		16.8		0.055		30	H3	154
		16.8		0.522		30	H5,120	250
		17.5		0.050		30	H8	154
		17.5	0.5	0.13	0.03	30	H12,144	45
		17.5	0.5	0.13	0.03	30	H54,144	45
		17.8		0.065		25	H7	624
		17.95	0.27	0.065	0.002	0/75	H7	690
		17.95		0.056		25	H7	624
		18.9		0.023		30	H2	154
		20.3		0.022			H61,Q	1387
		22.0	0.5	0.022	0.002	26	H5,120	1172
		23	3	0.046	0.006	30	H2/3	912
		26.0		0.043		30	H6	154
		26.0		0.040		30	H9	154
		28		0.040		70	H7,121	474
		28.0		0.036		70	H7	154
		33.36		0.032		30	H24	690
		33.4		0.032		0/75	H24	692
		35.3		0.027		25	H5,Q	1387
		37		0.027		70	H24,121	474
		61		0.016		70	H65,121	474
		76		0.013		70	H66,121	474
	1-PROPENE, 2-METHYL-	2.1		0.49		130-220	M1	656
	1-PROPENE, 3,3,3-TRIFLUORO-	0.18		0.8		70	M10	1181
		0.18		0.9		70	M10	1181
	VINYL ACETATE	0.16		1.12		60	M4	1366
		0.16		1.14		60	M4	246
		0.18		1.05		80	M4	1366
		0.28		1.14			Q	296
		0.45		1		60	M8	1366
		0.47		1		80	M8	1366
		0.49		1.02		135	M4	1366
		0.52		1.02		60	M10	1366
		0.67		0.95		80	M10	1366
		0.70		3.70		60	M9	246
		0.73		1.515	0.007	135	M10	1366
		0.743	0.005	1.02	0.02	62	I45,M6	1198
		0.77	0.04	1.02		166	I7,M2	953
		0.85		1.02		200	C,M10	1366
		0.89		1.02		130	C,M10	1366
		0.97	0.03	1.02	0.02	150	C,I7,M2	953
		1.07		1.08		90	M7,R	142
		1.07	0.06				M3	148
		1.2		1.1	0.19	160	C,M9	1366

M1	M2	R1	+/-	R2	+/-	TEMP.	REMARKS	REF.
ETHYLENE								
	VINYL ACETATE	1.2		1.1		210	C,M9	1366
		1.2		1.1		230	B,M14	1366
ETHYLENE, 1,1-BIS(P-CHLOROPHENYL)-								
	ACRYLIC ACID, METHYL ESTER	0		0.092	0.006	60	D	212
	ACRYLONITRILE	0		0.024	0.003	60	C	214
ETHYLENE, 1,1-BIS(P-METHOXYPHENYL)-								
	ACRYLIC ACID, METHYL ESTER	0		0.049	0.005	60	D	212
	ACRYLONITRILE	0		0.014	0.002	60	C	214
ETHYLENE, BROMO-								
	ACRYLAMIDE	0.018	0.1	70	16	50	C,I12,L	991
	ACRYLIC ACID, BUTYL ESTER	0.19		13.8		0	F,K	1098
		0.007		3.7		60		1098
	ACRYLONITRILE	0.055	0.003	0.94	0.02	60	F,K	1098
		0		2.25		50	C	991
	METHACRYLIC ACID	0.0024	0.1	90	15	0	F,K	1098
	METHYL METHACRYLATE	0.05	0.01	27.4	2	28	D	111
		0.05	0.015	25	2	60	A	111
		0.33		20		60		1098
	STYRENE	0.038		17.1		60		1097
		0.05		12.6	2	0	D	111
		0.06	0.015	23	2	28	A	111
	VINYL ACETATE	1.82		0.68	0.16	60		1098
		3	0.65	0.38	0.09	0	F,K	1098
		4.5	1.2	0.35		60		622
ETHYLENE, 1-BROMO-1-CHLORO-								
		2.38	0.06	0.83	0.08		I12,Q	1032
ETHYLENE, CHLORO-								
	ACENAPHTHYLENE	0.001		64		40	Q	391
	ACETIC ACID, ALLYL ESTER	1.16		0		40		654
	ACETIC ACID, CHLORO-, VINYL ESTER	1.2				60		620
	ACETIC ACID, DICHLORO-, VINYL ESTER	1.6		0.4		60		1213
	ACROLEIN	1.25		0.7		60	I5	1115
	ACRYLAMIDE	0.04		5.4		40		616
	ACRYLAMIDE, N-(HYDROXYMETHYL)-	0		19.6				617
	ACRYLIC ACID	0.025		23.5		50	I10	1236
		0.027		6.4		50	I7	935
		0.107		8.2		60	I57	541
	ACRYLIC ACID, BUTYL ESTER	0.07		6.8		45		963
	ACRYLIC ACID, ETHYL ESTER, STANNIC CHLORIDE COMPLEX	0.01		4.4		50	D	1428
	ACRYLIC ACID, 2-ETHYLHEXYL ESTER	0.002	0.002	8.25	0.75	-50	H38,I20	938
		0.16	0.03	4.15	0.6	63	C	938

M1	M2	R1	+/-	R2	+/-	TEMP.	REMARKS	REF.
ETHYLENE, CHLORO-								
	ACRYLIC ACID, METHYL ESTER	0				50		543
		0.06		5		60	P	620
		0.06		4		45		964
		0.083		4		45		963
	ACRYLIC ACID, OCTYL ESTER	0.12	0.01	9.0	0.5	50		165
	ACRYLONITRILE	0.12		4.4		50	C	600
		0.02	0.02	4.8	0.06	45	C	963
		0.04	0.02	3.28	0.5	60		560
		0.040	0.03	2.8	0.7	40		817
		0.052		2.7		60	H38	62
		0.074	0.009	4.0	0.2	38	E,I7	1008
	ACRYLONITRILE, TIN COMPLEX	0.01		3.7		50	I26	958
	ACRYLOYL CHLORIDE	0.017		0.5		50		165
	BENZOIC ACID, VINYL ESTER	0.72		2.65		35	D	1428
		1.7		0.28		50	I24	1236
		1.8		0.5		45		964
	BENZOIC ACID, P-(1-HYDROXYETHOXY)-, VINYL ESTER, ACETATE	1.06		0.3		40	C,I11,J	503
	BENZOIC ACID, P-MALEIMIDO-, ETHYL ESTER	3.3		1.18		60		1213
	BENZOIC ACID, P-(VINYLOXY)-, VINYL ESTER	0.61		0.05		50	I5	1250
	1,3-BUTADIENE	0.035		0.89		60		1343
	1-BUTENE	3.4		8.8		50	I5	1250
	1-BUTENE, 3,3-DIMETHYL-	5.0		0.21		60	C,K	962
	2-BUTENE, CIS-	8.8	0.7	0.001		70		936
	2-BUTENE, TRANS-	7.3		0.02		60		23
	3-BUTENOIC ACID, 3-HYDROXY-, BETA-LACTONE	5.74		0.2		60	I7	936
	3-BUTEN-2-ONE	0.10		8.3		70		936
	BUTYRIC ACID, VINYL ESTER	1.35		0.65		50		614
		1.75	0.05	0.55	0.04	60		22
		2	0.03	0.28	0.05	60	Q	799
	CARBAZOLE, 9-VINYL-	0.17		4.77		60	Q	176
	CARBONIC ACID, CYCLIC VINYLENE ESTER	5.2		0.09		60		1213
	CARBON MONOXIDE	13		0.15		80		391
	CROTONALDEHYDE	13.47		0.02		60		349
	CYCLOBUTANECARBOXYLIC ACID, 3-ACETYL-2,2-DIMETHYL-, VINYL ESTER *	1.93		0.02		60	I7	1403
		1.93		0.02		60	I7	750
	CYCLOHEXANE, 1,2-EPOXY-4-VINYL-	3.14		0.2		60	I5	1115
	DECANOIC ACID, VINYL ESTER	4.7		0.2		60	I5	1115
	ETHENESULFONIC ACID, BUTYL ESTER	0.35	0.05	0.30	0.05	40	Q	1107
	ETHENESULFONIC ACID, PROPYL ESTER	0.22		0.40		70	C,I11,J	503
	ETHER, ALLYL PHENYL	2.53		0.138		70	C	754
	ETHER, P-TERT-BUTYLPHENYL VINYL	1.08		0.024		70	I2	937
	ETHER, BUTYL VINYL	2.6		0.63		60		714
	ETHER, 2-CHLOROETHYL VINYL	2.59	0.2	0.15	0.001	60		1178
	ETHER, P-CHLOROPHENYL VINYL	1		0.02		60	I7	1340
	ETHER, DODECYL VINYL	1.93	0.15	0.02	0.2	60		1178
	ETHER, ISOBUTYL VINYL	2.0	0.2	0.1	0.01	50	C	9
	ETHER, OCTADECYL VINYL	2.0	0.2	0.1	0.01	50	Q	775
	ETHER, OCTYL VINYL	2.10	0.20	0.1	0.2	50	C	166
	ETHER, PHENYL VINYL	1.90	0.20	0.1	0.1	50	C	9
		1.059		0.161		60		1179

M1	M2	R1	+/-	R2	+/-	TEMP.	REMARKS	REF.
ETHYLENE, CHLORO-								
	ETHER, PHENYL VINYL	1.06		0.16		60		1178
		1.93		0.43		60		714
	ETHER, M-TOLYL VINYL	1.23		0.131		60		1178
	ETHER, O-TOLYL VINYL	1.33		0.138		60		1178
	ETHER, P-TOLYL VINYL	1.17		0.128		60		1178
	ETHERS, C8-C18 ALKYL VINYL	0		1.2		50	C,P	12
	ETHYLENE	1.13	0.32	0.285	0.075	60	F,JJ	744
		1.85		0.16		70	I13,M10	225
		1.85		0.2		60	Q	296
		1.85	0.2	0.16	0.02	70	M10	246
		1.85	0.2	0.20	0.02	70	M5	226
		1.85		0.20		70	M10	226
		2		0.2		70	M2	245
		2.0		0.2		60	L,M10	225
		2.0		0.3		50	C	1113
		3.21	0.22	0.21	0.02	60	H71	643
		3.60	0.30	0.24	0.07	60	C,I21	744
		3.60	0.30	0.24	0.07	50	C,M14	1046
		4.16		0.05		90	M3	1083
		4.70		0.08		90	M3	148
						0	H38,I15	642
						60	C	642
	ETHYLENE, CHLOROTRIFLUORO-	2.53		0.01		68	C	510
	ETHYLENE, 1,1-DICHLORO-	0.14		1		30	C	8
		0.15		0.60		50	H71	1063
		0.16	0.03	4.5	0.5	50	J	1441
		0.2		4.5		45	C,K	962
		0.2	0.2	1.8	0.5	70		799
		0.20		1.8		55	C	1063
		0.23		3.15		60		244
		0.3		3.2		30		620
		0.30		0.90		25	H71	1063
		0.5		0.001		47	G17,I9	526
		0.5		7.5		93	I7	93
	ETHYLENE, FLUORO-	9		0.07	0.02	30	H45,I3	879
		11	1	0.05	0.005	30	H38	1165
		11.0	1	0.05	0.005	30	H38,I38	879
	ETHYLENE:AGNO3 COMPLEX (1:1)	0.002		60		60	F,JJ	744
	FUMARIC ACID, DIETHYL ESTER	0.12	0.01	0.47	0.05	50	C,I11,J	131
		0.12	0.05	0.47	0.04	60	C,I11,J	557
	HEXANOIC ACID, VINYL ESTER	1.35		0.65		40		503
		1.8		0.1		40	C,I11,J	503
	IMIDAZOLE, 2-METHYL-1-VINYL-	2		0.28		60		1213
	LAURIC ACID, VINYL ESTER	0.22		2.13		60	C,I11,J	613
	LEVULINIC ACID, VINYL ESTER	7.4		0.2		40	C,K	503
	MALEIC ACID, BIS(2-ETHYLHEXYL) ESTER	1.40	0.004	0.419	0.002	60		591
	MALEIC ACID, DIBUTYL ESTER	0.42		0		68	C	8
		0.5		0		68	C	8
	MALEIC ACID, DIETHYL ESTER	1.4		0.0		40	C	500
		0.77	0.03	0.009	0.003	60	F,K	557
		0.8		0.0		40	F,K	502
		0.8		0.0		40		499
		0.9	0.1	0		70		23

M1	M2	R1	+/-	R2	+/-	TEMP.	REMARKS	REF.
ETHYLENE, CHLORO-								
	MALEIC ACID, DIISOPROPYL ESTER	0.65	0.05	0.1		40	C	501
	MALEIC ANHYDRIDE	0.037		0.40		60	I7,L	972
		0.104	0.07	0.668		60	I5	972
	MALEIMIDE, N-BUTYL-	0.296		0.008		75	C	1033
	MALEIMIDE, N-(P-CHLOROPHENYL)-	2.08		0.04		60	I7	1342
	MALEIMIDE, N-ETHYL-	3.6		0.03		60	I7	1343
	MALEIMIDE, N-HEXYL-	2.41		0.02		60	I7	1342
	MALEIMIDE, N-METHYL-	2.03		0.06		60	I7	1342
	MALEIMIDE, N-OCTYL-	3.12		0.01		60	I7	1342
	MALEIMIDE, N-PHENYL-	1.84		0.06		60	I7	1342
	MALEIMIDE, N-PROPYL-	3.82		0.03		60	I7	1343
	MALEIMIDE, N-P-TOLYL-	2.02		0.04		60	I7	1342
	MALONONITRILE, METHYLENE-	4.37		0.03		60		1343
		0.0093	0.01	0.72		60	Q	1343
	METHACRYLIC ACID	0.017		0.54	0.2	50		255
		0.027		36		50	I57	289
	METHACRYLIC ACID, BORNYL ESTER	0.034		23.8		60		935
	METHACRYLIC ACID, BUTYL ESTER	0.06		12.5		60	C	541
		0.025		30.5		80	C	388
	METHACRYLIC ACID, 2,3-EPOXYPROPYL ESTER	0.05		13.5		45		1194
		0.04		8.84			Q	964
	METHACRYLIC ACID, ISOBORNYL ESTER	0.3		2.3		60	C,I7,Q	391
	METHACRYLIC ACID, OCTYL ESTER	0.12		10.0		45		504
	METHYL METHACRYLATE	0.04		14.0		45	N,O	388
		0		12.5				964
	NONANOIC ACID, VINYL ESTER	0.02		15		68	C	620
	2,5-NORBORNADIENE	0.1		10		60	C	964
		1.16	0.06	0.282	0.035	50		8
	NORBORNANE, 2-VINYL-	0.74		0.35			C	591
	2-NORPINANE-2-ETHANOL, ACRYLATE	0.74		0.35			C	1082
	OCTADECANOIC ACID, 12-OXO-, VINYL ESTER	0.85		0.76		60	Q	1081
	OCTADECANOIC ACID, 4-OXO-, VINYL ESTER	0.14	0.01	4.3	0.2	60	C	387
	OCTANOIC ACID, VINYL ESTER	0.963	0.01	0.248	0.01	60	C	600
	2-OXAZOLIDINONE, 3-VINYL-	0.874	0.044	0.320	0.076	86	C	593
	1-PENTENE	3.2		0.2		50	C,I11,J	593
	1-PHENANTHRENEMETHANOL, 1,2,3,4,4A,4B,5,6,10,10A-DECAHYDRO- *	0.84	0.02	0.35	0.02	68	C	503
	7-ISOPROPYL-1,4A-DIMETHYL-, ACRYLATE							
	PHTHALIC ACID, DIALLYL ESTER	0.5		0.38		68	C	116
	1-PROPENE	0.5		2.57		60	C	8
		1.68		2.33		60	CC	8
		0.09		4.52		60	C,M14	8
		0.09		0.3		30	C,M9	1291
		1.12	0.2	0.29	0.5	60	H72	1337
		2.27	0.2	0.09	0.1	30	H71	1337
		2.35		0.24		60	C,M14	1062
	1-PROPENE, 1-CHLORO-	2.45		0.08		50	C,K	615
	1-PROPENE, 1-CHLORO-, CIS-	1.13		0.18		60		1062
	1-PROPENE, 1-CHLORO-, TRANS-	12.1		5.21		60		1206
	1-PROPENE, 2-CHLORO-	4.0		0.58		60		962
	1-PROPENE, 3-CHLORO-,OLIGOMER (DP 13-20)	0.18		0.33		60	C,K	745
	1-PROPENE, 2-CHLORO-3,3-DIMETHOXY-	0.75				50	Q	745
		0.63		0.04				962
		0.288	0.007	0.04	0.63	60		241

M1	M2	R1	+/-	R2	+/-	TEMP.	REMARKS	REF.
ETHYLENE, CHLORO-								
	1-PROPENE, 3-CHLORO-2-METHYL-	0.31		0		45		654
	1-PROPENE, 2-METHYL-	0.08		1.53		60	C,M14	1337
		0.08		1.76		60	C,M14	1337
		0.08	0.02	1.33		60	C,M9	1337
		1.54		0.08		60	C,M14	1206
		2.05		0.08		50		131
		2.05	0.3	0.08	0.1	60		560
		2.11		0.34		60		615
		4.3		0		65	N,O	292
	2-PROPENE-1,1-DIOL, DIACETATE	1.75		0.58			Q	1112
	1-PROPENE-1,2,3-TRICARBOXYLIC ACID, TRANS-, TRIMETHYL ESTER	0.15	0.10	0.00	0.50	60	C	595
	1-PROPEN-2-OL, ACETATE	2.2		0.25		65	C	772
	2-PROPEN-1-OL, 2-CHLORO-, ACETATE	0.70		0.65		100	C	484
	PROPIONIC ACID, VINYL ESTER	1.35	0.05	0.65	0.04	40	C,I11,J	503
		1.60	0.04	0.60	0.10		Q	176
		1.60	0.04	0.60	0.10		Q	176
	2H-PYRAN-3-CARBOXYLIC ACID, 5,6-DIHYDRO-2,6-DIMETHYL-, VINYL* ESTER	0.02		23.4		60	Q	613
	PYRIDINE, 4-VINYL-	0.38		0.53			Q	132
	2-PYRROLIDINONE, 1-VINYL-	0.53		0.38			Q	132
		0.55		0.34			Q,I24	1398
		0.74		0.73			Q,I15	1398
	SILANE, ALLYLTRIETHOXY-	2.0		0		50	C	957
	SILANE, 2-BUTENYLTRIETHOXY-	0.4		0		50	C	957
	SILANE, (1-CHLOROVINYL)TRIETHOXY-	0.2		0		50	C	957
	SILANE, (1-CYCLOHEXEN-4-YL)TRIETHOXY-	0.4		0		50	C	957
	SILANE, DIETHOXYETHYLVINYL-	1.0		0		50	C	957
	SILANE, DIETHOXYMETHYLVINYL-	1.0		0		50	C	957
	SILANE, DIETHOXYPHENYLVINYL-	0.7		0		50	C	957
	SILANE, TRIETHOXY(5-NORBORNEN-2-YL)-	1.6		0		50	C	957
	SILANE, TRIETHOXYPROPENYL-	8.0		0		50	C	957
	SILANE, TRIETHOXYVINYL-	0.9		0		50	C	957
	SILANE, TRIISOPROPOXYVINYL-	0.8		0		50	C	957
	SILANE, TRIMETHOXYVINYL-	0.8		0		50	C	957
	STEARIC ACID, VINYL ESTER	0.745	0.026	0.290	0.025	60	C	591
	STYRENE	0.02		17		60	P,Q,R	321
		0.040	0.001	17.9	3	60		210
		0.045		12.4	0.1	68		451
		0.067		35		50	Q	961
		0.077		35		50	I26	775
		0.08		28		50	K	165
		0.17		32		50	I26,M	395
		0.43		5.3		50	H81	990
	STYRENE, 2,3,4,5,6-PENTACHLORO-	0.06	0.01	5.65	0.25	50		678
	SUCCINIC ACID, METHYLENE-, DIETHYL ESTER	0.06	0.01	6.0	0.5	50		678
	SUCCINIC ACID, METHYLENE-, DIISOPROPYL ESTER	0.053	0.01	5.0	0.2	50		677
	SUCCINIC ACID, METHYLENE-, DIMETHYL ESTER	0.06	0.02	7.0	1.0	50		677
	SUCCINIC ACID, METHYLENE-, DIOCTYL ESTER	1.5		0		60		678
	TIN, BUTYLDICHLOROVINYL-	0.85		0.001		60		1443
	TIN, (MALEOYLDIOXY)BIS(TRIETHYL-	0.1		0		60		1195
	TIN, TRIBUTYL(METHACRYLOYLOXY)-	1.2		4.5		60		1195
	TIN, TRIETHYLVINYL-							1359
	TRIDECANEDIOIC ACID, 2-METHYLPENTYL VINYL ESTER	1.10	0.3	0.239		60	C,I7	164

COPOLYMERIZATION PARAMETER

M1	M2	R1	+/-	R2	+/-	TEMP.	REMARKS	REF.
ETHYLENE, CHLORO-								
	TRISILOXANE, 1,1,1,3,5,5-HEPTAMETHYL-3-VINYL-	0.90	0.13	0.50	0.07	50		773
	TRISILOXANE, 1,1,5,5,5-HEXAMETHYL-3-TRIMETHYLSILOXY-3-VINYL- *	0.90	0.13	0.50	0.07	50		773
	UNDECANOIC ACID, 11-IODO-, ALLYL ESTER	1.64		0.42		60	I7	177
	9-UNDECENOIC ACID, VINYL ESTER	1.06	0.05	0.358	0.065	60	C	591
	VALERIC ACID, VINYL ESTER	2		0.28		60		1213
	VINYL ACETATE	1.35	0.05	0.65	0.04	40	C,I11,J	503
		1.60		0.30		50	I26,M	395
		1.68	0.08	0.23	0.02	60		622
		1.8	0.6	0.6	0.2	40		596
		2		0.28		60		1213
		2.1		0.3		68		8
		2.45	0.15	0.30	0.08	60	C	304
		2.5		0.1		-40	H38	807
		3.74		0.033		-32	H7	80
ETHYLENE, CHLOROTRIFLUORO-								
	ETHER, ETHYL VINYL	0.008	0.002	0	0.0015	20	E	1400
	ETHYLENE	0.001	0.001	0.070	0.001	-78	H38	794
		0.009	0.001	0.116	0.001	-40	H38	794
		0.0285	0.001	0.173	0.0025	60	H38	794
		0.01		0.250		60	C	510
		0.02		2.53		60	C	510
		0.06		17.14		30	H38,I38	879
		0.06		0.18		30	H38	1165
		1.2		0.8		80	K,N	227
		0.04		25		-145	E,I1	145
		1.0		1.0		0.62	E	145
		1.0		1.0		60	K	227
		0.005		75.0		60	C	955
	ETHYLENE, CHLORO-	0		0.06		-35	E	1227
	ETHYLENE, 1,1-DICHLORO-	0.01		0.06		-35	E	1401
	ETHYLENE, FLUORO-	0.01	0.02	0.24	0.02	-78	E	1401
	ETHYLENE, TETRAFLUORO-	0.016	0.005	0.022	0.005	-40	E	1227
	METHYL METHACRYLATE	0.017	0.005	0.03	0.01	0	H38	1369
	1-PROPENE	0.028	0.015	0.06	0.02	40	H38	1369
	1-PROPENE, 2-METHYL-	0		0.06		-35	H38	1369
				0.04		0	E	1227
				0.04		0	E	1227
	STYRENE	0.004	0.001	0.038	0.007	60	E	1401
	TRISILOXANE, 1,1,1,3,5,5-HEPTAMETHYL-3-VINYL-	0.001		7.0		60	H38	1167
	TRISILOXANE, 1,1,5,5,5-HEXAMETHYL-3-TRIMETHYLSILOXY-3-VINYL- *	0.05		0.20		60	C	955
	VINYL ACETATE	0.05		0.20		60	C	773
		0.01		0.6		60	C	955
ETHYLENE, 1,1-DIBROMO-								
	ETHYLENE, 1,1-DICHLORO-	1.90	0.11	1.04	0.10	60	I12,Q	1032

M1 / M2	R1	+/-	R2	+/-	TEMP.	REMARKS	REF.
ETHYLENE, 1,1-DICHLORO-							
ACETIC ACID, ALLYL ESTER	6.6		0		60		620
ACRYLAMIDE	0.15	0.08	4.89	0.08	60		41
ACRYLAMIDE, N-OCTADECYL-	0.438	0.008	1.37	0.008	60	C,I14	454
ACRYLIC ACID	0.46		1.26		50	I10	1236
ACRYLIC ACID, BUTYL ESTER	0.84	0.2	0.46	0.13	70		575
	0.87	0.01	0.58	0.03	0		575
	0.88	0.10	0.83	0.02	50	K	453
ACRYLIC ACID, ETHYL ESTER	0.5		0.8		40	F,K	1053
ACRYLIC ACID, METHYL ESTER	0.005		38		50	G17,I21	526
	0.1		3.67		60	KK	1220
	0.3		1.1		60	I7,FF	373
	0.7		0.8		60	I7,FF	373
	0.85		0.65		60		1220
	0.99	0.10	0.6	0.06	60	I7,FF	373
	1		0.84		60	C	214
			1		60		620
					70		32
ACRYLIC ACID, OCTADECYL ESTER	0.91	0.05	1.01	0.01	60	C	453
	0.91	0.05	1.01	0.01	60		1239
ACRYLIC ACID, OCTYL ESTER	0.87	0.02	0.70	0.01	60	C	453
ACRYLONITRILE	0.37	0.10	0.91	0.10	60	C	558
	0.390		0.585		32,8	K	589
	0.40	0.05	0.44	0.05	40	I31/I60	215
	0.42	0.03	0.58	0.02	40	I11/I14	215
	0.49		1.20		45	C,I14,L	764
	0.69	0.03	0.65	0.03	40	I11/I59	215
	0.76	0.02	0.57	0.01	40	I11/I58	215
	0.95	0.03	0.50	0.03	40	I31/I45	215
	0.95	0.02	0.48	0.02	40	I14/I31	215
	1.30	0.03	0.65	0.03	40	I31/I61	215
	1.39	0.03	0.65	0.03	40	I31/I59	215
	1.43	0.03	0.73	0.03	40	I11/I12	215
	1.70	0.02	0.63	0.02	40	I31/I58	215
	1.74	0.02	0.63	0.02	40	I11/I15	215
	1.80	0.03	0.66	0.03	40	I12/I31	215
		0.03	0.70	0.03	40	I15/I31	215
ACRYLOYL CHLORIDE	0.5		1.12		50	I24	1236
ANISOLE, P-PROPENYL-/MALEIC ANHYDRIDE	0.006	0.0055	53.5	6.1	60	I7	1259
BENZENE, ETHYNYL-	0.1		1.4		60		211
BENZOIC ACID, VINYL ESTER	7.0	1	0.1	0.02	50	C	166
	7.0	1	0.1	0.02	50	C	775
1,3-BUTADIENE	0.05		1.9		5	K	1025
3-BUTEN-2-ONE	0.55	0.2	1.8		70		22
3-BUTEN-2-ONE, 3-METHYL-	0.15		4.5		60		775
CARBAZOLE, 9-VINYL-	0.020	0.02	4.5	0.1	60	Q	166
COUMARIN	1		3.7		75	C	331
CROTONALDEHYDE	17		0		70	I7	23
CROTONIC ACID	35		0		70		217
CROTONIC ACID, BUTYL ESTER	33	5	0.065	0.005	60	C	166
CROTONIC ACID, PROPYL ESTER	20		0		70		217
CYCLOBUTANECARBOXYLIC ACID, 3-ACETYL-2,2-DIMETHYL-, VINYL ESTER *	20		0		70		217

M1	M2	R1	+/-	R2	+/-	TEMP.	REMARKS	REF.
ETHYLENE, 1,1-DICHLORO-	CYCLOHEXENE, 4-VINYL-	1.8		0		70		217
	4-CYCLOPENTENE-1,3-DIONE	2.4	0.6	0.15	0.05	65	I7	1049
	ETHANESULFONIC ACID, 2-HYDROXY-, METHACRYLATE	0.22	0.03	3.6	0.5	60	I55	1241
	ETHENESULFONIC ACID, BUTYL ESTER	7.5	0.6	0.065	0.007	80	C	754
	ETHER, P-TERT-BUTYLPHENYL VINYL	2.01		0.048		60		1178
	ETHER, BUTYL VINYL	1.73		0.012		60		1178
	ETHER, P-CHLOROPHENYL VINYL	2.43		0.144		60		1178
	ETHER, DODECYL VINYL	1.30	0.015	0.00	0.2	50	C	10
	ETHER, ETHYL VINYL	3.2		0		60	N,O	620
	ETHER, OCTADECYL VINYL	1.50	0.15	0.0	0.3	50	C	10
	ETHER, OCTYL VINYL	1.35	0.15	0.0	0.2	50	C	10
	ETHER, PHENYL VINYL	1.63		0.041		60		1179
		1.634		0.041		60		1179
		2.37	0.42	0.38	0.12	60	Q	986
	ETHER, M-TOLYL VINYL	1.91		0.036		60		1178
	ETHER, O-TOLYL VINYL	2.08		0.041		60		1178
	ETHER, P-TOLYL VINYL	1.96		0.034		60		1178
	ETHERS, C8-C18 ALKYL VINYL	15.2		0.03		50	C,P	12
ETHYLENE	ETHYLENE	0.83	0.08	2.38	0.06	30	E,M2	1302
	ETHYLENE, 1-BROMO-1-CHLORO-	0.001		0.5		25	I12,Q	1032
	ETHYLENE, CHLORO-	0.60		0.15		30	G17,I9	526
		0.90		0.30		30	H71	1063
		1.8		0.14		68	H71	1063
		1.8	0.5	0.20	0.2	70	C	8
		3.15		0.2		45	C	1063
		3.2		0.23		55		799
		4.5		0.2		60		244
		4.5		0.2		50	C,K	620
		7.5		0.16		50	J	962
		17.14	0.5	0.5		47	I7	1441
	ETHYLENE, CHLOROTRIFLUORO-	1.04	0.10	1.90	0.11	60	C	93
	ETHYLENE, 1,1-DIBROMO-	12.2	2.0	0.046	0.015	60	I12,Q	510
	FUMARIC ACID, DIETHYL ESTER	11.7	0.07	0.15	0.014			1032
	FURAN, 2-VINYL-	0.40	0.06	0.33	0.05	70	C,Q	210
	5-HEXEN-3-YN-2-OL, 2-METHYL-	1.46	0.28	0.33	0.18	70	C,P	463
	INDENE	12.5	8	0.02		60		607
	ISOCYANIC ACID, VINYL ESTER	40				60		23
	MALEIC ACID, DIETHYL ESTER	9				60		420
	MALEIC ANHYDRIDE			0.0		60		211
	MALEIMIDE	0.71		0.48		60	N,O	620
	MALEIMIDE, N-2-BORNYL-	1.15		0.47		75	I13	1013
	MALONONITRILE, METHYLENE-	0.012		0.049		50	C,I5	1433
	METHACRYLIC ACID	0.15	0.02	3.0	0.4	22	R	289
	METHACRYLIC ACID, BENZYL ESTER	0.34	0.05	3.3	0.5	70	C	23
	METHACRYLIC ACID, BUTYL ESTER	0.35		0.22		60	C	405
	METHACRYLIC ACID, ETHYL ESTER	0.35		0.22		70	C	8
	METHACRYLOYL CHLORIDE	0.30	0.06	2.0	0.6	68	C	405
	METHYL METHACRYLATE	0.05	0.06	1.32	0.18	60	D,O,O	1458
		0.06		16.0		25	I36	386
		0.09		2.1		40	F,K	1053
		0.2	0.2	16	3	50	MM	1458

M1	M2	R1	+/-	R2	+/-	TEMP.	REMARKS	REF.
ETHYLENE, 1,1-DICHLORO-								
	METHYL METHACRYLATE	0.24	0.03	2.53	0.01	60	C	558
		0.34	0.05	2.5	0.3	60	C	405
		0.4	0.05	2.5	0.1	20	D	1458
		0.5	0.15	2.54	0.2	50		1458
		0.50		2.50		25		386
	NONANOIC ACID, VINYL ESTER	4.08	0.20	0.0	0.01	60	C	591
	2,5-NORBORNADIENE	1.41		0.08		50		1082
		1.41		0.08		50		1081
	OCTADECANOIC ACID, 12-OXO-, VINYL ESTER	4.0	0.5	0	0.16	60	C	593
	OCTADECANOIC ACID, 4-OXO-, VINYL ESTER	3.8	0.6	0	0.17	60	C	593
	OCTADECANOIC ACID, 9(10)-OXO-, VINYL ESTER	3.7	0.3	0.01	0.09	60	C	593
	2-OXAZOLIDINONE, 3-VINYL-	1.35		0.08		75		336
	1-PHENANTHRENEMETHANOL, 1,2,3,4,4A,4B,5,6,10,10A-DECAHYDRO- * 7-ISOPROPYL-1,4A-DIMETHYL-	1.35		0.08		75		336
	PHTHALIC ACID, DIALLYL ESTER	5.0		0.2		70		24
	1-PROPENE, 3-CHLORO-	3.8		0.26		68		8
		4.5		0		60	D	620
	1-PROPENE, 3-CHLORO-2-METHYL-	1.1		0		60	N,O	620
	1-PROPENE, 2-METHYL-	0		25		-78	E	860
		0.02		56		-79,5	E,L1	540
		0.11		1.5		-39	E,L1	540
		0.21		1.27		-40	E	860
		1.0		0.40		0	E,L1	540
		1.3		0.03		0	E	860
		1.5		0.05		60	N,O	620
		3.00		0.05		50		351
		3.30		0.01		30	D	597
	1-PROPENE-1,2,3-TRICARBOXYLIC ACID, TRANS-, TRIMETHYL ESTER	54	5	0.31	0.1	60	C,I7	419
	1-PROPEN-2-OL, ISOCYANATE	0.85	0.04	0.16	0.02	60	Q	620
	2-PROPEN-1-OL, 2-METHYL-, ACETATE	2.4		0.075		40	N,O	139
	2-PROPEN-1-OL, 2-METHYL-, OXALATE	4.8	0.2	0	0.01	60	C	591
	STEARIC ACID, VINYL ESTER	3.80	0.05	0.0054	0.025	80	C,I14	454
	STERAMIDE, N-ALLYL-	5.23	0.067	3.4	0.136	45		289
	STYRENE	0.001		0		26	G17,I9	526
		0.015		1.85		60	Q	637
		0.046	0.010	1.75	0.05	60		210
		0.085	0.02	2.0	0.10	60		318
		0.12	0.05	2.1	0.1	50	C	558
		0.14	0.009	1.80	0.2	25	C	958
	SUCCINIMIDE, N-VINYL-	0.145	0.02	0.32	0.27	60	E	149
	2H-TETRAZOLE, 2-PHENYL-5-(P-VINYLPHENYL)-	0.15		1.8	0.03	60	I7	365
	9-UNDECENOIC ACID, VINYL ESTER	2.58	0.09	0.054	0.030	60	C	591
	VINYL ACETATE	3.6	0.5	0.0		60		210
		5.0		0.05		50		447
		6		0.1		68	C	8
		6.7	0.5	0.05	0.03	60	C	405
ETHYLENE, 1,2-DICHLORO-, CIS-								
	MALONONITRILE, METHYLENE-	0		30		40		289
	STYRENE	0		56.1		50		133

COPOLYMERIZATION PARAMETER

M1	M2	R1	+/-	R2	+/-	TEMP.	REMARKS	REF.
ETHYLENE, 1,2-DICHLORO-, CIS-								
	STYRENE	0		210	31.5	60	D	557
		0		75	25	68	C	27
		0		47.5		90		133
		0		72.5		90		133
	VINYL ACETATE	0.018	0.003	2.8	0.2	68	C	27
				6.3		60		557
ETHYLENE, 1,2-DICHLORO-, TRANS-								
	MALONONITRILE, METHYLENE-	0		30		40		289
	STYRENE	0		44.9		50		133
		0		44.4		50		133
		0		37	3	60	D	557
		0		10		68	C	27
		0		25.9		90		133
		0		29.8		90		133
	VINYL ACETATE	0.063		0.85		68	C	27
		0.086	0.01	0.95	0.02	-45	H38	1177
				0.99		60		557
ETHYLENE, 1,1-DICHLORO-2,2-DIFLUORO-								
	STYRENE	0		1.6		45	K,N	626
	VINYL ACETATE	0		0.60	0.09	70		23
ETHYLENE, 1,1-DIFLUORO-								
	ETHYLENE, FLUORO-	0.17	0.03	5.5	0.5	25	H38	686
		0.17	0.03	5.5	0.5	30	H38,I38	879
		0.17	0.03	5.5	0.5	30	H38	1165
		0.18	0.02	4.2	0.4	30	H45,I3	879
ETHYLENE, 1,2-DIMETHOXY-								
	ACRYLIC ACID, METHYL ESTER	0.02	0.02	4.8	0.4	60		434
	ACRYLONITRILE	0.01	0.01	0.65	0.05	60		434
ETHYLENE, 1,2-DIMETHOXY-, CIS-								
	ETHYLENE, 1,2-DIMETHOXY-, TRANS-	4.45	0.38	0.27	0.12	~-70	H10,I3	1334
	MALEIMIDE	0.222		0			Q	665
ETHYLENE, 1,2-DIMETHOXY-, TRANS-								
	MALEIMIDE	0.27	0.12	4.45	0.38	-70	H10,I3	1334
		0.192		0			Q	665
ETHYLENE, 1,1-DIPHENYL-								
	ACRYLIC ACID, ETHYL ESTER	0.5		0.8	0.010	36	F,K	240
	ACRYLIC ACID, METHYL ESTER	0		0.092	0.006	60	Q	409
		0		0.102	0.006	60	D	212
	ACRYLONITRILE	0		0.028	0.003		C	211

M1	M2	R1	+/-	R2	+/-	TEMP.	REMARKS	REF.
ETHYLENE, 1,1-DIPHENYL-								
	ACRYLONITRILE	0		0.028	0.003	60	C	214
	ANISOLE, O-VINYL-	0		0		40	G17,I5	1449
		0		20		0	G17,I5	1449
	ANISOLE, P-VINYL-	0		0		0	G17,I5	1449
	1,3-BUTADIENE	0		0.3		40	G17,I7	1449
		0		0.09		0	G5,I5	1450
		0		0.13		0	G17,I5	1450
		0		0.09		0	G19,I5	1450
		0		0.71		40	G18,I7	1450
		0		0.1		40	G5,I7	1450
		0		54		40	G17,I7	1450
	1,3-BUTADIENE, 2-CHLORO-	0.00	0.05	3.17	0.16	60	C	214
	1,3-BUTADIENE, 2,3-DICHLORO-	0		4.5	0.45	60	C	214
	1,3-BUTADIENE, 2,3-DIMETHYL-	0		4.5	0.45	60	C	211
	ISOPRENE	0		0.23	0.04	40	G17,I7	1451
		0		0.12		0	G19,I5	1448
		0		0.11		0	G5,I5	1448
		0		0.11		0	G17,I5	1448
		0		0.04		25	G17,I10	1448
		0		0.5		30	G17,I20	1448
		0		29		40	G17,I7	1448
		0		37		40	G18,I7	1448
		0		0.38		40	G5,I7	1448
	METHACRYLONITRILE	0.09		0.05		60	R	573
	METHYL METHACRYLATE	0		0.33		36	F,K	240
	STYRENE	0		2.1	0.04	110	A,O	1189
		0		0.34		110	A	1188
		0		0.34		30	H17,I5	1075
		0		0.13		30	H19,I5	1075
		0.		0.29		30	H17,I2	1075
		0		0.44		30	H20,I5	1075
		0		0.16		30	H19,I5	1075
		0		0.23		30	H17,I9	1075
		0		0.63		30	H17,I7	1075
		0		0.71		30	H19,I2	1075
		0		0.42		30		1075
ETHYLENE, FLUORO-								
	ACRYLONITRILE	0.001		24	2	30	H38	1165
		0.001		24	2	30	H38,I38	879
	CYCLOBUTENE, HEXAFLUORO-	0.005	0.2	44.0	0.2	25	E,I7	1007
		3	0.6	0		30	H38	1165
	ETHYLENE	3	0.6	0		30	H38,I38	879
		0.16	0.05	4.39	0.77	160	M3	148
	ETHYLENE, CHLORO-	0.3	0.03	1.7	0.1	30	H38	1165
		0.3	0.03	1.7	0.1	30	H38,I38	879
		0.05	0.005	11.0	1	30	H38,I38	879
		0.05	0.005	11	1	30	H38	1165
		0.07	0.02	9		30	H45,I3	879
	ETHYLENE, CHLOROTRIFLUORO-	0.18	0.02	0.06	0.02	30	H38	1165
		0.18	0.02	0.06	0.02	30	H38,I38	879
		0.8		1.2		80	K,N	227

COPOLYMERIZATION PARAMETER

M1	M2	R1	+/-	R2	+/-	TEMP.	REMARKS	REF.
ETHYLENE, FLUORO-	ETHYLENE, 1,1-DIFLUORO-	4.2	0.4	0.18	0.02	30	H45,I3	879
		5.5	0.5	0.17	0.03	25	H38	686
		5.5	0.5	0.17	0.03	30	H38	1165
	ETHYLENE, TETRAFLUORO-	0.27	0.03	0.05	0.02	30	H38,I38	686
		0.27	0.03	0.05	0.02	30	H38	879
		0.27	0.03	0.05	0.02	30	H38	1165
	1-PROPENE, HEXAFLUORO-	1.01	0.01	0		30	H38,I38	1165
		1.1	0.01	0		30	H38	879
	1-PROPENE, 1,2,3,3,3-PENTAFLUORO-, CIS-	25	5	0	0.02	30	H45,I3	879
	VINYL ACETATE	0.09	0.05	0		30	H64,I3	1165
		0.16	0.01	2.9	0.2	30	H38,I38	1165
		0.16	0.01	2.9	0.2	30	H38,I38	879
ETHYLENE, IODO-	STYRENE	0.145	0.01	7.0	0.2	60		130
ETHYLENE, NITRO-	ACRYLONITRILE	65	15	0.01	0.01	78	E,I5	1060
ETHYLENE, 1-PHENYL-1-(P-VINYLPHENYL)-	STYRENE	4.4		0.4		80		1158
ETHYLENE, TETRACHLORO-	ACRYLIC ACID, METHYL ESTER	0		200		60	N,0	620
	ACRYLONITRILE	0		830		60	N,0	211
		0		470		60	N,0	210
				456	70	75	C	1310
	METHYL METHACRYLATE			240.	20	60	C	824
				200.	20	60	C	824
	STYRENE	0		165		50		133
		0		208		50		133
		0		185	20	60	0	210
		0		187		90		133
		0		66.4		90		133
	VINYL ACETATE	0		129	20	90		210
		0		6.8	0.5	60	C	133
		5		5		68		8
ETHYLENE, TETRAFLUORO-	ETHYLENE	0.024		0.61		25	D,I71	1240
		0.1		0.38		25	D	1240
	ETHYLENE, CHLOROTRIFLUORO-	0.85		0.15		80	K	396
		1.0		1.0		0.62	E	145
		1.0		1.0		60	K	227
		25		0.04		-145	E,I1	145

M1 / M2	R1	+/-	R2	+/-	TEMP.	REMARKS	REF.
ETHYLENE, TETRAFLUORO-							
ETHYLENE, FLUORO-	0.05	0.02	0.27	0.03	25	H38	686
	0.05	0.02	0.27	0.03	30	H38	1165
	0.05	0.02	0.27	0.03	30	H38,I38	879
1-PENTENE, 3,3,4,5,5,5-HEPTAFLUORO-	0.21		2.3		22	E,M13	1160
1-PROPENE	0.21		2.3		22	E,M14	1159
	0.06		1.0		-78	E	923
	0.06		1.0		-78	E	922
1-PROPENE, HEXAFLUORO-	3.5		0		25	D	1240
1-PROPENE, 2-METHYL-	0		0.2		-45	E	920
	0		0		-78	E	921
	0		0		-78	E	920
1-PROPENE, 1,3,3,3-TETRAFLUORO-, TRANS-	0.3		0.0		80	K	396
1-PROPENE, 2,3,3,3-TETRAFLUORO-	22		0.18		22	E,M13	1160
	0.37		5.4		22	E,M13	1160
1-PROPENE, 3,3,3-TRIFLUORO-	0.12		5		100	E,M13	141
	0.12		5		21	E,M14	141
	0.15		2.5		21	E,M	141
1-PROPENE, 3,3,3-TRIFLUORO-2-(TRIFLUOROMETHYL)-	0.58		0.09		22	E,M11	1160
STYRENE	0.01		5.2		80	E	1254
ETHYLENE, TRICHLORO-							
ACRYLIC ACID, METHYL ESTER	0		33		60	N,O	620
ACRYLONITRILE	0		67		60	N,O	620
1,3-BUTADIENE	-0.007	0.022	62.1	3.5	75	C	1311
METHYL METHACRYLATE	0		12.32	0.31	80	C	310
STYRENE	0		100		60	N,O	620
	0		17.1		50		133
	0		16.5		50		133
	0		12.1		90		133
	0		12.7		90		133
	0.0		16		60		210
VINYL ACETATE	0.0		0.67	2	68	C	27
	0.01	0.01	0.66	0.04	60		622
ETHYLENE:AGNO3 COMPLEX (1:1)							
ETHYLENE, CHLORO-	60		0.002		60	F,JJ	744
ETHYLENEDIAMINE, N,N'-BIS(P-HYDROXYBENZYLIDENE)-, DIMETHACRY*LATE							
STYRENE	2.4		0.235		50		1028
ETHYLENE GLYCOL, BIS(2-CHLOROACRYLATE)							
ACRYLIC ACID, METHYL ESTER						P,Q	900
METHYL METHACRYLATE						P,Q	900
STYRENE	0.1		0.6			P,Q	900
						Q	908

M1 / M2	R1	+/-	R2	+/-	TEMP.	REMARKS	REF.
ETHYLENE GLYCOL, DIMETHACRYLATE							
ACRYLONITRILE	1.24	0.10	0.22	0.01		Q	71
STYRENE	0.65		0.35		60		1044
STYRENE	0.65		0.40		61	C	1042
ETHYLENE OXIDE							
ACETALDEHYDE	1.05	0.40	0.40	0.12	-20	H10	728
ACETONITRILE	7.8		0		70	H20,R	230
ACRYLONITRILE	0		12		70	C,R	230
FURAN, TETRAHYDRO-	0.2		0		70	H11,R	230
	5.1		0		70	H20,R	230
	0.08		2.2		0	H10,14	673
GLYCIDAMIDE	0.13	0.03	14.2	1.5	70	H10	437
OXETANE, 3,3-BIS(CHLOROMETHYL)-	0.2		0.5		0	G13	347
OXETANE, 3-(CHLOROMETHYL)-3-METHYL-	0.027	0.09	67	10	-50	H10,13	47
PROPANE, 1-BROMO-2,3-EPOXY-	0.19	0.01	2.55	0.5	80	H10	283
	0.2	0.12	36.0	9	3	H11	1118
PROPANE, 1-CHLORO-2,3-EPOXY-	3.15	0.15	0.31	0.02	3	H13	283
PROPYLENE OXIDE	1.15	0.05	0.12	0.01	25	H13,12	1265
	6.5		0.5			H13,12	1265
						G10	1371
FATTY ACIDS, C6-C18, VINYL ESTERS							
ACRYLONITRILE	0.04	0.03	4.0	0.3	60	C,I6,S	58
FERROCENE, VINYL-							
ACRYLIC ACID, METHYL ESTER	0.82	0.05	0.63	0.03	70	I7	1268
ACRYLONITRILE	0.15	0.05	0.16	0.05	70	I7	1268
ETHER, ISOBUTYL VINYL	0.1	0.1	9.7	1	0	H10,I2	1139
METHYL METHACRYLATE	0.52	0.27	1.22	0.37	70	I7	1268
2-PYRROLIDINONE, 1-VINYL-	0.66		0.4		70	I7	1354
STYRENE	0.08	0.04	2.5	0.2	70	I7	1268
STYRENE	0.2	0.1	4	1	70	I7	1140
FERROCENEETHANOL, ACRYLATE							
ACRYLIC ACID, METHYL ESTER	0.76	0.027	0.687	0.043	65	I7	1358
STYRENE	0.408	0.138	1.06	0.051	65	I7	1358
VINYL ACETATE	3.43	0.313	0.074	0.094	65	I7	1358
FERROCENEETHANOL, METHACRYLATE							
METHYL METHACRYLATE	0.203	0.035	0.646	0.014	65	I7	1358
STYRENE	0.084	0.018	0.582	0.01	65	I7	1358
VINYL ACETATE	8.791	0.231	0.058	0.003	65	I7	1358
FERROCENEMETHANOL, ACRYLATE							
ACRYLIC ACID, METHYL ESTER	0.14	0.02	4.46	0.2	70	I7	1267
METHYL METHACRYLATE	0.08	0.03	2.9	0.5	70	I7	1267
STYRENE	0.02	0.01	2.5	0.1	60	I7	1356

M1	M2	R1	+/-	R2	+/-	TEMP.	REMARKS	REF.
FERROCENEMETHANOL, ACRYLATE								
	STYRENE	0.02	0.01	2.3	0.3	70	I7	1267
	VINYL ACETATE	1.44	0.38	0.46	0.07	70	I7	1267
FERROCENEMETHANOL, METHACRYLATE								
	ACRYLIC ACID, METHYL ESTER	0.08	0.03	0.82	0.07	70	I7	1267
	METHYL METHACRYLATE	0.12	0.02	3.27	0.06	70	I7	1267
	STYRENE	0.01	0.01	3.7	0.01	60	I7	1356
		0.03	0.02	3.7	0.2	70	I7	1267
	VINYL ACETATE	1.52	0.3	0.2	0.1	70	I7	1267
FORMALDEHYDE								
	KETENE	5		0.5	0.02	-30	G29,I20	1228
	PROPYLENE OXIDE	37	2	-0.13		-78	H10,I2	461
		40	13	0.21	0.25	-78	H10,I3	461
	STYRENE	30		0		-30	E	161
		52		0		-78	E,I3	161
FORMAMIDE, N-ALLYL-								
	ACRYLIC ACID	0.06		2.6		80	I12	857
	METHACRYLIC ACID	0.03		4.6		80	I12	857
FORMIC ACID, VINYL ESTER								
	ACRYLONITRILE	0.04	0.005	3.0	0.05	60	Q	166
	1,3-BUTADIENE	3.0		0.04		120	Q	775
	1,3-BUTADIENE, 2-CHLORO-	0.2	0.05	5.0	0.01	65	F	994
		0.01		30			C	1002
		-0.035		15.0			Q	493
	CARBAZOLE, 9-VINYL-	0.196	0.004	4.22	0.16	100	C	1001
	METHYL METHACRYLATE	0.05	0.01	28.6	2.5	60	C,I7	1135
	1-PROPENE, 3-CHLORO-	0.57	0.08	0.78	0.10	60	C	457
	STYRENE	0.01	0.01	17.6	0.6	70	C	858
	VINYL ACETATE	0.68		1.41		50	C	205
		0.95	0.07	0.94	0.07	60	C,I7	1135
FUMARAMIDE, N,N'-DIBUTYL-								
	STYRENE	0		10		70		1144
FUMARAMIDE, N,N,N',N'-TETRAETHYL-								
	STYRENE	0.69		0.69		70	I7	1144
FUMARANILIDE								
	STYRENE	0		1.4		100	I6	1144
FUMARIC ACID								
	ACRYLAMIDE	0.09		7.0			I6,Q	577

M1 / M2	R1	+/-	R2	+/-	TEMP.	REMARKS	REF.
FUMARIC ACID, BIS(2-CHLORO-1-METHYLPROPYL) ESTER							
STYRENE	-0.05	0.01	0.25	0.02	60		1010
FUMARIC ACID, 1,3-BUTYLENEGLYCOL POLYESTER							
ACETIC ACID, DIETHYLPHOSPHONYL-, ALLYL ESTER	10.0		0.075		80		906
ACRYLONITRILE	1.12	0.40	1.03	0.2	60		908
ALLYL ETHYL PHOSPHITE	5.5	2.50	0.035	0.035	80	Q	906
ETHER, BUTYL VINYL	5.50	0.5	0.035		60		604
ETHER, ISOBUTYL VINYL	1.8	0.7	0		60		908
ETHER, ISOPENTYL VINYL	2.0	0.7	0		60		908
ETHER, PENTYL VINYL	3.8	0.7	0		60		908
ETHER, PROPYL VINYL	2.7	0.5	0		60		908
METHYL METHACRYLATE	1.6	0.5	0		60		908
PHOSPHINIC ACID, DIETHYL-, ALLYL ESTER	0.5	0.5	2.1	0.3	60		908
PHOSPHONIC ACID, ALLYL-, DIETHYL ESTER	10.00	2.00	0.075	0.075	60	Q	604
STYRENE	9.25	3.00	0.12	0.008	60	Q	605
VINYL ACETATE	0.03	0.03	3.0	0.4	60	C	966
	0.2	0.2	0.15	0.07	60	C	908
FUMARIC ACID, SEC-BUTYL ESTER							
STYRENE	0	0.01	0.34	0.03	60		1010
FUMARIC ACID, DI-SEC-BUTYL ESTER							
STYRENE	0	0.01	0.50	0.03	60		1010
FUMARIC ACID, DIESTER WITH HYDRACRYLONITRILE							
ACRYLIC ACID, BUTYL ESTER	0.03	0.01	2.65	0.35	90	C	1137
ACRYLIC ACID, ETHYL ESTER	0.03		8.3	0.5	90	C	1137
STYRENE	0	0.0005	0.15	0.05	90	C	1137
VINYL ACETATE	0.02	0.01	0.02	0.02	90	C	1137
FUMARIC ACID, DIETHYL ESTER							
ACRYLONITRILE	0.25		8		60	N,O	620
1,3-BUTADIENE	0.027	0.010	2.13		60	Q,T	587
1,3-BUTADIENE, 2-CHLORO-	10		6.65	0.37	150	C	214
ETHYLENE	0.47		0.25		50	M7,R	142
ETHYLENE, CHLORO-	0.47	0.05	0.12	0.01	60		131
	0.046	0.015	0.12		60		557
ETHYLENE, 1,1-DICHLORO-	0.03	0.02	12.2	2.0	60		210
METHYL METHACRYLATE	0.05		40.4	0.2	60		92
	0		2.10	0.4	60	C,I10	153
1-PROPENE, 2-METHYL-	1.10	0.10	0.17		60	H	366
1-PROPENE, 3,3,3-TRICHLORO-	0.05	0.02	1.46	0.35	60	C	211
STYRENE	0.0697	0.0041	0.301	0.024	60		1010
	0.070	0.007	0.30	0.02	60		559
	0.0905	0.007	0.400	0.02	131	H20,I7	559
	0.1	0.02	0.36	0.04	100	C	592
	0		0		60	A	1219
SULFIDE, ISOBUTYL VINYL							1225

REMARKS PAGE II-366, REFERENCES PAGE II-368

M1 / M2	R1	+/-	R2	+/-	TEMP.	REMARKS	REF.
FUMARIC ACID, DIETHYL ESTER							
VINYL ACETATE	0.444	0.003	0.011	0.001	60		557
FUMARIC ACID, DIMETHYL ESTER							
STYRENE	0.025	0.015	0.21	0.02	60		557
FUMARIC ACID, DIMETHYLSILOXYLETHYLENEGLYCOL POLYESTER							
STYRENE	0.6	0.3	0.03	0.03		Q	146
FUMARIC ACID, DINONYL ESTER							
1,3-BUTADIENE	0.32		2.02			Q,T	587
FUMARIC ACID, ETHYLENEGLYCOLPHOSPHINATE POLYESTER							
ACETIC ACID, DIETHYLPHOSPHONYL-, ALLYL ESTER	1.73	0.03	0.15	0.06		Q	604
PHOSPHINIC ACID, DIETHYL-, ALLYL ESTER	2.07	1.12	0.09	0.05		Q	604
FUMARIC ACID, ETHYLENEGLYCOL POLYESTER							
METHYL METHACRYLATE	0.35	0.35	17.5	7.5	60	C	300
	0.50	0.20	1.05	0.25	60		151
STYRENE	2.16	0.50	0.062	0.030	60		151
VINYL ACETATE	0.2	0.1	0.020	0.02	60		908
FUMARIC ACID, ETHYL ESTER							
STYRENE	0.25	0.10	0.18	0.1	60	C	557
	0.25	0.10	0.18	0.1	60		501
FUMARIC ACID, PHENOLPHTHALEIN ISOPHTHALATE POLYESTER							
STYRENE	5.8	0.5	0		60	C	1015
FUMARIC ACID, DITHIO-, DIMETHYL ESTER							
1,3-BUTADIENE	-0.0014	0.027	0.0106	0.0175	50	C,17	598
STYRENE	0.0163	0.013	0.098	0.013	50	C,17	598
FUMARONITRILE							
ACRYLIC ACID, 2-CHLOROETHYL ESTER	0.11	0.03	10.2	0.3	60	17	436
ETHER, DODECYL VINYL	0.019	0.003	0.004	0.003	60	17	436
METHYL METHACRYLATE	0.01	0.01	3.5	0.5	79	128	780
2-PYRROLIDINONE, 1-VINYL-	0.041	0.007	0.030	0.033	76	C,17	460
STYRENE	0		0.3	0.3	50/70	R	659
	0		0.19	0.03	60		557
	0.00		0.09	0.005	79		262
	0.00	0.02	0.30	0.02	49,65		805
STYRENE, ALPHA-METHYL-	0.01	0.01	0.23	0.01	60	128	780
VINYL ETHER	0.00		0.022	0.005	50		262
	0				60	16,V,CC	156

COPOLYMERIZATION PARAMETER

M1	M2	R1	+/-	R2	+/-	TEMP.	REMARKS	REF.
FUMAROYL CHLORIDE								
	STYRENE	0.0		0.04		27		315
	VINYL ACETATE	0.00		0.14		70	C,M	315
2-FURALDEHYDE								
	STYRENE	0.25		2.0		70	C	234
		0.27		0.1		50	H10	234
FURAN, 2,5-DIHYDRO-								
	OXETANE, 3,3-BIS(CHLOROMETHYL)-	0.07		1.6		0	H10,I3	1130
	PROPANE, 1-CHLORO-2,3-EPOXY-	0.06		0.05		0	H10,I3	1130
	PROPYLENE OXIDE	0.08		0.12		0	H10,I3	1130
FURAN, 2-METHYL-								
	ACRYLONITRILE	0.044		1.893			E,Q	1005
FURAN, TETRAHYDRO-								
	ACETIC ACID, 2,3-EPOXYPROPYL ESTER	2.5		0.25		0	H10	738
	BUTANE, 1,2-EPOXY-	0.19		0.8		0	H83,I24	1154
		0.25		0.96		0	H82,I24	1154
	1,3-DIOXOLANE	0.58	0.02	0.58	0.02	0	H10,I4	1207
		25	3	0.71	0.10	0	H10,I2	1061
		28	4	0.25	0.05	0	H10	718
		40		0.245		0	H20,I3	1061
		195		0.24		0	H20,I2	1061
	ETHYLENE OXIDE	2.2	10	0.08	0.03	0	H10,I4	673
		2.2		0.08		0	H10	437
	B-D-GLUCOPYRANOSE, 1,6-ANHYDRO-2,3,4-TRI-O-METHYL-	0.1	0.02	0.5	0.05	24	H10,I3	1099
	HYDRACRYLIC ACID, BETA-LACTONE	2.9	0.5	0.4	0.2	0	H10,Q	925
		5.5	0.1	0.1	0.1	0	H13,Q	925
	METHACRYLIC ACID, 2,3-EPOXYPROPYL ESTER	5.5		0.25		0	H10	738
	OXETANE, 3-(ALLYLOXYMETHYL)-3-(CHLOROMETHYL)-	0.60	0.05	1.40	0.05	0	H10	707
	OXETANE, 3,3-BIS(CHLOROMETHYL)-	0.40		0.45		20	H13,I18	20
		0.5		0.3		120	H14	771
		1.00	0.05	0.80	0.05	25	H10,I2	771
		1.4		0.82		30	H14	822
		1.50	0.20	0.1		0	H14	771
		1.80	0.01	0.00	0.01	0	H13,I3	823
	OXETANE, 3-(CHLOROMETHYL)-3-ETHYL-	0.22	0.05	0.01		0	H13,I3	823
		0.72		4.3	0.3	0	H13	1128
		0.72		1.49		0	H13	1202
	OXETANE, 3-(CHLOROMETHYL)-3-METHYL-	0.17	0.1	1.43	0.1	50	H14,I2	20
		0.26	0.04	1.74	0.2	-50	H10	283
		2.5		6.2		20	H10	808
	PROPANE, 1-CHLORO-2,3-EPOXY-	2.5		0.25		20	H10	809
		3.85	0.05	0.00	0.05	0	H10,I4	399
		3.9		0.06		0	H10	437
		20	5	0.5	0.3	0	H13,I3	823
		80	20	2	1	0	H13,I3	823

REMARKS PAGE II-366, REFERENCES PAGE II-368

M1	M2	R1	+/-	R2	+/-	TEMP.	REMARKS	REF.
FURAN, TETRAHYDRO-								
	PROPANE, 1,2-EPOXY-3-PROPOXY-	0.46	0.03	0.27	0.03	0	H10,14	1207
	1-PROPANOL, 2,3-EPOXY-, NITRATE	0.88		0.07		0	H10,17	1086
	PROPYLENE OXIDE	0.09		0.68		0	H83,124	1154
		0.1		1.4		0	H10,124	110
		0.18		0.85		0	H82,124	1154
		0.29		2.9		30	H32,L	800
		0.37	0.09	2.30	0.9	0	H58	548
		0.5	0.10	1.3	0.12	0	H10,124	110
		0.6		1.8		0	H10,124	110
		0.63	0.05	0.34	0.05	0	H10,14	1207
		0.65	0.21	1.98	0.12	0	H59	548
		0.76	0.02	2.16	0.32	50	H60	548
		1.12	0.09	2.87	0.13	0	H13	548
FURAN, TETRAHYDRO-2-METHYL-								
	OXETANE, 3,3-BIS(CHLOROMETHYL)-	0.05	0.05	1.3	0.25	0	H10	977
	PROPANE, 1-CHLORO-2,3-EPOXY-	0.27	0.05	2.70	0.05	0	H10,14	399
				0.04		0	H10,14	399
FURAN, TETRAHYDRO-2-PHENOXYMETHYL-								
	OXETANE, 3,3-BIS(CHLOROMETHYL)-	0		3.8		0	H10,13	49
FURAN, TETRAHYDRO-2-PHENYL-								
	OXETANE, 3,3-BIS(CHLOROMETHYL)-	0		2.6		0	H10,13	49
FURAN, 2-VINYL-								
	ACRYLONITRILE	0.82	0.06	0.037	0.004	70		74
	ETHYLENE, 1,1-DICHLORO-	0.15	0.014	11.7	0.07		C,Q	463
	STYRENE	1.9	0.1	0.25	0.05	70		74
D-GALACTOPYRANOSE, 1,2:3,4-DI-O-ISOPROPYLIDENE-, 6-METHACRY * LATE								
	METHYL METHACRYLATE	1.30	0.19	0.93	0.03	50		108
GERMANE, DIALLYL-								
	METHYL METHACRYLATE						P,Q	517
	STYRENE						P,Q	517
GERMANE, TRIETHYLMETHACRYLOYL-								
	STYRENE	0.93	0.08	1.05	0.02	60		518
GERMANE, TRIMETHYLVINYL-								
	METHYL METHACRYLATE	0.05		19.98		50	I7	1300
	STYRENE	0.009		24.4		50	I7	1300

COPOLYMERIZATION PARAMETER

M1 M2	R1	+/-	R2	+/-	TEMP.	REMARKS	REF.
D-GLUCITOL, 1-ACRYLAMIDO-1-DEOXY-							
METHYL METHACRYLATE	0.206		4.22		50	C	1029
STYRENE	0.056		2.72		50	C	1029
VINYL ACETATE	2.41		0.18		50	C	1029
D-GLUCITOL, 1-DEOXY-1-METHACRYLAMIDO-							
METHYL METHACRYLATE	0.036		4.20		50	C	1029
STYRENE	0.005		2.09		50	C	1029
VINYL ACETATE	0.56		0.16		50	C	1029
D-GLUCITOL, 1,2:3,4:5,6-TRI-O-ALLYLIDENE-							
ACRYLONITRILE	0.26		0.7		80	C	1350
STYRENE	0.13		4.6		80	C	1350
D-GLUCITOL, 1,2:3,4:5,6-TRI-O-(2-BUTENYLIDENE)-							
ACRYLONITRILE	0.36		0.3		80	C	1350
STYRENE	0.39		2.31		80	C	1350
D-GLUCOFURANOSE, 1,2:5,6-DI-O-ISOPROPYLIDENE-, 3-METHACRYLA * TE							
ACRYLONITRILE	0		0.25		50	I6	498
B-D-GLUCOPYRANOSE, 1,6-ANHYDRO-2,3,4-TRI-O-METHYL-							
FURAN, TETRAHYDRO-	0.5	0.05	0.1	0.02	24	H10,I3	1099
PROPANE, 1-CHLORO-2,3-EPOXY-	0.15		1.4		0	H10,I3	1136
PROPYLENE OXIDE	0.35		0.7		0	H10,I3	1136
D-GLUCOSE, METHACRYLOYL-							
ACRYLONITRILE	0		0.25		50	I6	498
GLUTARONITRILE, 2-METHYLENE-							
1,3-BUTADIENE	0.00		0.00		25	F,K	789
	0.02		0.24		25	F,K,DD	789
3-BUTEN-2-ONE	1.24	0.76	5.05	0.95	0	Q	946
METHYL METHACRYLATE	0.10		1.25		25	F,K	789
	0.12		1.27		25	F,K,DD	789
STYRENE	0.00		0.45		25	F,K	789
	0.02		0.50		25	F,K,DD	789
	0.25	0.15	0.85	0.06	25	Q	946
GLYCIDAMIDE							
ETHYLENE OXIDE	14.2	1.5	0.13	0.03	70	G13	347
PROPANE, 1-CHLORO-2,3-EPOXY-	5.8	0.7	0.40	0.15	70	G13	347
PROPANE, 1,2-EPOXY-3-METHOXY-	12.0	1.0	0.14	0.04	70	G13	347
PROPANE, 1,2-EPOXY-3-PHENOXY-	10.0	0.8	0.35	0.10	70	G13	347
PROPYLENE OXIDE	23	3	0.10	0.10	70	G13	347

M1 / M2	R1	+/-	R2	+/-	TEMP.	REMARKS	REF.
GLYCINE, N-ALLYL-							
ACRYLONITRILE	0.07	0.03	3.9	0.15	50	C,136	1372
GLYCOLIC ACID, ETHYL ESTER, ACRYLATE							
STYRENE	0.08	0.02	0.68	0.03		Q	1446
GLYCOLONITRILE, METHACRYLATE							
STYRENE	0.36	0.02	0.45	0.02	70	C	1016
STYRENE, ALPHA-METHYL-	0.36	0.01	0.12	0.01	70	C	1016
3,5-HEPTADIEN-2-ONE							
ACRYLIC ACID, METHYL ESTER	2.3	0.2	0.35	0.06	60		272
STYRENE	0.59	0.12	0.18	0.08	60		272
1-HEPTANOL, 2,2,3,3,4,5,5,6,6,7,7-DODECAFLUORO-, ACRYLATE							
ACRYLIC ACID	1.8	0.3	3.8	0.6	30	E,128	1105
1-HEPTENE							
STYRENE	0.51	0.12	1.92	0.05	50	H7,I7	119
	0.95	0.14	0.43	0.05	30	H29,I7	79
	1.34	0.15	0.47	0.01	30	H12,I2	79
	5.70		0.61		35		43
1,5-HEXADIEN-3-YNE							
ACRYLONITRILE	4.1	0.125	0.03	0.025	50	16	840
HEXANOIC ACID, VINYL ESTER							
ACRYLONITRILE	0.04	0.03	4.0	0.3	60	C,16	58
ETHYLENE, CHLORO-	0.1		1.8		40	C,I11,J	503
	0.28		2		60		1213
	0.65	0.04	1.35	0.05	40	C,I11,J	503
HEXANOIC ACID, 2-ETHYL-, VINYL ESTER							
ACRYLONITRILE	0.01	0.01	12	2	30	C	166
1-PROPENE, 3-CHLORO-	0.71	0.18	1.52	0.13	60	C	457
HEXANOIC ACID, 6-HYDROXY-, EPSILON-LACTONE							
OXETANE, 3,3-BIS(CHLOROMETHYL)-	0.44		0.21		0	H10	980
			0.24		0	H10,I20	979
HEXANOIC ACID, 2-METHYLENE-							
STYRENE	0.32	0.04	0.52	0.01	70		1212

COPOLYMERIZATION PARAMETER

M1	M2	R1	+/-	R2	+/-	TEMP.	REMARKS	REF.
HEXANDIC ACID, 2-METHYLENE-, METHYL ESTER								
	STYRENE	0.20	0.05	0.80	0.05	65		174
HEXASILOXANE, 1,11-DIETHYL-1,3,3,5,5,7,7,9,9,11-DECAMETHYL-1,11-DIVINYL- *								
	ISOPRENE	0.47	0.47	3.34	0.1	120	C	1117
HEXASILOXANE, 1,1,3,5,7,9,11,11-OCTAMETHYL-3,5,7,9-TETRAPHENYL-1,11-DIVINYL- *								
	ISOPRENE	0.6	0.4	4.29	0.19	120	C	1117
1,3,5-HEXATRIENE, 2,3,4,5-TETRACHLORO-, TRANS-								
	ACRYLIC ACID, METHYL ESTER	3.17	0.27	0.27	0.04	70	Q	18
	ACRYLONITRILE	4.05		0.20		70	C	17
	1,3-BUTADIENE, 2-CHLORO-	0.2	0.07	3.6	0.17	70	C	542
	ISOPRENE	1.58	0.1	0.58	0.02	70	C	542
	METHYL METHACRYLATE	0.315	0.14	0.09	0.105		C	542
		1.82	0.075	0.465	0.05		Q	18
	STYRENE	0.84		0.21		70	C	17
	VINYL ACETATE	32		0.013		70	C	17
1-HEXENE								
	ACRYLIC ACID, METHYL ESTER	0		8.5	2	60	Q	212
	ACRYLONITRILE			12.2	2.4	60	C	212
	1-DECENE	1.3		0.90		23	H54,I21	565
	1-PENTENE, 4-METHYL-	5.4		0.62		50	H54	75
	STYRENE	9.75	0.81	0.19	0.04	35	H12,I2	43
1-HEXENE, 4-METHYL-								
	STYRENE	1.30	0.06	1.80	0.04	35	H12,I2	43
1-HEXENE, 5-METHYL-								
	STYRENE	4.00	0.28	0.591	0.057	35	H12,I2	44
1-HEXENE, 1-PHENYL-, CIS-								
	ACRYLONITRILE	-0.67	2.17	5.09	0.96	60		1065
1-HEXENE, 1-PHENYL-, TRANS-								
	ACRYLONITRILE	-0.09	0.29	1.09	0.03	60		1065
1-HEXEN-3-YNE								
	ACRYLONITRILE	0.63	0.4	0.17	0.01	60		784
	PYRIDINE, 2-VINYL-	0.6	0.1	1.5	0.5	60		784

M1	M2	R1	+/-	R2	+/-	TEMP.	REMARKS	REF.
5-HEXEN-3-YN-2-OL, 2-METHYL-	ACRYLONITRILE					70	C,P	607
	ETHYLENE, 1,1-DICHLORO-					70	C,P	607
	METHYL METHACRYLATE					70	C,P	607
	PROPANE, 1-((1,1-DIMETHYL-4-PENTEN-2-YNYL)OXY)-2,3-EPOXY-	1.40	0.10	0.54	0.16	70	B,Q	606
	STYRENE					70	C,P	607
	VINYL ACETATE					70	C,P	607
1-HEXYNE	ACRYLIC ACID, METHYL ESTER	0		11.2	2	60	D	212
	ACRYLONITRILE	0		5.4	0.3	60	C	212
HYDANTOIN, 5-METHYLENE-	STYRENE	24		0.03		70		91
HYDRACRYLIC ACID, BETA-LACTONE	ACRYLONITRILE	0		1.2		-60	G24,I6	1243
		0		0		25	G10,I2	1243
		0		0.84		50		1244
		3.64		0		30	G10,I6	1243
		5		0		40	G10,I6	1243
	AZIRIDINE, 1-PHENYL-	0.15		7.5			I63	459
	BENZENE, (EPOXYETHYL)-	15.0		0.1		50	H14	171
		15.0		0.1		50	H14	171
	BUTYRIC ACID, 3-HYDROXY-, BETA-LACTONE (COPOLYMERIC WITH ACRYLONITRILE) *	18	2	0.36	0.10	0	H13	924
	BUTYRIC ACID, 4-HYDROXY-, GAMMA-LACTONE	0.1	0.1	5.5	0.1		H13,Q	925
		0.4	0.2	2.9	0.5		H10,Q	925
	FURAN, TETRAHYDRO-	0.4		1.4			H10,I3	171
	7-OXABICYCLO(4.1.0)HEPTANE, 3-VINYL-	3.5		0.1	0.1	-30	H55,I2	171
	2,5-OXAZOLIDINEDIONE, 4-ISOPROPYL-	-0.16		11		20	G1	208
	OXETANE, 3,3-BIS(CHLOROMETHYL)-	0.03		25		35	H10	980
		0.04		10		20	H10	1440
		0.04		8		20	H20	1440
		0.05	0.04	30	10	-50	H13,I3	926
		0.05		3.5		0	H13,I9	926
		0.06	0.05	16	3	-50	H10	926
		0.1	0.1	38	3	0	H10	926
		0.15		30	10	0	H13,I3	978
	PROPANE, 1-CHLORO-2,3-EPOXY-	0.18	0.01	16.7	0.5	0	H10,EE	927
		0.5		10		0	H10,EE	927
		0.65		7		0	H10,EE	927
		1		0		0	H10,EE	927
	STYRENE	9.0	0.1	6.2	0.1	50	H14	171
		0		70		80	C	1244
		0		32		50		1244
	VALERIC ACID, 4,5-EPOXY-3-HYDROXY-, BETA-LACTONE	0.2		20		0	H10,I3	48
		0.2		20		0	H20,I3	48
		10.0		0.3		20	H40	715

M1 / M2	R1	+/-	R2	+/-	TEMP.	REMARKS	REF.
HYDRACRYLIC ACID, 2-METHYL-, BETA-LACTONE							
OXETANE, 3,3-BIS(CHLOROMETHYL)-	0.13	0.02	11.9	0.2	0	H10	978
HYDRACRYLONITRILE, ACRYLATE							
STYRENE	0.13		0.40		60		740
	0.13		0.40		60		742
HYDRACRYLONITRILE, METHACRYLATE							
STYRENE	0.32	0.01	0.45	0.01	70	C	1016
STYRENE, ALPHA-METHYL-	0.35	0.02	0.12	0.02	70	C	1016
HYDRAZINE, 1-ACRYLOYL-2-BENZOYL-							
ACRYLONITRILE	0.23	0.08	0.33	0.08	60	I6	950
STYRENE	0.65	0.05	1.40	0.05	60	I6	950
HYDRAZINE, 1-CROTONOYL-2-ISONICOTINOYL-							
2-PYRROLIDINONE, 1-VINYL-	0.58	0.02	0.04	0.04	60		1261
HYDRAZINIUM CHLORIDE, 2-ACRYLOYL-1,1,1-TRIMETHYL-							
STYRENE	0.46		0.58		70	I15	1175
HYDRAZINIUM CHLORIDE, 2-METHACRYLOYL-1,1,1-TRIMETHYL-							
STYRENE	0.23		0.51		70	I27	1175
HYDRAZINIUM HYDROXIDE, 1-(2,3-DIHYDROXYPROPYL)-2-METHACRYLO YL-1,1-DIMETHYL-, INNER SALT *							
METHYL METHACRYLATE	0.04		1.60		60	I6	192
HYDRAZINIUM HYDROXIDE, 1-(2-HYDROXYPROPYL)-2-METHACRYLOYL * -1,1-DIMETHYL-, INNER SALT							
METHYL METHACRYLATE	0.0		1.99		60	I6	192
HYDRAZINIUM HYDROXIDE, 2-METHACRYLOYL-1,1,1-TRIMETHYL-, IN * NER SALT							
ACRYLONITRILE	0.10	0.01	0.37	0.04	70	I63	193
HYDRAZINIUM HYDROXIDE, 1,1,1-TRIMETHYL-2-(P-VINYLBENZOYL) * -, INNER SALT							
STYRENE	0.63	0.07	0.47	0.05	70	I63	191
HYDROCINNAMIC ACID, ALPHA-METHYLENE-, METHYL ESTER							
STYRENE	0.17	0.03	0.59	0.05	65		174

REMARKS PAGE II-366, REFERENCES PAGE II-368

M1	M2	R1	+/-	R2	+/-	TEMP.	REMARKS	REF.
HYDROQUINONE, VINYL-, DIBENZOATE								
	ACRYLIC ACID	0.95	0.002	0.44	0.13	60	I2	482
	ACRYLIC ACID, METHYL ESTER	0.75		0.46		78	I7	465
	METHACRYLIC ACID	0.91	0.25	1.91	0.23	60	I2	482
	METHYL METHACRYLATE	0.41		0.34		78	I7	465
	PYRIDINE, 4-VINYL-	0.40		0.48		78	I7	465
	2-PYRROLIDINONE, 1-VINYL-	0.26		0.98		60		995
	STYRENE	0.43		0.22		78	I7	465
	STYRENE, ALPHA-METHYL-	0.30		0.11		78	I7	465
IMIDAZOLE, 2-METHYL-1-VINYL-								
	ACRYLAMIDE	0.22		2.50			C,I26,Q	578
	ACRYLIC ACID, METHYL ESTER	0.05		1.28		70	C,I10	579
	CROTONIC ACID, CIS-	0.09		0.6			Q	1273
	CROTONIC ACID, TRANS-	0.19		0.49			Q	1273
	ETHYLENE, CHLORO-	2.13		0.22		60		613
	METHYL METHACRYLATE	0.003		3.48			Q	667
	STYRENE	0.069		8.97			Q	667
	SUCCINIC ACID, METHYLENE-	0.17		1.02		70	C,I11	579
IMIDAZOLE, 2-PHENYL-1-VINYL-								
	METHYL METHACRYLATE	0.006		3.50			Q	667
1H-INDAZOLE, 1-ISOPROPENYL-								
	STYRENE	0.5		0.75		70		1218
INDENE								
	ACRYLONITRILE	0		0.03	0.03	0	Q	1070
	ANISOLE, P-PROPENYL-	1.6		0.3		-78	H23	1110
	BENZALDEHYDE	0.38	0.02	0.007	0.007	-78	H10	73
	BENZOFURAN	1.85	0.25	0.15	0.01	-78	H23,I3	1455
	BENZOFURAN, 4,7-DIMETHYL-	0.7	0.1	1.3	0.1	-78	H23,I3	1455
	BENZOTHIOPHENE	2.4	0.025	0.08	0.002	-78	H23,I3	1455
	1,1'-BIINDENE	2.07	0.04	0.045	0.008	-72	H23	586
	BUTANE, 1,4-DIINDEN-1-YL-	0.5	0.1	0.5	0.1	-78	H23,I3	1280
	2-BUTENE, 1,4-DIINDEN-1-YL-	0.2	0.05	0.75	0.05	-78	H23,I3	1280
	ETHANE, 1,2-DIINDEN-1-YL-	0.6	0.05	0.4	0.05	-78	H23,I3	1280
	ETHYLENE, 1,1-DICHLORO-	0.33	0.05	0.40	0.06	70		23
	INDENE, 4,6-DIMETHYL-	0.4	0.08	2.1	0.1	0	H23	1278
	INDENE, 4,7-DIMETHYL-	0.15	0.03	2.5	0.2	0	H10	1278
		0.4	0.05	3.7	0.3	0	H23	1278
	INDENE, 5,6-DIMETHYL-	0.5	0.1	4.2	0.4	-30	H23	1095
	INDENE, 5,7-DIMETHYL-	0.1	0.05	3.8	0.04	-30	H23	1278
		0.1	0.05	3.8	0.04	0	H23	1278
	INDENE, 1-METHYL-	3.18	0.05	0.14	0.02	-72	H23	883
	INDENE, 2-METHYL-	0.99	0.03	0.052	0.004	-72	H23	883
	INDENE, 3-METHYL-	0.25	0.05	0.72	0.15	-72	H23,R	883
	INDENE, 5-METHYL-	1.15	0.05	0.85	0.05	0	H23	1164
		1.15	0.05	0.85	0.05	0	H23	1278
	INDENE, 6-METHYL-	0.45	0.05	2.5	0.25	0	H23	1278

COPOLYMERIZATION PARAMETER

M1	M2	R1	+/-	R2	+/-	TEMP.	REMARKS	REF.
INDENE	INDENE, 6-METHYL-	0.45	0.05	2.5	0.25	0	H23	1164
	INDENE, 7-METHYL-	0.8	0.005	1.15	0.05	0	H23	1164
		0.8	0.005	1.15	0.05	0	H23	1278
	INDENE, 4,5,6,7-TETRAMETHYL-	0.2	0.04	7.5	0.05	0	H23	1278
		0.75	0.18	16	0.5	-30	H50	1278
	INDENE, 4,6,7-TRIMETHYL-	0.15	0.05	3.4	0.1	-72	H23	1278
		0.15	0.05	3.4	0.1	-78	H23	1095
	1-PROPENE, 2-METHYL-	2.2		1.1		10	H23,I3	585
	STYRENE	1.7		0.7		10	H10,I1	831
		1.8		0.5		-72	H10,I1	831
		2.16	0.15	0.97	0.07	20	H23	883
		2.3		0.5		20	H10,I1	831
		2.5	0.3	0.5	0.3	20	H10,I1	831
		2.7		0.4		20	H10,I1	831
		3.0	0.1	0.3	0.1		H10,I1	831
		3.3		0.3			H10,I1	831
		3.7	0.2	0.6	0.1	30	H10,124	653
		5.05	0.3	0.9	0.04	-72	H23	1094
	STYRENE, O-CHLORO-	0		3.5			H23	23
	STYRENE, METHYL-, 3:2 META:PARA MIXTURE	8.5	0.3	0.35	0.5	70	C,Q	1146
INDENE, 6-BROMO-	STYRENE	0.21	0.02	4.2	0.04	-30	H23,I3	1365
INDENE, 4,6-DIMETHYL-	INDENE, 4,7-DIMETHYL-	2.1	0.1	0.4	0.08	0	H10	1278
	INDENE, 4,5,6,7-TETRAMETHYL-	0.6	0.04	0.6	0.04	0	H23	1278
		0.53	0.08	2.2	0.3	0	H23	1278
INDENE, 4,7-DIMETHYL-	INDENE	2.5	0.2	0.15	0.03	0	H10	1278
	INDENE, 4,6-DIMETHYL-	3.7	0.3	0.4	0.05	0	H23	1278
	INDENE, 4,5,6,7-TETRAMETHYL-	0.6	0.04	0.6	0.04	0	H23	1278
		0.3	0.08	2	0.2	0	H23	1278
INDENE, 5,6-DIMETHYL-	INDENE	4.2	0.4	0.5	0.1	0	H23	1278
INDENE, 5,7-DIMETHYL-	INDENE	3.8	0.04	0.1	0.05	-30	H23	1278
		3.8	0.04	0.1	0.05	-30	H23	1095
INDENE, 4-METHOXY-	STYRENE	0.15	0.05	26	1	-78	H23	1417

REACTIVITY RATIOS

M1	M2	R1	+/-	R2	+/-	TEMP.	REMARKS	REF.
INDENE, 5-METHOXY-	STYRENE	0.35	0.05	6.3	0.5	-50	H23	1,417
		0.36	0.05	5.8	0.5	-30	H23	1,417
INDENE, 6-METHOXY-	STYRENE	0.1	0.05	4.1	0.5	-78	H23	1417
INDENE, 1-METHYL-	INDENE	0.14	0.02	3.18	0.05	-72	H23	883
	STYRENE	0.06	0.01	1.15	0.1	-72	H23	883
INDENE, 2-METHYL-	INDENE	0.052	0.004	0.99	0.03	-72	H23	883
	STYRENE	0.038	0.015	2.84	0.2	-72	H23	883
INDENE, 3-METHYL-	INDENE	0.72	0.15	0.25	0.05	-72	H23,R	883
INDENE, 5-METHYL-	INDENE	0.85	0.05	1.15	0.05	0	H23	1164
		0.85	0.05	1.15	0.05	0	H23	1278
INDENE, 6-METHYL-	INDENE	2.5	0.25	0.45	0.05	0	H23	1164
		2.5	0.25	0.45	0.05	0	H23	1278
INDENE, 7-METHYL-	INDENE	1.15	0.05	0.8	0.005	0	H23	1278
		1.15	0.05	0.8	0.005	0	H23	1164
INDENE, 4,5,6,7-TETRAMETHYL-	INDENE	7.5	0.5	0.2	0.04	0	H23	1278
		16	0.5	0.75	0.18	0	H50	1278
	INDENE, 4,6-DIMETHYL-	2.2	0.3	0.53	0.08	0	H23	1278
	INDENE, 4,7-DIMETHYL-	2	0.2	0.3	0.08	0	H23	1278
INDENE, 4,6,7-TRIMETHYL-	INDENE	3.4	0.1	0.15	0.05	-30	H23	1278
		3.4	0.1	0.15	0.05	-72	H23	1095
INDOLE, 1-VINYL-	METHACRYLIC ACID	1.43	0.05	0.096	0.03	60		873
	METHYL METHACRYLATE	1.9	0.5	0.42	0.1	60		873

COPOLYMERIZATION PARAMETER

M1	M2	R1	+/-	R2	+/-	TEMP.	REMARKS	REF.
ISOBUTYRALDEHYDE, DIVINYL ACETAL								
	VINYL ACETATE	0.985	0.05	1.002	0.047	60		608
ISOCYANIC ACID, BUTYL ESTER								
	BUTYRALDEHYDE	0		10		-75	G17,19	1331
		5		70		-75	G19,154	1331
		10		500		-75	G19,15	1331
		10		26		-75	G17,154	1331
	CHLORAL	0				-75	G19,154	1331
		0				-75	G19,19	1331
ISOCYANIC ACID, METHYL ESTER								
	SUCCINALDEHYDDNITRILE	0.01	0.01	8.3	0.3	-78	G23,15	1210
ISOCYANIC ACID, PHENYL ESTER								
	CHLORAL	0		0.5		-75	G19,19	1331
		0		0.48		-75	G19,154	1331
	PROPIONALDEHYDE, 2,3-DICHLORO-	0.002				-21	H40	1314
ISOCYANIC ACID, VINYL ESTER								
	ACRYLIC ACID, METHYL ESTER	0.14	1.0	1.38	0.27	60		420
	ACRYLONITRILE	0.16	0.06	0.19	0.03	60		420
	ETHYLENE, 1,1-DICHLORO-	0.33	0.18	1.46	0.28	60		420
	METHYL METHACRYLATE	0.01		3.3			Q	1027
		0.01	0.02	3.3	0.2	60		371
	STYRENE	0.16	0.08	5.57	0.33	60		420
		0.08	0.04	8.13	0.04	60		420
		0.1		6.9			Q	1027
		0.1	0.05	6.9	0.5	60		371
ISOCYANIC ACID, P-VINYLPHENYL ESTER								
	STYRENE	1.10	0.10	0.75	0.10	60		430
ISOPHTHALALDEHYDE								
	STYRENE	0		0.77		-78	H10,13	1142
		0		0.77		0	H10	1402
ISOPHTHALIC ACID, BIS(2-METHYLALLYL)ESTER								
	STYRENE	0.065	0.003	22.9	0.4	60	C	1294
ISOPHTHALIC ACID, DIALLYL ESTER								
	BENZOIC ACID, ALLYL ESTER	0.69		1.02		60	C	1296
	STYRENE	0.092	0.003	26.7	0.3	60	C	1294

REMARKS PAGE II-366, REFERENCES PAGE II-368

M1	M2	R1	+/-	R2	+/-	TEMP.	REMARKS	REF.
ISOPRENE								
	ACRYLIC ACID, METHYL ESTER	0.75		0.12	0.02	50		377
	ACRYLONITRILE	0.29	0.02	0.05	0.03	50		353
		0.45	0.05	0.03	0.05	50		352
	BENZALDEHYDE	0.65	0.23	0.04	0.05	0	H10	73
	1,3-BUTADIENE	0.11		3.6			G17,18,Q	135
		0.13		4.5			G17,15,Q	135
		0.47	0.03	3.38	0.14	50	G17,19	796
		0.55		1.88			H74,Q	898
		0.59		1.15			H27,17,Q	276
		0.6		2.0			H73,Q	898
		0.60		2.80			H74,Q	898
		0.80		4.50			H75,Q	898
		0.85		0.75		5	K	290
		0.9		1.0			H25,Q	898
		1		1			H73,Q	898
		1		1		-15	H43	765
		1		1		13	H43	765
		1		1		20	H43	765
		1		1		48	H43	765
		1.0	0.05	1.0	0.05	30	H24,L	136
		1.06		0.94		-18	K	732
		1.15	0.05	2.3	0.1	30	H25	136
		1.25		0.92			H25,17,Q	276
		1.37		0.99			H26,17,Q	276
		3.4		0.12			H4,Q	137
	1,3-BUTADIENE, 2-CHLORO-	0.063	0.051	2.82	0.22	50	K	353
		0.133	0.025	3.65	0.11	50		352
	1,3-BUTADIENE, 2,3-DIMETHYL-	1.18		0.84		-18		732
		17	5			50	H	797
	1,3-BUTADIENE, 2-FLUORO-	0.19	0.10	2.05	0.19	50	K	733
	1,3-BUTADIENE, HEXAFLUORO-	1.19	0.12	0.78	0.05	40		508
	1,3-CYCLOHEXADIENE	0.4	0.05	0.6	0.05	0	H23,12	203
		0.4		0.6		30	H17,17	203
		0.5		1.64		50	G30,17	1419
		1.5	0.1	0.2	0.05	30	G17,120	203
	ETHYLENE, 1,1-DIPHENYL-	0.04		0		25	G17,15	1448
		0.05		0		40	G5,17	1448
		0.11		0		0	G5,15	1448
		0.11		0		0	G17,15	1448
		0.12		0		0	G19,15	1448
		0.38		0		40	G18,17	1448
		0.5		0		30	G17,110	1448
		29		0		40	G17,120	1448
		37		0		40	G17,17	1448
		37		0		40	G17,17	1448
	HEXASILOXANE, 1,11-DIETHYL-1,3,3,5,5,7,9,9,11-DECAMETHYL * -1,11-DIVINYL-							
	HEXASILOXANE, 1,1,3,5,7,9,11,11-OCTAMETHYL-3,5,7,9-TETRAPHE * NYL-1,11-DIVINYL-	37		0		40	G17,17	1448
	1,3,5-HEXATRIENE, 2,3,4,5-TETRACHLORO-, TRANS-	0.58	0.02	1.58	0.1	70	C	542
	MALEIC ACID, METHYL ESTER	0.18	0.12	0.4			Q	1318
	METHYL METHACRYLATE	0.78	0.13	0.4	0.1	80	17	1333
	1,3-PENTADIENE	17	5			50	G	797
	PHOSPHONIC ACID, (DIMETHYLENEETHYLENE)DI-, TETRAETHYL ESTER	0.12	0.03	0.05	0.05	70		1395

COPOLYMERIZATION PARAMETER

M1	M2	R1	+/-	R2	+/-	TEMP.	REMARKS	REF.
ISOPRENE								
	1-PROPENE	0.50		0.23		-78	H46,I18	383
	1-PROPENE, 2-METHYL-	0.38		2.26		-95,6	H	776
		0.4	0.1	2.5	0.5	-103	H	956
		0.44		2.27		-90	H	776
		0.5		2.17		-100	H4,I3	483
	PYRIDINE, 2-VINYL-	0.58	0.05	0.46	0.07	50		536
		0.59	0.05	0.47	0.07	60		534
		0.59	0.05	0.47	0.07	60		533
	PYRIDINE, 4-VINYL-	0.32		2.49		50		377
	QUINOLINE, 2-VINYL-	0.534	0.001	1.882	0.002	50		536
		1.88	0.02	0.53	0.01	60		534
		1.88	0.02	0.53	0.01	60		533
	STILBENE, TRANS-	0		0		0	G17,I5	1452
	STYRENE	50		0		40	G17,I7	1452
		0	0.84	40	14	-35	H17,I5,N	909
		0.08	0.02	12	0.01	-15	G32,I5	1397
		0.08	0.00	9	0.00	27	H12,Q	1064
		0.1		2.2		20	H17,I5,N	909
		0.15		2.2		75	H4,I9,BB	537
		0.27		0.46		25	H12,Z	1076
		0.5		0.35		20	H14,I3	1271
		0.6		0.8		27	H4,I7,BB	537
		1.18		0.66		22	G17,I2/8	911
		1.30		0.48		-18	E	1038
		1.68	0.9	0.80	0.39	-18	K	290
		1.8		0.63		50	K	732
		1.92		0.50		100	K	353
		1.92		0.54		80	H,Q	1064
		1.98		0.44		50	C	1038
		2.02		0.42		30	C	1038
		2.05	0.45	1.38	0.54	30	K,Q	1085
		5.89		0.03		50	Q	1085
		7.0	0.6	0.14	0.02	27	G17	353
		7.6		0.25		40	G17,I7	1038
		9.5		0.046			G32,I7	531
		16.6					G17,I2	1397
							G17,I2I	911
								1054
ISOTHIOCYANIC ACID, VINYL ESTER								
	ACRYLAMIDE	0.47		0.68			Q	1070
	ACRYLONITRILE	1.4		0.36			Q	1070
	METHYL METHACRYLATE	0.6	0.02	0.85	0.03	60		371
	STYRENE	0.37	0.02	0.65	0.02	60		371
		0.5		0.8			C,Q	1070
ISOVALERALDEHYDE, DIVINYL ACETAL								
	VINYL ACETATE	0.987	0.057	1.04	0.08	60		608

M1 / M2	R1	+/-	R2	+/-	TEMP.	REMARKS	REF.
ISOXAZOLE, 3-ISOPROPENYL-5-METHYL-							
STYRENE	0.98		0.67		60		1233
ISOXAZOLE, 5-PHENYL-3-VINYL-							
STYRENE	0.48	0.04	0.72		60		426
KETENE							
FORMALDEHYDE	0.5		5		-30	G29,I20	1228
KETENE, DIMETHYL-							
SUCCINALDEHYDONITRILE	0.04	0.03	0.6	0.1	-78	G23,I5	1209
KETENE, DIPHENYL-							
1-PROPENE, 3-CHLORO-2-(CHLOROMETHYL)-	8.9		0.05		50		917
VINYL ACETATE	0.56	0.05	0.05	0.03	-10	H2O	675
KETONE, METHYL 5-NORBORNEN-2-YL							
MALEIC ANHYDRIDE	0.4		0		60		410
LAURIC ACID, ALLYL ESTER							
VINYL ACETATE	0.8		0.71		60	C	835
LAURIC ACID, VINYL ESTER							
ACETIC ACID, CHLORO-, VINYL ESTER	0.65		0.87		0	E	325
ACRYLONITRILE	0.04		4.0		60	C,I6	58
ETHYLENE, CHLORO-	0.2	0.03	7.4	0.3	40	C,I11,J	503
VINYL ACETATE	0.7		1.4		60	C	835
LEAD, TRIPHENYL(P-VINYLPHENYL)-							
STYRENE	0.76	0.01	4.7	0.1	60		535
	0.978	0.00	1.22	0.0	60	I2	535
LEVULINIC ACID, VINYL ESTER							
ETHYLENE, CHLORO-	0.419	0.002	1.40	0.004	60	C,K	591
LINOLEIC ACID, METHYL ESTER							
ACRYLIC ACID, ETHYL ESTER	0.026	0.015	14.7	0.9	60		619
ACRYLONITRILE	0		5		60		618
STYRENE	0		30		130		618
	0		36		130	C	618
	0		140		60		518
	0		28		70	C	330
	0		45		90	C	618

M1	M2	R1	+/-	R2	+/-	TEMP.	REMARKS	REF.
LINOLENIC ACID, METHYL ESTER								
	ACRYLIC ACID, ETHYL ESTER	0.16	0.02	7.4	0.4	60		619
	ACRYLONITRILE	0		5		60		618
	ACRYLONITRILE	0		60		60		618
	STYRENE	0		61		70	C	330
MALEALDEHYDE								
	STYRENE	0.30	0.05	0.08	0.03		Q	68
	STYRENE	0.30	0.05	0.08	0.03		Q	68
MALEAMIC ACID, N-CARBAMOYL-, METHYL ESTER								
	STYRENE	0		1.45		70	I10	705
	STYRENE	0		0.9		70	I26	704
MALEIC ACID								
	ACRYLAMIDE	0.01	0.015	2.2		25	C,I11	674
	ACRYLAMIDE	0.00		12.8		25	I6,Q	577
	VINYL ACETATE	0.00		0.045	0.015	60	F	341
MALEIC ACID, BIS(2-ETHYLHEXYL) ESTER								
	ETHYLENE, CHLORO-	0		0.42		68	C	8
	ETHYLENE, CHLORO-	0		0.5		68	C	8
MALEIC ACID, DIBENZYL ESTER								
	STYRENE	0.02	0.01	0.55	0.2	70	C	836
MALEIC ACID, DIBUTYL ESTER								
	ETHYLENE, CHLORO-	0.0		1.4		40	C	500
MALEIC ACID, DIETHYL ESTER								
	ACRYLONITRILE	0.11		12		60	N,O	620
	1,3-BUTADIENE	10		8.08		150	Q,T	587
	ETHYLENE			0.25		70	M7,R	142
	ETHYLENE, CHLORO-	0.0		0.9	0.1	40		23
	ETHYLENE, CHLORO-	0.0		0.8		40	F,K	502
	ETHYLENE, CHLORO-	0.0		0.8		60	F,K	499
	ETHYLENE, CHLORO-	0.009	0.003	0.77		60		557
	ETHYLENE, 1,1-DICHLORO-	0.0	0.04	12.5		60	C	211
	METHYL METHACRYLATE	0.0		40	8	60	N,O	620
	STYRENE	-0.10	0.11	20		60		92
	STYRENE	0.0		354	57	131	D	34
	STYRENE	0.005	0.1	5	1.5	70		559
	VINYL ACETATE	0.01	0.01	6.52	0.05	60		559
	VINYL ACETATE	0.043	0.005	0.17	0.01	60		557

M1	M2	R1	+/-	R2	+/-	TEMP.	REMARKS	REF.
MALEIC ACID, DIISOPROPYL ESTER								
	ETHYLENE, CHLORO-	0.1		0.65	0.05	40	C	501
	STEARIC ACID, VINYL ESTER	0.0075		0		70		1050
	VINYL ACETATE	0.043		0.17		60		1024
MALEIC ACID, DIMETHYL ESTER								
	STYRENE	0.03	0.01	8.5	0.2	60	E,Q	557
	SUCCINIMIDE, N-VINYL-	0.13		0.1		60		1392
	SULFIDE, ISOBUTYL VINYL	0.04		1.2		60	A	1217
	VINYL ACETATE	0.028		0.2		60	135	1225
				0.12				1055
MALEIC ACID, DINONYL ESTER								
	1,3-BUTADIENE	0.12		5.36			Q,T	587
MALEIC ACID, ETHYL ESTER								
	STYRENE	0.035	0.01	0.13	0.01	60		557
MALEIC ACID, METHYL ESTER								
	ISOPRENE	0.0		0.18	0.12	60	Q	1318
	VINYL ACETATE	0.0	0.007	0.09	0.005	60	135	1055
		0.015	0.007	0.09	0.005	60	135	1055
		0.035	0.01	0.13	0.03	75	135	1055
		0.4747		0.522		56		1000
		0.522		0.0468		56		998
		0.9996		0.0345		65		999
				0.1168	78			999
MALEIC ACID, 2-CHLORO-, DIETHYL ESTER								
	STYRENE	0		2.5		70		34
MALEIC ACID, 1,4-DITHIO-, CYCLIC ANHYDROSULFIDE								
	STYRENE	0.005		0.045		60		1393
MALEIC ANHYDRIDE								
	ACETIC ACID, ALLYL ESTER	<0.13		<0.0075		30	C	87
	ACRYLAMIDE	0		0.75		50		1125
		0.3		0.25		70	E	1125
	ACRYLIC ACID, 2-CHLOROETHYL ESTER	0.03	0.05	6.2	0.5	60		438
	ACRYLIC ACID, METHYL ESTER	0.02		2.5		60	N,O	620
				2.8	0.05	75	C	1033
	ACRYLONITRILE	0		6		60		620
	2-BUTENE, CIS-	0.016	0.005	0		60	17	669
		0.016	0.005	0		60	17	668
	2-BUTENE, TRANS-	0.030	0.009	0		60	17	669
	2-BUTENE, 2-METHYL-	0.030	0.009	0		60	17	668
		0.005	0.002	0		60		669

COPOLYMERIZATION PARAMETER

M1 = MALEIC ANHYDRIDE

M2	R1	+/-	R2	+/-	TEMP.	REMARKS	REF.
CARBAMIC ACID, DIETHYL-, VINYL ESTER	0		0.035		80	I7	316
CROTONONITRILE, CIS-	0	0.002	0		60	I7	669
CYCLOHEPTENE	0.068		0		60	I7	668
CYCLOHEXANE, VINYL-	0.22	0.005	0.008		65	I7	1148
CYCLOHEXENE	0.083	0.006	0		60	I7	668
CYCLOOCTENE	0.067	0.003	0		60	I7	668
CYCLOPENTENE	0.010		0		60	R	668
M-DIOXIN, 4-METHYL-	0.17		0.18		50	C	481
1,3-DIOXOLANE, 2-VINYL-	0.045		0.12		60	C,I26	1348
ETHER, BUTYL VINYL	0.046		0		50	A,I7	1328
ETHER, DODECYL VINYL	0.046	0.052	-0.046	0.054	50	C,P	15
ETHERS, C8-C16 ALKYL VINYL					50	E,O	12
ETHYLENE	0.008		0.296		40	C	576
ETHYLENE, CHLORO-	0.40		0.037	0.07	75	I7,L	1033
ETHYLENE, 1,1-DICHLORO-	0.668		0.104		60	I5	972
KETONE, METHYL 5-NORBORNEN-2-YL	0		9		60	N,O	972
MALONONITRILE, METHYLENE-	0		0.4		60		620
MALONONITRILE, METHYLENE-	0		45		50		410
METHACRYLIC ACID, 2,3-EPOXYPROPYL ESTER	0		4.5		40	I26,R	289
METHACRYLIC ACID, 2,3-EPOXYPROPYL ESTER	0		4.2		50	I26,R	1327
METHYL METHACRYLATE	0.01		0.50		60	I21,L	1327
METHYL METHACRYLATE	0.01		3.10		60	I2,L	975
METHYL METHACRYLATE	0.01		3.85		60	I28,L	975
METHYL METHACRYLATE	0.01		3.40		60	I5,L	975
METHYL METHACRYLATE	0.02		3.85		60	I26	975
METHYL METHACRYLATE	0.02		0.90		60	I32,L	975
METHYL METHACRYLATE	0.02		6.7	0.2	75	C	975
NAPHTHALENE, 1,2-DIHYDRO-	0.03		3.5		60		1033
NAPHTHALENE, 1,2-DIHYDRO-	0.5	0.3	4.63	0.4	60		109
NAPHTHALENE, 2-ISOPROPENYL-	-0.18	0.28	0	1.14	30	D	1407
NAPHTHALENE, 2-ISOPROPENYL-	0.026		0		30		708
5-NORBORNENE-2-CARBONITRILE	0.028		0		60	I26	664
5-NORBORNENE-2-CARBONITRILE	0.09		0.16		60	I7,L	664
5-NORBORNENE-2-CARBOXYLIC ACID, METHYL ESTER	0.012		1.1		60	I2	1205
5-NORBORNENE-2-CARBOXYLIC ACID, 2-METHYL-, METHYL ESTER	0		1.1		60		410
4-PENTENOIC ACID, 2-ALLYL-2-CYANO-, ETHYL ESTER	1.4		0.04		60		410
PHTHALIMIDE, N-VINYL-	0.001		0.20		65	Q	410
PROPANE, 1-(ALLYLOXY)-2,3-EPOXY-	0.003		0.3		90	I7	163
1-PROPENE, 1-CHLORO-, CIS-	0.002		0.01		60	I26	694
1-PROPENE, 1-CHLORO-, TRANS-	0.41		0.004		60		1217
1-PROPENE, 2-CHLORO-	0.26		0.05		60		702
1-PROPENE, 3-CHLORO-2-METHYL-	0.06		0.06		60		745
1-PROPEN-2-OL, ACETATE	0.002		0.18	0.05	155		745
2-PROPEN-1-OL, 2-CHLORO-, ACETATE			0.032	0.005	75	E	745
2-PYRROLIDINONE, 1-VINYL-	0.074	0.04	-0.027	0.02	120	C	281
SILANE, TRIETHOXYVINYL-	0.035		0.004		30	N	1033
STILBENE, CIS-	0.08	0.08	0.07	0.07	70	D	484
STILBENE, TRANS-	0.03	0.03	0.03	0.03	60		1407
STYRENE			0.019		60		1379
STYRENE	0		0.02		50		557
STYRENE	0				60		557
STYRENE							46
STYRENE							81

REMARKS PAGE II-366, REFERENCES PAGE II-368

M1	M2	R1	+/-	R2	+/-	TEMP.	REMARKS	REF.
MALEIC ANHYDRIDE	STYRENE	0		0.02		60	R	168
		0		0.01		60		620
		0	0.002	0.042	0.008	80		31
		0	0.001	0.097	0.002	50		368
		0.001	0.001	0.035	0.003	70	I26	973
		0.005		0.13	0.03	70	I26	973
		0.005		0.050	0.005	70	I5	973
		0.01		0.041		60	I2	861
		0.010	0.002	0.025	0.005	70	C	973
		0.015		0.040		50	I32	703
		0.018		0.043		70	I26	973
		0.035	0.001	0.012		40	I7	988
		0.049	0.022	0.040	0.009	70	I28	973
	STYRENE, ALPHA-METHYL-	0.08	0.03	0.038	0.003	60		787
		0.27	0.03	0.005	0.005	60		787
	SUCCINIMIDE, N-VINYL-	0.03		0.15		60		1217
	2,4,8,10-TETRAOXASPIRO(5.5)UNDECANE, 3,9-DIISOPROPENYL-	0.025		0.05		60	C,17	1347
	2,4,8,10-TETRAOXASPIRO(5.5)UNDECANE, 3,9-DIPROPENYL-	0.02		0.02		60	C,17	1347
	TIN, (METHACRYLOYLOXY)TRIMETHYL-	0		0.53		50	I6	1431
		1.57				60	I6	1431
	VINYL ACETATE	0.003		0.055	0.015	75	C	1033
MALEIC ANHYDRIDE, MONO(O-BENZYLOXIME)	STYRENE	0.09	0.06	0.22	0.08	70	I10	1088
MALEIC ANHYDRIDE/ 2-PYRROLIDINONE, 1-VINYL-	METHYL METHACRYLATE	555	240	0.0017	0.0005	30	D	1407
MALEIC ANHYDRIDE/SILANE, TRIETHOXYVINYL-	MALEIC ANHYDRIDE/STYRENE	0.081		1.24		70		1379
MALEIC ANHYDRIDE/STYRENE	2H-AZEPIN-2-ONE, HEXAHYDRO-1-VINYL-	1.73	0.06	0.21	0.01	65		1221
	MALEIC ANHYDRIDE/SILANE, TRIETHOXYVINYL-	1.24		0.081		70		1379
	2-PYRROLIDINONE, 1-VINYL-	1.16	0.03	0.23	0.01	65		1221
	SILANE, TRIETHOXYVINYL-	1.63		0.29		70		1379
	VINYL ACETATE	0.01		0.072	0.04	70	Q	384
MALEIMIDE	ACROLEIN	0.12		3.2		60	I17	1286
	P-DIOXIN, 2,3-DIHYDRO-	0		1.28		75	Q	665
	ETHYLENE, 1,1-DICHLORO-	0.48		0.71		75	I13	1013
	ETHYLENE, 1,2-DIMETHOXY-, CIS-	0		0.222			Q	665
	ETHYLENE, 1,2-DIMETHOXY-, TRANS-	0		0.192			Q	665
	METHYL METHACRYLATE	0.17		2.5		60		1432
		0.17		2.50		75	I13	1013
	STYRENE	0.1		0.1		75	I13	1013

II-230

COPOLYMERIZATION PARAMETER

M1 / M2	R1	+/-	R2	+/-	TEMP.	REMARKS	REF.
MALEIMIDE, N-(BENZYLOXY)-							
STYRENE	0.03		0.02		35	I10	1319
MALEIMIDE, N-2-BORNYL-							
ETHYLENE, 1,1-DICHLORO-	0.47		1.15		50	C,I5	1433
METHYL METHACRYLATE	0.16		2.02		50	C,I5	1433
STYRENE	0.05		0.13		50	C,I5	1433
MALEIMIDE, N-BUTYL-							
ETHYLENE, CHLORO-	0.04		2.08		60	I7	1342
METHYL METHACRYLATE	0.12	0.02	1.33	0.03	50		180
STYRENE	0.06		0.025		60	C	861
STYRENE	0.06	0.02	0.025	0.025	50		180
MALEIMIDE, N-BUTYL-2-METHYL-							
STYRENE	0.206		0.14		60	C	861
MALEIMIDE, N-(P-CHLOROPHENYL)-							
ETHYLENE, CHLORO-	0.03		3.6		60		1343
MALEIMIDE, N,2-DIMETHYL-							
ACRYLONITRILE	0.604		0.527		60	C	553
METHYL METHACRYLATE	0.15		3.24		70	C	553
NAPHTHALENE, 2-VINYL-	0.256		0.187		70	C	553
STYRENE	0.24		0.135		60	C	861
STYRENE	0.24		0.153		70	C	553
VINYL ACETATE	2.56	0.9	0.26	0.18	60	C	554
MALEIMIDE, N-ETHYL-							
ETHYLENE, CHLORO-	0.02		2.41		60	I7	1342
MALEIMIDE, N-HEXYL-							
ETHYLENE, CHLORO-	0.06		2.03		60	I7	1342
MALEIMIDE, N-HYDROXY-, ACETATE							
STYRENE	0.01		0.02		35	I10	1319
MALEIMIDE, N-(2-HYDROXYETHYL)-							
METHYL METHACRYLATE	0.26		1.5		60	I5	1432
2-PYRROLIDINONE, 1-VINYL-	0.07		0.02		60	I11	1432
STYRENE	0.01	0.01	0.04	0.02	60	I15	1432
VINYL ACETATE	0.86		0.05		60	I12	1432

REMARKS PAGE II-366, REFERENCES PAGE II-368

M1 / M2	R1	+/-	R2	+/-	TEMP.	REMARKS	REF.
MALEIMIDE, N-(2-HYDROXYETHYL)-, ACETATE							
VINYL ACETATE	0.85		0.25		70	I7	1056
MALEIMIDE, N-(2-HYDROXYETHYL)-, PROPIONATE							
VINYL ACETATE	1.02		0.03		70	I7	1057
MALEIMIDE, N-(HYDROXYMETHYL)-							
METHYL METHACRYLATE	0.27	0.03	1.66	0.01	60	I13	1432
STYRENE	0.06		0.01		60	I10	1432
VINYL ACETATE	1.4		0.05		60	I10	1432
MALEIMIDE, N-(P-HYDROXYPHENYL)-							
METHYL METHACRYLATE	0.35	0.05	1.35	0.05	60	I5	1432
STYRENE	0.12		0.08		60	I5	1432
VINYL ACETATE	1.78		0.02		60	I6	1432
MALEIMIDE, N-(3-HYDROXYPROPYL)-, ACETATE							
VINYL ACETATE	1.50		0.03		70	I7	1057
MALEIMIDE, N-(METHOXY)-							
STYRENE	0.02		0.02		35	I10	1319
MALEIMIDE, N-METHYL-							
ETHYLENE, CHLORO-	0.01		3.12		60	I7	1342
MALEIMIDE, 2-METHYL-N-PHENYL-							
STYRENE	0.196		0.145		60	C	861
MALEIMIDE, N-OCTYL-							
ETHYLENE, CHLORO-	0.06		1.84		60	I7	1342
MALEIMIDE, N-PHENYL-							
ETHYLENE, CHLORO-	0.03		3.82		60		1343
METHYL METHACRYLATE	0.3		0.98		60		1432
MALEIMIDE, N-PROPYL-							
ETHYLENE, CHLORO-	0.04		2.02		60	I7	1342
MALEIMIDE, N-P-TOLYL-							
ETHYLENE, CHLORO-	0.03		4.37		60		1343

M1	M2	R1	+/-	R2	+/-	TEMP.	REMARKS	REF.
MALEONITRILE								
	STYRENE	0		0.19	0.01	60		557
MALONIC ACID, DIALLYL ESTER								
	STYRENE	0.099	0.003	29.8	0.3	60	C	1293
MALONIC ACID, METHACRYLOYL-, DIETHYL ESTER								
	STYRENE	0.26	0.02	0.17	0.04	60	I7	1339
		0.88		0.41		100	I10	237
MALONIC ACID, METHYLENE-, DIETHYL ESTER								
	1-PROPENE, 2-NITRO-	0.9	0.7	0.15	0.07	0	G13,I5	842
		5.5	2	0.03	0.01	0	G13,I7	842
	STYRENE	0.08	2	0.3	0.2	0	G14,I5	842
		0.08		0.03		60		1429
				0.03		60		1430
MALONIC ACID, METHYLENE-, DIMETHYL ESTER								
	METHYL METHACRYLATE	0.35	0.05	0.43	0.02	60	I2	1235
MALONONITRILE, BENZYLIDENE-								
	ACRYLONITRILE	-0.5	0.4	6.2	1.8	70	C	1374
	STYRENE	-0		0.125		80	C	538
MALONONITRILE, METHYLENE-								
	ACETIC ACID, CHLORO-, VINYL ESTER	0.13	0.05	<0.004		40		289
	ACRYLIC ACID	0.29	0.08	0.26	0.06	50		289
	ACRYLIC ACID, 2-CHLORO-, METHYL ESTER	0.091	0.05	0.41	0.14	50	R	289
	BENZOIC ACID, VINYL ESTER	0.10		0.008		43		289
	1,3-BUTADIENE, 2-CHLORO-	0.0017		0.010		40		289
		0.0048		0.016		40		289
	ETHYLENE, CHLORO-	0.20	0.06	0.017		40		289
		0.54	0.2	0.0093	0.01	50		255
	ETHYLENE, 1,1-DICHLORO-	0.72		0.012		22	Q	289
	ETHYLENE, 1,2-DICHLORO-, CIS-	0.049		0		40	R	289
	ETHYLENE, 1,2-DICHLORO-, TRANS-	30		0		40		289
	MALEIC ANHYDRIDE	30		0		50		289
	METHYL METHACRYLATE	0.031		0.046		50	Q	254
		0.031		0.046		40		289
	1-PROPENE, 2-CHLORO-	0.20		1.40		45	Q	289
	STYRENE	0.001		0.005		45	Q	157
		0.0459		0		40		289
	STYRENE, 2,5-DICHLORO-	0.0092		0.031		45	R	637
	VINYL ACETATE	0.11		0.0054		45		289
	VINYL ETHER	0.12		0.23		50	I6	156

M1	M2	R1	+/-	R2	+/-	TEMP.	REMARKS	REF.
MANGANESE, TRICARBONYL-P1-(VINYLCYCLOPENTADIENYL)-								
	2-PYRROLIDINONE, 1-VINYL-	0.14		0.094		70	I7	1354
MELAMINE, N(2),N(4)-DIALLYL-								
	METHYL METHACRYLATE	0		28.9	2.3	60		812
		0		27.8	3.2	60		812
	STYRENE	0		103.0	11.8	60		812
		0		102.3	15.1	60		812
	VINYL ACETATE	0.19	0.004	1.41	0.06	60		812
		0.20	0.03	1.44	0.28	60		812
MENTHOL, ACRYLATE								
	STYRENE	0.28	0.05	0.84	0.05	50		844
MERCURY, DIACETATOBIS(5-VINYL-2-PICOLINE)-								
	STYRENE	0.4		0.5		60	I6	952
METANILYL FLUORIDE, N-ACRYLOYL-								
	METHYL METHACRYLATE	0.60		1.47			Q	337
METANILYL FLUORIDE, N-METHACRYLOYL-								
	METHYL METHACRYLATE	0.71		1.24			Q	337
	STYRENE	0.24		0.63			Q	337
METHACRYLALDEHYDE								
	ACRYLONITRILE	1.7	0.3	0.15	0.04	30	C,I11	525
		2.0	0.0	0.06		70	C	315
	METHACRYLONITRILE	1.78	0.06	0.40	0.04	50	I6	845
	STYRENE	0.88	0.02	0.22	0.02	60		899
METHACRYLAMIDE								
	ACRYLAMIDE	1.1	0.20	0.74	0.11	25	D,I11	194
	ACRYLIC ACID, METHYL ESTER	0.22	0.0	2.0		65		187
	METHACRYLIC ACID	0.22		2.0		50	C	188
		0.3		2.50		70		774
		0.30		1.55		78		774
	METHACRYLIC ACID, CALCIUM SALT	1.28		0.96		30	E,Q	1006
	METHACRYLIC ACID, 2-ETHYLHEXYL ESTER	0.12		2.42		80	F,I11	1052
	METHYL METHACRYLATE	0.66	0.03	1.68	0.08	70	C	967
		0.43	0.04	1.5	0.07	65	I12	1384
		0.47	0.04	1.65	0.02	70	C	187
		0.49	0.02	1.65	0.05	70	C	187
		0.49	0.02	1.55	0.05	70	I27	660
		1.27	0.19	1.2	0.22	48	I10	1384
		0.14						878
2-PICOLINIUM METHYL SULFATE, 1-METHYL-5-VINYL-								
	STYRENE	0.54	0.08	1.44	0.15	70	I12	1384
		1.29	0.08	1.46	0.09	70	I10	1384

M1	M2	R1	+/-	R2	+/-	TEMP.	REMARKS	REF.
METHACRYLANILIDE	ACRYLIC ACID, METHYL ESTER	1.05	0.19	0.66	0.22	60	I7,P	1445
	ACRYLONITRILE					70		16
	METHYL METHACRYLATE	0.46		0.54		70	C	772
METHACRYLIC ACID	ACETAMIDE, N-ALLYL-	16.9	0.03	0.04		80	I12	857
	ACETAMIDE, N-ALLYL-2-CHLORO-	12.2	0.02	0.05		80	I12	857
	ACRYLAMIDE			0.5	0.01	25	B,I11	1309
				1		60	C,I15	1309
	ACRYLIC ACID, BUTYL ESTER	0.25	0.2	0.12	0.02	60	I11	1416
	ACRYLONITRILE	1.36		0.18	0.01	60	I49	1416
		1.31	0.05	0.35		50	C,I12	766
	BENZAMIDE, N-ALLYL-	2.5	0.05	0.093	0.03	70	C,I62	886
	BENZENE, ETHYNYL-	13.3		0.03		80	I12	857
	1,3-BUTADIENE	2.3		0.07	0.01	60	C	291
	1,3-BUTADIENE, 2-CHLORO-	0.526		0.201		50	C,K	267
	BUTYRAMIDE, N-ALLYL-3-METHYL-	0.15		2.7	0.2	40	C,K	1030
	ETHYLENE, BROMO-	8.1		0.05		80	I12	857
	ETHYLENE, CHLORO-	90	15	0	0.1	50	C	991
	ETHYLENE, 1,1-DICHLORO-	23.8		0.034	0.02	60	I57	541
		36		0.027		60	I7	935
	FORMAMIDE, N-ALLYL-	3.0	0.4	0.15		50		23
	HYDROQUINONE, VINYL-, DIBENZOATE	4.6		0.03		70	I12	857
	INDOLE, 1-VINYL-	1.91	0.23	0.91	0.25	80	I2	482
	METHACRYLAMIDE	0.096	0.03	1.43	0.05	60		873
		1.55		1.28		60	E,Q	1006
		2.0		0.22		50	C	188
		2.0		0.3		70		774
		2.50				70		774
	METHACRYLIC ACID, BUTYL ESTER	0.53		0.30	0.15	80	I13	901
	METHACRYLIC ACID, 2-(DIETHYLAMINO)ETHYL ESTER	0.75	0.16	1.11	0.83	50	C,I12	766
		0.98		1.20	0.23	70		36
	METHACRYLIC ACID, 2,3-EPOXYPROPYL ESTER	0.85	0.15	0.90	0.13	80	I13	901
	METHACRYLIC ACID, ISOBUTYLAMINE COMPLEX	0.47	0.14	1.18	0.17	60	Q	1161
	METHACRYLIC ACID, PHENYL ESTER	0.21	0.08	1.52	0.08	60	I75	1454
		0.4	0.05	0.59	0.05	60	I7	1454
	METHACRYLONITRILE	1.63	0.1	0.59	0.15	80	C	159
		1.64	0.05	0.62		65		303
	METHYL METHACRYLATE	0.43	0.05	1.66	0.04	30		855
		0.63	0.1	2.61	0.04	55	D	1153
		0.68	0.027	1.18	0.07	60	LL	1378
		0.68	0.06	0.98	0.007	45	I6	820
		0.92		0.98	0.02	45		1377
		1.33		0.6	0.02	70	C,I62	886
		1.55	0.06	0.12	0.01	55	I10	1153
		1.55		0.55		45	C	1377
		1.6		0.55		45		820
		1.63	0.12	0.3		60		1378
				0.35		55		1153
	PHOSPHONIC ACID, ALLYL-, DIETHYL ESTER	6.1	0.4	0.04	0.04	60		401
	PHOSPHONIC ACID, 2-BUTENYLIDENEDI-, TETRAETHYL ESTER	2.0	1.0	0.0	0.10	60		401
	PHOSPHONIC ACID, ALPHA-PHENYLVINYL-	3.50	0.2	0.36	0.12	80	C	521

M1	M2	R1	+/-	R2	+/-	TEMP.	REMARKS	REF.
METHACRYLIC ACID								
	PHOSPHONIC ACID, ALPHA-PHENYLVINYL-	4.0	1	0.00		80	I49,J	522
	PHOSPHONIC ACID, VINYL-, BIS(2-CHLOROETHYL) ESTER	1.7	0.5	0.1	0.075		Q	729
	PHOSPHONIC ACID, VINYL-, DIETHYL ESTER	1.9	0.5	0.15	0.15	60		401
	2-PICOLINE, 5-VINYL-	0.43	0.2	0.85	0.03		Q	1149
	2-PICOLINIUM METHYL SULFATE, 1-METHYL-5-VINYL-	0.58		0.60		50		878
	1-PROPENE, 2,3-DICHLORO-	4.0		0.0		100		484
	2-PROPEN-1-OL, 2-CHLORO-	4.5		0.0		100		484
	PYRIDINE, 2-VINYL-	0.44	0.02	1.38	0.04	70	C,Q	819
		0.58	0.05	1.55	0.10	60	I11	35
	2-PYRROLIDINONE, 1-VINYL-	0.65	0.02	0.33	0.01	60	I49	1416
		0.9	0.04	0.3	0.03	60	I11	1416
		1.4	0.02	0.3	0.01	60	I11	1416
	STYRENE	0.28		0.38		60	LL	1378
		0.45	0.06	0.53	0.02	45	I6	820
		0.45	0.06	0.53	0.02	45	C,I6	1377
		0.47		0.45		40	GG,I19	695
		0.47		0.63		40	GG,I17	695
		0.51		0.37		40	GG,I56	695
		0.56		0.3		40	GG,I15	695
		0.62		0.2		60		1378
		0.64		0.22		40		1323
		0.64		0.22		45	Q	695
		0.66	0.08	0.2	0.02	45	C	1377
		0.66	0.08	0.2	0.05	70		820
		0.7	0.01	0.15	0.01	60		1212
		0.7	0.05	0.15		5		166
		0.70		0.15		70	C	218
	STYRENE, O-CHLORO-	0.7	0.1	0.12	0.02	70	F,K	23
	VINYL ACETATE	20		0.01				32
METHACRYLIC ACID, ALLYL ESTER								
	ACRYLONITRILE	0.515	0.10	9.55	0.47	-40	G17,I5	197
	METHYL METHACRYLATE	0.69	0.02	0.87	0.025	65		818
METHACRYLIC ACID, 3-ANILINO-2-HYDROXYPROPYL ESTER								
	STYRENE	0.46		0.23		60		992
		0.58		0.41		60	I10	992
		0.75		0.19		60	I7	992
METHACRYLIC ACID, 2-(1-AZIRIDINYL)ETHYL ESTER								
	METHYL METHACRYLATE	0.95		1.02		60		221
	STYRENE	0.63	0.04	0.53	0.02	60		221
METHACRYLIC ACID, BENZYL ESTER								
	ACRYLONITRILE	0.810	0.02	0.206	0.02	60		63
	BENZENE, ETHYNYL-	1.15		1.055		25	E	83
	ETHYLENE, 1,1-DICHLORO-	2.5	0.2	0.24	0.01	60		291
	METHACRYLIC ACID, 2-CHLOROETHYL ESTER	3.3	0.5	0.34	0.05	60	C	405
		0.96		1.04		60		743

M1 / M2	R1	+/-	R2	+/-	TEMP.	REMARKS	REF.
METHACRYLIC ACID, BENZYL ESTER							
METHACRYLIC ACID, PHENYL ESTER	0.65		1.42		30	G18,17	100
	0.65		1.42		60		100
METHYL METHACRYLATE	1.04	0.03	0.44	0.07	60	G16,12	1229
	1.05		0.93		0		100
	1.1		0.89		60	G19,15	1229
	1.14	0.10	0.85	0.10	0		408
	1.17		0.90		60	G18,17	100
	1.38	0.09	0.78	0.10	30		1067
	1.84	0.01	0.7	0.05	0	G17,12	1229
	1.98	0.07	0.97	0.07	0	G16,15	1229
	2.04	0.09	1.02	0.1	0	G17,15	1229
	2.71	0.05	1.16	0.06	0	H21,15	1229
	2.87	0.05	0.95	0.06	0	H21,12	1229
1-PROPENE, 3-CHLORO-	58.7	4.1	0.016	0.018	60	C	457
STYRENE	0.42	0.07	0.48	0.04	60		1067
	0.51		0.45		60		741
	0.51	0.01	0.44	0.01	60		743
	0.62	0.10	0.46	0.05	60		408
METHACRYLIC ACID, BENZYL-14C ESTER							
METHACRYLIC ACID, CYCLOHEXYL ESTER	0.93	0.05	1.15		60		99
METHACRYLIC ACID, PHENYL ESTER	0.65		1.42		60		99
METHYL METHACRYLATE-METHYL-T	1.05		0.93		60		99
METHACRYLIC ACID, P-(BENZYLIDENEAMINO)PHENYL ESTER							
STYRENE	0.25	0.05	0.24	0.01	60		801
METHACRYLIC ACID, BORNYL ESTER							
ETHYLENE, CHLORO-	12.5		0.06		60		388
STYRENE	0.44		0.49		60		388
METHACRYLIC ACID, 2-BROMOETHYL ESTER							
STYRENE	0.44	0.02	0.35	0.02	60		570
METHACRYLIC ACID, 2(TERT-BUTYLAMINO)ETHYL ESTER							
STYRENE	0.83	0.06	0.47	0.05	60		570
METHACRYLIC ACID, BUTYL ESTER							
ACRYLIC ACID	3.67		0.29		50	C,I12	766
ACRYLIC ACID, BUTYL ESTER	2.2		0.3		50	C,I12	766
ACRYLIC ACID, 2-CHLORO-, METHYL ESTER					60/80	P	916
ACRYLONITRILE	1.08		0.31		60	C	679
BENZENE, ETHYNYL-	0.21		1.7		60		1242
ETHYLENE, CHLORO-	13.5		0.05		45		964
ETHYLENE, 1,1-DICHLORO-	30.5		0.025		80	C	1194
METHACRYLIC ACID	0.22		0.35		70	C	8
	1.11		0.53		80	I13	901

REMARKS PAGE II-366, REFERENCES PAGE II-368

M1	M2	R1	+/-	R2	+/-	TEMP.	REMARKS	REF.
METHACRYLIC ACID, BUTYL ESTER								
	METHACRYLIC ACID	1.20		0.75		50	C,I12	766
	METHACRYLIC ACID, 2-CHLOROETHYL ESTER	0.85		1.10		60		743
	METHACRYLIC ACID, 2,3-EPOXYPROPYL ESTER	0.79		0.94		80	I13	901
	METHACRYLONITRILE	0.69		0.51		80	C	159
	METHYL METHACRYLATE	1.27		0.79		60		95
	1-PROPENE, 3-CHLORO-	46.3	2.6	0.058	0.024	60	C	457
	STYRENE	0.40		0.56		60		741
	STYRENE	0.40	0.03	0.56	0.03	60		743
	STYRENE	0.47	0.06	0.52	0.06	60		570
	STYRENE	0.64		0.63		50		1011
	STYRENE	0.64		0.54		70		1011
	STYRENE	0.67		0.97		30	D	152
	STYRENE, O-CHLORO-	0.324	0.11	1.24	0.33	40	C,K	1186
	VINYL ACETATE	28.8		0.023		60	C	679
		62.1	3.8	0.127	0.015	60	C	638
METHACRYLIC ACID, TERT-BUTYL ESTER								
	CARBONIC ACID, THIO-, O-TERT-BUTYL S-VINYL ESTER	1.7	0.05	0.15	0.05	70		1176
	METHACRYLIC ACID, 2-CHLOROETHYL ESTER	0.88		1.14		60		743
	METHACRYLONITRILE	0.70	0.03	0.37	0.03	80	C	159
	STYRENE	0.49		0.58		60		570
METHACRYLIC ACID, SEC-BUTYL ESTER								
	STYRENE	0.56	0.03	0.60	0.03	60		570
METHACRYLIC ACID, TERT-BUTYL ESTER								
	STYRENE	0.60		0.56		60		741
	STYRENE	0.67	0.04	0.59	0.03	60		743
METHACRYLIC ACID, 2-BUTYNYL ESTER								
	ACRYLIC ACID, METHYL ESTER	0.35	0.05	0.56	0.08	-40	G2,I5	196
	ACRYLONITRILE	0.04	0.01	1.10	0.20	-40	G2,I5	196
	METHYL METHACRYLATE	0.35	0.06	0.34	0.04	-40	G2,I5	196
	STYRENE	1.20	0.16	0.40	0.05	-40	G2,I5	196
METHACRYLIC ACID, CADMIUM SALT								
	ACRYLAMIDE	2.13		0.04		30	C,I11	1426
METHACRYLIC ACID, CALCIUM SALT								
	ACRYLAMIDE	0.71		0.24		30	C,I11	1426
	METHACRYLAMIDE	2.28		0.07		30	F,I11	1052
	METHACRYLAMIDE	0.96		0.12		30	F,I11	1052
METHACRYLIC ACID, 2-CHLOROETHYL ESTER								
	ACRYLONITRILE	1.3	0.2	0.14	0.02	60	C	64
	ANISOLE, P-PROPENYL-/MALEIC ANHYDRIDE	0.095	0.014	11.5	1.2	60	I7	1259

COPOLYMERIZATION PARAMETER

M1 / M2	R1	+/-	R2	+/-	TEMP.	REMARKS	REF.
METHACRYLIC ACID, 2-CHLOROETHYL ESTER							
METHACRYLIC ACID, BENZYL ESTER	1.04		0.96		60		743
METHACRYLIC ACID, BUTYL ESTER	1.10		0.85		60		743
METHACRYLIC ACID, TERT-BUTYL ESTER	1.14		0.88		60		743
METHACRYLIC ACID, ISOBUTYL ESTER	1.07		0.88		60		743
METHACRYLIC ACID, PHENYL ESTER	0.84		0.98		60		743
METHYL METHACRYLATE	0.48	0.2	0.55	0.07	60	C	64
	1.13		0.88		60		743
STYRENE	0.23	0.08	0.42	0.06	60	C	64
	0.46		0.33		60		742
STYRENE, O-CHLORO-	0.46	0.04	0.33	0.02	60		743
	0.885	0.107	0.909	0.005	40	C,K	1186
METHACRYLIC ACID, M-CHLOROPHENYL ESTER							
STYRENE	0.40		0.21		60		742
	0.40		0.21		60		739
METHACRYLIC ACID, P-CHLOROPHENYL ESTER							
STYRENE	0.45		0.22		60		742
	0.45		0.22		60		739
METHACRYLIC ACID, P-(N-(P-CHLOROPHENYL)FORMIMIDOYL)PHENYL ESTER *							
STYRENE	5.4	0.6	0.18	0.03	60		801
METHACRYLIC ACID, CINNAMYL ESTER, TRANS-							
METHYL METHACRYLATE	1.1	0.15	0.8	0.13	-78	G17,I2	1222
METHACRYLIC ACID, CYCLOHEXYL ESTER							
METHACRYLIC ACID, BENZYL-14C ESTER	1.15		0.93		60		99
METHYL METHACRYLATE	0.49		1.96		30		100
	0.8	0.1	1.4	0.1	30	G18,I7	1151
	1.15		0.86		60	G19,I10	100
METHYL METHACRYLATE-METHYL-14C	0.45		0.52		60		99
STYRENE	0.45	0.09	0.52	0.07	60		741
							743
METHACRYLIC ACID, DECYL ESTER							
2-OXAZOLIDINONE, 3-VINYL-	12.8	0.5	0.015	0.05	50		116
METHACRYLIC ACID, 2-(DIETHYLAMINO)ETHYL ESTER							
ACRYLIC ACID, ETHYL ESTER	1.6	0.1	0.15	0.1	70	P,Q	943
ACRYLIC ACID, METHYL ESTER	1.85	0.1	0.19	0.08	70	P,Q	943
ACRYLONITRILE	0.90	0.23	0.98	0.16	70	P,Q	943
						I12	1109
							1109
METHACRYLIC ACID						C	36

M1	M2	R1	+/-	R2	+/-	TEMP.	REMARKS	REF.
METHACRYLIC ACID, 2-(DIETHYLAMINO)ETHYL ESTER								
	METHACRYLIC ACID, ETHYL ESTER	0.65	0.03	0.08	0.015	70	P,Q	943
	METHACRYLIC ACID, SODIUM SALT						C	36
	METHYL METHACRYLATE						P,Q	943
	STYRENE						P,Q	943
		0.23		0.74		60	16	866
		0.34		0.78		60	112	866
		0.55	0.1	0.37	0.02	70	16	1109
		0.6	0.1	0.22	0.05	70		3
		0.66	0.05	0.35	0.1	70	16	1109
		3.00		0.07		80		866
		6.20		0.11		80		866
	VINYL ACETATE						P,Q	943
METHACRYLIC ACID, 2-(DIETHYLAMINO)ETHYL ESTER, ETHIODIDE								
	PYRIDINE, 4-VINYL-	0.61	0.09	0.30	0.02	60		755
METHACRYLIC ACID, 2-(DIETHYLAMINO)ETHYL ESTER, HYDROCHLORIDE								
	STYRENE	0.3	0.05	0.25	0.05	70	16	3
METHACRYLIC ACID, 2-(DIETHYLAMINO)ETHYL ESTER, METHIODIDE								
	STYRENE	0.25	0.05	0.34	0.05	70	16	3
METHACRYLIC ACID, 2-(DIMETHYLAMINO)ETHYL ESTER								
	STYRENE	0.37	0.01	0.53	0.02	60		570
METHACRYLIC ACID, 2,2-DINITROPROPYL ESTER								
	STYRENE	0.18		0.37		80	126	930
METHACRYLIC ACID, DODECYL ESTER								
	PHOSPHINE OXIDE, DIALLYLPHENYL-	5.59	0.24	-0.089	0.022	140	R	103
		11.1	0.4	-0.013	0.025	110	R	103
		18.1	0.4	-0.011	0.013	80	R	103
	PHOSPHONIC ACID, ALLYL-, DIETHYL ESTER	52.5	3.6	0.066	0.12	80	C,17	102
	PHOSPHONIC ACID, BUTYL-, DIALLYL ESTER	18.7	0.2	0.091	0.04	80	C,17	102
	PHOSPHONIC ACID, PHENYL-, DIALLYL ESTER	19.5	3.6	0.072	0.44	80	C,17	102
	STYRENE	0.36	0.04	0.56	0.02	60		743
		0.45	0.04	0.57	0.04	60		741
		0.57		0.63		60		570
	UREA, 1-ALLYL-	29.9	1.2	-0.22	0.06	100	C	104
	UREA, 1,1-DIALLYL-	27.4	0.8	-0.014	0.05	100	C	104
	UREA, 1,3-DIALLYL-	14.5	0.7	-0.12	0.04	100	C	104
METHACRYLIC ACID, 2,3-EPOXYPROPYL ESTER								
	ACRYLONITRILE	1.32	0.03	0.14	0.001	60	C,17,Q	414
		2.3		0.3				504
	ETHYLENE, CHLORO-	8.84		0.04			Q	391

COPOLYMERIZATION PARAMETER

M1	M2	R1	+/-	R2	+/-	TEMP.	REMARKS	REF.
METHACRYLIC ACID, 2,3-EPOXYPROPYL ESTER								
	FURAN, TETRAHYDRO-	0.25		2.5		0	H10	738
	MALEIC ANHYDRIDE	4.2		0		50	I26,R	1327
		4.5		0		40	I26,R	1327
	METHACRYLIC ACID	1.18		0.85		80	I13	901
	METHACRYLIC ACID, BUTYL ESTER	0.94		0.79		80	I13	901
	METHYL METHACRYLATE	0.88	0.10	0.76	0.07	60		412
		0.94		0.75		80	C,I2,I10	905
		1.05		0.80		80		293
	STYRENE	0.53		0.44		60		414
		0.544	0.07	0.441	0.001	60	C,I13	903
		0.55	0.004	0.45	0.007	60		570
		0.55	0.002	0.45	0.001	60	C,I13	902
		0.60		0.50		120		903
		0.63	0.1	0.34	0.05	65	I5	887
		0.68		0.37		80		866
METHACRYLIC ACID, ETHYL ESTER								
	ACETIC ACID, ALLYL ESTER	85	2			60		855
	ACRYLIC ACID, 2-CHLORO-, METHYL ESTER					60/80	P	916
	ALLYL ALCOHOL	108	5			60		855
	BENZENE, ETHYNYL-	0.23		2.1		60		1242
	ETHANESULFONIC ACID, 2-HYDROXY-, METHACRYLATE	1	0.1	2	0.4	68	I55	1241
	ETHYLENE, 1,1-DICHLORO-	0.22		0.35		80	C	8
	METHACRYLIC ACID, 2-(DIETHYLAMINO)ETHYL ESTER	0.83	0.1	0.46	0.1	60	P,Q	943
	METHACRYLONITRILE	0.98		1.09		60	C	159
	METHYL METHACRYLATE	1.08		0.92				305
		1.09		0.98			C	100
	1-PROPENE, 3-CHLORO-	57.9	15.0	0.082	0.083	30	G18,I7	100
	STYRENE	0.26		0.67		60	C	457
		0.29	0.03	0.65	0.06	50		1011
		0.33		0.55		70		1011
		0.41	0.03	0.53	0.03	60		570
		0.45	0.03	0.53		60		741
						60		743
	STYRENE, O-CHLORO-			1.34	0.032	40	C,K	1186
	VINYL ACETATE	142	2			60		855
METHACRYLIC ACID, ETHYL-1-14C ESTER								
	METHYL METHACRYLATE	1.08		0.92		60		99
METHACRYLIC ACID, 2-ETHYLHEXYL ESTER								
	METHACRYLAMIDE	2.42	0.08	0.66	0.03	80	C	967
	STYRENE, O-CHLORO-	0.394	0.07	2.74	0.04	40	C,K	1186
METHACRYLIC ACID, 2-(ETHYLTHIO)ETHYL ESTER								
	METHYL METHACRYLATE	1.1		1		50		1211
	STYRENE	0.51		0.42		50		1211

REMARKS PAGE II-366, REFERENCES PAGE II-368

M1	M2	R1	+/-	R2	+/-	TEMP.	REMARKS	REF.
METHACRYLIC ACID, FURFURYL ESTER								
	METHYL METHACRYLATE	1.19	0.037	0.75	0.111	60	I7,Q	95
	STYRENE	0.733	0.037	0.711	0.111		I7,Q	94
METHACRYLIC ACID, FURFURYL ESTER, PI COMPLEX WITH O2								
	STYRENE	0.336	0.069	0.258	0.069		I7,Q	94
METHACRYLIC ACID, HEPTYL ESTER								
	VINYL ACETATE	60.4	0.4	0.271	0.039	60	C	638
METHACRYLIC ACID, HEXADECYL ESTER								
	VINYL ACETATE	68.3	3.2	0.135	0.055	60	C	638
METHACRYLIC ACID, HEXYL ESTER								
	CARBONIC ACID, CYCLIC (HYDROXYMETHYL)ETHYLENE ESTER, ACRYLATE	68.3	3.2	0.135	0.055	60	C	638
	1,3-DIOXOLANE-4-METHANOL, 2,2-DIMETHYL-, ACRYLATE	2.37	0.01	0.47	0.06	50	C,12	195
	1,3-DIOXOLANE-4-METHANOL, 2,2-DIMETHYL-, 4-(METHYLENESUCCINATE)	2.37	0.01	0.47	0.06	50	C,12	195
	1,3-DIOXOLANE-4-METHANOL, 2,2-DIMETHYL-, METHYL FUMARATE	17.5	2.5	0		50	C,12	195
	METHACRYLONITRILE	0.56		0.75		80	C	159
	STYRENE	0.45	0.06	0.60	0.02	60		741
	STYRENE	0.65	0.02	0.60	0.03	60		743
	STYRENE			0.45				570
METHACRYLIC ACID, X-HYDROXYBUTYL ESTER, NITRATE								
	STYRENE	0.4		0.37		60		1307
METHACRYLIC ACID, 2-HYDROXYETHYL ESTER								
	ETHANESULFONIC ACID, 2-HYDROXY-, SODIUM SALT, METHACRYLATE	1.6	0.1	0.7	0.1	6	I11	1241
	1-PROPENE, 3-CHLORO-	35.6	1.7	0.033	0.048	60	C	457
	STYRENE	0.65		0.57		60		570
	TOLUENE-2,4-DICARBAMIC ACID, BIS(2-HYDROXYETHYL)ESTER, DIMETHACRYLATE	0.65		0.57		60		570
METHACRYLIC ACID, 2-HYDROXYETHYL ESTER, NITRATE								
	STYRENE	0.49		0.35		60		1307
METHACRYLIC ACID, 2-HYDROXY-3-(N-METHYLANILINO)PROPYL ESTER								
	STYRENE	0.44		0.30		60		992
	STYRENE	0.74		0.26		60	I7	992
METHACRYLIC ACID, 2-HYDROXY-3-(2-NAPHTHYLAMINO)PROPYL ESTER								
	STYRENE	0.37		0.21		60	I7	992

COPOLYMERIZATION PARAMETER

M1	M2	R1	+/-	R2	+/-	TEMP.	REMARKS	REF.
METHACRYLIC ACID, 2-HYDROXY-3-PHENOXYPROPYL ESTER								
	STYRENE	0.40		0.39		60		992
		0.82		0.58		60	I7	992
METHACRYLIC ACID, 2-HYDROXY-3(P-(PHENYLAZO)ANILINO)PROPYL ESTER *								
	STYRENE	0.28		0.39		60	I10	992
METHACRYLIC ACID, 2-HYDROXY-3((8-(PHENYLAZO)-2-NAPHTHYL)AMINO)PROPYL ESTER *								
	STYRENE	0.20		0.45		60	I10	992
METHACRYLIC ACID, 2-HYDROXYPROPYL ESTER								
	STYRENE	0.65	0.02	0.56	0.02	60		570
METHACRYLIC ACID, 2-HYDROXYPROPYL ESTER, NITRATE								
	STYRENE	0.5		0.36		60		1307
METHACRYLIC ACID, ISOBORNYL ESTER								
	ETHYLENE, CHLORO-	10.0		0.12		60		388
	STYRENE	0.32		0.70		60		388
METHACRYLIC ACID, ISOBUTYLAMINE COMPLEX								
	METHACRYLIC ACID	0		0.47	0.15		Q	1161
METHACRYLIC ACID, ISOBUTYL ESTER								
	ACRYLONITRILE	1.04		0.21		60	C	679
	BENZENE, ETHYNYL-	0.27		1.9		60		1242
	METHACRYLIC ACID, 2-CHLOROETHYL ESTER	0.88		1.07		60		743
	METHACRYLONITRILE	0.67		0.73		80	C	1070
	METHYL METHACRYLATE	1.09		0.91		60		95
	1-PROPENE, 3-CHLORO-	45.5	10.2	0.047	0.056	60	C	457
	STYRENE	0.40		0.55		60		741
		0.40	0.05	0.55	0.02	60		743
		0.58	0.02	0.56	0.02	60		570
	VINYL ACETATE	29.8		0.025		60	C	679
METHACRYLIC ACID, ISOPROPYL ESTER								
	METHACRYLONITRILE	0.92	0.20	0.43	0.21	80	C	159
	METHYL METHACRYLATE	1.20		0.89		60	I7	1068
	STYRENE	0.42	0.04	0.50	0.04	60		570
		0.74	0.05	0.47	0.06	60	I7	1068
METHACRYLIC ACID, LITHIUM SALT								
	ACRYLAMIDE	0.73		0.5		30	C,I11	1425

REMARKS PAGE II-366, REFERENCES PAGE II-368

M1	M2	R1	+/-	R2	+/-	TEMP.	REMARKS	REF.
METHACRYLIC ACID, MAGNESIUM SALT								
	ACRYLAMIDE	0.92		0.2		30	C,I11	1426
METHACRYLIC ACID, 5-(METHOXYCARBONYL)FURFURYL ESTER								
	STYRENE	0.143	0.077	0.352	0.022		I7,O	94
METHACRYLIC ACID, 2-METHOXYETHYL ESTER								
	METHYL METHACRYLATE	1.06	0.11	0.86	0.11	60	I7	1068
	STYRENE	0.58	0.04	0.50	0.01	60	I7	1068
METHACRYLIC ACID, P-METHOXYPHENYL ESTER								
	STYRENE	0.60		0.28		60		739
	STYRENE	0.60		0.26		60		742
METHACRYLIC ACID, 1-METHYL-2-NITROPROPYL ESTER								
	STYRENE	0.078		0.43		80	I26	930
METHACRYLIC ACID, 2-METHYL-2-NITROPROPYL ESTER								
	STYRENE	0.37		0.32		80	I26	930
METHACRYLIC ACID, 2-NITROBUTYL ESTER								
	STYRENE	0.34		0.25		80	I26	930
METHACRYLIC ACID, 2-NITROETHYL ESTER								
	STYRENE, ALPHA-METHYL-	0.46	0.03	0.52	0.03	70	C	1016
	STYRENE, ALPHA-METHYL-	0.26	0.01	0.12	0.01	70	C	1016
METHACRYLIC ACID, M-NITROPHENYL ESTER								
	STYRENE	0.36		0.18		60		739
	STYRENE	0.36		0.18		60		742
METHACRYLIC ACID, P-NITROPHENYL ESTER								
	STYRENE	0.22		0.19		60		739
	STYRENE	0.22		0.19		60		742
METHACRYLIC ACID, 2-NITROPROPYL ESTER								
	STYRENE	0.37		0.39		80	I26	930
METHACRYLIC ACID, NONYL ESTER								
	1,3-BUTADIENE	0.32		0.76			Q	588
METHYL METHACRYLATE-METHYL-T		0.86		1.10		60		99

COPOLYMERIZATION PARAMETER

M1 / M2	R1	+/-	R2	+/-	TEMP.	REMARKS	REF.
METHACRYLIC ACID, NOVOLAK RESIN ESTER							
ACRYLONITRILE	0		4.1		70		984
STYRENE	2.76		0.1		70	16	985
METHACRYLIC ACID, OCTADECYL ESTER							
METHACRYLONITRILE	1.13		0.90		80	C	159
METHACRYLIC ACID, OCTYL ESTER							
ETHYLENE, CHLORO-	14.0		0.04		45		964
METHACRYLONITRILE	0.58		0.75		80	C	159
STYRENE	0.55	0.07	0.67	0.03	60		741
STYRENE	0.68	0.04	0.56	0.03	60		743
STYRENE							570
METHACRYLIC ACID, PENTAMETHYLDISILOXANYL METHYL ESTER							
ACRYLONITRILE	1.44	0.15	0.19	0.04	50		632
METHYL METHACRYLATE	1.13	0.1	0.93	0.1	50		632
STYRENE	0.58	0.02	0.77	0.02	50		632
VINYL ACETATE	24	5	0.16	0.16	50		632
METHACRYLIC ACID, PENTYL ESTER							
METHACRYLONITRILE	0.51		0.55		80	C	159
STYRENE	0.40		0.55		60		741
STYRENE	0.40	0.05	0.55	0.02	60		743
METHACRYLIC ACID, PHENETHYL ESTER							
METHYL METHACRYLATE	1.33	0.50	1.09	0.23	60		1067
STYRENE	0.51	0.09	0.55	0.03	60		1067
METHACRYLIC ACID, PHENYL ESTER							
METHACRYLIC ACID	0.59	0.17	0.4	0.14	60	17	1454
METHACRYLIC ACID	1.52	0.13	0.21	0.02	60	I75	1454
METHACRYLIC ACID, BENZYL ESTER	1.42		0.65		30	G18,17	100
METHACRYLIC ACID, BENZYL ESTER	1.42		0.65		60		99
METHACRYLIC ACID, BENZYL-14C ESTER	0.98		0.84		60		743
METHACRYLIC ACID, 2-CHLOROETHYL ESTER	1.67		0.53		30		100
METHYL METHACRYLATE	1.69		0.58		30	G18,17	100
METHYL METHACRYLATE	1.72		0.56		60		100
STYRENE	0.51	0.01	0.26	0.01	60		570
STYRENE	0.60		0.30		60		741
STYRENE	0.60		0.30		60		739
STYRENE	0.60	0.05	0.30	0.03	60		743
METHACRYLIC ACID, PHENYL-14C ESTER							
METHYL METHACRYLATE-METHYL-T	1.30		0.56		110		99
METHYL METHACRYLATE-METHYL-T	1.67		0.53		30		99

M1	M2	R1	+/-	R2	+/-	TEMP.	REMARKS	REF.
METHACRYLIC ACID, PHENYL-14C ESTER								
	METHYL METHACRYLATE-METHYL-T	1.72		0.56		60		99
METHACRYLIC ACID, P-(N-PHENYLFORMIMIDOYL)PHENYL ESTER								
	STYRENE	2.4	0.5	0.25	0.03	60		801
METHACRYLIC ACID, PIPERIDIDE								
	ETHER, BUTYL VINYL	0.075		0			E,Q	1301
METHACRYLIC ACID, POTASSIUM SALT								
	ACRYLAMIDE	0.46		0.41		30	C,III	1425
METHACRYLIC ACID, PROPYL ESTER								
	BENZENE, ETHYNYL-	0.22		1.4		60		1242
	METHACRYLONITRILE	0.79		0.29		80	C	159
	1-PROPENE, 3-CHLORO-	52.0	3.2	0.067	0.027	60	C	457
	STYRENE	0.38		0.57		60		741
		0.38	0.04	0.57	0.01	60		743
	VINYL ACETATE	73.3	7.4	0.186	0.038	60	C	638
METHACRYLIC ACID, 2-PROPYNYL ESTER								
	ACRYLONITRILE	1.75		0.15		60		1306
	METHYL METHACRYLATE	0.79		0.96		60		1306
	STYRENE	0.45		0.3		60		1306
METHACRYLIC ACID, PYRIDINE COMPLEX								
	STYRENE	0.47		0.45			Q	1323
METHACRYLIC ACID, 2-PYRIDYL ESTER								
	METHYL METHACRYLATE	1.4	0.15	0.6	0.05	60		1101
	STYRENE	0.62	0.11	0.35	0.03	60		1101
METHACRYLIC ACID, QUINOLINE COMPLEX								
	STYRENE	0.51		0.37			Q	1323
METHACRYLIC ACID, SODIUM SALT								
	ACRYLAMIDE	0.59		0.42		30	C,III	1425
	METHACRYLIC ACID, 2-(DIETHYLAMINO)ETHYL ESTER	0.08	0.015	0.65	0.03	70	C	36
METHACRYLIC ACID, TETRAHYDROFURFURYL ESTER								
	PROPANE, 1-CHLORO-2,3-EPOXY-	0.72	0.15	0.14	0.01	0	H10	706

COPOLYMERIZATION PARAMETER

M1	M2	R1	+/-	R2	+/-	TEMP.	REMARKS	REF.
METHACRYLIC ACID, TETRAHYDROFURYL ESTER								
	ACRYLONITRILE	0.710	0.01	0.154	0.03	60		63
		1.01		0.477		25	E	83
METHACRYLIC ACID, 2,4,8,10-TETRAOXASPIRO(5.5)UNDEC-3,9-YLENE* DIETHYLENE ESTER								
	ACRYLONITRILE	0.4		0.35		60	C,CC	1344
	STYRENE	0.2		0.35		60	C,CC	1344
METHACRYLIC ACID, P-TOLYL ESTER								
	STYRENE	0.70		0.30		60		739
		0.70		0.30		60		742
METHACRYLIC ACID, TRIETHYLAMINE COMPLEX								
	STYRENE	0.36		2.1			O	1323,
METHACRYLIC ACID, 2-(VINYLOXY)ETHYL ESTER								
	ETHER, BUTYL VINYL	0.004		0.82	0.02	26	H37,I30	550
METHACRYLIC ACID, ZINC SALT								
	ACRYLAMIDE	1.2		0.04		30	C,I11	1426
METHACRYLIC ANHYDRIDE								
	ACETIC ACID, CHLORO-, ALLYL ESTER	42	2	0.16		60	C,I7	372
	ACRYLIC ACID, METHYL ESTER	4.75	0.15	1.94		60		1145
	ACRYLONITRILE	0.21	0.3	0		60		66
	ETHER, 2-CHLOROETHYL VINYL	6.1		0.27	0.02	60	C,I7	372
	METHACRYLONITRILE	1.6	0.07	0.22		36,6	I13	891
	METHYL METHACRYLATE	1.7		0.12	0.02	60	C,I7	1145
	STYRENE	0.26	0.03	0.1	0.01	36,6	I13	891
		0.33		0		60	C,I7	1145
	SULFIDE, BENZYL VINYL	0.8	0.1			60		372
	UREA, 1-ALLYL-	28	2			60		372
METHACRYLONITRILE								
	ACENAPHTHYLENE	0.15		2.38		60	I7	1326
	ACROLEIN	1.20	0.08	0.72	0.06	50	I10	845
		1.20	0.08	0.72	0.06	50		488
	ACRYLIC ANHYDRIDE	0.4		0.9		35	I13	891
	ACRYLONITRILE	0.05	0.02	1.0	0.1	-50	G	204
		0.50	0.05	0.43	0.05	-50	H21,I2	204
		1.69		0.33		70	C,J	803
		1.77	0.20	0.32	0.05	60	I6	354
	ANISOLE, P-PROPENYL-/MALEIC ANHYDRIDE	2.68		29.8	2.6	60	I7	724
	ATROPIC ACID, METHYL ESTER	0.028	0.008	0.25		60	R	1259
		0.19	0.04	0.36	0.07	60		573
	1,3-BUTADIENE	0.04				5	K	211

M1	M2	R1	+/-	R2	+/-	TEMP.	REMARKS	REF.
METHACRYLONITRILE								
	CINNAMIC ACID, VINYL ESTER	4.6		0.15	0.06	70	I7	806
	ETHER, METHYL 1-PHENYLVINYL-	0.55	0.4	0		60	R	573
	ETHYLENE, 1,1-DIPHENYL-	0.33		0		60	R	573
	METHACRYLALDEHYDE	0.40	0.04	1.78	0.06	50	I6	845
	METHACRYLIC ACID	0.59	0.08	1.63	0.08	80	C	159
	METHACRYLIC ACID	0.62	0.05	1.64	0.05	65	C	303
	METHACRYLIC ACID, BUTYL ESTER	0.51		0.69		80	C	159
	METHACRYLIC ACID, TERT-BUTYL ESTER	0.37		0.70		80	C	159
	METHACRYLIC ACID, ETHYL ESTER	0.46		0.83		80	C	159
	METHACRYLIC ACID, HEXYL ESTER	0.75		0.56		80	C	159
	METHACRYLIC ACID, ISOBUTYL ESTER	0.73		0.67		80	C	1070
	METHACRYLIC ACID, ISOPROPYL ESTER	0.43		0.92		80	C	159
	METHACRYLIC ACID, OCTADECYL ESTER	0.90		1.13		80	C	159
	METHACRYLIC ACID, OCTYL ESTER	0.75		0.58		80	C	159
	METHACRYLIC ACID, PENTYL ESTER	0.55		0.51		80	C	159
	METHACRYLIC ACID, PROPYL ESTER	0.29		0.79		80	C	159
	METHACRYLIC ANHYDRIDE	0.27		1.6		36,6	I13	891
	METHYL METHACRYLATE	0.05	0.02	5.0	0.5	-30	H21,I2	204
		0.30	0.05	2.0	0.5	-78	H21,I2	204
		0.65	0.06	0.67	0.1	60		560
		0.70		0.74		80	C	159
		0.80		0.68		80	G3	1070
	STILBENE, TRANS-	5.2	1.0	0.67	0.2	-55	R	264
	STYRENE	70		0		60		573
		0.16	0.06	0.30	0.10	60	C	560
		0.25	0.02	0.25	0.02	80	C	260
		0.26	0.05	0.38	0.05	80	C	159
		0.28		0.43		80	C	1070
		0.32		0.39		60	I2	1376
		0.41		0.37		90	I2	1376
		0.42		0.38		120	I2	1376
	STYRENE, O-CHLORO-	12.0	1.0	0.05	0.02	-78/-30	H21,I2	204
	STYRENE, P-CHLORO-	0.86		0.78		80	C	159
	STYRENE, ALPHA-METHYL-	0.73		0.45		80	C	1070
		0.21		0.15		80	C	1070
		0.35	0.02	0.12	0.02	80	C	260
	SUCCINIC ACID, METHYLENE-, DIMETHYL ESTER	0.383	0.04	0.545	0.15	60	I2	1238
	VINYL ACETATE	1.26		0.28			Q	169
	VINYL ACETATE	12	2	0.01	0.01	70	C	260
METHACRYLOPHENONE								
	STYRENE	0.34	0.16	0.23	0.11	75	C,R	663
METHACRYLOYL CHLORIDE								
	ACRYLONITRILE	2.8	0.4	0.35	0.05	50	I7	1375
	ETHYLENE, 1,1-DICHLORO-	2.0	0.6	0.30	0.06	60	C	405
METHANE, BIS(VINYLOXY)-								
	VINYL ACETATE	1.005	0.105	1.012	0.107	60		608

COPOLYMERIZATION PARAMETER

M1	M2	R1	+/-	R2	+/-	TEMP.	REMARKS	REF.
METHANE, BIS(P-VINYLPHENYL)-								
	STYRENE	0.93		1.01		61	C	1042
METHANE, ETHOXY(VINYLTHIO)-								
	STYRENE	0.26	0.05	6.4	0.05	60		1260
METHANE, (METHYLTHIO)(VINYLTHIO)-								
	STYRENE	0.23	8	4.25		60		1260
METHANOL, DIPHENYL(P-VINYLPHENYL)-								
	STYRENE	1.60	0.15	0.47	0.10	50	I7	129
METHYL METHACRYLATE								
	ACENAPHTHYLENE	0.44		2.25		60	I7	1326
	ACETANILIDE, N-VINYL-	8.90		0.005		60	Q	666
	ACETANILIDE, 4'-VINYL-	0.30	0.05	0.50	0.05	65		317
	ACETIC ACID, ALLYL ESTER	23		0		60		620
	ACETIC ACID, CHLORO-, ALLYL ESTER	98	8			60		855
		50		0		60		372
	ACETIC ACID, METHACRYLAMINO-, ETHYL ESTER	50		0		75		178
	ACETIC ACID, METHACRYLAMINO-, ETHYL ESTER	1.09		0.90		70	C	772
	ACETIC ACID, METHACRYLIMINODI-, DIMETHYL ESTER	16.67		0.013		70	C	772
	ACETIC ACID, PHENYL-, VINYL ESTER	26.4	2.5	0.03	0.01	60	C,I7	1135
	ACETOPHENONE, 4'-VINYL-	0.32		1.37		60		1078
	ACETOPHENONE, 4'-VINYL-, OXIME	0.98		0.50		60		1078
	ACROLEIN	1		0.5		60	C,H	1252
		1.2		0.8		50	I10	1251
	ACRYLAMIDE	10.0	0.2	0.2	0.05	20	C	843
		2.53	0.22	0.82	0.07	70	I72	1382
		2.55	0.4	2.45	0.35	70	I10	1382
		2.6	0.21	0.44	0.06	70	I12	1382
	ACRYLAMIDE, N,N-DIBUTYL-	1.85	0.05	0.42	0.04	50	I7	117
	ACRYLAMIDE, N,N-DIMETHYL-	1.8	0.18	0.45	0.08	50		710
		2.3	0.11	0.51	0.07	70	I10	1383
	ACRYLAMIDE, N,2-DIMETHYL-	2.04	0.24	0.42	0.1	70	I12	1383
		1.54		0.24		70	C	772
	ACRYLAMIDE, N-(1,1-DIMETHYL-3-OXOBUTYL)-	1.54	0.05	0.24	0.02	70	I127	660
	ACRYLAMIDE, N-ETHYL-N-(1,1-DIHYDROPERFLUOROBUTYL)-	1.68	0.06	0.57	0.03	60		179
	ACRYLAMIDE, N-ETHYL-2-METHYL-	0.77		0.89		66		513
	ACRYLAMIDE, N-HYDROXYMETHYL-2-METHYL-	1.75		0.11		70	C	772
	ACRYLAMIDE, N-METHOXYMETHYL-2-METHYL-, BUTYRATE	1.65	0.13	0.50	0.07	70	C,I7	51
	ACRYLAMIDE, N-METHOXYMETHYL-2-METHYL-	1.66	0.02	0.27	0.03	70	I127	660
	ACRYLAMIDE, N-OCTADECYL-	3.85	0.17	0.44	0.04	60		1239
	ACRYLAMIDE, N-OCTYL-	3.50	0.06	0.24	0.06	50	I7	117
	ACRYLAMIDE, N-TERT-OCTYL-	3.6	0.04	0.24	0.04	50	I7	117
	ACRYLANILIDE, 4'-CHLORO-2-METHYL-	0.61		0.24		70	C	772
	ACRYLANILIDE, 2,4'-DIMETHYL-	0.39		0.67		70	C	772
	ACRYLANILIDE, 4'-METHOXY-2-METHYL-	0.14		0.57		70	C	772
	ACRYLIC ACID	1.25		0.225		70		84
		1.5		0.25		50	I7	486

M1	M2	R1	+/-	R2	+/-	TEMP.	REMARKS	REF.
METHYL METHACRYLATE	ACRYLIC ACID	1.5		0.31	0.04	60		1378
		1.86	0.06	0.24		50		486
		2.02	0.04	0.11	0.05	60	LL	1378
		2.17	0.04	0.33	0.04	50	I63	1247
		2.3	0.04	0.31	0.03	50	I29	1247
		2.32		0.30	0.02	50	I10,L	486
		2.46		0.25		50	I5,L	486
		2.64		0.12		50	I17	1247
		2.89	0.05	0.24	0.03	50	I6,L	486
		9.40	1.0			30	D	855
	ACRYLIC ACID, BENZYL ESTER	1.7	0.17	0.34	0.03	60		1157
	ACRYLIC ACID, BUTYL ESTER	1.74		0.20		60		95
		1.8	0.1	0.37	0.1	60		305
	ACRYLIC ACID, 2-BUTYNYL ESTER	0.26	0.05	2.78	0.45	-40	G2,I5	196
		1.83		0.47		60		95
	ACRYLIC ACID, ETHYL ESTER	2.0	0.08	0.28	0.02	60	Q	590
		2.03	0.12	0.24	0.12	60		305
		2.03	0.12	0.24	0.12	60		305
	ACRYLIC ACID, 3-HYDROXY-2,2-BIS(HYDROXYMETHYL)PROPYL ESTER, * TRINITRATE	1.71	0.1	0.20	0.5	-30	G3	95
	ACRYLIC ACID, ISOBUTYL ESTER	0.1	0.005	4.5	0.01	65		551
	ACRYLIC ACID, METHYL ESTER	0.3	0.05	1.5	0.11	48	D,MM	187
		0.96	0.05	0.96	0.1	50	MM	1458
		0.99		0.93		20	H2O,I7	1458
		1		1		60	C,I7	551
		1.69	0.4	0.34		65		96
		1.8		0.35	0.1	130	A	305
		1.91		0.504		50		863
		2.23	0.13	0.36	0.1	48	D	1458
		2.3	0.15	0.45	0.15	130		1458
		2.3	0.04	0.5	0.1	130	A	532
	ACRYLIC ACID, METHYL-D ESTER	2.00	0.10	0.43	0.02	60	C	78
	ACRYLIC ACID, METHYL-14C ESTER	1.60		0.22		70	C	945
	ACRYLIC ACID, 2-NITROBUTYL ESTER	2.23	0.01	0.3	0.03	60	I7	1239
	ACRYLIC ACID, 9-OCTADECEN-1-YL ESTER	2.36	0.04	0.48	0.12	60	I7	1239
	ACRYLIC ACID, OCTADECYL ESTER	1.61	0.1	0.29	0.1	60		305
	ACRYLIC ACID, PROPYL ESTER	1.12		0.05		60		1306
	ACRYLIC ACID, 2-PROPYNYL ESTER	0.23		2.08		60		96
	ACRYLIC ACID, 2-BROMO-, METHYL ESTER	0.15		2.00		60	P,Q	900
	ACRYLIC ACID, 2-CHLORO-, 2-ETHOXYETHYL ESTER	0.15		2.00		60	P,Q	916
	ACRYLIC ACID, 2-CHLORO-, METHYL ESTER	0.3	0.04	1.2	0.2	70	C,I7	97
						60	C,I7	96
	ACRYLIC ACID, 2-CYANO-, METHYL ESTER	0.04	0.04	0.25	0.2	70	I7	23
						60	I7	507
	ACRYLIC ACID, 2-FLUOROMETHYL-, ETHYL ESTER	0.10	0.04	0.13	0.05	60	I2	747
	ACRYLIC ACID, 2-HYDROXY-, ETHYL ESTER, ACETATE	1.05	0.07	0.513	0.046	60		779
	ACRYLONITRILE	1.65		0.65	0.05	60	C	993
		0.02	0.01	6.5	0.5	-78	H21,I2	204
		0.03	0.01	4.0	0.5	-78	H22,I2	204
		0.03	0.01	3.0	0.1	-30	H21,I2	204
		0.05	0.01	5.0	0.5	-78	H21,I2	204
				7.0	0.5	-78	E,I8	636

COPOLYMERIZATION PARAMETER

M1	M2	R1	+/-	R2	+/-	TEMP.	REMARKS	REF.
METHYL METHACRYLATE	ACRYLONITRILE	0.05	0.02	1.3	0.1	-40	H22,I2	204
		0.06	0.02	2.0	0.2	20	H21,I2	204
		0.07	0.02	4.5	0.5	-50	H21,I2	204
		0.10	0.03	1.5	0.1	-15	H21,I2	204
		0.13		0.01		-78	E	918
		0.14		5		-12	G2	1084
		0.14		5		-12	H42	1084
		0.25	0.25	7.9	1.6	-30	G3	551
		0.30		10		-80	H42,I5	1084
		0.32		0.9		-60	G	65
		0.39		7		-8	G17	1084
		0.40		7		-12	H41	1084
		0.85		0.09		26	H40	1084
		1		1		-55	G3,P	265
		1.01		0.09		20	H	551
		1.05	0.07	0.15	0.05	26	I16,Q	609
		1.09	0.21	0.15	0.02	70	H14	1084
		1.14	0.02	0.46	0.08	100	C,I62	886
		1.186	0.22	0.160	0.05	100	C,I17	448
		1.20	0.120	0.03	0.100	26	H38	62
		1.224	0.14	0.150	0.07	60	C	456
		1.25	0.100	0.15	0.080	80	H36	1084
		1.30		0.14		-20/-52	C	558
		1.34		0.10		-12	C	456
		1.34		0.12		26	Q	871
		1.34		0.12		26	E	918
		1.34		0.12		70	H38,I24	1084
		1.34		0.12		90	H38	1084
		1.351	0.133	0.100	0.070	40	H39	1084
		1.45		0.15		-80	H38	1084
		1.46		0.37		70		147
		1.46		0.37		90		1083
		1.65	0.40	0.15		20	C	456
		2.01	0.03	0.45	0.03	70	H38,I2	1084
		2.01		0.43		90	M4	147
	ACRYLONITRILE, 2-CHLORO-	0.48		1.51		55	M4	302
	ACRYLOYL CHLORIDE	91	5			45		894
	ALLYL ALCOHOL	0.205		0.11		60	F,I11	855
	ALLYL ESTERS	2.0		0.01		60	M3	854
	ANILINE, N,N-DIMETHYL-P-VINYL-	0.07	0.02	0.75	0.02	60	M3	1024
	ANILINE, N,N-DIVINYL-	0.29		0.32		60		162
	ANILINE, N-VINYLOXY-	0.29	0.03	0.32	0.05	60		874
	ANISOLE, P-VINYL-	3.81		0.071		60		115
	ANTHRACENE 9-VINYL-	0.31	0.03	0.03	0.05	78	P,Q	1024
	ATROPIC ACID, ETHYL ESTER	0.21		0	0.02	65	C	478
	ATROPIC ACID, METHYL ESTER	0.3	0.1	0	0.03	60		175,571
	1-AZIRIDINECARBOXAMIDE, N-VINYL-	1.70	0.10	0.08	0.04	60	I5	418

M1	M2	R1	+/-	R2	+/-	TEMP.	REMARKS	REF
METHYL METHACRYLATE	BENZENE, M-DIVINYL-	0.41		0.61		70	C	1423
	BENZENE, P-DIVINYL-	0.62		1.3		70	C	1423
	BENZENE, ETHYNYL-	1.5		0.20		60		475
	BENZOIC ACID, VINYL ESTER	20.3	2.5	0.07	0.01	60	C,I7	1135
	BENZOIC ACID, P-VINYL-	1.20		0.40		60		1078
	BENZONITRILE, P-VINYL-	0.22	0.02	1.41	0.13	60		1024
	BENZOTHIAZOLE, 2-(ALLYLTHIO)-	3.15		0		60		1381
	BORAZINE, 2,4,6-TRIALLYL-1,3,5-TRIPHENYL-	8.6		0		80		768
	BORAZINE, 1,3,5-TRIPHENYL-2,4,6-TRIVINYL-	0.70		0		80		768
	1,3-BUTADIENE	0.06	0.03	0.53	0.05	5	K	1025
		0.25	0.03	0.75	0.05	90		560
		0.32		0.70				588
	1,3-BUTADIENE, 2-CHLORO-	0.080	0.007	6.12	0.2	60	Q	214
	1,3-BUTADIENE, 2,3-DICHLORO-	0.18	0.06	3.9	0.25	60	C	242
	1,3-BUTADIENE, 2-FLUORO-	0.64	0.015	10.3	1.5	60	K	214
	1,3-BUTADIENOLS, ACETATE, MIXED ISOMERS	0.35	0.08	1.54	0.08	50	K	733
	1-BUTANOL, 2,2,3,4,4-HEPTAFLUORO-, ACRYLATE	1.4		1.2		70	I2	1275
	1-BUTENE, 3,4-EPOXY-2-METHYL-	56		0.25		50	C,K	838
	3-BUTEN-2-OL, 1-CHLORO-3-METHYL-	90		0		60-140	C	630
	BUTYRIC ACID, VINYL ESTER	25	2.5	0		60-140	C	630
	BUTYRIC ACID, 2-METHYLENE-, METHYL ESTER	1.16		0.03	0.01	60	C,I7	1135
		2.03		0.96		30	G18	98
				0.1		30,60		98
	CARBAMIC ACID, ISOPROPENYL-, 2,3-EPOXYPROPYL ESTER	1.70		0.12		60	I5	418
	CARBAMIC ACID, VINYL-, 2,3-EPOXYPROPYL ESTER	2.75		0.10		60	I5	418
	CARBAZOLE, 9-VINYL-	2.0		0.04		75	I7	331
	CARBONIC ACID, CYCLIC (HYDROXYMETHYL)ETHYLENE ESTER, ACRYLATE	2.0	0.3	0.20	0.03	70	C	29
	CARBONIC ACID, CYCLIC (HYDROXYMETHYL)ETHYLENE ESTER, MALEATE	2.7	0.1	0.07	0.01	30	I5	1129
	CARBONIC ACID, CYCLIC (HYDROXYMETHYL)ETHYLENE ESTER, 4-(METHYLENESUCCINATE)	2.7	0.1	0.07	0.01	30	I5	1129
	CARBONIC ACID, CYCLIC VINYLENE ESTER	7.5	0.5	0		50	C,I2	199
		7.5	0.5	0		50	C,I2	199
	CARBONIC ACID, THIO-, O-TERT-BUTYL S-VINYL ESTER	67		0.01		20	I11	581
		70		0.005		70		349
	3-CARENE	1.4	0.05	0.17	0.05	70		1176
	CINNAMIC ACID	3.58		0.22		70	Q	813
	CINNAMIC ACID, ALPHA-CYANO-	3.8		0		70		84
	CITRACONIC ANHYDRIDE	17.0		0		70		84
	CROTONIC ACID	31.8	2.5	-0.06	0.08	30	D	709
	1,5,9-CYCLODODECATRIENE, TRANS-,TRANS-,CIS-	4.3		0.2		70		84
	4-CYCLOPENTENE-1,3-DIONE	4.84		0.2		80	Q,I2	1422
		4.84				65	I7	1415
	P-DIOXIN, 2,3-DIHYDRO-	7.4		0.083			I7	324
	1,3-DIOXOLANE, 2-VINYL-	15.3	0.6	0.02	0.02	80	I7	433
	DODECANAMIDE, N-(ACRYLAMIDOMETHYL)-	28.9		0.03		60		1351
	ETHENESULFONIC ACID	0.61	0.04	1.39	0.03	70	C	462
	ETHENESULFONIC ACID, 2,3-EPOXYPROPYL ESTER	1		0.01		70	I5	315
	ETHER, ALLYL PHENYL	14.4	2.4	0.06	0.13	60		415
	ETHER, BUTYL VINYL	66.		0.2		60		714
		1.6	0.2		0.1			493
	ETHER, DODECYL VINYL	1.0		0		60	Q	11
	ETHER, ETHYL VINYL	37		0.01		60	Q	372

COPOLYMERIZATION PARAMETER

M1	M2	R1	+/-	R2	+/-	TEMP.	REMARKS	REF.
METHYL METHACRYLATE	ETHER, METHYL 1-PHENYLVINYL	2.5	0.3	0		60		572
	ETHER, OCTADECYL VINYL	1		0		50		11
	ETHER, OCTYL VINYL	1				50		11
	ETHER, PHENYL VINYL	3.5		0.01			C,Q	376
	ETHERS, C8-C18 ALKYL VINYL	140		0.13		60	C,P	714
	ETHYLAMINE, N,N-DIMETHYL-2-(VINYLOXY)-	17		0.2		50	P,Q	12
	ETHYLENE	17.1		0.33		150	M7,R	170
		20	2	0.05	0.015	60	A	142
		25	2	0.05	0.01	28	D	1098
		27.4		0.0024		0	F,K	111
						0	C	111
	ETHYLENE, BROMO-	10		0.1		68	N,D	8
	ETHYLENE, CHLORO-	12.5		0.02		60	C	620
		15		0.005		45	D,OO	964
	ETHYLENE, CHLOROTRIFLUORO-	75.0	0.18	0.05	0.06	60	F,K	955
	ETHYLENE, 1,1-DICHLORO-	1.32		0.09	0.05	20	D	1458
		2.1	0.1	0.4	0.05	40	C	1053
		2.5	0.3	0.34		20	C	1458
		2.50	0.01	0.50	0.03	60		405
		2.53	0.2	0.24	0.15	25		386
		2.54		0.5	0.2	60	C	558
	ETHYLENE, 1,1-DIPHENYL-	16.0	3	0.2		50	MM	1458
	ETHYLENE, TETRACHLORO-	2.1		0.06		50	136	386
	ETHYLENE, TRICHLORO-	240.	20	0.09		25	F,K	240
						36	C	824
	ETHYLENE GLYCOL, BIS(2-CHLOROACRYLATE)	100		0		60	N,D	620
	FERROCENE, VINYL-	1.22	0.37	0.52	0.27	60	P,Q	900
	FERROCENEETHANOL, METHACRYLATE	0.646	0.014	0.203	0.035	70	17	1268
	FERROCENEMETHANOL, ACRYLATE	2.9	0.5	0.08	0.03	65	17	1358
	FERROCENEMETHANOL, METHACRYLATE	3.27	0.06	0.12	0.02	70	17	1267
	FORMIC ACID, VINYL ESTER	28.6	2.5	0.05	0.01	70	17	1267
	FUMARIC ACID, 1,3-BUTYLENEGLYCOL POLYESTER	2.1	0.3	0.5	0.5	60	C,17	1135
	FUMARIC ACID, DIETHYL ESTER	2.10	0.4	0.05	0.1	60	C,I10	908
	FUMARIC ACID, ETHYLENEGLYCOL POLYESTER	40.4	0.2	0.03	0.02	60	C	153
	FUMARONITRILE	1.05	0.25	0.50	0.20	60		92
		17.5	7.5	0.35	0.35	60	C	151
	D-GALACTOPYRANOSE, 1,2:3,4-DI-O-ISOPROPYLIDENE-, 6-METHACRYLATE *	3.5	0.5	0.01	0.01	79	128	300
		3.5	0.5	0.01	0.01	79	128	780
	GERMANE, DIALLYL-							780
	GERMANE, TRIMETHYLVINYL-	19.98		0.05		50	P,Q	517
	D-GLUCITOL, 1-ACRYLAMIDO-1-DEOXY-	4.22		0.206		50	17	1300
	D-GLUCITOL, 1-DEOXY-1-METHACRYLAMIDO-	4.20		0.036		50	C	1029
	GLUTARONITRILE, 2-METHYLENE-	1.25		0.10		50	C	1029
		1.27		0.12		25	F,K	789
	1,3,5-HEXATRIENE, 2,3,4,5-TETRACHLORO-, TRANS-	0.09	0.105	0.315	0.14	70	F,K,DD	542
	5-HEXEN-3-YN-2-OL, 2-METHYL-	0.465	0.05	1.82	0.075	70	C	18
	HYDRAZINIUM HYDROXIDE, 1-(2,3-DIHYDROXYPROPYL)-2-METHACRYLOYL-1,1-DIMETHYL-, INNER SALT *					70	C,P	607
						70	C,P	607

REMARKS PAGE II-366, REFERENCES PAGE II-368

M1: METHYL METHACRYLATE

M2	R1	+/-	R2	+/-	TEMP.	REMARKS	REF.
HYDRAZINIUM HYDROXIDE, 1-(2-HYDROXYPROPYL)-2-METHACRYLOYL * -1,1-DIMETHYL-, INNER SALT	0.34		0.41		70	C,P	607
HYDROQUINONE, VINYL-, DIBENZOATE	3.48		0.003		78	17	465
IMIDAZOLE, 2-METHYL-1-VINYL-	3.50		0.006			Q	667
IMIDAZOLE, 2-PHENYL-1-VINYL-	0.42	0.1	1.9	0.5		Q	667
INDOLE, 1-VINYL-	3.3		0.01		60		873
ISOCYANIC ACID, VINYL ESTER	3.3		0.01		60	Q	1027
	5.57	0.2	0.16	0.02	60		371
ISOPRENE	0.4	0.33	0.78	0.08	80	17	420
ISOTHIOCYANIC ACID, VINYL ESTER	0.85	0.1	0.6	0.13	60		1333
MALEIC ACID, DIETHYL ESTER	20	0.03	0	0.02	60	N,O	620
	354	57	-0.10	0.11	60		92
MALEIC ANHYDRIDE	0.50		0.01		60	121,L	975
	0.90		0.02		60	132,L	975
	1	0.4	0.5	0.3	30	D	1407
	3.10		0.01		60	I2,L	975
	3.40		0.01		60	I5,L	975
	3.5		0.03		60		109
	3.85		0.02		60	I26	975
	3.85		0.01		60	I28,L	975
	4.63	1.14	-0.18	0.28	30		708
	6.7	0.2	0.02		75	C	1033
MALEIC ANHYDRIDE/ 2-PYRROLIDINONE, 1-VINYL-	0.0017	0.0005	555	240	30	D	1407
MALEIMIDE	2.5		0.17		60	I13	1432
MALEIMIDE, N-2-BORNYL-	2.50	0.03	0.17		75	C,I5	1013
MALEIMIDE, N-BUTYL-	2.02	0.02	0.16	0.02	50	C	1433
MALEIMIDE, N,2-DIMETHYL-	1.33		0.12		50	I5	180
MALEIMIDE, N-(2-HYDROXYETHYL)-	3.24		0.15		70	I13	553
MALEIMIDE, N-(HYDROXYMETHYL)-	1.5		0.26		60	I5	1432
MALEIMIDE, N-(P-HYDROXYPHENYL)-	1.66		0.27		60		1432
MALEIMIDE, N-PHENYL-	1.35		0.35	0.05	60	I2	1432
MALONIC ACID, METHYLENE-, DIMETHYL ESTER	0.98		0.3		60	Q	1432
	0.43		0.35		60		1235
MALONONITRILE, METHYLENE-	0.046		0.031		50		254
	0.046		0.031		60		289
MELAMINE, N(2),N(4)-DIALLYL-	27.8	3.2	0		50		812
	28.9	2.3	0		60		812
METANILYL FLUORIDE, N-ACRYLOYL-	1.47		0.60		60	Q	337
METANILYL FLUORIDE, N-METHACRYLOYL-	1.24		0.71		60	Q	337
METHACRYLAMIDE	1.5	0.02	0.47	0.04	65	C	187
	1.55	0.22	1.27	0.19	70	I10	1384
	1.65	0.05	0.49	0.02	70	I27	660
	1.65	0.05	0.49	0.02	70	C	187
	1.68	0.07	0.43	0.04	70	I12	1384
METHACRYLANILIDE	0.54		0.46		70	C	772
METHACRYLIC ACID	0.12	0.007	1.33	0.027	55	I10	1153
	0.3		1.6		60		1378
	0.35	0.01	1.63	0.12	55		1153
	0.55	0.02	1.55	0.06	45		820
	0.55	0.02	1.55	0.06	45		1377
	0.6	0.07	0.92	0.1	70	C,I62	886
	0.98	0.04	0.68	0.05	45	I6	820
	0.98	0.04	0.68	0.05	45	C,I6	1377

COPOLYMERIZATION PARAMETER

M1: METHYL METHACRYLATE

M2	R1	+/-	R2	+/-	TEMP.	REMARKS	REF.
METHACRYLIC ACID	1.18		0.63	0.1	60	LL	1378
	1.66	0.15	0.43	0.02	30	D	855
	2.61	0.83	0.69		55	I11	1153
	0.87	0.025	0.95		65		.818
METHACRYLIC ACID, ALLYL ESTER	1.02		1.04	0.03	60		221
METHACRYLIC ACID, 2-(1-AZIRIDINYL)ETHYL ESTER	0.44	0.07	1.84	0.01	0	G16,12	1229
METHACRYLIC ACID, BENZYL ESTER	0.7	0.05	1.38	0.09	0	G17,12	1229
	0.78	0.10	1.14	0.10	60		1067
	0.85	0.10	1.1		60		408
	0.89		1.17		30	G19,15	1229
	0.90		1.05		60	G18,17	1229
	0.93		2.87				100
METHACRYLIC ACID, BUTYL ESTER	0.95	0.06	1.98			H21,12	100
	0.97	0.07	2.04		0	G16,15	1229
	1.02	0.1	2.71		0	G17,15	1229
	1.16	0.06	1.27		0	H21,15	1229
METHACRYLIC ACID, 2-BUTYNYL ESTER	0.79	0.04	0.35	0.06	-40	G2,15	95
METHACRYLIC ACID, 2-CHLOROETHYL ESTER	0.34	0.07	0.48	0.2	60	C	196
METHACRYLIC ACID, CINNAMYL ESTER, TRANS-	0.55		1.13		60		64
METHACRYLIC ACID, CYCLOHEXYL ESTER	0.88	0.13	1.1	0.15	-78	G17,12	743
METHACRYLIC ACID, 2-(DIETHYLAMINO)ETHYL ESTER	0.8		1.15		60		1222
METHACRYLIC ACID, 2,3-EPOXYPROPYL ESTER	0.86	0.1	0.8	0.1	30	G19,I10	100
	1.4		0.49		30	G18,I7	1151
	1.96		0.94			P,Q	100
METHACRYLIC ACID, ETHYL ESTER	0.75	0.07	0.88	0.10	80	C,12/10	943
	0.76		1.05		60		905
	0.80		1.08		60		412
	0.92		0.98		60		293
	0.98		1.08		30		100
	1.09	0.1	1.1	0.1	60	G18,I7	100
	0.92		1.19		60		305
METHACRYLIC ACID, ETHYL-1-14C ESTER	1		1.09		30		99
METHACRYLIC ACID, 2-(ETHYLTHIO)ETHYL ESTER	0.75	0.111	1.1	0.037	60	17,0	1211
METHACRYLIC ACID, FURFURYL ESTER	0.91		1.20		50		95
METHACRYLIC ACID, ISOBUTYL ESTER	0.89	0.21	1.06	0.20	60		95
METHACRYLIC ACID, ISOPROPYL ESTER	0.86	0.11	1.13	0.11	60	I7	1068
METHACRYLIC ACID, 2-METHOXYETHYL ESTER	0.93	0.1	1.33	0.1	60	I7	1068
METHACRYLIC ACID, PENTAMETHYLDISILOXANYL METHYL ESTER	1.09	0.23	1.67	0.50	50		632
METHACRYLIC ACID, PHENETHYL ESTER	0.53		1.72		60		1067
METHACRYLIC ACID, PHENYL ESTER	0.56		1.69		30		100
METHACRYLIC ACID, 2-PROPYNYL ESTER	0.58	0.05	0.79	0.15	60	G18,I7	100
	0.6	0.02	1.4	0.07	60		100
METHACRYLIC ACID, 2-PYRIDYL ESTER	0.22	0.1	1.7	0.06	60	C,I7	1306
METHACRYLIC ANHYDRIDE	0.67	0.2	0.65	1.0	60		1101
METHACRYLONITRILE	0.67		5.2		-55	G3	1145
	0.68		0.80		80	C	560
	0.74		0.70		80	C	264
	2.0	0.5	0.30	0.05	-78	H21,I2	1070
	5.0	0.5	0.05	0.02	-30	H21,I2	159
NAPHTHALENE, 4-CHLORO-1-VINYL-	0.7	0.2	0.7	0.2	60	C	204
NAPHTHALENE, 6-CHLORO-2-VINYL-	0.45	0.05	1.6	0.2	60	C	783

REMARKS PAGE II-366, REFERENCES PAGE II-368

M1: METHYL METHACRYLATE

M2	R1	+/-	R2	+/-	TEMP.	REMARKS	REF.
NAPHTHALENE, 2-ISOPROPENYL-	0.45	0.03	0.00		99	A	213
NAPHTHALENE, 2-VINYL-	0.4	0.05	1.0	0.15	60	C	783
2,5-NORBORNADIENE	10.0		0		50		1082
	10.0		0.0		50		1081
2,6-OCTADIENE, 2,6-DIMETHYL-	2.35		0.30		60	Q	813
3-OXA-1-AZASPIRO(4.5)DECAN-4-ONE, 2-ISOPROPENYL-	0.05		3.22		60		427
OXALIC ACID, ETHYL VINYL ESTER	6.0		0.1		60		479
4H-1,3-OXAZINE, 5,6-DIHYDRO-4,6,6-TRIMETHYL-2-VINYL-	1.81		0.19		60	I7	1418
2-OXAZOLIDINONE, 3-VINYL-	6.00		0.03		75		336
2,4-PENTADIENOIC ACID, 5,5-DICHLORO-4-HYDROXY-, GAMMA-LACTONE *	9.6	0.2	0.035	0.015	50		116
	9.6	0.2	0.035	0.015	50		116
4-PENTENOIC ACID, 2-ALLYL-2-CYANO-, ETHYL ESTER	5.4		0.0		60	Q	163
PHENETHYL ALCOHOL, ALPHA-METHYL-P-VINYL-	0.26	0.06	1.00	0.02	60	I5	42
PHENOL, M-VINYL-	0.43	0.08	0.4	0.06	60	I5	1245
PHENOL, O-VINYL-	0.33	0.06	0.21	0.04	60	I5	476
PHENOL, P-VINYL-	0.34	0.06	0.25		60	I5	1245
PHOSPHINE, DIPHENYLVINYL-	5.1	0.2	0		60	C	791
PHOSPHINE, DIPHENYL(P-VINYLPHENYL)-	0.32	0.02	0.91	0.12	60	I7	792
PHOSPHINE OXIDE, DIISOBUTYLVINYL-	30	10	0		60	C	791
PHOSPHINE OXIDE, DIPHENYLVINYL-	11	4	0		60	C	791
PHOSPHINE OXIDE, DIPHENYL(P-VINYLPHENYL)-	0.38	0.02	1.46	0.10	60	I7	792
PHOSPHINE SULFIDE, DIPHENYLVINYL-	13	3	0		60	C	791
PHOSPHINE SULFIDE, DIPHENYL(P-VINYLPHENYL)-	0.29	0.01	1.22	0.34	60	I7	792
PHOSPHINOTHIOIC ACID, DIPHENYL-, O-2-(METHACRYLOXY)ETHYL * ESTER	0.29	0.01	1.22	0.34	60	I7	792
PHOSPHONIC ACID, 1-HYDROXYETHYL-, DIETHYL ESTER, ACRYLATE	2.02	0.05	0.15	0.03	65		760
PHOSPHONIC ACID, HYDROXYMETHYL-, DIETHYL ESTER, ACRYLATE	1.88	0.03	0.27	0.01	75		760
PHOSPHONIC ACID, HYDROXYMETHYL-, DIETHYL ESTER, METHACRYLATE	1.43	0.05	0.50	0.03	68		760
PHOSPHONIC ACID, PHENYL-, DIALLYL ESTER	22.96		0.135		70		338
PHOSPHONIC ACID, ALPHA-PHENYLVINYL-	3.30	0.2	0.06	0.04	80		521
PHOSPHONIC ACID, VINYL-, BIS(2-CHLOROETHYL) ESTER	3.55	0.5	0.00		80	149,J	522
PHOSPHONIC DIFLUORIDE, (2-METHYL-1,3-BUTADIENYL)-	29.9		1.3		30	E	2271
PHOSPHORIC ACID, DIALLYL PHENYL ESTER	0.2	0.01	0.26	0.1	80	I2	339
PHTHALIC ACID, DIALLYL ESTER	9.78		0.225		70	C	612
	17.5	1.5	0.113	0.002	80	C	612
	20.8	0.6	0.074	0.008	70	C	612
	25.0	1.0	0.070	0.007	70	C	940
2-PICOLINE, 5-VINYL-	0.46	0.02	0.61	0.08	60	I12	1196
2-PICOLINIUM CHLORIDE, 5-VINYL-	0.46		0.94		70		878
2-PICOLINIUM METHYL SULFATE, 1-METHYL-5-VINYL-	0.12		1.8		50	Q	813
2-PINENE	6.60		0.07		70	C,I2	904
PROPANE, 1-(ALLYLOXY)-2,3-EPOXY-	40.7	5.0	0.035		70	C	457
1-PROPENE, 3-CHLORO-	48.1		0.048	0.038	60	N,O	620
	50		0		60	C	824
	56		0		60	N,O	620
1-PROPENE, 3-CHLORO-2-METHYL-	7.7		0		60	N,O	484
1-PROPENE, 2,3-DICHLORO-	0.5		0		40		29
1-PROPEN-2-OL, ACETATE	5.5	0.8	0.017	0.003	70	Q	334
1-PROPEN-2-OL, ISOCYANATE	30		0.017		75		419
2-PROPEN-1-OL, 2-CHLORO-	3.10	0.29	0.14	0.10	100	N,O	484
2-PROPEN-1-OL, 2-CHLORO-, ACETATE	1.0		0		50	N,O	484

M1 / M2	R1	+/-	R2	+/-	TEMP.	REMARKS	REF.
METHYL METHACRYLATE							
2-PROPEN-1-OL, 2-METHYL-, ACETATE	10		0		60	N,O	620
PROPIONIC ACID, VINYL ESTER	24	2.5	0.03	0.01	60	C,I7	1135
PYRIDINE, 5-ETHYL-2-VINYL-	0.11	0.003	0.69	0.03	60		939
PYRIDINE, 5-ETHYL-2-VINYL-, 1-OXIDE	0.12	0.01	5.5	0.5	60		942
PYRIDINE, 2-VINYL-	0.12	0.02	4.7	0.6	60	I12	942
	0.20	0.03	0.84	0.13	25	E	149
	0.33	0.05	0.70	0.10	70		23
	0.395	0.025	0.86	0.06	60		1023
	0.439	0.002	0.77	0.02	60		939
PYRIDINE, 2-VINYL-, 1-OXIDE	1.25		0.17		60	O	871
PYRIDINE, 4-VINYL-	0.13	0.03	3.9	0.8	60	I12	942
PYRIDINIUM FLUOROBORATE, 1-VINYL-	0.574	0.004	0.79	0.05	60		939
2-PYRROLIDINETHIONE, 1-VINYL-	4.75		0.008		60	I31	223
	0.44	0.06	1.72	0.09	60		888
2-PYRROLIDINONE, 1-VINYL-	1.85	0.04	0.32	0.02	70	I7	1171
	1.6	0.8	0.6	0.2	70	L3	1406
	4.6	0.4	0.02	0.02	70		1408
	4.6	0.4	0.02	0.02	70		1408
	4.7	0.5	0.005	0.05	50		116
	5.35	0.6	0	0.5	30		1407
SILANE, DIALLYL-	30.5	5	0.04	0.01	80	D,Q	517
SILANE, DIMETHYLPHENYLVINYL-	0.50	0.05	0.05	0.05	80	P,Q	685
SILANE, DIMETHYLPHENYL(P-VINYLPHENYL)-	0.76	0.02	0.18	0.02	80	C	307
SILANE, DIMETHYLVINYL(P-VINYLPHENYL)-	34	4	0.02	0.01	80	C	307
SILANE, TRIMETHYLVINYL-	0.005		20		80		685
STYRENE	0.005		20		-78	E,I18	1
	0.056	0.003	0.25	0.03	-78	E,I24	859
	0.1	0.05	10.5	0.2	50	MM	1458
	0.1	0.05	10.5	0.2	20	H,I7,Q	358
	0.23		0.55		30	H2O	552
	0.32	0.11	0.42		100	G9,I	288
	0.35	0.024	0.35	0.024	60		988
	0.39	0.054	0.44	0.054	60	I70	1230
	0.42	0.1	0.54	0.04	60	I69	1230
	0.422		0.485		60		787
	0.44		0.50		30	D,C	56
	0.44	0.02	0.50	0.02	35	I7	1020
	0.45	0.02	0.52	0.02	35	I7	1021
	0.45		0.47		50		59
	0.45		0.43		50		1458
	0.46	0.01	0.49	0.03	50	H36,Q	277
	0.46	0.016	0.48	0.013	50	I24	59
	0.46		0.52		60	C	1043
	0.46		0.52		60	E,I24	859
	0.46		0.52		60		743
	0.46	0.02	0.57	0.02	60		570
	0.47	0.02	0.45	0.02	60		559
	0.47		0.52		50	I7	1230
	0.48	0.026	0.58	0.026	60	I12	59
	0.48	0.032	0.38	0.032	50	I2	311
	0.49		0.56		60	I7,M4	59
					60	C,I29	506
					35	K	1020

REMARKS PAGE II-366, REFERENCES PAGE II-368

M1	M2	R1	+/-	R2	+/-	TEMP.	REMARKS	REF.
METHYL METHACRYLATE	STYRENE	0.49	0.03	0.53	0.03	60	C	1036
		0.49	0.045	0.54	0.045	99		213
		0.49	0.04	0.48	0.08	60	I68	1230
		0.5		0.44		60		787
		0.50		0.63		25	E,I24	859
		0.50		0.44	0.02	50	I19	1021
		0.50	0.02	0.56	0.02	35	K	558
		0.50	0.02	0.50	0.01	60	C	700
		0.50	0.03	0.56	0.10	40	H38	406
		0.50	0.06	0.54	0.02	60	C	634
		0.50	0.07	0.54	0.09	60	C	149
		0.52		0.63		25	E	59
		0.52		0.53		50	I24,M4	59
		0.52		0.49		50	I12,M4	506
		0.53		0.49		60	C,I38	506
		0.536	0.026	0.51	0.026	60	C,I7	559
		0.54		0.590		131	I10	992
		0.55		0.43		60	C	1043
		0.57		0.60		132	I19,M4	59
		0.57		0.49		50	I7	992
		0.58		0.51		60	I7,M3	59
		0.588	0.007	0.545	0.006	50	A,I53	411
		0.62		0.59		132	I12,M3	59
		0.66		0.63		50	I24,M3	59
		0.8		0.8		250	A	713
		0.80		0.56		50	I19,M3	59
		1		1		50	H,I24,O	358
		1		0		-78	H21,I2	204
		1.		0		-50/-20	H21,I2	204
		2.0	0.3	0.30	0.05	20	H21,I2	204
		4.5	1	0.07	0.02	20	H21,I2	204
		6.4	0.05	0.12	0.05	20	G3	552
		6.4	0.05	0.12	0.05	-30	G3	551
		14.0	2	0.05	0.02	-30	H21,I2	204
		17.1	2	0.31		-30	H47	278
		20.		0.02		30	H21,I2	204
		25.0		0.02		-78	H21,I2	204
	STYRENE, 2,5-BIS(TRIFLUOROMETHYL)-	0.57	0.07	1.35	0.01	60	C	181
	STYRENE, M-BROMO-	0.48	0.02	1.17	0.05	60		1024
	STYRENE, P-BROMO-	0.395	0.02	1.10	0.25	60		1024
	STYRENE, M-CHLORO-	0.47	0.075	0.91	0.11	60		1024
	STYRENE, O-CHLORO-	0.46	0.075	1.34	0.01	40	C,K	1186
	STYRENE, P-CHLORO-	0.50	0.03	1.37	0.1	30/40		1023
		0.4	0.02	0.8	0.4	60		599
		0.415	0.02	0.89	0.05	60		1024
	STYRENE, 2,5-DICHLORO-	0.434	0.056	0.468	0.068	60	D,OO	655
	STYRENE, 2,6-DICHLORO-	0.44	0.16	2.25	0.17	68	C	8
	STYRENE, 2,5-DIMETHOXY-	0.49	0.04	-0.3	0.02	20		1458
		2.05	0.1	0.12	0.03	50		1458
		3.68	0.01	0.09	0.04	50	MM	1458
	STYRENE, P-IODO-	0.25	0.03	0.72	0.20	70		466
		0.36		0.95		60		1024

M1	M2	R1	+/-	R2	+/-	TEMP.	REMARKS	REF.
METHYL METHACRYLATE								
	STYRENE, P-ISOPROPYL-	0.44		0.39		70	C	1424
	STYRENE, ALPHA-METHYL-	0.5	0.03	0.3	0.01	20		1051
		0.50		0.14		50		1023
		0.55		0.51		60		1051
		0.55		0.6		60		1051
		0.65		0.81		80		1051
		0.7		1.0		100	A	1051
	STYRENE, M-METHYL-	0.89		-0.01	0.01	99		213
	STYRENE, P-METHYL-	0.53	0.025	0.49	0.02	60		1024
	STYRENE, M-NITRO-	0.405	0.025	0.44	0.02	60		1024
	STYRENE, 2,3,4,5,6-PENTACHLORO-	0.35	0.05	0.85	0.2	75	C,17	895
	STYRENE, 2,3,4,5,6-PENTAFLUORO-	4.0	0.4	0.35	0.05	70	C	25
	STYRENE, 2,3,4,5-TETRAETHYL-	0.98		0.9		60		1364
	STYRENE, M-TRIFLUOROMETHYL-	0.56	0.06	0.33	0.08	50		364
	STYRENE, 2,4,6-TRIMETHYL-	0.60	0.10	0.98	0.15	60	C	181
		1.4		0.08		85		207
	SUCCINIC ACID, METHYLENE-	1.6	0.3	0.05	0.01	130	I10	247
		1.14		0.05		90	Q	1033
	SUCCINIC ACID, METHYLENE-, BIS(2-CHLOROETHYL) ESTER	1.23		0.10				677
	SUCCINIC ACID, METHYLENE-, DIBUTYL ESTER	2.63	0.2	0.4	0.05	50		169
	SUCCINIC ACID, METHYLENE-, DIMETHYL ESTER	0.8	1.3	0.3			C	169
	SUCCINIC ACID, METHYLENE-, 2,3-EPOXYPROPYL METHYL ESTER	1.2	0.05	0.20	0.02	60	C	54
	SUCCINIMIDE, N-VINYL-	1.73		0.064		60		365
	SULFIDE, P-ANISYL VINYL	9.5		0.04		60	C	976
	SULFIDE, P-BROMOPHENYL VINYL	0.77		0.20		60		976
	SULFIDE, P-CHLOROPHENYL VINYL	0.85		0.08		60		976
	SULFIDE, ETHYL VINYL	0.89		0.07		60		1338
		0.93		0.3				872
	SULFIDE, ISOBUTYL VINYL	2.7	1.5	0.04	0.1	60		1338
	SULFIDE, ISOPROPYL VINYL	0.94		0.04		60		1225
	SULFIDE, METHYL VINYL	0.94		0.07		60	A	1338
	SULFIDE, PHENYL VINYL	0.85		0.03		60		372
	SULFIDE, P-TOLYL VINYL	1.0		0.08		60		976
	SULFONE, METHYL VINYL	0.85		0.07		60		976
	SULFOXIDE, ETHYL VINYL	0.81		0.01		60	D	211
	SULFOXIDE, METHYL VINYL	14	2	0		6	O	1226
	TARTARIC ACID, DIVINYL ESTER	10	10	0		60		781
	TEREPHTHALIC ACID, 2,2-BIS(3-ALLYL-4-HYDROXYPHENYL)PROPANE -CO-PHENOLPHTHALEIN POLYESTER *	20		0.04		60	C	869
		0.97		0.04		60	C	869
	2,4,8,10-TETRAOXASPIRO(5.5)UNDECANE, 3,9-DIISOPROPENYL-	0.97		0.16		80	C	1349
	2,4,8,10-TETRAOXOSPIRO(5.5)UNDECANE, 3,9-DIVINYL-	2.88		0.18		60	C,17	1123
		4.58		0.05		60	C,16	1345
	2H-TETRAZOLE, 2-PHENYL-5-(P-VINYLPHENYL)-	19.6		0.4		60	17	1399
	THIANTHRENE, 2-VINYL-	0.7		0.07		60		363
	TIN, DIALLYL-	2.99		0.03		60		517
	TIN, TRIBUTYLVINYL-	27.9		0.705	0.025	50	P,Q	640
	TIN, TRIETHYL(METHACRYLOYLOXY)-	0.37	0.02	0.03		60	17	1329
	TIN, TRIMETHYLVINYL-	25.1		0.03		50	17	640
	P-TOLUENESULFONAMIDE, N-METHYL-N-VINYL-	4.68		0		60	C	280
	S-TRIAZINE, 2,4-DIMETHYL-6-VINYL-	0.37		1.75		60	C	279

M1	M2	R1	+/-	R2	+/-	TEMP.	REMARKS	REF.
METHYL METHACRYLATE								
	S-TRIAZINE, 2,4,6-TRIS(ALLYLOXY)-	46.3	5.3	0		60		812
		50.2	7.5	0		60		812
	S-TRIAZINE-2,4,6-(1H,3H,5H)-TRIONE, TRIALLYL-	45.0	5.1	0		60		812
		48.9	7.3	0		60		812
	UREA, 1-ETHYL-3-VINYL-	1.8		0.015		70		335
		1.8		0.015		75		335
	VINYL ACETATE	2.75		0.1		80	G31	1321
		2.77	0.02	0.11	0.01		G31,Q	1322
		3.2	1.1	0.4	0.2	-30	G3	551
		20		0.015	0.015	60	C,Q	1322
		20	3	0.015		60		622
		22.21	0.89	0.072	0.026	30	C	638
		26	2.5	0.03	0.01	60	C,17	1135
		28.6		0.035		30		76
		127	10			60		855
		181	10			60	F	162
	VINYL ETHER	10.0	0.05	0.006		30		855
	VINYL SULFIDE	0.85		0.13	0.05	60		850
	VINYL SULFONE	8.5		0.045		60		162
METHYL METHACRYLATE-METHYL-D								
	ACRYLIC ACID, METHYL ESTER	2.3	0.5	0.50	0.08	130	V	863
METHYL METHACRYLATE-METHYL-T								
	METHACRYLIC ACID, BENZYL-14C ESTER	0.93		1.05		60		99
	METHACRYLIC ACID, NONYL ESTER	1.10		0.86		60		99
	METHACRYLIC ACID, PHENYL-14C ESTER	0.53		1.67		30		99
		0.56		1.30		110		99
		0.56		1.72		60		99
METHYL METHACRYLATE-METHYL-14C								
	ACRYLIC ACID, METHYL ESTER	2.00	0.10	0.43	0.02	60	C	78
	METHACRYLIC ACID, CYCLOHEXYL ESTER	0.86		1.15		60		99
	STYRENE	0.46	0.02	0.52	0.03	60	C	78
MYRISTIC ACID, VINYL ESTER								
	ACRYLONITRILE	0.04	0.03	4.0	0.3	60	C,16	58
NAPHTHALENE, 1-(3-BUTENYL)-								
	ACRYLONITRILE	0		3.6		70		67
NAPHTHALENE, 4-CHLORO-1-VINYL-								
	METHYL METHACRYLATE	0.7	0.2	0.7	0.2	60	C	783
	STYRENE	0.8	0.1	0.85	0.1	60	C	783

COPOLYMERIZATION PARAMETER

M1 / M2	R1	+/-	R2	+/-	TEMP.	REMARKS	REF.
NAPHTHALENE, 6-CHLORO-2-VINYL-							
METHYL METHACRYLATE	1.6	0.2	0.45	0.05	60	C	783
STYRENE	1.5	0.2	0.4	0.1	60	C	783
NAPHTHALENE, 1-(DIAZOMETHYL)-							
TOLUENE, ALPHA-DIAZO-	2	0.05	0.22	0.03	40	I74	1396
NAPHTHALENE, 2-(DIAZOMETHYL)-							
TOLUENE, ALPHA-DIAZO-	3.55	0.3	0.2	0.1	40	I74	1396
NAPHTHALENE, 1,2-DIHYDRO-							
MALEIC ANHYDRIDE	0		0.026		60	I26	664
	0		0.028		60	I7,L	664
STYRENE	0.4	0.2	1.0	0.3	30	H10,I24	653
NAPHTHALENE, 2-ISOPROPENYL-							
ACRYLONITRILE	0.23	0.02	0.05	0.01	99	I2	213
MALEIC ANHYDRIDE	0		0.09	0.03	60	A	1205
METHYL METHACRYLATE	0.00		0.45		99		213
NAPHTHALENE, 1-VINYL-							
STYRENE	0.35	0.2	3.3	0.5	40	H12	880
	1.35	0.15	0.67	0.03	60		568
NAPHTHALENE, 2-VINYL-							
MALEIMIDE, N,2-DIMETHYL-	0.187	0.095	0.256		70		553
METHYL METHACRYLATE	1.0	0.02	0.4	0.05	60	C	783
STYRENE	1.4	0.1	0.5	0.1	60	C	783
NONANOIC ACID, VINYL ESTER							
ACRYLONITRILE	0.059	0.095	3.57	0.16	60	C	591
1,3-BUTADIENE	0.02	0.02	26.3	10.0	60	C	591
ETHYLENE, CHLORO-	0.282	0.035	1.16	0.86	60	C	591
ETHYLENE, 1,1-DICHLORO-	0.0	0.01	4.08	0.20	60	C	591
STYRENE	0.01	0.01	49.5	15	60	C	591
2,5-NORBORNADIENE							
ACRYLIC ACID, ETHYL ESTER	0.01		3.05		50		1081
	0.01		3.05		50		1082
ACRYLONITRILE	0.08		0.67		50		1081
	0.08		0.65	0.02	50		1082
	0.47	0.08	0.65		60	C	248
ETHYLENE, CHLORO-	0.35		0.74		50		1082
	0.35		0.74		50		1081
ETHYLENE, 1,1-DICHLORO-	0.08		1.41		50		1081
	0.08		1.41		50		1082

REMARKS PAGE II-366, REFERENCES PAGE II-368

REACTIVITY RATIOS

II-261

REACTIVITY RATIOS

M1	M2	R1	+/-	R2	+/-	TEMP.	REMARKS	REF.
2,5-NORBORNADIENE								
	METHYL METHACRYLATE	0		10.0		50		1082
		0.0		10.0		50		1081
	STYRENE, P-CHLORO-	0.01		85		60		769
	VINYL ACETATE	1.28		0.82		60		769
NORBORNANE, 2-VINYL-								
	ETHYLENE, CHLORO-	0.76	0.01	0.85	0.01		C	387
2-NORBORNANECARBOXYLIC ACID, ENDO-, VINYL ESTER								
	STYRENE	0.01		30		60		385
2-NORBORNANECARBOXYLIC ACID, EXO-, VINYL ESTER								
	STYRENE	0.01		30		60		385
5-NORBORNENE-2-CARBONITRILE								
	MALEIC ANHYDRIDE	0.16		0.012		60		410
5-NORBORNENE-2-CARBOXYLIC ACID, 2,3-EPOXYPROPYL ESTER								
	ACRYLONITRILE	0		1.4		100		416
5-NORBORNENE-2-CARBOXYLIC ACID, METHYL ESTER								
	ACRYLONITRILE	0.2	0.1	1.5	0.5	60		458
	MALEIC ANHYDRIDE	1.1		0		60		410
	VINYL ACETATE	0.45	0.07	1.5	0.24	60		458
5-NORBORNENE-2-CARBOXYLIC ACID, 2-METHYL-, 2,3-EPOXYPROPYL ESTER *								
	ACRYLONITRILE	0		1.5		100		416
5-NORBORNENE-2-CARBOXYLIC ACID, 2-METHYL-, METHYL ESTER								
	MALEIC ANHYDRIDE	1.1		0		60		410
2-NORPINANE-2-ETHANOL, ACRYLATE								
	ACRYLONITRILE	0.9	0.1	1.1	0.1	60	C	600
	1,3-BUTADIENE	0.2		1.1			C,O	601
	ETHYLENE, CHLORO-	4.3	0.2	0.14	0.01	60	C	600
	STYRENE	0.29	0.1	0.66	0.1	50	C	600
10,12-OCTADECADIENOIC ACID, TRANS,TRANS-, METHYL ESTER								
	ACRYLIC ACID, ETHYL ESTER	0.064	0.003	1.9	0.2	60		619
	ACRYLONITRILE	0		0.4		60		618
	STYRENE	0		11		130	C	618
		0		12		130	C	618

M1 / M2	R1	+/-	R2	+/-	TEMP.	REMARKS	REF.
10,12-OCTADECADIENOIC ACID, TRANS,TRANS-, METHYL ESTER							
STYRENE	0		12		60		618
	0		17		70	C	330
	0		15		90	C	618
9,12-OCTADECADIENOIC ACID							
STYRENE	0.29		7.16		150	C	868
OCTADECANOIC ACID, 12-OXO-, VINYL ESTER							
ACRYLONITRILE	0.0	0.02	3.11	0.05	60	C	593
ETHYLENE, CHLORO-	0.248	0.01	0.963	0.01	60	C	593
ETHYLENE, 1,1-DICHLORO-	0	0.16	4.0	0.5	60	C	593
VINYL ACETATE	1.26	0.08	1.07	0.02	60	C	593
OCTADECANOIC ACID, 4-OXO-, VINYL ESTER							
ETHYLENE, CHLORO-	0.320	0.076	0.874	0.044	60	C	593
ETHYLENE, 1,1-DICHLORO-		0.17	3.8	0.6	60	C	593
VINYL ACETATE	1.18	0.22	1.04	0.2	60	C	593
OCTADECANOIC ACID, 9(10)-OXO-, VINYL ESTER							
ACRYLONITRILE	0.0	0.091	2.96	0.12	60	C	593
ETHYLENE, 1,1-DICHLORO-	0.01	0.09	3.7	0.3	60	C	593
VINYL ACETATE	1.49	0.25	1.03	0.05	60	C	593
9,11,13-OCTADECATRIENOIC ACID							
STYRENE	4		0.46			C,I2,Q	814
1-OCTADECENE							
STYRENE	0.75	0.10	1.94	0.05	30	H29,I7	79
	0.87	0.09	1.32	0.14	30	H7,I7	79
2,6-OCTADIENE, 2,6-DIMETHYL-							
METHYL METHACRYLATE	0.30		2.35			Q	813
OCTANOIC ACID, VINYL ESTER							
ACRYLONITRILE	0.04	0.03	4.00	.3	60	C,I6	58
ETHYLENE, CHLORO-	0.2		3.2		86	C,I11,J	503
1-OCTENE							
ACRYLIC ACID, 2-CYANO-, METHYL ESTER	-0.17	0.04	0.63	0.06	60	I7	507
ETHYLENE			3.2		130/220	M1	656
			3.66	0.15	130/220	M5	112

REMARKS PAGE II-366, REFERENCES PAGE II-368

M1	M2	R1	+/-	R2	+/-	TEMP.	REMARKS	REF.
3-OXA-1-AZASPIRO(4.5)DECAN-4-ONE, 2-ISOPROPENYL-								
	METHYL METHACRYLATE	3.22		0.05		60		427
	STYRENE	0.82		0.12		60		427
7-OXABICYCLO(2.2.1)HEPTANE								
	OXETANE, 3-(CHLOROMETHYL)-3-METHYL-	0.125	0.025	1.22	0.05	0	H13	284
7-OXABICYCLO(4.1.0)HEPTANE								
	OXETANE, 3-(CHLOROMETHYL)-3-METHYL-	0.8	0.5	0.5	0.1	80	H11	1118
	OXETANE, 3-(CHLOROMETHYL)-3-METHYL-	0.95	0.15	0.35	0.05	0	H13	284
7-OXABICYCLO(4.1.0)HEPTANE, 3-VINYL-								
	HYDRACRYLIC ACID, BETA-LACTONE	0.1	0.1	3.5		20	H55,12	171
	HYDRACRYLIC ACID, BETA-LACTONE	1.4		0.4		-30	H10,13	171
1,3,4-OXADIAZOLE, 2-ISOPROPENYL-5-METHYL-								
	STYRENE	0.83		0.16		60		1233
1,3,4-OXADIAZOLE, 5-PHENYL-2-(4-VINYLPHENYL)-								
	STYRENE	4.5	0.4	0.17	0.05	70	I10	19
1-OXA-2,5-DISILACYCLOPENTANE, 2,2-DIMETHYL-5,5-DIPHENYL-								
	1-OXA-2,5-DISILACYCLOPENTANE, 2,2,5,5-TETRAMETHYL-	5.66	0.3	0.178	0.012	52	G	633
1-OXA-2,5-DISILACYCLOPENTANE, 2,5-DIMETHYL-2,5-DIPHENYL-								
	1-OXA-2,5-DISILACYCLOPENTANE, 2,2,5-TRIMETHYL-5-PHENYL-	2.23	0.36	0.33	0.04	25	G	633
1-OXA-2,5-DISILACYCLOPENTANE, 2-METHYL-2,5,5-TRIPHENYL-								
	1-OXA-2,5-DISILACYCLOPENTANE, 2,2,5-TRIMETHYL-5-PHENYL-	2.84	0.10	0.172	0.004	25	G	633
1-OXA-2,5-DISILACYCLOPENTANE, 2,2,5,5-TETRAMETHYL-								
	CYCLOTRISILOXANE, HEXAMETHYL-	4.10	0.77	0.31	0.07	100	G	633
	CYCLOTRISILOXANE, HEXAMETHYL-	8.11	0.13	0.146	0.014	40	G	633
	1-OXA-2,5-DISILACYCLOPENTANE, 2,2-DIMETHYL-5,5-DIPHENYL-	0.178	0.012	5.66	0.3	52	G	633
	1-OXA-2,5-DISILACYCLOPENTANE, 2,2,5-TRIMETHYL-5-PHENYL-	0.56	0.015	2.11	0.04	25	G	633
1-OXA-2,5-DISILACYCLOPENTANE, 2,2,5,5-TETRAPHENYL-								
	1-OXA-2,5-DISILACYCLOPENTANE, 2,2,5-TRIMETHYL-5-PHENYL-	11.3	3.2	0.091	0.007	25	G	633
1-OXA-2,5-DISILACYCLOPENTANE, 2,2,5-TRIMETHYL-5-PHENYL-								
	CYCLOTRISILOXANE, HEXAPHENYL-	0.58	0.16	1.47	0.25	25	G	633
	1-OXA-2,5-DISILACYCLOPENTANE, 2,5-DIMETHYL-2,5-DIPHENYL-	0.33	0.04	2.23	0.36	25	G	633
	1-OXA-2,5-DISILACYCLOPENTANE, 2-METHYL-2,5,5-TRIPHENYL-	0.172	0.004	2.84	0.10	25	G	633

COPOLYMERIZATION PARAMETER

M1	M2	R1	+/-	R2	+/-	TEMP.	REMARKS	REF.
1-OXA-2,5-DISILACYCLOPENTANE, 2,2,5-TRIMETHYL-5-PHENYL-								
	1-OXA-2,5-DISILACYCLOPENTANE, 2,2,5,5-TETRAMETHYL-	2.11	0.04	0.56	0.015	25	G	633
	1-OXA-2,5-DISILACYCLOPENTANE, 2,2,5,5-TETRAPHENYL-	0.091	0.007	11.3	3.2	25	G	633
OXALIC ACID, DIALLYL ESTER								
	STYRENE	0.104	0.003	27.5	0.4	60	C	1293
OXALIC ACID, ETHYL VINYL ESTER								
	ACRYLONITRILE	0.2		2.0		60		479
	METHYL METHACRYLATE	0.1		6.0		60		479
	STYRENE	0.1		8.0		60		479
	VINYL ACETATE	3.0		0.3		60		479
4H-1,3-OXAZINE, 5,6-DIHYDRO-4,6,6-TRIMETHYL-2-VINYL-								
	METHYL METHACRYLATE	0.19		1.81		60	I7	1418
3-OXAZIN-5-ONE, 2-(DICHLOROMETHYLENE)-4-ISOBUTYL-								
	STYRENE	0		0.1		60		1232
3-OXAZIN-5-ONE, 2-(DICHLOROMETHYLENE)-4-ISOPROPYL-								
	STYRENE	0		0.1		60		1232
3-OXAZIN-5-ONE, 2-(DICHLOROMETHYLENE)-4-METHYL-								
	STYRENE	0		0.1		60		1232
OXAZOLE, 5-(4-BIPHENYLYL)-2-(4-VINYLPHENYL)-								
	STYRENE	4.6	0.5	0.46	0.07	70	I10	19
OXAZOLE, 4,5-DIMETHYL-2-VINYL-								
	STYRENE	1.3		0.18		60		1233
		1.30	0.1	0.18	0.05	70		424
OXAZOLE, 4-ISOBUTYL-2-ISOPROPENYL-5-METHYL-								
	STYRENE	3.4		0.1		60		1233
OXAZOLE, 4-ISOBUTYL-5-METHYL-2-VINYL-								
	STYRENE	3		0.1		60		1233
OXAZOLE, 2-ISOPROPENYL-4,5-DIMETHYL-								
	STYRENE	2.2		0.15		60		1233
		2.20		0.15		60		426
		2.20	0.1	0.15	0.05	70		424

REMARKS PAGE II-366, REFERENCES PAGE II-368

M1 / M2	R1	+/-	R2	+/-	TEMP.	REMARKS	REF.
OXAZOLE, 5-(1-NAPHTHYL)-2-(4-VINYLPHENYL)-							
STYRENE	3.8	0.4	0.24	0.05	70	I10	19
OXAZOLE, 5-(2-NAPHTHYL)-2-(4-VINYLPHENYL)-							
STYRENE	4.5	0.5	0.36	0.05	70	I10	19
OXAZOLE, 5-PHENYL-2-(4-VINYLPHENYL)-							
STYRENE	3.5	0.4	0.49	0.08	70	I10	19
	4.5	0.4	0.68	0.04	70	I2	107
2,5-OXAZOLIDINEDIONE, 4-ISOBUTYL-							
4-OXAZOLIDINEPROPIONIC ACID, 2,5-DIOXO-, BENZYL ESTER	0.61	0.16	1.57	0.08	25	G7,I6	712
2,5-OXAZOLIDINEDIONE, 4-ISOPROPYL-							
HYDRACRYLIC ACID, BETA-LACTONE	11		-0.16		35	G1	208
4-OXAZOLIDINEPROPIONIC ACID, 2,5-DIOXO-, BENZYL ESTER							
2,5-OXAZOLIDINEDIONE, 4-ISOBUTYL-	1.57	0.08	0.61	0.16	25	G7,I6	712
2-OXAZOLIDINONE, 5-DECYL-3-VINYL-							
STYRENE	0.059		14.3		60		1174
2-OXAZOLIDINONE, 3-METHACRYLAMIDO-							
STYRENE	0.38		1.15		70	I6	1185
2-OXAZOLIDINONE, 3-METHACRYLOYL-							
STYRENE	0.27		1.1		70	I6	1184
2-OXAZOLIDINONE, 3-VINYL-							
ETHYLENE, CHLORO-	0.35	0.02	0.84	0.02	50		116
ETHYLENE, 1,1-DICHLORO-	0.08		1.35		75		336
METHACRYLIC ACID, DECYL ESTER	0.015	0.05	12.8	0.5	50		116
METHYL METHACRYLATE	0.03		6.00		75		336
	0.035	0.015	9.6	0.2	50		116
STYRENE	0.05	0.05	30	0.5	50		116
VINYL ACETATE	1.50		0.60		75		336
	1.90	0.10	0.52	0.08	50		116
2-OXAZOLIN-5-ONE, 4-ETHYL-2-ISOPROPENYL-4-METHYL-							
STYRENE	0.70		0.27		60		427
2-OXAZOLIN-5-ONE, 4-ISOBUTYL-2-ISOPROPENYL-4-METHYL-							
STYRENE	0.69		0.24		60		427

COPOLYMERIZATION PARAMETER

M1	M2	R1	+/-	R2	+/-	TEMP.	REMARKS	REF.
2-OXAZOLIN-5-ONE, 2-ISOPROPENYL-4,4-DIMETHYL-	STYRENE	1.00	0.09	0.31	0.06	60		427
2-OXAZOLIN-5-ONE, 2-ISOPROPENYL-4-ISOPROPYL-	STYRENE	1.12	0.10	0.31	0.03	60		426
3-OXAZOLIN-5-ONE, 4-ISOBUTYL-2-ISOPROPYLIDENE-	STYRENE	0.0		0.39	0.10	60		426
3-OXAZOLIN-5-ONE, 2-ISOPROPYLIDENE-4-METHYL-	STYRENE	0.0		0.36	0.07	60		426
3-OXAZOLIN-5-ONE, 4-ISOPROPYL-2-ISOPROPYLIDENE-	STYRENE	0.00	0.10	0.39	0.06	60		426
OXETANE, 3-(ALLYLOXYMETHYL)-3-(CHLOROMETHYL)-	FURAN, TETRAHYDRO-	1.40	0.01	0.60	0.01	0	H10	707
	VINYL ACETATE	0.22		1.36		65	C	832
OXETANE, 3-(ALLYLOXYMETHYL)-3-METHYL-	OXETANE, 3-(CHLOROMETHYL)-3-ETHYL-	10.6	3.5	0.12	0.03	0	H13	1202
OXETANE, 3,3-BIS(BROMOMETHYL)-	OXETANE, 3,3-BIS(CHLOROMETHYL)-	1.0		1.0		-78	E,I1	348
		1.0		1.0		-78	E,I1	1026
		1.0		1.0		0	E,I1	350
		1.0		1.0		30	H10,I3	350
		0.4		16.0		30	H10,I3	348
		0.4		16.0		30	H10,I3	350
	OXETANE, 3-(BROMOMETHYL)-3-ETHYL-	1.0		1.0		-78	E,I1	348
		1.0		1.0		-78	E,I1	348
		1.0		1.0		0	E,I1	1026
								350
OXETANE, 3,3-BIS(CHLOROMETHYL)-	ACETALDEHYDE	1.16	0.46	0.10	0.09	-5	H10	728
	BENZENE, (EPOXYETHYL)-	0.8		0.8		0	H10,I3	47
	BENZENE, (1,3-EPOXYPROPYL)-	0.05		2.7		0	H10,I3	49
	BENZENE, (2,3-EPOXYPROPYL)-	0.45		1.1		0	H10,I3	49
	BUTANE, 1,2-EPOXY-	0.3		2.0		0	H10,I3	49
	BUTANE, 1,3-EPOXY-	0.04		3.4		0	H10,I3	49
	BUTYRIC ACID, 4-HYDROXY-, GAMMA-LACTONE	1.4		0.04		0	H10,I20	979
		4.5				0	H10	980
	M-DIOXANE	13		0		0	H10,I3	50
	M-DIOXANE, 4-(CHLOROMETHYL)-	13		0		0	H10,I3	50
	M-DIOXANE, 4-METHYL-	13		0		0	H10,I3	50

REMARKS PAGE II-366, REFERENCES PAGE II-368

M1	M2	R1	+/-	R2	+/-	TEMP.	REMARKS	REF.
OXETANE, 3,3-BIS(CHLOROMETHYL)-								
	M-DIOXANE, 4-PHENYL-	12.5		0		0	H10	977
		46				0	H10,I3	50
	P-DIOXANE	1.85				0	H10	977
		7				0	H10,I3	50
	1,3-DIOXOLANE	1.5		0.65	0.05	0	H10	437
		1.5	0.1	0.65		0	H10	718
		2.5				0	H10	977
	1,3-DIOXOLANE, 4-METHYL-	0.5		0.2		0	H10,I3	47
	ETHYLENE OXIDE	1.6		0.07		0	H10,I3	1130
	FURAN, 2,5-DIHYDRO-	0.00		1.50	0.20	0	H13,I3	823
	FURAN, TETRAHYDRO-	0.01	0.01	1.80	0.01	0	H13,I3	823
		0.1		1.4		30	H14	771
		0.3		0.5		120	H14	771
		0.45		0.40	0.05	20	H13,I18	20
		0.80		1.0		25	H10,I2	771
	FURAN, TETRAHYDRO-2-METHYL-	0.82	0.05	1.00	0.05	0	H10	822
		1.3				0	H10	977
	FURAN, TETRAHYDRO-2-PHENOXYMETHYL-	2.70	0.25	0.05	0.05	0	H10,I4	399
	FURAN, TETRAHYDRO-2-PHENYL-	3.8		0		0	H10,I3	49
		2.6		0		0	H10,I3	49
	HEXANOIC ACID, 6-HYDROXY-, EPSILON-LACTONE	0.21				0	H10	980
	HYDRACRYLIC ACID, BETA-LACTONE	0.24				0	H10,I20	979
		3.5		0.44		0	H10,EE	927
		7		0.65		0	H13,I9	926
		8		0.5		20	H10,EE	927
		10		0.04			H20	1440
		10		0.18		20	H10,EE	927
		16	3	0.03		-50	H10	1440
		16.7	0.5	0.05			H10	926
		25		0.15		-50	H10	978
		30	10	0.04	0.04		H10	980
		30	10	0.1	0.1		H10	926
		38	3	0.06	0.05		H13,I3	926
		70		1			H13,I3	926
	HYDRACRYLIC ACID, 2-METHYL-, BETA-LACTONE	11.9	0.2	0.13	0.02	0	H10,EE	978
	OXETANE, 3,3-BIS(BROMOMETHYL)-	1.0		1.0		-78	E,I1	1026
		1.0		1.0		-78	E,I1	348
		1.0		1.0		0	E,I1	350
		1.0		1.0		30	H10,I3	348
	OXETANE, 3,3-BIS(FLUOROMETHYL)-	0.45		2.0		30	H10,I3	350
		1.0		1.0		20	H10,I3	348
	OXETANE, 3-(CHLOROMETHYL)-3-ETHYL-	0.05		6.5		-78	E,I1	348
		0.07	0.04	8.0	0.2	30	H10,I3	350
		0.07	0.04	8.0	0.2	26	E,L2	931
		0.07	0.04	8.0	0.2	30	H10,L2	931
		0.25		4.0		50	C,L2	931
		0.7		1.6		20	I3	348
	OXETANE, 3-(CHLOROMETHYL)-3-METHYL-	1.0		1.0		30	H10,I3	348
		1.0		1.0		-78	E,I1	1026
		1.0		1.0		-78	E,I1	348
	OXETANE, 3-(CHLOROMETHYL)-3-METHYL-	0.02	0.02	3.6	0.1	0	H13	283

COPOLYMERIZATION PARAMETER

M1	M2	R1	+/-	R2	+/-	TEMP.	REMARKS	REF.
OXETANE, 3,3-BIS(CHLOROMETHYL)-	OXETANE, 3-(CHLOROMETHYL)-3-METHYL-	0.09	0.05	10.85	0.05	50	H14,I2	20
	2-OXETANONE, 4,4-DIMETHYL-	3.2	0.1	0.15	0.01	0	H10,I2	1442
	PROPANE, 1-CHLORO-2,3-EPOXY-	2.5	0.5	0.4	0.1	0	H10,I3	47
	PROPANE, 1,2-EPOXY-3-METHOXY-	0.7		0.1		0	H10,I3	49
	PROPANE, 1,2-EPOXY-3-PHENOXY-	1.1		0.25		0	H10,I3	49
	PROPYLENE OXIDE	0.3		0.65		0	H10,I3	47
	PYRAN, TETRAHYDRO-	1.515		0		0	H10	977
	PYRAN, TETRAHYDRO-2-METHYL-	7.2		0		0	H10,I3	49
	PYRAN, TETRAHYDRO-2-PHENYL-	1				0	H10,I3	49
	TRIMETHYLENE OXIDE	0.01		10.8		0	H10,I3	49
	VALERIC ACID, 5-HYDROXY-, DELTA LACTONE	0.20		0		30	H10,I20	981
OXETANE, 3,3-BIS(ETHOXYMETHYL)-	OXETANE, 3-(CHLOROMETHYL)-3-METHYL-	19.0	1.5	0.21	0.05	0	H13	284
OXETANE, 3,3-BIS(FLUOROMETHYL)-	OXETANE, 3,3-BIS(CHLOROMETHYL)-	1.0		1.0		-78	E,I1	348
		2.0		0.45		20	H10,I3	348
OXETANE, 3,3-BIS(METHOXYMETHYL)-	OXETANE, 3-(CHLOROMETHYL)-3-METHYL-	8.5	0.5	0.14	0.04	0	H13	284
		9	1.5	0.125	0.125	80	H11	1118
OXETANE, 3-(BROMOMETHYL)-3-ETHYL-	OXETANE, 3,3-BIS(BROMOMETHYL)-	1.0		1.0		-78	E,I1	348
		1.0		1.0		-78	E,I1	1026
		1.0		1.0		0	E,I1	350
		16		0.4		30	H10,I3	350
		16.0		0.4		30	H10,I3	348
	OXETANE, 3-(CHLOROMETHYL)-3-ETHYL-	1.0		1.0		-78	E,I1	1026
		1.0		1.0		-78	E,I1	348
		1.6		0.7		0	H10,I3	350
		6.5		0.05		30	H10,I3	350
						30	H10,I3	348
OXETANE, 3-(CHLOROMETHYL)-3-ETHYL-	FURAN, TETRAHYDRO-	1.43		0.72		0	H13	1202
		1.49		0.72		0	H13	1128
		4.3	0.3	0.22	0.05	0	H13	283
	OXETANE, 3-(ALLYLOXYMETHYL)-3-METHYL-	0.12	0.03	10.6	3.5	0	H13	1202
	OXETANE, 3,3-BIS(CHLOROMETHYL)-	1.0		1.0		-78	E,I1	1026
		1.0		1.0		-78	E,I1	348
		1.0		1.0		0	H10,I3	350
		1.6		0.7		30	I3	348
		4.0		0.25		20	H10,I3	348
		6.5		0.05		30	H10,I3	350
		8.0	0.2	0.07	0.04	26	E,L2	931
		8.0	0.2	0.07	0.04	30	H10,L2	931

REMARKS PAGE II-366, REFERENCES PAGE II-368

M1	M2	R1	+/-	R2	+/-	TEMP.	REMARKS	REF.
OXETANE, 3-(CHLOROMETHYL)-3-ETHYL-								
	OXETANE, 3,3-BIS(CHLOROMETHYL)-	8.0	0.2	0.07	0.04	50	G,L2	931
	OXETANE, 3-(BROMOMETHYL)-3-ETHYL-	0.05		6.5		30	H10,I3	348
		0.7		1.6		30	H10,I3	350
		1.0		1.0		-78	E,I1	348
	OXETANE, 3-ETHYL-3-(FLUOROMETHYL)-	1.0		1.0		-78	E,I1	1026
		1.0		1.0		0	E,I1	350
						0-80	E,I1	647
	PROPYLENE OXIDE	8.4	1.3	0.27	0.12	0	H13	283
	TRIMETHYLENE OXIDE	0.64	0.22	3.0	0.4	0	H13	283
OXETANE, 3-(CHLOROMETHYL)-3-METHYL-								
	BUTANE, 1,3-EPOXY-	0.25	0.17	3.75	0.65	0	H13	284
	BUTANE, 2,3-EPOXY-2-METHYL-	1.77	0.2	0.75	0.1	80	H11	1118
		2.2	0.2	0.4	0.14	-50	H13	283
	ETHYLENE OXIDE	3.6	0.1	0.31	0.04	80	H10	1118
		2.55	0.5	0.19	0.01	0	H11	283
	FURAN, TETRAHYDRO-	36.0	9	0.2	0.12	-50	H13	283
		67	10	0.027	0.09	50	H10	20
		1.74	0.1	0.17	0.1	-50	H14,I2	283
	7-OXABICYCLO(2.2.1)HEPTANE	6.2	0.2	0.26	0.04	50	H10	284
	7-OXABICYCLO(4.1.0)HEPTANE	1.22	0.05	0.125	0.025	0	H13	284
		0.35	0.05	0.95	0.15	80	H11	1118
	OXETANE, 3,3-BIS(CHLOROMETHYL)-	0.5	0.1	0.8	0.5	80	H13	283
		3.6	0.1	0.02	0.02	0	H13	283
	OXETANE, 3,3-BIS(ETHOXYMETHYL)-	10.85	0.05	0.09	0.05	50	H14,I2	20
		0.21	0.05	19.0	1.5	0	H13	284
	OXETANE, 3,3-BIS(METHOXYMETHYL)-	0.125	0.125	9	1.5	80	H11	1118
	OXETANE, 3-(METHOXYMETHYL)-3-METHYL-	0.14	0.04	8.5	0.8	0	H13	284
	PROPANE, 1-CHLORO-2,3-EPOXY-	0.075	0.025	5.4	0.8	-50	H13	283
	PROPYLENE OXIDE	35	5	0.015	0.005	0	H13	283
		39.6		0.008		80	H11	1118
		2.1	0.5	0.185	0.025	50	H14,I2	20
		4.46	0.1	0.33	0.1	50	H13	283
	PYRAN, TETRAHYDRO-	8.3	0.75	0.2	0.05	-50	H10	283
	TRIMETHYLENE OXIDE	13.2	1.4	0.24	0.06	0	H13	284
		8.5	0.5	0.02	0.001	0	H13	284
		0.64	0.22	3.05	0.40	0	H13	284
OXETANE, 3-ETHYL-3-(FLUOROMETHYL)-								
	OXETANE, 3-(CHLOROMETHYL)-3-ETHYL-	1.0	0.8	1.0		0-80	E,I1	647
OXETANE, 3-(METHOXYMETHYL)-3-METHYL-								
	OXETANE, 3-(CHLOROMETHYL)-3-METHYL-	5.4	0.8	0.075	0.025	0	H13	284
2-OXETANONE, 4,4-DIMETHYL-								
	1,3-DIOXOLANE	0.94	0.05	1.5	0.1	0	H20,I2	1442
	OXETANE, 3,3-BIS(CHLOROMETHYL)-	0.15	0.01	3.2	0.1	0	H10,I2	1442

COPOLYMERIZATION PARAMETER

M1 / M2	R1	+/-	R2	+/-	TEMP.	REMARKS	REF.
2-OXETANONE, 4-METHYLENE-							
STYRENE	0.008		100		70		1190
PALMITIC ACID, VINYL ESTER							
ACRYLONITRILE	0.04	0.03	4.0	0.3	60	C,16	58
VINYL ACETATE	0.66	0.07	0.84	0.10	70	C	778
	0.78	1.0	1.15	0.13	70	C	778
1,3-PENTADIENE							
ISOPRENE			17	5	50	G	797
1,3-PENTADIENE, CIS-							
1,3-PENTADIENE, TRANS-	1.15		0.44			C,M11,Q	446
1,3-PENTADIENE, TRANS-							
1,3-PENTADIENE, CIS-	0.44		1.15			C,M11,Q	446
1,4-PENTADIENE							
ACRYLONITRILE	0.01		1.116		50		88
1,4-PENTADIENE, 3,3-DIMETHYL-							
ACRYLONITRILE	0	3	3.31		50	CC	155
2,4-PENTADIENENITRILE							
ACRYLONITRILE	40		0.10	0.03	50	16	840
1,3-BUTADIENE	1.70		0		50	K	1022
2,4-PENTADIENOIC ACID, 2,3-EPOXYPROPYL ESTER							
STYRENE	4.99	0.14	0.13	0.02	60	17	1231
2,4-PENTADIENOIC ACID, METHYL ESTER							
ACRYLIC ACID, METHYL ESTER	2.0	0.2	0.50	0.07	60		272
1,3-BUTADIENE	0.05		5.8		50	H76,17	101
STYRENE	0.80	0.12	0.12	0.08	60		272
2,4-PENTADIENOIC ACID, 5,5-DICHLORO-4-HYDROXY-, GAMMA-LACTO * NE							
METHYL METHACRYLATE	1.43	0.45	0.16	0.15	65	17	1047
2,4-PENTANEDIONE, 3-ETHYLIDENE-							
ACRYLONITRILE	0		15		60	17	1339
STYRENE	0		9		60	17	1339

REACTIVITY RATIOS

II-271

M1	M2	R1	+/-	R2	+/-	TEMP.	REMARKS	REF.
3-PENTANONE, 1-HYDROXY-5-(VINYLOXY)-, ACETATE	ETHANOL, 2-(VINYLOXY)-	0.83	0.2	0.93	0.05	60		1299
PENTASILOXANE, 1,1,3,3,5,5,7,9,9-UNDECAMETHYL-9-(P-VINYL* PHENYL)-	STYRENE	1.2	0.1	1.11	0.01	80		306
PENTASILOXANE, 1-(P-VINYLPHENYL)-	STYRENE	1.2	0.1	1.11	0.01	80		371
1-PENTENE	ETHYLENE	0.0145		33.2		-10	H45	1388
	ETHYLENE	0.0145		33.2		-10	H45	853
	ETHYLENE	0.015		41.0		-20	H61	852
	ETHYLENE	0.018		48.2		-20	H5	852
	ETHYLENE, CHLORO-	0.16	0.02	0.5		68	C	8
	1-PROPENE, 3-CHLORO-2-METHYL-					150	E	281
	STYRENE	1.51		0.448		50	H12	827
1-PENTENE, 5-CYCLOHEXYL-	ACRYLONITRILE	0		4.5		70		67
1-PENTENE, 4,4-DIMETHYL-2-PHENYL-	STYRENE			2		110	A,O	1189
1-PENTENE, 3,3,4,4,5,5,5-HEPTAFLUORO-	ETHYLENE, TETRAFLUORO-	2.3		0.21		22	E,M14	1159
	ETHYLENE, TETRAFLUORO-	2.3		0.21		22	E,M13	1160
1-PENTENE, 2-METHYL-	ACRYLONITRILE	0		1.29		60	C,J	497
1-PENTENE, 3-METHYL-	1-PENTENE, 4-METHYL-	0.1	0.1	6.2	0.2	80	H12	398
1-PENTENE, 4-METHYL-	1-HEXENE	0.62		5.4		50	H54	75
	1-PENTENE, 3-METHYL-	6.2	0.2	0.1	0.1	80	H12	398
	STYRENE	1.15	0.12	0.49	0.06	30	H7,I7	79
	STYRENE	1.23	0.15	0.55	0.10	30	H29,I7	79
	STYRENE	3.67	0.22	0.89	0.05	45	H12,I2	44
	STYRENE	3.92	0.15	0.98	0.1	70	H12	1253

M1 / M2	R1	+/-	R2	+/-	TEMP.	REMARKS	REF.
1-PENTENE, 4-METHYL-2,4-DIPHENYL- / STYRENE			1.9		110	A,O	1189
1-PENTENE, 3-METHYL-1-PHENYL-, CIS- / ACRYLONITRILE	-0.34	0.43	22.6	1.0	60		1065
1-PENTENE, 3-METHYL-1-PHENYL-, TRANS- / ACRYLONITRILE	-0.47	0.33	6.15	0.50	60		1065
1-PENTENE, 4-METHYL-2-PHENYL- / STYRENE			1.9		110	A,O	1189
1-PENTENE, 1-PHENYL- / STYRENE, P-CHLORO-					0	H2O133,V	761
1-PENTENE, 1-PHENYL-, CIS- / ACRYLONITRILE	-0.05	0.51	3.04	0.24	60		1065
1-PENTENE, 1-PHENYL-, TRANS- / ACRYLONITRILE	0.002	0.22	1.04	0.03	60		1065
1-PENTENE, 5-PHENYL- / ACRYLONITRILE	0		3.5		70		67
1-PENTENE, 2,4,4-TRIMETHYL- / ACRYLONITRILE	0		1.22		60	C,J	497
1-PENTENE, 2,4,4-TRIPHENYL- / STYRENE			1.6		110	A,O	1189
2-PENTENOIC ACID, CIS- / ACRYLAMIDE	7.1	0.5	0		20	F,173	1297
2-PENTENOIC ACID, TRANS- / ACRYLAMIDE	11	0.6	0		20	F,173	1297
2-PENTENOIC ACID, 4-HYDROXY-, GAMMA-LACTONE / ACRYLONITRILE	0		113	14		O	421
3-PENTENOIC ACID, 4-HYDROXY-, GAMMA-LACTONE / ACRYLONITRILE	0		14.5	0.5		O	421

M1	M2	R1	+/-	R2	+/-	TEMP.	REMARKS	REF.
4-PENTENOIC ACID, 2-ALLYL-2-CYANO-, ETHYL ESTER								
	MALEIC ANHYDRIDE	0.04		1.4			C	163
	METHYL METHACRYLATE	0.		5.4			C	163
	STYRENE	0.		16.1			C	163
	VINYL ACETATE	0.30		0.75			C	163
1-PENTEN-3-ONE								
	STYRENE	0.12	0.02	0.7	0.09	60		1114
3-PENTEN-2-ONE								
	STYRENE	0.01		13.7		60	C,17	641
	STYRENE	0.01		13.7		60	C,17	928
PEROXYFUMARIC ACID, DI-TERT-BUTYL ESTER								
	STYRENE	-0.02	0.02	0.67	0.02	40	C	580
PHENANTHRENE, 2-VINYL-								
	ACRYLIC ACID, METHYL ESTER	2.0	0.2	0.1	0.03	60	C	783
PHENANTHRENE, 3-VINYL-								
	ACRYLIC ACID, METHYL ESTER	1.75	0.25	0.8	0.05	60	C	783
PHENANTHRENE, 9-VINYL-								
	STYRENE	2.36		0.58		60	C	477
1-PHENANTHRENEMETHANOL, 1,2,3,4,4B,5,6,10,10A-DECAHYDRO- * 7-ISOPROPYL-1,4A-DIMETHYL-, ACRYLATE								
	ETHYLENE, CHLORO-	4.2	0.2	0.13	0.01	60	C	600
1-PHENANTHRENEMETHANOL, 1,2,3,4,4B,5,6,10,10A-DECAHYDRO- * 7-ISOPROPYL-1,4A-DIMETHYL-								
	ETHYLENE, 1,1-DICHLORO-	0		2.6		70		217
PHENETHYL ALCOHOL, ALPHA-METHYL-P-VINYL-								
	ACRYLONITRILE	0.53		0.10		60		42
	METHYL METHACRYLATE	1.00		0.26		60		42
	STYRENE	0.91		0.97		60		42
	STYRENE, P-CHLORO-	0.63		0.91		60		42
PHENETOLE, P-STYRYL-								
	STYRENE	0.038	0.15	5.33	0.64	70	16	1152

COPOLYMERIZATION PARAMETER

M1 / M2	R1	+/-	R2	+/-	TEMP.	REMARKS	REF.
PHENOL, M-VINYL-							
METHYL METHACRYLATE	0.4	0.02	0.43	0.06	60	I5	1245
STYRENE	1.21		0.91		60	I5	1245
	1.21		0.9		60		115
PHENOL, O-VINYL-							
1,3-CYCLOPENTADIENE	0.36	0.23	1.79	0.27	-25	H10,13	1224
	0.78	0.08	2.99	0.21	-25	H10,12	1224
METHYL METHACRYLATE	0.21	0.06	0.33	0.08	60	I5	476
STYRENE	1.32	0.25	0.72	0.10	60	I5	476
PHENOL, P-VINYL-							
METHYL METHACRYLATE	0.25	0.04	0.34	0.06	60	I5	1245
STYRENE	1.2	0.13	0.79	0.08	60	I5	1245
PHENOL, P-VINYL-, ACETATE							
STYRENE	1.35	0.20	0.85	0.15	60		201
PHOSPHINE, DIPHENYLVINYL-							
METHYL METHACRYLATE	0		5.1	0.2	60	C	791
STYRENE	0		7	1	60	C	791
PHOSPHINE, DIPHENYL(P-VINYLPHENYL)-							
METHYL METHACRYLATE	0.91	0.12	0.32	0.02	60	I7	792
STYRENE	1.11	0.12	0.46	0.07	65	I2	126
	1.43	0.25	0.52	0.05	60	I7	792
PHOSPHINE OXIDE, DIALLYLPHENYL-							
METHACRYLIC ACID, DODECYL ESTER	-0.011	0.013	18.1	0.4	80	R	103
	-0.013	0.025	11.1	0.4	110	R	103
	-0.089	0.022	5.59	0.24	140	R	103
PHOSPHINE OXIDE, DIISOBUTYLVINYL-							
METHYL METHACRYLATE	0		30	10	60	C	791
STYRENE	0		17	5	60	C	791
PHOSPHINE OXIDE, DIPHENYLVINYL-							
METHYL METHACRYLATE	0		11	4	60	C	791
STYRENE	0		5	1	60	C	791
PHOSPHINE OXIDE, DIPHENYL(P-VINYLPHENYL)-							
METHYL METHACRYLATE	1.46	0.10	0.38	0.02	60	I7	792
STYRENE	1.40	0.15	0.42	0.02	60	I7	792

REMARKS PAGE II-366, REFERENCES PAGE II-368

M1 / M2	R1	+/-	R2	+/-	TEMP.	REMARKS	REF.
PHOSPHINE SULFIDE, DIPHENYLVINYL-							
METHYL METHACRYLATE	0		13	3	60	C	791
STYRENE	0		2.1	0.3	60	C	791
PHOSPHINE SULFIDE, DIPHENYL(P-VINYLPHENYL)-							
METHYL METHACRYLATE	1.22	0.34	0.29	0.01	60	17	792
STYRENE	1.49	0.33	0.43	0.05	60	17	792
PHOSPHINIC ACID, DIETHYL-, ALLYL ESTER							
FUMARIC ACID, 1,3-BUTYLENEGLYCOL POLYESTER	0.075	0.075	10.00	2.00		Q	604
FUMARIC ACID, ETHYLENEGLYCOLPHOSPHINATE POLYESTER	0.09	0.05	2.07	1.12		Q	604
PHOSPHINOTHIOIC ACID, DIPHENYL-, O-2-(METHACRYLOXY)ETHYL ESTER *							
METHYL METHACRYLATE	0.450		0.875		70		340
STYRENE	0.380		0.467		70		340
PHOSPHONIC ACID, ALLYL-, DIETHYL ESTER							
ACRYLONITRILE	0.12	0.008	4.3	3.00	40		889
FUMARIC ACID, 1,3-BUTYLENEGLYCOL POLYESTER	0.04	0.04	9.25	0.4		Q	605
METHACRYLIC ACID			6.1		60		401
METHACRYLIC ACID, DODECYL ESTER	0.066	0.12	52.5	3.6	80	C,17	102
PHOSPHONIC ACID, 2-BUTENYLIDENEDI-, TETRAETHYL ESTER							
METHACRYLIC ACID	0.0	0.10	2.0	1.0	60		401
PHOSPHONIC ACID, BUTYL-, DIALLYL ESTER							
METHACRYLIC ACID, DODECYL ESTER	0.091	0.04	18.7	0.2	80	C,17	102
PHOSPHONIC ACID, (DIMETHYLENEETHYLENE)DI-, TETRAETHYL ESTER							
1,3-BUTADIENE	0.05	0.05	0.05	0.05	70		1395
ISOPRENE	0.05	0.05	0.12	0.03	70		1395
PHOSPHONIC ACID, 1-HYDROXYETHYL-, DIETHYL ESTER, ACRYLATE							
METHYL METHACRYLATE	0.15	0.03	2.02	0.05	65		760
PHOSPHONIC ACID, HYDROXYMETHYL-, DIETHYL ESTER, ACRYLATE							
ACRYLIC ACID, METHYL ESTER	0.29	0.05	0.88	0.09	75		760
METHYL METHACRYLATE	0.27	0.01	1.88	0.03	75		760
PHOSPHONIC ACID, HYDROXYMETHYL-, DIETHYL ESTER, METHACRYLATE							
METHYL METHACRYLATE	0.50	0.03	1.43	0.05	68		760

COPOLYMERIZATION PARAMETER

M1 / M2	R1	+/-	R2	+/-	TEMP.	REMARKS	REF.
PHOSPHONIC ACID, METHYL-, DIALLYL ESTER							
ETHYLENE			0.29	0.02	80	I7,M5	1412
			0.49	0.04	80	I7,M2	1412
			0.41	0.04	80	I7,M3	1412
PHOSPHONIC ACID, (2-METHYL-1,3-BUTADIENYL)-, DIISOPROPYL ESTER *							
STYRENE	1.28	0.11	0.67	0.03	75		1281
PHOSPHONIC ACID, (2-METHYL-1,3-BUTADIENYL)-, DIMETHYL ESTER							
STYRENE	1.14	0.11	0.4	0.02	75		1281
PHOSPHONIC ACID, PHENYL-, DIALLYL ESTER							
METHACRYLIC ACID, DODECYL ESTER	0.072	0.44	19.5	3.6	80	C,I7	102
METHYL METHACRYLATE	0.135		22.96		70		338
STYRENE	0.027		28.97		70		338
PHOSPHONIC ACID, ALPHA-PHENYLVINYL-							
ACRYLIC ACID	0.44	0.03	0.98	0.08	70	C	524
ACRYLONITRILE	0.32	0.07	0.69	0.18	70	C	524
1,3-BUTADIENE, 2-CHLORO-	0.1		7.5		60	I10	1091
METHACRYLIC ACID	0.00	0.12	4.0	1	80	I49,J	522
METHYL METHACRYLATE	0.36		3.50	0.2	80	C	521
	0.00	0.04	3.55	0.5	80	I49,J	522
	0.06		3.30	0.2	80	C	521
STYRENE	0.2	0.05	1.5	0.05	70	I10	1090
PHOSPHONIC ACID, VINYL-, BIS(2-CHLOROETHYL) ESTER							
METHACRYLIC ACID	0.1	0.075	1.7	0.5		Q	729
METHYL METHACRYLATE	0.26		29.9		30	E	271
STYRENE	0.03		2.43			Q	527
	0.16		2.3		30	E	271
	0.2	0.2	2.2	0.4		Q	729
PHOSPHONIC ACID, VINYL-, DIBUTYL ESTER							
STYRENE	0.3	0.3	5.4	1.5	60		519
PHOSPHONIC ACID, VINYL-, DIETHYL ESTER							
ETHYLENE	0.15	0.15	0.17	0.03	80	I7,M2	1411
METHACRYLIC ACID	0		1.9	0.5	60		401
STYRENE	0		3.25		116		52
	0	0.02	4.1	0.5	60		519
	0.06		8.87	0.14		O	982
PHOSPHONIC ACID, VINYL-, DIISOBUTYL ESTER							
STYRENE	0.5	0.4	4.4	1.0	60		519

REMARKS PAGE II-366, REFERENCES PAGE II-368

M1	M2	R1	+/-	R2	+/-	TEMP.	REMARKS	REF.
PHOSPHONIC ACID, VINYL-, DIISOPROPYL ESTER								
	STYRENE	0		2.39	0.37	60		519
PHOSPHONIC ACID, VINYL-, DIMETHYL ESTER								
	STYRENE	0.40	0.17	4.61	0.43	60		519
PHOSPHONIC ACID, VINYL-, DIPHENYL ESTER								
	ETHYLENE	0		0.055	0.007	140	I7,M3	1411
		0		0.11	0.02	80	I7,M2	1411
		0		0.04	0.004	80	I7,M3	1411
	STYRENE	0		2.03	0.39	60		519
PHOSPHONIC ACID, VINYL-, DIPROPYL ESTER								
	STYRENE	0.90	0.09	4.24	1.04	60		519
PHOSPHONIC DICHLORIDE, (2-METHYL-1,3-BUTADIENYL)-								
	STYRENE	1.35		0.3		75		1283
PHOSPHONIC DICHLORIDE, VINYL-								
	ACRYLONITRILE	0.05		6		25	E	1192
PHOSPHONIC DIFLUORIDE, (2-METHYL-1,3-BUTADIENYL)-								
	METHYL METHACRYLATE	1.3	0.1	0.2	0.01	80	I2	1274
	STYRENE	1.15		0.54		75	I2	1282
PHOSPHONOUS DICHLORIDE, PHENYL-								
	BENZOIC ACID, ALLYL ESTER	0.05		0.003		60		1368
	1,3-BUTADIENE	0.05		0.003		60	I39	793
PHOSPHORIC ACID, DIALLYL PHENYL ESTER								
	METHYL METHACRYLATE	0.225		9.78		70		339
	STYRENE	0.145		16.84		70		339
PHTHALIC ACID, BIS(2-METHYLALLYL) ESTER								
	STYRENE	0.059	0.005	25.4	0.7	60	C	1294
	VINYL ACETATE	1.08		0.99		60	C,CC	1295
PHTHALIC ACID, BIS(TRIETHYLENE GLYCOL) ESTER, DIMETHACRYLATE								
	STYRENE	0.54		0.36		0	AA,H23	564
		0.75		0.15		0	AA,H23	564
		0.75		0.15		0	H23,I18	563
		2.45		0.05		0	AA,H23	564

M1	M2	R1	+/-	R2	+/-	TEMP.	REMARKS	REF.
PHTHALIC ACID, DIALLYL ESTER								
	ACRYLIC ACID, METHYL ESTER	0.05		11.5		80	C,CC	1292
		0.058		12.9		60	C,CC	1292
	ACRYLONITRILE	0.049	0.004	2.54	0.16	80	C	612
		0.057	0.004	3.72	0.15	60	C	612
	ETHYLENE, CHLORO-	0.38		1.68		60	CC	1291
	ETHYLENE, 1,1-DICHLORO-	0.2		5.0		70		24
	METHYL METHACRYLATE	0.070	0.007	25.0	1.0	60	C	612
		0.074	0.008	20.8	0.6	70	C	612
		0.113	0.002	17.5	1.5	80	C	612
	STYRENE	0.053	0.005	21.4	0.2	100	C	611
		0.057	0.007	32.8	0.6	60	C	611
		0.066	0.005	19.4	0.5	110	C	611
		0.081	0.005	18.8	0.2	90	C	611
		0.088	0.007	27.5	0.6	70	C	611
		0.105	0.001	16.9	0.1	120	C	611
		0.105	0.009	23.8	0.2	80	C	611
		0.134	0.008	15.1	0.5	130	C	611
	VINYL ACETATE	0.93		0.83		80	C,CC	1292
		1.2		0.72		60	C,CC	1292
		2.0		0.72			C,Q	932
PHTHALIMIDE, N-1-(1,3-BUTADIENYL)-								
	STYRENE	1.48		0.32		60	C,I6	672
PHTHALIMIDE, N-2-(1,3-BUTADIENYL)-								
	STYRENE	5.2	0.5	0.11	0.02	60	C,I6	671
PHTHALIMIDE, N-(3-METHYLENE-1-BUTEN-4-YL)-								
	STYRENE	2	0.13	0.15	0.02	60	C,I6	1111
PHTHALIMIDE, 4-(TRIMETHYLSILYL)-N-VINYL-								
	STYRENE	0.08		6.1		60	I7	1217
PHTHALIMIDE, N-VINYL-								
	ACETIC ACID, THIO-, VINYL ESTER	0.23		3.2		60	I24	329
	ETHER, BUTYL VINYL	3.27		0.03		60	I7	694
	MALEIC ANHYDRIDE	0.20		0.001		65		694
		0.3	0.003	0.003		90		1217
	STYRENE	0.075	0.003	8.3	0.3	85	C	696
		0.09		6.3		60	C,I6	672
	STYRENE, ALPHA-METHYL-	0.24		0.92		75		693
		0.24		0.92		80		693
	VINYL ACETATE	2.4		0.07		95	C,Q	698
2-PICOLINE, 5-VINYL-								
	ACRYLAMIDE	0.01	0.09	0.56	0.09	60		41

M1	M2	R1	+/-	R2	+/-	TEMP.	REMARKS	REF.
2-PICOLINE, 5-VINYL-								
	ACRYLIC ACID, METHYL ESTER	0.70	0.10	0.35	0.05		Q	105
		0.88	0.10	0.172	0.007	60		940
	ACRYLONITRILE	0.27	0.04	0.16	0.003	60	Q	940
		1.10	0.2	0.10	0.05			1059
	ALLYL ETHER	80				70		72
	BENZENE, O-DIVINYL-	0.90	0.10	0.45	0.15	60	C,I7	69
	1,3-BUTADIENE	0.412	0.026	1.30	0.07	30	K,Q	189
		0.474	0.031	0.606	0.006		Q	562
		0.72	0.03	1.32	0.01		C	971
	CINNAMIC ACID, ETHYL ESTER	1.49	0.16	0.17	0.09	70	C	1121
	METHACRYLIC ACID	0.85	0.03	0.43	0.2		Q	1149
	METHYL METHACRYLATE	0.61	0.08	0.46	0.02	60		940
	2-PICOLINIUM METHYL SULFATE, 1-METHYL-5-VINYL-	0.3		0.01		50	I67	1197
	STYRENE	0.58	0.1	0.98	0.1	50	I15	1197
		0.68	0.030	0.6	0.010	70		830
		0.801	0.2	0.738	0.12	30		189
		0.88	0.02	1.19	0.005	60		737
		0.91	0.13	0.812	0.23	60		940
		1.20	0.3	0.72	0.1	60	I6	952
		2.9	0.3	0.5	0.1	70	H24	830
	SUCCINIC ACID, METHYLENE-, DIETHYL ESTER	0.51	0.11	0.17	0.06	70	G18	1089
	SUCCINIC ACID, METHYLENE-, DIMETHYL ESTER	0.875	0.05	0.225	0.03	70		1089
	TRIETHYLENE GLYCOL, DIMETHACRYLATE	0.32	0.065	0.59	0.06	70	Q	202
2-PICOLINE, 5-VINYL-, COBALT COMPLEX								
	STYRENE	0.3	0.2	1.8	0.2	50	I6	1410
2-PICOLINIUM CHLORIDE, 5-VINYL-								
	ACRYLONITRILE	2.3		0.15			F,I11,Q	235
	METHYL METHACRYLATE	0.94		0.46		70	I12	1196
	STYRENE	0.42		0.27		70	I12	1196
2-PICOLINIUM METHYL SULFATE, 1-METHYL-5-VINYL-								
	ACRYLAMIDE	2.7		0.19		48		878
	METHACRYLAMIDE	1.2		0.14		48		878
	METHACRYLIC ACID	0.60		0.58		50		878
	METHYL METHACRYLATE	1.8		0.12		50		878
	2-PICOLINE, 5-VINYL-	0.01		0.3		50	I67	1197
		0.98		0.58		50	I15	1197
2-PINENE								
	METHYL METHACRYLATE	0.07		6.60			Q	813
2-PIPERIDONE								
	2H-AZEPIN-2-ONE, HEXAHYDRO-	0.16		0.64			I7,Q	762

M1	M2	R1	+/-	R2	+/-	TEMP.	REMARKS	REF.
PIVALIC ACID, VINYL ESTER								
	VINYL ACETATE	0.43	0.05	0.75	0.04	60		392
		0.94	0.15	0.80	0.10	60	I7	954
PROPANE, 1-(ALLYLOXY)-2,3-EPOXY-								
	MALEIC ANHYDRIDE	0.01		0.002		60	I26	702
	METHYL METHACRYLATE	0.035		40.7		70	C,I2	904
	PROPANE, 1-CHLORO-2,3-EPOXY-	0.25	0.02	0.62	0.04	50	H85,I7	1264
	PROPYLENE OXIDE	0.64	0.03	1.89	0.09	50	H85,I7	1264
PROPANE, 1,3-BIS(P-VINYLPHENYL)-								
	STYRENE	0.89		1.11		61	C	1042
PROPANE, 1-BROMO-2,3-EPOXY-								
	ETHYLENE OXIDE	0.31	0.02	3.15	0.15	3	H13,I2	1265
PROPANE, 1-TERT-BUTOXY-2,3-EPOXY-								
	PROPANE, 1,2-EPOXY-3-ETHOXY-	1.4	0.1	0.7	0.1	0	H13,I43	1258
	PROPANE, 1,2-EPOXY-3-ISOPROPOXY-	1.5	0.1	0.8	0.1	0	H10,I43	1258
		1.1	0.1	0.9	0.1	0	H10,I43	1258
	PROPANE, 1,2-EPOXY-3-METHOXY-	1.2	0.1	0.8	0.1	0	H13,I43	1258
		1.5	0.1	0.7	0.1	0	H10,I43	1258
	SILANE, (2,3-EPOXYPROPOXY)TRIMETHYL-	1.7	0.1	0.6	0.1	0	H13,I43	1258
		1.25	0.1	0.79	0.1	0	H13,I43	1258
		1.28	0.1	0.84	0.1	0	H10,I43	1258
PROPANE, 1-CHLORO-2,3-EPOXY-								
	ACETONITRILE	0.2		0		70	H49,R	232
		2.5		0		70	H11,I7,R	232
		2.5		0		70	H11,R	232
		4.0		0		70	H11,I7,R	232
		4.2	0.4	0	0.02	70	H20	231
		6.1		0		70	H11,R	232
	ACRYLIC ACID, TETRAHYDROFURFURYL ESTER	0.12	0.01	0.56	0.05	0	H10	706
	BENZONITRILE	2.8	0.2	0	0.02	70	H20	231
	CYCLOBUTANE, 3-(EPOXYETHYL)-1,1,2,2-TETRAFLUORO-	1.45	0.15	0.408	0.007	20	H13	1360
	1,3-DIOXOLANE	1.2		0.08		0	H10	1061
	ETHYLENE OXIDE	0.12	0.01	1.15	0.05	3	H13,I2	1265
	FURAN, 2,5-DIHYDRO-	0.05		0.06		0	H10,I3	1130
	FURAN, TETRAHYDRO-	0.00	0.05	3.85	0.05	0	H10,I4	399
		0.06		3.9		0	H10	437
		0.25		2.5		20	H10	808
		0.25		2.5		20	H10	809
	FURAN, TETRAHYDRO-2-METHYL-	0.5	0.3	20	5	0	H13,I3	823
		2	1	80	20.	0	H13,I3	823
	B-D-GLUCOPYRANOSE, 1,6-ANHYDRO-2,3,4-TRI-O-METHYL-	0.04	0.05	0.27	0.05	0	H10,I4	399
	GLYCIDAMIDE	1.4		0.15		0	H10,I3	1136
	HYDRACRYLIC ACID, BETA-LACTONE	0.40	0.15	5.8	0.7	70	G13	347
		0.2	0.1	9.0		50	H14	171

M1	M2	R1	+/-	R2	+/-	TEMP.	REMARKS	REF.
PROPANE, 1-CHLORO-2,3-EPOXY-								
	METHACRYLIC ACID, TETRAHYDROFURFURYL ESTER	0.14	0.01	0.72	0.15	0	H10	706
	OXETANE, 3,3-BIS(CHLOROMETHYL)-	0.4	0.1	2.5	0.5	0	H10,I3	47
	OXETANE, 3-(CHLOROMETHYL)-3-METHYL-	0.008	0.005	39.6		0	H13	283
		0.015		35	5	-50	H10	283
	PROPANE, 1-(ALLYLOXY)-2,3-EPOXY-	0.62	0.04	0.25	0.02	50	H85,I7	1264
	PROPANE, 1,2-EPOXY-3-PHENOXY-	0.66	0.1	1.4	0.1	85	H13	1363
		1.07	0.1	1.05	0.2	85	H44	1363
	PROPYLENE OXIDE	0.09	0.02	4.65	0.15	80	H13	1118
		0.55	0.03	1.76	0.13	50	H85,I7	1264
		0.55	0.45	6.25	3.25	80	H11	1118
		1.8	0.3	0.60	0.05		H44,Q	375
	STYRENE	2.5		-0.115		70	H11	1100
		8.9	0.6	-0.1	0.1	70	H11	1100
PROPANE, 1-(P-CHLOROPHENOXY)-2,3-EPOXY-								
	PROPANE, 1,2-EPOXY-3-PHENOXY-	0.6	0.1	0.89	0.1	85	H13	1363
		0.68	0.1	1.09	0.1	85	H44	1363
		0.73	0.1	1.2	0.1	85	H40	1363
		1.45	0.1	0.65	0.05	30	G13,I17	1362
PROPANE, 1-((1,1-DIMETHYL-4-PENTEN-2-YNYL)OXY)-2,3-EPOXY-								
	5-HEXEN-3-YN-2-OL, 2-METHYL-	0.54	0.16	1.40	0.10		B,Q	606
PROPANE, 1,2-EPOXY-3-ETHOXY-								
	PROPANE, 1-TERT-BUTOXY-2,3-EPOXY-	0.7	0.1	1.4	0.1	0	H13,I43	1258
		0.8	0.1	1.5	0.1	0	H10,I43	1258
	STYRENE	0.75		0		50	H10,I10	443
PROPANE, 1,2-EPOXY-3-ISOPROPOXY-								
	PROPANE, 1-TERT-BUTOXY-2,3-EPOXY-	0.8	0.1	1.2	0.1	0	H13,I43	1258
		0.9	0.1	1.1	0.1	0	H10,I43	1258
	SILANE, (2,3-EPOXYPROPOXY)TRIMETHYL-	1.1	0.1	0.8	0.1	0	H13,I43	1258
PROPANE, 1,2-EPOXY-3-METHOXY-								
	ETHER, ISOBUTYL VINYL	1.03		7.3		50	H10,I10	443
	GLYCIDAMIDE	0.14	0.04	12.0	1.0	70	G13	347
	OXETANE, 3,3-BIS(CHLOROMETHYL)-	0.1		0.7		0	H10,I3	49
	PROPANE, 1-TERT-BUTOXY-2,3-EPOXY-	0.6	0.1	1.7	0.1	0	H13,I43	1258
		0.7	0.1	1.5	0.1	0	H10,I43	1258
	SILANE, (2,3-EPOXYPROPOXY)TRIMETHYL-	0.8	0.1	1	0.1	0	H13,I43	1258
	STYRENE	1.8		0		50	H10,I10	443
PROPANE, 1,2-EPOXY-3-PHENOXY-								
	BENZENE, 1-(2,3-EPOXYPROPOXY)-3-METHOXY-	0.9	0.05	1.05	0.15	30	G13,I17	1362
		0.9	0.2	0.6	0.2	85	H44	1363
		1	0.2	0.98	0.1	85	H13	1363
		1.24	0.2	0.88	0.1	85	H40	1363

COPOLYMERIZATION PARAMETER

M1	M2	R1	+/-	R2	+/-	TEMP.	REMARKS	REF.
PROPANE, 1,2-EPOXY-3-PHENOXY-	BENZENE, 1-(2,3-EPOXYPROPOXY)-4-METHOXY-	0.7	0.1	1.49	0.2	85	H44	1363
		0.71	0.1	1.37	0.2	85	H13	1363
		0.91	0.1	1.75	0.2	85	H40	1363
		1.19	0.02	0.85	0.05	30	G13,I17	1362
	GLYCIDAMIDE	0.35	0.10	10.0	0.8	70	G13	347
	OXETANE, 3,3-BIS(CHLOROMETHYL)-	0.25		1.1		0	H10,I3	49
	PROPANE, 1-CHLORO-2,3-EPOXY-	1.05	0.2	1.07	0.1	85	H44	1363
		1.4	0.1	0.66	0.1	85	H13	1363
	PROPANE, 1-(P-CHLOROPHENOXY)-2,3-EPOXY-	0.65	0.05	1.45	0.1	30	G13,I17	1362
		0.89	0.1	0.6	0.1	85	H13	1363
		1.09	0.1	0.68	0.1	85	H44	1363
		1.2	0.1	0.73	0.1	85	H40	1363
	PROPANE, 1,2-EPOXY-3-(P-TOLYLOXY)-	0.9	0.1	0.98	0.1	85	H13	1363
		1.06	0.1	0.92	0.1	85	H44	1363
		1.08	0.05	0.92	0.1	30	G13,I17	1362
	PROPYLENE OXIDE	1.11	0.1	1.52	0.2	85	H40	1363
		0.65	0.1	0.52	0.1	85	H13	1363
		1.32	0.2	0.2	0.2	85	H44	1363
		7	1	0.2		30	G13,I17	1362
PROPANE, 1,2-EPOXY-3-PROPOXY-	FURAN, TETRAHYDRO-	0.27	0.03	0.46	0.03	0	H10,I4	1207
	STYRENE	4.9		0		50	H10,I10	443
PROPANE, 1,2-EPOXY-3-(P-TOLYLOXY)-	PROPANE, 1,2-EPOXY-3-PHENOXY-	0.9	0.1	0.9	0.1	85	H13	1363
		0.92	0.1	1.08	0.05	30	G13,I17	1362
		0.92	0.1	1.11	0.1	85	H40	1363
		0.98	0.1	1.06	0.1	85	H44	1363
1-PROPANESULFONIC ACID, 3-HYDROXY-, ACRYLATE, SODIUM SALT	ACRYLONITRILE	0.55	0.01	1.8	0.02	50	I17	1373
1-PROPANESULFONIC ACID, 3-HYDROXY-, METHACRYLATE, SODIUM SALT *	ACRYLONITRILE	1.05	0.04	0.37	0.02	50	I17	1373
1-PROPANOL, 2,3-EPOXY-, NITRATE	FURAN, TETRAHYDRO-	0.07		0.88		0	H10,I7	1086
	PROPYLENE OXIDE	0.03		1.85		0	H10,I7	1086
	1,3,5-TRIOXANE	0.3		0.35		0	H10,I7	1086
2-PROPANONE, 1,3-DICHLORO-1,1,3-TETRAFLUORO-	VINYL ACETATE	5.6	0.3	0.41	0:02	60	C,R	566

M1	M2	R1	+/-	R2	+/-	TEMP.	REMARKS	REF.
2-PROPANONE, HEXAFLUORO-	VINYL ACETATE	4.1	0.5	0.12	0.01	60	C,R	566
1-PROPENE	ACRYLIC ACID	0		9.7		50	I?	749
	ACRYLONITRILE	0.1		0.8			H10/14,O	480
	1-BUTENE	0.7		0.7		-78	G,P,Q	625
		1.62		0.50		60	H61	1079
		2.4	0.4	0.5	0.2	60	H12,I20	346
		3.3		0.45		30	H24	1269
		4.04		0.252		0-75	H57	185
		4.3		0.8		30	H29	690
		4.39		0.227		0-75	H57,M12	690
		4.5		0.20		23	H28	185
		80		0.049	0.15	130/220	H54,I21	690
	CYCLOHEXANE, VINYL-						H54,Q	565
								1427
	ETHYLENE	0.013		3.43		30	M5	112
		0.016		28.0		30	H7	154
				76		70	H66,I21	474
				61		70	H65,I21	474
		0.022	0.002	10.7		30	H15	582
		0.022		20.3	0.5	26	H5,I20	1172
		0.023		20			H61,Q	1387
		0.025	0.005	10.3	2.3	26	H28,I20	1172
		0.027		35.3			H5,Q	1387
		0.027		37		70	H24,I21	474
		0.032		33.36		0/75	H24	690
		0.032		33.4		25	H24	692
		0.036		28		70	H7,I21	474
		0.037		14.8		30	H5	154
		0.04		16			H45,Q	1387
		0.04		15.0			H45	691
		0.040		26.0		-20	H6	154
		0.040		26.0		30	H9	154
		0.043	0.006	23	3	30	H2/3	912
		0.046		22.0		30	H1	154
		0.050		17.5		30	H8	154
		0.055		16.8		30	H3	154
		0.056		18.9		30	H2	154
		0.062		16		70	H67,I21	474
		0.062		16		70	H28,I21	474
		0.065		17.95		0/75	H7	690
		0.065	0.002	17.8	0.27	25	H7	624
		0.065		17.95		25	H7	624
		0.070		15.0		30	H4	154
		0.073		3		21	H61,I20	250
		0.08		7.08		25	H28	623
		0.088	0.003	7.08	0.20	0/75	H28	690
		0.11		15.72		0/75	H18	690
		0.11		15.72		0/75	H12	692
		0.110		15.7		75	H12	692
		0.110		15.7		75	H18	
		0.13	0.03	17.5	0.5	30	H54,I44	45

COPOLYMERIZATION PARAMETER

M1	M2	R1	+/-	R2	+/-	TEMP.	REMARKS	REF.
1-PROPENE	ETHYLENE	0.13	0.03	17.5	0.5	30	H12,I44	45
		0.145		5.61		0/75	H29	690
		0.145		5.61		75	H29	689
		0.18		8		30	H7,I44	45
		0.21	0.02	4.7	0.6	20/60	H2/3	912
		0.28	0.04	7.2	0.2	30	H24,I44	45
		0.522		16.8		30	H5,I20	250
		0.7		6.4		30	H53,I44	45
		1.9	0.2	0.9	0.2	120	H35,I20	273
		<0.62	0.2	3.2	0.1	130-220	M1	656
	ETHYLENE, CHLORO-	0.09		2.45	0.12	60	C,M14	1206
		0.29	0.1	2.35	0.2	30	H71	1062
		0.3		2.27		60		615
		2.33		0.09		60	C,M9	1337
		2.57		0.09		60	C,M14	1337
	ETHYLENE, CHLOROTRIFLUORO-	4.52	0.5	1.12	0.2	30	H72	1062
		0.022	0.005	0.016	0.005	-40	H38	1369
		0.03	0.01	0.017	0.005	0	H38	1369
		0.06		0		-35	E	1227
		0.06		0		-35	E	1401
	ETHYLENE, TETRAFLUORO-	0.06	0.02	0.028	0.015	40	H38	1369
		0.24		0.01		-78	E	1227
		0.24		0.01		-78	E	1401
	ISOPRENE	1.0		0.06		-78	E	922
		1.0		0.06		-78	E	923
	STYRENE	0.23		0.50		-78	H46,I18	383
		0.62		0.93		45	H12	173
		1.0		1.5		60	H7	344
		2.4		1.9		60	H28	344
		3.7		2.0		60	H24	344
		5.4		1.0		60	H29	344
		7.2	0.7	0.16	0.08	40	H29	61
		7.70		0.12		60	H12	343
	STYRENE, P-CHLORO-	20.0	3.5	0.20	0.13	40	H12	61
	STYRENE, P-METHYL-	20.5	3.0	0.30	0.15	40	H12	61
1-PROPENE, 1-CHLORO-	ETHYLENE, CHLORO-	0.1		8.80		60	H12,I20	345
		0.1		7.70		60	H12,I20	345
		0.24		1.13		50	C,K	962
1-PROPENE, 1-CHLORO-, CIS-	ACRYLONITRILE	0.005		29.5		60		745
	ETHYLENE, CHLORO-	0.08		12.1		60		745
	MALEIC ANHYDRIDE	0.004		0.41		60		745
	VINYL ACETATE	0.01		7.0		60		745
1-PROPENE, 1-CHLORO-, TRANS-	ACRYLONITRILE	0.01		18.0		60		745
	ETHYLENE, CHLORO-	0.18		4.0		60		745

REMARKS PAGE II-366, REFERENCES PAGE II-368

M1	M2	R1	+/-	R2	+/-	TEMP.	REMARKS	REF.
1-PROPENE, 1-CHLORO-, TRANS-	MALEIC ANHYDRIDE	0.05		0.26		60		745
	VINYL ACETATE	0.08		3.56		60		745
1-PROPENE, 2-CHLORO-	ACRYLONITRILE	0.01		1.4		60		745
	ETHYLENE, CHLORO-	0.58		0.75		50	C,K	962
	MALEIC ANHYDRIDE	5.21		0.18		60		745
	MALONONITRILE, METHYLENE-	0		0.06		40		289
	VINYL ACETATE	1.84		0.20		60		745
1-PROPENE, 3-CHLORO-	ACRYLIC ACID, ALLYL ESTER	0.08	0.01	10.4	0.3	60	C	1367
	ACRYLIC ACID, BENZYL ESTER	0.058	0.003	9.90	0.14	60	C	457
	ACRYLIC ACID, BUTYL ESTER	0.10	0.17	5.83	1.32	60	C	457
	ACRYLIC ACID, ETHYL ESTER	0.084	0.100	7.73	0.35	60	C	457
		0.48	0.05	2.3		70	C	30
	ACRYLIC ACID, 2-HYDROXYETHYL ESTER	0.016	0.030	8.85	1.32	60	C	457
	ACRYLIC ACID, METHYL ESTER	0.054	0.039	22	4	60	C	824
		0.071	0.011	8.450	0.370	60		456
		0.404	0.073	5.450	0.070	120		456
		-0.020	0.100	6.460	0.570	90		456
		-0.091	0.013	9.100	0.930	30		456
	ACRYLIC ACID, PROPYL ESTER	0		7.56	0.10	60	C	457
	ACRYLONITRILE	0.05		1.0		0	D,L3	389
		0.1		5.5		60		620
		1.25		3.0		0		166
		0.88	0.01	4.5	0.2	60		389
	BENZOIC ACID, ALLYL ESTER	0.1		2.5		60	C	834
	BENZOIC ACID, VINYL ESTER	1.15	0.47	0.46	0.16	60	C	457
	1-BUTENE	0	0.05	5.0	1.0	-78		404
	BUTYRIC ACID, VINYL ESTER		0.59	0.31	0.35	60	C	457
	CARBAZOLE, 9-VINYL-			1		70		30
	ETHYLENE, 1,1-DICHLORO-			4.5		60		620
		0.26		3.8		68		8
	FORMIC ACID, VINYL ESTER	0.78	0.10	0.57	0.08	60	C	457
	HEXANOIC ACID, 2-ETHYL-, VINYL ESTER	1.52	0.13	0.71	0.18	60	C	457
	METHACRYLIC ACID, BENZYL ESTER	0.016	0.018	58.7	4.1	60	C	457
	METHACRYLIC ACID, BUTYL ESTER	0.058	0.024	46.3	2.6	60	C	457
	METHACRYLIC ACID, ETHYL ESTER	0.082	0.083	57.9	15.0	60	C	457
	METHACRYLIC ACID, 2-HYDROXYETHYL ESTER	0.033	0.048	35.6	1.7	60	C	457
	METHACRYLIC ACID, ISOBUTYL ESTER	0.047	0.056	45.5	10.2	60	C	457
	METHACRYLIC ACID, PROPYL ESTER	0.067	0.027	52.0	3.2	60	C	457
	METHYL METHACRYLATE			56		60	C	824
				50		60		620
		0	0.038	48.1	5.0	60	C	457
	PROPIONIC ACID, VINYL ESTER	0.68	0.110	0.62	0.09	60	C	457
	STYRENE			30	4	70	N,D	824
		0.016	0.016	31.5		60	C	28
		0.029	0.006	36.8	0.60	60	C	457
	VALERIC ACID, VINYL ESTER	1.62	0.310	0.59	0.20	60	C	457

COPOLYMERIZATION PARAMETER

M1 / M2		R1	+/-	R2	+/-	TEMP.	REMARKS	REF.
1-PROPENE, 3-CHLORO-								
	VINYL ACETATE	0.67		1.30		60	P	854
		0.75	0.46	0.7		60	P	834
				0.34	0.26	60	C	824
						68	C	8
						60	C	457
1-PROPENE, 3-CHLORO-,OLIGOMER (DP 13-20)								
	ETHYLENE, CHLORO-	0.33		0.63			O	1119
1-PROPENE, 3-CHLORO-2-(CHLOROMETHYL)-								
	KETENE, DIPHENYL-	0.05		8.9		50		917
	STYRENE	0.01		3.8		50		917
1-PROPENE, 2-CHLORO-3,3-DIMETHOXY-								
	ETHYLENE, CHLORO-	0.04	0.63	0.288	0.007	60		241
	STYRENE	0.005	0.036	7.75	0.076	60		241
1-PROPENE, 3-CHLORO-2-METHYL-								
	ACETIC ACID, ALLYL ESTER	4.5	0.81	0.15	0.02	153	E	281
		11.1	2.6	0.058	0.014	133	E	281
	ACRYLONITRILE	0.0		1.26	0.04	60	I7,L	6
		0.0		1.28	0.04	60	I6	6
		0.0		1.14	0.04	80	I7,L	6
		0.0		0.31		45		654
	ETHYLENE, CHLORO-	0.18	0.05	1.1		60	N,O	620
	ETHYLENE, 1,1-DICHLORO-	0				155	E	281
	MALEIC ANHYDRIDE			7.7		60	N,O	620
	METHYL METHACRYLATE			0.16	0.02	150	E	281
	1-PENTENE	0				60		620
	STYRENE	0		22		73/90		654
	VINYL ACETATE	2.3	0.31	0.13		150	E	281
1-PROPENE, 1,1-DICHLORO-								
	STYRENE	0.025		37		50	C,Q	915
	VINYL ACETATE	0.06		0.98		50	M13	1457
				0.92				1457
1-PROPENE, 1,3-DICHLORO-								
	STYRENE	0.0		82.5			C,Q	915
1-PROPENE, 2,3-DICHLORO-								
	CARBAZOLE, 9-VINYL-	0		1		70		24
				1		70		30
	METHACRYLIC ACID	0.0		4.0		100		484
	METHYL METHACRYLATE	0		0.5		40	N,O	484
		0.017	0.003	5.5	0.8	70		29

REMARKS PAGE II-366, REFERENCES PAGE II-368

M1	M2	R1	+/-	R2	+/-	TEMP.	REMARKS	REF.
1-PROPENE, 2,3-DICHLORO-	STYRENE			2.5		40		484
		0.06	0.009	5.0	0.6	70	C	29
1-PROPENE, 3,3-DICHLORO-	VINYL ACETATE	3.6		0.07			C,Q	915
1-PROPENE, HEXAFLUORO-	ETHYLENE, FLUORO-	0		1.01	0.01	30	H38	1165
		0		1.01	0.01	30	H38,I38	879
		0		1.1	0.05	30	H45,I3	879
	ETHYLENE, TETRAFLUORO-	0.04	0.02	25	5	30	H64,I3	879
		0		3.5		25	D	1240
1-PROPENE, 2-METHYL-	ACRYLONITRILE	0.00		1.02		60		724
		0.02		0.80	0.2	60		865
	1,3-BUTADIENE	8.0	0.02	1.8		50	H38	60
		14.0		1.0		-78	E,L1	286
		30.0		1.2		-78	E,L1	286
		43		2.0		-100	H4,I3	483
		115				-103		956
	1-BUTEN-3-YNE	8.0	15	0.01	0.01	-100	H10,N	206
	ETHYLENE	0.49		0.13		130-220	M1	656
	ETHYLENE, CHLORO-	0		2.1		65	N,O	292
		0.08		4.3		50		131
		0.08	0.1	2.05	0.02	60	C,M14	1206
		0.08		1.54		60		560
		0.34		2.05		60		615
		1.33		2.11	0.3	60	C,M9	1337
		1.53		0.08		60	C,M14	1337
		1.76		0.08		60	C,M14	1337
	ETHYLENE, CHLOROTRIFLUORO-	0.038	0.007	0.004	0.001	0	H38	1167
		0.04		0		0	E	1401
		0.04		0		0	E	1227
	ETHYLENE, 1,1-DICHLORO-	0.06		1.5		-35	N,O	620
		0		1.3		60	E	860
		0.03		3.30		0	D	251
		0.05		3.00		30		351
		0.40		1.0		50	E,L1	540
		1.27		0.21		-40	E	860
		1.5		0.11		-39	E,L1	540
	ETHYLENE, TETRAFLUORO-	25		0		-78	E,L1	860
		56		0.02		-79,5	E	540
		0		0		-78	E	920
		0.0		0.3		-78	K	921
		0.2		0		80	E	396
				0		-45	E	920
	FUMARIC ACID, DIETHYL ESTER	0.17		0		60	H	366

COPOLYMERIZATION PARAMETER

M1: 1-PROPENE, 2-METHYL-

M2	R1	+/-	R2	+/-	TEMP.	REMARKS	REF.
INDENE	1.1		2.2		-78	H23,I3	585
ISOPRENE	2.17		0.5		-100	H4,I3	483
	2.26		0.38		-95,6	H	776
	2.27		0.44		-90	H	776
	2.5	0.5	0.4	0.1	-103	H	956
STYRENE	0.37	0.07	2.41	0.12	-78	H,P	721
	0.37	0.07	2.41	0.12	-78	H23,I9	381
	0.5		3.7		0	H23,I9	1223
	0.51	0.15	1.27	0.39	-78	E	777
	0.54	0.24	1.20	0.11	-20	H70,I9	722
	0.75	0.15	1.30	0.15	-78	H23,I52	722
	0.8		0.5		-78	H20,I16	722
	1.43	0.43	1.71	0.47	-20	H70,I9	379
	1.49	0.12	1.07	0.17	-20	H70,I9	722
	1.5		1.0		-78	H20,I16	722
	1.51	0.15	2.44	0.15	-78	H23,I50	379
	1.53	0.3	2.61	0.4	-78	H23,I51	722
	1.60	0.02	0.17	0.02	0	H20,I18	574
	1.78	0.10	0.42	0.10	-90	H46,I3	798
	1.79	0.02	1.20	0.02	-78	H23,I2	722
	1.9		0.24		-90	H46,I3	798
	2.2		0.6		-78	H20,I16	379
	2.36	0.06	1.1	0.13	-78	H20,I16	379
	2.51	0.05	0.76	0.06	-30	H46,I3	798
	2.63	0.52	1.21	0.55	-30	H46,I3	798
	3.1		5.50		-78	H23,I41	381
	3.25	0.25	2.75	0.25	-78	H20,I16	379
	3.3		0.2		-78	H23,I42	381
	3.5		0.33		0	E,L1	777
	3.75	0.25	2.75	0.25	-78	E,I18	1
	3.75	0.45	1.92	0.41	-78	H23,I42	1223
	4.11	0.19	1.70	0.07	-78	H20,I43	381
	4.48	0.28	1.08	0.07	-78	H23,I43	381
	4.48	0.28	1.08	0.07	-78	H23,I3	1223
	9.02	0.77	1.99	0.24	-91	H23,I3	381
	1.01		1.02		0	H46,I3	798
STYRENE, P-CHLORO-	1.1		1.04		0	H30,I9	757
	1.14	0.10	0.99	0.10	0	H30,I7	758
	2.80		0.89		0	H30,I24	758
	8.6	0.20	1.2	0.1	0	H20,I22	758
	8.6	1.0	1.25	0.03	0	H20,I22	757
	12.2		2.8		0	H20,I7	758
	14.7	0.2	0.15	0.04	0	H30,I22	757
	14.9		0.53		0	H30,I22	758
	18.0	0.5	0.75	0.05	0	H30,I25	758
	22.2		0.73		0	H30,I23	758
	22.5		0.7		0	H30,I23	757
STYRENE, P-CHLOROMETHYL-ALPHA-METHYL-	0.4		1		-100	H10	452
STYRENE, ALPHA-METHYL-	0.27	0.08	1.41	0.25	-78	H,P	721
	1.2		5.5		-78	H23,I2	722

M1	M2	R1	+/-	R2	+/-	TEMP.	REMARKS	REF.
1-PROPENE, 2-METHYL-								
	STYRENES, CHLOROMETHYL-	4.5		0.7	0.1	-100	H10	452
	VINYL ACETATE	0.31	1	2.15		60		865
1-PROPENE, 1-NITRO-								
	SORBIC ACID, 2-CYANO-, BUTYL ESTER	0.01	0.01	1.1	0.1	0	G13,I5	842
		0.03	0.03	1.50	0.2	20	G14	842
		0.4	0.4	1.7	0.4	-60	G13,I6	842
		0.6	0.3	0.9	0.3	0	G13,I6	842
		0.93	0.1	0.80	0.1	20	G13	842
		1.54	0.2	2.0	0.2	15	G13,I17	842
1-PROPENE, 2-NITRO-								
	MALONIC ACID, METHYLENE-, DIETHYL ESTER	0.03	0.01	5	2	0	G13,I7	842
		0.15	0.07	0.9	0.7	0	G13,I5	842
		0.3	0.2	5.5	2	0	G14,I5	842
	SORBIC ACID, 2-CYANO-, BUTYL ESTER	1.0	0.5	0.23	0.05	0	G13,I6	842
		1.1	0.5	0.93	0.05	-60	G13,I6	842
		1.3	0.5	0.30	0.05	90	G13,I6	842
		3.56		0.70	0.05	20	G14	842
		4.7	0.3	0.20	0.03	0	G13,I7	842
		4.7	0.5	0.04	0.05	0	G13,I5	842
				0.15		20	G13	842
1-PROPENE, 1,2,3,3,3-PENTAFLUORO-, CIS-								
	ETHYLENE, FLUORO-	0		0.9	0.05	30	H38	1165
		0		0.09	0.05	30	H38,I38	879
1-PROPENE, 1,3,3,3-TETRAFLUORO-, TRANS-								
	ETHYLENE, TETRAFLUORO-	0.18		22		22	E,M13	1160
1-PROPENE, 2,3,3,3-TETRAFLUORO-								
	ETHYLENE, TETRAFLUORO-	5.4		0.37		22	E,M13	1160
1-PROPENE, 3,3,3-TRICHLORO-								
	ACRYLONITRILE	0.100	0.015	12.2	1.2	60		211
	FUMARIC ACID, DIETHYL ESTER	1.46	0.35	1.10	1.10	60	C	211
	STYRENE	0.0	0.02	6.9	0.2	60		211
	VINYL ACETATE	0.19	0.03	0.19	0.04	60		211
1-PROPENE, 3,3,3-TRIFLUORO-								
	ETHYLENE	0.8		0.18		70	M10	1181
		0.9		0.18		70	M9	1181
		2.5		0.15		21	E,M	141
	ETHYLENE, TETRAFLUORO-	5		0.12		100	E,M13	141
		5		0.12		21	E,M14	141

COPOLYMERIZATION PARAMETER

M1	M2	R1	+/-	R2	+/-	TEMP.	REMARKS	REF.
1-PROPENE, 3,3,3-TRIFLUORO-2-(TRIFLUOROMETHYL)-	ETHYLENE, TETRAFLUORO-	0.09		0.58		22	E,M11	1160
2-PROPENE-1,1-DIOL, DIACETATE	ETHYLENE, CHLORO-	0.58		1.75				1112
	2-PYRROLIDINONE, 1-VINYL-	0.94		0.92		60	Q	639
	VINYL ACETATE	1.34	0.05	0.48	0.03	60		1380
2-PROPENE-1,1-DIOL, 2-CHLORO-, DIACETATE	STYRENE	0.14	0.018	4.6	0.21	60		241
2-PROPENE-1-SULFONAMIDE, N-(2-CHLOROETHYL)-	ACRYLONITRILE	0		2.8		60		417
2-PROPENE-1-SULFONIC ACID	ACRYLONITRILE	0.43	0.01	1.85	0.01	45	I17	441
2-PROPENE-1-SULFONIC ACID, ALLYL ESTER	ACRYLIC ACID, METHYL ESTER	0.37	0.09	5.3	0.7	60	I7	1201
	STYRENE	0.01	0.01	13	1	60	I7	1201
	VINYL ACETATE	1.54	0.08	0.5	0.15	60	I7	1201
2-PROPENE-1-SULFONIC ACID, 2,3-EPOXYPROPYL ESTER	ACRYLONITRILE	0		3.1		60		417
	STYRENE	0.02		50		5	60	417
2-PROPENE-1-SULFONIC ACID, SODIUM SALT	ACRYLONITRILE	0.07	0.06	4.94	0.06	30	F,J	648
		0.18	0.05	0.69	0.05	60	I17	648
		0.28	0.02	1.25	0.01	45	I40	441
		0.38	0.02	1.00	0.01	45	I17	441
1-PROPENE-1,2,3-TRICARBOXYLIC ACID	1,3-BUTADIENE					50	K,P	734
	STYRENE					60	C,R	734
1-PROPENE-1,2,3-TRICARBOXYLIC ACID, TRANS-, TRIETHYL ESTER	1,3-BUTADIENE	0.00	0.01	0.37	0.03	60	C	595
		0.02	0.01	0.65	0.05	50	K	734
	STYRENE	-0.10	0.10	1.10	0.10	60	C	595
1-PROPENE-1,2,3-TRICARBOXYLIC ACID, TRANS-, TRIMETHYL ESTER	ACRYLONITRILE	-0.10	0.10	5.50	0.50	60	C,16	595
	1,3-BUTADIENE	0.00	0.03	0.40	0.03	60	C	595

REMARKS PAGE II-366, REFERENCES PAGE II-368

M1	M2	R1	+/-	R2	+/-	TEMP.	REMARKS	REF.
1-PROPENE-1,2,3-TRICARBOXYLIC ACID, TRANS-, TRIMETHYL ESTER								
	ETHYLENE, CHLORO-	0.00	0.50	0.15	0.10	60	C	595
	ETHYLENE, 1,1-DICHLORO-	0.01	0.1	54	5	60	C,17	597
	STYRENE	0.00	0.01	1.10	0.01	60	C	595
1-PROPEN-2-OL, ACETATE								
	ETHYLENE, CHLORO-	0.25		2.2		65	C	772
	MALEIC ANHYDRIDE	0.032	0.005	0.002		75	C	1033
	METHYL METHACRYLATE	0.017		30		75		334
	STYRENE	0.5		2.6		30	H10,I22	374
	VINYL ACETATE	1.0		1.0		75	C	334
1-PROPEN-2-OL, ISOCYANATE								
	ACRYLIC ACID, ETHYL ESTER	0.15	0.07	0.79	0.11	60	C	419
	ACRYLIC ACID, METHYL ESTER	0.08		0.8				333
	ACRYLONITRILE	0.11	0.04	0.60	0.03			419
	ETHYLENE, 1,1-DICHLORO-	0.10	0.11	0.24	0.02			419
	METHYL METHACRYLATE	0.31	0.02	0.85	0.04			419
	METHYL METHACRYLATE	0.14	0.10	3.10	0.29			419
	STYRENE	0.07		8.12				419
	STYRENE	0.14		7.0		60	C	333
2-PROPEN-1-OL, 2-CHLORO-								
	METHACRYLIC ACID	0.0		4.5		100	N,O	484
	METHYL METHACRYLATE	0		4.4		100		484
	STYRENE	0		12.5		40		484
2-PROPEN-1-OL, 2-CHLORO-, ACETATE								
	ACRYLIC ACID	0.0		1.0		100	N,O	484
	ACRYLIC ACID, METHYL ESTER	0		0.7		100		484
	ETHYLENE, CHLORO-	0		0.70		100		484
	MALEIC ANHYDRIDE	0		0		120	N	484
	METHYL METHACRYLATE	0		1.0		50	N,O	484
	STYRENE	0		4.10		50		484
2-PROPEN-1-OL, 2-METHYL-, ACETATE								
	ACRYLIC ACID, 2-CHLOROETHYL ESTER	0		4	1	60	N,O	560
	ETHYLENE, 1,1-DICHLORO-	0		2.4		60	N,O	620
	METHYL METHACRYLATE	0		10		60		620
	STYRENE	0		71	10	60		620
2-PROPEN-1-OL, 2-METHYL-, OXALATE								
	ETHYLENE, 1,1-DICHLORO-	0.16	0.01	4.8	0.2	40		139
2-PROPEN-1-ONE, 1-(6-METHYL-3-PYRIDYL)-3-PHENYL-								
	STYRENE	-0.15	0.2	0.92	0.08	50	17	602

M1	M2	R1	+/-	R2	+/-	TEMP.	REMARKS	REF.
2-PROPEN-1-ONE, 1-PHENYL-3-(3-PYRIDYL)-								
	STYRENE	0.00	0.25	0.50	0.10	60	I7	602
2-PROPEN-1-ONE, 3-PHENYL-1-(3-PYRIDYL)-								
	STYRENE	0.09	0.10	0.85	0.05	50	I7	602
PROPIOLONITRILE								
	CARBAZOLE, 9-VINYL-	0.03	0.005	0.075	0.005	30	I5	1129
PROPIONALDEHYDE, 3-(ALLYLOXY)-								
	ACROLEIN, 2-CHLORO-	0.01		10		60		1287
	ACRYLONITRILE	0.01		3.9		-78	H21	1287
	STYRENE	0.015		5		60	H10	1287
		1		3		0		1287
PROPIONALDEHYDE, 2,3-DICHLORO-								
	ISOCYANIC ACID, PHENYL ESTER	0.48		0.002		-21	H40	1314
PROPIONIC ACID, VINYL ESTER								
	1,3-BUTADIENE, 2-CHLORO-	0.05	0.018	70	0.14	65	C	1002
	CARBAZOLE, 9-VINYL-	0.076	0.10	1.68	0.04	100	C	1001
	ETHYLENE, CHLORO-	0.60	0.04	1.60	0.05		O	176
		0.65	0.01	1.35	0.05	40	C,III,J	503
	METHYL METHACRYLATE	0.03	0.09	24	2.5	60	C,I7	1135
	1-PROPENE, 3-CHLORO-	0.62	0.07	0.68	0.110	60	C	457
	VINYL ACETATE	0.98	0.1	0.98	0.07	60	C,I7	1135
		1.0		0.9	0.2		O	378
PROPYLENE OXIDE								
	BUTANE, 1,2-EPOXY-	0.71		0.39		0	H83,I24	1154
		0.76		0.5		0	H82,I24	1154
		0.5		6.5		25	G10	1371
	ETHYLENE OXIDE	0.21	0.25	40	13	-78	H10,I3	461
	FORMALDEHYDE	-0.13	0.02	37	2	-78	H10,I2	461
	FURAN, 2,5-DIHYDRO-	-0.12		0.08		0	H10,I3	1130
	FURAN, TETRAHYDRO-	0.34	0.05	0.63	0.05	0	H10,I4	1207
		0.68		0.09		0	H83,I24	1154
		0.85		0.18		0	H82,I24	1154
		1.3		0.5		0	H10,I24	110
		1.4		0.1		0	H10,I24	110
		1.8		0.6		0	H10,I24	110
		1.98	0.12	0.65	0.21	0	H59	548
		2.16	0.32	0.76	0.02	50	H60	548
		2.30	0.12	0.37	0.10	0	H58	548
		2.87	0.13	1.12	0.09	0	H13	548
		2.9	0.9	0.29	0.09	30	H32,L	800
B-D-GLUCOPYRANOSE, 1,6-ANHYDRO-2,3,4-TRI-O-METHYL-								
		0.7		0.35		0	H10,I3	1136
	GLYCIDAMIDE	0.10	0.10	23	3	70	G13	347

M1 / M2	R1	+/-	R2	+/-	TEMP.	REMARKS	REF.
PROPYLENE OXIDE							
OXETANE, 3,3-BIS(CHLOROMETHYL)-	0.65		0.3		0	H10,I3	47
OXETANE, 3-(CHLOROMETHYL)-3-ETHYL-	0.27	0.12	8.4	1.3	0	H13	283
OXETANE, 3-(CHLOROMETHYL)-3-METHYL-	0.185	0.025	2.1	0.5	80	H11	1118
	0.2	0.05	8.3	0.75	0	H13	283
	0.24	0.06	13.2	1.4	-50	H10	283
PROPANE, 1-(ALLYLOXY)-2,3-EPOXY-	0.33	0.1	4.46	0.1	50	H14,I2	20
	1.89	0.09	0.64	0.03	50	H85,I7	1264
PROPANE, 1-CHLORO-2,3-EPOXY-	1.76	0.05	1.8	0.3	50	H44,0	375
	4.65	0.13	0.55	0.03	80	H85,I7	1264
	6.25	0.15	0.09	0.02	80	H13	1118
PROPANE, 1,2-EPOXY-3-PHENOXY-	0.2	3.25	7	0.45	80	G13,I17	1362
	0.52	0.2	1.32	1	30	H44	1363
	1.52	0.1	0.65	0.2	85	H13	1363
1-PROPANOL, 2,3-EPOXY-, NITRATE	1.85	0.2	0.03	0.1	85	H10,I7	1086
SUCCINIC ANHYDRIDE	0.19		0.0		150	H78,I22	828
	0.21		0.0		150	H79,I22	828
	0.28		0.0		150	H80,I22	828
VALERIC ACID, 5-HYDROXY-, DELTA LACTONE	0.25	0.1	5.3		20	H55	171
2-PROPYN-1-OL, ACETATE							
ACRYLONITRILE	0.05		6.15		60		1305
STYRENE	0		53		60		1305
PYRAN, TETRAHYDRO-							
OXETANE, 3,3-BIS(CHLOROMETHYL)-	0.02	0.001	1.515	0.5	0	H10	977
OXETANE, 3-(CHLOROMETHYL)-3-METHYL-			8.5		0	H13	284
PYRAN, TETRAHYDRO-2-METHYL-							
OXETANE, 3,3-BIS(CHLOROMETHYL)-	0		7.2		0	H10,I3	49
PYRAN, TETRAHYDRO-2-PHENYL-							
OXETANE, 3,3-BIS(CHLOROMETHYL)-	0		1		0	H10,I3	49
2H-PYRAN-3-CARBOXYLIC ACID, 5,6-DIHYDRO-2,6-DIMETHYL-, VINYL* ESTER							
ACRYLONITRILE	0		2.38		60	I6	1404
ETHYLENE, CHLORO-	1.03		0.93		60	I5	1404
STYRENE	0		13		60		1404
PYRAZOLE, 3,5-DIMETHYL-1-VINYL-							
STYRENE	0.08	0.02	8.2	0.1	60		523
3-PYRAZOLIDINONE, 1-(M-VINYLPHENYL)-							
STYRENE	0.8		1.2		60	I6	1298

M1 / M2	R1	+/-	R2	+/-	TEMP.	REMARKS	REF.
3,6-PYRIDAZINEDIONE, 1,2-DIHYDRO-							
ACROLEIN	0		16		60	I17	1286
ACRYLIC ACID, ETHYL ESTER	0		24		60	I17	1127
STYRENE	0		55		60	I17	1127
3(2H)-PYRIDAZINONE, 2(P-CHLOROPHENYL)-6-HYDROXY-							
ACRYLONITRILE	0		15		60	I17	1290
STYRENE	0		10		60	I17	1290
3(2H)-PYRIDAZINONE, 6-HYDROXY-							
ACROLEIN	0		16		60	I17	1289
ACROLEIN, 2-CHLORO-	0		30		60	I17	1289
STYRENE	0		55		60	I17	1289
3(2H)-PYRIDAZINONE, 6-HYDROXY-, ACETATE							
ACROLEIN	0		4		60	I17	1289
ACROLEIN, 2-CHLORO-	0	6	20		60	I17	1289
STYRENE	0		6		60	I17	1289
3(2H)-PYRIDAZINONE, 6-HYDROXY-2-PHENYL-							
ACRYLONITRILE	0		12		60	I17	1290
STYRENE	0		13		60	I17	1290
3(2H)-PYRIDAZINONE, 6-HYDROXY-2-(P-TOLYL)-							
ACRYLONITRILE	0		12		60	I17	1290
STYRENE	0		16		60	I17	1290
3(2H)-PYRIDAZINONE, 6-METHYL-							
ACROLEIN	0		8		60	I17	1289
ACROLEIN, 2-CHLORO-	0		48		60	I17	1289
STYRENE	0		25		60	I17	1289
PYRIDINE, 2-DIMETHYLAMINO-4-VINYL-							
STYRENE	1.4	0.1	0.35	0.02	60		759
PYRIDINE, 5-ETHYL-2-VINYL-							
ACRYLIC ACID, METHYL ESTER	1.16	0.08	0.179	0.006	60		939
ACRYLONITRILE	0.43	0.05	0.02	0.02	60		422
METHYL METHACRYLATE	0.69	0.03	0.395	0.003	60		939
STYRENE	1.2	0.2	0.79	0.03	60		939
PYRIDINE, 5-ETHYL-2-VINYL-, 1-OXIDE							
METHYL METHACRYLATE	4.7	0.6	0.12	0.02	60	I12	942
	5.5	0.5	0.11	0.01	60		942
STYRENE	2.6	0.3	0.10	0.01	60	I12	942

M1 / M2	R1	+/-	R2	+/-	TEMP.	REMARKS	REF.
PYRIDINE, 2-ISOPROPENYL-							
STYRENE	0.95		0.42		60		1233
PYRIDINE, 2-VINYL-							
ACROLEIN	0.097	0.04	4.0	0.05	50	I6	845
ACRYLIC ACID, BUTYL ESTER	0.23	0.05	2.51	0.06	60	C	955
ACRYLIC ACID, ETHYL ESTER	1.58	0.05	0.19	0.003	75		1012
ACRYLIC ACID, METHYL ESTER	2.03	0.49	0.168	0.09	60		939
			0.20		60		7
ACRYLONITRILE	0.47	0.03	0.113	0.002	60	P,Q	1080
	0.83		0.77		60		422
	2.13		0.086		60	Q	871
	21.9	5.52	0.05		60		1073
	21.9	5.52	0.05		60	C,111,Q	533
BENZENE, ETHYNYL-	4.0	0.07	0.2	0.01	60	C	534
1,3-BUTADIENE, 2-CHLORO-	0.06	0.01	5.19	0.05	50		782
	0.06	0.001	5.19	0.03	50		534
	0.064	0.005	5.195	0.003	50		533
1-BUTEN-3-YNE, 2-METHYL-	1.65		0.55	0.1	60		536
1,3-DIOXOLANE-4-METHANOL, 2,2-DIMETHYL-, ACRYLATE	0.836	0.18	0.12	0.1	50	C,12	784
1,3-DIOXOLANE-4-METHANOL, 2,2-DIMETHYL-, 4-(METHYLENESUCCINATE)	0.836	0.18	0.12	0.1	50	C,12	195
1,3-DIOXOLANE-4-METHANOL, 2,2-DIMETHYL-, METHYL FUMARATE	0.76	0.06	0.6	0.1	50	C,12	195
1-HEXEN-3-YNE	1.5	0.5	0.58	0.05	60		195
ISOPRENE	0.46	0.07	0.59	0.05	60		784
	0.47	0.07	0.59	0.05	70		536
	0.47	0.07	0.44	0.02	70		534
METHACRYLIC ACID	1.38	0.04	0.58	0.05	60		533
	1.55	0.10	1.25	0.05	25	C,Q	819
METHYL METHACRYLATE	0.17	0.10	0.33	0.002	60		35
	0.70	0.02	0.439	0.03	60	Q	871
	0.77	0.13	0.20	0.025	60		23
	0.84	0.06	0.395	0.02	60		939
	0.86	0.2			25	E	149
STYRENE	0.9	0.08	0.56	0.025	50		1023
	1.135	0.19	0.55	0.07	60		939
	1.14	0.04	0.55	0.03	60		759
	1.27	0.05	0.50	0.03	60	E	149
	1.33	0.05	0.57	0.03	70	C	821
	1.81	0.05	0.55	0.03	70		536
	1.81	0.07	0.55	0.07	50		534
	1.81		0.55		60		533
	0.63		0.11		60		7
STYRENE, DICHLORO-, MIXED ISOMERS	1.1		0.9		60	C	32
STYRENE, 2,5-DICHLORO-	0		0		70	131,Q	1389
VINYL ACETATE	10		0.3		60	Q	723
	30	15	0		70		23
PYRIDINE, 2-VINYL-, COBALT COMPLEX							
STYRENE	0.18	0.05	1.33	0.7	50	I6	1410

M1	M2	R1	+/-	R2	+/-	TEMP.	REMARKS	REF.
PYRIDINE, 2-VINYL-, 1-OXIDE								
	METHYL METHACRYLATE	3.9	0.8	0.13	0.03	60	I12	942
	STYRENE	2.1	0.6	0.11	0.01	60	I12	942
PYRIDINE, 4-VINYL-								
	ACRYLIC ACID, BUTYL ESTER	5.15	0.09	0.46	0.09	60	C	274
	ACRYLIC ACID, METHYL ESTER	1.7	0.2	0.22	0.01	60		939
	ACRYLONITRILE	2.14		0.15		50		377
	ETHYLENE, CHLORO-	0.41	0.09	0.113	0.005	60		422
	HYDROQUINONE, VINYL-, DIBENZOATE	23.4		0.02		60		613
	ISOPRENE	0.48		0.40		78		465
	METHACRYLIC ACID, 2-(DIETHYLAMINO)ETHYL ESTER, ETHIODIDE	2.49		0.32		50	I7	377
	METHYL METHACRYLATE	0.30	0.02	0.61	0.09	60		755
	METHYL METHACRYLATE	0.79	0.05	0.574	0.004	60		939
	STYRENE	0.52	0.06	0.62	0.02	80	C,I2	275
	VINYL ETHER	0.7	0.1	0.54	0.03	60	I7,CC	939
		32		0.032		60		156
PYRIDINE, 4-VINYL-, COBALT COMPLEX								
	STYRENE	0.21	0.1	1.9	0.25	50	I6	1410
PYRIDINE, 4,4'-VINYLENEDI-								
	ACRYLONITRILE	0.02	0.05	0.95	0.05	50	C,K	603
	STYRENE	0.17	0.1	1.85	0.1	50	C,K	603
PYRIDINIUM ACETATE, 2-VINYL-								
	STYRENE	0.36	0.04	0.16	0.02	22	D	821
PYRIDINIUM FLUOROBORATE, 1-VINYL-								
	ACRYLIC ACID, METHYL ESTER	0.2		1.5		60	I31	223
	ACRYLONITRILE	0.20		1.06		60	I31	223
	METHYL METHACRYLATE	0.008		4.75		60	I31	223
PYRIMIDINE, 4-VINYL-								
	STYRENE	1.2	0.1	0.17	0.02	60		759
2-PYRROLIDINETHIONE, 1-VINYL-								
	METHYL METHACRYLATE	0.32	0.02	1.85	0.04	60	I7	1171
	STYRENE	1.72	0.09	0.44	0.06	60		888
2-PYRROLIDINONE, 1-VINYL-								
	METHYL METHACRYLATE	1.50	0.30	0.13	0.02	60		888
	STYRENE	0.45	0.03	1.75	0.05	60	I7	1171
2-PYRROLIDINONE								
	2H-AZEPIN-2-ONE, HEXAHYDRO-	5.0		0.75		90	G18	514

REMARKS PAGE II-366, REFERENCES PAGE II-368

M1	M2	R1	+/-	R2	+/-	TEMP.	REMARKS	REF.
2-PYRROLIDINONE, 1-VINYL-	ACRYLIC ACID	0.15	0.1	1.3	0.2	75		1012
	ACRYLIC ACID, METHYL ESTER	0.041	0.024	0.27	0.16	60	C	1070
	BENZOIC ACID, VINYL ESTER	2.45	0.1	0.44	0.09	60		1019
	CARBAMIC ACID, VINYL-, ETHYL ESTER	2		0.42		65	I15	287
	CARBONIC ACID, CYCLIC VINYLENE ESTER	0.7		0.4		60		349
	CARBONIC ACID, THIO-, O-TERT-BUTYL S-VINYL ESTER	0.12	0.01	3.94	0.05	70		1176
	CINNAMIC ACID, VINYL ESTER	0.01		1.20		70	I7	806
	CROTONALDEHYDE	0.5		0.03		80		1004
	CROTONIC ACID	0.85		0.02		65		996
	ETHANOL, 2-(VINYLOXY)-, ACETATE	0.08	0.05	0.02	0.02	60		1394
	ETHER, BUTYL VINYL	2.97	0.01	1.02		60		882
	ETHER, CYCLOHEXYL VINYL	3.84		0		60		881
	ETHER, ISOPROPYL VINYL	1.68		0		60		882
	ETHER, PHENYL VINYL	4.43	0.001	0.22	0.001	60		881
	ETHYLENE, CHLORO-	0.34		0.55			Q,I24	1398
		0.38		0.53			Q	132
		0.53		0.38			Q	132
		0.73		0.74			Q,I15	1398
		0.4		0.66		70	I7	1354
	FERROCENE, VINYL-	0.030	0.033	0.041	0.007	76	C,I7	460
	FUMARONITRILE	0.04	0.04	0.58	0.02	60		1261
	HYDRAZINE, 1-CROTONOYL-2-ISONICOTINOYL-	0.98		0.26		60		995
	HYDROQUINONE, VINYL-, DIBENZOATE	-0.027		0.074		30	D	1407
	MALEIC ANHYDRIDE	0.23	0.02	1.16	0.04	65		1221
	MALEIC ANHYDRIDE/STYRENE	0.02	0.01	0.07	0.03	70		1432
	MALEIMIDE, N-(2-HYDROXYETHYL)-	0.094		0.14		60	I11	1354
	MANGANESE, TRICARBONYL-PI-(VINYLCYCLOPENTADIENYL)-	0.3	0.01	1.4	0.02	60	I7	1416
	METHACRYLIC ACID	0.3	0.03	0.9	0.04	60	I11	1416
		0.33	0.01	0.65	0.02	60	I49	1416
		0	0.01	5.35	0.02	30	I11	1407
	METHYL METHACRYLATE	0.005	0.5	4.7	0.6	50	D	116
		0.02	0.05	4.6	0.5	70		1408
		0.02	0.02	4.6	0.4	70		1406
		0.6	0.02	1.6	0.4	70		1408
		0.92	0.9	0.94	0.8	70	L3	639
	2-PROPENE-1,1-DIOL, DIACETATE	0.13	0.02	1.50	0.30	60		888
	2-PYRROLIDINETHIONE, 1-VINYL-	0.045		15.7		60		116
	STYRENE	0.11	0.05	9.0	0.5	50	C	1070
	SUCCINIMIDE, N-VINYL-	0.6		1.6		80	E	328
	TRISILOXANE, 1,1,1,5,5,5-HEXAMETHYL-3-TRIMETHYLSILOXY-3-VINYL- *	0.6		1.6		30	E	328
	VINYL ACETATE	0.44		0.38		30	Q	132
		0.44		0.38		70		349
		2.28	0.19	0.237	0.037	76	O,I7	460
		3.3		0.205		70		5
		3.30	0.15	0.205	0.015	50		116
QUINOLINE, 2-VINYL-	1,3-BUTADIENE, 2-CHLORO-	0.38	0.03	2.10	0.13	50		536
		0.38	0.03	2.10	0.13	60		534
		0.38	0.03	2.10	0.13	60		533
	ISOPRENE	0.53	0.01	1.88	0.02	60		534

M1	M2	R1	+/-	R2	+/-	TEMP.	REMARKS	REF.
QUINOLINE, 2-VINYL-								
	ISOPRENE	0.53	0.01	1.88	0.02	60		533
	STYRENE	1.882	0.002	0.534	0.001	50		536
		2.09	0.55	0.49	0.14	60		534
		2.09	0.55	0.49	0.14	60		533
		2.69	0.55	0.49	0.14	50		536
SALICYLALDEHYDE								
	STYRENE	0		4.2		0	H10	1402
SEBACIC ACID, DIALLYL ESTER								
	STYRENE	0.103	0.013	35	1.2	60	C	1293
SEMICARBAZIDE, 1-(METHACRYLOYLAMIDINO)-, HYDROCHLORIDE								
	ACRYLAMIDE, N,N'-METHYLENEBIS-	0.091	0.016	2.9	0.25	60		1420
SILANE, ALLYLDIMETHYLPHENYL-								
	ACRYLONITRILE	0.2		2.24		70	C	1317
SILANE, ALLYLDIMETHYL(P-VINYLPHENYL)-								
	STYRENE	0.96	0.10	0.69	0.10	80		307
SILANE, ALLYLMETHYL-								
	STYRENE	0		36		60		423
SILANE, ALLYLPHENYL-								
	STYRENE	0		29		60		423
SILANE, ALLYLTRIETHOXY-								
	ACRYLONITRILE	0		1.7		50	C	957
	ETHYLENE, CHLORO-	0		2.0		50	C	957
SILANE, ALLYLTRIMETHYL-								
	ACRYLONITRILE	0.1		3.98		70	C	1317
SILANE, 2-BUTENYLTRIETHOXY-								
	ACRYLONITRILE	0		10.0		50	C	957
	ETHYLENE, CHLORO-	0		0.4		50	C	957
SILANE, (CHLOROMETHYL)DIMETHYL(P-VINYLPHENYL)-								
	STYRENE	0.85	0.03	0.69	0.03	80	I2	308

REMARKS PAGE II-366, REFERENCES PAGE II-368

M1	M2	R1	+/-	R2	+/-	TEMP.	REMARKS	REF.
SILANE, (1-CHLOROVINYL)TRIETHOXY-								
	ACRYLONITRILE	0		0.7		50	C	957
	ETHYLENE, CHLORO-	0		0.2		50	C	957
SILANE, (1-CYCLOHEXEN-4-YL)TRIETHOXY-								
	ACRYLONITRILE	0		12.0		50	C	957
	ETHYLENE, CHLORO-	0		0.4		50	C	957
SILANE, DIALLYL-								
	METHYL METHACRYLATE						P,Q	517
	STYRENE						P,Q	517
SILANE, DIETHOXYETHYLVINYL-								
	ACRYLONITRILE	0		9.0		50	C	957
	ETHYLENE, CHLORO-	0		1.0		50	C	957
SILANE, DIETHOXYMETHYLVINYL-								
	ACRYLONITRILE	0		6.0		50	C	957
	ETHYLENE, CHLORO-	0		1.0		50	C	957
SILANE, DIETHOXYPHENYLVINYL-								
	ACRYLONITRILE	0		8.3		50	C	957
	ETHYLENE, CHLORO-	0		0.7		50	C	957
SILANE, DIMETHYLPHENYLVINYL-								
	ACRYLONITRILE	0.18	0.01	3.84		70	C	1316
	METHYL METHACRYLATE	0.04	0.01	30.5	5	80	C	685
	STYRENE	0.03		30.6	5	80	C	685
		1.30		0.54		20	G17	684
SILANE, DIMETHYLPHENYL(P-VINYLPHENYL)-								
	METHYL METHACRYLATE	0.05	0.05	0.50	0.05	80		307
	STYRENE	2.3	0.4	0.80	0.16	80		307
SILANE, DIMETHYLPHENYL(P-VINYLPHENYL)-, POLYMER								
	STYRENE, P-CHLORO-	0.51	0.10	0.51	0.10	80		307
SILANE, DIMETHYL(P-VINYLPHENYL)-								
	STYRENE	1.00	0.01	0.56	0.01	80		307
SILANE, DIMETHYLVINYL(P-VINYLPHENYL)-								
	METHYL METHACRYLATE	0.18	0.02	0.76	0.02	80		307
	STYRENE	1.8	0.45	0.9	0.2	80		307
	STYRENE, P-CHLORO-	0.84	0.08	0.76	0.04	80		307

M1	M2	R1	+/-	R2	+/-	TEMP.	REMARKS	REF.
SILANE, (2,3-EPOXYPROPOXY)TRIMETHYL-								
	PROPANE, 1-TERT-BUTOXY-2,3-EPOXY-	0.79	0.1	1.25	0.1	0	H13,I43	1258
		0.84	0.1	1.28	0.1	0	H10,I43	1258
	PROPANE, 1,2-EPOXY-3-ISOPROPOXY-	0.8	0.1	1.1	0.1	0	H13,I43	1258
	PROPANE, 1,2-EPOXY-3-METHOXY-	1	0.1	0.8	0.1	0	H13,I43	1258
SILANE, METHYL(P-VINYLPHENYL)-								
	STYRENE	1.1		0.91		60		423
SILANE, PHENYLVINYL-								
	STYRENE	0		5.7		60		423
SILANE, TETRAVINYL-								
	ACRYLIC ACID, METHYL ESTER	0.133	0.006	8.87	0.07	0	O	884
SILANE, TRIETHOXY(5-NORBORNEN-2-YL)-								
	ACRYLONITRILE	0		0.7		50	C	957
	ETHYLENE, CHLORO-	0		1.6		50	C	957
SILANE, TRIETHOXYPROPENYL-								
	ACRYLONITRILE	0		20.0		50	C	957
	ETHYLENE, CHLORO-	0		8.0		50	C	957
SILANE, TRIETHOXYVINYL-								
	ACRYLONITRILE	0		4.5		50	C	957
		0		5	1	60	C	849
	ETHYLENE, CHLORO-	0		0.9		50	C	957
	MALEIC ANHYDRIDE	0.004		0.035		70		1379
	MALEIC ANHYDRIDE/STYRENE	0.29		1.63		70		1379
	STYRENE	0.056		1.13		70		1379
SILANE, TRIISOPROPOXYVINYL-								
	ACRYLONITRILE	0		6.5		50	C	957
	ETHYLENE, CHLORO-	0		0.8		50	C	957
SILANE, TRIMETHOXYVINYL-								
	ACRYLONITRILE	0		6.0		50	C	957
	ETHYLENE, CHLORO-	0		0.8		50	C	957
	STYRENE	0		22	5	60	C	849
SILANE, TRIMETHOXY(P-VINYLPHENYL)-								
	STYRENE	1.4	0.1	0.71	0.02	70,3	C	556

M1	M2	R1	+/-	R2	+/-	TEMP.	REMARKS	REF.
SILANE, TRIMETHYLVINYL-								
	ACRYLONITRILE	0.07	0.03	3.9	3	60		849
		0.08		3.85		70	c	1316
	METHYL METHACRYLATE	0.02	0.01	34	4	80	c	685
	STYRENE	0		26	8	60		849
		0.01	0.01	24	4	80	c	685
		0.06	0.01	5.7	0.3	12	G17,I20	249
SILANE, TRIMETHYLVINYLOXY-								
	ETHER, BUTYL VINYL	1.04	0.09	0.32	0.09	-78	H19,I23	1330
		1.4	0.1	0.18	0.03	-78	H19,I2	1330
	ETHER, TERT-BUTYL VINYL	0.21	0.04	1.6	0.04	-78	H19,I2	1330
		0.24	0.05	1.45	0.09	-78	H19,I23	1330
	ETHER, ISOBUTYL VINYL	1.08	0.05	0.35	0.03	-78	H19,I23	1330
		1.37	0.06	0.34	0.02	-78	H19,I2	1330
SILANE, TRIMETHYL(P-VINYLPHENYL)-								
	STYRENE	1.0	0.2	1.0	0.2	70,3	c	556
SORBIC ACID, 2,3-EPOXYPROPYL ESTER								
	STYRENE	0.51	0.06	0.88	0.07	60	I7	1231
SORBIC ACID, METHYL ESTER								
	STYRENE	0.46	0.08	0.47	0.08	60		272
SORBIC ACID, 2-CYANO-, BUTYL ESTER								
	1-BUTENE, 2-NITRO-	0.6	0.1	0.84	0.1	90	G13,I6	1216
		0.6	0.2	0.85	0.2	0	G26,I45	1216
		0.65	0.25	0.8	0.25	0	G27,I45	1216
		0.65	0.25	0.8	0.25	0	G28,I6	1216
		0.75	0.1	0.42	0.1	-30	G13,I6	1216
		0.85	0.25	0.75	0.25	24	G13,I45	1216
		0.98	0.2	0.55	0.2	0	G24,I6	1216
		1	0.1	0.9	0.1	0	G25,I45	1216
		1.12	0.2	0.57	0.2	0	G13,I5	1216
		1.2		0.44		0	G13,I6	1216
		1.5	0.3	0.5	0.3	0	G13,I6	1216
		2.14	0.4	0.8	0.4	0	G13,I7	1216
	1-PROPENE, 1-NITRO-	0.80	0.1	1.04	0.1	20	G13	842
		0.9	0.3	0.93	0.3	0	G13,I6	842
		1.1	0.1	0.6	0.01	0	G13,I5	842
		1.50	0.2	0.01	0.03	20	G14	842
		1.7	0.4	0.03	0.4	-60	G13,I6	842
		2.0	0.2	0.4	0.2	15	G13,I17	842
	1-PROPENE, 2-NITRO-	0.04	0.03	1.54	0.3	0	G13,I15	842
		0.15	0.05	4.7	0.5	20	G13	842
		0.20		4.7		0	G13,I7	842
		0.23	0.05	3.56	0.5	0	G13,I6	842
				1.0		0		842

COPOLYMERIZATION PARAMETER

M1	M2	R1	+/-	R2	+/-	TEMP.	REMARKS	REF.
SORBIC ACID, 2-CYANO-, BUTYL ESTER	1-PROPENE, 2-NITRO-	0.30	0.05	1.1	0.5	90	G13,I6	842
		0.70	0.05	1.3	0.5	20	G14	842
		0.93	0.05	1.1	0.5	-60	G13,I6	842
	STYRENE, BETA-NITRO-	0.47	0.1	0.18	0.1	0	G13,I22	842
		0.52	0.1	0.27	0.1	74	G13,I5	842
		1.4	0.3	0.3	0.3	0	G13,I5	842
		2.9	0.4	0.5	0.4	15	G13,I7	842
		5.0	1	0.3	0.3	-60	G13,I5	842
SORBIC ACID, 2-CYANO-, METHYL ESTER	1-BUTENE, 2-NITRO-	1.5	0.15	0.4	0.15	0	G13,I6	1216
SORBONITRILE	ACRYLIC ACID, METHYL ESTER	2.5	0.3	0.30	0.06	60		272
	STYRENE	0.58	0.09	0.15	0.05	60		272
STEARIC ACID, VINYL ESTER	ACRYLIC ACID, METHYL ESTER	0.03	0.03	5.8	0.3	70	C,I6	1050
	ACRYLONITRILE	0.03		4.3		70		1050
		0.04	0.005	4.0	0.02			58
	1,3-BUTADIENE	0.064	0.034	4.20	6.0	60	C	591
	ETHYLENE, CHLORO-	0.034		34.5		60	C	591
	ETHYLENE, 1,1-DICHLORO-	0.290	0.025	0.745	0.026	60	C	591
		0.075	0.025	3.80	0.05	70		1050
	MALEIC ACID, DIISOPROPYL ESTER	0		0.0075		60	C	591
	STYRENE	0.01	0.01	68	30			
		0.73		0.90		70	C	1050
	VINYL ACETATE	1.00		0.97		50		4
STERAMIDE, N-ALLYL-	ACRYLONITRILE	0.118	0.084	3.61	0.087	80	C,I14	454
	ETHYLENE, 1,1-DICHLORO-	0	0.136	5.23	0.067	80	C,I14	454
	VINYL ACETATE	0.532	0.012	0.740	0.087	70	C,I14	454
STIBINE, DIPHENYL(P-VINYLPHENYL)-	STYRENE	1.3	0.1	0.6		65	I2	126
4-STILBENAMINE, N,N-DIMETHYL-	STYRENE	0.04	0.1	6.56	0.69	70	I6	1152
STILBENE, CIS-	MALEIC ANHYDRIDE	0.07	0.07	0.08	0.08	60		557
STILBENE, TRANS-	1,3-BUTADIENE	0		0		0	G17,I5	1452

M1	M2	R1	+/-	R2	+/-	TEMP.	REMARKS	REF.
STILBENE, TRANS-								
	1,3-BUTADIENE	0		50		40	G17,I7	1452
	1,3-BUTADIENE, 2,3-DIMETHYL-	0		0		0	G17,I5	1452
	ISOPRENE	0		8.5		40	G17,I7	1452
		0		0		0	G17,I5	1452
		0		50		40	G17,I7	1452
	MALEIC ANHYDRIDE	0.03	0.03	0.03	0.03	60		557
	METHACRYLONITRILE	0		70		60	R	573
	STYRENE	0		2.3		60	G17,I5	1126
		0		18		30	G17,I7	1126
		0.033	0.07	11.2	1.2	60	D	160
				5.17	0.3	70	I6	1152
STILBENE, 4-CHLORO-								
	STYRENE	0.036	0.05	4.9	0.2	70	I6	1152
STILBENE, 4-NITRO-								
	STYRENE	0.09	0.25	3.23	0.56	70	I6	1152
STILBENE, 4-VINYL-, TRANS-								
	STYRENE	5.5	0.5	0.36	0.02	60	I7	1156
2,2'-STILBENEDISULFONIC ACID, 4,4'-BIS(3-PHENYLUREIDO)-, DI * SODIUM SALT								
	ACRYLONITRILE	0		0.38	0.06	25	F	150
STYRENE								
	ACENAPHTHYLENE	0.3	0.1	4.4	0.3	30	H10,I7	841
		3.81		0.33		90	A	987
	ACETANILIDE, N-VINYL-	13.0	0.66	0.054	0.017	75		367
	ACETIC ACID, ALLYL ESTER	22.3	10	0.009		60	O	666
	ACETIC ACID, CHLORO-, VINYL ESTER	90	0.005	0.000		60		620
	ACETIC ACID, DICHLORO-, VINYL ESTER	1.0		0.05	0.02	80	C	870
	ACETIC ACID, THIO-, VINYL ESTER	45		0.03		80	C	1070
	ACETOPHENONE, 4'-HYDROXY-, METHACRYLATE	20		0.28		50		1070
	ACETOPHENONE, 4'-ISOPROPENYL-	4		0.25		60		1249
	ACETOPHENONE, 4'-ISOPROPENYL-, OXIME	0.15		0.49		60		739
	ACETOPHENONE, 4'-VINYL-	0.15		0.49		60		742
	ACETOPHENONE, 4'-VINYL-, OXIME	0.95		0.27		60		1078
		0.91		0.64		60		1078
		0.25	0.03	1.15	0.10	60		1078
		1.04		0.54		60		1078
	ACETYLENEDICARBONITRILE	1.40		0.33		55	I5	158
	ACROLEIN	0.22		0.25		50	I10	1346
		0.25		0.034		60	I10	1251
		0.32		0.18		50	C,H	1252
		0.06		0.18		60	I17	1288
	ACROLEIN, 2-CHLORO-	0.078	0.022	0.18	0.026	60		241
	ACROLEIN, 3-CYANO-	0.20		0		60	I5,R	913

M1	M2	R1	+/-	R2	+/-	TEMP.	REMARKS	REF.
STYRENE								
	ACRYLAMIDE	1.05		0.2		50	I15	1170
		1.13		0.59		70	I72	1270
		1.13	0.05	1.38	0.1	70	I72	1382
		1.27	0.17	0.3	0.09	70	I10	1382
		1.44	0.22	0.25	0.09	70	I12	1382
	ACRYLAMIDE, N-(1-ACETYL-3-METHYLBUTYL)-	1.87	0.1	0.48	0.05	60	I7	425
	ACRYLAMIDE, N-(1-ACETYL-3-METHYLBUTYL)-2-METHYL-	1.07	0.05	0.57	0.05	60	I7	425
	ACRYLAMIDE, N-(1-BENZYL-2-OXOPROPYL)-	2.06	0.05	0.32	0.1	60	I6	425
	ACRYLAMIDE, N-(1-BENZYL-2-OXOPROPYL)-2-METHYL-	1.30	0.1	0.23	0.02	60	I7	425
	ACRYLAMIDE, N,N-DIBUTYL-	1.65	0.03	0.44	0.13	50	I7	117
	ACRYLAMIDE, N,N-DIMETHYL-	1.23	0.43	0.42	0.08	70		710
	ACRYLAMIDE, N-(1,1-DIMETHYL-3-OXOBUTYL)-	1.28	0.15	0.49	0.05	70	I10	1383
	ACRYLAMIDE, N-(2-HYDROXY-1,1-BIS(HYDROXYMETHYL)ETHYL)-2-ME THYL- *	1.33	0.08	0.49	0.06	60	I12	1383
	ACRYLAMIDE, N,N'-ISOPROPYLICENEBIS-	1.77	0.08	0.565	0.06	60	I15	179
	ACRYLAMIDE, 2-METHYL-N-(1-METHYL-2-OXOPROPYL)-	1.77	0.08	0.01	0.02	60	I7	179
	ACRYLAMIDE, N-(1-METHYL-N-(2-OXOPROPYL)-	1.57	0.02	0.05	0.1	60	I7	1313
	ACRYLAMIDE, 2-METHYL-N-(2-OXOPROPYL)-	0.91	0.02	0.13	0.05	80	I7	425
	ACRYLAMIDE, N-OCTADECYL-	1.83	0.1	0.2	0.05	50	I7	425
	ACRYLAMIDE, N-OCTYL-	1.15	0.1	0.20	0.06	50	I7	425
	ACRYLAMIDE, N-TERT-OCTYL-	1.41	0.10	0.25	0.06	100	I7	1239
	ACRYLANILIDE, 4'-((2-CHLOROETHYL)SULFAMOYL)-	2.70	0.06	0.38	0.07	70	I5	117
	ACRYLANILIDE, 4'-((2-CHLOROETHYL)SULFAMOYL)-2-METHYL-	2.8	0.08	0.33	0.02	60	I6	117
	ACRYLIC ACID	0.69		0.25		70	I5	680
		1.02		0.25		50	I6	680
		0.15	0.01	0.35	0.02	45		168
		0.15		0.15		80	Q	1212
		0.22		0.07		50		299
		0.25		0.05		50		775
		0.25		0.15		50	C	485
		0.25		0.45		45	C	1377
		0.25		0.15		60	C	166
		0.70		0.13		60	I29	1070
		0.75		0.14		60	I10	485
		0.90		0.15		60	I5	485
		1.03		0.08		60	I6	485
		1.1	0.04	0.14	0.03	25	C,16	485
	ACRYLIC ACID, ALLYL ESTER	0.61	0.03	0.25	0.02	60	C	1377
	ACRYLIC ACID, BENZYL ESTER	0.5	0.05	0.20	0.02	60		1367
		0.55		0.17		60		1157
		0.55		0.15		60	C,D	742
	ACRYLIC ACID, 2-BUTOXYETHYL ESTER	0.76	0.13	0.19	0.06	60	I7	740
	ACRYLIC ACID, BUTYL ESTER	0.48	0.04	0.15	0.04	60		570
		0.66	0.02	0.21		60		55
		0.76		0.34		60		120
		0.8		0.28		60		120
		0.82	0.01	0.12	0.01	60	I7	1096
	ACRYLIC ACID, SEC-BUTYL ESTER	1.03		0.12		60		570
		0.97	0.05	0.10	0.02	60		120
	ACRYLIC ACID, 2-CHLOROETHYL ESTER	0.43				60		570
		0.43			60		742
		0.54	0.01	0.10	0.01	60		560

M1	M2	R1	+/-	R2	+/-	TEMP.	REMARKS	REF.
STYRENE								
	ACRYLIC ACID, 2-CHLOROETHYL ESTER	0.59	0.03	0.08	0.01	60	C	211
	ACRYLIC ACID, 2,3-EPOXYPROPYL ESTER	0.60		0.17		60		413
	ACRYLIC ACID, ETHYL ESTER	0.73		0.25		138		492
		0.77		0.17		60		742
		0.77		0.17		60		740
		0.79		0.19		50		1011
		0.80		0.20		70		1011
		0.80		0.48		80	C	1070
		1.01	0.14	0.16		60		570
	ACRYLIC ACID, 2-ETHYLHEXYL ESTER	0.91		0.29		138		492
	ACRYLIC ACID, 3-HYDROXY-2,2-BIS(HYDROXYMETHYL)PROPYL ESTER, * TRINITRATE	0.94	0.07	0.26	0.02	60		570
		0.94	0.07	0.26	0.02	60		570
	ACRYLIC ACID, 4-HYDROXYBUTYL ESTER	0.72		0.73		45	C,K	1018
	ACRYLIC ACID, 2-HYDROXYETHYL ESTER, NITRATE	0.48		0.08		60		1307
	ACRYLIC ACID, X-HYROXYBUTYL ESTER, NITRATE	0.41		0.2		60		1307
	ACRYLIC ACID, ISOPROPYL ESTER	0.76		0.26		60		742
	ACRYLIC ACID, 2-METHOXYETHYL ESTER	0.76		0.26		60		740
	ACRYLIC ACID, METHYL ESTER	0.72	0.03	0.21	0.02	60	E	570
		0.13		5.0		-78	E,I8,AA	816
		0.4		0.16		-78	E,I8,AA	816
		0.65	0.04	0.3		0	E,I8	816
		0.68	0.1	0.14	0.01	60		570
		0.7		0.15	0.05	60	C	600
		0.75	0.03	0.18	0.02	60		168
		0.75	0.1	0.18	0.05	60		559
		0.75	0.11	0.20	0.03	70		34
		0.825	0.0	0.18	0.02	25	E	149
		0.9		0.238		131		559
		0.90	0.60	0.07		-78	E,I18	815
		1.0	0.1	0.13	0.02	0	H10,I16	610
		2.2	0.2	0.16	0.2	60	C	407
		2.2	0.2	0.4	0.2	20	H,I7,Q	358
		>1		0.4			H2O,I7	551
				<1			Q,H,I24	358
	ACRYLIC ACID, METHYL-14C ESTER	0.74	0.04	0.21	0.01	60	C	78
	ACRYLIC ACID, 1-METHYL-2-NITROPROPYL ESTER	0.58		0.13		80	I26	930
	ACRYLIC ACID, 2-METHYL-2-NITROPROPYL ESTER	0.71		0.08		80	I26	930
	ACRYLIC ACID, 2-NITROBUTYL ESTER	0.58		0.115		70	C	945
	ACRYLIC ACID, OCTADECYL ESTER	0.44	0.07	0.18	0.45	60	I45	1239
		0.75	0.12	0.34	0.06	60		1239
		0.79	0.09	0.31	0.31	60		1239
	ACRYLIC ACID, 2-PROPYNYL ESTER	0.45		0.24		60	I7	1306
	ACRYLIC ACID, TETRAHYDROFURFURYL ESTER	0.501		0.485		70		701
	ACRYLIC ACID, 2-BROMO-, ETHYL ESTER	0.06		0.5		60	C	1429
	ACRYLIC ACID, 3-BUTOXY-, BUTYL ESTER	0.06		0.5		60		1430
	ACRYLIC ACID, 2-CHLORO-, 2-ETHOXYETHYL ESTER	78		0.01		60	P,Q	829
	ACRYLIC ACID, 2-CHLORO-, ETHYL ESTER	0.08		0.3		60		900
	ACRYLIC ACID, 2-CHLORO-, METHYL ESTER	0.08		0.3		60		259
	ACRYLIC ACID, 2-CYANO-, METHYL ESTER	0.25		0.30		70	C	259
	ACRYLIC ACID, 3-ETHOXY-, ETHYL ESTER	0.01		0.03		60	I7	507
		23.5	1	0		80	C	786

M1	M2	R1	+/-	R2	+/-	TEMP.	REMARKS	REF.
STYRENE	ACRYLIC ACID, 3-ETHOXY-, ETHYL ESTER	24		0.01		60		829
	ACRYLIC ACID, 2-FLUOROMETHYL-, ETHYL ESTER	0.341	0.013	0.091	0.017	60	I2	779
	ACRYLIC ACID, 2-HYDROXY-, ETHYL ESTER, ACETATE	0.57	0.05	0.20	0.05	60	C	993
	ACRYLIC ACID, 2-METHOXY-, METHYL ESTER	1.1		0.51		60		1430
		1.1		0.51		60		1341
	ACRYLIC ACID, 3-PROPOXY-, PROPYL ESTER	42		0.01		60		829
	ACRYLIC ACID, THIO-, BENZYL ESTER	0.15		0.2		60		1133
	ACRYLIC ACID, THIO-, BUTYL ESTER	0.19		0.34		60		1133
	ACRYLIC ACID, THIO-, TERT-BUTYL ESTER	0.21		0.45		60		1133
	ACRYLIC ACID, THIO-, ETHYL ESTER	0.19		0.28		60		1133
	ACRYLIC ACID, THIO-, PROPYL ESTER	0.17		0.32		60		1133
	ACRYLIC ANHYDRIDE			0.1		35	I13	891
	ACRYLONITRILE	0		0.28		75	R	319
		0		0		-20	E	919
		0.005		33		-78	E	919
		0.005		33.0		-78	E,16	2
		0.02	0.01	20	2	-78	E,16	635
		0.03	0.02	0.41	0.03	-70/0	H21,I2	204
		0.05	0.02	15	1	65	I15,L	285
		0.05	0.02	0.35	0.04	-45	H21,I2	204
		0.09		0.17		65	I15,L	285
		0.2		1.5		40	E,I15	833
		0.3	0.08	0.02	0.02	0	H12,I21	847
		0.33		0.03		40	H38	62
		0.33	0.03	0.00	0.02	0/15	E	919
		0.36		0.00		50		560
		0.37		0.07		20	E	167
		0.37	0.02	0.05	0.02	90		1083
		0.38	0.03	0.07	0.006	50		352
		0.38	0.03	0.05	0.02	50		958
		0.39	0.02	0.06	0.01	41,5		297
		0.398		0.058		99		213
		0.4	0.1	0.05	0.01	70	C,J	803
		0.40	0.05	0.04	0.04	30	D	1102
		0.41	0.08	0.04	0.04	60	C	212
		0.41	0.08	0.03	0.03	75	C	558
		0.43		0.13		90	C	259
		0.45	0.01	0.385	0.015	130	M4	1083
		0.45	0.03	0.02	0.02	65	G22/G21	1120
		0.47	0.03	0.02	0.02	86,5		297
		0.5		2		80		297
		0.52	0.04	0.03	0.03	67/80	H12,I21	847
		0.55		0.14		90	K	258
		0.62	0.02	0.21	0.01	110	M3	1083
		0.7		12		0	G22/G21	1120
		2		5		80	H7,I21	847
	ACRYLONITRILE, 2-CHLORO-	0.055	0.01	0.075	0.01	80	H7,I21	302
		0.06		0.13		60		1430
		0.06		0.13		60		1429
		0.1	0.03	0.08	0.02	80	C	134
	ACRYLONITRILE, 2-HYDROXY-, ACETATE	0.16		0.2		60		1430
		0.19	0.056	0.20	0.052	60		1070

REMARKS PAGE II-366, REFERENCES PAGE II-368

M1: STYRENE

M2	R1	+/-	R2	+/-	TEMP.	REMARKS	REF.
ACRYLONITRILE, 2-HYDROXY-, ACETATE	0.2	0.04	0.16	0.04		Q	727
ACRYLONITRILE, 2-METHOXY-	0.35		0.35	0.10	60		1430
	0.53		0.35	0.01	60		1341
ACRYLOPHENONE	0.107	0.01	1.10	0.007	60		746
ACRYLOYL CHLORIDE	0.10		0.03	0.01	67	I7	892
ADIPIC ACID, DIALLYL ESTER	35.5	0.9	0.087	0.05	60	C	1293
ADIPIC ACID, METHYL VINYL ESTER	49.5	0.7	0.02	0.1	80	C	268
AMMONIUM CHLORIDE, DIETHYL(2-HYDROXYETHYL)-, METHACRYLATE	0.3	0.02	0.25		70	I12	1109
	0.43		0.55	0.05	70	I12	1109
ANILINE, N,N-DIMETHYL-P-VINYL-	0.89	0.1	0.62			Q	1040
	0.89		0.62		60		1168
	1.015	0.06	0.84	0.08	60		1024
ANILINE, N,N-DIMETHYL-P-(P-VINYLPHENYLAZO)-	0.4		3.8	0.05	60		125
ANILINE, N,N-DIVINYL-	13.0		0.45		60		162
ANILINE, N-(ALPHA-PHENYL-P-VINYLBENZYLIDENE)-	0.36	0.02	1.88	0.08			125
O-ANISALDEHYDE	0.1	0.02	0	0.01	0	H10	1402
P-ANISALDEHYDE	0.1	0.02	0	0.2	0	H10	1402
ANISOLE, P-PROPENYL-	0.45	0.2	2.8	0.2	30	H10,I24	652
ANISOLE, P-PROPENYL-, CIS-	0.42	0.12	8.24	0.86	0	H10,I23	1215
ANISOLE, P-PROPENYL-, TRANS-	0.13	0.04	1.56	0.11	0	H10,I23	1215
ANISOLE, M-VINYL-	0.90	0.15	1.1	0.15	0	H20,I33	752
ANISOLE, O-VINYL-	0.11	0.04	3.6	0.8	-20	H10,I28	382
	0.20	0.02	3.9	0.7	30	H19,I28	382
ANISOLE, P-VINYL-	0.00	0.03	31	6	0	H23,I32	965
	0.01		72		0	H23,I2	965
	0.01		100		0	H20,I32	752
	0.05	0.04	46	3	0	H23,I32	965
	0.12	0.07	14	3	0	H23,I33	965
	0.22		7.6		0	H23,I22	965
	0.34	0.05	11	1	0	H46,I33	965
	0.38	0.04	11.5	0.7	0	H23,I33	965
	0.48	0.08	5.6	0.08	30	H23,I22	965
	0.5		38		25	H10,I28	382
	0.70	0.10	1.50	0.22	30	E	149
	1.0		0.85				505
	1.05		0.79		60	Q	1040
	1.05		0.79		60		1168
	1.13		0.93		60		631
	1.16	0.09	0.82	0.07	0		1024
	2.9	0.2	0.23	0.02	1	G4,I5	965
	4.1	0.5	0.13	0.02	0	G18,I5	965
	-0.02	0.04	29	5	0	H20,I33	965
	-0.04	0.04	19	3	0	H20,I33	965
	-0.12	0.012	35	5	0	H23,I33	965
	-0.33	0.03	12	2	0	H23,I2	965
	10.9	0.8	0.05		0	H17,I12	965
ANTHRACENE, 1-VINYL-	19		0.045		25	G18,I5	862
ANTHRACENE 9-VINYL-	0.81	0.03	0.57	0.04	65		477
	2.12	0.03	0.25	0.02	100		477
ARSINE, DIPHENYL(P-VINYLPHENYL)-	2.2	0.03	0.3	0.02	70	I2	172
ATROPIC ACID, BUTYL ESTER	0.46		1.18	0.04	65		126
ATROPIC ACID, 2-CHLOROETHYL ESTER	0.09	0.03	0.23	0.03	60		748
	0.06	0.03	0.30	0.03	60		748

COPOLYMERIZATION PARAMETER

M1	M2	R1	+/-	R2	+/-	TEMP.	REMARKS	REF.
STYRENE	ATROPIC ACID, ETHYL ESTER	0.04	0.03	0.19	0.03	60		748
	ATROPIC ACID, METHYL ESTER	0.04	0.03	0.19	0.04	65		175
		0.03	0.02	0.4	0.2	65		174
	ATROPIC ACID, PROPYL ESTER	0.055		1.0		60		571
	ATROPONITRILE	0.06	0.03	0.45	0.03	60		748
		0.08	0.03	0.19	0.03	60		748
	2H-AZEPIN-2-ONE, HEXAHYDRO-7-(1-CYCLOHEXENYL)-	0.02		0.7		80	C	538
	2H-AZEPIN-2-ONE, HEXAHYDRO-1-METHACRYLOYL-	2		0.36		60	I7	1332
		1.0		0		70/80		1031
	AZIRIDINE, 1-ACRYLOYL-	1.86	0.10	0.25		130	C,I	270
		0.56		0		70	I10	1421
	AZIRIDINE, 1-CROTONOYL-, TRANS-	20		0.41		70	I10	1421
	AZIRIDINE, 1-METHACRYLOYL-	0.47		0.01		70	I10	1421
	1-AZIRIDINECARBOXAMIDE, N-ISOPROPENYL-	15.7	0.8	0.05	0.03	60	I5	418
	1-AZIRIDINECARBOXAMIDE, N-VINYL-	14.5	0.9	0.0	0.04	60	I5	418
	1-AZIRIDINECARBOXYLIC ACID, ALLYL ESTER	50		0.01		70	I10	1421
	BENZALDEHYDE	0.27	0.06	0.01	0.01	0	H10	73
		0.3	0.02	0.0	0.02	0	H11	1402
		0.35		0.0		50	H10	228
	BENZALDEHYDE, O-CHLORO-	1.8		0.5		50	H11	228
	BENZALDEHYDE, P-CHLORO-	0.53	0.08	0		0	H10	1402
	BENZALDEHYDE, N-NITRO-	0.45		0		0	H10	1402
	BENZALDEHYDE, O-NITRO-	0.8		0		0	H10	1402
	BENZALDEHYDE, P-NITRO-	1.52		0		0	H10	1402
	BENZAMIDE, P-VINYL-	2.13		1.0		100	I5	57
	BENZAMIDE, P-VINYL-, LITHIUM CHLORIDE COMPLEX	0.47		1.6		100	G17,16	1041
	BENZENE, M-DIVINYL-	0.10		0.88		80	C,EE	1044
		0.60		0.60		60	C	1042
	BENZENE, O-DIVINYL-	0.65		0.60		61	C	1041
	BENZENE, P-DIVINYL-	0.65		1.08		80	C,EE	1039
		1.27		1.0		80	C	1034
	BENZENE, PROPENYL-, CIS-	0.92		0.5		70	C	1041
	BENZENE, PROPENYL-, TRANS-	0.14		1.46		80	H20,I21	650
	BENZENE, 2,3,5,6-TETRACHLORO-P-DIVINYL-	0.77		2.08		0	H20,I21	650
	BENZENE, 1,2,4-TRIVINYL-	0.77		0.2		0	H10,I24	652
	BENZENESULFONAMIDE, N-(2-CHLOROETHYL)-P-VINYL-	1.95	0.1	0.07		30	C	810
	BENZENESULFONAMIDE, P-VINYL-	1.4	0.15	1.12		49.2	I10	1034
	BENZENESULFONIC ACID, MALEIMIDO ESTER	1.8	0.2	1.25		70	B,I6	428
	BENZENESULFONIC ACID, P-VINYL-, POTASSIUM SALT	2.62		1.07		85	I10	1045
		1.80		0.03		90	B	1319
		0.25	0.1	0.56		35	B,I6	1045
	BENZENESULFONIC FLUORIDE, M-VINYL-	0.24		0.93		90	B	1045
	BENZENESULFONIC FLUORIDE, P-VINYL-	-0.02		0.54		110	I2	332
		-0.06		1.25		90	I29	332
	BENZOFURAN	0.8	0.1	0.95		30	H10,I24	653
	BENZOIC ACID, MALEIMIDO ESTER	0.03		0.01		35	I10	1319
	BENZOIC ACID, VINYL ESTER	38		0.05		80	C	82
	BENZOIC ACID, P-HYDROXY-, METHACRYLATE	0.13		0.84		105	C,I10	1150
		0.32		1.24		105	C	1150

M1	M2	R1	+/-	R2	+/-	TEMP.	REMARKS	REF.
STYRENE								
	BENZOIC ACID, P-VINYL-	1.04		0.28	0.10	60		1078
	BENZOIC ACID, P-VINYL-, 2,3-EPOXYPROPYL ESTER	0.40	0.02	0.95	0.10	70	C	429
		0.40	0.02	0.95		70		430
	BENZONITRILE, P-((4-AMINO-6-VINYL-S-TRIAZIN-2-YL)AMINO)-	0.64		0.31		60	I17	1453
	BENZONITRILE, P-VINYL-	0.19		1.2		30		505
		0.28		1.16		60		1168
	BENZOPHENONE, 4-VINYL-	0.28	0.025	1.16	0.13	60		1024
		0.18	0.08	3.0	0.3	60		129
	BENZOYL CHLORIDE, P-VINYL-	0.17	0.03	1.80	0.20	60		430
	BENZYL ALCOHOL, ALPHA, ALPHA-DIMETHYL-P-VINYL-	0.50	0.05	0.56	0.05	50		128
		0.79		1.25		60	I66	1092
	1,1'-BIINDENE	1.95	0.1	0.17	0.05	-72	H23,I3	1279
	BIPHENYL, 4-FLUORO-4'-VINYL-	0.50	0.15	0.88	0.07	70		106
	BIPHENYL, 4-VINYL-	0.89	0.1	1.4	0.1	60	I7	1248
		0.98	0.04	0.92	0.08	70		106
	BISMUTHINE, DIPHENYL(P-VINYLPHENYL)-	0.63	0.05	1.35	0.10	65	I2	126
	BORAZINE, 2,4,6-TRIALLYL-1,3,5-TRIPHENYL-	42.9		0		80		768
	BORAZINE, 1,3,5-TRIPHENYL-2,4,6-TRIVINYL-	0.26		0		80		768
	1,3-BUTADIENE	0.01	0.01	1.30	0.1	43	H38	60
		0.04		10		50	G32,I7	1397
		0.05		20		50	G17,I7	529
		0.06		3.0		29	G17,I7	657
		0.1		12.5		25	H17,I2,N	909
		0.23	0.07	1.48	0.08	50		352
		0.3		5.5		25	H17,I8,N	909
		0.38		1.37		-18	K	290
		0.38		1.37		-18	K	731
		0.38	0.08	1.40	0.08	5	K	730
		0.44		1.40		5		730
		0.44	0.03	1.59	0.05	50	G32,I5	353
		0.5		1.12		50	K	1397
		0.5		1.4		50	C,K	628
		0.58	0.1	1.35	0.2	50		353
		0.6	0.15	1.8	0.12	45	K	644
		0.64	0.1	1.38	0.4	5	K	290
		0.65		1.83		45	K	290
		0.65	0.16	1.83	0.32	45	K	644
		0.68		0.50		25	U	776
		0.78	0.01	1.39	0.03	60		560
		0.825	0.086	1.39	0.15	30		189
		1.68		0.6		20	G32,I5	1397
		4.6	0.5	0.13	0.015		H76,Q	1122
		8	1	0.2	0.1	-35	H17,I5,N	909
		9.8		0.44		-15	G32,I5	1397
		<0.04		27.5		40	G4,I21	449
		-0.1		7.0		30	G17,I20	547
	1,3-BUTADIENE, 2-CHLORO-	0.00		6.3		50		353
		0.00	0.02	5.22	0.1	50	K	353
		0.005	0.01	6.3	0.64	50	Q	266
		0.05		7	2	70	C	26
		0.052	0.02	8.11	0.34	60	C	211
		0.2	0.01	18		30	H71	1063
		6.9	0.5	0.04	0.01	0	H30,I7	758

M1 = STYRENE

M2	R1	+/-	R2	+/-	TEMP.	REMARKS	REF.
1,3-BUTADIENE, 2-CHLORO-	12.8	0.3	0.06	0.01	0	H10,I9	758
	15.6		2.4			H,Q	266
	15.6	0.05	0.24	0.01	-18	G3	263
	16.0	0.5	0.065	0.05	0	H30,I22	758
	33.0	0.012	0.15		0	H10,I23	758
	0.42	0.02	10.8	1.2	60	C	211
1,3-BUTADIENE, 2,3-DICHLORO-	0.16	0.08	0.92	0.02	-18	K	732
1,3-BUTADIENE, 2,3-DIMETHYL-	0.50	0.10	1.61	0.24	5	K	733
1,3-BUTADIENE, 2-FLUORO-	1		1.55	0.10	50	K	733
1,3-BUTADIENE, HEXACHLORO-	0.07		0		70		23
1,3-BUTADIENE, 1,1,2-TRICHLORO-	0.07		1.18		70		269
1,3-BUTADIENOLS, ACETATE, MIXED ISOMERS	0.3	0.03	1.18	0.08	60	K	520
BUTANE, 1,4-BIS(P-VINYLPHENYL)-	1.13		2.4		70	I2	1275
1-BUTANOL, 2,2,3,3,4,4,4-HEPTAFLUORO-, ACRYLATE	0.33		0.78		61	C	1042
1-BUTENE	1.98		0.07		50	C,K	838
1-BUTENE, 3,3-DIMETHYL-2-PHENYL-	30	3.7	5.51		50	H12	827
1-BUTENE, 2,4-DIPHENYL-	1.6				110	A,O	1189
1-BUTENE, 3,4-EPOXY-2-METHYL-	77		0		110	A,O	1189
2-BUTENE, 1,3-DICHLORO-	15		0		60-140	C	630
2-BUTENE, 2-METHYL-	1.55		0.65	0.03	70	C	229
3-BUTEN-2-OL, 1-CHLORO-3-METHYL-	1.65	0.31	1.7	0.19	-78	H23,I9	1223
3-BUTEN-2-ONE	1.85	0.21	0.8	0.2	-78	H23,I3	1223
3-BUTEN-2-ONE, 3-METHYL-	110	0.25		0.11	-78	H23,I42	1223
BUTYRALDEHYDE	0.29	0.04	0.35		60-140	C	630
	0.35	0.06	0.27		60		560
	0.32		0.66		60	C	1114
	0.44	0.10	0.29		80	C	1070
	0.3		0.6		80		1070
	0.4		0.5		-10	H11,I7,Q	1108
	1.8		0.017			H10	1182
	8.73		0.31			H10,I7,Q	1108
BUTYRIC ACID, HEPTAFLUORO-, VINYL ESTER	0.68	0.01	0.21	0.02	70	C,Q	1070
BUTYRIC ACID, 2-METHYLENE-	0.82	0.08	0.20	0.02	65		1212
BUTYRIC ACID, 2-METHYLENE-, METHYL ESTER	0.90	0.07	0		70		174
	1.85	0.07			65		175
BUTYRIC ACID, 3-METHYL-2-METHYLENE-, METHYL ESTER	1.94	0.05	0		70		174
	0.56	0.05			65		175
CADMIUM, DIACETATOBIS(5-VINYL-2-PICOLINE)-	32	5	0.47	0.04	60	I6	952
CARBAMIC ACID, DIETHYL-, VINYL ESTER	4.05	0.6	0.03	0.01	66		802
CARBAMIC ACID, DIETHYLDITHIO-, S-VINYL ESTER	4.4	3.0	0.143		60		681
CARBAMIC ACID, DIETHYLTHIO-, S-VINYL ESTER	13.0	0.02	0.14	0.03	66		802
CARBAMIC ACID, N-METHYL-N-VINYL THIO-, S-ETHYL ESTER	10.4	0.01	0.025	0.01	66		802
CARBAMIC ACID, VINYL-, 2,3-EPOXYPROPYL ESTER	0.01		0.02	0.02	60	I5	418
CARBAZOLE, 9-VINYL-	5.5	0.8	>7		30	H52,I5/6	949
CARBONIC ACID, CYCLIC (HYDROXYMETHYL)ETHYLENE ESTER, ACRYLATE	5.7		0.012	0.002	70	C	29
CARBONIC ACID, CYCLIC (HYDROXYMETHYL)ETHYLENE ESTER, MALEATE	5.7		0.035	0.05	75	I7	331
CARBONIC ACID, CYCLIC (HYDROXYMETHYL)ETHYLENE ESTER, 4-(METHYLENESUCCINATE)	0.74	0.03	0.035	0.05	75	I7	331
	0.74	0.03	0.10		50	C,I2	199
			0.10		50	C,I2	199
CARBONIC ACID, CYCLIC VINYLENE ESTER	<20.0		0		60		458
CARBONIC ACID, THIO-, O-TERT-BUTYL S-VINYL ESTER	3	0.1	0.2	0.05	70		1176

M1	M2	R1	+/-	R2	+/-	TEMP.	REMARKS	REF.
STYRENE								
	CHROMIUM, TRICARBONYL-P1-(BENZYL ACRYLATE)-	0.34	0.13	0.1	0.02	60	I38	1357
	CHROMIUM, TRICARBONYL-P1-(STYRENE)-	1.39		0		70	I7	1355
	CINNAMIC ACID	1.85	0.03	0.16	0.03	60	I7	86
	CINNAMIC ACID, BENZYL ESTER	1.45	0.03	0.10	0.03	60		748
	CINNAMIC ACID, BUTYL ESTER	0.87		0.08	0.03	80	C	1070
	CINNAMIC ACID, TERT-BUTYL ESTER	1.96	0.03	0.10	0.03	60		748
	CINNAMIC ACID, 2-CHLOROETHYL ESTER	1.83	0.03	0.16	0.03	60		748
	CINNAMIC ACID, ETHYL ESTER	1.48	0.03	0.05	0.03	60		748
		1.50	0.03	0.10	0.03	60		748
	CINNAMIC ACID, METHYL ESTER	2.7	0.3	0.08		75	I7	806
		1.9	0.2	0.05		60	D	211
	CINNAMIC ACID, PHENYL ESTER	2.03	0.03	0.10	0.03	60	I7	748
	CINNAMIC ACID, VINYL ESTER	1.53	0.03	0.08	0.03	70		806
	CINNAMIC ACID, ALPHA-CYANO-, BENZYL ESTER	1.25	0.1	0.25	0.1	80	C,R	1199
	CINNAMIC ACID, ALPHA-CYANO-, BUTYL ESTER	0.43		0		80	C,R	1199
	CINNAMIC ACID, ALPHA-CYANO-, CYCLOHEXYL ESTER	0.53		0		80	C,R	1199
	CINNAMIC ACID, ALPHA-CYANO-, ETHYL ESTER	0.44		0		80	C	538
		0.3		0		80	C,R	1199
	CINNAMIC ACID, ALPHA-CYANO-, 2-ETHYLHEXYL ESTER	0.51		0		80	C,R	1199
	CINNAMIC ACID, ALPHA-CYANO-, HEXYL ESTER	0.59		0		80	C,R	1199
	CINNAMIC ACID, ALPHA-CYANO-, METHYL ESTER	0.42		0		80	C,R	1199
		0.45		0		80		118
	CINNAMONITRILE	0.9	0.2	0.1	0.1	90	C	538
		2.2		0		80	C	861
	CITRACONIC ANHYDRIDE	0.15	0.02	0.01	0.01	60	C	166
		0.15		0.01		60	M3	393
	CROTONALDEHYDE	14.7		0.12		100	M2	393
		14.7		0.07		100		393
		14.7		0.03		100		211
	CROTONIC ACID	20		0.01		60	C,I7	641
	CROTONIC ACID, METHYL ESTER	26.0		0.01		60	C,I7	928
		26.0				60		1306
	CROTONIC ACID, 2-PROPYNYL ESTER	3.42	5.6	0		60		458
	CROTONIC ACID, 4-HYDROXY-, GAMMA-LACTONE	8.5	0.5	0		60	C,R	444
	CROTONONITRILE, TRANS-	20.0		0		60	C,I7	928
		24.0		0		60	C,I7	641
		24.0		0		60	C,I7	641
	CYCLOBUTANECARBOXYLIC ACID, 3-ACETYL-2,2-DIMETHYL-, VINYL * ESTER	2.54	0.05	0.22	0.05	90	C,I7	1414
	1,5,9-CYCLODODECATRIENE, TRANS-,TRANS-,CIS-	0.53	0.10	0.36	0.10	0	G17	1058
	1,3-CYCLOHEXADIENE	0.53		0.77		0	H10	1058
	1,3-CYCLOHEXADIENE, 1-METHOXY-	4.68	0.2	0.01	0.2	60		1434
	CYCLOHEXANE, 1,2-DIVINYL-	30		0.15		0	H4,I2	1138
	CYCLOHEXANE, 1,1,2,3,3,4,4-OCTAFLUORO-5,6-DIMETHYLENE-	0.34		0		50		1155
	CYCLOHEXANEACETIC ACID, ALPHA-METHYLENE-, METHYL ESTER	1.65	0.05	0		65		174
	3-CYCLOHEXENE-1-CARBOXYLIC ACID, 1-METHYL-, 2,3-EPOXYPROPYL * ESTER	1.65	0.05	0		65		174
	3-CYCLOHEXENE-1,5-DICARBOXYLIC ACID, 1,2-DIMETHYL-, BIS(2,3 * -EPOXYPROPYL) ESTER	1.65	0.05	0		65		174
	1,3-CYCLOOCTADIENE, CIS-,CIS-	1.63	0.05	0.48	0.05	0	H20	1304
		1.64	0.05	0.5	0.05	-20	H20	1304
		1.66	0.05	0.56	0.05	-20	H23	1304

M1	M2	R1	±	R2	±	TEMP.	REMARKS	REF.
STYRENE	1,3-CYCLOOCTADIENE, CIS-,CIS-	1.9	0.05	0.32	0.05	0	H23	1304
	4-CYCLOPENTENE-1,3-DIONE	0.024		0.415		50		1048
	CYCLOPROPENE, TETRACHLORO-	5.7	0.4	0		60		1214
	1-DECENE	5.0		0.16		50		827
	1,3-DIOXEPANE	0.37	0.07	24.5	2.7	50	H12	719
	M-DIOXIN, 4-METHYL-	2.5		0.05		0	H10,I2	720
	1,3-DIOXOLANE	0.35	0.05	1.9	0.2	25	R	719
	1,3-DIOXOLANE	0.65	0.07	6.5	0.85	0	H10,I2	481
	1,3-DIOXOLANE, 4-(CHLOROMETHYL)-	10.5	1.5	0.38	0.05	0	H10,I2	719
	1,3-DIOXOLANE, 2-VINYL-	44.9		0.02		60	I7	1351
	1,3-DIOXOLANE-4-METHANOL, 2,2-DIMETHYL-, ACRYLATE	0.389	0.04	0.17	0.02	50	C,7	195
	1,3-DIOXOLANE-4-METHANOL, 2,2-DIMETHYL-, 4-(METHYLENESUCCINA* TE)	0.389	0.04	0.17	0.02	50	C,I2	195
	1,3-DIOXOLANE-4-METHANOL, 2,2-DIMETHYL-, METHYL FUMARATE	0.17	0.05	0.03		50	C,I2	195
	1,3-DIOXOLANE-4-METHANOL, 2-VINYL-	20.4		0.22		60	I7	1351
	1,3-DIOXOLAN-4-OL, 2-VINYL-, ACRYLATE	0.81		0.51		60		1352
	1,3-DIOXOLAN-4-OL, 2-VINYL-, METHACRYLATE	0.49		0		60		1352
	DISILANE, PENTAMETHYLVINYL-	50	10	0		60	I7	837
	DISILOXANE, 1-(CHLOROMETHYL)-1,1,3,3-TETRAMETHYL-3-(P-VINYL * PHENYL)-	50	10	0		60	I7	837
	DISILOXANE, 1,1,1,3,3-PENTAMETHYL-3-VINYL-	60	9	0.10	0.01	50		773
	DISILOXANE, 1,1,1,3,3-PENTAMETHYL-3-(P-VINYLPHENYL)-	1.04	0.06	1.2	0.4	80		306
	DISILOXANE, 1-(P-VINYLPHENYL)-	1.04	0.07	1.2	0.4	80		371
	DODECANAMIDE, N-(ACRYLAMIDOMETHYL)-	0.24		0.88	0.02	80	I7	462
	ETHANE, 1,2-BIS(P-VINYLPHENYL)-	1.05	0.03	0.87	0.2	61	C	1042
	ETHANESULFONIC ACID, 2-HYDROXY-, METHACRYLATE	0.37		0.6		60	I55	1241
	ETHENESULFONAMIDE, N-(2-CHLOROETHYL)-	3		0.075		62		1234
	ETHENESULFONIC ACID, ALLYL ESTER	1.6	0.1	0.23	0.03	60	I7	209
	ETHENESULFONIC ACID, BUTYL ESTER	2.5	1.0	0.13	0.03	90	C	754
	ETHENESULFONIC ACID, 2,3-EPOXYPROPYL ESTER	1.19	0.11	0.03	0.08	70	I5	415
	ETHER, BIS 2-(VINYLOXY ETHYL)	16.8	0.5	0		60		1456
	ETHER, BIS(P-VINYLPHENYL)	10.94		1.06	0.02	61	C	1042
	ETHER, 2-BROMOVINYL ETHYL	37.5	2.0	0.02	0.4	80	C	786
	ETHER, BUTYL VINYL	0.34	0.03	14.4	0.5	-78	H84	1257
		0.44	0.07	18	0.15	20	H84	1257
	ETHER, 2-CHLOROETHYL VINYL	15.0	6.0	0.0		100		988
		0	0.01	51.0	15	25	H33,I7	143
		0.04	0.01	16.0	1.5	-30	H10,I24	143
		0.06	0.02	8.0	1.0	-30	H32,I24	143
		0.06	0.02	12.0	1.0	25	H34,I24	143
		0.08	0.03	24.0	2.0	25	H33,I24	143
		0.11	0.04	24.0	3.0	-30	H10,I22	143
		0.12	0.05	30.5	8.0	30	H33,I24	143
		3		36		100	H31,I24	470
	ETHER, DODECYL VINYL	27	2	0.0	0.05	70		14
		56	5	0.0	0.05	80		14
	ETHER, ETHYL VINYL	80	40	0.0		60	C	786
		90	20	0.0		0		560
	ETHER, ISOBUTYL VINYL	17.0	4.0	10.0	4.0	0	E,X	989
		50	0.1	0.1	0.1	0		775
	ETHER, METHYL 1-PHENYLVINYL	2.7	0.3	0.01		60	Q	572
	ETHER, METHYL VINYL	100	0.6	0.01			Q	775

REMARKS PAGE II-366, REFERENCES PAGE II-368

M1	M2	R1	+/-	R2	+/-	TEMP.	REMARKS	REF.
STYRENE	ETHER, OCTYL VINYL	65		0.0		70	C,Q	14
	ETHER, PHENYL VINYL	1.7	5	0.01	0.04	50	C,P	376
	ETHERS, C8-C18 ALKYL VINYL	12.6		0.038		60		12
	ETHYLENE, BROMO-	18	2	0.06	0.015	28	A	1097
		23	2	0.05		0	D	111
	ETHYLENE, CHLORO-	12.4		0.045		60		111
		17	3	0.02		68	P,O,R	321
		17.9	0.1	0.040	0.001	50	Q	961
		28		0.08		50		210
		32		0.17		50	I26,M	451
		35		0.077		60	H81	395
		35		0.067		45	K	990
	ETHYLENE, CHLOROTRIFLUORO-	7.0		0.001		60	I26	165
	ETHYLENE, 1,1-DICHLORO-	0		0.046		25	C	775
		0.0054		0.001		60	Q	955
		1.75	0.10	0.12	0.02	50	C	637
		1.80	0.27	0.15	0.02	26	E	289
		1.85	0.05	0.085	0.010	90		318
		2.0	0.1	0.14	0.05	50	C	149
		2.1	0.2	0.145	0.009	90		210
		3.4		0.015		68		558
	ETHYLENE, 1,2-DICHLORO-, CIS-	47.5		0		60	G17,I9	958
		56.1		0		68		526
		72.5		0		90		133
		75	25	0		90		133
		210	31.5	0		60		27
		10		0		50	C	557
	ETHYLENE, 1,2-DICHLORO-, TRANS-	25.9		0		50	C	27
		29.8		0				133
		37	3	0				133
		44.4		0				557
		44.9		0			D	133
	ETHYLENE, 1,1-DICHLORO-2,2-DIFLUORO-	1.6		0		45	K,N	626
	ETHYLENE, 1,1-DIPHENYL-	0.13		0		30	H17,I5	1075
		0.16		0		30	H20,I5	1075
		0.23		0		30	H19,I5	1075
		0.29		0		30	H19,I5	1075
		0.34	0.04	0		110	A,O	1189
		0.34		0		110	A	1188
		0.42		0		30	H17,I2	1075
		0.44		0		30	H17,I2	1075
		0.63		0		30	H17,I9	1075
		0.71		0		30	H17,I7	1075
	ETHYLENE, IODO-	7.0	0.2	0.145	0.01	60		130
	ETHYLENE, 1-PHENYL-1-(P-VINYLPHENYL)-	0.4		4.4		80		1158
	ETHYLENE, TETRACHLORO-	66.4		0		90		133
		129		0		90		133
		165		0		50		133
		185	20	0		60	D	210
		187	20	0		90		133
		200		0		60	C	824

M1	M2	R1	+/-	R2	+/-	TEMP.	REMARKS	REF.
STYRENE	ETHYLENE, TETRACHLORO-	208		0		50		133
	ETHYLENE, TETRAFLUORO-	5.2	2	0.01		80	E	1254
	ETHYLENE, TRICHLORO-	12.1		0		90		133
		12.7		0.0		90		133
		16.				60		210
		16.5		0		50		133
	ETHYLENEDIAMINE, N,N'-BIS(P-HYDROXYBENZYLIDENE)-, DIMETHACRYLATE	17.1		0		50		133
	ETHYLENE GLYCOL, BIS(2-CHLOROACRYLATE)	17.1		0		50		133
	ETHYLENE GLYCOL, DIMETHACRYLATE	0.6		0.1		60	P,Q	900
		0.35		0.65		60	Q	908
	FERROCENE, VINYL-	0.40	0.2	0.65	0.04	61	C	1044
		2.5	1	0.08	0.1	70	I7	1042
		4		0.2		70	I7	1268
	FERROCENEETHANOL, ACRYLATE	1.06	0.051	0.408	0.138	65	I7	1140
	FERROCENEETHANOL, METHACRYLATE	0.582	0.01	0.084	0.018	65	I7	1358
	FERROCENEMETHANOL, ACRYLATE	2.3	0.3	0.02	0.01	70	I7	1358
		2.5	0.01	0.02	0.01	60	I7	1267
	FERROCENEMETHANOL, METHACRYLATE	3.6	0.01	0.01	0.01	60	I7	1356
		3.7	0.2	0.03	0.02	70	I7	1267
	FORMALDEHYDE	0		30		-30	E	161
		0		52		-78	E,13	161
	FORMIC ACID, VINYL ESTER	17.6	0.6	0.01	0.01	70	C	858
	FUMARAMIDE, N,N'-DIBUTYL-	10		0.69		70		1144
	FUMARAMIDE, N,N,N',N'-TETRAETHYL-	0.69		0		70	I7	1144
	FUMARANILIDE	1.4		-0.05		100	I6	1144
	FUMARIC ACID, BIS(2-CHLORO-1-METHYLPROPYL) ESTER	0.25	0.02	0.03	0.01	60	C	1010
	FUMARIC ACID, 1,3-BUTYLENEGLYCOL POLYESTER	3.0	0.4	0	0.03	60		966
	FUMARIC ACID, SEC-BUTYL ESTER	0.34	0.03	0	0.01	60		1010
	FUMARIC ACID, DI-SEC-BUTYL ESTER	0.50	0.03	0.070	0.0005	60		1010
	FUMARIC ACID, DIESTER WITH HYDRACRYLONITRILE	0.15	0.05	0.0697	0.007	90		1137
	FUMARIC ACID, DIETHYL ESTER	0.30	0.02	0.05	0.0041	60	C	559
		0.301	0.024	0.1	0.02	60		559
	FUMARIC ACID, DIMETHYL ESTER	0.31	0.02	0.0905	0.02	100	C	1010
		0.36	0.04	0.025	0.007	131		1219
	FUMARIC ACID, DIMETHYLSILOXYLETHYLENEGLYCOL POLYESTER	0.400	0.02	0.6	0.015	60	H2O,I7	592
	FUMARIC ACID, ETHYLENEGLYCOL POLYESTER	0.21	0.02	2.16		60		557
	FUMARIC ACID, ETHYL ESTER	0.03	0.03	0.25		60	Q	146
		0.062	0.030			60		151
		0.18	0.1			60	C	501
			0.1			60	C	557
	FUMARIC ACID, PHENOLPHTHALEIN ISOPHTHALATE POLYESTER	0.098	0.013	5.8	0.5	60	C	1015
	FUMARIC ACID, DITHIO-, DIMETHYL ESTER	0.09	0.005	0.0163	0.013	50	C,I7	598
	FUMARONITRILE	0.19	0.03	0.00		79		262
		0.23	0.01	0		60		557
	FUMAROYL CHLORIDE	0.3	0.3	0.01	0.01	60	I2R	780
		0.30	0.02	0.00		50/70	R	659
	2-FURALDEHYDE	0.04		0.0		27		805
		0.1		0.27		49,65		315
	FURAN, 2-VINYL-	2.0		0.25	0.02	50	H10	234
		0.25	0.05	1.9	0.1	70	C	234
						70		74

M1	M2	R1	+/-	R2	+/-	TEMP.	REMARKS	REF.
STYRENE	GERMANE, DIALLYL-	1.05		0.93	0.08	60	P,Q	517
	GERMANE, TRIETHYLMETHACRYLOYL-	24.4	0.02	0.009		50	I7	518
	GERMANE, TRIMETHYLVINYL-	2.72		0.056		50	C	1300
	D-GLUCITOL, 1-ACRYLAMIDO-1-DEOXY-	2.09		0.005		50	C	1029
	D-GLUCITOL, 1-DEOXY-1-METHACRYLAMIDO-	4.6		0.13		80	C	1029
	D-GLUCITOL, 1,2:3,4:5,6-TRI-O-ALLYLIDENE-	2.31		0.39		80	C	1350
	D-GLUCITOL, 1,2:3,4:5,6-TRI-O-(2-BUTENYLIDENE)-	0.45		0.00		25	F,K	1350
	GLUTARONITRILE, 2-METHYLENE-	0.50		0.02		25	F,K,DD	789
	GLYCOLIC ACID, ETHYL ESTER, ACRYLATE	0.85	0.06	0.25	0.15	70	Q	789
	GLYCOLONITRILE, METHACRYLATE	0.68	0.03	0.08	0.02	60	Q	946
	3,5-HEPTADIEN-2-ONE	0.45	0.02	0.36	0.02	30		1446
	1-HEPTENE	0.18	0.05	0.59	0.12	30	H7,I7	1016
		0.43	0.05	0.95	0.14	35	H29,I7	272
		0.47	0.01	1.34	0.15	50	H12,I2	79
	HEXANOIC ACID, 2-METHYLENE-	0.61		5.70		50		79
	HEXANOIC ACID, 2-METHYLENE-, METHYL ESTER	1.92		0.51		70		43
	1,3,5-HEXATRIENE, 2,3,4,5-TETRACHLORO-, TRANS-	0.52	0.01	0.32	0.04	65		119
	1-HEXENE	0.80	0.05	0.20	0.05	70		1212
	1-HEXENE, 4-METHYL-	0.21	0.04	0.84	0.81	35	H12,I2	174
	1-HEXENE, 5-METHYL-	0.19	0.04	9.75	0.06	35	H12,I2	17
	5-HEXEN-3-YN-2-OL, 2-METHYL-	1.80		1.30		35	H12,I2	43
		0.591	0.057	4.00	0.28	70	C,P	607
	HYDANTOIN, 5-METHYLENE-			24		70		91
	HYDRACRYLIC ACID, BETA-LACTONE	0.03		0.2		50	H10,I3	1244
		6.2		0.2		0	H20,I3	48
		20				0		48
	HYDRACRYLONITRILE, ACRYLATE	20				80	C	1244
		32				60		742
		0.40	0.01	0.13	0.01	60	C	740
	HYDRACRYLONITRILE, METHACRYLATE	0.40	0.05	0.13	0.05	70	C	1016
	HYDRAZINE, 1-ACRYLOYL-2-BENZOYL-	0.45		0.32		60	I6	950
	HYDRAZINIUM CHLORIDE, 2-ACRYLOYL-1,1,1-TRIMETHYL-	1.40		0.65		70	I15	1175
	HYDRAZINIUM CHLORIDE, 2-METHACRYLOYL-1,1,1-TRIMETHYL-	0.58		0.46		70	I27	1175
	HYDRAZINIUM HYDROXIDE, 1,1,1-TRIMETHYL-2-(P-VINYLBENZOYL)- *, INNER SALT	0.51		0.23		70	I27	1175
	HYDROCINNAMIC ACID, ALPHA-METHYLENE-, METHYL ESTER	0.59	0.05	0.17	0.03	65	H10,II	174
	HYDROQUINONE, VINYL-, DIBENZOATE	0.22		0.43		78	I7	465
	IMIDAZOLE, 2-METHYL-1-VINYL-	8.97		0.069		70	Q	667
	1H-INDAZOLE, 1-ISOPROPENYL-	0.75		0.5		70		1218
	INDENE	0.3		2.7		20	H10,II	831
		0.3		3.3		20	H10,II	831
		0.3	0.1	3.0	0.1	0	H10,II	831
		0.4		2.5		20	H10,II	831
		0.5		1.8		10	H10,II	831
		0.5		2.3		20	H10,II	831
		0.5		3.7		0	H10,I24	653
		0.6	0.3	1.7	0.2	30	H10,II	831
		0.7	0.1			10	H10,II	831
		0.9	0.04	5.05	0.3	-72	H23	1094
		0.97	0.07	2.16	0.15	-72	H23	883
	INDENE, 6-BROMO-	4.2	0.04	0.21	0.02	-30	H23,I3	1365

M1	M2	R1	+/-	R2	+/-	TEMP.	REMARKS	REF.
STYRENE	INDENE, 4-METHOXY-	26	1	0.15	0.05	-78	H23	1417
	INDENE, 5-METHOXY-	5.8	0.5	0.36	0.05	-30	H23	1417
	INDENE, 6-METHOXY-	6.3	0.5	0.35	0.05	-50	H23	1417
	INDENE, 1-METHYL-	4.1	0.5	0.1	0.05	-78	H23	1417
	INDENE, 2-METHYL-	1.15	0.1	0.06	0.01	-72	H23	883
		2.84	0.2	0.038	0.015	-72	H23	883
	ISOCYANIC ACID, VINYL ESTER	6.9	0.5	0.1	0.05	60	Q	1027
		6.9		0.08		60		371
	ISOCYANIC ACID, P-VINYLPHENYL ESTER	8.13	0.04	1.10	0.10			420
	ISOPHTHALALDEHYDE	0.75	0.10	0		-78		430
		0.77		0			H10,I3	1142
		0.77		0		0	H10	1402
	ISOPHTHALIC ACID, BIS(2-METHYLALLYL)ESTER	22.9	0.4	0.065	0.003	60	C	1294
	ISOPHTHALIC ACID, DIALLYL ESTER	26.7	0.3	0.092	0.003	60	C	1294
	ISOPRENE	0.03		5.89		30	G17	1038
		0.046		16.6		40	G17,I21	1054
		0.14		7.6		50	G32,I7	1397
		0.14		7.0		30	G17,I7	531
		0.25		9.5		27	G17,I2	911
		0.35		0.6	0.6	20	H4,I7,BB	537
		0.42		2.02		25		1085
		0.44		1.98		-18	Q	1085
		0.46	0.02	0.5		-18	H14,I3	1271
		0.48		1.30		100	K	290
		0.48		1.30		80	K	732
		0.50	0.01	1.92	0.02		C	1038
		0.54		1.92			C	1038
		0.63	0.39	1.8	0.9	22	H,Q	1064
		0.66		1.18		27	E	1038
		0.8		1.68		50	G17,I2/8	911
		0.80	0.00	2.05		50	K	353
		1.38	0.54	0.15	0.45	20	H4,I9,BB	353
		2.2		0.27		75	H12,Z	537
		2.2		0.1		27	H17,I5,N	1076
		9		0.08		-15	H12,O	1064
		11		0.08		-35	G32,I5	909
		12		0.37			H17,I5,N	1397
		40	14	0.5	0.84			909
	ISOTHIOCYANIC ACID, VINYL ESTER	0.65	0.02	0.37	0.02	60	C,Q	371
		0.8		0.5				1070
	ISOXAZOLE, 3-ISOPROPENYL-5-METHYL-	0.67		0.98		60		1233
	ISOXAZOLE, 5-PHENYL-3-VINYL-	0.72	0.04	0.48	0.04	60		426
	LEAD, TRIPHENYL(P-VINYLPHENYL)-	1.22	0.0	0.978	0.00	60	I2	535
		4.7	0.1	0.76	0.01	70		535
	LINOLEIC ACID, METHYL ESTER	28		0		70	C	330
		30		0		130	C	618
		36		0		130	C	618
		45		0		90	C	618
		140		0		60	C	618
	LINOLENIC ACID, METHYL ESTER	60		0		60	C	330
	MALEALDEHYDE	0.08	0.03	0.30	0.05	60	Q	68
		0.08	0.03	0.30	0.05	70	Q	68

M1	M2	R1	+/-	R2	+/-	TEMP.	REMARKS	REF.
STYRENE	MALEAMIC ACID, N-CARBAMOYL-, METHYL ESTER	0.9		0		70	I26	704
		1.45		0		70	I10	705
	MALEIC ACID, DIBENZYL ESTER	0.55	0.2	0.02	0.01	70	C	836
	MALEIC ACID, DIETHYL ESTER	5	1.5	0.0	0.1	70		34
		5.48	0.56	0.01		131	D	559
	MALEIC ACID, DIMETHYL ESTER	6.52	0.05	0.005	0.01	60		559
		6.52	0.50	0.005		60		559
		0.1		0.13		60	E,Q	1392
	MALEIC ACID, ETHYL ESTER	8.5	0.2	0.03	0.01	60		557
		0.13	0.01	0.035	0.01	60		557
	MALEIC ACID, 2-CHLORO-, DIETHYL ESTER	2.5		0		70		34
	MALEIC ACID, 1,4-DITHIO-, CYCLIC ANHYDROSULFIDE	0.045		0.005		60		1393
	MALEIC ANHYDRIDE	0.01		0.035	0.022	40	R	620
		0.012	0.009	0		50		988
		0.019		0		60		46
		0.02		0		60		81
		0.02		0.010	0.002	60	I32	973
		0.025	0.005	0.001	0.001	70	I26	973
		0.035	0.003	0.015		70	I26	703
		0.040		0.049	0.005	50	I28	973
		0.040	0.005	0.01		70	C	861
		0.041		0.018		60		31
		0.042	0.008	0.005	0.001	80	I7	973
		0.043	0.005	0.005	0.001	70	I2	973
		0.050	0.005	0.09	0.002	50	I26	368
		0.097	0.002	0.1	0.001	70	I5	973
	MALEIC ANHYDRIDE, MONO(O-BENZYLOXIME)	0.13	0.03	0.03		70	I10	1088
	MALEIMIDE	0.22	0.08	0.05	0.06	75	I13	1013
	MALEIMIDE, N-(BENZYLOXY)-	0.1		0.06		35	I10	1319
	MALEIMIDE, N-2-BORNYL-	0.02		0.06		50	C,15	1433
	MALEIMIDE, N-BUTYL-	0.13		0.206		60	C	861
	MALEIMIDE, N-BUTYL-2-METHYL-	0.025	0.025	0.24	0.02	50		180
	MALEIMIDE, N,2-DIMETHYL-	0.14		0.24		60	C	861
	MALEIMIDE, N-HYDROXY-, ACETATE	0.135		0.01		60	C	861
	MALEIMIDE, N-(2-HYDROXYETHYL)-	0.153		0.01		70	C	553
	MALEIMIDE, N-(HYDROXYMETHYL)-	0.02		0.06		35	I10	1319
	MALEIMIDE, N-(P-HYDROXYPHENYL)-	0.04	0.02	0.12	0.01	60	I15	1432
	MALEIMIDE, N-(METHOXY)-	0.01	0.01	0.02	0.03	60	I10	1432
	MALEIMIDE, 2-METHYL-N-PHENYL-	0.08	0.05	0.196	0.05	60	I5	1432
	MALEONITRILE	0.02		0.099		35	I10	1319
	MALONIC ACID, DIALLYL ESTER	0.145		0.26	0.003	60	C	861
		0.19	0.01	0.88		60		557
	MALONIC ACID, METHACRYLOYL-, DIETHYL ESTER	29.8	0.3	0.08		60	C	1293
		0.17		0.08		60	I17	1339
	MALONIC ACID, METHYLENE-, DIETHYL ESTER	0.41	0.04	0.88	0.02	100	I10	237
	MALONONITRILE, BENZYLIDENE-	0.03		0		60		1429
		0.03		0.0459		60		1430
	MALONONITRILE, METHYLENE-	0.125		0.001		80	C	538
		0		0			Q	637
		0.005				45		289
	MELAMINE, N(2),N(4)-DIALLYL-	1.40					Q	157
		102.3	15.1	0		60		812

COPOLYMERIZATION PARAMETER

M1	M2	R1	+/-	R2	+/-	TEMP.	REMARKS	REF.
STYRENE	MELAMINE, N(2),N(4)-DIALLYL-	103.0	11.8	0		60		812
	MENTHOL, ACRYLATE	0.84	0.05	0.28	0.05	50	I6	844
	MERCURY, DIACETATUBIS(5-VINYL-2-PICOLINE)-	0.5		0.4		60	Q	952
	METANILYL FLUORIDE, N-METHACRYLOYL-	0.63		0.24		60		899
	METHACRYLALDEHYDE	0.22	0.02	0.88	0.02	70	I12	1384
	METHACRYLAMIDE	1.44	0.15	0.54	0.08	70	I10	1384
		1.46	0.09	1.29	0.08	5	F,K	218
	METHACRYLIC ACID	0.15	0.01	0.70	0.05	60	C	166
		0.15	0.05	0.7	0.01	70		1212
		0.15		0.7		60		1378
		0.2	0.02	0.62	0.08	45	C	1377
		0.2	0.02	0.66	0.08	45		820
		0.22		0.66		40	Q	1323
		0.22		0.64		40		695
		0.3		0.56		40	GG,I5	695
		0.37		0.51		60	GG,I56	695
		0.38		0.28		40	LL	1378
		0.45		0.47		45	GG,I19	695
		0.53	0.02	0.45	0.06	45	I6	820
		0.53	0.02	0.45	0.06	40	C,I6	1377
		0.63		0.47		60	GG,I17	695
	METHACRYLIC ACID, 3-ANILINO-2-HYDROXYPROPYL ESTER	0.19		0.75		60	I7	992
		0.23		0.46		60		992
		0.41		0.58		60	I10	992
	METHACRYLIC ACID, 2-(1-AZIRIDINYL)ETHYL ESTER	0.53	0.02	0.63	0.04	60		221
	METHACRYLIC ACID, BENZYL ESTER	0.44	0.01	0.51	0.01	60		743
		0.45		0.51		60		741
	METHACRYLIC ACID, P-(BENZYLIDENEAMINO)PHENYL ESTER	0.46	0.05	0.62	0.10	60		408
		0.48	0.04	0.42	0.07	60		1067
	METHACRYLIC ACID, BORNYL ESTER	0.24	0.01	0.25	0.05	60		801
	METHACRYLIC ACID, 2-BROMOETHYL ESTER	0.49		0.44		60		388
	METHACRYLIC ACID, 2(TERT-BUTYLAMINO)ETHYL ESTER	0.35	0.02	0.44	0.02	60		570
	METHACRYLIC ACID, BUTYL ESTER	0.47	0.05	0.83	0.06	60		570
		0.52	0.06	0.47	0.06	60		570
		0.54		0.64		70		1011
		0.56		0.40		60		741
		0.56	0.03	0.40	0.03	60		743
		0.63		0.64		50		1011
		0.97		0.67		30	D	152
	METHACRYLIC ACID, TERT-BUTYL ESTER	0.56	0.03	0.60	0.03	60		741
		0.58	0.03	0.49	0.04	60		570
	METHACRYLIC ACID, SEC-BUTYL ESTER	0.59	0.03	0.67	0.03	60		743
	METHACRYLIC ACID, 2-BUTYNYL ESTER	0.60	0.05	0.56	0.16	-40	G2,I5	570
	METHACRYLIC ACID, 2-CHLOROETHYL ESTER	0.40		1.20		60		196
		0.33		0.46		60		742
		0.33		0.46		60		743
	METHACRYLIC ACID, M-CHLOROPHENYL ESTER	0.42	0.02	0.23	0.04	60	C	64
		0.21	0.06	0.40	0.08	60		742
	METHACRYLIC ACID, P-CHLOROPHENYL ESTER	0.21		0.40		60		739
		0.22		0.45		60		742
		0.22		0.45		60		739

REMARKS PAGE II-366, REFERENCES PAGE II-368

M1	M2	R1	+/-	R2	+/-	TEMP.	REMARKS	REF.
STYRENE	METHACRYLIC ACID, P-(N-(P-CHLOROPHENYL)FORMIMIDOYL)PHENYL * ESTER	0.22		0.45		60		739
	METHACRYLIC ACID, CYCLOHEXYL ESTER	0.52		0.45		60		741
	METHACRYLIC ACID, CYCLOHEXYL ESTER	0.52	0.07	0.45	0.09	60		743
	METHACRYLIC ACID, 2-(DIETHYLAMINO)ETHYL ESTER	0.07		3.00		80	P,Q	943
	METHACRYLIC ACID, 2-(DIETHYLAMINO)ETHYL ESTER	0.11		6.20		80		866
	METHACRYLIC ACID, 2-(DIETHYLAMINO)ETHYL ESTER	0.22	0.05	0.6	0.1	70	I6	866
	METHACRYLIC ACID, 2-(DIETHYLAMINO)ETHYL ESTER, HYDROCHLORIDE	0.35	0.1	0.66	0.05	70		3
	METHACRYLIC ACID, 2-(DIETHYLAMINO)ETHYL ESTER, METHIODIDE	0.37	0.02	0.55	0.1	70	I12	1109
	METHACRYLIC ACID, 2-(DIMETHYLAMINO)ETHYL ESTER	0.74		0.23		60	I6	1109
	METHACRYLIC ACID, 2,2-DINITROPROPYL ESTER	0.78		0.34		60	I6	866
	METHACRYLIC ACID, DODECYL ESTER	0.25	0.05	0.3	0.05	70	I6	866
		0.34	0.05	0.25	0.05	60		3
		0.53	0.02	0.37	0.01	80		570
		0.37		0.18	0.04	60	I126	930
		0.56		0.36		60		743
	METHACRYLIC ACID, 2,3-EPOXYPROPYL ESTER	0.57		0.45		60		741
		0.63	0.04	0.57	0.04	65		570
		0.34	0.05	0.63	0.1	80	I5	887
		0.37		0.68		60		866
	METHACRYLIC ACID, ETHYL ESTER	0.441	0.001	0.53	0.07	60		414
		0.45	0.007	0.544	0.004	120	C,I13	903
		0.45		0.55		60		570
		0.50		0.55		60		902
		0.53	0.001	0.60	0.002	70	C,I13	903
	METHACRYLIC ACID, 2-(ETHYLTHIO)ETHYL ESTER	0.55	0.03	0.41	0.03	50		741
		0.65	0.06	0.41	0.03	50		743
		0.67		0.33		60		570
		0.42		0.29		60		1011
				0.26		60		1011
				0.51		60		1211
	METHACRYLIC ACID, FURFURYL ESTER	0.711	0.111	0.733	0.037	50	I7,O	94
	METHACRYLIC ACID, FURFURYL ESTER, PI COMPLEX WITH O2	0.258	0.069	0.336	0.069	50	I7,O	94
	METHACRYLIC ACID, HEXYL ESTER	0.45	0.03	0.65	0.02	60		570
		0.60		0.45		60		741
		0.60	0.02	0.45	0.06	60		743
	METHACRYLIC ACID, X-HYDROXYBUTYL ESTER, NITRATE	0.37		0.4		60		1307
	METHACRYLIC ACID, 2-HYDROXYETHYL ESTER	0.57		0.65		60		570
	METHACRYLIC ACID, 2-HYDROXYETHYL ESTER, NITRATE	0.35		0.49		60		1307
	METHACRYLIC ACID, 2-HYDROXY-3-(N-METHYLANILINO)PROPYL ESTER	0.26		0.74		60	I7	992
	METHACRYLIC ACID, 2-HYDROXY-3-(2-NAPHTHYLAMINO)PROPYL ESTER	0.30		0.44		60	I7	992
	METHACRYLIC ACID, 2-HYDROXY-3-PHENOXYPROPYL ESTER	0.21		0.37		60	I7	992
	METHACRYLIC ACID, 2-HYDROXY-3(P-(PHENYLAZO)ANILINO)PROPYL * ESTER	0.39		0.40		60	I7	992
	METHACRYLIC ACID, 2-HYDROXY-3((8-(PHENYLAZO)-2-NAPHTHYL)AMI NO)PROPYL ESTER	0.58		0.82		60	I7	992
	METHACRYLIC ACID, 2-HYDROXYPROPYL ESTER	0.58		0.82		60		570
	METHACRYLIC ACID, 2-HYDROXYPROPYL ESTER, NITRATE	0.58		0.82		60		1307
	METHACRYLIC ACID, ISOBORNYL ESTER	0.56	0.02	0.65	0.02	60	I7	388
	METHACRYLIC ACID, ISOBUTYL ESTER	0.36		0.5		60		741
		0.70		0.32		60		743
		0.55	0.02	0.40	0.05	60		
		0.55		0.40		60		

COPOLYMERIZATION PARAMETER

M1	M2	R1	+/-	R2	+/-	TEMP.	REMARKS	REF.
STYRENE	METHACRYLIC ACID, ISOBUTYL ESTER	0.56	0.02	0.58	0.02	60		570
	METHACRYLIC ACID, ISOPROPYL ESTER	0.47	0.06	0.74	0.05	60	17	1068
		0.50	0.04	0.42	0.04	60	17,Q	570
	METHACRYLIC ACID, 5-(METHOXYCARBONYL)FURFURYL ESTER	0.352	0.022	0.143	0.077	60	17	94
	METHACRYLIC ACID, 2-METHOXYETHYL ESTER	0.50	0.01	0.58	0.04	60		1068
	METHACRYLIC ACID, P-METHOXYPHENYL ESTER	0.26		0.60		60		742
		0.28		0.60		60		739
	METHACRYLIC ACID, 1-METHYL-2-NITROPROPYL ESTER	0.43		0.078		80	126	930
	METHACRYLIC ACID, 2-METHYL-2-NITROPROPYL ESTER	0.32		0.37		80	126	930
	METHACRYLIC ACID, 2-NITROBUTYL ESTER	0.25		0.34		80	126	930
	METHACRYLIC ACID, 2-NITROETHYL ESTER	0.52	0.03	0.46	0.03	70	C	1016
	METHACRYLIC ACID, M-NITROPHENYL ESTER	0.18		0.36		60		739
		0.18		0.36		60		742
	METHACRYLIC ACID, P-NITROPHENYL ESTER	0.19		0.22		60		739
		0.19		0.22		80	126	930
	METHACRYLIC ACID, 2-NITROPROPYL ESTER	0.39	0.03	0.37		70	16	985
	METHACRYLIC ACID, NOVOLAK RESIN ESTER	0.1		2.76		60		570
	METHACRYLIC ACID, OCTYL ESTER	0.56		0.68	0.04	60		741
	METHACRYLIC ACID, PENTAMETHYLDISILOXANYL METHYL ESTER	0.67	0.03	0.55	0.07	50		743
	METHACRYLIC ACID, PENTYL ESTER	0.77	0.02	0.58	0.02	60		632
	METHACRYLIC ACID, PHENETHYL ESTER	0.55		0.40		60		741
	METHACRYLIC ACID, PHENYL ESTER	0.55	0.02	0.51	0.05	60		743
		0.26	0.03	0.51	0.09	60		1067
		0.30	0.01	0.60	0.01	60		570
	METHACRYLIC ACID, P-(N-PHENYLFORMIMIDOYL)PHENYL ESTER	0.30		0.60		60		741
	METHACRYLIC ACID, PROPYL ESTER	0.25	0.03	0.60	0.05	60		739
	METHACRYLIC ACID, 2-PROPYNYL ESTER	0.57	0.03	2.4	0.5	60		743
	METHACRYLIC ACID, PYRIDINE COMPLEX	0.57	0.01	0.38		60		801
	METHACRYLIC ACID, 2-PYRIDYL ESTER	0.3		0.38	0.04	60		741
	METHACRYLIC ACID, QUINOLINE COMPLEX	0.45		0.45		60	Q	743
	2,4,8,10-TETRAOXASPIRO(5.5)UNDEC-3,9-YLENE* DIETHYLENE ESTER	0.35	0.03	0.47		60		1306
	METHACRYLIC ACID, P-TOLYL ESTER	0.37		0.62	0.11	60	Q	1323
		0.37		0.51			Q	1101
		0.30		0.51		60	Q	1323
	METHACRYLIC ACID, TRIETHYLAMINE COMPLEX	0.30		0.70		60		1323
	METHACRYLIC ACID ANHYDRIDE	2.1		0.70		60		742
	METHACRYLONITRILE	0.12	0.01	0.36		60		739
		0.05		0.33	0.03	60	C,17	1323
		0.25	0.02	0.26		36,6	I13	1145
		0.30	0.02	12.0	1.0	-78/-30	H21,12	891
		0.37	0.10	0.25	0.02	80		204
		0.38		0.16	0.06	60		260
		0.38		0.41		90	I2	560
		0.39	0.05	0.42	0.05	120	I2	1376
		0.43		0.26		80	C	1376
	METHACRYLOPHENONE	0.23	0.11	0.32		60	I2	159
				0.28		80	C	1376
	METHANE, BIS(P-VINYLPHENYL)-	1.01		0.34	0.16	75	G,R	1070
	METHANE, ETHOXY(VINYLTHIO)-	6.4		0.93		61	C	663
				0.26		60		1042

REMARKS PAGE II-366, REFERENCES PAGE II-368

M1	M2	R1	+/-	R2	+/-	TEMP.	REMARKS	REF.
STYRENE	METHANE, (METHYLTHIO)(VINYLTHIO)-	4.25		0.23		60		1260
	METHANOL, DIPHENYL(P-VINYLPHENYL)-	0.47	0.10	1.60	0.15	50	I7	129
	METHYL METHACRYLATE	0		1.		-50/-20		204
		0		1		-78		204
		0.01	0.01	25.0	2	-30	H21,I2	204
		0.02	0.01	20.	2	-78	H21,I2	204
		0.05	0.02	14.0	2	-30	H21,I2	204
		0.07	0.02	4.5	1	20	H21,I2	204
		0.12	0.05	6.4	0.05	-30	G3	551
		0.25	0.05	6.4	0.05	-30	G3	552
		0.30	0.03	0.056	0.003	50	MM	1458
		0.31	0.05	2.0	0.3	20	H21,I2	204
		0.35		17.1		30	H47	278
		0.38	0.024	0.35	0.024	60	I70	1230
		0.42	0.03	0.48	0.11	100	C,I29	506
		0.43		0.32		60	I10	988
		0.43	0.013	0.54	0.016	50	H36,Q	992
		0.44	0.054	0.45	0.054	60	I19	277
		0.44	0.08	0.50	0.04	60	I69	59
		0.45		0.39		50	I12	1230
		0.45	0.10	0.5	0.05	50	C	787
		0.47	0.03	0.47	0.01	60	C	59
		0.48		0.50		60	I68	406
		0.48	0.045	0.45	0.045	50	D,C	1458
		0.485		0.46		50	I24	1043
		0.49		0.49		50	I19,M4	1230
		0.49		0.422		60	I12,M4	56
		0.49		0.46		35	C,I38	59
		0.50		0.57		35	I7	59
		0.50	0.02	0.52	0.02	60	C	59
		0.50	0.02	0.52	0.02	60	I7	506
		0.51	0.02	0.44	0.02	60	C,I7	1021
		0.51		0.50		50	I7	1020
		0.52		0.57		60	I2	558
		0.52		0.53		60	E,I24	992
		0.52		0.45		60		506
		0.52		0.47		60		59
		0.53	0.02	0.46	0.02	60		311
		0.53	0.02	0.46	0.02	60	I24,M4	859
		0.54	0.026	0.46	0.026	60	C	743
		0.54		0.52		50	C	570
		0.545		0.49		60		559
		0.55	0.02	0.50	0.06	60		59
		0.56	0.03	0.49	0.03	99	A,I53	1036
		0.56	0.04	0.42	0.1	60	G9,I	634
		0.56	0.006	0.588	0.007	132	K	213
				0.23		30	I19,M3	787
				0.49		35	H38	411
				0.80		50	K	288
				0.50	0.01	40		1020
				0.50	0.02	35		59
								700
								1021

COPOLYMERIZATION PARAMETER

M1	M2	R1	+/-	R2	+/-	TEMP.	REMARKS	REF.
STYRENE	METHYL METHACRYLATE	0.57	0.032	0.46	0.032	60	I7	1230
		0.58		0.48		50	I7,M4	59
		0.59		0.62		50	I12,M3	59
		0.590	0.026	0.536	0.026	131		559
		0.60		0.55		132	C	1043
		0.63		0.50		25	E,I24	859
		0.63		0.66		50	I24,M3	59
		0.63	0.09	0.50	0.07	25	E	149
		0.72		0.58		50	I7,M3	59
		0.8		0.8		250	A	713
		1		1			H,I24,Q	358
		10.5	0.2	0.1	0.05		H,I7,Q	358
		10.5	0.2	0.1	0.05	20	H2O	552
		20		0.005		-78	E,I24	859
		20		0.005		-78	E,I8	1
	METHYL METHACRYLATE-METHYL-14C	0.52	0.03	0.46	0.02	60	C	78
	NAPHTHALENE, 4-CHLORO-1-VINYL-	0.85	0.1	0.8	0.1	60	C	783
	NAPHTHALENE, 6-CHLORO-2-VINYL-	0.4	0.1	1.5	0.2	60	C	783
	NAPHTHALENE, 1,2-DIHYDRO-	1.0	0.03	0.4	0.2	30	H10,I24	653
	NAPHTHALENE, 1-VINYL-	0.67	0.5	1.35	0.15	60		568
		3.3	0.1	0.35	0.2	40	H12	880
	NAPHTHALENE, 2-VINYL-	0.5		1.4		60		783
	NONANOIC ACID, VINYL ESTER	49.5	15	0.01	0.1	60	C	591
	2-NORBORNANECARBOXYLIC ACID, ENDO-, VINYL ESTER	30		0.01	0.01	60	C	385
	2-NORBORNANECARBOXYLIC ACID, EXO-, VINYL ESTER	30		0.01		60		385
	2-NORPINANE-2-ETHANOL, ACRYLATE	0.66	0.1	0.29	0.1	50		600
	10,12-OCTADECADIENOIC ACID, TRANS,TRANS-, METHYL ESTER	11		0		130	C	618
		12		0		130	C	618
	9,12-OCTADECADIENOIC ACID	12		0		60	C	618
	9,11,13-OCTADECATRIENOIC ACID	15		0		90		618
	1-OCTADECENE	17		0		70		330
	3-OXA-1-AZASPIRO(4,5)DECAN-4-ONE, 2-ISOPROPENYL-	7.16		0.29		150		868
	1,3,4-OXADIAZOLE, 2-ISOPROPENYL-5-METHYL-	0.46		4		30	C,I2,Q	814
		1.32	0.14	0.87	0.09	30	H7,I7	79
	1,3,4-OXADIAZOLE, 5-PHENYL-2-(4-VINYLPHENYL)-	1.94	0.05	0.75	0.10	60	H29,I7	79
		0.12		0.82		70		427
		0.16		0.83		60	I10	1233
		0.17		4.5		60		19
	OXALIC ACID, DIALLYL ESTER	27.5	0.05	0.104	0.4	60	C	1293
	OXALIC ACID, ETHYL VINYL ESTER	8.0	0.4	0.1	0.003	60		479
	3-OXAZIN-5-ONE, 2-(DICHLOROMETHYLENE)-4-ISOBUTYL-	0.1		0				1232
	3-OXAZIN-5-ONE, 2-(DICHLOROMETHYLENE)-4-ISOPROPYL-	0.1		0				1232
	3-OXAZIN-5-ONE, 2-(DICHLOROMETHYLENE)-4-METHYL-	0.1		0				1232
	OXAZOLE, 5-(4-BIPHENYLYL)-2-(4-VINYLPHENYL)-	0.46	0.07	4.6	0.5	60	I10	19
	OXAZOLE, 4,5-DIMETHYL-2-VINYL-	0.18	0.05	1.3	0.1	70		1233
		0.18	0.05	1.30	0.1	70		424
	OXAZOLE, 4-ISOBUTYL-2-ISOPROPENYL-5-METHYL-	0.1		3.4		60		1233
	OXAZOLE, 4-ISOBUTYL-5-METHYL-2-VINYL-	0.1		3		60		1233
	OXAZOLE, 2-ISOPROPENYL-4,5-DIMETHYL-	0.15		2.2		60		1233
		0.15		2.20		60		426
		0.15		2.20		70		424
	OXAZOLE, 5-(1-NAPHTHYL)-2-(4-VINYLPHENYL)-	0.24	0.05	3.8	0.4	70	I10	19
	OXAZOLE, 5-(2-NAPHTHYL)-2-(4-VINYLPHENYL)-	0.36	0.05	4.5	0.5	70	I10	19

REMARKS PAGE II-366, REFERENCES PAGE II-368

M1	M2	R1	+/-	R2	+/-	TEMP.	REMARKS	REF.
STYRENE	OXAZOLE, 5-PHENYL-2-(4-VINYLPHENYL)-	0.49	0.08	3.5	0.4	70	T,0	19
		0.68	0.04	4.5	0.4	70	I2	107
	2-OXAZOLIDINONE, 5-DECYL-3-VINYL-	14.3		0.059		60		1174
	2-OXAZOLIDINONE, 3-METHACRYLAMIDO-	1.15		0.38		70	I6	1185
	2-OXAZOLIDINONE, 3-METHACRYLOYL-	1.1	0.5	0.27	0.05	70	I6	1184
	2-OXAZOLIDINONE, 3-VINYL-	30		0.05		50		116
	2-OXAZOLIN-5-ONE, 4-ETHYL-2-ISOPROPENYL-4-METHYL-	0.27		0.70		60		427
	2-OXAZOLIN-5-ONE, 4-ISOBUTYL-2-ISOPROPENYL-4-METHYL-	0.24		0.69		60		427
	2-OXAZOLIN-5-ONE, 2-ISOPROPENYL-4,4-DIMETHYL-	0.31	0.06	1.00	0.09	60		427
	2-OXAZOLIN-5-ONE, 2-ISOPROPENYL-4-ISOPROPYL-	0.31	0.03	1.12	0.10	60		426
	3-OXAZOLIN-5-ONE, 4-ISOBUTYL-2-ISOPROPYLIDENE-	0.39	0.10	0.0		60		426
	3-OXAZOLIN-5-ONE, 2-ISOPROPYLIDENE-4-METHYL-	0.36	0.07	0.00	0.10	60		426
	3-OXAZOLIN-5-ONE, 4-ISOPROPYL-2-ISOPROPYLIDENE-	0.39	0.06	0.008		60		426
	2-OXETANONE, 4-METHYLENE-	100		4.99	0.14	70		1190
	2,4-PENTADIENOIC ACID, 2,3-EPOXYPROPYL ESTER	0.13	0.02	0.80	0.12	60	I7	1231
	2,4-PENTADIENOIC ACID, METHYL ESTER	0.12	0.08	0		60		272
	2,4-PENTANEDIONE, 3-ETHYLIDENE-	9		0		60	I7	1339
	PENTASILOXANE, 1,1,3,3,5,5,7,7,9,9-UNDECAMETHYL-9-(P-VINYL* PHENYL)-	9		0		60	I7	1339
	PENTASILOXANE, 1-(P-VINYLPHENYL)-	1.11	0.01	1.2	0.1	80	H12	371
	1-PENTENE	0.448		1.51		50	A,0	827
	1-PENTENE, 4,4-DIMETHYL-2-PHENYL-	2				110	H7,I7	1189
	1-PENTENE, 4-METHYL-	0.49	0.06	1.15	0.12	30	H29,I7	79
		0.55	0.10	1.23	0.15	30	H12,I2	79
		0.89	0.05	3.67	0.22	45		44
		0.98	0.1	3.92	0.15	70	H12	1253
	1-PENTENE, 4-METHYL-2,4-DIPHENYL-	1.9		0.12		110	A,0	1189
	1-PENTENE, 4-METHYL-2-PHENYL-	1.9		0.01		110	A,0	1189
	1-PENTENE, 2,4,4-TRIPHENYL-	1.6		0.01		110	A,0	1189
	4-PENTENOIC ACID, 2-ALLYL-2-CYANO-, ETHYL ESTER	16.1		-0.02		60	0	163
	1-PENTEN-3-ONE	0.7	0.09	0.12	0.02	60	C,I7	1114
	3-PENTEN-2-ONE	13.7		0.01		60	C,I7	928
	PEROXYFUMARIC ACID, DI-TERT-BUTYL ESTER	13.7	0.02	-0.02	0.02	40	C	641
	PHENANTHRENE, 9-VINYL-	0.67		2.36		60	C	580
	PHENETHYL ALCOHOL, ALPHA-METHYL-P-VINYL-	0.58		0.91		60		477
	PHENETOLE, P-STYRYL-	0.97		0.038		70	I6	42
	PHENOL, M-VINYL-	5.33	0.64	1.21	0.15	60		1152
	PHENOL, O-VINYL-	0.9		1.21		60	I5	115
	PHENOL, P-VINYL-	0.72	0.10	1.32	0.25	60	I5	1245
	PHENOL, P-VINYL-, ACETATE	0.79	0.08	1.2	0.13	60	I5	476
	PHOSPHINE, DIPHENYLVINYL-	0.85	0.15	1.35	0.20	60	C	1245
	PHOSPHINE, DIPHENYL(P-VINYLPHENYL)-	7	1	0		60	I2	201
	PHOSPHINE OXIDE, DIISOBUTYLVINYL-	0.46	0.07	1.11	0.12	60	I7	791
	PHOSPHINE OXIDE, DIPHENYLVINYL-	0.52	0.05	1.43	0.25	65	C	126
	PHOSPHINE OXIDE, DIPHENYL(P-VINYLPHENYL)-	17	5	0		60	C	792
	PHOSPHINE SULFIDE, DIPHENYLVINYL-	5	1	0		60	I7	791
	PHOSPHINE SULFIDE, DIPHENYL(P-VINYLPHENYL)-	0.42	0.02	1.40	0.15	60	C	791
	PHOSPHINOTHIOIC ACID, DIPHENYL-, O-2-(METHACRYLOXY)ETHYL* ESTER	2.1	0.3	0		60	I7	792
		0.43	0.05	1.49	0.33	60	I7	791
		0.43	0.05	1.49	0.33	60	I7	792

COPOLYMERIZATION PARAMETER

M1	M2	R1	+/-	R2	+/-	TEMP.	REMARKS	REF.
STYRENE	PHOSPHONIC ACID, (2-METHYL-1,3-BUTADIENYL)-, DIISOPROPYL ESTER *	0.43	0.05	1.49	0.33	60	17	792
	PHOSPHONIC ACID, (2-METHYL-1,3-BUTADIENYL)-, DIMETHYL ESTER	0.4	0.02	1.14	0.11	75		1281
	PHOSPHONIC ACID, PHENYL-, DIALLYL ESTER	28.97		0.027	0.05	70		338
	PHOSPHONIC ACID, ALPHA-PHENYLVINYL-	1.5	0.05	0.2	0.2	70	I10	1090
	PHOSPHONIC ACID, VINYL-, BIS(2-CHLOROETHYL) ESTER	2.2	0.4	0.16		30	Q	729
	PHOSPHONIC ACID, VINYL-, DIBUTYL ESTER	2.3		0.03		60	E	271
	PHOSPHONIC ACID, VINYL-, DIETHYL ESTER	2.43		0.3	0.3	116	Q	527
	PHOSPHONIC ACID, VINYL-, DIISOBUTYL ESTER	5.4	1.5	0		60		519
	PHOSPHONIC ACID, VINYL-, DIISOPROPYL ESTER	3.25		0		60		519
	PHOSPHONIC ACID, VINYL-, DIMETHYL ESTER	4.1		0.06	0.02	60	Q	982
	PHOSPHONIC ACID, VINYL-, DIPHENYL ESTER	8.87	0.5	0.5	0.4	60		519
	PHOSPHONIC ACID, VINYL-, DIPROPYL ESTER	4.4	1.0	0.40		60		519
	PHOSPHONIC DICHLORIDE, (2-METHYL-1,3-BUTADIENYL)-	2.39	0.37	0		60		1283
	PHOSPHONIC DIFLUORIDE, (2-METHYL-1,3-BUTADIENYL)-	4.61	0.43	0.90	0.17	75		1282
	PHOSPHORIC ACID, DIALLYL PHENYL ESTER	2.03	0.39	1.35	0.09	75	I2	339
	PHTHALIC ACID, BIS(2-METHYLALLYL) ESTER	4.24	1.04	1.15		70		1294
	PHTHALIC ACID, BIS(2-METHYLALLYL) ESTER	16.84	0.7	0.145	0.005	70	C	564
		25.4		-0.059		60	AA,H23	563
	PHTHALIC ACID, BIS(TRIETHYLENE GLYCOL) ESTER, DIMETHACRYLATE	0.05		2.45		0	H23,I18	564
		0.15		0.75		0	AA,H23	564
		0.15		0.75		0	AA,H23	564
		0.36		0.54			C	611
	PHTHALIC ACID, DIALLYL ESTER	15.1	0.5	0.134	0.008	130	C	611
		16.9	0.1	0.105	0.001	120	C	611
		18.8	0.2	0.081	0.005	90	C	611
		19.4	0.5	0.066	0.005	110	C	611
		21.4	0.2	0.053	0.005	100	C	611
		23.8	0.2	0.105	0.009	80	C	611
		27.5	0.6	0.088	0.007	70	C	611
		32.8	0.6	0.057	0.007	60	C	611
	PHTHALIMIDE, N-1-(1,3-BUTADIENYL)-	0.32		1.48		60	C,I6	672
	PHTHALIMIDE, N-2-(1,3-BUTADIENYL)-	0.11	0.02	5.2	0.5	60	C,I6	671
	PHTHALIMIDE, N-(3-METHYLENE-1-BUTEN-4-YL)-	0.15	0.02	2	0.13	60	C,I6	1111
	PHTHALIMIDE, 4-(TRIMETHYLSILYL)-N-VINYL-	6.1		0.08		60	I7	1217
	PHTHALIMIDE, N-VINYL-	6.3		0.09		60	C,I6	672
	PHTHALIMIDE, N-VINYL-	8.3	0.3	0.075	0.03	85	C	696
	2-PICOLINE, 5-VINYL-	0.3	0.1	2.9	0.3	70	G18	830
		0.5	0.1	2.9	0.3	70	H24	830
		0.6	0.1	0.68	0.1	70		830
	2-PICOLINE, 5-VINYL-, COBALT COMPLEX	0.72	0.23	1.20	0.13	60	I6	952
	2-PICOLINIUM CHLORIDE, 5-VINYL-	0.738	0.010	0.801	0.030	30		189
		0.812	0.005	0.91	0.02	60		940
	PROPANE, 1,3-BIS(P-VINYLPHENYL)-	1.19	0.12	0.88	0.2	60	I6	737
	PROPANE, 1-CHLORO-2,3-EPOXY-	1.8	0.2	0.3	0.2	50	I12	1410
	PROPANE, 1,2-EPOXY-3-ETHOXY-	0.27		0.42		70	C	1196
	PROPANE, 1,2-EPOXY-3-ETHOXY-	1.11		0.89		61	H11	1042
	PROPANE, 1,2-EPOXY-3-METHOXY-	-0.115		2.5	0.6	70	H11	1100
		-0.1	0.1	8.9		70	H10,I10	1100
		0		0.75		50	H10,I10	443
	PROPANE, 1,2-EPOXY-3-PROPOXY-	0		1.8		50	H10,I10	443
				4.9		50	H10,I10	443

M1	M2	R1	+/-	R2	+/-	TEMP.	REMARKS	REF.
STYRENE	1-PROPENE	0.12		7.70	0.7	60	H12	343
		0.16	0.08	7.2	3.5	40	H29	61
		0.20	0.13	20.0	3.0	40	H12	61
		0.30	0.15	20.5		40	H12	61
		0.93		0.62		45	H12	173
		1.0		5.4		60	H29	344
		1.5		1.0		60	H7	344
		1.9		2.4		60	H28	344
		2.0		3.7		60	H24	344
	1-PROPENE, 3-CHLORO-	30		0		60	C	824
		31.5	4	0.016	0.016	70	C	28
		36.8	0.60	0.029	0.006	60	C	457
	1-PROPENE, 3-CHLORO-2-(CHLOROMETHYL)-	3.8		0.01		50		917
	1-PROPENE, 2-CHLORO-3,3-DIMETHOXY-	7.75	0.076	0.005	0.036	60		241
	1-PROPENE, 3-CHLORO-2-METHYL-	22		0.025		60		620
	1-PROPENE, 1,1-DICHLORO-	37		0.0		40	C,Q	915
	1-PROPENE, 1,3-DICHLORO-	82.5		0		70	C,Q	915
	1-PROPENE, 2,3-DICHLORO-	2.5		0		40		484
		5.0	0.6	0.06	0.009	70	C	29
	1-PROPENE, 2-METHYL-	0.17		1.60		40	H,P	721
		0.2		3.3		0	H20,I18	574
		0.24	0.02	1.79	0.02	0	E,L1	777
		0.33		3.5		-90	H46,I3	798
		0.42	0.02	1.66	0.02	-78	E,I18	1
		0.5		0.8		-90	H46,I3	798
		0.6		1.9		-78	H20,I16	379
		0.76	0.13	2.36	0.06	-78	H20,I16	379
		1.0		1.5		-30	H46,I3	798
		1.07	0.17	1.49	0.12	-78	H20,I16	379
		1.08	0.07	4.48	0.28	-20	H70,I19	722
		1.08	0.07	4.48	0.28	-78	H23,I3	1223
		1.1		3.1		-78	H23,I3	381
		1.1		2.2		-78	H20,I16	379
		1.20	0.10	1.78	0.10	-78	H20,I16	379
		1.20	0.11	0.54	0.24	-78	H23,I2	722
		1.21	0.06	2.51	0.05	-20	H23,I9	722
		1.27	0.39	0.51	0.15	-30	H46,I3	798
		1.30	0.15	0.75	0.15	-78	H70,I19	722
		1.70	0.07	4.11	0.19	-78	H23,I9	381
		1.71	0.47	1.43	0.43	-78	H70,I19	722
		1.92	0.41	3.75	0.45	-20	H20,I43	381
		1.99	0.24	9.02	0.77	-78	H46,I3	722
		2.41	0.12	0.37	0.07	-91	H23,I9	381
		2.41	0.12	0.37	0.07	-78	H23,I9	798
		2.44	0.15	1.51	0.15	-78	H23,I50	381
		2.61	0.4	1.53	0.3	-78	H23,I51	1223
		2.75	0.25	3.25	0.25	-78	H23,I42	722
		2.75	0.25	3.75	0.25	-78	H23,I42	722
		3.7	0.55	0.5	0.52	0	E	777
	1-PROPENE, 3,3,3-TRICHLORO-	5.50	0.2	2.63	0.02	-78	H23,I41	381
		6.9		0.0		60		211
	2-PROPENE-1,1-DIOL, 2-CHLORO-, DIACETATE	4.6	0.21	0.14	0.018	60		241

COPOLYMERIZATION PARAMETER

M1	M2	R1	+/-	R2	+/-	TEMP.	REMARKS	REF.
STYRENE	2-PROPENE-1-SULFONIC ACID, ALLYL ESTER	13	1	0.01	0.01	60	I7	1201
	2-PROPENE-1-SULFONIC ACID, 2,3-EPOXYPROPYL ESTER	50	5	0.02		60		417
	1-PROPENE-1,2,3-TRICARBOXYLIC ACID	1.10	0.10	-0.10	0.10	60	C,R	734
	1-PROPENE-1,2,3-TRICARBOXYLIC ACID, TRANS-, TRIETHYL ESTER	1.10	0.01	0.00	0.01	60	C	595
	1-PROPENE-1,2,3-TRICARBOXYLIC ACID, TRANS-, TRIMETHYL ESTER	2.6		0.5		60	C	595
	1-PROPEN-2-OL, ACETATE	7.0		0.14		30	H10,I22	374
	1-PROPEN-2-OL, ISOCYANATE	8.12		0.07		60		333
	2-PROPEN-1-OL, 2-CHLORO-	12.5		0		40	Q	419
	2-PROPEN-1-OL, 2-CHLORO-, ACETATE	4.10		0		50		484
	2-PROPEN-1-OL, 2-METHYL-, ACETATE	71	10	0		50		620
	2-PROPEN-1-ONE, 1-(6-METHYL-3-PYRIDYL)-3-PHENYL-	0.92	0.08	-0.15	0.2	50	I7	602
	2-PROPEN-1-ONE, 1-PHENYL-3-(3-PYRIDYL)-	0.50	0.10	0.00	0.25	60	I7	602
	2-PROPEN-1-ONE, 3-PHENYL-1-(3-PYRIDYL)-	0.85	0.05	0.09	0.10	50	I7	602
	PROPIONALDEHYDE, 3-(ALLYLOXY)-	3		1		0	H10	1287
	2-PROPYN-1-OL, ACETATE	53		0		60		1305
	2H-PYRAN-3-CARBOXYLIC ACID, 5,6-DIHYDRO-2,6-DIMETHYL-, VINYL* ESTER	53		0		60		1305
	PYRAZOLE, 3,5-DIMETHYL-1-VINYL-	8.2	0.1	0.08	0.02	60	I6	523
	3-PYRAZOLIDINONE, 1-(M-VINYLPHENYL)-	1.2		0.8		60	I17	1298
	3,6-PYRIDAZINEDIONE, 1,2-DIHYDRO-	55		0		60	I17	1127
	3(2H)-PYRIDAZINONE, 2(P-CHLOROPHENYL)-6-HYDROXY-	10		0		60	I17	1290
	3(2H)-PYRIDAZINONE, 6-HYDROXY-	55		0		60	I17	1289
	3(2H)-PYRIDAZINONE, 6-HYDROXY-, ACETATE	6		0	6	60	I17	1289
	3(2H)-PYRIDAZINONE, 6-HYDROXY-2-PHENYL-	13		0		60	I17	1290
	3(2H)-PYRIDAZINONE, 6-HYDROXY-2-(P-TOLYL)-	16		0		60		1290
	3(2H)-PYRIDAZINONE, 6-METHYL-	25		0		60		1289
	PYRIDINE, 2-DIMETHYLAMINO-4-VINYL-	0.35	0.02	1.4	0.1	60		759
	PYRIDINE, 5-ETHYL-2-VINYL-	0.79	0.03	1.2	0.2	60		939
	PYRIDINE, 5-ETHYL-2-VINYL-, 1-OXIDE	0.10	0.01	2.6	0.3	60		942
	PYRIDINE, 2-ISOPROPENYL-	0.42		0.95		60	I12	1233
	PYRIDINE, 2-VINYL-	0.50	0.07	1.27	0.19	25	E	149
	PYRIDINE, 2-VINYL-	0.55	0.025	1.14	0.08	60		759
	PYRIDINE, 2-VINYL-	0.55	0.03	1.135	0.05	50		1023
	PYRIDINE, 2-VINYL-	0.55	0.03	1.81	0.05	60		536
	PYRIDINE, 2-VINYL-	0.55	0.02	1.81	0.05	60		534
	PYRIDINE, 2-VINYL-	0.56	0.03	1.81	0.2	60		533
	PYRIDINE, 2-VINYL-	0.57		0.9	0.04	50		939
	PYRIDINE, 2-VINYL-, COBALT COMPLEX	1.33	0.7	0.18	0.05	50	C	1410
	PYRIDINE, 2-VINYL-, 1-OXIDE	0.11	0.01	2.1	0.6	60	I6	942
	PYRIDINE, 4-VINYL-	0.54	0.03	0.7	0.1	60	I12	939
	PYRIDINE, 4-VINYL-	0.62	0.02	0.52	0.06	80		275
	PYRIDINE, 4-VINYL-, COBALT COMPLEX	1.9	0.25	0.21	0.1	50	C,I2	1410
	PYRIDINE, 4,4'-VINYLENEDI-	1.85	0.1	0.17	0.1	60	I6	603
	PYRIDINIUM ACETATE, 2-VINYL-	0.16	0.02	0.36	0.04	22	C,K	821
	PYRIMIDINE, 4-VINYL-	0.17	0.02	1.2	0.1	60	D	759
	2-PYRROLIDINETHIONE, 1-VINYL-	1.75	0.05	0.45	0.03	60	I7	1171
	2-PYRROLIDINONE, 1-VINYL-	9.0		0.11		80	C	1070
	2-PYRROLIDINONE, 1-VINYL-	15.7	0.5	0.045	0.05	50		116
	QUINOLINE, 2-VINYL-	0.49	0.14	2.69	0.55	50		536
		0.49	0.14	2.09	0.55	60		533
		0.49	0.14	2.09	0.55	60		534

REMARKS PAGE II-366, REFERENCES PAGE II-368

M1	M2	R1	+/-	R2	+/-	TEMP.	REMARKS	REF.
STYRENE	SALICYLALDEHYDE	4.2				0	H10	1402
	SEBACIC ACID, DIALLYL ESTER	35	1.2	0.103	0.013	60		1293
	SILANE, ALLYLDIMETHYL(P-VINYLPHENYL)-	0.69	0.10	0.96	0.10	80	C	307
	SILANE, ALLYLMETHYL-	36		0		60		423
	SILANE, ALLYLPHENYL-	29		0		80		423
	SILANE, (CHLOROMETHYL)DIMETHYL(P-VINYLPHENYL)-	0.69	0.03	0.86	0.03	20	I2	308
	SILANE, DIALLYL-	0.54		1.30		80	P,Q	517
	SILANE, DIMETHYLPHENYLVINYL-	30.6	5	0.03	0.01	80	G17	684
	SILANE, DIMETHYLPHENYL(P-VINYLPHENYL)-	0.80	0.16	1.00	0.4	80	C	685
	SILANE, DIMETHYL(P-VINYLPHENYL)-	0.56	0.01	1.8	0.01	80		307
	SILANE, DIMETHYLVINYL(P-VINYLPHENYL)-	0.9	0.2	1.1	0.45	80		307
	SILANE, METHYL(P-VINYLPHENYL)-	0.91		0.056		60		423
	SILANE, PHENYLVINYL-	5.7		0		60		423
	SILANE, TRIETHOXYVINYL-	1.13		1.4		70		1379
	SILANE, TRIMETHOXYVINYL-	22	5	0.06	0.1	60		849
	SILANE, TRIMETHOXY(P-VINYLPHENYL)-	0.71	0.02	0.01	0.01	70,3	G17,I20	556
	SILANE, TRIMETHYLVINYL-	5.7	0.3			12	C	249
	SILANE, TRIMETHYL(P-VINYLPHENYL)-	24	4	0		80	C	685
	SORBIC ACID, 2,3-EPOXYPROPYL ESTER	26	8	1.0	0.2	60		849
	SORBIC ACID, METHYL ESTER	1.0	0.2	0.51	0.06	70,3	C	1231
	SORBONITRILE	0.88	0.07	0.46	0.08	60	I7	272
	STEARIC ACID, VINYL ESTER	0.47	0.08	0.58	0.09	60		272
	STIBINE, DIPHENYL(P-VINYLPHENYL)-	0.15	0.05	0.01	0.01	60	C	591
	4-STILBENAMINE, N,N-DIMETHYL-	68	30	1.3		65	I2	126
	STILBENE, TRANS-	0.6		0.04	0.1	70	I6	1152
		6.56	0.69	0.033	0.07	70	G17,I15	1126
		2.3		0		70	I6	1152
		5.17	0.3			70	I6	1152
		11.2				60	D	160
		18	1.2			30		1126
	STILBENE, 4-CHLORO-	4.9	0.2	0.036	0.05	70	G17,I7	1152
	STILBENE, 4-NITRO-	3.23	0.56	0.09	0.25	70	I6	1152
	STILBENE, 4-VINYL-, TRANS-	0.36	0.02	5.5	0.5	60	I7	1156
	STYRENE, 2,5-BIS(1-ETHOXYETHOXY)-	0.9		0.5		60	I7	1103
	STYRENE, 2,5-BIS(1-METHYLBUTOXY)-	0.96		0.2		60	I7	1103
	STYRENE, 2,5-BIS(TRIFLUOROMETHYL)-	0.45	0.05	1.15	0.08	60	C	181
	STYRENE, M-BROMO-	0.50		1.6		30		699
	STYRENE, P-BROMO-	0.55	0.03	1.05	0.21	60		1024
		0.69		1.1		30		505
		0.695	0.02	0.99	0.07	60		1168
		0.71	0.02	1.05	0.05	60		1024
		0.8	0.3	0.4	0.1	60	H10,I16	490
	STYRENE, P-(2-BROMOETHYL)-	1.75	0.1	0.55	0.1	40	H12	880
	STYRENE, 4-TERT-BUTYL-2,5-DIMETHOXY-	1.8	0.6	0.3	0.2	-15/10	H10,I29	968
	STYRENE, M-CHLORO-	0.85	0.01	0.79	0.01	50	I7	127
		0.6		1.2		60		1103
		0.57		2.3		30		699
	STYRENE, O-CHLORO-	0.64	0.05	1.09	0.23	60	H20,I33	1024
	STYRENE, P-CHLORO-	3.3	0.4	0.3	0.05	0		752
		0.56	0.03	1.6	0.07	60	G17,P,Q	910

COPOLYMERIZATION PARAMETER

M1	M2	R1	+/-	R2	+/-	TEMP.	REMARKS	REF.
STYRENE	STYRENE, P-CHLORO-	0.66		1.1		30		505
		0.74	0.025	1.02	0.05	60		1168
		0.74	0.030	1.025	0.030	60		559
		0.742		1.032		60		560
		0.76		1.76		60	132	974
		0.816	0.03	1.042	0.05	131		559
		1.51	0.03	0.40	0.02	0	H30,I32	756
		1.7	0.2	0.55	0.03	0	H48,I33	756
		2.0	0.1	0.43	0.05	0	H44,I34	756
		2.0	0.2	0.34	0.05	0	H30,I33	756
		2.0	0.4	0.43	0.05	25	H51,I24	144
		2.10	0.2	0.50	0.02	0	H23,I37	756
		2.2	0.2	0.35	0.02	0	H20,I33	753
		2.2	0.2	0.45	0.02	0	H23,I22	756
		2.2	0.2	0.45	0.02	0	H20,I22	756
		2.2	0.2	0.45	0.02	0	H22,I33	756
		2.2	0.2	0.5	0.1	40	H12	687
		2.2	0.2	0.5	0.1	40	H12	880
		2.3	0.2	0.35	0.05	0	H20,I32	752
		2.5	0.5	0.36	0.05	0	H30,I33	756
		2.5	0.4	0.45		30	H31,I24	470
		2.7	0.4	0.30	0.03	0	H20,I32	752
		2.72	0.3	0.35	0.05	0	H20,I32	38
		2.89		0.48		-23	H19,I2	1284
		2.89		0.43		30	H19,I2	1284
	STYRENE, BETA-CHLORO-ALPHA,BETA-DIFLUORO-	2.099		0.402	0.03	60		238
	STYRENE, BETA-CHLORO-ALPHA,BETA-DIFLUORO-P-METHYL-	2.157		0.294	0.05	60		238
	STYRENE, 4-CHLORO-2,5-DIMETHOXY-	0.59		>1		60	I7,BB	1103
	STYRENE, ALPHA-CHLORO-BETA-FLUORO-	2.1		0.55		60	C	282
	STYRENE, DIBROMO-, MIXED ISOMERS	0.22	0.05	1.4	0.15	30	A	1173
	STYRENE, 2,5-DICHLORO-	0.18	0.07	0.25	0.09	41.5		280
		0.2		0.8		70		34
		0.23		2.2		70		355
		0.29		2.2		70		355
		0.30		1.8		70		355
		0.31		1.9		70		355
		0.32	0.06	0.08	0.05	65	C	297
		0.38		1.77		70		1037
		0.40	0.03	0.05		86.5	C	297
	STYRENE, 3,4-DICHLORO-	14.8	2	0.34	0.2	30	C	252
		2.8	0.2	0.45	0.10	0	H46	253
		3.0	0.5	0.20	0.15	0	H30	253
		3.1	0.1	0.48	0.08	0	H46	253
		3.5	0.5	0.10	0.2	0	H46	253
		4.2	0.2	0.27	0.05	30	H50	253
		5.9	0.8	0.0	0.07	0	H49	253
		6.8	0.5	0.0	0.2	30	H23	253
		7.2		0.38	0.20	0	H10	253
	STYRENE, ALPHA,BETA-DIFLUORO-	2.42		0.04		60	C	282
	STYRENE, BETA,BETA-DIFLUORO-	10.4		0.75		60	C	282
	STYRENE, 2,4-DIFLUORO-	1.05		2.195		60		238
	STYRENE, 2,5-DIFLUORO-	0.044		1.313		60	C	282
	STYRENE, ALPHA-(DIFLUOROMETHYL)-	0.54				60	C	238

REMARKS PAGE II-366, REFERENCES PAGE II-368

M1	M2	R1	+/-	R2	+/-	TEMP.	REMARKS	REF.
STYRENE	STYRENE, 2,5-DIMETHOXY-	0.77	0.07	1.13		70		466
	STYRENE, 2,5-DIMETHOXY-3,6-DIMETHYL-	3.5		0		60	I7	1103
	STYRENE, 2,5-DIMETHOXY-3-METHYL-	0.88		1.3		60	I7	1103
	STYRENE, 2,5-DIMETHOXY-4-METHYL-	0.88		1.3		60	I7	1103
	STYRENE, 2,5-DIMETHOXY-3,4,6-TRIMETHYL-	3		0		60	I7	1103
	STYRENE, X,ALPHA-DIMETHYL-, (X = 62:10:28 META:ORTHO:PARA)	0.63		0.39	0.04	75	C	39
	STYRENE, 2,4-DIMETHYL-	0.6	0.05	1.35	0.04	-30	H23	1094
		3	0.5	0.3		80	I7	1094
	STYRENE, ALPHA-ETHYL-	1.6		1.05		110	A,O	1189
	STYRENE, P-ETHYL-	0.95	0.1	1.0	0.1	40	H12	880
		1.0	0.2	0.01	0.2	40	H12	687
	STYRENE, BETA-FLUORO-	5.95		0.9		60	C	282
		0.7		0.9		60	C	202
	STYRENE, P-FLUORO-	0.7		0.60		60		1168
		1.5	0.1	0.7	0.1	40	H12	880
		1.5	0.2	0	0.1	40	H12	687
	STYRENE, P-FLUORO-ALPHA-TRIFLUOROMETHYL-	0.56	0.02	1.03	0.05	60		282
	STYRENE, P-IODO-	0.45		1.25		50	C	124
		0.62	0.05	1.09	0.30	60		1168
		0.62	0.01	0.54	0.01	60		1024
		1.11		0.89		50		127
	STYRENE, P-ISOPROPYL-	1.22		2.90		70	C	1424
	STYRENE, ALPHA-METHYL-	0.05	0.05	>20		-40	H20,I18	574
		0.05	0.06	10.1	1.5	-20	H10,I16	380
		0.14		2.5		20	H20	411
		0.2		3.0		20	H10,I1	831
		0.2		1.12	0.09	-78	H10,I1	831
		0.24		12	2	-20	H69,I3	722
		0.25	0.05	8.5	4.0	0	H10,I3	380
		0.25	0.25	1.19	0.73	-78	E	989
		0.49	0.12	0.297		90	H69,I2	722
		0.788		0.15		60		1237
		1.09		0.3		90		1187
		1.124		0.627		60	A	1187
		1.13		0.4		110		1237
		1.18	0.04	0.36		150	A	1187
		1.2		0.8		0	C,K,O	298
		1.3		0.3		100	A	1187
		1.3		0.3		60	C	713
		2.2	0.2	0.6		0	A	713
		2.3		0.38	0.2	0	C	713
							E,X	989
							Q	775
		17		0.015		25	G20,I5	969
		28		0.024		-20	G20,I5	969
		43		0.029			G20,I5	969
	STYRENE, M-METHYL-	2.0	0.2	0.5	0.1	40	H12	880
		2.0	0.2	0.5	0.1	40	H12	687
	STYRENE, O-METHYL-	12.1	1	0.1	0.1	40	H12	880
	STYRENE, P-METHYL-	0.20	0.10	4.2	0.14	30	H31,I24	470
		0.32	0.07	1.08	0.18	0	H23,I32	965
		0.48	0.04	1.18	0.1	0	H20,I33	965
		0.54		3.6		0	H23,I2	965

COPOLYMERIZATION PARAMETER

M1	M2	R1	+/-	R2	+/-	TEMP.	REMARKS	REF.
STYRENE	STYRENE, P-METHYL-	0.55	0.06	1.18	0.08	-78	H23,I2	965
		0.68	0.04	1.10	0.05	0	H23,I22	965
		0.82	0.1	1.15	0.05	40	H12	687
		0.83		0.96		63	C	1168
		0.90		0.96		40	H12	1035
		1	0.08	1.15	1.18	70	H12	880
		1.30	0.12	1	0.11	0	G4,I5	751
		1.97	0.15	0.91	0.11	0	G18,I5	965
		2.5	0.10	0.38	0.09	0	G17,I5	965
		5.3	0.1	0.26	0.03	0	G18,I5	965
				0.18		25		862
	STYRENE, BETA-NITRO-	0.4		0		80	C	539
	STYRENE, M-NITRO-	0.45	0.05	0.85	0.1	75	C,I7	895
		20	4	0.03	0.03	0	H20,I33	752
	STYRENE, P-NITRO-	0.19	0.02	1.15		60		1168
		0.19	0.2	0.10	0.20	60	C	1024
	STYRENE, 2,3,4,5,6-PENTACHLORO-	1.31		0.22	0.02	70		25
	STYRENE, 2,3,4,5,6-PENTAFLUORO-	0.43		0.6		60		1364
	STYRENE, 2,4,5-TRIBROMO-	0.05	0.15	0.07	0.35	30	I7	1173
	STYRENE, ALPHA,BETA,BETA-TRIFLUORO-	0.66		0.50		50	K	567
		0.81		0.15		60		238
		3.5		0.73		60	C	282
	STYRENE, ALPHA,BETA,BETA-TRIFLUORO-M-METHYL-	0.40		0.806		60		238
	STYRENE, M-TRIFLUOROMETHYL-	0.603		0.75		60		238
		0.62	0.05	1.05	0.05	60	C	282
	STYRENE, ALPHA,BETA,BETA-TRIFLUORO-O-METHYL-	0.70		0.403		60	C	181
	STYRENE, ALPHA,BETA,BETA-TRIFLUORO-P-METHYL-	1.177		0.63		60		238
	STYRENE, 2,4,6-TRIMETHYL-	0.40		0.004		60	G18,I5	238
	STYRENES, CHLOROMETHYL-	1060		1.08		25	P,Q	862
	SUCCINIC ACID, DIALLYL ESTER	0.72	1	0.106	0.009	60		216
	SUCCINIC ACID, METHYLENE-	35.9		0.2		60	C	41
		0.3		0		70	C,I10	1293
	SUCCINIC ACID, METHYLENE-, 1-BENZYL 2-METHYL ESTER	0.34		0.19		60	Q	261
	SUCCINIC ACID, METHYLENE-, BIS(2-CHLOROETHYL) ESTER	0.42	0.05	0.5	0.05	50		359
	SUCCINIC ACID, METHYLENE-, DIBUTYL ESTER	0	0.05	0.38	0.02	50		1444
	SUCCINIC ACID, METHYLENE-, DIETHYL ESTER	0.50	0.02	0.33	0.03	60		677
	SUCCINIC ACID, METHYLENE-, DIMETHYL ESTER	0.40		0.05		70	C	677
		0.30		0.25		60		123
	SUCCINIC ACID, METHYLENE-, DIOCTYL ESTER	1.4	0.02	0.14	0.02	60		123
	SUCCINIC ACID, METHYLENE-, DIPENTYL ESTER	0.32		0.06		60	Q	1391
	SUCCINIC ACID, METHYLENE-, DIPROPYL ESTER	0.48		0.60		60		123
	SUCCINIC ANHYDRIDE, METHYLENE-	0.5		0.52		60		169
		0.35	0.03	0.41	0.02	60		1444
		0.34	0.02	0.53	0.02	60		123
		0.25	0.03	0.60	0.04	60		123
		0.018		1.62		60		123
	SUCCINIMIDE, N-1-(1,3-BUTADIENYL)-	0.10		0.09		65	I7,Q	2
	SUCCINIMIDE, N-VINYL-	0.28		0.07		60	I5	222
		7.0		0.07		60	C,I6	672
		9.6		0.18		60	C	279
		10.5				60	C,I6	672
	SULFIDE, P-ANISYL VINYL	3.98						365
								976

M1	M2	R1	+/-	R2	+/-	TEMP.	REMARKS	REF.
STYRENE	SULFIDE, BENZYL VINYL	4.7		0.27		60		1338
	SULFIDE, BIS(P-VINYLPHENYL)	0.60		1.50		61	C	1042
	SULFIDE, P-BROMOPHENYL VINYL	3.42		0.31		60		976
	SULFIDE, BUTYL VINYL	4		0.2		60		1338
	SULFIDE, TERT-BUTYL VINYL	4.7		0.2		60		1338
	SULFIDE, P-CHLOROPHENYL VINYL	3.57		0.32		60		976
	SULFIDE, ETHYL VINYL	4.4		0.12		60		1338
		6		0.25		60		1260
		6.0	1.5	0.25	0.1	60		872
	SULFIDE, ISOBUTYL VINYL	4		0.2		60		1338
	SULFIDE, ISOPROPYL VINYL	4.3		0.2		60		1338
	SULFIDE, METHYL VINYL	4.5		0.15		60		785
		5		0.15		60		1338
	SULFIDE, PENTACHLOROPHENYL VINYL	5.1	1.0	0.12	0.05	60	C	788
	SULFIDE, PHENYL VINYL	3.9		0.24		80		360
		3.88		0.36		60		976
		4.5		0.15		60		785
	SULFIDE, P-TOLYL VINYL	3.65		0.25		60		976
	SULFONE, METHYL VINYL	1.40	0.5	0.01	0.01	60	C	788
		2.0	0.10	0.01	0.2	60	C	788
	SULFONE, PHENYL VINYL	2.4		0.01		60	I70	211
	SULFOXIDE, ETHYL VINYL	3.3		0.01	0.01	60	I112	785
		3.3		0.01	0.01	60		785
	SULFOXIDE, METHYL VINYL	2.8		0.2	0.01	60		1226
		6	0.2	0.1		60		1226
		7	0.1	0.02		60		1226
	SULFOXIDE, P-TOLYL VINYL	4.2	0.2	0.01		60		781
		5.77	0.1	0.1	0.07	70		1312
	TEREPHTHALALDEHYDE	0.6		0		-78	H10,I13	1142
		0.6		0		0	H10	1402
	TEREPHTHALIC ACID, DIALLYL ESTER	26.5		0.067	0.002	60	C	1294
	P-TERPHENYL, 3-VINYL-	0.94		1.75	0.1	60	I7	1248
	2,4,8,10-TETRAOXASPIRO(5.5)UNDECANE, 3,9-BIS(1-CHLOROVINYL)-	8.78		0.28		80	C,16	1353
	2,4,8,10-TETRAOXASPIRO(5.5)UNDECANE, 3,9-DIISOPROPENYL-	26		0.1		80	C	1349
	2,4,8,10-TETRAOXASPIRO(5.5)UNDECANE, 3,9-DIPROPENYL-	8.61		0.03		80	C,16	1349
	2,4,8,10-TETRAOXASPIRO(5.5)UNDECANE, 3,9-DISTYRYL-	18.8		0.11		80	C,I7	1353
	2,4,8,10-TETRAOXOSPIRO(5.5)UNDECANE, 3,9-DIVINYL-	55		0.02		80	C,16	1123
	TETRASILOXANE, 1-(CHLOROMETHYL)-1,1,3,5,5,7,7-OCTAMETHYL-7-(P-VINYLPHENYL)- *	55		0.01		60	C,16	1345
	TETRASILOXANE, 1,1,3,5,5,7,7-NONAMETHYL-7-(P-VINYLPHENYL)- *	55		0.01		60	C,16	1345
				0.01		60	C,16	1345
	TETRAZOLE, 1-(P-VINYLPHENYL)-	1.15	0.05	1.1	0.3	80		371
	2H-TETRAZOLE, 2-PHENYL-5-(P-VINYLPHENYL)-	0.95	0.02	1	0.01	60	I7	1399
	THIANTHRENE, 2-VINYL-	0.477		0.133		60		363
	4H-1,3-THIAZINE, 5,6-DIHYDRO-4,4-DIMETHYL-2-VINYL-	1.52		0.3		60	I7	1418
	THIAZOLE, 2-ISOPROPENYL-	0.09		3.8		60		1233
	THIAZOLE, 4-METHYL-2-VINYL-	0.15		2.8		60		1233
	THIAZOLE, 2-VINYL-	0.14		3.32		80		1106
	THIAZOLE, 4-VINYL-	0.66		0.82		80		1106
	THIOPHENE, 2-VINYL-	0.35	0.025	3.10	0.45	60		1023
	TIN, BIS(METHACRYLOYLOXY)DIMETHYL-	0.28		0.83		60	16	1431
	TIN, (1,3-BUTADIEN-2-YL)TRIETHYL-	0.23		0.68		70		1276

COPOLYMERIZATION PARAMETER

M1	M2	R1	+/-	R2	+/-	TEMP.	REMARKS	REF.
STYRENE								
	TIN, DIALLYL-	16.0		0.005		50	P,Q	517
	TIN, TRIBUTYLVINYL-	1.6	0.1	0.96	0.01	60	I7	640
	TIN, TRICYCLOHEXYL(P-VINYLPHENYL)-	24	0.5	0.004	0.0005	25	I2	535
	TIN, TRIETHYLVINYL-	49.0		0.001		50	G17,I7	1277
	TIN, TRIMETHYLVINYL-	4.7	0.1	0.897	0.004	60	I7	640
	TIN, TRIPHENYL(4'-VINYL-4-BIPHENYLYL)-	0.93		2.55		108	I2	535
	TIN, TRIPHENYL(P-VINYLPHENYL)-	2.67	0.07	0.71	0.01	60		839
	P-TOLUALDEHYDE	2.86	0.04	0.826	0.004	60		535
	TOLUENE, P-PROPENYL-	0.4		0		60	I2	1402
	P-TOLUENESULFONAMIDE, N-METHYL-N-VINYL-	0.75	0.1	0.36	0.05	0	H10	652
	1,3,5,2,4,6-TRIAZAPHOSPHORINE, 2-CHLORO-2,2,4,4,6,6-HEXAHYDRO-2,4,4,6,6-PENTAALLYLOXY- *	12.3		0		30	H10,I24	280
	1,3,5,2,4,6-TRIAZAPHOSPHORINE, 2-CHLORO-2,2,4,4,6,6-HEXAHYDRO-2,4,4,6,6-PENTAALLYLOXY- *	12.3		0		60	C	279
	S-TRIAZINE, 2-AMINO-4-ANILINO-6-ISOPROPENYL-	0.39		0.78		60	C	279
	S-TRIAZINE, 2-AMINO-4-ANILINO-6-VINYL-	0.6		0.24		60	I17	1453
	S-TRIAZINE, 2-AMINO-4-P-ANISIDINO-6-ISOPROPENYL-	0.43		0.81		60		1453
	S-TRIAZINE, 2-AMINO-4-P-ANISIDINO-6-VINYL-	0.78		0.29		60	I17	1193
	S-TRIAZINE, 2-AMINO-4-P-CHLOROANILINO-6-ISOPROPENYL-	0.48		0.66		60		1453
	S-TRIAZINE, 2-AMINO-4-(DIMETHYLAMINO)-6-ISOPROPENYL-	0.32		1.17		60	I17	1193
	S-TRIAZINE, 2-AMINO-4-(DIMETHYLAMINO)-6-VINYL-	0.7		0.81		60	I17	1453
	S-TRIAZINE, 2-AMINO-4-ISOPROPENYL-6-(N-METHYLANILINO)-	0.29		1.39		60	I17	1453
	S-TRIAZINE, 2-AMINO-4-ISOPROPENYL-6-P-TOLUIDINO-	0.43		0.61		60		1453
	S-TRIAZINE, 2-AMINO-4-(N-METHYLANILINO)-6-VINYL-	0.5		0.58		60	I17	1193
	S-TRIAZINE, 2-AMINO-4-O-TOLUIDINO-6-VINYL-	0.45		0.93		60	I17	1453
	S-TRIAZINE, 2-AMINO-4-P-TOLUIDINO-6-VINYL-	0.8		0.3		60	I17	1453
	S-TRIAZINE, 2,4-DIAMINO-6-ISOPROPENYL-	0.66		0.97		60	I17	1453
	S-TRIAZINE, 2,4-DIAMINO-6-VINYL-, HYDROCHLORIDE	0.09		0.14		60	I17	1453
	S-TRIAZINE, 4,6-DIAMINO-2-VINYL-	1.3		0.65		60	I17	1453
	S-TRIAZINE, 2,4-DIANILINO-6-VINYL-	0.53		0.4		60	I17	1453
	S-TRIAZINE, 2,4-DIMETHYL-6-VINYL-	0.12		0.92		60	I17	186
	S-TRIAZINE, 2,4-DIMETHYL-6-VINYL-	0.12		0.92		60		1453
	S-TRIAZINE, 2,4,6-TRIS(ALLYLOXY)-	8.7	9.3	0		60	I17	812
	S-TRIAZINE-2,4,6-(1H,3H,5H)-TRIONE, TRIALLYL-	90.6	13.2	0		60		812
	S-TRIAZINE-2,4,6-(1H,3H,5H)-TRIONE, TRIALLYL-	87.6	10.0	0		60		812
	S-TRIAZINE-2,4,6-(1H,3H,5H)-TRIONE, TRIALLYL-	90.1	13.1	0		60		488
	1,3,5-TRIOXANE	13.8	0.3	1.23	0.10	25		1131
	1,3,6-TRIOXOCANE	21		0.8		0	H10,I7	1132
	1,3,6-TRIOXOCANE	30		0.14		0	H23,I7	1132
	1,3,6-TRIOXOCANE	33		0.9		0	H10,I64	1132
	1,3,6-TRIOXOCANE	40		0.06		0	H10,I28	1132
	1,3,6-TRIOXOCANE	90		0		0	H20,I7	1132
	1,3,6-TRIOXOCANE			0		0	H10,I65	1132
	1,3,6-TRIOXOCANE					0	H10,I24	1132
	1,3,6-TRIOXOCANE					0	H10,I3	719
	TRISILANE, 1,1,1,2,3,3-HEPTAMETHYL-2-VINYL-	0.22	0.05	48	7	0	H10,I12	837
	TRISILOXANE, 1-(CHLOROMETHYL)-1,1,3,3,5,5-HEXAMETHYL-5-(P-VINYLPHENYL)-	38	5	0		60	I7	837
	TRISILOXANE, 1,1,1,3,5,5-HEPTAMETHYL-3-VINYL-	38	5	0		60	I7	773
	TRISILOXANE, 1,1,1,3,5,5-HEPTAMETHYL-5-(P-VINYLPHENYL)-	60	9	0.10	0.01	50		306
	TRISILOXANE, 1,1,1,3,5,5-HEPTAMETHYL-5-(P-VINYLPHENYL)-	0.90	0.03	1.2	0.2	80		

REMARKS PAGE II-366, REFERENCES PAGE II-368

M1	M2	R1	+/-	R2	+/-	TEMP.	REMARKS	REF.
STYRENE	TRISILOXANE, 1,1,1,5,5,5-HEXAMETHYL-3-TRIMETHYLSILOXY-3-VINYL- *	0.90	0.03	1.2	0.2	80		306
	TRISILOXANE, 1-(P-VINYLPHENYL)-	0.90	0.03	1.2	0.2	80		371
	9-UNDECENOIC ACID, VINYL ESTER	29.	9.	0.02	0.02	60		591
	UREA, 1-ETHYL-3-VINYL-	20.		0.020		70		335
		20.0		0.020		75		335
	VALERIC ACID, 2-METHYLENE-	0.725	0.005	0.68	0.08	70		1212
	VALERIC ACID, 2-METHYLENE-, METHYL ESTER	0.86	0.08	0.22	0.04	65		174
	VALERIC ACID, 3-METHYL-2-METHYLENE-, METHYL ESTER	2.25	0.05	0		65		174
	VALERIC ACID, 4-METHYL-2-METHYLENE-, METHYL ESTER	0.96	0.05	0.20	0.03	65		174
	VINYL ACETATE	0.01	0.01	0.1	0.1	-30		551
		2.64	0.35	0.30	0.15		H,I24,Q	358
		6.1	0.8	0.18	0.08		H,I22,Q	358
		6.5		0.15			H10,I22	374
		8.25	0.05	0.015	0.015	30	H2O	552
		10		0.05		20	C,I7	683
		16.0	5.6	0.0	0.30	0		988
		48		0.05		100	I7	683
		55	10	0.01		60		622
		56		0.01	0.01	60	I7	683
		60		0.16	0.02	60	C,I7	683
	VINYL ETHER	40		0.02		60		162
	VINYL SULFIDE	1.90	0.1	0.41	0.05	60		850
	VINYL SULFONE	1.3		0.01		60		162
	ZINC, DIACETATOBIS(5-VINYL-2-PICOLINE)-	0.35	0.05	2.0	0.2	50	16	951
	ZINC, DIACETATOBIS(2-VINYLPYRIDINE)-	0.75	0.045	0.32	0.03	60	16	952
	ZINC, DIACETATOBIS(4-VINYLPYRIDINE)-	0.55	0.15	3.35	0.3	50	16	951
	ZINC, DIBROMOBIS(5-VINYL-2-PICOLINE)-	0.08	0.03	2.7	0.5	50	16	952
	ZINC, DICHLOROBIS(5-VINYL-2-PICOLINE)-	0.33	0.05	0.38	0.18	60	16	952
	ZINC, DIIODOBIS(5-VINYL-2-PICOLINE)-	0.93	0.075	0.08	0.015	60	16	952
		0.60	0.10	0.55	0.06	60	16	952
STYRENE, 2,5-BIS(1-ETHOXYETHOXY)-	STYRENE	0.5		0.9		60	I7	1103
STYRENE, 2,5-BIS(1-METHYLBUTOXY)-	STYRENE	0.2		0.96		60	I7	1103
STYRENE, 2,5-BIS(TRIFLUOROMETHYL)-	METHYL METHACRYLATE	1.35	0.05	0.57	0.07	60	C	181
	STYRENE	1.15	0.08	0.45	0.05	60	C	181
STYRENE, M-BROMO-	ANISOLE, P-VINYL-	1.4		0.25		30		699
	BENZONITRILE, P-VINYL-	0.56		1.1		30		699
	METHYL METHACRYLATE	1.17	0.25	0.48	0.02	60		1024
	STYRENE	1.05	0.21	0.55	0.03	60		1024
	STYRENE, P-METHYL-	1.6		0.50		30		699
STYRENE, P-METHYL-		1.6		0.35		30		699

COPOLYMERIZATION PARAMETER

M1	M2	R1	+/-	R2	+/-	TEMP.	REMARKS	REF.
STYRENE, O-BROMO-	ACRYLIC ACID, METHYL ESTER	6.9		0		40		1186
STYRENE, P-BROMO-	ACRYLIC ACID, METHYL ESTER	6		0.01		40		1186
	ANISOLE, P-VINYL-	1.1		0.43		30		505
	METHYL METHACRYLATE	1.10	0.25	0.395	0.02	60		1024
	STYRENE	0.3	0.2	1.8	0.6	-15/10	H10,I29	968
	STYRENE	0.4	0.1	0.8	0.3	-20/10	H10,I16	968
	STYRENE	0.55	0.1	1.75	0.1	40	H12	880
	STYRENE	0.99	0.07	0.695	0.02	60		1024
	STYRENE	1		0.69		60		1168
	STYRENE	1.05	0.05	0.71	0.02	60		490
	STYRENE, P-CHLORO-	1.1		0.60		30		505
STYRENE, P-(2-BROMOETHYL)-	STYRENE, P-CHLORO-	1.0	0.1	1.0	0.1	0	H20,I34	752
STYRENE, 4-TERT-BUTYL-2,5-DIMETHOXY-	STYRENE	0.79	0.01	0.85	0.01	50		127
	STYRENE	1.2		0.6		60	I7	1103
STYRENE, M-CHLORO-	ACRYLIC ACID, METHYL ESTER	4.8		0.002		40		1186
	ANISOLE, P-VINYL-	1.9		0.20		30		699
	BENZONITRILE, P-VINYL-	0.60		1.20		30		699
	METHYL METHACRYLATE	0.91	0.11	0.47	0.075	60		1024
	STYRENE	0.3	0.05	3.3	0.4	0	H20,I33	752
	STYRENE	1.09	0.23	0.64	0.05	60		1024
	STYRENE	2.3		0.57		30		699
	STYRENE, P-METHYL-	2.2		0.20		30		699
STYRENE, O-CHLORO-	ACRYLIC ACID, BUTYL ESTER	2.25	0.2	0.2	0.05	40	C,K	1186
	ACRYLIC ACID, 2-CHLOROETHYL ESTER	1.71	0.08	0.533	0.015	40	C,K	1186
	ACRYLIC ACID, ETHYL ESTER	4.01	0.43	0.048	0.05	40	C,K	1186
	ACRYLIC ACID, 2-ETHYLHEXYL ESTER	2.65	0.23	0.17	0.1	40	C,K	1186
	ACRYLIC ACID, METHYL ESTER	6	0.4	0	0.03	40	C,K	1186
	ANISOLE, P-PROPENYL-	0.03	0.005	18	3	0	H20,I32	21
	INDENE	22	8	0		70	C	21
	METHACRYLIC ACID	3.5	0.5	0.7		70		23
	METHACRYLIC ACID, BUTYL ESTER	0.12	0.02	0.324	0.1	70		23
	METHACRYLIC ACID, 2-CHLOROETHYL ESTER	1.24	0.33	0.885	0.11	40	C,K	1186
	METHACRYLIC ACID, ETHYL ESTER	0.909	0.005	0.45	0.107	40	C,K	1186
	METHACRYLIC ACID, 2-ETHYLHEXYL ESTER	1.34	0.032	0.394	0.03	40	C,K	1186
	METHACRYLONITRILE	2.74	0.04	0.86	0.07	80	C,K	1186
	METHYL METHACRYLATE	0.78		0.46		40	C	159
	METHYL METHACRYLATE	1.34	0.01	0.50	0.075	40	C,K	1186
	METHYL METHACRYLATE	1.37	0.1		0.03	60		1023

REMARKS PAGE II-366, REFERENCES PAGE II-368

M1	M2	R1	+/-	R2	+/-	TEMP.	REMARKS	REF.
STYRENE, O-CHLORO-	STYRENE	1.6	0.07	0.56	0.03	60		1023
STYRENE, P-CHLORO-	ACRYLIC ACID, METHYL ESTER	3.9		0.01		40		1186
	ANISOLE, M-VINYL-	0.38	0.05	2.6	0.4	0	H20,I33	752
	ANISOLE, P-VINYL-	0.70		0.48		30		505
	BENZENE, PROPENYL-, CIS-	0.86	0.08	0.58	0.03	60		1024
	BENZENE, PROPENYL-, TRANS-	1.0	0.1	0.32	0.02	0	H20,I33	761
	BENZONITRILE, P-VINYL-	0.74	0.06	0.32	0.04	30	H20,I33	761
	BENZYL ALCOHOL, ALPHA, ALPHA-DIMETHYL-P-VINYL-	0.34		1.4		60	I66	505
	BENZYL ALCOHOL, ALPHA-ETHYL-ALPHA-METHYL-P-VINYL-	1.24		0.53		60	I66	1092
	1,3-BUTADIENE	1.43		0.33		50	K	1092
	1-BUTENE, 1-PHENYL-	0.42		1.07		0	H20,I33	1022
	CARBAZOLE, 9-VINYL-	0.88		0		30	I5	761
	CARBONIC ACID, DIVINYL ESTER	7	0.30	0.023	0.003	60	I7	1129
	CARBONIC ACID, ETHYL VINYL ESTER	27.5	0.2	0.036	0.017	60	I7	0496
	1,3-CYCLOOCTADIENE, CIS-,CIS-	39.8	0.7	0.025	0.05	-20	H20	0496
	4-CYCLOPENTENE-1,3-DIONE	0.32	0.05	0.84	0.05	0	H20	1304
	METHACRYLONITRILE	0.45	0.05	0.7	0.05	-20	H23	1304
	METHYL METHACRYLATE	0.52	0.05	0.52	0.05	50	H23	1304
		0.56	0.06	0.42	0.01	80	I7	1304
		0.32		0.02		60	C	1049
		0.45	0.068	0.73	0.056	30/40		1070
		0.468	0.4	0.434	0.2	60		655
		0.8	0.05	0.4		60		599
		0.89	0.05	0.415	0.02	60		1024
	2,5-NORBORNADIENE	85		0.01		60	H20I33,V	769
	1-PENTENE, 1-PHENYL-	0.91		0.63		60	H12,I20	761
	PHENETHYL ALCOHOL, ALPHA-METHYL-P-VINYL-	8.80		0.1		60	H30,I22	42
	1-PROPENE	0.15	0.04	14.7	0.2	0	H30,I22	345
		0.53		14.9		0	H30,I23	757
		0.7	0.05	22.5	0.5	0	H30,I23	757
		0.73		22.2		0	H30,I25	758
		0.75		18.0		0	H30,I24	758
		0.89	0.10	2.80	0.10	0	H30,I7	758
		0.99		1.14		0	H30,I9	758
		1.02		1.01		0	H30,I9	757
		1.04		1.1		0	H20,I22	758
		1.2	0.1	8.6	0.20	0	H20,I22	757
		1.25	0.03	8.6	1.0	0	H20,I7	758
		2.8		12.2		0		758
	1-PROPENE, 2-METHYL-	0.51	0.10	0.51	0.10	80		307
		0.76	0.04	0.84	0.08	80		307
	SILANE, DIMETHYLPHENYL(P-VINYLPHENYL)-, POLYMER	0.30	0.03	2.5	0.4	0	G17,P,Q	910
	SILANE, DIMETHYLVINYL(P-VINYLPHENYL)-	0.34	0.05	2.0	0.2	0	H20,I32	752
	STYRENE	0.35	0.02	2.10	0.2	0	H30,I33	756
		0.35	0.05	2.2	0.3	30	H20,I33	753
		0.36	0.05	2.7	0.4	0	H20,I32	752
		0.40	0.02	2.3	0.03	0	H20,I32	38

COPOLYMERIZATION PARAMETER

M1	M2	R1	+/-	R2	+/-	TEMP.	REMARKS	REF.
STYRENE, P-CHLORO-	STYRENE	0.43		2.82		30	H19,I2	1284
		0.43	0.03	2.0		0	H44,I34	756
		0.43	0.05	2.0		25	H51,I24	144
		0.45		2.5		30	H31,I24	470
		0.45	0.02	2.2	0.2	0	H23,I22	756
		0.45	0.02	2.2	0.2	0	H20,I22	756
		0.45	0.02	2.2	0.2	0	H22,I33	756
		0.48		2.72		-23	H19,I2	1284
		0.5	0.1	2.2	0.2	40	H12	880
		0.5	0.1	2.2	0.2	40	H12	687
		0.50	0.05	2.2	0.4	0	H23,I37	756
		0.55	0.05	1.7	0.2	0	H48,I33	756
	STYRENE, P-BROMO-	1.02		0.74		60		1168
		1.025	0.05	0.74	0.025	60		559
		1.032	0.030	0.742	0.030	60		560
		1.042	0.05	0.816	0.03	131		559
		1.1		0.66		30		505
	STYRENE, P-ETHYL-	1.76		0.76		60	I32	974
	STYRENE, ALPHA-METHYL-	1.0	0.1	1.0	0.1	0	H20,I34	752
		0.29	0.04	4.1	0.5	-78	H20,I33	761
	STYRENE, P-METHYL-	0.12	0.03	28.	2	0	G,P,Q	910
		0.35	0.05	15	1.5	0	H20	890
		0.35	0.05	15.5	1.5	74	H20,I32	752
	STYRENE, M-NITRO-	1.48	0.02	6.5	0.05	30	H20,I33	753
	STYRENE, P-NITRO-	0.19	0.05	4.5	0.7	0	C	890
		0.22	0.05	0.61	0.03	60	H31,I24	470
	STYRENE, 2,4,6-TRIMETHYL-	1.15	0.05	1.3	0.1	75	H20,I33	752
		0.25	0.05	0.91	0.37	60	C,I7	1024
		0.70	0.08	0.06		130	C,I7	895
		10		0.15				1024
		0.69						207
	SUCCINIC ACID, METHYLENE-, DIMETHYL ESTER						Q	169
STYRENE, BETA-CHLORO-ALPHA,BETA-DIFLUORO-	STYRENE	0.402		2.099		60		238
STYRENE, BETA-CHLORO-ALPHA,BETA-DIFLUORO-P-METHYL-	STYRENE	0.294		2.157		60		238
STYRENE, 4-CHLORO-2,5-DIMETHOXY-	STYRENE	>1		0.59		60	I7,BB	1103
STYRENE, ALPHA-CHLORO-BETA-FLUORO-	STYRENE	0.55		2.1		60	C	282
STYRENE, P-CHLOROMETHYL-ALPHA-METHYL-	1-PROPENE, 2-METHYL-	1		0.4		-100	H10	452

REMARKS PAGE II-366, REFERENCES PAGE II-368

M1	M2	R1	+/-	R2	+/-	TEMP.	REMARKS	REF.
STYRENE, DIBROMO-, MIXED ISOMERS								
	STYRENE	1.4	0.15	0.22	0.05	30	A	1173
STYRENE, DICHLORO-, MIXED ISOMERS								
	PYRIDINE, 2-VINYL-	0.11	0.07	0.63	0.07	60	C	7
STYRENE, 2,5-DICHLORO-								
	ACRYLIC ACID, METHYL ESTER	3.4	1.4	0.15	0.03	70	C	555
		4.27	0.28	0.25	0.04	60	C	7
	ACRYLONITRILE	0.07	0.05	0.22	0.05	67,5		297
		0.07	0.06	0.25	0.11	86,5		297
		0.09	0.02	0.26	0.02	38,5		297
	1,3-BUTADIENE	0.2	0.04	0.65	0.1	70	C	26
		0.46	0.01	0.46	0.01	50	K	1022
	3-BUTEN-2-ONE	2.0		0.5		70		22
	CARBAZOLE, 9-VINYL-	8	0.5	0.016	0.002	70	Q	32
		8.0		0.016				33
	MALONONITRILE, METHYLENE-	0.031		0.0092		40	B	289
	METHYL METHACRYLATE	2.25		0.44		68	C	8
	PYRIDINE, 2-VINYL-	0.9		1.1		70		32
	STYRENE	0.05	0.03	0.40	0.03	86,5		297
		0.08	0.05	0.32	0.06	65		297
		0.25	0.09	0.18	0.07	41,5		280
		0.34	0.2	14.8	2	30		252
		0.8		0.2		70	H46	34
		1.77		0.38		70		1037
		1.8		0.30		70	C	355
		1.9		0.31		70		355
		2.2		0.29		70		355
		2.2		0.23		70		355
	STYRENE, 2,5-DIMETHYL-	1.55		0.27		70	C	315
	STYRENE, ALPHA-METHYL-	3		0.14		70		32
	VINYL ACETATE			0.04		70		355
STYRENE, 2,6-DICHLORO-								
	METHYL METHACRYLATE	0.09	0.03	3.68	0.1	50	MM	1458
		0.12	0.02	2.05	0.04	50		1458
		-0.3	0.17	0.49	0.16	20	D,OO	1458
STYRENE, 3,4-DICHLORO-								
	STYRENE	0.0	0.2	6.8	0.8	0	H23	253
		0.0	0.2	3.5	0.5	0	H46	253
		0.10	0.05	4.2	0.2	0	H50	253
		0.20	0.15	3.0	0.5	0	H46	253
		0.27	0.07	5.9	0.2	30	H49	253
		0.38	0.20	7.2	0.5	0	H10	253
		0.45	0.10	2.8	0.2	0	H30	253
		0.48	0.08	3.1	0.1	0	H46	253

COPOLYMERIZATION PARAMETER

M1	M2	R1	+/-	R2	+/-	TEMP.	REMARKS	REF.
STYRENE, BETA,BETA-DIFLUORO-	STYRENE	0		10.4		60	C	282
STYRENE, ALPHA,BETA-DIFLUORO-	STYRENE	0.04		2.42		60	C	282
STYRENE, 2,4-DIFLUORO-	STYRENE	0.75		1.05		60	C	282
STYRENE, 2,5-DIFLUORO-	STYRENE	2.195		0.044		60		238
STYRENE, ALPHA-(DIFLUOROMETHYL)-	STYRENE	1.313		0.54		60		238
STYRENE, 2,5-DIMETHOXY-	METHYL METHACRYLATE	0.72	0.04	0.25	0.01	70		466
	STYRENE	1.13		0.77	0.07	70		466
STYRENE, 2,5-DIMETHOXY-3,6-DIMETHYL-	STYRENE	0		3.5		60	17	1103
STYRENE, 2,5-DIMETHOXY-3-METHYL-	STYRENE	1.3		0.88		60	17	1103
STYRENE, 2,5-DIMETHOXY-4-METHYL-	STYRENE	1.3		0.88		60	17	1103
STYRENE, 2,5-DIMETHOXY-3,4,6-TRIMETHYL-	STYRENE	0		3		60	17	1103
STYRENE, P,ALPHA-DIMETHYL-	ETHER, 2-CHLOROETHYL VINYL	0.51		1.7		RT	H20,17	592
		0.54		1.7		RT	H20,17	592
		0.64		1.7		RT	H20,I22	592
STYRENE, X,ALPHA-DIMETHYL-, (X = 62:10:28 META:ORTHO:PARA)	STYRENE	0.39		0.63		75	C	39
STYRENE, 2,4-DIMETHYL-	1,3-BUTADIENE	0.06		16.65	0.5	80	G17,19,Q	1017
	STYRENE	0.3	0.04	3			17	1094

REMARKS PAGE II-366, REFERENCES PAGE II-368

M1	M2	R1	+/-	R2	+/-	TEMP.	REMARKS	REF.
STYRENE, 2,4-DIMETHYL-								
	STYRENE	1.35	0.04	0.6	0.05	-30	H23	1094
STYRENE, 2,5-DIMETHYL-								
	STYRENE, 2,5-DICHLORO-	0.27		1.55		70	C	315
	STYRENE, ALPHA,BETA,BETA-TRIFLUORO-	0.32		0.67		60		238
	STYRENE, ALPHA,BETA,BETA-TRIFLUORO-M-METHYL-	0.20		0.50		60		238
	STYRENE, ALPHA,BETA,BETA-TRIFLUORO-P-METHYL-	0.44		0.7		60		238
STYRENE, ALPHA-ETHYL-								
	STYRENE			1.6		110	A,O	1189
STYRENE, P-ETHYL-								
	STYRENE	1.0	0.2	1.0	0.2	40	H12	687
		1.05	0.1	0.95	0.1	40	H12	880
	STYRENE, P-CHLORO-	4.1	0.5	0.29	0.04	0	H20,I33	761
STYRENE, BETA-FLUORO-								
	STYRENE	0.01		5.95		60	C	282
STYRENE, O-FLUORO-								
	ACRYLIC ACID, METHYL ESTER	3.8		0.01		40		1186
STYRENE, P-FLUORO-								
	ACRYLIC ACID, METHYL ESTER	2.6	0.1	0.04	0.1	40		1186
	STYRENE	0.60		1.5		40	H12	880
		0.7		1.5		40	H12	687
		0.9	0.1	0.7	0.2	60	C	282
		0.9		0.7		60		1168
STYRENE, P-FLUORO-ALPHA-TRIFLUOROMETHYL-								
	STYRENE	0		0.56		60	C	282
STYRENE, P-IODO-								
	METHYL METHACRYLATE	0.95	0.20	0.36	0.03	60		1024
	STYRENE	1.03	0.05	0.45	0.02	50		124
		1.09	0.30	0.62	0.05	60		1024
		1.25		0.62		60		1168
STYRENE, P-ISOPROPYL-								
	METHYL METHACRYLATE	0.39	0.01	0.44	0.01	70	C	1424
	STYRENE	0.54		1.11		50		127
		0.89		1.22		70	C	1424

COPOLYMERIZATION PARAMETER

M1 = STYRENE, ALPHA-METHYL-

M2	R1	+/-	R2	+/-	TEMP.	REMARKS	REF.
ACROLEIN, 3-CYANO-	0.033		0		80	C,I10	1405
ACRYLIC ACID, 2-CYANO-, METHYL ESTER	0.05		0.001		60	I7	507
ACRYLONITRILE	0.11	0.02	0.06	0.02	75	C,R	260
	0.17		0.03		20		1051
	0.2		0.04		50	A	713
	0.24		0.04		75		1051
	0.25		0.045		60		1051
ANILINE, N,N-DIMETHYL-P-VINYL-	0.035	0.015	0.08	0.03	70	I16,0	609
ANISOLE, M-VINYL-	5	1	31	19	5	H20,I32	752
ANISOLE, P-VINYL-	0.30	0.1	0.3	0.1	5	H20,I33	752
1,1'-BIINDENE	3.6	0.2	15	5	5	H20,I32	752
1,3-BUTADIENE	0.010	0.01	0.15	0.15	-72	H23,I3	1279
1,3-BUTADIENE, 2-FLUORO-	1.45	0.11	1.6	0.5	12,8		257
3-BUTEN-2-ONE, 3-METHYL-	0.38		0.12		20	K	790
ETHER, 2-CHLOROETHYL VINYL	0.03	0.01	1.71	0.19	50	K	733
	0.07	0.05	1.7		25	C,Q	1070
	0.31		4.5	0.5	-78	H10,I2	143
	0.33		5.72	0.7	RT	H20,I7	1285
	0.34	0.05	5.0	0.66	30	H20,I7	40
	0.42		5.03		RT	H20,I22	224
	0.42		2.6		40	H19,I2	40
	0.42		5		-23	H10,I3	1284
	0.46	0.07	6.02	0.25	-23	H20,I22	1284
	0.48	0.04	3.4	0.15	RT	H19,I2	592
	0.68	0.14	3.3	0.33	RT	H33,I24	40
	0.75	0.07	3.46	0.2	-78	H10,I3	1285
	0.76		3.9		0	H33,I24	143
			2.05		-78	H19,I3	1285
			2.0		-30	H19,I3	1284
			2.5		-23		1285
FUMARONITRILE	1	0.1	1.02	0.1	50		262
GLYCOLONITRILE, METHACRYLATE	0.022	0.005	0.00	0.01	70	C	1016
HYDRACRYLONITRILE, METHACRYLATE	0.12	0.01	0.36	0.02	70	C	1016
HYDROQUINONE, VINYL-, DIBENZOATE	0.11	0.02	0.35		78	I7	465
MALEIC ANHYDRIDE	0.005	0.005	0.30	0.03	60		787
METHACRYLIC ACID, 2-NITROETHYL ESTER	0.038	0.003	0.27		70		787
METHACRYLONITRILE	0.12	0.01	0.08	0.01	80	C	1016
	0.12	0.02	0.26	0.02	80	C	260
	0.15		0.35		80	C	1070
METHYL METHACRYLATE	0.545	0.15	0.21		60	I2	1238
	0.14	0.01	0.383	0.04	20		1023
	0.3		0.50	0.03	50		1051
	0.51		0.5		60		1051
	0.6		0.55		80		1051
	0.81		0.55		100		1051
	1.0		0.65		99		1051
PHTHALIMIDE, N-VINYL-	-0.01	0.01	0.89	0.03	75	A	213
1-PROPENE, 2-METHYL-	0.92		0.24		80		693
	0.92		0.24		95		693
	0.92		0.24		-78	H,P	721

REMARKS PAGE II-366, REFERENCES PAGE II-368

M1	M2	R1	+/-	R2	+/-	TEMP.	REMARKS	REF.
STYRENE, ALPHA-METHYL-	1-PROPENE, 2-METHYL-	1.41	0.25	0.27	0.08	-78	H23,13	722
		5.5		1.2		-78	H23,12	722
	STYRENE	0.015		17		25	G20,15	969
		0.024		28		0	G20,15	969
		0.029		43		-20	G20,15	969
		0.15		1		60		1187
		0.297		0.788		90		1237
		0.3		1.3		0	C	713
		0.3		1.3		100	A	713
		0.3		1.3		60	C	713
		0.36		1.09	0.04	90	A	1187
		0.38		1.18			C,K,Q	298
		0.4		2.3			Q	775
		0.6	0.2	2.2	0.2		A	989
		0.627		1.124		110	E,X	1187
		0.8		1.2		0	A	1237
		1.12	0.09	0.24	0.05	60		722
		1.19	0.73	0.49	0.12	150	H69,13	722
		2.5		0.2		-78	H69,12	831
		2.90		0.05		-78	H10,II	574
		3.0		0.2		20	H20,118	831
		8.5	4.0	0.25	0.25	20	H10,II	989
		10.1	1.5	0.14	0.06	0	E	411
		12	2	0.25	0.05	20	H20	380
		>20		0.05	0.05	-20	H10,13	910
	STYRENE, P-CHLORO-	0.25	0.05	1.48	0.02	74	G,P,Q	890
		15	1.5	0.35	0.05	0	C	752
		15.5	1.5	0.35	0.05	0	H20,132	753
		28.5	2	0.12	0.03	-78	H20,133	890
	STYRENE, 2,5-DICHLORO-	0.14		3		70	H20	32
	STYRENE, METHYL-, 1:2 ORTHO:PARA MIXTURE	0.68	0.07	1.016	0.006	80	C,K	1009
	STYRENE, 2,4,6-TRIMETHYL-	0.49	0.02	0.29	0.03	-78	G19,15	411
		0.72	0.07	0.2	0.1	0	G19,15	411
STYRENE, METHYL-, 3:2 META:PARA MIXTURE	INDENE	0.35		8.5			C,Q	1146
	TIN, TRIPHENYL(P-VINYLPHENYL)-	0.63		3.0		108		839
STYRENE, METHYL-, 1:2 ORTHO:PARA MIXTURE	ACRYLIC ACID, ETHYL ESTER	0.48	0.06	0.15	0.2	80		41
	ACRYLIC ACID, METHYL ESTER	0.59	0.06	0.15	0.20	80		41
	ACRYLIC ACID, OCTYL ESTER	0.49	0.06	0.16	0.13	80		41
	STYRENE, ALPHA-METHYL-	1.016	0.006	0.68	0.07	80	C,K	1009
STYRENE, M-METHYL-	ACRYLIC ACID, METHYL ESTER	1.65	0.1	0.14	0.04	40	Q	1186
	ACRYLONITRILE	0.43	0.02	0.07	0.025			40
	METHYL METHACRYLATE	0.49		0.53		60		1024

COPOLYMERIZATION PARAMETER

M1	M2	R1	+/-	R2	+/-	TEMP.	REMARKS	REF.
STYRENE, M-METHYL-	STYRENE	0.5	0.1	2.0	0.2	40	H12	687
		0.5	0.1	2.0	0.2	40	H12	880
STYRENE, O-METHYL-	ACRYLIC ACID, METHYL ESTER	0.98		0.27	0.05	40		1186
	ACRYLONITRILE	0.33	0.1	0.06			Q	40
	STYRENE	0.1	0.1	12.1	1	40	H12	880
STYRENE, P-METHYL-	ACRYLIC ACID, METHYL ESTER	1.54		0.17	0.02	40	Q	1186
	ACRYLONITRILE	0.33		0.05		60		40
		11.8		0.05		30		162
	ANILINE, N,N-DIVINYL-	0.1	0.14	10.	0.3	30		471
	ANISOLE, P-VINYL-	0.14		1.9		30	H31,I32	965
		0.3	0.06	4.3	0.10	0	H20,I33	471
		0.30	0.03	4.3	0.02	0	H31,I24	470
		0.52	0.1	1.54	0.06	0	H31,I24	965
		1.93	0.07	0.72	0.40	0	H23,I22	965
		2.2		0.68		0	G4,I5	965
		-0.03		3.7		0	G18,I5	965
		-0.4	0.03	18.2	0.45	-78	H23,I29	1284
		0.4	0.07	8.8	1.8	-78	H19,I2	1285
		0.5		10.1		0	H10,I3	1285
		0.88	0.06	8		-78	H19,I2	1284
		1.12		3.31	0.17	-78	H19,I3	1284
		1.12		2.31		-78	H19,I3	1285
		5		45		30	H19,I3	470
	ETHER, 2-CHLOROETHYL VINYL	0.44	0.02	0.405	0.025	60	H31,I24	1024
	METHYL METHACRYLATE	7.70		0.1		60	H12,I20	345
	1-PROPENE	0.18		5.3		25	G18,I5	862
	STYRENE	0.26	0.03	2.5	0.1	0	G17,I5	965
		0.38	0.09	1.97	0.10	0	G18,I5	965
		0.91	0.11	1.30	0.15	0	G4,I5	965
		0.96		0.83		60		1168
		0.96		0.83		63	C	1035
		1.08	0.12	1		70	H12	751
		1.10	0.14	0.32	0.12	0	H23,I32	965
		1.15	0.05	0.68	0.10	0	H23,I22	965
		1.15	0.05	0.82	0.04	40	H12	687
		1.18	1.18	0.90	0.1	40	H12	880
		1.18	0.08	0.55	0.08	-78	H23,I2	965
		3.6	0.18	0.48	0.06	0	H20,I33	965
		4.2	0.1	0.54	0.07	0	H23,I2	965
	STYRENE, M-BROMO-	0.35		0.20	0.04	30	H31,I24	470
	STYRENE, M-CHLORO-	0.20		1.6		30		699
	STYRENE, P-CHLORO-	0.61	0.03	2.2		60		699
		4.5	0.7	1.15	0.05	0	H20,I33	1024
		6.5		0.22	0.05	30	H31,I24	752
	TOLUENE, P-PROPENYL-	1.3	0.3	0.19	0.04	30	H10,I24	470
				0.04				652

M1	M2	R1	+/-	R2	+/-	TEMP.	REMARKS	REF.
STYRENE, BETA-NITRO-								
	SORBIC ACID, 2-CYANO-, BUTYL ESTER	0.18	0.1	0.47	0.1	0	G13,122	842
		0.27	0.1	0.52	0.1	74	G13,15	842
		0.3	0.3	5.0	1	-60	G13,15	842
		0.3	0.3	1.4	0.3	0	G13,15	842
	STYRENE	0.5	0.4	2.9	0.4	15	G13,117	842
		0		0.4		80	C	539
STYRENE, M-NITRO-								
	METHYL METHACRYLATE	0.85	0.2	0.35	0.05	75	C,17	895
	STYRENE	0.03	0.03	20	4	0	H20,133	752
	STYRENE, P-CHLORO-	0.85	0.1	0.45	0.05	75	C,17	895
		1.3	0.1	0.25	0.05	75	C,17	895
STYRENE, P-NITRO-								
	STYRENE	1.15		0.19		60		1168
		1.15	0.20	0.19	0.02	60		1024
	STYRENE, P-CHLORO-	0.91	0.37	0.70	0.08	60		1024
STYRENE, 2,3,4,5,6-PENTACHLORO-								
	ETHYLENE, CHLORO-	5.3		0.43			C	961
	METHYL METHACRYLATE	0.35	0.05	4.0	0.4	70	C	25
	STYRENE	0.10	0.02	1.31	0.2	70	C	25
STYRENE, 2,3,4,5,6-PENTAFLUORO-								
	METHYL METHACRYLATE	0.9		0.98		60		1364
	STYRENE	0.22		0.43		60		1364
STYRENE, 2,3,4,5-TETRAETHYL-								
	METHYL METHACRYLATE	0.33	0.08	0.56	0.06	50		364
STYRENE, 2,4,5-TRIBROMO-								
	STYRENE	0.6	0.35	0.05	0.15	30	17	1173
STYRENE, ALPHA,BETA,BETA-TRIFLUORO-								
	STYRENE	0.07		0.66		50	K	567
		0.15		3.5		60	C	282
	STYRENE, 2,5-DIMETHYL-	0.50		0.81		60		238
		0.67		0.32		60		238
STYRENE, M-TRIFLUOROMETHYL-								
	METHYL METHACRYLATE	0.98	0.15	0.60	0.10	60	C	181
STYRENE, ALPHA,BETA,BETA-TRIFLUORO-M-METHYL-								
	STYRENE	0.73		0.40		60		238

COPOLYMERIZATION PARAMETER

M1	M2	R1	+/-	R2	+/-	TEMP.	REMARKS	REF.
STYRENE, M-TRIFLUOROMETHYL-								
	STYRENE	0.75		0.62		60	C	282
		0.806	0.05	0.603	0.05	60		238
		1.05		0.70		60	C	181
STYRENE, ALPHA,BETA,BETA-TRIFLUORO-M-METHYL-								
	STYRENE, 2,5-DIMETHYL-	0.50		0.20		60		238
STYRENE, ALPHA,BETA,BETA-TRIFLUORO-O-METHYL-								
	STYRENE	0.403		1.177		60		238
STYRENE, ALPHA,BETA,BETA-TRIFLUORO-P-METHYL-								
	STYRENE, 2,5-DIMETHYL-	0.63		0.40		60		238
		0.7		0.44		60		238
STYRENE, 2,4,6-TRIMETHYL-								
	ACRYLONITRILE	0.16	0.02	0.98	0.2	138		207
	METHYL METHACRYLATE	0.05	0.01	1.6	0.3	130		207
		0.08		1.4		85		207
	STYRENE, P-CHLORO-	0.004		1060		25	G18,15	862
	STYRENE, ALPHA-METHYL-	0.06		10		130		207
		0.2	0.1	0.72	0.07	0	G19,15	411
		0.29	0.03	0.49	0.02	-78	G19,15	411
STYRENES, CHLOROMETHYL-								
	ACRYLONITRILE	0.67	0.1	0.06		60	H10	41
	1-PROPENE, 2-METHYL-	0.7		4.5	1	-100	P,Q	452
	STYRENE	1.08	0.1	0.72		60		216
								41
SUCCINALDEHYDONITRILE								
	ISOCYANIC ACID, METHYL ESTER	8.3	0.3	0.01	0.01	-78	G23,15	1210
	KETENE, DIMETHYL-	0.6	0.1	0.04	0.03	-78	G23,15	1209
SUCCINIC ACID, DIALLYL ESTER								
	STYRENE	0.106	0.009	35.9	1	60	C	1293
SUCCINIC ACID, METHYLENE-								
	ACRYLONITRILE	1.57		0.25		50	C,I11	676
	IMIDAZOLE, 2-METHYL-1-VINYL-	1.02		0.17		70	C,I11	579
	METHYL METHACRYLATE	0		1.23			0	1033
	STYRENE	0		1.14		90	I10	247
		0.2		0.34			Q	359
				0.3		70	C,I10	261

M1	M2	R1	+/-	R2	+/-	TEMP.	REMARKS	REF.
SUCCINIC ACID, METHYLENE-, 1-BENZYL 2-METHYL ESTER								
	STYRENE	0.19		0.42		60		1444
SUCCINIC ACID, METHYLENE-, BIS(2-CHLOROETHYL) ESTER								
	METHYL METHACRYLATE	0.10	0.05	2.63	0.2	50		677
	STYRENE	0.5	0.05	0.50	0.05	50		677
	STYRENE			0		50		677
SUCCINIC ACID, METHYLENE-, DIBUTYL ESTER								
	METHYL METHACRYLATE	0.4		0.8		60	C	169
	STYRENE	0.38	0.02	0.40	0.05	60		123
	SUCCINIC ACID, METHYLENE-, DIMETHYL ESTER	1.1		1.1			C	169
	VINYL ACETATE	6.3		0.02			C	169
SUCCINIC ACID, METHYLENE-, DIETHYL ESTER								
	ETHYLENE, CHLORO-	5.65	0.25	0.06	0.01	50		678
	2-PICOLINE, 5-VINYL-	0.17	0.06	0.51	0.11	70		1089
	STYRENE	0.05		1.4		70	C	1391
	STYRENE	0.33	0.03	0.30	0.02	60		123
SUCCINIC ACID, METHYLENE-, DIISOPROPYL ESTER								
	ETHYLENE, CHLORO-	6.0	0.5	0.06	0.01	50		678
SUCCINIC ACID, METHYLENE-, DIMETHYL ESTER								
	ACRYLIC ACID, BUTYL ESTER	0.94		0.40			C	169
	ETHYLENE, CHLORO-	5.0	0.2	0.053	0.01	50		677
	METHACRYLONITRILE	0.28		1.26			C	169
	METHYL METHACRYLATE	0.3		1.2	1.3		C	169
	2-PICOLINE, 5-VINYL-	0.225	0.03	0.875	0.05	70		1089
	STYRENE	0.06		0.5		60		1444
	STYRENE	0.14		0.48			C	169
	STYRENE, P-CHLORO-	0.25	0.02	0.32	0.02	60		123
	SUCCINIC ACID, METHYLENE-, DIBUTYL ESTER	0.15		0.69			C	169
		1.1		1.1			C	169
SUCCINIC ACID, METHYLENE-, DIOCTYL ESTER								
	ETHYLENE, CHLORO-	7.0	1.0	0.06	0.02	50		678
	STYRENE	0.60	0.02	0.35	0.03	60		123
SUCCINIC ACID, METHYLENE-, DIPENTYL ESTER								
	STYRENE	0.52	0.02	0.34	0.02	60		123
SUCCINIC ACID, METHYLENE-, DIPOTASSIUM SALT								
	ACRYLONITRILE	0.10		0.43		50	C, III	676

COPOLYMERIZATION PARAMETER

M1 / M2	R1	+/-	R2	+/-	TEMP.	REMARKS	REF.
SUCCINIC ACID, METHYLENE-, DIPROPYL ESTER							
STYRENE	0.41	0.04	0.25	0.03	60		123
SUCCINIC ACID, METHYLENE-, DISODIUM SALT							
VINYL ACETATE	0.198		-0.007		60	F,K	825
SUCCINIC ACID, METHYLENE-, 2,3-EPOXYPROPYL METHYL ESTER							
METHYL METHACRYLATE	0.20	0.02	1.73	0.05	60		54
SUCCINIC ANHYDRIDE							
BENZENE, (EPOXYETHYL)-	0.0		0.33		150	H80,I22	828
PROPYLENE OXIDE	0.0		0.28		150	H80,I22	828
	0.0		0.19		150	H78,I22	828
	0.0		0.21		150	H79,I22	828
SUCCINIC ANHYDRIDE, METHYLENE-							
ACRYLIC ACID, 2-CHLOROETHYL ESTER	2.25		0.34			I5,0	2
	2.7		0.065			I7,0	2
ACRYLONITRILE	4.8		0.13			I7,0	2
STYRENE	0.53		0.018			I7,0	2
VINYL ACETATE	0.60		0.10		65	I5	222
	3.50		0.08			I5,0	2
	5.75		0.08			I7,0	2
SUCCINIMIDE, N-1-(1,3-BUTADIENYL)-							
STYRENE	1.62		0.28		60	C,I6	672
SUCCINIMIDE, N-VINYL-							
ACETIC ACID, THIO-, VINYL ESTER	0.095		3.1		60		329
ACRYLIC ACID, METHYL ESTER	0.4		1.2		60		279
ACRYLONITRILE	0.16		0.54		60	C	279
CARBAZOLE, 9-VINYL-	1.05		0.3		60	C	1217
ETHER, BUTYL VINYL	15				60		279
ETHYLENE, 1,1-DICHLORO-	0.32		1.44		60	C	365
MALEIC ACID, DIMETHYL ESTER	1.2		0.04		60	C	1217
MALEIC ANHYDRIDE	0.15		0.03		60		365
METHYL METHACRYLATE	0.064		9.5		60		328
2-PYRROLIDINONE, 1-VINYL-	1.6		0.6		30	E	672
STYRENE	0.07		9.6		60	C,I6	365
	0.07		10.5		60	C	279
VINYL ACETATE	0.09		7.0		60	C	672
	5.1		0.175		65	C	365
	6.1		0.18		60	C	697
SULFIDE, P-ANISYL VINYL							
METHYL METHACRYLATE	0.04		0.77		60		976
STYRENE	0.18		3.98		60		976

M1	M2	R1	+/-	R2	+/-	TEMP.	REMARKS	REF.
SULFIDE, P-ANISYL VINYL								
	SULFIDE, P-BROMOPHENYL VINYL	0.53		0.93		60		976
	SULFIDE, P-CHLOROPHENYL VINYL	0.58		0.98		60		976
	SULFIDE, PHENYL VINYL	0.66		1.09		60		976
	SULFIDE, P-TOLYL VINYL	0.96		1.61		60		976
SULFIDE, BENZYL VINYL								
	METHACRYLIC ANHYDRIDE	0		0.8	0.1	60		372
	STYRENE	0.27		4.7		60		1338
SULFIDE, BIS(P-VINYLPHENYL)								
	STYRENE	1.50		0.60		61	C	1042
SULFIDE, P-BROMOPHENYL VINYL								
	METHYL METHACRYLATE	0.20		0.85		60		976
	STYRENE	0.31		3.42		60		976
	SULFIDE, P-ANISYL VINYL	0.93		0.53		60		976
	SULFIDE, P-CHLOROPHENYL VINYL	1.00		0.98		60		976
	SULFIDE, PHENYL VINYL	0.96		0.86		60		976
	SULFIDE, P-TOLYL VINYL	0.73		0.58		60		976
SULFIDE, BUTYL VINYL								
	ACRYLONITRILE	0.046		0.097		60		1338
	STYRENE	0.2		4		60		1338
SULFIDE, TERT-BUTYL VINYL								
	ACRYLONITRILE	0.01		0.13		60		1338
	STYRENE	0.2		4.7		60		1338
SULFIDE, P-CHLOROPHENYL VINYL								
	METHYL METHACRYLATE	0.08		0.89		60		976
	STYRENE	0.32		3.57		60		976
	SULFIDE, P-ANISYL VINYL	0.98		0.58		60		976
	SULFIDE, P-BROMOPHENYL VINYL	0.98		1.00		60		976
	SULFIDE, PHENYL VINYL	0.91		0.86		60		976
	SULFIDE, P-TOLYL VINYL	0.78		0.59		60		976
SULFIDE, ETHYL VINYL								
	ACRYLONITRILE	0.055		0.065		60	A	1225
	METHYL METHACRYLATE	0.07		0.93		60		1338
	STYRENE	0.3	0.1	2.7	1.5	60		872
		0.12		4.4		60		1338
		0.25		6		60		1260
		0.25	0.1	6.0	1.5	60		872

M1	M2	R1	+/-	R2	+/-	TEMP.	REMARKS	REF.
SULFIDE, ISOBUTYL VINYL								
	ACRYLIC ACID, ETHYL ESTER	0.05		0.35	0.04	60	A	1225
	ACRYLIC ACID, METHYL ESTER	0.05		0.35	0.04	60	A	1225
	ACRYLONITRILE	0.05		0.068		60		1338
	FUMARIC ACID, DIETHYL ESTER	0.2		0		60	A	1225
	MALEIC ACID, DIMETHYL ESTER			0		60		1338
	METHYL METHACRYLATE	0.04		0.94		60	A	1225
		0.04		0.94		60	A	1225
	STYRENE	0.1		4		60		1338
SULFIDE, ISOPROPYL VINYL								
	METHYL METHACRYLATE	0.07		0.85		60		1338
	STYRENE	0.2		4.3		60		1338
SULFIDE, METHYL VINYL								
	ACRYLIC ACID, METHYL ESTER	0.05	0.03	0.35	0.04	60		788
	CARBONIC ACID, CYCLIC VINYLENE ESTER	10.6	1.2	0.05	0.04	60		458
	METHYL METHACRYLATE	0.03		1.0		60		372
	STYRENE	0.12	0.05	5.1	1.0	60		788
		0.15		5		60	C	1338
		0.15		4.5		60		785
SULFIDE, PENTACHLOROPHENYL VINYL								
	ACRYLIC ACID, METHYL ESTER	0.25		0.89		80		360
	STYRENE	0.24		3.9		80		360
SULFIDE, PHENYL VINYL								
	ACRYLIC ACID, METHYL ESTER	0.05		0.40		60		785
	METHYL METHACRYLATE	0.08		0.85		60		976
	STYRENE	0.15	0.02	4.5		60		785
	SULFIDE, P-ANISYL VINYL	0.36		3.88		60		976
	SULFIDE, P-BROMOPHENYL VINYL	1.09		0.66		60		976
	SULFIDE, P-CHLOROPHENYL VINYL	0.86		0.96		60		976
		0.86		0.91		60		976
SULFIDE, P-TOLYL VINYL								
	METHYL METHACRYLATE	0.07		0.81		60		976
	STYRENE	0.25		3.65		60		976
	SULFIDE, P-ANISYL VINYL	1.61		0.96		60		976
	SULFIDE, P-BROMOPHENYL VINYL	0.58		0.73		60		976
	SULFIDE, P-CHLOROPHENYL VINYL	0.59		0.78		60		976
SULFONE, BUTYL VINYL								
	ACRYLONITRILE	0.2	0.1	1.1	0.2	-78	63	265
SULFONE, ALPHA-DIAZO-P-TOLYL METHYL								
	TOLUENE, ALPHA-DIAZO-	0.055	0.015	9.7	0.1	40	174	1396

M1	M2	R1	+/-	R2	+/-	TEMP.	REMARKS	REF.
SULFONE, METHYL VINYL	METHYL METHACRYLATE	0		14	2	60	D	211
	STYRENE	0.0	0.2	2.4	0.10	60		211
		0.01	0.01	1.40		60	C	788
		0.01	0.01	2.0	0.5	60	C	788
	VINYL ACETATE	0.35		3.3		60		785
		0.4	0.08	0.29		60	C	788
				0.0	0.01	60		211
SULFONE, PHENYL VINYL	STYRENE	0.01	0.01	3.3		60		785
	VINYL ACETATE	0.35		0.28		60		785
SULFOXIDE, ETHYL VINYL	ACRYLONITRILE	0.2		3		60		1226
	METHYL METHACRYLATE	0.01		10		6	C	1226
	STYRENE	0.02		7		60		1226
		0.1		6		60	I12	1226
		0.2		2.8		60	I70	1226
SULFOXIDE, METHYL VINYL	METHYL METHACRYLATE	0.01		20	10	60		781
	STYRENE	0.01	0.01	4.2	0.2	60		781
SULFOXIDE, P-TOLYL VINYL	STYRENE	0.1	0.07	5.77	0.1	70		1312
TARTARIC ACID, DIVINYL ESTER	METHYL METHACRYLATE	0.04		0.97		60	C	869
TAURINE, N-METHACRYLOYL-N-METHYL-, SODIUM SALT	ACRYLONITRILE	1.25		0.75		60	C,I17	1435
TEREPHTHALALDEHYDE	STYRENE	0		0.6		-78	H10,I3	1142
		0		0.6		0	H10	1402
TEREPHTHALIC ACID, 2,2-BIS(3-ALLYL-4-HYDROXYPHENYL)PROPANE * -CO-PHENOLPHTHALEIN POLYESTER	METHYL METHACRYLATE	0.25	0.05	2.8		70	C	1014
TEREPHTHALIC ACID, DIALLYL ESTER	BENZOIC ACID, ALLYL ESTER	0.77		0.89		60	C	1296
	STYRENE	0.067	0.002	26.5	0.2	60	C	1294

M1 / M2	R1	+/-	R2	+/-	TEMP.	REMARKS	REF.
P-TERPHENYL, 3-VINYL-							
STYRENE	1.75	0.1	0.94	0.1	60	I7	1248
2,4,8,10-TETRAOXASPIRO(5.5)UNDECANE, 3,9-BIS(1-CHLOROVINYL)-							
ACRYLONITRILE	0.8		0.58		80	C,I6	1353
STYRENE	0.28		0.95		80	C,I6	1353
2,4,8,10-TETRAOXASPIRO(5.5)UNDECANE, 3,9-DIISOPROPENYL-							
ACRYLONITRILE	0.04		0.89		80	C	1349
MALEIC ANHYDRIDE	0.05		0.025		60	C,I7	1347
METHYL METHACRYLATE	0.16		2.88		80	C	1349
STYRENE	0.1		8.78		80	C	1349
VINYL ACETATE	3		0.33		80	C	1349
2,4,8,10-TETRAOXASPIRO(5.5)UNDECANE, 3,9-DIPROPENYL-							
ACRYLONITRILE	0.04		4		80	C	1349
MALEIC ANHYDRIDE	0.02		0.02		60	C,I7	1347
STYRENE	0.03		26		80	C	1349
VINYL ACETATE	0.74		1.25		80	C	1349
2,4,8,10-TETRAOXASPIRO(5.5)UNDECANE, 3,9-DISTYRYL-							
ACRYLONITRILE	0.03		2.32		80	C,I6	1353
STYRENE	0.11		8.61		80	C,I6	1353
2,4,8,10-TETRAOXOSPIRO(5.5)UNDECANE, 3,9-DIVINYL-							
ACRYLIC ACID, METHYL ESTER	0.23		3.14		60	C,I7	1123
ACRYLONITRILE	0.02		7.11		60	C,I6	1345
	0.056		2.3		60	C,I6	1123
METHYL METHACRYLATE	0.05		19.6		60	C,I6	1345
	0.18		4.58		60	C,I7	1123
STYRENE	0.01		55		60	C,I6	1345
	0.02		18.8		60	C,I7	1123
VINYL ACETATE	0.8		1.15		60	C,I6	1345
	1		0.23		60	C,I10	1123
TETRASILOXANE, 1-(CHLOROMETHYL)-1,1,3,3,5,5,7,7-OCTAMETHYL-7-(P-VINYLPHENYL)- *							
STYRENE	1.08	0.08	0.98	0.04	80	I2	308
TETRASILOXANE, 1,1,1,3,3,5,5,7,7-NONAMETHYL-7-(P-VINYLPHENYL)- *							
STYRENE	1.1	0.3	1.15	0.05	80		306
TETRASILOXANE, 1-(P-VINYLPHENYL)-							
STYRENE	1.1	0.3	1.15	0.05	80		371

M1	M2	R1	+/-	R2	+/-	TEMP.	REMARKS	REF.
2H-TETRAZOLE, 2-PHENYL-5-(P-VINYLPHENYL)-								
	ETHYLENE, 1,1-DICHLORO-	1.8		0.25		60	17	1399
	METHYL METHACRYLATE	0.4	0.01	0.7		60	17	1399
	STYRENE	1		0.95	0.02	60	17	1399
THIANTHRENE, 2-VINYL-								
	ACRYLIC ACID, ETHYL ESTER	0.314		0.07		60		363
	METHYL METHACRYLATE	0.07		2.99		60		363
	STYRENE	0.133		0.477		60		363
4H-1,3-THIAZINE, 5,6-DIHYDRO-4,4-DIMETHYL-2-VINYL-								
	STYRENE	0.3		1.52		60	17	1418
THIAZOLE, 2-ISOPROPENYL-								
	STYRENE	3.8		0.09		60		1233
THIAZOLE, 4-METHYL-2-VINYL-								
	STYRENE	2.8		0.15		60		1233
THIAZOLE, 2-VINYL-								
	STYRENE	3.32		0.14		80		1106
THIAZOLE, 4-VINYL-								
	STYRENE	0.82		0.66		80		1106
THIOPHENE, 2,3-DIHYDRO-, 1,1-DIOXIDE								
	VINYL ACETATE	0.0	0.1	1.52	0.12	60	17	294
THIOPHENE, 2-VINYL-								
	STYRENE	3.10	0.45	0.35	0.025	60		1023
TIN, BIS(METHACRYLOYLOXY)DIMETHYL-								
	STYRENE	0.83		0.28		60	16	1431
TIN, (1,3-BUTADIEN-2-YL)TRIETHYL-								
	STYRENE	0.68		0.23		70		1276
TIN, BUTYLDICHLOROVINYL-								
	ETHYLENE, CHLORO-	0		1.5		60		1443
TIN, DIALLYL-								
	METHYL METHACRYLATE						P,Q	517

COPOLYMERIZATION PARAMETER

M1 M2	R1	+/-	R2	+/-	TEMP.	REMARKS	REF.
TIN, DIALLYL-							
STYRENE						P,Q	517
TIN, (MALEOYLDIOXY)BIS(TRIETHYL-							
ETHYLENE, CHLORO-	0.001		0.85		60		1195
TIN, (METHACRYLOYLOXY)TRIMETHYL-							
MALEIC ANHYDRIDE	0.2		0		50	I6	1431
	0.53		1.57		60	I6	1431
TIN, TRIBUTYL(METHACRYLOYLOXY)-							
ETHYLENE, CHLORO-	4.5		0.1		60		1195
TIN, TRIBUTYLVINYL-							
METHYL METHACRYLATE	0.03		27.9		50	I7	640
STYRENE	0.005		16.0		50	I7	640
TIN, TRICYCLOHEXYL(P-VINYLPHENYL)-							
STYRENE	0.96	0.01	1.6	0.1	60	I2	535
TIN, TRIETHYL(METHACRYLOYLOXY)-							
METHYL METHACRYLATE	0.705	0.025	0.37	0.02	60		1329
TIN, TRIETHYLUINYL-							
ETHYLENE, CHLORO-	0	0.0005	1.2	0.3	60	G17,I7	1359
STYRENE	0.004		24	0.5	25		1277
TIN, TRIMETHYLVINYL-							
METHYL METHACRYLATE	0.03		25.1		50	I7	640
STYRENE	0.001		49.0		50	I7	640
TIN, TRIPHENYL(4'-VINYL-4-BIPHENYLYL)-							
STYRENE	0.897	0.004	4.7	0.1	60	I2	535
TIN, TRIPHENYL(P-VINYLPHENYL)-							
STYRENE	0.71	0.01	2.67	0.07	60	I2	535
	0.826	0.004	2.86	0.04	60	I2	535
	2.55		0.93		108		839
	3.0		0.63		108		839
STYRENE, METHYL-, 3:2 META:PARA MIXTURE							
P-TOLUALDEHYDE							
STYRENE	0		0.4		0	H10	1402

M1	M2	R1	+/-	R2	+/-	TEMP.	REMARKS	REF.
TOLUENE, P-CHLORO-ALPHA-DIAZO-	TOLUENE, ALPHA-DIAZO-	0.3	0.05	1.2	0.1	67	I74	1396
TOLUENE, ALPHA-DIAZO-	ANISOLE, P-(DIAZOMETHYL)-	0.62	0.1	1.2	0.1	40	I74	1396
	NAPHTHALENE, 1-(DIAZOMETHYL)-	0.22	0.03	2	0.05	40	I74	1396
	NAPHTHALENE, 2-(DIAZOMETHYL)-	0.2	0.1	3.55	0.3	40	I74	1396
	SULFONE, ALPHA-DIAZO-P-TOLYL METHYL	9.7	0.1	0.055	0.015	40	I74	1396
	TOLUENE, P-CHLORO-ALPHA-DIAZO-	1.2	0.1	0.3	0.05	67	I74	1396
	TOLUENE, 2,4-DICHLORO-ALPHA-DIAZO-	2.13	0.05	0.21	0.02	40	I74	1396
	TOLUENE, 3,4-DICHLORO-ALPHA-DIAZO-	2.75	0.07	0.12	0.02	40	I74	1396
	P-XYLENE, ALPHA-DIAZO-	0.6	0.07	1.3	0.1	40	I74	1396
		0.65	0.15	2.3	0.2	40	I74	1396
TOLUENE, 2,4-DICHLORO-ALPHA-DIAZO-	TOLUENE, ALPHA-DIAZO-	0.21	0.02	2.13	0.05	40	I74	1396
TOLUENE, 3,4-DICHLORO-ALPHA-DIAZO-	TOLUENE, ALPHA-DIAZO-	0.12	0.02	2.75	0.07	40	I74	1396
TOLUENE, P-(EPOXYETHYL)-	BENZENE, (EPOXYETHYL)-	1.2	0.2	0.65	0.1	-20	H10,I20	717
TOLUENE, P-PROPENYL-	STYRENE	0.36	0.05	0.75	0.1	30	H10,I24	652
	STYRENE, P-METHYL-	0.04	0.04	1.3	0.3	30	H10,I24	652
TOLUENE-2,4-DICARBAMIC ACID, BIS(2-HYDROXYETHYL)ESTER, DIME* THACRYLATE	METHACRYLIC ACID, 2-HYDROXYETHYL ESTER	0.8	0.1	1.4	0.1	30	E	1266
P-TOLUENESULFONAMIDE, N-METHYL-N-VINYL-	ACRYLONITRILE	0		0.42		60	C	280
	ACRYLONITRILE	0		0.42		60	C	279
	METHYL METHACRYLATE	0		4.68		60	C	279
	METHYL METHACRYLATE	0		4.68		60	C	280
	STYRENE	0		12.3		60	C	279
	STYRENE	0		12.3		60	C	280
P-TOLUIC ACID, VINYL ESTER	BENZOIC ACID, P-BROMO-, VINYL ESTER	0.64		1.04		60		397
	BENZOIC ACID, P-CHLORO-, VINYL ESTER	0.74		0.89		60		397
	BENZOIC ACID, P-CYANO-, VINYL ESTER	0.65		0.86		60		397
	VINYL ACETATE	1.73		0.73		30		397

COPOLYMERIZATION PARAMETER

M1	M2	R1	+/-	R2	+/-	TEMP.	REMARKS	REF.
1,3,5,2,4,6-TRIAZAPHOSPHORINE, 2-CHLORO-2,2,4,4,6,6-HEXAHYDRO-2,4,6,6-PENTAALLYLOXY- *	STYRENE	0.221		0.510		70	C	528
S-TRIAZINE, 2-AMINO-4-ANILINO-6-ISOPROPENYL-	STYRENE	0.78		0.39		60	I17	1453
S-TRIAZINE, 2-AMINO-4-ANILINO-6-VINYL-	STYRENE	0.24		0.6		60	I17	1453
S-TRIAZINE, 2-AMINO-4-P-ANISIDINO-6-ISOPROPENYL-	STYRENE	0.81		0.43		60		1193
S-TRIAZINE, 2-AMINO-4-P-ANISIDINO-6-VINYL-	STYRENE	0.29		0.78		60	I17	1453
S-TRIAZINE, 2-AMINO-4-P-CHLOROANILINO-6-ISOPROPENYL-	STYRENE	0.66		0.48		60		1193
S-TRIAZINE, 2-AMINO-4-(DIMETHYLAMINO)-6-ISOPROPENYL-	STYRENE	1.17		0.32		60	I17	1453
S-TRIAZINE, 2-AMINO-4-(DIMETHYLAMINO)-6-VINYL-	STYRENE	0.81		0.7		60	I17	1453
S-TRIAZINE, 2-AMINO-4-ISOPROPENYL-6-(N-METHYLANILINO)-	STYRENE	1.39		0.29		60	I17	1453
S-TRIAZINE, 2-AMINO-4-ISOPROPENYL-6-P-TOLUIDINO-	STYRENE	0.61		0.43		60		1193
S-TRIAZINE, 2-AMINO-4-(N-METHYLANILINO)-6-VINYL-	STYRENE	0.58		0.5		60	I17	1453
S-TRIAZINE, 2-AMINO-4-O-TOLUIDINO-6-VINYL-	STYRENE	0.93		0.45		60	I17	1453
S-TRIAZINE, 2-AMINO-4-P-TOLUIDINO-6-VINYL-	STYRENE	0.3		0.8		60	I17	1453

REMARKS PAGE II-366, REFERENCES PAGE II-368

M1 / M2	R1	+/-	R2	+/-	TEMP.	REMARKS	REF.
S-TRIAZINE, 2,4-DIAMINO-6-ISOPROPENYL-							
STYRENE	0.97		0.66		60	I17	1453
S-TRIAZINE, 2,4-DIAMINO-6-VINYL-, HYDROCHLORIDE							
STYRENE	0.14		0.09		60	I17	1453
S-TRIAZINE, 4,6-DIAMINO-2-VINYL-							
3-BUTEN-2-ONE	1.2	0.15	0.26	0.04	60		759
STYRENE	0.65		1.3		60	I17	1453
S-TRIAZINE, 2,4-DIANILINO-6-VINYL-							
STYRENE	0.4		0.53		60	I17	1453
S-TRIAZINE, 2,4-DIMETHYL-6-VINYL-							
METHYL METHACRYLATE	1.75		0.37		60		186
STYRENE	0.92		0.12		60		186
STYRENE	0.92		0.12		60	I17	1453
S-TRIAZINE, HEXAHYDRO-1,3,5-TRIACRYLYLEYL-							
ACRYLIC ACID, CALCIUM SALT	1.71	0.03	0.56	0.02	20	F	295
S-TRIAZINE, 2,4,6-TRIS(ALLYLOXY)-							
METHYL METHACRYLATE	0		46.3	5.3	60		812
	0		50.2	7.5	60		812
STYRENE	0		90.6	13.2	60		812
	0		80.7	9.3	60		812
S-TRIAZINE-2,4,6-(1H,3H,5H)-TRIONE, TRIALLYL-							
METHYL METHACRYLATE	0		48.9	7.3	60		812
	0		45.0	5.1	60		812
STYRENE	0		87.6	10.0	60		812
	0		90.1	13.1	60		812
VINYL ACETATE	0.70	0.09	0.95	0.18	60		812
	0.75	0.06	0.91	0.03	60		812
TRIDECANEDIOIC ACID, 2-METHYLPENTYL VINYL ESTER							
ETHYLENE, CHLORO-	0.239		1.10		60	C,17	164
TRIETHYLENE GLYCOL, DIMETHACRYLATE							
2-PICOLINE, 5-VINYL-	0.59	0.06	0.32	0.065		Q	202
TRIMETHYLENE OXIDE							
1,3-DIOXOLANE	90		0.01		0	H2O,I3	1061
	100		0.03		0	H2O,I2	1061

M1 / M2	R1	+/-	R2	+/-	TEMP.	REMARKS	REF.
TRIMETHYLENE OXIDE							
1,3-DIOXOLANE	165		0.013		0	H10,I2	1061
OXETANE, 3,3-BIS(CHLOROMETHYL)-	10.8		0.01		0	H10,I3	49
OXETANE, 3-(CHLOROMETHYL)-3-ETHYL-	3.0	0.4	0.64	0.22	0	H13	283
OXETANE, 3-(CHLOROMETHYL)-3-METHYL-	3.05	0.40	0.64	0.22	0	H13	284
1,3,5-TRIOXANE							
BENZENE, (EPOXYETHYL)-	2.7	0.3	0		35		488
M-DIOXANE, 4-PHENYL-	9	2	0		35		488
1,3-DIOXOLANE	0.47	0.15	1.36	0.03	35	C,I22	489
	0.57		1.75		70	H63	544
1-PROPANOL, 2,3-EPOXY-, NITRATE	1.36		0.47	0.15	35		488
STYRENE	0.35	0.03	0.3		0	H10,I7	1086
	0		30		0	H10,I28	1132
	0		90		0	H10,I24	1132
	0		90		0	H10,I3	1132
	0		33		0	H20,I7	1132
	0.06		40		0	H10,I65	1132
	0.14		21		0	H23,I7	1131
	0.8		13.8		0	H10,I7	1131
	0.9		30		0	H10,I64	1132
	1.23	0.10	0.4	0.3	25		488
1,3,6-TRIOXOCANE							
1,3-DIOXOLANE, 4-(CHLOROMETHYL)-	75	15	0.08	0.02	0	H10	716
STYRENE	48	7	0.22	0.05	0	H10,I2	719
TRISILANE, 1,1,2,3,3-HEPTAMETHYL-2-VINYL-							
ACRYLONITRILE	0		2	1	60	I6	837
STYRENE	0		38	5	60	I7	837
TRISILOXANE, 1-(CHLOROMETHYL)-1,1,3,3,5,5-HEXAMETHYL-5-(P-VINYLPHENYL)- *							
STYRENE	0.97	0.10	1.02	0.04	80	I2	308
TRISILOXANE, 1,1,3,5,5,5-HEPTAMETHYL-3-VINYL-							
ACRYLONITRILE	0.1	0.01	8.0	1.2	50		773
ETHYLENE, CHLORO-	0.50	0.07	0.90	0.13	50		773
ETHYLENE, CHLOROTRIFLUORO-	0.20		0.05		60		773
STYRENE	0.10	0.01	60	9	50		773
VINYL ACETATE	0.010	0.001	0.99	0.15	70		773
TRISILOXANE, 1,1,3,5,5,5-HEPTAMETHYL-5-(P-VINYLPHENYL)-							
STYRENE	1.2	0.2	0.90	0.03	80	I2	306

M1	M2	R1	+/-	R2	+/-	TEMP.	REMARKS	REF.
TRISILOXANE, 1,1,1,5,5,5-HEXAMETHYL-3-TRIMETHYLSILOXY-3-VINYL- *								
	ACRYLONITRILE	0.1	0.01	8.0	1.2	50		773
	ETHYLENE, CHLORO-	0.50	0.07	0.90	0.13	50		773
	ETHYLENE, CHLOROTRIFLUORO-	0.20		0.05		60		773
	2-PYRROLIDINONE, 1-VINYL-	0.10	0.01	4.0	0.6	125		773
	STYRENE	0.10	0.01	60	9	50		773
	VINYL ACETATE	0.010	0.001	0.99	0.15	70		773
TRISILOXANE, 1-(P-VINYLPHENYL)-								
	STYRENE	1.2	0.2	0.90	0.03	80		371
UNDECANOIC ACID, 11-IODO-, ALLYL ESTER								
	ETHYLENE, CHLORO-	0.42		1.64		60	I7	177
9-UNDECENOIC ACID, VINYL ESTER								
	ACRYLIC ACID, METHYL ESTER	0.031	0.12	3.69	0.12	60	C	591
	ACRYLONITRILE	0.0	0.10	1.82	0.04	60	C	591
	1,3-BUTADIENE	0.015	0.015	37.9	4.0	60	C	591
	ETHYLENE, CHLORO-	0.358	0.065	1.06	0.05	60	C	591
	ETHYLENE, 1,1-DICHLORO-	0.054	0.030	2.58	0.09	60	C	591
	STYRENE	0.02	0.02	29.	9.	60	C	591
UREA, 1-ALLYL-								
	METHACRYLIC ACID, DODECYL ESTER	-0.22	0.06	29.9	1.2	100	C	104
	METHACRYLIC ANHYDRIDE	0		28	2	60		372
UREA, 1,1-DIALLYL-								
	METHACRYLIC ACID, DODECYL ESTER	0.014	0.05	27.4	0.8	100	C	104
UREA, 1,3-DIALLYL-								
	METHACRYLIC ACID, DODECYL ESTER	-0.12	0.04	14.5	0.7	100	C	104
UREA, 1-ETHYL-3-VINYL-								
	METHYL METHACRYLATE	0.015		1.8		70		335
		0.015		1.8		75		335
	STYRENE	0.020		20		70		335
	VINYL ACETATE	0.020		20.0		75		335
		0.63		0.45		70		335
		0.63		0.45		75		335
VALERIC ACID, VINYL ESTER								
	ETHYLENE, CHLORO-	0.28	0.20	2	0.310	60	C	1213
	1-PROPENE, 3-CHLORO-	0.59		1.62		60		457

M1	M2	R1	+/-	R2	+/-	TEMP.	REMARKS	REF.
VALERIC ACID, 4,5-EPOXY-3-HYDROXY-, BETA-LACTONE								
	HYDRACRYLIC ACID, BETA-LACTONE	0.3		10.0		20	H40	715
VALERIC ACID, 5-HYDROXY-, DELTA LACTONE								
	OXETANE, 3,3-BIS(CHLOROMETHYL)-	0		0.20		30	H10,I20	981
	PROPYLENE OXIDE	5.3		0.25	0.1	20	H55	171
VALERIC ACID, 4-METHYL-, VINYL ESTER								
	VINYL ACETATE	1.14		0.67		80	C	549
VALERIC ACID, 2-METHYLENE-								
	STYRENE	0.68	0.08	0.725	0.005	70		1212
VALERIC ACID, 2-METHYLENE-, METHYL ESTER								
	STYRENE	0.22	0.04	0.86	0.08	65		174
VALERIC ACID, 3-METHYL-2-METHYLENE-, METHYL ESTER								
	STYRENE	0		2.25	0.05	65		174
VALERIC ACID, 4-METHYL-2-METHYLENE-, METHYL ESTER								
	STYRENE	0.20	0.03	0.96	0.05	65		174
VERSATIC 10 ACID, VINYL ESTER								
	ACRYLIC ACID, ETHYL ESTER	0.10	0.10	6.46	0.50	60	I7	954
	ACRYLONITRILE	0	0.10	5.22	0.20	60	I7	954
	VINYL ACETATE	0.92	0.12	0.99	0.10	60	I7	954
VERSATIC 13 ACID, VINYL ESTER								
	VINYL ACETATE	0.93	0.13	1.25	0.14	60	I7	954
VERSATIC 9 ACID, VINYL ESTER								
	VINYL ACETATE	0.90	0.08	0.93	0.07	60	I7	954
VINYL ACETATE								
	ACETALDEHYDE, DIVINYL ACETAL	1.005	0.08	0.99	0.08	60		608
	ACETANILIDE, N-VINYL-	0.15	0.015	1.60	0.13	70		367
	ACETIC ACID, ALLYL ESTER	0.60	0.15	0.45	0.15	60		560
		1.0		0.7		60	C	834
	ACETIC ACID, CHLORO-, VINYL ESTER	0.73		1.20		60		256
	ACETIC ACID, PHENYL-, VINYL ESTER	0.96	0.07	0.92	0.07	60	C,I7	1135
	ACETIC ACID, THIO-, VINYL ESTER	0.05		5.5		50		1249
	ACETIC ACID, TRIFLUORO-, VINYL ESTER	0.25		0.2		70		1203
		0.6		0.32		60	C	312
ACROLEIN		0.1	0.05	3.33	0.1	20	C	843

M1	M2	R1	+/-	R2	+/-	TEMP.	REMARKS	REF.
VINYL ACETATE								
	ACROLEIN	0.1	0.05	3.3	0.1	50	L	845
	ACRYLAMIDE, N-OCTADECYL-	0.027	0.009	6.11	0.045	60	C,I14	454
	ACRYLIC ACID	0.01	0.003	10	1	70	C	119
		0.11		2		70		32
	ACRYLIC ACID, ALLYL ESTER	0.111	0.01	17.35	0.5	60	C	1367
	ACRYLIC ACID, BUTYL ESTER	0.06	0.01	3.07	0.3	70		1324
	ACRYLIC ACID, 2,3-EPOXYPROPYL ESTER	0.0029		7.6		60	15	1166
	ACRYLIC ACID, 2-(2-(2-ETHOXYETHOXY)ETHOXY)ETHYL ESTER	0.065		5		60	15	1166
	ACRYLIC ACID, 2-(2-ETHOXYETHOXY)ETHYL ESTER	0.032		1.25		60	15	1166
	ACRYLIC ACID, 2-ETHOXYETHYL ESTER	0.027		3.95		60	15	1166
	ACRYLIC ACID, 2-ETHYLHEXYL ESTER	0.04		7.5			C,I7,Q	1308
	ACRYLIC ACID, 2-HYDROXYPROPYL ESTER	0.28	0.1	2.1	0.15	60	K,Q	1143
	ACRYLIC ACID, METHYL ESTER	0.01		4.55		60	15	1166
	ACRYLIC ACID, SODIUM SALT	0.029	0.011	6.7	2.2	50	P,Q	763
		0.031	0.006	6.3	0.4	60	C	1263
		0.1	0.1	9	2.5	60		622
		0.5	0.06	3	0.3	60	C,K	77
	ACRYLIC ACID, 2-CYANO-, METHYL ESTER	0.01		2		60	C	119
	ACRYLIC ACID, 2-HYDROXY-, ETHYL ESTER, ACETATE	0.005		0.5		60	I7	507
	ACRYLONITRILE	0.08	0.03	5.4	0.5	60	C	993
		0.009		3.88		25	P,Q	1071
		0.02	0.02	6	2	60	D	947
		0.03	0.03	5.6	0.5	40	C	260
		0.05		4.2		50	H38	62
		0.061	0.013	4.2	0.3	50	C,16	219
		0.07	0.01	4.05		60	C,16	220
		0.07		6.0		60		622
		0.2	0.04	6.0	0.9	70	C,J	497
		0.57		6.0				23
		0.1		6.0		-196	E	40
	ADIPIC ACID, METHYL VINYL ESTER	0.78		0.66	0.04	80	C	268
	ANILINE, N,N-DIVINYL-	0.63		2.0		60		162
	P-ANISIC ACID, VINYL ESTER	500	100	1.45		30		397
	2H-AZEPIN-2-ONE, HEXAHYDRO-1-VINYL-	0.35		0.31		65		1183
	BENZENE	0.66		0		60	I7	767
	BENZOIC ACID, VINYL ESTER	0.7	0.09	0.99	0.13	30		1019
		0.7		1.74		80		397
	BENZOIC ACID, P-BROMO-, VINYL ESTER	0.60		1.70		60		82
	BENZOIC ACID, P-CHLORO-, VINYL ESTER	0.62		1.85		30		1135
	BENZOIC ACID, P-CYANO-, VINYL ESTER	0.61	0.07	1.63	0.07	30	C,17	397
	BORAZINE, 1,3,5-TRIPHENYL-2,4,6-TRIVINYL-	0		0		80		397
	1,3-BUTADIENE, 2-CHLORO-	0.01		50		65		768
	1,3-BUTADIEN-1-OL, ACETATE	0.0		>9		70		1002
	1-BUTENE	2.0		0.34		80	C	315
	2-BUTENE, CIS-	8.0		0.07		60	C	865
	2-BUTENE, TRANS-	7.0		0.07		60		865
	2-BUTENE, 1,3-DICHLORO-	0.0		4.8	0.9	50		233
	3-BUTENOIC ACID, 2-ETHYL-2-METHYL-, ETHYL ESTER	0.1	0.1	2.2	0.4	60	C	933
		0.3	0.2	3.2		60	C	933
	3-BUTEN-2-ONE	0.05		7.00	0.50	70	C	315

COPOLYMERIZATION PARAMETER

M1: VINYL ACETATE

M2	R1	+/-	R2	+/-	TEMP.	REMARKS	REF.
BUTYRALDEHYDE, DIVINYL ACETAL	1.005	0.015	1.06	0.01	60		608
BUTYRIC ACID, VINYL ESTER	1	0.07	0.97	0.07	60	C,I7	1135
CARBAMIC ACID, DIETHYL-, VINYL ESTER	1.35	0.15	1.25	0.15	60	I7	954
CARBAMIC ACID, DIETHYLTHIO-, S-VINYL ESTER	1.8	0.4	0.25	0.08	66		802
CARBAMIC ACID, N-METHYL-N-VINYL THIO-, S-ETHYL ESTER	0.16	0.08	1.5	0.3	66		802
CARBAMIC ACID, VINYL-, 2,3-EPOXYPROPYL ESTER	0.6	0.1	1.3	0.2	66		802
CARBAMIC ACID, VINYL-, ETHYL ESTER	0.26	0.04	3.2	0.3	60	I5	418
CARBAZOLE, 9-VINYL-	0.33		0.33		65	I15	287
	0.03	0.01	0.15	0.10	65	I31,Q	1389
CARBONIC ACID, CYCLIC (HYDROXYMETHYL)ETHYLENE ESTER, ACRYLATE	0.126	0.032	2.68		65	C	1001
CARBONIC ACID, CYCLIC (HYDROXYMETHYL)ETHYLENE ESTER, MALEATE	0.13	0.03	3.9	0.2	30	I5	1129
CARBONIC ACID, CYCLIC (HYDROXYMETHYL)ETHYLENE ESTER, 4-(METHYLENESUCCINATE)	0.152	0.018	3.02	0.24	100	C	1001
	0.152	0.018	3.02	0.24	100	C	1001
CARBONIC ACID, CYCLIC VINYLENE ESTER	0.05	0.01	0.09	0.02	50	C,I2	199
	0.05	0.01	0.09	0.02	50	C,I2	199
	0.036		3.075		25	F,I11	512
	3.0		0.27		70		349
	3.71		0.0579		55		584
	4.00		0.15		70	C	314
CARBONIC ACID, DIVINYL ESTER	7.3	0.7	0.13	0.1	60		458
CARBONIC ACID, ETHYL VINYL ESTER	0.77	0.05	1.33	0.08	60	I7	0496
CARBONIC ACID, THIO-, O-TERT-BUTYL S-VINYL ESTER	0.87	0.04	0.79	0.07	60	I7	0496
CARBON MONOXIDE	0.04	0.01	11	0.1	70		1176
CINNAMIC ACID, VINYL ESTER	0.24	0.05	0.33	0.05	60		636
CROTONAMIDE	0.04		1.20	0.10	110	I7	806
CROTONAMIDE, N-(HYDROXYMETHYL)-	0.01	0.01	2.0	0.05	110		1003
CROTONIC ACID	0.01	0.01	0.045	0.1	70		1003
	0.3		0.01		70	C	955
	0.3		0.01		70	C	997
	0.33		0		70	Q	319
CYANAMIDE, DIALLYL-	0.6	0.05	0.01	0.01	68	C	166
CYCLOHEXANE, 1,2-EPOXY-4-VINYL-	2.26	0.02	1.62	0.01	60	K,Q	1386
CYCLOPROPENE, TETRACHLORO-	0.72		0		80		1169
1,3-DIOXOLANE-4-METHANOL, 2-VINYL-	0.51	0.02	1.8		60		1214
DISILOXANE, 1,1,1,3-PENTAMETHYL-3-VINYL-	0.99	0.15	0.010	0.001	60	I7	1351
DODECANAMIDE, N-(ACRYLAMIDOMETHYL)-	0.05	0.03	10.0	0.8	70		773
ETHENESULFONIC ACID, ALLYL ESTER	0.38	0.08	3.60	0.20	80	I7	462
ETHENESULFONIC ACID, BUTYL ESTER	0.04	0.01	0.20	0.05	80	I7	209
	0.71	0.1	0.2	0.05	170	P,Y	960
ETHER, BUTYL VINYL	2.5		0.16		70	C	754
	3.7		0		60	Q	494
ETHER, DODECYL VINYL	3.67		0		50	I5	1166
ETHER, ETHYL VINYL	3.0		0		50	I7	1325
ETHER, OCTADECYL VINYL	3.67	0.45	0	0.23	60	C	12
ETHER, OCTYL VINYL	3.0	0.1	0		50	D	622
ETHERS, C8-C18 ALKYL VINYL	4.50	0.58	0	0.35	50	C	12
	3.47	0.51	0	0.31	50	C	12
ETHYLENE	0.95		0.67		80	C,P	12
	1		0.73		135	M10	1366
	1		1.01		150	M7,R	142

REMARKS PAGE II-366, REFERENCES PAGE II-368

M1	M2	R1	+/-	R2	+/-	TEMP.	REMARKS	REF.
VINYL ACETATE	ETHYLENE	1		0.89		200	C,M10	1366
		1		0.45		60	M8	1366
		1		0.47		80	M8	1366
		1.02		0.49		135	M4	1366
		1.02		0.85		166	C,M10	1366
		1.02		0.52		60	M10	1366
		1.02	0.02	0.97	0.03	130	C,I7,M2	953
		1.02	0.02	0.77	0.04	70	I7,M2	953
		1.05		0.18		80	M4	1366
		1.08	0.19	1.07	0.06	90	M3	148
		1.1		1.2		160	C,M9	1366
		1.1		1.2		210	C,M9	1366
		1.1		1.2		230	B,M14	1366
		1.12		0.16		60	M4	1366
		1.14		0.28			Q	296
				0.16			M4	246
		1.515	0.007	0.743	0.005	60	145,M6	1198
		3.70		0.70		62	H9	246
	ETHYLENE, BROMO-	0.35		4.5	1.2	60	F,K	622
		0.38	0.09	3	0.65	60		1098
		0.68	0.16	1.82				1098
	ETHYLENE, CHLORO-	0.033		3.74		-32	H7	8
		0.1		2.5		-40	H38	80
		0.23	0.02	1.68	0.08	60		622
		0.28		2		60		1213
		0.3		2.1		68	C	8
	ETHYLENE, CHLOROTRIFLUORO-	0.30	0.08	1.60	0.15	50	I26,M	395
		0.30	0.2	2.45	0.6	60		304
		0.6		1.8		40		596
		0.65	0.04	1.35	0.05	40	C,I11,J	503
	ETHYLENE, 1,1-DICHLORO-	0.6		0.01		60	C	955
		0.0	0.03	3.6	0.5	60		210
		0.05		5.0		50		447
		0.05	0.03	6.7	0.5	60		405
		0.1		6		68	C	8
	ETHYLENE, 1,2-DICHLORO-, CIS-	2.8		0		68	C	27
		6.3	0.2	0.018	0.003	60		557
	ETHYLENE, 1,2-DICHLORO-, TRANS-	0.85		0		68	C	27
		0.95	0.02	0.063		-45		1177
	ETHYLENE, 1,1-DICHLORO-2,2-DIFLUORO-	0.99	0.09	0.086	0.01	60	H38	557
		0.60		0		70		23
	ETHYLENE, FLUORO-	2.9	0.2	0.16	0.01	30		1165
		2.9	0.2	0.16	0.01	30	H38,I38	879
	ETHYLENE, TETRACHLORO-	5		0		68	C	8
		6.8	0.5	0	0.01	60		210
	ETHYLENE, TRICHLORO-	0.66	0.04	0.01		60		622
		0.67		0		68	C	27
	FERROCENEETHANOL, ACRYLATE	0.074	0.094	3.43	0.313	65	I7	1358
	FERROCENEETHANOL, METHACRYLATE	0.058	0.003	8.791	0.231	65	I7	1358
	FERROCENEMETHANOL, ACRYLATE	0.46	0.07	1.44	0.38	70	I7	1267
	FERROCENEMETHANOL, METHACRYLATE	0.2	0.1	1.52	0.3	70	I7	1267
	FORMIC ACID, VINYL ESTER	0.94	0.07	0.95	0.07	60	C,I7	1135
		1.41		0.68		50	C	205

COPOLYMERIZATION PARAMETER

M1	M2	R1	+/-	R2	+/-	TEMP.	REMARKS	REF.
VINYL ACETATE								
	FUMARIC ACID, 1,3-BUTYLENEGLYCOL POLYESTER	0.15	0.07	0.2	0.2	60		908
	FUMARIC ACID, DIESTER WITH HYDRACRYLONITRILE	0.02	0.02	0.02	0.01	90	C	1137
	FUMARIC ACID, DIETHYL ESTER	0.011	0.001	0.444	0.003	60		557
	FUMARIC ACID, ETHYLENEGLYCOL POLYESTER	0.020	0.02	0.2	0.1	60		908
	FUMAROYL CHLORIDE	0.14		0.00		70	C,W	315
	D-GLUCITOL, 1-ACRYLAMIDO-1-DEOXY-	0.18		2.41		50	C	1029
	D-GLUCITOL, 1-DEOXY-1-METHACRYLAMIDO-	0.16		0.56		50	C	1029
	1,3,5-HEXATRIENE, 2,3,4,5-TETRACHLORO-, TRANS-	0.013		32		70	C	17
	5-HEXEN-3-YN-2-OL, 2-METHYL-					70	C,P	607
	ISOBUTYRALDEHYDE, DIVINYL ACETAL	1.002	0.047	0.985	0.05	60		608
	ISOVALERALDEHYDE, DIVINYL ACETAL	1.04	0.08	0.987	0.057	60		608
	KETENE, DIPHENYL-	0.05	0.03	0.56	0.05	-10	H2O	67
	LAURIC ACID, ALLYL ESTER	0.71		0.8		60	C	835
	LAURIC ACID, VINYL ESTER	1.4		0.7		60	C	835
	MALEIC ACID	0.045	0.015	0.00		60	F	341
	MALEIC ACID, DIETHYL ESTER	0.17	0.01	0.043		60		557
	MALEIC ACID, DIISOPROPYL ESTER	0.12		0.028		60		1024
	MALEIC ACID, DIMETHYL ESTER	0.0345	0.005	0.522		60	135	1055
	MALEIC ACID, METHYL ESTER	0.0468	0.005	0.4747		65		999
	MALEIC ANHYDRIDE	0.09		0.0	0.007	56	135	998
		0.09		0.0	0.007	60	135	1055
		0.1168	0.03	0.9996	0.01	78	135	1055
		0.13		0.015		75		999
		0.522	0.015	0.035		56	C	1000
		0.055	0.04	0.003		75	Q	1033
		0.072	0.18	0.01			C	384
	MALEIMIDE, N,2-DIMETHYL-	0.26		2.56	0.9	60		554
	MALEIMIDE, N-(2-HYDROXYETHYL)-	0.05		0.86		60	I12	1432
	MALEIMIDE, N-(2-HYDROXYETHYL)-, ACETATE	0.25		0.85		70	I7	1056
	MALEIMIDE, N-(2-HYDROXYETHYL)-, PROPIONATE	0.03		1.02		70	I7	1057
	MALEIMIDE, N-(HYDROXYMETHYL)-	0.05		1.4		60	I10	1432
	MALEIMIDE, N-(P-HYDROXYPHENYL)-	0.02		1.78		60	I6	1432
	MALEIMIDE, N-(3-HYDROXYPROPYL)-, ACETATE	0.03		1.50		70	I7	1057
	MALONONITRILE, METHYLENE-	0.0054		0.11		45		289
	MELAMINE, N(2),N(4)-DIALLYL-	1.41	0.06	0.19	0.004	60		812
		1.44	0.28	0.20	0.03	60		812
	METHACRYLIC ACID	0.01		20		70		32
	METHACRYLIC ACID, BUTYL ESTER	0.023	0.015	28.8		60	C	679
				62.1	3.8	60	C	638
	METHACRYLIC ACID, 2-(DIETHYLAMINO)ETHYL ESTER	0.127	0.015	142	2	60	P,Q	943
	METHACRYLIC ACID, ETHYL ESTER	0.271	0.039	60.4	0.4	60		855
	METHACRYLIC ACID, HEPTYL ESTER	0.135	0.055	68.3	3.2	60	C	638
	METHACRYLIC ACID, HEXADECYL ESTER	0.025		29.8		60	C	638
	METHACRYLIC ACID, ISOBUTYL ESTER	0.16	0.16	24	5	60		679
	METHACRYLIC ACID, PENTAMETHYLDISILOXANYL METHYL ESTER	0.186	0.038	73.3	7.4	50		632
	METHACRYLIC ACID, PROPYL ESTER	0.01	0.01	12	2	70	C	638
						60	C	260
	METHACRYLONITRILE	1.012	0.107	1.005	0.105	60	C	608
	METHANE, BIS(VINYLOXY)-	0.015		181	10	30	F	855
				127	10	60		855
	METHYL METHACRYLATE	0.015	0.015	20	3	60	C,Q	1322
		0.015		20				622

M1	M2	R1	+/-	R2	+/-	TEMP.	REMARKS	REF.
VINYL ACETATE	METHYL METHACRYLATE	0.03	0.01	26	2.5	60	C,I7	1135
		0.035		28.6		30		76
		0.072	0.026	22.21	0.89	60	C	638
	2,5-NORBORNADIENE	0.1		2.75		80	G31	1321
		0.11	0.01	2.77	0.02		G31,Q	1322
		0.2	0.2	3.2	1.1	-30	G3	55
	5-NORBORNENE-2-CARBOXYLIC ACID, METHYL ESTER	0.4		1.28		60		769
		0.82		0.45		60		458
	OCTADECANOIC ACID, 12-OXO-, VINYL ESTER	1.5	0.24	1.26	0.07	60	C	593
	OCTADECANOIC ACID, 4-OXO-, VINYL ESTER	1.07	0.02	1.18	0.08	60	C	593
	OCTADECANOIC ACID, 9(10)-OXO-, VINYL ESTER	1.04	0.2	1.49	0.22	60	C	593
	OXALIC ACID, ETHYL VINYL ESTER	1.03	0.05	3.0	0.25	60		479
	2-OXAZOLIDINONE, 3-VINYL-	0.3		1.90	0.10	50		116
		0.52	0.08	1.50		75		336
	OXETANE, 3-(ALLYLOXYMETHYL)-3-(CHLOROMETHYL)-	0.60	0.01	0.22	0.01	65	C	832
	PALMITIC ACID, VINYL ESTER	0.84	0.10	0.66	0.07	70	C	778
		1.15	0.13	0.78	1.0	70	C	778
	4-PENTENOIC ACID, 2-ALLYL-2-CYANO-, ETHYL ESTER	0.75		0.30			C	163
	PHTHALIC ACID, BIS(2-METHYLALLYL) ESTER	0.99	0.04	1.08	0.05	60	C,CC	1295
	PHTHALIC ACID, DIALLYL ESTER	0.72	0.10	2.0	0.15		C,Q	932
		0.72	0.02	1.2	0.3	60	C,CC	1292
		0.83	0.01	0.93	0.5	80	C,CC	1292
	PHTHALIMIDE, N-VINYL-	0.07		2.4			C,Q	698
	PIVALIC ACID, VINYL ESTER	0.75		0.43		60	I7	392
		0.80		0.94		60		954
	2-PROPANONE, 1,3-DICHLORO-1,1,3,3-TETRAFLUORO-	0.41		5.6		60	C,R	566
	2-PROPANONE, HEXAFLUORO-	0.12		4.1		60	C,R	566
	1-PROPENE, 1-CHLORO-, CIS-	7.0		0.01		60		745
	1-PROPENE, 1-CHLORO-, TRANS-	3.56		0.08		60		745
	1-PROPENE, 2-CHLORO-	0.22		1.84		60		745
	1-PROPENE, 3-CHLORO-	0.34	0.26	0.75	0.46	60	P	834
		0.7		0.67		68	P	854
	1-PROPENE, 3-CHLORO-2-METHYL-	1.30		2.3	0.31	60	C	457
	1-PROPENE, 1,1-DICHLORO-	0.13		0			C	8
		0.92	0.04	0.06	0.03	60	C	824
	1-PROPENE, 3,3-DICHLORO-	0.98		3.6		150	E	281
	1-PROPENE, 2-METHYL-	0.07		0.31		73/90		654
	1-PROPENE, 3,3,3-TRICHLORO-	2.15	0.03	0.19	0.05	50	M13	1457
		0.19	0.15	1.34	0.08	50		1457
	2-PROPENE-1,1-DIOL, DIACETATE	0.48		1.54			C,Q	915
	2-PROPENE-1-SULFONIC ACID, ALLYL ESTER	0.5		1.0		60		865
		1.0		1.0		60		211
		0.9		0.98		60		1380
	1-PROPEN-2-OL, ACETATE	0.98		0		60	I7	1201
		1.0	0.2	1.0	0.1	75	C	334
	PROPIONIC ACID, VINYL ESTER						Q	378
		0.9	0.07	0.98	0.07	60	C,I7	1135
		0.98					I31,Q	1389
	PYRIDINE, 2-VINYL-	0		30				23
		0		10	15	70	Q	723
	2-PYRROLIDINONE, 1-VINYL-	0.3		3.3		70		5
		0.205	0.015	3.30	0.15	50		116
		0.237	0.037	2.28	0.19	76	Q17	460

COPOLYMERIZATION PARAMETER

M1	M2	R1	+/-	R2	+/-	TEMP.	REMARKS	REF.
VINYL ACETATE	2-PYRROLIDINONE, 1-VINYL-	0.38		0.44		70	Q	132
		0.38		0.44		70		349
	STEARIC ACID, VINYL ESTER	0.90		0.73		50		1050
		0.97		1.00		70	C	4
	STERAMIDE, N-ALLYL-	0.740	0.067	0.532	0.012	70	C,I14	454
	STYRENE	0.0	0.30	16.0	5.6	100		988
		0.01		56		60		683
		0.01	0.01	55	10	60	I7	622
		0.015	0.015	8.25	0.05	60		552
		0.05		48		20	H20	683
		0.05		10		0	I7	683
		0.1	0.1	0.01	0.01	0	C,I7	374
		0.15		6.5		-30	G3	55
		0.16		60		30	H10,I22	683
		0.18	0.08	6.1	0.8	60	H,I22,Q	358
		0.30	0.15	2.64	0.35		H,I24,Q	355
	STYRENE, 2,5-DICHLORO-	0.04		6.3		70	Q	169
	SUCCINIC ACID, METHYLENE-, DIBUTYL ESTER	0.02		0.198			F,K	825
	SUCCINIC ACID, METHYLENE-, DISODIUM SALT	-0.007		5.75			I7,O	2
	SUCCINIC ANHYDRIDE, METHYLENE-	0.08		3.50		60	I5,O	697
	SUCCINIMIDE, N-VINYL-	0.175		5.1		65	C	365
	SULFONE, METHYL VINYL	0.18	0.01	6.1	0.08	60	C	211
	SULFONE, PHENYL VINYL	0.0		0.4		60		788
	2,4,8,10-TETRAOXASPIRO(5.5)UNDECANE, 3,9-DIISOPROPENYL-	0.29		0.35		60	C	785
	2,4,8,10-TETRAOXASPIRO(5.5)UNDECANE, 3,9-DIPROPENYL-	0.28		0.35		80	C	1349
	2,4,8,10-TETRAOXASPIRO(5.5)UNDECANE, 3,9-DIVINYL-	0.33		3		80	C	1349
	THIOPHENE, 2,3-DIHYDRO-, 1,1-DIOXIDE	1.25		0.74		60	C,I10	1123
	P-TOLUIC ACID, VINYL ESTER	0.23		1		60	C,I16	1345
	S-TRIAZINE-2,4,6-(1H,3H,5H)-TRIONE, TRIALLYL-	1.15		0.8		60	I7	294
	TRISILOXANE, 1,1,3,5,5,5-HEPTAMETHYL-3-VINYL-	1.52	0.12	0.0	0.1	30		397
	TRISILOXANE, 1,1,5,5,5-HEXAMETHYL-3-TRIMETHYLSILOXY-3-VINYL- *	0.73	0.03	1.73	0.06	60		812
		0.91	0.18	0.75	0.09	60		812
	UREA, 1-ETHYL-3-VINYL-	0.95	0.15	0.70		60		773
		0.99	0.15	0.010	0.001	70		773
		0.99		0.010	0.001	70		773
	VALERIC ACID, 4-METHYL-, VINYL ESTER	0.45		0.63		70	C	335
		0.45		0.63		75		335
	VERSATIC 10 ACID, VINYL ESTER	0.67		1.14		80		549
	VERSATIC 13 ACID, VINYL ESTER	0.99	0.10	0.92	0.12	60	I7	954
	VERSATIC 9 ACID, VINYL ESTER	1.25	0.14	0.93	0.13	60	I7	954
		0.93	0.07	0.90	0.08	60	I7	954
VINYL ETHER	ACRYLONITRILE	0.024		0.938		50	I6,V,CC	88
	FUMARONITRILE					60		156
	MALONONITRILE, METHYLENE-	0.23		0.12		50	I6	156
	METHYL METHACRYLATE	0.006		10.0		60		162
	PYRIDINE, 4-VINYL-	0.032		32		60	I7	156
	STYRENE	0.02		40		60	I7,CC	162

M1	M2	R1	+/-	R2	+/-	TEMP.	REMARKS	REF.
VINYL SULFIDE								
	METHYL METHACRYLATE	0.13	0.05	0.85	0.05	60		850
	STYRENE	0.41	0.05	1.90	0.1	60		850
VINYL SULFONE								
	ACRYLONITRILE	0.364		1.94		50		155
	METHYL METHACRYLATE	0.045		8.5		60	CC	162
	STYRENE	0.01		1.3		60		162
P-XYLENE, ALPHA-DIAZO-								
	TOLUENE, ALPHA-DIAZO-	1.3	0.1	0.6	0.07	40	I74	1396
		2.3	0.2	0.65	0.15	40	I74	1396
ZINC, DIACETATOBIS(5-VINYL-2-PICOLINE)-								
	STYRENE	2.0	0.2	0.35	0.05	50	I6	951
ZINC, DIACETATOBIS(2-VINYLPYRIDINE)-								
	STYRENE	3.35	0.3	0.55	0.15	50	I6	951
ZINC, DIACETATOBIS(4-VINYLPYRIDINE)-								
	STYRENE	2.7	0.5	0.08	0.03	50	I6	951
ZINC, DIBROMOBIS(5-VINYL-2-PICOLINE)-								
	STYRENE	0.38	0.18	0.33	0.05	60	I6	952
ZINC, DICHLOROBIS(PYRIDINE)-								
	ZINC, DICHLOROBIS(QUINOLINE)-	0.63	0.04	2.7	0.4	350	H68	970
ZINC, DICHLOROBIS(QUINOLINE)-								
	ZINC, DICHLOROBIS(PYRIDINE)-	2.7	0.4	0.63	0.04	350	H68	970
ZINC, DICHLOROBIS(5-VINYL-2-PICOLINE)-								
	STYRENE	0.08	0.015	0.93	0.075	60	I6	952
ZINC, DIIODOBIS(5-VINYL-2-PICOLINE)-								
	STYRENE	0.55	0.06	0.60	0.10	60	I6	952

REMARKS

A. Thermal initiation

B. Hydroperoxide initiation

C. Peroxide initiation

D. Photoinitiation

E. γ-Ray initiation

F. Redox initiation

G. Anionic initiation

G 1.	$(C_2H_5)_3N$	G17.	R-Li
G 2.	R-Na	G18.	Na
G 3.	$NaNH_2$	G19.	Na Naphthenate
G 4.	Li	G20.	Na α-Methylstyrene dimer
G 5.	K	G21.	Co Naphthenate
G 6.	K Naphthenate	G22.	Pd-on-Zeolite
G 7.	$C_6H_{13}NH_2$	G23.	Li_2 Benzophenone
G 8.	$(C_6H_5)_2N$-AlR_2	G24.	NaCN
G 9.	SO_2	G25.	LiOR
G10.	K-OH	G26.	KOR
G11.	NaOH	G27.	CsOR
G12.	LiOH	G28.	KCN
G13.	NaOR	G29.	Co Acetylacetonate
G14.	Piperidine	G30.	R-NiI
G15.	Li Naphthenate	G31.	Mn Acetylacetonate
G16.	$LiAlH_4$	G32.	R-CaI

H. Cationic initiation

H 1.	$R_2AlCl/(RO)_3VO$	H36.	$R_2=Cd$
H 2.	$R_2AlCl/(RO)_2VOCl$	H37.	$Al_2(SO_4)_3/H_2SO_4$
H 3.	$R_2AlCl/ROVOCl_2$	H38.	R_3B
H 4.	$RAlCl_2/HCl$	H39.	R_2BCl
H 5.	$R_2AlCl/VOCl_3$	H40.	R_2Zn
H 6.	$R_3Al/VOCl_3/R_2AlCl$	H41.	R_2Be
H 7.	$R_3Al/VOCl_3$	H42.	R_2Mg
H 8.	$RAlCl_2/ROVOCl_2$	H43.	R_2AlCl/Co Acetylacetonate
H 9.	$RAlCl_2/(RO)_3VO$	H44.	$FeCl_3$
H10.	BF_3/OR_2	H45.	R_2AlCl/V Acetylacetonate
H11.	$(RO)_3Al$	H46.	$AlCl_3$
H12.	$R_3Al/TiCl_3$	H47.	$CaZn(C_2H_5)_4$
H13.	R_3Al/H_2O	H48.	$SbCl_5$
H14.	R_3Al	H49.	$ZnCl_2$
H15.	$HAlCl_2/VCl_3/OR_2$	H50.	H_2SO_4
H16.	$(CH_3CO)_2O/HClO_4$	H51.	$HClO_4$
H17.	$LiAlH_4/TiCl_4$	H52.	$CuCl_2$
H18.	$R_3Al/TiCl_2$	H53.	$R_2AlCl/TiCl_4$
H19.	$SnCl_4/Cl_3CCOOH$	H54.	$R_2AlCl/TiCl_3$
H20.	$SnCl_4$	H55.	R_2AlCl
H21.	R-MgBr (anionic init.?)	H56.	$RAlCl_2/H_2O$
H22.	R-MgI (anionic init.?)	H57.	$RAlCl_2/TiCl_3/(Me_2N)_3PO$
H23.	$TiCl_3$	H58.	$R_3Al/H_2O/epichlorohydrin$
H24.	$R_3Al/TiCl_3$	H59.	$R_3Al/H_2O/CoCl_2$
H25.	$R_2AlCl/CoCl_2$	H60.	R_2Zn/H_2O
H26.	$R_2AlCl/CoCl_2/Pyridine$	H61.	R_2AlCl/VCl_4
H27.	$R_2AlCl/NiCl_2/Pyridine$	H62.	$HAlCl_2/TiCl_4/AlJ_3/ether$
H28.	R_3Al/VCl_4	H63.	$HSiO_4$
H29.	R_3Al/VCl_3	H64.	$R_3Al/Ti(OR)_4$
H30.	$AlBr_3$	H65.	$R_3Al/ZrCl_4$
H31.	I_2	H66.	$R_3Al/HfCl_4$
H32.	Fullers earth	H67.	$AlBr_3/Sn(Ph)_4/VCl_4$
H33.	BF_3/CH_3COOH	H68.	HPO_4
H34.	$HgCl_2$	H69.	$TiCl_3/Cl_3CCOOH$
H35.	$Al/TiCl_4$	H70.	$(RO)_3Al/TiCl_4$

H. Cationic initiation   (Cont'd.)

H71.	$R_2AlCl/(RO)_4Ti$	H79.	$R_3SnOCOCH_3$
H72.	$R_2AlCl/(RO)_2TiCl_2$	H80.	R_3SnCl_2
H73.	$R_2AlCl/CoCl_2/R_2S$	H81.	$R_2AlOR/TiCl_4/dioxane$
H74.	R_3Al/TiI_4	H82.	$BF_3/propanediol$
H75.	$R_3Al/TiI_4/R_2S$	H83.	$BF_3/glycerol$
H76.	$Ni/NiCl_2/crotyl\ Cl$	H84.	R_2AlCl/R_2TiCl_2
H77.	$R_3Al/SnCl_4$	H85.	R_2Al-Acetylacetonate$/R_2Zn$
H78.	R_3SnCl		

I. In solution :

I 1.	Solid	I 39.	Xylene
I 2.	Toluene	I 40.	DMSO/H_2O (94 : 6)
I 3.	Methylene chloride	I 41.	CH_2Cl_2/hexane (1 : 3)
I 4.	Glycol	I 42.	CH_2Cl_2/hexane (1 : 1)
I 5.	Tetrahydrofuran	I 43.	CH_2Cl_2/hexane (3 : 1)
I 6.	Dimethylformamide	I 44.	Isooctane
I 7.	Benzene	I 45.	Butanol
I 8.	Triethylamine	I 46.	Mg Perchlorate/water
I 9.	Hexane	I 47.	NaSCN/water
I 10.	p-Dioxane	I 48.	KCl/water
I 11.	Water	I 49.	NaCl/water
I 12.	Ethanol	I 50.	Carbon disulfide
I 13.	Cyclohexanone	I 51.	Decalin
I 14.	tert-Butanol	I 52.	Methylcyclohexane
I 15.	Methanol	I 53.	o-Dichlorobenzene
I 16.	Sulfur dioxide	I 54.	Ether
I 17.	Dimethyl sulfoxide	I 55.	Glycol dimethyl ether
I 18.	Ethyl chloride	I 56.	Quinoline
I 19.	Pyridine	I 57.	Acetone/methanol (1 : 1)
I 20.	Heptene	I 58.	Isopropanol
I 21.	Cyclohexane	I 59.	Propanol
I 22.	Nitrobenzene	I 60.	Isopentanol
I 23.	Nitroethane	I 61.	Isobutanol
I 24.	1,2-Dichloroethane	I 62.	Glycol carbonate
I 25.	Benzene/nitrobenzene (1 : 1)	I 63.	Acetonitrile
I 26.	Acetone	I 64.	Benzene/dichloroethane (3 : 1)
I 27.	Glycol monoethyl ether	I 65.	Benzene/dichloroethane (1 : 3)
I 28.	Chloroform	I 66.	Dioxane/water (1 : 1)
I 29.	2-Butanone	I 67.	Methanol/water (1 : 1)
I 30.	Pentane	I 68.	Benzonitrile
I 31.	Acetic acid	I 69.	Benzyl alcohol
I 32.	Carbon tetrachloride	I 70.	Phenol
I 33.	CCl_4/nitrobenzene (1 : 1)	I 71.	Perfluorotriethylamine
I 34.	CCl_4/nitrobenzene (2 : 1)	I 72.	Dioxane/ethanol (70/30)
I 35.	Acetic anhydride	I 73.	DMF/water (1 : 1)
I 36.	$ZnCl_2$/water	I 74.	Toluene/methanol (4 : 1)
I 37.	Trichloroacetic acid	I 75.	Benzene/DMF (1 : 3)
I 38.	Ethyl acetate		

J. In suspension

K. In emulsion

L. In heterogeneous system :

L 1.	ZnO added	L 3.	$ZnCl_2$ added (1 : 1, Zn : M_2)
L 2.	Maleic anhydride added		

M. Under pressure :

M 1.	1020-1700 psi (70-117 bar)	M 6.	278 psi (19.2 bar)
M 2.	5680 psi (391 bar)	M 7.	10000 psi (689.5 bar)
M 3.	14700 psi (1013 bar)	M 8.	2000 psi (137.9 bar)
M 4.	1470 psi (101 bar)	M 9.	18000 psi (1241 bar)
M 5.	25000 psi (1723 bar)	M10.	4200 psi (289.5 bar)

REMARKS (Cont'd.)

M. Under pressure (Cont'd.)
    M11. 170000psi (11721 bar)   M13. 73500 psi (5068 bar)
    M12. H_2 added   M14. 26000-118000 psi (1793-8136 bar)

N. Single experiment

O. One r assumed

P. r-Values not in source reference

Q. Temperature not in source reference

R. Penultimate group effects applicable

S. r-Values independent of ester chain length

T. r-Values calculated by method of Gundin (CA 42, 5145c (1948)

U. At lower styrene, r-values increase; mechanism unknown

V. Sufficient data given, but r-values not calculated

W. r appears to vary with monomer composition

X. Extremely dry conditions

Y. $r_1 = r_2$

Z. r-Values not normal ($r_1 = k_{11}/k_{21}$: $r_2 = k_{22}/k_{12}$)

AA. r-Values vary with concentration

BB. r-Values may be in error

CC. Data fits cyclocopolymerization mechanism of Barton : r_c not shown

DD. Calcd. by author from source data

EE. r-Values vary with time (catalyst changes)

FF. Complex formation acrylate/$SnCl_4$

GG. Complex formation acid/solvent

HH. r-Values vary with pH (3-7)

II. r-Values vary with electric field strength

JJ. Complex formation ethylene/$AgNO_3$

KK. Complex formation acrylate/$SbCl_5$

LL. Complex formation acid/mercaptan

MM. Complex formation methacrylate/$ZnCl_2$

NN. Complex formation methacrylate/$AlBr_3$

OO. Complex formation methacrylate/$AlCl_3$.

REFERENCES

1. A. D. Abkin, A. P. Sheinker, L. P. Mezhirova, Zh. Fiz. Khim. 33, 2636 (1959); from J. Polymer Sci. 53, 39 (1961).

2. A. D. Abkin, A. P. Sheinker, M. K. Yakovleva, L. P. Mezhirova, J. Polymer Sci. 53, 39 (1961).

3. E. I. Ablyakimov, R. K. Gavurina, N. K. Shakalova, Reakts. Soposobnost Org. Soedin., Tartu. Gos. Univ. 4, 838 (1967); from CA 69, 27903H (1968).

4. A. Adicoff, A. Buselli, J. Polymer Sci. 21, 340 (1956).

5. V. A. Agasandyan, A. D. Litmanovich, V. Ya. Shtern, Kinet. Katal. 8, 773 (1967); from CA 67, 109019X (1967).

6. A. I. Ageev, A. I. Ezrielev, E. S. Roskin, Vysokomolekul. Soedin., Ser. B, 9, 571 (1967).

7. S. L. Aggarwal, F. A. Long, J. Polymer Sci. 11, 127 (1953).

8. P. Agron, T. Alfrey, jr., J. Bohrer, H. Haas, H. Wechsler, J. Polymer Sci. 3, 157 (1948).

9. G. Akasome, S. Sakai, Y. Choshi, K. Murai, Kobunshi Kagaku 17, 478 (1960); from Chem. Zentr. 135 (9), 119 (1964).

10. G. Akasome, S. Sakai, Y. Choshi, K. Murai, Kobunshi Kagaku 17, 558 (1960); from Makromol. Chem. 41, 246 (1960).

11. G. Akasome, S. Sakai, Y. Choshi, K. Murai, Kobunshi Kagaku 17, 627 (1960); from Makromol. Chem. 42, 174 (1960).

12. G. Akasome, S. Sakai, K. Murai, Kobunshi Kagaku 17, 449 (1960); from Chem. Zentr. 135(9), 119 (1964).

13. G. Akasome, S. Sakai, K. Murai, Kobunshi Kagaku 17, 452 (1960); from Chem. Zentr. 135(9), 119 (1964).

14. G. Akasome, S. Sakai, K. Murai, Kobunshi Kagaku 17, 482 (1960); from Chem. Zentr. 135(9), 119 (1964).

15. G. Akasome, S. Sakai, K. Murai, Kobunshi Kagaku 17, 618 (1960); from Makromol. Chem. 42, 174 (1960).

16. S. Akiyoshi, C. Aso, K. Sadakata, Kogyo Kagaku Zasshi 60, 1081 (1957); from CA 53, 10836H (1959).

17. A. N. Akopyan, G. Ye. Krbekyan, Vysokomolekul. Soedin. 5, 201 (1963).

18. A. N. Akopyan, G. Ye. Krbekyan, E. G. Sinanyan, Vysokomolekul Soedin. 5, 681 (1963).

19. T. A. Alekseeva, L. I. Dmitrievskaya, V. I. Grigor'eva, B. M. Krasovitskii, V. D. Bezuglyi, Vysokomolekul. Soedin., Ser. B, 10, 226 (1968).

20. L. V. Alferova, V. A. Kropachev, Vysokomolekul. Soedin. 7, 1065 (1965).

21. T. Alfrey, jr., L. Arond, C. G. Overberger, J. Polymer Sci. 4, 539 (1949).

22. T. Alfrey, jr., L. Arond, C. G. Overberger, quoted by Alfrey, Copolymerization, p. 35.

23. T. Alfrey, jr., J. Bohrer, H. Haas, C. Lewis, J. Polymer Sci. 5, 719 (1950).

24. T. Alfrey, jr., J. Bohrer, H. Mark, "Copolymerization", High Polymer Series, Vol. VIII, Interscience, New York, 1952, p. 40.

25. T. Alfrey, jr., W. H. Ebelke, J. Am. Chem. Soc. 71, 3235 (1949).

26. T. Alfrey, jr., A. I. Goldberg, W. P. Hohenstein, J. Am. Chem. Soc. 68, 2464 (1946).

27. T. Alfrey, jr., S. Greenberg, J. Polymer Sci. 3, 297 (1948).

28. T. Alfrey, jr., J. G. Harrison, jr., J. Am. Chem. Soc. 68, 299 (1946).

29. T. Alfrey, jr., S. L. Kapur, J. Polymer Sci. 4, 215 (1949).

30. T. Alfrey, jr., S. L. Kapur, quoted in Alfrey, "Copolymerization", Wiley, New York, p. 33, 39.

31. T. Alfrey, jr., E. Lavin, J. Am. Chem. Soc. 67, 2044 (1945).

32. T. Alfrey, jr., B. Magel, quoted by Alfrey, "Copolymerization", Wiley, New York.

33. T. Alfrey, jr., B. Magel, quoted by Hart, Makromol. Chem. 47, 143 (1961).

34. T. Alfrey, jr., E. Merz, H. Mark, J. Polymer Sci. 1, 37 (1946).

35. T. Alfrey, jr., H. Morawetz, J. Am. Chem. Soc. 74, 436 (1952).

36. T. Alfrey, jr., C. G. Overberger, S. H. Pinner, J. Am. Chem. Soc. 75, 4221 (1953).

37. T. Alfrey, jr., C. R. Pfeifer, J. Polymer Sci., A-1, 4, 2447 (1966).

38. T. Alfrey, jr., H. Wechsler, J. Am. Chem. Soc. 70, 4266 (1948).

39. S. M. Aliev, Z. A. Akhmedzade, R. G. Ismailov, G. M. Mamedaliev, Azerb. Khim. Zh. 1966(5), 70; from CA 67, 117333K (1967).

40. American Cyanamid Co., "Chemistry of Acrylonitrile", 2nd Ed., 1960, p. 34.

41. American Cyanamid Co., private communication.

42. K. Anda, S. Iwai, Kogyo Kagaku Zasshi 70, 557 (1967).

43. I. H. Anderson, G. M. Burnett, W. C. Geddes, European Polymer J. 3, 161 (1967).

44. I. H. Anderson, G. M. Burnett, P. J. Tait, J. Polymer Sci. 56, 391 (1962).

45. I. N. Andreeva, V. M. Zapletnyak, N. N. Severova, Z. V. Arkhipova, Plasticheskie Massy 1965(10), 4.

46. T. L. Ang, H. J. Harwood, Am. Chem. Soc., Div. Polymer Chem., Preprints 5(1), 306 (1964).

47. S. Aoki, K. Fujisawa, T. Otsu, M. Imoto, Kogyo Kagaku Zasshi 69, 131 (1966).

48. S. Aoki, Y. Harita, T. Otsu, M. Imoto, Bull. Chem. Soc. Japan 38, 1928 (1965).

49. S. Aoki, Y. Harita, Y. Tanaka, H. Mandai, T. Otsu, J. Polymer Sci., A-1, 6, 2585 (1968).

50. S. Aoki, T. Otsu, M. Imoto, Kogyo Kagaku Zasshi 67, 1958 (1964).

51. I. A. Arbuzova, I. K. Mosevich, Vysokomolekul. Soedin. 8, 1307 (1966).

52. C. L. Arcus, R. J. S. Matthews, J. Chem. Soc. 1956, 4607.

53. A. Ardis, Can. 516, 315 (1955).

54. N. Ariga, M. Uchiyama, T. Kurosaki, Y. Iwakura, Makromol. Chem. 99, 126 (1966).

55. E. J. Arlman, H. W. Melville, Proc. Roy. Soc., Ser. A 203, 301 (1950).

56. E. J. Arlman, H. W. Melville, L. Valentine, Rec. Trav. Chim. 68, 945 (1949).

57. T. Asahara, K. Ikeda, N. Yoda, J. Polymer Sci., A-1, 6, 2489 (1968).

58. T. Asahara, K. Mitsuhashi, Yukagaku 6, 331 (1957); from CA 54, 20303F (1960).

59. H. Asai, T. Imoto, J. Polymer Sci. B2, 553 (1964).

60. N. Ashikari, Bull. Chem. Soc. Japan 32, 1060 (1959).

61. N. Ashikari, T. Kanemitsu, K. Yanagisawa, K. Nakagawa, H. Okamoto, S. Kobayashi, A. Nishioka, J. Polymer Sci. A2, 3009 (1964).

62. N. Ashikari, A. Nishimura, J. Polymer Sci. 31, 250 (1958).

63. M. A. Askarov, A. S. Bank, Vysokomolekul. Soedin. 8, 1012 (1966).

64. M. A. Askarov, S. R. Pinkhasov, A. S. Bank, Vysokomolekul. Soedin., Ser. B, 9, 601 (1967).

65. M. A. Askarov, S. N. Trubitskyna, Vysokomolekul. Soedin. 5, 1235 (1963).

66. C. Aso, Kogyo Kagaku Zasshi 63, 363 (1960); from Resins, Rubbers, Plastics 14, 867 (1960).

67. C. Aso, T. Kunitake, K. Watanabe, Kogyo Kagaku Zasshi 70, 1001 (1967).

68. C. Aso, M. Miura, Kobunshi Kagaku 24, 178 (1967); from CA 68, 3211J (1968).

69. C. Aso, T. Nawata, Kogyo Kagaku Zasshi 68, 549 (1965); from CA 63, 8485B (1965).

70. C. Aso, O. Ohara, Makromol. Chem. 109, 161 (1967).

71. C. Aso, K. Sadakata, Kogyo Kagaku Zasshi 62, 1610 (1959); from CA 57, 16836E (1962).

72. C. Aso, M. Sugabe, Kogyo Kagaku Zasshi 68, 1970 (1965).

73. C. Aso, S. Tagami, T. Kunitake, Kobunshi Kagaku 23, 63 (1966); from CA 64, 17725F (1966).

74. C. Aso, Y. Tanaka, Kobunshi Kagaku 21, 373 (1964); from Makromol. Chem. 79, 236 (1964); CA 62, 9239B (1965).

75. Y. Atarashi, Kogyo Kagaku Zasshi 68, 2487 (1965).

76. J. N. Atherton, A. M. North, Trans. Faraday Soc. 58, 2049 (1962).

77. I. S. Avetisyan, V. I. Eliseeva, O. G. Larinovo, Vysokomolekul. Soedin., Ser. A, 9, 570 (1967).

78. F. C. Baines, J. C. Bevington, J. Polymer Sci., A-1, 6, 2433 (1968).

79. B. Baker, P. J. T. Tait, Polymer 8, 225 (1967).

80. W. P. Baker, J. Polymer Sci. 42, 578 (1960).

81. C. H. Bamford, W. G. Barb, Discussions Faraday Soc. 14, 208 (1953).

82. S. Banerjee, M. S. Muthana, J. Polymer Sci. 35, 292 (1959).

83. A. S. Bank, S. D. Savranskaya, M. A. Askarov, Khim. i Fiz.-Khim. Prirodn. i Sintetich. Polimerov, Akad. Nauk Uz. SSR, Inst. Khim. Polimerov No. 2, 110 (1964); from CA 61, 13425H (1964).

84. V. P. Barabanov, S. G. Sannikov, Tr. Kaz. Khim.-Tekhnol. Inst. 34, 235 (1965); from CA 68, 30093D (1968).

85. W. G. Barb, J. Polymer Sci. 11, 117 (1953).

86. C. A. Barson, J. Polymer Sci. 62, S-128 (1962).

87. P. D. Bartlett, K. Nozaki, J. Am. Chem. Soc. 68, 1495 (1946).

88. J. M. Barton, G. B. Butler, E. C. Chapin, J. Polymer Sci. A3, 501 (1965).

89. T. F. Baskova, O. M. Klimova, Vysokomolekul. Soedin., Ser. B, 10, 63 (1968); from CA 68, 69367W (1968).

90. T. F. Baskova, O. M. Klimova, L. G. Stulova, Vysokomolekul. Soedin., Ser. B, 10, 220 (1968).

91. H. S. Bender, J. Polymer Sci., B 4, 895 (1966).

92. W. I. Bengough, D. Goldrich, R. A. Young, European Polymer J. 3, 117 (1967).

93. W. I. Bengough, R. G. W. Norrish, Proc. Roy. Soc., Ser. A, 218, 155 (1953).

94. A. A. Berlin, Khr. Budevska, M. Mikhailov, Izv. Akad. Nauk SSSR, Ser. Khim. 1966, 943; from CA 65, 13833A (1966). see also Vysokomolekul. Soedin., Ser. B, 9, 309 (1967).

95. J. C. Bevington, D. O. Harris, J. Polymer Sci., B, 5, 799 (1967).

96. J. C. Bevington, D. O. Harris, M. Johnson, European Polymer J. 1, 235 (1965).

97. J. C. Bevington, M. Johnson, Makromol. Chem. 87, 66 (1965).

98. J. C. Bevington, B. W. Malpass, Trans. Faraday Soc. 60, 1268 (1964).

99. J. C. Bevington, B. W. Malpass, European Polymer J. 1, 19 (1965).

100. J. C. Bevington, B. W. Malpass, European Polymer J. 1, 85 (1965).

101. T. I. Bevza, N. A. Pokatilo, M. P. Teterina, B. A. Dolgoplosk, Vysokomolekul. Soedin., Ser. A, 10, 207 (1968).

102. K. I. Beynon, J. Polymer Sci. A-1, 3343 (1963).

103. K. I. Beynon, J. Polymer Sci. A-1, 3357 (1963).

104. K. I. Beynon, E. J. Hayward, J. Polymer Sci. A-3, 1793 (1965).

105. V. D. Bezuglyi, T. A. Alekseeva, Ukr. Khim. Zh. 31, 392 (1965); from CA 63, 5750G (1965).

106. V. D. Bezuglyi, T. A. Alekseeva, L. I. Dmitrievskaya, A. V. Chernobai, L. P. Kruglyak, Vysokomolekul. Soedin. 6, 125 (1964).

107. V. D. Bezuglyi, T. A. Alekseeva, L. I. Dmitrievskaya, V. I. Grigor'eva, Vysokomolekul. Soedin., Ser. A, 9, 889 (1967).

108. W. A. P. Black, J. A. Colquhoun, E. T. Dewar, Makromol. Chem. 102, 266 (1967).

109. D. C. Blackley, H. W. Melville, Makromol. Chem. 18, 16 (1956).

110. L. P. Blanchard, J. Singh, M. D. Baijal, Can. J. Chem. 44, 2679 (1966).

111. G. Blauer, L. Goldstein, J. Polymer Sci. 25, 19 (1957).

112. L. Bogetich, G. A. Mortimer, G. W. Daues, J. Polymer Sci. 61, 3 (1962).

113. V. Ya. Bogomol'nyi, B. A. Dolgoplosk, Vysokomolekul. Soedin., Ser. B, 10, 370 (1968); from CA 69, 28383A (1968).

114. V. J. Bogomol'nyi, B. A. Dolgoplosk, Z. P. Chirikova, Dokl. Akad. Nauk SSSR 159, 1069 (1964).

115. E. P. Bonsall, L. Valentine, H. W. Melville, Trans. Faraday Soc. 48, 763 (1952).

116. J. F. Bork, L. E. Coleman, J. Polymer Sci. 43, 413 (1960).

117. J. F. Bork, D. P. Wyman, L. E. Coleman, J. Appl. Polymer Sci. 7, 451 (1963).

118. E. T. Borrows, R. N. Haward, J. Porges, J. Street, J. Appl. Chem. (London) 5, 379 (1955).

119. J. Bourdais, Bull. Soc. Chim. France 1955, 485.

120. J. H. Bradbury, H. W. Melville, Proc. Roy. Soc., Ser. A, 222, 456 (1954).

121. E. E. Brando, A. I. Dintses, Neftekhim. 4, 68 (1964); from CA 61, 728 D (1964).

122. J. Brandrup, Faserforsch. Textiltech. 12, 135 (1961).

123. D. Braun, T.-O. Ahn, Kolloid Z. 188, 1 (1963).

124. D. Braun, T.-O. Ahn, W. Kern, Makromol. Chem. 53, 154 (1962).

125. D. Braun, G. Arcache, R. J. Faust, W. Neumann, Makromol. Chem. 114, 51 (1968).

126. D. Braun, H. Daimon, G. Becker, Makromol. Chem. 62, 183 (1963).

127. D. Braun, H.-G. Keppler, Makromol. Chem. 78, 100 (1964).

128. D. Braun, H.-G. Keppler, Makromol. Chem. 82, 132 (1965).

129. D. Braun, W. Neumann, J. Faust, Makromol. Chem. 85, 143 (1965).

130. D. Braun, E. Seelig, Kolloid Z. 201, 111 (1965).

131. D. Braun, M. Thallmaier, J. Polymer Sci., C, 16, 2351 (1967).

132. J. W. Breitenbach, H. Edelhäuser, Ric. Sci. 25, 242 (1955).

133. J. W. Breitenbach, A. Schindler, C. Pflug, Monatsh. Chem. 81, 21 (1950).

134. S. Bresadola, P. Canal, J. Polymer Sci. B1, 523 (1963).

135. L. S. Bresler, B. A. Dolgoplosk, M. F. Kolechkova, E. N. Kropacheva, Dokl. Akad. Nauk SSSR 144, 347 (1962); from CA 57, 8716H (1962).

136. L. S. Bresler, B. A. Dolgoplosk, M. F. Kolechkova, E. N. Kropacheva, Vysokomolekul. Soedin., 5, 357 (1963).

137. L. S. Bresler, B. A. Dolgoplosk, E. N. Kropacheva, Dokl. Akad. Nauk SSSR 149, 595 (1963); from CA 59, 4044H (1964).

138. D. S. Breslow, A. Kutner, J. Polymer Sci. 27, 295 (1958).

139. E. C. Britton, C. W. Davis, F. L. Taylor, U.S. Patent 2,160,940 (1939); calc. by F. R. Mayo and C. Walling, assuming only one dimethylal group reacts.

140. E. L. Brokhina, E. S. Roskin, A. I. Ezrielev, Khim. Volokna 1967, 26; from CA 68, 40206G (1968).

141. D. W. Brown, L. A. Wall, Am. Chem. Soc., Div. Polymer Chem., Preprints 7(2), 1116 (1966); see also J. Polymer Sci., A-1, 6, 1367 (1968).

142. F. E. Brown, G. E. Ham, J. Polymer Sci. A2, 3623 (1964).

143. G. R. Brown, D. C. Pepper, J. Chem. Soc. 1963, 5930.

144. G. R. Brown, D. C. Pepper, Polymer 6, 497 (1965).

145. M. A. Bruk, A. D. Abkin, P. M. Khomikovskii, G. A. Gol'der, H.-L. Chu, Dokl. Akad. Nauk SSSR 157, 1399 (1964); from CA 61, 14790F (1964).

146. M. A. Bulatov, S. S. Spasskii, Vysokomolekul. Soedin. 2, 658 (1960).

147. R. D. Burkhart, N. L. Zutty, J. Polymer Sci. 57, 783 (1962).

148. R. D. Burkhart, N. L. Zutty, J. Polymer Sci. A1, 1137 (1963).

149. W. J. Burlant, D. H. Green, J. Polymer Sci. 31, 227 (1958).

150. Z. I. Burlyuk, N. M. Beder, B. K. Kruptsov, Karbotsepnye Volakna 1966, 44; from CA 68, 30933C (1968).

151. G. M. Burnett, Am. Chem. Soc., Div. Polymer Chem., Preprints 8(1), 96 (1967).

152. G. M. Burnett, P. Evans, H. W. Melville, Trans. Faraday Soc. 49, 1096 (1953).

153. G. M. Burnett, J. M. Pearson, J. D. B. Smith, J. Polymer Sci., A-1, 4, 2024 (1966).

154. R. D. Bushick, J. Polymer Sci. A3, 2047 (1965).

155. G. B. Butler, R. B. Kasat, J. Polymer Sci. A3, 4205 (1965).

156. G. B. Butler, G. Vanhaeren, M.-F. Ramadier, J. Polymer Sci., A-1, 5, 1265 (1967).

157. N. R. Byrd, Nasa Accession No. N 6420601, Rept. No. Nasa-CR-56035; Rept. 166-F; from CA 62, 2831F (1965).

158. N. R. Byrd, F. D. Kleist, A. Rembaum, J. Macromol. Sci. (Chem.) A1, 627 (1967).

159. G. G. Cameron, D. H. Grant, N. Grassie, J. E. Lamb, I. C. McNeill, J. Polymer Sci. 36, 173 (1959).

160. G. G. Cameron, N. Grassie, Makromol. Chem. 51, 130 (1962).

161. Y. P. Castille, V. Stannett, J. Polymer Sci. A-1, 4, 2063 (1966).

162. E. Y. C. Chang, C. C. Price, J. Am. Chem. Soc. 83, 4650 (1961).

163. H.-C. Chang, T. Yin, W. H. Ts'ao, H. T. Feng, Ko Fen Tzu T'ung Hsun 7, 415 (1965); from CA 64, 14352C (1966).

164. S. Chang, T. K. Miwa, W. H. Tallent, Am. Chem. Soc., Div. Polym. Chem., Preprints 9(1), 48 (1968); see also J. Polymer Sci. A-1, 7, 471 (1969).

165. E. C. Chapin, G. E. Ham, R. G. Fordyce, J. Am. Chem. Soc. 70, 538 (1948).

166. E. C. Chapin, G. E. Ham, C. L. Mills, J. Polymer Sci. 4, 597 (1949); errata, J. Polymer Sci. 55, S-6 (1961).

167. A. Chapiro, A.-M. Jendrychowska-Bonamour, J. Polymer Sci. C1, 1211 (1963).

168. C. B. Chapman, L. Valentine, J. Polymer Sci. 34, 319 (1959).

169. Chas. Pfizer and Co., Product News, quoted by L. J. Young, 1961.

170. I. A. Chekulaeva, M. F. Shostakovskii, V. A. Gladshevskaya, I. V. Lipovich, Vysokomolekul. Soedin. 3, 901 (1961).

171. H. Cherdron, H. Ohse, Makromol. Chem. 92, 213 (1966).

172. A. V. Chernobai, Zh. S. Tirak'yants, R. Ya. Delyatitskays, Vysokomolekul. Soedin. 8, 997 (1966).

173. R. P. Chernovskaya, V. P. Lebedev, K. S. Minsker, G. A. Razuvaev, Vysokomolekul. Soedin. 6, 1313 (1964).

174. K. Chikanishi, T. Tsuruta, Makromol. Chem. 73, 231 (1964).

175. K. Chikanishi, T. Tsuruta, Makromol. Chem. 81, 211 (1965).

176. M. A. Chilingaryan, A. E. Akopyan, M. G. Barkudaryan, Arm. Khim. Zh. 20, 428 (1967); from CA 68, 22292C (1968).

177. R. C. L. Chow, C. S. Marvel, J. Polymer Sci., A-1, 6, 1515 (1968).

178. S. G. Cohen, D. B. Sparrow, J. Polymer Sci. 3, 693 (1948).

179. L. E. Coleman, J. F. Bork, D. P. Wyman, J. Polymer Sci. A3, 1601 (1965).

180. L. E. Coleman, J. A. Conrady, J. Polymer Sci. 38, 241 (1959).

181. L. E. Coleman, W. S. Durrell, J. Org. Chem. 23, 1211 (1958).

182. P. Colombo, L. E. Kukacka, J. Fontana, R. N. Chapman, M. Steinberg, J. Polymer Sci., A-1, 4, 29 (1966).

183. P. Colombo, M. Steinberg, J. Fontana, J. Polymer Sci. B1, 447 (1963).

184. W. Cooper, E. Catterall, Can. J. Chem. 34, 387 (1956).

185. H. W. Coover, jr., R. L. McConnell, F. B. Joyner, D. F. Slonaker, R. L. Combs, J. Polymer Sci., A-1, 4, 2563 (1966).

186. A. T. Coscia, R. L. Kugel, J. Pellon, J. Polymer Sci. 55, 303 (1961).

187. K. Crauwels, G. Smets, Bull. Soc. Chim. Belges 59, 443 (1950).

188. K. Crauwels, G. Smets, Bull. Soc. Chim. Belges 59, 182 (1950).

189. I. Crescentini, G. B. Gechele, A. Zanella, J. Appl. Polymer Sci. 9, 1323 (1965).

190. Z. Csuros, M. Gara, I. Gyukovics, Acta. Chim. Acad. Sci. Hung. 29, 207 (1961); from CA 56, 10385E (1962).

191. B. M. Culbertson, E. A. Sedor, S. Dietz, R. E. Freis, J. Polymer Sci., A-1, 6, 2197 (1968).

192. B. M. Culbertson, E. A. Sedor, R. C. Slagel, Macromolecules 1, 254 (1968).

193. B. M. Culbertson, R. C. Slagel, J. Polymer Sci., A-1, 6, 363 (1968).

194. F. S. Dainton, W. D. Sisley, Trans. Faraday Soc. 59, 1385 (1963).

195. G. F. D'Alelio, R. J. Caiola, J. Polymer Sci., A-1, 5, 287 (1967).

196. G. F. D'Alelio, R. C. Evers, J. Polymer Sci., A-1, 5, 813 (1967).

197. G. F. D'Alelio, T. R. Hoffend, J. Polymer Sci., A-1, 5, 323 (1967).

198. G. F. D'Alelio, T. Huemmer, J. Polymer Sci., A-1, 5, 77 (1967).

199. G. F. D'Alelio, T. Huemmer, J. Polymer Sci., A-1, 5, 307 (1967).

200. G. F. D'Alelio, L. X. Mallavarapu, Makromol. Chem. 37, 25 (1960).

201. F. Danusso, P. Ferruti, C. G. Marabelli, Chim. Ind. (Milan) 47, 585 (1965); from CA 63, 10149C (1965).

202. A. B. Davankov, L. B. Zubakova, A. A. Gurov, Vysokomolekul. Soedin. 6, 237 (1964).

203. F. Dawans, J. Polymer Sci. A2, 3297 (1964).

204. F. Dawans, G. Smets, Makromol. Chem. 59, 163 (1963).

205. L. E. De Millo, G. A. Shtraikhman, E. N. Rostovskii, Zh. Prikl. Khim. 39, 1360 (1966); from CA 65, 10675F (1966).

206. C. E. Denoon, U. S. Patent 2,384,731 (1945); calc. by F. R. Mayo and C. Walling.

207. A. De Pauw, G. Smets, Bull. Soc. Chim. Belges 59, 629 (1950).

208. O. C. Dermer, W. A. Ames, J. Polymer Sci. A2, 4151 (1964).

209. E. De Witte, E. J. Goethals, Makromol. Chem. 115, 234 (1968).

210. K. W. Doak, J. Am. Chem. Soc. 70, 1525 (1948).

211. K. W. Doak, quoted by Mayo and Walling, 1950.

212. K. W. Doak, J. Am. Chem. Soc. 72, 4681 (1950).

213. K. W. Doak, M. A. Deahl, I. H. Christmas, 137th ACS Meeting, Cleveland, Ohio, Abstr. Papers, Vol. 1, No. 1, 151 (April 1960).

214. K. W. Doak, D. L. Dineen, J. Am. Chem. Soc. 73, 1084 (1951).

215. T. Doi, M. Matsuki, A. Sugiyama, Kogyo Kagaku Zasshi 70, 1792 (1967).

216. A. F. Dokukina, Z. A. Smirnova, M. M. Koton, Vysokomolekul. Soedin. 2, 1247 (1960).

217. G. Dolgin, P. Gordon, quoted in Alfrey, "Copolymerization", Wiley, New York, p. 40.

218. B. A. Dolgoplosk, E. I. Tinyakova, V. N. Reikh, T. G. Zhuravleva, G. P. Belonovskaya, Kauchuk; Rezina 1957(3), 11; from CA 52, 5869C (1958).

219. I. S. Dorokhina, A. D. Abkin, V. S. Klimenkov, Khim. Volokna, 1962(1), 49; from CA 59, 2956H (1963).

220. I. S. Dorokhina, A. D. Abkin, V. S. Klimenkov, Vysokomolekul. Soedin. 5, 385 (1963).

221. Dow Chemical Co. Bulletin No. 125-913-67, Aminelogics.

222. J. Drougas, R. L. Guile, J. Polymer Sci. 55, 297 (1961).

223. I. N. Duling, C. C. Price, J. Am. Chem. Soc. 84, 578 (1962).

224. J. F. Dunphy, C. S. Marvel, J. Polymer Sci. 47, 1 (1960).

225. F. I. Duntov, B. L. Erusalimskii, Vysokomolekul. Soedin. 7, 1075 (1965).

226. F. I. Duntov, A. L. Gol'denberg, M. A. Litvinova, B. L. Erusalimskii, Vysokomolekul. Soedin., Ser. A, 9, 1920 (1967).

227. DuPont Co., British Patent 593,605 (1947); calc. by F. R. Mayo and C. Walling.

228. A. A. Durgaryan, A. O. Agumyan, Vysokomolekul. Soedin. 5, 1755 (1963).

229. A. A. Durgaryan, A. O. Agumyan, Izv. Akad. Nauk Arm. SSR, Khim. Nauk 18, 290 (1965); from CA 64, 1917H (1966).

230. A. A. Durgaryan, R. A. Arakelyan, Vysokomolekul. Soedin. 8, 1321 (1966).

231. A. A. Durgaryan, R. M. Beginyan, Vysokomolekul. Soedin. 5, 28 (1963).

232. A. A. Durgaryan, R. M. Beginyan, Vysokomolekul. Soedin. 8, 1326 (1966).

233. A. A. Durgaryan, A. S. Grigoryan, O. A. Chaltykyan, Izv. Akad. Nauk Arm. SSR, Khim. Nauk 15, 455 (1962); from CA 58, 14104G (1963).

234. A. A. Durgaryan, Zh. N. Terlemezyan, Z. A. Kirakosyan, G. S. Sarkisyan, Vysokomolekul. Soedin., Ser. A, 10, 303 (1968).

235. V. S. Dyumbaum, V. S. Klimenkov, Khim. Volokna 1963(4), 8; from CA 59, 11711B (1963).

236. F. H. C. Edgecombe, Nature 198, 1085 (1963).

237. G. Egle, Makromol. Chem. 86, 181 (1965).

238. E. Egorova, A. F. Dokukina, Vysokomolekul. Soedin., Karbotsepnye Vysokomolekul. Soedin., Sb. Statei 1963, 40; from CA 61, 4493H (1964).

239. D. D. Eley, J. Saunders, J. Chem. Soc. 1954, 1677.

240. B. G. Elgood, B. J. Sauntson, Chem. Ind. (London) 1965, 1558.
241. H. -G. Elias, W. Lengweiler, Makromol. Chem. 113, 155 (1968).
242. V. I. Eliseeva, N. G. Karapetyan, I. S. Boshnyakov, A. S. Margaryan, Lakokrasochnye Materialy i ikh Primenenie 1965(3), 15; from CA 63, 8495C (1965). Polymer Sci. USSR 7, 550 (1965).
243. W. H. Embree, J. M. Mitchell, H. L. Williams, Can. J. Chem. 29, 253 (1951).
244. S. Enomoto, J. Polymer Sci. 55, 95 (1961).
245. B. Erussalimsky, F. Duntoff, N. Tumarkin, Makromol. Chem. 66, 205 (1963).
246. B. Erussalimsky, N. Tumarkin, F. Duntoff, S. Lyubetzky, A. Goldenberg, Makromol. Chem. 104, 288 (1967).
247. J. Exner, M. Bohdanecky, Chem. Listy 48, 483 (1954); from CA 48, 8583G (1954).
248. E. S. Ferdinandi, W. P. Garby, D. G. L. James, Can. J. Chem. 42, 2568 (1964).
249. V. G. Filippova, N. S. Nametkin, S. G. Durgar'yan, Izv. Akad. Nauk SSSR, Ser. Khim. 1966, 1727; from CA 66, 18893R (1967).
250. A. P. Firsov, I. N. Meshkova, N. D. Kostrova, N. M. Chirkov, Vysokomolekul. Soedin. 8, 1860 (1966); from CA 66, 19518J (1967).
251. T. Fischer, J. B. Kinsinger, C. W. Wilson, III, J. Polymer Sci., B, 4, 379 (1966).
252. R. E. Florin, J. Am. Chem. Soc. 71, 1867 (1949).
253. R. E. Florin, J. Am. Chem. Soc. 73, 4468 (1951).
254. V. L. Folt, Canadian Patent 509,259 (1955).
255. V. L. Folt, Canadian Patent 510,354 (1955).
256. J. W. L. Fordham, G. H. McCain, L. E. Alexander, J. Polymer Sci. 39, 335 (1959).
257. J. W. L. Fordham, H. L. Williams, J. Phys. Chem. 57, 346 (1953).
258. R. G. Fordyce, J. Am. Chem. Soc. 69, 1903 (1947).
259. R. G. Fordyce, E. C. Chapin, J. Am. Chem. Soc. 69, 581 (1947).
260. R. G. Fordyce, E. C. Chapin, G. E. Ham, J. Am. Chem. Soc. 70, 2489 (1948).
261. R. G. Fordyce, G. E. Ham, J. Am. Chem. Soc. 69, 695 (1947).
262. R. G. Fordyce, G. E. Ham, J. Am. Chem. Soc. 73, 1186 (1951).
263. F. C. Foster, J. Polymer Sci. 5, 369 (1950).
264. F. C. Foster, J. Am. Chem. Soc. 72, 1370 (1950).
265. F. C. Foster, J. Am. Chem. Soc. 74, 2299 (1952).
266. F. C. Foster, U. S. Patent 2,666,045 (1954).
267. C. E. Frank, G. Kraus, A. J. Haefner, Ind. Eng. Chem. 44, 1600 (1952).
268. G. N. Freidlin, M. N. Adamova, Vysokomolekul. Soedin., Ser. A, 9, 1531 (1967).
269. R. Kh. Freidlina, G. S. Kolesnikov, G. L. Slonimskii, A. P. Suprun, T. A. Soboleva, A. B. Belyavskii, V. A. Ershova, Sintez i Svoistva Monomerov, Akad. Nauk SSSR, Inst. Neftekhim. Sinteza, Rabot 12-01 Konf. po Vysokomolekul. Soedin. 1962, 42; from CA 62, 7877B (1965).
270. T. M. Frunze, V. V. Korshak, E. L. Baranov, B. V. Lokshin, Vysokomolekul. Soedin. 8, 455 (1966).
271. S. Fujii, Radiation Res. 33, 249 (1968).
272. R. Fujio, H. Sato, T. Tsuruta, Kogyo Kagaku Zasshi 69, 2315 (1966).
273. K. Fukui, T. Shimidzu, T. Yagi, S. Fukumoto, T. Kagiya, S. Yuasa, J. Polymer Sci. 55, 321 (1961).
274. B. L. Funt, E. A. Ogryzlo, J. Polymer Sci. 25, 279 (1957).
275. R. M. Fuoss, G. I. Cathers, J. Polymer Sci. 4, 97 (1949).
276. J. Furukawa, T. Saegusa, T. Narvyama, S. Kurahashi, Kogyo Kagaku Zasshi 65, 2082 (1962); from CA 58, 14263B (1963).
277. J. Furukawa, T. Tsuruta, T. Fueno, R. Sakata, K. Ito, Makromol. Chem. 30, 109 (1959).
278. J. Furukawa, T. Tsuruta, S. Inoue, A. Kawasaki, N. Kawabata, J. Polymer Sci. 35, 269 (1959).
279. J. Furukawa, T. Tsuruta, N. Yamamoto, H. Fukutani, J. Polymer Sci. 37, 215 (1959).
280. J. Furukawa, T. Tsuruta, N. Yameda, Kogyo Kagaku Zasshi 61, 734 (1958).

281. L. H. Gale, J. Org. Chem. 31, 2475 (1966).
282. A. R. Gantmakher, Yu. L. Spirin, S. S. Medvedev, Vysokomolekul. Soedin. 1, 1526 (1959).
283. N. M. Geller, V. A. Kropachev, B. A. Dolgoplosk, Vysokomolekul. Soedin. 8, 450 (1966).
284. N. M. Geller, V. A. Kropachev, B. A. Dolgoplosk, Vysokomolekul. Soedin., Ser. A, 9, 575 (1967).
285. H. Gerrens, H. Ohlinger, R. Fricker, Makromol. Chem. 87, 209 (1965).
286. D. O. Geymer, Am. Chem. Soc., Div. Polymer Chem., Preprints 5(2), 942 (1964).
287. L. Ghosez, G. Smets, J. Polymer Sci. 35, 215 (1959).
288. P. Ghosh, K. F. O'Driscoll, J. Polymer Sci., B, 4, 519 (1966).
289. H. Gilbert, F. F. Miller, S. J. Averill, E. J. Carlson, V. L. Folt, H. J. Heller, F. D. Stewart, R. F. Schmidt, H. L. Trumbull, J. Am. Chem. Soc. 78, 1669 (1956).
290. R. D. Gilbert, H. L. Williams, J. Am. Chem. Soc. 74, 4114 (1952).
291. G. P. Gladyshev, R. G. Karzhaubaeva, S. R. Rafikov, Vysokomolekul. Soedin., Ser. B, 10, 351 (1968); from CA 69, 27870V (1968).
292. A. H. Gleason, U. S. Patent 2,379,292 (1945); calc. by F. R. Mayo and C. Walling.
293. M. S. Gluckman, M. J. Kampf, J. L. O'Brien, T. G. Fox, R. K. Graham, J. Polymer Sci. 37, 411 (1959).
294. E. J. Goethals, Makromol. Chem. 109, 132 (1967).
295. Ts. A. Goguadze, V. V. Korshak, N. E. Ogneva, Vysokomolekul. Soedin. 6, 1875 (1964).
296. A. L. Gol'denberg, S. G. Lyubetskii, Dokl. Nauk SSSR 179, 900 (1968); from CA 69, 3148Q (1968).
297. G. Goldfinger, M. Steidlitz, J. Polymer Sci. 3, 786 (1948).
298. A. V. Golubeva, N. F. Usmanova, A. A. Vansheidt, J. Polymer Sci. 52, 63 (1961).
299. M. Goodstein, quoted in Alfrey, "Copolymerization", Wiley, New York, p. 35.
300. M. Gordon, B. M. Grieveson, I. D. McMillan, J. Polymer Sci. 18, 497 (1955).
301. C. E. Grabiel, D. L. Decker, J. Polymer Sci. 59, 425 (1962).
302. N. Grassie, E. M. Grant, European Polymer J. 2, 255 (1966).
3o3. N. Grassie, I. C. McNeill, J. Polymer Sci. 27, 207 (1958).
304. N. Grassie, I. C. NcNeill, I. F. McLaren, J. Polymer Sci. B3, 897 (1965).
305. N. Grassie, B. J. D. Torrance, J. D. Fortune, J. D. Gemmell, Polymer 6, 653 (1965).
306. G. Greber, E. Reese, Makromol. Chem. 55, 96 (1962).
307. G. Greber, E. Reese, Makromol. Chem. 77, 13 (1964).
308. G. Greber, J. Tolle, Makromol. Chem. 67, 98 (1963).
309. V. F. Gromov, P. M. Khomikovskii, A. D. Abkin, Vysokomolekul. Soedin. 3, 1015 (1961).
310. E. Grzyma, Z. Jedlinski, Vysokomolekul. Soedin. 8, 1653 (1966).
311. A. Guyot, J. Guillot, J. Chim. Phys. 61, 1434 (1964).
312. H. C. Haas, E. S. Emerson, N. W. Schuler, J. Polymer Sci. 22, 291 (1956).
313. H. C. Haas, R.D. Moreau, N. W. Schuler, J. Polymer Sci., A-2, 5, 915 (1967).
314. H. C. Haas, N. W. Schuler, J. Polymer Sci. 31, 237 (1958).
315. H. C. Haas, M. S. Simon, J. Polymer Sci. 9, 309 (1952).
316. G. Hagele, H. Frohlich, D. Bischoff, K. Hamann, Makromol. Chem. 75, 98 (1964).
317. W. Hahn, A. Fischer, Makromol. Chem. 21, 77 (1956).
318. R. A. Haldron, J. N. Hay, J. Polymer Sci., A-1, 6, 951 (1968).
319. G. E. Ham, J. Polymer Sci. 14, 87 (1954).
320. G. E. Ham, J. Polymer Sci. 24, 349 (1957).
321. G. E. Ham, J. Polymer Sci. 38, 543 (1959).
322. G. E. Ham, J. Polymer Sci. 45, 177 (1960).
323. G. E. Ham, J. Polymer Sci. 45, 183 (1960).
324. F. L. Hamb, A. Winston, J. Polymer Sci. A2, 4475 (1964).
325. G. Hardy, J. Boros-Gyevi, Magy. Kem. Foly. 72, 116 (1966); from CA 64, 15994C (1966).

326. G. Hardy, J. Boros-Gyevi, L. Koronczay, Magy. Kem. Foly. 71, 447 (1965); from CA 64, 8321B (1966).

327. G. Hardy, L. Nagy, J. Polymer Sci., C, 16, 2667 (1967).

328. G. Hardy, J. Varga, G. Nagy, Makromol. Chem. 85, 58 (1965).

329. G. Hardy, J. Varga, K. Nytrai, I. Tsajlik, L. Zubonyai, Vysokomolekul. Soedin. 6, 758 (1964).

330. S. A. Harrison, W. E. Tolberg, J. Am. Oil Chem. Soc. 30 114 (1953); calc. by F. R. Mayo and C. W. Gould (1964).

331. R. Hart, Makromol. Chem. 47, 143 (1961).

332. R. Hart, Makromol. Chem. 49, 33 (1961).

333. R. Hart, A. E. van Dormael, Bull. Soc. Chim. Belges 65, 571 (1956).

334. R. Hart, G. Smets, J. Polymer Sci. 5, 55 (1950).

335. R. Hart, D. Timmerman, Bull. Soc. Chim. Belges 67, 123 (1958).

336. R. Hart, D. Timmerman, Makromol. Chem. 31, 223 (1959).

337. R. Hart, D. Timmerman, J. Polymer Sci. 48, 151 (1960).

338. S. Hashimoto, I. Furukawa, Kobunshi Kagaku 21, 647 (1964); from CA 62, 7874D (1965).

339. S. Hashimoto, I. Furukawa, Kobunshi Kagaku 22, 231 (1965); from CA 63, 8496G (1965).

340. S. Hashimoto, I. Furukawa, Kobunshi Kagaku 24, 152 (1967).

341. K. Hattori, Y. Komeda, Kogyo Kagaku Zasshi 68, 1729 (1965).

342. H. Hayashi, I. Ito, T. Saegusa, J. Furukawa, Kogyo Kagku Zasshi 65, 1634 (1962); from CA 58, 9237F (1963).

343. I. Hayashi, Kogyo Kagaku Zasshi 66, 1350 (1963); from CA 60, 13324A (1964).

344. I. Hayashi, Kogyo Kagaku Zasshi 67, 258 (1964); from CA 61, 1947E (1964).

345. I. Hayashi, Kogyo Kagaku Zasshi 68, 1126 (1965).

346. I. Hayashi, K. Ohno, Kobunshi Kagaku 22, 446 (1965); from Makromol. Chem. 90, 305 (1966).

347. J. Hayashi, S. Sakai, Y. Ishii, Kogyo Kagaku Zasshi 70, 1808 (1967).

348. K. Hayashi, M. Nishi, K. Moser, A. Shimizu, S. Okamura, Am. Chem. Soc., Div. Polymer Chem., Preprints 5(2), 951 (1964).

349. K. Hayashi, G. Smets, J. Polymer Sci. 27, 275 (1958).

350. K. Hayashi, H. Watanabe, S. Okamura, J. Polymer Sci. B1, 397 (1963).

351. K. H. Hellwege, U. Johnsen, K. Kolbe, Kolloid-Z. -Z. Polymere 214, 45 (1966).

352. K. R. Henery-Logan, R. V. V. Nicholls, quoted by Simha and Wall.

353. K. R. Henery-Logan, R. V. V. Nicholls, quoted by Mayo and Walling.

354. H. Herma, V. Groebe, R. Schmolke, Faserforsch. Textiltech. 17(2), 56 (1966).

355. R. Hess, quoted in Alfrey, "Copolymerization", Wiley, New York, p. 36, 37.

356. T. Higashimura, S. Kusudo, Y. Ohsumi, S. Okamura, J. Polymer Sci., A-1, 6, 2523 (1968).

357. T. Higashimura, Y. Ohsumi, K. Kuroda, S. Okamura, J. Polymer Sci., A-1, 5, 863 (1967).

358. T. Higashimura, S. Okamura, Kobunshi Kagaku 17, 635 (1960); from CA 55, 21654E (1961).

359. T. Higuchi, H. Imoto, Kogyo Kagaku Zasshi 61, 1053 (1958).

360. E. D. Holly, J. Polymer Sci. 36, 329 (1959).

361. E. D. Holly, unpublished data.

362. E. D. Holly, W. R. Nummy, unpublished data.

363. H. Hopff, H. Gutenberg, Makromol. Chem. 60, 129 (1963).

364. H. Hopff, F. Lochner, Makromol. Chem. 84, 261 (1965).

365. H. Hopff, P. Schlumbom, Makromol. Chem. 43, 173 (1961).

366. H. Hopff, D. Starck, Makromol. Chem. 48, 50 (1961).

367. Y. -W. Hsu, S. S. Skorokhodov, A. A. Vansheidt, Vysokomolekul. Soedin. 6, 1291 (1964).

368. M. B. Huglin, Polymer 3, 335 (1962).

369. A. Hunyar, H. Reichert, Faserforsch. Textiltech. 5, 204 (1954); from CA 48, 11106D (1954).

370. A. Hunyar, E. Roth, Faserforsch. Textiltech. 8, 99 (1957).

371. E. Husemann, Univ. Freiburg i. Br. Germany, private communication; see also Greber and Reese, 1962.

372. J. C. H. Hwa, L. Miller, J. Polymer Sci. 55, 197 (1961).

373. M. Ibonai, T. Kato, Kogyo Kagaku Zasshi 70, 2078 (1967).

374. M. Ibonai, T. Kato, Y. Yamashita, Kogyo Kagaku Zasshi 67, 1068 (1964); from CA 61, 12095 D (1964).

375. S. Ichida, Bull. Chem. Soc. Japan 33, 731 (1960).

376. F. Ida, K. Uemura, S. Abe, Kagaku to Kogyo (Osaka) 38, 215 (1964); from CA 61, 7105D (1964).

377. F. Ida, K. Uemura, S. Abe, Kagaku to Kogyo (Osaka) 39, 565 (1965); from CA 64, 3695A (1966).

378. F. Ide, K. Okano, S. Nakano, K. Nakstsuka, Shikizai Kyokaishi 40, 571 (1967); from CA 69, 3162Q (1968).

379. M. Iino, N. Tokura, Bull. Chem. Soc. Japan 37, 23 (1964).

380. M. Iino, N. Tokura, Bull. Chem. Soc. Japan 38, 1094 (1965).

381. Y. Imanishi, T. Higashimura, S. Okamura, J. Polymer Sci. A3, 2455 (1965).

382. Y. Imanishi, A. Mizote, T. Higashimura, S. Okamura, Kobunshi Kagaku 20, 58 (1963); from Makromol. Chem. 70, 77 (1964).

383. E. H. Immergut, G. Kollmann, A. Malatesta, Makromol. Chem. 41, 15 (1960).

384. E. Imoto, H. Horiuchi, Kobunshi Kagaku 8, 463 (1951); from CA 47, 9664A (1953).

385. M. Imoto, T. Otsu, W. Fukuda, J. Polymer Sci. B1, 225 (1963).

386. M. Imoto., T. Otsu, Y. Harada, Makromol. Chem. 65, 180 (1963).

387. M. Imoto., T. Otsu, A. Takada, Kogyo Kagaku Zasshi 68, 369 (1965); from CA 63, 5770A (1965).

388. M. Imoto, T. Otsu, K. Tsuda, T. Ito, J. Polymer Sci. A2, 1407 (1964).

389. M. Imoto, T. Otsu, B. Yamada, Kogyo Kagaku Zasshi 68, 1132 (1965).

390. M. Imoto, K. Saotome, J. Polymer Sci. 31, 208 (1958).

391. M. Imoto, S. Shimizu, Kobunshi Kagaku 18, 747 (1961); from Makromol. Chem. 53, 228 (1962).

392. S. Imoto, J. Ukida, T. Kominami, Kobunshi Kagaku 14, 101 (1957); from CA 52, 1669H (1958).

393. T. Imoto, Y. Ogo, S. Goto, T. Mitani, Kogyo Kagaku Zasshi 69, 1371 (1966).

394. T. Imoto, Y. Ogo, T. Mitani, Kogyo Kagaku Zasshi 70, 1217 (1967).

395. T. Imoto, Y. Ogo, H. Nakamoto, Bull. Chem. Soc. Japan 41, 543 (1968).

396. Imperial Chemical Industries Ltd., British Patent 594,249 (1947); calc. by F. R. Mayo and C. Walling.

397. T. Irie, M. Kinoshita, M. Imoto, Kogyo Kagaku Zasshi 69, 980 (1966); see also Makromol. Chem. 110, 47 (1967).

398. R. B. Isaacson, I. Kirshenbaum, I. Klein, J. Appl. Polymer Sci. 9, 933 (1965).

399. A. Ishigaki, T. Shono, Y. Hachihama, Makromol. Chem. 79, 170 (1964).

400. K. Ishigure, Y. Tabata, K. Oshima, J. Macromol. Sci. (Chem.) A1, 591 (1967).

401. O. A. Iskhakov, E. V. Kuznetsov, G. M. Eliseeva, Vysokomolekul. Soedin., Ser. B, 10, 32 (1968); from CA 68, 69425P (1968).

402. H. Ito, S. Suzuki, Kogyo Kagaku Zasshi 58, 627 (1955); from CA 50, 7501E (1956).

403. H. Ito, S. Suzuki, Kogyo Kagaku Zasshi 60, 341 (1957); from CA 53, 5732F (1959).

404. I. Ito, H. Hayashi, T. Saegusa, J. Furukawa, Makromol. Chem. 55, 15 (1962).

405. K. Ito, S. Iwase, Y. Yamashita, Makromol. Chem. 110, 233 (1967).

406. K. Ito, Y. Yamashita, J. Polymer Sci. B3, 625 (1965).

407. K. Ito, Y. Yamashita, J. Polymer Sci. B3, 637 (1965).

408. K. Ito, Y. Yamashita, Kogyo Kagaku Zasshi 68, 1469 (1965).

409. K. Ito, Y. Yamashita, J. Polymer Sci., A-1, 4, 631 (1966).

410. T. Ito, T. Otsu, M. Imoto, Mem. Fac. Eng., Osaka City Univ. 7, 87 (1965); from CA 66, 76325T (1967).

411. K. J. Ivin, R. H. Spensley, J. Macromol. Sci. (Chem.) A1, 653 (1967).

412. Y. Iwakura, T. Kurosaki, N. Ariga, T. Ito, Makromol. Chem. 97, 128 (1966).

413. Y. Iwakura, T. Kurosaki, N. Nakabayashi, Makromol. Chem. 46, 570 (1961).

414. Y. Iwakura, K. Matsuzaki, Kobunshi Kagaku 17, 187 (1960).

415. Y. Iwakura, N. Nakabayashi, Makromol. Chem. 66, 142 (1963).

416. Y. Iwakura, N. Nakabayashi, M. H. Lee, Makromol. Chem. 78, 157 (1964).

417. Y. Iwakura, N. Nakabayashi, M. H. Lee, Makromol. Chem. 104, 37 (1967).

418. Y. Iwakura, N. Nakabayashi, H. Suzuki, Makromol. Chem. 78, 168 (1964).

419. Y. Iwakura, M. Sato, T. Tamikado, S. Mimashi, Kobunshi Kagaku 13, 125 (1956); from CA 51, 4045D (1957).

420. Y. Iwakura, M. Sato, T. Tamikado, T. Mizoguchi, Kobunshi Kagaku 13, 390 (1956); from CA 51, 18694B (1957).

421. Y. Iwakura, T. Tamikado, Y. Fujimoto, S. Ikegami, M. Maruyama, Kobunshi Kagaku 15, 469 (1958); from CA 54, 11155A (1960).

422. Y. Iwakura, T. Tamikado, M. Yamaguchi, K. Takei, J. Polymer Sci. 39, 203 (1959).

423. Y. Iwakura, F. Toda, K. Hattori, J. Polymer Sci., A-1, 6, 1633 (1968).

424. Y. Iwakura, F. Toda, N. Kusakawa, H. Suzuki, J. Polymer Sci., B, 6, 5 (1968).

425. Y. Iwakura, F. Toda, H. Suzuki, J. Polymer Sci., A-1, 5, 1599 (1967).

426. Y. Iwakura, F. Toda, Y. Torii, J. Polymer Sci., A-1, 4, 2649 (1966).

427. Y. Iwakura, F. Toda, Y. Torii, R. Sekii, J. Polymer Sci., A-1, 6, 2681 (1968).

428. Y. Iwakura, K. Uno, N. Nakabayashi, W.-Y. Chiang, J. Polymer Sci., A-1, 5, 3193 (1967).

429. Y. Iwakura, K. Uno, N. Nakabayashi, T. Kojima, Bull. Chem. Soc. Japan 38, 1223 (1965).

430. Y. Iwakura, K. Uno, N. Nakabayashi, T. Kojima, Bull. Chem. Soc. Japan 41, 186 (1968).

431. T. Iwata, T. Saegusa, H. Fujii, J. Furukawa, Makromol. Chem. 97, 49 (1966).

432. T. Iwata, G. Wasai, T. Saegusa, J. Furukawa, Makromol. Chem. 77, 229 (1964).

433. S. Iwatsuki, S. Iguchi, Y. Yamashita, Kogyo Kagaku Zasshi 67, 1464 (1964); from CA 62, 5343B (1965).

434. S. Iwatsuki, S. Iguchi, Y. Yamashita, Kogyo Kagaku Zasshi 68, 2463 (1965).

435. S. Iwatsuki, M. Murakami, Y. Yamashita, Kogyo Kagaku Zasshi 68, 1967 (1965).

436. S. Iwatsuki, M. Shin, Y. Yamashita, Makromol. Chem. 102, 232 (1967).

437. S. Iwatsuki, N. Takigawa, M. Okada, Y. Yamashita, Y. Ishii, J. Polymer Sci. B2, 549 (1964).

438. S. Iwatsuki, Y. Yamashita, Kogyo Kagaku Zasshi 68, 1963 (1965).

439. Z. Izumi, H. Kitagawa, J. Polymer Sci., A-1, 5, 1967 (1967).

440. Z. Izumi, H. Kiuchi, M. Watanabe, J. Polymer Sci. A1, 705 (1963).

441. Z. Izumi, H. Kiuchi, M. Watanabe, J. Polymer Sci. A3, 2965 (1965).

442. Z. Izumi, H. Kiuchi, M. Watanabe, H. Uchiyama, J. Polymer Sci. A3, 2721 (1965).

443. H. Jahn, J. Neels, Z. Chem. 2(12), 370 (1962); from CA 59, 4044D (1963).

444. D. G. L. James, T. Ogawa, J. Polymer Sci. B2, 991 (1964).

445. M. Jelinek, Z. Laita, M. Kucera, J. Polymer Sci., C 16, 431 (1967).

446. G. Jenner, Bull. Soc. Chim. France 1966, 1127.

447. U. Johnsen, K. Kolbe, Kolloid-Z. Z. Polymer 216, 97 (1967).

448. U. Johnsen, K. Kolbe, Makromol. Chem. 116, 173 (1968).

449. A. F. Johnson, D. J. Worsfold, Makromol. Chem. 85, 273 (1965).

450. W. A. Johnson, L. J. Young, unpublished data.

451. N. W. Johnston, H. J. Harwood, Am. Chem. Soc., Div. Polymer Chem., Preprints 9(1), 36 (1968).

452. G. D. Jones, J. R. Runyon, J. Ong, Ind. Eng. Chem. 53, 297 (1961).

453. E. F. Jordan, K. M. Doughty, W. S. Port, J. Appl. Polymer Sci. 4, 203 (1960).

454. E. F. Jordan, jr., A. N. Wrigley, J. Appl. Polymer Sci. 8, 527 (1964).

455. R. M. Joshi, S. L. Kapur, J. Polymer Sci. 14, 508 (1954).

456. R. M. Joshi, S. L. Kapur, J. Sci. Ind. Res. 16B, 379 (1957).

457. R. M. Joshi, S. L. Kapur, J. Sci. Ind. Res. 16B, 441 (1957).

458. J. M. Judge, C. C. Price, J. Polymer Sci. 41, 435 (1959).

459. T. Kagiya, T. Kondo, S. Narisawa, K. Fukui, Bull. Chem. Soc. Japan 41, 172 (1968).

460. D. J. Kahn, H. H. Horowitz, J. Polymer Sci. 54, 363 (1961).

461. H. Kakiuchi, W. Fukuda, Asahi Garasu Kogyo Gijutsu Shorei-Kai Kenkyu Hokoku 12, 305 (1966); from CA 68, 10557U (1968).

462. H. Kakiuchi, T. Sakita, Kogyo Kagaku Zasshi 67, 2142 (1964); from CA 62, 16390D (1965).

463. S. Kamenar, I. Simek, E. Regensbogenova, Chem. Zvesti 14, 581 (1960); from CA 55, 154501 (1961).

464. N. Kamiya, H. Matsuura, Nippon Gomu Kyokaishi 40, 64 (1967); from CA 68, 50748Z (1968).

465. H. Kamogawa, H. G. Cassidy, J. Polymer Sci. A1, 1971 (1963).

466. H. Kamogawa, H. G. Cassidy, J. Polymer Sci. A2, 2409 (1964).

467. H. Kamogawa, R. Murase, T. Sekiya, Kogyo Kagaku Zasshi 62, 1749 (1959); from CA 58, 3511E (1963).

468. H. Kamogawa, T. Sekiya, Kogyo Kagaku Zasshi 62, 1117 (1959); from CA 58, 3511C (1963).

469. H. Kamogawa, T. Sekiya, Kogyo Kagaku Zasshi 63, 1631 (1960); from CA 56, 13084H (1962).

470. N. Kanoh, A. Gotoh, T. Higashimura, S. Okamura, Makromol. Chem. 63, 106 (1963).

471. N. Kanoh, K. Ikeda, A. Gotoh, T. Higashimura, S. Okamura, Makromol. Chem. 86, 200 (1965).

472. N. G. Karapetyan, I. S. Boshnyakov, A. S. Magaryan, Vysokomolekul. Soedin. 7, 1993 (1965).

473. V. A. Kargin, N. A. Plate, T. I. Patrikeyeva, Vysokomolekul. Soedin. 6, 2040 (1964).

474. F. J. Karol, W. L. Carrick, J. Am. Chem. Soc. 83, 2654 (1961).

475. R. G. Karzhaubaeva, G. P. Gladyshev, S. R. Rafikov, Vysokomolekul. Soedin., Ser. B, 9, 453 (1967).

476. M. Kato, H. Kamogawa, J. Polymer Sci., A-1, 4, 2771 (1966).

477. D. Katz, J. Polymer Sci. A1, 1635 (1963).

478. D. Katz, J. Relis, J. Polymer Sci., A-1, 6, 2079 (1968).

479. N. Kawabata, T. Tsuruta, J. Furukawa, Makromol. Chem. 48, 106 (1961).

480. W. Kawai, Kogyo Kagaku Zasshi 66, 249 (1963); from CA 59, 10243B (1963).

481. W. Kawai, J. Polymer Sci., A-1, 6, 1945 (1968).

482. V. F. Kazanskaya, O. M. Klimova, B. M. Khlebnikov, Vysokomolekul. Soedin. 6, 1799 (1964).

483. J. P. Kennedy, N. H. Canter, J. Polymer Sci., A-1, 5, 2455 (1967).

484. W. O. Kenyon, J. H. van Campen, U. S. Patent 2,419,221 (1947); calc. by F. R. Mayo and C. Walling.

485. R. Kerber, Makromol. Chem. 96, 30 (1966).

486. R. Kerber, H. Glamann, Makromol. Chem. 100, 290 (1967).

487. R. J. Kern, J. Polymer Sci. 43, 549 (1960).

488. W. Kern, Univ. Mainz, Germany, private communication.

489. W. Kern, Chemiker Ztg. 88, 623 (1964).

490. W. Kern, D. Braun, Makromol. Chem. 27, 23 (1958).

491. A. D. Ketley, J. Polymer Sci. B1, 121 (1963).

492. J. T. Khamis, Polymer 6, 98 (1965).

493. A. M. Khomutov, Izv. Akad. Nauk SSSR Otd. Khim. Nauk 1961, 352; Engl. Transl., p. 324.

494. A. M. Khomutov, Vysokomolekul. Soedin. 5, 1121 (1963).

495. A. M. Khomutov, A. P. Alimov, Vysokomolekul. Soedin. 8, 1068 (1966).

496. K. Kikukawa, S. Nozakura, S. Murahashi, Kobunshi Kagaku 24, 801 (1967); from CA 68, 115007D (1968).

497. C. S. Y. Kim, F. O. Hook, F. Veatch, E. C. Hughes, Kunststoff-Rundschau 12(2), 65 (1965).

498. T. Kimura, M. Imoto, Makromol. Chem. 50, 155 (1961).

499. T. Kimura, K. Yoshida, Kagaku to Kogyo (Osaka) 27, 288 (1953); from CA 49, 12034A (1955).

500. T. Kimura, K. Yoshida, Kagaku to Kogyo (Osaka) 28, 158 (1954); from CA 49, 12873B (1955).

501. T. Kimura, K. Yoshida, Kagaku to Kogyo (Osaka) 29, 43 (1955); from CA 49, 13688G (1955).

502. T. Kimury, K. Yoshida, Kagaku to Kogyo (Osaka) 29, 288 (1955); from CA 49, 13688G (1955).

503. T. Kimura, K. Yoshida, Kagaku to Kogyo (Osaka) 32, 223, 341 (1958); from CA 53, 4806G (1959).

504. T. Kimura, K. Yoshida, Kagaku to Kogyo (Osaka) 33, 413 (1959); from CA 58, 1538H (1963).

505. M. Kinoshita, M. Imoto, Kogyo Kagaku Zasshi 68, 2454 (1965); see also Makromol. Chem. 94, 328 (1966).

506. J. B. Kinsinger, J. S. Bartlett, N. H. Rauscher, J. Appl. Polymer Sci. 6, 529 (1962).

507. J. B. Kinsinger, J. R. Panchak, R. L. Kelso, J. S. Bartlett, R. K. Graham, J. Appl. Polymer Sci. 9, 429 (1965).

508. A. L. Klebanskii, O. A. Timofeev, Zh. Prikl. Khim. 32, 2294 (1959); from CA 54, 8587A (1960); J. Polymer Sci. 52, 23 (1961).

509. A. L. Klebanskii, O. A. Timofeev, Zh. Obshch. Khim. 30, 60 (1960); Engl. Transl., p. 62.

510. N. Kliman, M. Lazar, Chem. Prum. 9, 668 (1959); from CA 54, 10390D (1960).

511. O. M. Klimova, V. F. Kazanskaya, Zh. Prikl. Khim. 38, 434 (1965); from CA 62, 14831E (1965).

512. L. E. Klubikova, O. M. Klimova, Vysokomolekul. Soedin., Ser. B, 9, 528 (1967).

513. F. W. Knobloch, J. Polymer Sci. 25, 453 (1957).

514. F. Kobayashi, K. Matsuya, J. Polymer Sci. A1, 111 (1963).

515. Y. Kobuke, J. Furukawa, T. Fueno, J. Polymer Sci., A-1, 5, 2701 (1967).

516. S. Kohjiya, Y. Imanishi, S. Okamura, J. Polymer Sci., A-1, 6, 809 (1968).

517. G. S. Kolesnikov, S. L. Davydova, T. I. Ermolaeva, N. D. Shilova, M. B. Bykhovaskaya, Vysokomolekul. Soedin. 2, 567 (1960).

518. G. S. Kolesnikov, S. L. Davydova, N. V. Klimentova, Vysokomolekul. Soedin. 4, 1098 (1962).

519. G. S. Kolesnikov, E. F. Rodionova, I. G. Safaralieva, Izv. Akad. Nauk SSSR, Ser. Khim. 1963, 2028; from CA 60, 6933F (1964).

520. G. S. Kolesnikov, A. P. Suprun, T. A. Soboleva, V. A. Ershova, V. B. Bondareva, Vysokomolekul. Soedin. 4, 743 (1962).

521. G. D. Kolesnikov, A. S. Tevlina, A. B. Alovitdinov, Vysokomolekul. Soedin. 7, 1913 (1965).

522. G. S. Kolesnikov, A. S. Tevlina, A. B. Alovitdinov, Plasticheskie Massy 1966(2), 12.

523. G. N. Kolesnikov, A. S. Tevlina, I. I. Grandberg, S. E. Vasyukov, G. I. Sharova, Vysokomolekul. Soedin., Ser. A, 9, 2492 (1967).

524. G. S. Kolesnikov, A. S. Tevlina, S. P. Novikova, S. N. Sividova, Vysokomolekul. Soedin. 7, 2160 (1965).

525. A. R. Kol'k, A. A. Konkin, Z. A. Rogovin, Khim. Volokna 1963(4), 12; from CA 59, 11670D (1963).

526. A. Konishi, Bull. Chem. Soc. Japan 35, 395 (1962).

527. S. Konya, M. Yokoyama, Kogyo Kagaku Zasshi 68, 1080 (1965).

528. S. Konya, M. Yokoyama, Kogyo Kagaku Zasshi 68, 2450 (1965).

529. A. A. Korotkov, N. N. Chesnokova, Vysokomolekul. Soedin. 2, 365 (1960).

530. A. A. Korotkov, S. P. Mitsengendler, K. M. Aleev, Vysokomolekul. Soedin. 2, 1811 (1960).

531. A. A. Korotkov, G. V. Rakova, Vysokomolekul. Soedin. 3, 1482 (1961).

532. A. Kotera, M. Shima, K. Akiyama, M. Kume, M. Miyakawa, Bull. Chem. Soc. Japan 39, 758 (1966).

533. M. M. Koton, J. Polymer Sci. 30, 331 (1958).

534. M. M. Koton, Inst. Akad. Nauk Latv. SSR Riga 119, (1957); from CA 55, 16546G (1961).

535. M. M. Koton, L. F. Dokukina, Vysokomolekul. Soedin. 6, 1791 (1964).

536. M. M. Koton, O. K. Surnina, Dokl. Akad. Nauk SSSR 113, 1063 (1957); from Resins, Rubbers, Plastics 1958, 245.

537. H. Krauserova, I. Kossler, B. Matyska, N. G. Gaylord, J. Polymer Sci., C, 23, 327 (1968).

538. M. Kreisel, U. Garbatski, D. H. Kohn, J. Polymer Sci. A2, 105 (1964).

539. M. Kreisel, U. Garbatski, D. H. Kohn, J. Polymer Sci. B2, 81 (1964).

540. E. V. Kristal'nyi, S. S. Medvedev, Vysokomolekul. Soedin. 7, 1377 (1965).

541. A. G. Kronman, I. V. Pasmanik, B. I. Fedoseev, V. A. Kargin, Vysokomolekul. Soedin., Ser. A, 9, 2503 (1967).

542. G. E. Krybekyan, E. G. Sinanyan, A. N. Akopyan, Izv. Akad. Nauk Arm. SSR, Khim Nauki 15, 527 (1962); from CA 59, 1762H (1963).

543. Y. Kubouchi, T. Yamamoto, Y. Sono, Kogyo Kagaku Zasshi 57, 316, 678 (1954); from CA 49, 4324G (1955).

544. M. Kucera, J. Pichler, Polymer 5, 371 (1964).

545. G. I. Kudryavtsev, V. N. Odnoralova, M. V. Shablygin, Vysokomolekul. Soedin. 8, 821 (1966).

546. S. Kunichika, T. Katagiri, Kogyo Kagaku Zasshi 64, 929 (1961); from CA 57, 7480C (1962).

547. I. Kuntz, J. Polymer Sci. 54, 569 (1961).

548. T. N. Kuren'gina, L. V. Alferova, V. A. Kropachev, Vysokomolekul. Soedin. 8, 293 (1966).

549. C. J. Kurian, M. S. Muthana, Makromol. Chem. 29, 26 (1959).

550. J. Lal, J. E. McGrath, J. Polymer Sci. A2, 3369 (1964).

551. Y. Landler, J. Polymer Sci. 8, 63 (1952).

552. Y. Landler, Compt. Rend., Ser. A, 230, 539 (1950).

553. G. N. Larina, Z. V. Borisova, T. V. Sheremeteva, Vysokomolekul. Soedin. 3, 1664 (1961).

554. G. N. Larina, V. P. Sklizkova, Vysokomolekul. Soedin., Ser. B, 10, 204 (1968); from CA 69, 3166U (1968).

555. F. Leonard, W. P. Hohenstein, E. Merz, J. Am. Chem. Soc. 70, 1283 (1948).

556. C. W. Lewis, D. W. Lewis, J. Polymer Sci. 36, 325 (1959).

557. F. M. Lewis, F. R. Mayo, J. Am. Chem. Soc. 70, 1533 (1948).

558. F. M. Lewis, F. R. Mayo, W. F. Hulse, J. Am. Chem. Soc. 67, 1701 (1945).

559. F. M. Lewis, C. Walling, W. Cummings, Er. Briggs, F. R. Mayo, J. Am. Chem. Soc. 70, 1519 (1948).

560. F. M. Lewis, C. Walling, W. Cummings, E. R. Briggs, W. J. Wenisch, J. Am. Chem. Soc. 70, 1527 (1948).

561. F.-M. Li, Y.-L. Ch'en, L.-C. Wang, H.-C. Wei, L.-H. Yeh, Ko Fen Tzu T'ung Hsun 8, 1 (1966); from CA 65, 17054D (1966).

562. P. T. Li, K.-K. Wu, K'O Hsueh Ch'u Pan She 1963, 151; from CA 64, 2254H (1966).

563. T. E. Lipatova, V. M. Siderko, Vysokomolekul. Soedin. 6, 910 (1964).

564. T. E. Lipatova, V. M. Siderko, V. A. Budnikova, Vysokomolekul. Soedin. 7, 580 (1965).

565. R. D. A. Lipman, Am. Chem. Soc., Div. Polymer Chem., Preprints 8(1), 396 (1967).

566. M. Litt, F. W. Bauer, J. Polymer Sci., C, 16, 1551 (1967).

567. D. I. Livingston, P. M. Kamath, R. S. Corley, J. Polymer Sci. 20, 485 (1956).

568. S. Loshaek, E. Broderick, J. Polymer Sci. 39, 241 (1959).

569. G. G. Lowry, W. K. Carrington, unpublished data.

570. L. S. Luskin, R. J. Myers, "Encyclopedia of Polymer Science and Technology", Vol. 1, Interscience Publishers, 1964, p. 246.

571. H. Lüssi, Makromol. Chem. 103, 62 (1967).

572. H. Lüssi, Makromol. Chem. 103, 68 (1967).

573. H. Lüssi, Makromol. Chem. 110, 100 (1967).

574. E. B. Lyudvig, A. R. Gantmakher, S. S. Medvedev, Vysokomolekul. Soedin. 1, 1333 (1959).

575. Z. Machacek, Chem. Listy 48, 477 (1954); from CA 48, 8583E (1954).

576. S. Machi, T. Sakai, M. Gotoda, T. Kagiya, J. Polymer Sci., A-1, 4, 821 (1966).

577. S. Machida, H. Narita, Yuki Gosei Kagaku Kyokai Shi 24, 467 (1966); from CA 65, 9032F (1966).

578. S. Machida, H. Saito, Sen-i Gakkaishi 23, 301 (1967); from CA 68, 87619J (1968).

579. S. Machida, Y. Shimizu, T. Makita, Kamipa Gikyoshi 21, 352 (1967); from CA 67, 64764T (1967).

580. O. L. Mageli, R. E. Light, jr., R. B. Gallagher, Am. Chem. Soc., Div. Polymer Chem., Preprints 8(1), 714 (1967); see also J. Polymer Sci., C, 24, 57 (1968).

581. N. K. Maratova, O. M. Klimova, Vysokomolekul. Soedin., Ser. B, 10, 87 (1968); from CA 68, 96219F (1968).

582. W. Marconi, S. Cesca, G. Della Fortuna, Chim Ind. (Milan) 46, 1131 (1964); from CA 62, 5343F (1965).

583. W. Marconi, A. Mazzei, G. Lugli, M. Bruzzone, J. Polymer Sci., C, 16, 805 (1967).

584. H. L. Marder, C. Schuerch, J. Polymer Sci. 44, 129 (1960).

585. E. Marechal, J. P. Menissez, J. P. Richard, C. Zaffran, Compt. Rend., Ser. C, 266, 1427 (1968).

586. E. Marechal, P. Sigwalt, Bull. Soc. Chim. France 1966, 1071.

587. M. F. Margaritova, G. D. Berezhnov, Tr. Mosk. Khim.-Tekhnol. Inst. 4, 46 (1953); from CA 50, 1361F (1956).

588. M. F. Margaritova, V. A. Raiskaya, Tr. Mosk. Khim.-Tekhnol. Inst. 4, 37 (1953); from CA 49, 14372H (1955).

589. L. Marker, O. J. Sweeting, J. G. Wepsic, J. Polymer Sci. 57, 855 (1962).

590. G. Markert, Makromol. Chem. 103, 109 (1967).

591. C. S. Marvel, W. G. Depierri, J. Polymer Sci. 27, 39 (1958).

592. C. S. Marvel, J. F. Dunphy, J. Org. Chem. 25, 2209 (1960).

593. C. S. Marvel, T. K. Dykstra, F. C. Magne, J. Polymer Sci. 62, 369 (1962).

594. C. S. Marvel, S. L. Jacobs, W. K. Taft, B. G. Labbe, J. Polymer Sci. 19, 59 (1956).

595. C. S. Marvel, J. W. Johnson, J. P. Economy, G. P. Scott, W. K. Taft, B. G. Labbe, J. Polymer Sci. 20, 437 (1956).

596. C. S. Marvel, G. D. Jones, T. W. Mastin, G. L. Schertz, J. Am. Chem. Soc. 64, 2356 (1942).

597. C. S. Marvel, E. B. Mano, J. Polymer Sci. 31, 165 (1958).

598. C. S. Marvel, J. F. Porter, J. Org. Chem. 24, 137 (1959).

599. C. S. Marvel, G. L. Schertz, J. Am. Chem. Soc. 65, 2054 (1943); 66, 2135 (1944).

600. C. S. Marvel, R. Schwen, J. Am. Chem. Soc. 79, 6003 (1957).

601. C. S. Marvel, R. Schwen, R. W. Hobson, R. J. Coleman, J. Polymer Sci. 33, 27 (1958).

602. C. S. Marvel, V. Sziraky, J. P. Economy, J. Org. Chem. 21, 1314 (1956).

603. C. S. Marvel, A. T. Tweedie, J. P. Economy, J. Org. Chem. 21, 1420 (1956).

604. M. E. Mat'kova, S. S. Spasskii, Vysokomolekul. Soedin. 2, 879 (1960).

605. M. E. Mat'kova, S. S. Spasskii, Vysokomolekul. Soedin. 3, 93 (1961).

606. S. G. Matsoyan, L. A. Akopyan, A. A. Saakyan, S. B. Gevorkyan, Arm. Khim. Zh. 20, 902 (1967); from CA 69, 107810 (1968).

607. S. G. Matsoyan, N. M. Morlyan, Izv. Akad. Nauk Arm. SSR, Khim. Nauki 17(5), 522 (1964); from CA 62, 14832H (1965).

608. S. G. Matsoyan, M. G. Voskanyan, A. A. Cholakyan, Vysokomolekul. Soedin. 5, 1035 (1963).

609. M. Matsuda, M. Iino, N. Tokura, Makromol. Chem. 65, 232 (1963).

610. M. Matsuda, K. Ohshima, N. Tokura, J. Polymer Sci. A2, 4271 (1964).

611. A. Matsumoto, M. Oiwa, Kogyo Kagaku Zasshi 70, 360 (1967).

612. A. Matsumoto, S. Shoda, T. Harada, M. Oiwa, Kogyo Kagaku Zasshi 70, 1007 (1967).

613. K. Matsuoka, M. Otsuka, K. Takemoto, M. Imoto, Kogyo Kagaku Zasshi 69, 137 (1966).

614. K. Matsuoka, K. Takemoto, M. Imoto, Kogyo Kagaku Zasshi 68, 1135 (1965).

615. K. Matsuoka, K. Takemoto, M. Imoto, Kogyo Kagaku Zasshi 68, 1941 (1965).

616. K. Matsuoka, K. Takemoto, M. Imoto, Kogyo Kagaku Zasshi 69, 134 (1966).

617. K. Matsuoka, K. Takemoto, M. Imoto, Kogyo Kagaku Zasshi 69, 142 (1966).

618. F. R. Mayo, C. W. Gould, J. Am. Oil Chem. Soc. 41, 25 (1964).

619. F. R. Mayo, C. W. Gould, J. Am. Oil Chem. Soc. 44, 178 (1967).

620. F. R. Mayo, F. M. Lewis, C. Walling, J. Am. Chem. Soc. 70, 1529 (1948).

621. F. R. Mayo, C. Walling, Chem. Rev. 46, 191 (1950).

622. F. R. Mayo, C. Walling, F. M. Lewis, W. F. Hulse, J. Am. Chem. Soc. 70, 1523 (1948).

623. G. Mazzanti, A. Valvassori, G. Pajaro, Chim. Ind. (Milan) 39, 825 (1957).

624. G. Mazzanti, A. Valvassori, G. Pajaro, Chim. Ind. (Milan) 39, 743 (1957).

625. G. Mazzanti, A. Valvassori, G. Sartori, G. Pajaro, Chim. Ind. (Milan) 42, 468 (1960).

626. E. T. McBee, H. M. Hill, B. G. Bachman, Ind. Eng. Chem. 41, 70 (1949).

627. G. H. McCain, D. E. Hudgin, I. Rosen, J. Polymer Sci., A-1, 5, 975 (1967).

628. E. J. Meehan, J. Polymer Sci. 1, 318 (1946).

629. R. L. Meier, J. Chem. Soc. 1950, 3656.

630. N. V. Meiya, A. F. Nikolaev, G. A. Balaev, Vysokomolekul. Soedin. 7, 211 (1965).

631. H. W. Melville, L. Valentine, Trans. Faraday Soc. 51, 1474 (1955).

632. R. L. Merker, M. J. Scott, J. Polymer Sci. 25, 115 (1957).

633. R. L. Merker, M. J. Scott, J. Polymer Sci. 43, 297 (1960).

634. V. E. Meyer, J. Polymer Sci., A-1, 4, 2819 (1966).

635. L. P. Mezhirova, A. P. Sheinker, A. D. Abkin, Vysokomolekul. Soedin. 3, 99 (1961).

636. L. P. Mezhirova, Z. Smigasevich, A. P. Sheinker, A. D. Abkin, Vysokomolekul. Soedin. 5, 473 (1963).

637. F. Miller, Canadian Patent 516,532 (1955).

638. Min (Szu-Kwei) and Chen Ho Chu, Hua Hsueh Hsueh Pao 23, 262 (1957); from CA 52, 19232 (1958).

639. T. T. Minakova, F. P. Sidel'kovskaya, M. F. Shostakovskii, Izv. Akad. Nauk SSSR, Ser. Khim 1965, 1880; from CA 64, 5216D (1966).

640. Y. Minoura, Y. Suzuki, Y. Sakanaka, H. Doi, Kogyo Kagaku Zasshi 69, 345 (1966); see also J. Polymer Sci., A-1, 4, 2757 (1966).

641. Y. Minoura, Y. Tadokoro, Y. Suzuki, J. Polymer Sci., A-1, 5, 2641 (1967).

642. A. Misono, Y. Uchida, Bull. Chem. Soc. Japan 39, 2458 (1966).

643. A. Misono, Y. Uchida, K. Yamada, Bull. Chem. Soc. Japan 40, 2366 (1967).

644. J. M. Mitchell, H. L. Williams, Can. J. Res. 27, 35 (1949).

645. S. P. Mitsengendler, V. N. Krasulina, L. B. Trukhmanova, Izv. Akad. Nauk SSSR Otd. Khim. Nauk 1956, 1120; from CA 51, 3178D (1957).

646. A. Mitsutani, M. Yano, Kogyo Kagaku Zasshi 67, 935 (1964); from CA 61, 10794E (1964).

647. M. Miura, T. Hirai, Kogyo Kagaku Zasshi 67, 490 (1964); from CA 62, 2829H (1965).

648. K. Miyamichi, A. Suzuki, S. Harada, M. Katayama, Kobunshi Kagaku 21, 79 (1964); from CA 61, 7111C (1964).

649. A. Mizote, T. Higashimura, S. Okamura, Am. Chem. Soc., Div. Polymer Chem., Preprints 7(2), 409 (1966).

650. A. Mizote, T. Higashimura, S. Okamura, J. Polymer Sci., A-1, 6, 1825 (1968).

651. A. Mizote, S. Kusudo, T. Higashimura, S. Okamura, J. Polymer Sci., A-1, 5, 1727 (1967).

652. A. Mizote, T. Tanaka, T. Higashimura, S. Okamura, J. Polymer Sci. A3, 2567 (1965).

653. A. Mizote, T. Tanaka, T. Higashimura, S. Okamura, J. Polymer Sci., A-1, 4, 869 (1966).

654. E. W. Moffett, R. E. Smith, U. S. Patent 2,356,871 (1944); cald. by F. R. Mayo and C. Walling.

655. R. B. Mohite, S. Gundiah, S. L. Kapur, Makromol. Chem. 108, 52 (1967).

656. G. A. Mortimer, J. Polymer Sci. B3, 343 (1965).

657. M. Morton, F. R. Ells, J. Polymer Sci. 61, 25 (1962).

658. D. T. Mowry, U. S. Patent 2,398,321 (1946); calc. by F. R. Mayo and C. Walling.

659. D. T. Mowry, U. S. Patent 2,417,607 (1947); calc. by F. R. Mayo and C. Walling.

660. E. Müller, K. Dinges, W. Graulich, Makromol. Chem. 57, 27 (1962).

661. J. Müller, Chem. Listy 48, 1593 (1954); from CA 49, 5077D (1955).

662. J. Müller, Collection Czech. Chem. Commun. 20, 241 (1955).

663. J. E. Mulvaney, J. G. Dillon, J. L. Laverty, J. Polymer Sci., A-1, 6, 1841 (1968).

664. S. Murahashi, S. Nozakura, K. Emura, K. Yasufuku, Kobunshi Kagaku 23, 361 (1966); from CA 66, 65880U (1967).

665. S. Murahashi, S. Nozakura, Y. Imai, Kobunshi Kagaku 22, 739 (1965); from CA 64, 15993D (1966).

666. S. Nurahashi, S. Nozakura, A. Umehara, F. Nagoshi, Kobunshi Kagaku 22, 451 (1965); from CA 63, 18272D (1965).

667. S. Murahashi, S. Nozakura, A. Umehara, K. Obata, Kobunshi Kagaku 21, 625 (1964); from CA 62, 7878D (1965).

668. S. Murahashi, S. Nozakura, K. Yasufuku, Bull. Chem. Soc. Japan 38, 2082 (1965).

669. S. Murahashi, S. Nozakura, K. Yasufuku, Bull. Chem. Soc. Japan 39, 1338 (1966).

670. S. Murahashi, H. Yuki, K. Kosai, F. Doura, Bull. Chem. Soc. Japan 39, 1563 (1966).

671. K. Murata, A. Terada, J. Polymer Sci., A-1, 4, 2989 (1966).

672. K. Murata, A. Terada, Bull. Chem. Soc. Japan 39, 2494 (1966).

673. W. J. Murbach, A. Adicoff, Ind. Eng. Chem. 52, 772 (1960).

674. V. A. Myagchenkov, V. F. Kurenkov, E. V. Kuznetsov, S. Ya. Frenkel, Vysokomolekul. Soedin., Ser. B, 9, 251 (1967).

675. Sh. Nadzhimutdinov, E. P. Cherneva, V. A. Kargin, Vysokomolekul. Soedin. 7, 1173 (1965).

676. S. Nagai, Bull. Chem. Soc. Japan 36, 1459 (1963).

677. S. Nagai, T. Uno, K. Yoshida, Kobunshi Kagaku 15, 550 (1958); from CA 54, 11558B (1960).

678. S. Nagai, K. Yoshida, Kobunshi Kagaku 17, 77 (1960); from CA 55, 14973G (1961).

679. A. S. Nair, M. S. Muthana, Makromol. Chem. 47, 138 (1961).

680. N. Nakabayashi, Y. Iwakura, Makromol. Chem. 81, 180 (1965).

681. T. Nakai, K. Shioyo, M. Okawara, Makromol. Chem. 108, 95 (1967).

682. T. Nakata, N. Choumei, J. Macromol. Sci. (Chem.) A1, 1433 (1967).

683. T. Nakata, T. Otsu, M. Imoto, J. Polymer Sci. A3, 3383 (1965).

684. N. S. Nametkin, S. G. Durgar'yan, V. G. Filippova, Dokl. Akad. Nauk SSSR 172, 1090 (1967); from CA 66, 95413M (1967).

685. N. S. Nametkin, S. G. Durgar'yan, V. G. Filippova, Dokl. Akad. Nauk SSSR 177, 853 (1967); from CA 68, 50108R (1968).

686. G. Natta, G. Allegra, I. W. Bassi, D. Sianesi, G. Caporiccio, E. Torti, J. Polymer Sci. A3, 4263 (1965).

687. G. Natta, F. Danusso, D. Sianesi, Makromol. Chem. 30, 238 (1959).

688. G. Natta, G. Mazzanti, A. Valvassori, G. Pajaro, Chim. Ind. (Milan) 41, 764 (1959).

689. G. Natta, G. Mazzanti, A. Valvassori, G. Sartori, Chim. Ind. (Milan) 40, 717 (1958).

690. G. Natta, G. Mazzanti, A. Valvassori, G. Sartori, A. Barbagalio, J. Polymer Sci. 51, 429 (1961).

691. G. Natta, G. Mazzanti, A. Valvassori, G. Sartori, D. Fiumani, J. Polymer Sci. 51, 411 (1961).

692. G. Natta, A. Valvassori, G. Mazzanti, G. Sartori, Chim. Ind. (Milan) 40, 896 (1958).

693. A. F. Nikolaev, M. A. Andreeva, Vysokomolekul. Soedin., Ser. A, 9, 1720 (1967).

694. A. F. Nikolaev, M. A. Andreeva, Vysokomolekul. Soedin., Ser. A, 10, 502 (1968).

695. A. F. Nikolaev, V. M. Gaperin, Vysokomolekul. Soedin., Ser. A, 9, 2469 (1967).

696. A. F. Nikolaev, M. N. Tereshchenko, Vysokomolekul. Soedin. 6, 379 (1964).

697. A. F. Nikolaev, S. N. Ushakov, L. S. Mishkileeva, Vysokomolekul. Soedin. 6, 287 (1964).

698. A. F. Nikolaev, S. N. Ushakov, L. P. Vishnevetskaya, N. A. Voronova, E. I. Rodina, Vysokomolekul. Soedin. 4, 1053 (1962).

699. M. Nishigaki, M. Kinoshita, M. Imoto, Kogyo Kagaku Zasshi 70, 1938 (1967).

700. A. Nishioka, Y. Kato, N. Ashikari, J. Polymer Sci. 62, S-10 (1962).

701. K. Noma, M. Niwa, Kobunshi Kagaku 21, 17 (1964); from CA 61, 7109A (1964).

702. K. Noma, M. Niwa, Kobunshi Kagaku 24, 785 (1967); from CA 68, 115001X (1968).

703. K. Noma, M. Niwa, K. Iwasaki, Kobunshi Kagaku 20, 646 (1963); from Makromol. Chem. 73, 250 (1964).

704. K. Noma, M. Niwa, Y. Kato, Kobunshi Kagaku 22, 235 (1965); from CA 63, 8496F (1965).

705. K. Noma, M. Niwa, Y. Kato, S. Fujii, A. Kawade, Kobunshi Kagaku 23, 245 (1966); from CA 66, 18912W (1967).

706. K. Noma, Y. Ohfuji, I. Sakurada, Kobunshi Kagaku 22, 69 (1965).

707. K. Noma, Y. Ohfuji, I. Sakurada, Kobunshi Kagaku 24, 139 (1967).

708. A. M. North, D. Postlethwaite, Polymer 5, 237 (1964).

709. A. M. North, D. Postlethwaite, Trans. Faraday Soc. 62, 2843 (1966).

710. A. M. North, A. M. Scallan, Polymer 5, 447 (1964).

711. R. M. Nowak, P. L. Brissette, unpublished data.

712. R. E. Nylund, W. G. Miller, J. Am. Chem. Soc. 87, 3537 (1965).

713. K. F. O'Driscoll, F. P. Gasparro, J. Macromol. Sci. (Chem.) A1, 643 (1967).

714. E. Ohmori, Y. Ohi, T. Otsu, M. Imoto, Kogyo Kagaku Zasshi 68, 1600 (1965).

715. H. Ohse, H. Cherdron, Makromol. Chem. 95, 283 (1966).

716. M. Okada, S. Iwatsuki, Y. Yamashita, Kogyo Kagaku Zasshi 68, 2466 (1965).

717. M. Okada, K. Suyama, Y. Yamashita, Y. Ishii, Kogyo Kagaku Zasshi 68, 546 (1964).

718. M. Okada, N. Takikawa, S. Iwatsuki, Y. Yamashita, Y. Ishii, Makromol. Chem. 82, 16 (1965).

719. M. Okada, Y. Yamashita, Kogyo Kagaku Zasshi 69, 506 (1966).

720. M. Okada, Y. Yamashita, Y. Ishii, Kogyo Kagaku Zasshi 68, 364 (1965); from CA 62, 16389G (1965).

721. S. Okamura, T. Higashimura, Kobunshi Kagaku 18, 389 (1961); from CA 55, 24099I (1961).

722. S. Okamura, T. Higashimura, Y. Imanishi, R. Yamamoto, K. Kimura, J. Polymer Sci., C, 16, 2365 (1967).

723. S. Okamura, K. Uno, Kobunshi Kagaku 8, 467 (1951); from CA 47, 9663H (1953).

724. S. Okamura, T. Yamashita, J. Soc. Textile Cellulose Ind. Japan 9, 446 (1953); from CA 48, 1010C (1954).

725. T. Okuyama, T. Fueno, J. Furukawa, J. Polymer Sci., A-1, 6, 993 (1968).

726. T. Okuyama, T. Fueno, J. Furukawa, K. Uyeo, J. Polymer Sci., A-1, 6, 1001 (1968).

727. T. Oota, Enka Biniiru to Porima 8, 26 (1968); from CA 69, 36484Z (1968).

728. T. Oota, S. Masuda, Kogyo Kagaku Zasshi 69, 721 (1966).

729. V. A. Orlov, O. G. Tarakanov, Plasticheskie Massy 1964, 6.

730. R. J. Orr, Polymer 2, 79 (1961); from data of Furukawa.

731. R. J. Orr, H. L. Williams, Can. J. Chem. 29, 270 (1951).

732. R. J. Orr, H. L. Williams, Can. J. Chem. 30, 108 (1952).

733. R. J. Orr, H. L. Williams, Can. J. Chem. 33, 1328 (1955).

734. R. J. Orr, H. L. Williams, J. Polymer Sci. 32, 89 (1958).

735. E. Osawa, K. Wang, O. Kurihara, Makromol. Chem. 83, 100 (1965).

736. G. Oster, Y. Mizutani, J. Polymer Sci. 22, 173 (1956).

737. V. G. Ostroverkhov, I. S. Vakarchuk, V. G. Sinyavskii, Vysokomolekul. Soedin. 3, 1197 (1961).

738. T. Otsu, K. Goto, S. Aoki, M. Imoto, Makromol. Chem. 71, 150 (1964).

739. T. Otsu, T. Ito, Y. Fujii, M. Imoto, Bull. Chem. Soc. Japan 41, 204 (1968).

740. T. Otsu, T. Ito, T. Fukumizu, M. Imoto, Bull. Chem. Soc. Japan 39, 2257 (1966).

741. T. Otsu, T. Ito, M. Imoto, J. Polymer Sci. B3, 113 (1965).

742. T. Otsu, T. Ito, M. Imoto, J. Polymer Sci., A-1, 4, 733 (1966).

743. T. Otsu, T. Ito, M. Imoto, Kogyo Kagaku Zasshi 69, 986 (1966).

744. T. Otsu, Y. Kinoshita, A. Nakamachi, Makromol. Chem. 115, 275 (1968).

745. T. Otsu, A. Shimizu, M. Imoto, J. Polymer Sci. A3, 615 (1965).

746. T. Otsu, J. Ushirone, M. Imoto, Kogyo Kagaku Zasshi 69, 516 (1966).

747. T. Otsu, B. Yamada, Makromol. Chem. 110, 297 (1967).

748. T. Otsu, B. Yamada, T. Nozaki, Kogyo Kagaku Zasshi 70, 1941 (1967).

749. M. Otsuka, K. Takemoto, M. Imoto, Kobunshi Kagaku 23, 765 (1966); from CA 66, 65888C (1967).

750. M. Otsuka, Y. Yasuhara, K. Takemoto, M. Imoto, Makromol. Chem. 103, 291 (1967).

751. C. G. Overberger, F. Ang, J. Am. Chem. Soc. 82, 929 (1960).

752. C. G. Overberger, L. H. Arnold, D. Tanner, J. J. Taylor, T. Alfrey, jr., J. Am. Chem. Soc. 74, 4848 (1952).

753. C. G. Overberger, L. H. Arnold, J. J. Taylor, J. Am. Chem. Soc. 73, 5541 (1951).

754. C. G. Overberger, D. E. Baldwin, H. P. Gregor, J. Am. Chem. Soc. 72, 4864 (1950).

755. C. G. Overberger, H. Biletch, R. G. Nickerson, J. Polymer Sci. 27, 381 (1958).

756. C. G. Overberger, R. J. Ehrig, D. Tanner, J. Am. Chem. Soc. 76, 772 (1954).

757. C. G. Overberger, V. G. Kamath, J. Am. Chem. Soc. 81, 2910 (1959).

758. C. G. Overberger, V. G. Kamath, J. Am. Chem. Soc. 85, 446 (1963).

759. C. G. Overberger, F. W. Michelotti, J. Am. Chem. Soc. 80, 988 (1958).

760. C. G. Overberger, E. Sarlo, J. Polymer Sci. A2, 1017 (1964).

761. C. G. Overberger, D. Tanner, E. M. Pearce, J. Am. Chem. Soc. 80, 4566 (1958).

762. Yu. L. Pankratov, G. I. Kudryavtsev, Vysokomolekul. Soedin. 6, 1862 (1964).

763. G. S. Park, private communication, 1966; see T. A. Garrett, G. S. Park, J. Polymer Sci., A-1, 4, 2714 (1966).

764. R. B. Parker, jr., B. V. Mekler, J. Polymer Sci. B2, 19 (1964).

765. I. Pasquon, etal., Chim. Ind. (Milan) 43, 509 (1961).

766. T. R. Paxton, J. Polymer Sci. B1, 73 (1963).

767. L. H. Peebles, J. T. Clarke, W. H. Stockmayer, J. Am. Chem. Soc. 82, 4780 (1960).

768. J. Pellon, W. G. Deichert, W. M. Thomas, J. Polymer Sci. 55, 153 (1961).

769. J. Pellon, R. L. Kugel, R. Marcus, R. Rabinowitz, J. Polymer Sci. A2, 4105 (1964).

770. J. Pellon, L. H. Schwind, M. J. Guinard, W. M. Thomas, J. Polymer Sci. 55, 161 (1961).

771. I. Penczek, S. Penczek, J. Polymer Sci., B, 5, 367 (1967).

772. G. A. Petrova, G. A. Shtraikman, A. A. Vansheidt, Zh. Fiz. Khim. 33, 1246 (1959); from CA 54, 8613G (1960).

773. R. M. Pike, D. L. Bailey, J. Polymer Sci. 22, 55 (1956).

774. S. H. Pinner, J. Polymer Sci. 10, 379 (1953).

775. Platzer, Monsanto Chemical Co., private communication, based on unpublished data of E. C. Chapin, P. C. Hamm, R. G. Fordyce.

776. Polymer Corporation Ltd., Sarnia, Can., private communication.

777. A. I. Popova, A. P. Sheinker, A. D. Abkin, Dokl. Akad. Nauk SSSR 157, 1192 (1964); from CA 61, 13423H (1964).

778. W. S. Port, E. F. Jordan, J. E. Hansen, D. Swern, J. Polymer Sci. 9, 493 (1952).

779. J. A. Powell, R. K. Graham, J. Polymer Sci. A3, 3451 (1965).

780. C. C. Price, R. D. Gilbert, J. Polymer Sci. 8, 580 (1952).

781. C. C. Price, R. D. Gilbert, J. Am. Chem. Soc. 74, 2073 (1952).

782. C. C. Price, C. E. Greene, J. Polymer Sci. 6, 111 (1951).

783. C. C. Price, B. D. Halpern, S. T. Voong, J. Polymer Sci. 11, 575 (1953).

784. C. C. Price, T. F. McKeon, J. Polymer Sci. 41, 445 (1959).

785. C. C. Price, H. Morita, J. Am. Chem. Soc. 75, 4747 (1953).

786. C. C. Price, T. C. Schwan, J. Polymer Sci. 16, 577 (1955).

787. C. C. Price, J. G. Walsh, J. Polymer Sci. 6, 239 (1951).

788. C. C. Price, J. Zomlefer, J. Am. Chem. Soc. 72, 14 (1950).

789. E. G. Pritchett, P. M. Kamath, J. Polymer Sci., B, 4, 849 (1966).

790. Y. N. Prokofiev, M. I. Farberov, V. A. Shadricheva, Vysokomolekul. Soedin. 2, 185 (1960).

791. R. Rabinowitz, R. Marcus, J. Pellon, J. Polymer Sci. A2, 1233 (1964).

792. R. Rabinowitz, R. Marcus, J. Pellon, J. Polymer Sci. A2, 1241 (1964).

793. S. R. Rafikov, N. D. Kazakova, G. A. D'yachkov, Dokl. Akad. Nauk SSSR 176, 346 (1967); from CA 67, 117338R (1967).

794. M. Ragazzini, C. Garbuglio, D. Carcano, B. Minasso, Gb. Cevidalli, European Polymer J. 3, 129 (1967).

795. N. V. Rakityanskii, R. L. Rabinovich, Rept. Allunion Sci. Inst. Synthetic Rubber, 1951.

796. G. V. Rakova, A. A. Korotkov, Dokl. Akad. Nauk SSSR 119, 982 (1958); from CA 53, 4809I (1959).

797. G. V. Rakova, A. A. Korotkov, T. Chan-Li, Dokl. Akad. Nauk SSSR 126, 582 (1959); from CA 53, 19425C (1959).

798. J. Rehner, R. L. Zapp, W. J. Sparks, J. Polymer Sci. 11, 21 (1953).

799. R. C. Reinhardt, Ind. Eng. Chem. 35, 422 (1943); calc. by F. R. Mayo and C. Walling.

800. L. I. Reitburd, M. A. Markevich, M. S. Akutin, Vysokomolekul. Soedin., Ser. A, 9, 1144 (1967).

801. H. Ringsdorf, G. Greber, Makromol. Chem. 31, 27 (1959).

802. H. Ringsdorf, N. Weinshenker, C. G. Overberger, Makromol. Chem. 64, 126 (1963).

803. W. M. Ritchey, L. E. Ball, J. Polymer Sci., B, 4, 557 (1966).

804. A. F. Roche, G. Corey, unpublished data.

805. L. Rodriguez, Makromol. Chem. 12, 110 (1954).

806. J. Roovers, G. Smets, Makromol. Chem. 60, 89 (1963).

807. I. Rosen, W. E. Marshall, J. Polymer Sci. 56, 501 (1962).

808. B. A. Rosenberg, E. B. Lyudvig, N. V. Desyatova, A. R. Gantmakher, S. S. Medvedev, Vysokomolekul. Soedin. 7, 1010 (1965).

809. B. A. Rosenberg, E. B. Lyudvig, A. R. Gantmakher, S. S. Medvedev, J. Polymer Sci., C, 16, 1917 (1967).

810. S. D. Ross, M. Markarian, H. H. Young, jr., M. Nazzewski, J. Am. Chem. Soc. 72, 1133 (1950).

811. S. A. Rostovtsev, E. S. Roskin, A. I. Ezrielev, Vysokomolekul. Soedin., Ser. B, 9, 289 (1967).

812. R. W. Roth, R. F. Church, J. Polymer Sci. 55, 41 (1961).

813. A. M. Rozhkov, Izv. Sib. Otd. Akad. Nauk SSSR, Ser. Khim. Nauk 1963, 103; from CA 60, 15989C (1964).

814. N. N. Rozovskaya, A. D. Abkin, Lakokrasochnye Materiali i ikh Primenenie 1961, 9; from CA 55, 20500C (1961).

815. N. N. Rozovskaya, A. P. Sheinker, A. D. Abkin, Vysokomolekul. Soedin. 7, 1383 (1965).

816. N. N. Rozovskaya, A. P. Sheinker, A. D. Abkin, Vysokomolekul. Soedin. 7, 1500 (1965).

817. E. W. Rugeley, T. A. Field, jr., G. H. Fremon, Ind. Eng. Chem. 40, 1724 (1948).

818. B. N. Rutovskii, A. M. Shur, Zh. Prikl. Khim. 24, 1173 (1951).

819. A. V. Ryabov, Yu. D. Semchikov, N. N. Slavnitskaya, Tr. Khim. Khim. Tekhnol. 1963, 334; from CA 61, 5771E (1964).

820. A. V. Ryabov, Yu. D. Semchikov, N. N. Slavnitskaya, Dokl. Akad. Nauk SSSR 145, 822 (1962); from CA 57, 15344A (1962).

821. A. V. Ryabov, Yu. D. Semchikov, V. N. Vakhrusheva, Tr. Khim. Khim. Tekhnol. 1963, 188; from CA 60, 9448H (1964).

822. T. Saegusa, H. Imai, J. Furukawa, Makromol. Chem. 56, 55 (1962).

823. T. Saegusa, T. Ueshima, H. Imai, J. Furukawa, Makromol. Chem. 79, 221 (1964).

824. M. K. Saha, P. Ghosh, S. R. Palit, J. Polymer Sci. A2, 1365 (1964).

825. Y. Saheki, K. Negoro, Hiroshima Daigaku Kogakubu Kenkyu Hokoku 13, 91 (1965); from CA 64, 2231A (1966).

826. G. Saini, G. Polla-Mattiot, M. Meirone, J. Polymer Sci. 50, S-13 (1961).

827. F. Sakaguchi, R. Kitamaru, W. Tsuji, Bull. Inst. Chem. Res. Kyoto Univ. 43, 455 (1965); from CA 65, 862C (1966).

828. S. Sakai, H. Ito, Y. Ishii, Kogyo Kagaku Zasshi 71, 186 (1968).

829. N. Sakota, K. Nishihara, Kogyo Kagaku Zasshi 71, 276 (1968).

830. I. Sakurada, Kobunshi Kagaku 18, 496 (1961); from Makromol. Chem. 49, 252 (1961).

831. I. Sakurada, N. Ise, Y. Hayashi, M. Nakao, Macromolecules 1, 265 (1968).

832. I. Sakurada, K. Noma, Y. Ofuji, Kobunshi Kagaku 23, 224 (1966); from CA 66, 2843H (1967).

833. I. Sakurada, T. Okada, S. Hatakeyama, F. Kimura, J. Polymer Sci. C1, 1233 (1963).

834. I. Sakurada, G. Takahashi, Kobunshi Kagaku 11, 286 (1954); from CA 50, 602A (1956).

835. I. Sakurada, G. Takahashi, H. Mata, Kobunshi Kagaku 12, 362 (1955); from Chem. Zentr. 131, 7869 (1960).

836. F. Sakurai, C. Huang, Kogyo Kagaku Zasshi 61, 1629 (1958); from CA 55, 27964D (1961).

837. H. Sakurai, K. Tominaga, M. Kumada, Bull. Chem. Soc. Japan 39, 1279 (1966).

838. C. L. Sandberg, F. A. Bovey, J. Polymer Sci. 15, 553 (1955).

839. S. R. Sandler, J. Dannin, K. C. Tsou, J. Polymer Sci. A3, 3199 (1965).

840. B. Sandner, J. Ulbricht, Faserforsch. Textiltech. 17, 286 (1966).

841. K. Saotome, M. Imoto, Kobunshi Kagaku 15, 368 (1958); from J. Polymer Sci. 31, 208 (1958).

842. R. Sattelmeyer, K. Hamann, Makromol. Chem. 107, 1 (1967).

843. R. C. Schulz, H. Cherdron, W. Kern, Makromol. Chem. 28, 197 (1958).

844. R. C. Schulz, E. Kaiser, Makromol. Chem. 86, 80 (1965).

845. R. C. Schulz, E. Kaiser, W. Kern, Makromol. Chem. 58, 160 (1962).

846. R. C. Schulz, R. Wolf, Kolloid-Z. -Z. Polymere 220, 148 (1967).

847. T. C. Schwan, Proc. Indiana Acad. Sci. 74, 189 (1964).

848. T. C. Schwan, C. C. Price, Proc. Indiana Acad. Sci. 63, 103 (1953).

849. C. E. Scott, C. C. Price, J. Am. Chem. Soc. 81, 2670 (1959).

850. C. E. Scott, C. C. Price, J. Am. Chem. Soc. 81, 2672 (1959).

851. G. P. Scott, J. Org. Chem. 20, 736 (1955).

852. N. M. Seidov, M. A. Dalin, D. A. Koptev, Dokl. Akad. Nauk Azerb. SSR 23(3), 16 (1967); from CA 68, 69918B (1968).

853. N. M. Seidov, D. A. Koptev, Azerb. Khim. Zh. 1967, 101; from CA 69, 28386D (1968).

854. H. Senda, R. Oda, Kobunshi Kagaku 7, 150 (1951); from CA 47, 345C (1953).

855. P. K. Sengupta, A. R. Mukherjee, P. Ghosh, J. Macromol. Chem. 1, 481 (1966).

856. M. M. Sharabash, Dissertation Abstr. B27, 2301 (1967).

857. F. F. Shcherbina, I. P. Fedorova, A. P. Ivanova, Vysokomolekul. Soedin., Ser. A, 9, 2615 (1967).

858. I. A. Shefer, L. D. Budovskaya, E. N. Rostovskii, Vysokomolekul. Soedin., Ser. B, 9, 598 (1967).

859. A. P. Sheinker, A. D. Abkin, Tr. Tashkentsk. Konf. po Mirnomu Ispol Z. at. Energii, Akad. Nauk Uz. SSR 1, 395 (1961); from CA 57, 3608E (1962).

860. A. P. Sheinker, M. K. Yakovleva, B. V. Kristal'nyi, A. D. Abkin, Dokl. Adak. Nauk SSSR 124, 632 (1959); from Resins, Rubbers, Plastics 13, 659 (1959).

861. T. V. Sheremeteva, G. N. Larina, Dokl. Akad. Nauk SSSR 162, 1323 (1965); from CA 63, 11713C (1965).

862. M. Shima, D. N. Bhattacharyya, J. Smid, M. Szwarc, J. Am. Chem. Soc. 85, 1306 (1963).

863. M. Shima, A. Kotera, J. Polymer Sci. A1, 1115 (1963).

864. H. Shimasaki, Nippon Kagaku Zasshi 87, 472 (1966); from CA 65, 15590D (1966).

865. A. Shimizu, T. Otsu, M. Imoto, Bull. Chem. Soc. Japan 38, 1535 (1965).

866. T. Shimomura, Y. Kuwabara, E. Tsuchida, I. Shinohara, Kogyo Kagaku Zasshi 71, 283 (1968).

867. T. Shiota, Y. Goto, K. Hayashi, S. Okamura, J. Appl. Polymer Sci. 11, 791 (1967).

868. V.V. Shneiderova, Lakokrasochnye Materiali i ikh Primenenie 1962(4), 33; from CA 58, 579D (1963).

869. M. F. Shostakovskii, A. M. Khomutov, A. P. Alimov, Izv. Akad. Nauk SSSR, Ser. Khim. 1961, 706; from CA 55, 27021 (1961).

870. M. F. Shostakovskii, A. M. Khomutov, A. P. Alimov, Izv. Akad. Nauk SSSR, Ser. Khim. 1963, 1839; from CA 60, 4259G (1964).

871. M. F. Shostakovskii, A. M. Khomutov, F. P. Sidel'kovskaya, Izv. Akad. Nauk SSSR, Ser. Khim. 1961, 2222; from CA 56, 10387B (1962).

872. M. F. Shostakovskii, E. N. Prilezhaeva, J. M. Karavaeva, Vysoko-molekul. Soedin. 1, 781 (1959).

873. M. F. Shostakovskii, G. G. Skvortsova, E. S. Domnina, N. P. Glazkova, T. V. Kashik, Vysokomolekul. Soedin., Ser. A, 9, 2161 (1967).

874. M. F. Shostakovskii, G. G. Skvortsova, M. Ya. Samoilova, Vysoko-molekul. Soedin. 5, 966 (1963).

875. M. F. Shostakovskii, G. G. Skvortsova, K. V. Zapunnaya, V. G. Kozyrev, Vysokomolekul. Soedin., Ser. A, 9, 704 (1967).

876. G. A. Shtraikhman, A. A. Vansheidt, G. A. Petrova, Zh. Fiz. Khim. 32, 512 (1958); from CA 52, 14299A (1958).

877. A. M. Shur, M. M. Filimonova, B. F. Filimonov, Vysokomolekul. Soedin., Ser. A, 9, 2193 (1967).

878. W. P. Shyluk, J. Polymer Sci. A2, 2191 (1964).

879. D. Sianesi, G. Caporiccio, Am. Chem. Soc., Div. Polymer Chem., Preprints 7(2), 1104 (1966); see also J. Polymer Sci., A-1, 6, 335 (1968).

880. D. Sianesi, G. Pajaro, F. Danusso, Chim. Ind. (Milan) 41, 1176 (1959).

881. F. P. Sidel'kovskaya, M. A. Askarov, F. Ibragimov, Vysokomolekul. Soedin. 6, 1810 (1964).

882. F. P. Sidel'kovskaya, M. F. Shostakovskii, F. Ibragimov, M. A. Askarov, Vysokomolekul. Soedin 6, 1585 (1964).

883. P. Sigwalt, E. Marechal, European Polymer J. 2, 15 (1966).

884. I. Simek, L. Komora, Chem. Zvesti 17, 757 (1963); from CA 60, 9362G (1964).

885. R. Simha, L. A. Wall, J. Res. Natl. Bur. Std., A 41, 521 (1948).

886. C. Simionescu, N. Asandei, A. Liga, Makromol. Chem. 110, 278 (1967).

887. J. A. Simms, J. Appl. Polymer Sci. 5, 58 (1961).

888. S. N. Sividova, A. A. Avetisyan, G. S. Kolesnikov, F. P. Sidel'-kovskaya, A. S. Tevlina, Vysokomolekul. Soedin. 7, 2164 (1965).

889. T. Skwarski, T. Wodka, Zesz. Nauk. Politech. Lodz, Wlok. No. 16, 55 (1967); from CA 68, 115003Z (1968).

890. G. Smets, L. Dehaes, Bull. Soc. Chim. Belges 59, 13 (1950).

891. G. Smets, N. Deval, P. Hous, J. Polymer Sci. A2, 4835 (1964).

892. G. Smets, E. Dysseleer, Makromol. Chem. 91, 160 (1966).

893. G. Smets, A. M. Hesbain, J. Polymer Sci. 40, 217 (1959).

894. G. Smets, A. Poot, G. L. Duncan, J. Polymer Sci. 54, 65 (1961).

895. G. Smets, A. Reckers, Rec. Trav. Chim. 68, 983 (1949).

896. G. Smets, W. van Humbeeck, J. Polymer Sci. A1, 1227 (1963).

897. J. Smid, M. Szwarc, J. Polymer Sci. 61, 31 (1962).

898. I. N. Smirnova, V. A. Krol, B. A. Dolgoplosk, Dokl. Akad. Nauk SSSR 177, 647 (1967); from CA 68, 50755Z (1968).

899. W. E. Smith, G. E. Ham, H. D. Anspon, S. E. Gebura, D. W. Al-wani, J. Polymer Sci. A-1, 6, 2001 (1968).

900. L. B. Sokolov, A. D. Abkin, Zh. Fiz. Khim. 33, 1387 (1959); from CA 54, 10386D (1960).

901. M. F. Sorokin, M. M. Babkina, Vysokomolekul. Soedin. 7, 734 (1965).

902. M. F. Sorokin, I. M. Kochnov, Plasticheskie Massy 1963 (1), 7.

903. M. F. Sorokin, I. M. Kochnov, Vysokomolekul. Soedin. 6, 798 (1964).

904. M. F. Sorokin, V. K. Latov, Zh. T. Korkishko, Z. A. Kochnova, Plasticheskie Massy 1963(5), 11.

905. M. F. Sorokin, K. A. Lyalyushko, R. A. Dudakova, V. S. Vasil'ev, A. N. Shuvalova, Plasticheskie Massy 1963(3), 3.

906. S. S. Spasskii, M. E. Mat'kova, Zh. Obshch. Khim. 29, 3438 (1959).

907. S. S. Spasskii, A. I. Tarasov, Zh. Obshch. Khim. 30, 257 (1960).

908. S. S. Spasskii, A. V. Tokarev, S. A. Mikhailova, T. V. Molchanova, M. E. Mat'kova, Zh. Obshch. Khim. 30, 250 (1960).

909. Yu. L. Spirin, A. A. Arest-Yokubovich, D. K. Polyakova, A. R. Gantmakher, S. S. Medvedev, J. Polymer Sci. 58, 1181 (1962).

910. Yu. L. Spirin, A. R. Gantmakher, S. S. Medvedev, Dokl. Akad. Nauk SSSR 128, 1232 (1959); from CA 55, 26518E (1961).

911. Yu. L. Spirin, D. K. Polyakov, A. R. Gantmakher, S. S. Medvedev, Vysokomolekul. Soedin. 2, 1082 (1960).

912. H. M. Spurlin, Hercules Powder Co., private communication.

913. H. Sumitomo, K. Azuma, J. Polymer Sci., B, 4, 883 (1966).

914. I.-S. Sun, H.-T. Feng, Ko Fen Tzu T'ung Hsun 8, 59 (1966); from CA 65, 20220H (1966).

915. A. P. Suprin, V. V. Korshak, N. V. Klimentova, Vysokomolekul. Soedin., Ser. B, 9, 377 (1967).

916. S. Suzuki, H. Tatemichi, Kogyo Kagaku Zasshi 56, 870 (1953); from CA 48, 14288 (1954).

917. Y. Suzuki, Y. Minoura, S. Nishikawa, H. Kishimoto, Kogyo Kagaku Zasshi 70, 746 (1967).

918. Y. Tabata, Y. Hashizume, H. Sobue, J. Polymer Sci. A2, 2647 (1964).

919. Y. Tabata, Y. Hashizume, H. Sobue, J. Polymer Sci. A2, 3649 (1964).

920. Y. Tabata, K. Ishigure, K. Oshima, Makromol. Chem. 85, 91 (1965).

921. Y. Tabata, K. Ishigure, K. Oshima, H. Sobue, J. Polymer Sci. A2, 2445 (1964).

922. Y. Tabata, K. Ishigure, H. Sobue, Kogyo Kagaku Zasshi 65, 1626 (1962); from CA 58, 6931F (1963).

923. Y. Tabata, K. Ishigure, H. Sobue, J. Polymer Sci. A2, 2235 (1964).

924. K. Tada, Y. Numato, T. Saegusa, J. Furukawa, Makromol. Chem. 77, 220 (1964).

925. K. Tada, T. Saegusa, J. Furukawa, Kogyo Kagaku Zasshi 66, 1501 (1963); from CA 61, 727F (1964).

926. K. Tada, T. Saegusa, J. Furukawa, Makromol. Chem. 71, 71 (1964).

927. K. Tada, T. Saegusa, J. Furukawa, Makromol. Chem. 102, 47 (1967).

928. T. Tadokoro, H. Konishi, Kogyo Kagaku Zasshi 69, 511 (1966).

929. A. Takahashi, H. Tanaka, I. Kagawa, Kogyo Kagaku Zasshi 70, 988 (1967).

930. K. Takahashi, S. Abe, K. Namba, J. Appl. Polymer Sci. 12, 1683 (1968).

931. K. Takakura, K. Hayashi, S. Okamura, J. Polymer Sci., A-1, 4, 1747 (1966).

932. G. Takashi, Kobunshi Kagaku 14, 151 (1957); from CA 52, 1670C (1948).

933. M. Takebayashi, Y. Ito, Bull. Chem. Soc. Japan 29, 287 (1956).

934. K. Takemoto, M. Izubayashi, Makromol. Chem. 109, 81 (1967).

935. K. Takemoto, Y. Kikuchi, M. Imoto, Kogyo Kagaku Zasshi 69, 1367 (1966).

936. K. Takemoto, K. Matsuoka, M. Otsuka, M. Imoto, Kogyo Kagaku Zasshi 69, 2026 (1966).

937. K. Takemoto, M. Otsuka, K. Matsuoka, M. Imoto, Kobunshi Kagaku 23, 570 (1966); from CA 67, 64746P (1967).

938. G. Talamini, G. Vidotto, C. Garbuglio, Chim. Ind. (Milan) 47, 955 (1965).

939. T. Tamikado, J. Polymer Sci. 43, 489 (1960).

940. T. Tamikado, Makromol. Chem. 38, 85 (1960).

941. T. Tamikado, Y. Iwakura, J. Polymer Sci. 36, 529 (1959).

942. T. Tamikado, T. Sakai, K. Sagisaka, Makromol. Chem. 50, 244 (1961).

943. H. Tanabe, K. Nakano, Yakuzaigaku 22, 193 (1962); from CA 59 7656G (1963).

944. A. Tanaka, Y. Hozumi, K. Hatada, Kobunshi Kagaku 22, 216 (1965); from CA 63, 14993F (1965).

945. A. Tanaka, K. Sasaki, Y. Hozumi, O. Hashimoto, J. Appl. Polymer Sci. 8, 1787 (1964).

946. M. Tanaka, S. Asai, S. Takeya, Kogyo Kagaku Zasshi 62, 1786 (1959); from CA 57, 13972I (1962).

947. M. Taniyama, G. Oster, Bull. Chem. Soc. Japan 30, 856 (1957).

948. H. Tatemichi, S. Suzuki, Kogyo Kagaku Zasshi 63, 1843 (1960); from CA 56, 13084 (1962).

949. S. Tazuke, K. Nakagawa, S. Okamura, J. Polymer Sci. B3, 923 (1965).

950. S. Tazuke, A. Nakamura, Makromol. Chem. 95, 92 (1966).

951. S. Tazuke, S. Okamura, J. Polymer Sci., A-1, 5, 1083 (1967).

952. S. Tazuke, N. Sato, S. Okamura, J. Polymer Sci., A-1, 4, 2461 (1966).

953. R. A. Terteryan, A. I. Dintses, M. V. Rysakov, Neftekhimiya 3(5), 719 (1963); from CA 62, 1749D (1965).

954. R. W. Tess, W. T. Tsatsos, Am. Chem. Soc., Div. Org. Coatings Plast. Chem., Preprints 26(2), 276 (1966).

955. W. M. Thomas, M. T. O'Shaughnessy, J. Polymer Sci. 11, 455 (1953).

956. W. M. Thomas, W. J. Sparks, U. S. Patent 2,356,128 (1944); calc. by F. R. Mayo and C. Walling.

957. B. R. Thompson, J. Polymer Sci. 19, 373 (1956).

958. B. R. Thompson, R. H. Raines, J. Polymer Sci. 41, 265 (1959).

959. B. R. Thompson, R. H. Raines, J. Polymer Sci. 54, S-31 (1961).

960. J. R. Tichy, J. Polymer Sci. 33, 353 (1958).

961. G. V. Tkachenko, V. S. Etlis, L. V. Stupen, L. P. Kofman, Zh. Fiz. Khim. 33, 25 (1959); from CA 54, 11557E (1960).

962. G. V. Tkachenko, P. H. Khomikovskii, A. D. Abkin, S. S. Medvedev, Zh. Fiz. Khim. 31, 242 (1957).

963. G. V. Tkachenko, L. V. Stupen, L. P. Kofman, L. Z. Frolova, Zh. Fiz. Khim. 31, 2676 (1957); from CA 52, 8614D (1958).

964. G. V. Tkachenko, L. V. Stupen, L. P. Kofman, L. A. Karacheva, Zh. Fiz. Khim. 32, 2492 (1958).

965. A. V. Tobolsky, R. J. Boudreau, J. Polymer Sci. 51, S-53 (1961).

966. A. V. Tokarev, S. S. Spasskii, Zh. Fiz. Khim. 33, 544 (1959); from CA 53, 20900I (1959).

967. M. Tokarzewska, J. Golusinska, Vysokomolekul. Soedin. 6, 2093 (1964).

968. N. Tokura, M. Matsuda, M. Iino, Bull. Chem. Soc. Japan 36, 278 (1963).

969. J. Tolle, P. Wittmer, H. Gerrens, Makromol. Chem. 113, 23 (1968).

970. D. A. Topchiev, V. G. Popov, M. V. Shishkina, V. A. Kabanov, V. A. Kargin, Vysokomolekul. Soedin. 8, 1767 (1966).

971. V. L. Tsailingol'd, M. I. Farberov, G. A. Burgova, Vysokomolekul. Soedin. 1, 415 (1959).

972. E. Tsuchida, T. Kawagoe, Enka Biniiru to Porima 8, 21 (1968); from CA 69, 36467W (1968).

973. E. Tsuchida, Y. Ohtani, H. Nakadai, I. Shinohara, Kogyo Kagaku Zasshi 70, 573 (1967).

974. E. Tsuchida, Z. Okuno, T. Yao, I. Shinohara, Kogyo Kagaku Zasshi 69, 1230 (1966).

975. E. Tsuchida, T. Shimomura, K. Fujimori, Y. Ohtani, I. Shinohara, Kogyo Kagaku Zasshi 70, 566 (1967).

976. K. Tsuda, S. Kobayashi, T. Otsu, J. Polymer Sci., A-1, 6, 41 (1968).

977. T. Tsuda, T. Nomura, Y. Yamashita, Makromol. Chem. 86, 301 (1965).

978. T. Tsuda, T. Shimizu, Y. Yamashita, Kogyo Kagaku Zasshi 67, 1661, (1964); from CA 62, 10519C (1965).

979. T. Tsuda, T. Shimizu, Y. Yamashita, Kogyo Kagaku Zasshi 67, 2145 (1964).

980. T. Tsuda, T. Shimizu, Y. Yamashita, Kogyo Kagaku Zasshi 67, 2150 (1964); from CA 62, 14831H (1965).

981. T. Tsuda, T. Shimizu, Y. Yamashita, Kogyo Kagaku Zasshi 68, 2473 (1965).

982. T. Tsuda, Y. Yamashita, Kogyo Kagaku Zasshi 65, 811 (1962); from CA 57, 15344 G (1962).

983. Y. Tsuda, Kogyo Kagaku Zasshi 62, 1112 (1959); from CA 57, 13972E (1962).

984. A. Ya. Tsybul'ko, Yu. S. Lipatov, T. E. Lipatova, Vestsi Akad. Nauk Belarusk. SSR, Ser. Khim. Nauk 1965(4), 75; from CA 64, 19797A (1966).

985. A. Ya. Tsybul'ko, T. E. Lipatova, Yu. S. Lipatov, Vysokomolekul. Soedin. 7, 1626 (1965).

986. N. A. Tyukavkina, A. V. Kalabina, G. I. Deryabina, G. T. Zhikharev, A. D. Biryukova, Vysokomolekul. Soedin. 6, 1573 (1964).

987. K. Ueberreiter, W. Krull, Z. Physik. Chem. (Frankfurt) 12, 303 (1957).

988. K. Uehara, T. Nishi, T. Tsuyuri, F. Tamura, N. Murata, Kogyo Kagaku Zasshi 70, 750 (1967).

989. K. Ueno, K. Hayashi, S. Okamura, J. Polymer Sci. B3, 363 (1965).

990. J. Ulbricht, J. Giesemann, M. Gebauer, Angew. Makromol. Chem. 3, 69 (1968).

991. A. Ulinska, Z. Mankowski, Studia Soc. Sci. Torun., Sect. B 2(2), 16 (1960); from CA 59, 5270D (1963).

992. K. Uno, M. Makita, S. Doi, Y. Iwakura, J. Polymer Sci., A-1, 6, 257 (1968).

993. C. C. Unruh, T. M. Laakso, J. Polymer Sci. 33, 87 (1958); Ind. Eng. Chem. 50, 1124 (1958).

994. S. N. Ushakov, S. S. Ivanov, Izv. Akad. Nauk SSSR, Ser. Khim. 1957, 1465.

995. S. N. Ushakov, O. M. Klimova, O. S. Karchamarchik, E. M. Smul'skaya, Dokl. Akad. Nauk SSSR 143, 231 (1962); from CA 57, 3620I (1962).

996. S. M. Ushakov, V. A. Kropachev, L. B. Trukhmanova, R. I. Gruz, T. M. Markelova, Vysokomolekul. Soedin., Ser. A, 9, 1807 (1967).

997. S. N. Ushakov, E. M. Lavrent'eva, Zh. Prikl. Khim. 31, 1686 (1958); from CA 53, 4811A (1959).

998. S. N. Ushakov, S. P. Mitsengendler, B. M. Polyatskina, Zh. Prikl. Khim. 23, 512 (1950); from CA 46, 8893A (1952).

999. S. N. Ushakov, S. P. Mitsengendler, B. M. Polyatskina, Zh. Prikl. Khim. 24, 289 (1951); from CA 47, 7820D (1953).

1000. S. N. Ushakov, S. P. Mitsengendler, B. M. Polyatskina, Khim. i Fiz. Khim. Vysokomolekul. Soedin, Dokl. Konf. po Vysokomolekul. Soedin., 7-ya Konf. 1952, 59; from CA 47, 7820 (1953).

1001. S. N. Ushakov, A. F. Nikolaev, Izv. Akad. Nauk SSSR, Ser. Khim. 1956, 83; from CA 50, 13867 (1956).

1002. S. N. Ushakov, L. B. Trukhmanova, Izv. Akad. Nauk SSSR, Ser. Khim. 1957, 980; from CA 52, 42371 (1958).

1003. S. N. Ushakov, L. B. Trukhmanova, Vysokomolekul. Soedin. 1, 1754 (1959).

1004. S. N. Ushakov, L. B. Trukhmanova, T. M. Markelova, V. A. Kropachev, Vysokomolekul. Soedin., Ser. A, 9, 999 (1967).

1005. Kh. U. Usmanov, R. S. Tillaev, U. N. Musaev, Nauchn. Tr., Tashkentsk. Gos. Univ. No. 257, 3 (1964); from CA 63, 3049G (1965).

1006. Kh. U. Usmanov, R. S. Tillaev, U. N. Musaev, M. M. Ishanov, Nauchn. Tr., Tashkentsk. Gos. Univ. No. 257, 30 (1964); from CA 63, 3049D (1965).

1007. Kh. U. Usmanov, A. A. Yul'chibaev, Kh. Yuldasheva, Uzb. Khim. Zh. 11, 27 (1967); from CA 67, 32977Y (1967).

1008. Kh. U. Usmanov, A. A. Yul'chibaev, Kh. Yuldasheva, Uzb. Khim. Zh. 11, 41 (1967); from CA 67, 91133T (1967).

1009. N. F. Usmanova, V. M. Bulatova, V. N. Stepanova, V. V. Kerzhkovskaya, T. N. Kalacheva, A. V. Golubeva, Plasticheskie Massy 1967(6), 9; from CA 68, 59920G (1968).

1010. L. F. Vanderburgh, C. E. Brockway, J. Polymer Sci. A3, 575 (1965).

1011. J. W. Vanderhoff, in E. H. Riddle, "Monomeric Acrylic Esters", Reinhold, New York, 1954, p. 94, Ref. 154.

1012. G. van Paesschen, G. Smets, Bull. Soc. Chim. Belges 64, 173 (1955).

1013. G. van Paesschen, D. Timmerman, Makromol. Chem. 78, 112 (1964).

1014. S. V. Vinogradova, V. V. Korshak, M. G. Korchevei, Vysokomolekul. Soedin. 7, 1889 (1965).

1015. S. V. Vinogradova, V. V. Korshak, V. I. Kul'chitskii, B. V. Lokshin, G. Z. Mirkin, Vysokomolekul. Soedin., Ser. A, 10, 1108 (1968).

1016. A. I. Volkova, P. A. El'tsova, M. M. Koton, A. V. Mikhailova, Vysokomolekul. Soedin. 8, 1997 (1966); from CA 66, 29266Q (1967).

1017. E. L. Vollershtein, M. N. Shvarts, S. I. Beilin, B. A. Dolgoplosk, Dokl. Akad. Nauk SSSR 177, 1088 (1967); from CA 68, 50096K (1968).

1018. B. Vollmert, Angew. Makromol. Chem. 3, 1 (1968).

1019. M. Vrancken, G. Smets, Makromol. Chem. 30, 197 (1969).

1020. F. T. Wall, quoted by F. R. Mayo and C. Walling.

1021. F. T. Wall, R. E. Florin, C. J. Delbecq, J. Am. Chem. Soc. 72, 4769 (1950).

1022. F. T. Wall, R. W. Powers, G. D. Sands. G. S. Stent, J. Am. Chem. Soc. 70, 1031 (1948).

1023. C. Walling, E. R. Briggs, K. B. Wolfstirn, J. Am. Chem. Soc. 70, 1543 (1948).

1024. C. Walling, F. R. Briggs, K. B. Wolfstirn, F. R. Mayo, J. Am. Chem. Soc. 70, 1537 (1948).

1025. C. Walling, J. A. Davison, J. Am. Chem. Soc. 73, 5736 (1951).

1026. H. Watanabe, K. Hayashi, S. Okamura, Nippon Hoshasen Kobunshi Kenkyu Kyokai Nempo 4, 127 (1962); from CA 62, 9242B (1965).

1027. G. Welzel, G. Greber, Makromol. Chem. 31, 230 (1959).

1028. H. Wesslau, Angew. Makromol. Chem. 1, 56 (1967).

1029. R. L. Whistler, J. L. Goatley, J. Polymer Sci. 50, 127 (1961).

1030. G. S. Wich, N. Brodoway, J. Polymer Sci. A1, 2163 (1963).

1031. O. Wichterle, V. Gregor, J. Polymer Sci. 34, 309 (1959).

1032. O. Wichterle, J. Zelinka, Chem. Listy 51, 2146 (1957); from CA 52, 3395 (1958).

1033. M. C. de Wilde, G. Smets, J. Polymer Sci. 5, 253 (1950).

1034. R. H. Wiley, T.-O. Ahn, J. Polymer Sci., A-1, 6, 1293 (1968).

1035. R. H. Wiley, B. Davis, J. Polymer Sci. 46, 423 (1960).

1036. R. H. Wiley, B. Davis, J. Polymer Sci. 62, S-132 (1962).

1037. R. H. Wiley, B. Davis, J. Polymer Sci. 62, S-140 (1962).

1038. R. H. Wiley, B. Davis, J. Polymer Sci. A1, 2819 (1963).

1039. R. H. Wiley, B. Davis, J. Polymer Sci. B1, 463 (1963).

1040. R. H. Wiley, L. K. Heidemann, B. Davis, J. Polymer Sci. B1, 521 (1963).

1041. R. H. Wiley, W. K. Mathews, K. F. O'Driscoll, J. Macromol. Sci. (Chem.) A1, 503 (1967).

1042. R. H. Wiley, G. L. Mayberry, J. Polymer Sci. A1, 217 (1963).

1043. R. H. Wiley, E. E. Sale, J. Polymer Sci. 42, 497 (1960).

1044. R. H. Wiley, E. E. Sale, J. Polymer Sci. 42, 491 (1960).

1045. R. H. Wiley, W. A. Trinler, J. Polymer Sci. 28, 163 (1958).

1046. C. E. Wilkes, J. C. Westfahl, R. H. Backderf, Am. Chem. Soc., Div. Polymer Chem., Preprints 8(1), 386 (1967); see also J. Polymer Sci., A-1, 7, 23 (1969).

1047. A. Winston, D. A. Chapman, G. T. C. Li, Am. Chem. Soc., Div. Polymer Chem., Preprints 9(1), 41 (1968).

1048.   A. Winston, F. L. Hamb, J. Polymer Sci. A3, 583 (1965).

1049.   A. Winston, G. T. C. Li, J. Polymer Sci., A-1, 5, 1223 (1967).

1050.   L. P. Witnauer, E. Watkins, W. S. Port, J. Polymer Sci. 20, 213 (1956).

1051.   P. Wittmer, Makromol. Chem. 103, 188 (1967).

1052.   T. Wojnarowski, Polimery 12, 509 (1967); from CA 69, 3165T (1968).

1053.   D. M. Woodford, Chem. Ind. (London), 1966, 316.

1054.   D. J. Worsfold, J. Polymer Sci., A-1, 5, 2783 (1967).

1055.   M. Yamada, I. Takase, Kobunshi Kagaku 18, 85 (1961); from CA 55, 24100D (1961).

1056.   M. Yamada, I. Takase, Kobunshi Kagaku 22, 626 (1965); from Makromol. Chem. 93, 300 (1966).

1057.   M. Yamada, I. Takase, Kogyo Kagaku Zasshi 71, 572 (1968).

1058.   T. Yamaguchi, T. Ono, T. Yanoshita, R. Shishido, Kobunshi Kagaku 22, 835 (1965); from CA 64, 17721H (1966).

1059.   T. Yamamoto, Kogyo Kagaku Zasshi 62, 476 (1959).

1060.   H. Yamaoka, R. Uchida, K. Hayashi, S. Okamura, Kobunshi Kagaku 24, 79 (1967).

1061.   Y. Yamashita, T. Tsuda, M. Okada, S. Iwatsuki, J. Polymer Sci., A-1, 4, 2121 (1966).

1062.   N. Yamazaki, M. Aridomi, S. Maeda, S. Kambara, Kogyo Kagaku Zasshi 70, 1989 (1967).

1063.   N. Yamazaki, S. Kambara, J. Polymer Sci., C, 22, 75 (1968).

1064.   N. Yamazaki, T. Suminoe, T. Furuhama, S. Kambara, Kogyo Kagaku Zasshi 64, 1687 (1961); from CA 57, 3601B (1962).

1065.   K. Yasufuku, S. Hirose, S. Nozakura, S. Murahashi, Bull. Chem. Soc. Japan 40, 2139 (1967).

1066.   K. Yasufuku, S. Nozakura, S. Murahashi, Bull. Chem. Soc. Japan 40, 2146 (1967).

1067.   K. Yokota, Y. Ishii, Kogyo Kagaku Zasshi 69, 1057 (1966).

1068.   K. Yokota, M. Kani, Y. Ishii, J. Polymer Sci., A-1, 6, 1325 (1968).

1069.   M. Yoshida, I. Sakurada, Kobunshi Kagaku 7, 334 (1950).

1070.   L. J. Young, unpublished data, quoted in J. Polymer Sci. 54, 411 (1961).

1071.   S. Yuguchi, H. Kiuchi, M. Watanabe, Kobunshi Kagaku 18, 510 (1961); from Makromol. Chem. 49, 252 (1961).

1072.   S. Yuguchi, M. Watanabe, Kobunshi Kagaku 15, 129 (1958).

1073.   S. Yuguchi, M. Watanabe, Kobunshi Kagaku 18, 386 (1961); from Makromol. Chem. 49, 243 (1961).

1074.   S. Yuguchi, M. Watanabe, Kobunshi Kagaku 18, 613 (1961); from Makromol. Chem. 51, 246 (1962).

1075.   H. Yuki, K. Kosai, S. Murahashi, J. Hotta, J. Polymer Sci. B2, 1121 (1964).

1076.   E. V. Zabolotskaya, V. A. Khodzhemirov, A. R. Gantmakher, S. S. Medvedev, Izv. Akad. Nauk SSSR, Ser. Khim. 140, 964 (1961).

1077.   E. V. Zabolotskaya, V. A. Khodzhemirov, A. R. Gantmakher, S. S. Medvedev, Vysokomolekul. Soedin. 6, 81 (1964).

1078.   B. A. Zaitsev, G. A. Shtraikhman, Vysokomolekul. Soedin., Ser. A, 10, 438 (1968).

1079.   A. Zambelli, A. Lety, C. Tosi, I. Pasquon, Makromol. Chem. 115, 73 (1968).

1080.   M. A. Zharkova, G. I. Kudryavtsev, Khim. Volokna 3, 15 (1960); from CA 54, 23419H (1960).

1081.   N. L. Zutty, Union Carbide Chemicals Co., private communication.

1082.   N. L. Zutty, J. Polymer Sci. A1, 2231 (1963).

1083.   N. L. Zutty, R. D. Burkhart, Advances in Chem. Series No. 34, "Polymerization and Condensation Processes", ACS, Washington D.C., 1962, pp. 52-9.

1084.   N. L. Zutty, F. J. Welch, J. Polymer Sci. 43, 447 (1960).

1085.   M. P. Zverev, M. F. Margaritova, Ukr. Khim. Zh. 24, 626 (1958); from CA 53, 10823F (1959).

1086.   S. Abe, M. Ito, K. Namba, Makromol. Chem. 134, 121 (1970).

1087.   M. Ya. Agakishieva, N. M. Seidov, T. A. Kuliev, Azerb. Khim. Zh. 1969, 111; from CA 74, 31989 (1971).

1088.   M. Akiyama, Y. Yanagisawa, M. Okawara, J. Polymer Sci., A-1, 7, 1905 (1969).

1089.   D. D. Al-Khashimi, T. R. Abdurashidov, M. A. Askarov, Uzb. Khim. Zh. 14, 87 (1970); from CA 74, 76731N (1971).

1090.   A. B. Alovitdinov, D. K. Khamdamova, A. I. Kurbanov, A. B. Kuchkarov, Vysokomolekul. Soedin., Ser. B, 12, 886 (1970).

1091.   A. B. Alovitdinov, A. B. Kuchkarov, A. I. Kurbanov, Vysokomolekul. Soedin., Ser. B, 13, 657 (1971).

1092.   K. Anda, S. Iwai, J. Polymer Sci., A-1, 7, 2414 (1969).

1093.   K. Anda, S. Iwai, Kobunshi Kagaku 29, 212 (1972).

1094.   A. Anton, E. Marechal, Bull. Soc. Chim. France 1971, 3753.

1095.   A. Anton, J. Zwegers, E. Marechal, Bull. Soc. Chim. France 1970, 1466.

1096.   J. Aoyagi, K. Kitamura, I. Shinohara, Kogyo Kagaku Zasshi 73, 2045 (1970).

1097.   J. Aoyagi, I. Shinohara, Kogyo Kagaku Zasshi 74, 1191 (1971).

1098.   J. Aoyagi, I. Shinohara, Nippon Kagaku Kaishi 1972, 788.

1099.   B. Apsite, R. Pernikis, J. Suna, Latv. Psr Zinat. Akad. Vestis, Kim. Ser. 1970, (2), 233; from CA 73, 25924Z (1970).

1100.   A. A. Durgaryan, A. S. Grigoryan, Arm. Khim. Zh. 21, 137 (1968); from CA 69, 44222C (1968).

1101.   K. Yokota, M. Sasaki, Y. Ishii, J. Polymer Sci., A-1, 6, 2935 (1968).

1102.   S. Tazuke, S. Okamura, J. Polymer Sci., A-1, 6, 2907 (1968).

1103.   K. Uno, M. Ohara, H. G. Cassidy, J. Polymer Sci., A-1, 6, 2729 (1968).

1104.   Z. Jedlinski, J. Paprotny, J. Polymer Sci., C, 16, 3605 (1968).

1105.   J. E. Hinsch, J. G. Curro, Macromolecules 1, 405 (1968).

1106.   C. L. Schilling, jr., J. E. Mulvaney, Macromolecules 1, 445 (1968).

1107.   Y. Choshi, A. Tsuji, G. Akazome, K. Murai, Kogyo Kagaku Zasshi 71, 908 (1968).

1108.   A. A. Durgaryan, A. O. Agumyan, A. S. Grigoryan, Vysokomolekul. Soedin., Ser. B, 10, 483 (1968); from CA 69, 67731G (1968).

1109.   N. N. Loginova, R. K. Gavurina, S. I. Kulicheva, Vysokomolekul. Soedin., Ser. B, 10, 422 (1968); from CA 69, 52489K (1968).

1110.   E. Marechal, J. P. Richard, J. P. Menissez, C. Zaffran, Compt. Rend., Ser. C, 266, 1635 (1968).

1111.   K. Murata, A. Terada, J. Polymer Sci., A-1, 6, 2945 (1968).

1112.   T. Oota, T. Otsu, M. Imoto, Kogyo Kagaku Zasshi 71, 736 (1968); from CA 69, 67781Y (1968).

1113.   T. Otsu, T.-C. Lai, Y. Kinoshita, A. Nakamachi, M. Imoto, Kogyo Kagaku Zasshi 71, 904 (1968).

1114.   J. Ushirone, T. Otsu, Kogyo Kagaku Zasshi 71, 772 (1968).

1115.   K. Takemoto, S. Takahashi, M. Imoto, Kogyo Kagaku Zasshi 71, 742 (1968).

1116.   A. Mizote, T. Higashimura, S. Okamura, J. Macromol. Sci.-Chem., A2, 717 (1968).

1117.   K. A. Andrianov, L. A. Gavrikova, E. F. Rodionova, Izv. Akad. Nauk SSSR, Ser. Khim. 1968, 1786; from CA 69, 97223K (1968).

1118.   N. M. Geller, V. A. Kropachev, B. A. Dolgoplosk, Vysokomolekul. Soedin., Ser. A, 10, 1878 (1968).

1119.   T. Kagiya, S. Takayama, K. Fukui, Kogyo Kagaku Zasshi 71, 745 (1968).

1120.   K. Kurokawa, T. Kondo, Kogyo Kagaku Zasshi 71, 1087 (1968).

1121.   E. V. Kuznetsov, I. P. Prokhorova, N. I. Avvakumova, Tr. Kazan, Khim.-Tekhnol. Inst. 1967, No. 36, 441; from CA 69, 97219P (1968).

1122.   I. Ya. Ostrovskaya, K. L. Makovetskii, E. I. Tinyakova, B. A. Dolgoplosk, Dokl. Akad. Nauk SSSR 181, 892 (1968); from CA 69, 87822R (1968).

1123.   T. Ouchi, S. Yamamoto, Y. Akao, Y. Nagaoka, M. Oiwa, Kogyo Kagaku Zasshi 71, 1078 (1968).

1124.   N. M. Seidov, M. A. Dalin, S. M. Kyazimov, Azerb. Khim. Zh. 1968(1), 68; from CA 69, 87824T (1968).

1125.   M. K. Yakovleva, A. P. Sheinker, A. D. Abkin, Vysokomolekul. Soedin., Ser. A, 10, 1946 (1968).

1126. H. Yuki, M. Kato, Y. Okamoto, Bull. Chem. Soc. Japan 41, 1940 (1968).

1127. H. Nishiwaki, H. Sumitomo, T. Maeshima, Kogyo Kagaku Zasshi 71, 912 (1968).

1128. Yu. A. Gorin, K. N. Charskaya, E. I. Rodina, Vysokomolekul. Soedin., Ser. A, 10, 405 (1968).

1129. A. M. North, K. E. Whitelock, Polymer 9, 590 (1968).

1130. Y. Minoura, M. Mitch, Makromol. Chem. 119, 104 (1968).

1131. Y. Minoura, M. Mitch, Y. Mabuchi, Makromol. Chem. 119, 86 (1968).

1132. Y. Minoura, M. Mitch, Y. Mabuchi, Makromol. Chem. 119, 96 (1968).

1133. T. Otsu, K. Tsuda, T. Fukumizu, Makromol. Chem. 119, 140 (1968).

1134. K. Takemoto, M. Izubayashi, M. Imoto, Makromol. Chem. 119, 147 (1968).

1135. J. C. Bevington, M. Johnson, European Polymer J. 4, 669 (1968).

1136. B. Apsite, R. Pernikis, J. Surna, Latv. Psr Zinat. Akad. Vestis, Kim. Ser. 1971, 600; from CA 76, 72835T (1972).

1137. V. Arendt, S. Kaizerman, J. Polymer Sci., A-1, 7, 2741 (1969).

1138. C. Aso, T. Kunitake, R. K. Khattak, N. Sugi, Makromol. Chem. 134, 147 (1970).

1139. C. Aso, T. Kunitake, T. Nakashima, Makromol. Chem. 124, 232 (1969).

1140. C. Aso, T. Kunitake, T. Nakashima, Kogyo Kagaku Zasshi 72, 1411 (1969).

1141. C. Aso, O. Ohara, Makromol. Chem. 127, 78 (1969).

1142. C. Aso, S. Tagami, T. Kunitake, J. Polymer Sci., A-1, 8, 1323 (1970).

1143. I. S. Avetisyan, V. I. Eliseeva, Vysokomolekul. Soedin., Ser. B, 11, 316 (1969); from CA 71, 30719Z (1969).

1144. C. Azuma, N. Ogata, Kobunshi Kagaku 29, 254 (1972).

1145. F. C. Baines, J. C. Bevington, Polymer 11, 647 (1970).

1146. G. D. Ballova, M. M. Koton, S. G. Lyubetskii, E. I. Egorova, G. P. Fratkina, Vysokomolekul. Soedin., Ser. B, 13, 442 (1971).

1147. J. M. Barrales-Rienda, G. R. Brown, D. C. Pepper, Polymer 10, 327 (1969).

1148. S. T. Bashkatova, V. I. Kleiner, L. L. Stotskaya, B. A. Krentsel', Vysokomolekul. Soedin., Ser. A, 11, 2603 (1969).

1149. B. K. Basov, V. L. Tsailingol'd, E. G. Lazaryants, Prom. Sin. Kauch. 1967(1), 17; from CA 69, 77783Q (1968).

1150. A. A. Baturin, Yu. B. Amerik, B. A. Krentsel', Mol. Cryst. Liquid Cryst. 16, 117 (1972); from CA 76, 127543D (1972).

1151. J. C. Bevington, D. O. Harris, F. S. Rankin, European Polymer J. 6, 725 (1970).

1152. V. D. Bezuglyi, A. I. Shepeleva, T. A. Alekseeva, L. I. Smitrievskaya, L. Ya. Malkes, A. I. Nazarenko, Vysokomolekul. Soedin., Ser. A, 11, 1578 (1969).

1153. V. D. Bezuglyi, I. B. Voskresenskaya, T. A. Alekseeva, M. M. Gemer, Vysokomolekul. Soedin., Ser. A, 14, 540 (1972).

1154. L.-P. Blanchard, S. Kondo, J. Moinard, J. F. Pierson, F. Tahiani, J. Polymer Sci., A-1, 9, 3547 (1971).

1155. A. T. Blomquist, D. W. Durandetta, G. B. Robinson, J. Polymer Sci., A-1, 8, 2061 (1970).

1156. D. Braun, F.-J. Q. Lucas, W. Neumann, Makromol. Chem. 127, 253 (1969).

1157. D. Braun, G. Mott, Angew. Makromol. Chem. 18, 183 (1971).

1158. D. Braun, M. H. Tio, W. Neumann, Makromol. Chem. 123, 29 (1969).

1159. D. W. Brown, R. E. Lowry, L. A. Wall, J. Polymer Sci., A-1, 8, 2441 (1970).

1160. D. W. Brown, R. E. Lowry, L. A. Wall, J. Polymer Sci., A-1, 9, 1993 (1971).

1161. L. A. Budarina, E. V. Kuznetsov, Tr. Kazan. Khim.-Tekhnol. Inst. 40, 135 (1969); from CA 75, 98860B (1971).

1162. G. S. Buslaev, V. S. Ivanov, Vestn. Leningrad. Univ., Fiz. Khim. 1970(2), 155; from CA 73, 131761D (1970).

1163. W. R. Cabaness, T. Y.-C. Lin, C. Parkanyi, J. Polymer Sci., A-1, 9, 2155 (1971).

1164. P. Caillaud, J.-M. Huet, E. Marechal, Bull. Soc. Chim. France 1970(4), 1473.

1165. G. Caporiccio, D. Sianesi, Chim. Ind. (Milan) 52, 37 (1970).

1166. J. M. Carcamo, F. Arranz, Rev. Plast. Mod. 22 (182), 1,169 (1971).

1167. D. Carcano, M. Modena, M. Ragazzini, O. Pilati, Chim. Ind. (Milan) 53, 547 (1971).

1168. A. V. Chernobai, Zh. Kh. Zelichenko, Vysokomolekul. Soedin., Ser. A, 11, 1470 (1969).

1169. Y. Choshi, A. Tsuji, G. Akazome, K. Murai, Kogyo Kagaku Zasshi 72, 1031 (1969).

1170. K. H. Chung, Daehan Hwahak Hwoejee 14, 333 (1970); from CA 75, 118647T (1971).

1171. L. E. Coleman, J. F. Bork, J. Polymer Sci., A-1, 8, 2073 (1970).

1172. C. Cozewith, G. ver Strate, Macromolecules 4, 482 (1971).

1173. R. C. P. Cubbon, J. D. B. Smith, Polymer 10, 489 (1969).

1174. B. M. Culbertson, S. Dietz, R. D. Wilson, J. Polymer Sci., A-1, 9, 2727 (1971).

1175. B. M. Culbertson, R. E. Freis, Macromolecules 3, 715 (1970).

1176. W. H. Daly, C.-D. S. Lee, C. G. Overberger, J. Polymer Sci., A-1, 9, 1723 (1971).

1177. T. L. Dawson, R. D. Lundberg, F. J. Welch, J. Polymer Sci., A-1, 7, 173 (1969).

1178. G. I. Deryabina, Yu. L. Frolov, A. V. Kalabina, Vysokomolekul. Soedin., Ser. A, 13, 1162 (1971).

1179. G. I. Deryabina, G. I. Gershengorn, Vysokomolekul. Soedin., Ser. A, 11, 941 (1969).

1180. G. I. Deryabina, A. K. Khaliullin, S. Yu. Federova, L. A. Kron, Izv. Nauch.-Issled. Inst. Nefte-Uglekhim. Sin. Irkutsk. Univ. 12, 57 (1970); from CA 75, 64326S (1971).

1181. F. I. Duntov, B. L. Erusalimskii, Vysokomolekul. Soedin., Ser. B, 13, 539 (1971).

1182. A. A. Durgaryan, Zh. N. Terlemezyan, Arm. Khim. Zh. 23, 1057 (1970); from CA 74, 126164C (1971).

1183. M. Z. El-Sabban, S. I. Dmitrieva, A. I. Meos, Vysokomolekul. Soedin., Ser. B, 12, 243 (1970).

1184. T. Endo, R. Numazawa, M. Okawara, Kobunshi Kagaku 28, 260 (1971).

1185. T. Endo, M. Okawara, Makromol. Chem. 146, 237 (1971).

1186. J. W. H. Faber, W. F. Fowler, jr., J. Polymer Sci., A-1, 8, 1777 (1970).

1187. J. P. Fischer, Makromol. Chem. 155, 211 (1972).

1188. J. P. Fischer, Makromol. Chem. 155, 227 (1972).

1189. J. P. Fischer, W. Lüders, Makromol. Chem. 155, 239 (1972).

1190. T. M. Frunze, M. A. Surikova, V. V. Kurashev, L. Komarova, Vysokomolekul. Soedin., Ser. A, 12, 460 (1970).

1191. T. Fueno, T. Okuyama, I. Matsumura, J. Furukawa, J. Polymer Sci., A-1, 7, 1447 (1969).

1192. S. Fujii, K. Maeda, Nippon Kagaku Kaishi 1972, 215.

1193. M. Furuta, Y. Yuki, M. Nagano, Kobunshi Kagaku 28, 766 (1971).

1194. T. A. Gadzhiev, Z. M. Rzaev, S. G. Mamedova, Azerb. Khim. Zh. 1970, 91; from CA 75, 21146R (1971).

1195. G. A. Gadzhiev, Z. M. Rzaev, S. G. Mamedova, Vysokomolekul. Soedin., Ser. A, 13, 2386 (1971).

1196. R. K. Gavurina, V. G. Karkozov, O. Yu. Vybornov, Vysokomolekul. Soedin., Ser. B, 11, 589 (1969).

1197. V. R. Georgieva, V. P. Zubov, V. A. Kabanov, V. A. Kargin, Dokl. Akad. Nauk SSSR 190, 1128 (1970); from CA 72, 133263W (1970).

1198. A. L. German, D. Heikens, J. Polymer Sci., A-1, 9, 2225 (1971).

1199. A. Gilath, S. H. Ronel, M. Shmueli, D. H. Kohn, J. Appl. Polymer Sci. 14, 1491 (1970).

1200. K. L. Glazomitskii, T. R. Kirpichenko, B. E. Gol'tsin, E. S. Roskin, E. N. Rostovskii, Zh. Prikl. Khim. (Leningrad) 44, 1192 (1971); from CA 75, 49685Q (1971).

1201. E. J. Goethals, E. de Witte, J. Macromol. Sci., Chem. A5, 63 (1971).

1202. Yu. A. Gorin, E. I. Rodina, N. V. Kozlova, K. V. Nel'son, Vysokomolekul. Soedin., Ser. A, 11, 1477 (1969).

1203. H. C. Haas, R. L. MacDonald, C. K. Chiklis, J. Polymer Sci., A-1, 7, 633 (1969).

1204. H. C. Haas, R. L. MacDonald, A. N. Schuler, J. Polymer Sci., A-1, 9, 959 (1971).

1205. M. L. Hallensleben, I. Lumme, Makromol. Chem. 144, 261 (1971).

1206. K. Hamanoue, Rev. Phys. Chem. Japan 38, 120 (1968); from CA 71, 70972H (1969).

1207. J. M. Hammond, J. F. Hooper, W. G. P. Robertson, J. Polymer Sci., A-1, 9, 265 (1971).

1208. K. Hara, Y. Imanishi, T. Higashimura, M. Kamachi, J. Polymer Sci., A-1, 9, 2933 (1971).

1209. K. Hashimoto, H. Sumitomo, Polymer J. 1, 190 (1970).

1210. K. Hashimoto, H. Sumitomo, J. Polymer Sci., A-1, 9, 107 (1971).

1211. S. Hashimoto, I. Furukawa, Kobunshi Kagaku 27, 110 (1970).

1212. G. W. Hastings, J. Chem. Soc. D1969(18), 1039.

1213. K. Hayashi, T. Otsu, Makromol. Chem. 127, 54 (1969).

1214. J. K. Hecht, N. D. Ojha, Macromolecules 2, 94 (1969).

1215. T. Higashimura, K. Kawamura, T. Masuda, J. Polymer Sci., A-1, 10, 85 (1972).

1216. K. Höfelmann, R. Sattelmeyer, K. Hamann, Makromol. Chem. 130, 221 (1969).

1217. H. Hopff, G. Becker, Makromol. Chem. 133, 1 (1970).

1218. H. Hopff, P. Perlstein, Makromol. Chem. 125, 247 (1969).

1219. K. Horie, I. Mita, H. Kambe, J. Polymer Sci., A-1, 7, 2561 (1969).

1220. M. Ibonai, T. Kuramochi, Kogyo Kagaku Zasshi 74, 1002 (1971).

1221. F. Ibragimov, D. Mukhamadaliev, T. G. Gafurov, Vysokomolekul. Soedin., Ser. A, 12, 1475 (1970).

1222. T. Ichihashi, W. Kawai, Kobunshi Kagaku 28, 225 (1971).

1223. Y. Imanishi, H. Imamura, T. Higashimura, Kobunshi Kagaku 27, 242 (1970).

1224. Y. Imanishi, S. Kanagawa, T. Higashimura, Kobunshi Kagaku 28, 666 (1971).

1225. H. Inoue, T. Otsu, Makromol. Chem. 153, 21 (1972).

1226. H. Inoue, I. Umeda, T. Otsu, Makromol. Chem. 147, 271 (1971).

1227. K. Ishigure, Y. Tabata, K. Oshima, J. Macromol. Sci.-Chem. A5, 263 (1971).

1228. T. Ishii, T. Suzuki, T. Inoue, Kogyo Kagaku Zasshi 72, 2649 (1969).

1229. K. Ito, T. Sugie, Y. Yamashita, Makromol. Chem. 125, 291 (1969).

1230. T. Ito, T. Otsu, J. Macromol. Sci., Chem. 3, 197 (1969).

1231. Y. Iwakura, F. Toda, R. Iwata, Y. Torii, Bull. Chem. Soc. Japan 42, 837 (1969).

1232. Y. Iwakura, F. Toda, M. Kosugi, Y. Torii, J. Polymer Sci., A-1, 9, 2423 (1971).

1233. Y. Iwakura, F. Toda, H. Suzuki, N. Kusakawa, K. Yagi, J. Polymer Sci., A-1, 10, 1133 (1972).

1234. Y. Iwakura, K. Uno, N. Nakabayashi, W.-Y. Chiang, Bull. Chem. Soc. Japan 42, 741 (1969).

1235. M. Jabben, A. Chafik, A. Saupe, E. Klesper, H. J. Cantow, Fresenius' Z. Anal. Chem. 249, 1 (1970); from CA 72, 111960U (1970).

1236. Z. Janovic, C. S. Marvel, M. J. Diamond, D. J. Kertesz, G. Fuller, J. Polymer Sci., A-1, 9, 2639 (1971).

1237. H. K. Johnston, A. Rudin, J. Paint Technol. 42, 435 (1970).

1238. H. K. Johnston, A. Rudin, Macromolecules 4, 661 (1971).

1239. E. F. Jordon, jr., R. Bennett, A. C. Shuman, A. N. Wrigley, J. Polymer Sci., A-1, 8, 3113 (1970).

1240. A. S. Kabankin, S. A. Balabanova, A. M. Markevich, Vysokomolekul. Soedin., Ser. A, 12, 267 (1970).

1241. D. A. Kangas, R. R. Pelletier, J. Polymer Sci., A-1, 8, 3543 (1970).

1242. R. G. Karzhaubaeva, G. M. L'dokova, G. P. Gladyshev, S. R. Rafikov, Tr. Inst. Khim. Nauk, Akad. Nauk Kaz. SSR 28, 115 1970); from CA 73, 15305J (1970).

1243. S. Katayama, H. Horikawa, N. Masuda, J. Polymer Sci., A-1, 9, 1173 (1971).

1244. S. Katayama, H. Horikawa, O. Toshima, J. Polymer Sci., A-1, 9, 2915 (1971).

1245. M. Kato, J. Polymer Sci., A-1, 7, 2175 (1969).

1246. I. Katsura, T. Yamamoto, Kogyo Kagaku Zasshi 74, 752 (1971).

1247. R. Kerber, H. Glamann, Makromol. Chem. 144, 1 (1971).

1248. W. Kern, W. Heitz, M. Jäger, K. Pfitzner, H. O. Wirth, Makromol. Chem. 126, 73 (1969).

1249. M. Kinoshita, T. Irie, M. Imoto, Kogyo Kagaku Zasshi 72, 1210 (1969).

1250. M. Kinoshita, S. Kataoka, M. Imoto, Kogyo Kagaku Zasshi 72, 969 (1969).

1251. Y. Kinoshita, J. Kobayashi, F. Ide, K. Nakatsuka, Kobunshi Kagaku 27, 469 (1970).

1252. Y. Kinoshita, J. Kobayashi, F. Ide, K. Nakatsuka, Kobunshi Kagaku 28, 430 (1971).

1253. Yu. V. Kissin, Yu. Ya. Goldfarb, B. A. Krentsel, H. Uyliem, European Polymer J. 8, 487 (1972).

1254. K. Kitanaka, Y. Tabata, Kobunshi Kagaku 28, 206 (1971).

1255. Y. Kobuke, Y. Fukui, J. Furukawa, T. Fueno, J. Polymer Sci., A-1, 8, 3155 (1970).

1256. L. M. Kogan, A. I. Ezrielev, A. V. Lebedev, A. B. Peizner, Vysokomolekul. Soedin., Ser. B, 12, 316 (1970).

1257. A. Kogerman, H. Martinson, T. Evseev, A. Kongas, Esti Nsv Tead. Akad. Toim., Keem., Geol. 18, 232 (1969); from CA 71, 124986M (1969).

1258. H. Koinuma, S. Inoue, T. Tsuruta, Makromol. Chem. 136, 65 (1970).

1259. T. Kokubo, S. Iwatsuki, Y. Yamashita, Macromolecules 3, 518 (1970).

1260. R. Kroker, H. Ringsdorf, Makromol. Chem. 121, 240 (1969).

1261. V. A. Kropachev, T. M. Markelova, L. B. Trukhmanova, Vysokomolekul. Soedin., Ser. A, 12, 1091 (1970).

1262. V. A. Kruglova, A. V. Kalabina, L. V. Morozova, A. D. Biryukova, Vysokomolekul. Soedin., Ser. A, 12, 1600 (1970).

1263. N. G. Kulkarni, N. Krishnamurti, P. C. Chatterjee, M. A. Sivasamban, Makromol. Chem. 139, 165 (1970).

1264. I. Kuntz, C. Cozewith, H. T. Oakley, G. Via, H. T. White, Z. W. Wilchinsky, Macromolecules 4, 4 (1971).

1265. T. N. Kuren'gina, L. V. Alferova, V. A. Kropachev, Vysokomolekul. Soedin., Ser. A, 11, 1985 (1969).

1266. S. S. Labana, J. Polymer Sci., A-1, 8, 179 (1970).

1267. J. C. Lai, T. D. Rounsefell, C. U. Pittman, jr., Macromolecules 4, 155 (1971).

1268. J. C. Lai, T. Rounsfell, C. U. Pittman, jr., J. Polymer Sci., A-1, 9, 651 (1971).

1269. R. Laputte, A. Guyot, Makromol. Chem. 129, 234 (1969).

1270. A. Leoni, S. Franco, Macromolecules 4, 355 (1971).

1271. N. T. Lipscomb, W. K. Matthews, J. Polymer Sci., A-1, 9, 563 (1971).

1272. I. S. Lishanskii, O. S. Fomina, Vysokomolekul. Soedin., Ser. A, 11, 1398 (1969).

1273. S. Machida, K. Matsuo, Yuki Gosei Kagaku Kyokoi Shi 27, 759 (1969); from CA 71, 102295H (1969).

1274. K. A. Makarov, L. N. Mashlyakovskii, S. D. Shenkov, A. F. Nikolaev, Vysokomolekul. Soedin., Ser. B, 12, 894 (1970).

1275. K. A. Makarov, L. N. Vorob'ev, G. M. Tomashevskaya, A. F. Nikolaev, A. A. Petrov, Vysokomolekul. Soedin., Ser. B, 13, 222 (1971).

1276. K. A. Makarov, V. S. Zavgorodnii, E. N. Mal'tseva, A. F. Nikolaev, A. A. Petrov, A. I. Andreev, Vysokomolekul. Soedin., Ser. A, 12, 1429 (1970).

1277. V. A. Mal'tsev, N. A. Plate, Vysokomolekul. Soedin., Ser. A, 12, 182 (1970).

1278. E. Marechal, J. Polymer Sci., A-1, 8, 2867 (1970).

1279. E. Marechal, C. Bit, P. Sigwalt, Bull. Soc. Chim. France 1966(11), 3487.

1280. E. Marechal, G. Zaffran, C. Zaffran, A. Lepert, Compt. Rend., Ser. C, 268, 1350 (1969).

1281. L. N. Mashlyakovskii, K. A. Makarov, I. S. Okhrimenko, A. F. Nikolaev, Vysokomolekul. Soedin., Ser. B, 12, 451 (1970).

1282. L. N. Mashlyakovskii, K. A. Makarov, S. D. Shenkov, I. S. Okhrimenko, A. F. Nikolaev, Vysokomolekul. Soedin. Ser. B, 11, 913 (1969); from CA 72, 55932X (1970).

1283. L. N. Mashlyakovskii, K. A. Makarov, T. K. Solov'eva, Vysokomolekul. Soedin., Ser. B, 11, 712 (1969).

1284. T. Masuda, T. Higashimura, Polymer J. 2, 29 (1971).

1285. T. Masuda, T. Higashimura, S. Okamura, Polymer J. 1, 19 (1970).

1286. Y. Matsubara, J. Asakura, N. Yamashita, H. Sumitomo, T. Maeshima, Kogyo Kagaku Zasshi 72, 2658 (1969).

1287. Y. Matsubara, J. Asakura, M. Yoshihara, T. Maeshima, Kogyo Kagaku Zasshi 74, 1907 (1971).

1288. Y. Matsubara, P. R. Kavi, M. Yoshihara, T. Maeshima, Bull. Chem. Soc. Japan 44, 3127 (1971).

1289. Y. Matsubara, M. Yoshihara, T. Maeshima, Kogyo Kagaku Zasshi 74, 477 (1971).

1290. Y. Matsubara, M. Yoshihara, T. Maeshima, Kogyo Kagaku Zasshi 74, 2163 (1971).

1291. A. Matsumoto, T. Ise, M. Oiwa, Nippon Kagaku Kaishi 1972, 209.

1292. A. Matsumoto, H. Muraoka, M. Taniguchi, M. Oiwa, Kogyo Kagaku Zasshi 74, 1913 (1971).

1293. A. Matsumoto, M. Oiwa, Kogyo Kagaku Zasshi 71, 2063 (1968).

1294. A. Matsumoto, M. Oiwa, Kogyo Kagaku Zasshi 73, 228 (1970).

1295. A. Matsumoto, M. Oiwa, J. Polymer Sci., A-1, 9, 3607 (1971).

1296. A. Matsumoto, H. Sasaki, M. Oiwa, Nippon Kagaku Kaishi 1972, 166.

1297. K. Matsuo, S. Machida, J. Polymer Sci., A-1, 10, 187 (1972).

1298. T. Matsushita, M. Okawara, Kogyo Kagaku Zasshi 74, 1258 (1971).

1299. T. T. Minakova, A. S. Atavin, S. M. Maksimov, E. P. Kel'man, V. Kugeleviciene, L. V. Morozova, Vysokomolekul. Soedin., Ser. B, 14, 166 (1972).

1300. Y. Minoura, Y. Sakanaka, J. Polymer Sci., A-1, 7, 3287 (1969).

1301. M. M. Mirkhidoyatov, U. N. Musaev, A. A. Pulatova, R. S. Tillyaev, Kh. U. Usmanov, Nauch. Tr., Tashkent. Gos. Univ. 1970, No. 377, 282; from CA 76, 100165V (1972).

1302. H. Mitsui, K. Tsuneta, T. Kagiya, Kogyo Kagaku Zasshi 74, 1918 (1971).

1303. Z. Moniyama, A. Yamamoto, M. Kagiyama, R. Chujo, J. Macromol. Sci.-Chem. A4, 1649 (1970).

1304. N. A. S. Mondal, R. N. Young, European Polymer J. 7, 1575 (1971).

1305. M. Moriya, M. Kimura, T. Yamashita, Kobunshi Kagaku 28, 152 (1971).

1306. M. Moriya, S. Mano, T. Yamashita, Kobunshi Kagaku 28, 143 (1971).

1307. M. Moriya, T. Yamashita, Bull. Chem. Soc. Japan 43, 208 (1970).

1308. K. Moser, R. Signer, H. U. Stuber, Chimia 23, 393 (1969); from CA 72, 13098Z (1970).

1309. S. Mukhopadhyay, B. C. Mitra, S. R. Palit, J. Polymer Sci., A-1, 7, 2442 (1969).

1310. S. U. Mullik, A. R. Khan, Pak. J. Sci., Ind. Res. 12, 186 (1970); from CA 73, 4250M (1970).

1311. S. U. Mullik, M. A. Quddus, Pak. J. Sci., Ind. Res. 12, 181 (1970); from CA 73, 4249T (1970).

1312. J. E. Mulvaney, R. A. Ottaviani, J. Polymer Sci., A-1, 8, 2293 (1970).

1313. D. L. Murfin, K. Hayashi, L. E. Miller, J. Polymer Sci., A-1, 8, 1967 (1970).

1314. T. Nakagawa, H. Sumitomo, Kogyo Kagaku Zasshi 74, 1695 (1971).

1315. T. Nakaya, K. Sugiyama, M. Imoto, Makromol. Chem. 140, 275 (1970).

1316. N. S. Nametkin, I. N. Kozhukhova, S. G. Durgar'yan, V. G. Filippova, Vysokomolekul. Soedin., Ser. A, 11, 2523 (1969).

1317. N. S. Nametkin, I. N. Kozhukhova, V. G. Filippova, G. G. Durgar'yan, Vysokomolekul. Soedin., Ser. B, 12, 180 (1970).

1318. I. Nanu, C. Andrei, Rev. Roum. Chim. 16, 75 (1971); from CA 75, 6468D (1971).

1319. M. Narita, T. Teramoto, M. Okawara, Bull. Chem. Soc. Japan 44, 1084 (1971).

1320. S. K. Ngo, Yu. V. Korshak, B. A. Dolgoplosk, Vysokomolekul. Soedin., Ser. B, 13, 641 (1971).

1321. A. F. Nikolaev, K. V. Belogorodskaya, N. I. Duvakina, L. V. Chashnikova, Zh. Prikl. Khim. (Leningrad) 44, 1605 (1971).

1322. A. F. Nikolaev, K. V. Belogorodskaya, V. G. Shibalovich, Vysokomolekul. Soedin., Ser. B, 13, 608 (1971); from CA 76, 15007A (1972).

1323. A. F. Nikolaev, V. M. Gal'perin, L. N. Pirozhnaya, Vysokomolekul. Soedin., Ser. B, 11, 625 (1969); from CA 71, 124959E (1969).

1324. A. F. Nikolaev, L. P. Vishnevetskaya, O. A. Gromova, M. M. Grigor'eva, M. S. Kleshcheva, Vysokomolekul. Soedin., Ser. A, 11, 2418 (1969).

1325. K. Noma, M. Niwa, Kobunshi Kagaku 29, 52 (1972).

1326. K. Noma, M. Niwa, H. Norisada, Doshisha Daigaku Rikogaku Kenkyu Hokoku 10, 349 (1970); from CA 73, 77680M (1970).

1327. K. Noma, T. Saito, M. Niwa, T. Okuda, Kobunshi Kagaku 27, 406 (1970).

1328. K. Noma, H. Utsumi, M. Niwa, Kobunshi Kagaku 26, 889 (1969).

1329. I. S. Novoderezhkina, D. A. Kochkin, P. I. Zubov, G. I. Kuznetsova, Plasticheskie Massy 1969(3), 13; from CA 71, 3706M (1969).

1330. S. Nozakura, M. Kitamura, S. Murahashi, Polymer J. 1, 736 (1970).

1331. G. Odian, L. S. Hiraoka, J. Macromol. Sci.-Chem. A6, 109 (1972).

1332. N. Ogata, H. Tanaka, Kobunshi Kagaku 28, 738 (1971).

1333. E. Oikawa, K. Yamamoto, Polymer J. 1, 669 (1970).

1334. T. Okuyama, T. Fueno, J. Polymer Sci., A-1, 9, 629 (1971).

1335. T. Okuyama, T. Fueno, J. Furukawa, J. Polymer Sci., A-1, 7, 2433 (1969).

1336. T. Okuyama, T. Fueno, J. Furukawa, J. Polymer Sci., A-1, 7, 3045 (1969).

1337. J. Osugi, K. Hamanoue, Nippon Kagaku Zasshi 90, 552 (1969); from CA 71, 39441K (1969).

1338. T. Otsu, H. Inoue, J. Macromol. Sci., A-1, 4, 35 (1970).

1339. T. Otsu, T. Misaki, Nippon Kagaku Kaishi 1972, 211.

1340. T. Otsu, O. Nakatsuka, Kobunshi Kagaku 29, 168 (1972).

1341. T. Otsu, B. Yamada, H. Yoneno, Bull. Chem. Soc. Japan 42, 3207 (1969).

1342. M. Otsuka, K. Matsuoka, K. Takemoto, M. Imoto, Kogyo Kagaku Zasshi 72, 2505 (1969).

1343. M. Otsuka, K. Matsuoka, K. Takemoto, M. Imoto, Kogyo Kagaku Zasshi 73, 1062 (1970).

1344. T. Ouchi, Y. Imase, M. Oiwa, Bull. Chem. Soc. Japan 43, 2863 (1970).

1345. T. Ouchi, S. Kondo, M. Oiwa, Kogyo Kagaku Zasshi 73, 2508 (1970).

1346. T. Ouchi, M. Oiwa, Kogyo Kagaku Zasshi 72, 1587 (1969); from CA 71, 113282C (1969).

1347. T. Ouchi, M. Oiwa, Bull. Chem. Soc. Japan 43, 2858 (1970).

1348. T. Ouchi, M. Oiwa, Kogyo Kagaku Zasshi 73, 1717 (1970).

1349. T. Ouchi, S. Tatsuno, T. Nakayama, M. Oiwa, Kogyo Kagaku Zasshi 73, 607 (1970).

1350. T. Ouchi, Y. Yaguchi, M. Oiwa, Bull. Chem. Soc. Japan 44, 1623 (1971).

1351. T. Ouchi, K. Yokoi, M. Oiwa, Kogyo Kagaku Zasshi 73, 812 (1970).

1352. T. Ouchi, K. Yokoi, K. Yoshimura, M. Oiwa, Bull. Chem. Soc. Japan 44, 1339 (1971).

1353. T. Ouchi, M. Yoshida, M. Oiwa, Kogyo Kagaku Zasshi 73, 2511 (1970).

1354. C. U. Pittman, jr., P. L. Grube, J. Polymer Sci., A-1, 9, 3175 (1971).

1355. C. U. Pittman, jr., P. L. Brube, O. E. Ayers, S. P. McManus, M. D. Rausch, G. A. Moser, J. Polymer Sci., A-1, 10, 379 (1972).

1356. C. U. Pittman, jr., J. C. Lai, D. P. Vanderpool, Macromolecules 3 105 (1970).

1357. C. U. Pittman, jr., R. L. Voges, J. Elder, Macromolecules 4, 302 (1971).

1358. C. U. Pittman, jr., R. L- Voges, W. B. Jones, Macromolecules 4, 298 (1971).

1359. N. A. Plate, T. B. Zavarova, V. V. Mal'tsev, K. S. Minsker, G. T. Fedoseyeva, V. A. Kargin, Vysokomolekul. Soedin., Ser. A, 11, 803 (1969).

1360. V. A. Ponomarenko, S. P. Krukovskii, N. M. Khomutova, A. G. Kechina, Izv. Akad. Nauk SSSR, Ser. Khim. 1970, 2357.

1361. G. P. Popova, T. R. Kirpichenko, K. L. Glazomitskii, B. E. Gol'tsin, E. S. Roskin, E. N. Rostovskii, Izv. Vyssh. Ucheb. Zaved., Khim. Khim. Tekhnol. 13, 259 (1970); from CA 73, 15312J (1970).

1362. C. C. Price, Y. Atarashi, R. Yamamoto, J. Polymer Sci., A-1, 7, 569 (1969).

1363. C. C. Price, L. R. Brecker, J. Polymer Sci., A-1, 7, 575 (1969).

1364. W. A. Pryor, T. -L. Huang, Macromolecules 2, 70 (1969).

1365. J. P. Quere, E. Marechal, Bull. Soc. Chim. France 1971, 2983.

1366. M. Rätzsch, W. Schneider, D. Musche, J. Polymer Sci., A-1, 9, 785 (1971).

1367. M. Rätzsch, L. Stephan, Plaste Kautschuk 18, 572 (1971).

1368. S. R. Rafikov, N. D. Kazakova, G. A. D'yachkov, Izv. Akad. Nauk Kaz. SSR, Ser. Khim. 18, 71 (1968); from CA 69, 36839U (1968).

1369. M. Ragazzini, D. Carcano, M. Modena, G. C. Serboli, European Polymer J. 6, 763 (1970).

1370. E. A. Rassolova, M. A. Zharkova, G. I. Kudryavtsev, V. S. Klimenkov, Khim. Volokna 1969, 50; from CA 71, 92509Q (1969).

1371. A. K. Rastogi, L. E. St. Pierre, J. Appl. Polymer Sci. 14, 1179 (1970).

1372. H. Reichert, V. Gröbe, Faserforsch. Textiltech. 20, 317 (1969).

1373. H. Reichert. H. J. Schauder, Faserforsch. Textiltech. 21, 21 (1970).

1374. S. H. Ronel, D. H. Kohn, J. Polymer Sci., A-1, 7, 2209 (1969).

1375. S. A. Rostovtseva, Vysokomolekul. Soedin., Ser. B, 12, 588 (1970).

1376. A. Rudin, R. G. Yule, J. Polymer Sci., A-1, 9, 3009 (1971).

1377. A. V. Ryabov, Yu. D. Semchikov, N. N. Slvanitskaya, Vysoko-molekul. Soedin., Ser. A, 12, 553 (1970).

1378. A. V. Ryabov, L. A. Smirnova, V. A. Soldatov, Dokl. Akad. Nauk SSSR, 194, 1338 (1970).

1379. Z. M. Rzaev, L. V. Bryksina, Sh. K. Kyazimov, S. I. Sadikhzade, Vysokomolekul. Soedin., Ser. A, 14, 259 (1972).

1380. M. Sadamichi, K. Noro, J. Macromol. Sci., Chem. 3, 845 (1969).

1381. B. D. Saidov, A. S. Bank. M. A. Askarov, G. V. Leplyanin, Vysokomolekul. Soedin., Ser. B, 12, 845 (1970).

1382. G. Saini, A. Leoni, S. Franco, Makromol. Chem. 144, 235 (1971).

1383. G. Saini, A. Leoni, S. Franco, Makromol. Chem.146, 165 (1971).

1384. G. Saini, A. Leoni, S. Franco, Makromol. Chem. 147, 213 (1971).

1385. S. D. Sandomirskaya, G. N. Bondarenko, N. N. Stefanovskaya, M. P. Teterina, B. A. Dolgoplosk, Vysokomolekul. Soedin., Ser. B, 13, 855 (1971).

1386. A. G. Sayadyan, D. A. Simonyan, Arm. Khim. Zh. 22, 528 (1969); from CA 71, 102273 Z (1969).

1387. N. M. Seidov, R. M. Aliguliev, A. I. Abasov, Vysokomolekul. Soedin., Ser. A, 11, 2107 (1969).

1388. N. M. Seidov, D. A. Koptev, M. A. Dalin, D. M. Lisitsyn, N. M. Chirkov, Vysokomolekul. Soedin., Ser. A, 11, 844 (1969).

1389. Yu. D. Semchikov, A. V. Ryabov, V. N. Kashaeva, Vysokomolekul. Soedin., Ser. B, 12, 567 (1970).

1390. Yu. D. Semchikov, A. V. Ryabov, V. N. Kashaeva, Vysokomolekul. Soedin., Ser. B, 14, 138 (1972).

1391. L. N. Semenova, B. L. Gafurov, M. A. Askarov, Uzb. Khim. Zh. 13, 62 (1969); from CA 72, 111869W (1970).

1392. F. F. Shcherbina, O. I. Bakumenko, Vysokomolekul. Soedin., Ser. A, 11, 772 (1969).

1393. K. Shima, M. Kinoshita, M. Imoto, Kogyo Kagaku Zasshi 72, 1584 (1969).

1394. M. F. Shostakovskii, F. P. Sidel'kovskaya, A. S. Atavin, T. T. Minakova, E. P. Kel'man, L. V. Morozova, Vysokomolekul. Soedin., Ser. B, 13, 847 (1971).

1395. S. V. Shulyndin, S. V. Yakovleva, Kh. G. Sanatullin. V. Sh. Gurskaya, B. E. Ivanov, Vysokomolekul. Soedin., Ser. B, 12, 106 (1970).

1396. G. Smets, A. Bourtembourg, J. Polymer Sci., A-1, 8, 3251 (1970).

1397. N. A. Smirnyagina, E. I. Tinyakova, Dokl. Nauk SSSR 186, 1099 (1969); from CA 71, 81759V (1969).

1398. Yu. L. Spirin, T. S. Yatsimirskaya, Vysokomolekul. Soedin., Ser. B, 11, 515 (1969); from CA 71, 102246Z (1969).

1399. J. K. Stille, L. D. Gotter, Macromolecules 2, 468 (1969).

1400. Y. Tabata, T. A. Du Plessis, J. Polymer Sci., A-1, 9, 3425 (1971).

1401. Y. Tabata, K. Ishigure, H. Higaki, K. Oshima, J. Macromol. Sci.-Chem., A4, 801 (1970).

1402. S. Tagami, C. Aso. Kobunshi Kagaku 27, 922 (1970).

1403. S. Takahashi, K. Takemoto, M. Imoto, Kobunshi Kagaku 27, 156 (1970).

1404. K. Takemoto, H. Wada, M. Imoto, Kobunshi Kagaku 28, 185 (1971).

1405. I. Takemura, H. Sumitomo, Bull. Chem. Soc. Japan 43, 334 (1970).

1406. H. Tamura, M. Tanaka, N. Murata, Bull. Chem. Soc. Japan 42, 3042 (1969).

1407. H. Tamura, M. Tanaka, N. Murata, Kobunshi Kagaku 27, 652 (1970).

1408. H. Tamura, M. Tanaka, N. Murata, Kobunshi Kagaku 27, 736 (1970).

1409. S. Tazuke, S. Okamura, J. Polymer Sci., A-1, 7, 715 (1969).

1410. S. Tazuke, K. Shimada, S. Okamura, J. Polymer Sci., A-1, 7, 879 (1969).

1411. R. A. Terteryan, Vysokomolekul. Soedin., Ser. A, 11, 1798 (1969).

1412. R. A. Terteryan, Vysokomolekul. Soedin., Ser. A, 11, 1850 (1969).

1413. R. A. Terteryan, N. V. Fomicheva, V. N. Monastyrskii, Vysoko-molekul. Soedin., Ser. B, 13, 485 (1971).

1414. L. Tokarzewski, A. Wdowin, Polimery 14, 25 (1969).

1415. L. Tokarzewski, A. Wdowin, Polimery 14, 533 (1969).

1416. D. A. Topchiev, V. Z. Shakirov, L. P. Kalinina, T. M. Karaputadze, V. A. Kabanov, Vysokomolekul. Soedin., Ser. A, 14, 581 (1972).

1417. J. P. Tortai, E. Marechal, Bull. Soc. Chim. France 1971, 2673.

1418. N. Ueda, M. Shibata, K. Takemoto, Kobunshi Kagaku 28, 684 (1971); corr. Kobunshi Kagaku 28, 774 (1971).

1419. L. M. Vardanyan, S. Q. Ngo, Yu. V. Korshak, B. A. Dolgoplosk, Vysokomolekul. Soedin., Ser. B, 13, 19 (1971).

1420. V. A. Vvedenskaya, N. E. Ogneva, V. V. Korshak, Ts. A. Goguadze, Tr. Mosk. Khim-Tekhnol. Inst. 66, 170 (1970); from CA 75, 118656V (1971).

1421. N. Watanabe, M. Sakai, Y. Sakakibara, N. Uchino, Kogyo Kagaku Zasshi 73, 1056 (1970).

1422. A. S. Wdowin, P. Szewczyk, Vysokomolekul. Soedin., Ser. B, 11, 509 (1969); from CA 71, 102262V (1969).

1423. R. H. Wiley, J. I. Jin, J. Macromol. Sci.-Chem., A2, 1097 (1968).

1424. R. H. Wiley, J.-I. Jin, J. Macromol. Sci., Chem. 3, 835 (1969).

1425. T. J. Wojnarowski, Polimery 14, 155 (1969).

1426. T. J. Wojnarowski, Polimery 14, 207(1969).

1427. F. I. Yakobson, V. V. Amerik, D. V. Ivanyukov, V. F. Petrova, Yu. V. Kissin, B. A. Krentsel, Vysokomolekul. Soedin., Ser. A, 13, 2699 (1971); from CA 76, 100167X (1972).

1428. B. Yamada, Y. Kusuki, T. Otsu, Makromol. Chem. *137*, 29 (1970).

1429. B. Yamada, T. Otsu, J. Macromol. Sci., Chem. *3*, 1551 (1969).

1430. B. Yamada, T. Otsu, J. Polymer Sci., A-1, *7*, 2439 (1969).

1431. B. Yamada, H. Yoneno, T. Otsu, J. Polymer Sci., A-1, *8*, 2021 (1970).

1432. M. Yamada, I. Takase, T. Tsukano, Y. Ueda, N. Koutou, Kobunshi Kagaku *26*, 593 (1969).

1433. H. Yamaguchi, Y. Minoura, J. Polymer Sci., A-1, *8*, 1467 (1970).

1434. T. Yamaguchi, T. Ono, S. Kawagishi, Kobunshi Kagaku *26*, 187 (1969); from CA *71*, 3741U (1969).

1435. T. Yamamoto, S. Sumida, I. Katsura, Kogyo Kagaku Zasshi *74*, 987 (1971).

1436. Y. Yamamoto, S. Tsuge, T. Takeuchi, Macromolecules *5*, 325 (1972).

1437. N. Yamashita, T. Seita, M. Yoshihara, T. Maeshima, J. Polymer Sci., B, *9*, 641 (1971).

1438. N. Yamashita, T. Seita, M. Yoshihara, T. Maeshima, Kogyo Kagaku Zasshi *74*, 2157 (1971).

1439. T. Yamashita, M. Moriya, Kogyo Kagaku Zasshi *73*, 1252 (1970).

1440. Y. Yamashita, T. Asakura, M. Okada, K. Ito, Macromolecules *2*, 613 (1969).

1441. Y. Yamashita, K. Ito. H. Ishii, S. Hoshino, M. Kai, Macromolecules *1*, 529 (1968).

1442. Y. Yamashita, S. Kondo, K. Ito, Polymer J. *1*, 327 (1970).

1443. V. A. Yashkov, V. V. Mal'tsev, N. A. Plate, Vysokomolekul. Soedin., Ser. B, *13*, 866 (1971).

1444. K. Yokota, J. Macromol. Sci.-Chem. *A4*, 65 (1970).

1445. K. Yokoty, J. Oda, Kogyo Kagaku Zasshi *73*, 224 (1970).

1446. M. Yonezawa, K. Kimura, S. Suzuki, H. Ito, Yuki Gosei Kagaku Kyokai Shi *27*, 1218 (1969); from CA *74*, 54229C (1971).

1447. H. Yuki, K. Hatada, M. Takeshita, J. Polymer Sci., A-1, *7*, 667 (1969).

1448. H. Yuki, Y. Okamoto, Bull. Chem. Soc. Japan *42*, 1644 (1969).

1449. H. Yuki, Y. Okamoto, Polymer J. *1*, 13 (1970).

1450. H. Yuki, Y. Okamoto, Bull. Chem. Soc. Japan *43*, 148 (1970).

1451. H. Yuki, Y. Okamoto, K. Sadamoto, Bull. Chem. Soc. Japan *42*, 1754 (1969).

1452. H. Yuki, Y. Okamoto, K. Tsubota, K. Kosai, Polymer J. *1*, 147 (1970).

1453. Y. Yuki, T. Kakurai, T. Noguchi, Bull. Chem. Soc. Japan *43*, 2123 (1970).

1454. E. Yun, L. V. Lobanova, A. D. Litmanovich, N. A. Plate, M. V. Shishkina, T. A. Polikarpova, Vysokomolekul. Soedin., Ser. A, *12*, 2488 (1970).

1455. C. Zaffran, E. Marechal, Bull. Soc. Chim. France *1970*, 3523.

1456. B. A. Zhubanov, O. Sh. Kurmanaliev, E. M. Shaikutdinov, Izv. Akad. Nauk Kaz. SSR, Ser. Khim. *21*, 74 (1971).

1457. V. M. Zhulin, N. V. Klimentova, A. P. Suprun, M. G. Gonikberg, V. V. Korshak, Izv. Akad. Nauk SSSR, Ser. Khim. *1967*, 2447; from CA *68*, 22293D (1968).

1458. V. P. Zubov, L. I. Valuev, V. A. Kabanov, V. A. Kargin, J. Polymer Sci., A-1, *9*, 833 (1971).

TABULATION OF Q-e VALUES

Lewis J. Young
Koppers Company, Inc.
Law/Patent Department
Pittsburgh, Pennsylvania

These Q-e values are average values calculated by the use of the following two equations:

$$e_2 = e_1 \pm (-\ln r_1 r_2)^{1/2}$$

and

$$Q_2 = Q_1/r_1 \exp[-e_1(e_1 - e_2)]$$

The validity of the experimental r data was not normally a factor considered during the averaging process. Hence the values presented here may not agree with other sources.

The author wishes also to caution the reader to become familiar with the possible problems involved when the monomer is either a 1,2- or 1,1-disubstituted ethylene. (See G.E. Ham, "Copolymerization", Interscience Publisher, New York, 1964, Chapter II, Section III.) For systems which involve these hindered ethylenes, the Q-e values should be used only when no r data are available and then only with a forewarning of the possibility that the results may be widely at variance with the true experimental results.

For convenience, the data have been presented in two tables. Table I lists the monomers in order of increasing e-value, while Table II lists the monomers alphabetically.

TABLE I

Monomer	e	Q	Notes
1-Propene, 2-chloro-3,3-dimethoxy-	-2.250	0.125	b
1-Aziridinecarboxamide, N-isopropenyl-	-2.160	0.190	b
Ethylene, 1,1-bis(p-methoxyphenyl)-	-1.960	1.460	a
Benzyl alcohol, α,α-dimethyl-p-vinyl-	-1.930	4.940	b
1-Hexene, 1-phenyl-, trans-	-1.820	0.015	b
1,3-Butadiene, 2,3-dimethyl-	-1.810	5.860	
Ether, isobutyl vinyl	-1.770	0.023	
Benzene, m-divinyl-	-1.770	3.350	b
Ethenesulfonic acid, propyl ester	-1.760	0.270	b
Benzene, Propenyl-, cis-	-1.750	0.011	b
Sulfide, isobutyl vinyl	-1.700	0.530	b
Anthracene, 1-vinyl-	-1.680	2.500	b
Thianthrene, 2-vinyl-	-1.680	4.300	
1-Butene, 1-phenyl-, cis-	-1.670	0.008	b
2-Butene, 2-methyl-	-1.650	0.013	
Phthalimide, 4-(trimethylsilyl)-N-vinyl-	-1.650	0.320	b
1-Pentene, 1-phenyl-, cis-	-1.640	0.007	b
Cyclohexene, 4-vinyl-	-1.640	0.060	a,b
1-Butene, 3-methyl-1-phenyl-, trans-	-1.630	0.006	b
Carbamic acid, vinyl-, ethyl ester	-1.620	0.120	
Carbonic acid, cyclic phenylvinylene ester	-1.620	1.400	
2-Butene, 2,3-dimethyl-	-1.600	0.005	
Anthracene, 9-vinyl-	-1.600	0.900	b
Ether, tert-butyl vinyl	-1.580	0.150	b
Hydrazinium hydroxide, 1-(2-hydroxylpropyl)-2-methacryloyl-1,1-dimethyl-, inner salt	-1.580	0.170	b
Imidazole, 2-phenyl-1-vinyl-	-1.570	0.100	b
1-Hexene, 1-phenyl-, cis	-1.550	0.004	b
Ether, cyclohexyl vinyl	-1.550	0.081	a,b
Phthalimide, N-2-(1,3-butadienyl)-	-1.550	1.650	b
Acetanilide, N-vinyl-	-1.540	0.120	
Aniline, N,N-divinyl-	-1.540	0.190	
Carbamic acid, diethyldithio-, S-vinyl ester	-1.540	0.450	b
Benzene, propenyl-, trans-	-1.530	0.040	b
Urea, 1-ethyl-3-vinyl-	-1.530	0.130	
Ether, propyl vinyl	-1.520	0.014	a,b
2-Propene-1,1-diol, diacetate	-1.520	0.230	b
Phthalimide, N-vinyl-	-1.520	0.360	
Styrene, p-isopropyl-	-1.520	1.600	b
Cyclooctene	-1.510	0.004	
1-Pentene, 3-methyl-1-phenyl-, trans	-1.510	0.300	b
Carbamic acid, diethylthio-, S-vinyl ester	-1.460	0.310	b
Acetic acid, thio-, vinyl ester	-1.460	0.370	b
2-Propene-1,1-diol, 2-chloro-, diacetate	-1.450	0.200	b
Pyrazole, 3,5-dimethyl-1-vinyl-	-1.450	0.320	
Sulfide, methyl vinyl	-1.440	0.002	
Cyclohexene	-1.430	0.002	b
Cycloheptene			b

Table I (Cont'd.)

Monomer	e	Q	Notes
Styrene, p-(2-bromoethyl)-	-1.430	0.710	b
Sulfide, p-anisyl, vinyl	-1.420	0.430	
Ether, pentyl vinyl	-1.410	0.009	a, b
Sulfide, phenyl vinyl	-1.400	0.340	
Carbazole, 9-vinyl-	-1.400	0.410	
Terephthalic acid, 2,2-bis(3-allyl-4-hydroxyphenyl)propane-co- phenolphthalein polyester	-1.400	0.580	b
Ethanol, 2-(2-(vinyloxy)ethoxy)-	-1.390	0.046	b
Aniline, N,N-dimethyl-p-vinyl-	-1.370	1.510	
Cyclopentene	-1.350	0.008	
Ethylene, 1,1-diphenyl-	-1.350	1.500	c
Pyridine, 4,4'-vinylenedi-	-1.340	0.660	
2-Propen-1-ol, 2-methyl-, acetate	-1.330	0.037	a
Oxetane, 3-(allyloxymethyl)-3-(chloromethyl)-	-1.320	0.024	b
Silane, 2-butenyltriethoxy-	-1.320	0.040	a
1-Butene, 3-methyl-1-phenyl-, cis-	-1.310	0.002	b
Benzene, o-divinyl-	-1.310	1.640	
Ether, isopropyl vinyl	-1.310	45.400	b
Succinic anhydride	-1.300	0.002	b
Imidazole, 2-methyl-1-vinyl-	-1.300	0.150	
Carbamic acid, N-methyl-N-vinyl thio-, S-ethyl ester	-1.290	0.110	
1-Pentene, 1-phenyl-, trans-	-1.280	0.029	b
Vinyl ether	-1.280	0.037	
1-Pentene, 3-methyl-1-phenyl-, cis-	-1.270	0.001	b
Styrene, α-methyl-	-1.270	0.980	
Silane, dimethylphenyl(p-vinylphenyl)	-1.270	1.630	
Hydrazinium hydroxide, 1-(2,3-dihydroxypropyl)-2-methacryloyl- 1,1-dimethyl-, inner salt	-1.260	0.240	b
Carbonic acid, thio-, O-tert-butyl S-vinyl ester	-1.260	0.520	
Acetylene, diphenyl-	-1.230	0.003	a
Isoprene	-1.220	3.330	
Ether, phenyl vinyl	-1.210	0.082	
Ether, butyl vinyl	-1.200	0.087	c
Sulfide, isopropyl vinyl	-1.200	0.310	b
Sulfide, butyl vinyl	-1.200	0.330	b
Sulfide, p-tolyl vinyl	-1.200	0.400	
Ether, methyl 1-phenylvinyl	-1.200	0.650	a
1-Aziridinecarboxamide, N-vinyl-	-1.190	0.180	
2-Oxazolidinone, 5-methyl-3-vinyl	-1.190	0.210	
Phosphonic acid, vinyl-, bis(2-chloroethyl) ester	-1.190	0.760	
Acrylanilide, 4'-methoxy-2-methyl-	-1.190	2.800	b
Methacrylic acid, sodium salt	-1.180	1.360	
Ether, ethyl vinyl	-1.170	0.032	a
Butane, 1,4-bis(p-vinylphenyl)	-1.170	1.200	b
Carbamic acid, vinyl-, 2,3-epoxypropyl ester	-1.150	0.180	
2-Pyrrolidinone, 1-vinyl-	-1.140	0.140	b
Acetic acid, allyl ester	-1.130	0.028	c

Table I (Cont'd.)

Monomer	e	Q	Notes
Phenol, o-vinyl-	-1.130	1.410	
1,3-Cyclohexadiene	-1.130	1.610	
p-Dioxin, 2,3-dihydro-	-1.120	0.008	a
2-Propen-1-ol, 2-chloro-, acetate	-1.120	0.530	b
Naphthalene, 1-vinyl-	-1.120	1.940	
Vinyl sulfide	-1.110	0.580	b
Methane, bis(p-vinylphenyl)	-1.110	1.280	b
Anisole, p-vinyl-	-1.110	1.360	
p-Toluenesulfonamide, N-methyl-N-vinyl-	-1.100	0.082	a
Styrene, 2,4,6-trimethyl-	-1.100	0.220	
Sulfide, tert-butyl vinyl	-1.100	0.260	b
Sulfide, bis(p-vinylphenyl)	-1.090	2.070	b
Silane, triethoxypropenyl-	-1.080	0.003	a
Ethane, 1,2-bis(p-vinylphenyh)	-1.080	1.180	b
Ether, bis(p-vinylphenyl)	-1.070	1.360	
Carbamic acid, diethyl-, vinyl ester	-1.060	0.028	
2,5-Norbornadiene	-1.060	0.092	
1,3-Butadiene	-1.050	2.390	c
Naphthalene, 2-isopropenyl-	-1.040	0.920	c
Styrene, 2,5-dimethoxy-	-1.040	1.750	
Indene	-1.030	0.360	
1-Propen-2-ol, isocyanate	-1.020	0.230	c
Silane, dimethylvinyl(p-vinylphenyl)	-1.010	1.200	
s-Triazine, 2,4,6-tris(allyloxy)-	-1.000	0.020	c
Benzyl alcohol, α-ethyl-α-methyl-p-vinyl-	-1.000	0.820	
Ethylene, 1,2-dimethoxy-	-0.980	0.049	
Acrylanilide, 4'-chloro-2-methyl-	-0.980	0.700	b
Styrene, p-methyl-	-0.980	1.270	
Acetanilide, 4'-vinyl-	-0.980	1.420	b
Tridecanedioic acid, 2-methylpentyl vinyl ester	-0.960	0.032	b
1-Propene, 2-methyl-	-0.960	0.033	c
Styrene, 2,5-dimethyl-	-0.960	0.970	
Melamine, N(2),N(4)-diallyl-	-0.950	0.017	c
Silane, allyltriethoxy-	-0.940	0.024	a
1-Propene, 1-chloro-	-0.940	0.031	b
Benzothiazole, 2-vinylthio-	-0.920	1.680	b
1-Propene, 3-chloro-2-methyl-	-0.910	0.120	a
Sulfide, p-chlorophenyl vinyl	-0.910	0.310	
Propane, 1,3-bis(p-vinylphenyl)	-0.910	0.980	b
Styrene, 2,3,4,5-tetraethyl-	-0.900	0.790	b
Trisiloxane, 1-(chloromethyl)-1,1,3,5,5-hexamethyl-5- (p-vinylphenyl)-	-0.900	1.060	
Styrene, p-tert-butyl-	-0.900	1.370	b
Acrylamide, N-ethyl-2-methyl-	-0.880	0.250	b
Sulfide, p-bromophenyl vinyl	-0.880	0.350	
Silane, trimethoxy(p-vinylphenyl)-	-0.880	1.500	b
Silane, diethoxymethylvinyl-	-0.860	0.020	a

Table I (Cont'd.) Monomer	e	Q	Notes
Carbamic acid, isopropenyl-, 2,3-epoxypropyl ester	-0.860	0.260	b
Ethylene, 1,1-bis(p-chlorophenyl)-	-0.840	2.160	a
2-Butanol, 1-(p-vinylphenyl)-	-0.830	0.760	
Ethylene, fluoro-	-0.820	0.025	
Formic acid, vinyl ester	-0.820	0.170	
2,4-Pentadienoic acid, 5,5-dichloro-4-hydroxy-, γ-lactone	-0.820	2.840	b
Quinoline, 2-vinyl-	-0.820	3.790	
Disiloxane, 1,1,1,3,3-pentamethyl-3-vinyl-	-0.810	0.034	
2-Propene-1-sulfonic acid, 2,3-epoxypropyl ester	-0.800	0.020	b
1-Phenanthrenemethanol, 1,2,3,4,4A,4B,5,6,10,10A-decahydro-7-isopropyl-1,4A-dimethyl-	-0.800	0.056	a,b
2-Oxazolidinone, 3-vinyl-	-0.800	0.057	
Ethylene, iodo-	-0.800	0.140	b
Acenaphthylene	-0.800	0.260	b
2-Furaldehyde	-0.800	0.500	b
Benzyl alcohol, α,α-diethyl-p-vinyl-	-0.800	0.970	
Disiloxane, 1-(chloromethyl)-1,1,3,3-tetramethyl-3-(p-vinyl-phenyl)	-0.800	0.990	b
Styrene	-0.800	1.000	(Base)
Tetrasiloxane, 1-(chloromethyl)-1,1,3,5,5,7,7-octamethyl-7-(p-vinylphenyl)-	-0.800	1.020	b
Phenol, m-vinyl-	-0.800	1.100	b
Silane, methyl(p-vinylphenyl)-	-0.800	1.100	b
Styrene, m-(1,1-dichloro-2-cyclopropyl)-	-0.800	1.120	b
Phenol, p-vinyl-, acetate	-0.800	1.180	b
Phenanthrene, 9-vinyl-	-0.800	1.730	b
Phenanthrene, 3-vinyl-	-0.800	2.260	b
Thiophene, 2-vinyl-	-0.800	2.860	b
3(2H)-Pyridazinone, 6-hydroxy-	-0.790	0.010	
Ether, octyl vinyl	-0.790	0.061	c
1-Propene	-0.780	0.002	
Trisiloxane, 1,1,1,3,5,5,5-heptamethyl-3-vinyl-	-0.780	0.036	b
Methacrylanilide	-0.780	0.850	b
Styrene, o-methyl-	-0.780	0.900	b
Acrylanilide, 2,4'-dimethyl-	-0.760	1.200	b
Pivalic acid, vinyl ester	-0.750	0.037	b
Ether, dodecyl vinyl	-0.740	0.033	c
Valeric acid, 4-methyl-, vinyl ester	-0.740	0.043	b
Benzoic acid, p-vinyl-, sodium salt	-0.740	0.650	b
Pyridine, 5-ethyl-2-vinyl-	-0.740	1.370	
Benzenesulfonyl fluoride, m-vinyl-	-0.730	1.330	
Styrene, m-methyl-	-0.720	0.910	
Acrylic acid, 2-formamido-	-0.720	6.000	b
1-Propene, 2-chloro-	-0.710	0.035	c
3(2H)-Pyridazinone, 6-methyl-	-0.710	0.037	
1-Hexyne	-0.700	0.014	a
Tristlane, 1,1,1,2,3,3,3-heptamethyl-2-vinyl-	-0.700	0.020	a

Table I (Cont'd.) Monomer	e	Q	Notes
Isocyanic acid, vinyl ester	-0.700	0.160	
Phenethyl alcohol, α-methyl-p-vinyl-	-0.700	1.200	
Trisiloxane, 1,1,1,5,5,5-hexamethyl-3-trimethylsiloxy-3-vinyl-	-0.690	0.030	
Nonanoic acid, vinyl ester	-0.680	0.030	
Formamide, N-allyl-	-0.680	0.110	
Acetic acid, dichloro-, vinyl ester	-0.680	0.170	
Acrylamide, N-(1-benzyl-2-oxopropyl)-	-0.680	0.440	b
Phenanthrene, 2-vinyl-	-0.670	1.960	
Benzene, ethynyl-	-0.660	0.350	
Carbonic acid, cyclic vinylene ester	-0.650	0.007	
Acetic acid, chloro-, vinyl ester	-0.650	0.074	c
Benzyl alcohol, α-methylene-, acetate	-0.650	0.820	b
Versatic[9]acid, vinyl ester	-0.640	0.031	b
9-Undecenoic acid, vinyl ester	-0.640	0.035	c
Germane, triethylmethacryloyl-	-0.640	0.840	b
1-Butene, 3,3-dimethyl-	-0.630	0.007	a,b
Ether, octadecyl vinyl	-0.630	0.069	c
1-Pentene	-0.630	0.074	a,b
Hydrazinium hydroxide, 2-methacryloyl-1,1,1-trimethyl-, inner salt	-0.620	0.180	b
Urea, 1,1-diallyl-	-0.610	0.018	b
S-Triazine-2,4,6-(1H,3H,5H)-trione, triallyl-	-0.600	0.011	c
Disilane, pentamethylvinyl-	-0.600	0.020	a
Acrylamide, N,2-dimethyl-	-0.600	0.320	b
2H-Tetrazole, 2-phenyl-5-(p-vinylphenyl)-	-0.600	0.750	
3-Pyrazolidinone, 1-(m-vinylphenyl)-	-0.600	0.900	b
3(2H)-Pyridazinone, 6-hydroxy-2-(p-tolyl)-	-0.590	0.024	a
Benzenesulfonic acid, p-vinyl-, sodium salt	-0.590	2.490	b
Succinic acid, methylene-, dipotassium salt	-0.580	0.160	
Sulfide, pentachlorophenyl vinyl	-0.580	0.220	
2-Picoline, 5-vinyl-	-0.580	0.990	
Stearamide, N-allyl-	-0.570	0.040	c
1-Propene, 1-chloro-, cis-	-0.550	0.004	
Benzoic acid, vinyl ester	-0.550	0.061	
Allyl ether	-0.540	0.015	b
Styrene, 3,4-dibromo-	-0.540	1.300	b
Levulinic acid, vinyl ester	-0.530	0.027	b
2,4,8,10-Tetraoxaspiro[5,5]undecane, 3,9-distyryl-	-0.530	0.040	
Stearic acid, vinyl ester	-0.520	0.034	
p-Toluic acid, vinyl ester	-0.520	0.051	
Octadecanoic acid, 12-oxo-, vinyl ester	-0.500	0.032	c
1-Propen-2-ol, acetate	-0.500	0.045	
Silane, (1-cyclohexen-4-yl)triethoxy-	-0.500	0.050	a
Acrylamide, N-methoxymethyl-2-methyl-	-0.500	0.310	b
Styrene, p-tert-butyl-o-chloro-	-0.500	1.080	b
Pyridine, 2-vinyl-	-0.500	1.300	

Table I (Cont'd.)

Monomer	e	Q	Notes
1H-Indazole, 1-isopropenyl-	-0.490	0.430	b
Versatic [10] acid, vinyl ester	-0.480	0.037	
Octadecanoic acid, 4-oxo-, vinyl ester	-0.480	0.037	c
2,4,8,10-Tetraoxaspiro[5.5]undecane, 3,9-diisopropenyl-	-0.480	0.050	
Biphenyl, 4-vinyl-	-0.480	0.790	b
Hexanoic acid, vinyl ester	-0.470	0.025	
Norbornane, 2-vinyl-	-0.460	0.045	
Acetamide, 2-acrylamido-	-0.460	0.810	b
3(2H)-Pyridazinone, 6-hydroxy-2-phenyl-	-0.450	0.026	b
Hydrazine, 1-acryloyl-2-benzoyl-	-0.440	0.420	a
Methacrylic acid, zinc salt	-0.440	9.900	b
Acetic acid, chloro-, allyl ester	-0.430	0.011	a,b
1,3-Butadiene, 2-fluoro-	-0.430	2.080	
Acrolein, bis(2-butoxyethyl)acetal	-0.420	0.010	b
Silane, triethoxyvinyl-	-0.420	0.028	a
p-Anisic acid, vinyl ester	-0.420	0.048	
Undecanoic acid, 11-iodo-, allyl ester	-0.410	0.024	b
Acrolein, bis(2-ethoxyethyl)acetal	-0.400	0.010	b
1-Buten-3-yne	-0.400	0.690	b
Phosphonic acid, (2-methyl-1,3-butadienyl)-, diisopropyl ester	-0.400	1.090	
Bismuthine, diphenyl(p-vinylphenyl)-	-0.400	1.150	b
Styrene, p-iodo-	-0.400	1.170	b
1-Propene, 1-chloro-, trans-	-0.390	0.008	
Maleic acid, dinonyl ester	-0.390	0.220	b
Styrene, β-chloro-α,β-difluoro-	-0.390	0.340	b
Fumaric acid, dinonyl ester	-0.390	0.600	b
Silane, trimethoxyvinyl-	-0.380	0.031	a
Silane, diethoxyphenylvinyl-	-0.380	0.034	a
2-Oxazolidinone, 5-decyl-3-vinyl-	-0.380	0.050	
Butyramide, N-allyl-3-methyl-	-0.380	0.130	
Acetamide, N-allyl-	-0.380	0.180	
Naphthalene, 2-vinyl-	-0.380	1.250	
Phosphine sulfide, diphenyl(p-vinylphenyl)-	-0.380	1.530	
Ether, 2-bromovinyl ethyl	-0.370	0.012	c
Benzamide, N-allyl-	-0.370	0.095	
Tin, (methacryloyloxy)trimethyl-	-0.370	0.450	b
Silane, triisopropoxyvinyl-	-0.360	0.031	a
Isocyanic acid, p-vinylphenyl ester	-0.360	0.940	b
Styrene, m-chloro-	-0.360	1.030	
Styrene, o-chloro-	-0.360	1.280	
Silane, diethoxyethylvinyl-	-0.350	0.011	a,b
Cyclobutanecarboxylic acid, 3-acetyl-2,2-dimethyl-, vinyl ester	-0.350	0.034	
Silane, triethoxy(5-norbornen-2-yl)-	-0.350	0.072	a
Phosphine, diphenylvinyl-	-0.350	0.100	a
Acetamide, N-allyl-2-chloro-	-0.340	0.120	
Succinimide, N-vinyl-	-0.340	0.130	
Styrene, p-chloro-	-0.330	1.030	

Table I (Cont'd.)

Monomer	e	Q	Notes
2,4,8,10-Tetraoxaspiro[5.5]undecane, 3,9-dipropenyl-	-0.320	0.010	b
Versatic[13]acid, vinyl ester	-0.320	0.026	b
1-Butene, 1-phenyl-, trans-	-0.320	0.100	b
Styrene, p-bromo-	-0.320	1.040	
Phosphine, diphenyl(p-vinylphenyl)-	-0.320	1.290	
Octanoic acid, vinyl ester	-0.310	0.021	
Styrene, 2,4-difluoro-	-0.310	0.650	b
Naphthalene, 4-chloro-1-vinyl-	-0.310	0.740	b
Stibine, diphenyl(p-vinylphenyl)-	-0.300	1.120	b
Benzenesulfonic acid, p-vinyl-, potassium salt	-0.300	1.140	
Octadecanoic acid, 9(10)-oxo-, vinyl ester	-0.290	0.039	b
1-Hexen-3-yne	-0.290	0.600	
Styrene, m-trifluoromethyl-	-0.290	0.920	
1-Hexene	-0.280	0.019	b
3(2H)-Pyridazinone, 2(p-chlorophenyl)-6-hydroxy-	-0.280	0.074	a
Pyridine, 4-vinyl-	-0.280	1.000	
Methanol, diphenyl(p-vinylphenyl)-	-0.270	1.340	b
Methacrylic acid, cadmium salt	-0.270	8.600	b
Carbonic acid, ethyl vinyl ester	-0.260	0.003	
Butyric acid, vinyl ester	-0.260	0.042	
Oxalic acid, ethyl vinyl ester	-0.260	0.092	
Benzenesulfonic acid, p-vinyl-	-0.260	1.040	b
Ethylene, bromo-	-0.250	0.047	
D-Glucitol, 1,2:3,4:5,6-tri-O-(2-butenylidene)-	-0.250	0.100	b
Acrylamide, N-(1-benzyl-2-oxopropyl)-2-methyl-	-0.250	0.490	b
2-Propene-1-sulfonic acid, sodium salt	-0.240	0.150	b
Acetophenone, 4'-vinyl-, oxime	-0.240	0.530	
3(2H)-Pyridazinone, 6-hydroxy-, acetate	-0.230	0.058	
Vinyl acetate	-0.220	0.026	
Acetaldehyde, divinyl acetal	-0.220	0.026	b
Isovaleraldehyde, divinyl acetal	-0.220	0.026	b
Isobutyraldehyde, divinyl acetal	-0.220	0.026	b
Butyraldehyde, divinyl acetal	-0.220	0.026	b
Methane, bis(vinyloxy)-	-0.220	0.026	b
Phosphine oxide, diphenyl(p-vinylphenyl)-	-0.220	1.380	
Acrylamide, N-ethyl-N-(1,1-dihydroperfluorobutyl)-	-0.210	0.750	b
Styrene, m-bromo-	-0.210	1.070	
Styrene, α-(difluoromethyl)-	-0.210	1.160	b
Benzonitrile, p-vinyl-	-0.210	1.860	
Ethylene	-0.200	0.015	
1-Buten-3-yne, 2-methyl-	-0.180	0.540	
Butyric acid, 2-methylene-, ethyl ester	-0.180	0.610	b
Succinic acid, methylene-, 2-(4,6-dimethoxy-s-triazin-2-yl) hydrazide	-0.170	0.520	
Acrylic acid, 3-butoxy-, butyl ester	-0.160	0.008	b
Silane, allyldimethyl(p-vinylphenyl)-	-0.160	0.870	b
2-Picolinium methyl sulfate, 1-methyl-5-vinyl-	-0.160	2.490	b

Table I (Cont'd.)

Monomer	e	Q	Notes
m-Dioxane, 2-vinyl-	-0.150	0.030	b
Myristic acid, vinyl ester	-0.150	0.030	b
2-Propen-1-ol, 2-methyl-, oxalate	-0.150	0.038	b
Isoxazole, 3-isopropenyl-5-methyl-	-0.150	0.880	b
Acrylanilide, 4'-((2-chloroethyl)sulfamoyl)-2-methyl-	-0.140	0.980	
Methacrylic acid, calcium salt	-0.140	2.410	
Styrene, β-chloro-α,β-difluoro-p-methyl-	-0.130	0.270	b
s-Triazine, 2,4-diamino-6-isopropenyl-	-0.130	0.890	b
Naphthalene, 6-chloro-2-vinyl-	-0.130	1.250	
1,3-Dioxolane, 2-vinyl-	-0.120	0.020	
1,3-Dioxolane-4-methanol, 2-vinyl-	-0.120	0.040	
Sulfide, ethyl vinyl	-0.120	0.370	
Acrylic acid, sodium salt	-0.120	0.710	
Benzoic acid, p-vinyl-	-0.120	0.760	
Styrene, p-fluoro-	-0.120	0.830	
m-Dioxane, 5,5-dimethyl-2-vinyl-	-0.110	0.055	b
Acrylamide, N-(1,1-dimethyl-3-oxobutyl)-	-0.110	0.420	
Phosphonic difluoride, (2-methyl-1,3-butadienyl)-	-0.110	1.060	b
2-Pyrrolidinethione, 1-vinyl-	-0.110	1.480	
Decanoic acid, vinyl ester	-0.100	0.020	
Valeric acid, vinyl ester	-0.100	0.034	b
Acrylamide, N-tert-octyl-	-0.100	0.200	
Acetophenone, 4'-vinyl-	-0.100	0.370	
Pyridine, 5-ethyl-2-vinyl-, 1-oxide	-0.100	4.520	
Hexanoic acid, 2-ethyl-, vinyl ester	-0.080	0.024	
Stilbene, trans-	-0.080	0.030	
D-Glucitol, 1,2:3,4:5,6-tri-O-allylidene-	-0.080	0.050	
Silane, (chloromethyl)dimethyl(p-vinylphenyl)-	-0.080	0.820	b
Phosphonic acid, allyl-, diethyl ester	-0.070	0.019	b
Phosphine oxide, diallylphenyl-	-0.070	0.051	b
Phosphonic acid, phenyl-, diallyl ester	-0.070	0.051	b
Propionic acid, vinyl ester	-0.070	0.052	b
Phosphonic acid, butyl-, diallyl ester	-0.070	0.053	b
Acetophenone, 4'-isopropenyl-, oxime	-0.060	0.610	b
m-Dioxane, 4-methyl-2-vinyl-	-0.050	0.029	b
Acrylamide, N,N-dibutyl-	-0.050	0.320	
s-Triazine, 2-amino-4-(dimethylamino)-6-vinyl-	-0.050	0.780	b
Styrene, 2,5-bis(trifluoromethyl)-	-0.050	1.110	
2,4-Pentadienoic acid, 2,3-epoxypropyl ester	-0.040	4.790	b
Silane, dimethyl(p-vinylphenyl)-	-0.040	0.970	b
Stilbene, cis-	-0.030	0.017	b
Palmitic acid, vinyl ester	-0.020	0.026	
Benzoic acid, p-bromo-, vinyl ester	-0.020	0.053	c
Ethenesulfonic acid	-0.020	0.093	
Acrylamide, N-octyl-	-0.020	0.190	
Arsine, diphenyl(p-vinylphenyl)-	-0.020	1.160	b
Benzophenone, 4-vinyl-	-0.020	2.970	b

Table I (Cont'd.)

Monomer	e	Q	Notes
1,3-Butadiene, 2-chloro-	-0.020	7.260	
Lauric acid, vinyl ester	-0.010	0.018	b
Pyridine, 2-vinyl-, 1-oxide	-0.010	3.770	b
Methacrylic acid, magnesium salt	0.000	0.400	b
Methacrylic acid, potassium salt	0.010	0.540	b
s-Triazine, 2-amino-4-isopropenyl-6-(N-methylanilino)-	0.010	1.800	c
Silane, trimethylvinyl-	0.040	0.029	c
Acrylic acid, 2-methoxy-, methyl ester	0.040	0.470	b
Methacrylic acid, furfuryl ester	0.040	0.780	
Pyrimidine, 2-(dimethylamino)-4-vinyl-	0.040	1.450	b
Styrene, 2,4-dibromo-	0.050	1.530	
Glycine, N-allyl-	0.060	0.039	b
Styrene, α,β,β-trifluoro-o-methyl-	0.060	0.430	b
Phthalimide, N-1-(1,3-butadienyl)-	0.060	1.570	b
Acrylamide, N-(carbamoylmethyl)-2-methyl-	0.060	1.660	b
Acrylamide, N-(1-acetyl-3-methylbutyl)-	0.070	0.270	b
Benzamide, p-vinyl-	0.070	1.060	b
Phosphonic acid, (2-methyl-1,3-butadienyl)-, dimethyl ester	0.080	1.230	b
1,5,9-Cyclododecatriene, trans-, trans-, cis-	0.090	0.170	
4H-1,3-Thiazine, 5,6-dihydro-4,4-dimethyl-2-vinyl-	0.090	0.320	b
Styrene, 2,5-dichloro-	0.090	1.600	
Succinimide, N-1-(1,3-butadienyl)-	0.090	1.750	b
1-Propene, 3-chloro-	0.110	0.056	c
Biphenyl, 4-fluoro-4'-vinyl-	0.110	0.960	b
Benzoic acid, p-chloro-, vinyl ester	0.130	0.052	
Ethenesulfonic acid, allyl ester	0.130	0.150	
s-Triazine, 2-amino-4-o-toluidino-6-vinyl-	0.130	1.050	b
Thiazole, 4-methyl-2-vinyl-	0.130	3.200	b
Acrylic acid, 3-propoxy-, propyl ester	0.140	0.011	b
Benzoic acid, p-cyano-, vinyl ester	0.140	0.046	
m-Dioxane, 5,5-bis(bromomethyl)-2-vinyl-	0.140	0.058	
Acrylanilide, 4'-((2-chloroethyl)sulfamoyl)-	0.140	0.620	
Phosphonic dichloride, (2-methyl-1,3-butadienyl)-	0.150	1.560	b
Pyridine, 2-isopropenyl-	0.160	1.100	b
s-Triazine, 2-amino-4-(dimethylamino)-6-isopropenyl-	0.160	1.450	b
Methacrylic acid, 2(tert-butylamino)ethyl ester	0.170	0.980	d
2-Picolinium chloride, 5-vinyl-	0.170	1.160	
Acrylic acid, 3-ethoxy-, ethyl ester	0.180	0.015	c
Silane, (1-chlorovinyl)triethoxy-	0.180	0.230	a
Sorbic acid, 2,3-epoxypropyl ester	0.180	0.560	b
Methacrylic acid, octyl ester	0.180	1.140	
Benzoic acid, p-vinyl-, 2,3-epoxypropyl ester	0.180	1.140	
Methacrylic acid, pentamethyldisiloxanylmethyl ester	0.190	0.740	
Ethylene, chloro-	0.200	0.044	
Ethanesulfonic acid, 2-hydroxy-, sodium salt, methacrylate	0.200	0.530	b
Methacrylic acid, 2-hydroxypropyl ester	0.200	0.790	
Methacrylic acid, 2-methoxyethyl ester	0.200	0.790	d

Table I (Cont'd.)

Monomer	e	Q	Notes
Methacrylic acid, 2-hydroxyethyl ester	0.200	0.800	b, d
Benzenesulfonyl fluoride, p-vinyl-	0.200	1.640	
β-Alanine, N-allyl-	0.220	0.044	b
Acrylic acid, ethyl ester	0.220	0.520	
Styrene, α,β,β-trifluoro-	0.220	0.750	
Carbonic acid, divinyl ester	0.230	0.035	
Phosphonic acid, vinyl-, diethyl ester	0.230	0.160	c
1-Propanesulfonic acid, 3-hydroxy-, methacrylate, sodium salt	0.230	0.510	b
s-Triazine, 2-amino-4-p-anisidino-6-isopropenyl-	0.230	1.020	b
Cinnamic acid, vinyl ester	0.240	0.240	
Methacrylic acid, sec-butyl ester	0.240	0.720	b
Methacrylic acid, 2-(1-aziridinyl)ethyl ester	0.240	0.740	
Methacrylic acid, tert-butyl ester	0.240	0.760	
Ethylene glycol, dimethacrylate	0.240	0.880	
Styrene, α,β,β-trifluoro-p-methyl-	0.240	0.880	b
Oxazole, 4-isobutyl-2-isopropenyl-5-methyl-	0.240	4.100	b
Thiazole, 2-isopropenyl-	0.240	4.800	
Isothiocyanic acid, vinyl ester	0.250	0.540	
Methacrylaldehyde	0.250	1.450	
Oxazole, 2-isopropenyl-4,5-dimethyl-	0.250	2.880	b
5-Norbornene-2-carboxylic acid, methyl ester	0.260	0.059	
Silane, allyltrimethyl-	0.270	0.036	b
s-Triazine, 2-amino-4-p-chloroanilino-6-isopropenyl-	0.270	0.890	b
5-Norbornene-2-carbonitrile	0.280	0.070	d
2-Oxazolin-5-one, 2-isopropenyl-4,4-dimethyl-	0.280	1.360	b
Benzenesulfonamide, N-(2-chloroethyl)-p-vinyl-	0.280	1.690	b
2,4-Pentadienenitrile	0.280	5.980	a,b
s-Triazine, 2-amino-4-anilino-6-isopropenyl-	0.290	1.070	b
Benzoyl chloride, p-vinyl-	0.290	2.460	b
2-Norbornanecarboxylic acid, endo-, vinyl ester	0.300	0.014	b
2-Norbornanecarboxylic acid, exo-, vinyl ester	0.300	0.014	b
Silane, allyldimethylphenyl-	0.300	0.070	b
Methacrylic acid, lithium salt	0.300	0.640	b
Hydrazinium hydroxide, 1,1,1-trimethyl-2-(p-vinylbenzoyl)-, inner salt	0.300	0.880	b
Oxazole, 4-isobutyl-5-methyl-2-vinyl-	0.300	4.200	b
s-Triazine, 2-amino-4-(N-methylanilino)-6-vinyl-	0.310	0.820	b
Methacrylic acid, isobornyl ester	0.310	0.910	
3-Cyclohexene-1-carboxylic acid, 1-methyl-, 2,3-epoxypropyl ester	0.320	0.014	c
2,4,8,10-Tetraoxaspiro[5.5]undecane, 3,9-bis(1-chlorovinyl)-	0.340	0.200	b
Acrylic acid, sec-butyl ester	0.340	0.410	b
Methacrylic acid, hexyl ester	0.340	0.670	b
Hydrazinium chloride, 2-acryloyl-1,1,1-trimethyl-	0.340	0.690	
Methacrylic acid, isopropyl ester	0.340	0.850	
Methacrylic acid, isobutyl ester	0.340	0.870	
Benzoic acid, p-isopropenyl-, methyl ester	0.340	1.010	b

Table I (Cont'd.)

Monomer	e	Q	Notes
Phosphine oxide, diisobutylvinyl-	0.350	0.240	a
Acrylamide, N-(1-acetyl-3-methylbutyl)-2-methyl-	0.350	0.370	b
Methacrylic acid, tetrahydrofuryl ester	0.350	0.450	b
Methacrylic acid, dodecyl ester	0.350	0.700	
s-Triazine, 2-amino-4-isopropenyl-6-p-toluidino-	0.350	0.930	b
Crotonic acid, butyl ester	0.360	0.007	a,b
Crotonic acid, propyl ester	0.360	0.011	a,b
Crotonic acid, methyl ester	0.360	0.015	b
Angelonitrile	0.360	0.015	b
Phthalic acid, diallyl ester	0.360	0.044	b
Allyl alcohol	0.360	0.048	
4-Stilbenamine, N,N-dimethyl-	0.360	0.060	
Ethylene, 1,1-dichloro-	0.360	0.220	
Acetophenone, 4'-isopropenyl-	0.360	0.420	b
Acrylic acid, tetrahydrofurfuryl ester	0.360	0.540	
Methacrylic acid, phenethyl ester	0.360	0.740	
Benzenesulfonamide, p-vinyl-	0.370	1.620	b
1,3-Dioxolan-4-ol, 2-vinyl-, methacrylate	0.380	0.810	b
Acrylic acid, 2-ethylhexyl ester	0.390	0.410	d
Methacrylic acid, nonyl ester	0.390	0.820	
Styrene, p-nitro-	0.390	1.630	
Propane, 1-(allyloxy)-2,3-epoxy-	0.400	0.018	b
Cinnamic acid, 2-chloroethyl ester	0.400	0.260	b
s-Triazine, 2-amino-4-p-toluidino-6-vinyl-	0.400	0.480	b
Acrylonitrile, 2-methoxy-	0.400	0.720	b
Methyl methacrylate	0.400	0.740	
Methyl methacrylate-methyl-14C	0.400	0.740	
Methyl methacrylate-methyl-D	0.400	0.740	
Methyl methacrylate-methyl-T	0.400	0.740	
D-Galactopyranose, 1,2:3,4-di-O-isopropylidene-, 6-methacrylate	0.400	0.820	b
Oxazole, 4,5-dimethyl-2-vinyl-	0.400	2.130	
Ethenesulfonic acid, sodium salt	0.410	0.064	a
Cinnamic acid, benzyl ester	0.410	0.260	b
Tin, bis(methacryloyloxy)dimethyl-	0.410	1.360	b
Ethenesulfonamide, N-(2-chloroethyl)-	0.420	0.125	b
s-Triazine, 2-amino-4-p-anisidino-6-vinyl-	0.420	0.480	b
Methacrylic acid, benzyl ester	0.420	0.700	
Methacrylic acid, 2-(diethylamino)ethyl ester	0.420	2.080	
Germane, trimethylvinyl-	0.430	0.021	
Methacrylic acid, pentyl ester	0.430	0.680	b
Crotonaldehyde	0.440	0.016	c
Phenetole, p-styryl-	0.440	0.069	
1-Propene, 2,3-dichloro-	0.440	0.270	c
Acrylic acid, 2-hydroxy-, ethyl ester, acetate	0.440	0.440	
Methacrylic acid, propyl ester	0.440	0.650	
Sorbic acid, methyl ester	0.440	0.790	b

Table I (Cont'd.)

Monomer	e	Q	Notes
Styrene, α,β,β-trifluoro-m-methyl-	0.440	1.070	
Crotonic acid	0.450	0.013	c
s-Triazine, 2,4-dianilino-6-vinyl-	0.450	0.700	b
Methacrylic acid, cyclohexyl ester	0.450	0.820	
Pyrimidine, 4-vinyl-	0.450	2.180	b
Benzonitrile, p-(4-amino-6-vinyl-s-triazin-2-yl)amino-	0.470	0.570	b
Methacrylic acid, 2-(dimethylamino)ethyl ester	0.470	0.680	
1,3-Butadiene, hexafluoro-	0.470	0.930	
Cinnamic acid, methyl ester	0.480	0.140	c
Valeric acid, 4-methyl-2-methylene-, methyl ester	0.480	0.370	b
Aziridine, 1-methacryloyl-	0.480	0.760	b
Methacrylic acid, decyl ester	0.480	1.370	b
1,3-Butadiene, 2,3-dichloro-	0.480	12.860	c
Valeric acid, 2-methylene-, methyl ester	0.490	0.410	b
2-Oxazolin-5-one, 4-ethyl-2-isopropenyl-4-methyl-	0.490	1.320	b
Crotononitrile, trans-	0.500	0.013	c
Phosphine oxide, diphenylvinyl-	0.500	0.070	a
3-Butenoic acid, 2-ethyl-2-methyl-, ethyl ester	0.500	0.140	
Cinnamic acid	0.500	0.190	a,b
Cinnamic acid, tert-butyl ester	0.500	0.190	b
Succinic acid, methylene-	0.500	0.760	c
1-Propene, 3,3,3-trifluoro-	0.510	0.170	b
Phosphinothioic acid, diphenyl-, O-2-(methacryloxy) ethyl ester	0.510	0.750	b
Methacrylic acid, butyl ester	0.510	0.780	
Methacrylic acid, 2-hydroxypropyl ester, nitrate	0.510	0.980	b
Acrylic anhydride	0.510	1.270	
Styrene, 2,3,4,5,6-pentachloro-	0.520	0.220	
Butyric acid, 2-methylene-, methyl ester	0.520	0.420	b
Methacrylic acid, ethyl ester	0.520	0.730	
Succinic acid, methylene-, dipentyl ester	0.520	1.020	b
Lauric acid, allyl ester	0.520	0.031	b
Methacrylic acid, 2-chloroethyl ester	0.530	0.810	
Methacrylic acid, 2-hydroxyethyl ester, nitrate	0.530	0.990	b
3-Buten-2-one, 3-methyl-	0.530	1.490	
Acrylic acid, 2-chloroethyl ester	0.540	0.410	c
Acetic acid, methacrylamino-, ethyl ester	0.540	0.720	b
2-Oxazolin-5-one, 4-isobutyl-2-isopropenyl-4-methyl-	0.540	1.420	b
Acrylic acid, isopropyl ester	0.550	0.410	d
Hexanoic acid, 2-methylene-, methyl ester	0.550	0.420	b
Benzamide, p-vinyl-, lithium chloride complex	0.550	3.400	b
Ether, allyl phenyl	0.560	0.015	
2-Propen-1-ol, 2-chloro-	0.560	0.240	a
Methacrylic acid, octadecyl ester	0.560	1.070	b
Ethanesulfonic acid, 2-hydroxy-, methacrylate	0.560	1.340	
Acrylic acid, 2-methoxyethyl ester	0.570	0.410	b
Methacrylic acid, 2-bromoethyl ester	0.570	0.950	b
Methacrylic acid, 2,3-epoxypropyl ester	0.570	1.030	

Table I (Cont'd.)

Monomer	e	Q	Notes
Tin, (1,2-butadien-2-yl)triethyl-	0.570	1.500	b
Butyric acid, heptafluoro-, vinyl ester	0.580	0.038	b
Acrylamide, 2-methyl-N-(2-oxopropyl)-	0.580	0.290	b
Methacrylic acid, X-hydroxybutyl ester, nitrate	0.580	0.900	b
Acrylic acid, thio-, methyl ester	0.580	1.230	b
s-Triazine, 2-amino-4-anilino-6-vinyl-	0.590	0.550	b
Methacrylic acid, bornyl ester	0.590	0.790	
Methacrylic acid, 2-nitropropyl ester	0.590	0.840	b
Cinnamic acid, ethyl ester	0.600	0.170	
Acrylic acid, methyl ester	0.600	0.420	
Acrylic acid, methyl-14C ester	0.600	0.420	
Acrylic acid, methyl-D ester	0.600	0.420	
Carbonic acid, cyclic (hydroxymethyl)ethylene ester, acrylate	0.600	0.440	b
Aziridine, 1-acryloyl-	0.600	0.580	b
3-Penten-2-one	0.610	0.024	b
Sulfoxide, ethyl vinyl	0.610	0.130	b
Acrylamide, N-(hydroxymethyl)-	0.610	0.390	
1,3-Dioxolane-4-methanol, 2,2-dimethyl-, 4-(methylene-succinate)	0.610	0.390	
Methacrylic acid, 2-propynyl ester	0.620	1.070	b
1,3,4-Oxadiazole, 2-isopropenyl-5-methyl-	0.620	2.000	b
Acrylic acid, 2-butoxyethyl ester	0.630	0.420	b
Dodecanamide, N-(acrylamidomethyl)-	0.630	1.120	
2-Propen-1-one, 1-phenyl-3-(3-pyridyl)-	0.640	0.640	a,b
Cinnamic acid, p-chloro-, methyl ester	0.650	0.150	a,b
Cinnamic acid, phenyl ester	0.650	0.200	b
1,3-Dioxolan-4-ol, 2-vinyl-, acrylate	0.650	0.410	
Methacrylic acid	0.650	2.340	
Cinnamic acid, butyl ester	0.660	0.250	b
Metanilyl fluoride, N-acryloyl-	0.660	0.550	
Hydrazinium chloride, 2-methacryloyl-1,1,1-trimethyl-	0.660	0.610	
Methacrylic acid, 2-methyl-2-nitropropyl ester	0.660	0.970	b
2,4-Pentadienoic acid, methyl ester	0.660	1.640	
Acrylophenone	0.660	2.900	b
Styrene, α-chloro-β-fluoro-	0.670	0.150	b
2-Norpinane-2-ethanol, acrylate	0.670	0.500	
D-Glucitol, 1-acrylamido-1-deoxy-	0.680	0.150	
1,3,5,2,4,6-Triazophosphorine, 2-chloro-2,2,4,4,6,6-hexa-hydro-2,4,4,6,6-pentaallyloxy-	0.680	0.600	b
3-Buten-2-one	0.680	0.690	
Acrylic acid, 2-hydroxyethyl ester	0.680	0.840	b
Acrylic acid, 2-propynyl ester	0.690	0.680	b
Styrene, β,β-difluoro-	0.700	0.029	a
Succinic acid, methylene-, dipropyl ester	0.710	1.190	b
Triethylene glycol, dimethacrylate	0.710	1.460	b
Hytrocinnamic acid, α-methylene-, methyl ester	0.720	0.500	b
Acrylic acid, propyl ester	0.720	0.660	b

Table I (Cont'd.)

Monomer	e	Q	Notes
Acrylic acid, 3-hydroxy-2, 2-bis(hydroxymethyl) propyl ester trinitrate	0.720	0.780	
3-Oxa-1-azaspiro[4,5]decan-4-one, 2-isopropenyl-	0.720	2.470	b
Styrene, α, β-difluoro-	0.730	0.120o	b
Acrolein	0.730	0.850	
Methacrylic acid, phenyl ester	0.730	1.490	
Styrene, 2,5-difluoro-	0.730	6.700	b
Acrylamide, N-(1-methyl-2-oxopropyl)-	0.740	0.160	b
Styrene, 2,3,4,5,6-pentafluoro-	0.740	0.780	b
Acrylic acid, 2-hydroxypropyl ester	0.740	0.960	
Cinnamonitrile	0.750	0.320	b
1,3,5-Hexatriene, 2, 3, 4, 5-tetrachloro-, trans-	0.750	1.850	
Metanilyl fluoride, N-methacryloyl-	0.760	0.690	b
Methacrylic acid, furfuryl ester, π complex with O$_2$	0.760	0.850	b
1,3-Butadiene, hexachloro-	0.760	1.310	
1,3-Dioxolane-4-methanol, 2,2-dimethyl-, acrylate	0.770	0.500	
Methacrylic acid, 2-(diethylamino)ethyl ester, methiodide	0.770	0.840	b
Methacrylic acid, 2-nitrobutyl ester	0.770	1.140	b
Acrylic acid	0.770	1.150	
Acrylic acid, 2-chloro-, methyl ester	0.770	2.020	
Acrylic acid, X-hydroxybutyl ester, nitrate	0.780	0.690	b
Acrylic acid, 2-cyano-, methyl ester	0.780	2.140	
1,3-Butadiene, 1,1,2-trichloro-	0.780	4.040	
Succinic acid, methylene-, dioctyl ester	0.790	0.960	
Tin, tributylvinyl-	0.800	0.024	
Phosphonic acid, vinyl-, bis(2-bromoethyl) ester	0.800	0.060	c
2-Propen-1-one, 3-phenyl-1-(3-pyridyl)-	0.800	0.330	b
Methacrylophenone	0.800	1.330	b
Acrylic acid, 1-methyl-2-nitropropyl ester	0.810	0.470	b
Methacrylic acid, 2-(diethylamino)ethyl ester, hydrochloride	0.810	1.100	b
Methacrylonitrile	0.810	1.120	
Styrene, m-nitro-	0.810	2.470	
4-Pentenoic acid, 2-allyl-2-cyano-, ethyl ester	0.820	0.059	c
Cinnamic acid, o-chloro-, methyl ester	0.850	0.280	a
Methacrylic acid, 2,2-dinitropropyl ester	0.850	0.720	b
Cinnamic acid, α-cyano-, ethyl ester	0.870	1.240	a,b
s-Triazine, 2, 4-dimethyl-6-vinyl-	0.870	2.560	
Styrene, β-fluoro-	0.880	0.044	b
Ethylene glycol, bis(2-chloroacrylate)	0.880	0.430	b
Methacrylic acid, p-benzylideneamino)phenyl ester	0.880	1.080	b
3,5-Heptadien-2-one	0.880	1.620	
Acrylonitrile, 2-hydroxy-, acetate	0.880	1.820	
Succinic anhydride, methylene-	0.890	2.500	b
Acrylic acid, 2-methyl-2-nitropropyl ester	0.890	0.370	b
Pyridinium acetate, 2-vinyl-	0.890	1.620	b
Glutaric acid, 2-methylene-, diethyl ester	0.900	0.550	d
Acrylic acid, benzyl ester	0.900	0.680	

Table I (Cont'd.)

Monomer	e	Q	Notes
Methacrylic acid, 5-(methoxycarbonyl) furfuryl ester	0.900	0.720	b
2,4-Pentanedione	0.910	0.028	a
Glycolic acid, ethyl ester, acrylate	0.910	0.380	b
Crotononitrile, cis-	0.920	0.041	
Cyclohexane, 1,1,2,2,3,3,4,4-octafluoro-5,6-dimethylene-	0.930	0.730	
Tin, trimethylvinyl-	0.940	0.020	
Stilbene, 4-chloro-	0.940	0.051	
Sorbonitrile	0.940	1.910	
2-Propen-1-one, 1-(6-methyl-3-pyridyl)-3-phenyl-	0.960	0.260	b
Acrylic acid, 2,3-epoxypropyl ester	0.960	0.550	
Fumaric acid, ethyl ester	0.960	1.330	b
Sulfoxide, methyl vinyl	0.980	0.057	b
1-Phenanthrenemethanol, 1, 2, 3, 4, 4A, 4B, 5, 6, 10, 10A-decahydro-7-isopropyl-1, 4A-dimethyl-, acrylate	0.980	0.400	b
Phosphonic acid, hydroxymethyl-, diethyl ester, methacrylate	0.980	0.650	b
Succinic acid, methylene-, diethyl ester	0.980	0.940	b
Chalcone, 2-chloro-	0.990	0.370	b
Acrylic acid, 2-hydroxyethyl ester, nitrate	1.000	0.490	b
Acrylamide, N, N'-methylenebis-	1.000	0.740	b
Phosphinic acid, diethyl-, allyl ester	1.020	0.031	b
Maleimide, N-(2-hydroxyethyl)-, acetate	1.020	0.079	b
Fumaric acid, 1, 3-butyleneglycol polyester	1.020	0.290	c
Acryloyl chloride	1.020	1.780	
Methacrylic anhydride	1.030	1.600	
Ferrocenemethanol, acrylate	1.040	0.190	
Methacrylic acid, 1-methyl-2-nitropropyl ester	1.040	0.530	b
1-Propanol, 3-bromo-2, 2-bis(bromomethyl)-, acrylate	1.050	0.470	
Acetic acid, trifluoro-, vinyl ester	1.060	0.033	b
Acrylic acid, butyl ester	1.060	0.500	
Acrylic acid, octyl ester	1.070	0.350	
Methacryloyl chloride	1.070	0.950	b
Acrylic acid, 2-bromo-, ethyl ester	1.070	3.700	b
Stilbene, 4-nitro-	1.080	0.069	
Maleic acid, bis(2-ethylhexyl) ester	1.080	0.100	a, b
Succinic acid, methylene-, dibutyl ester	1.090	1.070	
1-Propanesulfonic acid, 3-hydroxy-, acrylate, sodium salt	1.100	0.300	b
Methacrylic acid, 2-(diethylamino)ethyl ester, ethiodide	1.100	2.100	b
Glutaronitrile, 2-methylene	1.110	0.570	
Acrylic acid, octadecyl ester	1.120	0.420	b
Acrylic acid, 2-fluoromethyl-, ethyl ester	1.120	0.820	b
Acrylamide, N-octadecyl-	1.130	0.660	
Malealdehyde	1.130	2.680	b
1-Propene-1, 2, 3-tricarboxylic acid, trans-, trimethyl ester	1.140	0.005	
Acrylic acid, 2-nitrobutyl ester	1.150	0.610	
1-Butanol, 2, 2, 3, 3, 4, 4, 4-heptafluoro-, acrylate	1.150	0.780	
Hydroquinone, vinyl-, dibenzoate	1.170	1.800	
Atropic acid, butyl ester	1.170	2.290	b

Table I (Cont'd.) Monomer	e	Q	Notes
Sulfone, phenyl vinyl	1.180	0.069	
Maleic anhydride, mono(O-benzyloxime)	1.180	0.960	
Ethenesulfonic acid, butyl ester	1.190	0.130	
Carbonic acid, cyclic (hydroxymethyl)ethylene ester, maleate	1.190	0.220	
Acrylamide	1.190	1.120	
Acrylonitrile	1.200	0.600	
Atropic acid, 2-chloroethyl ester	1.200	3.370	b
Atropic acid, methyl ester	1.200	4.780	
Succinic acid, methylene-, diisopropyl ester	1.210	0.890	b
Ethylene, tetrafluoro-	1.220	0.049	c
Acrylanilide	1.220	0.650	
Phosphonic acid, (dimethyleneethylene)di-, tetraethyl ester	1.220	2.720	
Methacrylamide	1.240	1.460	
Ethylene, 1,2-dichloro-, cis-	1.250	0.003	c
Fumaric acid, diethyl ester	1.250	0.610	c
Atropic acid, propyl ester	1.250	2.430	b
Ferrocenemethanol, methacrylate	1.260	0.130	
Acrylic acid, 2-bromo-, methyl ester	1.260	4.540	b
Atroponitrile	1.260	9.600	b
Maleic acid, dimethyl ester	1.270	0.090	
Carbonic acid, cyclic (hydroxymethyl)ethylene ester, 4-(methylenesuccinate)	1.270	1.120	
Acrolein, 2-chloro-	1.270	2.450	b
Ethylene, 1,2-dichloro-, trans-	1.280	0.010	c
Acetic acid, diethylphosphonyl-, allyl ester	1.290	0.014	b
Sulfone, methyl vinyl	1.290	0.110	c
s-Triazine, 2,4-diamino-6-vinyl-, hydrochloride	1.290	2.080	b
Maleic acid, dibenzyl ester	1.320	0.340	b
Maleimide, N-(hydroxymethyl)	1.320	0.490	
Vinyl sulfone	1.330	0.140	
2H-Azepin-2-one, hexahydro-1-methacryloyl-	1.340	0.180	a,b
Succinic acid, methylene-, dimethyl ester	1.340	1.030	
Phosphine sulfide, diphenylvinyl-	1.350	0.083	a
Maleimide	1.350	0.440	
Acrylamide, 2-methyl-N-(1-methyl-2-oxopropyl) -	1.360	0.200	b
1-Propene, 3,3,3-trichloro-	1.370	0.056	c
Ethenesulfonyl fluoride	1.390	0.120	b
Maleimide, N-(p-hydroxyphenyl) -	1.390	0.800	
4-Cyclopentene-1,3-dione	1.400	0.220	
Ethenesulfonic acid, 2,3-epoxypropyl ester	1.410	0.140	
Atropic acid, ethyl ester	1.410	4.280	
Maleimide, N-(2-hydroxyethyl) -	1.420	0.430	
4H-1,3-Oxazine, 5,6-dihydro-4,6,6-trimethyl-2-vinyl-	1.430	0.620	b
Succinic acid, methylene-, 2,3-epoxypropyl methyl ester	1.430	0.650	b
Fumaric acid, dioctyl ester	1.450	0.240	
Maleimide, N-2-bornyl-	1.470	0.480	
Ethylene, chlorotrifluoro-	1.480	0.020	

Table I (Cont'd.) Monomer	e	Q	Notes
D-Glucitol, 1-deoxy-1-methacrylamido-	1.480	0.170	
Acrylonitrile, 2-chloro-	1.480	2.160	
Maleimide, N,2-dimethyl-	1.480	2.350	
Maleic acid, diethyl ester	1.490	0.059	c
Phosphonic acid, 1-hydroxyethyl-, diethyl ester, acrylate	1.490	0.570	b
Fumaric acid, dimethyl ester	1.490	0.760	b
Phosphonic acid, hydroxymethyl-, diethyl ester, acrylate	1.500	0.760	
Succinic acid, methylene-, bis(2-chloroethyl) ester	1.520	0.380	c
Maleic acid, ethyl ester	1.520	1.230	b
1-Propene-1,2,3-tricarboxylic acid, cis-, triethyl ester	1.540	0.280	c
Maleimide, N-(3-hydroxypropyl)-, acetate	1.540	0.590	b
Maleic acid, dibutyl ester	1.600	0.042	a,b
1-Propene-1,2,3-tricarboxylic acid, cis-, trimethyl ester	1.620	0.320	
Acetic acid, methacryliminodi-, dimethyl ester	1.640	0.073	b
Maleic acid, 2-chloro-, diethyl ester	1.650	0.056	a,b
Maleimide, N-(2-hydroxyethyl)-, propionate	1.650	0.570	b
Maleimide, N-hexyl-	1.650	0.980	b
Ethylene:AgNO$_3$ complex (1:1)	1.660	29,400	b
Maleimide, N-octyl-	1.680	0.990	b
Phosphonic acid, α-phenylvinyl-	1.690	0.470	b
Methacrylic acid, 2-(ethylthio)ethyl ester	1.690	4.280	d
Maleic acid, 1,4-dithio-, cyclic anhydrosulfide	1.700	0.330	b
Acrylamide, N-benzyl-2-methyl-	1.710	1.140	b
Citraconic anhydride	1.750	0.870	b
Maleimide, N-butyl-	1.750	3.080	
Crotonamide	1.760	0.008	b
s-Triazine, 4,6-diamino-2-vinyl-	1.760	5.520	b
Fumaric acid, dithio-, dimethyl ester	1.780	1.230	b
Maleimide, N-propyl	1.790	1.510	b
Maleic acid, diisobutyl ester	1.850	0.094	b
Ethylene, trichloro-	1.860	0.019	c
Acrylonitrile, 2-bromo-	1.920	5.130	
Fumaric acid, ethyleneglycol polyester	1.930	0.420	
Benzenesulfonic acid, maleimido ester	1.930	5.650	b
Maleimide, N-(benzyloxy)-	1.930	5.650	b
Maleimide, N-ethyl-	1.940	3.110	b
Fumaronitrile	1.960	0.800	c
Maleimide, N-(methoxy)-	2.000	5.320	b
Ethylene, tetrachloro-	2.030	0.003	a
Benzoic acid, maleimido ester	2.050	3.420	b
Maleic acid, diisopropyl ester	2.060	0.084	c
Maleimide, N-methyl-	2.060	6.360	b
Crotonamide, N-(hydroxymethyl) -	2.100	0.035	b
Ethylene, 1,1-dichloro-2,2-difluoro-	2.100	0.041	a
Pyridinium fluoroborate, 1-vinyl-	2.120	1.120	
Maleimide, N-hydroxy-, acetate	2.120	4.850	b
Semicarbazide, 1-(methacryloylamidino)-, hydrochloride	2.140	0.570	b

Table I (Cont'd.)

Monomer	e	Q	Notes
Maleic acid, methyl ester	2.190	0.100	c
Maleic anhydride	2.250	0.230	c
1,3-Dioxolane-4-methanol, 2,2-dimethyl-, methyl fumarate	2.250	0.600	a,b
Maleonitrile	2.320	0.420	a,b
Cinnamic acid, α-cyano-, 2-ethylhexyl ester	2.360	0.130	a,b
Cinnamic acid, α-cyano-, butyl ester	2.420	0.140	a,b
Fumaric acid, diester with hydracrylonitrile	2.420	0.800	
Cinnamic acid, α-cyano-, methyl ester	2.490	0.140	a,b
Cinnamic acid, α-cyano-, cyclohexyl ester	2.560	0.150	a,b
Cinnamic acid, α-cyano-, benzyl ester	2.570	0.150	a,b
Cinnamic acid, α-cyano-, hexyl ester	2.580	0.160	c
Maleonitrile, methylene-	2.580	20.130	
Maleic acid	2.600	0.450	
Phosphonic dichloride, vinyl-	2.700	0.870	
Carbon monoxide	3.760	0.100	a

TABLE II

Monomer	e	Q	Notes
Acenaphthylene	-0.800	0.260	b
Acetaldehyde, divinyl acetal	-0.220	0.026	b
Acetamide, 2-acrylamido-	-0.460	0.810	b
Acetamide, N-allyl-	-0.380	0.180	
Acetamide, N-allyl-2-chloro-	-0.340	0.120	
Acetanilide, N-vinyl-	-1.540	0.120	
Acetanilide, 4'-vinyl-	-0.980	1.420	b
Acetic acid, allyl ester	-1.130	0.028	c
-",-, chloro-, allyl ester	-0.430	0.011	a,b
-",-, chloro-, vinyl ester	-0.650	0.074	c
-",-, dichloro-, vinyl ester	-0.680	0.170	
-",-, diethylphosphonyl-, allyl ester	1.290	0.014	b
-",-, methacrylamino-, ethyl ester	0.540	0.720	b
-",-, methacryliminodi-, dimethyl ester	1.640	0.073	b
-",-, thio-, vinyl ester	-1.460	0.310	b
-",-, trifluoro-, vinyl ester	1.060	0.033	b
Acetophenone, 4'-isopropenyl-	0.360	0.420	b
-",-, 4'-isopropenyl-, oxime	-0.060	0.610	b
-",-, 4'-vinyl-	-0.100	0.370	
-",-, 4'-vinyl-, oxime	-0.240	0.530	
Acetylene, diphenyl-	-1.230	0.003	a
Acrolein	0.730	0.850	
-",-, bis(2-ethoxyethyl) acetal	-0.400	0.010	b
-",-, bis(2-butoxyethyl) acetal	-0.420	0.010	b
-",-, 2-chloro-	1.270	2.450	b
Acrylamide	1.190	1.120	
-",-, N-(1-acetyl-3-methylbutyl)-	0.070	0.270	b
-",-, N-(1-acetyl-3-methylbutyl)-2-methyl-	0.350	0.370	b
-",-, N-benzyl-2-methyl-	1.710	1.140	b
-",-, N-(1-benzyl-2-oxopropyl)-	-0.680	0.440	b
-",-, N-(1-benzyl-2-oxopropyl)-2-methyl-	-0.250	0.490	b
Acrylamide (Cont'd.)			
-",-, N-(carbamoylmethyl)-2-methyl-	0.060	1.660	b
-",-, N,N-dibutyl-	-0.050	0.320	b
-",-, N,N-dimethyl-	-0.500	1.080	b
-",-, N,2-dimethyl-	-0.600	0.320	b
-",-, N-(1,1-dimethyl-3-oxobutyl)-	-0.110	0.420	b
-",-, N-ethyl-N-(1,1-dihydroperfluorobutyl)-	-0.210	0.750	b
-",-, N-ethyl-2-methyl-	-0.880	0.250	b
-",-, N-(hydroxymethyl)-	0.610	0.390	b
-",-, N-methoxymethyl-2-methyl-	-0.500	0.310	b
-",-, N,N'-methylenebis-	1.000	0.740	b
-",-, 2-methyl-N-(1-methyl-2-oxopropyl)-	1.360	0.200	b
-",-, 2-methyl-N-(2-oxopropyl)-	0.580	0.290	b
-",-, N-(1-methyl-2-oxopropyl)-	0.740	0.160	b
-",-, N-octadecyl-	1.130	0.660	b
-",-, N-octyl-	-0.020	0.190	b
-",-, N-tert-octyl-	-0.100	0.200	b
Acrylanilide	1.220	0.650	
-",-, 4'-((2-chloroethyl)sulfamoyl)-	0.140	0.620	b
-",-, 4'-((2-chloroethyl)sulfamoyl)-2-methyl-	-0.140	0.980	b
-",-, 4'-chloro-2-methyl-	-0.980	0.700	b
-",-, 2,4'-dimethyl-	-0.760	1.200	b
-",-, 4'-methoxy-2-methyl-	-1.190	2.800	b
Acrylic acid	0.770	1.150	
-",-, benzyl ester	0.900	0.680	
-",-, 2-butoxyethyl ester	0.630	0.420	b
-",-, butyl ester	1.060	0.500	b
-",-, sec-butyl ester	0.340	0.410	b
-",-, 2-chloroethyl ester	0.540	0.410	c
-",-, 2,3-epoxypropyl ester	0.960	0.550	
-",-, ethyl ester	0.220	0.520	

Table II (Cont'd.) Monomer	e	Q	Notes
Acrylic acid (Cont'd.)			
—, 2-ethylhexyl ester	0.390	0.410	d
—, 3-hydroxy-2,2-bis(hydroxymethyl)propyl ester			
trinitrate	0.720	0.780	
—, X-hydroxybutyl ester, nitrate	0.780	0.690	b
—, 2-hydroxyethyl ester	0.680	0.840	
—, 2-hydroxyethyl ester, nitrate	1.000	0.490	b
—, 2-hydroxypropyl ester	0.740	0.960	
—, isopropyl ester	0.550	0.410	d
—, 2-methoxyethyl ester	0.570	0.410	b
—, methyl ester	0.600	0.420	
—, methyl-14C ester	0.600	0.420	
—, methyl-D ester	0.600	0.420	
—, 1-methyl-2-nitropropyl ester	0.810	0.470	b
—, 2-methyl-2-nitropropyl ester	0.890	0.370	b
—, 2-nitrobutyl ester	1.150	0.610	
—, octadecyl ester	1.120	0.420	
—, octyl ester	1.070	0.350	
—, propyl ester	0.720	0.660	b
—, 2-propynyl ester	0.690	0.680	b
—, sodium salt	-0.120	0.710	
—, tetrahydrofurfuryl ester	0.360	0.540	b
—, 2-bromo-, ethyl ester	1.070	3.700	b
—, 2-bromo-, methyl ester	1.260	4.540	b
—, 3-butoxy-, butyl ester	-0.160	0.008	b
—, 2-chloro-, methyl ester	0.770	2.020	
—, 2-cyano-, methyl ester	0.780	2.140	
—, 3-ethoxy-, ethyl ester	0.180	0.015	c
—, 2-fluoromethyl-, ethyl ester	1.120	0.820	
—, 2-formamido-	-0.720	6.000	b
—, 2-methoxy-, ethyl ester, acetate	0.440	0.440	
—, 2-methoxy-, methyl ester	0.040	0.470	b
—, 3-propoxy-, propyl ester	0.140	0.011	b
—, thio-, methyl ester	0.580	1.230	b
Acrylic anhydride	0.510	1.270	
Acrylonitrile	1.200	0.600	
—, 2-bromo-	1.920	5.130	
—, 2-chloro-	1.480	2.160	
—, 2-hydroxy-, acetate	0.880	1.820	
—, 2-methoxy-	0.400	0.720	b
Acrylophenone	0.660	2.900	b
Acryloyl chloride	1.020	1.780	
β-Alanine, N-allyl-	0.220	0.044	b
Allyl alcohol	0.360	0.048	
Allyl ether	-0.540	0.015	b
Angelonitrile	0.360	0.015	b
Aniline, N,N-dimethyl-p-vinyl-	-1.370	1.510	

Table II (Cont'd.) Monomer	e	Q	Notes
Aniline, N,N-divinyl-	-1.540	0.190	
p-Anisic acid, vinyl ester	-0.420	0.048	
Anisole, p-vinyl-	-1.110	1.360	
Anthracene, 1-vinyl-	-1.680	2.500	b
—, 9-vinyl-	-1.600	0.900	b
Arsine, diphenyl(p-vinylphenyl)-	-0.020	1.160	b
Atropic acid, butyl ester	1.170	2.290	b
—, 2-chloroethyl ester	1.200	3.370	b
—, ethyl ester	1.410	4.280	
—, methyl ester	1.200	4.780	b
—, propyl ester	1.250	2.430	b
Atroponitrile	1.260	9.600	b
2H-Azepin-2-one, hexahydro-1-methacryloyl-	1.340	0.180	a, b
Aziridine, 1-acryloyl	0.600	0.580	b
—, 1-methacryloyl-	0.480	0.760	b
1-Aziridinecarboxamide, N-isopropenyl-	-2.160	0.190	b
—, N-vinyl-	-1.190	0.180	
Benzamide, N-allyl-	-0.370	0.095	
—, p-vinyl-	0.070	1.060	b
—, p-vinyl-, lithium chloride complex	0.550	3.400	b
Benzene, m-divinyl-	-1.770	3.350	
—, o-divinyl-	-1.310	1.640	b
—, ethynyl-	-0.660	0.350	b
—, propenyl-, cis-	-1.750	0.011	b
—, propenyl-, trans-	-1.530	0.040	b
Benzenesulfonamide, N-(2-chloroethyl)-p-vinyl-	0.280	1.690	b
—, p-vinyl-	0.370	1.620	b
Benzenesulfonic acid, maleimido ester	1.930	5.650	b
—, p-vinyl-	-0.260	1.040	b
—, p-vinyl-, potassium salt	-0.300	1.140	
—, p-vinyl-, sodium salt	-0.590	2.490	
Benzenesulfonyl fluoride, m-vinyl-	-0.730	1.330	
—, p-vinyl-	0.200	1.640	
Benzoic acid, maleimido ester	2.050	3.420	b
—, vinyl ester	-0.550	0.061	
—, p-bromo-, vinyl ester	-0.020	0.053	
—, p-chloro-, vinyl ester	0.130	0.052	
—, p-cyano-, vinyl ester	0.140	0.046	
—, p-isopropenyl-, methyl ester	0.340	1.010	b
—, p-vinyl-	-0.120	0.760	
—, p-vinyl-, 2,3-epoxypropyl ester	0.180	1.140	
—, p-vinyl-, sodium salt	-0.740	0.650	b
Benzonitrile, p-[(4-amino-6-vinyl-s-triazin-2-yl)amino]-	0.470	0.570	b
—, p-vinyl-	-0.210	1.860	b
Benzophenone, 4-vinyl-	-0.020	2.970	b
Benzothiazole, 2-vinylthio-	-0.920	1.680	b
Benzoyl chloride, p-vinyl-	0.290	2.460	b

Table II (Cont'd.) Monomer	e	Q	Notes
Benzyl alcohol, α,α-diethyl-p-vinyl-	-0.800	0.970	
-", α,α-dimethyl-p-vinyl-	-1.930	4.940	b
-", α-ethyl-α-methyl-p-vinyl-	-1.000	0.820	b
-", α-methylene-, acetate	-0.650	0.820	b
Biphenyl, 4-fluoro-4'-vinyl-	0.110	0.960	b
-", 4-vinyl-	-0.480	0.790	b
Bismuthine, diphenyl(p-vinylphenyl)-	-0.400	1.150	
1,3-Butadiene	-1.050	2.390	
-", 2-chloro-	-0.020	7.260	
-", 2,3-dichloro-	0.480	12.860	c
-", 2,3-dimethyl-	-1.810	5.860	
-", 2-fluoro-	-0.430	2.080	
-", hexachloro-	0.760	1.310	
-", hexafluoro-	0.470	0.930	
-", 1,1,2-trichloro-	0.780	4.040	
Butane, 1,4-bis(p-vinylphenyl)-	-1.170	1.200	b
1-Butanol, 2,2,3,3,4,4,4-heptafluoro-, acrylate	1.150	0.780	
2-Butanol, 1-(p-vinylphenyl)-	-0.830	0.760	
1-Butene, 3,3-dimethyl-	-0.630	0.007	a,b
-", 3-methyl-1-phenyl-, cis-	-1.310	0.002	b
-", 3-methyl-1-phenyl-, trans-	-1.630	0.006	b
-", 1-phenyl-, cis-	-1.670	0.008	b
-", 1-phenyl-, trans-	-0.320	0.100	b
2-Butene, 2,3-dimethyl-	-1.600	0.005	
-", 2-methyl-	-1.650	0.013	
3-Butenoic acid, 2-ethyl-2-methyl-, ethyl ester	0.500	0.140	
3-Buten-2-one	0.680	0.690	
-", 3-methyl-	0.530	1.490	
1-Buten-3-yne	-0.400	0.690	b
-", 2-methyl-	-0.180	0.540	
Butyraldehyde, divinyl acetal	-0.220	0.026	b
-", 2-methylene-	-0.330	1.525	
Butyramide, N-allyl-3-methyl-	-0.380	0.130	
Butyric acid, vinyl ester	-0.260	0.042	
-", heptafluoro-, vinyl ester	0.580	0.038	b
-", 2-methylene-, ethyl ester	-0.180	0.610	b
-", 2-methylene-, methyl ester	0.520	0.420	b
Carbamic acid, diethyl-, vinyl ester	-1.060	0.028	
-", diethyldithio-, S-vinyl ester	-1.540	0.450	b
-", diethylthio-, S-vinyl ester	-1.460	0.300	
-", isopropenyl-, 2,3-epoxypropyl ester	-0.860	0.260	b
-", N-methyl-N-vinyl thio-, S-ethyl ester	-1.290	0.110	
-", vinyl-, 2,3-epoxypropyl ester	-1.150	0.180	
-", vinyl-, ethyl ester	-1.620	0.120	
Carbazole, 9-vinyl-	-1.400	0.410	
Carbonic acid, cyclic (hydroxymethyl)ethylene ester, acrylate	0.600	0.440	
-", -", maleate	1.190	0.220	
-", -", 4-methylenesuccinate	1.270	1.120	

Table II (Cont'd.) Monomer	e	Q	Notes
Carbonic acid (Cont'd.)			
-", cyclic phenylvinylene ester	-1.620	1.440	
-", cyclic vinylene ester	-0.650	0.007	
-", divinyl ester	0.230	0.035	
-", ethyl vinyl ester	-0.260	0.003	
-", thio-, O-tert-butyl S-vinyl ester	-1.260	0.520	
Carbon monoxide	3.760	0.100	a
Chalcone, 2-chloro-	0.990	0.370	b
Cinnamic acid, 2-chloro-	0.500	0.190	a,b
-", benzyl ester	0.410	0.260	b
-", butyl ester	0.660	0.250	
-", tert-butyl ester	0.500	0.190	b
-", 2-chloroethyl ester	0.400	0.260	b
-", ethyl ester	0.600	0.170	
-", methyl ester	0.480	0.140	c
-", phenyl ester	0.650	0.200	b
-", vinyl ester	0.240	0.240	
-", o-chloro-, methyl ester	0.850	0.280	a
-", p-chloro-, methyl ester	0.650	0.150	a,b
-", α-cyano-, benzyl ester	2.570	0.150	a,b
-", α-cyano-, butyl ester	2.420	0.140	a,b
-", α-cyano-, cyclohexyl ester	2.560	0.150	a,b
-", α-cyano-, ethyl ester	0.870	1.240	a,b
-", α-cyano-, 2-ethylhexyl ester	2.360	0.130	a,b
-", α-cyano-, hexyl ester	2.580	0.160	a,b
-", α-cyano-, methyl ester	2.490	0.140	a,b
Cinnamonitrile	0.750	0.320	b
Citraconic anhydride	1.750	0.870	b
Crotonaldehyde	0.440	0.016	c
Crotonamide	1.760	0.008	b
-", N-(hydroxymethyl)-	2.100	0.035	b
Crotonic acid	0.450	0.013	c
-", butyl ester	0.360	0.007	a,b
-", methyl ester	0.360	0.015	b
-", propyl ester	0.360	0.011	a,b
Crotononitrile, cis-	0.920	0.041	b
-", trans-	0.500	0.013	c
Cyclobutanecarboxylic acid, 3-acetyl-2,2-dimethyl-, vinyl ester	-0.350	0.034	
1,5,9-Cyclododecatriene, trans, trans-, cis-	0.090	0.170	
Cycloheptene	-1.430	0.002	
1,3-Cyclohexadiene	-1.130	1.610	
Cyclohexane, 1,1,2,2,3,3,4,4-octafluoro-5,6-dimethylene-	0.930	0.730	
Cyclohexene	-1.440	0.001	
-", 4-vinyl-	-1.640	0.060	a,b
3-Cyclohexene-1-carboxylic acid, 1-methyl-, 2,3-epoxypropyl ester	0.320	0.014	c
Cyclooctene	-1.510	0.004	

Table II (Cont'd.)

Monomer	e	Q	Notes
Cyclopentene	-1.350	0.008	
4-Cyclopentene-1,3-dione	1.400	0.220	
Decanoic acid, vinyl ester	-0.100	0.020	
m-Dioxane, 5,5-bis(bromomethyl)-2-vinyl-	0.140	0.058	b
-", 5,5-dimethyl-2-vinyl-	-0.110	0.055	b
-", 4-methyl-2-vinyl-	-0.050	0.029	b
-", 2-vinyl-	-0.150	0.030	b
p-Dioxin, 2,3-dihydro-	-1.120	0.008	
1,3-Dioxolane, 2-vinyl-	-0.120	0.020	
1,3-Dioxolane-4-methanol, 2,2-dimethyl-, acrylate	0.770	0.500	
-", 2,2-dimethyl-, 4-(methylenesuccinate)	0.620	0.980	
-", 2,2-dimethyl-, methyl fumarate	2.250	0.600	
-", 2-vinyl-	-0.120	0.040	
1,3-Dioxolan-4-ol, 2-vinyl-, acrylate	0.650	0.410	b
-", 2-vinyl-, methacrylate	0.380	0.810	a
Disilane, pentamethylvinyl-	-0.600	0.020	
Disiloxane, 1-(chloromethyl)-1,1,3,3-tetramethyl-3-(p-vinylphenyl)-	-0.800	0.990	b
Disiloxane, 1,1,1,3,3-pentamethyl-3-vinyl-	-0.810	0.034	
Dodecanamide, N-(acrylamidomethyl)-	0.630	1.120	
Ethane, 1,2-bis(p-vinylphenyl)	-1.080	1.180	b
Ethanesulfonic acid, 2-hydroxy-, methacrylate	0.560	1.340	
-", 2-hydroxy-, sodium salt, methacrylate	0.200	0.530	b
Ethanol, 2-(2-(vinyloxy)ethoxy)-	-1.390	0.046	b
Ethenesulfonamide, N-(2-chloroethyl)-	0.420	0.125	b
Ethenesulfonic acid	-0.020	0.093	c
-", allyl ester	0.130	0.150	
-", butyl ester	1.190	0.130	
-", 2,3-epoxypropyl ester	1.410	0.140	
-", propyl ester	-1.760	0.270	b
-", sodium salt	0.410	0.064	a
Ethenesulfonyl fluoride	1.390	0.120	b
Ether, allyl phenyl	0.560	0.015	
-", bis(p-vinylphenyl)	-1.070	1.360	b
-", 2-bromovinyl ethyl	-0.370	0.012	c
-", butyl vinyl	-1.200	0.087	c
-", tert-butyl vinyl	-1.580	0.150	b
-", cyclohexyl vinyl	-1.550	0.081	a,b
-", dodecyl vinyl	-0.740	0.033	c
-", ethyl vinyl	-1.170	0.032	a
-", isobutyl vinyl	-1.770	0.023	
-", isopropyl vinyl	-1.310	45.400	b
-", methyl 1-phenylvinyl	-1.200	0.650	a
-", octadecyl vinyl	-0.630	0.069	c
-", octyl vinyl	-0.790	0.061	c
-", pentyl vinyl	-1.410	0.009	a,b
-", phenyl vinyl	-1.210	0.082	

Table II (Cont'd.)

Monomer	e	Q	Notes
Ether (Cont'd.)			
-", propyl vinyl	-1.520	0.014	a,b
Ethylene	-0.200	0.015	
-", 1,1-bis(p-chlorophenyl)-	-0.840	2.160	a
-", 1,1-bis(p-methoxyphenyl)-	-1.960	1.460	a
-", bromo-	-0.250	0.047	
-", chloro-	0.200	0.044	
-", chlorotrifluoro-	1.480	0.020	
-", 1,1-dichloro-	0.360	0.220	
-", 1,2-dichloro-, cis-	1.250	0.003	c
-", 1,2-dichloro-, trans-	1.280	0.010	c
-", 1,1-dichloro-2,2-difluoro-	2.100	0.041	a
-", 1,2-dimethoxy-	-0.980	0.049	
-", 1,1-diphenyl-	-1.350	1.500	c
-", fluoro-	-0.820	0.025	b
-", iodo-	-0.800	0.140	a
-", tetrachloro-	2.030	0.003	c
-", tetrafluoro-	1.220	0.049	b
-", trichloro-	1.860	0.019	b
Ethylene : AgNO$_3$ complex (1 : 1)	1.660	29.400	b
Ethylene glycol, bis(2-chloroacrylate)	0.880	0.430	
-", dimethacrylate	0.240	0.880	
Ferrocenemethanol, acrylate	1.040	0.190	
-", methacrylate	1.260	0.130	
Formamide, N-allyl-	-0.680	0.110	
Formic acid, vinyl ester	-0.820	0.170	
Fumaric acid, 1,3-butyleneglycol polyester	1.020	0.290	c
-", diester with hydracrylonitrile	2.420	0.800	
-", diethyl ester	1.250	0.610	c
-", dimethyl ester	1.490	0.760	b
-", dinonyl ester	-0.390	0.600	b
-", dioctyl ester	1.450	0.240	
-", ethyl ester	0.960	1.330	b
-", ethyleneglycol polyester	1.930	0.420	b
-", dithio-, dimethyl ester	1.780	1.230	
Fumaronitrile	1.960	0.800	c
2-Furaldehyde	-0.800	0.500	b
D-Galactopyranose, 1,2:3,4-di-O-isopropylidene-, 6-methacrylate	0.400	0.820	
Germane, triethylmethacryloyl-	-0.640	0.840	
-", trimethylvinyl-	0.430	0.021	
D-Glucitol, 1-acrylamido-1-deoxy-	0.680	0.150	c
-", 1-deoxy-1-methacrylamido-	1.480	0.170	
-", 1,2:3,4:5,6-tri-O-allylidene-	-0.080	0.050	b
-", 1,2:3,4:5,6-tri-O-(2-butenylidene)-	-0.250	0.100	b
Glutaric acid, 2-methylene-, diethyl ester	0.900	0.550	
Glutaronitrile, 2-methylene-	1.110	0.670	d

Table II (Cont'd.) Monomer	e	Q	Notes
Glycine, N-allyl-	0.060	0.039	b
Glycolic acid, ethyl ester, acrylate	0.910	0.380	b
3,5-Heptadien-2-one	0.880	1.620	
Hexanoic acid, vinyl ester	-0.470	0.025	
-", 2-ethyl-, vinyl ester	-0.080	0.024	
-", 2-methylene-, methyl ester	0.550	0.420	b
1,3,5-Hexatriene, 2,3,4,5-tetrachloro-, trans-	0.750	1.850	
1-Hexene	-0.280	0.019	
-", 1-phenyl-, cis-	-1.550	0.004	b
-", 1-phenyl-, trans-	-1.820	0.015	b
1-Hexen-3-yne	-0.290	0.600	b
1-Hexyne	-0.700	0.014	a
Hydrazine, 1-acryloyl-2-benzoyl-	-0.440	0.420	
Hydrazinium chloride, 2-acryloyl-1,1,1-trimethyl-	0.340	0.690	
-", 2-methacryloyl-1,1,1-trimethyl-	0.660	0.610	
Hydrazinium hydroxide, 1-(2,3-dihydroxypropyl)-2-methacryloyl-1,1,1-dimethyl-, inner salt	-1.260	0.240	b
Hydrazinium hydroxide, 1-(2-hydroxypropyl)-2-methacryloyl-1,1-dimethyl-, inner salt	-1.580	0.170	b
Hydrazinium hydroxide, 2-methacryloyl-1,1,1-trimethyl-, inner salt	-0.620	0.180	b
Hydrazinium hydroxide, 1,1,1-trimethyl-2-(p-vinylbenzoyl)-, inner salt	0.300	0.880	b
Hydrocinnamic acid, α-methylene-, methyl ester	0.720	0.500	b
Hydroquinone, vinyl-, dibenzoate	1.170	1.800	
Imidazole, 2-methyl-1-vinyl-	-1.300	0.150	
-", 2-phenyl-1-vinyl-	-1.570	0.100	b
1H-Indazole, 1-isopropenyl-	-0.490	0.430	b
Indene	-1.030	0.360	c
Isobutyraldehyde, divinyl acetal	-0.220	0.026	b
Isocyanic acid, vinyl ester	-0.700	0.160	
-", p-vinylphenyl ester	-0.360	0.940	b
Isoprene	-1.220	3.330	
Isothiocyanic acid, vinyl ester	0.250	0.540	
Isovaleraldehyde, divinyl acetal	-0.220	0.026	b
Isoxazole, 3-isopropenyl-5-methyl-	-0.150	0.880	b
Lauric acid, allyl ester	0.530	0.031	b
-", vinyl ester	-0.010	0.018	
Levulinic acid, vinyl ester	-0.530	0.027	b
Malealdehyde	1.130	2.680	b
Maleic acid	2.600	0.450	
-", bis(2-ethylhexyl) ester	1.080	0.100	a,b
-", dibenzyl ester	1.320	0.340	b
-", dibutyl ester	1.600	0.042	a,b
-", diethyl ester	1.490	0.059	c
-", diisobutyl ester	1.850	0.094	b
-", diisopropyl ester	2.060	0.084	c

Table II (Cont'd.) Monomer	e	Q	Notes
Maleic acid (Cont'd.) dimethyl ester	1.270	0.090	b
-", dinonyl ester	-0.390	0.220	b
-", ethyl ester	1.520	1.230	b
-", methyl ester	2.190	0.100	c
-", 2-chloro-, diethyl ester	1.650	0.056	a,b
-", 1,4-dithio-, cyclic anhydrosulfide	1.700	0.330	b
Maleic anhydride	2.250	0.230	c
Maleimide, mono(O-benzyloxime)	1.180	0.960	
Maleimide	1.350	0.440	
-", N-(benzyloxy)-	1.930	5.650	b
-", N-2-bornyl-	1.470	0.480	
-", N-butyl-	1.750	3.080	
-", N,2-dimethyl-	1.480	2.350	
-", N-ethyl-	1.940	3.110	b
-", N-hexyl-	1.650	0.980	b
-", N-hydroxy-, acetate	2.120	4.850	b
-", N-(2-hydroxyethyl)-	1.420	0.430	
-", N-(2-hydroxyethyl)-, acetate	1.020	0.079	b
-", N-(2-hydroxyethyl)-, propionate	1.650	0.570	b
-", N-(hydroxymethyl)-	1.320	0.490	
-", N-(hydroxyphenyl)-	1.390	0.800	
-", N-(3-hydroxypropyl)-, acetate	1.540	0.590	b
-", N-(methoxy)-	2.000	5.320	b
-", N-methyl-	2.060	6.360	b
-", N-octyl-	1.680	0.990	b
-", N-propyl-	1.790	1.510	b
Maleonitrile	2.320	0.420	a,b
Malononitrile, methylene-	2.580	20.130	b
Melamine, N(2),N(4)-diallyl-	-0.950	0.017	c
Metanilyl fluoride, N-acryloyl-	0.660	0.550	c
-", N-methacryloyl-	0.760	0.690	b
Methacrylaldehyde	0.250	1.450	
Methacrylamide	1.240	1.460	
Methacrylanilide	-0.780	0.850	b
Methacrylic acid	0.650	2.340	
-", 2-(1-aziridinyl)ethyl ester	0.240	0.740	
-", benzyl ester	0.420	0.700	
-", p-(benzylideneamino)phenyl ester	0.880	1.080	b
-", bornyl ester	0.590	0.790	
-", 2-bromoethyl ester	0.570	0.950	
-", butyl ester	0.510	0.780	
-", sec-butyl ester	0.240	0.720	
-", tert-butyl ester	0.240	0.760	
-", 2(tert-butylamino)ethyl ester	0.170	0.980	d
-", cadmium salt	-0.270	8.600	b
-", calcium salt	-0.140	2.410	b

Table II (Cont'd.)

Monomer	e	Q	Notes
Methacrylic acid (Cont'd.)			
", 2-chloroethyl ester	0.530	0.810	
", cyclohexyl ester	0.450	0.820	
", decyl ester	0.480	1.370	b
", 2-(diethylamino)ethyl ester	0.420	2.080	b
", 2-(diethylamino)ethyl ester, ethiodide	1.100	2.100	b
", 2-(diethylamino)ethyl ester, hydrochloride	0.810	1.100	b
", 2-(diethylamino)ethyl ester, methiodide	0.770	0.840	b
", 2-(dimethylamino)ethyl ester	0.470	0.680	
", 2,2-dinitropropyl ester	0.850	0.720	b
", dodecyl ester	0.350	0.700	
", 2,3-epoxypropyl ester	0.570	1.030	
", ethyl ester	0.520	0.730	
", 2-(ethylthio)ethyl ester	1.690	4.280	d
", furfuryl ester	0.040	0.780	
", furfuryl ester, π complex with O_2	0.760	0.850	b
", hexyl ester	0.340	0.670	b
", X-hydroxybutyl ester, nitrate	0.580	0.900	b
", 2-hydroxyethyl ester	0.200	0.800	b, d
", 2-hydroxyethyl ester, nitrate	0.530	0.990	b
", 2-hydroxypropyl ester	0.200	0.790	d
", 2-hydroxypropyl ester, nitrate	0.510	0.980	b
", isobornyl ester	0.310	0.910	
", isobutyl ester	0.340	0.870	
", isopropyl ester	0.340	0.850	
", lithium salt	0.300	0.640	b
", magnesium salt	0.000	0.400	b
", 5-(methoxycarbonyl)furfuryl ester	0.900	0.720	b
", 2-methoxyethyl ester	0.200	0.790	
", 1-methyl-2-nitropropyl ester	1.040	0.530	b
", 2-methyl-2-nitropropyl ester	0.660	0.970	b
", 2-nitrobutyl ester	0.770	1.140	b
", 2-nitropropyl ester	0.590	0.840	b
", nonyl ester	0.390	0.820	
", octadecyl ester	0.560	1.070	b
", octyl ester	0.180	0.810	
", pentamethyldisiloxanylmethyl ester	0.190	0.740	
", pentyl ester	0.430	0.680	b
", phenethyl ester	0.360	0.740	
", phenyl ester	0.730	1.490	
", potassium salt	0.010	0.540	b
", propyl ester	0.440	0.650	
", 2-propynyl ester	0.620	1.070	b
", sodium salt	-1.180	1.360	
", tetrahydrofuryl ester	0.350	0.450	b
", zinc salt	-0.440	9.900	b
Methacrylic anhydride	1.030	1.600	

Table II (Cont'd.)

Monomer	e	Q	Notes
Methacrylonitrile	0.810	1.120	b
Methacrylophenone	0.800	1.330	b
Methacryloyl chloride	1.070	0.950	b
Methane, bis(p-vinylphenyl)-	-1.110	1.280	b
", bis(vinyloxy)-	-0.220	0.026	b
Methanol, diphenyl(p-vinylphenyl)-	-0.270	1.340	b
Methyl methacrylate	0.400	0.740	
Methyl methacrylate-methyl-14C	0.400	0.740	
Methyl methacrylate-methyl-D	0.400	0.740	
Methyl methacrylate-methyl-T	0.400	0.740	
Myristic acid, vinyl ester	-0.150	0.030	b
Naphthalene, 4-chloro-1-vinyl-	-0.310	0.740	
", 6-chloro-2-vinyl-	-0.130	1.350	
", 2-isopropenyl-	-1.040	0.920	c
", 1-vinyl-	-1.120	1.940	b
", 2-vinyl-	-0.380	1.250	
Nonanoic acid, vinyl ester	-0.680	0.030	
2,5-Norbornadiene	-1.060	0.092	
Norbornane, 2-vinyl-	-0.460	0.045	b
2-Norbornanecarboxylic acid, endo-, vinyl ester	0.300	0.014	b
", exo-, vinyl ester	0.300	0.014	b
5-Norbornene-2-carbonitrile	0.280	0.070	d
5-Norbornene-2-carboxylic acid, methyl ester	0.260	0.059	
2-Norpinane-2-ethanol, acrylate	0.670	0.500	
Octadecanoic acid, 4-oxo-, vinyl ester	-0.480	0.037	c
", 9(10)-oxo-, vinyl ester	-0.290	0.039	c
", 12-oxo-, vinyl ester	-0.500	0.032	c
Octanoic acid, vinyl ester	-0.310	0.021	
3-Oxa-1-azaspiro[4.5]decan-4-one, 2-isopropenyl-	0.720	2.470	b
1,3,4-Oxadiazole, 2-isopropenyl-5-methyl-	0.620	2.000	b
Oxalic acid, ethyl vinyl ester	-0.260	0.092	
4H-1,3-oxazine, 5,6-dihydro-4,6,6-trimethyl-2-vinyl-	1.430	0.620	b
Oxazole, 4,5-dimethyl-2-vinyl-	0.400	2.130	
", 4-isobutyl-2-isopropenyl-5-methyl-	0.240	4.100	b
", 4-isobutyl-5-methyl-2-vinyl-	0.300	4.200	b
", 2-isopropenyl-4,5-dimethyl-	0.250	2.880	
2-Oxazolidinone, 5-decyl-3-vinyl-	-0.380	0.050	b
", 5-methyl-3-vinyl-	-1.190	0.210	b
", 3-vinyl-	-0.800	0.057	
2-Oxazolin-5-one, 4-ethyl-2-isopropenyl-4-methyl-	0.490	1.320	b
", 4-isobutyl-2-isopropenyl-4-methyl-	0.540	1.420	b
", 1-isopropenyl-4,4-dimethyl-	0.280	1.360	b
Oxetane, 3-(allyloxymethyl)-3-(chloromethyl)-	-1.320	0.024	
Palmitic acid, vinyl ester	-0.020	0.026	
2,4-Pentadienenitrile	0.280	5.980	a,b
2,4-Pentadienoic acid, 2,3-epoxypropyl ester	-0.050	4.790	b
", methyl ester	0.660	1.640	

Table II (Cont'd.)

Monomer	e	Q	Notes
2,4-Pentadienoic acid, 5,5-dichloro-4-hydroxy-, γ-lactone	-0.820	2.840	c
2,4-Pentanedione, 3-ethylidene-	0.910	0.028	a
1-Pentene	-0.630	0.074	a,b
-, 3-methyl-1-phenyl-, cis-	-1.270	0.001	b
-, 3-methyl-1-phenyl-, trans-	-1.510	0.004	b
-, 1-phenyl-, cis-	-1.640	0.007	b
-, 1-phenyl-, trans-	-1.280	0.029	b
4-Pentenoic acid, 2-allyl-2-cyano-, ethyl ester	0.820	0.059	b
3-Penten-2-one	0.610	0.024	b
Phenanthrene, 2-vinyl-	-0.670	1.960	b
-, 3-vinyl-	-0.800	2.260	b
-, 9-vinyl-	-0.800	1.830	b
1-Phenanthrenemethanol, 1,2,3,4,4A,4B,5,6,10,10A-deca-hydro-7-isopropyl-1,4A-dimethyl-	-0.800	0.056	a,b
1-Phenanthrenemethanol, 1,2,3,4,4A,4B,5,6,10,10A-deca-hydro-7-isopropyl-1,4A-dimethyl-, acrylate			
Phenethyl alcohol, α-methyl-p-vinyl-	0.980	0.400	b
Phenetole, p-styryl-	-0.700	1.200	b
Phenol, o-vinyl-	0.440	0.069	b
-, m-vinyl-	-1.130	1.410	a
-, p-vinyl-, acetate	-0.800	1.100	b
Phosphine, diphenylvinyl-	-0.800	1.180	b
-, diphenyl(p-vinylphenyl)-	-0.350	0.100	a
Phosphine oxide, diallylphenyl-	-0.320	1.290	
-, diisobutylvinyl-	-0.070	0.051	b
-, diphenylvinyl-	0.350	0.024	a
-, diphenyl(p-vinylphenyl)-	0.500	0.070	a
Phosphine sulfide, diphenylvinyl-	-0.220	1.380	
-, diphenyl(p-vinylphenyl)-	1.350	0.083	a
Phosphinic acid, diethyl-, allyl ester	-0.380	1.530	
Phosphinothioic acid, diphenyl-, O-2-(methacryloxy)ethyl ester	1.020	0.031	b
Phosphonic acid, allyl-, diethyl ester	0.510	0.750	b
-, butyl-, diallyl ester	-0.070	0.019	b
(dimethyleneethylene)di-, tetraethyl ester	-0.070	0.053	b
1-hydroxyethyl-, diethyl ester, acrylate	1.220	2.720	
hydroxymethyl-, diethyl ester, acrylate	1.490	0.570	b
hydroxymethyl-, diethyl ester, methacrylate	1.500	0.760	
(2-methyl-1,3-butadienyl)-, diisopropyl ester	0.980	0.650	b
(2-methyl-1,3-butadienyl)-, dimethyl ester	-0.400	1.090	b
phenyl-, diallyl ester	0.080	1.230	b
α-phenylvinyl-	-0.070	0.051	b
vinyl-, bis(2-bromoethyl) ester	1.690	0.470	
vinyl-, bis(2-chloroethyl) ester	0.800	0.060	
vinyl-, diethyl ester	-1.190	0.760	c
Phosphonic dichloride, (2-methyl-1,3-butadienyl)-	0.230	0.160	c
-, vinyl-	0.150	1.560	b
Phosphonic difluoride, (2-methyl-1,3-butadienyl)-	2.700	0.870	
-, vinyl-	-0.110	1.060	b

Table II (Cont'd.)

Monomer	e	Q	Notes
Phthalic acid, diallyl ester	0.360	0.044	b
Phthalimide, N-1-(1,3-butadienyl)-	0.060	1.570	b
-, N-2-(1,3-butadienyl)-	-1.550	1.650	b
-, 4-(trimethylsilyl)-N-vinyl-	-1.650	0.320	b
-, N-vinyl-	-1.520	0.360	
2-Picoline, 5-vinyl-	-0.580	0.990	
2-Picolinium chloride, 5-vinyl-	0.170	1.160	b
2-Picolinium methyl sulfate, 1-methyl-5-vinyl-	-0.160	2.490	
Pivalic acid, vinyl ester	-0.750	0.037	b
Propane, 1-(allyloxy)-2,3-epoxy-	0.400	0.018	b
-, 1,3-bis(p-vinylphenyl)-	-0.910	0.980	b
1-Propanesulfonic acid, 3-hydroxy-, acrylate, sodium salt	1.100	0.300	b
-, 3-hydroxy-, methacrylate, sodium salt	0.230	0.510	
1-Propanol, 3-bromo-2,2-bis(bromomethyl)-, acrylate	1.050	0.470	
1-Propene	-0.780	0.002	
-, 1-chloro-	-0.940	0.031	b
-, 1-chloro-, cis-	-0.550	0.004	
-, 1-chloro-, trans-	-0.390	0.008	c
-, 2-chloro-	-0.710	0.035	c
-, 2-chloro-3,3-dimethoxy-	-2.250	0.125	c
-, 3-chloro-	0.110	0.056	a
-, 3-chloro-2-methyl-	-0.910	0.120	c
-, 2,3-dichloro-	0.440	0.270	c
-, 2-methyl-	-0.960	0.033	c
-, 3,3,3-trichloro-	1.370	0.056	c
-, 3,3,3-trifluoro-	0.510	0.170	b
2-Propene-1,1-diol, diacetate	-1.520	0.230	b
-, 2-chloro-, diacetate	-1.460	0.370	b
2-Propene-1-sulfonic acid, 2,3-epoxypropyl ester	-0.800	0.020	b
-, sodium salt	-0.240	0.150	b
1-Propene-1,2,3-tricarboxylic acid, cis-, triethyl ester	1.540	0.280	c
-, cis-, trimethyl ester	1.620	0.320	
-, trans-, trimethyl ester	1.140	0.005	b
1-Propen-2-ol, acetate	-0.500	0.045	
-, isocyanate	-1.020	0.230	
2-Propen-1-ol, 2-chloro-	0.560	0.240	a
-, 2-chloro-, acetate	-1.120	0.530	a
-, 2-methyl-, acetate	-1.330	0.037	a
-, 2-methyl-, oxalate	-0.150	0.038	b
2-Propen-1-one, 1-(6-methyl-3-pyridyl)-3-phenyl-	0.960	0.260	b
-, 1-phenyl-3-(3-pyridyl)-	0.640	0.640	a,b
-, 3-phenyl-1-(3-pyridyl)-	0.800	0.330	b
Propionic acid, vinyl ester	-0.070	0.052	b
Pyrazole, 3,5-dimethyl-1-vinyl-	-1.450	0.200	b
3-Pyrazolidinone, 1-(m-vinylphenyl)-	-0.600	0.900	b
3(2H)-Pyridazinone, 2(p-chlorophenyl)-6-hydroxy-	-0.280	0.074	a
-, 6-hydroxy-	-0.790	0.010	

Table II (Cont'd.)

Monomer	e	Q	Notes
3(2H)-Pyridazinone (Cont'd.)			
- , 6-hydroxy-, acetate	-0.230	0.058	
- , 6-hydroxy-2-phenyl-	-0.450	0.026	a
- , 6-hydroxy-2-(p-tolyl)-	-0.590	0.024	a
- , 6-methyl-	-0.710	0.037	
Pyridine, 5-ethyl-2-vinyl-	-0.740	1.370	
- , 5-ethyl-2-vinyl-, 1-oxide	-0.100	4.520	
- , 2-isopropenyl-	0.160	1.100	
- , 2-vinyl-	-0.500	1.300	b
- , 2-vinyl-, 1-oxide	-0.010	3.770	
- , 4-vinyl-	-0.280	1.000	
- , 4,4'-vinylenedi-	-1.340	0.660	
Pyridinium acetate, 2-vinyl-	0.890	1.620	b
Pyridinium fluoroborate, 1-vinyl-	2.120	1.120	
Pyrimidine, 2-(dimethylamino)-4-vinyl-	0.040	1.450	b
- , 4-vinyl-	0.450	2.180	b
2-Pyrrolidinethione, 1-vinyl-	-0.110	1.480	
2-Pyrrolidinone, 1-vinyl-	-1.140	0.140	
Quinoline, 2-vinyl-	-0.820	3.790	
Semicarbazide, 1-(methacryloylamidino)-, hydrochloride	2.140	0.570	b
Silane, allyldimethylphenyl-	0.300	0.070	b
- , allyldimethyl(p-vinylphenyl)-	-0.160	0.870	b
- , allyltriethoxy-	-0.940	0.024	a
- , allyltrimethyl-	0.270	0.036	b
- , 2-butenyltriethoxy-	-1.320	0.040	a
- , (chloromethyl)dimethyl(p-vinylphenyl)-	-0.080	0.820	b
- , (1-chlorovinyl)triethoxy-	0.180	0.230	a
- , (1-cyclohexen-4-yl)triethoxy-	-0.500	0.050	a
- , diethoxyethylvinyl-	-0.350	0.011	a,b
- , diethoxymethylvinyl-	-0.860	0.020	a
- , diethoxyphenylvinyl-	-0.380	0.034	a
- , dimethylphenyl(p-vinylphenyl)-	-1.270	1.630	
- , dimethyl(p-vinylphenyl)-	-0.040	0.970	b
- , dimethylvinyl(p-vinylphenyl)-	-1.010	1.200	
- , methyl(p-vinylphenyl)-	-0.800	1.100	b
- , triethoxy(5-norbornen-2-yl)-	-0.350	0.072	a
- , triethoxypropenyl-	-1.080	0.003	a
- , triethoxyvinyl-	-0.420	0.028	a
- , triisopropoxyvinyl-	-0.360	0.031	a
- , trimethoxyvinyl-	-0.380	0.031	a
- , trimethoxy(p-vinylphenyl)-	-0.880	1.500	b
- , trimethylvinyl-	0.040	0.029	c
- , trimethyl(p-vinylphenyl)-	-0.800	1.000	b
Sorbic acid, 2,3-epoxypropyl ester	0.180	0.560	b
methyl ester	0.440	0.790	b
Sorbonitrile	0.940	1.910	
Stearamide, N-allyl-	-0.570	0.040	c

Table II (Cont'd.)

Monomer	e	Q	Notes
Stearic acid, vinyl ester	-0.520	0.034	b
Stilbene, diphenyl(p-vinylphenyl)-	-0.300	1.120	b
4-Stilbenamine, N,N-dimethyl-	0.360	0.060	
Stilbene, cis-	-0.030	0.017	
- , trans-	-0.080	0.030	
- , 4-chloro-	0.940	0.051	
- , 4-nitro-	1.080	0.069	
Styrene	-0.800	1.000	(Base)
- , 2,5-bis(trifluoromethyl)-	-0.050	1.110	
- , m-bromo-	-0.210	1.070	
- , p-bromo-	-0.320	1.040	
- , p-(2-bromoethyl)-	-1.430	0.710	b
- , p-tert-butyl-	-0.900	1.370	
- , p-tert-butyl-o-chloro-	-0.500	1.080	
- , m-chloro-	-0.360	1.030	
- , o-chloro-	-0.360	1.280	
- , p-chloro-	-0.330	1.030	
- , β-chloro-α,β-difluoro-	-0.390	0.340	b
- , β-chloro-α,β-difluoro-p-methyl-	-0.130	0.270	b
- , α-chloro-β-fluoro-	0.670	0.150	b
- , 2,4-dibromo-	0.050	1.530	b
- , 3,4-dibromo-	-0.540	1.300	b
- , 2,5-dichloro-	0.090	1.600	b
- , m-(1,1-dichloro-2-cyclopropyl)-	-0.800	1.120	b
- , 2,4-difluoro-	-0.310	0.650	b
- , 2,5-difluoro-	0.730	6.700	b
- , α,β-difluoro-	0.730	0.120	b
- , β,β-difluoro-	0.700	0.029	a
- , α-(difluoromethyl)-	-0.210	1.160	b
- , 2,5-dimethoxy-	-1.040	1.750	
- , 2,5-dimethyl-	-0.960	0.970	
- , β-fluoro-	0.880	0.044	b
- , p-fluoro-	-0.120	0.830	
- , p-iodo-	-0.400	1.170	
- , p-isopropyl-	-1.520	1.600	b
- , α-methyl-	-1.270	0.980	
- , m-methyl-	-0.720	0.910	
- , o-methyl-	-0.780	0.900	
- , p-methyl-	-0.980	1.270	b
- , m-nitro-	0.810	2.470	
- , p-nitro-	0.390	1.630	
- , 2,3,4,5,6-pentachloro-	0.520	0.220	
- , 2,3,4,5,6-pentafluoro-	0.740	0.780	
- , 2,3,4,5-tetraethyl-	-0.900	0.790	
- , α,β,β-trifluoro-	0.220	0.750	b
- , α,β,β-trifluoro-m-methyl-	0.440	1.070	
- , α,β,β-trifluoro-o-methyl-	0.060	0.430	b

COPOLYMERIZATION PARAMTER

Table II (Cont'd.)

Monomer	e	Q	Notes
Styrene (Cont'd.)			
-", α,β,β-trifluoro-p-methyl-	0.240	0.880	
-", m-trifluoromethyl-	-0.290	0.920	
-", 2,4,6-trimethyl-	-1.100	0.220	
Succinic acid, methylene-	0.500	0.760	c
-", bis(2-chloroethyl) ester	1.520	0.380	c
-", diethyl ester	0.980	0.940	b
-", diisopropyl ester	1.210	0.890	b
-", 2-(4,6-dimethoxy-s-triazin-2-yl) hydrazide	-0.170	0.520	b
-", dimethyl ester	1.340	1.030	
-", dioctyl ester	0.790	0.960	
-", dipentyl ester	0.520	1.020	b
-", dipotassium salt	-0.580	0.160	b
-", dipropyl ester	0.710	1.190	b
-", 2,3-epoxypropyl ester	1.430	0.650	b
Succinic anhydride	-1.300	0.002	b
-", methylene-	0.880	2.500	b
Succinimide, N-1-(1,3-butadienyl)-	0.090	1.750	b
-", N-vinyl-	-0.340	0.130	b
Sulfide, p-anisyl vinyl	-1.420	0.430	b
-", bis(p-vinylphenyl)	-1.090	2.070	b
-", p-bromophenyl vinyl	-0.880	0.350	b
-", butyl vinyl	-1.200	0.330	b
-", tert-butyl vinyl	-1.100	0.260	b
-", p-chlorophenyl vinyl	-0.910	0.310	b
-", ethyl vinyl	-0.120	0.370	b
-", isobutyl vinyl	-1.700	0.530	b
-", isopropyl vinyl	-1.200	0.310	b
-", methyl vinyl	-1.450	0.320	b
-", pentachlorophenyl vinyl	-0.580	0.220	b
-", phenyl vinyl	-1.400	0.340	b
-", p-tolyl vinyl	-1.200	0.400	b
Sulfone, methyl vinyl	1.290	0.110	c
-", phenyl vinyl	1.180	0.069	b
Sulfoxide, ethyl vinyl	0.610	0.130	b
-", methyl vinyl	0.980	0.057	b
Terephthalic acid, 2,2-bis(3-allyl-4-hydroxyphenyl)propane-co-phenolphthalein polyester	-1.400	0.580	b
2,4,8,10-tetraoxaspiro[5.5]undecane, 3,9-bis(1-chlorovinyl)-	0.340	0.200	b
-", 3,9-dipropenyl-	-0.320	0.010	b
-", 3,9-distyryl-	-0.530	0.040	b
2H-Tetrazole, 2-phenyl-5-(p-vinylphenyl)-	-0.600	0.750	b
Thianthrene, 2-vinyl-	-1.680	4.300	b
4H-1,3-Thiazine, 5,6-dihydro-4,4-dimethyl-2-vinyl-	0.090	0.320	b
Thiazole, 2-isopropenyl-	0.240	4.800	b

Table II (Cont'd.)

Monomer	e	Q	Notes
Thiazole, 4-methyl-2-vinyl-	0.130	3.200	b
Thiophene, 2-vinyl-	-0.800	2.860	b
Tin, bis(methacryloyloxy)dimethyl-	0.410	1.360	b
-", (1,3-butadien-2-yl)triethyl-	0.570	1.500	b
-", (methacryloyloxy)trimethyl-	-0.370	0.450	
-", tributylvinyl-	0.800	0.024	b
p-Toluenesulfonamide, N-methyl-N-vinyl-	-1.100	0.082	a
p-Toluic acid, vinyl ester	-0.520	0.051	
1,3,5,2,4,6-Triazaphosphorine, 2-chloro-2,2,4,4,6,6-hexahydro-2,4,4,6,6-pentaallyloxy-	0.680	0.600	b
s-Triazine, 2-amino-4-anilino-6-isopropenyl-	0.290	1.070	b
-", 2-amino-4-anilino-6-vinyl-	0.590	0.550	b
-", 2-amino-4-chloroanilino-6-isopropenyl-	0.270	0.890	b
-", 2-amino-4-(dimethylamino)-6-isopropenyl-	0.160	1.450	b
-", 2-amino-4-(dimethylamino)-6-vinyl-	-0.050	0.780	b
-", 2-amino-4-isopropenyl-6-(N-methylanilino)-	0.010	1.800	b
-", 2-amino-4-isopropenyl-6-p-toluidino-	0.350	0.930	b
-", 2-amino-4-(N-methylanilino)-6-vinyl-	0.310	0.820	b
-", 2,4-diamino-6-isopropenyl)-	-0.130	0.890	b
-", 2,4-diamino-6-vinyl-, hydrochloride	1.760	5.520	c
-", 2,4-dianilino-6-vinyl-	1.290	2.080	c
-", 2,4-dimethyl-6-vinyl-	0.450	0.700	b
-", 2,4,6-tris(allyloxy)-	0.870	2.560	b
s-Triazine-2,4,6-(1H,3H,5H)-trione, triallyl-	-1.000	0.020	c
Tridecanedioic acid, 2-methylpentyl vinyl ester	-0.600	0.011	c
Triethylene glycol, dimethacrylate	-0.960	0.032	b
Trisilane, 1,1,1,2,3,3,3-heptamethyl-2-vinyl-	0.710	1.460	b
Trisiloxane, 1,1,1,3,5,5,5-heptamethyl-3-vinyl-	-0.700	0.020	a
-", 1,1,1,5,5,5-hexamethyl-3-trimethylsiloxy-3-vinyl-	-0.780	0.036	
Undecanoic acid, 11-iodo-, allyl ester	-0.690	0.030	
9-Undecenoic acid, vinyl ester	-0.410	0.024	b
Urea, 1,1-diallyl-	-0.640	0.035	c
-", 1-ethyl-3-vinyl-	-0.610	0.018	b
Valeric acid, vinyl ester	-1.530	0.130	b
-", 4-methyl-, vinyl ester	-0.100	0.034	b
-", 2-methylene-, methyl ester	-0.740	0.043	b
-", 4-methyl-2-methylene-, methyl ester	0.490	0.410	b
Versatic[9]acid, vinyl ester	0.480	0.370	b
Versatic[10]acid, vinyl ester	-0.640	0.031	b
Versatic[13]acid, vinyl ester	-0.480	0.037	b
Vinyl acetate	-0.220	0.026	
Vinyl ether	-1.280	0.037	
Vinyl sulfide	-1.110	0.580	b
Vinyl sulfone	1.330	0.140	b

a = r₂ assumed zero - b = 1 set of data only - c = some r₂ assumed zero - d = Rohm and Haas values (no r₁ r₂).

RATES OF POLYMERIZATION AND DEPOLYMERIZATION, AVERAGE MOLECULAR WEIGHTS,
AND MOLECULAR WEIGHT DISTRIBUTION OF POLYMERS

L. H. Peebles, Jr.
Office of Naval Research
Boston, Mass.
and
Department of Chemical Engeneering
Massachusetts Institute of Technology
Cambridge, Mass.

Contents

A. INTRODUCTION

An attempt is made to present a systematic guide to the literature dealing with rates of polymerization, average molecular weights, and molecular weight distribution of polymers for various types of polymerization. This chapter is based on a review on molecular weight distributions (1) in which many of the equations are given in detail along with graphs showing the interrelationship among various distributions: here we present only references to the literature. In addition, sections have been added on the effects of degradation and reactor design on the reaction rates and the molecular weight distributions.

The theoretical description of the molecular weight distribution of a polymer and its rate of polymerization is dependent on the assumed mechanism of polymerization and on the mathematical simplifications used to obtain analytical expressions. As the number of distinct reactions is increased, such as the various transfer reactions, the mathematical expressions can become quite complex and unwieldy. In general, the equations for the rate of polymerization are the most difficult to describe, the distribution equations are somewhat easier, and the average molecular weights are the simplest. In condensation polymerization, many of the distribution formulas are derived by considering the probability of a given reaction instead of the kinetics of the reaction. Therefore, the emphasis is on the distribution functions and their averages, while the rates of polymerization are given only if they have been derived from the kinetics of the reaction.

The chapter is divided into several sections and tables, each treating various types of polymerization. The Stockmayer distribution function for condensation polymers is given in detail because of its general applicability and usefulness. Some general distribution functions are given in Section C. For all the other expressions, the reader must refer to the original literature. Many of the simpler functions are adequately described in textbooks of polymer chemistry. Flory (2), Bamford, Barb, Jenkins, and Onyon (3), Odian (4), Billmeyer (5) and Kuechler (6) give extended descriptions of many systems. Bagdasarian (7) gives an extended discussion on methods of determining absolute values of propagation and termination constants, the influence of cage effects on the rate of initiation, and the influence of retardation, inhibition, and of diffusion-controlled termination of the rate of polymerization. A review of various ways of deriving molecular weight distributions and the moments of distribution is given by Chappelear and Simon (8).

Section B presents a series of tables describing the main assumption or conditions imposed on the theoretical models and references to the articles where the equations may be found. Tables 1 and 2 present rate equations and the distribution formulas for addition polymerization by a variety of mechanisms. No distinction is made among free radical, cationic, anionic, or coordination-type polymerization. Table 1 treats those cases where termination reactions predominate and where steady-state assumptions are usually made. Table 2 treats those cases where termination reactions either do not exist, or may be considered as side reactions, having a minor to major control over the molecular weight distributions. The sequence length distributions for addition-type copolymers are omitted. However, see Kuechler (6) for an extended discussion of copolymerization distributions. Table 3 contains distribution formulas for linear condensation polymers, in which the polymer is assumed to be perfectly linear and to contain no rings. Table 4 treats equilibrium polymerization. Table 5 describes non-linear systems. Table 6 treats those cases where polymers are degraded (or altered) by the application of heat, light, or ionizing radiation. In the latter case the polymer may undergo scission, cross-linking, or both reaction simultaneously.

Several references on cross-linking reactions are included here but not in Table 5. Finally, Table 7 is concerned with the influence of reactor design on the molecular weight distribution; the kinetic equations for addition polymerization with and without termination and condensation polymerization are considered.

Section C lists a number of distribution functions and their properties. Among them is the generalized exponential function which is a good approximation to many real systems (Equation C29).

Section D presents the Stockmayer distribution function for condensation polymerization wherein molecules of various types of kind A react with molecules of various types of kind B.

B. TABLES OF REFERENCES FOR THE CALCULATION OF RATES OF POLYMERIZATION, AVERAGE MOLECULAR WEIGHTS,
AND MOLECULAR WEIGHT DISTRIBUTIONS OF POLYMERS FOR VARIOUS TYPES OF POLYMERIZATION

R_p rate of polymerization - R_d rate of depolymerization - R_1^* polymer radical containing one monomer unit - R_n^* and Q_n^* polymer radicals containing n monomer units - P molecular weight distribution - k_i rate constant for initiation - k_p rate constant for propagation - I initiator concentration - M monomer concentration - M_n and M_w number average and weight average molecular weights, respectively - 1° and 2° represent 1st- and 2nd- order respectively. No distinction is made among free-radical, cationic, anionic, or coordination-type polymerizations.

The rate of initiation may be held constant (const) throughout the polymerization, or it may depend on some function of the catalyst and monomer concentrations, or it may be instantaneous (instant) in which case only the total number of initiating species need be known.

Transfer reactions may occur to initiator, monomer, solvent, or to polymer.

Termination of active species may occur by a first-order deactivation, or by second-order combination (comb) or disproportionation (disprop) or not at all ("living polymers").

Confusion exists over the meaning of the transfer-to-catalyst reaction. In free-radical systems, it means transfer of the active species to the initiator by a second-order mechanism. In ionic polymerization, it means the expulsion of an active fragment from a growing center by a first-order mechanism to form polymer and an active initiator fragment. This first-order mechanism is called here the "catalyst expulsion reaction" (cat ex).

The nomenclature for distribution functions can be quite confusing. In this work, the Flory distribution (Equation C41) is also known as the Schulz-Flory distribution, the "most probable" distribution, and the exponential distribution. The Schulz distribution (Equation C36) is also known as the Schulz-Zimm distribution as well as the generalized Poisson distribution; at large values of $\underline{k}$ it approximates the Poisson distribution (Equation C48). The Pearson Type III distribution is a variation of the Schulz distribution. If an addition polymer is made at constant monomer concentration, no transfer reactions occur, and termination is only by second-order combination, the distribution of the polymer is described by the Schulz distribution with k = 2. This distribution is sometimes called the self-convolution distribution or the convoluted exponential distribution. In a uniform distribution, all molecules have the same size; it is monodisperse. A rectangular or box distribution has no molecules whose size is below r_a, an equal number (or weight) of molecules between r_a and r_b and no molecules whose size is above r_b.

	Set	Initiation	Monomer	Transfer	Termination	R_p	P_r	M_n, M_w
1.	**Addition Polymerization with Termination**							
1.1	**Invarient Monomer Concentration**							
	1.1.1	const	const	none	2° disprop or comb	2-4	1-4	1-4
	1.1.2	const	const	monomer, solvent	2° disprop	2, 3, 10-12	1-3	1-3
	1.1.3	const	const	monomer, solvent	2° disprop and comb	3	1, 3	1, 3
	1.1.4	const, redox	const	activator	2° comb	13	13	13
	1.1.5	const, initiation by activator	const	none	2° with catalyst; redox system	14	-	14
	1.1.6	$k_i M^2$	const	dimer, monomer initiator	2° term	15	-	15
	1.1.7	instant	const	none	2° term; 1° or 2° reactivation	16	-	16
1.2	**Varying Monomer Concentration, No Transfer-to-Monomer Reaction**							
	1.2.1	const	varies	none	2° comb or disprop	3	1, 3	1, 3
	1.2.2	const	varies	solvent	2° comb and disprop	3	1, 3	1, 3
	1.2.3	photosensitized	varies	none	2° disprop	39	39	-
	1.2.4	const	varies	none	2° disprop and comb pseudo 1° with scavenger	17 (Rate of Scavanger disappearance)	-	-
1.3	**Varying Monomer Concentration, Transfer-to-Monomer Reaction Occurs**							
	1.3.1	const	varies	monomer, solvent	2° disprop or comb	3, 18	1, 3, 18	1
	1.3.2	const	varies	monomer, solvent	2° disprop and comb	-	-	19
	1.3.3	$k_i M^2$	varies	monomer	2° comb and disprop	3, 4	1, 3, 4	1, 3, 4
	1.3.4	$k_i MI$	varies	monomer, initiator	2° disprop; degradative chain transfer	20	-	20
1.4	**1° Termination or Deactivation. Steady-State Kinetics**							
	1.4.1	const	const	monomer	1° deactivation	4	1, 4	1, 4
	1.4.2	$k_i MI$	const	monomer, solvent	1° deactivation	4	1, 4, 21, 22	1, 4, 21, 22
	1.4.3	$k_i M^2$	const	monomer	1° deactivation	4	1, 4	1, 4
	1.4.4	$k_i MI$	varies	solvent	1° plus solvent term	4, 23	1, 4	1, 4
	1.4.5	$k_i M^2$	varies	none	1° deactivation	4, 23	1, 4	1, 4
	1.4.6	$k_i MI$	const	none	deactivation by init. expulsion	21	1, 21	1, 21

References page II-418

Set	Initiation	Monomer	Transfer	Termination	R_p	P_r	M_n, M_w
1.5	**Termination by 2° Reaction with Monomer, Steady-State Kinetics**						
1.5.1	const	varies	none	2° with monomer	4,23	1,4	1,4
1.5.2	k_iMI	varies	monomer	2° with monomer	23	1	1
1.5.3	const	const	monomer	2° with monomer	4	1,4	1,4
1.6	**Two Active Ends per Chain**						
1.6.1	const	const	none	2° disprop	3	1,3	1,3
1.6.2	const	const	monomer	2° disprop	-	18	18
1.6.3	const	const	monomer	2° comb; two active chains couple to form a chain with two active ends	3	1,3	1,3
1.7	**Slow Exhaustion of Initiator. Nonsteady-State Kinetics**						
1.7.1	k_iI	const	none	2° disprop* or comb	25*,26	-	1,25*,26
1.7.2	k_iI	varies	none	2° comb	27	-	27
1.8	**Dead-end Polymerization, Rapid Decay of Initiator, Monomer Concentration Invarient**						
1.8.1	k_iI	const	monomer, solvent	2° disprop or comb	28	1,28	1,28
1.8.2	k_iI^2	const	none	2° disprop or comb	28	1,28	1,28
1.9	**Dead-end Polymerization, Slow Decay of Initiator, Monomer Concentration Varies**						
1.9.1	k_iI	varies	monomer, solvent	2° disprop* or comb	30	-	1,19*
1.9.2	k_iI	varies	monomer, solvent	2° comb	30	-	1
1.9.3	k_iIM	varies	none	1° term. bimolecular monomer addition. I* and R_1* do not terminate	31	-	-
1.9.4	instant	varies	monomer	1° term, or 2° with monomer	40	-	-
1.9.5	redox	varies	none	2° term	41	-	-
1.10	**Copolymerization: Two Different Monomers Present**						
1.10.1	const	const	monomer, solvent	2° disprop or comb	1,2,3,4,10, 32,33	34	1,35
1.11	**Diffusion - Controlled Termination**						
1.11.1	const	varies	none or solvent*	2° disprop or comb* and diffusion controlled	37*	1,36	-
1.11.2	k_iI	varies	none	2° comb, $R_t = k_t (M)^\nu \; \nu > 1$	27	-	27
1.11.3	const	varies	none	2° process	42	-	-
1.12	**Primary Radical Termination**						
1.12.1	const	varies	monomer	2° comb and disprop and primary radicals	38	-	1,38
1.12.2	const	const	none or init. and monomer*	2° term and primary radicals	33,43-45	-	44,45,15*
1.13	**Mixed Species Propagation**						
1.13.1	const	varies	monomer	propagation, transfer and termination rates depend on species type, i.e. free radical and cationic occuring simultaneously	46	46	-
1.13.2	const	const	none	zwitter-ion polymerization; distance between ions varies as $r^{3/2}$, equilibrium between free and solvated ions; 1° term.	47	-	47
1.14	**Emulsion Polymerization (see also Tables 2 and 7)**						
1.14.1	const	const within particles	none	general theory; particles act as independent units; 2° term within particles, number of particles remain constant; number of radicals per particle.	48-54	-	53
1.14.2	const	const within particles	none	general theory; slow term, rate within particles	55	-	55

---

* Data belong together

	Set	Initiation	Monomer	Transfer	Termination	R_p	P_r	M_n, M_w
1.14		Emulsion Polymerization (Cont'd.)						
	1.14.3	const	const within particles	none	inst. term within particles; number of particles varies	56	-	-
	1.14.4	const	const within particles	monomer	normal 2° term within particles; rapid interchange between phases of small-sized radicals; number of radicals per particle	57	-	-
	1.14.5	const	const within particles, varies outside	monomer	rapid interchange between phases of small-sized radicals; term in aq. phase and in particles	58	-	-
	1.14.6	slow and const	const	none	slow 2° comb; calculation of number of radicals per particle	-	59	59
	1.14.7	const	const within particles	none	constant number of particles; term by 2° comb	60	-	60
	1.14.8	const	const	none	inst. term on entering a particle; const number of particles	-	-	61
	1.14.9	varies	varies	minimize	adjust initiation and term rates while minimizing transfer to obtain "mono-disperse" polymers	72	72	72
1.15		Heterogeneous Polymerization						
	1.15.1	k_iMI, I varies	varies	monomer	polymerization in monomer rich and monomer poor phases; 2° disprop.	62	62	62
	1.15.2	instant	varies	none	term by precipitation onto growing solid particle.	63	-	63
	1.15.3	const	const	none	2° comb, 1° occlusion onto particle surfaces, and primary radical term	64	-	-
	1.15.4	const	const	none	2° term in liquid phase, 1° radical precipitation, propagation and 2° term at solid-liquid interface	65	-	-
1.16		Inhibition and Retardation						
	1.16.1	const	varies	none	2° term, 2° addition to inhibitor	3,66,67,73	-	-
	1.16.2	const	const	inhibitor	2° term, inhibitor term, inhibitor coupling	68	-	68
	1.16.3	const	const	retarder	2° term, retarder reinitiation and term	69	-	69
	1.16.4	const	const	retarder	2° term, retarder reinitiation, term and coupling	3,12,70,71,74	-	-
	1.16.5	const	const	retarder	2° term, rate of transfer equals rate of reinitiation	10	-	-
	1.16.6	const	const	retarder	2° term, copolymerization of retarder	11	-	11
	1.16.7	const	varies	none	pseudo 1° with scavenger	17	-	-
	1.16.8	const	const	inhibitor	2° disprop, 2° reaction with inhibitor	75	-	75
2.		Addition Polymerization -- "Living Polymers with Partial Deactivation"						
2.1		The Poisson Distribution: $k_i = k_p$						
	2.1.1	k_pMI	varies	none	none	2,101,102	1,2,101, 102	1,2,101, 102
	2.1.2	k_pMIt	varies	none	none; initiator added at a constant rate	103	-	103
2.2		The Gold Distribution: $k_i \neq k_p$						
	2.2.1	k_iMI	varies	none	none	102,104	1,102	1,102
2.3		Partial Deactivation by a 1° Process or a 2° Process with an Impurity						
	2.3.1	instant	varies	none	1° or 2°	105	1,106	1,106
	2.3.2	instant	varies	none	rate of term/R_p = constant	-	1,106,107	1,106 107
	2.3.3	instant, initiator has two active sizes	varies	none	rate of term/R_p = constant	-	1,107	1,107
	2.3.4	instant	varies	none	rate of term independent of R_p	1,106	-	1,106
	2.3.5	instant	varies	none	1° at infinite time	108,109	-	108,109

	Set	Initiation	Monomer	Transfer	Termination	R_p	P_r	M_n, M_w
2.4	Two Propagating Species: $R_n^\star$ can Transform into $Q_n^\star$							
	2.4.1	instant	const	none	none	-	76	76
	2.4.2	k_iMI	varies	none	none	1,77-79	1,77,78,80	1,77-80
2.5	Simultaneous Polymerization and Depolymerization							
	2.5.1	k_iMI	const	none	none	-	1,81	1,81
	2.5.2	k_iMI	varies	none	none	82,83	81	83
	2.5.3	k_pM^2	varies	none	none	84	-	-
2.6	The k_p Varies with Chain Length							
	2.6.1	k_iMI	varies	none	none			
	a.	all propagation constants (k_r) are different				-	1,85,86	-
	b.	$k_i : k_1 : k_2 : k_3 : \ldots \ldots k_n = m : (m-1) : (m-2) : (m-3) : \ldots$				-	1,85	-
	c.	$k_1 \neq k_2 \ldots \neq k_m = k_{m+1} = \ldots = k_n$				-	1,85	-
	d.	$k_i \neq k_1 \neq k_2 = k_m$				-	1,85	-
	2.6.2	instant, $k_1 = k_2 \ldots = k_m$, k_{m+1}, etc. = 0				86	86	-
	2.6.3	instant, k_p is a linear decreasing function of r				86	86	86
2.7	Deactivation by Transfer to Monomer							
	2.7.1	k_iMI	varies	monomer	none	1,87	1,87	1,87,88,74
	2.7.2	instant	varies	monomer	none	1,74,90	1,89,90	1,74,90
	2.7.3	instant	varies	monomer	none, two active ends per initiator	1,91	1,92,91	1,91
2.8	Deactivation by Slow 1° Termination							
	2.8.1	instant	varies	none	slow 1°	1,74	-	1,74
	2.8.2	instant	varies	monomer	slow 1°, at infinite time	1,74	-	1,74
2.9	Deactivation by Initiator Expulsion Reaction							
	2.9.1	instant	varies	none	none	-	-	1,93
	2.9.2	k_iMI	varies	monomer	1° term	67	-	67
2.10	Deactivation by Degradative Chain Transfer							
	2.10.1	k_iMI	varies	degradative transfer to polymer	none	94	-	95
2.11	Copolymerization							
	2.11.1	instant	varies	none	none	-	96	-
2.12	Heterogeneous Polymerization							
	2.12.1	instant	varies	none	none; diffusion of monomer through solid polymer	97	-	97
	2.12.2	instant	varies	none	2° term with monomer; diffusion of monomer through solid polymer	98	-	98
	2.12.3	const	const	none	sorption and desorption of chains from the surface; slow 1° term which depends on chain length	-	99	99
2.13	Spontaneous Polymerization							
	2.13.1	const	const	none	vinyl compound and activator form monomer; monomer both initiates and propagates; no term	100	-	100

Set	Conditions	References

3.   **Linear Condensation Polymerization without Ring Formation**

3.1	Condensation of bifunctional monomer AB. The Flory distribution. See equation C41.	1, 2, 110
3.2	Bifunctional monomer AA reacting with bifunctional monomer BB. The nylon case of hexamethylene diamine and adipic acid.	1, 24, 111
3.3	Deviations from the principle of equal reactivity.	1, 112
3.4	Rate of condensation proportional to chain size.	1, 113, 114, 134
3.5	Other simple linear condensation cases:	
	A. AA reacting with BC. BC is an anhydride. Within a given molecule, B must react before C.	1, 111
	B. AA reacting with BC. BC is an unsymmetrical acid or glycol. B reacts only with A at a different rate from that of C reacting with A.	1, 111
	C. AB reacting with C and itself. B and C react only with A. C is a terminator or capping material.	1, 111
	D. AA reacting with BB and C. The nylon case again with acetic acid as terminator.	1, 111
	E. AA reacting with BC. A and B react with C.	1, 111
	F. AB reacts with CC or DD kinetics.	115
3.6	Further polymerization of polymers with an initial geometric distribution:	
	A. Further polymerization of AB when the initial distribution is geometric.	1, 116
	B. Further polymerization of AB when the initial distribution is a superposition of two geometric distributions.	1, 116
	C. Further polymerization of AA with BB when the initial distribution of both is geometric.	1, 116
3.7	Copolymerization of Condensation Polymers:	
	A. AB reacting with CD. AB and CD are both hydroxy acids or similar materials.	1, 111
	B. AA reacting with BB and CC. A reacts with B and C only and vice versa. BB and CC could be adipic and sebacic acids, respectively.	111
	C. AA and BB reacting with CC and DD. A and B react only with C and D and vice versa.	111
	D. AA reacting with BC and DD. A reacts only with B, C, and D.	111
	E. AA and DD reacting with BC. A and B react only with C and D and vice versa.	111
3.8	Coupled Polymers:	
	A. AB polymerized to extent of conversion α, then coupled with CC.	1, 117, 118
	B. AB polymerized to extent of conversion α, then coupled with CD. A and B can react with C or D.	1, 117
	C. AA and BB polymerized to extent of conversion α, then coupled with an excess of CC. A and B to react completely on coupling.	1, 117
	D. Poisson-distribution polymer of AA coupled with BC.	1, 117
	E. Monomer AB polymerized to extent of conversion α, the coupled with excess CC to extent of conversion γ, the recoupled with excess DD.	1, 117
	F. Particularly narrow distributions via coupling reactions	
	I. AA and BB (great excess) → BBAABB, then remove excess BB. CC and BBAABB (great excess) → BBAABBCCBBAABB, then remove excess BBAABB, and continue in like manner.	119
	II. AA and 2BC → CBAABC - CBAABC and 2DE → EDCBAABCDE - etc.	119
	III. AB and CD → ABCD - ABCD and EF → ABCDEF - etc.	119
	G. Blocks of polymers of known distribution are coupled together.	
	I. A series of Poisson-type-polymers coupled together	
	II. Poisson-type-polymers coupled to "Most Probable"-type polymers.	
	III. A series of "Most Probable"-type polymers coupled together.	120, 118
3.9	AB reacts with CC or CD. Rate of reaction of A dependent upon whether or not B has reacted.	115

4.   **Equilibrium Polymerization**

4.1	The "most probable" distribution of Flory has been derived for condensation polymerization when all reactions are assumed to have the same probability, regardless of chain length, and whether or not exchange reactions occur:	Equation C41

$$P_r + P_s = P_{r+s-i} + P_i \qquad i < r + s$$

Here P_i can be a by-product such as water, or a polymer molecule whose size is smaller than $r + s$. The same distribution results when random scission occurs to infinitely long chains.   2, 121, 122

4.2	The theoretical equilibrium molecular weight distribution for the system	

$$rM = P_r$$

depends upon the thermodynamic definition of the final product. If the change in Gibbs free energy of formation and polymerization is a linear function of molecular size r, then two cases can be considered:

	a. The monomer is polymerized to a pure perfectly ordered state (solid). This gives the "most probable" distribution.	Equation C41
	b. The monomer is polymerized to a pure randomly ordered state (liquid). A completely different distribution results.	123

Set	Conditions	References
4.2 (Cont'd.)		
	If the free energy change upon disordering of the pure polymer is independent of the molecular weight distribution, a new distribution arises.	124
4.3	Thermodynamics of living equilibrium polymers.	125
4.4	Given an initial distribution, find the distribution as a function of time when chain ends react at random with all monomer units.	126
4.5	Initiation, propagation, and termination reactions are all equilibrium reactions. An initiator is required.	127
4.6	Initiation, propagation, and termination reactions are all equilibrium reactions. The system is self-initiating.	127
4.7	Equilibrium initiation and equilibrium propagation with a multifunctional initiator.	128
4.8	Equilibrium initiation and equilibrium propagation. No termination. Nonequilibrium solution.	129
4.9	The catalyst can form an active species with any polymer molecule, and the total number of moles of the system is kept constant (constant pressure and volume if all polymer molecules are ideal gases).	130
4.10	Addition polymerization, instantaneous initiation, no transfer, no termination, monomer concentration held constant, active species in equilibrium with inactive species.	77
4.11	Addition polymerization, rate of initiation equals $k_i MC$, no transfer, no termination, two active species in equilibrium, monomer concentration varies.	78
4.12	Addition polymerization, instantaneous initiation, no transfer, no termination, active species in equilibrium with inactive species, monomer concentration varies.	77
4.13	Poly(phosphate) equilibria.	131
4.14	Instantaneous initiation, polymerization and depolymerization, 1° termination.	132
4.15	Instantaneous initiation, polymerization and depolymerization, degradative transfer to monomer.	133
4.16	Copolymerization: bimolecular addition of one monomer, equilibrium addition of other monomer.	230

5. Non-Linear Polymerization Systems

In this table, systems are treated where branching reactions, ring formation or gelation may occur. See also Table 6 for references on gelation reactions during degradation. The symbol RA_f means a monomer containing f reactive A units.

Set	Conditions	References
5.1	Vinyl polymerization. Self-grafting.	
A.	Constant rate of initiation, monomer concentration invarient, transfer to monomer and to polymer, termination by 2° disproportionation.	1, 3, 18
B.	Constant rate of initiation, monomer concentration varies, transfer to polymer, termination by 2° combination.	135
5.2	Vinyl polymerization. Grafting onto a preformed polymer whose initial distribution is either (i) of constant length or (ii) a geometric distribution.	18
5.3	Vinyl polymerization. All segments of the backbone chains have equal probability to be grafted.	136
5.4	Vinyl polymerization. Terminal double bond polymerization. Constant rate of initiation, monomer concentration varies, transfer to monomer and to polymer, terminal double bond polymerization, termination predominantly by transfer. Batch or continuous polymerization.	1, 137-139
5.5	Vinyl polymerization. Long chain branching. Constant rate of initiation, monomer concentration invarient, transfer to monomer and to polymer, termination by 2° disproportionation.	1, 140
5.6	Vinyl polymerization. Living branched polymer. Polyfunctional initiator RA_f. Rate of initiation equals rate of propagation, no transfer, no termination.	141
5.7	Vinyl polymerization. R_p in a diffusion controlled gelling system. Termination by disproportionation or combination, no transfer.	142
5.8	Vinyl polymerization. Branching density as a function of conversion. Branches formed by polymerization through a vinyl group (diene polymers) or by transfer-to-polymer reaction.	
A.	Batch polymerization	1, 2
B.	Continuous polymerization	1, 138
5.9	Vinyl polymerization. Branching density as a function of conversion of (1) long chain branching, (2) short chain branching (backbiting), (3) very short chain branching (radical migration).	143
5.10	Emulsion polymerization. Branching density.	144
5.11	Cyclopolymerization: The concentration of pendent double bonds per molecule.	145-149
5.12	Gelation condition in a system when cyclopolymerization also occurs. Transfer to monomer reaction occurs. Molecular weight given.	150
5.13	General theory of gelation for polyaddition- and polycondensation-type polymers.	151
5.14	The general distribution function for various molecules of type A reacting with various molecules of type B. A can only react with B. All functional groups have equal reactivity, independent of position within the molecule or the size of the molecule. Ring formation is excluded.	Section D
5.15	Cross-linking or coupling of a polymer with a known primary distribution. Formation of rings excluded prior to gelation. See also - Table 6.	
A.	General case	1, 152-155
B.	All primary chains have the same size	1, 2, 152, 155-157
C.	Primary chains are of the Poisson type	155, 157

Set	Conditions	References
5.16	Homopolymer of ARB_{f-1}. A can only react with B. The B's may have different reactivities. Formation of rings excluded prior to gelation.	
	A. All B groups have the same reactivity	1, 2, 158
	B. Let B groups have different reactivities	1, 159, 160
	C. A controversy over the statistical approach	161-164
5.17	Copolymer of ARB_{f-1} and AB. A can only react wtih B. Formation of rings excluded prior to gelation.	1, 2, 158
5.18	Homopolymer of RA_f. Formation of rings excluded prior to gelation.	
	A. All A's equally reactive	1, 2, 121, 158, 165, 166
	B. Reactivity of A depends on number previously reacted	167
5.19	A. Copolymer of RA_f and AA. Formation of rings excluded prior to gelation.	1, 2, 165, 166
	B. Copolymer of RA_f, AA, and BB. A can react only with B. Formation of rings excluded prior to gelation. All branch points completely reacted. See set 5.16C for a controversy.	1, 168, 169
5.20	Homopolymer of RA_fB_g. A can only react with B. Formation of rings excluded prior to gelation.	170
5.21	Branching without gelation. Copolymer of RA_f and AB. A can only react with B.	1, 171
5.22	Gelation conditions. Formation of rings excluded prior to gelation.	
	A. AA reacting with B_fC_g. A can react with B and C, but with different velocities, B cannot react with C.	1, 172
	B. AA, BB, and C reacting with DDE and FF. A, B, and C individually can react with D, E, and F, but with different velocities.	1, 172
	C. AAB, RC_4, and GG reacting with DE and F. DE is an anhydride or similar material. D must react first and may have a different velocity from E. A, B, C, and G may react with different velocities.	1, 172
	D. AB and CD reacting with EEF and GG. AB and CD are anhydrides or similar materials, where A must react before B and C must react before D. A, B, C, and D may react with different velocities.	1, 172
	E. AA and BC reacting with DDE and FG. BC is an anhydride, B reacting first, and FG is like an unsymmetrical glycol.	1, 172
5.23	Ring formation in linear polymers.	
	A. Homopolymer of type AA.	1, 173, 174
	B. Homopolymer of type AB.	1, 173, 175
	C. Copolymer of type AABB.	1, 173, 176
	D. Distribution for chain fraction.	1, 2
5.24	Cross-linking of a polymer with a known distribution. Rings are permitted prior to gelation.	153
5.25	Gelation conditions for RA_f when rings are permitted prior to gelation.	1, 177
5.26	Homopolymer of RA_f when rings are permitted prior to gelation.	1, 178, 179
5.27	Copolymer of RA_f and BB when rings are permitted prior to gelation.	180
5.28	Gelation condition for RA_f, AA and BB when rings are permitted prior to gelation.	1, 181

6.   <u>Degradation of Polymers -- May be Accompanied by Cross-Linking</u>

Set	Initiation	Mass	Initial Distribution	Transfer	Termination	R_d	P_r	M_n, etc.
6.1	Random Scission Only							
6.1.1	proportional to time	const	arbitrary	none	none	-	183-186	182-184
6.1.2	proportional to time	const	infinite or uniform	none	none	-	183-187	183, 184, 188
6.1.3	random at all bonds	const	uniform	none	none	-	189	189
6.1.4	random at chain ends	const	uniform	none	none	-	190	-
6.1.5	random at all chains	const	uniform	none	none	-	191	191
6.1.6	random	const	rectangular	none	none	-	184, 187	184
6.1.7	random	const	Flory	none	none	-	184, 187	184, 188
6.1.8	random	const	Schulz(k=2)	none	none	-	184, 187	187
6.1.9	random	const	Poisson	none	none	-	187	-
6.1.10	random at normal and weak bonds	const	uniform	none	none	-	-	192
6.2	Unzipping-Type Reaction							
6.2.1	1° end	varies	uniform	none	1°	193	-	193
6.2.2	1° end	varies	uniform	polymer	2° disprop	194	-	194
6.2.3	1° end	varies	Schulz	none	1°	193	-	193
6.2.4	1° end	varies	Schulz	none	2° disprop	195	-	195
6.2.5	1° end	varies	Schulz (k=2)	none	1°	196	-	196
6.2.6	1° end	varies	Schulz (k=2)	none	2° comb	196	-	196
6.2.7	1° end; various rates of initiation	varies; monomer radicals may evaporate	Flory	none	1° or 2° disprop	197	-	-

	Set	Initiation	Mass	Initial Distribution	Transfer	Termination	R_d	P_r	M_n, etc.
6.2	Unzipping-Type Reaction (Cont'd.)								
	6.2.8	random scission only	varies	uniform	polymer	2° disprop	194	-	194
	6.2.9	random scission only	varies	Schulz	none	2° disprop	198	-	198
	6.2.10	1° end and random scission	varies	unspecified	none	1° or 2° process	199 (all molecules must terminate) 200 (some molecules can completely disappear)		
	6.2.11	1° end and random scission	varies	Flory	polymer	1° and 2° disprop	201	-	201
	6.2.12	1° end and random scission	varies	Flory	polymer	2° comb	201	-	201
	6.2.13	1° end and random scission	varies	Flory	polymer	1° and 2° process; comparison of random scission, weak link, and transfer theories.	202 203	- -	- 203
	6.2.14	1° end and random scission	varies	uniform	polymer	comparison of various term. mechanisms	204	-	-
	6.2.15	1° end and random scission	varies; units other than monomer may evaporate	uniform	polymer	2° disprop	205	-	205
	6.2.16	1° end and random scission	varies; units other than monomer may evaporate	Flory	polymer	1° and 2° disprop or 2° comb	206	206	206
	6.2.17	unzipping reactions	varies	various	various	various, a review	207	-	-

	Set	Initiation	Mass	Distribution	Transfer	Termination	P_r	M_n, etc.	Solfraction
6.3	Random Scission and Cross-Linking								
	6.3.1	proportional to time	const	arbitrary	none	unspecified term. with and without ring formation	153 208	153 208	153 208
	6.3.2	varies with depth of penetration	const	Flory	none	1° and 2° disprop without ring formation	209	210 209	210
	6.3.3	proportional to time	const	arbitrary	none	end-linking (grafting) rather than cross-linking; without ring formation	211 208	211 208	211 308
	6.3.4	proportional to time	const	arbitrary	none	end-linking; with ring formation	211	211	211
	6.3.5	proportional to time	const	Schulz	none	unspecified term. without ring formation	29	-	29
6.4	Cross-Linking Only								
	6.4.1	proportional to time	const	arbitrary	none	ring formation prohibited	153	153	153
	6.4.2	proportional to time	const	uniform	none	ring formation prohibited	182	182	182
	6.4.3	proportional to time	const	Schulz	none	ring formation prohibited	182, 212, 213	182	182
	6.4.4	proportional to time	const	Beasley (140)	none	ring formation prohibited	-	182	-
	6.4.5	proportional to time	const	arbitrary	none	ring formation permitted	153	153	153

7. Influence of Reactor Conditions and Design on Molecular Weight Distributions
(CFSTR: Continuous-Flow, Stirred-Tank Reactor. BR: Batch Reactor. MWD: Molecular Weight Distribution)

7.1 Comparison of 1° term., 2° term. with monomer, 2° disprop, 2° comb and no term. for 1° and 2° self-initiation and initiation by light in a CFSTR. R_p, P_r and M_n is given. ... 214

7.2 Emulsion polymerization in a CFSTR. Instantaneous 2° term. R_p and M_n's. ... 215

7.3 Effect of inadequate stirring in a CFSTR. Comparison with BR and normal CFSTR. Term. by 2° or no term. R_p, P_r, M_n's. ... 216

7.4 Control of MWD by use of temperature variation. Term. by 2° disprop or 2° comb. M_n's. ... 217

7.5 Effect of expanding drop of monomer in a catalyst solution on MWD of "Living Polymers". ... 218

7.   Influence of Reactor Conditions and Design on Molecular Weight Distribution (Cont'd.)

C.   SOME DISTRIBUTION FUNCTIONS AND THEIR PROPERTIES

The frequency function, $F(r)$, is the fraction of molecules of size r. Furthermore, $F(r)$ is normalized

$$\sum_1^\infty F(r) = \int_0^\infty F(r)\, dr = 1 \tag{C1}$$

the weight fraction of molecules of size r is

$$W(r) = rF(r)/\sum_1^\infty rF(r) = rF(r)/\int_0^\infty rF(r)\, dr \tag{C2}$$

Averages of any distribution are defined by

$$\bar{r}_i = \sum_1^\infty r^i F(r)/\sum_1^\infty r^{i-1} F(r) = \int_0^\infty r^i F(r)\, dr / \int_0^\infty r^{i-1} F(r)\, dr \tag{C3}$$

The number-average degree   polymerization is defined by

$$\bar{r}_n = \sum_1^\infty rF(r)/\sum_1^\infty F(r) = \sum_1^\infty W(r)/\sum_1^\infty [W(r)/r] \tag{C4}$$

Thus, by the definitions of $F(r)$ and $W(r)$

$$W(r) = rF(r)/\bar{r}_n \tag{C5}$$

The weight-average of polymerization is

$$\bar{r}_W = \sum_1^\infty r^2 F(r)/\sum_1^\infty rF(r) = \sum_1^\infty r W(r)/\sum_1^\infty W(r) \tag{C6}$$

The "z" average and the "z+1" averages are defined by equation (C3) with i set equal to 3 and 4. There is no need to restrict averages of a distribution to positive integers-- any useful average can be defined; such as the (-5/2) average. If   intrinsic viscosity of a polymer is related to the degree of polymerization through the equation

$$[\eta] = k\bar{r}_V^a \tag{C7}$$

where $\bar{r}_V$ is the "viscosity-average degree of polymerization", then it is related to the frequency function by

$$\bar{r}_V = \left[\sum_1^\infty r^{1+a} F(r)/\sum_1^\infty rF(r) \right]^{1/a} \tag{C8}$$

As a approaches unity, $\bar{r}_V$ approaches $\bar{r}_W$. In principle, a distribution function can be determined if sufficient averages of the distribution can be determined. In practice, only the number, weight, and perhaps "z" averages can be found, which are insufficient to define any distribution without making further assumptions.

The degree of polymerization is a useful concept as long as one is describing polymers made with a single monomer or with monomers of equal molecular weight. When considering copolymers, one must work with the actual molecular weight of the reacted unit. This is done in condensation polymerization. The molecular weight distribution of addition copolymers is not included here because of the extreme complexity of these systems.

When performing the summation of a distribution equation to find an average of the distribution, use is frequently made of the following sums

$$\sum_{x=1}^\infty p^{x-1} = 1/(1-p) \qquad\qquad p < 1$$

$$\sum_{x=1}^\infty x p^{x-1} = 1/(1-p)^2 \qquad\qquad p < 1$$

$$\sum_{x=1}^\infty x^2 p^{x-1} = (1+p)/(1-p)^3 \qquad p < 1 \tag{C9}$$

$$\sum_{x=1}^\infty x^n p^{x-1} = \frac{d}{dp}\left[p \sum_{x=1}^\infty x^{n-1} p^{x-1} \right] p < 1$$

$$\sum_{i=0}^\infty a^{n-i} b^i n!/(n-i)!\, i! = (a+b)^n \qquad \text{n is an integer}$$

$$\sum_{i=0}^{\infty} a^i i n!/(n-i)!i! = na(1+a)^{n-1} \tag{C9}$$

$$\sum_{i=0}^{\infty} a^i i^2 n!/(n-i)!i! = na(1+na)(1+a)^{n-2}$$

$$\sum_{n=1}^{\infty} A^{n-1}/(n-1)! = \exp A$$

$$\sum_{n=1}^{\infty} n A^{n-1}/(n-1)! = (1+A)\exp A$$

$$\sum_{n=1}^{\infty} n^2 A^{n-1}/(n-1)! = (1+3A+A^2)\exp A$$

$$\sum_{n=1}^{\infty} n^p A^{n-1}/(n-1)! = \frac{d}{dA}\left[A\sum_{n=1}^{\infty} n^{p-1} A^{n-1}/(n-1)!\right]$$

1. Normal Distribution Function (Gaussian Distribution)

$$F(r) = \frac{\exp\left\{-(r-\bar{r})^2/2\sigma^2\right\}}{(2\pi)^{1/2}\sigma} \tag{C10}$$

$$W(r) = (r/\bar{r})F(r) \tag{C11}$$

$$\bar{r} = \int_{-\infty}^{+\infty} \frac{r\exp\left\{-(r-\bar{r})^2/2\sigma^2\right\}}{(2\pi)^{1/2}\sigma} = \bar{r}_n \tag{C12}$$

$$\sigma^2 = \int_{-\infty}^{+\infty} \frac{(r-\bar{r})^2\exp\left\{-(r-\bar{r})^2/2\sigma^2\right\}}{(2\pi)^{1/2}\sigma} \tag{C13}$$

$$\bar{r}_w = \frac{\sigma^2}{\bar{r}} + \bar{r} \tag{C14}$$

Values of F(r) and $\int_0^t F(r)dr$ are found in many statistical tables (223).

2. Logarithmic Normal Distribution Function.
    Under the assumption that the weight distribution of the logarithm of molecular size is normally distributed, we can replace r in the Normal Distribution Function by $\ln r$ and $\bar{r}$ by $\ln \bar{r}_m$

$$W(\ln r) = \frac{\exp\left\{-(\ln r - \ln \bar{r}_m)^2/2\sigma^2\right\}}{(2\pi)^{1/2}\sigma} \tag{C15}$$

where

$$\int_1^{\infty} W(\ln r)\,d(\ln r) = 1 \tag{C16}$$

or in the alternate form

$$W(r) = \frac{\exp\left\{-(\ln r - \ln \bar{r}_m)^2/2\sigma^2\right\}}{r(2\pi)^{1/2}\sigma} \tag{C17}$$

$$\int_0^{\infty} W(r)\,dr = 1 \tag{C18}$$

and

$$\ln \bar{r}_m = \int_0^{\infty} (\ln r)\,W(r)\,dr \tag{C19}$$

here $\bar{r}_m$ is the median value of the distribution, that is, one-half of the values of r are less than $\bar{r}_m$.

$$\sigma^2 = \int_0^{\infty} (\ln r - \ln \bar{r}_m)^2\,W(r)\,dr \tag{C20}$$

$$\bar{r}_n = \bar{r}_m \exp(-\sigma^2/2) \tag{C21}$$

$$\bar{r}_w = \bar{r}_m \exp(+\sigma^2/2) \tag{C22}$$

$$\bar{r}_z = \bar{r}_m \exp(+3\sigma^2/2) \tag{C23}$$

$$\bar{r}_i = \bar{r}_m \exp\{(2i - 3)\, \sigma^2/2\} \tag{C24}$$

Note that
$$\bar{r}_m = (\bar{r}_n \cdot \bar{r}_w)^{1/2} \tag{C25}$$

$$\bar{r}_w/\bar{r}_n = \bar{r}_z/\bar{r}_w = \bar{r}_{z+1}/\bar{r}_z = \exp \sigma^2 \tag{C26}$$

The Logarithmic Normal Distribution sometimes is given in a "generalized" form (224)

$$W_s(r) = \frac{r^s \exp[-(\ln r/\bar{r}_s)^2/2\sigma^2]}{(2\pi)^{1/2}\sigma \bar{r}_s^{s+1} \exp\{(s+1)^1\sigma^2/2\}} \tag{C27}$$

but Honig (225) has shown that
$$\ln \bar{r}_m = \ln \bar{r}_s + (s+1)\,\sigma^2. \tag{C28}$$

Thus, the normalized Lansing–Kramer function (s = 0) (226) is identical to the normalized Wesslau function (s = -1) (227).
The generalized Logarithmic Normal Distribution is skewed to large values of r, but Kotliar (237, 238) shows that this distribution is not a good representation of a polymer after either low or high molecular weight material is removed or degradation has occurred.

3.   Generalized Exponential Distribution (1) (228)

$$F_{m,k,y}(r) = m\,y^{k/m}\,r^{k-1}\,[\exp(-y r^m)]/\Gamma(k/m) \tag{C29}$$

$$W_{m,k,y}(r) = m\,y^{(k+1)/m}\,r^k\,[\exp(-y r^m)]/\Gamma[(k+1)/m] \tag{C30}$$

$$\bar{r}_n = \Gamma[(k+1)/m]/y^{1/m}\,\Gamma(k/m) \tag{C31}$$

$$\bar{r}_w = \Gamma[(k+2)/m]/y^{1/m}\,\Gamma[k+1]/m] \tag{C32}$$

$$r_z = \Gamma[(k+3)/m]/y^{1/m}\,\Gamma[(k+2)/m] \tag{C33}$$

$$\bar{r}_i = \Gamma(k+i)/m]/y^{1/m}\,\Gamma[k+i-1)/m] \tag{C34}$$

and
$$r_v = \{\Gamma[(k+a+1)/m]/y^{a/m}\,\Gamma[(k+1)/m]\}^{1/a} \tag{C35}$$

where a is defined in equation (C7)

a) The distribution function of Schulz (9) and Zimm (229) is obtained by setting m = 1 and requiring that k > 0. This distribution is equivalent to the Pearson Type III distribution

$$F(r) = y^k\,r^{k-1}\,[\exp(-yr)]/\Gamma(k) \tag{C36}$$

$$W(r) = y^{(k+1)}\,r^k\,[\exp(-yr)]/\Gamma(k+1) \tag{C37}$$

The distribution reduces to the "most probable" distribution where k = 1.
The cumulative number or weight fraction of the Schulz distribution may be computed from

$$\int_0^r F_{1,k,y}(r) = \frac{k\,y^k}{\Gamma(1+k)} \sum_{i=0}^{\infty} \frac{(-1)^i\,y^i\,r^{k+i}}{i!\,(k+i)} \tag{C38}$$

$$\int_0^r W_{1,k,y}(r) = \frac{y^{k+1}}{\Gamma(1+k)} \sum_{i=0}^{\infty} \frac{(-1)^i\,y^i\,r^{k+i+1}}{i!\,(a+i+1)} \tag{C39}$$

b) The Weibull (231) or Tung (232) distribution is obtained by setting m = 1+k and k > 0. This distribution is usually seen in the form

$$\int_0^r W(r)\,dr = 1 - \exp[-y r^{(1+k)}] \tag{C40}$$

Kotliar (233) shows that evaluation of y and k by the Tung method can lead to erroneous values of $\bar{r}_w/\bar{r}_n$.

c) The Flory distribution, as noted above, occurs when k = m = 1, provided that ln p can be replaced by -(1-p), i.e., p ≈ 1. The Flory distribution is also written

$$F(r) = p^{r-1}(1-p) \tag{C41}$$

$$W(r) = r\,p^{r-1}(1-p)^2 \tag{C42}$$

$$\bar{r}_n = 1/(1-p) \tag{C43}$$

$$\bar{r}_w = (1 + p)/(1 - p) \tag{C44}$$

$$\bar{r}_v = \bar{r}_n \left([\Gamma (2 + a)]^{1/a} \right) \tag{C45}$$

and with the approximation $\ln p = p - 1$

$$\int_0^r W(r)\, dr \cong (1/p) - [1 + (1 - p)\, r]\, p^{r-1} \tag{C46}$$

for high molecular weights $p \approx 1$ hence $r_n : r_w : r_z = 1 : 2 : 3$ \tag{C47}

4. Poisson Distribution

$$F(r) = \exp(-v) \cdot v^{r-1}/(r-1)! \tag{C48}$$

$$W(r) = [v/v + 1)]\, r \exp(-v) \cdot v^{r-2}/(r-1)! \tag{C49}$$

$$\bar{r}_n = 1 + v \tag{C50}$$

$$\bar{r}_w = 1 + v + v/(1 + v) \tag{C51}$$

D. MOLECULAR WEIGHT DISTRIBUTION IN CONDENSATION POLYMERS: THE STOCKMAYER DISTRIBUTION FUNCTION

Stockmayer (234) has presented a generalized distribution formula for a variety of monomers containing end groups of type A which can only react with a variety of monomers containing end groups of type B. In the original mixture there are A_1, A_2, A_3, ... A_i, ... moles of reactants bearing respectively f_1, f_2, f_3, ... f_i, ... functional groups of type A each, together with B_1, B_2, ... B_j, ... moles of reactants of functionalities g_1, g_2, ... g_j, ... in groups of type B. All functional groups of a given type are equally reactive and ring formation does not occur appreciably; which obviously is not true near the gel point. The system reacts until a fraction α of the A groups and a fraction β of the B groups have reacted. Further

$$\alpha \Sigma_i f_i A_i = \beta \Sigma_j g_j B_j$$

Now $N\{m_i, n_j\}$ represents the number of moles of that species which consists of m_1, m_2, ... m_i, ... monomer units of the A type combined with n_1, n_2, ... n_j units of the B type.

$$N\{m_i, n_j\} = \frac{K\,(\Sigma_i f_i m_i - \Sigma_i m_i)!\,(\Sigma_j g_j n_j - \Sigma_j n_j)!\,\prod_i (x_i^{m_i}/m_i!)\,\prod_j (y_j^{n_j}/n_j!)}{(\Sigma_i f_i m_i - \Sigma_i m_i - \Sigma_j n_j + 1)!\,(\Sigma_j g_j n_j - \Sigma_j n_j - \Sigma_i m_i + 1)!} \tag{D1}$$

where

$$x_i = \left[\frac{f_i A_i}{(\Sigma_i f_i A_i)} \right] \left[\frac{\beta\,(1 - \alpha)^{f_i - 1}}{(1 - \beta)} \right]; \quad y_j = \left[\frac{g_j B_j}{(\Sigma_j g_j B_j)} \right] \left[\frac{\alpha\,(1 - \beta)^{g_j - 1}}{(1 - \alpha)} \right]$$

$$K = (\Sigma_i f_i A_i)\,(1 - \alpha)\,(1 - \beta)/\beta = (\Sigma_j g_j B_j)\,(1 - \alpha)\,(1 - \beta)/\alpha \tag{D2}$$

an example will illustrate the use of Eq. (D1).

Suppose we have the monomers acetic acid (CH_3COOH, A_1, $f_1 = 1$) and adipic acid ($HOOC(CH_2)_4COOH$, A_2, $f_2 = 2$) reacting with ethylene glycol ($HOCH_2CH_2OH$, B_1, $g_1 = 2$) and glycerol ($HOCH_2CHOHCH_2OH$, B_2, $g_2 = 3$) (all hydroxyl groups of the glycerol are considered equally reactive). What is the number of molecules which contain exactly 1 acetic acid unit, 4 adipic acid units, 3 glycol units and 2 glycerol units? It is $N\{1, 4, 3, 2\}$ and

$$N\{1, 4, 3, 2\} = \frac{K\,(1 + 8 - 1 - 4)!\,(6 + 6 - 3 - 2)!}{(1 + 8 - 1 - 4 - 3 - 2 + 1)!\,(6 + 6 - 3 - 2 - 1 - 4 + 1)!} \tag{D3}$$

$$\cdot \frac{x_1^1}{1!} \cdot \frac{x_2^4}{4!} \cdot \frac{y_1^3}{3!} \cdot \frac{y_2^2}{2!}$$

$$x_1 = \left[\frac{A_1}{A_1 + 2A_2} \right] \left[\frac{\beta}{1 - \beta} \right] \qquad x_2 = \left[\frac{2A_2}{A_1 + 2A_2} \right] \left[\frac{\beta(1 - \alpha)}{(1 - \beta)} \right]$$

$$y_1 = \left[\frac{2B_1}{2B_1 + 3B_2} \right] \left[\frac{\alpha(1 - \beta)}{(1 - \alpha)} \right] \qquad y_2 = \left[\frac{3B_2}{2B_1 + 3B_2} \right] \left[\frac{\alpha(1 - \beta)^2}{1 - \alpha} \right]$$

$$K = (A_1 + 2A_2)\,(1 - \alpha)\,(1 - \beta)/\beta \tag{D4}$$

If each species $\underline{i}$ has an effective molecular weight M_i, which is lower than the original molecular weight by the term $W_o f_i/2$, where W_o is the molecular weight of the by-product, then

$$W = \Sigma_i M_i A_i + \Sigma_j B_j M_j \qquad (D5)$$

The number of molecules of the end of the reaction is, neglecting by-product,

$$N = \Sigma_i A_i + \Sigma_j B_j - \alpha \Sigma_i f_i A_i \qquad (D6)$$

The number-average molecular weight is

$$\bar{M}_n = W/N \qquad (D7)$$

The weight-average molecular weight is

$$\bar{M}_w = \left[\beta \frac{\Sigma_i M_i^2 A_i}{\Sigma_i f_i A_i} + \alpha \frac{\Sigma_j M_j^2 B_j}{\Sigma_j g_j B_j} + \frac{\alpha \beta \left[\alpha (f_e - 1) M_b^2 + \beta (g_e - 1) M_a^2 + 2 M_a M_b \right]}{1 - \alpha \beta (f_e - 1)(g_e - 1)} \right]$$

$$\left[\beta \frac{\Sigma_i M_i A_i}{\Sigma_i f_i A_i} + \alpha \frac{\Sigma_j M_j B_j}{\Sigma_j g_j B_j} \right]^{-1} \qquad (D8)$$

where
$$f_e = (\Sigma_i f_i^2 A_i)/(\Sigma_i f_i A_i)$$

$$g_e = (\Sigma_j g_j^2 B_j)/(\Sigma_j g_j B_j)$$

$$M_a = (\Sigma_i M_i f_i A_i)/(\Sigma_i f_i A_i)$$

$$M_b = (\Sigma_j M_j g_j B_j)/(\Sigma_j g_j B_j) \qquad (D9)$$

Simpler expressions for M_n and M_w are given by Ziegler, et al. (235).

The gel point is
$$(\alpha \beta)_c = 1/(f_e - 1)(g_e - 1) \qquad (D10)$$

Extensive computations of $N\{m_i, n_j\}$ for the case $f_1 = 1$, $f_2 = 2$, $f_4 = 4$ and $g_2 = 2$ have appeared (236).

E. REFERENCES

1. L. H. Peebles, Jr., "Molecular Weight Distributions in Polymers", Interscience, New York, 1971.

2. P. J. Flory, "Principles of Polymer Chemistry", Cornell University Press, Ithaca, N.Y., 1953.

3. C. H. Bamford, W. G. Barb, A. D. Jenkins, P. F. Onyon, "The Kinetics of Vinyl Polymerization by Radical Mechanisms", Academic Press, New York, 1958.

4. G. Odian, "Principles of Polymerization", McGraw-Hill, New York, 1970.

5. F. W. Billmeyer, Jr., "Textbook of Polymer Science", Second Edition, Interscience, New York, 1971.

6. L. Kuechler, "Polymerisationskinetik", Springer-Verlag, Berlin, 1951.

7. Kh. S. Bagdasarian, "Theory of Free Radical Polymerization", Israel Program for Scientific Translations, Jerusalem, 1968.

8. D. C. Chappelear, R. H. M. Simon, "Addition and Condensation Polymerization Processes", N. Platzer, Ed., (Adv. in Chem. Ser. 91), Am. Chem. Soc., Washington, D.C. 1969, p.1.

9. G. V. Schulz, Z. Physik. Chem. B43, 25 (1939).

10. G. M. Burnett, L. D. Loan, Trans. Faraday Soc. 51, 214 (1955).

11. L. H. Peebles, Jr., J. T. Clarke, W. H. Stockmayer, J. Am. Chem. Soc. 82, 4780 (1960).

12. J. L. Kice, J. Am. Chem. Soc. 76, 6274 (1954).

13. L. H. Peebles, Jr., J. Appl. Polymer Sci. 17, 113 (1973).

14. S. Saccubai, M. Santappa, Makromol. Chem. 117, 50 (1968).

15. W. A. Pryor, J. H. Coco, Macromolecules 3, 300 (1970).

16. L. S. Bresler, V. A. Grechanovsky, A. Muzsay, I. Ya. Poddubnyi, Makromol. Chem. 133, 111 (1970).

17. C. H. Bamford, A. D. Jenkins, R. Johnson, Trans. Faraday Soc. 58, 1212 (1962).

18. C. H. Bamford, H. Tompa, Trans. Faraday Soc. 50, 1097 (1954).

19. M. Litt, J. Polymer Sci., A2, 8, 2105 (1970).

20. M. Litt, F. R. Eirich, J. Polymer Sci. 45, 379 (1960).

21. P. H. Plesch, Ed., "The Chemistry of Cationic Polymerization", MacMillan Co., New York, 1963.

22. A. A. Korotkov, S. P. Mitsengendler, V. N. Krasulina, J. Polymer Sci. 53, 217 (1961).

23. D. O. Jordan, A. R. Mathieson, J. Chem. Soc. 1952, 2358. (Note: equation 13 et seq. contain errors owing to an integration error).

24. H. E. Grethlein, Ind. Eng. Chem. Fundamentals 8, 206 (1969).

25. J. C. W. Chien, J. Polymer Sci. A1, 1839 (1963).

26. J. C. W. Chien, J. Am. Chem. Soc. 81, 86 (1959).

27. S. Katz, G. M. Saidel, Am. Inst. Chem. Eng. J. 13, 319 (1967).

28. C. H. Bamford, Polymer 6, 63 (1965).

29. M. Inokuti, J. Chem. Phys. 38, 2999 (1963).

30. A. V. Tobolsky, J. Am. Chem. Soc. 80, 5927 (1958); A. V. Tobolsky, C. E. Rogers, R. D. Brinkman, ibid. 82, 1277 (1960).

31. C. B. Wooster, Macromolecules 1, 324 (1968).

32. K. F. O'Driscoll, R. Knorr, Macromolecules 1, 367 (1968).

33. J. C. Bevington, "Radical Polymerization", Academic Press, New York, 1961.

34. W. H. Ray, T. L. Douglas, E. W. Goodsolve, Macromolecules 4, 166 (1971).

35. W. H. Stockmayer, J. Chem. Phys. 13, 199 (1945).

36. K. Ito, J. Polymer Sci., A2, 7, 241 (1969).

37. S. W. Benson, A. M. North, J. Am. Chem. Soc. 84, 935 (1962).

38. S. Okamura, T. Manabe, Polymer 2, 83 (1961).

39. K. Venkatarao, M. Santappa, J. Polymer Sci., A1, 8, 1785, 3429 (1970).

40. S. Kohjiya, Y. Imanishi, T. Higashimura, J. Polymer Sci., A1, 9, 747 (1971).

41. K. F. O'Driscoll, S. A. McArdle, J. Polymer Sci. 40, 557 (1959).

42. K. Ito, J. Polymer Sci., A1, 7, 827, 2995 (1969).

43. C. H. Bamford, A. D. Jenkins, R. Johnson, Trans. Faraday Soc. 55, 1451 (1959).

44. K. Ito, J. Polymer Sci., A1, $\underline{7}$, 2247, 3387 (1969); $\underline{8}$, 1823 (1970); $\underline{10}$, 1481 (1972); K. Ito, T. Matsuda, J. Appl. Polymer Sci. $\underline{14}$, 311 (1970).

45. T. Nagabhushanam, M. Santappa, J. Polymer Sci., A1, $\underline{10}$, 1511 (1972).

46. J. F. Westlake, R. Y. Huang, J. Polymer Sci., A1, $\underline{10}$, 1429, 1442 (1972).

47. M. A. Markevich, Ye. V. Kochetov, N. S. Yenikolopyan, Vysoko-molekul. Soedin.. A13, 1013 (1971); Polymer Sci. USSR 13, 1153 (1971).

48. R. N. Haward, J. Polymer Sci. $\underline{4}$, 273 (1949).

49. W. V. Smith, R. H. Ewart, J. Chem. Phys. $\underline{16}$, 592 (1948).

50. W. H. Stockmayer, J. Polymer Sci. $\underline{24}$, 314 (1957).

51. J. T. O'Toole, J. Appl. Polymer Sci. $\underline{9}$, 1291 (1965).

52. A. G. Parts, D. E. Moore, J. G. Watterson, Makromol. Chem. $\underline{89}$, 156 (1965).

53. J. L. Gardon, J. Polymer Sci., A1, $\underline{6}$, 623 (1968).

54. L. M. Pisman, S. I. Kuchanov, Polymer Sci. USSR 13, 1187 (1971); Vysokomolekul. Soedin. A13, 1055 (1971).

55. J. L. Gardon, J. Polymer Sci., A1, $\underline{6}$, 665 (1968).

56. J. L. Gardon, J. Polymer Sci., A1, $\underline{9}$, 2763 (1971).

57. P. Harriot, J. Polymer Sci., A1, $\underline{9}$, 1153 (1971).

58. M. Litt, R. Patsiga, V. Stannett, J. Polymer Sci., A1, $\underline{8}$, 3607 (1970).

59. S. Katz, R. Shinnar, G. M. Saidel, "Addition and Condensation Poly-merization Processes", N. Platzer, Ed., (Adv. in Chem. Ser. $\underline{91}$), Am. Chem. Soc., Washington, D.C. 1969, p. 145.

60. G. M. Saidel, S. Katz, J. Polymer Sci. $\underline{37}$, 149 (1969).

61. J. G. Watterson, J. G. Parts, Makromol. Chem. $\underline{146}$, 1 (1971).

62. A. H. Abdeh-Alim, A. E. Hamielec, J. Appl. Polymer Sci. $\underline{16}$, 783 (1972).

63. M. Iguchi, J. Polymer Sci., A1, $\underline{8}$, 1013 (1970).

64. O. G. Lewis, R. M. King, Jr., "Addition and Condensation Polymeriza-tion Processes", N. Platzer, Ed., (Adv. in Chem. Ser. $\underline{91}$), Am. Chem. Soc. Washington, D.C. 1969, p. 25.

65. R. Wessling, I. R. Harrison, J. Polymer Sci., A1, $\underline{9}$, 3471 (1971).

66. P. D. Barlett, H. Kwart, J. Am. Chem. Soc. $\underline{72}$, 1051 (1950).

67. Th. Deleanu, M. Dimonie, J. Polymer Sci., A1, $\underline{8}$, 95 (1970).

68. I. Kar, B. M. Mandal, S. R. Palit, Makromol. Chem. $\underline{127}$, 196 (1969).

69. W. H. Atkinson, C. H. Bamford, G. C. Eastmond, Trans. Faraday Soc. $\underline{66}$, 1446 (1970).

70. A. D. Jenkins, Trans. Faraday Soc. $\underline{54}$, 1885 (1958).

71. A. D. Jenkins, Trans. Faraday Soc. $\underline{54}$, 1895 (1958).

72. J. P. Bianchi, F. P. Price, B. H. Zimm, J. Polymer Sci. $\underline{25}$, 27 (1957).

73. Ff. Williams, K. Hayashi, K. Ueno, K. Hayashi, S. Okamura, Trans. Faraday Soc. $\underline{63}$, 1501 (1967).

74. P. W. Allen, F. M. Merrett, J. Scanlan, Trans. Faraday Soc. $\underline{51}$, 95 (1955).

75. C. H. Bamford, A. D. Jenkins, R. Johnson, Proc. Roy. Soc. (Lon-don) A239, 214 (1957).

76. B. D. Coleman, T. G. Fox, J. Am. Chem. Soc. $\underline{85}$, 1241 (1963).

77. M. Szwarc, J. J. Hermans, Polymer Letters, B2, 815 (1964).

78. R. V. Figini, Makromol. Chem. $\underline{71}$, 193 (1964).

79. R. V. Figini, Makromol. Chem. $\underline{107}$, 170 (1967).

80. R. V. Figini, Makromol. Chem. $\underline{88}$, 272 (1965).

81. A. Miyake, W. H. Stockmayer, Makromol. Chem. $\underline{88}$, 90 (1965).

82. C. R. Huang, H. H. Wang, J. Polymer Sci., A1, $\underline{10}$, 791 (1972).

83. V. S. Nanda, S. C. Jain, European Polymer J. $\underline{6}$, 1517 (1970).

84. J. Leonard, Macromolecules $\underline{2}$, 661 (1969).

85. H. J. R. Maget, J. Polymer Sci. A2, 1281 (1964).

86. I. Mita, J. Macromol. Sci.-Chem. A5, 883 (1971).

87. W. T. Kyner, J. R. M. Radok, M. Wales, J. Chem. Phys. $\underline{30}$, 363 (1959).

88. V. S. Nanda, Trans. Faraday Soc. $\underline{60}$, 949 (1964).

89. L. H. Peebles, Jr., Polymer Letters $\underline{7}$, 75 (1969) (see ref. 91).

90. S. C. Jain, V. S. Nanda, Polymer Letters $\underline{8}$, 843 (1970).

91. L. H. Peebles, Jr., J. Polymer Sci., A2, $\underline{8}$, 1235 (1970): (Note eqn. 13 of ref 89 should be written $x = (\tau - r)/r^{1/2}$; equations there and in this paper containing x should be corrected. The figures are correct.

92. V. S. Nanda, R. K. Jain, Trans. Faraday Soc. $\underline{64}$, 1022 (1968).

93. J. C. W. Chien, J. Polymer Sci. A1, 425 (1963).

94. S. Penzek, P. Kubisa, Makromol. Chem. $\underline{130}$, 186 (1969).

95. V. I. Irzhak, N. S. Enikolopyan, Dokl. Akad. Nauk SSSR 185, 862 (1969); Dokl. Phys. Chem. $\underline{185}$, 226 (1969).

96. W. H. Ray, Macromolecules $\underline{4}$, 162 (1971).

97. W. R. Schmeal, J. R. Street, Am. Inst. Chem. Eng. J. $\underline{17}$, 1188 (1971).

98. D. Singh, R. P. Merrill, Macromolecules $\underline{4}$, 599 (1971).

99. M. Gordon, R. J. Roe, Polymer $\underline{2}$, 41 (1961).

100. M. I. Mustafayev, K. V. Aliev, V. A. Kabanov, Vysokomolekul. Soedin. A12, 855 (1970); Polymer Sci. USSR 12, 968 (1970).

101. P. J. Flory, J. Am. Chem. Soc. $\underline{62}$, 1561 (1940).

102. L. Gold, J. Chem. Phys. $\underline{28}$, 91 (1958).

103. M. Litt, D. H. Richards, Polymer Letters $\underline{5}$, 867 (1967).

104. L. F. Beste, H. K. Hall, Jr., J. Phys. Chem. $\underline{68}$, 269 (1964).

105. H. N. Friedlander, J. Polymer Sci. A2, 3885 (1964).

106. B. D. Coleman, F. Gornick, G. Weiss, J. Chem. Phys. $\underline{39}$, 3233 (1963).

107. T. A. Orofino, F. Wenger, J. Chem. Phys. $\underline{35}$, 532 (1961).

108. R. Chiang, J. J. Hermans, J. Polymer Sci., A1, $\underline{4}$, 2843 (1966).

109. A. Guyot, J. Chim. Phys. $\underline{1964}$, 548.

110. P. J. Flory, J. Am. Chem. Soc. $\underline{58}$, 1877 (1936).

111. L. C. Case, J. Polymer Sci. $\underline{29}$, 455 (1958).

112. G. Challa, Makromol. Chem. $\underline{38}$, 105 (1960).

113. V. S. Nanda, S. C. Jain, J. Chem. Phys. $\underline{49}$, 1318 (1968).

114. A. Ninagawa, I. Ijichi, M. Imoto, Makromol. Chem. $\underline{107}$, 196 (1967).

115. L. C. Case, J. Polymer Sci. $\underline{48}$, 27 (1960).

116. J. J. Hermans, Makromol. Chem. $\underline{87}$, 21 (1965).

117. L. C. Case, J. Polymer Sci. $\underline{37}$, 147 (1959).

118. C. H. Bamford, A. D. Jenkins, Trans. Faraday Soc. $\underline{56}$, 907 (1960).

119. L. C. Case, J. Polymer Sci. $\underline{39}$, 175 (1959).

120. L. C. Case, J. Polymer Sci. $\underline{39}$, 183 (1959).

121. P. J. Flory, Chem. Rev. $\underline{39}$, 137 (1946).

122. W. Kuhn, Chem. Ber. $\underline{63}$, 1503 (1930).

123. J. L. Lundberg, J. Polymer Sci. A2, 1121 (1964).

124. F. E. Harries, J. Polymer Sci. $\underline{18}$, 351 (1955).

125. M. Szwarc, Adv. Polymer Sci. $\underline{4}$, 457 (1967).

126. J. J. Hermans, J. Polymer Sci. C12, 345 (1966).

127. A. V. Tobolsky, J. Polymer Sci. $\underline{25}$, 220 (1957); J. Polymer Sci. $\underline{31}$, 126 (1958); A. V. Tobolsky, A. Eisenberg, J. Am. Chem. Soc. $\underline{81}$, 2302 (1959); J. Am. Chem. Soc. $\underline{82}$, 289 (1960).

128. M. E. Baur, A. Eisenberg, J. Chem. Phys. $\underline{42}$, 85 (1965).

129. F. P. Adams, J. B. Carmichael, R. J. Zeman, J. Polymer Sci., A1, $\underline{5}$, 741 (1967).

130. C. M. Fontana, "The Chemistry of Cationic Polymerization", P. H. Plesch, Ed., MacMillan Co., New York (1963).

131. J. R. Van Wazer, "Phosphorus and Its Compounds", Vol. 1, Inter-science Publishers, Inc., New York (1958).

132. T. Saegusa, T. Shiota, S. Matsumoto, H. Fujii, Macromolecules $\underline{5}$, 34 (1972).

133. R. Z. Greenley, J. C. Stauffer, J. E. Kurz, Macromolecules $\underline{2}$, 561 (1969).

134. V. S. Nanda, S. C. Jain, J. Polymer Sci., A1, $\underline{8}$, 1871 (1970).

135. G. M. Saidel, S. Katz, J. Polymer Sci., A2, $\underline{6}$, 1149 (1968).

136. L. H. Tung, R. M. Wiley, Polymer Preprints 13, 1060 (1972).

137. W. W. Graessley, H. Mittelhauser, R. Maramba, Makromol. Chem. $\underline{86}$, 129 (1965).

138. W. W. Graessley, Am. Inst. Chem. Eng. - Inst. Chem. Eng. Joint Meeting, London, June, 1965, $\underline{3}$, 16.

139. O. Saito, K. Najasubramanian, W. W. Graessley, J. Polymer Sci. A2, $\underline{7}$, 1937 (1969).

140. J. K. Beasley, J. Am. Chem. Soc. $\underline{75}$, 6123 (1953).

141. K. Fukii, T. Yamabe, J. Polymer Sci. A2, 3743 (1964).

142. M. Gordon, R. J. Roe, J. Polymer Sci. 21, 57 (1956).

143. A. Rigo, G. Palma, G. Talamini, Makromol. Chem. 153, 219 (1972).

144. L. E. Dannals, J. Polymer Sci., A1, 8, 2989 (1970).

145. M. Gordon, J. Chem. Phys. 22, 610 (1954).

146. C. Aso. T. Nawata, H. Kamao, Makromol. Chem. 68, 1 (1963).

147. Y. Minoura, M. Mitoh, J. Polymer Sci. A3, 2149 (1965).

148. G. B. Butler, M. A. Raymond, J. Polymer Sci. A3, 3413 (1965).

149. L. Trassarelli, M. Guaita, J. Macromol. Sci.-Chem. 1, 471 (1966).

150. M. Gordon, R. J. Roe, J. Polymer Sci. 21, 75 (1956).

151. M. Gordon, Proc. Roy. Soc. (London) A268, 240 (1962).

152. W. H. Stockmayer, J. Chem. Phys. 12, 125 (1944).

153. O. Saito, J. Phys. Soc. (Japan) 13, 198 (1958).

154. A. A. Miller, J. Polymer Sci. 42, 441 (1960).

155. G. R. Dobson, M. Gordon, J. Chem. Phys. 43, 705 (1965).

156. P. J. Flory, J. Am. Chem. Soc. 63, 3096 (1941).

157. A. Amemiya, J. Phys. Soc. (Japan) 23, 1394, 1402 (1967).

158. P. J. Flory, J. Am. Chem. Soc. 74, 2718 (1952).

159. E. S. Allen, J. Polymer Sci. 21, 349 (1956).

160. S. Erlander, D. French, J. Polymer Sci. 20, 7 (1956).

161. M. Gordon, M. Judd, Nature 234, 96 (1971).

162. C. R. Masson, I. B. Smith, S. G. Whiteway, Nature 234, 97 (1971).

163. C. R. Masson, I. B. Smith, S. G. Whiteway, Can. J. Chem. 48, 1456 (1970).

164. S. G. Whiteway, I. B. Smith, C. R. Masson, Can. J. Chem. 48, 33 (1970).

165. P. J. Flory, J. Am. Chem. Soc. 63, 3091 (1941).

166. W. H. Stockmayer, J. Chem. Phys. 11, 45 (1943).

167. M. Gordon, G. R. Scantlebury, Trans. Faraday Soc. 60, 604 (1964).

168. A. R. Shultz, J. Polymer Sci. A3, 4211 (1965).

169. H. L. Berger, A. R. Shultz, J. Polymer Sci. A3, 4227 (1965).

170. L. M. Pismen, S. I. Kuchanov, Polymer Sci. USSR 13, 890 (1971); Vysokomolekul. Soedin. A13, 791 (1971).

171. J. R. Schaefgen, P. J. Flory, J. Am. Chem. Soc. 70, 2709 (1948).

172. L. C. Case, J. Polymer Sci. 26, 333 (1957).

173. H. Jacobson, W. H. Stockmayer, J. Chem. Phys. 18, 1600 (1950).

174. P. J. Flory, J. A. Semlyen, J. Am. Chem. Soc. 88, 3209 (1966).

175. H. Morawetz, N. Goodman, Macromolecules 3, 699 (1970).

176. M. Gordon, W. B. Temple, Makromol. Chem. 152, 277 (1972).

177. F. E. Harris, J. Chem. Phys. 23, 1518 (1955).

178. C. A. J. Hoeve, J. Polymer Sci. 21, 11 (1956).

179. M. Gordon, G. R. Scantlebury, J. Polymer Sci. C16, 3933 (1968).

180. M. Gordon, G. R. Scantlebury, J. Chem. Soc., B 1967, 1.

181. R. W. Kilb, J. Phys. Chem. 62, 969 (1958).

182. A. Charlesby, S. H. Pinner, Proc. Roy. Soc. (London) A249, 367 (1959).

183. D. Mejzler, J. Schmorak, M. Lewin, J. Polymer Sci. 46, 289 (1960).

184. V. S. Nanda, R. K. Pathria, Proc. Roy. Soc. (London) A270, 14 (1962).

185. L. Monnerie, J. Neel, J. Chim. Phys. 62, 53 (1965).

186. S. W. Lee, J. Polymer Sci., A2, 7, 77 (1969).

187. L. Monnerie, J. Neel, J. Chim. Phys. 62, 510 (1965).

188. J. Durup, J. Chim. Phys. 51, 64 (1954).

189. R. Simha, J. Appl. Phys. 12, 569 (1941).

190. B. J. Coyne, J. Polymer Sci., A2, 5, 633 (1967).

191. E. W. Montroll, R. Simha, J. Chem. Phys. 8, 721 (1940).

192. J. R. MacCallum, Makromol. Chem. 83, 129 (1965).

193. M. Gordon, L. R. Shenton, J. Polymer Sci. 38, 157 (1959).

194. R. Simha, L. A. Wall, J. Phys. Chem. 56, 707 (1952).

195. R. H. Boyd, T. P. Lin, J. Chem. Phys. 45, 773 (1966).

196. M. Gordon, L. R. Shenton, J. Polymer Sci. 38, 179 (1959).

197. M. Gordon, J. Phys. Chem. 64, 19 (1960).

198. R. H. Boyd, T. P. Lin, J. Chem. Phys. 45, 778 (1966).

199. G. G. Cameron, Makromol. Chem. 100, 255 (1967).

200. G. G. Cameron, G. P. Kerr, Makromol. Chem. 115, 268 (1968).

201. R. H. Boyd, J. Chem. Phys. 31, 321 (1959).

202. M. Gordon, Trans. Faraday Soc. 53, 1662 (1957).

203. R. Simha, Trans. Faraday Soc. 54, 1345 (1958).

204. L. A. Wall, S. Strauss, Polymer Preprints 10, 1472 (1969).

205. L. A. Wall, S. Strauss, J. H. Flynn, D. McIntyre, J. Phys. Chem. 70, 53 (1966).

206. R. H. Boyd, J. Polymer Sci., A1, 5, 1573 (1967).

207. M. Gordon, Soc. Chem. Ind. (London) Monograph 13, 163 (1961).

208. O. Saito, J. Phys. Soc. (Japan) 13, 1465 (1958).

209. A. R. Shultz, J. Appl. Polymer Sci. 10, 353 (1966).

210. A. R. Shultz, J. Chem. Phys. 29, 200 (1958); J. Phys. Chem. 65, 967 (1961).

211. O. Saito, J. Phys. Soc. (Japan) 13, 1451 (1958).

212. T. Kimura, J. Phys. Soc. (Japan) 17, 1884 (1962). See also ref. 213.

213. D. I. C. Kells, J. E. Guillet, J. Polymer Sci., A2, 7, 1895 (1969).

214. K. G. Denbigh, Trans. Faraday Soc. 43, 648 (1947).

215. A. W. DeGraff, G. W. Poehlein, J. Polymer Sci., A2, 9, 1955 (1971).

216. Z. Tadmor, J. A. Biesenberger, Ind. Eng. Chem. Fundamentals 5, 336 (1966).

217. M. E. Sacks, S. Lee, J. A. Biesenberger, Chem. Eng. Sci. 28, 241 (1973).

218. R. V. Figini, G. V. Schulz, Z. Physik. Chem. (Frankfurt) 23, 233 (1960); Makromol. Chem. 41, 1 (1960).

219. M. Litt, J. Polymer Sci. 58, 429 (1962).

220. L. F. Beste, H. K. Hall, Jr., J. Macromol. Chem. 1, 121 (1966).

221. A. Eisenberg, D. A. McQuarrie, J. Polymer Sci., A1, 4, 737 (1966).

222. T. E. Corrigan, M. J. Dean, J. Macromol. Sci.-Chem. A2, 645 (1968).

223. "Handbook of Chemistry and Physics", Vol. 34 et seq. Chemical Rubber Publishing Co., Cleveland.

224. W. F. Espenscheid, M. Kerker, E. Matijevic, J. Phys. Chem. 68, 3093 (1964).

225. E. P. Honig, J. Phys. Chem. 69, 4418 (1965).

226. W. D. Lansing, E. O. Kramer, J. Am. Chem. Soc. 57, 1369 (1935).

227. H. Wesslau, Makromol. Chem. 20, 111 (1956).

228. L. T. Muus, W. H. Stockmayer, quoted by F. W. Billmeyer, Jr., in "Textbook of Polymer Science", Interscience, New York-London (1962).

229. B. H. Zimm, J. Chem. Phys. 16, 1099 (1948).

230. M. Matsuda, M. Iino. T. Hirayama, T. Miyashita, Macromolecules 5, 240 (1972).

231. W. Weibull, J. Appl. Mech. 18, 293 (1951).

232. L. H. Tung, J. Polymer Sci. 20, 495 (1956).

233. A. M. Kotliar, J. Polymer Sci. A2, 1373 (1964).

234. W. H. Stockmayer, J. Polymer Sci. 9, 69 (1952); 11, 424 (1953).

235. K. D. Ziegal, A. W. Fogiel, R. Pariser, Macromolecules 5, 95 (1972).

236. I. Nakamura, R. Yokouchi, T. Ito, D. Miura, K. Fujii, Kobunshi Kagaku 21, 553 (1964).

237. A. M. Kotliar, J. Polymer Sci. A2, 4304 (1964).

238. A. M. Kotliar, J. Polymer Sci. A2, 4327 (1964).

HEATS AND ENTROPIES OF POLYMERIZATION, CEILING
TEMPERATURES, EQUILIBRIUM MONOMER CONCENTRATIONS;
AND POLYMERIZABILITY OF HETEROCYCLIC COMPOUNDS

K. J. Ivin
Department of Chemistry
The Queen's University of Belfast,
N. Ireland

CONTENTS

TABLE A - HEATS OF POLYMERIZATION

Symbols: The subscripts xx to ΔH denote the state of the monomer (first letter) and the state of the polymer (second letter), as follows:

   g   gaseous state (hypethetical in case of polymer)
   c   condensed amorphous state
   c'   crystalline or partially crystalline state
   l   liquid state
   s   in solution; solvent specified in sixth column (ls denotes polymer dissolved in monomer).

For emulsion polymerization the subscript is given lc and a note added in the sixth column. In all cases where monomer or polymer is present in solution, the value of ΔH will depend to some extent on the composition. Where the polymer is crystalline ΔH will depend on the degree of crystallinity. ΔH values are in kJ per mol of monomer and are generally the limiting values for high degree of polymerization n [1 kcal = 4.187 kJ]. Where more than one value is available in the literature for given states of monomer and polymer and a given method, only one value is shown in the tables. This corresponds to one of the references indicated.

Example: C_2H_4 (g) $\rightarrow \frac{1}{n} (C_2H_4)_n$ (g), ΔH_{gg} = -93.5 kJ

Precision: generally 0.5 - 4.0 kJ mol^{-1} (for details see Ref. 1).

Methods of determination: These are summarized below using the numbering system in Ref. 1.

   2   Combustion of monomer or polymer or both
   3   Reaction calorimetry
   4a   Thermodynamic (van' t Hoff Equation).
   4b   Semi-empirical rules applied to evaluate heat of formation of polymer or monomer or both. Such rules may be found in Refs. 25,
           143, 223, 224.

Monomer	State of Monomer and Polymer xx	$-\Delta H_{xx}$ [kJ mol^{-1}]	Temp. [°C]	Method	Solvent/Notes	References

1. MONOMERS GIVING POLYMERS CONTAINING MAIN CHAIN ACYCLIC CARBON ONLY

1.1 DIENES

Monomer	State	$-\Delta H_{xx}$	Temp.	Method	Solvent/Notes	References
1,3-Butadiene	gg	73	25	4b	1:2 polymerization	10
	gg	78	25	4b	1:4 polymerization	10
	lc	73	25	2	N_1	45
Chloroprene	lc	68	61.3	3		41
Cyclopentadiene	ss	59	-70	3	methylene chloride	190
Isoprene	gg	70.5	25	4b		10
	lc	75	25	2		46
	ls	71	74.5	3		15
	ls	65.5	34.6	3	N_2	47

1.2 MONOMERS GIVING POLYMERS WITHOUT OR WITH ALIPHATIC SIDE CHAINS WHICH CONTAIN ONLY C, H

1.2.1 OLEFINS
(listed by increasing carbon number), N_4

Monomer	State	$-\Delta H_{xx}$	Temp.	Method	Solvent/Notes	References
Ethylene	gg	93.5	25	4b		2
	gc'	108.5	25	2		2,3
	gc'	107.5	25	3		183
	gc	101.5	25	2	ΔH_{fus} (Polymer) taken as 5.0	2,4
Propene	gg	86.5	25	4b		5
	gc'	104	25	2		3
	lc	84	25	4b		5
	sc	69	-78	3	n-butane	6
1-Butene	gg	86.5	25	4b		5
	lc	83.5	25	4b		5
2-Butene, cis	gg	80	25	4b		5
	lc	75	25	4b		5
2-Butene, trans	gg	75.5	25	4b		5
	lc	71	25	4b		5
Isobutene	gc	72	25	2		7
	lc	48	25	2	low polymer	3,8
	ss	53.5	25	?	solvent? low polymer	8
	ss	54	-35	3	CH_2Cl_2, polymer swollen, N_3	9
2-Pentene, cis	gg	80.5	25	4b		10
2-Pentene, trans	gg	75.5	25	4b		10
1-Hexene	lc	83	25	4b		5
1-Heptene	gg	86	25	4b		5

1.2.2 CYCLOALKANES
(listed by increasing ring size)

Monomer	State	$-\Delta H_{xx}$	Temp.	Method	Solvent/Notes	References
Cyclopropane	lc	113	25	2		11
--, methyl-	lc	105	25	4b		11
--, 1,1-dimethyl-	lc	97.5	25	4b		11
Cyclobutane	lc	105	25	2		12
--, methyl-	lc	100	25	4b		11
--, 1,1-dimethyl-	lc	93.5	25	4b		11
Cyclopentane	lc	22	25	2		12
--, methyl-	lc	17	25	4b		11
--, 1,1-dimethyl-	lc	13.5	25	4b		11
Cyclohexane	lc	-3	25	2		12
--, methyl-	lc	-9	25	4b		11
--, 1,1-dimethyl-	lc	-7.5	25	4b		11
Cycloheptane	lc	21.5	25	2		12
Cyclooctane	lc	34.5	25	2		12
Cyclononane	lc	47	25	2		13
Cyclodecane	lc	48	25	2		13
Cycloundecane	lc	45	25	2		13
Cyclododecane	lc	14	25	2		13

Monomer	State of Monomer and Polymer xx	$-\Delta H_{xx}$ [kJ mol^{-1}]	Temp. [$^{\circ}$C]	Method	Solvent/Notes	References
Cyclotridecane	lc	22	25	2		13
Cyclotetradecane	lc	7	25	2		13
Cyclopentadecane	lc	12	25	2		13
Cyclohexadecane	lc	8	25	2		13
Cycloheptadecane	lc	8.5	25	2		13

1.3 MONOMERS GIVING POLYMERS WITH ALIPHATIC SIDE CHAINS WHICH CONTAIN HETEROATOMS

1.3.1 ACRYLIC DERIVATIVES
(listed alphabetically)

Monomer	State of Monomer and Polymer xx	$-\Delta H_{xx}$ [kJ mol^{-1}]	Temp. [$^{\circ}$C]	Method	Solvent/Notes	References
Acrolein	lc	80	74.5	3		15
	ss	57.5	74.5	3	hexane	17
	ss	81.5	74.5	3	water	17
Acrylamide	ss	81.5	74.5	3	water	17,14
	ss?	70.5	74.5	3	acetone	17
	sc	60	74.5	3	benzene	17
	sc	57.5	74.5	3	hexane	17
Acrylic acid	lc	67	74.5	3		15
	ss	77.5	20	3	water	36,37
	sc	73.5	74.5	3	benzene	15
	sc	72	74.5	3	carbon tetrachloride	15
	sc	74.5	74.5	3	hexane	15
--, n-butyl ester	lc	78	74.5	3		17,37
	ss	77	74.5	3	acetone	17
	ss	79	74.5	3	butanone	17
	ss	76	74.5	3	benzene	17
	ss	75.5	74.5	3	carbon tetrachloride	17
	ss	78	74.5	3	hexane	17
--, ethyl ester	lc	78	74.5	3		34,37
--, methyl ester	lc	78	76.8	3		21,34,37
	ss	84.5	20	3	ethyl alcohol	36
	ss	81	74.5	3	hexane	34
Acrylonitrile	lc'	76.5	74.5	3		17,21,28,201
	sc'	77.5	74.5	3	benzene	17

1.3.2. METHACRYLIC DERIVATIVES
(listed alphabetically)

Monomer	State of Monomer and Polymer xx	$-\Delta H_{xx}$ [kJ mol^{-1}]	Temp. [$^{\circ}$C]	Method	Solvent/Notes	References
Methacrolein	lc	65.5	74.5	3		15
Methacrylamide	ss	56	74.5	3	water	17
	ss	42.5	74.5	3	chloroform	17
	ss	39.5	74.5	3	acetone	17
	ss	35	74.5	3	benzene	17
Methacrylic acid	lc	42.5	74.5	3		15
	lc	64.5	25	2		206
	ss	66	20	3	water	36
	ss	56.5	25	3	water	37
	ss	54	25	3	dimethylformamide, N_7	203
	sc	57	74.5	3	methanol, N_5	15
--, benzyl ester	lc	56	76.8	3	N_6	39
--, n-butyl ester	lc	57.5	74.5	3		34,37,39
	ls	60	26.9	3		14
--, tert-butyl ester	ls	54.5	26.9	3		14
--, cyclohexyl ester	lc	51	76.8	3		39
	ls	53	26.9	3		14
--, 2-ethoxyethyl ester	lc	57.5	74.5	3		34
	ls	62	26.9	3		14
--, ethyl ester	lc	60	120	4a		40
	lc	59.5	74.5	3		34,37
	ls	57.5	26.9	3		14
--, β-(N-piperidyl)ethyl ester	lc	57	40	3		188

Monomer	State of Monomer and Polymer xx	$-\Delta H_{xx}$ [kJ mol^{-1}]	Temp. [$^{\circ}$C]	Method	Solvent/Notes	References
Methacrylic acid (Cont'd.)						
--, n-hexyl ester	lc	58.5	25	3	emulsion(aq)	37
	ls	60	26.9	3		14
--, 2-hydroxyethyl ester	lc	50	25	3	emulsion (aq)	37
--, 2-hydroxypropyl ester	lc	50.5	25	3	emulsion (aq)	37
--, isobutyl ester	lc	60		3		34
--, isopropyl ester	lc	60	74.5	3		34
--, methyl ester	lc	56	130	4a		42,43,29
	lc	55.5	74.5	3		17,34,36,37,41,201,235
	lc	55	25	2		206
	ls	57.5	26.9	3		14
	ss	54	130	4a	o-dichlorobenzene	42,29
	ss	58.5	74.5	3	acetonitrile	34
	ss	57.5	74.5	3	tetrahydrofuran	34
	ss	58.5	74.5	3	hexane	34
--, n-octyl ester	lc	57	65	3		188
--, phenyl ester	lc	51.5	76.8	3	N_6	39
--, n-propyl ester	lc	57.5	74.5	3		34
Methacrylonitrile	lc	56.5	74.5	3		15
	ss	64	130	4a	benzonitrile	44,29

1.3.3 VINYL ETHERS

Monomer	State	$-\Delta H_{xx}$	Temp.	Method	Solvent/Notes	References
Vinyl n-butyl ether	lc?	60	50	3	N_6	216

1.3.4 VINYL ALCOHOL, KETONES, HALOGENS

Monomer	State	$-\Delta H_{xx}$	Temp.	Method	Solvent/Notes	References
Vinyl alcohol					see acetaldehyde	195
Methyl vinyl ketone	ls	74	74.5	3		17
Vinyl chloride	gc	132	25	2		18,19,20
	lc	71	25	4b		10
	lc	111.5	25	2	ΔH_{fus} (monomer) taken as 20	18
	lc	96	74.5	3		215

1.3.5 VINYL ESTERS

Monomer	State	$-\Delta H_{xx}$	Temp.	Method	Solvent/Notes	References
Vinyl acetate	lc	88	74.5	3		17,34,21
	ls	89.5	25	3		35
	ss	90	74.5	3	acetone	17
	ss	86.5	74.5	3	hexane	17
	ss	86	74.5	3	benzene	17
Vinyl propionate	lc	86	74.5	3		17,34
Vinyl 2-ethylhexoate	lc	88	74.5	3		34

1.3.6 OTHERS
(listed alphabetically)

Monomer	State	$-\Delta H_{xx}$	Temp.	Method	Solvent/Notes	References
Allyl chloride	ls	77.5	74.5	3		15
Ethylene, nitro-	lc	88	20	2		214
	ss	91	20	3	dimethylformamide, tetrahydrofuran	214
--, tetrafluoro-	gg	155	25	2		22
	gc'	172	25	2		22,23
	lc'	163±17	25	2		1
Fumaric acid, diethyl ester	lc	65	100	3		235
Itaconic acid, dimethyl ester	ss	60.5	26.9	3	o-dichlorobenzene	14
Maleic anhydride	ls	59	74.5	3		15
Maleimide	ss	67.5	74.5	3	chlorobenzene	24
	ss	89.5	74.5	3	dioxane	24
	ss	88.5	74.5	3	acetonitrile	24
	ss	87.5	74.5	3	dimethylformamide	24
Vinylidene chloride	lc'	60	76.8	3		21
	lc'	75.5	25	2		18

Monomer	State of Monomer and Polymer xx	$-\Delta H_{xx}$ [kJ mol^{-1}]	Temp. [°C]	Method	Solvent/Notes	References
Vinylidene chloride	lc'	73	74.5	3		15

1.4 MONOMERS GIVING POLYMERS WITH AROMATIC SIDE CHAINS WHICH CONTAIN ONLY C, H
(listed alphabetically)

Monomer	State of Monomer and Polymer xx	$-\Delta H_{xx}$ [kJ mol^{-1}]	Temp. [°C]	Method	Solvent/Notes	References
Acenaphthylene	ss	98.5	26.9	3	o-dichlorobenzene	14
	ss	67	74.5	3	o-dichlorobenzene	15
	ss	67	74.5	3	benzene	15
	c' c	82	26.9	3		14
Biphenyl, p-isopropenyl-	ss	34	-15	4a	tetrahydrofuran	16
Indene	ss	58	-30	3	methylene chloride indep. of T [-70° to 10°]	190
--, 5-methyl-	ss	58	-50	3	methylene chloride	175
--, 6-methyl-	ss	57.5	-60	3	methylene chloride	175, 169
--, 7-methyl-	ss	58.5	-70	3	methylene chloride	175, 169
--, 4,6-dimethyl-	ss	77.5	-40	3	methylene chloride	175
--, 4,7-dimethyl-	ss	65	-40	3	methylene chloride	175
--, 5,6-dimethyl-	ss	79	-40	3	methylene chloride	175
--, 4,5,6,7-tetramethyl-	ss	75	-40	3	methylene chloride	175
Naphthalene, 2-isopropenyl-	ss	36.5	-5	4a	tetrahydrofuran	16
Styrene	gg	74.5	25	4b		25
	lc	70	25	2		26
	lc	68.5	26.9	3		14, 17, 27, 201, 235
	lc	73	127	4a		28, 29
	ls	73	25	2		26
	ls	73	26.9	3		14
	ss	66.5	-60	3	methylene chloride	30, 190
--, o-chloro-	lc	68.5	76.8	3		27
--, p-chloro-	lc	67	76.8	3		27
--, 2,5-dichloro-	lc	69	76.8	3		27
--, ar-ethyl-	lc	68	76.8	3		27
--, α-methyl-	lc	35	25	2		31
	lc	34.5	-20	4a		32, 29
	lc	39	25		N$_9$	123
	ss	00.5	-20	4a	tetrahydrofuran	32, 29, 33, 10
--, 2,4,6-trimethyl-	lc	70	26.9	3		14

1.5 MONOMERS GIVING POLYMERS WITH AROMATIC SIDE CHAINS AND WHICH CONTAIN HETEROATOMS

Monomer	State of Monomer and Polymer xx	$-\Delta H_{xx}$ [kJ mol^{-1}]	Temp. [°C]	Method	Solvent/Notes	References
Atropic acid, methyl ester	ss	28	-40	4b	toluene, N$_{10}$	149
Atroponitrile	ss	39	50	4b	toluene, N$_{10}$	149
Benzoic acid, vinyl ester	lc	84.5	74.5	3		34
Carbazole, N-vinyl-	sc'	63.5	74.5	3	hexane	15
Oxazole, 2-isopropenyl-	ss	39.5	20	4a	tetrahydrofuran	284
Pyridine						
--, 2-isopropenyl-	ss	26	20	4a	tetrahydrofuran	284
--, 2-vinyl-	lc	71.5	74.5	3		15
	ls	75.5	74.5	3		15
	sc	73.5	74.5	3	benzene	15
--, 4-vinyl-	lc	78	74.5	3		17
	ss	78	74.5	3	benzene, hexane	17
Thiazole, 2-isopropenyl-	ss	28.5	20	4a	tetrahydrofuran	284

2. MONOMERS GIVING POLYMERS CONTAINING HETEROATOMS IN THE MAIN CHAIN

2.1. MONOMERS GIVING POLYMERS CONTAINING O IN THE MAIN CHAIN, BONDED TO CARBON ONLY

2.1.1 ETHERS AND ACETALS
(listed by increasing ring size of monomer)

2-RINGS (ALDEHYDES AND KETONES)

Monomer	State of Monomer and Polymer xx	$-\Delta H_{xx}$ [kJ mol^{-1}]	Temp. [°C]	Method	Solvent/Notes	References
Acetaldehyde	lc	0	25	4b	N$_{11}$	48
	lc'	64.5	25	2	to poly(vinyl alcohol)	195

Monomer	State of Monomer and Polymer xx	$-\Delta H_{xx}$ [kJ mol^{-1}]	Temp. [°C]	Method	Solvent/Notes	References
Acetaldehyde (Cont' d.)	lc	62.5	25	2	to poly(vinyl alcohol)	195
Acetone	gg	-12	25	4b		234
	gc	10	25	4b		144,148
	lc	-25	25	4b		48,234
n-Butyraldehyde	lc	20.5	20	3		161
	sc	35.5	18	4a	n-hexane [9 kbar], N_{12}	161
	sc	21.5	20	3	n-hexane [1 bar, 1M], N_{12}	161
Chloral	gc'	71±8	50	4a		50
	gc'	65.5	-50	4a		166
	gc	51	-50	4a		166
	lc'	38±8	50	4a		50
	lc'	34.5	-50	4a		166
	lc	20	-50	4a		166
	ss	14.5	-50	4a	tetrahydrofuran	167
	sc'	33.5	13	4a	pyridine	50
	sc'	39	10	4a	n-heptane	166
Fluoral	gc	64.5	45	4a		129
Formaldehyde	gc'	66	25	4a	N_{13}	49,164,165,132
	gc'	55	25	2		7,49
Isobutyraldehyde	gc	46	-65	4a		166
	lc	16	-65	4a		166
	ss	15.5	-65	4a	tetrahydrofuran	167,172
	ss	16.5	-65	4a	diethyl ether	166
	ss	22	-65	4a	n-pentane	166
--, 1-chloro-	gc	47	-65	4a		172
	lc	19.5	-65	4a		172
	ss	19.5	-65	4a	tetrahydrofuran	172
o-Phthalaldehyde	ss	22	-70	4a	methylene chloride, cyclopolymer	154
Propionaldehyde, 1,1-dichloro-	gc	50	-65	4a		172
	lc	20.5	-65	4a		172
	ss	17	-65	4a	tetrahydrofuran	172

3-RINGS

Monomer	State of Monomer and Polymer xx	$-\Delta H_{xx}$ [kJ mol^{-1}]	Temp. [°C]	Method	Solvent/Notes	References
Ethylene oxide	gg	104	25	4b		14
	gc'	140	25	3		168
	lc'	94.5	25	2		51
Propylene oxide	gg	75.5	25	4b		14,52
Styrene oxide	lc	101.5	26.9	3		14
--, 3-nitro-	lc	101	26.9	3		14

4-RINGS

Monomer	State of Monomer and Polymer xx	$-\Delta H_{xx}$ [kJ mol^{-1}]	Temp. [°C]	Method	Solvent/Notes	References
Oxetane $\overline{OCH_2CH_2CH_2}$	ss	81	-9	3	methyl + ethyl chloride	53
--, 3,3-di(chloromethyl)-	lc	84.5	26.9	3		14
--, 3,3-dimethyl-	ss	67.5	-9	3	methyl + ethyl chloride	53
--, 3,3-di(phenoxymethyl)-	ss	83.5	26.9	3	o-dichlorobenzene	14

5-RINGS

Monomer	State of Monomer and Polymer xx	$-\Delta H_{xx}$ [kJ mol^{-1}]	Temp. [°C]	Method	Solvent/Notes	References
1,3-Dioxolane $\overline{OCH_2OCH_2CH_2}$	gg	26	20	4b		54
	gc	50	25	4a		292
	lc	24	100	4b	N_{14}	198
	lc	15	25	4a	N_{17}	292
	lc	23	40	4a	N_{15}	159
	ls	6.5	55	4a		219
	ss	22	0	4a/3	methylene chloride	133,159,220
	ss	27	20	4a	ethyl chloride	293
	ss	15	30	4a	benzene	134
	ss	17.5	70	3	methylene chloride	220
Tetrahydrofuran	gg	21	20	4b		55
	gg	12	25	4b		56
	gc	39.5	25	4a		292

Monomer	State of Monomer and Polymer xx	$-\Delta H_{xx}$ [kJ mol^{-1}]	Temp. [$^\circ$C]	Method	Solvent/Notes	References
Tetrahydrofuran (Cont'd.)	lc'	38	25	2		56,57
	lc	7.5	25	4a	N[17]	292
	lc	12.5	50	4a		130
	lc	15	30	4a	N[16]	231
	lc	25.5	25	2		140,208
	ls	23	40	4a		57,58,211
	ss	23	40	4a	diethyl ether	211

6-RINGS

Monomer	State of Monomer and Polymer xx	$-\Delta H_{xx}$ [kJ mol^{-1}]	Temp. [$^\circ$C]	Method	Solvent/Notes	References
m-Dioxane	gg	0	20	4b		54,59,60
Tetrahydropyran	gg	1.5	20	4b		55,56,59
Trioxane	sc'	17.5	40	4a	1,2-dichloroethane	178
	sc'	26.5	58	3	1,2-dichloroethane	212
	sc'	19.5	30	3	methylene chloride	182,202
	sc'	12.5	30	4a	nitrobenzene	182
	sc'	21.5	30	3	nitrobenzene	202
	c'c'	6	30	3	N[18]	202
	c'c'	11.5	58	3	N[19]	212
	c'c'	4.5	25	4a		164
	c'c'	5.5	25	2		see 202

7-RINGS

Monomer	State of Monomer and Polymer xx	$-\Delta H_{xx}$ [kJ mol^{-1}]	Temp. [$^\circ$C]	Method	Solvent/Notes	References
1,3-Dioxepane $\overline{OCH_2O(CH_2)_4}$	gg	19.5	20	4b		54
	ss	13.5	60	4a	benzene	134,87
	ss	15	-30	4a/3	methylene chloride	136
1-Oxa-4,5-dithiepane $\overline{O(CH_2)_2S_2(CH_2)_2}$	lc	7.5	26.9	3		61
	ss	9	26.9	3	dioxane	61
	ss	8	26.9	3	benzene	61

8-RINGS

Monomer	State of Monomer and Polymer xx	$-\Delta H_{xx}$ [kJ mol^{-1}]	Temp. [$^\circ$C]	Method	Solvent/Notes	References
1,3-Dioxocane $\overline{OCH_2O(CH_2)_5}$	gg	53.5	20	4b		54
Tetraoxane $(OCH_2)_4$	sc'	26	30	3	nitrobenzene	202
	c'c'	3.5	30	3		131,202
	c'c'	4	25	2		see 202
1,3,6-Trioxocane $\overline{OCH_2O(CH_2)_2O(CH_2)_2}$	ss	22	60	4a	benzene	134

2.1.2 ESTERS

Monomer	State of Monomer and Polymer xx	$-\Delta H_{xx}$ [kJ mol^{-1}]	Temp. [$^\circ$C]	Method	Solvent/Notes	References
β-Propiolactone	lc'	80.5	25	2/4b		137

2.2 MONOMERS GIVING POLYMERS CONTAINING O IN THE MAIN CHAIN, BONDED TO OTHER HETEROATOMS (S, Si)

Monomer	State of Monomer and Polymer xx	$-\Delta H_{xx}$ [kJ mol^{-1}]	Temp. [$^\circ$C]	Method	Solvent/Notes	References
Cyclotrisiloxane, hexamethyl-	lc	14.5	25	3		84
Sulfur trioxide	gc'	56	13	4a		86
	lc'	12.5	25	4a		86

2.3 MONOMERS GIVING POLYMERS CONTAINING S IN THE MAIN CHAIN, BONDED IN THE CHAIN TO CARBON ONLY
(listed by increasing ring size)
(Polysulfones listed under 3. Copolymers)

Monomer	State of Monomer and Polymer xx	$-\Delta H_{xx}$ [kJ mol^{-1}]	Temp. [$^\circ$C]	Method	Solvent/Notes	References
Carbon disulfide	lc'	23	25	2		179
Thiirane	gg	81	25	4b	N[20]	69,70,71
	lc	73±4	20	3		173
	sc'	80±4	20	3	various solvents	173
--, 2,2-dimethyl-	gg	70.5	25	4b	N[20]	69,70,71
--, cis-2,3-dimethyl-	gg	63	25	4b	N[20]	69,70,71
--, trans-2,3-dimethyl-	gg	55.5	25	4b	N[20]	69,70,71
--, 2-methyl-	gg	71	25	4b	N[20]	69,70,71
Thietane	gg	80	25	4b	N[20]	69,70,71
Thiolane	gg	7.5	25	4b	N[20]	69,70,71

Monomer	State of Monomer and Polymer xx	$-\Delta H_{xx}$ [kJ mol^{-1}]	Temp. [°C]	Method	Solvent/Notes	References
1,2-Dithiolane SSCH$_2$CH$_2$CH$_2$	ss	26.5	30	4a	ethyl alcohol	72
Thiane	gg	-2.5	25	4b	N$_{20}$	69, 70, 71
o-Dithiane SS(CH$_2$)$_3$CH$_2$	lc	2	26.9	3		14
Thiepane	gg	14.5	25	4b	N$_{20}$	69, 70, 71
1,2-Dithiepane SS(CH$_2$)$_4$CH$_2$	lc	10.5	26.9	3		14
	sc	11.5	26.9	3	dioxane	14
1,2-Dithiocane SS(CH$_2$)$_5$CH$_2$	lc	16	26.9	3		14
Sulfur S$_8$	ls	13.5	200	4a		73, 74

2.4 MONOMERS GIVING POLYMERS CONTAINING N IN THE MAIN CHAIN, BONDED IN THE CHAIN TO CARBON ONLY

(listed by increasing ring size)

5-RINGS

Monomer	State of Monomer and Polymer xx	$-\Delta H_{xx}$ [kJ mol^{-1}]	Temp. [°C]	Method	Solvent/Notes	References
2-Pyrollidinone	lc	4.5	75	2/4b		62
	lc	3.5	43	3		135
	lc'	12	43	3		135
	c' c'	-5.5	25	2/4b		63
--, 1-methyl-	lc	3.5	25	4b		64

6-RINGS

Monomer	State of Monomer and Polymer xx	$-\Delta H_{xx}$ [kJ mol^{-1}]	Temp. [°C]	Method	Solvent/Notes	References
2-Piperidone	lc	9	75	2/4b		62
	c' c'	4.5	25	2/4b		63
--, 1-methyl-	lc	-2	25	4b		64

7-RINGS

Monomer	State of Monomer and Polymer xx	$-\Delta H_{xx}$ [kJ mol^{-1}]	Temp. [°C]	Method	Solvent/Notes	References
ε-Caprolactam	lc	16	75	2		62
	lc	15.5	200	3		65, 135, 138
	ls	16.5	250	4a		68, 67, 66
	c' c'	12.5	25	2		63
--, 1-methyl-	lc	9.5	25	4b		64
--, 3-methyl-	ls	12.5	260	4a		194
--, 3-ethyl-	ls	16	240	3		191
--, 3-propyl-	ls	16.5	240	3		191
--, 4-ethyl-	ls	14.5	240	3		171
--, 5-methyl-	lc	16	75	4b		62
	ls	11	200	4a		225
--, 7-methyl-	lc	16	75	4b		62
	ls	16.5	225	4a		226
--, 7-ethyl-	ls	14.5	240	3		191
--, 7-propyl-	ls	14.5	240	3		191

8-RINGS

Monomer	State of Monomer and Polymer xx	$-\Delta H_{xx}$ [kJ mol^{-1}]	Temp. [°C]	Method	Solvent/Notes	References
2-Oxo-hexamethyleneimine	lc	22	75	4b		62
	lc	22	230	3		138, 207
	c' c'	24	25	2		63
--, 1-methyl-	lc	16.5	25	4b		64
--, 4-ethyl-	ls	22	240	3		171
--, 7-ethyl-	ls	21.5	240	3		171

9-, 10-, 11-, 12- and 13-RINGS

Monomer	State of Monomer and Polymer xx	$-\Delta H_{xx}$ [kJ mol^{-1}]	Temp. [°C]	Method	Solvent/Notes	References
2-Oxo-heptamethyleneimine	lc	32.5	230	3		138, 189
2-Oxo-octamethyleneimine	c' c'	23.5	25	2/4b		138
2-Oxo-nonamethyleneimine	c' c'	11.5	25	2/4b		138

Monomer	State of Monomer and Polymer xx	$-\Delta H_{xx}$ [kJ mol^{-1}]	Temp. [°C]	Method	Solvent/Notes	References
2-Oxo-decamethyleneimine	c' c'	-2	25	2/4b		138
2-Oxo-undecamethyleneimine	lc	~0	230	3		138,189
	ls	13	290	4a		139

2.5 MONOMERS GIVING POLYMERS CONTAINING N IN THE MAIN CHAIN, BONDED TO OTHER HETEROATOMS (P)

Monomer	State	$-\Delta H_{xx}$	Temp.	Method	Solvent/Notes	References
Phosphonitrile chloride,						
--, cyclic trimer	gc	61	230	3		125
	lc	6	230	3		125
--, cyclic tetramer	gc	68	230	3		125
	lc	4	230	3		125
--, cyclic pentamer	gc	79.5	230	3		125
	lc	3.5	230	3		125
--, cyclic hexamer	gc	88	230	3		125
	lc	1.5	230	3		125
--, cyclic heptamer	gc	96.5	230	3		125
	lc	0	230	3		125

2.6 OTHER MONOMERS GIVING POLYMERS NOT LISTED ABOVE

Monomer	State	$-\Delta H_{xx}$	Temp.	Method	Solvent/Notes	References
Carbon C (graphite)	c' c'	-39.5	25	2	linear polymer (carbyne)	187
Selenium Se$_8$	ls	-9.5	400	4a		85
Sulfur S$_8$	ls	-13.5	200	4a		73,74

Monomer A	Monomer B	State of Monomer and Polymer xx	$-\Delta H_{xx}$ [kJ mol^{-1}]	Temp. [°C]	Method	Solvent/Notes	References

3. COPOLYMERS
(listed alphabetically under Monomer A)   N$_{21}$

Monomer A	Monomer B	State	$-\Delta H_{xx}$	Temp.	Method	Solvent/Notes	References
Acrylonitrile	methacrylic acid, methyl ester	sc		25	3	emulsion (aq)	38
		ss		30.5	3		76
	styrene	ss		20	3		76,201
	vinyl acetate	ss		20	3		76
	vinylidene chloride	sc		25	3	emulsion (aq), N$_{22}$	75
Allyl chloride	maleic anhydride	ss	74	74.5	3		34
1,3-Butadiene	styrene	lc		25	2		45
	sulfur dioxide						197
1-Butene	sulfur dioxide	ss	44.5	26.9	3	excess B	5
		ss	43.5	55	4a	excess B	77
--, 3-methyl-	sulfur dioxide	lc'	46	35	4a	from B-rich mixture	199,200
		lc	22	25	4a	from A-rich mixture	199,200
2-Butene, cis	sulfur dioxide	ss	42.5	26.9	3	excess B	5
		ss	43.5	25	4a	excess B	5
--, trans	sulfur dioxide	ss	39.5	26.9	3	excess B	5
		ss	40.5	25	4a	excess B	5
Cyclohexene	sulfur dioxide	ss	38	25	4a	excess B	78
Cyclopentene	sulfur dioxide	ss	45	26.9	3	excess B	5
Fumaric acid, diethyl ester	vinyl acetate	ss?	78	76.8	3	excess B, N$_{22}$	79
Fumaroyl chloride	styrene	ss	80	74.5	3	hexane	34
	styrene, α-methyl-	ss	71.5	74.5	3	excess B	34
Isobutene	sulfur dioxide	lc	31	25	2		82
		lc	39.5	0	4a	N$_{23}$	83
1-Hexadecene	sulfur dioxide	ss	42	26.9	3	chloroform	5
		ss	40	30	4a	chloroform	80
1-Hexene	sulfur dioxide	ss	43.5	26.9	3	excess B	5
Isopropenyl acetate	maleic anhydride	sc	74.5	76.8	3		79
Maleic acid, diethyl ester	vinyl acetate	ss?	83.5	76.8	3	excess B, N$_{22}$	79
Maleic anhydride	allyl chloride	ss	74	74.5	3		34

Monomer A	Monomer B	State of Monomer and Polymer xx	$-\Delta H_{xx}$ [kJ mol^{-1}]	Temp. [°C]	Method	Solvent/Notes	References
Maleic anhydride (Cont'd.)	isopropenyl acetate	sc	74.5	76.8	3		79
	styrene	ss	81	74.5	s	benzene	34
		ss	82.5	74.5	3	acetonitrile	34
	styrene, α-methyl-	ss	72.5	74.5	3	excess B	34
	vinyl acetate	sc	84.5	76.8	3	excess B	79
	vinyl n-butyl ether	ss	90	74.5	3	benzene	34
Maleimide	styrene	ss	87.5	74.5	3	acetonitrile	34
	styrene, α-methyl-	ss	72	74.5	3	acetonitrile	34
Methacrylic acid	methacrylic acid, methyl ester	lc		25	2		206
--, methyl ester	acrylonitrile	sc		25	3	emulsion (aq)	38
		ss		30.5	3		76
	methacrylic acid	lc		25	2		206
	styrene	ss		24	3		81
	vinyl acetate	ss		24	3		81
Propene	sulfur dioxide	sc	42.5	26.9	3	excess B	5
Styrene	acrylonitrile	ss		20	3		76, 201
	1,3-butadiene	lc		25	2		45
	fumaroyl chloride	ss	80	74.5	3	hexane	34
	maleic anhydride	ss	81	74.5	3	benzene	34
		ss	82.5	74.5	3	acetonitrile	34
	maleimide	ss	87.5	74.5	3	acetonitrile	34
	methacrylic acid, methyl ester	ss		24	3		81
	vinyl acetate	ss		35	3		81
--, α-methyl-	fumaroyl chloride	ss	71.5	74.5	3	excess A	34
	maleic anhydride	ss	72.5	74.5	3	excess A	34
	maleimide	ss	72	74.5	3	acetonitrile	34
Sulfur dioxide	butadiene						197
	1-butene	ss	44.5	26.9	3	excess A	5
		ss	43.5	55	4a	excess A	77
	--, 3-methyl-	lc'	92	35	4a	from A-rich mixture	199
		lc	44.5	25	4a	from B-rich mixture	199
	2-butene, cis	ss	42.5	26.9	3	excess A	5
		ss	43.5	25	4a	excess A	5
	--, trans	ss	39.5	26.9	3	excess A	5
		ss	40.5	25	4a	excess A	5
	cyclohexene	ss	38	25	4a	excess A	78
	cyclopentene	ss	45	26.9	3	excess A	5
	1-hexadecene	ss	42	26.9	3	chloroform	5
		ss	40	30	4a	chloroform	80
	1-hexene	ss	43.5	26.9	3	excess A	5
	isobutene	lc	31	25	2		82
		lc	39.5	0	4a	N_{24}	83
	propene	sc	42.5	26.9	3	excess A	5
Vinyl acetate	acrylonitrile	ss		20	3		76
	fumaric acid, diethyl ester	ss?	78	76.8	3	excess A, N_{22}	79
	maleic acid, diethyl ester	ss?	83.5	76.8	3	excess A, N_{22}	79
	maleic anhydride	sc	84.5	76.8	3	excess A	79
	methacrylic acid, methyl ester	ss		24	3		81
	styrene	ss		35	3		81
Vinyl n-butyl ether	maleic anhydride	ss	90	74.5	3	benzene	34
Vinylidene chloride	acrylonitrile	sc		25	3	emulsion (aq), N_{22}	75

ENTROPIES OF POLYMERIZATION

TABLE B - ENTROPIES OF POLYMERIZATION

Symbols: The subscripts to ΔS^o denote the state of the monomer (first letter) and the state of the polymer (second letter) as in Table A. The standard state of the monomer is 1 atm for the gaseous state and 1 [mol litre^{-1}] in solution, unless otherwise stated. In all cases where monomer or polymer is present in solution, the value of ΔS^o will depend to some extent on the composition. Where the polymer is crystalline ΔS^o will depend on the degree of crystallinity. ΔS values are in [J K^{-1}] per mol of monomer and are generally the limiting values for high degree of polymerization [1 kcal = 4.187 kJ].

Precision: generally 0.5 - 8.0 [J K^{-1} mol^{-1}]

Methods of determination: (numbering conforms with that in Table A)

1.   Third law or statistical.
4a.   Thermodynamic (van't Hoff Equation).
4b.   Semi-empirical rules applied to evaluate entropy of monomer or polymer or both.

Monomer	State of Monomer and Polymer xx	$-\Delta S^o_{xx}$ [J K^{-1} mol^{-1}]	Temp. [^{o}C]	Method	Solvent/Notes	References

1. MONOMERS GIVING POLYMERS CONTAINING MAIN CHAIN ACYCLIC CARBON ONLY

1.1 DIENES

Monomer	State	$-\Delta S^o_{xx}$	Temp.	Method	Solvent/Notes	References
1,3-Butadiene	lc	89	25	1		95,96
	lc	84	25	1	cis-1,4-polymer	97
Isoprene	lc	101	25	1		100,101

1.2 MONOMERS GIVING POLYMERS WITHOUT OR WITH ALIPHATIC SIDE CHAINS WHICH CONTAIN ONLY C, H

1.2.1 OLEFINS
(listed by increasing carbon number)

Monomer	State	$-\Delta S^o_{xx}$	Temp.	Method	Solvent/Notes	References
Ethylene	gg	142	25	4b		2,87
	gc	155	25	4b		11
	gc	158	25	1		88-91
	gc'	172	25	4b		11
	gc'	174	25	1	100% cryst. polymer	88-91
Propene	gg	167	25	4b		5
	gc'	191	25	1	syndiotactic polymer	186
	gc'	205	25	1	isotactic polymer	207
	lc	113	25	4b		5
	lc	116	25	1	isotactic polymer	88
	lc	116	25	1	atactic polymer	88,92
	lc'	136	25	1	100% cryst. isotactic	88,92
1-Butene	gg	166	25	4b		5
	gc	190	25	1	isotactic polymer	88
	gc'	219	25	1	100% cryst. isotactic	88
	lc	113	25	4b		5
	lc	112	25	1	isotactic polymer	88
	lc'	141	25	1	100% cryst. isotactic	88
2-Butene, cis	gg	163	25	4b		5
	lc	104	25	4b		5
--, trans	gg	159	25	4b		5
	lc	100	25	4b		5
Isobutene	gg	172	25	4b		5
	lc	112	25	4b		5
	lc	121	25	1		11,93
1-Pentene, 4-methyl-	gc'	216	25	1	isotactic polymer	196
1-Hexene	lc	113	25	4b		5
1-Heptene	gg	168	25	4b		5

Monomer	State of Monomer and Polymer xx	$-\Delta S^o_{xx}$ [J K^{-1} mol^{-1}]	Temp. [^{o}C]	Method	Solvent/Notes	References

1.2.2 CYCLOALKANES
(listed by increasing ring size)

Monomer	State	$-\Delta S$	Temp	Method	Solvent/Notes	Refs
Cyclopropane	lc	69	25	4b		11
--, methyl-	lc	85	25	4b		87
--, 1,1-dimethyl-	lc	93	25	4b		87
Cyclobutane	lc	55	25	4b		11
--, methyl-	lc	72	25	4b		87
--, 1,1-dimethyl-	lc	75	25	4b		87
Cyclopentane	lc	43	25	4b		11
--, methyl-	lc	64	25	4b		87
--, 1,1-dimethyl-	lc	66	25	4b		87
Cyclohexane	lc	11	25	4b		11,87
--, methyl-	lc	32	25	4b		87
--, 1,1-dimethyl-	lc	36	25	4b		87
Cycloheptane	lc	16	25	4b	N_{25}	11,94
Cyclooctane	lc	3	25	4b	N_{25}	11,94

1.3 MONOMERS GIVING POLYMERS WITH ALIPHATIC SIDE CHAINS WHICH CONTAIN HETEROATOMS

Monomer	State	$-\Delta S$	Temp	Method	Solvent/Notes	Refs
Acrylonitrile	lc'	109	25	1		193
Ethylene, tetrafluoro-	gc'	197	-75.7	1		98,99
	lc'	112	-75.7	1		98,99
Methacrylic acid						
--, ethyl ester	lc	126	120	4a		40
--, methyl ester	lc	117	127	4a		42,43,102
	c'c'	40	-63	1		102,103
	ss	130	127	4a	o-dichlorobenzene	42,29
Methacrylonitrile	ss	151	127	4a	benzonitrile	44,29
Vinylidene chloride	lc'	89	-73	1		213

1.4 MONOMERS GIVING POLYMERS WITH AROMATIC SIDE CHAINS WHICH CONTAIN ONLY C, H

Monomer	State	$-\Delta S$	Temp	Method	Solvent/Notes	Refs
Biphenyl, p-isopropenyl-	ss	118	-20	4a	tetrahydrofuran	16
Naphthalene, 2-isopropenyl-	ss	122	-5	4a	tetrahydrofuran	16
Styrene	gg	149	25	4b		25
	lc	90	-23	1		104
	lc	104	25	1		104
	lc	116	127	1		104
	lc	104	127	4a	N_{26}	28
	lc	105	25	1	isotactic polymer	88
	lc	112	25	1		105
	lc'	111	25	1	100% cryst. isotactic	88
--, α-methyl-	lc	110	-20	4a		32
	ss	130	-20	4a	tetrahydrofuran	32,29,16

1.5 MONOMERS GIVING POLYMERS WITH AROMATIC SIDE CHAINS AND WHICH CONTAIN HETEROATOMS

Monomer	State	$-\Delta S$	Temp	Method	Solvent/Notes	Refs
Oxazole, 2-isopropenyl-4,5-dimethyl-	ss	96	20	4a	tetrahydrofuran	284
Pyridine, 2-isopropenyl-	ss	70	20	4a	tetrahydrofuran	284
Thiazole, 2-isopropenyl-	ss	69	20	4a	tetrahydrofuran	284

2. MONOMERS GIVING POLYMERS CONTAINING HETEROATOMS IN THE MAIN CHAIN

2.1 MONOMERS GIVING POLYMERS CONTAINING O IN THE MAIN CHAIN, BONDED TO CARBON ONLY
(listed by increasing ring size)

2-RINGS (ALDEHYDES AND KETONES)

Monomer	State	$-\Delta S$	Temp	Method	Solvent/Notes	Refs
Acetone	gc	188	25	4b		144,148
n-Butyraldehyde	sc	122	10	4a	n-hexane [9 kbar], N_{12}	161
Chloral	gc'	190±30	50	4a		50

Monomer	State of Monomer and Polymer xx	$-\Delta S^o_{xx}$ [J K^{-1} mol^{-1}]	Temp. [^{o}C]	Method	Solvent/Notes	References
Chloral (Cont'd.)	lc'	95±30	50	4a		50
	sc'	117	13	4a	pyridine, N$_{27}$	50
	ss	52	-50	4a	tetrahydrofuran	167
	sc'	142	10	4a	n-heptane	166
Fluoral	gc	187	45	4a		129
Formaldehyde	gg	124	25	1		141,142
	gc'	174	25	1		106
	gc'	169	25	4a	N$_{28}$	164,165
Isobutyraldehyde	ss	74	-65	4a	tetrahydrofuran	172
	ss	78	-65	4a	diethyl ether	166
	ss	94	-65	4a	n-pentane	166
--, 1-chloro-	ss	90	-65	4a	tetrahydrofuran	172
o-Phthalaldehyde	ss	96	-70	4a	methylene chloride	154
Propionaldehyde, 1,1-dichloro-	ss	69	-65	4a	tetrahydrofuran	172

3-RINGS

Ethylene oxide	gc'	174	25		100 % cryst. polymer	210
Propylene oxide	gc'	189	25		estimated for 100% cryst. polymer	210

4-RINGS

Oxetane, 3,3-di(chloromethyl)-	lc	83	25	1		107

5-RINGS

1,3-Dioxolane	gc'	205	25	1		177
	gc	167	25	1		177
	gc	139	25	4a		292
	lc'	100	25	1		177
	lc	37	25	4a	N$_{17}$	292
	lc	67	25	1		177
	lc	63	40	4a	N$_{15}$	159
	lc	76	100	4a	N$_{14}$	198
	ss	78	0	4a	methylene chloride	133,159
	ss	94	20	4a	ethyl chloride	293
	ss	59	30	4a	benzene	134
Tetrahydrofuran	gc'	177	25	1		140
	gc	112	25	4a		292
	gc	139	25	1		140
	lc'	100	25	1		140
	lc	16	25	4a	N$_{17}$	292
	lc	62	25	1		140
	lc	41	50	4a		130
	lc	49	30	4a	N$_{16}$	231
	ls	67	40	4a		57,58
	ss	87	40	4a	diethyl ether	211

6-RINGS

Paraldehyde	gc'	201	25	1	isotactic polymer	184
	gc'	159	25	1	syndiotactic polymer	184
Trioxane	gg	64	25	1		141
	gc'	156	25	1		141
	ss	41	40	4a	1,2-dichloroethane	178
	sc'	42	25	4a	nitrobenzene	182
	c' c'	18±16	25	4a		164

7-RINGS

1,3-Dioxepane	gc'	181	25	1		170
	gc	144	25	1		170

Notes page II-446, References page II-447

Monomer	State of Monomer and Polymer xx	$-\Delta S^O_{xx}$ [J K^{-1} mol^{-1}]	Temp. [^{O}C]	Method	Solvent/Notes	References
1,3-Dioxepane	lc'	77	25	1		170
	lc	39	25	1		170
	ss	48	-30	4a	methylene chloride	136
	ss	39	60	4a	benzene	134
1-Oxa-4,5-dithiepane	ss	>12	25	4a	benzene	61

8-RINGS

Tetraoxane	gg	51	25	1		142
	c' c'	-3	27	1		131
1,3,6-Trioxocane	ss	39	60	4a	benzene	134

2.2 MONOMERS GIVING POLYMERS CONTAINING O IN THE MAIN CHAIN, BONDED TO OTHER HETEROATOMS (S)

Sulfur trioxide	gc'	178	41	4a		86

2.3 MONOMERS GIVING POLYMERS CONTAINING N IN THE MAIN CHAIN, BONDED IN THE CHAIN TO CARBON ONLY

(listed by increasing ring size)

5-RINGS

2-Pyrrolidinone	c' c'	31	25	1/4b		63

6-RINGS

2-Piperidone	c' c'	25	25	1/4b		63

7-RINGS

ε-Caprolactam	ls	29	250	4a		68
	c' c'	-5	25	1		63
--, 3-methyl-	ls	11	260	4a		194
--, 5-methyl-	ls	16	200	4a		225
--, 7-methyl-	ls	21	225	4a		226

8-RINGS

2-Oxo-hexamethyleneimine	c' c'	-17	25	1		63

N.B. Values for 9-13 rings by probably unreliable extrapolation of $\Delta S_{c' c'}$ for 5-8 rings are given in Ref. 138

13-RINGS

2-Oxo-undecamethyleneimine	ls	16	290	4a		139

2.4 OTHER MONOMERS GIVING POLYMERS NOT LISTED ABOVE

Selenium Se$_8$	ls	-27	200	4a	from estimated T$_c$	88,111
Sulfur S$_8$	ls	-31	159	4a		73,74,111

Monomer A	Monomer B	State of Monomer and Polymer xx	$-\Delta S^O_{xx}$ [J K^{-1} mol^{-1}]	Temp. [^{O}C]	Method	Solvent/Notes	References

3. COPOLYMERS

(listed alphabetically under Monomer A)

Monomer A	Monomer B	State of Monomer and Polymer xx	$-\Delta S^O_{xx}$	Temp. [^{O}C]	Method	Solvent/Notes	References
1,3-Butadiene	styrene	lc		30	1		108
1-Butene	sulfur dioxide	lc	116	25	1		109

Monomer A	Monomer B	State of Monomer and Polymer xx	$-\Delta S^o_{xx}$ [J K^{-1} mol^{-1}]	Temp. [^{o}C]	Method	Solvent/Notes	References
1-Butene (Cont'd.)	sulfur dioxide (Cont'd.)	ss	145	64	4a	excess B	110
--, 3-methyl-	sulfur dioxide	lc	162	35	4a	excess B	199,200
		lc	81	25	4a	excess A	199,200
2-Butene, cis	sulfur dioxide	ss	146	25	4a	excess B	5
--, trans	sulfur dioxide	ss	140	25	4a	excess B	5
Cyclopentene	sulfur dioxide	ss	134	102	4a	excess B	110
1-Hexadecene	sulfur dioxide	ss	139	30	4a	chloroform	110
1-Hexene	sulfur dioxide	lc	116	25	1		109
		ss	145	60	4a	excess B	110
Isobutene	sulfur dioxide	lc	134	0	4a	N$_{23}$	83
Methacrylic acid	methacrylic acid, methyl ester						206
Propene	sulfur dioxide	lc	117	25	1		109
		sc	130	90	4a	excess B	110
	ethylene	gc	172	25	1	31 mol% A in copolymer, N$_{30}$	222

TABLE C - CEILING TEMPERATURES AND EQUILIBRIUM MONOMER CONCENTRATIONS

Most addition polymerization reactions are exothermic and exentropic. The free energy of polymerization per monomer unit therefore becomes less negative as the temperature is raised. At the ceiling temperature T_c the free energy of polymerization under the prevailing conditions is zero and above this temperature polymerization to long-chain polymer is impossible (just as in physical aggregation a liquid connot form a solid when the temperature is above the melting point). The reverse phenomenon of a floor temperature is also known, e.g. for sulfur.

In general a pure liquid monomer which gives an insoluble polymer will have a single well-defined ceiling temperature, given by $T_c = \Delta H_{1c}/\Delta S_{1c}$. A pure liquid monomer which gives a soluble polymer will have a series of ceiling temperatures corresponding to different percentage conversions of monomer to polymer. The condition for equilibrium is then

$$-\Delta \overline{G}_1 + \Delta G_{1c} + \Delta \overline{G}_2 = 0$$

The partial molar free energy per mol of monomer, $\Delta \overline{G}_1$ and per base-mol of polymer, $\Delta \overline{G}_2$, are then functions of composition and may be evaluated from an appropriate equation for mixing of monomer and polymer, e.g. the Flory-Huggins equation. For a monomer dissolved in a solvent the situation is more complex and the ceiling temperature at a given monomer concentration (or the equilibrium concentration of monomer at a given temperature) is dependent on the nature of the solvent and the composition of the medium (Refs. 28, 29, 112, 113, 114, 160). For the case where both monomer and polymer are in solution the variation of T_c with concentration is given to a first approximation by

$$T_c = \Delta H^o_{ss}/(\Delta S^o_{ss} + R \ln[M])$$

where [M] is the concentration of monomer and ΔH^o_{ss} and ΔS^o_{ss} refer to the heat and entropy changes in an appropriate standard state. A more general expression may be derived from the free energy condition by insertion of suitable expressions for ΔG_1 and ΔG_2. These will contain the various interaction parameters appropriate to the polymer-monomer-solvent system.

The values of T_c quoted in the Table are mostly obtained from experimental values by interpolation or short extrapolation. Some unpolymerizable monomers are included where these are structurally closely related to monomers which do polymerize and where the cause of non-polymerization appears to be thermodynamic. This is amplified for cyclic monomers in Table D.

Table C is divided into three sections:

        1 - Equilibria involving pure liquid monomers (lc, lc', ls)
        2 - Equilibria involving gaseous monomers (gc, gc')
        3 - Equilibria involving monomers in solution (sc, sc', ss) and copolymerizations

For meaning of symbols see Table A.

Monomer	State of Monomer and Polymer	T_c [°C]	Wt. Fraction Monomer at Equilibrium	Notes	References

1. EQUILIBRIA INVOLVING PURE LIQUID MONOMERS

1.1 MONOMERS GIVING POLYMERS CONTAINING MAIN CHAIN ACYCLIC CARBON ONLY

Monomer	State	T_c	Wt. Fraction	Notes	References
2-Butene					
--, 2-methyl-	ll	-29	1	calculated, N_{31}	143
--, 2,3-dimethyl-	ll	-223	1	calculated, N_{31}	143
Atropic acid, methyl ester	ls	-8	1		149
Styrene, α-methyl-	ls	61	1	1 bar	119
	ls	170	1	6.57 kbar	119

N.B. The following compounds are not polymerizable for thermodynamic reasons but can sometimes be copolymerized with other monomers (see Ref. 205): α-trifluoromethyl vinyl acetate, α-methoxystyrene, 1,1-diphenylethylene, trans-crotonitrile, trans-stilbene, trans-1,2-di(2-pyridyl)-ethylene, trans-1,2-dibenzoylethylene, trans-1,2-diacetylethylene, α-stilbazole, methyl 2-tert-butylacrylate, 1-isopropenylnaphthalene, 2,4-dimethyl-α-methyl-styrene

1.2 MONOMERS GIVING POLYMERS CONTAINING HETEROATOMS IN THE MAIN CHAIN

1.2.1 MONOMERS GIVING POLYMERS CONTAINING O IN THE MAIN CHAIN, BONDED TO CARBON ONLY

Monomer	State	T_c	Wt. Fraction	Notes	References
Acetaldehyde	ls	-31	1	atactic polymer (1 bar)	115
	ls	-39	1	isotactic polymer (1 bar)	115
	ls	20	1	< 10 kbar	158
Chloral	lc'	58	1		233
1,3-Dioxepane	ls	100	0.1	N_{22}	116
--, 2-phenyl-	ls	20	0.36	N_{22}	116
(1,3-Dioxolane)$_n$ n = 1	ls	100	0.3		116,219
n = 1	ls	165	1		198
n = 1-8	ls	60		K_n determined for cyclic monomers	232
1,3-Dioxolane, 4-phenyl-	ls	20	1		285
1,2-Oxathiolane-2,2-dioxide	ls	95	0.74	N_{22}	152
Oxepane	ls	41.5	1		147
Propionaldehyde	ls	-31	1	atactic polymer	115
	ls	-39	1	isotactic polymer	115
Tetrahydrofuran	ls	80	1	1 bar	58,130
	ls	129	1	2.5 kbar	146
Tri(ethylene terephthalate)	ls	297	0.014		see 174

N.B. The following compounds are not polymerizable for thermodynamic reasons but can sometimes be copolymerized with other monomers (see Ref. 205): benzaldehyde (Refs. 297-299), hexafluoroacetone, 1,1-dimethylpropionaldehyde; also numerous cyclic ethers - see Table D.

1.2.2 MONOMERS GIVING POLYMERS CONTAINING O IN THE MAIN CHAIN, BONDED TO OTHER HETEROATOMS (S, P, Si)

Monomer	State	T_c	Wt. Fraction	Notes	References
Sulfur trioxide	lc'	30.4	1		86
Sodium metaphosphate (Na$_n$(OPO$_2$)$_n$)					
n = 3	ls	800	0.043	K_n determined for cyclic monomers, n = 3-7	300,301
n = 4	ls	800	0.026		300,301
n = 5	ls	800	0.008		300,301
Siloxanes (R$_1$R$_2$SiO)$_n$					
--, R$_1$, R$_2$					
Me, H n = 3-5	ls	0	0.045	K_n determined for cyclic monomers, n = 4-15	176
n = 6-18	ls	0	0.034		176
n = 19-∞	ls	0	0.046		176
n = 3-00	ls	0	0.125		176
Me, Me n = 3-5	ls	110	0.100	K_n determined for cyclic monomers, n = 4-40	176
n = 6-18	ls	110	0.036		176
n = 19-∞	ls	110	0.047		176
n = 3-∞	ls	110	0.183		176
n = 4-40	ls	110		K_n determined for values of n indicated	180
n = 4-30	ls	145		K_n determined for values of n indicated	180
Me, Et n = 3-5	ls	110	0.170	K_n determined for cyclic monomers, n = 4-20	176
n = 6-18	ls	110	0.049		176
n = 19-∞	ls	110	0.039		176
n = 3-∞	ls	110	0.258		176

Monomer	State of Monomer and Polymer	T_c [°C]	Wt. Fraction Monomer at Equilibrium	Notes	References
Siloxanes $(R_1 R_2 SiO)_n$ (Cont'd.)					
--, R_1, R_2					
Me, Pr	n = 3-5 ls	110	0.270	K_n determined for cyclic monomers, n = 4-8	176
	n = 3-8 ls	110	0.310		176
Me, $CF_3(CH_2)_2$	n = 3-5 ls	110	0.711	K_n determined for cyclic monomers, n = 4-20	176
	n = 6-18 ls	110	0.089		176
	n = 19-∞ ls	110	0.027		176
	n = 3-∞ ls	110	0.827		176

1.2.3 MONOMERS GIVING POLYMERS CONTAINING S IN THE MAIN CHAIN, BONDED IN THE CHAIN TO CARBON ONLY

Monomer	State of Monomer and Polymer	T_c [°C]	Wt. Fraction Monomer at Equilibrium	Notes	References
Thioacetone	lc	95	1		148
Thioacetophenone	ls	~ 40	1		288

N.B. Isothiocyanates (Ref. 295) are not polymerizable for thermodynamic reasons, but can be copolymerized with thiiranes; carbon disulfide (Ref. 296) can also be copolymerized with thiiranes.

1.2.4 MONOMERS GIVING POLYMERS CONTAINING N IN THE MAIN CHAIN, BONDED IN THE CHAIN TO CARBON ONLY

Monomer	State of Monomer and Polymer	T_c [°C]	Wt. Fraction Monomer at Equilibrium	Notes	References
(ε-Caprolactam)$_n$	n = 1 ls	220	0.055		68
	ls	225	0.067		192
	ls	277	0.084		see 181
	n = 2 ls	277	0.0078	cyclic monomer	see 181
	n = 3 ls	277	0.0052	cyclic monomer	see 181
	n = 4 ls	277	0.0056	cyclic monomer	see 181
	n = 5 ls	277	0.0048	cyclic monomer	see 181
	n = ≥ 6 ls	277	0.0076	cyclic monomer	see 181
--, 3-methyl-	ls	225	0.086		192,194
--, 3-ethyl-	ls	240	0.1		191
--, 3-propyl-	ls	240	0.1		191
--, 4-ethyl-	ls	240	0.35		171
--, 5-methyl-	ls	172	0.18		225,302
--, 7-methyl-	ls	225	0.107		226,302
--, 7-ethyl-	ls	240	0.1		191,302
--, 7-propyl-	ls	240	0.1		191,302
2-Oxo-hexamethylenimine	ls	240			302
--, 4-ethyl-	ls	240	0.03		171
--, 7-ethyl-	ls	240	0.06		171
--, 8-propyl-	ls	240			302
2-Oxo-undecamethylenimine	ls	290	0.024		139
2-Piperidone	ls	60	0.32		118

1.2.5 OTHER MONOMERS GIVING POLYMERS NOT LISTED ABOVE

Monomer	State of Monomer and Polymer	T_c [°C]	Wt. Fraction Monomer at Equilibrium	Notes	References
Selenium Se$_8$	ls	83	1	N_{32}	85
Sulfur S$_8$	ls	159	1	N_{33} effect of pressure	73,111 145

Monomer	State of Monomer and Polymer	T_c [°C]	Equilibrium pressure [mbar]	Notes	References

2. EQUILIBRIA INVOLVING GASEOUS MONOMERS
(listed alphabetically)

Monomer	State of Monomer and Polymer	T_c [°C]	Equilibrium pressure [mbar]	Notes	References
Acetone	gc	(-220)		estimated, N_{36}	144
Chloral	gc'	98	1013		50
1,3-Dioxolane	gc	87	1013		292
Ethylene, tetrafluoro-	gc	560	128	N_{34}	23,120
Fluoral	gc	73	1013		129

Monomer	State of Monomer and Polymer	T_c [°C]	Equilibrium pressure [mbar]	Notes	References
Formaldehyde	gc'	119	1013	N_{28}	49,164,165
Methacrylic acid					
--, ethyl ester	gs	173	1013	N_{35}	40
--, methyl ester	gs	164	1013	N_{35}	43
Sulfur trioxide	gc'	27.0	372		86
		0.0	40.8		86
Tetrahydrofuran	gc	83	1013		292

Monomer	State of Monomer and Polymer	T_c [°C]	M [mol litre^{-1}]	Solvent/Notes	References

3. EQUILIBRIA INVOLVING MONOMERS IN SOLUTION

3.1 MONOMERS GIVING POLYMERS CONTAINING MAIN CHAIN ACYCLIC CARBON ONLY

Monomer	State	T_c [°C]	M	Solvent/Notes	References
Acrylonitrile	-	25	3×10^{-8}	estimated from $\Delta G_{1c'}$	193
Atropic acid					
--, methyl ester	ss	-40	1.0	toluene	149
--, ethyl ester	ss	-67	1.0	toluene	150
--, n-propyl ester	ss	-72	1.0	toluene	150
--, n-butyl ester	ss	-80	1.0	toluene	150
--, p-methyl-, methyl ester	ss	-37	1.0	toluene	150
Atroponitrile	ss	50	1.0	toluene	149
Biphenyl, p-isopropenyl-	ss	0	0.515	tetrahydrofuran	16
Cyclo-octene	ss	20		benzene N_{38}	229,230
Methacrylic acid, methyl ester	ss	155.5	0.82	o-dichlorobenzene	42
Methacrylonitrile	ss	145	0.27	benzonitrile	44
Naphthalene, 2-isopropenyl-	ss	0	0.284	tetrahydrofuran	16
Oxazole					
--, 2-isopropenyl-4,5-dimethyl-	ss	0	0.0025	tetrahydrofuran	284
Pyridine, 2-isopropenyl-	ss	0	0.043	tetrahydrofuran	284
Styrene	ss	150	9.1×10^{-4}	benzene	28
	ss	150	6.5×10^{-4}	cyclohexane	28
--, α-methyl-	ss	0	0.76	tetrahydrofuran N_{37}	16,32,33
--, --, o-methoxy-	ss	-25	~ 2	methylene chloride	155
Thiazole, 2-isopropenyl-	ss	0	0.010	tetrahydrofuran	284

3.2 MONOMERS GIVING POLYMERS CONTAINING HETEROATOMS IN THE MAIN CHAIN

3.2.1 MONOMERS GIVING POLYMERS CONTAINING O IN THE MAIN CHAIN, BONDED TO CARBON ONLY

Monomer	State	T_c [°C]	M	Solvent/Notes	References
Acetaldehyde	sc	~ -80	6	toluene, N_{22}	228
n-Butyraldehyde	sc	-18	~ 2	toluene, isotactic polymer	227
	sc	12	1.0	n-hexane [8 kbar], N_{12}	161
	sc	27	1.0	n-hexane [10 kbar], N_{12}	161
Chloral	ss	11	1.0	tetrahydrofuran	167
	sc'	12.5	(0.1)	pyridine, N_{39}, N_{40}	50
	sc'	35	1.0	n-heptane	166
1,3-Dioxepane	ss	78	1.0	benzene	116,134
	ss	27	1.0	methylene chloride	136
(1,3-Dioxolane)$_n$	n = 1	ss	1	1.0	methylene chloride, N_{41}	133,159
	n = 1	ss	-8	1.0	benzene, N_{41}	134,159
	n = 1	ss	20	2.0	1,3-dioxane, N_{41}	159
	n = 1	ss	20	1.0	ethyl chloride	293
	n = 1	ls	60	2.0	a few percent methylene chloride present	198
	n = 1	ls	145	10.0	a few percent methylene chloride present	198
	n = 1-8	ss	60		dichloroethane, N_{42}	232
Formaldehyde	sc'	30	0.06	methylene chloride	122
	sc'	25	0.004	nitrobenzene	182
Isobutyraldehyde	ss	-63	1.0	tetrahydrofuran	172
	ss	-61.6	1.0	diethyl ether	166

Monomer	State of Monomer and Polymer	T_c [°C]	M [mol litre^{-1}]	Solvent/Notes	References
Isobutyraldehyde (Cont'd.)	ss	-36.6	1.0	n-pentane	166
--, 1-chloro-	ss	-54	1.0	tetrahydrofuran	172
Oxa-4,5-dithiepane	ss	25	<0.005	benzene	61
Oxepane	ss	30	0.08	methylene chloride	236
o-Phthalaldehyde	ss	-43	1.0	methylene chloride, cyclopolymer	154
Propionaldehyde					
--, 1,1-dichloro-	ss	-24	1.0	tetrahydrofuran	172
--, 1,2-dichloro-	ss	~ 5	(0.25)	methylene chloride, N_{22}, N_{40}	153
Tetrahydrofuran	ss	25	4.45	benzene, N_{37}, N_{44}	231
	ss	20	2.8	diethyl ether	211
Tetraoxane	sc'	50	<0.008	1,2-dichloroethane	178
	sc'	25	0.12	methylene chloride	182
	sc'	25	0.10	nitrobenzene	182
Trioxane	sc'	50	0.20	1,2-dichloroethane	178
	sc'	30	0.13	1,2-dichloroethane	178
	sc'	30	0.05	benzene	178
	sc'	30	0.19	nitrobenzene, N_{45}	178
	sc'	25	1.5	nitrobenzene, N_{45}	182
	sc'	25	2.5	methylene chloride	182
1,3,6-Trioxocane	ss	297	1.0	benzene, N_{43}	134

3.2.2 MONOMERS GIVING POLYMERS CONTAINING O IN THE MAIN CHAIN, BONDED TO OTHER HETEROATOMS (Si)

Siloxanes $(R_1 R_2 SiO)_n$
--, R_1, R_2

	State of Monomer and Polymer	T_c [°C]	M [mol litre^{-1}]	Solvent/Notes	References
H, Me	ss	0	(0.2)	toluene, N_{46}	176
Me, Me	ss	110	0.3	toluene, N_{47}	217, 218
Me, Et	ss	110	(0.4)	toluene, N_{46}	176
Me, $(CH_2)_2$	ss	110	(0.9)	cyclohexanone, N_{46}	176

3.2.3 MONOMERS GIVING POLYMERS CONTAINING S IN THE MAIN CHAIN, BONDED IN THE CHAIN TO CARBON ONLY

1,3-Dithiolane, 2-phenylimino-4-methyl-	ss	65	2.7	benzene	290

3.2.4 MONOMERS GIVING POLYMERS CONTAINING N IN THE MAIN CHAIN, BONDED IN THE CHAIN TO CARBON ONLY

n-Hexyl isocyanate	sc	-22	2	dimethylformamide	127

Monomer A	Monomer B	State of Monomer and Polymer	T_c [°C]	[A][B] [mol^2 litre^{-2}]	Solvent/Notes	References

3.5 COPOLYMERS

3.5.1 1:1 COPOLYMERS

Monomer A	Monomer B	State	T_c	[A][B]	Solvent/Notes	References
Allyl acetate	sulfur dioxide	ss	45	27	excess B	110
Allyl alcohol	sulfur dioxide	sc	76	27	excess B	110
Allyl ethyl ether	sulfur dioxide	ss	68	27	excess B	110
Allyl formate	sulfur dioxide	sc	45	27	excess B	110
1-Butene	sulfur dioxide	ss	64	27	excess B	110
--, 2-ethyl-	sulfur dioxide		<-80		all compositions	110
--, 3-methyl-	sulfur dioxide	sc	36	27	excess B	110,199
2-Butene, cis	sulfur dioxide	ss	46	27	excess B, N_{48}	110
--, trans	sulfur dioxide	ss	38	27	excess B, N_{48}	110
2-Butene (50% cis)	sulfur dioxide	ss	34.6	27	excess B, N_{49}	110
--, 2-methyl-	sulfur dioxide		<-80		all compositions	110
Cycloheptene	sulfur dioxide	ss	11	27	excess B	124
Cyclohexene	sulfur dioxide	ss	24	27	excess B	110
1,3-Cyclooctadiene	sulfur dioxide	ss	37	10	excess B (1,2 addition)	156
Cyclopentene	sulfur dioxide	ss	103	27	excess B	110

Notes page II-446, References page II-447

Monomer A	Monomer B	State of Monomer and Polymer	T_c [°C]	[A] [B] [$mol^2 litre^{-2}$]	Solvent/Notes	References
Ethylene	sulfur dioxide	sc'	> 135	27	excess B	110
2-Heptene (88% cis)	sulfur dioxide	ss	-38	33	excess B, N[49]	110
1-Hexadecene	sulfur dioxide	sc	69	27	excess B	110
		ss	30	1.15	chloroform	80
1-Hexene	sulfur dioxide	ss	60	27	excess B	110
--, 2-ethyl-	sulfur dioxide		< -80		all compositions	110
1-Pentene	sulfur dioxide	ss	63	27	excess B	110
--, 2-methyl-	sulfur dioxide	ss	-34	27	excess B	110
--, 4,4-dimethyl-	sulfur dioxide	sc	14	27	excess B	110
--, 2,4,4-trimethyl-	sulfur dioxide		< -80		all compositions	110
2-Pentene (50% cis)	sulfur dioxide	ss	8.5	32	N[49]	110
--, 4-methyl-			< -80		all compositions	110
4-Pentenoic acid	sulfur dioxide	sc	66	27	excess B	110
Propene	sulfur dioxide	sc	90	27	excess B	110
Isobutene	sulfur dioxide	sc	5	27	excess B	110

3.5.2 GENERAL COPOLYMERS

ε-Caprolactam	ε-Caprolactam, 3-methyl-	ls	225		N[50]	192,194

TABLE D - POLYMERIZABILITY OF 5-, 6- AND 7-MEMBERED HETEROCYCLIC RING COMPOUNDS

The free energy of polymerization is generally negative for 3-, 4-, 8- and higher-membered ring compounds. With heterocyclic rings there is generally an ionic mechanism available which allows these compounds to polymerize. With 5-, 6- and 7-membered ring compounds the sign of ΔG, and hence the polymerizability, is critically dependent on the nature of the ring and on the extent and position of substitution. Substitution generally makes ΔG less negative (more positive) and reduces polymerizability in the thermodynamic sense.

Tables D 1, 2, 3 summarize the polymerizability of 5-, 6- and 7-membered rings, respectively, as a function of the nature of the ring and extent of substitution. Table D 4 provides a comparison of polymerizability of unsubstituted 5-, 6- and 7-membered rings.

1. 5-MEMBERED RING COMPOUNDS

Ring	Polymerized	Not Polymerized	References
C_2N_2O	N[51]	R = Me, Ph	237
C_3NO	N[52] R = Me, Ph etc.		238,239
C_3NS			239,240

Ring	Polymerized	Not Polymerized	References
C_3N_2		R = Me, MeCO	239
C_3OS			152
C_3O_2	R = Me, CH_2Cl, Ph R = H, Me	R = Me, Ph R = Me, CH_2Cl	116,133,134,159, 177,239,241-244, 285-287
C_3S_2			245,246,290
C_4N			239,240,243
			240,247-249
C_4O		R = Me, CH_2Cl	51,58,117,130, 140,146,239, 243,250-256, 291,303,304
	R' = H, R = H R' = H, R = endo-Me R' = H, R = exo-Me R' = endo-Me, R = exo-Me	R' = exo-Me R = exo-Me	
C_4S			185,256

2. 6-MEMBERED RING COMPOUNDS

C_3N_2O		Bu, Bu	239

page II-446, References page II-447

Ring	Polymerized	Not Polymerized	References
C_3N_3			239
C_3O_3			141,178, 184,257
C_3S_3			148,258, 259,260
C_4NO			185,239,240
C_4NS			240
C_4N_2		R = Me, Ph	239,240,261
C_4OS			152
C_4O_2			116,136,239, 243,262
		R = Me, Ph R = Me, Ph	239,243, 262,263
C_4S_2			246,265
C_5N		R = H, Et, Ph	118,209,239, 240,243,266

Ring	Polymerized	Not Polymerized	References
C_5O			153, 239, 243, 264, 226-270, 283
C_2S			185, 256

3. 7-MEMBERED RING COMPOUNDS

Ring	Polymerized	Not Polymerized	References
C_4OS_2			61
C_5NS			240
C_5N_2			239, 240
C_5O_2			116, 136, 170, 241, 262, 271
C_5S_2			246
C_6N	R = 3-Me, 4-Me, 5-Me, 5-Et, 5-n-Propyl, 5-i-Propyl 5-Cyclohexyl 6-Me, 7-Me / R = H, Me	R = Ph, PhCH$_2$ n-Heptyl / R = Me, Ph, CH$_2$OH, EtSCH$_2$	63, 68, 240, 249, 272-279
			240, 243, 276, 277, 280 305

Ring	Polymerized	Not Polymerized	References
C_6O			153,239,243, 281,282
C_6S			256

4. COMPARISON OF POLYMERIZABILITY (+ or -) OF UNSUBSTITUTED 5-, 6- AND 7-MEMBERED RING COMPOUNDS
(The formulae denote rings of any size. Thus in the first line $1,3$-C_4NO is drawn as and represents

Type of ring		Name of 5-ring	5-ring	6-ring	7-ring
$1,3$-C_nNO		2-oxazolidinone	-	+	
$1,3$-C_nN_2		2-imidazolidinone	+	-	+
$1,2$-C_nOS		1,2-oxathiolane-2,2-dioxide	+	-	
$1,3$-C_nO_2		1,3-dioxolane	+	-	+
		ethylene carbonate	-	+	
$1,2$-C_nS_2		1,2-dithiolane	+	+	+
C_nN		pyrrolidine	-	-	+
		2-pyrrolidinone	+	+	+
		succinimide	-	-	+
		2-pyrrolidinethione	-		+
C_nO		tetrahydrofuran (oxolane)	+	-	+
		oxolane-2-one (γ-butyrolactone)	+	+	+

Type of ring	Name of 5-ring	5-ring	6-ring	7-ring
$\overset{C\ O}{n}$	oxolane-2,5-dione (succinic anhydride)	-	-	+
$\overset{C\ S}{n}$	thiolan-2-one (γ-thiobutyrolactone)	-	+	+

E. NOTES

N_1 Corrected for end-group effects

N_2 Is assumed

N_3 partial allowance for unreacted monomer

N_4 semi-empirical values for a number of other olefins are given in Ref. 143

N_5 Value given is for dilute solution. Values determined for complete range of composition; maximum (14.4) at 50 mol % monomer.

N_6 No allowance for unreacted monomer.

N_7 Strongly dependent on both monomer concentration and temperature.

N_9 Corrected for enthalpy of glassy state.

N_{10} $-\Delta S_{ss}$ assumed to be 120.3 J K^{-1} mol^{-1}.

N_{11} Zero heat not necessarily in conflict with observed polymerizability below -40°C. Additional loss of free energy may be provided by the crystallization of the polymer.

N_{12} Polymer largely insoluble and remains in suspension.

N_{13} After correction for species in vapour other than formaldehyde.

N_{14} In the presence of a few per cent methylene chloride

N_{15} From measurements in methylene chloride, benzene and 1,4-dioxane.

N_{16} From measurements in benzene.

N_{17} From measurements on vapour-solid equilibrium.

N_{18} From measurements in both methylene chloride and nitrobenzene

N_{19} From measurements in 1,2-dichloroethane

N_{20} Value calculated by the compiler of this table from data in Ref. cited.

N_{21} Numerical values for heats of copolymerization are listed only for those systems yielding 1:1 copolymers. The values refer to the copolymerization of 0.5 mol of each monomer. In all other cases listed the copolymers have a range of composition; details of the corresponding heats of copolymerization are given in the Ref. cited.

Where no solvent is specified in the seventh column, the symbol ΔH denotes that the measured heat is for a liquid mixture of monomers going to a solution of copolymer. ΔH_{lc} denotes the heat change for pure liquid monomers going to condensed amorphous polymer. (This symbolism differs from that in Ref. 1)

N_{22} The states of monomer and/or polymer are not stated in Ref. but are likely to be the ones given.

N_{23} Value calculated from measurements on mixtures containing excess B.

N_{24} Value calculated from measurements on mixtures containing excess A.

N_{25} Value in Ref. 11 corrected using entropy of monomer in Ref. 94.

N_{26} From measurements in both benzene and cyclohexane.

N_{27} Standard state: mol fraction of monomer = 0.1.

N_{28} After correction for species in vapour other than formaldehyde.

N_{29} Numerical values for entropies of polymerization listed are for systems yielding 1:1 copolymers. The values refer to the copolymerization of 0.5 mol of each monomer. The symbolism and standard states are the same as used for the heats of copolymerization (see N_{21}).

N_{30} Value of ΔS^o for the composition of polymer indicated.

N_{31} No allowance for free energy of mixing of polymer and monomer.

N_{32} Floor temperature (hypothetical for supercooled liquid).

N_{33} Floor temperature.

N_{34} Calculated value. Not measurable experimentally because of side reactions.

N_{35} Small amount of vapour dissolved in the polymer; approximately gc.

N_{36} Hypothetical value for acetone as a gas at 1013 mbar pressure.

N_{37} Decreases with increasing polymer concentration (112,160,289).

N_{38} Recognised that polymerization over WCl$_6$/EtAlCl$_2$/EtOH yields an equilibrium mixture of unsaturated cyclic rings (C$_8$H$_{14}$)$_n$. Species up to n = 15 have been identified and higher cyclic polymers are also likely to be present. No firm equilibrium data exist as yet.

N_{39} dT$_c$/dP = 19 deg kbar^{-1}.

N_{40} Value in [M] column is the mol fraction of monomer.

N_{41} Decreases with increasing polymer concentration in methylene chloride, in benzene and in 1,4-dioxane (Ref. 159).

N_{42} K$_n$ values determined for cyclic monomers, n = 1-8.

N_{43} Extrapolated value

N_{44} For equilibrium concentrations of linear oligomeric species see Ref. 294.

N_{45} Note the substantial discrepancy between the two sets of results with nitrobenzene as solvent.

N_{46} Value in [M] column is the volume fraction of cyclic monomer (rings of all sizes).

N_{47} Concentration of rings, n = 13-∞, taking mol.wt = 74 (value for n = 1); K_n determined for cyclic monomers, n = 3-400.
N_{48} Corrected for isomerization effect.
N_{49} Uncorrected for isomerization effect.
N_{50} Equilibrium concentrations of A and B lower than expected from values for homopolymerizations.
N_{51} Polymer is ┤CONHN(COMe)├.
N_{52} Polymer is ┤CH$_2$CH$_2$N(COR)├.

F. REFERENCES

1. F. S. Dainton, K. J. Ivin, in H. A. Skinner, Ed. "Experimental Thermochemistry," Vol. II, Interscienece, New York-London, 1962. p. 251.

2. R. S. Jessup, J. Chem. Phys. 16, 661 (1948).

3. J. W. Richardson, G. S. Parks, J. Am. Chem. Soc. 61, 3545 (1939).

4. F. A. Quinn, L. Mandelkern, J. Am. Chem. Soc. 80, 3178 (1958).

5. F. S. Dainton, J. Diaper, K. J. Ivin, D. R. Sheard, Trans. Faraday Soc. 53, 1269 (1957).

6. C. M. Fontana, G. A. Kidder, J. Am. Chem. Soc. 70, 3745 (1948).

7. G. S. Parks, H. P. Mosher, J. Polymer Sci. A1, 1979 (1963).

8. A. G. Evans, M. Polanyi, Nature 152, 738 (1943).

9. R. H. Biddulph, P. H. Plesch, P. P. Rutherford, Polymer 1, 521 (1960).

10. D. E. Roberts, J. Res. Natl. Bur. Std. 44, 221 (1950).

11. F. S. Dainton, T. R. E. Devlin, P. A. Small, Trans. Faraday Soc. 51, 1710 (1955).

12. S. Kaarsemaker, J. Coops, Rec. Trav. Chim. 71, 261 (1952).

13. H. van Kamp, J. Coops, W. A. Lambregts, B. J. Visser, H. Dekker, Rec. Trav. Chim. 79, 1226 (1960).

14. F. S. Dainton, K. J. Ivin, D. A. G. Walmsley, Trans. Faraday Soc. 56, 1784 (1960).

15. R. M. Joshi, Makromol. Chem. 55, 35 (1962).

16. H. Hopff, H. Luessi, Makromol. Chem. 62, 31 (1963).

17. R. M. Joshi, J. Polymer Sci. 56, 313 (1962).

18. G. C. Sinke, D. R. Stull, J. Phys. Chem. 62, 397 (1958).

19. J. R. Lacher, E. E. Merz, E. Bohmfalk, J. D. Park, J. Phys. Chem. 60, 492 (1956).

20. J. R. Lacher, H. B. Gottlieb, J. D. Park, Trans. Faraday Soc. 58, 2348 (1962).

21. L. K. J. Tong, W. O. Kenyon, J. Am. Chem. Soc. 69, 2245 (1947).

22. W. M. D. Bryant, J. Polymer Sci. 56, 277 (1962).

23. C. R. Patrick, Tetrahedron 4, 26 (1958).

24. R. M. Joshi, Makromol. Chem. 62, 140 (1963).

25. F. S. Dainton, K. J. Ivin, Trans. Faraday Soc. 46, 331 (1950).

26. D. E. Roberts, W. W. Walton, R. S. Jessup, J. Res. Natl. Bur. Std. 38, 627 (1947).

27. L. K. J. Tong, W. O. Kenyon, J. Am. Chem. Soc. 69, 1402 (1947).

28. S. Bywater, D. J. Worsfeld, J. Polymer Sci. 58, 571 (1962).

29. S. Bywater, Makromol. Chem. 52, 120 (1962).

30. R. H. Biddulph, W. R. Longworth, J. Penfold, P. H. Plesch, P. P. Rutherford, Polymer 1, 521 (1960).

31. D. E. Roberts, R. S. Jessup, J. Res. Natl. Bur. Std. 46, 11 (1951).

32. D. J. Worsfold, S. Bywater, J. Polymer Sci. 26, 299 (1957).

33. H. W. McCormick, J. Polymer Sci. 25, 488 (1957).

34. R. M. Joshi, Makromol. Chem. 66, 114 (1963).

35. W. I. Bengough, Trans. Faraday Soc. 54, 1560 (1958).

36. A. G. Evans, E. Tyrrall, J. Polymer Sci. 2, 387 (1947).

37. K. G. McCurdy, K. J. Laidler, Can. J. Chem. 42, 818 (1964).

38. J. H. Baxendale, G. W. Madaras, J. Polymer Sci. 19, 171 (1956).

39. L. K. J. Tong, W. O. Kenyon, J. Am. Chem. Soc. 68, 1355 (1946).

40. R. E. Cook, K. J. Ivin, Trans. Faraday Soc. 53, 1132 (1957).

41. S. Ekegren, S. Oehrn, K. Granath, P. Kinell, Acta Chem. Scand. 4, 126 (1950).

42. S. Bywater, Trans. Faraday Soc. 51, 1267 (1955).

43. K. J. Ivin, Trans. Faraday Soc. 51, 1273 (1955).

44. S. Bywater, Can. J. Chem. 35, 552 (1957).

45. R. A. Nelson, R. S. Jessup, D. E. Roberts, J. Res. Natl. Bur. Std. 48, 275 (1952).

46. R. S. Jessup, A. D. Cummings, J. Res. Natl. Bur. Std. 13, 357 (1934).

47. A. A. Korotkov, E. N. Marandzheva, Russ. J. Phys. Chem. (English Transl.) 37, 135 (1963).

48. V. A. Kargin, V. A. Kabanov, V. P. Zubov, I. M. Papisov, Dokl. Akad. Nauk SSSR, 134, 1098 (1960). Chem. Abst. 55, 8282 (1961).

49. F. S. Dainton, K. J. Ivin, D. A. G. Walmsley, Trans. Faraday Soc. 55, 61 (1959).

50. W. K. Busfield, E. Whalley, Trans. Faraday Soc. 59, 679 (1963).

51. H. C. Raine, R. B. Richards, H. Ryder, Trans. Faraday Soc. 41, 56 (1945).

52. P. Gray, A. Williams, Trans. Faraday Soc. 55, 760 (1959).

53. J. B. Rose, J. Chem. Soc. 1946, 546.

54. S. M. Skuratov, A. A. Strepikheev, S. M. Shtekher, A. V. Volokhina, Dokl. Akad. Nauk.SSSR, 117, 263 (1957).

55. S. M. Skuratov, A. A. Strepikheev, M. P. Kozina, Dokl. Akad. Nauk. SSSR, 117, 452 (1957).

56. R. C. Cass, S. E. Fletcher, C. T. Mortimer, H. D. Springall, T. R. White, J. Chem. Soc. 1958, 1406.

57. D. Sims, J. Chem. Soc. 1964, 864.

58. C. E. H. Bawn, R. M. Bell, A. Ledwith, Polymer 6, 95 (1965).

59. A. Snelson, H. A. Skinner, Trans. Faraday Soc. 57, 2125 (1961).

60. S. E. Fletcher, C. T. Mortimer, H. D. Springall, J. Chem. Soc. 580 (1959).

61. F. S. Dainton, J. A. Davies, P. P. Manning, S. A. Zahir, Trans. Faraday Soc. 53, 813 (1957).

62. A. A. Strepikheev, S. M. Skuratov, O. N. Kachinskaya, R. S. Muramova, E. P. Brykina, S. M. Shtekher, Dokl. Akad. Nauk. SSSR, 102, 105 (1955).

63. V. P. Kolesov, I. E. Paukov, S. M. Skuratov, Zh. Fiz. Khim. 36, 770 (1962); Russ. J. Phys. Chem. 36, 400 (1962).

64. M. P. Kozina, S. M. Skuratov, Dokl. Akad. Nauk. SSSR, 127, 561 (1959).

65. S. M. Skuratov, A. A. Strepikheev, Y. N. Kanarskaya, Kolloidn. Zh. 14, 185 (1952).

66. A. B. Meggy, J. Chem. Soc. 1953, 796.

67. P. F. van Velden, G. M. van der Want, D. Heikens, C. A. Kruissink, P. H. Hermans, A. J. Staverman, Rec. Trav. Chim. 74, 1376 (1955).

68. A. V. Tobolsky, A. Eisenberg, J. Am. Chem. Soc. 81, 2302 (1959).

69. H. Mackle, P. A. G. O'Hare, Tetrahedron 19, 961 (1963).

70. S. Sunner, Acta Chem. Scand. 17, 728 (1963).

71. H. Mackle, R. G. Mayrick, Trans. Faraday Soc. 58, 230 (1962).

72. J. A. Barltrop, P. M. Hayes, M. Calvin, J. Am. Chem. Soc. 76, 4348 (1954).

73. F. Fairbrother, G. Gee, G. T. Merrall. J. Polymer Sci. 16, 459 (1955).

74. A. V. Tobolsky, A. Eisenberg, J. Am. Chem. Soc. 81, 780 (1959).

75. H. Nagao. T. Yamaguchi, J. Chem. Soc. Japan, Ind. Eng. Chem. Sect. 59, 1363 (1956).

76. H. Miyama, S. Fujimoto, J. Polymer Sci. 54, S 32 (1961).

77. F. S. Dainton, K. J. Ivin, Proc. Roy. Soc. (London), A 212, 207 (1952).

78. J. E. Hazell, K. J. Ivin, Trans. Faraday Soc. 58, 342 (1962).

79. L. K. J. Tong, W. O. Kenyon, J. Am. Chem. Soc. 71, 1925 (1949).

80. F. S. Dainton, K. J. Ivin, D. R. Sheard, Trans. Faraday Soc. 52, 414 (1956).

81. M. Suzuki, H. Miyama, S. Fujimoto, J. Polymer Sci. 31, 212 (1958).

82. K. J. Ivin, W. A. Keith, H. Mackle, Trans. Faraday Soc. 55, 262 (1959).

83. R. E. Cook, K. J. Ivin, J. H. O'Donnell, Trans. Faraday Soc. 61, 1881 (1965).

84. W. A. Piccoli, G. G. Haberland, R. L. Merker, J. Am. Chem. Soc. 82, 1883 (1960).

85. A. Eisenberg, A. V. Tobolsky, J. Polymer Sci. 46, 19 (1960).

86. D. C. Abercromby, R. A. Hyne, P. F. Tiley, J. Chem. Soc. 5832 (1963).

87. F. S. Dainton, K. J. Ivin, Quart. Rev. 12, 61 (1958).

88. F. S. Dainton, D. M. Evans, F. E. Hoare, T. P. Melia, Polymer 3, 277, 286 (1962).

89. E. Passaglia, H. K. Kevorkian, J. Appl. Polymer Sci. 7, 119 (1963).

90. R. W. Warfield, M. C. Petree, Makromol. Chem. 51, 113 (1962).

91. B. Wunderlich, J. Chem. Phys. 37, 1203 (1962).

92. E. Passaglia, H. K. Kevorkian, J. Appl. Phys. 34, 90 (1963).

93. G. T. Furukawa, M. L. Reilly, J. Res. Natl. Bur. Std. 56, 285 (1956).

94. H. L. Finke, D. W. Scott, M. E. Gross, J. F. Messerly, G. Waddington, J. Am. Chem. Soc. 78, 5469 (1956).

95. R. B. Scott, C. H. Meyers, R. D. Rands, F. G. Brickwedde, N. Bekkedahl, J. Res. Natl. Bur. Std. 35, 39 (1945).

96. G. T. Furukawa, R. E. McCoskey, J. Res. Natl. Bur. Std. 51, 321 (1953).

97. F. S. Dainton, D. M. Evans, F. E. Hoare, T. P. Melia, Polymer 3, 297 (1962).

98. G. T. Furukawa, R. E. McCoskey, M. L. Reilly, J. Res. Natl. Bur. Std. 51, 69 (1953).

99. G. T. Furukawa, R. E. McCoskey, G. J. King. J. Res. Natl. Bur. Std. 49, 273 (1952).

100. N. Bekkedahl, L. A. Wood, J. Res. Natl. Bur. Stod. 19, 551 (1937).

101. N. Bekkedahl, H. Matheson, J. Res. Natl. Bur. Std. 15, 503 (1935).

102. T. P. Melia, Polymer 3, 317 (1962).

103. R. W. Warfield, M. C. Petree, J. Polymer Sci. A1, 1701 (1963).

104. R. H. Boundy, R. F. Boyer, "Styrene," Reinhold, New York, 1952, p. 67.

105. R. W. Warfield, M. C. Petree, J. Polymer Sci. 55, 497 (1961).

106. F. S. Dainton, D. M. Evans, F. E. Hoare, T. P. Melia, Polymer 3, 263 (1962).

107. F. S. Dainton, D. M. Evans, F. E. Hoare, T. P. Melia, Polymer 3, 271 (1962).

108. R. J. Orr, Polymer 2, 74 (1961).

109. F. S. Dainton, D. M. Evans, F. E. Hoare, T. P. Melia, Polymer 3, 310 (1962).

110. R. E. Cook, F. S. Dainton, K. J. Ivin, J. Polymer Sci. 26, 351 (1957).

111. G. Gee, Chemical Society (London) Special Publication 15, 67 (1961).

112. A. Vrancken, J. Smid, M. Szwarc, Trans. Faraday Soc. 58, 2036 (1962).

113. A. V. Tobolsky, A. Rembaum, A. Eisenberg, J. Polymer Sci. 45, 347 (1960).

114. K. J. Ivin, Pure Appl. Chem. 4, 271 (1962).

115. A. M. North, D. Richardson, Polymer 6, 333 (1965).

116. A. A. Strepikheev, A. V. Volokhina, Dokl. Akad. Nauk. SSSR, 99, 407 (1954).

117. C. L. Hamermesh, V. E. Haury, J. Org. Chem. 26, 4748 (1961).

118. N. Yoda, A. Miyake, J. Polymer Sci. 43 117 (1960).

119. J. G. Kilroe, K. E. Weale, J. Chem. Soc. 1960, 3849.

120. H. H. G. Jellinek, H. Kachi, Makromol. Chem. 85, 1 (1965).

121. R. B. Whitney, M. Calvin, J. Chem. Phys. 23, 1750 (1955).

122. W. Kern, V. Jaacks, J. Polymer Sci. 48, 399 (1960).

123. B. J. Cottam, J. M. G. Cowie, S. Bywater, Makromol. Chem. 86, 116 (1965).

124. J. E. Hazell, K. J. Ivin, Trans. Faraday Soc. 58, 176 (1962).

125. J. K. Jacques, M. F. Mole, N. L. Paddock, J. Chem. Soc. 1965, 2112.

126. J. W. C. Crawford, J. Chem. Soc. 1953, 2658.

127. V. E. Shashoua, W. Sweeny, R. E. Tietz, J. Am. Chem. Soc. 82, 866 (1960).

128. R. C. P. Cubbon, Makromol. Chem. 80, 44 (1964).

129. W. K. Busfield, Polymer 7, 541 (1966).

130. K. J. Ivin, J. Leonard, Polymer 6, 621 (1965).

131. K. Nakatsuka, H. Suga, S. Seki, J. Polymer Sci. B7, 361 (1969).

132. J. B. Thompson, W. M. D. Bryant, Am. Chem. Soc. Polymer Preprints, 11, 204 (1970).

133. P. H. Plesch, P. H. Westermann, J. Polymer Sci. C16, 3837 (1968).

134. Y. Yamashita, M. Okada, K. Suyama, H. Kasahara, Makromol. Chem. 114, 146 (1968).

135. O. Riedel, P. Wittmer, Makromol. Chem. 97, 1 (1966).

136. P. H. Plesch, P. H. Westermann, Polymer 10, 105 (1969).

137. B. Boerjesson, Y. Nakase, S. Sunner, Acta Chem. Scand. 20, 803 (1966).

138. A. K. Bonetskaya, S. M. Skuratov, Vysokomol. Soedin. A11, 532 (1969).

139. H. G. Elias, A. Fritz, Makromol. Chem. 114, 31 (1968); 120, 238 (1968).

140. G. A. Clegg, D. R. Gee, T. P. Melia, A. Tyson, Polymer 9, 501 (1968).

141. G. A. Clegg, T. P. Melia, A. Tyson, Polymer 9, 75 (1968).

142. G. A. Clegg, T. P. Melia, Makromol. Chem. 123, 184 (1969).

143. R. M. Joshi, B. J. Zwolinski, C. W. Hayes, Macromolecules 1, 30 (1968).

144. D. R. Waywell, J. Polymer Sci B8, 327 (1970).

145. G. C. Vezzoli, F. Dachille, R. Roy, J. Polymer Sci. A-1 7, 1557 (1969).

146. M. Rahman, K. E. Weale, Polymer 11, 122 (1970).

147. F. P. Jones, Ph. D. Thesis, Keele University (1970).

148. V. C. E. Burnop, K. G. Latham, Polymer 8, 589 (1967).

149. H. Hopff, H. Luessi, L. Borla, Makromol. Chem. 81, 268 (1965).

150. L. Borla, Promotionsarbeit Nr. 3668, E. T. H. Zuerich (1965).

151. H. Luessi, Makromol. Chem. 103, 68 (1967).

152. S. Hashimoto, T. Yamashita, Kobunshi Kagaku, 27, 400 (1970).

153. H. Sumimoto, T. Nakagawa, J. Polymer Sci. B7, 739 (1969).

154. C. Aso, S. Tagami, T. Kunitake, J. Polymer Sci, A-1, 7, 497 (1969).

155. Y. Okamoto, H. Takano, H. Yuki, Polymer J. $\underline{1}$, 403 (1970).

156. T. Yamaguchi, K. Nagai, Kobunshi Kagaku $\underline{26}$, 809 (1969).

157. D. J. Stein, P. Wittmer, J. Toelle, Angew. Makromol. Chem. $\underline{8}$, 61 (1969).

158. A. Novak, E. Whalley, Can. J. Chem. $\underline{37}$, 1710 (1959).

159. L. I. Kuzub, M. A. Markevich, A. A. Berlin, N. S. Yenikolopyan, Vysokomol. Soedin. $\underline{A10}$, 2007 (1968); Polymer Sci. SSSR, $\underline{10}$, 2332 (1968).

160. K. J. Ivin, J. Leonard, Europ. Polymer J. $\underline{6}$, 331 (1970).

161. Y. Ohtsuka, C. Walling, J. Am. Chem. Soc. $\underline{88}$, 4167 (1966).

162. C. Walling, T. A. Augurt, J. Am. Chem. Soc. $\underline{88}$, 4163 (1966).

163. G. Natta, G. Mazzanti, P. Corradini, I. W. Bassi, Makromol. Chem. $\underline{37}$, 156 (1960).

164. W. K. Busfield, D. Merigold, Makromol. Chem. $\underline{138}$, 65 (1970).

165. Y. Iwasa, T. Imoto, J. Chem. Soc. Japan, Pure Chem. Sect. (Nippon Kagaku Zasshi) $\underline{84}$, 29 (1963).

166. I. Mita, I. Imai, H. Kambe, Makromol. Chem. $\underline{137}$, 169 (1970).

167. I. Mita, I. Imai, H. Kambe, Makromol. Chem. $\underline{137}$, 143 (1970).

168. V. E. Ostrovskii, V. A. Khodzhemirov, A. P. Barkova, Dokl. Akad. Nauk. SSSR, $\underline{191}$, 1095 (1970).

169. P. Caillaud, J. M. Huet, E. Maréchal, Bull. Soc. Chim. France 1473 (1970).

170. G. A. Clegg, T. P. Melia, Polymer $\underline{11}$, 245 (1970).

171. A. K. Bonetskaya, T. V. Sopova, O. B. Salamatina, S. M. Skuratov, B. P. Fabrichuyi, I. F. Shalavina, V. L. Gol' dfarb, Vysokomol. Soedin. $\underline{B11}$, 894 (1969).

172. I. Mita, I. Imai, H. Kambe, Makromol. Chem. $\underline{137}$, 155 (1970).

173. A. Nicco, B. Boucheron, Europ. Polymer J. $\underline{6}$, 1477 (1970).

174. G. R. Walker, J. A. Semlyen, Polymer $\underline{11}$, 472 (1970).

175. E. Maréchal, J. Polymer Sci. A-1, $\underline{8}$, 2867 (1970).

176. P. V. Wright, J. A. Semlyen, Polymer $\underline{11}$, 462 (1970).

177. G. A. Clegg, T. P. Melia, Polymer $\underline{10}$, 912 (1969).

178. T. Miki, T. Higashimura, S. Okamura, J. Polymer Sci. A-1, $\underline{8}$, 157 (1970).

179. E. G. Butcher, H. Mackle, D. V. McNally, Trans. Faraday Soc. $\underline{65}$, 2331 (1969).

180. J. A. Semlyen, P. V. Wright, Polymer $\underline{10}$, 543 (1969).

181. J. A. Semlyen, G. R. Walker, Polymer $\underline{10}$, 597 (1969).

182. A. A. Berlin, K. A. Bogdanova, I. P. Kravchuk, G. P. Rakova, N. S. Yenikolopyan, Dokl. Akad. Nauk. SSSR, $\underline{184}$, 1128 (1969).

183. V. E. Ostrovskii, V. A. Khodzhemirov, S. P. Kostareva, Dokl. Akad. Nauk. SSSR, $\underline{184}$, 103 (1969).

184. G. A. Clegg, T. P. Melia, Makromol. Chem. $\underline{123}$, 194 (1969).

185. G. Nabi, Pakistan J. Sci. $\underline{20}$, 29 (1968).

186. D. R. Gee, T. P. Melia, Makromol. Chem. $\underline{116}$, 122 (1968).

187. V. I. Kasatochkin, M. E. Kazakov, Y. P. Kudryavtsev, A. M. Sladkov, V. V. Korshak, Dokl. Akad. Nauk. SSSR, $\underline{183}$, 109 (1968).

188. E. M. Morazova, V. I. Yeliseyeva, M. A. Korshunov, Vysokomol. Soedin. $\underline{A10}$, 2354 (1968); Polymer Sci. SSSR, $\underline{A10}$, 2736 (1968).

189. A. K. Bonetskaya, S. M. Skuratov, N. A. Lukina, A. A. Strel' tsova, K. E. Kuznetsova, M. P. Lazareva, Vysokomol. Soedin. $\underline{B10}$, 75 (1968).

190. H. Cheradame, J. P. Vairon, P. Sigwalt, Europ. Polymer J. $\underline{4}$, 13 (1968).

191. O. B. Salamatina, A. K. Bonetskaya, S. M. Skuratov, B. P. Fabrichnyi, I. F. Shalavina, Y. L. Gol' dfarb, Vysokomol. Soedin. $\underline{B10}$, 10 (1968).

192. P. Čefelín, P. Schmidt, J. Šebenda, J. Polymer Sci. $\underline{C23}$, 175 (1966).

193. B. V. Lebedev, I. B. Rabonovich, L. Y. Martynenko, Vysokomol. Soedin. $\underline{A9}$, 1640 (1967).

194. P. Čefelín, A. Frydrychová, J. Labský, P. Schmidt, J. Šebenda, Coll. Czech Chem. Comm. $\underline{32}$, 2787 (1967).

195. M. Furue, S. Nozakura, S. Murahashi, T. Tanaka, Bull. Chem. Soc. Japan $\underline{40}$, 2700 (1967).

196. T. P. Melia, A. Tyson, Makromol. Chem. $\underline{109}$, 87 (1967).

197. S. Iwatsuki, S. Amano, Y. Yamashita, Kogyo Kagaku Zasshi $\underline{70}$, 2027 (1967).

198. L. A. Kharitonova, G. V. Rakova, A. A. Shaginyan, N. S. Yenikolopyan, Vysokomol. Soedin. $\underline{A9}$, 2586 (1967).

199. B. H. G. Brady, J. H. O' Donnell, Trans. Faraday Soc. $\underline{64}$, 29 (1968).

200. B. H. G. Brady, J. H. O' Donnell, Europ. Polymer J. $\underline{4}$, 537 (1968).

201. J. Chiu, Proc. 2nd Symp. on Anal. Calorimetry (1970) 171.

202. K. A. Bogdanova, A. K. Bonetskaya, A. A. Berlin, G. V. Rakova, N. S. Yenikolopyan, Dokl. Akad. Nauk. SSSR $\underline{197}$, 618 (1971).

203. F. Delben, V. Crescenzi, Ann. Chim. (Rome) $\underline{60}$, 782 (1970).

204. G. Natta, G. Mazzanti, P. Corradini, I. W. Bassi, Makromol. Chem. $\underline{37}$, 156 (1968).

205. K. J. Ivin, Thermodynamics of Addition Polymerzation Processes, Ch. 16 in "Reactivity, Mechanism and Structure in Polymer Chemistry", edited by A. D. Jenkins, A. Ledwith, (Wiley, New York, 1973).

206. I. B. Rabinovich, L. I. Pavlinov, Vysokomol. Soedin. $\underline{A10}$, 416 (1968).

207. S. M. Skuratov, V. V. Voevodskii, A. A. Strepikheev, Y. N. Kanarskaya, R. S. Muramova, N. V. Fok, Dokl. Akad. Nauk. SSSR $\underline{95}$, 591 (1954).

208. J. H. S. Green, Quart. Rev. Chem. Soc. $\underline{15}$, 125 (1961).

209. F. Korte, W. Glet, J. Polymer Sci. $\underline{B4}$, 685 (1966).

210. R. H. Beaumont, B. Clegg, G. Gee, J. B. M. Herbert, D. J. Marks, R. C. Roberts, D. Sims, Polymer $\underline{7}$, 401 (1966).

211. B. A. Rozenberg, O. M. Chekhuta, Y. B. Lyudvig, A. R. Gantmakher, S. S. Medvedev, Vysokomol. Soedin. $\underline{6}$, 2030 (1964); Polymer Sci. SSSR $\underline{6}$, 2246 (1964).

212. L. Leese, M. W. Baumber, Polymer $\underline{6}$, 269 (1965).

213. R. W. Warfield, M. C. Petree, J. Polymer Sci. A-2, $\underline{4}$, 532 (1966).

214. J. Grodzinsky, A. Katchalsky, D. Vofsi, Makromol. Chem. $\underline{44\text{-}46}$, 591 (1961).

215. R. M. Joshi, Indian. J. Chem. $\underline{2}$, 215 (1964).

216. M. F. Shostakovskii, I. F. Bogdanov, J. Appl. Chem. Russ. $\underline{15}$, 249 (1942).

217. J. F. Brown, G. M. J. Slusarczuk, J. Am. Chem. Soc. $\underline{87}$, 931 (1965).

218. P. J. Flory, J. A. Semlyen, J. Am. Chem. Soc. $\underline{88}$, 3209 (1966).

219. M. Kučera, Y. Pichler, Vysokomol. Soedin. $\underline{7}$, 10 (1965); Polymer Sci. SSSR $\underline{7}$, 9 (1965).

220. G. M. Chil'-Gevorgyan, A. K. Bonetskaya, S. M. Skuratov, N. S. Yenikolopyan, Vysokomol. Soedin. $\underline{A9}$, 1363 (1967).

221. I. Mita, S. Yabe, I. Imai, H. Kambe, Makromol. Chem. $\underline{137}$, 133 (1970).

222. T. P. Melia, G. A. Clegg, A. Tyson, Makromol. Chem. $\underline{112}$, 84 (1968).

223. O. E. Grikina, V. M. Tatevskii, N. F. Stepanov, S. S. Yarovoi, Vestn. Mosk. Univ. Ser. II, $\underline{22}$, 8 (1967).

224. O. E. Grikina, N. F. Stepanov, V. M. Tatevskii, S. S. Yarovoi, Vysokomol. Soedin. $\underline{A13}$, 575 (1971); Polymer Sci. SSSR $\underline{13}$, 653 (1971).

225. P. Čefelín, D. Doskočilova, A. Frydrychová, J. Šebenda, Coll. Czech. Chem. Comm. 29, 485 (1964).

226. P. Čefelín, A. Frydrychová, P. Schmidt, J. Šebenda, Coll. Czech. Chem. Comm. 32, 1006 (1967).

227. O. Vogl, J. Polymer Sci. A2, 4607 (1964).

228. H. Takida, K. Noro, Chemistry of High Polymers, Japan 21, 452 (1964).

229. E. Wasserman, D. A. Ben-Efraim, R. Wolovsky, J. Am. Chem. Soc. 90, 3286 (1968).

230. N. Calderon, J. Macromol. Sci. Rev. Macromol. Chem. C7, 105 (1972).

231. J. Leonard, D. Maheux, Am. Chem. Soc. Polymer Preprints 13, 78 (1972).

232. J. M. Andrews, J. A. Semlyen, Polymer 13, 142 (1972).

233. O. Vogl, personal communication.

234. O. Vogl, W. M. D. Bryant, J. Polymer Sci. A2, 4633 (1964).

235. H. Kambe, I. Mita, K. Horie, Proc. 2nd Int. Conf. on Thermal Analysis (1969) 1071.

236. T. Saegusa, T. Shiota, S. Matsumoto, H. Fujii, quoted in Macromolecules 5, 34 (1972).

237. T. Endo, Bull. Chem. Soc. Japan 44, 870 (1971).

238. T. Kagiya, S. Narisawa, T. Maeda, K. Fukui, Kogyo Kagaku Zasshi 69, 732 (1969); J. Polymer Sci. B4, 441 (1966).

239. H. K. Hall, A. K. Schneider, J. Am. Chem. Soc. 80, 6409 (1958).

240. H. K. Hall, J. Am. Chem. Soc. 80, 6404 (1958).

241. J. W. Hill, W. H. Carothers, J. Am. Chem. Soc. 57, 925 (1935).

242. M. Okada, Y. Yamashita, Y. Ishii, Makromol. Chem. 80, 196 (1964).

243. W. H. Carothers, Collected papers of, Ed. H. Mark, G. S. Whitby, Interscience, New York, 1940.

244. I. Maruyama, M. Nakaniwa, T. Saegusa, J. Furukawa, J. Chem. Soc. Japan Ind. Chem. Sect. 68, 1149 (1965).

245. R. B. Whitney, M. Calvin. J. Chem. Phys. 23, 1750 (1955).

246. J. G. Affleck, G. Dougherty, J. Org. Chem 15, 865 (1950).

247. W. O. Ney, W. R. Nummy, C. E. Bames, USP. 2638463 (1953).

248. W. O. Ney, M. Crowther, USP. 2739959 (1956).

249. M. P. Kozina, S. M. Skuratov, Dokl. Akad. Nauk. SSSR 127, 561 (1959).

250. H. Meerwein, Angew. Chem. 59, 168 (1947).

251. H. Meerwein, D. Delfs, H. Morschel, Angew. Chem. 72, 927 (1960).

252. M. P. Dreyfuss, P. Dreyfuss, J. Polymer Sci. A4, 2179 (1966).

253. T. Saegusa, S. Matsumoto, J. Macromol. Sci. Chem. A4, 873 (1970).

254. J. B. Rose, J. Stuart-Webb, unpublished results quoted in The Chemistry of Cationic Polymerization, Ed. P. H. Plesch, Pergamon, Oxford 1963.

255. R. Chiang, J. H. Rhodes, J. Polymer Sci. B7, 643 (1969).

256. C. G. Overberger, J. K. Weise, J. Am. Chem. Soc. 90, 3533 (1968).

257. W. Kern, H. Cherdron, V. Jaacks, Angew. Chem. 73, 177 (1961).

258. E. Gipstein, E. Wellisch, O. J. Sweeting, J. Polymer Sci. B1, 237 (1963).

259. J. B. Lando, V. Stannett, J. Polymer Sci. B2, 375 (1964).

260. O. G. von Ettinghausen, E. Kendrick, Polymer 7, 469 (1966).

261. A. B. Meggy, J. Chem. Soc. 1956, 1444.

262. H. K. Hall, Am. Chem. Soc. Polymer Preprints 6, 535 (1965).

263. W. Dittrich, R. C. Schulz, Angew. Makromol. Chem. 15, 109 (1971).

264. E. Hollo, Ber. 61, 895 (1928).

265. A. Schoeberl, G. Wiehler, Ann. 595, 101 (1955).

266. P. A. Small, Trans. Faraday Soc. 51, 1717 (1955).

267. F. Fichter, A. Beisswenger, Ber. 36, 1200 (1903).

268. K. Saotome, Y. Kodaira, Makromol. Chem. 82, 41 (1965).

269. H. Batzer, G. Fritz, Makromol. Chem. 14, 179 (1954).

270. H. D. K. Drew, W. N. Haworth, J. Chem. Soc. 1927, 775.

271. M. H. Palomaa, V. Toukola, Ber. 66B, 1629 (1933).

272. D. D. Coffman, N. L. Cox, E. L. Martin, W. E. Mochel, F. J. van Natta, J. Polymer Sci. 3, 85 (1948).

273. W. E. Hanford, R. M. Joyce, J. Polymer Sci. 3, 167 (1948).

274. H. R. Mighton, USP. 2647105 (1953).

275. F. N. S. Carver, B. L. Hollingsworth, Makromol. Chem 95, 135 (1966).

276. A. Schäffler, W. Ziegenbein, Ber. 88, 1374 (1955).

277. W. Ziegenbein, A. Schäffler, R. Kaufhold, Ber. 88, 1906 (1955).

278. L. E. Wolinski, H. R. Mighton, J. Polymer Sci. 49, 217 (1961).

279. M. Imoto, H. Sakurai, T. Kono, J. Polymer Sci. 50, 467 (1961).

280. N. L. Cox, W. E. Hanford, USP. 2276164 (1942); Swiss P. 270546 (1951), 276924 (1952).

281. F. P. Jones, Ph. D. Thesis, Univers. of Keele, 1970.

282. R. Gehm, Angew. Makromol. Chem. 18, 159 (1971).

283. T. Saegusa, T. Hodaka, H. Fujii, Polymer J. 2, 670 (1971).

284. K. Yagi, T. Miyazaki, H. Okitsu, F. Toda, Y. Iwakura, J. Polymer Sci. A-1, 10, 1149 (1972).

285. B. Krummenacher, H.G. Elias, Makromol. Chem. 150, 271 (1971).

286. P. C. Wollwage, P. A. Seib, J. Polymer Sci, A-1, 9, 2877 (1971).

287. C. C. Tu, C. Schuerch, J. Polymer Sci. B1, 163 (1963).

288. T. Kunitake, M. Yasumatsu, C. Aso, J. Polymer Sci. A-1, 9, 3678 (1971).

289. I. Mita, H. Okuyama, J. Polymer Sci. A-1, 9, 3437 (1971).

290. G. Belonovskaya, Z. Tchernova, B. Dolgoplosk, Europ. Polymer J. 8, 35 (1972).

291. J. Kops, H. Spanggaard, Makromol. Chem. 151, 21 (1972).

292. W. K. Busfield, R. M. Lee, D. Merigold, Makromol. Chem. 156, 183 (1972).

293. Y. B. Lyudvig, Y. E. Berman, Z. N. Nysenko, V. A. Ponomarenko, S. S. Medvedev, Vysokomol. Soedin. A13, 1375 (1971); Polymer Sci. SSSR 13, 1546 (1971).

294. J. M. Andrews, J. A. Semlyen, Polymer 12, 642 (1971).

295. V. S. Etlis, G. A. Razuvaev, A. P. Sineokof, Khim. Geterosicl. Soedin. 2, 223 (1967).

296. G. A. Razuvaev, V. S. Etlis, L. N. Gribov, Zh. Obshch. Khim. 33, 1366 (1963).

297. C. Aso, S. Tagami, T. Kunitake, Kobunshi Kagaku 23, 63 (1966).

298. A. A. Dugaryan, A. V. Agumyan, Vysokomol. Soedin. 5, 1755 (1963).

299. R. Raff, J. L. Cook, B. V. Etting, J. Polymer Sci. A3, 3511 (1965).

300. E. Thilo, U. Schuelke, Z. Anorg. Allgem. Chem. 341, 293 (1965).

301. D. R. Cooper, J. A. Semlyen, Polymer 13, 414 (1972).

302. O. B. Salamatina, A. K. Bonetskaya, S. M. Skuratov, B. P. Fabrichnyi, I. F. Shalavina, Y. L. Gol'dfarb, Vysokomol. Soedin. 7, 485 (1965).

303. K. Weissermel, E. Noelken, Makromol. Chem. 68, 140 (1963).

304. T. Saegusa, M. Matoi, S. Matsumoto, H. Fujii, Macromolecules, 5, 233, 236 (1972).

305. T. Kodaira, J. Stehlicek, J. Šebenda, European Polymer J. 6, 1451 (1970).

ACTIVATION ENERGIES OF PROPAGATION AND TERMINATION IN FREE RADICAL POLYMERIZATION

R. Korus and K. F. O Driscoll
Department of Chemical Engineering
University of Waterloo
Waterloo, Ontario, Canada

Where the rate constants k_p and k_t for a given monomer have been determined (cf. preceding chapter) at different temperatures, the activation energies E_p and E_t may be determined from the Arrhenius expression, $k = A \exp(-E/RT)$ where A is the frequency factor, R the gas constant, and T the absolute temperature.

Monomer	E_p [kcal/mol]	E_p [kJ/mol]	E_t [kcal/mol]	E_t [kJ/mol]	Remarks	References
1.1 Dienes						
Butadiene	2.6	10.9	-	-	in gas phase	13
	9.3	38.9	-	-	in emulsion	14
	5.8;5.0	24.3;20.9	-	-	in gas phase	16
	4.9	20.5	-	-	in gas phase	21
Chloroprene	9.7	40.6	-	-	in emulsion	35
Dimethylbutadiene	9.0	37.7	-	-	in emulsion	45
Isoprene	9.8	41.0	-	-	in emulsion	15
1.2 Olefins						
Acetylene	5.1	21.4	-	-	in gas phase	21
Ethylene	8.2	34.3	-	-	in gas phase	21
	4.4	18.4	0.3	1.3	calculated using results of ref. 24	32
	5.7	23.9	-	-	in gas phase	44
Propene	5.6	23.4	-	-	in gas phase	21
1.3.1 Acrylic derivatives						
Acrylic acid,						
--, butyl ester	2.1	8.8	0	0		8
	12.5	52.3	17.6	73.7	at 20% conversion	26
--, methyl ester	7.1	29.7	5.3	22.2		12
	4.7	19.7	~ 0	~ 0		23
Acrylonitrile	4.1	17.2	5.4	22.6	solvent: water	25
	3.88	16.2	3.70	15.5	solvent: DMF	32
	6.4	26.8	~ 0	~ 0	solvent: DMF	37
1.3.2 Methacrylic derivatives						
Methacrylamide	3.7	15.5	4.0	16.7	solvent: water	34
Methacrylic acid,						
--, t-butyl ester	4.4	18.4	1.1	4.6		27
--, methyl ester	6.31	26.4	2.84	11.9		4
	4.4	18.4	0	0		7
	5.8	24.3	0.5	2.1		28
	5.0	20.9	0 to 4.0	0 to 16.7	termination by combination or disproportionation	30
	5.35	22.4	-	-		47
	4.5	18.8	-	-	in emulsion	31
	5.77±1.12	24.2±4.7	6.42±1.97	26.9±8.2	average (with 95% confidence limits) of all data in previous chapter	-
Methacrylonitrile	11.5	48.1	5.0	20.9		20
1.3.3 Vinyl Halogens						
Tetrafluoroethylene	4.16	17.4	3.25	13.6		46
	1.3	5.4	0.2	0.8		43
Vinyl bromide	7.1	30	3.9	16.3		42
Vinyl chloride	3.7	16	4.2	17.6		17
Vinylidene chloride	25	104	40	167	solvent: hexane	9
1.3.4 Vinyl esters						
Vinyl acetate	4.4	18	0	0		1
	7.32	30,6	5.24	21.9		5
	3.2	13	0	0		6
	4.2	18	< 1	< 4		18
	4.2	18	0	0		22
	9.4±0.8	39±3	-	-		38

Monomer	E_p [kcal/mol]	[kJ/mol]	E_t [kcal/mol]	[kJ/mol]	Remarks	References
1.4 Styrene derivatives						
Styrene	11.7	49	-	-	in emulsion	2
	6.5+1	27+4	2.8+1	12+4		3
	6.3	26	1.9	8.0		10
	7.76	32.5	2.37	9.92		11
	7.4;8.4	31;35	-	-	in emulsion	15
	13.0;14.1	54;59	-		in emulsion	19
	5.9	25	0.5	2		29
	17.6	73.7	-	-	in emulsion	36
	6.9	29	-	-		42
	10.42+1.41	43.6+5.9	8.87+7.72	37.1+32.3	average (with 95% confidence limits) of all data in previous chapter	-
--, p-bromo-	7.6	32	-	-		39
--, o-methyl-	13.9	58.2	-	-	in emulsion	36
--, p-methyl-	7.7	32	-	-	in emulsion	36
	7.7	32	-	-		39
1.5 Vinyl Heteroaromatics						
N-Vinylcarbazole	6.9	29	-	-		42
2-Vinylpyridine	8.0	33	5.0	21		40
N-Vinylpyrrolidone	7.1	30	1.6	6.7		41

References

1. G. M. Burnett, H.W. Melville; Nature 156, 661 (1945).
2. W. V. Smith, J. Am. Chem. Soc. 70, 3695 (1948).
3. C. H. Bamford, M. J. S. Dewar Proc. Roy. Soc. (London) A 192, 309 (1948).
4. M. S. Matheson, E. E. Auer, E. B. Bevilacqua, E. J. Hart, J. Am. Chem. Soc. 71, 497 (1949).
5. M. S. Matheson, E. E. Auer, E. B. Bevilacqua, E. J. Hart, J. Am. Chem. Soc. 71, 2610 (1949).
6. G. Dixon-Lewis, Proc. Roy. Soc. (London) A 198, 510 (1949).
7. M. H. Mackay, H. W. Melville; Trans. Faraday Soc. 45, 323 (1949).
8. H. W. Melville, A. F. Bickel, Trans. Faraday Soc. 45, 1049 (1949).
9. J. D. Burnett, H. W. Melville, Trans. Faraday Soc. 46, 976 (1950).
10. G. M. Burnett. Trans. Faraday Soc. 47, 772 (1950).
11. M. S. Matheson, E. E. Auer, E. B. Bevilacqua, E. J. Hart, J. Am. Chem. Soc. 73, 1700 (1951).
12. M. S. Matheson, E. E. Auer, E. B. Bevilacqua, E. J. Hart, J. Am. Chem. Soc. 73, 5395 (1951).
13. D. H. Volman. J. Chem. Phys. 19, 668 (1951).
14. M. Morton, P. P. Salatiello, H. Landfield, J. Polymer Sci. 8, 215 (1952).
15. M. Morton, P. P. Salatiello, H. Landfield, J. Polymer Sci. 8, 279 (1952).
16. D. H. Volman, W. M. Graven, J. Am. Chem. Soc. 75, 3111 (1953).
17. G. M. Burnett, W. W. Wright, Proc. Roy. Soc. (London) A 221, 41 (1954).
18. W. J. Bengough, H. W. Melville, Proc. Roy. Soc. (London) A 230, 429 (1955).
19. E. Bartholome, H. Gerrens, R. Herbeck, H. Weitz, Z. Elektrochem. 60, 334 (1956).
20. N. Grassie E. Vance, Trans. Faraday Soc. 52, 727 (1956).
21. L. C. Landers, D. H. Volman. J. Am. Chem. Soc. 79, 2996 (1957).
22. W. J. Bengough, Trans. Faraday Soc. 54, 868 (1958).
23. Z. A. Sinitsyna, Kh. S. Bagdasaryan, Zh. Fiz. Khim. 32, 1319 (1958).
24. Z. Laita, Z. Machacek, J. Polymer Sci. 38, 459 (1959).
25. F. S. Dainton, R. S. Eaton J. Polymer Sci. 39, 313 (1959).
26. W. J. Bengough, H. W. Melville, Proc. Roy. Soc. (London) A 249, 445 (1959).
27. D. H. Grant, N. Grassie, Trans. Faraday Soc. 55, 1042 (1959).
28. P. Hyden, H. Melville. J. Polymer Sci. 43, 201 (1960).
29. G. Henrici-Olive, S. Olive, Makromol. Chem. 37, 71 (1960).
30. G. V. Schulz, G. Henrici-Olive, S. Olive, Z. Physik. Chem. (Frankfurt), 27, 1 (1960).
31. P. E. M. Allen, G. M. Burnett, J. M. Downer, J. R. Majer, Makromol. Chem. 58, 169 (1962).
32. W. Rabel, K. Ueberreiter, Ber. Bunsenges. 67, 710 (1963).
33. N. Colebourne, E. Collinson, D. J. Currie, F. S. Dainton, Trans. Faraday Soc. 59, 1357 (1963).
34. F. S. Dainton, W. D. Sisley, Trans Faraday Soc. 59, 1369 (1963).
35. M. Morton, W. E. Gibbs, J. Polymer Sci. A 1, 2679 (1963).
36. K- P. Paoletti, F. W. Billmeyer, Jr., J. Polymer. Sci. A 2, 2049 (1964).
37. A. F. Revzin, Kh. S. Bagdasaryan, Zh. Fiz. Khim. 38, 1020 (1964).
38. T. A. Berezsnich-Földes, F. Tüdös, Vysokomolekul. Soedin. 6, 1529 (1964).
39. M. Imoto, M. Kinoshita, M. Nishigaki, Makromol. Chem. 86, 217 (1965).
40. W. I. Bengough, ,W. Henderson, Trans. Faraday Soc. 61, 141 (1965).
41. V. A. Agasandyan, E. A. Trosman, Kh. S. Bagdasaryan, A. D. Litmanovich, V. Ya. Shtern, Vysokomolekul. Soedin. 8, 1580 (1966).
42. J. Hughes, A. M. North, Trans. Faraday Soc. 62, 1866 (1966).
43. A. N. Plyusnin, N. M. Chirkov, Teoret. i Eksp. Khimiya 2, 777 (1966).
44. S. Machi, S. Kise, M. Hagiwara, T. Kagiya, J. Polymer Sci. A-1, 5, 3115 (1967).
45. F. Hrabák, M. Bezděk, V. Hynková, Z. Pelzbauer, J. Polymer Sci. C 16, 1345 (1967).
46. É. F. Nosov, Kinetika i Kataliz 8, 680 (1967).
47. J. P. Fischer G. V. Schulz, Ber. Bunsenges. Physik. Chem. 74, 1077 (1970).

ACTIVATION VOLUMES OF POLYMERIZATION REACTIONS

G. Luft

Institut für Chemische Technologie

Technische Hochschule

Darmstadt, Germany

Contents

A. INTRODUCTION

The activation volume is included in the pressure dependence of the reaction rate constant (1-4):

$$(\partial \ln k/\partial p)_T = - \Delta v^\star/RT + \frac{1}{V} (\partial V/\partial p)_T \tag{1}$$

k = reaction rate constant; p = pressure; T = temperature; V = volume; $\Delta v^\star$ = activation volume (analogous to the activation energy).

It is the difference between the partial molar volume of the activated complex, as formed in the transition state theory by the efficient collision of molecules, and those of the initial reactants. The sign of the activation volume depends on the type of the chemical reaction. If new bonds are formed in the transition state, the activated complex is larger than the initial species, and hence the activation volume is positive. Inversely the activation volume is negative if bonds are stretched and broken.

For a rough calculation of the activation volume, we can assume that the formation of new bonds as well as stretching occurs along the axis of a cylinder, whose constant cross section is determined by the van der Waal's radii $dm_{A/2}$, $dm_{B/2}$, and $dm_{C/2}$ of the atoms A, B, C:

$$\Delta v^\star = \frac{\pi}{8} N [(d_{mA}^2 + d_{mB}^2)(\Delta_{AB}) + (d_{mB}^2 + d_{mC}^2)(\Delta_{B,C})] \tag{2}$$

with Δ_{AB} = (0.10 to 0.35) · d_{mAB} and $\Delta_{B,C}$ = (1.10 to 1.35) d_{mBC} - $d_{mB,C}$ (see (2, 5, 6)).

N = Avogadro number; d_{mAB}, d_{mBC} = bond length (values see (7)); $d_{mB,C}$ = interatomic distance at the minimum of potential (sum of van der Waal's radii); $d_{mA/2}$, $d_{mB/2}$, see (8).

The stretching of the bonds in the transition state is about 0.01 to 0.1 nm, whereas the change in the distance $d_{mB,C}$ between the unbonded atoms is usually larger (up to 0.25 nm).

If the decomposition of the activated complex causes only small changes in the bond length or in the distance between the atoms, its partial molar volume differs from that of the activated complex only slightly. Hence we can assume:

$$\Delta v^\star \cong \Delta v$$

where Δv = excess of the partial molar volume of the reaction products over the partial molar volumes of the initial species.

This simplification is valid for reactions in which products of cyclic structure are formed, as in the dimerization of cyclopentadiene with $\Delta v^\star \approx \Delta v$ = - 30 cm³/mol (9, 10).

A second term is added to the volume-change of the reacting molecules in the transition state if a polar solvent is involved. This term takes into account the change in the packing density of the surrounding solvent molecules, due to the arising or disappearing of electrostatic charges between the solvent and the reactant species. This volume change is negative in bimolecular association, positive in unimolecular dissociation and tends therefore to counteract the effect of volume change during the reaction. The activation volume can be determined according to eqn. (1) from rate measurements at different pressures and constant temperature using a semi-log plot of reaction rate constant versus pressure. In this evaluation the value of $(\partial V/\partial p)_T/V$ can be often neglected, because the compressibility of fluid reactants and compressed gases is generally small at pressures above 1 to 2 kbar. Experimentally determined activation volumes of some polymerization reactions are listed in Section B. The values are more or less negative, hence the rate of polymerization increases with pressure.

The overall activation volume is composed of the activation volumes of the different polymerization steps, initiation or initiator decomposition, chain propagation and chain termination:

$$\Delta v^\star = \Delta v^\star_p + \Delta v^\star_d/2 - \Delta v^\star_t/2 \tag{3}$$

$\Delta v^\star_p$ = activation volume of chain propagation; $\Delta v^\star_d$ = activation volume of initiation resp. initiator decomposition; $\Delta v^\star_t$ = activation volume of chain termination.

The activation volumes of initiator decomposition (values in section C) are always positive because this reaction is a unimolecular dissociation, in which a bond (e.g. the O-O bond in peroxides, peresters or the N=N bond in azo-compounds) is stretched in the transition state and finally broken. Aliphatic or alicyclic hydrocarbons, when used as solvents, can influence the activation volume.

In the chain propagation reaction, the decrease in the distance between the radical and the monomer molecule is greater than the increase in length of the double bond of the monomer. Hence the activation volume is negative (e.g. $\Delta v^\star_p$ = - 16 to -20 cm^3/mol in the free radical polymerization of ethylene).

The activation volumes of chain termination are also negative, those of the combination reaction being larger (ethylene polymerization: $\Delta v^\star_{t\ comb.}$ = -15 to -20 cm^3/mol) than those of the disproportionation reaction (ethylene polymerization: $\Delta v^\star_{t\ disp.}$ = -4 to -10 cm^3/mol).

The chain transfer of a radical to the monomer, to a "dead" polymer or to a modifier molecule is favoured by the pressure, which means that the activation volumes should be negative, whereas the activation volume of the intramolecular transfer (by "back biting"), which initiates the formation of short side chains, can be negative or positive. The activation volumes of these reactions are determined from the change in short and long chain branching or from the change with pressure of the mean molecular weight of the formed polymers.

Under the assumption that the monomer and the polymer concentration do not change over a wide range and that the temperature is kept constant, the following relation can be derived from kinetic considerations:

$$\Delta v^\star_p - \Delta v^\star_{tr} = RT \cdot \left(\frac{\partial \ln \cdot VZG}{\partial p} \right)_T \tag{4}$$

VZG = number of branches; $\Delta v^\star_{tr}$ = activation volume of the chain transfer reaction.

Analogously the influence of the pressure on the chain transfer to a modifier can be described by the expression

$$\Delta v^\star_p - \Delta v^\star_{tr} = RT \cdot \left(\frac{\partial \ln (k_{tr}/k_p)}{\partial p} \right)_T = RT \cdot \left(\frac{\partial \ln(1/\overline{P} - 1/\overline{P}_{no})}{\partial p} \right)_T \tag{5}$$

P = number average molecular weight obtained in the presence of a transfer agent; P_{no} = number average molecular weight obtained in the absence of a transfer agent.

Because of the mostly negative sign of the activation volumes (listed in Section D), the radical transfer to the modifier molecule is often slightly decreased at high pressures.

The activation volume of degradation reactions which can occur at high temperatures, should be positive because bond cleavage increases the volume of the activated complex more than the formation of double bonds from single bonds decreases it.

The difference in the activation volumes of the various component reactions results in a different influence of the pressure on the rate constants and leads to modified polymers. Because of its larger negative activation volume the chain propagation reaction is more favoured by high pressures than the termination. The chain length and consequently the tensile strength as well as the tensile modulus increase with pressure:

$$\left(\frac{\partial \ln \nu}{\partial p} \right)_T = - \frac{1}{RT} [\Delta v^\star_p - \frac{1}{4} (3 \Delta v^\star_d + \Delta v^\star_t)] \tag{6}$$

ν = chain length

The small negative activation volume of the intramolecular chain transfer shows, that this reaction step, which determines the short chain branching, is favoured by high pressures; that means that polymers with high density and crystallinity can be synthesized at high pressures. The increase with pressure of chain termination by combination relative to disproportionation, which can be deduced from the ratio of the respective activation volumes, shifts the molecular weight distribution toward higher molecular weights.

Similar considerations also show that the composition of copolymers and thus their molecular structure is influenced by the synthesis pressure. The pressure dependence of the copolymerization parameters r_1 and r_2, which determine the composition of a copolymer, is expressed by the relationship

$$\left(\frac{\partial \ln(r_1/r_2)}{\partial p} \right)_T = - \frac{1}{RT} (\Delta v^\star_{P11} - \Delta v^\star_{P12} - \Delta v^\star_{P22} + \Delta v^\star_{P21}) \tag{7}$$

$$\approx - \frac{1}{RT} \cdot (\Delta v^\star_{P11} - \Delta v^\star_{P22})$$

$\Delta v^\star_{P11}, \Delta v^\star_{P22}$ = activation volume of the propagation reaction between a monomer molecule and a radical formed from the same monomer

$\Delta v^\star_{P12}, \Delta v^\star_{P21}$ = activation volume of the propagation reaction between a monomer molecule and a radical formed from the comonomer.

Eqn. (7) is valid, if the composition of the initial reaction mixture does not change appreciably with pressure. According to eqn. (7), an increase in pressure favours the inclusion of monomers, which in homopolymerization show a large negative activation volume (e.g. substituted olefins). Values of copolymerization parameters obtained at different pressures and activation volumes of some copolymerization reactions are listed in Section E.

At very high pressures the increase of the raction rate constant at negative activation volume is retarded by the lower mobility of the molecules, due to the increased viscosity. This effect was first measured by Hamann (11) in an examination of the alkaline etherification of ethyl bromide at pressures up to 40 kbar. It was also found in the polymerization of styrene by Nicholson and Norrish (12).

In order to obtain the reaction rate as an explicit function of pressure, eqn. (1) can be integrated, neglecting the compressibility and assuming that the activation volume does not change greatly with pressure (13):

$$k_p = k_o \cdot e - \frac{\Delta v^\star}{RT} (p - p_o) \tag{8}$$

k_p = reaction rate constant at pressure p; k_o = reaction rate constant at reference pressure p_o.

References page II-458

According to eqn. (8) the reaction rate constant increases exponentially with pressure if the activation volume is negative and decreases if $\Delta v^\star$ is positive. Eqn. (7) is valid only in the high pressure range ($p > 500$ bar). At low pressures the compressibility cannot be neglected anymore, and the activation volume changes noticeably with pressure, especially in the neighbourhood of the critical point. Simmons and Mason (14) studied the dimerization of chlorotrifluoroethylene at pressures up to 100 bar. They found that the value of the negative activation volume decreases first with pressure and then increases rapidly. It has a maximum at the critical point and after decreasing again it approaches asymptoticly a constant value.

The authors describe the pressure dependence of the rate constant, taking account of the partial molar volume of the initial reactants as well as that of the activated complex by the use of suitable equations of state (e.g. Redlich-Kwong or virial coefficient equations).

In order to appreciate the influence of pressure on the reaction rate constant one can compare it with the influence of the temperature: An increase in pressure (activation volume $\Delta v^\star = -25$ cm^3/mol) at a temperature of 50°C from 1 bar to 4.5 kbar, corresponds to a temperature increase (activation energy $E = 84$ kJ/mol) from 50 to 105°C.

B. ACTIVATION VOLUMES OF SOME POLYMERIZATION REACTIONS

Monomer	Solvent	Initiator	Pressure abs. [bar]	Temperature [$^\circ$C]	Activation Volume $\Delta v^\star$ [cm^3/mol]	Ref.
Acenaphthylene	toluene	AIBN[**]	1 - 3900	60	-6	15
Acrylic acid anhydride	dimethylformamide	AIBN	2500 - 4000	50	-14	50
Acrylonitrile	dimethylformamide	AIBN	1 - 2000	50	-22	51
Allyl acetate	-	benzoyl peroxide	1 - 8500	80	-13.4	16
3,3-Bis(chloromethyl) oxetane	toluene	borontrifluoride/diethyl ether	1 - 5000	60	-16	54
2-Cyclopropylpropene-1	-	AIBN	1 - 14000	70	-23	52
	-	stannic chloride	1 - 14000	20	-15	52
Diallylcyanamide	toluene	tert.-butyl perbenzoate	1000 - 6000	80	-14	53
3,3-Dichloropropene-1	-	dicyclohexyl peroxydicarbonate	1 - 1000	40	0	56
	-	dicyclohexyl peroxydicarbonate	1000 - 10000	40	-8	56
1,1-Dichloro-2-vinylcyclopropane	-	AIBN	1 - 10000	80	-35	55
	-	stannic chloride	1 - 10000	20	-52	55
Diethyl fumarate	-	AIBN	1 - 3000	60	-14.7	17
3,3,3-Difluorochloropropene-1	-	dicyclohexyl peroxydicarbonate	1 - 14000	30	-15	57
1,1-Difluoro-2-vinylcyclopropane	-	AIBN	1 - 10000	50	-15	58
	-	stannic chloride	1 - 10000	20	-45	59
Ethylene	-	AIBN	3000 - 7600	50 - 70	-3 to -6	18
	-	di-tert-butyl peroxide	750 - 2500	129	-20 to -23	19
	-	tert-butyl peroctoate	1100 - 2200	190	-18 to -22	60
	-	peroxide	800 - 2000	225 - 255	-20 to -26	72
	-	oxygen	1100 - 5500	170 - 240	-12 to -18	20
	-	dicyclohexyl peroxydicarbonate	400 - 1900	80	-10	73
1,2-Epoxycyclohexane	-	benzoyl peroxyde	8400 - 12500	60	-13	54
Indene	heptane	benzoyl peroxide		64	-21.2	21
Isoprene	-	oxygen	8000 - 14000		-11.0	22
	-	benzoyl peroxide	7000 - 14000	20 - 30	-18.0 to -7.9	22
	ethyl acetate	hydrogen iodide/water[***]		20	-9.8	23
	dichloromethane	hydrogen iodide/water		20	-10.8	23
	toluene	hydrogen iodide/water		20	-7.9	23
Methyl methacrylate	-	peroxide	1 - 3000	40	-19	24, 25
	acetone	peroxide	1 - 1000	50	-26	26
	benzene	peroxide	1 - 1000	40	-17	26
	benzene	peroxide	1 - 1000	50	-19	26
	benzene	peroxide	1 - 1000	60	-21	26
	benzene	peroxide	1 - 1000	70	-23	26
	butyraldehyde	AIBN	1 - 4000	50	-25	61
	carbon tetrachloride	AIBN	1 - 1000	50	-19	26
	1,2-dichloroethane	AIBN	1 - 1000	50	-21	26
	ethanol	AIBN	1 - 1000	50	-20	26
	ethyl acetate	peroxide	1 - 1000	50	-24	26
	n-hexane	peroxide	1 - 1000	50	-19	26
	trichloromethane	peroxide	1 - 1000	50	-22	26
α-Methylstyrene	-		2500 - 3500	60	-18	17

Monomer	Solvent	Initiator	Pressure abs. [bar]	Temperature [°C]	Activation Volume $\Delta v^\star$ [cm³/mol]	Ref.
Propylene	-	thermal and radiation	5000 - 16000	21 - 83	-9.6 to -12.2	27
Styrene	-	AIBN and peroxide	1 - 2930	60	-18	28
	butyraldehyde	AIBN	1 - 4000	50	-22	62, 64
	benzene	peroxide	1 - 1000	50	-18	29
	benzene	AIBN	1 - 3000	50	-23	62, 63
	para-butyraldehyde	AIBN	1 - 4000	50	-25	64
	tetrachloroethylene	peroxide	1 - 2680	60	-16	30
	toluene	peroxide	1 - 2930	50	-17	28
	triethylamine	peroxide	1 - 4400	60	-17	28
	triethylamine	benzoyl peroxide	1 - 3000	30	-18	31
	triethylamine	benzoyl peroxide	1 - 3000	40	-11.5	32
	triethylamine	thermal	1000 - 6000	72	-17.5	33
	water****	200	2000 - 5000	60	-11.1	34
Tetrahydrofuran	-	borontrifluoride/diethyl ether	2500 - 5400	60	-22	54
Vinyl acetate	acetone	AIBN	1 - 4000	40	-19	61
	butyraldehyde	AIBN	1 - 4000	40	-21	61
	carbon tetrachloride	AIBN	1 - 3000	40	-13	35
	trichloroethylene	AIBN	1 - 4000	65	-10.5	64

** Azobisisobutyronitrile.   *** Cationic polymerization.   **** Emulsion polymerization.

C.   ACTIVATION VOLUMES OF INITIATOR DECOMPOSITION

Initiator	Solvent	Pressure abs. [bar]	Temperature [°C]	Activation Volume $\Delta v_d^\star$ [cm³/mol]	Ref.
AIBN**	toluene	1 - 1500	70	13.1	36
	toluene	1 - 10000	62.5	3.8*** to 9.4****	36
Benzoyl peroxide	acetophenone	1 - 6500	80	4.8	37
	allyl acetate	1 - 5500	80	4.7	16
	carbon tetrachloride	1 - 3000	60	9.7	38
	carbon tetrachloride	1 - 3000	70	8.6	38
di-tert-butyl peroxide	benzene	1 - 7300	120	12.6	39
	carbon tetrachloride	1 - 7300	120	13.3	39
	cyclohexane	1 - 7300	120	6.7	39
	toluene	1 - 7300	120	5.4	39
Pentaphenylethane	toluene	1 - 1500	60	7	36
	toluene	1000 - 2700	50	5	36

** Azo-bisisobutyronitrile.   *** Photometric analysis.   **** By I_2 scavenger technique.

D.   ACTIVATION VOLUMES OF CHAIN TRANSFER REACTIONS

Monomer	Transfer Agent	Pressure abs. [bar]	Temperature [°C]	Activation Volume [cm³/mol] $\Delta v_{tr}^\star$	$\Delta v_p^\star - \Delta v_{tr}^\star$	Ref.
Acrylonitrile	butyraldehyde	1 - 4000	50		-6.6	61
3,3-Dichloropropene-1	3,3-dichloropropene-1	1	50		-8	61
Ethylene	acetone	1360 - 2380	130		-1.1	40
	butane	1360 - 2380	130		-3.2	40
	butanone	1360 - 2380	130		-2.9	40
	cyclohexane	1360 - 2380	130		-1.5	40

Monomer	Transfer Agent	Pressure abs. [bar]	Temperature [°C]	$\Delta v^{\star}_{tr}$	$\Delta v^{\star}_{p} - \Delta v^{\star}_{tr}$	Ref.
				Activation Volume [cm³/mol]		
Ethylene (Cont'd.)	cyclopentane	1360 - 2380	130		-4.7	40
	ethanol	1360 - 2380	130		-3.1	40
	ethylbenzene	1360 - 2380	130		-6.7	40
	isopropanol	1360 - 2380	130		-6.1	40
	propane	1360 - 2380	130		-3.1	40
	p-xylene	1360 - 2380	130		7.8	40
	toluene	1360 - 2380	130		4.7	40
Methyl methacrylate	butyralodehyde	1 - 4000	50		-6.6	66
Styrene	butyraldehyde	1 - 4000	50		-15	62, 64, 67
	carbon tetrachloride	6000	60	-11		32
	isobutyraldehyde	1 - 4000	50		-6.6	67
	trithylamine	1810 - 4400	60	-2 to -3		41
Vinyl acetate	butyraldehyde	1 - 4000	50		-3.2	66
Vinyl chloride	butyraldehyde	1 - 4000	50		-3.8	66

E.   INFLUENCE OF PRESSURE ON COPOLYMERIZATION
1.   COPOLYMERIZATION PARAMETERS

Monomer 1	Monomer 2	Solvent	Pressure abs. [bar]	Temp. [°C]	r_1	r_2	Ref.
					Copolymerization Parameters		
Acrylonitrile	ethyl cinnamate	-	1	60	2.4	0.17	68
			800		2.8	0.21	
			2800		3.9	0.36	
Ethylene	vinyl acetate	-	510	120	0.6	0.95	40
			1020 - 2040		0.82	0.99	
	vinyl chloride	-	340	74	0.16	2.7	40
			1020		0.16	2.1	
			2040		0.21	2.0	
Maleic anhydride	acenaphthylene	acetic anhydride	1	60	0.07	0.43	42
			2880		0.10	0.32	
Methyl methacrylate	acenaphthylene	toluene	1	60	0.57	3.1	42
			1900		0.60	1.9	
			3900		0.58	1.6	
	acrylonitrile	toluene	1	70	1.34	0.12	43
			100		1.46	0.37	
			1000		2.01	0.45	
Styrene	acrylic acid	-	1	60	0.25	0.07	68
			1000		0.27	0.08	
			2500		0.31	0.095	
	acrylonitrile	toluene	1	70	0.37	0.07	43
			100		0.43	0.13	
			1000		0.55	0.14	
	diethyl fumarate	benzene	1	60	0.26	0.06	69
			100		0.29	0.09	
			1000		0.32	0.15	
	diethyl maleate	benzene	1	60	6.53	0.01	69
			100		7.08	0.02	
			1000		9.50	0.02	
	ethyl methacrylate	benzene	1	50	0.56	0.50	44
			100		0.62	0.55	
			1000		0.72	0.67	
	methyl methacrylate	-	1	60	0.704	0.159	68
			1000		0.710	0.163	
			3000		0.718	0.171	
	vinyl acetate	benzene	1	60	44	0.01	45
			100		46	0.01	
			1000		47	0.01	

Monomer 1	Monomer 2	Solvent	Pressure abs. [bar]	Temp. [°C]	Copolymerization Parameters		Ref.
					r_1	r_2	
Styrene (Cont'd.)	vinylidene chloride	-	1	60	2.6	0.04	46
			1500		2.5	0.04	
			2000		2.5	0.01	
	vinyl-2-pyridine	-	1	60	0.540	1.135	68
			2000		0.560	1.145	
			3000		0.605	1.165	
Vinyl acetate	vinylidene chloride	-	1	60	0.025	3.2	46
			1500		0.025	3.1	
			2500		0.03	3.1	
Vinylidene chloride	indene	-	1	60	0.475	0.205	68
			2000		0.500	0.215	
			4000		0.525	0.225	

2. ACTIVATION VOLUMES

Monomer 1	Monomer 2	Solvent	Pressure abs. [bar]	Temp. [°C]	Activation Volume [cm³/mol]			Ref.
					$\Delta v^\star$	$\Delta v^\star_{11} - \Delta v^\star_{12}$	$\Delta v^\star_{22} - \Delta v^\star_{11}$	
Ethylene[★★]	vinyl acetate	-	1100 - 1900	240	-36			71
Indene[★★★]	acrylonitrile	heptane		50	-17.5	0	-4.3	47
	acrylonitrile	dioxane		50	-18.0	0	-3.6	47
	methyl vinyl ketone	heptane		50	-24.0	-2.5	-1.0	48
	methyl methacrylate	heptane		50	-20.1	0	0	47
	vinylidene chloride				-17.8 to -19.1			70
Isoprene[★★★]	acrylonitrile	heptane		50	-27.0	-	-	49
Methyl methacrylate	maleic anhydride				-18 to -25			70
Styrene	diethyl fumarate				-18 to -25			70
	methyl acrylate				-13.8 to -14.5			70
	methyl metnacyrlate				-18 to -25			70
	vinyl-2-pyridine				-23.5 to -26			70

[★★] Initiator: Oxygen.   [★★★] Initiator: Benzoyl peroxide

F. REFERENCES

1. M. G. Evans, M. Polanyi, Trans. Faraday Soc. 32, 1333 (1936).
2. S. D. Hamann, "High Pressure Physics and Chemistry," London 1963.
3. M. G. Gonikberg, "Chemical Equilibrium and Reaction Rates at High Pressures," Jerusalem 1963.
4. G. Luft, Dissertation, Darmstadt 1967.
5. A. E. Stearn, H. Eyring, Chem. Rev. 29, 509 (1941).
6. G. Luft, Chem. Ing. Techn. 41, 712 (1969).
7. Landolt-Boernstein, "Zahlenwerte und Funktionen aus Physik und Chemie," Berlin 1951.
8. W. Kleinpaul, Z. Physik. Chem. (Frankfurt) 29, 201 (1961), 30, 262 (1961).
9. B. Raistrick, R. H. Sapiro, D. M. Newitt, J. Chem. Soc. (London) 1761 (1939).
10. M. G. Gonikberg, L. F. Vereshchagin, Zh. Fiz. Khim. 23, 1447 (1949).
11. S. D. Hamann, Trans. Faraday Soc. 54, 507 (1958).
12. A. E. Nicholson, R. G. W. Norrish, Discussions Faraday Soc. 22, 104 (1956).
13. R. Steiner, G. Luft, Chem. Eng. Sci. 22, 537 (1967).

14. G. M. Simmons, D. M. Mason, Chem. Eng. Sci. 27, 89 (1972).
15. M. N. Romanii, K. E. Weale, Trans. Faraday Soc. 62, 2264 (1966).
16. C. Walling, J. Pellon, J. Am. Chem. Soc. 79, 4782 (1957).
17. R. L. Hemmings, K. E. Weale, "Chemical Reactions at High Pressures," Spon. London 1967.
18. A. L. Shriers, B. F. Dodge, R. H. Bretton, "Free Radical Polymerization of Ethylene at High Pressures," Paper presented on the A. I. Ch. E./J. Chem. Eng. Symposium, London 1965.
19. R. O. Symrox, P. Ehrlich, J. Am. Chem. Soc. 84, 531 (1961).
20. D. Lierse, G. Luft, unpublished
21. T. Yala, G. Jenner, unpublished
22. G. Jenner, Thesis, Strasbourg 1966 and Bull. Soc. Chim. France 1, 344 (1967).
23. H. Abdi, G. Jenner, C. Brun, Makromol. Chem., in press
24. J. A. Lamb, K. E. Weale, Symp. Phys. Chem. High Pressures 229 (1963), (Soc. Chem. Ind., London).
25. A. A. Zharov, N. S. Enikolopyan, Zh. Fiz. Khim. 38, 2727 (1964).
26. H. Asai, T. Imoto, J. Chem. Soc. (Japan) 66, 863 (1963).

27. D. W. Brown, L. A. Wall, J. Phys. Chem. 67, 1016 (1963).
28. A. C. Toohey, K. E. Weale, Trans. Faraday Soc. 58, 2439 (1962).
29. H. Asai, T. Imoto, J. Chem. Soc. (Japan) 66, 871 (1963).
30. K. Salahuddin, K. E. Weale in K. E. Weale, "Chemical Reactions at High Pressures," Spon, London 1967.
31. A. E. Nicholson, R. G. W. Norrish, Discussions Faraday Soc. 22, 104 (1956).
32. C. Walling, J. Pellon, J. Am. Chem. Soc. 79, 4776 (1957).
33. P. P. Kobeko, E. U. Kuvshinskii, A. S. Semonova, Zh. Fiz. Khim. 24, 345 (1950).
34. F. M. Merrett, R. G. W. Norrish, Proc. Roy. Soc. (London) A 206, 309 (1951).
35. V. M. Zhulin, M. G. Gonikberg, R. F. Barkova, Izv. Akad. Nauk. SSSR 432 , 1965).
36. A. H. Ewald, Discussions Faraday Soc. 22, 138 (1956).
37. C. Walling, J. Pellon, J. Am. Chem. Soc. 79, 4786 (1957).
38. A. H. Ewald, Discussions Faraday Soc. 22, 146 (1956).
39. C. Walling, G. Metzger, J. Am. Chem. Soc. 81, 5365 (1959).
40. P. Ehrlich, G. Mortimer, in "Advances in Polymer Science," Vol. 7, No. 3, 420 (1970).
41. A. C. Toohey, K. E. Weale, Trans. Faraday Soc. 58, 2446 (1962).
42. M. N. Romani, K. E. Weale, Thesis, London University 1964.
43. R. D. Burkhart, N. L. Zutti, J. Polymer Sci. 57, 793 (1962).
44. H. Asai, T. Imoto, J. Chem. Soc. (Japan) 85, 155 (1964).
45. ib. 85, 152 (1964).
46. B. S. Maulik, K. E. Weale, in K. E. Weale, "Chemical Reactions at High Pressures," Spon. London 1967.
47. T. Yala, G. Jenner, Communication 10. Meeting EHPRG, Strasbourg 1972.
48. T. Yala, Thesis 3. Cycle, Strasbourg 1972, will be published
49. G. Jenner, J. Rimmelin, unpublished
50. J. P. J. Higgins, K. E. Weale, J. Polymer Sci. A-1, 8, 1705 (1970).
51. R. I. Baikova, V. M. Zhulin, M. G. Gonikberg, Izv. Akad. Nauk SSSR, Ser. Khim. 154 (1966).
52. V. M. Zhulin, A. R. Volchek, M. G. Gonikberg, A. S. Shashkov, S. V. Zotova, Vysokomolekul. Soedin. 7A, 1513 (1972).

53. J. P. J. Higgins, K. E. Weale, unpublished
54. S. A. Mehdi, K. E. Weale, unpublished
55. A. R. Volchek, V. M. Zhulin, A. S. Shashkov, O. M. Nefedov, A. A. Ivashenko, Vysokomolekul. Soedin., in press
56. V. M. Zhulin, N. V. Klimentova, M. G. Gonikberg, V. V. Korshak, A. P. Suprun, Vysokomolekul. Soedin. 11A, 101 (1969). V. M. Zhulin, N. V. Klimentova, A. P. Suprun, V. N. Zagorbinina, Vysokomolekul. Soedin., in press
57. V. M. Zhulin, A. P. Suprun, G. P. Lopatina, T. A. Soboleva, A. S. Shashkov, M. G. Gonikberg, G. P. Shakhovskoi, Vysokomolekul. Soedin. 13A, 2518 (1971).
58. V. M. Zhulin, M. G. Gonikberg, A. R. Volchok, O. M. Nefedov, A. S. Shashkov, Vysokomolekul. Soedin. 9A, 2153 (1971).
59. A. R. Volchek, V. M. Zhulin, A. S. Shashkov, O. M. Nefedov, A. A. Ivashenko, Vysokomolekul. Soedin., in press
60. G. Luft, H. Bitsch, unpublished
61. V. M. Zhulin, Thesis for Doctor of Science degree, Moscow (1971).
62. V. M. Zhulin, M. G. Gonikberg, V. N. Zagorbinina, Izv. Akad. Nauk, SSSR, Ser. Khim. 827 (1966).
63. V. M. Zhulin, R. I. Baikova, M. G. Gonikberg, G. P. Shakhovskoi, Vysokomolekul. Soedin. 13, 1071 (1971).
64. V. M. Zhulin, M. G. Gonikberg, V. N. Zagorbinina, Izv. Akad. Nauk SSSR.
65. M. G. Gonikberg, R. I. Baikova, V. M. Zhulin, Izv. Akad. Nauk SSSR, 1164 (1962).
66. V. M. Zhulin, M. G. Gonikberg, A. L. Goff, V. N. Zagorbinina, Vysokomolekul. Soedin. 11A, 777 (1969).
67. V. M. Zhulin, M. G. Gonikberg, V. N. Zagorbinina, Dokl. Akad. Nauk SSSR, 163, 106 (1965).
68. W. F. Dellsperger, K. E. Weale, ACS Polymer Preprints 11, 645 (1970).
69. H. Asai, J. Chem. Soc. (Japan) 85, 247 (1964).
70. J. A. Lamb, Thesis, London University, 1961.
71. H. Bitsch, G. Luft, Angew. Makromol. Chem. 32, 17 (1973).
72. Th. J. van der Molen, Preprint IUPAC Symp. on Macromolecular Chemistry, Budapest 1969.
73. Th. J. van der Molen, International Symp. on Macromolecules, Helsinki 1972.

ACTIVATION ENTHALPIES AND ENTROPIES OF STEREOCONTROL IN FREE RADICAL POLYMERIZATIONS

H. -G. Elias

Midland Macromolecular Institute

Midland, Mich.

Contents

A. INTRODUCTION

The tacticity of the resulting polymer depends in free radical polymerization on the propagation step, i.e. on stereocontrol by the propagating end. The following simple cases exist:

(1)   The last monomer unit does not control the stereospecifity of the propagation step. Isotactic and syndiotactic additions thus exhibit the same probabilities ($p_i = p_s$). The mole fractions of isotactic and syndiotactic diads are equal ($X_i = X_s$). If all steps occur at random, the four probabilities for the formation of triads are equal ($p_{i/i} = p_{i/s} = p_{s/i} = p_{s/s}$) and so are the probabilities for tetrad, pentad etc. formations. This polymer is a true atactic polymer. It follows

$$p_i = X_i = 0.5; \qquad p_s = X_s = 0.5 \tag{1}$$

(2)   The last monomer unit, but not the last diad, controls the propagation step. The probabilities for the formation of isotactic and dyndiotactic diads are thus different ($p_i \neq p_s$). The process is Bernoullian with respect to the formation of diads. Consequently, the probability of forming an isotactic diad at an existing isotactic diad equals the probability of forming an isotactic diad at an existing syndiotactic diad ($p_{i/i} = p_{s/i}$), and by analogy, $p_{s/s} = p_{i/s}$. It follows (6, 8)

$$p_i = X_i \neq 0.5; \qquad p_s = X_s \neq 0.5 \tag{2}$$

and

$$X_{ii} = p_i^2 = X_i^2 \tag{3}$$

$$X_{ss} = p_s^2 = X_s^2 \tag{4}$$

$$X_{is} = p_i p_s + p_s p_i = 2 X_i X_s = 2 X_i (1-X_i) \tag{5}$$

where X_{is} is the mole fraction of heterotactic triads, regardless of whether formed by the formation of an isotactic unit at a syndiotactic one (probability $p_{s/i}$) or vice versa (probability $p_{i/s}$).

(3)   The last two monomer units, i.e. the last diad, regulate the stereocontrol of the propagation step. There are consequently four different probabilities, $p_{i/i} \neq p_{i/s} \neq p_{s/i} \neq p_{s/s}$. The mole fractions of isotactic (X_{ii}), syndiotactic (X_{ss}) and heterotactic triads (X_{is}) are thus given (11, 8)

$$X_{ii} = X_i p_{i/i} \tag{6}$$

$$X_{ss} = X_s p_{s/s} \tag{7}$$

$$X_{is} = X_i p_{i/s} + X_s p_{s/i} \tag{8}$$

This process corresponds to a Markov trial of first order.

(4)   Second order Markov statistics are present if the last two diads (i.e. the last triad or the last three monomer units) control the propagation step. In this case

$$X_{iii} = X_{ii} p_{i/i} \tag{9}$$

$$X_{sss} = X_{ss} p_{s/s} \tag{10}$$

$$X_{iis} = X_{ii}P_{i/s} + X_{si}P_{i/i} \tag{11}$$

$$X_{iss} = X_{is}P_{s/s} + X_{ss}P_{s/i} \tag{12}$$

$$X_{isi} = X_{is}P_{s/i} \tag{13}$$

$$X_{sis} = X_{si}P_{i/s} \tag{14}$$

In order to distinguish between Markov second order and Markov first order mechanisms, at least tetrads must be known.

The probability of the addition of a b diad to an existing end a is given by the corresponding rates

$$P_{a/b} = v_{a/b}/(v_{a/a} + v_{a/b}) \tag{15}$$

with b = i or s, a = i or s in the Markov first order case, and a = ii, is, si or ss in the Markov second order case.

Using these definitions and assuming stationary states and the equality of instantaneous and final diad and triad fractions, one can express ratios of diads, triads, and tetrads in terms of rate constants (Table 1) or ratios of rate constants in terms of diad, triad etc. fractions (Table 2). The rate constants themselves can be calculated if the rate constants of propagation are known for the corresponding experimental conditions (1).

From the temperature dependence of the expressions given in Table 2, activation enthalpies $(\Delta H_A^\star - \Delta H_B^\star)$ and activation entropies $(\Delta S_A^\star - \Delta S_B^\star)$ can be calculated via

$$\frac{k_A}{k_B} = \exp\left(\frac{\Delta S_A^\star - \Delta S_B^\star}{R}\right) \exp\left(\frac{\Delta H_A^\star - \Delta H_B^\star}{RT}\right) \tag{16}$$

At the present time, it is not clear whether the quantities $(\Delta H_A^\star - \Delta H_B^\star)$ and $(\Delta S_A^\star - \Delta S_B^\star)$ really reflect activation parameters (2,3) or conformation parameters (4,5) or combinations thereof.

Table 3 contains the differences $\Delta\Delta H^\star = \Delta H_A^\star - \Delta H_B^\star$ and $\Delta\Delta S^\star = \Delta S_A^\star - \Delta S_B^\star$ calculated from the temperature dependence of diad ratios X_i/X_s. In the case of Bernoullian trials, they represent $\Delta H_i^\star - \Delta H_s^\star$ and $\Delta S_i^\star - \Delta S_s^\star$ resp. For first order Markov trials, they stand for $\Delta H_{s/i}^\star - \Delta H_{i/s}^\star$ and $\Delta S_{s/i}^\star - \Delta S_{i/s}^\star$ resp. Both $\Delta\Delta H^\star$ and $\Delta\Delta S^\star$ in general depend on the solvent (9,10) and consequently on the monomer/solvent ratio (12), but are nearly independent of conversion (7).

Table 4 contains differences of activation enthalpies for other modes of addition than s/i vs i/s, and Table 5 contains the corresponding differences of activation entropies. Only a few of these activation parameters could be calculated from literature data, because of the very few reported triad fractions.

For any two different modes of addition, compensation effects exist between $\Delta\Delta H^\star$ and $\Delta\Delta S^\star$ (10-12, 43) for the polymerization of a given monomer in different solvents and/or in different monomer/solvent ratios. The compensation effects can be described by

$$\Delta H_A^\star - \Delta H_B^\star = \Delta\Delta H_o^\star - T_o(\Delta S_A^\star - \Delta S_B^\star) \tag{17}$$

The compensation temperature T_o has been found to be independent of the mode of addition (e.g. A = i/i vs B = i/s) within the limits of error (12,43). The compensation enthalpy $\Delta\Delta H_o^\star$ depends however on the mode of addition.

B. TABLES

1. Ratios of Some i-ads for Different Mechanisms (from (42))

i-ad ratio	Mechanism		
	Bernoulli	Markov 1st order	Markov 2nd order
X_i/X_s	$\dfrac{k_i}{k_s}$	$\dfrac{k_{s/i}}{k_{i/s}}$	$\dfrac{k_{ss/i}(k_{si/i} + k_{ii/s})}{k_{ii/s}(k_{is/s} + k_{ss/i})}$
X_{ii}/X_{ss}	$\dfrac{k_i^2}{k_s^2}$	$\dfrac{k_{i/i}k_{s/i}}{k_{s/s}k_{i/s}}$	$\dfrac{k_{ss}k_{si/i}}{k_{ii/s}k_{is/s}}$
X_{iii}/X_{sss}	$\dfrac{k_i^3}{k_s^3}$	$\dfrac{k_{i/i}^2 k_{s/i}}{k_{s/s}^2 k_{i/s}}$	$\dfrac{k_{ss/i}k_{si/i}k_{ii/i}}{k_{ii/s}k_{is/s}k_{ss/s}}$
X_{is}/X_{ss}	$\dfrac{k_i}{k_s}$	$\dfrac{2k_{s/i}}{k_{s/s}}$	

2. Ratios of Rate Constants for Markov First Order Mechanisms (from (43))

Ratio of rate constants	Expressions giving the ratios of rate constants in column 1
$\dfrac{k_{i/i}}{k_{i/s}}$	$\dfrac{2X_{ii}}{X_{is}}$
$\dfrac{k_{i/i}}{k_{s/i}}$	$\dfrac{2X_sX_{ii}}{X_iX_{is}} = \dfrac{1 + (2X_{ss}/X_{is})}{1 + (0.5X_{is}/X_{ii})}$
$\dfrac{k_{i/i}}{k_{s/s}}$	$\dfrac{X_sX_{ii}}{X_iX_{ss}} = \dfrac{1 + (0.5X_{is}/X_{ss})}{1 + (0.5X_{is}/X_{ii})}$
$\dfrac{k_{i/s}}{k_{s/i}}$	$\dfrac{X_s}{X_i} = \dfrac{X_{ss} - 0.5X_{is}}{X_{ii} - 0.5X_{is}}$
$\dfrac{k_{i/s}}{k_{s/s}}$	$\dfrac{X_sX_{is}}{2X_iX_{ss}} = \dfrac{1 + (0.5X_{is}/X_{ss})}{1 + (2X_{ii}/X_{is})}$
$\dfrac{k_{s/i}}{k_{s/s}}$	$\dfrac{X_{is}}{2X_{ss}}$

B. 3. $\Delta H^{\star}_{s/i} - \Delta H^{\star}_{i/s}$ and $\Delta S^{\star}_{s/i} - \Delta S^{\star}_{i/s}$ of Free Radical Polymerizations in Different Solvents

a) Mole ratio solvent/monomer; b) Mol-%; c) mol/l; d) g/ml; e) data directly from literature, mechanistic assumptions unknown; f) literature data, calculated under the assumption of Bernoulli statistics. In all cases, except those marked by IR, diad fractions were determined via NMR.

Monomer Compound	Vol.-%	Solvent	$\Delta H^{\star}_{s/i} - \Delta H^{\star}_{i/s}$ [J/mol]	$\Delta S^{\star}_{s/i} - \Delta S^{\star}_{i/s}$ [J mol^{-1} K^{-1}]	Ref.	Calculation Remarks
Acrylonitrile	100	bulk	0	0	13	-
	30	toluene	3768± 59	2.68±0.193	10, 14	-
Acrylic acid, isopropyl ester	?	?	0	-5.9	15	e)
Acrylic acid, methyl ester	?	?	0	-5.9	15	e)
α-Chloroacrylic acid, methyl ester	?	toluene	3768±188	2.51±0.59	16	-
Methacrylonitrile	100	bulk	251± 50	3.02±0.142	17	-
	100	bulk	-(4103± 71)	-(10.84±0.209)	18	-
Methacrylonitrile/SnCl$_4$	3:1 a)	toluene	-(628± 54)	-(0.67±0.176)	18	monomer/SnCl$_4$ = 2/1 mol/mol
	2:1 a)	toluene	168± 63	-(0.13±0.180)	18	monomer/SnCl$_4$ = 1/1 mol/mol
Methacrylonitrile/ZnCl$_2$	1:1 a)	benzene	251	0.75±0.088	18	monomer/ZnCl$_2$ = 1/1 mol/mol
	1:1 a)	benzene	2345± 38	9.76±0.096	18	monomer/ZnCl$_2$ = 2/1 mol/mol
Methacrylic acid	20	methanol	3140± 42	-(4.10±0.134)	19	-
	20	propanol-1	3852± 80	-(1.80±0.276)	19	-
	20	propanol-2	7830±176	10.5 ±0.63	19	-
	10	propanol-1	3978± 92	-(2.18±0.322)	19	-
Methacrylic acid anhydride	?	benzene	13315±209	44.4 ±0.63	20	-
Methacrylic acid, benzyl ester	50	toluene	2303±126	-(1.7 ±3.77)	21	e)
Methacrylic acid, sec-butyl ester	50	toluene	3433± 63	1.97±0.218	10, 22	-
Methacrylic acid, tert-butyl ester	50	toluene	2094± 80	-(2.64±0.289)	10, 22	-
Methacrylic acid, p-carboxyphenyl ester	?	dimethylformamide	4313	2.09	23	f)
	?	p-cetyloxybenzoic acid	0	-6.87	10, 23	-
Methacrylic acid, cyclohexyl ester	100	bulk	1214± 80	-(4.98±0.204)	10, 24	-
Methacrylic acid, glycidyl ester	100	bulk	9840±109	17.46±0.283	10, 25	-
	57 b)	toluene	8583±130	14.19±0.042	10, 25	-
	45 b)	acetone	7034±100	9.50±0.360	10, 25	-
	50 b)	dimethylformamide	5778± 42	4.98±0.151	10, 25	-
Methacrylic acid, α-methylbenzyl ester	50	toluene	3643± 59	2.18±0.218	10, 22	-
Methacrylic acid, methyl ester	25	ethanol	4983± 50	3.60±0.138	9	IR
	25	acetone	7034± 75	9.04±0.230	9	IR
	25	n-heptane	5192±209	3.69±0.059	9	IR
	25	n-hexane	6699± 29	8.50±0.092	9	IR
	25	carbon tetrachloride	5862± 21	5.36±0.067	9	IR
	25	ethyl acetate	6281± 29	6.99±0.092	9	IR
	25	chloroform	9379± 71	16.04±0.205	9	IR
	25	dichloroethylene	6992± 46	8.37±0.134	9	IR
	28.5	toluene	5653± 80	3.94±0.255	10, 26	-
	20	paraffin	2722± 34	1.59±0.130	27	-
	50	toluene	2931±419	1.26±1.26	21	e)
	100	bulk	6197±113	9.21±3.77	27	-
	100	bulk	4480	4.15	7	e)
	100	bulk	3810±226	-(0.04±0.63)	25	e)
	100	bulk	2303± 54	-(1.80±0.163)	28	-
	100	bulk	3559± 84	3.31±0.239	12	-
	9.1 b)	acetone	6281± 75	10.22±0.226	12	-
	9.1 b)	acetonitrile	5359± 75	7.33±0.255	12	-
	9.1 b)	dioxane	5317± 8	7.37±0.301	12	-
	9.1 b)	tetrachloroethane	5150±100	7.16±0.293	12	-
	9.1 b)	chloroform	5108± 71	7.54±0.230	12	-
	9.1 b)	pyridine	4857± 75	7.66±0.218	12	-
	9.1 b)	tetrahydrofuran	3894±113	4.40±0.331	12	-
	9.1 b)	butanone	3852±159	4.19±0.46	12	-
	9.1 b)	dimethylformamide	3475±147	3.43±0.46	12	-
	9.1 b)	benzene	3475± 71	2.26±0.226	12	-
	9.1 b)	dimethyl sulfoxide	3350± 96	2.47±0.281	12	-
	9.1 b)	γ-butyrolactone	3140± 88	1.38±0.268	12	-

Monomer	Vol. -%	Solvent	$\Delta H^{\star}_{s/i} - \Delta H^{\star}_{i/s}$ [J/mol]	$\Delta S^{\star}_{s/i} - \Delta S^{\star}_{i/s}$ [J mol^{-1} K^{-1}]	Ref.	Calculation Remarks
Methacrylic acid, methyl ester (Cont' d.)	9.1 [b)	methylene chloride	3098± 67	2.01±0.218	12	-
	9.1 [b)	toluene	2805± 96	0.75±0.264	12	-
Methacrylic acid, methyl ester/ZnCl$_2$	3.2 [c)	ethyl acetate	4103± 71	5.23±0.222	10, 29	MMA/ZnCl$_2$ = 1/1 mol/mol
	1.1 [c)	water	2847± 71	2.18±0.197	10, 29	MMA/ZnCl$_2$ = 1/8.7 mol/mol
	100	bulk	2638± 187	1.13±0.054	10, 29	MMA/ZnCl$_2$ = 1/0.04 mol/mol
	100	bulk	-(126± 80)	-(9.80±0.067)	27	MMA/ZnCl$_2$ = 2/1 mol/mol
	100	bulk	-(42± 126)	-(6.7 ±0.46)	27	MMA/ZnCl$_2$ = 1/1 mol/mol
Methacrylic acid, methyl ester/SnCl$_4$	100	bulk	796± 100	-(8.21±0.331)	27	MMA/SnCl$_4$ = 2/1 mol/mol
	100	bulk	419± 13	-(14.07±0.054)	27	MMa/SnCl$_4$ = 1/1 mol/mol
Methacrylic acid, L-menthyl ester	50	toluene	963± 46	0.335±0.180	30	-
Methacrylic acid, phenetyl ester	50	toluene	5652±1675	4.6 ±4.6	21	e)
Trifluorochloroethylene	0.4 [d)	carbon tetrachloride	0	-5.78	31	e)
Vinyl acetate	100	bulk	712	2.5	32	e)
	?	n-butyl acetate	251	2.1	32	e)
Vinyl bromide	100	bulk	2135± 17	5.99±0.080	34	-
	6.5 [b)	o-dichlorobenzene	1926± 54	5.28±0.167	34	-
Vinyl chloride	33.3	water/ethanol	5024±1047	18.0 ±4.2	35	-
	33.3	water	1465± 628	2.93±2.09	35	-
	10	cyclohexane	1298± 84	2.51±0.42	36	e)
	100	bulk	2931± 159	6.78±0.067	37	IR
	100	bulk	2261± 50	5.73±0.142	34	-
Vinyl chloride β, β-d$_2$	100	bulk	-(2650± 4)	-(6.28±0.04)		-
Vinyl formate	100	bulk	1047± 54	3.31±0.172	38	-
	100 [b)	bulk	879± 25	3.39±0.080	41	-
	50 [b)	dimethylformamide	251± 38	0.50±0.12	41	-
	50 [b)	chloroform	293± 50	0.08±0.155	41	-
	50 [b)	acetone	335± 25	0.21±0.080	41	-
Vinyl trichloroacetone	100	bulk	2931	84	32	e)
	?	n-butyl acetone	1800	4.6	32	e)
Vinyl trifluoroacetate	100	bulk	1591	4.2	32	e)
	?	n-butyl acetate	879	2.1	32	e)

4. Activation Enthalpy Differences ($\Delta H^{\star}_A - \Delta H^{\star}_B$), Calculated from Data of Various Authors

Monomer	Solvent	$\Delta H^{\star}_A - \Delta H^{\star}_B$ [J/mol] for					Ref.
		A = i/i; B = i/s	s/i; i/i	i/i; s/s	i/s; s/s	s/i; s/s	
Methyl methacrylate/ZnCl$_2$							
9.34/0.369 mol/mol	bulk	2776± 117	-(126± 109)	2428± 59	-(339± 59)	2311± 54	29
1.10/9.6 mol/mol	70% ZnCl$_2$ in water	674± 239	2169± 180	1386± 151	703± 92	3551± 38	29
3.19/3.19 mol/mol	ethyl acetate	2265± 234	1842± 184	2638± 172	373± 63	4484± 54	29
1.00/1.00 mol/mol	bulk	-(21± 163)	-(21± 109)	-(29± 126)	-(8± 17)	-(46± 134)	27
2.00/1.00 mol/mol	bulk	-(13± 113)	-(100± 75)	-(42± 88)	-(29± 29)	-(142± 75)	27
Methacrylic acid (20 Vol.%)	methanol	24661±1382	37599±2470	6490± 754	419± 80	3278± 109	19
	propanol-1	14403±1465	20474±1759	310± 155	-(1244± 75)	4007± 96	19
	propanol-2	26839±1842	32784±2559	5556± 8	151± 29	7955± 218	19
Glycidyl methacrylate	toluene	-	-	-	-	10091±1005	25
	dimethylformamide	-	-	-	-	578± 33	25
	acetone	32910±2303	-	-	-	7126± 113	25
	bulk	-	-	-	-	9274± 142	25
Vinyl formate	bulk	2847± 84	-(1645± 100)	649± 25	-(1897± 105)	-(909± 126)	41
	acetone	666± 25	-(1055± 88)	-(419± 4)	-(1063± 92)	-(2152± 92)	41
	chloroform	1658± 180	-(3165± 193)	-(523± 46)	-(339± 193)	-(3668± 247)	41
	dimethylformamide	0± 151	-(247± 188)	-(176± 46)	-(184± 197)	-(435± 234)	41
Vinyl chloride	water	423	1206	172	984	2504	35
	water/methanol	3597± 63	-(1227± 54)	4421± 29	829± 84	5644± 138	35
Vinyl chloride β, β-d$_2$	bulk	3182± 46	540± 46	2856± 21	-(327± 29)	2320± 25	39

5. Activation Entropy Differences $(\Delta S^{\star}_A - \Delta S^{\star}_B)$. Calculated from Data of Various Authors

Monomer	Solvent	$\Delta S^{\star}_A - \Delta S^{\star}_B$ [J mol^{-1} K^{-1}]					Ref.
		A = i/i B = i/s	s/i i/i	i/i s/s	i/s s/s	s/i s/s	
Methyl methacrylate/ZnCl$_2$							
9.34/0.369 mol/mol	bulk	3.98±0.34	-(2.89±0.29)	1.76±0.17	-(2.22±0.17)	-(1.13±0.13)	29
1.10/9.6 mol/mol	70% ZnCl$_2$ in water	-(3.06±0.71)	5.23±0.54	-(1.51±0.42)	1.55±0.25	3.77±0.13	29
3.19/3.19 mol/mol	ethyl acetate	1.38±0.88	3.85±0.71	1.80±0.63	0.38±0.25	5.65±0.13	29
1.00/1.00 mol/mol	bulk	-(5.19±0.63)	-(1.72±0.38)	-(5.99±0.46)	-(0.84±0.21)	-(7.70±0.42)	27
2.00/1.00 mol/mol	bulk	-(7.03±0.46)	-(2.76±0.29)	-(8.00±0.38)	-(1.51±0.13)	-(10.76±0.29)	27
Methacrylic acid (20 Vol.%)	methanol	54.0 ±4.2	125.2 ±8.4	-(1.30±2.26)	2.30±0.25	-(2.81±0.34)	19
	propanol-1	31.8 ±7.1	72.0 ±8.0	-(15.79±0.54)	-(3.39±0.25)	-(0.88±0.38)	19
	propanol-2	74.5 ±7.5	99.7±10.9	-(1.84±0.04)	0.71±0.13	11.14±0.75	19
Glycidyl methacrylate	toluene	-	-	-	-	17.00±3.89	25
	dimethylformamide	-	-	-	-	5.82±0.13	25
	acetone	85.4 ±8.8	-	-	-	10.34±0.42	25
	bulk	-	-	-	-	16.08±0.54	25
Vinyl formate	bulk	11.14±0.25	-(7.62±0.25)	2.55±0.08	-(8.58±0.34)	-(5.02±0.38)	41
	acetone	5.11±0.25	-(5.57±0.29)	-(0.75±0.04)	-(5.78±0.29)	-(9.25±0.29)	41
	chloroform	7.66±0.54	-(12.98±0.63)	-(1.05±0.13)	-(13.90±0.59)	-(13.94±0.75)	41
	dimethylformamide	4.56±0.46	-(3.89±0.59)	-(0.29±0.13)	-(4.31±0.39)	-(3.85±0.71)	41
Vinyl chloride	water	-1.30	4.61	4.48	3.77	6.78	35
	water/methanol	11.72±0.25	-(4.06±0.21)	15.32±0.54	3.64±0.34	20.56±0.54	35
Vinyl chloride β,β-d$_2$	bulk	8.33±0.17	2.09±0.17	7.03±0.08	-(1.30±0.08)	4.98±0.08	39

6. Compensation Temperature T_o and Compensation Enthalpies $\Delta\Delta H^{\star}_o$ for Various Monomers and Modes of Addition, Calculated under the Assumption of a Markov First Order Process (data from 43, 40)

Monomer	Modes of Addition		T_o [K]	$\Delta\Delta H^{\star}_o$ [J/mol]
	A	B		
Acrylonitrile	s/i	i/s	1406	0
Methacrylic acid	s/i	i/s	455	2780
Methacrylic acid in alcohols	i/i	i/s	294± 26	6281±1256
	s/i	i/i	323± 5	-(2286± 511)
	i/i	s/s	403±137	6662± 205
	s/i	i/s	323± 10	4476± 21
	i/s	s/s	302± 74	-(184± 63)
	s/i	s/s	333± 23	4254± 25
Methacrylic acid, iso-butyl ester	s/i	i/s	294± 3	3304± 4
Methacrylic acid, n-butyl ester	s/i	i/s	236± 4	3576± 8
Methacrylic acid, p-carboxyphenyl ester	s/i	i/s	481	3308
Methacrylic acid, glycidyl ester	s/i	i/s	345± 15	3764± 201
	s/i	s/s	373± 11	3471± 121
Methacrylic acid, hexyl ester	s/i	i/s	303± 11	3257± 29
Methacrylic acid, methyl ester	i/i	i/s	315± 49	121± 155
	s/i	i/i	310± 20	2571± 126
	i/i	s/s	323±124	1440± 84
	s/i	i/s	362± 15	2470± 50
	i/s	s/s	338± 51	1398± 92
	s/i	s/s	334± 54	4007± 75
Methacrylic acid, methyl ester/ZnCl$_2$	i/i	i/s	314± 39	1650± 67
	s/i	i/i	321± 9	540± 33
	i/i	s/s	331± 24	1930± 42
	s/i	i/s	336± 21	2248± 50
	i/s	s/s	304± 10	243± 4
	s/i	s/s	328± 42	2529± 80
Methacrylic acid, iso-propyl ester	s/i	i/s	230± 54	3421± 67

Monomer	Modes of Addition		T_o [K]	$\Delta\Delta H_o^{\star}$ [J/mol]
	A	B		
Methacrylic acid, n-propyl ester	s/i	i/s	248± 41	3262±176
Vinyl acetate	s/i	i/s	1100	-2052
Vinyl chloride	i/i	i/s	283± 25	821±130
	s/i	i/i	282± 6	-(75± 17)
	i/i	s/s	235± 28	900±159
	s/i	i/s	305±124	-(222±419)
	i/s	s/s	248± 9	-(13± 33)
	s/i	s/s	219± 10	1130± 59
Vinyl formate	i/i	i/s	362± 23	-(1361±126)
	s/i	i/i	315± 16	854± 80
	i/i	s/s	323± 27	-(201± 25)
	s/i	i/s	330± 95	-(289± 42)
	i/s	s/s	320± 18	971± 96
	s/i	s/s	316± 5	741± 21
Vinyl chloroacetate	s/i	i/s	300	419
Vinyl trifluoroacetate	s/i	i/s	340	167

C. REFERENCES

1. H.-G. Elias, Makromol. Chem. 137, 277 (1970).
2. J. W. L. Fordham, J. Polymer Sci. 39, 321 (1959).
3. C. E. H. Hawn, W. H. Janes, A. M. North, J. Polymer Sci. C 4, 427 (1963).
4. H. Fischer, Kolloid Z. 206, 131 (1965).
5. P. L Luisi, R. M. Mazo J. Polymer Sci. A-2 7, 775 (1969).
6. F. A. Bovey, G. V. D. Tiers, J. Polymer Sci. 44, 173 (1960).
7. T. G. Fox, H. W. Schnecko, Polymer 3, 575 (1962).
8. F. A. Bovey, "Polymer Conformation and Configuration", Academic Press, New York, 1969.
9. H Watanabe, Y. Sono, Kogyo Kagaku Zasshi 65, 273 (1962).
10. H.-G. Elias, P. Goeldi, V. S. Kamat, Makromol. Chem. 117, 269 (1968).
11. R. L. Miller, SPE Transactions 3, 1 (1963).
12. P. Goeldi, H.-G. Elias, Makromol. Chem. 153, 81 (1972).
13. G. Svegliado, G. Talamini, G. Vidotto, J. Polymer Sci. A-1 5, 2875 (1967).
14. H. Murano, R. Yamadera, J. Polymer Sci. B 5, 333 (1967).
15. C. Schuerch, W. Fowells, F. P. Hood, F. A. Bovey, cited in 36).
16. K. Matsuzaki, T. Uryu, K. Ito, Makromol. Chem. 126, 292 (1969).
17. K. Matsuzaki, T. Uryu, J. Polymer Sci. B 4, 255 (1966).
18. H. Hirai T. Ikegami, S. Makishima, J. Polymer Sci. A-1 7, 2059 (1969).
19. J. B. Lando, J. Semen, B. Farmer, Polymer Preprints 10, 586 (1969).
20. W. L. Miller, W. S. Brey, G. B. Butler, J. Polymer Sci. 54, 329 (1961).
21. K. Yokota, Y. Ishi, Kogyo Kagaku Zasshi 69, 1053 (1966).
22. K. Matsuzaki, A. Ishida, N. Tateno, T. Asakura, A. Hasegawa, T. Tameda, Kogyo Kagaku Zasshi 68 852 (1965).
23. Y. B. Amerik, I. I. Konstantinov, B. A. Krentsel, IUPAC, International Symposium, Macromolecular Chemistry, Tokyo 1966, Preprint 1.1.09.
24. K. Matsuzaki, A. Ishida, N. Tateno, IUPAC, International Symposium, Macromolecular Chemistry, Prague 1965, Preprint 265.
25. Y. Iwakura, F. Toda, T. Ito, K. Aoshima, Makromol. Chem. 104, 26 (1967).
26. Y. Kato, A. Nishioka, Bull. Chem. Soc. Japan 37, 1614 (1964).
27. S. Okuzawa, H. Hirai, S. Makishima, J. Polymer Sci. A-1, 7, 1039 (1969).
28. F. A. Bovey, J. Polymer Sci. 46, 59 (1960).
29. T. Otsu, B. Yamada, M. Imoto, J. Macromol. Chem. 1, 61 (1966).
30. H. Sobue, K. Matsuzaki, S. Nakano. J. Polymer Sci. A 2, 0000 (1964).
31. F. A. Bovey, G. V. D. Tiers, Fortschr. Hochpolymer.-Forsch. 3, 139 (1963).
32. M. Uoi, M. Sumi, S. Nozakura, S. Murahashi, cited in 33).
33. S. Murahashi in IUPAC, International Symposium, Macromolecular Chem. 3, 435 (1967) (= Pure Appl. Chem. 15, nos. 3-4 (1967).)
34. G. Talamini, G. Vidotto, Makromol. Chem. 100, 48 (1967).
35. J. Bargon, K. H. Hellwege, U. Johnsen, Makromol. Chem. 95, 187 (1966).
36. F. A. Bovey, F. D. Hood, E. W. Anderson, R. L. Kornegay, J. Phys. Chem. 71, 312 (1967).
37. H. U. Pohl, D. O. Hummel, Makromol. Chem. 113, 203 (1968).
38. K. C. Ramey, D. C. Lini, G. Statton, J. Polymer Sci. A-1, 5, 257 (1967).
39. L. Cavalli, G. C. Borsini, G. Carraro, G. Confalonieri, J. Polymer Sci. A-1, 8, 801 (1970).
40. H.-G. Elias, P. Goeldi, Makromol. Chem. 144, 85 (1971).
41. H.-G. Elias, M. Riva, P. Goeldi, Makromol. Chem. 145, 163 (1971).
42. B. L. Johnson, H.-G. Elias, Makromol. Chem. 155, 121 (1972).
43. H.-G. Elias, P. Goeldi, B. L. Johnson, Adv. Chem. Ser., in print.

ACTIVATION ENERGIES FOR THE THERMAL DEGRADATION OF POLYMERS

N. Grassie and A. Scotney
Chemistry Department
The University
Glasgow, Scotland

The rate constant of a chemical reaction, k, is given by the Arrhenius equation, k = A exp (-E/RT) in which A is a constant, R and T are the gas constant and the absolute temperature respectively and E is the energy of activation of the reaction expressed in kJ/mol. Since log k = log A - 2.303 E/RT, E is obtained from the slope of the log k vs. 1/T plot. The rate of a reaction is given quite generally by, $kC_A^a C_B^b C_C^c$ . . . . . in which C_A, C_B, C_C, ... are the concentrations of the reactants and a, b, c, ... are constants. Thus provided rate measurements at different temperatures are made under the same concentration conditions, the energy of activation may be obtained simply from the slope of the plot of log rate vs. 1/T.

The majority of the data presented refers to the rate of production of volatile material as measured by the loss in weight of the polymer. Other less common methods of rate measurement are referred to in column 4 of the table.

Polymer	Temp. Range [°C]	Energy of Activation [kJ/mol]	Type of Measurement and Remarks	References
1. Main-chain Acyclic Carbon Polymers				
1.1 Poly(dienes)				
Poly(butadiene)	380-395	259.6		4
Poly(isoprene) (natural rubber)	291-311	234.5-263.8		30
	60-100	108.0	Chain scission in cyclohexane and transdecalin solution measured by decrease in M_w	31
Poly(perfluoro-4-chloro-1,6-heptadiene)		237.0		40
1.2 Poly(alkenes)				
Poly(cyclohexylethylene)	321-336	205.2		1
Poly(ethylene)	375-436	192.6	MW = 11.000	13
		220.2	MW = 16.000	13
		276.8	MW = 23.000	13
	360-392	263.8	MW = 20.000	14,15,16
	350.9-372.6	268.0	Branched	14,15,16
	345-396	301.5	Polymethylene	14,15,16
Poly(isobutene)	306-326	205.2		14
Poly(4-methyl-1-pentene)	291-341	224.0		37
Poly(propylene)	336-366	242.8		14
	320-341	230.3	Volatile material produced	48
	250-300	272.2	Rate of random bond scission	49
1.3 Poly(acrylics), Poly(methacrylics)				
Poly(acrylic acid)				
--, methyl ester	271-286	142.4		1
--, tert-butyl ester	161-222	163.3	Initially	2
		121.4	Crystalline	2
		129.8	Amorphous	2
--, α-phenyl, methyl ester	210-280	234.5-238.7	M_n = 140.000 (by anionic polymerization)	3
Poly(acrylonitrile)	218-260	129.8		4
	286-456	96.3	Initial rate of formation of cyanogen	5
	286-456	62.8	Initial rate of formation of HCN	5
Poly(methacrylic acid)	153-192	154.9±12.6	Water produced	32
--, methyl ester	220-280	134.0-175.9	Monomer produced - E increasing from 134.0 to 175.9 with extent of degradation from 0-100 % (benzoyl peroxide initiated polymer)	33
	170-210	123.5±4.2	Rate of production of volatile material measured mass spectrometrically	34
	240-270	125.6	Benzoyl peroxide initiated polymer	1
	310-325	217.7	Thermally polymerized at room temperature	1

THERMAL DEGRADATION OF POLYMERS

Polymer	Temp. Range [°C]	Energy of Activation [kJ/mol]	Type of Measurement and Remarks	References
Poly(methacrylic acid)				
--, methyl ester (Cont'd.)	110-180		Increase of pressure in a closed system	35
		201.0	$MW = 10^4$	
		129.8	$MW = 10^6$	
	300-375	131.5-75.4	Rate of monomer production; for chain-end initiated reaction, E falls from 131.5 to 75.4 as MW increases from 40.000 to 1,611.000.	36
	300-375	322.0-264.6	Rate of monomer production; for randomly initiated reaction at 40 % conversion, E falls from 322.0 to 264.6 as MW increases from 40.000 to 1,611.000.	36
	300-375	333.3-284.7	Rate of monomer production; for randomly initiated reaction at 60 % conversion, E falls from 333.3 to 284.7 as MW increases from 40.000 to 1,611.000.	36

1.4   Poly(vinyl halides), Poly(vinyl esters)

Polymer	Temp. Range [°C]	Energy of Activation [kJ/mol]	Type of Measurement and Remarks	References
Poly(chlorotrifluoroethylene)	331-371	234.5		18
Poly(chlorotrifluoroethylene-co-vinylidene fluoride)	340-380	221.9	Volatile material produced	23
Poly(1,1-dichloro-2,2-difluoroethylene)	170-220	192.6		17
Poly(perfluoroheptene)	-	263.8		40
Poly(perfluoropropylene)	-	237.0		40
Poly(tetrafluoroethylene)	423.5-513	337.1	Loss in weight and pressure of volatile products	19
	480.5-508.5	318.2		20
	450-550	314.0±16.7		21
	500-570	293.1	Less than 20 % conversion	22
Poly(vinyl acetate)	242-264	224.4	Production of acetic acid	60
Poly(vinyl chloride)	150-190	83.7	Production of HCl	61
	200-250	138.2	Production of HCl in nitrogen	62
		100.5	in oxygen	
	235-260	108.9-134.0		63
Poly(vinyl fluoride)	365-382	230.3		64
Poly(vinylidene chloride)	175-225	131.1		65
Poly(vinylidene fluoride)	371-420	201.0		53
	340.5-368	297.3		20

1.5   Poly(styrenes)

Polymer	Temp. Range [°C]	Energy of Activation [kJ/mol]	Type of Measurement and Remarks	References
Poly(styrene)	250-301	201.0	Rate of bond scission. $M_n = 268.000$; $M_w = 285.000$	51
	299-348	230.3		52
--, (cross-linked)	330-390	221.9-242.8	E increases from 221.9 to 242.8 with divinyl benzene content increasing from 2 % to 56 %	53,54
	360-390	272.2	Containing 25 % trivinyl benzene	53,54
--, (divinyl benzene)	360-390	272.2		53,54
--, (trivinyl benzene)	394-440	305.7		53,54
--, α-deutero-	321-341	230.3		1,14
--, β-deutero-	326-346	234.5		1,14
--, α-methyl-	228.8-275.5	230.3		1,16
	214-240	260.4	Rate of monomer production	55
	253-289	272.2		56
--, 3-methyl-	318-338	234.5		1
--, 2,3,4,5,6-pentafluoro-	395-415	272.2		57
--, α-β-β-trifluoro-	311-326	268.0		18

1.6   Others

Polymer	Temp. Range [°C]	Energy of Activation [kJ/mol]	Type of Measurement and Remarks	References
Poly(phenylvinylene)	300-340	102.2	In argon	46
	300-330	102.2	In air	46
Poly(pyridylvinylene)	320-340	102.2	In argon	46

Polymer	Temp. Range [°C]	Energy of Activation [kJ/mol]	Type of Measurement and Remarks	References
2. **Main-chain Carbocyclic Polymers**				
Poly(perfluoro-m-phenylene)	360-470	108.9	In oxygen (267 mbar pressure)	41
Poly(perfluoro-p-phenylene)	360-470	163.3	In oxygen (267 mbar pressure); pre-heated in vacuum for 1 hour at 400°C. MW = 15.000	41
	360-470	129.8	In oxygen (267 mbar pressure); pre-heated in vacuum for 1 hour at 400°C. MW = 3.700	41
Poly(1,4-phenyleneethylene) (poly(p-xylylene))	401-411	305.7		8
	-	242.8	Rate of random bond scission	66
	408-450	309.8		67
	475-515	194.7		67
Poly(1,4-phenylenemethylene)	386-416	209.4		8
3. **Main-chain Heteroatom Polymers**				
3.1 **Poly(oxides)**				
Poly(oxybutylethylene)	320-430	209.4±8.4		12
Poly(oxyethylene)	320-335	192.6		24
Poly(oxymethylene) hydroxyl end groups	100-120	113.5	Pressure of volatile product	26
	170-285	108.9	Pressure of volatile product	27
	135-190	83.7	Pressure of volatile product, E initial - increases to 217.7 after 20 % volatilization	28
acetate end groups	240-340	134.0	Pressure of volatile product	27
	178-190	234.5	Pressure of volatile product	28
Poly(oxymethyleneoxyethylene) (Poly(dioxolane))	320-430	209.4±8.4		12
Poly(oxypropylene)	265-285	83.7	Atactic	24
	285-300	188.4	Isotactic	24
Poly(oxytetramethylene)	265-343	188.4	Bulk polymer degraded	58
	262-340	206.8	Viscometric molecular weight fall used to determine E; bulk polymer degraded	58
	320-430	209.4±8.4		12
	160-190	121.4	Bulk polymer degraded in air	58
	20-147	55.7	Viscometric molecular weight fall used to determine E; solution of polymer in xylene degraded in air	58
3.2 **Poly(esters)**				
Poly(oxycarbonyloxy-1,4-phenyleneisopropylidene-1,4-phenylene)	300-389	117.2	Rate of gas evolution	47
Poly(oxyethyleneoxyterephthaloyl) (poly(ethylene terephthalate))	336-356	159.1		25
3.3 **Poly(sulfides)**				
Poly(thioacetone)	105-145	111.4	M_w = 20.000	59
Poly(thioperfluorophenylene)	355-396.5	226.1	In vacuum	42
	336.5-408	138.2	In oxygen	42
Poly(thiophenylene)	398.5-442	201.0		42
	390-454	154.9	In oxygen	42
3.4 **Poly(amides)**				
Poly(2,5-dimethyl-1,4-piperazinediylisophthaloyl)	390-439	221.9		29
Poly(2,5-dimethyl-1,4-piperazinediyloxalyl)	397-445	268.0	MW = 25.000-30.000	38
Poly(2,5-dimethyl-1,4-piperazinediylterephthaloyl)	400-462	268.0-284.7	MW = 25.000-30.000	38
Poly(iminoisophthaloylimino-1,3-phenylene)	330-390	141.5	Rate of increase of pressure of volatiles	45
Poly(iminoisophthaloylimino-1,4-phenylene)	390-470	131.5	Rate of increase of pressure of volatiles	45
Poly(imino(1-oxohexamethylene)) (Nylon 6)	355-365	142.4-180.0	Increases with purification of monomer	6
Poly(iminoterephthaloylimino-1,3-phenylene)	410-480	177.5	Rate of increase of pressure of volatiles	45
Poly(iminoterephthaloylimino-1,4-phenylene)	390-460	48.2	Rate of increase of pressure of volatiles	45
	470-500	222.7	Rate of increase of pressure of volatiles	45

Polymer	Temp. Range [°C]	Energy of Activation [kJ/mol]	Type of Measurement and Remarks	References
3.4 Poly(amides) (Cont'd.)				
Poly(2-methyl-1,4-piperazinediylterephthaloyl)	403-450	280.5		29
Poly(piperazinediylterephthaloyl)	418-462	297.3		29
4. Main-chain Heterocyclic Polymers				
Poly(2,5-benzimidazolediyl)	530-780	123.5-125.6	In air	7
Poly(2,5-benzimidazolediyl-5,2-benzimidazolediyl-1,3-phenylene)	530-780	142.4-167.5	In air	7
Poly(2,5-benzimidazolediyl-5,2-benzimidazolediyl-1,4-phenylene)	500-700	188.4	In vacuum	44
	450-550	148.2	In air	44
Poly(2,5-benzimidazolediyl-5,2-benzimidazolediyl-1,4-phenyleneoxy-1,4-phenylene)	530-780	114.7	In air	7
Poly(2,6-benzoxazolediyl)	530-780	140.7-144.5	In air	7
Poly(2,6-benzoxazolediyl-6,2-benzoxazolediyl-1,3-phenylene)	530-780	151.2-155.8	In air	7
Poly(2,6-benzoxazolediyl-6,2-benzoxazolediyl-1,4-phenyleneoxy-1,4-phenylene)	530-780	144.5	In air	7
Poly(2,6-benzothiazolediyl)	530-780	118.9	In air	7
Poly(2,6-benzothiazolediyl-6,2-benzothiazolediyl-1,3-phenylene)	530-780	147.4	In air	7
Poly(2,6-benzothiazolediyl-6,2-benzothiazolediyl-1,4-phenyleneoxy-1,4-phenylene)	530-780	155.8	In air	7
Poly(5,7-dihydro-1,3,5,7-tetraoxobenzo[1,2-c : 4,5-c'] dipyrrole-2,6[1H,3H]-diyl-1,4-phenyleneoxy-1,4-phenylene) (Poly(pyromellitimide))	521-660 435-485	309.8 138.2	In vacuum In air	50 50
Poly(3-perfluoropropyl-1,5-triazinediylperfluorotrimethylene)	491-513	180.0		39
Poly(5-phenyl-as-3.4-triazinediyl-2,6-pyridinediyl-5-phenyl-as-3.4-triazinediyl-1,4-phenyleneoxy-1,4-phenylene)	330-380	94.2		43
Poly(1,3,5-triazinetriylperfluorotrimethylene) (poly(perfluoroglutaro-diamidine))	491-513	163.3		39
5. Cellulose and Derivatives				
Cellulose	251-291	209.4	Cotton	9,10
		205.2	Viscose	9,10
		196.8	Hydrocellulose	9,10
		192.6	Fortisan	9,10
	180-225	73.3	Untreated	11
	225-340	90.0	Untreated	11
	270-340	117.2	Untreated	11
	340-370	239.5	Untreated	11
	180-225	72.0	Mercerised	11
	225-340	85.4	Mercerised	11
	270-340	146.5	Mercerised	11
	340-370	251.2	Mercerised	11
Cellulose acetate	180-200	62.8		11
--, (D.S. = 0.9)	200-270	29.3		11
	270-285	146.5		11
	285-360	87.9		11
--, (D.S. = 1.6)	265-295	135.7		11
	280-295	190.5		11
	295-360	275.1		11
--, (D.S. = 2.65)	275-285	12.6		11
	285-315	100.5		11
	315-365	226.1		11
--, (D.S. = 2.94)	280-300	222.3		11
	300-318	177.9		11
	318-360	286.8		11

Polymer	Temp. Range [°C]	Energy of Activation [kJ/mol]	Type of Measurement and Remarks	References
5. Cellulose and Derivatives (Cont'd.)				
Cellulose triacetate	283-306	188.4		10
	280-304	38.5		11
	304-322	126.4		11
	322-360	139.4		11

REFERENCES

1. S. L. Madorsky, J. Polymer Sci. 11, 491 (1953).
2. J. R. Schaefgen, I. M. Sarasohn, J. Polymer Sci. 58, 1049 (1962).
3. G. G. Cameron, G. P. Kerr, J. Polymer Sci., A-1, 7, 3067 (1969).
4. S. Straus, S. L. Madorsky, J. Res. Nat. Bur. Std. 61, 77 (1958).
5. A. R. Monahan, J. Polymer Sci., A-1, 4, 2391 (1966).
6. S. Straus, L. A. Wall, J. Res. Nat. Bur. Std., 63A, 269 (1959).
7. W. Wrasidlo. R. Empey, J. Polymer Sci., A-1, 5, 1513 (1967).
8. L. A. Wall, R. E. Florin, J. Res. Nat. Bur. Std. 60, 451 (1958).
9. S. L. Madorsky, V. E. Hart. S. Straus, J. Res. Nat. Bur. Std. 56, 343 (1956).
10. S. L. Madorsky, V. E. Hart, S. Straus, J. Res. Nat. Bur. Std. 60, 343 (1958).
11. K. S. Patel, K. C. Patel, R. D. Patel, Makromol. Chem. 132, 7 (1970).
12. A. B. Blyumenfel'd, B. M. Kovarskaya, Polymer Sci. USSR 12, 710 (1970).
13. H. H. G. Jellinek, J. Polymer Sci. 4, 13 (1949).
14. S. L. Madorsky, S. Straus, J. Res. Nat. Bur. Std. 53, 361 (1954).
15. S. L. Madorsky, J. Polymer Sci. 9, 133 (1952).
16. S. L. Madorsky, J. Res. Nat. Bur. Std. 62, 219 (1959).
17. J. L. Cotter, G. J. Knight, W. W. Wright, J. Polymer Sci., B, 6, 763 (1968).
18. S. L. Madorsky, S. Straus, J. Res. Nat. Bur. Std. 55, 223 (1955).
19. S. L. Madorsky, V. E. Hart, S. Straus, V. A. Sedlek, J. Res. Nat. Bur. Std. 51, 327 (1953).
20. J. M. Cox, B. A. Wright, W. W. Wright, J. Appl. Polymer Sci. 8, 2935 (1964).
21. H. C. Anderson, Makromol. Chem. 51, 233 (1962).
22. L. Reich, Makromol. Chem. 105, 223 (1967).
23. T. G. Degteva, I. M. Sedova, A. S. Kuz'minskii, Polymer Sci. USSR 4, 1036 (1963).
24. S. L. Madorsky, S. Straus, J. Polymer Sci. 36, 183 (1959).
25. S. Straus, L. A. Wall, J. Res. Nat. Bur. Std. 60, 39 (1958).
26. Y. Iwasa, T. Imoto, Nippon Kagaku Zasshi 84, 31 (1963).
27. L. A. Dudina, N. S. Enikolopyan, Polymer Sci. USSR 5, 36 (1964).
28. N. Grassie, R. S. Roche, unpublished.
29. S. D. Bruck, Polymer 7, 231 (1966).
30. S. Straus, S. L. Madorsky, Ind. Eng. Chem. 48, 1212 (1956).
31. P. S. Sarfare, H. L. Bhatnagar, A. B. Biswas, J. Appl. Polymer Sci. 7, 2199 (1963).
32. D. H. Grant, N. Grassie, Polymer 1, 125 (1960).
33. N. Grassie, H. W. Melville, Proc. Roy. Soc. A190, 1 (1949).
34. P. D. Zemany, Nature 171, 391 (1953).
35. S. Bywater, J. Phys. Chem. 57, 879 (1953).
36. H. H. G. Jellinek, Ming Dean Luh, Makromol. Chem. 115, 89 (1968).
37. L. Reginato, Makromol. Chem. 132, 113 (1970).
38. S. D. Bruck, Polymer 6, 483 (1965).
39. L. A. Wall, S. Straus, J. Res. Nat. Bur. Std. 65A, 227 (1961).
40. S. Straus, L. A. Wall, Soc. Plastics Eng. J. 56 (January 1964).
41. J. L. Cotter, G. J. Knight, J. M. Lancaster, W. W. Wright, J. Appl. Polymer Sci. 12, 2481 (1968).
42. N. S. J. Christopher, J. L. Cotter, G. J. Knight, W. W. Wright, J. Appl. Polymer Sci. 12, 863 (1968).
43. W. Wrasidlo, P. M. Hergenrother, Macromolecules 3, 548 (1970).
44. V. V. Rode, N. M. Kotsoyeva, G. M. Cherkasova, D. S. Tugushi, G. M. Tseitlin, A. L. Rusanov, V. V. Korshak, Polymer Sci. USSR 12, 2103 (1970).
45. Ye. P. Krasnov, V. M. Savinov, L. B. Sokolov, V. I. Logunova, V. K. Belyakov, T. A. Polyakova, Polymer Sci. USSR 8, 413 (1966).
46. A. A. Berlin, G. V. Belova, A. I. Sherle, N. A. Markova, Polymer Sci. USSR 11, 194 (1969).
47. A. Davis, J. H. Golden, J. Chem. Soc., B1, 45 (1968).
48. V. D. Moiseev, M. B. Neiman, A. I. Kriukova, Polymer Sci. USSR 2, 55 (1961).
49. T. E. Davis, R. L. Tobias, E. B. Peterli, J. Polymer Sci. 56, 485 (1962).
50. S. D. Bruck, Polymer 5, 435 (1964).
51. J. Wegner, F. Patat, J. Polymer Sci., C, 31, 121 (1970).
52. S. L. Madorsky, D. McIntyre, J. H. O'Mara, S. Straus, J. Res. Nat. Bur. Std. 66A, 307 (1962).
53. S. L. Madorsky, S. Straus, J. Res. Nat. Bur. Std. 63A, 261 (1959).
54. S. Straus, S. L. Madorsky, J. Res. Nat. Bur. Std. 65A, 243 (1961).
55. D. H. Grant, E. Vance, S. Bywater, Trans. Faraday Soc. 56, 1697 (1960).
56. D. W. Brown, L. A. Wall, J. Phys. Chem. 62, 848 (1958).
57. L. A. Wall, J. M. Antonucci, S. Straus, M. Tryon, Soc. Chem. Ind. (London), Monograph 13, 295 (1961).
58. A. Davis, J. H. Golden, Makromol. Chem. 81, 38 (1965).
59. V. C. E. Burnop, K. G. Latham, Polymer 8, 589 (1967).
60. N. Grassie, Trans. Faraday Soc. 49, 835 (1953).
61. A. Guyot, J. P. Benevise, Y. Trambouze, J. Appl. Polymer Sci. 6, 103 (1962).
62. G. Talamini, G. Pezzin, Makromol. Chem. 39, 26 (1960).
63. R. R. Stromberg, S. Straus, B. G. Achhammer, J. Polymer Sci. 35, 355 (1959).
64. L. A. Wall, S. Straus, R. E. Florin, J. Polymer Sci., A-1, 4, 349 (1966).
65. R. M. Aseyeva, A. A. Berlin, V. I. Kasatochkin, Z. S. Smutkina, Polymer Sci. USSR 8, 2404 (1966).
66. J. R. Schaefgen, J. Polymer Sci. 41, 133 (1959).
67. H. H. G. Jellinek, S. N. Lipovac, J. Polymer Sci., A-1, 8, 2517 (1970).

PRODUCTS OF THERMAL DEGRADATION OF POLYMERS

N. Grassie and A. Scotney
Chemistry Department
The University
Glasgow, Scotland

The chemical nature of the thermal decomposition of polymers varies widely from one material to another. At its simplest it may consist of complete breakdown to a single, and readily identifiable product which is often monomer. On the other hand, a complex mixture of products may be obtained together with a relatively stable intractable residue.

From the point of view of identification there are three types of products of thermal degradation. Firstly, substances of the molecular dimensions of monomer or less, of which a detailed analysis can usually be made. Secondly, substances are often produced which are volatile at the degradation temperatures but involatile at ordinary temperatures. These are usually polymer chain fragments which are larger than monomer. Although they may be referred to as dimeric, trimeric, etc., their precise chemical structures have usually not been determined. Finally, the involatile residue which often remains is frequently insoluble although a knowledge of the mechanism of the degradation reaction combined with spectral data can often give some information about its structure. The relative amounts of these three kinds of products can vary with the temperature of degradation although not a great deal of information of this kind is available. However, the temperature ranges in which the experiments have been carried out are quoted in the table since they give some idea of the relative stability of the polymers.

Most of the data are concerned with the volatile small molecular products of degradation. Information about larger chain fragments and involatile residues is given where it is available and relevant.

Polymer	Temp. Range [^{o}C]	Degradation Products	References
1. Main-chain Acyclic Carbon Polymers			
1.1 Poly(dienes)			
Poly(butadiene)	325-475	14.1 % of products are volatile at 25°C including 1.5 % of monomer among other saturated and unsaturated hydrocarbons; 85.9 % of products are larger fragments involatile at 25°C.	16
Poly(butadiene-co-acrylonitrile) (70/30)	310-400	14.5 % of products are volatile at 25°C, consisting of saturated hydrocarbons.	10
Poly(butadiene-co-styrene) (75/25)	327-430	11.8 % of products are volatile at 25°C - 1.9 % butadiene with other saturated and unsaturated hydrocarbons.	10, 16
Poly(chloroprene)	377	Hydrogen chloride.	19
Poly(isoprene), synthetic	287-400	3.4 % isoprene, 8.8 % dipentene, small amounts of p-menthene.	2, 16, 31
natural rubber	287-400	3.9 % isoprene, 13.2 % dipentene, small amounts of p-menthene.	2, 16, 31
	450-800	Dipentene main product at 450°C, optimum yields of isoprene in range 675-800°C, e.g. 58 % at 750°C and 13 mbar.	32
gutta percha	287-400	3.0 % isoprene, 15.6 % dipentene, small amounts of p-menthene.	2, 16, 31
Poly(perfluoro-4-chloro-1,6-heptadiene)	320-400	Completely volatilized - products unknown.	49
1.2 Poly(alkenes)			
Poly(cyclohexylethylene)	335-391	Small amounts of cyclohexene, cyclohexane, methylcyclohexene, methylcyclohexane, vinylcyclohexene, vinylcyclohexane, ethylcyclohexane with larger chain fragments.	2, 3
Poly(ethylene)	335-450	Continuous spectrum of saturated and unsaturated hydrocarbons from C_2-C_{90} - lower temperature favours larger fragments.	13, 16
Poly(isobutene)	288-425	18.1 % monomer together with methane, isobutane and C_5 and higher saturated and unsaturated hydrocarbons.	16
	up to 1200	As temperature is increased the yields of fragments smaller than monomer increase at the expense of larger fragments.	28
	325	At 4.9 % weight loss, products are isobutene (64.3 %), CH_4 (13.6 %) neopentane (10.3 %), remainder C_2-C_{12} hydrocarbons.	29
	345	At 15.7 % weight loss, products are isobutene (78.9 %), CH_4 (5.9 %) neopentane (4.7 %), remainder C_2-C_{12} hydrocarbons.	29
	365	At 46.8 % weight loss, products are isobutene (81.6 %), CH_4 (3.9 %) neopentane (3.1 %), remainder C_2-C_{12} hydrocarbons.	29
Poly(4-methyl-1-pentene)	291-341	After 22 hours at 341°C, 20 % residue, 71 % saturated and unsaturated chain fragments; 9 % of products volatile at 25°C, comprising isobutene (56 %), propane (34 %), traces of monomer and hydrocarbons.	43

Polymer	Temp. Range [°C]	Degradation Products	References
1.2 Poly(alkenes) (Cont'd.)			
Poly(propylene)	328-410	Saturated and unsaturated hydrocarbons from C_2 upwards, monomer yield 0.17 %.	13
	400-1200	As temperature is raised, yield of small fragments increases at expense of large fragments.	28
1.3 Poly(acrylics), Poly(methacrylics)			
Poly(acrylic acid)			
--, benzyl ester	260-300	Major products are benzyl alcohol, chain fragments, partially crosslinked residue; minor product is CO_2, with traces of toluene, CO, CH_4, H_2.	1
--, methyl ester	292-399	26 % of products are volatile at 25 °C, mainly methyl alcohol and carbon dioxide with traces of monomer and methyl methacrylate and C_4-C_6 oxygenated compounds. 74 % of products are larger chain fragments involatile at 25 °C.	2,3
--, tert-butyl ester	> 160	86 % isobutylene, 11 % water, 3 % carbon dioxide.	4
--, α-bromo-, methyl ester	110-150	Methyl bromide, hydrogen bromide.	5
--, α-cyano-, methyl ester	> 180	Yellows and some monomer formed.	6
--, α-chloro-, sec-butyl ester	190	sec-Butyl chloride, butylene, hydrogen chloride.	7
--, α-phenyl-, methyl ester	210-280	Monomer is the sole product.	8
Poly(acrylonitrile)	< 200	Colours through yellow, orange, red and black.	9
	250-280	12 % of products are volatile at 25 °C, consisting of hydrogen cyanide, acrylonitrile and vinyl acetonitrile; 88 % of products are involatile at room temperature.	10,11
	280-450	Five major volatile products are cyanogen, HCN, acrylonitrile, acetonitrile, vinyl acetonitrile; involatile residue remains.	12
Poly(methacrylic acid)	200	Almost quantitative yields of H_2O, traces of monomer, residue of poly(methacrylic anhydride).	33
--, methyl ester	170-300	100 % monomer.	34,35
	246-1200	As temperature is raised fragmentation increases to give complex series of products and monomer yields correspondingly decreases.	2
	160	100 % monomer under 253.7 nm radiation.	36
--, ethyl ester	250	Monomer.	37
--, n-propyl ester	250	Monomer.	37
--, i-propyl ester	250	Monomer.	37
--, n-butyl ester	250	40 % monomer and traces of 1-butene.	38
	170	100 % monomer under 253.7 nm radiation.	38
--, i-butyl ester	250	Monomer.	37
--, sec-butyl ester	250	Monomer and small amount of olefin by cracking of side chain.	37
--, tert-butyl ester	180-200	High yields of isobutylene and water, 1 % monomer, trace of methacrylic acid, residue of poly(methacrylic anhydride).	39
	< 180	100 % monomer under 253.7 nm radiation.	
--, n-amyl ester	250	Monomer.	37
--, i-amyl ester	250	Monomer.	37
--, 1,2-dimethylpropyl ester	250	Monomer and small amount of olefin by side chain cracking.	37
--, neopentyl ester	250	Monomer.	37
--, 3,3-dimethylbutyl ester	250	Monomer.	37
--, 1,3-dimethylbutyl ester	250	Monomer and small amount of olefin by side chain cracking.	37
Poly(methacrylonitrile)	< 200	No volatile material, coloration through yellow, orange and red.	40
	220-270	50-100 % monomer depending upon pretreatment and purity of polymer.	40
1.4 Poly(vinyl halides), Poly(vinyl alcohol), Poly(vinyl esters)			
Poly(chlorotrifluoroethylene) (KEL-F)	347-418	25 % of products volatile at 25 °C - monomer with traces of C_3F_5Cl and $C_3F_4Cl_2$; 72.1 % of larger chain fragments involatile at 25 °C.	14
Poly(1,1-dichloro-2,2-difluoroethylene)	240	20 % black involatile residue; 80 % monomer.	22
Poly(perfluoroheptene)	210-270	100 % monomer.	49
Poly(perfluoropropylene)	280-400	100 % monomer.	49
Poly(tetrafluoroethylene)	504-538	> 95 % monomer, 2-3 % C_3F_6, no larger fragments (in vacuum).	14
	1200	Monomer yield drops, larger fragments appear (in vacuum).	23
	600-700	At 6.66 mbar pressure: pure monomer; at 1013 mbar pressure, 15.9 % monomer, 25.7 % C_3F_6, 58.4 % C_4F_8.	24

	Polymer	Temp. Range [$^{\circ}$C]	Degradation Product	References

1.4   Poly(vinyl halides), Poly(vinyl alcohol), Poly(vinyl esters)   (Cont'd.)

	Poly(trifluoroethylene)	380-800	High yields of HF and products involatile at 25°C.	23
	Poly(vinyl acetate)	213-235	Quantitative yields of acetic acid.	71
		300	Small amounts of aromatics including benzene.	71
	Poly(vinyl alcohol)	250	Quantitative yields of H_2O.	72
		240	Main products H_2O and C_2H_5OH, with aldehydes $CH_3(CH=CH)_nCHO$ and ketones $CH_3(CH=CH)_nCO \cdot CH_3$ where n = 0, 1, 2, etc.	73
	Poly(vinyl butyrate)	300-325	Butyric acid.	74
	Poly(vinyl chloride)	200-300	Quantitative yields of HCl.	75
		400	Saturated and unsaturated, aliphatic and aromatic hydrocarbons are produced with benzene and toluene in high yield.	75
		600	In helium; quantitative yield of HCl, remainder residue and hydrocarbons; benzene is major volatile hydrocarbon product.	76
	Poly(vinyl fluoride)	372-480	High yields of HF and products involatile at 25°C - little carbonization.	23
	Poly(vinylidene chloride)	225-275	High yields of HCl.	77
	Poly(vinylidene cyanide)	> 160	High yields of monomer	78
	Poly(vinylidene fluoride)	400-530	35 % HF and high yields of products involatile at 25°C - some carbonization.	23

1.5   Poly(vinyl ketones), Poly(vinyl ethers)

	Poly(methyl isopropenyl ketone)	270-360	H_2O.	41
		150-190	Monomer under 313 nm radiation.	42
	Poly(methyl vinyl ketone)	270-360	H_2O, 3-methyl-2-cyclohexene-1-one and other six membered ring ketones.	44
	Poly(trifluorovinyl phenyl ether)	275-500	Maximum of 75 % volatilization - products unknown.	49

1.6   Poly(styrenes)

	Poly(styrene)	300-400	40.6 % monomer, 2.0 % toluene, 0.1 % CO, remainder dimer, trimer and tetramer - monomer yield increases with pressure of nitrogen - 62 % at 1013 mbar.	62, 63
		500-1200	Small hydrocarbon fragments appear (C_1-C_6) - fragmentation is greater the higher the temperature and the greater the pressure of inert gas.	28
--,	(cross-linked)	346-450	Cross-linking with increasing quantities of divinyl or trivinyl benzene progressively decreases the styrene yield - the yield of larger chain fragments and the amount of carbonization also increase.	11, 64
--,	(poly(divinyl benzene))	385-450	Volatile products include toluene, benzene, styrene and xylene.	11, 64
--,	(poly(trivinyl benzene))	470-500	Mixture of aliphatic and aromatic hydrocarbons.	11, 64
--,	m-amino-	340-500	Gaseous fraction CO_2, CH_4, C_2H_6, C_3H_6, C_3H_8; liquid fraction m-aminostyrene, m-toluidine; gum fraction mainly dimer and trimer; cross-linked residue remains.	65
		320-500	Monomer yield rises from 41 % (320°C) to 61 % (500°C); minor product is m-toluidine, with traces of aniline, m-ethylaniline, α-methyl-m-aminostyrene, H_2, CO_2, CH_4; remainder comprises chain fragments and residue.	66
--,	α-deutero-	334-387	68.4 % monomer, 1.5 % α-deuterostyrene, 0.6 % α-methylstyrene, 29.5 % larger chain fragments.	2, 13
--,	β-deutero-	345-384	39.7 % monomer, 1.2 % toluene, 0.1 % deuterotoluene, 59 % larger chain fragments.	2, 13
--,	3-methyl-	309-399	44.4 % monomer, 7.3 % xylene, 48.3 % larger chain fragments.	3
--,	4-methyl-	200-350	Ratio of monomer to oligomer rises from 40 % at 200°C to 95 % at 350°C.	79
--,	α-methyl-	200-500	95-100 % monomer.	28
		500-1200	Fragments both larger and smaller than monomer appear in increasing amounts the higher the temperature, particularly CH_4, C_2H_4 and C_6H_6 until at 1200°C the monomer yield is only 33.9 %.	28
--,	2, 3, 4, 5, 6-pentafluoro-	390-446	63 % of the products are volatile at 25°C - contains some monomer.	67
--,	α, β, β-trifluoro-	333-382	72 % monomer, 28 % larger chain fragments.	14
	Poly(styrene-co-SO_2) (1.85:1)	200	At 26.7 mbar pressure; 40 % weight loss; main product is 2,4-diphenylthiophene; at least 11 unidentified minor products.	68

2.   Main-chain Carbocyclic Polymers

	Poly(perfluoro-m-phenylene)	700	Volatiles comprise SiF_4 (from silica vessel), CO_2, with traces of C_6F_6 and C_6F_5H; white sublimate on vessel walls; residue contains 1 % fluorine.	50

Polymer	Temp. Range [$^\circ$C]	Degradation Product	References
2. Main-chain Carbocyclic Polymers (Cont'd.)			
Poly(perfluoro-p-phenylene)			
high molecular weight (15.000)	700	Volatiles comprise SiF_4 (from silica vessel), CO_2, with traces of C_6F_6 and C_6F_5H; white sublimate on vessel walls; residue contains 1 % fluorine.	50
low molecular weight (3.700)	700	Volatiles comprise SiF_4 (from silica vessel), CO_2, with traces of C_6F_6 and C_6F_5H; no white sublimate; residue contains 33-41% fluorine.	50
Poly(1,4-phenylene)	250-620	79 % residue, 10 % chain fragments, 11 % volatiles comprising H_2, CH_4, H_2O, HCl (from catalyst).	55
--, 2-hydroxy-	300-620	60 % residue, 24 % chain fragments, 16 % volatiles comprising CO, CO_2, H_2, CH_4.	55
Poly(1,4-phenyleneethylene) (Poly(p-xylylene))	420-465	3.6 % of products are volatile at 25°C and consist of 2.83 % xylene, 0.29 % toluene, 0.28 % methylethyl benzene, 0.14 % methylstyrene, 0.06 % benzene - products involatile at 25°C consist of dimeric-octameric fragments.	13, 14, 80
	408-515	Mainly chain fragments (dimer - pentamer); traces of H_2, monomer.	81
Poly(1,4-phenylenemethylene)	386-416	7.4 % of products are volatile at 25°C; 5.9 % toluene, 1.4 % benzene, 0.1 % xylene.	13, 14
3. Main-chain Heteroatom Polymers			
3.1 Poly(oxides)			
Poly(oxybutylethylene)	321-365	Mixture of saturated and unsaturated hydrocarbons (C_1-C_6), aldehydes (C_1-C_6), H_2, H_2O.	20
Poly(oxy-2,5-dimethoxy-1,4-phenylene)	100-550	41 % residue, 29 % fragments, 30 % volatiles comprising H_2, CH_4, CO, CO_2, CH_3OH, traces of H_2O, C_2H_6, other hydrocarbons.	55
Poly(oxy-2,6-dimethyl-1,4-phenylene)	100-550	26 % residue, 66 % chain fragments, 8 % volatiles comprising H_2, CH_4, H_2O, CO, CO_2.	55
Poly(oxyethylene)	324-363	9.7 % of products volatile at 25°C - 3.9 % monomer with smaller amounts of CO_2, formaldehyde, ethanol and saturated and unsaturated C_1-C_7 compounds.	25
Poly(oxymethylene)	222	100 % monomer.	27
Poly(oxymethyleneoxyethylene)	314-338	Major volatile products CH_4, C_2H_6, C_2H_4, C_2H_2, $H \cdot CO_2C_2H_5$.	20
Poly(oxy-1,3-phenylene)	300-620	34 % residue, 53 % chain fragments, 13 % volatiles comprising H_2O, CO, H_2, CH_4, CO_2, trace of C_6H_6.	55
Poly(oxypropylene)			
atactic	270-330	12.8 % of products volatile at 25°C including 4.00 % acetaldehyde, 2.22 % acetone, 1.43 % dipropyl ether and 0.75 % propylene.	25
isotactic	275-355	20 % of products volatile at 25°C including 6.34 % acetaldehyde, 2.39 % acetone, 2.19 % dipropyl ether and 2.22 % propylene.	25
Poly(oxytetramethylene)	347	Volatiles comprise a mixture of CH_4, C_2H_6, C_3H_8, C_4H_{10}, C_2H_4, C_3H_6, $CH_3 \cdot CHO$, $C_2H_5 \cdot CHO$, $C_3H_7 \cdot CHO$.	20
3.2 Poly(esters)			
Poly(oxycarbonyloxy-1,4-phenyleneisopropylidene-1,4-phenylene)	300-389	Major products CO_2, bisphenol A; minor products CO, CH_4, 4-alkyl phenols.	60
Poly(oxyethyleneoxyterephthaloyl) (Poly(ethylene terephthalate))	283-306	Acetaldehyde major gaseous product with CO_2, CO, C_2H_4, H_2O, CH_4, benzene, 2-methyl-dioxolane, terephthalic acid and more complex chain fragments.	26
3.3 Poly(sulfides)			
Poly(thioacetone)	145	Main product is cyclic trimer, $[(CH_3)_2CS]$.	69
Poly(thioperfluorophenylene)	500	73 % residue, remainder, chain fragments, H_2, SiF_4 (from silica vessel), CO_2.	51
Poly(thiophenylene)	460	72 % residue, remainder chain fragments, H_2, dibenzthiophene.	51
	300-620	38 % residue, 47 % chain fragments, 15 % volatiles comprising mainly H_2S and H_2.	59

References page II-478

Polymer	Temp. Range [$^{\circ}$C]	Degradation Product	References
3. Main-chain Heteroatom Polymers (Cont'd.)			
3.4 Poly(amides)			
Poly(2,5-dimethyl-1,4-piperazinediylisophthaloyl)	475	After 1 hour, 5.1 % residue, 73.8 % chain fragments, 20.5 % volatiles comprising mainly CO, H_2O, CH_4, CO_2, H_2, with traces of hydrocarbons, pyrazines, pyrroles.	30
Poly(2,5-dimethyl-1,4-piperazinediyloxalyl)	475	After 1 hour, 1 % residue, 73.5 % chain fragments, comprising mainly CO, with traces of H_2, CO_2, H_2O, hydrocarbons, pyrazines, pyrroles.	47
Poly(2,5-dimethyl-1,4-piperazinediyltere-phthaloyl)	475	After 1 hour, 9.4 % residue, 3.2 % chain fragments, 87.4 % volatiles comprising mainly CO, H_2O, CO_2, CH_4, NH_3, with traces of hydrocarbons, pyrazines and pyrroles.	47
Poly(iminoisophthaloylimino-1,3-phenylene)	300-500	Mixture of H_2, CO, CO_2, H_2O, HCN, benzene, toluene, benzonitrile.	57
Poly(iminoisophthaloylimino-1,4-phenylene)	300-500	Mixture of H_2, CO, CO_2, H_2O, HCN, benzene, toluene, benzonitrile.	57
Poly(iminoisophthaloyl-co-terephthaloylimino-1,4-phenylene)	250-620	36 % residue, 24 % chain fragments, 40 % volatiles comprising H_2, CO, CO_2, CH_4.	58
Poly(imino(1-oxohexamethylene) (Nylon 6) Poly(iminohexamethyleneiminoadipoyl) (Nylon 66)	310-380	H_2O, CO_2, cyclopentanone, traces of saturated and unsaturated hydrocarbons; purification from water and acid polymerization catalysts increases stability and decreases yield of CO_2.	45, 46
Poly(iminoterephthaloylimino-1,3-phenylene)	300-500	Mixture of H_2, CO, CO_2, H_2O, HCN, benzene, toluene, benzonitrile.	57
Poly(iminoterephthaloylimino-1,4-phenylene)	300-500	Mixture of H_2, CO, CO_2, H_2O, HCN, benzene, toluene, benzonitrile.	57
Poly(2-methylpiperazinediylterephthaloyl)	475	After 1 hour, 11.3 % residue, 46.2 % chain fragments, 42.5 % volatiles comprising mainly CO, H_2O, CO_2, CH_4, NH_3, H_2, with traces of hydrocarbons, pyrazines and pyrroles.	30
Poly(1,4-piperazinediylterephthaloyl)	475	After 1 hour, 21.1 % residue, 45.6 % chain fragments, 33.3 % volatiles comprising mainly H_2O, CO, H_2, CO_2, NH_3, with traces of hydrocarbons, pyrazines and pyrroles.	30
3.5 Poly(siloxanes)			
Poly(borontri(dimethyl siloxane)), non-linear	250-350	Molecular weight increases; traces of cyclic products $[(CH_3)_2SiO]_3$ and $[(CH_3)_2SiO]_4$.	15
	350-500	Extensive weight loss (96 % at 500°C); major products $[(CH_3)_2SiO]_3$ and $[(CH_3)_2SiO]_4$; traces of H_2 and CH_4.	15
Poly(oxy-3,3,3-trifluoropropyl methyl silylene)	400	Main products are low molecular weight cyclic oligomers; some 1,1-difluoropropene, fluoroform, CH_4.	70
4. Main-chain Heterocyclic Polymers			
Phenol-Formaldehyde Resin	250-400	Volatiles comprise xylene (76 %), traces of phenol, cresol, benzene.	53
Poly(2,5-benzimidazolediyl-5,2-benzimidazolediyl-1,4-phenylene)	400-700	Mixture of phenol, terephthalodinitrile, benzonitrile, 4,4'-diaminodiphenyl, H_2, CO, NH_3, HCN, traces of aniline; carbonaceous residue remains.	56
Poly(5,7-dihydro-1,3,5,7-tetraoxobenzo[1,2-c:4,5-c'] dipyrrole-2,6[1H,3H]-diyl) (Poly(pyromellitimide))			
--, 1,4-phenylenecarbonyl-1,4-phenylene	700	Nitrogen 1013mbar pressure; 60 % residue, remainder H_2, CO, CH_4, CO_2, H_2O, C_6H_6, benzonitrile, chain fragments.	61
--, 1,4-phenylenemethylene-1,4-phenylene	700	Nitrogen 1013 mbar pressure; 60 % residue, remainder H_2, CO, CH_4, CO_2, H_2O, C_6H_6, toluene, benzonitrile, chain fragments.	61
--, 1,4-phenyleneoxy-1,4-phenylene	700	Nitrogen 1013 mbar pressure; 60 % residue, remainder H_2, CO, CH_4, CO_2, H_2O, C_6H_6, benzonitrile, phenol, chain fragments.	61
Poly(3-perfluoropropyl-1,5-triazinediylperfluoro-trimethylene)	415-505	C_2F_4, CF_4, C_2F_6 and smaller amounts of C_3F_6 and C_3F_8 and larger chain fragments.	48
Poly(perfluoropyridinediyl)	600	Volatiles comprise CO_2, CO, SiF_4, (all from silica vessel); carbonaceous residue remains.	52
Poly(5-phenyl-as-3,4-triazinediyl-2,6-pyridinediyl-5-phenyl-as-3,4-triazinediyl-1,4-phenylene-oxy-1,4-phenylene)	305-465	Only volatile product is benzonitrile.	54

Polymer	Temp. Range [^{o}C]	Degradation Product	References
Poly(5-phenyl-as-3,4-triazinediyl-2,6-pyridine-diyl-5-phenyl-as-3,4-triazinediyl-1,4-phenyleneoxy-1,4-phenylene) (Cont' d.)	480-660	Volatile products HCN, NH_3.	54
	690-760	Major volatile product is H_2; carbonaceous residue remains.	54
Poly(1,3,5-triazinetriylperfluorotrimethylene) (Poly(perfluoroglutarodiamidine))	430-503	C_2F_4 with traces of CF_4, C_3F_6 and larger chain fragments.	48

5. Cellulose and Derivatives

Cellulose	250-397	H_2O with smaller amounts of CO_2 and CO and a tar containing principally levoglucosan.	17, 18
Cellulose (oxidised)	180-331	Mainly H_2O and CO_2, smaller amounts of CO, formaldehyde, methanol, acetic acid, ethanol and acetaldehyde, and very little tar.	
Cellulose triacetate	250-310	Product fraction volatile at 25°C contains acetic acid, CO_2, CO, CH_4, H_2, acetaldehyde and acetone; heavier fractions do not contain levoglucosan acetate.	18

REFERENCES

1. G. G. Cameron, D. R. Kane, Polymer 9, 461 (1968).
2. S. Straus, S. L. Madorsky, J. Res. Natl. Bur. Std. 50, 165 (1953).
3. S. L. Madorsky, J. Polymer Sci. 11, 491 (1953).
4. J. R. Schaefgen, I. M. Sarasohn, J. Polymer Sci. 58, 1049 (1962).
5. C. S. Marvel, J. C. Cowan, J. Am. Chem. Soc. 61, 3156 (1939).
6. A. J. Canale, W. E. Goode, J. B. Kinsinger, J. R. Panchak, R. L. Kelso, R. K. Graham, J. Appl. Polymer Sci. 4, 231 (1960).
7. J. W. C. Crawford, D. Plant, J. Chem. Soc. 1952, 4492.
8. G. G. Cameron, G. P. Kerr, J. Polymer Sci., A1, 7, 3067 (1969).
9. W. J. Burlant, J. L. Parsons, J. Polymer Sci. 22, 249 (1956).
10. S. Straus, S. L. Madorsky, J. Res. Natl. Bur. Std. 61, 77 (1958).
11. S. L. Madorsky, S. Straus, J. Res. Natl. Bur. Std. 63A, 261 (1959).
12. A. R. Monahan, J. Polymer Sci., A1, 4, 2391 (1966).
13. S. L. Madorsky, S. Straus, J. Res. Natl. Bur. Std. 53, 361 (1954).
14. S. L. Madorsky, S. Straus, J. Res. Natl. Bur. Std. 55, 223 (1955).
15. M. A. Verkhotin, K. A. Andrianov, M. N. Yermakova, S. R. Rafikov, V. V. Rode, Polymer Sci. USSR 8, 2369 (1966).
16. S. L. Madorsky, S. Straus, D. Thompson, L. Williamson, J. Res. Natl. Bur. Std. 42, 499 (1949).
17. S. L. Madorsky, V. E. Hart, S. Straus, J. Res. Natl. Bur. Std. 56, 343 (1956).
18. S. L. Madorsky, V. E. Hart, S. Straus, J. Res. Natl. Bur. Std. 60, 343 (1958).
19. R. F. Schwenker, Jr., L. R. Beck, Textile Research J. 624, (August, 1960).
20. A. B. Blyumenfel'd, B. M. Kovarskaya, Polymer Sci. USSR 12, 710 (1970).
21. L. G. Kaufman, P. T. Funke, A. A. Volpe, Macromolecules 3, 358 (1970).
22. J. L. Cotter, G. J. Knight, W. W. Wright, J. Polymer Sci., B, 6, 763 (1968).
23. S. L. Madorsky, V. E. Hart, S. Straus, V. A. Sedlek, J. Res. Natl. Bur. Std. 51, 327 (1953).
24. E. E. Lewis, M. A. Naylor, J. Am. Chem. Soc. 69, 1968 (1947).
25. S. L. Madorsky, S. Straus, J. Polymer Sci. 36, 183 (1959).
26. E. P. Goodings, Soc. Chem. Ind. (London), Monograph 13, 211 (1961).
27. C. E. Schweitzer, R. N. MacDonald, J. O. Punderson, J. Appl. Polymer Sci. 1, 158 (1959).
28. S. Straus, S. L. Madorsky, J. Res. Natl. Bur. Std. 66A, 401 (1962).
29. Y. Tsuchiya, K. Sumi, J. Polymer Sci., A1, 7, 813 (1969).
30. S. D. Bruck, Polymer 7, 231 (1966).
31. S. Straus, S. L. Madorsky, Ind. Eng. Chem. 48, 1212 (1956).
32. B. S. T. Boonstra, G. J. Van Amerongen, Ind. Eng. Chem. 41, 161 (1949).
33. D. H. Grant, N. Grassie, Polymer 1, 125 (1960).
34. N. Grassie, H. W. Melville, Proc. Roy. Soc. A190, 1 (1949).
35. P. D. Zemany, Nature 171, 391 (1953).
36. P. R. E. J. Cowley, H. W. Melville, Proc. Roy. Soc. A210, 461 (1952).
37. J. W. C. Crawford, J. Soc. Chem. Ind. 68, 201 (1949).
38. N. Grassie, J. R. MacCallum, J. Polymer Sci. 2, 983 (1964).
39. D. H. Grant, N. Grassie, Polymer 1, 445 (1960).
40. N. Grassie, I. C. McNeill, J. Chem. Soc. 1956, 3929.
41. C. S. Marvel, E. H. Riddle, J. O. Corner, J. Am. Chem. Soc. 64, 92 (1942).
42. K. F. Wissbrun, J. Am. Chem. Soc. 81, 58 (1959).
43. L. Reginato, Makromol. Chem. 132, 113 (1970).
44. C. S. Marvel, C. L. Levesque, J. Am. Chem. Soc. 60, 280 (1938).
45. S. Straus, L. A. Wall, J. Res. Natl. Bur. Std. 60, 39 (1958).
46. S. Straus, L. A. Wall, J. Res. Natl. Bur. Std. 63A, 269 (1959).
47. S. D. Bruck, Polymer 6, 483 (1965).
48. L. A. Wall, S. Straus, J. Res. Natl. Bur. Std. 65A, 227 (1961).
49. S. Straus, L. A. Wall, Soc. Plastics Eng. J. 56 (January 1964).
50. J. L. Cotter, G. J. Knight, J. M. Lancaster, W. W. Wright, J. Appl. Polymer Sci. 12, 2481 (1968).
51. N. S. J. Christopher, J. L. Cotter, G. J. Knight, W. W. Wright, J. Appl. Polymer Sci. 12, 863 (1968).
52. J. L. Cotter, H. Dickinson, G. J. Knight, W. W. Wright, J. Appl. Polymer Sci. 15, 317 (1971).
53. A. A. Berlin, V. V. Yarkina, A. P. Firsov, Polymer Sci. USSR 10, 2219 (1968).
54. W. Wrasidlo, P. M. Hergenrother, Macromolecules 3, 548 (1970).
55. G. F. L. Ehlers, K. R. Fisch, W. R. Powell, J. Polymer Sci., A1, 7, 2931 (1969).
56. V. V. Rode, N. M. Kotsoyeva, G. M. Cherkasova, D. S. Tugushi, G. M. Tseitlin, A. L. Rusanov, V. V. Korshak, Polymer Sci. USSR 12, 2103 (1970).
57. Ye. P. Krasnov, V. M. Savinov, L. B. Sokolov, V. I. Logunova, V. K. Belyakov, T. A. Polyakova, Polymer Sci. USSR 8, 413 (1966).
58. G. F. L. Ehlers, K. R. Fisch, W. R. Powell, J. Polymer Sci., A1, 7, 2969 (1969).
59. G. F. L. Ehlers, K. R. Fisch, W. R. Powell, J. Polymer Sci., A1, 7, 2955 (1969).
60. A. Davis, J. H. Golden, J. Chem. Soc., B1, 45 (1968).
61. D. P. Bishop, D. A. Smith, J. Appl. Polymer Sci. 14, 345 (1970).
62. S. L. Madorsky, S. Straus, J. Res. Natl. Bur. Std. 40, 417 (1948).
63. S. L. Madorsky, "Thermal Degradation of Organic Polymers", Wiley, New York (1964).

64. S. Straus, S. L. Madorsky, J. Res. Natl. Bur. Std. 65A, 243 (1961).

65. R. H. Still, P. B. Jones, A. L. Mansell, J. Appl. Polymer Sci. 13, 401 (1969).

66. R. H. Still, P. B. Jones, J. Appl. Polymer Sci. 13, 2033 (1969).

67. L. A. Wall, J. M. Antonucci, S. Straus, M. Tryon, Soc. Chem. Ind. (London), Monograph 13, 295 (1961).

68. D. C. Allport, Polymer 8, 492 (1967).

69. V. C. E. Burnop, K. G. Latham, Polymer 8, 589 (1967).

70. S. N. Novikov, Ye. G. Kagan, A. N. Pravednikov, Polymer Sci. USSR 8, 1114 (1966).

71. N. Grassie, Trans. Faraday Soc. 48, 379 (1952).

72. J. B. Gilbert, J. J. Kipling, Fuel 12, 249 (1962).

73. Y. Tsuchiya, K. Sumi, J. Polymer Sci., A1, 7, 3151 (1969).

74. J. B. Gilbert, J. J. Kipling, B. McEnaney, J. N. Sherwood, Polymer 3, 1 (1962).

75. R. R. Stromberg, S. Straus, B. G. Achhammer, J. Polymer Sci. 35, 355 (1959).

76. M. M. O'Mara, J. Polymer Sci., A1, 8, 1887 (1970).

77. F. H. Winslow, W. O. Baker, W. A. Yager, quoted in ref. 63, Chapter V.

78. H. Gilbert, F. F. Miller, S. J. Averill, R. F. Schmidt, F. D. Stewart, H. L. Trumbull, J. Am. Chem. Soc. 76, 1074 (1954).

79. Y. A. Glagoleva, V. R. Regel, Polymer Sci. USSR 12, 1078 (1970).

80. P. Brandt, V. H. Dibeler, F. L. Mohler, J. Res. Natl. Bur. Std. 50, 201 (1953).

81. H. H. G. Jellinek, S. N. Lipovac, J. Polymer Sci., A1, 8, 2517 (1970).

RADIATION CHEMICAL YIELDS: "G-VALUES"

A. Chapiro

Laboratoire de Chimie des Radiations du C. N. R. S.
92-Bellevue, France

Figures compiled from the following books and review articles: "Radiation Chemistry of Polymeric Systems" Interscience Publishers, 1962, "Selected Constants--Radiolytic Yields", Tables des Constantes, Vol. 13, IUPAC Pergamon Press, 1963. "Polymérisation en phase solide amorcées par les radiations" in M. Haissinsky's Ed. "Actions Chimiques et Biologiques des Radiations", Masson et Cie, Paris, 10e Série (1966) pp. 187-312. "Radiation-Induced Reactions" in "Encyclopedia of Polymer Science and Technology", John Wiley & Sons Inc., Vol. 11 (1969) pp. 702-60.

"G-Values" are defined as radiation-chemical yields of individual atomic or molecular events for 100 eV absorbed in the system. $G(R^{\cdot})$ stands for the yields of free radicals per 100 eV; $G(\text{init.})$ for yield of initiating centers; $G(\text{c.l.})$ for yield of cross-links; $G(\text{breaks})$ for yield of main chain scissions.

Table I: Yields of free radicals for monomers

Monomer	$G(R^{\cdot})$	Monomer	$G(R^{\cdot})$
Styrene	0.66 - 0.69	Vinyl acetate	9.6 - 12
Ethylene	4.4	Acrylonitrile	2.4 - 5.6
Methyl methacrylate	5.5 - 11.5	Ethyl acrylate	10.9
Methyl acrylate	6.3 - 15	n-Butyl acrylate	17.4

Table II: Yields of free radicals for solvents

Solvent	$G(R^{\cdot})$	Solvent	$G(R^{\cdot})$
Benzene	0.76	Methyl acetate	10.9
Toluene	1.15	Ethyl acetate	11.5 - 12.0
Ethylbenzene	4.0	Acetone	9.4
Chlorobenzene	8.0	Ethyl bromide	11.8

Table III: Estimated yields of free radicals for polymers

Polymer	$G(R^{\cdot})$	Polymer	$G(R^{\cdot})$
Poly(butadiene) } Poly(isoprene) }	2 - 4	Silicones	3.6 or 7.2
Poly(styrene)	1.5 - 3	Cellulose } Poly(vinyl alcohol) }	10
Poly(ethylene)	6 - 8	Poly(vinyl chloride) }	
Poly(isobutylene)	6 - 8	Poly(vinylidene chloride) }	10 - 15
Poly(vinyl acetate) } Poly(methyl acrylate) }	6 or 12		

Table IV: Yields of ionic initiation in liquid monomers

Monomer (and solvent)	$G(\text{init.})$	Monomer (and solvent)	$G(\text{init.})$
Isobutylene	0.15 - 0.30	Styrene/CH_2Cl_2 (1/3)	0.5
Butadiene	0.2		

Table V: Yields of initiation and of trapped radicals for crystalline monomers

Monomer	Temperature [°C]	$G(\text{init.})$	$G(\text{trapped } R^{\cdot})$
Acrylamide	20	1.8 - 2.0	1.0
	30	0.8 - 1.0	-
Acrylonitrile	-196	0.15 - 0.28	0.5

Monomer	Temperature [°C]	G (init.)	G (trapped R·)
Acrylonitrile (Cont'd.)	-110	0.71	-
Acrylic acid	-196	-	1.2
Styrene	-80	0.06	-
	-35	0.61	-
Vinyl acetate (vitreous)	-150	0.18	0.67
Tetrafluoroethylene	-196	-	0.4
Formaldehyde	-196	-	3.7
Acetaldehyde	-196	-	1.6

Table VI: Yields of chemical changes in poly(ethylene)

Chemical Process	at 20°C	$G_{below\ -\ 40°C}$
Crosslinking	2.0	1.0
Hydrogen evolution	4.1	3.2
Unsaturation	1.8	1.8

Table VII: Yields of crosslinking (c.l.) and ratio β/α of degradation to crosslinking for polymers of the "crosslinking type"

Polymer	G(c.l.) at 20°C	β/α
Poly(propylene)	0.6	0.8 - 1.0
Poly(styrene)	0.04 - 0.06	0 - 0.2
Natural rubber	1.3	0.14
Poly(butadiene)	2.0	
Poly(acrylonitrile)	1.4	
Poly(methyl acrylate)	0.5 - 1.1	0.17
Poly(vinyl chloride)	0.2 - 0.5	
Poly(vinyl acetate)	0.28	0.1
Poly(dimethyl siloxane)	2.5	0
Poly(methylphenyl siloxane)	0.8	
Poly(amides)	0.3	

Table VIII: Yields of degradation for polymers of the "degrading type"

Polymer	G (breaks) at 20°C	Polymer	G (breaks) at 20°C
Poly(isobutene)	3.0	Cellulose	10.0
Poly(methyl methacrylate)	1.9	Poly(α-methylstyrene)	0.25

CRITICAL MICELLE CONCENTRATION

H. Gerrens and G. Hirsch

BASF AG
Ludwigshafen, Germany

Contents

A. INTRODUCTION

In an aqueous solution of a surface active agent (surfactant) the surfactant is molecularly dispersed at low concentrations. At higher concentrations, however, when a certain critical concentration is reached the molecules form micelles. These micelles are in equilibrium with the free surfactant molecules. In the case of ionic surfactants the micelles contain about 50 - 200 molecules. The concentration that must be reached in order that micelles are formed is called the critical micelle concentration (CMC). Many physical properties of the surfactant solution when plotted against the concentration show more or less sudden changes at the CMC. By measuring such properties as electrical conductivity, interfacial tension, surface tension, refractive index, viscosity and light scattering as a function of the concentration of the surfactant the CMC is determined as the concentration at which the property versus concentration curve shows a change in slope. The hydrophobic part of the surfactant molecule is situated at the inside of the micelle, the hydrophilic part at the outside. Inside the micelles lipophilic substances may be solubilized. Certain dyes show a spectral change when they are solubilized in the micelles. This effect too, is used for the measurement of the CMC.

In emulsion polymerization the monomer is emulsified in the solution of a surfactant (emulsifier) in water. Part of the monomer is solubilized in the micelles. As free radicals enter the micelles the monomer in them is polymerized and polymer particles (latex particles) are formed. The surfactant molecules are adsorbed at the surface of the particles and protect them from coagulation. According to the theory of Smith and Ewart [1], the number N of latex particles formed in emulsion polymerization is given by

$$N = 0.43 \cdot \left(\frac{\rho}{\mu}\right)^{2/5} \cdot \left(a_s [S]\right)^{3/5} \qquad (1)$$

N = number of latex particles per cm^3 of emulsion $[cm^{-3}]$
ρ = rate of formation of free radicals $[cm^{-3} s^{-1}]$
μ = rate of volume increase of a particle $[cm^3 \cdot s^{-1}]$
a_s = interfacial area occupied by one surfactant molecule in micelles or at the surface of latex particles $[cm^2 \cdot molecule^{-1}]$
$[S]$ = concentration of surfactant $[molecules \cdot cm^{-3}]$

Colloidal and technological properties of a polymer are to a large extent determined by the size and the number of the latex particles. The rate of polymerization R_p in emulsion is also governed by the number N of latex particles and can be expressed as

$$R_p = \frac{1}{2 N_A} \cdot N \cdot k_p \cdot [M] \qquad (2)$$

R_p = overall rate of polymerization $[mol \cdot cm^{-3} \cdot s^{-1}]$
1/2 = average number of radicals per latex particle
N_A = Avogadro's number $[mol^{-1}]$
k_p = propagation constant $[cm^3 \cdot mol^{-1} \cdot s^{-1}]$
$[M]$ = monomer concentration $[mol \cdot cm^{-3}]$

In the derivation of Equation 1 the molecularly dissolved portion of the emulsifier is neglected. Technical emulsion polymerizations, however, frequently use very low emulsifier concentrations. Here the simplification is no longer permissible and the factor $[S]$ in equation 1 must be substituted by $([S] - CMC)$ [2]. Generally, with decreasing CMC the number of latex particles N and the rate of polymerization R_p will increase.

In a homologous series, there is found a linear dependence of the logarithm of the CMC on the number n of carbon atoms in the paraffin chain of the surfactant [3] or on the number m of ethylene oxide units in the poly(oxyethylene) chain of ethylene oxide adducts.

$$\log CMC = A - B \cdot n \tag{3}$$

and

$$\log CMC = A' - B' \cdot m \tag{4}$$

Literature values of the CMC were plotted according to Equations 3 or 4 and the best straight line was drawn through the points. The constants A, B, A' and B' of Equations 3 and 4 are listed in Tables 1-4 together with the numerical values of the CMC calculated with the aid of these constants. Equations 3 and 4 may be used for interpolation and within reasonable boundaries also for extrapolation of the reported values. Comparing the values for ionic surfactants obtained by different authors, it was observed that the deviations from the straight line given by Equation 3 were always greater than the differences caused by different temperatures. Therefore, only the temperature range of the measurements is given in Tables 1-3. Table 5 shows the temperature dependence of the CMC for some selected examples. For ionic surfactants the CMC increases slightly with increasing temperature, for nonionics it decreases. Table 5 includes values of the conventional heat of micelle formation ΔH_m, which according to Stainsby and Alexander (4), is given by the relation:

$$\Delta H_m = -RT^2 \ (d \ln CMC/dT) \tag{5}$$

Addition of salts lowers the CMC of ionic surfactants (5-8). Equation 3 changes to

$$\log CMC = K_1 - K_2 \ n - K_3 \cdot \log c_G \tag{3a}$$

where c_G is the total concentration of counterions ("gegenionen"), K_1, K_2 and K_3 are constants. K_3 is generally of the order 0.5. For theoretical foundations of Equations 3, 3a and 4 see ref. (9). Nonionics show a similar behaviour on the addition of salts. However, certain organic compounds such as urea and formamide are known to considerably increase the CMC of ionic (10) and especially of nonionic surfactants (11-13).

Mixing of different surfactants results in the formation of mixed micelles with a specific CMC different from that of the pure components (10, 14-18, 83, 114).

The Krafft point (19) is the triple point where three phases are in equilibrium: the molecularly dissolved surfactant, the micellar surfactant and the crystalline or gel-like surfactant. The Krafft point can be determined by measurements of the solubility of the surfactant with increasing temperature. The solubility suddenly increases when the temperature of the Krafft point is reached. This temperature is listed in Tables 1 and 2. The Krafft points show a pronounced alternating between odd and even members of a homologous series (20, 21). This must be kept in mind for interpolation. Most nonionic surfactants exhibit no Krafft points. For emulsion polymerizations the polymerization temperature should be above the Krafft point, or not all of the emulsifier will participate in the reaction. This effect may be desirable in certain special cases, for instance the formation of monodisperse latices (22). Table 6 shows the influence of counterions on the Krafft point (21a, 23). Addition of salt results in a decrease of surfactant solubility and in an increase in Krafft point temperature (21a).

When the aqueous solution of a nonionic surface active agent is heated, it suddenly becomes cloudy within a narrow temperature range. This is called the Cloud Point. Further heating causes separation into two phases (24). No micelles are present above the cloud point. Emulsion polymerization therefore should be carried out at temperatures below the cloud point. Temperatures of cloud points are given in Table 4. The cloud point can be influenced by electrolytes and by solubilized organic substances. At low surfactant concentrations its temperature is practically independent of concentration (25).

The surface area a_s occupied by a surfactant molecule adsorbed on a latex particle is most conveniently determined by soap titration (26-37), but measurements have been published for only a few surfactants. The nature of the polymer that forms the latex particles appears to have little influence on the adsorption area a_s. In Tables 1-4 values for a_s determined by soap titration are marked with an asterisk. Another method for the determination of approximate values of a_s is the evaluation of measurements of the interfacial tension γ of a surfactant solution against air or better against an organic solvent immiscible with water (35). In the presence of an excess of salt (constant ionic strength) Gibbs' Equation (38,39) for the interfacial concentration Γ of the surfactant is valid

$$\Gamma = -1/RT \ (d\gamma/d \ln C) \quad [mol \ cm^{-2}] \tag{6}$$

$$a_s = 1/\Gamma \cdot N_A \qquad [cm^2/molecule] \tag{6a}$$

In the absence of salt the factor 1/2 must be added to the right side of Equation 6 (40,41).

According to recipe and polymerization technique, e.g. batchwise, semicontinuous (42,43) or continuous (44-46) emulsion polymerization, the surface of the resulting latex particles may be more or less covered with surfactant molecules, saturation being reached with the formation of a monolayer. The degree of coverage can be determined by means of soap titration and has an influence on colloidal and technological properties such as latex stability, rheology after addition of pigments, and film formation.

Table 1 - Critical Micelle Concentration, Krafft Point and Adsorption Area of Anionic Surfactants

Surfactant	Mol.Wt.	Temp. [°C]	CMC [mol/l]	CMC [g/l]	Krafft Point [°C]	Adsorption Area $a_s \cdot 10^{16}$ [cm²/molecule]	References
Na- and K-Salts of Saturated Fatty Acids $C_nH_{2n-1}OONa(K)$			$\log CMC = 1.96 - 0.296n$ [x]				
$C_5H_9O_2$ Na	124.12	20	3.0	3.7×10^2			47
$C_5H_9O_2$ K	140.22	25		4.2×10^2			48
$C_6H_{11}O_2$ Na	138.14	20	1.5	2.1×10^2			47
$C_6H_{11}O_2$ K	154.24	25		2.3×10^2			48,49,50
$C_7H_{13}O_2$ Na	152.17	20	7.7×10^{-1}	1.2×10^2			47
$C_7H_{13}O_2$ K	168.27	25		1.3×10^2			49,50
$C_8H_{15}O_2$ Na	166.19	20	3.9×10^{-1}	6.5×10^1			47,51
$C_8H_{15}O_2$ K	182.30	25		7.1×10^1			48,49,50
$C_9H_{17}O_2$ Na	180.22	20	1.95×10^{-1}	3.5×10^1			47,51
$C_9H_{17}O_2$ K	196.32	25		3.8×10^1			49
$C_{10}H_{19}O_2$ Na	194.25	20	1.0×10^{-1}	1.9×10^1			47,51
$C_{10}H_{19}O_2$ K	210.35	25		2.1×10^1			48,49,50,51
$C_{11}H_{21}O_2$ Na	208.28	-	5.0×10^{-2}	1.0×10^1			-
$C_{11}H_{21}O_2$ K	224.38	25		1.1×10^1			49
$C_{12}H_{23}O_2$ Na	222.30	20-70	2.5×10^{-2}	5.6	36	41.4*, 45	23,26,31,47,51,52,53
$C_{12}H_{23}O_2$ K	238.40	20-25		6.0	< 0	44	23,26,48,49,51,53
$C_{13}H_{25}O_2$ Na	236.33	-	1.3×10^{-2}	3.1			-
$C_{13}H_{25}O_2$ K	252.43	25		3.3			49
$C_{14}H_{27}O_2$ Na	250.35	50-70	6.5×10^{-3}	1.6	53	34.1*	23,26,31,52
$C_{14}H_{27}O_2$ K	266.45	20-25		1.7	8		23,26,48,49,51
$C_{15}H_{29}O_2$ Na	264.38	-	3.3×10^{-3}	8.7×10^{-1}			-
$C_{15}H_{29}O_2$ K	280.48	-		9.3×10^{-1}			-
$C_{16}H_{31}O_2$ Na	278.40	50-70	1.7×10^{-3}	4.7×10^{-1}	62	25.1*	23,26,31,51,54
$C_{16}H_{31}O_2$ K	294.50	35		5.0×10^{-1}	30		23,26,48
$C_{17}H_{33}O_2$ Na	292.43	-	8.6×10^{-4}	2.5×10^{-1}			-
$C_{17}H_{33}O_2$ K	308.53	-		2.65×10^{-1}			-
$C_{18}H_{35}O_2$ Na	306.45	50-60	4.4×10^{-4}	1.3×10^{-1}	71	23.4*	23,26,31,51
$C_{18}H_{35}O_2$ K	322.55			1.4×10^{-1}	46		23,26
Sodium n-Alkyl Sulfates $C_nH_{2n+1}OSO_3Na$			$\log CMC = 1.43 - 0.290n$				
$C_6H_{13}SO_4Na$	204.22	-	4.9×10^{-1}	9.9×10^1		-	-
$C_8H_{17}SO_4Na$	232.27	25-50	1.3×10^{-1}	3.0×10^1	< 0		23,40,54,55
$C_9H_{19}SO_4Na$	236.30	20-25	6.5×10^{-2}	1.5×10^1		55.7	21,53
$C_{10}H_{21}SO_4Na$	260.33	25-50	3.4×10^{-2}	8.85	8	50;52	21,23,40,54,56
$C_{11}H_{23}SO_4Na$	274.36	25	1.7×10^{-2}	4.7	7.0		21
$C_{12}H_{25}SO_4Na$	288.38	25-60	9.0×10^{-3}	2.6	20;8.0;16	49;52;53.4	21,23,40,53,54,55 56,57,58,59,60
$C_{13}H_{27}SO_4Na$	302.41	25	4.6×10^{-3}	1.4	20.8		21
$C_{14}H_{29}SO_4Na$	316.43	40-60	2.4×10^{-3}	7.5×10^{-1}	33;20.5;30	50;51;44.2	21,21a,23,40 54,55,57,59
$C_{15}H_{31}SO_4Na$	330.46	31.5	1.2×10^{-3}	4.0×10^{-1}	31.5		21
$C_{16}H_{33}SO_4Na$	344.48	40-60	6.2×10^{-4}	2.15×10^{-1}	46;31.0;45	54	21,23,40,53,54, 55,57,59,60,61
$C_{17}H_{35}SO_4Na$	358.51	38.2	3.3×10^{-4}	1.2×10^{-1}	38.2		21
$C_{18}H_{37}SO_4Na$	372.54	40-60	1.65×10^{-4}	6.2×10^{-2}	58;40.5;56		21,23,40,54,55,57,59
Sodium n-Alkyl Sulfonates $C_nH_{2n+1}SO_3Na$			$\log CMC = 1.53 - 0.290n$				
$C_6H_{13}SO_3Na$	188.22	-	6.2×10^{-1}	1.2×10^2	-		
$C_8H_{17}SO_3Na$	216.27	25	1.6×10^{-1}	3.5×10^1	(15)		23,48,62,63
$C_{10}H_{21}SO_3Na$	244.33	25-80	4.3×10^{-2}	1.0×10^1	24;22.5		23,48,62-66
$C_{11}H_{23}SO_3Na$	258.36	20	2.2×10^{-2}	5.6	21.5	54	20,53
$C_{12}H_{25}SO_3Na$	272.38	35-80	1.1×10^{-2}	2.3	33;31.5;38		20,23,48,59, 60,62-66,67
$C_{13}H_{27}SO_3Na$	286.41	25-50	5.8×10^{-3}	1.7	35.5		20,59
$C_{14}H_{29}SO_3Na$	300.43	40-80	2.9×10^{-3}	8.7×10^{-1}	42;39.5;48	63*	20,23,32,48,59, 60,62-66,67

[x] No significant differences between the CMC of Na- and K-compounds.

* Determined by soap titration, all other values of a_s derived from measurements of surface or interfacial tension.

CRITICAL MICELLE CONCENTRATION

Surfactant	Mol.Wt.	Temp. [°C]	CMC [mol/l]	CMC [g/l]	Krafft Point [°C]	Adsorption Area $a \cdot 10^{16}$ [cm²/molecule]	References

Sodium n-Alkyl Sulfonates (Cont'd.) $C_nH_{2n+1}SO_3Na$ log CMC = 1.53 - 0.290n

Surfactant	Mol.Wt.	Temp.	CMC [mol/l]	CMC [g/l]	Krafft Point	Ads. Area	References
$C_{15}H_{31}SO_3Na$	314.46	25-50	1.5×10^{-3}	4.7×10^{-1}	48		20,59
$C_{16}H_{33}SO_3Na$	328.48	50	7.7×10^{-4}	2.5×10^{-1}	51;47.5;57		20,23,48,59 60,63,64,67
$C_{17}H_{35}SO_3Na$	342.51	25-50	4.0×10^{-4}	1.4×10^{-1}	62		20,59
$C_{18}H_{37}SO_3Na$	356.53	57	2.0×10^{-4}	7.2×10^{-2}	60;57.0;70		20,23,64

Alkylsulfonic Acids $C_nH_{2n+1}SO_3H$ log CMC = 0.928 - 0.274n

Surfactant	Mol.Wt.	Temp.	CMC [mol/l]	CMC [g/l]	Krafft Point	Ads. Area	References
$C_{12}H_{25}SO_3H$	250.44	25-50	4.4×10^{-3}	1.1			59
$C_{13}H_{27}SO_3H$	264.47	25-50	2.4×10^{-3}	6.3×10^{-1}			59
$C_{14}H_{29}SO_3H$	278.50	25-50	1.3×10^{-3}	3.6×10^{-1}	12.5		59
$C_{15}H_{31}SO_3H$	292.53	25-50	6.6×10^{-4}	1.9×10^{-1}	27.0		59
$C_{16}H_{33}SO_3H$	306.56	25-50	3.5×10^{-4}	1.1×10^{-1}	34.0		59
$C_{17}H_{35}SO_3H$	320.59	25-50	1.9×10^{-4}	6.1×10^{-2}	42.5		59
$C_{18}H_{37}SO_3H$	334.62	25-50	1.0×10^{-4}	3.3×10^{-2}	50.0		59

Sodium n-Alkylbenzene Sulfonates $C_nH_{2n+1}C_6H_4SO_3Na$ log CMC = 0.084 - 0.253n

Surfactant	Mol.Wt.	Temp.	CMC [mol/l]	CMC [g/l]	Krafft Point	Ads. Area	References
$C_6H_{13}C_6H_4SO_3Na$	264.31	75	3.7×10^{-2}	9.8			68
$C_7H_{15}C_6H_4SO_3Na$	278.34	75	2.1×10^{-2}	5.85			68
$C_8H_{17}C_6H_4SO_3Na$	292.37	20-75	1.15×10^{-2}	3.4			68-72
$C_9H_{19}C_6H_4SO_3Na$	306.39	75	6.5×10^{-3}	2.0			68
$C_{10}H_{21}C_6H_4SO_3Na$	320.42	50-75	3.6×10^{-3}	1.2	47		68,69,73,74
$C_{11}H_{23}C_6H_4SO_3Na$	334.44	-	2.0×10^{-3}	6.7×10^{-1}	50		74
$C_{12}H_{25}C_6H_4SO_3Na$	348.47	50-75	1.15×10^{-3}	4.0×10^{-1}	64	52	53,68,69,71-74
$C_{14}H_{29}C_6H_4SO_3Na$	376.52	70-75	3.45×10^{-4}	1.3×10^{-1}	75		68,73,74
$C_{16}H_{33}C_6H_4SO_3Na$	404.57	50-75	1.1×10^{-4}	4.4×10^{-2}			68,72,73
$C_{18}H_{37}C_6H_4SO_3Na$	432.63	-	3.4×10^{-5}	1.5×10^{-2}			-

Sodium p-1-Methyl Alkylbenzene Sulfonates $C_nH_{2n+1}CH(CH_3)C_6H_4SO_3Na$ log CMC = -0.456 - 0.215n

Surfactant	Mol.Wt.	Temp.	CMC [mol/l]	CMC [g/l]	Krafft Point	Ads. Area	References
$C_{10}H_{21}$-CH(CH$_3$)-$C_6H_4SO_3Na$	348.47	19-40	2.5×10^{-3}	8.7×10^{-1}	19.0		69
$C_{12}H_{25}$-CH(CH$_3$)-$C_6H_4SO_3Na$	376.52	28-40	9.3×10^{-4}	3.5×10^{-1}	27.7		69
$C_{14}H_{29}$-CH(CH$_3$)-$C_6H_4SO_3Na$	404.57	33-40	3.45×10^{-4}	1.4×10^{-1}	32.6		69
$C_{16}H_{33}$-CH(CH$_3$)-$C_6H_4SO_3Na$	432.63	45-50	1.3×10^{-4}	5.6×10^{-2}	45.5		69

Sodium Benzene Sulfonates, p-substituted $R-C_6H_4SO_3Na$

Surfactant	Mol.Wt.	Temp.	CMC [mol/l]	CMC [g/l]	Krafft Point	Ads. Area	References
R = n-$C_{11}H_{23}$O-	350.50	70	3.2×10^{-3}	1.1	75		74
R = n-$C_{11}H_{23}$NH-	349.52	60	3.2×10^{-3}	1.1	< 10		74
R = n-$C_{11}H_{23}$S-	366.56	75	2.8×10^{-3}	1.0	84		74
R = n-$C_{11}H_{23}$SO-	382.66	60	6.5×10^{-3}	2.5	32		74
R = n-$C_{11}H_{23}$SO$_2$-	398.56	72	8.5×10^{-3}	3.4	72		74
R = n-$C_{10}H_{21}$OOC-	364.48	60	4.1×10^{-3}	1.5	52		74
R = n-$C_{10}H_{21}$COO-	364.48	60	5.0×10^{-3}	1.8	24		74
R = n-C_9H_{19}OOCCH$_2$-	364.48	60	8.5×10^{-3}	3.1	< 10		74
R = n-C_8H_{17}OOCCH$_2$O-	366.51	60	2.8×10^{-2}	1.0	32		74
R = n-$C_{10}H_{21}$CONH-	363.50	60	7.3×10^{-3}	2.65	41		74
(n-$C_{11}H_{23}$NHC$_6$H$_4$SO$_3$)$_2$Ca	693.12	60	7×10^{-4} x	9.7×10^{-1}	50		74

Sodium 2-Alkene Sulfonates $R_{n-3}-CH=CH-CH_2SO_3Na$ log CMC = 1.819 - 0.310n

Surfactant	Mol.Wt.	Temp.	CMC [mol/l]	CMC [g/l]	Krafft Point	Ads. Area	References
$C_{12}H_{23}SO_3Na$	270.41	40	1.2×10^{-2}	3.25	< 6		67
$C_{14}H_{27}SO_3Na$	298.47	40	2.9×10^{-3}	8.7×10^{-1}	< 6		67
$C_{16}H_{31}SO_3Na$	326.53	40	7.0×10^{-4}	2.3×10^{-1}	36		67
$C_{18}H_{35}SO_3Na$	354.59	40	1.7×10^{-4}	6.0×10^{-2}	54		67

---

x mequiv/l.

Surfactant	Mol.Wt.	Temp. [°C]	CMC [mol/l]	CMC [g/l]	Krafft Point [°C]	Adsorption Area $a_s \cdot 10^{16}$ [cm²/molecule]	References

Sodium 3-Oxoalkane Sulfonates R_{n-3}-CO-CH$_2$-CH$_2$SO$_3$Na   log CMC = 2.294 - 0.309n

Surfactant	Mol.Wt.	Temp.	CMC [mol/l]	CMC [g/l]	Krafft Pt.	References
$C_{12}H_{23}(O)SO_3Na$	286.41	40	2.8×10^{-2}	8.0	28	67
$C_{14}H_{27}(O)SO_3Na$	314.47	40	7.2×10^{-3}	2.3	40	67
$C_{16}H_{31}(O)SO_3Na$	342.53	40	1.9×10^{-3}	6.5×10^{-1}	51	67
$C_{18}H_{35}(O)SO_3Na$	370.59	60	4.9×10^{-4}	1.8×10^{-1}	59	67

Sodium 3-Hydroxyalkane Sulfonates R_{n-3}-CHOH-CH$_2$-CH$_2$SO$_3$Na   log CMC = 2.065 - 0.305n

Surfactant	Mol.Wt.	Temp.	CMC [mol/l]	CMC [g/l]	Krafft Pt.	References
$C_{12}H_{24}(OH)SO_3Na$	288.43	40	2.5×10^{-2}	7.2	< 6	67
$C_{14}H_{28}(OH)SO_3Na$	316.49	40	6.3×10^{-3}	2.0	21	67
$C_{16}H_{32}(OH)SO_3Na$	344.55	40	1.5×10^{-3}	5.2×10^{-1}	37	67
$C_{18}H_{36}(OH)SO_3Na$	372.61	40	3.8×10^{-4}	1.4×10^{-1}	51	67

Sodium 1-Hydroxy-2-Alkylsulfonates R-CH(SO$_3$Na)CH$_2$OH   log CMC = 2.529 - 0.364n[x]

Surfactant	Mol.Wt.	Temp.	CMC [mol/l]	CMC [g/l]	Krafft Pt.	References
$C_{10}H_{21}CH(SO_3Na)CH_2OH$	288.43	25	1.5×10^{-2}	4.3	59	59
$C_{12}H_{25}CH(SO_3Na)CH_2OH$	316.49	25	2.7×10^{-3}	8.6×10^{-1}	73	59
$C_{14}H_{29}CH(SO_3Na)CH_2OH$	344.55	25	5.1×10^{-4}	1.8×10^{-1}	84	59
$C_{16}H_{31}CH(SO_3Na)CH_2OH$	370.59	25	1.0×10^{-4}	3.7×10^{-2}	93	59

1-Hydroxy-2-Alkylsulfonic Acids R-CH(SO$_3$H)CH$_2$OH   log CMC = 1.763 - 0.305n[x]

Surfactant	Mol.Wt.	Temp.	CMC [mol/l]	CMC [g/l]	Krafft Pt.	References
$C_{10}H_{21}CH(SO_3H)CH_2OH$	266.44	25-50	1.3×10^{-2}	3.5		59
$C_{12}H_{25}CH(SO_3H)CH_2OH$	294.50	25-50	3.2×10^{-3}	9.4×10^{-1}		59
$C_{14}H_{29}CH(SO_3H)CH_2OH$	322.56	25-50	7.9×10^{-4}	2.6×10^{-1}		59
$C_{16}H_{31}CH(SO_3H)CH_2OH$	348.60	25-50	1.9×10^{-4}	6.6×10^{-2}	28.5	59

Disodium α-Sulfoalkanoates R-CH(SO$_3$Na)COONa   log CMC = 1.514 - 0.240n[x]

Surfactant	Mol.Wt.	Temp.	CMC [mol/l]	CMC [g/l]	Krafft Pt.	References
$C_{12}H_{25}CH(SO_3Na)COONa$	352.46	25	3.6×10^{-2}	12.7		59
$C_{14}H_{29}CH(SO_3Na)COONa$	380.52	25	8.8×10^{-3}	3.35	76	59
$C_{16}H_{31}CH(SO_3Na)COONa$	406.56	25	2.1×10^{-3}	8.5×10^{-1}	91	59

Sodium n-Alkyl Sulfoacetates $C_nH_{2n+1}OOCCH_2SO_3Na$ log CMC = 1.263 - 0.303n

Surfactant	Mol.Wt.	Temp.	CMC [mol/l]	CMC [g/l]	Krafft Pt.	References
$C_8H_{17}OOCCH_2SO_3Na$	274.35	40	7.0×10^{-2}	19.2		75,76
$C_{10}H_{21}OOCCH_2SO_3Na$	302.41	40	1.7×10^{-2}	5.1	6.5	75,76
$C_{12}H_{25}OOCCH_2SO_3Na$	330.47	40	4.2×10^{-3}	1.4	17.5	75,76
$C_{14}H_{29}OOCCH_2SO_3Na$	358.53	40	1.1×10^{-3}	3.9×10^{-1}	28.5	75,76

Sodium n-Alkyl β-Sulfopropionates $C_nH_{2n+1}OOC(CH_2)_2SO_3Na$ log CMC = 1.108 - 0.301n

Surfactant	Mol.Wt.	Temp.	CMC [mol/l]	CMC [g/l]	Krafft Pt.	References
$C_8H_{17}OOC(CH_2)_2SO_3Na$	288.38	40	5.0×10^{-2}	14.4	-	75,77
$C_{10}H_{21}OOC(CH_2)_2SO_3Na$	316.44	40	1.3×10^{-2}	4.1	12.5	75,77
$C_{12}H_{25}OOC(CH_2)_2SO_3Na$	344.50	40	3.1×10^{-3}	1.1	26.5	75,77
$C_{14}H_{29}OOC(CH_2)_2SO_3Na$	372.56	40	7.9×10^{-4}	2.9×10^{-1}	39.0	75,77

Sodium Salts of Fatty Acid Sulfoalkyl Esters $C_nH_{2n+1}COO(CH_2)_mSO_3Na$ log CMC = 1.314 - 0.293 (n+0.5m)

Surfactant	Mol.Wt.	Temp.	CMC [mol/l]	CMC [g/l]	References
$C_9H_{19}COO(CH_2)_2SO_3Na$	302.41	40	2.5×10^{-2}	7.5	75
$C_9H_{19}COO(CH_2)_3SO_3Na$	316.44	40	1.8×10^{-2}	5.7	75
$C_9H_{19}COO(CH_2)_4SO_3Na$	330.47	40	1.25×10^{-2}	4.2	75
$C_{10}H_{21}COO(CH_2)_3SO_3Na$	330.47	40	9.0×10^{-3}	3.0	75
$C_{11}H_{23}COO(CH_2)_2SO_3Na$	330.47	40	6.5×10^{-3}	2.15	75
$C_{11}H_{23}COO(CH_2)_3SO_3Na$	344.50	40	4.7×10^{-3}	1.6	75
$C_{11}H_{23}COO(CH_2)_4SO_3Na$	358.53	40	3.3×10^{-3}	1.2	75

Sodium di-n-Alkyl Sulfosuccinates NaO$_3$S-CHCOOR / CH$_2$COOR   R = C_nH_{2n+1} log CMC = 2.08 - 0.681n

Surfactant	Mol.Wt.	Temp.	CMC [mol/l]	CMC [g/l]	References
R = C_4H_9	332.34	50	2.25×10^{-1}	7.5×10^{1}	72,78
R = C_5H_{11}	360.40	50	4.7×10^{-2}	1.7×10^{1}	72,78
R = C_6H_{13}	388.44	50	9.7×10^{-3}	3.8	72,78
R = C_7H_{15}	416.50	50	2.0×10^{-3}	8.3×10^{-1}	72
R = C_8H_{17}	444.55	50	4.3×10^{-4}	1.9×10^{-1}	72,78

[x] n = total number of C atoms

Surfactant	Mol.Wt.	Temp. [°C]	C M C [mol/l]	C M C [g/l]	Krafft Point [°C]	Adsorption Area $a_s \cdot 10^{16}$ [cm²/molecule]	References

Sodium di-n-Alkyl Sulfosuccinates (Cont'd.) $\text{NaO}_3\text{S-CHCOOR}$ CH_2COOR $R = C_n H_{2n+1}$ log CMC = 2.08 - 0.681n

Surfactant	Mol.Wt.	Temp.	CMC [mol/l]	CMC [g/l]	Krafft	Ads.	Ref.
$R = C_9 H_{19}$	472.60	50	9.0×10^{-5}	4.25×10^{-2}			72
$R = C_{10} H_{21}$	500.65	50	1.85×10^{-5}	9.3×10^{-3}			72
$R = C_{11} H_{23}$	528.71	-	3.8×10^{-6}	2.0×10^{-3}			-
$R = C_{12} H_{25}$	556.76	50	8.0×10^{-7}	4.45×10^{-4}			72

Potassium n-Alkyl Malonates $\text{RCH} \genfrac{}{}{0pt}{}{\text{COOK}}{\text{COOK}}$ $R = C_n H_{2n+1}$ log CMC = 1.30 - 0.219n

Surfactant	Mol.Wt.	Temp.	CMC [mol/l]	CMC [g/l]	Krafft	Ads.	Ref.
$R = C_8 H_{17}$	292.44	20-25	3.5×10^{-1}	1.0×10^{2}			79
$R = C_{10} H_{21}$	320.50	25	1.3×10^{-1}	4.2×10^{1}			79
$R = C_{12} H_{25}$	348.55	20-25	4.7×10^{-2}	1.6×10^{1}			79
$R = C_{14} H_{29}$	376.60	20-25	1.7×10^{-2}	6.4			79
$R = C_{16} H_{33}$	404.65	20-25	6.2×10^{-3}	2.5			79
$R = C_{18} H_{37}$	432.70	25	2.3×10^{-3}	1.0			79

Potassium Alkyl Tricarboxylates $\text{RCH-CH} \genfrac{}{}{0pt}{}{\text{COOK}}{\text{COOK}}$ COOK $R = C_n H_{2n+1}$ log CMC = 1.26 - 0.227n

Surfactant	Mol.Wt.	Temp.	CMC [mol/l]	CMC [g/l]	Krafft	Ads.	Ref.
$R = C_6 H_{13}$	360.52	25	7.9×10^{-1}	2.8×10^{2}			80
$R = C_8 H_{17}$	388.57	25	2.8×10^{-1}	1.1×10^{2}			-
$R = C_{10} H_{21}$	416.62	25	9.7×10^{-2}	4.0×10^{1}			80
$R = C_{12} H_{25}$	444.67	25	3.4×10^{-2}	1.5×10^{1}			-
$R = C_{14} H_{29}$	472.73	25	1.2×10^{-2}	5.7			80

Ether Alcohol Sulfates $R \cdot (OCH_2CH_2)_m \cdot OSO_3Na$

$R = C_{12}H_{25}$ log CMC = -2.12 - 0.198m

Surfactant	Mol.Wt.	Temp.	CMC [mol/l]	CMC [g/l]	Krafft	Ads.	Ref.
$C_{12}H_{25}(OCH_2CH_2)_0OSO_3Na$	288.38	50	7.6×10^{-3}	2.2	12		81
$C_{12}H_{25}(OCH_2CH_2)_1OSO_3Na$	332.43	50	4.9×10^{-3}	1.6	< 0		81
$C_{12}H_{25}(OCH_2CH_2)_2OSO_3Na$	376.48	50	3.05×10^{-3}	1.1	-		81
$C_{12}H_{25}(OCH_2CH_2)_3OSO_3Na$	420.53	50	1.95×10^{-3}	8.2×10^{-1}	-		81
$C_{12}H_{25}(OCH_2CH_2)_4OSO_3Na$	404.59	50	1.2×10^{-3}	5.0×10^{-1}	-		81

$R = C_{14}H_{29}$ log CMC = -2.61 - 0.200m

Surfactant	Mol.Wt.	Temp.	CMC [mol/l]	CMC [g/l]	Krafft	Ads.	Ref.
$C_{14}H_{29}(OCH_2CH_2)_0OSO_3Na$	316.43	50	2.45×10^{-3}	7.7×10^{-1}	29		81
$C_{14}H_{29}(OCH_2CH_2)_1OSO_3Na$	360.48	50	1.55×10^{-3}	5.6×10^{-1}	14		81
$C_{14}H_{29}(OCH_2CH_2)_2OSO_3Na$	404.53	50	9.9×10^{-4}	4.0×10^{-1}	~ 0		81
$C_{14}H_{29}(OCH_2CH_2)_3OSO_3Na$	448.59	50	6.2×10^{-4}	2.8×10^{-1}	-		81

$R = C_{16}H_{33}$ log CMC = -3.42 - 0.219m (25°C)
   log CMC = -3.21 - 0.188m (50°C)

Surfactant	Mol.Wt.	Temp.	CMC [mol/l]	CMC [g/l]	Krafft	Ads.	Ref.
$C_{16}H_{33}(OCH_2CH_2)_0OSO_3Na$	344.48	25	3.8×10^{-4}	1.3×10^{-1}	46		82
		50	6.2×10^{-4}	2.1×10^{-1}			81
$C_{16}H_{33}(OCH_2CH_2)_1OSO_3Na$	388.53	25	2.3×10^{-4}	8.9×10^{-2}	31		82
		50	4.0×10^{-4}	1.55×10^{-1}			81
$C_{16}H_{33}(OCH_2CH_2)_2OSO_3Na$	432.58	25	1.4×10^{-4}	6.1×10^{-2}	26		82
		50	2.6×10^{-4}	1.1×10^{-1}			81
$C_{16}H_{33}(OCH_2CH_2)_3OSO_3Na$	476.63	25	8.5×10^{-5}	4.05×10^{-2}	19.4		82
		50	1.7×10^{-4}	8.1×10^{-2}			81
$C_{16}H_{33}(OCH_2CH_2)_4OSO_3Na$	520.68	25	(5.1×10^{-5})	(2.7×10^{-2})	16		82
		50	1.1×10^{-4}	5.7×10^{-2}			81

$R = C_{18}H_{37}$ log CMC = -3.78 - 0.165m (25°C)
   log CMC = -3.65 - 0.187m (50°C)

Surfactant	Mol.Wt.	Temp.	CMC [mol/l]	CMC [g/l]	Krafft	Ads.	Ref.
$C_{18}H_{37}(OCH_2CH_2)_0OSO_3Na$	372.53	25	1.65×10^{-4}	6.15×10^{-2}	56		82
		50	2.25×10^{-4}	8.4×10^{-2}			81
$C_{18}H_{37}(OCH_2CH_2)_1OSO_3Na$	416.58	25	1.1×10^{-4}	4.6×10^{-2}	47.5		82
		50	1.45×10^{-4}	6.0×10^{-2}			81
$C_{18}H_{37}(OCH_2CH_2)_2OSO_3Na$	460.63	25	7.8×10^{-5}	3.6×10^{-2}	40		82
		50	9.5×10^{-5}	4.3×10^{-2}			81

Surfactant	Mol.Wt.	Temp. [°C]	C M C [mol/l]	C M C [g/l]	Krafft Point [°C]	Adsorption Area $a_s \cdot 10^{16}$ [cm²/molecule]	References
Ether Alcohol Sulfates (Cont'd.) $R \cdot (OCH_2CH_2)_m \cdot OSO_3Na$							
$R = C_{18}H_{37}$ log CMC = -3.78 - 0.165m (25°C)							
log CMC = -3.65 - 0.187m (50°C)							
$C_{18}H_{37}(OCH_2CH_2)_2OSO_3Na$	504.68	25	5.3×10^{-5}	2.7×10^{-2}	34		82
		50	6.1×10^{-5}	3.1×10^{-2}			81
$C_{18}H_{37}(OCH_2CH_2)_4OSO_3Na$	548.73	25	3.6×10^{-5}	2.0×10^{-2}	30		82
		50	4.0×10^{-5}	2.2×10^{-2}			81
$R = C_8H_{17}CH=CHC_8H_{16}$ log CMC = -2.45 - 0.173m							
$R(OCH_2CH_2)_0OSO_3Na$	370.52	50	3.6×10^{-3}	1.3			81
$R(OCH_2CH_2)_1OSO_3Na$	414.57	50	2.4×10^{-3}	9.9×10^{-1}			81
$R(OCH_2CH_2)_2OSO_3Na$	458.62	50	1.6×10^{-3}	7.3×10^{-1}			81
$R(OCH_2CH_2)_3OSO_3Na$	502.67	50	1.1×10^{-3}	5.5×10^{-1}			81
Alkyl Trimethylammonium Alkyl Sulfates $C_nH_{2n+1}N(CH_3)_3 O SOC_mH_{2m+1}$ log CMC = 2.634 - 0.299(n+m)							
$C_6H_{13}N(CH_3)_3 O SOC_6H_{13}$	325.57	25	1.1×10^{-1}	35.8			83
$C_6H_{13}N(CH_3)_3 O SOC_8H_{17}$	353.63	25	2.8×10^{-2}	9.9			83
$C_8H_{17}N(CH_3)_3 O SOC_8H_{17}$	381.69	25	7.1×10^{-3}	2.7			83
$C_8H_{17}N(CH_3)_3 O SOC_{10}H_{21}$	409.75	25	1.8×10^{-3}	7.4×10^{-1}			83
$C_8H_{17}N(CH_3)_3 O SOC_{12}H_{25}$	437.81	25	4.5×10^{-4}	2.0×10^{-1}			83
$C_{10}H_{21}N(CH_3)_3 O SOC_{10}H_{21}$	437.81	25	4.5×10^{-4}	2.0×10^{-1}			83
$C_{12}H_{25}N(CH_3)_3 O SOC_8H_{17}$	437.81	25	4.5×10^{-4}	2.0×10^{-1}			83
Alkyl Trimethylammonium Alkyl Sulfonates $C_nH_{2n+1}N(CH_3)_3 O SC_mH_{2m+1}$ log CMC = 2.713 - 0.277(n+m)							
$C_6H_{13}N(CH_3)_3 O SC_6H_{13}$	309.57	25	2.5×10^{-1}	77.4			83
$C_8H_{17}N(CH_3)_3 O SC_8H_{17}$	365.69	25	1.9×10^{-2}	6.95			83
$C_{10}H_{21}N(CH_3)_3 O SC_{10}H_{21}$	421.81	25	1.5×10^{-3}	6.3×10^{-1}			83

Table 2 - Critical Micelle Concentration, Krafft Point and Adsorption Area of Cationic Surfactants

Surfactant	Mol.Wt.	Temp. [°C]	C M C [mol/l]	C M C [g/l]	Krafft Point [°C]	Adsorption Area $a_s \cdot 10^{16}$ [cm²/molecule]	References
n-Alkylamine Hydrochlorides $RNH_2 \cdot HCl$ $R = C_nH_{2n+1}$ log CMC = 1.51 - 0.286n							
$R = C_6H_{13}$	137.65	-	6.2×10^{-1}	8.5×10^{1}			-
$R = C_8H_{17}$	165.71	-	1.65×10^{-1}	2.7×10^{1}			-
$R = C_{10}H_{21}$	193.76	25	4.4×10^{-2}	8.5			48
$R = C_{12}H_{25}$	221.81	30-50	1.2×10^{-2}	2.7	26		23,30,48,51
$R = C_{14}H_{29}$	249.86	40	3.2×10^{-3}	8.0×10^{-1}			48
$R = C_{16}H_{33}$	277.91	50	8.6×10^{-4}	2.4×10^{-1}			48
$R = C_{18}H_{37}$	305.97	60	2.3×10^{-4}	7.0×10^{-2}	56		23,48
n-Alkyltrimethylammonium Bromides $RN(CH_3)_3Br$ $R = C_nH_{2n+1}$ log CMC = 1.98 - 0.311n							
$R = C_8H_{17}$	252.24	25	3.1×10^{-1}	7.8×10^{1}			48
$R = C_{10}H_{21}$	280.29	25;60	7.5×10^{-2}	2.1×10^{1}			48
$R = C_{12}H_{25}$	308.35	25;60	1.75×10^{-2}	5.4			48
$R = C_{14}H_{29}$	336.40	60	4.2×10^{-3}	1.4			48
$R = C_{16}H_{33}$	364.45	60	1.0×10^{-3}	3.6×10^{-1}			48
$R = C_{18}H_{37}$	392.50	-	2.4×10^{-4}	9.4×10^{-2}			-
di-n-Alkyldimethylammonium Chlorides $R_2N(CH_3)_2Cl$ $R = C_nH_{2n+1}$ log CMC = 2.77 - 0.548n							
$R = C_8H_{17}$	305.97	30	2.7×10^{-2}	8.3			84
$R = C_{10}H_{21}$	362.07	30	2.1×10^{-3}	7.8×10^{-1}			84
$R = C_{12}H_{25}$	418.17	30	1.8×10^{-4}	7.5×10^{-2}			84

CRITICAL MICELLE CONCENTRATION

Table 3 - Critical Micelle Concentration, Krafft Point and Adsorption Area of Zwitterionic Surfactants

Alkyl Betaines $R\overset{\ominus}{C}HCOO$, $\overset{\oplus}{N}(CH_3)_3$ $R = C_nH_{2n+1}$ $\log CMC = 2.75 - 0.469n$

Surfactant	Mol.Wt.	Temp. [°C]	CMC [mol/l]	CMC [g/l]	Krafft Point [°C]	Adsorption Area $a_s \cdot 10^{16}$ [cm²/molecule]	References
$R = C_6H_{13}$	201.30	-	8.5×10^{-1}	1.7×10^{2}			-
$R = C_8H_{17}$	229.35	27-60	9.7×10^{-2}	2.2×10^{1}	x	60.0	85-88
$R = C_{10}H_{21}$	257.41	27-60	1.15×10^{-2}	3.0	x	59.5	85-88
$R = C_{12}H_{25}$	285.46	27-60	1.3×10^{-3}	3.7×10^{-1}	x	53.9	85-88
$R = C_{14}H_{29}$	313.51	-	1.5×10^{-4}	4.7×10^{-2}			-

N-Alkyl N,N'-Dimethylglycines $RN^{\oplus}(CH_3)_2CH_2COO^{\ominus}$ $R = C_nH_{2n+1}$ $\log CMC = 3.109 - 0.485n$

Surfactant	Mol.Wt.	Temp. [°C]	CMC [mol/l]	CMC [g/l]	Krafft Point [°C]	Adsorption Area $a_s \cdot 10^{16}$ [cm²/molecule]	References
$R = C_8H_{17}$	215.38	20	1.7×10^{-1}	36.6		-	89
$R = C_{10}H_{21}$	243.44	20	1.8×10^{-2}	4.4		36	89
$R = C_{11}H_{23}$	257.47	20	6.0×10^{-3}	1.55		43	89
$R = C_{12}H_{25}$	271.50	20	2.0×10^{-3}	5.4×10^{-1}		48	89
$R = C_{14}H_{29}$	299.56	20	2.1×10^{-4}	6.3×10^{-2}		60	89
$R = C_{16}H_{33}$	327.62	20	2.2×10^{-5}	7.2×10^{-3}		70	89

Alkyl Dimethylamine Oxides $C_nH_{2n+1}(CH_3)_2N \to O$ $\log CMC = 3.626 - 0.446n$

Surfactant	Mol.Wt.	Temp. [°C]	CMC [mol/l]	CMC [g/l]	Krafft Point [°C]	Adsorption Area $a_s \cdot 10^{16}$ [cm²/molecule]	References
$C_8H_{17}(CH_3)_2N \to O$	173.34	27	1.4×10^{-1}	24.3			90
$C_{10}H_{21}(CH_3)_2N \to O$	201.40	27	1.7×10^{-2}	3.4			90
$C_{12}H_{25}(CH_3)_2N \to O$	229.46	27	2.1×10^{-3}	4.8×10^{-1}			90
$C_{14}H_{29}(CH_3)_2N \to O$	257.52	27	2.6×10^{-4}	6.7×10^{-2}			90
$C_{12}H_{25}(CH_3)_2NOH^{\oplus}$ xx	230.47	27	8.2×10^{-3}	1.9			90

Alkyl Dimethylammoniopropane Sulfonate $RN^{\oplus}(CH_3)_2C_3H_6SO_3^{\ominus}$ $R = C_nH_{2n+1}$ $\log CMC = 3.757 - 0.517n$

Surfactant	Mol.Wt.	Temp. [°C]	CMC [mol/l]	CMC [g/l]	Krafft Point [°C]	Adsorption Area $a_s \cdot 10^{16}$ [cm²/molecule]	References
$R = C_{10}H_{21}$	307.55	30	3.9×10^{-2}	12.0			91
$R = C_{12}H_{25}$	335.61	30	3.6×10^{-3}	1.2			91
$R = C_{14}H_{29}$	363.67	-	3.4×10^{-4}	1.2×10^{-1}			-
$R = C_{16}H_{33}$	391.73	30	3.1×10^{-5}	1.2×10^{-2}			91

Various Zwitterion Surfactants

Surfactant	Mol.Wt.	Temp. [°C]	CMC [mol/l]	CMC [g/l]	Krafft Point [°C]	Adsorption Area $a_s \cdot 10^{16}$ [cm²/molecule]	References
$C_{12}H_{25}N^{\oplus}(CH_3)_2C_3H_6COO^{\ominus}$	299.56	30	5.3×10^{-3}	1.6			91
$C_{12}H_{25}P^{\oplus}(CH_3)_2C_3H_6SO_3^{\ominus}$	352.58	30	2.4×10^{-3}	8.5×10^{-1}			91
$C_{12}H_{25}N^{\oplus}(C_2H_5)_2CH_2COO^{\ominus}$	299.56		2.8×10^{-3}	8.4×10^{-1}			92
$C_{12}H_{25}N^{\oplus}(CH_3)_2CH_2CH(OH)CH_2SO_3^{\ominus}$	351.61	30	2.0×10^{-3}	7.0×10^{-1}			91
$C_{12}H_{25}N^{\oplus}(n-C_3H_7)_2C_3H_6SO_3^{\ominus}$	391.73	30	1.8×10^{-3}	7.0×10^{-1}			91
$C_{12}H_{25}CH[N^{\oplus}(CH_3)_3]COO^{\ominus}$	285.53		1.6×10^{-3}	4.5×10^{-1}			93
$C_{12}H_{25}N^{\oplus}(CH_3)_2C_3H_6OSO_3^{\ominus}$	351.61	30	5.7×10^{-4} xxx	2.0×10^{-1} xxx			91

---

x   Krafft Point and Cloud Point do not seem to exist between 0 and 100 °C.
xx 1×10^{-3} m Cl⁻
xxx   In 0.20 m sodium p-toluene sulfonate

References page II-496

Table 4 - Critical Micelle Concentration, Cloud Point and Adsorption Area of Nonionic Surfactants

Surfactant	Mol.Wt.	Temp. [°C]	CMC [mol/l]	CMC [g/l]	Cloud Point [°C]	Adsorption Area $a_s \cdot 10^{16}$ [cm²/molecule]	References
Glucosyloxy Alkanes $R\,C_6H_{11}O_6$ $R = C_nH_{2n+1}$ $\log CMC = 1.64 - 0.531n$							
$R = C_8H_{17}$	292.37	25	2.5×10^{-2}	7.3		41	94
$R = C_{10}H_{21}$	320.42	25	2.2×10^{-3}	7.05×10^{-1}		47	95
$R = C_{12}H_{25}$	348.47	25	1.9×10^{-4}	6.6×10^{-2}		36	95
Saccharose Monoesters of Fatty Acids $ROOC_{12}H_{21}O_{10}$ $R = C_nH_{2n-1}$ $\log CMC = 0.514 - 0.348n$ (25-27.5°C) $\log CMC = -0.668 - 0.350n$ (50°C)							
$R = C_{12}H_{23}$	524.59	20-27.5	2.2×10^{-4}	1.2×10^{-1}			96,97
		50	1.35×10^{-5}	7.1×10^{-3}			
$R = C_{14}H_{27}$	552.64	20	4.5×10^{-5}	2.5×10^{-2}			97
		50	2.7×10^{-6}	1.5×10^{-3}			
$R = C_{16}H_{31}$	580.70	20	9.1×10^{-6}	5.3×10^{-3}			97
		50	5.5×10^{-7}	3.2×10^{-4}			
$R = C_{18}H_{35}$	608.75	20-27.5	1.85×10^{-6}	1.1×10^{-3}			96,97
		50	1.1×10^{-7}	6.7×10^{-5}			
Saccharose Dipalmitate							
$C_{12}H_{20}O_9(C_{16}H_{31}O_2)_2$	819.10	20	1.34×10^{-5}	1.1×10^{-2}			97
Poly(oxyethylene) Monododecyl Ethers $C_{12}H_{25}(OCH_2CH_2)_mOH$ $\log CMC = -4.51 + 0.056m$ (23°C) $\log CMC = -4.40 + 0.009m$ (25°C) $\log CMC = -4.80 + 0.013m$ (55°C)							98 / 99,100 / 99,100
$C_{12}H_{25}(OCH_2CH_2)_4OH$	362.54	25	4.3×10^{-5}	1.6×10^{-2}		44	99
		55	1.8×10^{-5}	6.5×10^{-3}		44	99
$C_{12}H_{25}(OCH_2CH_2)_5OH$	406.59	23	5.9×10^{-5}	2.4×10^{-2}	25	54	98,101
$C_{12}H_{25}(OCH_2CH_2)_6OH$	450.64				48		101,102
		23	7.6×10^{-5}	3.8×10^{-2}		59	98
$C_{12}H_{25}(OCH_2CH_2)_7OH$	494.69	25	4.6×10^{-5}	2.3×10^{-2}		48	100
		55	1.95×10^{-5}	9.6×10^{-3}		46	100
$C_{12}H_{25}(OCH_2CH_2)_9OH$	582.80	23	9.9×10^{-5}	5.8×10^{-2}		71	98
$C_{12}H_{25}(OCH_2CH_2)_{12}OH$	670.90	23	1.45×10^{-4}	9.7×10^{-2}		77	98
$C_{12}H_{25}(OCH_2CH_2)_{14}OH$	803.06	25	5.35×10^{-5}	4.3×10^{-2}		65	100
		55	2.4×10^{-5}	1.9×10^{-2}		65	100
$C_{12}H_{25}(OCH_2CH_2)_{23}OH$	1199.53	25	6.5×10^{-5}	7.8×10^{-2}		82	100
		55	3.3×10^{-5}	4.0×10^{-2}		80	100
$C_{12}H_{25}(OCH_2CH_2)_{30}OH$	1507.89	25	7.5×10^{-5}	1.1×10^{-1}		107	100
		55	3.9×10^{-5}	5.9×10^{-2}		105	100
Poly(oxyethylene) Monotridecyl Ethers $C_{13}H_{27}(OCH_2CH_2)_mOH$ $\log CMC = -4.04 + 0.011m$ (25°C) $\log CMC = -4.23 + 0.008m$ (55°C)							
$C_{13}H_{27}(OCH_2CH_2)_5OH$	420.62	25	1.05×10^{-4}	4.4×10^{-2}		60	99
		55	6.6×10^{-5}	2.8×10^{-2}		58	99
$C_{13}H_{27}(OCH_2CH_2)_{9.5}OH$	618.85	25	1.15×10^{-4}	7.1×10^{-2}		74	99
		55	7.2×10^{-5}	4.5×10^{-2}		71	99
$C_{13}H_{27}(OCH_2CH_2)_{14}OH$	817.09	25	1.3×10^{-4}	1.1×10^{-1}			99
		55	7.8×10^{-5}	6.3×10^{-2}			99
$C_{13}H_{27}(OCH_2CH_2)_{20}OH$	1081.40	25	1.55×10^{-4}	1.7×10^{-1}			99
		55	8.8×10^{-5}	9.5×10^{-2}			99
$C_{13}H_{27}(OCH_2CH_2)_{30}OH$	1521.92	25	2.0×10^{-4}	3.0×10^{-1}		89	99
		55	1.05×10^{-4}	1.6×10^{-1}		87	99
Poly(oxyethylene) Monohexadecyl Ethers $C_{16}H_{33}(OCH_2CH_2)_mOH$ $\log CMC = -5.95 + 0.028m$							
$C_{16}H_{33}(OCH_2CH_2)_6OH$	506.74	25	1.65×10^{-6}	8.4×10^{-4}	32		101,103

CRITICAL MICELLE CONCENTRATION

Surfactant	Mol.Wt.	Temp. [°C]	CMC [mol/l]	CMC [g/l]	Cloud Point [°C]	Adsorption Area $a_s \cdot 10^{16}$ [cm²/molecule]	References
Poly(oxyethylene) Monohexadecyl Ethers (Cont'd.) $C_{16}H_{33}(OCH_2CH_2)_mOH$ $\quad$ log CMC = -5.95 + 0.028m							
$C_{16}H_{33}(OCH_2CH_2)_7OH$	550.80	25	1.75×10^{-6}	9.6×10^{-4}			103
$C_{16}H_{33}(OCH_2CH_2)_9OH$	638.90	25	2.0×10^{-6}	1.3×10^{-3}			103
$C_{16}H_{33}(OCH_2CH_2)_{12}OH$	771.06	25	2.4×10^{-6}	1.85×10^{-3}			103
$C_{16}H_{33}(OCH_2CH_2)_{15}OH$	903.21	25	2.9×10^{-6}	2.6×10^{-3}			103
$C_{16}H_{33}(OCH_2CH_2)_{21}OH$	1167.52	25	4.3×10^{-6}	5.0×10^{-3}			103
Poly(oxyethylene) Monooctadecyl Ethers $C_{18}H_{37}(OCH_2CH_2)_mOH$							
$C_{18}H_{37}(OCH_2CH_2)_{14}OH$	887.21	25	6.0×10^{-5}	5.3×10^{-2}		94	99
		55	2.0×10^{-5}	1.8×10^{-2}		92	99
$C_{18}H_{37}(OCH_2CH_2)_{100}OH$	4675.68	25	2.0×10^{-5}	9.35×10^{-2}		150	99
		55				145	99
Methoxy Poly(oxyethylene) Decanoates $C_9H_{19}CO(OCH_2CH_2)_mOCH_3$ $\quad$ log CMC = -3.29 + 0.036m							
$C_9H_{19}CO(OCH_2CH_2)_5OCH_3$	406.55	27	7.7×10^{-4}	3.1×10^{-1}			-
$C_9H_{19}CO(OCH_2CH_2)_7OCH_3$	494.65	27	9.2×10^{-4}	4.55×10^{-1}	44		104
$C_9H_{19}CO(OCH_2CH_2)_{10.3}OCH_3$	640.03	27	1.2×10^{-3}	7.7×10^{-1}	65		104
$C_9H_{19}CO(OCH_2CH_2)_{11.9}OCH_3$	710.51	27	1.38×10^{-3}	9.8×10^{-1}	74		104
$C_9H_{19}CO(OCH_2CH_2)_{16}OCH_3$	891.12	27	1.9×10^{-3}	1.7			104
Methoxy Poly(oxyethylene) Dodecanoates $C_{11}H_{23}CO(OCH_2CH_2)_mOCH_3$ $\quad$ log CMC = -3.97 + 0.038m							
$C_{11}H_{23}CO(OCH_2CH_2)_6OCH_3$	478.65	27	1.8×10^{-4}	8.7×10^{-2}	31		104
$C_{11}H_{23}CO(OCH_2CH_2)_{8.4}OCH_3$	584.38	27	2.3×10^{-4}	1.3×10^{-1}	53		104
$C_{11}H_{23}CO(OCH_2CH_2)_{11.2}OCH_3$	707.72	27	2.9×10^{-4}	2.05×10^{-1}	74		104
$C_{11}H_{23}CO(OCH_2CH_2)_{12.5}OCH_3$	764.99	27	3.25×10^{-4}	2.5×10^{-1}	79		104
$C_{11}H_{23}CO(OCH_2CH_2)_{15}OCH_3$	875.12	27	4.0×10^{-4}	3.5×10^{-1}			-
Poly(oxyethylene) Nonyl Phenyl Ethers $C_9H_{19}C_6H_4(OCH_2CH_2)_mOH$ $\quad$ log CMC = -4.54 + 0.014m (55°C)							
$C_9H_{19}C_6H_4(OCH_2CH_2)_{10}OH$	660.86	25	$7.5-9.0 \times 10^{-5}$	$5.0-6.0 \times 10^{-2}$	60	55-60	100,105,106
		55	4.0×10^{-5}	2.6×10^{-2}			100
$C_9H_{19}C_6H_4(OCH_2CH_2)_{15}OH$	881.12	25	$1.1-1.3 \times 10^{-4}$	$9.7-11 \times 10^{-2}$	95	72	100,105,106
		55	4.7×10^{-5}	4.1×10^{-2}			100
$C_9H_{19}C_6H_4(OCH_2CH_2)_{20}OH$	1101.38	25	$1.3-1.8 \times 10^{-4}$	$1.4-2.0 \times 10^{-1}$		82;89[*];78[*]	36,100,105
		55	5.6×10^{-5}	6.2×10^{-2}			100
$C_9H_{19}C_6H_4(OCH_2CH_2)_{30}OH$	1541.90	25	$1.8-3.0 \times 10^{-4}$	$2.8-4.6 \times 10^{-1}$		101	100,105
		55	7.8×10^{-5}	1.2×10^{-1}			100
$C_9H_{19}C_6H_4(OCH_2CH_2)_{32}OH$	1630.00	26	1.5×10^{-4}	2.4×10^{-1}		116[*];103[*]	36
$C_9H_{19}C_6H_4(OCH_2CH_2)_{50}OH$	2422.94	25	2.8×10^{-4}	6.8×10^{-1}			100
		55	1.5×10^{-4}	3.6×10^{-1}			100
$C_9H_{19}C_6H_4(OCH_2CH_2)_{100}OH$	4625.54	25	1.0×10^{-3}	4.6		173	105

[*] determined by soap titration.

Table 5 - Temperature Dependence of Critical Micelle Concentration

Surfactant	Temp. [$^\circ$C]	C M C [mol/l]	ΔH_m [J/mol]	References
$C_{12}H_{23}OOK$	25	2.55×10^{-2}	-2.1×10^3	48
	30	2.60	-4.6×10^3	
	35	2.70	-7.5×10^3	
	45	3.05	-1.1×10^4	
	55	3.50	-1.4×10^4	
	65	4.20×10^{-2}	-2.3×10^4	
$C_{14}H_{27}OOK$	25	6.6×10^{-3}	-4.2×10^3	48
	35	7.0	-4.2×10^3	
	45	7.4	-5.4×10^3	
	55	7.9	-6.7×10^3	
	65	8.6×10^{-3}	-9.2×10^3	
$C_{12}H_{25}SO_4Na$	20	6.3×10^{-3}	-1.7×10^3	57
	40	6.7	-3.8×10^3	
	60	7.9	-9.6×10^3	
	75	9.5×10^{-3}	-1.2×10^4	
$C_8H_{17}SO_3Na$	25	1.55×10^{-1}	-8.4×10^2	48
	40	1.62	-4.2×10^3	
	50	1.77×10^{-1}	-8.8×10^3	
$C_{10}H_{21}SO_3Na$	25	4.1×10^{-2}	-1.3×10^3	48
	35	4.2	-3.4×10^3	
	45	4.5	-6.3×10^3	
	55	4.9	-8.8×10^3	
	65	5.5×10^{-2}	-1.5×10^4	
$C_{12}H_{25}SO_3Na$	35	1.0×10^{-2}	-6.3×10^3	48
	45	1.1	-8.0×10^3	
	55	1.2	-1.05×10^4	
	65	1.4×10^{-2}	-1.6×10^4	
$C_8H_{17}-\overset{\displaystyle N^{\oplus}(CH_3)_3}{\underset{\,}{CH}}-COO^{\ominus}$	6	10.4×10^{-2}	2.9×10^3	87
	27	9.7	2.9×10^3	
	45	9.1	2.9×10^3	
	60	8.6×10^{-2}	2.9×10^3	
$C_{12}H_{25}-\overset{\displaystyle N^{\oplus}(CH_3)_3}{\underset{\,}{CH}}-COO^{\ominus}$	10	1.39×10^{-3}	2.5×10^3	87
	27	1.32	2.5×10^3	
	45	1.25	2.5×10^3	
	60	1.20×10^{-3}	2.5×10^3	
$C_{10}H_{21}N(CH_3)_3Br$	25	6.8×10^{-2}	2.5×10^3	48
	60	7.5×10^{-2}	2.5×10^3	
$C_{12}H_{25}N(CH_3)_3Br$	10	3.52×10^{-3}	-8.4×10^2	107
	20	3.56	-1.7×10^3	
	30	3.72	-4.6×10^3	
	40	4.02	-8.0×10^3	
	50	4.48	-1.1×10^4	
	60	5.12	-1.25×10^4	
	70	5.97	-1.3×10^4	
	80	6.88×10^{-3}	-1.5×10^4	
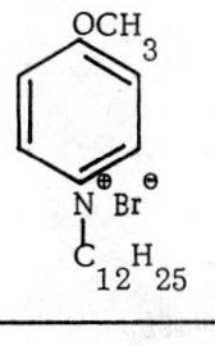 (see structure below)	5	8.70×10^{-3}	1.1×10^4	108
	10	8.43	5.0×10^3	
	15	8.38	$+3.3 \times 10^2$	
	20	8.40	-3.3×10^3	
	25	8.52×10^{-3}	-6.3×10^3	

Structure for the last surfactant: a benzene ring with an OCH_3 substituent and an $\overset{\oplus}{N}\,Br^{\ominus}$ group bearing a $C_{12}H_{25}$ chain.

Surfactant	Temp. [°C]	CMC [mol/l]	ΔH_m [J/mol]	References
(benzene ring, OCH$_3$; N$^\oplus$Br$^\ominus$; C$_{12}$H$_{25}$) (Cont'd.)	30	8.76×10^{-3}	-8.4×10^{3}	108
	35	9.04	-1.1×10^{4}	
	40	9.39	-1.4×10^{4}	
	45	9.84×10^{-3}	-1.8×10^{4}	
	50	1.04×10^{-2}	-2.3×10^{4}	
	55	1.12×10^{-2}	-3.1×10^{4}	
(benzene ring, OCH$_3$; N$^\oplus$Cl$^\ominus$; C$_{12}$H$_{25}$)	5	1.34×10^{-2}	1.3×10^{4}	108
	10	1.29	9.2×10^{3}	
	15	1.25	5.9×10^{3}	
	20	1.23	3.0×10^{3}	
	25	1.22	$+5.9 \times 10^{2}$	
	30	1.23	-1.5×10^{3}	
	35	1.24	-3.4×10^{3}	
	40	1.25	-5.0×10^{3}	
	45	1.28	-6.7×10^{3}	
	50	1.30	-8.8×10^{3}	
	55	1.34	-1.1×10^{4}	
	60	1.39×10^{-2}	-1.4×10^{4}	
(benzene ring, OC$_{12}$H$_{25}$; N$^\oplus$Br$^\ominus$; CH$_3$)	15	3.44×10^{-3}	$+6.7 \times 10^{2}$	108
	20	3.46	-3.2×10^{3}	
	25	3.52	-7.1×10^{3}	
	30	3.62	-1.05×10^{4}	
	35	3.76	-1.4×10^{4}	
	40	3.95	-1.7×10^{4}	
	45	4.18	-2.0×10^{4}	
	50	4.44	-2.2×10^{4}	
	55	4.73	-2.4×10^{4}	
	60	5.08×10^{-3}	-2.6×10^{4}	
(benzene ring, OCH$_3$; N$^\oplus$Br$^\ominus$; C$_{14}$H$_{29}$)	20	2.07×10^{-3}	-8.4×10^{-3}	108
	25	2.13	-1.3×10^{4}	
	30	2.24	-1.5×10^{4}	
	35	2.36	-1.6×10^{4}	
	40	2.49	-1.7×10^{4}	
	45	2.61	-1.8×10^{4}	
	50	2.75	-2.1×10^{4}	
	55	2.96	-2.7×10^{4}	
	60	3.22×10^{-3}	-3.8×10^{4}	
$C_{10}H_{21}(OCH_2CH_2)_4OH$	1	1.45×10^{-3}	2.93×10^{4}	109
	6.8	1.12	2.89×10^{4}	
	10	0.98	2.5×10^{4}	
	15	0.90	1.8×10^{4}	
	20	0.73	1.4×10^{4}	
	25	0.68×10^{-3}	9.6×10^{3}	
$C_{10}H_{21}(OCH_2CH_2)_5OH$	0.8	1.75×10^{-3}	3.4×10^{4}	109
	5	1.41	2.9×10^{4}	
	10	1.18	2.3×10^{4}	
	15	0.97	1.8×10^{4}	
	20	0.90	1.4×10^{4}	
	25	0.81	1.1×10^{4}	
	30	0.76	9.2×10^{3}	
	35	0.72	8.0×10^{3}	
	40	0.68×10^{-3}	6.3×10^{3}	
$C_{10}H_{21}(OCH_2CH_2)_8OH$	5	2.1×10^{-3}	3.0×10^{4}	109
	10	1.72	3.0×10^{4}	
	15	1.4	3.0×10^{4}	
	20	1.1×10^{-3}	3.0×10^{4}	

References page II-496

Surfactant	Temp. [°C]	CMC [mol/l]	ΔH_m [J/mol]	References
$C_{10}H_{21}(OCH_2CH_2)_8OH$ (Cont'd.)	25	0.9×10^{-3}	2.1×10^4	109
	35	0.77×10^{-3}	9.2×10^3	
$C_{10}H_{21}(OCH_2CH_2)_{12}OCH_3$	9.7	17×10^{-4}	1.6×10^4	110
	29.0	11	1.5×10^4	
	50.7	7.8	1.3×10^4	
	58.5	7.0	1.1×10^4	
	69.7	6.2	9.6×10^3	
	73.4	6.0	2.5×10^3	
	75.0	6.0×10^{-4}	1.7×10^3	
$C_{12}H_{25}(OCH_2CH_2)_7OH$	1	1.0×10^{-4}		111
	5	9.0×10^{-5}		
	10	8.0×10^{-5}		
	25	5.0×10^{-5}	2.1×10^4	
	40	3.2×10^{-5}		
	55	2.0×10^{-5}		
$C_{12}H_{25}(OCH_2CH_2)_{30}OH$	1	1.25×10^{-4}		111
	5	1.0×10^{-4}		
	10	9.0×10^{-5}		
	25	8.0×10^{-5}	1.55×10^4	
	40	5.0×10^{-5}		
	55	4.0×10^{-5}		
$C_{16}H_{33}(OCH_2CH_2)_{30}OH$	1	3.0×10^{-5}		111
	5	2.5×10^{-5}		
	10	2.0×10^{-5}		
	25	1.2×10^{-5}	2.8×10^{-4}	
	40	6.0×10^{-6}		
	55	4.0×10^{-6}		
$C_7H_{15}CO(OCH_2CH_2)_{7,6}OCH_3$	11	12.5×10^{-3}	1.4×10^4	112
	25	9.8	1.05×10^4	
	40	8.2	9.2×10^3	
	43	7.8×10^{-3}	9.2×10^3	
$C_9H_{19}C_6H_4(OCH_2CH_2)_{15}OH$	21	3.10×10^{-5}	5.4×10^3	113
	25	2.85×10^{-5}	5.4×10^3	
	28	2.71×10^{-5}	5.4×10^3	
	33.5	2.69×10^{-5}	5.4×10^3	
	41.5	2.36×10^{-5}	5.4×10^3	
	45	2.27×10^{-5}	5.4×10^3	

Table 6 - Krafft Points [°C] for Various Counterions
(References 21, 21a, 23, 115)

Counterion \ Surfactant	Li^+	Na^+	K^+	Rb^+	Cs^+
$C_{12}H_{23}OO^-$	124	36	<0	-	-
$C_{14}H_{27}OO^-$	134	53	8	-	-
$C_{16}H_{31}OO^-$	140	62	30	-	-
$C_{18}H_{35}OO^-$	(148)	71	46	51	48
$C_{12}H_{25}OSO_3^-$	-	8	-	-	-
$C_{14}H_{29}OSO_3^-$	-	21	41		

Counterion \ Surfactant	Mg^{2+}	Ca^{2+}	Sr^{2+}	Ba^{2+}	Al^{3+}
$(C_{12}H_{25}OSO_3)_2^{2-}$	25	50	64	100	-
$(C_{14}H_{29}OSO_3)_2^{2-}$	38.5	67	-	>100	-
$(C_{14}H_{29}OSO_3)_3^{3-}$	-	-	-	-	>100

Counterion \ Surfactant	F^-	Cl^-	Br^-	J^-
$C_{12}H_{25}NH_3^+$	(22)	26	34	39
$C_{18}H_{37}NH_3^+$	(50)	56	(62)	(68)

Table 6 (Cont'd.) - Influence of Added Salt on Krafft Point and CMC of Sodium Tetradecyl Sulfate

Krafft-Point [°C]	ΔK [°C]	CMC [mol/l]	Na^+ [mol/l]
20.5	0.0	1.9×10^{-3}	1.9×10^{-3}
23.8	3.3	1.2×10^{-3}	4.7×10^{-3}
24.2	3.7	0.9×10^{-3}	6.3×10^{-3}
25.2	4.7	0.7×10^{-3}	10.4×10^{-3}
26.9	6.4	0.5×10^{-3}	19.4×10^{-3}
29.7	9.2	0.45×10^{-3}	70.4×10^{-3}

C. References

1. W. V. Smith, R. H. Ewart, J. Chem. Phys. 16, 592 (1948).
2. H. Gerrens, Fortschr. Hochpolymer. Forsch. 1, 234 (1959).
3. J. Stauff, Z. Physik. Chem. (Leipzig) A183, 55 (1938/39).
4. G. Stainsby, A. E. Alexander, Trans. Faraday Soc. 46, 587 (1950).
5. V. V. Subba Rao, R. J. Fix, A. C. Zettlemoyer, J. Am. Oil Chemists Soc. 45, 449 (1968).
6. M. U. Oko, R. L. Venable, J. Colloid Interface Sci. 35, 53 (1971).
7. J. Steigman, I. Cohen, F. Spingola, J. Colloid Sci. 20, 732 (1965).
8. E. K. Goette, J. Colloid Sci. 4, 459 (1949).
9. K. Shinoda, T. Nakagawa, B. Tamamushi, T. Isemura, "Colloidal Surfactants", Academic Press, New York - London 1963, p. 25-48

10. W. U. Malik, S. P. Verma, Kolloid-Z.u.Z. Polymere 229, 985 (1969).
11. W. U. Malik, S. M. Saleem, J. Am. Oil Chemists Soc. 45, 670 (1968).
12. M. J. Schick, A. H. Gilbert, J. Colloid Sci. 20, 464 (1965).
13. W. U. Malik, O.P. Jhamb, Kolloid-Z.u.Z. Polymere 242, 1209 (1970).
14. M. J. Schick, D. J. Manning, J. Am. Oil Chemists Soc. 43, 133 (1966).
15. K. J. Mysels, R. J. Otter, J. Colloid Sci. 16, 462 (1961).
16. A. M. Mankowich, J. Am. Oil Chemists Soc. 41, 449 (1964).
17. J. M. Corkill, J. F. Goodman, J. R. Tate, Trans Faraday Soc, 60, 986 (1964).
18. K. J. Mysels, R. J. Otter, J. Colloid Sci. 16, 474 (1961).

a

19. F. Krafft, H. Wiglow, Ber. 28, 2566 (1895), F. Krafft, Ber. 32, 1596 (1899).

20. J. K. Weil, F. D. Smith, A. J. Stirton, J. Org. Chem. 27, 2950 (1962).

21. H. Lange, M. J. Schwuger, Kolloid-Z.u.Z. Polymere 223, 145 (1968).

21a. M. J. Schwuger, Kolloid-Z.u.Z. Polymere 233, 979 (1969).

22. A. H. Loranger, T. T. Serafini, W. v. Fischer, E. G. Bobalek, Off. Dig. Federation Paint and Varnish Prod. Clubs 31, 482 (1959).

23. D. G. Dervichian, Proceedings, 3rd International Congress of Surface Activity, Cologne (Koeln), 1960, Vol. 1, Sect. A, p. 182.

24. B. Wurzschmitt, Z. Anal. Chem. 130, 105 (1950).

25. W. N. Maclay, J. Colloid Sci. 11, 272 (1956).

26. S. H. Maron, M. E. Elder, J. N. Ulevitch, J. Colloid Sci. 9, 89 (1954).

27. S. H. Maron, M. E. Elder, C. Moore, J. Colloid Sci. 9, 104 (1954).

28. S. H. Maron, M. E. Elder, J. Colloid Sci. 9, 263 (1954).

29. S. H. Maron, M. E. Elder, J. Colloid Sci. 9, 347 (1954).

30. S. H. Maron, M. E. Elder, J. Colloid Sci. 9, 353 (1954).

31. S. H. Maron, M. E. Elder, J. N. Ulevitch, J. Colloid Sci. 9, 382 (1954).

32. B. Jacobi, Angew. Chem. 64, 539 (1952).

33. E. A. Willson, J. R. Miller, E. H. Rowe, J. Phys. and Colloid Chem. 53, 357 (1949).

34. M. Morton, J. A. Cala, M. W. Altier, J. Polymer Sci. 19, 547 (1956).

35. J. G. Brodnyan, G. L. Brown, J. Colloid. Sci. 15, 76 (1960).

36. R. J. Orr, L. Breitman, Can. J. Chem. 38, 668 (1960).

37. R. J. Orr, Rubber Plastics Age 41, 1027 (1960).

38. J. W. Gibbs, Trans. Conn. Acad. Arts Sci. 3, 108 and 343 (1874-1878).

39. C. P. Roe, P. D. Brass, J. Am. Chem. Soc. 76, 4703 (1954).

40. W. Kling, H. Lange, Proceedings, 2nd International Congress of Surface Activity, London 1957, Vol 1, p. 295 and 308.

41. B. A. Pethica, Trans. Faraday Soc. 50, 413 (1954).

42. J. J. Krackeler, H. Naidus, J. Polymer Sci. C, 27, 207 (1969).

43. H. Gerrens, J. Polymer Sci. C, 27, 77 (1969).

44. D. B. Gershberg, J. E. Longfield, Symp. Polymerization Kinetics Catalyst Systems: Part I, 54th Meeting, A.I.Ch.E., Dec. 2-7, 1961, Prepr. 10.

45. G. Beckmann, H. Matis, Chemie-Ing.-Techn. 38, 209 (1966).

46. H. Gerrens, K. Kuchner, Brit. Polymer J. 2, 18 (1970).

47. K. Hess, W. Philippoff, H. Kiessig, Kolloid-Z. 88, 40 (1939).

48. H. B. Klevens, J. Phys. Colloid Chem. 52, 130 (1948).

49. S. H. Herzfeld, J. Phys. Chem. 56, 953 (1952).

50. K. Shinoda, J. Phys. Chem. 58, 541 (1954).

51. K. Tyuzyo, Bull. Chem. Soc. Japan 31, 117 (1958).

52. J. Powney, C. C. Addison, Trans. Faraday Soc. 34, 372 (1938).

53. F. Van Voorst Vader, Trans. Faraday Soc. 56, 1067 (1960).

54. E. K. Goette, J. Colloid Sci. 4, 459 (1949).

55. H. C. Evans, J. Chem. Soc, 579 (1956).

56. B. D. Flockhart, J. Colloid Sci. 16, 484 (1961).

57. J. Powney, C. C. Addison, Trans. Faraday Soc. 33, 1243 (1937).

58. H. Suzuki, J. Am. Oil Chemists Soc. 47, 273 (1970).

59. J. K. Weil, F. D. Smith, A. J. Stirton, R. G. Bistline, J. Am. Oil Chemists Soc. 40, 538 (1963).

60. M. J. Rosen, J. Solash, J. Am. Oil Chemists Soc, 46, 399 (1969).

61. O. R. Howell, H. G. B. Robinson, Proc. Roy. Soc. A. 155, 386 (1936).

62. H. V. Tartar, A. L. M. Lelong, J. Phys. Chem. 59, 1185 (1955).

63. H. B. Klevens, J. Phys. Colloid Chem. 51, 1143 (1947).

64. H. V. Tartar, K. A. Wright, J. Am. Chem. Soc. 61, 539 (1939).

65. K. A. Wright, A. D. Abbott, V. Sivertz, H. V. Tartar, J. Am. Chem. Soc. 61, 549 (1939).

66. E. C. Lingafelter, O. L. Wheeler, H. V. Tartar, J. Am. Chem. Soc. 68, 1490 (1946).

67. F. Pueschel, Tenside 4, 320 (1967).

68. W. Griess, Fette, Seifen, Anstrichmittel 57, 24 (1955).

69. J. W. Gershman, J. Phys. Chem. 61, 581 (1957).

70. R. G. Paquette, E. C. Lingafelter, H. V. Tartar, J. Am. Chem. Soc. 65, 686 (1943).

71. M. S. Schick, F. M. Fowkes, J. Phys. Chem. 61, 1062 (1957).

72. H. Schuller, private communication.

73. H. Lange, 4th International Congress of Surface Activity, Brussels 1964, Preprints, paper No. B/IV.1.

74. F. Pueschel, O. Todorov, Tenside 5, 193 (1968).

75. T. Hikota, K. Meguro, J. Am. Oil Chemists Soc. 47, 197 (1970).

76. T. Hikota, K. Meguro, J. Am. Oil Chemists Soc. 46, 579 (1969).

77. T. Hikota, K. Meguro, J. Am. Oil Chemists Soc. 47, 158 (1970).

78. E. F. Williams, N. T. Woodberry, J. K. Dixon, J. Colloid Sci. 12, 452 (1957).

79. K. Shinoda, J. Phys. Chem. 59, 432 (1955).

80. K. Shinoda, J. Phys. Chem. 60, 1439 (1956).

81. E. Goette, Proceedings, 3rd International Congress of Surface Activity, Cologne (Koeln), 1960, Vol. 1, Sect. A, p. 45.

82. J. K. Weil, R. G. Bistline, A. J. Stirton, J. Phys. Chem. 62, 1083 (1958).

83. J. M. Corkill, J. F. Goodman, S. P. Harrold, J. R. Tate, Trans. Faraday Soc. 62, 994 (1966).

84. A. W. Ralston, D. N. Eggenberger, P. L. Du Brow, J. Am. Chem. Soc. 70, 977 (1948).

85. K. Tori, T. Nakagawa, Kolloid.-Z. 180, 47 (1962).

86. K. Tori, T. Nakagawa, Kolloid.-Z. 187, 44 (1963).

87. K. Tori, T. Nakagawa, Kolloid.-Z. 189, 50 (1963).

88. K. Tori, T. Nakagawa, Kolloid.-Z. 191, 42 (1963).

89. P. Molyneux, C. T. Rhodes, J. Swarbrick, Trans. Faraday Soc. 61, 1043 (1965).

90. K. W. Herrmann, J. Phys. Chem. 66, 295 (1962).

91. K. W. Herrmann, J. Colloid Interface Sci. 22, 352 (1966).

92. T. Kusano, J. Mikumo, Kogyo Kagaku Zasshi 59, 458 (1956).

93. K. Tori, T. Nakagawa, Kolloid.-Z.u.Z. Polymere 191, 48 (1963).

94. K. Shinoda, T. Yamanaka, K. Kinoshita, J. Phys. Chem. 63, 648 (1959).

95. K. Shinoda, T. Yamaguchi, R. Hori, Bull. Chem. Soc. Japan 34, 237 (1961).

96. L. Ossipow, F. D. Snell, J. Hickson, Proceedings, 2nd International Congress of Surface Activity, London 1957, Vol. I, p. 50.

97. W. Wachs, J. Hayano, Kolloid.-Z. 181, 139 (1962).

98. H. Lange, Proceedings, 3rd International Congress of Surface Activity, Cologne (Koeln) 1960, Vol. I, Sect. A, p. 279.

99. M. S. Schick, J. Colloid Sci. 17, 801 (1962).

100. M. S. Schick, S. M. Atlas, F. R. Eirich, J. Phys. Chem. 66, 1326 (1962).

101. B. A. Mulley, Proceedings, 3rd International Congress of Surface Activity, Cologne (Koeln) 1960, Vol. I. Sect. A, p. 31.

102. R. R. Balmbra, J. S. Clunie, J. M. Corkill, J. F. Goodman, Trans. Faraday Soc. 58, 1661 (1962).

103. P. H. Elworthy, C. B. Macfarlane, J. Chem. Soc. 1963, 537.

104. T. Nakagawa, K. Kuriyama, K. Tori, J. Chem. Soc. Japan, Pure Chem. Sect. 78, 1573 (1957), through Ref. (9).

105. L. Hsiao, H. V. Dunning, P. B. Lorenz, J. Phys. Chem. 60, 657 (1956).

106. C. F. Jelinek, R. L. Mayhew, Textile Res. J. 24, 765 (1954).

107. J. E. Addison, H. Taylor, 4th International Congress of Surface Activity, Brussels, 1964, Preprints, paper No. B/IV. 17.

108. J. A. Stead, H. Taylor, J. Colloid Interface Sci. 30, 482 (1969).

109. R. A. Hudson, B. A. Pethica, 4th International Congress of Surface Activity, Brussels, 1964, Preprints, paper No. B/IV. 21.

110. K. Kuriyama, H. Inoue, T. Nakagawa, Ann. Rept. Shionogi Res. Lab. 9, 1061 (1959), through Ref. (9).

111. M. J. Schick, J. Phys. Chem. 67, 1796 (1963).

112. T. Nakagawa, H. Inoue,, K. Tori, K. Kuriyama, J. Chem. Soc. Japan, Pure Chem. Sect. 79, 1194 (1958), through Ref. (9).

113. A. M. Mankowich, J. Am. Oil Chemists Soc. 43, 615 (1966).

114. M. J. Schick, J. Am. Oil Chemists Soc. 43, 681 (1966).

115. S. Miyamoto, Bull. Chem. Soc. Japan 33, 371 (1960).

III. SOLID STATE PROPERTIES

CRYSTALLOGRAPHIC DATA FOR VARIOUS POLYMERS

Robert L. Miller
Midland Macromolecular Institute
Midland, Michigan

Contents

A. INTRODUCTION - NOMENCLATURE

These tables contain data for approximately 2000 polymers - roughly twice the number contained in the 1st edition of the Polymer Handbook. In general, the format and organization of the tables follow that previously used. To the two tables of previous versions (Section B: Crystallographic Data and Section C: Melting Points) has been added a third table (Section E: Formula Index) for this edition. Section D contains references for the data contained in the tables.

Nomenclature: Contrary to the rest of the Polymer Handbook the ACS-IUPAC Nomenclature proposals (1, 2) for naming regular single-strand organic polymers have not yet been adopted for these tables.
The primary reason for the compilers reluctance to use these nomenclature recommendations for these tables was the tentative nature of these proposals which probably will be changed significantly before a reasonably definitive set of rules is finally obtained. Clearly, the inordinate amount of work involved in changing names for the approximately 2000 polymer structures already existing in these tables is not justified if the results have only a limited lifetime. The compiler would like to await a final set of rules. Thus, the compiler has chosen to remain (at least temporarily) with the nomenclature of more common usage, in spite of the inconsistencies in that usage. As an aid to the searcher, he has provided a structural index to polymer names with this edition of the tables. This section permits one to determine the polymer name used in the tables from the structural line formula for the polymer of interest (see Section E: Formula Index).

Basically, the nomenclature used in this portion of the Polymer Handbook is that of the 1st edition and of the 1967 recommendations of the IUPAC Nomenclature Committee (3). The majority of polymers were named as the bivalent radical(s) comprising the constitutional repeating unit (CRU) with priority in naming given to incorporation of important structural features (amide, ester, oxide, etc.) and to emphasis on symmetry of the CRU (when present). In the past, many common polymers (poly(styrene), poly(vinyl chloride), etc.) were named according to source names.

In the majority of cases, the polymer was named according to its structure, particularly whenever the CRU was composed of two or more subunits following each other in a regular fashion. An example is poly(ethylene terephthalate). With more complicated structures, the CRU was divided into its possible subunits each of which was treated as a bivalent radical and so named. One of the subunits was chosen as the parent compound, e.g., the "terephthalic acid" of the above example. For poly(amides) and poly(esters), the acid was chosen as the parent compound; for poly(urethanes) it was the isocyanate.

Whenever the CRU (or a portion thereof) exhibited a center of symmetry, naming began with the central subunit using Chemical Abstracts/IUPAC names for the constituent bivalent radicals (see under Examples). Conjunctive names were used as needed in constructing names. For symmetrical units, this process of naming could be continued indefinitely.

Unsymmetrical bivalent units were named with replacement nomenclature or from end-to-end, depending upon the complexity of the bivalent radical and the ease of naming by the alternate choices. Replacement nomenclature was used, for example, to name poly(hexamethylene 6-oxaheptadecanediamide). The polymer from the symmetrical oxyacid would be named poly(hexamethylene oxydicaprylamide).

More complicated structures without symmetry were named end-to-end. This method was particularly appropriate for the poly(oxides) of the form $(-R-O-R'-O-)_n$ (and for the analogous sulfides, sulfones, ureas, and anhydrides) and for the poly(ω-aminoacids) and the poly(ω-hydroxyacids). The constituent hydrocarbon subunits were ordered alphabetically in naming the polymer as, for example, in poly(methyleneoxypentamethylene oxide).

For silicon-containing polymers, bivalent radical names replaced the source-based silane/siloxane names. For example, the source name poly[(tetramethylsilphenylene)siloxane] appears in these tables as poly(p-phenylene tetramethyldisiloxanylene).

For the regular polymers of multi-peptides, a cyclic permutation of the individual peptide residues in the constitutional base unit was performed to permit alphabetical ordering without loss of directional or sequential sense. Thus, poly(alanylglycylproline) is the name given to polymers prepared from (ala-gly-pro), (gly-pro-ala), and (pro-ala-gly), but not from (ala-pro-gly) or its cyclic permutations.

Examples of Polymer Names:

1. <u>Polymer Names Based on Source.</u>

   <u>Olefin, Vinyl, and Acyclic Polymers</u>

$[-CH_2-CH_2-]_n$ poly(ethylene)

$[-CH-CH_2-]_n$ with CH_2-CH_3 poly(1-butene)

$[-CH-CH_2-]_n$ with CN poly(acrylonitrile)

$[-CH-CH_2-]_n$ with Cl poly(vinyl chloride)

$[-CH-CH_2-]_n$ with $COOCH_3$ poly(methyl acrylate)

   <u>Poly(amides)</u>*

$[-NH-CH_2-COO-]_n$ poly(glycine)

$[-NH-(CH_2)_2-COO-]_n$ poly(3-aminopropionic acid) (poly(β-alanine))

$[-NH-(CH_2)_3-COO-]_n$ poly(4-aminobutyric acid)

$[-NH-(CH_2)_4-COO-]_n$ poly(5-aminovaleric acid)

$[-NH-(CH_2)_5-COO-]_n$ poly(6-aminocaproic acid)

$[-NH-(CH_2)_6-COO-]_n$ poly(7-aminoenanthic acid)

$[-NH-(CH_2)_7-COO-]_n$ poly(8-aminocaprylic acid)

$[-NH-(CH_2)_8-COO-]_n$ poly(9-aminopelargonic acid)

$[-NH-(CH_2)_9-COO-]_n$ poly(10-aminocapric acid)

$[-NH-(CH_2)_{10}-COO-]_n$ poly(11-aminoundecanoic acid)

$[-NH-(CH_2)_{11}-COO-]_n$ poly(12-aminolauric acid)

   <u>Poly(oxides)</u>

$[-CH_2-O-]_n$ poly(formaldehyde)

$[-CH_2-O-]_n$ with CH_3 poly(acetaldehyde)

$[-CH_2-CH_2-O-]_n$ poly(ethylene oxide)

$[-CH-CH_2-O-]_n$ with CH_3 poly(propylene oxide)

$[-CH_2-CH_2-CH_2-O-]_n$ poly(trimethylene oxide)

2. <u>Polymer Names Based on Structure</u>

a) <u>Linking Radicals</u>

$-CH_2-CH_2-$ ethylene

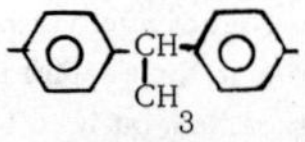

$-CH_2-CH_2-CH_2-$ trimethylene

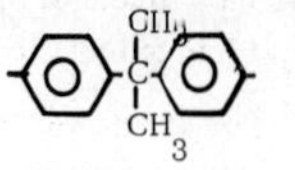

$-CH_2-CH_2-O-CH_2-CH_2-$ oxydiethylene

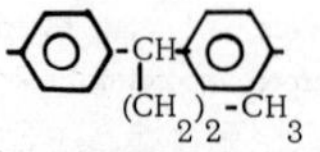

$-CH_2-CH_2-O-CH_2-CH_2-O-CH_2-CH_2-$ ethylenedioxy-diethylene

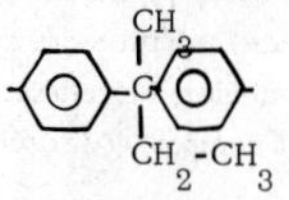

$-CH_2-CH_2-\overset{O}{\underset{O}{S}}-CH_2-CH_2-$ sulfonyldiethylene

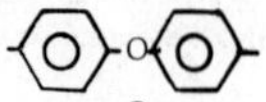

$-CH_2-CH_2-CH_2-\overset{CH_3}{P}-CH_2-CH_2-CH_2-$ methylphosphinidene-ditri-methylene

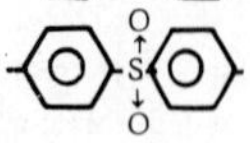

$-CH_2-CH_2-⬡-CH_2-CH_2-$ p-phenylenediethylene

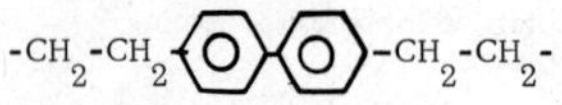

$-CH_2-CH_2-⬡⬡-CH_2-CH_2-$ 4,4'-biphenylenediethylene

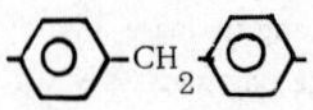

$⬡-CH_2-⬡-$ 4,4'-methylenediphenylene

$⬡-\overset{CH_3}{CH}-⬡$ 4,4'-ethylidenediphenylene

$⬡-\overset{CH_3}{\underset{CH_3}{C}}-⬡$ 4,4'-isopropylidenediphenylene

$⬡-\overset{}{\underset{(CH_2)_2-CH_3}{CH}}-⬡$ 4,4'-butylidenediphenylene

$⬡-\overset{CH_3}{\underset{CH_2-CH_3}{C}}-⬡$ 4,4'-(2,2-butylidene)diphenylene

$⬡-O-⬡$ 4,4'-oxydiphenylene

$⬡-\overset{O}{\underset{O}{S}}-⬡$ 4,4'-sulfonyldiphenylene

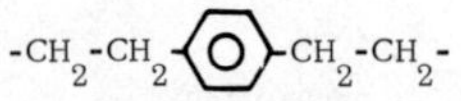

---

* Poly(esters) are treated similarly

b) <u>Parent Compounds</u>

-OOC-COO- (-NHOC-CONH-)	oxalate (oxamide)	-OOC⟨O⟩-CH$_2$-⟨O⟩-COO-	4, 4'-methylenedibenzoate
-OOC-CH$_2$-COO-	malonate (malonamide)	-OOC⟨O⟩-O-⟨O⟩-COO-	4, 4'-oxydibenzoate
-OOC-(CH$_2$)$_2$-COO-	succinate (succinamide)		
-OOC-(CH$_3$)$_3$-COO-	glutarate (glutaramide)	-OOC⟨O⟩-S-⟨O⟩-COO-	4, 4'-sulfonyldibenzoate
-OOC-(CH$_2$)$_4$-COO-	adipate (adipamide)		
-OOC-(CH$_2$)$_5$-COO-	pimelate (pimelamide)	-OOC-CH$_2$-⟨O⟩-O-⟨O⟩-CH$_2$-COO-	4, 4'-oxydiphenylenediacetate
-OOC-(CH$_2$)$_6$-COO-	suberate (suberamide)		
-OOC-(CH$_2$)$_7$-COO-	azelaate (azelaamide)	-OOC⟨O⟩-O-CH$_2$-CH$_2$-O-⟨O⟩-COO-	4, 4'-(ethylenedioxy)dibenzoate
-OOC-(CH$_2$)$_8$-COO-	sebacate (sebacamide)		
-OOC-(CH$_2$)$_9$-COO-	nonanedioate (nonanediamide)	-OOC-CH$_2$-O-⟨O⟩-O-CH$_2$-COO-	p-(phenylenedioxy)diacetate
-OOC-CH$_2$-CH$_2$-O-CH$_2$-CH$_2$-COO-	oxydipropionate		
		-OOC⟨O⟩-O-CH$_2$-COO-	(p-carboxyphenoxy)acetate
-OOC-CH$_2$-⟨O⟩-CH$_2$-COO-	p-phenylene diacetate		
		-OOC-CH$_2$-N⟨O⟩N-CH$_2$-COO-	 1, 4-piperazinediacetate
-OOC-(CH$_2$)$_2$-⟨O⟩-(CH$_2$)$_2$-COO-	p-phenylenedipropionate		
-OOC-⟨O⟩⟨O⟩-COO-	4, 4'-dibenzoate		

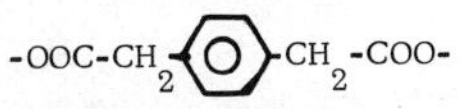

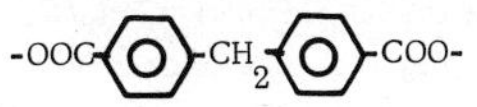

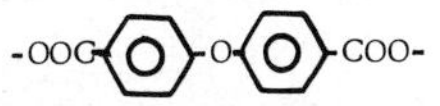

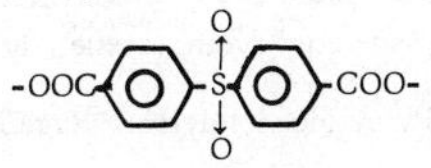

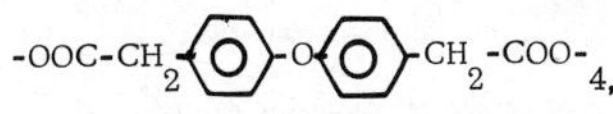

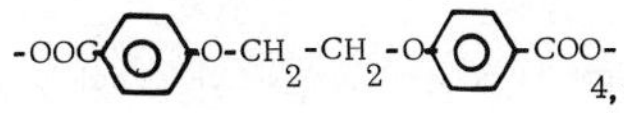

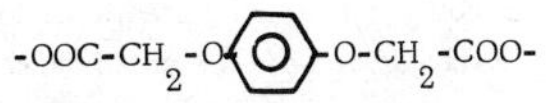

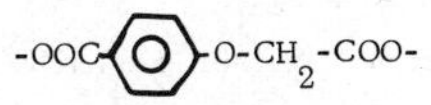

c) <u>Polymer Names</u>

<u>Poly(esters)</u>*

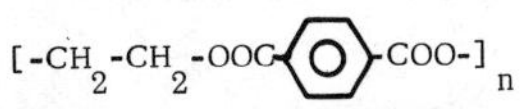

[-CH$_2$-CH$_2$-OOC⟨O⟩-COO-]$_n$	poly(ethylene terephthalate)	[-CH$_2$-CH$_2$-OOC⟨O⟩-C(CH$_3$)$_2$⟨O⟩-COO-]$_n$	poly(ethylene 4, 4'-isopropyli-denedibenzoate)
[-CH$_2$-CH$_2$-OOC⟨O⟩⟨O⟩-COO-]$_n$	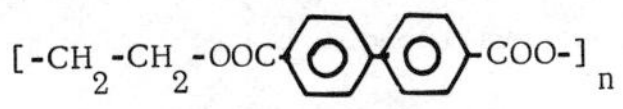 poly(ethylene 4, 4'-dibenzoate)		
[-CH$_2$-CH$_2$-OOC-CH$_2$-⟨O⟩-CH$_2$-COO-]$_n$	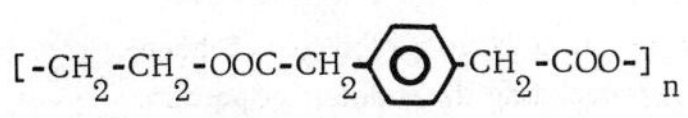 poly(ethylene p-phenylene-diacetate)	[-CH$_2$-CH$_2$-OOC⟨O⟩-S-⟨O⟩-COO-]$_n$	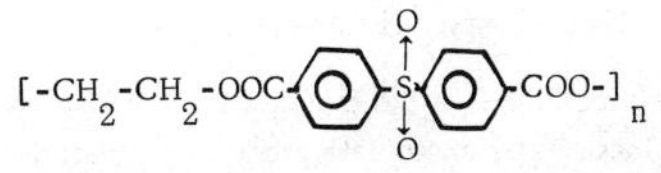 poly(ethylene 4, 4'-sulfonyl-dibenzoate)

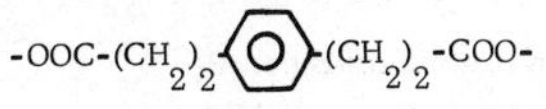

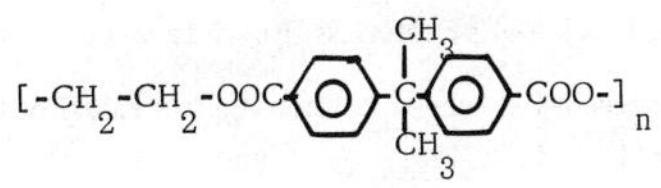

<u>Poly(urethanes)</u>

[-CH$_2$-CH$_2$-CH$_2$-O-C(=O)-NH-CH$_2$-CH$_2$-NH-C(=O)-O-]$_n$

poly(trimethylene ethylene-urethane)

<u>Poly(oxides), Poly(sulfides), Poly(amines)</u>

[-(CH$_2$)$_2$-O-(CH$_2$)$_3$-]$_n$	poly(ethylene trimethylene oxide)	[-(CH$_2$)$_2$-NH-(CH$_2$)$_3$-]$_n$	poly(ethylene trimethylene amine)
[-(CH$_2$)$_2$-S-(CH$_2$)$_3$-]$_n$	poly(ethylene trimethylene sulfide)		

REFERENCES - INTRODUCTION

1. ACS Structure-based Nomenclature for Linear Polymers, Macromolecules <u>1</u>, 193 (1968).
2. IUPAC Appendices on Tentative Nomenclature, Symbols, Units, and Standards - Number 29, Nomenclature of Regular Single-strand Organic Polymers, November 1972 - Macromolecules <u>6</u>, 149 (1973). (See this Handbook page I-1).
3. 1967 IUPAC Report on Nomenclature, J. Polymer Sci., Pt. B, <u>6</u>, 257 (1968).

* Poly(amides) are treated similarly.

B. CRYSTALLOGRAPHIC DATA FOR VARIOUS POLYMERS

The following table presents crystallographic data for about 400 polymers (and for their many polymorphs) grouped, according to the generic structure of the constitutional base unit of the chain, into poly(olefins), poly(vinyls) and poly(vinylidenes), poly(aromatics), poly(dienes), poly(peptides), poly(amides), poly(esters), poly(urethanes) and poly(ureas), poly(ethers), poly(oxides), poly(sulfides) and poly(sulfones), poly(saccharides), and other polymers. Where a polymer might be included in more than one group the following group definitions were used:

(1) Poly(olefins): Those olefin polymers not containing aromatic rings; all substituted derivatives of these, excluding the vinyl and vinylidene polymers. For example, poly(ethylene), poly(tetrafluoroethylene), and poly(cyclopentene), but not poly(styrene) or poly(vinyl chloride).

(2) Poly(vinyls) and Poly(vinylidenes): Those vinyl and vinylidene polymers containing atoms other than carbon and hydrogen, excepting the poly(ethers). Thus, poly(vinyl chloride), but not poly(methylvinyl ether).

(3) Poly(aromatics): Those hydrocarbon polymers containing aromatic rings. Thus, poly(styrene) and poly(p-xylylene).

(4) Poly(peptides): Those polymers whose basic structure, ignoring substituents, is poly(2-aminoacetic acid), i. e., poly(glycine). This group contains mainly the synthetic poly(α-amino acids).

(5) Poly(amides): All amide polymers except the poly(α-amino acids) separately grouped under poly(peptides).

(6) The remaining categories contain polymers according to functional groups as, for example, the poly(esters). Poly(ethers) and poly(oxides) are differentiated according to whether the ether linkage is in the side-group or in the chain backbone, respectively. For example, poly(1-butoxy-2-chloroethylene) is a poly(ether) and poly(oxymethylene) is a poly(oxide).

(7) Those polymers not otherwise categorized, such as poly(phosphonitrile chloride) and the poly(siloxanes), are grouped in the "other" category.

To avoid confusion which might arise from nests of parentheses (the only bracket symbol available to the compiler), the recommendation that the structural repeating unit of the polymer be enclosed in parentheses or brackets has been ignored. Instead, a hyphen is used throughout the tables to separate the prefix "Poly" from the structural repeating unit of the chain which is to be taken as the entire name following "Poly"-.

Within each group, polymers are listed alphabetically according to the basic structure ignoring substituents (excepting the polypeptides all of which have the same basic structure). Substituted polymers are listed alphabetically according to the substituent under the entry for the unsubstituted polymer. Thus, poly(tetrafluoroethylene) and poly(4-methyl-1-pentene) appear, respectively, under poly(ethylene) and poly(1-pentene).

Included as part of the polymer name column is the molecular weight of the chemical repeat unit of the chain, i. e., of the constitutional base unit. This appears below the polymer name in the table and is bracketed by asterisks (*). The values reported are in $[\text{g mol}^{-1}]$.

Unless otherwise indicated, the reference cited in the second column applies to all of the data in that line of the table. Where an entry in a line has been taken from a source different from that listed in column 2, a solidus (/) separates the value of the entry from the reference citation. For example, one value of the melting point of poly(1-butene) is given as "126/12" which is to be read as $126\,^{\circ}$C according to reference 12.

The crystal system, where known, is given in column 3 of the table, using the abbreviations below. The next column contains the space group symbol (the Schoenflies notation is used because of the limitations of electronic data processing symbolisms) with the subscript portion of the symbol preceeding the superscript portion. Thus, according to reference 14, the poly(ethylene) unit cell is orthorhombic (ORTHO) with space group D_{2h}^{16} (D2H-16). For a fuller discussion of space group symbols, see International Tables for X-Ray Crystallography, Vol. I, Kynoch Press, Birmingham, England, 1952.

The dimensions of the unit cell (a, b, and c) are given next in Angstroms (1 $\text{\AA}$ = 0.1 nm). Unless otherwise indicated (by an *), c is the fiber axis. Also, unless otherwise indicated, the unit cell angles are all 90°. Where only one angle differs from 90°, it is identified according to the abbreviations below and its value given. When all three angles differ from 90°, their values are listed in the order: α, β, γ. The next column contains the number of constitutional base units in the unit cell described.

Columns 10 and 11 contain the densities of the crystalline and of the completely amorphous polymer, respectively. The values given, normally appropriate to room temperature, are in $[\text{g cm}^{-3}]$ $(1\ \text{g cm}^{-3} = 10^3\ \text{kg m}^{-3})$. For uniformity, crystallographic densities (column 10) have been recalculated on the basis of the 1969 IUPAC atomic weight values. Melting points in $[^{\circ}\text{C}]$ and heats of fusion in $[\text{kJ mol}^{-1}]$ are listed in the next two columns. The heats of fusion correspond to the amount of substance represented by the molecular weight associated with each polymer name.

The last column of the table indicates briefly the conformation of the polymer chain in the crystal (in a helical notation). The designation, n*p/q, specifies the number (n) of skeletal atoms in the asymmetric unit of the chain and the number of such asymmetric units (p) per q turns of the helix in the crystallographic repeat. Thus, poly(ethylene) as listed has two carbon atoms in the backbone with one such unit per turn in the repeat - it is designated as a 2*1/1 helix. Alternatively, poly(ethylene) considered as poly(methylene) would be designated as a 1*2/1 helix - an entirely equivalent description of the conformation. Note that n may differ from the number of chain atoms in the constitutional base unit. On the one hand, isotactic poly(propene) has two skeletal atoms in the asymmetric unit and three units per turn (2*3/1). On the other hand, syndiotactic poly(propene) with the same constitutional base unit as has isotactic poly(propene) has four skeletal atoms in the asymmetric unit and two units per turn (4*2/1). In this case, n signifies the number of skeletal atoms in the stereobase unit. A fuller discussion of this helical notation is given in Hughes and Lauer, J. Chem. Phys. 30, 1165 (1959) and in Nagai and Kobayashi, J. Chem. Phys. 36, 1268 (1961).

This list of polymers and data is not considered to be exhaustive. The compiler expands it as rapidly as new information is unearthed and he is quite receptive to notification of ommissions and/or corrections and to submission of unpublished data. All of the data in this table cannot be considered to have the same validity - the number of polymers for which detailed and accurate crystal structure analyses have been conducted is still quite limited. The reader may find it desirable to refer to the original literature cited to gain a fuller appreciation of the reasons for apparent discrepancies within the tables.

Abbreviations used in this table: TRI-triclinic; MONO-monoclinic; ORTHO-orthorhombic; TET-tetragonal; RHO-rhombohedral (trigonal); HEX-hexagonal; P-pseudo; A-α; B-β; and G-γ.

POLYMER	REF	CRYST SYST.	SPACE GROUP	UNIT CELL PARAMETERS A	B	C	ANGLES	MON/ UNIT CELL	DENSITY (G./CC.) CRYSTAL	AMORPH.	MELT. POINT	HEAT OF FUSION KJ/MOL	CHAIN CONF. N*P/Q
POLY-OLEFINS													
-HCH-CHR-													
POLY-ACETYLENE *26.04*	69	PHEX		4.2	4.2	2.43		1	1.16				2*1/1
	76			4.22									
POLY-ALLENE *40.06* I.	497	ORTHO	D2H-6	8.20	7.81	3.88		4	1.071		122/536		2*2/1
II.	497	MONO OR	C2-2 / C2H-2	6.37	3.88*	5.12	B=96.6	2	1.058				2*2/1
III.	497					3.88							2*2/1
--, TETRAFLUORO- *112.03*	38	TET OR	C4-2 / D4-3	6.88	6.88	15.4		8	2.042		126		2*4/1
POLY-ALLYLBENZENE, SEE POLY-PROPENE, 3-PHENYL-													
POLY-1-BUTENE *56.11* I.	35 391 252	RHO	D3D-6	17.7	17.7	6.50		18	0.951 0.95/12 0.950	0.87/45 0.860 0.868	126/12 135/345 136/277 132/250 142/313 135/320 138/380	13.9/82 7.16/345 6.07/277 7.03/320	2*3/1
II.	207 350 442	TET TET	S4-1	14.89 14.85	14.89 14.85	20.87 20.6 76		44 44	0.886 0.902		122/345 126/313 122/320 130/380 124/277	4.06/320 6.28/277	2*11/3 2*11/3 2*40/ 11
III.	207 461 531	ORTHO		12.49 12.38	8.96 8.92	7.6 7.45		8 8	(0.876) 0.906		106/277	6.49/277	
--, 3-METHYL- *70.14*	9 231 355 431 280 549	MONO MONO PORT TET MONO MONO	C4V-12 C2H-5	9.55 19.1 19.25 34.3 19.25 9.55	8.54 17.8 17.20 34.3 17.20 17.08	6.84 6.85 6.85 6.85 6.63 6.84	G=116.5 G=116 G=116.5 G=116.5 G=116.5	4 16 16 64 16 8	0.933 0.890 0.918 0.925 0.948 0.933		300/48 310/90 300/282 306/632	17.3/82	2*4/1 2*4/1 2*4/1 2*4-1

POLYMER	REF	CRYST SYST.	SPACE GROUP	UNIT CELL PARAMETERS A	B	C	ANGLES	MON/ UNIT CELL	DENSITY (G./CC.) CRYSTAL	AMORPH.	MELT. POINT	HEAT OF FUSION KJ/MOL	CHAIN CONF. N*P/Q
---, 3-METHYL- (VIA HYDRIDE SHIFT) *70.14*	332			5.4		7.8					55 66/319		
---, 4-PHENYL- *132.21*	483 67	PORT	CS-2	10.4	18.0	6.61 6.55		6	1.064	0.962	158 160/90 168/187 159/282	4.6	2*3/1 2*3/1
POLY-2-BUTENE-ALT-ETHYLENE *84.16*	324	MONO	C2H-5	10.92	7.73	9.10	A=130	4	0.950	0.87	135		4*2/1
POLY-CHLOROTRIFLUORO- ETHYLENE-ALT-ETHYLENE *144.52*	635 637	HEX		9.86	9.86	5.02 5		3	1.703		262 264	18.8 13.30	4*1/1
POLY-CYCLODECENE TRANS- III. *138.25*	477	TRI	CI-1	4.40	5.39	12.30	66104118	1	0.985				10*1/1
IV.	530	MONO	C2H-5	7.42	5.00	12.40	B=94.2	2	1.001		80/562	32.9/562	10*1/1
POLY-CYCLODODECENE TRANS- III. *166.31*	477	TRI	CI-1	4.40	5.39	14.78	66104118	1	0.986				12*1/1
IV.	530	MONO	C2H-5	7.43	5.00	14.85	B=93.5	2	1.003		80/439 84/562	41.2/562	12*1/1
POLY-CYCLOHEPTENE TRANS- I. *96.17*	97 530	ORTHO OR	D2H-16 C2V-9	7.40	5.00	17.10		4	1.010		51/439		7*2/1
POLY-CYCLOHEPTENE-ALT- ETHYLENE *124.23*	259					9.0					74		
POLY-CYCLOOCTENE TRANS- III. *110.20*	126 476	TRI	CI-1	4.34	5.41	9.78 9.78	64105119	1	1.008		67/439 73/284	20.1/284	8*1/1
IV.	530	MONO	C2H-5	7.43	5.00	9.90	B=95.2	2	0.999		77/562 62/439	23.8/562	8*1/1
POLY-CYCLOPENTENE TRANS- I. *68.12*	147	ORTHO OR	C2V-9 D2H-16	7.28	4.97	11.90		4	1.051		23 34/682	12.0/682	5*2/1

POLYMER	REF	CRYST SYST.	SPACE GROUP	UNIT CELL A	B	C	ANGLES	MON/ UNIT CELL	DENSITY CRYSTAL	DENSITY AMORPH.	MELT. POINT	HEAT OF FUSION KJ/MOL	CHAIN CONF. N*P/Q
POLY-CYCLOPENTENE-ALT-ETHYLENE *96.17*	247	ORTHO		8.76	7.83	9.02		4	1.032				
	412	ORTHO OR	D2H-17 D2H-14	8.76	8.05	9.00		4	1.006		185		4*2/1
POLY-1,3-CYCLOPENTYLENE-METHYLENE *82.15*	267	ORTHO		7.69	6.21	4.80		2	1.190		146		
	340	ORTHO		13.30	15.52	4.80		8	1.101		128		
POLY-1-DECENE *140.27* I.	150					13.2					34/250 40/537		
POLY-1-DODECENE *168.32* I.	150					13.2					45/48 49/250		
POLYETHYLENE *28.05* I.	14	ORTHO	D2H-16	7.40	4.93	2.534		2	1.008		137/85	7.87/85	2*1/1
	15	ORTHO		7.36	4.94	2.534		2	1.011		141/206	7.70/86	
	535	ORTHO		7.418	4.946	2.546		2	0.9972		145/351	8.67/700	
	311	ORTHO		7.406	4.939	2.547		2	0.9998		141/286	7.79/156	
	406								0.991	0.811	142/315	8.12/262	
	138									0.855	146/362	7.62/343	
	363								0.9988	0.8866		8.37/260	
(SINGLE CRYSTAL)	72	TRI		7.84	5.56	120	63,71,82	96	1.004				
II.	58	PMONO		4.05	4.85	2.54	G=105	1	0.966				2*1/1
	233	MONO	C2H-3	8.09	2.53*	4.79	B=107.9	1	0.998				
	254	TRI		4.285	4.820	2.54	90 110 108	1	1.002				
	557	PMONO		7.11	5.46	2.534	G=96.5	2	0.953				
---, CHLOROTRIFLUORO- *116.47*	208	HEX		6.34	6.34	35		14	2.222	2.08/49	220/49	5.02/88	2*14/1
	21	HEX		6.5	6.5	35		14	2.11		210/46		2*17/
	296	HEX		6.385	6.385	42		17	2.217				2*17/1
	629	HEX		6.4	6.4	42.5		17	2.18				2*16/1
	205					43		16					
	139								2.19/49	1.925	222/341		
	116								2.192	2.032	215		
---, ETHYLSILYL-ISOTACTIC *86.21*	489	HEX	C3I-2	21.60	21.60	6.50		18	0.981				2*3/1
---, TETRAFLUORO- *50.01* II.	11	PHEX		5.54	5.54	16.8	G=119.5	13	2.406				1*13/1
	266	TRI		4.882	4.875	5.105	90,87,87	4	2.744				
	209	TRI		5.59	5.59	16.88	G=119.3	13	2.347				1*13/6
	602	MONO	C2-2	5.59	9.76	16.88	B=90	26	2.344		327/46		

References page III-92

CRYSTALLOGRAPHIC DATA FOR VARIOUS POLYMERS

POLYMER	REF	CRYST SYST.	SPACE GROUP	UNIT CELL A	B	C	PARAMETERS ANGLES	MON/UNIT CELL	DENSITY CRYSTAL	AMORPH.	MELT. POINT	HEAT OF FUSION KJ/MOL	CHAIN CONF. N*P/Q
ABOVE 20C I AND IV.	11	HEX		5.61	5.61	16.8		13	2.358		330		1*13/1
	209	HEX		5.66	5.66	19.50		15	2.302		330/101	5.74/91	1*15/7
III. (AT 12 KBAR)									2.304/66				
---, TRIFLUORO- *82.02*	681	MONO	C2H-3	9.50	5.05	2.62	G=105.5	4	2.742				1*2/1
	389	HEX		5.59	5.59	2.50		1	2.013				
POLY-ETHYLIDENE *28.05*	178	ORTHO		12.38	6.28	2.5		4	0.958		100/617 200/618 195/619		1*
POLY-1-HEPTENE *98.19* I.	150					6.45					17/250		2*3/1
	359												2*3/1
---, 5-METHYL- (S) *112.22* (S),(R)	67	MONO	C2-2	18.40	10.62	6.40	B=90	6	0.900		52/129		2*3/1
	685	TET	S4-1	20.00	20.00	6.36		76	0.913				2*3/1
	685					38.76							2*19/6
POLY-1-HEXADECENE *224.43* II.	150	ORTHO		7.5	63.2	6.6		8	0.95		68/250		2*4/1
POLY-1-HEXENE *84.16* OR	150	MONO		22.2	8.89	13.7	G=94.5	14	0.726		-55/48		2*7/2
	150	ORTHO		11.7	26.9	13.7		28	0.908				2*7/2
---, 4-METHYL- *98.19*	67	TET		19.64	19.64	14.00		28	0.845		200/9		2*7/2
	569	TET	C1-1	19.85	19.85	13.50		28	0.858		190/632		2*7/2
---, 5-METHYL- *98.19*	9	HEX		10.2	10.2	6.50		3	0.835		130		2*3/1
	685	MONO	C2-2	17.62	10.17	6.33	B=90	6	0.862		110/282		2*3/1
POLY-ISOBUTENE *56.11*	34	ORTHO	D2-4	6.94	11.96	18.63		16	0.964	0.915	44/46	12.0/82	2*8/5
	690	ORTHO	D2-4	6.88	11.91	18.60		16	0.978	0.912/53			2*8/3
	139									0.842			
POLY-1-OCTADECENE *252.49* II	150	ORTHO		7.5	70.4	6.6		8	0.96		80/6		2*4/1
III.	493	HEX		4.24	4.24						100/90		

POLYMER	REF	CRYST SYST.	SPACE GROUP	UNIT CELL A	CELL B	PARAMETERS C	ANGLES	MON/UNIT CELL	DENSITY CRYSTAL	(G./CC.) AMORPH.	MELT. POINT	HEAT OF FUSION KJ/MOL	CHAIN CONF. N*P/Q
POLY-OCTAMETHYLENE													
---, 1-METHYL- *126.24*	639										-5	10.9	
POLY-1-PENTENE													
70.13 IA.	355	MONO		11.35	20.85	6.49	B=99.6	12	0.923		130/276		2*3/1
			OR	21.15	11.20	6.49	B=99.6	12	0.922		130/380		
									0.96/408	0.85/408	111/501	6.28/501	
IB.	355					6.5							2*3/1
IIA.	152			19.30	16.90	6.60	G=116	16	0.898		75/48		2*4/1
	355	PORT				7.08					80/9		
											78/282		
											79/501	4.00/501	
IIB.	355	PORT		19.60	16.75	7.08	G=115.3	16	0.887				2*4/1
5-AMINO-													
---, N,N-DIISOBUTYL- *197.37*	191					6.85					112		
---, N,N-DIISOPROPYL- *169.31*	191					7.32					130		
---, 4,4-DIMETHYL- *98.19*	431	TET	S4-1	20.3	20.3	13.8		28	0.803		231/282		2*7/2
	442	TET		20.35	20.35	7.01		16	0.899		350/90		2*4/1
											380/441		
5-HYDROXY- *86.13*	191					6.5							
3-METHYL- *84.16*	622	TET	C4-6	13.35	13.35	6.80		8	0.922		362/632		2*4/1
											273/129		
											200/441		
---, 4-METHYL- *84.16* I.	55	TET	S4-1	18.66	18.66	13.80		28	0.814		235/48	19.7/82	
	452	TET		18.54	18.54	13.84		28	0.822		250/94		IRREG.
	333	TET		18.50	18.50	13.76		28	0.831		238/632	11.9/370	2*7/2
	67	TET		18.60	18.60	13.84		28	0.817		228/282		2*7/2
	689	TET	S4-1	18.63	18.63	13.85		28	0.814 / 0.828/94	0.838/94			
II.	8			19.16	19.16	7.12		16	0.855		125		2*4/1
III.	8			19.36	19.36	7.05		16	0.846		75		2*4/1

POLYMER	REF	CRYST SYST.	SPACE GROUP	UNIT CELL PARAMETERS A	B	C	ANGLES	MON/UNIT CELL	DENSITY (G./CC.) CRYSTAL	AMORPH.	MELT. POINT	HEAT OF FUSION KJ/MOL	CHAIN CONF. N*P/Q
---, 5-TRIMETHYLSILYL- *142.32*	67					6.55					133/71		2*3/1
POLY-PROPENE ISOTACTIC I. *42.08*	127	MONO	C2H-6	6.65	20.96	6.50	B=99.3	12	0.938	0.85/45	176/10	9.92/82	2*3/1
	308	MONO		6.66	20.78	6.495	B=99.62	12	0.946		189/392	7.96/478	2*3/1
	170	TRI	CI-1	13.36	6.50*	10.99	87,10899	12	0.935		186/394	8.8/394	
	519	MONO	C2H-5	6.65	20.73	6.50	B=99.0	12	0.947		208/676	5.80/676	2*3/1
	626	MONO		6.63	20.78	6.504	B=99.5	12	0.949				2*3/1
	131	MONO		6.64	20.88	6.51	B=98.7	12	0.940		178/349	10.0/358	
	136	MONO	C2H-5	6.69	20.98	6.504	B=99.5	12	0.931	0.8535	183/380	6.16/182	2*3/1
	330	MONO	C2H-3					12			180/250	7.91/251	
												10/474	
II.	166	HEX	D3-4	12.74	12.74	6.35		12	0.939	.907/112			2*3/1
	196	RHO	OR -6	6.38	6.38	6.33		3	0.939				
	308	ORTHO	OR	19.08	11.01	6.490		18	0.88/167				2*3/1
	308	HEX		22.03	22.03	6.490		36	0.922				2*3/1
	485	RHO		19.08	19.08	6.49		27	0.922				2*3/1
III.	308	TRI		6.47	10.71	6.50	G=99.07	12	0.946				2*3/1
	592	TRI		6.54	21.40		8910099						2*3/1
SYNDIOTACTIC I.	169	ORTHO	D2-2	14.5	5.8	7.4		8	0.898	.858/393	161/415		4*2/1
	430	ORTHO	D2-5	14.50	5.60	7.40		8	0.930		138/545	1.88/545	4*2/1
	409								0.898	0.858			
II.	306	PHEX				5.05							4*1/1
---, 3-CYCLOHEXYL- *124.23* I.	431	TET OR	C4-5 S4-2	21.06	21.06	20.09		40	0.926		230/90 215/282 214/328		2*10/3
II.	431	RHO OR	C4-3 C3I-2	19.12	19.12	6.33		9	0.926				2*3/1
---, 3-CYCLOPENTYL- *110.20*	431	TET OR	C4-5	20.34	20.34	47.49		96	0.894		210/328		2*24/7
	625	TET	S4-2	20.30	20.30	47.40		96	0.899		225/90		2*24/7
---, HEXAFLUORO- *150.02*	268										160		2*4/1
---, 3-PHENYL- (ALLYLBENZENE) *118.18*	67					6.40					230/90 208/187 185/282		2*3/1

POLYMER	REF	CRYST. SYST.	SPACE GROUP	UNIT CELL A	CELL B	PARAMETERS C	ANGLES	MON/ UNIT CELL	DENSITY (G./CC.) CRYSTAL	AMORPH.	MELT. POINT	HEAT OF FUSION KJ/MOL	CHAIN CONF. N*P/Q
---, 3-SILYL- *72.18	67					6.45					128/71		2*3/1
---, TRIMETHYLSILYL- *114.26*	67					6.50					360/71		2*3/1
POLY-1-TETRADECENE II. *196.38*	150	ORTHO		7.5	56.0	6.6		8	0.94		57/250		2*4/1
POLY-TETRAFLUOROETHYLENE-ALT-ETHYLENE *128.07*	722	MONO		9.6	9.25	5.0	G=96	4	1.93				
POLY-VINYLCYCLOBUTANE *82.15*	431	TET		34.12	34.12	6.6		64	1.14		228/437		2*4/1
POLY-VINYLCYCLOHEPTANE *124.23*	431	TET	C4H-6	23.4	23.4	6.5		16	0.927		>300/437		2*4/1
POLY-VINYLCYCLOHEXANE *110.20* I.	431	TET	C4H-6	21.99	21.99	6.43		16	0.942		305/90		2*4/1
	67	TET		21.76	21.76	6.50		16	0.951		300/282		2*4/1
	95	TRI		11.6	7.8	6.6	92,10898	3	0.98		383/328		
											372/441		
II.	431	TET OR	C4-5 S4-2	20.48	20.48	44.58		96	0.940				2*24/7
POLY-VINYLCYCLOPENTANE *96.17* I.	95	TRI		10.5	7.4	6.6	92,10899	3	1.00		292/328		2*4/1
	431	TET	C4H-6	20.14	20.14	6.50		16	0.969		270/441		
II.	431	TET OR	C4-5 S4-2	20.14	20.14	19.5		48	0.969				2*12/3
III.	431	TET OR	D4H-1 D4H-7	37.3	37.3	19.8			0.927				2*10/3
POLY-VINYLCYCLOPROPANE *68.12* I.	175	TRI	C3-2	13.6	13.6	6.5		9	0.981		230		2*3/1
	95	RHO OR	C3-3			6.48							
	526												
II.	526	TET	S4-2	15.21	15.21	20.85		40	0.9380				2*10/3
POLY-1-VINYLENE-3-CYCLOPENTYLENE *94.16*	620					11.8					202		7*2/

POLY-VINYLS AND POLY-VINYLIDENES
-HCH-CHR-

POLYMER	REF	CRYST SYST.	SPACE GROUP	UNIT CELL A	CELL B	PARAMETERS C	ANGLES	MON/UNIT CELL	DENSITY CRYSTAL	DENSITY AMORPH.	MELT. POINT	HEAT OF FUSION KJ/MOL	CHAIN CONF. N*P/Q
POLY-ACRYLAMIDE													
---, N,N-DIBUTYL- *183.30*	93	HEX		26.3	26.3	6.3		12	0.97 (1.06)				2*3/1
---, N-ISOPROPYL- *113.16*	102								1.118	1.070	200		
POLY-ACRYLIC ACID													
---, ALLYL ESTER *112.13*	278					6.5					90		2*3/1
---, SEC-BUTYL ESTER *128.17*	401	ORTHO		17.92	10.34	6.49		6	1.062		130		2*3/1
---, T-BUTYL ESTER *128.17*	67 / 401	ORTHO		17.92	10.50	6.45 / 6.48		6	1.047		193/16 / 200		2*3/1 / 2*3/1
---, ISOBUTYL ESTER *128.17*	401	ORTHO		17.92	17.92	6.42		12	1.239		81		2*3/1
---, ISOPROPYL ESTER ISOTACTIC *114.14*	67 / 149					6.5 / 5.32			1.08		162/120 / 162 / 178/541	5.9/541	2*3/1
SYNDIOTACTIC	113					5.18			1.18		115		
POLY-ACRYLONITRILE SYNDIOTACTIC *53.06*	76	HEX	C2V-16	5.99	5.99	5.10	G=120	4	1.111		317/77	4.86/77	4*1/1
	133	ORTHO		10.20	6.10	5.00		8	1.273		341/499	5.23/77	
	162	ORTHO		18.1	6.12	5.08		4	1.134				
	210	ORTHO		10.55	5.80	(5.1)		2	1.07				
	322	PHEX		6.1	6.11	(5.1)		16	1.125				
	486	ORTHO		21.18	11.60	5.04		8	1.137				
	575	ORTHO		10.6	11.6								
ISOTACTIC	133	TET		4.74	4.74	2.55		1	1.538				
POLY-ISOPROPENYLMETHYL KETONE *84.12*	347	TET		15.08	15.08	8.54		16	1.151		240 / 240/255 / 200/379		2*4/1

POLYMER	REF	CRYST SYST.	SPACE GROUP	UNIT CELL A	CELL B	PARAMETERS C	ANGLES	MON/UNIT CELL	DENSITY (G./CC.) CRYSTAL	AMORPH.	MELT. POINT	HEAT OF FUSION KJ/MOL	CHAIN CONF. N*P/Q
POLY-METHACRYLIC ACID													
---, METHYL ESTER													
ISOTACTIC *100.12*	30	PORT		21.08	12.17	10.55		20	1.228	1.22/31	160/31		2*5/2
	520	ORTHO		21.08	12.17	10.50		20	1.234				2*5/1
SYNDIOTACTIC	30									1.19/31	200/31		4*5/2
---, OCTADECYL ESTER													
SYNDIOTACTIC *338.58*	500	HEX		4.73	4.73	30.2		1	0.961		36		
POLY-METHACRYLONITRILE *67.09* I.	427	PHEX		9.03	9.03	6.87		4	0.918		250/436		
	729			7.87	8.97	6.87		4	0.918				
II.	427	MONO		13.5	7.71	7.62	B=97.7	8	1.134				2*4/1
POLY-THIOLACRYLIC ACID													
---, SEC-BUTYL ESTER *144.23*	401	RHO	C3-2			6.35							2*3/1
---, ISOBUTYL ESTER *144.23*	401	MONO	CS-2			6.42							
---, ISOPROPYL ESTER *130.20*	401	RHO	C3-2			6.42							
---, PROPYL ESTER *130.20*	401	MONO	CS-2			6.4							2*3/1
POLY-VINYL ACETATE													
---, TRIFLUORO- *140.06*	638	ORTHO		9.54	12.44	4.8		4	1.633				
POLY-VINYL ALCOHOL *44.05*	29	MONO	C2H-2	7.81	2.52*	5.51	B=91.7	2	1.350	1.291/1	232/310		2*1/1
	132	MONO		7.805	2.533	5.485	B=92.2	2	1.350	1.26/53	265/323		
	184	MONO	C2H-2	7.81	2.52*	5.50	B=92	2	1.345	1.269	228/433	6.87/433	
	261	MONO		7.81		5.43	B=91.5		1.352		267/450		
	316								1.34/410	1.27/410	243/480	6.99/480	
QUENCHED	261	ORTHO	C2V-20	7.42	2.52	5.25		2	1.490				
SINGLE CRYSTAL	151	HEX		5.45	5.45	2.51		1	1.133				2*1/1
POLY-VINYL CHLORIDE *62.50*	7	ORTHO	D2H-11	10.6	5.4	5.1		4	1.42	1.41/413	273/372	11/372	4*1/1
	64	ORTHO		10.11	5.27	5.12		4	1.522		212/143	2.8/145	
	140	MONO		10.65	5.15*	5.20	B=90	4	1.456		310/423	3.3/423	
	235	ORTHO	D2H-11	10.40	5.30	5.10		4	1.477				
	718	ORTHO	D2H-11	10.24	5.24	5.08		4	1.523				

CRYSTALLOGRAPHIC DATA FOR VARIOUS POLYMERS

POLYMER	REF	CRYST. SYST.	SPACE GROUP	UNIT CELL PARAMETERS A	B	C	ANGLES	MON/ UNIT CELL	DENSITY (G./CC.) CRYSTAL	AMORPH.	MELT. POINT	HEAT OF FUSION KJ/MOL	CHAIN CONF. N*P/Q
POLY-VINYL FLUORIDE *46.04*													
	62	HEX		4.93	4.93	2.53		1	1.436		200/79	7.54/79	2*1/1
	236	ORTHO	C2V-14	8.57	4.95	2.52		2	1.430		230/434		2*1/1
POLY-VINYL FORMATE ISOTACTIC *72.06*													
	232	RHO	C3V-6- D3D-6	15.9	15.9	5.55		18	1.502				
SYNDIOTACTIC													
	232					5.0							
POLY-VINYLIDENE BROMIDE *185.85*													
	33	MONO		25.88	4.77*	13.87	B=70.2	16	3.065				
POLY-VINYLIDENE CHLORIDE *96.94*													
	33	MONO		22.54	4.68*	12.53	B=84.2	16	1.958		200/721		4*1/1
	245	MONO	C2-2	6.73	4.68*	12.54	B=123.6	4	1.957				
	32	MONO		13.69	4.67*	5.296	B=55.2	4	1.948	1.66/43	190/46		2*2/1
	462									1.7754			
	231												
POLY-VINYLIDENE FLUORIDE (ALPHA) I. *64.04*													
	171	MONO		5.02	25.4*	4.62	B=107	10	1.888		185/434		4*1/1
	604	MONO	C2H-5	4.96	9.64	4.62	B=90	4	1.926		178		4*1/1
	417	MONO	C2-2	17.72	4.57*	11.68	B=87.3	16	1.801		220/443		2*2/1
	418	ORTHO	C2V-9	9.66	4.96	4.64		4	1.913		171/373		
	168	MONO	C2-2,1	9.64	4.64*	5.02	B=91.1	4	1.895		178/726	8.91/726	2*2/1
	563	PORT	C2-2	5.02	9.63	4.62		4	1.904				
	597	ORTHO		9.60	5.00	4.655		4	1.904				
(BETA) II.													
	417	ORTHO	C2V-14	8.45	4.88	2.55		2	2.022		212/443		2*1/1
	418	ORTHO	C2V-14	8.47	4.90	2.56		2	2.002				2*1/1
	597	ORTHO		8.60	4.97	2.57		2	1.936				
	604	ORTHO	C2V-14	8.58	4.91	2.56		2	1.972		191		2*1/1
III.													
	604	MONO	C2-3	8.66	4.93	2.58	B=97	2	1.945		197		2*1/1
POLY-VINYLMETHYL KETONE *70.09*													
	164	TET	S4-1	14.52	14.52	14.40		28	1.073		170		2*7/2
	379	TET		14.56	14.56	14.10		28	1.090				2*7/2

POLYMER	REF	CRYST. SYST.	SPACE GROUP	UNIT CELL PARAMETERS A	B	C	ANGLES	MON/ UNIT CELL	DENSITY (G./CC.) CRYSTAL	AMORPH.	MELT. POINT	HEAT OF FUSION KJ/MOL	CHAIN CONF. N*P/Q
POLY-AROMATICS													
-HCH-CHR-													
POLY-2,5-DISTYRYLPYRAZINE *284.36*	558	ORTHO		10.8	18.4	7.4		4	1.28		321		
	564	ORTHO	D2H-15	18.36	10.88	7.52		4	1.257		343/723		
POLY-METHYLENE-P-PHENYLENE *90.12*	745	ORTHO		8.16	6.32	9.90		4	1.172		170		
											142/744		
POLY-4,4'-OXYDIPHENYLENE 2,2',3,3'-OXYDIPHENYLENE- TETRACARBOXIMIDE *474.43*	684	MONO		5.17	10.85	38.2	G=103.7	4	1.514				*2/1
POLY-4,4'-OXYDIPHENYLENE P-PHENYLENE- DITRIMELLITATEDIIMIDE *622.54*	658	MONO		7.95	5.38	57.4	G=85	4	1.691				
	750	MONO		7.95	5.56	58	G=85	4	1.62				*2/1
POLY-4,4'-OXYDIPHENYLENE PYROMELLITIMIDE *382.33*	642	ORTHO		6.35	4.05	32.6		2	1.514				
	658	ORTHO				31.9							
	684	ORTHO		12.62	7.94	32		8	1.58				
POLY-P-PHENYLENE *76.10*	475	ORTHO		7.81	5.53	4.20		2	1.393		<550/735		6*1/1
	508												6*2/1
POLY-P-PHENYLENECYCLO- BUTYLENE ---, DIPYRIDYL- *284.36*	202	ORTHO	D2H-15	19.16	10.69	7.45		4	1.238				
	723	ORTHO	D2H-15	18.9	10.5	7.53		4	1.26		340		
POLY-4,4'-(P-PHENYLENEDIOXY)- DIPHENYLENE 2,2',3,3'-OXY- DIPHENYLENETETRACARBOXIMIDE *566.52*	684	TRI		10.52	9.45	26.1	90,95,98	4	1.472				
POLY-4,4'-(P-PHENYLENEDIOXY)- DIPHENYLENE P-PHENYLENE- DITRIMELLITATEDIIMIDE *714.64*	658	ORTHO		5.64	8.05	68.0		4	1.537				
	750	ORTHO		5.64	8.05	34		2	1.54				*2/1

CRYSTALLOGRAPHIC DATA FOR VARIOUS POLYMERS

POLYMER	REF	CRYST SYST.	SPACE GROUP	UNIT CELL A	CELL B	PARAMETERS C	ANGLES	MON/ UNIT CELL	DENSITY (G./CC.) CRYSTAL	AMORPH.	MELT. POINT	HEAT OF FUSION KJ/MOL	CHAIN CONF. N*P/Q
POLY-4,4'-(P-PHENYLENEDIOXY)-DIPHENYLENE PYROMELLITIMIDE *474.43*	658 684	MONO		5.64	8.30	21.2 21.8	B=98.7	2	1.562				*2/1
POLYSTYRENE *104.15*	6 128 411 139	RHO RHO	D3D-6 D3D-6	22.08 21.9	22.08 21.9	6.626 6.65		18 18	1.113 1.127 1.114 1.12/45	1.04 TO 1.065/17 1.052 1.024	240/10 250/90 235/282	9.00/89 8.37/174	2*3/1
---, P-TERT-BUTYL- *160.26*	198									0.950	300		
---, O-FLUORO- *122.14*	123	RHO	C3V-6	22.15	22.15	6.63		18	1.296		270/75		2*3/1
---, P-FLUORO- *122.14*	67					8.30					265/75		2*4/1
---, A-METHYL- *118.18*	357	RHO				6.6							
---, O-METHYL- *118.18*	125	TET	C4V-12	19.01	19.01	8.10		16	1.073		360/74		2*4/1
---, P-FLUORO- *136.17*	67					8.05					360/75		2*4/1
---, M-METHYL- *118.18*	80 163 431 103	TET TET OR	S4-1 C4-5 S4-2	19.81 19.9	19.81 19.9	21.74 57.1 78.9 57.0		44	1.012 1.005		215 215/74		2*11/3 2*29/8 2*40/ 11
---, P-METHYL- *118.18*	103					22.9				1.04/67			
---, TRIMETHYLSILYL- *176.34*	103					60.4					284		
POLY-N-VINYL CARBAZOLE *193.25*	126 607	HEX PHEX		12.30 12.3	12.30 12.3	7.44		3	0.988		>320/228		2*3/1 2*3/1
POLY-1-VINYL NAPHTHALENE *154.21*	122	TET	C4V-12	21.20	21.20	8.10		16	1.125		360/75		2*4/1
POLY-2-VINYLPYRIDINE *105.14*	185					6.7					212		

POLYMER	REF	CRYST SYST.	SPACE GROUP	UNIT CELL A	B	C	PARAMETERS ANGLES	MON/ UNIT CELL	DENSITY (G./CC.) CRYSTAL	AMORPH.	MELT. POINT	HEAT OF FUSION KJ/MOL	CHAIN CONF. N*P/Q
POLY-P-XYLYLENE													
ALPHA *104.15*	223	MONO		11.68	6.10	9.16	B=102.5	4	1.086	1.05	375/142		8*
	426	ORTHO		21.3	33.6	6.58		31	1.138		420/219	30.2/82	
	595	MONO		5.92	10.64	6.55	B=134.7	2	1.179		375/82		
											435/323		
											400/395		
											440/546		
BETA	223	MONO		8.10	5.25	6.53	B=95	2	1.250		420		
	426	HEX		20.52	20.52	6.58		16	1.153		412		
	224					6.55							
POLY-DIENES --------													
1,2-POLY-1,3-BUTADIENE													
SYNDIOTACTIC *54.09*	7	ORTHO	D2H-11	10.98	6.60	5.14		4	0.964		154		4*1/1
ISOTACTIC	12	RHO	D3D-6	17.3	17.3	6.5		18	0.96		120		2*3/1
											125/44		
---, 4,4-DIMETHYL-													
ISOTACTIC II. *82.15*	488	TET	D2D-10	17.80	17.80	36.50		72	0.849		167/707		2*18/5
1,4-POLY-1,3-BUTADIENE													
TRANS- I. *54.09*	37	PHEX	C2H-5	4.54	4.54	4.9		1	1.03		100/352	10.0/352	4*1/1
	496	MONO		8.63	9.11	4.83	B=114	4	1.036		96/451	13.8/451	
											55/624	6.31/624	
II. (ABOVE 65)	154	PHEX		4.88	4.88	4.68		1	0.930		141/352	4.61/352	4*1/1
	550	HEX		4.95	4.95	4.66		1	0.908		148/44	5.99/342	
											145/451	4.61/451	
											142/624	3.62/624	
CIS-	60	MONO	C2H-5	8.53	8.16	12.66	B=83.33	8	0.821		6.3/335	9.2/353	8*1/1
	124	MONO	CS-4 - C2H-6	4.60	9.50	8.60	B=109	4	1.011		1 1/287		
---, 2-TERT-BUTYL-													
CIS- *110.20*	367	TRI		13.95	20.78	15.3		22	0.908		106		4*11/3

CRYSTALLOGRAPHIC DATA FOR VARIOUS POLYMERS

POLYMER	REF	CRYST SYST.	SPACE GROUP	UNIT CELL A	B	PARAMETERS C	ANGLES	MON/ UNIT CELL	DENSITY (G./CC.) CRYSTAL	AMORPH.	MELT. POINT	HEAT OF FUSION KJ/MOL	CHAIN CONF. N*P/Q
---, 2-CHLORO- TRANS- (CHLOROPRENE) *88.54*	40	ORTHO	D2-4	8.84	10.24	4.79		4	1.356		115/229		
	109	ORTHO		9.0	8.23	4.79		4	1.658		80/81	8.37/81	
											107/556		
---, 1-CYANO- TRANS- *79.10*	275					4.8							
---, 2,3-DICHLORO- TRANS- I, *122.98*	221	MONO	C2H-5	5.34	9.95	4.86	B=93.5	2	1.604				4*1/1
	565					4.80							
---, II.	699	ORTHO	C2V-9	8.81	12.34	4.80		4	1.565				4*1/1
---, THIOUREA COMPLEX	565	MONO	C2H-5	9.87	15.83	12.53	B=114.1		1.44				
---, 2,3-DIMETHYL- TRANS- *82.15*	104	ORTHO	C2V-9	9.13	13.00	4.35		4	0.978		260		4*1/1
	699					4.70					272/221		
---, CIS-	220					7.0					198		
---, 2-METHYL- (ISOPRENE) TRANS- ALPHA *68.12*	288	MONO	C2H-5	7.98	6.29	8.70	B=102	4	1.051				4*2/1
	502					8.76							8*1/1
	715					8.77							
BETA	23	ORTHO	D2-4	7.78	11.78	4.72		4	1.046		65/46	12.7/81	4*1/1
	288	ORTHO		11.9	4.8*	7.85		4	1.009		74/81		4*1/1
	502	ORTHO		7.83	11.87	4.75		4	1.025		68/287		
											87/568		
											78/630		
DELTA	502	MONO		5.9	7.9	9.2	G=94	4	1.06				
	492			7.84	5.99								
EPSILON	492			7.80	6.29								
CIS-	40	MONO	C2H-5	12.46	8.89	8.10	B=92	8	1.009	0.906/53	28/81	4.40/81	8*1/1
	288	ORTHO		8.97	8.20*	25.2		16	0.976	0.910/47	14/287		8*1/1
	124	ORTHO	D2H-15								36/50		8*1/1
---, 3-CHLORO- TRANS- *102.56*	165					4.9							

POLYMER	REF	CRYST. SYST.	SPACE GROUP	UNIT CELL PARAMETERS				MON/ UNIT CELL	DENSITY (G./CC.)		MELT. POINT	HEAT OF FUSION KJ/MOL	CHAIN CONF. N*P/Q
				A	B	C	ANGLES		CRYSTAL	AMORPH.			
--, HYDROCHLORINATED *104.58*	229 70	PORTH ORTHO	C2H-5	5.83 11.9	10.38 9.1*	8.95 10.4	B=90	4 8	1.282 1.23		115 110		
--, 2-METHYLACETOXY- TRANS- *140.18*	165	ORTHO		16.2	9.3	4.75		4	1.30		135		
--, 2-PROPYL- TRANS- *96.17*	368	ORTHO		10.95	6.65	9.2		4	0.95		42 42/368		8*1/1
POLY-CHLOROPRENE, SEE POLY-BUTADIENE, 2-CHLORO-													
1,4-POLY-1,3-HEPTADIENE TRANS- ISOTACTIC *96.17*	375	MONO	C2-2	8.62	7.95	4.85	B=99	2	0.973		85/371		4*1/1
--, 6-METHYL- TRANS- ISOTACTIC *110.20*	371					4.85					119		
1,4-POLY-1,3-HEXADIENE TRANS- ISOTACTIC *82.15*	274	ORTHO	D2-4	14.02	8.02	4.85		4	1.000		82/371		4*1/1
--, 5-METHYL- TRANS- ISOTACTIC *96.17*	371					4.85					88		
2,5-POLY-2,4-HEXADIENE TRANS-ERYTHRO-DI-ISOTACTIC *82.15*	525					2.3					84		
--, 2,5-DIMETHYL- TRANS- *110.20*	183					4.8					265 265/376		
2,5-POLY-2,4-HEXADIENEDIOIC ACID --, DIISOPROPYL ESTER TRANS-ERYTHRO-DI-ISOTACTIC *226.27*	420	ORTHO		14.16	10.28	9.70		4	1.064		230/420		8*1/1

References page III-92

CRYSTALLOGRAPHIC DATA FOR VARIOUS POLYMERS

POLYMER	REF	CRYST SYST.	SPACE-GROUP	UNIT CELL PARAMETERS A	B	C	ANGLES	MON/UNIT CELL	DENSITY (G./CC.) CRYSTAL	AMORPH.	MELT. POINT	HEAT OF FUSION KJ/MOL	CHAIN CONF. N*P/Q
2,5-POLY-2,4-HEXADIENOIC ACID TRANS-ERYTHRO-DI-ISOTACTIC													
--, BUTYL ESTER *168.24*	384	ORTHO		11.36	9.70	4.80		2	1.056				
---, ETHYL ESTER *140.18*	384					4.80							
---, ISOAMYL ESTER *182.26*	384					4.80							
---, ISOBUTYL ESTER *168.24*	384					4.80							
---, ISOPROPYL ESTER *154.21*	384					4.80							
---, METHYL ESTER *126.16*	384					4.80							
2,5-POLY-2,4-HEXADIYNEDIOL													
---, BIS-PHENYLURETHANE *348.36* I. (0.5 DIOXANE/UNIT)	716	TRI	CI-1	4.89*	12.53	16.78	69,97,96	2	1.367				4*1/1
II.	749			4.9*									
III.	749			4.9*									
POLY-ISOPRENE, SEE POLY-BUTADIENE, 2-METHYL-													
1,4-POLY-1,3-OCTADIENE TRANS-													
ISOTACTIC *110.20*	371					4.85					87		
1,2-POLY-1,3-PENTADIENE TRANS-													
SYNDIOTACTIC *68.12*	230					5.1					10		
1,4-POLY-1,3-PENTADIENE TRANS-													
ISOTACTIC *68.12*	189	ORTHO		19.73	4.85	4.8		4	0.98		95		4*1/1
	200	ORTHO	D2-4	19.80	4.86	4.85		4	0.969				4*1/1
CIS- ISOTACTIC	263					3.15					44		4*2/1

POLYMER	REF	CRYST SYST.	SPACE GROUP	UNIT CELL PARAMETERS A	B	C	ANGLES	MON/ UNIT CELL	DENSITY (G./CC.) CRYSTAL	AMORPH.	MELT. POINT	HEAT OF FUSION KJ/MOL	CHAIN CONF. N*P/Q
SYNDIOTACTIC	234					8.50					53		8*1/1
2,5-POLY-5-PHENYL-2,4-PENTADIENOIC ACID TRANS-ERYTHRO-DI-ISOTACTIC ---, BUTYL ESTER *230.31*	384					4.80							
---, METHYL ESTER *188.23*	384					4.80							
POLY-PEPTIDES -NH-CHR-CO-													
POLY-L-ALANINE ALPHA HELIX *71.08*	507	HEX		8.55	8.55	70.3		47	1.246				3*47/ 13
BETA	507 509	ORTHO ORTHO		4.79 4.734	10.7 10.54	6.88 6.89		4 4	1.339 1.373				3*2/1
POLY-BETA-ALANINE, SEE POLY-3-AMINOPROPIONIC ACID													
POLY-L-ALA-L-ALA-GLY *200.22*	753	ORTHO		9.44	21.0*	9.96		8	1.351				9*2/1
POLY-L-ALA-L-ALA-GLY-L-PRO-L-PRO-GLY *450.49*	755	ORTHO		10.5		29.5							
POLY-ALA-GLU-GLY ---, GAMMA-ETHYL- *285.30*	756	MOND		9.54	21.4*	8.54	B=106		1.133				9*2/1
POLY-L-ALA-GLY *128.13*	591 754	ORTHO ORTHO		9.42 9.44	6.95* 6.94*	8.87 8.96		4 4	1.466 1.450				6*1/1 6*1/1
II.	741	ORTHO		4.72	14.4	9.6		4	1.30				6*2/1
POLY-L-ALA-GLY-L-ALA-GLY-L-SER-GLY *400.39*	754	ORTHO		9.39	20.6*	9.05		4	1.522				

POLYMER	REF	CRYST SYST.	SPACE GROUP	UNIT CELL A	B	PARAMETERS C	ANGLES	MON/ UNIT CELL	DENSITY (G./CC.) CRYSTAL	AMORPH.	MELT. POINT	HEAT OF FUSION KJ/MOL	CHAIN CONF. N*P/Q
POLY-L-ALA-GLY-L-ALA- L-PRO-GLY-L-PRO *450.49*	755	ORTHO		10.5		29.5							
POLY-L-ALA-GLY-GLY II. (1 H2O/RESIDUE) *185.18*	596	MONO		8.86	22.0	9.42	B=90	9	1.507				9*1/1
	740	MONO	C2-2	11.0	4.8	9.45	B=90	2	1.35				
POLY-L-ALA-GLY-GLY-GLY II. *242.24*	740	HEX		4.89	4.89	36.60		3	1.592				12*1/1
POLY-L-ALA-GLY-L-PRO DRY	614	HEX		11.9	11.9	28.8		12	1.27				9*
WET *225.25*	614	HEX		11.4	11.4	9.3							9*1/1
POLY-L-ALA-GLY-L-PRO- L-PRO-GLY-L-PRO *476.53*	755	ORTHO		10.6		28.7							
POLY-L-ALA-L-PRO-GLY WITH H2O *225.25*	616	MONO	C2-2	8.6	7.2	9.4	B=90	2	1.4				9*
POLY-L-ALA-L-PRO-GLY- L-PRO-L-PRO-GLY *476.53*	755	HEX		11.9	11.9	29.5		5	1.093				
POLY-L-ARGININE ---, HYDROCHLORIDE ALPHA HELIX (2.5 H2O/RESIDUE) *192.65*	641	HEX		14.7	14.7	27.0		18	1.406				3*18/
BETA (>5 H2O/RESIDUE)	641	MONO		9.26	22.05	6.76	G=108.9	4	1.438				3*2/1
POLY-L-ASPARTIC ACID ---, BETA-BENZYL- OMEGA *205.21*	582	TET		13.85	13.85	5.30		4	1.341				3*4/1
	703	TET		13.80	13.80	5.42		4	1.320				
---, ---, P-CHLORO- ALPHA *239.66*	606	HEX	D6-3	14.9	14.9	27.0		18	1.380				3*18/5

POLYMER	REF	CRYST SYST.	SPACE GROUP	UNIT CELL A	B	C	PARAMETERS ANGLES	MON/ UNIT CELL	DENSITY (G./CC.) CRYSTAL	AMORPH.	MELT. POINT	HEAT OF FUSION KJ/MOL	CHAIN CONF. N*P/Q
OMEGA	606	TET	D4-4	23.3	23.3	5.20		8	1.13				3*4/1
---, BETA-N-PROPYL- BETA *157.17*	579	MONO	C2-2	9.57	6.79*	25.08	B=96	8	1.288				3*4/1
POLY-L-CYSTEINE ---, S-CARBOBENZOXY- BETA *237.27* I.	578	ORTHO		4.76	6.95*	32.4		4	1.470				
II.	578	ORTHO		4.89	6.89*	32.8		4	1.426				
POLY-2,4-DIAMINOBUTYRIC ACID ---, N-CARBOBENZOXY- *234.26*	720	RHO		27.5	27.5	27.0		56	1.232				
POLY-L-GLUTAMIC ACID BETA *129.12*	585	MONO		9.66	6.98*	9.10	B=105	4	1.447				
---, GAMMA-BENZYL- ALPHA HELIX (ENANTIOMORPHIC)	587	TRI		15.25	15.25	27.0	8484122	18	1.257				3*18/5
(RACEMIC)	584	PHEX		14.95	14.95	27.1	G=56	18	1.305				3*71/
	58.	HEX		14.95	14.95	106		71	1.260				20
219.24	702	MONO		25.2	15.0	26.8	G=122.5	30	1.278				3*43/
(DRY)	757	ORTHO		14.76	25.40	64.11		86	1.303				12
---, GAMMA-CALCIUM (WET) BETA	585	MONO	C2-2	9.40	6.83*	12.82	B=100.3						
---, GAMMA-METHYL- ALPHA HELIX *143.14*	588	HEX		11.96	11.96	27.5		18	1.256				3*18/5
	576	HEX		11.95	11.95	27.0		18	1.281				3*29/8
	589	ORTHO		11.95	20.70	43.2		58	1.290				*65/18
	583					97.26		65	(1.284)				
BETA	702			4.725		6.83							
POLY-GLYCINE 2 *57.05* I.	581			4.77	4.77	7.0	G=66	2	1.30				
II.	273	HEX	C3-2	4.8	4.8	9.3		3	1.53				3*3/1
POLY-GLY-GLY-L-PRO WITH H2O *211.22*	615	MONO	C2-2	12.2	4.9	9.3	B=90	2	1.4				9*
W/FORMIC ACID	615	MONO	C2-2	13.5	4.9	9.3	B=94	2					9*

CRYSTALLOGRAPHIC DATA FOR VARIOUS POLYMERS

POLYMER	REF	CRYST SYST.	SPACE GROUP	UNIT CELL PARAMETERS A	B	C	ANGLES	MON/ UNIT CELL	DENSITY (G./CC.) CRYSTAL	AMORPH.	MELT. POINT	HEAT OF FUSION KJ/MOL	CHAIN CONF. N*P/Q
POLY-GLY-PRO-HYP *267.28*	612	HEX		13.7	13.7	28.2		10	0.97				9*
POLY-GLY-L-PRO-L-PRO DRY *251.29*	613	HEX		12.5	12.5	28.7		10	1.07				9*
POLY-L-HISTIDINE ---, 1-BENZYL- *227.27*	719	PHEX		17.4	17.4	27.0	G=130	18	1.085				
POLY-L-LYSINE ALPHA HELIX *128.18*	586	HEX		19.55	19.55								
BETA	586	ORTHO		9.44	6.8*	17.16		62	1.159				
E-CARBOBENZOXY- ALPHA HELIX *262.31*	577	HEX		16.69	16.69	26.90		18	1.208				3*18/5
HYDROCHLORIDE DRY BETA *164.64*	666	ORTHO		4.62	15.20	6.66		4	2.338				
WET 65% R.H. (BETA)	666	ORTHO		4.71	16.67	6.66		4					
>84% R.H. (ALPHA HELIX)	666	HEX		16.80	16.80								
POLY-L-ORNITHINE ---, N-CARBOBENZOXY- *248.28*	720	MONO		23.3	22.7	16.2	G=119.2	22	1.213				
POLY-L-PROLINE *97.12* I.	590	HEX	C2-2	9.05	9.05	19.0		10	1.197				3*10/3
II.	510	RHO	C3-3	6.62	6.62	9.36		3	1.362				3*3/1
	511	RHO	C3-3	6.68	6.68	9.36		3	1.338				3*3/1
	512	RHO	C3-3	6.62	6.62	9.31		3	1.369				3*3/1
ACID COMPLEXES ---, ACETIC	701	TET		9.13	9.13	19.0		10					3*10/3
---, ---, FORMIC	701	TET		8.92	8.92	19.0		10					3*10/3
---, ---, PROPIONIC	701	TET		9.13	9.13	19.0		10					3*10/3
HIGHLY SOLVATED	701	ORTHO		9.00	25.1	19.0							3*10/3

POLYMER	REF	CRYST. SYST.	SPACE GROUP	UNIT CELL PARAMETERS A	B	C	ANGLES	MON/ UNIT CELL	DENSITY (G./CC.) CRYSTAL	AMORPH.	MELT. POINT	HEAT OF FUSION KJ/MOL	CHAIN CONF. N*P/Q
---, HYDROXY- A. *113.12*	580	HEX	C3-3	12.3	12.3	9.15		9	1.410				
---, O-ACETYL- II. *155.15*	737	HEX		11.4	11.4	9.3		6	1.10				3*3/1
POLY-L-SERINE ---, O-ACETYL- BETA *129.12*	706	ORTHO		9.6*	6.48	9.71							

POLY-AMIDES   SEE ALSO POLY-PEPTIDES
-NH-X-NH-CO-Y-CO-   OR   -NH-X-CO-

POLY-2-AMINOACETIC ACID, SEE POLY-GLYCINE

POLYMER	REF	CRYST. SYST.	SPACE GROUP	UNIT CELL PARAMETERS A	B	C	ANGLES	MON/ UNIT CELL	DENSITY CRYSTAL	AMORPH.	MELT. POINT	HEAT OF FUSION KJ/MOL	CHAIN CONF. N*P/Q
POLY-3-AMINOBUTYRIC ACID *85.11*	503	ORTHO		10.9	9.6	4.6		4	1.17		330		4*
POLY-4-AMINOBUTYRIC ACID ALPHA	272	MONO		9.44	12.1*	8.22	B=64	8	1.340		250/337		
85.11	457	MONO	C2-2	9.29	12.2*	7.97	B=114.5	8	1.375		260/244		5*2/1
BETA	458	MONO			12.2*			2					
	65												
DELTA	458	HEX		4.65	4.65								
POLY-10-AMINOCAPRIC ACID *169.27*	65	PHEX	C2-2	4.9	4.9	26.5		2	1.02		192/177		
	61	HEX						2			177/146		11*2/1
											188/337		
POLY-6-AMINOCAPROIC ACID ALPHA	3	MONO	C2-2	9.56	17.2*	8.01	B=67.5	8	1.235		215	20.8/155	
(CAPROLACTAM)	26	MONO	C2-2	4.81	17.10	7.61	B=79.5	4	1.221		223/153	18.1/216	
113.16	135	MONO		9.65	17.2*	8.11	B=66.3	8	1.220		223/244	23.0/405	
	212	MONO	C2-2	9.45	8.02	17.08	G=68	8	1.252		226/293	21.6/343	
	334	MONO		9.66	8.32	17.0	G=65	8	1.214		214/304	17.8/346	7*2/1
									1.23/213	1.10/213	228/346	18.5/534	
									1.25/449	1.11/449	250/532	24.1/697	

POLYMER	REF	CRYST SYST.	SPACE GROUP	A	B	C	ANGLES	MON/ UNIT CELL	CRYSTAL	AMORPH.	MELT. POINT	HEAT OF FUSION KJ/MOL	CHAIN CONF. N*P/Q
BETA	222	HEX		4.8	4.8	8.6		1	1.10				
GAMMA	336	MONO	C2H-5	9.35	16.6*	4.81	B=120	4	1.162				7*2/1
	272	HEX		4.79	4.79	16.7		2	1.132				
	279	ORTHO		4.82	7.82	16.70		4	1.194				
	246	MONO	C2H-5	9.33	16.9*	4.78	B=121	4	1.163				
	697	MONO	OR	9.14	4.84	16.68	G=121	4	1.188				
	697	ORTHO		4.83	7.83	16.68		4	1.191				
ABOVE 150	26	RHO		4.90	16.28	8.22		4	1.146				
---, D-(-)-3-METHYL- *127.19*	440	MONO	C2-2	9.15	16.96	4.84	B=90	4	1.125		225/19 226/36		7*2/1
POLY-6-AMINOCAPROIC ACID- ALT-11-AMINOUNDECANOIC ACID *296.46*	714					22.5					184		
POLY-8-AMINOCAPRYLIC ACID ALPHA *141.21*	272	MONO	C2-2	9.8	22.4*	3.3	B=65	8	1.14		185/146 202/153 200/244		9*2/1
	65	PHEX	C2-2						1.18/153				
BETA	121	MONO											
GAMMA	272	HEX		4.79	4.79	21.7		2	1.088				
	61	HEX		4.9	4.9	21.7		2	1.04	1.04/495			
POLY-7-AMINOENANTHIC ACID *127.19*	61	TRI	CI-1	4.9	5.4	9.85	49,77,63	1	1.21		225/146 233/215 223/291 217/153		8*1/1
	65	OR	CI-1					1	1.20/153				
			CI-1										
---, (R)-3-METHYL- *141.21*	479					9.24					179		
---, (S)-4-METHYL- *141.21*	479					9.43					176		
---, (R)-5-METHYL- *141.21*	479					9.11					182		
---, (R)-6-METHYL- *141.21*	479					9.57					188		
POLY-12-AMINOLAURIC ACID GAMMA *197.32*	660	HEX	C2H-5	4.70	4.70	31		2	1.10		179/177		13*2/1
	680	PHEX		4.79	31.9*	9.58	B=120	4	1.034				

POLYMER	REF	CRYST. SYST.	SPACE GROUP	UNIT CELL PARAMETERS A	B	C	ANGLES	MON/ UNIT CELL	DENSITY (G./CC.) CRYSTAL	AMORPH.	MELT. POINT	HEAT OF FUSION KJ/MOL	CHAIN CONF. N*P/Q
POLY-9-AMINOPELARGONIC ACID													
155.24	421	TRI	CI-1	4.9	5.4	12.5	49,77,64	1	1.15		210		10*1/1
	65		CI-1								194/146		
		OR	CI-1								198/292		
											209/177		
POLY-2-AMINOPROPIONIC ACID, SEE POLY-ALANINE													
POLY-3-AMINOPROPIONIC ACID													
ALPHA	491	MONO	C2-2	9.33	4.78*	8.73	B=60	4	1.400		340		
71.08											330/337		
(BETA ALANINE)													
---, 2,2-DIMETHYL- I.											189/73		
99.13													
II.	73					8.4					273	13	
											270/402		
POLY-11-AMINOUNDECANOIC ACID													
ALPHA	61	TRI	CI-1	4.9	5.4	14.9	49,77,63	1	1.15		182/146		12*1/1
183.30	59	TRI		9.6	4.2	15.0	72,90,64	2	1.19		194/101	41/343	
	243	TRI		4.78	4.13	14.9	82,75,66	1	1.174		186		
	365	TRI		4.78	4.13	13.1	90,75,66	1	1.343		220		
	65		OR CI-1 / CI-1								188/444		
											183/292		
GAMMA	529	MONO		9.48	29.4*	4.51	B=118.5	4	1.10				
POLY-N-BUTYL ISOCYANATE													
99.13	522	PHEX		13.3	13.3	15.4		16	1.12		175/528		2*8/3
POLY-DECAMETHYLENE ADIPAMIDE													
10.6 *282.43*	110					20.0					230		18*1/1
											236/244		
											240/544		
POLY-DECAMETHYLENE AZELAAMIDE													
10.9 *324.51*											214/137	68.2/405	
GAMMA	65	PHEX						2				36.7/159	

References page III-92

CRYSTALLOGRAPHIC DATA FOR VARIOUS POLYMERS

POLYMER	REF	CRYST SYST.	SPACE GROUP	UNIT CELL PARAMETERS A	B	C	ANGLES	MON/ UNIT CELL	DENSITY (G./CC.) CRYSTAL	AMORPH.	MELT. POINT	HEAT OF FUSION KJ/MOL	CHAIN CONF. N*P/Q
POLY-4,4'-DECAMETHYLENE-DIPIPERAZINE SEBACAMIDE *476.75*											130/159	69.1/159	
POLY-DECAMETHYLENE 4-OCTENEDIAMIDE TRANS- *308.47*	657					24.1					243		
POLY-DECAMETHYLENE SEBACAMIDE 10.10 *338.54*	65 110	TRI				25.6		1			203/244 197 198/454 216/137	51.1/118 34.7/159 72.0/405 32.7/160	22*1/1
POLY-DODECAMETHYLENE OXAMIDE 12.2 *254.37*	743					15.5					230		
POLY-4,4'-(ETHYLENEDIOXY)-DIPHENYLENEDITRIMETHYLENE ADIPAMIDE *438.57*	516					26.3							
POLY-HEPTAMETHYLENE PIMELAMIDE 7.7 *254.37*	63	PHEX	CS-1	4.82	19.0*	4.82	B=60	1	1.105		214/244 205/2 196/339		16*1/1
GAMMA	65	PHEX			18.95			1					
POLY-HEXAMETHYLENE ADIPAMIDE 6.6 *226.32* ALPHA	25 402 407	TRI MONO TRI	CI-1	4.9 15.7 5.00	5.4 10.5 4.17	17.2 17.3 17.3	48,77,63 B=73 81,76,63	1 9 1	1.24 1.240 1.204 1.220/54	1.09/52 1.12/495 1.069/54 1.09/495	265/2 270/289	46.5/82 40/155 36.8/216 46.9/405	14*1/1
BETA	25	TRI	CI-1	4.9	8.0	17.2	90,77,67	2	1.25				14*1/1
POLY-HEXAMETHYLENE AZELAAMIDE 6.9 *268.40*	402	MONO		7.8	40.15	5.3	B=87	4	1.08		226/244 185/339		
POLY-HEXAMETHYLENE 4-OCTENEDIAMIDE TRANS- *252.36*	657					19.1					259		

POLYMER	REF	CRYST. SYST.	SPACE GROUP	UNIT CELL PARAMETERS				MON/ UNIT CELL	DENSITY (G./CC.)		MELT. POINT	HEAT OF FUSION KJ/MOL	CHAIN CONF. N*P/Q
				A	B	C	ANGLES		CRYSTAL	AMORPH.			
POLY-HEXAMETHYLENE SEBACAMIDE 6.10 ALPHA *282.43*	25	TRI	CI-1	4.95	5.4	22.4	49,76,63	1	1.16 1.152/54 1.17/153 1.189/52	1.041/54	228/51 233/244 216/153 215/291 225/454	30.6/160 58.6/405 56.5/137	18*1/1
BETA	25	TRI	CI-1	4.9	8.0	22.4	90,77,67	2	1.20				18*1/1
GAMMA	65	PHEX						1					
POLY-NONAMETHYLENE AZELAAMIDE 9.9 *310.48*	110					24.0					177 189/2 165/339		20*1/1
POLY-PENTAMETHYLENE AZELAAMIDE 5.9 *254.37*	2					19.5					179/291 178/339		16*1/1
POLY-M-PHENYLENE ISOPHTHALAMIDE *238.25*	735			6.7	4.71	11.0		1	1.14		375/735 390/748		
POLY-P-PHENYLENE PHTHALAMIDE *238.25*	396	ORTHO	D2H-14	22.8	5.5	8.1		4	1.56				
POLY-PIPERAZINE ADIPAMIDE *196.25*	386					9.2					355		
POLY-PIPERAZINE SEBACAMIDE *252.36*											180/81	26.0/81	
POLY-1,4-PIPERIDINEDIYL- TRIMETHYLENE ---,2,6-DIOXO- *153.18*	481	TRI		9.64	11.32	15.80	9396114	8	1.326		281		9*2/1
POLY-TETRAMETHYLENE 4-OCTENEDIAMIDE TRANS- *224.30*	657					17.0					294		

| POLYMER | REF | CRYST SYST. | SPACE GROUP | UNIT CELL PARAMETERS | | | | MON/ UNIT CELL | DENSITY (G./CC.) | | MELT. POINT | HEAT OF FUSION KJ/MOL | CHAIN CONF. N*P/Q |
				A	B	C	ANGLES		CRYSTAL	AMORPH.			
POLY-M-XYLYLENE ADIPAMIDE *246.31*	84	MONO		5.10	4.70	15.2	G=69.6	1	1.198		246/326		16*2/1
	459	TRI	CI-1	12.01	4.83	29.8	75,26,65	2	1.250		244/544		
POLY-P-XYLYLENE SEBACAMIDE *302.42*	204	TRI		5.74	4.87	20.6	76,55,65	1	1.168		300/290 268/291 281/385 291/454		20*1/1
POLY-ESTERS -O-X-O-CO-Y-CO- OR -O-X-CO-													
POLY-1,3-CYCLOBUTYLENE CARBONATE ---,2,2,4,4-TETRAMETHYL- CIS- *170.21*	364	ORTHO		9.16	8.22	12.9		4	1.164		253		
TRANS-	364	TRI		9.25	8.28		G=96.5		1.08		360		
POLY-1,4-CYCLOHEXYLENEDIMETHYLENE TEREPHTHALATE CIS- *274.32*	199	TRI		6.02	6.01	13.7	89,53112	1	1.319		256/547		
TRANS-	199	TRI		6.37	6.63	14.2	89,47114	1	1.266		318/547		
POLY-DECAMETHYLENE ADIPATE 10.6 *284.40*	106	MONO		5.0	7.4	22.1		2	1.16		80/81 77/300	42.7/81 45.6/405	16*1/1
POLY-DECAMETHYLENE AZELAATE 10.9 *326.48*	106	MONO		5.0	7.4	51.7		4	1.13		69/137	41.9/81 50.7/405	
POLY-DECAMETHYLENE GLUTARATE 10.5 *270.37*	106	MONO		5.0	7.4	41.6		4	1.17				

POLYMER	REF	CRYST. SYST.	SPACE GROUP	UNIT CELL PARAMETERS				MON/ UNIT CELL	DENSITY (G./CC.)		MELT. POINT	HEAT OF FUSION KJ/MOL	CHAIN CONF. N*P/Q
				A	B	C	ANGLES		CRYSTAL	AMORPH.			
POLY-DECAMETHYLENE OCTADECANEDIOATE 10.18 *452.72	521	MONO		5.47	7.38	37	B=115	2	1.11		93		
POLY-DECAMETHYLENE OXALATE 10.2 *228.29*	106	MONO		5.28	7.00	17.0		2	1.207		79/305		14*1/1
POLY-DECAMETHYLENE SEBACATE 10.10 *340.50*	106	MONO		5.0	7.4	27.1		2	1.13		80/137 73/291	50.2/81 51.5/160 56.5/405 30.2/301	2<*1/1
POLY-DECAMETHYLENE SUBERATE 10.8 *312.45*	106	MONO		5.0	7.4	24.6		2	1.14				20*1/1
POLY-DECAMETHYLENE SUCCINATE 10.4 *256.34*	106	MONO		5.0	7.4	19.6		2	1.17		68/290		16*1/1
POLY-DECAMETHYLENE TEREPHTHALATE 10.T *304.39*	68	TRI		4.62	6.30	20.10	1079 6113	1	1.022		138/81 129/99 131/453	46.1/81 44.0/405	20*1/1
POLY-DIKETENE *84.07*	217	MONO		5.50	7.78	9.06	B=92	4	1.441		115 115/190		
POLY-ETHYLENE ADIPATE 2.6 *172.18*	211 203 105 523	MONO MONO MONO MONO	C2H-5	7.26 5.47 25.7 5.47	5.40 7.23 30.7 7.24	10.85 11.72 11.71 11.55	A=67.7 B=113.5 B=103.8 B=113.5	2 2 40 2	1.453 1.345 1.275 1.363		47/265 54/291 52/27 50/46 55/543 65/571	15.9/265	10*1/1
POLY-ETHYLENE AZELAATE 2.9 *214.26*	108 109 105	ORTHO ORTHO MONO		7.45 5.0 25.7	4.97 7.4 30.7	31.5 31.2 31.2	B=103.8	4 4 80	1.220 1.23 1.190		46/543	21.0/571	
POLY-ETHYLENE 2,5-FURAN- DICARBOXYLATE *166.13*	742	TRI		5.75	5.35	20.1	13490112	2	1.436				

POLYMER	REF	CRYST SYST.	SPACE GROUP	UNIT CELL PARAMETERS A	B	C	ANGLES	MON/UNIT CELL	DENSITY CRYSTAL	DENSITY AMORPH	MELT. POINT	HEAT OF FUSION KJ/MOL	CHAIN CONF. N*P/Q
POLY-ETHYLENE ISOPHTHALATE 2.I *192.17* I.	398												
	111					14.8							
	573	TRI	CI-1	5.20	7.08	14.8	10913696	2	1.358	1.346	143		12*1/1
									2.034		240		
II.	398												
	572	TRI	CI-1	5.41	6.35	21.0	11613684	2	1.478				12*2/1
						2L.2							
POLY-ETHYLENE OXALATE 2.2 *116.07*	675	ORTHO	D2H-14	6.44	6.22	11.93		4	1.613		172/678		6*2/1
POLY-ETHYLENE-P-OXYBENZOATE ALPHA *164.16*	494	ORTHO	D2-4	10.52	4.75	15.68		4	1.392		203/327		
	356					15.7					220/354		
BETA	494			11.2	14.7	18.99		16	1.395				11*2/1
POLY-ETHYLENE 1,4-PIPERAZINEDICARBOXYLATE 2.00.19*	386					13.4					245		
POLY-ETHYLENE SEBACATE 2.10 *228.29*	100	MONO		5.5	15	16.9	B=65	4	1.20		72/265	13.8/265	
	109	MONO		5.52	7.4	16.9	B=65	2	1.21		76/137	29.1/137	
	105	MONO		25.7	30.7	16.67	B=103.8	40	1.187		78/302	25.6/158	
	521	MONO	C2H-5	5.58	7.31	16.76	B=115.5	2	1.229		79/46	35/405	
	523	MONO		5.52	7.30	16.65	B=115.0	2	1.247		73/539	32.0/571	14*1/1
											83/571		
POLY-ETHYLENE SUBERATE 2.8 *200.23*	203	MONO		5.51	7.25	14.28	B=114.5	2	1.281		75/571	26.5/571	
	108	MONO	C2H-5	5.0	7.4	14.1		2	1.27		55/27		
	523	MONO		5.50	7.25	14.10	B=114.5	2	1.300				14*1/1
POLY-ETHYLENE SUCCINATE 2.4 *144.13*	108	MONO		5.0	7.4	8.32		2	1.55		108/290		8*1/1
	105	MONO		9.05	11.09	8.32	B=102.8	4	1.176		103/302		8*1/1
	675	ORTHO	D2H-10	7.60	10.75	8.33		4	1.407				
POLY-ETHYLENE TEREPHTHALATE 2.T *192.17*	27	TRI	CI-1	4.56	5.94	10.75	98118112	1	1.457	1.335	265	24.1/87	12*1/1
	195	TRI		5.54	4.14	10.86	10711292	1	1.472	1.337	284/264	22.6/157	
	400	TRI		4.52	5.98	10.77	101,1811	1	1.477		267/265	9.2/265	
											270/374	25/405	
											265/290	16.7/155	
											245/472	27.8/472	
											278/724		

POLYMER	REF	CRYST. SYST.	SPACE GROUP	UNIT CELL PARAMETERS A	B	C	ANGLES	MON/ UNIT CELL	DENSITY (G./CC.) CRYSTAL	AMORPH.	MELT. POINT	HEAT OF FUSION KJ/MOL	CHAIN CONF. N*P/Q
POLY-HEPTAMETHYLENE SEBACATE 7.10 *298.42*												25/301	
POLY-HEXAMETHYLENE SEBACATE 6.10 *284.40*	521	MONO		5.52	7.40	22.15	B=115	2	1.152		78 67/290	22/301	
POLY-HEXAMETHYLENE TEREPHTHALATE 6.T *248.28*	68	TRI		4.57	6.10	15.40	105 98 114	1	1.146		160/81 161/405 154/99	34.8/81 33.5/405 35.3/159	16*1/1
POLY-2-HYDROXYACETIC ACID *58.04* (GLYCOLIC ACID)	312 490 465 605	ORTHO ORTHO	D2-4 D2H-16	6.36 5.22	5.13 6.19	7.04 7.02 7		4 4 2	1.678 1.700 1.707 1.69	1.50	223/305 233 230	12/464	3*2/1
POLY-3-HYDROXYBUTYRIC ACID *86.09*	419	ORTHO	D2-4	5.76	13.20	5.96		4	1.262		176/422 168/570		4*2/1
POLY-10-HYDROXYCAPRIC ACID *170.25*	108	ORTHO		7.45	4.96	27.1		4	1.129		80/309		
POLY-6-HYDROXYCAPROIC ACID *114.14*	559 561	ORTHO ORTHO	D2-4 D2-4	7.496 7.47	4.974 4.98	17.30 17.05		4 4	1.1753 1.195	1.09/712	55/560 64/623	15.4/623	7*2/1 7*2/1
POLY-2-HYDROXYPROPIONIC ACID *72.06* (LACTIC ACID)	663	PORT		10.7	6.45	27.8		20	1.247				
POLY-3-HYDROXYPROPIONIC ACID ALPHA *72.06* (HYDRACRYLIC ACID)	314					7.02					122/317 84/623	8.6/623	4*2/1
BETA	314 689			7.76	4.50	4.82 4.76		2	1.440				4*1/1

POLYMER	REF	CRYST SYST.	SPACE GROUP	UNIT CELL PARAMETERS A	B	C	ANGLES	MON/ UNIT CELL	DENSITY (G./CC.) CRYSTAL	AMORPH.	MELT. POINT	HEAT OF FUSION KJ/MOL	CHAIN CONF. N*P/Q
---, 2,2-DIMETHYL- *100.12*	321	MONO	C2H-5	9.02	11.64	6.02	B=121.5	4	1.234	1.10/551	240/551	14.9/551	4*2/1
	683	MONO	C2H-5	9.05	11.58	6.03	B=121.5	4	1.234				
---, 3-ISOPROPYL- ISOTACTIC *114.14*	689	ORTHO		10.63	18.13	6.49		8	1.212		89		
POLY-4,4'-ISOPROPYLIDENE- DIPHENYLENE ADIPATE ---, 3,3',5,5'-TETRACHLORO- *476.18*	713										280	8.1	
POLY-4,4'-ISOPROPYLIDENE- DIPHENYLENE CARBONATE *254.28*	5	ORTHO	D2-2-	11.9	10.1	21.5		8	1.307	1.20	267/57	27.9/566	16*2/1
	435	MONO	D2-3	12.3	10.1	20.8	G=84	8	1.314 1.30/283	1.20/283	240 263/370 230/438 260/736	36.8/299	16*2/1
POLY-4,4'-METHYLENE- DIPHENYLENE CARBONATE *226.23*	435	ORTHO	C2V-9	5.0	10.5	22.0		4	1.301	1.26	230 300/438 278/736		16*2/1
POLY-NONAMETHYLENE AZELAATE 9.9 *312.45*											65/405	43.1/159 49.0/405	
POLY-OXYDIETHYLENE SEBACATE *272.34*	105	TET		17.6	17.6	38.0		32	1.229		44/325		
POLY-TETRAMETHYLENE ISOPHTHALATE 4.I *220.22*	111								1.309	1.268	152.5	42.3	14*2/1
POLY-TETRAMETHYLENE SEBACATE 4.10 *256.34*											60/325 62/425	17/301 2.08/425	
POLY-TETRAMETHYLENE SUCCINATE 4.4 *172.18*	689	MONO	C2H-5	5.21	9.14	10.94	B=124	2	1.324		34/696		

POLYMER	REF	CRYST. SYST.	SPACE GROUP	UNIT CELL PARAMETERS A	B	C	ANGLES	MON/ UNIT CELL	DENSITY (G./CC.) CRYSTAL	AMORPH.	MELT. POINT	HEAT OF FUSION KJ/MOL	CHAIN CONF. N*P/Q
POLY-TETRAMETHYLENE TEREPHTHALATE 4.T *220.22*									1.08/111		232/99 221/453	32/111 31/405	14*1/1
POLY-4,4'-THIODIPHENYLENE CARBONATE *244.27*	435	ORTHO	C2V-9	5.6	8.7	22.2		4	1.50	1.35	220 240/540		16*2/1
POLY-TRIMETHYLENE ADIPATE 3.6 *186.21*	107	MONO		5.0	7.4	21.5		4	1.55		38 45/300 46/609		
POLY-TRIMETHYLENE AZELAATE 3.9 *228.29*	107	MONO		5.0	7.4	27.7		4	1.48		50 60/609		
POLY-TRIMETHYLENE DODECANEDIOATE 3.12 *270.37*	107	MONO		5.0	7.4	35.8		4	1.36		61		
POLY-TRIMETHYLENE GLUTARATE 3.5 *172.18*	107	MONO		5.0	7.4	15.4		2	1.00		39 53/609		10*1/1
POLY-TRIMETHYLENE OCTADECANEDIOATE 3.18 *354.53*	107	MONO		5.0	7.4	51.6		4	1.23		76 76/307		
POLY-TRIMETHYLENE PIMELATE 3.7 *200.23*	107	MONO		5.0	7.4	23.6		4	1.52		37 51/609		
POLY-TRIMETHYLENE SEBACATE 3.10 *242.32*	107 105	MONO PTET		5.0 31.2	7.4 31.2	31.3 33.5	G=90	4 96	1.39 1.184		53 56/305 58/609		
POLY-TRIMETHYLENE SUBERATE 3.8 *214.26*	107	MONO		5.0	7.4	26.1		4	1.47		41 52/609		

POLYMER	REF	CRYST SYST.	SPACE GROUP	UNIT CELL PARAMETERS A	B	C	ANGLES	MON/ UNIT CELL	DENSITY (G./CC.) CRYSTAL	AMORPH.	MELT. POINT	HEAT OF FUSION KJ/MOL	CHAIN CONF. N*P/Q
POLY-TRIMETHYLENE SUCCINATE 3.4 *158.15*	107	MONO		5.0	7.4	15.2		4	1.87		47		
POLY-TRIMETHYLENE UNDECANEDIOATE 3.11 *256.34*	107	ORTHO		5.0	7.4	32.4		4	1.42		59		

POLY-URETHANES AND POLY-UREAS

-NH-X-NH-CO-O-Y-O-CO- OR -NH-X-O-CO- AND
-NH-X-NH-CO-NH-Y-NH-CO- OR -NH-X-NH-CO-

POLYMER	REF	CRYST SYST.	SPACE GROUP	UNIT CELL PARAMETERS A	B	C	ANGLES	MON/ UNIT CELL	DENSITY (G./CC.) CRYSTAL	AMORPH.	MELT. POINT	HEAT OF FUSION KJ/MOL	CHAIN CONF. N*P/Q
POLY-ETHYLENE DECAMETHYLENEDIURETHANE *286.37* ALPHA	414					21.8					174		
BETA	414					18.9							
POLY-ETHYLENE 4,4'-ETHYLENE-DIPHENYLENEDIURETHANE *326.35*	414	TRI				19.7					312		
POLY-ETHYLENE HEXAMETHYLENEDIURETHANE *230.26* ALPHA	627	TRI	CI-1	4.93	4.58	16.8	13,03,09	1	1.266		170 184/402 166/473		14*1/1
GAMMA	627	TRI	CI-1	4.59	5.14	13.9	G=119	1	1.333		170		14*1/1
POLY-ETHYLENE 4,4'-METHYLENE-DIPHENYLENEDIURETHANE *312.32*	414	HEX				15.7					239		
POLY-ETHYLENE NONAMETHYLENEDIURETHANE *272.34*	414					36.2					168		
POLY-HEPTAMETHYLENEUREA *156.23*												10.6/404	

POLYMER	REF	CRYST SYST.	SPACE GROUP	UNIT CELL PARAMETERS A	B	C	ANGLES	MON/ UNIT CELL	DENSITY (G./CC.) CRYSTAL	AMORPH.	MELT. POINT	HEAT OF FUSION KJ/MOL	CHAIN CONF. N*P/Q
POLY-HEXAMETHYLENE HEXAMETHYLENEDIURETHANE *286.36*	627	TRI	CI-1	5.05	4.54	21.9	12,08,08	1	1.226		165 150/291 171/473 164/634		18*1/1
POLY-HEXAMETHYLENEUREA *142.20*											300/402	13.9/404	17*2/1
POLY-PENTAMETHYLENE HEXAMETHYLENEDIURETHANE *272.34*	627	MONO	C2H-3	4.70	8.36	39.0	G=115	4	1.302		158 151/633 157/473 235/402		16*1/1
POLY-TETRAMETHYLENE HEXAMETHYLENEDIURETHANE *258.32*	662 334 627	TRI TRI TRI	CI-1	4.95 9.05 4.98	8.69 19.1* 4.71	19.17 8.38 19.4	9010460 90,63,65 16,05,09	2 4 1	1.248 1.510 1.258		180/291 173/101 184 182/473 184/402		15*2/1
POLY-TRIMETHYLENE HEXAMETHYLENEDIURETHANE *244.29*	627	MONO	C2H-3	4.70	8.36	33.9	G=115	4	1.344		168 167/633 163/473		
POLY-ETHERS -HCH-CHOR-													
POLY-BENZYLVINYL ETHER *134.18*	253					6.30					162/114		2*3/1
POLY-BUTYLVINYL ETHER *100.16*	360	RHO	C3I-2	23.7	23.7	6.50		18	0.947	0.92	64/114		2*3/1
---, 2-METHYL- *114.19*	382					6.50					140 128/608		2*3/1

CRYSTALLOGRAPHIC DATA FOR VARIOUS POLYMERS

POLYMER	REF	CRYST SYST.	SPACE GROUP	UNIT CELL PARAMETERS A	B	C	ANGLES	MON/ UNIT CELL	DENSITY (G./CC.) CRYSTAL	AMORPH.	MELT. POINT	HEAT OF FUSION KJ/MOL	CHAIN CONF. N*P/Q
POLY-SEC-BUTYLVINYL ETHER, SEE POLY-PROPYLVINYL ETHER, 1-METHYL-													
POLY-TERT-BUTYLVINYL ETHER *100.16*	269	TET	C4H-6	18.84	18.84	7.65		16	0.980		160 260/114 238/281		2*4/1
POLY-ETHYLENE --, 1-BUTOXY-2-CHLORO- CIS- *134.61*	242					8.6							2*4/1
TRANS-	242					6.5							2*3/1
--, 1-CHLORO-2-ISOBUTOXY- TRANS- *134.61*	242					20.8							2*10/3
--, 1-ISOBUTOXY-2-METHYL- TRANS- *114.19*	67 141					13.77 13.8					226		2*7/2
POLY-ISOBUTYLVINYL ETHER *100.16*	161	ORTHO		16.8	9.70	6.50		6	0.942 0.94/269		117 117/269 165/114 170/281 115/46		2*3/1
	152												2*3/1
POLY-ISOPROPYLVINYL ETHER *86.13*	161	TET		17.2	17.2	35.5		68	0.926 0.93/269		191/281 98/269 190/114		2*17/5
POLY-METHYLVINYL ETHER *58.08*	176	RHO	D3D-6	16.20	16.20	6.50		18	1.175		144/114		2*3/1
	429	RHO	D3D-6	16.25	16.25	6.50		18	1.168				2*3/1
POLY-A-METHYLVINYL METHYL ETHER SYNDIOTACTIC *72.11*	383	TET		15.2	15.2	16.4		32	1.011				2*8/3
	514	PHEX		9.02	9.02	16.0		10	1.062				4*5/2
	554	HEX		9.02	9.02	16.6		10	1.024				4*5/2
POLY-NEOPENTYLVINYL ETHER *114.19*	161	ORTHO		18.2	10.51	6.50		6	0.915 0.91/269		216/281 155/269 216/114		2*3/1

POLYMER	REF	CRYST. SYST.	SPACE GROUP	UNIT CELL PARAMETERS A	B	C	ANGLES	MON/ UNIT CELL	DENSITY (G./CC.) CRYSTAL	AMORPH.	MELT. POINT	HEAT OF FUSION KJ/MOL	CHAIN CONF. N*P/Q
POLY-PROPYLVINYL ETHER													
---, 1-METHYL- *100.16*	424	TET		18.25	18.25	35.5		68	0.956		177/608		2*17/5
	382					35.3					170		2*17/5
POLY-OXIDES													
-X-O-Y-O- OR -R-O-													
POLY-ACETALDEHYDE *44.05*	92	TET	C4H-6	14.63	14.63	4.79		16	1.142		165/329		2*4/1
---, 2-CHLORO- *73.50*	387	TET	C4H-6			4.80							2*4/1
---, 2,2-DICHLORO- *112.94*	387	TET	C4H-6			5.22							2*4/1
---, 2,2,2-TRICHLORO *147.39*	387	TET	C4H-6			6.45					>220/388		2*4/1
POLY-ACETONE *58.08*	214	TET	S4-1	14.65	14.65	10.22		28	1.231		60		2*7/2
POLY-2-BUTENE OXIDE CIS- *72.11*	397	ORTHO		11.20	10.44	7.01		8	1.169		162/399		3*2/1
TRANS-	397	ORTHO	D2-4	13.72	4.60	6.90		4	1.100		114/399 100/527		3*2/1
POLY-BUTYRALDEHYDE *72.11*	92	TET	C4H-6	20.01	20.01	4.78		16	1.001		225/329		2*4/1
POLY-1,3-CYCLOBUTYLENEOXY- METHYLENE OXIDE													
---, 2,2,4,4-TETRAMETHYL- *156.22* CIS- ALPHA	361					11.5					285		
BETA	361					5.75					285/369		
TRANS- ALPHA	361					11.5					260 260/369		

POLYMER	REF	CRYST SYST.	SPACE GROUP	UNIT CELL A	B	PARAMETERS C	ANGLES	MON/ UNIT CELL	DENSITY (G./CC.) CRYSTAL	AMORPH.	MELT. POINT	HEAT OF FUSION KJ/MOL	CHAIN CONF. N*P/Q
POLY-DECAMETHYLENE OXIDE *156.27*													
BETA	361					5.75							
	482	ORTHO		7.40	4.94	27.49		4	1.033		79		11*2/1
	689	ORTHO	D2H-16	7.40	4.93	27.29		4	1.042		79/181		11*2/1
											72/180		
											60/291		
POLY-DODECAMETHYLENE OXIDE *184.32*													
	482	ORTHO		7.40	4.94	32.53		4	1.030				13*2/1
POLY-ETHYLENE OXIDE *44.05* I.													
	109	MONO		9.5	19.5*	12.0	B=101	36	1.207	1.13/366	66/81	8.29/81	3*7/2
	194	MONO	CS-2	8.03	13.09	19.52	B=125.1	28	1.220		62/180	8.04/498	3*7/2
	188	MONO		7.95	13.11	19.39	B=124.6	28	1.231		72/318	11.7/466	
	227	MONO		8.02	13.4	19.25	B=126.9	30	1.326		69/728	9.41/728	
	348	ORTHO		12.83	12.83	19.3		56	1.289		66		3*7/2
	460	MONO		8.16	12.99	19.30	B=126.1	28	1.239		66		3*7/2
	303							4	1.234	1.124	70/538	7.33	3*7/2
	390								1.235		76/567		
	190					19.3			1.227	1.123	75/665	9.0/665	
II.	690	TRI	CI-1	4.69	4.44	7.05	62,93111	2	1.214				3*2/1
---, TERT-BUTYL- *100.16*	717	TET	D2D-8	15.42	15.42	24.65		36	1.022		135/691		3*9/4
											140/692		
											150/693		
											152/527		
---, CHLOROMETHYL- (EPICHLOROHYDRIN) *92.52*	119	ORTHO OR	D2-4	12.14	4.90	7.05			1.461		117		
	194		C2V-9			7.07					121		
	555	ORTHO	C2V-9	12.16	4.90	7.03		4	1.467		135/318		3*2/1
	601	ORTHO		12.24	4.92	6.96		4	1.466				
---, MERCURIC COMPLEXES ---, HGBR2 I. (1 HG/4 RESIDUES)	732	ORTHO		13.73	8.66	11.80		16	2.540				
---, HGCL2 I. (1 HG/4 RESIDUES)	731	ORTHO		13.5	17.1	11.6		32	2.22				6*2/1
	732	ORTHO		13.55	8.58	11.75		16	2.177				
II. (1 HG/RESIDUE)	733	ORTHO		7.75	12.09	5.88		4	3.804				6*1/1
---, HGI2	732					13.52							

POLYMER	REF	CRYST SYST.	SPACE GROUP	UNIT CELL A	B	PARAMETERS C	ANGLES	MON/ UNIT CELL	DENSITY (G./CC.) CRYSTAL	AMORPH.	MELT. POINT	HEAT OF FUSION KJ/MOL	CHAIN CONF. N*P/Q
---, UREA COMPLEX (2 UREA/RESIDUE)	730	RHO		10.43	10.43	9.12					143/734		
POLY-ETHYLENEOXYMETHYLENE OXIDE *74.08* I. (1,3-DIOXOLANE)	690 505	TRI	C1-1	12.32	4.66	24.7 36.6	G=100.9	15	1.325		55/506		5*5/1
II.	688	ORTHO	D2H-15	9.07	7.79	9.85		8	1.414				5*2/1
III.	690	HEX		8.07	8.07	29.5		18	1.331				5*6/1
POLY-FORMALDEHYDE, SEE POLY-OXYMETHYLENE													
POLY-HEXAMETHYLENE OXIDE *100.16*	482 689	MOND MOND	C2H-6 C2H-6	5.65 5.64	9.01 8.98	17.28 17.32	B=134.5 B=134.5	4 4	1.060 1.063		58 58/181 62/542		7*2/1 7*2/1
POLY-HEXAMETHYLENEOXY- METHYLENE OXIDE *130.19*	690	ORTHO	C2V-9	8.4	4.85	18.8		4	1.13		38/695		9*2/1
POLY-ISOBUTYLENE OXIDE *72.11*	594	ORTHO	D2-4	10.76	5.76	7.00		4	1.104		158/593 155/318 177/548 160/664		3*2/1
POLY-ISOBUTYRALDEHYDE *72.11*	96	TET				5.2					>260/329		2*4/1
POLY-ISOVALERALDEHYDE *86.13*	96	TET		20.6	20.6	5.2		16	1.04				
POLY-4,4'-METHYLENEDIPHENYLENE OXIDE *182.22*	745	ORTHO		8.10	5.60	10.00		2	1.334		140		
POLY-METHYLENEOXY- PENTAMETHYLENE OXIDE *116.16*	690	TRI	CI-1	8.36	4.84	8.15	90,90,90	2	1.170		39/695		8*1/1
POLY-METHYLFNEOXY- TETRAMETHYLENE OXIDE *102.13*	686	ORTHO	C2V-9	8.50	4.79	13.50		4	1.234				14*1/1

POLYMER	REF	CRYST SYST.	SPACE GROUP	UNIT CELL PARAMETERS A	B	C	ANGLES	MON/ UNIT CELL	DENSITY (G./CC.) CRYSTAL	AMORPH.	MELT. POINT	HEAT OF FUSION KJ/MOL	CHAIN CONF. N*P/Q
POLY-NONAMETHYLENE OXIDE *142.24*	482	ORTHO		7.36	4.94	12.45		2	1.044		73		10*1/1
POLY-OCTAMETHYLENE OXIDE *128.22* I.	482	MONO	C2H-6	5.67	9.04	22.45	B=134.5	4	1.038		67 74/727	29.8/727	9*2/1
II.	482	ORTHO		7.36	4.93	22.43		4	1.046				9*2/1
POLY-OXACYCLOBUTANE (TRIMETHYLENE OXIDE) (HYDRATE) I. *58.08*	446	MONO	C2H-3	12.3	7.27	4.80	B=91	4	1.178		36/180 34/181 35/447		4*1/1
II.	446	RHO	C3V-6	14.13	14.13	8.41		18	1.194				8*1/1
III.	446	ORTHO	D2-5	9.23	4.82	7.21		4	1.203				8*1/1
---, 3,3-BISCHLOROMETHYL- *155.02* ALPHA	344	ORTHO	D2H-2	17.85	8.16	4.8		4	1.47		190/173	32.2/271	3*1/1
	378	ORTHO	D2H-2	8.16	17.85	4.82		4	1.466				
	689	ORTHO	D2H-16	17.85	8.16	4.67		4	1.514 1.47/343	1.39/343	180/148	23.0/343	3*1/1
BETA	172	MONO	CS-1	6.85	11.42	4.75	B=109.8	2	1.472				3*1/1
	378	OR MONO	CS-2	11.42	7.06	4.82	G=114.5	2	1.456				3*1/1
	689	ORTHO	C2V-12	13.01	11.71	4.67		4	1.447				
POLY-OXYMETHYLENE *30.03* I. (FORMALDEHYDE)	42	RHO	C3-2	4.46	4.46	17.30		9	1.506	1.25	181	7.45/91	2*9/5
	134	OR	C3-3										
	258	RHO		4.43	4.43	17.25		9	1.531	1.32/640	198/217 178/329	6.66/186 7.37/343	2*9/5 2*9/5
	362	RHO	D6H-1	4.471	4.471	17.39		9	1.491		200/455	10.0/628	2*29/ 16
	513	RHO		4.470	4.470	56.00		29	1.492				
	270												
II.	270	ORTHO	D2-4	4.767	7.660	3.563		4	1.533				2*2/1
	249	ORTHO		7.75	4.46	17.30		18	1.501				
---, CYANOETHYL- *83.09*	240	PMONO		9.44	5.32	4.95	G=102	2	1.135		176		2*4/1
POLY-P-PHENYLENE OXIDE *92.10*	487	ORTHO	D2H-14	5.54	8.07	9.72		4	1.408	1.27	298		7*2/1

POLYMER	REF	CRYST. SYST.	SPACE GROUP	UNIT CELL PARAMETERS A	B	C	ANGLES	MON/ UNIT CELL	DENSITY CRYSTAL	DENSITY AMORPH.	MELT. POINT	HEAT OF FUSION KJ/MOL	CHAIN CONF. N*P/Q
---, 2,6-DIMETHYL- *120.15*	226 621	TET		8.45 11.92	6.02 11.92	17.10	G=91	16	1.314		261 272/471 262/484	5.86/524 3.8/471 5.0/484 5.40/598 5.08/574	7*
---, 2,6-DIPHENYL- *244.29* I.	57	TET	D4-4	12.51	12.51	17.08		8	1.2140		484/201	12.2/201	7*4/1
POLY-PROPIONALDEHYDE *58.08*	92	TET	C4H-6	17.52	17.52	4.78		16	1.052		185/329		2*4/1
---, 3-CARBOMETHOXY- *116.12*	704					4.56					150		
POLY-PROPYLENE OXIDE *58.08*	13	ORTHO OR	C2V-9 D2-4	10.52	4.67	7.16		4	1.097	.998/139	75/18 75/377	8.4/498 8.4/377	3*2/1
	41	ORTHO	D2-4	10.52	4.68	7.10		4	1.104		75/377		3*2/1
	78	ORTHO	D2-4	10.40	4.64	6.92		4	1.155		73/285		
	448	ORTHO	D2-4	10.46	4.66	7.03		4	1.126		72/664		3*2/1
---, 3-PHENOXY- *150.18*	238	ORTHO		17.0	8.2	5.48		4	1.30	1.27	215 210/297 208/318 203/527		
---, O-CHLORO- *184.62*	445	ORTHO		12.6	9.90	6.93		4	1.418		200		
---, 3,3,3-TRIFLUORO *112.05*	603	ORTHO	D2-4	11.42	6.26	6.26		4	1.663				3*2/1
POLY-TETRAHYDROFURAN (TETRAMETHYLENE OXIDE) *72.11*	348	MONO	C2H-6	5.48	8.73	12.07	B=134.2	4	1.157		35	12.6/425	5*2/1
	460	MONO	C2H-6	5.59	8.90	12.07	B=134.2	4	1.112		43		5*2/1
	533	ORTHO	D2-4	12.25	8.75	7.22		8	1.238		37/180		
	403	MONO	C2H-6	5.61	8.92	12.25	B=134.5	4	1.095	0.982	60/366	12.4/600	5*2/1
	599				8.89	12.15			1.116			14.4	
POLY-4,4'-THIODIPHENYLENE OXIDE *200.26*	745	ORTHO		8.16	5.55	10.20		2	1.440		340		
POLY-TRIMETHYLENE OXIDE, SEE, POLY-OXACYCLOBUTANE													

POLYMER	REF	CRYST SYST.	SPACE GROUP	UNIT CELL PARAMETERS A	B	C	ANGLES	MON/ UNIT CELL	DENSITY (G./CC.) CRYSTAL	AMORPH.	MELT. POINT	HEAT OF FUSION KJ/MOL	CHAIN CONF. N*P/Q
POLY-SULFIDES AND POLY-SULFONES													
POLYMER OF SELENIUM *78.96*	751	RHO	D3-6	4.355	4.355	4.949		3	4.839		219/752	5.20/752	1*3/1
POLYMER OF SULFUR *32.06*	288	MONO	C2H-2	26.4	9.26*	12.32	B=79.25		2.34				INDET.
	469	ORTHO		8.11	9.20								
	192	MONO	C2-1	17.6	9.25	13.8	B=113	80	2.059				1*10/3
POLY-ETHYLENE DISULFIDE *92.17*	134					8.8					130/294		
											113/295		
	468					8.8					145/725		4*3/1
POLY-ETHYLENE SULFIDE *60.11*	241	HEX		4.92	4.92	6.74		2	1.413		210		3*2/1
	504	ORTHO	D2H-6	8.50	4.95	6.70		4	1.416		190/181		3*2/1
---, TERT-BUTYL- *116.22*	690	RHO	C3-2	16.91	16.91	6.50		9	1.079		205/694		3*3/1
POLY-ETHYLENE TETRASULFIDE *156.29*	256	ORTHO		8.57	5.0	4.27		1	1.42				
	257	MONO		8.68	5.03	4.32	G=87	1	1.378				
	134					4.32							4*1/1
POLY-HEXAMETHYLENESULFONYL- PENTAMETHYLENE SULFONE *282.42*	39	MONO		9.88	9.26	34.00	B=121.7	8	1.418		223		
POLY-HEXAMETHYLENE SULFONE *148.22*	39	MONO		9.88	9.26	18.24	B=121.7	8	1.387		220		
POLY-HEXAMETHYLENESULFONYL- TETRAMETHYLENE SULFONE *268.39*	39	MONO		9.88	9.26	15.68	B=121.7	4	1.460		246		
POLY-METHYLENE DISULFIDE *78.15*	467					4.18							3*2/1
POLY-METHYLENE SELENIDE *92.99* I.	416	HEX		5.22	5.22	46.25		21	2.971		190		2*21/ 11

POLYMER	REF	CRYST SYST.	SPACE GROUP	UNIT CELL A	B	PARAMETERS C	ANGLES	MON/ UNIT CELL	DENSITY (G./CC.) CRYSTAL	AMORPH.	MELT. POINT	HEAT OF FUSION KJ/MOL	CHAIN CONF. N*P/Q
	II.												
POLY-METHYLENE SULFIDE *46.09*	463	ORTHO	D2-4	5.37	9.03	4.27		4	2.983		170		2*2/1
POLY-PENTAMETHYLENE SULFIDE	237	ORTHO		12.7	12.0	5.10		16	1.575		260		2*17/9
	331	HEX		5.07	5.07	36.52		17	1.600		245/181 260/298		
POLY-PENTAMETHYLENE SULFIDE *102.20*	687	MOND	C2H-5	9.61	9.78	7.84	B=131	4	1.221		65/295		6*1/1
POLY-PENTAMETHYLENE SULFONE *134.19*	39	MOND		9.88	9.26	7.76	B=121.7	4	1.476		243		
POLY-PENTAMETHYLENESULFONYL-TETRAMETHYLENE SULFONE *254.36*	39	MOND		9.88	9.26	28.33	B=121.7	8	1.532		247		
POLY-P-PHENYLENE SULFIDE *108.16*	677	ORTHO	D2H-14	8.67	5.61	10.26		4	1.440		295 290/679		7*2/1
POLY-PROPYLENE SULFIDE ISOTACTIC (S) *74.14*	515	ORTHO	D2-4	9.95	4.89	8.20		4	1.234				3*2/1
POLY-1,4-B-D-GALACTO-1,4-B-D-MANNOSE (1:2) (GUAR GALACTOMANNAN) DRY *486.42*	643			13.5	10.3*	8.66		2	1.34				
16.5% H2O	643	MONO	C2-2	15.45	10.3*	8.65	B=90	6					
POLY-1,3-B-D-GALACTOSAMINE-ALT-1,4-B-D-GLUCURONIC ACID													
---, N-ACETYL-	705	ORTHO			18.6*								14*2/1
---, 6-SULFATE	705	OR		12.1	28.5*	14.4							14*3/1
---, 6-SULFATE *459.38*	705	MONO		12.1	28.5*	9.3	B=93						14*3/1

POLY-SACCHARIDES*

* NOTE: FOR THE POLY-SACCHARIDES, A = ALPHA AND B = BETA.

POLYMER	REF	CRYST. SYST.	SPACE GROUP	A	B	C	ANGLES	MON/UNIT CELL	DENSITY CRYSTAL	DENSITY AMORPH.	MELT. POINT	HEAT OF FUSION KJ/MOL	CHAIN CONF. N*P/Q
POLY-1,4-B-D-GLUCOSAMINE													
(CHITIN) ALPHA	652	ORTHO	D2-4	4.76	10.28	18.85		4	1.463				7*2/1
203.19	669	ORTHO	D2-4	4.69	19.13	10.43		4	1.442				7*2/1
BETA	653	MONO	C2-2	4.7	10.5	10.3		2	1.44				
(1 H2O/RESIDUE)	670	ORTHO		9.32	10.2*	22.15	B=90	8	1.400				
POLY-1,3-B-D-GLUCOSAMINE-ALT-1,4-B-D-GLUCURONIC ACID													
---, N-ACETYL-													
(HYALURONIC ACID)													
II.	710	ORTHO		11.4	9.8	33.7		8	1.34				14*4/1
(AS K SALT)	711	ORTHO		11.0	9.9	33.0		8	1.54				14*4/1
(AS NA SALT) III. *379.32*	711	ORTHO		10.4	9.0	37.1		8	1.54				14*4/1
(AS NA SALT) IV.	659	HEX		11.7	11.7	28.5		6	1.183				14*3/1
(AS NA SALT) V.	659	HEX		18.7	18.7	28.5		18	1.39				14*3/1
VII.	659	MONO				19.6							14*2/1
POLY-1,4-A-D-GLUCOSE													
(AMYLOSE) B.	644	ORTHO	D2-1	16.0	10.6*	9.2		8	1.38				7*2/1
(1 H2O/RESIDUE) *162.14*	645	ORTHO		15.6	10.6*	9.0		8	1.61				7*4/1
(1 H2O/RESIDUE) V.	646	ORTHO	D2-4	13.7	23.8	8.05		12	1.37				7*6/1
	647	PHEX		13.2	23.3	7.93							7*7/1
	668			14.94	14.94								
---, COMPLEXES WITH DMSO													
(8 DMSO/CELL)	649	TET		19.17	19.17	24.4		36					7*18/
	667	PTET	D2-4	19.21	19.21	8.12		12					7*6/1
FATTY ACIDS	646	ORTHO		13.0	23.0	8.05							
HYDROXIDES													
LIOH	650			12.1	22.6	8.8		12					
NAOH	650			12.3	22.6	8.9		12					
KOH	650	ORTHO	D2-4	12.7	22.6	9.0		12					
NH4OH	650			12.7	22.6	9.0		12					
CSOH	650			12.4	22.6	8.9		12					
C(NH2)3OH	650			13.1	22.6	9.0		12					
IODINE *204.44*	651	HEX		12.97	12.97	7.91		6	1.77				
(1 1/3 RESIDUES)													

POLYMER	REF	CRYST. SYST.	SPACE GROUP	UNIT CELL A	B	C	ANGLES	MON/UNIT CELL	DENSITY CRYSTAL	AMORPH.	MELT. POINT	HEAT OF FUSION KJ/MOL	CHAIN CONF. N*P/Q
POTASSIUM SALTS DRY BROMIDE	661	TET	D4-8	10.2	10.2	16.4		8					7*4/1
FORMATE	661	TET	D4-8	10.2	10.2	16.4		8					7*4/1
IODIDE	661	TET	D4-8	10.5	10.5	15.9		8					7*4/1
WET (10% H2O) ACETATE	661	TET	D4-8	10.8	10.8	16.1		8					7*4/1
BICARBONATE	661	TET	D4-8	10.8	10.8	15.8		8					7*4/1
BROMIDE	661	TET	D4-8	10.7	10.7	16.1		8					7*4/1
FORMATE	661	TET	D4-8	10.8	10.8	16.1		8					7*4/1
IODIDE	661	TET	D4-8	10.7	10.7	16.1		8					7*4/1
ATM. DRIED ACETATE	661	ORTHO		11.0	18.1	17.9		16					
PROPIONATE	661	ORTHO		11.4	18.0	17.6		16					
---, TRIACETATE *288.25*	648	ORTHO		10.87	18.83	52.33		28	1.251				7*14/3
POLY-1,4-B-D-GLUCOSE (CELLULOSE) I. *162.14*	22	MONO		8.35	10.3*	7.9	B=84	4	1.59				
	98	MONO		8.20	10.3*	7.90	B=83.3	4	1.625				
	20	MONO	C2-2										
	28	MONO		8.171	10.34	7.846	B=83.6	4	1.635				
	709	MONO		10.85	10.3*	12.08	B=93.2	8	1.598				
II.	22	MONO		8.14	10.3*	9.14	B=62	4	1.592				
	98	MONO		8.02	10.3*	9.03	B=62.8	4	1.623				
	28	MONO		7.917	10.34	9.083	B=62.7	4	1.630				
	708	TRI		8.97	10.34	7.31	B=99.4	4	1.610				
III.	28	MONO		7.74	10.3*	9.9	B=58	4	1.61				
IV.	98	MONO		8.12	10.3*	7.99	B=90	4	1.612				
	239	MONO		7.9	10.3*								
	28	MONO		8.068	10.3*	7.946	B=90	4	1.63				
X.	98	MONO		8.10	10.3*	8.16	B=78.3	4	1.615				
	218	MONO		8.12	10.3*	7.99	B=90.0	4	1.612				
---, TRIACETATE *288.25*	4	PORT	C2-2	24.5	11.6*	10.43		8	1.296		306/144		
---, TRIBUTYRATE *372.41*											206/46 207/118	12.6/118	
---, TRICAPRYLATE *540.74*											116/117	13.0/117	

CRYSTALLOGRAPHIC DATA FOR VARIOUS POLYMERS

POLYMER	REF	CRYST SYST.	SPACE GROUP	UNIT CELL PARAMETERS A	B	C	ANGLES	MON/ UNIT CELL	DENSITY (G./CC.) CRYSTAL	AMORPH.	MELT. POINT	HEAT OF FUSION KJ/MOL	CHAIN CONF. N*P/Q
---, TRINITRATE *297.13*	24	ORTHO		12.25	25.4*	9.0		8	1.409		697/77 700/81	3.8 - 6.3/77	
---, 2.44-NITRATE											617/81	5.65/81	
POLY-1,4-A-L-GULURONIC ACID *176.12* (1 H2O/RESIDUE)	674 673	ORTHO ORTHO	D2-4 D2-4	7.75 8.6	8.7* 8.72*	10.6 10.74		4 4	1.64 1.60				7*2/1 7*2/1
POLY-1,4-B-D-MANNOSE I. *162.41*	636			7.21	10.27	8.82		4	1.65				
II.	636	MONO		18.8	10.2*	18.7	B=57.5	16	1.43				
---, TRIACETATE *288.25*	672					15.24							7*3/1
POLY-1,4-B-D-MANNURONIC ACID *176.12*	673	ORTHO	D2-4	7.58	10.4*	8.58		4	1.738				7*2/1
POLY-1,3-B-D-XYLOSE *148.11* (XYLAN)	654	HEX		13.7	5.85*	13.7		6	1.55				7*3/1
POLY-1,4-B-D-XYLOSE DRY (HARDWOOD XYLAN) *148.11*	671	HEX		8.8	8.8	14.85		6	1.48				7*3/1
---, DIACETATE *232.19*	656	MONO	C2-2	7.64	12.44	10.31	G=85	4	1.580				7*2/1
---, DIHYDRATE *184.14*	671	HEX		9.64	9.64	14.95		6	1.525				7*3/1
---, HYDRATE *166.13*	671 655	HEX PHEX	D3D-6 C1-1	9.16 9.16	9.16 9.16	14.85 14.84	G=120	6 6	1.534 1.535				7*3/1 7*3/1

POLYMER	REF	CRYST SYST.	SPACE GROUP	UNIT CELL A	CELL B	PARAMETERS C	ANGLES	MON/ UNIT CELL	DENSITY (G./CC.) CRYSTAL	AMORPH.	MELT. POINT	HEAT OF FUSION KJ/MOL	CHAIN CONF. N*P/Q
OTHER POLYMERS													
POLY-CARBON MONOXIDE-ALT-ETHYLENE *56.06*	193	ORTHO	D2H-16	7.97	4.76	7.57		4	1.296		185		3*2/1
POLY-DIMETHYL KETENE (KETONE) I. ALPHA *70.09*	432 179	ORTHO	C2V-9	12.85	6.53	8.80 8.8		8	1.261		255/130 250		8*1/1
BETA	456					4.40							2*2/1
(ESTER) II. *140.18*											170/179 180/248		
POLY-ETHYLENE AMINE *43.07*											58/738		
---, N-BENZOYL- *147.18*											210/747		
---, P-CHLORO- *181.62*	746	TRI		4.74	14.8	6.55	88,86,99	2	1.334		285/747		2*3/1
---, N-BUTYRYL- *113.16*											155/747		
---, 4-(4-METHYLTHIO-PHENOXY)- *251.34*	739	TRI		4.35	24.0	12.7		4	1.259		105		3*4/1
---, N-HEPTANOYL- *155.24*	746	TRI		5.0	17.7	6.4	85,62,98	2	1.06		175/747		2*3/1
---, N-HEXANOYL- *141.21*	746	TRI		4.9	15.9	6.4	87,64,95	2	1.06		175/747		2*3/1
---, N-ISOBUTYRYL- *113.16*	746					6.4					210/747		
---, N-ISOVALERYL- *127.19*	746	TRI		4.7	13.0	6.4	80,82,88	2	1.11		210/747		2*3/1
---, N-OCTANOYL- *169.27*	746	TRI		5.03	19.9	6.3	91,60,97	2	1.04		165/747		2*3/1

POLYMER	REF	CRYST SYST.	SPACE GROUP	UNIT CELL A	B	PARAMETERS C	ANGLES	MON/ UNIT CELL	DENSITY CRYSTAL	(G./CC.) AMORPH.	MELT. POINT	HEAT OF FUSION KJ/MOL	CHAIN CONF. N*P/Q
---, N-P-TOLUOYL- *161.20*	746	TRI		4.84	14.8	6.61	91,78,99	2	1.172		260/747		2*3/1
---, N-VALERYL- *127.19*	746	TRI		4.7	14.4	6.4	86,73,97	2	1.03		172/747		2*3/1,
POLY-HYDROXYMETHYLENE *30.03*	338					2.5							
POLY-6-MERCAPTOCAPROIC ACID *130.21*	428					17.8					106		7*2/1
POLY-OCTAMETHYLENE-5,5'- DIBENZIMIDAZOLE *344.46*	470					21							
POLY-2,5-OCTAMETHYLENE- 1,3,4-TRIAZOLE ---, 1-AMINO- *194.28*	610					25.5							
POLY-P-PHENYLENE- DISILOXANYLENE ---, TETRAMETHYL- *208.41*	115	TET		9.08	9.08	15.38		4	1.092		148/225	18.2/225	9*1/1
POLY-PHOSPHAZENE ---, BISPHENOXY- *231.19*	552					4.9							
---, BIS-2,2,2- TRIFLUOROETHOXY- *243.04*	552					4.8					240/553		
POLY-PHOSPHONITRILE CHLORIDE *115.89*	56	ORTHO	C2V-9	11.07	4.92*	12.72		8	2.222	1.91			2*2/1
POLY-SILOXANE ---, DIETHYL- *102.21*	197										17/518	1.34/518	
---, DIMETHYL- *74.16*	631	MONO		13.0 / 13	8.3* / 8.3	7.75	B=60	6	1.02	0.98	-40/517 / -40	1.34/517	
---, DIPROPYL- *130.26*	518	TET	C4-2	9.52	9.52	9.40		4	1.016		74	1.51	2*4/1

References page III-92

C. MELTING POINTS OF VARIOUS POLYMERS

In the following table of melting points, entries are alphabetical according to the basic structure of the polymer ignoring substituents. Substituted polymers are listed alphabetically (according to the substituent) under the entry for the unsubstituted polymer. The molecular weight is that of the chemical repeat (the constitutional base unit) in the polymer. Melting points in $^\circ$C are taken from the reference cited.

POLYMER	MOLECULAR WEIGHT	MELTING POINT	REF.
CELLULOSE, SEE POLY-1,4-B-D-GLUCOSE			
POLY-ACETALDEHYDE	44.05	165	329
---, 2,2,2-TRICHLORO-	147.39	>220	388
POLY-ACETONE	58.08	60	214
POLY-ACRYLAMIDE	71.08		
---, N-DOCOSYL-	379.67	68	906
---, N-DODECYL-	239.40	-8	906
---, N-HEXADECYL-	295.51	45	906
---, N-ISOPROPYL-	113.16	200	102
---, N-OCTADECYL-	323.56	68	906
---, N-TETRADECYL-	267.46	18	906
POLY-ACRYLIC ACID	72.06		
---, ALLYL ESTER	112.13	90	278
---, BUTYL ESTER	128.17	47	401
---, SEC-BUTYL ESTER	128.17	130	401
---, TERT-BUTYL ESTER	128.17	193	16
		200	401
---, DECYL ESTER	212.33		
---, ---, 1,1-DIHYDROPERFLUORO-	554.15	100	898
---, DOCOSYL ESTER	380.66	72	906
---, DODECYL ESTER	240.39	12	906
---, HEXADECYL ESTER	296.50	43	906
---, ISOBUTYL ESTER	128.17	81	401
---, ISOPROPYL ESTER	114.14		
ISOTACTIC		162	120
		162	149
		178	541
		115	113
SYNDIOTACTIC		56	906
---, OCTADECYL ESTER	324.55		
---, OCTYL ESTER	184.28		
---, ---, 1,1-DIHYDROPERFLUORO-	454.14	35	898
---, TETRADECYL ESTER	268.44	32	906
POLY-ACRYLONITRILE	53.06		
SYNDIOTACTIC		317	77
		341	499
---, ALPHA-ETHYL-	81.12	200	436
---, ALPHA-ISOPROPYL-	95.14	310	436
---, ALPHA-PROPYL-	95.14	210	436
POLY-ADIPIC ANHYDRIDE	128.13	85	840
		98	841
		77	826
POLY-2,2'-(ADIPYLDIAMINO)-DIBENZOIC ISOPHTHALIC ANHYDRIDE	514.49	225	871
POLY-4,4'-(ADIPYLDIAMINO)-DIBENZOIC ISOPHTHALIC ANHYDRIDE	514.49	273	871
POLY-2,2'-(ADIPYLDIAMINO)-DIBENZOIC TEREPHTHALIC ANHYDRIDE	514.49	235	871
POLY-4,4'-(ADIPYLDIAMINO)-DIBENZOIC TEREPHTHALIC ANHYDRIDE	514.49	285	871
POLY-ADIPYL DITHIONISOPHTHALOYLDIHYDRAZIDE	336.43	280	889
POLY-ADIPYL ETHYLENEDIUREA	256.26	241	402
POLY-ADIPYL OCTAMETHYLENEDIUREA	340.42	290	402
POLY-ADIPYL PENTAMETHYLENEDIUREA	298.34	222	402

CRYSTALLOGRAPHIC DATA FOR VARIOUS POLYMERS

POLYMER	MOLECULAR WEIGHT	MELTING POINT	REF.
POLY-ALLENE	40.06	122	536
---, TETRAFLUORO-	112.03	126	38
POLY-ALLYLBENZENE, SEE POLY-PROPENE, 3-PHENYL-			
POLY-M-AMINOBENZOIC ACID	119.12	425	735
POLY-P-AMINOBENZOIC ACID	119.12	550	735
POLY-3-AMINOBUTYRIC ACID	85.11	330	503
POLY-4-AMINOBUTYRIC ACID	85.11	260	337
		260	244
POLY-10-AMINOCAPRIC ACID	169.27	177	146
		188	337
		192	177
POLY-6-AMINOCAPROIC ACID	113.16	215	3
		214	304
		223	153
		223	244
		226	293
		228	346
		250	532
---, D-(-)-3-METHYL-	127.19	225	19
		226	36
---, 6-METHYL-	127.19	185	402
POLY-6-AMINOCAPROIC ACID-ALT-11-AMINOUNDECANOIC ACID	296.46	184	714
POLY-8-AMINOCAPRYLIC ACID	141.21	185	146
		202	153
		200	244
POLY-(3-AMINOCYCLOBUTYLENE)-PROPIONIC ACID	125.17		
---, 2,2-DIMETHYL-	153.22	358	816
POLY-4-AMINOCYCLOHEXYLENEACETIC ACID	139.20	400	402
POLY-AMINODIHEXAMETHYLENE SUBERAMIDE	353.55	160	402
POLY-AMINODITRIMETHYLENE OXAMIDE	185.23		
---, N-METHYL-	199.25	202	291
POLY-22-AMINODOCOSANOIC ACID	337.59	145	177
POLY-7-AMINOENANTHIC ACID	127.19	225	146
		217	153
		233	215
		223	291
---, (R)-3-METHYL-	141.21	179	479
---, (S)-4-METHYL-	141.21	176	479
---, (R)-5-METHYL-	141.21	182	479
---, (R)-6-METHYL-	141.21	188	479
---, N-METHYL-	141.21	65	828
POLY-P-(AMINOETHYLENE)-PHENYLENEACETIC ACID	161.20	283	402
POLY-P-(AMINOETHYLENE)-PHENYLENEBUTYRIC ACID	189.26	224	402
POLY-P-(AMINOETHYLENE)-PHENYLENEPROPIONIC ACID	175.23	382	402
POLY-P-(AMINOETHYLENE)-PHENYLENEVALERIC ACID	203.28	275	402
POLY-2-AMINOETHYLENESULFONIC ACID	107.13	0275	938
POLY-17-AMINOHEPTADECANOIC ACID	267.46	150	292
POLY-P-AMINOHYDROCINNAMIC ACID	147.18	310	402
POLY-12-AMINOLAURIC ACID	197.32	179	177
---, N-METHYL-	211.35	52	929
POLY-(3-AMINOMETHYLENE)-CYCLOHEXYLENEACETIC ACID	153.22		
---, 1,3-DIMETHYL-	181.28	297	816
POLY-P-(AMINOMETHYLENE)-PHENYLENEACETIC ACID	147.18	355	402
POLY-P-(AMINOMETHYLENE)-PHENYLENEBUTYRIC ACID	175.23	267	402
POLY-P-(AMINOMETHYLENE)-PHENYLENEPROPIONIC ACID	161.20	300	402
POLY-P-(AMINOMETHYLENE)-PHENYLENEVALERIC ACID	189.26	233	402
POLY-6-AMINO-4-OXACAPROIC ACID	115.13		
---, 3,5-DIMETHYL-	143.19	210	902
POLY-5-AMINO-3-OXAVALERIC ACID	101.10	148	402
POLY-9-AMINOPELARGONIC ACID	155.24	194	146
		198	292
		209	177
		210	421
POLY-3-AMINOPROPIONIC ACID (BETA-ALANINE)	71.08	340	491
		330	337
---, 2,2-DIMETHYL I.	99.13	189	73
II.		270	402
		273	73

References page III-92

POLYMER	MOLECULAR WEIGHT	MELTING POINT	REF.
---, 2,3-DIMETHYL-	99.13		
ERYTHRO		405	503
THREO		355	503
---, 3,3-DIMETHYL-	99.13	296	503
---, N-ISOPROPYL-	113.16	130	869
---, N-METHYL-	85.11	225	863
		202	869
---, N-PHENYL-	147.18	205	869
POLY-13-AMINO-11-THIATRIDECANOIC ACID	229.38	150	402
POLY-13-AMINO-12-THIATRIDECANOIC ACID	229.38	150	402
POLY-6-AMINOTHIOCAPROIC ACID	129.22	120	402
POLY-7-AMINOTHIOENANTHIC ACID	143.25	235	402
POLY-ALPHA-AMINOTOLUIC ACID	133.15	300	444
POLY-13-AMINOTRIDECANOIC ACID	211.35	183	177
POLY-3-AMINOTRIMETHYLENESULFONIC ACID	107.15	260	932
POLY-11-AMINOUNDECANOIC ACID	183.30	194	101
		182	146
		183	292
		186	243
		188	444
		220	365
---, N-ALLYL-	223.36	350	402
---, N-ETHYL-	212.36	-30	402
---, 2-METHYL-	197.32	130	402
---, N-METHYL-	197.32	80	402
---, N-PHENYL-	259.39	-30	402
---, N-PIPERAZINYL-	267.42	142	402
POLY-5-AMINOVALERIC ACID	99.13	258	402
POLY-L-ASPARTIC ACID	115.09		
---, BETA-ISOBUTYL ESTER	171.20	286	924
POLY-AZELAIC ANHYDRIDE	170.21	54	823
POLY-BENZALDEHYDE-ALT-DIMETHYLKETENE	176.22	290	817
---, P-CHLORO-	210.66	260	817
---, P-METHOXY-	206.24	240	817
---, M-NITRO-	221.21	240	817
POLY-BENZYLIDENETHIODECAMETHYLENE SULFIDE	294.52	135	923
POLY-BENZYLIDENETHIOHEXAMETHYLENE SULFIDE	238.42	115	923
---, P-METHOXY-	268.44	130	923
POLY-BENZYLVINYL ETHER	134.18	162	114
POLY-4,4'-BIPHENYLDICARBOXALDEHYDE	210.23	250	801
POLY-4,4'-BIPHENYLENE ADIPAMIDE	294.35	400	402
POLY-4,4'-BIPHENYLENE ADIPATE	296.32	300	901
POLY-4,4'-BIPHENYLENE CARBODIIMIDE	192.22	>300	907
---, 3,3'-DIMETHOXY-	220.28	190	907
POLY-4,4'-(4,4'-BIPHENYLENEDIOXY)-DIPHENYLENE CARBONATE	396.40	250	915
POLY-4,4'-BIPHENYLENE DISILOXANYLENEDIPROPIONAMIDE	370.56		
---, TETRAMETHYL- (SI)	426.66	300	402
POLY-4,4'-(2,2'-2,2'-BIPHENYLENEDITHIAZOLE)-OXY-P-PHENYLENE	486.61	240	810
POLY-4,4'-(2,2'-2,2'-BIPHENYLENEDITHIAZOLE)-P-PHENYLENE	394.51	250	810
POLY-4,4'-BIPHENYLENE DITHIOLISOPHTHALATE	348.44	340	922
POLY-4,4'-BIPHENYLENE DITHIOLTEREPHTHALATE	348.44	>460	922
POLY-4,4'-BIPHENYLENE SEBACAMIDE	350.46	435	402
POLY-4,4'-BIPHENYLENE TEREPHTHALAMIDE	314.34	500	402
---, 3,3'-DIMETHYL-	342.40	440	402
---, ---, N,N'-DIETHYL-	398.51	254	870
---, N,N'-DIETHYL-	370.45	316	870
POLY-4,4'-BIPHENYLENE TEREPHTHALATE	316.31		
---, 2,2'-DIPROPYL-	400.47	350	901
1,2-POLY-1,3-BUTADIENE	54.09		
ISOTACTIC		120	12
		125	44
		154	7
SYNDIOTACTIC			
---, 4,4-DIMETHYL-	82.15		
ISOTACTIC		167	707
1,4-POLY-1,3-BUTADIENE	54.09		
CIS-		1	287
		1	124
		6	335

CRYSTALLOGRAPHIC DATA FOR VARIOUS POLYMERS

POLYMER		MOLECULAR WEIGHT	MELTING POINT	REF.
TRANS-	I.		100	352
			96	451
			55	624
	II.		141	352
			148	44
			145	451
			142	624
---, 2-TERT-BUTYL-		110.20		
CIS-			106	367
---, 2-CHLORO- (CHLOROPRENE)		88.54		
CIS-			70	893
TRANS-			115	229
			107	556
			80	81
---, 2,3-DIMETHYL-		82.15		
CIS-			198	220
TRANS-			260	104
			272	221
---, 1-METHOXY-		84.12		
TRANS-			118	329
---, 2-METHYL- (ISOPRENE)		68.12		
CIS-			28	81
			14	287
			36	50
TRANS-			65	46
			68	287
			74	81
			78	630
			87	568
---, ---, HYDROCHLORINATED		104.58	115	229
			110	70
---, 2-METHYLACETOXY-		140.18		
TRANS-			135	165
---, 2-PROPYL-		96.17		
TRANS-			42	165
			42	368
1,4-POLY-1,3-BUTADIENE-ALT-METHACRYLONITRILE		121.18		
---, 1,1,4,4-TETRADEUTERO-		125.22	75	825
1,4 POLY 1,3 BUTADIENE ALT-METHYL METHACRYLATE		154.21	56	849
POLY-BUTADIENE OXIDE		70.09	74	809
POLY-1-BUTENE	I.	56.11	126	12
			132	250
			136	277
			135	345
			142	313
			135	320
			135	380
	II.		124	277
			122	345
			126	313
			122	320
			130	380
	III.		106	277
---, 4-CYCLOHEXYL-		138.25	170	282
			138	328
---, 4-N,N-DIISOPROPYLAMINO-		155.28	315	191
---, 3,3-DIMETHYL-		84.16	260	282
---, 3-METHYL-		70.14	300	48
			306	632
			310	90
			300	282
---, 3-METHYL- (VIA HYDRIDE SHIFT)		70.14	55	332
			66	319
---, 3-PHENYL-		132.21	360	90
---, 4-PHENYL-		132.21	160	90
			168	187
			159	282
			158	483

POLYMER	MOLECULAR WEIGHT	MELTING POINT	REF.
---, 4-O-TOLYL-	146.23	239	187
---, 4-P-TOLYL-	146.23	196	187
---, 4,4,4-TRIFLUORO-	110.08	263	813
---, 3-TRIFLUOROMETHYL-	124.11	300	813
POLY-2-BUTENE-ALT-ETHYLENE	84.16	135	324
POLY-2-BUTENE 4-OCTENEDIAMIDE	222.29		
TRANS-, TRANS-		301	657
POLY-2-BUTENE OXIDE	72.11		
CIS-		162	399
TRANS-		114	399
		100	527
---, 2-METHYL-	86.13	196	879
POLY-2-BUTENE SULFIDE	88.17		
CIS-		156	548
TRANS-		192	548
POLY-2-BUTENYLENE HEXAMETHYLENEDIURETHAN	256.30		
CIS-		136	402
TRANS-		177	402
POLY-4,4'-BUTYLIDENEDIPHENYLENE CARBONATE	268.31	170	438
POLY-4,4'-(2,2-BUTYLIDENE)-DIPHENYLENE CARBONATE	268.31	222	438
POLY-BUTYLISOCYANATE	99.13	175	528
---, 2-METHYL-	113.16	220	909
POLY-BUTYLVINYL ETHER	100.16	64	114
---, 2-METHYL-	114.19	140	382
POLY-SEC-BUTYLVINYL ETHER, SEE POLY- PROPYLVINYL ETHER, METHYL-			
POLY-TERT-BUTYLVINYL ETHER	100.16	160	269
		260	114
		238	281
POLY-BUTYRALDEHYDE	72.11	225	329
POLY-CAPRYLALDEHYDE	128.22	35	329
POLY-CARBON MONOXIDE-ALT-ETHYLENE	56.06	185	193
POLY-M-(CARBOXYPHENOXY)-ACETIC ANHYDRIDE	178.14	134	827
POLY-P-(CARBOXYPHENOXY)-ACETIC ANHYDRIDE	178.14		
---, 3-BROMO-	257.04	179	827
POLY-CHLOROPRENE, SEE POLY-1,3-BUTADIENE, CHLORO-			
POLY-CHLOROTRIFLUOROETHYLENE-ALT-ETHYLENE	144.52	262	635
POLY-CYCLOBUTENE	54.09		
I.		210	864
II.		150	864
POLY-1,3-CYCLOBUTYLENE CARBONATE	114.10		
---, 2,2,4,4-TETRAMETHYL-	170.21		
CIS-		253	364
TRANS-		360	364
POLY-1,3-CYCLOBUTYLENEOXYMETHYLENE OXIDE	110.12		
---, 2,2,4,4-TETRAMETHYL-	156.22		
CIS-		285	361
		285	369
		260	361
TRANS-		260	369
POLY-CYCLODECENE	138.25	80	562
POLY-CYCLODODECENE	166.31	80	439
		84	562
POLY-CYCLOHEPTENE	96.17	51	439
POLY-CYCLOHEPTENE-ALT-ETHYLENE	124.23	74	259
1,4-POLY-1,3-CYCLOHEXADIENE	80.13		
TRANS-		380	221
POLY-CYCLOHEXENE	82.15	61	439
POLY-CYCLOHEXENE OXIDE	98.14	126	527
POLY-1,3-CYCLOHEXYLENE ADIPAMIDE	224.30		
CIS-		170	402
TRANS-		300	402
POLY-1,4-CYCLOHEXYLENE ADIPAMIDE	224.30	400	402
POLY-1,3-CYCLOHEXYLENE AZELAAMIDE	266.38		
CIS-		125	402
TRANS-		300	402
POLY-1,4-CYCLOHEXYLENE 3,3'-DIBENZAMIDE	320.39	390	402
POLY-1,4-CYCLOHEXYLENEDIMETHYLENE ADIPAMIDE	252.36		
CIS-		246	385
		225	878

POLYMER	MOLECULAR WEIGHT	MELTING POINT	REF.
TRANS-		345	385
		342	806
POLY-1,4-CYCLOHEXYLENEDIMETHYLENE ADIPATE	254.33	347	878
CIS-			
TRANS-		55	547
POLY-1,4-CYCLOHEXYLENEDIMETHYLENE AZELAAMIDE	294.44	124	547
CIS-			
TRANS-		195	385
POLY-1,4-CYCLOHEXYLENEDIMETHYLENE AZELAATE	296.41	275	385
CIS-			
TRANS-		41	547
POLY-1,4-CYCLOHEXYLENEDIMETHYLENE 1,4-CYCLOHEXYLENEDICARBOXYLATE	280.36	50	547
CIS-, TRANS-			
TRANS-, TRANS-		205	547
POLY-1,4-CYCLOHEXYLENEDIMETHYLENE DITHIOL-1,4-CYCLOHEXYLENEDICARBOXYLATE	312.49	246	547
TRANS-, TRANS-			
POLY-1,4-CYCLOHEXYLENEDIMETHYLENE DODECANEDIAMIDE	336.52	310	922
CIS-			
		215	385
TRANS-		190	878
		278	385
POLY-1,4-CYCLOHEXYLENEDIMETHYLENE DODECANEDIOATE	338.49	280	878
CIS-			
TRANS-		46	547
POLY-1,4-CYCLOHEXYLENEDIMETHYLENE (ETHYLENEDITHIO) DIETHYLENEDIURETHAN	376.53	85	547
TRANS-			
POLY-1,4-CYCLOHEXYLENEDIMETHYLENE GLUTARAMIDE	238.33	121	897
CIS-			
TRANS-		167	385
POLY-1,4-CYCLOHEXYLENEDIMETHYLENE GLUTARATE	240.30	290	385
TRANS-			
POLY-1,4-CYCLOHEXYLENEDIMETHYLENE (HEXAMETHYLENE-DITHIO)-DIETHYLENEDIURETHAN	432.64	50	547
TRANS-			
POLY-1,4-CYCLOHEXYLENEDIMETHYLENE ISOPHTHALAMIDE	272.35	124	897
TRANS-			
POLY-1,4-CYCLOHEXYLENEDIMETHYLENE ISOPHTHALATE	274.32	310	385
TRANS-			
POLY-1,4-CYCLOHEXYLENEDIMETHYLENE 2,6-NAPHTHALATE	324.38	197	547
CIS-			
TRANS-		287	547
POLY-1,4-CYCLOHEXYLENEDIMETHYLENE OCTAMETHYLENE-DIURETHAN	340.46	341	547
TRANS-			
		160	402
POLY-1,4-CYCLOHEXYLENEDIMETHYLENE OXALATE	198.22	160	291
TRANS-			
POLY-(1,4-CYCLOHEXYLENEDIMETHYLENE)-OXYDIMETHYLENE OXIDE	156.22	215	547
TRANS-			
POLY-1,4-CYCLOHEXYLENEDIMETHYLENE (PENTAMETHYLENE-DITHIO)-DIETHYLENEDIURETHAN	418.61	78	369
TRANS-			
POLY-1,4-CYCLOHEXYLENEDIMETHYLENE P-PHENYLENE-DIACETATE	302.37	102	897
CIS-			
POLY-1,4-CYCLOHEXYLENEDIMETHYLENE PIMELAMIDE	266.38	55	547
CIS-			
TRANS-		191	385
POLY-1,4-CYCLOHEXYLENEDIMETHYLENE PIMELATE	268.35	293	385
TRANS-			
POLY-1,4-CYCLOHEXYLENEDIMETHYLENE SEBACAMIDE	308.47	42	547
CIS-			
		208	385
		200	878

POLYMER	MOLECULAR WEIGHT	MELTING POINT	REF.
TRANS-		300	385
		296	878
POLY-1,4-CYCLOHEXYLENEDIMETHYLENE SEBACATE	310.43		
CIS-		50	547
TRANS-		78	547
POLY-1,4-CYCLOHEXYLENEDIMETHYLENE SUBERAMIDE	280.41		
CIS-		215	385
TRANS-		311	385
POLY-1,4-CYCLOHEXYLENEDIMETHYLENE SUBERATE	282.38		
CIS-		50	547
TRANS-		96	547
POLY-1,4-CYCLOHEXYLENEDIMETHYLENE SUCCINATE	226.27		
CIS-		62	547
TRANS-		147	547
POLY-1,4-CYCLOHEXYLENEDIMETHYLENE TEREPHTHALATE	274.32		
CIS-		256	547
TRANS-		318	547
POLY-1,4-CYCLOHEXYLENEDIMETHYLENE (TETRAMETHYLENE-DITHIO)-DIETHYLENEDIURETHAN	404.58		
TRANS-		129	897
POLY-1,4-CYCLOHEXYLENEDIMETHYLENE (TRIMETHYLENE-DITHIO)-DIETHYLENEDIURETHAN	390.56		
TRANS-		105	897
POLY-(1,4-CYCLOHEXYLENEDIOXY)-DITRIMETHYLENE ADIPAMIDE	340.46	196	402
POLY-(1,4-CYCLOHEXYLENEDIOXY)-DITRIMETHYLENE 4,4'-(ETHYLENEDIOXY)-DIBENZAMIDE	496.60	250	402
POLY-(1,4-CYCLOHEXYLENEDIOXY)-DITRIMETHYLENE 4,4'-(HEXAMETHYLENEDIOXY)-DIBENZAMIDE	552.71	215	402
POLY-(1,4-CYCLOHEXYLENEDIOXY)-DITRIMETHYLENE OXAMIDE	284.36	246	402
POLY-(1,4-CYCLOHEXYLENEDIOXY)-DITRIMETHYLENE 4,4'-(OXYDIETHYLENEDIOXY)-DIBENZAMIDE	540.66	125	402
POLY-(1,4-CYCLOHEXYLENEDIOXY)-DITRIMETHYLENE (P-PHENYLENEDIOXY)-DIACETAMIDE	420.51	160	402
POLY-(1,4-CYCLOHEXYLENEDIOXY)-DITRIMETHYLENE TEREPHTHALAMIDE	360.45	384	402
POLY-(1,4-CYCLOHEXYLENEDIOXY)-DITRIMETHYLENE 4,4'-(TETRAMETHYLENEDIOXY)-DIBENZAMIDE	524.66	224	402
POLY-1,4-CYCLOHEXYLENE DITHIOLSUCCINATE	230.35		
TRANS-		302	922
POLY-1,4-CYCLOHEXYLENE (ETHYLENEDITHIO)-DIETHYLENEDIURETHAN	348.48		
TRANS-		214	897
POLY-1,4-CYCLOHEXYLENE (HEXAMETHYLENEDITHIO)-DIETHYLENEDIURETHAN	404.58		
TRANS-		204	897
POLY-1,4-CYCLOHEXYLENE 3,3'-METHYLENEDIBENZAMIDE	334.42	174	402
POLY-1,4-CYCLOHEXYLENE OCTAMETHYLENEDIURETHAN	312.41		
TRANS-		255	633
		255	402
		221	291
POLY-1,4-CYCLOHEXYLENEOXYMETHYLENE OXIDE	128.17		
TRANS-		209	369
POLY-1,4-CYCLOHEXYLENE (PENTAMETHYLENEDITHIO)-DIETHYLENEDIURETHAN	390.56		
TRANS-		189	897
POLY-1,3-CYCLOHEXYLENE SEBACAMIDE	280.41		
CIS-		120	402
TRANS-		290	402
POLY-1,2-CYCLOHEXYLENE SULFIDE	114.21	130	917
POLY-1,3-CYCLOHEXYLENE SULFIDE	114.21	139	917
POLY-1,2-CYCLOHEXYLENE SULFONE	146.21	284	917
POLY-1,3-CYCLOHEXYLENE SULFONE	146.21	309	917
POLY-1,4-CYCLOHEXYLENE (TETRAMETHYLENEDITHIO)-DIETHYLENEDIURETHAN	376.53		
TRANS-		214	897
POLY-1,4-CYCLOHEXYLENE (TRIMETHYLENEDITHIO)-DIETHYLENEDIURETHAN	362.50		

POLYMER	MOLECULAR WEIGHT	MELTING POINT	REF.
TRANS-		202	897
POLY-1,4-CYCLOHEXYLENE URETHAN	141.17	>355	444
POLY-CYCLOHEXYLIDENEDIMETHYLENE OXIDE	126.20	152	935
---, 3,4-DIHYDRO-	124.18	102	935
---, ---, 4-METHYL-	138.21	155	935
---, 4-METHYL-	140.23	165	935
POLY-4,4'-CYCLOHEXYLIDENEDIPHENYLENE CARBONATE	294.35	260	438
---, 2,2'-DIMETHYL-	322.40	200	915
---, 3,3',5,5'-TETRACHLORO-	427.09	270	540
POLY-CYCLOHEXYLIDENETHIOHEXAMETHYLENE SULFIDE	230.44	75	923
POLY-CYCLOOCTENE	110.20	67	439
		73	284
		77	562
POLY-CYCLOPENTENE	68.12		
TRANS-		23	147
POLY-CYCLOPENTENE-ALT-ETHYLENE	96.17	185	412
POLY-1,3-CYCLOPENTYLENE-METHYLENE	82.15	146	267
		128	340
POLY-4,4'-CYCLOPENTYLIDENEDIPHENYLENE CARBONATE	280.32	250	438
POLY-CYCLOPROPYLENE CYCLOPROPYLENEDICARBOXAMIDE	166.18	285	865
POLY-CYCLOPROPYLENE (CYCLOPROPYLENEDICARBOXOYL)- DIURETHAN	254.20		
TRANS-, TRANS-		245	865
POLY-CYCLOPROPYLENEDIMETHYLENE CYCLOPROPYLENE- DICARBOXAMIDE	194.23	220	865
POLY-CYCLOPROPYLENEDIMETHYLENE CYCLOPROPYLENE- DIURETHAN	226.23		
CIS-		175	865
TRANS-		210	865
POLY-CYCLOPROPYLENEDIMETHYLENE HEXAMETHYLENE- DIURETHAN	270.33		
CIS-		120	860
TRANS-		165	860
POLY-CYCLOPROPYLENEDIMETHYLENE ISOPHTHALAMIDE	230.27	220	865
POLY-CYCLOPROPYLENEDIMETHYLENE ISOPHTHALATE	232.24	100	860
POLY-CYCLOPROPYLENEDIMETHYLENE 4,4'-(METHYLENE- DIPHENYLENE)-DIURETHAN	352.39		
CIS-		230	860
TRANS-		290	860
POLY-CYCLOPROPYLENEDIMETHYLENE PIPERAZINEDIURETHAN	240.26		
CIS-		70	860
TRANS-		100	860
POLY-CYCLOPROPYLENEDIMETHYLENE SEBACAMIDE	266.38	223	865
POLY-CYCLOPROPYLENEDIMETHYLENE TEREPHTHALATE	232.24	130	860
POLY-CYCLOPROPYLENEDIMETHYLENE TOLUYLENEDIURETHAN	276.29		
CIS-		170	860
TRANS-		200	860
POLY-CYCLOPROPYLENE HEXAMETHYLENEDIUREA	240.31	180	865
POLY-CYCLOPROPYLENE ISOPHTHALAMIDE	202.21	250	865
POLY-CYCLOPROPYLENE PIPERAZINEDIUREA	210.24	260	865
POLY-CYCLOPROPYLENE SEBACAMIDE	238.33	220	865
POLY-CYCLOPROPYLIDENEDIMETHYLENE OXIDE	84.12	45	935
POLY-DECAMETHYLENE	140.27		
---, 1,2-DICHLORO-	209.16		
TRANS-		47	925
POLY-DECAMETHYLENE ADIPAMIDE	282.43	230	110
		236	244
		240	544
POLY-DECAMETHYLENE ADIPATE	284.40	80	81
		77	300
POLY-DECAMETHYLENE ADIPYLDIURETHAN	370.45	164	402
POLY-DECAMETHYLENEAMINOHEXAMETHYLENE AMINE	254.46	102	829
POLY-DECAMETHYLENE AZELAAMIDE	324.51	214	137
POLY-DECAMETHYLENE AZELAATE	326.48	69	137
POLY-DECAMETHYLENE CARBONATE	200.28	55	820
POLY-DECAMETHYLENE M-CARBOXYCARBANILATE	319.40	127	891
POLY-DECAMETHYLENE P-CARBOXYCARBANILATE	319.40	158	891
POLY-DECAMETHYLENE (DECAMETHYLENEDISULFONYL)- DICAPROAMIDE	634.98	207	402

POLYMER	MOLECULAR WEIGHT	MELTING POINT	REF.
POLY-DECAMETHYLENE DECAMETHYLENEDIURETHAN	398.59	145	291
		142	634
POLY-(DECAMETHYLENEDIOXY)-DIHEXAMETHYLENE OXIDE	356.59	68	542
POLY-4,4'-DECAMETHYLENEDIPHENYLENE CARBONATE	352.47	110	915
POLY-DECAMETHYLENE 4,4'-DIPHENYLENEDIURETHAN			
---, 3,3'-DIMETHYL- (DIISOCYANATE)	438.57	219	633
POLY-4,4'-DECAMETHYLENEDIPIPERAZINE SEBACAMIDE	476.75	130	159
		129	402
POLY-DECAMETHYLENE DISILOXANYLENEDIPROPIONAMIDE	358.63		
---, TETRAMETHYL- (SI)	414.74	50	402
POLY-DECAMETHYLENE DISULFIDE	204.39	45	868
		65	725
POLY-DECAMETHYLENEDITHIOETHYLENE DISULFIDE	296.56	80	837
POLY-DECAMETHYLENE DITHIOLADIPATE	316.53	75	922
POLY-DECAMETHYLENE DITHIOLSEBACATE	372.63	103	922
POLY-DECAMETHYLENE DITHIOLTEREPHTHALATE	336.52	201	922
POLY-DECAMETHYLENEDITHIOTETRAMETHYLENE DISULFIDE	324.62	56	837
POLY-DECAMETHYLENE DOCOSANEDIAMIDE	506.86	169	454
POLY-DECAMETHYLENE DODECAMETHYLENEDIUREA	424.67	190	634
POLY-DECAMETHYLENE DODECANEDIAMIDE	366.59	191	244
		192	454
POLY-DECAMETHYLENE EICOSANEDIAMIDE	478.80	171	244
POLY-DECAMETHYLENE 4,4'-(ETHYLENEDIOXY)-DIBENZOATE	440.54	135	290
POLY-DECAMETHYLENE 4,4'-(ETHYLENEDIPHENYLENE)-DIOXYDIACETAMIDE	466.62	220	402
POLY-DECAMETHYLENE N,N'-ETHYLENEDITEREPHTHALAMATE	494.59	277	886
POLY-DECAMETHYLENE 4,4'-ETHYLIDENEDIBENZAMIDE	406.57	150	402
POLY-DECAMETHYLENE FUMARAMIDE	252.36	50	402
---, 2,3-DIMETHYL- (DIACID)	280.41		
TRANS-		267	931
---, METHYL- (DIACID)	266.38	166	931
POLY-DECAMETHYLENE 2,5-FURANDICARBOXAMIDE	292.38	130	905
POLY-DECAMETHYLENE 2,5-FURANDIPROPIONAMIDE	348.49	140	402
POLY-DECAMETHYLENE GLUTARAMIDE	268.40		
---, 3-CARBOXYL- (DIACID)	312.41	70	402
POLY-DECAMETHYLENE (HEXAMETHYLENEDISULFONYL)-DICAPROAMIDE	578.87	218	402
POLY-DECAMETHYLENE N,N'-HEXAMETHYLENE-DITEREPHTHALAMATE	550.70	235	886
POLY-DECAMETHYLENE HEXAMETHYLENEDIUREA	340.51	210	402
		216	634
POLY-DECAMETHYLENE HEXAMETHYLENEDIURETHAN	342.48	148	634
		161	473
POLY-DECAMETHYLENE ISOPHTHALAMIDE	302.42	186	868
POLY-DECAMETHYLENE 4,4'-(ISOPROPYLIDENEDI-PHENYLENE)-DIOXYDIACETAMIDE	480.65	105	402
POLY-DECAMETHYLENE 3,3'-METHYLENEDIBENZAMIDE	392.54	65	402
POLY-DECAMETHYLENE 4,4'-METHYLENEDIBENZAMIDE	392.54	100	402
POLY-DECAMETHYLENE 4,4'-METHYLENEDIPHENYLENEDIUREA	422.57	246	634
POLY-DECAMETHYLENE 4,4'-(METHYLENEDIPHENYLENE)-DIURETHAN	424.54	166	634
		194	473
POLY-DECAMETHYLENE (METHYLENE-2,5-FURAN)-DICARBOXAMIDE	306.41	111	402
POLY-DECAMETHYLENE (METHYLENE-2,5-TETRAHYDROFURAN)-DICARBOXAMIDE	310.44	85	402
POLY-DECAMETHYLENE OCTADECAMETHYLENEDIUREA	508.84	172	634
POLY-DECAMETHYLENE OCTADECANEDIAMIDE	450.75	170	454
POLY-DECAMETHYLENE OCTADECANEDIOATE	452.72	93	521
POLY-DECAMETHYLENE OCTAMETHYLENEDIUREA	368.57	201	634
POLY-DECAMETHYLENE OCTAMETHYLENEDIURETHAN	370.53	146	402
POLY-DECAMETHYLENE 4-OCTENDIAMIDE	308.47		
TRANS-		243	657
POLY-2,5-DECAMETHYLENE-1,3,4-OXADIAZOLE	208.30	100	921
POLY-DECAMETHYLENE OXALATE	228.29	79	305
POLY-DECAMETHYLENE OXAMIDE	226.32	229	291
		252	743
		290	402

POLYMER	MOLECULAR WEIGHT	MELTING POINT	REF.
POLY-DECAMETHYLENE OXIDE	156.27	79	181
		72	180
		60	291
		79	482
POLY-DECAMETHYLENEOXYMETHYLENE OXIDE	186.30	57	695
POLY-DECAMETHYLENEOXY-P-PHENYLENE OXIDE	248.37	<100	868
POLY-DECAMETHYLENE (PENTAMETHYLENEDISULFONYL)- DICAPROAMIDE	564.84	223	402
POLY-DECAMETHYLENE N,N'-PENTAMETHYLENE- DITEREPHTHALAMATE	536.67	205	886
POLY-DECAMETHYLENE P-PHENYLENEDIACETAMIDE	330.47	242	291
		265	454
POLY-DECAMETHYLENE (P-PHENYLENEDIOXY)-DIACETAMIDE	362.47	188	402
POLY-DECAMETHYLENE P-PHENYLENEDIPROPIONAMIDE	358.53	265	402
		271	454
POLY-DECAMETHYLENE 3,3'-PHOSPHINYLIDENEDIBENZAMIDE	426.50		
---, METHYL- (P)	440.52	155	402
POLY-DECAMETHYLENE 4,4'-PHOSPHINYLIDENEDIBENZAMIDE	426.50		
---, METHYL- (P)	440.52	186	402
---, PHENYL- (P)	502.59	192	402
POLY-DECAMETHYLENE PHTHALAMIDE	302.42	115	868
POLY-DECAMETHYLENE 2,4-PYRIDINEDICARBOXAMIDE	303.41	190	905
POLY-DECAMETHYLENE 2,5-PYRIDINEDICARBOXAMIDE	303.41	232	905
POLY-DECAMETHYLENE 2,6-PYRIDINEDICARBOXAMIDE	303.41	200	905
---, 1,4-DIHYDRO-4-OXO-	319.41	137	905
POLY-DECAMETHYLENE 3,5-PYRIDINEDICARBOXAMIDE	303.41	210	905
POLY-DECAMETHYLENE SEBACAMIDE	338.54	197	110
		198	454
		203	244
		216	137
POLY-DECAMETHYLENE SEBACATE	340.50	80	137
		73	291
POLY-DECAMETHYLENE SEBACYLDIURETHAN	426.55	153	402
POLY-DECAMETHYLENE SUBERAMIDE	310.48	217	244
		208	291
POLY-DECAMETHYLENE SUCCINATE	256.34	68	290
---, D-1,2-DIHYDROXY- (TARTRATE)	288.34	66	607
POLY-DECAMETHYLENE SULFIDE	172.33	78	819
		91	181
POLY-DECAMETHYLENE TEREPHTHALAMIDE	302.42	276	868
POLY-DECAMETHYLENE TEREPHTHALATE	304.39	138	81
		129	99
POLY-DECAMETHYLENE TETRADECAMETHYLENEDIUREA	452.73	180	634
POLY-DECAMETHYLENE TETRADECANEDIAMIDE	394.64	189	2
POLY-DECAMETHYLENE 2,5-TETRAHYDROFURAN- DIPROPIONAMIDE	352.52	178	402
POLY-DECAMETHYLENE (TETRAMETHYLENEDISULFONYL)- DICAPROAMIDE	550.81	236	402
POLY-DECAMETHYLENE N,N'-TETRAMETHYLENE- DITEREPHTHALAMATE	522.64	261	886
POLY-DECAMETHYLENE TETRAMETHYLENEDIURETHAN	314.43	171	633
POLY-DECAMETHYLENETHIOHEXAMETHYLENE SULFIDE	288.55	78	819
POLY-DECAMETHYLENE 2,5-THIOPHENEDICARBOXAMIDE	308.44	250	905
POLY-2,5-DECAMETHYLENE-1,3,4-TRIAZOLE	207.32		
---, 1-AMINO-	222.34	242	610
POLY-DECAMETHYLENE TRIDECANEDIAMIDE	380.62	175	339
POLY-DECAMETHYLENE N,N'-TRIMETHYLENE- DITEREPHTHALAMATE	508.61	222	886
POLY-DECAMETHYLENE UREA	198.31	200	402
		197	634
		210	291
POLY-DECAMETHYLENE P-XYLYLENEDIUREA	360.50	260	634
POLY-DECAMETHYLENE P-XYLYLENEDIURETHAN	362.47	184	634
POLY-1-DECENE	140.27	34	250
		40	537
POLY-DECYLVINYL ETHER	184.32	7	900
POLY-DIKETENE	84.07	115	190
		115	217

POLYMER		MOLECULAR WEIGHT	MELTING POINT	REF.
POLY-DIMETHYL KETENE	I.	70.09	250	179
			255	130
	II.	140.18	170	179
			180	248
POLY-2,5-DISTYRYLPYRAZINE		284.36	321	558
			343	723
POLY-DITHIODIETHYLENE SEBACAMIDE		318.49	155	402
POLY-4,4'-DITHIODIPHENYLENE HEXAMETHYLENEDIUREA		416.56	262	402
POLY-DIVINYLBENZAL		176.22	100	830
---, 2-METHYL-		190.24	95	830
---, 4-METHYL-		190.24	115	830
POLY-DIVINYLFURFURAL		166.18	145	830
POLY-DODECAMETHYLENE		168.32		
---, 1,2-DICHLORO-		237.21		
TRANS-			40	925
---, 1-METHYL-		182.35	30	926
POLY-DODECAMETHYLENE ADIPAMIDE		310.48	210	339
			230	544
POLY-DODECAMETHYLENE 4,4'-BIPHENYLENEDIURETHAN		438.57		
---, 3,3'-DIMETHYL- (DIISOCYANATE)		466.62	219	402
POLY-DODECAMETHYLENE DECAMETHYLENEDIURETHAN		426.64	133	634
POLY-DODECAMETHYLENE DOCOSANEDIAMIDE		534.92	164	454
POLY-DODECAMETHYLENE DODECAMETHYLENEDIURETHAN		454.70	128	633
POLY-DODECAMETHYLENE DODECANEDIAMIDE		394.64	183	454
POLY-DODECAMETHYLENE N,N'-ETHYLENE-DITEREPHTHALAMATE		522.64	264	886
POLY-DODECAMETHYLENE N,N'-HEXAMETHYLENE-DITEREPHTHALAMATE		578.75	231	886
POLY-DODECAMETHYLENE HEXAMETHYLENEDIUREA		368.57	205	634
			238	892
POLY-DODECAMETHYLENE HEXAMETHYLENEDIURETHAN		370.53	139	634
POLY-DODECAMETHYLENE 4,4'-METHYLENEDIPHENYLENE-DIUREA		450.63	240	634
POLY-DODECAMETHYLENE 4,4'-METHYLENEDIPHENYLENE DIURETHAN		452.59	164	634
POLY-DODECAMETHYLENE OCTADECANEDIAMIDE		478.80	167	454
POLY-DODECAMETHYLENE OXAMIDE		254.37	230	743
POLY-DODECAMETHYLENE N,N'-PENTAMETHYLENE-DITEREPHTHALAMATE		564.72	206	886
POLY-DODECAMETHYLENE P-PHENYLENEDIACETAMIDE		358.53	256	454
POLY-DODECAMETHYLENE P-PHENYLENEDIPROPIONAMIDE		386.58	263	454
POLY-DODECAMETHYLENE SEBACAMIDE		366.59	173	339
			192	454
POLY-DODECAMETHYLENE TEREPHTHALAMIDE		330.47	296	454
			301	544
POLY-DODECAMETHYLENE N,N'-TETRAMETHYLENE-DITEREPHTHALAMATE		550.70	258	886
POLY-DODECAMETHYLENE N,N'-TRIMETHYLENE-DITEREPHTHALAMATE		536.67	216	886
POLY-DODECAMETHYLENE P-XYLYLENEDIUREA		388.56	258	634
POLY-DODECAMETHYLENE P-XYLYLENEDIURETHAN		390.52	178	634
POLY-DODECANEDIOIC ANHYDRIDE		212.29	87	823
POLY-1-DODECENE		168.32	45	48
			49	250
POLY-DODECYLVINYL ETHER		212.38	30	900
POLY-EICOSAMETHYLENE MALONATE		382.58	69	868
POLY-ENANTHALDEHYDE		114.19	150	329
POLY-ETHYLENE		28.05	137	85
			141	206
			141	286
			142	315
			145	472
			146	362
---, 1-BUTOXY-2-METHYL-		114.19		
CIS- (ERYTHRO-DI-ISOTACTIC)			100	880
TRANS- (THREO-DI-ISOTACTIC)			100	880
---, 1-TERT-BUTOXY-2-METHYL-		114.19		
CIS- (ERYTHRO-DI-ISOTACTIC)			>250	880

CRYSTALLOGRAPHIC DATA FOR VARIOUS POLYMERS

POLYMER	MOLECULAR WEIGHT	MELTING POINT	REF.
---, CHLOROTRIFLUORO-	116.47	210	46
		220	49
		215	116
		222	341
---, 1-ETHOXY-2-METHOXY-	102.13		
TRANS-		217	141
---, 1-ETHOXY-2-METHYL-	86.13		
CIS- (ERYTHRO-DI-ISOTACTIC)		191	880
TRANS- (THREO-DI-ISOTACTIC)		207	880
		230	804
		243	141
---, 1-ISOBUTOXY-2-METHYL-	114.19		
TRANS-		226	141
---, 1-ISOPROPOXY-2-METHYL-	100.16		
CIS- (ERYTHRO-DI-ISOTACTIC)		204	880
TRANS- (THREO-DI-ISOTACTIC)		211	880
---, 1-METHOXY-2-METHYL-	72.11		
CIS- OR TRANS- (THREO-DI-ISOTACTIC)		230	880
		287	804
		210	141
---, 1-METHYL-2-PROPOXY-	100.16		
TRANS-		168	804
---, TETRAFLUORO-	100.02	327	46
		330	11
		330	101
POLY-ETHYLENE ADIPAMIDE	170.21	310	402
POLY-ETHYLENE ADIPATE	172.18	52	27
		47	265
		54	291
		50	46
		55	543
		65	571
POLY-ETHYLENE ADIPYLDIURETHAN	258.23	210	402
POLY-ETHYLENE AMINE	43.07	58	738
---, N-ACETYL-	85.11	200	747
---, N-BENZOYL-	147.18	210	747
---, ---, P-CHLORO-	181.62	285	747
---, N-BUTYRYL-	113.16	155	747
---, ---, 4-(4-METHYLTHIOPHENOXY)-	251.34	105	739
---, N-CYCLOHEXANECARBONYL-	153.22	285	747
---, N-DODECANOYL-	225.38	155	747
---, N-HEPTANOYL-	155.24	175	747
---, N-HEXANOYL-	141.21	175	747
---, N-ISOBUTYRYL-	113.16	210	747
---, N-ISOVALERYL-	127.19	210	747
---, N-2-NAPHTHOYL-	197.24	258	747
---, N-OCTADECANOYL-	309.54	145	747
---, N-OCTANOYL-	169.27	165	747
---, N-PERFLUOROOCTANOYL-	439.12	245	747
---, N-PERFLUOROPROPIONYL-	189.08	216	747
---, N-PIVALOYL-	127.19	320	747
---, N-P-TOLUOYL-	161.20	260	747
---, N-VALERYL-	127.19	172	747
POLY-ETHYLENE AZELAATE	214.26	46	543
POLY-4,4'-ETHYLENEBIPHENYLENE	180.25	>550	832
POLY-ETHYLENE P-(CARBOXYPHENOXY)-ACETATE	222.20	140	290
POLY-ETHYLENE P-(CARBOXYPHENOXY)-BUTYRATE	250.25	85	290
POLY-ETHYLENE P-(CARBOXYPHENOXY)-CAPROATE	278.30	45	290
POLY-ETHYLENE P-(CARBOXYPHENOXY)-HEPTANOATE	292.33	55	290
POLY-ETHYLENE P-(CARBOXYPHENOXY)-UNDECANOATE	348.44	65	290
POLY-ETHYLENE P-(CARBOXYPHENOXY)-VALERATE	264.28	55	290
POLY-ETHYLENE P-(CARBOXYPHENYLENE)-ACETAMIDE	204.23	221	904
POLY-ETHYLENE 1,4-CYCLOHEXYLENEDICARBOXYLATE	198.22		
CIS-		< 30	912
TRANS-		120	912
POLY-ETHYLENE CYCLOPROPYLENEDICARBOXAMIDE	154.17		
TRANS-		350	860
POLY-ETHYLENE DECAMETHYLENEDIURETHAN	286.37	174	414

POLYMER	MOLECULAR WEIGHT	MELTING POINT	REF.
POLY-ETHYLENE 4,4'-DIBENZOATE	268.27	346	291
POLY-4,4'-ETHYLENEDIBENZOIC ANHYDRIDE	252.27	340	854
POLY-4,4'-ETHYLENEDICYCLOHEXYLENE DODECANEDIAMIDE	418.66	316	926
POLY-4,4'-ETHYLENEDICYCLOHEXYLENE SEBACAMIDE	390.61	333	926
POLY-3,3'-(ETHYLENEDIOXY)-DIBENZOIC ANHYDRIDE	284.27	237	854
POLY-4,4'-(ETHYLENEDIOXY)-DIBENZOIC ANHYDRIDE	284.27	208	854
		215	827
		222	872
---, 3,3'-DIMETHOXY-	344.32	220	855
POLY-4,4'-(ETHYLENEDIOXY)-DIBENZOIC ISOPHTHALIC ANHYDRIDE	432.38	140	871
POLY-4,4'-(ETHYLENEDIOXY)-DIBENZOIC PHTHALIC ANHYDRIDE	432.38	185	872
POLY-4,4'-(ETHYLENEDIOXY)-DIBENZOIC TEREPHTHALIC ANHYDRIDE	432.38	209	871
POLY-(ETHYLENEDIOXY)-DIETHYLENE ADIPAMIDE	258.32	160	291
		190	402
POLY-(ETHYLENEDIOXY)-DIETHYLENE 1,4-PIPERAZINE-DIACETAMIDE	314.39	115	291
POLY-(ETHYLENEDIOXY)-DIETHYLENE SEBACATE	316.39	29	696
POLY-(ETHYLENEDIOXY)-DIETHYLENE UNDECANEDIAMIDE	328.45		
---, 6-HYDROXY- (DIACID)	344.45	156	402
POLY-4,4'-ETHYLENEDIPHENYLENE ADIPAMIDE	322.41	400	402
POLY-4,4'-ETHYLENEDIPHENYLENE CARBONATE	240.26	300	438
POLY-4,4'-(ETHYLENEDIPHENYLENE)-METHYLENE	194.28	255	832
POLY-4,4'-ETHYLENEDIPHENYLENE SEBACAMIDE	378.52	360	402
POLY-ETHYLENE DISILOXANYLENEDIPROPIONAMIDE	246.42		
---, TETRAMETHYL- (SI)	302.52	55	402
POLY-ETHYLENE DISULFIDE	92.17	130	294
		113	295
		145	725
POLY-(ETHYLENEDISULFONYL)-DIACETIC ANHYDRIDE	256.24	185	856
POLY-(ETHYLENEDISULFONYL)-DIPROPIONIC ANHYDRIDE	284.30	255	856
POLY-4,4'-(2,2'-ETHYLENEDITHIAZOLE)-P-PHENYLENE	270.37	265	810
POLY-(ETHYLENEDITHIO)-DIACETIC ANHYDRIDE	192.25	83	827
POLY-(ETHYLENEDITHIO)-DIPROPIONIC ANHYDRIDE	220.30	75	827
POLY-ETHYLENEDITHIOHEXAMETHYLENE DISULFIDE	240.56	102	837
POLY-ETHYLENE DITHIOLADIPATE	204.31	125	922
POLY-ETHYLENE DITHIOLISOPHTHALATE	224.30	185	922
POLY-ETHYLENE DITHIOL-2,5-PYRIDINEDICARBOXYLATE	252.29	D280	922
POLY-ETHYLENE DITHIOLTEREPHTHALATE	224.30	340	922
POLY-ETHYLENE DITHIONISOPHTHALATE	224.29	80	889
POLY-ETHYLENEDITHIOTETRAMETHYLENE DISULFIDE	212.40	96	837
POLY-ETHYLENE DODECANEDIAMIDE	254.37	261	244
POLY-ETHYLENE (ETHYLENEDIAMINO)-DIPROPIONAMIDE	228.30		
---, N,N'-DIISOPROPYL-, N,N'-DIMETHYL-	340.51	102	875
POLY-ETHYLENE 4,4'-ETHYLENEDIBENZOATE	296.32	212	903
POLY-ETHYLENE (ETHYLENEDIOXY)-DIACETAMIDE	202.21	168	894
POLY-ETHYLENE 3,3'-(ETHYLENEDIOXY)-DIBENZOATE	328.32	141	111
POLY-ETHYLENE 4,4'-(ETHYLENEDIOXY)-DIBENZOATE	328.32	240	290
		240	291
---, 3,3'-DIMETHOXY-	388.37	210	914
POLY-ETHYLENE 4,4'-(ETHYLENEDIPHENYLENE)-DIURETHAN	326.35	312	414
POLY-ETHYLENE 4,4'-(ETHYLENEDITHIO)-DIBENZOATE	360.44	190	291
POLY-ETHYLENE (ETHYLENEDITHIO)-DIETHYLENEDIURETHAN	294.38	168	897
POLY-ETHYLENE FUMARAMIDE	140.14	50	402
POLY-ETHYLENE 2,4-HEXADIENEDIAMIDE	166.18	61	402
POLY-ETHYLENE (HEXAMETHYLENEDITHIO)-DIETHYLENE-DIURETHAN	350.49	130	897
POLY-ETHYLENE HEXAMETHYLENEDIUREA	228.30	293	892
POLY-ETHYLENE HEXAMETHYLENEDIURETHAN	230.26	184	402
		170	627
		166	473
POLY-ETHYLENE ISOPHTHALATE	192.17	240	111
		143	398
POLY-ETHYLENE 4,4'-METHYLENEDIBENZOATE	282.30	220	291
POLY-ETHYLENE 4,4'-(METHYLENEDIPHENYLENE)-DIURETHAN	312.32	239	414

POLYMER	MOLECULAR WEIGHT	MELTING POINT	REF.
POLY-ETHYLENE 1,5-NAPHTHALATE	242.23	230	291
POLY-ETHYLENE 2,6-NAPHTHALATE	242.23	260	291
POLY-ETHYLENE 2,7-NAPHTHALATE	242.23	270	911
POLY-ETHYLENE NONAMETHYLENEDIURETHAN	272.34	168	414
POLY-ETHYLENE OXALATE	116.07	172	678
POLY-ETHYLENE OXIDE	44.05	66	81
		62	180
		72	318
		66	348
		70	538
		76	567
		66	480
		75	665
---, 1,1-BISCHLOROMETHYL-	141.00	180	816
---, BROMOMETHYL- (EPIBROMOHYDRIN)	136.98	112	809
---, TERT-BUTYL-	100.16		
ISOTACTIC		152	527
		150	693
		140	692
		135	691
SYNDIOTACTIC		63	691
---, CHLOROMETHYL- (EPICHLOROHYDRIN)	92.52	117	119
		121	194
		135	318
---, FLUOROMETHYL- (EPIFLUOROHYDRIN)	76.07	68	119
---, NEOPENTYL-	114.19	82	527
---, TETRAFLUORO-	116.02	42	858
---, TETRAMETHYL-	100.16	<300	664
---, UREA COMPLEX		143	734
POLY-ETHYLENE-P-OXYBENZOATE	164.16	220	354
		203	327
POLY-ETHYLENE 4,4'-OXYDIBENZOATE	284.27	152	291
		134	939
POLY-ETHYLENE 4,4'-(OXYDIETHYLENE)-DIOXYDIBENZOATE	372.37		
---, 3,3'-DIMETHOXY-	432.42	118	914
POLY-ETHYLENE 4,4'-(OXYDIMETHYLENE)-DI-2-(1,3-DIOXOLANE)-CAPRYLATE	500.63	19	908
POLY-ETHYLENEOXYMETHYLENE OXIDE (DIOXOLANE)	74.08	55	506
POLY-ETHYLENE P-OXYPHENYLENEACETATE	178.19	172	041
POLY-ETHYLENEOXY-P-PHENYLENE OXIDE	136.15	270	868
POLY-ETHYLENE 4,4'-(PENTAMETHYLENEDIOXY)-DIBENZOATE	370.40	150	912
POLY-ETHYLENE (PENTAMETHYLENEDITHIO)-DIETHYLENE-DIURETHAN	336.46	126	897
POLY-ETHYLENE P-PHENYLENEDIACETATE	220.22	107	325
		137	341
POLY-ETHYLENE PHTHALAMIDE	190.20	250	833
POLY-ETHYLENE 1,4-PIPERAZINEDICARBOXYLATE	200.19	245	386
POLY-ETHYLENE 1,4-PIPERAZINEDIPROPIONAMIDE	254.33	D231	875
---, N,N'-DIISOPROPYL-	338.50	210	875
POLY-ETHYLENE 1,4-PIPERAZINEDIPROPIONATE	256.30	93	875
POLY-ETHYLENE SEBACAMIDE	226.32	276	244
		254	291
		280	402
POLY-ETHYLENE SEBACATE	228.29	78	302
		72	265
		76	137
		79	46
		73	539
		83	571
POLY-ETHYLENE SEBACYLDIUREA	312.37	228	402
POLY-ETHYLENE SEBACYLDIURETHAN	314.34	198	402
POLY-ETHYLENE SUBERATE	200.23	55	27
		75	571
POLY-ETHYLENE SUCCINATE	144.13	103	302
		108	290
POLY-ETHYLENE SULFIDE	60.11	190	181
		210	241

POLYMER	MOLECULAR WEIGHT	MELTING POINT	REF.
---, TERT-BUTYL-	116.22	205	694
POLY-ETHYLENE 4,4'-SULFONYLDIBENZAMIDE	330.36	380	835
POLY-ETHYLENE TEREPHTHALAMIDE	190.20	455	834
---, N,N'-DIMETHYL-	218.26	379	834
POLY-ETHYLENE TEREPHTHALATE	192.17	265	27
		265	290
		267	265
		270	374
		284	264
		245	472
		278	724
---, CHLORO-	226.62	92	913
---, 2,5-DICHLORO-	261.06	165	913
---, 2,5-DIMETHYL-	220.22	180	913
---, METHYL-	206.20	70	913
POLY-ETHYLENE 2,5-TETRAHYDROFURANDIPROPIONAMIDE	240.30	218	402
POLY-ETHYLENE 4,4'-TETRAMETHYLENEDIBENZOATE	324.38	170	291
POLY-ETHYLENE 3,3'-(TETRAMETHYLENEDIOXY)-DIBENZOATE	356.37	100	111
POLY-ETHYLENE 4,4'-(TETRAMETHYLENEDIOXY)-DIBENZOATE	356.37	252	841
---, 3,3'-DIMETHOXY-	416.43	117	914
POLY-ETHYLENE 4,4'-(TETRAMETHYLENEDIPHENYLENE)-DIURETHAN	354.41	274	414
POLY-ETHYLENE (TETRAMETHYLENEDITHIO)-DIETHYLENE-DIURETHAN	322.44	126	897
POLY-ETHYLENE 4,4'-THIODIBENZOATE	300.33	200	291
POLY-ETHYLENE THIODIENANTHAMIDE	314.49	210	402
POLY-ETHYLENE THIODIVALERAMIDE	258.38	220	402
POLY-ETHYLENETHIOHEXAMETHYLENE SULFIDE	176.34	86	819
POLY-ETHYLENETHIOTETRAMETHYLENE SULFIDE	148.29	89	181
POLY-2,5-ETHYLENE-1,3,4-TRIAZOLE	95.10		
---, 1-AMINO-	110.12	360	610
POLY-ETHYLENE 4,4'-(TRIMETHYLENEDIOXY)-DIBENZOATE	342.35	190	912
---, 3,3'-DIMETHOXY-	403.00	105	914
POLY-ETHYLENE 4,4'-(TRIMETHYLENEDIPHENYLENE)-DIURETHAN	340.38	207	414
POLY-ETHYLENE N,N'-TRIMETHYLENEDITEREPHTHALAMATE	396.40	205	886
POLY-ETHYLENE UREA	86.09	400	402
POLY-ETHYLIDENE	28.05	100	617
		200	618
		195	619
POLY-4,4'-ETHYLIDENEDIPHENYLENE CARBONATE	240.26	195	438
---, PHENYL-	316.36	230	438
POLY-ETHYLIDENE P-PHENYLENE	104.15	205	876
POLY-ETHYLVINYL ETHER	72.11	86	114
---, 2-CHLORO-	106.55	150	114
---, 2-METHOXY-	102.13	73	114
---, 2,2,2-TRIFLUORO-	126.08	128	114
POLY-FORMALDEHYDE, SEE POLY-OXYMETHYLENE			
POLY-2,5-FURANDIPROPIONIC ANHYDRIDE	194.19	67	827
POLY-FURFURAL-ALT-DIMETHYLKETENE	166.18	180	817
POLY-1,4-B-D-GLUCOSE (CELLULOSE)	162.14		
---, TRIACETATE	288.25	306	144
---, TRIBUTYRATE	372.41	207	118
		206	46
---, TRICAPRATE	624.90	88	144
---, TRICAPROATE	456.58	94	144
---, TRICAPRYLATE	540.74	116	117
---, TRIHEPTYLATE	498.66	88	144
---, TRILAURATE	709.06	91	144
---, TRIMYRISTATE	793.22	106	144
---, TRINITRATE	297.11	697	77
		700	81
---, 2.44-TRINITRATE		617	81
---, TRIPALMITATE	877.38	105	144
---, TRIPROPIONATE	330.33	234	144
---, TRIVALERATE	414.50	122	144

POLYMER	MOLECULAR WEIGHT	MELTING POINT	REF.
1,4-POLY-1,3-HEPTADIENE	96.17		
TRANS-			
ISOTACTIC		85	371
---, 5-METHYL-	110.20		
TRANS-			
ISOTACTIC		180	896
---, 6-METHYL-	110.20		
TRANS-			
ISOTACTIC		119	371
POLY-HEPTAMETHYLENE	98.19		
---, 1,2-DICHLORO-	167.08		
TRANS-		53	925
POLY-HEPTAMETHYLENE ADIPAMIDE	240.35	250	244
		209	291
		226	339
		245	859
POLY-HEPTAMETHYLENE AZELAAMIDE	282.43	201	244
POLY-HEPTAMETHYLENE DISULFIDE	162.31	130	868
POLY-HEPTAMETHYLENE N,N'-ETHYLENE-			
DITEREPHTHALAMATE	452.51	275	886
POLY-HEPTAMETHYLENE N,N'-HEXAMETHYLENE-			
DITEREPHTHALAMATE	508.61	227	886
POLY-HEPTAMETHYLENE HEXAMETHYLENEDIUREA	298.43	243	892
POLY-HEPTAMETHYLENE HEXAMETHYLENEDIURETHAN	300.40	151	473
POLY-HEPTAMETHYLENE 4,4'-(METHYLENEDIPHENYLENE)-			
DIURETHAN	382.46	198	473
POLY-HETAMETHYLENE 4-OCTENDIAMIDE	266.38		
TRANS- I.		261	657
II.		249	657
POLY-HEPTAMETHYLENE N,N'-PENTAMETHYLENE-			
DITEREPHTHALAMATE	494.59	199	886
POLY-HEPTAMETHYLENE P-PHENYLENEDIACETAMIDE	288.39	234	402
POLY-HEPTAMETHYLENE PIMELAMIDE	254.37	214	244
		196	339
		205	2
POLY-HEPTAMETHYLENE SEBACAMIDE	296.46	208	244
		187	339
POLY-HEPTAMETHYLENE SUBERAMIDE	268.40	230	244
POLY-HEPTAMETHYLENE 4,4'-SULFONYLDIBENZAMIDE	400.49		
---, 4,4-DIMETHYL-	428.55	268	835
POLY-HEPTAMETHYLENE TEREPHTHALAMIDE	260.34	328	895
		341	834
POLY-HEPTAMETHYLENE TEREPHTHALATE	262.30	98	453
POLY-HEPTAMETHYLENE 2,5-TETRAHYDROFURAN-			
DIPROPIONAMIDE	310.44	148	402
POLY-HEPTAMETHYLENE N,N'-TETRAMETHYLENE-			
DITEREPHTHALAMATE	480.56	269	886
POLY-2,5-HEPTAMETHYLENE-1,3,4-TRIAZOLE	165.24		
---, 1-AMINO-	180.26	237	610
POLY-HEPTAMETHYLENE N,N'-TRIMETHYLENE			
DITEREPHTHALAMATE	466.53	222	886
POLY-HEPTAMETHYLENE UNDECANEDIAMIDE	310.48	195	2
POLY-1-HEPTENE	98.19	17	250
---, 6,6-DIMETHYL-	126.24	104	282
---, 5-METHYL-	112.22	52	129
POLY-4,4'-(4,4-HEPTYLIDENE)-DIPHENYLENE CARBONATE	310.39	200	438
POLY-HEXADECAMETHYLENE DECAMETHYLENEDIURETHAN	482.75	128	634
POLY-HEXADECAMETHYLENE HEXAMETHYLENEDIURETHAN	426.64	134	634
POLY-HEXADECAMETHYLENE 4,4'-METHYLENE-			
DIPHENYLENEDIURETHAN	508.70	152	634
POLY-HEXADECAMETHYLENE P-XYLYLENEDIURETHAN	446.63	168	634
POLY-1-HEXADECENE	224.43	68	250
POLY-1,5-HEXADIENE, SEE POLY-1-METHYLENE-3-			
CYCLOPENTYLENE			
1,4-POLY-1,3-HEXADIENE	82.15		
TRANS-			
ISOTACTIC		82	371
---, 5-METHYL-	96.17		
TRANS-			

POLYMER	MOLECULAR WEIGHT	MELTING POINT	REF.
ISOTACTIC		88	371
2,5-POLY-2,4-HEXADIENE	82.15		
TRANS- (ERYTHRO-DI-ISOTACTIC)		84	525
---, 2,5-DIMETHYL-	110.20		
TRANS-		265	376
		265	183
2,5-POLY-2,4-HEXADIENEDIOIC ACID	142.11		
---, DIISOPROPYL ESTER	226.27		
TRANS- (ERYTHRO-DI-ISOTACTIC)		230	420
POLY-HEXAMETHYLENE ADIPAMIDE	226.32	265	2
		270	289
---, 2,5-DIHYDROXY- (DIACID)	258.32	163	402
---, N,N'-DIMETHYL-	254.37	75	402
---, 3-METHYL- (DIACID)	240.35	216	291
		230	859
---, 3-METHYL- (DIAMINE)	240.35	180	291
---, N-METHYL-	240.35	145	291
POLY-HEXAMETHYLENE ADIPATE	228.29	56	290
		57	539
POLY-HEXAMETHYLENE ADIPYLDIURETHAN	314.34	206	402
POLY-HEXAMETHYLENE AZELAAMIDE	268.40	226	244
		185	339
POLY-HEXAMETHYLENE AZELAATE	270.37	53	539
POLY-HEXAMETHYLENE CARBONATE	144.17	60	820
POLY-HEXAMETHYLENE M-CARBOXYCARBANILATE	263.29	96	891
POLY-HEXAMETHYLENE P-(CARBOXYPHENOXY)-ACETATE	278.30	50	290
POLY-HEXAMETHYLENE P-(CARBOXYPHENOXY)-CAPROATE	334.41	60	290
POLY-HEXAMETHYLENE P-(CARBOXYPHENOXY)-UNDECANOATE	404.55	72	290
POLY-HEXAMETHYLENE P-(CARBOXYPHENOXY)-VALERATE	320.38	60	290
POLY-HEXAMETHYLENE P-(CARBOXYPHENYLENE)-ACETAMIDE	260.34	227	904
POLY-HEXAMETHYLENE 1,2-CYCLOHEXYLENEDIACETAMIDE	280.41	255	402
POLY-HEXAMETHYLENE 1,2-CYCLOHEXYLENEDICARBOXAMIDE	252.36		
TRANS-		242	402
POLY-HEXAMETHYLENE 1,3-CYCLOHEXYLENEDICARBOXAMIDE	252.36		
TRANS-		312	402
POLY-HEXAMETHYLENE 1,4-CYCLOHEXYLENEDICARBOXAMIDE	252.36		
TRANS-		360	402
POLY-HEXAMETHYLENE CYCLOPROPYLENEDICARBOXAMIDE	210.28		
CIS-		180	860
TRANS-		300	860
---, 1-METHYL- (DIACID)	224.30		
TRANS-		115	860
---, 3-METHYL- (DIACID)	224.30		
TRANS-		270	860
POLY-HEXAMETHYLENE (DECAMETHYLENEDISULFONYL)-			
DICAPROAMIDE	578.87	210	402
POLY-HEXAMETHYLENE DECAMETHYLENEDIURETHAN	342.48	154	291
		144	634
POLY-HEXAMETHYLENE 2,2'-DIBENZAMIDE	322.41	175	402
POLY-HEXAMETHYLENE 3,3'-DIBENZAMIDE	322.41	142	402
POLY-HEXAMETHYLENE 3,4'-DIBENZAMIDE	322.41	140	402
POLY-HEXAMETHYLENE 4,4'-DIBENZAMIDE	322.41	360	402
POLY-HEXAMETHYLENE 4,4'-DIBENZOATE	324.38	214	291
		230	903
POLY-4,4'-HEXAMETHYLENEDIBENZOIC ANHYDRIDE	308.38	151	854
POLY-3,3'-(HEXAMETHYLENEDIOXY)-DIBENZOIC ANHYDRIDE	340.38	157	854
POLY-4,4'-(HEXAMETHYLENEDIOXY)-DIBENZOIC ANHYDRIDE	340.38	157	854
POLY-(HEXAMETHYLENEDIOXY)-DIDECAMETHYLENE OXIDE	412.70	76	542
POLY-HEXAMETHYLENE 4,4'-DISILOXANYLENEDIBENZAMIDE	398.61		
---, TETRAMETHYL- (SI)	454.72	145	402
POLY-HEXAMETHYLENE DISILOXANYLENEDIPROPIONAMIDE	302.52		
---, DIETHYL-DIMETHYL- (SI)	386.68	50	402
---, TETRAETHYL- (SI)	414.74	70	402
---, TETRAMETHYL- (SI)	358.63	95	402
POLY-HEXAMETHYLENE DISULFIDE	148.28	57	868
		46	725
POLY-HEXAMETHYLENE DITHIODICARBOXYLATE	236.30	55	936
POLY-HEXAMETHYLENE DITHIOLADIPATE	260.42	115	922

POLYMER	MOLECULAR WEIGHT	MELTING POINT	REF.
POLY-HEXAMETHYLENE DITHIOLAZELAATE	302.50	110	922
POLY-HEXAMETHYLENE DITHIOL-1,4-CYCLOHEXYLENE-DICARBOXYLATE	286.46		
TRANS-		215	922
POLY-HEXAMETHYLENE DITHIOLISOPHTHALATE	280.41	113	922
POLY-HEXAMETHYLENE DITHIOL-P-PHENYLENEDIACETATE	308.46	105	922
POLY-HEXAMETHYLENE DITHIOLPIMELATE	274.44	70	922
POLY-HEXAMETHYLENE DITHIOLSEBACATE	316.53	140	922
POLY-HEXAMETHYLENE DITHIOLSUBERATE	288.47	110	922
POLY-HEXAMETHYLENE DITHIOL-4,4'-SULFONYLDIBENZOATE	420.57		
---, 2-ETHYL- (DIOL)	448.62	208	922
POLY-HEXAMETHYLENE DITHIOLTEREPHTHALATE	280.41	300	922
POLY-HEXAMETHYLENE DITHIOTEREPHTHALAMIDE	278.43	190	402
POLY-HEXAMETHYLENEDITHIOTETRAMETHYLENE DISULFIDE	268.51	38	837
POLY-HEXAMETHYLENE DOCOSANEDIAMIDE	450.75	180	454
POLY-HEXAMETHYLENE DODECANEDIAMIDE	310.48	217	244
		219	454
POLY-HEXAMETHYLENE EICOSANEDIAMIDE	422.70	189	244
POLY-HEXAMETHYLENE (ETHYLENEDIOXY)-DIACETAMIDE	258.32	126	894
POLY-HEXAMETHYLENE 4,4'-(ETHYLENEDIOXY)-DIBENZOATE	384.43	175	290
POLY-HEXAMETHYLENE 4,4'-(ETHYLENEDIPHENYLENE)-DIOXYDIACETAMIDE	410.51	220	402
POLY-HEXAMETHYLENE N,N'-ETHYLENE-DITEREPHTHALAMATE	438.48	274	886
POLY-HEXAMETHYLENE (ETHYLENEDITHIO)-DIETHYLENE-DIURETHAN	350.49	140	897
POLY-HEXAMETHYLENE 4,4'-ETHYLIDENEDIBENZAMIDE	350.46	175	402
POLY-HEXAMETHYLENE 2,5-FURANDICARBOXAMIDE	236.27	175	905
POLY-HEXAMETHYLENE 2,5-FURANDIPROPIONAMIDE	292.38	190	402
POLY-HEXAMETHYLENE GLUTARAMIDE	212.29	241	402
---, PERFLUORO- (DIACID)			
---, ---, N,N'-DIBUTYL-	432.45	20	402
---, ---, N,N'-DIETHYL-	376.34	5	402
---, ---, N,N'-DIISOPROPYL-	404.40	20	402
---, ---, N,N'-DIMETHYL-	348.29	30	402
POLY-HEXAMETHYLENE (HEXAMETHYLENEDIOXY)-DIPROPIONAMIDE	342.48	105	402
POLY-HEXAMETHYLENE (HEXAMETHYLENEDISULFONYL)-DICAPROAMIDE	522.76	222	402
POLY-HEXAMETHYLENE N,N'-HEXAMETHYLENE-DITEREPHTHALAMATE	494.59	265	886
POLY-HEXAMETHYLENE (HEXAMETHYLENEDITHIO)-DIETHYLENEDIURETHAN	406.60	106	897
POLY-HEXAMETHYLENE HEXAMETHYLENEDIURETHAN	286.37	150	291
		164	634
		165	627
		171	473
POLY-HEXAMETHYLENE ISOPHTHALAMIDE	246.31	220	402
		198	868
---, 5-TERT-BUTYL- (DIACID)	302.42	210	402
POLY-HEXAMETHYLENE ISOPHTHLATE	248.28	140	111
POLY-HEXAMETHYLENE 4,4'-ISOPROPYLIDENEDIBENZAMIDE	364.49	180	402
POLY-HEXAMETHYLENE 4,4'-KETODIBENZAMIDE	350.42	350	402
POLY-HEXAMETHYLENE KETODIPROPIONAMIDE	254.33	180	402
POLY-HEXAMETHYLENE MALONAMIDE	184.24		
---, DIMETHYL- (DIACID)	212.29	117	402
POLY-HEXAMETHYLENE 3,3'-METHYLENEDIBENZAMIDE	336.44	113	402
POLY-HEXAMETHYLENE 4,4'-METHYLENEDIBENZAMIDE	336.44	132	402
POLY-HEXAMETHYLENE 4,4'-(METHYLENEDIPHENYLENE)-DIOXYDIACETAMIDE	396.49	174	402
POLY-HEXAMETHYLENE 4,4'-(METHYLENEDIPHENYLENE)-DIUREA	366.46	250	402
---, 3,3'-DIMETHYL-	394.52	290	402
---, ---, N,N'-DIETHYL-	450.63	120	402
---, ---, N,N'-DIPROPYL-	478.68	90	402
POLY-HEXAMETHYLENE 4,4'-(METHYLENEDIPHENYLENE)-DIURETHAN	368.43	179	634
		200	473

References page III-92

POLYMER	MOLECULAR WEIGHT	MELTING POINT	REF.
POLY-HEXAMETHYLENE (METHYLENE-2,5-FURAN)-DICARBOXAMIDE	250.30	115	402
POLY-HEXAMETHYLENE (P-METHYLENEPHENOXY)-DIACETAMIDE	304.39	220	402
POLY-HEXAMETHYLENE (METHYLENE-2,5-TETRAHYDROFURAN)-DICARBOXAMIDE	254.33	78	402
POLY-HEXAMETHYLENE NAPHTHALENEDICARBOXAMIDE	298.39	200	402
POLY-HEXAMETHYLENE NONAMETHYLENEDIUREA	326.48	243	892
POLY-HEXAMETHYLENE OCTADECAMETHYLENEDIUREA	452.73	192	634
POLY-HEXAMETHYLENE OCTADECANEDIAMIDE	394.64	192	454
---, 9,10-DIHYDROXY-	426.64	172	402
---, N,N'-DIMETHYL-	422.70	52	877
POLY-HEXAMETHYLENE OCTAMETHYLENEDITHIOUREA	344.58	160	402
POLY-HEXAMETHYLENE OCTAMETHYLENEDIUREA	312.46	239	634
		253	892
		255	402
POLY-HEXAMETHYLENE OCTAMETHYLENEDIURETHAN	342.48	153	633
POLY-HEXAMETHYLENE 4-OCTENDIAMIDE	252.36		
TRANS-		259	657
POLY-HEXAMETHYLENE 5-OXADODECANEDIAMIDE	312.45	149	847
POLY-HEXAMETHYLENE 6-OXAHEPTADECANEDIAMIDE	382.59	160	847
POLY-HEXAMETHYLENE 5-OXAHEXADECANEDIAMIDE	368.56	159	847
POLY-HEXAMETHYLENE 6-OXAPENTADECANEDIAMIDE	354.53	154	847
POLY-HEXAMETHYLENE OXALATE	172.18	66	305
POLY-HEXAMETHYLENE OXAMIDE	170.21	320	402
		320	743
---, 3-OXYPROPYL-	228.29	240	402
POLY-HEXAMETHYLENE 5-OXASEBACAMIDE	284.40	160	847
POLY-HEXAMETHYLENE 5-OXATETRADECANEDIAMIDE	340.51	155	847
POLY-HEXAMETHYLENE 5-OXAUNDECANEDIAMIDE	298.43	152	847
POLY-HEXAMETHYLENE OXIDE	100.16	58	181
		58	482
		62	542
POLY-HEXAMETHYLENE OXYDIACETAMIDE	214.26	143	402
		172	894
POLY-HEXAMETHYLENE 4,4'-OXYDIBENZOATE	340.38	74	939
POLY-HEXAMETHYLENE OXYDIBUTYRAMIDE	270.37	187	847
POLY-HEXAMETHYLENE OXYDICAPROAMIDE	326.48	175	847
POLY-HEXAMETHYLENE OXYDIENANTHAMIDE	354.53	170	847
POLY-HEXAMETHYLENE 4,4'-(OXYDIMETHYLENE)-DI-2-(1,3-DIOXOLANE)-CAPRYLAMIDE	554.77	145	908
POLY-HEXAMETHYLENE OXYDIPELARGONAMIDE	410.64	158	847
POLY-HEXAMETHYLENE 4,4'-(OXYDIPHENYLENE)-DIOXY-DIACETAMIDE	398.46	220	402
POLY-HEXAMETHYLENE OXYDIVALERAMIDE	298.43	180	847
POLY-HEXAMETHYLENEOXYMETHYLENE OXIDE	130.19	38	695
POLY-HEXAMETHYLENEOXY-P-PHENYLENE OXIDE	192.26	170	868
POLY-HEXAMETHYLENE (PENTAMETHYLENEDISULFONYL)-DICAPROAMIDE	508.73	226	402
POLY-HEXAMETHYLENE N,N'-PENTAMETHYLENE-DITEREPHTHALAMATE	480.56	238	886
POLY-HEXAMETHYLENE (PENTAMETHYLENEDITHIO)-DIETHYLENEDIURETHAN	392.57	98	897
POLY-HEXAMETHYLENE PENTAMETHYLENEDIUREA	270.38	251	892
POLY-HEXAMETHYLENE (P-PHENOXY)-DIACETAMIDE	290.36	86	402
POLY-HEXAMETHYLENE M-PHENYLENEDIACETAMIDE	274.36	182	402
POLY-HEXAMETHYLENE P-PHENYLENEDIACETAMIDE	274.36	300	454
POLY-HEXAMETHYLENE (P-PHENYLENEDIOXY)-DIACETAMIDE	306.36	237	402
POLY-HEXAMETHYLENE P-PHENYLENEDIPROPIONAMIDE	302.42	290	291
		295	402
POLY-HEXAMETHYLENE M-PHENYLENEDIUREA	276.34		
---, 4-METHYL- (DIISOCYANATE)	290.37	235	402
POLY-HEXAMETHYLENE 3,3'-PHOSPHINYLIDENEDIBENZAMIDE	370.39		
---, METHYL- (P)	384.42	172	402
POLY-HEXAMETHYLENE 4,4'-PHOSPHINYLIDENEDIBENZAMIDE	370.39		
---, HYDROXY- (P)	386.39	245	402
---, METHYL- (P)	384.42	213	402
---, PHENYL- (P)	445.48	231	402

POLYMER	MOLECULAR WEIGHT	MELTING POINT	REF.
POLY-HEXAMETHYLENE PHTHALAMIDE	246.31	150	833
POLY-HEXAMETHYLENE PIMELAMIDE	240.35	228	244
		202	339
POLY-HEXAMETHYLENE PIMELATE	242.32	50	539
POLY-HEXAMETHYLENE 1,4-PIPERAZINEDIACETAMIDE	282.39	168	291
POLY-HEXAMETHYLENE (1,4-PIPERAZINEDITHIO)- DICARBOXYLATE	322.44	144	934
POLY-HEXAMETHYLENE PIPERAZINEDIUREA	254.33	245	402
		265	838
POLY-HEXAMETHYLENE 4(H)-PYRAN-2,6-DICARBOXAMIDE	250.30		
---, 4-OXO- (DIACID)	264.28	77	905
POLY-HEXAMETHYLENE 2,3-PYRIDINEDICARBOXAMIDE	247.30	103	905
POLY-HEXAMETHYLENE 2,4-PYRIDINEDICARBOXAMIDE	247.30	220	905
POLY-HEXAMETHYLENE 2,5-PYRIDINEDICARBOXAMIDE	247.30	272	905
POLY-HEXAMETHYLENE 2,6-PYRIDINEDICARBOXAMIDE	247.30	220	905
---, 1,4-DIHYDRO-4-OXO-	263.30	178	905
POLY-HEXAMETHYLENE 3,4-PYRIDINEDICARBOXAMIDE	247.30	101	905
POLY-HEXAMETHYLENE 3,5-PYRIDINEDICARBOXAMIDE	247.30	257	905
---, 2,6-DIMETHYL- (DIACID)	275.35	140	905
POLY-HEXAMETHYLENE 2,5-PYRROLEDIPROPIONAMIDE	291.40		
---, 1-METHYL- (DIACID)	305.42	180	402
POLY-HEXAMETHYLENE 2,5-PYRROLIDINEDIPROPIONAMIDE	295.43		
---, 1-METHYL- (DIACID)	309.45	200	402
POLY-HEXAMETHYLENE 5-PYRROLIDONYLIDENE DIPROPIONAMIDE	295.40	250	402
POLY-HEXAMETHYLENE SEBACAMIDE	282.43	228	51
		216	153
		225	454
		233	244
		215	291
---, 3-METHYL- (DIAMINE)	296.46	153	402
POLY-HEXAMETHYLENE SEBACATE	284.40	67	290
		78	521
POLY-HEXAMETHYLENE SEBACYLDIURETHAN	370.45	158	402
POLY-HEXAMETHYLENE 4,4'-SILYLENEDIBENZAMIDE	352.51		
---, DIMETHYL- (SI)	380.56	190	402
POLY-HEXAMETHYLENE SUBERAMIDE	254.37	232	244
		235	291
POLY-HEXAMETHYLENE SUBERATE	256.34	58	291
POLY-HEXAMETHYLENE SUCCINAMIDE	198.27	212	402
POLY-HEXAMETHYLENE SUCCINATE	200.23	57	305
POLY-HEXAMETHYLENE SULFIDE	116.22	79	181
		68	295
		76	819
POLY-HEXAMETHYLENE SULFONE	148.22	220	538
POLY-HEXAMETHYLENE 4,4'-SULFONYLDIBENZAMIDE	386.46	310	835
POLY-HEXAMETHYLENE SULFONYLDIVALERAMIDE	346.49	215	402
POLY-HEXAMETHYLENESULFONYLPENTAMETHYLENE SULFONE	282.41	223	538
POLY-HEXAMETHYLENESULFONYLTETRAMETHYLENE SULFONE	268.39	246	538
POLY-HEXAMETHYLENE TEREPHTHALAMIDE	246.31	371	834
		D350	291
		>400	841
---, 2,5-DIHYDROXY- (DIACID)	278.31	334	402
---, 2,5-DIMETHYL- (DIACID)	274.36	143	402
---, N,N'-DIMETHYL-	274.36	260	834
---, METHYL- (DIACID)	260.34	248	402
POLY-HEXAMETHYLENE TEREPHTHALATE	248.28	160	81
		154	99
		161	405
POLY-HEXAMETHYLENE TETRADECAMETHYLENEDIUREA	396.62	198	634
POLY-HEXAMETHYLENE TETRADECANEDIAMIDE	338.54	209	2
POLY-HEXAMETHYLENE 2,5-TETRAHYDROFURAN- DIPROPIONAMIDE	296.41	182	402
POLY-HEXAMETHYLENE (TETRAMETHYLENEDIOXY)- DIPROPIONAMIDE	314.43	110	402
POLY-HEXAMETHYLENE (TETRAMETHYLENEDISULFONYL)- DICAPROAMIDE	494.70	241	402
POLY-HEXAMETHYLENE N,N'-TETRAMETHYLENE- DITEREPHTHALAMATE	466.53	290	886

POLYMER	MOLECULAR WEIGHT	MELTING POINT	REF.
POLY-HEXAMETHYLENE (TETRAMETHYLENEDITHIO)-			
DIETHYLENEDIURETHAN	378.55	120	897
POLY-HEXAMETHYLENE TETRAMETHYLENEDIUREA	256.35	283	892
POLY-HEXAMETHYLENE 4,4'-TETRASILOXANYLENE-			
DIBENZAMIDE	490.81		
---, TETRAMETHYL-TETRAPHENYL- (SI)	847.28	140	402
POLY-HEXAMETHYLENE THIODIBUTYRAMIDE	286.43	200	291
POLY-HEXAMETHYLENE THIODIENANTHAMIDE	370.60	170	402
POLY-HEXAMETHYLENE THIODIPROPIONAMIDE	258.38	216	859
		219	402
POLY-HEXAMETHYLENE THIODIVALERAMIDE	314.49	180	402
		185	853
POLY-HEXAMETHYLENETHIO-4,4'-(METHYLENEDIPHENYLENE)			
DIETHYLENE SULFIDE	370.61	88	819
POLY-HEXAMETHYLENETHIOPENTAMETHYLENE SULFIDE	218.42	65	819
POLY-HEXAMETHYLENE 2,5-THIOPHENEDIACETAMIDE	280.39	230	905
POLY-HEXAMETHYLENE 2,5-THIOPHENEDICARBOXAMIDE	252.33	315	905
POLY-HEXAMETHYLENE 2,5-THIOPHENEDIPROPIONAMIDE	308.44	232	402
		210	905
POLY-HEXAMETHYLENETHIOTETRAMETHYLENE SULFIDE	204.39	67	819
		75	181
POLY-HEXAMETHYLENETHIOUNDECAMETHYLENE SULFIDE	302.59	78	923
POLY-HEXAMETHYLENE THIOUREA	158.26	160	402
POLY-2,5-HEXAMETHYLENE-1,3,4-TRIAZOLE	151.21		
---, 1-AMINO-	166.23	275	610
POLY-HEXAMETHYLENE 4,4'-(TRIMETHYLENEDIPHENYLENE)-			
DIOXYDIACETAMIDE	424.54	80	402
POLY-HEXAMETHYLENE (TRIMETHYLENEDISULFONYL)-			
DIIMINODICAPROAMIDE	446.59	185	402
POLY-HEXAMETHYLENE N,N'-TRIMETHYLENE-			
DITEREPHTHALAMATE	452.51	256	886
POLY-HEXAMETHYLENE TRIMETHYLENEDIUREA	242.32	266	892
POLY-HEXAMETHYLENE UNDECANEDIAMIDE	296.46		
---, 6-HYDROXY- (DIACID)	312.45	165	402
POLY-HEXAMETHYLENE UREA	142.20	300	402
		270	802
		268	634
		>300	291
POLY-HEXAMETHYLENE VINYLENEDIBENZAMIDE			
---, METHYL- (DIACID)	362.47	160	402
POLY-HEXAMETHYLENE P-XYLYLENEDIUREA	304.39	306	634
POLY-HEXAMETHYLENE P-XYLYLENEDIURETHAN	306.36	209	634
1,6-POLY-1,3,5-HEXATRIENE	80.13	250	848
POLY-1-HEXENE	84.16	-55	48
---, 4,4-DIMETHYL-	112.22	350	90
---, 4-ETHYL-	112.22	234	632
---, 3-METHYL-	98.19	285	441
---, 4-METHYL	98.19	200	9
		190	632
---, 5-METHYL-	98.19	130	9
		110	282
POLY-4,4'-(2,2-HEXYLIDENE)-DIPHENYLENE CARBONATE	296.36	200	438
POLY-HYDRAZO ADIPAMIDE	142.16	320	402
POLY-HYDRAZO AZELAAMIDE	184.24	237	402
POLY-HYDRAZO DOCOSANEDIAMIDE	366.59	210	402
POLY-HYDRAZO DODECANEDIAMIDE	226.32	242	402
POLY-HYDRAZO GLUTARAMIDE	128.13	140	402
POLY-HYDRAZO PIMELAMIDE	156.18	260	402
POLY-HYDRAZO SEBACAMIDE	198.27	260	402
POLY-HYDRAZO SUBERAMIDE	170.21	275	402
POLY-HYDRAZO SUCCINAMIDE	114.10	>360	402
POLY-HYDRAZO THIODIVALERAMIDE	230.33	180	402
POLY-2-HYDROXYACETIC ACID (GLYCOLIC ACID)	58.04	223	305
		230	465
		233	490
---, DIETHYL-	114.14	200	928
---, ISOPROPYL-	100.12	205	930
POLY-P-HYDROXYBENZOIC ACID	120.11	>320	839
		>350	911

POLYMER	MOLECULAR WEIGHT	MELTING POINT	REF.
---, 3,5-DI-TERT-BUTYL-	232.32	480	814
POLY-3-HYDROXYBUTYRIC ACID	86.09	176	422
		168	570
POLY-10-HYDROXYCAPRIC ACID	170.24	80	309
POLY-6-HYDROXYCAPROIC ACID	114.14	55	560
		64	623
POLY-5-HYDROXY-2-(1,3-DIOXANE)-CAPRYLIC ACID	228.29		
CIS-		120	908
TRANS-		150	908
POLY-7-HYDROXYENANTHIC ACID	128.17		
---, 4-METHYL- (R+)	142.20	36	836
POLY-4-HYDROXYMETHYLENE-2-(1,3-DIOXOLANE)-CAPRYLIC ACID	228.29		
CIS-		53	908
TRANS-		53	908
POLY-5-HYDROXYMETHYLENE-5-METHYL-2-(1,3-DIOXANE)-CAPRYLIC ACID	256.34	81	908
POLY-5-HYDROXY-3-OXAVALERIC ACID	102.09	89	305
POLY-3-HYDROXYPROPIONIC ACID (HYDRACRYLIC ACID)	72.06	122	317
		84	623
---, 2,2-DIMETHYL-	100.12	240	551
---, 3-ISOPROPYL-	114.14		
ISOTACTIC		89	698
POLY-4-HYDROXYTETRAMETHYLENE-2-(1,3-DIOXOLANE)-CAPRYLIC ACID	270.37		
CIS-		23	908
TRANS-		23	908
POLY-5-HYDROXYVALERIC ACID	100.12	55	844
		53	305
---, 2,3,4-TRIMETHOXY-	190.20	138	305
POLY-2-IMINO-1,3-DITHIAPENTAMETHYLENE	119.20		
---, N-ETHYL-	147.25	68	927
---, N-METHYL-	133.23	89	927
---, N-PHENYL-	195.30	128	927
---, ---, 4-METHYL-	209.32	89	927
POLY-ISOBUTENE	56.11	44	746
POLY-ISOBUTYLENE OXIDE	72.11	158	593
		160	664
		155	318
		177	548
POLY-ISOBUTYLENE SULFIDE	88.17	187	548
POLY-4,4'-ISOBUTYLIDENEDIPHENYLENE CARBONATE	268.31	180	438
POLY-ISOBUTYLVINYL ETHER	100.16	115	161
		117	269
		165	114
		170	281
		115	46
POLY-ISOBUTYRALDEHYDE	72.11	D260	329
POLY-ISOPHTHALALDEHYDE	134.13	80	801
POLY-ISOPHTHALIC ANHYDRIDE	148.12	259	827
POLY-ISOPHTHALIC 2,2'-(ISOPHTHALOYLDIAMINO)-DIBENZOIC ANHYDRIDE	534.48	259	871
POLY-ISOPHTHALIC (4,4'-ISOPROPYLIDENEDIPHENYLENE-DIOXY)-DIACETIC ANHYDRIDE	474.46	65	871
POLY-ISOPHTHALIC 4,4'-(METHYLENEDIOXY)-DIBENZOIC ANHYDRIDE	418.36		
---, 3,3'-DIMETHOXY-	478.41	228	871
POLY-ISOPHTHALIC (P-PHENYLENEDIOXY)-DIACETIC ANHYDRIDE	356.29	297	871
POLY-ISOPHTHALIC 2,2'-(TEREPHTHALOYLDIAMINO)-DIBENZOIC ANHYDRIDE	534.48	291	871
POLY-ISOPHTHALIC 4,4'-(TETRAMETHYLENEDIOXY)-DIBENZOIC ANHYDRIDE	460.44	100	871
POLY-ISOPHTHALIC 4,4'-(TRIMETHYLENEDIOXY)-DIBENZOIC ANHYDRIDE	446.41	165	871
---, 3,3'-DIMETHOXY-	506.46	216	871
POLY-ISOPHTHALOYL (M-CARBOXYPHENOXYACETYL)-DIHYDRAZIDE	354.32	125	833

POLYMER	MOLECULAR WEIGHT	MELTING POINT	REF.
POLY-ISOPHTHALOYL (M-CARBOXYPHENOXYBUTYRYL)-DIHYDRAZIDE	382.38	265	833
POLY-ISOPHTHALOYL (M-CARBOXYPHENOXYCAPRYL)-DIHYDRAZIDE	410.43	249	833
POLY-ISOPHTHALOYL (M-CARBOXYPHENOXYPROPIONYL)-DIHYDRAZIDE	368.35	238	833
POLY-2,2'-(ISOPHTHALOYLDIAMINO)-DIBENZOIC TEREPHTHALIC ANHYDRIDE	534.48	264	871
POLY-ISOPHTHALOYL DITHIONISOPHTHALOYLDIHYDRAZIDE	356.42	324	889
POLY-ISOPRENE, SEE POLY-BUTADIENE, --- 2-METHYL-			
POLY-ISOPROPENYLMETHYL KETONE	84.12	240	347
		240	255
		200	379
POLY-4,4'-ISOPROPYLIDENEDIBENZOIC ANHYDRIDE	266.30	235	854
		240	855
POLY-ISOPROPYLIDENEDIMETHYLENE 4,4'-(METHYLENEDIPHENYLENE)-DIURETHAN	354.41	190	402
POLY-ISOPROPYLIDENEDIMETHYLENE (1,4-PIPERAZINEDITHIO)-DICARBOXYLATE	308.41		
---, 2,5-DIMETHYL-	336.46	135	934
POLY-4,4'-ISOPROPYLIDENEDIPHENYLENE ADIPATE	338.40		
---, 3,3',5,5'-TETRACHLORO-	476.18	283	713
POLY-4,4'-ISOPROPYLIDENEDIPHENYLENE CARBONATE	254.29	267	57
		260	736
		230	438
		263	370
		240	435
---, 3,3'-DICHLORO-	323.18	210	540
		230	736
---, 3,3'-DIISOPROPYL-	338.45	110	915
---, 3,3'-DIMETHYL-	282.34	170	540
---, 3,3',5,5'-TETRABROMO-	569.87	260	540
		267	736
---, 3,3',5,5'-TETRACHLORO-	392.06	260	540
		275	736
POLY-4,4'-ISOPROPYLIDENEDIPHENYLENE CYCLOPROPYLENEDICARBOXYLATE	322.36		
CIS-		130	860
TRANS-		180	860
---, 1-METHYL- (DIACID)	336.39		
TRANS-		90	860
---, 3-METHYL- (DIACID)	336.39		
TRANS-		130	860
POLY-(4,4'-ISOPROPYLIDENEDIPHENYLENEDIOXY)-DIACETIC ANHYDRIDE	326.35	202	855
POLY-(4,4'-ISOPROPYLIDENEDIPHENYLENEDIOXY)-DIACETIC 3,4-PYRIDINEDICARBOXYLIC ANHYDRIDE	475.45	120	872
POLY-(4,4'-ISOPROPYLIDENEDIPHENYLENEDIOXY)-DIACETIC 3,4-PYRIDENEDICARBOXYLIC ANHYDRIDE	475.45	120	872
POLY-(4,4'-ISOPROPYLIDENEDIPHENYLENEDIOXY)-DIACETIC TEREPHTHALIC ANHYDRIDE	474.47	270	871
POLY-4,4'-ISOPROPYLIDENEDIPHENYLENE DITHIOLADIPATE	370.53	198	922
POLY-4,4'-ISOPROPYLIDENEDIPHENYLENE DITHIOL-1,3-CYCLOHEXYLENEDICARBOXYLATE	396.57	220	922
POLY-4,4'-ISOPROPYLIDENEDIPHENYLENE DITHIONISOPHTHALATE	390.52	152	889
POLY-4,4'-ISOPROPYLIDENEDIPHENYLENE HEXAMETHYLENE-DIURETHAN	396.49	130	402
POLY-4,4'-ISOPROPYLIDENEDIPHENYLENE MALONATE	296.32	96	868
POLY-4,4'-ISOPROPYLIDENEDIPHENYLENE 4,4'-(METHYLENEDIPHENYLENE)-DIURETHAN	478.55	193	402
POLY-4,4'-ISOPROPYLIDENEDIPHENYLENE 1,4-PIPERAZINEDIPROPIONATE	422.52	145	875
POLY-ISOPROPYLIDENE SULFIDE	74.14	125	882
POLY-ISOPROPYLVINYL ETHER	86.13	191	281
		98	269
		190	114

POLY-KETONE, SEE POLY-ETHYLENE-ALT-CARBON MONOXIDE

POLYMER	MOLECULAR WEIGHT	MELTING POINT	REF.
POLY-2-MERCAPTOACETIC ACID	74.10	169	922
---, 2-METHYL-	88.13	152	922
POLY-6-MERCAPTOCAPROIC ACID	130.21	106	428
---, 4-METHYL-	144.24	62	922
POLY-3-MERCAPTOPROPIONIC ACID	88.13	145	922
---, 2-PHTHALIMIDO-	233.24	260	922
---, 2-(P-TOLUENESULFONAMIDO)-	257.33	250	922
POLY-METHACRYLIC ACID	86.09		
---, TERT-BUTYL ESTER	142.12		
ISOTACTIC		104	851
SYNDIOTACTIC		165	851
---, HEXADECYL ESTER	310.52	22	900
---, METHYL ESTER	100.12		
ISOTACTIC		160	31
SYNDIOTACTIC		>200	31
---, OCTADECYL ESTER	338.58		
SYNDIOTACTIC		36	500
POLY-METHACRYLONITRILE	67.09	250	436
POLY-METHYLENE (SEE ALSO POLY-ETHYLENE)	14.03		
---, DIPHENYL-	166.22	220	831
---, DI-P-TOLYL-	194.28	180	831
POLY-METHYLENE ADIPAMIDE	156.18	306	402
POLY-4,4'-METHYLENEDIBENZOIC ANHYDRIDE	238.24	332	854
POLY-4,4'-(METHYLENEDIOXY)-DIBENZOIC ANHYDRIDE	270.24	220	854
POLY-4,4'-(METHYLENEDIOXY)-DIBENZOIC TEREPHTHALIC ANHYDRIDE	418.36		
---, 3,3'-DIMETHOXY-	478.41	267	871
POLY-4,4'-METHYLENEDIPHENYLENE ADIPAMIDE	308.38	356	402
---, 3,3'-DIMETHYL- (DIAMINE)	336.44	326	402
---, ---, N,N'-DIBUTYL-	448.65	57	402
---, ---, N,N'-DIETHYL-	392.54	79	402
---, ---, N,N'-DIHEXYL-	504.76	45	402
---, ---, N,N'-DIISOAMYL-	476.70	45	402
---, ---, N,N'-DIISOPROPYL-	420.60	61	402
---, N,N'-DIETHYL-	364.49	62	402
---, N,N'-DIMETHYL-	336.44	120	402
POLY-4,4'-METHYLENEDIPHENYLENE ADIPATE	310.35	140	901
POLY-4,4'-METHYLENEDIPHENYLENE AZELAAMIDE	350.46	275	402
		268	866
---, N,N'-DIETHYL-	406.57	41	402
---, N,N'-DIMETHYL-	378.52	58	402
POLY-4,4'-METHYLENEDIPHENYLENE CARBONATE	226.23	300	438
		278	736
		230	435
---, 3,3'-DIMETHYL-	254.29	250	736
---, DIPHENYL-	378.43	230	540
---, PHENYL-	302.33	215	438
POLY-4,4'-(METHYLENEDIPHENYLENE)-DIETHYLENE SULFIDE	254.39	108	819
POLY-4,4'-METHYLENEDIPHENYLENE DISILOXANYLENE-DIPROPIONAMIDE	384.58		
---, TETRAMETHYL- (SI)	440.69	90	402
---, ---, 3,3'-DIMETHYL- (DIAMINE)	454.72	265	402
POLY-3,3'-(4,4'-METHYLENEDIPHENYLENE-N,N'-DISUCCINIMIDEDIYL)-IMINO-4,4'-METHYLENE-DIPHENYLENE AMINE	556.62	295	876
POLY-4,4'-METHYLENEDIPHENYLENE DODECANEDIAMIDE	392.54	256	866
POLY-4,4'-METHYLENEDIPHENYLENE OCTADECAMETHYLENE-DIUREA	534.79	225	634
POLY-4,4'-METHYLENEDIPHENYLENE OCTAMETHYLENEDIUREA	394.52	263	634
POLY-4,4'-METHYLENEDIPHENYLENE OXIDE	182.22	140	745
POLY-4,4'-METHYLENEDIPHENYLENE M-PHENYLENEDIUREA	358.40		
---, 3,3'-DIMETHYL-, 4-METHYL-	400.48	307	402
POLY-4,4'-METHYLENEDIPHENYLENE SEBACAMIDE	364.49	280	402
		270	866
---, N,N'-DIETHYL-	420.60	32	402
---, 3,3'-DIMETHYL- (DIAMINE)	392.54	227	402
---, N,N'-DIMETHYL-	392.54	55	402

POLYMER	MOLECULAR WEIGHT	MELTING POINT	REF.
POLY-4,4'-METHYLENEDIPHENYLENE TEREPHTHALAMIDE	328.37	420	402
---, N,N'-DIBUTYL-	440.59	195	870
---, N,N'-DIETHYL-	384.48	182	870
---, 3,3'-DIMETHYL- (DIAMINE)	356.42	380	402
---, ---, N,N'-DIBUTYL-	468.64	159	870
---, ---, N,N'-DIETHYL-	412.53	178	870
---, ---, N,N'-DIMETHYL-	384.48	229	870
---, ---, N,N'-DIPROPYL-	440.59	190	870
---, N,N'-DIMETHYL-	356.42	264	870
---, N,N'-DIPROPYL-	412.53	156	870
POLY-4,4'-METHYLENEDIPHENYLENE TETRADECAMETHYLENEDIUREA	478.68	229	634
POLY-N,N'-METHYLENEDITEREPHTHALAMIC ANHYDRIDE	324.29	330	827
POLY-METHYLENEOXYNONAMETHYLENE OXIDE	172.27	55	695
POLY-METHYLENEOXYOCTADECAMETHYLENE OXIDE	298.51	72	695
POLY-METHYLENEOXYPENTAMETHYLENE OXIDE	116.16	39	695
POLY-METHYLENEOXYTETRADECAMETHYLENE OXIDE	242.40	69	695
POLY-METHYLENE P-PHENYLENE	90.12	142	744
		170	745
---, 2,5-DIMETHYL-	118.18	300	744
POLY-METHYLENE SEBACAMIDE	212.29	268	887
		260	402
POLY-METHYLENE SELENIDE	92.99	190	416
		170	463
POLY-METHYLENE SULFIDE	46.09	245	181
		260	298
		260	237
---, DIFLUORO-	82.07	35	842
POLY-METHYLENETHIOTETRAMETHYLENE SULFIDE	134.26	73	181
POLY-METHYLVINYL ETHER	58.08	144	114
POLY-NEOPENTYLVINYL ETHER	114.19	216	281
		155	269
		216	114
POLY-NONAMETHYLENE ADIPAMIDE	268.40	205	339
		242	818
POLY-NONAMETHYLENE AZELAAMIDE	310.48	177	110
		165	339
		189	2
POLY-NONAMETHYLENE AZELAATE	312.45	65	405
POLY-NONAMETHYLENE DISILOXANYLENEDIPROPIONAMIDE	344.60		
---, DIETHYL-DIMETHYL- (SI)	428.76	10	402
---, TETRAETHYL- (SI)	456.82	10	402
POLY-NONAMETHYLENE DISULFIDE	190.36	55	868
		60	725
POLY-NONAMETHYLENE N,N'-ETHYLENE-DITEREPHTHALAMATE	480.56	267	886
POLY-NONAMETHYLENE N,N'-HEXAMETHYLENE-DITEREPHTHALAMATE	536.67	219	886
POLY-NONAMETHYLENE HEXAMETHYLENEDIURETHAN	328.45	147	633
		154	473
POLY-NONAMETHYLENE 4,4'-(METHYLENEDIPHENYLENE)-DIURETHAN	410.51	194	473
POLY-NONAMETHYLENE 4-OCTENEDIAMIDE	294.44		
TRANS- II.		248	657
POLY-2,5-NONAMETHYLENE-1,3,4-OXADIAZOLE	194.28	82	921
POLY-NONAMETHYLENE OXIDE	142.24	73	482
POLY-NONAMETHYLENE N,N'-PENTAMETHYLENE-DITEREPHTHALAMATE	522.64	194	886
POLY-NONAMETHYLENE 3,3'-PHOSPHINYLIDENEDIBENZAMIDE	412.47		
---, METHYL- (P)	426.50	141	402
POLY-NONAMETHYLENE 4,4'-PHOSPHINYLIDENEDIBENZAMIDE	412.47		
---, METHYL- (P)	426.50	167	402
---, PHENYL- (P)	488.57	179	402
POLY-NONAMETHYLENE PIMELAMIDE	282.43	196	2
POLY-NONAMETHYLENE SEBACAMIDE	324.51	176	339
		202	402
POLY-NONAMETHYLENE TEREPHTHALATE	290.36	85	99
		90	453

POLYMER	MOLECULAR WEIGHT	MELTING POINT	REF.
POLY-NONAMETHYLENE 2,5-TETRAHYDROFURAN- DIPROPIONAMIDE	338.49	149	402
POLY-NONAMETHYLENE N,N'-TETRAMETHYLENE- DITEREPHTHALAMATE	508.61	259	886
POLY-NONAMETHYLENE TRIDECANEDIAMIDE	366.59	183	2
POLY-NONAMETHYLENE N,N'-TRIMETHYLENE- DITEREPHTHALAMATE	494.59	211	886
POLY-NONAMETHYLENE UNDECANEDIAMIDE	338.54	196	2
POLY-NONAMETHYLENE UREA	184.28	220	611
POLY-1-NONENE	126.24	19	250
POLY-4,4'-(2,2-NONYLIDENE)-DIPHENYLENE CARBONATE	338.47	190	438
POLY-NORBORNENE, SEE POLY-1-VINYLENE-3- CYCLOPENTYLENE			
POLY-OCTADECAMETHYLENE DOCOSANEDIAMIDE	619.08	146	454
POLY-OCTADECAMETHYLENE DODECANEDIAMIDE	478.80	170	454
POLY-OCTADECAMETHYLENE 4,4'-(ETHYLENEDIOXY)- DIBENZOATE	552.77	122	290
POLY-OCTADECAMETHYLENE OCTADECANEDIAMIDE	562.97	152	454
POLY-OCTADECAMETHYLENE P-PHENYLENEDIACETAMIDE	442.69	225	454
POLY-OCTADECAMETHYLENE P-PHENYLENEDIPROPIONAMIDE	470.74	232	454
POLY-OCTADECAMETHYLENE SEBACAMIDE	450.75	171	454
POLY-OCTADECAMETHYLENE TEREPHTHALAMIDE	414.63	255	454
POLY-OCTADECAMETHYLENE TEREPHTHALATE	416.60	116	290
POLY-OCTADECAMETHYLENE P-XYLYLENEDIUREA	472.72	226	634
POLY-OCTADECANEDIOIC ANHYDRIDE	296.45	95	823
POLY-1-OCTADECENE	252.49	80	6
		100	90
1,4-POLY-1,3-OCTADIENE	110.20		
TRANS-			
ISOTACTIC		87	371
POLY-OCTAMETHYLENE	112.22		
---, 1-METHYL-	126.24	-5	639
POLY-OCTAMETHYLENE ADIPAMIDE	254.37	250	244
		235	291
		254	544
POLY-OCTAMETHYLENE AZELAAMIDE	296.46	206	244
POLY-OCTAMETHYLENE DECAMETHYLENEDIURETHAN	370.53	137	634
POLY-OCTAMETHYLENE DISILOXANYLENEDIPROPIONAMIDE	330.58		
---, TETRAMETHYL- (SI)	386.68	95	402
POLY-OCTAMETHYLENE DISULFIDE	176.34	58	725
POLY-4,4'-(2,2'-OCTAMETHYLENEDITHIAZOLE)- P-PHENYLENE	354.53	164	810
POLY-OCTAMETHYLENE DOCOSANEDIAMIDE	478.80	175	454
POLY-OCTAMETHYLENE DODECANEDIAMIDE	338.54	202	244
		200	454
		194	291
POLY-OCTAMETHYLENE DODECANEDIOATE	340.50	73	291
POLY-OCTAMETHYLENE EICOSANEDIAMIDE	450.75	179	244
POLY-OCTAMETHYLENE N,N'-ETHYLENE- DITEREPHTHALAMATE	466.53	282	886
POLY-OCTAMETHYLENE 2,5-FURANDICARBOXAMIDE	264.32	125	905
POLY-OCTAMETHYLENE N,N'-HEXAMETHYLENE- DITEREPHTHALAMATE	522.64	238	886
POLY-OCTAMETHYLENE HEXAMETHYLENEDIURETHAN	314.43	162	473
		152	634
POLY-OCTAMETHYLENE ISOPHTHALAMIDE	274.36	186	868
POLY-OCTAMETHYLENE 4,4'-(METHYLENEDIPHENYLENE)- DIURETHAN	396.49	172	634
		201	473
POLY-OCTAMETHYLENE OCTADECANEDIAMIDE	422.70	179	454
POLY-OCTAMETHYLENE OCTAMETHYLENEDITHIOUREA	372.63	190	402
POLY-OCTAMETHYLENE OCTAMETHYLENEDIURETHAN	342.48	144	402
POLY-OCTAMETHYLENE 4-OCTENEDIAMIDE	280.41		
TRANS-		256	657
POLY-2,5-OCTAMETHYLENE-1,3,4-OXADIAZOLE	180.25	110	921
		100	862
POLY-OCTAMETHYLENE OXAMIDE	198.27	276	743
POLY-OCTAMETHYLENE OXIDE	128.22	67	482

POLYMER	MOLECULAR WEIGHT	MELTING POINT	REF.
POLY-OCTAMETHYLENE N,N'-PENTAMETHYLENE- DITEREPHTHALAMATE	508.61	207	886
POLY-OCTAMETHYLENE P-PHENYLENEDIACETAMIDE	302.42	280	454
POLY-OCTAMETHYLENE P-PHENYLENEDIPROPIONAMIDE	330.47	289	454
POLY-OCTAMETHYLENE 3,3'-PHOSPHINYLIDENEDIBENZAMIDE	398.44		
---, METHYL- (P)	412.47	158	402
POLY-OCTAMETHYLENE 4,4'-PHOSPHINYLIDENEDIBENZAMIDE	398.44		
---, METHYL- (P)	412.47	172	402
---, PHENYL- (P)	474.54	202	402
POLY-OCTAMETHYLENE PHTHALAMIDE	274.36	123	868
POLY-OCTAMETHYLENE 2,4-PYRIDINEDICARBOXAMIDE	275.35	190	905
POLY-OCTAMETHYLENE 2,5-PYRIDINEDICARBOXAMIDE	275.35	232	905
POLY-OCTAMETHYLENE 2,6-PYRIDINEDICARBOXAMIDE	275.35	192	905
---, 1,4-DIHYDRO-4-OXO-	291.35	86	905
POLY-OCTAMETHYLENE 3,5-PYRIDINEDICARBOXAMIDE	275.35	214	905
POLY-OCTAMETHYLENE SEBACAMIDE	310.48	207	153
		206	454
		210	244
POLY-OCTAMETHYLENE SEBACYLDIUREA	396.53	212	402
POLY-OCTAMETHYLENE SUBERAMIDE	282.43	225	244
		216	291
POLY-OCTAMETHYLENE TEREPHTHALAMIDE	274.36	315	868
POLY-OCTAMETHYLENE TEREPHTHALATE	276.33	132	99
POLY-OCTAMETHYLENE TETRADECANEDIAMIDE	366.59	196	2
POLY-OCTAMETHYLENE 2,5-TETRAHYDROFURAN- DIPROPIONAMIDE	324.46	180	402
POLY-OCTAMETHYLENE N,N'-TETRAMETHYLENE- DITEREPHTHALAMATE	494.59	276	886
POLY-OCTAMETHYLENE TETRAMETHYLENEDITHIOUREA	316.53	160	402
POLY-OCTAMETHYLENE 2,5-THIOPHENEDICARBOXAMIDE	280.39	304	905
POLY-2,5-OCTAMETHYLENE-1,3,4-TRIAZOLE	179.27		
---, 1-AMINO-	194.28	258	610
POLY-OCTAMETHYLENE N,N'-TRIMETHYLENE- DITEREPHTHALAMATE	480.56	231	886
POLY-OCTAMETHYLENE UREA	170.26	260	291
POLY-OCTAMETHYLENE P-XYLYLENEDIUREA	332.45	278	634
POLY-OCTAMETHYLENE P-XYLYLENEDIURETHAN	334.42	196	634
POLY-1-OCTENE	112.22	-38	48
POLY-OXACYCLOBUTANE (TRIMETHYLENE OXIDE)	58.08	36	180
		35	447
		34	181
---, 3,3-BISBROMOMETHYL-	243.93	220	807
---, 3,3-BISCHLOROMETHYL-	155.02	190	173
		180	148
---, 3,3-BISETHOXYMETHYL-	174.24	83	808
---, 3,3-BISFLUOROMETHYL-	122.12	135	805
---, 3,3-BISHYDROXYMETHYL-	118.13	>280	280
---, 3,3-BISIODOMETHYL-	337.93	290	807
---, 3,3-DIMETHYL-	86.13	47	447
POLY-4,4'-OXYDIBENZOIC ANHYDRIDE	240.21	296	854
POLY-OXYDIETHYLENE AZELAATE	258.31	0	696
POLY-(OXYDIETHYLENE)-DIOXYDIBENZOIC ANHYDRIDE	328.32	190	854
POLY-OXYDIETHYLENE HEXAMETHYLENEDIURETHAN	274.32	120	633
POLY-(OXYDIETHYLENE)-OXY-P-PHENYLENE OXIDE	180.20	136	868
POLY-OXYDIETHYLENE SEBACATE	272.34	44	325
POLY-(OXYDIMETHYLENEDIOXY)-DIBENZOIC ANHYDRIDE	300.27	192	854
POLY-OXYDIPENTAMETHYLENE 5-OXADODECANEDIAMIDE	384.56	102	847
POLY-OXYDIPENTAMETHYLENE 6-OXADODECANEDIAMIDE	384.56	100	847
POLY-OXYDIPENTAMETHYLENE 6-OXAHEPTADECANEDIAMIDE	454.70	128	847
POLY-OXYDIPENTAMETHYLENE 5-OXAHEXADECANEDIAMIDE	440.67	122	847
POLY-OXYDIPENTAMETHYLENE 6-OXAPENTADECANEDIAMIDE	426.64	117	847
POLY-OXYDIPENTAMETHYLENE 5-OXASEBACAMIDE	356.51	127	847
POLY-OXYDIPENTAMETHYLENE 5-OXATETRADECANEDIAMIDE	412.61	118	847
POLY-OXYDIPENTAMETHYLENE 6-OXATRIDECANEDIAMIDE	398.59	106	847
POLY-OXYDIPENTAMETHYLENE 5-OXAUNDECANEDIAMIDE	370.53	108	847
POLY-OXYDIPENTAMETHYLENE OXYDIBUTYRAMIDE	342.48	138	847
POLY-OXYDIPENTAMETHYLENE OXYDICAPROAMIDE	398.59	128	847
POLY-OXYDIPENTAMETHYLENE OXYDIENANTHAMIDE	426.64	129	847

POLYMER	MOLECULAR WEIGHT	MELTING POINT	REF.
POLY-OXYDIPENTAMETHYLENE OXYDIPELARGONAMIDE	482.75	129	847
POLY-OXYDIPENTAMETHYLENE OXYDIVALERAMIDE	370.53	134	847
POLY-4,4'-OXYDIPHENYLENE CARBONATE	228.20	235	540
POLY-4,4'-OXYDIPHENYLENEDITHIODECAMETHYLENE DISULFIDE	436.70	80	837
POLY-4,4'-OXYDIPHENYLENEDITHIOETHYLENE DISULFIDE	324.49	158	837
POLY-4,4'-OXYDIPHENYLENEDITHIOTETRAMETHYLENE DISULFIDE	352.54	94	837
POLY-OXYDITETRAMETHYLENE HEXAMETHYLENEDIURETHAN	330.42	124	633
POLY-OXYDITRIMETHYLENE ADIPAMIDE	242.32	190	291
		213	895
POLY-OXYDITRIMETHYLENE TEREPHTHALAMIDE	262.31	281	895
POLY-OXYMETHYLENE (FORMALDEHYDE)	30.03	178	329
		181	42
		198	217
		200	455
---, CYANOETHYL-	83.09	176	240
1,2-POLY-1,3-PENTADIENE	68.12		
TRANS-			
SYNDIOTACTIC		10	230
1,4-POLY-1,3-PENTADIENE	68.12		
CIS-			
ISOTACTIC		44	263
SYNDIOTACTIC		53	234
TRANS-			
ISOTACTIC		95	189
POLY-PENTAMETHYLENE	70.14		
---, 1,2-DICHLORO-	139.02		
TRANS-		75	925
POLY-PENTAMETHYLENE ADIPAMIDE	212.29	223	339
		225	402
		251	852
		258	818
---, (L)-1-CARBOXY- (DIAMINE)	256.30	113	883
---, (L)-1-CARBOXY- (DIAMINE)	512.60	102	883
---, 2,2,3,3,4,4-HEXAFLUORO- (DIAMINE)			
---, ---, N,N'-DIBUTYL-	432.45	15	402
---, ---, N,N'-DIETHYL-	376.34	20	402
---, ---, N,N'-DIISOPROPYL-	404.40	35	402
---, ---, N,N'-DIMETHYL-	348.29	30	402
POLY-PENTAMETHYLENE ADIPATE	214.26		
---, 2,2,3,3,4,4-HEXAFLUORO- (DIAMINE)	322.20	34	933
POLY-PENTAMETHYLENE AZELAAMIDE	254.37	178	339
		179	291
POLY-PENTAMETHYLENE AZELAATE	256.34	41	696
POLY-PENTAMETHYLENE CARBONATE	130.14	46	820
POLY-PENTAMETHYLENE M-CARBOXYCARBANILATE	249.27	80	891
POLY-PENTAMETHYLENE P-CARBOXYCARBANILATE	249.27	160	891
POLY-PENTAMETHYLENE P-(CARBOXYPHENYLENE)-ACETAMIDE	246.31	216	904
POLY-PENTAMETHYLENE (CYCLOPROPYLENEDICARBOXOYL)-DIURETHAN	284.27		
TRANS-		250	865
POLY-PENTAMETHYLENE CYCLOPROPYLENEDIURETHAN	228.25	170	865
POLY-PENTAMETHYLENE (DECAMETHYLENEDISULFONYL)-DICAPROAMIDE	564.84	202	402
POLY-4,4'-PENTAMETHYLENEDIBENZOIC ANHYDRIDE	294.35	118	854
POLY-3,3'-(PENTAMETHYLENEDIOXY)-DIBENZOIC ANHYDRIDE	326.35	176	854
POLY-4,4'-(PENTAMETHYLENEDIOXY)-DIBENZOIC ANHYDRIDE	326.35	188	854
POLY-(PENTAMETHYLENEDIOXY)-DIDECAMETHYLENE OXIDE	398.67	72	542
POLY-(PENTAMETHYLENEDIOXY)-DIHEXAMETHYLENE OXIDE	286.46	46	542
POLY-PENTAMETHYLENE DISULFIDE	134.26	44	295
POLY-PENTAMETHYLENE DITHIOLISOPHTHALATE	266.38	175	922
POLY-PENTAMETHYLENE DITHIOLTEREPHTHALATE	266.38	232	922
POLY-PENTAMETHYLENE N,N'-ETHYLENEDITEREPHTHALAMATE	424.45	312	886
POLY-PENTAMETHYLENE (ETHYLENEDITHIO)-DIETHYLENE-DIURETHAN	336.46	130	897

POLYMER	MOLECULAR WEIGHT	MELTING POINT	REF.
POLY-PENTAMETHYLENE GLUTARAMIDE	198.27	198	291
POLY-PENTAMETHYLENE (HEXAMETHYLENEDISULFONYL)-DICAPROAMIDE	508.73	210	402
POLY-PENTAMETHYLENE N,N'-HEXAMETHYLENE-DITEREPHTHALAMATE	480.56	255	886
POLY-PENTAMETHYLENE (HEXAMETHYLENEDITHIO)-DIETHYLENEDIURETHAN	392.57	94	897
POLY-PENTAMETHYLENE HEXAMETHYLENEDIURETHAN	272.34	151	633
		157	473
		158	627
		235	402
POLY-PENTAMETHYLENE MALONAMIDE	170.21	191	402
POLY-PENTAMETHYLENE 4,4'-(METHYLENEDIPHENYLENE)-DIURETHAN	354.41	192	473
POLY-PENTAMETHYLENE OCTADECANEDIAMIDE	380.62	167	339
POLY-PENTAMETHYLENE 4-OCTENDIAMIDE	238.33		
TRANS- II.		258	657
POLY-PENTAMETHYLENE OXYDIACETAMIDE	200.24	130	291
POLY-PENTAMETHYLENE 4,4'-OXYDIBENZOATE	326.35	70	939
POLY-PENTAMETHYLENEOXY-P-PHENYLENE OXIDE	178.23	164	868
POLY-PENTAMETHYLENE (PENTAMETHYLENEDISULFONYL)-DICAPROAMIDE	494.70	212	402
POLY-PENTAMETHYLENE N,N'-PENTAMETHYLENE-DITEREPHTHALAMATE	466.53	198	886
POLY-PENTAMETHYLENE (PENTAMETHYLENEDITHIO)-DIETHYLENEDIURETHAN	378.55	87	897
POLY-PENTAMETHYLENE PIMELAMIDE	226.32	183	291
POLY-PENTAMETHYLENE SEBACAMIDE	268.40	195	339
POLY-PENTAMETHYLENE SEBACATE	270.37	51	696
POLY-PENTAMETHYLENE SEBACYLDIUREA	354.45	205	402
POLY-PENTAMETHYLENE SUBERAMIDE	240.35	202	339
POLY-PENTAMETHYLENE SULFIDE	102.20	65	295
POLY-PENTAMETHYLENE SULFONE	134.19	243	538
POLY-PENTAMETHYLENESULFONYLTETRAMETHYLENE SULFONE	254.36	247	538
POLY-PENTAMETHYLENE TEREPHTHALAMIDE	232.28	353	834
POLY-PENTAMETHYLENE TEREPHTHALATE	234.25	134	99
		116	291
POLY-PENTAMETHYLENE TETRADECANEDIAMIDE	324.51	178	339
POLY-PENTAMETHYLENE 2,5-TETRAHYDROFURAN-DIPROPIONAMIDE	282.38	153	402
POLY-PENTAMETHYLENE N,N'-TETRAMETHYLENE-DITEREPHTHALAMATE	452.51	284	886
POLY-PENTAMETHYLENE (TETRAMETHYLENEDITHIO)-DIETHYLENEDIURETHAN	364.52	95	897
POLY-PENTAMETHYLENETHIOTETRAMETHYLENE SULFIDE	190.37	67	181
POLY-2,5-PENTAMETHYLENE-1,3,4-TRIAZOLE	137.19		
---, 1-AMINO-	152.20	260	610
POLY-PENTAMETHYLENE TRIDECANEDIAMIDE	310.48	176	339
POLY-PENTAMETHYLENE N,N'-TRIMETHYLENE-DITEREPHTHALAMATE	438.48	235	886
POLY-PENTAMETHYLENE UNDECAMETHYLENEDIURETHAN	342.48	123	633
POLY-PENTAMETHYLENE UNDECANEDIAMIDE	282.43	176	291
POLY-PENTAMETHYLENE URETHAN	129.16	150	291
		155	444
POLY-1-PENTENE IA.	70.14	130	276
		130	280
		111	501
IIA.		75	48
		80	9
		78	282
		79	501
---, 5-CYCLOHEXYL-	152.28	123	328
---, 5-DIISOBUTYLAMINO-	197.37	112	191
---, 5-DIISOPROPYLAMINO-	169.31	130	191
---, 4,4-DIMETHYL-	98.19	350	90
		231	282
		380	441
---, 3-ETHYL-	98.19	425	632

CRYSTALLOGRAPHIC DATA FOR VARIOUS POLYMERS

POLYMER	MOLECULAR WEIGHT	MELTING POINT	REF.
---, 3-METHYL-	84.16	273	129
		200	441
		362	632
---, 4-METHYL-	84.16	235	48
		228	282
		238	632
		250	94
---, 5,5,5-TRIFLUORO-	124.11	225	813
---, 4-TRIFLUOROMETHYL-	138.13	255	813
---, 5-TRIMETHYLSILYL-	142.32	133	71
POLY-4,4'-(2,2-PENTYLIDENE)-DIPHENYLENE CARBONATE	282.34	220	438
---, 4-METHYL-	296.37	220	438
POLY-4,4'-(3,3-PENTYLIDENE)-DIPHENYLENE CARBONATE	282.34	195	915
---, 2,4-DIMETHYL-	310.39	220	915
POLY-P-PHENYLENE	76.10	<550	735
POLY-M-PHENYLENE ADIPAMIDE	218.26	296	402
		250	868
---, 3-METHYL- (DIAMINE)	232.28	225	402
POLY-O-PHENYLENE ADIPAMIDE	218.26	179	868
POLY-P-PHENYLENE ADIPAMIDE	218.26	262	402
		>340	868
POLY-PHENYLENECYCLOBUTYLENE	130.19		
---, DIPYRIDYL-	284.36	340	723
POLY-P-PHENYLENE CYCLOBUTYLENEDICARBOXAMIDE	216.24	405	723
POLY-P-PHENYLENE CYCLOBUTYLENEDICARBOXYLIC ACID	218.21	290	723
---, DIBUTYL ESTER	330.42		
---, ---, DICYANO-	380.44	330	723
---, DIETHYL ESTER	274.32	347	723
---, ---, DICYANO-	324.34	340	723
---, DIISOPROPYL ESTER	302.37	320	723
---, ---, DICYANO-	352.39	320	723
---, DIMETHYL ESTER	246.26	415	723
---, ---, DICYANO-	296.28	290	723
---, DIPHENYL ESTER	370.40	420	723
---, DIPROPYL ESTER	302.37	360	723
---, ---, DICYANO-	352.39	335	723
POLY-M-PHENYLENE CYCLOPROPYLENEDICARBOXYLATE	204.18		
CIS-		65	860
TRANS-		105	860
POLY-P-PHENYLENE CYCLOPROPYLENEDICARBOXYLATE	204.18		
CIS-		160	860
TRANS-		280	860
POLY-P-PHENYLENEDIACETIC ANHYDRIDE	176.17	92	827
		152	855
POLY-P-PHENYLENEDIETHYLENE ADIPAMIDE	274.36	310	402
POLY-P-PHENYLENEDIETHYLENE AZELAAMIDE	316.44	250	402
		290	454
POLY-P-PHENYLENEDIETHYLENE DOCOSANEDIAMIDE	498.80	230	454
POLY-P-PHENYLENEDIETHYLENE DODECANEDIAMIDE	358.53	280	454
POLY-P-PHENYLENEDIETHYLENE HEPTADECANEDIAMIDE	428.66	249	454
POLY-P-PHENYLENEDIETHYLENE HEXADECANEDIAMIDE	414.63	258	454
POLY-P-PHENYLENEDIETHYLENE OCTADECANEDIAMIDE	442.69	248	454
POLY-P-PHENYLENEDIETHYLENE OCTAMETHYLENEDIURETHAN	362.47	212	402
		212	291
POLY-P-PHENYLENEDIETHYLENE PENTADECANEDIAMIDE	400.61	248	454
POLY-P-PHENYLENEDIETHYLENE M-PHENYLENEDIACETAMIDE	322.41	222	402
POLY-P-PHENYLENEDIETHYLENE SEBACAMIDE	330.47	285	841
		285	402
		300	454
POLY-P-PHENYLENEDIETHYLENE TEREPHTHALATE	296.32	330	881
POLY-P-PHENYLENEDIETHYLENE TETRADECANEDIAMIDE	386.58	267	454
POLY-P-PHENYLENEDIETHYLENE THIODICAPROAMIDE	390.58	251	824
POLY-P-PHENYLENEDIETHYLENE THIODIENANTHAMIDE	418.64	234	824
POLY-P-PHENYLENEDIETHYLENE THIODIPELARGONAMIDE	474.75	222	824
POLY-P-PHENYLENEDIETHYLENE THIODIUNDECANOAMIDE	530.86	207	824
POLY-P-PHENYLENEDIETHYLENE THIODIVALERAMIDE	362.53	252	824
POLY-P-PHENYLENEDIETHYLENE TRIDECANEDIAMIDE	372.55	262	454
POLY-P-PHENYLENEDIETHYLENE UNDECANEDIAMIDE	344.50	275	454

POLYMER	MOLECULAR WEIGHT	MELTING POINT	REF.
POLY-(M-PHENYLENEDIOXY)-DIACETIC ANHYDRIDE	208.17	130	827
POLY-(P-PHENYLENEDIOXY)-DIACETIC ANHYDRIDE	208.17	158	855
		152	827
		160	872
POLY-(P-PHENYLENEDIOXY)-DIACETIC 3,4-PYRIDINE- DICARBOXYLIC ANHYDRIDE	357.27	312	872
POLY-(P-PHENYLENEDIOXY)-DIACETIC TEREPHTHALIC ANHYDRIDE	356.29	324	871
POLY-(P-PHENYLENEDIOXY)-DIACETIC 4,4'-(TRIMETHYLENEDIOXY)-DIBENZOIC ANHYDRIDE	474.46	260	872
POLY-(P-PHENYLENEDIOXY)-DIETHYLENE OCTAMETHYLENE- DIURETHAN	394.47	212	633
		212	402
POLY-4,4'-(P-PHENYLENEDIOXY)-DIPHENYLENE CARBONATE	320.30	215	915
POLY-P-PHENYLENEDIPENTAMETHYLENE TEREPHTHALATE	380.48	116	881
POLY-P-PHENYLENEDIPROPIONIC ANHYDRIDE	204.22	92	827
POLY-P-PHENYLENE DISILOXANYLENE	152.30		
---, TETRAMETHYL- (SI)	208.41	148	225
POLY-M-PHENYLENE DISILOXANYLENEDIPROPIONAMIDE	294.46		
---, 4-METHYL- (DIAMINE)			
---, ---, DIETHYL-DIMETHYL- (SI)	392.65	50	402
---, ---, TETRAETHYL- (SI)	420.70	90	402
---, ---, TETRAMETHYL-	364.59	70	402
---, TETRAETHYL- (SI)	406.67	70	402
---, TETRAMETHYL- (SI)	350.57	90	402
POLY-O-PHENYLENE DISILOXANYLENEDIPROPIONAMIDE	294.46		
---, DIETHYL-DIMETHYL- (SI)	378.62	40	402
---, TETRAETHYL- (SI)	406.67	40	402
---, TETRAMETHYL- (SI)	350.57	40	402
POLY-P-PHENYLENE DISILOXANYLENEDIPROPIONAMIDE	294.46		
---, DIETHYL-DIMETHYL- (SI)	378.62	300	402
---, TETRAETHYL- (SI)	406.67	290	402
---, TETRAMETHYL- (SI)	350.57	300	402
POLY-P-PHENYLENEDITETRAMETHYLENE TEREPHTHALATE	352.43	217	881
POLY-4,4'-(2,2'-M-PHENYLENEDITHIAZOLE)-OXY- P-DIPHENYLENE	410.51	340	810
POLY-(P-PHENYLENEDITHIO)-DIPROPIONIC ANHYDRIDE	268.34	55	827
		50	857
POLY-P-PHENYLENEDITRIMETHYLENE HEXAMETHYLENE- DIURETHAN	362.47	158	633
		240	291
POLY-P-PHENYLENEDITRIMETHYLENE TEREPHTHALATE	324.38	162	881
POLY-M-PHENYLENE HEXAMETHYLENEDIURETHAN	278.31	150	402
POLY-P-PHENYLENE HEXAMETHYLENEDIURETHAN	278.31	150	402
POLY-M-PHENYLENE ISOPHTHALAMIDE	238.25	390	748
		375	735
POLY-P-PHENYLENE ISOPHTHALAMIDE	238.25	500	748
POLY-M-PHENYLENE ISOPHTHALATE	240.21	245	837
POLY-P-PHENYLENE MALONATE	178.14	233	868
POLY-P-PHENYLENE OXIDE	92.10	298	487
---, 2,6-DIMETHYL-	120.15	261	226
		262	484
		272	471
---, 2,6-DIPHENYL-	244.29	484	201
POLY-P-PHENYLENEOXYTRIMETHYLENE OXIDE	150.18	166	868
POLY-M-PHENYLENE 3,3'-PHOSPHINYLIDENEDIBENZAMIDE	362.32		
---, METHYL- (P)	376.35	260	402
---, ---, 4-METHYL- (DIAMINE)	390.38	244	402
POLY-M-PHENYLENE 4,4'-PHOSPHINYLIDENEDIBENZAMIDE	362.32		
---, METHYL- (P)	376.35	340	402
---, ---, 4-METHYL- (DIAMINE)	390.38	252	402
---, PHENYL- (P)	438.42	340	402
---, ---, 4-METHYL- (DIAMINE)	452.45	280	402
POLY-O-PHENYLENE 3,3'-PHOSPHINYLIDENEDIBENZAMIDE	362.32		
---, METHYL- (P)	376.35	181	402
POLY-O-PHENYLENE 4,4'-PHOSPHINYLIDENEDIBENZAMIDE	362.32		
---, METHYL- (P)	376.35	193	402
---, PHENYL- (P)	438.42	207	402

POLYMER	MOLECULAR WEIGHT	MELTING POINT	REF.
POLY-P-PHENYLENE 3,3'-PHOSPHINYLIDENEDIBENZAMIDE	362.32		
---, METHYL- (P)	376.35	264	402
POLY-P-PHENYLENE 4,4'-PHOSPHINYLIDENEDIBENZAMIDE	362.32		
---, METHYL- (P)	376.35	340	402
---, PHENYL- (P)	438.42	340	402
POLY-P-PHENYLENE 1,4-PIPERAZINEDIPROPIONATE	304.35	189	875
---, METHYL- (DIACID)	318.37	164	875
POLY-M-PHENYLENE SEBACAMIDE	274.36	205	868
		256	402
---, 3-METHYL- (DIAMINE)	288.39	200	402
POLY-O-PHENYLENE SEBACAMIDE	274.36	125	868
POLY-P-PHENYLENE SEBACAMIDE	274.36	145	402
		325	868
POLY-M-PHENYLENE SUBERAMIDE	246.31	196	868
POLY-O-PHENYLENE SUBERAMIDE	246.31	150	868
POLY-P-PHENYLENE SUBERAMIDE	246.31	>340	868
POLY-P-PHENYLENE SUCCINATE	192.17	300	841
POLY-P-PHENYLENE SULFIDE	108.16	290	679
		295	677
---, 2-METHYL-	122.18	100	679
POLY-M-PHENYLENE TEREPHTHALAMIDE	238.25	500	748
POLY-P-PHENYLENE TEREPHTHALAMIDE	238.25	500	748
		600	735
POLY-M-PHENYLENE UREA	134.14		
---, 4-METHYL-	148.16	284	402
POLY-PHENYLISOCYANATE	119.12	275	528
POLY-PHOSPHAZENE	47.00		
---, BIS-BETA-NAPHTHOXY-	331.31	>350	553
---, BIS-PHENOXY-	231.19		
---, ---, P-CHLORO-	300.08	405	553
---, ---, 2,4-DICHLORO-	368.97	210	553
---, ---, P-PHENYL-	383.39	398	553
---, ---, M-TRIFLUOROMETHYL-	367.19	330	553
---, BIS-2,2,2-TRIFLUOROETHOXY-	243.04	240	553
POLY-PHOSPHINIDENEDITRIMETHYLENE ADIPAMIDE	258.30		
---, OCTYL-	370.52	135	859
POLY-PIMELIC ANHYDRIDE	142.15	55	823
POLY-PIPERAZINE ADIPAMIDE	196.25	355	386
---, 3-METHYL- (DIACID)	210.28	300	402
POLY-PIPERAZINE AZELAAMIDE	238.33	148	402
POLY-PIPERAZINE CYCLOPROPYLENEDICARBOXAMIDE	180.21		
TRANS-		330	860
---, 1-METHYL- (DIACID)	194.23		
TRANS-		130	860
---, 3-METHYL- (DIACID)	194.23		
TRANS-		280	860
POLY-1,4-PIPERAZINEDIETHYLENE HEXAMETHYLENE- DIURETHAN	342.44	165	291
POLY-1,4-PIPERAZINEDIETHYLENE SULFONE	204.29	200	875
POLY-PIPERAZINE (ETHYLENEDIIMINO)-DIPROPIONAMIDE			
---, N,N'-DIISOPROPYL-	338.50	97	875
---, N,N'-DIMETHYL-	282.39	113	875
POLY-PIPERAZINE (ETHYLENEDIOXY)-DIACETAMIDE	228.25	165	894
POLY-PIPERAZINE ISOPHTHALAMIDE	216.24	340	833
---, 2,5-DIMETHYL- (DIAMINE)	244.29	>350	833
---, METHYL- (DIAMINE)	230.27	280	833
POLY-PIPERAZINE 4-OCTENDIAMIDE	222.29		
TRANS-		252	657
POLY-PIPERAZINE OXYDIACETAMIDE	184.20	258	894
POLY-PIPERAZINE PHTHALAMIDE	216.24	325	833
---, 2,5-DIMETHYL- (DIAMINE)	244.29	350	833
---, METHYL- (DIAMINE)	230.27	350	833
POLY-PIPERAZINE 1,4-PIPERAZINEDIPROPIONAMIDE	280.37	D270	875
---, METHYL- (DIACID)	294.40	218	875
---, METHYL- (DIAMINE)	294.40	217	875
POLY-PIPERAZINE SEBACAMIDE	252.36	180	81
POLY-PIPERAZINE SUBERAMIDE	224.30	300	402
POLY-PIPERAZINE TEREPHTHALAMIDE	216.24	350	833
		380	402

POLYMER	MOLECULAR WEIGHT	MELTING POINT	REF.
---, 2,5-DIMETHYL- (DIAMINE)	244.29	350	833
---, METHYL- (DIAMINE)	230.27	350	833
POLY-PIPERAZINE (TRIMETHYLENEDITHIO)-DIPROPIONAMIDE	302.45	100	402
POLY-1,4-PIPERIDINEDIYLTRIMETHYLENE	125.22		
---, 2,6-DIOXO-	153.18	281	481
POLY-1,4-PIPERIDENE URETHAN	127.14	270	444
POLY-PROPENE	42.08		
ISOTACTIC		176	10
		180	250
		165	282
		178	349
		183	380
		186	394
		189	392
		208	676
SYNDIOTACTIC		161	415
		138	545
---, 3-CYCLOHEXYL-	124.23	230	90
		215	282
		214	328
---, 3-CYCLOPENTYL-	110.20	225	90
		210	328
---, HEXAFLUORO-	150.02	160	268
---, 3-PHENYL- (ALLYLBENZENE)	118.18	230	90
		208	187
		185	282
---, ---, 2,5-DIMETHYL-	146.23	338	187
---, ---, 3,4-DIMETHYL-	146.23	275	187
---, ---, 3,5-DIMETHYL-	146.23	252	187
---, 3-SILYL-	72.18	128	71
---, 3-M-TOLYL-	132.21	180	187
---, 3-O-TOLYL-	132.21	290	187
---, 3-P-TOLYL-	132.21	240	187
---, 3-TRIMETHYLSILYL-	114.26	360	71
POLY-PROPIONALDEHYDE	58.08	185	329
---, 3-CARBOMETHOXY-	116.12	150	704
POLY-PROPYLENE OXALATE	130.10	180	305
POLY-PROPYLENE OXIDE	58.08	75	18
		75	377
		72	664
		73	285
---, 2-CHLOROMETHYL-	106.55	126	816
		126	593
---, 3-(1-NAPHTHOXY)-	200.24	235	445
---, 3-(2-NAPHTHOXY)-	200.24	297	445
---, 3-PHENOXY-	150.18	215	238
		210	297
		208	318
		203	527
---, ---, O-CHLORO-	184.62	200	445
---, ---, P-CHLORO-	184.62	176	445
---, ---, O-ISOPROPYL-	192.26	128	445
---, ---, M-METHYL-	164.20	169	445
---, ---, O-METHYL-	164.20	191	445
---, ---, P-METHYL-	164.20	212	445
---, ---, O-PHENYL-	226.28	293	445
---, ---, 2,4,6-TRICHLORO-	253.51	130	445
POLY-PROPYLENE 4,4'-SULFONYLDIBENZAMIDE	344.38	335	835
POLY-PROPYLIDENE	42.08	90	617
POLY-PROPYLVINYL ETHER	86.13	76	114
---, 1-METHYL-	100.16	177	608
		170	382
POLY-2,5-PYRROLEDIPROPIONIC ANHYDRIDE	193.20		
---, N-METHYL-	207.23	188	827
POLY-SEBACIC ANHYDRIDE	184.24	80	826
		82	822
POLY-SEBACYL DITHIONISOPHTHALOYLDIHYDRAZIDE	392.54	294	889

POLYMER	MOLECULAR WEIGHT	MELTING POINT	REF.
POLY-SELENIUM	78.96	219	752
POLY-SILOXANE	46.10		
---, DIETHYL-	102.21	17	518
---, DIMETHYL-	74.16	-40	517
---, DIPROPYL-	130.26	74	518
POLY-4,4'-(SILYLENEDIPHENYLENE)-DIMETHYLENE 4,4'-DISILOXANYLENEDIBENZAMIDE	524.80		
---, DIMETHYL- (DIAMINE), TETRAMETHYL- (SI)	608.96	165	402
POLY-4,4'-(SILYLENEDIPHENYLENE)-DIMETHYLENE 4,4'-SILYLENEDIBENZAMIDE	478.70		
---, DIMETHYL- (DIAMINE), DIMETHYL- (SI)	550.81	215	402
POLY-4,4'-(SILYLENEDIPHENYLENE)-DIMETHYLENE 4,4'-TETRASILOXANYLENEDIBENZAMIDE	617.00		
---, DIMETHYL- (DIAMINE)			
---, ---, TETRAMETHYL-TETRAPHENYL- (SI)	1001.52	150	402
POLY-STYRENE	104.15	240	10
		250	90
		235	282
---, P-TERT-BUTYL-	160.26	300	198
---, 2,4-DIMETHYL-	132.21	310	75
		350	74
---, 2,5-DIMETHYL-	132.21	330	75
		340	74
---, 3,4-DIMETHYL-	132.21	240	75
---, 3,5-DIMETHYL-	132.21	290	75
---, M-FLUORO-	122.14	250	910
---, O-FLUORO-	122.14	270	75
---, P-FLUORO-	122.14	265	75
---, O-METHYL-	118.18	>360	74
---, ---, P-FLUORO-	136.17	360	75
---, M-METHYL-	118.18	215	74
		215	80
---, BETA-NITRO-	149.15	285	937
---, P-TRIMETHYLSILYL-	176.34	284	103
POLY-STYRENE OXIDE	120.15	149	809
		140	318
		162	693
POLY-STYRENE SULFONE	168.21	215	916
POLY-STYRYL PYRIDINE-ALT-VINYL PYRIDINE	206.30	350	920
POLY-SUBERIC ANHYDRIDE	156.18	66	823
POLY-4,4'-SULFINYLDIPHENYLENE CARBONATE	260.26	250	540
POLY-4,4'-SULFONYLDIPHENYLENE CARBONATE	276.26	210	540
POLY-SULFONYLDIPROPIONIC ANHYDRIDE	192.18	237	856
POLY-SULFUR TRIOXIDE	80.06	32	288
POLY-TEREPHTHALALDEHYDE	134.13	120	801
---, 2,5-DIMETHYL-	162.19	140	801
POLY-TEREPHTHALIC ANHYDRIDE	148.12	410	826
		400	827
POLY-TEREPHTHALIC 2,2'-(TEREPHTHALOYLDIAMINO)-DIBENZOIC ANHYDRIDE	534.48	312	871
POLY-TEREPHTHALIC 4,4'-(TETRAMETHYLENEDIOXY)-DIBENZOIC ANHYDRIDE	460.44	215	871
POLY-TEREPHTHALIC 4,4'-(TRIMETHYLENEDIOXY)-DIBENZOIC ANHYDRIDE	446.41	285	871
---, 3,3'-DIMETHOXY-	506.46	255	871
POLY-TEREPHTHALOYL (M-CARBOXYPHENOXYACETYL)-DIHYDRAZIDE	354.32	218	833
POLY-TEREPHTHALOYL DITHIONISOPHTHALOYLDIHYDRAZIDE	356.42	309	889
POLY-TETRADECAMETHYLENE DOCOSANEDIAMIDE	562.97	153	454
POLY-TETRADECAMETHYLENE DODECANEDIAMIDE	422.70	175	454
POLY-TETRADECAMETHYLENE N,N'-ETHYLENE-DITEREPHTHALAMATE	550.70	249	886
POLY-TETRADECAMETHYLENE N,N'-HEXAMETHYLENE-DITEREPHTHALAMATE	606.80	222	866
POLY-TETRADECAMETHYLENE OCTADECANEDIAMIDE	506.36	158	454
POLY-TETRADECAMETHYLENE N,N'-PENTAMETHYLENE-DITEREPHTHALAMATE	592.78	200	866
POLY-TETRADECAMETHYLENE P-PHENYLENEDIACETAMIDE	386.58	235	454

POLYMER	MOLECULAR WEIGHT	MELTING POINT	REF.
POLY-TETRADECAMETHYLENE P-PHENYLENEDIPROPIONAMIDE	414.63	246	454
POLY-TETRADECAMETHYLENE SEBACAMIDE	394.64	175	454
POLY-TETRADECAMETHYLENE TEREPHTHALAMIDE	358.53	265	454
POLY-TETRADECAMETHYLENE N,N'-TETRAMETHYLENE-DITEREPHTHALAMATE	578.75	248	886
POLY-TETRADECAMETHYLENE N,N'-TRIMETHYLENE-DITEREPHTHALAMATE	564.72	208	886
POLY-TETRADECAMETHYLENE P-XYLYLENEDIUREA	416.61	248	634
POLY-TETRADECANEDIOIC ANHYDRIDE	240.34	91	823
POLY-1-TETRADECENE	196.38	57	250
POLY-TETRAHYDROFURAN (TETRAMETHYLENE OXIDE)	72.11	37	180
		35	348
		60	366
		43	460
POLY-2,5-TETRAHYDROFURANDIPROPIONIC ANHYDRIDE	198.22	135	827
POLY-2,5-TETRAHYDROPYRROLEDIPROPIONIC ANHYDRIDE			
---, N-METHYL-	211.26	103	827
POLY-TETRAMETHYLENE ADIPAMIDE	198.27	295	291
		308	51
POLY-TETRAMETHYLENE ADIPATE	200.23	48	325
		54	539
POLY-TETRAMETHYLENE AZELAAMIDE	240.35	223	339
		253	402
POLY-TETRAMETHYLENE AZELAATE	242.32	37	325
		39	696
POLY-TETRAMETHYLENE CARBONATE	116.12	59	820
POLY-TETRAMETHYLENE M-CARBOXYCARBANILATE	235.24	60	891
POLY-TETRAMETHYLENE P-CARBOXYCARBANILATE	235.24	186	891
POLY-TETRAMETHYLENE P-(CARBOXYPHENYLENE)-ACETAMIDE	232.28	221	904
POLY-TETRAMETHYLENE 1,4-CYCLOHEXYLENEDIURETHAN	256.30	260	633
POLY-TETRAMETHYLENE (CYCLOPROPYLENEDICARBOXOYL)-DIURETHAN	270.24	226	865
POLY-TETRAMETHYLENE CYCLOPROPYLENEDIURETHAN	214.22	180	865
POLY-TETRAMETHYLENE (DECAMETHYLENEDISULFONYL)-DICAPROAMIDE	550.81	219	402
POLY-TETRAMETHYLENE 4,4'-DIBENZOATE	296.32	280	903
POLY-4,4'-TETRAMETHYLENEDIBENZOIC ANHYDRIDE	280.32	263	854
POLY-3,3'-(TETRAMETHYLENEDIOXY)-DIBENZOIC ANHYDRIDE	312.32	199	854
POLY-4,4'-(TETRAMETHYLENEDIOXY)-DIBENZOIC ANHYDRIDE	312.32	204	854
---, 3,3'-DIMETHOXY-	372.37	172	855
POLY-(TETRAMETHYLENEDIOXY)-DIDECAMETHYLENE OXIDE	384.64	73	542
POLY-(TETRAMETHYLENEDIOXY)-DIHEXAMETHYLENE OXIDE	272.43	50	542
POLY-(TETRAMETHYLENEDIOXY)-DITRIMETHYLENE OXAMIDE	258.32	160	291
		167	402
POLY-TETRAMETHYLENE DISILOXANYLENEDIPROPIONAMIDE	274.47		
---, TETRAMETHYL- (SI)	358.63	120	402
---, ---, N,N'-DIMETHYL-	120.23	39	295
POLY-TETRAMETHYLENE DISULFIDE			
POLY-4,4'-(2,2'-TETRAMETHYLENEDITHIAZOLE)-P-PHENYLENE	298.42	250	810
POLY-4,4'-(TETRAMETHYLENEDITHIO)-DIBENZOIC ANHYDRIDE	344.44	206	854
POLY-TETRAMETHYLENE DITHIODICARBOXYLATE	208.25	115	936
POLY-TETRAMETHYLENE DITHIOLADIPATE	232.36	128	922
POLY-TETRAMETHYLENE DITHIOLISOPHTHALATE	252.35	174	922
POLY-TETRAMETHYLENE DITHIOLSEBACATE	288.47	98	922
POLY-TETRAMETHYLENE DITHIOLTEREPHTHALATE	252.35	310	922
POLY-TETRAMETHYLENE DITHIONISOPHTHALATE	252.35	118	889
POLY-TETRAMETHYLENE DODECANEDIAMIDE	282.43	245	153
POLY-TETRAMETHYLENE 4,4'-(ETHYLENEDIOXY)DIBENZOATE	356.37	180	290
POLY-TETRAMETHYLENE (ETHYLENEDITHIO)-DIETHYLENE-DIURETHAN	322.44	177	897
POLY-TETRAMETHYLENE (HEXAMETHYLENEDISULFONYL)-DICAPROAMIDE	494.70	228	402
POLY-TETRAMETHYLENE N,N'-HEXAMETHYLENE-DITEREPHTHALAMATE	466.53	300	886

POLYMER	MOLECULAR WEIGHT	MELTING POINT	REF.
POLY-TETRAMETHYLENE (HEXAMETHYLENEDITHIO)-			
DIETHYLENEDIURETHAN	378.55	117	897
POLY-TETRAMETHYLENE HEXAMETHYLENEDIURETHAN	258.32	180	291
		173	101
		182	473
		184	402
		184	627
---, 1,4-DIMETHYL- (DIOL)	286.37	104	633
POLY-TETRAMETHYLENE ISOPHTHALATE	220.22	152	111
POLY-TETRAMETHYLENE (4,4'-METHYLENEDIPHENYLENE)-			
DIAMINODITHIODICARBOXYLATE	404.50	104	934
POLY-TETRAMETHYLENE 4,4'-(METHYLENEDIPHENYLENE)-			
DIURETHAN	340.38	194	634
		248	473
POLY-TETRAMETHYLENE NONAMETHYLENEDIURETHAN	300.40	140	633
POLY-TETRAMETHYLENE OCTAMETHYLENEDIURETHAN	286.37	160	633
POLY-TETRAMETHYLENE 4-OCTENDIAMIDE	224.30		
TRANS-		294	657
POLY-TETRAMETHYLENE OXIDE, SEE			
POLY-TETRAHYDROFURAN			
POLY-TETRAMETHYLENE 4,4'-OXYDIBENZOATE	312.32	88	939
POLY-TETRAMETHYLENE (PENTAMETHYLENEDISULFONYL)-			
DICAPROAMIDE	480.68	232	402
POLY-TETRAMETHYLENE (PENTAMETHYLENEDITHIO)-			
DIETHYLENEDIURETHAN	364.52	113	897
POLY-TETRAMETHYLENE PENTAMETHYLENEDIURETHAN	244.29	159	633
POLY-TETRAMETHYLENE P-PHENYLENEDIACETATE	248.28	63	325
POLY-TETRAMETHYLENE 4,4'-PHOSPHINYLIDENE-			
DIBENZAMIDE			
---, METHYL- (P)	356.36	208	402
---, PHENYL- (P)	418.43	230	402
POLY-TETRAMETHYLENE PIMELAMIDE	212.29	233	339
		238	402
POLY-TETRAMETHYLENE (1,4-PIPERAZINEDITHIO)-			
DICARBOXYLATE	294.38	D188	934
---, 2,5-DIMETHYL-	322.44	113	934
POLY-TETRAMETHYLENE 4,4'-PROPYLENEDIBENZAMIDE	336.44	180	402
POLY-TETRAMETHYLENE 4,4'-(PROPYLENEDIOXY)-			
DIBENZAMIDE	368.43	140	402
POLY-TETRAMETHYLENE SEBACAMIDE	254.33	254	153
		239	291
POLY-TETRAMETHYLENE SEBACATE	256.34	60	325
		62	425
POLY-TETRAMETHYLENE SUBERAMIDE	226.32	250	291
POLY-TETRAMETHYLENE SUBERATE	228.29	56	291
POLY-TETRAMETHYLENE SUCCINAMIDE	170.21	287	402
---, N,N'-DIMETHYL-	198.27	123	402
POLY-TETRAMETHYLENE SUCCINATE	172.18	34	696
---, D-1,2-DIMETHOXY- (DIACID)	232.23	92	609
POLY-TETRAMETHYLENE SULFIDE	88.17	67	181
POLY-TETRAMETHYLENE SULFONE	120.17	100	845
POLY-TETRAMETHYLENE 4,4'-SULFONYLDIBENZAMIDE	358.41	358	835
POLY-TETRAMETHYLENE TEREPHTHALAMIDE	218.26	436	834
---, N,N'-DIMETHYL-	246.31	272	834
POLY-TETRAMETHYLENE TEREPHTHALATE	220.22	232	99
		221	453
POLY-TETRAMETHYLENE 2,5-TETRAHYDROFURAN-			
DIPROPIONAMIDE	268.36	210	402
POLY-TETRAMETHYLENE (TETRAMETHYLENEDITHIO)-			
DIETHYLENEDIURETHAN	350.49	119	897
POLY-TETRAMETHYLENE TETRAMETHYLENEDIURETHAN	230.26	193	291
		193	633
POLY-TETRAMETHYLENE TETRATHIODICARBOXYLATE	272.37	92	936
POLY-TETRAMETHYLENE THIODICARBOXYLATE	176.19	90	936
POLY-TETRAMETHYLENE THIODITRIMETHYLENEDIURETHAN	290.38	133	633
POLY-TETRAMETHYLENETHIOTRIMETHYLENE SULFIDE	162.32	64	181
POLY-TETRAMETHYLENE THIOUREA	130.21	213	402
POLY-2,5-TETRAMETHYLENE-1,3,4-TRIAZOLE	123.16		

POLYMER	MOLECULAR WEIGHT	MELTING POINT	REF.
---, 1-AMINO-	138.17	300	610
POLY-TETRAMETHYLENE 1,1'-(4,4'-TRIMETHYLENE-DIPIPERIDINE)-DITHIODICARBOXYLATE	418.61	80	934
POLY-TETRAMETHYLENE N,N'-TRIMETHYLENE-DITEREPHTHALAMATE	424.45	193	886
POLY-TETRAMETHYLENE TRITHIODICARBOXYLATE	240.31	66	936
POLY-TETRAMETHYLENE UNDECAMETHYLENEDIURETHAN	328.45	146	866
POLY-TETRAMETHYLENE UNDECANEDIAMIDE	268.40	208	339
		208	402
POLY-TETRAMETHYLENE UREA	114.15	400	402
POLY-TETRAMETHYLENE P-XYLYLENEDIURETHAN	278.31	227	634
POLY-TETRAMETHYL-P-SILPHENYLENE SILOXANE, SEE POLY-P-PHENYLENE DISILOXANYLENE, ---TETRAMETHYL-			
POLY-THIODIETHYLENE DITHIOLADIPATE	264.43	70	922
POLY-THIODIETHYLENE DITHIOLTEREPHTHALATE	284.42	146	922
POLY-THIODIETHYLENE HEXAMETHYLENEDIURETHAN	290.38	134	633
POLY-THIODIPHENYLENE CARBONATE	244.26	220	435
		240	540
POLY-4,4'-THIODIPHENYLENE OXIDE	200.26	340	745
POLY-THIODIPROPIONIC ANHYDRIDE	160.19	55	857
POLY-THIODITETRAMETHYLENE HEXAMETHYLENEDIURETHAN	346.49	125	633
POLY-2,5-THIOPHENEDIPROPIONIC ANHYDRIDE	210.25	78	827
POLY-TRIDECAMETHYLENE N,N'-ETHYLENE-DITEREPHTHALAMATE	536.67	257	886
POLY-TRIDECAMETHYLENE N,N'-HEXAMETHYLENE-DITEREPHTHALAMATE	592.78	222	886
POLY-TRIDECAMETHYLENE N,N'-PENTAMETHYLENE-DITEREPHTHALAMATE	578.75	194	886
POLY-TRIDECAMETHYLENE N,N'-TETRAMETHYLENE-DITEREPHTHALAMATE	564.72	248	886
POLY-TRIDECAMETHYLENE TRIDECANEDIAMINE	422.70	172	899
POLY-TRIDECAMETHYLENE N,N'-TRIMETHYLENE-DITEREPHTHALAMATE	550.70	206	886
POLY-TRIDECANEDIOIC ANHYDRIDE	226.32	78	823
POLY-TRIMETHYLENE ADIPATE	186.21	45	300
		38	107
		46	609
---, 2,2-BISBROMOMETHYL- (DIOL)	372.05	120	874
---, 2,2-BISCHLOROMETHYL- (DIOL)	283.15	108	874
---, 2-BROMOMETHYL-2-CHLOROMETHYL- (DIOL)	327.60	111	874
---, 2,2-DIMETHYL- (DIOL)	214.26	37	609
POLY-TRIMETHYLENE ADIPYLDIURETHAN	272.26		
---, 2-HYDROXY- (DIOL)	288.26	208	402
POLY-TRIMETHYLENE AZELAATE	228.29	50	107
		60	609
---, 2,2-DIMETHYL- (DIOL)	256.34	0	609
POLY-TRIMETHYLENE P-(CARBOXYPHENYLENE)-ACETAMIDE	218.26	281	904
POLY-TRIMETHYLENE 1,4-CYCLOHEXYLENEDICARBOXYLATE TRANS-	212.24	110	918
POLY-TRIMETHYLENE CYCLOPROPYLENEDICARBOXAMIDE TRANS-	168.20	310	860
POLY-TRIMETHYLENE CYCLOPROPYLENEDIURETHAN	200.19	170	865
POLY-4,4'-TRIMETHYLENEDIBENZOIC ANHYDRIDE	266.30	215	854
POLY-3,3'-(TRIMETHYLENEDIOXY)-DIBENZOIC ANHYDRIDE	298.29	197	854
POLY-4,4'-(TRIMETHYLENEDIOXY)-DIBENZOIC ANHYDRIDE	298.29	267	854
		263	872
---, 3,3'-DIMETHOXY-	358.35	175	855
POLY-TRIMETHYLENE DISULFIDE	106.20	67	295
		84	725
POLY-TRIMETHYLENE DODECANEDIOATE	270.37	61	107
POLY-TRIMETHYLENE (ETHYLENEDITHIO)-DIETHYLENE-DIURETHAN	308.41	157	897
POLY-TRIMETHYLENE GLUTARATE	172.18	39	107
		53	609
POLY-TRIMETHYLENE (HEXAMETHYLENEDITHIO)-DIETHYLENEDIURETHAN	364.52	115	897
POLY-TRIMETHYLENE HEXAMETHYLENEDIURETHAN	244.29	167	633
		163	473
		168	627

CRYSTALLOGRAPHIC DATA FOR VARIOUS POLYMERS

POLYMER	MOLECULAR WEIGHT	MELTING POINT	REF.
---, 2-BUTYL-2-ETHYL- (DIOL)	314.43	60	402
---, 2,2-DIMETHYL- (DIOL)	272.35	120	402
---, 2-METHYL- (DIOL)	258.32	50	402
POLY-TRIMETHYLENE ISOPHTHALATE	206.20	132	111
POLY-TRIMETHYLENE MALONATE	144.13	33	609
---, 2,2-DIMETHYL- (DIOL)	172.18	67	609
POLY-TRIMETHYLENE 4,4'-(METHYLENEDIPHENYLENE)- DIURETHAN	326.35	241	473
---, 2-BUTYL-2-ETHYL- (DIOL)	410.51	120	402
---, 2,2-DIETHYL- (DIOL)	382.46	150	402
---, 1-ETHYL-2-PROPYL- (DIOL)	396.49	140	402
POLY-TRIMETHYLENE OCTADECANEDIOATE	354.53	76	307
		76	107
POLY-TRIMETHYLENE OCTAMETHYLENEDIURETHAN	272.35		
---, 1-METHYL- (DIOL)	286.37	82	633
		82	402
POLY-TRIMETHYLENE 4-OCTENDIAMIDE	210.28		
TRANS-		271	657
POLY-TRIMETHYLENE OXALATE	130.10	86	678
		88	305
		89	609
---, 2,2-BISCHLOROMETHYL-	227.04	122	874
---, 2,2-DIMETHYL-	158.15	111	609
POLY-TRIMETHYLENE OXIDE, SEE POLY-OXACYCLOBUTANE			
POLY-TRIMETHYLENE-P-OXYBENZOATE	178.19	185	841
		211	327
POLY-TRIMETHYLENE 4,4'-OXYDIBENZOATE	298.30	90	939
POLY-TRIMETHYLENE (PENTAMETHYLENEDITHIO)- DIETHYLENEDIURETHAN	350.49	112	897
POLY-TRIMETHYLENE P-PHENYLENEDIACETATE	234.25	58	918
POLY-TRIMETHYLENE PIMELATE	200.23	37	107
		51	609
POLY-TRIMETHYLENE SEBACATE	242.32	56	305
		53	107
		58	609
---, 2,2-DIMETHYL- (DIOL)	270.37	26	609
POLY-TRIMETHYLENE SEBACYLDIURETHAN	328.36		
---, 2-HYDROXY- (DIOL)	344.36	189	402
POLY-TRIMETHYLENE SUBERATE	214.26	41	107
		52	609
---, 2,2-DIMETHYL- (DIOL)	242.32	17	609
POLY-TRIMETHYLENE SUCCINATE	158.15	47	107
		52	300
---, 2,2-BISCHLOROMETHYL- (DIOL)	255.10	74	874
---, 2,2-DIMETHYL- (DIOL)	186.21	86	609
POLY-TRIMETHYLENE SULFIDE	74.14	100	181
---, 2,2-DIMETHYL-	102.20	140	917
POLY-TRIMETHYLENE SULFONE	106.14	300	890
---, 2,2-DIETHYL-	162.25	364	890
---, 2,2-DIMETHYL-	134.19	266	890
		303	917
---, 2,2-DIPENTYL-	246.41	385	890
POLY-TRIMETHYLENE 4,4'-SULFONYLDIBENZAMIDE	344.38	298	835
---, 2,2-DIMETHYL- (DIAMINE)	372.44	284	835
---, 1-METHYL- (DIAMINE)	358.41	272	835
POLY-TRIMETHYLENE TEREPHTHALAMIDE	204.23	399	834
---, N,N'-DIMETHYL- (DIAMINE)	232.28	220	834
POLY-TRIMETHYLENE TEREPHTHALATE	206.20	233	99
		227	453
		221	291
---, 2,2-DIMETHYL-	234.25	140	291
POLY-TRIMETHYLENE (TETRAMETHYLENEDITHIO)- DIETHYLENEDIURETHAN	336.46	113	897
POLY-TRIMETHYLENE N,N'-TRIMETHYLENE- DITEREPHTHALAMATE	410.43	235	886
POLY-TRIMETHYLENE UNDECANEDIOATE	256.34	59	107
POLY-UNDECAMETHYLENE ADIPAMIDE	296.46	218	544
POLY-UNDECAMETHYLENE N,N'-ETHYLENE- DITEREPHTHALAMATE	508.61	251	886

POLYMER	MOLECULAR WEIGHT	MELTING POINT	REF.
POLY-UNDECAMETHYLENE N,N'-HEXAMETHYLENE- DITEREPHTHALAMATE	564.72	217	886
POLY-UNDECAMETHYLENE N,N'-PENTAMETHYLENE- DITEREPHTHALAMATE	550.70	199	886
POLY-UNDECAMETHYLENE SEBACAMIDE	352.56	169	339
POLY-UNDECAMETHYLENE TEREPHTHALAMIDE	316.44	292	544
POLY-UNDECAMETHYLENE N,N'-TETRAMETHYLENE- DITEREPHTHALAMATE	536.67	248	886
POLY-UNDECAMETHYLENE N,N'-TRIMETHYLENE- DITEREPHTHALAMATE	522.64	198	886
POLY-UNDECANEDIOIC ANHYDRIDE	198.26	70	823
POLY-VALERALDEHYDE	86.13	155	329
POLY-VINYL ALCOHOL	44.05		
ISOTACTIC		212	450
SYNDIOTACTIC		228	433
		232	310
		243	480
		265	323
		267	450
---, 2-METHYL-	58.08	140	885
POLY-VINYL BENZOATE	148.16		
---, P-TERT-BUTYL-	204.27	327	821
POLY-VINYL-TERT-BUTYL KETONE	112.17	150	812
		>240	843
POLY-N-VINYL CARBAZOLE	193.25	>320	228
POLY-VINYL CHLORIDE	62.50	212	143
		273	372
		310	423
POLY-VINYLCYCLOBUTANE	82.15	228	437
POLY-VINYLCYCLOHEPTANE	124.23	>300	437
POLY-VINYLCYCLOHEXANE	110.20	305	90
		300	282
		383	328
		372	441
---, O-METHOXY-	140.23	195	815
---, 3-METHYL-	124.23	355	867
---, 4-METHYL-	124.23	250	867
POLY-VINYLCYCLOHEXENE	108.18	418	850
POLY-VINYLCYCLOHEXYL KETONE	138.21	>240	843
POLY-VINYLCYCLOPENTANE	96.17	292	28
		270	441
POLY-VINYLCYCLOPROPANE	68.12	230	175
POLY-N-VINYLDIPHENYLAMINE	195.26	320	811
POLY-1-VINYLENE-3-CYCLOPENTYLENE	94.16	202	620
POLY-VINYL FLUORIDE	46.04	200	79
		230	434
POLY-VINYLIDENE CHLORIDE	96.94	190	46
POLY-VINYLIDENE FLUORIDE	64.04	171	373
		185	434
		220	443
POLY-VINYLISOPROPYL KETONE	98.14	220	843
POLY-VINYL LAURATE	226.36	16	906
POLY-VINYLMETHYL KETONE	70.09	170	379
POLY-VINYL MYRISTATE	254.41	28	888
POLY-1-VINYLNAPHTHALENE	154.21	360	75
POLY-VINYL PALMITATE	282.47	41	888
		46	906
POLY-VINYLPHENYL KETONE	132.16		
---, O-HYDROXY-	148.16	195	884
POLY-3-VINYL PYRENE	228.29	214	873
POLY-2-VINYLPYRIDINE	105.14	212	185
POLY-VINYL STEARATE	310.52	54	846
		52	888
		58	906
POLY-M-XYLYLENE	104.15	80	546
---, 4,6-DIMETHYL-	132.21	135	832
POLY-P-XYLYLENE	104.15	425	142
		420	219
		420	426
		375	82
		435	323
		412	224
		400	395
		440	546

POLYMER	MOLECULAR WEIGHT	MELTING POINT	REF.
---, BROMO-	183.05	270	395
---, CHLORO-	138.60	290	395
---, CYANO-	129.16	270	395
---, 2,5-DICHLORO-	173.04	183	831
		>300	395
---, 2,5-DIMETHYL-	132.21	350	832
---, ETHYL-	132.21	170	395
---, METHYL-	118.18	210	395
		230	142
POLY-M-XYLYLENE ADIPAMIDE	246.31	246	326
		244	544
POLY-P-XYLYLENE ADIPAMIDE	246.31	333	385
		340	402
POLY-P-XYLYLENE ADIPATE	248.28	70	325
		80	547
		81	919
POLY-M-XYLYLENE AZELAAMIDE	288.39	172	402
POLY-P-XYLYLENE AZELAAMIDE	288.39	263	385
		282	454
POLY-P-XYLYLENE AZELAATE	290.36	59	325
		74	547
		79	919
POLY-P-XYLYLENE CARBONATE	164.16	185	820
POLY-P-XYLYLENE 1,4-CYCLOHEXYLENEDICARBOXYLATE TRANS-	274.32		
		106	547
POLY-P-XYLYLENE DISULFIDE	168.27	175	861
POLY-(P-XYLYLENEDISULFONYL)-DIACETIC ANHYDRIDE	332.34	170	856
POLY-(P-XYLYLENEDISULFONYL)-DIPROPIONIC ANHYDRIDE	360.40	260	856
POLY-(P-XYLYLENEDITHIO)-DIACETIC ANHYDRIDE	268.34	88	827
		87	857
POLY-(P-XYLYLENEDITHIO)-DIPROPIONIC ANHYDRIDE	296.40	91	827
		55	857
POLY-P-XYLYLENE DITHIOLADIPATE	280.41	170	922
---, 2,3,5,6-TETRAMETHYL-	336.52	255	922
POLY-P-XYLYLENE DITHIOLISOPHTHALATE	300.40	210	922
POLY-P-XYLYLENE DITHIOL-4,4'-METHYLENEDIBENZOATE	390.52	274	922
POLY-P-XYLYLENE DITHIOL-P-PHENYLENEDIACETATE	328.45		
---, 2,3,5,6-TETRAMETHYL- (DIOL)	384.56	262	922
POLY-P-XYLYLENE DITHIOLSEBACATE	336.52		
---, 2,3,5,6-TETRAMETHYL- (DIOL)	392.62	180	922
POLY-P-XYLYLENE DOCOSANEDIAMIDE	470.74	240	402
		225	454
POLY-P-XYLYLENE DODECANEDIAMIDE	330.47	272	385
		270	454
POLY-P-XYLYLENE DODECANEDIOATE	332.44	94	547
POLY-P-XYLYLENE GLUTARAMIDE	232.28	280	402
POLY-P-XYLYLENE GLUTARATE	206.20	58	919
POLY-P-XYLYLENE HEPTADECANEDIAMIDE	400.61	239	454
POLY-P-XYLYLENE HEXADECANEDIAMIDE	386.58	248	454
---, N,N'-DIMETHYL-	414.63	85	877
POLY-M-XYLYLENE ISOPHTHALAMIDE	266.30	225	402
POLY-P-XYLYLENE ISOPHTHALAMIDE	266.30	290	385
POLY-P-XYLYLENE ISOPHTHALATE	268.27	100	547
POLY-P-XYLYLENE MALONAMIDE	204.23	110	402
POLY-P-XYLYLENE OCTADECANEDIAMIDE	414.63	235	402
		242	454
---, N,N'-DIETHYL-	470.74	50	877
---, N,N'-DIMETHYL-	442.69	80	877
POLY-P-XYLYLENE OCTAMETHYLENEDIURETHAN	334.42	168	402
POLY-P-XYLYLENE 5-OXADODECANEDIAMIDE	332.44	200	847
POLY-P-XYLYLENE 6-OXADODECANEDIAMIDE	332.44	206	847
POLY-P-XYLYLENE 6-OXAHEPTADECANEDIAMIDE	402.58	223	847
POLY-P-XYLYLENE 5-OXAHEXADECANEDIAMIDE	388.55	220	847
POLY-P-XYLYLENE OXALATE	164.12	214	919
POLY-P-XYLYLENE 6-OXAPENTADECANEDIAMIDE	374.52	217	847
POLY-P-XYLYLENE 5-OXASEBACAMIDE	304.39	234	847
POLY-P-XYLYLENE 5-OXATETRADECANEDIAMIDE	360.50	215	847
POLY-P-XYLYLENE 6-OXATRIDECANEDIAMIDE	346.47	207	847

References page III-92

POLYMER	MOLECULAR WEIGHT	MELTING POINT	REF.
POLY-P-XYLYLENE 5-OXAUNDECANEDIAMIDE	318.42	190	847
POLY-P-XYLYLENE OXYDIBUTYRAMIDE	290.36	241	847
		243	853
POLY-P-XYLYLENE OXYDICAPROAMIDE	346.47	234	847
POLY-P-XYLYLENE OXYDIENANTHAMIDE	374.52	229	847
POLY-P-XYLYLENE OXYDIPELARGONAMIDE	430.63	215	847
POLY-P-XYLYLENE OXYDIVALERAMIDE	318.42	243	847
POLY-P-XYLYLENE PENTADECANEDIAMIDE	372.55	241	454
POLY-P-XYLYLENE P-PHENYLENEDIACETATE	296.32	146	291
POLY-P-XYLYLENE PHTHALAMIDE	266.30	230	402
POLY-M-XYLYLENE PIMELAMIDE	260.34	192	402
POLY-P-XYLYLENE PIMELAMIDE	260.34	284	385
POLY-P-XYLYLENE PIMELATE	262.30	67	547
		66	919
POLY-M-XYLYLENE SEBACAMIDE	302.42	193	402
POLY-P-XYLYLENE SEBACAMIDE	302.42	300	290
		291	454
		281	385
		268	291
POLY-P-XYLYLENE SEBACATE	304.39	84	325
		88	919
		93	547
POLY-M-XYLYLENE SUBERAMIDE	274.36	213	402
POLY-P-XYLYLENE SUBERAMIDE	274.36	305	385
POLY-P-XYLYLENE SUBERATE	276.33	82	547
POLY-P-XYLYLENE SUCCINAMIDE	218.26	360	402
POLY-P-XYLYLENE SUCCINATE	220.22	115	547
POLY-M-XYLYLENE TEREPHTHALAMIDE	266.30	300	402
POLY-P-XYLYLENE TEREPHTHALAMIDE	266.30	350	402
POLY-P-XYLYLENE TEREPHTHALATE	268.27	242	547
		267	881
POLY-M-XYLYLENE TETRADECANEDIAMIDE	358.53	192	402
POLY-P-XYLYLENE TETRADECANEDIAMIDE	358.53	257	454
POLY-P-XYLYLENE THIODICAPROAMIDE	362.53	236	853
POLY-P-XYLYLENE THIODIENANTHAMIDE	390.58	228	853
POLY-P-XYLYLENE THIODIPELARGONAMIDE	446.69	214	824
POLY-P-XYLYLENE THIODIUNDECANOAMIDE	502.80	200	824
POLY-P-XYLYLENE THIODIVALERAMIDE	334.48	242	853
POLY-P-XYLYLENE TRIDECANEDIAMIDE	344.50	247	454
POLY-P-XYLYLENE UNDECANEDIAMIDE	316.44	264	454

(1)   H. TADOKORO, K. KOZAI, S. SEKI, AND I. NITTA, J. POLYMER SCI. 26, 379 (1957).

(2)   W.P. SLICHTER, J. POLYMER SCI. 35, 77 (1959).

(3)   D.R. HOLMES, C.W. BUNN, AND D.J. SMITH, J. POLYMER SCI. 17, 159 (1955).

(4)   W.J. DULMAGE, J. POLYMER SCI. 26, 277 (1957).

(5)   A. PRIETZSCHK, KOLLOID-Z. 156, 8 (1958).

(6)   R.L. MILLER, UNPUBLISHED RESULTS.

(7)   G. NATTA AND P. CORRADINI, <A> ATTI ACCAD. NAZL. LINCEI, REND., CLASSE SCI. FIS., MAT. NAT. 19, 229 (1955), <B> J. POLYMER SCI. 20, 251 (1956).

(8)   Y. TANADA, K. IMADA, AND M. TAKAYANAGI, KOGYO KAGAKU ZASSHI 69, 1971 (1966).

(9)   G. NATTA, P. CORRADINI, AND I.W. BASSI, ATTI ACCAD. NAZL. LINCEI, REND., CLASSE SCI. FIS., MAT. NAT. 19, 404 (1955).

(10)   G. NATTA, SPE (SOC. PLASTICS ENGRS.) J. 15, 373 (1959).

(11)   C.W. BUNN AND E.R. HOWELLS, NATURE 174, 549 (1954).

(12)   G. NATTA, L. PORRI, P. CORRADINI, AND D. MORERO, ATTI ACCAD. NAZL. LINCEI, REND., CLASSE SCI. FIS., MAT. NAT. 20, 560 (1956).

(13)   G. NATTA, P. CORRADINI, AND G. DALL'ASTA, ATTI ACCAD. NAZL., LINCEI, REND., CLASSE SCI. FIS., MAT. NAT. 20, 408 (1956).

(14)   C.W. BUNN, TRANS. FARADAY SOC. 35, 482 (1939).

(15)   E.R. WALTER AND F.P. REDING, J. POLYMER SCI. 21, 561 (1956).

(16)   M.L. MILLER AND C.E. RAUHUT, J. POLYMER SCI. 38, 63 (1959).

(17)   G. NATTA, J. POLYMER SCI. 16, 143 (1955).

(18)   C.C. PRICE, M. OSGAN, R.E. HUGHES, AND C. SHAMBELAN, J. AM. CHEM. SOC. 78, 690 (1956).

(19)   C.G. OVERBERGER AND H. JABLONER, J. AM. CHEM. SOC. 85, 3431 (1963).

(20)   C. LEGRAND, ACTA CRYST. 5, 800 (1952).

(21)   H.S. KAUFMAN, J. AM. CHEM. SOC. 75, 1477 (1953).

(22)   P.H. HERMANS, PHYSICS AND CHEMISTRY OF CELLULOSE FIBRES, ELSEVIER PUBLISHING CO., INC., NEW YORK, 1949.

(23)   C.W. BUNN, CHEMICAL CRYSTALLOGRAPHY, CLARENDON PRESS, OXFORD, 1946.

(24)   H.S. PEISER, H.P. ROOKSBY, AND A.J.C. WILSON, X-RAY DIFFRACTION BY POLY-CRYSTALLINE MATERIALS, CHAPMAN AND HALL LTD., LONDON, 1955.

(25)   C.W. BUNN AND E.V. GARNER, PROC. ROY. SOC. (LONDON) A 189, 39 (1947).

(26)   A. OKADA, KOBUNSHI KAGAKU 7, 122 (1950).

(27)   R. DE P. DAUBENY, C.W. BUNN, AND C.J. BROWN, PROC. ROY. SOC. (LONDON) A226, 531 (1954).

(28)   H.J. WELLARD, J. POLYMER SCI. 13, 471 (1954).

(29)   C.W. BUNN, NATURE 161, 929 (1948).

(30)   J.D. STROUPE AND R.E. HUGHES, J. AM. CHEM. SOC. 80, 1768 (1958).

(31)   T.G. FOX, B.S. GARRETT, W.E. GOODE, S. GRATCH, J.F. KINCAID, A. SPELL, AND J.D. STROUPE, J. AM. CHEM. SOC. 80, 1768 (1958).

(32)   R.C. REINHARDT, IND. ENG. CHEM. 35, 422 (1943).

(33)   S. NARITA AND K. OKUDA, J. POLYMER SCI. 38, 270 (1959).

(34)   <A> C.S. FULLER, S.J. FROSCH, AND N.R. PAPE, J. AM. CHEM. SOC. 62, 1905 (1940), <B> A.M. LIQUORI, ACTA CRYST. 8, 345 (1955).

(35)   G. NATTA, P. CORRADINI, AND I.W. BASSI, <A> MAKROMOL. CHEM. 21, 240 (1956), <B> NUOVO CIMENTO, SUPPL. 15, 52 (1960).

(36)   V.M. COIRO, P. DE SANTIS, L. MAZZARELLA, AND L. PICOZZI, CHIM. IND. (MILAN) 47, 1236 (1965).

(37)   G. NATTA, P. CORRADINI, AND L. PORRI, ATTI ACCAD. NAZL. LINCEI, REND., CLASSE SCI. FIS., MAT. NAT. 20, 728 (1956).

(38)   J.D. MCCULLOUGH, R.S. BAUER, AND T.L. JACOBS, CHEM. IND. (LONDON) 1957, 706.

(39)   H.D. NOETHER, <A> J. POLYMER SCI. 25, 217 (1957), <B> TEXTILE RES. J. 28, 533 (1958).

(40)   C.W. BUNN, PROC. ROY. SOC. (LONDON) A180, 40 (1942).

(41)   <A> C. SHAMBELAN, PH.D. THESIS, UNIV. OF PENNA., 1959 – DISSERTATION ABSTR. 20, 120 (1959), <B> M. FUJISAKA, K. IMADA, AND M. TAKAYANAGI, REPTS. PROGR. POLYMER PHYS. JAPAN 11, 169 (1968).

(42)   C.F. HAMMER, T.A. KOCH, AND J.F. WHITNEY, J. APPL. POLYMER SCI. 1, 169 (1959).

(43)   W. GOGGIN AND R. LOWRY, IND. ENG. CHEM. 34, 327 (1942).

(44)   G. NATTA, CHEM. IND. (LONDON) 1957, 1520.

(45)   G. NATTA, P. PINO, P. CORRADINI, F. DANUSSO, E. MANTICA, G. MAZZANTI, AND G. MORAGLIO, J. AM. CHEM. SOC. 77, 1708 (1955).

(46)   R. BOYER, COMPT. REND. DE LA 2E REUNION DE CHIMIE PHYSIQUE (2-7 JUIN 1952, PARIS) P. 383.

(47)   D.E. ROBERTS AND L. MANDELKERN, J. AM. CHEM. SOC. 80, 1289 (1958).

(48)   F.P. REDING, J. POLYMER SCI. 21, 547 (1956).

(49)   J.D. HOFFMAN AND J.J. WEEKS, J. RES. NATL. BUR. STD. 60, 465 (1958).

(50)   L. WOOD, N. BEKKEDAHL, AND R.E. GIBSON, J. CHEM. PHYS. 13, 475 (1945).

(51)   R. BEAMAN AND F. CRAMER, J. POLYMER SCI. 21, 223 (1956).
(52)   H. STARKWEATHER, JR., G. MOORE, J. HANSEN, T.RODER, AND R. BROOKS, J.
        POLYMER SCI. 21, 189 (1956).
(53)   R. WILEY, IND. ENG. CHEM. 38, 959 (1946).
(54)   H.W. STARKWEATHER, JR., AND R.E. MOYNIHAN, J. POLYMER SCI. 22, 363 (1956).
(55)   F.C. FRANK, A. KELLER, AND A. O'CONNOR, PHIL. MAG. 4, 200 (1959).
(56)   <A> K.H. MEYER, W. LOTMAR, AND G.W. PANKOW, HELV. CHIM. ACTA 19, 930
        (1936), <B> E. GIGLIO, F. POMPA, AND A. RIPAMONTI, J. POLYMER SCI. 59, 293
        (1960).
(57)   J. BOON AND E.P. MAGRE', MAKROMOL. CHEM. 136, 267 (1970).
(58)   P.W. TEARE AND D.R. HOLMES, J. POLYMER SCI. 24, 496 (1957).
(59)   K. LITTLE, BRIT. J. APPL. PHYS. 10, 225 (1959).
(60)   C.J.B. CLEWS, PROC. RUBBER TECHNOLOGY CONFERENCE, W. HEFFER AND SONS LTD.,
        CAMBRIDGE, 1938, P. 955.
(61)   W.P. SLICHTER, J. POLYMER SCI. 36, 259 (1959).
(62)   R.C. GOLIKE, J. POLYMER SCI. 42, 583 (1960).
(63)   Y. KINOSHITA, MAKROMOL. CHEM. 33, 21 (1959).
(64)   P.H. BURLEIGH, J. AM. CHEM. SOC. 82, 749 (1960).
(65)   Y. KINOSHITA, MAKROMOL. CHEM. 33, 1 (1959).
(66)   R.E. MOYNIHAN, J. AM. CHEM. SOC. 81, 1045 (1959).
(67)   G. NATTA, MAKROMOL. CHEM. 35, 93 (1960).
(68)   J. BATEMAN, R.E. RICHARDS, G. FARROW, AND I.M. WARD, POLYMER 1, 63 (1960).
(69)   P. CORRADINI, ATTI ACCAD. NAZL. LINCEI, REND., CLASSE SCI. FIS., MAT. NAT.
        25, 517 (1958).
(70)   S.D. GEHMAN, J.E. FIELD, AND R.P. DINSMORE, PROC. RUBBER TECHNOLOGY
        CONFERENCE, W. HEFFER AND SONS LTD., CAMBRIDGE, 1938, P. 961.
(71)   G. NATTA, G. MAZZANTI, P. LONGI, AND F. BERNARDINI, CHIM. IND. (MILAN) 40,
        813 (1958).
(72)   W.D. NIEGISCH AND P.R. SWAN, J. APPL. PHYS. 31, 1906 (1960).
(73)   E. MARTUSCELLI, R. GALLO, AND G. PAIARO, MAKROMOL. CHEM. 103, 295 (1967).
(74)   D. SIANESI, G. NATTA, AND P. CORRADINI, GAZZ. CHIM. ITAL. 89, 775 (1959).
(75)   <A> G. NATTA, F. DANUSSO, AND D. SIANESI, MAKROMOL. CHEM. 28, 253 (1958),
        <B> D. SIANESI, M. RAMPICHINI, AND F. DANUSSO, CHIM. IND. (MILAN) 41, 287
        (1959).
(76)   G. NATTA, G. MAZZANTI, AND P. CORRADINI, ATTI ACCAD. NAZL. LINCEI, REND.,
        CLASSE SCI. FIS., MAT. NAT. 25, 3 (1958).
(77)   W.R. KRIGBAUM AND N. TOKITA, J. POLYMER SCI. 43, 467 (1960).
(78)   E. STANLEY AND M. LITT, J. POLYMER SCI. 43, 453 (1960).
(79)   D.I. SAPPER, J. POLYMER SCI. 43, 383 (1960).
(80)   P. CORRADINI AND P. GANIS, J. POLYMER SCI. 43, 311 (1960).
(81)   L. MANDELKERN, CHEM. REVS. 56, 903 (1956).
(82)   J.R. SCHAEFGEN, J. POLYMER SCI. 38, 549 (1959).
(83)   F. DANUSSO, G. MORAGLIO, AND E. FLORES, ATTI ACCAD. NAZL. LINCEI, REND.,
        CLASSE SCI. FIS., MAT. NAT. 25, 520 (1958).
(84)   N. YODA AND I. MATSUBARA, J. POLYMER SCI., PT. A, 2, 253 (1964).
        (UNIT CELL CONSTANTS GIVEN REFER TO A PROJECTED BASE).
(85)   F.A. QUINN, JR., AND L. MANDELKERN, J. AM. CHEM. SOC. 80, 3178 (1958).
(86)   F.W. BILLMEYER, JR., J. APPL. PHYS. 28, 1114 (1957).
(87)   M. DOLE, J. POLYMER SCI. 19, 347 (1956).
(88)   A.M. BUECHE, J. AM. CHEM. SOC. 74, 65 (1952).
(89)   F. DANUSSO AND G. MORAGLIO, ATTI ACCAD. NAZL. LINCEI, REND., CLASSE SCI.
        FIS., MAT. NAT. 27, 381 (1959).
(90)   T.W. CAMPBELL AND A.C. HAVEN, JR., J. APPL. POLYMER SCI. 1, 73 (1959).
(91)   H.W. STARKWEATHER, JR., AND R.H. BOYD, J. PHYS. CHEM. 64, 410 (1960).
(92)   <A> G. NATTA, G. MAZZANTI, P. CORRADINI, P. CHINI, AND I.W. BASSI, ATTI
        ACCAD. NAZL. LINCEI, REND., CLASSE SCI. FIS., MAT. NAT. 28, 8 (1960),
        <B> G. NATTA, G. MAZZANTI, P. CORRADINI, AND I.W. BASSI, MAKROMOL. CHEM.
        37, 156 (1960), <C> I.W. BASSI, REND. IST. LOMBARDO SCI. LETTERE A94, 579
        (1960), <D> G. NATTA, P. CORRADINI, AND I.W. BASSI, J. POLYMER SCI. 51,
        505 (1961).
(93)   D.V. BADAMI, POLYMER 1, 273 (1960).
(94)   J.H. GRIFFITH AND B.G. RANBY, J. POLYMER SCI. 44, 369 (1960).
(95)   C.G. OVERBERGER, A.E. BORCHERT, AND A. KATCHMAN, J. POLYMER SCI. 44, 491
        (1960).
(96)   G. NATTA, G. MAZZANTI, P. CORRADINI, A. VALVASSORI, AND I.W. BASSI, ATTI
        ACCAD. NAZL. LINCEI, REND., CLASSE SCI. FIS., MAT. NAT. 28, 18 (1960).
(97)   G. NATTA AND I.W. BASSI, EUROPEAN POLYMER J. 3, 33 (1967).
(98)   O. ELLEFSEN, NORELCO REPTR. 7, 104 (1960).
(99)   G. FARROW, J. MCINTOSH, AND I.M. WARD, MAKROMOL. CHEM. 38, 147 (1960).
(100)   N.G. ESIPOVA, L. PAN-TUN, N.S. ANDREEVA, AND P.V. KOZLOV, VYSOKOMOLEKUL.
        SOEDIN. 2, 1109 (1960).
(101)   A.G.M. LAST, J. POLYMER SCI. 39, 543 (1959).

(102)   D.J. SHIELDS AND H.W. COOVER, JR., J. POLYMER SCI. 39, 532 (1959).
(103)   S. MURAHASHI, S. NOZAKURA, AND H. TADOKORO, BULL. CHEM. SOC. JAPAN 32, 534
        (1959).
(104)   T.F. YEN, J. POLYMER SCI. 38, 272 (1959).
(105)   C.S. FULLER AND C.L. ERICKSON, J. AM. CHEM. SOC. 59, 344 (1937).
(106)   C.S. FULLER AND C.J. FROSCH, J. AM. CHEM. SOC. 61, 2575 (1939).
        (EXCEPT FOR POLY(DECAMETHYLENE OXALATE), UNIT CELL CONSTANTS GIVEN REFER
        TO A PROJECTED BASE).
(107)   C.S. FULLER, C.J. FROSCH, AND N.R. PAPE, J. AM. CHEM. SOC. 64, 154 (1942).
        (UNIT CELL CONSTANTS GIVEN REFER TO A PROJECTED BASE).
(108)   C.S. FULLER AND C.S. FROSCH, J. PHYS. CHEM. 43, 323 (1939).
(109)   C.S. FULLER, CHEM. REVS. 26, 143 (1940).
(110)   W.O. BAKER AND C.S. FULLER, J. AM. CHEM. SOC. 64, 2399 (1942).
(111)   A. CONIX AND R. VAN KERPEL, J. POLYMER SCI. 40, 521 (1959).
(112)   J.A. GAILEY AND R.H. RALSTON, SPE (SOC. PLASTICS ENGRS.) TRANS. 4, 29
        (1964).
(113)   H.S. YANI, QUOTED IN C.F. RYAN AND J.J. GORMLEY, MACROMOL. SYNTH. 1, 30
        (1963).
(114)   E.J. VANDENBERG, R.F. HECK, AND D.S. BRESLOW, J. POLYMER SCI. 41, 519
        (1960).
(115)   J.H. MAGILL, J. POLYMER SCI., PT. A-2, 5, 89 (1967).
(116)   S. FURUYA AND M. HONDA, J. POLYMER SCI. 28, 232 (1958).
(117)   P. GOODMAN, J. POLYMER SCI. 24, 307 (1960).
(118)   L. MANDELKERN AND P.J. FLORY, J. AM. CHEM. SOC. 73, 3206 (1951).
(119)   S. ISHIDA AND S. MURAHASHI, J. POLYMER SCI. 40, 571 (1959).
(120)   B.S. GARRETT, W.E. GOODE, S. GRATCH, J.F. KINCAID, C.L. LEVESQUE, A.
        SPELL, J.D. STROUPE, AND W.H. WATANABE, J. AM. CHEM. SOC. 81, 1007 (1959).
(121)   D.C. VOGELSONG AND E.M. PEARCE, J. POLYMER SCI. 45, 546 (1960).
(122)   P. CORRADINI AND P. GANIS, NUOVO CIMENTO, SUPPL. 15, 104 (1960).
(123)   G. NATTA, P. CORRADINI, AND I.W. BASSI, NUOVO CIMENTO, SUPPL. 15, 83
        (1960).
(124)   G. NATTA AND P. CORRADINI, <A> ANGEW. CHEM. 68, 615 (1956), <B> NUOVO
        CIMENTO, SUPPL. 15, 111 (1960).
(125)   P. CORRADINI AND P. GANIS, NUOVO CIMENTO, SUPPL. 15, 96 (1960).
(126)   R.G. CRYSTAL, MACROMOLECULES 4, 379 (1971).
(127)   G. NATTA AND P. CORRADINI, NUOVO CIMENTO, SUPPL. 15, 40 (1960).
(128)   G. NATTA, P. CORRADINI, AND I.W. BASSI, NUOVO CIMENTO, SUPPL. 15, 68
        (1960).
(129)   P. PINO AND G.P. LORENZI, J. AM. CHEM. SOC. 82, 4745 (1960).
(130)   G. NATTA, G. MAZZANTI, G. PREGLIA, M. BINAGHI, AND M. PERALDO, J. AM.
        CHEM. SOC. 82, 4742 (1960).
(131)   Z.W. WILCHINSKY, J. APPL. PHYS. 31, 1969 (1960).
(132)   T. MOCHIZUKI, NIPPON KAGAKU ZASSHI 81, 15 (1960).
(133)   R. STEFANI, M. CHEVRETON, M. GARNIER, AND C. EYRAUD, COMPT. REND. 251,
        2174 (1960).
(134)   M.L. HUGGINS, J. CHEM. PHYS. 13, 37 (1945).
(135)   C. RUSCHER AND H.J. SCHRODER, FASERFORSCH. TEXTILTECH. 11, 165 (1960).
(136)   Z. MENCIK, CHEM. PRUMYSL 10, 377 (1960).
(137)   M. DOLE AND B. WUNDERLICH, MAKROMOL. CHEM. 34, 29 (1959).
(138)   G. ALLEN, G. GEE, AND G.J. WILSON, POLYMER 1, 456 (1960).
(139)   G. ALLEN, G. GEE, D. MANGARAJ, D. SIMS, AND G.J. WILSON, POLYMER 1, 466
        (1960).
(140)   M. ASAHINA AND K. OKUDA, KOBUNSHI KAGAKU 17, 607 (1960).
(141)   G. NATTA, M. FARINO, M. PERALDO, P. CORRADINI, G. BRESSAN, AND P. GANIS,
        ATTI ACCAD. NAZL. LINCEI, REND., CLASSE SCI. FIS., MAT. NAT. 28, 442
        (1960).
(142)   L.A. AUSPOS, C.W. BURNHAM, L. HALL, J.K. HUBBARD, W. KIRK, J.R.
        SCHAEFGEN, AND S.B. SPECK, J. POLYMER SCI. 15, 9 (1955).
(143)   A.T. WALTER, J. POLYMER SCI. 13, 207 (1954).
(144)   C.J. MALM, J.W. MENCH, D.L. KENDALL, AND G.D. HIATT, IND. ENG. CHEM. 43,
        688 (1951).
(145)   C.E. ANAGNOSTOPOULOS, A.Y. CORAN, AND H.R. GAMRATH, J. APPL. POLYMER SCI.
        4, 181 (1960).
(146)   C.F. HORN, B.T. FREURE, H. VINEYARD, AND H.J. DECKER, ANGEW. CHEM. 74, 531
        (1962).
(147)   <A> G. NATTA, G. DALL'ASTA, AND G. MAZZANTI, ANGEW. CHEM. 76, 765 (1964) -
        ANGEW. CHEM., INTERN. ED. ENGL. 3, 723 (1964), <B> G. NATTA AND I.W.
        BASSI, ATTI ACCAD. NAZL. LINCEI, REND., CLASSE SCI. FIS., MAT. NAT. 38,
        315 (1965), <C> G. NATTA AND I.W. BASSI, J. POLYMER SCI., PT. C, NO. 16,
        2551 (1967).
(148)   A.C. FARTHING AND W.J. REYNOLDS, J. POLYMER SCI. 12, 503 (1954).
(149)   H.S. YANAI, QUOTED IN W.E. GOODE, R.P. FELLMAN, AND F.H. OWENS, MACROMOL.
        SYNTH. 1, 25 (1963).

(150)   A. TURNER-JONES, MAKROMOL. CHEM. 71, 1 (1964).
(151)   M. NIINOMI, T. FUKUDA, AND M. TAKAYANAGI, PRIVATE COMMUNICATION.
(152)   G. NATTA, ANGEW. CHEM. 68, 393 (1956).
(153)   G.F. SCHMIDT AND H.A. STUART, Z. NATURFORSCHUNG 13A, 222 (1958).
(154)   G. NATTA AND P. CORRADINI, NUOVO CIMENTO, SUPPL. 15, 9 (1960).
(155)   F. RYBNIKAR, CHEM. LISTY 52, 1042 (1958).
(156)   B. WUNDERLICH, AND M. DOLE, J. POLYMER SCI. 24, 201 (1957).
(157)   C.W. SMITH AND M. DOLE, J. POLYMER SCI. 20, 37 (1956).
(158)   B. WUNDERLICH AND M. DOLE, J. POLUMER SCI. 32, 125 (1958).
(159)   P.J. FLORY, H.D. BEDON, AND E.H. KEEFER, J. POLYMER SCI. 28, 151 (1958).
(160)   R.D. EVANS, M.R. MIGHTON, AND P.J. FLORY, J. AM. CHEM. SOC. 72, 2018
         (1950).
(161)   G. DALL'ASTA AND N. ODDO, CHIM. IND. (MILAN) 42, 1234 (1960).
(162)   Z. MENCIK, VYSOKOMOLEKUL. SOEDIN. 2, 1635 (1960).
(163)   Y. CHATANI, J. POLYMER SCI. 47, 491 (1960).
(164)   G. WASAI, T. TSURUTA, AND J. FURUKAWA, KOGYO KAGAKU ZASSHI 66, 1339 (1963)
(165)   M. CESARI, PRIVATE COMMUNICATION.
(166)   H.D. KEITH, F.J. PADDEN, JR., N.M. WALTER, AND H.W. WYCKOFF, J. APPL.
         PHYS. 30, 1485 (1959).
(167)   G. NATTA, M. PERALDO, AND P. CORRADINI, ATTI ACCAD. NAZL. LINCEI, REND.,
         CLASSE SCI. FIS., MAT. NAT. 26, 14 (1959).
(168)   K. OKUDA, T. YOSHIDA, M. SUGITA, AND M. ASAHINA, J. POLYMER SCI., PT. B,
         5, 465 (1967).
(169)   G. NATTA, I. PASQUON,. P. CORRADINI, M. PERALDO, M. PEGORARO, AND A.
         ZAMBELLI, ATTI ACCAD. NAZL. LINCEI, REND., CLASSE SCI. FIS., MAT. NAT. 28,
         539 (1960).
(170)   N.M. WALTER, QUOTED IN C.Y. LIANG AND F.G. PEARSON, J. MOL. SPECTRY. 5,
         290 (1960).
(171)   S.S. LESHCHENKO, V.L. KARPOV, AND V.A. KARGIN, <A> VYSOKOMOLEKUL. SOEDIN.
         1, 1538 (1959), <B> IBID., 5, 953 (1963) – POLYMER SCI. (USSR) (ENGLISH
         TRANSL.) 5, 1 (1964).
(172)   D.J.H. SANDIFORD, J. APPL. CHEM. 8, 188 (1958).
(173)   M. HATANO, AND S. KAMBARA, POLYMER 2, 1 (1961).
(174)   R. DEDEURWAERDER AND J.F.M. OTH, BULL. SOC. CHIM. BELGES 70, 37 (1961).
(175)   G. NATTA, D. SIANESI, D. MORERO, I.W. BASSI, AND G. CAPORICCIO, ATTI
         ACCAD. NAZL. LINCEI, REND., CLASSE SCI. FIS., MAT. NAT. 28, 551 (1960).
(176)   I.W. BASSI, ATTI ACCAD. NAZL. LINCEI, REND., CLASSE SCI. FIS., MAT. NAT.
         29, 193 (1960).
(177)   G. CHAMPETIER, M. LAUALOV, AND J.P. PIED, BULL. SOC. CHIM. FRANCE 1958,
         708.
(178)   A.G. NASINI, L. TROSSARELLI, AND G. SAINI, MAKROMOL. CHEM. 44-46, 550
         (1961).
(179)   G. NATTA, G. MAZZANTI, G.F. PREGAGLIA, AND M. BINAGHI, MAKROMOL. CHEM.
         44-46, 537 (1961).
(180)   J.C. SWALLOW, PROC. ROY. SOC. (LONDON) A238, 1 (1956).
(181)   J. LAL AND G.S. TRICK, J. POLYMER SCI. 50, 13 (1961).
(182)   R.J. WILKINSON AND M. DOLE, J. POLYMER SCI. 58, 1089 (1962).
(133)   F.B. MOODY, AM. CHEM. SOC. POLYMER PREPRINTS 2, 285 (1961).
(184)   I. SAKURADA, K. NUKUSHINA, AND Y. SONE, KOBUNSHI KAGAKU 12, 506 (1955).
(185)   G. NATTA, G. MAZZANTI, P. LONGI, G. DALL'ASTA, AND F. BERNARDINI, J.
         POLYMER SCI. 51, 487 (1961).
(186)   M. INOUE, J. POLYMER SCI. 51, S18 (1961).
(187)   J.A. PRICE, M.R. LYTTON, AND B.G. RANBY, J. POLYMER SCI. 51, 541 (1961).
(188)   F.P. PRICE AND R.W. KILB, J. POLYMER SCI. 57, 395 (1962).
(189)   G. NATTA, L. PORRI, P. CORRADINI, G. ZANINI, AND F. CIAMPELLI, J. POLYMER
         SCI. 51, 463 (1961).
(190)   <A> J.P. ARLIE AND A. SKOULIOS, COMPT. REND. 258, 2570 (1964), <B> J.P.
         ARLIE, P. SPEGT, AND A. SKOULIOS, MAKROMOL. CHEM. 104, 212 (1967).
(191)   U. GIANNINI, G. BRUCKNER, E. PELLINO, AND A. CASSATA, J. POLYMER SCI., PT.
         B, 5, 527 (1967).
(192)   <A> S. GELLER, SCIENCE 152, 644 (1966), <B> M.D. LIND AND S. GELLER, J.
         CHEM. PHYS. 51, 348 (1969).
(193)   Y. CHATANI, T. TAKIZAWA, S. MURAHASHI, Y. SAKATA, AND Y. NISHIMURA, J.
         POLYMER SCI. 55, 811 (1961).
(194)   J.R. RICHARDS, PH.D. THESIS, UNIV. OF PENNA., 1961, DISSERTATION ABSTR.
         22, 1029 (1961).
(195)   H.G. KILIAN, H. HABOTH, AND E. JENCKEL, KOLLOID-Z. 172, 166 (1960).
(196)   E.J. ADDINK AND J. BEINTEMA, POLYMER 2, 185 (1961).
(197)   G. DAMASCHUN, KOLLOID-Z. 180, 65 (1962).
(198)   F.L. SAUNDERS, J. POLYMER SCI., PT. A-1, 5, 2187 (1967).
(199)   C.A BOYE, J. POLYMER SCI. 55, 275 (1961).
(200)   I.W. BASSI, G. ALLEGRA, AND R. SCORDAMAGLIA, MACROMOLECULES 4, 575 (1971).

(201)   W. WRASIDLO, MACROMOLECULES 4, 642 (1971).
(202)   R.H. BAUGHMAN, J. APPL. PHYS. 42, 4579 (1971).
(203)   A. TURNER-JONES AND C.W. BUNN, ACTA CRYST. 15, 105 (1962).
(204)   D.C. VOGELSONG, J. POLYMER SCI. 57, 895 (1962).
(205)   C.Y. LIANG AND S. KRIMM, J. CHEM. PHYS. 25, 563 (1956).
(206)   M.G. BROADHURST, J. RES. NATL. BUR. STD. 66A, 241 (1962), J. CHEM. PHYS.
        36, 2578 (1962).
(207)   R.L. MILLER AND V.F. HOLLAND, J. POLYMER SCI., PT. B, 2, 519 (1964),
        V.F. HOLLAND AND R.L. MILLER, J. APPL. PHYS. 35, 3241 (1964).
(208)   A.V. ERMOLINA, G.S. MARKOVA, AND V.A. KARGIN, KRISTALLOGRAFIYA 2, 623
        (1957).
(209)   E.S. CLARK, AND L.T. MUUS, Z. KRIST. 117, 11. (1962).
(210)   V.F. HOLLAND, S.B. MITCHELL, W.L. HUNTER, AND P.H. LINDENMEYER, J. POLYMER
        SCI. 62, 145 (1962).
(211)   J.J. POINT, BULL. CLASSE SCI. ACAD. ROY. BELG. 30, 435 (1953).
(212)   L.G. WALLNER, MONATSH. CHEM. 79, 279 (1948).
(213)   H. HENDUS, K. SCHMIEDER, G. SCHNELL, AND K.A. WOLF. FESTSCHRIFT CARL
        WURSTER DER BASF VOM 2.12.1960.
(214)   J. FURUKAWA, T. SAEGUSA, T. TSURUTA, S. OTHA, AND G. WASAI, MAKROMOL.
        CHEM. 52, 230 (1962).
(215)   G. CHAMPETIER AND J.P. PIED, MAKROMOL. CHEM. 44-46, 64 (1961).
(216)   F. RYBNIKAR, COLLECTION CZECH. CHEM. COMMUN. 24, 2861 (1959).
(217)   S. OKAMURA, K. HAYASHI, AND Y. KITANISHI, J. POLYMER SCI. 58, 925 (1962).
(218)   O. ELLEFSEN AND N. NORMAN, J. POLYMER SCI. 58, 769 (1962).
(219)   L.A. ERREDE AND R.S. GREGORIAN, J. POLYMER SCI. 60, 21 (1962).
(220)   T.F. YEN, J. POLYMER SCI. 35, 533 (1959).
(221)   J.F. BROWN, JR., AND D.M. WHITE, J. AM. CHEM. SOC. 82, 5671 (1960).
(222)   A. ZIABICKI, KOLLOID-Z. 167, 132 (1959).
(223)   C.J. BROWN AND A.C. FARTHING, J. CHEM. SOC. 1953, 3270.
(224)   M.H. KAUFMAN, H.F. MARK, AND R.R. MESROBIAN, J. POLYMER SCI. 13, 3 (1954).
(225)   R.L. MERKER AND M.J. SCOTT, J. POLYMER SCI., PT. A, 2, 15 (1964).
(226)   W.A. BUTTE, C.C. PRICE, AND R.E. HUGHES, J. POLYMER SCI. 61, S28 (1962).
(227)   E.R. WALTER AND F.P. REDING, 133RD NATL. AM. CHEM. SOC. MEETING, SAN
        FRANCISCO, APRIL, 1958.
(228)   O.F. SOLOMON, M. DIMONIE, K. AMBROZH, AND M. TOMESKU, J. POLYMER SCI. 52,
        205 (1961).
(229)   C.W. BUNN AND E.V. GARNER, J. CHEM. SOC. 1942, 654.
(230)   G. NATTA, L. PORRI, AND G. SOVARZI, EUROPEAN POLYMER J. 1, 81 (1965).
(231)   F. SAKAGUCHI, R. KITAMARU, AND W. TSUJI, BULL. INST. CHEM. RES., KYOTO
        UNIV. 44, 155 (1966).
(232)   K. FUJII, T. MOCHIZUKI, S. IMOTO, J. UKIDA, AND M. MATSUMOTO, MAKROMOL.
        CHEM. 51, 225 (1962).
(233)   K. TANAKA, T. SETO, AND T. HARA, J. PHYS. SOC. JAPAN 17, 873 (1962).
        T. SETO, T. HARA, AND K. TANAKA, JAPAN. J. APPL. PHYS. 7, 31 (1968).
(234)   G. NATTA, L. PORRI, A. CARBONARO, F. CIAMPELLI, AND G. ALLEGRA, MAKROMOL.
        CHEM. 51, 229 (1962).
(235)   G. NATTA, I.W. BASSI, AND P. CORRADINI, ATTI ACCAD. NAZL. LINCEI, REND.,
        CLASSE SCI. FIS., MAT. NAT. 31, 17 (1961).
(236)   G. NATTA, I.W. BASSI, AND G. ALLEGRA, ATTI ACCAD. NAZL. LINCEI, REND.,
        CLASSE SCI. FIS., MAT. NAT. 31, 350 (1961).
(237)   J.B. LANDO AND V. STANNETT, J. POLYMER SCI., PT. B. 2, 375 (1964), AM.
        CHEM. SOC. POLYMER PREPRINTS 5, 969 (1964).
(238)   A. TAKAHASHI AND S. KAMBARA, MAKROMOL. CHEM. 72, 92 (1964).
(239)   T. PETITPAS, AND J. MERING, COMPT. REND. 244, 2611 (1962).
(240)   H. SUMITOMO AND K. KOBAYASHI, J. POLYMER SCI., PT. A-1, 5, 2247 (1967).
        (UNIT CELL CONSTANTS GIVEN REFER TO A PROJECTED BASE).
(241)   S. BOILEAU, J. COSTE, J.-M. RAYNAL, AND P. SIGWALT, COMPT. REND. 254, 2774
        (1962).
(242)   G. NATTA, M. PERALDO, M. FARINA, AND G. BRESSAN, MAKROMOL. CHEM. 55, 139
        (1962).
(243)   R. AELION, ANN. CHIM. (PARIS) 3, 5 (1948).
(244)   K. DACHS AND E. SCHWARTZ, ANGEW. CHEM. 74, 540 (1962), AGNEW. CHEM.,
        INTERN. ED. ENGL. 1, 430 (1962).
(245)   K. OKUDA, J. POLYMER SCI., PT. A, 2, 1749 (1964).
(246)   H. ARIMOTO, J. POLYMER SCI., PT. A, 2, 2283 (1964).
(247)   <A> G. NATTA, P. CORRADINI, P. GANIS, I.W. BASSI, AND G. ALLEGRA, CHIM.
        IND. (MILAN) 44, 532 (1962), <B> G. NATTA, G. ALLEGRA, I.W. BASSI, P.
        CORRADINI, AND P. GANIS, MAKROMOL. CHEM. 58, 242 (1962).
(248)   Y. YAMASHITA AND S. NUNOMOTO, MAKROMOL. CHEM. 58, 244 (1962).
(249)   L. BECKER, WISS. Z. KARL-MARX-UNIV. LEIPZIG, MATH.-NATURW. REIHE 11, 3
        (1962).
(250)   K.J. CLARK, A. TURNER-JONES, AND D.J.H. SANDIFORD, CHEM. IND. (LONDON)
        1962, 2010.

(251)   E. PASSAGLIA AND H.R. KEVORKIAN, J. APPL. PHYS. 34, 90 (1963).
(252)   A. NISHIOKA AND K. YANIGISAWA, KOBUNSHI KAGAKU 19, 667 (1962).
(253)   S. MURAHASHI, J. YUKI, T. SANO, U. YONEMURA, H. TADOKORO, AND Y. CHATANI,
        J. POLYMER SCI. 62, S77 (1962).
(254)   A. TURNER-JONES, J. POLYMER SCI. 62, S53 (1962).
(255)   H. WATANABE, R. KOYAMA, H. NAGAI, AND A. NISHIOKA, J. POLYMER SCI. 62, S74
        (1962).
(256)   J.-J. TRILLAT AND R. TERTIAN, COMPT. REND. 219, 395 (1944).
(257)   L. ULICKY, CHEM. ZVESTI.16, 818 (1962).
(258)   E. SAUTER, Z. PHYSIK. CHEM. B18, 417 (1932).
(259)   G. NATTA, G. DALL'ASTA, AND G. MAZZANTI, CHIM. IND. (MILAN) 44, 1212
        (1962).
(260)   M.G. BROADHURST, J. RES. NATL. BUR. STD. 67A, 233 (1963).
(261)   L. BECKER, PLASTE KAUTSCHUK 8, 557 (1961).
(262)   B. WUNDERLICH AND C.M. CORMIER, J. POLYMER SCI., PT. A-2, 5, 987 (1967).
(263)   G. NATTA, L. PORRI, G. STOPPA, G. ALLEGRA, AND F. CIAMPELLI, J. POLYMER
        SCI., PT. B, 1, 67 (1963).
(264)   G.W. TAYLOR, POLYMER 3, 543 (1962).
(265)   O.B. EDGAR AND E. ELLERY, J. CHEM. SOC. 1952, 2633.
(266)   H.-G. KILIAN, KOLLOID-Z. 185, 13 (1962).
(267)   H.S. MAKOWSKI, K.C. SHIM, AND Z.W. WILCHINSKY, J. POLYMER SCI., PT. A, 2,
        (1964).
(268)   D. SIANESI AND G. CAPORICCIO, MAKROMOL. CHEM. 60, 213 (1963).
(269)   I.W. BASSI, G. DALL'ASTA, U. CAMPIGLI, AND E. STREPPAROLA, MAKROMOL. CHEM.
        60, 202 (1963).
(270)   <A> G. CARAZZOLO, GAZZ. CHIM. ITAL. 92, 1345 (1962), <B> G. CARAZZOLO AND
        G. PUTTI, CHIM. IND. (MILAN) 45, 771 (1963), <C> G. CARAZZOLO AND M.
        MAMMI, J. POLYMER SCI., PT. A, 1, 965 (1963), <D> G. CARAZZOLO, J. POLYMER
        SCI., PT. A, 1, 1573 (1963), <E> G. CARAZZOLO, S. LEGHISSA, AND M. MAMMI,
        MAKROMOL. CHEM. 63, 171 (1963).
(271)   E. BAER, J.R. COLLIER, AND D.R. CARTER, SPE (SOC. PLASTICS ENGRS.) TRANS.
        5, 22 (1965).
(272)   D.C. VOGELSONG, J. POLYMER SCI., PT. A, 1, 1055 (1963).
(273)   F.H.C. CRICK AND A. RICH, NATURE 176, 780 (1955).
(274)   G. PEREGO.AND I.W. BASSI, MAKROMOL. CHEM. 61, 198 (1963).
(275)   U. GIANNINI, M. CAMBINI, AND A. CASSATA, MAKROMOL. CHEM. 61, 246 (1963).
(276)   F. DANUSSO AND G. GIANNOTTI, MAKROMOL. CHEM. 61, 164 (1963).
(277)   F. DANUSSO AND G. GIANNOTTI, MAKROMOL. CHEM. 61, 139 (1963).
(278)   M. DONATI AND M. FARINA, MAKROMOL. CHEM. 60, 233 (1963).
(279)   E.M. BRADBURY AND A. ELLIOTT, POLYMER 4, 47 (1963).
(280)   H. UTSUNOMIYA, N. KAWASAKI, M. NIINOMI, AND M. TAKAYANAGI, J. POLYMER
        SCI., PT. B, 5, 907 (1967).
(281)   E.J. VANDENBERG, J. POLYMER SCI., PT. C, NO. 1, 207 (1963).
(282)   K.R. DUNHAM, J. VANDENBERGHE, J.W.H. FABER, AND L.E. CONTOIS, J. POLYMER
        SCI., PT. A, 1, 751 (1963).
(283)   M. TOMIKA, KOBUNSHI KAGAKU 20, 145 (1963).
(284)   N. CALDERON AND M.C. MORRIS, J. POLYMER SCI., PT. A-2, 5, 1283 (1967).
(285)   N.S. CHU AND C.C. PRICE, J. POLYMER SCI., PT. A, 1, 1105 (1963).
(286)   K.H. MEYER AND A. VAN DER WYK, HELV. CHIM. ACTA. 20, 1313 (1937).
(287)   W. COOPER AND R.K. SMITH, J. POLYMER SCI., PT. A, 1, 159 (1963).
(288)   K.H. MEYER, NATURAL AND SYNTHETIC HIGH POLYMERS, INTERSCIENCE PUBLISHERS,
        NEW YORK, 1950.
(289)   J.R. WHINFIELD, NATURE 158, 930 (1946).
(290)   E.F. IZARD, J. POLYMER SCI. 8, 503 (1952).
(291)   R. HILL AND E.E. WALKER, J. POLYMER SCI. 3, 609 (1948).
(292)   D.D. COFFMAN, M.L. COX, F.L. MARTIN, W.E. MOCHEL, AND F.J. VAN NATTA, J.
        POLYMER SCI. 3, 85 (1948).
(293)   J.R. SCHAEFGEN AND P.J. FLORY, J. AM. CHEM. SOC. 70, 2709 (1948).
(294)   J.C. PATRICK, TRANS. FARADAY SOC. 32, 347 (1936).
(295)   C.W. BUNN, J. POLYMER SCI. 16, 323 (1955).
(296)   E.L. GAL'PERIN, S.S. DUBOV, E.V. VOLKOVA, AND M.P. MLENIK,
        KRISTALLOGRAFIYA 9, 102 (1964) - SOVIET PHYS. CRYST. (ENGLISH TRANSL.) 9,
        81 (1964).
(297)   A. NOSHAY AND C.C. PRICE, J. POLYMER SCI. 34, 165 (1959).
(298)   E. GIPSTEIN, E. WELLISCH, AND O.J. SWEETING, J. POLYMER SCI., PT. B, 1,
        237 (1963).
(299)   L.D. JONES AND F.E. KARASZ, J. POLYMER SCI., PT. B, 4, 803 (1966).
(300)   W.J. CAROTHERS AND J.A. ARVIN, J. AM. CHEM. SOC. 51, 2560 (1929).
(301)   K. UEBERREITER AND N. STEINER, MAKROMOL. CHEM. 74, 158 (1964).
(302)   W.H. CAROTHERS AND G.L. DOROUGH, J. AM. CHEM. SOC. 52, 711 (1930).
(303)   H. TADOKORO, Y. CHATANI, T. YOSHIHARA, S. TAHARA, AND S. MURAHASHI,
        MAKROMOL. CHEM. 73, 109 (1964).

(304) W.H. CAROTHERS AND G.J. BERCHET, J. AM. CHEM. SOC. 52, 5289 (1930).
(305) W.H. CAROTHERS, CHEM. REVS. 8, 353 (1931).
(306) G. NATTA, M. PERALDO, AND G. ALLEGRA, MAKROMOL. CHEM. 75, 215 (1964).
(307) W.H. CAROTHERS AND J.W. HILL, J. AM. CHEM. SOC. 54, 1559 (1932).
(308) A. TURNER-JONES, J.M. AIZLEWOOD, AND D.R. BECKETT, MAKROMOL. CHEM. 75, 134 (1964). (UNIT CELL CONSTANTS GIVEN FOR POLY(PROPENE)-FORM III REFER TO A PROJECTED BASE).
(309) W.H. CAROTHERS, J. AM. CHEM. SOC. 55, 4714 (1933).
(310) M.I. BESSONOV AND A.P. RUDAKOV, FIZ. TVERD. TELA 6, 1333 (1964), THROUGH CHEM. ABSTR. 61, 5793D (1964).
(311) P.R. SWAN, J. POLYMER SCI. 56, 403 (1962).
(312) K. HIRONO, G. WASAI, T. SAEGUSA, AND J. FURUKAWA, KOGYO KAGAKU ZASSHI 67, 604 (1964).
(313) J. BOOR, JR., AND J.C. MITCHELL, J. POLYMER SCI., PT. A, 1, 59 (1963).
(314) G. WASAI, T. SAEGUSA, AND J. FURUKAWA, KOGYO KAGUKU ZASSHI 67, 601 (1964).
(315) B. WUNDERLICH AND T. ARAKAWA, J. POLYMER SCI., PT. A, 2, 3697 (1964).
(316) K. TSUBOI AND T. MOCHIZUKI, J. POLYMER SCI., PT. B, 1, 531 (1963), KOBUNSHI KAGAKU 23, 645 (1966).
(317) K. HAYASHI, Y. KITANISHI, M. NISHII, AND S. OKAMURA, MAKROMOL. CHEM. 47, 237 (1961).
(318) S. KAMBARA AND A. TAKAHASHI, MAKROMOL. CHEM. 63, 89 (1963).
(319) J.P. KENNEDY, J.J. ELLIOTT, AND B. GROTEN, MAKROMOL. CHEM. 77, 26 (1964).
(320) J. WILSKY AND T. GREWER, J. POLYMER SCI., PT. C, NO. 6, 33 (1964).
(321) G. CARAZZOLO, CHIM. IND. (MILAN) 46, 525 (1964).
(322) G.W. URBANCZYK, ZESTY NAUK. POLITECH. LODZ, WLOKIENNICTWO 9, 79 (1962) - THROUGH CHEM. ABSTRS. 61, 5836B (1964).
(323) K. FUJII, T. MOCHIZUKI, S. IMOTO, J. UKIDA, AND M. MATSUMOTO, J. POLYMER SCI., PT. A, 2, 2327 (1964).
(324) <A> G. NATTA, G. DALL'ASTA, G. MAZZANTI, I. PASQUON, A. VALVASSORI, AND A. ZAMBELLI, J. AM. CHEM. SOC. 83, 3343 (1961), <B> P. CORRADINI AND P. GANIS, MAKROMOL. CHEM. 62, 97 (1963).
(325) E.N. ZILBERMAN, A.E. KULIKOVA, AND N.M. TEPLYAKOV, J. POLYMER SCI. 56, 417 (1962).
(326) G. ALLEGRA, A. PONOGLIO, AND I. PASQUON, REND. IST. LOMBARDO SCI. LETTERE A95, 335 (1961).
(327) M. ISHIBASHI, J. POLYMER SCI., PT. A, 2, 4361 (1964).
(328) A.D. KETLEY AND R.J. EHRIG, J. POLYMER SCI., PT. A, 2, 4461 (1964).
(329) O. VOGL, J. POLYMER SCI., PT. A, 2, 4621 (1964).
(330) A. CHIBA, H. FUTAMA, AND J. FURUICHI, REPTS. PROGR. POLYMER PHYS. JAPAN 7, 51 (1964).
(331) <A> G. CARAZZOLO AND M. MAMMI, J. POLYMER SCI., PT. B, 2, 1057 (1964), <B> G. CARAZZOLO AND G. VALLE, MAKROMOL. CHEM. 90, 66 (1966).
(332) J.P. KENNEDY AND R.M. THOMAS, MAKROMOL. CHEM. 64, 1 (1963).
(333) M. LITT, J. POLYMER SCI., PT. A, 1, 2219 (1963).
(334) R. BRILL, Z. PHYSIK. CHEM. B53, 61 (1943).
(335) J.C. MITCHELL, J. POLYMER SCI., PT. B, 1, 285 (1963).
(336) T. OTA, O. YOSHIZAKI, AND E. NAGAI, <A> J. POLYMER SCI., PT. B, 1, 57 (1963), <B> KOBUNSHI KAGAKU 20, 225 (1963).
(337) E. MUELLER, MELLIAND TEXTILBER. 44, 484 (1963).
(338) J.R. SCHAEFGEN AND R. ZBINDEN, J. POLYMER SCI., PT. A, 2, 4865 (1964).
(339) D.D. COFFMAN, G.J. BERCHET, U.R. PETERSON, AND E.W. SPANAGEL, J. POLYMER SCI. 2, 306 (1947).
(340) H.S. MAKOWSKY, B.K.C. SHIM, AND Z.W. WILCHINSKY, J. POLYMER SCI., PT. A, 2, 4973 (1964).
(341) F. RYBNIKAR, COLLECT. CZECH. CHEM. COMMUN. 27, 2864 (1962).
(342) L. MANDELKERN, M. TRYON, AND F.A. QUINN, JR., J. POLYMER SCI. 19, 77 (1956).
(343) M. INOUE, J. POLYMER SCI., PT. A, 1, 2697 (1963).
(344) I. HEBER, KOLLOID-Z. 189, 110 (1963).
(345) H. WILSKI AND T. GREWER, J. POLYMER SCI., PT. C, NO. 6, 33 (1964).
(346) G.B. GECHELE AND L. CRESCENTINI, J. APPL. POLYMER SCI. 7, 1349 (1963).
(347) <A> R. KOYAMA, BULL. INST. CHEM. RES., KYOTO UNIV. 41, 207 (1963), <B> A. NISHIOKA, H. WATANABE, R. KOYAMA, AND H. NAGAI, REPTS. PROGR. POLYMER PHYS. JAPAN 6, 311 (1963).
(348) <A> H. TADOKORO, Y. CHATANI, M. KOBAYASHI, T. YOSHIHARA, AND S. MURAHASHI, REPTS. PROGR. POLYMER PHYS. JAPAN 6, 303 (1963), <B> K. IMADA, T. MIYAKA, Y. CHATANI, H. TADOKORO, AND S. MURAHASHI, MAKROMOL. CHEM. 83, 113 (1965).
(349) H.W. WYCKOFF, J. POLYMER SCI. 62, 83 (1962).
(350) A. TURNER-JONES, J. POLYMER SCI., PT. B, 1, 455 (1963).
(351) M.G. BROADHURST, J. RES. NATL. BUR. STD. 70A, 481 (1966).
(352) G. NATTA AND G. MORAGLIO, RUBBER PLASTICS AGE 44, 42 (1962).
(353) G. NATTA AND G. MORAGLIO, MAKROMOL. CHEM. 66, 218 (1963).

(354) M. ISHIBASHI, POLYMER 5, 305 (1964).
(355) A. TURNER-JONES AND J.M. AIZLEWOOD, J. POLYMER SCI., PT. B, 1, 471 (1963).
(356) M. ISHIBASHI, J. POLYMER SCI., PT. B, 1, 629 (1963).
(357) Y. SAKURADA, M. MATSUMOTO, K. IMAI, A. NISHIODA, AND Y. KATO, J. POLYMER SCI., PT. B, 1, 633 (1963).
(358) I. KIRSHENBAUM, Z.W. WILCHINSKY, AND B. GROTEN, J. APPL. POLYMER SCI. 8, 2723 (1964).
(359) N.P. BORISOVA AND T.M. BIRSHTEIN, VYSOKOMOLEKUL. SOEDIN. 5, 279 (1963) - POLYMER SCI. (USSR) (ENGLISH TRANSL.) 5, 907 (1963).
(360) G. DALL'ASTA AND I.W. BASSI, CHIM. IND. (MILAN) 43, 999 (1961).
(361) C.A. BOYE, BULL. AM. PHYS. SOC. 8, 266 (1963).
(362) P.J. FLORY AND A. VRIJ, J. AM. CHEM. SOC. 85, 3548 (1963).
(363) P.R. SWAN, J. POLYMER SCI. 42, 525 (1960).
(364) A. TURNER-JONES AND R.P. PALMER, POLYMER 4, 525 (1963).
(365) M. GENAS, ANGEW. CHEM. 74, 535 (1962).
(366) N.G. GAYLORD, EDITOR, POLYETHERS II., INTERSCIENCE PUBLISHERS, NEW YORK, 1963.
(367) <A> M. CESARI, J. POLYMER SCI., PT. B, 2, 453 (1964), <B> W. MARCONI, A. MAZZEI, S. CUCINELLA, AND M. CESARI, J. POLYMER SCI., PT. A, 2, 4261 (1964). (A AND B OBTAINED FROM EQUATORIAL REFLECTIONS ALONE - TRUE CELL IS PROBABLY TRICLINIC).
(368) W. MARCONI, A. MAZZEI, A. CUCINELLA, M. CESARI, AND E. PAULEZZI, J. POLYMER SCI., PT. A, 3, 123 (1965).
(369) W.J. JACKSON, JR. AND J.R. CALDWELL, J. APPL. POLYMER SCI. 7, 1975 (1963).
(370) R.B. ISAACSON, I. KIRSHENBAUM, AND W.C. FEIST, J. APPL. POLYMER SCI. 8, 2789 (1964).
(371) G. NATTA, L. PORRI, AND M.C. GALLAZZI, CHIM. IND. (MILAN) 46, 1158 (1964).
(372) D.C. KOCKOTT, KOLLOID-Z. 198, 17 (1964).
(373) F.S. INGRAHAM AND D.F. WOOLEY, JR., IND. ENG. CHEM. 56, NO. 9, 53 (1964).
(374) R. JANSSEN, H. RUYSSCHAERT, AND R. VROOM, MAKROMOL. CHEM. 77, 153 (1964).
(375) G. NATTA, I.W. BASSI, AND G. PEREGO, ATTI ACCAD. NAZL. LINCEI, REND., CLASSE SCI. FIS., MAT. NAT. 36, 291 (1964).
(376) F.B. MOODY, MACROMOL. SYNTH. 1, 67 (1963).
(377) G. ALLEN, C. BOOTH, M.N. JONES, D.J. MARKS, AND W.D. TAYLOR, POLYMER 5, 547 (1964).
(378) G. WASAI, T. SAEGUSA, AND J. FURUKAWA, KOGYO KAGAKU ZASSHI 67, 1428 (1964)
(379) T. TSURUTA, R. FUJIO, AND J. FURUKAWA, MAKROMOL. CHEM. 80, 172 (1964).
(380) F. DANUSSO AND G. GIANOTTI, MAKROMOL. CHEM. 80, 1 (1964).
(381) F. DANUSSO, G. GIANOTTI, AND G. POLIZZOTTI, MAKROMOL. CHEM. 80, 13 (1964).
(382) G.P. LORENZI, E. BENEDETTI, AND F. CHIELLINI, CHIM. IND. (MILAN) 46, 1474 (1964).
(383) M. GOODMAN AND Y.-L. FAN, J. AM. CHEM. SOC. 86, 4922 (1964).
(384) G. NATTA, P. CORRADINI, AND P. GANIS, J. POLYMER SCI., PT. A, 3, 11 (1965)
(385) A. BELL, J.G. SMITH, AND C.J. KIBLER, J. POLYMER SCI., PT. A, 3, 19 (1965)
(386) E.L. WITTBECKER, W.S. SPLIETHOFF, AND G.R. STINE, J. APPL. POLYMER SCI. 9, 213 (1965).
(387) G. WASAI, T. IWATA, K. HIRONO, M. KURAGANO, T. SAEGUSA, AND J. FURUKAWA, KOGYO KAGAKU ZASSHI 67, 1920 (1964).
(388) D.E. ILYINA, B.A. KRENTSEL, AND G.E. SEMENIDO, J. POLYMER SCI., PT. C, NO. 4, 999 (1964).
(389) E.L. GAL'PERIN AND YU.V. STROGALIN, VYSOKOMOLEKUL. SOEDIN. 7, 16 (1965) - POLYMER SCI. (USSR) (ENGLISH TRANSL.) 7, 15 (1965).
(390) F.T. SIMON AND J.M. RUTHERFORD, JR., J. APPL. PHYS. 35, 82 (1964).
(391) J. POWERS, J.D. HOFFMAN, J.J. WEEKS, AND F.A. QUINN, JR., J. RES. NATL. BUR. STD. 69A, 335 (1965).
(392) G. FARROW, POLYMER 4, 191 (1963).
(393) F. DANUSSO, G. MORAGLIO, W. CHIGLIA, L. MOTTO, AND G. TALAMINI, CHIM. IND. (MILAN) 41, 748 (1959).
(394) W.R. KRIGBAUM AND I. UEMATSU, J. POLYMER SCI., PT. A, 3, 767 (1965).
(395) W.F. GORHAM, AM. CHEM. SOC. POLYMER PREPRINTS 6, 73 (1965).
(396) H. MORAWETZ, S.Z. ZAKABHAZY, J.B. LANDO, AND B. POST, PROC. NATL. ACAD. SCI. US 49, 789 (1963).
(397) M. BARLOW, J. POLYMER SCI., PT. A-2, 4, 121 (1966).
(398) R. YAMADERA AND C. SONODA, J. POLYMER SCI., PT. B, 3, 411 (1965).
(399) E.J. VANDENBERG, J. POLYMER SCI. 47, 489 (1960).
(400) YU. YA. TOMASHPOLSKII AND G.S. MARKOVA, VYSOKOMOLEKUL. SOEDIN. 6, 27 (1964) - POLYMER SCI. (USSR) (ENGLISH TRANSL.) 6, 316 (1964).
(401) <A> A. KAWASAKI, J. FURUKAWA, T. TSURUTA, Y. NAKAYAMA, AND G. WASAI, MAKROMOL. CHEM. 49, 112 (1961), <B> IBID., P. 136, <C> T. MAKIMOTO, T. TSURUTA, AND J. FURUKAWA, IBID., 50, 116 (1961), <D> G. WASAI, J. FURUKAWA, AND A. KAWASAKI, KOGYO KAGAKU ZASSHI 68, 210 (1965).
(402) V.V. KORSHAK AND T.M. FRUNZE, SYNTHETIC HETERO-CHAIN POLYAMIDES, TRANSLATED BY N. KANER, DANIEL DAVEY AND CO., NEW YORK, 1964 (TABLES I, II, XXI, XXVIII, AND XXX).

(403) M. CESARI, G. PEREGO, AND A. MAZZEI, MAKROMOL. CHEM. 83, 196 (1965).
(404) H. IIYAMA, M. ASAKURA, AND K. KIMOTO, KOGYO KAGAKU ZASSHI 68, 243 (1965).
(405) I. KIRSHENBAUM, J. POLYMER SCI., PT. A, 3, 1869 (1965).
(406) H. KOJIMA AND K. YAMAGUCHI, KOBUNSHI KAGAKU 19, 715 (1962).
(407) E. ECHOCHARD, J. CHIM. PHYS. 43, 113 (1946), THROUGH REF. 402, P. 363.
(408) F.A. QUINN, JR., AND J. POWERS, J. POLYMER SCI., PT. B, 1, 341 (1963).
(409) G. NATTA AND M. PEGORARO, ATTI ACCAD. NAZL. LINCEI, REND., CLASSE SCI.
FIS., MAT. NAT. 34, 110 (1963).
(410) H. TADOKORO, S. SEKI, AND I. NITTA, BULL. CHEM. SOC. JAPAN 28, 559 (1955).
(411) G. NATTA, F. DANUSSO, AND G. MORAGLIO, MAKROMOL. CHEM. 28, 166 (1958).
(412) <A> G. NATTA, G. DALL'ASTA, G. MAZZANTI, I. PASQUON, A. VALVASSORI, AND A.
ZAMBELLI, MAKROMOL. CHEM. 54, 95 (1962), <B> G. NATTA, G. ALLEGRA, I.W.
BASSI, P. CORRADINI, AND P. GANIS, ATTI ACCAD. NAZL. LINCEI, REND., CLASSE
SCI. FIS., MAT. NAT. 36, 433 (1964).
(413) V.P. LEBEDEV, N.A. OKLADNOV, K.S. MINSKER, AND B.P. SHTARKMAN,
VYSOKOMOLEKUL. SOEDIN. 7, 655 (1965).
(414) D.J. LYMAN, J. HELLER, AND M. BARLOW, MAKROMOL. CHEM. 84, 64 (1965).
(415) J. BOOR, JR., AND E.A. YOUNGMAN, J. POLYMER SCI., PT. B, 3, 577 (1965).
(416) <A> G. CARAZZOLO, L. MORTILLARO, L. CREDALI, AND S. BEZZI, J. POLYMER
SCI., PT. B, 2, 997 (1964), <B> L. MORTILLARO, L. CREDALI, M. RUSSO, AND
C. DE CHEECHI, J. POLYMER SCI., PT. B, 3, 581 (1965), <C> G. CARAZZOLO AND
G. VALLE, J. POLYMER SCI., PT. A, 3, 4013 (1965).
(417) E.L. GAL'PERIN, YU.V. STROGALIN, AND M.P. MLENIK, VYSOKOMOLEKUL. SOEDIN.
7, 933 (1965) - POLYMER SCI. (USSR) (ENGLISH TRANSL.) 7, 1031 (1965).
(418) J.B. LANDO, H.G. OLF, AND A. PETERLIN, J. POLYMER SCI., PT. A-1, 4, 941
(1966).
(419) <A> K. OKAMURA AND R.H. MARCHESSAULT, IN G.N. RAMACHANDRAN, ED.,
CONFORMATION OF BIOPOLYMERS, V. 2, ACADEMIC PRESS, NEW YORK, 1967, P. 709,
<B> J. CORNIBERT AND R.H. MARCHESSAULT, J. MOL. BIOL. 71, 735 (1972).
(420) M. DONATI, G. PEREGO, AND M. FARINA, MAKROMOL. CHEM. 85, 301 (1965).
(421) H. KOMOTO AND K. SAOTOME, KOBUNSHI KAGAKU 22, 337 (1965).
(422) W.G.C. FORSYTH, A.C. HAYWARD, AND J.B. ROBERTS, NATURE 182, 800 (1958).
(423) A. NAKAJIMA, H. HAMADA, AND S. HAYASHI, MAKROMOL. CHEM. 95, 40 (1966).
(424) G. NATTA, I.W. BASSI, AND G. ALLEGRA, MAKROMOL. CHEM. 89, 81 (1965).
(425) K. MIKI AND R. NAKATSUKA, REPTS. PROGR. POLYMER PHYS. JAPAN 8, 115 (1963).
(426) W.D. NIEGISCH, <A> J. APPL. PHYS. 37, 4041 (1966), <B> J. POLYMER SCI.,
PT. B, 4, 531 (1966).
(427) <A> Y. JOH, T. YOSHIHARA, Y. KOTAKE, F. IDE, AND K. NAKATSUKA, J. POLYMER
SCI., PT. B, 3, 933 (1965), <B> T. YOSHIHARA, Y. KOTAKE, AND Y. JOH,
IBID., 5, 459 (1967).
(428) C.G. OVERBERGER AND J.K. WEISE, <A> J. POLYMER SCI., PT. B, 2, 329 (1964),
<B> J. AM. CHEM. SOC. 90, 3533 (1960).
(429) P. CORRADINI AND I.W. BASSI, J. POLYMER SCI., PT. C, NO. 16, 3233 (1968).
(430) G. NATTA, P. CORRADINI, P. GANIS, AND P.A. TEMUSSI, J. POLYMER SCI., PT.
C, NO. 16, 2477 (1967).
(431) <A> M.G. HUGUET, MAKROMOL. CHEM. 94, 205 (1966), <B> H.D. NOETHER, J.
POLYMER SCI., PT. C, NO. 16, 725 (1967).
(432) I.W. BASSI, P. GANIS, AND P.A. TEMUSSI, J. POLYMER SCI., PT. C, NO. 16,
2867 (1967).
(433) R.K. TUBBS, J. POLYMER SCI., PT. A, 3, 4181 (1965).
(434) G. NATTA, G. ALLEGRA, I.W. BASSI, D. SIANESI, G. CAPORICCIO, AND E. TORTI,
J. POLYMER SCI., PT. A, 3, 4263 (1965).
(435) R. BONART, MAKROMOL. CHEM 92, 149 (1966).
(436) G. NATTA AND G. DALL'ASTA, CHIM. IND. (MILAN) 46, 1429 (1964).
(437) C.G. OVERBERGER, H. KAYE, AND G. WALSH, J. POLYMER SCI., PT. A, 2, 755
(1964).
(438) H. SCHNELL, ANGEW. CHEM. 68, 633 (1956).
(439) G. NATTA, G. DALL'ASTA, I.W. BASSI, AND G. CARELLA, MAKROMOL. CHEM. 91, 87
(1966).
(440) V.M. COIRO, P. DE SANTIS, L. MAZZARELLA, AND L. PICOZZI, J. POLYMER SCI.,
PT. A, 3, 4001 (1965).
(441) J.A. FAUCHER AND F.P. REDING, CRYSTALLINE OLEFIN POLYMERS, PART I, P. 677,
R.A.V. RAFF AND K.W. DOAK, EDITORS, INTERSCIENCE PUBLISHERS, NEW YORK 1965
(442) A. TURNER-JONES, POLYMER 7, 23 (1966).
(443) N.I. MAKAREVICH, ZH. PRIKL. SPEKTROSKOPII, AKAD. NAUK BELORUSSK. SSR 2,
341 (1965) - THROUGH CHEM. ABSTRS. 63, 8509A (1965).
(444) J.R. SCHAEFGEN, F.H. KOONTZ, AND R.F. TIETZ, J. POLYMER SCI. 40, 377
(1959).
(445) T.R. GIBB, JR., R.A. CLENDINNING, AND W.D. NIEGISCH, J. POLYMER SCI.,
PT. A-1, 4, 917 (1966).
(446) <A> H. TADOKORO, Y. TAKAHASHI, Y. CHATANI, AND H. KAKIDA, MAKROMOL. CHEM.
109, 96 (1967), <B> H. KAKIDA, D. MAKINO, Y. CHATANI, M. KOBAYASHI, AND H.
TADOKORO, MACROMOLECULES 3, 569 (1970).

(447)   J.B. ROSE, J. CHEM. SOC. 1956, 542, 546.
(448)   M. CESARI, G. PEREGO, AND W. MARCONI, MAKROMOL. CHEM. 94, 194 (1966).
(449)   I.I. NOVAK AND V.I. VETTEGREN, VYSOKOMOLEKUL. SOEDIN. 7, 1027 (1965) -
        POLYMER SCI. (USSR) (ENGLISH TRANSL.) 7, 1136 (1965).
(450)   F. HAMADA AND A. NAKAJIMA, KOBUNSHI KAGAKU 23, 395 (1966).
(451)   G. MORAGLIO, G. POLIZZOTTI, AND F. DANUSSO, EUROPEAN POLYMER J. 1, 183
        (1965).
(452)   K.-S. CHAN, PH.D. THESIS, STATE UNIV. COLLEGE OF FORESTRY AT SYRACUSE
        UNIV., DISSER. ABSTR. 26, 4260 (1966).
(453)   J.G. SMITH, C.J. KIBLER, AND B.J. SUBLETT, J. POLYMER SCI., PT. A-1, 4,
        1851 (1966).
(454)   K. SAOTOME AND H. KOMOTO, J. POLYMER SCI., PT. A-1, 4, 1463 (1966).
(455)   K.F. WISSBRUN, J. POLYMER SCI., PT. A-2, 4, 827 (1966).
(456)   P. GANIS AND P.A. TEMUSSI, EUROPEAN POLYMER J. 2, 401 (1966).
(457)   R.J. FREDERICKS, T.H. DOYNE, AND R.S. SPRAQUE, J. POLYMER SCI., PT. A-2,
        4, 899 (1966).
(458)   R.J. FREDERICKS, T.H. DOYNE, AND R.S. SPRAQUE, J. POLYMER SCI., PT. A-2,
        4, 913 (1966).
(459)   T. OTA, M. YAMASHITA, O. YOSHIZAKI, AND E. NAGAI, J. POLYMER SCI., PT.
        A-2, 4, 959 (1966).
(460)   H. TADOKORO, J. POLYMER SCI., PT C, NO. 15, 1 (1966).
(461)   H. YASUDA, Y. TANADA, AND M. TAKAYANAGI, KOGYO KAGAKU ZASSHI 69, 304
        (1966).
(462)   R.W. EYKAMP, A.M. SCHNEIDER, AND E.W. MERRILL, J. POLYMER SCI., PT. A-2,
        4, 1025 (1966).
(463)   G. CARAZZOLO AND M. MAMMI, MAKROMOL. CHEM. 100, 28 (1967).
(464)   K. CHUJO, H. KOBAYASHI, J. SUZUKI, S. TOKUHARA, AND M. TANABE, MAKROMOL.
        CHEM. 100, 262 (1967).
(465)   K. CHUJO, H. KOBAYASHI, J. SUZUKI, AND S. TOKUHARA, MAKROMOL. CHEM. 100,
        267 (1967).
(466)   W. BRAUN, K.-H. HELLWEGE, AND W. KNAPE, KOLLOID-Z. 215, 10 (1967).
(467)   M. HAYASHI, Y. SHIRO, AND H. MURATA, BULL. CHEM. SOC. JAPAN 39, 1857
        (1966).
(468)   M. HAYASHI, Y. SHIRO, AND H. MURATA, BULL. CHEM. SOC. JAPAN 39, 1861
        (1966).
(469)   F. TUINSTRA, ACTA CRYST. 20, 341 (1966).
(470)   B.M. GINZBURG, L.M. KORZHAVIN, S.YA. FRENKEL, L.A. LAIUS, AND M.B. ADROVA,
        VYSOKOMOLEKUL. SOEDIN. 8, 278 (1966) - POLYMER SCI. (USSR) (ENGLISH
        TRANSL.) 8, 302 (1966).
(471)   F.E. KARASZ, J.M. O'REILLY, H.E. BAIR, AND R.A. KLUGE, AM. CHEM. SOC.
        POLYMER PREPRINTS 9, 822 (1968).
(472)   P.E. SLADE AND T.A. OROFINO, AM. CHEM. SOC. POLYMER PREPRINTS 9, 825
        (1968).
(473)   W.J. MACKNIGHT, M. YANG, AND T. KAJIMA, AM. CHEM. SOC. POLYMER PREPRINTS
        9, 860 (1968).
(474)   V.G. BARANOV, BU. ZHU-CHAN, T.I. VOLKOV, AND S.YA. FRENKEL, VYSOKOMOL.
        SOEDIN., SER. A, 9, 81 (1967) - POLYMER SCI. (USSR) (ENGLISH TRANSL.) 9,
        87 (1967).
(475)   P. KOVACIC, M.B. FELDMAN, J.P. KOVACIC, AND J.B. LANDO, J. APPL. POLYMER
        SCI. 12, 1735 (1968).
(476)   I.W. BASSI AND G. FAGHERAZZI, EUROPEAN POLYMER J. 4, 123 (1968).
(477)   G. FAGHERAZZI AND I.W. BASSI, EUROPEAN POLYMER J. 4, 151 (1968).
(478)   F. DANUSSO AND G. GIANOTTI, EUROPEAN POLYMER J. 4, 165 (1968).
(479)   C.G. OVERBERGER AND T. TAKEKOSHI, MACROMOLECULES 1, 7 (1968).
(480)   K. KIKUKAWA, S. NOZAKURA, AND S. MURAHASHI, KOBUNSHI KAGAKU 25, 19 (1968).
(481)   <A> H.K. REIMSCHUESSEL, L.G. ROLDAN, AND J.P. SIBILIA, J. POLYMER SCI.,
        PT. A-2, 6, 559 (1968), <B> H.K. REIMSCHUESSEL, TRANS. N.Y. ACAD. SCI. 33,
        219 (1971).
(482)   S. KOBAYASHI, H. TADOKORO, AND Y. CHATANI, MAKROMOL. CHEM. 112, 225 (1968)
(483)   F.J. GOLEMBA, J.E. GUILLET, AND S.C. NYBURG, J. POLYMER SCI., PT. A-1, 6,
        1341 (1968).
(484)   F.E. KARASZ, H.E. BAIR, AND J.M. O'REILLY, J. POLYMER SCI., PT. A-2, 6,
        1141 (1968).
(485)   <A> A. TURNER-JONES AND A.J. COBBOLD, J. POLYMER SCI., PT. B, 6, 538
        (1968), <B> R.J. SAMUELS AND R.H. YEE, J. POLYMER SCI., PT. A-2, 10, 385
        (1972).
(486)   J.J. KLEMENT AND P.H. GEIL, J. POLYMER SCI., PT. A-2, 6, 1381 (1968).
(487)   <A> H.M. VAN DORT, C.A.M. HOEFS, E.P. MAGRE', A.J. SCHOPF, AND K. YNTEMA,
        EUROPEAN POLYMER J. 4, 275 (1968), <B> J. BOON AND E.P. MAGRE', MAKROMOL.
        CHEM. 126, 130 (1969).
(488)   G. NATTA, P. CORRADINI, I.W. BASSI, AND G. FAGHERAZZI, EUROPEAN POLYMER J.
        4, 297 (1968).

(489)   <A> A. CARBONARO, A. GRECO, AND I.W. BASSI, EUROPEAN POLYMER J. 4, 445 (1968), <B> I.W. BASSI AND G. CHIOCCOLA, IBID., 5, 163 (1969).

(490)   Y. CHATANI, K. SUEHIRO, Y. OKITA, H. TADOKORO, AND K. CHUJO, MAKROMOL. CHEM. 113, 215 (1968).

(491)   <A> J. MASAMOTO, K. SASAGURI, C. OHIZUMI, AND H. KOBAYASHI, J. POLYMER SCI., PT. A-2, 8, 1703 (1970), <B> J. MASAMOTO AND H. KOBAYASHI, KOBUNSHI KAGAKU 27, 220 (1970).

(492)   H. UTSUNOMIYA, T. MORI, K. IMADA, AND M. TAKAYANAGI, REPTS. PROGR. POLYMER PHYS. JAPAN 11, 153 (1968).

(493)   <A> S. MINAMI, S. MANABE, AND M. TAKAYANAGI, REPTS. PROGR. POLYMER PHYS. JAPAN 11, 155 (1968), <B> S. MANABE, S. MINAMI, AND M. TAKAYANAGI, KOGYO KAGAKU ZASSHI 73, 1577 (1970).

(494)   <A> S. TAKAMUKU, K. IMADA, AND M. TAKAYANAGI, REPTS. PROGR. POLYMER PHYS. JAPAN 11, 159 (1968), <B> M. KUROISHI, M. FUJISAKI, T. KAJIYAMA, AND M. TAKAYANAGI, NIPPON KAGAKU KAISHI 1972, 1281.

(495)   K.-H. ILLERS AND H. HABERKORN, MAKROMOL. CHEM. 146, 267 (1971).

(496)   S. IWAYANAGI, I. SAKURAI, T. SAKURAI, AND T. SETO, <A> REPTS. PROGR. POLYMER PHYS. JAPAN 10, 167 (1967), <B> J. MACROMOL. SCI., PHYS., B2, 163 (1968).

(497)   <A> H. TADOKORO, Y. TAKAHASHI, S. OTSUKA, K. MORI, AND F. IMAISUMI, J. POLYMER SCI., PT. B, 3, 697 (1965), <B> H. TADOKORO, M. KOBAYASHI, K. MORI, Y. TAKAHASHI, AND S. TANIYAN, REPTS. PROGR. POLYMER PHYS. JAPAN 10, 181 (1967), <C> J. POLYMER SCI., PT. C, NO. 22, 1031 (1969).

(498)   C. BOOTH, C.J. DEVOY, AND G. GEE, POLYMER 12, 327 (1971).

(499)   T. YUBAYASHI, H. SATO, AND N. YAMADA, REPTS PROGR. POLYMER PHYS. JAPAN 10, 271 (1967).

(500)   H. AILHAUD, Y. GALLOT, AND A. SKOULIOS, COMPT. REND. 267C, 139 (1968).

(501)   G. GIANNOTTI AND A. CAPIZZI, EUROPEAN POLYMER J. 4, 677 (1968).

(502)   D. FISHER, PROC. PHYS. SOC. (LONDON) B66, 7 (1963).

(503)   H. BESTIAN, UNPUBLISHED WORK OF H. SCHERER, ANGEW. CHEM., INTERN. ED. ENGL. 7, 278 (1968).

(504)   Y. TAKAHASHI, H. TAKOKORO, AND Y. CHATANI, J. MACROMOL. SCI., PHYS., B2, 361 (1968).

(505)   E.F. OLEINIK AND N.S. YENIKOLOPYAN, VYSOKOMOL. SOEDIN., SER. A, 9, 2609 (1967) - POLYMER SCI. (USSR) (ENGLISH TRANSL.) 9, 2951 (1967).

(506)   M. OKADA, Y. YAMASHITA, AND Y. ISHII, MAKROMOL. CHEM. 80, 196 (1964).

(507)   <A> C.H. BAMFORD, L. BROWN, A. ELLIOTT, W.E. HANBY, AND I.F. TROTTER, NATURE 173, 27 (1954), <B> L. BROWN AND I.F. TROTTER, TRANS. FARADAY SOC. 52, 537 (1956), <C> A. ELLIOTT AND B.R. MALCOLM, PROC. ROY. SOC. (LONDON), SER. A, 249, 30 (1959).

(508)   C.S. MARVEL AND G.E. HARTZELL, J. AM. CHEM. SOC. 81, 448 (1959).

(509)   S. ARNOTT, S.D. DOVER, AND A. ELLIOTT, J. MOL. BIOL. 30, 201 (1967).

(510)   P.M. COWAN AND S. MCGAVIN, NATURE 176, 501 (1955).

(511)   V. SASISEKHARAN, ACTA CRYST. 12, 897 (1959).

(512)   S. ARNOTT AND S.D. DOVER, ACTA CRYST. B24, 599 (1968).

(513)   T. UCHIDA AND H. TADOKORO, J. POLYMER SCI., PT. A-2, 5, 63 (1967).

(514)   M. GOODMAN, AM. CHEM. SOC. POLYMER PREPRINTS 10, 36 (1969). (PSEUDOHEXAGONAL CELL BASED ON EQUATORIAL REFLECTIONS).

(515)   H. SAKAKIHARA, Y. TAKAHASHI, H. TADOKORO, P. SIGWALT, AND N. SPASSKY, MACROMOLECULES 2, 515 (1969).

(516)   N. YASUOKA, N. KASAI, M. KAKUDO, T. ANDO, AND S. KURIBAYASHI, REPTS. PROGR. POLYMER PHYS. JAPAN 11, 149 (1968).

(517)   C.L. LEE, O.K. JOHANNSON, O.L. FLANINGAM, AND P. HAHN, AM. CHEM. SOC. POLYMER PREPRINTS 10, 1311 (1969).

(518)   <A> C.L. LEE, O.K. JOHANNSON, O.L. FLANINGAM, AND P. HAHN, AM. CHEM. SOC. POLYMER PREPRINTS 10, 1319 (1969), <B> D.R. PETERSEN, D.R. CARTER, AND C.L. LEE, J. MACROMOL. SCI., PHYS., B3, 519 (1969).

(519)   M. HIKOSAKA AND T. SETO, REPTS. PROGR. POLYMER PHYS. JAPAN 12, 153 (1969).

(520)   H. TADOKORO, Y. CHATANI, H. KUSANAGI, AND M. YOKAYAMA, MACROMOLECULES 3, 441 (1970).

(521)   <A> T. KANAMOTO, H. NAGAI, AND K. TANAKA, REPTS. PROGR. POLYMER PHYS. JAPAN 9, 135 (1966), <B> T. KANAMOTO, K. TANAKA, AND H. NAGAI, J. POLYMER SCI., PT. A-2, 9, 2043 (1971).

(522)   U. SCHMUELI, W. TRAUB, AND K. ROSENHECK, J. POLYMER SCI., PT. A-2, 7, 515 (1969).

(523)   S.Y. HOBBS AND F.W. BILLMEYER, JR., J. POLYMER SCI., PT. A-2, 7, 1119 (1969).

(524)   A.R. SCHULTZ AND C.R. MCCULLOUGH, J. POLYMER SCI., PT. A-2, 7, 1577 (1969)

(525)   S. MURAHASHI, M. KAMACHI, AND N. WAKABAYASHI, J. POLYMER SCI., PT. B, 7, 135 (1969).

(526)   H.D. NOETHER, C.G. OVERBERGER, AND G. HALEK, J. POLYMER SCI., PT. A-1, 7, 201 (1969).

(527)   E.J. VANDENBERG, J. POLYMER SCI., PT. A-1, 7, 525 (1969).
(528)   G. NATTA, J. DI PIETRO, AND M. CAMBINI, MAKROMOL. CHEM. 56, 200 (1962).
(529)   A. KAWAGUCHI, T. IKAWA, Y. FUJIWARA, AND K. MONOBE, REPTS. PROGR. POLYMER
        PHYS. JAPAN 9, 157 (1966).
(530)   <A> G. NATTA AND I.W. BASSI, EUROPEAN POLYMER J. 3, 43 (1967), <B> G.
        NATTA, I.W. BASSI, AND G. FAGHERAZZI, IBID., 339, <C> G. NATTA, I.W.
        BASSI, AND G. FAGHERAZZI, IBID., 5, 239 (1969).
(531)   E.W. FISCHER, F. KLOOS, AND G. LIESER, J. POLYMER SCI., PT. B, 7, 845
        (1969).
(532)   K. ISHIKAWA, K. MIYASAKA, AND T. OKABE, MAKROMOL. CHEM. 122, 123 (1969).
(533)   E.F. VAINSHTEIN, M.YA. KUSHEREV, A.A. POPOV, AND S.G. ENTELIS, VYSOKOMOL.
        SOEDIN., SER. A, 11, 1606 (1969).
(534)   A. MATTIUSSI, G.B. GECHELE, M.A. PARESI, AND A. RONDANELLI, CHIM. IND.
        (MILAN) 51, 376 (1969).
(535)   P. ZUGENMAIER AND H.-J. CANTOW, KOLLOID-Z. Z. POLYMERE 230, 229 (1969).
(536)   W.P. BAKER, JR., J. POLYMER SCI., PT. A, 1, 655 (1963).
(537)   A. TURNER-JONES, POLYMER 6, 249 (1965).
(538)   A. KOVACS, PRIVATE COMMUNICATION.
(539)   P.W.T. WILLMOTT AND F.W. BILLMEYER, JR., OFFIC. DIG. J. PAINT TECHNOL.
        ENG. 35, 847 (1963).
(540)   H. SCHNELL, IND. ENG. CHEM. 51, 157 (1959).
(541)   R.A. WESSLING, J.E. MARK, AND R.E. HUGHES, J. PHYS. CHEM. 70, 1909 (1966).
(542)   T.P. HOBIN, POLYMER 7, 367 (1966).
(543)   Y. IWAKURA, Y. TANEDA, AND S. UCHIDA, J. APPL. POLYMER SCI. 5, 108 (1961).
(544)   A.J. YU AND R.D. EVANS, J. POLYMER SCI. 62, 249 (1960).
(545)   J. BOOR, JR., AND E.A. YOUNGMAN, J. POLYMER SCI., PT. A-1, 4, 1861 (1966).
(546)   H.E. LUNK AND E.A. YOUNGMAN, J. POLYMER SCI., PT. A, 3, 2983 (1965).
(547)   C.J. KIBLER, A. BELL, AND J.G. SMITH, J. POLYMER SCI., PT. A, 2, 2115
        (1964).
(548)   E.J. VANDENBERG, J. POLYMER SCI., PT. A-1, 10, 329 (1972).
(549)   P. CORRADINI, P. GANIS, AND V. PETRACCONE, EUROPEAN POLYMER J. 6, 281
        (1970).
(550)   K. SUEHIRO AND M. TAKAYANAGI, J. MACROMOL. SCI., PHYS., B4, 39 (1970).
(551)   C. BORRI, S. BRUCKNER, V. CRESCENZI, G. DELLA FORTUNA, A. MARIANO, AND
        P. SCARAZZOTO, EUROPEAN POLYMER J. 7, 1515 (1971).
(552)   H.R. ALLCOCK, R.L. KUGEL, AND K.J. VALAN, INORG. CHEM. 10, 1709 (1966).
(553)   G. ALLEN, C.J. LEWIS, AND S.M. TODD, POLYMER 11, 44 (1970).
(554)   V.Y. CHEN, G. ALLEGRA, P. CORRADINI, AND M. GOODMAN, MACROMOLECULES 3, 274
        (1970).
(555)   G. PEREGO AND M. CESARI, MAKROMOL. CHEM. 133, 133 (1970).
(556)   W.R. KRIGBAUM AND J.H. O MARA, J. POLYMER SCI., PT. A-2, 8, 1011 (1970).
(557)   C. GIENIEWSKI AND R.S. MOORE, MACROMOLECULES 3, 97 (1970).
(558)   H. KANETSUNA, M. HASEGAWA, S. MITSUHASHI, T. KURITA, K. SASAKI, K. MAEDA,
        H. OBATA, AND T. HATAKEYAMA, J. POLYMER SCI., PT. A-2, 8, 1027 (1970).
(559)   H. BITTIGER, R.H. MARCHESSAULT, AND W.D. NIEGISCH, ACTA CRYST. B26, 1923
        (1970).
(560)   F.J. VAN NATTA, J.W. HILL, AND W.H. CAROTHERS, J. AM. CHEM. SOC. 56, 455
        (1934).
(561)   Y. CHATANI, Y. OKITA, H. TADOKORO, AND Y. YAMASHITA, POLYMER J. 1, 555
        (1970).
(562)   G. GIANOTTI AND A. CAPIZZI, EUROPEAN POLYMER J. 6, 743 (1970).
(563)   W.W. DOLL AND J.B. LANDO, J. MACROMOL. SCI., PHYS., B4, 309 (1970).
(564)   <A> H. NAKANISHI, N. NAKANO, AND M. HASEGAWA, J. POLYMER SCI., PT. B, 8,
        755 (1970), <B> H. NAKANISHI, M. HASEGAWA, AND Y. SASADA, J. POLYMER SCI.,
        PT. A-2, 10, 1537 (1972).
(565)   Y. CHATANI, S. NAKATANI, AND H. TADOKORO, MACROMOLECULES 3, 481 (1970).
(566)   J.P. MERCIER AND R. LEGRAS, J. POLYMER SCI., PT. B, 8, 645 (1970).
(567)   D.R. BEECH AND C. BOOTH, J. POLYMER SCI., PT. B, 8, 731 (1970).
(568)   E.G. LOVERING AND D.C. WOODEN, J. POLYMER SCI., PT. A-2, 9, 175 (1971).
(569)   I.W. BASSI, O. BONSIGNORI, G.P. LORENZI, P. PINO, P. CORRADINI, AND P.A.
        TEMUSSI, J. POLYMER SCI., PT. A-2, 9, 193 (1971).
(570)   J.R. SHELTON, J.B. LANDO, AND D.E. AGOSTINI, J. POLYMER SCI., PT. B, 9,
        173 (1971).
(571)   S.Y. HOBBS AND F.W. BILLMEYER, JR., J. POLYMER SCI., PT. A-2, 8, 1387
        (1970).
(572)   <A> M. HACHIBOSHI, T. FUKUDA, AND S. KOBAYASHI, J. MACROMOL. SCI., PHYS.,
        B3, 525 (1969), <B> S. KOBAYASHI AND M. HACHIBOSHI, REPTS. PROGR. POLYMER
        PHYS. JAPAN 13, 157 (1970).
(573)   S. KOBAYASHI AND M. HACHIBOSHI, REPTS. PROGR. POLYMER PHYS. JAPAN 13, 161
        (1970).
(574)   A.R. SHULTZ AND C.R. MCCULLOUGH, J. POLYMER SCI., PT. A-2, 10, 307 (1972).
(575)   G. HINRICHSEN AND H. ORTH, KOLLOID-Z. Z. POLYMERE 247, 844 (1971).

(576)　C.H. BAMFORD, L. BROWN, A. ELLIOTT, W.E. HANBY, AND I.F. TROTTER, NATURE 169, 357 (1952).
(577)　H.L. YAKEL, ACTA CRYST. 6, 724 (1953).
(578)　A. ELLIOTT, R.D.B. FRASER, T.P. MACRAE, I.W. STAPLETON, AND E. SUZUKI, J. MOL. BIOL. 9, 10 (1964).
(579)　E.M. BRADBURY, L. BROWN, A.R. DOWNIE, A. ELLIOTT, W.E. HANBY, AND T.R.R. MCDONALD, J. MOL. BIOL. 2, 276 (1960).
(580)　V. SASISEKHARAN, ACTA CRYST. 12, 903 (1959).
(581)　W.J. ASTBURY, NATURE 163, 722 (1949).
(582)　E.M. BRADBURY, L. BROWN, A.R. DOWNIE, A. ELLIOTT, AND W.E. HANBY, J. MOL. BIOL. 5, 230 (1962).
(583)　Y. MITSUI, ACTA CRYST. 20, 694 (1966).
(584)　A. ELLIOTT, R.D.B. FRASER, AND T.P. MACRAE, J. MOL. BIOL. 11, 821 (1965).
(585)　H.D. KEITH, F.J. PADDEN, JR., AND G. GIANNONI, J. MOL. BIOL. 43, 423 (1969).
(586)　F.J. PADDEN, JR., H.D. KEITH, AND G. GIANNONI, BIOPOLYMERS 7, 793 (1969).
(587)　C.H. BAMFORD, A. ELLIOTT, AND W.E. HANBY, SYNTHETIC POLY-PEPTIDES. PREPARATION, STRUCTURE, AND PROPERTIES, ACADEMIC PRESS, NEW YORK, 1956, P. 272.
(588)　L. PAULING AND R.B. COREY, PROC. NATL. ACAD. SCI. 37, 241 (1951).
(589)　C.H. BAMFORD, L. BROWN, A. ELLIOTT, W.E. HANBY, AND I.F. TROTTER, PROC. ROY. SOC. (LONDON) B141, 49 (1953).
(590)　W. TRAUB AND U. SHMUELI, NATURE 198, 1165 (1963).
(591)　R.D.B. FRASER, T.P. MACRAE, F.H.C. STEWART, AND E. SUZUKI, J. MOL. BIOL. 11, 706 (1965).
(592)　D.R. MORROW AND B.A. NEWMAN, J. APPL. PHYS. 39, 4944 (1968).
(593)　S. KAMBARA AND A. TAKAHASHI, MAKROMOL. CHEM. 58, 226 (1962).
(594)　<A> K. KAJI, THESIS, DEPT. OF POLYMER CHEM., KYOTO UNIV., 1970, <B> K. KAJI AND I. SAKURADA, MAKROMOL. CHEM. 148, 261 (1971), <C> K. KAJI, KYOTO DAIGAKU NIPPON KAGAKUSENI KENKYUSHO KOENSHU 28, 1 (1971) - THROUGH CHEM. ABSTR. 78, 136889Q (1973).
(595)　S. KUBO AND B. WUNDERLICH, <A> J. APPL. PHYS. 42, 4565 (1971), <B> J. POLYMER SCI., POLYMER CHEM. ED., 10, 1949 (1972), <C> MAKROMOL. CHEM. 162, 1 (1972).
(596)　J.C. ANDRIES, J.M. ANDERSON, AND A.G. WALTON, BIOPOLYMERS 10, 1049 (1971).
(597)　YE.L. GAL'PERIN AND B.P. KOSMYNIN, VYSOKOMOL. SOEDIN., SER, A, 11, 1432 (1969) - POLYMER SCI. (USSR) (ENGLISH TRANSL.) 11, 1624 (1969).
(598)　F.E. KARASZ AND D. MANGARAJ, AM. CHEM. SOC. POLYMER PREPRINTS 12, 317 (1971).
(599)　I. BOWMAN, D.S. BROWN, AND R.E. WETTON, POLYMER 10, 715 (1969).
(600)　G.A. CLEGG, D.R. GEE, T.P. MELIA, AND A. TYSON, POLYMER 9, 501 (1968).
(601)　S. MITSUHASHI, KOGYO KAGAKU ZASSHI 72, 2160 (1969).
(602)　F.J. BOERIO AND J.L. KOENIG, J. CHEM. PHYS. 54, 3667 (1971).
(603)　I.V. KUMPANENKO, K.S. KAZANSKII, N.V. PTITSYNA, AND M.YA. KUSHNEREV, VYSOKOMOL. SOEDIN., SER. A, 12, 822 (1970) - POLYMER SCI. (USSR) (ENGLISH TRANSL.) 12, 930 (1970).
(604)　<A> R. HASEGAWA, M. KOBAYASHI, AND H. TADOKORO, POLYMER J. 3, 591 (1972), <B> R. HASEGAWA, Y. TAKAHASHI, Y. CHATANI, AND H. TADOKORO, IBID., 600.
(605)　D.G. GRABER, MICROSCOPE 18, 203 (1970).
(606)　Y. TAKEDA, Y. IITAKA, AND M. TSUBOI, J. MOL. BIOL. 51, 101 (1970).
(607)　A. KIMURA, S. YOSHIMOTO, Y. AKANA, H. HIRATA, S. KUSABAYASHI, H. MIKAWA, AND N. NASAI, J. POLYMER SCI., PT. A-2, 8, 643 (1970).
(608)　P. PINO, G.P. LORENZI, AND E. CHIELLINI, RIC. SCI. REND. 34, 193 (1964).
(609)　K.W. DOAK AND H.N. CAMPBELL, J. POLYMER SCI. 18, 215 (1955).
(610)　J.W. FISHER, J. APPL. CHEM. 4, 212 (1954).
(611)　N.G. GAYLORD, ED., MACROMOLECULAR SYNTHESES, V. 3, JOHN WILEY AND SONS, INC., NEW YORK, 1969.
(612)　V.N. ROGULENKOVA, M.I. MILLIONOVA, AND N.S. ANDREEVA, J. MOL. BIOL. 9, 253 (1964).
(613)　<A> W. TRAUB AND A. YONATH, J. MOL. BIOL. 16, 404 (1966), <B> A. YONATH AND W. TRAUB, IBID., 43, 461 (1969).
(614)　W. TRAUB AND A. YONATH, J. MOL. BIOL. 25, 351 (1967).
(615)　W. TRAUB, J. MOL. BIOL. 43, 479 (1969).
(616)　D.M. SEGAL AND W. TRAUB, J. MOL. BIOL. 43, 487 (1969).
(617)　G.D. BUCKLEY, L.H. CROSS, AND N.H. RAY, J. CHEM. SOC. 1950, 2714.
(618)　G. SAINI, E. CAMPI, AND S. PARODI, GAZZ. CHIM. ITAL. 87, 342 (1957).
(619)　A. NASINI, J. POLYMER SCI. 34, 106 (1959).
(620)　W.L. TRUETT, D.R. JOHNSON, I.M. ROBINSON, AND B.A. MONTAGUE, J. AM. CHEM. SOC. 82, 2337 (1960).
(621)　S. HORIKIRI, J. POLYMER SCI., PT. A-2, 10, 1167 (1972).
(622)　V. PETRACCONE, P. GANIS, P. CORRADINI, AND G. MONTAGNOLI, EUROPEAN POLYMER J. 8, 99 (1972).

(623) V. CRESCENZI, G. MANZINI, G. CALZOLARI, AND C. BORRI, EUROPEAN POLYMER J. 8, 449 (1972).

(624) S.F. BERMUDEZ AND J.M.G. FATOU, EUROPEAN POLYMER J. 8, 575 (1972).

(625) V. PETRACCONE, B. PIROZZI, AND P. CORRADINI, EUROPEAN POLYMER J. 8, 107 (1972).

(626) Z. MENCIK, J. MACROMOL. SCI., PHYS., B6, 101 (1972).

(627) Y. SAITO, S. NANSAI, AND S. KINOSHITA, POLYMER J. 3, 113 (1972).

(628) H. WILSKI, KOLLOID-Z. Z. POLYMERE 248, 867 (1971).

(629) Y. MIYAMOTO, C. NAKAFUKU, AND T. TAKEMURA, POLYMER J. 3, 122 (1972).

(630) E. MARTUSCELLI, MAKROMOL. CHEM. 151, 159 (1972).

(631) K.A. ANDRIANOV, G.L. SLONIMSKII, A.A. ZHDANOV, V.YU. LEVIN, YU.K. GODOVSKII, AND V.A. MOSKALENKO, J. POLYMER SCI., PT. A-1, 10, 1 (1972).

(632) R. BACKSAI, J.E. GOODRICH, AND J.B. WILKES, J. POLYMER SCI., PT. A-1, 10, 1529 (1972).

(633) O. BAYER, ANGEW. CHEM. A59, 257 (1947).

(634) K. SAOTOME AND H. KOMOTO, J. POLYMER SCI., PT. A-1, 5, 119 (1967).

(635) J.P. SIBILIA, L.G. ROLDAN, AND S. CHANDRASEKARAN, J. POLYMER SCI., PT. A-2, 10, 549 (1972).

(636) H. BITTIGER AND E. HUSEMANN, J. POLYMER SCI., PT. B, 10, 367 (1972).

(637) <A> M. RAGAZZINI, C. GARBUGLIO, D. CARCANO, B. MINASSO, AND G. CEVIDALLI, EUROPEAN POLYMER J. 3, 129 (1967), <B> C. GARBUGLIO, M. RAGAZZINI, O. PILATI, D. CARCANO, AND G. CEVIDALLI, IBID., 137.

(638) <A> T. TAKIZAWA, T. SANO, Y. CHATANI, AND I. TAGUCHI, PREPRINTS OF THE 6TH POLYMER SYMP., POLYMER PHYS. DIV. (NAGOYAN), P. 83, OCT. 1957, <B> Y. CHATANI, I. TAGUCHI, T. SANO, AND T. TAKIZAWA, ANN. REPT. INST. TEXTILE SCI. JAPAN 13, 37 (1960), THROUGH K. FUJII, J. POLYMER SCI., PT. D, NO. 5, 431 (1971).

(639) <A> G. GIANOTTI, G. DALL'ASTA, A. VALVASSORI, AND V. ZAMBONI, MAKROMOL. CHEM. 149, 117 (1971), <B> G. DALL'ASTA, IBID., 154, 1 (1972).

(640) H. WILSKI, MAKROMOL. CHEM. 150, 209 (1971).

(641) M. SUWALSKY AND W. TRAUB, BIOPOLYMERS 11, 623 (1972).

(642) <A> L.G. KAZARYAN, YE.G. LUR'E, AND L.A. IGONIN, VYSOKOMOL. SOEDIN., SER. B, 11, 779 (1969), <B> YE.G. LUR'E, L.G. KAZARYAN, E.L. UCHASTKINA, V.V. KOVRIGA, K.N. VLASOVA, M.L. DOBROKHOTOVA, AND L.M. YEMEL'YANOVA, VYSOKOMOL. SOEDIN., SER. A, 13, 606 (1971) - POLYMER SCI. (USSR) (ENGLISH TRANSL.) 13, 685 (1971).

(643) K.J. PALMER AND M. BALLANTYNE, J. AM. CHEM. SOC. 72, 736 (1950).

(644) R.E. RUNDLE, L. DAASCH, AND D. FRENCH, J. AM. CHEM. SOC. 66, 130 (1944).

(645) <A> D.R. KREGER, BIOCHIM. BIOPHYS. ACTA 6, 406 (1951), <B> L.C. SPARK, IBID., 8, 101 (1952).

(646) <A> R.E. RUNDLE AND F.C. EDWARDS, J. AM. CHEM. SOC. 65, 2200 (1943), <B> F.F. MIKUS, R.M. HIXON, AND R.E. RUNDLE, IBID., 68, 1115 (1946).

(647) R.S.J. MANLEY, J. POLYMER SCI., PT. A-2, 4503 (1964).

(648) A. SARKO AND R.H. MARCHESSAULT, J. AM. CHEM. SOC. 89, 6454 (1967).

(649) W. WINTER AND A. SARKO, BIOPOLYMERS 11, 849 (1972).

(650) F.R. SENTI AND L.P. WITNAUER, J. AM. CHEM. SOC. 70, 1438 (1948).

(651) R.E. RUNDLE AND D. FRENCH, J. AM. CHEM. SOC. 65, 1707 (1943).

(652) D. CARLSTROM, J. BIOPHYS. BIOCHEM. CYTOL. 3, 669 (1957), QUOTED IN R.H. MARCHESSAULT AND A. SARKO, ADVAN. CARBOHYDRATE CHEM. 22, 421 (1967).

(653) N.E. DWELTZ, BIOCHIM. BIOPHYS. ACTA 51, 283 (1961).

(654) I.M. MACKIE AND E.E. PERCIVAL, J. CHEM. SOC. 1959, 1151, QUOTED IN R.H. MARCHESSAULT AND A. SARKO, ADVAN. CARBOHYDRATE CHEM. 22, 421 (1967).

(655) <A> R.H. MARCHESSAULT AND W. SETTINERI, J. POLYMER SCI., PT. B-2, 1047 (1964), <B> W. SETTINERI AND R.H. MARCHESSAULT, J. POLYMER SCI., PT. C, NO. 11, 253 (1965).

(656) S.M. GABBAY, P.R. SUNDARARAJAN, AND R.H. MARCHESSAULT, BIOPOLYMERS 11, 79 (1972).

(657) <A> N. LANZETTA, G. MAGLIO, C. MARCHETTA, AND R. PALUMBO, PREPRINT III-36, IUPAC SYMPOSIUM ON MACROMOLECULES, HELSINKII, JULY 2-7, 1972, <B> G. MAGLIO, C. MARCHETTA, R. PALUMBO, F. RIVA, AND M. DE SIMONE, MAKROMOL. CHEM. 156, 321 (1972), <C> J. POLYMER SCI., POLYMER CHEM. ED., 11, 913 (1973).

(658) A.I. KOLTSOV, IU.G. BAKLAGINA, B.M. GINZBURG, L.N. KORZHAVIN, M.M. KOTON, N.V. MIKHAILOVA, V.N. NIKITIN, AND A.V. SIDOROVICH, PREPRINT III-77, IUPAC SYMPOSIUM ON MACROMOLECULES, HELSINKII, JULY 2-7, 1972.

(659) <A> E.D.T. ATKINS, C.F. PHELPS, AND J.K. SHEEHAN, BIOCHEM. J. 128, 1255 (1972), <B> E.D.T. ATKINS AND J.K. SHEEHAN, NATURE, NEW BIOLOGY 235, 253 (1972).

(660) M.G. NORTHOLT, B.J. TABOR, AND J.J. VAN AARTSEN, J. POLYMER SCI., PT. A-2, 10, 191 (1972).

(661) F.R. SENTI AND L.P. WITNAUER, J. POLYMER SCI. 9, 115 (1952).

(662) W.Z. BORCHERT, ANGEW. CHEM. 63, 31 (1951).

(663)   P. DE SANTES AND A. KOVACS, BIOPOLYMERS 6, 299 (1968).
(664)   S. ISHIDA, BULL. CHEM. SOC. JAPAN 33, 924 (1960).
(665)   A.M. AFIFI-EFFAT, J. CHEM. SOC., FARADAY TRANS. II, 68, 656 (1972).
(666)   U. SHMUELI AND W. TRAUB, J. MOL. BIOL. 12, 205 (1965).
(667)   A.D. FRENCH AND H.F. ZOBEL, BIOPOLYMERS 5, 457 (1967).
(668)   B. ZASLOW, BIOPOLYMERS 1, 165 (1963).
(669)   N.E. DWELTZ, BIOCHIM. BIOPHYS. ACTA 44, 416 (1960).
(670)   W. LOTMAR AND L.E.R. PICKEN, EXPERIENTIA 6, 58 (1950).
(671)   I.A. NIEDUSZYNSKI AND R.H. MARCHESSAULT, <A> NATURE 232, 46 (1971),
         <B> BIOPOLYMERS 11, 1335 (1972).
(672)   H. BITTIGER AND R.H. MARCHESSAULT, CARBOHYDRATE RES. 18, 469 (1971).
(673)   <A> E.D.T. ATKINS, W. MACKIE, AND E.E. SMOLKO, NATURE 225, 626 (1970),
         <B> E.D.T. ATKINS, W. MACKIE, K.D. PARKER, AND E.E. SMOLKO, J. POLYMER
         SCI., PT. B, 9, 311 (1971).
(674)   W.T. ASTBURY, NATURE 155, 667 (1945).
(675)   A.S. UEDA, Y. CHATANI, AND H. TADOKORO, POLYMER J. 2, 387 (1971).
(676)   J.G. FATOU, EUROPEAN POLYMER J. 7, 1057 (1971).
(677)   B.J. TABOR, E.P. MAGRE', AND J. BOON, EUROPEAN POLYMER J. 7, 1127 (1971).
(678)   W.H. CAROTHERS, J.A. ARVIN, AND G.L. DOROUGH, J. AM. CHEM. SOC. 5, 3292
         (1930).
(679)   S. TSUNAWAKI AND C.C. PRICE, J. POLYMER SCI., PT. A, 2, 1511 (1964).
(680)   G. COJAZZI, A. FICHERA, C. GARBUGLIO, V. MALTA, AND R. ZANNETTI, CHIM.
         IND. (MILAN) 54, 40 (1972).
(681)   H.D. FLACK, J. POLYMER SCI., PT. A-2, 10, 1799 (1972).
(682)   A. CAPIZZI AND G. GIANOTTI, MAKROMOL. CHEM. 157, 123 (1972).
(683)   G. PEREGO, A. MELIS, AND M. CESARI, MAKROMOL. CHEM. 157, 269 (1972).
(684)   L.G. KAZARYAN, D.YA. TSVANKIN, B.M. GINZBURG, S. TAYCHIEV, L.N. KORZHAVIN,
         AND S.YA. FRENKEL', VYSOKOMOL. SOEDIN., SER. A, 14, 1199 (1972).
(685)   P. CORRADINI, E. MARTUSCELLI, G. MONTAGNOLI, AND V. PETRACCONE, EUROPEAN
         POLYMER J. 6, 1201 (1970).
(686)   S. SASAKI, Y. TAKAHASHI, AND H. TADOKORO, POLYMER J. 4, 172 (1973).
(687)   Y. GOTOH, H. SAKAKIHARA, AND H. TADOKORO, POLYMER J. 4, 68 (1973).
(688)   S. SASAKI, Y. TAKAHASHI, AND H. TADOKORO, J. POLYMER SCI., POLYMER PHYS.
         ED., 10, 2363 (1972).
(689)   H. TADOKORO, IN G. ALLEN, ED., MOLECULAR STRUCTURE AND PROPERTIES
         (PHYSICAL CHEMISTRY, SERIES ONE, V.2, P.45) (MTP INTERNATIONAL REVIEW OF
         SCIENCE) BUTTERWORTHS, LONDON, 1972.
(690)   H. TADOKORO, PRIVATE COMMUNICATION, 30 SEPTEMBER 1972.
(691)   N. DODDI, W.C. FORSMAN, AND C.C. PRICE, MACROMOLECULES 4, 648 (1971).
(692)   H. TANI AND N. OGUNI, J. POLYMER SCI., PT. B, 7, 803 (1969).
(693)   G. ALLEN, C. BOOTH, S.J. HURST, C. PRICE, F. VERNON, AND R.F. WARREN,
         POLYMER 8, 406 (1967).
(694)   P. DUMAS, N. SPASSKY, AND P. SIGWALT, MAKROMOL. CHEM. 156, 55 (1972).
(695)   J.W. HILL AND W.H. CAROTHERS, J. AM. CHEM. SOC. 57, 925 (1935).
(696)   R. NAKATSUKA, BULL. CHEM. SOC. JAPAN 36, 1294 (1963).
(697)   H.K. ILLERS, H. HABERKORN, AND P. SIMAK, MAKROMOL. CHEM. 158, 285 (1972).
(698)   K. TERANISHI, T. ARAKI, AND H. TANI, MACROMOLECULES 5, 660 (1972).
(699)   Y. CHATANI, QUOTED IN Y. CHATANI AND S. NAKATANI, MACROMOLECULES 5, 597
         (1972).
(700)   M.J. RICHARDSON, J. POLYMER SCI., PT. C, 38, 251 (1972).
(701)   W. TRAUB, U. SHMUELI, M. SUWALSKY, AND A. YONATH, IN G.N. RAMACHANDRAN,
         ED., CONFORMATION OF BIOPOLYMERS, V. 2, ACADEMIC PRESS, NEW YORK, 1967,
         P. 449.
(702)   B.K. VAINSHTEIN AND L.I. TATARINOVA, IN G.N. RAMACHANDRAN, ED.,
         CONFORMATION OF BIOPOLYMERS, V. 2, ACADEMIC PRESS, NEW YORK, 1967, P. 569.
(703)   D.B. GREEN, F. HAPPEY, AND B.M. WATSON, IN G.N. RAMACHANDRAN, ED.,
         CONFORMATION OF BIOPOLYMERS, V. 2, ACADEMIC PRESS, NEW YORK, 1967, P. 617.
(704)   H. SUMITOMO, K. KOBAYASHI, AND T. SAJI, J. POLYMER SCI., POLYMER CHEM.
         ED., 10, 3421 (1972).
(705)   E.D.T. ATKINS, R. GAUSSEN, D.H. ISAAC, V. NANDANWAR, AND J.K. SHEEHAN, J.
         POLYMER SCI., POLYMER LETTERS ED., 10, 863 (1972).
(706)   I. YAHARA, K. IMAHORI, Y. IITAKA, AND M. TSUBOI, J. POLYMER SCI., PT. B,
         1, 47 (1963).
(707)   L. PORRI AND M.C. GALLAZZI, EUROPEAN POLYMER J. 2, 189 (1966).
(708)   B.J. POPPLETON AND A.M. MATHIESON, NATURE 219, 1046 (1968).
(709)   K.C. ELLIS AND J.O. WARWICKER, J. POLYMER SCI. 56, 339 (1962).
(710)   I.C.M. DEA, R. MOORHOUSE, D.A. REES, S. ARNOTT, J.M. GUSS, AND E.A.
         BALAZS, SCIENCE 179, 560 (1973).
(711)   E.D.T. ATKINS AND J.K. SHEEHAN, SCIENCE 179, 562 (1973).
(712)   R. PERRET AND A. SKOULIOS, MAKROMOL. CHEM. 156, 157 (1972).
(713)   E. LANZA, H. BERGHMANS, AND G. SMETS, J. POLYMER SCI., POLYMER PHYS. ED.,
         11, 75 (1973).

(714)   J. PISANCHYN, J. POLYMER SCI., POLYMER CHEM. ED., 11, 135 (1973).
(715)   Y. TAKAHASHI, T. SATO, H. TADOKORO, AND Y. TANAKA, J. POLYMER SCI.,
        POLYMER PHYS. ED., 11, 233 (1973).
(716)   E. HADICKE, E.C. MEZ, C.H. KRAUCH, G. WEGNER, AND J. KAISER, ANGEW. CHEM.
        83, 253 (1971) – ANGEW. CHEM., INTERN. ED. ENGL. 10, 226 (1971).
(717)   H. SAKAKIHARA, Y. TAKAHASHI, H. TADOKORO, N. OGUNI, AND H. TANI,
        MACROMOLECULES 6, 205 (1973).
(718)   C.E. WILKES, V.L. FOLT, AND S. KRIMM, MACROMOLECULES 6, 235 (1973).
(719)   A. DEL PRA, M. MAMMI, AND E. PEGGION, BIOPOLYMERS 12, 937 (1973).
(720)   A. DEL PRA, P. SPADON, AND G. VALLE, BIOPOLYMERS 12, 941 (1973).
(721)   R.A. WESSLING, J.H. OSWALD, AND I.R. HARRISON, J. POLYMER SCI., POLYMER
        PHYS. ED., 11, 875 (1973).
(722)   F.C. WILSON AND H.W. STARKWEATHER, JR., J. POLYMER SCI., POLYMER PHYS.
        ED., 11, 919 (1973).
(723)   M. HASEGAWA, Y. SUZUKI, H. NAKANISHI, AND F. NAKANISHI, IN K. IMAHORI AND
        S. MURAHASHI, EDS., PROGRESS IN POLYMER SCIENCE JAPAN, V.5, P.143, HALSTED
        PRESS (JOHN WILEY AND SONS), NEW YORK, 1973.
(724)   A. WLOCHOWICZ AND W. PRZYGOCKI, J. APPL. POLYMER SCI. 17, 1197 (1973).
(725)   N. KOBAYASHI, A. OSAWA, AND T. FUJISAWA, J. POLYMER SCI., POLYMER LETTERS
        ED., 11, 225 (1973).
(726)   G. WELCH AND R.L. MILLER, UNPUBLISHED RESULTS
(727)   M. IKEDA, H. SUGA, AND S. SEKI, REPTS. PROGR. POLYMER PHYS. JAPAN 15, 255
        (1972).
(728)   M. IKEDA, H. SUGA, AND S. SEKI, REPTS. PROGR. POLYMER PHYS. JAPAN 15, 257
        (1972).
(729)   S. MINAMI, K. MURASE, AND T. YOSHIHARA, REPTS. PROGR. POLYMER PHYS. JAPAN
        15, 339 (1972).
(730)   H. TADOKORO, T. YOSHIHARA, Y. CHATANI, AND S. MURAHASHI, J. POLYMER SCI.,
        PT. B, 2, 363 (1964).
(731)   A.A. BLUMBERG, S.S. POLLACK, AND C.A.J. HOEVE, J. POLYMER SCI., PT. A, 2,
        2499 (1964).
(732)   R. IWAMOTO, Y. SAITO, H. ISHIHARA, AND H. TADOKORO, J. POLYMER SCI., PT.
        A-2, 6, 1509 (1968).
(733)   M. YOKOYAMA, H. ISHIHARA, R. IWAMOTO, AND H. TADOKORO, MACROMOLECULES 2,
        184 (1969).
(734)   F.E. BAILEY, JR., AND H.G. FRANCE, J. POLYMER SCI. 49, 397 (1961).
(735)   H. HERLINGER, H.-P. HORNER, F. DRUSCHKE, H. KNOLL, AND F. HAIBER, ANGEW.
        MAKROMOL. CHEM. 29/30, 229 (1973).
(736)   P.V. KOZLOV AND A.N. PEREPELKIN, POLYMER SCI. (USSR) (ENGLISH TRANSL.) 9,
        414 (1967) – VYSOKOMOL. SOEDIN., SER. A, 9, 370 (1967).
(737)   J.C. ANDRIES AND A.G. WALTON, BIOPOLYMERS 8, 465 (1969).
(738)   T. SAEGUSA, H. IKEDA, AND H. FUJII, MACROMOLECULES 5, 108 (1972).
(739)   J.W. SUMMERS AND M.H. LITT, J. POLYMER SCI., POLYMER CHEM. ED., 11, 1353
        (1973).
(740)   B. LOTZ AND H.D. KEITH, J. MOL. BIOL. 61, 195 (1971).
        THE AUTHORS REPORT C = 9.15A FOR THREE PEPTIDE RESIDUES IN A 'GLYCINE II'
        3/1 HELIX. CRYSTALLOGRAPHIC IDENTITY, THEN, OCCURS AT FOUR TIMES THIS.
(741)   B. LOTZ AND H.D. KEITH, J. MOL. BIOL. 61, 201 (1971).
(742)   L.G. KAZARYAN AND F.M. MEDVEDVA, VYSOKOMOL. SOEDIN., SER. B, 10, 305
        (1968).
(743)   S.W. SHALABY, E.M. PEARCE, R.J. FREDERICKS, AND E.A. TURI, J. POLYMER SCI.
        POLYMER CHEM. ED., 11, 1 (1973).
(744)   J.P. KENNEDY AND R.B. ISAACSON, J. MACROMOL. CHEM. 1, 541 (1966).
(745)   G. MONTAUDO, G. BRUNO, P. MARAVIGNA, P. FINOCCHIARO, AND P. CENTINEO, J.
        POLYMER SCI., POLYMER CHEM. ED., 11, 65 (1973).
(746)   M. LITT, F. RAHL, AND L.G. ROLDAN, J. POLYMER SCI., PT. A-2, 7, 463 (1969)
(747)   T.G. BASSIRI, A. LEVY, AND M. LITT, J. POLYMER SCI., PT. B, 5, 871 (1967).
(748)   H.F. MARK, S.M. ATLAS, AND N. OGATA, J. POLYMER SCI. 61, S49 (1962).
(749)   J. KAISER, G. WEGNER, AND E.W. FISCHER, KOLLOID-Z. Z. POLYMERE 250, 1158
        (1972).
(750)   N.A. ADROVA, A.I. ARTYUKHOV, YU.G. BAKLAGINA, T.I. BORISOVA, M.M. KOTON,
        N.V. MIKHAILOVA, V.N. NIKITIN, AND A.V. SIDOROVICH, VYSOKOMOL. SOEDIN.,
        SER. A, 15, 153 (1973).
(751)   M. STRAUMANIS, Z. KRIST. 102, 432 (1940).
(752)   R.G. CRYSTAL, J. POLYMER SCI., PT. A-2, 8, 1755 (1970).
(753)   B.B. DOYLE, W. TRAUB, G.P. LORENZI, F.R. BROWN III, AND E.R. BLOUT, J.
        MOL. BIOL. 51, 47 (1970). THE AUTHORS REPORT B = 6.98A FOR TWO PEPTIDE
        RESIDUES IN AN 'ANTIPARALLEL PLEATED SHEET' STRUCTURE. CRYSTALLOGRAPHIC
        IDENTITY, THEN, OCCURS AT THREE TIMES THIS.
(754)   R.D.B. FRASER, T.P. MACRAE, AND F.H.C. STEWART, J. MOL. BIOL. 19, 580
        (1966). THE AUTHORS REPORT B = 6.85A FOR TWO PEPTIDE RESIDUES. CRYSTAL-
        LOGRAPHIC IDENTITY OCCURS AT THREE TIMES THIS.

(755) D.M. SEGAL, W. TRAUB, AND A. YONATH, J. MOL. BIOL. 43, 519 (1969).
(756) J.C. ANDRIES AND A.G. WALTON, J. MOL. BIOL. 56, 515 (1971).
(757) J.M. SQUIRE AND A. ELLIOTT, J. MOL. BIOL. 65, 291 (1972).
(801) YU.V. MITIKIN, YU.N. SAZANOV, G.P. VLASOV, AND M.M. KOTON, VYSOKOMOLEKUL.
       SOEDIN. 2, 716 (1960 - POLYMER SCI. (USSR) (ENGLISH TRANSL.) 2, 423
       (1961).
(802) P.J. FLORY, PRINCIPLES OF POLYMER CHEMISTRY, CORNELL UNIVERSITY PRESS,
       ITHACA, NEW YORK, 1963).
(803) R.F. HECK AND D.S. BRESLOW, J. POLYMER SCI. 41, 521 (1960).
(804) R.F. HECK AND D.S. BRESLOW, J. POLYMER SCI. 41, 520 (1960).
(805) Y. ETIENNE, IND. PLASTIQUES MOD. (PARTS) 9, 37 (1957).
(806) T.I. SHEIN, G.I. KUDRYAVTSEW, AND G.P. LYUBIMTSEVA, VYSOKOMOLEKUL. SOEDIN.
       7, 1447 (1965) - POLYMER SCI. (USSR) (ENGLISH TRANSL.) 7, 1603 (1965).
(807) T.W. CAMPBELL, J. ORG. CHEM. 22, 1029 (1957).
(808) A.C. FARTHING, J. CHEM. SOC. 1955, 3648.
(809) E.J. VANDENBERG, J. POLYMER SCI. 47, 486 (1960).
(810) W.C. SHEEHAN, T.B. COLE, AND L.G. PICKLESIMER, J. POLYMER SCI., PT. A, 3,
       1443 (1965).
(811) P. LONGI, ATTI ACCAD. NAZL. LINCEI, REND., CLASSE SCI. FIS., MAT. NAT. 31,
       273 (1961).
(812) C.G. OVERBERGER AND A.M. SCHILLER, J. POLYMER SCI. 54, S30 (1961).
(813) C.G. OVERBERGER AND E.B. DAVIDSON, J. POLYMER SCI. 62, 23 (1962).
(814) D.R. STEVENSON AND J.E. MULVANEY, J. POLYMER SCI., PT. A-1, 10, 2713
       (1972).
(815) G. NATTA, G. DALL'ASTA, G. MAZZANTI, AND A. CASALE, MAKROMOL. CHEM. 58,
       217 (1962).
(816) H.K. HALL, JR., J. ORG. CHEM. 28, 3213 (1963).
(817) G. NATTA, G. MAZZANTI, G.F. PREGAGLIA, AND G. POZZI, J. POLYMER SCI. 58,
       1201 (1962).
(818) J.H. MAGILL, J. POLYMER SCI., PT. A, 3, 1195 (1965).
(819) C.S. MARVEL AND R.R. CHAMBERS, J. AM. CHEM. SOC. 70, 993 (1948).
(820) W.H. CAROTHERS AND F.J. VAN NATTA, J. AM. CHEM. SOC. 52, 314 (1930).
(821) M.A. OSMAN AND H.-G. ELIAS, J. MACROMOL. SCI., CHEM., A5, 805 (1971).
(822) J.W. HILL AND W.H. CAROTHERS, J. AM. CHEM. SOC. 54, 1569 (1932).
(823) J.W. HILL AND W.H. CAROTHERS, J. AM. CHEM. SOC. 55, 5023 (1933).
(824) H. KOMOTO, REV. PHYS. CHEM. JAP. 37, 112 (1967).
(825) T.L. ANG, R.C. CHANG, E.R. SANTEE, JR., AND H.J. HARWOOD, J. POLYMER SCI.,
       POLYMER LETTERS ED., 10, 791 (1972).
(826) N. YODA AND A. MIYAKE, BULL. CHEM. SOC. JAPAN 32, 1120 (1959).
(827) N. YODA, J. POLYMER SCI., PT. A, 1, 1323 (1963).
(828) R.S. MUROMOVA, A.A. STREPIKHEYEV, AND Z.A. ROGOVIN, VYSOKOMOLEKUL. SOEDIN.
       5, 1096 (1963) - POLYMER SCI. (USSR) (ENGLISH TRANSL.) 5, 157 (1964).
(829) H. ZAHN AND G.B. GLEITSMANN, MAKROMOL. CHEM. 63, 129 (1963).
(830) S.G. MATSOYAN AND L.M. AKOPYAN, VYSOKOMOLEKUL. SOEDIN. 3, 1311 (1961) -
       POLYMER SCI. (USSR) (ENGLISH TRANSL.) 3, 915 (1962).
(831) V.V. KORSHAK, S.L. SOSIN, AND V.P. ALEKSEYEVA, VYSOKOMOLEKUL. SOEDIN. 3,
       1332 (1961) - POLYMER SCI. (USSR) (ENGLISH TRANSL.) 3, 925 (1962).
(832) E.P. MELNIKOVA, A.A. VANSHEIDT, M.G. KRAKOVYAK, AND L.V. KUKHAREVA,
       VYSOKOMOLEKUL. SOEDIN. 2, 1817 (1960) - POLYMER SCI. (USSR) (ENGLISH
       TRANSL.) 3, 494 (1962).
(833) N. OGATA, K. SANUI, AND K. SHIRAISHI, KOBUNSHI KAGAKU 30, 51 (1973).
(834) V.E. SHASHOUA AND W.M. EARECKSON, III, J. POLYMER SCI. 40, 343 (1959).
(835) C.W. STEPHENS, J. POLYMER SCI. 40, 359 (1959).
(836) C.G. OVERGERGER, S. OZAKI, AND D.M. BRAUNSTEIN, MAKROMOL. CHEM. 93, 13
       (1966).
(837) N. KOBAYASHI AND T. FUJISAWA, J. POLYMER SCI., POLYMER CHEM. ED., 11, 545
       (1973).
(838) D.J. LYMAN AND S.L. JUNG, J. POLYMER SCI. 40, 407 (1959).
(839) J.R. CALDWELL AND R. GILKEY, 134TH NATIONAL AMERICAN CHEMICAL SOCIETY
       MEETING, CHICAGO, SEPTEMBER 1958.
(840) J.W. HILL, J. AM. CHEM. SOC. 52, 4110 (1930).
(841) O.B. EDGAR AND R. HILL, J. POLYMER SCI. 8, 1 (1952).
(842) <A> W.J. MIDDLETON, H.W. JACOBSON, R.E. PUTNAM, H.C. WALTER, D.G. PYE, AND
       W.H. SHARKEY, J. POLYMER SCI., PT. A, 3, 4115 (1965), <B> A.L. BARNEY,
       J.M. BRUCE, JR., J.N. COKER, H.W. JACOBSON, AND W.H. SHARKEY, IBID., PT.
       A, 4, 2617 (1966).
(843) P.R. THOMAS, G.J. TYLER, T.E. EDWARDS, A.T. RADCLIFFE, AND R.C.P. CUBBON,
       POLYMER 5, 525 (1964).
(844) W.H. CAROTHERS, G.L. DOROUGH, AND F.J. VAN NATTA, J. AM. CHEM. SOC. 54,
       761 (1932).
(845) E. WELLISCH, E. GIPSTEIN, AND O.J. SWEETING, J. POLYMER SCI., PT. B, 2, 35
       (1964).

(846) D.A. LUTZ AND L.P. WITNAUER, J. POLYMER SCI., PT. B, 2, 31 (1964).
(847) K. SAOTOME AND K. SATO, J. POLYMER SCI., PT. A-1, 4, 1303 (1966).
(848) V.L. BELL, J. POLYMER SCI., PT. A, 2, 5291 (1964).
(849) F. SHEPHERD, J. POLYMER SCI., POLYMER LETTERS ED., 10, 799 (1972).
(850) W. MARCONI, S. CESCA, AND G.D. FORTUNA, J. POLYMER SCI., PT. B, 2, 301 (1964).
(851) K. MATSUZAKI, T. OKAMOTO, A. ISHIDA, AND H. SOBUE, J. POLYMER SCI., PT. A, 2, 1105 (1964).
(852) B. KE AND A.W. SISKO, J. POLYMER SCI. 50, 87 (1961).
(853) K. SAOTOME AND H. KOMOTO, J. POLYMER SCI., PT. A-1, 4, 1475 (1966).
(854) A. CONIX, J. POLYMER SCI. 29, 343 (1958).
(855) N. YODA, MAKROMOL. CHEM. 32, 1 (1959).
(856) N. YODA, KOBUNSHI KAGAKU 19, 495 (1962).
(857) N. YODA, KOBUNSHI KAGAKU 19, 553 (1962).
(858) P. BARNABA, D. CORDISCHI, M. LENZI, AND A. MELE, CHIM. IND. (MILAN) 47, 1060 (1965).
(859) J. PELLON, J. POLYMER SCI., PT. A, 1, 3561 (1963).
(860) R. ODA, T. SHONO, A. OKU, AND H. TAKAO, MAKROMOL. CHEM. 67, 124 (1963).
(861) K. MURAYAMA AND S. MORIMOTO, CHEM. IND. (LONDON) 1967, 402.
(862) M. HASEGAWA AND T. UNISHI, J. POLYMER SCI., PT. B, 2, 237 (1964).
(863) T. KAGIYA, H. KISHIMOTO, S. NARISAWA, AND K. FUKUI, J. POLYMER SCI., PT. A, 3, 145 (1965).
(864) G. NATTA, G. DALL'ASTA, G. MAZZANTI, AND G. MOTRONI, MAKROMOL. CHEM. 69, 163 (1963).
(865) T. SHONO, T. MORIKAWA, R.-I. OKAYAMA, AND R. ODA, MAKROMOL. CHEM. 81, 142 (1965).
(866) D.A. HOLMER, O.A. PICKETT, JR., AND J.H. SAUNDERS, J. POLYMER SCI., PT. A-1, 10, 1547 (1972).
(867) C.G. OVERBERGER AND J.E. MULVANEY, J. AM. CHEM. SOC. 81, 4697 (1959).
(868) P.W. MORGAN, CONDENSATION POLYMERS, BY INTERFACIAL AND SOLUTION METHODS (POLYMER REVIEWS, V. 10), INTERSCIENCE PUBLISHERS, NEW YORK, 1965 (PP. 243, 366, 418, 424, AND 427).
(869) A. LEONI, M. GUAITA, AND G. SAINI, CHIM. IND. (MILAN) 47, 373 (1965).
(870) O.YA. FEDOTOVA, M.L. KERBER, AND I.P. LOSEV, VYSOKOMOLEKUL. SOEDIN. 6, 452 (1964) - POLYMER SCI. (USSR) (ENGLISH TRANSL.) 6, 502 (1964).
(871) H. SAWADA AND A. YASUE, KOGYO KAGAKU ZASSHI 67, 1444 (1964), THROUGH CHEM. ABSTR. 62, 10525B (1965).
(872) H. SAWADA AND A. YASUE, KOGYO KAGAKU ZASSHI 67, 1449 (1964), THROUGH CHEM. ABSTR. 62, 10525F (1965).
(873) K. TANIKAWA, S. KUSABAYASHI, H. HIRATA, AND H. MIKAWA, J. POLYMER SCI., PT. B, 6, 275 (1968).
(874) H. MIYAKE, K. HAYASHI, AND S. OKAMURA, J. POLYMER SCI., PT A, 3, 2731 (1965).
(875) <A> F. DANUSSO, P. FERRUTI, AND G. FERRONI, CHIM. IND. (MILAN) 49, 453 (1967), <B> F. DANUSSO AND P. FERRUTI, POLYMER 11, 88 (1970).
(876) J.V. CRIVELLO, J. POLYMER SCI., POLYMER CHEM. ED., 11, 1185 (1973).
(877) K. SAOTOME AND H. KOMOTO, J. POLYMER SCI., PT. A-1, 5, 107 (1967).
(878) F.R. PRINCE, E.A. TURI, AND E.M. PEARCE, J. POLYMER SCI., PT. A-1, 10, 465 (1972).
(879) N.D. FIELD, J.A. KIERAS, AND A.E. BORCHERT, J. POLYMER SCI., PT. A-1, 5, 2179 (1967).
(880) <A> T. HIGASHIMURA, S. KUSUDO, Y. OHSUMI, A. MIZOTE, AND S. OKAMURA, J. POLYMER SCI., PT. A-1, 6, 2511 (1968), <B> A. MIZOTE, T. HIGASHIMURA, AND S. OKAMURA, PURE APPL. CHEM. 16, 457 (1968).
(881) A.A. NISHIMURA, POLYMER 8, 446 (1967).
(882) V.C.E. BURNOP AND K.G. LATHAM, POLYMER 8, 589 (1967).
(883) K. SAOTOME AND R.C. SCHULZ, MAKROMOL. CHEM. 109, 239 (1967).
(884) M.M. KOTON, I.V. ANDREEVA, A.I. TURBINA, AND V.G. SINYAUSKII, DOKL. AKAD. NAUK SSSR 159, 602 (1964).
(885) V.B. LUSKCHIK AND S.S. SKOROKHODOV, VYSOKOMOL. SOEDIN., SER. B, 9, 797 (1967).
(886) J.L.R. WILLIAMS, T.M. LAAKSO, AND L.E. CONTOIS, J. POLYMER SCI. 61, 353 (1962).
(887) A. CANNEPIN, G. CHAMPETIER, AND A. PARISOT, J. POLYMER SCI. 8, 35 (1952).
(888) W.S. PORT, J.E. HANSEN, E.F. JORDAN, JR., T.T. DIETZ, AND D. SWERN, J. POLYMER SCI. 7, 207 (1951).
(889) Y. KISHIMOTO, K. SANUI, AND N. OGATA, POLYMER J. 4, 18 (1973).
(890) V.S. FOLDI AND W. SWEENEY, MAKROMOL. CHEM. 72, 208 (1964).
(891) Y. IWAKURA, K. HAYASHI, S. KANG, AND K. INAGAKI, MAKROMOL. CHEM. 95, 205 (1966).
(892) H.V. BOENIG, N. WALKER, AND E.H. MYERS, J. APPL. POLYMER SCI. 5, 384 (1961).

(893)   C.A. AUFDERMARSH AND R. PARISER, J. POLYMER SCI., PT. A, 2, 4727 (1964).
(894)   W.J. PEPPEL, J. POLYMER SCI. 51, S64 (1961).
(895)   F.B. CRAMER AND R.G. BEAMAN, J. POLYMER SCI. 21, 237 (1956).
(896)   L. PORRI AND D. PINI, CHIM. IND. (MILAN) 55, 196 (1973).
(897)   <A> Y. IWAKURA AND M. SAKAMOTO, J. POLYMER SCI. 47, 277 (1960), <B> Y.
         IWAKURA, M. SAKAMOTO, AND Y. AWATA, IBID., PT. A, 2, 881 (1964).
(898)   F.A. BOVEY, J.F. ABERE, G.B. RATHMANN, AND C.L. SANDBERG, J. POLYMER SCI.
         15, 520 (1955).
(899)   J.L. GREENE, JR., E.L. HUFFMAN, R.E. BURKS, JR., W.C. SHEEHAN, AND I.A.
         WOLFF, J. POLYMER SCI., PT. A-1, 5, 391 (1967).
(900)   J. LAL AND G.S. TRICK, J. POLYMER SCI., PT. A, 2, 4559 (1964).
(901)   M. LEVINE AND S.C. TEMIN, J. POLYMER SCI. 28, 179 (1958).
(902)   N. OGATA, T. ASAHARA, AND S. TOHYAMA, J. POLYMER SCI., PT. A-1, 4, 1359
         (1966).
(903)   E.J. IZARD, J. POLYMER SCI. 9, 35, (1952).
(904)   D.A. HOLMER, J. POLYMER SCI., PT. A-1, 6, 3177 (1968).
(905)   H. HOPFF AND A. KRIEGER, MAKROMOL. CHEM. 47, 93 (1961).
(906)   E.F. JORDAN, JR., D.W. FELDEISEN, AND A.N. WRIGLEY, J. POLYMER SCI., PT.
         A-1, 9, 1835 (1971). (SIDE CHAIN CRYSTALLIZATION ONLY).
(907)   T.W. CAMPBELL AND K.Z. SMELTZ, J. ORG. CHEM. 28, 2069 (1963).
(908)   R.W. LENZ, J. NELSON, AND P. GUIMOND, AM. CHEM. SOC. POLYMER PREPRINTS 9,
         1301 (1968).
(909)   M. GOODMAN AND S.-C. CHEN, MACROMOLECULES 5, 625 (1971).
(910)   W.C. WOOTEN AND H.W. COOVER, J. POLYMER SCI. 37, 560 (1959).
(911)   R.E. WILFONG, J. POLYMER SCI. 54, 385 (1961).
(912)   A.S. CARPENTER, J. SOC. DYERS COLOURISTS 65, 469 (1949), THROUGH REF. 911.
(913)   P.M. CACHIA, ANN. CHIM. PARIS 4, 5 (1959), THROUGH REF. 911.
(914)   L.H. BOCK AND J.K. ANDERSON, J. POLYMER SCI. 17, 553 (1955).
(915)   W.F. CHRISTOPHER AND D.W. FOX, POLYCARBONATES, REINHOLD PUBLISHING CO.,
         NEW YORK, 1962.
(916)   A. NOSHAY AND C.C. PRICE, J. POLYMER SCI. 54, 533 (1961).
(917)   J.K. STILLE AND J.A. EMPEN, J. POLYMER SCI., PT. A-1, 5, 273 (1967).
(918)   V.V. KORSHAK, S.V. VINOGRADOVA, AND V.M. BELYAKOV, BULL. ACAD. SCI. USSR,
         DIV. CHEM. SCI. ENGLISH TRANSL. 1957, 1029, THROUGH REF. 547.
(919)   V.V. KORSHAK AND S.V. VINOGRADOVA, BULL. ACAD. SCI. USSR, DIV. CHEM. SCI.
         ENGLISH TRANSL. 1959, 138, THROUGH REF. 547.
(920)   G. NATTA, P. LONGI, AND V. NORDIO, MAKROMOL. CHEM. 83, 161 (1965).
(921)   T. UNISHI AND M. HASEGAWA, J. POLYMER SCI., PT. A, 3, 3191 (1965).
(922)   H.G. BUHRER AND H.-G. ELIAS, ADVAN. CHEM. SER., SUBMITTED.
(923)   N.G. GAYLORD, POLYETHERS, PART III, INTERSCIENCE PUBLISHING CO., NEW YORK,
         1962.
(924)   H. YUKI AND Y. TAKETANI, J. POLYMER SCI., PT. B, 10, 373 (1972).
(925)   G. DALL'ASTA, P. MENEGHINI, AND U. GENNARD, MAKROMOL. CHEM. 154, 279
         (1972).
(926)   H. KOMOTO, F. HAYANO, T. TAKAMI, AND S. YAMATO, J. POLYMER SCI., PT. A-1,
         9, 2983 (1971).
(927)   G. BELONOVSKAYA, ZH. TCHERNOVA, AND B. DOLGOPLOSK, EUROPEAN POLYMER J. 8,
         35 (1972).
(928)   G.P. BLACKBURN AND B.J. TIGHE, J. POLYMER SCI., PT. A-1, 8, 3591 (1970).
(929)   S.W. SHALABY, AM. CHEM. SOC. POLYMER PREPRINTS 13, 316 (1971).
(930)   Y. IWAKURA, K. IWATA, S. MATSUO, AND A. TOHARA, MAKROMOL. CHEM. 146, 21
         (1971).
(931)   V. GUIDOTTI, M. RUSSO, AND L. MORTILLARO, MAKROMOL. CHEM. 147, 111 (1971).
(932)   A.D. BLISS, W.K. CLINE, C.E. HAMILTON, AND O.J. SWEETING, J. ORG. CHEM.
         28, 3537 (1963).
(933)   E.V. GOUINLOCK, JR., C.J. VERBANIC, AND G.C. SCHWEIKER, J. APPL. POLYMER
         SCI. 1, 361 (1959).
(934)   N. KOBAYASHI AND T. FUJISAWA, J. POLYMER SCI., POLYMER CHEM. ED., 10, 3165
         (1972).
(935)   T.W. CAMPBELL AND V.S. FOLDI, J. ORG. CHEM. 26, 4654 (1961).
(936)   N. KOBAYASHI AND T. FUJISAWA, J. POLYMER SCI., POLYMER CHEM. ED., 10, 3317
         (1972).
(937)   R.W.H. BERRY AND R.J. MAZZA, POLYMER 14, 172 (1973).
(938)   H. HIRUKAWA AND Y. IMAI, J. POLYMER SCI., POLYMER LETTERS ED., 11, 271
         (1973).
(939)   S. NISHIZAKI, S. ETO, AND T. MORIWAKI, KOGYO KAGAKU ZASSHI 71, 265 (1968).

E. FORMULA INDEX TO THE TABLES

As the number of crystallizable polymers and their complexity increased, it became necessary to devise a simple, reasonably unambiguous scheme to relate polymer structure to the specific name given that polymer.

Some time ago, the compiler designed a shorthand notation to represent the structure of linear polymers which has proved to be useful. This scheme requires one to write the line formula for the constitutional base unit (without concern for priority or direction rules) and to count the groups/atoms comprising the unit in a specified order. In a sense, this scheme is analogous to the Chemical Abstracts Formula Index but emphasizes groups commonly occurring in linear organic polymers rather than atoms exclusively. This scheme has the additional advantage of ensuring that data for a given polymer are not entered into the tables under differing polymer names.

The rules of this scheme are simple - count the number of groups/atoms in the constitutional base unit in the order:

1) $-CH_2-$ (methylenes as such, but not the $-CH_2-$ portion of a $-CH_3$ group)
2) $-CO-$ (carbonyls),
3) $-C_6H_4-$ (disubstituted benzenes; phenylenes; monosubstituted benzenes with the fifth proton counted under '5' below),
4) Carbon atoms not already counted (as in $-CH_3$, $-CH=$, $-C_6H_3=$, etc.),
5) Protons (-H) not already counted (as in $-CH_3$, $-NH-$, etc.),
6) Oxygen atoms (exclusive of those in carbonyls),
7) Nitrogen atoms,
8) Sulfur atoms,
9) Other atoms (F, Cl, Br, Si, etc.).

The string of nine integers thus generated, arranged in the above order, is the required structure code. By looking up this code in the Formula Index Table one can find the name under which the polymer in question appears in the data tables or, alternatively, by its absence, that no data has been found for the polymer in question. The Formula Index Table is arranged in ascending order of the structural code in columns 2-10 with the assigned polymer name in column 1. Note that some very long, complicated names have been truncated. This should cause little difficulty in the use of this table. Examples of the application of this scheme are shown below.

Experience has shown that this scheme generates unique codes for the vast majority of linear polymers capable of crystallizing. Where two or more polymers do have the same structural code, the choice of correct name is usually obvious from the structure. Of course, certain classes of polymers do tend to generate many members all having a single code, e.g., all of the condensation poly(amides) having a given total number of methylene groups and differing only in the proportioning of these between the diol and diacid constituents. Again, no ambiguities arise.

An advantage of this structural coding scheme is its adaptability to electronic data processing equipment. The logical step beyond this scheme could be a strict sequential notation detailing each group and its position (similar to the ACS-IUPAC Nomenclature Recommendation) in a short-hand notation for conciseness and for computer compatibility.

EXAMPLES★

Structure	Code	Name

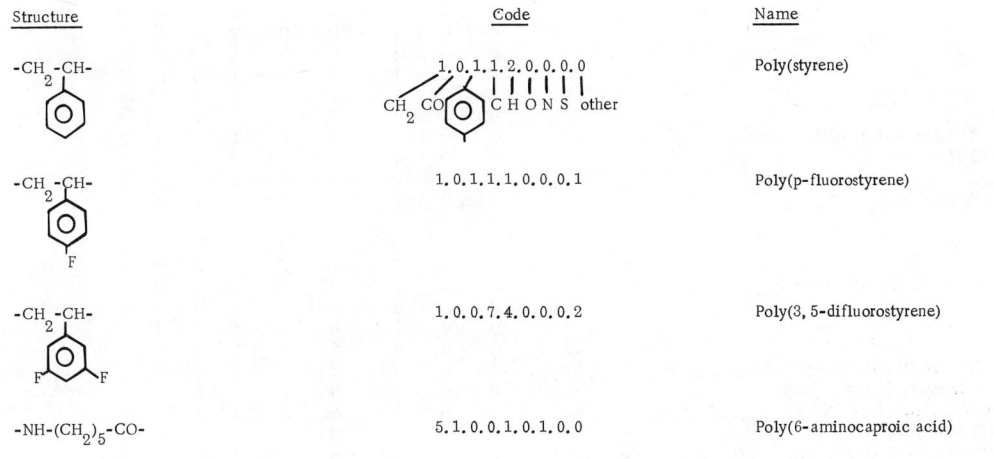

In the first example, there is one methylene, one phenylene, one carbon atom not otherwise counted (in -CH=), and two protons not otherwise counted (one in -CH= and one in $-C_6H_4H-$). In the second example, the fluorine atom replaces a proton from the substituted benzene ring - no other change in code occurs. In the third example, the trisubstituted benzene ring is removed from category 3 ($-C_6H_4-$) and is placed in category 4 (carbon atoms not otherwise counted) as 6 carbon atoms and in category 5 (protons) as 3 protons. In the last example, there are five methylene groups, one carbonyl, no other carbons, one proton (from -NH-), no uncounted oxygens, and one nitrogen. In similar fashion, the approximately 2000 entries in the Formula Index Table were generated.

★ Periods (.) are used to delineate the structural code integers generated.

POLY-	CH2	CO	C6H4	C	H	O	N	S	X
----- O									
SELENIUM	0	0	0	0	0	0	0	0	0
SULFUR	0	0	0	0	0	0	0	0	1
PHOSPHONITRILE CHLORIDE	0	0	0	0	0	0	0	1	0
SULFUR TRIOXIDE	0	0	0	0	0	0	1	0	3
PHOSPHAZENE	0	0	0	0	0	3	0	1	0
SILOXANE	0	0	0	0	2	0	1	0	1
DIFLUOROMETHYLENE SULFIDE	0	0	0	0	2	1	0	0	1
HYDROXYMETHYLENE	0	0	0	1	0	0	0	1	2
CHLOROTRIFLUOROETHYLENE	0	0	0	1	2	1	0	0	0
TETRAFLUOROETHYLENE	0	0	0	2	0	0	0	0	4
TETRAFLUOROETHYLENE OXIDE	0	0	0	2	0	1	0	0	4
TRIFLUOROETHYLENE	0	0	0	2	1	0	0	0	3
2,2,2-TRICHLOROACETALDEHYDE	0	0	0	2	1	1	0	0	3
ACETYLENE	0	0	0	2	2	0	0	0	0
2,2-DICHLOROACETALDEHYDE	0	0	0	2	2	1	0	0	0
ETHYLIDENE	0	0	0	2	4	0	0	0	0
ACETALDEHYDE	0	0	0	2	4	1	0	0	0
DIMETHYLSILOXANE	0	0	0	2	6	1	0	0	1
TETRAFLUOROALLENE	0	0	0	3	0	0	0	0	4
HEXAFLUOROPROPENE	0	0	0	3	0	0	0	0	6
ISOPROPYLIDENE SULFIDE	0	0	0	3	6	0	0	1	0
ACETONE	0	0	0	3	6	1	0	0	0
2-METHYLVINYLALCOHOL	0	0	0	3	6	1	0	0	0
2-BUTENE SULFIDE	0	0	0	4	8	0	0	1	0
2-BUTENE OXIDE	0	0	0	4	8	1	0	0	0
ISOBUTYRALDEHYDE	0	0	0	4	8	1	0	0	0
1-METHOXY-2-METHYLETHYLENE	0	0	0	4	8	1	0	0	0
2-METHYL-2-BUTENE OXIDE	0	0	0	5	10	1	0	0	0
2,4-HEXADIENE, 2,5-	0	0	0	6	10	0	0	0	0
1-ISOPROPOXY-2-METHYLETHYLENE	0	0	0	6	12	1	0	0	0
TETRAMETHYLETHYLENE OXIDE	0	0	0	6	12	1	0	0	0
2-METHYL-P-PHENYLENE SULFIDE	0	0	0	7	6	0	0	1	0
1-T-BUTOXY-2-METHYLETHYLENE	0	0	0	7	14	1	0	0	0
2,6-DIMETHYL-1,4-PHENYLENE OXIDE	0	0	0	8	8	1	0	0	0
2,5-DIMETHYL-2,4-HEXADIENE, 2,5-	0	0	0	8	14	0	0	0	0
BIS-2,4-DICHLOROPHENOXYPHOSPHAZENE	0	0	0	12	6	2	1	0	5
3,3'-DIMETHOXY-4,4'-BIPHENYLENE CARBODIIMIDE	0	0	0	15	12	0	2	0	0
BIS-BETA-NAPHTHOXYPHOSPHAZENE	0	0	0	20	14	2	1	0	1
P-PHENYLENE	0	0	1	0	0	0	0	0	0
P-PHENYLENE SULFIDE	0	0	1	0	0	0	0	1	0
P-PHENYLENE OXIDE	0	0	1	0	0	1	0	0	0
PHENYLENEDISILOXANYLENE	0	0	1	0	4	1	0	0	2
BETA-NITROSTYRENE	0	0	1	2	3	2	1	0	0
ETHYLIDENE P-PHENYLENE	0	0	1	2	4	0	0	0	0
P-PHENYLENE-(TETRAMETHYL)DISILOXANYLENE	0	0	1	4	12	1	0	0	2
P-PHENYLENEDIPYRIDYLCYCLOBUTYLENE	0	0	1	14	12	0	2	0	0
4,4'-THIODIPHENYLENE OXIDE	0	0	2	0	0	1	0	1	0
BIS-P-CHLOROPHENOXYPHOSPHAZENE	0	0	2	0	0	2	1	0	3
BISPHENOXYPHOSPHAZENE	0	0	2	0	2	2	1	0	1
4,4'-BIPHENYLENE CARBODIIMIDE	0	0	2	1	0	0	2	0	0
DIPHENYLMETHYLENE	0	0	2	1	2	0	0	0	0
BIS(M-TRIFLUOROMETHYLPHENOXY)PHOSPHAZENE	0	0	2	2	0	2	1	0	7
DI-P-TOLYLMETHYLENE	0	0	2	3	6	0	0	0	0
2,6-DIPHENYL-1,4-PHENYLENE OXIDE	0	0	2	6	4	1	0	0	0
2,5-DISTYRYLPYRAZINE	0	0	2	8	8	0	2	0	0
4,4'-ISOPROPYLIDENEDIPHENYLENE DITHIONISOPHTHALATE	0	0	3	5	6	2	0	2	0
2,2'-BIPHENYLENEDITHIAZOLE-P-PHENYLENE	0	0	3	6	2	0	2	2	0
M-PHENYLENEDITHIAZOLE-OXY-P-DIPHENYLENE	0	0	3	6	2	1	2	2	0
BIS(P-PHENYLPHENOXY)PHOSPHAZENE	0	0	4	0	2	2	1	0	1
2,2'-BIPHENYLENEDITHIAZOLE-OXY-P-DIPHENYLENE	0	0	4	6	2	1	2	2	0
2-MERCAPTO-2-METHYLACETIC ACID	0	1	0	2	4	0	0	1	0
2-HYDROXYPROPIONIC ACID (LACTIC ACID)	0	1	0	2	4	1	0	0	0
L-ALANINE	0	1	0	2	5	0	1	0	0
DIMETHYL KETENE I.	0	1	0	3	6	0	0	0	0
ISOPROPYLHYDROXYACETIC ACID	0	1	0	4	8	1	0	0	0
2,3-DIMETHYL-3-AMINOPROPIONIC ACID	0	1	0	4	9	0	1	0	0
B-D-GULURONIC ACID	0	1	0	5	8	5	0	0	0
B-D-MANNURONIC ACID	0	1	0	5	8	5	0	0	0
2,5-METHYL-2,4-HEXADIENOATE	0	1	0	6	10	1	0	0	0
4-METHYL-1,3-PHENYLENEUREA	0	1	0	7	8	0	2	0	0
DIMETHYL KETENE II.	0	1	0	7	12	1	0	0	0
FURFURAL-ALT-DIMETHYLKETENE	0	1	0	8	10	2	0	0	0
2,5-ISOPROPYL-2,4-HEXADIENOATE	0	1	0	8	14	1	0	0	0
TETRAMETHYLCYCLOBUTYLENE CARBONATE	0	1	0	8	14	2	0	0	0
DI-T-BUTYL-4-HYDROXYBENZOIC ACID	0	1	0	14	20	1	0	0	0
4,4'-ISOPROPYLIDENE-3,3',5,5'-TETRACHLORODIPHENYLENE CARBONATE	0	1	0	15	10	2	0	0	4

POLY-	CH2	CO	C6H4	C	H	O	N	S	X
4,4'-ISOPROPYLIDENE-3,3',5,5'-TETRABROMODIPHENYLENE CARBONATE	0	1	0	15	10	2	0	0	4
4,4'-ISOPROPYLIDENE-3,3'-DICHLORODIPHENYLENE CARBONATE	0	1	0	15	12	2	0	0	2
4,4'-ISOPROPYLIDENE-3,3'-DIMETHYLDIPHENYLENE CARBONATE	0	1	0	17	18	2	0	0	0
4,4'-CYCLOHEXYLIDENE-3,3',5,5'-TETRACHLORODIPHENYLENE CARBONATE	0	1	0	18	9	2	0	0	4
4,4'-ISOPROPYLIDENE-3,3'-DIISOPROPYLDIPHENYLENE CARBONATE	0	1	0	21	26	2	0	0	0
P-HYDROXYBENZOIC ACID	0	1	1	0	0	1	0	0	0
M-AMINOBENZOIC ACID	0	1	1	0	1	0	1	0	0
P-AMINOBENZOIC ACID	0	1	1	0	1	0	1	0	0
PHENYLISOCYANATE	0	1	1	0	1	0	1	0	0
M-PHENYLENE UREA	0	1	1	0	2	0	2	0	0
P-CHLOROBENZALDEHYDE-ALT-DIMETHYLKETENE	0	1	1	4	7	1	0	0	1
M-NITROBENZALDEHYDE-ALT-DIMETHYLKETENE	0	1	1	4	7	3	1	0	0
BENZALDEHYDE-ALT-DIMETHYLKETENE	0	1	1	4	8	1	0	0	0
2,5-METHYL-5-PHENYL-2,4-PENTADIENOATE	0	1	1	5	8	1	0	0	0
P-METHOXYBENZALDEHYDE-CO-DIMETHYLKETENE	0	1	1	5	10	2	0	0	0
4,4'-THIODIPHENYLENE CARBONATE	0	1	2	0	0	2	0	1	0
4,4'-OXYDIPHENYLENE CARBONATE	0	1	2	0	0	3	0	0	0
4,4'-SULFINYLDIPHENYLENE CARBONATE	0	1	2	0	0	3	0	1	0
4,4'-SULFONYLDIPHENYLENE CARBONATE	0	1	2	0	0	4	0	1	0
4,4'-ETHYLIDENEDIPHENYLENE CARBONATE	0	1	2	2	4	2	0	0	0
4,4'-ISOPROPYLIDENEDIPHENYLENE CARBONATE	0	1	2	3	6	2	0	0	0
4,4'-ISOBUTYLIDENEDIPHENYLENE CARBONATE	0	1	2	4	8	2	0	0	0
4,4'-(DIMETHYL-3,3-PENTYLIDENE)DIPHENYLENE CARBONATE	0	1	2	7	14	2	0	0	0
(P-PHENYLENEDIOXY)DIPHENYLENE CARBONATE	0	1	3	0	0	4	0	0	0
4,4'-(PHENYLMETHYLENE)DIPHENYLENE CARBONATE	0	1	3	1	2	2	0	0	0
4,4'-PHENYLETHYLYIDENEDIPHENYLENE CARBONATE	0	1	3	2	4	2	0	0	0
4,4'-(BIPHENYLENEDIOXY)DIPHENYLENE CARBONATE	0	1	4	0	0	4	0	0	0
4,4'-(DIPHENYLMETHYLENE)DIPHENYLENE CARBONATE	0	1	4	1	2	2	0	0	0
2,4-HEXADIENEDIOIC ACID, 2,5-	0	2	0	4	6	2	0	0	0
2,5-DIISOPROPYL-2,4-HEXADIENEDIOATE	0	2	0	10	18	2	0	0	0
ISOPHTHALIC ANHYDRIDE	0	2	1	0	0	1	0	0	0
TEREPHTHALIC ANHYDRIDE	0	2	1	0	0	1	0	0	0
P-XYLYLENE OXALATE	0	2	1	0	0	2	0	0	0
P-PHENYLENE CYCLOBUTYLENEDICARBOXYLIC ACID	0	2	1	4	6	2	0	0	0
P-PHENYLENE CYCLOBUTYLENEDICARBOXAMIDE	0	2	1	4	8	0	2	0	0
P-PHENYLENE DIMETHYLCYCLOBUTYLENEDICARBOXYLATE	0	2	1	6	10	2	0	0	0
P-PHENYLENE DICYANODIMETHYLCYCLOBUTYLENEDICARBOXYLATE	0	2	1	8	8	2	2	0	0
P-PHENYLENE DIISOPROPYLCYCLOBUTYLENEDICARBOXYLATE	0	2	1	10	18	2	0	0	0
P-PHENYLENE DICYANODIISOPROPYLCYCLOBUTYLENEDICARBOXYLATE	0	2	1	12	16	2	2	0	0
3,3'-DIMETHYL-4,4'-BIPHENYLENE TEREPHTHALAMIDE	0	2	1	14	14	0	2	0	0
4,4'-OXYDIBENZOIC ANHYDRIDE	0	2	2	0	0	2	0	0	0
M-PHENYLENE ISOPHTHALATE	0	2	2	0	0	2	0	0	0
M-PHENYLENE ISOPHTHALAMIDE	0	2	2	0	2	0	2	0	0
P-PHENYLENE ISOPHTHALAMIDE	0	2	2	0	2	0	2	0	0
P-PHENYLENE PHTHALAMIDE	0	2	2	0	2	0	2	0	0
M-PHENYLENE TEREPHTHALAMIDE	0	2	2	0	2	0	2	0	0
P-PHENYLENE TEREPHTHALAMIDE	0	2	2	0	2	0	2	0	0
ISOPHTHALOYL DITHIONISOPHTHALOYLDIHYDRAZIDE	0	2	2	2	4	0	4	2	0
TEREPHTHALOYL DITHIONISOPHTHALOYLDIHYDRAZIDE	0	2	2	2	4	0	4	2	0
4,4'-ISOPROPYLIDENEDIBENZOIC ANHYDRIDE	0	2	2	3	6	1	0	0	0
4,4'-ISOPROPYLIDENEDIPHENYLENE 3-METHYLCYCLOPROPYLENEDICARBOXYLA	0	2	2	7	12	2	0	0	0
4-METHYL-1,3-PHENYLENE 3,3'-METHYLPHOSPHINYLIDENEDIBENZAMIDE	0	2	2	8	11	1	2	0	1
4-METHYL-1,3-PHENYLENE 4,4'-METHYLPHOSPHINYLIDENEDIBENZAMIDE	0	2	2	8	11	1	2	0	1
4,4'-BIPHENYLENE DITHIOLISOPHTHALATE	0	2	3	0	0	0	0	2	0
4,4'-BIPHENYLENE DITHIOLTEREPHTHALATE	0	2	3	0	0	0	0	2	0
4,4'-BIPHENYLENE TEREPHTHALATE	0	2	3	0	0	2	0	0	0
4,4'-BIPHENYLENE TEREPHTHALAMIDE	0	2	3	0	2	0	2	0	0
PHENYLENE PHOSPHINYLIDENEDIBENZAMIDE	0	2	3	0	3	1	2	0	1
M-PHENYLENE 3,3'-(METHYLPHOSPHINYLIDENE)DIBENZAMIDE	0	2	3	1	5	1	2	0	1
M-PHENYLENE 4,4'-(METHYLPHOSPHINYLIDENE)DIBENZAMIDE	0	2	3	1	5	1	2	0	1
O-PHENYLENE 3,3'-(METHYLPHOSPHINYLIDENE)DIBENZAMIDE	0	2	3	1	5	1	2	0	1
O-PHENYLENE 4,4'-(METHYLPHOSPHINYLIDENE)DIBENZAMIDE	0	2	3	1	5	1	2	0	1
P-PHENYLENE 3,3'-(METHYLPHOSPHINYLIDENE)DIBENZAMIDE	0	2	3	1	5	1	2	0	1
P-PHENYLENE 4,4'-(METHYLPHOSPHINYLIDENE)DIBENZAMIDE	0	2	3	1	5	1	2	0	1
P-PHENYLENE DIPHENYLCYCLOBUTYLENEDICARBOXYLATE	0	2	3	4	6	2	0	0	0
4-METHYL-1,3-PHENYLENE 4,4'-PHENYLPHOSPHINYLIDENEDIBENZAMIDE	0	2	3	7	9	1	2	0	1
M-PHENYLENE 4,4'-(PHENYLPHOSPHINYLIDENE)DIBENZAMIDE	0	2	4	0	3	1	2	0	1
O-PHENYLENE 4,4'-(PHENYLPHOSPHINYLIDENE)DIBENZAMIDE	0	2	4	0	3	1	2	0	1
P-PHENYLENE 4,4'-(PHENYLPHOSPHINYLIDENE)DIBENZAMIDE	0	2	4	0	3	1	2	0	1
4,4'-OXYDIPHENYLENE PYROMELLITIMIDE	0	4	2	6	2	1	2	0	0
4,4'-OXYDIPHENYLENE 2,2',3,3'-OXYDIPHENYLENETETRACARBOXIMIDE	0	4	2	12	6	2	2	0	0
(P-PHENYLENEDIOXY)DIPHENYLENE PYROMELLITIMIDE	0	4	3	6	2	2	2	0	0
(P-PHENYLENEDIOXY)DIPHENYLENE OXYDIPHENYLENETETRACARBOXIMIDE	0	4	3	12	6	3	2	0	0
4,4'-OXYDIPHENYLENE P-PHENYLENEDITRIMELLITATEDIIMIDE	0	6	3	12	6	3	2	0	0
ISOPHTHALIC 2,2'-(ISOPHTHALOYLDIAMINO)DIBENZOIC ANHYDRIDE	0	6	4	0	2	2	2	0	0
ISOPHTHALIC 2,2'-(TEREPHTHALOYLDIAMINO)DIBENZOIC ANHYDRIDE	0	6	4	0	2	2	2	0	0
2,2'-(ISOPHTHALOYLDIAMINO)DIBENZOIC TEREPHTHALIC ANHYDRIDE	0	6	4	0	2	2	2	0	0

POLY-	CH2	CO	C6H4	C	H	O	N	S	X
TEREPHTHALIC 2,2'-(TEREPHTHALOYLDIAMINO)DIBENZOIC ANHYDRIDE	0	6	4	0	2	2	2	0	0
(P-PHENYLENEDIOXY)DIPHENYLENE P-PHENYLENEDITRIMELLITATEDIIMIDE	0	6	4	12	6	4	2	0	0
------ 1	1	0	0	0	0	0	0	0	0
METHYLENE	1	0	0	0	0	0	0	0	0
METHYLENE SELENIDE	1	0	0	0	0	0	0	0	1
METHYLENE SULFIDE	1	0	0	0	0	0	0	1	0
METHYLENE DISULFIDE	1	0	0	0	0	0	0	2	0
OXYMETHYLENE	1	0	0	0	0	1	0	0	0
VINYLIDENE BROMIDE	1	0	0	1	0	0	0	0	2
VINYLIDENE CHLORIDE	1	0	0	1	0	0	0	0	2
VINYLIDENE FLUORIDE	1	0	0	1	0	0	0	0	2
VINYL CHLORIDE	1	0	0	1	1	0	0	0	1
VINYL FLUORIDE	1	0	0	1	1	0	0	0	1
2-CHLOROACETALDEHYDE	1	0	0	1	1	1	0	0	1
VINYL ALCOHOL	1	0	0	1	2	1	0	0	0
ACRYLONITRILE	1	0	0	2	1	0	1	0	0
3,3,3-TRIFLUOROPROPYLENE OXIDE	1	0	0	2	1	1	0	0	3
PROPENE	1	0	0	2	4	0	0	0	0
PROPYLIDENE	1	0	0	2	4	0	0	0	0
PROPYLENE SULFIDE	1	0	0	2	4	0	0	1	0
METHYLVINYL ETHER	1	0	0	2	4	1	0	0	0
PROPIONALDEHYDE	1	0	0	2	4	1	0	0	0
PROPYLENE OXIDE	1	0	0	2	4	1	0	0	0
METHACRYLONITRILE	1	0	0	3	3	0	1	0	0
ISOBUTENE	1	0	0	3	6	0	0	0	0
ISOBUTYLENE SULFIDE	1	0	0	3	6	0	0	1	0
ISOBUTYLENE OXIDE	1	0	0	3	6	1	0	0	0
ALPHA-METHYLVINYLMETHYL ETHER	1	0	0	3	6	1	0	0	0
1-CYANO-1,3-BUTADIENE, 1,4-	1	0	0	4	3	0	1	0	0
3-TRIFLUOROMETHYL-1-BUTENE	1	0	0	4	5	0	0	0	3
1,3-PENTADIENE, 1,2-	1	0	0	4	6	0	0	0	0
1,3-PENTADIENE, 1,4-	1	0	0	4	6	0	0	0	0
1-METHOXY-1,3-BUTADIENE, 1,4-	1	0	0	4	6	1	0	0	0
1,3-D-XYLOSE (XYLAN)	1	0	0	4	6	5	0	0	0
1,4-D-XYLOSE (HARDWOOD XYLAN)	1	0	0	4	6	5	0	0	0
3-METHYL-1-BUTENE	1	0	0	4	8	0	0	0	0
1-ETHOXY-2-METHYLETHYLENE	1	0	0	4	8	1	0	0	0
ISOPROPYLVINYL ETHER	1	0	0	4	8	1	0	0	0
ISOVALERALDEHYDE	1	0	0	4	8	1	0	0	0
1-ETHOXY-2-METHOXYETHYLENE	1	0	0	4	8	2	0	0	0
1,4-D-XYLOSE HYDRATE	1	0	0	4	8	6	0	0	0
1,4-D-XYLOSE DIHYDRATE	1	0	0	4	10	7	0	0	0
B-D-GLUCOTRINITRATE	1	0	0	5	5	11	3	0	0
ALPHA-ISOPROPYLACRYLONITRILE	1	0	0	5	7	0	1	0	0
4,4-DIMETHYL-1,3-BUTADIENE, 1,2-	1	0	0	5	8	0	0	0	0
A-D-GLUCOSE (AMYLOSE)	1	0	0	5	8	5	0	0	0
B-D-GLUCOSE (CELLULOSE)	1	0	0	5	8	5	0	0	0
B-D-MANNOSE	1	0	0	5	8	5	0	0	0
1-CHLORO-2-ISOBUTOXYETHYLENE	1	0	0	5	9	1	0	0	1
3,3-DIMETHYL-1-BUTENE	1	0	0	5	10	0	0	0	0
TERT-BUTYLETHYLENE SULFIDE	1	0	0	5	10	0	0	1	0
TERT-BUTYLETHYLENE OXIDE	1	0	0	5	10	1	0	0	0
TERT-BUTYLVINYL ETHER	1	0	0	5	10	1	0	0	0
2-VINYL PYRIDINE	1	0	0	6	5	0	1	0	0
5-METHYL-1,3-HEXADIENE, 1,4-	1	0	0	6	10	0	0	0	0
1-ISOBUTOXY-2-METHYLETHYLENE	1	0	0	6	12	1	0	0	0
1,4-TETRADEUTEROBUTADIENE-ALT-METHACRYLONITRILE	1	0	0	7	5	0	1	0	4
P-FLUORO-O-METHYLSTYRENE	1	0	0	8	7	0	0	0	1
METHYLENE 2,5-DIMETHYL-P-PHENYLENE	1	0	0	8	8	0	0	0	0
TETRAMETHYLCYCLOBUTYLENEOXYMETHYLENE OXIDE	1	0	0	8	14	2	0	0	0
2,4-DIMETHYLSTYRENE	1	0	0	9	10	0	0	0	0
2,5-DIMETHYLSTYRENE	1	0	0	9	10	0	0	0	0
3,4-DIMETHYLSTYRENE	1	0	0	9	10	0	0	0	0
3,5-DIMETHYLSTYRENE	1	0	0	9	10	0	0	0	0
1-VINYL NAPHTHALENE	1	0	0	11	8	0	0	0	0
3-VINYL PYRENE	1	0	0	17	10	0	0	0	0
METHYLENE P-PHENYLENE	1	0	1	0	0	0	0	0	0
M-FLUOROSTYRENE	1	0	1	1	1	0	0	0	1
O-FLUOROSTYRENE	1	0	1	1	1	0	0	0	1
P-FLUOROSTYRENE	1	0	1	1	1	0	0	0	1
STYRENE	1	0	1	1	2	0	0	0	0
STYRENE OXIDE	1	0	1	1	2	1	0	0	0
STYRENE SULFONE	1	0	1	1	2	2	0	1	0
ALPHA-METHYLSTYRENE	1	0	1	2	4	0	0	0	0
M-METHYLSTYRENE	1	0	1	2	4	0	0	0	0
O-METHYLSTYRENE	1	0	1	2	4	0	0	0	0
P-METHYLSTYRENE	1	0	1	2	4	0	0	0	0

POLY-	CH2	CO	C6H4	C	H	O	N	S	X
N-PHENYL-2-IMINO-1,3-DITHIA-4-METHYLPENTAMETHYLENE	1	0	1	3	5	0	1	2	0
3-PHENYL-1-BUTENE	1	0	1	3	6	0	0	0	0
P-TRIMETHYLSILYLSTYRENE	1	0	1	4	10	0	0	0	1
TERT-BUTYLSTYRENE	1	0	1	5	10	0	0	0	0
STYRYL PYRIDINE-ALT-VINYL PYRIDINE	1	0	1	13	12	0	2	0	0
4,4'-METHYLENEDIPHENYLENE OXIDE	1	0	2	0	0	1	0	0	0
N-VINYL CARBAZOLE	1	0	2	1	1	0	1	0	0
N-VINYL DIPHENYLAMINE	1	0	2	1	3	0	1	0	0
2-MERCAPTOACETIC ACID	1	1	0	0	0	0	0	1	0
2-HYDROXYACETIC ACID (GLYCOLIC ACID)	1	1	0	0	0	1	0	0	0
GLYCINE	1	1	0	0	1	0	1	0	0
THIOLACRYLIC ACID	1	1	0	1	2	0	0	1	0
ACRYLIC ACID	1	1	0	1	2	1	0	0	0
VINYL FORMATE	1	1	0	1	2	1	0	0	0
ACRYLAMIDE	1	1	0	1	3	0	1	0	0
L-SERINE	1	1	0	1	3	1	1	0	0
VINYL TRIFLUOROACETATE	1	1	0	2	1	1	0	0	3
VINYLMETHYL KETONE	1	1	0	2	4	0	0	0	0
3-HYDROXYBUTYRIC ACID	1	1	0	2	4	1	0	0	0
METHACRYLIC ACID	1	1	0	2	4	1	0	0	0
VINYL ACETATE	1	1	0	2	4	1	0	0	0
3-AMINOBUTYRIC ACID	1	1	0	2	5	0	1	0	0
ISOPROPENYLMETHYL KETONE	1	1	0	3	6	0	0	0	0
2,2-DIMETHYL-3-HYDROXYPROPIONIC ACID	1	1	0	3	6	1	0	0	0
METHYLMETHACRYLATE	1	1	0	3	6	1	0	0	0
2,2-DIMETHYL-3-AMINOPROPIONIC ACID	1	1	0	3	7	0	1	0	0
3,3-DIMETHYL-3-AMINOPROPIONIC ACID	1	1	0	3	7	0	1	0	0
HISTIDINE	1	1	0	4	5	0	3	0	0
VINYLISOPROPYL KETONE	1	1	0	4	8	0	0	0	0
ISOPROPYLTHIOLACRYLATE	1	1	0	4	8	0	0	1	0
ISOPROPYL ACRYLATE	1	1	0	4	8	1	0	0	0
3-ISOPROPYL-3-HYDROXYPROPIONIC ACID	1	1	0	4	8	1	0	0	0
N-ISOPROPYL ACRYLAMIDE	1	1	0	4	9	0	1	0	0
VINYL-TERT-BUTYL KETONE	1	1	0	5	10	0	0	0	0
TERT-BUTYL ACRYLATE	1	1	0	5	10	1	0	0	0
2,5-ETHYL-2,4-HEXADIENOATE	1	1	0	6	10	1	0	0	0
B-D-2-GLUCOSAMINE (CHITIN)	1	1	0	6	11	4	1	0	0
TERT-BUTYLMETHACRYLATE	1	1	0	6	12	1	0	0	0
2,3,4-TRIMETHOXY-5-HYDROXYVALERIC ACID	1	1	0	6	12	4	0	0	0
2,5-DIMETHYLTEREPHTHALALDEHYDE	1	1	0	8	8	1	0	0	0
2,5-ISOBUTYL-2,4-HEXADIENOATE	1	1	0	8	14	1	0	0	0
1,3-D-GALACTOSAMINE-ALT-1,4-D-GLUCURONIC ACID	1	1	0	10	17	9	1	0	0
3,3'-DIMETHYL-4,4'-METHYLENEDIPHENYLENE CARBONATE	1	1	0	14	12	2	0	0	0
ISOPHTHALALDEHYDE	1	1	1	0	0	1	0	0	0
TEREPHTHALALDEHYDE	1	1	1	0	0	1	0	0	0
ALPHA-AMINO-P-TOLUIC ACID	1	1	1	0	1	0	1	0	0
VINYLPHENYLKETONE	1	1	1	1	2	0	0	0	0
VINYL BENZOATE	1	1	1	1	2	1	0	0	0
VINYL-O-HYDROXYPHENYL KETONE	1	1	1	1	2	1	0	0	0
3-MERCAPTO-2-(P-TOLUENESULFONAMIDE)PROPIONIC ACID	1	1	1	2	5	2	1	2	0
VINYL-P-TERT-BUTYLBENZOATE	1	1	1	5	10	1	0	0	0
4,4'-BIPHENYLDICARBOXALDEHYDE	1	1	2	0	0	1	0	0	0
4,4'-METHYLENEDIPHENYLENE CARBONATE	1	1	2	0	0	2	0	0	0
4,4'-BUTYLIDENEDIPHENYLENE CARBONATE	1	1	2	3	6	2	0	0	0
4,4'-(METHYL-2,2-PENTYLIDENE)DIPHENYLENE CARBONATE	1	1	2	5	10	2	0	0	0
ASPARTIC ACID	1	2	0	1	3	1	1	0	0
PROPYLENE OXALATE	1	2	0	2	4	2	0	0	0
O-ACETYL-L-SERINE	1	2	0	2	5	1	1	0	0
ALANYLGLYCINE	1	2	0	2	6	0	2	0	0
(3-BROMO-4-CARBOXYPHENOXY)ACETIC ANHYDRIDE	1	2	0	6	3	2	0	0	1
B-D-XYLODIACETATE	1	2	0	6	10	5	0	0	0
B-D-2-GLUCOSAMINE-ALT-1,4-B-D-GLUCURONIC ACID	1	2	0	11	19	9	1	0	0
N-ACETYLGALACTOSAMINE-6-SULFATE-ALT-1,4-D-GLUCURONIC ACID	1	2	0	11	19	12	1	1	0
3,3'-DIMETHYL(METHYLENEDIPHENYLENE) METHYL-1,3-PHENYLENEDIUREA	1	2	0	21	22	0	4	0	0
(M-CARBOXYPHENOXY)ACETIC ANHYDRIDE	1	2	1	0	0	2	0	0	0
(P-CARBOXYPHENOXY)ACETIC ANHYDRIDE	1	2	1	0	0	2	0	0	0
P-PHENYLENE MALONATE	1	2	1	0	0	2	0	0	0
M-PHENYLENE CYCLOPROPYLENEDICARBOXYLATE	1	2	1	2	2	2	0	0	0
P-PHENYLENE CYCLOPROPYLENEDICARBOXYLATE	1	2	1	2	2	2	0	0	0
CYCLOPROPYLENE ISOPHTHALAMIDE	1	2	1	2	4	0	2	0	0
3,3'-DIMETHYL-4,4'-METHYLENEDIPHENYLENE TEREPHTHALAMIDE	1	2	1	14	14	0	2	0	0
N,N'-DIMETHYL-4,4'-(DIMETHYLMETHYLENE)DIPHENYLENE TEREPHTHALAMID	1	2	1	16	18	0	2	0	0
4,4'-METHYLENEDIBENZOIC ANHYDRIDE	1	2	2	0	0	1	0	0	0
4,4'-(METHYLENEDIOXY)DIBENZOIC ANHYDRIDE	1	2	2	0	0	3	0	0	0
PROPYLENE 4,4'-SULFONYLDIBENZAMIDE	1	2	2	2	6	2	2	1	0
4,4'-ISOPROPYLIDENEDIPHENYLENE MALONATE	1	2	2	3	6	2	0	0	0
4,4'-ISOPROPYLIDENEDIPHENYLENE CYCLOPROPYLENEDICARBOXYLATE	1	2	2	5	8	2	0	0	0

CRYSTALLOGRAPHIC DATA FOR VARIOUS POLYMERS

POLY-	CH2	CO	C6H4	C	H	O	N	S	X
4,4'-ISOPROPYLIDENEDIPHENYLENE 1-METHYLCYCLOPROPYLENEDICARBOXYLA	1	2	2	6	10	2	0	0	0
4,4'-METHYLENEDIPHENYLENE TEREPHTHALAMIDE	1	2	3	0	2	0	2	0	0
4,4'-METHYLENEDIPHENYLENE M-PHENYLENEDIUREA	1	2	3	0	4	0	4	0	0
N,N'-DIMETHYL-4,4'-METHYLENEDIPHENYLENE TEREPHTHALAMIDE	1	2	3	2	6	0	2	0	0
4,4'-ISOPROPYLIDENEDIPHENYLENE METHYLENEDIPHENYLENEDIURETHAN	1	2	4	3	8	2	2	0	0
ALANYLALANYLGLYCINE	1	3	0	4	12	0	3	0	0
A-D-GLUCOTRIACETATE	1	3	0	8	14	5	0	0	0
B-D-GLUCOTRIACETATE	1	3	0	8	14	5	0	0	0
B-D-MANNOTRIACETATE	1	3	0	8	14	5	0	0	0
3-MERCAPTO-2-PHTHALIMIDOPROPIONIC ACID	1	3	1	1	1	0	1	1	0
ISOPHTHALIC 4,4'-METHYLENEDIOXY-3,3'-DIMETHOXYDIBENZOIC ANH	1	4	1	14	12	6	0	0	0
4,4'-(METHYLENEDIOXY)DIMETHOXYDIBENZOIC TEREPHTHALIC ANHYDRIDE	1	4	1	14	12	6	0	0	0
N,N'-METHYLENEDITEREPHTHALAMIC ANHYDRIDE	1	4	2	0	2	1	2	0	0
ISOPHTHALOYL (M-CARBOXYPHENOXYACETYL)DIHYDRAZIDE	1	4	2	0	4	1	4	0	0
TEREPHTHALOYL (M-CARBOXYPHENOXYACETYL)DIHYDRAZIDE	1	4	2	0	4	1	4	0	0
ISOPHTHALIC 4,4'-(METHYLENEDIOXY)DIBENZOIC ANHYDRIDE	1	4	3	0	0	4	0	0	0
4,4'-(METHYLENEDIOXY)DIBENZOIC TEREPHTHALIC ANHYDRIDE	1	4	3	0	0	4	0	0	0
------ 2	2	0	0	0	0	0	0	0	0
ETHYLENE	2	0	0	0	0	0	0	0	0
ETHYLENE SULFIDE	2	0	0	0	0	0	0	1	0
ETHYLENE DISULFIDE	2	0	0	0	0	0	0	2	0
ETHYLENE TETRASULFIDE	2	0	0	0	0	0	0	4	0
ETHYLENE OXIDE	2	0	0	0	0	1	0	0	0
ETHYLENE AMINE	2	0	0	0	1	0	1	0	0
2-AMINOETHYLENESULFONIC ACID	2	0	0	0	1	2	1	1	0
ALLENE	2	0	0	1	0	0	0	0	0
2-IMINO-1,3-DITHIAPENTAMETHYLENE	2	0	0	1	1	0	1	2	0
BROMOMETHYLETHYLENE OXIDE	2	0	0	1	1	1	0	0	1
CHLOROMETHYLETHYLENE OXIDE	2	0	0	1	1	1	0	0	1
FLUOROMETHYLETHYLENE OXIDE	2	0	0	1	1	1	0	0	1
3-SILYLPROPENE	2	0	0	1	4	0	0	0	1
2,3-DICHLORO-1,3-BUTADIENE, 1,4-	2	0	0	2	0	0	0	0	2
CHLOROTRIFLUOROETHYLENE-ALT-ETHYLENE	2	0	0	2	0	0	0	0	4
TETRAFLUOROETHYLENE-ALT-ETHYLENE	2	0	0	2	0	0	0	0	4
BIS-2,2,2-TRIFLUOROETHOXYPHOSPHAZENE	2	0	0	2	0	2	1	0	7
2-CHLORO-1,3-BUTADIENE, 1,4-	2	0	0	2	1	0	0	0	1
4,4,4-TRIFLUORO-1-BUTENE	2	0	0	2	1	0	0	0	3
2,5-ETHYLENE-1,3,4-TRIAZOLE	2	0	0	2	1	0	3	0	0
2,2,2-TRIFLUOROETHYLVINYL ETHER	2	0	0	2	1	1	0	0	3
CYANOETHYLOXYMETHYLENE	2	0	0	2	1	1	1	0	0
1,3-BUTADIENE, 1,2-	2	0	0	2	2	0	0	0	0
1,3-BUTADIENE, 1,4-	2	0	0	2	2	0	0	0	0
CYCLOBUTENE	2	0	0	2	2	0	0	0	0
2,5-ETHYLENE-1-AMINO-1,3,4-TRIAZOLE	2	0	0	2	2	0	4	0	0
BUTADIENE OXIDE	2	0	0	2	2	1	0	0	0
N-METHYL-2-IMINO-1,3-DITHIAPENTAMETHYLENE	2	0	0	2	3	0	1	2	0
2-CHLOROMETHYLPROPYLENE OXIDE	2	0	0	2	3	1	0	0	1
1-BUTENE	2	0	0	2	4	0	0	0	0
BUTYRALDEHYDE	2	0	0	2	4	1	0	0	0
ETHYLVINYL ETHER	2	0	0	2	4	1	0	0	0
ETHYLSILYLETHYLENE	2	0	0	2	6	0	0	0	1
DIETHYLSILOXANE	2	0	0	2	6	1	0	0	1
3-CHLORO-2-METHYL-1,3-BUTADIENE, 1,4-	2	0	0	3	3	0	0	0	1
ALPHA-ETHYLACRYLONITRILE	2	0	0	3	3	0	1	0	0
2-METHYL-1,3-BUTADIENE, 1,4- (ISOPRENE)	2	0	0	3	4	0	0	0	0
3-METHYL-1-BUTENE, 1,3-	2	0	0	3	6	0	0	0	0
2,2-DIMETHYLTRIMETHYLENE SULFIDE	2	0	0	3	6	0	0	1	0
3,3-DIMETHYLOXACYCLOBUTANE	2	0	0	3	6	1	0	0	0
2,2-DIMETHYLTRIMETHYLENE SULFONE	2	0	0	3	6	2	0	1	0
2,4-HEXADIYNE-1,6-DIOL	2	0	0	4	2	2	0	0	0
1,3-CYCLOHEXADIENE, 1,4-	2	0	0	4	4	0	0	0	0
1,3,5-HEXATRIENE, 1,6-	2	0	0	4	4	0	0	0	0
4-TRIFLUOROMETHYL-1-PENTENE	2	0	0	4	5	0	0	0	3
2,3-DIMETHYL-1,3-BUTADIENE, 1,4-	2	0	0	4	6	0	0	0	0
1,3-HEXADIENE, 1,4-	2	0	0	4	6	0	0	0	0
2-BUTENE-ALT-ETHYLENE	2	0	0	4	8	0	0	0	0
3-METHYL-1-PENTENE	2	0	0	4	8	0	0	0	0
4-METHYL-1-PENTENE	2	0	0	4	8	0	0	0	0
ISOBUTYLVINYL ETHER	2	0	0	4	8	1	0	0	0
1-METHYL-2-PROPOXYETHYLENE	2	0	0	4	8	1	0	0	0
1-METHYLPROPYLVINYL ETHER	2	0	0	4	8	1	0	0	0
3-TRIMETHYLSILYLPROPENE	2	0	0	4	10	0	0	0	1
4,4-DIMETHYL-1-PENTENE	2	0	0	5	10	0	0	0	0
NEOPENTYLETHYLENE OXIDE	2	0	0	5	10	1	0	0	0
NEOPENTYLVINYL ETHER	2	0	0	5	10	1	0	0	0
2,5-DICHLORO-P-XYLYLENE	2	0	0	6	2	0	0	0	2
BROMO-P-XYLYLENE	2	0	0	6	3	0	0	0	1

POLY-	CH2	CO	C6H4	C	H	O	N	S	X
CHLORO-P-XYLYLENE	2	0	0	6	3	0	0	0	1
2-T-BUTYL-1,3-BUTADIENE, 1,4-	2	0	0	6	10	0	0	0	0
5-METHYL-1,3-HEPTADIENE	2	0	0	6	10	0	0	0	0
6-METHYL-1,3-HEPTADIENE, 1,4-	2	0	0	6	10	0	0	0	0
CYANO-P-XYLYLENE	2	0	0	7	3	0	1	0	0
2,4,6-TRICHLORO-3-PHENOXYPROPYLENE OXIDE	2	0	0	7	3	2	0	0	3
METHYL-P-XYLYLENE	2	0	0	7	6	0	0	0	0
DIVINYLFURFURAL	2	0	0	7	6	3	0	0	0
4,6-DIMETHYL-M-XYLYENE	2	0	0	8	8	0	0	0	0
2,5-DIMETHYL-P-XYLYENE	2	0	0	8	8	0	0	0	0
2,5-DIMETHYL-3-PHENYLPROPENE	2	0	0	9	10	0	0	0	0
3,4-DIMETHYL-3-PHENYLPROPENE	2	0	0	9	10	0	0	0	0
3,5-DIMETHYL-3-PHENYLPROPENE	2	0	0	9	10	0	0	0	0
3-(1-NAPHTHOXY)PROPYLENE OXIDE	2	0	0	11	8	2	0	0	0
3-(2-NAPHTHOXY)PROPYLENE OXIDE	2	0	0	11	8	2	0	0	0
M-XYLYLENE	2	0	1	0	0	0	0	0	0
P-XYLYLENE	2	0	1	0	0	0	0	0	0
P-XYLYLENE DISULFIDE	2	0	1	0	0	0	0	2	0
ETHYLENEOXY-P-PHENYLENE OXIDE	2	0	1	0	0	2	0	0	0
N-PHENYL-2-IMINO-1,3-DITHIAPENTAMETHYLENE	2	0	1	1	1	0	1	2	0
O-CHLORO-3-PHENOXYPROPYLENE OXIDE	2	0	1	1	1	2	0	0	1
P-CHLORO-3-PHENOXYPROPYLENE OXIDE	2	0	1	1	1	2	0	0	1
3-PHENYLPROPENE (ALLYLBENZENE)	2	0	1	1	2	0	0	0	0
BENZYLVINYL ETHER	2	0	1	1	2	1	0	0	0
3-PHENOXYPROPYLENE OXIDE	2	0	1	1	2	2	0	0	0
ETHYLENE DITHIONISOPHTHALATE	2	0	1	2	0	2	0	2	0
P-PHENYLENECYCLOBUTYLENE	2	0	1	2	2	0	0	0	0
3-M-TOLYLPROPENE	2	0	1	2	4	0	0	0	0
3-O-TOLYLPROPENE	2	0	1	2	4	0	0	0	0
3-P-TOLYLPROPENE	2	0	1	2	4	0	0	0	0
M-METHYL-3-PHENOXYPROPYLENE OXIDE	2	0	1	2	4	2	0	0	0
O-METHYL-3-PHENOXYPROPYLENE OXIDE	2	0	1	2	4	2	0	0	0
P-METHYL-3-PHENOXYPROPYLENE OXIDE	2	0	1	2	4	2	0	0	0
DIVINYLBENZAL	2	0	1	3	4	2	0	0	0
2-METHYLDIVINYLBENZAL	2	0	1	4	6	2	0	0	0
4-METHYLDIVINYLBENZAL	2	0	1	4	6	2	0	0	0
O-ISOPROPYL-3-PHENOXYPROPYLENE OXIDE	2	0	1	4	8	2	0	0	0
2,2'-ETHYLENEDITHIAZOLE-P-PHENYLENE	2	0	1	6	2	0	2	2	0
4,4'-ETHYLENEBIPHENYLENE	2	0	2	0	0	0	0	0	0
4,4'-OXYDIPHENYLENEDITHIOETHYLENE DISULFIDE	2	0	2	0	0	1	0	4	0
O-PHENYL-3-PHENOXYPROPYLENE OXIDE	2	0	2	1	2	2	0	0	0
CARBON MONOXIDE-ALT-ETHYLENE	2	1	0	0	0	0	0	0	0
3-MERCAPTOPROPIONIC ACID	2	1	0	0	0	0	0	1	0
3-HYDROXYPROPIONIC ACID (HYDRACRYLIC ACID)	2	1	0	0	0	1	0	0	0
3-AMINOPROPIONIC ACID (BETA-ALANINE)	2	1	0	0	1	0	1	0	0
ETHYLENE UREA	2	1	0	0	2	0	2	0	0
DIKETENE	2	1	0	1	0	1	0	0	0
N-ACETYLETHYLENE AMINE	2	1	0	1	3	0	1	0	0
N-METHYL-3-AMINOPROPIONIC ACID	2	1	0	1	3	0	1	0	0
2,4-DIAMINOBUTYRIC ACID	2	1	0	1	4	0	2	0	0
N-(PERFLUOROPROPIONYL)ETHYLENE AMINE	2	1	0	2	0	0	1	0	5
1,3-CYCLOBUTYLENE CARBONATE	2	1	0	2	2	2	0	0	0
HYDROXYPROLINE	2	1	0	2	3	1	1	0	0
3-CARBOMETHOXYPROPIONALDEHYDE	2	1	0	2	4	2	0	0	0
DIETHYL-2-HYDROXYACETIC ACID	2	1	0	3	6	1	0	0	0
N-ISOBUTYRYLETHYLENE AMINE	2	1	0	3	7	0	1	0	0
N-ISOPROPYL-3-AMINOPROPIONIC ACID	2	1	0	3	7	0	1	0	0
2-METHYLBUTYLISOCYANATE	2	1	0	3	7	0	1	0	0
SEC-BUTYLTHIOLACRYLATE	2	1	0	4	8	0	0	1	0
ISOBUTYLTHIOLACRYLATE	2	1	0	4	8	0	0	1	0
SEC-BUTYL ACRYLATE	2	1	0	4	8	1	0	0	0
ISOBUTYL ACRYLATE	2	1	0	4	8	1	0	0	0
N-PIVALOYLETHYLENE AMINE	2	1	0	4	9	0	1	0	0
3,5-DIMETHYL-6-AMINO-4-OXACAPROIC ACID	2	1	0	4	9	1	1	0	0
N-(PERFLUOROOCTANOYL)ETHYLENE AMINE	2	1	0	7	0	0	1	0	15
1,1-DIHYDROPERFLUOROOCTYL ACRYLATE	2	1	0	8	1	1	0	0	15
2,5-ISOAMYL-2,4-HEXADIENOATE	2	1	0	8	14	1	0	0	0
1,1-DIHYDROPERFLUORODECYL ACRYLATE	2	1	0	10	1	1	0	0	19
N-2-NAPHTHOYLETHYLENE AMINE	2	1	0	10	7	0	1	0	0
N-(P-CHLOROBENZYL)ETHYLENE AMINE	2	1	1	0	0	0	1	0	1
ETHYLENE-P-OXYBENZOATE	2	1	1	0	0	2	0	0	0
P-XYLYLENE CARBONATE	2	1	1	0	0	2	0	0	0
P-AMINOHYDROCINNAMIC ACID	2	1	1	0	1	0	1	0	0
P-(AMINOMETHYLENE)PHENYLENEACETIC ACID	2	1	1	0	1	0	1	0	0
N-BENZOYLETHYLENE AMINE	2	1	1	0	1	0	1	0	0
N-PHENYL-3-AMINOPROPIONIC ACID	2	1	1	0	1	0	1	0	0
N-(P-TOLUOYL)ETHYLENE AMINE	2	1	1	1	3	0	1	0	0

POLY-	CH2	CO	C6H4	C	H	O	N	S	X
1-BENZYLHISTIDINE	2	1	1	4	5	0	3	0	0
4,4'-ETHYLENEDIPHENYLENE CARBONATE	2	1	2	0	0	2	0	0	0
4,4'-BUTYLIDENEDIPHENYLENE CARBONATE	2	1	2	2	4	2	0	0	0
4,4'-PENTYLIDENEDIPHENYLENE CARBONATE	2	1	2	3	6	2	0	0	0
ETHYLENE OXALATE	2	2	0	0	0	2	0	0	0
HYDRAZO SUCCINAMIDE	2	2	0	0	2	0	2	0	0
GLUTAMIC ACID	2	2	0	1	3	1	1	0	0
ETHYLENE FUMARAMIDE	2	2	0	2	4	0	2	0	0
METHYL GLUTAMATE	2	2	0	2	5	1	1	0	0
O-ACETYLHYDROXYPROLINE	2	2	0	3	5	1	1	0	0
2,2-DIMETHYLTRIMETHYLENE OXALATE	2	2	0	3	6	2	0	0	0
ETHYLENE 2,5-FURANDICARBOXYLATE	2	2	0	4	2	2	0	0	0
CYCLOPROPYLENE CYCLOPROPYLENEDICARBOXAMIDE	2	2	0	4	6	0	2	0	0
ETHYLENE 2,4-HEXADIENEDIAMIDE	2	2	0	4	6	0	2	0	0
ISOBUTYL ASPARTATE	2	2	0	4	9	1	1	0	0
ETHYLENE DITHIOL-2,5-PYRIDINEDICARBOXYLATE	2	2	0	5	3	0	1	2	0
ETHYLENE 2,5-DICHLOROTEREPHTHALATE	2	2	0	6	2	2	0	0	2
ETHYLENE CHLOROTEREPHTHALATE	2	2	0	6	3	2	0	0	1
ETHYLENE METHYLTEREPHTHALATE	2	2	0	7	6	2	0	0	0
ETHYLENE 2,5-DIMETHYLTEREPHTHALATE	2	2	0	8	8	2	0	0	0
ETHYLENE 1,5-NAPHTHALATE	2	2	0	10	6	2	0	0	0
ETHYLENE 2,6-NAPHTHALATE	2	2	0	10	6	2	0	0	0
ETHYLENE 2,7-NAPHTHALATE	2	2	0	10	6	2	0	0	0
4,4'-ETHYLENEDIOXY-3,3'-DIMETHOXYDIBENZOIC ANHYDRIDE	2	2	0	14	12	5	0	0	0
2,5-DIMETHYLTEREPHTHALALDEHYDE	2	2	0	16	16	2	0	0	0
ETHYLENE DITHIOLISOPHTHALATE	2	2	1	0	0	0	0	2	0
ETHYLENE DITHIOLTEREPHTHALATE	2	2	1	0	0	0	0	2	0
P-PHENYLENEDIACETIC ANHYDRIDE	2	2	1	0	0	1	0	0	0
ETHYLENE ISOPHTHALATE	2	2	1	0	0	2	0	0	0
ETHYLENE TEREPHTHALATE	2	2	1	0	0	2	0	0	0
P-PHENYLENE SUCCINATE	2	2	1	0	0	2	0	0	0
(M-PHENYLENEDIOXY)DIACETIC ANHYDRIDE	2	2	1	0	0	3	0	0	0
(P-PHENYLENEDIOXY)DIACETIC ANHYDRIDE	2	2	1	0	0	3	0	0	0
ETHYLENE PHTHALAMIDE	2	2	1	0	2	0	2	0	0
ETHYLENE TEREPHTHALAMIDE	2	2	1	0	2	0	2	0	0
P-CHLOROBENZYL-L-ASPARATE	2	2	1	1	2	1	1	0	1
BENZYL ASPARATE	2	2	1	1	3	1	1	0	0
S-CARBOBENZOXY CYSTEINE	2	2	1	1	3	1	1	1	0
N,N'-DIMETHYLETHYLENE TEREPHTHALAMIDE	2	2	1	2	6	0	2	0	0
2,2-DIMETHYLTRIMETHYLENE TEREPHTHALATE	2	2	1	3	6	2	0	0	0
2,5-DIMETHYLPIPERAZINE ISOPHTHALAMIDE	2	2	1	4	8	0	2	0	0
2,5-DIMETHYLPIPERAZINE PHTHALAMIDE	2	2	1	4	8	0	2	0	0
2,5-DIMETHYLPIPERAZINE TEREPHTHALAMIDE	2	2	1	4	8	0	2	0	0
P-PHENYLENE DIETHYLCYCLOBUTYLENEDICARBOXYLATE	2	2	1	6	10	2	0	0	0
P-PHENYLENE DICYANODIETHYLCYCLOBUTYLENEDICARBOXYLATE	2	2	1	8	8	2	2	0	0
N,N'-DIETHYL-3,3'-DIMETHYLBIPHENYLENE TEREPHTHALAMIDE	2	2	1	16	18	0	2	0	0
P-XYLYLENE DITHIOLISOPHTHALATE	2	2	2	0	0	0	0	2	0
4,4'-ETHYLENEDIBENZOIC ANHYDRIDE	2	2	2	0	0	1	0	0	0
ETHYLENE 4,4'-DIBENZOATE	2	2	2	0	0	2	0	0	0
ISOPHTHALALDEHYDE	2	2	2	0	0	2	0	0	0
TEREPHTHALALDEHYDE	2	2	2	0	0	2	0	0	0
P-XYLYLENE ISOPHTHALATE	2	2	2	0	0	2	0	0	0
P-XYLYLENE TEREPHTHALATE	2	2	2	0	0	2	0	0	0
ETHYLENE 4,4'-THIODIBENZOATE	2	2	2	0	0	2	0	1	0
3,3'-(ETHYLENEDIOXY)DIBENZOIC ANHYDRIDE	2	2	2	0	0	3	0	0	0
4,4'-(ETHYLENEDIOXY)DIBENZOIC ANHYDRIDE	2	2	2	0	0	3	0	0	0
ETHYLENE 4,4'-OXYDIBENZOATE	2	2	2	0	0	3	0	0	0
4,4'-(OXYDIMETHYLENE)DIOXYDIBENZOIC ANHYDRIDE	2	2	2	0	0	4	0	0	0
M-XYLYLENE ISOPHTHALAMIDE	2	2	2	0	2	0	2	0	0
P-XYLYLENE ISOPHTHALAMIDE	2	2	2	0	2	0	2	0	0
P-XYLYLENE PHTHALAMIDE	2	2	2	0	2	0	2	0	0
M-XYLYLENE TEREPHTHALAMIDE	2	2	2	0	2	0	2	0	0
P-XYLYLENE TEREPHTHALAMIDE	2	2	2	0	2	0	2	0	0
ETHYLENE 4,4'-SULFONYLDIBENZAMIDE	2	2	2	0	2	2	2	1	0
1-METHYLTRIMETHYLENE 4,4'-SULFONYLDIBENZAMIDE	2	2	2	2	6	2	2	1	0
(4,4'-ISOPROPYLIDENEDIPHENYLENE)DIOXYDIACETIC ANHYDRIDE	2	2	2	3	6	3	0	0	0
2,2-DIMETHYLTRIMETHYLENE SULFONYLDIBENZAMIDE	2	2	2	3	8	2	2	1	0
BISPHENYL-2,4-HEXADIYNE-1,6-DIURETHAN	2	2	2	4	2	2	2	0	0
N,N'-DIETHYLBIPHENYLENE TEREPHTHALAMIDE	2	2	3	2	6	0	2	0	0
4,4'-BIPHENYLDICARBOXALDEHYDE	2	2	4	0	0	2	0	0	0
4,4'-(SILYLENEDIPHENYLENE)DIMETHYLENE 4,4'-SILYLENEDIBENZAMIDE	2	2	4	0	6	0	2	0	2
4,4'-(SILYLENEDIPHENYLENE)DIMETHYLENE 4,4'-DISILOXANYLENEDIBENZAM	2	2	4	0	8	1	2	0	3
4,4'-(SILYLENEDIPHENYLENE)DIMETHYLENE 4,4'-TETRASILOXANYLENEDIBEN	2	2	4	0	12	3	2	0	5
4,4'-DIMETHYLSILYLENEDIPHENYLENEDIMETHYLENE DIMETHYLSILYLENEDIBE	2	2	4	4	14	1	2	0	2
4,4'-DIMETHYLSILYLENEDIPHENYLENEDIMETHYLENE TETRAMETHYLDISILOXAN	2	2	4	6	20	1	2	0	3
4,4'-DIMETHYLSILYLDIPHENYLENEDIMETHYLENE TETRAMETHYLTETRAPHENYLT	2	2	8	6	20	3	2	0	5
ALANYLGLYCYLGLYCINE	2	3	0	2	7	0	3	0	0

POLY-	CH2	CO	C6H4	C	H	O	N	S	X
CYCLOPROPYLENE (CYCLOPROPYLENEDICARBOXOYL)DIURETHAN	2	4	0	4	6	2	2	0	0
(P-PHENYLENEDIOXY)DIACETIC 3,4-PYRIDINEDICARBOXYLIC ANHYDRIDE	2	4	1	5	3	4	1	0	0
ISOPHTHALIC (P-PHENYLENEDIOXY)DIACETIC ANHYDRIDE	2	4	2	0	0	4	0	0	0
(P-PHENYLENEDIOXY)DIACETIC TEREPHTHALIC ANHYDRIDE	2	4	2	0	0	4	0	0	0
ISOPHTHALOYL (M-CARBOXYPHENOXYPROPIONYL)DIHYDRAZIDE	2	4	2	0	4	1	4	0	0
4,4'-ISOPROPYLIDENEDIPHENYLENEDIOXYDIACETIC PYRIDINEDICARBOXYLIC	2	4	2	8	9	4	1	0	0
4,4'-(ETHYLENEDIOXY)DIBENZOIC ISOPHTHALIC ANHYDRIDE	2	4	3	0	0	4	0	0	0
4,4'-(ETHYLENEDIOXY)DIBENZOIC PHTHALIC ANHYDRIDE	2	4	3	0	0	4	0	0	0
4,4'-(ETHYLENEDIOXY)DIBENZOIC TEREPHTHALIC ANHYDRIDE	2	4	3	0	0	4	0	0	0
ISOPHTHALIC (ISOPROPYLIDENEDIPHENYLENE)DIACETIC ANHYDRIDE	2	4	3	3	6	4	0	0	0
4,4'-ISOPROPYLIDENEDIPHENYLENEDIOXYDIACETIC TEREPHTHALIC ANHYDRI	2	4	3	3	6	4	0	0	0
------ 3	3	0	0	0	0	0	0	0	0
TRIMETHYLENE SULFIDE	3	0	0	0	0	0	0	1	0
TRIMETHYLENE DISULFIDE	3	0	0	0	0	0	0	2	0
OXACYCLOBUTANE (TRIMETHYLENE OXIDE)	3	0	0	0	0	1	0	0	0
ETHYLENEOXYMETHYLENE OXIDE (DIOXOLANE)	3	0	0	0	0	2	0	0	0
TRIMETHYLENE SULFONE	3	0	0	0	0	2	0	1	0
3-AMINOTRIMETHYLENESULFONIC ACID	3	0	0	0	1	2	0	1	0
1,1-BISCHLOROMETHYLETHYLENE OXIDE	3	0	0	1	0	1	0	0	2
2-CHLOROETHYLVINYLETHER	3	0	0	1	1	1	0	0	1
5,5,5-TRIFLUORO-1-PENTENE	3	0	0	2	1	0	0	0	3
CYCLOPENTENE	3	0	0	2	2	0	0	0	0
VINYL CYCLOPROPANE	3	0	0	2	2	0	0	0	0
1,2-DICHLOROPENTAMETHYLENE	3	0	0	2	2	0	0	0	2
1,3-CYCLOBUTYLENEOXYMETHYLENE OXIDE	3	0	0	2	2	2	0	0	0
HYDROCHLORINATED 2-METHYL-1,3-BUTADIENE, 1,4-	3	0	0	2	3	0	0	0	1
N-ETHYL-2-IMINO-1,3-DITHIAPENTAMETHYLENE	3	0	0	2	3	0	1	2	0
1-PENTENE	3	0	0	2	4	0	0	0	0
PROPYLVINYL ETHER	3	0	0	2	4	1	0	0	0
VALERALDEHYDE	3	0	0	2	4	1	0	0	0
2-METHOXYETHYLVINYL ETHER	3	0	0	2	4	2	0	0	0
ALPHA-PROPYLACRYLONITRILE	3	0	0	3	3	0	1	0	0
1-BUTOXY-2-CHLOROETHYLENE	3	0	0	3	5	1	0	0	1
1-VINYLENE-3-CYCLOPENTYLENE	3	0	0	4	4	0	0	0	0
1,3-HEPTADIENE, 1,4-	3	0	0	4	6	0	0	0	0
3-ETHYL-1-PENTENE	3	0	0	4	8	0	0	0	0
3-METHYL-1-HEXENE	3	0	0	4	8	0	0	0	0
4-METHYL-1-HEXENE	3	0	0	4	8	0	0	0	0
5-METHYL-1-HEXENE	3	0	0	4	8	0	0	0	0
1-BUTOXY-2-METHYLETHYLENE	3	0	0	4	8	1	0	0	0
2-METHYLBUTYLVINYL ETHER	3	0	0	4	8	1	0	0	0
1,4-BUTADIENE-ALT-METHACRYLONITRILE	3	0	0	5	5	0	1	0	0
4,4-DIMETHYL-1-HEXENE	3	0	0	5	10	0	0	0	0
ETHYL-P-XYLYLENE	3	0	0	7	6	0	0	0	0
4-DIISOPROPYLAMINO-1-BUTENE	3	0	0	7	15	0	1	0	0
B-D-GALACTO-D-MANNOSE (GUAR GALLACTOMANNAN)	3	0	0	15	24	15	0	0	0
P-PHENYLENEOXYTRIMETHYLENE OXIDE	3	0	1	0	0	2	0	0	0
4-PHENYL-1-BUTENE	3	0	1	1	2	0	0	0	0
4-O-TOLYL-1-BUTENE	3	0	1	2	4	0	0	0	0
4-P-TOLYL-1-BUTENE	3	0	1	2	4	0	0	0	0
4,4'-(ETHYLENEDIPHENYLENE)METHYLENE	3	0	2	0	0	0	0	0	0
5-HYDROXY-3-OXAVALERIC ACID	3	1	0	0	0	2	0	0	0
4-AMINOBUTYRIC ACID	3	1	0	0	1	0	1	0	0
5-AMINO-3-OXAVALERIC ACID	3	1	0	0	1	1	1	0	0
L-PROLINE	3	1	0	1	1	0	1	0	0
BUTYLISOCYANATE	3	1	0	1	3	0	1	0	0
ORNITHINE	3	1	0	1	4	0	2	0	0
ALLYL ACRYLATE	3	1	0	2	2	1	0	0	0
PROPYLTHIOLACRYLATE	3	1	0	2	4	0	0	1	0
ARGININE	3	1	0	2	6	0	4	0	0
ARGININE HYDROCHLORIDE	3	1	0	2	7	0	4	0	1
N-ISOVALERYLETHYLENE AMINE	3	1	0	3	7	0	1	0	0
1,4-BUTADIENE-ALT-METHYL METHACRYLATE	3	1	0	5	8	1	0	0	0
(2,2-DIMETHYL-3-AMINOCYCLOBUTYLENE)PROPIONIC ACID	3	1	0	5	9	0	1	0	0
2,5-BUTYL-2,4-HEXADIENOATE	3	1	0	6	10	1	0	0	0
ETHYLENE P-OXYPHENYLENEACETATE	3	1	1	0	0	2	0	0	0
TRIMETHYLENE-P-OXYBENZOATE	3	1	1	0	0	2	0	0	0
P-(AMINOETHYLENE)PHENYLENEACETIC ACID	3	1	1	0	1	0	1	0	0
P-(AMINOMETHYLENE)PHENYLENEPROPIONIC ACID	3	1	1	0	1	0	1	0	0
2,5-BUTYL-5-PHENYL-2,4-PENTADIENOATE	3	1	1	5	8	1	0	0	0
4,4'-HEXYLIDENEDIPHENYLENE CARBONATE	3	1	2	3	6	2	0	0	0
TRIMETHYLENE OXALATE	3	2	0	0	0	2	0	0	0
HYDRAZO GLUTARAMIDE	3	2	0	0	2	0	2	0	0
ETHYLENE CYCLOPROPYLENEDICARBOXAMIDE	3	2	0	2	4	0	2	0	0
N-PROPYL ASPARATE	3	2	0	2	5	1	1	0	0
2,2-DIMETHYLTRIMETHYLENE MALONATE	3	2	0	3	6	2	0	0	0
CYCLOPROPYLENEDIMETHYLENE TOLUYLENEDIURETHAN	3	2	0	9	10	2	2	0	0

CRYSTALLOGRAPHIC DATA FOR VARIOUS POLYMERS

POLY-	CH2	CO	C6H4	C	H	O	N	S	X
4,4'-TRIMETHYLENEDIOXY-3,3'-DIMETHOXYDIBENZOIC ANHYDRIDE	3	2	0	14	12	5	0	0	0
TRIMETHYLENE ISOPHTHALATE	3	2	1	0	0	2	0	0	0
TRIMETHYLENE TEREPHTHALATE	3	2	1	0	0	2	0	0	0
P-XYLYLENE GLUTARATE	3	2	1	0	0	2	0	0	0
ETHYLENE P-(CARBOXYPHENOXY)ACETATE	3	2	1	0	0	3	0	0	0
ETHYLENE P-(CARBOXYPHENYLENE)ACETAMIDE	3	2	1	0	2	0	2	0	0
TRIMETHYLENE TEREPHTHALAMIDE	3	2	1	0	2	0	2	0	0
P-XYLYLENE MALONAMIDE	3	2	1	0	2	0	2	0	0
BENZYL GLUTAMATE	3	2	1	1	3	1	1	0	0
N-CARBOBENZOXY-2,4-DIAMINOBUTYRIC ACID	3	2	1	1	4	1	2	0	0
CYCLOPROPYLENEDIMETHYLENE ISOPHTHALATE	3	2	1	2	2	2	0	0	0
CYCLOPROPYLENEDIMETHYLENE TEREPHTHALATE	3	2	1	2	2	2	0	0	0
CYCLOPROPYLENEDIMETHYLENE ISOPHTHALAMIDE	3	2	1	2	4	0	2	0	0
METHYLPIPERAZINE ISOPHTHALAMIDE	3	2	1	2	4	0	2	0	0
METHYLPIPERAZINE PHTHALAMIDE	3	2	1	2	4	0	2	0	0
METHYLPIPERAZINE TEREPHTHALAMIDE	3	2	1	2	4	0	2	0	0
N,N'-DIMETHYLTRIMETHYLENE TEREPHTHALAMIDE	3	2	1	2	6	0	2	0	0
N,N'-DIETHYL-3,3'-DIMETHYL(METHYLENEDIPHENYLENE) TEREPHTHALAMIDE	3	2	1	16	18	0	2	0	0
4,4'-TRIMETHYLENEDIBENZOIC ANHYDRIDE	3	2	2	0	0	1	0	0	0
ETHYLENE 4,4'-METHYLENEDIBENZOATE	3	2	2	0	0	2	0	0	0
3,3'-(TRIMETHYLENEDIOXY)DIBENZOIC ANHYDRIDE	3	2	2	0	0	3	0	0	0
4,4'-(TRIMETHYLENEDIOXY)DIBENZOIC ANHYDRIDE	3	2	2	0	0	3	0	0	0
TRIMETHYLENE 4,4'-OXYDIBENZOATE	3	2	2	0	0	3	0	0	0
ETHYLENE 4,4'-(METHYLENEDIPHENYLENE)DIURETHAN	3	2	2	0	2	2	2	0	0
TRIMETHYLENE 4,4'-SULFONYLDIBENZAMIDE	3	2	2	0	2	2	2	1	0
ISOPROPYLIDENEDIMETHYLENE METHYLENEDIPHENYLENEDIURETHAN	3	2	2	3	8	2	2	0	0
P-XYLYLENE DITHIOL-4,4'-METHYLENEDIBENZOATE	3	2	3	0	0	0	0	2	0
N,N'-DIETHYL(METHYLENEDIPHENYLENE) TEREPHTHALAMIDE	3	2	3	2	6	0	2	0	0
ALANYLGLYCYLGLYCYLGLYCINE	3	4	0	2	8	0	4	0	0
ALANYLGLUTAMYLGLYCINE	3	4	0	3	9	1	3	0	0
ISOPHTHALIC TRIMETHYLENEDIOXYDIMETHOXYDIBENZOIC ANHYDRIDE	3	4	1	14	12	6	0	0	0
TEREPHTHALIC TRIMETHYLENEDIOXYDIMETHOXYDIBENZOIC ANHYDRIDE	3	4	1	14	12	6	0	0	0
ISOPHTHALOYL (M-CARBOXYPHENOXYBUTYRYL)DIHYDRAZIDE	3	4	2	0	4	1	4	0	0
ISOPHTHALIC 4,4'-(TRIMETHYLENEDIOXY)DIBENZOIC ANHYDRIDE	3	4	3	0	0	4	0	0	0
TEREPHTHALIC 4,4'-(TRIMETHYLENEDIOXY)DIBENZOIC ANHYDRIDE	3	4	3	0	0	4	0	0	0
------ 4	4	0	0	0	0	0	0	0	0
TETRAMETHYLENE SULFIDE	4	0	0	0	0	0	0	1	0
TETRAMETHYLENE DISULFIDE	4	0	0	0	0	0	0	2	0
TETRAHYDROFURAN (TETRAMETHYLENE OXIDE)	4	0	0	0	0	1	0	0	0
TETRAMETHYLENE SULFONE	4	0	0	0	0	2	0	1	0
CYCLOPROPYLIDENEDIMETHYLENE OXIDE	4	0	0	1	0	1	0	0	0
3,3-BISBROMOMETHYLOXACYCLOBUTANE	4	0	0	1	0	1	0	0	2
3,3-BISCHLOROMETHYLOXACYCLOBUTANE	4	0	0	1	0	1	0	0	2
3,3-BISFLUOROMETHYLOXACYCLOBUTANE	4	0	0	1	0	1	0	0	2
3,3-BISIODOMETHYLOXACYCLOBUTANE	4	0	0	1	0	1	0	0	2
TETRAMETHYLENE THIOUREA	4	0	0	1	2	0	2	1	0
5-HYDROXY-1-PENTENE	4	0	0	1	2	1	0	0	0
3,3-BISHYDROXYMETHYLOXACYCLOBUTANE	4	0	0	1	2	3	0	0	0
2,5-TETRAMETHYLENE-1,3,4-TRIAZOLE	4	0	0	2	1	0	3	0	0
CYCLOHEXENE	4	0	0	2	2	0	0	0	0
1,3-CYCLOPENTYLENEMETHYLENE	4	0	0	2	2	0	0	0	0
VINYL CYCLOBUTANE	4	0	0	2	2	0	0	0	0
1,2-CYCLOHEXYLENE SULFIDE	4	0	0	2	2	0	0	1	0
1,3-CYCLOHEXYLENE SULFIDE	4	0	0	2	2	0	0	1	0
2,5-TETRAMETHYLENE-1-AMINO-1,3,4-TRIAZOLE	4	0	0	2	2	0	4	0	0
CYCLOHEXENE OXIDE	4	0	0	2	2	1	0	0	0
1,2-CYCLOHEXYLENE SULFONE	4	0	0	2	2	2	0	1	0
1,3-CYCLOHEXYLENE SULFONE	4	0	0	2	2	2	0	1	0
1-HEXENE	4	0	0	2	4	0	0	0	0
BUTYLVINYL ETHER	4	0	0	2	4	1	0	0	0
DIPROPYLSILOXANE	4	0	0	2	6	1	0	0	1
2-PROPYL-1,3-BUTADIENE, 1,4-	4	0	0	3	4	0	0	0	0
2,2-DIETHYLTRIMETHYLENE SULFONE	4	0	0	3	6	2	0	1	0
1,3-OCTADIENE, 1,4-	4	0	0	4	6	0	0	0	0
4-ETHYL-1-HEXENE	4	0	0	4	8	0	0	0	0
5-METHYL-1-HEPTENE	4	0	0	4	8	0	0	0	0
5-TRIMETHYLSILYL-1-PENTENE	4	0	0	4	10	0	0	0	1
6,6-DIMETHYL-1-HEPTENE	4	0	0	5	10	0	0	0	0
5-DIISOPROPYLAMINO-1-PENTENE	4	0	0	7	15	0	1	0	0
OXYDIETHYLENEOXY-P-PHENYLENE OXIDE	4	0	1	0	0	3	0	0	0
TETRAMETHYLENE DITHIONISOPHTHALATE	4	0	1	2	0	2	0	2	0
2,2'-TETRAMETHYLENEDITHIAZOLE-P-PHENYLENE	4	0	1	6	2	0	2	2	0
4,4'-OXYDIPHENYLENEDITHIOTETRAMETHYLENE DISULFIDE	4	0	2	0	0	1	0	4	0
5-HYDROXYVALERIC ACID	4	1	0	0	0	1	0	0	0
TETRAMETHYLENE CARBONATE	4	1	0	0	0	2	0	0	0
5-AMINOVALERIC ACID	4	1	0	0	1	0	1	0	0
6-AMINO-4-OXACAPROIC ACID	4	1	0	0	1	1	1	0	0

POLY-	CH2	CO	C6H4	C	H	O	N	S	X
TETRAMETHYLENE UREA	4	1	0	0	2	0	2	0	0
1,4-PIPERIDINE URETHAN	4	1	0	1	1	1	1	0	0
N-BUTYRYLETHYLENE AMINE	4	1	0	1	3	0	1	0	0
LYSINE	4	1	0	1	4	0	2	0	0
LYSINE HYDROCHLORIDE	4	1	0	1	5	0	2	0	1
(3-AMINOCYCLOBUTYLENE)PROPIONIC ACID	4	1	0	2	3	0	1	0	0
1,4-CYCLOHEXYLENE URETHAN	4	1	0	2	3	1	1	0	0
6-MERCAPTO-4-METHYLCAPROIC ACID	4	1	0	2	4	0	0	1	0
BUTYL ACRYLATE	4	1	0	2	4	1	0	0	0
3-METHYL-6-AMINOCAPROIC ACID	4	1	0	2	5	0	1	0	0
6-METHYL-6-AMINOCAPROIC ACID	4	1	0	2	5	0	1	0	0
2-METHYLACETOXY-1,3-BUTADIENE	4	1	0	3	4	1	0	0	0
P-(AMINOETHYLENE)PHENYLENEPROPIONIC ACID	4	1	1	0	1	0	1	0	0
P-(AMINOMETHYLENE)PHENYLENEBUTYRIC ACID	4	1	1	0	1	0	1	0	0
4,4'-CYCLOPENTYLIDENEDIPHENYLENE CARBONATE	4	1	2	1	0	2	0	0	0
4,4'-HEPTYLIDENEDIPHENYLENE CARBONATE	4	1	2	3	6	2	0	0	0
ADIPIC ANHYDRIDE	4	2	0	0	0	1	0	0	0
THIODIPROPIONIC ANHYDRIDE	4	2	0	0	0	1	0	1	0
(ETHYLENEDITHIO)DIACETIC ANHYDRIDE	4	2	0	0	0	1	0	2	0
ETHYLENE SUCCINATE	4	2	0	0	0	2	0	0	0
TRIMETHYLENE MALONATE	4	2	0	0	0	2	0	0	0
TETRAMETHYLENE THIODICARBOXYLATE	4	2	0	0	0	2	0	1	0
TETRAMETHYLENE DITHIODICARBOXYLATE	4	2	0	0	0	2	0	2	0
TETRAMETHYLENE TRITHIODICARBOXYLATE	4	2	0	0	0	2	0	3	0
TETRAMETHYLENE TETRATHIODICARBOXYLATE	4	2	0	0	0	2	0	4	0
SULFONYLDIPROPIONIC ANHYDRIDE	4	2	0	0	0	3	0	1	0
(ETHYLENEDISULFONYL)DIACETIC ANHYDRIDE	4	2	0	0	0	5	0	2	0
HYDRAZO ADIPAMIDE	4	2	0	0	2	0	2	0	0
2,2-BISCHLOROMETHYLTRIMETHYLENE OXALATE	4	2	0	1	0	2	0	0	2
TRIMETHYLENE CYCLOPROPYLENEDICARBOXAMIDE	4	2	0	2	4	0	2	0	0
TRIMETHYLENE CYCLOPROPYLENEDIURETHAN	4	2	0	2	4	2	2	0	0
2,2-DIMETHYLTRIMETHYLENE SUCCINATE	4	2	0	3	6	2	0	0	0
2,5-THIOPHENEDIPROPIONIC ANHYDRIDE	4	2	0	4	2	1	0	1	0
2,5-FURANDIPROPIONIC ANHYDRIDE	4	2	0	4	2	2	0	0	0
2,5-PYRROLEDIPROPIONIC ANHYDRIDE	4	2	0	4	3	1	1	0	0
CYCLOPROPYLENEDIMETHYLENE CYCLOPROPYLENEDICARBOXAMIDE	4	2	0	4	6	0	2	0	0
PIPERAZINE (3-METHYLCYCLOPROPYLENE)DICARBOXAMIDE	4	2	0	4	6	0	2	0	0
CYCLOPROPYLENEDIMETHYLENE CYCLOPROPYLENEDIURETHAN	4	2	0	4	6	2	2	0	0
TETRAMETHYLENE 1,2-DIMETHOXYSUCCINATE	4	2	0	4	8	4	0	0	0
N-METHYL-2,5-PYRROLEDIPROPIONIC ANHYDRIDE	4	2	0	5	5	1	1	0	0
3-METHYL-M-PHENYLENE ADIPAMIDE	4	2	0	7	8	0	2	0	0
ISOPROPYLIDENEDIMETHYLENE DIMETHYLPIPERAZINEDITHIODICARBOXY	4	2	0	7	16	2	2	2	0
4-METHYL-1,3-PHENYLENE TETRAMETHYLDISILOXANYLENEDIPROPIONAMID	4	2	0	11	20	1	2	0	2
4,4'-TETRAMETHYLENEDIOXY-3,3'-DIMETHOXYDIBENZOIC ANHYDRIDE	4	2	0	14	12	5	0	0	0
ETHYLENE 3,3'-DIMETHOXY-4,4'-(ETHYLENEDIOXY)DIBENZOATE	4	2	0	14	12	6	0	0	0
4,4'-TETRACHLOROISOPROPYLIDENEDIPHENYLENE ADIPATE	4	2	0	15	10	2	0	0	4
3,3'-DIMETHYLMETHYLENEDIPHENYLENE TETRAMETHYLDISILOXANYLENEDIPRO	4	2	0	18	26	1	2	0	2
TETRAMETHYLENE DITHIOLISOPHTHALATE	4	2	1	0	0	0	0	2	0
TETRAMETHYLENE DITHIOLTEREPHTHALATE	4	2	1	0	0	0	0	2	0
THIODIETHYLENE DITHIOLTEREPHTHALATE	4	2	1	0	0	0	0	3	0
PIPERAZINE ISOPHTHALAMIDE	4	2	1	0	0	0	2	0	0
PIPERAZINE PHTHALAMIDE	4	2	1	0	0	0	2	0	0
PIPERAZINE TEREPHTHALAMIDE	4	2	1	0	0	0	2	0	0
P-PHENYLENEDIPROPIONIC ANHYDRIDE	4	2	1	0	0	1	0	0	0
(P-PHENYLENEDITHIO)DIPROPIONIC ANHYDRIDE	4	2	1	0	0	1	0	2	0
(P-XYLYLENEDITHIO)DIACETIC ANHYDRIDE	4	2	1	0	0	1	0	2	0
ETHYLENE P-PHENYLENEDIACETATE	4	2	1	0	0	2	0	0	0
TETRAMETHYLENE ISOPHTHALATE	4	2	1	0	0	2	0	0	0
TETRAMETHYLENE TEREPHTHALATE	4	2	1	0	0	2	0	0	0
P-XYLYLENE SUCCINATE	4	2	1	0	0	2	0	0	0
(P-XYLYLENEDISULFONYL)DIACETIC ANHYDRIDE	4	2	1	0	0	5	0	2	0
TETRAMETHYLENE M-CARBOXYCARBANILATE	4	2	1	0	1	2	1	0	0
TETRAMETHYLENE P-CARBOXYCARBANILATE	4	2	1	0	1	2	1	0	0
M-PHENYLENE ADIPAMIDE	4	2	1	0	2	0	2	0	0
O-PHENYLENE ADIPAMIDE	4	2	1	0	2	0	2	0	0
P-PHENYLENE ADIPAMIDE	4	2	1	0	2	0	2	0	0
TETRAMETHYLENE TEREPHTHALAMIDE	4	2	1	0	2	0	2	0	0
TRIMETHYLENE P-(CARBOXYPHENYLENE)ACETAMIDE	4	2	1	0	2	0	2	0	0
P-XYLYLENE SUCCINAMIDE	4	2	1	0	2	0	2	0	0
PHENYLENE DISILOXANYLENEDIPROPIONAMIDE	4	2	1	0	6	1	2	0	2
N-CARBOBENZOXYORNITHINE	4	2	1	1	4	1	2	0	0
ADIPYL DITHIONISOPHTHALOYLDIHYDRAZIDE	4	2	1	2	4	0	4	2	0
N,N'-DIMETHYLTETRAMETHYLENE TEREPHTHALAMIDE	4	2	1	2	6	0	2	0	0
M-PHENYLENE (TETRAMETHYLDISILOXANYLENE)DIPROPIONAMIDE	4	2	1	4	14	1	2	0	2
O-PHENYLENE (TETRAMETHYLDISILOXANYLENE)DIPROPIONAMIDE	4	2	1	4	14	1	2	0	2
P-PHENYLENE (TETRAMETHYLDISILOXANYLENE)DIPROPIONAMIDE	4	2	1	4	14	1	2	0	2
P-PHENYLENE DIPROPYLCYCLOBUTYLENEDICARBOXYLATE	4	2	1	6	10	2	0	0	0

POLY-	CH2	CO	C6H4	C	H	O	N	S	X
P-PHENYLENE DICYANODIPROPYLCYCLOBUTYLENEDICARBOXYLATE	4	2	1	8	8	2	2	0	0
TETRAMETHYL-P-XYLYLENE DITHIOL-P-PHENYLENEDIACETATE	4	2	1	10	12	0	0	2	0
2,2'-DIPROPYL-4,4'-BIPHENYLENE TEREPHTHALATE	4	2	1	14	12	2	0	0	0
P-XYLYLENE DITHIOL-P-PHENYLENEDIACETATE	4	2	2	0	0	0	0	2	0
4,4'-TETRAMETHYLENEDIBENZOIC ANHYDRIDE	4	2	2	0	0	1	0	0	0
4,4'-(TETRAMETHYLENEDITHIO)DIBENZOIC ANHYDRIDE	4	2	2	0	0	1	0	2	0
4,4'-BIPHENYLENE ADIPATE	4	2	2	0	0	2	0	0	0
ETHYLENE 4,4'-ETHYLENEDIBENZOATE	4	2	2	0	0	2	0	0	0
P-PHENYLENEDIETHYLENE TEREPHTHALATE	4	2	2	0	0	2	0	0	0
TETRAMETHYLENE 4,4'-DIBENZOATE	4	2	2	0	0	2	0	0	0
P-XYLYLENE P-PHENYLENEDIACETATE	4	2	2	0	0	2	0	0	0
ETHYLENE 4,4'-(ETHYLENEDITHIO)DIBENZOATE	4	2	2	0	0	2	0	2	0
TETRAMETHYLENE 4,4'-OXYDIBENZOATE	4	2	2	0	0	3	0	0	0
3,3'-(TETRAMETHYLENEDIOXY)DIBENZOIC ANHYDRIDE	4	2	2	0	0	3	0	0	0
4,4'-(TETRAMETHYLENEDIOXY)DIBENZOIC ANHYDRIDE	4	2	2	0	0	3	0	0	0
ETHYLENE 3,3'-(ETHYLENEDIOXY)DIBENZOATE	4	2	2	0	0	4	0	0	0
ETHYLENE 4,4'-(ETHYLENEDIOXY)DIBENZOATE	4	2	2	0	0	4	0	0	0
(OXYDIETHYLENEDIOXY)DIBENZOIC ANHYDRIDE	4	2	2	0	0	4	0	0	0
4,4'-BIPHENYLENE ADIPAMIDE	4	2	2	0	2	0	2	0	0
ETHYLENE 4,4'-(ETHYLENEDIPHENYLENE)DIURETHAN	4	2	2	0	2	2	2	0	0
TRIMETHYLENE 4,4'-(METHYLENEDIPHENYLENE)DIURETHAN	4	2	2	0	2	2	2	0	0
TETRAMETHYLENE 4,4'-SULFONYLDIBENZAMIDE	4	2	2	0	2	2	2	1	0
4,4'-BIPHENYLENE DISILOXANYLENEDIPROPIONAMIDE	4	2	2	0	6	1	2	0	2
TETRAMETHYLENE 4,4'-(METHYLPHOSPHINYLIDENE)DIBENZAMIDE	4	2	2	1	5	1	2	0	1
1,4-CYCLOHEXYLENE 3,3'-DIBENZAMIDE	4	2	2	2	4	0	2	0	0
CYCLOPROPYLENEDIMETHYLENE 4,4'-METHYLENEDIPHENYLENEDIURETHA	4	2	2	2	4	2	2	0	0
4,4'-ISOPROPYLIDENEDIPHENYLENE DITHIOLADIPATE	4	2	2	3	6	0	0	2	0
4,4'-ISOPROPYLIDENEDIPHENYLENE ADIPATE	4	2	2	3	6	2	0	0	0
4,4'-BIPHENYLENE (TETRAMETHYLDISILOXANYLENE)DIPROPIONAMIDE	4	2	2	4	14	1	2	0	2
4,4'-ISOPROPYLIDENEDIPHENYLENEDIOXYDITHIOL-1,3-CYCLOHEXYLENEDICARBOX	4	2	2	5	8	0	0	2	0
TETRAMETHYLENE 4,4'-(PHENYLPHOSPHINYLIDENE)DIBENZAMIDE	4	2	3	0	3	1	2	0	1
ALANYLGLYCYLPROLINE	4	3	0	3	7	0	3	0	0
ALANYLPROLYLGLYCINE	4	3	0	3	7	0	3	0	0
B-D-GLUCOTRIPROPIONATE	4	3	0	8	14	5	0	0	0
ALANYL(ETHYLGLUTAMYL)GLYCINE	4	4	0	4	11	1	3	0	0
ISOPHTHALIC 4,4'-(TETRAMETHYLENEDIOXY)DIBENZOIC ANHYDRIDE	4	4	3	0	0	4	0	0	0
TEREPHTHALIC 4,4'-(TETRAMETHYLENEDIOXY)DIBENZOIC ANHYDRIDE	4	4	3	0	0	4	0	0	0
4,4'-ISOPROPYLIDENEDIPHENYLENEDIOXYDIACETIC P-PHENYLENEDIOXYDIAC	4	4	3	3	6	4	0	0	0
METHYLENEDIPHENYLENEDISUCCINIMIDEDIYLIMINOMETHYLENEDIPHENYL	4	4	4	2	4	0	4	0	0
ALANYLGLYCYLALANYLGLYCYLSERYLGLYCINE	4	6	0	5	16	1	6	0	0
2,2'-(ADIPYLDIAMINO)DIBENZOIC ISOPHTHALIC ANHYDRIDE	4	6	3	0	2	2	2	0	0
4,4'-(ADIPYLDIAMINO)DIBENZOIC ISOPHTHALIC ANHYDRIDE	4	6	3	0	2	2	2	0	0
2,2'-(ADIPYLDIAMINO)DIBENZOIC TEREPHTHALIC ANHYDRIDE	4	6	3	0	2	2	2	0	0
4,4'-(ADIPYLDIAMINO)DIBENZOIC TEREPHTHALIC ANHYDRIDE	4	6	3	0	2	2	2	0	0
------ 5	5	0	0	0	0	0	0	0	0
PENTAMETHYLENE	5	0	0	0	0	0	0	0	0
PENTAMETHYLENE SULFIDE	5	0	0	0	0	0	0	1	0
METHYLENETHIOTETRAMETHYLENE SULFIDE	5	0	0	0	0	0	0	2	0
PENTAMETHYLENE DISULFIDE	5	0	0	0	0	0	0	2	0
METHYLENEOXYTETRAMETHYLENE OXIDE	5	0	0	0	0	2	0	0	0
PENTAMETHYLENE SULFONE	5	0	0	0	0	2	0	1	0
6-AMINOTHIOCAPROIC ACID	5	0	0	1	1	0	1	1	0
2,5-PENTAMETHYLENE-1,3,4-TRIAZOLE	5	0	0	2	1	0	3	0	0
CYCLOHEPTENE	5	0	0	2	2	0	0	0	0
CYCLOPENTENE-ALT-ETHYLENE	5	0	0	2	2	0	0	0	0
VINYL CYCLOPENTANE	5	0	0	2	2	0	0	0	0
1,2-DICHLOROHEPTAMETHYLENE	5	0	0	2	2	0	0	0	2
2,5-PENTAMETHYLENE-1-AMINO-1,3,4-TRIAZOLE	5	0	0	2	2	0	4	0	0
1,4-CYCLOHEXYLENEOXYMETHYLENE OXIDE	5	0	0	2	2	2	0	0	0
1-HEPTENE	5	0	0	2	4	0	0	0	0
ENANTHALDEHYDE	5	0	0	2	4	1	0	0	0
VINYL CYCLOHEXENE	5	0	0	3	2	0	0	0	0
3,4-DIHYDROCYCLOHEXYLIDENEDIMETHYLENE OXIDE	5	0	0	3	2	1	0	0	0
3,4-DIHYDRO-4-METHYLCYCLOHEXYLIDENEDIMETHYLENE OXIDE	5	0	0	4	4	1	0	0	0
3-METHYLVINYL CYCLOHEXANE	5	0	0	4	6	0	0	0	0
4-METHYLVINYL CYCLOHEXANE	5	0	0	4	6	0	0	0	0
O-METHOXYVINYL CYCLOHEXANE	5	0	0	4	6	1	0	0	0
PENTAMETHYLENEOXY-P-PHENYLENE OXIDE	5	0	1	0	0	2	0	0	0
4,4'-(METHYLENEDIPHENYLENE)DIETHYLENE SULFIDE	5	0	2	0	0	0	0	1	0
6-MERCAPTOCAPROIC ACID	5	1	0	0	0	0	0	1	0
6-HYDROXYCAPROIC ACID	5	1	0	0	0	1	0	0	0
PENTAMETHYLENE CARBONATE	5	1	0	0	0	2	0	0	0
6-AMINOCAPROIC ACID	5	1	0	0	1	0	1	0	0
PENTAMETHYLENE URETHAN	5	1	0	0	1	1	1	0	0
N-VALERYLETHYLENE AMINE	5	1	0	1	3	0	1	0	0
(4-AMINOCYCLOHEXYLENE)ACETIC ACID	5	1	0	2	3	0	1	0	0
4-METHYL-7-HYDROXYHEPTANOIC ACID	5	1	0	2	4	1	0	0	0

POLY-	CH2	CO	C6H4	C	H	O	N	S	X
3-METHYL-7-AMINOENANTHIC ACID	5	1	0	2	5	0	1	0	0
4-METHYL-7-AMINOENANTHIC ACID	5	1	0	2	5	0	1	0	0
5-METHYL-7-AMINOENANTHIC ACID	5	1	0	2	5	0	1	0	0
6-METHYL-7-AMINOENANTHIC ACID	5	1	0	2	5	0	1	0	0
4,4'-CYCLOHEXYLIDENE-2,2'-DIMETHYLDIPHENYLENE CARBONATE	5	1	0	15	12	2	0	0	0
P-(AMINOETHYLENE)PHENYLENEBUTYRIC ACID	5	1	1	0	1	0	1	0	0
P-(AMINOMETHYLENE)PHENYLENEVALERIC ACID	5	1	1	0	1	0	1	0	0
N-4-(METHYLTHIO)PHENOXYBUTYRYLETHYLENE AMINE	5	1	1	1	3	1	1	1	0
4,4'-CYCLOHEXYLIDENEDIPHENYLENE CARBONATE	5	1	2	1	0	2	0	0	0
PIMELIC ANHYDRIDE	5	2	0	0	0	1	0	0	0
TRIMETHYLENE SUCCINATE	5	2	0	0	0	2	0	0	0
HYDRAZO PIMELAMIDE	5	2	0	0	2	0	2	0	0
METHYLENE ADIPAMIDE	5	2	0	0	2	0	2	0	0
(2,6-DIOXO-1,4-PIPERIDINEDIYL)TRIMETHYLENE	5	2	0	1	1	0	1	0	0
PIPERAZINE CYCLOPROPYLENEDICARBOXAMIDE	5	2	0	2	2	0	2	0	0
CYCLOPROPYLENE PIPERAZINEDIUREA	5	2	0	2	4	0	4	0	0
TETRAMETHYLENE CYCLOPROPYLENEDIURETHAN	5	2	0	2	4	2	2	0	0
PIPERAZINE (1-METHYLCYCLOPROPYLENE)DICARBOXAMIDE	5	2	0	3	4	0	2	0	0
ETHYLENE 3,3'-DIMETHOXY-4,4'-(TRIMETHYLENEDIOXY)DIBENZOATE	5	2	0	14	12	6	0	0	0
3,3'-DIMETHYL-4,4'-METHYLENEDIPHENYLENE ADIPAMIDE	5	2	0	14	14	0	2	0	0
N,N'-DIISOPROPYL-3,3'-DIMETHYL-METHYLENEDIPHENYLENE ADIPAMIDE	5	2	0	20	26	0	2	0	0
PENTAMETHYLENE DITHIOLISOPHTHALATE	5	2	1	0	0	0	0	2	0
PENTAMETHYLENE DITHIOLTEREPHTHALATE	5	2	1	0	0	0	0	2	0
PENTAMETHYLENE TEREPHTHALATE	5	2	1	0	0	2	0	0	0
TRIMETHYLENE P-PHENYLENEDIACETATE	5	2	1	0	0	2	0	0	0
ETHYLENE P-(CARBOXYPHENOXY)BUTYRATE	5	2	1	0	0	3	0	0	0
PENTAMETHYLENE M-CARBOXYCARBANILATE	5	2	1	0	1	2	1	0	0
PENTAMETHYLENE P-CARBOXYCARBANILATE	5	2	1	0	1	2	1	0	0
PENTAMETHYLENE TEREPHTHALAMIDE	5	2	1	0	2	0	2	0	0
TETRAMETHYLENE P-(CARBOXYPHENYLENE)ACETAMIDE	5	2	1	0	2	0	2	0	0
P-XYLYLENE GLUTARAMIDE	5	2	1	0	2	0	2	0	0
CARBOBENZOXY LYSINE	5	2	1	1	4	1	2	0	0
N,N'-DIPROPYLMETHYLENE-4,4'-DIMETHYLDIPHENYLENE TEREPHTHALAMIDE	5	2	1	16	18	0	2	0	0
4,4'-PENTAMETHYLENEDIBENZOIC ANHYDRIDE	5	2	2	0	0	1	0	0	0
4,4'-METHYLENEDIPHENYLENE ADIPATE	5	2	2	0	0	2	0	0	0
PENTAMETHYLENE 4,4'-OXYDIBENZOATE	5	2	2	0	0	3	0	0	0
3,3'-(PENTAMETHYLENEDIOXY)DIBENZOIC ANHYDRIDE	5	2	2	0	0	3	0	0	0
4,4'-(PENTAMETHYLENEDIOXY)DIBENZOIC ANHYDRIDE	5	2	2	0	0	3	0	0	0
ETHYLENE 4,4'-(TRIMETHYLENEDIOXY)DIBENZOATE	5	2	2	0	0	4	0	0	0
4,4'-METHYLENEDIPHENYLENE ADIPAMIDE	5	2	2	0	2	0	2	0	0
ETHYLENE 4,4'-(TRIMETHYLENEDIPHENYLENE)DIURETHAN	5	2	2	0	2	2	2	0	0
TETRAMETHYLENE 4,4'-(METHYLENEDIPHENYLENE)DIURETHAN	5	2	2	0	2	2	2	0	0
TETRAMETHYLENE METHYLENEDIPHENYLENEDIAMINODITHIODICARBOXYLA	5	2	2	0	2	2	2	2	0
4,4'-METHYLENEDIPHENYLENE DISILOXANYLENEDIPROPIONAMIDE	5	2	2	0	6	1	2	0	2
1,4-CYCLOHEXYLENE 3,3'-METHYLENEDIBENZAMIDE	5	2	2	2	4	0	2	0	0
N,N'-DIMETHYL-4,4'-METHYLENEDIPHENYLENE ADIPAMIDE	5	2	2	2	6	0	2	0	0
TETRAMETHYLENE 4,4'-PROPYLENEDIBENZAMIDE	5	2	2	2	6	0	2	0	0
TETRAMETHYLENE 4,4'-(PROPYLENEDIOXY)DIBENZAMIDE	5	2	2	2	6	2	2	0	0
2,2-DIETHYLTRIMETHYLENE 4,4'-(METHYLENEDIPHENYLENE)DIURETHAN	5	2	2	3	8	2	2	0	0
1-ETHYL-2-PROPYLTRIMETHYLENE (METHYLENEDIPHENYLENE)DIURETHAN	5	2	2	4	10	2	2	0	0
4,4'-METHYLENEDIPHENYLENE TETRAMETHYLDISILOXANYLENEDIPROPIONAMID	5	2	2	4	14	1	2	0	2
N,N'-DIPROPYL-4,4'-METHYLENEDIPHENYLENE TEREPHTHALAMIDE	5	2	3	2	6	0	2	0	0
GLYCYLGLYCYLPROLINE	5	3	0	1	3	0	3	0	0
TETRAMETHYLENE (CYCLOPROPYLENEDICARBOXOYL)DIURETHAN	5	4	0	2	4	2	2	0	0
ETHYLENE N,N'-TRIMETHYLENEDITEREPHTHALAMATE	5	4	2	0	2	2	2	0	0
ISOPHTHALOYL (M-CARBOXYPHENOXYCAPRYL)DIHYDRAZIDE	5	4	2	0	4	1	4	0	0
P-PHENYLENEDIOXYDIACETIC TRIMETHYLENEDIOXYDIBENZOIC ANHYDRIDE	5	4	3	0	0	4	0	0	0
------ 6	6	0	0	0	0	0	0	0	0
HEXAMETHYLENE SULFIDE	6	0	0	0	0	0	0	1	0
ETHYLENETHIOTETRAMETHYLENE SULFIDE	6	0	0	0	0	0	0	2	0
HEXAMETHYLENE DISULFIDE	6	0	0	0	0	0	0	2	0
ETHYLENEDITHIOTETRAMETHYLENE DISULFIDE	6	0	0	0	0	0	0	4	0
HEXAMETHYLENE OXIDE	6	0	0	0	0	1	0	0	0
METHYLENEOXYPENTAMETHYLENE OXIDE	6	0	0	0	0	2	0	0	0
HEXAMETHYLENE SULFONE	6	0	0	0	0	2	0	1	0
7-AMINOTHIOENANTHIC ACID	6	0	0	1	1	0	1	1	0
HEXAMETHYLENE THIOUREA	6	0	0	1	2	0	2	1	0
2,5-HEXAMETHYLENE-1,3,4-TRIAZOLE	6	0	0	2	1	0	3	0	0
CYCLOOCTENE	6	0	0	2	2	0	0	0	0
3-CYCLOPENTYLPROPENE	6	0	0	2	2	0	0	0	0
VINYL CYCLOHEXANE	6	0	0	2	2	0	0	0	0
2,5-HEXAMETHYLENE-1-AMINO-1,3,4-TRIAZOLE	6	0	0	2	2	0	4	0	0
1-OCTENE	6	0	0	2	4	0	0	0	0
CAPRYLALDEHYDE	6	0	0	2	4	1	0	0	0
4-METHYLCYCLOHEXYLIDENEDIMETHYLENE OXIDE	6	0	0	3	4	1	0	0	0
3,3-BISETHOXYMETHYLOXACYLCLOBUTANE	6	0	0	3	6	3	0	0	0
5-DIISOBUTYLAMINO-1-PENTENE	6	0	0	7	15	0	1	0	0

POLY-	CH2	CO	C6H4	C	H	O	N	S	X
HEXAMETHYLENEOXY-P-PHENYLENE OXIDE	6	0	1	0	0	2	0	0	0
BENZYLIDENETHIOHEXAMETHYLENE SULFIDE	6	0	1	1	2	0	0	2	0
HEXAMETHYLENE DITHIOTEREPHTHALAMIDE	6	0	1	2	2	0	2	2	0
P-METHOXYBENZYLIDENETHIOHEXAMETHYLENE SULFIDE	6	0	1	2	4	1	0	2	0
7-HYDROXYENANTHIC ACID	6	1	0	0	0	1	0	0	0
HEXAMETHYLENE CARBONATE	6	1	0	0	0	2	0	0	0
7-AMINOENANTHIC ACID	6	1	0	0	1	0	1	0	0
HEXAMETHYLENE UREA	6	1	0	0	2	0	2	0	0
N-HEXANOYLETHYLENE AMINE	6	1	0	1	3	0	1	0	0
N-METHYL-7-AMINOENANTHIC ACID	6	1	0	1	3	0	1	0	0
VINYLCYCLOHEXYL KETONE	6	1	0	2	2	0	0	0	0
(3-AMINOMETHYLENE)CYCLOHEXYLENEACETIC ACID	6	1	0	2	3	0	1	0	0
(1,3-DIMETHYL-3-AMINOMETHYLENE)CYCLOHEXYLENEACETIC ACID	6	1	0	4	7	0	1	0	0
P-(AMINOETHYLENE)PHENYLENEVALERIC ACID	6	1	1	0	1	0	1	0	0
4,4'-NONYLIDENEDIPHENYLENE CARBONATE	6	1	2	3	6	2	0	0	0
ETHYLENE DITHIOLADIPATE	6	2	0	0	0	0	0	2	0
SUBERIC ANHYDRIDE	6	2	0	0	0	1	0	0	0
(ETHYLENEDITHIO)DIPROPIONIC ANHYDRIDE	6	2	0	0	0	1	0	2	0
PIPERAZINE OXYDIACETAMIDE	6	2	0	0	0	1	2	0	0
ETHYLENE ADIPATE	6	2	0	0	0	2	0	0	0
HEXAMETHYLENE OXALATE	6	2	0	0	0	2	0	0	0
TETRAMETHYLENE SUCCINATE	6	2	0	0	0	2	0	0	0
TRIMETHYLENE GLUTARATE	6	2	0	0	0	2	0	0	0
HEXAMETHYLENE DITHIODICARBOXYLATE	6	2	0	0	0	2	0	2	0
ETHYLENE 1,4-PIPERAZINEDICARBOXYLATE	6	2	0	0	0	2	2	0	0
(ETHYLENEDISULFONYL)DIPROPIONIC ANHYDRIDE	6	2	0	0	0	5	0	2	0
ETHYLENE ADIPAMIDE	6	2	0	0	2	0	2	0	0
HEXAMETHYLENE OXAMIDE	6	2	0	0	2	0	2	0	0
HYDRAZO SUBERAMIDE	6	2	0	0	2	0	2	0	0
PENTAMETHYLENE MALONAMIDE	6	2	0	0	2	0	2	0	0
TETRAMETHYLENE SUCCINAMIDE	6	2	0	0	2	0	2	0	0
ETHYLENE (ETHYLENEDIOXY)DIACETAMIDE	6	2	0	0	2	2	2	0	0
AMINODITRIMETHYLENE OXAMIDE	6	2	0	0	3	0	3	0	0
ETHYLENE DISILOXANYLENEDIPROPIONAMIDE	6	2	0	0	6	1	2	0	2
2,2-BISCHLOROMETHYLTRIMETHYLENE SUCCINATE	6	2	0	1	0	2	0	0	2
(METHYLAMINO)DITRIMETHYLENE OXAMIDE	6	2	0	1	5	0	3	0	0
1,4-CYCLOHEXYLENE DITHIOLSUCCINATE	6	2	0	2	2	0	0	2	0
1,4-CYCLOHEXYLENEDIMETHYLENE OXALATE	6	2	0	2	2	2	0	0	0
ETHYLENE 1,4-CYCLOHEXYLENEDICARBOXYLATE	6	2	0	2	2	2	0	0	0
2,5-TETRAHYDROFURANDIPROPIONIC ANHYDRIDE	6	2	0	2	2	2	0	0	0
PENTAMETHYLENE CYCLOPROPYLENEDIURETHAN	6	2	0	2	4	2	2	0	0
N,N'-DIMETHYLTETRAMETHYLENE SUCCINAMIDE	6	2	0	2	6	0	2	0	0
HEXAFLUOROPENTAMETHYLENE ADIPATE	6	2	0	3	0	2	0	0	6
N-METHYL-2,5-TETRAHYDROPYRROLEDIPROPIONIC ANHYDRIDE	6	2	0	3	5	1	1	0	0
2,2-DIMETHYLTRIMETHYLENE ADIPATE	6	2	0	3	6	2	0	0	0
HEXAMETHYLENE 2,2-DIMETHYLMALONAMIDE	6	2	0	3	8	0	2	0	0
ISOPROPYLIDENEDIMETHYLENE PIPERAZINEDITHIODICARBOXYLATE	6	2	0	3	8	2	2	2	0
HEXAMETHYLENE 2,5-THIOPHENEDICARBOXAMIDE	6	2	0	4	4	0	2	1	0
HEXAMETHYLENE 2,5-FURANDICARBOXAMIDE	6	2	0	4	4	1	2	0	0
2-BUTENE 4-OCTENEDIAMIDE	6	2	0	4	6	0	2	0	0
HEXAMETHYLENE (3-METHYLCYCLOPROPYLENE)DICARBOXAMIDE	6	2	0	4	8	0	2	0	0
TETRAMETHYLENE (2,5-DIMETHYLPIPERAZINE)DITHIODICARBOXYLATE	6	2	0	4	10	2	2	2	0
ETHYLENE (TETRAMETHYLDISILOXANYLENE)DIPROPIONAMIDE	6	2	0	4	14	1	2	0	2
HEXAMETHYLENE 2,3-PYRIDINEDICARBOXAMIDE	6	2	0	5	5	0	3	0	0
HEXAMETHYLENE 2,4-PYRIDINEDICARBOXAMIDE	6	2	0	5	5	0	3	0	0
HEXAMETHYLENE 2,5-PYRIDINEDICARBOXAMIDE	6	2	0	5	5	0	3	0	0
HEXAMETHYLENE 2,6-PYRIDINEDICARBOXAMIDE	6	2	0	5	5	0	3	0	0
HEXAMETHYLENE 3,4-PYRIDINEDICARBOXAMIDE	6	2	0	5	5	0	3	0	0
HEXAMETHYLENE 3,5-PYRIDINEDICARBOXAMIDE	6	2	0	5	5	0	3	0	0
N,N'-DIMETHYL(HEXAFLUOROPENTAMETHYLENE) ADIPAMIDE	6	2	0	5	6	0	2	0	6
N,N'-DIMETHYLHEXAMETHYLENE PERFLUOROGLUTARAMIDE	6	2	0	5	6	0	2	0	6
HEXAMETHYLENE 2,5-DIHYDROXYTEREPHTHALAMIDE	6	2	0	6	6	2	2	0	0
HEXAMETHYLENE METHYLTEREPHTHALAMIDE	6	2	0	7	8	0	2	0	0
HEXAMETHYLENE 2,6-DIMETHYL-3,5-PYRIDINEDICARBOXAMIDE	6	2	0	7	9	0	3	0	0
HEXAMETHYLENE 4-METHYL-1,3-PHENYLENEDIUREA	6	2	0	7	10	0	4	0	0
HEXAMETHYLENE 2,5-DIMETHYLTEREPHTHALAMIDE	6	2	0	8	10	0	2	0	0
N,N'-DIISOPROPYL(HEXAFLUOROPENTAMETHYLENE) ADIPAMIDE	6	2	0	9	14	0	2	0	6
N,N'-DIISOPROPYLHEXAMETHYLENE PERFLUOROGLUTARAMIDE	6	2	0	9	14	0	2	0	6
TETRAMETHYL-P-XYLYLENE DITHIOLADIPATE	6	2	0	10	12	0	0	2	0
HEXAMETHYLENE 5-TERT-BUTYLISOPHTHALAMIDE	6	2	0	10	14	0	2	0	0
4-METHYLPHENYLENE DIETHYLDIMETHYLDISILOXANYLENEDIPROPIONAMIDE	6	2	0	11	20	1	2	0	2
1,4-CYCLOHEXYLENEDIMETHYLENE 2,6-NAPHTHALATE	6	2	0	12	8	2	0	0	0
ETHYLENE 3,3'-DIMETHOXY-4,4'-(TETRAMETHYLENEDIOXY)DIBENZOAT	6	2	0	14	12	6	0	0	0
ETHYLENE 3,3'-DIMETHOXY-4,4'-(OXYDIETHYLENE)DIOXYDIBENZOATE	6	2	0	14	12	7	0	0	0
HEXAMETHYLENE DITHIOLISOPHTHALATE	6	2	1	0	0	0	0	2	0
HEXAMETHYLENE DITHIOLTEREPHTHALATE	6	2	1	0	0	0	0	2	0
P-XYLYLENE DITHIOLADIPATE	6	2	1	0	0	0	0	2	0

POLY-	CH2	CO	C6H4	C	H	O	N	S	X
(P-XYLYLENEDITHIO)DIPROPIONIC ANHYDRIDE	6	2	1	0	0	1	0	2	0
HEXAMETHYLENE ISOPHTHALATE	6	2	1	0	0	2	0	0	0
HEXAMETHYLENE TEREPHTHALATE	6	2	1	0	0	2	0	0	0
TETRAMETHYLENE P-PHENYLENEDIACETATE	6	2	1	0	0	2	0	0	0
P-XYLYLENE ADIPATE	6	2	1	0	0	2	0	0	0
ETHYLENE P-(CARBOXYPHENOXY)VALERATE	6	2	1	0	0	3	0	0	0
(P-XYLYLENEDISULFONYL)DIPROPIONIC ANHYDRIDE	6	2	1	0	0	5	0	2	0
HEXAMETHYLENE M-CARBOXYCARBANILATE	6	2	1	0	1	2	1	0	0
HEXAMETHYLENE ISOPHTHALAMIDE	6	2	1	0	2	0	2	0	0
HEXAMETHYLENE PHTHALAMIDE	6	2	1	0	2	0	2	0	0
HEXAMETHYLENE TEREPHTHALAMIDE	6	2	1	0	2	0	2	0	0
PENTAMETHYLENE P-(CARBOXYPHENYLENE)ACETAMIDE	6	2	1	0	2	0	2	0	0
M-PHENYLENE SUBERAMIDE	6	2	1	0	2	0	2	0	0
O-PHENYLENE SUBERAMIDE	6	2	1	0	2	0	2	0	0
P-PHENYLENE SUBERAMIDE	6	2	1	0	2	0	2	0	0
M-XYLYLENE ADIPAMIDE	6	2	1	0	2	0	2	0	0
P-XYLYLENE ADIPAMIDE	6	2	1	0	2	0	2	0	0
OXYDITRIMETHYLENE TEREPHTHALAMIDE	6	2	1	0	2	1	2	0	0
M-PHENYLENE HEXAMETHYLENEDIURETHAN	6	2	1	0	2	2	2	0	0
P-PHENYLENE HEXAMETHYLENEDIURETHAN	6	2	1	0	2	2	2	0	0
TETRAMETHYLENE P-XYLYLENEDIURETHAN	6	2	1	0	2	2	2	0	0
HEXAMETHYLENE M-PHENYLENEDIUREA	6	2	1	0	4	0	4	0	0
1,4-CYCLOHEXYLENEDIMETHYLENE ISOPHTHALATE	6	2	1	2	2	2	0	0	0
1,4-CYCLOHEXYLENEDIMETHYLENE TEREPHTHALATE	6	2	1	2	2	2	0	0	0
P-XYLYLENE 1,4-CYCLOHEXYLENEDICARBOXYLATE	6	2	1	2	2	2	0	0	0
1,4-CYCLOHEXYLENEDIMETHYLENE ISOPHTHALAMIDE	6	2	1	2	4	0	2	0	0
N,N'-DIMETHYLHEXAMETHYLENE TEREPHTHALAMIDE	6	2	1	2	6	0	2	0	0
HEXAMETHYLENE NAPHTHALENEDICARBOXAMIDE	6	2	1	4	6	0	2	0	0
O-PHENYLENE (DIETHYLDIMETHYLDISILOXANYLENE)DIPROPIONAMIDE	6	2	1	4	14	1	2	0	2
P-PHENYLENE (DIETHYLDIMETHYLDISILOXANYLENE)DIPROPIONAMIDE	6	2	1	4	14	1	2	0	2
P-PHENYLENE DIBUTYLCYCLOBUTYLENEDICARBOXYLATE	6	2	1	6	10	2	0	0	0
P-PHENYLENE DICYANODIBUTYLCYCLOBUTYLENEDICARBOXYLATE	6	2	1	8	8	2	2	0	0
4,4'-HEXAMETHYLENEDIBENZOIC ANHYDRIDE	6	2	2	0	0	1	0	0	0
ETHYLENE 4,4'-TETRAMETHYLENEDIBENZOATE	6	2	2	0	0	2	0	0	0
HEXAMETHYLENE 4,4'-DIBENZOATE	6	2	2	0	0	2	0	0	0
P-PHENYLENEDITRIMETHYLENE TEREPHTHALATE	6	2	2	0	0	2	0	0	0
HEXAMETHYLENE DITHIOL-4,4'-SULFONYLDIBENZOATE	6	2	2	0	0	2	0	3	0
HEXAMETHYLENE 4,4'-OXYDIBENZOATE	6	2	2	0	0	3	0	0	0
3,3'-(HEXAMETHYLENEDIOXY)DIBENZOIC ANHYDRIDE	6	2	2	0	0	3	0	0	0
4,4'-(HEXAMETHYLENEDIOXY)DIBENZOIC ANHYDRIDE	6	2	2	0	0	3	0	0	0
ETHYLENE 3,3'-(TETRAMETHYLENEDIOXY)DIBENZOATE	6	2	2	0	0	4	0	0	0
ETHYLENE 4,4'-(TETRAMETHYLENEDIOXY)DIBENZOATE	6	2	2	0	0	4	0	0	0
TETRAMETHYLENE 4,4'-(ETHYLENEDIOXY)DIBENZOATE	6	2	2	0	0	4	0	0	0
ETHYLENE 4,4'-(OXYDIETHYLENE)DIOXYDIBENZOATE	6	2	2	0	0	5	0	0	0
4,4'-ETHYLENEDIPHENYLENE ADIPAMIDE	6	2	2	0	2	0	2	0	0
HEXAMETHYLENE 2,2'-DIBENZAMIDE	6	2	2	0	2	0	2	0	0
HEXAMETHYLENE 3,3'-DIBENZAMIDE	6	2	2	0	2	0	2	0	0
HEXAMETHYLENE 3,4'-DIBENZAMIDE	6	2	2	0	2	0	2	0	0
HEXAMETHYLENE 4,4'-DIBENZAMIDE	6	2	2	0	2	0	2	0	0
P-PHENYLENEDIETHYLENE M-PHENYLENEDIACETAMIDE	6	2	2	0	2	0	2	0	0
ETHYLENE 4,4'-(TETRAMETHYLENEDIPHENYLENE)DIURETHAN	6	2	2	0	2	2	2	0	0
PENTAMETHYLENE 4,4'-(METHYLENEDIPHENYLENE)DIURETHAN	6	2	2	0	2	2	2	0	0
HEXAMETHYLENE 4,4'-SULFONYLDIBENZAMIDE	6	2	2	0	2	2	2	1	0
HEXAMETHYLENE PHOSPHINYLIDENEDIBENZAMIDE	6	2	2	0	3	1	2	0	1
HEXAMETHYLENE 4,4'-(HYDROXYPHOSPHINYLIDENE)DIBENZAMIDE	6	2	2	0	3	2	2	0	1
HEXAMETHYLENE 4,4'-SILYLENEDIBENZAMIDE	6	2	2	0	4	0	2	0	1
4,4'-DITHIODIPHENYLENE HEXAMETHYLENEDIUREA	6	2	2	0	4	0	4	2	0
HEXAMETHYLENE 4,4'-DISILOXANYLENEDIBENZAMIDE	6	2	2	0	6	1	2	0	2
HEXAMETHYLENE 4,4'-TETRASILOXANYLENEDIBENZAMIDE	6	2	2	0	10	3	2	0	4
HEXAMETHYLENE 3,3'-(METHYLPHOSPHINYLIDENE)DIBENZAMIDE	6	2	2	1	5	1	2	0	1
HEXAMETHYLENE 4,4'-(METHYLPHOSPHINYLIDENE)DIBENZAMIDE	6	2	2	1	5	1	2	0	1
2-ETHYLHEXAMETHYLENE DITHIOL-4,4'-SULFONYLDIBENZOATE	6	2	2	2	4	2	0	3	0
HEXAMETHYLENE 4,4'-ETHYLIDENEDIBENZAMIDE	6	2	2	2	6	0	2	0	0
HEXAMETHYLENE 4,4'-(DIMETHYLSILYLENE)DIBENZAMIDE	6	2	2	2	8	0	2	0	1
HEXAMETHYLENE 4,4'-(METHYLVINYLENE)DIBENZAMIDE	6	2	2	3	6	0	2	0	0
HEXAMETHYLENE 4,4'-ISOPROPYLIDENEDIBENZAMIDE	6	2	2	3	8	0	2	0	0
4,4'-ISOPROPYLIDENEDIPHENYLENE HEXAMETHYLENEDIURETHAN	6	2	2	3	8	2	2	0	0
4,4'-DIMETHYLHEPTAMETHYLENE SULFONYLDIBENZAMIDE	6	2	2	3	8	2	2	1	0
HEXAMETHYLENE 4,4'-(TETRAMETHYLDISILOXANYLENE)DIBENZAMIDE	6	2	2	4	14	1	2	0	2
HEXAMETHYLENE 4,4'-(PHENYLPHOSPHINYLIDENE)DIBENZAMIDE	6	2	3	0	2	1	2	0	1
HEXAMETHYLENE TETRAMETHYLTETRAPHENYLTETRASILOXANYLENEDIBENZ	6	2	6	4	14	3	2	0	4
GLYCYLPROLYLHYDROXYPROLINE	6	3	0	3	5	1	3	0	0
HEXAMETHYLENE 4-OXO-4(H)-PYRAN-2,6-DICARBOXAMIDE	6	3	0	4	4	1	2	0	0
HEXAMETHYLENE 1,4-DIHYDRO-4-OXO-2,6-PYRIDINEDICARBOXAMIDE	6	3	0	4	5	0	3	0	0
HEXAMETHYLENE 4,4'-KETODIBENZAMIDE	6	3	2	0	2	0	2	0	0
ETHYLENE ADIPYLDIURETHAN	6	4	0	0	2	2	2	0	0
ADIPYL ETHYLENEDIUREA	6	4	0	0	4	0	4	0	0

POLY-	CH2	CO	C6H4	C	H	O	N	S	X
2-HYDROXYTRIMETHYLENE ADIPYLDIURETHAN	6	4	0	1	4	3	2	0	0
PENTAMETHYLENE (CYCLOPROPYLENEDICARBOXOYL)DIURETHAN	6	4	0	2	4	2	2	0	0
TRIMETHYLENE N,N'-TRIMETHYLENEDITEREPHTHALAMATE	6	4	2	0	2	2	2	0	0
------ 7									
HEPTAMETHYLENE	7	0	0	0	0	0	0	0	0
HEPTAMETHYLENE DISULFIDE	7	0	0	0	0	0	0	2	0
TETRAMETHYLENETHIOTRIMETHYLENE SULFIDE	7	0	0	0	0	0	0	2	0
HEXAMETHYLENEOXYMETHYLENE OXIDE	7	0	0	0	0	2	0	0	0
CYCLOHEXYLIDENEDIMETHYLENE OXIDE	7	0	0	1	0	1	0	0	0
1,4-PIPERIDINEDIYLTRIMETHYLENE	7	0	0	1	1	0	1	0	0
2,5-HEPTAMETHYLENE-1,3,4-TRIAZOLE	7	0	0	2	1	0	3	0	0
CYCLOHEPTENE-ALT-ETHYLENE	7	0	0	2	2	0	0	0	0
3-CYCLOHEXYLPROPENE	7	0	0	2	2	0	0	0	0
VINYL CYCLOHEPTANE	7	0	0	2	2	0	0	0	0
2,5-HEPTAMETHYLENE-1-AMINO-1,3,4-TRIAZOLE	7	0	0	2	2	0	4	0	0
1,4-CYCLOHEXYLENEDIMETHYLENEOXYMETHYLENE OXIDE	7	0	0	2	2	2	0	0	0
1-METHYLOCTAMETHYLENE	7	0	0	2	4	0	0	0	0
1-NONENE	7	0	0	2	4	0	0	0	0
8-AMINOCAPRYLIC ACID	7	1	0	0	1	0	1	0	0
HEPTAMETHYLENE UREA	7	1	0	0	2	0	2	0	0
N-CYCLOHEXANECARBONYLETHYLENE AMINE	7	1	0	1	1	0	1	0	0
N-HEPTANOYLETHYLENE AMINE	7	1	0	1	3	0	1	0	0
N,N-DIBUTYLACRYLAMIDE	7	1	0	3	7	0	1	0	0
AZELAIC ANHYDRIDE	7	2	0	0	0	1	0	0	0
TRIMETHYLENE ADIPATE	7	2	0	0	0	2	0	0	0
HEXAMETHYLENE MALONAMIDE	7	2	0	0	2	0	2	0	0
HYDRAZO AZELAAMIDE	7	2	0	0	2	0	2	0	0
PENTAMETHYLENE 2,2'-OXYDIACETAMIDE	7	2	0	0	2	1	2	0	0
TRIMETHYLENE 1,4-CYCLOHEXYLENEDICARBOXYLATE	7	2	0	2	2	2	0	0	0
CYCLOPROPYLENEDIMETHYLENE PIPERAZINEDIURETHAN	7	2	0	2	2	2	2	0	0
HEXAMETHYLENE CYCLOPROPYLENEDICARBOXAMIDE	7	2	0	2	4	0	2	0	0
PIPERAZINE 3-METHYLADIPAMIDE	7	2	0	2	4	0	2	0	0
TRIMETHYLENE 4-OCTENDIAMIDE	7	2	0	2	4	0	2	0	0
CYCLOPROPYLENE HEXAMETHYLENEDIUREA	7	2	0	2	6	0	4	0	0
3-OXYPROPYLHEXAMETHYLENE OXAMIDE	7	2	0	2	6	1	2	0	0
HEXAMETHYLENE (1-METHYLCYCLOPROPYLENE)DICARBOXAMIDE	7	2	0	3	6	0	2	0	0
HEXAMETHYLENE (METHYLENE-2,5-FURAN)DICARBOXAMIDE	7	2	0	4	4	1	2	0	0
HEXAMETHYLENE 4(H)-PYRAN-2,6-DICARBOXAMIDE	7	2	0	4	4	1	2	0	0
HEXAMETHYLENE 3,3'-DIMETHYL-4,4'-METHYLENEDIPHENYLENEDIUREA	7	2	0	14	16	0	4	0	0
N,N'-DIETHYL-3,3'-DIMETHYL-4,4'-METHYLENEDIPHENYLENE ADIPAMIDE	7	2	0	16	18	0	2	0	0
HEPTAMETHYLENE TEREPHTHALATE	7	2	1	0	0	2	0	0	0
P-XYLYLENE PIMELATE	7	2	1	0	0	2	0	0	0
ETHYLENE P-(CARBOXYPHENOXY)CAPROATE	7	2	1	0	0	3	0	0	0
HEXAMETHYLENE P-(CARBOXYPHENOXY)ACETATE	7	2	1	0	0	3	0	0	0
HEPTAMETHYLENE TEREPHTHALAMIDE	7	2	1	0	2	0	2	0	0
HEXAMETHYLENE P-(CARBOXYPHENYLENE)ACETAMIDE	7	2	1	0	2	0	2	0	0
M-XYLYLENE PIMELAMIDE	7	2	1	0	2	0	2	0	0
P-XYLYLENE PIMELAMIDE	7	2	1	0	2	0	2	0	0
P-PHENYLENE (METHYL-1,4-PIPERAZINE)DIPROPIONATE	7	2	1	2	4	2	2	0	0
N,N'-DIBUTYL-4,4'-(DIMETHYLMETHYLENE)DIPHENYLENE TEREPHTHALAMIDE	7	2	1	16	18	0	2	0	0
ETHYLENE 4,4'-(PENTAMETHYLENEDIOXY)DIBENZOATE	7	2	2	0	0	4	0	0	0
HEXAMETHYLENE 3,3'-METHYLENEDIBENZAMIDE	7	2	2	0	2	0	2	0	0
HEXAMETHYLENE 4,4'-METHYLENEDIBENZAMIDE	7	2	2	0	2	0	2	0	0
HEXAMETHYLENE 4,4'-(METHYLENEDIPHENYLENE)DIURETHAN	7	2	2	0	2	2	2	0	0
HEPTAMETHYLENE 4,4'-SULFONYLDIBENZAMIDE	7	2	2	0	2	2	2	1	0
HEXAMETHYLENE 4,4'-(METHYLENEDIPHENYLENE)DIUREA	7	2	2	0	4	0	4	0	0
N,N'-DIETHYL(METHYLENEDIPHENYLENE) ADIPAMIDE	7	2	2	2	6	0	2	0	0
2-BUTYL-2-ETHYLTRIMETHYLENE METHYLENEDIPHENYLENEDIURETHAN	7	2	2	3	8	2	2	0	0
N,N'-DIBUTYL(METHYLENEDIPHENYLENE) TEREPHTHALAMIDE	7	2	3	2	6	0	2	0	0
GLYCYLPROLYLPROLINE	7	3	0	2	3	0	3	0	0
B-D-GLUCOTRIBUTYRATE	7	3	0	8	14	5	0	0	0
TRIMETHYLENE ADIPYLDIURETHAN	7	4	0	0	2	2	2	0	0
PENTAMETHYLENE N,N'-ETHYLENEDITEREPHTHALAMATE	7	4	2	0	2	2	2	0	0
TETRAMETHYLENE N,N'-TRIMETHYLENEDITEREPHTHALAMATE	7	4	2	0	2	2	2	0	0
------ 8									
OCTAMETHYLENE	8	0	0	0	0	0	0	0	0
ETHYLENETHIOHEXAMETHYLENE SULFIDE	8	0	0	0	0	0	0	2	0
OCTAMETHYLENE DISULFIDE	8	0	0	0	0	0	0	2	0
ETHYLENEDITHIOHEXAMETHYLENE DISULFIDE	8	0	0	0	0	0	0	4	0
OCTAMETHYLENE OXIDE	8	0	0	0	0	1	0	0	0
1,4-PIPERAZINEDIETHYLENE SULFONE	8	0	0	0	0	2	2	1	0
2,5-OCTAMETHYLENE-1,3,4-OXADIAZOLE	8	0	0	2	0	1	2	0	0
2,5-OCTAMETHYLENE-1,3,4-TRIAZOLE	8	0	0	2	1	0	3	0	0
CYCLODECENE	8	0	0	2	2	0	0	0	0
4-CYCLOHEXYL-1-BUTENE	8	0	0	2	2	0	0	0	0
1,2-DICHLORODECAMETHYLENE	8	0	0	2	2	0	0	0	2
2,5-OCTAMETHYLENE-1-AMINO-1,3,4-TRIAZOLE	8	0	0	2	2	0	4	0	0

POLY-	CH2	CO	C6H4	C	H	O	N	S	X
1-DECENE	8	0	0	2	4	0	0	0	0
OCTAMETHYLENE 5,5'-DIBENZIMIDAZOLE	8	0	0	14	8	0	4	0	0
2,2'-OCTAMETHYLENEDITHIAZOLE-P-PHENYLENE	8	0	1	6	2	0	2	2	0
9-AMINOPELARGONIC ACID	8	1	0	0	1	0	1	0	0
OCTAMETHYLENE UREA	8	1	0	0	2	0	2	0	0
N-OCTANOYLETHYLENE AMINE	8	1	0	1	3	0	1	0	0
OCTYL ACRYLATE	8	1	0	2	4	1	0	0	0
TETRAMETHYLENE DITHIOLADIPATE	8	2	0	0	0	0	0	2	0
THIODIETHYLENE DITHIOLADIPATE	8	2	0	0	0	0	0	3	0
PIPERAZINE ADIPAMIDE	8	2	0	0	0	0	2	0	0
SEBACIC ANHYDRIDE	8	2	0	0	0	1	0	0	0
ETHYLENE SUBERATE	8	2	0	0	0	2	0	0	0
HEXAMETHYLENE SUCCINATE	8	2	0	0	0	2	0	0	0
TETRAMETHYLENE ADIPATE	8	2	0	0	0	2	0	0	0
TRIMETHYLENE PIMELATE	8	2	0	0	0	2	0	0	0
PIPERAZINE (ETHYLENEDIOXY)DIACETAMIDE	8	2	0	0	0	2	2	0	0
HEXAMETHYLENE SUCCINAMIDE	8	2	0	0	2	0	2	0	0
HYDRAZO SEBACAMIDE	8	2	0	0	2	0	2	0	0
OCTAMETHYLENE OXAMIDE	8	2	0	0	2	0	2	0	0
PENTAMETHYLENE GLUTARAMIDE	8	2	0	0	2	0	2	0	0
TETRAMETHYLENE ADIPAMIDE	8	2	0	0	2	0	2	0	0
HYDRAZO THIODIVALERAMIDE	8	2	0	0	2	0	2	1	0
HEXAMETHYLENE OXYDIACETAMIDE	8	2	0	0	2	1	2	0	0
ETHYLENE HEXAMETHYLENEDIURETHAN	8	2	0	0	2	2	2	0	0
TETRAMETHYLENE TETRAMETHYLENEDIURETHAN	8	2	0	0	2	2	2	0	0
ETHYLENE (ETHYLENEDITHIO)DIETHYLENEDIURETHAN	8	2	0	0	2	2	2	2	0
TETRAMETHYLENE (1,4-PIPERAZINEDITHIO)DICARBOXYLATE	8	2	0	0	2	2	2	2	0
ETHYLENE (ETHYLENEDIAMINO)DIPROPIONAMIDE	8	2	0	0	4	0	4	0	0
ETHYLENE HEXAMETHYLENEDIUREA	8	2	0	0	4	0	4	0	0
TETRAMETHYLENE DISILOXANYLENEDIPROPIONAMIDE	8	2	0	0	6	1	2	0	2
2,2-BISBROMOMETHYLTRIMETHYLENE ADIPATE	8	2	0	1	0	2	0	0	2
2,2-BISCHLOROMETHYLTRIMETHYLENE ADIPATE	8	2	0	1	0	2	0	0	2
2,2-(BROMOMETHYLCHLOROMETHYL)TRIMETHYLENE ADIPATE	8	2	0	1	0	2	0	0	2
PIPERAZINE 4-OCTENDIAMIDE	8	2	0	2	2	0	2	0	0
1,4-CYCLOHEXYLENEDIMETHYLENE SUCCINATE	8	2	0	2	2	2	0	0	0
1,3-CYCLOHEXYLENE ADIPAMIDE	8	2	0	2	4	0	2	0	0
1,4-CYCLOHEXYLENE ADIPAMIDE	8	2	0	2	4	0	2	0	0
TETRAMETHYLENE 4-OCTENDIAMIDE	8	2	0	2	4	0	2	0	0
ETHYLENE 2,5-TETRAHYDROFURANDIPROPIONAMIDE	8	2	0	2	4	1	2	0	0
2-BUTENYLENE HEXAMETHYLENEDIURETHAN	8	2	0	2	4	2	2	0	0
TETRAMETHYLENE 1,4-CYCLOHEXYLENEDIURETHAN	8	2	0	2	4	2	2	0	0
HEXAMETHYLENE 2,5-DIHYDROXYADIPAMIDE	8	2	0	2	6	2	2	0	0
2-METHYLTRIMETHYLENE HEXAMETHYLENEDIURETHAN	8	2	0	2	6	2	2	0	0
2,2-DIMETHYLTRIMETHYLENE SUBERATE	8	2	0	3	6	2	0	0	0
2,2-DIMETHYLTRIMETHYLENE HEXAMETHYLENEDIURETHAN	8	2	0	3	8	2	2	0	0
HEXAMETHYLENE 2,5-THIOPHENEDIACETAMIDE	8	2	0	4	4	0	2	1	0
OCTAMETHYLENE 2,5-THIOPHENEDICARBOXAMIDE	8	2	0	4	4	0	2	1	0
OCTAMETHYLENE 2,5-FURANDICARBOXAMIDE	8	2	0	4	4	1	2	0	0
1,4-DIMETHYLTETRAMETHYLENE HEXAMETHYLENEDIURETHAN	8	2	0	4	10	2	2	0	0
OCTAMETHYLENE 2,4-PYRIDINEDICARBOXAMIDE	8	2	0	5	5	0	3	0	0
OCTAMETHYLENE 2,5-PYRIDINEDICARBOXAMIDE	8	2	0	5	5	0	3	0	0
OCTAMETHYLENE 2,6-PYRIDINEDICARBOXAMIDE	8	2	0	5	5	0	3	0	0
OCTAMETHYLENE 3,5-PYRIDINEDICARBOXAMIDE	8	2	0	5	5	0	3	0	0
N,N'-DIETHYL(HEXAFLUOROPENTAMETHYLENE) ADIPAMIDE	8	2	0	5	6	0	2	0	6
N,N'-DIETHYLHEXAMETHYLENE PERFLUOROGLUTARAMIDE	8	2	0	5	6	0	2	0	6
N,N'-DIMETHYLTETRAMETHYLENE TETRAMETHYLDISILOXANYLENEDIPROPIONAM	8	2	0	6	18	1	2	0	2
3-METHYL-M-PHENYLENE SEBACAMIDE	8	2	0	7	8	0	2	0	0
N,N'-DIISOPROPYLETHYLENE DIMETHYL(ETHYLENEDIAMINO)DIPROPIONAMIDE	8	2	0	8	20	0	4	0	0
4-METHYL-1,3-PHENYLENE TETRAETHYLDISILOXANYLENEDIPROPIONAMIDE	8	2	0	11	20	1	2	0	2
HEXAMETHYLENE DITHIOL-P-PHENYLENEDIACETATE	8	2	1	0	0	0	0	2	0
OCTAMETHYLENE TEREPHTHALATE	8	2	1	0	0	2	0	0	0
P-XYLYLENE SUBERATE	8	2	1	0	0	2	0	0	0
P-PHENYLENE 1,4-PIPERAZINEDIPROPIONATE	8	2	1	0	0	2	2	0	0
ETHYLENE P-(CARBOXYPHENOXY)HEPTANOATE	8	2	1	0	0	3	0	0	0
HEXAMETHYLENE M-PHENYLENEDIACETAMIDE	8	2	1	0	2	0	2	0	0
HEXAMETHYLENE P-PHENYLENEDIACETAMIDE	8	2	1	0	2	0	2	0	0
OCTAMETHYLENE ISOPHTHALAMIDE	8	2	1	0	2	0	2	0	0
OCTAMETHYLENE PHTHALAMIDE	8	2	1	0	2	0	2	0	0
OCTAMETHYLENE TEREPHTHALAMIDE	8	2	1	0	2	0	2	0	0
P-PHENYLENEDIETHYLENE ADIPAMIDE	8	2	1	0	2	0	2	0	0
M-PHENYLENE SEBACAMIDE	8	2	1	0	2	0	2	0	0
O-PHENYLENE SEBACAMIDE	8	2	1	0	2	0	2	0	0
P-PHENYLENE SEBACAMIDE	8	2	1	0	2	0	2	0	0
M-XYLYLENE SUBERAMIDE	8	2	1	0	2	0	2	0	0
P-XYLYLENE SUBERAMIDE	8	2	1	0	2	0	2	0	0
HEXAMETHYLENE P-PHENOXYDIACETAMIDE	8	2	1	0	2	1	2	0	0
P-XYLYLENE OXYDIBUTYRAMIDE	8	2	1	0	2	1	2	0	0

POLY-	CH2	CO	C6H4	C	H	O	N	S	X
HEXAMETHYLENE (P-PHENYLENEDIOXY)DIACETAMIDE	8	2	1	0	2	2	2	0	0
HEXAMETHYLENE P-XYLYLENEDIURETHAN	8	2	1	0	2	2	2	0	0
HEXAMETHYLENE P-XYLYLENEDIUREA	8	2	1	0	4	0	4	0	0
1,4-CYCLOHEXYLENEDIMETHYLENE P-PHENYLENEDIACETATE	8	2	1	2	2	2	0	0	0
SEBACYL DITHIONISOPHTHALOYLDIHYDRAZIDE	8	2	1	2	4	0	4	2	0
M-PHENYLENE (TETRAETHYLDISILOXANYLENE)DIPROPIONAMIDE	8	2	1	4	14	1	2	0	2
O-PHENYLENE (TETRAETHYLDISILOXANYLENE)DIPROPIONAMIDE	8	2	1	4	14	1	2	0	2
P-PHENYLENE (TETRAETHYLDISILOXANYLENE)DIPROPIONAMIDE	8	2	1	4	14	1	2	0	2
P-PHENYLENEDITETRAMETHYLENE TEREPHTHALATE	8	2	2	0	0	2	0	0	0
HEXAMETHYLENE (ETHYLENEDIOXY)DIBENZOATE	8	2	2	0	0	4	0	0	0
4,4'-BIPHENYLENE SEBACAMIDE	8	2	2	0	2	0	2	0	0
4,4'-METHYLENEDIPHENYLENE AZELAAMIDE	8	2	2	0	2	0	2	0	0
HEPTAMETHYLENE 4,4'-(METHYLENEDIPHENYLENE)DIURETHAN	8	2	2	0	2	2	2	0	0
HEXAMETHYLENE 4,4'-(OXYDIPHENYLENE)DIOXYDIACETAMIDE	8	2	2	0	2	3	2	0	0
OCTAMETHYLENE PHOSPHINYLIDENEDIBENZAMIDE	8	2	2	0	3	1	2	0	1
OCTAMETHYLENE 3,3'-(METHYLPHOSPHINYLIDENE)DIBENZAMIDE	8	2	2	1	5	1	2	0	1
OCTAMETHYLENE 4,4'-(METHYLPHOSPHINYLIDENE)DIBENZAMIDE	8	2	2	1	5	1	2	0	1
N,N'-DIMETHYL-4,4'-METHYLENEDIPHENYLENE AZELAAMIDE	8	2	2	2	6	0	2	0	0
4,4'-ISOPROPYLIDENEDIPHENYLENE 1,4-PIPERAZINEDIPROPIONATE	8	2	2	3	6	2	2	0	0
OCTAMETHYLENE 4,4'-(PHENYLPHOSPHINYLIDENE)DIBENZAMIDE	8	2	3	1	3	1	2	0	1
L-1-CARBOXYPENTAMETHYLENE ADIPAMIDE	8	3	0	1	4	1	2	0	0
OCTAMETHYLENE 1,4-DIHYDRO-4-OXO-2,6-PYRIDINEDICARBOXAMIDE	8	3	0	4	5	0	3	0	0
HEXAMETHYLENE N,N'-ETHYLENEDITEREPHTHALAMATE	8	4	2	0	2	2	2	0	0
PENTAMETHYLENE N,N'-TRIMETHYLENEDITEREPHTHALAMATE	8	4	2	0	2	2	2	0	0
ALANYLALANYLGLYCYLPROLYLPROLYLGLYCINE	8	6	0	6	14	0	6	0	0
ALANYLGLYCYLALANYLPROLYLGLYCYLPROLINE	8	6	0	6	14	0	6	0	0
------ 9	9	0	0	0	0	0	0	0	0
NONAMETHYLENE DISULFIDE	9	0	0	0	0	0	0	2	0
PENTAMETHYLENETHIOTETRAMETHYLENE SULFIDE	9	0	0	0	0	0	0	2	0
NONAMETHYLENE OXIDE	9	0	0	0	0	1	0	0	0
PENTAMETHYLENESULFONYLTETRAMETHYLENE SULFONE	9	0	0	0	0	4	0	2	0
2,5-NONAMETHYLENE-1,3,4-OXADIAZOLE	9	0	0	2	0	1	2	0	0
5-CYCLOHEXYL-1-PENTENE	9	0	0	2	2	0	0	0	0
10-HYDROXYCAPRIC ACID	9	1	0	0	0	1	0	0	0
10-AMINOCAPRIC ACID	9	1	0	0	1	0	1	0	0
NONAMETHYLENE UREA	9	1	0	0	2	0	2	0	0
5-HYDROXY-2-(1,3-DIOXANE)CAPRYLIC ACID	9	1	0	2	2	3	0	0	0
4-HYDROXYMETHYLENE-2-(1,3-DIOXOLANE)CAPRYLIC ACID	9	1	0	2	2	3	0	0	0
2-METHYL-11-AMINOUNDECANOIC ACID	9	1	0	2	5	0	1	0	0
UNDECANEDIOIC ANHYDRIDE	9	2	0	0	0	1	0	0	0
ETHYLENE AZELAATE	9	2	0	0	0	2	0	0	0
PENTAMETHYLENE ADIPATE	9	2	0	0	0	2	0	0	0
TRIMETHYLENE SUBERATE	9	2	0	0	0	2	0	0	0
HEXAMETHYLENE GLUTARAMIDE	9	2	0	0	2	0	2	0	0
METHYLENE SEBACAMIDE	9	2	0	0	2	0	2	0	0
PENTAMETHYLENE ADIPAMIDE	9	2	0	0	2	0	2	0	0
TETRAMETHYLENE PIMELAMIDE	9	2	0	0	2	0	2	0	0
TETRAMETHYLENE PENTAMETHYLENEDIURETHAN	9	2	0	0	2	2	2	0	0
TRIMETHYLENE HEXAMETHYLENEDIURETHAN	9	2	0	0	2	2	2	0	0
TRIMETHYLENE (ETHYLENEDITHIO)DIETHYLENEDIURETHAN	9	2	0	0	2	2	2	2	0
HEXAMETHYLENE TRIMETHYLENEDIUREA	9	2	0	0	4	0	4	0	0
1,4-CYCLOHEXYLENEDIMETHYLENE GLUTARATE	9	2	0	2	2	2	0	0	0
1,4-CYCLOHEXYLENEDIMETHYLENE GLUTARAMIDE	9	2	0	2	4	0	2	0	0
CYCLOPROPYLENE SEBACAMIDE	9	2	0	2	4	0	2	0	0
PENTAMETHYLENE 4-OCTENDIAMIDE	9	2	0	2	4	0	2	0	0
HEXAMETHYLENE (METHYLENE-2,5-TETRAHYDROFURAN)DICARBOXAMIDE	9	2	0	2	4	1	2	0	0
CYCLOPROPYLENEDIMETHYLENE HEXAMETHYLENEDIURETHAN	9	2	0	2	4	2	2	0	0
HEXAMETHYLENE 3-METHYLADIPAMIDE	9	2	0	2	6	0	2	0	0
3-METHYLHEXAMETHYLENE ADIPAMIDE	9	2	0	2	6	0	2	0	0
2,2-DIMETHYLTRIMETHYLENE AZELAATE	9	2	0	3	6	2	0	0	0
3,3'-DIMETHYL-4,4'-METHYLENEDIPHENYLENE SEBACAMIDE	9	2	0	14	14	0	2	0	0
HEXAMETHYLENE (DIETHYLDIMETHYL)METHYLENEDIPHENYLENEDIUREA	9	2	0	16	20	0	4	0	0
N,N'-DIISOAMYL-3,3'-DIMETHYL-4,4'-METHYLENEDIPHENYLENE ADIPAMIDE	9	2	0	20	26	0	2	0	0
NONAMETHYLENE TEREPHTHALATE	9	2	1	0	0	2	0	0	0
P-XYLYLENE AZELAATE	9	2	1	0	0	2	0	0	0
HEPTAMETHYLENE P-PHENYLENEDIACETAMIDE	9	2	1	0	2	0	2	0	0
M-XYLYLENE AZELAAMIDE	9	2	1	0	2	0	2	0	0
P-XYLYLENE AZELAAMIDE	9	2	1	0	2	0	2	0	0
HEXAMETHYLENE (METHYLENE-P-PHENOXY)DIACETAMIDE	9	2	1	0	2	1	2	0	0
P-XYLYLENE 5-OXASEBACAMIDE	9	2	1	0	2	1	2	0	0
4,4'-METHYLENEDIPHENYLENE SEBACAMIDE	9	2	2	0	2	0	2	0	0
HEXAMETHYLENE 4,4'-(METHYLENEDIPHENYLENE)DIOXYDIACETAMIDE	9	2	2	0	2	2	2	0	0
OCTAMETHYLENE 4,4'-(METHYLENEDIPHENYLENE)DIURETHAN	9	2	2	0	2	2	2	0	0
NONAMETHYLENE PHOSPHINYLIDENEDIBENZAMIDE	9	2	2	0	3	1	2	0	1
4,4'-METHYLENEDIPHENYLENE OCTAMETHYLENEDIUREA	9	2	2	0	4	0	4	0	0
NONAMETHYLENE 3,3'-(METHYLPHOSPHINYLIDENE)DIBENZAMIDE	9	2	2	1	5	1	2	0	1
NONAMETHYLENE 4,4'-(METHYLPHOSPHINYLIDENE)DIBENZAMIDE	9	2	2	1	5	1	2	0	1

POLY-	CH2	CO	C6H4	C	H	O	N	S	X
N,N'-DIMETHYL-4,4'-METHYLENEDIPHENYLENE SEBACAMIDE	9	2	2	2	6	0	2	0	0
NONAMETHYLENE 4,4'-(PHENYLPHOSPHINYLIDENE)DIBENZAMIDE	9	2	3	0	3	1	2	0	1
ADIPYL PENTAMETHYLENEDIUREA	9	4	0	0	4	0	4	0	0
HEPTAMETHYLENE N,N'-ETHYLENEDITEREPHTHALAMATE	9	4	2	0	2	2	2	0	0
HEXAMETHYLENE N,N'-TRIMETHYLENEDITEREPHTHALAMATE	9	4	2	0	2	2	2	0	0
PENTAMETHYLENE N,N'-TETRAMETHYLENEDITEREPHTHALAMATE	9	4	2	0	2	2	2	0	0
------10	10	0	0	0	0	0	0	0	0
DECAMETHYLENE	10	0	0	0	0	0	0	0	0
DECAMETHYLENE SULFIDE	10	0	0	0	0	0	0	1	0
DECAMETHYLENE DISULFIDE	10	0	0	0	0	0	0	2	0
HEXAMETHYLENETHIOTETRAMETHYLENE SULFIDE	10	0	0	0	0	0	0	2	0
HEXAMETHYLENEDITHIOTETRAMETHYLENE DISULFIDE	10	0	0	0	0	0	0	4	0
DECAMETHYLENE OXIDE	10	0	0	0	0	1	0	0	0
METHYLENEOXYNONAMETHYLENE OXIDE	10	0	0	0	0	2	0	0	0
HEXAMETHYLENESULFONYLTETRAMETHYLENE SULFONE	10	0	0	0	0	4	0	2	0
2,5-DECAMETHYLENE-1,3,4-OXADIAZOLE	10	0	0	2	0	1	2	0	0
2,5-DECAMETHYLENE-1,3,4-TRIAZOLE	10	0	0	2	1	0	3	0	0
CYCLODODECENE	10	0	0	2	2	0	0	0	0
1,2-DICHLORODODECAMETHYLENE	10	0	0	2	2	0	0	0	2
2,5-DECAMETHYLENE-1-AMINO-1,3,4-TRIAZOLE	10	0	0	2	2	0	4	0	0
1-DODECENE	10	0	0	2	4	0	0	0	0
DECYLVINYL ETHER	10	0	0	2	4	1	0	0	0
2,2-DIPENTYLTRIMETHYLENE SULFONE	10	0	0	3	6	2	0	1	0
DECAMETHYLENEOXY-P-PHENYLENE OXIDE	10	0	1	0	0	2	0	0	0
BENZYLIDENETHIODECAMETHYLENE SULFIDE	10	0	1	1	2	0	0	2	0
4,4'-OXYDIPHENYLENEDITHIODECAMETHYLENE DISULFIDE	10	0	2	0	0	1	0	4	0
DECAMETHYLENE CARBONATE	10	1	0	0	0	2	0	0	0
11-AMINOUNDECANOIC ACID	10	1	0	0	1	0	1	0	0
DECAMETHYLENE UREA	10	1	0	0	2	0	2	0	0
N-METHYL-11-AMINOUNDECANOIC ACID	10	1	0	1	3	0	1	0	0
DECYL ACRYLATE	10	1	0	2	4	1	0	0	0
5-HYDROXYMETHYLENE-5-METHYL-2-(1,3-DIOXANE)CAPRYLIC ACID	10	1	0	3	4	3	0	0	0
N-PHENYL-11-AMINOUNDECANOIC ACID	10	1	1	0	1	0	1	0	0
4,4'-DECAMETHYLENEDIPHENYLENE CARBONATE	10	1	2	0	0	2	0	0	0
HEXAMETHYLENE DITHIOLADIPATE	10	2	0	0	0	0	0	2	0
PIPERAZINE SUBERAMIDE	10	2	0	0	0	0	2	0	0
DODECANEDIOIC ANHYDRIDE	10	2	0	0	0	1	0	0	0
DECAMETHYLENE OXALATE	10	2	0	0	0	2	0	0	0
ETHYLENE SEBACATE	10	2	0	0	0	2	0	0	0
HEXAMETHYLENE ADIPATE	10	2	0	0	0	2	0	0	0
TETRAMETHYLENE SUBERATE	10	2	0	0	0	2	0	0	0
TRIMETHYLENE AZELAATE	10	2	0	0	0	2	0	0	0
ETHYLENE 1,4-PIPERAZINEDIPROPIONATE	10	2	0	0	0	2	2	0	0
DECAMETHYLENE OXAMIDE	10	2	0	0	2	0	2	0	0
ETHYLENE SEBACAMIDE	10	2	0	0	2	0	2	0	0
HEXAMETHYLENE ADIPAMIDE	10	2	0	0	2	0	2	0	0
HYDRAZO DODECANEDIAMIDE	10	2	0	0	2	0	2	0	0
PENTAMETHYLENE PIMELAMIDE	10	2	0	0	2	0	2	0	0
TETRAMETHYLENE SUBERAMIDE	10	2	0	0	2	0	2	0	0
ETHYLENE THIODIVALERAMIDE	10	2	0	0	2	0	2	1	0
HEXAMETHYLENE THIODIPROPIONAMIDE	10	2	0	0	2	0	2	1	0
ETHYLENE 1,4-PIPERAZINEDIPROPIONAMIDE	10	2	0	0	2	0	4	0	0
HEXAMETHYLENE PIPERAZINEDIUREA	10	2	0	0	2	0	4	0	0
OXYDITRIMETHYLENE ADIPAMIDE	10	2	0	0	2	1	2	0	0
(ETHYLENEDIOXY)DIETHYLENE ADIPAMIDE	10	2	0	0	2	2	2	0	0
HEXAMETHYLENE (ETHYLENEDIOXY)DIACETAMIDE	10	2	0	0	2	2	2	0	0
TETRAMETHYLENE HEXAMETHYLENEDIURETHAN	10	2	0	0	2	2	2	0	0
(TETRAMETHYLENEDIOXY)DITRIMETHYLENE OXAMIDE	10	2	0	0	2	2	2	0	0
TETRAMETHYLENE (THIODITRIMETHYLENE)DIURETHAN	10	2	0	0	2	2	2	1	0
THIODIETHYLENE HEXAMETHYLENEDIURETHAN	10	2	0	0	2	2	2	1	0
ETHYLENE (TETRAMETHYLENEDITHIO)DIETHYLENEDIURETHAN	10	2	0	0	2	2	2	2	0
HEXAMETHYLENE (1,4-PIPERAZINEDITHIO)DICARBOXYLATE	10	2	0	0	2	2	2	2	0
TETRAMETHYLENE (ETHYLENEDITHIO)DIETHYLENEDIURETHAN	10	2	0	0	2	2	2	2	0
OXYDIETHYLENE HEXAMETHYLENEDIURETHAN	10	2	0	0	2	3	2	0	0
PHOSPHINIDENEDITRIMETHYLENE ADIPAMIDE	10	2	0	0	3	0	2	0	1
HEXAMETHYLENE TETRAMETHYLENEDIUREA	10	2	0	0	4	0	4	0	0
HEXAMETHYLENE DISILOXANYLENEDIPROPIONAMIDE	10	2	0	0	6	1	2	0	2
N-METHYLHEXAMETHYLENE ADIPAMIDE	10	2	0	1	4	0	2	0	0
HEXAMETHYLENE DITHIOL-1,4-CYCLOHEXYLENEDICARBOXYLATE	10	2	0	2	2	0	0	2	0
1,4-CYCLOHEXYLENEDIMETHYLENE ADIPATE	10	2	0	2	2	2	0	0	0
1,4-CYCLOHEXYLENEDIMETHYLENE ADIPAMIDE	10	2	0	2	4	0	2	0	0
DECAMETHYLENE FUMARAMIDE	10	2	0	2	4	0	2	0	0
HEXAMETHYLENE 1,2-CYCLOHEXYLENEDICARBOXAMIDE	10	2	0	2	4	0	2	0	0
HEXAMETHYLENE 1,3-CYCLOHEXYLENEDICARBOXAMIDE	10	2	0	2	4	0	2	0	0
HEXAMETHYLENE 1,4-CYCLOHEXYLENEDICARBOXAMIDE	10	2	0	2	4	0	2	0	0
HEXAMETHYLENE 4-OCTENDIAMIDE	10	2	0	2	4	0	2	0	0
TETRAMETHYLENE 2,5-TETRAHYDROFURANDIPROPIONAMIDE	10	2	0	2	4	1	2	0	0

POLY-	CH2	CO	C6H4	C	H	O	N	S	X
(1,4-CYCLOHEXYLENEDIOXY)DITRIMETHYLENE OXAMIDE	10	2	0	2	4	2	2	0	0
1,4-CYCLOHEXYLENE (ETHYLENEDITHIO)DIETHYLENEDIURETHAN	10	2	0	2	4	2	2	2	0
DECAMETHYLENE 1,2-DIHYDROXYSUCCINATE	10	2	0	2	4	4	0	0	0
N,N'-DIMETHYLHEXAMETHYLENE ADIPAMIDE	10	2	0	2	6	0	2	0	0
PIPERAZINE N,N'-DIMETHYL(ETHYLENEDIIMINO)DIPROPIONAMIDE	10	2	0	2	6	0	4	0	0
1-METHYLTRIMETHYLENE OCTAMETHYLENEDIURETHAN	10	2	0	2	6	2	2	0	0
DECAMETHYLENE METHYLFUMARAMIDE	10	2	0	3	6	0	2	0	0
2,2-DIMETHYLTRIMETHYLENE SEBACATE	10	2	0	3	6	2	0	0	0
1,4-CYCLOHEXYLENEDIMETHYLENE DITHIOL-1,4-CYCLOHEXYLENEDICARBOXY	10	2	0	4	4	0	0	2	0
DECAMETHYLENE 2,5-THIOPHENEDICARBOXAMIDE	10	2	0	4	4	0	2	1	0
HEXAMETHYLENE 2,5-THIOPHENEDIPROPIONAMIDE	10	2	0	4	4	0	2	1	0
DECAMETHYLENE 2,5-FURANDICARBOXAMIDE	10	2	0	4	4	1	2	0	0
HEXAMETHYLENE 2,5-FURANDIPROPIONAMIDE	10	2	0	4	4	1	2	0	0
1,4-CYCLOHEXYLENEDIMETHYLENE 1,4-CYCLOHEXYLENEDICARBOXYLATE	10	2	0	4	4	2	0	0	0
HEXAMETHYLENE 2,5-PYRROLEDIPROPIONAMIDE	10	2	0	4	5	0	3	0	0
DECAMETHYLENE 2,3-DIMETHYLFUMARAMIDE	10	2	0	4	8	0	2	0	0
HEXAMETHYLENE (TETRAMETHYLDISILOXANYLENE)DIPROPIONAMIDE	10	2	0	4	14	1	2	0	2
DECAMETHYLENE 2,4-PYRIDINE DICARBOXAMIDE	10	2	0	5	5	0	3	0	0
DECAMETHYLENE 2,5-PYRIDINE DICARBOXAMIDE	10	2	0	5	5	0	3	0	0
DECAMETHYLENE 2,6-PYRIDINE DICARBOXAMIDE	10	2	0	5	5	0	3	0	0
DECAMETHYLENE 3,5-PYRIDINE DICARBOXAMIDE	10	2	0	5	5	0	3	0	0
HEXAMETHYLENE (N-METHYL-2,5-PYRROLE)DIPROPIONAMIDE	10	2	0	5	7	0	3	0	0
N,N'-DIISOPROPYLETHYLENE 1,4-PIPERAZINEDIPROPIONAMIDE	10	2	0	6	14	0	4	0	0
PIPERAZINE N,N'-DIISOPROPYL(ETHYLENEDIIMINO)DIPROPIONAMIDE	10	2	0	6	14	0	4	0	0
TETRAMETHYL-P-XYLYLENEDITHIOL SEBACATE	10	2	0	10	12	0	0	2	0
DECAMETHYLENE 3,3'-DIMETHYL-4,4'-DIPHENYLENEDIURETHAN	10	2	0	14	14	2	2	0	0
DECAMETHYLENE DITHIOLTEREPHTHALATE	10	2	1	0	0	0	0	2	0
P-XYLYLENE DITHIOLSEBACATE	10	2	1	0	0	0	0	2	0
DECAMETHYLENE TEREPHTHALATE	10	2	1	0	0	2	0	0	0
P-XYLYLENE SEBACATE	10	2	1	0	0	2	0	0	0
HEXAMETHYLENE P-(CARBOXYPHENOXY)VALERATE	10	2	1	0	0	3	0	0	0
DECAMETHYLENE M-CARBOXYCARBANILATE	10	2	1	0	1	2	1	0	0
DECAMETHYLENE P-CARBOXYCARBANILATE	10	2	1	0	1	2	1	0	0
DECAMETHYLENE ISOPHTHALAMIDE	10	2	1	0	2	0	2	0	0
DECAMETHYLENE PHTHALAMIDE	10	2	1	0	2	0	2	0	0
DECAMETHYLENE TEREPHTHALAMIDE	10	2	1	0	2	0	2	0	0
HEXAMETHYLENE 3,3'-P-PHENYLENEDIPROPIONAMIDE	10	2	1	0	2	0	2	0	0
OCTAMETHYLENE P-PHENYLENEDIACETAMIDE	10	2	1	0	2	0	2	0	0
M-XYLYLENE SEBACAMIDE	10	2	1	0	2	0	2	0	0
P-XYLYLENE SEBACAMIDE	10	2	1	0	2	0	2	0	0
P-XYLYLENE THIODIVALERAMIDE	10	2	1	0	2	0	2	1	0
P-XYLYLENE 5-OXAUNDECANEDIAMIDE	10	2	1	0	2	1	2	0	0
P-XYLYLENE OXYDIVALERAMIDE	10	2	1	0	2	1	2	0	0
OCTAMETHYLENE P-XYLYLENEDIURETHAN	10	2	1	0	2	2	2	0	0
P-XYLYLENE OCTAMETHYLENEDIURETHAN	10	2	1	0	2	2	2	0	0
OCTAMETHYLENE P-XYLYLENEDIUREA	10	2	1	0	4	0	4	0	0
(1,4-CYCLOHEXYLENEDIOXY)DITRIMETHYLENE TEREPHTHALAMIDE	10	2	1	2	4	2	2	0	0
P-PHENYLENEDIPENTAMETHYLENE TEREPHTHALATE	10	2	2	0	0	2	0	0	0
4,4'-ETHYLENEDIPHENYLENE SEBACAMIDE	10	2	2	0	2	0	2	0	0
HEXAMETHYLENE 4,4'-(ETHYLENEDIPHENYLENE)DIOXYDIACETAMIDE	10	2	2	0	2	2	2	0	0
NONAMETHYLENE 4,4'-(METHYLENEDIPHENYLENE)DIURETHAN	10	2	2	0	2	2	2	0	0
DECAMETHYLENE PHOSPHINYLIDENEDIBENZAMIDE	10	2	2	0	3	1	2	0	1
DECAMETHYLENE 3,3'-(METHYLPHOSPHINYLIDENE)DIBENZAMIDE	10	2	2	1	5	1	2	0	1
DECAMETHYLENE 4,4'-(METHYLPHOSPHINYLIDENE)DIBENZAMIDE	10	2	2	1	5	1	2	0	1
DECAMETHYLENE 4,4'-ETHYLIDENEDIBENZAMIDE	10	2	2	2	6	0	2	0	0
N,N'-DIETHYL(METHYLENEDIPHENYLENE) AZELAAMIDE	10	2	2	2	6	0	2	0	0
DECAMETHYLENE 4,4'-(PHENYLPHOSPHINYLIDENE)DIBENZAMIDE	10	2	3	0	3	1	2	0	1
HEXAMETHYLENE KETODIPROPIONAMIDE	10	3	0	0	2	0	2	0	0
DECAMETHYLENE 1,4-DIHYDRO-4-OXO-2,6-PYRIDINEDICARBOXAMIDE	10	3	0	4	5	0	3	0	0
ß-D-GLUCOTRIVALERATE	10	3	0	8	14	5	0	0	0
ETHYLENE SEBACYLDIURETHAN	10	4	0	0	2	2	2	0	0
HEXAMETHYLENE ADIPYLDIURETHAN	10	4	0	0	2	2	2	0	0
ETHYLENE SEBACYLDIUREA	10	4	0	0	4	0	4	0	0
2-HYDROXYTRIMETHYLENE SEBACYLDIURETHAN	10	4	0	1	4	3	2	0	0
HEPTAMETHYLENE N,N'-TRIMETHYLENEDITEREPHTHALAMATE	10	4	2	0	2	2	2	0	0
HEXAMETHYLENE N,N'-TETRAMETHYLENEDITEREPHTHALAMATE	10	4	2	0	2	2	2	0	0
OCTAMETHYLENE N,N'-ETHYLENEDITEREPHTHALAMATE	10	4	2	0	2	2	2	0	0
PENTAMETHYLENE N,N'-PENTAMETHYLENEDITEREPHTHALAMATE	10	4	2	0	2	2	2	0	0
TETRAMETHYLENE N,N'-HEXAMETHYLENEDITEREPHTHALAMATE	10	4	2	0	2	2	2	0	0
------11	11	0	0	0	0	0	0	0	0
HEXAMETHYLENETHIOPENTAMETHYLENE SULFIDE	11	0	0	0	0	0	0	2	0
DECAMETHYLENEOXYMETHYLENE OXIDE	11	0	0	0	0	2	0	0	0
HEXAMETHYLENESULFONYLPENTAMETHYLENE SULFONE	11	0	0	0	0	4	0	2	0
CYCLOHEXYLIDENETHIOHEXAMETHYLENE SULFIDE	11	0	0	1	0	0	0	2	0
1-METHYLDODECAMETHYLENE	11	0	0	2	4	0	0	0	0
HEXAMETHYLENETHIO-4,4'-METHYLENEDIPHENYLENEDIETHYLENESULFID	11	0	2	0	0	0	0	2	0
12-AMINOLAURIC ACID	11	1	0	0	1	0	1	0	0

POLY-	CH2	CO	C6H4	C	H	O	N	S	X
13-AMINO-11-THIATRIDECANOIC ACID	11	1	0	0	1	0	1	1	0
13-AMINO-12-THIATRIDECANOIC ACID	11	1	0	0	1	0	1	1	0
N-METHYL-12-AMINOLAURIC ACID	11	1	0	1	3	0	1	0	0
N-ETHYL-11-AMINOUNDECANOIC ACID	11	1	0	1	4	0	1	0	0
VINYL LAURATE	11	1	0	2	4	1	0	0	0
HEXAMETHYLENE DITHIOLPIMELATE	11	2	0	0	0	0	0	2	0
PIPERAZINE AZELAAMIDE	11	2	0	0	0	0	2	0	0
PIPERAZINE (TRIMETHYLENEDITHIO)DIPROPIONAMIDE	11	2	0	0	0	0	2	2	0
TRIDECANEDIOIC ANHYDRIDE	11	2	0	0	0	1	0	0	0
HEXAMETHYLENE PIMELATE	11	2	0	0	0	2	0	0	0
TETRAMETHYLENE AZELAATE	11	2	0	0	0	2	0	0	0
TRIMETHYLENE SEBACATE	11	2	0	0	0	2	0	0	0
OXYDIETHYLENE AZELAATE	11	2	0	0	0	3	0	0	0
HEPTAMETHYLENE ADIPAMIDE	11	2	0	0	2	0	2	0	0
HEXAMETHYLENE PIMELAMIDE	11	2	0	0	2	0	2	0	0
PENTAMETHYLENE SUBERAMIDE	11	2	0	0	2	0	2	0	0
TETRAMETHYLENE AZELAAMIDE	11	2	0	0	2	0	2	0	0
ETHYLENE NONAMETHYLENEDIURETHAN	11	2	0	0	2	2	2	0	0
PENTAMETHYLENE HEXAMETHYLENEDIURETHAN	11	2	0	0	2	2	2	0	0
TRIMETHYLENE OCTAMETHYLENEDIURETHAN	11	2	0	0	2	2	2	0	0
ETHYLENE (PENTAMETHYLENEDITHIO)DIETHYLENEDIURETHAN	11	2	0	0	2	2	2	2	0
PENTAMETHYLENE (ETHYLENEDITHIO)DIETHYLENEDIURETHAN	11	2	0	0	2	2	2	2	0
TRIMETHYLENE (TETRAMETHYLENEDITHIO)DIETHYLENEDIURETHAN	11	2	0	0	2	2	2	2	0
HEXAMETHYLENE PENTAMETHYLENEDIUREA	11	2	0	0	4	0	4	0	0
1,4-CYCLOHEXYLENEDIMETHYLENE PIMELATE	11	2	0	2	2	2	0	0	0
1,3-CYCLOHEXYLENE AZELAAMIDE	11	2	0	2	4	0	2	0	0
1,4-CYCLOHEXYLENEDIMETHYLENE PIMELAMIDE	11	2	0	2	4	0	2	0	0
CYCLOPROPYLENEDIMETHYLENE SEBACAMIDE	11	2	0	2	4	0	2	0	0
HEPTAMETHYLENE 4-OCTENDIAMIDE	11	2	0	2	4	0	2	0	0
METHYLPIPERAZINE 1,4-PIPERAZINEDIPROPIONAMIDE	11	2	0	2	4	0	4	0	0
PIPERAZINE (METHYL-1,4-PIPERAZINE)DIPROPIONAMIDE	11	2	0	2	4	0	4	0	0
PENTAMETHYLENE 2,5-TETRAHYDROFURANDIPROPIONAMIDE	11	2	0	2	4	1	2	0	0
1,4-CYCLOHEXYLENE (TRIMETHYLENEDITHIO)DIETHYLENEDIURETHAN	11	2	0	2	4	2	2	2	0
2-BUTYL-2-ETHYLTRIMETHYLENE HEXAMETHYLENEDIURETHAN	11	2	0	3	8	2	2	0	0
DECAMETHYLENE (METHYLENE-2,5-FURAN)DICARBOXAMIDE	11	2	0	4	4	1	2	0	0
N,N'-DIBUTYL-3,3'-DIMETHYL-4,4'-METHYLENEDIPHENYLENE ADIPAMIDE	11	2	0	16	18	0	2	0	0
HEXAMETHYLENE (DIMETHYLDIPROPYL)METHYLENEDIPHENYLENEDIUREA	11	2	0	16	20	0	4	0	0
HEXAMETHYLENE P-(CARBOXYPHENOXY)CAPROATE	11	2	1	0	0	3	0	0	0
P-PHENYLENEDIETHYLENE AZELAAMIDE	11	2	1	0	2	0	2	0	0
UNDECAMETHYLENE TEREPHTHALAMIDE	11	2	1	0	2	0	2	0	0
P-XYLYLENE UNDECANEDIAMIDE	11	2	1	0	2	0	2	0	0
P-XYLYLENE 5-OXADODECANEDIAMIDE	11	2	1	0	2	1	2	0	0
P-XYLYLENE 6-OXADODECANEDIAMIDE	11	2	1	0	2	1	2	0	0
DECAMETHYLENE 3,3'-METHYLENEDIBENZAMIDE	11	2	2	0	2	0	2	0	0
DECAMETHYLENE 4,4'-METHYLENEDIBENZAMIDE	11	2	2	0	2	0	2	0	0
4,4'-METHYLENEDIPHENYLENE DODECANEDIAMIDE	11	2	2	0	2	0	2	0	0
DECAMETHYLENE 4,4'-(METHYLENEDIPHENYLENE)DIURETHAN	11	2	2	0	2	2	2	0	0
HEXAMETHYLENE 4,4'-(TRIMETHYLENEDIPHENYLENE)DIOXYDIACETAMID	11	2	2	0	2	2	2	0	0
DECAMETHYLENE 4,4'-(METHYLENEDIPHENYLENE)DIUREA	11	2	2	0	4	0	4	0	0
N,N'-DIETHYL(METHYLENEDIPHENYLENE) SEBACAMIDE	11	2	2	2	6	0	2	0	0
TRIMETHYLENE SEBACYLDIURETHAN	11	4	0	0	2	2	2	0	0
HEPTAMETHYLENE N,N'-TETRAMETHYLENEDITEREPHTHALAMATE	11	4	2	0	2	2	2	0	0
HEXAMETHYLENE N,N'-PENTAMETHYLENEDITEREPHTHALAMATE	11	4	2	0	2	2	2	0	0
NONAMETHYLENE N,N'-ETHYLENEDITEREPHTHALAMATE	11	4	2	0	2	2	2	0	0
OCTAMETHYLENE N,N'-TRIMETHYLENEDITEREPHTHALAMATE	11	4	2	0	2	2	2	0	0
PENTAMETHYLENE N,N'-HEXAMETHYLENEDITEREPHTHALAMATE	11	4	2	0	2	2	2	0	0
ALANYLGLYCYLPROLYLPROLYLGLYCYLPROLINE	11	6	0	5	10	0	6	0	0
ALANYLPROLYLGLYCYLPROLYLPROLYLGLYCINE	11	6	0	5	10	0	6	0	0
------12	12	0	0	0	0	0	0	0	0
DODECAMETHYLENE	12	0	0	0	0	0	0	0	0
DECAMETHYLENEDITHIOETHYLENE DISULFIDE	12	0	0	0	0	0	0	4	0
DODECAMETHYLENE OXIDE	12	0	0	0	0	1	0	0	0
1-TETRADECENE	12	0	0	2	4	0	0	0	0
OCTAMETHYLENE TETRAMETHYLENEDITHIOUREA	12	0	0	2	4	0	4	2	0
DODECYLVINYL ETHER	12	0	0	2	4	1	0	0	0
13-AMINOTRIDECANOIC ACID	12	1	0	0	1	0	1	0	0
N-ALLYL-11-AMINOUNDECANOIC ACID	12	1	0	1	1	0	1	0	0
N-DODECANOYLETHYLENE AMINE	12	1	0	1	3	0	1	0	0
4-HYDROXYTETRAMETHYLENE-2-(1,3-DIOXOLANE)CAPRYLIC ACID	12	1	0	2	2	3	0	0	0
DODECYL ACRYLATE	12	1	0	2	4	1	0	0	0
N-DODECYL ACRYLAMIDE	12	1	0	2	5	0	1	0	0
HEXAMETHYLENE DITHIOLSUBERATE	12	2	0	0	0	0	0	2	0
TETRAMETHYLENE DITHIOLSEBACATE	12	2	0	0	0	0	0	2	0
PIPERAZINE SEBACAMIDE	12	2	0	0	0	0	2	0	0
PIPERAZINE 1,4-PIPERAZINEDIPROPIONAMIDE	12	2	0	0	0	0	4	0	0
TETRADECANEDIOIC ANHYDRIDE	12	2	0	0	0	1	0	0	0
DECAMETHYLENE SUCCINATE	12	2	0	0	0	2	0	0	0

POLY-	CH2	CO	C6H4	C	H	O	N	S	X
HEXAMETHYLENE SUBERATE	12	2	0	0	0	2	0	0	0
PENTAMETHYLENE AZELAATE	12	2	0	0	0	2	0	0	0
TETRAMETHYLENE SEBACATE	12	2	0	0	0	2	0	0	0
TRIMETHYLENE UNDECANEDIOATE	12	2	0	0	0	2	0	0	0
OXYDIETHYLENE SEBACATE	12	2	0	0	0	3	0	0	0
DODECAMETHYLENE OXAMIDE	12	2	0	0	2	0	2	0	0
ETHYLENE DODECANEDIAMIDE	12	2	0	0	2	0	2	0	0
HEPTAMETHYLENE PIMELAMIDE	12	2	0	0	2	0	2	0	0
HEXAMETHYLENE SUBERAMIDE	12	2	0	0	2	0	2	0	0
OCTAMETHYLENE ADIPAMIDE	12	2	0	0	2	0	2	0	0
PENTAMETHYLENE AZELAAMIDE	12	2	0	0	2	0	2	0	0
TETRAMETHYLENE SEBACAMIDE	12	2	0	0	2	0	2	0	0
HEXAMETHYLENE THIODIBUTYRAMIDE	12	2	0	0	2	0	2	1	0
DITHIODIETHYLENE SEBACAMIDE	12	2	0	0	2	0	2	2	0
HEXAMETHYLENE 1,4-PIPERAZINEDIACETAMIDE	12	2	0	0	2	0	4	0	0
HEXAMETHYLENE OXYDIBUTYRAMIDE	12	2	0	0	2	1	2	0	0
ETHYLENE DECAMETHYLENEDIURETHAN	12	2	0	0	2	2	2	0	0
HEXAMETHYLENE HEXAMETHYLENEDIURETHAN	12	2	0	0	2	2	2	0	0
TETRAMETHYLENE OCTAMETHYLENEDIURETHAN	12	2	0	0	2	2	2	0	0
ETHYLENE (HEXAMETHYLENEDITHIO)DIETHYLENEDIURETHAN	12	2	0	0	2	2	2	2	0
HEXAMETHYLENE (ETHYLENEDITHIO)DIETHYLENEDIURETHAN	12	2	0	0	2	2	2	2	0
TETRAMETHYLENE (TETRAMETHYLENEDITHIO)DIETHYLENEDIURETHAN	12	2	0	0	2	2	2	2	0
TRIMETHYLENE (PENTAMETHYLENEDITHIO)DIETHYLENEDIURETHAN	12	2	0	0	2	2	2	2	0
(ETHYLENEDIOXY)DIETHYLENE 1,4-PIPERAZINEDIACETAMIDE	12	2	0	0	2	2	4	0	0
OCTAMETHYLENE DISILOXANYLENEDIPROPIONAMIDE	12	2	0	0	6	1	2	0	2
1,4-CYCLOHEXYLENEDIMETHYLENE SUBERATE	12	2	0	2	2	2	0	0	0
1,4-CYCLOHEXYLENEDIMETHYLENE SUBERAMIDE	12	2	0	2	4	0	2	0	0
1,3-CYCLOHEXYLENE SEBACAMIDE	12	2	0	2	4	0	2	0	0
HEXAMETHYLENE 1,2-CYCLOHEXYLENEDIACETAMIDE	12	2	0	2	4	0	2	0	0
OCTAMETHYLENE 4-OCTENEDIAMIDE	12	2	0	2	4	0	2	0	0
HEXAMETHYLENE 2,5-TETRAHYDROFURANDIPROPIONAMIDE	12	2	0	2	4	1	2	0	0
1,4-CYCLOHEXYLENE OCTAMETHYLENEDIURETHAN	12	2	0	2	4	2	2	0	0
1,4-CYCLOHEXYLENEDIMETHYLENE ETHYLENEDITHIODIETHYLENEDIURETHAN	12	2	0	2	4	2	2	2	0
1,4-CYCLOHEXYLENE (TETRAMETHYLENEDITHIO)DIETHYLENEDIURETHAN	12	2	0	2	4	2	2	2	0
HEXAMETHYLENE 2,5-PYRROLIDENEDIPROPIONAMIDE	12	2	0	2	5	0	3	0	0
HEXAMETHYLENE (N-METHYL-2,5-PYRROLIDINE)DIPROPIONAMIDE	12	2	0	3	7	0	3	0	0
HEXAMETHYLENE (DIETHYLDIMETHYLDISILOXANYLENE)DIPROPIONAMIDE	12	2	0	4	14	1	2	0	2
OCTAMETHYLENE (TETRAMETHYLDISILOXANYLENE)DIPROPIONAMIDE	12	2	0	4	14	1	2	0	2
N,N'-DIBUTYL(HEXAFLUOROPENTAMETHYLENE) ADIPAMIDE	12	2	0	5	6	0	2	0	6
N,N'-DIBUTYLHEXAMETHYLENE PERFLUOROGLUTARAMIDE	12	2	0	5	6	0	2	0	6
DODECAMETHYLENE 3,3'-DIMETHYL-4,4'-BIPHENYLENEDIURETHAN	12	2	0	14	14	2	2	0	0
P-XYLYLENE DODECANEDIOATE	12	2	1	0	0	2	0	0	0
ETHYLENE P-(CARBOXYPHENOXY)UNDECANOATE	12	2	1	0	0	3	0	0	0
DECAMETHYLENE P-PHENYLENEDIACETAMIDE	12	2	1	0	2	0	2	0	0
DODECAMETHYLENE TEREPHTHALAMIDE	12	2	1	0	2	0	2	0	0
OCTAMETHYLENE P-PHENYLENEDIPROPIONAMIDE	12	2	1	0	2	0	2	0	0
P-PHENYLENEDIETHYLENE SEBACAMIDE	12	2	1	0	2	0	2	0	0
P-XYLYLENE DODECANEDIAMIDE	12	2	1	0	2	0	2	0	0
P-PHENYLENEDIETHYLENE THIODIVALERAMIDE	12	2	1	0	2	0	2	1	0
P-XYLYLENE THIODICAPROAMIDE	12	2	1	0	2	0	2	1	0
P-XYLYLENE 6-OXATRIDECANDEDIAMIDE	12	2	1	0	2	1	2	0	0
P-XYLYLENE OXYDICAPROAMIDE	12	2	1	0	2	1	2	0	0
DECAMETHYLENE (P-PHENYLENEDIOXY)DIACETAMIDE	12	2	1	0	2	2	2	0	0
DECAMETHYLENE P-XYLYLENEDIURETHAN	12	2	1	0	2	2	2	0	0
P-PHENYLENEDIETHYLENE OCTAMETHYLENEDIURETHAN	12	2	1	0	2	2	2	0	0
P-PHENYLENEDITRIMETHYLENE HEXAMETHYLENEDIURETHAN	12	2	1	0	2	2	2	0	0
(P-PHENYLENEDIOXY)DIETHYLENE OCTAMETHYLENEDIURETHAN	12	2	1	0	2	4	2	0	0
DECAMETHYLENE P-XYLYLENEDIUREA	12	2	1	0	4	0	4	0	0
(1,4-CYCLOHEXYLENEDIOXY)DITRIMETHYLENE PHENYLENEDIOXYDIACETAMIDE	12	2	1	2	4	4	2	0	0
DECAMETHYLENE 4,4'-(ETHYLENEDIOXY)DIBENZOATE	12	2	2	0	0	4	0	0	0
DODECAMETHYLENE 4,4'-BIPHENYLENEDIURETHAN	12	2	2	0	2	2	2	0	0
4,4'-(ETHYLENEDIOXY)DIPHENYLENEDITRIMETHYLENE ADIPAMIDE	12	2	2	0	2	2	2	0	0
(1,4-CYCLOHEXYLENEDIOXY)DITRIMETHYLENE ETHYLENEDIOXYDIBENZAMIDE	12	2	2	2	4	4	2	0	0
DECAMETHYLENE ISOPROPYLIDENEDI-P-PHENYLENEDIOXYDIACETAMIDE	12	2	2	3	8	2	2	0	0
HEXAMETHYLENE 5-PYRROLIDONYLIDENEDIPROPIONAMIDE	12	3	0	1	3	0	2	0	0
DECAMETHYLENE 3-CARBOXYLGLUTARAMIDE	12	3	0	1	4	1	2	0	0
ADIPYL OCTAMETHYLENEDIUREA	12	4	0	0	4	0	4	0	0
DECAMETHYLENE N,N'-ETHYLENEDITEREPHTHALAMATE	12	4	2	0	2	2	2	0	0
HEPTAMETHYLENE N,N'-PENTAMETHYLENEDITEREPHTHALAMATE	12	4	2	0	2	2	2	0	0
HEXAMETHYLENE N,N'-HEXAMETHYLENEDITEREPHTHALAMATE	12	4	2	0	2	2	2	0	0
NONAMETHYLENE N,N'-TRIMETHYLENEDITEREPHTHALAMATE	12	4	2	0	2	2	2	0	0
OCTAMETHYLENE N,N'-TETRAMETHYLENEDITEREPHTHALAMATE	12	4	2	0	2	2	2	0	0
------13	13	0	0	0	0	0	0	0	0
VINYL MYRISTATE	13	1	0	2	4	1	0	0	0
HEXAMETHYLENE DITHIOLAZELAATE	13	2	0	0	0	0	0	2	0
DECAMETHYLENE GLUTARATE	13	2	0	0	0	2	0	0	0
HEXAMETHYLENE AZELAATE	13	2	0	0	0	2	0	0	0

POLY-	CH2	CO	C6H4	C	H	O	N	S	X
PENTAMETHYLENE SEBACATE	13	2	0	0	0	2	0	0	0
TRIMETHYLENE DODECANDIOATE	13	2	0	0	0	2	0	0	0
DECAMETHYLENE GLUTARAMIDE	13	2	0	0	2	0	2	0	0
HEPTAMETHYLENE SUBERAMIDE	13	2	0	0	2	0	2	0	0
HEXAMETHYLENE AZELAAMIDE	13	2	0	0	2	0	2	0	0
NONAMETHYLENE ADIPAMIDE	13	2	0	0	2	0	2	0	0
PENTAMETHYLENE SEBACAMIDE	13	2	0	0	2	0	2	0	0
TETRAMETHYLENE UNDECANEDIAMIDE	13	2	0	0	2	0	2	0	0
HEXAMETHYLENE 5-OXASEBACAMIDE	13	2	0	0	2	1	2	0	0
HEPTAMETHYLENE HEXAMETHYLENEDIURETHAN	13	2	0	0	2	2	2	0	0
TETRAMETHYLENE NONAMETHYLENEDIURETHAN	13	2	0	0	2	2	2	0	0
PENTAMETHYLENE (TETRAMETHYLENEDITHIO)DIETHYLENEDIURETHAN	13	2	0	0	2	2	2	2	0
TETRAMETHYLENE (PENTAMETHYLENEDITHIO)DIETHYLENEDIURETHAN	13	2	0	0	2	2	2	2	0
TRIMETHYLENE (HEXAMETHYLENEDITHIO)DIETHYLENEDIURETHAN	13	2	0	0	2	2	2	2	0
HEPTAMETHYLENE HEXAMETHYLENEDIUREA	13	2	0	0	4	0	4	0	0
NONAMETHYLENE DISILOXANYLENEDIPROPIONAMIDE	13	2	0	0	6	1	2	0	2
1,4-CYCLOHEXYLENEDIMETHYLENE AZELAATE	13	2	0	2	2	2	0	0	0
1,4-CYCLOHEXYLENEDIMETHYLENE AZELAAMIDE	13	2	0	2	4	0	2	0	0
NONAMETHYLENE 4-OCTENEDIAMIDE	13	2	0	2	4	0	2	0	0
DECAMETHYLENE (METHYLENE-2,5-TETRAHYDROFURAN)DICARBOXAMIDE	13	2	0	2	4	1	2	0	0
HEPTAMETHYLENE 2,5-TETRAHYDROFURANDIPROPIONAMIDE	13	2	0	2	4	1	2	0	0
1,4-CYCLOHEXYLENEDIMETHYLENE TRIMETHYLENEDITHIODIETHYLENEDIURET	13	2	0	2	4	2	2	2	0
1,4-CYCLOHEXYLENE (PENTAMETHYLENEDITHIO)DIETHYLENEDIURETHAN	13	2	0	2	4	2	2	2	0
3-METHYLHEXAMETHYLENE SEBACAMIDE	13	2	0	2	6	0	2	0	0
P-PHENYLENEDIETHYLENE UNDECANEDIAMIDE	13	2	1	0	2	0	2	0	0
P-XYLYLENE TRIDECANEDIAMIDE	13	2	1	0	2	0	2	0	0
P-XYLYLENE 5-OXATETRADECANEDIAMIDE	13	2	1	0	2	1	2	0	0
DODECAMETHYLENE 4,4'-(METHYLENEDIPHENYLENE)DIURETHAN	13	2	2	0	2	2	2	0	0
DODECAMETHYLENE 4,4'-(METHYLENEDIPHENYLENE)DIUREA	13	2	2	0	4	0	4	0	0
B-D-GLUCOTRICAPROATE	13	3	0	8	14	5	0	0	0
PENTAMETHYLENE SEBACYLDIUREA	13	4	0	0	4	0	4	0	0
DECAMETHYLENE N,N'-TRIMETHYLENEDITEREPHTHALAMATE	13	4	2	0	2	2	2	0	0
HEPTAMETHYLENE N,N'-HEXAMETHYLENEDITEREPHTHALAMATE	13	4	2	0	2	2	2	0	0
NONAMETHYLENE N,N'-TETRAMETHYLENEDITEREPHTHALAMATE	13	4	2	0	2	2	2	0	0
OCTAMETHYLENE N,N'-PENTAMETHYLENEDITEREPHTHALAMATE	13	4	2	0	2	2	2	0	0
UNDECAMETHYLENE N,N'-ETHYLENEDITEREPHTHALAMATE	13	4	2	0	2	2	2	0	0
------14	14	0	0	0	0	0	0	0	0
DECAMETHYLENEDITHIOTETRAMETHYLENE DISULFIDE	14	0	0	0	0	0	0	4	0
1-HEXADECENE	14	0	0	2	4	0	0	0	0
HEXAMETHYLENE OCTAMETHYLENEDITHIOUREA	14	0	0	2	4	0	4	2	0
N-PIPERAZINYL-11-AMINOUNDECANOIC ACID	14	1	0	0	1	0	3	0	0
TETRADECYL ACRYLATE	14	1	0	2	4	1	0	0	0
N-TETRADECYL ACRYLAMIDE	14	1	0	2	5	0	1	0	0
DECAMETHYLENE DITHIOLADIPATE	14	2	0	0	0	0	0	2	0
HEXAMETHYLENE DITHIOLSEBACATE	14	2	0	0	0	0	0	2	0
DECAMETHYLENE ADIPATE	14	2	0	0	0	2	0	0	0
HEXAMETHYLENE SEBACATE	14	2	0	0	0	2	0	0	0
(ETHYLENEDIOXY)DIETHYLENE SEBACATE	14	2	0	0	0	4	0	0	0
DECAMETHYLENE ADIPAMIDE	14	2	0	0	2	0	2	0	0
HEPTAMETHYLENE AZELAAMIDE	14	2	0	0	2	0	2	0	0
HEXAMETHYLENE SEBACAMIDE	14	2	0	0	2	0	2	0	0
NONAMETHYLENE PIMELAMIDE	14	2	0	0	2	0	2	0	0
OCTAMETHYLENE SUBERAMIDE	14	2	0	0	2	0	2	0	0
PENTAMETHYLENE UNDECANEDIAMIDE	14	2	0	0	2	0	2	0	0
TETRAMETHYLENE DODECANEDIAMIDE	14	2	0	0	2	0	2	0	0
ETHYLENE THIODIENANTHAMIDE	14	2	0	0	2	0	2	1	0
HEXAMETHYLENE THIODIVALERAMIDE	14	2	0	0	2	0	2	1	0
HEXAMETHYLENE 5-OXAUNDECANEDIAMIDE	14	2	0	0	2	1	2	0	0
HEXAMETHYLENE OXYDIVALERAMIDE	14	2	0	0	2	1	2	0	0
DECAMETHYLENE TETRAMETHYLENEDIURETHAN	14	2	0	0	2	2	2	0	0
HEXAMETHYLENE (TETRAMETHYLENEDIOXY)DIPROPIONAMIDE	14	2	0	0	2	2	2	0	0
OCTAMETHYLENE HEXAMETHYLENEDIURETHAN	14	2	0	0	2	2	2	0	0
HEXAMETHYLENE SULFONYLDIVALERAMIDE	14	2	0	0	2	2	2	1	0
THIODITETRAMETHYLENE HEXAMETHYLENEDIURETHAN	14	2	0	0	2	2	2	1	0
HEXAMETHYLENE (TETRAMETHYLENEDITHIO)DIETHYLENEDIURETHAN	14	2	0	0	2	2	2	2	0
PENTAMETHYLENE (PENTAMETHYLENEDITHIO)DIETHYLENEDIURETHAN	14	2	0	0	2	2	2	2	0
TETRAMETHYLENE (HEXAMETHYLENEDITHIO)DIETHYLENEDIURETHAN	14	2	0	0	2	2	2	2	0
1,4-PIPERAZINEDIETHYLENE HEXAMETHYLENEDIURETHAN	14	2	0	0	2	2	4	0	0
OXYDITETRAMETHYLENE HEXAMETHYLENEDIURETHAN	14	2	0	0	2	3	2	0	0
HEXAMETHYLENE OCTAMETHYLENEDIUREA	14	2	0	0	4	0	4	0	0
DECAMETHYLENE DISILOXANYLENEDIPROPIONAMIDE	14	2	0	0	6	1	2	0	2
HEXAMETHYLENE 6-HYDROXYUNDECANEDIAMIDE	14	2	0	1	4	1	2	0	0
(ETHYLENEDIOXY)DIETHYLENE 6-HYDROXYUNDECANEDIAMIDE	14	2	0	1	4	3	2	0	0
1,4-CYCLOHEXYLENEDIMETHYLENE SEBACATE	14	2	0	2	2	2	0	0	0
1,4-CYCLOHEXYLENEDIMETHYLENE SEBACAMIDE	14	2	0	2	4	0	2	0	0
DECAMETHYLENE 4-OCTENDIAMIDE	14	2	0	2	4	0	2	0	0
OCTAMETHYLENE 2,5-TETRAHYDROFURANDIPROPIONAMIDE	14	2	0	2	4	1	2	0	0

POLY-	CH2	CO	C6H4	C	H	O	N	S	X
1,4-CYCLOHEXYLENEDIMETHYLENE OCTAMETHYLENEDIURETHAN	14	2	0	2	4	2	2	0	0
(1,4-CYCLOHEXYLENEDIOXY)DITRIMETHYLENE ADIPAMIDE	14	2	0	2	4	2	2	0	0
1,4-CYCLOHEXYLENEDIMETHYLENE TETRAMETHYLENEDITHIODIETHYLENEDIUR	14	2	0	2	4	2	2	2	0
1,4-CYCLOHEXYLENE (HEXAMETHYLENEDITHIO)DIETHYLENEDIURETHAN	14	2	0	2	4	2	2	2	0
HEXAMETHYLENE OCTAMETHYLENEDIURETHAN	14	2	0	2	6	2	2	0	0
DECAMETHYLENE 2,5-FURANDIPROPIONAMIDE	14	2	0	4	4	1	2	0	0
DECAMETHYLENE (TETRAMETHYLDISILOXANYLENE)DIPROPIONAMIDE	14	2	0	4	14	1	2	0	2
HEXAMETHYLENE (TETRAETHYLDISILOXANYLENE)DIPROPIONAMIDE	14	2	0	4	14	1	2	0	2
DECAMETHYLENE P-PHENYLENEDIPROPIONAMIDE	14	2	1	0	2	0	2	0	0
DODECAMETHYLENE P-PHENYLENEDIACETAMIDE	14	2	1	0	2	0	2	0	0
P-PHENYLENEDIETHYLENE DODECANEDIAMIDE	14	2	1	0	2	0	2	0	0
TETRADECAMETHYLENE TEREPHTHALAMIDE	14	2	1	0	2	0	2	0	0
M-XYLYLENE TETRADECANEDIAMIDE	14	2	1	0	2	0	2	0	0
P-XYLYLENE TETRADECANEDIAMIDE	14	2	1	0	2	0	2	0	0
P-PHENYLENEDIETHYLENE THIODICAPROAMIDE	14	2	1	0	2	0	2	1	0
P-XYLYLENE THIODIENANTHAMIDE	14	2	1	0	2	0	2	1	0
P-XYLYLENE 6-OXAPENTADECANEDIAMIDE	14	2	1	0	2	1	2	0	0
P-XYLYLENE OXYDIENANTHAMIDE	14	2	1	0	2	1	2	0	0
DODECAMETHYLENE P-XYLYLENEDIURETHAN	14	2	1	0	2	2	2	0	0
DODECAMETHYLENE P-XYLYLENEDIUREA	14	2	1	0	2	2	2	0	0
DECAMETHYLENE 4,4'-(ETHYLENEDIPHENYLENE)DIOXYDIACETAMIDE	14	2	2	0	2	2	2	0	0
1,4-CYCLOHEXYLENEDIOXYDITRIMETHYLENE TETRAMETHYLENEDIOXYDIBENZA	14	2	2	2	4	4	2	0	0
1,4-CYCLOHEXYLENEDIOXYDITRIMETHYLENE OXYDIETHYLENEDIOXYDIBENZAM	14	2	2	2	4	5	2	0	0
DECAMETHYLENE ADIPYLDIURETHAN	14	4	0	0	2	2	2	0	0
HEXAMETHYLENE SEBACYLDIURETHAN	14	4	0	0	2	2	2	0	0
DECAMETHYLENE N,N'-TETRAMETHYLENEDITEREPHTHALAMATE	14	4	2	0	2	2	2	0	0
DODECAMETHYLENE N,N'-ETHYLENEDITEREPHTHALAMATE	14	4	2	0	2	2	2	0	0
NONAMETHYLENE N,N'-PENTAMETHYLENEDITEREPHTHALAMATE	14	4	2	0	2	2	2	0	0
OCTAMETHYLENE N,N'-HEXAMETHYLENEDITEREPHTHALAMATE	14	4	2	0	2	2	2	0	0
UNDECAMETHYLENE N,N'-TRIMETHYLENEDITEREPHTHALAMATE	14	4	2	0	2	2	2	0	0
------15	15	0	0	0	0	0	0	0	0
METHYLENEOXYTETRADECAMETHYLENE OXIDE	15	0	0	0	0	2	0	0	0
VINYL PALMITATE	15	1	0	2	4	1	0	0	0
HEPTAMETHYLENE SEBACATE	15	2	0	0	0	2	0	0	0
6-AMINOCAPROIC ACID-ALT-11-AMINOUNDECANOIC ACID	15	2	0	0	2	0	2	0	0
HEPTAMETHYLENE SEBACAMIDE	15	2	0	0	2	0	2	0	0
HEXAMETHYLENE UNDECANEDIAMIDE	15	2	0	0	2	0	2	0	0
OCTAMETHYLENE AZELAAMIDE	15	2	0	0	2	0	2	0	0
UNDECAMETHYLENE ADIPAMIDE	15	2	0	0	2	0	2	0	0
HEXAMETHYLENE 5-OXADODECANEDIAMIDE	15	2	0	0	2	1	2	0	0
(ETHYLENEDIOXY)DIETHYLENE UNDECANEDIAMIDE	15	2	0	0	2	2	2	0	0
NONAMETHYLENE HEXAMETHYLENEDIURETHAN	15	2	0	0	2	2	2	0	0
TETRAMETHYLENE UNDECAMETHYLENEDIURETHAN	15	2	0	0	2	2	2	0	0
HEXAMETHYLENE (PENTAMETHYLENEDITHIO)DIETHYLENEDIURETHAN	15	2	0	0	2	2	2	2	0
PENTAMETHYLENE (HEXAMETHYLENEDITHIO)DIETHYLENEDIURETHAN	15	2	0	0	2	2	2	2	0
HEXAMETHYLENE NONAMETHYLENEDIUREA	15	2	0	0	4	0	4	0	0
NONAMETHYLENE 2,5-TETRAHYDROFURANDIPROPIONAMIDE	15	2	0	2	4	1	2	0	0
1,4-CYCLOHEXYLENEDIMETHYLENE PENTAMETHYLENEDITHIODIETHYLENEDIUR	15	2	0	2	4	2	2	2	0
TETRAMETHYLENE TRIMETHYLENEDIPIPERIDENEDITHIODICARBOXYLATE	15	2	0	2	4	2	2	2	0
NONAMETHYLENE (DIETHYLDIMETHYLDISILOXANYLENE)DIPROPIONAMIDE	15	2	0	4	14	1	2	0	2
N,N'-DIHEXYL-3,3'-DIMETHYL-4,4'-METHYLENEDIPHENYLENE ADIPAMIDE	15	2	0	16	18	0	2	0	0
P-PHENYLENEDIETHYLENE TRIDECANEDIAMIDE	15	2	1	0	2	0	2	0	0
P-XYLYLENE PENTADECANEDIAMIDE	15	2	1	0	2	0	2	0	0
P-XYLYLENE 5-OXAHEXADECANEDIAMIDE	15	2	1	0	2	1	2	0	0
4,4'-METHYLENEDIPHENYLENE TETRADECAMETHYLENEDIUREA	15	2	2	0	4	0	4	0	0
DECAMETHYLENE N,N'-PENTAMETHYLENEDITEREPHTHALAMATE	15	4	2	0	2	2	2	0	0
DODECAMETHYLENE N,N'-TRIMETHYLENEDITEREPHTHALAMATE	15	4	2	0	2	2	2	0	0
NONAMETHYLENE N,N'-HEXAMETHYLENEDITEREPHTHALAMATE	15	4	2	0	2	2	2	0	0
TRIDECAMETHYLENE N,N'-ETHYLENEDITEREPHTHALAMATE	15	4	2	0	2	2	2	0	0
UNDECAMETHYLENE N,N'-TETRAMETHYLENEDITEREPHTHALAMATE	15	4	2	0	2	2	2	0	0
------16	16	0	0	0	0	0	0	0	0
DECAMETHYLENETHIOHEXAMETHYLENE SULFIDE	16	0	0	0	0	0	0	2	0
(TETRAMETHYLENEDIOXY)DIHEXAMETHYLENE OXIDE	16	0	0	0	0	3	0	0	0
DECAMETHYLENEAMINOHEXAMETHYLENE AMINE	16	0	0	0	2	0	2	0	0
1-OCTADECENE	16	0	0	2	4	0	0	0	0
OCTAMETHYLENE OCTAMETHYLENEDITHIOUREA	16	0	0	2	4	0	4	2	0
17-AMINOHEPTADECANOIC ACID	16	1	0	0	1	0	1	0	0
HEXADECYL ACRYLATE	16	1	0	2	4	1	0	0	0
N-HEXADECYL ACRYLAMIDE	16	1	0	2	5	0	1	0	0
HEXADECYL METHACRYLATE	16	1	0	3	6	1	0	0	0
OCTADECANEDIOIC ANHYDRIDE	16	2	0	0	0	1	0	0	0
DECAMETHYLENE SUBERATE	16	2	0	0	0	2	0	0	0
NONAMETHYLENE AZELAATE	16	2	0	0	0	2	0	0	0
DECAMETHYLENE SUBERAMIDE	16	2	0	0	2	0	2	0	0
DODECAMETHYLENE ADIPAMIDE	16	2	0	0	2	0	2	0	0
HEPTAMETHYLENE UNDECANEDIAMIDE	16	2	0	0	2	0	2	0	0
HEXAMETHYLENE DODECANEDIAMIDE	16	2	0	0	2	0	2	0	0

POLY-	CH2	CO	C6H4	C	H	O	N	S	X
NONAMETHYLENE AZELAAMIDE	16	2	0	0	2	0	2	0	0
OCTAMETHYLENE SEBACAMIDE	16	2	0	0	2	0	2	0	0
PENTAMETHYLENE TRIDECANEDIAMIDE	16	2	0	0	2	0	2	0	0
HEXAMETHYLENE OXYDICAPROAMIDE	16	2	0	0	2	1	2	0	0
DECAMETHYLENE HEXAMETHYLENEDIURETHAN	16	2	0	0	2	2	2	0	0
HEXAMETHYLENE DECAMETHYLENEDIURETHAN	16	2	0	0	2	2	2	0	0
HEXAMETHYLENE (HEXAMETHYLENEDIOXY)DIPROPIONAMIDE	16	2	0	0	2	2	2	0	0
OCTAMETHYLENE OCTAMETHYLENEDIURETHAN	16	2	0	0	2	2	2	0	0
OXYDIPENTAMETHYLENE OXYDIBUTYRAMIDE	16	2	0	0	2	2	2	0	0
PENTAMETHYLENE UNDECAMETHYLENEDIURETHAN	16	2	0	0	2	2	2	0	0
HEXAMETHYLENE (HEXAMETHYLENEDITHIO)DIETHYLENEDIURETHAN	16	2	0	0	2	2	2	2	0
DECAMETHYLENE HEXAMETHYLENEDIUREA	16	2	0	0	4	0	4	0	0
1,4-CYCLOHEXYLENEDIMETHYLENE DODECANEDIOATE	16	2	0	2	2	2	0	0	0
1,4-CYCLOHEXYLENEDIMETHYLENE DODECANEDIAMIDE	16	2	0	2	4	0	2	0	0
DECAMETHYLENE 2,5-TETRAHYDROFURANDIPROPIONAMIDE	16	2	0	2	4	1	2	0	0
1,4-CYCLOHEXYLENEDIMETHYLENE HEXAMETHYLENEDITHIODIETHYLENEDIURE	16	2	0	2	4	2	2	2	0
HEXAMETHYLENE P-(CARBOXYPHENOXY)UNDECANOATE	16	2	1	0	0	3	0	0	0
DODECAMETHYLENE P-PHENYLENEDIPROPIONAMIDE	16	2	1	0	2	0	2	0	0
P-PHENYLENEDIETHYLENE TETRADECANEDIAMIDE	16	2	1	0	2	0	2	0	0
TETRADECAMETHYLENE P-PHENYLENEDIACETAMIDE	16	2	1	0	2	0	2	0	0
P-XYLYLENE HEXADECANEDIAMIDE	16	2	1	0	2	0	2	0	0
P-PHENYLENEDIETHYLENE THIODIENANTHAMIDE	16	2	1	0	2	0	2	1	0
P-XYLYLENE 6-OXAHEPTADECANEDIAMIDE	16	2	1	0	2	1	2	0	0
TETRADECAMETHYLENE P-XYLYLENEDIUREA	16	2	1	0	4	0	4	0	0
N,N'-DIMETHYL-P-XYLYLENE HEXADECANEDIAMIDE	16	2	1	2	6	0	2	0	0
1,4-CYCLOHEXYLENEDIOXYDITRIMETHYLENE HEXAMETHYLENEDIOXYDIBENZAM	16	2	2	2	4	4	2	0	0
B-D-GLUCOTRIHEPTYLATE	16	3	0	8	14	5	0	0	0
OCTAMETHYLENE SEBACYLDIUREA	16	4	0	0	4	0	4	0	0
DECAMETHYLENE N,N'-HEXAMETHYLENEDITEREPHTHALAMATE	16	4	2	0	2	2	2	0	0
DODECAMETHYLENE N,N'-TETRAMETHYLENEDITEREPHTHALAMATE	16	4	2	0	2	2	2	0	0
TETRADECAMETHYLENE N,N'-ETHYLENEDITEREPHTHALAMATE	16	4	2	0	2	2	2	0	0
TRIDECAMETHYLENE N,N'-TRIMETHYLENEDITEREPHTHALAMATE	16	4	2	0	2	2	2	0	0
UNDECAMETHYLENE N,N'-PENTAMETHYLENEDITEREPHTHALAMATE	16	4	2	0	2	2	2	0	0
L-1-CARBOXYPENTAMETHYLENE ADIPAMIDE	16	6	0	2	8	2	4	0	0
------17	17	0	0	0	0	0	0	0	0
HEXAMETHYLENETHIOUNDECAMETHYLENE SULFIDE	17	0	0	0	0	0	0	2	0
(PENTAMETHYLENEDIOXY)DIHEXAMETHYLENE OXIDE	17	0	0	0	0	3	0	0	0
VINYL STEARATE	17	1	0	2	4	1	0	0	0
DECAMETHYLENE AZELAATE	17	2	0	0	0	2	0	0	0
DECAMETHYLENE AZELAAMIDE	17	2	0	0	2	0	2	0	0
NONAMETHYLENE SEBACAMIDE	17	2	0	0	2	0	2	0	0
PENTAMETHYLENE TETRADECANEDIAMIDE	17	2	0	0	2	0	2	0	0
HEXAMETHYLENE 5-OXATETRADECANEDIAMIDE	17	2	0	0	2	1	2	0	0
OXYDIPENTAMETHYLENE 5-OXASEBACAMIDE	17	2	0	0	2	2	2	0	0
OCTYLPHOSPHINIDENEDITRIMETHYLENE ADIPAMIDE	17	2	0	1	5	0	2	0	1
NONAMETHYLENE (TETRAETHYLDISILOXANYLENE)DIPROPIONAMIDE	17	2	0	4	14	1	2	0	2
P-PHENYLENEDIETHYLENE PENTADECANEDIAMIDE	17	2	1	0	2	0	2	0	0
P-XYLYLENE HEPTADECANEDIAMIDE	17	2	1	0	2	0	2	0	0
HEXADECAMETHYLENE 4,4'-(METHYLENEDIPHENYLENE)DIURETHAN	17	2	2	0	2	2	2	0	0
DODECAMETHYLENE N,N'-PENTAMETHYLENEDITEREPHTHALAMATE	17	4	2	0	2	2	2	0	0
TETRADECAMETHYLENE N,N'-TRIMETHYLENEDITEREPHTHALAMATE	17	4	2	0	2	2	2	0	0
TRIDECAMETHYLENE N,N'-TETRAMETHYLENEDITEREPHTHALAMATE	17	4	2	0	2	2	2	0	0
UNDECAMETHYLENE N,N'-HEXAMETHYLENEDITEREPHTHALAMATE	17	4	2	0	2	2	2	0	0
------18	18	0	0	0	0	0	0	0	0
N-OCTADECANOYLETHYLENE AMINE	18	1	0	1	3	0	1	0	0
OCTADECYL ACRYLATE	18	1	0	2	4	1	0	0	0
N-OCTADECYL ACRYLAMIDE	18	1	0	2	5	0	1	0	0
OCTADECYL METHACRYLATE	18	1	0	3	6	1	0	0	0
DECAMETHYLENE DITHIOLSEBACATE	18	2	0	0	0	0	0	2	0
DECAMETHYLENE SEBACATE	18	2	0	0	0	2	0	0	0
OCTAMETHYLENE DODECANEDIOATE	18	2	0	0	0	2	0	0	0
DECAMETHYLENE SEBACAMIDE	18	2	0	0	2	0	2	0	0
HEXAMETHYLENE TETRADECANEDIAMIDE	18	2	0	0	2	0	2	0	0
NONAMETHYLENE UNDECANEDIAMIDE	18	2	0	0	2	0	2	0	0
OCTAMETHYLENE DODECANEDIAMIDE	18	2	0	0	2	0	2	0	0
HEXAMETHYLENE THIODIENANTHAMIDE	18	2	0	0	2	0	2	1	0
HEXAMETHYLENE OXYDIENANTHAMIDE	18	2	0	0	2	1	2	0	0
HEXAMETHYLENE 6-OXAPENTADECANEDIAMIDE	18	2	0	0	2	1	2	0	0
DECAMETHYLENE OCTAMETHYLENEDIURETHAN	18	2	0	0	2	2	2	0	0
DODECAMETHYLENE HEXAMETHYLENEDIURETHAN	18	2	0	0	2	2	2	0	0
OCTAMETHYLENE DECAMETHYLENEDIURETHAN	18	2	0	0	2	2	2	0	0
OXYDIPENTAMETHYLENE 5-OXAUNDECANEDIAMIDE	18	2	0	0	2	2	2	0	0
OXYDIPENTAMETHYLENE OXYDIVALERAMIDE	18	2	0	0	2	2	2	0	0
AMINODIHEXAMETHYLENE SUBERAMIDE	18	2	0	0	3	0	3	0	0
DECAMETHYLENE OCTAMETHYLENEDIUREA	18	2	0	0	4	0	4	0	0
DODECAMETHYLENE HEXAMETHYLENEDIUREA	18	2	0	0	4	0	4	0	0
4,4'-ETHYLENEDICYCLOHEXYLENE SEBACAMIDE	18	2	0	4	6	0	2	0	0

CRYSTALLOGRAPHIC DATA FOR VARIOUS POLYMERS

POLY-	CH2	CO	C6H4	C	H	O	N	S	X
OCTADECAMETHYLENE TEREPHTHALATE	18	2	1	0	0	2	0	0	0
OCTADECAMETHYLENE TEREPHTHALAMIDE	18	2	1	0	2	0	2	0	0
P-PHENYLENEDIETHYLENE HEXADECANEDIAMIDE	18	2	1	0	2	0	2	0	0
TETRADECAMETHYLENE P-PHENYLENEDIPROPIONAMIDE	18	2	1	0	2	0	2	0	0
P-XYLYLENE OCTADECANEDIAMIDE	18	2	1	0	2	0	2	0	0
P-XYLYLENE THIODIPELARGONAMIDE	18	2	1	0	2	0	2	1	0
P-XYLYLENE OXYDIPELARGONAMIDE	18	2	1	0	2	1	2	0	0
HEXADECAMETHYLENE P-XYLYLENEDIURETHAN	18	2	1	0	2	2	2	0	0
N,N'-DIMETHYL-P-XYLYLENE OCTADECANEDIAMIDE	18	2	1	2	6	0	2	0	0
DECAMETHYLENE SEBACYLDIURETHAN	18	4	0	0	2	2	2	0	0
DODECAMETHYLENE N,N'-HEXAMETHYLENEDITEREPHTHALAMATE	18	4	2	0	2	2	2	0	0
TETRADECAMETHYLENE N,N'-TETRAMETHYLENEDITEREPHTHALAMATE	18	4	2	0	2	2	2	0	0
TRIDECAMETHYLENE N,N'-PENTAMETHYLENEDITEREPHTHALAMATE	18	4	2	0	2	2	2	0	0
------19	19	0	0	0	0	0	0	0	0
METHYLENEOXYOCTADECAMETHYLENE OXIDE	19	0	0	0	0	2	0	0	0
TRIMETHYLENE OCTADECANEDIOATE	19	2	0	0	0	2	0	0	0
UNDECAMETHYLENE SEBACAMIDE	19	2	0	0	2	0	2	0	0
HEXAMETHYLENE 5-OXAHEXADECANEDIAMIDE	19	2	0	0	2	1	2	0	0
OXYDIPENTAMETHYLENE 5-OXADODECANEDIAMIDE	19	2	0	0	2	2	2	0	0
OXYDIPENTAMETHYLENE 6-OXADODECANEDIAMIDE	19	2	0	0	2	2	2	0	0
TETRAMETHYLENE (PENTAMETHYLENEDISULFONYL)DICAPROAMIDE	19	2	0	0	2	4	2	2	0
HEXAMETHYLENE (TRIMETHYLENEDISULFONYL)DIAMINODICAPROAMIDE	19	2	0	0	4	4	4	0	0
P-PHENYLENEDIETHYLENE HEPTADECANEDIAMIDE	19	2	1	0	2	0	2	0	0
4,4'-METHYLENEDIPHENYLENE OCTADECAMETHYLENEDIUREA	19	2	2	0	4	0	4	0	0
B-D-GLUCOTRICAPRYLATE	19	3	0	8	14	5	0	0	0
TETRADECAMETHYLENE N,N'-PENTAMETHYLENEDITEREPHTHALAMATE	19	4	2	0	2	2	2	0	0
TRIDECAMETHYLENE N,N'-HEXAMETHYLENEDITEREPHTHALAMATE	19	4	2	0	2	2	2	0	0
------20	20	0	0	0	0	0	0	0	0
DECAMETHYLENE DODECANDIAMIDE	20	2	0	0	2	0	2	0	0
DODECAMETHYLENE SEBACAMIDE	20	2	0	0	2	0	2	0	0
HYDRAZO DOCOSANEDIAMIDE	20	2	0	0	2	0	2	0	0
NONAMETHYLENE TRIDECANEDIAMIDE	20	2	0	0	2	0	2	0	0
OCTAMETHYLENE TETRADECANEDIAMIDE	20	2	0	0	2	0	2	0	0
HEXAMETHYLENE 6-OXAHEPTADECANEDIAMIDE	20	2	0	0	2	1	2	0	0
DECAMETHYLENE DECAMETHYLENEDIURETHAN	20	2	0	0	2	2	2	0	0
OXYDIPENTAMETHYLENE 6-OXATRIDECANEDIAMIDE	20	2	0	0	2	2	2	0	0
OXYDIPENTAMETHYLENE OXYDICAPROAMIDE	20	2	0	0	2	2	2	0	0
HEXAMETHYLENE (TETRAMETHYLENEDISULFONYL)DICAPROAMIDE	20	2	0	0	2	4	2	2	0
PENTAMETHYLENE (PENTAMETHYLENEDISULFONYL)DICAPROAMIDE	20	2	0	0	2	4	2	2	0
TETRAMETHYLENE (HEXAMETHYLENEDISULFONYL)DICAPROAMIDE	20	2	0	0	2	4	2	2	0
HEXAMETHYLENE TETRADECAMETHYLENEDIUREA	20	2	0	0	4	0	4	0	0
HEXAMETHYLENE 9,10-DIHYDROXYOCTADECANEDIAMIDE	20	2	0	2	6	2	2	0	0
ETHYLENE 4,4'-(OXYDIMETHYLENE)DI 2 (1,3 DIOXOLANE)CAPRYLATE	20	2	0	4	4	7	0	0	0
4,4'-ETHYLENEDICYCLOHEXYLENE DODECANEDIAMIDE	20	2	0	4	6	0	2	0	0
OCTADECAMETHYLENE P-PHENYLENEDIACETAMIDE	20	2	1	0	2	0	2	0	0
P-PHENYLENEDIETHYLENE OCTADECANEDIAMIDE	20	2	1	0	2	0	2	0	0
P-PHENYLENEDIETHYLENE THIODIPELARGONAMIDE	20	2	1	0	2	0	2	1	0
OCTADECAMETHYLENE P-XYLYLENEDIUREA	20	2	1	0	4	0	4	0	0
N,N'-DIETHYL-P-XYLYLENE OCTADECANEDIAMIDE	20	2	1	2	6	0	2	0	0
OCTADECAMETHYLENE 4,4'-(ETHYLENEDIOXY)DIBENZOATE	20	2	2	0	0	4	0	0	0
TETRADECAMETHYLENE N,N'-HEXAMETHYLENEDITEREPHTHALAMATE	20	4	2	0	2	2	2	0	0
------21	21	0	0	0	0	0	0	0	0
22-AMINODOCOSANOIC ACID	21	1	0	0	1	0	1	0	0
EICOSAMETHYLENE MALONATE	21	2	0	0	0	2	0	0	0
DECAMETHYLENE TRIDECANEDIAMIDE	21	2	0	0	2	0	2	0	0
PENTAMETHYLENE OCTADECANEDIAMIDE	21	2	0	0	2	0	2	0	0
OXYDIPENTAMETHYLENE 5-OXATETRADECANEDIAMIDE	21	2	0	0	2	2	2	0	0
HEXAMETHYLENE (PENTAMETHYLENEDISULFONYL)DICAPROAMIDE	21	2	0	0	2	4	2	2	0
PENTAMETHYLENE (HEXAMETHYLENEDISULFONYL)DICAPROAMIDE	21	2	0	0	2	4	2	2	0
------22	22	0	0	0	0	0	0	0	0
(DECAMETHYLENEDIOXY)DIHEXAMETHYLENE OXIDE	22	0	0	0	0	3	0	0	0
DOCOSYL ACRYLATE	22	1	0	2	4	1	0	0	0
N-DOCOSYL ACRYLAMIDE	22	1	0	2	5	0	1	0	0
DECAMETHYLENE TETRADECANEDIAMIDE	22	2	0	0	2	0	2	0	0
DODECAMETHYLENE DODECANEDIAMIDE	22	2	0	0	2	0	2	0	0
HEXAMETHYLENE OCTADECANEDIAMIDE	22	2	0	0	2	0	2	0	0
TETRADECAMETHYLENE SEBACAMIDE	22	2	0	0	2	0	2	0	0
HEXAMETHYLENE OXYDIPELARGONAMIDE	22	2	0	0	2	1	2	0	0
DODECAMETHYLENE DECAMETHYLENEDIURETHAN	22	2	0	0	2	2	2	0	0
HEXADECAMETHYLENE HEXAMETHYLENEDIURETHAN	22	2	0	0	2	2	2	0	0
OXYDIPENTAMETHYLENE 6-OXAPENTADECANEDIAMIDE	22	2	0	0	2	2	2	0	0
OXYDIPENTAMETHYLENE OXYDIENANTHAMIDE	22	2	0	0	2	2	2	0	0
HEXAMETHYLENE (HEXAMETHYLENEDISULFONYL)DICAPROAMIDE	22	2	0	0	2	4	2	2	0
DECAMETHYLENE DODECAMETHYLENEDIUREA	22	2	0	0	4	0	4	0	0
N,N'-DIMETHYLHEXAMETHYLENE OCTADECANEDIAMIDE	22	2	0	2	6	0	2	0	0
OCTADECAMETHYLENE P-PHENYLENEDIPROPIONAMIDE	22	2	1	0	2	0	2	0	0
P-XYLYLENE DOCOSANEDIAMIDE	22	2	1	0	2	0	2	0	0

POLY-	CH2	CO	C6H4	C	H	O	N	S	X
P-XYLYLENE THIODIUNDECANOAMIDE	22	2	1	0	2	0	2	1	0
------23	23	0	0	0	0	0	0	0	0
OXYDIPENTAMETHYLENE 5-OXAHEXADECANEDIAMIDE	23	2	0	0	2	2	2	0	0
------24	24	0	0	0	0	0	0	0	0
(TETRAMETHYLENEDIOXY)DIDECAMETHYLENE OXIDE	24	0	0	0	0	3	0	0	0
HEXAMETHYLENE EICOSANEDIAMIDE	24	2	0	0	2	0	2	0	0
OCTAMETHYLENE OCTADECANEDIAMIDE	24	2	0	0	2	0	2	0	0
TETRADECAMETHYLENE DODECANEDIAMIDE	24	2	0	0	2	0	2	0	0
TRIDECAMETHYLENE TRIDECANEDIAMIDE	24	2	0	0	2	0	2	0	0
DODECAMETHYLENE DODECAMETHYLENEDIURETHAN	24	2	0	0	2	2	2	0	0
OXYDIPENTAMETHYLENE 6-OXAHEPTADECANEDIAMIDE	24	2	0	0	2	2	2	0	0
DECAMETHYLENE (TETRAMETHYLENEDISULFONYL)DICAPROAMIDE	24	2	0	0	2	4	2	2	0
TETRAMETHYLENE (DECAMETHYLENEDISULFONYL)DICAPROAMIDE	24	2	0	0	2	4	2	2	0
DECAMETHYLENE TETRADECAMETHYLENEDIUREA	24	2	0	0	4	0	4	0	0
HEXAMETHYLENE OCTADECAMETHYLENEDIUREA	24	2	0	0	4	0	4	0	0
HEXAMETHYLENE 4,4'-(OXYDIMETHYLENE)DI-DIOXOLANECAPRYLAMIDE	24	2	0	4	6	5	2	0	0
P-PHENYLENEDIETHYLENE DOCOSANEDIAMIDE	24	2	1	0	2	0	2	0	0
P-PHENYLENEDIETHYLENE THIODIUNDECANOAMIDE	24	2	1	0	2	0	2	1	0
------25	25	0	0	0	0	0	0	0	0
(PENTAMETHYLENEDIOXY)DIDECAMETHYLENE OXIDE	25	0	0	0	0	3	0	0	0
DECAMETHYLENE (PENTAMETHYLENEDISULFONYL)DICAPROAMIDE	25	2	0	0	2	4	2	2	0
PENTAMETHYLENE (DECAMETHYLENEDISULFONYL)DICAPROAMIDE	25	2	0	0	2	4	2	2	0
B-D-GLUCOTRICAPRATE	25	3	0	8	14	5	0	0	0
------26	26	0	0	0	0	0	0	0	0
(HEXAMETHYLENEDIOXY)DIDECAMETHYLENE OXIDE	26	0	0	0	0	3	0	0	0
4,4'-DECAMETHYLENEDIPIPERAZINE SEBACAMIDE	26	2	0	0	0	0	4	0	0
DECAMETHYLENE OCTADECANEDIOATE	26	2	0	0	0	2	0	0	0
DECAMETHYLENE OCTADECANEDIAMIDE	26	2	0	0	2	0	2	0	0
HEXAMETHYLENE DOCOSANEDIAMIDE	26	2	0	0	2	0	2	0	0
OCTADECAMETHYLENE SEBACAMIDE	26	2	0	0	2	0	2	0	0
OCTAMETHYLENE EICOSANEDIAMIDE	26	2	0	0	2	0	2	0	0
HEXADECAMETHYLENE DECAMETHYLENEDIURETHAN	26	2	0	0	2	2	2	0	0
OXYDIPENTAMETHYLENE OXYDIPELARGONAMIDE	26	2	0	0	2	2	2	0	0
DECAMETHYLENE (HEXAMETHYLENEDISULFONYL)DICAPROAMIDE	26	2	0	0	2	4	2	2	0
HEXAMETHYLENE (DECAMETHYLENEDISULFONYL)DICAPROAMIDE	26	2	0	0	2	4	2	2	0
------28	28	0	0	0	0	0	0	0	0
DECAMETHYLENE EICOSANEDIAMIDE	28	2	0	0	2	0	2	0	0
DODECAMETHYLENE OCTADECANEDIAMIDE	28	2	0	0	2	0	2	0	0
OCTADECAMETHYLENE DODECANEDIAMIDE	28	2	0	0	2	0	2	0	0
OCTAMETHYLENE DOCOSANEDIAMIDE	28	2	0	0	2	0	2	0	0
DECAMETHYLENE OCTADECAMETHYLENEDIUREA	28	2	0	0	4	0	4	0	0
------30	30	0	0	0	0	0	0	0	0
DECAMETHYLENE DOCOSANEDIAMIDE	30	2	0	0	2	0	2	0	0
TETRADECAMETHYLENE OCTADECANEDIAMIDE	30	2	0	0	2	0	2	0	0
DECAMETHYLENE (DECAMETHYLENEDISULFONYL)DICAPROAMIDE	30	2	0	0	2	4	2	2	0
B-D-GLUCOTRILAURATE	31	3	0	8	14	5	0	0	0
DODECAMETHYLENE DOCOSANEDIAMIDE	32	2	0	0	2	0	2	0	0
OCTADECAMETHYLENE OCTADECANEDIAMIDE	34	2	0	0	2	0	2	0	0
TETRADECAMETHYLENE DOCOSANEDIAMIDE	34	2	0	0	2	0	2	0	0
B-D-GLUCOTRIMYRISTATE	37	3	0	8	14	5	0	0	0
OCTADECAMETHYLENE DOCOSANEDIAMIDE	38	2	0	0	2	0	2	0	0
B-D-GLUCOTRIPALMITATE	43	3	0	8	14	5	0	0	0

THE GLASS TRANSITION TEMPERATURES OF POLYMERS

W. A. Lee
Royal Aircraft Establishment
Farnborough, Hants, England

and

R. A. Rutherford
Rubber and Plastics Research Association of Great Britain
Shawbury, Shrewsbury, England

Contents

A. DEFINITIONS OF THE GLASS TEMPERATURE (T_g)

A wide range of polymer transition temperatures are reported in the literature. They appear under a variety of names, e.g., T_g (sometimes as glass-to-rubber transition temperature, or just glass temperature), dynamic T_g, mechanical T_g, glass-to-glass transition temperature (T_{gg}), low temperature TT, first-order TT, second-order TT, brittle temperature (T_b), melting point (T_m), polymer melt temperature (PMT), softening point (SP), flow temperature (T_f), freezing temperature, stiffness temperature, temperature of 10% tensile retraction, heat deflection and distortion temperatures (HDT), and α, β, γ etc. relaxations and dispersion temperatures. Unfortunately, some of these terms mean different things to different people. Thus one SP may lie very close to T_g and another to T_m while T_b may be a T_{gg} in one case yet analogous to T_g in another. All of these TT appear in the literature so an account of the meanings of the amorphous phase transitions and of factors governing them is relevant here.

Many properties are affected at the T_g so definitions in terms of the property changes are numerous; the principal definitions are usually expressed in thermodynamic terms or in terms of physical property changes.

The T_g of a polymer is a characteristic of the amorphous (non-crystalline) region; this region possesses the amorphous X-ray diffraction pattern of a liquid above and below T_g (44). At the T_g, a discontinuity occurs in the temperature (and pressure) derivatives of the energy, heat content, entropy, and volume, but these quantities are continuous functions of temperature (46); for this reason T_g has been referred to as a second-order transition though it only has some of the characteristics of such a transition. There is an abrupt increase in the coefficient of expansion and specific heat, but no absorption of latent heat, when a polymer is heated through the T_g region.

In terms of physical property changes, the T_g is often taken as the mid-point of the temperature interval over which the discontinuity takes place, but this is not necessarily so as some authors adopt other conventions.

The T_g is interpreted in terms of molecular behaviour as the temperature above which the polymer has acquired sufficient thermal energy for isomeric rotational motion, or for significant torsional oscillation to occur about most of the bonds in the main chain which are capable of such motion in the undegraded molecule. This definition, in referring to "most" of the bonds instead of "all" of the bonds in the main chain places so-called liquid-liquid transitions, of the type which occur in polystyrene (47) at $160^{\circ}C$, above T_g. Constraints to motion can arise from the barrier to rotation which in turn is raised by increasing the magnitude of cohesive forces, including hydrogen bonding, or from steric interactions. Below T_g, segmental motion takes place relatively infrequently and the majority of in-chain groups have fixed conformations. The T_g is usually unambiguous because considerable cooperation between neighbouring segments is required to effect rotational motion about a long sequence of in-chain bonds, the situation being analogous to the interaction of a set of engaging cog-wheels.

Unfortunately, these definitions, while serving to locate transitions, do not provide adequate criteria for T_g. For example, on a rising temperature scale, T_g could be the lowest or the highest amorphous transition, or the one which accompanied the biggest change in properties. It could further be specified that the transition should be of the glass to liquid type. None of these definitions find general acceptance and therefore when multiple transitions occur, but no clear T_g emerges, it is best to report all of the significant amorphous transitions.

The molecular mechanism of amorphous transitions (T_{gg} below T_g involves rotational motion but to a greatly limited degree. Short-range co-operative motions can occur along the main-chains of polymers, but the commonest form of motion is that involving side-group rotational isomerism.

The terms α, β, or γ, etc are often used to discriminate between multiple relaxations in polymers. The transition occurring at the highest temperature is usually denoted as α, the next β and so on. Frequently, therefore, α-relaxations are T_m , β-relaxations are T_g, and γ-relaxations are T_{gg} . With increasing usage of β for T_g a T_g may be referred to as a β relaxation even though a higher transition is not reported. T_g is, moreover, associated with the α relaxation in some works.

Other terms for transitions, notably SP , PMT , T_f, and HDT could be T_g or T_m ; freezing temperatures and stiffness temperatures are usually in the vicinity of T_g. Further test results are usually required to define these transitions in terms of their relevance to the amorphous or crystalline regions but in the absence of more specific information some such data is included in the tables which follow with a clear indication of its origin in such measurements.

B. DATA ASSESSMENT AND FACTORS AFFECTING T_g

Though the factors which govern T_g have been known for some years, there is still a wide spread in reported data on particular polymers. The distinction between polymers containing diluents, and/or having an unknown molecular weight distribution, and well-characterised polymers, with persistent (or limiting) properties (see section (6)) is too infrequently drawn. Polymer T_gs are frequently sensitive to parameters which are varied, so the provision of reliable data requires an analysis of the published values in terms of all the factors which affect T_g. The main factors are now outlined so that reliability may be judged independently; more detailed treatments are published in reviews (6, 48, 49).

(1)   Structure

Most polymer structures would be difficult to prove in detail and thus they are idealised and presumed from the characteristics of the reactants rather than proved by chemical and structural analysis. Structural uncertainties can arise both from the multiplicity of chemical reactions which may occur during polymerization processes, especially those taken to high conversions, and from the structure of the reactants themselves. For example, if a polymer possesses an asymmetric in-chain tetravalent atom, then the structure may be stereoregular to some degree in one or more of a variety of possible forms.

Alternatively, a polymer containing a residual double bond in the repeating unit could be in a cis, or trans, conformation. Polymers of 1,3-dienes can have various combinations of cis, trans, 1,2- or 1,4-structures. Variations in these structural features can have a large effect on T_g data. Despite these uncertainties, the data are often useful if they can be regarded as reliably pertaining to a polymer with reproducible properties and so data are reported here on many polymers of uncertain structure.

(2)   Crystallinity

The T_g of a semicrystalline polymer is a characteristic of the amorphous region and this has led some workers to suggest that the presence of crystallinity does not appreciably affect T_g. However, it seems possible that the steric constraint imposed by crystalline regions on neighbouring, otherwise-mobile polymer segments might result in an increase in T_g. It is found that with increasing crystallinity polymer T_gs may be little affected (50), may increase (51, 52) (at least for isothermal crystallisation (53)) or may decrease (51, 54). Wherever possible therefore the effect of variations in the degree of crystallinity on T_g should be measured and reported. T_g values selected in this work are the highest quoted on the sample with the lowest degree of crystallinity, other factors being equal.

(3)   <u>Dynamic Measurements and Rate Effects</u>

Rate effects are of three kinds. The first is associated with the rate of heat transfer to or from a sample, the second with the rate of attainment of equilibrium at the temperature imposed, and the third with the rate of response to an applied electrical or mechanical stress. These effects are considered in turn.

Fast heating or cooling rates, obviously may not permit the whole of a polymer sample to attain temperature equilibrium with its environment, particularly near and below the T_g where relaxation processes are very slow. In practice, the best criteria for a suitable heat transfer rate is that halving, or doubling it has no significant effect on the results. Reliable results are obtained using a rising temperature scale of about 1°C/h and for many polymers the same results, within 2°C, can be obtained using heating rates of 3°C/h. True equilibrium results are very difficult to attain near T_g and very slow, isothermal drifts in volume, heat content etc. occur over long time periods. In the interests of time-saving, it is nearly always necessary to extrapolate results in the vicinity of T_g but corrections for rate may be applied (55-57). To put T_g values on a comparable basis it is clear that many published T_gs need correcting for rate. In this study, where it is expected that rate effects might have caused a shift in T_g value, the heating rate is given specifically or as that rate to be associated with a particular method of measurement.

Methods of determining T_g which involve high rates of application of electrical or mechanical stress often provide T_g values 20° or more above those determined dilatometrically. Various approximations to correct for frequency effects are available (58-60), but the correct form of extrapolation to low frequencies is open to question and is certainly structure-dependent. Dynamic measurements made at frequencies near 1Hz often correspond to dilatometric measurements made at heating rates of 3°C/h. In assessing T_g values, priority has been given to data obtained at the slowest heating rates and lowest frequencies but no corrections for rate variation have been applied. In the data tables which follow, however, the use of a dynamic method is recorded when its use is reported.

(4)   <u>Static Methods of Measurement</u>

Clearly, no measurement can be conducted at zero rate but at least twenty relatively static, as opposed to dynamic, or high frequency, methods have been used in the determination of T_gs. Many of these are variants of the same principle, for example, the β-ray transmission method (61) is based on the change in dimensions of the sample on heating rather than on a change in β-ray transmission characteristics. Other techniques include dilatometry (10-12), dilatometry by differential pressure measurements (62), dilatometry under pressure (63, 64), buoyancy (13, 14), expansivity (15, 16), optical (65), x-ray diffraction (66-68), refractometry (27, 29), (laser) (69), calorimetry and differential scanning calorimetry (DSC) (57, 70-72), differential thermal analysis (DTA) (73-75), resistivity (76, 77), penetrometry (78, 79), T_b measurements (18, 80-82), SP measurements (83), infrared spectrometry (84, 85), radiothermoluminescence (86, 87), and o-positronium decay (88, 89). Recent methods claimed as relevant for T_g and/or TT measurement, include Rayleigh scattering (90), Landau-Placzek ratio (69, 91, 49), craze relaxation (37), chromatography (92, 93), current glow (94) and depolarisation current (95). Length/-temperature measurements (15) are less reproducible than volume methods, especially when the polymer is orientated, or when internal stresses are present. T_gs from refractive index and dilatometric measurements agree well. Mechanical techniques, such as penetrometer, T_b etc. measurements can give high or low results, but correlations with T_gs determined dilatometrically can be established (72, 83, 96). In the following tables, when it is known that mechanical techniques have been used for measuring T_g data recorded the fact is noted against the data.

A further disparity in published data, mentioned earlier arises from variations in the criteria used for selecting the T_g when, as usually occurs, results are obtained in the form of a curve showing an inflection region. In dynamic mechanical loss and dielectric loss measurements T_g is often assumed to coincide with the peaks of graphs showing the variation of these parameters with temperature; this temperature usually coincides with the mid-point of the region in a graph of modulus or dielectric constant against temperature where a significant fall in the magnitude of these quantities occurs. In the case of DSC and DTA data some authors prefer to take the inception of the inflection as T_g (97). Frequently, the criteria adopted are not stated.

(5)   <u>Diluents</u>

Much of the divergence in published T_g data is caused by the use of impure samples. Common impurities are unpolymerised monomer, low molecular weight polymer, solvents, and water. Great care should be taken to remove such impurities because their presence in small concentrations can lead to a shift (48, 98-103) in T_g of over 40°C and sometimes the occurrence of "diluent transitions" for example "water peaks". A full description in the original publication of the precautions taken to exclude diluents in the preparation of samples is necessary if the validity of published results is to be properly assessed. As these descriptions have usually been omitted there is some uncertainty attached to such reported data and they should strictly be regarded as only provisional.

(6)   <u>Molecular Weight</u>

The T_g of a regular homopolymer generally increases up to a limiting value, known as the limiting or persistent T_g value, with increase in molecular weight (48, 104, 105) but the reverse may hold for polymers with particular end-groups (106), or where crystallinity decreases with increasing molecular weight (107). For some polymers, T_gs are said to be independent of molecular weight (108).

Many data, especially on condensation polymers, relate to polymers of rather low molecular weight and it seems likely that higher T_g values would be obtained were higher molecular weight polymers available. Many polymers are not properly characterised with respect to molecular weight and the only measure of molecular weight is a viscosity value which itself can be very dependent on solvent/polymer interactions and to some extent temperature. In the absence of better criteria it has been assumed, other factors being equal, that the highest viscosities and the highest molecular weight polymers are associated with the most reliable data.

(7)   Thermal History

Ideally, polymer samples for T_g measurements should be amorphous, free from internal stress, and unoriented. T_g measurements should be made on samples which have been melted, rapidly quenched, and, if practicable, annealed for at least thirty minutes at some $20\,^{\circ}C$ above the T_g, but below the crystallisation temperature range. It is important that the T_g should be shown to be reproducible after repeated annealing and that the period between measurements at all temperatures, including room temperature should be recorded. In certain polyamides, heat treatment can cause the disappearance of a TT but the TT can reappear after storage for several days at RT (75).

(8)   Pressure

Increasing pressure increases T_g at about $0.025\,^{\circ}C$ per bar (48) and therefore the effect is negligible up to imposed pressures of about 40 bars.

The foregoing remarks emphasize the need for considerable caution in the determination of T_g and show why many literature values should be regarded as the most reliable values available rather than fully authenticated; extensive experimentation is required (6) to fully satisfy the criteria for T_g. In this study, the T_g values quoted have been selected using the criteria outlined in the previous discussion but it should be noted that these values are not necessarily comparable. Consideration must be given to all factors which govern T_g and suitable corrections applied before data become comparable. Details of all the data relevant to such an assessment, and covering all TT data will be published in the fuller study mentioned earlier (43). The references cited here indicate all the papers which have been considered in allocating T_g values but not all of the papers quote the value adopted.

C.   CLASSIFICATION AND NOMENCLATURE

A vast amount of data has been published in over 10,000 papers which form the subject of a special study (43) of all TTs to be published separately. This present Section of the Polymer Handbook represents a preliminary survey only of the T_g field.

The data in the tables which follow is restricted to undiluted linear homopolymers. This definition excludes nearly all polymers (44) and therefore requires qualification. The term "undiluted" here means that no extraneous materials of any kind have been added to the polymer and the term 'linear homopolymer' refers to a homopolymer prepared from difunctional reagent(s), which is expected from its mode of synthesis to have a predominantly regular, not unintentionally branched, repeating unit (44). Such a polymer is essentially represented by its chemical structural formula and/or the polymer name.

Polymers are subdivided into principle classes of increasing seniority substantially in accordance with the recommendations of the editors. Heterocyclics being most senior are found at the end of the table. All polymers are placed in the most senior class their structure commands and appear in only one class. Polymers containing heterocyclic rings in the main-chain are arranged in decreasing order of seniority of the most senior main-chain heterocyclic ring (109, 110). This results in the order represented in the table of contents. With the exception of the most common polymers with accepted trivial names, the polymers are named substantially according to the ACS recommendations for polymer nomenclature (110) in conjunction with IUPAC rules (109); less common polymers are cross-referenced from the trivial to the systematic name. Systematic names are not given for all the polymers in order to save space. Substitutive nomenclature is generally used for simple radicals, but for long combinations of radicals, replacement nomenclature has been used to provide a much shorter name (as for some fluoroacrylates with ether side chains). When sequences of radicals have repeated, the repeating sequence has been written once and prefixed "di", "tri" etc. as appropriate, for example di(oxyethylene) for the sequence $-O-CH_2-CH_2-O-CH_2-CH_2-$. Note that the diradical "di(oxyethylene)" must be distinguished from the diradical "dioxyethylene" which has the structure $-O-O-CH_2-CH_2-$, and also the diradical "ethylenedioxy" which has the structure $-O-CH_2-CH_2-O-$ (IUPAC rule C006.9). The principle underlying the last-named diradical has not generally been extended to the naming of polymers in this section, i.e. diradicals of structure $-X-Y-X-$ are not named YdiX, with the exception of alkanedioyl diradicals, because of the difficulty of locating indexed polymer names in which the diradicals are not named from left to right. Many polymers are derivatives of the diradical "propylene" $-CH(CH_3)-CH_2-$; the substituted diradical has been employed in naming polymers instead of "1-methylethylene" which could be preferred.

Polymer names are tabulated in alphabetical order within each subsection, but

1.   multiplying prefixes and prefixes like sec-, tert-, including designated atoms and the numbers showing locations of substituents are ignored except as secondary and tertiary indicators of order thus "8-ferrocenyl-3-methyl" comes before "1,2-dimethyl-1-butenylene",

2.   monosubstituents are placed before polysubstituents, thus "methyl" comes before "dimethyl", but "tetramethyl" is alphabetically placed before "trimethyl",

3.   the locations of substituents in otherwise identical polymers are taken as tertiary indicators of order: the numbers are arranged in increasing order at the first point of difference. Thus, 2,3,8- comes before 2,4,1-.

The authors gratefully acknowledge the assistance of Miss S. A. Watts.

D. TABLE OF GLASS TRANSITION TEMPERATURES OF POLYMERS

Polymer	$T_g(K)$	Remarks	References
1. Main-chain Acyclic Carbon Polymers			
1.1 Poly(dienes)			
Natural rubber see Poly(isoprene) cis			
Neoprene see Poly(1-chloro-1-butenylene)			
Poly(butadiene)s see Poly(1-butenylene)s & poly(vinylethylene)			
Poly(1-butenylene) cis	171		26, 61, 574-597
trans	215	Wide spread in published data	580, 583, 585-587, 589 594, 595, 598, 599
Poly(1-butyl-1-butenylene)	192	Dynamic method	600
Poly(1-tert-butyl-1-butenylene)	293		80, 348, 601, 602
Poly(2-chloro-1,3-butadiene)			
see Poly(1-chloro-1-butenylene)			
Poly(1-chloro-1-butenylene) cis	253		603
trans	233		1, 24, 604-611
Poly(2-chloro-1,4,4-trifluoro-1-butenylene)	256	No experimental details	587
Poly(1-decyl-1-butenylene)	220		1
Poly(1-ethyl-1-butenylene)	197		600, 612
Poly(1,4,4-trifluoro-1-butenylene)	238	Slightly crystalline	587
Poly(octafluoro-4-methyl-1-butenylene)	270	DSC heating rate	97
Poly(1-heptyl-1-butenylene)	190		80, 601
Poly(isoprene) cis	200		349, 388, 394, 464, 469, 574, 591, 594, 595, 613-623
trans	215		514, 581, 589, 594, 608, 617, 624
Poly(1-isopropyl-1-butenylene)	221		600, 601, 612
Poly(4-methoxy-1-butenylene)	256		591
Poly(4-methoxycarbonyl-3-methyl-1-butenylene)	326	Dynamic method	625
Poly(1,2-dimethyl-1-butenylene)	262		576
Poly(methyl sorbate) see Poly(4-methoxycarbonyl-3-methyl-1-butenylene)			
Poly(1-pentenylene) cis	159	Heating rate 16 K/min	626
trans	183	DTA heating rate	594
Poly(1-phenyl-1-butenylene)	~283	Low molecular weight	584
Poly(1-propyl-1-butenylene)	196		600
Poly(vinylethylene) atactic	269	Published values, range	585, 586, 592, 597
syndiotactic	245	from 245 to 283 K	627, 628
1.2 Poly(alkenes)			
Poly(butene-1) see Poly(ethylethylene)			
Poly(butylethylene)	223		1, 272, 574, 629-633
Poly(tert-butylethylene)	337	Softening point, highly crystalline sample	631
Poly(cyclohexylethylene) atactic	393		634-639
isotactic	406	Dynamic method	640
Poly(2-cyclohexylethylethylene)	313	Mechanical method	634
Poly[(cyclohexylmethyl)ethylene]	348	Mechanical method	634
Poly(3-cyclohexylpropylethylene)	248	Softening point, comparative data reported	631
Poly(cyclopentylethylene)	348	Dynamic method	634
Poly[(cyclopentylmethyl)ethylene]	333	Mechanical method	634
Poly(decylethylene)	237		629, 632, 641
Poly(dodecylethylene)	241		629, 632, 641
Poly(ethylene)	148	Conflicting interpretations of data; branch point transition at 252 K	1, 6, 16, 28, 61, 80, 191, 223, 238, 261, 262, 272, 317, 318, 344, 349, 395, 396, 469, 521-523, 574, 589, 592, 608, 629, 641-676

Polymer	T_g(K)	Remarks	References
1.2 <u>Poly(alkenes)</u> (Cont' d.)			
Poly(ethylethylene)	249	Wide spread in reported values	1, 61, 67, 273, 349, 395, 397, 574, 629, 632, 641, 645, 646, 677-680
Poly(1-ethyl-1-methyltetramethylene)	∼250	DTA heating rate	681
Poly(ethyl-2-propylene)	268	Dynamic method	682
Poly(heptylethylene)	226	Dynamic method	629
Poly(hexadecylethylene)	328	Dynamic method, stereo-regular sample; may be first-order transition	629, 641, 683
Poly(hexylethylene)	In range 208 to 228	Conflicting data	1, 574, 629, 632
Poly(isobutylene) see Poly(1,1-dimethylethylene)			
Poly(isobutylethylene)	302	Reported values range from 297-333 K	1, 51, 675 684-692
Poly(isohexylethylene)	239	Softening point	631
Poly(isopentylethylene)	259	Softening point	631
Poly(isopropylethylene) atactic	323	Dynamic method	634, 641, 685, 691, 692,
isotactic	367	Dilatometry; suggested transition crystal/crystal type	693
Poly(3,3-dimethylbutylethylene)	326	Softening point	631
Poly(methylene)	155		215, 238, 262, 398, 574, 694
Poly(1,1-dimethylethylene)	200		1, 23, 24, 58, 61, 216, 223, 695-704
Poly(1-methyloctamethylene)	215	Heating rate 4-8K/min	705
Poly(1,1-dimethyltetramethylene)	253	Dynamic method	706
Poly(1,1-dimethyltrimethylene)	263	Dynamic method	706-709
Poly(1,1,2-trimethyltrimethylene)	310		682
Poly(4-methylpentene-1) see Poly(isobutylethylene)			
Poly(4,4-dimethylpentene-1)			
see Poly(neopentylethylene)			
Poly(4,4-dimethylpentylethylene)	313	Softening point, crystalline sample	631
Poly(neopentylethylene)	332	Softening point, crystalline sample	631
Poly(nonylethylene)	236		632
Poly(octylethylene)	232		629, 632, 641
Poly(pentylethylene)	242	Dynamic method	629
Poly(propylene) atactic (a)	∼260	Conflicting data; most values reported range (a) 258 to 270 (b) 238 to 260 (c) 263 to 267 K	1, 6, 15, 61, 80, 122, 191, 223, 261, 272, 273, 282, 318, 326, 394-397, 469, 574, 615, 629, 632, 641, 645, 646, 659, 666, 668, 675, 677, 678, 684, 710-726
isotactic (b)	∼265		
syndiotactic (c)	∼265		
Poly(propylethylene)	∼233	Conflicting data	1, 272, 574, 595, 629, 630, 632, 645, 685, 727
Poly(propyl-2-propylene)	300	Dynamic method	682
Poly(tetradecylethylene)	246		629, 632, 641
1.3 <u>Poly(acrylics) and Poly(methacrylics)</u>			
1.3.1 <u>Poly(acrylic acid) and Poly(acrylic acid esters)</u>			
Poly(acrylic acid)	379		720, 811-818
Poly(benzyl acrylate)	279		746
Poly(4-biphenylyl acrylate)	∼383		819
Poly(4-butoxycarbonylphenyl acrylate)	286		746
Poly(butyl acrylate)	219	Mechanical method	1, 23, 634, 775, 820-822

Polymer	$T_g(K)$	Remarks	References

1.3.1   Poly(acrylic acid) and Poly(acrylic acid esters)   (Cont'd.)

Polymer	$T_g(K)$	Remarks	References
Poly(sec-butyl acrylate) conventional	251		823,824
syndiotactic	253		
isotactic	250		
Poly(tert-butyl acrylate)	380,316,346	Conflicting data	746,824,825
Poly(2-tert-butylphenyl acrylate)	345		826
Poly(4-tert-butylphenyl acrylate)	344		826
Poly(cesium acrylate)	447	Extrapolated from DSC data on water plasticised samples	817
Poly[3-chloro-2,2-bis(chloromethyl)propyl acrylate]	319		746
Poly(2-chlorophenyl acrylate)	326		746
Poly(4-chlorophenyl acrylate)	331		826
Poly(2,4-dichlorophenyl acrylate)	333		746
Poly(pentachlorophenyl acrylate)	420		746
Poly(4-cyanobenzyl acrylate)	317		746
Poly(4-cyanobutyl acrylate)	233-238	No experimental details	827
Poly(2-cyanoethyl acrylate)	277		746,820
Poly(cyanomethyl acrylate)	296	No experimental details	820
Poly(5-cyano-3-oxapentyl acrylate)	250	No measurement details	820
Poly(4-cyanophenyl acrylate)	363		746
Poly(4-cyano-3-thiabutyl acrylate)	249		828
Poly(6-cyano-3-thiahexyl acrylate)	215		828
Poly(6-cyano-4-thiahexyl acrylate)	215		828
Poly(8-cyano-7-thiaoctyl acrylate)	214		828
Poly(5-cyano-3-thiapentyl acrylate)	223		828
Poly(cyclohexyl acrylate) conventional	292		824
syndiotactic	289		
isotactic	285		
Poly(dodecyl acrylate)	270	Brittle point	821,829
Poly(2-ethoxycarbonylphenyl acrylate)	303		746
Poly(3-ethoxycarbonylphenyl acrylate)	297		746
Poly(4-ethoxycarbonylphenyl acrylate)	310		746
Poly(2-ethoxyethyl acrylate)	223		830
Poly(3-ethoxypropyl acrylate)	218		830,831
Poly(ethyl acrylate) conventional	249		23,634,775
syndiotactic	249		820,821,824,
isotactic	248		832,833
Poly(2-ethylbutyl acrylate)	223	Brittle point	823
Poly(2-ethylhexyl acrylate)	223	Brittle point	821
Poly(ferrocenylethyl acrylate)	430	No experimental details	834
Poly(ferrocenylmethyl acrylate)	470-483	DSC heating rate	835
Poly(1H,1H-heptafluorobutyl acrylate)	243		155,836,837
Poly(1H,1H,3H-hexafluorobutyl acrylate)	251		836
Poly(2,2,2-trifluoroethyl acrylate)	263		836
Poly[2,2-difluoro-2-(2-heptafluorotetrahydrofuranyl)ethyl acrylate]	275	Brittle temperature	830
Poly(1H,1H-undecafluorohexyl acrylate)	234		836
Poly(fluoromethyl acrylate)	288	Estimated T_g	838
Poly(1H,1H-pentadecafluorooctyl acrylate)	256	Crystalline	836
Poly(5,5,6,6,7,7,7-heptafluoro-3-oxaheptyl acrylate)	228		830
Poly(1H,1H-undecafluoro-4-oxaheptyl acrylate)	205		830
Poly(1H,1H-nonafluoro-4-oxahexyl acrylate)	224		830
Poly(7,7,8,8-tetrafluoro-3,6-dioxaoctyl acrylate)	233		830
Poly(1H,1H-tridecafluoro-4-oxaoctyl acrylate)	205		830
Poly(2,2,3,3,5,5,5-heptafluoro-4-oxapentyl acrylate)	218		830,837
Poly(4,4,5,5-tetrafluoro-3-oxapentyl acrylate)	251		830
Poly(5,5,5-trifluoro-3-oxapentyl acrylate)	235		830
Poly(1H,1H-nonafluoropentyl acrylate)	236		836
Poly(1H,1H,5H-octafluoropentyl acrylate)	238		836
Poly(heptafluoro-2-propyl acrylate)	278-283	No details on sample or measurement	839
Poly(1H,1H-pentafluoropropyl acrylate)	247		836
Poly(heptyl acrylate)	213	Brittle point	821
Poly(2-heptyl acrylate)	235	Brittle point	823

	Polymer		T_g(K)	Remarks	References
1.3.1	Poly(acrylic acid) and Poly(acrylic acid esters) (Cont' d.)				
	Poly(hexadecyl acrylate)		308	Brittle point	23, 821, 840, 841
	Poly(hexyl acrylate)		216	Brittle point	823
	Poly(isobornyl acrylate)	conventional	367		824
		syndiotactic	369		
		isotactic	363		
	Poly(isobutyl acrylate)		249	Brittle point	823
	Poly(isopropyl acrylate)	conventional	267-270		746, 823, 824
		syndiotactic	271-284		
		isotactic	262		
	Poly(1,2:3,4-di-O-isopropylidene-α-D-galactopyranos-6-O-yl acrylate)		371		11, 842
	Poly(magnesium acrylate)		673	Estimated from copolymer data	843
	Poly(3-methoxybutyl acrylate)		217		844
	Poly(2-methoxycarbonylphenyl acrylate)		319		746
	Poly(3-methoxycarbonylphenyl acrylate)		311		746
	Poly(4-methoxycarbonylphenyl acrylate)		340		746
	Poly(2-methoxyethyl acrylate)		223		830
	Poly(4-methoxyphenyl acrylate)		324		826
	Poly(3-methoxypropyl acrylate)		198		830
	Poly(methyl acrylate)	conventional	283		18, 22, 23, 81
		head to tail	278		576, 720, 775-777
		head to head	304		821, 824, 831, 841, 845-848
	Poly(3,5-dimethyladamantyl acrylate)		379	DSC heating rate; data corrected (sic)	849, 850
	Poly(3-dimethylaminophenyl acrylate)		320		746
	Poly(2-methylbutyl acrylate)		241	Brittle point	823
	Poly(3-methylbutyl acrylate)		228	Brittle point	823
	Poly(1,3-dimethylbutyl acrylate)		258	Brittle point	823
	Poly(2-methyl-7-ethyl-4-undecyl acrylate)		253	Brittle point	823
	Poly(2-methylpentyl acrylate)		235	Brittle point	823
	Poly(2-naphthyl acrylate)		358		826
	Poly(neopentyl acrylate)		295		746
	Poly(nonyl acrylate)		215	Brittle point	821
	Poly(octyl acrylate)		208	Brittle point	821, 841
	Poly(2-octyl acrylate)		228	Brittle point	823
	Poly(3-pentyl acrylate)		267		746, 823
	Poly(phenethyl acrylate)		270		746
	Poly(phenyl acrylate)		330		746, 826
	Poly(potassium acrylate)		467	Extrapolated from data on water plasticised samples	817
	Poly(propyl acrylate)		236	Brittle point	23, 592, 832, 836, 851, 852
	Poly(sodium acrylate)		503	Estimated from copolymer data	817, 843
	Poly(tetradecyl acrylate)		297	Brittle point; probably 1st order transition	23, 821
	Poly(3-thiabutyl acrylate)		213		831
	Poly(4-thiahexyl acrylate)		197		831
	Poly(5-thiahexyl acrylate)		203		831
	Poly(3-thiapentyl acrylate)		202		831
	Poly(4-thiapentyl acrylate)		208		831
	Poly(m-tolyl acrylate)		298		826
	Poly(o-tolyl acrylate)		325		826
	Poly(p-tolyl acrylate)		316		826
1.3.2	Poly(acrylamides)				
	Poly(acrylamide)		438	No experimental details	820
	Poly(N-butylacrylamide)		319	Mechanical method	853
	Poly(N-sec-butylacrylamide)		390	No experimental details	820
	Poly(N-tert-butylacrylamide)		401	No experimental details	820
	Poly(N,N-dibutylacrylamide)		333	Softening point, amorphous	854
	Poly(N-dodecylacrylamide)		198-320	Conflicting data, 198K more probable in "amorphous sample"	820, 853

Polymer		T_g(K)	Remarks	References
1.3.2 Poly(acrylamides) (Cont'd.)				
Poly(isodecylacrylamide)		313	No experimental details	820
Poly(isohexylacrylamide)		344	No experimental details	820
Poly(isononylacrylamide)		325	No experimental details	820
Poly(isooctylacrylamide)		339	No experimental details	820
Poly(N-isopropylacrylamide)		358,403	Conflicting data	820,855
Poly(N,N-diisopropylacrylamide)		~393	Softening point, almost amorphous	854
Poly(N,N-dimethylacrylamide)		362		746
Poly[N-(1-methylbutyl)acrylamide]		380	No experimental details	820
Poly(N-methyl-N-phenylacrylamide)		~453	Softening point, amorphous	854
Poly(morpholylacrylamide) [Poly(morpholinocarbonylethylene)]		420		9,856
Poly(N-octadecylacrylamide)		162	Mechanical method	853
Poly(N-octylacrylamide)		220		853
Poly(piperidylacrylamide)		381		9,856
1.3.3 Poly(methacrylic acid) and Poly(methacrylic acid esters)				
Poly(adamantyl methacrylate)		414	DSC heating rate	849
Poly(benzyl methacrylate)		327		746
Poly(2-bromoethyl methacrylate)		325		746
Poly(2-N-tert-butylaminoethyl methacrylate)		306		746,857
Poly(butyl methacrylate)	atactic	293	Range from ~286 to 308K	22,69,272,695,720, 775,822 824,847, 858-862
	isotactic	249		824
Poly(sec-butyl methacrylate)		333		746,863
Poly(tert-butyl methacrylate)	atactic	391	Maximum value	746,847,864
	syndiotactic	387	Maximum value	
	isotactic	280		
Poly(2-chloroethyl methacrylate)		~365	Likely to be slightly high	262
Poly(2-cyanoethyl methacrylate)		364		746,865
Poly(4-cyanomethylphenyl methacrylate)		~401		746
Poly(4-cyanophenyl methacrylate)		428		746
Poly(cyclohexyl methacrylate)	atactic	356		263,353,820,824,
	isotactic	324		847,862
Poly(decyl methacrylate)		203		23,821,840,846
Poly(dodecyl methacrylate)		208	Conflicting data	1,821,858,866,867
Poly(diethylaminoethyl methacrylate)		~289-297		857
Poly(2-ethylhexyl methacrylate)		263	Brittle point	821
Poly(ethyl methacrylate)	atactic	338	Data covers range 320-343K	22,69,286,352,353, 521,695,821,824, 857,858,860,862, 868-874
	isotactic	285		824
	syndiotactic	339		847
Poly(2-ethylsulfinylethyl methacrylate)		298		746
Poly(ferrocenylethyl methacrylate)		482	No experimental details	834
Poly(ferrocenylmethyl methacrylate)		~458-468	DSC heating rate	835
Poly(1H,1H-heptafluorobutyl methacrylate) syndiotactic		~330		875
Poly(1H,1H,7H-dodecafluoroheptyl methacrylate)		286	Mechanical method	876
Poly(1H,1H,9H-hexadecafluorononyl methacrylate)		258	Mechanical method	876
Poly(1H,1H,5H-octafluoropentyl methacrylate)		309	Mechanical method	876
Poly(1,1,1-trifluoro-2-propyl methacrylate)		354	Vicat softening temperature	863
Poly(hexadecyl methacrylate)		288	Brittle point, sample probably crystalline - may be T_m	821,866
Poly(hexyl methacrylate)		268		695,846,858,877
Poly(2-hydroxyethyl methacrylate)		328,359	Conflicting data	746,878-880
Poly(2-hydroxypropyl methacrylate)		349		846,878
Poly(isobornyl methacrylate)		383		824
Poly(isobutyl methacrylate)	random	326		746,821,824
	isotactic	281		846,881
	80% syndiotactic, 20% isotactic	326		

Polymer		T_g(K)	Remarks	References
1.3.3 Poly(methacrylic acid) and Poly(methacrylic acid esters) (Cont'd.)				
Poly(isopropyl methacrylate)	atactic	354		746,824,862
	isotactic	300		
	syndiotactic	358		
Poly(1,2:3,4-di-O-isopropylidene-α-D-galactopyranos-6-O-yl methacrylate)		399		11,842
Poly(2,3-O-isopropylidene-DL-glyceritol-1-O-yl methacrylate)		335	Heating rate 20K/min	842
Poly(magnesium methacrylate)		~763	Extrapolated value	843
Poly(methacrylic acid)		501	Extrapolated data from plasticised samples	882
Poly(methacrylic anhydride) see Section 4.27				
Poly(4-methoxycarbonylphenyl methacrylate)		379		746
Poly(3,5-dimethyladamantyl methacrylate)		469	Temperature reported as "corrected"	849,850
Poly(dimethylaminoethyl methacrylate)		292		746
Poly(3,3-dimethylbutyl methacrylate)		~318	Vicat softening point 332K	863
Poly(3,3-dimethyl-2-butyl methacrylate)		~381	Vicat softening point 396K	863
Poly(3,5,5-trimethylhexyl methacrylate)		274		867
Poly(methyl methacrylate)	atactic	378		1,17,22,25-27,69,78, 79,81,190,201,263, 286,287,317,318,352- 354,400,614,684,698, 720,775,777-779,789, 804,821,824,846,858, 860,862,880,883-895
	isotactic	311		6,122,720,824,847,884. 886,890,895-901
	syndiotactic	378		122,720,824,847,890, 895,896,898,900,901
Poly(trimethylsilyl methacrylate)	isotactic	341	Heating rate 15K/min, weak T_g for syndiotactic polymer	902
	syndiotactic	400		
Poly[(2-nitratoethyl) methacrylate]		328		903
Poly(octadecyl methacrylate)		173		1,720,904
Poly(octyl methacrylate)		203,253	Conflicting data	23,695,821,840, 846,858
Poly(3-oxabutyl methacrylate)		289		846,857
Poly(3-oxa-5-hydroxypentyl methacrylate)		278-280	Mechanical method	880
Poly(pentyl methacrylate)		268	Brittle point	821
Poly(phenethyl methacrylate)		299		746
Poly(phenyl methacrylate)		383		353,746,820,847, 863,875
Poly(propyl methacrylate)		308	Conflicting data, 308-345K reported	22,262,272,821,847, 857,858,860,862, 877,878,880
Poly(sodium methacrylate)		~583	Extrapolated value	843
Poly(tetradecyl methacrylate)		201-264	Conflicting data	23,821,840,866
1.3.4 Poly(methacrylamides)				
Poly(4-butoxycarbonylphenylmethacrylamide)		401	Softening point	905
Poly(N-tert-butylmethacrylamide)		433	No experimental details	820
Poly(4-carboxyphenylmethacrylamide)		473	Softening point	905
Poly(4-ethoxycarbonylphenylmethacrylamide)		441	Softening point	905
Poly(4-methoxycarbonylphenylmethacrylamide)		453	Softening point	905
1.3.5 Other α- and β-substituted Poly(acrylics) and Poly(methacrylics)				
Poly(butyl butoxycarbonylmethacrylate)		298		906
Poly(butyl chloroacrylate)		330	Vicat softening point	863
Poly(sec-butyl chloroacrylate)		347	Vicat softening point	863
Poly(butyl cyanoacrylate)		358		907
Poly(dibutyl itaconate) see Poly(butyl butoxycarbonylmethacrylate)				
Poly(cyclohexyl chloroacrylate)		387	Vicat softening point	863

Polymer		T_g(K)	Remarks	References
1.3.5 **Other α- and β-substituted Poly(acrylics) and Poly(methacrylics)** (Cont'd.)				
Poly(ethyl chloroacrylate)		366	Vicat softening point	832,863,908,909
	100% isotactic	308	Calculated for infinite Mn; heating rate 20K/min	
	100% syndiotactic	404	Calculated for infinite Mn; heating rate 20K/min	
Poly(ethyl ethoxycarbonylmethacrylate)		325	Intrinsic viscosity only 0.24 dl/g	906
Poly(ethyl ethacrylate)		300		746
Poly(ethyl fluoromethacrylate)		316		910
Poly(hexyl hexyloxycarbonylmethacrylate)		269	Intrinsic viscosity only 0.24 dl/g	906
Poly(iosbutyl chloroacrylate)		363		832,911
Poly(isopropyl chloroacrylate)		363	Vicat softening point	832,863
Poly(methyl chloroacrylate)		413	Vicat softening point	863
Poly(methyl β-chloroacrylate)		416	No measurement details	820
Poly(methyl fluoroacrylate)		404	No details on samples or measurement	820,912
Poly(methyl fluoromethacrylate)		357	No experimental details	820,910
Poly(methyl phenylacrylate)	atactic	391		913
	isotactic	397		
Poly(propyl chloroacrylate)		344	Vicat softening point	832,863
1.4 **Poly(vinyl ethers) and Poly(vinyl thioethers)**				
Poly(butoxyethylene)		218		1,832,866,914-918
Poly(sec-butoxyethylene)		253		866,914
Poly(tert-butoxyethylene)		361		1,918
Poly(butylthioethylene)		253		866
Poly(butyl vinyl ether) see Poly(butoxyethylene)				
Poly(cyclohexyloxyethylene)		354	Softening point	918
Poly(decyloxyethylene)		~183,211	Conflicting data, independent DSC data supports higher value	866,914,916,917
Poly(ethoxyethylene)		230		1,223,521,573,832, 866,914-920
Poly[(2-ethylhexyloxy)ethylene]		207		866,915
Poly(ethylthioethylene)		266	Viscosity only 0.2 dl/g	866
Poly(dodecafluorobutoxyethylene)		263-273	T_g estimated from copolymer data	457
Poly(2,2,2-trifluoroethoxytrifluoroethylene)		308	DTA heating rate	457,921
Poly[1,1-bis(trifluoromethoxy)difluoroethylene]		213	Poorly defined DTA endotherms	457
Poly(1,1-difluoro-2-trifluoromethoxyethylene)		263-273	Estimated from copolymer data	457
Poly(1,2-difluoro-1-trifluoromethoxyethylene)		263-273	Estimated from copolymer data	457
Poly(hexafluoromethoxyethylene)		268	DTA heating rate	457
Poly[heptafluoro-2-propoxy)ethylene]		~328-338		839
Poly(hexyloxyethylene)		199		866,914-918
Poly(isobutoxyethylene)		254		1,223,573,832,866, 914,916-918,920, 922
Poly(isopropenyl methyl ether) see Poly(2-methoxypropylene)				
Poly(isopropoxyethylene)		270		832,866,914,918
Poly(methoxyethylene)		242		1,223,805,866, 914-918, 920
Poly(2-methoxypropylene)		340		866
Poly(2,2-dimethylbutoxyethylene)		282	Mechanical method	918
Poly(methylthioethylene)		272		866
Poly(neopentyloxyethylene)		~424	Softening point	918
Poly(octyloxyethylene)		194		866,914-917
Poly(pentyloxyethylene)		207		866,915
Poly(propoxyethylene)		224		832,915
Poly(vinyl methyl ether) see Poly(methoxyethylene)				

Polymer	T_g(K)	Remarks	References

1.5   Poly(vinyl alcohol) and Poly(vinyl ketones)

Poly(1-acetyl-1-fluoroethylene)	415		923
Poly(benzoylethylene)	314,347	Conflicting data	924,925
Poly(4-bromo-3-methoxybenzoylethylene)	317	Mechanical method	924
Poly(4-tert-butylbenzoylethylene)	377		925
Poly(4-chlorobenzoylethylene)	310,362	Conflicting data	924,925
Poly(4-ethylbenzoylethylene)	325		924,925
Poly(4-isopropylbenzoylethylene)	336		925
Poly(4-methoxybenzoylethylene)	319	Mechanical method	924
Poly(3,4-dimethylbenzoylethylene)	315	Mechanical method	924
Poly(phenyl vinyl ketone) see Poly(benzoylethylene)			
Poly(4-propylbenzoylethylene)	317		925
Poly(p-toluoylethylene)	344		924,925
Poly(vinyl alcohol) [Poly(hydroxyethylene)]	358	Dynamic method	1,191,202,216,230, 261,573,926,927

1.6   Poly(vinyl halides) and Poly(vinyl nitriles)

Poly(1-acetoxy-1-cyanoethylene)	420	Heating rate 8K/min	255
Poly(acrylonitrile) [poly(cyanoethylene)]	370		1,7,26,80,191,202, 393,400,521,525, 576,752,772,779, 787,928-941
Poly(1,1-dichloroethylene)	255		1,6,62,80,216,521, 524,648,942-944
Poly(chlorotrifluoroethylene)	~325	By 'static' methods	66,80,191,261,349,355,
	373	By mechanical methods even at low frequencies	396,401,521,647,893, 945-961
Poly(1,1-dichloro-2-fluoroethylene)	~320	[η] only 0.035 dl/g	962
Poly(1,2-dichloro-1,2-difluoroethylene)	350	Heating rate 5-20K/min, high pressure, radiation synthesis	97
Poly(1,1-difluoroethylene)	~233		401,963-970
Poly(1,2-difluoroethylene)	371	DTA heating rate	114,457,971
Poly(tetrafluoroethylene) [poly(difluoromethylene)]	160,400	Much data, some conflicting	1,6,66,190,191,253,261, 355,356,396,521,522, 641,651,659,891,893, 914,921,952,972-1005
Poly(trifluoroethylene)	304	Quoted value, DTA?	457
Poly[(pentafluoroethyl)ethylene]	314	Heating rate 5-20 K/min, high pressure, radiation synthesis	97
Poly(tetradecafluoropentylethylene)	503?	Heating rate 5-20K/min, high pressure, radiation synthesis, value uncertain	97
Poly(hexafluoropropylene)	425	Heating rate 5-20K/min	97,984,1006
Poly(2,3,3,3-tetrafluoropropylene)	315	Heating rate 5-20K/min, high pressure, radiation synthesis	97
Poly(3,3,3-trifluoropropylene)	300		97,914,1007,1008
Poly[(heptafluoropropyl)ethylene]	331	Heating rate 5-20K/min, high pressure, radiation synthesis	97
Poly(2-iodoethylethylene)	343	Dynamic method	1009
Poly(9-iodononylethylene)	267	Dynamic method	1009
Poly(3-iodopropylethylene)	~303	Dynamic method	1009
Poly(methacrylonitrile)	393		1
Poly(vinyl chloride)	354	Increasing syndiotactic content increases T_g to 371K	1,21,52,78,80,104,188, 191,230,286,349,353, 354,356,469,608,696, 720,772,777,787,798, 861,926,942,944,972, 1010-1031
Poly(vinyl fluoride)	314	Mechanical method	521,1032,1033
Poly(vinylidene chloride) see Poly(1,1-dichloroethylene)			

Polymer	T_g(K)	Remarks	References

1.6 Poly(vinyl halides) and Poly(vinyl nitriles)   (Cont' d.)

Poly(vinylidene fluoride)   see   Poly(1,1-difluoroethylene)

1.7 Poly(vinyl esters)

Polymer	T_g(K)	Remarks	References
Poly[(2-acetoxybenzoyloxy)ethylene]	333		29,925
Poly(4-acetoxybenzoyloxyethylene)	∼349		29,925
Poly(acetoxyethylene) (Poly(vinyl acetate))	305		1,9,12,19,22,62,187,216, 223,230,286,352,354,399, 400,573,593,695,775,777, 792,816,840,845,888,926, 928,932,1016,1034-1044
Poly [(1-acetylindazol-3-ylcarbonyloxy)ethylene]	423	Heating rate 20K/min - "onset" value	1045
Poly(4-benzoylbutyryloxyethylene)	318	DTA heating rate	1046
Poly(benzoyloxyethylene)	344		29,354,755,820,925, 1047,1048
Poly(3-bromobenzoyloxyethylene)	331		29,925
Poly(4-bromobenzoyloxyethylene)	365		29,925,1047,1048
Poly[(tert-butoxycarbonylamino)ethylene]	393	DTA heating rate	1049
Poly(4-tert-butylbenzoyloxyethylene)	374		29,925,1050
Poly(4-butyryloxybenzoyloxyethylene)	334		29,925
Poly[(2-chlorobenzoyloxy)ethylene]	335		29,925,1048
Poly(3-chlorobenzoyloxyethylene)	338		29,925,1048
Poly(4-chlorobenzoyloxyethylene)	357		29,925,1047,1048
Poly(cyclohexanoyloxyethylene)	349	Mechanical method, heating rate 30K/h	1043
Poly(cyclohexylacetoxyethylene)	298	Mechanical method, heating rate 30K/h	1043
Poly(4-cyclohexylbutyryloxyethylene)	∼263	Mechanical method, heating rate 30K/h	1043
Poly(cyclopentanoyloxyethylene)	309	Mechanical method, heating rate 30K/h	1043
Poly(cyclopentylacetoxyethylene)	270	Mechanical method, heating rate 30K/h	1043
Poly(4-ethoxybenzoyloxyethylene)	343		29,925
Poly(4-ethylbenzoyloxyethylene)	326		29,925
Poly[(2-ethyl-2,3,3-trimethylbutyryloxy)ethylene]	388	Mechanical method	1043
Poly[(trifluoroacetoxy)ethylene]	∼319,>348	Conflicting data	1051,1052
Poly[(heptafluorobutyryloxy)ethylene]	300		1051
Poly[(undecafluorocyclohexylcarbonyloxy)ethylene]	327		1051
Poly[(nonadecafluorodecanoyloxy)ethylene]	253-255		1051
Poly[(undecafluorohexanoyloxy)ethylene]	264	Plasticiser may be present	1051
Poly[(pentadecafluorooctanoyloxy)ethylene]	258-263		1051
Poly[(pentafluoropropionyloxy)ethylene]	315		1051
Poly[(nonafluorovaleryloxy)ethylene]	288-293		1051
Poly(formyloxyethylene)	310	∼60% syndiotactic diads	593
	306	50% syndiotactic diads	
Poly(isonicotinoyloxyethylene)	372	From polyvinyl alcohol	29,288,1053
Poly(4-isopropylbenzoyloxyethylene)	342		29,925
Poly[(2-isopropyl-2,3-dimethylbutyryloxy)ethylene]	393	Mechanical method	1043
Poly[(2-methoxybenzoyloxy)ethylene]	338		29,925
Poly[(3-methoxybenzoyloxy)ethylene]	∼317		29,925
Poly[(4-methoxybenzoyloxy)ethylene]	360		29,757,925,1047,1048
Poly[(2-methylbenzoyloxy)ethylene]	321		29,820,925,1048
Poly[(3-methylbenzoyloxy)ethylene]	324		29,925,1048
Poly[(4-methylbenzoyloxy)ethylene]	343		29,925,1047,1048
Poly[(1-methylcyclohexanoyloxy)ethylene]	359	Mechanical method, heating rate 30K/h	1043
Poly(3,3-dimethyl-3-phenylpropionyloxyethylene)	293	Mechanical method, heating rate 30K/h	1043
Poly[(3-trimethylsilylbenzoyloxy)ethylene]	353	DTA heating rate	1054
Poly[(4-trimethylsilylbenzoyloxy)ethylene]	408	DTA heating rate	1054,1055

	Polymer	T_g(K)	Remarks	References
1.7	Poly(vinyl esters)　(Cont'd.)			
	Poly[(2,2-dimethylvaleryloxy)ethylene]	283	Mechanical method	1043
	Poly[(2,2,3,3-tetramethylvaleryloxy)ethylene]	363	Mechanical method	1043
	Poly[(2,2,3,4-tetramethylvaleryloxy)ethylene]	323	Mechanical method	1043
	Poly[(2,2,4,4-tetramethylvaleryloxy)ethylene]	328	Mechanical method	1043
	Poly(nicotinoyloxyethylene)	360	From polyvinyl alcohol	29,1053
	Poly(nitratoethylene)	307		903
	Poly[(3-nitrobenzoyloxy)ethylene]	366		29,925
	Poly[(4-nitrobenzoyloxy)ethylene]	395		29,925
	Poly[(4-phenylbenzoyloxy)ethylene]	358		29,925
	Poly(pivaloyloxyethylene)	359	Mechanical method	1043
	Poly[(4-propionyloxybenzoyloxy)ethylene]	345		29,925
	Poly(propionyloxyethylene)	283	Mechanical method	1043
	Poly[(4-p-toluoylbutyryloxy)ethylene]	313	DTA heating rate	1046
	Poly(vinyl acetate)　see　Poly(acetoxyethylene)			
	Poly(vinyl formate)　see　Poly(formyloxyethylene)			
	Poly(vinyl-4-isopropylbenzoate)			
	see　Poly(4-isopropylbenzoyloxyethylene)			
1.8	Poly(styrenes)			
	Poly(4-acetylstyrene)	389	Mechanical method	728
	Poly(4-p-anisoylstyrene)	376	Mechanical method	728
	Poly(4-benzoylstyrene)	371	Mechanical method	728
	Poly[(2-benzyloxymethyl)styrene]	345	Mechanical method	729
	Poly[3-(4-biphenylyl)styrene]	~471	Softening point	730
	Poly[4-(4-biphenylyl)styrene]	593	Softening point	730
	Poly(5-bromo-2-butoxystyrene)	320	Mechanical method	731
	Poly(5-bromo-2-ethoxystyrene)	353	Mechanical method	731
	Poly(5-bromo-2-isopentyloxystyrene)	310	Mechanical method, low viscosity	731
	Poly(5-bromo-2-isopropoxystyrene)	308	Mechanical method	731
	Poly(5-bromo-2-methoxystyrene)	359	Mechanical method	731
	Poly(5-bromo-2-pentyloxystyrene)	322	Mechanical method; low viscosity	731
	Poly(5-bromo-2-propoxystyrene)	327	Mechanical method; low viscosity	731
	Poly(1-bromostyrene)	391		732-734
	Poly(2-butoxycarbonylstyrene)	339	Mechanical method	735
	Poly(4-butoxycarbonylstyrene)	349	Mechanical method	728
	Poly(4-[(2-butoxyethoxy)methyl]styrene)	< 235		736
	Poly(2-butoxymethylstyrene)	340	Mechanical method	729
	Poly(4-butoxymethylstyrene)	< 283		736
	Poly[4-(sec-butoxymethyl)styrene]	313	Dynamic method	736
	Poly(4-butoxystyrene)	~320	Mechanical method	736,737
	Poly(5-tert-butyl-2-methylstyrene)	360		732
	Poly(4-butylstyrene)	279		736,738,739
	Poly(4-sec-butylstyrene)	359	Softening point	739
	Poly(4-tert-butylstyrene)	403		83,739-741
	Poly(4-butyrylstyrene)	347	Mechanical method	728
	Poly(2-carboxystyrene)	450	Mechanical method	742
	Poly(4-carboxystyrene)	386	Mechanical method	728
	Poly(4-chloro-3-fluorostyrene)	395		732
	Poly(4-chloro-2-methylstyrene)	418		732
	Poly(4-chloro-3-methylstyrene)	387		732
	Poly(2-chlorostyrene)	392		732
	Poly(3-chlorostyrene)	363		732
	Poly(4-chlorostyrene)	383		83,732-734
	Poly(2,4-dichlorostyrene)	406		732,743
	Poly(2,5-dichlorostyrene)	379		44,732,744
	Poly(2,6-dichlorostyrene)	440		83,569,745
	Poly(3,4-dichlorostyrene)	401		732,744
	Poly(4-cyanostyrene)	393	Sample thought to be crosslinked	746
	Poly(4-decylstyrene)	208		738
	Poly(4-dodecylstyrene)	221		738
	Poly(2-ethoxycarbonylstyrene)	391	Mechanical method	735

Polymer	T_g(K)	Remarks	References
1.8 Poly(styrenes) (Cont'd.)			
Poly(4-ethoxycarbonylstyrene)	367	Mechanical method	728
Poly[4-(2-ethoxyethoxymethyl)styrene]	273	Dynamic method	736
Poly(2-ethoxymethylstyrene)	347	Mechanical method; viscosity low	729
Poly(4-ethoxystyrene)	~359	Mechanical method	737
Poly[4-(2-diethylaminoethoxycarbonyl)styrene hydrochloride]	347	Mechanical method	728
Poly(4-diethylcarbamoylstyrene)	375	Mechanical method	728
Poly[4-(1-ethylhexyloxymethyl)styrene]	250		736
Poly(2-ethylstyrene)	376	Softening point	739
Poly(3-ethylstyrene)	~303	Softening point	739
Poly(4-ethylstyrene)	300,<351	Conflicting data	736,738,739
Poly[4-(pentadecafluoroheptyl)styrene]	320	Heating rate 32K/min, T_g thermal history dependent	747
Poly(2-fluoro-5-methylstyrene)	384	Mechanical method	748
Poly(4-fluorostyrene)	368		732-734,749-751
Poly(2,5-difluorostyrene)	374	Softening point	83
Poly(2,3,4,5,6-pentafluorostyrene)	378	Heating rate 5-20K/min	97
Poly(perfluorostyrene)	467	Heating rate 5-20K/min	97
Poly(α,β,β-trifluorostyrene)	475	Heating rate 5-20K/min values up to 513K reported	83,97,114,752,753
Poly(4-hexadecylstyrene)	278		738
Poly(4-hexanoylstyrene)	339	Mechanical method	728
Poly(2-hexyloxycarbonylstyrene)	318	Mechanical method	735
Poly(4-hexyloxycarbonylstyrene)	339	Mechanical method	728
Poly(4-hexyloxymethylstyrene)	253	Dynamic method	736
Poly(4-hexylstyrene)	246		738
Poly[4-(4-hydroxybutoxymethyl)styrene]	293	Dynamic method	736
Poly[4-(2-hydroxyethoxymethyl)styrene]	319	Dynamic method	736
Poly[4-(1-hydroxyiminoethyl)styrene]	407	Mechanical method	728
Poly(4-[(1-hydroxyimino)-2-phenethyl]styrene)	384	Mechanical method	728
Poly[4-(1-hydroxy-3-dimethylaminopropyl)styrene]	316	Mechanical method; viscosity low	754
Poly[4-(1-hydroxy-1-methylbutyl)styrene]	~403	Softening point	754
Poly[4-(1-hydroxy-1-methylethyl)styrene]	~438	Softening point	754
Poly[4-(1-hydroxy-1-methylhexyl)styrene]	~364	Softening point	754
Poly[4-(1-hydroxy-1-methylpentyl)styrene]	~356	Softening point	754
Poly[4-(1-hydroxy-1-methylpropyl)styrene]	~459	Softening point	754
Poly(2-hydroxymethylstyrene)	433		1
Poly(3-hydroxymethylstyrene)	398	Dynamic method	1
Poly(4-hydroxymethylstyrene)	413		1
Poly[4-(1-hydroxy-3-morpholinopropyl)styrene]	323	Mechanical method; viscosity low	754
Poly[4-(1-hydroxy-3-piperidinopropyl)styrene]	327	Mechanical method; viscosity low	754
Poly(4-iodostyrene)	429		733,734
Poly(2-isobutoxycarbonylstyrene)	400	Mechanical method	735
Poly(4-isobutoxycarbonylstyrene)	363	Mechanical method	728
Poly(2-isopentyloxycarbonylstyrene)	341	Mechanical method	735
Poly(2-isopentyloxymethylstyrene)	351	Mechanical method; viscosity low	729
Poly(4-isopentyloxystyrene)	~330	Mechanical method	737
Poly(2-isopropoxycarbonylstyrene)	419	Mechanical method	735
Poly(4-isopropoxycarbonylstyrene)	368	Mechanical method	728
Poly(2-isopropoxymethylstyrene)	361	Mechanical method, viscosity low	729
Poly(4-isopropylstyrene)	360	Softening point	739
Poly(2,4-diisopropylstyrene)	~435	Softening point	83
Poly(2,5-diisopropylstyrene)	441	Softening point	83
Poly(2-methoxycarbonylstyrene)	403	Mechanical method, viscosity low	742
Poly(4-methoxycarbonylstyrene)	386	Mechanical method	728
Poly(2-methoxymethylstyrene)	362	Mechanical method; viscosity low	729
Poly(4-methoxymethylstyrene)	350	Dynamic method	736
Poly(4-methoxy-2-methylstyrene)	~358	Softening point	755
Poly(2-methoxystyrene)	~348	Softening point	755
Poly(4-methoxystyrene)	~362	Mechanical method	737,755-757
Poly(2-methylaminocarbonylstyrene)	462	Mechanical method	742
Poly(2-dimethylaminocarbonylstyrene)	463	Mechanical method; viscosity low	742
Poly(4-dimethylaminocarbonylstyrene)	398	Mechanical method	728

Polymer		T_g(K)	Remarks	References
1.8	**Poly(styrenes)** (Cont'd.)			
Poly[2-(2-dimethylaminoethoxycarbonyl)styrene]		342	Mechanical method	735
Poly[4-(2-dimethylaminoethoxycarbonyl)styrene]		373	Mechanical method	728
Poly[4-(2-dimethylaminoethoxycarbonyl)styrene hydrochloride]		355	Mechanical method	728
Poly(α-methylstyrene)		441		6,83,758-765
Poly(2-methylstyrene)		409		63,732,739,749,755
Poly(3-methylstyrene)		370	DTA heating rate	732,739,749,766
Poly(4-methylstyrene)		366,374	Conflicting data	732,739,756,766,767
Poly(2,4-dimethylstyrene)		385		122,732,739,749,768
Poly(2,5-dimethylstyrene)		416	DTA heating rate	732
Poly(3,4-dimethylstyrene)		384	DTA heating rate	732
Poly(3,5-dimethylstyrene)		377		749
Poly(2,4,5-trimethylstyrene)		~409	Softening point	83
Poly(2,4,6-trimethylstyrene)		~435	Softening point	83
Poly(4-morpholinocarbonylstyrene)		400	Mechanical method	728
Poly[4-(3-morpholinopropionyl)styrene]		314	Mechanical method	754
Poly(4-nonadecylstyrene)		305		1
Poly(4-nonylstyrene)		220		738
Poly(4-octadecylstyrene)		305		738
Poly(4-octanoylstyrene)		323	Mechanical method	728
Poly[4-(octyloxymethyl)styrene]		231	Dynamic method	736
Poly(2-octyloxystyrene)		286		769
Poly(4-octylstyrene)		228		738
Poly(2-pentyloxycarbonylstyrene)		365	Mechanical method	735
Poly(2-pentyloxymethylstyrene)		320	Mechanical method	729
Poly(2-phenethyloxymethylstyrene)		336	Mechanical method; viscosity low	729
Poly(2-phenoxycarbonylstyrene)		397	Mechanical method; viscosity low	742
Poly(4-phenoxystyrene)		~373	Softening point	83
Poly(4-phenylacetylstyrene)		351	Mechanical method	728
Poly(2-phenylaminocarbonylstyrene)		464	Mechanical method; viscosity low	742
Poly(4-phenylstyrene)		434	Extrapolated to zero rate	763,770
Poly(4-piperidinocarbonylstyrene)		387	Mechanical method	728
Poly[4-(3-piperidinopropionyl)styrene]		311	Mechanical method	754
Poly(4-propionylstyrene)		375	Mechanical method	728
Poly(2-propoxycarbonylstyrene)		381	Mechanical method	735
Poly(4-propoxycarbonylstyrene)		365	Mechanical method	728
Poly(2-propoxymethylstyrene)		370	Mechanical method; viscosity low	729
Poly(4-propoxymethylstyrene)		295	Dynamic method	736
Poly(4-propoxystyrene)		343	Mechanical method	737
Poly(4-propoxysulfonylstyrene)	isotactic	490	DTA heating rate	771
Poly(styrene)	isotactic and atactic	373		9,17,21,22,25,47,51,57, 63,64,72,78,79,188,190, 191,261,263,317,318,344, 350-352,394,397,399,469, 524,569,576,619,628,630, 635,637,640,646,647,684, 699,731-734,746,750,766, 768,772-810.
Poly(4-tetradecylstyrene)		237		738
Poly(4-p-toluoylstyrene)		372	Mechanical method	728
Poly(4-valerylstyrene)		343	Mechanical method	728
1.9	**Others**			
Poly(benzylethylene)		333		631,634,1056
Poly(N-carbazolylethylene)		357,423,481	Conflicting values reported	1,272,573,845
Poly(ferrocenylethylene)		457-467	DSC heating rate	835
Poly(indazol-2-ylethylene)		331,298	Two values for different preparations, heating rate 20K/min	1045
Poly[dimethylamino(ethoxy)phosphinylethylene]		305		1057
Poly[dimethylamino(phenoxy)phosphinylethylene]		300		1057
Poly(4,4-dimethyl-5-oxazolonylethylene)		365	DTA heating rate, viscosity low	1058
Poly(4,4-dimethyl-5-oxazolonyl-2-propylene)		380,438	Conflicting data	1058,1059

Polymer	T_g(K)	Remarks	References

1.9   Others   (Cont' d.)

Poly[(2-methyl-5-pyridyl)ethylene]	403	Mechanical method	1060-1062
Poly[(2-methyl-6-pyridyl)ethylene]	365	Heating rate 20K/min	1061
Poly(2,4-dimethyl-1,3,5-triazinylethylene)	350		820
Poly(1-naphthylethylene)	432?	Values range from 323-453K	83,755,763,1063
Poly(2-naphthylethylene)	424	Extrapolated to zero heating rate	770
Poly(phenethylethylene)	283	Crystalline sample	631,634,1056,1064
Poly(phenethylmethylethylene)	245	Softening Point	631
Poly(phenylacetylene) see Poly(phenylvinylene)			
Poly(diphenylphosphinylethylene)	453		1065
Poly(phenylvinylene)	393	Crystallline	1066
Poly(phthalimidoethylene)	497	No experimental details	820
Poly(2-pyridylethylene)	377		1061,1062,1067
Poly(4-pyridylethylene)	415		1061,1062
Poly(N-pyrrolidinylethylene)	327		573
Poly(N-pyrrolidonylethylene)	359		1,4,907
Poly(m-tolylmethylethylene)	313	Crystalline sample	1056
Poly(o-tolylmethylethylene)	353	Crystalline sample	1056
Poly(p-tolylmethylethylene)	338	Crystalline sample	1056
Poly(vinyl carbazole) see Poly(N-carbazolylethylene)			
Poly(vinyltrimethylgermanium)	463	Mechanical method	1068
Poly(vinylpyridine) see Poly(pyridylethylene)			
Poly(vinyl pyrrolidine) see Poly(N-pyrrolidinylethylene)			

2.   Main-chain Carbocyclic Polymers

2.1   Poly(phenylenes)

Poly(3,3'-biphenylylenehexafluorotrimethylene)	324	Heating rate 32K/min, η_{inh} only 0.05 dl/g	563
Poly(2-bromo-1,4-phenyleneethylene)	353	Mechanical method	564
Poly(2-chloro-1,4-phenyleneethylene)	343	Mechanical method, heating rate 1.5K/min	300,564,565
Poly(2,5-dichloro-1,4-phenyleneethylene)	613	Softening point, no measurement details	565
Poly(2-cyano-1,4-phenyleneethylene)	363	Mechanical method	564
Poly(2-ethyl-1,4-phenyleneethylene)	298	Mechanical method	564
Poly(2-methyl-1,4-phenyleneethylene)	328	Mechanical method, heating rate 1.5K/min	564,565
Poly(2,5-dimethyl-1,4-phenyleneethylene)	373	Mechanical method, heating rate 1.5K/min	566
Poly(1,4-phenylenetrichloroethylene)	433	Polymer contains small amount of different structure, mechanical method; heating rate 1.5K/min	565
Poly(1,4-phenyleneethylene)	~353	Mechanical method, variable data	300,564,565,567
Poly(1,3-phenylenehexafluorotrimethylene)	303	Heating rate 32K/min, η_{inh} only 0.1 dl/g	148
Poly(1,4-phenylene-1-phenylethylene)	428	Softening point, no experimental details	565
Poly(2',3',6',2''',3''',5'''-hexaphenyl-p,p,m,p,p-quinquephenylylenedeca-methylene)	453	Heating rate 20K/min, structure may contain more m-links	568
Poly(2',3',6',2''',3''',5'''-hexaphenyl-4,4''''-p-quinquephenylylenedeca-methylene)	508	Heating rate 20K/min, structure may contain m-links	568
Poly(2',3',6',2''',3''',5'''-hexaphenyl-p,p,m,p,p,quinquephenylylenehexa-methylene)	488	Heating rate 20K/min, structure may contain more m-links	568
Poly(2',3',6',2''',3''',5'''-hexaphenyl-4,4''''-p-quinquephenylylenehexa-methylene)	523	Heating rate 20K/min, structure may contain m-links	568
Poly(2',3',6',2''',3''',5'''-hexaphenyl-p,p,m,p,p-quinquephenylylenetetra-decamethylene)	433	Heating rate 20K/min, structure may contain more m-links	568
Poly(2',3',6',2''',3''',5'''-hexaphenyl-4,4''''-p-quinquephenylylenetetra-decamethylene)	458	Heating rate 20K/min, structure may contain m-links	568

	Polymer	$T_g(K)$	Remarks	References
2.1	Poly(phenylenes) (Cont' d).			
	Poly(2',3',6',2''',3''',5'''-hexaphenyl-p,p,m,p,p-quinquephenylylene-tetramethylene)	453	Heating rate 20K/min, structure may contain more m-links	568
	Poly(2',3',6',2''',3''',5'''-hexaphenyl-4,4''''-p-quinquephenylylene-tetramethylene)	478	Heating rate 20K/min, structure may contain m-links	568
	Poly(2',3',6',2''',3''',5'''-hexaphenyl-p,p,m,p,p-quinquephenylylenetri-methylene)	513	Heating rate 20K/min, structure may contain more m-links	568
	Poly(2',3',6',2''',3''',5'''-hexaphenyl-4,4''''-quinquephenylylenetri-methylene)	563	Heating rate 20K/min, structure may contain m-links	568
2.2	Others			
	Poly(acenaphthylene)	487-618	Conflicting data reported.	31,83,566,569-571
	Poly(5-chlorononafluoro-1,3-cyclohexylenedifluoromethylene)	420	Heating rate 5-20K/min, high pressure, radiation synthesis, ring structure unproven	97
	Poly(cyclobutene) erythro di-isotactic	293	Degree of polymerization 150	572
	erythro di-syndiotactic	273	Degree of polymerization 15	
	Poly(hexafluoro-1,3-cyclobutylenedifluoromethylene)	395	Heating rate 5-20K/min, high pressure, radiation synthesis, ring structure unproven, $[\eta]$ only 0.03 dl/g	97
	Poly(dodecafluoro-1,3-cycloheptylenedifluoromethylene)	471?	Heating rate 5-20K/min, high pressure, radiation synthesis, data queried in original paper, ring structure unproven.	97
	Poly(decafluoro-1,3-cyclohexylenedifluoromethylene)	374?	Heating rate 5-20K/min, ring structure unproven, crystalline, high pressure, radiation synthesis	97
	Poly(octafluoro-1,3-cyclopentylenedifluoromethylene)	390	Heating rate 5-20K/min, high pressure, radiation synthesis, ring structure unproven	97
	Poly(indenylene)	358		573
	Poly(1,4-naphthyleneethylene)	433	Mechanical method	565

3. Main-chain Acyclic Heteroatom Polymers

     3.1   Main-chain -C-O-C-Polymers

        3.1.1   Poly(oxides)

	Polymer	$T_g(K)$	Remarks	References
	Poly(acetaldehyde) see Poly(oxyethylidene)			
	Poly[3,3-bis(chloromethyl)oxacyclobutane]			
	see Poly[oxy-2,2-bis(chloromethyl)trimethylene]			
	Poly(epichlorhydrin) see Poly[oxy(chloromethyl)ethylene]			
	Poly(ethylene oxide)s see Poly(oxyethylene)s			
	Poly(formaldehyde) see Poly(oxymethylene)			
	Poly(tetrahydrofuran) see Poly(oxytetramethylene)			
	Poly(methylene oxide) see Poly(oxymethylene)			
	Poly(2,6-dimethyl-1,4-phenylene oxide) [PPO]			
	see Poly(oxy-2,6-dimethyl-1,4-phenylene)			
	Poly(oxacyclobutane) see Poly(oxytrimethylene)			
	Poly(oxetane)s see Poly(oxytrimethylene)s			
	Poly(oxy-2-acetoxytrimethyleneoxy-2,6-dichloro-1,4-phenyleneisopropylidene-3,5-dichloro-1,4-phenylene)	373		219
	Poly(oxy-2-acetoxytrimethyleneoxy-1,4-phenylene)	322		219
	Poly(oxy-2-acetoxytrimethyleneoxy-1,4-phenylene-1-ethyl-1,3-cyclohexylene-1,4-phenylene)	380		219
	Poly(oxy-2-acetoxytrimethyleneoxy-1,4-phenyleneisopropylidene-1,4-phenylene)	333		219
	Poly[oxy-2-acetoxytrimethyleneoxy-1,4-phenylenemethyl(phenyl)methylene-1,4-phenylene]	383		219
	Poly[oxy(allyloxymethyl)ethylene]	195		220
	Poly(oxy-2-benzoyloxytrimethyleneoxy-1,4-phenyleneisopropylidene-1,4-phenylene)	338		219
	Poly[oxy-2-benzoyloxytrimethyleneoxy-1,4-phenylenemethyl(phenyl)methylene-1,4-phenylene]	399		219
	Poly[oxy-2-(2-biphenylyl)-6-phenyl-1,4-phenylene]	484	DSC heating rate; low viscosity	221

References page III-179

Polymer	T_g(K)	Remarks	References
3.1.1 Poly(oxides) (Cont'd.)			
Poly[oxy(bromomethyl)ethylene]	259	Heating rate 20K/min	222
Poly(oxybutadiene)	198	Structure not given	223,224
Poly[oxy(butoxymethyl)ethylene]	194	Heating rate 10K/min	220
Poly(oxybutylethylene)	203	Heating rate 10K/min	220
Poly(oxy-tert-butylethylene)	308		223,225-227
Poly[oxy-2-(4-tert-butylphenyl)-6-phenyl-1,4-phenylene]	513	DSC heating rate	221
Poly[oxy-2-(chloroacetoxy)trimethyleneoxy-1,4-phenyleneisopropylidene-1,4-phenylene]	338		219
Poly[oxy-2-(2-chlorobenzoyloxy)trimethyleneoxy-1,4-phenyleneisopropylidene-1,4-phenylene]	339		219
Poly(oxy-2,2,2-trichloroethylethylene)	271	No measurement details	229
Poly[oxy(chloromethyl)ethylene]	251	Heating rate 20K/min	222
Poly[oxy-2,2-bis(chloromethyl)trimethylene]	265		1,122,230-236
Poly(oxydecylethylene)	232		220,223
Poly[oxy(ethoxymethyl)ethylene]	212		220
Poly(oxy-2-ethyl-2-chloromethyltrimethylene)	~293	Dynamic method	237
Poly(oxyethylene)	232?	Conflicting data, values range from 158-233K	107,223,234,238-244
Poly(oxyethylethylene)	203		220
Poly(oxyethylidene)	243		250,251
Poly[oxy-1-(2,2,3,3-tetrafluorocyclobutyl)ethylene]	248	Mechanical method, heating rate 2K/min	252
Poly(oxytetrafluoroethylene)	225	Low molecular weight	253,254
Poly[oxy-2,2,3,3,4,4-hexafluoropentamethyleneoxy-4,4'-octafluorobiphenylylene)	314	Heating rate 32K/min	255
Poly(oxy-2,2,3,3,4,4-hexafluoropentamethyleneoxymethylene-1,4-phenylene-methylene)	247	Heating rate 32K/min	256
Poly(oxyoctafluorotetramethylene)	~208	Heating rate 10K/min	254,257
Poly[oxy-1-(heptafluoro-2-propoxymethyl)ethylene]	~230	DTA heating rate	258
Poly(oxyhexylethylene)	206	Heating rate 10K/min	220
Poly[oxy(hexyloxymethyl)ethylene]	190		220
Poly(oxy-2-hydroxytrimethyleneoxy-2-chloro-1,4-phenyleneisopropylidene-3-chloro-1,4-phenylene)	358		259
Poly(oxy-2-hydroxytrimethyleneoxy-2,6-dichloro-1,4-phenyleneisopropylidene-3,5-dichloro-1,4-phenylene)	388		259
Poly(oxy-2-hydroxytrimethyleneoxy-1,4-phenylene)	333		259
Poly(oxy-2-hydroxytrimethyleneoxy-1,4-phenylene-1-ethyl-1,4-cyclohexylene-1,4-phenylene)	413		259
Poly(oxy-2-hydroxytrimethyleneoxy-1,4-phenyleneisobutylidene-1,4-phenylene)	368		259
Poly(oxy-2-hydroxytrimethyleneoxy-1,4-phenyleneisopropylidene-1,4-phenylene)	373		259
Poly(oxy-2-hydroxytrimethyleneoxy-1,4-phenylene-5-isopropyl-2-methyl-1,2-cyclohexylene-1,4-phenylene)	448		259
Poly(oxy-2-hydroxytrimethyleneoxy-1,4-phenylene-1-methyl-1,4-cyclohexylene-isopropylidene-1,4-phenylene)	408		259
Poly(oxy-2-hydroxytrimethyleneoxy-1,4-phenylenemethylene-1,4-phenylene)	353		259
Poly(oxy-2-hydroxytrimethyleneoxy-1,4-phenylene-3,3-dimethyl-1,6-indanylene)	393		259
Poly(oxy-2-hydroxytrimethyleneoxy-1,4-phenylene-1,3,3-trimethyltrimethylene-1,4-phenylene)	348		259
Poly(oxy-2-hydroxytrimethyleneoxy-1,4-phenylenemethyl(phenyl)methylene-1,4-phenylene)	388		219
Poly[oxy(methoxymethyl)ethylene]	211	Heating rate 10K/min	220
Poly(oxy-2,6-dimethoxy-1,4-phenylene)	440		260
Poly(oxymethylene)	191	Values range from 188-263K	1,6,68,191,230,233,238,239,261-271
Poly(oxytetramethylene)	189		237,238,243,272-278
Poly(oxytrimethylene)	195		238,272,273,278-281
Poly(oxymethyleneoxy-2,2,3,3,4,4-hexafluoropentamethylene)	220	Heating rate 32K/min	284
Poly[di(oxymethylene)oxy-2,2,3,3,4,4-hexafluoropentamethylene]	218		285
Poly(oxymethyleneoxy-2,2,3,3,4,4,5,5-octafluorohexamethylene)	220	Heating rate 32K/min	284
Poly(oxymethylene-1,3-phenylenemethyleneoxy-2,2,3,3,4,4-hexafluoropentamethylene)	238		256

Polymer	T_g(K)	Remarks	References
3.1.1 Poly(oxides) (Cont'd.)			
Poly(oxymethylene-1,4-phenyleneoxy-1,4-phenylenemethyleneoxy-2,2,3,3,4,4-hexafluoropentamethylene)	279	Heating rate 32K/min; high rate effect	256
Poly(oxy-1,1-dimethylethylene)	264		223
Poly(oxy-1,2-dimethylethylene) trans	277		223
Poly(oxy-1-methyltrimethylene)	223	Dynamic method	281
Poly(oxy-2-methyltrimethylene)	218	Dynamic method	281
Poly(oxy-2,6-dimethyl-1,4-phenylene)	482		102,114,202,262, 286-300
Poly(oxy-2-methyl-6-phenyl-1,4-phenylene)	428		299
Poly[oxy-2-(1-naphthyl)-6-phenyl-1,4-phenylene]	507	DSC heating rate; low viscosity	221
Poly(oxy-3-nitro-1,3-phenyleneoxy-1,4-phenyleneisopropylidene-1,4-phenylene)	423	Mechanical method	116
Poly(oxydiphenoxymethyleneoxy-4,4' biphenylylene)	385	DSC heating rate	379
Poly(oxydiphenoxymethyleneoxy-3,3'-dimethyl-4,4'-biphenylylene)	398	DSC heating rate	379
Poly(oxydiphenoxymethyleneoxy-1,4-phenyleneisopropylidene-1,4-phenylene)	365	DSC heating rate	379
Poly(oxy-1,3-phenylene)	~318	DTA heating rate; low viscosity	114,295
Poly(oxy-1,4-phenylene)	358	DTA heating rate	288,305-307
Poly[di(oxy-1,4-phenylene)carbonyl-1,4-phenylene]	433	Mechanical method	116
Poly[tri(oxy-1,4-phenylene)carbonyl-1,4-phenylene]	423	Mechanical method	116
Poly(oxy-1,4-phenylenecarbonyl-1,4-phenyleneoxy-1,4-phenyleneisopropylidene-4-methyl-1,4-cyclohexylene-1,4-phenylene)	493	Mechanical method	116
Poly(oxy-1,4-phenylenecarbonyl-1,4-phenyleneoxy-1,4-phenyleneisopropylidene-1,4-phenylene)	428	Mechanical method	116
Poly(oxy-1,4-phenylene-2,2-dicyanotrimethylene-1,4-phenylene)	420	DTA heating rate	310
Poly[oxy-1,4-phenylene-(2-cyano)-2-phenyltrimethylene-1,4-phenylene]	416	Proposed structure; DTA heating rate	310
Poly[oxy-1,4-phenylene-2,2-di(ethoxycarbonyl)trimethylene-1,4-phenylene]	327	DTA heating rate	310
Poly(oxy-1,4-phenylenehexafluoro-2,2-propylidene-1,4-phenyleneoxy-1,4-phenylenecarbonyl-1,4-phenylene)	448	Mechanical method	116
Poly(oxy-1,4-phenyleneisopropylidene-1,4-phenyleneoxy-1,4-phenyleneazo-1,4-phenylene)	448	Mechanical method	116
Poly(oxy-1,4-phenyleneisopropylidene-1,4-phenyleneoxy-1,4-phenylenetere-phthaloyl-1,4-phenylene)	438	Mechanical method	116
Poly(oxy-1,4-phenylenemethylenefluoren-9-ylidenemethylene-1,4-phenylene)	411	Proposed structure; DTA heating rate	310
Poly(oxyphenylethylene)	313		220,223,225,324
Poly(oxy-2,6-diphenyl-1,4-phenylene)	493		56,299,305,325
Poly[oxy-6-phenyl-2-(3'-o-terphenylyl)-1,4-phenylene]	485	DSC heating rate; low viscosity	221
Poly[oxy-6-phenyl-2-(m-tolyl)-1,4-phenylene]	492	DSC heating rate; low viscosity	221
Poly[oxy-6-phenyl-2-(p-tolyl)-1,4-phenylene]	491	DSC heating rate; low viscosity	221
Poly(oxy-2-propionyloxytrimethyleneoxy-1,4-phenyleneisopropylidene-1,4-phenylene)	333		219
Poly(oxypropylene)	198	Conflicting data	220,223,243,251,269, 273,276,281,326-332
Poly(phenylene oxide)s see Poly(oxyphenylene)s			
3.1.2 Poly(carbonates)			
Polycarbonate of Bisphenol A see Poly(oxycarbonyloxy-1,4-phenyleneiso-propylidene-1,4-phenylene)			
Poly(oxycarbonyloxy-2,6-dibromo-1,4-phenyleneisopropylidene-3,5-dibromo-1,4-phenylene)	430		333,334
Poly(oxycarbonyloxy-2-chloro-6-methyl-1,4-phenyleneisopropylidene-3-chloro-5-methyl-1,4-phenylene)	427	Mechanical method, heating rate 1K/min	335
Poly(oxycarbonyloxy-2-chloro-1,4-phenylenecyclohexylidene-3-chloro-1,4-phenylene)	443-452	Mechanical method, heating rate 1K/min	335
Poly(oxycarbonyloxy-2,6-dichloro-1,4-phenylenecyclohexylidene-3,5-di-chloro-1,4-phenylene)	436		334
Poly(oxycarbonyloxy-2-chloro-1,4-phenyleneisopropylidene-3-chloro-1,4-phenylene)	420	Mechanical method	334,335
Poly(oxycarbonyloxy-2,6-dichloro-1,4-phenyleneisopropylidene-3,5-dichloro-1,4-phenylene)	453,493,504	Conflicting data	334,336,337
Poly(oxycarbonyloxy-2-cyclohexyl-1,4-phenyleneisopropylidene-1,4-phenylene)	405		338

Polymer	T_g(K)	Remarks	References

3.1.2 Poly(carbonates)   (Cont'd.)

Polymer	T_g(K)	Remarks	References
Poly(oxycarbonyloxy-2,2,3,3,4,4,5,5-octafluorohexamethylene)	232	DTA heating rate	339
Poly(oxycarbonyloxy-2-isopropyl-1,4-phenyleneisopropylidene-1,4-phenylene)	385	No experimental details	338
Poly(oxycarbonyloxy-2-methoxy-1,4-phenyleneisopropylidene-1,4-phenylene)	418	No experimental details	338
Poly[oxycarbonyloxy-1,3-(2,2,4,4-tetramethylcyclobutylene)]	∼500,<433,418	Conflicting data	340-342
Poly(oxycarbonyloxyhexamethylene)	230	Low molecular weight sample	339
Poly(oxycarbonyloxy-2-methyl-1,4-phenylenecyclohexylidene-3-methyl-1,4-phenylene)	408	Mechanical method	335,343
Poly(oxycarbonyloxy-2-methyl-1,4-phenyleneisopropylidene-3-methyl-1,4-phenylene)	368,373,418 363-383	Conflicting data	334-336,343
Poly(oxycarbonyloxy-2-methyl-1,4-phenyleneisopropylidene-1,4-phenylene)	413		338
Poly(oxycarbonyloxy-2-methyl-1,4-phenylenemethylene-3-methyl-1,4-phenylene)	324	Mechanical method	335
Poly[oxycarbonyloxy-4,6-dimethyl-1,2-phenylenemethylene-3,5-dimethyl-1,2-phenylene]	410	No experimental details	338
Poly(oxycarbonyloxy-3-methyl-1,4-phenylenebenzylidene-2-methyl-1,4-phenylene)	455	Mechanical method	335
Poly(oxycarbonyloxy-1,4-phenylenebenzylidene-1,4-phenylene)	394		335,338
Poly(oxycarbonyloxy-1,4-phenylene-1,1-butylidene-1,4-phenylene)	396		344
Poly(oxycarbonyloxy-1,4-phenylene-2,2-butylidene-1,4-phenylene)	407		335,344
Poly(oxycarbonyloxy-1,4-phenylene-1,3-dichloro-1,1,3,3-tetrafluoro-2,2-propylidene-1,4-phenylene)	456	DTA heating rate	337
Poly[oxycarbonyloxy-1,4-phenylene(cyano)phenylmethylene-1,4-phenylene]	411		345
Poly(oxycarbonyloxy-1,4-phenylenecyclohexylidene-1,4-phenylene)	448		334,335,344
Poly(oxycarbonyloxy-1,4-phenylenecyclopentylidene-1,4-phenylene)	440		334,346
Poly(oxycarbonyloxy-1,4-phenyleneethylidene-1,4-phenylene)	403		344
Poly(oxycarbonyloxy-1,4-phenylenefluoren-9-ylidene-1,4-phenylene)	653	Heating rate 10K/min	347
Poly(oxycarbonyloxy-1,3-phenylenehexafluorotrimethylene-1,3-phenylene)	319		148
Poly(oxycarbonyloxy-1,4-phenylenehexafluoro-2,2-propylidene-1,4-phenylene)	449	DTA heating rate	337
Poly(oxycarbonyloxy-1,4-phenylene-4,4-heptylidene-1,4-phenylene)	421		334
Poly(oxycarbonyloxy-1,4-phenyleneisobutylidene-1,4-phenylene)	422		344
Poly(oxycarbonyloxy-1,4-phenyleneisopropylidene-1,3-phenyleneisopropylidene-1,4-phenylene)	393	No experimental details	338
Poly(oxycarbonyloxy-1,4-phenyleneisopropylidene-1,4-phenylene)	418		78,102 131,145,230, 233,262,263,287, 317-319,321,335- 337,344,348-378
Poly[oxycarbonyloxy-1,4-phenylenedi(isopropylidene-1,4-phenylene)]	435	No experimental details	338
Poly(oxycarbonyloxy-1,4-phenylenemethylene-1,4-phenylene)	420	No experimental details	338,367
Poly[oxycarbonyloxy-1,4-phenylene(methyl)phenylmethylene-1,4-phenylene]	449		334,336
Poly(oxycarbonyloxy-1,4-phenylene-2,2-pentylidene-1,4-phenylene)	410		334
Poly(oxycarbonyloxy-1,4-phenylenediphenylmethylene-1,4-phenylene)	394		334
Poly(oxyethyleneoxy-1,4-phenylenecarbonyloxycarbonyl-1,4,-phenylene)	318		247
Poly[di(oxyethylene)oxy-1,4-phenylenecarbonyloxycarbonyl-1,4-phenylene]	314		247
Poly(oxymethyleneoxy-1,4-phenylenecarbonyloxycarbonyl-1,4-phenylene)	357		247
Poly[di(oxymethylene)oxy-1,4-phenylenecarbonyloxycarbonyl-1,4-phenylene]	325 335	Amorphous Crystalline	247
Poly(oxypentamethyleneoxy-1,4-phenylenecarbonyloxycarbonyl-1,4-phenylene)	326		247
Poly(oxytetramethyleneoxy-1,4-phenylenecarbonyloxycarbonyl-1,4-phenylene)	348		247
Poly(oxytrimethyleneoxy-1,3-phenylenecarbonyloxycarbonyl-1,3-phenylene)	326		247
Poly(oxytrimethyleneoxy-1,4-phenylenecarbonyloxycarbonyl-1,4-phenylene)	368		247
Poly(oxy-1,3-phenylenecarbonyloxycarbonyl-1,3-phenyleneoxypentamethylene)	334		247
Poly(oxy-1,3-phenylenecarbonyloxycarbonyl-1,3-phenyleneoxytetramethylene)	< 293		247

3.1.3 Poly(esters)

Polymer	T_g(K)	Remarks	References
Poly(bisphenol A terephthalate) see Poly(oxyterephthaloyloxy-1,4-phenyleneiso-propylidene-1,4-phenylene)			
Poly(ethylene adipate) see Poly(oxyethyleneoxyadipoyl)			
Poly(ethylene terephthalate) see Poly(oxyethyleneoxyterephthaloyl)			
Poly(1,4,7-trioxa-3,3,5,5-tetrafluoroheptamethylenecarbonyl-1,3-phenylene-hexafluorotrimethylene-1,3-phenylenecarbonyl)	303	Heating rate 32K/min	218

Polymer	T_g(K)	Remarks	References

3.1.3   Poly(esters)   (Cont' d.)

Polymer	T_g(K)	Remarks	References
Poly(1,4,7-trioxa-3,3,5,5-tetrafluoroheptamethyleneisophthaloyl)	301	Heating rate 32K/min; η_{inh} 0.23 only	218
Poly(1,4,7-trioxa-3,3,5,5-tetrafluoroheptamethylene-5-pentyloxyisophthaloyl)	287	Heating rate 32K/min; [η] only 0.24 dl/g	218
Poly(oxyadipoyloxy-2,6-dichloro-1,4-phenyleneisopropylidene-3,5-dichloro-1,4-phenylene)	383	No measurement details	380
Poly(oxyadipoyloxy-3,3',5,5'-tetramethyl-4,4'-biphenylylene)	381	Heating rate 8K/min	371
Poly(oxyadipoyloxydecamethylene)	217		96,381
Poly(oxyadipoyloxy-2,6-dimethyl-1,4-phenyleneisopropylidene-3,5-dimethyl-1,4-phenylene)	366	Heating rate 8K/min	371
Poly(oxyadipoyloxy-1,4-phenyleneisopropylidene-1,4-phenylene)	341		382
Poly(oxyadipoyloxy-2,6-diphenyl-1,4-phenylenemethylene-3,5-diphenyl-1,4-phenylene)	388	Heating rate 8K/min	305,371
Poly(oxy-2-butenyleneoxysebacoyl) cis	232		383
trans	233		
Poly(oxy-5-butyl-1,3-phenyleneoxyisophthaloyl)	359		384
Poly(oxydibutyltinoxyadipoyl)	383	Heating rate 20K/min	385
Poly(oxydibutyltinoxyfumaroyl)	459	Heating rate 20K/min	385
Poly(oxydibutyltinoxyterephthaloyl)	588	Heating rate 20K/min	385
Poly(oxy-2-butynyleneoxysebacoyl)	246		383,429
Poly(oxycarbonyl-3,3'-biphenylylenecarbonyloxy-1,4-phenyleneisopropylidene-1,4-phenylene)	460	Heating rate 8K/min	386
Poly(oxycarbonyl-1,4-cyclohexylenecarbonyloxy-1,4-phenyleneisopropylidene-1,4-phenylene) trans	423	Heating rate 20K/min	42
Poly(oxycarbonyl-2,6-naphthylenecarbonyloxydecamethylene)	287	No measurement details	387
Poly(oxyethyleneoxyadipoyl)	210	Values range from 163-233K	1,96,243,354, 381,388-390
Poly[di(oxyethylene)oxyadipoyl]	227		96
Poly[di(oxyethylene)oxyazelaoyl]	205		96
Poly(oxyethyleneoxycarbonyl-1,4-cyclohexylenecarbonyl) trans	291	Heating rate 20K/min	42
Poly(oxyethyleneoxycarbonyl-2,2'-dimethyl4,4'-biphenylylenecarbonyl)	346		391
Poly(oxyethyleneoxycarbonyl-1,1,3-trimethylindan-3,5-ylene-1,4-phenylene-carbonyl)	427	DTA heating rate	392
Poly(oxyethyleneoxycarbonyl-1,4-naphthylenecarbonyl)	337		393
Poly(oxyethyleneoxycarbonyl-1,5-naphthylenecarbonyl)	344	Amorphous	393
	351	Crystalline	
Poly(oxyethyleneoxycarbonyl-2,6-naphthylenecarbonyl)	386		387,393
Poly(oxyethyleneoxycarbonyl-2,7-naphthylenecarbonyl)	392		393
Poly(oxyethyleneoxycarbonyl-1,4-phenylene-sec-butylidene-1,4-phenylene-carbonyl)	~380		165
Poly[di(oxyethylene)oxydodecanedioyl]	202		96
Poly[di(oxyethylene)oxyglutaryl]	226		96
Poly[di(oxyethylene)oxyheptylmalonyl]	215		96
Poly(oxyethyleneoxyisophthaloyl)	324		393
Poly[di(oxyethylene)oxymalonyl]	244		96
Poly[di(oxyethylene)oxymethylmalonyl)]	244		96
Poly[di(oxyethylene)oxynonylmalonyl]	214		96
Poly[di(oxyethylene)oxyoctadecanedioyl]	205		96
Poly[di(oxyethylene)oxyoxalyl]	265		96
Poly[di(oxyethylene)oxypentylmalonyl]	226		96
Poly[di(oxyethylene)oxypimeloyl]	213		96
Poly[di(oxyethylene)oxypropylmalonyl]	235		96
Poly[di(oxyethylene)oxysebacoyl]	199		96
Poly[di(oxyethylene)oxysuberoyl]	212		96
Poly[di(oxyethylene)oxysuccinyl]	244		96
Poly(oxyethyleneoxyterephthaloyl)	342		1,42,61,78,94,183,202,233, 261,262,272,319,344,349, 352-354,360,362,371,381, 388,389,393-418
Poly(oxycarbonyl-1,3-phenyleneoxy-1,3-phenylenecarbonyloxy-2,2,3,3,4,4-hexafluoropentamethylene)	293		148
Poly(oxy-5-ethyl-1,3-phenyleneoxyisophthaloyl)	395		384

Polymer	T_g(K)	Remarks	References

3.1.3 Poly(esters) (Cont'd.)

Polymer	T_g(K)	Remarks	References
Poly[oxy-5-(pentadecafluoroheptyl)isophthaloyloxy-2,2,3,3,4,4-hexa-fluoropentamethylene]	300	Heating rate 32K/min; η_{inh} only 0.10 dl/g	386
Poly(oxy-2,2,3,3,4,4-hexafluoropentamethyleneoxyadipoyl)	216		419-423
Poly(oxy-2,2,3,3,4,4-hexafluoropentamethyleneoxycarbonyl-3,3'-bi-phenylylenecarbonyl)	318		148
Poly(oxy-2,2,3,3,4,4-hexafluoropentamethyleneoxycarbonyl-1,3-phenylene-decafluoropentamethylene-1,3-phenylenecarbonyl)	301	Heating rate 32K/min; η_{inh} low	148
Poly(oxy-2,2,3,3,4,4-hexafluoropentamethyleneoxycarbonyl-1,3-phenylene-hexafluorotrimethylene-1,3-phenylenecarbonyl)	290		148
Poly(oxy-2,2,3,3,4,4-hexafluoropentamethyleneoxy-3,6-dithiaoctanedioyl)	~233	Brittle point	422
Poly(oxy-2,2,3,3,4,4,5,5-octafluorohexamethyleneoxy-3,6-dithiaoctanedioyl)	~235	Brittle point	422
Poly(oxy-3-heptafluoropropylglutaryloxy-2,2,3,3,4,4-hexafluoropenta-methylene)	243-248	Brittle point	422
Poly(oxy-3-heptafluoropropylglutaryloxy-2,2,3,3,4,4,5,5-octafluorohexa-methylene)	~248	Brittle point	422
Poly(oxyglutaryloxy-2,2,3,3,4,4-hexafluoropentamethylene)	218-223	Brittle point	422
Poly(oxyglutaryloxy-2,2,3,3,4,4,5,5-octafluorohexamethylene)	218-223	Brittle point	422
Poly(oxy-5-hexyl-1,3-phenyleneoxyisophthaloyl)	335		384
Poly(oxyisophthaloyloxy-4,4'-biphenylylene)	437,583	Conflicting data	313,424
Poly(oxyisophthaloyloxy-2,2,3,3,4,4-hexafluoropentamethylene)	298	Heating rate 32K/min; η_{inh} 0.14 dl/g only	386,425
Poly(oxyisophthaloyloxy-2-methyl-1,4-phenyleneisopropylidene-3-methyl-1,4-phenylene)	438	Dynamic method	366
Poly(oxyisophthaloyloxy-2,6-dimethyl-1,4-phenyleneisopropylidene-3,5-dimethyl-1,4-phenylene)	498	Heating rate 8K/min	371
Poly(oxyisophthaloyloxy-2-methyl-1,4-phenylenemethylene-3-methyl-1,4-phenylene)	418	Mechanical method	369
Poly(oxyisophthaloyloxy-2,6-dimethyl-1,4-phenylenemethylene-3,5-di-methyl-1,4-phenylene)	461	Mechanical method	369
Poly(oxyisophthaloyloxy-1,4-phenylenebenzylidene-1,4-phenylene)	433		426
Poly(oxyisophthaloyloxy-1,4-phenylenecyclohexylidene-1,4-phenylene)	400		145
Poly(oxyisophthaloyloxy-1,4-phenyleneisopropylidene-1,4-phenylene)	462	Heating rate 8K/min	145,369,371,427
Poly(oxyisophthaloyloxy-1,4-phenylenemethylene-1,4-phenylene)	423	Heating rate 20K/min	313
Poly(oxy-3,3',5,5'-tetramethyl-4,4'-biphenylyleneoxysebacoyl)	330	Heating rate 8K/min	371
Poly(oxy-2,2,4,-tetramethyl-1,3-cyclobutyleneoxycarbonyl-trans-1,4-cyclohexylenecarbonyl)	442	Heating rate 20K/min	42
Poly(oxy-2,2,4,4-tetramethyl-1,3-cyclobutyleneoxyterephthaloyl) 60-68% trans	457	Heating rate 20K/min	42,319,428
Poly(oxymethylene-1,4-cyclohexylenemethyleneoxycarbonyl-trans-1,4-cyclohexylenecarbonyl)	325	Heating rate 20K/min	42
Poly(oxypentamethyleneoxyadipoyl)	204		96
Poly[di(oxy-1,4-phenylene)oxyisophthaloyl]	446	Heating rate 20K/min	313
Poly[di(oxy-1,4-phenylene)oxy-5-pentyloxyisophthaloyl]	399	Heating rate 20K/min	313
Poly(oxytetramethyleneoxyadipoyl)	155	Transition at 194-206K, 1st order (ref. 243), rate 10K/min.	80,96,243
Poly(oxytrimethyleneoxyadipoyl)	214		96
Poly(oxytetramethyleneoxycarbonyl-1,4-cyclohexylenecarbonyl)	263	Heating rate 20K/min	42
Poly(oxytrimethyleneoxycarbonyl-1,4-cyclohexylenecarbonyl) trans	267	Heating rate 20K/min	42
Poly(oxyhexamethyleneoxycarbonyl-2,6-naphthylenecarbonyl)	317	No measurement details	387
Poly(oxypentamethyleneoxycarbonyl-2,6-naphthylenecarbonyl)	311	No measurement details	387
Poly(oxytetramethyleneoxycarbonyl-2,6-naphthylenecarbonyl)	349	No measurement details	387
Poly(oxytrimethyleneoxycarbonyl-2,6-naphthylenecarbonyl)	346	No measurement details	387
Poly(oxytetramethyleneoxysebacoyl)	216	T_g decreased as intrinsic viscosity increased	1,383,429
Poly(oxypentamethyleneoxyterephthaloyl)	283	DTA heating rate	411
Poly(oxytetramethyleneoxyterephthaloyl)	290,353	Conflicting data	6,42,403,411
Poly(oxytrimethyleneoxyterephthaloyl)	368,308	Conflicting data	42,183,403,411
Poly(oxy-2,6-dimethyl-1,4-phenyleneisopropylidene-3,5-dimethyl-1,4-phenyleneoxysebacoyl)	318	Heating rate 8K/min	371
Poly(oxy-5-methyl-1,3-phenyleneoxyisophthaloyl)	426		384
Poly(oxy-2,5-dimethylterephthaloyloxy-1,4-phenyleneisopropylidene-1,4-phenylene)	457	Mechanical method	366

Polymer	T_g(K)	Remarks	References
3.1.3 Poly(esters) (Cont'd.)			
Poly(oxyneopentyleneoxycarbonyl-1,4-cyclohexylenecarbonyl) trans	303	Heating rate 20K/min	42
Poly(oxyneopentyleneoxyterephthaloyl)	341	Heating rate 20K/min	42
Poly(oxy-5-nonyl-1,3-phenyleneoxy-2-fluoroisophthaloyl)	307		432
Poly(oxy-5-nonyl-1,3-phenyleneoxy-5-fluoroisophthaloyl)	293		255
Poly(oxy-5-nonyl-1,3-phenyleneoxyisophthaloyl)	304		384
Poly(oxy-5-octyl-1,3-phenyleneoxyisophthaloyl)	314		384
Poly(oxy-5-pentyloxyisophthaloyloxy-4,4'-biphenylene)	411	Heating rate 20K/min	313
Poly(oxy-2-pentyloxyisophthaloyloxy-2,2,3,3,4,4-hexafluoropentamethylene)	290	DTA heating rate; η_{inh} only 0.12 dl/g	425
Poly(oxy-4-pentyloxyisophthaloyloxy-2,2,3,3,4,4-hexafluoropentamethylene)	282	DTA heating rate; η_{inh} only 0.25 dl/g	425
Poly(oxy-5-pentyloxyisophthaloyloxy-2,2,3,3,4,4-hexafluoropentamethylene)	243	DTA heating rate	425
Poly(oxy-5-pentyloxyisophthaloyloxy-1,4-phenylenemethylene-1,4-phenylene)	383	Heating rate 20K/min; η_{inh} only 0.13	313
Poly(oxy-3,3',5,5'-tetraphenyl-4,4'-biphenylyleneoxysebacoyl)	371	Heating rate 8K/min.	371
Poly(oxy-1,4-phenylenefluoren-9-ylidene-1,4-phenyleneoxysebacoyl)	424	Heating rate 10K/min	347
Poly(oxy-1,3-phenylenehexafluorotrimethylene-1,3-phenyleneoxycarbonyl-1,3-phenylenehexafluorotrimethylene-1,3-phenylenecarbonyl)	345		148
Poly(oxy-1,4-phenyleneisopropylidene-1,4-phenyleneoxycarbonyl-1,4-phenylene-sec-butylidene-1,4-phenylenecarbonyl)	~480		165
Poly(oxy-1,4-phenyleneisopropylidene-1,4-phenyleneoxycarbonyl-1,3-phenylene-hexafluorotrimethylene-1,3-phenylenecarbonyl)	389		433
Poly(oxy-1,4-phenyleneisopropylidene-1,4-phenyleneoxysebacoyl)	280	Heating rate 8K/min	363,371,409
Poly[oxy-1,3-phenyleneoxy-5-(10H-eicosafluorodecyl)isophthaloyl]	356		432
Poly[oxy-1,3-phenyleneoxy-5-(pentadecafluoroheptyl)isophthaloyl]	374		432
Poly(oxy-1,3-phenyleneoxy-2-fluoroisophthaloyl)	393	Heating rate 32K/min	255
Poly[oxy-1,3-phenyleneoxy-5-(heptafluoropropyl)isophthaloyl]	394	Sample of low η_{inh}	434
Poly(oxy-1,3-phenyleneoxyisophthaloyl)	411	Values range from 411-463K	313,384,427, 435-437
Poly(oxy-1,3-phenyleneoxy-5-pentyloxyisophthaloyl)	369	Heating rate 20K/min; η_{inh} only 0.13	313
Poly(oxy-2,6-diphenyl-1,4-phenylenemethylene-3,5-diphenyl-1,4-phenylene-oxysebacoyl)	365	Heating rate 8K/min	371
Poly(oxypimeloyloxy-3,3',5,5'-tetramethyl-4,4'-biphenylylene)	369	Heating rate 8K/min	371
Poly(oxypimeloyloxy-2,6-dimethyl-1,4-phenyleneisopropylidene-3,5-dimethyl-1,4-phenylene)	357	Heating rate 8K/min	371
Poly(oxypimeloyloxy-2,6-diphenyl-1,4-phenylenemethylene-3,5-diphenyl-1,4-phenylene)	384	Heating rate 8K/min	371
Poly(oxypropyleneoxycarbonyl-2,6-naphthylenecarbonyl)	361	No measurement details	387
Poly(oxypropyleneoxyterephthaloyl)	341		389
Poly(oxy-5-propyl-1,3-phenyleneoxyisophthaloyl)	394		384
Poly(oxyterephthaloyloxy-2-butyl-1,4-phenyleneisopropylidene-3-butyl-1,4-phenylene)	433		439
Poly(oxyterephthaloyloxy-2-sec-butyl-1,4-phenyleneisopropylidene-3-sec-butyl-1,4-phenylene)	373	Mechanical method	439
Poly(oxyterephthaloyloxy-2-chloro-1,4-phenyleneisopropylidene-3-chloro-1,4-phenylene)	463	Mechanical method	439
Poly(oxyterephthaloyloxy-2-isopropyl-1,4-phenyleneisopropylidene-3-isopropyl-1,4-phenylene)	403		439
Poly(oxyterephthaloyloxydecamethylene)	298, 268	Conflicting data	6,403,411
Poly(oxyterephthaloyloxyheptamethylene)	276	DTA heating rate	411
Poly(oxyterephthaloyloxyhexamethylene)	318, 264	Conflicting data	6,403,411
Poly(oxyterephthaloyloxynonamethylene)	308, 270	Conflicting data	6,403,411
Poly(oxyterephthaloyloxyoctamethylene)	318	Dynamic method	403
Poly(oxyterephthaloyloxypentamethylene)	318	Dynamic method	6,403
Poly(oxyterephthaloyloxymethylene-1,4-cyclohexylenemethylene) 70% trans	358	Heating rate 20K/min	42,440
Poly(oxyterephthaloyloxy-2-methyl-1,4-phenyleneisopropylidene-3-methyl-1,4-phenylene)	~428	Dilatometric method	366,439,441
	461	Dynamic method	
Poly(oxyterephthaloyloxy-2,6-dimethyl-1,4-phenyleneisopropylidene-3,5-di-methyl-1,4-phenylene)	498	Heating rate 8K/min	371
Poly(oxyterephthaloyloxy-1,4-phenylene-9,9-anthronylidene-1,4-phenylene)	570	Heating rate 10K/min	347
Poly(oxyterephthaloyloxy-1,4-phenylenebenzylidene-1,4-phenylene)	473		426

	Polymer	T_g(K)	Remarks	References
3.1.3	Poly(esters) (Cont'd.)			
	Poly(oxyterephthaloyloxy-1,4-phenylenecyclohexylmethylene-1,4-phenylene)	543	No measurement details	442
	Poly(oxyterephthaloyloxy-1,4-phenylenefluoren-9-ylidene-1,4-phenylene)	654	Heating rate 10K/min	347
	Poly(oxyterephthaloyloxy-1,4-phenylenehexafluoroisopropylidene-1,4-phenylene)	534	Softening point	443
	Poly(oxyterephthaloyloxy-1,4-phenylene-1,1-indanylidene-1,4-phenylene)	608	Heating rate 10K/min	347
	Poly(oxyterephthaloyloxy-1,4-phenyleneisopropylidene-1,4-phenylene)	478	DTA heating rate	363,369,371,382, 427,439,441,443
	Poly(oxy-5-tridecyl-1,3-phenyleneoxyisophthaloyl)	291		384
	Poly(oxy-5-undecyl-1,3-phenyleneoxyisophthaloyl)	295		384
	Poly(resorcinol isophthalate) see Poly(oxy-1,3-phenyleneoxyisophthaloyl)			
3.1.4	Poly(anhydrides)			
	Poly(oxycarbonyl-1,4-phenylenehexafluorotrimethylene-1,4-phenylene-carbonyl)	371		433
	Poly(oxycarbonyl-1,4-phenyleneisopropylidene-1,4-phenylenecarbonyl)	333		247
	Poly(oxycarbonyl-1,4-phenylenemethylene-1,4-phenylenecarbonyl)	395		247
	Poly(oxycarbonyl-1,4-phenylenepentamethylene-1,4-phenylenecarbonyl)	312	Amorphous	247
		319	Crystalline	
	Poly(oxycarbonyl-1,4-phenylenetetramethylene-1,4-phenylenecarbonyl)	319		247
	Poly(oxycarbonyl-1,4-phenylenethiotetramethylenethio-1,4-phenylenecarbonyl)	335		247
	Poly(oxyisophthaloyl)	403	Dynamic mechanical method	398,444
3.1.5	Poly(urethanes)			
	Poly(oxy-2-butenyleneoxycarbonyliminohexamethyleneiminocarbonyl) cis	234		383
	trans	229		
	Poly(oxy-2-butynyleneoxycarbonyliminohexamethyleneiminocarbonyl)	228		429
	Poly(oxycarbonyliminodecamethyleneiminocarbonyloxyhexadecamethylene)	307	Heating rate 1K/min	445
	Poly(oxycarbonyliminohexamethyleneiminocarbonyloxydodecamethylene)	309	Heating rate 1K/min	445
	Poly(oxycarbonyliminohexamethyleneiminocarbonyloxyhexadecamethylene)	312	Heating rate 1K/min	445
	Poly(oxycarbonyliminomethylene-1,4-phenylenemethyleneiminocarbonyloxy-decamethylene)	322	Heating rate 1K/min	445
	Poly(oxycarbonyliminomethylene-1,4-phenylenemethyleneiminocarbonyloxy-dodecamethylene)	318	Heating rate 1K/min	445
	Poly(oxycarbonyliminomethylene-1,4-phenylenemethyleneiminocarbonyloxy-hexadecamethylene)	320	Heating rate 1K/min	445
	Poly(oxycarbonylimino-4-methyl-1,3-phenyleneiminocarbonyloxydecamethylene)	291	Heating rate 10K/min	446
	Poly(oxycarbonylimino-4-methyl-1,3-phenyleneiminocarbonyloxynona-methylene)	335	Heating rate 10K/min	446
	Poly(oxycarbonylimino-4-methyl-1,3-phenyleneiminocarbonyloxyoctamethylene)	337	Heating rate 10K/min	446
	Poly(oxycarbonylimino-1,4-phenylenemethylene-1,4-phenyleneiminocarbonyl-oxyhexadecamethylene)	313	Heating rate 1K/min	445
	Poly(oxyethyleneoxycarbonyliminodecamethyleneiminocarbonyl)	334		447
	Poly(oxyethyleneoxycarbonyliminohexamethyleneiminocarbonyl)	329	Heating rate 10K/min	446
	Poly(oxyethyleneoxycarbonyliminononamethyleneiminocarbonyl)	317		447
	Poly(oxyethyleneoxycarbonylimino-4-methyl-1,3-phenyleneiminocarbonyl)	326	Heating rate 10K/min	446,448
	Poly(oxyethyleneoxycarbonylimino-1,4-phenyleneethylene-1,4-phenyleneimino-carbonyl)	390		447
	Poly(oxyethyleneoxycarbonylimino-1,4-phenylenemethylene-1,4-phenylene-iminocarbonyl)	412,366	Conflicting data	446,447
	Poly(oxyethyleneoxycarbonylimino-1,4-phenylenetetramethylene-1,4-phenylene-iminocarbonyl)	379		447
	Poly(oxyethyleneoxycarbonylimino-1,4-phenylenetrimethylene-1,4-phenylene-iminocarbonyl)	347		447
	Poly(oxy-2,2-diethyltrimethyleneoxycarbonylimino-4-methyl-1,3-phenylene-iminocarbonyl)	213		448
	Poly(oxy-2,2,3,3,4,4-hexafluoropentamethyleneoxycarbonylimino-2,2,3,3-4,4,5,5-octafluorohexamethyleneiminocarbonyl)	298	DTA heating rate	449
	Poly(oxy-2,2,3,3,4,4,5,5-octafluorohexamethyleneoxycarbonylimino-2,2,3,3-4,4,5,5-octafluorohexamethyleneiminocarbonyl)	293	DTA heating rate	449
	Poly(oxy-2,2,3,3-tetrafluorotetramethyleneoxycarbonylimino-2,2,3,3,4,4,5,5-octafluorohexamethyleneiminocarbonyl)	311	DTA heating rate	450

	Polymer	T_g(K)	Remarks	References
3.1.5	Poly(urethanes) (Cont'd.)			
	Poly(oxy-2,2,3,3,4,4-hexafluoropentamethyleneoxycarbonyliminohexamethylene-iminocarbonyl)	289	DTA heating rate	449
	Poly(oxy-2,2,3,3,4,4,5,5-octafluorohexamethyleneoxycarbonyliminohexa-methyleneiminocarbonyl)	271	DTA heating rate	451,452
	Poly(oxy-2,2,3,3-tetrafluorotetramethyleneoxycarbonyliminohexamethylene-iminocarbonyl)	282	DTA heating rate	450
	Poly(oxymethylene-5-tert-butyl-1,3-phenylenemethyleneoxycarbonylimino-1,4-phenylenemethylene-1,4-phenyleneiminocarbonyl)	387	DTA heating rate	453
	Poly(oxyhexamethyleneoxycarbonylimino-2,2,3,3,4,4,5,5-octafluorohexa-methyleneiminocarbonyl)	278	DTA heating rate	451,452
	Poly(oxypentamethyleneoxycarbonylimino-2,2,3,3,4,4,5,5-octafluorohexa-methyleneiminocarbonyl)	306	DTA heating rate	449
	Poly(oxytetramethyleneoxycarbonylimino-2,2,3,3,4,4,5,5-octafluorohexa-methyleneiminocarbonyl)	306	DTA heating rate	450
	Poly(oxydecamethyleneoxycarbonyliminohexamethyleneiminocarbonyl)	328	Heating rate 10K/min	446
	Poly(oxydodecamethyleneoxycarbonyliminodecamethyleneiminocarbonyl)	307	Heating rate 1K/min	445
	Poly(oxyheptamethyleneoxycarbonyliminohexamethyleneiminocarbonyl)	328	Heating rate 10K/min	446
	Poly(oxyhexamethyleneoxycarbonyliminohexamethyleneiminocarbonyl)	332	Heating rate 10K/min	446
	Poly(oxynonamethyleneoxycarbonyliminohexamethyleneiminocarbonyl)	331	Heating rate 10K/min	446
	Poly(oxyoctamethyleneoxycarbonyliminohexamethyleneiminocarbonyl)	331	Heating rate 10K/min	446
	Poly(oxypentamethyleneoxycarbonyliminohexamethyleneiminocarbonyl)	331,281	Conflicting data	446,449
	Poly(oxytetramethyleneoxycarbonyliminohexamethyleneiminocarbonyl)	215,253,303, 273,332	Conflicting data	160,261,344,429,446
	Poly(oxytrimethyleneoxycarbonyliminohexamethyleneiminocarbonyl)	328	Heating rate 10K/min	446
	Poly(oxyhexamethyleneoxycarbonyliminomethylene-1,4-phenylenemethylene-iminocarbonyl)	329	Heating rate 1K/min	445
	Poly(oxyheptamethyleneoxycarbonylimino-4-methyl-1,3-phenyleneimino-carbonyl)	334	Heating rate 10K/min	446
	Poly(oxyhexamethyleneoxycarbonylimino-4-methyl-1,3-phenyleneiminocarbonyl)	305	Heating rate 10K/min	446
	Poly(oxypentamethyleneoxycarbonylimino-4-methyl-1,3-phenyleneimino-carbonyl)	325	Heating rate 10K/min	446
	Poly(oxytetramethyleneoxycarbonylimino-4-methyl-1,3-phenyleneiminocarbonyl)	315	Heating rate 10K/min	446,454
	Poly(oxytrimethyleneoxycarbonylimino-4-methyl-1,3-phenyleneiminocarbonyl)	345,213	Conflicting data	446,448
	Poly(oxydecamethyleneoxycarbonylimino-1,4-phenylenemethylene-1,4-phenyleneiminocarbonyl)	321	Heating rate 1K/min	445,446
	Poly(oxydodecamethyleneoxycarbonylimino-1,4-phenylenemethylene-1,4-phenyleneiminocarbonyl)	316	Heating rate 1K/min	445
	Poly(oxyheptamethyleneoxycarbonylimino-1,4-phenylenemethylene-1,4-phenyleneiminocarbonyl)	357	Heating rate 10K/min	446
	Poly(oxyhexamethyleneoxycarbonylimino-1,4-phenylenemethylene-1,4-phenyleneiminocarbonyl)	324	Heating rate 1K/min	445,446
	Poly(oxynonamethyleneoxycarbonylimino-1,4-phenylenemethylene-1,4-phenyleneiminocarbonyl)	345	Heating rate 10K/min	446
	Poly(oxyoctamethyleneoxycarbonylimino-1,4-phenylenemethylene-1,4-phenyleneiminocarbonyl)	352	Heating rate 10K/min	446
	Poly(oxypentamethyleneoxycarbonylimino-1,4-phenylenemethylene-1,4-phenyleneiminocarbonyl)	368	Heating rate 10K/min	446
	Poly(oxytetramethyleneoxycarbonylimino-1,4-phenylenemethylene-1,4-phenyleneiminocarbonyl)	382	Heating rate 10K/min	446
	Poly(oxytrimethyleneoxycarbonylimino-1,4-phenylenemethylene-1,4-phenylene-iminocarbonyl)	392	Heating rate 10K/min	446
	Poly(oxymethylene-1,3-phenylenemethyleneoxycarbonylimino-1,4-phenylene-methylene-1,4-phenyleneiminocarbonyl)	379	Viscosity only 0.20	453
	Poly(oxyneopentyleneoxycarbonylimino-4-methyl-1,3-phenyleneiminocarbonyl)	241	DTA heating rate	448
3.2	Main-chain O-Heteroatom Polymers			
3.2.1	Poly(sulfonates)			
	Poly(oxysulfonyl-1,3-phenylenecarbonyl-1,3-phenylenesulfonyloxy-1,4-phenyleneisopropylidene-1,4-phenylene)	395	Mechanical method	361
	Poly(oxysulfonyl-1,3-phenylenesulfonyloxy-1,4-phenyleneisopropylidene-1,4-phenylene)	~385	Not well defined; mechanical method	361

Polymer	T_g(K)	Remarks	References
3.2 **Main-chain O-Heteroatom Polymers** (Cont'd.)			
3.2.2 **Nitroso-polymers**			
Poly(oxy-2-bromotetrafluoroethyliminotetrafluoroethylene)	256	Heating rate 15K/min	455
Poly(oxy-1,1-difluoroethyliminotetrafluoroethylene)	238	No measurement details	456
Poly(oxytrifluoromethyliminotetrafluoroethylene)	219		423,457-462
Polymer from trifluoronitrosomethane and bromotrifluoroethylene	276	Heating rate 15K/min	455,461
Polymer from trifluoronitrosomethane and chlorotrifluoroethylene	248-253		423
Polymer from trifluoronitrosomethane and perfluorobutadiene	266,230	Heating rate 15K/min; conflicting data	455,461
Polymer from trifluoronitrosomethane and 1,1,2-trifluorobutadiene	218	No measurement details	456
Polymer from trifluoronitrosomethane and trifluoromethyl trifluorovinyl ether	268	DSC heating rate	457
Polymer from trifluoronitrosomethane and trifluoromethyl trifluorovinyl sulfide	225	No measurement details	461
Polymer from trifluoronitrosomethane and hexafluoropropylene	264	No experimental details	456
3.2.3 **Poly(siloxanes)**			
Poly(dimethylsiloxane) see Poly(oxydimethylsilylene)			
Poly[oxydi(pentafluorophenyl)silylenedi(oxydimethylsilylene)]	231	No experimental details	463
Poly[oxymethylchlorotetrafluorophenylsilylenedi(oxydimethylsilylene)]	198	No experimental details	463
Poly(oxymethylpentafluorophenylsilylene)	248	No measurement details; [η] only 0.2 dl/g	463
Poly(oxymethylpentafluorophenylsilyleneoxydimethylsilylene)	208	No measurement details; [η] only 0.24 dl/g	463
Poly[oxymethylpentafluorophenylsilylenedi(oxydimethylsilylene)]	190	No measurement details	463
Poly(oxymethyl-3,3,3-trifluoropropylsilylene)	< 193		421,423,464, 465,466
Poly[oxy(methyl)-3,3,3-trifluoropropylsilyleneethylenedodecafluorohexamethylene-ethylene(methyl)-3,3,3-trifluoropropylsilylene]	249	DSC heating rate	466
Poly[oxy(methyl)-3,3,3-trifluoropropylsilyleneethyleneeicosafluorodeca-methyleneethylene(methyl)-3,3,3-trifluoropropylsilylene]	262	DSC heating rate	466
Poly[oxy(methyl)-3,3,3-trifluoropropylsilyleneethylenehexadecafluoroocta-methyleneethylene(methyl)-3,3,3-trifluoropropylsilylene]	245	DSC heating rate	466
Poly[oxy(methyl)-3,3,3-trifluoropropylsilyleneethyleneoctafluorotetra-methyleneethylene(methyl)-3,3,3-trifluoropropylsilylene]	248	DSC heating rate	466
Poly[oxy(methyl)-3,3,3-trifluoropropylsilylene-3,3-difluoropentamethylene-(methyl)-3,3,3-trifluoropropylsilylene]	235	DSC heating rate	466
Poly[oxy(methyl)-3,3,3-trifluoropropylsilylene-3,3,4,4-tetrafluorohexa-methylene(methyl)-3,3,3-trifluoropropylsilylene]	246	DSC heating rate	466
Poly[oxy(methyl)phenylsilylene]	187		467
Poly[oxy(methyl)phenylsilyleneoxy-1,4-phenylene]	368	Low viscosity	468
Poly[oxy(methyl)phenylsilyleneoxy-1,4-phenyleneisopropylidene-1,4-phenylene]	331	Low viscosity	468
Poly(oxydimethylsilylene)	146		1,6,223,243, 467,469-474
Poly(oxydimethylsilylene-2,4,5,6-tetrafluorophenylenedimethylsilylene)	245	No measurement details	463
Poly(oxydimethylsilyleneneopentyleneoxyisophthaloyloxyneopentylene-dimethylsilylene)	221		430
Poly(oxydimethylsilyleneneopentyleneoxyterephthaloyloxyneopentylene-dimethylsilylene)	238		431
Poly[tri(oxydimethylsilylene)oxy(methyl)trimethylsiloxysilylene]	148		475
Poly[tri(oxydimethylsilylene)oxy(methyl)-2-phenylethylsilylene]	171		475
Poly[tri(oxydimethylsilylene)oxy(methyl)phenylsilylene]	201		475
Poly[oxytri(dimethylsilyleneoxy)(methyl)phenylsilylene-1,3-phenylene-(methyl)phenylsilylene]	231	DTA heating rate	302
Poly(oxydimethylsilyleneoxydimethylsilylene-2,4,5,6-tetrafluorophenylene-dimethylsilylene)	212	No measurement details	463
Poly(oxydimethylsilyleneoxy-1,4-phenylene)	363	Mechanical method	468,476
Poly(oxydimethylsilyleneoxy-1,4-phenyleneisopropylidene-1,4-phenylene)	318	Mechanical method	468,476
Poly(oxydimethylsilylene-1,3-phenylenetetrafluoroethylene-1,3-phenylene-dimethylsilylene)	271	Heating rate 32K/min; crystalline samples	477
Poly(oxydimethylsilylene-1,3-phenylenehexafluorotrimethylene-1,3-phenylene-dimethylsilylene)	265	Heating rate 32K/min	148
Poly[di(oxydimethylsilylene)-1,4-phenylenedimethylsilylene]	210	DSC heating rate	478
Poly[penta(oxydimethylsilylene)-1,4-phenylenedimethylsilylene]	193	DTA heating rate	302
Poly[tetra(oxydimethylsilylene)-1,3-phenylenedimethylsilylene]	198	DTA heating rate	302

Polymer	$T_g(K)$	Remarks	References
3.2.3 Poly(siloxanes) (Cont' d.)			
Poly[tetra(oxydimethylsilylene)-1,4-phenylenedimethylsilylene]	201	DTA heating rate	302
Poly[tri(oxydimethylsilylene)-1,4-phenylenedimethylsilylene]	211	DTA heating rate	302
Poly(oxydimethylsilylene-1,4-phenyleneoxy-1,4-phenylenedimethylsilylene)	293	Heating rate 8K/min	301
Poly[penta(oxydimethylsilylene)-1,4-phenyleneoxy-1,4-phenylenedimethyl-silylene]	208	DTA heating rate	302
Poly[tetra(oxydimethylsilylene)-1,4-phenyleneoxy-1,4-phenylenedimethyl-silylene]	221	DTA heating rate	302
Poly[tri(oxydimethylsilylene)-1,4-phenyleneoxy-1,4-phenylenedimethylsilylene]	236	DTA heating rate	302
Poly[oxy-1,3-phenyleneoxy-5-(triphenylsiloxydimethylsilyl)isophthaloyl]	376	Heating rate 32K/min	438
Poly(oxydiphenylsilyleneoxydimethylsilylene-1,4-phenylenedimethylsilylene)	~273	DSC heating rate	478
Poly(oxydiphenylsilylene-1,3-phenylene)	~331	DTA heating rate	114
3.3 Main-chain -C-(S)-C- and -C-S-N-Polymers			
3.3.1 Poly(sulfides)			
Poly(oxycarbonyloxy-2-methyl-1,4-phenylenethio-3-methyl-1,4-phenylene)	340	No experimental details	338
Poly(oxycarbonyloxy-1,4-phenylenethio-1,4-phenylene)	~383	DTA heating rate	367
Poly(oxyethylenedithioethylene)	220	Mechanical method	248,249
Poly(oxyethylenetetrathioethylene)	233		248
Poly(oxymethyleneoxyethylenedithioethylene)	214		248,249,283
Poly(oxymethyleneoxytetramethylenedithiotetramethylene)	197		248
Poly(oxytetramethylenedithiotetramethylene)	197		248,249
Poly[thio-1-(allyloxymethyl)ethylene]	213		220
Poly(thiocarbonyl fluoride) see Poly(thiodifluoromethylene)			
Poly(thio-1,2-cyclohexylene)	~256,~228	Conflicting data	479,480
Poly(thio-1,3-cyclohexylene)	221	DTA heating rate	480
Poly(thioethylene) possibly	~253	Interpretation difficult	481-483
Poly(dithioethylene)	246	Mechanical method	248,249
Poly(tetrathioethylene)	249		248,249
Poly(thio-1-ethylethylene)	218		220
Poly(thio-2-ethyl-2-methyltrimethylene)	~223		160
Poly(thiodifluoromethylene)	155		484-486
Poly(thiomethylene)	218	No measurement details	482
Poly(thiotrimethylene)	~228	Conflicting data	480,487,488
Poly(dithiodecamethylene)	208	DTA heating rate	489
Poly(dithiohexamethylene)	199		248
Poly(dithiopentamethylene)	201		248
Poly(trithiodecamethylene)	203	DTA heating rate	489
Poly(tetrathiodecamethylene)	197	DTA heating rate	489
Poly(dithiomethylene-1,4-phenylenemethylene)	296	DTA heating rate	249,489
Poly(tetrathiomethylene-1,4-phenylenemethylene)	276	DTA heating rate	489
Poly(trithiomethylene-1,4-phenylenemethylene)	291	DTA heating rate	489
Poly(thio-1-methyltrimethylene)	214	DSC heating rate	490
Poly(thioneopentylene)	233	Heating rate 20K/min	160,480
Poly(thio-1,4-phenylene)	370	DSC heating rate	305,307,493,494
Poly(thiopropylene)	226		220,482,483, 495,496
Thiokols see Poly(thiomethylene)s			
3.3.2 Poly(thioesters)			
Poly(thio-3-methyl-6-oxohexamethylene)	293	DTA heating rate	497
Poly(thio-1-methyl-3-oxotrimethylene)	285		498
Poly(thio-6-oxohexamethylene)	292	DTA heating rate	499
3.3.3 Poly(sulfones), Poly(sulfonamides)			
Poly(oxy-2-acetoxytrimethyleneoxy-1,4-phenylenesulfonyl-1,4-phenylene)	403		219
Poly(oxy-4,4'-biphenylyleneoxy-1,4-phenylenesulfonyl-1,4-phenylene)	503	Mechanical method	116
Poly(oxycarbonylneopentylenesulfonylneopentylene)	323		160

Polymer	T_g(K)	Remarks	References

3.3.3 Poly(sulfones), Poly(sulfonamides) (Cont'd.)

Polymer	T_g(K)	Remarks	References
Poly[oxycarbonyldi(oxy-1,4-phenylene)sulfonyl-1,4-phenyleneoxy-1,4-phenylene]	~478	Mechanical method	228
Poly(oxy-2-hydroxytrimethyleneoxy-1,4-phenylenesulfonyl-1,4-phenylene)	428		219
Poly(oxytetramethyleneoxy-1,4-phenyleneisopropylidene-1,4-phenyleneoxy-1,4-phenylenesulfonyl-1,4-phenyleneoxy-1,4-phenyleneisopropylidene-1,4-phenylene)	413	Mechanical method	116
Poly(oxyneopentylenesulfonylneopentyleneoxycarbonyliminohexamethyleneiminocarbonyl)	303		160
Poly(oxy-5-pentyloxyisophthaloyloxy-1,4-phenylenesulfonyl-1,4-phenylene)	443	Heating rate 20K/min; η_{inh} only 0.11	313
Poly(oxy-1,4-phenyleneisopropylidene-1,4-phenyleneoxy-1,4-phenylenesulfonylmethyliminotetramethylenemethyliminosulfonyl-1,4-phenylene)	393	Mechanical method	116
Poly(oxy-1,4-phenylenesulfinyl-1,4-phenyleneoxy-1,4-phenylenecarbonyl-1,4-phenylene)	478	Mechanical method	116
Poly(oxy-1,4-phenylenesulfinyl-1,4-phenyleneoxy-1,4-phenyleneisopropylidene-1,4-phenylene)	438	Mechanical method	116
Poly(oxy-1,4-phenylenesulfonyl-4,4'-biphenylylenesulfonyl-1,4-phenylene)	533		184
Poly(oxy-1,4-phenylenesulfonyl-2,7-naphthylenesulfonyl-1,4-phenylene)	523		184
Poly(oxy-1,4-phenylenesulfonyl-1,4-phenylene)	487	Calculated T_g for infinite molecular weight: 511 K (ref.316)	116,184,314-316
Poly[di(oxy-1,4-phenylene)sulfonyl-1,4-phenylene]	483	Mechanical method	116
Poly[tri(oxy-1,4-phenylene)sulfonyl-1,4-phenylene]	453	Mechanical method	116
Poly(oxy-1,4-phenylenesulfonyl-1,4-phenyleneoxy-2-chloro-1,4-phenyleneisopropylidene-3-chloro-1,4-phenylene)	478	Mechanical method	116
Poly(oxy-1,4-phenylenesulfonyl-1,4-phenyleneoxy-2,6-dimethyl-1,4-phenyleneisopropylidene-3,5-dimethyl-1,4-phenylene)	508	Mechanical method	116
Poly(oxy-1,4-phenylenesulfonyl-1,4-phenyleneoxy-1,4-phenylenecarbonyl-1,4-phenylene)	478	Mechanical method	116
Poly(oxy-1,4-phenylenesulfonyl-1,4-phenyleneoxy-1,4-phenylenecyclohexylidene-1,4-phenylene)	478	Mechanical method	116
Poly(oxy-1,4-phenylenesulfonyl-1,4-phenyleneoxy-1,4-phenyleneethylidene-1,3-cyclohexylene-1,4-phenylene)	503	Mechanical method	116
Poly(oxy-1,4-phenylenesulfonyl-1,4-phenyleneoxy-1,4-phenylenehexafluoro-2,2-propylidene-1,4-phenylene)	478	Mechanical method	116
Poly(oxy-1,4-phenylenesulfonyl-1,4-phenyleneoxy-1,4-phenyleneisobutylidene-1,4-phenylene)	473	Mechanical method	116
Poly(oxy-1,4-phenylenesulfonyl-1,4-phenyleneoxy-1,4-phenyleneisopropylidene-1,4-phenylene)	449		102,116,262,287,316-323
Poly(oxy-1,4-phenylenesulfonyl-1,4-phenyleneoxy-1,4-phenylenemethylene-1,4-phenylene)	453	Mechanical method	116
Poly(oxy-1,4-phenylenesulfonyl-1,4-phenyleneoxy-1,4-phenylenemethylphenylmethylene-1,4-phenylene)	473	Mechanical method	116
Poly(oxy-1,4-phenylenesulfonyl-1,4-phenyleneoxy-1,4-phenylene-2,2-norbornylene-1,4-phenylene)	523	Mechanical method	116
Poly(oxy-1,4-phenylenesulfonyl-1,4-phenyleneoxy-1,4-phenylenediphenylmethylene-1,4-phenylene)	503	Mechanical method	116
Poly(oxy-1,4-phenylenesulfonyl-1,4-phenyleneoxy-1,4-phenylenethio-1,4-phenylene)	448	Mechanical method	116
Poly(oxy-1,4-phenylenesulfonyl-1,4-phenyleneoxyterephthaloyl)	522	Heating rate 20K/min	313
Poly(oxyterephthaloyloxyneopentylenesulfonylneopentylene)	378	? DTA heating rate; η low	160
Poly(sulfonyl-1,2-cyclohexylene)	401	Heating rate 20K/min	480
Poly(sulfonyl-1,3-cyclohexylene)	381	Heating rate 20K/min	480
Poly(sulfonylneopentylene)	386	Heating rate 20K/min	160,480
Poly(sulfonyl-1,4-phenylenemethylene-1,4-phenylene)	497	Heating rate 40K/min; calculated T_g at infinite molecular weight: 511K	316

3.4 Main-chain -C-N-C-Polymers

3.4.1 Poly(amides)

Poly(ε-caprolactam)   see   Nylon 6

Poly(hexamethylene adipamide)

                see   Nylon 6,6

	Polymer	T_g(K)	Remarks	References
3.4.1	Poly(amides) (Cont'd.)			
	Nylon 3 [Poly(imino-1-oxotrimethylene)]	384	Crystalline, DTA heating rate	502
	Nylon 4,6 [Poly(iminotetramethyleneiminoadipoyl)]	316	Dynamic method	503
	Nylon 5,6 [Poly(iminopentamethyleneiminoadipoyl)]	318	Heating rate 2K/min	504,505
	Nylon 6 [Poly(imino-1-oxohexamethylene)]	313-360	Conflicting data, most values ~313-325K	1,75,101,122,187, 191,261,262,318, 394,396,399,416, 506-520
	Nylon 6,6 [Poly(iminoadipoyliminohexamethylene)]	~323	Conflicting data, most values about 323K	6,75,78,161,170,171, 191,234,261,262,272, 317,326,344,381,388, 394,396,399,506.510, 521,522-533
	Nylon 6,7 [Poly(iminohexamethyleneiminopimeloyl)]	331		492,506
	Nylon 6,8 [Poly(iminohexamethyleneiminosuberoyl)]	330	Heating rate 0.5K/min	506
	Nylon 6,9 [Poly(iminohexamethyleneiminoazelaoyl)]	331		521
	Nylon 6,10 [Poly(iminohexamethyleneiminosebacoyl)]	323	Variable data below 323K	75,161,166,167,191, 399,503,505.506, 528,534,535
	Nylon 6,12 [Poly(iminohexamethyleneiminododecanedioyl)]	319	Heating rate 0.5K/min	506,535,536
	Nylon 7 [Poly(imino-1-oxoheptamethylene)]	325		191,506,537,538
	Nylon 7,6 [Poly(iminoadipoyliminoheptamethylene)]	318,333	Conflicting data	282,492,505,506
	Nylon 7,7 [Poly(iminopimeloyliminoheptamethylene)]	328	Heating rate 0.5K/min	506
	Nylon 8 [Poly(imino-1-oxooctamethylene)]	323		538
	Nylon 8,6 [Poly(iminoadipoyliminooctamethylene)]	318	Heating rate 2K/min	505
	Nylon 8,10 [Poly(iminooctamethyleneiminodecanedioyl)]	333	Heating rate 2K/min	505
	Nylon 8,12 [Poly(iminooctamethyleneiminododecanedioyl)]	323	Dynamic method	535
	Nylon 8,22 [Poly(iminooctamethyleneiminodocosanedioyl)]	321	Dynamic method	535
	Nylon 9 [Poly(imino-1-oxononamethylene)]	319		506,537,538
	Nylon 9,6 [Poly(iminoadipoyliminononamethylene)]	318	Estimated	504
	Nylon 10 [Poly(imino-1-oxodecamethylene)]	315		538
	Nylon 10,6 [Poly(iminoadipoyliminodecamethylene)]	313	Heating rate 2K/min	505
	Nylon 10,10 [Poly(iminosebacoyliminodecamethylene)]	333,319	Conflicting data	505,506
	Nylon 10,12 [Poly(iminodecamethyleneiminododecanedioyl)]	322	Dynamic method	535
	Nylon 11 [Poly(imino-1-oxoundecamethylene)]	316,365	Transitions affected by thermal history and relaxation effects	75,191,506,519, 538
	Nylon 12 [Poly(imino-1-oxododecamethylene)]	314	Affected by thermal history and relaxation effects	75,520,538
	Nylon 12,18 [Poly(iminododecamethyleneiminooctadecanedioyl)]	323	Dynamic method	535
	Nylon 13 [Poly(imino-1-oxotridecamethylene)]	314		538
	Nylon 14,18 [Poly(iminotetradecamethyleneiminooctadecanedioyl)]	321	Dynamic method	535
	Nylon 18,18 [Poly(iminooctadecanedioyliminooctadecamethylene)]	323	Dynamic method	535,536
	Poly(γ-benzyl L-glutamate)	~288	Dynamic method	273
	Poly(butyliminohexafluoroglutarylbutyliminohexamethylene)	~293	Brittle point	539
	Poly(butylimino-2,2,3,3,4,4-hexafluoropentamethylenebutyliminoadipoyl)	283-288	Brittle point	539
	Poly(ethyliminohexafluoroglutarylethyliminohexamethylene)	~278	Brittle point	539
	Poly(ethylimino-2,2,3,3,4,4-hexafluoropentamethyleneethyliminoadipoyl)	293	Brittle point	539
	Poly(iminoadipoylimino-1,4-cyclohexylenemethylene-1,4-cyclohexylene) 98% trans/trans, 2% cis/trans	458	T_g depends on cis/trans isomer content	540
	Poly(iminoadipoyliminotrimethylenefluoren-9-ylidenetrimethylene)	393	Mechanical method, heating rate 1K/min	78,161
	Poly(iminoadipoyliminotrimethyleneisobutylphosphinidenetrimethylene)	344	DTA heating rate	492
	Poly(iminoadipoyliminotrimethylenemethyliminotrimethylene)	278		492
	Poly(iminoadipoyliminomethylene-2,5-dimethyl-1,4-phenylenemethylene)	343	DTA heating rate	303
	Poly(iminoadipoyliminotrimethylenemethylphosphinylidenetetramethylene)	332	DTA heating rate	492
	Poly(iminoadipoyliminotrimethylenemethylphosphinylidenetrimethylene)	332	DTA heating rate	492
	Poly(iminoadipoyliminotrimethyleneoctylphosphinylidenetrimethylene)	285	DTA heating rate	492
	Poly[iminoadipoyliminotrimethylene(phenylphosphinidene)trimethylene]	322		492
	Poly[iminoadipoyliminotrimethylene(phenylphosphinylidene)trimethylene]	328		492
	Poly(imino-5-tert-butylisophthaloyliminohexamethylene)	436	DTA heating rate	453
	Poly(imino-5-tert-butylisophthaloyliminomethylene-1,4-cyclohexylenemethylene)	509	DTA heating rate	453

Polymer	T_g(K)	Remarks	References

3.4.1   Poly(amides)   (Cont' d.)

Polymer	T_g(K)	Remarks	References
Poly(imino-5-tert-butylisophthaloyliminomethylene-1,3-phenylenemethylene)	465	DTA heating rate	453
Poly(imino-5-tert-butylisophthaloyliminomethylene-1,4-phenylenemethylene)	477	DTA heating rate	453
Poly[imino-5-tert-butylisophthaloylimino(2,5-dimethylhexamethylene)]	422	DTA heating rate	453
Poly[imino-5-tert-butylisophthaloylimino(3,4-dimethylhexamethylene)]	446	DTA heating rate	453
Poly(iminocarbonyl-1,3-adamantylenecarbonyliminohexamethylene)	~385	Values from 378-393K obtained	255
Poly(iminocarbonyl-2,2'-biphenylylenecarbonyliminotrimethylenefluoren-9-ylidenetrimethylene)	438	Mechanical method, heating rate 1K/min	78,161
Poly(iminocarbonyl-1,4-cyclohexylenemethylene)	466	DTA heating rate	517
Poly(iminocarbonyl-1,4-phenylene-2-oxoethyleneiminohexamethylene)	377	Heating rate 20K/min	541
Poly(imino-1,4-cyclohexylenemethylene-1,4-cyclohexyleneiminododecanedioyl) 98% trans/trans, 2% cis/trans	408	T_g depends on cis/trans isomer content	540
Poly(imino-1,4-cyclohexylenemethylene-1,4-cyclohexyleneiminosebacoyl) 98% trans/trans, 2% cis/trans	417	T_g depends on cis/trans isomer content	540
Poly(iminoethylene-1,4-phenyleneethyleneimino-1,16-dioxohexadecamethylene)	358	Dynamic method	536
Poly(iminoethylene-1,4-phenyleneethyleneimino-1,18-dioxooctadecamethylene)	348	Dynamic method	536
Poly(iminoethylene-1,4-phenyleneethyleneimino-1,14-dioxotetradecamethylene)	366	Dynamic method	536
Poly(iminoethylene-1,4-phenyleneethyleneimino-1,11-dioxoundecamethylene)	369	Dynamic method	536
Poly(iminoethylene-1,4-phenyleneethyleneiminosebacoyl)	378	Dynamic method	536
Poly(iminoglutarylimino-2,2-dimethylpentamethylene)	355	DTA heating rate	542
Poly(iminoisophthaloylimino-4,4'-biphenylylene)	558	DTA heating rate	311,312
Poly(iminoisophthaloyliminohexamethylene)	390		78,161,255,543
Poly(iminoisophthaloyliminooctamethylene)	388	Dynamic method	544
Poly(iminoisophthaloyliminomethylene-1,4-cyclohexylenemethylene)	481	DSC heating rate	453
Poly(iminoisophthaloyliminotrimethylenefluoren-9-ylidenetrimethylene)	448	Mechanical method	78,161
Poly(iminoisophthaloyliminomethylene-1,3-phenylenemethylene)	438	Mechanical method, heating rate 1K/min	78,161,453
Poly(iminoisophthaloylimino-3,4-dimethylhexamethylene)	398	DSC heating rate	453
Poly(iminoisophthaloylimino-2,2-dimethylpentamethylene)	426	DTA heating rate	308
Poly(iminoisophthaloylimino-1,4-phenylenemethylene-1,4-phenylene)	> 500	Dynamic method	311
Poly(iminomesaconoyliminodecamethylene)	347	Heating rate 16K/min	545
Poly(iminomesaconoyliminohexamethylene)	367	Heating rate 16K/min	545
Poly(iminomesaconoyliminomethylene-1,3-phenylenemethylene)	408	Heating rate 16K/min	545
Poly(imino-3-methyladipoyliminohexamethylene)	290	DTA heating rate	492
Poly(iminomethylene-5-tert-butyl-1,3-cyclohexylenemethyleneiminoisophthaloyl)	482	DSC heating rate	453
Poly(iminomethylene-5-tert-butyl-1,3-phenylenemethyleneiminoadipoyl)	382	DTA heating rate	453
Poly(iminomethylene-5-tert-butyl-1,3-phenylenemethyleneiminoisophthaloyl)	461	DSC heating rate	453
Poly(iminomethylene-1,3-cyclohexylenemethyleneimino-5-tert-butylisophthaloyl)	473	DTA heating rate	453
Poly(iminomethylene-1,3-cyclohexylenemethyleneiminoisophthaloyl)	453	DSC heating rate	453
Poly(iminotrimethylenefluoren-9-ylidenetrimethyleneiminosebacoyl)	358	Mechanical method, heating rate 1K/min	78,161
Poly(iminohexamethyleneiminocarbonyl-2,2'-biphenylylenecarbonyl)	400	Mechanical method, heating rate 1K/min	78,161
Poly(iminohexamethyleneiminocarbonyl-1,4-phenylene-2,2-butylidene-1,4-phenylenecarbonyl)	427-437		165
Poly(iminotetramethyleneiminocarbonyl-1,4-phenylene-2,2-butylidene-1,4-phenylenecarbonyl)	446-455		165
Poly(iminopentamethyleneiminocarbonyl-1,4-phenylene-2-oxoethylene)	376	Heating rate 20K/min	541
Poly(iminotetramethyleneiminocarbonyl-1,4-phenylene-2-oxoethylene)	357	Heating rate 20K/min	541
Poly(iminotrimethyleneiminocarbonyl-1,4-phenylene-2-oxoethylene)	382	Heating rate 20K/min	541
Poly(iminohexamethyleneimino-4-methylpimeloyl)	323		521
Poly(iminooctamethyleneimino-1-oxoethylene-1,4-phenylene-2-oxoethylene)	383	Dynamic method	536
Poly(iminohexamethyleneimino-1-oxotrimethylenefluoren-9-ylidene-3-oxotrimethylene)	395	Mechanical method, heating rate 1K/min	78,161
Poly(iminohexamethyleneimino-1-oxotrimethylenephenylphosphinothioylidene-3-oxotrimethylene)	316		492
Poly(iminohexamethyleneimino-1-oxotrimethylenephenylphosphinylidene-3-oxotrimethylene)	302		492
Poly(iminopentamethyleneiminoterephthaloyl)	~500	Dynamic method	518
Poly(iminomethylene-1,3,3-trimethyl-1,5-cyclohexyleneiminoadipoyl)	433	No experimental details	381
Poly(iminomethylene-1,3,3-trimethyl-1,5-cyclohexyleneiminoterephthaloyl)	473	No experimental details	381
Poly[iminomethylene(2,5-dimethyl-1,4-phenylene)methyleneiminosuberoyl]	351	DTA heating rate	303
Poly(iminomethylene-1,3-phenylenemethyleneiminoadipoyl)	346	Mechanical method, heating rate 1K/min	78

Polymer	T_g (K)	Remarks	References

3.4.1　Poly(amides)　(Cont'd.)

Polymer	T_g (K)	Remarks	References
Poly(iminomethylene-1,3-phenylenemethyleneiminocarbonyl-2,2'-biphenylylene-carbonyl)	432	Mechanical method, heating rate 1K/min	78
Poly[iminomethylene-1,4-phenylenemethyleneiminocarbonyl(1,3,3-trimethyl-5-oxopentamethylene)]	398	No experimental details	381
Poly(iminomethylene-1,4-phenylenemethyleneiminododecanedioyl)	378		491,536
Poly(iminomethylene-1,4-phenylenemethyleneiminooctadecanedioyl)	348	Dynamic method	536
Poly(iminomethylene-1,3-phenylenemethyleneimino-1-oxotrimethylenefluoren-9-ylidene-3-oxotrimethylene)	423	Mechanical method, heating rate 1K/min	78,161
Poly(iminomethylene-1,4-phenylenemethyleneiminopentadecanedioyl)	363	Dynamic method	536
Poly(iminomethylene-1,4-phenylenemethyleneiminosebacoyl)	388	Dynamic method	536
Poly(iminomethylene-1,4-phenylenemethyleneiminotridecanedioyl)	373	Dynamic method	536
Poly(iminomethylene-1,4-phenylenemethyleneiminoundecanedioyl)	380		491,536
Poly(imino-5-methylisophthaloyliminohexamethylene)	393		255
Poly(imino-1-methyl-3-oxotrimethylene)	369	Crystalline; DTA heating rate	502
Poly(imino-2,2-dimethylpentamethyleneiminoadipoyl)	350	DTA heating rate	542
Poly(imino-2,4,4-trimethylhexamethyleneiminoadipoyl)	338	No measurement details	381
Poly(imino-2,2-dimethylpentamethyleneiminoazelaoyl)	336	DTA heating rate	542
Poly(imino-2,2-dimethylpentamethyleneiminopimeloyl)	344	DTA heating rate	542
Poly(imino-2,2-dimethylpentamethyleneiminoterephthaloyl)	430	DTA heating rate	308
Poly(imino-1,5-naphthyleneiminoisophthaloyl)	598	DTA heating rate	312
Poly(imino-1,5-naphthyleneiminoterephthaloyl)	578	Heating rate 10K/min	312
Poly(iminooxalylimino-2,2-dimethylpentamethylene)	382	DTA heating rate	542
Poly(imino-1-oxoethylene-1,4-phenylene-2-oxoethyleneiminooctadecamethylene)	351	Dynamic method	536
Poly(imino-1-oxotrimethylenefluoren-9-ylidene-3-oxotrimethyleneiminotri-methylenefluoren-9-ylidenetrimethylene)	438	Mechanical method, heating rate 1K/min	78
Poly(imino-1-oxotrimethylene-1,4-phenylene-3-oxotrimethyleneiminododeca-methylene)	358	Dynamic method	536
Poly(imino-1-oxotrimethylene-1,4-phenylene-3-oxotrimethyleneiminooctadeca-methylene)	338	Dynamic method	536
Poly(imino-1,3-phenyleneiminocarbonyl-1,3-adamantylenecarbonyl)	580	Heating rate 32K/min	255
Poly(imino-1,3-phenyleneiminoisophthaloyl)	553	Mechanical method, heating rate 2K/min	311,518,546
Poly(imino-1,4-phenyleneiminoisophthaloyl)	> 500	Dynamic method	311
Poly(imino-1,3-phenyleneimino-5-methylisophthaloyl)	570	Heating rate 02K/min	255
Poly(imino-1,3-phenyleneiminosebacoyl)	383	Dynamic method	544
Poly(imino-1,3-phenyleneiminoterephthaloyl)	> 500	Dynamic method	311
Poly(imino-1,4-phenyleneiminoterephthaloyl)	618	Heating rate 10K/min	311,312
Poly(iminoterephthaloylimino-4,4'-biphenylylene)	613	DTA heating rate	311,312
Poly(iminoterephthaloylimino-3-ethylhexamethylene)	403	DSC heating rate	381
Poly(iminoterephthaloylimino-3-isopropylhexamethylene)	416	DSC heating rate	381
Poly(iminoterephthaloyliminododecamethylene)	393	Dynamic method	543,547
Poly(iminoterephthaloyliminoheptamethylene)	470	Dynamic method	518
Poly(iminoterephthaloyliminohexamethylene)	413	No experimental details	381
Poly(iminoterephthaloyliminononamethylene)	388	No experimental details	381
Poly(iminoterephthaloyliminomethylene-2,5-dimethyl-1,4-phenylenemethylene)	498	DTA heating rate	303
Poly(iminoterephthaloylimino-4,4-dimethylheptamethylene)	425	No experimental details	381
Poly(iminoterephthaloylimino-1,4,4-trimethylheptamethylene)	423	DTA heating rate	548
Poly(iminoterephthaloylimino-2,2,4-trimethylhexamethylene)	418	DSC heating rate	262,381
Poly(iminoterephthaloylimino-2,4,4-trimethylhexamethylene)	421,432	Two different samples; DSC heating rate	381
Poly(iminoterephthaloylimino-1,4-phenylenemethylene-1,4-phenylene)	> 500	Dynamic method	311
Poly(isopropyliminohexafluoroglutarylisopropyliminohexamethylene)	~ 293	Brittle point	539
Poly(isopropylimino-2,2,3,3,4,4-hexafluoropentamethyleneisopropylimino-adipoyl)	303-308	Brittle point	539
Poly(methyliminohexafluoroglutarylmethyliminohexamethylene)	298-303	Brittle point	539
Poly(methylimino-2,2,3,3,4,4-hexafluoropentamethylenemethyliminoadipoyl)	298-303	Brittle point	539
Poly[di(oxyethylene)oxycarbonyliminohexamethyleneiminocarbonyl]	272	Heating rate 1-2K/min	245
Poly[tri(oxyethylene)oxycarbonyliminohexamethyleneiminocarbonyl]	260		246
Poly(oxytrimethyleneiminoadipoyliminotrimethylene)	307		282
Poly(oxy-3-oxotrimethyleneiminomethylene-2,5-dimethyl-1,4-phenylene-methyleneimino-1-oxotrimethylene)	353	DTA heating rate	303

	Polymer	T_g(K)	Remarks	References
3.4.1	Poly(amides) (Cont' d.)			
	Poly(oxy-5-oxopentamethyleneiminomethylene-1,4-phenylenemethyleneimino-1-oxopentamethylene)	343	Estimated from copolymer data	304
	Poly(oxy-1,4-phenylenecarbonylimino-2,2-dimethylpentamethyleneiminocarbonyl-1,4-phenylene)	428	DTA heating rate	308
	Poly(oxy-1,4-phenylenecarbonylimino-1,4-phenylenesulfonyl-1,4-phenyleneiminocarbonyl-1,4-phenylene)	571	Softening point; mechanical method	309
	Poly(oxy-1,4-phenyleneiminoisophthaloylimino-1,4-phenylene)	463,554	Conflicting data	130,311,312
	Poly(oxy-1,4-phenyleneiminoterephthaloylimino-1,4-phenylene)	613	Heating rate 10K/min	311,312
	Poly(sulfonyl-1,3-phenyleneiminoadipoylimino-1,3-phenylene)	413	DTA heating rate	500,501
	Poly(sulfonyl-1,4-phenyleneiminoadipoylimino-1,4-phenylene)	467	DTA heating rate	500,501
	Poly(sulfonyl-1,3-phenyleneiminoazelaoylimino-1,3-phenylene)	398	DTA heating rate	500,501
	Poly(sulfonyl-1,4-phenyleneiminoazelaoylimino-1,4-phenylene)	451	DTA heating rate	500,501
	Poly(sulfonyl-1,3-phenyleneiminocarbonyl-1,4-naphthylenecarbonylimino-1,3-phenylene)	> 573	Softening point; mechanical method	309
	Poly(sulfonyl-1,4-phenyleneiminocarbonyl-1,4-naphthylenecarbonylimino-1,4-phenylene)	> 573	Softening point; mechanical method	309
	Poly(sulfonyl-1,4-phenyleneiminocarbonyl-1,4-phenylenemethylene-1,4-phenylenecarbonylimino-1,4-phenylene)	> 573	Softening point	309
	Poly(sulfonyl-1,3-phenyleneiminododecanedioylimino-1,3-phenylene)	380	DTA heating rate	500,501
	Poly(sulfonyl-1,4-phenyleneiminododecanedioylimino-1,4-phenylene)	433	DTA heating rate	500,501
	Poly(sulfonyl-1,3-phenyleneiminoisophthaloylimino-1,3-phenylene)	573-593	Softening point; mechanical method	309
	Poly(sulfonyl-1,4-phenyleneimino-2-methoxyisophthaloylimino-1,4-phenylene)	568-583	Softening point	309
	Poly(sulfonyl-1,4-phenyleneiminopimeloylimino-1,4-phenylene)	436	DTA heating rate	500,501
	Poly(sulfonyl-1,3-phenyleneiminosebacoylimino-1,3-phenylene)	385	DTA heating rate	500,501
	Poly(sulfonyl-1,4-phenyleneiminosebacoylimino-1,4-phenylene)	444	DTA heating rate	500,501
	Poly(sulfonyl-1,3-phenyleneiminosuberoylimino-1,3-phenylene)	398	DTA heating rate	500,501
	Poly(sulfonyl-1,4-phenyleneiminosuberoylimino-1,4-phenylene)	453	Softening point	500,501
	Poly(sulfonyl-1,4-phenyleneiminoterephthaloyl-1,4-phenylenecarbonylimino-1,4-phenylene)	590	Softening point; mechanical method	309
	Poly(thio-11-oxoundecamethyleneiminoethylene-1,4-phenyleneethyleneimino-1-oxoundecamethylene)	331	Dynamic method; low viscosity	491
	Poly(thio-3-oxotrimethyleneiminohexamethyleneimino-1-oxotrimethylene)	287-300		492
	Poly(thio-7-oxoheptamethyleneiminomethylene-1,4-phenylenemethyleneimino-1-oxoheptamethylene)	338	Dynamic method	491
	Poly(thio-5-oxopentamethyleneiminomethylene-1,4-phenylenemethyleneimino-1-oxopentamethylene)	343	Dynamic method	491
3.4.2	Poly(imines)			
	Poly(acetyliminoethylene)	~353	Mechanical method	549
	Poly(acetyliminotrimethylene)	~303	Mechanical method; low viscosity	550
	Poly(benzoyliminoethylene)	378	Mechanical method	549
	Poly(benzoyliminotrimethylene)	345	Mechanical method	550
	Poly(butyryliminoethylene)	~303	Mechanical method	549
	Poly(dodecanoyliminoethylene)	~283	Mechanical method - apparent T_g	549
	Poly[(pentadecafluorooctanoylimino)trimethylene]	298	Mechanical method	550
	Poly(heptanoyliminoethylene)	~283	Mechanical method - apparent T_g	549
	Poly(hexanoyliminoethylene)	~283	Mechanical method - apparent T_g	549
	Poly(hexanoyliminotrimethylene)	257	Mechanical method	550
	Poly(isobutyryliminoethylene)	~303	Mechanical method	549
	Poly(isovaleryliminoethylene)	~303	Mechanical method	549
	Poly(3-methoxycarbonylpropionyliminoethylene)	304	Mechanical method	551
	Poly(2-naphthoyliminoethylene)	403	Mechanical method	549
	Poly(octadecanoyliminoethylene)	~283	Mechanical method - apparent T_g	549
	Poly(octanoyliminoethylene)	~283	Mechanical method - apparent T_g	549
	Poly(propionyliminoethylene)	~343	Mechanical method	549
	Poly(propionyliminotrimethylene)	281	Mechanical method; low viscosity	550
	Poly(valeryliminoethylene)	~286	Mechanical method	549
3.4.3	Poly(ureas)			
	Poly(ureylenemethylene-1,4-phenylenemethyleneureylenedecamethylene)	328	Heating rate 1K/min	445
	Poly(ureylenemethylene-1,4-phenylenemethyleneureylenehexamethylene)	345	Heating rate 1K/min	445

Polymer	T_g(K)	Remarks	References

3.4.3 Poly(ureas) (Cont'd.)

Polymer	T_g(K)	Remarks	References
Poly(ureylenemethylene-1,4-phenylenemethyleneureyleneoctadecamethylene)	325	Heating rate 1K/min	445
Poly(ureylenemethylene-1,4-phenylenemethyleneureyleneoctamethylene)	341	Heating rate 1K/min	445
Poly(ureylenemethylene-1,4-phenylenemethyleneureylenetetradecamethylene)	327	Heating rate 1K/min	445
Poly(ureylenedecamethyleneureylenetetradecamethylene)	313	Heating rate 1K/min	445
Poly(ureylenehexamethyleneureylenedodecamethylene)	322	Heating rate 1K/min	445
Poly(ureylenehexamethyleneureylenetetradecamethylene)	320	Heating rate 1K/min	445
Poly(ureylenehexamethyleneureylene-1,4-phenylenemethylene-1,4-phenylene)	328	Heating rate 1K/min	445
Poly(ureyleneoctamethyleneureylene-1,4-phenylenemethylene-1,4-phenylene)	323	Heating rate 1K/min	445
Poly(ureylene-1,4-phenylenemethylene-1,4-phenyleneureylenedecamethylene)	319	Heating rate 1K/min	445
Poly(ureylene-1,4-phenylenemethylene-1,4-phenyleneureylenedodecamethylene)	324	Heating rate 1K/min	445
Poly(ureylene-1,4-phenylenemethylene-1,4-phenyleneureyleneoctadecamethylene)	321	Heating rate 1K/min	445

3.5 Poly(phosphazenes)

Polymer	T_g(K)	Remarks	References
Poly(chloro-2,2,2-trifluoroethoxyphosphazene) (average structure)	213	Heating rate 20K/min; probably branched structure	552
Poly(dichlorophosphazene)	210	DTA heating rate	553
Poly(diethoxyphosphazene)	189	DTA heating rate	553
Poly[bis(ethylamino)phosphazene]	303	DTA heating rate	554
Poly[bis(2,2,2-trifluoroethoxy)phosphazene]	203		552,555,556
Poly[bis(3-trifluoromethylphenoxy)phosphazene]	238	Heating rate 20K/min	552
Poly[bis(1H,1H-pentadecafluorooctyloxy)phosphazene]	233	Heating rate 20K/min	552
Poly[bis(4-fluorophenoxy)phosphazene]	259	Heating rate 20K/min	552
Poly[bis(1H,1H-pentafluoropropoxy)phosphazene]	198,218		557
Poly(dimethoxyphosphazene)	197	DTA heating rate	553
Poly[bis(dimethylamino)phosphazene]	269	Dynamic mechanical method	554
Poly(diphenoxyphosphazene)	265	DTA heating rate	553
Poly[bis(phenylamino)phosphazene]	364	DTA heating rate	554
Poly[bis(piperidino)phosphazene]	292		554

3.6 Poly(silanes) and Poly(silazanes)

Polymer	T_g(K)	Remarks	References
Poly[(4-dimethylaminophenyl)methylsilylenetrimethylene]	267		558
Poly[(4-dimethylaminophenyl)phenylsilylenetrimethylene]	225		558
Poly[(methyl)phenylsilylenetrimethylene]	243	No measurement details	559
Poly(1,1-dimethylsilazane)	191		560,561
Poly(dimethylsilylenemethylene)	173	No measurement details	562
Poly(dimethylsilylenetrimethylene)	203		559
Poly(di-p-tolylsilylenetrimethylene)	311	Mechanical method	558

4. Main-chain Heterocyclic Polymers

4.1 Poly(furan tetracarboxylic acid diimides)

Polymer	T_g(K)	Remarks	References
Poly(4,5,6,8,9,10-hexahydro-1,3,6,8-tetraoxofuro[3,2-c:4,5-c']dipyrrole-2,7[1H,3H]-diyl-1,4-phenylene-3-phenylquinoxaline-2,7-diylcarbonyl-3-phenylquinoxaline-7,2-diyl-1,4-phenylene)	528	Dynamic method	111

Polymer	T_g(K)	Remarks	References
Poly(4,5,6,8,9,10-hexahydro-1,3,6,8-tetraoxofuro[3,2-c:4,5-c']dipyrrole-2,7[1H,3H]-diyl-1,4-phenylene-3-phenylquinoxaline-2,7-diyloxy-3-phenylquinoxaline-7,2-diyl-1,4-phenylene)	583	Dynamic method	111

Polymer	T_g(K)	Remarks	References

4.1   <u>Poly(furan tetracarboxylic acid diimides)</u>   (Cont'd.)

Poly(4,5,6,8,9,10-hexahydro-1,3,6,8-tetraoxofuro[3,2-c:4,5-c']dipyrrole-2,7[1H,3H]-diyl-1,4-phenylene-3-phenylquinoxaline-2,7-diyl-3-phenylquinoxaline-7,2-diyl-1,4-phenylene)   503   Dynamic method   111

4.2   <u>Poly(benzoxazoles)</u>

Polymer	T_g(K)	Remarks	References
Poly(2,6-benzoxazolediyl-6,2-benzoxazolediyloctamethylene)	623	No method	112
Poly(benzoxazole-5,2-diyloctamethylenebenzoxazole-2,5-diyliminoadipoylimino)	358	Heating rate 20K/min	113
Poly(benzoxazole-5,2-diylhexamethylenebenzoxazole-2,5-diyliminosebacoylimino)	359	Heating rate 20K/min	113
Poly(benzoxazole-5,2-diyloctamethylenebenzoxazole-2,5-diyliminosebacoylimino)	366	Heating rate 20K/min	113
Poly(benzoxazole-5,2-diylhexamethylenebenzoxazole-2,5-diyliminosuberoylimino)	350	Heating rate 20K/min	113
Poly(benzoxazole-5,2-diyl-1,3-phenylenebenzoxazole-2,5-diyliminosebacoylimino)	393	Heating rate 20K/min	113

4.3   <u>Poly(oxadiazoles)</u>

Polymer	T_g(K)	Remarks	References
Poly(1,3,4-oxadiazolediyl-1,3-phenylene-1,3,4-oxadiazolediyl-1,4-phenylene)	551	DTA heating rate	114
Poly(1,3,4-oxadiazolediyl-1,4-phenyleneoxy-1,4-phenylene)	513	Mechanical method	115
Poly(1,3,4-oxadiazolediyl-1,4-phenyleneoxy-1,4-phenyleneisopropylidene-1,4-phenyleneoxy-1,4-phenylene)	453	Mechanical method	116
Poly(1,3,4-oxadiazolediyl-1,4-phenylene-3,3-phthalidylidene-1,4-phenylene)	653	Mechanical method	115

4.4   <u>Poly(benzothiazinophenothiazines)</u>

Polymer	T_g(K)	Remarks	References
Poly(7,7'-[1",4"-dihydro(2,3H)benzothiazino(2',3'-b)hexahydro(1,2,3,4H)-phenothiazine]diyl)	483	Dynamic method	117

4.5   <u>Poly(benzothiazoles)</u>

Polymer	T_g(K)	Remarks	References
Poly(2,6-benzothiazolediyl-6,2-benzothiazolediyloctamethylene)	511	No method	112
Poly(2,6-benzothiazolediyl-6,2-benzothiazolediyl-1,3-phenylene)	768	Dynamic method	118

4.6   <u>Poly(pyrazinoquinoxalines)</u>

Polymer	T_g(K)	Remarks	References
Poly[3,7-diphenylpyrazino(2,3-g)quinoxaline-2,8-diyl-1,3-phenylene]	638	Heating rate 20K/min	119
Poly[3,7-diphenylpyrazino(2,3-g)quinoxaline-2,8-diyl-1,4-phenylene]	668	Heating rate 20K/min	119
Poly[pyrazino(2,3-g)quinoxaline-2,8-diyl-1,4-phenylene]	665	Heating rate 20K/min	119
Poly[pyrazino(2,3-g)quinoxaline-2,8-diyl-1,4-phenyleneoxy-1,4-phenylene]	626	Heating rate 20K/min	119

4.7   <u>Poly(pyromellitimides)</u>

Polymer	T_g(K)	Remarks	References
Poly(5,7-dihydro-1,3,5,7-tetraoxobenzo[1,2-c:4,5-c']dipyrrole-2,6[1H,3H]-diyl-2,7-fluorenylene)	~623	Heating rate 3K/min	120
Poly(5,7-dihydro-1,3,5,7-tetraoxobenzo[1,2-c:4,5-c']dipyrrole-2,6[1H,3H]-diylnonamethylene)	383	No experimental details	121
Poly(5,7-dihydro-1,3,5,7-tetraoxobenzo[1,2-c:4,5-c']dipyrrole-2,6[1H,3H]-diyl-3-methylheptamethylene)	408	No experimental details	121
Poly(5,7-dihydro-1,3,5,7-tetraoxobenzo[1,2-c:4,5-c']dipyrrole-2,6[1H,3H]-diyl-4,4-dimethylheptamethylene)	408	No experimental details	121
Poly(5,7-dihydro-1,3,5,7-tetraoxobenzo[1,2-c:4,5-c']dipyrrole-2,6[1H,3H]-diyl-1,4-phenyleneoxy-1,4-phenylene) probably >773		Conflicting data, transitions from 523K upwards reported as T_g	119,120,122-131

Polymer	T_g(K)	Remarks	References

4.7   Poly(pyromellitimides)   (Cont'd.)

Polymer	T_g(K)	Remarks	References
Poly(5,7-dihydro-1,3,5,7-tetraoxobenzo[1,2-c:4,5-c'] dipyrrole-2,6[1H,3H] - diyl-1,4-phenylene-3-phenylquinoxaline-2,7-diylcarbonyl-3-phenyl-quinoxaline-7,2-diyl-1,4-phenylene)	513	Dynamic method, heating rate 5K/min	132

(chemical structure)

| Poly(5,7-dihydro-1,3,5,7-tetraoxobenzo[1,2-c:4,5-c'] dipyrrole-2,6[1H,3H] -diyl-1,4-phenylene-3-phenylquinoxaline-2,7-diyloxy-3-phenylquinoxaline-7,2-diyl-1,4-phenylene) | 547 | Dynamic method, heating rate 5K/min | 132 |
| Poly(5,7-dihydro-1,3,5,7-tetraoxobenzo[1,2-c:4,5-c'] dipyrrole-2,6[1H,3H] - diyl-1,4-phenylene-3-phenylquinoxaline-2,7-diyl-3-phenylquinoxaline-7,2-diyl-1,4-phenylene) | 490 | Dynamic method, heating rate 5K/min | 132 |

(chemical structure)

Poly(5,7-dihydro-1,3,5,7-tetraoxobenzo[1,2-c:4,5-c'] dipyrrole-2,6[1H,3H] -diyl-1,4-phenylene-3-phenylquinoxaline-2,7-diylsulfonyl-3-phenylquinoxaline-7,2-diyl-1,4-phenylene)	459	Dynamic method, heating rate 5K/min	132
Poly(5,7-dihydro-1,3,5,7-tetraoxobenzo[1,2-c:4,5-c']dipyrrole-2,6[1H,3H] -diyl-1,4-phenylenethio-1,4-phenylene)	618	Dynamic method	125
Poly[N,N'(pp'-oxydiphenylene)pyromellitimide] see Poly(5,7-dihydro-1,3,5,7-tetraoxobenzo[1,2-c:4,5-c'] -dipyrrole-2,6[1H,3H] -diyl-1,4-phenyleneoxy-1,4-phenylene)			

4.8   Poly(quinoxalines)

Polymer	T_g(K)	Remarks	References
Poly(3-phenylquinoxaline-2,6-diylcarbonyl-3-phenylquinoxaline-7,2-diyl-1,4-phenylene)	583	Dynamic method	133
Poly(3-phenylquinoxaline-2,7-diylcarbonyl-3-phenylquinoxaline-7,2-diyl-1,4-phenylene)	595	Values range from 531-651K	119,134,135
Poly(3-phenylquinoxaline-2,7-diylcarbonyl-3-phenylquinoxaline-7,2-diyl-1,4-phenylene-1,3-dioxoisoindoline-2,5-diylcarbonyl-1,3-dioxoisoindoline-5,2-diyl-1,4-phenylene)	528	Dynamic method, heating rate 5K/min	111
Poly(3-phenylquinoxaline-2,7-diylcarbonyl-3-phenylquinoxaline-7,2-diyl-1,4-phenyleneoxy-1,4-phenylene)	516,544	Conflicting data	119,134
Poly(3-phenylquinoxaline-2,6-diyloxy-3-phenylquinoxaline-6,2-diyl-4,4'-bi-phenylene)	683	Dynamic method	136
Poly(3-phenylquinoxaline-2,6-diyloxy-3-phenylquinoxaline-7,2-diyl-1,4-phenylylene)	565	Dynamic method	133
Poly(3-phenylquinoxaline-2,7-diyloxy-3-phenylquinoxaline-7,2-diyl-1,3-phenylene)	526,543	Conflicting data possibly arising from different rate effects	119,134,137
Poly(3-phenylquinoxaline-2,7-diyloxy-3-phenylquinoxaline-7,2-diyl-1,4-phenylene)	571	Dynamic method	119,134,137-139
Poly(3-phenylquinoxaline-2,7-diyloxy-3-phenylquinoxaline-7,2-diyl-1,4-phenylene-1,3-dioxoisoindoline-2,5-diylcarbonyl-1,3-dioxoisoindoline-5,2-diyl-1,4-phenylene)	527	Dynamic method	111
Poly(3-phenylquinoxaline-2,6-diyloxy-3-phenylquinoxaline-6,2-diyl-1,4-phenyleneoxy-1,4-phenylene)	710	Dynamic method	136
Poly(3-phenylquinoxaline-2,6-diyloxy-3-phenylquinoxaline-7,2-diyl-1,4-phenyleneoxy-1,4-phenylene)	516	Dynamic method	133

Polymer	T_g(K)	Remarks	References
4. 8 <u>Poly(quinoxalines)</u> (Cont'd.)			
Poly(3-phenylquinoxaline-2,7-diyloxy-3-phenylquinoxaline-7,2-diyl-1,4-phenyleneoxy-1,4-phenylene)	541		119,134,137
Poly(3-phenylquinoxaline-2,6-diyl-3-phenylquinoxaline-6,2-diyl-4,4'-biphenylylene)	677	Dynamic method	136
Poly(3-phenylquinoxaline-2,6-diyl-3-phenylquinoxaline-7,2-diyl-1,4-phenylene)	639-698	Dynamic methods	133
Poly(3-phenylquinoxaline-2,7-diyl-3-phenylquinoxaline-7,2-diyl-1,3-phenylene)	593	Dynamic method	134,137
Poly(3-phenylquinoxaline-2,7-diyl-3-phenylquinoxaline-7,2-diyl-1,4-phenylene) range	590-638	Conflicting data; dilatometric T_g 638K	119,134,139
Poly(3-phenylquinoxaline-2,7-diyl-3-phenylquinoxaline-7,2-diyl-1,4-phenylene-1,3-dioxoisoindoline-2,5-diylcarbonyl-1,3-dioxoisoindoline-5,2-diyl-1,4-phenylene)	533	Dynamic method, heating rate 5K/min	111
Poly(3-phenylquinoxaline-2,6-diyl-3-phenylquinoxaline-6,2-diyl-1,4-phenyleneoxy-1,4-phenylene)	567	Values range from 558-693K	119,134,136,140
Poly(3-phenylquinoxaline-2,6-diyl-3-phenylquinoxaline-7,2-diyl-1,4-phenyleneoxy-1,4-phenylene)	558	Dynamic method	133
Poly(3-phenylquinoxaline-2,7-diyl-3-phenylquinoxaline-7,2-diyl-1,4-phenylenethio-1,4-phenylene)	561,533	Conflicting data	119,134
Poly(3-phenylquinoxaline-2,6-diylsulfonyl-3-phenylquinoxaline-7,2-diyl-1,4-phenylene)	543	Dynamic method	133
Poly(3-phenylquinoxaline-2,7-diylsulfonyl-3-phenylquinoxaline-7,2-diyl-1,4-phenylene)	485-618	Conflicting data	119,134,135
Poly(3-phenylquinoxaline-2,7-diylsulfonyl-3-phenylquinoxaline-7,2-diyl-1,4-phenylene-1,3-dioxoisoindoline-2,5-diylcarbonyl-1,3-dioxoisoindoline-5,2-diyl-1,4-phenylene)	513	Dynamic method, heating rate 5K/min	111
Poly(3-phenylquinoxaline-2,7-diylsulfonyl-3-phenylquinoxaline-7,2-diyl-1,4-phenyleneoxy-1,4-phenylene)	468,563	Conflicting data	119,134
Poly(quinoxaline-2,7-diylcarbonylquinoxaline-7,2-diyl-1,4-phenylene)	591	Heating rate 20K/min	119
Poly(quinoxaline-2,6-diyloxyquinoxaline-6,2-diyl-4,4'-biphenylylene)	663	Dynamic method	136
Poly(quinoxaline-2,7-diyloxyquinoxaline-7,2-diyl-1,4-phenylene)	579	Heating rate 20K/min	119
Poly(quinoxaline-2,6-diyloxyquinoxaline-6,2-diyl-1,4-phenyleneoxy-1,4-phenylene)	655	Dynamic method	136
Poly(quinoxaline-2,6-diyloxyquinoxaline-7,2-diyl-1,4-phenyleneoxy-1,4-phenylene)	543		137
Poly(quinoxaline-2,6-diylquinoxaline-6,2-diyl-4,4'-biphenylylene)	659	Dynamic method	136,141
Poly(quinoxaline-2,6-diylquinoxaline-6,2-diyl-1,4-phenylene)	623	Dynamic method	141
Poly(quinoxaline-2,7-diylquinoxaline-7,2-diyl-1,4-phenylene)	649	Heating rate 20K/min	119
Poly(quinoxaline-2,6-diylquinoxaline-6,2-diyl-1,4-phenyleneoxy-4,4'-biphenylyleneoxy-1,4-phenylene)	486	Dynamic method	141
Poly(quinoxaline-2,6-diylquinoxaline-6,2-diyl-1,4-phenyleneoxy-1,4-phenylene)	661,553	Dynamic method, conflicting data	136,137,141
Poly(quinoxaline-2,7-diylquinoxaline-7,2-diyl-1,4-phenyleneoxy-1,4-phenylene)	580		119
Poly[quinoxaline-2,6-diylquinoxaline-6,2-diyldi(1,4-phenyleneoxy)-1,4-phenylene]	468	Dynamic method	141
Poly[quinoxaline-2,7-diylquinoxaline-7,2-diyldi(1,4-phenyleneoxy)-1,4-phenylene]	508	Heating rate 20K/min	119
Poly[quinoxaline-2,6-diylquinoxaline-6,2-diyltri(1,4-phenyleneoxy)-1,4-phenylene]	406	Dynamic method	141
Poly[quinoxaline-2,7-diylquinoxaline-7,2-diyltri(1,4-phenyleneoxy)-1,4-phenylene]	489	Heating rate 20K/min	119
Poly(quinoxaline-2,7-diylquinoxaline-7,2-diyl-p-terphenyl-4,4"-ylene)	578	Heating rate 20K/min	119
Poly(quinoxaline-2,7-diylsulfonylquinoxaline-7,2-diyl-1,4-phenylene)	615	Heating rate 20K/min	119
4. 9 <u>Poly(benzimidazoles)</u>			
Poly(2,6-benzimidazolediyl-6,2-benzimidazolediyloctamethylene)	507	No experimental details	112
Poly(2,5-benzimidazolediyl-6,2-benzimidazolediylpentamethyleneimino-4-diethylamino-1,3,5-triazinediyliminopentamethylene)	463	Heating rate 2K/min	142
Poly(2,6-benzimidazolediyl-6,2-benzimidazolediyltrimethylene)	548	No experimental details	112
Poly(2,5-benzimidazolediyl-5,2-benzimidazolediyl-1,3-phenylene)	703	T_g increased after storage at 773K	143

	Polymer	T_g(K)	Remarks	References
4.9	Poly(benzimidazoles) (Cont'd.)			
	Poly(2,5-benzimidazolediyl-6,2-benzimidazolediyl-1,4-phenyleneimino-4-di-phenylamino-1,3,5-triazinediylimino-1,4-phenylene)	602	Heating rate 2K/min; η_{inh} only 0.21 dl/g	142
	Poly(5,5'-bibenzimidazole-2,2'-diyl-2,2'-biphenylylene)	663	Mechanical method	144
	Poly(2,6-benzimidazolediylsulfonyl-5,2-benzimidazolediyl-1,3-phenylene)	~560	DTA heating rate	114
4.10	Poly(oxindoles)			
	Poly(5,7-dichloro-3,3-oxindolylidene-1,4-phenyleneoxyisophthaloyloxy-1,4-phenylene)	543		145
	Poly(3,3-oxindolylidene-1,4-phenyleneoxyisophthaloyloxy-1,4-phenylene)	529		145
4.11	Poly(oxoisoindolines)			
	Poly(N-methyl-3,3-oxoisoindolylidene-1,4-phenyleneoxyisophthaloyloxy-1,4-phenylene)	~558	Transition indistinct	145
	Poly(N-methyl-3,3-oxoisoindolylidene-1,4-phenyleneoxyterephthaloyloxy-1,4-phenylene)	~555	Transition indistinct	145
	Poly(3,3-oxoisoindolylidene-1,4-phenyleneoxyisophthaloyloxy-1,4-phenylene)	598		145
	Poly(3,3-oxoisoindolylidene-1,4-phenyleneoxyterephthaloyloxy-1,4-phenylene)	600		145
4.12	Poly(dioxoisoindolines)			
	Poly(1,3-dioxoisoindoline-2,5-diylcarbonyl-1,3-dioxoisoindoline-5,2-diyl-1,3-phenylene)	~638	Dynamic method	124,125,146,147
	Poly(1,3-dioxoisoindoline-2,5-diylcarbonyl-1,3-dioxoisoindoline-5,2-diyl-1,3-phenylenedecafluoropentamethylene-1,3-phenylene)	533-543	Softening point; viscosity only 0.11 dl/g	148
	Poly(1,3-dioxoisoindoline-2,5-diylcarbonyl-1,3-dioxoisoindoline-5,2-diyl-1,3-phenylenehexafluorotrimethylene-1,3-phenylene)	~393	Sample hygroscopic	149
	Poly(1,3-dioxoisoindoline-2,5-diylcarbonyl-1,3-dioxoisoindoline-5,2-diyl-1,4-phenyleneoxy-1,4-phenylene)	>469	Heating rate 32K/min	124,125,150
	Poly(1,3-dioxoisoindoline-2,5-diylcarbonylimino-1,4-phenyleneiminocarbonyl-1,3-dioxoisoindoline-5,2-diyl-1,4-phenylenesulfonyl-1,4-phenylene)	641	Dynamic method	151
	Poly(1,3-dioxoisoindoline-2,5-diylhexafluorotrimethylene-1,3-dioxoisoindoline-5,2-diyl-1,3-phenylenedecafluoropentamethylene-1,3-phenylene)	371	Heating rate 32K/min	148
	Poly(1,3-dioxoisoindoline-2,5-diylhexafluorotrimethylene-1,3-dioxoisoindoline-5,2-diyl-1,3-phenylenehexafluorotrimethylene-1,3-phenylene)	418		148
	Poly(1,3-dioxoisoindoline-2,5-diyloctafluorotetramethylene-1,3-dioxoiso-indoline-5,2-diyl-1,3-phenylenedecafluoropentamethylene-1,3-phenylene)	417	Heating rate 32K/min	148
	Poly(1,3-dioxoisoindoline-2,5-diyloctafluorotetramethylene-1,3-dioxoiso-indoline-5,2-diyl-1,3-phenylenehexafluorotrimethylene-1,3-phenylene)	432	Heating rate 32K/min	148
	Poly(1,3-dioxoisoindoline-2,5-diyltetradecafluoroheptamethylene-1,3-dioxoiso-indoline-5,2-diyl-1,3-phenylenedecafluoropentamethylene-1,3-phenylene)	385	Heating rate 32K/min	148
	Poly(1,3-dioxoisoindoline-2,5-diyltetradecafluoroheptamethylene-1,3-dioxoiso-indoline-5,2-diyl-1,3-phenylenehexafluorotrimethylene-1,3-phenylene)	406	Heating rate 32K/min	148
	Poly(1,3-dioxoisoindoline-2,5-diylhexafluorotrimethylene-1,3-dioxoisoindoline-5,2-diyl-1,4-phenylenemethylene-1,4-phenylene)	498	Heating rate 32K/min	152
	Poly(1,3-dioxoisoindoline-2,5-diyldodecafluorohexamethylene-1,3-dioxoiso-indoline-5,2-diyl-1,4-phenyleneoxy-1,4-phenylene)	458	Heating rate 32K/min	153
	Poly(1,3-dioxoisoindoline-2,5-diylhexafluorotrimethylene-1,3-dioxoisoindoline-5,2-diyl-1,3-phenyleneoxy-1,3-phenylene)	451	Heating rate 32K/min	148
	Poly(1,3-dioxoisoindoline-2,5-diylhexafluorotrimethylene-1,3-dioxoisoindoline-5,2-diyl-1,4-phenyleneoxy-1,4-phenylene)	483		148
	Poly(1,3-dioxoisoindoline-2,5-diylhexadecafluorooctamethylene-1,3-dioxoiso-indoline-5,2-diyl-1,4-phenyleneoxy-1,4-phenylene)	457	Heating rate 32K/min; highly crystalline	152
	Poly(1,3-dioxoisoindoline-2,5-diyloctafluorotetramethylene-1,3-dioxoisoindoline-5,2-diyl-1,4-phenyleneoxy-1,4-phenylene)	485	Heating rate 32K/min	148
	Poly(1,3-dioxoisoindoline-2,5-diyltetradecafluoroheptamethylene-1,3-dioxoiso-indoline-5,2-diyl-1,4-phenyleneoxy-1,4-phenylene)	460	Heating rate 32K/min	148
	Poly(1,3-dioxoisoindoline-2,5-diylhexafluorotrimethylene-1,3-dioxoisoindoline-5,2-diyl-1,4-phenylenesulfonyl-1,4-phenylene)	533	Heating rate 32K/min	154
	Poly(1,3-dioxoisoindoline-2,5-diyloxy-1,3-dioxoisoindoline-5,2-diyl-1,3-phenylenedecafluoropentamethylene-1,3-phenylene)	393		148
	Poly(1,3-dioxoisoindoline-2,5-diyloxy-1,3-dioxoisoindoline-5,2-diyl-1,3-phenylenehexafluorotrimethylene-1,3-phenylene)	457	Heating rate 32K/min	148

Polymer	T_g(K)	Remarks	References
4.12　Poly(dioxoisoindolines) (Cont'd.)			
Poly(1,3-dioxoisoindoline-2,5-diyloxy-1,3-dioxoisoindoline-5,2-diyl-1,3-phenyleneimino-4-dimethylamino-1,3,5-triazinediylimino-1,3-phenylene)	567	Heating rate 2K/min	142
Poly(1,3-dioxoisoindoline-2,5-diyloxy-1,3-dioxoisoindoline-5,2-diyl-1,3-phenyleneimino-4-diphenylamino-1,3,5-triazinediylimino-1,3-phenylene)	563	Heating rate 2K/min	142
Poly(1,3-dioxoisoindoline-2,5-diyloxy-1,3-dioxoisoindoline-5,2-diyl-1,3-phenyleneimino-4-phenyl-1,3,5-triazinediylimino-1,3-phenylene)	565	Heating rate 2K/min	142
Poly(1,3-dioxoisoindoline-2,5-diyloxy-1,3-dioxoisoindoline-5,2-diyl-1,3-phenyleneimino-1,3,5-triazinediylimino-1,3-phenylene)	563	Heating rate 2K/min	142
Poly(1,3-dioxoisoindoline-2,5-diyloxy-1,3-dioxoisoindoline-5,2-diyl-1,3-phenyleneoxy-1,3-phenylene)	480	Heating rate 32K/min	148
Poly(1,3-dioxoisoindoline-2,5-diyloxy-1,3-dioxoisoindoline-5,2-diyl-1,4-phenyleneoxy-1,4-phenylene)	523	Heating rate 2K/min	129
Poly(1,3-dioxoisoindoline-2,5-diyloxy-1,3-dioxoisoindoline-5,2-diyl-1,4-phenylene-4-phenyl-1,3,5-triazinediyl-1,4-phenylene)	628	Heating rate 2K/min	142
Poly(1,3-dioxoisoindoline-5,2-diyl-1,4-phenylenesulfonyl-1,4-phenylene-1,3-dioxoisoindoline-2,5-diylcarbonylimino-4,4'-biphenylyleneimino-carbonyl)	643	Dynamic method	151
Poly(1,3-dioxoisoindoline-5,2-diyl-1,4-phenylenesulfonyl-1,4-phenylene-1,3-dioxoisoindoline-2,5-diylcarbonylimino-1,3-phenyleneiminocarbonyl)	584	Dynamic method	151
Poly(1,3-dioxoisoindoline-5,2-diyl-1,4-phenylenesulfonyl-1,4-phenylene-1,3-dioxoisoindoline-2,5-diylcarbonylimino-1,4-phenylenemethylene-1,4-phenyleneiminocarbonyl)	597	Dynamic method	151
Poly(1,3-dioxoisoindoline-5,2-diyl-1,4-phenylenesulfonyl-1,4-phenylene-1,3-dioxoisoindoline-2,5-diylcarbonylimino-1,4-phenyleneoxy-1,4-phenylene-iminocarbonyl)	583	Dynamic method	151
Poly(1,3-dioxoisoindoline-5,2-diyl-1,4-phenylenesulfonyl-1,4-phenylene-1,3-dioxoisoindoline-2,5-diylcarbonylimino-1,4-phenylenesulfonyl-1,4-phenyleneiminocarbonyl)	580	Dynamic method	151
4.13　Poly(triazines)			
Poly[(4-H-octafluorobutyl)-1,3,5-triazinediyltrimethylene]	253-255	No measurement details	155
Poly(pentafluoroethyl-1,3,5-triazinediyltrimethylene)	253-255	No measurement details	155
Poly(2-H-tetrafluoroethyl-1,3,5-triazinediyltrimethylene)	253-255	No measurement details	155
Poly(heptafluoropropyl-1,3,5-triazinediyltrimethylene)	253-255	No measurement details	155
Poly[perfluoro(propyl-1,3,5-triazinediylhexamethylene)]	261	DTA heating rate	156
Poly(6-phenyl-1,3,5-triazinediylphenylimino-1,3,5-triazinediylphenylimino)	541-552	DTA heating rate	114
Poly(6-phenyl-1,2,4-triazine-5,3-diyl-2,6-pyridinediyl-6-phenyl-1,2,4-triazine-3,5-diyl-1,4-phenylene)	533	Dynamic method; different structural isomers may be present	157
Poly(6-phenyl-1,2,4-triazine-5,3-diyl-2,6-pyridinediyl-6-phenyl-1,2,4-triazine-3,5-diyl-1,4-phenylenemethylene-1,4-phenylene)	478	Dynamic method; different structural isomers may be present	157
Poly(6-phenyl-1,2,4-triazine-5,3-diyl-2,6-pyridinediyl-6-phenyl-1,2,4-triazine-3,5-diyl-1,4-phenyleneoxy-1,4-phenylene)	488	Dynamic method; different structural isomers may be present	157,158
Poly(1,2,4-triazine-5,3-diyl-2,6-pyridinediyl-1,2,4-triazine-3,6-diyl-1,4-phenyleneoxy-1,4-phenylene)	496	Heating rate 20K/min	159
4.14　Poly(pyridazines)			
Poly(3,6-pyridazinediyloxy-1,4-phenyleneisopropylidene-1,4-phenyleneoxy)	453	Mechanical method	116
4.15　Poly(piperazines)			
Poly(2,5-dimethyl-1,4-piperazinediylcarbonyloxyneopentylenesulfonylneopentylene-oxycarbonyl)	358		160
Poly(2,5-dimethyl-1,4-piperazinediylcarbonyloxyneopentylenethioneopentyleneoxy-carbonyl)	313		160
Poly(1,4-piperazinediyladipoyl)	399	No experimental details	161,162
Poly(1,4-piperazinediylcarbonyl-2,2'-biphenylylenecarbonyl)	466	Mechanical method, heating rate 1K/min; low viscosity	78,161
Poly(1,4-piperazinediylcarbonyloxyethyleneoxycarbonyl)	333		163
Poly(1,4-piperazinediylcarbonyloxyneopentyleneoxycarbonyl)	343		160
Poly(1,4-piperazinediylcarbonyloxy-9-oxabicyclo[3,3,1]nonan-2,6-yleneoxy-carbonyl)	386	DTA heating rate	164

	Polymer	T_g(K)	Remarks	References
4.15	Poly(piperazines) (Cont' d.)			
	Poly(1,4-piperazinediylcarbonyl-1,4-phenylene-2,2-butylidene-1,4-phenylene-carbonyl)	492-505		165
	Poly(1,4-piperazinediylisophthaloyl)	465	Mechanical method, heating rate 1K/min	78,161
	Poly(1,4-piperazinediyl-1-oxotrimethylenefluoren-9-ylidene-3-oxotrimethylene)	418	Mechanical method, heating rate 1K/min	78,161
	Poly(1,4-piperazinediylsebacoyl)	~355	Rather ill-defined transition	161,166-168
4.16	Poly(pyridines)			
	Poly(2,3,5-trifluoropyridinediyloxy-2,2,3,3,4,4-hexafluoropentamethyleneoxy)	260	Polymer structure may contain o-linkages	169
	Poly(2,5-pyridinediylcarbonyliminohexamethyleneiminocarbonyl)	322		170
4.17	Poly(piperidines)			
	Poly(2,6-dioxopiperidine-1,4-diyltrimethylene)	363	Heating rate 10K/min	172
4.18	Poly(triazoles)			
	Poly(4-phenyl-1,2,4-triazolediyl-1,3-phenyleneiminoterephthaloylimino-1,3-phenylene)	623	DTA heating rate	173
	Poly(4-phenyl-1,2,4-triazolediyl-1,3-phenylene-4-phenyl-1,2,4-triazolediyl-1,4-phenylene)	538		174
4.19	Poly(pyrazoles)			
	Poly(1,3-pyrazolediyl-1,3-phenylene-3,1-pyrazolediylhexamethylene)	343	DTA heating rate; viscosity low	175
	Poly(1,3-pyrazolediyl-1,4-phenylene-3,1-pyrazolediylhexamethylene)	353	DTA heating rate; high crystallinity	175
	Poly(1,3-pyrazolediyl-1,3-phenylene-3,1-pyrazolediyl-1,4-phenylene)	353	DTA heating rate; high crystallinity	175
	Poly(1,3-pyrazolediyl-1,4-phenylene-3,1-pyrazolediyl-1,4-phenylene)	373	DTA heating rate; high crystallinity	175
4.20	Poly(pyrrolidines)			
	Poly(2,5-dioxo-1,3-pyrrolidinediylethylene)	408	Heating rate 20K/min	176,177
4.21	Poly(carboranes)			
	Poly(C$_2$B$_5$H$_5$-carboranylenedimethylsilyleneoxydimethylsilylene)	277, 211	Conflicting data	178,179
	Poly(C$_2$B$_5$H$_5$-carboranylenedi(dimethylsilyleneoxy)dimethylsilylene)	< 188	Dynamic method	180
	Poly(C$_2$B$_{10}$H$_{10}$-carboranylenedimethylsilyleneoxydimethylsilylene)	298	Mechanical method	181
	Poly(C$_2$B$_{10}$H$_{10}$-carboranylenedi(dimethylsilyleneoxy)dimethylsilylene)	239		180-182
	Poly(C$_2$B$_{10}$H$_{10}$-carboranylenetri(dimethylsilyleneoxy)dimethylsilylene)	213	Mechanical method	180-182
4.22	Poly(fluoresceins)			
	Poly(fluorescein-3',6'-diyloxyisophthaloyloxy)	549		183
4.23	Poly(oxabicyclononanes)			
	Poly[9-oxabicyclo(3,3,1)nonane-2,6-diyloxycarbonylimino-1,4-cyclohexylene-methylene-1,4-cyclohexyleneiminocarbonyloxy]			
	53% trans, trans	494	DTA heating rate	164
	70% trans, trans	502		
	Poly[9-oxabicyclo(3,3,1)nonane-2,6-diyloxycarbonylimino-1,4-phenylene-methylene-1,4-phenyleneiminocarbonyloxy]	473		164
	Poly[9-oxabicyclo(3,3,1)nonane-2,6-diyloxycarbonyloxy-1,4-phenyleneiso-propylidene-1,4-phenyleneoxycarbonyloxy]	467	DTA heating rate	164
4.24	Poly(dibenzofurans)			
	Poly(3,6-dibenzofurandiylsulfonyl)	633		184
	Poly(3,6-dibenzofurandiylsulfonyl-1,4-phenyleneoxy-1,4-phenylenesulfonyl)	563		184
4.25	Poly(phthalides)			
	Poly(3,3-phthalidylidene-1,4-phenyleneoxy-5-tert-butylisophthaloyloxy-1,4-phenylene)	552		145
	Poly(3,3-phthalidylidene-1,4-phenyleneoxycarbonyloxy-1,4-phenylene)	513,538	Conflicting data	145,185

Polymer	$T_g(K)$	Remarks	References

4.25   Poly(phthalides)   (Cont'd.)

Polymer	$T_g(K)$	Remarks	References
Poly(3,3-phthalidylidene-1,4-phenyleneoxy-5-chloroisophthaloyloxy-1,4-phenylene)	586		145
Poly(3,3-phthalidylidene-1,4-phenyleneoxyisophthaloyloxy-1,4-phenylene)	591		145

4.26   Poly(acetals)

Polymer	$T_g(K)$	Remarks	References
Poly(2-ethyl-1,3-dioxa-4,6-cyclohexylenemethylene)	345		1
Poly(2-isopropyl-1,3-dioxa-4,6-cyclohexylenemethylene)	329		1
Poly(2-methyl-1,3-dioxa-4,6-cyclohexylenemethylene)	355		1
Poly(1,3-dioxa-4,6-cyclohexylenemethylene)	378		1,187
Poly(1,3-dioxa-2-propyl-4,6-cyclohexylenemethylene)	322		1,188
Poly(vinyl acetal) see Poly(2-methyl-1,3-dioxa-4,6-cyclohexylene-methylene)			
Poly(vinyl butyral) see Poly(1,3-dioxa-2-propyl-4,6-cyclohexylene-methylene)			

4.27   Poly(anhydrides)

Polymer	$T_g(K)$	Remarks	References
Poly(methacrylic anhydride)	432	Vicat softening point	186

4.28   Carbohydrates

Polymer	$T_g(K)$	Remarks	References
Amylose triacetate	440	Heating rate 5-10K/min	189
Amylose tributyrate	365	Heating rate 5-10K/min	189
Amylose hexanoate (2.9)	315	Heating rate 5-10K/min	189
Amylose tripropionate	406	Heating rate 5-10K/min	189
Amylose valerate (2.8)	330	Heating rate 5-10K/min	189
Cellulose	range 243-433	Conflicting data	190-200
Cellulose triacetate	range 322-751	Conflicting data; depend on acetate and water content and degree of crystallinity	1,21,93,143,188, 191,197,201-214
Cellulose tributyrate	388		1,206,215
Cellulose tridecanoate	321		215
Cellulose triheptanoate	320		215
Cellulose trihexanoate	237		215
Cellulose nitrate	326,339	Conflicting data	18,21,188,201,216,217
Cellulose tripropionate	400		215
Cellulose trivalerate	338		215
Cyanoethyl cellulose	453	Dynamic method	197
Ethyl cellulose	316		1,201
Methyl cellulose	423	Dynamic method	197
Triphenylmethyl cellulose	426	Dynamic method	197

E.   REFERENCES

1. L.E. Nielsen, "Mechanical Properties of Polymers", Reinhold, 1962.
2. C. A. Kumins, J. Roteman, J. Polymer Sci., A, 1, 527 (1963).
3. K. Kanamaru, M. Sugiura, Kolloid. Z. - Z. Polymere 194, 110 (1964).
4. M. Sugiura, E. Fujii, Tokyo Kogyo Shikensho Hokoku 58, 534 (1963).
5. K. Kanamaru, M. Sugiura, J. Chem. Soc. Japan 65, 1434 (1962).
6. R. F. Boyer, Rubber Chem. Technol. 36, 1303 (1963).
7. T. M. A. Hossain, T. Iijima, Z. Morita, H. Maeda, J. Appl. Polymer Sci. 13, 541 (1969).
8. T. R. Manley, Polymer 10, 148 (1969).
9. A. J. Kovacs, J. Polymer Sci. 30, 131 (1958).
10. N. Bekkedahl, J. Res. Nat. Bur. Std. 43, 145 (1949).
11. J. H. Sewell, Royal Aircraft Establishment, Farnborough, Hants, U. K. Unpublished report.
12. T. Holleman, Rheol. Acta 10, No. 2, 194 (1971).
13. A. Eisenberg, A. V. Tobolsky, J. Polymer Sci. 61, 483 (1962).
14. S. Nakamura, A. V. Tobolsky, Rept. Progr. Polymer Phys. Japan 12, 303 (1969).
15. J. J. Maurer, H. C. Tsien, J. Appl. Polymer Sci. 8, 1719 (1964).
16. F. C. Stehling, L. Mandelkern, Macromolecules 3, 242 (1970).
17. G. M. Martin, S. S. Rogers, L. Mandelkern, J. Polymer Sci. 20, 579 (1956).
18. R. F. Boyer, R. S. Spencer, "Advances in Colloid Science", Vol. II, Interscience, New York, 1946, p. 1.
19. R. H. Wiley, J. Polymer Sci. 2, 10 (1947).
20. L. Turunen, Kunststoffe 52, 672 (1962).
21. B. D. Sully, "Science of Surface Coatings", Ernest Benn, 1962, p. 281.
22. R. H. Wiley, G. M. Brauer, J. Polymer Sci. 3, 455 (1948).
23. R. H. Wiley, G. M. Brauer, J. Polymer Sci. 3, 647 (1948).
24. R. H. Wiley, G. M. Brauer, A. R. Bennett, J. Polymer Sci. 5, 609 (1950).
25. R. B. Beevers, Trans. Faraday Soc. 58, 1465 (1962).
26. R. B. Beevers, E. F. T. White, Trans. Faraday Soc. 56, 1529 (1960).
27. R. B. Beevers, E. F. T. White, Trans. Faraday Soc. 56, 744 (1960).
28. K. H. Illers, Kolloid. Z. - Z. Polymere 190, 16 (1963).

29. P. L. Magagnini, Chim. Ind. (Milan) 49, 1041 (1967).

30. R. Kaneko, Chem. High Polymers (Tokyo) 24, 272 (1967).

31. K. R. Dunham, J. Vandenberghe, J. W. H. Faber, W. F. Fowler,
     J. Appl. Polymer Sci. 7, 143 (1963).

32. B. Maxwell, L. F. Rahm, SPE. J. 6, No. 9, 7 (1950).

33. S. Newman, S. Strella, J. Appl. Polymer Sci. 9, 2297 (1965).

34. S. L. Rosen, Polymer Eng. Sci. 7, 115 (1967).

35. R. J. Seward, Rubber Chem. Technol. 43, 1 (1970).

36. I. Momiyama, H. Iwai, S. Matsuzaki, Shikizai Kyokaishi 43, 427
     (1970); C.A. 74, 4067c (1971).

37. G. Salee, J. Appl. Polymer Sci. 15, 2049 (1971).

38. L. E. Nielsen, R. Buchdahl, R. Levreault, J. Appl. Phys. 21, 607
     (1950).

39. F. S. Conant, J. W. Liska, J. Appl. Phys. 15, 767 (1944).

40. U. S. 3,514,427 - 26 May 1970 (F. S. Owens).

41. J. A. Gorbatkina, V. G. Ivanova-Mumjieva, IUPAC Preprints, Intern.
     Symp. Macromol. Chem., Toronto, Section A9.9 (1968).

42. W. J. Jackson, T. F. Gray, J. R. Caldwell, J. Appl. Polymer Sci. 14,
     685 (1970).

43. RAPRA Data Handbook Series on Transition Temperatures to be published
     (1973) by Rubber & Plastics Research Association of Great Britain,
     Shawbury, Shrewsbury, SY4 4NR, England.

44. O. G. Lewis, "Physical Constants of Linear Homopolymers", Springer-
     Verlag, 1968.

45. W. A. Lee, J. Polymer Sci, A-2, 8, No. 4, 555 (1970).

46. J. H. Gibbs, "Modern Aspects of the Vitreous State", Ed. J. D. Macken-
     zie, Butterworths, 1960, p. 152.

47. S. Krimm, A. V. Tobolsky, J. Polymer Sci. 6, 667 (1951).

48. M. C. Shen, A. Eisenberg, "Progress in Solid State Chemistry", Vol. 3,
     Ed. H. Reiss, Pergamon Press, 1966, p. 407; reprint in Rubber Chem.
     Technol. 43, 95 (1970).

49. A. Eisenberg, M. Shen, Rubber Chem. Technol. 43, 156 (1970).

50. G. Allen, SCI Plastics Polymer Group Symp., London, Sept. 1962.

51. B. G. Ranby, K. S. Chan, H. Brumberger, J. Polymer Sci. 58, 545
     (1962).

52. S. A. Iobst, J. A. Manson, ACS Polymer Preprints 11, 765 (1970).

53. W. Schermann, H. G. Zachmann, Kolloid. Z. - Z. Polymere 241,
     921 (1970).

54. V. P. Roshchupkin, V. V. Kochervinskii, Vysokomolekul. Soedin.,B,
     13, 194 (1971).

55. J. M. Barton, Polymer 10, 151 (1969).

56. W. Wrasidlo, Macromolecules 4, 642 (1971).

57. G. W. Miller, J. Appl. Polymer Sci. 15, 2335 (1971).

58. M. L. Williams, R. F. Landel, J. D. Ferry, J. Am. Chem. Soc. 77,
     3701 (1955).

59. A. F. Lewis, J. Polymer Sci., B, 1, 649 (1963).

60. Yu. V. Zelenev, V. I. Abramova, Vysokomolekul. Soedin., A, 11,
     920 (1969); (transl'n in Polymer Sci. USSR 11, 1040 (1969)).

61. R. Zannetti, P. Manaresi, L. Baldi, J. Polymer Sci. 62, S33 (1962).

62. J. Heller, D. J. Lyman, J. Polymer Sci., B, 1, 317 (1963).

63. A. Quach, R. Simha, J. Appl. Phys. 42, 4592 (1971).

64. S. Ichihara, A. Komatsu, Y. Tsujita, T. Nose, T. Hata, Polymer J.
     (Japan) 2, 530 (1971).

65. A. J. Kovacs, S. Y. Hobbs, J. Appl. Polymer Sci. 16, 301 (1972).

66. H. G. Kilian, E. Jenckel, "Struktur und Physikalisches Verhalten der
     Kunststoffe", Ed. K. A. Wolf, Springer-Verlag, 1962, p. 176.

67. E. W. Fischer, F. Kloos, G. Lieser, J. Polymer Sci., B, 7, 845 (1969).

68. Y. Aoki, A. Nobuta, A. Chiba, M. Kaneko, Polymer J. (Japan) 2,
     502 (1971).

69. A. B. Romberger, D. P. Eastman, J. L. Hunt, J. Chem. Phys. 51,
     3723 (1969).

70. G. A. Clegg, T. P. Melia, Polymer 11, 245 (1970).

71. R. Hoffman, W. Knappe, Kolloid - Z. - Z. Polymere 247, 763 (1971).

72. A. Lambert, Polymer 10, 319 (1969).

73. J. J. Maurer, Rubber. Chem. Technol. 42, 110 (1969).

74. S. M. Wolpert, A. Weitz, B. Wunderlich, J. Polymer Sci., A-2, 9,
     1887 (1971).

75. G. A. Gordon, J. Polymer Sci., A-2, 9, 1693 (1971).

76. R. W. Warfield, SPE J. 15, 625 (1959).

77. A. Rembaum, J. Polymer Sci., C, 157 (1970).

78. S. C. Temin, J. Appl. Polymer Sci. 9, 471 (1965).

79. J. Periard, A. Banderet, G. Reiss, Angew. Makromol. Chem. 15, 37
     (1971).

80. R. F. Boyer, Changements de Phases, Soc. Chim. Phys., Paris,
     p. 383 (1952).

81. E. H. Riddle, "Monomeric Acrylic Esters", Reinhold, New York,
     1954, p. 59.

82. A. L. Machek, J. Elastoplastics 1, 213 (1969).

83. W. G. Barb, J. Polymer Sci. 37, 515 (1959).

84. H. Aida, M. Urushizaki, H. Takeuchi, Fukui Daigaku Kogakubu
     Kenkyu Hokoku 18, 173 (1970); C.A. 74, 42806a (1971).

85. M. Tsuge, S. Tanimoto, S. Tanaka, Kogyo Kagaku Zasshi 73, 440
     (1970); C. A. 72, 133340u (1970).

86. A. Charlesby, R. H. Partridge, Proc. Roy. Soc. A, 271, 170 (1963).

87. M. Mozisek, Intern. J. Appl. Radiation Isotopes 21, 11 (1970).

88. J. R. Stevens, A. C. Mao, J. Appl. Phys. 41, 4273 (1970).

89. S. Y. Chuang, S. J. Tao, J. M. Wilkenfeld, J. Appl. Phys. 43, 737
     (1972).

90. J. E. Guillet, E. Dan. R. S. Mitchell, J. P. Valleau, Nature Phys.
     Sci. 234, 135 (1971).

91. J. N. Gayles, W. L. Peticolas, Light Scattering Spectra Solids, Proc.
     Intern. Conf., 1968, Ed. G. B. Wright, Springer-Verlag, New York,
     p. 715 (1969); C.A. 72, 22089z (1970).

92. A. Lavoie, J. E. Guillet, Macromolecules 2, 443 (1969).

93. S. Nakamura, S. Shindo, K. Matsuzaki, J. Polymer Sci., B, 9, 591
     (1971).

94. E. Sacher, J. Macromol. Sci., B4, 449 (1970).

95. T. Takamatsu, E. Fukada, Rika Gaku Kenkyusho Hokoku 46, 47 (1970);
     C.A. 74, 23281j (1971).

96. B. M. Grieveson, Polymer 1, 499 (1960).

97. D. W. Brown, L. A. Wall, J. Polymer Sci., A-2, 7, 601 (1969).

98. F. Jenckel, R. Heusch, Kolloid -Z. 130, 89 (1953).

99. H. Jacobs, E. Jenckel, Makromol. Chem. 47, 72 (1961).

100. E. A. DiMarzio, J. H. Gibbs, J. Polymer Sci., A, 1, 1417 (1963).

101. J. Kolarik, J. Janacek, J. Polymer Sci., C, 16, Pt. 1, 441 (1967).

102. G. Allen, J. McAinsh, G. M. Jeffs, Polymer 12, 85 (1971).

103. R. D. McCammon, R. G. Saba, R. N. Work, J. Polymer Sci., A-2,
     7, 1721 (1969).

104. G. Pezzin, F. Zilio-Grandi, P. Sanmartin, European Polymer J. 6,
     1053 (1970).

105. F. M. Smekhov, A. I. Nepomnyashchii, A. T. Sanzharovskii, S. V.
     Yakubovich, Vysokomolekul. Soedin. A, 13, 2102 (1971); (transl'n
     in Polymer Sci. USSR 13, 2362 (1971)).

106. K. Ueberreiter, U. Rohde-Liebenau, Makromol. Chem. 49, 164 (1961).

107. J. A. Faucher, J. V. Koleske, E. R. Santee, J. J. Stratta, C. W.
     Wilson, J. Appl. Phys. 37, 3962 (1966).

108. J. A. Faucher, J. Polymer Sci., B, 3, 143 (1965).

109. IUPAC, "Nomenclature of Organic Chemistry", Sections A & B,
     Third Edit., Sect. C, Second Edit., London, Butterworths, 1969.

110. Macromolecules 1, 193 (1968) and IUPAC Information Bulletin Ap-
     pendix No. 29, 1972.

111. J. M. Augl, J. V. Duffy, J. Polymer Sci., A-1, 9, 1343 (1971).

112. F. D. Trischler, K. J. Kjoller, H. H. Levine, ACS Org. Coatings
     Plast. Chem. 27, 381 (1967).

113. W. De Winter, R. Stein, H. Uwents, C. Masquelier, J. Polymer Sci.,
     A-1, 8, 1955 (1970).

114. G. F. L. Ehlers, K. R. Fisch, Appl. Polymer Symp. 8, 171 (1969).

115. V. V. Korshak, V. M. Mamedov, G. E. Golubkov, D. R. Tur, Vyso-
     komolekul. Soedin. B, 12, 57 (1970).

116. R. N. Johnson, A. G. Farnham, R. A. Glendinning, W. F. Hale, C.
     N. Merriam, J. Polymer Sci., A-1, 5, 2375 (1967).

117. J. M. Augl, W. J. Wrasidlo, J. Polymer Sci., A-1, 8, 63 (1970).

118. P. M. Hergenrother, H. H. Levine, J. Polymer Sci., A-1, 4, 2341 (1966).

119. W. Wrasidlo, AD 718834 (1971); in part in ACS Polymer Preprints 12, 755 (1971).

120. J. K. Gillham, M. B. Roller, Polymer Eng. Sci. 11, 295 (1971).

121. U.S. 2,710,853, 14 June 1955 (W. M. Edwards, I. M. Robinson); via ref. 128.

122. K. W. Doak, 1968 Modern Plastics Encyclopedia 45, 1A, 14 (1967).

123. J. M. Barton, J. P. Critchley, Polymer 11, 212 (1970).

124. S. L. Cooper, A. D. Mair, A. V. Tobolsky, Textile Res. J. 35, 1110 (1965).

125. J. H. Freeman, L. W. Frost, G. M. Bower, E. J. Traynor, Conf. Structural Plastics, Adhesives, Filament Wound Composites, WPAFB, Dayton, Ohio 1, 30 (1962).

126. G. A. Bernier, D. E. Kline, J. Appl. Polymer Sci. 12, 593 (1968).

127. A. D. Mair, M. C. Shen, A. V. Tobolsky, AD 604010 (1964) "High temperature polymers: H-Film and SP-Polymer", ONR Tech. Rept. RLT-83; via ref. 126.

128. C. E. Sroog, J. Polymer Sci., C, 16, 1191 (1967); data quoted from W. E. Tatum, L. E. Amborski, C. W. Gerow, J. F. Heacock, R. S. Mallouk, Electrical Insulation Conf., Chicago 1963.

129. A. V. Sidorovich, Ye. V. Kuvshinskii, Polymer Sci. USSR 10, 1627 (1968) (Transl'n of Vysokomolekul. Soedin. A 10, 1401 (1968)).

130. W. Wrasidlo, J. M. Augl, J. Polymer Sci., A-1, 7, 321 (1969).

131. G. A. Pogany, Polymer 11, 66 (1970).

132. J. M. Augl, J. Polymer Sci., A-1, 8, 3145 (1970).

133. W. Wrasidlo, ACS Org. Coatings Plast. Chem. 30, 97 (1970).

134. P. M. Hergenrother, AD 710504 (1970). Includes values of N. J. Johnston and W. J. Wrasidlo.

135. W. Wrasidlo, J. M. Augl, Macromolecules 3, 544 (1970).

136. P. M. Hergenrother, H. H. Levine, J. Polymer Sci., A-1, 5, 1453 (1967).

137. W. Wrasidlo, J. Polymer Sci., A-1, 8, 1107 (1970).

138. W. Wrasidlo, J. M. Augl, J. Polymer Sci., B, 7, 281 (1969).

139. W. Wrasidlo, J. M. Augl, J. Polymer Sci., A-1, 7, 3393 (1969).

140. P. M. Hergenrother, SAMPE Quarterly 3, 1 (1971).

141. P. M. Hergenrother, D. E. Kiyohara, Macromolecules 3, 387 (1970).

142. R. J. Kray, R. Seltzer, R. A. E. Winter, ACS Org. Coatings Plast. Chem. 31, 569 (1971).

143. J. K. Gillham, Polymer Eng. Sci. 7, 225 (1967).

144. J. K. Gillham, Science 139, 494 (1963).

145. P. W. Morgan, J. Polymer Sci., A, 2, 437 (1964).

146. G. M. Bower, J. H. Freeman, E. J. Traynor, L. W. Frost, H. A. Burgman, C. R. Ruffing, J. Polymer Sci., A-1, 6, 877 (1968).

147. J. H. Freeman, L. W. Frost, G. M. Bower, E. J. Traynor, H. A. Burgman, C. R. Ruffing, Polymer Eng. Sci. 9, 56 (1969).

148. J. P. Critchley, V. C. R. McLoughlin, J. Thrower, I. M. White, Brit. Polymer J. 2, 288 (1970).

149. R. Dine-Hart, W. W. Wright, Royal Aircraft Establishment, Farnborough, Hants, U.K. - Unpublished results.

150. J. Thrower, W. A. Lee, Royal Aircraft Establishment, Farnborough, Hants, U.K. - Unpublished results.

151. W. Wrasidlo, J. M. Augl, J. Polymer Sci., A-1, 7, 1589 (1969).

152. J. P. Critchley, M. A. White, Royal Aircraft Establishment, Farnborough, Hants, U.K. - Unpublished results.

153. J. P. Critchley, M. A. White, Royal Aircraft Establishment, Farnborough, Hants, U.K., J. Polymer Sci., A-1, 10, 1809 (1972).

154. J. P. Critchley, Royal Aircraft Establishment, Farnborough, Hants, U.K. - Unpublished results.

155. I. M. Dolgopol'skii, Kauch. i Rezina 30, No. 2, 31 (1971) (transl'n in Soviet Rubber Technol. 30, No. 2, 28 (1971)).

156. E. Dorfman, W. E. Emerson, R. J. Gruber, A. A. Lemper, B. M. Rushton, T. L. Graham, Angew. Makromol. Chem. 16/17, 75 (1971).

157. P. M. Hergenrother, D. E. Kiyohara, J. Macromol. Sci. A5, 365 (1971).

158. W. J. Wrasidlo, P. M. Hergenrother, Macromolecules 3, 548 (1970).

159. W. Wrasidlo, J. Polymer Sci., A-2, 9, 1603 (1971).

160. G. L. Brode, ACS Polymer Preprints 6, 626 (1965).

161. R. G. Beaman, J. Appl. Polymer Sci. 9, 3949 (1965).

162. W. W. Moseley, R. G. Parrish; private communication to ref. 161.

163. E. L. Wittbecker, Chem. Eng. News 41, 41 (1963).

164. A. H. Frazer, Macromolecules 1, 199 (1968).

165. A. Schiller, J. C. Petropoulos, C. S. H. Chen, J. Appl. Polymer Sci. 8, 1699 (1964).

166. M. E. Baird, G. T. Goldsworthy, C. J. Creasey, J. Polymer Sci., B, 6, 737 (1968).

167. M. E. Baird, J. Polymer Sci., A-2, 8, 739 (1970).

168. P. J. Flory, L. Mandelkern, H. K. Hall, J. Am. Chem. Soc. 73, 2532 (1951).

169. P. Johncock, P. A. Grattan, M. A. H. Hewins, Royal Aircraft Establishment, Farnborough, Hants, U. K. - Unpublished report.

170. O. Ishizuka, A. Okada, S. Ueda, T. Muroi, I. Ikoma, Chem. High Polymers (Tokyo) 17, 143 (1960); C.A. 55, 15997c (1961).

171. F. D. Hartley, F. W. Lord, L. B. Morgan, Intern. Symp. Macromol. Chem., Ric. Sci. 25, 577 (1955).

172. H. K. Reimschuessel, L. G. Roldan, J. P. Sibilia, J. Polymer Sci., A-2, 6, 559 (1968).

173. U.S. 3,376,268, 2 April 1968 (J. Preston).

174. M. R. Lilyquist, J. R. Holsten, J. Polymer Sci., C, 19, 77 (1967).

175. J. K. Stille, M. A. Bedford, J. Polymer Sci., A-1, 6, 2331 (1968).

176. H. K. Reimschuessel, K. P. Klein, G. J. Schmitt, Macromolecules 2, 567 (1969).

177. H. K. Reimschuessel, K. P. Klein, J. Polymer Sci., A-1, 9, 3071 (1971).

178. R. E. Kesting, K. F. Jackson, E. B. Klusmann, F. J. Gerhart, J. Appl. Polymer Sci. 14, 2525 (1970).

179. R. E. Kesting, K. F. Jackson, J. M. Newman, J. Appl. Polymer Sci. 15, 1527 (1971).

180. J. M. Augl, AD 718329 (1970).

181. E. J. Zaganiaris, L. H. Sperling, A. V. Tobolsky, J. Macromol. Sci., A-1, 6, 1111 (1967).

182. L. H. Sperling, S. L. Cooper, A. V. Tobolsky, J. Appl. Polymer Sci. 10, 1725 (1966).

183. A. B. Thompson, D. W. Woods, Trans. Faraday Soc. 52, 1383 (1956).

184. H. A. Vogel, J. Polymer Sci., A-1, 8, 2035 (1970).

185. Z. Dobkowski, B. Krajewski, Z. Wielgosz, Polim. Tworz. Wielk. 15, 428 (1970).

186. J. C. H. Hwa, W. A. Fleming, L. Miller, J. Polymer Sci., A, 2, 2385 (1964).

187. K. Kanamaru, M. Hirata, Kolloid-Z. - Z. Polymere 230, 206 (1969).

188. R. F. Clash, L. M. Runkiewiez, Ind. Eng. Chem. 36, 279 (1944).

189. J. M. G. Cowie, P. M. Toporowski, F. Costaschuk, Makromol. Chem. 121, 51 (1969).

190. C. Klason, J. Kubat, A. de Ruvo, Rheol. Acta 6, 390 (1967).

191. K. E. Perepelkin, Mekh. Polim. 5, 790 (1971).

192. T. Hatakeyama, H. Kanetsuna, Chem. High Polymer (Tokyo) 26, 76 (1969).

193. J. Kubat, C. Pattyranie, Nature 215, 390 (1967).

194. E. Fukada, M. Date, N. Hirai, J. Polymer Sci., C, 23, Pt. 2, 509 (1968).

195. V. A. Kargin, P. V. Kozlov, Nai-Ch'ang Wang, Dokl. Akad. Nauk SSSR 130, 356 (1960); C.A. 56, 605b (1962).

196. K. Tsuge, Y. Wada, J. Phys. Soc. (Japan) 17, 156 (1962); via ref. 199.

197. G. P. Mikhailov, A. I. Artyukhov, V. A. Shevelev, Polymer Sci. USSR 11, 628 (1969) (transl'n of Vysokomolekul. Soedin. A 11, 553 (1969)).

198. E. L. Back, I. E. Dioriksson, Svensk Paperstid. 72, 687 (1969); via Chem. Titles, No. 26 (1969).

199. Y. Ogiwara, H. Kubota, S. Hayashi, N. Mitomo, J. Appl. Polymer Sci. 14, 303 (1970).

200. Y. Nakamura, J. C. Arthur, M. Negishi, K. Doi, E. Kageyama, K. Kudo, J. Appl. Polymer Sci. 14, 929 (1970).

201. F. E. Wiley, Ind. Eng. Chem. 34, 1052 (1942).
202. J. K. Gillham, R. F. Schwenker, Appl. Polymer Symp. 2, 59 (1966).
203. A. Sharples, F. L. Swinton, J. Polymer Sci. 50, 53 (1961).
204. J. H. Daane, R. E. Barker, J. Polymer Sci., B, 2, 343 (1964).
205. K. Nakamura, Chem. High Polymer (Tokyo) 13, 47 (1956).
206. L. Mandelkern, P. J. Flory, J. Am. Chem. Soc. 73, 3206 (1951).
207. J. Russell, R. G. Van Kerpel, J. Polymer Sci. 25, 77 (1957).
208. R. E. Barker, C. R. Thomas, J. Appl. Phys. 35, 87 (1964).
209. B. S. Sprague, J. Polymer Sci., C, 20, 159 (1967); via ref. 7.
210. N. N. Druzhinina, M. P. Pen'kova, R. M. Livshits, Z. A. Rogovin, Polymer Sci. USSR 10, 3181 (1968) (transl'n of Vysokomolekul. Soedin. A 10, 2743 (1968)).
211. K. S. Patel, K. C. Patel, R. D. Patel, Makromol. Chem. 132, 7 (1970).
212. J. K. Gillham, Appl. Polymer Symp. 2, 45 (1966).
213. J. M. G. Cowie, R. J. Ranson, Makromol. Chem. 143, 105 (1971).
214. K. Ueberreiter, Z. Physik. Chem. B48, 197 (1941); C.A. 36, 1489 (1942).
215. A. F. Klarman, A. V. Galanti, L. H. Sperling, J. Polymer Sci., A-2, 7, 1513 (1969).
216. R. Houwink, "Fundamentals of Synthetic Polymer Technology", Elsevier, New York, 1949, p. 49.
217. G. A. Sorokin, I. V. Tishunin, E. N. Fominykh, Vysokomolekul. Soedin. B, 11, 522 (1969).
218. W. J. Feast, W. K. R. Musgrave, N. Reeves, J. Polymer Sci., A-1, 9, 2733 (1971).
219. N. H. Reinking, A. E. Barnabeo, W. F. Hale, J. Appl. Polymer Sci. 7, 2153 (1963).
220. J. Lal, G. S. Trick, J. Polymer Sci., A-1, 8, 2339 (1970).
221. A. S. Hay, R. F. Clark, Macromolecules 3, 533 (1970).
222. A. R. Blythe, G. M. Jeffs, J. Macromol. Sci., B3, 141 (1969).
223. D. J. Marks, Thesis, Manchester Univ. 1961.
224. G. Allen, J. Hurst - Unpublished work.
225. G. Allen, C. Booth, S. J. Hurst, C. Price, F. Vernon, R. F. Warren, Polymer 8, 406 (1967).
226. P. G. Wapner (supervisor: W. C. Forsman), Ph.D. Thesis, Univ. Pennsylvania 1971; via Diss. Abstr. Intern., B, 32, 2332 (1971).
227. N. Doddi, W. C. Forsman, C. C. Price, Macromolecules 4, 648 (1971).
228. Brit. 1,140,300 - 15. Jan. 1969 (H. Schnell, L. Bottenbruch, G. Darsow, K. Weirauch)
229. P. E. Wei, P. E. Butler, J. Polymer Sci., A-1, 6, 2461 (1968).
230. V. Frosini, E. Butta, M. Calamia, J. Appl. Polymer Sci. 11, 527 (1967).
231. F. S. Dainton, D. M. Evans, F. E. Hoare, T. P. Melia, Polymer 3, 271 (1962).
232. D. J. H. Sandiford, J. Appl. Chem. 8, 188 (1958).
233. K.-H. Hellwege, J. Hennig, W. Knappe, Kolloid Z. - Z. Polymere 186, 29 (1962).
234. K.-H. Hellwege, R. Hoffmann, W. Knappe, Kolloid Z. - Z. Polymere 226, 109 (1968).
235. M. Baccaredda, E. Butta, V. Frosini, European Polymer J. 2, 423 (1966).
236. A. Nagai, T. Ishibashi, M. Takayanagi, Rept. Progr. Polymer Phys. Japan 10, 341 (1967).
237. A. I. Marei, Ye. A. Sidorovich, G. Ye. Novikova, E. I. Rodina, Polymer Sci. USSR 10, 630 (1968) (transl'n of Vysokomolekul. Soedin. A 10, 542 (1968)).
238. J. A. Faucher, J. V. Koleske, Polymer 9, 44 (1968).
239. J. C. Swallow, Proc. Roy. Soc., London, A238, 1 (1957).
240. L. Mandelkern, N. L. Jain, H. Kim, J. Polymer Sci., A-2, 6, 165 (1968); from results of ref. 241.
241. Y. Ishida, M. Matsuo, M. Takayanagi, J. Polymer Sci., B, 3, 321 (1965).
242. A. Aoki, S. Sawada, Repts. Progr. Polymer Phys. (Japan) 10, 261,509 (1967).
243. G. W. Miller, J. H. Saunders, J. Appl. Polymer Sci. 13, 1277 (1969).
244. E. J. Vandenberg, R. H. Ralston, B. J. Kocher, Rubber Age 102, 47 (1970).
245. Yu. K. Godovskii, Yu. S. Lipatov, Polymer Sci. USSR 10, 34 (1968) (transl'n of Vysokomolekul. Soedin. A 10, 32 (1968)).
246. Yu. K. Godovskii, Yu. S. Lipatov, Vysokomolekul. Soedin. B, 10, 323 (1968).
247. A. Conix, J. Polymer Sci. 29, 343 (1958).
248. R. H. Gobran, M. B. Berenbaum, ACS Meeting 133, April 1958; via E. R. Bertozzi, Rubber Chem. Technol. 41, 114 (1968).
249. A. V. Tobolsky, W. J. MacKnight, "Polymer Reviews", Vol. 13: Polymeric Sulfur and Related Polymer, Eds. H. F. Mark, E. H. Immergut, Interscience, p. 35 (1965).
250. G. Williams, Trans. Faraday Soc. 59, 1397 (1963).
251. B. E. Read, Polymer 3, 529 (1962).
252. V. A. Ponomarenko, N. M. Khomutova, Izv. Akad. Nauk SSSR, Ser. Khim. 5, 1153 (1969).
253. A. Mele, A. Delle Site, C. Bettinali, A. Di Domenico, J. Chem. Phys. 49, 3297 (1968).
254. J. L. Zollinger, J. R. Throckmorton, S. T. Ting, R. A. Mitsch, D. E. Elrick, J. Macromol. Sci. A3, 1443 (1969).
255. W. A. Lee, Royal Aircraft Establishment, Farnborough, Hants, U.K. Unpublished results.
256. P. Johncock, M. A. H. Hewins, Royal Aircraft Establishment, Farnborough, Hants, U.K. - Unpublished report.
257. D. E. Rice, J. Polymer Sci., B, 6, 335 (1968).
258. A. G. Pittman, D. L. Sharp, J. Polymer Sci., B, 3, 379 (1965).
259. N. H. Reinking, A. E. Barnabeo, W. F. Hale, J. Appl. Polymer Sci. 7, 2135 (1963).
260. H. M. Van Dort, C. R. H. I. De Jonge, W. J. Mijs, J. Polymer Sci., C, 22, 431 (1968).
261. K. Fujimoto, Chem. High Polym. (Japan) 18, 415 (1961).
262. J. Heijboer, British Polymer J. 1, 3 (1969).
263. S. G. Turley, ACS Org. Coatings Plast. Chem. 27, 453 (1967); ACS Polymer Preprints 8, 1524 (1967).
264. B. E. Read, G. Williams, Polymer 2, 239 (1961).
265. W. H. Linton, H. H. Goodman, J. Appl. Polymer Sci. 1, 179 (1959).
266. N. G. McCrum, J. Polymer Sci. 54, 561 (1961).
267. F. S. Dainton, D. M. Evans, F. E. Hoare, T. P. Melia, Polymer 3, 263 (1962).
268. G. Williams, Polymer 4, 27 (1963).
269. R. E. Wetton, G. Allen, Polymer 7, 331 (1966).
270. H. Fischer, W. Langbein, Kolloid Z. - Z. Polymere 216-7, 329 (1967).
271. L. Bohn, Kolloid Z. - Z. Polymere 201, 20 (1965).
272. A. H. Willbourn, Trans. Faraday Soc. 54, 717 (1958).
273. R. G. Saba, J. A. Sauer, A. E. Woodward, J. Polymer Sci., A, 1, 1483 (1963).
274. D. Sims; private communication: Ministry of Aviation, E.R.D.E., Waltham Abbey, Essex, U. K.
275. G. A. Clegg, D. R. Gee, T. P. Melia, A. Tyson, Polymer 9, 501 (1968).
276. J. A. Faucher, J. Polymer Sci., B, 3, 143 (1965).
277. R. E. Wetton, G. Williams, Trans. Faraday Soc. 61, 2132 (1965).
278. R. E. Wetton, G. S. Fielding-Russell, K. U. Fulcher, J. Polymer Sci., C, 30, 219 (1968).
279. S. Yoshida, M. Sakiyama, S. Seki, Rept. Progr. Polymer Phys. (Japan) 12, 247 (1969).
280. S. Yoshida, M. Sakiyama, S. Seki, Polymer J. (Japan) 1, 573 (1970).
281. J. J. Stratta, F. P. Reding, J. A. Faucher, J. Polymer Sci., A, 2, 5017 (1964).
282. F. B. Cramer, R. G. Beaman, J. Polymer Sci. 21, 237 (1956).
283. W. W. Schwarz, R. D. Lowrey, J. Appl. Polymer Sci. 11, 553 (1967).
284. P. Johncock, P. A. Grattan, Royal Aircraft Establishment, Farnborough, Hants, U.K. - Unpublished report.
285. P. Johncock, J. D. Lee, Royal Aircraft Establishment, Farnborough, Hants, U. K. - Unpublished report.

286. O. G. Lewis, L. V. Gallacher, AD 649869 (1966).

287. A. T. DiBenedetto, K. L. Trachte, J. Appl. Polymer Sci. $\underline{14}$, 2249 (1970).

288. E. Butta, Mater. Plastics Elastomers $\underline{35}$, 1411 (1969).

289. S. de Petris, V. Frosini, E. Butta, M. Baccaredda, Makromol. Chem. $\underline{109}$, 54 (1967).

290. F. E. Karasz, J. M. O'Reilly, J. Polymer Sci., B, $\underline{3}$, 561 (1965).

291. F. E. Karasz, H. E. Bair, J. M. O'Reilly, J. Polymer Sci., A-2, $\underline{6}$, 1141 (1968).

292. G. A. Bernier, R. P. Kambour, Macromolecules $\underline{1}$, 393 (1968).

293. M. H. Litt, A. V. Tobolsky, J. Macromol. Sci. B1, 3, 433 (1967).

294. A. J. Chalk, A. S. Hay, J. Polymer Sci., A-1, $\underline{7}$, 691 (1969).

295. G. F. L. Ehlers, K. R. Fisch, W. R. Powell, J. Polymer Sci., A-1, $\underline{7}$, 2931 (1969).

296. G. Allen. M. W. Coville, R. M. John, R. F. Warren, Polymer $\underline{11}$, 492 (1970).

297. F. E. Karasz, W. J. MacKnight, J. Stoelting, J. Appl. Phys. $\underline{41}$, 4357 (1970).

298. J. M. Barrales-Rienda, J. M. G. Fatou, Kolloid Z. - Z. Polymere $\underline{244}$, 317 (1971).

299. A. Eisenberg, B. Cayrol, J. Polymer Sci., C, $\underline{35}$, 129 (1971).

300. C. I. Chung, J. A. Sauer, J. Polymer Sci., A-2, $\underline{9}$, 1097 (1971); T_g values mostly from manufacturers' data.

301. P. A. Grattan, V. C. R. McLoughlin, W. A. Lee, Royal Aircraft Establishment, Farnborough, Hants, U.K. - Unpublished data.

302. L. W. Breed, R. L. Elliot, M. E. Whitehead, J. Polymer Sci., A-1, $\underline{5}$, 2745 (1967).

303. L. T. C. Lee, E. M. Pearce, J. Polymer Sci., A-1, $\underline{9}$, 557 (1971).

304. K. Saotome, H. Komoto, J. Polymer Sci., A-1, $\underline{4}$, 1475 (1966).

305. H. G. Weyland, P. J. Hoftyzer, D. W. Van Krevelen, Polymer $\underline{11}$, 79 (1970).

306. H. M. van Dort, C. A. M. Hoefs, E. P. Magre, A. J. Schopf, K. Yntema, European Polymer J. $\underline{4}$, 275 (1968).

307. B. J. Tabor, E. P. Magré, J. Boon, European Polymer J. $\underline{7}$, 1127 (1971).

308. J. C. Mileo, B. Sillion, G. De Gaudemaris, Compt. Rend., C, $\underline{268}$, 1949 (1969).

309. Brit. 1,128,807, 2. Oct. 1968 (M. E. B. Jones, I. Goodman).

310. D. Brown, M. E. B. Jones, W. R. Maltman, J. Polymer Sci., B, $\underline{6}$, 635 (1968).

311. V. Frosini, M. Pasquini, E. Butta, Chim. Ind. $\underline{53}$, 140 (1971).

312. E. Butta, S. de Petris, V. Frosini, M. Pasquini, European Polymer J. $\underline{7}$, 387 (1971).

313. G. F. L. Ehlers, R. C. Evers, K. R. Fisch, J. Polymer Sci., A-1, $\underline{7}$, 3413 (1969).

314. B. E. Jennings, M. E. B. Jones, J. B. Rose, J. Polymer Sci., C, $\underline{16}$, 715 (1967).

315. M. E. A. Cudby, R. G. Feasey, B. E. Jennings, M. E. B. Jones, J. B. Rose, Polymer $\underline{6}$, 589 (1965).

316. J. E. Kurz, J. C. Woodbrey, M. Ohta, J. Polymer Sci., A-2, $\underline{8}$, 1169 (1970).

317. G. W. Miller, ACS Polymer Preprints $\underline{8}$, 1072 (1967).

318. G. W. Miller, "Thermal Analysis", Academic Press Inc., New York, Vol. 1, 1969, p. 435.

319. T. F. Gray, R. L. Combs, D. F. Slonaker, W. C. Wooten, SPE Tech. Papers $\underline{13}$, 370 (1967).

320. M. Baccaredda, E. Butta, V. Frosini, S. de Petris, J. Polymer Sci., A-2, $\underline{5}$, 1296 (1967).

321. L. M. Robeson, J. A. Faucher, J. Polymer Sci., B, $\underline{7}$, 35 (1969).

322. G. F. L. Ehlers, K. R. Fisch, W. R. Powell, J. Polymer Sci., A-1, $\underline{7}$, 2955 (1969).

323. N. J. Mills, A. Nevin, J. McAinsh, J. Macromol. Sci. B4, 863 (1970).

324. E. J. Vandenberg, J. Polymer Sci. $\underline{47}$, 486,489a (1960).

325. A. S. Hay, Macromolecules $\underline{2}$, 107 (1969).

326. J. M. Crissman, J. A. Sauer, A. E. Woodward, J. Polymer Sci., A, $\underline{2}$, 5075 (1964).

327. W. P. Slichter, Makromol. Chem. $\underline{34}$, 67 (1959).

328. L. E. St. Pierre, C. C. Price, J. Am. Chem. Soc. $\underline{78}$, 3432 (1956).

329. T. M. Connor, A. Hartland, Polymer $\underline{9}$, 591 (1968).

330. G. Williams, Trans. Faraday Soc. $\underline{61}$, 1564 (1965); data used by ref. 329.

331. G. Allen, C. Booth, M. N. Jones, D. J. Marks, W. D. Taylor, Polymer $\underline{5}$, 547 (1964).

332. I. Kuntz, C. Cozewith, H. T. Oakley, G. Via, H. T. White, Z. W. Wilchinsky, Macromolecules $\underline{4}$, 4 (1971).

333. I. P. Losev, O. V. Smirnova, Ye. V. Korovina, Vysokomolekul. Soedin. $\underline{5}$, 1603 (1963); transl'n in Polymer Sci. USSR $\underline{5}$, 707 (1964).

334. H. Schnell, Ind. Eng. Chem. $\underline{51}$, 157 (1959).

335. A. N. Perepelkin, P. V. Kozlov, Polymer Sci. USSR $\underline{8}$, 57 (1966); (transl'n of Vysokomolekul. Soedin. $\underline{8}$, 56 (1966)).

336. F. P. Reding, J. A. Faucher, R. D. Whitman, J. Polymer Sci. $\underline{54}$, S56 (1961).

337. L. J. Garfield, J. Polymer Sci., C, $\underline{30}$, 561 (1970).

338. H. Schnell, "Chemistry and Physics of Polycarbonates", Interscience, New York, 1964, Chap. III.

339. B. F. Malichenko, V. I. Feoktistova, A. E. Nesterov, Vysokomolekul. Soedin. B, $\underline{11}$, 543 (1969).

340. U.S. 3,313,777 - 11. April 1967 (Brit. 962,913) (E. N. Elam, J. C. Martin, R. Gilkey).

341. W. Gawlak, R. P. Palmer, J. B. Rose, D. J. H. Sandiford, A. Turner-Jones, Chem. Ind. $\underline{25}$, 1148 (1962).

342. Brit. 1,011,283 - 24. Nov. 1965 (A. A. D'Onofrio).

343. O. V. Smirnova, El Said Ali Khasan, I. P. Losev, G. S. Kolesnikov, Polymer Sci. USSR $\underline{7}$, 557 (1965) (transl'n of Vysokomolekul. Soedin. $\underline{7}$, 503 (1965)).

344. H. Schnell, Angew. Chem. $\underline{68}$, 633 (1956).

345. H. Schnell, Plastics Inst. Trans. J. $\underline{28}$, 143 (1960).

346. O. V. Smirnova, Ye. V. Korovina, G.S. Kolesnikov, A. M. Lipkin, S. I. Kuzina, Polymer Sci. USSR $\underline{8}$, 776 (1966) (transl'n of Vysoko-molekul. Soedin. $\underline{8}$, 708 (1966)).

347. U. S. 3,546,165 - 8. Dec. 1970 (P.W. Morgan).

348. D. Stefan. H. L. Williams, D. R. Renton, M. M. Pintar, J. Macromol. Sci. B4, 853 (1970).

349. D. W. McCall, D. R. Falcone, Trans. Faraday Soc. $\underline{66}$, 262 (1970).

350. R. E. Robertson, General Elec. Co., Res. & Dev. Center, Schenectady, N.Y., Gen. Chem. Lab. Report No. 67-C-353, 1967; also Appl. Polymer Symp. $\underline{7}$, 201 (1968).

351. G. W. Miller, Appl. Polymer Symp. $\underline{10}$, 35 (1969).

352. M. S. Ali, R. P. Sheldon, J. Appl. Polymer Sci. $\underline{14}$, 2619 (1970).

353. J. van Turnhout, Polymer J. (Japan) $\underline{2}$, 173 (1971).

354. S. Matsuoka, Y. Ishida, J. Polymer Sci., C, $\underline{14}$, 247 (1966).

355. F. Krum, F. H. Muller, Kolloid Z. $\underline{164}$, 81 (1959).

356. I. I. Perepechko, L. A. Bodrova, Soviet Plastics $\underline{7}$, 58 (1967). (transl'n of Plasticheskie Massy $\underline{7}$, 56 (1967)).

357. J. P. Mercier, J. J. Aklonis, A. V. Tobolsky, AD 410922 (1963), "The Viscoelastic Behaviour of Polycarbonates of Bisphenol A", ONR Tech. Report RLT-60.

358. J. H. Golden, B. L. Hammant, E. A. Hazell, J. Appl. Polymer Sci. $\underline{11}$, 1571 (1967).

359. J. P. Mercier, J. J. Aklonis, M. Litt, A. V. Tobolsky, J. Appl. Polymer Sci. $\underline{9}$, 447 (1965).

360. W. J. Jackson, J. R. Caldwell, Ind. Eng. Chem. Prod. Res. Develop. $\underline{2}$, 246 (1963).

361. J. L. Work, J. E. Herweh, J. Polymer Sci., A-1, $\underline{6}$, 2022 (1968).

362. G. W. Miller, ACS Polymer Preprints $\underline{9}$, 832 (1968).

363. G. P. Mikhailov, M. P. Eidel' nant, Vysokomolekul. Soedin. $\underline{2}$, 287 (1960).

364. J. M. O'Reilly, F. E. Karasz, H. E. Bair, J. Polymer Sci., C, $\underline{6}$, 109 (1964).

365. A. Conix, L. Jeurissen, Adv. Chem. Ser. $\underline{48}$, 172 (1964).

366. J. Bussink, J. Heijboer, "Physics of Non-Crystalline Solids", Ed. J.A. Prins, Proc. Intern. Conf., Delft 1964, p. 388, North Holland Publishing Co., Amsterdam, 1965.

367. B. von Falkai, W. Rellensmann, Makromol. Chem. 75, 112 (1964).
368. T. Sulzberg, R. J. Cotter, Macromolecules 2, 146 (1969).
369. E. V. Gouinlock, E. J. Quinn, H. W. Marciniak, R. R. Hindersinn, IUPAC Preprints, Intern. Symp. Macromol. Chem., Toronto, Sect. A13.3 (1968).
370. Du Pont trade literature, A-49783; Instruments: 940 Thermomechanical Analyzer.
371. H. G. Weyland, C. A. M. Hoefs, K. Yntema, W. J. Mijs, European Polymer J. 6, 1339 (1970).
372. P. W. Morgan, Macromolecules 3, 536 (1970).
373. I. V. Yannas, A. C. Lunn, J. Macromol. Sci. B4, 603 (1970).
374. I. I. Perepechko, L. A. Kvacheva, I. I. Levantovskaya, Vysokomolekul. Soedin. A, 13, 702 (1971) (trans'n in Polymer Sci. USSR 13, 796 (1971)).
375. Y. Yamamoto, S. Tsuge, T. Takeuchi, Bull. Chem. Soc. (Japan) 44, 1145 (1971).
376. A. Conix, J. Polymer Sci., C, 16, 3821 (1968).
377. H. Van Hoorn, J. Appl. Polymer Sci. 12, 871 (1968).
378. I. I. Perepechko, L. A. Kvacheva, Soviet Plastics 8, 40 (1971). (transl'n of Plasticheskie Massy 8, 42 (1971)).
379. T. Takekoshi, ACS Polymer Preprints 10, 103 (1969).
380. Brit. 1,052,757 - 3o. Dec. 1966.
381. G. Bier, Adv. Chem. Ser. 91, 612 (1969).
382. A. Conix, IUPAC Preprints, Intern. Symp. Macromol. Chem., Montreal, Sect. D39 (1961); via ref. 44.
383. C. S. Marvel, C. H. Young, J. Am. Chem. Soc. 73, 1066 (1951).
384. W. A. Lee, B. Stagg, Royal Aircraft Establishment, Farnborough, Hants, U.K., TR 71223 (1971).
385. S. D. Bruck, J. Polymer Sci., B, 4, 933 (1966).
386. V. C. R. McLoughlin, J. Thrower, M. A. H. Hewins, J. S. Pippett, M. A. White, Royal Aircraft Establishment, Farnborough, Hants, U. K., TR 70160 (1970).
387. U. S. 3,436,376 - 1. April 1969 (I. N. Duling).
388. R. G. Beaman, J. Polymer Sci. 9, 470 (1952).
389. O. B. Edgar, J. Chem. Soc. 1952, 2638.
390. O. B. Edgar, R. Hill, J. Polymer Sci. 8, 1 (1952).
391. E. F. Izard, J. Polymer Sci. 9, 35 (1952).
392. A. Steitz, J. O. Knobloch, J. Paint Technol. 40, 384 (1968).
393. H. J. Kolb, E. F. Izard, J. Appl. Phys. 20, 564 (1949).
394. H. Mark, Kolloid Z. - Z. Polymere 216/7, 126 (1967).
395. N. G. Gaylord, H. F. Mark, "Linear and Stereoregular Addition Polymers", Interscience, New York, 1959.
396. Y. Wada, J. Phys. Soc. (Japan) 16, 1226 (1961).
397. R. Zannetti, P. Manaresi, L. Baldi, Chim. Ind. 43, 1310 (1961).
398. I. M. Ward, Textile Res. J. 13, 650 (1961).
399. A. Anton, J. Appl. Polymer Sci. 12, 2117 (1968); also ACS Polymer Preprints 8, 873 (1967).
400. S. Saito, T. Nakajima, J. Appl. Polymer Sci. 2, 93 (1959).
401. L. Mandelkern, G. M. Martin, F. A. Quinn, J. Res. Nat. Bur. Std. 58, 137 (1957).
402. I. M. Ward, Polymer 5, 59 (1964).
403. G. Farrow, J. McIntosh, I. M. Ward, Makromol. Chem. 38, 147 (1960).
404. J. Bateman, R. E. Richards, G. Farrow, I. M. Ward, Polymer 1, 63 (1960).
405. B. Ke, J. Appl. Polymer Sci. 6, 624 (1962).
406. J. H. Dumbleton, T. Murayama, Kolloid Z. - Z. Polymere 220, 41 (1967).
407. W. R. Vieth, E. S. Matulevicius, S. R. Mitchell, Kolloid Z. - Z. Polymere 220, 49 (1967).
408. G. S. Y. Yeh, P. H. Geil, J. Macromol. Sci. B1, 235 (1967).
409. G. P. Mikhailov, V. A. Shevelev, Polymer Sci. USSR 8, 840 (1966) (transl'n of Vysokomolekul. Soedin. 8, 763 (1966)).
410. J. H. Dumbleton, J. P. Bell, T. Murayama, J. Appl. Polymer Sci. 12, 2491 (1968).

411. J. G. Smith, C. J. Kibler, B. J. Sublett, J. Polymer Sci., A-1, 4, 1851 (1966). Values taken from data of ref. 412.
412. R. M. Schulken, R. E. Boy, R. H. Cox, J. Polymer Sci., C, 6, 17 (1964).
413. I. Ito, S. Okajima, F. Shibata, J. Appl. Polymer Sci. 14, 551 (1970).
414. J. B. Jackson, G. W. Longman, Polymer 10, 873 (1969).
415. J. H. Dumbleton, T. Murayama, J. Appl. Polymer Sci. 14, 2921 (1970).
416. E. A. Egorov, V. V. Zhizhenkov, Mekh. Polim. 192, 24 (1971).
417. T. R. Manley, J. Thermal Anal. 2, 411 (1970).
418. H. Sasabe, K. Sawamura, S. Saito, K. Yoda, Polymer J. (Japan) 2, 518 (1971).
419. I. M. Dolgopol'skii, Kh. A. Dobina, V. S. Fikhtengol'ts, S. A. Kamysheva, M. I. Sinaiskaya, L. G. Balashova, R. V. Zolotareva, Polymer Sci. USSR 9, 1723 (1967) (transl'n of Vysokomolekul. Soedin. A 9, 1536 (1967)).
420. E. V. Gouinlock, C. J. Verbanic, G. C. Schweiker, J. Appl. Polymer Sci. 1, 361 (1959).
421. W. R. Griffin, Rubber World 136, 687 (1957).
422. G. C. Schweiker, P. Robitschek, J. Polymer Sci. 24, 33 (1957).
423. J. H. Sewell, Royal Aircraft Establishment, Farnborough, Hants, U.K. - Unpublished results.
424. I. J. Goldfarb, R. McGuchan, AD 697987 (1969).
425. R. C. Evers, G. F. L. Ehlers, J. Polymer Sci., A-1, 7, 3020 (1969).
426. A. Conix, Ind. Eng. Chem. 51, 147 (1959).
427. V. Frosini, G. Vallebona, Chim. Ind. 52, 499 (1970).
428. N. H. Shearer, SPE Tech. Papers 12, 22nd A.N.T.E.C., Montreal 1966, Session XIV, paper 1.
429. C. S. Marvel, J. H. Johnson, J. Am. Chem. Soc. 72, 1674 (1950).
430. S. M. Somerville, I. M. White, Royal Aircraft Establishment, Farnborough, Hants, U. K., TR 66345 (1966).
431. I. M. White, M. A. H. Hewins, S. M. Somerville, Paper presented at Second Intern. Symp. Organosilicon Chem., Univ. Bordeaux, 9-12 July 1968.
432. W. A. Lee, J. H. Sewell, Royal Aircraft Establishment, Farnborough, Hants, U. K. - Unpublished results.
433. J. H. Sewell, B. Stagg, Royal Aircraft Establishment, Farnborough, Hants, U. K. - Unpublished report.
434. W. A. Lee, F. Pinchin, Royal Aircraft Establishment, Farnborough, Hants, U. K. - Unpublished results.
435. F. P. 1,175,362 (1959) - Brit. 863,704 (1961); via ref. 313.
436. W. M. Eareckson, J. Polymer Sci. 40, 399 (1959).
437. V. V. Korshak, S. V. Vinogradova, C. L. Slonimskii, Ya. S. Vygodskii, S. N. Salazkin, A. A. Askadskii, A. I. Mzhel'skii, V. P. Sidorova, Polymer Sci. USSR 10, 2395 (1968) (transl'n of Vysokomolekul. Soedin. A 10, 2058 (1968)).
438. W. A. Lee, I. M. White, F. Pinchin, Royal Aircraft Establishment, Farnborough, Hants, U. K. - Unpublished results.
439. K. A. Gol'dgammer, G. G. Pimenov, A. I. Maklakov, V. V. Korshak, S. V. Vinogradova, P. M. Valetskii, A. N. Baskakov, Vysokomolekul. Soedin. A, 10, 821 (1968) (transl'n in Polymer Sci. USSR 10, 953 (1968)).
440. R. L. Combs, R. G. Nations, J. Polymer Sci., C, 30, 407 (1970). Unpublished result from T. F. Gray.
441. V. V. Korshak, G. L. Berestneva, S. V. Vinogradova, A. N. Baskakov, P. M. Valetskii, Polymer Sci. USSR 10, 2300 (1968) (transl'n of Vysokomolekul. Soedin. A 10, 1984 (1968)).
442. Brit. 956,206 - 22. April 1964 (W. J. Jackson, J. R. Caldwell).
443. S. S. Karapetyan, A. Ya. Yakubovich, I. L. Knunyants, Polymer Sci. USSR 6, 1718 (1964) (transl'n of Vysokomolekul. Soedin. 6, 1550 (1964)).
444. Brit. 838,986 (1960) (J. E. McIntyre, E. C. Pugh).
445. K. Saotome, H. Komoto, J. Polymer Sci., A-1, 5, 119 (1967).
446. W. J. Macknight, M. Yang, T. Kajiyama, ACS Polymer Preprints 9, 860 (1968).

447. D. J. Lyman, J. Heller, M. Barlow, Makromol. Chem. 84, 64 (1965).

448. A. Anton, H. C. Beachell, J. Polymer Sci., B, 7, 215 (1969).

449. B. F. Malichenko, E. V. Shelud'ko, Yu. Yu. Kercha, R. L. Savchenko, Vysokomolekul. Soedin. A, 11, 377 (1969) (transl'n in Polymer Sci. USSR 11, 423 (1969)).

450. B. F. Malichenko, E. V. Shelud'ko, Yu. Yu. Kercha, R. L. Savchenko, Vysokomolekul. Soedin. A, 11, 1518 (1969) (transl'n in Polymer Sci. USSR 11, 1721 (1969)).

451. B. F. Malichenko, E. V. Shelud'ko, Yu. Yu. Kercha, Polymer Sci. USSR 9, 2808 (1967) (transl'n of Vysokomolekul. Soedin. A 9, 2482 (1967)).

452. Yu. Yu. Kercha, L. I. Ryabokon, B. F. Malichenko, "Synthesis and Physico-Chemistry of Polyurethanes", Ed. Yu. S. Lipatov, Naukova Dumka, 1968, p. 198.

453. M. A. McCall, J. R. Caldwell, H. G. Moore, H. M. Beard, J. Macromol. Sci. A3, 911 (1969).

454. A. Ye. Nesterov, Yu. Yu. Kercha, Yu. S. Lipatov, L. M. Sushko, L. I. Ryabokon, Polymer Sci. USSR 9, 2782 (1967) (transl'n of Vysokomolekul. Soedin. A 9, 2459 (1967)).

455. C. B. Griffis, C. W. Shurtleff, AD 635114 (1966).

456. M. C. Henry, C. B. Griffis, AD 632196 (1966). Quoted values.

457. P. D. Schuman, E. C. Stump, G. Westmoreland, J. Macromol. Sci. B1, 815 (1967).

458. D. A. Barr, R. N. Haszeldine, C. J. Willis, J. Chem. Soc. 1961 1351.

459. G. H. Crawford, D. E. Rice, Chem. Eng. News 38, 107 (1960).

460. A. Heslinga, Plastica 21, 206 (1968).

461. E. C. Stump, C. D. Padgett, AD 666801 (1967).

462. G. L. Ball, I. O. Salyer, J. V. Pustinger, H. S. Wilson, AD 672523 (1967).

463. Yu. A. Yuzhelevskii, E. G. Kagan, A. L. Klebanskii, I. A. Zevakin, A. V. Kharlamova, Vysokomolekul. Soedin. B, 11, 854 (1969).

464. M. S. Paterson, J. Appl. Phys. 35, 176 (1964).

465. J. B. Alexopoulos, J. A. Barrie, J. C. Tye, M. Fredrickson, Polymer 9, 56 (1968).

466. O. R. Pierce, Y. K. Kim, J. Elastoplastics 3, 82 (1971).

467. K. E. Polmanteer, M. J. Hunter, J. Appl. Polymer Sci. 1, 3 (1959).

468. K. A. Andrianov, T. K. Dzhashiashvili, V. V. Astakhin, G. N. Shumakova, Vysokomolekul. Soedin. B, 10, 766 (1968).

469. W. P. Gergen. T. L. Keelen, 11th Annual Meeting Proc. Intern. Inst. Synth. Rubber Producers, Inc., p. 81 (1970).

470. C. E. Weir, W. H. Leser, L. A. Wood, J. Res. Nat. Bur. Std. 44, 367 (1950).

471. L. A. Wood, J. Polymer Sci. 28, 319 (1958).

472. M. Sh. Yagfarov, V. S. Ionkin, Polymer Sci. USSR 10, 1867 (1968) (transl'n of Vysokomolekul. Soedin. A 10, 1613 (1968)).

473. A. Yim, L. E. St. Pierre, J. Polymer Sci., B, 7, 237 (1969).

474. J. D. Helmer, K. E. Polmanteer, J. Appl. Polymer Sci. 13, 2113 (1969).

475. K. A. Andrianov, S. E. Yakushkina, Vysokomolekul. Soedin. 4, 1193 (1962).

476. K. A. Andrianov, T. K. Dzhashiashvili, V. V. Astakhin, G. N. Shumakova, Soviet Plastics 2, 47 (1968) (transl'n of Plasticheskie Massy 2, 44 (1968)).

477. V. C. R. McLoughlin, P. A. Grattan, Royal Aircraft Establishment, Farnborough, Hants, U. K., TR 71224 (1971).

478. R. E. Burks, J. L. Greene, N68-16509; Southern Res. Inst. Rept. 25 (1967).

479. J. A. Empen, J. K. Stille, "Macromolecular Syntheses", 3, Ed. N. G. Gaylord, Wiley, New York, 1969, p. 56.

480. J. K. Stille, J. A. Empen, J. Polymer Sci., A-1, 5, 273 (1967).

481. E. H. Catsiff, J. Appl. Polymer Sci. 15, 1641 (1971).

482. A. Nicco, J. P. Machon, H. Fremaux, J. Ph. Pied, B. Zindy, M. Thiery, European Polymer J. 6, 1427 (1970).

483. R. S. Nevin, E. M. Pearce, J. Polymer Sci., B, 3, 487 (1965).

484. W. J. Middleton, H. W. Jacobson, R. E. Putnam, H. C. Walter, D. G. Pye, W. H. Sharkey, J. Polymer Sci., A, 3, 4115 (1965).

485. V. A. Engelhardt; report in Chem. Eng. News 43, 80 (1965).

486. W. H. Sharkey, W. J. Middleton, H. W. Jacobson, O. S. Acker, H. C. Walter, Chem. Eng. News 41, 46 (1963).

487. J. A. Empen, J. K. Stille, "Macromolecular Syntheses", 3, Ed. N. G. Gaylord, Wiley, New York, 1969, p. 53.

488. C. C. Price, E. A. Blair, J. Polymer Sci., A-1, 5, 171 (1967).

489. R. M. Fitch, D. C. Helgeson, J. Polymer Sci., C, 22, 1101 (1969).

490. M. Morton, R. F. Kammereck, L. J. Fetters, Brit. Polymer J. 3, 120 (1971).

491. H. Komoto, Rev. Phys. Chem. Japan 37, 112 (1967).

492. J. Pellon, J. Polymer Sci., A, 1, 3561 (1963).

493. R. L. Fyans; private communication to ref. 114.

494. M. Russo, Mater. Plastics Elastomers 34, 61 (1968); via RAPRA Abstr. 1968, abstr. 4571.

495. E. A. Peterson; private communication to ref. 481.

496. S. Adamek, B. B. J. Wood, R. T. Woodhams, Rubber Plastics Age Intern. 46, 56 (1965); Rubber Age 96, 581 (1965).

497. C. G. Overberger, J. K. Weise, J. Am. Chem. Soc. 90, 3538 (1968).

498. H. G. Buhrer, H.-G. Elias, Makromol. Chem. 140, 41 (1970).

499. C. G. Overberger, J. K. Weise, J. Am. Chem. Soc. 90, 3533 (1968).

500. M. E. B. Jones, SPE Tech. Papers 15, 27th A.N.T.E.C., Chicago 1969, p. 453.

501. G. Jarrett, M. E. B. Jones, Brit. Polymer J. 2, 229 (1970).

502. H. Wexler, Makromol. Chem. 115, 262 (1968).

503. R. G. Beaman, F. B. Cramer, J. Polymer Sci. 21, 223 (1956).

504. J. H. Magill, J. Polymer Sci., A, 3, 1195 (1965).

505. B. Ke, A. W. Sisko, J. Polymer Sci. 50, 87 (1961).

506. Y. Nishijima, J. Seki, T. Kawai, Rept. Progr. Polymer Phys. (Japan) 10, 473 (1967).

507. M. J. Forster, Textile Res. J. 38, 474 (1968).

508. K. Neki, M. Takayanagi, Rept. Progr. Polymer Phys. (Japan) 8, 281 (1965).

509. W. H. Charch, W. W. Moseley, Textile Res. J. 29, 552 (1959); via ref. 507.

510. F. Rybnikar, J. Polymer Sci. 28, 633 (1958).

511. M. Takayanagi, Memoirs Faculty Eng. Kyushu Univ. 23, 2 (1963); via ref. 507.

512. S. R. Urzendowski, A. H. Guenther, J. R. Asay, ACS Polymer Preprints 9, 878 (1968).

513. J. Rubin, R. D. Andrews, Polymer Eng. Sci. 8, 302 (1968).

514. E. R. Neuhausl, Plastics Polymer 36, 93 (1968).

515. D. R. Gee, T. P. Melia, Polymer 11, 192 (1970).

516. T. Shibusawa, J. Appl. Polymer Sci. 14, 1553 (1970).

517. F. R. Prince, E. M. Pearce, R. J. Fredericks, J. Polymer Sci., A-1, 8, 3533 (1970).

518. V. Frosini, E. Butta, J. Polymer Sci., B, 9, 253 (1971).

519. V. Zilvar, I. Boukal, J. Hell, Plastics Inst. Trans. J. 35, 403 (1967).

520. D. C. Prevorsek, R. H. Butler, H. K. Reimschuessel, J. Polymer Sci., A-2, 9, 867 (1971).

521. K. Schmieder, K. Wolf, Kolloid Z. 134, 149 (1953).

522. K. H. Illers, H. G. Kilian, R. Kosfeld, Ann. Rev. Phys. Chem. 12, 49 (1961).

523. K. C. Dao, J. H. Percy, Inst. Physics and Physical Soc., Proc. Conf. on the Science of Materials, p. 151, Aug. 1969; New Zealand, DSIR, Information Series 71.

524. R. F. Boyer, R. S. Spencer, J. Appl. Phys. 15, 398 (1944).

525. R. Meredith, B. Hsu, J. Polymer Sci. 61, 271 (1962).

526. G. F. D'Alelio, "Fundamental Principles of Polymerisation", Wiley, New York, 1952, p. 124.

527. D. W. McCall, E. W. Anderson, J. Chem. Phys. 32, 237 (1960).

528. A. E. Woodward, J. M. Crissman, J. A. Sauer, J. Polymer Sci. 44, 23 (1960).

529. A. M. Thomas, Nature 179, 862 (1957).

530. F. P. Chappel, M. F. Culpin, R. G. Gosden, T. C. Tranter, Conference on Adv. Polymer Sci. Technol. London, 1963.
531. J. S. Ridgeway, J. Polymer Sci., A-1, 7, 2195 (1969).
532. J. S. Ridgeway, J. Polymer Sci., A-1, 8, 3089 (1970).
533. J. P. Bell, J. Appl. Polymer Sci. 12, 627 (1968).
534. E. A. Tippetts, J. Zimmerman, J. Appl. Polymer Sci. 8, 2465 (1964).
535. H. Komoto, Rev. Phys. Chem. (Japan) 37, 105 (1967).
536. K. Saotome, H. Komoto, J. Polymer Sci., A-1, 4, 1463 (1966).
537. H. Komoto, K. Saotome, Chem. High Polymers (Japan) 22, 337 (1965); C.A. 63, 15045f (1965).
538. G. Champetier, J. P. Pied, Makromol. Chem. 44-6, 64 (1961); C.A. 56, 1584c (1962).
539. B. S. Marks, G. C. Schweiker, J. Polymer Sci. 43, 229 (1960).
540. F. R. Prince, E. M. Pearce, Macromolecules 4, 347 (1971).
541. D. A. Holmer, J. Polymer Sci., A-1, 6, 3177 (1968).
542. J. C. Mileo, B. Sillion, G. De Gaudemaris, Compt. Rend., C, 268, 2007 (1969).
543. J. B. Jackson, Polymer 10, 159 (1969).
544. L. C. Glover, B. J. Lyons, ACS Polymer Preprints 9, 243 (1968).
545. V. Guidotti, M. Russo, L. Mortillaro, Makromol. Chem. 147, 111 (1971).
546. G. A. Kuznetsov, V. D. Gerasimov, L. N. Fomenko, A. I. Maklakov, G. G. Pimenov, L. B. Sokolov, Polymer Sci. USSR 7, 1763 (1965) (transl'n of Vysokomolekul. Soedin. 7, 1592 (1965)).
547. G. S. Kolesnikov, O. Ya. Fedotova, V. V. Trezvov, V. N. Kuz'-micheva, Polymer Sci. USSR 10, 2612 (1968) (transl'n of Vysokomolekul. Soedin. A 10, 2248 (1968)).
548. E. Bessler, G. Bier, Makromol. Chem. 122, 30 (1969).
549. T. G. Bassiri, A. Levy, M. Litt, J. Polymer Sci., B, 5, 871 (1967).
550. A. Levy, M. Litt, J. Polymer Sci., B, 5, 881 (1967).
551. A. Levy, M. Litt, J. Polymer Sci., A-1, 6, 1883 (1968).
552. G. Allen, C. J. Lewis, S. M. Todd, Polymer 11, 44 (1970).
553. H. R. Allcock, R. L. Kugel, K. J. Valan, Inorg. Chem. 5, 1709 (1966).
554. H. R. Allcock, R. L. Kugel, Inorg. Chem. 5, 1716 (1966).
555. G. Allen, Brit. Polymer J. 1, 168 (1969).
556. H. R. Allcock, R. L. Kugel, J. Am. Chem. Soc. 87, 4216 (1965).
557. S. H. Rose, K. A. Reynard, J. R. Cable, AD 704332 (1970).
558. N. S. Nametkin, N. V. Ushakov, V. M. Vdovin, Polymer Sci. USSR 13, 31 (1971) (transl'n of Vysokomolekul. Soedin. A 13, 29 (1971)).
559. N. S. Nametkin, V. M. Vdovin, V. I. Zav'yalov, Dokl. Akad. Nauk SSSR 162, 824 (1965).
560. F. S. Model, G. Redl, E. G. Rochow, J. Polymer Sci., A-1, 4, 639 (1966).
561. F. S. Model, E. G. Rochow, J. Polymer Sci., A-2, 8, 999 (1970).
562. N. S. Nametkin, V. M. Vdovin, V. I. Zav'yalov, Polymer Sci. USSR 7, 836 (1965) (transl'n of Vysokomolekul. Soedin. 7, 757 (1965)).
563. V. C. R. McLoughlin, J. Thrower, P. A. Grattan, M. A. H. Hewins, Royal Aircraft Establishment, Farnborough, Hants, U. K, TR 71111 (1971).
564. W. F. Gorham, J. Polymer Sci., A-1, 4, 3027 (1966).
565. H. G. Gilch, W. L. Wheelwright, J. Polymer Sci., A-1, 4, 1337 (1966).
566. I. N. Markevich, S. I. Beilin, M. P. Teterina, G. P. Karpacheva, B. A. Dolgoplosk, Dokl. Akad. Nauk SSSR 191, 362 (1970).
567. L. A. Auspos, C. W. Bumham, L. A. R. Hall, J. K. Hubbard, W. Kirk, J. R. Schaefgen, S. B. Speck, J. Polymer Sci. 15, 19 (1955).
568. J. K. Stille, R. O. Rakutis, H. Mukamal, F. W. Harris, Macromolecules 1, 431 (1968).
569. M. Morton, J. Elastoplastics 3, 112 (1971).
570. R. J. Schaffhauser, M. C. Shen, A. V. Tobolsky, J. Appl. Polymer Sci. 8, 2825 (1964).
571. M. Kaufman, A. F. Williams, J. Appl. Chem. 1, 489 (1951).
572. A. Chierico, G. Del Nero, G. Lanzi, E. R. Mognaschi, European Polymer J. 3, 245 (1967).
573. E. Jenckel, Kolloid Z. 100, 163 (1942).
574. M. L. Dannis, J. Appl. Polymer Sci. 1, 121 (1959).

575. M. L. Dannis, J. Appl. Polymer Sci. 7, 231 (1963).
576. R. H. Gerke, J. Polymer Sci. 13, 295 (1954).
577. M. Baccaredda, E. Butta, Chim. Ind. (Milan) 42, 978 (1960).
578. E. Pedemonte, U. Bianchi, J. Polymer Sci., B, 2, 1025 (1964).
579. A. W. Meyer, R. R. Hampton, J. A. Davison, J. Am. Chem. Soc. 74, 2294 (1952).
580. F. S. Dainton, D. M. Evans, F. E. Hoare, T. P. Melia, Polymer 3, 297 (1962).
581. R. J. Morgan, L. E. Nielsen, R. Buchdahl, J. Appl. Phys. 42, 4653 (1971).
582. R. N. Kienle, E. S. Dizon, T. J. Brett, C. F. Eckert, Rubber Chem. Technol. 44, 996 (1971).
583. G. Dall'asta, G. Motroni, Angew. Makromol. Chem. 16/17, 51 (1971).
584. W. Marconi, A. Mazzei, G. Lugli, M. Bruzzone, J. Polymer Sci.,C, 16, 805 (1967).
585. W. S. Bahary, D. I. Sapper, J. H. Lane, Rubber Chem. Technol. 40, 1529 (1967).
586. G. Kraus, C. W. Childers, J. T. Gruver, J. Appl. Polymer Sci. 11, 1581 (1967); reprint in Rubber Chem. Technol. 42, 520 (1969).
587. D. I. Relyea, H. P. Smith, A. N. Johnson, AD 657675 (1967).
588. K. A. Wolf, Z. Electrochem. 65, 604 (1961); via R. Havinga, A. Schors, J. Macromol. Sci., A2, 12 (1968) and Sci. Abstr., A, 65, No. 773, abs. 8141 (1962).
589. E. N. Tinyakova, B. A. Dolgoplosk, T. G. Zhuravleva, R. N. Kovalevskaya, T. N. Kuren'gina, J. Polymer Sci. 52, 159 (1961).
590. M. Sh. Yagfarov, Polymer Sci. USSR 10, 1465 (1968) (transl'n of Vysokomolekul. Soedin. A 10, 1264 (1968)).
591. G. S. Trick, J. Appl. Polymer Sci. 3, 253 (1960).
592. K.-H. Illers, Europeam Polymer J. (Supplt.) 133 (Aug. 1969).
593. W. R. Brown, G. S. Park, J. Paint Technol. 42, 16 (1970).
594. F. Haas, K. Nutzel, G. Pampus, D. Theisen, Rubber Chem. Technol. 43, 1116 (1970). See also Kautschuk Gummi Kunststoffe 23, 502 (1970).
595. M. Takeda, K. Tanaka, R. Nagao, J. Polymer Sci. 57, 517 (1962).
596. Shell Chem. Tech. Bull., RB/70/14.
597. N. K. Kalfoglou, H. L. Williams, J. Appl. Polymer Sci. 14, 2481 (1970).
598. M. Baccaredda, E. Butta, J. Polymer Sci. 51, S39 (1961).
599. R. Nagao, Nippon Gomu Kyokaishi 41, 509 (1968).
600. I. A. Livshits, L. M. Korobova, E. A. Sidorovich, Mekh. Polim. 4, 596 (1967).
601. C. G. Overberger, L. H. Arond, R. H. Wiley, R. R. Garrett, J. Polymer Sci. 7, 431 (1951).
602. W. Marconi, A. Mazzei, S. Cucinella, M. Cesari, J. Polymer Sci., A, 2, 4261 (1964).
603. C. A. Aufdermarsh, R. Pariser, J. Polymer Sci., A, 2, 4727 (1964).
604. R. M. Kell, B. Bennett, P. B. Stickney, J. Appl. Polymer Sci. 2, 8 (1959).
605. J. B. Campbell, Science 141, 329 (1963).
606. B. L. Dolgoplosk, S. P. Erusalimskii, A. V. Merkur'eva, Vysokomolekul. Soedin. 4, 1333 (1962); C.A. 59, 4045d (1963).
607. R. R. Garrett. Unpublished results quoted by C.A. Aufdermarsh, R. Pariser, J. Polymer Sci., A, 2, 4727 (1964).
608. R. Nagao, Polymer 9, 517 (1968).
609. R. F. Robbins, R. P. Reed, Adv. Cryogenic Eng. 13, 252 (1968).
610. U. Eisele, Kautschuk Gummi Kunststoffe 23, 486 (1970).
611. R. R. Garrett, C. A. Hargreaves, D. N. Robinson, J. Macromol. Sci., A4, 1679 (1970).
612. I. A. Livshits, L. M. Korobova, Dokl. Akad. Nauk SSSR 121, 474 (1958).
613. N. V. Zakharenko, R. A. Gavrilina, Soviet Rubber Technol. 25, No.11,44 (1966) (transl'n of Kauchuk i Rezina 25, No.11, 47 (1966)).
614. J. P. Mercier, Rev. Gen. Caoutchouc,Edn. Plast. 5, 25 (1968).
615. J. J. Maurer, ACS Polymer Preprints 9, 866 (1968).
616. D. Heinze, K. Schmieder, G. Schnell, K. A. Wolf, Kautschuk Gummi 14, WT208 (1961) (transl'n in Rubber Chem. Technol. 35, 776 (1962)); via R. Havinga, A. Schors, J. Macromol. Sci., A2, 12 (1968), and ref. 262.

617. H. S. Gutowsky, A. Saika, M. Takeda, D. E. Woessner, J. Chem. Phys. 27, 534 (1957).

618. W. V. C. van Beek, W. F. Koch, B. F. Steggerda, O. F. K. Bussemaker, Assoc. Francaise Ing. Caoutchouc, Plastiques Conf. Intern. Caoutchouc, Paris, June 1970, paper 55; also Rev. Gen. Caoutchouc 48, 929 (1971).

619. M. Girolamo, J. R. Urwin, European Polymer J. 7, 225 (1971).

620. D. Reichenbach, W. Eckelmann, Angew. Makromol. Chem. 16/17, 157 (1971).

621. S. S. Chang, A. B. Bestul, J. Res. Nat. Bur. Std. 75A, 113 (1971).

622. R. T. Humpidge, A. Maclean, S. H. Morrell, Rubber Chem. Technol. 44, 479 (1971).

623. Z. R. Glaser, F. R. Eirich, J. Polymer Sci., C, 31, 275 (1970).

624. K. Ueberreiter, Z. Physik. Chem. B45, 361 (1940).

625. E. R. Mognaschi, A. Chierico, Trans. Faraday Soc. 66, 86 (1970).

626. G. Dall'asta, P. Scaglione, Rubber Chem. Technol. 42, 1235 (1969).

627. J. N. Short, G. Kraus, R. P. Zelinski, F. E. Naylor, Rubber Chem. Technol. 32, 614 (1959).

628. R. M. Ikeda, M. L. Wallach, R. J. Angelo, "Block Copolymers", Ed. S. L. Aggarwal, Plenum Press, 1970, p. 43; ACS Polymer Preprints 10, 1446 (1969).

629. K. J. Clark, A. T. Jones, D. H. Sandiford, Chem. Ind. 47, 2010 (1962).

630. G. Natta, F. Danusso, G. Moraglio, J. Polymer Sci. 25, 119 (1957).

631. K. R. Dunham, J. Vandenberghe, J. W. H. Faber, L. E. Contois, J. Polymer Sci., A, 1, 751 (1963).

632. G. Natta, Rend. Accad. Nazl. Lincei, Ser. 8, 24 (1958); via ref. 223.

633. S. F. Kurath, E. Passaglia, R. Pariser, J. Appl. Phys. 28, 499 (1957).

634. F. P. Reding, J. A. Faucher, R. D. Whitman, J. Polymer Sci. 57, 483 (1962).

635. H.-G. Elias, O. Etter, J. Macromol. Sci. A1, 943 (1967).

636. W. H. McCarty, G. Parravano, J. Polymer Sci., A, 3, 4029 (1965).

637. G. L. Taylor, S. Davison, J. Polymer Sci., B, 6, 699 (1968).

638. S. T. Barsamyan, A. S. Apresyan, V. I. Kleiner, L. L. Stotskaya, Soviet Plastics 3, 17 (1968) (transl'n of Plasticheskie Massy 3, 13 (1968)).

639. R. E. Kelchner, J. J. Aklonis, J. Polymer Sci., A-2, 8, 799 (1970).

640. A. Abe, T. Hama, J. Polymer Sci., B, 7, 427 (1969).

641. M. Takayanagi, Pure Appl. Chem. 23, 151 (1970).

642. P. R. Swan, J. Polymer Sci. 42, 525 (1960).

643. T. Shibukawa, V. D. Gupta, R. Turner, J. H. Dillon, A. V. Tobolsky, Textile Res. J. 32, 810 (1962).

644. F. S. Dainton, D. M. Evans, F. E. Hoare, T. P. Melia, Polymer 3, 277 (1962).

645. M. Baccaredda, E. Butta, Chim. Ind. (Milan) 44, 1228 (1962).

646. F. S. Dainton, D. M. Evans, F. E. Hoare, T. P. Melia, Polymer 3, 286 (1962).

647. A. V. Tobolsky, "Properties and Structure of Polymers", Wiley, New York, 1960.

648. N. L. Zutty, C. J. Whitworth, J. Polymer Sci., B, 2, 709 (1964).

649. F. Danusso, G. Moraglio, G. Talamini, J. Polymer Sci. 21, 139 (1956).

650. H. Thurn, Kolloid Z. 173, 72 (1960).

651. R. Simha, R. F. Boyer, J. Chem. Phys. 37, 1003 (1962).

652. A. Odajima, J. A. Sauer, A. E. Woodward, J. Phys. Chem. 66, 718 (1962).

653. W. G. Oakes, D. W. Robinson, J. Polymer Sci. 14, 505 (1954).

654. O. D. Frampton, J. F. Nobis, Ind. Eng. Chem. 45, 404 (1953).

655. K. Wolf, K. Schmieder, Intern. Symp. Macromol. Chem., Ric. Sci. 25, 732 (1955).

656. F. Wuerstlin, Kunststoffe Plastiks 40, 158 (1950).

657. A. E. Woodward, J. A. Sauer, Adv. Polymer Sci. 1, 114 (1958).

658. B. Wunderlich, J. Polymer Sci., C, 1, 41 (1963).

659. N. G. McCrum, Makromol. Chem. 34, 50 (1959).

660. P. Manaresi, V. Giannela, J. Appl. Polymer Sci. 4, 251 (1960).

661. R. Nakane, J. Appl. Polymer Sci. 3, 125 (1960).

662. E. Hunter, E. W. Oakes, Trans. Farad y Soc. 41, 49 (1945)

663. F. A. Quinn, L. Mandelkern, J. Am. Chem. Soc. 80, 3178 (1958).

664. D. E. Kline, J. A. Sauer, A. E. Woodward, J. Polymer Sci. 22, 455 (1956).

665. E. G. Kontos, W. P. Slichter, J. Polymer Sci. 61, 61 (1962).

666. H. A. Floecke, Kolloid Z. - Z. Polymere 180, 118 (1962).

667. W. Pechhold, S. Blasenbrey, S. Woerner, Kolloid Z. - Z. Polymere 189, 14 (1963).

668. C. A. F. Tuijnman, J. Polymer Sci., C, 16, 2379 (1967).

669. J. Heijboer, J. Polymer Sci., C, 16, 3755 (1968).

670. H. Schlein, M. Shen, Rev. Sci. Instrum. 40, 587 (1969).

671. I. I. Perepechko, L. A. Kvacheva, Vysokomolekul. Soedin. B, 12, 484 (1970).

672. B. Wunderlich, J. Chem. Phys. 37, 1203 (1962).

673. E. W. Fischer, F. Kloos, J. Polymer Sci., B, 8, 685 (1970).

674. S. Matsuoka, J. H. Daane, H. E. Bair, T. K. Kwei, J. Polymer Sci., B, 6, 87 (1968).

675. R. A. Jackson, S. R. D. Oldland, A. Pajaczkowski, J. Appl. Polymer Sci. 12, 1297 (1968).

676. T. K. Kwei, T. T. Wang, H. E. Bair, J. Polymer Sci., C, 31, 87 (1970).

677. G. Natta, Angew. Chem. 68, 393 (1956).

678. F. P. Reding, J. Polymer Sci. 21, 547 (1956).

679. R. W. Warfield, R. Brown, J. Polymer Sci., A-2, 5, 791 (1967).

680. G. Vidotto, A. J. Kovacs, Kolloid Z. - Z. Polymere 220, 1 (1967).

681. J. P. Kennedy, W. Naegele, J. J. Elliott, J. Polymer Sci., B, 3, 729 (1965).

682. O. E. Van Lohuizen, K. S. De Vries, J. Polymer Sci., C, 16, 3943 (1968).

683. D. W. Aubrey, A. Barnatt, J. Polymer Sci., A-2, 6, 241 (1968).

684. A. E. Woodward, Trans. N. Y. Acad. Sci. 24, 250 (1962).

685. A. E. Woodward, J. A. Sauer, R. A. Wall, J. Polymer Sci. 50, 117 (1961).

686. J. H. Griffith, B. G. Ranby, J. Polymer Sci. 44, 369 (1960).

687. R. W. Penn, Trans. Soc. Rheol. 3, 416 (1963).

688. W. A. Hewett, F. E. Weir, J. Polymer Sci., A, 1, 1239 (1963).

689. F. E. Karasz, H. E. Bair, J. M. O'Reilly, Polymer 8, 547 (1967).

690. T. P. Melia, A. Tyson, Makromol. Chem. 109, 87 (1967).

691. I. Kirshenbaum, R. B. Isaacson, M. Druin, J. Polymer Sci., B, 3, 525 (1965).

692. N. Kawasaki, M. Takayanagi, Rept. Progr. Polymer Phys. (Japan) 10, 337 (1967).

693. R. G. Quynn, B. S. Sprague, J. Polymer Sci., A-2, 8, 1971 (1970).

694. A. A. Miller, J. Polymer Sci., A-2, 6, 249 (1968).

695. J. D. Ferry, "Viscoelastic Properties of Polymers", Wiley, New York, 1961.

696. N. Hirai, H. Eyring, J. Polymer Sci. 37, 51 (1959).

697. J. D. Ferry, G. S. Parks, J. Chem. Phys. 4, 70 (1936).

698. T. G. Fox, S. Loshaek, J. Polymer Sci. 15, 371 (1955).

699. L. A. Wood, "Synthetic Rubber", Ed. G. S. Wiley, Wiley New York, 1954, Chap. 10.

700. G. T. Furukawa, R. E. McCoskey, M. L. Reilly, J. Res. Nat. Bur. Std. 55, 127 (1955).

701. G. M. Zhidomirov, U. D. Tsvetkov, Y. S. Lebedev, Zh. Strukt. Khim. 2, 696 (1961).

702. E. Butta, P. Giusti, Ric. Sci. Rend. Sec. A, 2, 362 (1962).

703. T. L. Smith, J. Polymer Sci., C, 16, 841 (1967).

704. J. J. Maurer, ACS Division of Rubber Chem., Spring Meeting, May 1967, paper 26 (33).

705. G. Gianotti, G. Dall'asta, A. Valvassori, V. Zamboni, Makromol. Chem. 149, 117 (1971).

706. A. Turner, F. E. Bailey, J. Polymer Sci., B, 1, 601 (1963).

707. J. P. Kennedy, R. M. Thomas, Makromol. Chem. 64, 1 (1963).

708. J. P. Kennedy, R. M. Thomas, Makromol. Chem. 53, 28 (1962).

709. J. P. Kennedy, J. J. Elliott, B. Groten, Makromol. Chem. 77, 26 (1964).

710. D. L. Beck, A. A. Hiltz, J. R. Knox, SPE Trans. 3, 279 (1963).

711. B. Ke, J. Polymer Sci., B, 1, 167 (1963).

712. L. T. Muus, N. G. McCrum, F. C. McGrew, SPE J. 15, 368 (1959).

713. N. G. McCrum, J. Polymer Sci., B, 2, 495 (1964).

714. R. W. Wilkinson, M. Dole, J. Polymer Sci. 58, 1089 (1962).

715. H. Wilski, Kunststoffe 54, 90 (1964).

716. J. A. Sauer, R. A. Wall, N. Fuschillo, A. E. Woodward, J. Appl. Phys. 29, 1385 (1958).

717. V. A. Kargin, I. Yu. Marchenko, Polymer Sci. USSR 2, 370 (1961); (transl'n of Vysokomolekul. Soedin. 2, 549 (1960)).

718. E. Passaglia, G. M. Martin, J. Res. Nat. Bur. Std. 68A, 273 (1964).

719. S. Nara, H. Kashiwabara, J. Sohma, J. Polymer Sci., A-2, 5, 929 (1967).

720. J. P. Mercier, Ind. Chim. Belge 30, 813 (1965).

721. D. R. Gee, T. P. Melia, Makromol. Chem. 116, 122 (1968).

722. D. R. Gee, T. P. Melia, Polymer 10, 239 (1969).

723. D. R. Gee, T. P. Melia, Makromol. Chem. 132, 195 (1970).

724. I. Abu-Isa, V. A. Crawford, A. R. Haly, M. Dole, J. Polymer Sci., C, 6, 149 (1964).

725. E. Passaglia, H. K. Kevorkian, J. Appl. Phys. 34, 90 (1963).

726. M. S. Akutin, Z. I. Salina, L. Yu. Zlatkevich, B. V. Andrianov, L. E. Vlaskina, Soviet Plastics, No. 2, p. 19 (1971) (transl'n of Plasticheskie Massy 2, 23 (1971)).

727. G. Gianotti, A. Capizzi, European Polymer J. 4, 677 (1968).

728. G. M. Pogosyan, G. A. Zhamkochyan, S. G. Matsoyan, Arm. Khim. Zh. 22, 364 (1969).

729. G. M. Pogosyan, L. M. Akopyan, E. V. Vanyan, S. G. Matsoyan, Vysokomolekul. Soedin. B, 13, 242 (1971).

730. W. Kern, W. Heitz, M. Jager, K. Pfitzner, H. O. Wirth, Makromol. Chem. 126, 73 (1969).

731. G. M. Pogosyan, L. M. Akopyan, E.V. Vanyan, S. G. Matsoyan, Vysokomolekul. Soedin. B, 12, 142 (1970).

732. K. R. Dunham, J. W. H. Faber, J. Vandenberghe, W. F. Fowler, J. Appl. Polymer Sci. 7, 897 (1963).

733. R. Kosfeld, Kolloid Z. 172, 182 (1960).

734. K. H. Illers, Z. Elektrochem. 65, 679 (1961).

735. G. M. Pogosyan, T. G. Karapetyan, S. G. Matsoyan, Vysokomolekul. Soedin. B, 13, 228 (1971).

736. E. C. Chapin, J. G. Abrams, V. L. Lyons, J. Org. Chem. 27, 2595 (1962).

737. G. S. Kolesnikov, G. M. Pogosyan, Izv. Akad. Nauk SSSR Otd. Khim. Nauk 2, 227 (1958).

738. C. G. Overberger, C. Frazier, J. Mandelman, H. F. Smith, J. Am. Chem. Soc. 75, 3326 (1953).

739. T. E. Davies, British Plastics 32, 283 (1959).

740. U. S. 2,723,261 - 8. Nov. 1955.

741. L. L. Ferstandig, J. C. Butler, A. E. Straus, J. Am. Chem. Soc. 76, 5779 (1954).

742. G. M. Pogosyan, T. G. Karapetyan, S. G. Matsoyan, Vysokomolekul. Soedin. B, 12, 463 (1970).

743. G. S. Kolesnikov, Izv. Akad. Nauk SSSR, Otd. Khim. Nauk, 1333 (1959).

744. K. R. Dunham, J. Vandenberghe, J. W. H. Faber, W. F. Fowler, J. Appl. Polymer Sci. 7, 1531 (1963).

745. Brit. 609,482 - 1. Oct. 1948.

746. S. Krause, J. J. Gormley, N. Roman, J. A. Shetter, W. H. Watanabe, J. Polymer Sci., A, 3, 3573 (1965).

747. J. Thrower, M. A. H. Hewins, Royal Aircraft Establishment, Farnborough, Hants, U. K., TR 70056 (1970).

748. R. A. Abdrashitov, N. M. Bazhenov, M. V. Vol'kenshtein. A. I. Kol'tsov, A. S. Khachaturov, Vysokomolekul. Soedin. 5, 405 (1963); (transl'n in Polymer Sci. USSR 4, 1066 (1963)); Z. Phys. Chem. 220, 413 (1962).

749. F. Danusso, G. Polizzotti, Makromol. Chem. 61, 157 (1963).

750. K. Tanaka, O. Yano, S. Maru, R. Nagao, Rept. Progr. Polymer Phys. (Japan) 12, 379 (1969).

751. R. A. Abdrashitov, N. M. Bazhenov, M. V. Vol'kenshtein, A. I. Kol'tsov, A. S. Khachaturov, Polymer Sci. USSR 6, 2074 (1964); (transl'n of Vysokomolekul. Soedin. 6, 1871 (1964)).

752. U. S. 2,651,627 - 8. Sept. 1953.

753. D. I. Livingstone, P. M. Kamath, R. S. Corley, J. Polymer Sci. 20, 485 (1956).

754. G. M. Pogosyan, G. A. Zhamkochyan, R. A. Stepanyan S. G. Matsoyan, Arm. Khim. Zh. 22, 915 (1969).

755. Polaroid Corp., Offic. Sci. Res. Dev. Rept. No. 4417, 246p, Feb. 1945.

756. V. Frosini, P. L. Magagnini, European Polymer J. 2, 129 (1966).

757. V. Frosini, E. Butta, Mater. Sci. Eng. 6, 274 (1970).

758. T. Fujimoto, N. Ozaki, M. Nagasawa, J. Polymer Sci., A-2, 6, 129 (1968).

759. H.-G. Elias, V. S. Kamat, Makromol. Chem. 117, 61 (1968).

760. J. M. G. Cowie, P. M. Toporowski, European Polymer J. 4, 621 (1968).

761. M. Baccaredda, E. Butta, V. Frosini, P. Magagnini, F. Andruzzi, J. Polymer Sci., C, 16, 3581 (1968).

762. J. M. G. Cowie, P. M. Toporowski, J. Macromol. Sci. B3, 81 (1969).

763. E. Rovira, A. Eisenberg, Unpublished results quoted by ref. 48

764. H. Endo, T. Fujimoto, M. Nagasawa, J. Polymer Sci., A-2, 7, 1669 (1969).

765. S. Ichihara, A. Komatsu, T. Hata, Polymer J. (Japan) 2, 650 (1971).

766. G. T. Kennedy, F. Morton, J. Chem. Soc. 1949, 2383.

767. L. C. Corrado, J. Chem. Phys. 50, 2260 (1969).

768. W. Hodes, American Cyanamid Co. Unpublished results quoted by ref. 44.

769. G. S. Kolesnikov, G. M. Pogosyan, Izv. Akad. Nauk SSSR, Otd. Khim. Nauk, 2098 (1962); C.A. 58, 8047a (1963).

770. L. A. Utracki, R. Simha, Makromol. Chem. 117, 94 (1968).

771. N. N. Aylward, J. Polymer Sci., B, 8, 377 (1970).

772. J. J. Keavney, E. C. Eberlin, J. Appl. Polymer Sci. 3, 47 (1960).

773. R. S. Spencer, J. Colloid Sci. 4, 229 (1949).

774. T. G. Fox, P. J. Flory, J. Appl. Phys. 21, 581 (1950).

775. L. J. Hughes, G. L. Brown, J. Appl. Polymer Sci. 5, 580 (1961).

776. E. Jenckel, K. Ueberreiter, Z. Physik. Chem. (Leipzig) A182, 361 (1938).

777. F. Wuerstlin, H. Thurn, "Die Physik der Hochpolymeren", Ed. H. A. Stuart, Springer-Verlag, Berlin 1956.

778. "Die Physik der Hochpolymeren", III, Ed. H. A. Stuart, Springer-Verlag, Berlin 1955.

779. R. B. Beevers, E. F. T. White, J. Polymer Sci., B, 1, 171 (1963).

780. S. Newman, W. P. Cox, J. Polymer Sci. 46, 29 (1960).

781. R. S. Spencer, R. F. Boyer, J. Appl. Phys. 17, 398 (1946).

782. T. G. Fox, P. J. Flory, J. Polymer Sci. 14, 315 (1954).

783. T. G. Fox, P. J. Flory, J. Am. Chem. Soc. 70, 2384 (1948).

784. T. G. Fox, P. J. Flory, J. Phys. Colloid Chem. 55, 221 (1951).

785. R. H. Boundy, R. F. Boyer, "Styrene", Reinhold, New York, 1952.

786. W. Patnode, W. J. Scheiber, J. Am. Chem. Soc. 61, 3449 (1939).

787. W. J. Roff, "Fibres, Plastics and Rubbers", Academic Press, New York, 1956.

788. Yu. V. Zelenev, Vysokomolekul. Soedin. 4, 1486 (1962); (transl'n in Polymer Sci. USSR 4, 457 (1963)).

789. M. S. Parker, V. J. Krasnansky, B. G. Achhammer, J. Appl. Polymer Sci. 8, 1825 (1964).

790. A. D. Pasquino, M. N. Pilsworth, J. Polymer Sci., B, 2, 253 (1964).

791. A. E. Martin, H. F. Rase, IEC Prod. Res. Dev. 6, 104 (1967).

792. S. N. Kolesov, Vysokomolekul. Soedin. A, 9, 1860 (1967); (transl'n in Polymer Sci. USSR 9, 2098 (1967)).

793. G. Rehage, H. Breuer, J. Polymer Sci., C, 16, 2299 (1967).

794. W. Simpson, Chem. Ind. 5, 215 (1965).

795. S. G. Turley. Appl. Polymer Symp. 7, 237 (1968).

796. K. C. Rusch, J. Macromol. Sci. B2, 421 (1968).

797. J. Stoelting, F. E. Karasz, W. J. MacKnight, Polymer Eng. Sci. 10, 133 (1970).

798. K. Thinius, B. Hoesselbarth, Plaste Kautschuk 17, 653 (1970).

799. N. Nemoto, Polymer J. (Japan) 1, 485 (1970).
800. F. E. Karasz, H. E. Bair, J. M. O'Reilly, J. Phys. Chem. 69, 2657 (1965).
801. I. Abu-Isa, M. Dole, J. Phys. Chem. 69, 2668 (1965).
802. E. H. Merz, L. E. Nielsen, R. Buchdahl, Ind. Eng. Chem. 43, 1396 (1951).
803. D. J. Plazek, V. M. O'Rourke, J. Polymer Sci., A-2, 9, 209 (1971).
804. C. H. Bamford, G. C. Eastmond, D. Whittle, Polymer 12, 247 (1971).
805. M. Bank, J. Leffingwell, C. Thies, Macromolecules 4, 43 (1971).
806. S. Hozumi, T. Wakabayashi, K. Sugihara, Polymer J. (Japan) 1, 632 (1970)
807. G. V. Vinogradov, E. A. Dzyura, A. Ya. Malkin, V. A. Grechanovskii, J. Polymer Sci., A 2, 9, 1153 (1971).
808. R. A. Wallace, J. Polymer Sci., A-2, 9, 1325 (1971).
809. S. Ichihara, A. Komatsu, T. Hata, Polymer J. (Japan) 2, 644 (1971).
810. O. Yamamoto, H. Kambe, Polymer J. (Japan) 2, 623 (1971)
811. E. Jenckel, E. Braucker, Z. Physik. Chem. (Leipzig) A185, 465 (1940).
812. L. J. T. Hughes, D. B. Fordyce, J. Polymer Sci. 22, 509 (1956).
813. K. H. Illers, Z. Elektrochem. 70, 353 (1966); via ref. 815.
814. L. E. Nielsen, ACS Polymer Preprints 9, 596 (1968).
815. A. Eisenberg, T. Yokoyama, E. Sambalido, J. Polymer Sci., A-1, 7, 1717 (1969).
816. A. Chifor, Mater. Plast. 7, 575 (1970).
817. A. Eisenberg, H. Matsuura, T. Yokoyama, AD 715093 (1970).
818. M. Pegoraro, L. Szilagyi, A. Penati, G. Alessandrini, European Polymer J. 7, 1709 (1971).
819. M. Baccaredda, P. L. Magagnini, G. Pizzirani, P. Giusti, J. Polymer Sci., B, 9, 303 (1971).
820. E. C. Eberlin - Unpublished values quoted by ref. 44.
821. C. E. Rehberg, C.H. Fisher, Ind. Eng. Chem. 40, 1429 (1948).
822. J. Hrouz, J. Janacek, J. Macromol. Sci. B5, 245 (1971).
823. C. E. Rehberg, W. A. Faucette, C. H. Fisher, J. Am. Chem. Soc. 66, 1723 (1944).
824. J. A. Shetter, J. Polymer Sci., B, 1, 209 (1963).
825. A. R. Monahan, J. Polymer Sci., A-1, 4, 2381 (1966).
826. G. Pizzirani, P. L. Magagnini, Chim. Ind. (Milan) 50, 1218 (1968).
827. Brit. 1,116,933 - 12. June 1968 (B. Vollmert, H. Haas).
828. J. H. Prager, R. M. McCurdy, C. B. Rathmann, J. Polymer Sci., A, 2, 1941 (1964).
829. V. P. Shibayev, B. S. Petrukhin, Yu. A. Zubov, N. A. Plate, V. A. Kargin, Polymer Sci. USSR 10, 258 (1968) (transl'n of Vysokomolekul. Soedin. A 10, 216 (1968)).
830. F. A. Bovey, J. F. Abere, J. Polymer Sci. 15, 537 (1955).
831. R. M. McCurdy, J. H. Prager, J. Polymer Sci., A, 2, 1185 (1964).
832. G. P. Mikhailov, V. A. Shevelev, Polymer Sci. USSR 9, 2762 (1967). (transl'n of Vysokomolekul. Soedin. A 9, 2442 (1967)).
833. J. Lawler, D. C. Chalmers, J. Timar, ACS Div. Rubber Chem., Spring Meeting, May 1967, paper 42, (10).
834. C. U. Pittman, R. L. Voges - Unpublished results quoted by ref. 835.
835. C. U. Pittman, J. C. Lai, D. P. Vanderpool, M. Good, R. Prado, Macromolecules 3, 746 (1970).
836. F. A. Bovey, J. F. Abere, G. B. Rathmann, C. L. Sandberg, J. Polymer Sci. 15, 520 (1955).
837. A. S. Khachaturov, N. M. Bazhenov, M. V. Vol'kenshtein, I. M. Dolgopol'skii, A. I. Kol'tsov, Soviet Rubber Technol. 24, No. 12, 9 (1965); (transl'n of Kauchuk i Rezina 24, No. 12, 6 (1965)).
838. G. K. Dyvik, Polymer 2, 449 (1961).
839. A. G. Pittman, B. A. Ludwig, D. L. Sharp, J. Polymer Sci., A-1, 6, 1741 (1968).
840. R. H. Wiley, G. M. Brauer, J. Polymer Sci. 4, 351 (1949).
841. W. R. Sorenson, T. W. Campbell, "Preparative Methods of Polymer Chemistry", 2nd edn., 1968, p. 248.
842. W. A. P. Black, J. A. Colquhoun, E. T. Dewar, Makromol. Chem. 122, 244 (1969).
843. E. P. Otocka, T. K. Kwei, Macromolecules 1, 401 (1968).

844. J. R. Constanza, J. A. Vona, J. Polymer Sci., A-1, 4, 2659 (1966).
845. F. Wuerstlin, "Die Physik der Hochpolymeren", Ed. H. A. Stuart, Springer-Verlag, Berlin 1955, Chap. 11.
846. R. A. Haldon, R. Simha, J. Appl. Phys. 39, 1890 (1968).
847. J. Heijboer, "Physics of Non-Crystalline Solids", Ed. J. A. Prins, Proc. Intern. Conf., Delft 1964, p. 231, North Holland Publishing Co., Amsterdam, 1965.
848. T. Otsu, S. Aoki, R. Nakatani, Makromol. Chem. 134, 331 (1970).
849. M. E. Hoagland, I. N. Duling, ACS Div. Petrol. Chem. Preprints 15, B85 (1970).
850. U. S. 3,518,241 - 30. June 1970 (I.N. Duling, A. Schneider, R. E. Moore).
851. C. E. Rehberg, C. H. Fisher, J. Am. Chem. Soc. 66, 1203 (1944).
852. P. Weitz, Paint Varn. Prod. 57, 99 (1967).
853. E. F. Jordan, G. R. Riser, B. Artymyshyn, W. E. Parker, J. W. Pensabene, A. N. Wrigley, J. Appl. Polymer Sci. 13, 1777 (1969).
854. K. Butler, P. R. Thomas, G. J. Tyler, J. Polymer Sci. 48, 357 (1960).
855. O. Smidsrod, J. E. Guillet, Macromolecules 2, 272 (1969).
856. J. Elles, Chim. Mod. 4, 53 (1959).
857. A. V. Tobolsky, M. C. Shen, J. Phys. Chem. 67, 1886 (1963).
858. S. S. Rogers, L. Mandelkern, J. Phys. Chem. 61, 985 (1957).
859. A. F. Wilde, J. J. Ricca, AD 675463, Proc. Army Symp. on Solid Mechanics, p. 39 (1968).
860. T. Hata, T. Nose, J. Polymer Sci., C, 16, 2019 (1967).
861. N. W. Johnston, ACS Polymer Preprints 10, 608 (1969).
862. Z. G. Gardlund, J. J. Laverty, J. Polymer Sci., B, 7, 719 (1969).
863. E. A. W. Hoff, D. W. Robinson, A. H. Willbourn, J. Polymer Sci. 18, 161 (1955).
864. Z. A. Azimov, S. P. Mitsengendler, A. A. Korotkov, Vysokomolekul. Soedin. 7, 843 (1965); (transl'n in Polymer Sci. USSR 7, 929 (1965)).
865. G. P. Mikhailov, A. I. Artyukhov, T. I. Borisova, Vysokomolekul. Soedin. A, 10, 1755 (1968); (transl'n in Polymer Sci. USSR 10, 2033 (1968)).
866. J. Lal, G. S. Trick, J. Polymer Sci. A, 2, 4559 (1964).
867. G. Williams, D. C. Watts, Trans. Faraday Soc. 67, 2793 (1971).
868. K. M. Sinnott, SPE Trans. 2, 65 (1962).
869. M. O. Samsoen, Ann. Phys. 9, 35 (1928).
870. E. Baer, Offic. Dig. J. Paint Technol. Eng. 36, 464 (1964).
871. L. de Brouckère, G. Offergeld, Bull. Soc. Chim. Belges 67, 96 (1958).
872. J. D. Ferry, W. C. Child, R. Zard, D. M. Stern, M. L. Williams, R. F. Landel, J. Colloid Sci. 12, 53 (1957).
873. S. Saito, Kolloid Z. - Z. Polymere 189, 116 (1963) and Res. Electrotech. Lab. (Tokyo), No. 648 (1964); via ref. 286.
874. W. L. Peticolas, G. I. A. Stegeman, B. P. Stoicheff, Phys. Rev. Lett. 18, 1130 (1967).
875. W. M. Lee (advisers : F. R. Eirich, B. R. McGarvey), Dissertation, Polytechnic. Inst., Brooklyn 1967; via Diss. Abs., B, 28, 1897 (1967).
876. E. N. Rostovskii, L. D. Rubinovich, Vysokomolekul. Soedin. Karb. Vys. Soedin. Sb. Statei, p. 140 (1963).
877. M. Ilavsky, G. L. Slonimskii, J. Janacek, J. Polymer Sci., C, 16, 329 (1967).
878. M. C. Shen, J. D. Strong, F. J. Matusik, J. Macromol. Sci. B1, 15 (1967).
879. J. Janacek, J. D. Ferry, J. Macromol. Sci. B5, 219 (1971).
880. F. Lednicky, J. Janacek, J. Macromol. Sci. B5, 335 (1971).
881. H. Ochiai, H. Shindo, H. Yamamura, J. Polymer Sci., A-2, 9, 431 (1971).
882. I. N. Razinskaya, N. E. Kharitonova, B. P. Shtarkman, Vysokomolekul. Soedin. B, 11, 892 (1969).
883. R. W. Warfield, M. C. Petree, P. Donovan, SPE J. 15, 1055 (1959).
884. T. G. Fox, B. S. Garrett, W. E. Goode, S. Gratch, J. F. Kincaid, A. Spell, J. D. Stroupe, J. Am. Chem. Soc. 80, 1768 (1958).
885. S. Loshaek, J. Polymer Sci. 15, 391 (1955).
886. A. Odajima, A. E. Woodward, J. A. Sauer, J. Polymer Sci. 55, 181 (1961).

887. W. G. Gall, N. G. McCrum, J. Polymer Sci. 50, 489 (1961).

888. G. Allen, D. Sims, G. J. Wilson, Polymer 2, 375 (1961).

889. B. E. Read, J. Polymer Sci., C, 16, 1887 (1967).

890. M. F. Margaritova, S. D. Stavrova, S. N. Trubitsyna, S. D. Medvedev, J. Polymer Sci., C, 16, 2251 (1967).

891. S. R. Urzendowski, A. H. Guenther, J. R. Asay, ACS Polymer Preprints 9, 878 (1968).

892. A. A. Berlin, E. F. Samarin, R. V. Kronman, I. G. Sumin, Soviet Plastics 2, 22 (1971);
(transl'n of Plasticheskie Massy 2, 26 (1971)).

893. A. K. Schulz, J. Chim. Phys. 53, 933 (1956).

894. F. P. Wolf, K. Ueberreiter, Kolloid Z. - Z. Polymere 245, 399 (1971).

895. S. Bywater, P. M. Toporowski, Polymer 13, 94 (1972).

896. G. V. Schulz, W. Wunderlich, R. Kirste, Makromol. Chem. 75, 23 (1964).

897. J. P. Mercier, H. Berghmans, G. Smets, Paper given at VIII FATIPEC Congress, Scheveningen (The Hague), p. 26 (1966).

898. E. V. Thompson, J. Polymer Sci., A-2, 4, 199 (1966).

899. E. V. Thompson, J. Polymer Sci., A-2, 6, 433 (1968).

900. S. Havriliak, Polymer 9, 289 (1968).

901. H. Shindo, I. Murakami, H. Yamamura, J. Polymer Sci., A-1, 7, 297 (1969).

902. N. N. Aylward, J. Polymer Sci., A-1, 8, 319 (1970).

903. T. I. Borisova, Vysokomolekul. Soedin. A, 12, 932 (1970);
(transl'n in Polymer Sci. USSR 12, 1060 (1970)).

904. M. J. Bowden, J. H. O'Donnell, J. Polymer Sci., A-1, 7, 1665 (1969).

905. G. M. Chetyrkina, T. A. Sokolova, M. M. Koton, Vysokomolekul. Soedin., Vsesoyuz. Khim. Obshchestvo im. D. I. Mendeleeva 1, 248 (1959); C.A. 53, 23059i (1959).

906. B. E. Tate, Adv. Polymer Sci. 5, 214 (1967) - Unpublished results of Charles Pfizer and Co. Inc.

907. J. B. Kinsinger, J. R. Panchak, R. L. Kelso, J. S. Bartlett, R. K. Graham, J. Appl. Polymer Sci. 9, 429 (1965).

908. B. Wesslen, R. W. Lenz, ACS Polymer Preprints 11, 105 (1970).

909. B. Wesslen, R. W. Lenz, W. J. MacKnight, F. E. Karasz, Macromolecules 4, 24 (1971).

910. J. A. Powell, R. K. Graham, J. Polymer Sci., A, 3, 3451 (1965).

911. V. A. Kargin, T. I. Sogolova, Zh. Fiz. Khim. 23, 540 (1949).

912. H. D. Anspon, J. J. Baron, WADC Tech' l Report 57-24, Pt. I, 1957, Pt. II, 1958; via Ref. 910.

913. H. Yuki, K. Hatada, T. Nijnomi, M. Hashimoto, J. Ohshima, Polymer J. (Japan) 2, 629 (1971).

914. R. A. Haldon, W. J. Schell, R. Simha, J. Macromol. Sci. B1, 759 (1967).

915. D. S. Otto, Vinyl Technol. Newsletter 5, 20 (1968).

916. W. J. Schell, R. Simha, J. J. Aklonis, J. Macromol. Sci. A3, 1297 (1969).

917. W. J. Schell, Thesis, Univ. S. California 1969; via Diss. Abstr. Intern. B, 30, 2120 (1969).

918. L. Fishbein, B. F. Crowe, Makromol. Chem. 48, 221 (1961).

919. H. Thurn, K. Wolf, Kolloid-Z. 148, 16 (1956).

920. E. J. Vandenberg, R. F. Heck, D. S. Breslow, J. Polymer Sci. 41, 519 (1959).

921. W. S. Durrell, E. C. Stump, P. D. Schuman, J. Polymer Sci., B, 3, 831 (1965).

922. C. E. Schildknecht, S. T. Gross, H. R. Davidson, J. M. Lambert, A. O. Zoss, Ind. Eng. Chem. 40, 2104 (1948).

923. J. A. Sedlak, K. Matsuda, J. Polymer Sci., A, 3, 2329 (1965).

924. G. M. Pogosyan, E. S. Avanesyan, S. G. Matsoyan, Arm. Khim. Zh. 24, 694 (1971).

925. G. Pizzirani, P. Magagnini, P. Giusti, J. Polymer Sci., A-2, 9, 1133 (1971).

926. J. B. Clark, Polymer Eng. Sci. 7, 137 (1967).

927. T. N. Kalinina, G. N. Afanas'eva, L. A. Vol'f, A. I. Meos, E. B. Kremer, S. Ya. Frenkel', S. S. Mnatsakanov, Vysokomolekul. Soedin. B, 12, 661 (1970).

928. W. H. Howard, J. Appl. Polymer Sci. 5, 303 (1961).

929. L. V. Holroyd, R. S. Codrington, B. A. Mrowca, E. Guth, Rubber Chem. Technol. 25, 767 (1952).

930. W. R. Krigbaum, N. Tokita, J. Polymer Sci. 43, 467 (1960).

931. T. G. Fox, Bull. Am. Phys. Soc. 1, 123 (1956).

932. R. J. Kokes, F. A. Long, J. L. Hoard, J. Chem. Phys. 20, 1711 (1952).

933. Rohm & Haas Co., Philadelphia. Leaflet SP-251 (1/66) of Spec. Prod. Dept.

934. S. Okajima, A. Takeuchi, J. Polymer Sci., A-2, 5, 1317 (1967).

935. K.-H. Illers, Makromol. Chem. 124, 278 (1969).

936. R. D. Andrews, R. M. Kimmel, J. Polymer Sci., B, 3, 167 (1965).

937. S. H. Ronel, D. H. Kohn, J. Polymer Sci., A-1, 2221 (1969).

938. C. R. Bohn, J. R. Schaefgen, W. O. Statton, J. Polymer Sci. 55, 531 (1961).

939. R. B. Beevers, J. Polymer Sci., A, 2, 5257 (1964).

940. T. M. A. Hossain, H. Maeda, T. Iijima, Z. Morita, J. Polymer Sci., B, 5, 1069 (1967).

941. K. Ogura, S. Kawamura, H. Sobue, Macromolecules 4, 79 (1971).

942. H. Kakutani, M. Asahina, J. Polymer Sci., A-1, 5, 1717 (1967).

943. M. Kryszewski, M. Mucha, J. Appl. Polymer Sci. 15, 2687 (1971).

944. V. Heidingsfeld, J. Zelinger, V. Kuska, J. Appl. Polymer Sci. 15, 2447 (1971).

945. J. D. Hoffman, J. Am. Chem. Soc. 74, 1696 (1952).

946. R. F. Boyer, J. Appl. Phys. 25, 825 (1954).

947. F. P. Reding; private communication to ref. 946.

948. J. D. Hoffman, J. J. Weeks, J. Polymer Sci. 28, 472 (1958); J. Res. Nat. Bur. Std. 60, 465 (1958).

949. T. Nakajima, S. Saito, J. Polymer Sci. 31, 423 (1958).

950. A. Nishioka, J. Polymer Sci. 37, 163 (1959).

951. F. Krum, Kolloid-Z. 165, 77 (1959).

952. K.-H. Illers, E. Jenckel, Kolloid-Z. 165, 84 (1959).

953. M. Baccaredda, E. Butta, J. Polymer Sci. 44, 421 (1960).

954. A. W. Myers, V. Tammela, V. Stannett, M. Szwarc, Mod. Plast. 37, 139 (1960).

955. N. G. McCrum, J. Polymer Sci. 60, S3 (1962).

956. A. H. Scott, D. J. Scheiber, A. J. Curtis, J. I. Lauritzen, J. D. Hoffman, J. Res. Nat. Bur. Std. 66A, 269 (1962).

957. E. Passaglia, J. M. Crissman, R. R. Stromberg, ACS Polymer Preprints 6, 590 (1965).

958. J. D. Hoffman, G. Williams, E. Passaglia, J. Polymer Sci., C, 14, 173 (1966).

959. J. M. Crissman, E. Passaglia, J. Polymer Sci., C, 14, 237 (1966).

960. I. I. Perepechko, L. A. Bodrova, Vysokomolekul. Soedin., B, 10, 148 (1968).

961. I. I. Perepechko, L. A. Kvacheva, L. A. Ushakov, A. Ya. Svetov, V. A. Grechishkin, Soviet Plastics 8, 39 (1970);
(transl'n of Plasticheskie Massy 8, 43 (1970)).

962. G. S. Kolesnikov, M. G. Avetyan, Izv. Akad. Nauk SSSR, Otd. Khim. Nauk, 331 (1959).

963. L. A. Wood, Army Elastomer Conference 1955.

964. R. N. Haszeldine, U.M.I.S.T., Manchester, U. K.; private communication.

965. A. Peterlin, J. D. Holbrook, Kolloid Z. - Z. Polymere 203, 68 (1965).

966. N. Koizumi, S. Yano, K. Tsunashima, J. Polymer Sci., B, 7, 59 (1969).

967. H. Sasabe, S. Saito, M. Asahina, H. Kakutani, J. Polymer Sci., A-2, 7, 1405 (1969).

968. A. Miglierina, G. Ceccato, SRS 4 (4th Intern. Synth. Rubber Symp. London 1969) 2, 65 (1969).

969. S. Yano, J. Polymer Sci., A-2, 8, 1057 (1970).

970. H. Kakutani, J. Polymer Sci., A-2, 8, 1177 (1970).

971. W. S. Durrell, G. Westmoreland, M. G. Moshonas, J. Polymer Sci., A, 3, 2975 (1965).

972. G. T. Furukawa, R. E. McCoskey, G. J. King, J. Res. Nat. Bur. Std. 49, 273 (1952).

973. A. J. Warner; private communication to ref. 80.

974. H. Sack; private communication to ref. 946.

975. J. A. S. Smith, Discussions Faraday Soc. 19, 207 (1955).

976. S. Nohara, Chem. High Polymer (Japan) 14, 318 (1957).

977. S. P. Kabin, Zh. Tekh. Fiz. 26, 2628 (1956); (transl'n in Soviet Phys. -Tech. Phys. 1, 2542 (1957)).

978. Y. Maeda, Chem. High Polymer (Japan) 14, 442 (1957); C.A. 52, 5022f (1958).

979. K. Nagamatsu, T. Yoshitomi, T. Takemoto, J. Colloid. Sci. 13, 257 (1958).

980. M. Baccaredda, E. Butta, J. Polymer Sci. 31, 189 (1958).

981. K.-H. Illers, E. Jenckel, Kolloid-Z. 160, 97 (1958).

982. N. G. McCrum, J. Polymer Sci. 27, 555 (1958).

983. N. G. McCrum, J. Polymer Sci. 34, 355 (1959).

984. A. V. Tobolsky, J. Polymer Sci. 35, 555 (1959).

985. L. Mandelkern; private communication to ref. 659.

986. G. W. Becker, Kolloid-Z. 167, 44 (1959).

987. M. Baccaredda, E. Butta, Ann. Chim. (Rome) 49, 559 (1959); C.A. 53, 18538a (1959).

988. J. A. Sauer, A. E. Woodward, Rev. Mod. Phys. 32, 88 (1960).

989. C. A. Sperati, H. W. Starkweather, Adv. Polymer Sci. 2, 465 (1961).

990. R. K. Eby, K. M. Sinnott, J. Appl. Phys. 32, 1765 (1961).

991. T. Satokawa, S. Koisumi, Kogyo Kagaku Zasshi 65, 1211 (1962); C.A. 58, 1544h (1963).

992. A. V. Tobolsky, D. Katz, M. Takahashi, J. Polymer Sci., A, 1, 483 (1963).

993. Y. Ohzawa, Y. Wada, Rept. Progr. Polymer Phys. (Japan) 6, 147 (1963).

994. Y. Ohzawa, Y. Wada, Japan J. Appl. Phys. 3, 436 (1964).

995. Y. Araki, J. Appl. Polymer Sci. 9, 421 (1965).

996. Y. Araki, J. Appl. Polymer Sci. 9, 1515 (1965).

997. K.-L. Hsu, D. E. Kline, J. N. Tomlinson, J. Appl. Polymer Sci. 9, 3567 (1965).

998. Y. Araki, J. Appl. Polymer Sci. 9, 3575 (1965).

999. Y. Araki, J. Appl. Polymer Sci. 9, 3585 (1965).

1000. Y. Araki, J. Appl. Polymer Sci. 11, 953 (1967).

1001. E. G. Howard, P. B. Sargeant, J. Macromol. Sci. A1, 1011 (1967).

1002. K. Frigge, G. Dube, H. Kriegsman, Plaste Kautschuk 15, 470 (1968).

1003. S. Miyakawa, T. Takemura, Japan J. Appl. Phys. 7, 814 (1968); via ref. 641.

1004. A. W. Neumann, W. Tanner, J. Colloid Interface Sci. 34, 1 (1970).

1005. D. W. Brown, L. A. Wall, ACS Polymer Preprints 12, 302 (1971).

1006. H. S. Eleuterio, E. P. Moore, Chem. Eng. News 40, 44 (1962).

1007. D. W. Brown, L. A. Wall, ACS Polymer Preprints 7, 1116 (1966).

1008. U. S. 3,110,705 - 12. Nov. 1962 (E. M. Sullivan, E. W. Wise, F. P. Reding).

1009. J. A. E. Kail, Polymer 6, 535 (1965).

1010. F. P. Reding, E. R. Walter, F. J. Welch, J. Polymer Sci. 56, 225 (1962).

1011. L. Bohn, Kunststoffe Plastiks 53, 93 (1963).

1012. F. E. Bailey, J. P. Henry, R. D. Lundberg, J. M. Whelan, J. Polymer Sci., B, 2, 447 (1964).

1013. R. M. Barrer, R. Mallinder, P. S.-L. Wong, Polymer 8, 321 (1967).

1014. G. Pezzin, G. Sanmartin, F. Zilio-Grandi, J. Appl. Polymer Sci. 11, 1539 (1967).

1015. K. S. Minsker, Yu. A. Sangalov, G. A. Razuwayev, J. Polymer Sci., C, 16, 1489 (1967).

1016. U. Bianchi, A. Turturro, G. Basile, J. Phys. Chem. 71, 3555 (1967).

1017. K. Nakamura, T. Nakagawa, Rept. Progr. Polymer Phys. (Japan) 10, 277 (1967).

1018. A. M. Jendrychowska-Bonamour, European Polymer J. 4, 627 (1968).

1019. A. M. Sharetskii, S. V. Svetozarskii, I. B. Kotlyar, E. N. Zil'berman, Soviet Plastics 3, 10 (1968); (transl'n of Plasticheskie Massy 3, 6 (1968)).

1020. I. I. Perepechko, L. I. Trepelkova, L. A. Bodrova, E. K. Kosikova, L. O. Bunina, Soviet Plastics 5, 7 (1968); (transl'n of Plasticheskie Massy 5, 11 (1968)).

1021. J. Zelinger, J. Polymer Sci., C, 16, 4259 (1969).

1022. J. V. Koleske, R. D. Lundberg, J. Polymer Sci., A-2, 7, 795 (1969).

1023. J. Petersen, B. Ranby, Makromol. Chem. 133, 251 (1970).

1024. M. Kryszewski, M. Mucha, J. Polymer Sci., B, 5, 1095 (1967).

1025. D. E. Witenhafer, J. Macromol. Sci. B4, 915 (1970).

1026. G. Pezzin, G. Ajroldi, C. Garbuglio, J. Appl. Polymer Sci. 11, 2553 (1967).

1027. S. A. Liebman, C. R. Foltz, J. F. Reuwer, R. J. Obremski, Macro-molecules 4, 134 (1971).

1028. J. Barton, M. Pegoraro, L. Szilagyi, G. Pagani, Makromol. Chem. 144, 245 (1971).

1029. Kh. U. Usmanov, A. A. Yulchibaev, M. K. Asamov, A. Valiev, J. Polymer Sci., A-1, 9, 1459 (1971).

1030. J. Furukawa, Pure Appl. Chem. 26, 153 (1971).

1031. A. M. Sharetskii, S.V. Svetozarskii, E. N. Zil'berman, I. B. Kotlyar, Soviet Plastics 9, 1 (1971); (transl'n of Plasticheskie Massy 9, 5 (1971)).

1032. E. Sacher, J. Polymer Sci., A-2, 6, 1813 (1968).

1033. Kh. U. Usmanov, A. A. Yulchibaev, T. Sirlibaev, J. Polymer Sci., A-1, 9, 1779 (1971).

1034. J. M. O'Reilly, J. Polymer Sci. 57, 429 (1962).

1035. D. M. Chackraburtty, J. Chem. Phys. 26, 427 (1957).

1036. E. W. Merrill, D. A. Gibbs, Chem. Eng. News 41, 41 (1963).

1037. P. Meares, Trans. Faraday Soc. 53, 31 (1957).

1038. P. L. Magagnini, F. Sardelli, G. Pizzirani, Ann. Chim. (Rome) 57, 805 (1967).

1039. S. Saito, Denki Sikense Kenkyu Khakoku, Res. Electrotech. Lab. 1: 648 (1964); via ref. 792.

1040. M. Scatena, P. Sanmartin, F. Zilio-Grandi, FATIPEC Congr. 9, Section 3, p. 68 (1968), also Ind. Vernice 23(2), 17 (1969); C.A. 71, 50752h (1969).

1041. R. M. Holsworth, J. Paint Technol. 41, 167 (1969).

1042. J. M. Carcamo, F. Arranz, Rev. Plast. Mod. 185, 1683 (1971).

1043. H. van Hoorn, Rheol. Acta 10, 208 (1971).

1044. S. Ichihara, A. Komatsu, T. Hata, Polymer J. (Japan) 2, 640 (1971).

1045. H. Hopff, P. Perlstein, Makromol. Chem. 125, 247 (1969).

1046. H. Hopff, M. A. Osman, Angew. Makromol. Chem. 6, 39 (1969).

1047. M. Kinoshita, T. Irie, M. Imoto, Makromol. Chem. 110, 47 (1967).

1048. P. L. Magagnini, V. Frosini, European Polymer J. 2, 139 (1966).

1049. A. R. Monahan, Macromolecules 1, 408 (1968).

1050. M. A. Osman, H.-G. Elias, J. Macromol. Sci. A5, 805 (1971).

1051. T. S. Reid, D. W. Codding, F. A. Bovey, J. Polymer Sci. 18, 417 (1955).

1052. H. C. Haas, E. S. Emerson, N. W. Schuler, J. Polymer Sci. 22, 291 (1956).

1053. G. Pizzirani, P. L. Magagnini, J. Appl. Polymer Sci. 11, 1173 (1967).

1054. H. Hopff, M. A. Osman, Makromol. Chem. 135, 175 (1970).

1055. H. Hopff, M. A. Osman, Makromol. Chem. 143, 289 (1971).

1056. J. A. Price, M. R. Lytton, B. G. Ranby, J. Polymer Sci. 51, 541 (1961).

1057. G. S. Kolesnikov, E. F. Rodionova, L. S. Fedorova, T. Ya. Medved, M. I. Kabacknik, Vysokomolekul. Soedin. 4, 1385 (1962); C.A. 58, 14108e (1963).

1058. L. D. Taylor, C. K. Chiklis, T. E. Platt, J. Polymer Sci., B, 9, 187 (1971).

1059. K. Hubner, F. Kollinsky, G. Markert, H. Pennewiss, Angew. Makromol. Chem. 11, 109 (1970).

1060. L. D. Lovyagina, N. V. Meiya, A. F. Nikolaev, Zh. Prikl. Khim. 44, 2056 (1971).

1061. C. Noel, L. Monnerie, J. Chim. Phys. Phys.-Chim. Biol. 65, 2089 (1968).

1062. V. Frosini, S. de Petris, Chim. Ind. (Milan) 49, 1178 (1967).

1063.   L. A. Utracki, R. Simha, Makromol. Chem. 117, 94 (1968).

1064.   F. J. Golemba, J. E. Guillet, S. C. Nyburg, J. Polymer Sci., A-1,
        6, 1341 (1968).

1065.   B. L. Tsetlin, T. Ya. Medved, Yu. G. Chikishev, Yu. M. Polikar-
        pov, S. R. Rafikov, M. I. Kabuchnik, Vysokomolekul. Soedin. 3,
        1117 (1961);C. A. 56, 2568h (1962).

1066.   P. Ehrlich, R. J. Kern, E. D. Pierron. T. Provder, J. Polymer Sci.,
        B, 5, 911 (1967).

1067.   C. Noel, Compt. Rend. 258, 3702 (1964).

1068.   N. S. Nametkin, S. G. Durgar'yan, L. I. Tikhonova, V. G. Fillip-
        pova, Vysokomolekul. Soedin. A, 13, 672 (1971);
        (transl'n in Polymer Sci. USSR 13, 765 (1971)).

RATE OF CRYSTALLIZATION OF POLYMERS

J. H. Magill
Department of Metallurgical and Materials Engineering
University of Pittsburgh
Pittsburgh, PA/USA

Contents

A. INTRODUCTION

Polymers characteristically crystallize in two distinct morphological forms. Spontaneous crystallization occurs on cooling a supersaturated dilute solution to produce single crystals (1). From the supercooled liquid state and concentrated solutions, spherulites normally form. The crystalline texture and morphology depends upon experimental conditions (2, 3, 19).

Growth occurs outwards from a central nucleus until impingement produces polygonized structures and termination of the radial growth results. This growth[*] (28) which is extremely temperature sensitive, has been called primary crystallization. Crystallization which occurs after the radial growth is completed is sometimes termed secondary crystallization.

The overall isothermal transformation kinetics have been described by the Avrami theory (29) which has been simplified by Evans (32) and Morgan (33) and others. The equation

$$\theta = \exp(-Kt^n)$$

relates the amount or fraction θ of uncrystallized material remaining after some time t at the crystallization temperature, to the growth rate parameter K and mode of nucleation n. The overall transformation rate, K, embodies both nucleation and growth presumed by Avrami to occur under constant volume conditions and modified by Mandelkern (34) to include incomplete crystallization. A very recent development by Misra and Stein (38) also described mathematically the embryonic stages of spherulitic development.

Generally, the change in some physical property unambiguously associated with the crystalline transformation, can be employed to study the rate of transformation under isothermal conditions. A variety of experimental techniques have been used to obtain the parameter θ for this purpose. The most commonly used parameter is specific volume (34) where

$$\theta = (v_\infty - v_t)/(v_\infty - v_a)$$

and v_a, v_t, and v_∞ are measured parameters corresponding to the initial, intermediate and final stages of the isothermal crystallization. The first example of this procedure dates back many decades to the classical work of Wood and Bekkedahl on natural rubber (39, 40) which clearly established that the overall rate process goes through a maximum as the temperature is lowered. It was later demonstrated experimentally (41) and shown theoretically (42) that the spherulitic growth rate proceeds similarly.

Calorimetry (43), dilatometry (34), density balance (44) and density gradient (45), light depolarization (46, 47) (for overall rate studies) and/or small-angle light scattering (48, 49) (for average spherulite growth rate) infrared (50), x-ray (51), NMR (52) and electrical resistivity (53) are all properties which have been used to study the overall crystallization process since they all, to a degree, reflect the ordering of chain molecules on passing from the liquid to crystalline phase.

On the whole, bulk crystallization rates are probably of limited value because they depend markedly on the concentration of nucleation centers (31, 35-38, 55, 56-58), whatever their origin. There is really no satisfactory way to quantify the concentration of nuclei since so much is dependent on adventitious impurities. Therefore, only literature references are tabulated in part B.

Spherulitic growth rates are a more fundamental and reproducible property for a given polymeric system and are retained in this article. At a given temperature such rates are independent of the origin of the primary nucleation site and depend primarily on the polymer molecular weight and distribution. Under isothermal conditions in the absence of non-crystallizable impurities, spherulites grow at a constant rate which is typical of a nucleation controlled process (28).

In general, theories for isothermal growth rates G(T) in polymeric and other less complex systems can be formulated by the equation:

$$G(T) = f(T)\, D(T)$$

which is approximated by

$$G(T) = f(T)/\eta(T)$$

wherein it is postulated that the temperature dependence of the diffusive term $D(T)$ is proportional to the reciprocal zero shear viscosity $\eta(T)$. This assumption is open to question in condensed systems of anisotropic molecules. The form of $f(T)$ differentiates between various growth rate models. Different formalisms (42) in-

---

[*] Another definition is frequently used by kineticists. Here the term primary crystallization is used to describe that part of the transformation process which extends from zero time at the crystallization temperature up to the point where deviations occur from Avrami kinetics (29). At this point, a very much slower and more complex process (secondary crystallization) ensues within the boundaries of the spherulites and further densification (30, 31) usually follows.

voking chain folding (59) and diverse geometries (60, 61) can be set up for growing polymer crystals or spherulites. Among the most successful to date is the Hoffman-Lauritzen model which is similar to others (60) in the field. In this model the equation for two-dimensional growth can be represented by

$$G = G_o \exp\left[-\frac{A}{R(T-T_\infty)}\right] \quad \exp\left[\frac{-BT_M}{T\Delta T}\right]$$

G_o is a constant for a given molecular weight, and the first exponential is a Williams-Landel-Ferry-type of expression (42, 55) describing the transport of macroscopic molecules. In this term, A is an activation energy associated with segmental transport und T_∞ is some temperature below the glass transition temperature where segmental motions fade away. The undercooling $\Delta T = T_M^{\neq} - T$, and B is a constant (for a given molecular weight) and is related to the product of the lamellae surface (of interfacial) energies $\sigma\sigma_e$. Here σ and σ_e refer to the crystal lateral and end surfaces respectively.

With this formalism*, many analyses of spherulitic growth rates have been made over the last decade. It has been shown that $\sigma\sigma_e$ is molecular weight dependent (67-72). Overall transformation rates (70, 71), growth rates of spherulites (12, 55, 69) and monocrystals (11, 12) are dependent on chain length. The spherulitic growth rate has also been found to vary inversely as the molecular weight (48, 55, 69) for three different polymers.

More recently (163) crystallization in polymeric materials has been characterized by a single empirical relation. Non-isothermal crystallization has also been treated theoretically in articles (4, 5) published within the last few years.

Experimental data** selected from the literature are presented below. Only spherulitic growth rates* are tabulated. The reader is referred to the Polymer Handbook 1st edition, Wiley Interscience, 1966, for data on bulk crystallization studies.

B. OVERALL RATES OF CRYSTALLIZATION (Literature References)

Poly(1-butene)	78-82	Poly(oxydimethylsilylene-1, 4-phenylene-dimethylsilylene)	73, 131
Poly(chloroprene)	92	Poly(oxyethylene)	115, 116, 124
Poly(chlorotrifluoroethylene)	105, 106	Poly(oxyethyleneoxyadipoyl)	125, 127
Poly(ethylene)	93-95	Poly(oxyethyleneoxysuccinoyl)	71
Poly(iminohexamethyleneiminoadipoyl) (Nylon 66)	46, 132, 133	Poly(oxyethyleneoxyterephthaloyl)	
Poly(iminomethylene-1, 3-phenylenemethyleneimino-		(Poly(ethylene terephthalate))	30, 50, 128, 129
adipoyl)	141	Poly(oxymethylene)	111-113
Poly(imino(1-oxohexamethylene)) (Nylon 6)	134-138	Poly(oxypropylene)	117, 118
Poly(imino(1-oxoundecamethylene)) (Nylon 11)	139	Poly(oxyterephthaloyloxydecamethylene)	124
Poly(isobutene)	82	Poly(oxytetramethylene)	119
Poly(isoprene) cis	14, 39, 40, 85-89	Poly(oxytetramethylene) glycol	119
trans	91	Poly(1-pentene)	83, 84
Poly(4-methyl-1-pentene)	90	Poly(1, 4-piperazinediylsebacoyl)	140
Poly(oxyadipoyloxydecamethylene)	114, 115	Poly(propylene)	96-100, 103
Poly(oxy-2, 2-bischloromethyltrimethylene)		Poly(styrene)	57, 107-110
(Poly(3, 3-bischloromethyloxacyclobutane))	120, 121	Poly(vinyl chloride)	104
Poly(oxydecamethyleneoxysebacoyl)	122, 123	Selenium	21, 149
Poly(oxydimethylsilylene) (Poly(dimethyl siloxane))	130		

$\neq$ By far the most serious deficiency in this analysis is the absence of precise thermodynamic melting points (60-66).

*   More complex theories (75, 76) have been evolved to account for rates of crystallization of multicomponent system wherein fractionation and other effects occur during crystallization (77).

**   The influence of additives or nucleants (6-8), stress (13, 14), shear (9, 10) and solvents (15-18, 57, 125, 127-130) on crystallization behavior have been investigated.

x   It is apt to point out that other materials also form spherulites; for example, sulfur (20), selenium (21), antimony (22), proteins (23), graphite (24), silicates (25), inorganic(26) and low molecular weight organic molecules (27).

C. RATES OF RADIAL SPHERULITIC GROWTH

Polymer	T_f [°C]	T [°C]	G [nm/s]	Remarks	References
1. Poly(dienes), Poly(alkenes), Poly(vinyls)					
Poly(1-butene)					
M_v = 400,000	160 (10 min)	90.0	338.	Supercooled liquid, form 2	81
		96.0	168.		
		100.0	60.4		
		105.0	19.6		
Poly(chlorotrifluoroethylene)		197.4	2.217		105
		200	1.009		
		202	0.3201		
	305	163	450.1		106
	277	169.0	440.1		
	305	169.1	430.1		
	277	171.7	360.1		
	277	180.1	195.0		
	267	185.5	110.0		
	267	191.0	40.01		
		195	7.251		
		197	2.501		
(Kel-F NST 300)	270 (10 min)	170	406.7		147
		175	310.0		
		180	216.7		
		185	124.2		
		190	36.34		
Poly(ethylene)					
D* = 0 MR	292	120.7	117		142
	192	122.9	103.4		
	193	125.1	18.2		
	196	127.0	1.08		
	168	122.4	88.4		
	153	122.5	100.4		
D = 40MR	195	118.9	41.68		
	195	113.1	291.7		
	195	114.6	216.7		
	195	115.8	65.01		
	195	117.8	2.334		
D = 100MR	195	100.0	71.69		
	195	100.0	61.68		
	195	101.8	31.67		
	195	104.0	7.835		
M_v = 19,000	200 (20 min)	125.0	126.7		143
		126.0	36.67		
		127.0	16.67		
		128.0	3.334		
M_n = 4,000		81.0	510.0	$CH_3/100CH_2$ = 5.32	144
		83.0	256.7		
		85.0	143.4		
		88.0	63.35		144
7,000		91.0	215.0	$CH_3/100CH_2$ = 5.32	
		93.0	86.68		
		95.2	35.01		
		97.0	14.00		
		98.0	7.835		
10,000		92.0	175.0	$CH_3/100CH_2$ = 4.70	
		93.0	101.2		
		95.0	40.01		
		97.0	20.01		
		98.0	11.50		
12,000		90.3	316.7	$CH_3/100CH_2$ = 4.51	
		92.0	185.0		

* D = Radiation dose, MR = megarad

Polymer	T_f [$^{\circ}$C]	T [$^{\circ}$C]	G [nm/s]	Remarks	References
Poly(ethylene) (Cont'd.)					
M_n = 12,000		98.0	12.17	$CH_3/100CH_2 = 4.51$	
		100.3	3.334		
M_n = 5,140	160 (30 min)	121.39	422.1★	Fractionated samples from	70
M_w = 6,290		125.12	1.069	N.B.S. material S.R.M. ★★★★ 1475	
M_z = 7,620					
M_n = 10,510		125.32	93.39		
M_w = 11,740		127.81	0.4328		
M_z = 13,220					
M_n = 56,350		124.46	25.10		
M_w = 68,570		128.65	0.3770		
M_z = 83,930					
M_n = 76,760		123.07	117.3		
M_w = 134,300		127.74	1.039		
M_z = 276,000					
M_n = 195,000		120.25	196.3		
M_w = 323,200		126.30	1.062		
M_z = 500,800					
Poly(isoprene), trans (HMF)★★					
M_n = 7,000	100	33.6	72.18		69
		35.6	63.01		
		37.5	48.01		
		37.6	53.18		
		4o.4	39.01		
		43.0	22.67		
12,000	100	37.6	44.01	Data for M_n 18,500 and 37,000	69
		39.5	36.01	see ref. 69	
		43.0	21.34		
		45.6	14.34		
45,000	100	34.0	24.67		69
		38.0	11.67		
		40.0	9.34		
		43.0	5.67		
		46.0	2.83		
91,000	100	34.0	16.50		
		38.0	7.92		
		40.0	5.85		
		43.0	3.52		
		46.0	1.71		
165,000	100	38.0	5.83		
		40.0	4.25		
		43.0	2.10		
		46.0	1.05		
400,000	100	34.0	14.00		
		38.0	6.67		
		40.0	4.38		
		43.0	2.33		
		46.0	0.95		
(LMF)★★★					72
M_n = 12,000	100	25.8	173.37	Data for M_n 18,500; 37,000 see	
		29.6	111.69	ref. 72	
		33.6	66.68		
		37.6	34.01		
		43.0	9.17		
45,000	100	26.0	51.01		
		30.0	34.84		
		34.0	16.50		
		38.0	8.17		
		40.0	4.83		
		43.0	1.83		

★ From least square fit to experimental data;   ★★ HMF denotes high melting form spherulites;   ★★★LMF denotes low melting form spherulites
★★★★ Standard reference material

Polymer	T_f [°C]	T [°C]	G [nm/s]	Remarks	References
Poly(isoprene) (Cont'd.)					
(LMF)★					
M_n = 91,000	100	26.0	36.01		
		30.0	20.83		
		34.0	11.16		
		38.0	4.33		
		40.0	2.50		
		43.0	1.17		
Poly(propylene)	190-195	125	87.52		36
		130	61.51		
		134	28.67		
		138	8.668		
		140	4.501		
	180 (15 min)	122.0	300.1		97
		125.0	200.1		
		130.0	71.68		
		135.0	26.67		
		140.0	9.835		
		145.0	4.501		
M_v = 178,000		120	490.1		145
		125	216.7		
		131	64.68		
		135	27.17		
	220	120.5	480.1	Pro-fax 6513E	98
		129	120.0		
		145	4.501		
		121	408.4	Pro-fax 6501	98
		132	46.67		
		145	2.334		
Poly(styrene) (isotactic)					
M_v = 60,000		140	0.333		145
		155	1.500		
		177	4.334		
		190	3.667		
		200	1.667		
		210	0.500		
190,000	270	140	0.83		148
		150	0.333		
		170	1.334		
		180	1.334		
		190	0.500		
M_v = 330,000	250	120	0.0400		107
M_n = 185,000		130	0.162		
		140	0.600		
		150	1.43		
		160	2.50		
		170	3.42		
		180	3.42		
		190	2.50		
		200	1.43		
		210	0.45		
M_v = 1,250,000		155	2.167		145
		170	4.834		
		184	4.501		
		200	2.167		
		210	0.6668		
M_v = 1,380,000		130	0.333		148
		150	1.834		
		170	4.501		
		180	7.502		
		190	3.167		

★ LMF denotes low melting form spherulites

Polymer	T_f [$^\circ$C]	T [$^\circ$C]	G [nm/s]	Remarks	References
Poly(styrene) (isotactic) (Cont'd.)					
M_v = 2,200,000		112.01	0.004544		146
		119.62	0.02582		
		129.70	0.1518		
		150.26	1.526		
		159.22	2.532		
		172.15	3.844		
		179.70	3.980		
		195.06	2.529		
		200.52	1.832		
2. Poly(oxides)					
Poly(oxy-2,2-bischloromethyltrimethylene)					
(Poly(3,3-bis(chloromethyl)-oxacyclobutane))		75	208.4		121
		80	183.4		
		90	150.0		
		100	125.0		
		110	100.0		
		120	66.68		
		130	33.34		
		140	25.01		
Poly(oxyethylene)					
M_v = 5,470		42.0	13769.4		115
		44.2	4734.2		
		45.2	2217.1		
		47.2	395.1		
12,400		47.2	11752.3		
		48.8	6034.5		
		51.2	1514.0		
		55.2	55.51		
33,100		47.6	1784.0		
		48.2	1516.9		
		51.0	421.8		
		53.8	94.19		
		56.4	1.60		
Poly(oxymethylene)					
M_n = 45,000		156.0	346.7		149(112, 113)
		157.0	180.0		
		158.0	99.19		
		159.8	77.02		
		160.8	56.61		
		163.0	20.84		
		163.9	14.25		
		164.5	4.801		
		166.1	2.951		
		169.1	0.4751		
Poly(oxypropylene)		42.3	325.89		117
		46.3	132.52		
		46.7	39.1		
		51.0	16.67		
		40.5	165.86		
		42.3	104.18		
		44.2	66.68		
		46.3	73.3		
		39.0	8.3		
		40.5	40.0		
		42.3	33.34		
		44.2	8.3		
	85	45	1.37	Mean isotactic sequence	119
		50	0.75	length = 122	
		55	0.14		
		38	1.00	Mean isotactic sequence	119
		43	0.42	length = 14	
		48	0.15		

Polymer	T_f [°C]	T [°C]	G [nm/s]	Remarks	References
Poly(oxypropylene) (Cont'd.)					
M_n = 10,300	120 (5 min)	0	500.1	Optically active	151
		5	746.8		
		10	723.4		
		15	838.5		
		25	476.8		
		30	403.4		
		35	240.0		
		40	145.7		
		45	56.68		
		50	23.50		
		60	2.050		
135,000	120 (5 min)	25	255.2		151
		30	157.5		
		35	103.3		
		40	41.7		
		45	23.6		
		50	6.01		
		55	2.02		
		60	0.467		

3.   Poly(carbonates)

Polymer	T_f [°C]	T [°C]	G [nm/s]	Remarks	References
Poly(oxacarbonyloxy-1,4-phenyleneisopropylidene-1,4-phenylene)					
(Polycarbonate of Bisphenol A)		190	0.08335		152
		195	0.04168		
Poly(oxycarbonyloxy-1,4-phenylenemethylene-1,4-phenylene) (Polycarbonate-F)		175	16.67		152
		180	10.00		
		185	5.00		
		190	3.33		
Poly(oxycarbonyloxy-1,4-phenylenethio-1,4-phenylene) (Polycarbonate-S)		145	20.0		152
		150	15.00		
		160	8.33		
		170	5.85		
		180	4.167		

4.   Poly(esters)

Polymer	T_f [°C]	T [°C]	G [nm/s]	Remarks N[nuclei/cc/min]	References
Poly(oxydecamethyleneoxysebacoyl)					123
M_v = 10,300	100 (24 hours)	67.1	87.35	1.50×10^5	
		68.1	21.671	3.72×10^4	
		69.1	6.451	8.95×10^3	
		70.1	1.884	1.12×10^3	
		71.1	0.5833	1.65×10^2	
		72.0	0.08301	1.65×10^1	
Poly(oxyethyleneoxyadipoyl)					127
M_v = 9,900		20	66.346		
		40	18.00		
		44	13.169		
		47	9.002		
Poly(oxyethyleneoxyterephthaloyl)		39.6	147.0		153
(Poly(ethylene terephthalate))		40.9	140.0		
		42.3	158.5		
		50.3	166.0		
		53.4	150.2		
		57.0	153.4		
		57.6	120.0		
		58.7	116.0		
		61.4	76.81		
		67.6	74.18		
		72.0	51.34		

Polymer	T_f [°C]	T [°C]	G [nm/s]	Remarks	References
Poly(oxyethyleneoxyterephthaloyl)					
(Poly(ethylene terephthalate)) (Cont'd.)		75.2	26.17		
		81.8	5.801		
		85.0	3.001	N [nuclei/cc/min]	
M_n = 16,800	294	220	41.08	3.2×10^8	128
		240	3.100	4.93×10^6	
19,000		120	6.84	Data for M_n 22,000; 24,800;	48★
		130	15.0	30,100; 35,400 in ref. 48	
		140	47.4		
		150	75.2		
		160	93.2		
		170	112.2		
		180	112.8		
		190	117.4		
		200	85.0		
27,400		120	3.28		48★
		130	8.50		
		140	19.9		
		150	36.8		
		160	56.2		
		180	72.1		
		190	63.6		
		200	43.9		
		210	20.2		
39,100		120	1.54		48★
		130	4.40		
		140	8.26		
		150	14.2		
		160	18.0		
		170	27.8		
		180	26.6		
		190	19.1		
		200	13.4		
Poly(oxymethyleneoxysuccinoyl)					153
M_w = 5,980		16.4	12.00		
		23.0	31.17		
		26.3	62.34		
		28.5	44.18		
		35.6	91.35		
		36.3	91.83		
Poly(oxyterephthaloyloxydecamethylene)				N [nuclei/cc/min]	125
M_n = 10,000	160 (5 min)	120.0	5.534	54.6	
		121.0	3.451	22.6	
		123.0	1.592	3.33	
		125.0	0.6217		
		126.0	0.2550		
5. Poly(amides)					
Poly(iminohexamethyleneiminoadipoyl) (Nylon 66)				I [nuclei/cc]	
M_n = 11,600	295	241	166.7	1.06×10^6	132, (156,157)
	295	247	58.35	7.60×10^5	
	295	250	13.84	6.08×10^5	
	295	252	10.50	$N = 6.99 \times 10^3$	
	285	247	66.08	9.13×10^5	
Negative spherulites		251	14.24		158
		256	83.35		
		257±1	13.34		159
		259±1	9.168		
		261±1	6.668		
		263±1	4.167		
		265±0.5	2.500		

★ Growth rates with liquid additives have also been measured.

Polymer	T_f [°C]	T [°C]	G [nm/s]	Remarks	References
Poly(iminohexamethyleneiminoadipoyl) (Nylon 66) (Cont'd.)				N[nuclei/cc/min]	160
M_n = 12,900	300 (30 min)	246.0	106.68	1.16×10^6	
		248.0	56.34	0.88×10^6	
		253.0	10.84		
13,700	300 (30 sec)	141.0	13502.7		154
		160.0	13669.4		
		180.0	12119.1		
		199.0	8901.8		
		215.0	5167.7		
		230.0	2117.1		
		234.0	1530.3		
		237.0	920.18		
		239.5	765.15		
		241.0	471.76		
		244.0	368.40		
14,600				N[nuclei/cc/min]	160
	280	241.5	283.39		
		243	230.05	22.5×10^6	
		245	180.86	5.73×10^6	
		248	33.685	2.26×10^6	
		252	14.66	0.33×10^6	
	300	241.5	204.4	14.4×10^6	
		243	175.0	10.5×10^6	
		245	128.3	6.2×10^6	
		248	58.34	2.80×10^6	
		252	5.501		
	315	241.5	280.0	45.0×10^6	
		243	168.4	10.5×10^6	
		245	113.4	6.7×10^6	
		248	56.68	2.89×10^6	
		252	6.335	0.09×10^6	
Positive spherulites		50	3650.7		161
		100	4706.6		
		142	6751.3		
		160	6101.2		
		178	5201.0		
		198	3700.7		
		200	2900.6		
		228	466.7		
Positive spherulites	300 (30 sec)	180.0	11435.6		154
M_n = 25,500		200.0	7951.5		
		211.0	5284.4		
		222.0	2733.8		
		230.0	1615.3		
		235.5	680.13		
		240	483.4		
Poly(iminohexamethyleneiminosebacoyl)					
Positive spherulites		200.0	253.4		161
		205.0	120.02		
		212.0	36.67		
		217.0	13.34		
Negative spherulites		205.0	186.7		161
		212.0	130.0		
		217.0	73.3		
Poly(imino(1-oxohexamethylene)) (Nylon 6)					
M_n = 27,400	300(30 sec)	102.3	900.2		154(155)
		112.5	1478.6		
		122.0	2033.7		
		130.0	2767.2		
		140.5	3133.9		
		150.5	2933.9		
		159.0	2450.5		
		170.0	1431.9		

Polymer	T_f [^{o}C]	T [^{o}C]	G [nm/s]	Remarks	References
Poly(imino(1-oxohexamethylene) (Nylon 6) (Cont'd.)					
M_n = 27,400	300 (30 sec)	180.0	710.1		
		182.0	610.1		
M_v = 24,700		90	375.1		156
		95	610.1		
		101	640.1		
		107	1108.5		
		117	1571.9		
		124.5	1933.7		
		135	2438.8		
		141	2438.8		
		148	2105.4		
		157	1790.3		
		172	983.5		
		184	305.1		

6. Others

Polymer	T_f [^{o}C]	T [^{o}C]	G [nm/s]	Remarks	References
Poly(oxydimethylsilylene-1,4-phenylenedimethylsilylene) 200 (5 min)					
M_v = 8,700		23	316	Data for M_v 15,800; 56,000;	55
		30	530	see ref.	
		40	900		
		50	1450		
		60	1640		
		70	1740		
		80	1460		
		90	959		
		100	417		
		110	161		
		120	14.5		
		130	0.593		
372,000	200 (5 min)	24	80.0		55
		30	113.0		
		40	166		
		60	328		
		70	342.0		
		80	246		
		90	146		
		100	65.0		
		110	20.0		
		130	0.124		
		134	0.0365		
1,400,000	200 (5 min)	24	41.5		55
		40	165		
		51	280		
		60	328		
		70	302		
		80	240		
		90	146		
		100	70.6		
		110	23.4		
Selenium	600 (2 hours)	80	0.1717	99.9999 %	21
	and ice water	90	0.8084		
	quenched	100	1.5353		
		110	2.3838		
		120	3.0839		
		130	5.8345		
		140	4.8843		
		150	2.8505		

D. REFERENCES

1. P. H. Geil, "Polymer Single Crystals", Wiley Interscience Publishers Inc., New York (1963).

2. J. H. Magill, J. Polymer Sci., A2, 4, 243 (1966); A2, 7, 123 (1969); A2, 9, 815 (1971).

3. H. D. Keith, J. Polymer Sci., A, 2, 4339 (1964); J. Appl. Phys. 35, 3115 (1964).

4. J. Tomka, J. Sebenda, Collection Czech. Chem. Commun. 32, 2779 (1967).

5. T. Ozawa, Polymer 12, 150 (1971).

6. H. N. Beck, J. Appl. Polymer Sci. 11, 673 (1967).

7. A. G. M. Last, J. Polymer Sci. 34, 543 (1959).

8. V. L. Folt, Rubber Chem. Technol. 44, 29 (1971).

9. A. Wereta, Jr., C. G. Gogos, Polymer Eng. 11, 20 (1971).

10. A. K. Fritzke, F. P. Price, Bull. Am. Phys. Soc., B 28, March 27-30, Atlantic City, N. J. 1972.

11. M. J. Nardini, F. P. Price, J. Phys. Chem. Solids (Suppl.), Proceedings Intern. Conf. Crystal Grwoth, Boston, June 20-24, 1966.

12. V. F. Holland, P. H. Lindenmeyer, J. Polymer Sci. 57, 589 (1962).

13. H. Kim, L. Mandelkern, J. Polymer Sci., A2, 6, 181 (1968).

14. A. N. Gent, Trans. Faraday Soc. 50, 521 (1954).

15. L. Mandelkern, "Growth and Perfection of Crystals", R. H. Doremus, B. W. Roberts, D. Turnbull (Editors), John Wiley and Sons, 467 (1958).

16. M. I. Kashmiri, R. P. Sheldon, Brit. Polymer J. 1, 65 (1969).

17. C. Devoy, L. Mandelkern, L. Bouland, J. Polymer Sci., A2, 8, 869 (1970).

18. J. B. Helms, G. Challa, J. Polymer Sci., A2, 11, 761 (1972).

19. A. Keller, Kolloid-Z.-Z. Polymere 231, 386 (1969).

20. C. Briske, N. H. Hawthorne, Trans. Faraday Soc. 63, 1546 (1967).

21. R. G. Crystal, J. Polymer Sci., A2, 8, 2153 (1970).

22. I. Ye. Bolotov, S. B. Fisheleva, Physics Metals and Metallog. (USSR) (English Transl.) 20, 147 (1966).

23. J. E. Coleman, B. J. Allan, B. L. Allee, Science 131, 350 (1960).

24. I. Ezhek, I. Koritta, K. Leble, Soviet Phys. Cryst. (English Transl.) 2, 653 (1957).

25. J. G. Morley, Glass Technol. 6, 69, 77 (1965).

26. H. W. Morse, J. D. H. Donnay, J. Am. Mineralogist 21, 391 (1936).

27. J. H. Magill, D. J. Plazek, J. Chem. Phys. 46, 3757 (1967).

28. P. J. Flory, A. D. McIntyre, J. Polymer Sci. 18, 592 (1955).

29. M. Avrami, J. Chem. Phys. 7, 1103 (1939); 8, 212 (1940); 9, 177 (1941).

30. H. G. Zachmann, H. A. Stuart, Makromol. Chem. 41, 131 (1960).

31. A. J. Kovacs, Ric. Sci. 25A, 668 (1955).

32. U. R. Evans, Trans. Faraday Soc. 41, 365 (1945).

33. L. B. Morgan, Phil. Trans. Roy. Soc., London, Ser. A, 247, 13 (1954).

34. L. Mandelkern, "Crystallization of Polymers", Chap. 8, McGraw Hill, New York (1964).

35. I. H. Hillier, J. Polymer Sci. A3, 3067 (1965).

36. S. Hoshino, F. Meineck, J. Powers, R. S. Stein, J. Polymer Sci. A3, 3041 (1965).

37. F. P. Price, J. Appl. Phys. 36, 3014 (1965).

38. A. Misra, R. S. Stein, Polymer Letters, B, 10, 473 (1972).

39. L. A. Wood, N. Bekkedahl, J. Res. Natl. Bur. Std. 36, 487 (1946).

40. N. Bekkedahl, Rubber Chem. Technol. 40, 25 (1967).

41. B. B. Burnett, W. F. McDevit, J. Appl. Phys. 28, 1107 (1957).

42. J. D. Hoffman, Soc. Plastics Engrs. Trans. 4, 1 (1964).

43. F. H. Mueller, H. Martin, J. Polymer Sci., C, 6, 83 (1964).

44. J. H. Magill, Research Develop. (London) 11, 30 (1962).

45. A. Keller, G. R. Lester, L. B. Morgan, Phil. Trans. Roy. Soc. (London), Ser. A, 247, 1 (1954).

46. J. H. Magill, Polymer 2, 221 (1961).

47. A. Ziabicki, Kolloid-Z.-Z. Polymere 219, 1 (1967).

48. F. van Antwerpen, Thesis "Kinetics of Crystallization Phenomena of Spherulites in Polyethyleneterephthalate", Delft Technical Univ. 1971.

49. V. G. Baranov, Discussions Faraday Soc. 49, 137 (1970).

50. W. H. Cobbs, R. L. Burton, J. Polymer Sci. 10, 276 (1953).

51. K. P. Mamedov, A. D. Nurieva, Soviet Phys. Cryst. (English Transl.) 12, 605 (1968).

52. A. Peterlin, E. Roechl, J. Appl. Phys. 34, 102 (1963).

53. R. W. Warfield, Makromol. Chem. 89, 269 (1965).

54. R. P. Sheldon, J. Polymer Sci., B1, 655 (1963).

55. J. H. Magill, J. Appl. Phys. 35, 3249 (1964); J. Polymer Sci., A2, 5, 89 (1967).

56. W. Banks, M. Gordon, A. Sharples, Polymer 4, 289 (1963).

57. J. N. Hay, J. Polymer Sci. A3, 433 (1965).

58. A. Sharples, Polymer 3, 250 (1962).

59. J. I. Lauritzen, Jr., J. D. Hoffmann, J. Res. Natl. Bur. Std. 64A, 73 (1960); 65A, 297 (1961).

60. See for example review by F. P. Price in "Nucleation", Chapt. 8 (Editor A. C. Zettlemoyer) Marcel Dekker Inc., New York, 1969.

61. L. Mandelkern, N. L. Jain, H. Kim, J. Polymer Sci., A2, 6 (1968).

62. R. L. Miller, Kolloid-Z.-Z. Polymere 225, 62 (1969).

63. P. J. Flory, A. Vrij, J. Am. Chem. Soc. 85, 3548 (1963).

64. M. G. Broadhurst, J. Res. Natl. Bur. Std. 67A, 233 (1963); 70A, 481 (1966).

65. T. Arakawa, B. Wunderlich, J. Polymer Sci., C, 16, 653 (1967).

66. C. M. L. Atkinson, M. J. Richardson, Trans. Faraday Soc. 65, 1749, 1764 (1969).

67. C. Devoy, L. Mandelkern, J. Polymer Sci., A2, 7, 1883 (1969).

68. J. H. Magill, J. Polymer Sci., A2, 7, 1187 (1969).

69. E. G. Lovering, J. Polymer Sci., C, 30, 329 (1970).

70. J. D. Hoffman, G. S. Ross, L. J. Frolen, Private communication.

71. K. Ueberreiter, G. Kanig, A. S. Brenner, J. Polymer Sci. 16, 53 (1955).

72. E. G. Lovering, J. Polymer Sci., A2, 8, 747 (1970).

73. J. H. Magill, J. Polymer Sci., B, 6, 853 (1968).

74. L. Mandelkern, J. G. Fatou, K. Ohno, J. Polymer Sci., B, 6, 615 (1968).

75. E. A. DiMarzio, J. Chem. Phys. 47, 3451 (1971).

76. J. I. Lauritzen, Jr., E. A. DiMarzio, E. Passaglia, Jr., J. Chem. Phys. 45, 4444 (1966).

77. H. D. Keith, F. J. Padden, Jr., J. Appl. Phys. 35, 1270 (1964).

78. K. S. Dastry, European Polymer J. 8, 63 (1972).

79. J. Boor, Jr., J. C. Mitchell, J. Polymer Sci., A, 1, 59 (1963).

80. G. Videtto, A. J. Kovacs, Kolloid-Z.-Z. Polymere 220, 1 (1967).

81. J. Powers, J. D. Hoffmann, J. J. Weeks, F. A. Quinn, Jr., J. Res. Natl. Bur. Std. 69A, 335 (1965).

82. R. M. Kell, B. Bennett, P. B. Stickney, Rubber, Chem. Technol. 31, 499 (1958).

83. F. A. Quinn, Jr., J. Powers, J. Polymer Sci., C1, 341 (1963).

84. A. Capizzi, G. Gianotti, Makromol. Chem. 157, 123 (1972).

85. E. W. Russell, Trans. Faraday Soc. 47, 539 (1951).

86. A. N. Gent, I. R. I. Trans. 30, 139, 144 (1954).

87. A. N. Gent, J. Polymer Sci. 18, 321 (1955); 18, 257 (1955).

88. H. G. Kim, L. Mandelkern, J. Polymer Sci. A2, 6, 181 (1968).

89. J. C. Mitchell, D. J. Meier, J. Polymer Sci., A2, 6, 1689 (1968).

90. J. H. Griffith, B. G. Ranby, J. Polymer Sci. 44, 369 (1960).

91. E. C. Lovering, J. Polymer Sci., A2, 8, 2197 (1970).

92. A. N. Gent, J. Polymer Sci., A, 3, 3787 (1965).

93. L. Mandelkern, "Growth and Perfection of Crystals", R. H. Doremus, B. W. Roberts, D. Turnbull (Editors), John Wiley, New York, 467-497 (1958).

94. J. J. Weeks, J. Res. Natl. Bur. Std. 67A, 44 (1963).

95. W. Banks, M. Gordon, R. J. Roe, A. Sharples, Polymer 4, 61 (1963).

96. J. H. Griffith, B. G. Ranby, J. Polymer Sci. 38, 107 (1959).

97. B. von Falkai, H. A. Stuart, Kolloid-Z. 162, 138 (1959).

98. L. Marker, P. M. Hays, G. P. Tilley, R. M. Early, O. J. Sweeting, J. Polymer Sci. 38, 33 (1959).

99. P. Parrini, G. Corrieri, Makromol. Chem. 62, 83 (1963).

100. J. H. Magill, Polymer 3, 35 (1962).

101. F. Rybnikár, M. Mozísek, O. Jelínek, Plaste Kautschuk 6, 324 (1963).

102. H. N. Beck, H. D. Ledbetter, J. Appl. Polymer Sci. 9, 2131 (1965).

103. F. L. Binsberger, B. G. M. de Lange, Polymer 11, 309 (1970).

104. F. Rybnikár, Makromol. Chem. 140, 91 (1970).

105. F. Rybnikár, Collection Czech. Chem. Commun. 27, 2307 (1962).

106. J. D. Hoffman, J. J. Weeks, J. Chem. Phys. 37, 1723 (1962).

107. J. Boon, G. Challa, D. W. van Krevelen, J. Polymer Sci., A2, 6, 1835 (1968).

108. A. S. Kenyon, R. L. Gross, A. L. Wurstner, J. Polymer Sci. 15, 159 (1959).

109. J. L. Talen, G. Challa, J. Polymer Sci., B, 4, 407 (1966).

110. J. Boon, J. M. Azcue, J. Polymer Sci., A2, 6, 885 (1968).

111. M. Inoue, T. Takayanaki, J. Polymer Sci. 47, 498 (1960).

112. F. Rybnikár, Collection Czech. Chem. Commun. 31, 4080 (1966).

113. F. De Candia, V. Vittoria, Makromol. Chem. 155, 17 (1972).

114. L. Mandelkern, F. A. Quinn, Jr., P. J. Flory, J. Appl. Phys. 25, 830 (1954).

115. W. J. Barnes, W. G. Luetzel, F. P. Price, J. Phys. Chem. 65, 80 (1961).

116. J. P. Arlie, P. Spegt, A. Skoulios, Makromol. Chem. 104, 212 (1967).

117. C. Booth, D. V. Dodgson, I. H. Hillier, Polymer 11, 11 (1970).

118. S. Aggarwal, L. Marker, W. L. Kollar, R. Geroch, J. Polymer Sci., A2, 4, 715 (1966).

119. G. S. Trick, J. M. Ryan, J. Polymer Sci., C, 18, 93 (1967).

120. M. Hatano, S. Kambara, Polymer 2, 1 (1961).

121. M. Inoue, J. Polymer Sci. 61, 343 (1962).

122. A. D. McIntyre, Thesis, Cornell Univ., Univ. Microfilm 16, 256 (1956).

123. P. J. Flory, A. D. McIntyre, J. Polymer Sci. 18, 592 (1955).

124. F. T. Simon, J. M. Rutherford, Jr., J. Appl. Phys. 35, 82 (1964).

125. A. Sharples, F. L. Swinton, Polymer 4, 119 (1963).

126. H. A. Lanceley, A. Sharples, Makromol. Chem. 94, 30 (1966).

127. M. Takayanagi, Mem. Fac. Eng. Kyushu. Univ. 16 (3) 111 (1957).

128. F. D. Hartley, F. W. Lord, L. B. Morgan, Proc. Roy. Soc. (London) A247, 23 (1954).

129. R. P. Sheldon, Polymer 3, 27 (1962).

130. L. Cottram, R. P. Sheldon, Advan. Polymer Sci. Technol. Soc. Chem. Ind. Monogr. 26, 65 (1967).

131. Yu. K. Godovskii, V. Yu. Levin, G. L. Slonimskii, A. A. Zhdanov, K. A. Andrianov, Polymer Sci. (USSR) (English Transl.) 11, 2778 (1969).

132. F. D. Hartley, F. W. Lord, L. B. Morgan, Ric. Sci. Suppl. A25, 577 (1955).

133. P. W. Allen, Trans. Faraday Soc. 48, 1178 (1952).

134. M. Inoue, J. Polymer Sci. 55, 753 (1961).

135. P. Cefelin, M. Chmelir, O. Wichterle, Collection Czech. Chem. Commun. 25, 1267 (1960).

136. F. Rybnikár, Collection Czech. Chem. Commun. 27, 106 (1962).

137. M. Inoue, J. Polymer Sci., A1, 2013 (1963).

138. J. Tomka, J. Sebenda, O. Wichterle, Collection Czech. Chem. Commun. 31, 4341 (1966).

139. B. Kahle, Z. Electrochem. 61, 1318 (1957).

140. L. Mandelkern, F. A. Quinn, Jr., P. J. Flory, J. Appl. Phys. 25, 830 (1954).

141. G. Allegra, A. Pontoglio, I. Pasquon, Rend. Ist. Lombardo Sci. Lettere A95, 335 (1961).

142. F. P. Price, J. Phys. Chem. 64, 169 (1960).

143. W. Banks, J. N. Hay, A. Sharples, G. Thompson, Polymer 5, 163 (1964).

144. T. Naono, J. Sci. Hiroshima Univ. A24, 653 (1960).

145. H. D. Keith, F. J. Padden, Jr., J. Appl. Phys. 35, 1286 (1964).

146. T. Suzuki, A. J. Kovacs, Polymer J. 1, 82 (1970).

147. F. P. Price, J.   Chem. Soc. 74, 311 (1952).

148. A. S. Kenyon, R. C. Gross, A. L. Wurstner, J. Polymer Sci. 40, 159 (1959).

149. S. F. Dzhaliov, J. Phys. Chem. (USSR) 39, 10, 1376 (1965).

150. S. L. Aggarwal, L. Marker, W. L. Kollar, R. Geroch, J. Polymer Sci., A2, 4, 715 (1966).

151. J. H. Magill, Makromol. Chem. 86, 283 (1965).

152. B. von Falkai, W. Rellensmann, Makromol. Chem. 75, 112 (1964).

153. M. Takayanagi, N. Kusumoto, J. Chem. Soc. Japan (Ind. Chem. Sect.) 62, 587 (1959).

154. B. B. Burnett, W. F. McDevit, J. Appl. Phys. 28, 1101 (1957).

155. J. Tomka, J. Sebenda, Collection Czech. Chem. Commun. 37, 453 (1972).

156. J. H. Magill, Polymer 6, 367 (1965).

157. F. P. Price, "Fundamental Phenomena in the Materials Sciences", Surface Phenomena Chem. Biolog. 3, 85 (1966).

158. E. H. Boasson, I. M. Woestenenk, J. Polymer Sci. 24, 57 (1957).

159. F. Khoury, J. Polymer Sci. 38, 389 (1958).

160. J. V. McLaren, Polymer 4, 175 (1963).

161. C. R. Lindegren, J. Polymer Sci. 50, 181 (1961).

162. S. Hamada, T. Sato, T. Shirai, Bull. Chem. Soc. (Japan) 40, 864 (1967).

163. A. Gaudica, J. H. Magill, Polymer 13, 595 (1972).

ISOMORPHOUS POLYMER PAIRS

G. Allegra
Istituto di Chimica dell'Università
Trieste, Italy

and

I. W. Bassi
Montecatini Edison S.p.A., Centro Ricerche di Milano
Milano, Italy

Contents

A. INTRODUCTION

The following chapter is intended to provide a list of all known pairs of isomorphous monomer units in crystalline polymer systems, together with some basic additional information. This consists of: a) the range of relative concentrations where isomorphism is observed, whenever given; b) the chain conformation, whenever known; c) the literature references.

The isomorphous polymer systems are subdivided according to the scheme already proposed in a review article on the subject (1). Two basic types of macromolecular isomorphism were distinguished, namely that corresponding to the isomorphous monomer units being more or less randomly distributed within the same chains (Type 1), and that where mixed crystals of different homopolymer chains are formed (Type 2). Type 1 was further subdivided into three different classes: 1.1) the monomer units differ in chemical constitution (copolymers in the usual sense) (Table 1.1); 1.2) the monomer units have randomly opposite, non-superimposable configurations (Table 1.2); 1.3) the monomer units, although being stereochemically identical, assume different conformations at random (Table 1.3). For sake of greater generality, we have included in the class 1.2) (Table 1.2) all the cases where the isomorphous units are related either by geometrical stereoisomerism or by random head-to-head, head-to-tail enchainment. As to Type 2, two classes were considered: 2.1) stereochemically different homopolymer chains are present (Table 2.1); 2.2) stereochemically identical chains assume different orientations at random within the crystals (Table 2.2).

As for the subdivision of the data within each separate heading (1.1, 1.2 etc.), we have tried to conform as far as possible to the standard criterion adopted in this Handbook. Each monomer pair has been assigned to the class corresponding to the higher ranking monomer.

Several vinyl copolymers show macromolecular isomorphism when the side groups are sterically ordered. If, as it always happens in the examples known hitherto, the order is of the isotactic type, we have called them "isotactic copolymers".

Whenever the conformation of the chain skeleton may be described as a regular helix, we have used the symbolism adopted by R. L. Miller in the chapter "Crystallographic Data for Various Polymers" of this Handbook. As an example, the helical conformation of the crystalline isotactic copolymers of 3-methyl-1-butene and 4-methyl-1-pentene (see ref. 14), corresponding to 4 monomer units in one turn, each unit having two skeletal atoms, is designated as $2^{\star}4\text{-}1$. However, referring to the general notation $n^{\star}p\text{-}q$, we have not considered as helical those conformations where both p and q are equal to unity (i.e. one repetitive unit in one turn). In these instances, the crystalline polymer approaching at best the conformation under examination is mentioned, with the appropriate reference.

It may happen that, for a given relative concentration of the two monomer units, more than one crystalline phase is observed. Under these circumstances, the symbols of all the corresponding chain conformations are given together.

For a full discussion of macromolecular isomorphism, see ref. 1) and some of the papers quoted therein.

B.   TABLES OF ISOMORPHOUS PAIRS OF MONOMER UNITS

Monomers Giving Isomorphous Copolymers	Mol % of First Component	Chain Conformation	References
1. Isomorphous Units within the Same Macromolecule			
1.1 Isomorphous Units with Different Chemical Constitution			
1.1.1 Main-chain acyclic carbon copolymers			
Copoly(dienes)			
1,3-Butadiene/1,3-n-pentadiene (trans-1,4)	100-0	Similar to that present in trans-1,4-poly-(butadiene), high temp. modification (see ref. 3)	4
trans-1-Pentenamer/trans-1-octenamer	100-0	Mainly planar, with skew rotations adjacent to the double bonds (see ref. 5)	6,7
trans-1-Octenamer/trans-1-dodecenamer	100-0	Mainly planar, with skew rotations adjacent to the double bonds (see ref. 5)	6,7
Copoly(alkenes)			
Ethylene/propylene	100-~75	Planar zig-zag	8,9,10,11
n-1-Butene/3-methyl-1-butene	100-0	Isotactic copolymer - (2★4-1) and (2★29-8)	12
n-1-Butene/n-1-hexene	100-~30	Isotactic copolymer - (2★3-1) and (2★29-8)	13
n-1-Butene/4-methyl-1-pentene	100-~75	Isotactic copolymer - (2★3-1) and (2★29-8)	13
	~75-~50	(2★29-8) and (2★7-2)	
	~50-0	2★7-2	
n-1-Butene/n-1-octene	100-~50	Isotactic copolymer - (2★3-1) and (2★29-8)	13
n-1-Butene/n-1-decene	100-~90	Isotactic copolymer - (2★3-1) and (2★29-8)	13
n-1-Butene/n-1-dodecene	100-~90	Isotactic copolymer - (2★3-1) and (2★29-8)	13
3-Methyl-1-butene/4-methyl-1-pentene	100-~50	Isotactic copolymer - 2★4-1	14
	~50-0	(2★4-1) and (2★7-2)	
n-1-Pentene/4-methyl-1-pentene	100-~50	Isotactic copolymer - 2★4-1	15,16,17
	~50-0	2★7-2	
4-Methyl-1-pentene/n-1-hexene	100-~20	Isotactic copolymer - 2★7-2	15,16,17
4-Methyl-1-pentene/4-methyl-1-hexene	100-0	Isotactic copolymer - 2★7-2	18
4-Methyl-1-pentene/n-1-octene	100-~70	Isotactic copolymer - 2★7-2	15
4-Methyl-1-pentene/n-1-decene	100-~85	Isotactic copolymer - 2★7-2	15
Copoly(vinyls)			
Ethylene/vinyl alcohol	(not given)	Planar zig-zag	19
Ethylene/(ethylene + CO)	100-0	Planar zig-zag	20,21
Randomly chlorinated poly(ethylene)	up to 5 Cl atoms per 100 C atoms	Planar zig-zag	22
Randomly chlorinated isotactic poly(propylene)	up to 5 Cl atoms per 100 C atoms	2★3-1	22
Vinyl fluoride/vinylidene fluoride	100-0	Planar zig-zag	23
Vinyl fluoride/tetrafluoroethylene	100-20	Planar zig-zag	23
Isopropyl vinyl ether/sec-butyl vinyl ether	100-0	Isotactic copolymer - 2★17-5	18
Copoly(styrenes)			
Styrene/p-methylstyrene	100-~50	Isotactic copolymer - 2★3-1	24,25
Styrene/o-methylstyrene	100-~80	Isotactic copolymer - 2★3-1	24,25
Styrene/o-fluorostyrene	100-0	Isotactic copolymer - 2★3-1	24,25
Styrene/p-fluorostyrene	100-~50	Isotactic copolymer - 2★3-1	24,25
	~50-0	- 2★4-1	24,25
Styrene/p-ethylstyrene	100-~85	Isotactic copolymer - 2★3-1	24,25
Styrene/p-chlorostyrene	100-~85	Isotactic copolymer - 2★3-1	24,25
Styrene/p-bromostyrene	100-~85	Isotactic copolymer - 2★3-1	24,25
1.1.2 Main-chain heteroatom copolymers			
Copoly(aldehydes)			
Acetaldehyde/propionaldehyde	100-0	Isotactic copolymer - 2★4-1	26

Monomers Giving Isomorphous Copolymers	Mol % of First Component	Chain Conformation	References
Copoly(aldehydes) (Cont'd.)			
Acetaldehyde/n-butyraldehyde	100-0	Isotactic copolymer - 2★4-1	26
Acetaldehyde/isobutyraldehyde	100-0	Isotactic copolymer - 2★4-1	26
Propionaldehyde/n-butyraldehyde	100-0	Isotactic copolymer - 2★4-1	26
n-Butyraldehyde/isobutyraldehyde	100-0	Isotactic copolymer - 2★4-1	26
n-Butyraldehyde/n-heptanal	100-50	Isotactic copolymer - 2★4-1	26
Copoly(carbonates)			
Hydroquinone/chlorohydroquinone/$COCl_2$	100-0	Essentially extended	27
Toluhydroquinone/chlorohydroquinone/$COCl_2$	100-0	Essentially extended	27
Copoly(esters) and copoly(ethers)			
Diethylene glycol/decamethylene glycol/adipic acid	20-0	Essentially extended (planar zig-zag for the $-CH_2$-sequences)	28
Tetramethylene glycol/decamethylene glycol/adipic acid	20-0	Essentially extended (planar zig-zag for the $-CH_2$-sequences)	28
Pentamethylene glycol/decamethylene glycol/adipic acid	20-0	Essentially extended (planar zig-zag for the $-CH_2$-sequences)	28
Ethylene glycol/terephthalic acid/p-(3-hydroxypropoxy) benzoic acid	100-0	Essentially extended (planar zig-zag for the $-CH_2$-sequences)	29
Hexamethylene glycol/decamethylene glycol/adipic acid	100-0	Essentially extended (planar zig-zag for the $-CH_2$-sequences)	30
Hydroquinone/decamethylene glycol/adipic acid	20-0	Essentially extended (planar zig-zag for the $-CH_2$-sequences)	28
Decamethylene glycol/adipic acid/sebacic acid	100-80	Essentially extended (planar zig-zag for the $-CH_2$-sequences)	28
Hexamethylene glycol/decamethylene glycol/adipic acid/ sebacic acid	100-0	Essentially extended (planar zig-zag for the $-CH_2$-sequences)	30
Ethylene glycol/sebacic acid/undecandioic acid/dodecan-dioic acid	Equimolar proportions of the three acids	Essentially extended (planar zig-zag for the $-CH_2$-sequences)	31
Copoly(amides) and copoly(peptides)			
γ-Methyl-L-glutamate/γ-benzyl-L-glutamate	100-0	3★18-5 (α-helix)	32
Caprolactam/4-aminomethylbenzoic acid	100-∼70	Essentially extended (planar zig-zag for the $-CH_2$-sequences)	33
Caprolactam/4-aminomethylcyclohexanecarboxylic acid	100-0	Essentially extended (planar zig-zag for the $-CH_2$-sequences)	33,34
n-Methylenediamine/adipic acid/terephthalic acid n-between 6 and 12	100-0	Essentially extended (planar zig-zag for the $-CH_2$-sequences)	35,36,37,38
Heptamethylenediamine/adipic acid/terephthalic acid	100-0	Essentially extended (planar zig-zag for the $-CH_2$-sequences)	39
Bis(3-aminopropyl)ether/heptamethylenediamine/adipic acid	100-0	Essentially extended (planar zig-zag for the $-CH_2$-sequences)	39
Bis(3-aminopropyl)ether/heptamethylenediamine/terephthalic acid	100-0	Essentially extended (planar zig-zag for the $-CH_2$-sequences)	39
Bis(3-aminopropyl)ether/adipic acid/terephthalic acid	100-0	Essentially extended (planar zig-zag for the $-CH_2$-sequences)	39
Hexamethylenediamine/N,N'-diisobutylhexamethylenedi-amine/sebacic acid	100-0	Essentially extended (planar zig-zag for the $-CH_2$-sequences)	40
Hexamethylenediamine/p-phenylenedipropionic acid/(p-carb-oxymethoxy)phenylpropionic acid	100-0	Essentially extended (planar zig-zag for the $-CH_2$-sequences)	41
Bisaminomethylcyclohexane, cis/trans/adipic acid	100-30	Essentially extended (planar zig-zag for the $-CH_2$-sequences)	42
Bisaminomethylcyclohexane, cis/trans/sebacic acid	100-30	Essentially extended (planar zig-zag for the $-CH_2$-sequences)	42
Bisaminocyclohexane, cis/trans/decandioic acid	100-30	Essentially extended (planar zig-zag for the $-CH_2$-sequences)	42

	Polymer	Type of Stereoisomerism	Chain Conformation	References
1.2	Isomorphous Units with Different Configuration			
	Poly(vinyl alcohol)	Atactic polymer (i.e. side groups not placed in steric order)	Planar zig-zag	19,32,43
	Poly(vinyl fluoride)	Atactic polymer	Planar zig-zag	44
	Poly(trifluorochloroethylene)	Atactic polymer	2★14-1 (see ref. 45); 2★16-1 (see ref. 46)	32
	Poly(ethylene-co-trifluorochloroethylene) (alternating copolymer)	Atactic polymer + head-to-head random inversions	Nearly planar zig-zag ("kinked" conformation) (see ref. 47)	47

See also the last three examples of the preceding table.

	Polymer	Type of Stereoisomerism	Chain Conformation	References
1.3	Isomorphous Units with Different Conformation			
	trans-1,4-Poly(butadiene)	Rotational disorder arount C-C single bonds adjacent to double bonds	Essentially extended (trans rotations for CH_2-CH_2 bonds, skew and	3,48,49,50
	cis-1,4-Poly(2-methylbutadiene)	Rotational disorder around C-C single bonds adjacent to double bonds	cis rotations for CH-CH_2 bonds)	50
	Poly(S-3-methyl-1-pentene)	Rotational disorder involving the side groups	2★4-1	50,51

Monomer Giving Isomorphous Copolymers	Mol % of First Component	Chain Conformation	References

2. Isomorphism of Macromolecules

	Monomer Giving Isomorphous Copolymers	Mol % of First Component	Chain Conformation	References
2.1	Isomorphism of Macromolecules with Different Chemical Constitution			
	Poly(4-methylpentene)/poly(4-methylhexene)	100-~75 and ~20-0	Isotactic polymers - 2★7-2	18
	Poly(isopropyloxyethylene)/poly(2-butyloxyethylene) (poly(i-propyl vinyl ether)/poly(sec-butyl vinyl ether))	100-0	Isotactic polymers - 2★17-5	18
	Poly(vinyl fluoride), atactic/poly(vinylidene fluoride)	100-0	Planar zig-zag	23
	Poly(styrene)/poly(styrene-co-p-methylstyrene) (70:30 mol %)	50 % of polystyrene	Isotactic copolymer - 2★3-1	25

	Polymer	Chain Conformation	References
2.2	Isomorphism of Identical Macromolecules with Randomly Different Orientations in the Crystals		

In all the cases where this case of isomorphism has been either proven or postulated, the possible orientations of the macromolecules (mostly "up" and "down" with reference to the orientation of the side groups) are inherently equivalent. Therefore, since they must be present in the same amount, no indication about the percentage composition is given.

Polymer	Chain Conformation	References
Poly(propylene)	Isotactic - 2★3-1	52
Poly(n-butene-1)	Isotactic - 2★3-1	53
1,2-Poly(4-methylpentadiene)	Isotactic - 2★18-5	54
Poly(ethylene-co-cyclopentene), alternating copolymer	Isotactic, all skeletal rotations are trans except those around C-C bonds included in the ring, which are gauche	55
Poly(oxyethylidene), (poly(acetaldehyde))	Isotactic - 2★4-1	56
Poly(oxypropylidene), (poly(propionaldehyde))	Isotactic - 2★4-1	56
Poly(oxy-n-butylidene), (poly(n-butyraldehyde))	Isotactic - 2★4-1	56

C. REFERENCES

1. G. Allegra, I. W. Bassi, Adv. Polymer Sci. 6, 549 (1969).

2. Macromolecules, Appendix 1, 1, 193 (1968).

3. P. Corradini, Polymer Letters 7, 211 (1969).

4. G. Natta, L. Porri, A. Carbonaro, P. Lugli, Makromol. Chem. 53, 52 (1969).

5. G. Natta, I. W. Bassi, G. Fagherazzi, European Polymer J. 5, 239 (1969).

6. G. Dall' Asta, G. Motroni, G. Carella, Ital. Pat. 773,657 (1966).

7. G. Motroni, G. Dall'Asta, I. W. Bassi, European Polymer J. in press

8. M. J. Richardson, P. J. Flory, J. B. Jackson, Polymer 4, 221 (1963).

9. C. H. Baker, L. Mandelkern, Polymer 7, 7 (1966).

10. C. H. Baker, L. Mandelkern, Polymer 7, 71 (1966).

11. I. W. Bassi, P. Corradini, G. Fagherazzi, A. Valvassori, European Polymer J. 6, 709 (1970).

12. A. Turner Jones, Polymer Letters 3, 591 (1965).

13. A. Turner Jones, Polymer 7, 23 (1966).

14. F. P. Reding, E. R. Walter, J. Polymer Sci. 37, 555 (1959).

15. A. Turner Jones, Polymer 6, 249 (1965).

16. T. W. Campbell, J. Appl. Polymer Sci. 5, 184 (1961).

17. W. A. Hewett, F. E. Weir, J. Polymer Sci., A, 1, 1239 (1963).

18. G. Allegra, I. W. Bassi, C. Carlini, E. Chiellini, G. Montagnoli, Macromolecules 2, 311 (1969).

19. C. W. Bunn, S. Peiser, Nature 159, 161 (1947).

20. M. M. Brubaker, D. D. Coffman, H. H. Hochn, J. Am. Chem. Soc. 74, 1509 (1952).

21. Y. Chatani, J. Tazikawa, S. Murahashi, Y. Sakata, Y. Nishimura, J. Polymer Sci. 55, 811 (1961).

22. N. A. Platè, Tran Kheu, U. P. Shibaev, J. Polymer Sci., C, 16, 1133 (1967).

23. G. Natta, G. Allegra, I. W. Bassi, D. Sianesi, G. Caporiccio, E. Torti, J. Polymer Sci., A, 3, 4263 (1965).

24. G. Natta, Makromol. Chem. 35, 93 (1960).

25. G. Natta, P. Corradini, D. Sianesi, D. Morero, J. Polymer Sci. 51, 527 (1961).

26. A. Tanaka, Y. Hozumi, K. Hatada, S. Endo, R. Fujishije, Polymer Letters 2, 181 (1964).

27. I. D. Rubin, J. Polymer Sci., A, 1, 1645 (1963).

28. R. B. Evans, H. R. Mighton, P. J. Flory, J. Am. Chem. Soc. 72, 2018 (1950).

29. M. Ishibashi, Polymer 5, 103 (1964).

30. G. J. Howard, S. Knutton, Polymer 9, 527 (1969).

31. C. J. Fuller, J. Am. Chem. Soc. 70, 421 (1948).

32. C. W. Bunn, J. Appl. Phys. 25, 820 (1954).

33. M. Levine, S. C. Temin, J. Polymer Sci. 49, 241 (1961).

34. F. R. Prince, E. M. Pearce, R. J. Fredericks, J. Polymer Sci., A-1, 8, 3533 (1970).

35. H. Plimmer, R. J. W. Reynolds, L. Wood, H. Hargreaves, I.C.I. Ltd. British Patent Appl. 604/49.

36. O. B. Edgar, R. H. Hill, J. Polymer Sci. 8, 1 (1952).

37. E. D. Harvey, F. J. Hybart, Polymer 12, 711 (1971).

38. A. J. Yu, R. D. Evans, J. Polymer Sci. 42, 249 (1960).

39. F. B. Cramer, R. G. Beaman, J. Polymer Sci. 21, 237 (1956).

40. E. L. Wittbecker, R. C. Houtz, W. W. Watkins, Ind. Eng. Chem. 40, 875 (1948).

41. F. C. Tranter, J. Polymer Sci., A, 2, 4289 (1964).

42. F. R. Prince, R. J. Fredericks, Macromolecules 5, 168 (1972).

43. C. W. Bunn, Nature 161, 929 (1948).

44. G. Natta, I. W. Bassi, G. Allegra, Acc. Naz. Lincei Rend. 31, 350 (1961).

45. H. S. Kaufman, J. Am. Chem. Soc. 75, 1477 (1953).

46. C. Y. Liang, S. Krimm, J. Chem. Phys. 25, 563 (1956).

47. J. P. Sibilia, L. G. Roldan, S. Chandrasekaran, J. Polymer Sci. 10, A-2, 549 (1972).

48. G. Natta, P. Corradini, L. Porri, Acc. Naz. Lincei Rend. 20, 728 (1956).

49. S. Iwayanagi, H. Nakane, I. Sakurai, International Symp. Macromol. Chem. Tokyo 1966, Preprint 2-4-11.

50. P. Corradini, V. Petraccone, "Interactions between Polymer Molecules", Lecture held at Ciba Foundation Symposium, London 14-16 March, 1972.

51. V. Petraccone, P. Ganis, P. Corradini, G. Montagnoli, European Polymer J. 8, 99 (1972).

52. G. Natta, P. Corradini, Nuovo Cimento Suppl. 15, 40 (1960).

53. G. Natta, P. Corradini, I. W. Bassi, Nuovo Cimento Suppl. 15, 52 (1960).

54. G. Natta, P. Corradini, I. W. Bassi, G. Fagherazzi, European Polymer J. 4, 297 (1968).

55. G. Natta, G. Allegra, I. W. Bassi, P. Corradini, P. Ganis, Accad. Naz. Lincei Rend. 36, 433 (1964).

56. G. Natta, P. Corradini, I. W. Bassi, J. Polymer Sci. 51, 505 (1961).

COMPATIBLE POLYMERS

L. Bohn

Hoechst AG

Frankfurt/Main, Germany

Compatibility

Compatibility as reviewed in the following tables will refer to the miscibility on an intimate molecular scale of polymers in the solid state. Such miscibility will only take place if the Gibbs free energy of mixing $\Delta F = \Delta H - T\Delta S$ is negative. The entropy term $T\Delta S$ is unsignificant in the mixing of high molecular weight species. The enthalpy of mixing ΔH is normally positive to such an extent as to overcompensate for the entropy term, resulting generally in an unfavorable energy of mixing for polymer blends. Real compatibility is therefore a rare event, especially in the solid state. For more details see the recently published comprehensive review by Krause (1) and the earlier review by Bohn (2).

Criteria for Compatibility

Two experimental methods seem to be most reliable:

(1) The "Transition Temperature Method"

   With polymeric components having glass transition temperatures or melting points far enough apart to be resolvable by DTA, DSC, dynamic mechanical, dilatometric or other suitable methods, compatibility is indicated by the shift or disappearance of the single components. Compatible blends show generally only one glass transition. This method seems to be the most unambiguous.

   Polymer pairs listed in Table I have been shown to be compatible by this method.

(2) The "Optical Method"

   Clarity of a film cast from a homogeneous solution of a polymer mixture which exhibits no heterogeneity under considerable magnification. This method is not so unambiguous, for incompatible blends may form transparent films when both components have the same refractive index. Blends with different refractive indices may give transparency by two-layered films or due to a highly dispersed internal phase having dimensions well below those of the wavelength of light.

   Table II contains polymer mixtures which gave clear films but have not been studied by method (1).

Method of Mixing

There are three principal methods to prepare a solid polymeric mixture:

(a) Mixing of polymers directly in the condensed state - usually in the melt - with the aid of mixers, mills, extruders and the like.

(b) Mixture of polymers in solution, from which the solid mixture is obtained by precipitation or drying.

(c) Mixture of one polymer with another monomer or other active low molecular weight preproduct, leading to a high molecular polymeric composite after polymerization. The polymeric components may be linked by primary valences in this process giving graft or block copolymers.

Table I: Polymers Compatible as Solids

Table I contains only polymer mixtures the compatibility of which has been demonstrated by the "transition temperature method". Most of these mixtures show low differences in the solubility parameters of the components (37). At the outset are listed 8 rubber blends, thereafter rigid composites roughly in order of increasing polarity.

Component 1	Component 2	Weight-% of Component 1	Method of Mixing[+)]	$\Delta \delta$[x)]	Notes	References
Poly(butadiene) (cis-1,4)	Poly(butadiene-co-styrene) (75/25)	20-80	S, vulc.	0.16	a	3,4,32,29
(emulsion)	Poly(butadiene-co-styrene) (75/25)	20-80	S, vulc.	0.34	a	3
(ca. 30 % cis)	Poly(butadiene-co-styrene) (75/25)	20-80	M, vulc.	0.27	b	5,38
Poly(isoprene) (natural rubber)	Poly(butadiene-co-styrene) (75/25)	50	M, vulc.	0.1		5
	Poly(butadiene) (ca. 50 % -1,2-)	25-75	M, vulc.	≤ 0.71		6
	Poly(butadiene)	ca. 50	B	≤ 0.31	c	7,32
Poly(butadiene-co-styrene)	Same, but diff. in composition ≤ 20 % styrene	50	S, vulc.	≤ 0.41		8
Poly(butadiene-co-acrylonitrile) (ca. 82/18)	Same (60/40)	25-75	M, vulc.	1.6-2.0		6
Poly(styrene-co-acrylonitrile) (ca. 80/20)	Poly(butadiene-co-acrylonitrile (ca. 65/35)	70-100	M	0.6		9,10,11
Poly(styrene)	Poly(butadiene-co-styrene) (75/25)	≤ 40	M	1.6-2.0		9
	Poly(vinyl methyl ether)	20-75	S	-		12
	Poly(α-methylstyrene)	44-50	B	-	d	13
	Poly(2,6-dimethylphenylene oxide)	25-75	M	-	e	14,15
Poly(α-methylstyrene)	Poly(2,6-dimethylphenylene oxide)	all comp.	-	-		16
Poly(methyl methacrylate) (isotactic)	Same (syndiotactic)	all comp.	S	-	f	17,18
	Poly(vinylidene fluoride)	≥ 66	M	-		19
Poly(ethyl methacrylate)	Poly(vinylidene fluoride)	≥ 50	M	-		19
Poly(methyl methacrylate-co-butyl-acrylate)	Same, diff. in composition ≤ 15-20 mol % methyl methacrylate	50	S	≤ 0.41		20

Component 1	Component 2	Weight-% of Component 1	Method of Mixing [+]	Δ δ [x]	Notes	References

Table I (Cont'd.)

Component 1	Component 2	Weight-% of Component 1	Method of Mixing	Δ δ	Notes	References
Poly(isopropyl methacrylate)	Poly(isopropyl acrylate)	all comp.	S	ca. 0.41		17
Poly(vinyl acetate)	Poly(methyl acrylate)	50	S, M	0.6-1.4		21,22,35
	Poly(vinyl chloride-co-vinyl acetate (ca. 90/10)	40-50	M	0.4		23
Poly(vinyl chloride)	Poly(butadiene-co-acrylonitrile) (ca. 30 % acrylonitrile)	90-20	S, M	0.8-1.2		9,11,24,25, 26,27
	Poly(ε-caprolactone)	≥ 50	M	-		28
	Poly(ethylene-co-vinyl acetate-co-sulfur dioxide) (72.7/18.5/8.8 molar comp.)	≥ 60	M	-		40
Poly(ethylene), chlorinated (62 % Cl)	Same (66 % Cl)	all comp.	S, M	0.2-0.4		29
Poly(butadiene-co-acrylonitrile) (60/40)	Cellulose acetate butyrate	< 20-90 <	M	ca. 2.0		30
Ebonite	Poly(sulfide)	95-80	M	-	k	31

[+] M = mixing in the melt; S = mixture from solution; B = block copolymer; vulc. = vulcanized
[x] Δ δ = difference in solubility parameters in $[J/cm^3]^{1/2}$; see table "Solubility Parameters" this Handbook.

Table II: Polymer Mixtures Giving Clear Films

Table II contains mixtures which gave clear films from solution. Compositions were in all cases about 1/1. The statement of compatibility only from the criterion 2, the "optical method", is not unambiguous, as mentioned above. Therefore the differences of solubility parameter Δ δ and of refractive index Δn have been estimated. General experience shows that transparency by matching the refractive indices requires $\Delta n \times 10^2 \leq 0.5$ to 1.

Component 1	Component 2	Δ δ [x]	$\Delta n \times 10^2$ [+]	Notes	References
Poly(styrene-co-acrylonitrile)	Same, but diff. in composition ca. 4 % acrylonitrile	0.41	0.3	i	33
Nitrocellulose	Cellulose acetate propionate	0-0.8	3-5	g	34
	Ethyl cellulose	≥ 0.41	2-3.5	g	34
	Poly(vinyl acetate)	≥ 2.3	3-4.5		34,36
	Poly(methyl methacrylate)	≥ 2.5	1-2.5		34,36
	Poly(styrene-co-acrylonitrile) (80/20)	0-0.8	6-7	g	34
	Poly(ester urethane)	(ca. 1.0)	-		34
	Poly(styrene-co-methyl methacrylate) (25/75)	> 2.5	0	h	34
Poly(styrene-co-methyl methacrylate) (25/75)	Poly(methyl methacrylate)	ca. 0	2.5		34
	Poly(epichlorohydrin)	ca. 0	-		34
Poly(vinyl chloride-co-vinyl acetate)	Poly(ester urethane)	ca. 1.0 (?)	-		34
	Poly(ether urethane)	ca. 1.0 (?)	-		34
	Poly(butadiene-co-acrylonitrile)	0-1.0	0-1		34
Poly(vinyl methyl ether-co-maleic anhydride)	Poly(methyl methacrylate)	-	-		34
	Poly(vinyl acetate)	-	-		34
Benzyl cellulose	Poly(styrene)	ca. 4.1	0-2	h	36
Poly(styrene)	Poly(o-methylstyrene)	(small)	0.3		35
Poly(o-methylstyrene)	Poly(p-methylstyrene)	0.6-1.4	(small)		35

[x] Difference of solubility parameters in $[J/cm^3]^{1/2}$; see table "Solubility Parameters" this Handbook.
[+] Difference of refractive indices; see corresponding table in this Handbook.

Notes

a) According to (3) and (4) compatibility only after vulcanization, but indication of partial compatibility in unvulcanized mixtures by (32) and (39).
b) Component 2 only compatible with ≤ 25 % styrene (38).
c) Rubbery sequences of block copolymers and block terpolymers with polystyrene.
d) High molecular weight homopolymers incompatible (13).
e) Mixed powders only moulded in compression (14). This may be the reason for some differences between the results from DSC and dynamic mechanical methods.

f) Mixture of isotactic poly(methyl methacrylate)/normal poly(methyl methacrylate) precipitated from solution showed two phases (18).
g) Mixtures seem compatible since high Δn and low Δ δ.
h) Mixtures seem to be transparent only due to the similarity in the refractive indices of the components.
i) Mixtures blended in melt.
k) Component 1: Poly(butadiene-co-styrene) (75/25) vulcanized with 22 weight-% sulfur.
Component 2: Thiokol FA.

REFERENCES

1. S. Krause, J. Macromol. Sci.-Rev. Macromol. Chem. C 7 (2), 251 (1972).

2. L. Bohn, Rubber Chem. Technol. 41, 495 (1968); Kolloid-Z. - Z. Polymere 213, 55 (1966).

3. N. Yoshimura, K. Fujimoto, Rubber Chem. Technol. 42, 1009 (1969).

4. K. Fujimoto, N. Yoshimura, Rubber, Chem. Technol. 41, 669 (1968).

5. P. J. Corish, Rubber Chem. Technol. 40, 324 (1967).

6. G. M. Bartenev, G. S. Kongarov, Vysokomolekul. Soedin. 2, 1692 (1960); Rubber Chem. Technol. 36, 668 (1963).

7. R. J. Angelo, R. M. Ikeda, M. L. Wallach, Polymer 6, 141 (1965).

8. G. Kraus, K. W. Rollmann, ACS Polymer Preprints 11, 377 (1970); ACS Adv. Chem. Ser. 99, 189, Washington 1971.

9. W. Breuers, W. Hild, H. Wolff, Plaste Kautschuk 1, 170 (1954).

10. E. B. Atkinson, R. F. Eagling, SCI-Monograph No. 5, 197, London 1959.

11. N. E. Davenport, L. W. Hubbard, M. R. Pettit, Brit. Plastics 32, 549 (1959).

12. M. Bank, J. Leffingwell, C. Thies, ACS Polymer Preprints 10, 622 (1969).

13. M. Baer, J. Polymer Sci., A, 2, 417 (1964).

14. J. Stoelting, F. E. Karasz, W. J. Macknight, Polymer Eng. Sci. 10, 133 (1970); ACS Adv. in Chem. Ser. 99, 29, Washington 1971.

15. H. E. Bair, Polymer Eng. Sci. 10, 222 (1970).

16. E. P. Cizek, U.S. P. 3,383,435 - 14.5.1968 - General Electric Co.

17. S. Krause, N. Roman, J. Polymer Sci., A, 3, 1631 (1965).

18. R. C. Bauer, N. C. Bletso, ACS Polymer Preprints 10, 632 (1969).

19. J. S. Noland, N. N. Hsu. R. Saxon, J. M. Schmitt, ACS Polymer Preprints 11, 355 (1970).

20. F. Kollinsky, G. Markert, Makromol. Chem. 121, 117 (1969); ACS Adv. Chem. Ser. 99, 175, Washington 1971.

21. L. J. Hughes, G. L. Brown, J. Appl. Polymer Sci. 5, 580 (1961).

22. L. E. Nielsen, "Mechanical Properties of Polymers", John Wiley, N. Y. 1962, p. 177.

23. H. Wolff, Plaste Kautschuk 4, 244 (1957).

24. L. E. Nielsen, J. Am. Chem. Soc. 75, 1435 (1953).

25. M. Takayanagi, Mem. Fac. Eng. Kyushu Univ. 23 (1), 11 (1963).

26. M. Matsuo, A. Ueda, Y. Kondo, Polymer Eng. Sci. 10, 253 (1970).

27. R. A. Reznikova, A. D. Zaionchkovski, S. S. Voyutsky, Zh. Tekh. Fiz. 25, 1045 (1955).

28. J. V. Koleske, R. D. Lundberg, J. Polymer Sci., A-2, 7, 795 (1969).

29. H. J. Oswald, E. T. Kubu, SPE-Trans. 3, 168 (1963).

30. R. M. Asimova, P. V. Koslov, V. A. Kargin, S. M. Vtorygin, Vysokomolekul. Soedin. 4, 554 (1962).

31. T. H. Meltzer, W. J. Dermody, A. V. Tobolsky, J. Appl. Polymer Sci. 8, 765 (1964).

32. M. C. Morris, Rubber Chem. Technol. 40, 341 (1967).

33. G. E. Molau, J. Polymer Sci. B-3, 1007 (1965).

34. R. J. Petersen, R. D. Corneliussen, L. T. Rozelle, ACS Polymer Preprints 10, 385 (1969).

35. R. J. Kern, R. J. Slocombe, J. Polymer Sci. 15, 183 (1955).

36. A. Dobry, F. Boyer-Kawenoki, J. Polymer Sci. 2, 90 (1947).

37. See chapter "Solubility Parameter Values" in this Handbook

38. K. Fujimoto, N. Yoshimura, Rubber Chem. Technol. 41, 1109 (1968).

39. D. Reichenbach, Kautschuk, Gummi, Kunststoffe 24, 387 (1971).

40. J. J. Hickman, R. M. Ikeda, J. Polymer Sci. 11, 1713 (1973).

HEAT CAPACITY OF HIGH POLYMERS

H. Wilski

Hoechst AG

Frankfurt/Main, Germany

The table contains all data found in the literature for the heat capacity (at constant pressure) at 25°C. Since the table is short, an alphabetical order war preferred. The unit used throughout this article for the heat capacity is kJ/kgK. The value obtained for the heat capacity of a polymeric substance is dependent on the state of the polymer. Whenever it was possible, the state of the polymer was characterized by the density at 25°C and by the following abbreviations:

am = amorphous; ann = annealed; c = (partially) crystalline; q = quenched.

In general stereo-regularity has only an indirect effect on heat capacity through its influence on crystallinity. The following abbreviations were used:

a = atactic; i = isotactic; s = syndiotactic.

The tabulated data for the heat capacity were always taken from the literature cited first under "references". The literature cited does not claim to be complete. An introduction to the theory of heat capacity is given by Wunderlich and Baur (85), the special problems at very low temperatures are discussed by Reese (58).

Because of the steadily increasing importance of differential thermal analysis some typical examples for the temperature dependence of the heat capacity are given graphically. To facilitate a comparision of the different results the same scale was mostly used. As a result, in many cases the melting peak is not completely visible in the diagram.

Polymer		Density $[\text{Mg m}^{-3}] \equiv [\text{g cm}^{-3}]$	Heat Capacity [kJ/kgK]	References
Cellulose fibers			1.25	21
Poly(acrylonitrile)			1.286	47
Poly(allyl methacrylate)			1)	33
Poly(cis-1,3-butadiene)			1.854	16
Poly(trans-1,3-butadiene)			2.402	16
Poly(1-butene)	(i, c)	0.9268	1.783	80,15
Poly(1-butene)	(a, am)	0.8772	2.093	80
Poly(1-butene sulfone)			1.219	17
Poly(n-butyl acrylate)			1.823	37
Poly(n-butyl methacrylate)			1.862	37
Poly(chlorotrifluoroethylene)	(ann)		0.875	35,59
Poly(chlorotrifluoroethylene)	(q)		0.923	
Poly(cyclohexyl methacrylate)			1)	33
Poly(dimethylaminoethyl methacrylate)			1)	33
Poly(dimethyl siloxane) see Poly(oxydimethylsilylene)				
Poly(di(oxyethylene)oxycarbonyliminohexamethyleneiminocarbonyl) (poly(urethane))			2.3 (0°C)	29
Poly(ethyl acrylate)			1.806	37,33
Poly(ethylene) (linear)		0.964	1.855	14,31,34,39,55,59, 60,70,84,86
Poly(ethylene) (branched)		0.921	2.315	14,19,34,55,59,70
Poly(ethylene terephthalate) see Poly(oxyethyleneoxyterephthaloyl)				
Poly(ethyl methacrylate)			1.487	37
Poly(1-hexene sulfone)			1.413	17
Poly(iminoadipoyliminohexamethylene) (ann) (Nylon 6-6)			1.419	75
Poly(imino(1-oxoheptamethylene)) (Nylon 7)			1.840	44
Poly(imino(1-oxohexamethylene)) (Nylon 6)			1.599	49,44
Poly(iminosebacoyliminohexamethylene) (ann) (Nylon 6-10)			1.601	75
Poly(isobutylene)			1.948	26,22
Poly(cis-1,4-isoprene) (natural rubber)			1.917	5,10,81
Poly(methacrylic acid)			1.068	56,68
Poly(methyl acrylate)			1.790	37,68
Poly(methyl methacrylate)	(i)		1.373	52
	(s)		1.381	52
	(a)		1.388	52,37,50,57,72
Poly(4-methyl-1-pentene) (i, c)		0.8325	1.725	42,51
Poly(α-methylstyrene)			1)	38
Poly(oxycarbonyloxy-1,4-phenyleneisopropylidene-1,4-phenylene)			1.181	18,53
(Polycarbonate from bisphenol A) (cryst. and amorphous same value for heat capacity)				
Poly(oxy-1,1-dichloromethyltrimethylene)			1.152	13
Poly(oxy-2,6-dimethyl-1,4-phenylene) (poly(phenylene oxide))			1.241	43
(cryst. and amorphous same value for heat capacity)				

Polymer		Density [Mg m^{-3}] ≡ [g cm^{-3}]	Heat Capacity [kJ/kgK]		References
Poly(oxydimethylsilylene)			1.48		30
Poly(oxyethylene)			1.499		4, 6
Poly(oxyethyleneoxysebacoyl)	(c, ann)	1.167	2.064		87
Poly(oxyethyleneoxyterephthaloyl) (c, ann) (poly(ethylene terephthalate))		1.423	1.103		63, 61
	(am, q)	1.340	1.146		63, 61
Poly(oxy-2-hydroxytrimethyleneoxy-1,4-phenyleneisopropylidene-1,4-phenylene) (Epikote 828)			1.86		2
Poly(oxymethylene)			1.214		34
Poly(oxymethylene) (trioxane copolymer with ethylene oxide)		1.412	1.369		12, 79
Poly(oxypropylene)	(am)		1.913		4
Poly(oxytetramethylene)		1.053	1.721	(0°C)	11
Poly(oxytetramethyleneoxycarbonyliminohexamethyleneiminocarbonyl) (poly(urethane))			1.767	(0°C)	29
Poly(oxytrimethylene)	(c)		1.843	(0°C)	88
Poly(3-phenylquinoxaline-2,7-diylcarbonyl-3-phenylquinoxaline-7,2-diyl-1,4-phenylene-oxy-1,4-phenylene)			1.229		82, 83
Poly(propene sulfone)			1.159		17
Poly(propylene)	(i, c)	0.9103	1.789		76, 15, 54
	(s, c)	0.880	1.805		27, 28
	(a, am)	0.8667	2.349		76, 15, 54
Poly(styrene)	(i, c)	1.072	1.220		1, 15, 41
	(i, am)	1.056	1.195		1, 41
	(a, am)	1.051	1.194		1, 7, 9, 34, 41, 45, 67, 69, 74
Poly(tetrafluoroethylene)	(ann)		0.938	(0°C)	23, 20, 48
Poly(trifluoroethylene)			2)		64
Poly(vinyl acetate)			1.331		71, 69
Poly(vinyl alcohol)			1.51		8, 66
Poly(vinyl butyral)		1.090	1.394		77, 40
Poly(vinyl carbazole)					
Poly(vinyl chloride) (annealed and quenched same value for heat capacity)			0.934		32, 3, 36, 46, 78
Poly(vinylidene chloride)			0.857		46, 65, 73

1) Values only above 400 K;   2) Values only below 120 K

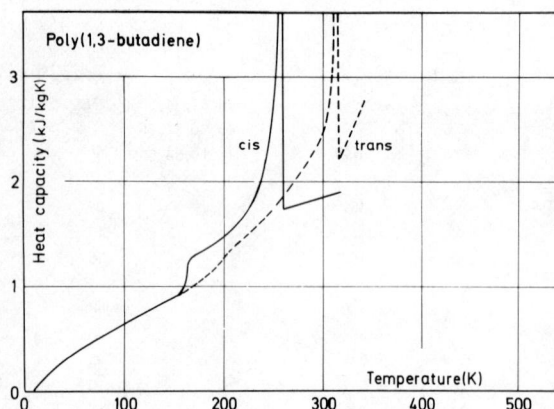

Figure 1: Poly(1,3-butadiene) (16)

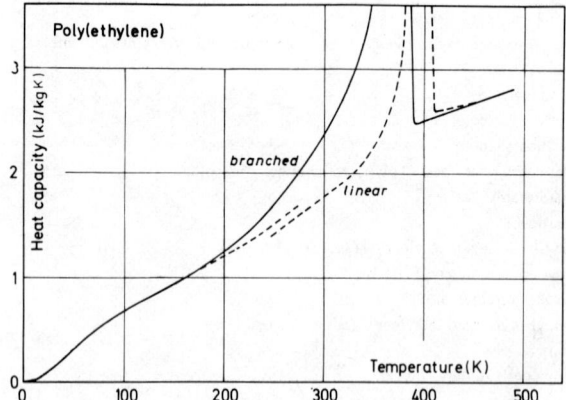

Figure 2: Poly(ethylene).
Low temp. (14), high temp. branched (19), linear (86).
Note the good agreement between the different measurements for branched PE; the difference for the linear PE is a result of different crystallinities of the two resp. samples.

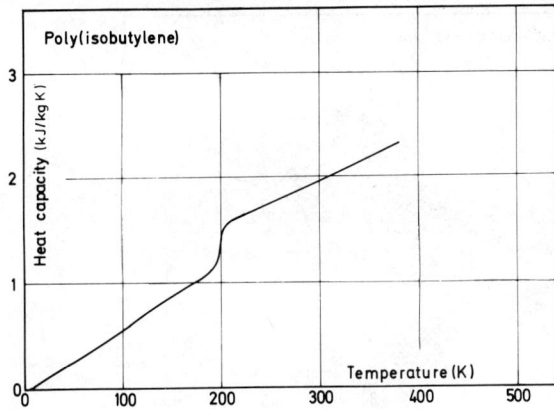

Figure 3 : Poly(isobutylene) (26).

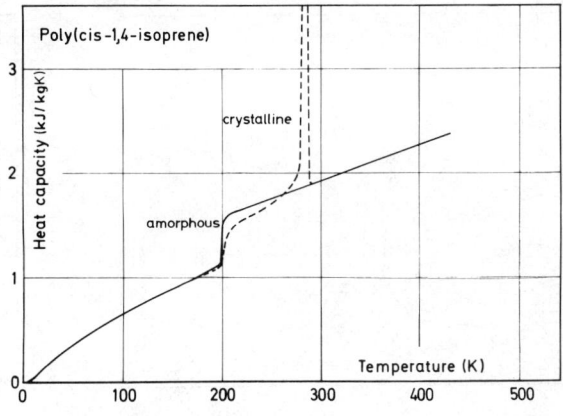

Figure 4 : Poly(cis-1,4-isoprene), natural rubber (5,10,81).

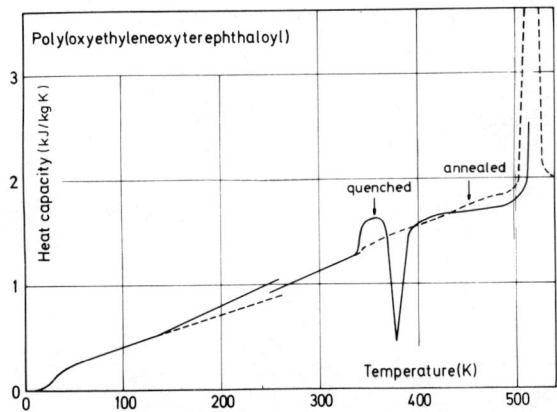

Figure 5 : Poly(oxyethyleneoxyterephthaloyl).
Low temp. (61), high temp. (63). The agreement between the
low temperature data and the high temperature data resp. is not
good.

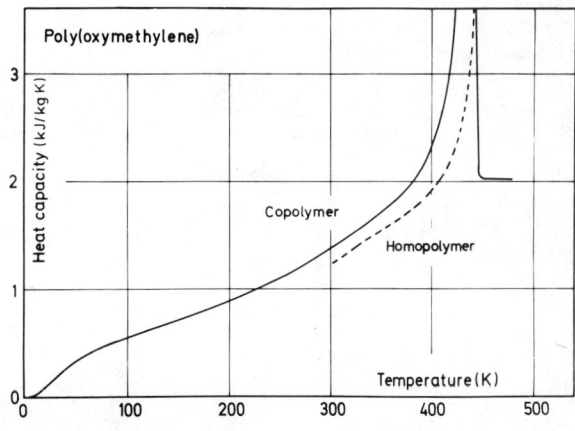

Figure 6 : Poly(oxymethylene).
Homopolymer (34), copolymer low temp. (12), high temp. (79).

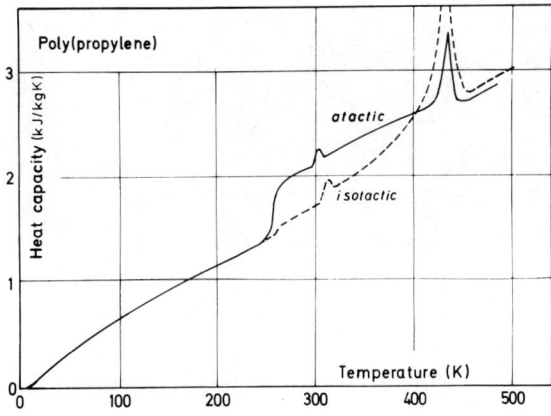

Figure 7 : Poly(propylene).
Low temp. (15), high temp. (54).
Note : The atactic material contains a small fraction of isotactic
(partly crystallized) poly(propylene).

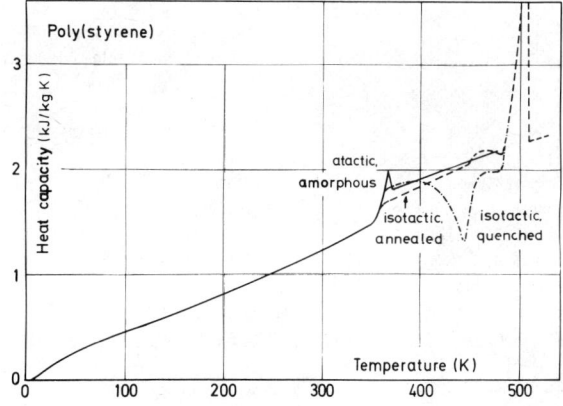

Figure 8 : Poly(styrene).
Low temp. (15), high temp. (41).

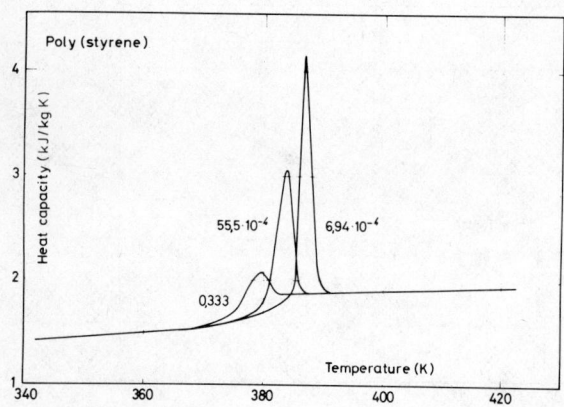

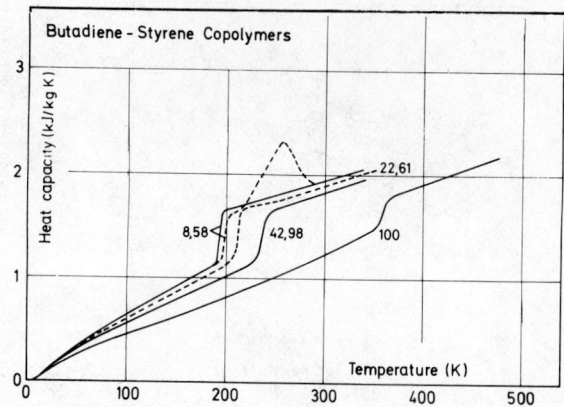

Figure 9: Poly(styrene).

Before the measurement of the heat capacity (with a heating rate of 8 K/min) the different samples were cooled from the melt with different cooling rates. The figures next to the curves give the cooling rates in K/min.

Note: The glass transition imitates a first order transition as a consequence of enthalpy relaxation (38 a).

Figure 10: Poly(butadiene-co-styrene).

Influence of copolymerization on the heat capacity. The figures next to the curves give the weight percentage of styrene in the copolymers. Dotted lines: Polymerization temperature 5°C, full lines: polymerization temperature 50°C. 8.58%-polymers (24), 22.61 and 42.98%-polymers (25), 100% polystyrene (15, 41).

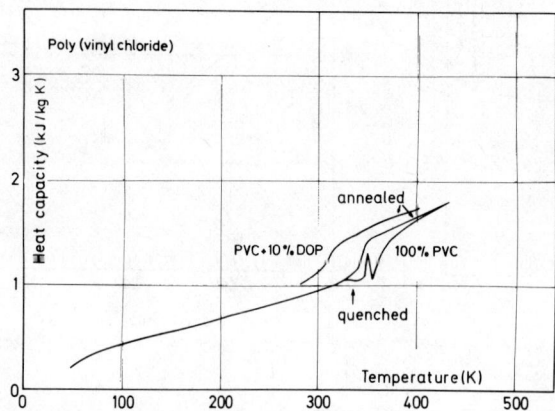

Figure 11: Poly(vinyl chloride).

Low temp. (46), high temp. (32).

Plastizised sample with 10% dioctylphthalate (48).

REFERENCES

1.   I. Abu-Isa, M. Dole, J. Phys. Chem. 69, 2668 (1965).
2.   J. V. Aleman, J. Polymer Sci., A-1, 9, 3501 (1971).
3.   S. Alford, M. Dole, J. Am. Chem. Soc. 77, 4774 (1955).
4.   R. H. Beaumont, B. Clegg, G. Gee, J. B. M. Herbert, D. J. Marks, R. C. Robert, D. Sims, Polymer 7, 401 (1966).
5.   N. Bekkedahl, H. Matheson, J. Res. Nat. Bur. Std. 15, 503 (1935).
6.   W. Braun, K. H. Hellwege, W. Knappe, Kolloid Z.-Z. Polymere 215, 10 (1967).
7.   F. G. Brickwedde, in: R. H. Boundy, R. F. Boyer, "Styrene". New York: Reinhold Publ. Corp. 1952.
8.   T. V. Burdzhanadze, P. L. Privalov, N. N. Tarvkhelidze, Bull. Acad. Sci. Georgian SSR 31, 277 (1963).
9.   S. S. Chang, A. B. Bestul, J. Polymer Sci., A-2, 6, 849 (1968).
10. S. S. Chang, A. B. Bestul, J. Res. Nat. Bur. Std. 75 A, 113 (1971).
11. C. A. Clegg, D. R. Gee, T. P. Melia, A. Tyson, Polymer 9, 501 (1968).
12. F. S. Dainton, D. M. Evans, F. E. Hoare, T. P. Melia, Polymer 3, 263 (1962).
13. Idem. ibid. 12, p. 271.
14. Idem. ibid. 12, p. 277.
15. Idem. ibid. 12, p. 286.
16. Idem. ibid. 12, p. 297.
17. Idem. ibid. 12, p. 310.
18. Idem. ibid. 12, p. 316.
19. M. Dole, W. P. Hettinger, N. R. Larson, J. A. Wethington, J. Chem. Phys. 20, 718 (1952).
20. T. B. Douglas, A. W. Harman, J. Res. Nat. Bur. Std. 69 A, 149 (1965).

21. E. Z. Fainberg, N. V. Mikhailov, Vysokomolekul. Soedin. A, 9, 920 (1967).

22. J. D. Ferry, G. S. Parks, J. Chem. Phys. 4, 70 (1936).

23. G. T. Furukawa, R. E. Mc Coskey, G. J. King, J. Res. Nat. Bur. Std. 49, 273 (1952).

24. G. T. Furukawa, R. E. Mc Coskey, G. J. King, J. Res. Nat. Bur. Std. 50, 357 (1953).

25. G. T. Furukawa, R. E. Mc Coskey, M. L. Reilly, J. Res. Nat. Bur. Std. 55, 127 (1955).

26. G. T. Furukawa, M. L. Reilly, J. Res. Nat. Bur. Std. 56, 285 (1956).

27. D. R. Gee, T. P. Melia, Makromol. Chem. 116, 122 (1968).

28. D. R. Gee, T. P. Melia, Polymer 10, 239 (1969).

29. Yu. K. Godovski, Yu. S. Lipatov, Vysokomolekul. Soedin. A, 10, 32 (1968).

30. Yu. K. Godovski, V. Yu. Levin, G. L. Slonimski, A. A. Shdanov, K. A. Andrianov, Vysokomolekul. Soedin. A, 11, 2444 (1969).

31. A. P. Gray, N. Brenner, Am. Chem. Soc. Div. Polymer Sci. Preprint 6, 956 (1965).

32. Th. Grewer, H. Wilski, Kolloid Z.-Z. Polymere 226, 46 (1968).

33. R. G. Griskey, D. O. Hubbel, J. Appl. Polymer Sci. 12, 853 (1968).

34. K. H. Hellwege, W. Knappe, W. Wetzel, Kolloid Z. 180, 126 (1962).

35. J. D. Hoffmann, J. Am. Chem. Soc. 74, 1696 (1952).

36. R. Hoffmann, W. Knappe, Kolloid Z.- Z. Polymere 240, 784 (1970).

37. R. Hoffmann, W. Knappe, Kolloid Z.- Z. Polymere 247, 763 (1971).

38. S. Ichihara, A. Komatsu, T. Hata, Polymer J. 2, 650 (1971).

38a. Idem. ibid. 38, p. 644.

39. L. L. Isaacs, C. W. Garland, J. Phys. Chem. Solids 23, 311 (1962).

40. A. N. Karasev, Plasticheskie Massy (1967) 52.

41. F. E. Karasz, H. E. Bair, J. M. O'Reilly, J. Phys. Chem. 69, 2657 (1965).

42. F. E. Karasz, H. E. Bair, J. M. O'Reilly, Polymer 8, 547 (1967).

43. F. E. Karasz, H. E. Bair, J. M. O'Reilly, J. Polymer Sci., A-2, 6 1141 (1968).

44. V. P. Kolesov, I. E. Paukov, S. M. Skuratov, Zh. Fiz. Khim. 36, 770 (1962).

45. H. Koplin, Thesis Techn. Hochschule Aachen 1962.

46. B. V. Lebedev, I. B. Rabinovich, V. A. Budarina, Vysokomolekul. Soedin., A, 9, 488 (1967).

47. B. V. Lebedev, I. B. Rabinovich, L. Ya. Martynenko, Vysokomolekul. Soedin. A, 9, 1640 (1967).

48. P. Marx, M. Dole, J. Am. Chem. Soc. 77, 4771 (1955).

49. P. Marx, C. W. Smith, A. E. Worthington, M. Dole, J. Phys. Chem. 59, 1015 (1955).

50. T. Melia, Polymer 3, 317 (1962).

51. T. P. Melia, A. Tyson, Makromol. Chem. 109, 87 (1967).

52. J. M. O'Reilly, F. E. Karasz, J. Polymer Sci., C, 14, 49 (1966).

53. J. M. O'Reilly, F. E. Karasz, H. E. Bair, J. Polymer Sci., C, 6, 109 (1963).

54. E. Passaglia, H. K. Kevorkian, J. Appl. Phys. 34, 90 (1963).

55. E. Passaglia, H. K. Kevorkian, J. Appl. Polymer Sci. 7, 119 (1963).

56. L. I. Pavlinov, I. B. Rabinovich, N. K. Okladnov, S. A. Arzhakov, Vysokomolekul. Soedin. A, 9, 483 (1967).

57. W. Reese, J. Appl. Phys. 37, 864, 3959 (1966).

58. W. Reese, J. Macromol. Sci. Chem., A, 3, 1257 (1969).

59. W. Reese, J. E. Tucker, J. Chem. Phys. 43, 105 (1965).

60. M. J. Richardson, Trans. Faraday Soc. 61, 1876 (1965).

61. E. Yu. Roinishvili, N. N. Tavkhelidze, V. B. Akopyan, Vysokomolekul. Soedin. B, 9, 254 (1967).

62. W. Schermann, G. Wegner, Makromol. Chem. in press

63. C. W. Smith, M. Dole, J. Polymer Sci. 20, 37 (1956).

64. I. V. Sochava, Dokl. Akad. Nauk USSR 130, 126 (1960).

65. I. V. Sochava, Vestn. Leningr. Univ. 19 (10), Ser. Fiz. i Khim. 2, 56 (1964).

66. I. V. Sochava, O. N. Trapeznikova, Dokl. Akad. Nauk USSR 113, 784 (1957).

67. I. V. Sochava, O. N. Trapeznikova, Vestn. Leningr. Univ. 16, Ser. Fiz. i Khim. 2, 70 (1961).

68. I. V. Sochava, O. N. Trapeznikova, Vestn. Leningr. Univ. 20, Ser. Fiz. i Khim. 4, 71 (1965).

69. H. Tautz, M. Glueck, G. Hartmann, R. Leuteritz, Plaste Kautschuk 10, 648 (1963).

70. J. E. Tucker, W. Reese, J. Chem. Phys. 46, 1388 (1967).

71. M. V. Volkenshtein, Yu. A. Sharanov, Vysokomolekul. Soedin. 3, 1739 (1961).

72. R. W. Warfield, M. C. Petree, J. Polymer Sci. A 1, 1701 (1963).

73. R. W. Warfield, M. C. Petree, J. Polymer Sci. 4, 532 (1966).

74. R. W. Warfield, M. C. Petree, P. Donovan, SPE-Journal 15, 1055 (1959).

75. R. C. Wilhoit, M. Dole, J. Phys. Chem. 57, 14 (1953).

76. R. W. Wilkinson, M. Dole, J. Polymer Sci. 58, 1089 (1962).

77. H. Wilski, Angew. Makromol. Chem. 6, 101 (1969).

78. H. Wilski, Kolloid Z.-Z. Polymere 238, 426 (1970).

79. H. Wilski, Kolloid Z.-Z. Polymere 248, 867 (1971).

80. H. Wilski, T. Grewer, J. Polymer Sci., C, 6, 33 (1964).

81. L. A. Wood, N. Bekkedahl, Polymer Letters 5, 169 (1967).

82. W. Wrasidlo, in: C. D. Craver (editor) "Polymer Characterization Interdisziplinary Approaches" (ACS-Meeting Chicago 1970), Plenum Press, New York 1971, p. 157.

83. W. Wrasidlo, in: H. G. Wiedemann (editor) "Thermal Analysis", Vol. 3, p. 397, Birkhäuser Verlag, Basel 1972.

84. B. Wunderlich, J. Phys. Chem. 69, 2078 (1965).

85. B. Wunderlich, H. Baur, Adv. Polymer Sci. 7, 151 (1970).

86. B. Wunderlich, M. Dole, J. Polymer Sci. 24, 201 (1957).

87. B. Wunderlich, M. Dole, J. Polymer Sci. 32, 125 (1958).

88. S. Yoshida, M. Sakiyama, S. Seki, Polymer J. 1, 573 (1970).

CRITICAL SURFACE TENSIONS OF POLYMERS

E. G. Shafrin
Laboratory for Chemical Physics
U. S. Naval Research Laboratory
Washington, D. C.

Contents

A. INTRODUCTION

The critical surface tension for spreading (γ_c) defines the wettability of a solid surface by noting the lowest surface tension (γ_L) a liquid can have and still exhibit a contact angle (θ) greater than zero degrees on that solid. The constant is expressed in units of milli-Newton per meter (formerly, dynes per centimeter) for a specific temperature. The values which follow are based on measurements made at $20\text{-}25^{\circ}$C.

The value of γ_c for a given solid is determined by observing the spreading behavior and the angle θ on it of a series of related liquids of decreasing γ_L. A rectilinear relation exists between $\cos \theta$ and γ_L; the intercept of this line with the $\cos \theta = 1$ line (i.e., θ = zero) gives the value of γ_c for that series of liquids (15). The closeness of the intercepts for various liquid series on a single solid makes γ_c a valuable parameter characteristic of the solid surface (58). In general, values of γ_c cited here were determined from measurements on a variety of liquids. Wherever a value of γ_c was available based on data for the homologous series of normal alkanes, however, this was the value preferred for inclusion in the following table, and such entries have been postdesignated by the letter "a" in the column of critical surface tension values. Where the γ_c values were derived solely from data for liquids (including solutions) capable of hydrogen-bond formation, the entries are marked with the letter "b" to emphasize that such liquids may induce or reflect conformational changes at the polymer surface (2,59).

B. TABLE OF CRITICAL SURFACE TENSIONS OF POLYMERS

Polymer		Critical Surface Tension (γ_c) $[\text{mN m}^{-1}] \equiv [\text{dyn cm}^{-1}]$		References
1. Main-chain Acyclic Carbon Polymers				
1.1 Poly(dienes)				
Poly(butadiene)		31	b	26
	cis	32	b	26
	trans	31	b	26
	1,2-structure	25	b	26
Poly(chloroprene)		38	b	26
Poly(isoprene)	chlorinated	37	b	26
	cis	32	b	25
		31	b	26
	cyclized	34	b	26
	trans	31	b	25
		30	b	26
	hydrochloride	36	b	26
1.2 Poly(alkenes)				
Poly(ethylene)		31		17
		36.0		41

Polymer	Critical Surface Tension (γ_c) [mN m^{-1}] $\equiv$ [dyn cm^{-1}]		References
Poly(ethylene) (Cont' d.)	28	b	14
	31.5	b	40
	31.0	b	38
	25.5-31	b	10
	29	b	23
high density	34.1	b	9
	28-32	b	54
(ρ = 0.95)	31	b	28
low density	26		24
	31.5	b	37
	34	b	50
	27	b	54
(ρ = 0.91)	31	b	28
Poly(ethylene-co-acrylic acid) see poly(acrylic acid-co-ethylene)			
Poly(ethylene-co-propylene) see poly(propylene-co-ethylene)			
Poly(ethylene-co-tetrafluoroethylene) see poly(tetrafluoroethylene-co-ethylene)			
Poly(ethylene-co-vinyl acetate) see poly(vinyl acetate-co-ethylene)			
Poly(isobutene)	27	b	28
Poly(isobutene-co-isoprene)	27	b	26
Poly(4-methyl-1-pentene) atactic	25	b	26
Poly(propylene)	34		1
	29	b	26
	32.0	b	38
Poly(propylene-co-ethylene)	28	b	26

1.3 Poly(acrylic acid) and Derivatives

Poly(acrylamide)	35-40		22
Poly(acrylate) (Plexiglas)	29	b	36
(Primal K 3)	35	b	13
Poly(acrylic acid-co-ethylene) 20.7:79.3 mol %	59	b	50
14.3:85.7	55	b	50
8.2:91.8	44	b	50
3.2:96.8	41	b	50
Poly(butyl acrylate)	28	b	28
Poly[(1-(chlorodifluoromethyl)-2-chloro-1,2,2-trifluoroethyl acrylate)]	20.3	a	43
Poly(1-chlorodifluoromethyl-tetrafluoroethyl acrylate)	18.9	a	43
Poly(2,2,3,3,4,4,4-heptafluorobutyl acrylate)	15.2	a	43
Poly(methyl acrylate)	41	b	28
Poly(perfluoroisobutyl acrylate)	13.5	a	48
Poly(perfluoroisobutyl acrylate-co-1H,1H-heptafluorobutyl acrylate) 95:5 mol%	13.9	a	48
80:20	14.1	a	48
50:50	14.3	a	48
20:80	14.9	a	48
5:95	13.4	a	48
Poly[2-(N-propyl-N-heptadecafluorooctylsulfonyl)aminoethyl acrylate]	11.1	a	8
Poly(2,2,3,3,4,4,5,5,6,6,7,7,8,8,8-pentadecafluorooctyl acrylate)	10.4	a	43
Poly(1-trifluoromethyltetrafluoroethyl acrylate)	14.1	a	43
Poly[2-(2-trifluoromethyl)tetrafluoroethoxy)ethyl acrylate]	15.5	a	43
Poly[5-(1-trifluoromethyl)tetrafluoroethoxy)pentyl acrylate]	17.7	a	43
Poly[11-(1-trifluoromethyl)tetrafluoroethoxy)undecyl acrylate]	20.3	a	43
Poly[(1-trifluoromethyl)-2,2,2-trifluoroethyl acrylate]	15.4	a	43
	15.0-15.4	a	44

1.4 Poly(methacrylic acid) and Derivatives

Poly(perfluoro-tert-butyl methacrylate)	14.7	a	48
Poly[1-(chlorodifluoromethyl)tetrafluoroethyl methacrylate]	19.1	a	43
Poly(ethyl methacrylate)	33	b	28
Poly(methyl methacrylate)	33-44		57
	39		22
	41.2		18
	26.5-31.5	b	10
	21	b	36
	38	b	28

Polymer		Critical Surface Tension (γ_c) $[mN\ m^{-1}] \equiv [dyn\ cm^{-1}]$		References
Poly(methyl methacrylate) (Cont'd.)		39.0	b	38
		37-39	b	54
Poly(2,2,3,3,4,4,5,5,6,6,7,7,8,8,8-pentadecafluorooctyl methacrylate)		10.6	a	8
Poly(2,2,3,3-tetrafluoropropyl methacrylate)		19	a	52
Poly(1-trifluoromethyltetrafluoroethyl methacrylate)		15.0	a	43
Poly(1-trifluoromethyl-2,2,2-trifluoroethyl methacrylate)		14.8-15.4	a	43,44
1.5 Poly(vinyl ethers)				
Poly(ethoxyethylene)		36		24
Poly(methoxyethylene)		29	b	28
Poly[(1-trifluoromethyl)tetrafluoroethoxyethylene]		14.2-14.6	a	42
Poly[(1-trifluoromethyl)tetrafluoroethoxymethylethylene]		21	a	42
Poly[(1-trifluoromethyl)tetrafluoroethoxymethyl-1-methylethylene-co-maleic acid]		18.8	a	56
Poly(vinyl ethyl ether) see poly(ethoxyethylene)				
Poly(vinyl methyl ether) see poly(methoxyethylene)				
1.6 Poly(vinyl alcohols), Poly(vinyl esters), Poly(vinyl halides)				
Poly(1-chlorotrifluoroethylene)		31		16
Poly(1-chlorotrifluoroethylene-co-tetrafluoroethylene)	40:60 mol%	24		16
	20:80	20		16
Poly(heptafluoropropylethylene)		15.5	a	21
Poly(heptafluoropropylethylene-co-tetrafluoroethylene)	79:21 mol%	16.2	a	21
	52:48	16.5	a	21
	25:75	16.5	a	21
Poly(hexafluoropropylene) see poly(trifluoromethyltrifluoroethylene)				
Poly(tetrafluoroethylene)		18.5	a	15
		19		1
		19.5		41
		16	b	55
		18-19	b	54
		19-22	b	10
Poly(tetrafluoroethylene-co-ethylene)		26-27		16
Poly(tetrafluoroethylene-co-hexafluoropropylene) see poly(trifluoromethyltri-fluoroethylene-co-tetrafluoroethylene)				
Poly(trifluoroethylene)		22		11
Poly(trifluoromethylethylene)		21.5	a	21
Poly(trifluoromethylethylene-co-tetrafluoroethylene)	58:42 mol%	21.1	a	21
	39:61	18.0	a	21
Poly(trifluoromethyltrifluoroethylene)		16.2-17.1	a	7
Poly(trifluoromethyltrifluoroethylene-co-tetrafluoroethylene)	23:77 mol%	17.8	a	6
	16:84	18	a	6
	14:86	18.2	a	6
	8:92	18.3	a	6
	6:94	19.0	a	6
Poly(vinyl acetate)		37	b	28
Poly(vinyl acetate-co-ethylene)		37		1
Poly(vinyl alcohol)		37		46
Poly(vinyl chloride)		39		11
		26-31.5	b	10
Poly(vinyl fluoride)		28		11
Poly(vinylidene chloride)		40		11
Poly(vinylidene fluoride)		25		11
1.7 Poly(styrenes)				
Poly(styrene)		32.8 (35)	(b)	12
		30-33		22
		33		52
		33.0		41
		30	b	36
		36	b	26
		30.5	b	38

Polymer		Critical Surface Tension (γ_c) [mN m^{-1}] $\equiv$ [dyn cm^{-1}]		References
Poly(styrene) (Cont'd.)		27–31.5	b	10
		32.7	b	39
		33	b	54
atactic		36	b	28
Poly(styrene-co-acrylonitrile)	32:68 mol %	43	b	27
	25:75	42	b	27
	16:84	40	b	27
	6:94	37	b	27
Poly(styrene-co-butadiene)	25:75 random copolymer	33	b	26
	25:75 block copolymer	33	b	26
Poly(styrene-co-dimethyl siloxane)	see Section 2.2			
Poly(styrene-co-2,2,3,3-tetrafluoropropyl methacrylate)	99:1 mol %	18	a	52
	98:2 to 90:10	19	a	52
	80:20	20	a	52
	70:30	21	a	52
	60:40 to 50:50	25		52
	40:60	26		52
	30:70	27		52
	20:10	28		52
	10:90 to 8:92	29		52
	6:94 to 4:96	30		52
	2:98 to 1:99	31		52
Poly(styrene), 4-bromo-		39.4		53
--, 2-chloro-		42	b	28
--, 4-isobutyl-		29	b	28
--, methyl-		35	b	28

1.8 Others

Polymer		Critical Surface Tension (γ_c)		References
Chlorosulfonated polyethylene		37	b	25
Poly(acrylonitrile)		44	b	28
Poly(acrylonitrile-co-butadiene)		37	b	26
Poly(acrylonitrile-co-vinylidene chloride)	20:80	38–44		22
Poly(vinyl formal)		38–40	b	28
Poly(vinyl butyral)		28	b	28
with 12 % residual poly(vinyl alcohol)		24–25	b	37
with 30 % residual poly(vinyl alcohol)		32 (25)	(b)	37

2. Main-chain Heteroatom Polymers

2.1 Poly(oxides), Poly(esters), Poly(urethanes)

Polymer		Critical Surface Tension (γ_c)		References
Poly(acetal) see poly(oxymethylene)				
Poly(carbonate) see poly(oxycarbonyloxy-1,4-phenyleneisopropylidene-1,4-phenylene)				
Poly(ethylene terephthalate) see poly(oxyethyleneoxyterephthaloyl)				
Polymer from (2,3-epoxy)propyleneoxycarbonyltetratriacontamethylenecarbonyloxy-(2,3-epoxy)propylene) (Epon resin 871)		37		20
Polymer from (2,3-epoxy)propyleneoxy-1,4-phenyleneisopropylidene-1,4-phenylene-oxy(2,3-epoxy)propylene		34–44		20
Poly[oxy-1,2-bis(perfluoroisobutoxymethyl)-ethylene]		11.4	a	45
Poly[oxy-1-(3,5-bis(trifluoromethyl)phenyl)-1-trifluoromethyltrifluoroethoxymethylethylene)]		19.1–19.2	a	47
Poly(oxycarbonyloxy-1,4-phenyleneisopropylidene-1,4-phenylene)		34.5		41
		45		1
		45		20
Poly(oxyethylene)		43	b	28
Poly(oxyethyleneoxyterephthaloyl) (poly(ethylene terephthalate))		43.0		12
		40.0		41
		43 (27–36.5)	(b)	10
		42.5	b	40
		40	b	54
Poly(oxymethylene)		38.0		41
		36	b	28
		29	b	23

Polymer	Critical Surface Tension (γ_c) [mN m^{-1}] $\equiv$ [dyn cm^{-1}]		References
2.1 Poly(oxides), Poly(esters), Poly(urethanes) (Cont' d.)			
Poly(oxymethylene-1,4-cyclohexylenemethyleneoxyterephthaloyl)	44.6 (27-33)	(b)	10
Poly(oxy-1-pentafluorophenyl-1-trifluoromethyltrifluoroethoxymethyl-ethylene)	20.2	a	47
Poly(oxyphenylene) (poly(phenylene oxide))	41	b	28
Poly(oxy-1-phenyl-1-trifluoromethyltrifluoroethoxymethylethylene)	20.2	a	47
Poly(oxypropylene)	32	b	28
Poly(oxy-3-trifluoromethylphenoxymethylethylene)	18.5	a	47
Poly[oxy-1-(3-trifluoromethyl)phenyl-1-trifluoromethyltrifluoroethoxymethyl-ethylene]	19.4-21.5	a	47
Poly[(1-trifluoromethyltetrafluoroethoxymethyl)ethylene-co-maleic anhydride]	15.7	a	56
Poly[1-(1-trifluoromethyltetrafluoroethoxymethyl)-1-methylethylene-co-maleic anhydride]	16.8	a	56
2.2 Poly(silanes), Poly(silazanes)			
Poly(dimethyl siloxane)	24		49
Poly(dimethylsiloxane-co-styrene) see poly(styrene-co-dimethyl siloxane)			
Polymer from 3-acetoxypropyltrimethoxysilane	37.5		30
Polymer from 2-aminoethyliminopropyltrimethoxysilane	33.5	b	31
Polymer from 3-aminopropyltriethoxysilane	35	b	31
Polymer from 4-bromomethylphenyltrimethoxysilane	42.0		30
Polymer from (bromo)(methyl)phenyltrimethoxysilane	39.5		29
Polymer from bromophenyltrimethoxysilane	43.5		29
Polymer from 2-(4-chlorophenyl)ethyltrichlorosilane	42		4
Polymer from 2-(4-chlorophenyl)ethyltriethoxysilane	43		4
Polymer from 2-(4-chlorophenyl)ethyltrimethoxysilane	39-43		4
	44-47		5
Polymer from 3-chloropropyltrichlorosilane	42		5
Polymer from 3-chloropropyltrimethoxysilane	43-45		5
	40.5	b	31
Polymer from 2-cyanoethyltrimethoxysilane	34.0		30
Polymer from dibromophenyltrimethoxysilane	41.5		30
Polymer from (2-(3,4-epoxy)cyclohexyl)ethyltrimethoxysilane	41.5		30
Polymer from ethyltriethoxysilane	26-33		4
Polymer from 3-fluoropropyltrimethoxysilane	20-25		5
Polymer from 3-glycidoxypropyldimethylsilane	35		20
Polymer from 3-glycidoxypropyltrimethoxysilane	38.5	b	31
Polymer from 4-isopropylphenyltrimethoxysilane	41.5		30
Polymer from 3-mercaptopropyltrimethoxysilane	41	b	31
Polymer from 3-methacryloxypropyltrimethoxysilane	28	b	31
Polymer from 4-methylphenyltrimethoxysilane	34.0		29
Polymer from methyltrimethoxysilane	22.5		29
Polymer from 3-(2,2,3,3,4,4,5,5,6,6,7,7,8,8,8-pentadecafluorooctoxy)propyltri-ethoxysilane	14	a	5
Polymer from phenyltrimethoxysilane	40.0		29
Polymer from propyltrimethoxysilane	28.5		30
Polymer from (3-(1-trifluoromethyl)tetrafluoroethoxy)propyltrichlorosilane	14.9-15.9	a	44
Polymer from (3-(1-trifluoromethyl)tetrafluoroethoxy)propyltrimethoxysilane	16-18	a	5
Polymer from 3,3,3-trifluoropropyltrimethoxysilane	33.5		30
Polymer from vinyltriacetoxysilane	26.0		30
Polymer from vinyltriethoxysilane	24-36		4
Polymer from vinyltrimethoxysilane	25	b	31
Poly(styrene-co-dimethyl siloxane-co-styrene), ABA block copolymer with approximate monomer units as indicated:			
Poly(styrene) + 1% 38-27-38 copolymer	28.3	b	39
+ 1% 29-48-29 copolymer	27.5	b	39
+ 1% 19-77-19 copolymer	22.0	b	39
+ 1% 10-96-10 copolymer	22.0	b	39
Silicone rubber	22		1
	24	b	32

CRITICAL SURFACE TENSION OF POLYMERS

Polymer	Critical Surface Tension (γ_c) [mN m^{-1}] $\equiv$ [dyn cm^{-1}]		References

2. Main-chain Heteroatom Polymers (Cont'd.)

2.3 Poly(sulfones)

Polymer	γ_c		References
Poly(oxy-1,4-phenylenesulfonyl-1,4-phenylene-oxy-1,4-phenyleneisopropylidene-1,4-phenylene) ((Poly(sulfone))	41	b	28

2.4 Poly(amides)

Polymer	γ_c		References
Nylon	41.0	b	38
Nylon 2, see poly[imino(1-oxoethylene)]			
Nylon 6, see poly[imino(1-oxohexamethylene)]			
Nylon 11, see poly[imino(1-oxoundecamethylene)]			
Nylon 6,6 see poly(iminohexamethyleneiminoadipoyl)			
Nylon 7,7 see poly(iminoheptamethyleneiminopimeloyl)			
Nylon 8,8 see poly(iminooctamethyleneiminosuberoyl)			
Nylon 9,9 see poly(iminononamethyleneiminoazelaoyl)			
Nylon 10,10 see poly(iminodecamethyleneiminosebacoyl)			
Poly(aminoacetic acid) see poly[imino(1-oxoethylene)]			
Poly(γ-benzyl L-glutamate) see poly[imino(1-oxo(2-(3-oxo-3-benzyloxy)-propyl)ethylene)]			
Poly(iminodecamethyleneiminosebacoyl)	32	b	14
Poly(iminoheptamethyleneiminopimeloyl)	43	b	14
Poly(iminohexamethyleneiminoadipoyl)	42.5 (46.0)	(b)	12
	46 (28-37.5)	(b)	10
	42	b	14
Poly(iminononamethyleneiminoazelaoyl)	36	b	14
Poly(iminooctamethyleneiminosuberoyl)	34	b	14
Poly[imino-(1-oxoethylene)]	44 (51)	(b)	2
Poly imino(1-oxohexamethylene)] (Nylon 6)	42	b	14
cast from dichloroacetic acid	42 (46)	(b)	3
cast from formic acid	38 (38)	(b)	3
Poly[imino(1-oxo(2-(3-oxo-3-benzyloxy)propyl)ethylene)]	40		2
Poly[imino(1-oxo(2-(3-oxo-3-methoxy)propyl)ethylene)], α-helix	41 (49)	(b)	2
random	42 (48)	(b)	2
extended chain	37 (37)	(b)	2
Poly[imino(1-oxoundecamethylene)]	33	b	14
	42 (46)	(b)	2
Poly(γ-methyl L-glutamate) see poly[imino(1-oxo(2-(3-methoxy)propyl)-ethylene)]			

2.5 Poly(imines)

Polymer	γ_c		References
Poly[(benzoylimino)ethylene]	25.5	b	33
Poly[(butyrylimino)ethylene]	24.5 (24.5)	(b)	33
Poly[(dodecanoylimino)ethylene]	22		34
Poly[(dodecanoylimino)ethylene-co-(acetylimino)trimethylene] 75:25 weight %	22		34
50:50 weight %	22		34
25:75 weight %	22		34
Poly[(heptanoylimino)ethylene]	22 (21)	(b)	33
Poly[(hexanoylimino)ethylene]	23 (21)	(b)	33
Poly[((3-methyl)butyrylimino)ethylene]	24 (22)	(b)	33
Poly[(octadecanoylimino)ethylene]	22		33
Poly[(12,12,13,13,14,14,15,15,16,16,17,17,18,18,18-pentadecafluoroocta-decanoylimino)ethylene]	11		33
Poly[(pentanoylimino)ethylene]	23 (22)	(b)	33

2.6 Others

Polymer	γ_c		References
Amylopectin	35		46
Amylose	37		46
Casein	43	b	13
Cellulose (regenerated)	44		46
from cotton linters	41.5 (42)	(b)	35
from wood pulp	35.5 (42)	(b)	35
Cellulose acetate	39.0	b	38

Polymer	Critical Surface Tension (γ_c) $[mN\ m^{-1}] \equiv [dyn\ cm^{-1}]$		References
2.6 <u>Others</u> (Cont'd.)			
Hemicellulose, arabinogalactan	33		35
galactoglucomannan	36.5		35
hardwood xylan	33-36.5		35
softwood xylan	35.0		35
Natural rubber/rosin adhesive	36		24
Phenol/resorcinol adhesive	52.0		19
Polyamide-epichlorohydrin resin	52	b	13
Poly(epichlorohydrin)	35	b	26
Resorcinol adhesive	51.0	b	19
Starch	39		46
Urea-formaldehyde-resin	61	b	13
Wool	45	b	13

C. REFERENCES

1. R. E. Baier, "Surface Properties of Materials for Prosthetic Implants", Cornell Aeronautical Laboratory CAL Report # VH-2801-P-2, Feb. 15, 1970.

2. R. E. Baier, W. A. Zisman, Macromolecules 3, 70 (1970).

3. R. E. Baier, W. A. Zisman, Macromolecules 3, 462 (1970).

4. W. D. Bascom, "The Wettability of Ethyl- and Vinyltriethoxysilane Films Formed at Organic Liquid Silica Interfaces", in R. F. Gould, Ed., Advances in Chemistry Series, 87, "Interaction of Liquids at Solid Substrates", Am. Chem. Soc., Washington, D.C., 1968, p. 38.

5. W. D. Bascom, J. Colloid Interface Sci. 27, 789 (1968).

6. M. K. Bernett, W. A. Zisman, J. Phys. Chem. 64, 1292 (1960).

7. M. K. Bernett, W. A. Zisman, J. Phys. Chem. 65, 2266 (1961).

8. M. K. Bernett, W. A. Zisman, J. Phys. Chem. 66, 1207 (1962).

9. K. Bright, B. A. W. Simmons, European Polymer J. 3, 219 (1967)

10. J. R. Dann, J. Colloid Interface Sci. 32, 302 (1970).

11. A. H. Ellison, W. A. Zisman, J. Phys. Chem. 58, 260 (1954).

12. A. H. Ellison, W. A. Zisman, J. Phys. Chem. 58, 503 (1954).

13. H. D. Feltman, J. R. McPhee, Textile Research J. 34, 634 (1964).

14. T. Fort, Jr., "The Wettability of a Homologous Series of Nylon Polymers" in R. F. Gould, Ed., Advances in Chemistry Series. 43. "Contact Angle, Wettability, and Adhesion", Am. Chem. Soc., Washington, D.C., 1964, p. 302.

15. H. W. Fox, W. A. Zisman, J. Colloid Sci. 5, 514 (1950).

16. H. W. Fox, W. A. Zisman, J. Colloid Sci. 7, 109 (1952).

17. H. W. Fox, W. A. Zisman, J. Colloid Sci. 7, 428 (1952).

18. P.-O. Glantz, J. Colloid Interface Sci. 37, 281 (1971).

19. V. R. Gray, "Contact Angles, Their Significance and Measurement" in Monograph No. 25, Soc. Chem. Ind. (London), London, 1967, p. 99.

20. A. Herczeg, G. S. Ronay, W. C. Simpson, Proc. National Soc. Aerospace Materials and Process Engineers Conf., Dallas, Texas, Vol. 2, 221 (1970).

21. W. K. H. Hu, W. A. Zisman, Macromolecules 4, 688 (1971).

22. N. L. Jarvis, R. B. Fox, W. A. Zisman, "Surface Activity at Organic Liquid-Air Interfaces. V. The Effect of Partially Fluorinated Additives on the Wettability of Solid Polymers", in R. F. Gould, Ed., Advances in Chemistry Series. 43. "Contact Angle, Wettability, and Adhesion", Am. Chem. Soc., Washington, D. C., 1964, p. 317.

23. D. H. Kaelble, E. H. Cirlin, J. Polymer Sci., A-2, 9, 363 (1971).

24. Y. Kitazaki, A. Watanabe, M. Toyama, Preprints, Fourth Conference on Adhesion, Osaka, Japan, June, 1966, p. 31; through M. Toyama, T. Ito, H. Moriguchi, J. Appl. Polymer Sci. 14, 2039 (1970).

25. L.-H. Lee, Polymer Preprints, Am. Chem. Soc. Div. Polymer Chem. 7, 916 (1966).

26. L.-H. Lee, J. Polymer Sci., A-2, 5, 1103 (1967).

27. L.-H. Lee, "Adhesion of High Polymers. III. Mechanisms of Adhesion at the Rubber-Resin Interface in Heterophase Systems", in R. F. Gould, Ed., Advances in Chemistry Series. 87. "Interaction of Liquids at Solid Substrates", Am. Chem. Soc., Washington, D. C. 1968, p. 85.

28. L.-H. Lee, "Adhesion of High Polymers. IV. Relationships between Surface Wetting and Bulk Properties of High Polymers", in R. F. Gould, Ed., Advances in Chemistry Series. 87. "Interactions of Liquids at Solid Substrates" Am. Chem. Soc., Washington, D. C. 1968, p. 106.

29. L.-H. Lee, private communication cited by W. A. Zisman, I & EC Product Res. and Dev. 8, 97 (June 1969).

30. L.-H. Lee, private communication.

31. L.-H. Lee, J. Colloid Interface Sci. 27, 751 (1968).

32. L.-H. Lee, J. Adhesion 4, 39 (1972).

33. M. Litt, J. Herz, J. Colloid Interface Sci. 31, 248 (1969).

34. M. Litt, J. Herz, Polymer Preprints, Am. Chem. Soc. Div. Polymer Chem. 10, 905 (1969).

35. P. Luner, M. Sandell, J. Polymer Sci., C, 28, Proc. 6. Cellulose Conf., 115 (1969).

36. J. E. Marian, "Surface Texture in Relation to Adhesive Bonding", in Symposium on Properties of Surfaces, Special Technical Publ. No. 340, American Society for Testing and Materials, 1962, p. 122.

37. S. Newman, J. Colloid Interface Sci. 25, 341 (1967).

38. D. A. Olsen, A. J. Osteraas, J. Appl. Polymer Sci. 13, 1523 (1969).

39. M. J. Owen, T. C. Kendrick, Macromolecules 3, 458 (1970).

40. D. K. Owens, J. Appl. Polymer Sci. 8, 1465 (1964) and private communication.

41. F. D. Petke, B. R. Ray, J. Colloid Interface Sci. 31, 216 (1969); erratum ibid 33, 195 (1970).

42. A. G. Pittman, B. A. Ludwig, D. L. Sharp, J. Polymer Sci., A-1, 6, 1741 (1968).

43. A. G. Pittman, D. L. Sharp, B. A. Ludwig, J. Polymer Sci., A-1, 6, 1729 (1968).

44. A. G. Pittman, W. L. Wasley, cited by A. G. Pittman, "Surface Properties of Fluorocarbon Polymers", in L. A. Wall, Ed., "Fluoropolymers" (High Polymers, Vol. XXV), Wiley-Interscience, New York, 1972, p. 419.

45. A. G. Pittman, W. L. Wasley, J. N. Roitman, Polymer Letters 8, 873 (1970).

46. B. R. Ray, J. R. Anderson, J. J. Scholz, J. Phys. Chem. 62, 1220 (1958).

47. J. P. Reardon, J. G. O'Rear, J. R. Griffith, Ind. Eng. Chem. Prod. Res. Dev. 11 (Sept. 1972).

48. J. N. Roitman, A. G. Pittman, Polymer Letters 10, 499 (1972).

49. E. G. Shafrin, W. A. Zisman, "Upper Limits to the Contact Angles of Liquids on Solids", in R. F. Gould, Ed., Advances in Chemistry Series 43. "Contact Angles, Wettability, and Adhesion", Am. Chem. Soc., Washington, D. C., 1964, p. 145.

50. W. H. Smarook, S. Bonotto, Polymer Eng. Sci. 41 (Jan. 1968).

51. C. A. Sutula, private communication cited in F. D. Petke, B. R. Ray, J. Colloid Interface Sci. 31, 216 (1969).

52. K. Tamaribuchi, Polymer Preprints, Am. Chem. Soc. Div. Polymer Chem. 8, 631 (1967).

53. J. Thrower, private communication cited in J. H. Sewell, Modern Plastics 48, 66 (1971).

54. M. Toyama, T. Ito, H. Moriguchi, J. Appl. Polymer Sci. 14, 2039 (1970).

55. F. Vergara, M. Rose, B. Lespinasse, Compt. Rend. Acad. Sci., C, 269, 441 (1969).

56. W. L. Wasley, A. G. Pittman, Polymer Letters 10, 279 (1972).

57. E. Wolfram, Kolloid Z. 182, 75 (1962).

58. W. A. Zisman, "Relation of the Equilibrium Contact Angle to Liquid and Solid Constitution", in R. F. Gould, Ed., Advances in Chemistry Series. 43. "Contact Angle, Wettability, and Adhesion", Am. Chem. Soc., Washington, D. C., 1964, p. 1

59. W. A. Zisman, Ind. Eng. Chem. Prod. Res. Dev. 8, 97 (June 1969).

PERMEABILITY COEFFICIENTS

H. Yasuda
Camille Dreyfus Laboratory
Research Triangle Institute
Research Triangle Park, North Carolina

and

V. Stannett
Department of Chemical Engineering
North Carolina State University
Raleigh, N. C.

<u>Contents</u>   Page

A. INTRODUCTION

When small molecules permeate through a polymer membrane, the rate of permeation can be expressed by parameters which may be characteristic of the polymer. The general concept of the ease with which a permeant passes through a barrier is often referred to as "permeability". This general term "permeability" does not refer to the mechanism of the permeation but only to the rate of the transmission. In the literature, the rates of transmission in several different units are often cited as permeability. For example, the total rate of transmission, (amount of penetrant)/(time); the rate of transmission per unit area, (amount of penetrant)/(area)-(time); the rate of transmission per unit area and unit film thickness, (amount of penetrant)(film thickness)/(area)(time); and also the rate of transmission per unit area, unit thickness, unit time and unit pressure-(or concentration)-drop across the film, (amount of penetrant)(film thickness)/(area)(time)(pressure-drop); all appear under the general term "permeability".

The coefficient which has the dimensions

$$P = \frac{(\text{amount of permeant})(\text{film thickness})}{(\text{area})(\text{time})(\text{pressure-drop across the film})}$$

or, in the more general form,

$$P = \frac{(\text{amount of permeant})}{(\text{area})(\text{time})(\text{driving force gradient across the film})}$$

may best be defined as the permeability coefficient P.

When a permeant does not interact with the polymer, this permeability coefficient P is usually characteristic for the permeant-polymer system. Such is the case with the permeation of many gases, such as H_2, He, N_2, O_2 and CO_2, through many polymers. On the other hand, if a permeant interacts with polymer molecules, P is no longer a constant, and may depend on the pressure, thickness and other environmental conditions. In such cases, a single value of P does not represent the characteristic permeability of the polymer membrane and it is necessary to know the dependency of P on all possible variables in order to obtain the complete profile of the permeability of the polymer. In these cases, the transmission rate Q, which has the dimensions

$$Q = (\text{amount of permeant})(\text{film thickness})/(\text{area})(\text{time})$$

is often used for practical purposes, when the saturated vapor pressure of the permeant at a specified temperature is applied across the film. Permeabilities of films to water and organic compounds are often presented in this way.

The amount of permeant, in both P and Q, can be expressed by weight, moles or gaseous volume at standard temperature and pressure. These can readily be converted from one unit into another. The preferred metric units of the permeability coefficient used in this handbook are;

$$\text{units of P: } [\text{cm}^3 (\text{STP})\ \text{cm}\ \text{cm}^{-2}\ \text{s}^{-1}\ (\text{cm Hg})^{-1}]$$

This composite unit sometimes appears in the literature as

$$[\text{cm}^2\ \text{s}^{-1}\ (\text{cm Hg})^{-1}]$$

or

$$[\text{cm}^3 (\text{STP})\ \text{cm}^{-2}\ \text{cm}^{-1}\ \text{s}^{-1}\ (\text{cm Hg})^{-1}]$$

Since permeability coefficients given in this composite unit are often in the range of 10^{-6} to 10^{-13} for many polymers and permeants, many other large composite units are used in the literature. Some of those composite units and their conversion factors are listed in Table B. A number of different composite units for Q are also used in the literature and some of them are listed in Table C together with their converstion factors.

The permeation of small molecules through flawless polymer films occurs by the consecutive steps of solution of a permeant in the polymer and diffusion of the dissolved permeant. Consequently, the permeability coefficient P can be given by

$$P = D \cdot S$$

where D is the diffusion constant and S is the solubility coefficient. Units of D and S used in these tables are:

$$\text{units of } D: \quad [cm^2 s^{-1}]$$

$$\text{units of } S: \quad [cm^3 (STP) cm^{-3} atm^{-1}]$$

Values of S cited in the following tables are based on the value calculated by $S = (P/D) \times 76$. The solubility coefficients in these units for condensable vapors such as water are hypothetical values extrapolated to vapor pressure of 1.013 bar (1 atm.).

The temperature dependence of the permeability coefficient, the diffusion constant, and the solubility coefficient can be represented by

$$P = P_o \exp (- E_p /RT)$$

and

$$D = D_o \exp (- E_D /RT)$$

and

$$S = S_o \exp (- \Delta H_s /RT)$$

consequently

$$E_p = E_D + \Delta H_s$$

where E_p is the activation energy of permeation, E_D is the activation energy of diffusion and ΔH_s is the heat of solution. Values of E_p, E_D and ΔH_s are given in kJ/mol in the following tables. R is the gas constant in joule (8.3144 joules per $^\circ$C per mol).

Since the transmission rate Q is not a real constant which is characteristic for a polymer, it should only be used as a means of comparing orders of magnitude. In the following tables, the parameter Q is presented in the units of

$$\text{units of } Q: \quad [g \; mil \; m^{-2} day^{-1}]$$

since most of the hydrophobic polymers, such as polyolefins and fluorocarbon polymers, have water transmission rates near unity at around room temperature when expressed in these units. This offers a convenient reference point for relative comparision of permeabilities of vapors in various polymers.

Although permeability coefficients are listed for many polymers in the following tables, the permeability coefficient, in a strict sense, is not solely a function of the chemical structure of polymers. The permeability coefficient varies with the morphology of polymers and depends on many physical factors such as density, degree of crystallinity, degree of orientation, etc. However, the chemical structure of a polymer can be considered as the predominant factor which controls the magnitude of the permeability coefficient.

Since $P = D \cdot S$, a high permeability coefficient does not always mean a high diffusivity of the permeant. The permeability coefficient defined previously becomes essentially equal to the diffusion constant only when the amount of permeant and the driving force can be expressed by the concentration of the penetrant inside the polymer.

Since the transmission rate Q includes neither pressure nor concentration of permeant in its dimensions, it is necessary to know either the vapor pressure at saturation or the concentration of permeant under the conditions of the measurement in order to correlate Q to P or D of the permeant in the polymer. Knowledge of the following general trends in permeability as related to some influencing factors may be useful for the proper interpretation of the tables:

Density can be regarded as a measure of "looseness" of the polymer structure and, in general, the lower the density the higher the permeability.

Molecular weight of a polymer has been found to have little effect on the permeability of polymer, except in the very low range of molecular weights.

Crystallinity of a semi-crystalline polymer reduces the permeability significantly from the value of the corresponding amorphous polymer, i.e., the higher the degree of crystallinity, the lower the permeability.

Orientation of polymer molecules reduces the permeability.

Crosslinking decreases the permeability, especially for large molecular size permeants.

Plasticizers usually, but not always, increase the permeability.

Humidity increases the permeability of some hydrophilic polymers.

Liquid permeants have slightly higher permeabilities than the corresponding saturated vapor under many practical conditions.

Solution-cast films have variable permeabilities depending upon the kind of solvent and the drying technique. Poor solvents tend to yield films of higher permeability.

The method of vulcanization has a significant effect on the permeability of elastomers.

Fillers generally decrease the permeability; however, the effect is complicated by the type, shape and amount of filler, and its interaction with the permeant.

Thickness of film does not, in principle, affect the permeability coefficient, the diffusion constant, and the solubility coefficient; however, different values may be obtained by films of identical sample but of various thickness due to the difference of morphology and to the effect of asymmetry introduced in the preparation of films of various thickness.

Details of these factors will be found in the following general references:

V. Stannett, M. Szwarc, R. L. Bhargava, J. A. Meyer, A. W. Myers and C. E. Rogers:
"Permeability of Plastic Films and Coated Papers to Gases and Vapors", Tappi Monograph Series No. 23, Technical Association of the Pulp and Paper Industry, New York, 1962.

V. Stannett and H. Yasuda:
"Permeability" in "Crystalline Olefin Polymers", Part II, High Polymers, Volume XX, edited by R. A. V. Raff and K. W. Doak, Wiley-Interscience, New York, 1964.

V. Stannett and H. Yasuda:
"The Measurement of Gas and Vapor Permeation and Diffusion in Polymers" in "Testing of Polymers" edited by J. V. Schmitz, Wiley-Interscience, New York, 1965.

C. E. Rogers:
"Solubility and Diffusivity" in "Physics and Chemistry of the Organic Solid State" edited by D. Fox, M. M. Labes and A. Weissberger, Wiley-Interscience, New York, 1965.

H. Yasuda and V. Stannett:
"Barriers, Vapor" in "Encyclopedia of Polymer Science and Technology", Vol. 2, edited by H. F. Mark, N. G. Gaylord and N. M. Bikales, Wiley-Interscience, New York, 1965.

H. Yasuda, H. G. Clark, and V. Stannett:
"Permeability" in "Encyclopedia of Polymer Science and Technology", Vol. 9, edited by H. F. Mark, N. G. Gaylord and N. M. Bikales, Wiley-Interscience, New Aork, 1968.

V. Stannett: "Simple Gases", and J. A. Barrie "Water in Polymer" in "Diffusion in Polymers", edited by J. Crank and G. S. Park, Academic Press, New York, 1968.

B. CONVERSION FACTORS FOR VARIOUS UNITS OF PERMEABILITY COEFFICIENT P

Multiplication Factors to Obtain P in

From	$\dfrac{[cm^3]\ [cm]}{[s]\ [cm^2]\ [cm\ Hg]}$	$\dfrac{[cm^3]\ [cm]}{[s]\ [cm^2]\ [Pa]}$	$\dfrac{[cm^3]\ [cm]}{[day]\ [m^2]\ [atm]}$
$\dfrac{[cm^3]\ [cm]}{[s]\ [cm^2]\ [cm\ Hg]}$	1	7.5×10^{-4}	6.57×10^{10}
$\dfrac{[cm^3]\ [mm]}{[s]\ [cm^2]\ [cm\ Hg]}$	10^{-1}	7.5×10^{-5}	6.57×10^{9}
$\dfrac{[cm^3]\ [cm]}{[s]\ [cm^2]\ [atm]}$	1.32×10^{-2}	9.9×10^{-6}	8.64×10^{8}
$\dfrac{[cm^3]\ [mil]}{[day]\ [m^2]\ [atm]}$	3.87×10^{-14}	2.90×10^{-17}	2.54×10^{-3}
$\dfrac{[in^3]\ [mil]}{[day]\ [100\ in^2]\ [atm]}$	9.82×10^{-12}	7.37×10^{-15}	6.46×10^{-1}
$\dfrac{[cm^3]\ [cm]}{[day]\ [m^2]\ [atm]}$	1.52×10^{-11}	1.14×10^{-14}	1

C. CONVERSION FACTORS FOR VARIOUS UNITS OF TRANSMISSION RATE Q

Multiplication Factors to Obtain Q in

From	$\dfrac{[g]\ [cm]}{[m^2]\ [day]}$	$\dfrac{[g]\ [mil]}{[m^2]\ [day]}$	$\dfrac{[g]\ [mil]}{[100\ m^2]\ [hr]}$	$\dfrac{[g]\ [mil]}{[100\ in^2]\ [hr]}$
$\dfrac{[g]\ [cm]}{[m^2]\ [day]}$	1	3.94×10^{2}	1.64×10^{3}	1.06
$\dfrac{[g]\ [mil]}{[m^2]\ [day]}$	2.54×10^{-3}	1	4.17	2.69×10^{-3}
$\dfrac{[g]\ [mil]}{[100\ m^2]\ [hr]}$	6.10×10^{-4}	2.40×10^{-1}	1	6.45×10^{-4}
$\dfrac{[g]\ [mil]}{[100\ in^2]\ [hr]}$	9.45×10^{-1}	3.72×10^{2}	1.55×10^{3}	1

D. PERMEABILITY COEFFICIENTS, DIFFUSION CONSTANTS AND SOLUBILITY COEFFICIENTS OF POLYMERS

Units used in Tables are

$$P \text{ in } [cm^3 \text{(STP)} \, cm \, cm^{-2} \, s^{-1} \, (cm \, Hg)^{-1}]$$

$$D \text{ in } [cm^2 \, s^{-1}]$$

$$S \text{ in } [cm^3 \text{(STP)} \, cm^{-3} \, atm^{-1}]$$

$$E_p, \, E_D, \, \Delta H_s \text{ in } [kJ \, mol^{-1}]$$

To obtain corresponding coefficients in suggested new international units the following factors should be used to multiply:

$$P \times 7.5 \times 10^{-4} = [cm^3 \text{(STP)} \, cm \, cm^{-2} \, s^{-1} \, Pa^{-1}]$$

$$S \times 0.987 \times 10^{-5} = [cm^3 \text{(STP)} \, cm^{-3} \, Pa^{-1}]$$

Polymer	Permeant	$T[^oC]$	$P \times 10^{10}$	E_p	$D \times 10^6$	E_D	$S \times 10^2$	ΔH_s	Ref.
1. Poly(dienes)									
Poly(1,3-butadiene)	H_2	25	41.9	27.6	9.6	21.3	3.3	6.2	35
	N_2	25	6.42	34.3	1.1	30.1	4.5	4.2	35
	O_2	25	19.0	29.7	1.5	28.4	9.7	1.3	35
	CO_2	25	138.0	21.8	1.05	30.5	100.0	- 8.8	35
	H_2O	37.5	5,070.0						12
cis	He	24.6	32.6	20.2	15.7	17.3	1.58	2.90	25
	Ne	24.6	19.2	21.7	6.55	17.4	2.23	4.4	25
	Ar	24.6	41.0	19.4	4.06	21.3	7.68	-1.8	25
	N_2	24.6	19.2	21.3★	2.96	25.0★	4.94	-3.7★	25
gutta percha	H_2	25	14.4	42.3★	5.0	31.8★	2.3	13.0★	34
	O_2	25	6.16	49.8★	0.7	38.5★	6.7	13.8★	34
	N_2	25	2.17	52.3★	0.5	41.0★	3.3	13.8★	34
	CO_2	25	35.4	39.3	0.4	41.0	56.0	0.84	34
	H_2O	25	510.0						31
hydrogenated (Hydropol) ρ = 0.894									
	He	25	15.7	35.1	15.1	24.6	0.795	10.4	16
	O_2	25	11.3	41.0	1.2	38.9	7.17	2.0	16
	Ar	25	11.0	43.0	0.96	40.1	8.75	2.9	16
	CO_2	25	48.2	36.4	0.91	36.8	40.4	- 0.4	16
	CO	25	6.16	44.8	0.82	37.2	5.73	7.5	16
	N_2	25	3.98	46.8	0.74	39.3	4.11	6.6	16
	CH_4	25	13.0	45.6	0.54	43.9	18.3	1.6	16
	C_2H_6	25	33.1	41.8	0.24	49.3	105.0	- 7.5	16
	C_3H_4	25	224.0	35.9	0.31	45.1	549	- 9.2	16
	C_3H_6	25	82.5	40.1	0.2	48.1	315.0	1.0	16
	C_3H_8	25	53.7	43.5	0.1	52.3	408	- 8.7	16
	SF_6	25	1.13	56.9	0.056	60.2	15.4	- 3.3	16
Poly(butadiene-co-acrylonitrile) 80/20 (Perbunan 18)									
	He	25	16.9	28.4	15.5	17.5	0.83	10.8	35
	H_2	25	25.2	30.1	6.43	25.9	3.0	4.1	35
	N_2	25	2.52	41.4	0.51	35.5	3.8	5.8	35
	O_2	25	8.16	35.9	0.79	33.8	7.8	2.0	35
	CO_2	25	63.1	29.2	0.425	38.4	113.0	- 9.2	35
73/27 (Perbunan)									
	He	25	12.2	29.2	11.7	21.7	0.80	7.5	35
	H_2	25	15.9	33.1	4.50	28.8	2.7	4.1	35
	N_2	25	1.06	47.6	0.25	43.5	3.2	4.1	35
	O_2	25	3.85	40.5	0.43	38.4	6.8	2.0	35
	CO_2	25	30.8	33.8	0.190	44.7	124.0	-10.8	35
	C_2H_2	25	24.8	41.0	0.0764	51.5	248.0	-10.5	35
	C_3H_8	50	77.7		0.141		420.0		35
61/39 (Hycar OR15)									
	He	25	6.81	32.2	7.92	23.0	0.66	9.2	35
	H_2	25	7.10	36.8	2.43	31.7	2.2	5.0	35
	N_2	25	0.234	57.7	0.064	53.1	2.8	4.6	35
	O_2	25	0.961	50.2	0.136	45.6	5.4	4.6	35

★ Values for below 40°C. Melting point 40∼50°C.

Polymer	Permeant	$T[^{o}C]$	$P \times 10^{10}$	E_p	$D \times 10^6$	E_D	$S \times 10^2$	ΔH_s	Ref.
Poly(butadiene-co-acrylonitrile) 61/39 (Hycar OR15) (Cont'd.)									
	CO_2	25	7.43	50.2	0.038	45.6	149.0	4.6	35
68/32 (Hycar OR25)									
	He	25	9.83	30.9	11.2	21.7	0.67	9.2	35
	H_2	25	11.8	34.3	3.85	29.2	2.3	5.2	35
	N_2	25	0.603	51.4	0.152	48.9	3.0	2.5	35
	O_2	25	2.33	43.9	0.28	43.0	6.4	0.8	35
	CO_2	25	18.5	37.6	0.107	50.2	132	-12.5	35
Poly(butadiene-co-styrene) 92/8 (Ameripol 1502)									
	He	24.0	22.9	26.7	15.7	23.6	1.11	3.0	25
	Ne	24.0	9.70	29.1	5.47	29.3	1.34	0.1	25
	Ar	24.0	12.7	33.6	1.39	33.2	6.95	0.4	25
	N_2	24.0	5.11	36.2	1.05	35.9	3.70	0.3	25
80/20 (Hycar 2001)									
	He	25.4	13.4	29.4	16.0	18.0	0.636	11.4	25
	Ne	25.4	5.01	29.4	4.23	25.6	0.897	3.8	25
	Ar	25.4	4.49	42.1	0.58	41.2	5.88	0.9	25
	N_2	25.4	1.71	41.5	0.43	42.1	3.04	-0.5	25
Poly(chloroprene) (Neoprene G)									
	H_2	25	13.6	33.8	4.31	27.6	2.6		33,36
	O_2	25	4.0	41.4	0.43	39.3	7.5		33,36
	N_2	25	1.2	44.3	0.29	43.0	3.6		33,36
	Ar	25	3.79		0.16	48.9			36
	CO_2	25	25.8	35.5	0.27	45.1	83.0		33,36
	CH_4	25	3.27						36
	H_2O	25	910						31
	NH_3	25					880		36
	SO_2	25					1810		36
Poly(dimethylbutadiene) (Methyl rubber)									
	He	25	14.4						
	H_2	25	17.0	33.4	3.9	31.3	3.3	2.0	35
	N_2	25	0.472	55.6	0.079	51.8	4.6	3.8	35
	O_2	25	2.10	47.3	0.14	46.4	11.4	0.8	35
	CO_2	25	7.47	46.8	0.063	53.5	91.0	- 6.6	36
Poly(isoprene) (Natural rubber)									
	He	25			21.6	19.7			16
	O_2	25	23.3	29.3	1.73	33.5	10.3	- 4.2	16
	Ar	25	22.8	32.6	1.36	33.1	12.8	- 0.4	16
	CO_2	25	153	21.8	1.25	34.3	93.6	-12.5	16
	CO	25	15.7	31.0	1.35	31.0	8.88	0	16
	N_2	25	9.43	35.6	1.17	33.5	6.15	2.1	16
	CH_4	25	30.1	31.0	0.89	36.4	25.8	- 5.4	16
	C_2H_6	25			0.40	42.7			16
	C_3H_4	25	550	22.6	0.50	37.7	840.0	-15.0	16
	C_3H_6	25	204	28.9	0.31	42.7	503.0	-13.8	16
	C_3H_8	25	168	23.0	0.21	46.4	610.0	-23.4	16
	SF_6	25	3.59	35.6	0.115	50.2	23.8	-14.6	16
	C_2H_2	25	98.9	30.5	0.467	39.7	162.0	- 9.2	35
	H_2O	25	2290						31
Poly(isoprene), hydrochloride (Pliofilm)									
	He	20	19						10
	N_2	30	0.14	41.8					37
	O_2	30	0.54	35.2					37
	CO_2	30	1.3	35.9					37
	H_2O	25	16	28.5					39
	H_2S	30	0.1	73.6					8
	ethylene oxide	0	7.3						37
	CH_3Br	60	34						37
Poly(isoprene-co-acrylonitrile) 74/26									
	He	25	7.77	31.8	8.01	20.5	0.74	11.3	35
	H_2	25	7.41	38.1	2.47	31.0	2.3	7.1	35
	N_2	25	0.181	62.8	0.045	60.7	3.1	2.1	35
	O_2	25	0.852	53.1	0.092	50.6	7.1	2.5	35
	CO_2	25	4.32	52.3	0.031	60.2	107.0	- 7.9	35

Polymer	Permeant	$T[^oC]$	$P \times 10^{10}$	E_p	$D \times 10^6$	E_D	$S \times 10^2$	ΔH_s	Ref.
Poly(isoprene-co-methacrylonitrile) 74/26									
	H_2	25	13.6	33.9	3.55	28.9	2.9	5.0	35
	N_2	25	0.596	53.6	0.123	48.5	3.7	5.0	35
	O_2	25	2.34	46.0	0.24	40.2	7.5	5.9	35
	CO_2	25	14.1	42.3	0.091	51.0	120.0	-8.7	35
Poly(2-methyl-co-4-methyl-1,3-pentadiene) 85/15									
	H_2		42.4	29.7					35
	N_2		2.74	46.9	0.30	46.4	7.0	0.4	35
	O_2		9.98	39.3	0.55	41.0	13.8	-1.7	35
	CO_2		44.8	34.7					35

2. Poly(alkenes)

Polymer	Permeant	$T[^oC]$	$P \times 10^{10}$	E_p	$D \times 10^6$	E_D	$S \times 10^2$	ΔH_s	Ref.
Poly(ethylene) (density 0.914)									
	He	25	4.9	34.7	6.8	24.6	0.551	10.0	16
	O_2	25	2.88	42.6	0.46	40.1	4.78	2.5	16
		25	2.72	45.1	0.36	42.2	5.78	2.9	16
	CO_2	25	12.6	38.9	0.37	38.4	25.8	0.4	16
	CO	25	1.48	46.4	0.33	39.7	3.40	6.6	16
	N_2	25	0.969	49.3	0.32	41.4	2.31	7.9	16
	CH_4	25	2.88	47.2	0.19	45.6	11.4	1.6	16
	C_2H_6	25	6.81	47.2	0.068	53.5	76.5	-6.2	16
	C_3H_4	25	42.2	38.9	0.11	49.7	306	-10.8	16
	C_3H_6	25	14.4	43.5	0.058	52.3	190.0	-8.8	16
	C_3H_8	25	9.43	46.8	0.032	55.6	216	-8.8	16
	SF_6	25	0.170	59.8	0.014	61.9	9.63	-2.1	16
	H_2O	25	90	33.4					21
	ethylene oxide	0	49						37
	CH_3Br	60	472						37
Poly(ethylene) (density 0.964)									
	He	25	1.14	29.7	3.07	23.4	0.283	6.3	16
	O_2	25	0.403	35.1	0.170	36.8	1.81	-1.7	16
	Ar	25	1.69	37.6	0.116	38.9	11.1	-1.3	16
	CO_2	25	0.36	30.2	0.124	35.5	2.21	-5.3	16
	CO	25	0.193	39.3	0.096	36.8	1.53	2.5	16
	N_2	25	0.140	39.7	0.090	37.6	1.17	0.1	16
	CH_4	25	0.388	40.5	0.057	43.5	5.19	-3.0	16
	C_2H_6	25	0.590	42.6	0.015	52.3	30.8	-9.7	16
	C_3H_4	25	4.01	33.1	0.025	47.3	123.0	-14.2	16
	C_3H_6	25	1.15	38.9	0.011	52.3	83.0	-13.4	16
	C_3H_8	25	0.537	44.7	0.0049	56.9	83.7	-12.2	16
	SF_6	25	0.0083	55.2	0.0016	62.8	4.0	-7.6	16
	H_2O	25	12.0						21
Poly(ethylene-co-propylene) 40/60									
	He	24.5	31.4	27.6	14.4	15.0	1.66	12.6	25
	Ne	24.5	12.0	31.8	4.50	28.5	2.02	3.3	25
	Ar	24.5	12.9	39.6	1.06	37.0	9.27	2.5	25
	N_2	24.5	4.87	43.7	0.67	45.8	5.52	-2.0	25
50/50									
	He	24.2	29.4	28.1	17.6	24.8	1.27	3.3	25
	Ne	24.2	11.3	32.7	5.30	31.8	1.62	0.8	25
	Ar	24.2	14.5	37.7	1.30	42.0	8.44	-4.3	25
	N_2	24.2	5.50	42.1	0.938	42.8	4.45	-0.7	25
60/40									
	He	26.7	21.4	29.7	9.42	27.1	1.73	2.6	25
	Ne	26.7	9.23	34.1	3.48	31.8	2.01	2.1	25
	Ar	26.7	12.2	37.2	1.06	39.5	8.74	-2.3	25
	N_2	26.7	4.89	41.2	0.79	43.9	4.70	-3.1	25
	H_2O	37.5	450						12
Poly(isobutene-co-isoprene) 98/2 (Butyl rubber)									
	He	25	8.38	31.8	5.93	24.3	1.1	7.5	35
	H_2	25	7.20	36.4	1.52	33.9	3.6	2.5	35
	N_2	25	0.324	52.3	0.045	50.6	5.5	1.7	35
	O_2	25	1.30	44.8	0.081	49.8	12.2	-5.0	35
	CO_2	25	5.16	41.4	0.0578	50.2	68.0	-8.8	35

Polymer	Permeant	$T[^oC]$	$P \times 10^{10}$	E_p	$D \times 10^6$	E_D	$S \times 10^2$	ΔH_s	Ref.
Poly(isobutene-co-isoprene) 98/2 (Butyl rubber) (Cont'd.)									
	C_3H_8	50	13.6		0.0224		470.0		35
	H_2O	37.5	110						12
Poly(4-methylpentene-1)									
	He	25	101						41
	H_2	25	136						41
	N_2	25	7.83		0.55		10.8		41
	O_2	25	32.3		1.01		24.3		41
	CO_2	25	92.6		0.684		103.0		41
Poly(propylene)	He	20	38	32.2	19.5	30.5			13
	H_2	20	41	38.5	2.12	34.7			13
	N_2	30	0.44	55.7					19
	O_2	30	2.3	47.7					19
	CO_2	30	9.2	38.1					19
	H_2O	25	51	42.3					21
	H_2S	20	0.33						5
	NH_3	20	9.2						5

3. Poly(methacrylates), Poly(nitriles), Poly(styrenes), Others

Polymer	Permeant	$T[^oC]$	$P \times 10^{10}$	E_p	$D \times 10^6$	E_D	$S \times 10^2$	ΔH_s	Ref.
Poly(ethyl methacrylate)									
	He	25	6.82	26.8	42.3	15.5	0.12	11.3	29
	Ne	25	2.98	28.7	1.51	23.9	1.5	5.1	29
	O_2	25	1.15	36.4	0.103	31.8	8.6	4.6	29
	A	25	0.565	38.1	0.020	43.1	21.5	- 5.1	29
	N_2	25	0.220	40.6	0.022	42.7	7.5	- 2.1	29
	CO_2	25	5.00	28.9	0.033	33.1	112.0	- 4.2	29
	Kr	25	0.377	43.1	0.0065	46.0	44.0	- 2.9	29
	H_2S	25	3.83	39.3	0.0044	47.7	66.2	- 8.4	29
	SF_6	25	1.65×10^{-6}	44.8	0.00022	64.4	0.0057	-19.7	29
	H_2O	25	3200	2.1	0.0989	36.4	24600	-34.3	29

Poly(nitriles)

Polymer	Permeant	$T[^oC]$	$P \times 10^{10}$	E_p	$D \times 10^6$	E_D	$S \times 10^2$	ΔH_s	Ref.
Barex (Sohio)	CO_2	25	0.016						42
	O_2	25	0.0054						42
	H_2O	25	660						42
	CO_2	38	0.031						42
	O_2	38	0.018						42
	H_2O	38	690						42
Lopac (Monsanto)	CO_2	25	0.011						42
	O_2	25	0.0035						42
	H_2O	25	340						42
	CO_2	38	0.021						42
	O_2	38	0.0070						42
	H_2O	38	370						42
Poly(acrylonitrile)	CO_2	25	0.0008						42
	O_2	25	0.0002						42
	H_2O	25	300						42
Poly(acrylonitrile-co-methyl acrylate-co-butadiene) 79/15/6									
	CO_2	25	0.0160						42
	O_2	25	0.0054						42
	H_2O	25	1,300						42
Poly(acrylonitrile-co-styrene) 86/14									
	CO_2	25	0.0150						42
	O_2	25	0.0042						42
	H_2O	25	850						42
66/34									
	CO_2	25	0.21						42
	O_2	25	0.048						42
	H_2O	25	2,000						42
57/43									
	CO_2	25	0.365						42
	O_2	25	0.180						42
	H_2O	25	2,500						42

Polymer	Permeant	T[°C]	$P \times 10^{10}$	E_p	$D \times 10^6$	E_D	$S \times 10^2$	ΔH_s	Ref.
Poly(methacrylonitrile)									
	CO_2	25	0.0032						42
	O_2	25	0.0012						42
	H_2O	25	410						42
Poly(methacrylonitrile-co-styrene) 97/3									
	CO_2	25	0.0078						42
	O_2	25	0.0024						42
	H_2O	25	490						42
82/18	CO_2	25	0.050						42
	O_2	25	0.013						42
	H_2O	25	930						42
61/39	CO_2	25	0.275						42
	O_2	25	0.090						42
	H_2O	25	1,700						42
53/47	CO_2	25	0.51						42
	O_2	25	0.16						42
	H_2O	25	1,900						42
38/62	CO_2	25	1.2						42
	O_2	25	0.39						42
	H_2O	25	2,100						42
18/82	CO_2	25	3.2						42
	O_2	25	1.1						42
	H_2O	25	2,000						42
Poly(methacrylonitrile-co-styrene-co-butadiene) 88/7/5									
	CO_2	25	0.014						42
	O_2	25	0.0048						42
	H_2O	25	600						42
83/7/10	CO_2	25	0.019						42
	O_2	25	0.0069						42
	H_2O	25	670						42
78/7/15	CO_2	25	0.032						42
	O_2	25	0.0096						42
	H_2O	25	770						42
Poly(styrene)	He	25	18.7		10.4	12.9			13,41
	H_2	25	23.3		43.6	16.7			13,41
	N_2	25	0.788						41
	O_2	25	2.63		0.11	34.7			30,41
	CO_2	25	10.5		0.058	36.4			30,41
	H_2O	25	1,200						20
		25	1,130		0.14		5092		27
Poly(tetrafluoroethylene)									
	H_2	25	9.8	21.3					23
	N_2	25	1.4	24.4	0.088	30.0	0.159	- 5.4	23
	O_2	25	4.2	20.1	0.152	26.3	0.276	- 7.2	23
	CO_2	25	11.7	13.9	0.095	28.6	1.230	-14.7	23
	NO_2	25	15.7	10.6	0.0037	58.6	43.0	-47.7	23
	N_2O_4	25	37.5	-2.1	0.025		15.0		23
Poly(tetrafluoroethylene-co-hexafluoropropene) (Teflon FEP)									
	N_2	25	1.59	30.5	0.0948	38.5	12.7	- 7.9	24
	O_2	25	4.9	25.5	0.184	34.7	20.5	- 9.2	24
	CO_2	25	12.7	20.9	0.105	36.6	92.0	-15.7	24
	CH_4	25	0.865	34.7	0.0298	45.6	22.0	-10.8	24
	C_2H_6	25	0.439	36.6	0.00470	51.8	71.0	-15.1	24
	C_3H_8	25	0.142	43.1	0.00077	60.3	140.0	-17.1	24
	C_2H_4	25	0.635	34.9	0.0098	49.4	49.4	-14.3	24
Poly(trifluorochloroethylene) (Kel-F)									
	He	20	6.8	23.4					10
	H_2	20	0.94	30.2					10
	N_2	25	0.003	49.8					20
	O_2	40	0.025	45.6					20
	CO_2	40	0.048	46.4					20
	H_2O	25	0.29						20
	ethylene oxide	0	1.2						37

Polymer	Permeant	T[°C]	$P \times 10^{10}$	E_p	$D \times 10^6$	E_D	$S \times 10^2$	ΔH_s	Ref.
Poly(trifluorochloroethylene) (Kel-F) (Cont'd.)									
	CH_3Br	60	4.6	★		★		★	37
Poly(vinyl acetate)	He	30	12.6	31.2★	9.55	22.4★	1.05	8.8★	14
	H_2	30	8.9	41.7★	2.63	31.4★	2.63	10.3★	14
	Ne	30	2.64	39.8★	1.66	35.4★	1.26	4.4★	14
	O_2	30	0.50	56.1★	0.055	60.6★	6.3	- 4.6★	14
	Ar	30	0.19	61.1★	0.016	69.1★	9.8	- 7.8★	14
	Kr	30	0.078	76.2★	0.0030	81.2★	20.0	- 5.1★	15
	CH_4	30	0.050	78.6	0.0028	80.7	15.0	- 2.1	15
Poly(vinyl alcohol)	He	20	0.001						10
	H_2	25★★	0.009						28
	N_2	14★★★★	< 0.001						11
		14	0.33		0.045		5.32		11
	O_2	25	0.0089						28
	CO_2	25★★	0.012						28
		23★★★★★	0.001						11
		23	11.9		0.0476		190		11
	H_2S	25	0.007						28
	ethylene oxide	0	0.002						37
Poly(vinyl chloride)	He	25	2.05	29.9	2.80	20.6	0.556	9.2	32
	H_2	25	1.70	34.4	0.500	34.4	2.58	0	32
	Ne	25	0.392	34.1	0.250	31.5	1.19	2.5	32
	N_2	25	0.0118	69.1	0.00378	61.9	2.37	7.1	32
	Ar	25	0.0115	57.7	0.00115	51.5	7.60	6.3	32
	O_2	25	0.0453	55.6	0.0118	54.4	2.92	1.3	32
	CO_2	25	0.157	56.9	0.00250	64.4	47.7	- 7.9	32
	CH_4	25	0.0286	66.1	0.00126	70.3	17.3	- 4.2	32
	H_2O	25	275	22.9	0.0238	41.7	8,780.0	-15.1	32
Poly(vinylidene chloride) (Saran)									
	He	34	0.31						10
	N_2	30	0.00094	70.2					37
	O_2	30	0.0053	66.5					37
	CO_2	30	0.03	51.4					37
	H_2O	25	0.5	46.0					21
	H_2S	30	0.03	74.4					8
	CH_3Br	60	0.008						37

4. Poly(amides), Poly(esters), Poly(oxides), Poly(siloxanes)

Polymer	Permeant	T[°C]	$P \times 10^{10}$	E_p	$D \times 10^6$	E_D	$S \times 10^2$	ΔH_s	Ref.
Poly(iminohexamethyleneiminoadipoyl) (Nylon 66)									
	CO_2	25	0.0069	8.4	0.00083	28.0	64.0	-19.7	4
Poly(imino(1-oxohexamethylene) (Nylon 6)									
	He	20	0.53	36.4					10
	N_2	30	0.0095	46.9					37
	O_2	30	0.038	43.5					37
	CO_2	20★★	0.088	40.6					37
		30★★★★★	0.10						11
		30	0.29						11
	H_2O	25	177						20
	H_2S	30	0.33	58.2					8
	NH_3	20	1.2						5
	CH_3Br	60	0.84						37
Poly(imino(1-oxoundecamethylene) (Nylon 11)									
	He	30	1.95	28.0	3.50	22.2	0.43	5.8	1
	Ne	30	0.344	35.4	0.437	41.4	0.60	2.5	1
	Ar	40	0.189	41.8	0.037	44.8	3.9	- 2.9	1
	H_2	30	1.78	40.0	0.984	30.0	1.38	- 1.7	1
	CO_2	40	1.00	33.9	0.019	51.9	40	-18.0	1

★ above 20°C.   ★★ R.H. 0%.   ★★★★ R.H. 90%.   ★★★★★ R.H. 94%.   ★★★★★ R.H. 95%.

Polymer	Permeant	T[$^{\circ}$C]	P x 10^{10}	E_p	D x 10^6	E_D	S x 10^2	ΔH_s	Ref.
Poly(oxy-2,6-dimethyl-1,4-phenylene)									
	He	25	78.1						41
	H_2	25	112.8						41
	N_2	25	3.81						41
	O_2	25	15.8						41
	CO_2	25	75.7		0.0601		958.0		41
	H_2O	25	4,060		0.170		18,200.0		40
Poly(oxydimethylsilylene) 10 weight% filler, vulcanized									
	He	0	233		41	7.9	4.3		2
		25	355						41
	Ne	0	191		16.1	8.4	9.0		2
	Ar	0	550		12.2	10.3	34		2
	Kr	0	1,020		8.0	12.2	98		2
	Xe	0	2,550		4.8	13.9	400		2
	H_2	0	464		47	9.3	7.4		2
		25	649						41
	N_2	0	227		8.5	10.0	20.3		2
		25	281						41
	O_2	0	489		12.0	11.3	31.1		2
		25	605						41
	CO_2	25	3,240						41
	H_2O	35	43,000						3
	n-butane	0	19,000		2.40	14.2	6,100		2
Poly(oxyethyleneoxyterephthaloyl) (Poly(ethylene terephthalate))									
crystalline									
	He	25	1.32	19.7	2.0	20	0.53	0.9	17,18
	O_2	25	0.035	32.3	0.0035	46.1	7.5	-13.0	17,18
	N_2	25	0.0065	32.7	0.0014	44.0	5.0	-18.4	17,18
	CO_2	25	0.17	18.4	0.0006	50.2	200	-31.4	17,18
	CH_4	25	0.0032	36.9	0.00011	52.4	22.5	-22.2	17,18
	H_2O	25	130	2.9					21
	CH_3Br	60	0.08						37
amorphous									
	He	25	3.28	21.3	3.0	19.3	0.8	1.1	17,18
	O_2	25	0.059	37.6	0.005	48.5	10.0	-14.6	17,18
		25	0.029	28.8	0.0023	48.1	8.0	-15.0	17,18
	N_2	25	0.013	26.4	0.002	47.4	6.0	-23.8	17,18
	CO_2	25	0.30	27.6	0.00085	52.3	300.0	-28.5	17,18
	CH_4	25	0.009	24.7	0.0003	51.1	25.0	-27.2	17,18
Poly(oxymethylene)	CO_2	25	180	31.4	0.014	49.0	0.14		38
	H_2O	25	910	13.0	0.027		0.42		38
Poly(oxycarbonyloxy-1,4-phenyleneisopropylidene-1,4-phenylene) (Lexan)									
	H_2	25	12.0	22.3	0.640	20.9	0.14	1.7	22
	Ar	25	0.8	22.3	0.015	25.1	0.42	-2.5	22
	O_2	25	1.4	19.3	0.021	32.2	0.51	-12.9	22
	CO_2	25	8.0	15.8	0.0048	37.6	12.6	-21.7	22
	SF_6	25	6.5×10^{-6}		1×10^{-7}	83.7	0.50		22
	H_2O	25	1400.0		0.680	25.9	169		22
	N_2	25	0.30	25.1	0.200				22
	Ne	100	14.5						22
5. Cellulose and Derivatives									
Cellulose (Cellophane)									
	He	20	0.0005	41.4					10
	H_2	25	0.0065						28
	N_2	25	0.0032						28
	O_2	25	0.0021						28
	CO_2	25	0.0047						28
	H_2O	25	1,900						7
	H_2S	45	0.006	89.5					8
	SO_2	25	0.0017						28

References page III-240

Polymer	Permeant	T[°C]	$P \times 10^{10}$	E_p	$D \times 10^{6}$	E_D	$S \times 10^{2}$	ΔH_s	Ref.
Cellulose acetate	He	20	13.6						10
	H_2	20	3.5						10
	N_2	30	0.28	27.1					37
	O_2	30	0.78	20.9					37
	CO_2	30	22.7	17.9					37
	H_2O	25	5,500	0					21
	H_2S	30	3.5	21.3					8
	ethylene oxide	0	17						37
	CH_3Br	60	6.8						37
Cellulose nitrate	He	25	6.9	21.7					9
	H_2	20	2.0	23.8					10
	N_2	25	0.12						9
	O_2	25	1.95						9
	CO_2	25	2.12						9
	H_2O	25	6,300						9
	H_2S	25							
	NH_3	25	57.1						9
	SO_2	25	1.76						9
Ethyl cellulose	He	30	400	13.3					26
	H_2	20	87	-					10
	N_2	30	8.4	17.5					37
	O_2	30	26.5	17.5					37
	CO_2	30	41.0	5.4					37
	H_2O	20	12,000	1.6					39
	NH_3	25	705						9
	SO_2	25	264						9
	ethylene oxide	0	420						37

E. PERMEABILITY COEFFICIENTS OF SOME ORGANIC SOLVENTS THROUGH POLYMERS[*]

in $[cm^3\ cm\ cm^{-2}\ s^{-1}\ (cm\ Hg)^{-1}] \times 10^{10}$ at 35°C

Polymer	Benzene	n-Hexane	carbon tetrachloride	Ethyl alcohol	Ethyl acetate
Cellulose	1.4	0.912	0.836	85.8	13.4
Cellulose acetate	512	2.80	3.74	2,980	3,595
Poly(hexamethylene adipate-co-sebacate-co-aminocaproate) (Nylon 66/610/6)					
	3.88	4.42	1.13	3,080	2.90
Poly(ethylene) low density	5,300	2,910	3,810	55.9	513
Poly(acrylonitrile)	2.61	1.59	1.47	0	1.34
Poly(styrene)	10,600		6,820	0	sol.
Poly(vinyl alcohol)	3.58	2.34	1.61	32.7	2.53

[*] data from V. L. Simril and A. Hershberger, Modern Plastics 27, No. 10, 97 (1950).

F.   TRANSMISSION RATE Q OF VARIOUS ORGANIC COMPOUNDS THROUGH LOW DENSITY POLY(ETHYLENE)[*]

in $[\text{g mil m}^{-2} \text{ day}^{-1}]$

Permeant	0°C	21.1°C	54.4°C	Permeant	0°C	21.1°C	54.4°C
Acetic acid	5.43	48.1	1,020	Ethyl acetate	29.5	256	5,860
Acetic anhydride	2.02	12.6	459	Ethyl alcohol	-	10.9	-
Acetone	21.7	105	2,850	Formaldehyde	-	8.22	-
Allyl alcohol	2.48	22.5	357	n-Heptane	753	4,190	41,100
Amyl acetate	8.68	134	4,170	Formic acid	3.88	10.4	186
i-Amyl alcohol	-	3.10	-	n-Hexane	744	5,430	140,000
Aniline	3.88	26.4	924	Methyl alcohol	3.88	18.9	431
Benzaldehyde	5.74	105	3,190	Methyl ethyl ketone	57.4	195	5.050
Benzene	700	6,820	69,800	Nitrobenzene	5.74	75.9	1,550
n-Butyl alcohol	1.55	7.13	316	Octyl alcohol	1.55	7.75	397
sec-Butyl alcohol	1.86	9.61	583	n-Pentane	1,500	8,150	233,000
t-Butyl alcohol	0.775	4.03	419	i-Pentane	744	4,190	77,500
Butyraldehyde	13.9	155	9,050	2-Pentane	2,790	10,800	248,000
Carbon tetrachloride	790	9,460	121,000	Phenol	1.55	7.75	372
Chlorobenzene	896	7,050	68,400	n-Propyl alcohol	1.09	7.75	347
p-Chlorotoluene	465	4,710	64,600	i-Propylamine	35.7	632	10,900
Cyclohexane	490	3,890	57,800	Tetradecane	26.4	226	6,260
Decane	147	1,100	18,900	Toluene	899	7,830	89,500
Diethyl ether	744	4,850	12,400	o-Xylene	558	3,970	55,800
Ethanethiol	403	6,200	155,000	p-Xylene	1,330	7,530	74,400

[*] data from M. Salame, S. P. E. Trans. **1**, 153 (1961).

G.   REFERENCES

1.   R. Ash, R. M. Barrer, D. G. Palmer, Polymer **11**, 421 (1970).

2.   R. M. Barrer, H. T. Chio, J. Polymer Sci. C **10**, 111 (1966).

3.   J. A. Barrie, B. Plott, Polymer **4**, 303 (1963).

4.   W. W. Brandt, J. Polymer Sci. **41**, 415 (1959).

5.   V. L. Braumisch, H. Lenhart, Kolloid-Z. **177**, 24 (1961).

6.   J. Crank, C. Robinson, Proc. Roy. Soc. (London) A **204**, 549 (1951).

7.   P. M. Hauser, A. D. Mclaren, Ind. Eng. Chem. intern. Edition **40**, 112 (1948).

8.   W. Heilman, V. Tammela, J. A. Meyer, V. Stannett, M. Szwarc, Ind. Eng. Chem. **48**, 821 (1956).

9.   P. Y. Hsieh, J. Appl. Polymer Sci. **7**, 1743 (1963).

10.   Y. Ito, Kobunshi Kagaku **18**, 124 (1961).

11.   Y. Ito, Kobunshi Kagaku **18**, 158 (1961).

12.   Y. Iyengar, J. Polymer Sci. B **3**, 663 (1965).

13.   D. Jeschke, Ph. D. Thesis, Mainz (1960).

14.   P. Mears, J. Am. Chem. Soc. **76**, 3415 (1954).

15.   P. Mears, Trans. Faraday Soc. **53**, 101 (1957).

16.   A. S. Michaels, H. J. Bixler, J. Polymer Sci. **50**, 413 (1961).

17.   A. S. Michaels, W. R. Vieth, J. A. Barrie, J. Appl. Phys. **34**, 1 (1963).

18.   A. S. Michaels, W. R. Vieth, J. A. Barrie, J. Appl. Phys. **34**, 13 (1963).

19.   A. W. Myers, V. Stannett, M. Szwarc, J. Polymer Sci. **35**, 285 (1959).

20.   A. W. Myers, V. Tammela, V. Stannett, M. Szwarc, Modern Plastics **37**, No. 10, 139 (1960).

21.   A. W. Myers, J. A. Myer, C. E. Rogers, V. Stannett, M. Szwarc, Tappi **44**, 58 (1961).

22.   F. J. Norton, J. Appl. Polymer Sci. **7**, 1649 (1963).

23.   R. A. Pasternak, M. V. Christenson, J. Heller, Macromolecules **3**, 366 (1970).

24.   R. A. Pasternak, G. L. Burns, J. Heller, Macromolecules **4**, 470 (1971).

25.   D. R. Paul, A. T. DiBenedetto, J. Polymer Sci. C **10**, 17 (1965).

26.   R. W. Roberts, K. Kammermeyer, J. Appl. Polymer Sci. **7**, 2183 (1963).

27.   G. Rust, F. Herrero, Materialpruefung **11**, No. 5, 166 (1969).

28.   V. L. Simril, A. Hershberger, Modern Plastics **27**, No. 11, 95 (1950).

29.   V. Stannett, J. L. Williams, J. Polymer Sci. C **10**, 45 (1966).

30.   V. Stannett, unpublished data

31.   R. L. Taylor, D. B. Herrmann, A. R. Kemp, Ind. Eng. Chem. intern. Edition **28**, 1255 (1936).

32.   B. P. Tikhomirov, H. B. Hopfenberg, V. T. Stannett, J. L. Williams, Makromol. Chem. **118**, 177 (1968).

33.   G. J. van Amerongen, J. Appl. Phys. **17**, 972 (1946).

34.   G. J. van Amerongen, J. Polymer Sci. **2**, 381 (1947).

35.   G. J. van Amerongen, J. Polymer Sci. **5**, 307 (1950).

36.   G. J. van Amerongen, Rubb. Chem. Technol. **37**, 1065 (1964).

37.   R. Waack, N. H. Alex, H. L. Frisch, V. Stannett, M. Szwarc, Ind. Eng. Chem. **47**, 2524 (1955).

38.   J. L. Williams, V. Stannett, J. Appl. Polymer. Sci. **14**, 1949 (1970).

39.   H. Yasuda, V. Stannett, J. Polymer Sci. **57**, 907 (1962).

40.   H. Yasuda, V. Stannett, J. Macromol. Sci. Phys. B **3** (4), 589 (1969).

41.   H. Yasuda, Kj. Rosengren. J. Appl. Polymer Sci. **14**, 2839 (1970).

42.   M. Salame, Polymer Symposia **41**, 46 (1973).

REFRACTIVE INDICES OF POLYMERS

L. Bohn
Hoechst AG
Frankfurt/Main, Germany

The table presents the refractive indices of polymers in order of increasing n_D. It contains mainly homopolymers but a few copolymers of special technical interest have been included. Temperatures are in $^\circ$C.

The following general remarks may be helpful for the calculation of approximate refractive indices of other polymers and copolymers.

Calculation of Refractive Indices

The Lorentz-Lorenz Equation

$$(n^2 - 1)/(n^2 + 2) = R/M \; \rho = r \, \rho \qquad (1)$$

correlates the refractive index n to the density ρ, the molecular weight M and the molar refractivity R, where $R/M = r$ is the specific refractivity. It seems to be reasonable to assume that the molar refractivity R of a polymer molecule is the sum of the molar refractivities of the monomer subunits R_M, i. e. $R = DP \cdot R_M$, where DP is the degree of polymerization. With $M = DP \cdot M_M$ (M_M = molecular weight of the monomer unit) the Lorentz-Lorenz Equation for polymers may be written

$$(n^2 - 1)/(n^2 + 2) = R_M/M_M \; \rho = r_M \, \rho \qquad (2)$$

containing only the refractivity of the repeating unit.

Using incremental atomic or bond refractivities (1-4) the refractive indices of various kinds of polymers have been calculated according to the above Equation (1) to give fairly good agreement with measured values.

Refractive Indices of Copolymers (Monomer A and B)

If only the refractive indices n_A and n_B of the corresponding homopolymers are known, a rough estimation of the refractive index n_{AB} of a random copolymer may be obtained by interpolation with respect to the composition by weight (5): $n_{AB} = c_A n_A + (1 - c_A) n_B$.

A more exact calculation is possible, if the densities ρ_A and ρ_B of the homopolymers as well as the density ρ_{AB} of the copolymer are known. It is advisable to calculate first the specific refractivities r_A and r_B from n_A, ρ_A and n_B, ρ_B by use of the Lorentz-Lorenz Equation. By interpolation with regard to the weight contents $r_{AB} = c_A r_A + (1 - c_A) r_B$ is obtained. With this r_{AB} and the known density ρ_{AB} the Lorentz-Lorenz Equation gives a fairly accurate refractive index n_{AB}.
The calculation methods described are applicable in the same way to multi-phase systems like blends, block or graft copolymers or partially crystalline polymers, if the phase dimensions are well below the wavelength of light.

Temperature Dependence (dn/dT)

From the Lorentz-Lorenz Equation the following expression can easily be derived:

$$\frac{dn}{dT} = \frac{(n^2 + 2)(n^2 - 1)}{6 n} \cdot \frac{1}{\rho} \frac{d\rho}{dT} = - q(n)\alpha \qquad (3)$$

where $\alpha = - d\rho/\rho dT$ is the thermal expansion coefficient. $q(n)$ increases from 0.46 ($n = 1.4$) to 0.59 ($n = 1.5$) to 0.74 ($n = 1.6$) in the refractive index range of interest for polymers. With $\alpha = 1.5 \ldots 2.5 \times 10^{-4}/^\circ$C for hard polymers below their glass transition temperature and $\alpha = 5 \ldots 7 \times 10^{-4}/^\circ$C for amorphous polymers above their glass transition temperature the following rough approximation may be obtained:

$$- dn/dT = 1 \ldots\ldots 2 \times 10^{-4}/^\circ C \quad \text{for glassy polymers,}$$
$$- dn/dT = 3 \ldots\ldots 5 \times 10^{-4}/^\circ C \quad \text{for amorphous polymers above the glass transition.}$$

Partially crystalline polymers which are ductile in character like polyolefins, aliphatic polyesters and cellulosic resins, have intermediate values.

Influence of Molecular Weight

End-group refractivities are always more or less different from the refractivity of the repeating unit of the chain. Therefore the refractive index of the whole polymer may strongly depend on the chain length at very low degrees of polymerization. Knowing the refractivities of end-group and repeating unit the degree of polymerization can be obtained from the refractive index (5). In polymers with more than 500 to 1000 chain atoms the refractive index has approached a limiting value (5).

Polymer	n_D	$T\,^\circ C$	Polymer	n_D	$T\,^\circ C$
Poly(tetrafluoroethylene-co-hexafluoropropylene)	1.338		Poly(heptafluorobutyl acrylate)	1.367	25
Poly(pentadecafluorooctyl acrylate)	1.339	25	Poly(trifluorovinyl acetate)	1.375	25
Poly(tetrafluoro-3-(heptafluoropropoxy)propyl acrylate)	1.346	25	Poly(octafluoropentyl acrylate)	1.380	25
Poly(tetrafluoro-3-(pentafluoroethoxy)propyl acrylate)	1.348	25	Poly(pentafluoropropyl acrylate)	1.385	25
			Poly(2-(heptafluorobutoxy)ethyl acrylate)	1.390	25
			Poly(2,2,3,4,4,4-hexafluorobutyl acrylate)	1.392	25
Poly(tetrafluoroethylene)	1.35 (-1.38)		Poly(trifluoroethyl acrylate)	1.407	25
Poly(undecafluorohexyl acrylate)	1.356		Poly(2-(1,1,2,2-tetrafluoroethoxy)ethyl acrylate)	1.412	25
Poly(nonafluoropentyl acrylate)	1.360	25	Poly(trifluoroisopropyl methacrylate)	1.4177	20
Poly(tetrafluoro-3-(trifluoromethoxy)propyl acrylate)	1.360	25	Poly(2,2,2-trifluoro-1-methylethyl methacrylate)	1.4185	
			Poly(2-(trifluoroethoxy)ethyl acrylate)	1.419	25
Poly(pentafluorovinyl propionate)	1.364	25	Poly(trifluorochloroethylene)	1.42-1.43	

Polymer	n_D	$T\,^{o}C$	Polymer	n_D	$T\,^{o}C$
Poly(vinylidene fluoride)	1.42	25	Poly(3,3,5-trimethylcyclohexyl methacrylate)	1.485	20
Poly(dimethylsilylene) (poly(dimethyl siloxane))	1.43		Poly(ethyl methacrylate)	1.485	20-25
Poly(trifluoroethyl methacrylate)	1.437		Poly(2-nitro-2-methylpropyl methacrylate)	1.4868	20
Poly(oxypropylene)	1.4495		Poly(triethylcarbinyl methacrylate)	1.4889	20
Poly(vinyl isobutyl ether)	1.4507	30	Poly(1,1-diethylpropyl methacrylate)	1.4889	20
Poly(vinyl ethyl ether)	1.4540	30	Poly(methyl methacrylate)	1.4893	23
Poly(oxyethylene)	1.4563	30		1.490	20
Poly(vinyl butyl ether)	1.4563	30	Poly(2-decyl-1,3-butadiene)	1.4899	20.5
Poly(vinyl pentyl ether)	1.4581	30	Poly(vinyl alcohol)	1.49-1.53	
Poly(vinyl hexyl ether)	1.4591	30	Poly(ethyl glycolate methacrylate)	1.4903	20
Poly(4-methyl-1-pentene)	1.459-1.465		Poly(3-methylcyclohexyl methacrylate)	1.4947	20
Cellulose acetate butyrate	1.46-1.49		Poly(cyclohexyl α-ethoxyacrylate)	1.4969	20
Poly(4-fluoro-2-trifluoromethylstyrene)	1.46		Methyl cellulose (low viscosity)	1.497	25
Poly(vinyl octyl ether)	1.4613	30	Poly(4-methylcyclohexyl methacrylate)	1.4975	20
Poly(vinyl 2-ethylhexyl ether)	1.4626	30	Poly(decamethylene glycol dimethacrylate)	1.4990	
Poly(vinyl decyl ether)	1.4628	30	Poly(urethanes)	1.5-1.6	
Poly(2-methoxyethyl acrylate)	1.463	25	Poly(1,2-butadiene)	1.5000	20
Poly(butyl acrylate)	1.4631	30	Poly(vinyl formal)	1.50	
Poly(butyl acrylate)	1.466	20	Poly(2-bromo-4-trifluoromethylstyrene)	1.5	25
Poly(tert-butyl methacrylate)	1.4638	20	Cellulose nitrate	1.50-1.514	
Poly(vinyl dodecyl ether)	1.4640	30	Poly(sec-butyl α-chloroacrylate)	1.500	25
Poly(3-ethoxypropyl acrylate)	1.465	25	Poly(2-heptyl-1,3-butadiene)	1.5000	
Poly(oxycarbonyl tetramethylene)	1.465	50	Poly(ethyl α-chloroacrylate)	1.502	25
Poly(vinyl propionate)	1.4665	20	Poly(2-isopropyl-1,3-butadiene)	1.5028	30
Poly(vinyl acetate)	1.4665	20	Poly(2-methylcyclohexyl methacrylate)	1.5028	20
Poly(vinyl methyl ether)	1.467	20	Poly(propylene) (density 0.9075 g/cm^3)	1.5030	20
Poly(ethyl acrylate)	1.4685	20	Poly(isobutene)	1.505-1.51	
Poly(ethylene-co-vinyl acetate)	1.47-1.50		Poly(bornyl methacrylate)	1.5059	20
(80% - 20% vinyl acetate)			Poly(2-tert-butyl-1,3-butadiene)	1.5060	24.6
Cellulose propionate	1.47-1.49		Poly(ethylene glycol dimethacrylate)	1.5063	20
Cellulose acetate propionate	1.47		Poly(cyclohexyl methacrylate)	1.5066	20
Benzyl cellulose	1.47-1.58		Poly(cyclohexanediol-1,4-dimethacrylate)	1.5067	20
Phenol-formaldehyde resins	1.47-1.70		Butyl rubber (unvulcanized)	1.508	
Cellulose triacetate	1.47-1.48		Poly(tetrahydrofurfuryl methacrylate)	1.5096	20
Poly(vinyl methyl ether) (isotactic)	1.4700	30	Gutta percha (β)	1.509	
Poly(3-methoxypropyl acrylate)	1.471	25	Poly(ethylene) Ionomer	1.51	
Poly(2-ethoxyethyl acrylate)	1.471	25	Poly(oxyethylene) (high molecular weight)	1.51-1.54	
Poly(methyl acrylate)	1.472-1.480		Poly(ethylene) (density 0.914 g/cm^3)	1.51	20
Poly(isopropyl methacrylate)	1.4728	20	--, (density 0.94-0.945 g/cm^3)	1.52-1.53	20
Poly(1-decene)	1.4730		--, (density 0.965 g/cm^3)	1.545	20
Poly(propylene) (atactic, density 0.8575 g/cm^3)	1.4735	20	Poly(1-methylcyclohexyl methacrylate)	1.5111	20
Poly(vinyl sec-butyl ether) (isotactic)	1.4740	30	Poly(2-hydroxyethyl methacrylate)	1.5119	20
Poly(dodecyl methacrylate)	1.4740	20	Poly(vinyl chloroacetate)	1.512	25
Poly(oxyethyleneoxysuccinoyl)	1.4744	25	Poly(butene) (isotactic)	1.5125	
(poly(ethylene succinate))			Poly(vinyl methacrylate)	1.5129	20
Poly(tetradecyl methacrylate)	1.4746	30	Poly(N-butyl·methacrylamide)	1.5135	20
Poly(ethylene-co-propylene) (EPR-rubber)	1.4748-1.48		Gutta percha (α)	1.514	50
Poly(hexadecyl methacrylate)	1.4750	30	Terpene resin	1.515	25
Poly(vinyl formate)	1.4757	20	Poly(1,3-butadiene)	1.5154	25
Poly(2-fluoroethyl methacrylate)	1.4768	20	Shellac	1.51-1.53	
Poly(isobutyl methacrylate)	1.477	20	Poly(methyl α-chloroacrylate)	1.517	20
Ethyl cellulose	1.479	21	Poly(2-chloroethyl methacrylate)	1.517	20
Poly(vinyl acetal)	1.48-1.50		Poly(2-diethylaminoethyl methacrylate)	1.5174	20
Cellulose acetate	1.48-1.50		Poly(2-chlorocyclohexyl methacrylate)	1.5179	20
Cellulose tripropionate	1.48-1.49		Poly(1,3-butadiene) (35% cis; 56% trans; 7% 1,2-	1.5180	
Poly(oxymethylene)	1.48	20	content)		
Poly(vinyl butyral)	1.48-1.49		Natural rubber	1.519-1.52	
Poly(n-hexyl methacrylate)	1.4813	20	Poly(allyl methacrylate)	1.5196	20
Poly(n-butyl methacrylate)	1.483	20-25	Poly(vinyl chloride) + 40% dioctyl phthalate	1.52	
Poly(ethylidene dimethacrylate)	1.4831	20	Poly(acrylonitrile)	1.52	
Poly(2-ethoxyethyl methacrylate)	1.4833	20		1.5187	25
Poly(oxyethyleneoxymaleoyl)	1.4840	25	Poly(methacrylonitrile)	1.52	
(poly(ethylene maleate))			Poly(1,3-butadiene) (high cis-type)	1.52	
Poly(n-propyl methacrylate)	1.484	25	Poly(butadiene-co-acrylonitrile)	1.52	

Polymer	n_D	T °C	Polymer	n_D	T °C
Poly(methyl isopropenyl ketone)	1.5200	20	Poly(o-methoxyphenyl methacrylate)	1.5705	20
Poly(isoprene)	1.521	20	Poly(phenyl methacrylate)	1.5706	20
Poly(ester) resin, rigid (ca. 50 % styrene)	1.523-1.54		Poly(o-cresyl methacrylate)	1.5707	20
Poly(N-(2-methoxyethyl)methacrylamide)	1.5246	20	Poly(diallyl phthalate)	1.572	20
Poly(2,3-dimethylbutadiene) (methyl rubber)	1.525		Poly(2,3-dibromopropyl methacrylate)	1.5739	20
Poly(vinyl chloride-co-vinyl acetate) (95/5-90/10)	1.525-1.535		Poly(oxycarbonyloxy-1,4-phenylene-1-methyl-	1.5745	
Poly(acrylic acid)	1.527	25	butylidene-1,4-phenylene)		
Poly(1,3-dichloropropyl methacrylate)	1.5270	20	Poly(oxy-2,6-dimethylphenylene)	1.575	
Poly(2-chloro-1-(chloromethyl)ethyl methacrylate)	1.5270	20	Poly(oxyethyleneoxyterephthaloyl) (amorphous)	1.5750	20
Poly(acrolein)	1.529		(poly(ethylene terephthalate))		
Poly(1-vinyl-2-pyrrolidone)	1.53		(crystalline fiber: 1.51 transverse;		
Hydrochlorinated rubber	1.53-1.55		1.64 in fiber direction)		
Nylon 6; Nylon 6,6; Nylon 6,10 (moulding)	1.53		Poly(vinyl benzoate)	1.5775	20
(Nylon-6-fiber: 1.515 transverse,			Poly(oxycarbonyloxy-1,4-phenylenebutylidene-1,4-	1.5792	
1.565 in fiber direction)			phenylene)		
Poly(butadiene-co-styrene) (ca. 30 % styrene)	1.53		Poly(1,2-diphenylethyl methacrylate)	1.5816	20
black copolymer			Poly(o-chlorobenzyl methacrylate)	1.5823	20
Poly(cyclohexyl α-chloroacrylate)	1.532	25	Poly(oxycarbonyloxy-1,4-phenylene-sec-butylidene-	1.5827	
Poly(2-chloroethyl α-chloroacrylate)	1.533	25	1,4-phenylene)		
Poly(butadiene-co-styrene) (ca. 75/25)	1.535		Poly(oxypentaerythritoloxyphthaloyl)	1.584	20
Poly(2-aminoethyl methacrylate)	1.537	20	Poly(m-nitrobenzyl methacrylate)	1.5845	20
Poly(furfuryl methacrylate)	1.5381		Poly(oxycarbonyloxy-1,4-phenyleneisopropylidene-	1.5850	
Proteins	1.539-1.541		1,4-phenylene)		
Poly(butylmercaptyl methacrylate)	1.5390	20	Poly(N-(2-phenylethyl)methacrylamide)	1.5857	20
Poly(1-phenyl-n-amyl methacrylate)	1.5396	20	Poly(4-methoxy-2-methylstyrene)	1.5868	20
Poly(N-methyl-methacrylamide)	1.5398	20	Poly(o-methylstyrene)	1.5874	20
Cellulose	1.54		Poly(styrene)	1.59-1.592	20
Poly(vinyl chloride)	1.54-1.55		Poly(oxycarbonyloxy-1,4-phenylenecyclohexylidene-	1.5900	
Urea formaldehyde resin	1.54-1.56		1,4-phenylene)		
Poly(sec-butyl α-bromoacrylate)	1.542	25	Poly(o-methoxystyrene)	1.5932	20
Poly(cyclohexyl α-bromoacrylate)	1.542	25	Poly(diphenylmethyl methacrylate)	1.5933	20
Poly(2-bromoethyl methacrylate)	1.5426	20	Poly(oxycarbonyloxy-1,4-phenyleneethylidene-1,4-	1.5937	
Poly(dihydroabietic acid)	1.544		phenylene)		
Poly(abietic acid)	1.546	25	Poly(p-bromophenyl methacrylate)	1.5964	20
Poly(ethylmercaptyl methacrylate)	1.547	20	Poly(N-benzyl methacrylamide)	1.5965	20
Poly(N-allyl methacrylamide)	1.5476	20	Poly(p-methoxystyrene)	1.5967	20
Poly(1-phenylethyl methacrylate)	1.5487	20	Hard rubber (32 % S)	1.6	
Poly(vinylfuran)	1.55	20	Poly(vinylidene chloride)	1.60-1.63	
Poly(2-vinyltetrahydrofuran)	1.55	20	Poly(sulfides) ("Thiokol")	1.6-1.7	
Poly(vinyl chloride) + 40 % tricresyl phosphate	1.55		Poly(o-chlorodiphenylmethyl methacrylate)	1.6040	20
Epoxy resins	1.55-1.60		Poly(oxycarbonyloxy-1,4-(2,6-dichloro)phenylene-	1.6056	
Poly(p-methoxybenzyl methacrylate)	1.552	20	isopropylidene-1,4-(2,6-dichloro)phenylene))		
Poly(isopropyl methacrylate)	1.552		Poly(oxycarbonyloxybis(1,4-(3,5-dichlorophenylene))	1.6056	
Poly(p-isopropylstyrene)	1.554	20	Poly(pentachlorophenyl methacrylate)	1.608	20
Poly(chloroprene)	1.554-1.558		Poly(o-chlorostyrene)	1.6098	20
Poly(oxyethylene)-α-benzoate-ω-methacrylate)	1.555	20	Poly(phenyl α-bromoacrylate)	1.612	25
Poly(p,p'-xylylenyl dimethacrylate)	1.5559	20	Poly(p-divinylbenzene)	1.6150	20
Poly(1-phenylallyl methacrylate)	1.5573	20	Poly(N-vinylphthalimide)	1.6200	20
Poly(p-cyclohexylphenyl methacrylate)	1.5575	20	Poly(2,6-dichlorostyrene)	1.6248	20
Poly(2-phenylethyl methacrylate)	1.5592	20	Poly(β-naphthyl methacrylate)	1.6298	20
Poly(oxycarbonyloxy-1,4-phenylene-1-propyl-	1.5602		Poly(α-naphthyl carbinyl methacrylate)	1.63	
butylidene-1,4-phenylene)			Poly(sulfone)	1.633	
Poly(1-(o-chlorophenyl)ethyl methacrylate)	1.5624	20	Poly(2-vinylthiophene)	1.6376	20
Poly(styrene-co-maleic anhydride)	1.564	21	Poly(α-naphthyl methacrylate)	1.6410	20
Poly(1-phenylcyclohexyl methacrylate)	1.5645	20	Poly(oxycarbonyloxy-1,4-phenylenediphenyl-	1.6539	
Poly(oxycarbonyloxy-1,4-phenylene-1,3-dimethyl-	1.5671		methylene-1,4-phenylene)		
butylidene-1,4-phenylene)			Poly(vinyl phenyl sulfide)	1.6568	20
Poly(methyl α-bromoacrylate)	1.5672	20	Butylphenol formaldehyde resin	1.66	
Poly(benzyl methacrylate)	1.5680	20	Urea-thiourea-formaldehyde resin	1.660	25
Poly(2-(phenylsulfonyl)ethyl methacrylate)	1.5682		Poly(vinylnaphthalene)	1.6818	20
Poly(m-cresyl methacrylate)	1.5683	20	Poly(vinylcarbazole)	1.683	20
Poly(styrene-co-acrylonitrile) (ca. 75/25)	1.57		Naphthalene-formaldehyde resin	1.696	
Poly(oxycarbonyloxy-1,4-phenyleneisobutylidene-	1.5702		Phenol-formaldehyde resin	1.70	
1,4-phenylene)			Poly(pentabromophenyl methacrylate)	1.71	20

REFRACTIVE INDICES

REFERENCES

1.   S. Nose, Polymer Letters 2, 1127 (1964).

2.   "Handbook of Chemistry and Physics", Ed., R.C. Weast, The Chemical
     Rubber Co., Cleveland 1971, E-203.

3.   A. F. Forziati in "High Polymers", Vol. XII, Ed. G. M. Kline, part II,
     chapter III, p. 137, table IV - Interscience   N.Y. 1962.

4.   D'Ans-Lax, "Taschenbuch fuer Chemiker und Physiker", Ed. K. Schaefer
     und D. Synowietz, Springer 1970, Vol. III, Chapter 4, p. 295.

5.   R. Albert, W. M. Malone, J. Macromol. Sci. Chem., A-6 (2), 347
     (1972).

IV. SOLUTION PROPERTIES

VISCOSITY-MOLECULAR WEIGHT RELATIONSHIPS AND
UNPERTURBED DIMENSIONS OF LINEAR CHAIN MOLECULES

M. Kurata, Y. Tsunashima, M. Iwama, K. Kamada
Institute for Chemical Research,
Kyoto University,
Kyoto, Japan

CONTENTS

A. INTRODUCTION

1. THE VISCOSITY-MOLECULAR WEIGHT RELATIONSHIP

The limiting viscosity number $[\eta]$ of a solution which has long been called the intrinsic viscosity is defined as

$$[\eta] = \lim_{c \to 0} \frac{\eta - \eta_o}{\eta_o \, c} \qquad (1)$$

in terms of the solvent viscosity η_o, the solution viscosity η and the solute concentration c. The concentration c is expressed in grams of solute per milliliter of solution or, more frequently, in grams of solute per 100 milliliters of solution, the limiting viscosity number being given in the reciprocal of these units, i. e. in milliliters per gram or in deciliters per gram. Here, following the IUPAC 1952-recommendations (1), we adopt the former unit. The quantity $[\eta]$ of a polymer solution is a measure of the capacity of a polymer molecule to enhance the viscosity, which depends on the size and the shape of the polymer molecule. Within a given series of polymer homologs, $[\eta]$ increases with the molecular weight M; hence it is a measure of M.

Table C gives the limiting viscosity number-molecular weight relationships for polymers in various solvents and at various temperatures. The table contains the constants of the equation

$$[\eta] = KM^a \qquad (2)$$

which is known as the Mark-Houwink-Sakurada equation.

It is now well established that for linear, flexible polymers, under special condition of temperature or solvent, (usually known as the Flory "theta" temperature or solvent (2)), the above equation becomes

$$[\eta]_{\Theta} = K_{\Theta} M^{0.50} \qquad (3)$$

The sign Θ in front of the temperature data in the table indicates that the viscosity constants were obtained under the Θ condition. Since Eq. (3) is approximately valid over the whole molecular weight range, K_{Θ} and $a = 0.50$ may be used, without modification outside of the molecular weight range in which they were determined. However, it must be noted that $[\eta]$ is rather sensitive to temperature in the vicinity of Θ, especially when M is higher than 5×10^{5}.

In ordinary good solvents, the constants K and a obtained are valid only within a rather limited range of M (3, 4). It is, therefore, quite probable that the tabulated relationships are in error outside the indicated range of M (see eighth column in the table). As for the effect of temperature, however, both K and a mostly become insensitive to the temperature when a exceeds about 0.70, and they may be used in a ten-degree range on either side of temperature at which the constants were determined.

The method of determination of the molecular weight and the number of fractionated samples (Fr.) or whole polymer samples (W.P.) used to determine the $[\eta]$-M relationship are also given in the ninth and the sixth or seventh columns, respectively. The abbreviations used are as follows.

(A) Methods yielding the number-average molecular weight, M_n.

CR,	cryoscopy.	EB,	ebullioscopy.
EG,	end-group titration.	OS,	osmotic pressure
VOS,	vapor pressure osmometry.		

(B) Methods yielding the weight-average molecular weight, M_w.

LS,	light scattering.	SA,	approach to the sedimentation equilibrium
SE,	sedimentation equilibrium.		(Archibald's method).

(C) Empirical or semi-empirical methods.

EM,	electron microscopy.	DV,	diffusion and viscosity
LV,	limiting viscosity number-molecular weight relationship.	MV,	melt viscosity-molecular weight relationship.
PR,	analysis of polymerization rate (yielding M_n).	SD,	sedimentation and diffusion.
		SV,	sedimentation and viscosity.

Thus, for example, the constants tabulated are for the $[\eta]$-M relationships expressed in terms of M_n or M_w if the method is specified as OS or LS, respectively; i.e.

$$[\eta] = K_n M_n^{a} \qquad (4)$$

or

$$[\eta] = K_w M_w^{a} \qquad (5)$$

The values of K_n and K_w, especially the former, are greatly influenced by the molecular weight distribution (MWD) of polymer samples, and caution must be taken in using these relationships.

To illustrate this effect, let us assume that:

(i) Eq. (2) is applicable to the molecule i with molecular weight M_i over the whole range of M; i.e.

$$[\eta]_i = K M_i^{a} \qquad (6)$$

(ii) The weight fraction w_i of the molecules i in a given sample can be represented by a continuous exponential function,

$$w_i (M_i) = [y^{h+1}/\Gamma(h+1)] M_i^{h} \exp(-yM_i) \qquad (7)$$

$$y = h/M_n = (h+1)/M_w \qquad (8)$$

or by the log-normal function,

$$w_i (M_i) = A M_i \exp[-p^2 (\ln M_i/M_o)^2] \qquad (9)$$

where h, A, p and M_o are constants, and Γ represents the gamma function.

Then, since $[\eta] = \Sigma_i w_i [\eta]_i$, we obtain

$$K_n = K\Gamma(a+h+1)/h^a\Gamma(h+1) \qquad (10)$$

$$K_w = K\Gamma(a+h+1)/(h+1)^a\,\Gamma(h+1) \tag{11}$$

for the exponential MWD, and

$$K_n = K(M_w/M_n)^{0.5a(a+1)} \tag{12}$$

$$K_w = K(M_w/M_n)^{0.5a(a-1)} \tag{13}$$

for the log-normal MWD (5). The values of K_n/K and K_w/K calculated by these equations are shown in Table B. This table may be used for estimating an error due to MWD in determination of M.

As an example, let us assume that a given polymer sample has the exponential MWD with $M_w/M_n = 2.0$, while an available $[\eta]$-M_n equation has been obtained for samples with a narrow MWD, e.g. $M_w/M_n = 1.1$. Further, let a be 0.70. Then, to find the correct value of M_n of the given sample from $[\eta]$, we must use the Equation (4) with $K_n = 1.54K$, instead of the available equation with $K_n = 1.06K$. Use of the latter would lead to an overestimate M_n' which is related to the correct M_n by

$$[\eta] = 1.54K\,M_n^{0.70} = 1.06K\,M_n'^{0.70} \tag{14}$$

The error amounts to about 70%, i.e. $M_n' = 1.7M_n$. Thus, application of the viscosity equation written in M_n is to be restricted to within a narrow class of samples, unless an appropiate correction is made. On the other hand, if an $[\eta]$-M_w equation is available for the same pair of working and reference samples as above, we have

$$[\eta] = 0.951K\,M_w^{0.70} = 0.991K\,M_w'^{0.70} \tag{15}$$

instead of Eq. (14). Hence, the error in M_w amounts to only 6% ($M_w' = 0.94M_w$), which will be negligible for most practical purposes.

Based on the above consideration, we classify the heterogeneity of polymers in four classes, A to D, as shown in the last column of Table B, and indicate it in the tenth column of Table C as a measure of the heterogeneity of the reference samples used.

It is desirable that readers select their own relationship by inspecting these data on heterogeneity as well as those on the number of samples and molecular weight range. Generally speaking, a "good" $[\eta]$-M relationship is one that has been obtained on the basis of M_w for at least four samples of classes A and B (exceptionally C) or on the basis of M_n for those of class A (exceptionally B), whose molecular weights range over at least one half orders of magnitude.

In the "Remarks" column of Table C, we have occasionally indicated by the letter R a "recommended" relationship for the convenience of readers. In the range of low molecular weight (mostly less than 10^4), the constant a becomes 0.50 irrespective of solvent. This type of relationship can not be used, even approximately, at higher molecular weights. This case is noted by the letter L. High conversion polymers are also marked by the letter H, where the $[\eta]$-M relationships are less reproducible due to chain branching than are ordinary ones. The abbreviations used are as follows.

A, narrow MWD polymers, or well-fractionated polymers, $M_w/M_n \leq 1.25$.
B, ordinary fractionated polymers, $1.30 \leq M_w/M_n \leq 1.75$.
C, poorly-fractionated polymers or most probable MWD polymers, $1.8 \leq M_w/M_n \leq 2.4$.
D, wide MWD polymers, $M_w/M_n > 2.5$.
H, high conversion polymers, including branches.
L, limited to low molecular weight polymers.
R, recommended relationship.

In this table, polymers are arranged according to their structure in subgroups. Within each subgroup, the polymers are, in principle, given in alphabetical order. Within each polymer, the solvents are also arranged in alphabetical order, followed by the mixed solvents.

Chain configurational data are occasionally given in the first column. The data given in parentheses refer to only one set of viscosity constants listed in the same row, while the data given without parentheses refer to a series of sets listed in the same and succeeding rows. Thus, for example, the data "N content, 13.9 wt%" are effective only for the sixth row of cellulose trinitrate, and the data "95%-cis, 1%-trans, 4%-1,2" are effective for the fourth to eighth rows of polybutadiene.

Table C is essentially based on the table published by Kurata and Stockmayer (3). Data were also taken from tables published by Peterlin (7), Meyerhoff (8), Elias (9) and Krause (10), the last one including a number of unpublished data on acrylic and methacrylic polymers. We are grateful to these authors. Thanks are tendered also to J. Brandrup and K. Kamide for their help with this compilation.

2. UNPERTURBED DIMENSIONS OF LINEAR CHAIN MOLECULES

The mean-square end-to-end distance $<r^2>$ of a linear chain molecule in solution is usually expressed in terms of two basic quantities, the unperturbed mean-square end-to-end distance $<r^2>_o$ and the expansion factor α; i.e.

$$<r^2> = <r^2>_o\, \alpha^2 \tag{16}$$

The latter quantity α represents the effect of "long-range interactions" which can be described as an osmotic swelling of the chain by the solvent-polymer interactions, while the unperturbed dimension $<r^2>_o$ represents the effect of "short-range interactions" such as bond angle restrictions and steric hindrances to internal rotation. The steric hindrances are also influenced by the torques exerted on the chain by solvent molecules, but the effect is rather small in many cases (11).

For sufficiently long chain, $<r^2>_o$ becomes proportional to $\Sigma_i n_i l_i^2$, where n_i is the number of the ith-kind bond of length l_i. The quantity C_∞ defined by

$$C_\infty = \lim_{n\to\infty} <r^2>_o / \Sigma_i n_i l_i^2 \tag{17}$$

is often called the characteristic ratio and it serves as a measure of the effect of short-range interactions.

The freely rotating state is a hypothetical state of the chain in which the bond angle restrictions are retained, but the steric hindrances to internal rotation are released. The mean-square end-to-end distance of the freely rotating chain $<r^2>_{of}$ can be readily calculated from the given basic structure of the chain. For instance, if the chain consists of only one kind of bond of length 1, we obtain

$$<r^2>_{of} = nl^2\,[(1 + \cos\,\theta)/(1 - \cos\,\theta)] \tag{18}$$

where n is the number of bonds and θ is the supplement of the valence bond angle. For vinyl polymer chains, $1 = 0.154$ [nm], $\cos\theta = 1/3$, and $n = M/m = 2M/M_u$; and hence,

$$(<r^2>_{of}/M)^{1/2} = 0.308/M_u^{1/2} = 0.218/m^{1/2}\ \text{[nm]} \tag{19}$$

where M_u is the molar weight of the repeating unit and m is the average molar weight per skeletal link. Similar expressions for r_{of} ($= <r^2>_{of}^{1/2}$) can be obtained also for more complicated chains. The results are summarized in Table D.

The ratio of $<r^2>_o$ to $<r^2>_{of}$, then, represents the effect of steric hindrance on the average chain dimension:

$$\sigma = r_o/r_{of} = (<r^2>_o/<r^2>_{of})^{1/2} \tag{20}$$

The quantity σ is independent of n. Table E gives a list of the unperturbed dimensions of linear chain molecules which were obtained under various conditions of solvent and temperature. The values of $r_o/M^{1/2}$, $r_{of}/M^{1/2}$, σ and C_∞ are given, together with the experimental values of S_{oz}/M_w, a_p or K_o from which r_o was computed. S_{oz} which is the abbreviation of $<S^2>_{oz}$ is the z-average value of the unperturbed radius of gyration, a_p is the persistence length and K_o is the viscosity constant corresponding to K_θ in Eq. (3). The methods used to determine these quantities are also indicated in the tenth column of the table by using the following abbreviations.

(A) Light scattering

   LT, Zimm's plot in a theta solvent yielding $S_{oz}/M_w^{1/2}$. After a heterogeneity correction is made, the tabulated value of $r_o/M^{1/2}$ (= $6^{1/2}S_{ow}/M_w^{1/2}$) is obtained.

   LD, dissymmetry method in a theta solvent. Less reliable for heterogeneous samples than the former method.

   LG, Zimm's plot in good solvents yielding $S_z/M_w^{1/2}$. After corrections for the excluded volume effect and heterogeneity are made, the tabulated value of $r_o/M^{1/2}$ is obtained (3, 12).

(B) X-ray small angle scattering

   XS, the persistence length a_p is obtained irrespective of the solvent nature. The tabulated values of $r_o/M^{1/2}$ are the asymptotic values for infinitely high molecular weight. (13, 14).

(C) Limiting viscosity number

   VT, viscosity-molecular weight relationship in a theta solvent, Eq. (3). $r_o/M^{1/2}$ is calculated by the Flory and Fox relation, $K_o = \Phi_o(r_o/M^{1/2})^3$. The following values of Φ_o were used:

      2.7×10^{23} for well fractionated polymers (class A in Table C);
      2.5×10^{23} for ordinary fractionated polymers (class B);
      2.1×10^{23} for poorly fractionated or unfractionated polymers (class C or D).

   VG, viscosity-molecular weight relationship in good solvents. K_o was estimated by using the Kurata-Stockmayer-Fixman plot (3, 4) or other analogous plots (12).

VA, viscosity in good solvents. The correction of excluded volume effect is made by using the Flory-Krigbaum-Orofino theory of the second virial coefficient A_2 or other analogous theories (12).

(D) Method yielding the temperature dependence of r_0.

ST, stress-temperature coefficient of undiluted or swollen samples.

The polymers are arranged in Table E in the same order as in Table C. For each polymer, smoothed values of $r_0/M^{1/2}$, σ and C_∞, which were mostly obtained by VT or VG, are given in the first line, followed by some typical values obtained by more direct methods such as LT or XS. The listed values of $r_0/M^{1/2}$ sometimes scatter appreciably, reflecting the difficulty, both experimental and theoretical, involved in determination of this quantity. Especially in the case of cellulosic chains, the right magnitude of r_0 is yet in controversy (542, 549, 3, 691, 696, 688, 678, 686, 12). In recent papers, emphasis has often been put on the effect of temperature or solvent on the unperturbed dimensions. These data are put together at the end of the tabulation for each polymer. Table E is also based on the tables published by Kurata and Stockmayer (3).

B. EFFECT OF MOLECULAR WEIGHT DISTRIBUTION ON VISCOSITY CONSTANT, K

M_w/M_n	a = 0.5		a = 0.6		a = 0.7		a = 0.8		a = 0.9		a = 1.0		Class
	K_n/K	K_w/K	K_n/K	K_w/K	K_n/K	K_w/K	K_n/K	K_w/K	K_n/K	K_w/K	K_n/K	K_w/K	
1 - MOLECULAR WEIGHT DISTRIBUTION: EXPONENTIAL TYPE, EQ. (7)													
30	4.87	0.890	6.91	0.897	9.85	0.911	14.18	0.933	20.56	0.963	30	1	D
15	3.46	0.893	4.57	0.900	6.08	0.914	8.16	0.935	11.02	0.964	15	1	D
10	2.83	0.896	3.59	0.903	4.59	0.917	5.91	0.937	7.67	0.965	10	1	D
5	2.03	0.907	2.40	0.913	2.85	0.925	3.42	0.943	4.12	0.968	5	1	D
3	1.60	0.921	1.79	0.926	2.02	0.936	2.29	0.952	2.62	0.973	3	1	D
2	1.33	0.940	1.43	0.943	1.54	0.951	1.68	0.963	1.83	0.979	2	1	C
1.75	1.25	0.948	1.33	0.951	1.42	0.958	1.51	0.968	1.63	0.982	1.75	1	B
1.50	1.18	0.959	1.23	0.961	1.28	0.967	1.35	0.975	1.42	0.986	1.50	1	B
1.25	1.09	0.975	1.12	0.977	1.15	0.980	1.18	0.985	1.21	0.991	1.25	1	A
1.10	1.04	0.989	1.05	0.989	1.06	0.991	1.07	0.993	1.09	0.996	1.10	1	A
2 - MOLECULAR WEIGHT DISTRIBUTION: LOG. NORMAL TYPE, EQ. (9)													
30	3.58	0.654	5.12	0.665	7.57	0.700	11.58	0.762	18.32	0.858	30	1	D
15	2.76	0.713	3.67	0.723	5.01	0.753	7.03	0.805	10.13	0.885	15	1	D
10	2.37	0.750	3.02	0.759	3.94	0.785	5.25	0.832	7.16	0.902	10	1	D
5	1.83	0.818	2.17	0.824	2.61	0.845	3.19	0.879	3.96	0.930	5	1	D
3	1.51	0.872	1.69	0.877	1.92	0.891	2.21	0.916	2.56	0.952	3	1	D
2	1.30	0.917	1.39	0.920	1.51	0.930	1.65	0.946	1.81	0.969	2	1	C
1.75	1.23	0.932	1.31	0.935	1.40	0.943	1.50	0.956	1.61	0.975	1.75	1	B
1.50	1.16	0.951	1.21	0.953	1.27	0.958	1.34	0.968	1.41	0.982	1.50	1	B
1.25	1.09	0.973	1.11	0.974	1.14	0.977	1.17	0.982	1.21	0.990	1.25	1	A
1.10	1.04	0.988	1.05	0.989	1.06	0.990	1.07	0.992	1.08	0.996	1.10	1	A

C. TABLES OF VISCOSITY-MOLECULAR WEIGHT RELATIONSHIPS, $[\eta] = KM^a$

Polymer	Solvent	Temp. [°C]	$K \times 10^3$ [ml/g]	a	Fr.	W.P.	Mol. Wt. Range $M \times 10^{-4}$	Method	Remarks	Ref.
1. MAIN-CHAIN ACYCLIC-CARBON POLYMERS										
1.1 POLY(DIENES)										
Poly(butadiene)										
98%-cis, 2%-1,2	benzene	30	33.7	0.715	9	--	5 - 50	OS	A,R	15
	isobutyl acetate	θ 20.5	185	0.50	6	--	5 - 50	OS	A	15
	toluene	30	30.5	0.725	9	--	5 - 50	OS	A	15
95%-cis, 1%-trans										
4%-1,2	benzene	30	8.5	0.78	4	--	15 - 50	LS	A	16
	cyclohexane	30	11.2	0.75	4	--	15 - 50	LS	A	16
	5-methyl-2-hexanone	θ 12.6	150	0.50	4	--	15 - 35	LS	B	17
	3-pentanone	θ 10.3	152	0.50	4	--	10 - 25	LS	B	17
	toluene	30	33.9	0.688	8	--	10 - 65	OS	A	18
94%-cis, 4%-trans,										
2%-1,2	benzene	25	41.4	0.70	8	--	9 - 120	OS	A	19
	dioxane	θ 20.2	205	0.50	8	--	9 - 120	OS	A	19
92%-cis, 3%-trans,										
5%-1,2	benzene	32	10	0.77	13	--	10 - 160	LS	B,R	20
51%-trans, 43%-cis,										
6%-1,2	toluene	30	39	0.713	6	--	11 - 25	OS	A	21
71%-trans, 4%-cis,										
25%-1,2	cyclohexane	25	12	0.77	8	--	230 - 880	LS	C	22
79%-trans, 21%-cis,	cyclohexane	20	36	0.70	12	--	23 - 130	LS	B,R	23
97%-trans, 3%-1,2	cyclohexane	40	28.2	0.70	7	--	4 - 17	LS	B	24
	toluene	30	29.4	0.753	6	--	5 - 16	OS	A	25
ca. 100%-cis	benzene	32	14.5	0.76	8	--	18 - 50	LS	A	26
	heptane/hexane (1/1 vol)	20	138	0.53	5	--	?	SD	A	27
65%-1,2, 25%-trans,										
10%-cis	toluene	25	110	0.60	8	--	7 - 70	OS	B	28
5°C-emulsion, randomly branched	3-pentanone	θ 24								
50°C-emulsion, randomly branched	benzene	θ 5								

(see equations)

$M^{2/3}/[\eta]^{4/3} = 7.15 + 3.47M$ — 3-pentanone, θ 24, 10 Fr., 10 - 100, OS, C, 29

$M^{2/3}/[\eta]^{4/3} = 4.61 + 0.328M$ — benzene, θ 5, 16 Fr., 5 - 124, OS, C, 29

Polymer	Solvent	Temp. [°C]	$K \times 10^3$	a	Fr.	W.P.	Range	Method	Remarks	Ref.
Poly(butadiene-co-acrylonitrile), Buna-N rubber	acetone	25	50	0.64	5	--	2.5 - 10	OS	B	28
	benzene	25	13	0.55	5	--	2.5 - 10	OS	B	28
	chloroform	25	54	0.68	5	--	2.5 - 10	OS	B	28
	toluene	25	49	0.64	7	--	2.5 - 40	OS	B	28
Poly(butadiene-co-styrene), Buna-S, GR-S, or SBR rubber	benzene	25	52.5	0.66	24	--	1 - 160	OS		45
		25	54	0.66	8	--	1 - 165	OS	B	46
	cyclohexane	30	31.6	0.70	6	--	5 - 25	OS	A	47
	2-pentanone	θ 21	185	0.50	6	--	5 - 25	OS	A	47
	toluene	25	52.5	0.667	25	--	2.5 - 50	OS	B	28
		30	16.5	0.78	--	9	3 - 35	OS		48
		30	37.9	0.71	6	--	5 - 25	OS	A	47
linear fraction	toluene	30	21.4	0.74	15	--	3 - 20	OS	A,R	41
branched fraction	toluene	30	535	0.48	20	--	20 - 100	OS	B	41
Poly(2-tert-butylbutadiene)	benzene	21	4.2	0.80	--	8	6 - 90	SD	A	30
	octane	21	4.2	0.80	--	7	6 - 35	SD	A	30
Poly(chloroprene)										
Neoprene CG	benzene	25	2.02	0.89	10	--	6 - 150	OS	B	31
Neoprene GN	benzene	25	14.6	0.73	16	--	2 - 96	OS	B	32
Neoprene W	benzene	25	15.5	0.71	8	--	5 - 100	OS	B	33
		25	15.5	0.72	9	--	5 - 80	LS	B,R	34
	butanone	θ 25	113	0.50	7	--	15 - 300	LS	A	35

Polymer	Solvent	Temp. [°C]	K x 10³ [ml/g]	a	No. of samples Fr.	W.P.	Mol. Wt. Range M x 10⁻⁴		Method	Remarks	Ref.
Poly(chloroprene) (Cont'd.)											
Neoprene W (Cont'd.)	butyl acetate	25	37.8	0.62	7	--	15	- 300	LS	A	35
	carbon tetrachloride	25	22.1	0.69	7	--	15	- 300	LS	A	35
	cyclohexane	θ 45.5	107	0.50	7	--	15	- 70	LS	B	34
type, unspecified	toluene	25	50	0.615	13	--	4	- 120	OS	B	28
Poly(isoprene)											
natural rubber	benzene	30	18.5	0.74	--	4	8	- 28	OS	C	37
	cyclohexane	27	30	0.70	--	1	ca 185		LS,SD	C	38
	2-pentanone	θ 14.5	119	0.50	--	4	8	- 28	OS	C	37
	toluene	25	50.2	0.667	20	--	7	- 100	OS	B,R	39
synthetic cis	hexane	20	68.4	0.58	5	--	5	- 80	SD	A	40
	toluene	30	8.51	0.77	5	--	20	- 100	LS	A	41
85-91%-cis	toluene	30	20.0	0.728	--	12	14	- 580	LS	A,R	42
		30	15	0.74	--	16	2	- 15	PR	A	43
	2,2,4-trimethylpentane	30	22.2	0.683	--	8	23	- 580	LS	A	42
	heptane/propanol (78/22 vol)	30	37	0.63	--	6	43	- 580	LS	A	42
84%-cis, 14%-trans, 2%-1,2	benzene	25	13.3	0.78	20	--	2	- 80	OS	B	44
		25	11.2	0.78	25	--	2	- 60	OS	B	44
	dioxane	θ 34	145	0.50	30	--	2	- 50	OS	B	44
gutta percha	benzene	25	35.5	0.71	9	--	0.2	- 5	OS	A,R	19
	dioxane	θ 47.7	191	0.50	9	--	0.2	- 5	OS	A	19
	propyl acetate	θ 60	232	0.50	--	3	10	- 20	OS	C	37
synthetic trans,	benzene	32	43.7	0.65	24	--	8	- 140	LS	C	26
Poly(1,1,2-trichloro-butadiene)	benzene	25	31.6	0.66	11	--	25	- 130	LS		36

1.2 POLY(ALKENES)

Polymer	Solvent	Temp. [°C]	K x 10³ [ml/g]	a	No. of samples Fr.	W.P.	Mol. Wt. Range M x 10⁻⁴		Method	Remarks	Ref.
Poly(alkene) C₁₀-C₁₈	toluene	25	12.7	1.04	12	--	2	- 18	LS	B	86
Poly(alkene) C₁₂-C₁₈	cetane	38	21	0.61	10	--	4	- 700	LS	B	87
Poly(1-butene),											
atactic	anisole	θ 86.2	123	0.50	3	--	10	- 130	LS	C	81
	benzene	30	22.4	0.72	11	--	0.03	- 0.5	EG	B.L	82
	ethylcyclohexane	70	7.34	0.80	5	--	4	- 130	LS	C	81
isotactic	ethylcyclohexane	70	7.34	0.80	4	--	8	- 94	LS	A	81
	decalin	115	9.49	0.73	6	--	4.5	- 90	LS		83
	heptane	35	4.73	0.80	6	--	4.5	- 90	LS		83
		60	15.0	0.69	6	--	4.5	- 90	LS		83
	nonane	80	5.85	0.80	4	--	11	- 94	LS	A	81
Poly(ethylene)											
low pressure	biphenyl	θ 127.5	323	0.50	4	--	2	- 30	LV	B	58
	1-chloronaphthalene	125	138	0.58	?	?	?		LS	?	59
		125	18.4	0.78	10	--	5	- 100	LS		60
		125	43	0.67	10	--	5	- 100	LS	C,D	61
		129	27.1	0.71	26	--	5	- 100	LS	D	62
	decalin	135	67.7	0.67	--	>10	3	- 100	LS	D	63
		135	46	0.73	23	--	3	- 64	LS		64
		135	62	0.70	7	--	2	- 105	LS	B,R	65,66
		135	58.5	0.725	9	--	0.4	- 50	OS	B	67,68
	decanol	θ 153.3	302	0.50	?	--	2	- 105	LV	B	58
	diphenyl ether	θ 161.4	295	0.50	6	--	2	- 105	LS	B	65
	diphenylmethane	θ 142.2	315	0.50	?	--	2	- 105	LV	B	58
	dodecanol	θ 137.3	307	0.50	5	--	2	- 105	LV	B	58
		θ 138	316	0.50	--	8	8	- 32	LS	D	69
	octanol	θ 180.1	286	0.50	?	--	2	- 105	LV	B	58
	tetralin	105	16.2	0.83	4	--	13	- 57	LS	C	70
		120	23.6	0.78	36	--	5	- 100	LS		60
		120	32.6	0.77	20	--	0.3	- 50	LS	B	71
		130	43.5	0.76	6	--	2	- 30	OS	B	71

Polymer	Solvent	Temp. [°C]	$K \times 10^3$ [ml/g]	a	No. of samples Fr.	No. of samples W.P.	Mol. Wt. Range $M \times 10^{-4}$		Method	Remarks	Ref.
Poly(ethylene) (Cont' d.)											
low pressure	tetralin (Cont' d.)	130	51	0.725	9	--	0.4	- 50	OS	B,R	72
		130	37.8	0.72	--	10	8	- 17	LS	D	73
	p-xylene	105	16.5	0.83	4	--	13	- 50	LS	C	70
		105	17.6	0.83	8	--	1	- 18	OS	C	74
		105	51	0.725	?	--	0.4	- 50	LV	B,R	75
	paraffin wax ($M_n = 390 \pm 10$)	150	(42)	(0.65)	9	--	0.04	- 11	LS	D	76
high pressure	decalin	70	38.73	0.738	8	--	0.2	- 3.5	OS	B	77
	p-xylene	75	135	0.63	--	22	0.2	- 7.6	OS	D	78
		81	105	0.63	7	--	1	- 10	OS	D	79
Poly(ethylene) (normal paraffin)	carbon tetrachloride	20	\multicolumn{2}{l}{$[\eta] = -1.14+0.104\ M$}	--	7	0.024	- 0.048	CR	A	80
Poly(ethylene-co-propylene-co-diene),											
EPDM rubber	cyclohexane	40	53.1	0.75	20	--	3	- 30	OS	A	41
Poly(isobutene)	anisole	θ 105	91	0.50	--	--	18	- 188	LV	B	49
	benzene	θ 24	107	0.50	15	--	18	- 188	LV	B	49
		25	83	0.53	9	--	0.05	- 126	OS,CR	B,R	50
		30	61	0.56	9	--	0.05	- 126	OS,CR	B	50
		40	43	0.60	9	--	0.05	- 126	OS,CR	B	50
		60	26	0.66	9	--	0.05	- 126	OS,CR	B	50
	carbon tetrachloride	30	29	0.68	12	--	0.05	- 126	OS,CR	B	50
	cyclohexane	25	40	0.72	6	--	14	- 34	OS	B	51
		30	27.6	0.69	7	--	4	- 71	OS	A,R	52
		30	26.5	0.69	12	--	0.05	- 126	OS,CR	B	50
	decalin	25	22	0.70	6	--	530	- 1680	LS	A-B	53,54
	diisobutylene	20	36	0.64	23	--	1	- 130	OS	A,R	55,52
		25	130	0.50	5	--	0.4	- 2.5	OS	A,L	56
	phenetol	θ 86	91	0.50	4	--	5	- 188	LV	B	49
	toluene	0	40	0.60	8	--	1	- 146	LV	B	50
		15	24	0.65	6	--	1	- 146	LV	B	50
		25	07	0.66	6	--	14	- 34	OS	B	51
		30	20	0.67	5	--	1	- 146	LV	B,R	50
		50	20	0.68	6	--	1	- 146	LV	B	50
		60	13.5	0.71	4	--	11	- 146	LV	B	50
		90	12.6	0.72	3	--	46	- 146	LV	B	50
Poly(isobutene-co-isoprene), butyl rubber	carbon tetrachloride	25	10.7	0.78	6	--	10	- 30	OS	A	57
	toluene	25	66	0.60	5	--	15	- 30	OS	A	57
		30	21.4	0.678	8	--	10	- 30	OS	A	57
Poly(4-methyl-1-pentene)	diisobutylene	20	42	0.63	6	--	1	- 20	LS	A	85
Poly(1-octene)	bromobenzene	25	2.90	0.78	5	--	25	- 400	LS	A	84
	cyclohexane	30	5.75	0.78	6	--	25	- 400	LS	A	84
	phenetol	θ 50.4	65.5	0.50	4	--	60	- 400	LS	A	84
Poly(propylene)											
atactic	benzene	25	27.0	0.71	6	--	6	- 31	OS	A	88
		30	33.8	0.67	6	--	2	- 34	OS	A	89
	1-chloronaphthalene	θ 74	182	0.50	3	--	4	- 33	OS	A	90
	cyclohexane	25	16.0	0.80	6	--	6	- 31	OS	A	88
		30	20.9	0.76	6	--	2	- 34	OS	A	89
	cyclohexanone	θ 92	172	0.50	4	--	1.5	- 33	OS	A	90
	decalin	135	15.8	0.77	6	--	2	- 39	OS	A	91
		135	11.0	0.80	6	--	2	- 62	LS	A,R	88
		135	54.3	0.65	--	10	2	- 72	LS	D	92
	isopentyl acetate	θ 34	168.5	0.50	6	--	2	- 34	OS	A	89
	phenyl ether	145	192	0.47	3	--	3.7	- 21	OS	A	90
		θ 153	120	0.50	3	--	3.7	- 21	OS	A	90
	tetralin	130	1.24	0.96	--	--	?		?		93
	toluene	30	21.8	0.725	7	--	2	- 34	OS	A	89
isotactic	biphenyl	θ 125.1	152	0.50	4	--	5	- 42	LV	A	94
	1-chloronaphthalene	139	21.5	0.67	11	--	10	- 170	LS		95

Polymer	Solvent	Temp. [°C]	K x 10³ [ml/g]	a	No. of samples Fr.	W.P.	Mol. Wt. Range M x 10⁻⁴			Method	Remarks	Ref.
Poly(propylene) Cont'd.)												
isotactic (Cont'd.)	1-chloronaphthalene	145	4.9	0.80	9	--	5	-	63	LS	A,R	96
	decalin	135	11.0	0.80	6	--	2	-	62	LS	A,R	88
		135	10.0	0.80	4	--	10	-	100	LS	A,R	97
	dibenzyl ether	θ 183.2	106	0.50	4	--	5	-	42	LV	A	94
	diphenyl ether	θ 142.8	137	0.50	4	--	5	-	42	LV	A	94
		θ 145	132	0.50	4	--	3	-	48	OS	A	90
		153	112	0.54	4	--	3	-	48	OS	A	90
	tetralin	135	2.5	1.0	5	--	2	-	11	OS		98
		135	9.17	0.80	9	--	4	-	54	OS	A,R	96
	p-xylene	85	96	0.63	12	--	?			OS		99
syndiotactic	heptane	30	31.2	0.71	5	--	9	-	45	LS	A	100

1.3 POLY(ACRYLIC ACID) AND DERIVATIVES

Polymer	Solvent	Temp. [°C]	K x 10³ [ml/g]	a	No. of samples Fr.	W.P.	Mol. Wt. Range M x 10⁻⁴			Method	Remarks	Ref.
Poly(acrylamide)	water	30	6.31	0.80	7	--	2	-	50	SD	B	101
		30	68	0.66	--	21	1	-	20	PR	C	102
Poly(acrylic acid)	1,4-dioxane	θ 30	76	0.50	--	4	13	-	82	OS	B	104
--, sodium salt	aqueous NaOH (2M)	25	42.2	0.64	12	--	4	-	50	OS	C	105
	aqueous NaCl (0.012M)	20	--	0.93	7	--	7	-	180	LV	B	106
	(1M)	25	15.47	0.90	12	--	4	-	50	OS	C	105
	aqueous NaBr (1.5M) θ	15	165	0.50	5	--	6	-	64	LV	C	107
	θ	15	124	0.50	4	--	12	-	83	LS	C	108
	(0.5M)	15	52.7	0.628	7	--	1	-	50	LV	C	109
		25	50.6	0.656	7	--	2	-	80	LV	C,R	110
	(0.1M)	15	25.4	0.755	7	--	1	-	50	LV	C	109
		25	31.2	0.755	7	--	2	-	80	LV	C	110
	(0.05M)	15	28.1	0.77	7	--	1	-	50	LV	C	109
	(0.025M)	15	16.3	0.84	7	--	1	-	50	LV	C	109
		25	17.6	0.85	7	--	2	-	80	LV	C	110
	(0.01M)	15	13.6	0.89	7	--	1	-	50	LV	C	109
		25	13.2	0.91	7	--	2	-	80	LV	C	110
	(0.005M)	15	(44.2)	0.83	7	--	1	-	50	LV	C	109
	(0.0025M)	15	(24.9)	0.89	7	--	1	-	50	LV	C	109
	aqueous NaSCN											
	(1.25M) θ	30	154	0.50	5	--	6	-	64	LV	C	107
	θ	30	121	0.50	4	--	12	-	83	LS	C	111
Poly(acrylonitrile)	γ-butyrolactone	20	34.3	0.730	5	--	4	-	40	LV(LS)	A,R	134
(polymerized at -30°C)		30	57.2	0.67	6	--	4	-	30	SA	B	135
(polymerized at 60°C)		30	34.2	0.70	5	--	6	-	30	SA	B	135
		30	40.0	0.69	--	5	15	-	53	LS	D	136
		50	28.7	0.740	5	--	4	-	40	LS	A	134
	dimethylformamide	20	17.7	0.78	5	--	7	-	30	LS	B	137
		25	16.6	0.81	5	--	5	-	27	SD	B	138
		25	24.3	0.75	--	4	3	-	25	LS	C	139
		25	39.2	0.75	--	16	3	-	100	OS	C	140
	(deionized DMF)	25	15.5	0.80	3	5	3	-	10	LS,SD	B-C	141
		25	57.4	0.73	--	8	0.3	-	1.5	EG	L	142
		25	39.6	0.75	--	7	4	-	30	OS	C	143
		25	44.3	0.70	--	7	2	-	20	LS	C	143
		25	69.8	0.65	--	21	8	-	140	LS	C	144
(polymerized at -30°C)		30	29.6	0.74	7	--	4	-	30	SA	B	135
(polymerized at 60°C)		30	20.9	0.75	7	--	6	-	30	SA	B	135
		30	33.5	0.72	--	6	16	-	48	LS	D	136
		35	27.8	0.76	9	--	3	-	58	DV	B	145
		35	31.7	0.746	12	--	9	-	76	LS	A,R	134
		50	30.0	0.752	22	--	4	-	102	LV	A	134
	dimethylacetamide	20	30.7	0.761	6	--	2	-	40	LV	A	134

VISCOSITY-MOLECULAR WEIGHT RELATIONSHIPS

Polymer	Solvent	Temp. [°C]	$K \times 10^3$ [ml/g]	a	No. of samples Fr.	W.P.	Mol. Wt. Ranges $M \times 10^{-4}$		Method	Remarks	Ref.
Poly(acrylonitrile)	dimethylacetamide	35	27.5	0.767	6	--	2	- 40	LV	A	134
(Cont'd.)	(Cont'd.)	50	27.4	0.764	6	--	2	- 40	LV	A	134
	dimethyl sulfoxide	20	32.1	0.750	9	--	9	- 40	LV	A	134
		50	28.3	0.758	9	--	9	- 40	LV	A	134
		140	20.9	0.75	--	6	4	- 40	LS		146
	ethylene carbonate	50	29.5	0.718	13	--	7	- 40	LV	A	134
	hydroxyacetonitrile	20	40.9	0.697	8	--	4	- 34	LV	A	134
		50	35.4	0.707	8	--	4	- 34	LV	A	134
	aqueous HNO₃ 60%)	0	33.9	0.740	6	--	2	- 40	LV	A	134
		20	30.7	0.747	5	--	4	- 40	LV	A	134
Poly(benzyl acrylate)	butanone	35	0.587	0.883	?		?		OS		337
Poly(butyl acrylate)	acetone	25	6.85	0.75	--	6	5	- 27	LS	C	112
Poly(1,1-dihydroper-											
fluorobutyl acrylate)	benzofluoride	26.6	13	0.56	7	3	20	- 200	LS	B	113
	methyl perfluorobutyrate	26.6	12	0.60	7	3	20	- 200	LS	B	113
Poly(N,N-dimethylacryl-											
amide)	methanol	25	17.5	0.68	--	8	5	- 122	LS	C	103
	water	25	23.2	0.81	--	6	5	- 122	LS	C	103
		40	20.0	0.65	--	4	11	- 122	LS	C	103
Poly(ethyl acrylate)	acetone	25	51	0.59	7	--	35	- 450	LS	B,R	114
		30	20.0	0.66	5	--	16	- 50	OS	B,R	115
	benzene	30	27.7	0.67	--	7	5	- 67	OS	C	116
	butanone	30	2.68	0.80	5	--	48	- 700	LS	B-C	117
	chloroform	30	31.4	0.68	--	5	9	- 54	OS	C	116
	ethyl acetate	30	26.0	0.66	--	5	9	- 54	OS	C	116
Poly(hexadecyl acrylate)	methanol	30	48.7	0.55	--	6	6	- 70	OS	C	116
	heptane	20	1.74	0.82	6	--	1	- 10	LS	B	118
Poly(isopropyl acrylate)	acetone	30	13.0	0.69	6	--	6	- 30	LS	B	119
	benzene	25	14.9	0.70	9	--	7	- 70	OS	B	120
		25	12.4	0.701	20	--	4	- 100	LS	B,R	121
		30	11.8	0.71	4	--	7	- 20	LS	B	119
	bromobenzene	25	11.3	0.704	20	--	4	- 100	LS	B	121
		60	11.6	0.698	20	--	4	- 100	LS	B	121
	chloroform	30	14.1	0.72	5	--	7	- 30	LS	B	122
(isotactic)	2,2,3,3-tetrafluoro-										
	propanol	25	19.7	0.697	7	--	10	- 65	LS	B	121
(atactic)		25	17.3	0.703	6	--	8	- 110	LS	B	121
(syndiotactic)		25	15.9	0.708	6	--	20	- 110	LS	B	121
(isotactic)		60	17.9	0.693	4	--	10	- 65	LS	B	121
(atactic and syndiotactic)		60	14.7	0.704	6	--	20	- 110	LS	B	121
Poly(methyl acrylate)	acetone	20	(7.40)	(0.76)	--	4	7	- 32	OS		123
		25	5.5	0.77	8	--	28	- 160	LS	B,R	124
		25	19.8	0.66	9	--	30	- 250	LS	B	125
		30	28.2	0.52	7	--	4	- 45	OS	B	126
	benzene	25	2.58	0.85	4	--	20	- 130	OS		127
		30	4.5	0.78	7	7	7	- 160	LS		128
		30	3.56	0.798	6	--	25	- 190	LS	B,R	129
		30	4.59	0.795	6	--	15	- 140	OS	B	129
		35	12.8	0.71	--	5	5	- 30	OS	C	130
	butanone	20	3.5	0.81	13	--	6	- 240	LS	A-B,R	128
		25	14.1	0.67	4	--	17	- 68	LS	B	131
		30	3.97	0.772	6	--	25	- 190	LS	B	129
		35	(34)	(0.61)	--	3	5	- 47	LV	C	132
	diethyl malonate	30	3.51	0.793	4	--	50	- 190	LS	B	129
	ethyl acetate	35	11	0.69	--	8	24	- 148	LS	A	133
	isopentyl acetate	θ 62.5	68	0.50	6	--	20	- 160	LS	B	129
	2-methylcyclohexanol	θ 56.0	68	0.50	4	--	40	- 105	LS	B	129
	toluene	30	7.79	0.697	6	--	25	- 190	LS	B	129
		35	21	0.60	--	7	12	- 69	LS	A	133
	butanone/2-propanol										
	(42/58 vol)	θ 20	81	0.50	5	--	29	- 140	LS	B	124

Polymer	Solvent	Temp. [°C]	$K \times 10^3$ [ml/g]	a	No. of samples Fr.	W.P.	Mol. Wt. Range $M \times 10^{-4}$		Method	Remarks	Ref.
Poly(methyl acrylate) (Cont'd.)	butanone/2-propanol (1/1 vol)	θ 27.5	54.4	0.50	4	--	14	- 83	LS	C	108
		θ 30	72	0.50	4	--	50	- 190	LS	B	129
(branched)	(42/58 vol)	θ 20	290	0.40	6	--	37	- 250	LS	B	125
Poly(1-methylphenyl acrylate)	butyl acetate	25	14.7	0.63	8	--	2	- 110	SD	A	346
Poly(morpholinocarbonyl-ethylene)	dimethylformamide	25	18	0.65	?		?		LS	C	338
	aqueous NaCl (0.1M)	20	64	0.68	?		?		LS	C	338
Poly(piperidinocarbonyl-ethylene)	dimethylformamide	25	32	0.56	?		?		LS	C	338
Poly(propyl acrylate)	butanone	30	15.0	0.687	4	--	71	- 181	LS	A	117

1.4 POLY(α-SUBSTITUTED ACRYLIC ACID) AND DERIVATIVES

Polymer	Solvent	Temp. [°C]	$K \times 10^3$ [ml/g]	a	No. of samples Fr.	W.P.	Mol. Wt. Range $M \times 10^{-4}$		Method	Remarks	Ref.
Poly(benzyl methacrylate)	benzene	30	1.03	0.82	--	9	17	- 120	LS		339
Poly(butyl methacrylate)	acetone	25	18.4	0.62	5	--	100	- 600	LS	A	150
	benzene	30	(4.0)	(0.77)	--	3	8	- 300	LS		151
	butanone	23	1.56	0.81	10	--	25	- 260	LS	B	152
		25	9.7	0.68	5	--	11	- 670	LS	A	150
		30	(1.15)	(0.89)	3	--	67	- 132	OS	C	153
	chloroform	20	2.9	0.78	8	--	4	- 800	LS	B,R	154
		25	4.37	0.80	6	--	8	- 80	OS		155
	2-propanol	θ 21.5	29.5	0.50	8	--	30	- 260	LS	B	152
		θ 21.5	38	0.50	9	--	4	- 800	LS	B,R	154
		θ 23.7	36.6	0.50	5	--	40	- 170	LS	B	156
Poly(tert-butyl methacrylate)	butyl acetate	25	22.0	0.63	6	--	46	- 870	LS	A	157
Poly(4-tert-butylphenyl methacrylate)	acetone	20	5.75	0.68	15	--	6	- 350	LS		340
	bromobenzene	20	4.1	0.71	7	--	15	- 2500	LS		341
	carbon tetrachloride	20	4.1	0.71	7	--	20	- 2500	LS		341
Poly[1-(N-carbethoxy-phenyl)-methacrylamide]	chloroform	20	2.4	0.78	15	--	6	- 300	LS	A-B	342
	acetone	unc.	0.00115	1.35	4	--	26	- 74	LS		369
	dimethylformamide	unc.	This relation not followed		5	--	48	- 140	LS		369
	ethyl acetate	unc.	0.00446	1.25	5	--	26	- 11	LS		369
Poly(4-chlorophenyl methacrylate)	benzene		9.2	0.66	8	--	10	- 610	LS	A	343
	carbon tetrachloride		20.0	0.58	8	--	10	- 610	LS	A	343
	dioxane		6.1	0.70	8	--	10	- 610	LS	A	343
Poly(cyclohexyl methacrylate)	benzene	30	8.4	0.69	5	--	80	- 200	LS		344
	butanol	θ 23	33.7	0.50	5	--	57	- 445	LS	B	345
	butanone	25	5.79	0.68	6	--	57	- 560	LS	B	345
		30	7.0	0.66	5	--	80	- 200	LS		344
Poly(dodecyl methacrylate)	butyl acetate	23	8.64	0.64	8	--	26	- 360	LS	A	158
	isopropyl acetate	θ 13	32.2	0.50	7	--	26	- 360	LS	A	158
	pentanol	θ 29.5	34.8	0.50	7	--	27	- 240	LS	A	159
Poly(2-ethylbutyl methacrylate)	butanone	25	2.21	0.77	8	--	48	- 332	LS	A	160
	2-propanol	θ 27.4	33.7	0.50	8	--	48	- 332	LS	A	160
Poly(ethyl methacrylate)	butanone	23	2.83	0.79	10	--	20	- 263	LS	A	161
	ethyl acetate	35	8.6	0.71	--	11	65	- 1200	LS	C	162
	2-propanol	θ 36.9	47.5	0.50	4	--	22	- 130	LS	B	156
	butanone/2-propanol (1/7 vol)	θ 23	47.3	0.50	10	--	20	- 263	LS	A	161
	ethyl acetate/ethanol (2/9 vol)	35	47.6	0.53	6	--	78	- 500	LS	A	162
	(1/6 vol)	θ 35	56.4	0.50	6	--	60	- 420	LS	A	162

VISCOSITY-MOLECULAR WEIGHT RELATIONSHIPS

Polymer	Solvent	Temp. [°C]	$K \times 10^3$ [ml/g]	a	No. of samples Fr.	No. of samples W.P.	Mol. Wt. Range $M \times 10^{-4}$		Method	Remarks	Ref.
Poly(hexadecyl methacrylate)	benzene	21	5.9	0.71	3	--	130	- 440	SD	B	163
	carbon tetrachloride	21	2.37	0.78	5	--	130	- 440	SD	B	163
	heptane	21	3.92	0.75	5	--	130	- 440	SD	B	163
		25	35.1	0.56	9	--	20	- 110	LS		164
Poly(hexyl methacrylate)	butanone	23	2.12	0.78	8	--	6	- 41	LS	A	165
	2-propanol	θ 32.6	43.0	0.50	8	--	6	- 41	LS	A	165
Poly(isobutyl methacrylate)	acetone	25	0.199	0.94	6	--	300	- 1100	LS	C	166
	butanone	20	5.56	0.73	6	--	300	- 1100	LS	C	166
		25	8.61	0.70	7	--	300	- 1100	LS	C	166
		30	7.47	0.71	6	--	300	- 1100	LS	C	166
		44	2.18	0.79	6	--	300	- 1100	LS	C	166
Poly(methacrolein)	dimethylformamide	20	2.8	0.97	--	?	0.5	- 2	OS, CR	?	204
Poly(methacrylic acid)	methanol	26	242	0.51	6	--	4	- 20	OS	B	147
	aqueous HCl (0.002M)	30	66	0.50	7	--	10	- 90	LV	C	148
	aqueous NaNO₃ (2M)	25	44.9	0.65	6	--	8	- 70	OS	B	149
Poly(methacrylonitrile)	acetone	20	95.5	0.56	--	4	35	- 100	OS	C	202
	dimethylformamide	29.2	306	0.503	--	15	0.6	- 8	LV	C, H	203
Poly(methyl butacrylate)	butanol	θ 13	57.0	0.50	4	--	6	- 60	LS	A	168
	butanone	30	5.43	0.73	10	--	7	- 430	LS	A	168
Poly(methyl ethacrylate)	benzene	30	2.35	0.82	6	--	16	- 110	LS	A	168
	butanone	30	4.29	0.75	10	--	4	- 200	LS	A	168
	2,6-dimethyl-4-heptanone	θ 11.4	67.6	0.50	10	--	4	- 200	LS	A	168
Poly(methyl methacrylate) atactic	acetone	20	5.5	0.73	7	--	7	- 700	SD	A-B,R	169
		20	3.90	0.76	7	--	7	- 700	SD	A-B	169
		25	7.5	0.70	9	--	8	- 137	LS	B	170
		25	6.76	0.71	10	--	3	- 700	SD	A-B	171
		25	7.5	0.70	14	--	2	- 740	LS, SD	A-B	172
		25	5.3	0.73	7	--	2	- 780	LS	A-B,R	173
		25	9.6	0.69	4	--	180	- 350	LS	A-B	174
		25	7.5	0.70	4	6	3	- 98	LS	B-C	175
		25	2.45	0.80	9	--	6	- 210	OS	B-C	176
		25	6.59	0.71	6	--	5	- 41	OS	B	177
		30	7.7	0.70	6	--	6	- 263	LS	A-B	178
		39	6.40	0.72	6	--	5	- 41	OS	B	177
		46	6.18	0.72	6	--	5	- 41	OS	B	177
	acetonitrile	30	39.3	0.50	6	--	10	- 86	LV	A-B	178
		θ 45	48	0.50	6	--	10	- 260	LV	A-B,R	179
		50	29	0.54	6	--	10	- 260	LV	A-B	180
		65	9.8	0.64	5	--	10	- 260	LV	A-B	180
	benzene	20	8.35	0.73	7	--	7	- 700	SD	A-B	169
		20	15.1	0.70	7	--	8	- 90	SD		181
		25	7.24	0.76	10	--	6	- 100	OS	B	182
		25	5.5	0.76	11	--	2	- 740	LS	A-B,R	173
		25	3.80	0.79	5	--	24	- 450	LS		183
		25	83	0.52	7	--	0.03	- 1	EB	A,L	184
		30	5.2	0.76	9	--	6	- 250	LS	A-B,R	178
		30	6.27	0.76	5	--	4	- 73	OS	A	185
		30	104	0.50	9	--	0.02	- 2	OS	A,L	185
		30	195	0.41	5	--	0.3	- 2	LS	A-B,L	178
		39	6.74	0.75	6	--	5	- 41	OS	B	177
		53	6.52	0.76	6	--	5	- 41	OS	B	177
	butanone	25	6.8	0.72	9	--	8	- 137	LS	B,R	170
		25	7.1	0.72	7	--	41	- 330	LS	A-B	174
		25	6.8	0.72	4	6	3	- 98	LS	B-C	175
		25	9.39	0.68	15	--	16	- 910	LS	A-B	186
	butyl chloride	θ 35.4	50.5	0.50	4	--	13	- 68	SA	A-B	187
	chloroform	20	9.6	0.78	18	--	1.4	- 60	OS		188

Polymer	Solvent	Temp. [°C]	$K \times 10^3$ [ml/g]	a	No. of samples Fr.	W.P.	Mol. Wt. Range $M \times 10^{-4}$	Method	Remarks	Ref.
Poly(methyl meth- acrylate) (Cont'd.) atactic	chloroform (Cont'd.)	20	4.88	0.82	8	--	6 - 100	OS	B	182
		20	4.85	0.80	9	--	8 - 200	SD	A-B,R	169
		20	6.0	0.79	12	--	3 - 780	LS	A-B	173,189
		25	4.8	0.80	9	--	8 - 137	LS	B	170
		25	3.4	0.83	6	--	40 - 330	LS	A-B	174
		25	5.81	0.79	6	--	5 - 41	OS	B	177
(living type)		30	4.3	0.80	--	8	13 - 263	LS	A-B	178
		39	5.02	0.80	6	--	5 - 41	OS	B	177
		53	3.90	0.79	6	--	5 - 41	OS	B	177
		unc.	5.1	0.79	13	--	7 - 400	LS	B	190
	p-cymene	θ 159.7	57.5	0.50	4	--	6.6 - 171	LV	A-B	191
	1,2-dichloroethane	25	17.0	0.68	4	6	3 - 98	LS	B-C	175
		30	5.3	0.77	--	7	6 - 263	LS	A-B,R	178
	ethyl acetate	20	21.1	0.64	8	34	6 - 110	SD		192
	3-heptanone	θ 33.7	63.1	0.50	4	--	6.6 - 171	LV	A-B	191
	4-heptanone	θ 33.8	48	0.50	5	--	1 - 172	LS	A-B,R	179
	methyl isobutyrate	30	9.9	0.67	6	--	19 - 260	LV	A-B	178
	methyl methacrylate	30	6.75	0.72	3	--	13 - 170	LV	A-B	178
	nitroethane	25	5.70	0.74	2	6	10 - 200	LS	C	193
	3-octane	θ 72	50	0.50	3	--	13 - 260	LV	A-B	179
	propanol	θ 84.4	67.9	0.50	4	--	6.6 - 171	LV	A-B	191
	tetrachloroethane	25	12.8	0.73	6	--	5 - 41	OS	B	177
		53	12.2	0.73	6	--	5 - 41	OS	B	177
	2,2,3,3-tetrafluoro- propanol	25	7.2	0.79	7	--	7 - 95	LV	A	194
	toluene	25	7.1	0.73	7	--	4 - 330	LS	A-B	174
		25	8.12	0.71	6	--	5 - 41	OS	B	177
		25	78	0.50	10	--	0.2 - 7	OS	A,L	195
		30	7.0	0.71	6	--	19 - 263	LV	A-B	178
		39	7.24	0.72	6	--	5 - 41	OS	B	177
		53	6.63	0.73	6	--	5 - 41	OS	B	177
	butanone/2-propanol (55/45 vol)	23	47.0	0.55	6	--	40 - 300	LS	A-B	174
	(50/50 vol)	θ 25	59.2	0.50	7	--	30 - 280	LS	A-B	196
		θ 25	42.8	0.50	5	--	77 - 490	LS	A-B	186
	methanol/toluene (9/5 vol)	θ 26.2	55.9	0.50	3	--	60 - 300	LS	A-B	156
isotactic	acetone	30	23.0	0.63	7	--	5 - 128	LS	A-B	199
	acetonitrile	20	130	0.448	5	--	3 - 19	LV	A	198
		θ 27.6	75.5	0.500	5	--	3 - 19	LV	A	198
		35	46	0.546	5	--	3 - 19	LV	A	198
		50	26.2	0.602	5	--	3 - 19	LV	A	198
	benzene	30	5.2	0.76	5	--	5 - 128	LS	A-B	199
	p-cymene	θ 152.1	56.6	0.50	4	--	7 - 131	LV	A-B	191
	3-heptanone	θ 40.0	87.0	0.50	4	--	7 - 131	LV	A-B	191
	propanol	θ 75.9	76.1	0.50	4	--	7 - 131	LV	A-B	191
	2,2,3,3-tetrafluoro- propanol	25	7.05	0.78	11	--	2 - 100	LV	B	194
	butanone/2-propanol (1/1 vol)	θ 30.3	90.0	0.50	4	--	7 - 131	LV	A-B	191
Poly(octadecyl meth- acrylate)	tetrahydrofuran	30	2.5	0.75	--	4	20 - 170	LS	C,H	200
Poly(octyl methacrylate)	butanol	θ 16.8	26.8	0.50	10	--	33 - 1250	LS	B	201
	butanone	23	4.47	0.69	10	--	33 - 1250	LS	B	201
Poly(N-phenyl meth- acrylamide)	acetone	20	28.2	0.75	8	--	10 - 320	LS		370

Polymer	Solvent	Temp. [^{o}C]	$K \times 10^{3}$ [ml/g]	a	No. of samples Fr.	No. of samples W.P.	Mol. Wt. Range $M \times 10^{-4}$		Method	Remarks	Ref.

1.5 POLY(VINYL ETHERS)

Polymer	Solvent	Temp.	$K \times 10^3$	a	Fr.	W.P.	Range		Method	Remarks	Ref.
Poly[(hexadecyloxy) ethylene]	heptane	21	70.8	0.50	6	--	0.5	- 3	SD	B,L	205
Poly(methoxyethylene)	benzene	30	76	0.60	13	--	1	- 45	LS	B	206
	butanone	30	137	0.56	13	--	1	- 45	LS	B	206
Poly[(octadecyloxy) ethylene]	benzene	25	170	0.47	--	7	0.1	- 1.5	LS	D,H	200
	tetrahydrofuran	30	224	0.35	--	7	9.4	- 11	LS	D,H	200

Poly(vinyl methyl ether), see Poly(methoxyethylene)

1.6 POLY(VINYL ALCOHOL), POLY(VINYL HALIDES)

Polymer	Solvent	Temp.	$K \times 10^3$	a	Fr.	W.P.	Range		Method	Remarks	Ref.
Poly(chlorotrifluoro-ethylene)	2,5-dichlorobenzotri-fluoride	130	6.15	0.74	7	--	7	- 51	OS	B	234
Poly(vinyl alcohol)	water	25	20	0.76	6	--	0.6	- 2.1	OS	B	208
		25	300	0.50	4	--	0.9	- 17	SD		209
		25	140	0.60	3	--	1	- 7	SD	B	210
		30	66.6	0.64	8	--	0.6	- 16	OS	B	212
		30	42.8	0.64	--	14	1	- 80	LS	C	213
		30	45.3	0.64	--	--	1	- 80	LS	A,R	213
		80	94	0.56	--	5	10	- 46	LS	B	214
	phenol/water (85/15 vol)	30	24.6	0.80	--	21	3	- 12	LV	B	215
Poly(vinyl bromide)	cyclohexanone	25	32.8	0.55	7	--	2	- 10	LS	B	217
	tetrahydrofuran	25	15.9	0.64	7	--	2	- 10	LS	B	217
	methanol/tetrahydro-furan (17/83 vol)	20	38.8	0.50	7	--	2	- 10	LS	B	218
Poly(vinyl chloride)	benzyl alcohol	θ 155.4	156	0.50	9	--	4	- 35	LS	B	219
	chlorobenzene	30	71.2	0.59	7	--	3	- 19	SA	B	220
	cyclohexanone	20	11.6	0.85	--	6	2	- 10	OS	C	221
		20	13.7	1.0	7	5	7	- 13	OS	C,D	222
		20	112.5	0.63	5	3	9	- 15	OS	D H	222
		25	12.3	0.83	11	--	2	- 14	OS		223
		25	24	0.77	13	--	3	- 14	OS		224
		25	204	0.56	?	--	2	- 15	OS	C	225
		25	174	0.55	6	--	6	- 22	LS	C	226
		25	8.5	0.75	5	--	4	- 20	LS	B	227
		25	13.8	0.78	28	--	1	- 12	LS	A,B,R	228
		30	16.3	0.77	6	--	3	- 19	SA	B	220
	tetrahydrofuran	20	3.63	0.92	20	--	2	- 17	OS	B	229
		25	15.0	0.77	22	--	1	- 12	LS	A,B	228
		25	16.3	0.766	23	--	2	- 30	LS	A,B,R	230
		25	49.8	0.69	5	--	4	- 40	LS	A-B	231
		30	63.8	0.65	9	--	3	- 32	LS		232
		30	83.3	0.83	7	-	3	- 19	SA	B	220
		30	219	0.54	16	--	5	- 30	LS		233
Poly(vinyl fluoride)	dimethylformamide	90	6.42	0.80	--	9	14	- 66	SV	D	235

1.7 POLY(VINYL ESTERS)

Polymer	Solvent	Temp.	$K \times 10^3$	a	Fr.	W.P.	Range		Method	Remarks	Ref.
Poly(allyl acetate)	benzene	27	66	0.53	8	--	0.1	- 0.3	CR		216
Poly(vinyl acetate)	acetone	6	$[\eta] = 0.104 M^{0.50} + 0.00725 M^{0.90}$		21	--	0.3	- 150	LS	A	236
		18	24.5	0.67	6	--	4	- 34	OS	B	237
		20	15.8	0.69	6	--	19	- 72	LS		238
		25	21.4	0.68	6	--	4	- 34	OS	B	237
		25	18.8	0.69	?	?	?		LS		239
		25	14.6	0.72	--	6	0.7	- 1.3	EG	C,L	240
		25	10.8	0.72	10	--	0.9	- 2.5	EG	B,L	240
		30	17.6	0.68	16	--	2	- 163	OS	A-B	241
		30	8.6	0.74	8	--	8	- 66	LS	A-B	242
		30	17.4	0.70	?	--	7	- 68	OS		243

Polymer	Solvent	Temp. [°C]	$K \times 10^3$ [ml/g]	a	No. of samples Fr.	No. of samples W.P.	Mol. Wt. Range $M \times 10^{-4}$	Method	Remarks	Ref.
Poly(vinyl acetate) (Cont'd.)	acetone (Cont'd.)	30	10.2	0.72	--	8	3 - 126	LS	C	244
		30	10.1	0.73	11	--	6 - 150	LS	A	236
		30	$[\eta] = 0.097 M^{0.50} + 0.00723 M^{0.90}$		22	--	0.3 - 150	LS	A	236
		46	13.8	0.71	6	--	4 - 34	OS	A	236
	acetonitrile	25	16.2	0.71	--	--	24 - 215	LS	B	246
		30	41.5	0.62	4	--	97 - 153	LS	A-B	247
	benzene	30	22	0.65	5	--	34 - 102	LS	A-B	248
		30	56.3	0.62	24	--	3 - 86	OS	B	249
		30	56.3	0.62	12	--	7 - 54	LS	B	250
		35	21.6	0.675	14	--	5 - 40	LS	A-B	251
	butanone	25	13.4	0.71	6	--	25 - 346	LS	A	252
		25	42	0.62	15	--	2 - 120	SD,LS	A,B	253
		30	10.7	0.71	--	13	3 - 120	LS	C	244
	chlorobenzene	25	110	0.50	9	--	0.15 - 7	OS	A	195
		25	94.4	0.56	6	--	4 - 34	OS	A	236
		53	53.7	0.60	6	--	4 - 34	OS	A	236
		67	28.9	0.65	6	--	4 - 34	OS	A	236
	chloroform	20	15.8	0.74	?	?	7 - 68	OS		243
		25	20.3	0.72	5	--	4 - 34	OS	A	236
		53	14.7	0.74	5	--	4 - 34	OS	A	236
	dioxane	25	11.4	0.74	5	--	4 - 34	OS	B	237
		53,60	10.2	0.75	5	--	4 - 34	OS	B	237
	ethanol	θ 56.9	90	0.50	5	--	4 - 150	OS,LS	A	236
	ethyl formate	30	32	0.65	4	--	16 - 154	LS	A-B	247
	3-heptanone	θ 26.8	82.0	0.50	5	--	4 - 150	OS,LS	A	236
		θ 29	92.9	0.50	18	--	5 - 83	LS	A-B	255
	methanol	θ 6	101	0.50		--	0.3 - 150	OS,LS,VOS		236,245
		25	38.0	0.59	5	--	4 - 22	OS	B	237
		30	31.4	0.60	--	13	3 - 120	LS	C	244
		53	36.6	0.59	5	--	4 - 22	OS	B	237
	6-methyl-3-hepta-none	θ 66	82.0	0.50	9	--	14 - 83	LS	A-B	255
		θ 66	78.0	0.50	3	--	9 - 150	OS,LS	A	236
	4-methyl-2-pentanone	30	44.9	0.60	5	--	12 - 69	LS		247
	toluene	25	108	0.53	4	--	4 - 15	OS	B	237
		67	156	0.49	4	--	4 - 15	OS	B	237
	1,2,4-trichlorobenzene	35	33.0	0.623			5 - 40	LS		251
	heptane/3-methyl-2-butanone (27.3/72.7 vol)	25	92	0.50	6	--	25 - 287	LS	C	244
Poly(vinyl benzoate)	xylene	θ 32.5	62.0	0.50	5	--	10 - 24	OS	B	334
Poly(vinyl butyrate)	benzene	30	11.15	0.735	--	4	3 - 15	OS	C	256
Poly(vinyl caproate)	benzene	30	15.47	0.689	--	4	3 - 126	OS	C	256
Poly(vinyl 4-chloro-benzoate)	water	30	64.0	0.64	7	--	6 - 35	LV	B	336
	butanol/butanone (47/53 vol)	θ 60	73	0.50	7	--	6 - 35	LV	B	336
Poly(vinyl formate)	acetone	30	29.3	0.63	--	9	3 - 41	LV	C	257
	acetonitrile	30	14.1	0.717	--	9	3 - 41	LV	C	257
	dioxane	30	20.7	0.68	--	8	3 - 41	LV	C	257
	methyl acetate	30	37.6	0.61	--	7	3 - 24	LV	C	257
	methyl formate	30	14.1	0.722	--	7	3 - 24	LV	C	257
Poly(vinyl isobutyrate)	benzene	30	11.05	0.711	--	4	5 - 20	OS	C	256
Poly(vinyl isocaproate)	benzene	30	51.0	0.575	--	4	3 - 17	OS	C	256
Poly(vinyl pivalate)	acetone	25	2.88	0.77	4	--	40 - 217	LS	C	258
	butanone/methanol (0.897g/ml)	20	53	0.50	2	--	222 - 344	LS	C	258
Poly(vinyl sulfate)	aqueous NaCl(0.5M)	20	0.55	1.06	6	--	1 - 6	LV	C	261

Polymer	Solvent	Temp. [°C]	$K \times 10^3$ [ml/g]	a	No. of samples Fr.	No. of samples W.P.	Mol.Wt. Range $M \times 10^{-4}$	Method	Remarks	Ref.

1.8 POLY(STYRENE) AND DERIVATIVES

Polymer	Solvent	Temp.	K×10³	a	Fr.	W.P.	Mol.Wt. Range	Method	Remarks	Ref.
Poly(4-bromostyrene)	benzene	θ 20	95.5	0.53	10	--	3 - 30	OS	B	347
		θ 26.3	50.0	0.50	5	--	84 - 250	LS	A,R	348
	chlorobenzene	30	7.43	0.69	5	--	59 - 400	LS	A	348
	toluene	30	18.2	0.57	5	--	63 - 400	LS	A	349
Poly(2-chlorostyrene)	toluene	30	14.3	0.65	10	--	23 - 143	LS	A	350
Poly(4-chlorostyrene)	benzene	30	30.6	0.56	--	8	10 - 200	LS	C	351
	butanone	25	29	0.59	7	--	3 - 140	LS	B,R	352
		30	3.52	0.75	6	--	17 - 270	OS	B	353
	chlorobenzene	30	2.19	0.80	6	--	17 - 270	OS	B	353
	chloroform	30	14.8	0.65	--	8	10 - 200	LS	C	351
	dioxane	30	17.6	0.62	--	8	10 - 200	LS	C	351
	toluene	20	24.1	0.605	--	7	2 - 40	LS	B	354
		25	13.2	0.645	--	7	1 - 244	LS	B	355
		30	13.0	0.64	6	--	3 - 140	LS	B,R	352
		30	11.8	0.65	7	--	21 - 140	LS	A	349
		30	5.37	0.71	7	--	17 - 270	OS	B	353
Poly(4-cyclohexylstyrene)	heptane	30	32.3	0.54	6	--	4 - 30	OS	A-B	266
	toluene	30	10.6	0.69	7	--	2 - 30	OS	A-B	266
Poly(2,5-dichlorostyrene)	toluene	21	12.6	0.69	9		7 - 66	LS		356
	ethanol/ethyl acetate (1/15 wt)	θ 30.5	35.5	0.50	8	--	50 - 130	LS		357
Poly(3,4-dichlorostyrene)	chlorobenzene	30	4.39	0.72	7	--	8 - 51	OS	A	358
	o-dichlorobenzene	30	4.11	0.73	7	--	8 - 51	OS	A	358
	butanol/butyl acetate (1/13 wt)	θ 32.9		0.50	8	--	40 - 540	LS		359
Poly(2,4-dimethylstyrene)	toluene	30	9.52	0.70	--	9	5 - 120	LS	C	333
Poly(4-iodostyrene)	dioxane	20	33	0.51	10	6	10 - 118	LV	B-C	360
Poly(p-isopropylstyrene)	toluene	25	12.3	0.69	--	5	14 - 75	LS	B,C	265
Poly(o-methoxystyrene)	butanone	30	18.6	0.59	5	--	13 - 35	LS	A-B	362
	toluene	30	6.40	0.71	5	--	13 - 35	LS	A-B	362
	methanol/toluene (25/75 vol)	θ 30	57.5	0.50	4	--	15 - 30	LS	A-B	362
Poly(p-methoxystyrene)	butanone	30	3.75	0.73	5	--	13 - 75	LS	A-B	362
		35	8.6	0.68	6	--	1 - 100	LS	B	352
	chlorocyclohexane	25	17.7	0.63	16	--	22 - 220	LS	A	363
	pentyl acetate	25	55	0.52	16	--	22 - 220	LS	A	363
	toluene	25	10.5	0.70	16	--	22 - 220	LS	A	363
		30	5.28	0.73	5	--	13 - 75	LS	B	362
		30	18.0	0.62	6	--	1 - 100	LS	B	352
	methanol/toluene (28.1/71.9 vol)	θ 30	62.1	0.50	5	--	7 - 180	LS	B	362
Poly(α-methylstyrene) anionic, (ca. 50%-hetero, ca. 40%-syndio)	benzene	30	10.3	0.72	--	9	4 - 170	LS	A	319
	cyclohexane	θ 34.5	73	0.50	--	10	4 - 750	LS,OS	A	320
		θ 37	78	0.50	--	9	9 - 400	LS	A	321
		θ 38	76	0.50	--	6	2 - 66	LS	A	322
		θ 38.6	76.0	0.50	--	9	4 - 170	LS	A	323
		39	71.3	0.51	--	9	3 - 140	LS	A	324
	trans-decalin	θ 9.5	67	0.50	--	9	8 - 750	LS,OS	A	320
	toluene	25	7.06	0.744	--	9	8 - 750	LS,OS	A	320
		25	7.81	0.73	--	6	3 - 60	SD	A	325
		30	10.8	0.71	--	13	2 - 66	LS	A	322,326
cationic (10%-hetero, 90%-syndio)	benzene	30	24.9	0.647	4	--	14 - 91	OS	B	327
(19%-hetero, 80%-syndio)	cyclohexane	θ 32.5	66.0	0.50	5	--	2 - 370	LS	B	328
		θ 33.3	72.7	0.50	8	--	2 - 18	LS	B	328
	toluene	θ 30	2.2	0.80	6	--	1 - 100	LS	B	329
	benzene methanol (79.4/20.6 vol)	30	76.8	0.50	4	--	14 - 91	OS	B	327

Polymer	Solvent	Temp. [°C]	$K \times 10^3$ [ml/g]	a	No. of samples Fr.	No. of samples W.P.	Mol. Wt. Range $M \times 10^{-4}$		Method	Remarks	Ref.
Poly(m-methylstyrene)	benzene	30	7.36	0.76	9	--	8	- 115	OS	A	330
	cyclohexane	30	11.76	0.70	7	--	15	- 83	OS	A	330
	ethyl acetate	30	17.42	0.64	7	--	15	- 83	OS	A	330
Poly(p-methylstyrene)	diethyl succinate	θ 16.4	70	0.50	6	--	16	- 200	LS	A	331
	toluene	30	8.86	0.74	9	--	19	- 180	LS	A	331
Poly(methylstyrene), position of substituent, unspecified											
	cyclohexane	20	22	0.68	6	--	11	- 133	SV	A	332
Poly[(2,3,4,5,6-pentafluorostyrene)]											
	4-methyl-2-pentanone	20	4.37	0.736	--	21	10	- 260	OS	C	364
Poly(styrene)											
atactic	benzene	20	6.3	0.78	18	--	1	- 300	SD	A	270
		20	12.3	0.72	7	--	0.6	- 520	SD	A,R	271
		25	22.7	0.72	--	7	0.2	- 0.8	CR	C,L	272
		25	41.7	0.60	9	--	0.1	- 1	CR	B,L	272
		25	34.0	0.65	11	--	0.04	- 0.8	EG	A.L	273
		25	9.52	0.744	6	--	3	- 61	OS	A	274
		25	9.18	0.743	6	--	3	- 70	LS	A	275
		25	11.3	0.73	10	--	7	- 180	OS	A	276
		34	9.8	0.737	10	--	8	- 80	DV	A	277
	butanone	25	39	0.58	16	--	1	- 180	LS	A,R	278
		25	30.5	0.60	5	--	7	- 150	OS	A	276
		25	19.5	0.635	7	--	12	- 280	LS	A	279
		30	23	0.62	7	--	40	- 370	LS	B	280
		34	28.9	0.60	10	--	8	- 80	DV	A	281,282
	butyl chloride	40.8	15.1	0.659	5	--	29	- 106	LS	B	283
	chlorobenzene	25.7	7.4	0.749	4	--	62	- 424	LS	B	283
	chloroform	25	7.16	0.76	8	--	12	- 280	LS	A	279
		25	11.2	0.73	5	--	7	- 150	OS	A	276
		30	4.9	0.794	4	--	19	- 373	OS	B	284
	cyclohexane	28	108.0	0.479	7	--	0.6	- 69	OS	A	285
		θ 34	82	0.50	15	--	1	- 70	LV	A	274
		θ 34	90.2	0.503	9	--	0.6	- 69	OS	A	285
		θ 34.5	84.6	0.50	8	--	14	- 200	LS	A,R	286
		θ 35	80	0.50	3	--	8	- 42	LS	A	287
		θ 35	70	0.50	8	--	3	- 200	SD	B	288
		θ 35	76	0.50	10	--	4	- 137	LS	B	283
		40	41.6	0.554	10	--	4	- 137	LS	B	283
		45	34.7	0.575	10	--	4	- 137	LS	B	283
		50	26.9	0.599	10	--	4	- 137	LS	B	283
		50	36.4	0.584	7	--	4	- 52	LS	A	289
	decalin (100%-trans)	20	149	0.44	7	--	14	- 200	LS	A	290
		23	98	0.48	7	--	14	- 200	LS	A	290
		θ 23.8	--	0.50	--	--		--	LS	A	290
		25	67	0.52	7	--	14	- 200	LS	A	290
		30	61	0.53	6	--	14	- 200	LS	A	290
		60	22	0.63	4	--	14	- 200	LS	A	290
	decalin(73%-trans)	θ 18	77	0.50	4	--	14	- 140	LS	A	290
		30	36	0.58	4	--	14	- 140	LS	A	290
		40	37	0.58	4	--	14	- 140	LS	A	290
		60	22	0.64	4	--	14	- 140	LS	A	290
		100	15.7	0.67	6	--	14	- 200	LS	A	290
	dichloroethane	25	21.0	0.66	7	--	1	- 180	LS	A	278
		35	14.3	0.69	11	--	10	- 500	LS	A	689
	diethyl malonate	θ 34.2	71.8	0.50	3	--	39	- 400	LV	B	291
	diethyl oxalate	θ 55.8	73.0	0.50	3	--	39	- 400	LV	B	291
	dioxane	34	15.0	0.694	10	--	8	- 80	DV	A	282
	ethylbenzene	25	17.6	0.68	5	--	7	- 150	OS	A	276
	ethylcyclohexane	θ 70	75	0.50	2	--	36	- 127	LV	B	292
	methylcyclohexane	θ 70	76	0.50	?	?	?		?		293
		θ 70.5	69.6	0.50	3	--	39	- 400	LV	B	291
	toluene	20	4.16	0.788	10	--	4	- 137	LS	B	283

Polymer	Solvent	Temp. [°C]	K x 10³ [ml/g]	a	No. of samples Fr.	No. of samples W.P.	Mol. Wt. Range M x 10⁻⁴	Method	Remarks	Ref.
Poly(styrene) (Cont'd.)										
atactic	toluene (Cont'd.)	25	7.5	0.75	8	--	12 - 280	LS	A	279
		25	8.48	0.748	7	--	4 - 52	LS	A	289
		25	10.5	0.73	6	--	16 - 100	LS	A,R	294
		25	17	0.69	9	--	1 - 160	LS	A	278
		25	7.54	0.783	?	?	5 - 80	OS		295
		25	13.4	0.71	5	--	7 - 150	OS	A	276
		25	44	0.65	--	9	0.5 - 4.5	OS		296
		25	(a increases with M)		10	--	0.08 - 3.7	CR	L	297
		25	100	0.50	8	--	0.05 - 0.5	CR	A,R,L	298
		30	9.2	0.72	9	--	4 - 146	LS	A	299
		30	12.0	0.71	8	--	40 - 370	LS	B	280
		30	11.0	0.725	7	--	8 - 85	OS	A-B	300
		34	9.7	0.733	10	--	8 - 80	DV	A	282
	trichloro-benzene	135	1.75	0.67						697
	benzene methanol (74/26 vol)	θ 34	89	0.50	10	--	8 - 80	DV	A	277
	butanone/methanol (97.5/2.5 vol)	25	22.4	0.62	8	--	12 - 280	LS	A	279
	(95.0/5.0 vol)	25	26.3	0.60	8	--	12 - 280	LS	A	279
	(92.5/7.5 vol)	25	35.7	0.57	8	--	12 - 280	LS	A	279
	(89/11 vol)	θ 25	73	0.50	8	--	12 - 280	LS	A	279
	butanone/2-propanol (6/1 vol)	θ 23	73	0.50	9	--	4 - 146	LS	A	299
	(82.6/17.4 vol)	θ 34	71.8	0.50	10	--	8 - 80	DV	A	282
	chloroform/methanol (90/10 vol)	25	7.7	0.75	8	--	12 - 280	LS	A	279,278
	(80/20 vol)	25	12	0.68	8	--	12 - 280	LS	A	279,278
	(75/25 vol)	25	46	0.54	8	--	12 - 280	LS	A	279,278
	(74.7/24.3 vol)	θ 25	73	0.50	8	--	12 - 280	LS	A	279,278
	dioxane/methanol (65.1/34.9 vol)	θ 34	72.6	0.50	10	--	8 - 80	DV	A	282
	toluene/methanol (90/10 vol)	25	10.4	0.715	8	--	12 - 280	LS	A	279
	(80/20 vol)	25	26	0.612	8	--	12 - 280	LS	A	279
	(76.9/23.1 vol)	25	92	0.50	12	--	0.07 - 3.5	DV	A,L	298,297
	(75.2/24.8 vol)	θ 34	88	0.50	10	--	8 - 80	DV	A	282
atactic, anionic	benzene	25	100	0.50	--	7	0.04 - 1	VOS,EB	A,L	301
		30	8.5	0.75	--	12	2.5 - 150	VOS	A	301
		30	11.5	0.73	--	5	25 - 300	LS	A	302
		30	9.50	0.74	--	6	31 - 500	LS	A	649
	cyclohexane	θ 34	74.5	0.50	--	?	?	LS	B	304
		θ 34.5	85	0.50	--	12	0.04 - 150	LS	A,R	301,303
		θ 34.5	88	0.50	--	9	31 - 970	LS	A	649
		θ 34.6	91	0.50	--	4	25 - 300	LS	A	302
		θ 35	86	0.50	--	7	2 - 50	LS	A	305
	cyclohexene	25	16.3	0.68	--	3	20 - 107	LS	A	306
	decalin (66%-cis)	θ 12.2	80	0.50	--	6	2 - 50	LS	A	303
	decalin(99%-trans)	θ 20.4	81	0.50	--	8	31 - 760	LS	A	649
	dichloroethane	30	8.38	0.74	--	5	25 - 300	LS	A	302
	dioctyl phthalate	θ 22.0	80	0.50	--	4	40 - 160	LS	A	303
	toluene	20	11.2	0.72	--	6	3 - 24	SD		307
		25	9.77	0.73	--	12	1 - 104	SD	A,R	308
		25	34.5	0.62	--	25	0.4 - 230	SD	B	309
		30	8.81	0.75	--	5	25 - 300	LS	A	302
		30.3	10.4	0.73	--	15	2.6 - 50	OS,LS	A	310
isotactic	benzene	30	9.5	0.77	6	--	4 - 75	OS		311
		30	10.6	0.735	7	--	4 - 37	OS	A-B,R	312
	chloroform	30	25.9	0.734	3	--	9 -- 32	OS	C-D	284
	o-dichlorobenzene	25	17.9	0.677	5	--	2 - 100	LV	C	313
	toluene	30	11.0	0.725	7	--	3 - 37	OS	A-B	312
		30	9.3	0.72	5	--	15 - 71	LS	A-B,R	314

Polymer	Solvent	Temp. [°C]	K x 10³ [ml/g]	a	No. of samples Fr.	W.P.	Mol. Wt. Range M x 10⁻⁴		Method	Remarks	Ref.
Poly(styrene) (Cont'd.)											
branched, random type	butanone	25	(a decreases with M)		5	--	30	- 200	LS	B-C	315
	cyclohexane	θ 35	(a decreases with M)		9	--	8	- 300	LS	A	316
	toluene	30	(a decreases with M)		9	--	8	- 300	LS	A	316
star type, anionic	cyclohexane	θ 34	g'=0.94 (3 branches) ★								304
			g'=0.82 (4 branches) ★								304
	decalin	15	g'=0.48 (9 branches) ★								318
	toluene	25	g'=0.90 (3 branches) ★								304
		34	g'=0.84 (4 branches) ★								304
Poly(styrenesulfonic acid)	aqueous HCl (0.52M)	25	(0.344)	(1.0)	3	--	18	- 46	LV		365
	aqueous NaCl (0.52M)	25	(0.312)	(1.0)	3	--	18	- 46	LV		365
--, sodium salt	aqueous NaCl (4.17M)	θ 25	20.4	0.50	4	--	49	- 228	LS	B	366
	(0.5M)	25	18.6	0.64	6	--	39	- 234	LS	B,R	366
	(0.1M)	25	17.8	0.68	6	--	39	- 234	LS	B	366
	(0.05M)	25	13.9	0.72	6	--	39	- 234	LS	B	366
	(0.02M)	25	10.1	0.78	6	--	39	- 234	LS	B	366
	(0.01M)	25	2.8	0.89	5	--	39	- 234	LS	B	366
	(0.005M)	25	2.3	0.93	5	--	49	- 234	LS	B	366
	aqueous KCl (3.1M)	25	20.4	0.50	4	--	49	- 234	LS	B	366

1.9 OTHERS

Polymer	Solvent	Temp. [°C]	K x 10³ [ml/g]	a	No. of samples Fr.	W.P.	Mol. Wt. Range M x 10⁻⁴		Method	Remarks	Ref.
Poly[(biphenyl-4-yl)-ethylene]											
	benzene	20	21.4	0.619	5	--	7	- 170	LS	B	264
		30	29.5	0.59	6	--	1	- 110	LV	B	264
		75	27.7	0.589	5	--	7	- 170	LS	B	264
Poly(carbanilinoxyethylene), (Poly(vinyl carbanilate))											
	dioxane	20	13.7	0.68	11	--	6	- 200	LS	A	335
	dioxane/methanol (28/72 vol)	θ 20	64.5	0.51	5	--	6	- 200	LS	A	335
Poly(diphenylmethylene)	benzene		218	0.328	?		1	- 90	?		267
Poly(1-methoxycarbonyl-1-phenylethylene)											
	benzene	30	35.6	0.566	8	--	6	- 40	LS	A	361
	chloroform	30	12.7	0.661	8	--	6	- 40	LS	A	361
	ethylbenzene	θ 15	51.4	0.507	8	--	6	- 40	LS	A	361
Poly(vinylcarbazole)	benzene	25	30.5	0.58	11	--	0.7	- 45	LS	A	367
	chloroform	25	13.6	0.67	8	--	3	- 45	LS	A	367
	cyclohexanone	25	20.0	0.61	9	--	2	- 45	LS	A	367
	tetrachloroethane	25	12.9	0.68	9	--	2	- 45	LS	A	367
	tetrahydrofuran	25	14.4	0.65	10	--	1	- 45	LS	A	367
	toluene	θ 37	76.2	0.50	7	--	4	- 107	OS	A	368
Poly(5-vinyl-2-methylpyridine)											
	butanone	25	13.9	0.65	5	--	13	- 88	LS	A	375
		25	19	0.64	15	--	6	- 100	LS	A	376
	dimethylformamide	25	13.0	0.76	6	--	4	- 40	OS	A-B	377
	methanol	25	18.0	0.83	8	--	4	- 40	OS	A-B	377
		25	18.6	0.70	9	--	7	- 80	LS	A	376
		25	8.0	0.76	9	--	13	- 88	LS	A	375
Poly(1-vinylnaphthalene)	benzene	20	2.20	0.82	4	--	4	- 17	LS	B	264
		75	1.03	0.88	4	--	4	- 17	LS	B	264
Poly(2-vinylnaphthalene)	benzene	17	1.7	0.80	11	--	10	- 100	LS		268
		20	6.90	0.719	6	--	6	- 68	LS	B	264
		75	8.69	0.695	6	--	6	- 69	LS	B	264
	decalin/toluene (13/10 wt)	θ 30.2		0.50	8	--	10	- 100	LS		269
Poly(2-vinylpyridine)	benzene	25	17.0	0.64	14	--	3	- 93	LS	B,C	371
	butanone	25	97.2	0.47	14	--	3	- 93	LS	B,C	207
	dimethylformamide	25	14.7	0.67	14	--	3	- 93	LS	B,C	371
	dioxane	25	30.9	0.58	14	--	3	- 93	LS	B,C	371
	methanol	25	11.3	0.73	14	--	3	- 93	LS	B,C	371
	pyridine	25	13.8	0.69	14	--	3	- 93	LS	B,C	207

★ g' = [η] of branched molec. / [η] of linear molec. with same mol. wt.

Polymer	Solvent	Temp. [°C]	$K \times 10^3$ [ml/g]	a	No. of samples Fr.	W.P.	Mol. Wt. Range $M \times 10^{-4}$			Method	Remarks	Ref.
Poly(2-vinylpyridine) (Cont'd.)	ethanol/water (92/8 wt)	25	12.2	0.73	14	--	3	-	93	LS	B,C	371
Poly(4-vinylpyridine)	ethanol	25	(1.51)	(0.52)	--	3	1	-	4	SD	C	372
		25	25.0	0.68	8	--	10	-	185	LS	A-B	373
	water	25	22.0	0.687	8	--	10	-	185	LS	A-B	373
	butanone/2-propanol	25	38.0	0.57	7	--	7	-	224	LS	B	374
	ethanol/water (92/8 wt)	25	12.0	0.73	7	--	7	-	224	LS	B	374
Poly(vinylpyrrolidone)	chloroform	25	19.4	0.64	4	2	2	-	23	LS	B	378
	methanol	30	23	0.65	--	6	2	-	23	LS	B	378
	water	20	64	0.58	3	--	1	-	9	SD	B	379
		25	67.6	0.55	15	--	0.7	-	10	LS	B,R	378
		25	4.1	0.85	--	5	1	-	4	SD	C,D	211
		30	14	0.70	9	--	1	-	20	SD	B	381
		30	39.3	0.59	6	--	8	-	110	OS	A,R	383
	acetone/water (66.8/33.2 vol) θ	25	75.0	0.50	--	3	1.2	-	108	LS	B	384
Poly(vinylsulfonic acid)	aqueous KBr (0.347M) θ	5.7	68.8	0.50	5	--	4	-	39	LS	B	259
		15	30.8	0.61	5	--	8	-	39	LS	B	259
		30	24.5	0.75	5	--	8	-	39	LS	B	259
		50	26.6	0.76	5	--	8	-	39	LS	B	259
	aqueous KCl (0.349M) θ	5.5	68.2	0.50	5	--	4	-	39	LS	B	259
		25	16.7	0.79	5	--	4	-	39	LS	B	259
	(0.650M) θ	26.0	79.5	0.50	5	--	4	-	39	LS	B	259
	(1.001M) θ	44.5	80.3	0.50	5	--	4	-	39	LS	B	259
	aqueous NaBr (0.346M) θ	-0.6	95.5	0.50	5	--	4	-	39	LS	B	259
		10	26.8	0.73	5	--	8	-	39	LS	B	259
		20	25.1	0.76	5	--	8	-	39	LS	B	259
		30	22.0	0.79	5	--	8	-	39	LS	B	259
	(1.008M) θ	40.1	94.5	0.50	5	--	4	-	39	LS	B	259
	aqueous NaCl (1.003M) θ	32.4	96.1	0.50	5	--	4	-	39	LS	B	259
	(0.5M)	20	21.5	0.65	--	6	0.3	-	3	SD	C	260
Poly(vinyltrimethylsilane)	cyclohexane	25	8.2	0.71	5	--	59	-	213	LS	B	610

1.10 COPOLYMERS

Polymer	Solvent	Temp. [°C]	$K \times 10^3$ [ml/g]	a	No. of samples Fr.	W.P.	Mol. Wt. Range $M \times 10^{-4}$			Method	Remarks	Ref.
Poly(acrylonitrile-co-butadiene), see also Poly(butadiene-co-acrylonitrile) in group 1.1												
18/82 wt, random	toluene	25	251	0.50	7	--	0.06	-	1.26	OS	A	590
26/74 wt, random	toluene	25	260	0.50	5	--	0.15	-	0.40	OS	A	590
Poly(acrylonitrile-co-glycidyl methacrylate)												
	dimethylformamide	30	175	0.65	?	?		?		?		591
Poly(acrylonitrile-co-methyl acrylate)												
	dimethylformamide	20	17.9	0.79	6	--	2	-	21	LS	B	592
Poly(acrylonitrile-co-styrene), 38.3/61.7 mol, azeotropic												
	butanone	30	36	0.62	16	--	15	-	120	LS	B	593
	tetrahydrofuran	25	21.5	0.68	4	--	10	-	78	LS	B	594
62.6/37.4 mol, random												
	butanone	30	53	0.61	11	--	19	-	56	LS	B	595
	dimethylformamide	30	12	0.77	11	--	19	-	56	LS	B	595
Poly(butadiene-co-methacrylamide), 90/10 wt, random												
	toluene	25	437	0.50	5	--	0.09	-	0.11	OS	A	590
Poly(butadiene-co-2-methyl-5-vinylpyridine)												
	toluene	25	309	0.50	5	--	0.08	-	1.04	OS	A	590
Poly(butadiene-co-styrene), see also Poly(butadiene-co-styrene) in group 1.1												
84/16 mol, random	benzene	25	39.4	0.70	4	--	2	-	51	OS	A	596
	dibutyl phthalate	56	472	0.40	6	--	2	-	51	OS	A	596
	2-pentanone θ	23.8	167	0.50	5	--	7	-	51	OS	A	596
Poly(butyl itaconate-co-dibutyl itaconate), 40/60 mol, random												
	acetone	25	575	0.32	6	--	9	-	70	LS	B	597
	methanol	25	354	0.32	7	--	11	-	110	LS	B	597

Polymer	Solvent	Temp. [°C]	$K \times 10^3$ [ml/g]	a	No. of samples Fr.	W.P.	Mol. Wt. Range $M \times 10^{-4}$		Method	Remarks	Ref.
Poly(butyl itaconate-co-dibutyl itaconate), 40/60 mol, random (Cont'd.)											
	m-xylene	25	1040	0.21	7	--	11	- 110	LS	B	597
Poly(p-chlorostyrene-co-methyl methacrylate), 52/48 mol, random											
	benzene	27	7.94	0.72	9	--	15	- 120	LS	C	598
	benzene/hexane										
	(60/40 vol)	θ 22.3	64	0.50	8	--	15	- 120	LS	C	598
Poly(diethyl fumarate-co-isobutene), 50/50 mol											
	benzene/petrol ether	20	340	0.44	4	--	1	- 14	SD	C	599
Poly(p-diethylphosphono-methylstyrene-co-styrene), 1/4 mol, random											
	benzene	20	1.95	0.90	5	--	15	- 50	SV	B	600
	tetrachloroethane	20	0.0836	1.18	11	--	9	- 51	SV	B	600
Poly(dimethyl itaconate-co-styrene)											
75/25 wt	toluene	25	6.6	0.68	6	--	6	- 22	LS	A	683
67/33 wt		25	9.0	0.67	6	--	4	- 19	LS	A	683
59/41 wt		25	9.7	0.67	8	--	5	- 38	LS	A	683
49/51 wt		25	11.7	0.67	7	--	6	- 24	LS	A	683
29.5/70.5 wt		25	12.8	0.67	8	--	7	- 36	LS	A	683
27/73 wt		25	10.9	0.69	8	--	6	- 40	LS	A	683
0/100 wt		25	11.45	0.712	12	--	3	- 58	LS	A	683
Poly(dimethyl siloxane-co-diphenyl siloxane)											
55/45 mol	benzene	25	40.7	0.60	5	--	7	- 57	OS	A	596
	dimethyl phthalate	82.5	512	0.31	5	--	11	- 57	OS	A	596
	ethanol/toluene										
	(37/63 wt)	θ 29.5	78	0.50	5	--	7	- 35	OS	A	596
66/34 mol	benzene	25	15.6	0.68	4	--	3.7	- 100	LS	A	596
	hexane	θ 36	141	0.44	4	--	3.7	- 100	LS	A	596
	benzene/2-propanol										
	(44/56 wt)	θ 42	74	0.50	4	--	3.7	- 100	LS	A	596
Poly(divinylstyrene-co-styrene), see also poly(styrene), branched, random type, in group 1.8.											
	benzene	25	37.2	0.70	5	--	5	- 80	SD		602
	octane	21	162	0.50	6	--	5	- 80	SD		602
Poly(ethyl acrylate-co-methyl methacrylate), 80/20 mol, random											
	acetone	25	62	0.57	10	--	65	- 800	LS	B	114
Poly(ethylene-co-α-methylstyrene), $[(ET)_m(MS)_n]_p$											
m/n=3/4	cyclohexane	30	92	0.56	5	--	0.7	- 6	SA	B	604
	dioxane	30	76	0.58	5	--	0.7	- 6	SA	B	604
	toluene	30	32	0.68	5	--	0.7	- 6	SA	B	604
	butanone/cyclohexane										
	(60/40 vol)	θ 30	135	0.50	5	--	0.7	- 6	SA	B	604
m/n=5/4	cyclohexane	30	65	0.60	5	--	0.8	- 7	SA	B	604
	dioxane	30	89	0.56	5	--	0.8	- 7	SA	B	604
	toluene	30	37	0.66	5	--	0.8	- 7	SA	B	604
	butanone/cyclohexane										
	(75/25 vol)	θ 30	140	0.50	5	--	0.8	- 7	SA	B	604
m/n=5/7	cyclohexane	θ 30	112	0.50	4	--	1.5	- 7	SA	B	604
	dioxane	30	123	0.49	4	--	1.5	- 7	SA	B	604
	toluene	30	56	0.58	4	--	1.5	- 7	SA	B	604
Poly(hexadecyl methacrylate-co-methyl methacrylate)											
25/75 mol, random	heptane	25	85.0	0.38	8	--	4	- 47	LS	A	605
	propyl acetate	25	17.1	0.62	8	--	4	- 47	LS	A	605
38/62 mol, random	chloroform	25	36.3	0.57	8	--	16	- 195	LS	A	605
	heptane	25	6.9	0.65	8	--	4	- 47	LS	A	605
	propyl acetate	25	53.6	0.50	8	--	4	- 47	LS	A	605
50/50 mol, random	chloroform	25	53	0.54	8	--	10	- 154	LS	A	605
	heptane	25	32.0	0.52	8	--	4	- 47	LS	A	605
	propyl acetate	25	91.3	0.43	8	--	4	- 47	LS	A	605
Poly(1-hexene-co-sulfur dioxide), see poly[sulfonyl(butylethylene)] in group 3.10.											
Poly(isobutene-co-isoprene), see group 1.2.											
Poly(methylacrylic acid-co-methyl methacrylate)											
7.4/92.6 wt	acetone	20	3.4	0.74	9	--	26	- 105	LS		607

VISCOSITY-MOLECULAR WEIGHT RELATIONSHIPS

Polymer	Solvent	Temp. [°C]	$K \times 10^3$ [ml/g]	a	No. of samples Fr.	No. of samples W.P.	Mol. Wt. Range $M \times 10^{-4}$	Method	Remarks	Ref.
Poly(methyl acrylate-co-methyl methacrylate)										
17/83 wt, random	butanone	25	11.7	0.70	4	--	56 - 208	LS	B	131
29/71 wt, random	butanone	25	11.1	0.63	4	--	37 - 137	LS	B	131
67/33 wt, random	butanone	25	36.6	0.60	4	--	71 - 187	LS	B	131
Poly(methyl acrylate-co-styrene)										
22/78 mol, random	benzene	30	8.93	0.744	16	--	2.6 - 80	LS	A	129
	butanone	30	21.1	0.640	12	--	2.6 - 80	LS	A	129
	2-methylcyclohexanol	θ 43.5	77	0.50	8	--	2.6 - 80	LS	A	129
33/67 mol, random	benzene	30	7.18	0.759	9	--	6.6 - 36	LS	A	129
	butanone	30	11.4	0.696	4	--	6.6 - 36	LS	A	129
	2-methylcyclohexanol	θ 35.0	76	0.50	7	--	6.6 - 36	LS	A	129
47/53 mol, random	butanone	30	10.7	0.724	9	--	6.7 - 24.4	OS	A	129
50/50 mol, random	ethyl acetate	35	41.6	0.57	9	--	18 - 116	LS	A	133
59/41 mol, random	benzene	30	6.15	0.780	6	--	7 - 40	LS	A	129
	butanone	30	11.3	0.703	4	--	12 - 40	LS	A	129
	2-methylcyclohexanol	θ 36.6	76	0.50	6	--	7 - 40	LS	A	129
76/24 mol, random	benzene	30	7.42	0.766	6	--	7.2 - 28	LS	A	129
	butanone	30	9.16	0.728	5	--	8.9 - 28	LS	A	129
	2-methylcyclohexanol	θ 29.4	75	0.50	5	--	6.5 - 24	LS	A	129
Poly(methyl methacrylate-co-p-isopropylstyrene)										
2/3 mol, graft	butanone	25	0.021	1.11	--	6	31 - 65	LS		611
Poly(methyl methacrylate-co-2-methyl-5-vinylpyridine)										
85/15 mol, random	acetic acid	25	170	0.51	3	--	37 - 150	LV	B	612
Poly(methyl methacrylate-co-styrene)										
10/90 mol, random	1-chlorobutane	40.8	16.6	0.609	5	--	20 - 82	LS	B	613
30/70 mol, random	1-chlorobutane	30	17.6	0.67	9	--	5 - 55	LS	B	614
	cyclohexanol	θ 64.0	71.6	0.51	4	--	5 - 55	LS	B	614
	toluene	30	8.32	0.75	10	--	5 - 55	LS	B	614
44/56 mol, random	1-chlorobutane	30	24.9	0.63	10	--	5 - 81	LS	B	614
	cyclohexanol	θ 64.0	70.0	0.51	4	--	10 - 81	LS	B	614
	toluene	30	13.2	0.71	11	--	4.8 - 81	LS	B	614
50/50 mol, random	butanone	25	15.4	0.675	11	--	5 - 227	LS	B	615
52/48 mol, random	1-chlorobutane	40.8	49.0	0.575	5	--	18 - 115	LS	B	613
71/29 mol, random	1-chlorobutane	30	24.9	0.63	10	--	4.8 - 81	LS	B	614
	cyclohexanol	θ 68.0	97.3	0.47	5	--	15 - 106	LS	B	614
	toluene	30	11.4	0.70	8	--	7 - 106	LS	B	614
94/6 mol, random	1-chlorobutane	40.8	27.6	0.617	5	--	20 - 100	LS	B	613
nearly equimolar, three blocks (MSM)	cyclohexanol	θ 81.0	$\lim\limits_{M_w \to 0} [\eta]_\theta / M^{1/2} = 63$		--	7	3.4 - 147	LS	B	616

PS%, 86.1-90.4, graft (S on M; M_{PS}=0.7-1.0 x 10⁴)

Polymer	Solvent	Temp. [°C]	$K \times 10^3$ [ml/g]	a	No. of samples Fr.	No. of samples W.P.	Mol. Wt. Range $M \times 10^{-4}$	Method	Remarks	Ref.
	benzene	25	$[\eta] = 0.00918 M_{PS}^{0.743} g^{0.77}$				83 - 121	LV		617
	butanone	25	$[\eta] = 0.0390 M_{PS}^{0.58} g^{0.59}$				53 - 285	LV		617
graft	bromoform	--	56	0.6	6	--	180 - 320	LV		618
Poly(2-methyl-1-pentene-co-sulfur dioxide), see Poly[sulfonyl(1-methyl-1-propylethylene)], group 3.10.										
Poly(α-methylstyrene-co-styrene)										
	benzene	20	14.4	0.69	5	--	18 - 80	SV	A	332
Poly(styrene-co-sulfur dioxide), see Poly[sulfonyl(phenylethylene)], group 3.10.										
Poly(styrene-co-monoethyl maleate)										
	acetone	θ 26.4	51.1	0.50	9	--	20 - 180	LS	A	317
	dioxane	25	11.2	0.702	9	--	20 - 180	LS	A	317
	tetrahydrofuran	25	7.50	0.695	9	--	20 - 180	LS	A	317
--, sodium salt	aqueous NaCl (0.005M)	25	5.8	0.87	4	--	40 - 130	LS	A	317
	(0.01M)	25	5.5	0.85	5	--	40 - 180	LS	A	317
	(0.03M)	25	6.3	0.80	5	--	40 - 180	LS	A	317
	(0.05M)	25	11	0.73	5	--	40 - 180	LS	A	317
	(0.075M)	25	10	0.71	5	--	40 - 180	LS	A	317
	(0.15M)	25	15	0.65	5	--	40 - 180	LS	A	317
	(0.3M)	25	21	0.60	5	--	40 - 180	LS	A	317
	(0.6M)	θ 25	55	0.50	5	--	40 - 180	LS	A	317

Polymer	Solvent	Temp. [°C]	$K \times 10^3$ [ml/g]	a	No. of samples Fr.	W.P.	Mol. Wt. Range $M \times 10^{-4}$		Method	Remarks	Ref.

2. MAIN-CHAIN CARBOCYCLIC POLYMERS

Polymer	Solvent	Temp.	K	a	Fr.	W.P.	Range		Method	Remarks	Ref.
Poly(acenaphthenylene)	benzene	25	30.04	0.594	11	--	2	- 100	OS	B	262
		25	2.82	0.74	4	--	4	- 100	LS	A,B	263
	ethylene chloride	25	20.0	0.54	6	--	6	- 125	LS	A,B	263
	dioxane	25	11.5	0.61	7	--	6	- 145	LS	A,B	263
	methylene chloride	25	6.92	0.66	5	--	6	- 145	LS	A,B	263
	toluene	25	6.76	0.66	17	--	3	- 175	LS	A,B	263

3. MAIN-CHAIN HETEROATOM POLYMERS

3.1 POLY(OXIDES)

Poly(butene oxide), see Poly[oxy(ethylethylene)]
Poly(ethylene oxide), see Poly(oxyethylene)
Poly[oxy(tert-butyl-ethylene)]

Polymer	Solvent	Temp.	K	a	Fr.	W.P.	Range		Method	Remarks	Ref.
	benzene	25	39.7	0.686	9	--	8	- 520	LS	A-B	385
Poly(oxy-1,2-cyclohexylene)											
	toluene	35	3.5	0.83	22	--	2	- 50	OS	B	472
Poly(oxydecamethylene)	benzene	35	195	0.53	7	--	0.1	- 0.9	SE	B	386
	chloroform	30	172	0.56	9	--	0.05	- 0.9	SE	B	386
Poly(oxy-2,6-dimethyl-1,4-phenylene)											
	benzene	25	26.0	0.69	8	--	3	- 17	LS	B	473
	carbon tetrachloride	25	75.5	0.585	5	--	7	- 17	LS	B	473
	chlorobenzene	25	37.8	0.66	7	--	2	- 42	LS	B	474
		90	51.4	0.63	7	--	3	- 18	LS	B	473
	chloroform	25	48.3	0.64	8	--	2	- 42	LS	B	474
	toluene	25	28.5	0.68	15	--	2	- 42	LS	B	474
Poly(dioxolane), see Poly(oxymethyleneoxyethylene)											
Poly(oxy-2,6-diphenyl-1,4-phenylene)											
	chlorobenzene	25	13.9	0.68	--	10	4	- 145	LS	C	473
		90	15.5	0.67	--	10	4	- 145	LS	C	473
	toluene	25	21.4	0.635	--	10	4	- 145	LS	C	473
Poly(oxyethylene)	acetone	25	32	0.67	5	--	7	- 100	LV	A,R	387
		25	156	0.50	7	--	0.02	- 0.3	EG	A,L	388
	benzene	20	48	0.68	12	--	0.01	- 1.9	EG	A	389
		25	39.7	0.686	9	--	8	- 520	LS	A,R	385
		25	129	0.50	12	--	0.02	- 0.8	EG	A,L	388
	carbon tetrachloride	20	69	0.61	9	--	0.02	- 1.1	EG	A	389
		25	62	0.64	5	--	7	- 100	LV	A	387
	chloroform	25	206	0.50	6	--	0.02	- 0.15	EG	A,L	388
	cyclohexane	20	$[\eta]=0.5+0.035M^{0.64}$		11	--	0.006	- 1.1	EG	A	389
	diethylene glycol diethyl ether	50	140	0.51	5	--	7	- 100	LV	A	387
	dimethylformamide	25	$[\eta]=2.0+0.024M^{0.73}$		10	--	0.1	- 3	LS,SD	A	390
	dioxane	20	$[\eta]=0.75+0.035M^{0.71}$		13	--	0.006	- 1.1	EG	A	389
		25	138	0.50	7	--	0.02	- 0.15	EG	A,L	388
	methanol	20	$[\eta]=2.0+0.033M^{0.72}$		12	--	0.006	- 1.9	EG	A	389
		25	85.2	0.57	?		?		LS,SD	A	391
	4-methylpentan-2-one	50	120	0.52	5	--	7	- 100	LV	A	387
	toluene	35	14.5	0.70	--	4	0.04	- 0.4	EG	C L	392
	water	20	$[\eta]=2.0+0.016M^{0.76}$		11	--	0.006	- 1.1	EG	A	391
		25	156	0.50	5	--	0.019	- 0.1	EG	A,L	393
		30	12.5	0.78	--	6	2	- 500	LS,SD	C	394
		35	6.4	0.82	--	5	3	- 700	LV	C,R	395
		35	16.6	0.82	--	4	0.04	- 0.4	EG	C,L	392
		45	6.9	0.81	--	5	3	- 700	LV	C	395

Polymer	Solvent	Temp. [°C]	K x 10³ [ml/g]	a	No. of samples Fr.	No. of samples W.P.	Mol. Wt. Range M x 10⁻⁴		Method	Remarks	Ref.
Poly(oxyethylene) (Cont'd.)											
	aqueous K₂SO₄ (0.45M)	θ 35	130	0.50	--	5	3 -	700	LV	C	395
		35	280	0.45	--	5	7 -	100	LV	A	387
	aqueous MgSO₄ (0.39M)	θ 45	100	0.50	--	5	3 -	700	LV	C	395
Poly[oxy(ethylethylene)]	benzene	25	15.9	0.75	10	--	5 -	120	LS	B-C	396
		30	3.39	0.84	9	--	20 -	210	LS	B	397
	butanol	25	19.6	0.69	10	--	5 -	120	LS	B-C	396
	butanone	30	4.08	0.79	9	--	20 -	210	LS	B	397
	hexane	25	14.3	0.73	10	--	5 -	120	LS	B-C	396
	2-propanol	θ 30	86.5	0.50	9	--	20 -	210	LS	B	397
		θ 30	111	0.50	9	--	5 -	120	LS	B-C	396
Poly(oxyhexamethylene)	benzene	25	86.9	0.62	1	8	0.01 -	1.5	SE,CR	C	398
	dioxane	25	131	0.55	1	10	0.01 -	1.5	SE,CR	C	398
Poly(oxymethylene)	dimethylformamide	130	22.4	0.71	7	--	0.15 -	1.5	EG	B	399
		140	18.1	0.73	7	--	0.15 -	1.5	EG	B	399
	hexafluoroacetone-ses-quihydrate (1/1.7 mol, with triethyl-amine 1% vol)	25	46.0	0.74	7	--	0.15 -	1.5	EG	B	399
		25	87	0.69	--	5	2 -	16	LS	C	402
	phenol/tetrachloroethane (1/3 wt)	90	27.5	0.80	--	18	0.8 -	10	EG	C,D	400
	(1/3 vol)	90	5.22	0.93	?		?		OS		403
Poly(oxymethyleneoxyethylene)											
	chlorobenzene	25	200	0.50	4	14	9 -	100	LS	D	404
	p-chlorophenol	60	41.3	0.724	--	3	7 -	13	LS	C	405
	1H,1H,5H-octafluoro-pentanol-1	110	13.35	0.810	--	3	7 -	13	LS	C	405
Poly[oxy(phenylethylene)]											
	benzene	30	92.2	0.758	10	--	1.4 -	81	LS	B,C	385
	toluene	25	67.9	0.766	10	--	1.4 -	81	LS	B,C	385
Poly(oxypropylene)	acetone	25	75.5	0.56	5	--	0.1 -	0.4	LS	A	406
	benzene	20	11.1	0.79	5	--	0.07 -	0.33	SE	A	407
		25	11.2	0.77	3	--	3 -	70	LS	A-B	408
		25	14	0.8	?		?		?		409
isotactic		25	38.5	0.73	--	8	0.5 -	92	LV	C	410
		25	41.3	0.64	11	--	1 -	8	LS	A	411
		25	41.5	0.65	5	--	0.05 -	0.4	LS	A	406
	hexane	46	19.7	0.67	6	10	3.4 -	367	LS	A-B	408
	methanol	20	40.6	0.64	6	--	0.05 -	0.33	SE	A	407
		25	76.9	0.55	10	--	1 -	7	LS		411
	tetrahydrofuran	20	55.0	0.62	6	--	0.05 -	0.33	SE	A	407
	toluene	20	20.8	0.72	5	--	0.07 -	0.33	SE	A	407
		25	12.9	0.75	3	--	3 -	70	LS	A-B	408
	toluene/2,2,4-trimethyl-pentane (5/7 vol)	θ 39.5	107.5	0.50	7	--	1 -	7	LS	A	411
Poly(oxytetramethylene)	benzene	30	131	0.60	--	12	2.6 -	113	LS	A	412
	ethyl acetate	30	42.2	0.65	--	12	2.6 -	113	LS	A	412
	toluene	28	25.1	0.78	10	--	3 -	12	OS	A-B	413
	ethyl acetate/hexane (22.7/77.3 wt)	θ 31.8	206	0.49	--	11	2.6 -	113	LS	A	412
Poly(oxytrimethylene)	acetone	30	76.0	0.59	--	7	2.8 -	20	LS	A	414
	benzene	30	21.9	0.78	--	15	2.8 -	30	LS	A	414
	carbon tetrachloride	30	26.7	0.75	--	11	2.8 -	25	LS	A	414

Poly(propylene oxide), see Poly(oxypropylene)
Poly(tetrahydrofuran), see Poly(oxytetramethylene)

Polymer	Solvent	Temp. [°C]	$K \times 10^3$ [ml/g]	a	No. of samples Fr.	No. of samples W.P.	Mol. Wt. Range $M \times 10^{-4}$			Method	Remarks	Ref.

3.2 POLY(ESTERS), POLY(CARBONATES)

Bisphenol A poly(carbonates), see Poly[oxycarbonyloxy-1,4-phenyleneisopropylidene-1,4-phenylene]
Poly(ethylene terephthalate), see Poly(oxyethyleneoxyterephthaloyl)

Polymer	Solvent	Temp.	K×10³	a	Fr.	W.P.	Range lo		hi	Method	Remarks	Ref.
Poly(oxyadipoyloxydecamethylene)												
	chlorobenzene	25	11.7	0.84	--	7	0.3	-	3	LV	C	415
	diethyl succinate	79	5.8	0.86	--	12	1	-	3	LV	C	415
Poly[oxycarbonyl(bicyclo[2,2,2]octan-2,5-dion)carbonyloxyhexamethylene]												
	chloroform	20	--	--	--	4	1.4	-	3.9	OS	C	475
Poly(oxybutynedioyloxyhexamethylene)												
	benzene	20	151	0.55	?	--	0.1	-	0.5	OS	B	416
	chloroform	20	91	0.61	?	--	0.1	-	0.5	OS	B	416
Poly[oxycarbonyloxy-1,4-phenyleneisopropylidene-1,4-phenylene]												
	butyl benzyl ether	θ 170	210	0.50	8	--	4	-	31	LS	B	476
	chloroform	20	277	0.50			1.5	-	6	LS		477
		25	12.0	0.82	8	--	1	-	7	LS	A	478
	ethylene chloride	25	20.4	0.76	8	--	1	-	7	LS	A	478
	methylene chloride	25	11.1	0.82	6	--	1	-	27	SD	B	479
		25	11.9	0.80	12	--	1	-	76	LS	B,R	476
		25	38.9	0.70	8	--	1	-	7	LS	A	478
	tetrachloroethane	25	13.4	0.82	8	--	1	-	7	LS	A	478
	tetrahydrofuran	25	38.9	0.70	8	--	1	-	7	LS	A	478
		25	39.9	0.70	--	6	1	-	27	SD	C,R	479
	cyclohexane/dioxane (36.1/63.9 wt)	25	210	0.50	4	--	30	-	75	LS	B	476
Poly(oxycarbonylpentamethylene)												
	benzene	30	9.94	0.82	9	--	1.4	-	15	SV	B	447
	dimethylformamide	30	19.1	0.73	9	--	1.4	-	15	SV	B	447
Poly(oxycarbonylpropylene)												
	chloroform	30	7.7	0.82	--	5	2	-	78	SD	C,D	660
	2,2,2-trifluoroethanol	30	25.1	0.74	--	6	2	-	101	LS	C,D	661
Poly(oxy-1,4-cyclohexyleneoxysebacoyl)												
cis	chloroform	20	27.8	0.78	--	5	2.1	-	4.6	OS	C	480
trans	chloroform	20	18.3	0.86	--	9	1.1	-	3.7	OS	C	480
Poly(oxyethyleneoxyterephthaloyl)												
	o-chlorophenol	25	17	0.83	--	7	0.8	-	2.0	EG	C	481
		25	19	0.81	6	--	1.5	-	3.8	EG	B	482
		25	30	0.77	--	34	1.1	-	2.9	EG	C	483
		25	42.5	0.69	7	--	2	-	15	SD	A	484
		25	6560	0.73	--	5	1.2	-	2.5	OS	C	485
		55	26	0.77	6	--	1.5	-	3.8	EG	B	482
	m-cresol	25	0.77	0.95	--	5	0.04	-	1.2	EG	A,L	486
	dichloroacetic acid	45	400	0.50	7	--	1.5	-	3.8	EG	B	482
	tetrachloroethane	50	13.8	0.87	--	6	0.04	-	0.1	EG	A,L	487
	trifluoroacetic acid	25	140	0.64	7	--	1.5	-	3.8	EG	B	482
		30	43.3	0.68	--	9	2.5	-	12	LS	C	488
		35	130	0.66	7	--	1.5	-	3.8	EG	B	482
		55	105	0.69	6	--	1.5	-	3.8	EG	B	482
	dichloroethane/phenol (6/4 vol)		9.2	0.8						EG		489
	phenol/tetrachloroethane (40/60 wt)	25	140	0.64	6	--	1.5	-	3.8	EG	B	482
		35	125	0.65	6	--	1.5	-	3.8	EG	B	482
	(3/5 vol)	30	22.9	0.73	--	9	2.5	-	12	LS	C	488
	(50/50 vol)	20	75.5	0.685	--	38	0.3	-	3	EG	C	490
		25	21	0.82	--	9	0.5	-	3	EG	C	491
			12.7	0.86								492
	phenol/tetrachlorophenol	25	46.8	0.68						LS		493
	phenol/trichlorophenol (10/7 vol)	29.8	28.0	0.775	--	4	0.1	-	0.4	EG	C	494
		30	630	0.47	--	8	1.1	-	4	OS	C	495

Polymer	Solvent	Temp. [°C]	$K \times 10^3$ [ml/g]	a	No. of samples Fr.	W.P.	Mol. Wt. Range $M \times 10^{-4}$		Method	Remarks	Ref.
Poly(oxyfumaroyloxyhexamethylene)											
	chloroform	20	27.1	0.80	5	--	2	- 4.3	OS	B	417
Poly[oxy(hexahydro-3,6-endomethylenephthaloyl)oxyhexamethylene]											
cis	benzene	20	4.64	0.86	13	--	2.3	- 7.5	OS	B	496
	chloroform	20	9.33	0.83	13	--	2.3	- 7.5	OS	B	496
trans	benzene	20	17.4	0.75	10	--	3.3	- 11	OS	B	496
	chloroform	20	17.9	0.77	11	--	3.3	- 15	OS	B	496
Poly[oxy(hexahydroterephthaloyl)oxyoctamethylene]											
cis	chloroform	20	22.9	0.79	6	--	3.3	- 5.5	OS	B	480
trans	chloroform	20	18.9	0.84	6	--	2.4	- 4.4	OS	B	480
Poly(oxyhexamethyleneoxy-2,9-dibutylsebacoyl)											
	benzene	20	37.4	0.74	?	--	0.9	- 2.4	OS	B	418
Poly(oxyhexamethyleneoxysebacoyl)											
	benzene	20	62.7	0.69	9	--	0.6	- 1.8	OS	B	418
	chloroform	20	72.5	0.70	9	--	2	- 10	OS	B	419
Poly(oxymaleoyloxyhexamethylene)											
	benzene	20	76.3	0.60	7	--	1.3	- 6.6	OS	B	417
	chloroform	20	36.2	0.73	7	--	1.3	- 6.6	OS	B	417
	tetrahydrofuran	20	43.7	0.66	7	--	1.3	- 6.6	OS	B	417
Poly(oxysebacoyloxyhexadecamethylene)											
	chloroform	20	74.7	0.70	4	--	2	- 10	OS	B	419
Poly(oxysuccinyloxyhexamethylene)											
	benzene	20	43.3	0.70	22	--	1.5	- 5	OS	B	417
	chloroform	20	24.4	0.79	18	--	1.5	- 5	OS	B	417
	tetrahydrofuran	20	44.3	0.69	13	--	1.5	- 5	OS	B	417
Poly[oxytetra(ethyleneoxy)carbonyl(1-methylethylene)thio(2-methylethylene)carbonyl]											
	chloroform	?	34.7	0.714	--	?	< 1.5		EG	?	421
Poly(oxyundecanoyl)	chloroform	20	21.4	0.60	7	--	3	- 49	OS	B	419
		25	36.3	0.82	--	6	0.5	- 1.3	EG	C	420

3.3 POLY(AMIDES)

Polymer	Solvent	Temp. [°C]	$K \times 10^3$ [ml/g]	a	No. of samples Fr.	W.P.	Mol. Wt. Range $M \times 10^{-4}$		Method	Remarks	Ref.
Poly[(butylimino)carbonyl], (poly(butyl isocyanate))											
	benzene	20	1.10	1.11	--	7	1.8	- 21	SD	A,R	441
	tetrachloromethane	20			--	2	6.6	- 16	SD	D	442
	tetrahydrofuran	20	0.457	1.18	--	7	1.8	- 21	SD	A	441
Poly(iminoadipoyliminohexamethylene), (Nylon 66)											
	o-chlorophenol	25	168	0.62	--	2	1.4	- 5	LS,EG	C	443
	m-cresol	25	240	0.61	--	2	1.4	- 5	LS,EG	C	443
		25	$[\eta]=0.5+0.0353M^{0.792}$		13	--	0.015	- 5	LS,EG	B	444
	dichloroacetic acid	25	$[\eta]=0.5+0.352M^{0.551}$		13	--	0.015	- 5	LS,EG	B	444
	2,2,3,3-tetrafluoropropanol, CF_3COONa (0.1M)	25	114	0.66	--	2	1.4	- 5	LV	C	443
	aqueous HCOOH (90% vol)	25	35.3	0.786	3	11	0.6	- 6.5	LS,EG	C	443
		25	110	0.72	--	20	0.5	- 2.5	EG	C	446
		25	$[\eta]=2.5+0.0132M^{0.873}$		13	--	0.015	- 5	LS,EG	B	444
	aqueous HCOOH (90% vol), HCOONa (0.1M)	25	32.8	0.74	--	19	1	- 5	EG	C,R	445
		25	87.7	0.65	--	2	1.4	- 5	LS,EG	C	443
		25	$[\eta]=1.0+0.0516M^{0.687}$		6	--	0.015	- 5	LS,EG	B,R	444
	aqueous HCOOH (90% vol), KCl (2.3M)	θ 25	227	0.50	--	2	1.4	- 5	LS,EG	C	443
		θ 25	253	0.50	7	--	0.015	- 5	LS,EG	B-C	444
	aqueous H_2SO_4 (95% vol)	25	$[\eta]=2.5+0.0249M^{0.832}$		12	--	0.015	- 5	LS,EG	B	444
	aqueous H_2SO_4 (96% vol)	25	115	0.67	--	2	1.4	- 5	LV	C	443

Polymer	Solvent	Temp. [°C]	$K \times 10^3$ [ml/g]	a	No. of samples		Mol. Wt. Range $M \times 10^{-4}$		Method	Remarks	Ref.
					Fr.	W.P.					
Poly(iminohexamethyleneiminosebacoyl), (Nylon 6 10)											
	m-cresol	25	13.5	0.96	--	5	0.8	- 2.4	SD	B	454
Poly[imino(1-oxohexamethylene)], (Nylon 6)											
	m-cresol	25	320	0.62	6	--	0.05	- 0.5	EG	B	448
	trifluoroethanol	-20	53.3	0.74	5	--	1.3	- 10	LS	B	450
		25	53.6	0.75	5	--	1.3	- 10	LS	B	450
		50	58.2	0.73	5	--	1.3	- 10	LS	B	450
	aqueous HCOOH (85%)	-10	26.8	0.82	6	--	0.7	- 12	LS	B	450
		0	24.8	0.82	6	--	0.7	- 12	LS	B	450
		10	23.4	0.82	6	--	0.7	- 12	LS	B	450
		20	75	0.70			0.45	- 1.6	EG		451
		25	22.6	0.82	11	--	0.7	- 12	LS	B,R	450
	aqueous HCOOH (65%)	25	229	0.50	5	--	0.7	- 12	LS	B	450
	aqueous H_2SO_4 (40%)	25	59.2	0.69			0.3	- 1.3	EG		452
ring oligomer	m-cresol	25	2100	0.22	--	4	0.02	- 0.06	VOS	A	449
	ethylene chlorohydrin	25	870	0.27	--	3	0.03	- 0.06	VOS	A	449
monochain, polymerized with stearic acid	conc. H_2SO_4	25	63	0.76	--	7	0.2	- 1.4	EG	B	453
dichain, polymerized with sebacic acid	conc. H_2SO_4	25	42	0.79	--	14	0.2	- 2.3	EG	B	453
tetrachain, polymerized with a tetrabasic acid	conc. H_2SO_4	25	55	0.74	--	11	0.2	- 1.9	EG	B	453
octachain, polymerized with a octabasic acid	conc. H_2SO_4	25	13.5	0.86	--	5	0.4	- 2.6	EG	B	453
Poly(iminoterephthaloylimino-1,4-phenylenefluoren-9-ylidene-1,4-phenylene)											
	dimethylformamide	25	110	0.66	7	--	1.5	- 7.8	LS	B	503
Poly(iminoterephthaloylimino-1,4-phenylene phthalidylidene-1,4-phenylene)											
	dimethylformamide	25	277	0.59	8	--	1.4	- 6.9	LS	B	503

3.4 POLY(AMINO ACIDS)

Poly(β-benzyl-L-aspartate), see Poly(iminocarbonyl-L-benzyloxycarbonylethylidene)
Poly(γ-benzyl-L-glutamate), see Poly(iminocarbonyl-L-benzyloxycarbonylpropylidene)

Polymer	Solvent	Temp. [°C]	$K \times 10^3$ [ml/g]	a	No. of samples		Mol. Wt. Range $M \times 10^{-4}$		Method	Remarks	Ref.
					Fr.	W.P.					
Poly[(benzylimino)carbonylethylene], (Poly(N-benzyl-β-alanine))											
	dichloroacetic acid	25	120	0.525	--	6	0.15	- 1.8	EG	B,L	455
Poly(iminocarbonyl-L-benzyloxycarbonylethylidene), (Poly(β-benzyl-L-aspartate))											
	m-cresol	15	--	1.15	5	--	0.8	- 24	LS	B	456
		70	--	0.74	5	--	0.8	- 24	LS	B	456
	hexamethylphosphoramide	25	--	0.80	4	--	2	- 24	LS	B	456
	chloroform/dichloroacetic acid (98/2 vol)	25	--	1.30	5	--	0.8	- 24	LS	B	456
Poly(iminocarbonyl-L-benzyloxycarbonylpropylidene), (Poly(γ-benzyl-L-glutamate))											
	dichloroacetic acid	25	2.78	0.87	--	6	2	- 34	LS	C	457
	dimethylformamide	25	0.00029	1.70	--	5	7	- 34	LS	C	457
	dichloroacetic acid/heptane										
	(55/45 vol)	21	116	0.53	--	4	1.5	- 10	LS	C	458
	(90/10 vol)	21	25.4	0.68	--	4	1.5	- 10	LS	C	458
D,L	dichloroacetic acid	25	2.85	0.85	--	6	1.5	- 10	LS	C	459
	dimethylformamide	25	37.7	0.55	--	6	1.5	- 10	LS	C	459
Poly[iminocarbonyl-L-(N-hydroxypropyl-carbamoylpropylidene)], (Poly(N[5]-(3-hydroxypropyl)-L-glutamine))											
	methanol	25	--	1.6	4	--	20	- 40	LS	B	460
	water	25	--	0.6 ~ 1.0	5	--	2	- 33	LS	B	460
Poly(iminocarbonyl-L-methoxycarbonylpropylidene), (Poly(γ-methyl-L-glutamate))											
	m-cresol	25		>1			3	- 21	OS	C	462
	dichloroacetic acid	25	29	0.74	--	6	3	- 21	OS	C	463
D,L	m-cresol	25	11	0.78	--	6	3.2	- 8.2	OS	C	464
	dichloroacetic acid	25	5.9	0.85	--	6	3.2	- 8.2	OS	C	464

Polymer	Solvent	Temp. [°C]	$K \times 10^3$ [ml/g]	a	No. of samples Fr.	No. of samples W.P.	Mol. Wt. Range $M \times 10^{-4}$		Method	Remarks	Ref.
Poly(iminocarbonyl-L-methoxyethylideneiminocarbonyl-L-hydroxyethylideneiminocarbonylmethylene), (Poly(Asp(OCH$_3$)-Ser(H)-Gly))											
	dichloroacetic acid	30	868	0.367	--	9	0.25	- 1.1	SA	B,C	467
Poly(iminocarbonyl-L-p-nitrobenzyloxycarbonylpropylidene), (Poly(γ-p-nitrobenzyl-L-glutamate))											
	dichloroacetic acid	25	11.5	0.72	--	10	1	- 5	LS	B	608
	dimethylformamide	25	0.0170	1.36	--	10	1	- 5	LS	B	608
Poly(iminocarbonyl-L-phenylethylidene), Poly(L-phenyl alanine))											
	chloroform	25	0.00346	1.48	--	11	2.2	- 14	LS	B	465
D, L	chloroform/dichloro-acetic acid (2/3 vol)	21	118	0.55					LS	B	466
Poly[(methylimino)carbonylmethylene], (Poly(sarcosine))											
	water	20	56	0.88	--	5	0.7	- 1.6	EG	C	468
Poly(1-proline), see Poly[(L-1,2pyrrolidindiyl)carbonyl] group 3.9.											

3.5 POLY(UREAS), POLY(URETHANES), POLY(IMINES)

Polymer	Solvent	Temp. [°C]	$K \times 10^3$ [ml/g]	a	No. of samples Fr.	No. of samples W.P.	Mol. Wt. Range $M \times 10^{-4}$		Method	Remarks	Ref.
Poly(iminoethylene)	water	25	$[\eta] = 2.14 P^{0.35}$ P: number of N atoms		4	--	P=4-13		CR	D	470
Poly(oxytetramethyleneoxycarbonylimino-2,4-tolyleneiminocarbonyl)											
	dimethylformamide	30	54	0.74	--	5	0.35	- 1.6	LS	C	497
Poly[oxytetramethyleneoxycarbonylimino-(6-pentyloxy-1,3-phenylene)iminocarbonyl]											
	dimethylformamide	20	8.1	0.86	--	5	0.9	- 4.3	SV	C	498
Poly(oxytetramethyleneoxycarbonylimino-[6-(αH,ωH,ωH-perfluoroalkylene)oxy-1,3-phenylene]-iminocarbonyl)											
number of F atoms											
4	acetone	20	7.1	0.81	--	5	0.5	- 4	SV	C	498
8	acetone	20	4.3	0.785	--	5	2	- 16	SV	C	498
12	acetone	20	13.5	0.67	--	5	1.7	- 28	SV	C	498
16	acetone	20	25.6	0.615	--	5	0.9	- 9	SV	C	498
Poly(ureyleneheptamethylene)											
	dichloroacetic acid	46	338	0.505	--	10	0.3	- 2.4	LS	C	471
	sulfuric acid (90%)	25	500	0.714	--	14	0.13	- 2.4	LS	C	471
		46	223	0.506	--	7	0.06	- 2.4	LS	C	471
	(96%)	25	37.5	0.757	--	5	0.4	- 2.4	LS	C	471
	(98%)	46	240	0.53	--	7	0.2	- 2.4	LS	C	471

3.6 POLY(SULFIDES)

Polymer	Solvent	Temp. [°C]	$K \times 10^3$ [ml/g]	a	No. of samples Fr.	No. of samples W.P.	Mol. Wt. Range $M \times 10^{-4}$		Method	Remarks	Ref.
Poly(thiopropylene)	benzene	20	3.3	0.86	7	--	3.8	- 20.4	LS	B	436

3.7 POLY(PHOSPHATES)

Polymer	Solvent	Temp. [°C]	$K \times 10^3$ [ml/g]	a	No. of samples Fr.	No. of samples W.P.	Mol. Wt. Range $M \times 10^{-4}$		Method	Remarks	Ref.
Poly[oxy(hydroxyphosphinylidene)]											
	aqueous NaBr										
	(0.35M)	25	6.5	0.69	--	16	1	- 125	LS	C	422
	(0.415M)	θ 25	49.4	0.50	--	9	1	- 125	LS	C	422
Poly[oxy(hydroxyphosphinylidene)], sodium salt											
	aqueous NaBr										
	(0.035M)	25.5	69	0.61	--	5	0.09	- 1	EG	C	423
Poly(phosphoric acid), see Poly[oxy(hydroxyphosphinylidene)]											

3.8 POLY(SILOXANES), POLY(SILSESQUIOXANES)

Polymer	Solvent	Temp. [°C]	$K \times 10^3$ [ml/g]	a	No. of samples Fr.	No. of samples W.P.	Mol. Wt. Range $M \times 10^{-4}$		Method	Remarks	Ref.
Poly(dimethyl siloxane), see Poly[oxy(dimethylsilylene)]											
Poly[(1-isobutyl-3-phenylsilsesquioxane)											
	benzene	21	1.4	0.90	?	--	1.2	- 15	SD	?	619
		21	110	0.54	?	--	20	- 230	SD	?	619
	butyl acetate	24	same as above two data								619
Poly(3-methylbutenesilsesquioxane)											
	benzene	21	5.4	0.88	5	--	9	- 60	SD	B	603
		21	1.6	0.90	?	--	0.35	- 74	SD	B	619
	butyl acetate	24	5.4	0.88	13	--	9	- 60	SD	B	603

Polymer	Solvent	Temp. [°C]	$K \times 10^3$ [ml/g]	a	No. of samples Fr.	No. of samples W.P.	Mol. Wt. Range $M \times 10^{-4}$	Method	Remarks	Ref.
Poly[oxy(dimethylsilylene)]										
	benzene	20	12	0.68	4	--	5.5 - 12	LV	A-B	424
	bromobenzene	θ 78.7	76	0.50	3	--	8 - 106	LS	A	425
	bromocyclohexane	θ 28	78	0.50	5	--	10 - 92	SD	A	426
		θ 29.0	74	0.50	5	--	3.3 - 106	LS	A,R	425
	butanone	θ 20	81	0.50	5	--	5 - 66	OS	A	427
		30	48	0.55	8	--	5 - 66	OS	A	427
	ethyl iodide	θ 2.1	70	0.50	2	--	34 - 106	LS	A	425
	phenetole	θ 83	79	0.50	--	2	5 - 66	OS	A	427
		θ 89.5	73	0.50	4	--	4.5 - 106	LS	A,R	425
	toluene	20	20.0	0.66	--	?	0.3 - 20	OS,LS	C	428
		25	2.43	0.84	--	7	1.9 - 13	LS	C	429
		25	8.28	0.72	5	--	10 - 92	SD	A	426
		25	21.5	0.65	--	?	2 - 130	OS		430
		25	75	0.50	5	--	0.2 - 1.0	OS		56
	bromocyclohexane/ phenetole (6/7 vol)	θ 36.3	75.5	0.50	4	--	4.5 - 106	LS	A	425
	chlorobenzene/dimethyl phthalate (45/6 vol)	θ 57.5	76	0.50	3	--	8 - 106	LS	A	425
	$C_8F_{18}/C_2Cl_4F_2$ (33.17/66.83 wt), low cohesive energy density mixture	θ 22.5	106	0.50	4	--	55 - 120	LS	A-B	424
star type, 3 branches	toluene	20	23.9	0.64	10	--	4 - 35	LS	A	431
styr type, 4 branches	toluene	20	64.5	0.54	10	--	0.8 - 25	LS	A	431
Poly[oxy(dimethylsilylene)-1,4-phenylene-dimethylsilylene]										
	toluene	25	11.2	0.75	6	--	7 - 40	LS	B	499
Poly[oxy(dipropylsilylene)]										
	2-pentanone	θ 76	87.1	0.50	4	--	2.5 - 27	OS	A	433
	toluene	θ 10	109	0.50	6	--	2.5 - 30	OS	A	433
		25	43.5	0.58	15	--	1.7 - 43	OS	A	433
Poly[oxy(methylsilylene)]										
Me/Si=1.5	chlorobenzene	θ 20	326	0.21	12	--	0.1 - 500	LS		432
Me/Si=1.8	chlorobenzene/dimethylphthalate (90.7/9.3 wt)	θ 20	240	0.28	3	--	5 - 100	LS		432
Poly[oxy(methylphenylsilylene)]										
	cyclohexane	25	5.52	0.72	13	--	6 - 124	LS	A	434
	diisobutylamine	θ 30.4	51.5	0.50	9	--	6 - 124	LS	A	434
	toluene	25	3.90	0.78	20	--	6 - 124	LS	A	434
Poly[oxy(γ-trifluoropropylmethylsilylene)]										
	cyclohexyl acetate	θ 25.0	41.0	0.50	12	--	12 - 451	LS	A	435
	ethyl acetate	25	5.92	0.70	9	--	20 - 451	LS	A	435
	methyl hexanoate	θ 72.8	44.5	0.50	7	--	44 - 451	LS	A	435
Poly(phenylsilsesquioxane)										
	benzene	--	--	0.92	--	--	--	LS		500
		21	0.77	0.90	7	--	1.7 - 6.1	SD	B	501
		21	2.38	0.85	14	--	0.4 - 88	SD	B	502
		21	0.13	1.10	8	--	3.7 - 15	SD	B,L	501,603
		21	7.6	0.70	5	--	10 - 31	SD	B	603
	bromoform	21	0.13	1.09	8	--	3.7 - 15	SD	B	501
		21	2.38	0.85	12	--	3.6 - 88	SD	B	502
	benzene/bromoform (60/40 wt)	θ 21	2.38	0.85	5	--	14 - 71	SD	B	502
		θ 21	220	0.50	?	--	60 - 340	SD	?	619
cis-syndiotactic	1,2-dichloroethane	θ 50.5	2.12	0.87	4	--	5 - 30	OS		601

Polymer	Solvent	Temp. [°C]	$K \times 10^3$ [ml/g]	a	No. of samples Fr.	W.P.	Mol Wt. Range $M \times 10^{-4}$		Method	Remarks	Ref.

3.9 POLY(HETEROCYCLICS)

Poly[(1,3-dihydro-3-oxoisobenzofuran-1-ylidene)-1,4-phenyleneiminoterephthaloylimino-1,4-phenylene]

| | dimethylformamide | 25 | 277 | 0.59 | 7 | -- | 1.4 | - 5.5 | LS | B | 503 |

Poly[(1,3-dihydro-3-oxo-2-phenylisoindole-1-ylidene)-1,4-phenyleneoxyterephthaloyloxy-1,4-phenylene]

| | tetrachloroethane | 20 | 41.0 | 0.684 | 10 | -- | 0.9 | - 3 | LS | B | 514 |
| | tetrahydrofuran | 20 | 259 | 0.488 | 5 | -- | 1 | - 3 | LS | B | 514 |

Poly[(5,7-dihydro-1,3,5,7-tetraoxobenzo[1,2-c:4,5-c']-dipyrrole-2,6(1H,3H)diyl)-1,4-phenylene-(1,3-dihydro-3-oxoisobenzofuran-1-ylidene)-1,4-phenylene]

| | dimethylformamide | 20 | 328 | 0.516 | 26 | -- | 0.4 | - 17 | LS | B | 515 |

Poly(1-isobutyl-2,5-oxopyrrolidin-3,4-diyl)

| | butyl acetate | 21 | 22 | 0.65 | 13 | -- | 19 | - 340 | SD | A | 512 |

Poly[(4-phenyl-1,2,4-triazol-3,5-diyl)-1,3(or 1,4)-phenylene]

| | phenol/water (90/10 wt) | -- | 845 | 0.56 | -- | 5 | 1.3 | - 2.7 | OS | -- | 516 |

Poly[(L-1,2-pyrrolidindiyl)carbonyl]

| | water, acetic acid | 25 | no simple relation | | | -- | 5 | 1 | - 5 | OS | C | 507 |

Poly(1-p-tolyl-2,5-oxopyrrolidin-3,4-diyl)

| | dimethylformamide | 21 | 15.5 | 0.7 | 6 | -- | 4 | - 56 | SD | B | 513 |

3.10 COPOLYMERS (MALEIC ANHYDRIDE, SULFONES)

Poly[(tetrahydro-2,5-dioxo-3,4-furandiyl(1-isobutyloxyethylene)]

	acetone	30	124.7	0.506	5	--	21	- 111	LS	B	504
	butanone	30	119.4	0.512	5	--	21	- 111	LS	B	504
	tetrahydrofuran	30	75.6	0.552	5	--	21	- 111	LS	B	504

Poly[(tetrahydro-2,5-dioxo-3,4-furandiyl(1-methoxycarbonyl-1-methylethylene)]

	acetone	30	12.4	0.69	6	--	20	- 71	LS	B	505
	dimethylsulfoxide	30	7.5	0.77	6	--	20	- 71	LS	B	505
	dioxane	30	26.1	0.64	6	--	20	- 71	LS	B	505
	tetrahydrofuran	30	13.4	0.69	6	--	20	- 71	LS	B	505

Poly[(tetrahydro-2,5-dioxo-3,4-furandiyl) (1-phenylethylene)]

| | acetone | 30 | 8.69 | 0.74 | 6 | -- | 13 | - 75 | OS | A | 506 |
| | tetrahydrofuran | 30 | 5.07 | 0.81 | 6 | -- | 13 | - 75 | OS | A | 506 |

Poly[sulfonyl(butylethylene)]

	acetone	20	5.9	0.74	7	--	5	- 60	LS,SD	B	437
	benzene	25	8.9	0.70	5	--	9	- 107	OS	A,R	438
	chloroform	25	5.8	0.75	6	--	7	- 54	OS	A,R	439
	dioxane	25	6.2	0.76	5	--	9	- 107	OS	A	438
	hexylchloride	θ 13	33	0.55	5	--	10	- 60	LS,SD	B	437
	butanone/2-propanol										
	(29.8/70.2 vol)	θ 8	53	0.50	6	--	7	- 54	OS	A	439
	(37/63 vol)	θ 24	53	0.50	6	--	7	- 54	OS	A	439
	dioxane/hexane										
	(40/60 vol)	θ 20	65	0.50	7	--	9	- 107	OS	A	438

Poly[sulfonyl(1-methyl-1-propylethylene)]

	chloroform	20	5.9	0.81	6	--	4	- 50	OS	A	439
	butanone/2-propanol										
	(39.5/60.5 vol)	θ 22.5	91	0.50	6	--	4	- 50	OS	A	439
	butanone/hexane										
	(35.4/64.6 vol)	θ 11.5	91	0.50	6	--	4	- 50	OS	A	439

Poly[sulfonyl(phenylethylene)]

| | tetrahydrofuran | 30 | 3.89 | 0.78 | 5 | -- | 15 | - 40 | OS | A | 440 |

4. CELLULOSE AND DERIVATIVES

Amylose	dimethyl sulfoxide	20	3.97	0.82	--	14	2	- 217	LS	C	517
		25	1.25	0.87	9	--	22	- 310	LS	B	518
		25	15.1	0.70			8	- 180	LS	B	519
		25	30.6	0.64	6	--	27	- 220	LS	B	520
	ethylenediamine	25	15.5	0.70	6	--	31	- 310	LS	B	518
	formamide	20	22.6	0.67	--	12	2	- 157	LS	C	517

Polymer	Solvent	Temp. [°C]	$K \times 10^3$ [ml/g]	a	No. of samples Fr.	No. of samples W.P.	Mol. Wt. Range $M \times 10^{-4}$	Method	Remarks	Ref.
Amylose (Cont'd.)	formamide (Cont'd.)	25	30.5	0.62	14	--	8 - 180	LS	B	519
	water	20	13.2	0.68	--	12	36 - 217	LS	C	517
	acetone/dimethyl sulfoxide (43.5/56.5 vol) θ	20	83.1	0.51	--	10	2 - 157	LS	C	517
	aqueous KCl									
	(0.33M)	22.5	33.9	0.59	5	--	16 - 230	LS	B	521
	θ	25	112	0.50	5	--	16 - 230	LS	B	522
	θ	25	115	0.50	6	--	27 - 220	LS	B	520
	(0.50M) θ	25	61.1	0.50					B	523
	aqueous KOH									
	(0.15M)	25	8.36	0.77	?	--	8 - 180	LS	B	519
	(0.2M)	25	6.92	0.78	5	--	16 - 229	LS	B	522
	(0.5M)	25	8.50	0.76	6	--	27 - 220	LS	B	520
	(1M)	25	1.18	0.89	5	--	31 - 310	LS	B	518
	aqueous NaOH (0.5M)	20	3.65	0.85	--	16	2 - 217	LS	C	517
Amylose triacetate	chloroform	30	1.06	0.92	12	--	12 - 480	LS	B	524
		30	4.90	0.85	4	--	21 - 102	LS	A	525
		50	5.20	0.83	4	--	21 - 102	LS	A	525
	methyl acetate	25	5.60	0.80	--	3	7 - 19	SD	D	526
	nitromethane	22.5	8.50	0.73	12	--	14 - 310	LS	B	519,527
		30	9.93	0.76	4	--	21 - 102	LS	A	525
		50	8.71	0.76	4	--	21 - 102	LS	A	525
	chloroform/cyclohexane									
	(80/20 vol)	30	4.64	0.85	4	--	21 - 102	LS	A	525
	(50/50 vol)	30	7.41	0.79	4	--	21 - 102	LS	A	525
	methanol/nitromethane									
	(70.7/29.3 vol) θ	30	98.4	0.51	4	--	21 - 102	LS	A	525
	(50/50 vol)	30	6.49	0.75	4	--	21 - 102	LS	A	525
	(25/75 vol)	30	10.23	0.76	4	--	21 - 102	LS	A	525
	nitromethane/propanol									
	(43.3/56.7 vol) θ	25	91.6	0.50	12	--	14 - 310	LS	B	519
	(50/50 vol)	25	17.0	0.66	12	--	14 - 310	LS	B	519
Amylose tricarbanilate	acetone	20	0.814	0.90	--	26	4 - 490	LS	B	528
	dioxane	20	0.906	0.92	--	25	4 - 360	LS	B	528
	pyridine	20	0.589	0.92	--	20	4 - 360	LS	B	528
Amylose tricarbethoxymethylcarbamate										
	acetone	20	27.6	0.63	13	--	9 - 380	LV	B	529
Carboxymethyl amylose, sodium salt										
	aqueous NaCl									
	(0.35M)	37.5	25.2	0.64				LS	A	609
	(0.5M, pH 8)	35	209	0.53	6	--	5 - 21	OS	B	530
	(0.78M; 0.02% NaN$_3$)	35	37.1	0.61	6	--	7 - 29	LS	B	531
Diethylaminoethyl amylose hydrochloride										
	aqueous NaCl									
	(0.78M; 0.02% NaN$_3$)	35	82.8	0.55	5	--	4 - 23	LS	B	531
Arginic acid, sodium salt										
	aqueous NaCl (0.2M)	25	7.97	1.0	--	7	5 - 19	OS	C	532
Cellulose, see also table "Properties of Cellulose Materials."										
	cadoxen	25	33.8	0.77	5	--	20 - 100	SD	C,R	533
		25	38.5	0.76	4	--	1.0 - 3.4	SE,LS	B-C,R	534
	cuprammonium	20	105	0.66	9	--	2 - 25	OS	C	535
		25	8.5	0.81	--	5	8 - 96	OS	C	536
	cupriethylene	25	13.3	0.905	32	--	1 - 54	OS	B-C	537
Cellulose acetate butyrate										
	acetic acid	25	14.6	0.83	--	5	1 - 21	OS	B-C	538
	acetone	25	13.7	0.85	--	11	1 - 21	OS	B-C	538
Cellulose triacetate	acetone	20	2.38	1.0	5	--	2 - 14	SD	B	539
		25	14.9	0.82	8	--	2 - 39	OS	A(?)	540
		25	8.97	0.90	14	--	1 - 18	OS	B,R	541
		25	33.0	0.760	9	--	2 - 30	OS	C	535
	chloroform	30	4.5	0.9	5	--	3 - 18	LV	C	542

Polymer	Solvent	Temp. [°C]	$K \times 10^3$ [ml/g]	a	No. of samples Fr.	W.P.	Mol. Wt. Range $M \times 10^{-4}$	Method	Remarks	Ref.
Cellulose triacetate (Cont'd.)										
	o-cresol	30	6.15	0.9	5	--	3 - 18	LV	C	542
	acetone/water									
	(80/20 vol)	20	2.65	1.0	9	--	2 - 11	SD	B	539
		25	21.0	0.803		--	2 - 30	OS	C	535
	ethanol/methylene									
	chloride (20/80 vol)	25	13.9	0.834		--	2 - 30	OS	C	535
Cellulose tributyrate	butanone	30	4.3	0.87	7	--	6 - 32	LS	B,R	543
		30	18.2	0.80	7	--	8 - 22	OS	C-D	544,2
	tributyrin	0	5.3	0.87	4	--	6 - 32	LS	B	543
		25	5.6	0.85	4	--	6 - 32	LS	B	543
		50	6.1	0.82	4	--	6 - 32	LS	B	543
		70	6.2	0.80	4	--	6 - 32	LS	B	543
	dodecane/tetralin									
	(75/25 vol)	θ130	82	0.50	3	--	11 - 21	OS	C-D	544,2
Cellulose tricarbanilate	acetone	0	1.10	0.93	6	--	31 - 220	LS	B-C	545
		20	4.66	0.84	--	16	7 - 270	LS	B	528
		25	1.43	0.91	6	--	31 - 220	LS	B-C,R	545
		35	1.31	0.90	6	--	31 - 220	LS	B-C	545
	anisol	θ 94	130	0.50	4	--	31 - 220	LS	B-C	546
	cyclohexanone	25	1.91	0.86	5	--	31 - 220	LS	B-C	545
		35	2.02	0.85	5	--	31 - 220	LS	B-C	545
	dioxane	20	4.20	0.88	--	15	7 - 270	LS	B	528
		25	0.813	0.97	5	--	31 - 220	LS	B-C	545
		35	0.865	0.96	5	--	31 - 220	LS	B-C	545
		50	0.849	0.95	4	--	31 - 94	LS	B-C	545
	pyridine	20	3.46	0.86	--	12	7 - 270	LS	B	528
Cellulose trihexanoate	dimethylformamide	θ 41	245	0.50	7	--	6 - 130	LS	C-D	547
	dioxane	35	125	0.57	7	--	4 - 130	LS	C-D	547
Cellulose trinitrate	acetone	20	2.80	1.00	13	--	1 - 250	SD	B	548
		25	1.69	1.00	11	--	8 - 265	LS	B-C	549
		25	1.66	0.86	6	--	68 - 250	LS	C	550
		25	10.8	0.89	4	--	4 - 32	LS	C-D	551
(N content, 12.9 wt%)		25	5.70	0.90	4	--	15 - 200	LS	A,R	552
(N content, 13.9 wt%)		25	6.93	0.91	6	--	8 - 400	LS	A,R	552
		25	7.00	0.933	9	--	5 - 50	OS	B-C	535
		25	11.0	0.91	33	--	3 - 100	OS	B-C	537
		25	23.5	0.78	6	--	7 - 26	OS	B-C	553
	butyl acetate	25	5.68	0.969	9	--	5 - 50	OS	B-C	535
	butyl formate	25	23	0.81	6	--	7 - 26	OS	B-C	553
	cyclohexanone	25	2.24	0.810	6	--	7 - 22	OS	B-C	554
	ethyl acetate	25	3.8	1.03	33	--	3 - 100	OS	B-C	537
		25	8.3	0.90	6	--	7 - 26	OS	B-C	553
		25	1.66	0.86	7	--	68 - 250	LS	C	550
		30	2.50	1.01	6	--	4 - 57	LS	B-C	555
	ethyl butyrate	25	3.64	1.0	?	--	5 - 50	OS	B-C	556
	ethyl formate	25	30	0.79	6	--	7 - 26	OS	B-C	553
	ethyl lactate	25	12.2	0.92	10	--	3 - 65	OS	B-C	537
	2-heptanone	25	5.0	0.93	6	--	7 - 26	OS	B-C	553
	methyl acetate	25	18.3	0.835	6	--	7 - 22	OS	B-C	554
	nitrobenzene	25	6.1	0.945	6	--	7 - 22	OS	B-C	554
	pentyl acetate	25	1.1	1.04	6	--	7 - 26	OS	B-C	553
Cellulose trioctanoate	dimethylformamide	θ140	113	0.50	3	--	10 - 32	OS	B-C	544
	γ-phenylpropanol	θ 48	129	0.50	3	--	8 - 32	OS	B-C	544
	toluene	30	17.3	0.70	6	--	8 - 35	OS	B-C	544
Ethyl cellulose	acetone	20	1.51	1.05	5	--	1.1 - 8	SD	A	557
	benzene	20	1.34	1.07	5	--	1.1 - 8	SD	A	557
		25	29.2	0.81	6	--	4 - 14	OS	B-C	558
		60	35.8	0.78	6	--	4 - 14	OS	B-C	558
	butanone	25	18.2	0.84	6	--	4 - 14	OS	B-C	558
		60	26.7	0.79	6	--	4 - 14	OS	B-C	558

Polymer	Solvent	Temp. [°C]	$K \times 10^3$ [ml/g]	a	No. of samples Fr.	W.P.	Mol. Wt. Range $M \times 10^{-4}$		Method	Remarks	Ref.
Ethyl cellulose (Cont'd.)	butyl acetate	25	14.0	0.87	6	--	4	- 14	OS	B-C	558
		60	18.1	0.83	6	--	4	- 14	OS	B-C	558
	chloroform	25	11.8	0.89	6	--	4	- 14	OS	B-C	558
		46	9.3	0.90	6	--	4	- 14	OS	B-C	558
	ethyl acetate	25	10.7	0.89	6	--	4	- 14	OS	B-C	558
		60	14.0	0.85	6	--	4	- 14	OS	B-C	558
	methanol	25	52.3	0.65	6	--	10	- 41	LS	B-C	559
	nitroethane	25	4.2	0.96	6	--	4	- 14	OS	B-C	558
		60	22.6	0.79	6	--	4	- 14	OS	B-C	558
Ethyl hydroxyethyl cellulose	water	25	37	0.80	4	--	5	- 18	SD, LS	B	560
Hydroxyethyl cellulose	cadoxen	25	17.4	0.79	4	--	8	- 61	LS	B	561
	water	25	9.53	0.87	5	--	8	- 63	LS	B	561
D.S.* 0.88	aqueous HCl (4M)	25	$[\eta]=1.2DP_w^{0.87}$ (DP_w; weight-average degree of polymerization)							B	562
Methyl cellulose D.S.* 1.74	water	25	316	0.55	--	5	12	- 57	LS	C-D	563
	aqueous HCl (4M)	25	$[\eta]=1.6DP_w^{0.86}$							LS	B	562
Sodium carboxymethylcellulose D.S.* 0.2-1.0	cadoxen	25	33.4	0.73	--	5	5	- 106	LS	C	564
D.S.* 0.96	aqueous HCl (4M)	25	$[\eta]=0.97DP_w^{0.83}$							B		562
D.S.* 0.62-0.74	aqueous NaCl										
	(0.001M)	25	0.100	1.40	8	--	4.5	- 35	SD	C-D	565
	(0.01M)	25	0.646	1.20	3	--	4.5	- 35	SD	C-D	565
	(0.1M)	25	12.3	0.91	8	--	4.5	- 35	SD	C-D	565
D.S.* 1.06	(0.005M)	25	7.2	0.95	4	--	14	- 106	LS	C-D	566
	(0.01M)	25	8.1	0.92	4	--	14	- 106	LS	C-D	566
	(0.05M)	25	19	0.82	4	--	14	- 106	LS	C-D	566
	(0.2M)	25	43	0.74	4	--	14	- 106	LS	C-D	566
Sodium cellulose xanthate	aqueous NaOH (1M)	0	$[\eta]=1.67DP_w^{0.82}+0.62DS_{2,3}-0.20DS_6 \, DP_w^{0.94}$							C		567
			DS_i, degree of substitution at the i(=2,3 or 6) positions in glucose unit									
Dextran, linear fraction	formamide	25	16.5	0.49	5	--	0.2	- 3.2	OS	C	568
	water	25	97.8	0.50	10	--	2	- 10	LS	C,R	569
		25	49.3	0.60	10	--	0.04	- 4.5	EA	C	570
		50	39.3	0.61	6	--	0.2	- 3.2	OS	C	568
branched fraction	water	25	--	0.20	9	--	80		LS	C	569
		32.7			6	--					571
Guaran triacetate	acetonitrile	25	2.62	0.87	4	--	7	- 85	LS	A	572
		25	311	0.52	5	--	206	- 534	LS	B	572
Hyaluronic acid	aqueous HCl (0.1M)	25	27.9	0.763	5	--	7	- 103	LS	A	573
	aqueous NaCl (0.2M)	25	22.8	0.816	8	--	7	- 103	LS	A	573
	(0.5M)	25	31.8	0.777	5	--	11	- 103	LS	A	573
Salep glucomannan triacetate	nitroethane	30	this relation not followed		11	--	0.06	- 0.4	LS	A-B	574

D. CALCULATED UNPERTURBED DIMENSIONS OF FREELY-ROTATING CHAINS

Chain Type	$r_{of}/M^{1/2}$ [nm mol$^{1/2}$ g$^{-1/2}$]		Reference
Polymethylene chain	$0.308/M_u^{1/2}$	$0.218/m^{1/2}$	2
Amylosic chain	$0.426/M_u^{1/2}$	$0.191/m^{1/2}$	518
Cellulosic chain	$0.790/M_u^{1/2}$	$0.353/m^{1/2}$	620
Gutta-percha (trans polydiene)	$0.580/M_u^{1/2}$	$0.290/m^{1/2}$	620,621
Natural rubber (cis polydiene)	$0.402/M_u^{1/2}$	$0.201/m^{1/2}$	620
Polypeptide	$0.383/M_u$	$0.221/m$	620

*D.S. - Degree of Substitution

E. UNPERTURBED DIMENSIONS OF LINEAR POLYMER MOLECULES

(References in parenthesis give data which were used for calculation of end-to-end distance in Ref. 3.)

1. MAIN-CHAIN ACYCLIC-CARBON POLYMERS

1.1 POLY(DIENES)

Polymer	Solvent	Temp. [°C]	$S_{oz}/M_w^{1/2}$ or $a_p \times 10^4$ [nm]	$K_o \times 10^3$ [ml/g]	$r_o/M^{1/2} \times 10^4$ [nm]	$r_{of}/M^{1/2} \times 10^4$ [nm]	$\sigma = r_o/r_{of}$	$C_\infty = r_o^2/nl^2$	Method	References
Poly(butadiene)										
100%-cis	dioxane	20,2	--	205	920	547	1.68	5.15	VT	19
98%-cis, 2%-1,2	isobutyl acetate	20.5	--	185	880	547	1.61	4.75	VT	15
95%-cis, 4%-1,2	2-pentanone	59.7	--	157	835	546	1.53	4.3	VT	17
95%-cis, 4%-1,2	3-pentanone	10.3	--	152	825	546	1.51	4.2	VT	17
92%-cis, 5%-1,2	benzene	32	--	150± 20	820± 40	545	1.50±0.08	4.15	VG	3(20)
71%-trans, 25%-1,2	cyclohexane	25	--	300± 40	1030± 50	702	1.45±0.08	7.3	VG	3(22)
79%-trans, 21%-1,2	cyclohexane	20	--	280± 25	1010± 30	742	1.36±0.05	6.9	VG	3(23)
97%-trans, 3%-1,2	cyclohexane	40	--	200± 30	935± 40	768	1.22±0.07	5.4	VG	24
100%-cis	various solvents	50					1.63	4.9	VT	622
100%-cis	undiluted	50~90 $d\ln r_o^2/dT = 0.4 \times 10^{-3}$ [deg^{-1}]							ST	622
100%-trans	decalin	55					1.23	5.8	VA	623
100%-trans	undiluted	$d\ln r_o^2/dT = -0.6 \times 10^{-3}$ [deg^{-1}]							ST	623
Poly(chloroprene)										
85%-trans	benzene	25	--	115± 20	750± 80	535	1.40±0.15	5.6	VG	3(32,33)
	butanone	25	--	113	750	535	1.40	5.6	VT	35
	cyclohexane	45.5	--	107	755	535	1.41	5.65	VT	34
	butanone	25	313	--	750	535	1.40	5.6	LT	624
Poly(isoprene)										
100%-cis	benzene; 2-pentanone	~20	--	130± 20	810± 45	485	1.67±0.09	5.0	VT,VG	3,37
	diisopropyl ether	22	0.76	--	847	485	1.74	5.5	XS	625
	2-pentanone	14.5		119				4.7	VT	626
	undiluted	-10~70 $d\ln r_o^2/dT = 0.41 \times 10^{-3}$ [deg^{-1}]							ST	626
100%-trans	propyl acetate	60	--	232	970	703	1.38	7.2	ST,VT	627
	dioxane	47.7		191	910	703	1.30	6.35	VT	37
	undiluted	30~70 $d\ln r_o^2/dT = 0.56 \times 10^{-3}$ [deg^{-1}]							VT	19
	undiluted	~60 $d\ln r_o^2/dT = -0.27 \times 10^{-3}$ [deg^{-1}]							ST	623

1.2 POLY(ALKENES)

Polymer	Solvent	Temp. [°C]	$S_{oz}/M_w^{1/2} \times 10^4$ or a_p [nm]	$K_\theta \times 10^3$ [ml/g]	$r_0/M^{1/2} \times 10^4$ [nm]	$r_{of}/M^{1/2} \times 10^4$ [nm]	$\sigma = r_0/r_{of}$	$C_\infty = r_0^2/nl^2$	Method	References
Poly(1-butene)										
atactic	anisole; ethylcyclohexane	~70	--	123± 10	775± 25	427	1.82±0.05	6.6	VT,VG	3(81)
	nonane	35	590± 50	--	1180± 70	427	2.76±0.20	15.1	LT	81
	undiluted	140~200 dln r_o^2/dT = (0.5±0.04) x 10⁻³ [deg⁻¹]	--	--					ST	633
isotactic	nonane	80	510± 50	--	1290± 90	427	3.00±0.20	18.0	LT	81
	undiluted	140~200 dln r_o^2/dT = (0.09±0.07) x 10⁻³ [deg⁻¹]							ST	634
	undiluted	160 dln r_o^2/dT = -0.1 x 10⁻³ [deg⁻¹]							ST	634
Poly(ethylene)	1-chloronaphthalene; tetralin;									
	p-xylene	~100	--	230	950± 40	582	1.63±0.08	5.3	VG	3(61, 67, 74)
	decalin	140	--	--	1070	582	1.84	6.8	VA	65
	bis-2-ethylhexyl adipate	145	--	225	940± 40	582	1.61	5.2	VT	630
		145	690±100	--	1320±150	582	2.27±0.26	10.3	LT	630
	biphenyl	127.5	--	323	1085	582	1.87	7.0	VT	58
	dodecanol	137.3	--	307	1070	582	1.84	6.8	VT	58
		138	--	316± 7	1080	582	1.86	6.9	VT	69
	diphenylmethane	142.2	--	315	1080	582	1.86	6.9	VT	58
	decanol	153.3	--	302	1065	582	1.83	7.6	VT	65
	diphenyl ether	161.4	--	295	1060	582	1.79	6.4	VT	58
	octanol	180.1	--	286	1040	582	1.79	6.4	VT	58
	biphenyl	127.5	--	330	1095	582	1.88	7.1	VT	631
	diphenylmethane	142.2	--	322	1085	582	1.87	7.0	VT	631
	diphenyl ether	163.9	--	309	1070	582	1.84	6.8	VT	631
	undiluted; diluted with tri-acontane; dotriacontane	140~190 dln r_o^2/dT = -(1.15±0.1) x 10⁻³ [deg⁻¹]							ST	629
	hexadecane	140 dln r_o^2/dT = -1.2 x 10⁻³ [deg⁻¹]							VA	632
Poly(isobutene)	anisole	105	--	91± 5	700± 20	412	1.70±0.05	5.8	VT	49
	benzene	24	--	107± 5	740± 20	412	1.80±0.05	6.5	VT	49
	phenetole	86	--	91± 5	700± 20	412	1.70±0.05	5.8	VT	49
	heptane/propanol (80/20 vol)	25	390	166	780	412	1.9	7.2	LT	628
	undiluted; diluted with hexa-decane	60 dln r_o^2/dT = -(0.1±0.05) x 10⁻³ [deg⁻¹]							ST	629
Poly(1-octene)	bromobenzene	25	--	50	570± 50	291	1.96±0.15	7.7	VG	84
	cyclohexane	30	--	100	710± 60	291	2.44±0.20	11.9	VG	84
	phenetole	50.4	--	65	625± 30	291	2.14±0.10	9.1	VT	84
Poly(1-pentene)										
atactic	undiluted	40-140 dln r_o^2/dT = (0.53±0.05) x 10⁻³ [deg⁻¹]							ST	633
isotactic	2-pentanol	62.4	--	121	790	368	2.14	9.2	VT	633
	undiluted	80~140 dln r_o^2/dT = (0.34±0.04) x 10⁻³ [deg⁻¹]							ST	633

Polymer	Solvent	Temp. [°C]	$S_{oz}/M_w^{1/2} \times 1$ or a_p [nm]	$K_o \times 10^3$ [ml/g]	$r_o/M^{1/2} \times 10^4$ [nm]	$r_{of}/M^{1/2} \times 10^4$ [nm]	$\sigma = r_o/r_{of}$ of	$C_\infty = r_o^2/nl^2$	Method	References
Poly(1-pentene) (Cont'd.)										
isotactic	undiluted	~90 $\mathrm{dln}\,r_o^2/dT = -0.3 \times 10^{-3}$ [deg^{-1}]							ST	634
		~60 $\mathrm{dln}\,r_o^2/dT = -0.2 \times 10^{-3}$ [deg^{-1}]							ST	635
Poly(propylene)										
atactic	isoamyl acetate; benzene; cyclohexane; toluene	30	--	156±15	835±25	475	1.76±0.05	6.2	VT,VG	3(88,89)
	decalin	135	--	125±20	775±35	475	1.63±0.08	5.3	VG	3(88,91)
	1-chloronaphthalene	74	--	182	880	475	1.85	6.85	VT	90
	cyclohexanone	92	--	172	870	475	1.83	6.7	VT	90
	diphenyl ether	153	--	120	765	475	1.61	5.2	VT	90
isotactic	1-chloronaphthalene; decalin; tetralin	~140	--	120±20	765±40	475	1.61±0.08	5.2	VG	3(88,96)
	diphenyl ether	145	--	132	790	475	1.66	5.5	VT	90
		145	--	94	710	475	1.49	4.45	VT	630
		145	370±30	--	685±30	475	1.44±0.07	4.15	LT	630
	biphenyl	125.1	--	152	809	475	1.70	5.8	VT	94
	diphenyl ether	142.8	--	137	782	475	1.62	5.25	VT	94
	dibenzyl ether	183.2	--	106	718	475	1.51	4.55	VT	94
syndiotactic	heptane	30	--	164	830	475	1.75	6.1	VG	100
	isoamyl acetate	45	--	172	843	475	1.77	6.25	VT	100

1.3 POLY(ACRYLIC ACID) AND DERIVATIVES

Polymer	Solvent	Temp. [°C]	$S_{oz}/M_w^{1/2} \times 1$ or a_p [nm]	$K_o \times 10^3$ [ml/g]	$r_o/M^{1/2} \times 10^4$ [nm]	$r_{of}/M^{1/2} \times 10^4$ [nm]	$\sigma = r_o/r_{of}$ of	$C_\infty = r_o^2/nl^2$	Method	References
Poly(acrylamide)	water	30	--	260±40	1000±50	367	2.72±0.10	14.8	VG	3(101)
Poly(acrylic acid)	1,4-dioxane	30	--	76	665	363	1.83	6.7	VT	104
--, sodium salt	aqueous NaBr (1.5M)	15	--	124	756	318	2.38	11.3	VT	108,109
		15	--	--	1030	318	3.24	21	LD	108
Poly(acrylonitrile)	aqueous NaSCN (1.25M)	30	--	121	752	318	2.36	11.1	VT	111
(polymd. at -30°C)	dimethylformamide	25	--	210±15	930±20	422	2.20±0.05	9.7	VG	3(138,139)
(polymd. at 60°C)	γ-butyrolactone; dimethylformamide	30	--	250	970	422	2.30	10.6	VG	135
		30	--	200	900	422	2.13	9.1	VG	135
Poly(butyl acrylate)	undiluted	60 $\mathrm{dln}\,r_o^2/dT = -0.2 \times 10^{-3}$ [deg^{-1}]							ST	635
		76 $\mathrm{dln}\,r_o^2/dT = 0$ [deg^{-1}]							ST	634
Poly(sec-butyl acrylate)	undiluted	60 $\mathrm{dln}\,r_o^2/dT = -0.2 \times 10^{-3}$ [deg^{-1}]							ST	635
Poly(tert-butyl acrylate)	undiluted	60 $\mathrm{dln}\,r_o^2/dT = -0.2 \times 10^{-3}$ [deg^{-1}]							ST	635
Poly(N,N'-dimethylacrylamide)	methanol; water	25	--	78±15	670±40	309	2.17±0.14	9.15	VG	3(103)
Poly(dodecyl acrylate)	undiluted	60 $\mathrm{dln}\,r_o^2/dT = 1.0 \times 10^{-3}$ [deg^{-1}]							ST	635
Poly(ethyl acrylate)	acetone; methanol	30	--	90±10	720±30	308	2.34±0.10	10.9	VG	3(115,116)
	acetone	25	--	--	856	308	2.78	15.4	VG	114

Polymer	Solvent	Temp. [°C]	$S_{oz}/M_w^{1/2} \times 10^4$ or a_p [nm]	$K_o \times 10^3$ [ml/g]	$r_o/M^{1/2} \times 10^4$ [nm]	$r_{of}/M^{1/2} \times 10^4$ [nm]	$\sigma = r_o/r_{of}$	$C_\infty = r_o^2/nl^2$	Method	References
Poly(ethyl acrylate) (Cont' d.)	undiluted	60	dln r_o^2/dT = -0.2 x 10^{-3} [deg^{-1}]						ST	635
		76	dln r_o^2/dT = -0.4 x 10^{-3} [deg^{-1}]						ST	634
Poly(hexyl acrylate)	undiluted	60	dln r_o^2/dT = -0.3 x 10^{-3} [deg^{-1}]						ST	635
Poly(isopentyl acrylate)	undiluted	60	dln r_o^2/dT = -0.2 x 10^{-3} [deg^{-1}]						ST	635
Poly(isopropyl acrylate)	benzene	25	--	--	540± 25	287	1.88±0.08	7.1	VA	120
	bromobenzene	60	--	--	540± 25	287	1.88±0.08	7.1	VA	120
	2,2,3,3-tetrafluoropropanol	25	--	92	700± 30	287	2.42±0.10	11.7	VG	121
	undiluted	60	dln r_o^2/dT = -0.3 x 10^{-3} [deg^{-1}]						ST	635
isotactic	bromobenzene	60	--	--	630± 30	287	2.20±0.10	9.7	VA	120
syndiotactic	bromobenzene	60	--	--	546± 35	287	1.90±0.12	7.2	VA	120
Poly(methyl acrylate)	various solvents	30	--	81± 10	680± 30	332	2.05±0.10	8.4	VG	3(123,132)
	isopentyl acetate	62.5	--	68	650	332	1.96	7.7	VT	129
	2-methylcyclohexanol	56.0	--	68	650	332	1.96	7.7	VT	129
	butanone/2-propanol (42/58 vol)	20	--	81	680	332	2.05	8.4	VT	124
	(50/50 vol)	30	--	72	665	332	2.00	8.0	VT	129
		27.5	--	--	720	332	2.17	9.4	LD	108
	undiluted	60	dln r_o^2/dT = -0.2 x 10^{-3} [deg^{-1}]						ST	635
Poly(octyl acrylate)	undiluted	60	dln r_o^2/dT = -0.2 x 10^{-3} [deg^{-1}]						ST	635
Poly(morpholinocarbonylethylene)	dimethylformamide	25	--	70± 10	630± 40	260	2.42±0.15	11.7	VG	3(338)
Poly(piperidinocarbonylethylene)	dimethylformamide	25	--	58± 10	600± 40	261	2.30±0.15	10.6	VG	3(338)
Poly(propyl acrylate)	undiluted	60	dln r_o^2/dT = -0.3 x 10^{-3} [deg^{-1}]						ST	635
1.4 POLY(α-SUBSTITUTED ACRYLIC ACID) AND DERIVATIVES										
Poly(butyl methacrylate)	butanone; 2-propanol	23	--	34± 5	510± 20	258	1.98±0.10	7.85	VT, VG	3(152,154,156)
	2-propanol	23	--	--	530	258	2.06	8.5	LD	152
	undiluted	60	dln r_o^2/dT = 2.5 x 10^{-3} [deg^{-1}]						ST	635
Poly(sec-butyl methacrylate)	undiluted	60	dln r_o^2/dT = -0.2 x 10^{-3} [deg^{-1}]						ST	635
Poly(tert-butyl methacrylate)	butyl acetate	25	--	--	894	258	3.45	23.8	VA	157
Poly(tert-butylphenyl methacrylate)	acetone	20	--	35± 5	515± 20	208	2.48±0.10		VG	3(340)
Poly(cyclohexyl methacrylate)	butanol	23	--	34	510	237	2.15	9.25	VT	345
Poly(dodecyl methacrylate)	isopropyl acetate	13	--	32.2	490	193	2.54	12.9	VT	158
	pentanol	29.5	--	34.8	500	193	2.59	13.4	VT	159
		29.5	222	--	500	193	2.59	13.4	LD	159
	undiluted	60	dln r_o^2/dT = 2.6 x 10^{-3} [deg^{-1}]						ST	635
Poly(2-ethylbutyl methacrylate)	butanone; 2-propanol	25	--	36± 5	510± 30	236	2.16±0.13	9.3	VG, VT	3(160)
	2-propanol	27.4	--	33.7	500	236	2.12	8.95	VT	160

References page IV-52

Polymer	Solvent	Temp. [°C]	$S_{oz}/M_w^{1/2} \times 10^4$ or a_p [nm]	$K_o \times 10^3$ [ml/g]	$r_o/M^{1/2} \times 10^4$ [nm]	$r_{of}/M^{1/2} \times 10^4$ [nm]	$\sigma = r_o/r_{of}$	$C_\infty = r_o^2/nl^2$	Method	References
Poly(ethyl methacrylate)	butanone	23	--	49.3	565± 15	288	1.96±0.05	7.7	VG	3(161)
	2-propanol	36.5	--	47.5	575	288	2.00	8.0	VT	156
	butanone/2-propanol (1/7 vol)	23	--	47.3	560	288	1.94	7.55	VT	161
Poly(hexadecyl methacrylate)	heptane	23	--	--	560	288	1.94	7.55	LD	161
Poly(hexyl methacrylate)	heptane	21	--	60	620	175	3.54	25.1	VG	163
	butanone	30	--	41± 4	530± 20	236	2.25±0.08	10.1	VT,VG	3(165)
	2-propanol	32.6	--	43	540	236	2.29	10.5	VT	165
	undiluted	32.6	--		580	236	2.46	12.1	LD	165
		60 dln $r_o^2/dT = 2.2 \times 10^{-3}$ [deg^{-1}]							ST	635
Poly(isopentyl methacrylate)	undiluted	60 dln $r_o^2/dT = 1.4 \times 10^{-3}$ [deg^{-1}]							ST	635
Poly(isopropyl methacrylate)	undiluted	60 dln $r_o^2/dT = 2.5 \times 10^{-3}$ [deg^{-1}]							ST	635
Poly(methacrylic acid)	aqueous NaCl	25	--	200	900	334	2.7	14.6	VG	110
Poly(methyl butacrylate)	butanol	13	--	57	590	258	2.28	10.4	VT	168
Poly(methyl ethacrylate)	2,6-dimethyl-4-heptanone	11.4	--	67.6	620	288	2.15	9.25	VT	168
Poly(methyl methacrylate) atactic	various solvents	25	--	70± 20	640± 60	308	2.08±0.20	8.65	VT,VG	3(170,173,174 181,193,196)
	butyl chloride	35.4	219	--	537	308	1.74	6.05	LT	636
		40.8	292± 6	--	620± 15	308	2.01±0.05	8.1	LT	283
	benzene, toluene	21	0.72±0.05	--	653± 25	308	2.12±0.08	9.0	XS	637
	2-methyl-4-pentanone	-42	--	36.0	500	308	1.62	5.25	VT	636
	methyl isovaleriate	-37	--	41.5	525	308	1.70	5.8	VT	636
	butyl acetate	-20	--	40.6	520	308	1.69	5.7	VT	636
	butanone/2-propanol (58.2/41.8 vol)	4.0	--	47.2	550	308	1.78	6.35	VT	636
	(55/45 vol)	12.8	--	49.8	560	308	1.82	6.65	VT	636
	(50/50 vol)	22.8	--	50.4	610	308	1.98	7.85	VT	636
	(46.8/53.2 vol)	28.5	--	50.8	610	308	1.99	7.95	VT	636
	butyl chloride	35.4	--	52.6	620	308	2.00	8.0	VT	636
	4-heptanone	40.4	--	53.2	620	308	2.01	8.1	VT	636
	isoamyl acetate	57.5	--	53.5	620	308	2.01	8.1	VT	636
	4-heptanone	33	--	47± 4	550± 15	308	1.78±0.05	6.35	VT	179
	acetonitrile	45	--	49± 5	555± 15	308	1.80±0.05	6.5	VT	179
	3-octanone	72	--	50± 3	560± 10	308	1.82±0.03	6.65	VT	179
	undiluted	168 dln $r_o^2/dT = 0.1 \times 10^{-3}$ [deg^{-1}]							ST	634
isotactic	acetonitrile	27.6	--	75.5	670	308	2.17	9.4	VT	198
	butanone/2-propanol (50/50 vol)	30.3	--	90	715	308	2.32	10.8	VT	191
	3-heptanone	40	--	87	710	308	2.30	10.6	VT	191
	propanol	75.9	--	76.1	680	308	2.21	9.75	VT	191
	p-cymene	152.1	--	56.6	610	308	1.98	7.85	VT	191
Poly(octyl methacrylate)	butanol; butanone	20	--	30± 5	480± 20	219	2.19±0.09	9.6	VT,VG	3(201)
	butanol	16.8	--	--	500	219	2.28	10.4	LD	201

Polymer	Solvent	Temp. [°C]	$S_{oz}/M_w^{1/2}$ or $a_p \times 10^4$ [nm]	$K_\theta \times 10^3$ [ml/g]	$r_o/M^{1/2} \times 10^4$ [nm]	$r_{of}/M^{1/2} \times 10^4$ [nm]	$\sigma = r_o/r_{of}$	$C_\infty = r_o^2/nl^2$	Method	References
Poly(octyl methacrylate) (Cont'd.)	undiluted								ST	635
		60	dln $r_o^2/dT = 2.2 \times 10^{-3}$ [deg^{-1}]							
Poly(N-phenylmethacryl-amide)	acetone	20	--	38± 9	520± 40	242	2.15±0.16	9.25	VG	3(370)

1.5 POLY(VINYL ETHERS), POLY(VINYL ALCOHOL), POLY(VINYL ESTERS), POLY(VINYL HALIDES)

Polymer	Solvent	Temp. [°C]	$S_{oz}/M_w^{1/2}$ or $a_p \times 10^4$ [nm]	$K_\theta \times 10^3$ [ml/g]	$r_o/M^{1/2} \times 10^4$ [nm]	$r_{of}/M^{1/2} \times 10^4$ [nm]	$\sigma = r_o/r_{of}$	$C_\infty = r_o^2/nl^2$	Method	References
Poly(chlorotrifluoroethylene)	2,5-dichlorobenzotrifluoride	130	--	52± 3	580± 15	286	2.03±0.07	8.25	VG	3(234)
Poly(methoxyethylene)	benzene; butanone	30	--	195± 30	900± 50	404	2.23±0.13	9.95	VG	3(206)
Poly(tetrafluoroethylene)	perfluorokerosene	300	--	~300	1070	308	~3.5	24	VG	690
Poly(vinyl acetate)	various solvents	25	--	93± 10	705± 10	332	2.12±0.09	9.0	VT,VG	3(242,244, 252,255)
	3-heptanone	29	0.95±0.05	--	790± 20	332	2.38±0.07	11.3	XS	641
	heptane/3-methyl-2-butanone (26.8/73.2 vol)	25	318± 10	--	745± 20	332	2.24±0.07	10.0	LT	252
	methanol	6	--	101	720	332	2.17	9.4	VT	236
	3-heptanone	26.8	--	82.0	670	332	2.02	8.15	VT	236
	ethanol	56.9	--	90	690	332	2.08	8.65	VT	236
	6-methyl-3-heptanone	66	--	78	660	332	1.99	7.9	VT	236
Poly(vinyl alcohol)	water	30	--	222± 25	950± 40	464	2.04±0.10	8.3	VG	3(208,210,212)
	undiluted								ST	638
		50	dln $r_o^2/dT = 0.0$ [deg^{-1}]						ST	639
		90	dln $r_o^2/dT = 0.5 \times 10^{-3}$ [deg^{-1}]							
Poly(vinyl benzoate)	xylene	32.5	--	62± 8	620± 25	252	2.46±0.10	12.1	VT	334
Poly(vinyl bromide)	cyclohexane; tetrahydrofuran; methanol/THF (17/88 vol)	20	--	40± 5	540± 20	298	1.82±0.07	6.6	VT,VG	3(217,218)
	1-methylnaphthalene	~20	1.09	--	763	298	2.56	13.1	XS	640
Poly(vinyl butyrate)	benzene	30	--	80± 10	670± 35	288	2.32±0.12	6.7	VG	3(256)
Poly(vinyl chloride)	cyclohexanone; tetrahydrofuran	~25	--	100± 30	670± 35		1.83±0.15	6.7	VG	3(221,223,224,229)
	benzyl alcohol	155.4	--	156	820	393	2.08	8.65	VT	219
Poly(vinyl 4-chlorobenzoate)	butanol/butanone (41/53 vol)	60	--	73	665	228	2.92	17.1	VT	336
Poly(vinyl fluoride)	dimethylformamide	90	--	128	787	457	1.72	5.9	VG	235
Poly(vinyl hexanoate)	benzene	30	--	91± 10	700± 30	258	2.71±0.12		VG	3(256)
Poly(vinyl isobutyrate)	benzene	30	--	80± 10	670± 35	288	2.32±0.12		VG	3(256)
Poly(vinyl methyl ether), see Poly(methoxyethylene)										
Poly(vinyl pivalate)	butanone/methanol (0.897 g/ml)	20	--	53± 5	580± 20	253	2.29±0.08		VT	258

1.6 POLY(STYRENE) AND DERIVATIVES

Polymer	Solvent	Temp. [°C]	$S_{oz}/M_w^{1/2}$ or $a_p \times 10^4$ [nm]	$K_\theta \times 10^3$ [ml/g]	$r_o/M^{1/2} \times 10^4$ [nm]	$r_{of}/M^{1/2} \times 10^4$ [nm]	$\sigma = r_o/r_{of}$	$C_\infty = r_o^2/nl^2$	Method	References
Poly(4-bromostyrene)	benzene	26.3	--	50	570	228	2.50	12.5	VT	348
	toluene	30	--	45	564	228	2.48	12.3	VG	349
		30	237	--	554	228	2.43	11.8	LG	349
Poly(4-chlorostyrene)	butanone; chlorobenzene; toluene	30	--	50± 5	560± 20	261	2.15±0.07	9.25	VG	3(352,355,353)

References page IV-52

Polymer	Solvent	Temp. [°C]	$S_{oz}/M_w^{1/2}$ or a_p × 10⁴ [nm]	K_o × 10³ [ml/g]	$r_o/M^{1/2}$ × 10⁴ [nm]	$r_{of}/M^{1/2}$ × 10⁴ [nm]	$\sigma = r_o/r_{of}$	$C_\infty = r_o^2/nl^2$	Method	References
Poly(4-chlorostyrene) (Cont'd.)	toluene	30	--	58	615	261	2.36	11.1	VG	349
	toluene	30	272	---	615	261	2.36	11.1	LG	349
Poly(4-cyclohexylstyrene)	heptane; toluene	30	--	53± 3	570± 20	226	2.52±0.07	12.7	VG	266
	toluene	60	--	51± 3	560± 20	226	2.48±0.07	12.2	VG	266
Poly(2,5-dichlorostyrene)	ethanol/ethyl acetate (1/15 wt)	30.5	--	35.5	510	234	2.18	9.5	VT	3(357)
Poly(3,4-dichlorostyrene)	butanol/butyl acetate (1/13 wt) chlorobenzene; o-dichlorobenzene	32.9	--	71	640	234	2.7	14.6	VT	359
Poly(2,4-dimethylstyrene)	benzene	30	--	38± 5	510± 20	234	2.18±0.08	9.5	VG	358
	toluene	30	--	60± 5	630± 15	268	2.35±0.07	11.0	VG	333
Poly(o-methoxystyrene)	methanol/toluene (25/75 vol)	30	--	57.5	600	266	2.26	10.2	VT	362
Poly(p-methoxystyrene)	methanol/toluene (28.1/71.9 vol)	30	--	62.1	630	266	2.37	11.2	VT	362
Poly(α-methylstyrene) anionic (atactic)	benzene; cyclohexane	~30	--	76± 5	650± 15	284	2.29±0.05	10.5	VT, VG	319,323
	trans-decalin	9.5	--	67± 2	625± 5	284	2.20±0.02	9.7	VT	320
	cyclohexane	~38	--	76± 2	650± 10	284	2.29±0.03	10.5	VT	320,321,322,323
cationic (syndiotactic)	toluene; benzene/methanol (79.4/20.6 vol)	30	--	74± 10	670± 25	284	2.36±0.10	11.1	VT, VG	3(329,327)
	cyclohexane	~33	--	69± 3	655± 10	284	2.31±0.03	10.7	VT	328
Poly(m-methylstyrene)	benzene; cyclohexane; ethyl acetate	30	--	84.0	664	284	2.34	11.0	VG	330
		40	--	86.8	671	284	2.37	11.2	VG	330
		50	--	89.7	678	284	2.39	11.4	VG	330
Poly(p-methylstyrene)	butanone; cyclohexane; toluene	30	--	68± 5	620± 15	284	2.18±0.05	9.5	VG	651,331
	diethyl succinate	16.4	--	70	655	284	2.31	10.7	VT	331
		16.4	291	--	680	284	2.39	11.4	LT	331
Poly(styrene) atactic	various solvents	~30	--	82± 5	670± 15	302	2.22±0.05	9.85	VT, VG	3(274,277,279, 282,286,299,301,304)
	ethylcyclohexane: methyl-cyclohexane	~70	--	75± 5	650± 15	302	2.15±0.05	9.25	VT	3(292,293)
	cyclohexane	34	282± 5	--	690± 10	302	2.28±0.04	10.4	LT	643
		35	306	--	730	302	2.42	11.7	LT	278
		35	300	--	670	302	2.22	9.85	LT	644
		35	0.92±0.03	--	705± 15	302	2.33±0.05	10.9	XS	645
	benzene; toluene	25	0.91±0.02	--	700± 15	302	2.32±0.04	10.8	XS	645
	benzene/ethanol (71.5/28.5 vol)	25	296	--	645	302	2.14	9.15	LT	628
	butanone/2-propanol (87/13 vol)	67	317	--	757	302	2.50	12.5	LT	278
	1-chloroundecane	32.8	--	--	775	302	2.56	13.1	VA	646
	cyclohexane	34.8	--	--	768	302	2.54	12.9	VA	646
	dimethyl malonate	35.9	--	--	762	302	2.52	12.7	VA	646

Polymer	Solvent	Temp. [°C]	$S_{oz}/M_w^{1/2} \times 10^4$ or a_p [nm]	$K_o \times 10^3$ [ml/g]	$r_o/M^{1/2} \times 10^4$ [nm]	$r_{of}/M^{1/2} \times 10^4$ [nm]	$\sigma = r_o/r_{of}$	$C_\infty = r_o^2/nl^2$	Method	References
Poly(styrene) (Cont'd.)	73%-trans-decalin	18	--	77	655	302	2.17	9.4	VT	290
	100%-trans-decalin	24	--	82	670	302	2.22	9.85	VT	290
	butyl formate	-9	--	77.4	655	302	2.17	9.4	VT	647
	hexyl-m-xylol	12.5	--	77.0	655	302	2.17	9.4	VT	647
	decalin	29.5	--	77.9	655	302	2.17	9.4	VT	647
	diethyl malonate	31	--	70.5	635	302	2.10	8.8	VT	647
	cyclohexane	34	--	79.5	660	302	2.18	9.5	VT	647
	diethyl oxalate	51.5	--	72.2	640	302	2.12	9.0	VT	647
	methylcyclohexane	68	--	78.0	655	302	2.17	9.4	VT	647
	cyclohexanol	83.5	--	50.8	575	302	1.90	7.2	VT	647
	1-chlorodecane	6.6	--	78.0	655	302	2.17	9.4	VT	648
	1-chloroundecane	32.8	--	78.7	660	302	2.18	9.5	VT	648
	1-chlorododecane	58.6	--	80.7	665	302	2.20	9.7	VT	648
	cyclohexane/methylcyclohexane									
	(1/0)	34.5	--	77.9	655	302	2.17	9.4	VT	291
	(2/1 vol)	43.0	--	77.6	655	302	2.17	9.4	VT	291
	(1/1 vol)	48.0	--	74.8	650	302	2.15	9.25	VT	291
	(1/2 vol)	54.0	--	73.0	645	302	2.14	9.15	VT	291
	(0/1)	70.5	--	69.6	635	302	2.10	8.8	VT	291
	diethyl malonate	34.2	--	71.8	640	302	2.12	9.0	VT	291
	diethyl oxalate	55.8	--	73.0	645	302	2.14	9.15	VT	291
	undiluted	150	dln $r_o^2/dT = 0.4 \times 10^{-3}$ [deg^{-1}]							634
atactic, anionic	cyclohexane	~34.5	--	88± 3	685± 10	302	2.27±0.03	10.3	VT	301,302,303 305,649
isotactic	decalin; dioctyl phthalate	12~22	--	80± 1	665± 5	302	2.20±0.02	9.7	VT	303,649
	benzene; toluene	30	--	90± 10	695± 25	302	2.30±0.08	10.5	VG	3(311,312,314)
	chlorobenzene	25.3	--	176	890	302	2.94	17.3	VG	650
Poly(styrene-p-sulfonic acid) --, sodium salt	aqueous NaCl (4.17M); aqueous KCl (31.M)	25	--	20.4	425	214	1.98	7.85	VT	366

1.7 OTHERS

Polymer	Solvent	Temp. [°C]	$S_{oz}/M_w^{1/2} \times 10^4$ or a_p [nm]	$K_o \times 10^3$ [ml/g]	$r_o/M^{1/2} \times 10^4$ [nm]	$r_{of}/M^{1/2} \times 10^4$ [nm]	$\sigma = r_o/r_{of}$	$C_\infty = r_o^2/nl^2$	Method	References
Poly[(biphenyl-4-yl)ethylene]	benzene	30	--	63.0	605	230	2.63	13.8	VG	264
		20~75	dln $r_o^2/dT = (0.23\pm0.01) \times 10^{-3}$ [deg^{-1}]						VG	264
Poly(carbanilinoxyethylene)	dioxane/methanol (28/72 vol)	20	--	75	680	241	2.82	15.9	VT	335
Poly(1-methoxycarbonyl-1-phenylethylene)	benzene;chloroform	30	--	54± 1	585± 5	242	2.42±0.02	11.7	VG	361
	ethylbenzene	15	--	54± 1	585± 5	242	2.42±0.02	11.7	VT	361
Poly(vinyl carbazole)	toluene	37	--	76.2±5	633	222	2.85	16.2	VT	368
	benzene; chloroform; tetra- chloroethane; tetrahydrofuran	25	--	68± 2	619	222	2.82	15.9	VG	367

References page IV-52

Polymer	Solvent	Temp. [°C]	$S_{oz}/M_w^{1/2} \times 10^4$ or a_p [nm]	$K_o \times 10^3$ [ml/g]	$r_o/M^{1/2} \times 10^4$ [nm]	$r_{of}/M^{1/2} \times 10^4$ [nm]	$\sigma = r_o/r_{of}$	$C_\infty = r_o^2/nl^2$	Method	References
Poly(1-vinylnaphthalene)	benzene	30	--	24.2	435	248	1.76	6.2	VG	264
		75	--	--	405	248	1.63	5.3	VG	264
		20~75	dln r_o^2/dT = -(1.87±0.04) x 10^{-3} [deg^{-1}]						VG	264
Poly(2-vinylnaphthalene)	benzene	30	--	64.7	610	248	2.45	12.0	VG	264
		65	--	--	595	248	2.40	11.5	VG	264
		20~75	dln r_o^2/dT = -(0.83±0.03) x 10^{-3} [deg^{-1}]						VG	264
	decalin/toluene (13/10 wt)	30.2	--	--	--	248	ca. 3.1	ca. 19.2	VT	269
Poly(2-vinylpyridine)	various solvents	25	--	82±10	660±30	300	2.20±0.10	9.7	VG	371
	benzene	15	--	72	635	300	2.12	9.0	VG	652
		25	--	71	633	300	2.11	8.9	VG	652
		30	--	59	595	300	1.98	7.85	VG	652
		40	--	52	570	300	1.90	7.2	VG	652
		50	--	57	590	300	1.96	7.65	VG	652
		60	--	62	605	300	2.02	8.15	VG	652
	chloroform	0	--	88	690	300	2.24	10.0	VG	652
		25	--	87.5	689	300	2.24	10.0	VG	652
Poly(4-vinylpyridine)	ethanol; water	25	--	94±10	710±30	300	2.37±0.10	11.2	VG	3(373,374)
Poly(5-vinyl-2-methylpyridine)	butanone; methanol	25	--	69±5	652±15	282	2.31±0.05	10.6	VG	375
	butyl acetate	21.8	--	83	675	282	2.39	11.4	VT	381
	4-methyl-2-pentanone	37.4	--	83	675	282	2.39	11.4	VT	381
	pentyl acetate	48.2	--	80	665	282	2.36	11.1	VT	381
Poly(vinylpyrrolidone)	water	~25	--	100±15	720±40	292	2.48±0.12	12.3	VG	3(380,382)
	acetone/water (66.8/33.2 vol)	25	--	75	650	292	2.22	9.85	VT	384
	butanone/2-propanol (96/4 vol)	25	--	61	630	292	2.16	9.3	VT	653
Poly(vinyl sulfate)	aqueous NaCl (0.5M)	20	--	25±15	460±80 ★	278	1.65±0.30 ★	5.45	VG	3(261)
Poly(vinylsulfonic acid)	aqueous KBr (0.349M)	5.7	--	68.8	650(788) ★	296	2.19(2.66) ★	9.6	VT	259,642
	aqueous KCl (0.349M)	5.5	--	68.2	650(786) ★	296	2.19(2.66) ★	9.6	VT	259,642
	(0.650M)	26.0	--	79.5	685(830) ★	296	2.31(2.80) ★	10.6	VT	259,642
	(1.001M)	44.5	--	80.3	690(832) ★	296	2.33(2.81) ★	10.8	VT	259,642
	aqueous NaBr (0.347M)	-0.6	--	95.5	730(882) ★	296	2.46(2.98) ★	12.1	VT	259,642
	aqueous NaCl (1.003M)	32.4	--	96.1	730(880) ★	296	2.46(2.97) ★	12.1	VT	259,642
	aqueous NaBr (1.008M)	40.1	--	94.5	725(875) ★	296	2.45(2.96) ★	12.0	VT	259,642

1.8 COPOLYMERS

Polymer	Solvent	Temp. [°C]	$S_{oz}/M_w^{1/2} \times 10^4$ or a_p [nm]	$K_o \times 10^3$ [ml/g]	$r_o/M^{1/2} \times 10^4$ [nm]	$r_{of}/M^{1/2} \times 10^4$ [nm]	$\sigma = r_o/r_{of}$	$C_\infty = r_o^2/nl^2$	Method	References
Poly(acrylonitrile-co-styrene) 38.3/61.7 mol, azeotropic	butanone	30	--	124	770	335	2.30±0.05	10.6	VG	593,595

★ The values of $r_o/M^{1/2}$ and σ given in parenthesis were obtained by using $\Phi_o = 1.39 \times 10^{23}$, while those given outside of it by using $\Phi_o = 2.5 \times 10^{23}$

Polymer	Solvent	Temp. [°C]	$S_{oz}/M_w^{1/2} \times 10^4$ or a_p [nm]	$K_\theta \times 10^3$ [ml/g]	$r_o/M^{1/2} \times 10^4$ [nm]	$r_{of}/M^{1/2} \times 10^4$ [nm]	$\sigma = r_o/r_{of}$	$C_\infty = r_o^2/nl^2$	Method	References
Poly(acrylonitrile-co-styrene) (Cont'd.)										
62.6/37.4 mol, random	butanone; dimethylformamide	30	--	170	840	362	2.32	10.8	VG	595
Poly(butadiene-co-styrene)										
84/16 mol, random	2-pentanone	23.8	--	23.8	460				VT	596
Poly(butyl itaconate-co-dibutyl itaconate)										
40.5/59.5 mol, random	acetone	25	--	83			3.2		VG	682
	methanol/m-xylene									
	(100/0 vol)	25	--	51			2.7		VG	682
	(80/20 vol)	25	--	62			2.8		VG	682
	(65/35 vol)	25	--	78			3.1		VG	682
	(50/50 vol)	25	--	98			3.4		VG	682
	(30/70 vol)	25	--	101			3.4		VG	682
	(10/90 vol)	25	--	82			3.2		VG	682
	(0/100 vol)	25	--	42			2.5		VG	682
Poly(p-chlorostyrene-co-methyl methacrylate)										
51.6/48.4 mol, random	benzene/hexane (60/40 vol)	22.3	--	64	660				VT	598
Poly(dimethyl itaconate-co-styrene)										
100/0 wt	benzene	25	--	30.4	495	245	2.02		VT	683
75/25 wt	toluene	25	--	35.4	508	260	1.96		VT	683
67/33 wt		25	--	40.3	544	266	2.04		VT	683
59/41 wt		25	--	45.7	553	270	2.05		VT	683
49/51 wt		25	--	55.3	590	278	2.12		VT	683
29.5/70.5 wt		25	--	63.4	617	287	2.15		VT	683
27/73 wt		25	--	63.7	618	288	2.15		VT	683
0/100 wt		25	--	78.0	661	302	2.17		VT	683
Poly(ethyl acrylate-co-methyl methacrylate)										
80/20 mol, random	acetone	25	--	--	823	308	2.67		VG	114
Poly(ethylene-co-α-methylstyrene), $[(ET)_m (MS)_n]_p$										
m/n = 3/4	butanone/cyclohexane (60/40 vol)	30	--	135	820	345	2.38		VT	604
m/n = 5/4	butanone/cyclohexane (75/25 vol)	30	--	140	830	373	2.22		VT	604
m/n = 5/7	cyclohexane	30	--	112	770	343	2.24		VT	604
Poly(methyl acrylate-co-styrene)										
50/50 mol, random	ethyl acetate	35	--	104± 2	1010± 20	314	3.22		VG	133
22/78 mol, random	various solvents	~30	--	75	650				VT,VG	129
33/67 mol, random		~30	--	76	650				VT	129
47/53 mol, random		~30	--	77	655				VG	129
59/41 mol, random		~30	--	76	650				VT,VG	129
76/24 mol, random		~30	--	75	650				VT	129
Poly(methyl methacrylate-co-styrene)										
100/0 mol, random	1-chlorobutane	40.8	--	50	583	308	1.89		VT	613
94/6 mol, random		40.8	--	59	616	308	2.00		VG	613

Polymer	Solvent	Temp. [°C]	$S_{oz}/M_w^{1/2} \times 10^4$ or a_p [nm]	$K_o \times 10^3$ [ml/g]	$r_o/M^{1/2} \times 10^4$ [nm]	$r_{of}/M^{1/2} \times 10^4$ [nm]	$\sigma = r_o/r_{of}$	$C_\infty = r_o^2/nl^2$	Method	References
Poly[methyl methacrylate-co-styrene] (Cont'd.)										
52/48 mol, random	1-chlorobutane	40.8	--	95	728	305	2.39		VG	613
10/90 mol, random		40.8	--	89	707	302	2.34		VG	613
0/100 mol, random		40.8	--	80	685	302	2.27		VG	613
71/29 mol, random	various solvents	~30	--	66± 2	625± 95		2.05		VG,VT	614
44/56 mol, random	various solvents	~30	--	75± 2	655± 95		2.15		VG,VT	614
30/70 mol, random	various solvents	~30	--	77± 2	660± 95		2.17		VG,VT	614
three blocks (MSM) nearly equimolar	cyclohexanol	81	--	63	617	305	2.04		VT	616
Poly(styrene-co-vinylpyrrolidone)										
87/13 wt, random	butanone	25	--	96					VT	684
	butanone/2-propanol (75/25 vol)	25	--	76					VT	684
13/87 wt, random	butanone/2-propanol (97/3 vol)	25	341	75					LT,VT	684
Poly(styrene-co-monomethyl maleate)										
	acetone	26.4	--	51.1	575	285	2.02	8.15	VT	317
	aqueous NaCl (0.6M)	25	--	55	585	285	2.05	8.4	VT	317
Poly(trifluoronitrosomethane-co-tetrafluoroethylene)		35	--	38	510± 25	304	1.68±0.08		VT	3(685)

2. MAIN-CHAIN CARBOCYCLIC POLYMERS

Polymer	Solvent	Temp. [°C]	$S_{oz}/M_w^{1/2} \times 10^4$ or a_p [nm]	$K_o \times 10^3$ [ml/g]	$r_o/M^{1/2} \times 10^4$ [nm]	$r_{of}/M^{1/2} \times 10^4$ [nm]	$\sigma = r_o/r_{of}$	$C_\infty = r_o^2/nl^2$	Method	References
Poly(1,2-acenaphthenylene)										
trans	various solvents	25	--	36± 3	520± 20	354	1.47±0.05		VG	263

3. MAIN-CHAIN HETEROATOM POLYMERS

3.1 POLY(OXIDES)

Polymer	Solvent	Temp. [°C]	$S_{oz}/M_w^{1/2} \times 10^4$ or a_p [nm]	$K_o \times 10^3$ [ml/g]	$r_o/M^{1/2} \times 10^4$ [nm]	$r_{of}/M^{1/2} \times 10^4$ [nm]	$\sigma = r_o/r_{of}$	$C_\infty = r_o^2/nl^2$	Method	References
Poly(butene oxide), see Poly[oxy(ethylethylene)]										
Poly(ethylene oxide), see Poly(oxyethylene)										
Poly[oxy(tert-butylethylene)]	benzene	25	--	230	930	377 ★	2.47	13.6	VG	385
Poly(oxy-1,2-cyclohexylene)	toluene	35	--	53	592	359	1.65		VG	472
Poly(oxydecamethylene)	benzene; chloroform	~30	--	240	960	570	1.68	7.5	VG	386
Poly[oxy(2,6-dimethyl-1,4-phenylene)]	chlorobenzene; toluene	25	--	166± 5	833± 10	715	1.16±0.02	2.7	VG	474
	benzene; carbon tetrachloride	25	--	175± 8	850± 10	715	1.13±0.02	2.6	VG	473
Poly[oxy(2,6-diphenyl-1,4-phenylene)]	chlorobenzene; toluene	25	--	80± 5	660± 20	500	1.32±0.04	3.5	VG	473

★ These values of r_{of} of poly(epoxide) chains were calculated by $0.377/M_u^{1/2}$ [nm. mol$^{1/2}$ gram$^{-1/2}$], while those given without asterisk were calculated by $0.360/M_u^{1/2}$. The former is due to Allen et al. (Ref. 695). The latter is based on the assumption that all valence angles of skeleton are tetrahedral.

Polymer	Solvent	Temp. [°C]	$S_{oz}/M_w^{1/2} \times 10^4$ or a_p [nm]	$K_o \times 10^3$ [ml/g]	$r_o/M^{1/2} \times 10^4$ [nm]	$r_{of}/M^{1/2} \times 10^4$ [nm]	$\sigma = r_o/r_{of}$	$C_\infty = r_o^2/nl^2$	Method	References
Poly(oxyethylene)	various solvents	~20	--	110± 10	750± 30	541	1.38±0.06	3.8	VG	3(389,390)
	aqueous K$_2$SO$_4$ (0.45M); aqueous MgSO$_4$ (0.39M)	~40	--	115± 15	775± 30	541	1.43±0.06	4.1	VT	395,655
	benzene	25	--	129	790	541	1.46	4.25	VG	656,388
	acetone	25	--	170	840	541	1.55	4.8	VG	387
	various poor solvents	50	--	170	840	541	1.55	4.8	VG	387
	undiluted	--	--	--					ST	655
	60 dln r_o^2/dT= $(0.23\pm0.02) \times 10^{-3}$ [deg^{-1}]									
Poly[oxy(ethylethylene)]	benzene; butanone; 2-propanol	30	--	87	700	423	1.66±0.05	5.5	VG,VT	397
	2-propanol	30	--	110	730	427★	1.71	5.85	VT	396
Poly(oxyhexamethylene)	benzene; dioxane	25	--	185	910	565	1.61	5.15	VG	398
Poly(oxymethylene)	hexafluoroacetone sesquihydrate (1/1.7 mol)	25	--	430± 40	1200± 80	522	2.3 ±0.2	10.5	VG	402
Poly[oxy(2-methyl-6-phenyl-1,4-phenylene)]	dioxane/methylcyclohexane (1/1 vol)	25	--	110± 5	790± 20	580	1.36±0.04	3.7	VT	673
Poly[oxy(phenylethylene)]	toluene	25	--	75	640	344★	1.85	6.85	VG	385
Poly(oxypropylene)	benzene; methanol	25	--	115± 10	750± 25	472	1.59±0.05	5.05	VG	3(411)
	toluene/2,2,4-trimethylpentane (5/7 vol)	39.5	--	107.5	735	472	1.56	4.85	VT	411
	2,2,4-trimethylpentane	50	375	--	800	494★	1.62	5.75	LT	654
Poly(oxytetramethylene)	ethyl acetate/hexane (22.7/77.3 wt)	31.8	--	210	900± 20	556	1.62±0.03	5.25	VT	412
	ethyl acetate	30	--	180± 20	860± 30	556	1.55±0.05	4.8	VG	412
	ethyl acetate/hexane (22.7/77 3 wt)	30.4	--	267	975	556	1.75	6.1	VT	658
	diethyl malonate	33.5	--	243	945	556	1.70	5.8	VT	658
	2-propanol	44.6	--	231	930	556	1.67	5.6	VT	658
	undiluted	--	--						ST	659
	60 dln r_o^2/dT = -1.33×10^{-3} [deg^{-1}]									
Poly(oxytrimethylene)	acetone; benzene; carbon tetra-chloride	30	--	126	795	550	1.45	4.2	VG	414
Poly(propylene oxide), see Poly(oxypropylene)										
Poly(tetrahydrofuran), see Poly(oxytetramethylene)										

3.2 POLY(ESTERS), POLY(CARBONATES)

Polymer	Solvent	Temp. [°C]	$S_{oz}/M_w^{1/2} \times 10^4$ or a_p [nm]	$K_o \times 10^3$ [ml/g]	$r_o/M^{1/2} \times 10^4$ [nm]	$r_{of}/M^{1/2} \times 10^4$ [nm]	$\sigma = r_o/r_{of}$	$C_\infty = r_o^2/nl^2$	Method	References
Bisphenol A poly(carbonate), see Poly[oxycarbonyloxy-1,4-phenyleneisopropylidene-1,4-phenylene]										
Poly(ethylene terephthalate), see Poly(oxyethyleneoxyterephthaloyl)										
Poly(oxyadipoyloxydecamethylene)	chlorobenzene	25	--	100± 10	720± 25	540	1.33±0.05	3.55	VT	3(415)
Poly(oxybutynedioyloxyhexamethylene)	benzene; chloroform	20	--	180± 20	870± 30	627	1.39±0.05	3.9	VG	3(416)

★ These values of r_{of} of poly(epoxide) chains were calculated by $0.377/M_u^{1/2}$ [nm. mol$^{1/2}$ gram$^{-1/2}$], while those given without asterisk were calculated by $0.360/M_u^{1/2}$. The former is due to Allen et al. (Ref. 695). The latter is based on the assumption that all valence angles of skeleton are tetrahedral.

References page IV-52

Polymer	Solvent	Temp. [°C]	$S_{oz}/M_w^{1/2}$ or a_p [nm]	$K_o \times 10^3$ [ml/g]	$r_o/M^{1/2} \times 10^4$ [nm]	$r_{of}/M^{1/2} \times 10^4$ [nm]	$\sigma = r_o/r_{of}$ of	$C_\infty = r_o^2/nl^2$	Method	References
Poly[oxycarbonyloxy-1,4-phenyleneisopropylidene-1,4-phenylene methylene chloride; tetrahydro-										
	furan	25	--	180± 20	880± 20	796	1.10±0.05		VG	3(479)
	butyl benzyl ether	170	--	210	940	796	1.18		VT	476
	cyclohexane/dioxane (36.1/63.9 wt)	25	--	210	940	796	1.18		VT	476
	chloroform; tetrahydrofuran	25	--	150± 13	840	796	1.05		VG	478
	hexane/tetrachloroethane (54/46 vol)	30	--	230	930	796	1.16		VT	694
Poly(oxy-1,4-cyclohexyleneoxysebacoyl)										
cis	chloroform	20	--	140± 20	800± 30	495	1.62±0.05		VG	3(480)
trans	chloroform	20	--	160± 20	840± 30	633	1.33±0.05		VG	3(480)
Poly(oxyethyleneoxyterephthaloyl)										
	phenol/tetrachloroethane (1/1 vol)	25	--	160± 15	840± 25	687	1.22±0.03	3.15	VG	3(491)
	o-chlorophenol	25	--	210	910	687	1.33	3.7	VG	484
	trifluoroacetic acid	30	--	242	975	687	1.42	4.25	VG	488
Poly(oxyfumaroyloxyhexamethylene)										
	chloroform	20-50	--	180± 20	870± 30	592	1.47±0.05	4.3	VG	3(417)
Poly[oxy(hexahydroterephthaloyl)oxyoctamethylene]										
cis	chloroform	20	--	140± 20	800± 30	495	1.62±0.05		VG	3(480)
trans	chloroform	20	--	160± 20	840± 30	633	1.33±0.05		VG	3(480)
Poly(oxyhexamethyleneoxy-2,9-dibutylsebacoyl)										
	benzene	20	--	155± 25	835± 70	457	1.82±0.15	6.6	VG	3(418)
Poly(oxyhexamethyleneoxysebacoyl)										
	benzene; chloroform	20	--	215± 60	910±100	540	1.70±0.17	5.8	VG	3(418, 419)
Poly[oxyisophthaloyloxy-1,4-phenylene(fluoren-9-ylidene)-1,4-phenylene] tetrachloroethan; tetrahydro-										
	furan	20	--	210	902				VG	674
Poly(oxymaleoyloxyhexamethylene)										
	benzene; chloroform; tetra- hydrofuran	20-50	--	135± 15	790± 30	510	1.55±0.05	4.8	VG	3(417)
Poly(oxysebacoyloxyhexadecamethylene)										
	chloroform	20	--	270± 40	1000± 50	555	1.80±0.10	6.5	VG	3(419)
Poly(oxysuccinyloxyhexamethylene)										
	benzene; chloroform; tetra- hydrofuran	20-60	--	165± 30	850± 60	522	1.62±0.14	5.25	VG	3(417)
Poly(oxyundecanoyl)										
	chloroform	20	--	185± 60	880±100	550	1.60±0.16	5.1	VG	3(419, 420)

3.3 POLY(AMIDES)

Polymer	Solvent	Temp. [°C]	$S_{oz}/M_w^{1/2}$ or a_p [nm]	$K_o \times 10^3$ [ml/g]	$r_o/M^{1/2} \times 10^4$ [nm]	$r_{of}/M^{1/2} \times 10^4$ [nm]	$\sigma = r_o/r_{of}$ of	$C_\infty = r_o^2/nl^2$	Method	References
Poly(iminoadipoyliminohexamethylene) (Nylon 66)										
	aqueous HCOOH (90% vol)	25	--	190± 20	890± 40	545	1.63±0.08	5.3	VG	3(445,446)

Polymer	Solvent	Temp. [°C]	$S_{oz}/M_w^{1/2} \times 10^4$ of a_p [nm]	$K_0 \times 10^3$ [ml/g]	$r_0/M^{1/2} \times 10^4$ [nm]	$r_{0f}/M^{1/2} \times 10^4$ [nm]	$\sigma = r_0/r_{0f}$ of	$C_\infty = r_0^2/nl^2$	Method	References
Poly(iminoadipoyliminohexamethylene) (Cont'd.)										
	aqueous HCOOH (90% vol), KCl (2.3M),	25	--	253	1010	545	1.85	6.85	VT	444
		25	--	192	935	545	1.72	5.95	VT	668, 669
Poly[imino(1-oxohexamethylene)] (Nylon 6)										
	conc. H_2SO_4	25	--	190±10	890±20	545	1.63±0.04	5.3	VG	3(453)
	aqueous HCOOH (65~85%)	25	--	229	970	545	1.78	6.35	VT,VG	450
Poly[iminoterephthaloylimino-1,4-phenylene(fluoren-9-ylidene)-1,4-phenylene]										
	dimethylformamide	25	--	409	~1200				VG	503

3.4 POLY(AMINO ACIDS)

Polymer	Solvent	Temp. [°C]	$S_{oz}/M_w^{1/2} \times 10^4$ of a_p [nm]	$K_0 \times 10^3$ [ml/g]	$r_0/M^{1/2} \times 10^4$ [nm]	$r_{0f}/M^{1/2} \times 10^4$ [nm]	$\sigma = r_0/r_{0f}$ of	$C_\infty = r_0^2/nl^2$	Method	References
Poly(β-benzyl-L-aspartate), see Poly(iminocarbonyl-L-benzyloxycarbonylethylidene)										
Poly(γ-benzyl-L-glutamate), see Poly(iminocarbonyl-L-benzyloxycarbonylpropylidene)										
Poly(iminocarbonyl-L-benzyloxycarbonylethylidene)										
	m-cresol	100	--	--	600	268	2.24±0.1	9.6	VA	670
Poly(iminocarbonyl-L-benzyloxycarbonylpropylidene)										
	dichloroacetic acid	25	--	58± 5	600± 20	259	2.32±0.08	10.3	VG	3(457)
		25	--				2.14	8.8	VA	670(457)
D,L	dichloroacetic acid; dimethylformamide	25	--	58± 5	600± 20	259	2.32±0.08	10.3	VG	3(459)
Poly(iminocarbonyl-L-carboxypropylidene), (Poly(L-glutamic acid))	phosphate buffer (Na^+, 0.3M; pH, 7.85)	17	--	--	720	337	2.14±0.1	8.8	VA	670
Poly[(methylimino)carbonylmethylene], (Poly(sarcosine))										
	water	20	--	50± 20	570± 90	455	1.25±0.20	3.0	VG	3(468)

3.6 POLY(URETHANES)

Polymer	Solvent	Temp. [°C]	$S_{oz}/M_w^{1/2} \times 10^4$ of a_p [nm]	$K_0 \times 10^3$ [ml/g]	$r_0/M^{1/2} \times 10^4$ [nm]	$r_{0f}/M^{1/2} \times 10^4$ [nm]	$\sigma = r_0/r_{0f}$ of	$C_\infty = r_0^2/nl^2$	Method	References
Poly(oxytetramethyleneoxycarbonylimino-2,4-tolyleneiminocarbonyl)										
	dimethylformamide	30	--	--	1030	515	2.0		VG	497

3.6 POLY(SULFIDES)

Polymer	Solvent	Temp. [°C]	$S_{oz}/M_w^{1/2} \times 10^4$ of a_p [nm]	$K_0 \times 10^3$ [ml/g]	$r_0/M^{1/2} \times 10^4$ [nm]	$r_{0f}/M^{1/2} \times 10^4$ [nm]	$\sigma = r_0/r_{0f}$ of	$C_\infty = r_0^2/nl^2$	Method	References
Poly(thiopropylene)										
	benzene	20	--	60	600				VG	436

3.7 POLY(PHOSPHATES)

Polymer	Solvent	Temp. [°C]	$S_{oz}/M_w^{1/2} \times 10^4$ of a_p [nm]	$K_0 \times 10^3$ [ml/g]	$r_0/M^{1/2} \times 10^4$ [nm]	$r_{0f}/M^{1/2} \times 10^4$ [nm]	$\sigma = r_0/r_{0f}$ of	$C_\infty = r_0^2/nl^2$	Method	References
Poly[oxy(hydroxyphosphinylidene)]										
	aqueous NaBr (0.35-0.415M)	25	--	50± 3	560± 20	370	1.51±0.04	6.6	VT,VG	3(422)
	aqueous LiBr									662
	aqueous CsCl (0.96M)	30	--				3.93	7.1	LT	663
	aqueous LiCl (2.9M)	30	--				2.25		LT	663

References page IV-52

Polymer	Solvent	Temp. [°C]	$S_{oz}/M_w^{1/2}$ or a_p $\times 10^{1/2}$ [nm]	$K_o \times 10^3$ [ml/g]	$r_o/M^{1/2} \times 10^4$ [nm]	$r_{of}/M^{1/2} \times 10^4$ [nm]	$\sigma = r_o/r_{of}$	$C_\infty = r_o^2/nl^2$	Method	References
Poly[oxy(hydroxyphosphinylidene)] (Cont'd.)	aqueous NaCl (0.52M)	30					2.79		LT	663

3.8 POLY(SILOXANES), POLY(SILSESQUIOXANES), POLY(SILMETHYLENES)

Polymer	Solvent	Temp. [°C]	$S_{oz}/M_w^{1/2}$ or a_p $\times 10^{1/2}$ [nm]	$K_o \times 10^3$ [ml/g]	$r_o/M^{1/2} \times 10^4$ [nm]	$r_{of}/M^{1/2} \times 10^4$ [nm]	$\sigma = r_o/r_{of}$	$C_\infty = r_o^2/nl^2$	Method	References
Poly(dimethyl siloxane), see Poly[oxy(dimethylsilylene)]										
Poly(dimethylsilmethylene)	heptane/propanol (68.8/31.2 vol)	25	540	--	988	450	2.2	--	LT	671
Poly(dimethylsiltrimethylene)	heptane	25	--	--	1200	480	2.5	--	LG	671
Poly(diphenylsiltrimethylene)	cyclohexanol/toluene (63.5/36.5 vol)	25	580	--	1160	322	3.6	--	LT	671
Poly[oxy(dimethylsilylene)]	butanone; toluene	~25	266±10	80±5	670±20	482	1.39±0.05	6.25	VT,VG	3(427,428),664
	various theta solvents	2-90		--	612±13	482	1.27±0.03	5.2	LT	425
	$C_8F_{18}/C_2Cl_4F_2$ (33/67 wt)	22.5		106	740	482	1.54	7.6	VT	424
	ethyl iodide	2		70	640	482	1.33	5.7	VT	425
	bromocyclohexane	29		74	655	482	1.36	6.0	VT	425
	bromocyclohexane/phenetole (6/7 vol)	36		75	660	482	1.37	6.1	VT	425
	chlorobenzene/dimethyl phthalate (45/6 vol)	57.5		76	660	482	1.37	6.1	VT	425
	bromobenzene	78.5		76	660	482	1.37	6.1	VT	425
	phenetole	89.5		73	650	482	1.35	5.9	VT	425
	undiluted	40-100	$\mathrm{dln}\, r_o^2 = (0.78\pm0.06)\times10^{-3}\ [\deg^{-1}]$						ST	665
	diluted with liquid silicon	30-105	$\mathrm{dln}\, r_o^2 = (0.71\pm0.13)\times10^{-3}\ [\deg^{-1}]$						VT	665
Poly[oxy(dipropylsilylene)]	2-pentanone	76	--	87.1	703	372	1.89	12.0	VT	433
	toluene	10	--	109	759	372	2.04	14.0	VT	433
Poly[oxy(methylphenylsilylene)]	diisobutylamine	30.4	--	51.5	575	363	1.58	8.35	VT	434
Poly[oxy(γ-trifluoropropylmethylsilylene)]	cyclohexyl acetate	25.0	--	41.0	550(648)★	341	1.61(1.90)★	6.3	VT	435
	methyl hexanoate	72.8	--	44.5	565(667)★	341	1.66(1.96)★	6.65	VT	435
Poly(phenylsilsesquioxane)	1,2-dichloroethane	50.5	--	--	1160	--	--	--	VT	675

★ The values of $r_o/M^{1/2}$ and σ given in parenthesis were obtained by using $\Phi_o = 1.5 \times 10^{23}$, while those given outside by using $\Phi_o = 2.5 \times 10^{23}$.

3.9 POLY(HETEROCYCLICS)

Polymer	Solvent	Temp. [°C]	$S_{oz}/M_w^{1/2} \times 10^4$ or a_p [nm]	$K_o \times 10^3$ [ml/g]	$r_o/M^{1/2} \times 10^4$ [nm]	$r_{of}/M^{1/2} \times 10^4$ [nm]	$\sigma = r_o/r_{of}$	$C_\infty = r_o^2/nl^2$	Method	References
Poly[(1,3-dihydro-3-oxo-isobenzofuran-1-ylidene)-1,4-phenyleneoxy-isophthaloyloxy-1,4-phenylene-tetrachloroethane: tetrahydro- furan		20	--	195	880	--	--		VG	674
Poly[(1,3-dihydro-3-oxoisobenzofuran-1-ylidene)-1,4-phenyleneiminoterephthaloylimino-1,4-phenylene] dimethylformamide		25	--	558	~1500	--	--		VG	503
Poly[(D,L-1,2-pyrrolidindiyl)carbonyl] water		25	--	ca. 25	ca. 570	390	ca. 1.5		VT	3(507)
Poly(1-isobutyl-2,5-dioxopyrrolidin-3,4-diyl) butyl acetate		21	--	132	790	--	--		VG	512
Poly(1-p-tolyl-2,5-dioxopyrrolidin-3,4-diyl) dimethylformamide		21	--	75	670	--	--		VG	513

3.10 COPOLYMERS (MALEIC ANHYDRIDE, SULFONES, SILOXANES)

MALEIC ANHYDRIDE COPOLYMERS

Polymer	Solvent	Temp. [°C]	$S_{oz}/M_w^{1/2} \times 10^4$ or a_p [nm]	$K_o \times 10^3$ [ml/g]	$r_o/M^{1/2} \times 10^4$ [nm]	$r_{of}/M^{1/2} \times 10^4$ [nm]	$\sigma = r_o/r_{of}$	$C_\infty = r_o^2/nl^2$	Method	References
Poly[(tetrahydro-2,5-dioxo-3,4-furandiyl)-1-phenylethylene] tetrahydrofuran		30	--	82.6	732	383	1.91		VG	506

SULFONES

Polymer	Solvent	Temp. [°C]	$S_{oz}/M_w^{1/2} \times 10^4$ or a_p [nm]	$K_o \times 10^3$ [ml/g]	$r_o/M^{1/2} \times 10^4$ [nm]	$r_{of}/M^{1/2} \times 10^4$ [nm]	$\sigma = r_o/r_{of}$	$C_\infty = r_o^2/nl^2$	Method	References
Poly[sulfonyl(butylethylene)] hexyl chloride		13	--	67± 2	625± 10	350	1.79±0.02		VT	666
		13	296± 9	--	725± 22	350	2.07±0.07		LT	666
benzene/cyclohexane (43/57 vol)		25	--	66± 2	642± 7	350	1.84±0.02		VG	348
butanone/2-propanol (30~37/70~63 vol)		8~24	--	53	580	350	1.66		VT	439
dioxane/hexane (40/60 vol)		20	--	65± 2	638± 7	350	1.83±0.02		VT	438
Poly[sulfonyl(1-methyl1-1-propylethylene)] butanone/hexane (35.4/64.6 vol)		11.5	--	91	700				VT	439
Poly[sulfonyl-(phenylethylene)] tetrahydrofuran		30	--	57.4	649	425	1.53		VG	440,667

SILOXANES

Polymer	Solvent	Temp. [°C]	$S_{oz}/M_w^{1/2} \times 10^4$ or a_p [nm]	$K_o \times 10^3$ [ml/g]	$r_o/M^{1/2} \times 10^4$ [nm]	$r_{of}/M^{1/2} \times 10^4$ [nm]	$\sigma = r_o/r_{of}$	$C_\infty = r_o^2/nl^2$	Method	References
Poly(dimethyl siloxane-co-diphenyl siloxane)										
95/5 mol bromobenzene		9	--	100	742	--	1.56		VT	596
66/34 mol benzene/2-propanol (44/56 wt)		42	--	74	670	--	1.70		VT	596
55/45 mol dimethylphthalate		82.5	--	78	675	--	1.81		VT	596
ethanol/toluene (37/63 wt)		29.5	--	78	675	--	1.81		VT	596

References page IV-52

4. CELLULOSE AND DERIVATIVES

Polymer	Solvent	Temp. [°C]	$S_{oz}/M_w^{1/2} \times 10^2$ or a_p [nm]	$K_o \times 10^3$ [ml/g]	$r_o/M^{1/2} \times 10^3$ [nm]	$r_{of}/M^{1/2} \times 10^4$ [nm]	$\sigma = r_o/r_{of}$ of	$C_\infty = r_o^2/nl^2$	Method	References
Amylose	dimethyl sulfoxide; ethylene diamine	25	--	56± 12	600± 50	335	1.79±0.15		VG	3(518)
	various solvents	25	--	--	700	335	2.08		VT,VG	517
	aqueous KCl (0.33M); dimethyl sulfoxide	25	--	110± 5	750± 25	335	2.24±0.08		VT,VG	3(520)
	aqueous KCl (0.5M)	25	--	61	625	335	1.87		VT	523
	aqueous KOH (0.15M)	25	--	164					VG	519
	aqueous KCl (0.33M)	25	--	--	--	--	--	5.2★	VT	530
	nitromethane	22.5	--	--	920	335	2.75		LG	527
Amylose triacetate	chloroform; nitromethane	30	--	47± 10	580± 60	250	2.32±0.24		VG	3(524)
		30	--	48	580	250	2.32		VG	676
		30	--		800± 15	250	3.2 ±0.06		VG	677
Amylose tricarbanilate	acetone; dioxane; pyridine	20	--	27± 5	470± 30	187	2.51±0.16		VG	3(528)
	dioxane/methanol (49/51 vol)	20	--	--	2180	--	11.7		LT	678
Carboxymethyl amylose, sodium salt	aqueous NaCl (0.65M)	37.5					2.62	7.9★	VT	609
	aqueous NaCl (0.5M; pH 8)	35					2.95	10.0★	VA	530
	aqueous NaCl (0.78M; 0.02% NaN₃)	35					2.15	5.3★	VA	531
Diethylaminoethyl amylose hydrochloride	aqueous NaCl (0.78M; 0.02% NaN₃)	35						6.4★	VA	531
Cellulose	cupriethylene diamine	25	--	180± 80	900±150	620	1.45±0.25		VA	3(537)
	cadoxen	25	0.24	495	1250	620	2.0		VG	691,534
Cellulose triacetate	acetone; chloroform; o-cresol ethylacetate; dioxane; methyl acetate; tetrahydrofuran	25-30	--	108± 10	750± 30	465	1.61±0.07		VG	3(541,542)
Cellulose tributyrate	butanone	25	--	97± 15	730~740	465	1.57~1.59		VG	680
	dodecane/tetralin (75/25 vol)	30	--	82	730± 40	408	1.79±0.10		VG	3(544)
		130	--		690	408	1.69		VT	544
Cellulose tricarbanilate	acetone; dioxane; pyridine	20	--	130± 30	810± 70	346	2.34±0.20		VG	3(528)
	acetone	~25	--	65± 3	635	346	1.83		VG	693
	cyclohexane	~25	--	83.5±3	690	346	1.99		VG	693
	dioxane	~25	--	44± 3	560	346	1.61		VG	693
	dioxane/methanol (42.5/57.5 vol)	20	--	--	1120	346	3.24		LT	678
	anisol	94	ca. 1000	130	805	346	2.32		LT,VT	546
	cyclohexanol	73	ca. 1050	--	875	346	2.52		LT	546

★ These values of the characteristic ratio C_∞ of cellulosic chains were obtained by $C_\infty = r_o^2/DP\, l^2$, where DP is the degree of polymerization and l = 0.425 [nm].

Polymer	Solvent	Temp. [°C]	$S_{oz}/M_w^{1/2} \times 10^4$ or a_p [nm]	$K_o \times 10^3$ [ml/g]	$r_o/M^{1/2} \times 10^4$ [nm]	$r_{of}/M^{1/2} \times 10^4$ [nm]	$\sigma = r_o/r_{of}$	$C_\infty = r_o^2/nl^2$	Method	References
Cellulose trihexanoate	dimethylformamide	41	--	240	980	370	2.65		VT	547
	dioxane/water (100/7 vol)	63	--	224	960	370	2.60		VT	547
Cellulose trinitrate	acetone; ethyl acetate	~25	--	130±30	810±50	458	1.77±0.11		VG	3(535,537,548,549)
	acetone	25	360	--	720	458	1.57		LG	3(549)
		22	0.26±0.01	--	930±15	458	2.03±0.03		XS	641
		~20	0.40~0.70	--	1180~1590	458	2.58~3.46		XS	679
		20	0.22~0.28	--	850~975	458	1.85~2.12		XS	679
(Nitrogen %, 13.9)	acetone	25	970	--	2410	462	4.7		LG	686
(Nitrogen %, 12.9)	acetone	25	530	--	1780	475	3.4		LG	686
(Nitrogen %, 13.5)	ethyl acetate	30	700	--	2050	467	3.9		LG	555
(Nitrogen %, 12.0)	acetone	20	0.48	--		486	2.5		XS	687
Cellulose trioctanoate	toluene	30	--	127±15	800±40	340	2.35±0.12		VG	3(544)
	dimethylformamide	140	--	113	770	340	2.27		VT	544
	γ-phenylpropanol	48	--	129	805	340	2.37		VT	544
Ethyl cellulose	methanol	25	--	232±10	970±20	520	1.87±0.03		VG	3(559)
Ethyl hydroxyethyl cellulose	water	25	--	550±80	1300±50	545	2.38±0.09		VG	3(560)
Hydroxyethyl cellulose	cadoxen; water	25	--	335	1100	545	2.0		VG	691,560,692
				250	1000	514	1.9		VG	691,561
Methyl cellulose	water	25	--	920±100	1500±60	581	2.58±0.10		VG	3(563)
Sodium carboxymethyl cellulose	cadoxen	25	--	357	1130	563	2.0		VG	691,565
	aqueous NaCl (0.005~0.2M)	25	--	420	1190	502	2.4		VG	691,565
Sodium cellulose xanthate	aqueous NaOH (1M)	30	--	290	1380				VG	681
Hyaluronic acid	aqueous HCl (0.1M)	25	--	--	--	--	2.5	--	VA	688

F. REFERENCES

1. "Report on Nomenclature in the Field of Macromolecules, International Union of Pure and Applied Chemistry", J. Polymer Sci. 8, 257 (1952).

2. P. J. Flory, "Principles of Polymer Chemistry," Cornell University Press, Ithaca, New York, (1953).

3. M. Kurata, W. H. Stockmayer, Fortschr. Hochpolymer Forsch. 3, 196 (1963).

4. W. H. Stockmayer, M. Fixman, J. Polymer Sci. C 1, 137 (1963).

5. R. Koningsveld, C. A. Tuijnman, Makromol. Chem. 38, 39, 44 (1960).

6. "A Structure-Based Nomenclature for Linear Polymers," Macromolecules 1, 193 (1968).

7. A. Peterlin, Viskosität und Form. In "Die Physik der Hochpolymeren," edited by H. A. Stuart, Vol. II, Springer-Verlag, Berlin, (1953).

8. G. Meyerhoff, Fortschr. Hochpolymer. Forsch. 3, 59 (1961).

9. H.-G. Elias, Kunststoffe-Plastics 4, 1 (1961).

10. S. Krause, "Dilute Solution Properties of Acrylic and Methacrylic Polymers," Part I, Revision 1, Rohm & Haas Co., Philadelphia, Pennsylvania, Feb. 1961.

11. P. J. Flory, "Statistical Mechanics of Chain Molecules," Interscience Pub., John Wiley & Sons, Inc., New York, 1969.

12. H. Yamakawa, "Modern Theory of Polymer Solutions," Harper & Row, Pub., New York, 1971.

13. A. Peterlin, J. Polymer Sci. 47, 403 (1960).

14. S. Heine, O. Kratky, G. Porod, P.J. Schmitz, Makromol. Chem. 44-46, 682 (1961); S. Heine, ibid. 48, 205 (1961); S. Heine, O. Kratky, J. Roppert, ibid. 56, 150 (1962).

15. F. Danusso, G. Moraglio, G. Gianott, J. Polymer Sci. 51, 475 (1961).

16. H. Fujita, N. Takeguchi, K. Kawahara, M. Abe, H. Utiyama, M. Kurata, Paper given at 12th Polymer Symposium, Nagoya, Japan (Nov. 1963).

17. M. Abe, Y. Murakami, H. Fujita, J. Appl. Polymer Sci. 9, 2549 (1965). See also Ref. 254.

18. R. Endo, Nippon Gomu Kyokaishi (J. Rubber Ind. Japan) 34, 522 (1961). See also, M. Takeda, R. Endo, Rept. Progr. Polymer Phys. Japan 6, 37 (1963).

19. I. Ya. Poddubnyi, Ye. G. Erenberg, M. A. Yeremina, Vysokomolekul. Soedin. Ser. A 10, 1381 (1968).

20. W. G. Cooper, G. Vaughan, D. E. Eaves, R. W. Madden, J. Polymer Sci. 50, 159 (1961).

21. R. Endo, Nippon Gomu Kyokaishi (J. Rubber Ind. Japan) 35, 658 (1962).

22. R. L. Cleland, J. Polymer Sci. 27, 349 (1958). See also Ref. 26.

23. P. L. Ribeyrolles, A. Guyot, H. Benoit, J. Chim. Phys, 56, 377 (1959). See also Ref. 26.

24. M. Kurata, H. Utiyama, K. Kajitani, T. Koyama, H. Fujita, Paper given at 12th Polymer Symposium, Nagoya, Japan (Nov. 1963).

25. R. Endo, Nippon Gomu Kyokaishi (J. Rubber Ind. Japan) 34, 527 (1961).

26. W. Cooper, D. E. Eaves, G. Vaughan, J. Polymer Sci. 59, 241 (1962).

27. I. Ya. Poddubnyi, V. A. Grechanovskii, Vysokomolekul. Soedin. 6, 64 (1964).

28. R. L. Scott, W. C. Carter, M. Magat, J. Am. Chem. Soc. 71, 220 (1949).

29. D. J. Pollock, L. J. Elyash, T. W. Dewitt, J. Polymer Sci. 15, 335 (1955).

30. V. S. Skazka, M. Kozhokaru, G. A. Fomin, L.F. Roguleva, Vysokomolekul. Soedin. Ser. A 9, 177 (1967).

31. W. E. Mochel, J. B. Nichols, J. Am. Chem. Soc. 71, 3435 (1949).

32. W. E. Mochel, J. B. Nichols, C. J. Mighton, J. Am. Chem. Soc. 70, 2185 (1948).

33. W. E. Mochel, J. B. Nichols, Ind. Eng. Chem. 43, 154 (1951).

34. K. Hanafusa, A. Teramoto, H. Fujita, J. Phys. Chem. 70, 4004 (1966).

35. K. Kawahara, T. Norisuye, H. Fujita, J. Chem. Phys. 49, 4339 (1968).

36. S. A. Pavlova, T. A. Soboleva, A. P. Suprun, Vysokomolekul. Soedin. 6, 122 (1964).

37. H. L. Wagner, P. J. Flory, J. Am. Chem. Soc. 74, 195 (1952).

38. K. Altgelt, G. V. Schulz, Makromol. Chem. 36, 209 (1960).

39. W. C. Carter, R. L. Scott, M. Magat, J. Am. Chem. Soc. 68, 1480 (1946).

40. I. Ya. Poddubnyi, V. A. Grechanovskii, A. V. Podalinskii, Vysokomolekul. Soedin. 5, 1588 (1964).

41. M. Abe, M. Iwama, T. Homma, Kogyo Kagaku Zasshi (J. Chem. Soc. Japan, Ind. Chem. Sec.) 72, 2313 (1969).

42. W. H. Beattie, C. Booth, J. Appl. Polymer Sci. 7, 507 (1963).

43. H. Brody, M. Ladaeki, R. Milkovitch, M. Szwarc, J. Polymer Sci. 25, 221 (1959).

44. I. Y. Poddubnyi, E. G. Ehrenberg, J. Polymer Sci. 57, 545 (1962). K=13.3 in benzene is for Li-type polymers, and 11.2 for Ziegler-types. No difference is found in dioxane.

45. H. C. Tingey, R. H. Ewart, G. E. Hulse, unpublished work; cited in Ref. 28. See also, D. M. French, R. H. Ewart, Anal. Chem. 19, 165 (1947).

46. J. A. Yanko, J. Polymer Sci. 3, 576 (1958).

47. T. Homma, H. Fujita, J. Appl. Polymer Sci. 9, 1701 (1965). Bound styrene, 24%.

48. J.-Y. Chien, W. Chin, Y.-S. Cheng, Kolloidn. Zh. 19, 515 (1957).

49. T. G. Fox, P. J. Flory, J. Am. Chem. Soc. 73, 1909 (1951).

50. T. G. Fox, P. J. Flory, J. Phys. Colloid Chem. 53, 197 (1949).

51. C. E. H. Bawn, E. S. Hill, M. A. Wajid, Trans. Faraday Soc. 52, 1651 (1956).

52. W. R. Krigbaum, P. J. Flory, J. Am. Chem. Soc. 75, 1775 (1953).

53. A. Ram, Thesis, Mass. Inst. Tech., Cambridge, Mass., USA, 1961.

54. F. R. Cottrell, E. W. Merrill, K. A. Smith, J. Polymer Sci. A-2 7, 1415 (1969).

55. P. J. Flory, J. Am. Chem. Soc. 65, 372 (1943).

56. U. Bianchi, M. Dalpiaz, E. Patrone, Makromol. Chem. 80, 112 (1964).

57. N. M. Tret'yakova, L. V. Kosmodem'yanskii, R. G. Romanova, E. G. Lazaryants, Vysokomolekul. Soedin. Ser. A 12, 2754 (1970).

58. R. Chiang, J. Phys. Chem. 70, 2348 (1966).

59. R. W. Wheatcraft, cited in H. Wesslau, Makromol. Chem. 20, 111 (1956).

60. E. Duch, L. Kuechler, Z. Elektrochem. 60, 218 (1956).

61. J. T. Atkins, L. T. Muus, C. W. Smith, E. T. Pieski, J. Am. Chem. Soc. 79, 5089 (1957).

62. A. Kotera, T. Saito, K. Takamizawa, Y. Miyazawa, Rept. Progr. Polymer Phys. Japan 3, 58 (1960).

63. P. S. Francis, R. Cooke, jr, J. H. Elliott, J. Polymer Sci. 31, 453 (1957).

64. P. M. Henry, J. Polymer Sci. 36, 3 (1959).

65. R. Chiang, J. Phys. Chem. 69, 1645 (1965).

66. R. Chiang, J. Polymer Sci. 36, 91 (1959).

67. L. H. Tung, J. Polymer Sci. 24, 333 (1957).

68. L. H. Tung, J. Polymer Sci. 36, 287 (1959).

69. C. J. Stacy, R. L. Arnett, J. Phys. Chem. 69, 3109 (1965). See also Ref. 73.

70. Q. A. Trementozzi, J. Polymer Sci. 36, 113 (1959).

71. H. Wesslau, Makromol. Chem. 26, 96 (1958).

72. H. S. Kaufmann, E. K. Walsh, J. Polymer Sci. 26, 124 (L) (1957).

73. C. J. Stacy, R. L. Amett, J. Polymer Sci. A 2, 167 (1964).

74. W. R. Krigbaum, Q. A. Trementozzi, J. Polymer Sci. 28, 295 (1958).

75. R. A. Mendelson, E. E. Drott, J. Polymer Sci. B 11, 795 (1968).

76. W. F. Busse, R. Longworth, J. Polymer Sci. 58, 49 (1962).

77. K. Ueberreiter, H.-J. Orthman, S. Sorge, Makromol. Chem. 8, 21 (1952).

78. I. Harris, J. Polymer Sci. 8, 353 (1952).

79. Q. A. Trementozzi, J. Polymer Sci. 23, 887 (1957).

80. K. H. Meyer, A. van der Wyk, Helv. Chim. Acta 18, 1067 (1935).

81. W. R. Krigbaum, J. E. Kurz, P. Smith, J. Phys. Chem. 65, 1984 (1961).

82. R. Endo, K. Iimura, M. Takeda, Bull. Chem. Soc. Japan, 37, 950 (1964).

83. S. S. Stivala, R. J. Valles, D. W. Levi, J. Appl. Polymer Sci. 7, 97 (1963).

84. J. B. Kinsinger, L. E. Ballard, J. Polymer Sci. A 3, 3963 (1965).

85. I. H. Billick, J. P. Kennedy, Polymer 6, 175 (1965).

86. J. L. Jungnickel, F. T. Weiss, J. Polymer Sci. 49, 437 (1961).

87. D. L. Flowers, W. A. Hewett, R. D. Mullineaux, J. Polymer Sci. A 2, 2305 (1964).

88. J. B. Kinsinger, R. E. Hughes, J. Phys. Chem. 63, 2002 (1959).

89. F. Danusso, G. Moraglio, Rend. Acad. Naz. Lincei 25, 509 (1958).

90. J. B. Kinsinger, R. E. Hughes, J. Phys. Chem. 67, 1922 (1963).

91. F. Danusso, G. Moraglio, Makromol. Chem. 28, 250 (1958).

92. L. Westerman, J. Polymer Sci. A 1, 411 (1963).

93. H. J. L. Schuurmans, R. A. Mendelson, unpublished work; cited in E. Kohn, H. J. L. Schuurmans, J. V. Cavender, R. A. Mendelson, J. Polymer Sci. 58, 681 (1962).

94. A. Nakajima, A. Saijo, J. Polymer Sci. A-2 6, 735 (1968).

95. A. Kotera, K. Takamizawa, T. Kamata, H. Kawaguchi, Rept. Progr. Polymer Phys. Japan 4, 131 (1961).

96. P. Parrini, F. Sebastiano, G. Messina, Makromol. Chem. 38, 27 (1960).

97. R. Chiang, J. Polymer Sci. 38, 235 (L) (1958).

98. G. Ciampa, Chim. Ind. (Milan) 38, 298 (1956).

99. F. Ang, H. Mark, Monatsh. Chem. 88, 427 (1957).

100. H. Inagaki, T. Miyamoto, S. Ohta, J. Phys. Chem. 70, 3420 (1966).

101. W. Scholtan, Makromol. Chem. 14, 169 (1954).

102. E. Collinson, F. S. Dainton, G. S. McNaughton, Trans. Faraday Soc. 53, 489 (1957).

103. L. Trossarelli, M. Meirone, J. Polymer Sci. 57, 445 (1962).

104. S. Newman, W. R. Krigbaum, C. Laugier, P. J. Flory, J. Polymer Sci. 14, 451 (1954).

105. A. Takahashi, N. Hayashi, I. Kagawa, Kogyo Kagaku Zasshi (J. Chem. Soc. Japan, Ind. Chem. Sec.) 60, 1059 (1957). See also Ref. 126.

106. V. N. Tsvetkov, S. Ya Lyubina, T. V. Barskaya, Vysoko-molekul. Soedin, 6, 806 (1964); V. N. Tsvetkov, V. S. Skazka, G. V. Tarasova, V. M. Yamshchikov, S. Ya Lyubina, ibid. Ser. A 10, 74 (1968).

107. A. Takahashi, S. Yamori, I. Kagawa, Nippon Kagaku Zasshi (J. Chem. Soc. Japan, Pure Chem. Sec.) 83, 11 (1962).

108. A. Takahashi, T. Kamei, I. Kagawa, Nippon Kagaku Zasshi (J. Chem. Soc. Japan, Pure Chem. Sec.) 83, 14 (1962).

109. A. Takahashi, M. Nagasawa, J. Am. Chem. Soc. 86, 543 (1964).

110. I. Noda, T. Tsuge, M. Nagasawa, J. Phys. Chem. 74, 710 (1970).

111. A. Soda, I. Kagawa, Nippon Kagaku Zasshi (J. Chem. Soc. Japan, Pure Chem. Sec.) 83, 412 (1962).

112. G. Saini, L. Trossarelli, Atti Accad. Sci. Torino, Classe Sci. Fis., Mat. Nat. 90, 410 (1955-56). cf. Ref. 10.

113. G. B. Rathmann, F. A. Bovey, J. Polymer Sci. 15, 544 (1955).

114. M. Giurgea, C. Ghita, I. Baltog, A. Lupu, J. Polymer Sci. A-2 4, 529 (1966).

115. H. Sumitomo, Y. Hachihama, Kobunshi Kagaku (Chem. High Polymers (Tokyo)) 10, 544 (1953). See also Y. Hachihama, H. Sumitomo, Technol. Rept. Osaka Univ. 3, 385 (1953).

116. H. Sumitomo, Y. Hachihama, Kobunshi Kagaku (Chem. High Polymers (Tokyo)) 12, 479 (1955). See also, Y. Hachihama, H. Sumitomo, Technol. Rept. Osaka Univ. 5, 485 (1956).

117. D. Mangaraj, S. K. Patra, Makromol. Chem. 107, 230 (1967).

118. I. G. Soboleva, N. V. Makletsova, S. S. Medvedev, Dokl. Akad. Nauk. SSSR, 94, 289 (1954).

119. E. S. Cohn, T. A. Orofino, I. L. Scogna, unpublished work; cited in Ref. 10.

120. J. E. Mark, R. A. Wessling, R. E. Hughes, J. Phys. Chem. 70, 1895 (1966).

121. R. A. Wessling, J. E. Mark, E. Hamori, J. E. Mark, J. Phys. Chem. 70, 1903 (1966).

122. S. Krause, unpublished work; cited in Ref. 10.

123. H. Staudinger, H. Warth, Z. Prakt. Chem. 155, 261 (1940).

124. L. Trossareli, G. Saini, Atti Accad. Sci. Torino, Classe Sci. Fis., Mat. Nat. 90, 419 (1955-56); cited in Ref. 10.

125. L. Trossareli, G. Saini, Atti Accad. Sci. Torino, Classe Sci. Fis., Mat. Nat. 90, 431 (1955-56); cited in Ref. 10.

126. H. Ito, S. Shimizu, S. Suzuki, Kogyo Kagaku Zasshi (J. Chem. Soc. Japan, Ind. Chem. Sec.) 59, 930 (1956).

127. G. M. Guzman, Anales Real Soc. Espan. Fis. Quim. (Madrid) Ser. B, 52, 377 (1956) ; cited in Ref. 10.

128. Rohm and Haas, old data; reported in Ref. 10.

129. H. Matsuda, K. Yamano, H. Inagaki, J. Polymer Sci. A-2 7, 609 (1969).

130. J. N. Sen, S. R. Chatterjee, S. R. Palit, J. Sci. Ind. Res. (India) 11 B, 90 (1952); cited in Ref. 10.

131. A. Kotera, T. Saito, Y. Watanabe, M. Ohama, Makromol. Chem. 87, 195 (1965).

132. N. T. Sririvasan, M. Santappa, Makromol. Chem. 27, 61 (1958).

133. K. Karunakaran, M. Santappa, J. Polymer Sci. A-2 6, 713 (1968).

134. Y. Fujisaki, H. Kobayashi, Kobunshi Kagaku (Chem. High Polymers (Tokyo)) 19, 73, 81 (1962).

135. H. Inagaki, K. Hayashi, T. Matsuo, Makromol. Chem. 84, 80 (1965).

136. T. Shibukawa, M. Sone, A. Uchida, K. Iwahori, J. Polymer Sci. A-1 6, 147 (1968).

137. W. Scholtan, H. Marzolph, Makromol. Chem. 57, 52 (1962).

138. J. Bisschops, J. Polymer Sci. 17, 81 (1955).

139. R. L. Cleland, W. H. Stockmayer, J. Polymer Sci. 17, 473 (1955).

140. R. F. Onyon, J. Polymer Sci. 22, 13 (1956).

141. W. R. Krigbaum, A. M. Kotliar, J. Polymer Sci. 32, 323 (1958).

142. C. H. Bamford, A. D. Jenkins, R. Johnston, E. F. T. White, Trans. Faraday Soc. 55, 168 (1959).

143. P. F. Onyon, J. Polymer Sci. 37, 315 (1959).

144. L. H. Peebles, J. Polymer Sci. A <u>3</u>, 361, (1965).

145. H. Kobayashi, J. Polymer Sci. <u>39</u>, 369 (1959).

146. R. Chiang, J. C. Stauffer, J. Polymer Sci. A-2 <u>5</u>, 101 (1967).

147. N. M. Wiederhorn, A. R. Brown, J. Polymer Sci. <u>8</u>, 651 (1952).

148. A. Katchalsky, H. Eisenberg, J. Polymer Sci. <u>6</u>, 145 (1951).

149. R. Arnold, S. R. Caplan, Trans. Faraday Soc. <u>51</u>, 857 (1955).

150. R. van Leemput, R. Stein, J. Polymer Sci. A <u>1</u>, 985 (1963). K values are given for monodisperse polymers.

151. Rohm and Haas, old data; reported in Ref. 10.

152. S. N. Chinai, R. A. Guzzi, J. Polymer Sci. <u>21</u>, 417 (1956).

153. A. S. Nair, M. S. Muthana, Makromol. Chem. <u>47</u>, 114 (1961).

154. V. N. Tsvetkov, S. I. Klenin, Zh. Tekhn. Fiz. <u>29</u>, 1393 (1959).

155. Z. Menčik, Chem. Listy <u>46</u>, 407 (1952).

156. S. N. Chinai, R. J. Valles, J. Polymer Sci. <u>39</u>, 363 (1959).

157. M. Kozhokaryu, V. S. Skazka, K. G. Berdnikova, Vysoko-molekul. Soedin. <u>8</u>, 1063 (1967).

158. S. N. Chinai, R. A. Guzzi, J. Polymer Sci. <u>41</u>, 475 (1959).

159. H. T. Lee, D. W. Levi, J. Polymer Sci. <u>47</u>, 449 (1960).

160. F. E. Didot, S. N. Chinai, D. W. Levi, J. Polymer Sci. <u>43</u>, 557 (1960).

161. S. N. Chinai, R. J. Samuels, J. Polymer Sci. <u>19</u>, 463 (1956).

162. K. Karunakaran, M. Santappa, Makromol. Chem. <u>111</u>, 20 (1968).

163. V. N. Tsvetkov, D. Khardi, I. N. Shtennikova, Ye. V. Komeyeva, G. F. Pirogova, K. Nitrai, Vysokomolekul. Soedin. Ser. A <u>11</u>, 349 (1969).

164. J.-Y. Chieng, M.-H. Shih, Z. Physik. Chem. (Leipzig) <u>207</u>, 60 (1957).

165. S. N. Chinai, J. Polymer Sci. <u>25</u>, 413 (1957).

166. R. J. Valles, J. Polymer Sci. A <u>3</u>, 3853 (1965); see also R. J. Valles, E. C. Schramm, ibid. <u>3</u>, 3664 (1965), and Ref. 167.

167. R. J. Valles, Makromol. Chem. <u>100</u>, 167 (1967).

168. M. Iwama, H. Utiyama, M. Kurata, J. Makromol. Chem. <u>1</u>, 701 (1966).

169. G. Meyerhoff, G. V. Schulz, Makromol. Chem. <u>7</u>, 294 (1951).

170. J. Bischoff, V. Desreux, Bull Soc. Chim. Belges <u>61</u>, 10 (1952).

171. G. V. Schulz, G. Meyerhoff, Z. Elektrochem. <u>56</u>, 904 (1952).

172. G. V. Schulz, H.-J. Cantow, G. Meyerhoff, J. Polymer Sci. <u>10</u>, 79 (1953).

173. H.-J. Cantow, G. V. Schulz, Z. Physik. Chem. (Frankfurt) <u>2</u>, 117 (1954); see also, H.-J. Cantow, G. V. Schulz, ibid. <u>1</u>, 365 (1954).

174. S. N. Chinai, J. D. Matlack, A. L. Resnick, R. J. Samuels, J. Polymer Sci. <u>17</u>, 391 (1955).

175. F. W. Billmeyer jr, C. B. de Than, J. Am. Chem. Soc. <u>77</u>, 4763 (1955).

176. S. L. Kapur, J. Sci. Ind. Res. (India) <u>15 B</u>, 239 (1956).

177. W. R. Moore, R. J. Fort, J. Polymer Sci. A <u>1</u>, 929 (1963).

178. E. Cohn-Ginsberg, T. G. Fox, H. F. Mason, Polymer <u>3</u>, 97 (1962).

179. T. G. Fox, Polymer <u>3</u>, 111 (1962).

180. E. S. Cohn, T. G. Fox, unpublished work; cited in Ref. 10.

181. A. F. V. Eriksson, Acta Chem. Scand. <u>7</u>, 623 (1953).

182. J. H. Baxendale, S. Bywater, M. G. Evans, J. Polymer Sci. <u>1</u>, 237 (1946).

183. J.-Y. Chien, L.-H. Shin, K.-I. Shin, Hua Hsueeh Pao, Acta Chim. Sinica <u>23</u>, 215 (1957). cf. Ref. 10.

184. K. G. Schoen, G. V. Schulz, Z. Physik. Chem. (Frankfurt) <u>2</u>, 197 (1954).

185. T. G. Fox, J. B. Kinsinger, H. F. Mason, E. M. Schuele, Polymer <u>3</u>, 71 (1962).

186. J. Hakozaki, Nippon Kagaku Zasshi (J. Chem. Soc. Japan, Pure Chem. Sec.) <u>82</u>, 155 (1961).

187. H. Inagaki, S. Kawai, Makromol. Chem. <u>79</u>, 42 (1964). see also Ref. 197.

188. G. V. Schulz, A. Dinglinger, Z. Prakt. Chem. <u>158</u>, 137 (1941).

189. H. J. Cantow, J. Pouget, C. Wippler, Makromol. Chem. <u>14</u>, 110 (1954).

190. V. N. Tsvetkov, S. I. Klenin, J. Polymer Sci. <u>30</u>, 187 (1958).

191. I. Sakurada, A. Nakajima, O. Yoshizaki, K. Nakamae, Kolloid-Z. <u>186</u>, 41 (1962).

192. A. F. V. Eriksson, Acta Chem. Scand. <u>10</u>, 378 (1956).

193. E. F. Casassa, W. H. Stockmayer, Polymer <u>3</u>, 53 (1962).

194. E. Hamori, L. R. Prusinowski, P. G. Sparks, R. E. Hughes, J. Phys. Chem. <u>69</u>, 1101 (1965).

195. E. Patrone, U. Bianchi, Makromol. Chem. <u>94</u>, 52 (1966).

196. S. N. Chinai, C. W. Bondurant, J. Polymer Sci. <u>22</u>, 555 (1956).

197. G. V. Schulz, R. Kirste, Z. Physik. Chem. (Frankfurt) <u>30</u>, 171 (1961).

198. S. Krause, E. Cohn-Ginsberg, J. Phys. Chem. <u>67</u>, 1479 (1963).

199. S. Krause, E. Cohn-Ginsberg, Polymer <u>3</u>, 565 (1962).

200. J. G. Fee, W. S. Port, L. P. Whitnauer, J. Polymer Sci. <u>33</u>, 95 (1958).

201. S. N. Chinai, A. L. Resnick, H. T. Lee, J. Polymer Sci. <u>33</u>, 471 (1958).

202. N. Fuhrman, R. B. Mesrobian, J. Am. Chem. Soc. <u>76</u>, 3281 (1954).

203. C. G. Overberger, E. M. Pearce, N. Mayes, J. Polymer Sci. <u>34</u>, 109 (1959).

204. R. C. Schulz, S. Suzuki, H. Cherdron, W. Kern, Makromol. Chem. <u>53</u>, 145 (1962).

205. Ye. V. Korneyeva, V. N. Tsvetkov, P. N. Lavrenko, Vysokomolekul. Soedin. Ser. A <u>12</u>, 1369 (1970).

206. J. A. Manson, G. J. Arquette, Makromol. Chem. <u>37</u>, 187 (1960).

207. S. Arichi, J. Sci. Hiroshima Univ. Ser. A-II <u>29</u>, 97 (1965).

208. P. J. Flory, F. Leutner, J. Polymer Sci. <u>3</u>, 880 (1948).

209. K. Dialer, K. Vogler, F. Patat. Helv. Chim. Acta <u>35</u>, 869 (1952).

210. H. A. Dieu, J. Polymer Sci. <u>12</u>, 417 (1954).

211. L. E. Miller, F. A. Hamm, J. Phys. Chem. <u>57</u>, 110 (1953).

212. A. Nakajima, K. Furutate, Kobunshi Kagaku (Chem. High Polymers, Tokyo) <u>6</u>, 460 (1949).

213. M. Matsumoto, Y. Ohyanagi, Kobunshi Kagaku (Chem. High Polymers, Tokyo), <u>17</u>, 191 (1960). The second line is the constant corrected for monodisperse sample.

214. T. Matsuo, H. Inagaki, Makromol. Chem. <u>55</u>, 151 (1962).

215. M. Matsumoto, K. Imai, J. Polymer Sci. <u>24</u>, 125 (1957).

216. M. Litt, F. R. Eirich, J. Polymer Sci. <u>45</u>, 379 (1960).

217. A. C. Ciferri, M. Kryszewski, G. Weil, J. Polymer Sci. <u>27</u>, 167 (1958).

218. A. Ciferri, M. Lauretti, Ann. Chim. (Rome) <u>48</u>, 198 (1958).

219. M. Sato, Y. Koshiishi, M. Asahina, J. Polymer Sci. B <u>1</u>, 233 (1963).

220. H. Inagaki, J. Nakazawa, Private communication.

221. J. W. Breitenbach, E. L. Forster, A. J. Renner, Kolloid-Z. <u>127</u>, 1 (1952).

222. C. Bier, H. Kramer, Makromol. Chem. <u>18-19</u>, 151 (1955).

223. G. Ciampa, H. Schwindt, Makromol. Chem. <u>21</u>, 169 (1954).

224. F. Danusso, G. Moraglio, S. Cazzera, Chim. Ind. (Milan) 36, 883 (1954).

225. Z. Mencik, Collection Czech. Chem. Commun. 21, 517 (1956).

226. W. R. Moore, R. J. Hutchinson, Nature 200, 1095 (1963).

227. J. Vavra, J. Lapcik, J. Sabados, J. Polymer Sci. A-2 5, 1305 (1967).

228. M. Bohdanecky, K. Solc, P. Kratochvil, M. Kolinsky, M. Pyska, D. Lim, J. Polymer Sci. A-2 5, 343 (1967).

229. H. Batzer, A. Nisch, Makromol. Chem. 22, 131 (1957).

230. M. Freeman, P. P. Manning, J. Polymer Sci. A 2, 2017 (1964).

231. A. Takahashi, M. Ohara, I. Kagawa, Kogyo Kagaku Zasshi (J. Chem. Soc. Japan, Ind. Chem. Sec.) 66, 960 (1963).

232. M. Sato, Y. Koshiishi, M. Asahina, cited in A. Nakajima, K. Kato, Makromol. Chem. 95, 52 (1966).

233. T. Kobayashi, Bull. Chem. Soc. Japan 35, 726 (1962).

234. E. K. Walsh, H. S. Kaufman, J. Polymer Sci. 26, 1 (1957).

235. M. L. Wallach, M. A. Kabayama, J. Polymer Sci. A-1 4, 2667 (1966).

236. M. Ueda, K. Kajitani, Makromol. Chem. 108, 138 (1967). and private communication.

237. W. R. Moore, M. Murphy, J. Polymer Sci. 56, 519 (1962).

238. V. N. Tsvetkov, S. Ya. Kotlyar, Zh. Fiz. Khim. 30, 1100 (1956).

239. G. Saini, L. Maldifassi, L. Trossarelli, Ann. Chim. (Rome) 44, 1954 (1954); cited in H.-G. Elias, Kunststoffe-Plastics 4, 8 (1961).

240. G. S. Misra, V. P. Gupta, Makromol. Chem. 71, 110 (1964).

241. R. H. Wagner, J. Polymer Sci. 2, 21 (1947).

242. S. N. Chinai, P. C. Scherer, D. W. Levi, J. Polymer Sci. 17, 117 (1955).

243. K. Z. Fattakhov, E. S. Pisarenko, L. N. Verkotina, Kolloidn. Zh. 18, 101 (1956).

244. M. Matsumoto, Y. Ohyanagi, J. Polymer Sci. 46, 441 (1960). Kobunshi Kagaku (Chem. High Polymers, Tokyo) 17, 1 (1960).

245. R. Naito, Kobunshi Kagaku (Chem. High Polymers, Tokyo) 16, 7 (1959).

246. Bevak, Thesis, Mass. Inst. Tech., Cambridge, Mass., USA 1955.

247. V. Kalpagam, R. Rao, J. Polymer Sci. A 1, 233 (1963).

248. M. R. Rao, V. Kalpagam, J. Polymer Sci. 49, S 14 (1961).

249. A. Nakajima, Kobunshi Kagaku (Chem. High Polymers, Tokyo) 11, 142 (1954).

250. V. V. Varadiah, J. Polymer Sci. 19, 477 (1956).

251. G. C. Berry, L. M. Hobbs, V. C. Long, Polymer 5, 31 (1964).

252. A. R. Shultz, J. Am. Chem. Soc. 76, 3423 (1954). See also, R. O. Howard, Thesis, Mass. Inst. Tech., Cambridge, Mass. USA, 1952.

253. H.-G. Elias, F. Patat, Makromol. Chem. 25, 13 (1957).

254. M. Abe, H. Fujita, J. Phys. Chem. 69, 3263 (1965).

255. M. Matsumoto, Y. Ohyanagi, J. Polymer Sci. 50, S 1 (1961). See also Ref. 244.

256. C. J. Kurian, M. S. Muthana, Makromol. Chem. 29, 1 (1959).

257. K. Fujii, S. Imoto, J. Ukida, M. Matsumoto, Kobunshi Kagaku (Chem. High Polymers, Tokyo) 19, 581 (1962).

258. H. Hopff, J. Dohany, Makromol. Chem. 69, 131 (1963).

259. H. Eisenberg, D. Woodside, J. Chem. Phys. 36, 1844 (1962).

260. K. Dialer, R. Kerber, Makromol. Chem. 17, 56 (1955).

261. F. Patat, K. Vogler, Helv. Chim. Acta 35, 128 (1952).

262. J. Springer, K. Ueberreiter, R. Wenzel, Makromol. Chem. 96, 122 (1966).

263. J. M. Barrales-Rienda, D. C. Pepper, Polymer 8, 337 (1967).

264. L. A. Utracki, R. Simha, Makromol. Chem. 117, 94 (1968). See also, L. A. Utracki, R. Simha, N. Eliezer, Polymer 10, 43 (1969).

265. F. S. Holahan, S. S. Stivala, D. W. Levi, J. Polymer Sci. A 3, 3987 (1965).

266. N. Kuwahara, K. Ogino, M. Konuma, N. Iida, M. Kaneko, J. Polymer Sci. A-2 4, 173 (1966).

267. V. V. Korshak, S. L. Sosin, V. P. Alekseeva, J. Polymer Sci. 52, 213 (1961).

268. V. E. Eskin, O. Z. Korotkina, Vysokomolekul. Soedin. 1, 1580 (1959).

269. V. E. Eskin, O. Z. Korotkina, Vysokomolekul. Soedin. 2, 272 (1960).

270. M. Cantow, G. Meyerhoff, G. V. Schulz, Makromol. Chem. 49, 1 (1961).

271. G. Meyerhoff, Z. Phys. Chem. (Frankfurt) 4, 335 (1955).

272. D. C. Pepper, J. Polymer Sci. 7, 347 (1951). See also D. C. Pepper, Proc. Roy. Dublin Soc. 25, 239 (1951).

273. G. S. Misra, R. C. Rastogi, V. P. Gupta, Makromol. Chem. 50, 72 (1961).

274. W. R. Krigbaum, P. J. Flory, J. Polymer Sci. 11, 37 (1953).

275. T. A. Orofino, F. Wenger, J. Phys. Chem. 67, 566 (1963).

276. C. E. H. Bawn, C. Freeman, A. Kamaliddin, Trans. Faraday Soc. 46, 1107 (1950).

278. C. Rossi, Chimica delle Macromolecole (Sept. 1961); Consiglio Nazionalle delle Ricerche, Roma, (1963) p. 153.

278. P. Outer, C. I. Carr, B. H. Zimm, J. Chem. Phys. 18, 830 (1950).

279. J. Oth, V. Desreux, Bull. Soc. Chim. Belges 63, 285 (1954).

280. T. Oyama, K. Kawahara, M. Ueda, Nippon Kagaku Zasshi (J. Chem. Soc. Japan, Pure Chem. Sec.) 79, 727 (1958).

281. U. Bianchi, V. Magnasso, C. Rossi, Ric. Sci. 58, 1412 (1958).

282. U. Bianchi, V. Magnasso, C. Rossi, Chim Ind. (Milan) 40, 263 (1958); see also, Ref. 276.

283. H. Utiyama, Thesis, Kyoto Univ., Kyoto, Japan, 1962.

284. R. Endo, M. Takeda, J. Polymer Sci. 56, 28 (1962). The tabulated constants are calculated from the original figure.

285. L. Utracki, R. Simha, J. Phys. Chem. 67, 1052 (1963).

286. H. Inagaki, H. Suzuki, M. Fujii, T. Matsuo, J. Phys. Chem. 70, 1718 (1966). See also, H. Inagaki, H. Suzuki, M. Kurata, J. Polymer Sci. C 15, 409 (1966).

287. T. Homma, K. Kawahara, H. Fujita, M. Ueda, Makromol. Chem. 67, 132 (1963).

288. H. J. Cantow, Makromol. Chem. 30, 169 (1959).

289. L. A. Papazian, Polymer 10, 399 (1969).

290. H. Inagaki, private communication. Based on the data given in Ref. 286, and also on unpublished data.

291. M. Abe, H. Fujita, J. Phys. Chem. 69, 3263 (1965).

292. T. G. Fox, P. J. Flory, J. Am. Chem. Soc. 73, 1915 (1951).

293. E. T. Dimitru, L. H. Cragg, unpublished work; cited in P. J. Flory, "Principles of Polymer Chemistry", Cornell Univ. Press, Ithaca, New York, 1953, p. 615, Table XXVII.

294. J. W. Breitenbach, H. Gabler, O. F. Olaj, Makromol. Chem. 81, 32 (1964).

295. R. H. Ewart, H. C. Tingey, M. Wales, unpublished work; cited in W. V. Smith, J. Am. Chem. Soc. 68, 2063 (1946).

296. C. H. Bamford, M. J. S. Dewar, Proc. Roy. Soc. London A 192, 329 (1948).

297. H. Marzolph, G. V. Schulz, Makromol. Chem. 13, 120 (1954).

298. C. Rossi, U. Bianchi, E. Bianchi, Makromol. Chem. 41, 31 (1960). See also, Ref. 301.

299. S. N. Chinai, P. C. Scherer, C. W. Bondurat, D. W. Levi, J. Polymer Sci. 22, 527 (1956).

300. F. Danusso, G. Moraglio, J. Polymer Sci. 24, 161 (1957).

301. T. Altares, D. P. Wyman, V. R. Allen, J. Polymer Sci. A 2, 4533 (1964).

302. A. Yamamoto, M. Fujii, G. Tanaka, H. Yamakawa, Polymer J. 2, 799 (1971).

303. G. C. Berry, J. Chem. Phys. 46, 1338 (1967).

304. M. Morton, T. E. Helminiak, S. D. Gadkary, F. Bueche, J. Polymer Sci. 57, 471 (1962).

305. J. M. G. Cowie, S. Bywater, Polymer 6, 197 (1965).

306. J. M. G. Cowie, E. L. Cussler, J. Chem. Phys. 46, 4886 (1967).

307. G. Meyerhoff, Z. Physik. Chem. (Frankfurt) 23, 100 (1960).

308. H. W. McCormick, J. Polymer Sci. 36, 341 (1959).

309. R. N. Mukherjea, P. Remmp, J. Chim. Phys. 56, 94 (1959).

310. J. M. G. Cowie, D. J. Worsfold, S. Bywater, Trans. Faraday Soc. 57, 537 (1961).

311. F. Ang. J. Polymer Sci. 25, 126 (1957).

312. G. Natta, F. Danusso, G. Moraglio, Makromol. Chem. 20, 37 (1956). See also Ref. 300.

313. W. R. Krigbaum, D. K. Carpenter, S. Newman, J. Phys. Chem. 62, 1586 (1958).

314. L. Trossarelli, E. Campi, G. Saini, J. Polymer Sci. 35, 205 (1959).

315. C. D. Thurmond, B. H. Zimm, J. Polymer Sci. 8, 477 (1952).

316. M. Kurata, M. Abe, M. Iwama, M. Matsushima, Polymer J. 3, 729 (1972).

317. S. Minatono, Thesis, Kyoto Univ., Kyoto, Japan, 1966.

318. Y. Mitsuda, K. Osaki, L. Schrag, J. D. Ferry, Polymer J. 4, 24 (1973).

319. K. Sakato, M. Kurata, Polymer J. 1, 260 (1970).

320. I. Noda, K. Mizutani, T. Kato, T. Fujimoto, M. Nagasawa, Macromolecules 3, 787 (1970).

321. J. M. G. Cowie, S. Bywater, D. J. Worsfold, Polymer 8, 105 (1967).

322. A. F. Sirianni, D. J. Worsfold, S. Bywater, Trans. Faraday Soc. 55, 2124 (1959).

323. M. Abe, K. Sakato, T. Kageyama, M. Fukatsu, M. Kurata, Bull. Chem. Soc. Japan 41, 2000 (1000).

324. I. Noda, S. Saito, T. Fujimoto, M. Nagasawa, J. Phys. Chem. 71, 4048 (1967).

325. H. W. McCormick, J. Polymer Sci. 41, 327 (1959).

326. B. J. Cottam, J. M. G. Cowie, S. Bywater, Makromol. Chem. 86, 116 (1965).

327. S. Okamura, T. Higashimura, Y. Imanishi, Kobunshi Kagaku (Chem. High Polymers, Tokyo) 16, 244 (1959).

328. J. M. G. Cowie, S. Bywater, J. Polymer Sci. A-2 6, 499 (1968).

329. A. Kotera, T. Saito, H. Matsuda, R. Kamata, Rept. Progr. Polymer Phys. Japan 3, 51 (1960).

330. A. K. Chaudhuri, D. K. Sarkar, S. Palit, Makromol. Chem. 111, 36 (1968).

331. G. Tanaka, S. Imai, H. Yamakawa, J. Chem. Phys. 52, 2639 (1970).

332. V. P. Budtov, V. M. Belyayev, Vysokomolekul. Soedin. Ser. A 12, 1909 (1970).

333. C. S. H. Chen, R. F. Stamm, J. Polymer Sci. 58, 369 (1962).

334. I. Sakurada, Y. Sakaguchi, S. Kokuryo, Kobunshi Kagaku, (Chem. High Polymers, Tokyo) 17, 227 (1960).

335. W. Burchard, M. Nosseir, Makromol. Chem. 82, 109 (1964).

336. Y. Sakaguchi, J. Nishino, K. Tsugawa, Kobunshi Kagaku, (Chem. High Polymer, Tokyo) 20, 661 (1963).

337. S. K. Patra, Thesis, I. I. T., Kharagpur, India, 1965; cited in S. K. Patra, D. Mangaraj, Makromol. Chem. 111, 168 (1968).

338. J. Parrod, J. Elles, J. Polymer Sci. 29, 411 (1958).

339. E. S. Cohn, unpublished work, cited in Ref. 10.

340. V. N. Tsvetkov, O. V. Kalisov, Zh. Fiz. Khim. 33, 710 (1959).

341. O. V. Kalisov, I. N. Shtennikova, Vysokomolekul. Soedin. 1, 842 (1959).

342. V. N. Tsvetkov, S. I. Klenin, J. Polymer Sci. 30, 187 (1958).

343. M. Tricot, V. Desreux, Makromol. Chem. 149, 185 (1970).

344. E. S. Cohn, I. L. Scogna, T. A. Orofino, unpublished work; cited in Ref. 10.

345. J. Hakozaki, Nippon Kagaku Zasshi (J. Chem. Soc, Japan, Pure Chem. Sec.) 82, 158 (1961).

346. V. N. Tsvetkov, V. S. Skazka, N. A. Nikitin, I. B. Steparenko, Vysokomolekul. Soedin. 6, 69 (1964).

347. W. Kern, D. Brawn, Makromol. Chem. 27, 23 (1958).

348. K. Takashima, G. Tanaka, H. Yamakawa, Polymer J. 2, 245 (1971).

349. Y. Noguchi, A. Aoki, G. Tanaka, H. Yamakawa, J. Chem. Phys. 52, 2651 (1970).

350. K. Matsumura, Makromol. Chem. 124, 204 (1969).

351. R. B. Mohite, S. Gundiah, S. L. Kapur, Makromol. Chem. 116, 280 (1968).

352. A. Kotera, T. Saito, H. Matsuda, R. Kamata, Rept. Progr. Polymer Phys. Japan 3, 51 (1960).

353. N. Kuwahara, K. Ogino, A. Kasai, S. Ueno, M. Kaneko, J. Polymer Sci. A 3, 985 (1965).

354. G. Greber, J. Tolle, W. Burchard, Makromol. Chem. 71, 47 (1964).

355. J. E. Davis, Thesis, Mass. Inst. Tech., Cambridge, Mass., USA, 1960.

356. E. F. Frisman, L. F. Shalaeva, Dokl. Akad. Nauk SSSR, 101, 970 (1955).

357. V. E. Eskin, K. Z. Gumargalieva, Vysokomolekul. Soedin. 2, 265 (1960).

358. N. Kuwahara, K. Ogino, M. Konuma, N. Iida, M. Kaneko, J. Polymer Sci. A-2 4, 173 (1966).

359. V. E. Eskin, L. N. Andreeva, Vysokomolekul. Soedin. 3, 435 (1961).

360. D. Braun, T.-O. Ahn, W. Kern, Makromol. Chem. 53, 154 (1962).

361. S. Minatono, Thesis, Kyoto Univ., Kyoto, Japan, 1967.

362. Y. Imanishi, T. Higashimura, S. Okamura, Kobunshi Kagaku (Chem. High Polymer Japan, Tokyo) 22, 241 (1965).

363. G. Ceccorulli, M. Pizzoli, G. Stea, Makromol. Chem. 142, 153 (1971).

364. W. A. Pryor, T.-L. Huang, Macromolecules, 2, 70 (1969).

365. M. Kato, T. Nakagawa, H. Akamatsu, Bull. Chem. Soc. Japan 33, 322 (1960).

366. A. Takahashi, T. Kato, M. Nagasawa, J. Phys. Chem. 71, 2001 (1967).

367. G. Sitaramaiah, D. Jacobs, Polymer 11, 165 (1970).

368. N. Kuwahara, S. Higashide, M. Nakata, M. Kaneko, J. Polymer Sci. A-2 7, 285 (1969).

369. G. M. Chetyrkina, V. G. Aldoshin, S. Y. Frenkel, Vysokomolekul. Soedin. 1, 1133 (1959).

370. V. N. Tsvetkov, V. G. Aldoshin, Zh. Fiz. Khim. 33, 2767 (1959).

371. S. Arichi, Bull. Chem. Soc. Japan 39, 439 (1966).

372. D. O. Jordan, A. R. Mathieson, M. R. Porter, J. Polymer Sci. 21, 473 (1956).

373. J. B. Berkowitz, M. Yamin, R. M. Fuoss, J. Polymer Sci. 28, 69 (1958).

374. A. G. Boyes, U- P. Strauss, J. Polymer Sci. 22, 463 (1956).

375. M. Miura, Y. Kubota, T. Masuzukawa, Bull. Chem. Soc. Japan 38, 316 (1965).

376. C. Garbuglio, L. Crescentini, A. Mula, G. B. Gechele, Makromol. Chem. 97, 97 (1966); see also Ref. 381.

377. H. Sato, T. Yamamoto, Nippon Kagaku Zasshi (J. Chem. Soc. Japan, Pure Chem. Sec.) 80, 1393 (1959).

378. G. B. Levy, H. P. Frank, J. Polymer Sci. 17, 247 (1955); see also Ref. 380.

379. K. Dialer, K. Vogler, Makromol. Chem. 6, 191 (1951).

380. G. B. Levy, H. P. Frank, J. Polymer Sci. 10, 371 (1953).

381. G. B. Gechele, L. Crescentini, J. Polymer Sci. A3, 3599 (1965).

382. W. Scholtan, Makromol. Chem. 7, 209 (1951).

383. L. C. Cerney, T. E. Helminiak, J. F. Meier, J. Polymer Sci. 44, 539 (1960).

384. H.-G. Elias, Makromol. Chem. 50, 1 (1961).

385. G. Allen, C. Booth, S. J. Hurst, M. N. Jones, C. Price, Polymer 8, 391 (1967).

386. K. Yamamoto, H. Fujita, Polymer 8, 517 (1967).

387. D. R. Beech, C. Booth, J. Polymer Sci., A-2, 7, 575 (1969).

388. C. Rossi, C. Cuniberti, J. Polymer Sci., B, 2, 681 (1964).

389. C. Sadron, P. Rempp, J. Polymer Sci. 29, 127 (1958).

390. T. A. Ritscher, H.-G. Elias, Makromol. Chem. 30, 48 (1959).

391. H.-G. Elias, Kunststoffe-Plastics 4, 1 (1961).

392. D. K. Thomas, A. Charlesby, J. Polymer Sci. 42, 195 (1960).

393. C. Rossi, E. Bianchi, G. Conio, Chim. Ind. (Milan) 45, 1498 (1963).

394. F. E. Bailey, jr., J. L. Kucera, L. G. Imhof, J. Polymer Sci. 32, 517 (1958).

395. F. E. Bailey, jr., R. W. Callard, J. Appl. Polymer Sci. 1, 56 (1959).

396. C. Booth, R. Orme, Polymer 11, 626 (1970).

397. M. Matsushima, M. Fukatsu, M. Kurata, Bull. Chem. Soc. Japan 41, 2570 (1968).

398. K. Yamamoto, H. Fujita, Polymer 7, 557 (1966).

399. L. Hoehr, V. Jaacks, H. Cherdron, S. Iwabuchi, W. Kern, Makromol. Chem. 103, 279 (1967).

400. J. Majer, Makromol. Chem. 86, 253 (1965).

401. V. Kokle, F. W. Billmeyer, jr., J. Polymer Sci., B, 3, 47 (1965).

402. W. H. Stockmayer, L.-L. Chen, J. Polymer Sci., A-2, 4, 437 (1966).

403. J. Majer, unpublished; cited in M. Kučera, E. Spousta, Makromol. Chem. 76, 183 (1964).

404. N. A. Pravikova, Ye. B. Berman, Ye. B. Lyudvig, A. G. Davtyan, Vysokomolekul. Soedin., A, 12, 580 (1970).

405. H. L. Wagner, K. F. Wissbrun, Makromol. Chem. 81, 14 (1964).

406. G. Meyerhoff, U. Moritz, Makromol. Chem. 109, 143 (1967).

407. W. Scholtan, S. Y. Lie, Makromol. Chem. 108, 104 (1967).

408. G. Allen, A. C. Booth, M. M. Jones, Polymer 5, 195 (1964).

409. R. E. Hughes, J. Richards, unpublished work; cited in P. E. Ebert, C. C. Price, J. Polymer Sci. 34, 157 (1959).

410. R. J. Valles, Makromol. Chem. 113, 147 (1968). The constants are expressed in terms of M_n.

411. J. Moacanin, J. Appl. Polymer Sci. 1, 272 (1959).

412. M. Kurata, H. Utiyama, K. Kamada, Makromol. Chem. 88, 281 (1965).

413. S. M. Ali, M. B. Huglin, Makromol. Chem. 84, 117 (1965).

414. K. Yamamoto, A. Teramoto, H. Fujita, Polymer 7, 267 (1966).

415. P. J. Flory, P. B. Stickney, J. Am. Chem. Soc. 62, 3032 (1940).

416. H. Batzer, G. Weisenberger, Makromol. Chem. 11, 83 (1953).

417. H. Batzer, B. Mohr, Makromol. Chem. 8, 217 (1952).

418. H. Batzer, Makromol. Chem. 10, 13 (1953).

419. H. Batzer, Makromol. Chem. 5, 5 (1950).

420. W. O. Baker, C. S. Fuller, J. H. Heiss, jr., J. Am. Chem. Soc. 63, 3316 (1941).

421. J. G. Erickson, J. Polymer Sci., A-1, 4, 519 (1966).

422. U. P. Strauss, P. L. Wineman, J. Am. Chem. Soc. 80, 2366 (1958).

423. M. Nakagaki, S. Ohashi, F. Minato, Bull. Chem. Soc. Japan 36, 341 (1963).

424. V. Crescenzi, P. J. Flory, J. Am. Chem. Soc. 86, 141 (1964).

425. G. V. Schulz, A. Haug, Z. Physik. Chem. (Frankfurt) 34, 328 (1962).

426. Von A. Haug, G. Meyerhoff, Makromol. Chem. 53, 91 (1962).

427. P. J. Flory, L. Mandelkern, J. B. Kinsinger, W. B. Schulz, J. Am. Chem. Soc. 74, 3364 (1952).

428. A. J. Barry, J. Appl. Phys. 17, 1020 (1946).

429. H. H. Takimoto, C. T. Forbes, R. K. Laudenslager, J. Appl. Polymer Sci. 5, 153 (1961).

430. A. Ya. Korolev, K. A. Andrianov, L. S. Utesheva, T. E. Vredenskaya, Dokl. Akad. Nauk. SSSR 89, 65 (1953).

431. K. A. Andrianov, B. G. Zavin, N. V. Pertsova, Vysokomolekul. Soedin., A, 10, 46 (1968).

432. F. P. Price, S. G. Martin, J. P. Bianchi, J. Polymer Sci. 22, 41 (1956).

433. C.-L. Lee, F. A. Emerson, J. Polymer Sci., A-2, 5, 829 (1967).

434. R. R. Buch, H. M. Klimisch, O. K. Johannson, J. Polymer Sci., A-2, 8, 541 (1970).

435. R. R. Buch, H. M. Klimisch, O. K. Johannson, J. Polymer Sci., A-2, 7, 563 (1969).

436. V. Ye. Eskin, A. Ye. Nesterov, Vysokomolekul. Soedin. 8, 141 (1966).

437. K. J. Ivin, H. A. Ende, G. Meyerhoff, Polymer 3, 129 (1962).

438. T. W. Bates, K. J. Ivin, Polymer 8, 263 (1967).

439. T. W. Bates, J. Biggins, K. J. Ivin, Makromol. Chem. 87, 180 (1965).

440. R. Endo, T. Manago, M. Takeda, Bull. Chem. Soc. Japan 39, 733 (1966).

441. W. Buchard, Makromol. Chem. 67, 182 (1963).

442. V. N. Tsvetkov, I. N. Shtennikova, Ye. I. Ryumtsev, L. N. Andreyeva, Yu. P. Getmanchuk, Yu. L. Spirin, R. I. Dryagileva, Vysokomolekul. Soedin., A, 10, 2132 (1968); see also V. N. Tsvetkov, European Polymer J. Suppl. 237 (1969).

443. J. J. Burke, T. A. Orofino, J. Polymer Sci., A-2, 7, 1 (1969).

444. H.-G. Elias, R. Schumacher, Makromol. Chem. 76, 23 (1964).

445. G. J. Howard, J. Polymer Sci. 37, 310 (1959).

446. G. B. Taylor, J. Am. Chem. Soc. 69, 635 (1947).

447. J. V. Koleske, R. D. Lundberg, J. Polymer Sci., A-2, 7, 897 (1967).

448. K. Hoshino, M. Watanabe, Nippon Kagaku Zasshi (J. Chem. Soc. Japan, Pure Chem. Sec.) 70, 24 (1949).

449. R. Okada, Y. Fukumura, H. Tanzawa, J. Polymer Sci., B, 4, 971 (1963).

450. A. Mattiussi, G. B. Gechele, R. Francesconi, J. Polymer Sci., A-2, 7, 411 (1969); see also G. B. Gechele, A. Mattiussi, European Polymer J. 1, 47 (1965); A. Mattiussi, G. B. Gechele, N. Pizzoli, ibid. 2, 283 (1966).

451. R. Bennewitz, Faserforsch. Textiltech. 5, 155 (1954).

452. J.-J. Chien, L.-H. Shih, K.-I. Shih, Acta Chem. Sinica 21, 50 (1955).

453. J. R. Schaefgen, P. J. Flory, J. Am. Chem. Soc. 70, 2709 (1948).

454. P. W. Morgan, S. L. Kwolek, J. Polymer Sci., A, 1, 1147 (1963).

455. A. Zilkha, Y. Burstein, Biopolymers 2, 147 (1964).

456. Y. Hayashi, A. Teramoto, K. Kawahara, H. Fujita, Biopolymers 8, 403 (1969).

457. P. Doty, J. H. Bradbury, A. M. Holtzer, J. Am. Chem. Soc. 78, 947 (1956).

458. E. Marchal, C. Dufour, J. Polymer Sci., C, 30, 77 (1970).

459. G. Spach, Compt. Rend. 249, 543 (1959).

460. K. Okita, A. Teramoto, H. Fujita, Biopolymers 9, 717 (1970).

461. N. Lupu-Lotan, A. Yaron, A. Berger, M. Sela, Biopolymers 3, 625 (1965).

462. A. Nakajima, T. Hayashi, Kobunshi Kagaku (Chem. High Polymer, Tokyo) 24, 230 (1967).

463. S. Tanaka, Thesis, Kyoto Univ., Kyoto, Japan, 1972.

464. A. Nakajima, S. Tanaka, K. Itoh, Polymer J. 3, 398 (1972).

465. J. Marchal, C. Lapp, J. Chim. Phys. 60, 756 (1963).

466. C. Lapp, J. Marchal, J. Chim. Phys. 62, 1032 (1965).

467. D. F. Detar, F. F. Rogers, jr., H. Bach, J. Am. Chem. Soc. 89, 3039 (1967).

468. J. H. Fessler, A. G. Ogston, Trans. Faraday Soc. 47, 667 (1951).

469. C. Tanford, K. Kawahara, S. Lapanje, J. Am. Chem. Soc. 89, 729 (1967).

470. G. Thomson, S. A. Rice, M. Nagasawa, J. Am. Chem. Soc. 85, 2537 (1963).

471. J. Feisst, H.-G. Elias, Makromol. Chem. 82, 78 (1964).

472. A. Mele, R. Rufo, L. Santonato, Polymers 10, 233 (1969).

473. P. J. Akers, G. Allen, M. J. Bethell, Polymer 9, 575 (1968).

474. J. M. Barrales-Rienda, D. C. Pepper, J. Polymer Sci., B, 4, 939 (1966).

475. H. Batzer, G. Benzing, Makromol. Chem. 62, 66 (1963).

476. G. C. Berry, H. Nomura, K. G. Mayhan, J. Polymer Sci., A-2, 5, 1 (1967).

477. A. D. Chirico, Chim. Ind. (Milan) 42, 248 (1960).

478. G. Sitaramaiah, J. Polymer Sci., A, 3, 2743 (1965).

479. G. V. Schulz, A. Horbach, Makromol. Chem. 29, 93 (1959).

480. H. Batzer, G. Fritz, Makromol. Chem. 14, 179 (1955).

481. D. A. S. Ravens, I. M. Ward, Trans. Faraday Soc. 57, 150 (1961).

482. W. R. Moore, D. Sanderson, Polymer 9, 153 (1968).

483. I. M. Ward, Nature 180, 141 (1957).

484. G. Meyerhoff, S. Shimotsuma, Makromol. Chem. 135, 195 (1970).

485. I. Marshall, A. Todd, Trans. Faraday Soc. 49, 67 (1953).

486. H. Zahn, C. Borstlap, G. Valk, Makromol. Chem. 64, 18 (1963).

487. B. Seidel, J. Polymer Sci. 55, 411 (1961).

488. M. L. Wallach, Makromol. Chem. 103, 19 (1967).

489. E. V. Kuznetsov, A. O. Wisel, I. M. Shermergorn, C. C. Tjulenjeu, Vysokomolekul. Soedin. 2, 205 (1960).

490. H. M. Koepp, H. Werner, Makromol. Chem. 32, 79 (1959).

491. A. Conix, Makromol. Chem. 26, 226 (1958).

492. W. Griehl, S. Neue, Faserforsch. Textiltech. 5, 423 (1954).

493. L. D. Moore, ACS Meeting, Cleveland, Polymer Preprints 1, 234 (1960).

494. N. G. Gaylord, S. Rosenbaum, J. Polymer Sci. 39, 545 (1959).

495. W. A. Lanke, reported in W. R. Krigbaum, J. Polymer Sci. 28, 213 (1958).

496. H. Batzer, H. Juergen, Makromol. Chem. 44-49, 179 (1961).

497. B. F. Malichenko, O. N. Tsypina, A. Ye. Nesterov, A. S. Shevlya-kov, Yu. S. Lipatov, Vysokomolekul. Soedin., A, 9, 2624 (1967).

498. B. F. Malichenko, A. V. Yazlovitskii, A. Ye. Nesterov, Vysokomolekul. Soedin., A, 12, 1700 (1970).

499. R. J. Merker, M. J. Scott, J. Polymer Sci., A, 2, 15 (1964).

500. J. F. Brown, jr., L. H. Vogt, jr., A. Katchman, J. W. Eustance, K. M. Kiser, K. W. Krantz, J. Am. Chem. Soc. 82, 6194 (1960).

501. V. N. Tsvetkov, K. A. Andrianov, G. I. Okhrimenko, I. N. Shtennikova, G. A. Fomin, M. G. Vitovskaya, V. I. Pakhomov, A. A. Yarosh, D. N. Andreyev, Vysokomolekul. Soedin., A, 12, 1892 (1970).

502. V. N. Tsvetkov, K. A. Andrianov, I. N. Shtennikova, G. I. Okhri-menko, L. N. Andreyeva, G. A. Fomin, V. I. Pakhomov, Vysoko-molekul. Soedin., A, 10, 547 (1968).

503. V. V. Rode, P. N. Gribkova, V. V. Korshak, Vysokomolekul. Soedin., A, 11, 57 (1969).

504. R. Endo, H. Iimura, M. Takeda, Kobunshi Kagaku (Chem. High Polymer, Tokyo) 29, 44 (1972).

505. R. Endo, M. Hattori, M. Takeda, Kobunshi Kagaku (Chem. High Polymer, Tokyo) 29, 48 (1972).

506. R. Endo, T. Hinokuma, M. Takeda, J. Polymer Sci., A-2, 6, 665 (1968).

507. I. Z. Steinberg, W. F. Harrington, A. Berger, M. Sela, E. Katchals-ki, J. Am. Chem. Soc. 82, 5263 (1960).

508. T. Nishihara, P. Doty, Proc. Natl. Acad. Sci. U. S. 44, 411 (1958).

509. J. Pouradier, A. M. Venet, J. Chim. Phys. 47, 391 (1950).

510. J. W. Williams, W. M. Saunders, J. S. Cicirelli, J. Phys. Chem. 58, 774 (1954).

511. M. Tamura, H. Odani, S. Imai, S. Nishida, Private communication.

512. V. N. Tsvetkov, G. A. Fomin, P. N. Lavrenko, I. N. Shtennikova, T. V. Sheremeteva, I. I. Godunova, Vysokomolekul. Soedin., A, 10, 903 (1968).

513. V. N. Tsvetkov, N. N. Kupriyanova, G. V. Tarasova, P. N. Lavren-ko, I. I. Migunova, Vysokomolekul. Soedin., A, 12, 1974 (1970).

514. L. V. Dubrovina, S. A. Pavlova, V. V. Korshak, Vysokomolekul. Soedin. 8, 752 (1966).

515. V. V. Korshak, S. A. Pavlova, L. V. Boiko, T. M. Babchinitser, S. V. Vinogradova, Ya. S. Vygodskii, N. A. Golubeva, Vysokomole-kul. Soedin., A, 12, 56 (1970).

516. J. R. Holsten, M. R. Lilyquist, J. Polymer Sci., A, 3, 3905 (1965).

517. W. Burchard, Makromol. Chem. 64, 110 (1963).

518. J. M. G. Cowie, Makromol. Chem. 42, 230 (1961).

519. W. Banks, C. T. Greenwood, Polymer 10, 257 (1969).

520. W. W. Everett, J. F. Foster, J. Am. Chem. Soc. 81, 3459, 3464 (1959).

521. W. Banks, C. T. Greenwood, Makromol. Chem. 67, 49 (1963).

522. W. Banks, C. T. Greenwood, European Polymer J. 5, 649 (1969).

523. J. M. G. Cowie, Makromol. Chem. 53, 13 (1962).

524. J. M. G. Cowie, J. Polymer Sci. 49, 455 (1961).

525. R. S. Patel, R. D. Patel, Makromol. Chem. 90, 262 (1966).

526. B. A. Dombrow, C. O. Beckmann, J. Phys. Colloid Chem. 51, 107 (1947).

527. W. Banks, C. T. Greenwood, D. J. Hourston, Trans. Faraday Soc. 64, 363 (1968); see also Makromol. Chem. 111, 226 (1968).

528. W. Burchard, E. Husemann, Makromol. Chem. 44-46, 358 (1961).

529. H. Husemann, R. Resz, R. Werner, Makromol. Chem. 47, 48 (1961).

530. D. A. Brant, B. K. Min, Macromolecules 2, 1 (1969).

531. K. D. Goebel, D. A. Brant, Macromolecules 3, 634 (1970).

532. F. G. Donnan, R. C. Rose, Can. J. Research B28, 105 (1950).

533. D. Henley, Arkiv Kemi 18, 327 (1961).

534. W. Brown, R. Wikstroem, European Polymer J. 1, 1 (1965).

535. W. G. Harland, in "Recent Advances in Chemistry of Cellulose and Starch" edited by Honeyman, Interscience Pub. Inc., New York, 1959, pp. 265-284.

536. N. Gralen, T. Svedberg, Nature 152, 625 (1943).

537. E. H. Immergut, B. G. Ranby, H. Mark, Ind. Eng. Chem. 45, 2383 (1953).

538. J. W. Tamblyn, D. R. Morey, R. H. Wagner, Ind. Eng. Chem. 37, 573 (1945).

539. F. H. Holmes, D. J. Smith, Trans. Faraday Soc. 53, 67 (1957).

540. W. J. B. Badgley, H. Mark, J. Phys. Colloid Chem. 51, 58 (1947).

541. H. J. Philipp, C. F. Bjork, J. Polymer Sci. 6, 549 (1951).

542. P. J. Flory, O. K. Spurr, jr., D. K. Carpenter, J. Polymer Sci. 27, 231 (1958).

543. L. Huppenthal, S. Claesson, Roczniki Chem. 39, 1867 (1965).

544. L. Mandelkern, P. J. Flory, J. Am. Chem. Soc. 74, 2517 (1952).

545. V. P. Shanbhag, Arkiv Kemi 29, 1 (1968).

546. V. P. Shanbhag, J. Oehman, Arkiv Kemi 29, 163 (1968).

547. W. R. Krigbaum, L. H. Sperling, J. Phys. Chem. 64, 99 (1960).

548. G. Meyerhoff, J. Polymer Sci. 29, 399 (1958).

549. A. M. Holtzer, H. Benoit, P. Doty, J. Phys. Chem. 58, 624 (1954).

550. M. M. Huque, D. A. I. Goring, S. G. Mason, Can. J. Chem. 36, 952 (1958).

551. R. M. Badger, R. H. Blaker, J. Phys. Colloid Chem. 53, 1056 (1949).

552. E. Penzel, G. V. Schulz, Makromol. Chem. 113, 64 (1968).

553. W. R. Moore, J. A. Epstein, J. Appl. Chem. (London) 6, 168 (1956).

554. W. R. Moore, G. D. Edge, J. Polymer Sci. 47, 469 (1960).

555. M. L. Hunt, S. Newman, H. A. Scheraga, P. J. Flory, J. Phys. Chem. 60, 1278 (1956).

556. W. G. Harland, Nature 170, 667 (1952).

557. G. Meyerhoff, N. Suetterlin, Makromol. Chem. 87, 258 (1965).

558. W. R. Moore, A. M. Brown, J. Colloid Sci. 14, 1, 343 (1959).

559. V. N. Tsvetkov, S. Ya. Kotlyar, Zh. Fiz. Khim. 30, 1100 (1956).

560. R. St. J. Manley, Arkiv Kemi 9, 519 (1956).

561. W. Brown, D. Henley, J. Oehman, Makromol. Chem. 64, 49 (1963); see also W. Brown, Arkiv Kemi 18, 227 (1961).

562. H. Vink, Makromol. Chem. 94, 1 (1966).

563. W. B. Neely, J. Polymer Sci., A, $\underline{1}$, 311 (1963).

564. W. Brown, D. Henley, J. Oehman, Makromol. Chem. $\underline{62}$, 164 (1963).

565. G. Sitaramaiah, D. A. I. Goring, J. Polymer Sci. $\underline{58}$, 1107 (1962).

566. W. Brown, D. Henley, Makromol. Chem. $\underline{79}$, 68 (1964).

567. H. Elmgren, Arkiv Kemi $\underline{24}$, 237 (1965).

568. K. Gekko, Makromol. Chem. $\underline{148}$, 229 (1971).

569. F. R. Senti, N. N. Hellman, N. H. Ludwig, G. E. Babcock, R. Tobin, C. A. Class, B. L. Lamberts, J. Polymer Sci. $\underline{17}$, 527 (1955); see also Ref. 8.

570. K. Gekko, H. Noguchi, Biopolymers $\underline{10}$, 1513 (1971).

571. L. H. Arond, H. P. Frank, J. Phys. Chem. $\underline{58}$, 953 (1954).

572. J. V. Koleska, S. F. Kurath, J. Polymer Sci., A, $\underline{2}$, 4123 (1964).

573. R. L. Cleland, J. L. Wang, Biopolymers $\underline{9}$, 799 (1970).

574. D. H. Juers, H. A. Swenson, S. F. Kurath, J. Polymer Sci., A-2, $\underline{5}$, 361 (1967).

575. P. Doty, B. B. McGill, S. A. Rice, Proc. Natl. Acad. Sci. U. S. $\underline{44}$, 432 (1958); see also Ref. 577.

576. J. T. Lett, K. A. Stacey, Makromol. Chem. $\underline{38}$, 204 (1960).

577. J. Eigner, P. Doty, J. Mol. Biol. $\underline{12}$, 549 (1965).

578. D. M. Crothers, B. H. Zimm, J. Mol. Biol. $\underline{12}$, 525 (1965). This $[\eta]$ -M_w relationship is based on their own data on phage T2 and T7 DNA, and on the earlier data on calf thymus, herring sperm and T2 DNA; see Refs. 575 and 579.

579. Y. Kawade, I. Watanabe, Biochim. Biophys. Acta $\underline{19}$, 513 (1956); K. Iso, I. Watanabe, Nippon Kagaku Zasshi (J. Chem. Soc. Japan, Pure Chem. Sec.) $\underline{78}$, 1268 (1957); and E. Burgi, A. D. Hershey, J. Mol. Biol. $\underline{3}$, 458 (1961).

580. P. D. Ross, R. L. Scruggs, Biopolymers $\underline{6}$, 1005 (1968).

581. K. E. Reinert, J. Strassburger, H. Triebel, Biopolymers $\underline{10}$, 285 (1971).

582. R. J. Douthart, V. A. Bloomfield, Biopolymers $\underline{6}$, 1297 (1968).

583. J. R. Fresco, P. Doty, J. Am. Chem. Soc. $\underline{79}$, 3928 (1957).

584. R. Haselkorn, unpublished work; cited in J. Eigner, Thesis, Harvard Univ., Mass., USA, 1960.

585. E. G. Richards, C. P. Flessel, J. R. Fresco, Biopolymers $\underline{1}$, 431 (1963).

586. B. D. Hall, P. Doty, "Microsomal Particles and Protein Synthesis", Washington Acad. Sci. p. 27 (1958).

587. Y. Kawade, Ann. Rept. Inst. Virus Research, Kyoto Univ. B2, 219 (1959); see also Ref. 589.

588. H. Boedtker, J. Mol. Biol. $\underline{2}$, 171 (1960).

589. C. G. Kurland, J. Mol. Biol. $\underline{2}$, 83 (1960).

590. N. M. Tret'yakova, L. V. Kosmodem'yanoskii, R. G. Romanova, E. G. Lazaryants, Vysokomolekul. Soedin., A, $\underline{12}$, 2754 (1970).

591. Houtz, cited in Y. Iwakura, T. Kurosaki, N. Nakabayashi, Makromol. Chem. $\underline{44\text{-}46}$, 570 (1961).

592. W. Scholtan, H. Marzolph, Makromol. Chem. $\underline{57}$, 52 (1962).

593. Y. Shimura, I. Mita, H. Kambe, J. Polymer Sci., B, $\underline{2}$, 403 (1964).

594. H. Gerrens, H. Ohlinger, R. Fricker, Makromol. Chem. $\underline{87}$, 209 (1965).

595. Y. Shimura, J. Polymer Sci., A-2, $\underline{4}$, 423 (1966).

596. Ye. G. Erenburg, G. G. Kartasheva, M. A. Yeremina, I. Ya. Poddubnyi, Vysokomolekul. Soedin., A, $\underline{9}$, 2709 (1967).

597. Ye. A. Bekturov, L. A. Bimendina, S. V. Bereza, Vysokomolekul. Soedin., A, $\underline{12}$, 2179 (1970).

598. R. B. Mohite, S. Gundiah, S. I. Kapur, Makromol. Chem. $\underline{108}$, 52 (1967).

599. H. Hopff, D. Starck, Makromol. Chem. $\underline{48}$, 50 (1961).

600. S. R. Refikov, S. A. Pavlova, Sh. Shayakhmotov, Vysokomolekul. Soedin., A, $\underline{11}$, 1990 (1969).

601. T. E. Helminiak, C. L. Benner, W. E. Gibbs, Polymer Preprints $\underline{8}$, 284 (1967); see also C.A. $\underline{1967}$, 116045k; cited by W. H. Stockmayer, IUPAC Symposium Leiden, Netherlands, 1970.

602. I. Ya. Poddubnyi, V. A. Grechanovskii, M. I. Mosevitskii, Vysokomolekul. Soedin. $\underline{5}$, 1042 (1964).

603. V. N. Tsvetkov, K. A. Andrianov, G. I. Okhrimenko, M. G. Vitovskaya, European Polymer J. $\underline{7}$, 1215 (1971).

604. H. Tanzawa, T. Tanaka, A. Soda, J. Polymer Sci., A-2, $\underline{7}$, 929 (1969).

605. S. V. Bereza, Ye. A. Bekturov, S. R. Rafikov, Vysokomolekul. Soedin., A, $\underline{11}$, 1681 (1969).

606. S. V. Bereza, S. R. Rafikov, Ye. A. Bekturov, R. Ye. Legkunets, Vysokomolekul. Soedin., A, $\underline{10}$, 2536 (1968).

607. V. A. Myagchenkov, E. V. Kuznetsov, O. A. Ishhakov, V. M. Lichkina, Vysokomolekul. Soedin. $\underline{5}$, 724 (1963).

608. J. Feyen, K. deGroot, A. C. deVisser, A. Bantjes, Biopolymers $\underline{10}$, 2509 (1971).

609. J. R. Patel, C. K. Patel, R. D. Patel, Stärke $\underline{19}$, 330 (1967); cited in J. R. Patel R. D. Patel, Biopolymers $\underline{10}$, 839 (1971).

610. A. B. Duchkova, N. S. Nametkin, V. S. Khotimskii, L. Ye. Gusel'-nikov, S. G. Durgar'yan, Vysokomolekul. Soedin. $\underline{8}$, 1814 (1966).

611. J. D. Matlack, S. N. Chinai, R. A. Guzzi, D. W. Levi, J. Polymer Sci. $\underline{49}$, 533 (1961).

612. S. M. Kochergin, V. P. Barabanov, I. G. Fedorova, Vysokomolekul. Soedin. $\underline{8}$, 916 (1966).

613. H. Utiyama, K. Kajitani, M. Kurata, unpublished work. Cf. Rept. Progr. Polymer Phys. Japan $\underline{6}$, 29 (1963).

614. T. Kotaka, Y. Murakami, H. Inagaki, J. Phys. Chem. $\underline{72}$, 829 (1968); see also T. Kotaka, H. Ohnuma, Y. Murakami, ibid. $\underline{70}$, 4099 (1966).

615. W. H. Stockmayer, L. D. Moore, jr., M. Fixman, B. N. Epstein, J. Polymer Sci. $\underline{16}$, 517 (1955).

616. T. Kotaka, H. Ohmuma, H. Inagaki, Polymers $\underline{10}$, 517 (1969); see also H. Ohnuma, T. Kotaka, H. Inagaki, ibid. $\underline{10}$, 501 (1969).

617. S. Polowinski, J. Polymer Sci., A-2, $\underline{5}$, 891 (1967).

618. V. N. Tsvetkov, S. Ya. Magarik, T. Kadyrov, G. A. Andreyeva, Vysokomolekul. Soedin., A, $\underline{10}$, 943 (1968).

619. V. N. Tsvetkov, Makromol. Chem. $\underline{160}$, 1 (1972).

620. H. Benoit, J. Polymer Sci. $\underline{3}$, 376 (1948).

621. H. Markovitz, J. Chem. Phys. $\underline{20}$, 868 (1952).

622. J. E. Mark, J. Am. Chem. Soc. $\underline{88}$, 4354 (1966).

623. J. E. Mark, J. Am. Chem. Soc. $\underline{89}$, 6829 (1967).

624. T. Norisuye, K. Kawahara, A. Teramoto, H. Fujita, J. Chem. Phys. $\underline{49}$, 4330 (1968).

625. O. Kratky, H. Sand, Kolloid-Z. $\underline{172}$, 18 (1960).

626. J. E. Mark, Ref. 622; based on the stress-temperature data of L. A. Wood and F. L. Roth, J. Appl. Phys. $\underline{15}$, 781 (1944).

627. J. E. Mark, Ref. 622; based on the stress-temperature data of A. Ciferri, Makromol. Chem. $\underline{43}$, 152 (1961).

628. E. D. Kunst, Rec. Trav. Chim. $\underline{69}$, 125 (1950).

629. A. Ciferri, C. A. J. Hoeve, P. J. Flory, J. Am. Chem. Soc. $\underline{83}$, 1015 (1961); A. Ciferri, J. Polymer Sci. $\underline{54}$, 149 (1961).

630. A. Kotera, H. Matsuda, A. Wada, Rept. Progr. Polymer Phys. Japan $\underline{8}$, 5 (1965).

631. A. Nakajima, F. Hamada, S. Hayashi, J. Polymer Sci., C, $\underline{15}$, 285 (1966).

632. P. J. Flory, A. Ciferri, R. Chiang, J. Am. Chem. Soc. $\underline{83}$, 1023 (1961).

633. J. E. Mark, P. J. Flory, J. Am. Chem. Soc. $\underline{87}$, 1423 (1965).

634. Unpublished data from the Chemstrand Research Center; cited in A. Ciferri, J. Polymer Sci., A, $\underline{2}$, 3089 (1964).

635. A. V. Tobolsky, D. Carlson, N. Indictor, J. Polymer Sci. $\underline{54}$, 175 (1961).

636. R. Kirste, G. V. Schulz, Z. Physik. Chem. (Frankfurt) $\underline{30}$, 171 (1961).

637. R. Kirste, O. Kratky, Z. Phys. Chem. (Frankfurt) $\underline{31}$, 363 (1962).

638. A. Nakajima, H. Yamakawa, J. Phys. Chem. $\underline{67}$, 654 (1963).

639. H. Abe, W. Prins, J. Polymer Sci., C, $\underline{2}$, 527 (1963).

640. O. Kratky, G. Porod, Rec. Trav. Chim. $\underline{68}$, 1106 (1949).

641. E. Wada, K. Okano, Rept. Progr. Polymer Phys. Japan $\underline{6}$, 1 (1963).

642. E. Eisenberg, E. F. Casassa, J. Polymer Sci. $\underline{47}$, 29 (1960).

643. N. T. Notley, P. J. W. Debye, J. Polymer Sci. $\underline{17}$, 99 (1955).

644. W. R. Krigbaum, D. K. Carpenter, J. Phys. Chem. $\underline{59}$, 1166 (1955).

645. E. Wada, K. Okano, Rept. Progr. Polymer Phys. Japan $\underline{7}$, 19 (1964).

646. T. A. Orofino, J. W. Mickey, jr., J. Chem. Phys. 38, 2513 (1963).

647. G. V. Schulz, H. Baumann, Makromol. Chem. 60, 120 (1963).

648. T. A. Orofino, A. Ciferri, J. Phys. Chem. 68, 3136 (1964).

649. M. Fukuda, M. Fukutomi, Y. Kato, T. Hashimoto, private communication.

650. H. Utiyama, J. Phys. Chem. 69, 4138 (1965).

651. N. Kuwahara, K. Ogino, A. Kasai, S. Ueno, M. Kaneko, J. Polymer Sci., A, 3, 985 (1965).

652. A. Dondos, H. Benoit, Makromol. Chem. 135, 181 (1970).

653. V. Ye. Eskin, O. Z. Korotkina, Vysokomolekul. Soedin., A, 12, 2216 (1970).

654. G. Allen, C. Booth, C. Price, Polymer 8, 397, 414 (1967).

655. J. E. Mark, P. J. Flory, J. Am. Chem. Soc. 87, 1415 (1965).

656. C. Rossi, A. Perico, J. Chem. Phys. 53, 1223 (1970).

657. W. Silberszyc, J. Polymer Sci., B, 1, 577 (1963).

658. J. M. Evans, M. B. Huglin, Makromol. Chem. 127, 141 (1969).

659. K. Bak, E. Elefante, J. E. Mark, J. Phys. Chem. 71, 4007 (1967).

660. R. H. Marchessault, K. Okamura, C. J. Su, Macromolecules 3, 735 (1970).

661. J. Cornibert, R. H. Marchessault, H. Benoit, G. Weill, Macromolecules 3, 741 (1970).

662. U. P. Strauss, P. Ander, J. Phys. Chem. 66, 2235 (1962).

663. J. K. Peterson, Thesis, Ohio State Univ., Ohio, U. S. A., 1961.

664. Ye. G. Erenburg, G. G. Kartasheva, M. A. Yeremina, I. Ya. Poddubnyi, Vysokomolekul. Soedin., A, 9, 2709 (1967).

665. J. E. Mark, P. J. Flory, J. Am. Chem. Soc. 86, 138 (1964).

666. K. J. Ivin, H. A. Ende, J. Polymer Sci. 54, S17 (1961).

667. R. Endo, T. Hinokuma, M. Takeda, J. Polymer Sci., A-2, 6, 665 (1968).

668. P. R. Saunders, J. Polymer Sci., A, 2, 3765 (1964).

669. P. J. Flory, A. D. Williams, J. Polymer Sci., A-2, 5, 399 (1967).

670. D. A. Brant, P. J. Flory, J. Am. Chem. Soc. 87, 2788 (1965).

671. A. Yu. Koshevnik, M. M. Kusakov, E. A. Razumovskaya, Vysokomolekul. Soedin., A, 10, 2795 (1968).

672. A. E. Tonelli, P. J. Flory, Macromolecules 2, 225 (1969).

673. A. R. Shultz, J. Polymer Sci., A-2, 8, 883 (1970).

674. S. A. Pavlova, V. V. Korshak, L. V. Dubrovina, G. I. Timofeyeva, Vysokomolekul. Soedin., A, 9, 2624 (1967).

675. W. H. Stockmayer, IUPAC Symposium, Leiden, Netherlands, 1970; This estimate of $r_0/M^{1/2}$ was based on the viscosity data given in Ref. 601.

676. J. M. G. Cowie, P. M. Toporowski, Polymer 5, 601 (1964).

677. R. S. Patel, R. D. Patel, J. Polymer Sci., A, 3, 2123 (1965); A-2, 4, 835 (1966).

678. W. Burchard, Makromol. Chem. 88, 11 (1965).

679. S. Heine, O. Kratky, G. Porod, P. Schmitz, Makromol. Chem. 44-46, 682 (1961).

680. A. M. Rijke, J. Polymer Sci., B, 4, 131 (1966).

681. B. Das, A. K. Ray, P. K. Choudhury, J. Phys. Chem. 73, 3413 (1969); see also B. Das, P. K. Choudhury, J. Polymer Sci., A-1, 5, 769 (1967).

682. Ye. A. Bekturov, L. A. Bimendina, S. V. Bereza, Vysokomolekul. Soedin., A, 12, 2179 (1970).

683. J. Velickovic, D. Jovanovic, J. Vukajlovic, Makromol. Chem. 129, 203 (1969).

684. V. Ye. Eskin, O. Z. Korotkina, Vysokomolekul. Soedin., A, 12, 2216 (1970).

685. G. A. Morneau, P. I. Roth, A. R. Shultz, J. Polymer Sci. 55, 609 (1961).

686. G. V. Schulz, E. Penzel, Makromol. Chem. 112, 260 (1968).

687. O. Kratky, H. Leopold, G. Puchwein, Kolloid-Z.-Z. Polymere 216-217, 255 (1967).

688. R. L. Cleland, Biopolymers 10, 1925 (1971).

689. M. Nakata, Makromol. Chem. 149, 99 (1971).

690. T. W. Bates, W. H. Stockmayer, J. Chem. Phys. 45, 2321 (1966); Macromolecules 1, 12 (1968).

691. W. Brown, D. Henley, Makromol. Chem. 75, 179 (1964).

692. R. St. John Manley, Svensk Papperstidn. 61, 96 (1958).

693. V. P. Shanbhag, Arkiv Kemi 29, 139 (1968).

694. S. Teramachi, A. Takahashi, I. Kagawa, Kogyo Kagaku Zasshi (J. Chem. Soc. Japan, Ind. Chem. Sec.) 69, 685 (1966).

695. G. Allen, C. Booth, C. Price, Polymer 8, 414 (1967).

696. P. J. Flory, Makromol. Chem. 98, 128 (1966).

697. Recommendation of IUPAC Macromolecular Division working party "molecular characterization of commercial polymers", 12.3.1971 Strassbourg.

Note: References 508-511 and 575-589 are for biological polymers such as collagen, gelatin and poly(nucleotides), which do not appear in the present tables.

SEDIMENTATION COEFFICIENTS, DIFFUSION COEFFICIENTS, PARTIAL SPECIFIC VOLUMES, FRICTIONAL RATIOS

and

SECOND VIRIAL COEFFICIENTS OF POLYMERS IN SOLUTION

P. E. O. Klärner[+]
Institut für Physikalische Chemie
Universität Mainz, Germany

and

H. A. Ende
BASF, Ludwigshafen, Germany

Contents

A. SEDIMENTATION COEFFICIENTS, DIFFUSION COEFFICIENTS, PARTIAL SPECIFIC VOLUMES AND FRICTIONAL RATIOS OF POLYMERS IN SOLUTION
(Detailed bibliography see Ref. 2, 349-357)

1. Introduction

1.1 Sedimentation Coefficients

According to Svedberg (2) the sedimentation coefficient s is defined as the sedimentation velocity in a unit field (1 dyn; 10^{-5} N)

$$s = (dr/dt)(1/\omega^2 r) \qquad (1)$$

For a given molecule in a given solvent the sedimentation coefficient is dependent on temperature, polymer concentration and pressure.

(a) Temperature Dependence

The Svedberg Equation (2) (to be discussed in A.1.3), requires that the sedimentation coefficient s, the diffusion coefficient D, the partial specific volume $\bar{v}_2$, and the density ρ, are given at equal temperatures. The temperature commonly used is $20\,^{\circ}C$. The sedimentation coefficient, s_{20}, can either be measured at $20\,^{\circ}C$, or be calculated from s_T by equation (2)

$$s_{20} = s_T \frac{\eta_T}{\eta_{20}} \frac{1 - (\bar{v}_2)_{20}\, \varrho_{20}}{1 - (\bar{v}_2)_T\, \varrho_T} \qquad (2)$$

where η_T and η_{20} are the viscosities of the solvent at $T\,^{\circ}C$ and $20\,^{\circ}C$, respectively. Especially with aqueous solvents (buffers, salt solutions) it is often desirable to reduce s_{20} to a reference solvent (e.g. water). Equation (2) becomes then (2)

+) Present address: BASF, Kunststoff-Laboratorium, Ludwigshafen, Germany.

$$s_{20}^0 = s_T \frac{\eta_T}{\eta_{20}^0} \frac{1 - (\bar{v}_2)_{20}\, \varrho_{20}^0}{1 - (\bar{v}_2)_T\, \varrho_T} \tag{3}$$

with η_{20}^0 = viscosity of the reference solvent at 20°C.

(b) Concentration Dependence

For the majority of systems studied by ultracentrifugation it is found that $1/s$ or s is a linear function of concentration:

and

$$\frac{1}{s} = \frac{1}{s_0} (1 + k_s c) \tag{4a}$$

$$s = s_0 (1 - k_s' c) \tag{4b}$$

where s_o is the sedimentation constant at $c = 0$.

The concentration dependence according to Equation (4a) is found more frequently than that according to Equation (4b), which especially holds for protein solutions. The dimensions of k_s and k_s' depend on those of c; usually, cm^3/g or 100 cm^3/g are used. In Table A.2. a concentration dependence according to Equation (4a) is denoted by c.a. and that according to Equation (4b) by c.b. In rare cases special extrapolation procedures are used. In these cases Table A.2. refers to the appropriate reference. For special treatments see e.g. Gehatia (358,359).

The sector-shape of the ultracentrifugation cell causes a uniform decrease of the concentration during sedimentation (360). In many cases this decrease in concentration has to be taken into account. At time t

$$c_t = c_0 (r_0/r_t)^2 \tag{5}$$

where r_0 and r_t are the distances of the meniscus and of the boundary from the axis of rotation, c_o and c_t the concentrations at zero time and at time t, respectively.

(c) Pressure Dependence

The high pressures encountered within the cell due to the high centrifugal fields applied in ultracentrifugation change appreciably the viscosity and the density of the solvent; likewise, the partial specific volume is influenced by high pressures. Thus, the sedimentation coefficient, s_p, measured at a pressure p, differs from the sedimentation coefficient, s_1, measured at 1 atm. The difference between s_p and s_1 is especially large in organic solvents. Mosimann and Signer (361) derived an equation analogous to Equation (3) for calculating s_1 from s_p:

$$s_1 = s_p \frac{\eta_p}{\eta_1} \frac{(1 - (\bar{v}_2)_1\, \varrho_1)}{(1 - (\bar{v}_2)_p\, \varrho_p)} \tag{6}$$

The calculation requires knowledge of the viscosity η_p, density of the solvent ρ_p, and partial specific volume $(\bar{v}_2)_p$, each at pressure p. Values for η_p are given by Bridgman (362), and may be calculated from the compressibilities found in a number of handbooks. Recently, values for $(\bar{v}_2)_p$ have become available through the work of Andersson (148). More precise equations for the calculation of the pressure influence on the sedimentation constant have been worked out by Oth and Desreux (363), Wales (364), and Fujita (365). An elegant and simple extrapolation procedure, which does not require the knowledge of any of the pressure dependent quantities, was suggested by Elias (366).

(d) Averages of Sedimentation Coefficients

Most ultracentrifuges are equipped with schlieren optics which measure the refractive index gradient dn/dr versus r. The increment dn/dr is proportional to the concentration gradient dc/dr. Monodisperse polymers yield a symmetrical curve with a maximum. The gradient curve of a polydisperse polymer is a summation of an infinite number of single curves from the single fractions contained in the polymer. Each of the fractions has its own sedimentation coefficient so that a distribution of sedimentation coefficients results. In the majority of cases a continuous curve with one maximum is obtained. It can be shown (14) that with certain assumptions the gradient curve represents to a fair approximation the distribution function of s, f(s) = dc/ds. Since

$$\frac{dn}{dr} = const \frac{dc}{ds} \tag{7}$$

it should in principle be possible to evaluate the various averages of sedimentation coefficients, e.g. $<s>_o = s_n$, $<s>_1 = s_w$, and $<s>_2 = s_z$, from the distribution of the concentration gradient within the cell. The common practice, however, is to determine simply the migration of the maximum which leads to a sedimentation coefficient s_t, not readily defined with respect to its average. A less common method of evaluating sedimentation coefficients is by observing the movement of the median, that is the line dividing the gradient curve in two equal areas, yielding, likewise, a rather undefined average s_m. According to Jullander (259) s_z and s_m can be related to s_w by

$$s_w = \frac{3 s_m - s_t}{2} \tag{8}$$

provided the skewness of the curve is not very pronounced. In Table A.2., we have occasionally listed instead of s_o, the intrinsic sedimentation coefficient $[s_o] = s_o\, \eta/(1 - v_2\rho)$.

(e) Sedimentation Coefficient - Molecular Weight Relationship

In analogy to the Mark-Houwink Equation (367,368) the dependence of the sedimentation coefficient on molecular weight can often be expressed as follows:

$$s = k' \cdot M^{a'} \tag{9}$$

where k' and a' are constants for a given polymer homologous series in a given solvent. Whenever quoted in the literature, relationship (9) is listed in the table rather than single s-values.

1.2   Diffusion Coefficients

The diffusion coefficient D is defined by Fick's first law which states that in the process of translational diffusion the amount of material (in solution) crossing a plane of unit area per unit time is proportional to the concentration gradient $\partial c/\partial x$ through the diffusion coefficient D. Thus, the rate of flow F is

$$F = -D \, (\partial c/\partial x) \tag{10}$$

The negative sign in Equation (10) indicates that diffusion takes place in the direction of decreasing concentration. For practical reasons Fick's second law is preferred

$$\partial c/\partial t = D \, (\partial^2 c/\partial x^2) \tag{11}$$

which, when solved with consideration of certain initial and boundary conditions, allows the determination of D measuring dc/dx as a function of the time t and the distance x.

The diffusion coefficient is dependent on temperature and polymer concentration. Constancy of pressure is generally assumed during measurements.

(a)   Temperature Dependence
The diffusion coefficient D_{20} for $20\,^{\circ}C$ can be calculated (2) from that measured at $T\,^{\circ}C$ from Equation (12)

$$D_{20} = D_T \frac{293 \, \eta_T}{(273 + T)\eta_{20}} \tag{12}$$

The diffusion coefficient D_{20}^{o}, reduced to a reference solvent (usually water), is calculated from

$$D_{20}^0 = D_T \frac{293 \, \eta_T}{(273 + T)\eta_{20}^0} \tag{13}$$

(b)   Concentration Dependence
In general, it is possible to describe the concentration dependence by a linear equation:

$$D = D_0 \, (1 + k_D c) \tag{14}$$

The direct measurement of the concentration dependence according to Equation (14) is the method which is used most frequently nowadays. There are, however, methods which determine D_0 and k_D from a measurement at only one concentration. Another method is that of Gralén (14), and that of Beckmann and Rosenberg (13). Whenever one of the latter two methods was used the remarks "Gralén" or "B. + R." are inserted in Table A.2.

(c)   Averages of Diffusion Coefficients
Similar considerations as made under 1.1.d. on sedimentation coefficients hold for diffusion coefficients also. Moments over the concentration gradient curve obtained by schlieren optics yield certain averages of diffusion coefficients. The schlieren curve has usually one maximum. The abscissa is conveniently chosen such that the zero point is at the maximum of the concentration gradient $dn/dx \sim dc/dx = g(x)$. The moments over the concentration gradient Λ_s are then defined as

$$\Lambda_s = \int\limits_{-\infty}^{\infty} |x|^s \, g \, (x) \, \mathrm{d}x \tag{15}$$

with s = 0, 1, 2, and 4. The following diffusion coefficients D can then be obtained:

$$D_{0,h} = \frac{1}{t} \, \frac{\Lambda_0^2}{4 \, \pi h_{max}^2} = D_A \tag{16}$$

$$h_{max} = [g \, (x)]_{max}$$

$$D_{2,0} = \frac{1}{t} \, \frac{\Lambda_2}{2 \, \Lambda_0} = D_w \tag{17}$$

$$D_{4,2} = \frac{1}{t} \, \frac{\Lambda_4}{6 \, \Lambda_2} = D_z \tag{18}$$

$$D_\sigma = \frac{\sigma^2}{2 \, t} \, \frac{\Lambda_2 \Lambda_0 - \Lambda_1^2}{2 \, t \Lambda_0^2} \tag{19}$$

A somewhat unusual diffusion coefficient is that defined by Equation (20)

$$D_h = \frac{x_1^2 - x_2^2}{4 \, t \ln \dfrac{h_2}{h_1}} = \frac{x_1^2}{4 \, t \ln \dfrac{h_{max}}{h_1}} \tag{20}$$

determined from two points $h_1 = [g(x)]_1$ and $h_2 = [g(x)]_2$ at x_1 and x_2, h_{max} at the latter value being equal to zero.

The averages of the diffusion coefficients are defined as follows:

$$\langle D \rangle_a = \frac{\displaystyle\int\limits_0^{\infty} f \, (D) \, D^a \, \mathrm{d}D}{\displaystyle\int\limits_0^{\infty} f \, (D) \, D^{a-1} \, \mathrm{d}D} \tag{21}$$

Thus, for a = 0, 1, and 2 one obtains D_n, D_w, and D_z, respectively. The diffusion coefficient $D_A = D_{0,h}$ is larger than D_n. It is defined by Equation (22):

$$D_A = \left[\frac{\int_0^\infty f(D)\,dD}{\int_0^\infty f(D)\,D^{-1/2}\,dD} \right]^2 \tag{22}$$

The diffusion coefficients most frequently found in the literature are D_A and D_w.

(d) Diffusion Coefficient - Molecular Weight Relationship

The diffusion coefficient molecular weight dependence frequently takes the form

$$D = k'' M^{-a''} \tag{23}$$

with k'' and a'' being constants for a given polymer homologous series in a given solvent. In Table A.2. the relationship (23) is listed whenever quoted in the literature, in preference to single D-values.

1.3 Molecular Weight Averages Determined from Sedimentation and Diffusion Coefficients

For calculation of the molecular weight from sedimentation and diffusion data the Svedberg Equation (2) is used:

$$M = \frac{RTs}{(1 - \bar{v}_2 \varrho)\,D} \qquad (T = K) \tag{24}$$

Here s and D are the corrected and standardized coefficients for zero polymer concentration, $\bar{v}_2$ is the partial specific volume of the polymer, ρ is the density of the solution, R and T are the gas constant and absolute temperature, respectively. For polydisperse polymers the various averages of the sedimentation and diffusion coefficient $<s> = s_i$ and $<D> = D_i$, are inserted; certain molecular weight averages $<M> = M_{i,j}$ are then obtained. Thus, Equation (24) acquires the more general form:

$$M_{i,j} = \frac{RTs_i}{D_j(1 - \bar{v}_2 \varrho)} \tag{25}$$

For i = n, w, z, and j = n, w, z, Equation (25) defines nine different molecular weight averages. $M_{n,n}$, $M_{w,n}$, $M_{w,w}$, etc. (The averages $M_{n,n}$ and $M_{w,w}$ are different from M_n and M_w, respectively). As described in the preceding sections, the coefficients with i = n and z, and j = n and z are determinable either not at all, or only with large error. Instead, the coefficients with i = t and m, and j = A, w, σ (and h) are usually evaluated. Thus, rather peculiar averages, e. g. $M_{t,A}$, $M_{m,\sigma}$, and $M_{w,h}$, result. The more straightforward molecular weight averages, $M_{n,w}$, $M_{w,w}$, $M_{z,w}$, etc., in relation to the familiar averages $\bar{M}_n$, $\bar{M}_w$, and $\bar{M}_z$, are found elsewhere in the literature (369-371).

1.4 Partial Specific Volumes

The volume, V_{12}^{id}, of an ideal two component system can be expressed in terms of the masses m_1 and m_2, the specific volumes v_1 and v_2 of the two components by the equation

$$V_{12}^{id} = m_1 v_1 + m_2 v_2 \tag{26}$$

Most components do not behave ideally upon mixing; i.e., they react with each other in a way so that the total volume deviates from V_{12}^{id} owing to concentration effects or to expansion of component 1 and/or 2. The total volume can then be written:

$$V_{12} = m_1 \bar{v}_1 + m_2 \bar{v}_2 \tag{27}$$

where $\bar{v}_1$ and $\bar{v}_2$ are the partial specific volumes of components 1 and 2, respectively. When component 2 is polymer and component 1 solvent, then, for practical reasons, it is convenient to introduce the so-called apparent partial specific volume $\bar{v}_2^\star$ which is defined by

$$V_{12} = m_1 \bar{v}_1^0 + m_2 \bar{v}_2^\star \tag{28}$$

where $\bar{v}_1^0 = v_1$, i.e. the specific volume of the solvent. The quantity $\bar{v}_2^\star$ now contains the parameters of non-ideal mixing of both the solvent and the polymer, in practice, however, $\bar{v}_2^\star$ differs not much from $\bar{v}_2$ if the polymer concentration is kept low (up to 1 %).

Dividing Equation (28) by the total mass $m_1 + m_2$ leads to Equation (29)

$$V_{12} = g_1 v_1^0 + g_2 \bar{v}_2^\star \tag{29}$$

where V_{12} is the specific volume of the solution and $g_1 = m_1/(m_1 + m_2)$, $g_2 = m_2/(m_1 + m_2)$ are the weight fractions of the solvent and the polymer, respectively. With $V_{12} = 1/\rho_{12}$ and $\bar{v}_1^0 = 1/\rho_1$, ρ_{12} and ρ_1 being the density of the solution and the solvent, respectively, it is readily found that

$$\bar{v}_2^\star = \frac{1}{g_2}\frac{\varrho_1 - g_1 \varrho_{12}}{\varrho_{12}\varrho_1} \tag{30}$$

Equation (30) shows that $\bar{v}_2^\star$ can be determined by measuring ρ_1 and ρ_{12}. Numerous methods for determining densities are described in the literature (268, 372-379).

In order to determine the partial specific volume from the apparent specific volume, Equation (29) yields

$$\left(\frac{\partial (m_2 \bar{v}_2^\star)}{\partial m_2}\right)_{m_1} = m_2 \left(\frac{\partial \bar{v}_2^\star}{\partial m_2}\right)_{m_1} + \bar{v}_2^\star = \left(\frac{\partial V_{12}}{\partial m_2}\right)_{m_1} \tag{31}$$

where $(\partial V_{12}/\partial m_2)_{m_1}$ is, according to definition, equal to $\bar{v}_2$. In terms of weight fractions, Equation (31) can finally be written

$$\bar{v}_2 = \bar{v}_2^{\star} + g_1 g_2 \frac{\partial \bar{v}_2^{\star}}{\partial g_2} \qquad (32)$$

Most values reported in the literature are $\bar{v}_2^{\star}$ - values rather than $\bar{v}_2$ - values, since extrapolation according to Equation (32) is usually omitted. The differences between $\bar{v}_2^{\star}$ and $\bar{v}_2$ are, however, often small.

1.5   Frictional Ratios

The molar frictional coefficient f_0 of an unsolvated spherical molecule may be computed by the formula based on Stokes law

$$f_0 = 6 \pi \eta \left(\frac{3 \, N^2 \, M \bar{v}_2}{4 \, \pi} \right)^{1/3} \qquad (33)$$

where η is the viscosity of the solvent and N is the Avogadro number. When the shape of a molecule deviates from a sphere, or when it is solvated, then the molar frictional coefficient f of such a molecule is larger than that of the spherical molecule. The frictional ratio f/f_0 thus permits to draw conclusions concerning either solvation or shape of the molecule. It is possible to calculate the dimensions of the non-spherical molecule provided a particular model (ellipsoid, cylinder, etc.) for the molecule is adopted and either the degree of solvation is known or assumed to be negligible. The molar coefficient f can be determined (2) either from sedimentation velocity data, provided the molecular weight is known from independent measurements according to

$$f = \frac{M \, (1 - \bar{v}_2 \varrho)}{s} \qquad (34)$$

or from diffusion measurements, using the relation

$$f = \frac{R \, T}{D} \qquad (T = K) \qquad (35)$$

Combining Equations (24), (33) and (35) the following relationship is found:

$$f/f_0 = \frac{1}{\eta} \left(\frac{R^2 \, T^2 \, (1 - \bar{v}_2 \varrho)}{162 \, \pi^2 \, N^2 \, D^2 \, \bar{v}_2 \, s} \right)^{1/3} \qquad (36)$$

For aqueous solutions, Equation (36) reduces to the approximated relation (valid at $20\,^{\circ}\mathrm{C}$):

$$f/f_0 = 10^{-8} \left(\frac{1 - \bar{v}_2 \varrho}{D^2 \, s \, \bar{v}_2} \right)^{1/3} \qquad (37)$$

Equations (36) and (37) are most frequently used for calculating the frictional ratio; they are termed a and b, respectively, in the present table. Other relationships for the determination of f/f_0 are quoted in the literature (2). In these cases special reference to the literature is made in the table.

In a few cases, only the molar frictional coefficient f, rather than f/f_0, was quoted in the literature. These values are inserted in the same column as the values for the frictional ratio.

1.6   Miscellaneous

(a)   In Table A.2. all concentrations are given in g/liter, if not otherwise stated.

(b)   Molecular weights determined from viscosity data, via a Mark-Houwink relation, are called M_v.

(c)   With certain assumptions it is possible to determine the molecular weight from a combination of the intrinsic viscosity $[\eta]$ and the limiting sedimentation coefficient s_0. Mandelkern and Flory (128) derived an expression

$$M = \left[\frac{s_0 \, [\eta]^{1/3} \, \eta \, N}{p^{-1} \, \Phi^{-1/3} \, (1 - \bar{v}_2 \varrho)} \right]^{3/2} \qquad (38)$$

in which the parameter $\Phi^{-1/3} \, p^{-1}$ is assumed to be a constant, equal to approximately 2.5×10^6. An expression similar to that of Equation (38) was derived by combining relationships involving D_0 and $[\eta]$ (128).

In addition, Wales and Van Holde (134) derived the following expression

$$M = 9.5 \times 10^{24} \, (1.66 \, [\eta])^{0.5} \left(\frac{s_0 \eta}{1 - \bar{v}_2 \varrho} \right)^{3/2} \qquad (39)$$

Equations (38) and (39) are referred to as F + M and W + H, respectively, in Table A.2., and the molecular weights determined from either one of these or similar relationships are termed M_η.

1.7   Symbols

c	concentration
c_0	initial polymer concentration
c_t	polymer concentration at time t
D	diffusion coefficient $[\mathrm{cm}^2/\mathrm{s}]$
D_n	number average diffusion coefficient
$D_{0,h} = D_A$	diffusion coefficient determined from zeroth moment of concentration gradient curve and its maximum height (so-called "area method").
D_σ	diffusion coefficient determined from zeroth, first, and second moment.

$D_{2,0} = D_w$ diffusion coefficient determined from second and zeroth moment of concentration gradient, is equal to the weight average of diffusion coefficient D_w. This is so-called "moment method" (frequently the symbol D_m is used in literature).

$D_{4,2} = D_z$ diffusion coefficient determined from fourth and second moment of concentration gradient, it is equal to the "z" - average of diffusion coefficient D_z.

D_h diffusion coefficient determined from two heights, h_1 and h_2 at distances x_1 and x_2 of the concentration gradient curve.

D_{20} diffusion coefficient at 20°C.

D_T diffusion coefficient at T^{o}C

D_{20}^{o} diffusion coefficient at 20°C, corrected for water.

η viscosity of solvent

η_1 viscosity of solvent at 1 atm. = 1.013 bar

η_p viscosity of solvent at p atm.

η_T viscosity of solvent at T^{o}C.

η_{20} viscosity of solvent at 20°C.

η_{20}^{o} viscosity of reference solvent at 20°C.

f_o molar frictional coefficient of spherical molecules

f molar frictional coefficient of any polymer molecule

k_D concentration coefficient defined by Equation (14).

k_s concentration coefficient defined by Equation (4a).

k'_s concentration coefficient defined by Equation (4b).

M molecular weight of polymer

$\overline{M}_n$ number average molecular weight.

$\overline{M}_w$ weight average molecular weight.

$\overline{M}_z$ "z" – average molecular weight.

M_v molecular weight determined by Mark-Houwink equation.

M_η molecular weight determined from Equations (38) or (39) or from similar relations.

$M_{i,j}$ molecular weight average determined by Equation (25) from s_i and D_j

$M_{t,w}$ molecular weight determined from s_t and D_w

$M_{t,A}$ molecular weight determined from s_t and D_A

$M_{t,\sigma}$ molecular weight determined from s_t and D_σ

$M_{s,w}$ molecular weight determined with undefined sedimentation coefficient and D_w

$M_{s,A}$ molecular weight determined with undefined sedimentation coefficient and D_A

$M_{s,D}$ molecular weight determined from undefined sedimentation coefficient and undefined diffusion constant

$M_{w,w}$ molecular weight determined from s_w and D_w

n refractive index of polymer solution

ω angular velocity

P_n number average degree of polymerization

P_w weight average degree of polymerization

r distance from center of rotation

r_o distance of menicus from center of rotation

r_t distance of boundary (maximum) from center of rotation at time t

ρ density of solution [g/cm^3]

ρ_T density of solvent at T^{o}C

ρ_{20} density of reference solvent at 20°C

ρ_1 density of solvent at 1 atm. = 1.013 bar

ρ_p density of solvent at pressure p atm.

R gas constant

s sedimentation coefficient [s]

s_o sedimentation coefficient at zero polymer concentration

$[s_o]$ intrinsic sedimentation coefficient, see Section A.1.1.d.

s_{20} sedimentation coefficient at 20°C.

s_{20}^{o} sedimentation coefficient at 20°C reduced to reference solvent.

s_1 sedimentation coefficient at 1 atm. pressure = 1.013 bar

s_p sedimentation coefficient at pressure p atm.

s_t sedimentation coefficient determined from migration of the gradient curve maximum

s_T sedimenation coefficient at T^{o}C

$s_n = <s>_o$ number average of sedimentation coefficient

$s_w = <s>_1$ weight average of sedimentation coefficient

$s_z = <s>_2$ "z" - average of sedimentation coefficient

s_m sedimentation coefficient determined from migration of gradient curve median

T temperature in ^{o}C, unless specified as K

t time

v_2 specific volume of polymer

$\overline{v}_2$ partial specific volume of polymer

$(\overline{v}_2)_{20}$ partial specific volume of polymer at 20°C.

$(\overline{v}_2)_T$ partial specific volume of polymer at T^{o}C.

$\overline{v}_2^{\star}$ apparent partial specific volume of polymer

$(\overline{v}_2)_p$ partial specific volume of polymer at pressure p

$(\overline{v}_2)_1$ partial specific volume of polymer at 1 atm. = 1.013 bar

$v_1^{o} = v_1$ specific volume of solvent.

1.8   Abbreviations

A   Archibald method

LS   Light scattering method

OS   Osmometry

SE   Sedimentation equilibrium

SV   Sedimentation velocity

V   number of single values given in Ref. cited.

c.a.   concentration dependence according to Equation (4a) [100 cm^3/g]

c.b.   concentration dependence according to Equation (4b) [100 cm^3/g] or [cm^3/g]

a   frictional ratio determined through Equation (36)

b   frictional ratio determined through Equation (37)

"Gralén"   concentration dependence of diffusion coefficient determined by Gralén method (14)

"B.+R."   concentration dependence of diffusion coefficient determined by the method of Beckmann and Rosenberg (13)

F+M   see Section A.1.6.c.

W+H   see Section A.1.6.c.

This table includes the litereature until June 1972.

2. Table of Sedimentation Coefficients, Diffusion Coefficients, Partial Specific Volumes and Frictional Ratios of Polymers in Solution

2.1. POLY(DIENES)

Polymer	Solvent	T[°C]	$s_o \times 10^{13}$	k_s	$D_o \times 10^7$	k_D	f/f_o	$\bar{v}_2$	$M \times 10^{-4}$	Remarks	Ref.
Poly(butadiene)	diethyl ketone	10.3	1.76						6.0	M_η	590
			2.76						18.7		
			3.45						35.0		
			4.28						43.6		
			4.52						77.8		
			5.15						138		
95%, 1,4-cis	diethyl ketone θ-cond.	10.3	$s_o=0.530 \times 10^{-15} \times M^{0.5}$							$s_n, s_w, s_v, s_{m,o}$ given in Ref.	574
90%, 1,4-cis (Al(isobutyl)$_3$/TiI$_4$)	hexane/heptane (1:1)	20	$s_o=2.80 \times 10^{-15} \times M^{0.48}$	0.48					5.5 -108	$M_{s,D}$	323
90%, 1,4-cis (Al(isobutyl)$_3$/CoCl$_2$)	hexane/heptane (1:1)		$s_o=2.33 \times 10^{-15} \times M^{0.50}$	0.50					3.47-104		323
Poly(butadiene-co-acrylonitrile) statistical	methyl ethyl ketone/cyclo-hexane 47.5/52.5 (θ cond.)	22	$s_o=1.28 \times 10^{-2} \times M^{0.5}$		$D_o=1.38 \times 10^{-4} \times M^{-0.5}$				3.48- 78	D_A;$M_{s,D}$ c.a. [g/100 ml] k_s given in Ref.	552
	methyl ethyl ketone/isopropanol 60/40 (θ-cond.)	20	7.58	0.197					47.5	D_A;$M_{s,D}$; c.a. [g/100 ml]	
			4.65	0.160					20.5		
			3.43	0.206					10.1		
Poly(butadiene-co-styrene) branched	benzene	25	$4.63 - 1,21$					0.995±0.005	16.8 - 8,19	see Ref.	406
	methyl n-propyl ketone (θ-cond.)	21	$s_o=1.04 \times 10^{-15} \times M^{0.5}$						4.9 - 51.4	$\bar{M}_w$	573
			$s_o=0.83 \times 10^{-15} \times M^{0.5}$						4.9 - 51.4	$\bar{M}_n$	344
21% weight styrene	methyl ethyl ketone	25	$s_o=3.1 \times 10^{-15} \times M^{0.5}$				0.773		5.6 - 38	$\bar{M}_w$;OS;SE	444
Poly(chloroprene)											
Poly(isoprene) natural rubber, crepe	chloroform	20	15.5		2.24		3.32		27.0	$\bar{M}_n$ Determ. of f/f_o see Ref.	337
			15.5		1.26		5.10		48.5		
			27.5		1.16		5.26		93.0		
					2.63		3.82		12.5		
					1.64		4.71		27.5	M_n;OS;Determ. of f/f_o see Ref.	337
					1.55		4.67		33.0		
					1.41		4.64		45.0		
					1.44		3.82		76.0		
	cyclohexane	20	4.6		0.48				160.0	$M_{s,D}$	508
			4.76		0.46				175.0		
			4.35		0.41				180		
			4.6		0.48			1.10	160	s_t;c.a.;$M_{t,A}$	29

Polymer	Solvent	$T[^\circ C]$	$s_o \times 10^{13}$	k_s	$D_o \times 10^7$	k_D	f/f_o	$\bar{v}_2$	$M \times 10^{-4}$	Remarks	Ref.
Poly(isoprene) (Cont'd.)											
natural rubber	hexane	20	9		3				27	M_s,D	337
1,4-cis	hexane	20	21		1.01 $\times 10^{-2} \times M^{-0.55}$			1.054	166	M_s,D	324
			$s_o = 5.01 \times 10^{-15} \times M^{0.45}$		$D_o = 3.98 \times 10^{-2} \times M^{-0.55}$						
					**2.2. POLY(ALKENES)**						
Poly(1-butene)	ethyl octanoate	22 (θ-temp.)	1.30	55.5				1.106★	80.0	M_η;c.a. [g/ml] ★ corrected for pressure influence	409, 410
Poly(ethylene)											
linear, high density	1-bromonaphthalene	110	$s_o = 7.74 \times 10^{-15} \times M^{0.344}$						0.35- 27.4	M	38
		120	-4.09						6.9	M_η^V	509
branched		120	-5.14						13.6	M_η	509
linear	1-chloronaphthalene	120	-5.18						8.8	$[\eta] = 0.51$ (100 ml/g)	27
			-1.67							0.90	
			-2.37							1.69	
			-3.31								
branched, fractions											
branches per molec.									$[\eta] = 0.18$ [100 ml/g]	27
1.9			-1.06							0.33	
2.6			-1.91							0.56	
3.0			-2.85							0.60	
2.9			-2.77							0.66	
4.6			-3.57							0.74	
5.7			-4.11							0.88	
6.4			-4.52							1.18	
13.7			-7.30							1.29	
16.1			-7.87							1.39	
26.3			-11.37				★ ★				
	diphenyl	110			9.80				2.6	★ see Ref. for various methods applied and values obtained	532
		123			6.42				6.0	$D_A.\bar{M}_w;D_{0.2}$ given in Ref.	593
		130			6.42				7.9		
					4.73				9.9		
					3.13				26.1		
					2.15				49.3		
Poly(isobutene)	cyclohexane (θ-cond.)	34	6.4★	721±159					130 ±20	c.a.:[ml/g] M_v;Oppanol B100; see Ref. for def. of s ★	576
		20	0.925	0.54					3.09	c.a.[100 ml/g];$\bar{M}_n$	32
			1.49	1.28					8.67		
			1.94	1.91					17.2		
			3.33	5.1					67.2		
			4.45	8.36					142		
	ethyl octanoate	22	1.3	55.5				1.106	75	s;[ml/g] $D_A;\bar{M}_w$;c from 0.295 g/100 ml to 0.053 g/100 ml;$D_{0.1};D$; $\bar{M}_n$ given in Ref.	237
	n-heptane	25							14.5		511

Polymer	Solvent	$T[°C]$	$s_o \times 10^{13}$	k_s	$D_o \times 10^7$	k_D	f/f_o	$\bar{v}_2$	$M \times 10^{-4}$	Remarks	Ref.
Poly(isobutene) (Cont'd.)	heptane	25			$D_o=5.01 \times 10^{-4} \times M^{-0.555}$				6.52-30.8	$\bar{M}_w$;LS	449
	octane	20.9						1.072	5.2	M_v	148
		23.2						1.069	160		
		20.7						1.075	160		
Poly(1-hexene sulfone)	acetone	20	11.5		16.2				4.68	$s;D;c.a.;M_{t,A}$	53
			21.5		6.4				22.2		
			33.8		3.0				57.1		
Poly(methylene)	1-bromonaphthalene	20	1.42					0.793	0.47	M_v	54
		110	2.03						1.42		38
			2.39						2.18		
			3.39						5.50		

2.3. POLY(ACRYLIC ACID), POLY(METHACRYLIC ACID) and DERIVATIVES

Polymer	Solvent	$T[°C]$	$s_o \times 10^{13}$	k_s	$D_o \times 10^7$	k_D	f/f_o	$\bar{v}_2$	$M \times 10^{-4}$	Remarks	Ref.
Poly(acrylamide)	water	20	$s=8.17 \times 10^{-15} \times M^{0.31}$		$D_A=8.46 \times 10^{-4} \times M^{-0.69}$			0.769	1.94-53.4	$s;c.a.;D$ measured in 0.2 M KCl;$M_{t,A}$ ★see Ref.	266, 154
Poly(acrylic acid)	aqueous NaCl mol/l	20 to 22 ★	s_t ★					0.73			555
	0.2			5.6		4.77			110	$D;M;c.a.;[ml/g]$ $\sigma_{s,D}$	
	0.02			35.0		20.0					
	0.012		7.3±0.5	81.0	0.65±0.02	28.2					
	0.006			114.0		49.0					
	1.0			2.4		0.2					
	0.2			4.2		1.0					
	0.012		6.5±0.5	32.0	1.35±0.05	10.0			42	$D;M;c.a.;[ml/g]$ $\sigma_{s,D}$	
	0.006			60.0		27.4					
	0.0012			-		146					
	1.0			-		22.3					
	0.2			0.0		22.3					
	0.012		2.4±0.1	2.75	3.2 ±0.1	150.0			7	$D;M;c.a.;[ml/g]$ $\sigma_{s,D}$	
	0.006			8.0		286					
Poly(benzyl acrylate-co-methyl acrylate) different comp.	methyl ethyl ketone	20						0.944		$\bar{v}_2$ and M given	506
Poly(n-butyl acrylate)	methyl isobutyl ketone	23.3	17.7	0.94					4.6	M_v	148
Poly(tert-butyl acrylate)	methyl ethyl ketone	25							30.6		512
	acetone		0.207	0.84							
	methanol		0.237	0.53							
	pentane		0.266	0.22							
	methyl ethyl ketone	25	0.160	0.88					23.7	$[s_o];\bar{M}_w$ in methyl ethyl ketone	
	acetone		0.194	0.74							
	methanol		0.230	0.49							
	pentane		0.244	0.21							

References page IV-106

Polymer	Solvent	T[°C]	$s_o \times 10^{13}$	k_s	$D_o \times 10^7$	k_D	f/f_o	$\bar{v}_2$	$M \times 10^{-4}$	Remarks	Ref.
Poly(tert-butyl acrylate) (Cont'd.)											
	acetone	25	0.182	0.64					18.0	$[s_o]; \bar{M}_w$ in methyl ethyl ketone	512
	methanol		0.203	0.43							
	pentane		0.228	0.20							
	methyl ethyl ketone	25	0.139	0.65					14.9	$[s_o]; \bar{M}_w$ in methyl ethyl ketone	
	acetone		0.162	0.55							
	methanol		0.191	0.39							
	pentane		0.109	0.19							
	methyl ethyl ketone	25	0.109	0.59					7.5	$[s_o]; \bar{M}_w$ in methyl ethyl ketone	
	acetone		0.117	0.32							
	methanol		0.127	0.22							
	pentane		0.137	0.13							
Poly(butyl methacrylate)	methyl isobutyl ketone	21.2						0.922	82	M_v	148
	2-propyl alcohol	21.5			$6.3 \times 10^{-5} \times M^{-0.50}$				4-800		399
Poly(tert-butyl methacrylate)											
	butyl acetate	20 (s)	$s = 3.02 \times 10^{-5} \bar{M}_w^{0.41}$							$\bar{M}_w; D; M_A; s, D$ given in Ref. $c = 0.5$ g/l for s	551
		21(D)			$D = 2.51 \times 10^{-4} \times M_w^{-0.55}$						
Poly(p-carbethoxyphenylmethacrylamide)											
	acetone				7.05				47.9	$D; \bar{M}_w$	267
					7.59				57.5		
					3.90				95.5		
					2.52				106.0		
	dimethylformamide (DMF)				2.18				100	$D; \bar{M}_w$	267
	DMF/10% formamide				1.16				74	$\bar{M}_w; w$	267
					1.11				110		
	ethyl acetate				$D_A = 2.8 \times 10^{-4} \times \bar{M}_w^{-0.69}$				26-74		267
Poly(cetyl methacrylate)	heptane	21	$s_o = 1.42 \times 10^{-14} \times M^{0.36}$		$D = 1.74 \times 10^{-3} \times M^{-0.64}$					$M_s, D; D_A$	560
Poly(cholesteryl acrylate)											
	benzene	21	$s_o = 1.9 \times 10^{-15} \times M^{0.46}$		$D_o = 3.2 \times 10^{-4} \times M^{0.54}$				14-78	M_s, D	580
Poly(ethyl acrylate)	methyl isobutyl ketone	24.3						0.951	3.4	M_v^s, D	148
	n-propanol	28	$s_o = 1.05 \times 10^{-2} \bar{M}_n^{0.47}$						1.8-15.5	M_η	456
Poly(ethyl acrylate-co-methyl acrylate) different comp.	methyl ethyl ketone	20	$s_o \sim M^{0.43}$		$D_o \sim M^{-0.57}$		$f_o \sim M^{0.57}$		7.7-744	$\bar{v}_2$ and M given	506
Poly(ethyl methacrylate)	acetone	20	8.95-120		17.4-0.85			0.798	7.7-744	$M; t_r A = M_r; s$; see Ref.	59
								0.798	3.41-935	$t_r A = M_{t,w}$; extrapolated value	18
			7.45-107	0.023-0.61	20.0-0.95	0-0.063		0.7942	2.46-744	$s; [1/g]; D_A [1/g]; M; t_{t,w}$; see Ref.	268
											52

Polymer / Solvent	T[°C]	$s_o \times 10^{13}$	k_s	$D_o \times 10^7$	k_D	f/f_o	$\bar{v}_2$	$M \times 10^{-4}$	Remarks	Ref.
Poly(ethyl methacrylate) (Cont'd.)										
methyl isobutyl ketone	22.9						0.863	41	M_v	148
Poly(isooctyl methacrylate)										
θ-solvent★	25	63.0-67.5					0.900	103.6 / 87.5	A; see Ref. $c=2,5$ g/ml;s_w (Trautmann)	427
Poly(N-isopropylacrylamide)										
water	15		0.99★				0.887	29.0	$\bar{M}_n$ in chloroform; for all temperatures	498
	20	2.48								
	25	3.04								
	27.5	3.55					0.896	100.0	$\bar{M}_w$ in chloroform	
	30	4.17								
	32	4.76								
	33	5.08					0.902			
Poly(methacrylic acid), sodium salt										
NaCl/water, Y = 4	20	2.90		1.150				23.7	s_t;apparent M	124
5		3.87		1.072				33.9		
7		4.36		0.923				44.2		
9		4.73		0.838				53.0		
14		5.46		0.742				69.0		
19		5.66		0.692				76.8		
29		6.04		0.643				88.0		
39		6.28		0.621				94.3		
49		6.43		0.611				98.6		
59		6.50		0.602				101.4		
∞		7.22				3.45		111.2	M of uncharged polymer;b	
0.01 M HCl pH 2.0	21.3						0.712	6.2	M_v	148
0.01 M NaOH pH 12.0	21.4						0.385	7.8	M_v	148
0.01 M NaOH/1 M NaCl pH 12.0	21.8						0.412	7.8	M_v	148
Poly(methyl acrylate-co-styrene) different comp.										
methyl ethyl ketone	20								$\bar{v}_2$ and M given	506
Poly(methyl methacrylate)										
acetone	20	20.3		6.8	10.25	$[f]=1,19 \times 10^{-5}$		11.5	M_v;[f] in [cm]	460
			72		18	$[f]=1,97 \times 10^{-5}$	0.798	20.1	s;c.a.[ml/g];D;k[ml/g];D_A D[ml/g] t,A	15
acetonitrile		46.1	224	2.2		$[f]=5,35 \times 10^{-5}$		138	s;c.a.[ml/g];D;k[ml/g];D_A D t,A	15
		88.5	614	0.92		$[f]=14,1 \times 10^{-5}$		635	$M_{t,A}$	15
	30						0.825	494	see Ref. for range of $\bar{M}_w$	494
	38	$s_o = 0.186 \times M^{0.51}$						11.5	M_η;[f] in [cm]	570
benzene	20					$[f]=1,25 \times 10^{-5}$		11.5	M_η;[f] in [cm]	460
	25				5.05		0.8273 to 0.8077	0.19-7.7	M;$\bar{v}=[\bar{v}_2]_{M=\infty}+50/M$ (AIBN)	268

★ - total molarity of sodium ions divided by molarity of poly(methacrylic acid).

Poly(methyl methacrylate) (Cont' d.)

Polymer	Solvent	T[°C]	$s_o \times 10^{13}$	k_s	$D_o \times 10^7$	k_D	f/f_o	$\bar{v}_2$	$M \times 10^{-4}$	Remarks	Ref.
Poly(methyl methacrylate) (Cont' d.)	benzene	25						0.8069	∞	$\bar{v}_2=[\bar{v}_2]_{M=\infty}+39.4/M$ $(B_2^o)_2$	268
								0.818			211
	butyl acetate	30						0.879			494
	n-butyl chloride (θ-cond.)	25	40.1★	251±53				0.8080	∞	extrapolated value	208
		35								c.a.,[ml/g]; see Ref. for def. of s	576
		35.6	15.7	22	7.18	-30	1.65×10^{-5}★	0.82	20	★[η]=kT/$D_o\eta$	6, 15
			39.1	41	2.91	-55	4.09×10^{-5}★		130	★[η];c.a,[ml/g];	6, 15
			86.0	107	1.15	-35	9.38×10^{-5}★		655	★[η];k,[ml/g];M_t, A	6, 15
	carbon tetrachloride							0.8038	∞	extrapolated value	268
	chlorobenzene							0.8118	∞	extrapolated value	268
	chloroform							0.7942	∞	extrapolated value	•268
	o-dichlorobenzene							0.8187	∞	extrapolated value	268
	diethyl ketone							0.8015	∞	extrapolated value	268
	dioxane	25						0.8181	∞	extrapolated value	268
		25	2.11	134			2.17×10^{-5}★	0.818	20	★[η] in [cm];s;c.a,[ml/g]; $\bar{M}_w$	15
			4.5	412			6.35×10^{-5}★		130		15
			8.55	1040			[η]=17.6×10^{-5}		655		15
	ethyl acetate							0.7963	∞	extrapolated value	268
		20	7.6	2.5	7.21	0		0.787	7.9	c.a.,[ml/g];$D_2;D_2^o$; s_o $s=-0.06+0.028\sqrt{M}$	211
			8.8	0.8	5.66	0			12.7		
			11.7	1.8	4.36				20.7		
			14.1	1.9	3.73				31.5		
			16.3	2.4	3.79				38.1		
			19.4	5.2	3.38				47.3		
			21.5	5.2	2.54				70.1		
			25.0	6.0	2.34				93.1		
			9.86	0.59					18		212
				3.1					196.5	c.a.[ml/g];$M_{s,w}$	
			$34.0 \times 10^{15} \times M^{0.48}$						18-196.5		
	methyl ethyl ketone							0.7993	∞	extrapolated value	268
	methyl isobutyl ketone	21.8						0.816	48	$\bar{M}_v$	148
	tetrahydrofuran							0.8085	∞	extrapolated value	268
	toluene	20			5.80		1.19×10^{-5}	0.807	11.5	$\bar{M}_n$	460
								0.8101	∞	extrapolated value	268
	m-xylene	17						0.8063	19	$\bar{M}_v$	148
		22.15						0.8019	19	$\bar{M}_v$	
		29.5						0.8190	19	$\bar{M}_v$	
		45.0						0.8143	19	$\bar{M}_v$	

$s = 1.51 \times 10^{15} \times M^{0.48}$

Poly(methyl methacrylate-co-ethylene dimethylacrylate)

Polymer	Solvent	T[°C]	$s_o \times 10^{13}$	k_s	$D_o \times 10^7$	k_D	f/f_o	$\bar{v}_2$	$M \times 10^{-4}$	Remarks	Ref.
randomly branched	ethyl acetate	20								$\bar{M}_w$; see Ref. for polymerization conditions and s_o, k_s and M values	549

2.4. VINYL POLYMERS

Polymer	Solvent	T[°C]	$s_o \times 10^{13}$	k_s	$D_o \times 10^7$	k_D	f/f_o	$\bar{v}_2$	$M \times 10^{-4}$	Remarks	Ref.
Poly(acrylonitrile)	dimethylformamide	20	1.72						7.87	M_η	569
			1.65-4.68						6.7-103.0	see Ref. for s and M_{w+H}	533
		25	1.52		3.6			0.83	4.8	$s;D_t;M_{2,0};M_{t,w}$	115
			2.08		2.1				11.3		
			2.36		1.75				15.5		
			2.95		1.25				27.0		74
			2.04					0.830	10.5	$s;\overline{M}_{t,w}$	
			2.48						21.0		
					7.14				1.6	$D_{2,0};D_n$ given in Ref.	206
					5.72				2.8		
					5.08				6.3		
					4.03				19.3		
		35			3.91				21.4		201
			$D_o=2.19 \times 10^{-4} \times M^{-0.58}$			$k_D=2.31 \times 10^{-9} \times D_o^{-1.29}$			2.8-57.5		
	dimethylformamide/benzene 30%	R.T.	3.08						32	$M_{s,D}$	462
	dimethylformamide/benzene 37%		138						700,000	M;calculated from $s_o = 0.025\, M_{s,D}^{0.38}$	462
Poly(methoxyethylene)	amyl alcohol	23.5						0.983	1.5	M_v	148
Poly(vinyl acetal)	toluene	20.9						0.848		M_v	148
Poly(vinyl acetate)	cyclohexanone	23.8						0.849	4.8	M_v	148
	diethyl ketone	13						0.821			60
		40						0.833			
	methyl ethyl ketone	20	$s_o=9.8 \times 10^{-15} \times \overline{M}_w^{0.38}$		$D_o=7.8 \times 10^{-4} \times \overline{M}_w^{-0.63}$				1.7-120	$s;c.a.;D;M_{t,A};M_{t,A}$	110,112
	1,2,3-trichloropropane	13						0.83			60
		40						0.841			
Poly(vinyl alcohol)	water	20	0.96		7.46			0.765	2.24	$s;c.a.[100\ ml/g];M_{t,w}$	129
			1.26		3.77				1.3		129
			1.96		2.68				3.4		
			$s_o=4.4 \times 10^{-15} \times M^{0.32}$ 1.86		$D_o=5.5 \times 10^{-10} \times M^{-0.68}$ 2.16				7.4		
									9		
		21.6						0.765	2.1	M_v	130
		25	0.61		6.89	2.28		0.750	0.85	$s;c.a.;D;M$ see Ref.	148
			0.96		3.97	2.82			2.33	D_A given in Ref.	119
			1.54		2.3	3.47	$1.04 \times M^{-0.166}$		6.47		119
			2.52		1.45	4.00			16.7		
		30						0.769	0.64	M_v	151
								0.766	1.98		
								0.763	5.4		
	0.2 M NaCl	25	2.14					0.79		c.a.	152
	0.2 M NaCl (D_2O)		1.22								

References page IV-106

Polymer	Solvent	T[°C]	$s_0 \times 10^{13}$	k_s	$D_0 \times 10^7$	k_D	f/f_0	$\bar{v}_2$	$M \times 10^{-4}$	Remarks	Ref.
Poly(4-vinylbiphenyl)		25						0.82		see Ref. for solvent	480
Poly(4-vinyl-N-butylpyridinium bromide)											
	0.2 M NaCl	20	3.0		2.8		2.38	0.77	11.5	a;c=1.0 g/l	135
		25	1.10						2.5		131
Poly(vinyl butyral)	amyl alcohol	20	33.3					0.883	455	$[\eta]=122$ ml/g	148
Poly(N-vinylcarbazole)	benzene	20	16.3					0.786	84.7	LS	457
Poly(vinylchloride)	chlorobenzene	23.1						0.793		LS	148
	cyclohexanone	23.4						0.711	94	$[\eta]=122$ ml/g ; M_v	148
	cyclohexanone	25	0.622					0.6926	1.9	$\bar{M}_n$	602
			1.075					0.705	5.3		75
			1.18						6.6		
			1.56						12.0		
	tetrahydrofuran	20	7.69-5.32	2.47--.32				0.7429	9.7-10.5	c.a. [100 ml/g]; $\bar{M}_w$; M(F+M)	544
		25						0.6951		M(F+M)	181
fractionated			6.8						9.0	M(F+M)	602
			$5.41 \times 10^{-2} \times M^{0.41}$						4.6		602
			$6.50 \times 10^{-2} \times M^{0.41}$	0.55					2.02-18.3	c.a. [g/100ml]; M_v	596
Poly(vinyl chloride-co-vinyl acetate)	cyclohexanone	23.5						0.723		$[\eta]=77.2$ [ml/g]	148
Poly(vinyl chloride-co-vinylidene chloride)	cyclohexanone	23.5						0.703		$[\eta]=77.4$ [ml/g]	148
Poly(vinyl fluoride)	0.1 n LiBr in dimethylformamide	100	$s=0.101 \times M^{0.40}$						11.4-52.1	$\bar{M}_n$;F+M;W+H	421
			$9.29 \times 10^{-2} \times M^{0.4}$					0.72	14.3-65.4	s from $s=s_0(1+1.66[\eta]c)$; F+M	441
Poly(vinyl methyl ether) see Poly(methoxyethylene)											
Poly(1-vinylnaphthalene)	toluene	25						0.82			480
Poly(2-vinylnaphthalene)	toluene	25						0.84			480
Poly(4-vinylpyridine)	ethyl alcohol	25	3.6	1.79				0.682	9.7	c.a.[100 ml/g];W+H; Ref. 134	132
Poly(vinylpyrrolidone)	methanol	25	6.45	6.29	9.60			0.689	43.0	c.a.;From SV;$M_{t,D}$	150
			3.41					0.730	2.3		144
	water	20						0.776	2.0		144
								0.783	50.0		144
			$s_o=8.81 \times 10^{-16} \times M^{0.50}$		$D_o=1.0 \times 10^{-4} \times M^{-0.50}$			0.802	1.2-7.4	D_A;M_A ; s;D_A;$M_{t,A}$; $\bar{v}_2$ at 24.85°C;D_A given in Ref.	144,143
			0.82 to 1.42		7.55 to 4.14		1.70 to 2.13		1.3-4.15		40
			10		1.17				9.45	s;D_A;$M_{t,A}$	153
			12.20		0.50				270.0	$M_{t,A}$	153
			0.85		8.66		1.66		1.06	c=5%;$M_{s,D}$	178

Polymer	Solvent	T[°C]	$s_o \times 10^{13}$	k_s	$D_o \times 10^7$	k_D	f/f_o	$\bar{v}_2$	$M \times 10^{-4}$	Remarks	Ref.
Poly(vinylpyrrolidone) (Cont'd.)	water	20	1.66		4.31		2.12		4.2	c=5% M, D	178
	water	20	1.94		2.44		2.94		8.6	c=5% M_s, D	178
unfractionated		23.1						0.775			178
		25						0.781	2.9	M_v	148
		25						0.820	ca. 3.2		150
		30						0.780	2.0	M_η	487
		30						0.784	5.0		144
		30						0.801			144
	1 M $CaBr_2$	20			5.5				2.45	c=1 g/100ml	611, 613
	1 M KBr				4.78				2.45	c=1 g/100ml	611, 613
	1 M LiBr				4.06				2.45	c=1 g/100ml	611, 613
	1 M $MgBr_2$				4.01				2.45	c=1.003 g/100ml	611, 613
	0.2 M NaCl		1.0				1.46		1.2	c. a. [1/g] ; $M_{t,A}$	144
			1.43	0.068			2.09		2.5		144
			1.59	0.085			2.38		3.92		144
			1.76	-			-		5.56		144
			4.77	0.391			2.81		31.0		144
			7.13	-			4.36		58.4		144
			7.13	0.720			-		112.9		144
			12.5	1.310			-		190		144
			16.7	1.63			-		342		144
			$s_t = 8.81 \times 10^{-16} \times M^{0.5}$								
(Luviskol K30)	0.6 M NaSCN	25						0.772		M_η	487
	0.2 M Na_2SO_4							0.771		M_η	487
Poly(vinylsulfonic acid)	0.1 M phosphate buffer pH=7	20	7.15						94.5	$M_{t,A}$; s	153
sodium salt	0.5 M NaCl	20	1.57		21.3			0.410	0.3	s ; D ; $M_{t,\sigma}$	116
			1.96		15.0			0.428	0.5		
			2.32		11.0			0.430	0.9		
			2.94		9.75			0.424	1.25		
			4.27		5.95			0.408	2.95		
			4.46		6.40			0.424	2.95		

2.5. POLY(STYRENES)

Polymer	Solvent	T[°C]	$s_o \times 10^{13}$	k_s	$D_o \times 10^7$	k_D	f/f_o	$\bar{v}_2$	$M \times 10^{-4}$	Remarks	Ref.
Poly(o-chlorostyrene)	benzene	30						0.792		$\bar{M}_w$ from SE LS; k_s; D; k_D given in Ref.	585
Poly(p-chlorostyrene)	benzene	30						0.797		$\bar{M}_w$; D given in Ref.	585
Poly(α-methylstyrene)	benzene	30	$s_o = 3.61 \times 10^{-15} \times M^{0.43}$						4.84-167		522
monodisperse	chloroform	21			3.1				50		558
monodisperse	cyclohexane	39	$s_o = 2.0 \times 10^{-15} \times M^{0.5}$						4.84-167	$\bar{M}_w$ from SE; k_s given in Ref. 522	522

References page IV-106

Polymer	Solvent	T[°C]	$s_o \times 10^{13}$	k_s	$D_o \times 10^7$	k_D	f/f_o	$\bar{v}_2$	$M \times 10^{-4}$	Remarks	Ref.
Poly(α-methylstyrene)	cyclohexane θ-cond.	35	$s_o=1.72\times10^{-15}\times M_m^{0.49}$						3.7-62.7	$\bar{M}_w$; A; $\bar{M}_n$; given in Ref; s_m	321
			13.89						73		616
		37	19.47						152	$\bar{M}_w$; D given in Ref. $\bar{M}_w^o$	68
		38.2	$s_o=1.86\times10^{-15}\times \bar{M}_w^{0.5}$				$f_o=4.1\times10^{-8}\times M_w^{0.5}$			$\bar{M}_w$ dep. of k_s given in Ref.	594
		35-125	$s_o=1.93\times10^{-15}\times M_w^{0.5}$					0.889	650	c=1.64 g/100 ml; see Ref. for s-values	550
	cyclohexane	39	$s_o=2.0\times10^{-15}\times M_w^{0.5}$; $k_s=6.2\times10^{-4}\times M_w^{0.5}$						4.27-143	k_s [100 ml/g]; $\bar{M}_w$; pressure dep. of s given in Ref.	601
monodisperse		39	$s_o=2.04\times10^{-15}\times M_w^{0.50}$					0.886	4.25-117.0	$\bar{M}_\eta$; LS; k_s given in Ref.	472
monodisperse	toluene	25	$s_o=3.01\times10^{-15}\times M_w^{0.44}$					0.873	4.25-117.0	$\bar{M}_n$; A; LS; c. a. [100 ml/g]	472
		35	11.37						1.98	$\bar{M}_n$	616
		35	17.78						4.01	$\bar{M}_v$	
		25-120	2.38-8.0						650	c=1.64 g/100 ml; see Ref. for s values	550
		37	$s_o=4.02\times10^{-15}\times M_w^{0.43}$							$\bar{M}_w^o$; relationship valid for $\bar{M}_w>2\times10^4$; D and [f] given in Ref.	68
Poly(α-methylstyrene-co-butadiene) block and statistical different comp.											
Poly(styrene)	cyclohexane	37	11.9-16.7		1.18-0.80				120-245	s_o and $\bar{M}_n$ given	567
	acetone	20								$s; D=D_{2,0}; \bar{M}_{t,w}$	50
		25						0.899		extrapolated value	263
	benzene	20						0.911	15.0	$\bar{M}_w; \bar{M}=45\times10^4$	606
		25								see Ref. for $\bar{v}_2$	532
		25			27.9				0.92		480
					25.0				0.132		180
					21.1				0.195		
					17.2				0.28		
					11.7				0.39		
		25	2.5		4.1				1.06	c. a.; SV; $\bar{M}_w$	150
			7.14		1.50				6.7		
									60.6		
		25						0.91	∞	extrapolated value	150
			$s_o \sim M^{0.47}$		$D_o \sim M^{-0.53}$			0.9175	0.118-52	extrapolated value	268
								0.9078-0.9177	0.23-22.2	$\bar{M}_w; \bar{v}_2=[\bar{v}_2]_{M=\infty}=+(-23/M); B_o$	43
										$\bar{M}_w$; c. a. [100 ml/g]; D; A	268
NBS 705		35			3.29	0.22			17.9	$\bar{M}_w$; c. a. [100 ml/g]; D; A	577
NBS 419					1.57	0.59			41.1	$\bar{M}_w$; c. a. [100 ml/g]; D; A	577
Styron 683-7					3.26	0.45			27.7		577
					$D_o=3.05\times10^{-4}\times M^{-0.55}$						577
		40	19.68	100.4				0.917	400	$\bar{M}_w$; LS; c. a.	172

Polymer	Solvent	T[°C]	$s_o \times 10^{13}$	k_s	$D_o \times 10^7$	k_D	f/f_o	$\bar{v}_2$	$M \times 10^{-4}$	Remarks	Ref.
Poly(styrene) (Cont'd.)	benzene/methanol 1/3.27 vol	25						0.660	∞	extrapolated value	150
	butyl acetate	25						0.9162	∞		268
	carbon tetrachloride	27			4.43	-1.31			8.2	$D_{2,0}$;B+Rmethod;$\overline{M}_n$	12
					2.27	0			14	Gralén-method given in Ref.	
					1.41	0			48		
					1.04	0.5			110		
					$D_{2,0}=k \times M_{t,w}^{-0.59}$						
								0.9087	∞	extrapolated value	268
									∞		12
	chlorobenzene	25						0.9207	∞	extrapolated value	268
	chloroform	20	9.2	1.8					25	c.a.[100 ml/g]	125
			13.1	3.2					55		
			15.4	4.0					80		
			17.8	5.0					130		
		20	-15.1		(1.84)		$[f]=3.86 \times 10^{-5}$	0.908	54	$s;\overline{M}_t;[f]=kT/D_o\eta$ in cm	3
											5
standard sample No 706	cyclohexane	21	$\log[s_o]=17.922 + 0.415 \log \overline{M}_w$		5.8				27.8	$\overline{M}_w;D_w$	558
		25						0.9100	∞	extrapolated value	268
									11.5-280	$[s_o]=-1.595 \times 10^{-2}$	268
		25	$s_o=3.85 \times 10^{-15} \times \overline{M}_w^{0.42}$					0.9293	∞	extrapolated value	268
		25							18.1-104	$\overline{M}_s,D$	605
		28			ca. 2.2				18.0	$\overline{M}_s,D$	414
		30			ca. 2.5				18.0	$\overline{M}_n$	414
					ca. 4.0				9.0	$\overline{M}_n$	587
					ca. 3.1				23.7	c=0.0178 g/ml	587
	(θ-cond.)	34	16.9★	72±22			2.95×10^{-5}		115	c.a.;[I ml/g];$\overline{M}_v$★; from 0.00147 to 0.0179 g/ml; see Ref.★ for def. of s	576
		35	12.1		1.92				3.2-164	$\overline{M}_w;[f]=kT/D_o\eta$ in [cm] given in Ref.	6
									54	$\overline{M}_t;[f]=kT/D_o\eta$ in [cm] given in Ref.	3
					$D_o=1.21 \times 10^{-4} \times \overline{M}_w^{-0.49}$						
			$s_o=1.69 \times 10^{-15} \times M_w^{0.48}$					0.934	1.14-104	$\overline{M}_w;A$	5
			$s_o=1.35 \times 10^{-15} \times \overline{M}_w^{0.51}$	5.06						$\overline{M}_w;A$	107
			28.1								96
								0.928	400	$\overline{M}_w$;LSc.a.; see Ref. for $\bar{v}_2$	172
										$\overline{M}_w;\overline{M}_n$;given in Ref.	532
Styron 683-7					3.86	-0.07			12.5	$\overline{M}_w;\overline{M}_n$;given in Ref.	592
					(2.71)					$D_{0,2}$;$\overline{k}$[100 ml/g];(D_A)	592
NBS 705			8.43		3.19	0.05			30.9	$D_{0,2}$;$\overline{k}$[100 ml/g];(D_A)	592
			16.2		(2.88)			0.943	97.5	$\overline{M}_n$	486
			34.5						490		

Poly(styrene) (Cont'd.)

Polymer	Solvent	T[°C]	$s_0 \times 10^{13}$	k_s	$D_0 \times 10^7$	k_D	f/f_0	$\bar{v}_2$	$M \times 10^{-4}$	Remarks	Ref.
Styron 683-7	cyclohexane	35			3.86	-0.06				27.7	$\bar{M}_w$;c.a.[100 ml/g]:D_A	577
NBS 705					3.19	-0.04				17.9	$\bar{M}_w$;c.a.[100 ml/g]:D_A	605
NBS 419					2.67	-0.04				41.1	$\bar{M}_w$;c.a.[100 ml/g]:D_A	606
					$s_0 = 1.50 \times 10^{-15} \times M^{0.5}$ $s_0 = 1.50 \times 10^{-15} \times M^{0.5}$ $D_0 = 1.21 \times 10^{-4} \times M^{-0.5}$ 0.928				19.4-92.9	$\bar{M}$ dep. of k_s given in Ref. 594	543
Waters sample No. 25170		13	2.28	-0.043				0.933	15.0	$\bar{M}_w$; $\bar{M}_w^s$;$\bar{M}_n$=45 x 10^4 $\bar{M}_n$;$\bar{M}_w$=5.1 x 10^4; $\bar{M}_n$=4.9 x 10^4	543
		16	2.42	-0.0_4					5.1		
		22	2.61	+0.008							
		28	2.88	+0.058							
		35	3.26	+0.1_1							
		32	16.8	+0.2_							
		34	17.2	0.4_							
		36	17.6	0.7_							
		38	18.0	0.8_							
		40	18.6	1.0_							
		42	19.0	1.2_							
		28-40			ca. 2.9				150	M_η;c.a.[100 ml/g]	543
									18.0	$\bar{M}$;LS;c.a.	414
		45	29.0	15.7				0.934	400	$\bar{M}_w$;LS;c.a.	172
		55	34.0	35.7				0.936	400	$\bar{M}_w$;LS;c.a.	172
Standard sample Polymer Society Japan	trans-decalin	18.2			0.85				15.0	$\bar{M}_n$;D;$D_{2,0}$;$\bar{M}_w$ and M_{F+M} given in Ref.	502
		30			1.16						
		40			0.93						
		50			1.43						
		60			1.52						
		70			1.82						
		80			1.96						
		90			2.69						
		100			3.30						
		110			3.59						
Standard sample Polymer Society Japan		18.2			0.69				45.0	$\bar{M}_n$;D;$D_{2,0}$;$\bar{M}_w$ and M_{F+M} given in Ref.	502
		40			0.76						
		60			1.10						
		80			1.45						
		110			2.23						
	o-dichlorobenzene	27	2.73	0.73	0.50	-0.15		0.9289	77	$D_{2,0}$;B+R method;$\bar{M}_n$	12
	diethyl ketone	25							∞	extrapolated value	268
		25						0.9106	∞	extrapolated value	268

Polymer	Solvent	T[°C]	$s_o \times 10^{13}$	k_s	$D_o \times 10^7$	k_D	f/f_o	$\bar{v}_2$	$M \times 10^{-4}$	Remarks	Ref.
Poly(styrene) (Cont'd.)											
	dioxane	25						0.9270	∞	extrapolated value	268
	ethyl acetate	25						0.9132	∞	extrapolated value	268
		27			2.64	0.12				77	$D_{2,0}$;B+R method;$\bar{M}$	12
	ethyl benzene	27			0.96	1.3				77	$D_{2,0}$;B+R method;$\bar{M}$	12
		30	$s_o=3.28 \times 10^{-2} \times M^{0.42}$	3.55					4.25-143	$12\bar{v}$;M;see Ref. for k_s	612
	mesitylene	25						0.9260	∞	extrapolated value	268
	methyl ethyl ketone	20	7.13						11.0	LS;M: av. of 6 observers	11
			13.41						33.1	LS;M: av. of 6 observers	
			26.0						98.5	LS;M: av. of 6 observers	
			23.8						72.0	OS;M: av. of 6 observers	
			18.4		3.25				52.8	s;D;$\bar{M}$	3
			12.6	0.73					25	t_c,a, [100 ml/g]	125
			18.2	1.3					55		
			21.3	1.7					80		
			26.0	2.3					130		
			12		12		1.38		9	s;$D_{2,0}$;$\bar{M}_w$	126
			21		6.4		2.05		18		
			21		2.6		3.18		75		
			22		2.04		3.75		96		
			31		1.70		3.71		170		
			45		1.48		3.59		280		
			30		0.83		6.05		340		
			45		0.84		5.24		500		
			48		0.81		5.25		550		
		23.4	17.9		3.4				45	D_A; av. of 4 observers	4
		25			3.75				52.8	D_A	4
		25					$[f]=3.12 \times 10^{-5}$			$[f]=kT/D_o\eta$	3
									52.8	LS	568
									180	LS	568
								0.9078	∞	extrapolated value	268
NBS 706			$\log s_o=17{,}845+0.455 \log \bar{M}_w$		$D_o=3.1 \times 10^{-4} \times M^{-0.53}$				32.5-280	$[s]=1.385 \times 10^{-2} \times s_o$	169
NBS 705		25			3.88				25.78	$\bar{M}^o$;LS	571
					$D_o=5.4 \times 10^{-4} \times M^{-0.49}$				17.93	$\bar{M}_w$;LS	571
					$D_o=1.96 \times 10^{-4} \times M^{-0.49}$					$\bar{M}_w$;LS	571
Waters sample W-61970					$D_o=1.96 \times 10^{-4} \times M^{-0.49}$				214.5	$\bar{M}_w$;LS	571
	θ-solvent	ca. 25	35.76-37.82						12.7	for solvents see Ref.	427
		27	9.8	0.43	9.4	-0.86			8.2	s_w(Trautmann);$A_2c=2.0$ g/ml	12
			13.3	0.48	5.14	-0.12			14	[100 ml/g];$D_{2,0}$ Gralén	
			20.0	1.11	3.32	0.49			48	$\bar{M}_n$ $D_{2,0}=k \times M_{t,w}^{-0.53}$	
					2.51	0			77		
			25.0	1.75	2.64	0.40			110		

References page IV-106

Polymer	Solvent	$T[^{\circ}C]$	$s_o \times 10^{13}$	k_s	$D_o \times 10^7$	k_D	f/f_o	$\bar{v}_2$	$M \times 10^{-4}$	Remarks	Ref.
Poly(styrene) (Cont'd.)											
	methyl ethyl ketone / butyl alcohol ϕ=0.26★	25	18.05						37.25	$\bar{M}_w$;LS ★volume fraction of butyl alcohol	169
			26.6	1.23					83		
	methyl isopropyl ketone	20	32.0	1.62					106.5		
			47.1	2.72					230		
			7.2-34		6.95-2.5					7V;s;D=$D_{2,0}$;$\bar{M}_w$ t_A extrapolated value	66
									∞		268
	tetrahydrofuran	25	8.75		1.95		$[f]=3.55 \times 10^{-5}$	0.9102	53.6	s;D;M;$[f]=kT/D$ in [cm] o_{η}	3
	toluene	20	7.92		2.02				45	t_A av. of 3 observers	4
								0.910	5		5
								0.917	15.0	$\bar{M}_n$;$\bar{M}_w$=45 $\times 10^4$	606
			$s \sim M^{0.47}$		$D \sim M^{-0.53}$				0.12-52		43
			6.25		2.50				30	s;D;$\bar{M}$	1
			7.20		2.85				4.5	s;D;A_w^w	
			4.75		4.03				14	w;w;w;n	16
			9.65		1.17				96	t_A;s;D;M	
			18.1		0.74				285		
Szwarc "living polymer"		22.7	5.9-11.2		8.05-3.29				25-130	4V;c,a;k_s given in Ref.	125
		23.5	3.7-5.20	36.4				0.918	5.75-22.6	4V;M_t;c=1,0 g/l	44
			3.24	36.4					19	$\bar{M}_v$	148
			3.24							s;[ml/g]	237
		25			$D=2.15 \times 10^{-4} \times \bar{M}_w^{-0.53}$			0.917	7.50	M_{η};c.a. [g/ml]	409
								0.916	∞	extrapolated value	268
					4.13					see Ref. for $\bar{v}_2$	591
					3.78				7.4	D_A;D_D and see Ref. c=0,079 g/100 ml	532
										c=0,119 g/100 ml	511
		27			1.22	0.96		0.917	77	$D_{2,0}$;B+R method;Ref. 13 $\bar{M}_n$	511
										s_t	57
			0.66						0.12		12
			0.44						0.2		17
									4		17
		30						0.948	1.33	$\bar{M}_n$	1, 237 / 467
	toluene/methyl ethyl ketone 3:1	20	9.9		2.11		$[f]=3.65 \times 10^{-5}$		52.8	s;D;$[f]=kT/D$ o_{η}	3
	1:1		12.9		2.45		$[f]=3.43 \times 10^{-5}$		55.3	t_A;s;D;$[f]=kT/D$ o_{η}	3
	1:3		15.6		2.82		$[f]=3.29 \times 10^{-5}$		54.5	t_A;s;D;$[f]=kT/D$ o_{η}	3
	m-xylene	20							∞	extrapolated value	268
	p-xylene	25						0.9205		see Ref. for $\bar{v}_2$	532
halomethylated, containing 5.66% P	tetrachloroethane	20	$s_o = 9.77 \times 10^{-2} \times M^{0.266}$						9.07-50.7	M;F+M; for preparation see Ref.	561

Polymer	Solvent	T[°C]	$s_o \times 10^{13}$	k_s	$D_o \times 10^7$	k_D	f/f_o	$\bar{v}_2$	$M \times 10^{-4}$	Remarks	Ref.
Poly(styrene-co-butadiene)											
	methyl n-propyl ketone	21	$s_o=3.31 \times 10^{-16} \times M^{0.62}$						12.5-98.5	$\bar{M}_n$	590
24 wt.% St											
	(θ-cond.)	21	$s_o=0.83 \times 10^{-15} \times M^{0.5}$						4.9-51.4		573
Poly(styrene-co-divinylbenzene)											
30/70	n-octane	21	$s_o=1.59 \times 10^{-15} \times M^{0.50}$		$D_o=1.49 \times 10^{-4} \times M^{-0.50}$			1.05	6.5-71.0	$M_{s,D}$	322
Poly(styrene-co-isooctyl methacrylate)★											
20/80%	θ-solvent★	25	25.87-28.42					0.761	59.6 / 53.5	A / c=1.67 g/ml;s_w (Trautmann) ★ see Ref.	427
Poly(styrenesulfonic acid)											
	0.2 M KCl	20	8.6-11.5						67-115	F+M;see Ref. for single data	127
	0.005-0.2 M NaCl	25	4.95-16.4						35-265	M;see Ref. for single data	452

2.6. POLY(OXIDES), POLY(CARBONATES), POLY(ESTERS)

Poly(carbonate) see Poly(oxycarbonyloxy......)
Poly(ethylene terephthalate) see Poly(oxyethyleneoxyterephthaloyl)
Poly(formaldehyde) see Poly(oxymethylene)

Polymer	Solvent	T[°C]	$s_o \times 10^{13}$	k_s	$D_o \times 10^7$	k_D	f/f_o	$\bar{v}_2$	$M \times 10^{-4}$	Remarks	Ref.
Poly(oxycarbonyloxy-1,4-phenyleneisopropylidene-1,4-phenylene)											
	methylene chloride	20				$1/D_o=1.569 \times 10^{-3} \times M_{s,D}^{0.638}$		0.8099	0.85-26.6	$M_{t,w}$;s;c.a. $_t$ [ml/g]	45
	tetrahydrofuran	20	$s_o=1.33 \times 10^{-14} \times M_{s,D}^{0.362}$					0.7738			45
Poly(oxycarbonylpropylene)											
	chloroform	20	-9.95		37.8			0.812	78.0	$\bar{M}_w$ from Scheraga-Mandel-kern-Eq.;Ref. 536	535, 537
			-8.16		5.76				37.0		
			-4.96		23.5				15.6		
			-3.50		10.7				8.35		
			-2.00		10.7				2.11		
Poly(oxyethylene)											
	acetone	25							0.43	$\bar{M}$;LS	184
	formamide	25							0.43	$\bar{M}_w$;LS	184
	methanol	25							0.43	$\bar{M}_w$;LS	184
									2.38		
	water	20	3.18					0.785	1.92	s;$M_{t,D}$ $D_{2.0}$	150
					37.0				0.029		177
					29.2				0.0625		
					24.0				0.125		
					13.3				0.33		
					11.6				0.58		
					10.3				0.88		
					7.2				1.06		
					23.6				0.1426		180
					22.6				0.1470		
					19.7				0.1778		
					20.1				0.1822		

References page IV-106

Polymer	Solvent	$T[°C]$	$s_o \times 10^{13}$	k_s	$D_o \times 10^7$	k_D	f/f_o	$\bar{v}_2$	$M \times 10^{-4}$	Remarks	Ref.
Poly(oxyethylene) (Cont'd.)											
	water	22.8						0.834	47	M_v	148
		23.0						0.846	0.024		184
		25.0							0.43		184
Poly(oxyethyleneoxyterephthaloyl) (Poly(ethylene terephthalate))											
			0.59		11.5				23.8		
			0.93		4.85				1.2		
			1.09		7.35				2.38		
					4.85						150
					3.95				3.73	$\overline{M}_w$;LS	
	m-cresol	25	0.103		0.120				11.1	M_s, D	526
			0.153		0.155				7.73	A_2 given in Ref.	
			0.138		0.186				6.10		
			0.132		0.263				4.13		
			0.125		0.280				3.68		
				180	0.365	40			2.30		
					0.205				6.02		
unfractionated		50			0.130				8.73		
			$s_o \sim K \times M^{0.26}$		$D_o \sim K \times M^{-0.74}$			0.678		M_s, D	526
			0.435						11.1		
			0.405						7.73		
			0.391						6.10		
			0.372	140	0.992	80			4.13		
			0.338						3.68		
			0.439						2.30		
unfractionated			0.408		0.381				6.02		
			$s_o \sim K \times M^{0.22}$						8.73		
Poly(oxypropylene)	acetone	20			$D_o = K \times M^{-0.52}$				0.0074–	D ;$[f]=4.4 \times 10^{-7} \times M^{0.52}$	465
									0.3375	$[f]=kT/D_o$	
	benzene	20			$D_o = K \times M^{-0.55}$				0.0134–	D ;M from condensation and boiling temperature elevation	465
									0.3375	$[f]=3.2 \times 10^{-8} \times M^{0.55}$	

2.7. POLY(AMIDES), POLY(UREAS), POLY(URETHANES)

Polymer	Solvent	$T[°C]$	$s_o \times 10^{13}$	k_s	$D_o \times 10^7$	k_D	f/f_o	$\bar{v}_2$	$M \times 10^{-4}$	Remarks	Ref.
Poly(γ-benzyl-L-glutamate)											
	dichloroacetic acid 0 to 100 Vol.% ★	25	ca. 3.8–4.2						53	M_v ; $[s]$ ★ helix to coil transition between 60 to 80 Vol.%	566
	dimethylformamide	25		1.0	1.1			0.779	30	LS;k_s[100 ml/g]	581
				1.8	0.75				55	LS;k_s[100 ml/g]	
		21	$s_c =2.8 \times 10^{-14} \times M^{0.2}$		$D_c =2.8 \times 10^{-3} \times M^{-0.8}$			0.786	5.9–36.4	M_s, D ;for s_c and D_c see Ref.	559
		25	1.62–3.91	0.09–1.41					5.1–41	c. a.[100 ml/g] see Ref. for s_o and k_s	434

Polymer	Solvent	T[°C]	$s_o \times 10^{13}$	k_s	$D_o \times 10^7$	k_D	f/f_o	$\bar{v}_2$	$M \times 10^{-4}$	Remarks	Ref.
Poly(γ-benzyl-L-glutamate) (Cont' d.)											
	dimethylformamide	25	1.77–3.95	0.15–1.41					3.64–435	LS;c.a. [100 ml/g] see Ref. for s_o, k_s, M_A-values	450
Poly(n-butyliminocarbonyl) (Poly(butylisocyanate)											
	tetrahydrofuran	20	$s_o=3.05 \times 10^{-14} \times M^{0.16}$		$D_o=1.69 \times 10^{-4} \times M^{-0.85}$				1.8–21.1	$M_{s,D}$	325
Poly(iminohexamethyleneiminoadipoyl) (Nylon 66)											
	formic acid (98–100%)	22.4						0.890	1.6	M_v	148
Poly(iminoisophthaloyliminio-1,3-phenylene)											
	dimethylformamide	25	$s_o=1.9 \times 10^{-2} \times M^{0.44}$					0.678	3.02–15.6		589
	LiCl (0.25 g/100 ml + 98% sulfuric acid) in dimethylformamide		$s_o=2.8 \times 10^{-2} \times M^{0.39}$					0.682	2.07–14.2		589
Poly(imino(1-oxohexamethylene)) (Nylon 6)											
	formic acid (85%) + 2 M KCl	25						0.883		from dn/dc see Ref.	532
Poly(iminoterephthaloylimino-1,4-phenyleneanthron-9-ylidene-1,4-phenylene)											
	dimethylformamide	25	$s_o=3.0 \times 10^{-2} \times M^{0.40}$					0.747			579
Poly(L-lysine) branched by benzyl asparate and benzyl glutamate											
	dimethylformamide	20								s_o, k_s, D given dependent on composition and M c, a,	448
Poly(oxy-1,4-phenyleneiminocarbonyl-2,5-dicarboxy-1,4-phenylenecarbonylimino-1,4-phenylene)											
	dimethylacetamide + 0.1 n LiBr	15	0.99–3.22						0.99–26.6	17V;LS;	455
		15	$s_o=3.72 \times 10^{-2} \times \bar{M}_w^{0.36}$						1.84–4.54	$\bar{M}_n$	403
Poly(urethane) from Poly(oxycaproyl)(A), diphenylmethanediisocyanate, ethylenediamine											
linear											
A: $\bar{M}_n$ 1.3 x 10³ (I)*	dimethylformamide	25	$s_o=1.36 \times 10^{-15} \times M^{0.44}$					0.848	2.62–12.8	M_{F+M};c.a. [100 ml/g]	607, 608
A: $\bar{M}_n$ 2.8 x 10³ (II)*			$s_o=9.86 \times 10^{-16} \times M^{0.46}$					0.8745	26.6–3.06	7V;k_s given in Ref.	
A: $\bar{M}_n$ 1.8 x 10³ (III)*			$s_o=1.18 \times 10^{-15} \times M^{0.45}$					0.8581	14.64–1.79	7V;k_s given in Ref.	
branched (I)*			2.91						20.0	M_{F+M};c.a. [100 ml/g]	
(II)*			2.10						9.67	M_{F+M};c.a. [100 ml/g]	
			2.92						13.7	M_{F+M};c.a. [100 ml/g]	
			1.76						7.12	M_{F+M};c.a. [100 ml/g]	
Poly(ureyleneheptamethylene)											
	dimethylformamide	55	2.50					0.955			183
		90	2.43					0.949			183
	formic acid (97%)	25	1.85					0.915			183
	sulfonic acid (98%)	60	1.59					0.974	2.35	$\bar{M}_w$;LS	183
								0.965			
Poly(ureylene-2,4,4-trimethylpentamethylene)											
	dimethylformamide	55						0.955			183
		90						0.949			
	formic acid (97%)	25						0.915			

2.8. POLY(SILOXANES)

Polymer	Solvent	T[°C]	$s_o \times 10^{13}$	k_s	$D_o \times 10^7$	k_D	f/f_o	$\bar{v}_2$	$M \times 10^{-4}$	Remarks	Ref.
Poly(3-methyl-1-butene silsesquioxane), cyclolinear											
	butyl acetate	24						0.863±0.003	9.2-60	$M_{s,D}$	583
Poly(oxydimethylsilylene) (Poly(dimethyl siloxane))			$s_o=1.1 \times 10^{-14} \times M^{0.31}$		$D_o=1.1 \times 10^{-3} \times M^{-0.69}$						
	bromocyclohexane	28	-2.40		1.66				10.1	$M_{s,D}$	326
			-2.70		1.15				16.5		
			-4.60		0.59				55.0		
	methyl ethyl ketone	10						1.025	1.6	M_v	148
		20						1.018			
		35						1.0282			
								1.0468			
	toluene	20	4.09		3.00				30.1	$M_{s,D}$	326
			6.07		1.60				92.0		
		22.4						1.024	0.94	M_v	148
Poly(oxydiphenylsilylene) (Poly(phenyl siloxane))											
	benzene	21	$s_o=9.16 \times 10^{-15} \times M^{0.37}$		$D_o=6.16 \times 10^{-4} \times M^{-0.63}$			0.807	0.4-335	$D;M_{s,D}$	556
Poly(oxymethylphenylsilylene)											
	toluene	22						0.996		$[\eta]=132$ [ml/g]	148

2.9. OTHERS

Polymer	Solvent	T[°C]	$s_o \times 10^{13}$	k_s	$D_o \times 10^7$	k_D	f/f_o	$\bar{v}_2$	$M \times 10^{-4}$	Remarks	Ref.
Poly(acenaphthylene)	benzene	25						0.77			480
Poly(metaphosphoric acid)											
magnesium complex	0.4 M NaCl (water)	38	1.10		18.5	1.39	2.8	0.370	100	$s;D_{2.0};a;M_{t,w}$	139
potassium salt	0.1 M NaCl (water)	14	1.36		12.5	1.05	5.2	0.311	48		139
	0.4 M NaCl (water)	32	1.55		11.1	1.08	3.8	0.316	110		
	0.1 M NaCL (5% ethanol)	20	1.73		11.0	1.25	4.1	0.315	57		
	0.1 M NaCl (8% ethanol)	25	2.20		12.1	1.36	3.6	0.318	65		
Poly(o-methylphenylene)	toluene	20	2.35		9.41		1.68	0.841	0.56	$D_{2.0};M_{t,w};\bar{v}_2$ independ. of M	140
			2.87		8.38		1.94		0.93	$D_{2.0}$	
			4.40		5.34		2.04		1.26	D_A	
			8.25		3.52		2.06		1.58	$D_A;M_{t,A}$	
									1.68	$D_{2.0}$	
							2.04		2.34	$D_A;M_{t,w}$	
							2.26		3.67	$D_{2.0};M_{t,w}$	
							2.42		7.72	$D_{2.0};M_{t,w}$	
							2.97		27.0	$D_{2.0}$	
Poly(1-p-tolyl-2,5-dioxopyrrolidin-3,4-diyl) (Poly(p-tolyl-maleimide))											
	dimethylformamide	21	$[s_o]=2.76 \times 10^{-14} \times M^{0.42}$		$D_o=3.47 \times 10^{-3} \times M^{-0.58}$			0.74	4-56	$M_{s,D}$	563

2.10. POLY(SACCHARIDES)

Polymer	Solvent	T[°C]	$s_o \times 10^{13}$	k_s	$D_o \times 10^7$	k_D	f/f_o	$\bar{v}_2$	$M \times 10^{-4}$	Remarks	Ref.
Cellulose											
absorbent cotton			4.28		2.64					s; c=1.0 g/l; t	63
aged, alkali cell.	cuprammonium hydroxide	20	3.8	1.1	0.84	0.3	5.3	0.508	15	s; c. a, [100 ml/g]: $D_{2,0}$; a; t $M_{t,w}$	14
overaged			1.9	0.2	1.54	0.45	4.6	0.508	4.4		14
cotton (Georgia)			10.4	9.0	0.2	3.5	9.8	0.508	175		14
holo cellulose (spruce)			7.2	3.9	0.72	1.45	4.7	0.508	34		14
linters, unbleached			10.3	6.0	0.23	3.2	9.0	0.508	150		14
--, bleached			7.2	5.2	0.43	1.3	6.1	0.508	49		14
--, cotton filter paper			9.2	6.9	0.26	2.5	8.6	0.508	120		14
flax fiber			3.94		2.83					s; c=1.0 g/l; t	63
			17.5	17.0	0.1	12.0	13.1	0.508	590	s; c. a, [100 ml/g]: $D_{2,0}$; a; t $M_{t,w}$	14
nettle fiber			14.0	10.0	0.25	3.2	7.7	0.508	190		14
ramie fiber			10.8	8.6	0.18	2.5	10.4	0.508	200		14
sulfate cell.			6.5	3.9	0.54	1.05	5.9	0.508	40	s; c. a, [100 ml/g]: $D_{2,0}$; a; t $M_{t,w}$	14
--, acetylated			6.3		0.37	1.9	7.7	0.508	57		14
--, precipet.			6.3		0.66	1.65	5.2	0.508	32		14
sulfite cell.			6.3	3.8	0.45	1.15		0.508	50		14
--, bleached			4.28		2.72					s; constituent A	63
			2.73		3.67					constituent B	63
viscose staple fiber (sulfate cellulose)											
--, (spruce)			2.6	0.5	1.04	0	4.8	0.508	7.5	s; c. a.; $D_{2,0}$; a; $M_{t,w}$	14
--, (pine)			4.1	1.7	0.54	0.75	6.4	0.508	23		
wood cellulose (cross and Bevan aspen)			4.28		2.70					s; constituent A	63
			2.73		3.53					constituent B	63
α-Cellulose											
cotton linters		25	4.6		5.4			0.642	5.3	s; t	62
holo cellulose			6.5	3.2	0.32	2.95	8.4	0.508	68	s; c. a, [100 ml/g]: $D_{2,0}$; a; t $M_{t,w}$	14
β-Cellulose											
Cellulose	cupriethylenediamine	25	1.6	0.5	1.97	0.5	4.0	0.508	2.7		14
			5.5		1.2				17.5	c, a,; $D_{2,0}$; $M_{s,w}$	234
			8.3		0.95			0.65	0.95		97
Cellulose	cadmium ethylenediamine	12	1.25						3.36	$\overline{M}_w$; SE	425
			1.13						2.45		
			1.04						1.88		
			0.74						1.01		
linters hydrol.		25	1.80	1.53	1.77			0.50	4.3	s; [100 ml/g]: $D_{2,0}$; $M_{t,w}$	97
			2.75	3.70	0.75				15.5		97

References page IV-106

Polymer	Solvent	T[°C]	$s_o \times 10^{13}$	k_s	$D_o \times 10^7$	k_D	f/f_o	$\bar{v}_2$	$M \times 10^{-4}$	Remarks	Ref.
Cellulose (Cont'd.)											
linters hydrol.	cadmium ethylenediamine	25	2.75	3.70	0.75				15.5		97
			3.17	4.65	0.64				21.0		
			3.80	7.10	0.58				27.5		
			4.53	10.0	0.44				43.5		
			5.49	14.1	0.31				74.5		
sulfite pulp			2.87	3.67	1.06				11.5		97
	FeTNa*	20	1.75		0.89			0.654	11.1	s;c.a.;D_A;$M_{s,A}$; *FeTNa = Iron-Tartaric acid complex Na salt	235
			2.17		0.60				20,14		
			2.98		0.46				36.4		
			3.35		0.35				53.9		
Cellulose acetate	acetone	20	4.07	6	5.4	0			10	$D_{2,0}$;B·R	13
					20.7			0.68	1.035	s;c.a.[100 ml/g];$D_{2,0}$; $M_{t,w}$	39
			7.5	56	7.7				5.1		39
			10.9	143	4.0				14.3		
			12.5	191	3.4				19.4		
from cotton linters	acetone	25	16.7					0.687	15.0	$\bar{M}_n$;OS;M_{F+M} given in Ref.	554, 557
			16.0					0.687	13.6		
			10.5					0.692	6.6		
			14.7					0.698	14.5		
			12.8					0.697	11.0		
			7.0					0.690	2.6		
from wood cellulose			$s_o = 1.2 \times 10^{-14} \times M^{0.39}$					0.700	-		
			10.0					0.700	4.0		
			8.4								
Cellulose triacetate	sym-tetrachloroethane	21.9						0.744	25	M_v	148
Cellulose acetate-co-butyrate	sym-tetrachloroethane	22.7	$s_o = 2.35 \times 10^{-15} \times M^{0.36}$					0.777	9.7	M_v	148
Cellulose carbanilate	dioxane	20						0.72	258-0.135	M;D;c.a.[g/l]:k_s and k_o given in Ref.	68
Cellulose nitrate	acetone	20	12.0				1.7	0.51	0.62	M;D	223
							3.2		3.0	SE;M,D	
							4.6		8.02	SE	
							5.4		19.9	M,D	
							6.9		61.3	M,D	
			30.0				3.74		10	a;$M_{s,D}$	276
(N=13,8)			13.9		6.5					s;c.a.;$D_A=D_{2,0}$	64
			14.0		3.29			0.57	19.0		
			14.2		2.60				24.2	$M_{t,w}$	65

Polymer	Solvent	T[°C]	$s_o \times 10^{13}$	k_s	$D_o \times 10^7$	k_D	f/f_o	$\bar{v}_2$	$M \times 10^{-4}$	Remarks	Ref.
Cellulose nitrate (Cont'd.) (N=13.8%)	acetone	20	6.5		20.9				1.38	$s;D_{w\,2.0};M_{w,w}$	259
			7.1		19.3			0.572	1.63		
			8.3		13.3			0.570	2.78		
			12.6		5.6			0.566	10.0		
			15.3		4.5				15.3		
			14.5		3.5				18.7		
			16.8		2.8				26.7		
			25.3		2.3				49.5		
			26.2		1.0				117.5		
cotton (N=13.8%)			7.2		25.0				1.3	$s;c.a.;D_A=D_{2.0};M_{t,w}$	64
			8.8		16.5				2.3		
			11.1		10.8				4.6		
			10.8		6.45				7.4		
			12.8		6.30				9		
			15.2		3.80				17.8		
			17.1		2.90				26.2		
			20.5		2.00				45.6		
			21.5		1.50				63.0		58, 64
			37.5		0.66				249.0		
cotton (N=13.5-14%)			10.8		6.45				7.4	$s;D=D_{A\,2.0};M_{t,w}$	49, 58
			15.2		3.80				17.8		58
			17.1		2.90				26.2		
			20.5		2.00				45.6		
			20.0		1.50				59.3		
			30.5		1.20				113.0		
holo cellulose --, (spruce)			21.6	4.2	2.57	1.75	6.3	0.51	34	$s;c.a.;D_{2.0};a;M_{t,w}$	14
			13.4	2.1	2.4	0.65			78	$s;c.a.;D_{2.0};a;M_{t,w}$	14
--, (α-cellulose)			16				7.7	0.51	23	$s;c.a.;D_2 0=D_{A\,2.0};M_{t,w}$	14
linters			15.8		2.8				25	$s;c.a.;D_{A\,2.0};M_{t,w}$	65
			19.0	4.2	2.05	2.1	12.2	0.510	34.4		
					1.0				78		
--, bleached			14.0	2.8	1.44	1.2	10.6	0.510	40	$s;c.a.;D_{2.0};a;M_{t,w}$	14
			19.0	4.2	1.11	2.25	11.5	0.510	60	$s;c.a.;a;M_{t,w}$	14
									68		
--, bleached and unbleached			5.2		35				0.6	M_s,D	223
			8.7						3.0	SE,D	
			12.0						8.0	SE	
			18.0		3.7				19.9	M_s,D	
			30.0		2.0				61.3	M_s,D	
rayon pulp			17.6	4.8	3.01	2.35	6.0	0.51	24	$s;c.a.$ [100 ml/g]$;D_{2.0};M_{t,w}$	14

References page IV-106

Polymer	Solvent	T[°C]	$s_0 \times 10^{13}$	k_s	$D_0 \times 10^7$	k_D	f/f_0	$\bar{v}_2$	$M \times 10^{-4}$	Remarks	Ref.
Cellulose nitrate (Cont'd.)											
rayon pulp	acetone	20	15.5	2.6	2.15	1.0	7.9	0.51	30	s;c.a. [100 ml/g]:$D_{t,2.0}$; a;$M_{t,w}$	14
			16.7	3.6	1.94		8.2	0.51	35		
			18.3	4.8	1.48		9.6	0.51	51		
sulfate pulp			23.3	11.5	1.66	2.8	8.2	0.51	57	s;c.a. [100 ml/g]:$D_{t,2.0}$; a;$M_{t,w}$	14
			16.2	3.5	1.56		9.6	0.51	41		
sulfite pulp			16.4	3.6	1.56	2.65	9.6	0.51	43	s;c.a. [100 ml/g]:$D_{t,2.0}$; a;$M_{t,w}$	14
	butyl acetate		1.1					0.584	8.02	SE:[s] given in Ref.	259
	cyclohexanone		2.0					0.59	61.3	M_o	223
chemical cotton	ethyl acetate	21.3	11.4		1.49			0.556	35	M_s,D	148
		20	13.1		0.91				37	s;$D_{2.0}$;$M_{t,w}$	91
raw cotton (Delfos variety)					0.47				69	s;c.a.;$D = D_{2.0}$;$M_{t,w}$	91
linters			9.3		2.40				150	s;$D_{2.0}$;$M_{t,w}$	65
viscose rayon	methanol	20	9.26		4.77			0.545	19.5	$M_{t,w}$	91
									9.3	SE_n	91
									1.75	SE	223
	pentyl acetate	20			13.3			0.54	2.7	$M_{w,w}$	259
					5.8				9.1		
					5.6				10		
					4.5				15.3		
					3.5				18.6		
					2.9				28.2		
			4.0					0.55	8.02	SE	223
			6.5						61.3	M_s,D	
Cellulose trinitrate											
purified chemical linters	ethyl acetate	30	6.34						4.13	$\overline{M}_w$:$\overline{M}_n = 3.46 \times 10^4$	90
			7.10						7.6	$\overline{M}_w$:$\overline{M}_n = 6.06 \times 10^4$	
			8.65						12.8	$\overline{M}_w$:$\overline{M}_n = 8.9 \times 10^4$	
purified cotton linters			9.62						11.0	$\overline{M}_w$;$s_n = 0.304 \times 10^{-13} \times M_w^{0.29}$	90
			11.13						24.8	$\overline{M}_o$	
			12.66						57.3	$\overline{M}_w$:$\overline{M}_n = 25.7$	
Cellulose xanthate	1M NaOH	0	2.10				1.16		1460 ★	see Ref. for degree of subst. $f=[g/sec \times mol] \times 10^6$ ★ degree of polymerization	437
			2.25				1.01		1060 ★		
			1.68				0.73		590 ★		
			1.81				0.57		450 ★		
			1.22				0.52		360 ★		
			1.05				0.49		350 ★		
Dextran	water	20	1.1-6.0						1-20	s vs M diagram	23

Polymer	Solvent	T[°C]	$s_o \times 10^{13}$	k_s	$D_o \times 10^7$	k_D	f/f_o	$\bar{v}_2$	$M \times 10^{-4}$	Remarks	Ref.
Dextran (Cont'd.)	water	20	2.02						2	s;M determ. by Ogston Method (Ref. 24,25)	24
			3.04						4.4		
			3.47						6.5		
			5.47						15.0		
			9.6						46.0		
		20.9						0.60	16.0	$\bar{M}_v$	148
		20							1.8–40		26
									1.3–20		26
				$s_o = 2.45 \times 10^{-16} \times \bar{M}_w^{0.44}$							
				$s_o = 2.36 \times 10^{-16} \times \bar{M}_n^{0.45}$							
	0.025 M Na$_2$HPO$_4$ + 0.025 M NaH$_2$PO$_4$	20	1.64		7.50		1.88	0.62	1.4	s;D A;a;$M_{t,A}$	92
			2.64		4.69		2.19		3.6		
			4.1		3.12		2.49		8.4		
			5.3		2.40		2.79		14.1		
			6.55		1.73		3.15		24		
			23.3		0.75		3.61		199		
			26.4		0.55		4.25		305		
			150		0.25		4.03		3800		
leuconostoc mesenteroides											
ca. 5% branched	0.05 Na$_2$HPO$_4$ + 0.05 M NaH$_2$PO$_4$	20	1.57–9.82		8.80–1.40		1.74–3.22	0.611	1.12–43.7	s;D;M;see Ref. for f/f_o and D_o $s = 1.45 \times 10^{-15} \times M_{t,A}^{0.50}$	167
highly branched	0.05 M Na$_2$HPO$_4$ + 0.05 M NaH$_2$PO$_4$	20						0.603	2.88	s;D;M;see Ref. for f/f_o and D_o $s = 3.43 \times 10^{-15} \times M_{t,A}^{0.42}$	167
β-Dextrin	water	20	0.48					0.627	0.1134	D;c=1.0 g/l	171
Ethyl cellulose	acetone	20	5.53		5.7			0.885	7.88	$\bar{M}_v$, D A	412
			4.81		7.5				5.21		
			4.33		9.8				3.59		
			3.48		14.6				1.94		
			3.02		21.4				1.145		
					3.46				2.0	D;c=6.1 g/l	21
					1.43				5.3	c=5.0 g/l	
					1.05				9.0	c=5.16 g/l	
	sym-tetrachloroethane	21.9						0.864	19.0	$\bar{M}_v$	148
Ethyl hydroxyethylcellulose											
ca. 18% ethoxyl	water	20	2.0		1.9	0.3	3.8	0.721	9.2	s;D$_{2,0}$;D A given in Ref.;b $M_{t,A}^{0.36}$	94
ca. 30% ethoxyl+hydroxyethyl			2.32		1.4	0.3	4.1	0.711	14.0	$s = 3.1 \times 10^{-15} \times M_{t,w}^{0.7}$	
			2.63		1.1	0.3	5.1	0.703	20.0	$D_{2,0} = 10^{-4} \times M^{-0.7}$	
			3.0		0.8	0.3	6.0	0.705	30.0		
Hydroxyethyl cellulose	water	19.2						0.678	47.0	$\bar{M}_v$	148
		25	2.82	3.85	1.86		3.96	0.701	12.5	s;c.a.;b;$\bar{v}_2$ at 20°C;$M_{t,w}^{0.46}$	98
			4.45	10.2	1.11		4.78		33.0	$s = 1.0 \times 10^{-15} M_{t,w}$	
			4.90	11.9	0.96		5.12		42		

Polymer	Solvent	T[°C]	$s_o \times 10^{13}$	k_s	$D_o \times 10^7$	k_D	f/f_o	$\bar{v}_2$	$M \times 10^{-4}$	Remarks	Ref.
Hydroxyethyl cellulose (Cont'd.)	water	25	5.65	15.8	0.85		5.29		54.5	$D=8.2 \times 10^{-5} \times M^{-0.54}$	98
Methyl cellulose											
22.6 % methoxyl	water	20	0.83				3.0	0.68	1.4	$\bar{M}_w$;SE	222
27.05% methoxyl			1.29		1.49				7.06	$s;c.a.;D_t;M_{t,A}$	19
27.44% methoxyl			1.67		0.92			0.704	15.1		
28.45% methoxyl			2.50		0.71			0.710	30.2		
28.8 % methoxyl			0.79		3.05			0.683	2.26		19,20
			0.79				3.9	0.717	2.43	$\bar{M}_w$;SE	222
31.7 % methoxyl			0.89		2.34			0.720	3.43	$s;c.a.;D_A;M_{t,A}$	19,20
31.8 % methoxyl			0.89				4.5	0.73	3.81	$\bar{M}_w$;SE	222
ca. 28% methoxyl										$\bar{v}_2=0.0560+0.0542 \times$ (% OCH$_3$)	19,20

$$s_o=0.85 \times 10^{-15} \times M_{t,A}^{0.45}$$
$$D_o=0.79 \times 10^{-4} \times M_{t,A}^{-0.56}$$

Polymer	Solvent	T[°C]	$s_o \times 10^{13}$	k_s	$D_o \times 10^7$	k_D	f/f_o	$\bar{v}_2$	$M \times 10^{-4}$	Remarks	Ref.
	0.2 M NaCl	20			4.45				1.41	D; av. of M not specified	21
					3.05				2.43	D^o_o;c=5.0 g/l	
					2.47				3.81	D^o_σ	
Sodium carboxymethylcellulose	0.01 M NaOH;pH=12.0	20,6	2.40					0.505	70	M_v	148
	0.001 M NaCl	25	2.30						4.47	$M^v_{t,w}$	314
			2.50						4.6		
			2.58						9.06		
			2.65						9.42		
			2.62						16.3		
			2.85						19.4		
			2.96						28.0		
	0.005 M NaCl		2.54						34.6		
			3.00						7	$\bar{M}_n$	277,278
			3.14						10.4		
			3.97						17.5		
	0.01 M NaCl		2.76						29.0		
			3.18						7	$\bar{M}_n$	277,278
			4.25						10.4		
			4.28						17.5		
			2.70						29		
			3.11						4.47		314
			4.53						16.3		
									34.6		
Sodium cellulose glycolate	0.2 M NaCl	20			3.08	1.85			5.5	$D_{2.0}$ [100 ml/g];$M_{t,w}$	14
	0.5 M NaCl				3.30	2.15			5.5	$D_{2.0}$ [100 ml/g];$M_{t,w}$	14
	1.0 M NaCl		3.4	1.3	3.22		3.2	0.53	5.5	c.a. [100 ml/g];s;$D_{2.0}$; a;$M_{t,w}$	
Sodium cellulose xanthate	0.2 M NaOH + 1.0 M NaCl		2.30-4.4	0.7-2.3	3.35-1.83	0.20-0	3.4-4.3	0.530	3.5-12.6	$D_{2.0}$;a;$M_{t,w}$	14

B. SECOND VIRIAL COEFFICIENTS OF POLYMERS IN SOLUTION

1. Introduction

The osmotic pressure π of polymer solutions cannot be described by van't Hoff's Law except in a θ-solvent. Thus, the expression:

$$\pi/c = RT/M \qquad (1)$$

where c, R, T and M are the solute concentration, the gas constant, the absolute temperature and the molecular weight of the solute, respectively, only holds under theta conditions, or at infinite dilution, i.e. lim c = 0. The deviation from ideality is strongly dependent on the polymer concentration. The osmotic pressure is thus conveniently developed in a power series of c such that in the limit of zero polymer concentration, Equation (1) results. The forms which are most frequently encountered are those as in Equations (2), (3), and (4).

$$\pi/c = RT [A_1 + A_2 c + A_3 c^2 + \ldots] \qquad (2)$$
$$A_1 = 1/M$$

$$\pi/c = (\pi/c)_{c=0} [1 + \Gamma_2 c + \Gamma_3 c^2 + \ldots] \qquad (3)$$

$$\pi/c = \frac{RT}{M} + Bc + Cc^2 + \ldots \qquad (4)$$

The parameters A_2, Γ_2, and B are the so-called second virial coefficients, A_3, Γ_3, and C the third virial coefficients, and so on. With certain assumptions it is possible to express the third virial coefficient in terms of the second virial coefficient (379-381). Especially when the second virial coefficient is small it is found (382) that, for example, Equation (3) can be approximately expressed as:

$$\pi/c = (\pi/c)_{c=0} [1 + (\Gamma_2/2) c]^2 \qquad (5)$$

a form particularly convenient in plots of $(\pi/c)^{1/2}$ versus c.

The second virial coefficients defined by Equations (2), (3), and (4) are related to each other through Equation (6).

$$B = RTA_2 = (RT/M) \Gamma_2 \qquad (6)$$

The dimensions of the virial coefficients follow from Equations (2), (3), and (4). Depending on whether the dimensions [g/mol] or just [g] are accepted for M, A_2 has the dimension [mol cm^3/g^2], when c is expressed in g/cm^3. For Γ_2 the dimension depends only on that of c; frequently one finds [cm^3/g]. A large variety of dimensions are found for B since its dimension depends on those of R, c, and M. In this case, for example [erg. cm^3/g^2], [Joule cm^3/g^2], [l^2 bar /g^2], etc. are found in the literature. Table B.2. lists the second virial coefficients in the dimension [mol cm^3/g^2] unless otherwise indicated under "Remarks" and by an asterisk.

Based on the fluctuation theory of v. Smoluchowski (383) and of Einstein (384), which links osmotic pressure with light scattering, Debye (385) established the fundamental equation for the light scattering of polymers in solution

$$\frac{Kc}{R_\theta} RT = \frac{RT}{M} + 2Bc + 3Cc^2 + \ldots \qquad (7)$$

which makes it possible to determine the second virial coefficient (and the third virial coefficient) from light scattering measurements.

In Equation (7), K is a constant and R_θ is the intensity of the (excess) scattered light at angle θ at unit distance when the primary intensity is unity. Variants of Equation (7), corresponding to those of Equations (2) and (3) are also frequent. The determination of second virial coefficients, usually as a "by-product" of molecular weight measurements, by both osmotic pressure and light scattering, is the most common practice now. It should, however, be mentioned that according to Schulz (386), the second virial coefficient can also be related to the coefficients of the concentration dependence, k_D and k_s, of the diffusion and sedimentation coefficients.

$$B = \frac{RT}{M} (k_D + k_s) \qquad (8)$$

Later Flory (387) established an equation, by means of which the separate determination of k_D and k_s becomes unnecessary:

$$(D/s) (1 - \bar{v}_2 \varrho) = \frac{RT}{M} + 2Bc + 3Cc^2 + \ldots \qquad (9)$$

Equation (9) is based on previous considerations of Flory and Mandelkern (388).

In the early 1940's the first attempts were made to find theoretical explanations for the nonideal behavior of polymer solutions. Since then the literature on polymer solution theory has increased vastly. It would go far beyond the frame of the "Polymer Handbook" to go into this subject, especially since numerous excellent books and review articles are available (272, 387, 389-398, 618).

The following table gives second virial coefficients for different polymer solutions. The polymers are divided into certain classes and arranged alphabetically keeping similar polymers together. The tremendous amount of data reported in the literature for homologous series of polymers prevented a listing of the second virial coefficients for every molecular weight. Following the suggestion of the Editors, only ranges of values for the second virial coefficients are reported here, taking the lowest and the highest values from the reference cited. If no trend was apparent in the data reported, an average value is given. In order to indicate the number of values reported in the original reference, a number, e.g. 5V, is given in the "Remarks" column. Values for solvent mixtures of different ratios, data for different pH etc. are treated similarly. The reader is asked to refer to the original reference to find the detailed data and conditions.

2. Table of Second Virial Coefficients of Polymers in Solution

Polymer	Solvent	T[°C]	$M \times 10^{-4}$	Sec. Virial Coeff. $[\text{mol cm}^3/\text{g}^2] \times 10^4$	Remarks	Ref.
				2.1. POLY(DIENES)		
Poly(butadiene)						
1,4-cis	benzene	28.6	6.0 to 29.3	ca. 15.3	11V;OS	297
			13.8	27.9	OS	
unfractionated	cyclohexane	28.6	84 to 435	ca. 2.92	11V;LS	297
fractionated		28.6	14.3 to 164	7.50 to -1.63	7V;LS	297
1,4-trans						
69%		28.6	107	1.67	LS	297
60-70% unfractionated		25	540 to 1900	ca. 0.52	7V;LS	288
fractionated		25	340 to 1800	2.6 to -0.4	12V;LS	288
			230 to 1100	ca. 2.67	5V;LS	
		34.2	1.74 to 37.0	★ (18.3 to 9.48) x 10^{-4}	4V;OS;(cgs)	497
	dioxane	34.2	1.79 to 43.4	★ (3.81 to 1.49) x 10^{-4}	4V;OS;(cgs)	497
	toluene	34.2	1.82 to 46.6	★ (19.4 to 11.5) x 10^{-4}	7V;OS;(cgs)	497
Poly(butadiene-co-styrene)						
25-30% styrene	benzene	28.6	5.8 to 11.2	ca. 13.5	4V;OS	297
25-30%	cyclohexane	28.6	80 to 2670	ca. 0.9	4V;LS	297
24%		30	4.9 to 51.4	1.7 to 1.0 x 10^{-4}	8V;S;(cgs)	344
diblock, 32.9 to 3.59 weight% PST						
		34.2	4.26 to 62.5	★ (10.0 to 7.55) x 10^{-4}	4V;OS;(cgs)	497
	dioxane	34.2	4.36 to 61.0	★ (5.8 to 1.64) x 10^{-4}	4V;OS(cgs)	
	toluene	34.2	4.30 to 62.5	(12.6 to 8.17) x 10^{-4}	4V;OS(cgs)	
triblock, 45.9 to 4.05 weight% PST						
	cyclohexane	34.2	3.58 to 51.7	★ (10.1 to 8.69) x 10^{-4}	8V;OS(cgs)	497
	dioxane	34.2	3.4 to 17.0	★ (6.63 to 2.16) x 10^{-4}	6V;OS(cgs)	
	toluene	34.2	3.46 to 51.6	★ (12.5 to 9.57) x 10^{-4}	8V;OS;(cgs)	
Poly(chloroprene)	n-butyl acetate	25	14.9 to 302	2.52 to 1.24	7V;LS	493
	carbon tetrachloride	25	15.6 to 318	4.59 to 2.03	7V;LS	493
	trans decalin	0.4 to 48.7	58.7	-0.10 to +1.49	12V;LS	493
		1.1 to 48.7	86.5	-0.074 to +1.43	13V;LS	
		0.4 to 48.7	166	-0.05 to +1.12	12V;LS	
	methyl ethyl ketone	25	14.7 to 292	0	7V;LS	493
Chlorinated rubber	methyl ethyl ketone	29	15.6	3.60	OS	478
		35	441	-0.18	LS	478
Chlorinated rubber-co-poly(ethyl methacrylate)						
20.72%	methyl ethyl ketone	29	19.7	4.80	OS	478
Poly(isoprene)						
cis	cyclohexane	20	160	★ 6.5 x 1.013	$M_{s,D}$ [bar cm^6/g^2]	29
		25	62.0±10%	5.0±20%	OS	508
natural rubber		7	170	★ 14.2 x 1.013	LS [bar cm^6/g^2]	30
		27		★ 14.3 x 1.013		
		7	130	★ 11.7 x 1.013	LS [bar cm^6/g^2]	28
		27	130	★ 12.7 x 1.013		
		25	30	6.2	OS;values for LS in Ref.	508
brominated, 0% bromine	cyclohexane	20	180	★ 14.3 x 1.013	LS; [bar cm^6/g^2]	31
5.8% bromine			185	★ 9 x 1.013		
7.3% bromine			190	★ 4.5 x 1.013		
25.8% bromine			290	★ 2.6 x 1.013		
41.5% bromine			310	★ 0.4 x 1.013		
43.8% bromine			380	★ 0		
Poly(isoprene-co-styrene)						
block copolymer	cyclohexane	30	22.1 to 37.0	5.51 to 4.40^x	LS; Values for OS given in Ref. 495^x	
	methyl isobutyl ketone	30	22.1 to 37.0	1.25 to 1.74^x	LS; Values for OS given in Ref. 495^x	
	toluene	30	22.1 to 37.0	9.04 to 10.2^x	LS; Values for OS given in Ref. 495^x	

References page IV-106

Polymer	Solvent	$T[^{o}C]$	$M \times 10^{-4}$	Sec. Virial Coeff. $[mol\ cm^{3}/g^{2}] \times 10^{4}$	Remarks	Ref.

2.2. POLY(ALKENES)

Polymer	Solvent	$T[^{o}C]$	$M \times 10^{-4}$	Sec. Virial Coeff.	Remarks	Ref.
Poly(1-butene)						
atactic	n-nonane	35	4.41 to 130	2.4	LS	109
isotactic		80	10.5 to 93.5	6.05 to 1.05		
atactic	toluene	45	2.63 to 55.8	10.8 to 4.10	OS	109
isotactic		45	9.01 to 77.5	6.57 to 3.73		
atactic	toluene	45		$M^{-0.32}$	OS	316
isotactic				$M^{-0.25}$		
Poly(ethylene)	1-chloronaphthalene	125	11.5 to 216	★ $(12.0\ to\ 0.78) \times 10^{-4}$	9V;LS; $[cm^{3}/g]$	291
low pressure			12.1 to 52.6	ca. 10.5	8V;LS	296
Marlex			31	11.6	LS	296
			62.5	10.6	LS	296
high pressure			14.4 to 72.0			
linear, fractions		135	2 to 100	$6.3 \times 10^{-3} \times M_{v}^{-0.15}$		231a
	n-decane	115	14.5 to 70.0	5.9 to 1.6	2V;LS; $\overline{M}_{n}$ given in Ref.	231
	1,2,3,4-tetrahydro-naphthalene	105	14.5 to 69.0	21.8 to 6.6	2V;LS; $\overline{M}_{n}$ given in Ref.	231
		125	10.5 to 219	★ 26.8 to 1.68	8V;LS; $[cm^{3}/g]$	291
branched, varying degree		81	3.8 to 73.5	3.6 to 0.9	6V;LS	292
			260 to 610	ca. 0		
high pressure		81	57.3	0.92	LS	293
			198	0.82		
low pressure		105	12.5 to 46.5	23.1 to 15.9	3V;LS	293
high pressure, fractionated	p-xylene	81	0.9 to 21.5	27.8 to 6.54	12V;OS	292
low pressure, fractionated		105	1.12 to 16	33.5 to 26.1	4V;OS	292
				$26.2 \times 10^{-3} \times M^{-0.24}$		292a
Poly(isobutene)	benzene	24-30	126	★ 78.1×10^{-4}	$(1-(298/T));LS;M_{v}\ [cm^{3}/g^{2}]$	308, 311
		30	55.6	★ 0.45	$[100\ cm^{3}/g]$	160
			72.2	★ 0.52		
		40		$A_{2}=0.75 \times 10^{-3} \times M^{-0.12}$	OS	317
	cyclohexane	25	55	★ 2.6	LS; $[100\ cm^{3}/g^{2}]$	202
			240	5.0	LS	454
	dibutyl ether	0 ˣ	900±100	ca. 1.0	LS	438
	ethyl caprylate	25	67.4	★ 0.41	LS; $[100\ cm^{3}/g^{2}]$	202
	n-heptane		5.0 to 2.2	$A_{2}=1.49 \times 10^{-2} \times \overline{M}_{w}^{-0.28}$	17V;LS ★★	600
			40 to 190	2.85 to 2.75	3V;LS	56
		60	190	1.75		56
	heptane/propyl alcohol					
	90/10	25	190	1.95	LS	56
	80/20		190	0		56
		60	190	0.85	LS	56
	72/28	60	190	-0.25		56
	isooctane	25	240.0	1.1	LS	454
	n-pentane	0 ˣ	900±100	ca. 1.2	LS	438
Poly(1-octene)	bromobenzene	25	25 to 400	3.12 to 1.62	LS	420
Poly(propylene)						
isotactic	α-chloronaphthalene	147±1	ca. 48.4	ca. 4.3	3V;LS;three different wavelengths used	446
			ca. 11.6	ca. 3.65	3V;LS;three different wavelengths used	
atactic	benzene	25		$3.2 \times 10^{-3} \times M^{-0.20}$	LS	318
	bromobenzene	100	0.216	1.21	OS	518, 520
				10.8	LS	518, 520
		37-130	0.216	★ $(-2\ to\ +46.5) \times 1.013$	4V;OS; $[bar\ cm^{6}/g^{2}]$	518, 520
	carbon tetrachloride	37	0.125 to 0.250	★ $(54\ to\ 69.5) \times 1.013$	6V;OS; $[bar\ cm^{6}/g^{2}]$ values for 4 more solvents given in Ref.	518, 520
isotactic	1-chloronaphthalene	125		$4.3 \times 10^{-3} \times M^{-0.16}$	LS	318
		135		$16.5 \times 10^{-3} \times M^{-0.27}$	LS	318
		140	10.7 to 111	ca. 3.9	4V;LS	315

ˣ graphical representation fo T-dependance up to 76°C and 140°C resp. given in Ref.   ★★ different methods of polymer preparation. More details given in Ref.

Polymer	Solvent	T[°C]	$M \times 10^{-4}$	Sec. Virial Coeff. $[\text{mol cm}^3/\text{g}^2] \times 10^4$	Remarks	Ref.
Poly(propylene) (Cont'd.)						
atactic	cyclohexane	25		$20 \times 10^{-3} \times M^{-0.26}$		318
Poly(1-pentene)						
isotactic	toluene	30	54.9 to 2.52	★ $(0.86 \text{ to } 2.96) \times 10^{-4}$	7V;OS [cm^4/g]	604

2.3. POLY(ACRYLIC ACID), POLY(METHACRYLIC ACID) and DERIVATIVES

Polymer	Solvent	T[°C]	$M \times 10^{-4}$	Sec. Virial Coeff. $[\text{mol cm}^3/\text{g}^2] \times 10^4$	Remarks	Ref.
Poly(acrylamide)	water	25-60	39	★ $217 \times (1-(235/T))$	LS; [cm^3/g^2]	308
		25	470	★ 0.64	LS; [J·cm^6/g^2]	93
Poly(acrylic acid)	aqueous NaCl					
	IS[x]:0.102 to 0.994	27.5	77	5.95 to 196	6V;LS	320
	1,4-dioxane	25	13.4 to 122	★ 0.28 to 0.18	5V;LS; [100 cm^3/g]	202
	0.2 M hydrogen chloride	20-68	110	★ $49.9 \times (1-(287/T))$	LS; [cm^3/g^2]	308
sodium salt	aqueous NaCl 0.01 M	25	6.4	★ $7.98 \times 10^{-4} \times 1.013$	OS; [bar 1/g^2]	254
	0.05 M		5.87	★ $1.22 \times 10^{-4} \times 1.013$	OS;	
	0.1 M		3.0-64.4	★ca. $0.68 \times 10^{-4} \times 1.013$	6V; [	
	1.0 M		5.8	★ $0.07 \times 10^{-4} \times 1.013$		
Poly(n-butyl acrylate)	acetone	20	3.2 to 92	★ 5.0 to 2.9	7V;LS; [cm^3/g^2]	525
Poly(n-butyl acrylate-co-methyl methacrylate)						
weight% 79.5/20.5	acetone	20	2.2 to 74	★ 8.0 to 2.6	7V;LS; [cm^3/g^2]	524
56/44			2.4 to 71	★ 11.0 to 4.0	7V;LS; [cm^3/g^2]	
29.5/70.5			2.7 to 80	★ 9.5 to 3.6	7V;LS; [cm^3/g^2]	
Poly(tert-butyl acrylate)	methyl ethyl ketone	25	30.6-7.5	2.37 to 4.27	5V;LS	512
Poly(n-butyl methacrylate)						
	acetone	20	5.7 to 49	★ 2.3 to 1.5	5V;LS; [cm^3/g^2]	525
		r.t.	10.8 to 58.2	3.61 to 1.69	5V;OS	280
			11.6 to 61.3	3.04 to 1.33	5V;LS★★	
	isopropyl alcohol	20	10.8 to 50.6	-0.7 to -0.39	4V;LS	279
		23.7	10.7 to 65	-0.15 to 0.1	5V	
		25.0	10.7 to 65.3	+0.03 to 0.2	5V	
		31.0	10.7 to 65.3	ca. 0.54	5V	
		45.0	10.9 to 26.2	ca. 1.44	3V	
	methyl ethyl ketone	23	25 to 258	2.08 to 1.69	7V;LS	165,205
			11.7 to 66.6	3.65 to 1.95	5V;LS★★	280
Poly(butyl methacrylate-co-methyl methacrylate)						
50/50 weight%	acetone	20	7.0 to 43.0	★ 3.2 to 2.2	5V;LS; [cm^3/g^2]	525
Poly(tert-butylphenyl methacrylate)						
	acetone	r.t.	6 to 352	1.4 to 0.48	15V;LS	307
Poly(carbethoxyphenylmethacrylamide)						
	dimethylformamide	20	99.5	2.2	LS	267
	DMF/formamide 90/10			1.2		
	ethyl acetate			6.2		
Poly(cetyl methacrylate)	heptane	25	20 to 107	3.38 to 1.9	9V;LS	263
Poly(1,1-dihydroperfluorobutyl acrylate)						
	benzotrifluoride	25	200 to 3800	ca. 0.2	10V;LS	203
	methyl perfluorobutyrate		520 to 1410	0.5	2V;LS	203
Poly(ethyl acrylate)	acetone	20	5.5 to 86	★ 5.0 to 3.1	6V;LS; [cm^3/g^2]	525
		28	32 to 800	★ca. 10.52×1.013	10V;OS; [bar dl^2/g^2]	208
		30	14.5 to 69.1	★ca. 14.6×1.013	13V;OS; [bar dl^2/g^2]	233
			13.2 to 69.1	★ $(18.1 \text{ to } 14.1) \times 1.013$	7V;OS; [bar dl^2/g^2]	271
			14.5 to 71.1	★ca. 13.3×1.013	5V;OS; [bar dl^2/g^2]	270
Poly(ethyl acrylate-co-methyl methacrylate)						
50/50 weight%	acetone	20	22.0 to 66.0	★ 4.3 to 3.5	5V;LS; [cm^3/g^2)	525
	dimethylformamide	20		see Ref.	LS	515
	methyl ethyl ketone	20		see Ref.	OS	515
Poly(2-ethylhexyl acrylate)						
	toluene	37	60.0 to 7.2	$A_2 = 0.0045 \, M_n^{-0.25}$	4V;OS	586
Poly(ethyl methacrylate)	ethyl acetate	35	92.6 to 1162	1.45 to 1.01	9V;LS	471
				(1.38 to 1.20)	calculated acc. to Orofino, Flory	
	methyl ethyl ketone	23	20 to 263	4.9 to 1.83	9V;LS	162
Poly(n-hexyl methacrylate)						
	methyl ethyl ketone	23	64 to 405	1.77 to 1.17	8V;LS	295

---

[x] IS= ionic strength   ★★ Third virial coefficient reported

Polymer	Solvent	T[°C]	$M \times 10^{-4}$	Sec. Virial Coeff. $[\mathrm{mol\ cm^3/g^2}] \times 10^4$	Remarks	Ref.
Poly(isopropyl acrylate)						
highly isotactic	bromobenzene	60	7.43 to 50.9	6.21 to 3.52	6V;OS	429
moderately isotactic	bromobenzene	60	10.9 to 4.6	5.83 to 4.12	4V;OS	
syndiotactic	bromobenzene	60	14.2 to 56.4	7.06 to 3.71	4V;OS	
atactic	benzene	25	7.17 to 70.3	7.27 to 3.60	5V;OS	
Poly(lauryl methacrylate)	n-butyl acetate	23	26 to 360	1.10 to 0.47	8V;LS	205
Poly(methacrylamide)	water	22-56	32	★ 4.49 x (1-(279/T))	LS; $[\mathrm{cm^3/g^2}]$	308
Poly(methacrylic acid)	hydrochloric acid 0.02 M	27-53	59	★ 14.8 x ((329/T)-1)	LS; $[\mathrm{cm^3/g}]$; T=K	308
	0.1 M		14.6 to 68.2	ca. 1.76	3V;OS	240
	0.114 M	25	20 to 42.7	ca. 16.9	5V;LS	240
	1.0 M		8.4 to 16.9	2.0 to 0.4	3V;OS	240
	LiBr in ethanol 0.125 M	25	4.2 to 38.3	5.2 to 2.3	4V;LS	249
		30	3.4 to 42.1	4.9 to 1.8	5V;OS	249
	methanol	20-40	ca. 51.7	★ca. 2.42	3V;LS; $[\mathrm{cm^3/g^2}]$	346
	0.05 M NaCl	20-60	ca. 50.5	★ 1.465 to 0	4V;LS; $[\mathrm{cm^3/g^2}]$	346
	0.5 M NaCl	20-40	51.0	★ 0.96 to 0.1	3V;LS	
	water + x M NaCl x=0.18 to 0.0007	25	19.6	0 to ~780	5V;LS	355
Poly(methyl acrylate)	acetone	20	7.7 to 88	★ 4.5 to 2.8	5V;LS; $[\mathrm{cm^3/g^2}]$	525
		25	28 to 250	4.2 to 2.4	17V;LS	246,247
	ethyl acetate	35	36.2 to 148	ca. 1.92	5V;LS	482
	methyl ethyl ketone/ isopropyl alcohol 58/42	20	29 to 172	0.1 to 0.06	8V;LS	246
Poly(methyl acrylate-co-chlorinated rubber) 86.9/13.1						
graft copolymer	methyl ethyl ketone	35	3394	0.079	LS	478
Poly(methyl acrylate-co-methyl methacrylate)						
50/50 weight%	acetone	20	8.0 to 52.5	★ 5.5 to 3.5	10V;LS; $[\mathrm{cm^3/g^2}]$	525
Poly(methyl butacrylate)	methyl ethyl ketone	30	7.52 to 428	2.77 to 1.00	10V;LS	445
Poly(methyl ethacrylate)	methyl ethyl ketone	30	4.61 to 191	4.78 to 1.72	10V;LS	445
Poly(methyl methacrylate)	acetone	20		see Ref.	small angle x-ray scat., varying tacticities measured	474
			0.02 to 51	30 to 1.5	10V;small angle x-ray scat.	492
			2.3 to 530	★(12.2 to 1.55) x 10^{-6} x1.013	10V;SD; $[\mathrm{bar\ 1^2/g^2}]$	52
			20 to 655	★ 2.25 to 0.68	3V;SD; $[\mathrm{cm^3/g}]$	15
		22	55.0	1.9	LS;identical samples;	469
			51.0	1.5	small angle x-ray scat.	469
		25	0.0203 to 0.0385	★(-24.2 to +113.0)x10^{-6} x1.013	4V; $[\mathrm{bar\ 1^2/g^2}]$; $\overline{M}_n$	408
			0.0691 to 0.2480	★(627 to 47.5) x 10^{-6} x 1.013	3V; $[\mathrm{bar\ 1^2/g^2}]$; $\overline{M}_n$	408
			5.1±0.43	★ (8.2±0.61) x 1.013	LS; $[\mathrm{bar\ cm^6/g^2}]$	419
			367±5.0	★ (2.18±0.05) x 1.013	LS; $[\mathrm{bar\ cm^3/g}]$	419
				2.9 x 10^{-3} x $M^{-0.22}$	LS	273
			6.8 to 137	★ (7.1 to 3.5) x 10^{12}	7V;LS; $[981\ \mathrm{cm^3/g}]$	232
			24 to 447	2.3 to 1.1	5V;LS	264
			2.77 to 780	★ 0.975 to 0.29	12V;LS; $[\mathrm{J \cdot cm^3/g^2}]$	99
			145	★ 1.9	LS; $[100\ \mathrm{cm^3/g}]$	202
			0.269 to 250	★0.0097 to 3.90	13V; $[100\ \mathrm{cm^3/g}]$	275
			8.6 to 105	2.90 to 1.58	7V;OS	202
			9.2 to 79.8	1.89 to 1.52	9V;LS	202
atactic			48.6 to 662	★ 2.38 to 0.82	4V;LS; $[\mathrm{cm^3/g^2}]$	55
		30		★ 2.63 x 10^{-5} x $\overline{M}_n^{0.78}$	OS; $[100\ \mathrm{cm^3/g}]$	274
			16.1	1.85	OS	494
unfractionated		30.5	16.2 to 64.7	★ 28.8 to 88	12V;OS; $[\mathrm{cm^3/g}]$	274
fractionated			6.4 to 72.6	★ 12.3 to 95	5V;OS; $[\mathrm{cm^3/g}]$	274
	acetonitrile	40	1.0	-0.1	small angle x-ray scat.	492
		43.3	1.0	0	small angle x-ray scat.	492
		50	1.0	+0.2	small angle x-ray scat.	492
	benzene	25	0.0188 to 0.0935	★(-222.5 to +8.9) x 10^{-6} x1.013	8V; $[\mathrm{bar\ 1^2/g}]$; $\overline{M}_n$	408
		30	43.3	2.63	OS	494
	butanone/isopropanol 58.2/41.8	4-36	21	★ +0.001 to +0.369	9V;LS; $[\mathrm{J \cdot cm^3/g^2}]$	13
	55/45	4-36	21	★ -0.128 to +0.293	9V;LS; $[\mathrm{J \cdot cm^3/g^2}]$	13
	50/50	8-40	21	★ -0.365 to +0.237	9V;LS; $[\mathrm{J \cdot cm^3/g^2}]$	13

Polymer	Solvent	T[$^{\circ}$C]	M x 10^{-4}	Sec. Virial Coeff. [mol cm^3/g^2] x 10^4	Remarks	Ref.
Poly(methyl methacrylate) (Cont' d.)						
fractionated	butanone/isopropanol					
	46.8/53.2	18-40	21	★ -0.248 to +0.174	7V;LS; [J · cm^3/g^2]	13
	butyl acetate	25	48.6 to 662	★ 0.96 to 0.36	4V;LS; [cm^3/g^2]	55
		-18 to +60	21	★ 0 to 1.20	6V;LS; [cm^3/g^2]	210, 215
	butyl chloride	14-48	21	★ -0.242 to +0.088	10V;LS [J · cm^3/g^2]	13
	carbon tetrachloride	25	0.0202 to 0.0755	★(-400.0 to +6.7) x 10^{-6} x 1.013	3V; [bar l/g^2];$\overline{M}_n$	408
			0.1160 to 0.4625	★(-23.2 to 0.0) x 10^{-6} x 1.013	4V; [bar l/g^2];$\overline{M}_n$	408
	chlorobutane	4-48	3	★ -1.6 to +0.365	12V;LS; [cm^3/g^2]	210, 215
		28-48	110	★ -0.3 to +0.325	7V;LS; [cm^3/g^2]	210, 215
		28-48	460	★ -0.23 to +0.31	7V;LS; [cm^3/g^2]	210, 215
		35.6	20 to 655	★ -0.24 to +0.07	3V;SD; [cm^3/g^2]$_{12}$	15
	chloroform	25	83 to 137	★ (12 to 10) x 10^3	2V;LS; [981 cm^3/g]	232
			0.0204 to 0.0379	★(-53.9 to +81.0) x 10^{-6} x 1.013	4V; [bar l/g^2];$\overline{M}_n$	408
			0.0710 to 0.5240	★(+49.6 to +21.9) x 10^{-6} x 1.013	3V; [bar l/g^2];$\overline{M}_n$	408
	dimethylformamide		3.22±0.03	★ (10.9 to 0.16) x 1.013	OS; [bar cm^6/g^2]	419
			218.0±6.0	★ (6.2±0.08) x 1.013	OS; [bar cm^3/g]	419
	dioxane		48.6 to 662	3.24 to 1.43	4V;LS; [cm^3/g^2]	5
			122 to 200	3.1 to 3.0	2V;LS	294
	ethyl acetate		17.2 to 310	1.9 to 1.2	3V;OS	196
	4-heptanone	16-62	210	★ -0.76 to 0.52	10V;LS; [cm^3/g^2]	210, 215
	isoamyl acetate	20-60	21	★ -0.276 to 0.014	11V;LS; [J · cm^3/g^2]	13
	isopentyl acetate	20-60	210	★ -1.13 to 0.055	12V;LS; [cm^3/g^2]	210, 215
	methyl ethyl ketone	25	0.5 to 5.28	7.38 to 2.18	6V;LS	418
			7.55	2.70	LS	418
unfractionated	methyl ethyl ketone	25	2.0 to 173	★ (4.02 to 1.39) x 10^{-6}	7V;LS; [cm^3/g^2] ★★	273
			13.0 to 276	(3.04 to 1.07) x 10^{-6}	4V;LS; [cm^3/g^2]$_{12}$	273
			76 to 140	★ (5.4 to 4.2) x 10^3	4V;LS; [981 cm^3/g]	232
			122	1.6	LS	294
			3.4 to 980	2.78 to 1.46	10V;LS	253
			41 to 326	1.92 to 1.66	7V;LS	163
	methyl ethyl ketone/isopropyl					
	alcohol 55/45	23	41 to 326	0.08 to 0.02	6V;LS	163
	58.2/41.8	4-36	21	★ 0.005 to 1.44	9V;LS; [cm^3/g^2]	210, 215
	55/45	4-36	21	★ -0.57 to 1.14	10V; [cm^3/g^2]	210, 215
	50/50	8-50	21	★ -1.56 to 0.91	10V; [cm^3/g^2]	210, 215
	46.8/53.2	8-40	21	★ -1.02 to 0.68	8V; [cm^3/g^2]	210, 215
linear	methyl ethyl ketone	40	4.59 to 94.0	4.43 to 2.34	8V;LS;OS results also given in Ref.	570
randomly branched			4.89 to 449	5.02 to 1.95	12V;LS;OS results also given in Ref.	570
			147 to 699	1.44 to 0.61	3V;LS;OS results also given in Ref.	570
unfractionated	nitroethane	25	2.2 to 171	★ 5.51 to 1.68	6V;LS; [cm^3/g^2] ★★	273
			11.4 to 56.4	★ 3.38 to 1.73	2V;LS; [cm^3/g^2] ★★	273
				0.67 x 10^{-3} x M$^{-0.257}$	LS	273
			122 to 200	ca. 2.6	2V;LS	294
	tetrahydrofuran		4.7 to 930	5.3 to 1.1	6V;LS	121
	toluene	30.5	1.23 to 2.0	★ 5.7 to 8.5	3V;OS; [cm^3/g^2]	274
Poly(methyl methacrylate-co-chlorinated rubber) (45.9/54.1)						
graft copolymer	methyl ethyl ketone	35	815	0.36	LS	478
Poly(methyl methacrylate-co-ethylene dimethacrylate)						
randomly branched	methyl ethyl ketone	40		see Ref.	11V	549
Poly(n-octyl methacrylate)						
	methyl ethyl ketone	23	326 to 1252	0.77 to 0.35	7V;LS	209
Poly(N-phenylmethacrylamide)						
	acetone	20	10 to 320	★ 4.1 to 3.2 x 10^{-4}	LS; [cm^3/g^2]	229

2.4 VINYL POLYMERS

Polymer	Solvent	T[$^{\circ}$C]	M x 10^{-4}	Sec. Virial Coeff. [mol cm^3/g^2] x 10^4	Remarks	Ref.
Poly(acrylonitrile)	dimethylformamide	20	9.8 to 12.0	★ 22.9 to 21.4	LS; [cm^3/g^2]	603
				4.94 x 10^{-2} x M$_w^{-0.24}$	2 different wavelenghts used	

★★ third virial coefficient reported.

Polymer	Solvent	$T[^{o}C]$	$M \times 10^{-4}$	Sec. Virial Coeff. $[mol\ cm^{3}/g^{2}] \times 10^{4}$	Remarks	Ref.
Poly(acrylonitrile) (Cont' d.)						
	dimethylformamide	20	0.9 to 6.9 ★	32.2 to 7.0	5V;OS; $[cm^{4}/g]$	248
			9.1 to 76.2	$2.43 \times 10^{-2} \times M^{-22}$	LS	95
		25	4.3 to 29.8 ★ca.	21	7V;OS; $[cm^{3}/g^{2}]$	199
			2.7 to 15.9	16 to 20	OS	200
		25-40	3.3 to 10.1 ca.	19.1	13V;OS	74
		35	2.8 to 57.5	$2.74 \times 10^{-2} \times M^{-0.24}$	$9V;M_{\eta}$	201
Poly(n-butoxyethylene)	methyl ethyl ketone	25	21.5 to 86.1	2.3 to 0.7	2V;LS	149
Poly(ethoxyethylene)	methyl ethyl ketone	25	34.7 to 40.9	3.4 to 2.3	2V;LS	149
Poly(isopropoxyethylene)	methyl ethyl ketone	25	53.6 to 89.4	1.9 to 1.7	2V;LS	149
Poly(methoxyethylene)	methyl ethyl ketone	25	1.3 to 45 ca.	3.89	12V;LS	149
Poly(2-methyl-5-vinylpyridine)						
	methyl ethyl ketone	25	6.7 to 120.0	2.8 to 1.2	13V;LS	432
Poly(1-trimethylsilyl ethylene)						
	cyclohexane	25	213 to 59 ★	$(4.85\ to\ 12.00) \times 10^{-4}$	6V;LS; $[cm^{3}/g^{2}]$	553
Poly(vinyl acetate)	acetone	ca. 30	2.7 to 84.5	8.80 to 3.34	9V;LS	330
			34.3 to 72.2	3.66 to 3.50	LS	331
			7.8 to 66	6.5 to 2.5	18V;LS	289
branched	benzene	30	25.8 to 4.52 ca.	4.76	10V;OS	453
			520.3 to 33.1	0.208 to 2.066	8V;LS	
	methanol	ca. 30	37.0 to 62.0 ca.	3.5	2V;LS	331
linear		ca. 36	162 to 190	1.14 to 1.3	2V;LS	333
branched		ca. 36	292	0.99	3V;LS	333
		ca. 36	358	1.06		
	methyl ethyl ketone	25	87 to 342	3.51 to 2.43	11V;LS	111
			5.8 to 36.7 ★ca.	10×1.013	8V;OS; $[bar\ cm^{6}/g^{2}]$	112
			9.3 to 110 ★ca.	5×1.013	LS; $[bar\ cm^{6}/g^{2}]$	112
		30	4.0 to 126.8	7.00 to 2.95	12V;LS	330
			30.0 to 77.2	3.93 to 2.90	LS	331
linear	methyl ethyl ketone	ca. 36	44 to 358	3.2 to 2.74	10V;LS	333
branched		ca. 36	43 to 1,260	2.99 to 0.66	9V;LS	333
linear	trichlorobenzene	ca. 36	138 to 376	2.29 to 1.75	12V;LS	333
				$5.94 \times 10^{-3} \times M^{-0.230}$		333
branched	trichlorobenzene	ca. 36	161 to 411	1.69 to 1.18	19V;LS	333
Poly(vinyl alcohol)	water	30	18 to 19.6 ★	3.9 to 5.2	2V;LS; $[cm^{3}/g^{2}]$	282, 283
		73.5	24.5 ★	1.119	LS; $[cm^{3}/g^{2}]$	282, 283
		80	10 to 45.8 ★	2.189 to 0.775	5V;LS; $[cm^{3}/g^{2}]$	282, 283
Poly(vinyl bromide)	tetrahydrofuran	r.t.	0.6 to 7.6	6 to 0	14V;LS	287
			10.4 to 27	2.7 to 2.0	2V;LS	
Poly(vinyl butyral)	dioxane	37	54.1 to 8.95	5.53 to 10.3	8V;OS	597
			20.8 to 7.39	9.15 to 11.3	5V;OS	
			18.1 to 5.75	9.45 to 12.2	7V;OS	
Poly(vinyl-iso-butyral)			46.2 to 20.7	6.9 to 9.6	6V;OS	
Poly(vinyl carbanilate)	cyclohexanol	ca. 40 to 90	110	★ca. -1.2 to +1.2	8V;LS $[cm^{3}/g^{2}]$	504
	cyclohexanone	ca. 15 to 90		★ca. 2.2 to 1.9	6V;LS $[cm^{3}/g^{2}]$	504
	diethylene glycol/					
	diethyl ether	ca. 5 to 95		★ca. 2.7 to 1.5	7V;LS $[cm^{3}/g^{2}]$	504
	diethyl ketone	ca. 20 to 90		★ca. -0.5 to +1.8	8V;LS $[cm^{3}/g^{2}]$	
	dioxane	20	0.72 to 18.3 ★	$(3.4\ to\ 1.71) \times 10^{-4}$	11V;LS $[cm^{2}/g^{2}]$	401
		ca. 5 to 95	110	★ca. 2.7 to 2.2	7V;LS $[cm^{3}/g^{2}]$	504
	pyridine	ca. 10 to 90		★ca. 2.5 to 2.0	4V;LS $[cm^{3}/g^{2}]$	504
Poly(N-vinylcarbazole)	benzene	25	455 to 13.6	0.57 to 1.32	LS	457
			163	0.61	LS	499
			229	0.54	LS	499
	dioxane	37	4.24 to 107	2.35 to 0.91	8V	499
				$4.26 \times 10^{-3} \times \overline{M}_{n}^{-0.27}$		
	toluene	37 to 43	29.8	0 to 0.27	3V;OS	499
		37 to 43	52.2	0.01 to 0.26	3V;OS	499
Poly(vinyl chloride)	cyclohexanone	25	11.8 (11.9)	11	A;(LS)	602
fractionated			5.9 (5.0)	12 (18)	A;(LS)	602
		30	3.23 to 9.74 ★	3.7×10^{2}	OS; $[cm^{4}/g]$	108
		30 to 70	9.0 ★	$3.9 \times (1+(1613/T))$	LS;M_{v}; $[cm^{3}/g^{2}]$	308, 309

Polymer	Solvent	$T[^\circ C]$	$M \times 10^{-4}$	Sec. Virial Coeff. $[\text{mol cm}^3/\text{g}^2] \times 10^4$	Remarks	Ref.
Poly(vinyl chloride) (Cont'd.)						
prepared in 2,4,6-trichloroheptane		see Ref.	0.5 to 2.1	★ (16 to 27) $\times 10^{-4}$	7V;LS; $[\text{cm}^3/\text{g}]$	451
	tetrahydrofuran	see Ref.	8.7 to 1.65	★ (15 to 18.5) $\times 10^{-4}$	11V;LS; $[\text{cm}^3/\text{g}]$	451
		25	4.88	★ 12.8×10^{-4}	OS; $[\text{cm}^3/\text{g}]$	451
			4.49	★ 16.2×10^{-4}	OS; $[\text{cm}^3/\text{g}]$;more values for 7 different PVC-preparations given in Ref.	451
			2.48	★ 6.85	OS;$[\text{J} \cdot \text{cm}^3/\text{g}^2]$	75
			3.7 to 11.0	★ca. 2.5	5V;OS; $[\text{J} \cdot \text{cm}^3/\text{g}^2]$	75
			12.2	15	A	602
			5.6 (5.3)	14 (17)	A;(LS)	602
		29	13.6	9.03	OS	478
Poly(vinyl chloride-co-poly(ethyl methacrylate))						
84.55%/15.45%	tetrahydrofuran	29	16.1	4.80	OS	478
Poly(vinyl chloride-co-poly(methyl methacrylate))						
45.1%/54.9%, graft copolymer						
	tetrahydrofuran	29	30.2	9.55	OS	478
Poly(vinyl methyl ether) see Poly(methoxyethylene)						
Poly(4-vinylpyridine)	ethyl alcohol	r.t.	10.2 to 185	ca. 4.13	10V;LS	286
Poly(vinylpyrrolidone)	cyclohexanone	25	2.45 to 3.79	★ca. 2.6×10^{-7}	3V;OS; $[1/g^2]$	191
	water	25	1.16 to 7.54	★ca. 3.39×10^{-7}	12V;LS; $[1/g^2]$	191
			1.95 to 93.3	★ 64.7 to 2.52×10^{-7}	6V;LS; $[1/g^2]$	

2.5. POLY(STYRENES)

Polymer	Solvent	$T[^\circ C]$	$M \times 10^{-4}$	Sec. Virial Coeff. $[\text{mol cm}^3/\text{g}^2] \times 10^4$	Remarks	Ref.
Poly(p-bromostyrene)	toluene	30	386 to 63.0	0.076 to 0.113	5V;LS	516
Poly(m-chlorostyrene)	butanone	30	106.4 to 32.2	1.44 to 1.78 $A_2 = 1.54 \times 10^{-3} \times \overline{M}_w^{-0.17}$	5V;LS	595
	toluene	35	107.5 to 14.5	2.42 to 5.08 $A_2 = 26.5 \times 10^{-3} \times \overline{M}_w^{-0.33}$	6V;LS	
Poly(o-chlorostyrene)	n-butyl acetate	5	41.5	0.570	OS	585
		30	40.3	0.449	OS	585
		55	40.1	0.337	OS	585
	chlorobenzene	30	40.6	1.72	OS	585
	toluene	25	100.3 to 14.2	1.78 to 2.70	6V;LS	527
		30	40.7	1.73	OS	585
		35	108.7 to 43.5	2.18 to 2.70	6V;LS	500
			142.9 to 235	1.96 to 3.00 $5.22 \times 10^{-3} \times \overline{M}_w^{-0.23}$	4V;LS	500
Poly(p-chlorostyrene)	n-butyl acetate	5	18	2.96	OS	585
		30	17.6	2.39	OS	585
		50	16.5	1.90	OS	585
	chlorobenzene	30	31.6	2.17	OS	585
	methyl ethyl ketone	21	48 to 422	1.29 to 0.752	10V;LS	300
	toluene	30	47.2 to 19.2	0.96 to 1.69 $A_2 = K \times \overline{M}_n^{-0.6}$	5V;OS; A_3 given in Ref.	609
			137 to 21.4	0.570 to 0.895	6V;LS	516
			20.6	0.824	LS	516
			46.3	0.625	OS	585
Poly(2,5-dichlorostyrene)	dioxane	20	59.0 to 1960	0.88 to 0.20	7V;LS	250
Poly(p-methoxystyrene)	butanone	25	87 to 22	$A_2 \sim \overline{M}_w^{-0.20}$	LS;6V	610
	toluene	30	85 to 22	$A_2 \sim \overline{M}_w^{-0.40}$	LS;6V	
Poly(α-methylstyrene)	benzene	25	0.145	★ 17.4×1.013	LS; $[\text{bar cm}^6/\text{g}^2]$	407
			0.676	★ 13.7×1.013	LS; $[\text{bar cm}^6/\text{g}^2]$	407
		30		see Ref.		522
monodisperse	cyclohexane	40	122	0.218		484
		50	122	0.509		484
		50 to 32	31	0.575 to -0.142	7V;LS;A_2-values for mixtures of these samples given in Ref.	484
			1.73	0.461 to -0.116	7V;LS;A_2-values for mixtures of these samples given in Ref.	484
			300	0.406 to -0.109	7V;LS;A_2-values for mixtures of these samples given in Ref.	484

References page IV-106

Polymer	Solvent	T[°C]	M x 10^{-4}	Sec. Virial Coeff. [mol cm^3/g^2] x 10^4	Remarks	Ref.
Poly(α-methylstyrene) (Cont'd.)						
	toluene	25	0.145	★ 49 x 1.013	LS [bar cm^6/g^2]	407
			0.676	★ 34 x 1.013	M$_{s,D}$;SE; [bar cm^6/g^2]	407
		37	27.8 to 4.69	4.2 to 5.9	8V;OS	584
Poly(α-methylstyrene-co-styrene)						
50/50	toluene	37	25.3 to 10.6	5.9 to 5.2	4V;OS	584
Poly(o-methylstyrene)	toluene	30	43 to 10.2	2.64 to 4.80 $A_2 = 3.89 \times 10^{-2} \times \overline{M}_w^{-0.38}$	7V;LS	588
Poly(p-methylstyrene)	cyclohexane	30	84.7 to 26.3	1.56 to 1.92	3V;LS	517
	dichloroethane	30	118 to 50.0	1.70 to 2.28	3V;LS	517
	diethyl succinate	60 to 16	197	0.589 to -0.026	8V;LS	517
			89.6	0.819 to -0.001	8V;LS	
			76.2	0.900 to -0.013	8V;LS	
			68.4	0.830 to -0.046	8V;LS	
	methyl ethyl ketone	30	121 to 39.1	1.00 to 1.28	4V;LS	517
	toluene	30	180 to 19.2	2.18 to 4.37	9V;LS	517
Poly(styrene)	benzene	20	133 to 710	★ (3.6 to 3.3) x 10^{-6}	4V;LS; [100 cm^3/g]	514
			270	★ 3.4 x 10^{-6}	LS; [100 cm^3/g];A$_3$ given in Ref.	514
IUPAC-sample			7.9	6.2	OS	11
IUPAC-sample			23.3	5.0	OS	11
IUPAC-sample			63	5.1	OS	11
			0.338±0.008	23±3	M from isothermal destillation	463
branched			17	4.9	LS	312
linear				5.7		312
branched			25.7	3.3		312
linear				5.4		312
		25	10.9 to 500	ca. 3.83	8V;LS	56
			320 to 6.2	2.9 to 6.0	5V;LS	454
atactic			68	★ 4.9 x 10^{-4}	M$_\eta$; [cm^4/g]	562
		40	400	3.2	LS	172
		60		2.30		56
	benzene/ethyl alcohol					
80/20		25	175	1.75	LS	56
		60		1.55		56
70/30		25		0		56
		60		1.1		56
	benzene/heptane					
heptane 0-56 Vol.%		35.0±0.2	2.34 to 237	3.06 to 0	9V;LS;A$_2$-values corrected for preferential adsorption given in Ref.	440
	benzene/hexane 50/50	25	175	1.0		56
		60		0.9		56
	40/60	25		0		56
		60		0.65		56
	30/70	60		0		56
	benzene/isopropanol					
isopropanol 0-39 Vol.%		35.0 ±0.2	234 to 417	3.06 to 0	8V;LS	440
	n-butyl acetate	5	41.1	1.88	OS	585
		30	40.4	2.08		
		55	40.2	2.00		
	carbon tetrachloride	25	92	2.6	LS	294
	chlorobenzene	25	9.84	5.41	OS	404
		30	38.8	3.77	OS	585
	chloroform	20	37.2 to 280	8.0 to 6.6	4V;LS	176
	chloroform/methyl alcohol					
100/0		25	33	7.5	LS	176
90/10				5.4		
80/20				1.9		
	chloroform	30	10	5.4 to 5.1 ★★	5V;LS;A$_{2,S}$;A$_{2,H}$ given in Ref.	528
isotactic	1-chloro-4-methyl benzene					
		r.t.	41.0 to 105	ca. 2.4	2V;LS	328
	1-chloronaphthalene	130	ca. 34.5	★ca. 1.33 x 10^2	OS; [cm^3/g];see Ref. for various extrapolation methods	523

★★
pressure dependance from 0-810 bar measured.

SECOND VIRIAL COEFFICIENTS

Polymer	Solvent	T[$^\circ$C]	M x 10^{-4}	Sec. Virial Coeff. [mol cm^3/g^2] x 10^4	Remarks	Ref.
Poly(styrene) (Cont' d.)						
atactic	1-chloro-n-undecane	27.96 to 43.99	40.6 ★	-6.0 to 9.0	9V;LS; [cm^3/g];$\overline{M}_n$ given in Ref.	133
	cyclohexane	23 to 45	6.87 ★	-0.410 to 0.226	6V;LS; [J cm^3/g^2]	105
		27 to 41.5	161	-0.37 to 0.258	5V;LS	166
		30	5 to 20.3	-0.276 to -0.445	4V;OS	106
		40		0.634 to 0.445	4V	106
		50		1.33 to 0.901	4V	106
		34.2	7.8 to 21.0 ★	(-0.061) to 0.071) x 10^{-4}	3V;OS;(cgs)	497
		32.5 to 55	see Ref.	-0.153 to 0.538	5V;LS	329
		35 to 55	320	0 to 0.54	3V;LS	334
			400	0 to 0.88	3V;LS	172
		31.94 to 43.99	40.6 ★	-131 to 23.1	5V;LS; [cm^3/g]	133
		29.92 to 44.03	★	-16.3 to 22.7	5V;LS; [cm^3/g]	133
		28	50	0.07	LS	454
		33		0		
		50		0.5		
		40	10	0.4 to 0.25 ★★	3V;LS	528
		45		0.79 to 0.87 ★★	3V;LS	528
		50		1.11 to 1.19 ★★	3V;LS	528
		50	51.3 to 4.59	0.95 to 1.29	7V;LS	503
	decalin	25	50 ★	0.8 x 10^{-4}	M_η; [cm^4/g]	562
		25	180	0.19 to -0.01 ★★★	4V;LS	528
		30		0.40 to 0.20 ★★★★	5V;LS	528
		40		0.72 to 0.45 ★★★★	5V;LS	528
isotactic	o-dichlorobenzene	r.t.	19.6 to 26.4	2.9 to 1.9	3V;OS	328
atactic	dichloroethane	22	0.3 to 1.7	7 to 2.54	6V;LS	166
		25	0.3	6.8	LS	166
		35	43.9 to 536	see Ref.	11V;LS	564
		67	1.6	2.7	LS	166
	dichloroethane/cyclohexane 65/35	22	161	3.98	LS	166
		67		3.32		166
	35/65	22		2.76		166
		67		2.70		166
	6/94	22		0.78		166
		67		1.27		166
	dioxane	25	90 to 92	2.9 to 2.8	2V;LS	294
			9.73	3.08	OS;A_3 given in Ref.	404
		34.2	7.76 to 19.6 ★	(5.11 to 4.48) x 10^{-4}	3V;OS;(cgs)	497
	dioxane/heptane heptane 0-58.5 Vol.%	35±0.2	224 to 263	2.51 to 0	9V;LS	440
	ethyl acetate	25	24.3	0.7	OS	196
	ethylbenzene		68 ★	4.7 x 10^{-4}	M_η; [cm^4/g]	562
	hexyl-m-xylene	16	57.0 to 600	0	3V;LS; small angle x-ray scat.	492
		20	1.49 to 3.98	0.2 to 0.05	2V;LS; small angle x-ray scat.	492
	poly(isobutene)/cyclohexane	25 to 40	8.42	3.59	5V;LS	599
		30	3.88	1.81		599
		35		0.93		599
		40		0.46		599
	methyl ethyl ketone	7.5 to 45	6.87 ★	0.586	LS; [J cm^3/g^2]	105
IUPAC-sample		r.t.	103	1.29	LS	253
		20	11.5 to 280	2.2 to 0.72	8V;LS	176
		22	0.2 to 177	4.3 to 0.86	9V;LS	166
		25	0.32 to 98	3.5 to 1.05	8V;LS	166
			24.5 to 54	1.2 to 1.0	2V;OS	196
			9.63 to 78.5	166 to 5.3	5V;OS	176
			17 to 25.7	1.9	2V;LS;OS	312
			90	1.3	LS	294
		35	4.51 to 635	see Ref.	13V;LS	564
		67	51.7 to 98	0.89 to 0.81	3V;LS	166

★★ pressure dependance from 0-405 bar measured.   ★★★ pressure dependance from 0-570 bar measured.   ★★★ pressure dependance from 0-759 bar measured.

References page IV-106

Polymer	Solvent	T[°C]	M x 10⁻⁴	Sec. Virial Coeff. [mol cm³/g²] x 10⁴	Remarks	Ref.
Poly(styrene) (Cont'd.)	methyl ethyl ketone/ isopropyl alcohol 87/13	22	1.63 to 1.74	-0.2 to -0.3	2V;LS	166
		67		0	2V;LS	166
	tetrahydrofuran	30	2.44 to 570	8.32 to 2.105	6V;LS;$A_{2,H}$;$A_{2,S}$ given in Ref.	481
	tetralin	25	68	★ 4.0×10^{-4}	M_η;[cm⁴/g]	562
	toluene	12	5.4 to 440	★ 30 to 792	8V;LS;[cm³/g]	548
star branched			38.4	★ 115	LS;[cm³/g]	548
			187	★ 380	LS;[cm³/g]	548
comb branched			232	★ 404	LS;[cm³/g]	548
			388	★ 520	LS;[cm³/g]	548
		15	180	ca. 2.24 to 2.39★	5V;LS	528
		20	0.585 to 12.5	9.4 to 4.5	5V;small angle x-ray scat.	492
		25	180	2.27 to 2.41★	5V;LS	528
			20.9 to 392	4.20 to 1.84	4V;LS	492
			68	★ 5.0×10^{-4}	M_η;[cm⁴/g]	562
			9.9 to 95.2	6.5 to 3.5	7V;LS;OS	176
			23.7	★ 1.23	see Ref.;[J cm³/g²]	56
			7.2 to 180	★ (11 to 7.0) x 10³	10V;LS;[cm⁴/g]	245
		30	10	4.6	5V;LS;$A_{2,H}$;$A_{2,S}$ given in Ref.	528
			37.3	3.45	OS	585
			1.33	4.49	OS	467
			41.8 to 3.76	4.54 to 5.88	7V;OS	503
			3.09 to 61.2	★ $\Gamma_2 = 7.41 \times 10^{-5} \times M^{0.75}$	[100 cm³/g];$\Gamma_3 = \frac{1}{4}\Gamma_2^2$	169
			15.5 to 71.0	≈ 4.17 to 1.56	4V;LS	290
				$A_2 = k \times M^{-0.22}$	OS	317
				$= 4.6 \times 10^{-3} \times \overline{M}_n^{-0.22}$	OS	319
				$= 2.1 \times 10^{-3} \times \overline{M}_n^{-0.146}$	OS	319
		30.27	5.6 to 45.2	★ 29 to 145	5V;OS;[cm³/g]	327
		34.2	7.78 to 19.6	★ $(5.43 \text{ to } 4.46) \times 10^{-4}$	3V;OS;(cgs)	497
		35	3.31 to 646	see Ref.	16V;LS	564
		37	36.6 to 5.09	5.5 to 6.9	8V;OS	584
		40	20.9 to 160	4.28 to 2.46	3V;LS	121
		22 to 67	161	3.12 to 2.43	2V;LS	166
		28 to 67	98	★ 20.4 x ((479/T)-1)	M_v;LS;[cm³/g]	308,310

Poly(styrene-co-maleic anhydride)/poly(ethyl acrylate)

A		B				

graft copolymer

A	B						
56.5/43.5		acetone	25	5.34	★ 2.1×10^{-4}	OS[cm⁴/g]	489
77 /23				7.12	★ 2.1×10^{-4}		
0 /100				35.0	★ 5.7×10^{-4}		

Poly(styrene-co-p-methoxystyrene)
random
mol% p-methoxystyrene

	Solvent	T[°C]	M x 10⁻⁴	Sec. Virial Coeff.	Remarks	Ref.
26.0	toluene	25	7.1 to 35	5.0 to 3.0	7V;OS	617
53.8			3.55 to 70.1	4.1 to 3.2	12V;OS	617
75.6			4.4 to 42.1	5.9 to 3.5	8V;OS	617
26.4			6.1 to 66.5	5.34 to 4.29 $A_2 = 2.0 \times 10^{-3} \times \overline{M}_w^{-0.115}$	9V;LS	617
53.0			6.6 to 178.3	5.46 to 3.41 $A_2 = 2.4 \times 10^{-3} \times \overline{M}_w^{-0.135}$	11V;LS	617
75.6			7.9 to 171.7	4.54 to 3.00 $A_2 = 2.3 \times 10^{-3} \times \overline{M}_w^{-0.145}$	8V;LS	617

Poly(styrene-co-methyl acrylate)

	Solvent	T[°C]	M x 10⁻⁴	Sec. Virial Coeff.	Remarks	Ref.
	dimethylformamide	20			see Ref. OS	515
	methyl ethyl ketone	20			see Ref. LS	515

Poly(styrene-co-methyl methacrylate)

	Solvent	T[°C]	M x 10⁻⁴	Sec. Virial Coeff.	Remarks	Ref.
ca. 50/50	carbon tetrachloride	25	133 to 190	2.3 to 2.4	2V;LS	294
	dioxane		133 to 190	3.8 to 3.6	2V;LS	294
random, mol% St. 29.3	2-ethoxyethanol	40.0(68.0)	35.4	0	OS	443
56.2	(cyclohexanol)	58.4(61.3)	35.0	0		
70.2		72.8(63.0)	34.2	0		

★ pressure depandance from 1 to 759 bar measured

SECOND VIRIAL COEFFICIENTS

Polymer	Solvent	T[$^{\circ}$C]	M x 10^{-4}	Sec. Virial Coeff. $[\text{mol cm}^3/\text{g}^2]$ x 10^4	Remarks	Ref.
Poly(styrene-co-methyl methacrylate) (Cont'd.)						
block copolymer						
mol% St 49.6	2-ethoxyethanol	81.0	31.7	0	OS	443
	(cyclohexanol)	81.3	31.7	0		
73.2		84.0	39.2	0		
85.1		84.0	19.3	0		
100		81.8	20.6	0		
fractionated	methyl ethyl ketone	25	4.9 to 227	3.2 to 1.45	11V;LS	294
unfractionated			97	1.55		294
	isoamyl acetate				see Ref.	507
	nitroethane		133 to 190	1.5 to 1.2	2V;LS	294
	toluene				see Ref.	507
random, copolymer						
29±1% St	toluene	30	59.2 to 4.67	2.6 to 3.7	7V;OS	479
56±1% St		30	50.0 to 9.66	3.3 to 4.1	8V;OS	479
70±1% St		30	43.2 to 6.67	2.9 to 4.1	6V;OS	479
29±1% St	1-chloro-n-butane	30	59.2 to 10.0	1.5 to 2.1	7V;OS	479
56±1% St		30	50.0 to 3.42	2.2 to 5.6	11V;OS	479
70'1% St		30	43.2 to 4.00	2.1 to 5.2	10V;OS	479
Poly(styrene-p-sulfonic acid), sodium salt				OS LS		
	aqueous NaCl 0.5 mol/l					
	to 0.005 mol/l	25	32.04	1.6 to 35.53 2.47 to 37.4	6V;$\overline{M}_n$;A_3 given in Ref.	513
Poly(styrene-p-sulfonic acid), potassium salt						
	aqueous KCl	27.5	48	4.68	LS	320

2.6. POLY(OXIDES), POLY(CARBONATES), POLY(ESTERS)

Polymer	Solvent	T[$^{\circ}$C]	M x 10^{-4}	Sec. Virial Coeff. $[\text{mol cm}^3/\text{g}^2]$ x 10^4	Remarks	Ref.
Poly(carbonate) see Poly(oxycarbonyloxy......)						
Poly(ethylene terephthalate) see Poly(oxyethyleneoxyterephthaloyl)						
Poly(formaldehyde) see Poly(oxymethylene)						
Poly(oxybutene)	methyl ethyl ketone	30	210 to 34.3	2.3 to 1.8	6V;LS	598
Poly(oxycarbonyloxy-1,4-phenyleneisopropylidene-1,4-phenylene)						
	chloroform	25	7.69 to 2.85	11.9 to 14.4	6V;OS	496
		30	8.8 to 3.73	12.5 to 18.0	6V;OS	496
	cyclohexanone	25	7.69 to 3.16	8.8 to 10.1	5V;OS	496
	dioxane			10.9 to 12.4	5V;OS	496
	dioxane/cyclohexanone		7.69 to 2.85	0	4V;OS	496
	ethylene chloride		7.69 to 3.16	10.6 to 12.7	5V;OS	496
	methylene chloride		7.69 to 2.85	12.1 to 14.4	4V;OS	496
			6.1 to 12.2 ★	3.6 to 2.5	2V;LS;$[\text{J cm}^3/\text{g}^2]$	45
		27	2.3 to 18.2 ★	3.9 to 2.9	6V;OS;$[\text{J cm}^3/\text{g}^2]$	45
	tetrahydrofuran	20	0.8 to 26.6 ★	4 to 1.29	6V;$M_{s,D}$;$]\text{J cm}^3/\text{g}^2]$	45
		25	7.69 to 3.16	11.0 to 13.0	5V;OS	496
		27	2.3 to 9.2 ★	2.5 to 1.7	6V;OS;$[\text{J cm}^3/\text{g}^2]$	45
Poly(oxy-2,6-dimethyl-1,4-phenylene)						
	toluene	25	41.5 to 2.56	7.80 to 15.4	15V;LS	466
			20.9 to 1.27	7.90 to 15.7	15V;OS	
Poly(oxyethylene)	benzene	25	0.770 ★	27.4 x 1.013	LS;$[\text{bar cm}^6/\text{g}^2]$	407
			0.379 ★	78 x 1.013	OS;$[\text{bar cm}^6/\text{g}^2]$	407
	dimethylformamide		0.379 ★	30 x 1.013	OS;$[\text{bar cm}^6/\text{g}^2]$	407
		25 to 120 ca.	0.35 ★	(37 to 47) x 1.013	4V;OS;$[\text{bar cm}^6/\text{g}^2]$	188
	methanol	25	0.770 ★	66.0 x 1.013	OS;$[\text{bar cm}^6/\text{g}^2]$	407
			0.379 ★	56.0 x 1.013	OS;$[\text{bar cm}^6/\text{g}^2]$	407
			0.0316 to 0.675	18.0 to 16.4	3V;OS	518
			see Ref.	170 to 34.8	3V;LS	423
			0.0062 to 3.73	★(1220 to 46) x 1.013	LS;$[\text{bar cm}^6/\text{g}^2]$;data for seven solvents and for OS, SE given in Ref.	423
			0.1 to 3.1 ★	(102.5 to 39) x 1.013	11V;LS;$[\text{bar cm}^6/\text{g}^2]$	192
			0.1 to 1.0 ★	(84.5 to 47.5) x 1.013	4V;LS;$[\text{bar cm}^6/\text{g}^2]$	193
			0.4 to 2.3 ★	(87 to 46) x 1.013	3V;LS;$[\text{bar cm}^6/\text{g}^2]$	184
			0.3 to 4.8 ★	(48 to 27.5) x 1.013	2V;LS;$[\text{bar cm}^6/\text{g}^2]$	185

References page IV-106

Polymer	Solvent	$T[^{\circ}C]$	$M \times 10^{-4}$	Sec. Virial Coeff. $[\text{mol cm}^3/\text{g}^2] \times 10^4$	Remarks	Ref.
Poly(oxyethylene) (Cont'd.)						
	water	25	1.09 to 80	★ (116 to 30.4) x 1.013	5V;[bar cm^6/g^2]	189
			1.01	★ 62 x 1.013	LS;[bar cm^6/g^2]	184
Poly(oxyethyleneoxyterephthaloyl)						
fractionated	o-chlorophenol	25	10.7 to 1.81	★ 14 to 25.5	8V;OS;[cm^3/g^2]	526
				★ca.40	different types of osmometers used, see Ref.	
unfractionated		25	3.9	★ 26	different types of osmometers used, see Ref.	
fractionated		50	7.4 to 2.62	★ca.15	4V;OS;[cm^3/g^2]	
	o-chlorophenol/tetra-chloroethylene 60/40	25	1.22	19	OS	542
Poly(oxyhexamethyleneoxyadipoyl)						
	chloroform	25	1.72	53	OS	542
Poly(oxyhexamethyleneoxyterephthaloyl)						
	o-chlorophenol/tetra-chloroethylene 60/40	25	1.13	35	OS	542
Poly(oxymethylene)	hexafluoroacetone hydrate	25	2.3 to 17.0	★ (5.6 to 2.0) x 10^{-3}	5V;LS;[cm^3/g^2];★ see Ref.	430
--, diacetate	hexafluoroacetone	25	2.3 to 18.5	56 to 25	5V;LS	236
unfractionated	p-chlorophenol	70	4.1	32	OS	510
fractionated			8.65 to 2.7	35 av.	5V;OS	
unfractionated			3.5 to 5.8	31 to 28	3V;OS	
fractionated			4.6 to 6.65	32 to 31	2V	
Poly(oxymethylene-co-oxyethylene)						
	1H, 1H, 5H-octafluoro-pentanol-1	110	3.4 to 5.6	see Ref.	OS;[cm^3/mol];LS-measurements also given in Ref.	402
Poly(oxypentamethyleneoxyterephthaloyl)						
	o-chlorophenol/tetra-chloroethylene 60/40	25	1.06	47	OS	542
Poly(oxypropylene)	acetone	25	0.0067	-90	LS	475
			0.0125	0	LS	
			0.045 to 0.385	ca. 15.2	6V;LS	
	hexane	46	78.3 to 90.1	0.46 to 4.50	5V;LS	428
	methanol	20	0.0535 to 0.331	★ (10.75 to 0.95) x 10^{-3}	6V;A;[cm^3/g];see Ref.	470
	iso-octane	48 to 85	90.1	0 to 1.58	9V;LS	428
		50 to 89	78.3	-0.25 to 1.72	7V;LS	
	hexane	46	3.42 to 441	3.16 to 0.523	16V;LS	207
Poly(oxytetramethylene), Poly(tetrahydrofuran)	ethyl acetate	30	3.46 to 103.0	6.14 to 2.47	10V;LS	413
Poly(oxytetramethyleneoxyadipoyl)						
	chloroform	25	1.47	40	OS	542
Poly(oxytetramethyleneoxyisophthaloyl)						
	o-chlorophenol/tetra-chloroethylene 60/40	25	1.28	27	OS	542
Poly(oxytetramethyleneoxyterephthaloyl)						
	o-chlorophenol/tetra-chloroethylene 60/40	25	1.34	30	OS	542

2.7. POLY(AMIDES), POLY(UREAS)

Polymer	Solvent	$T[^{\circ}C]$	$M \times 10^{-4}$	Sec. Virial Coeff. $[\text{mol cm}^3/\text{g}^2] \times 10^4$	Remarks	Ref.
Poly(γ-benzyl-L-glutamate)						
	chlorobenzene	25	92	5.1	LS	488
	chloroform		13 to 35.8	5.0 to 4.7	4V;LS	299
	chloroform/formamide		39.2	★ 2.0	(cgs)	435
			40.8	★ 2.2	(cgs)	435
	cyclohexanone		23	16	LS	488
	dichloroacetic acid		2.14 to 33.6	15.7 to 7.5	3V;LS	299
	dichloroethane		49	5.0	LS	488
			55.5	★ 23.2 x 1.013	LS	476
	dimethylformamide	15 to 55	2.74	-6±1.3 to 2.5±0.27	5V;OS	521

Polymer	Solvent	T[°C]	M x 10⁻⁴	Sec. Virial Coeff. [mol cm³/g²] x 10⁴	Remarks	Ref.

Polymer	Solvent	$T[^oC]$	$M \times 10^{-4}$	Sec. Virial Coeff. $[\text{mol cm}^3/\text{g}^2] \times 10^4$	Remarks	Ref.
Poly(γ-benzyl-L-glutamate) (Cont'd.)						
	dimethylformamide	20 to 55	14.0	−0.9±0.44 to 2.6±0.15	5V;OS	521
		25	1.85 to 41.4	★ca. 2.5	13V;LS;(cgs)	435
			35	4.7	LS	488
			13.2 to 41.7	see Ref.	LS;(cgs)	450
		37	6.9 to 25.0	★ 3.5 to 2.8	OS	450
			7.6 to 17.6	★ 3.8 to 2.0	5V;OS;(cgs)	435
		25 to 70	0.96	★ 16.0 x 1.013	3V;OS;LS;M ;[bar cm⁶/g²]	476
			3.34	★ 9.1 x 1.013	5V;[bar cm⁶/g²]	476
			15.4	★ 9.3 x 1.013	4V;[bar cm⁶/g²]	476
			22.2	★ 9.1 x 1.013	4V;[bar cm⁶/g²]; for measurements in 5 more solvents see Ref.	476
		45	2.74 to 49	ca. 2.2	8V;OS	521
	dioxane	25	40	5.3	LS	488
	tetrahydrofuran		44	5.8	LS	488
Poly((n-butylimino)carbonyl) (Poly(n-butyl isocyanate))						
	chloroform	22	6.1 to 1000	25.1	31V;LS	565
	toluene	37	2.3±0.05 to 53.0±6.0		7V;OS	433
				21.2±2.7		433
		37	2.3 to 52.3	21.3	21V;OS	565
Poly(N-(3-hydroxypropyl)-L-glutamine)						
	methanol	25	36.6±0.5 to 10.0±0.5		5V;LS;(cgs)	534
				★ 2.40 to 4.1±0.6		534
			5.5±0.1	3.3	LS;(cgs)	
Poly(iminohexamethyleneiminoadipoyl)						
(Nylon 66)	m-cresol	60	1.8	★ 183 x 1.013	OS;[bar cm⁶/g²]	217
	formic acid (90%)	25	1.8	★ 840 x 1.013	OS;[bar cm⁶/g²]	217
	formic acid (90%)/KCl 0.2-2.5 M KCl	25	3.1	★ (59.2 to -7.0) x 1.013	7V;LS;[bar cm⁶/g²]	219
	formic acid (90%)/KCl 2.3 M KCl	25	3.1	★ 0		219
	formic acid/2 M KCl 82.5-98% acid	25	3.1	★ (-9.4 to 36.5) x 1.013	5V;LS;[bar cm⁶/g²]	219
	formic acid (90%)/ 2 M KCl	25	0.2 to 5.2	★ (312 to 10.1) x 1.013	10V;LS;[bar cm⁶/g²]	218
	formic acid (75-98%), 0.5 M sodium formiate, 2,2,3,3,-tetrafluoropropanol	25	3.2	ca. 1.0 to 4.0	6V;LS	405
	0.1 M sodium trifluoroacetate/2,2,3,3-tetrafluoropropanol	25	6.2	★ 57.1 x 1.013	LS;[bar cm⁶/g²]	220,221
Poly(n-hexyliminocarbonyl)						
	toluene	37	6.1 to 60.6	ca. 20	3V;OS; see Ref. for measurements in THF	565
Poly((n-octylimino)carbonyl)						
	toluene	37	6.8 to 20	ca. 20	3V;OS; see Ref. for measurements in THF	565
Poly(oxy-1,4-phenyleneiminocarbonyl-2,5-dicarboxy-1,4-phenylenecarbonylimino-1,4-phenylene)						
	dimethylacetamide, 0.1 M LiBr	15	2.25 to 26.6	★ (21.2 to 14.3) x 10⁻⁴	13V;LS;[cm³/g];OS;	403
Poly(ureyleneheptamethylene)						
	dichloroacetic acid	25	2.85	★ 0	LS;[bar cm⁶/g²]	400
		45	1.75	★ 6.5 x 1.013	LS;[bar cm⁶/g²]	400
		73	1.78	★ 38 x 1.013	LS;[bar cm⁶/g²]	400
		98	1.32	★ 52 x 1.013	LS;[bar cm⁶/g²]	400
		45	0.31 to 1.78	★ (32 to 6.2) x 1.013	6V;LS;[bar cm⁶/g²]	183
		73	0.26 to 2.35	★ (82 to 8.0) x 1.013	6V;LS;[bar cm⁶/g²]	183
		25 to 98	ca. 1.75	★ (o to 52) x 1.013	4V;LS;[bar cm⁶/g²]	183
	formic acid 98%/KCl 96/4	45	0.05	★ 5 x 1.013	LS;[bar cm⁶/g²]	183

References page IV-106

Polymer	Solvent	T[°C]	$M \times 10^{-4}$	Sec. Virial Coeff. $[mol\ cm^3/g^2] \times 10^4$	Remarks	Ref.
Poly(ureyleneheptamethylene) (Cont'd.)						
	formic acid 98%/KCl					
	96/4	45	0.3 to 1.39	★ (104 to 59) x 1.013	6V;LS;[bar cm^6/g^2]	183
	98/2	73	0.46 to 1.62	★ (105 to 64) x 1.013	3V;LS;[bar cm^6/g^2]	183
	formic acid 85.8-99.5%/					
	KCl 3.0 to 10.0%	73	0.57 to 1.25	★ (512 to 30) x 1.013	8V;LS;[bar cm^6/g^2]	400
		83	1.95 to 2.63	★ (83 to 0.0) x 1.013	6V;LS;[bar cm^6/g^2]	400
	formic acid 98%/					
	KCl 4%	25	1.06	★ -190 x 1.013	LS;[bar cm^6/g^2]	400

2.8. POLY(SILOXANES)

Polymer	Solvent	T[°C]	$M \times 10^{-4}$	Sec. Virial Coeff. $[mol\ cm^3/g^2] \times 10^4$	Remarks	Ref.
Poly(oxydimethylsilylene)	bromobenzene	59.1-110.8	8	★ -2.025 to 1.750	9V;LS;[cm^3/g^2]	216
		70-108.3	34	★ -0.587 to 0.890	8V;LS	
		73.7-111.1	106	★ -0.272 to 0.772	7V;LS	
	bromocyclohexane	16.7-60.2	3.3	★ -1.3 to 2.260	8V;LS;[cm^3/g^2]	216
		18.2-60.2	8.5	★ -0.782 to 2.012	8V;LS	
		29.2-60.2	34	★ -0.136 to 1.329	8V;LS	
		24.1-90.3	78	★ -0.257 to 1.985	10V;LS	
		25.5-60.2	106	★ -0.190 to 0.985	6V;LS	
	bromocyclohexane/					
	phenetol (6/7)	23.9-88.8	8	★ -1.23 to 2.14	10V;LS;[cm^3/g^2]	216
		26.8-93.9	34	★ -0.536 to 1.490	10V;LS	
		31.8-90	106	★ -0.239 to 1.280	8V;LS	
	chlorobenzene	20	10.7	1.88	OS	501
			23.2	1.70	OS	501
	phenetol	80.8-122.8	4.5	★ -2.15 to 3.823	8V;LS;[cm^3/g^2]	216
		81.7-123.1	8.0	★ -0.93 to 3.23	7V	
		79.5-122.8	34	★ -0.54 to 1.796	7V	
		90-112.8	106	★ 0.015 to 1.466	6V	
	toluene	20	10.0	3.53	OS	501
			23.5	3.18	OS	501
		25	0.066 to 0.2015	25.0 to 4.05	3V;OS	518
				-2.66 to 10.6	3V;LS	519
Poly(phenyl silsesquioxane)						
	toluene	37	48.85 to 2.64	1.123 to 2.434	15V;OS	459

2.9. OTHERS

Polymer	Solvent	T[°C]	$M \times 10^{-4}$	Sec. Virial Coeff. $[mol\ cm^3/g^2] \times 10^4$	Remarks	Ref.
Poly(acenaphthylene)	toluene	25	172 to 3.3	0.26 to 2.42	17V;LS	458
			114 to 2.3	0.28 to 2.10	17V;OS	

2.10. POLY(SACCHARIDES)

Polymer	Solvent	T[°C]	$M \times 10^{-4}$	Sec. Virial Coeff. $[mol\ cm^3/g^2] \times 10^4$	Remarks	Ref.
Cellulose						
hydrolized linters	cadmium ethylenediamine	25	22.5 to 94.5	ca. 16.1	5V;LS	97
sulfite pulp			21.5	12.1	LS	
Cellulose acetate	acetone	r.t.	6.0 to 17.3	9.4 to 5.8	4V;LS	302
Cellulose carbanilate	dioxane	20	258 to 8.08	★ 2.45 to 11.5	7V;[cm^3/g^2]	68
			4.20 to 0.135	★ 7.5 to 20.0	6V;[cm^3/g^2]	68
Cellulose nitrate						
(13.9 % N)	acetone	25	8.1 to 385	10.8 to 8.2	6V;LS	477
(12.9% N)		25	14.1 to 170	13.3 to 12.5	4V;LS	
(ca. 13.55 % N)		r.t.	6.16 to 248.2	★ca. 0.24 x 1.013	14V;OS;[bar l^2/g^2]	61
		25	7.7 to 264	ca. 6.10	11V;LS	22
from row cotton						
(ca. 13.66% N)		15	2.28 to 41.7	★ca. 0.24 x 1.013	8V;OS;[bar l^2/g^2]	195
from cotton		20	3.1 to 66.1	★ca. 0.28 x 1.013	11V;OS;[bar l^2/g^2]	194
	butyl acetate	20	15 to 40	★ (1.0 to 0.5) x 10^{-6} x 1.013	LS;[bar l^2/g^2]	239
		25	3.0 to 36	★ (3.5 to 0.3) x 10^{-6} x 1.013	OS;[bar l^2/g^2]	239
from viscose rayon	ethyl acetate	29.7	7.15	★ 4.41	OS;[100 cm^3/g]	91
from chemical cotton			29.5 to 45	★ 2.85 to 2.57	2V;OS;[100 cm^3/g]	91
(12.4% N)	methyl ethyl ketone	25	13	10.8	OS	196

Polymer	Solvent	T[^oC]	M x 10^{-4}	Sec. Virial Coeff. [mol cm^3/g^2] x 10^4	Remarks	Ref.
Cellulose tricaproate	1-chloronaphthalene	24	6.25 to 131	6.5 to 1.1	4V;LS	334
	dimethylformamide	30-53.5	13.2	-0.58 to 0.65	3V;OS	334
	dioxane/water 100/7	37.1-49.4	20.6	-0.19 to 0.37	3V;OS	334
		43-61	131	0 to 1.2	3V;LS	334
		63	19.5 to 148	7.0 to 3.0	3V;LS	334
Cellulose tricarbanilate (8.19-8.41% N)	acetone	7	6.7 to 267	★ 3.14 to 1.36	16V;LS;[cm^3/g^2]	281
		17	6.7 to 267	★ 3.16 to 1.32	16V;LS;[cm^3/g^2]	281
		27	6.7 to 267	★ 3.16 to 1.27	16V;LS;[cm^3/g^2]	281
	acetone/water 0-7.8% water	20	84	★ 2.02 to 0	7V;LS;[cm^3/g^2]	197
	cyclohexanone	10-80	50	★ca. 4.8 to 3.8	6V;LS;[cm^3/g^2]	504
	dibutyl ketone	50	ca. +3-43	★ca. 2.8 to 0	6V;LS;[cm^3/g^2]	504
	diethylene glycol diethyl ether	ca. +5-ca. +95	50	★ca. 4.7 to 3.1	6V;LS;[cm^3/g^2]	504
	diethyl ketone/methanol 0-52% methanol	20	84	★ 4.26 to 0	7V;LS;[cm^3/g^2]	197
	dioxane/methanol 0-57.5% methanol	20	84	★ 6.18 to 0	7V;LS;[cm^3/g^2]	197
	dioxane	20-95	50	★ca. 5.5 to 3.7	5V;LS;[cm^3/g^2]	504
	methyl ethyl ketone/ methanol 0-57% methanol	20	84	★ 3.63 to 0	8V;LS;[cm^3/g^2]	197
	pyridine	ca. +5-ca. +95	50	★ca. 5.5 to 4.7	6V;LS;[cm^3/g^2]	504
Cellulose trinitrate from cotton linters (13.58% N)	acetone	25	ca. 7.5	7.3 to 7.4	2V;LS	89
from purified chemical linters			ca. 6	23 to 79	3V;OS; [cm^3/g^2]	90
from bleached ramie; (ca. 13.8% N)			9.9 to 22.8	ca. 4.1	5V;LS	89
from unbleached row ramie; (ca. 13.7% N)			ca.24.5	5.1 to 3.8	2V;LS	89
from linters	ethyl acetate	25	4.1 to 57.3	★ 32 to 367	5V;LS;[cm^3/g]	90
			6.8	7.3	LS	89
from bleached ramie; (13.83% N)		25	9.8 to 23.2	3.1 to 4.8	4V;LS	89
from unbleached ramie, (13.70% N)		25	25	3.6	LS	89
from purified cotton linters	methyl ethyl ketone	r.t.	11 to 25.7	★ 92 to 177	2V;OS;[cm^3/g]	90
Cellulose xanthate	1 M NaOH	1	$\overline{P}_n$ 510 to 130	ca. 30	4V;OS	437
		5	$\overline{P}_w$ 2400	24	LS	437
			2100	30		437
			1540	32		437
			1400	37		437
			580	36		437
			540	43		437
			440	38		437
			430	50		437
Ethyl hydroxyethyl cellulose	water	25	7.7 to 17.6	9.8 to 5.45	7V;LS and OS	94
Hydroxypropyl cellulose	ethanol	25	3.9 to 143	23.2 to 5.6	14V;LS	578

C. REFERENCES

1. G. Meyerhoff, Z. Electrochem. 61, 1250 (1957).
2. T. Svedberg, K. O. Pederson, "Die Ultrazentrifuge," Th. Steinkopf, Dresden und Leipzig, 1940.
3. G. Meyerhoff, Makromol. Chem. 37, 97 (1960).
4. H. P. Frank, H. F. Mark, J. Polymer Sci. 17, 1 (1955).
5. G. Meyerhoff, in "Conference on the Ultracentrifuge," Academic Press, New York (1963).
6. G. Meyerhoff, Makromol. Chem. 72, 214 (1964).
7. J. G. Kirkwood, J. Riseman, J. Chem. Phys. 16, 565 (1948).
8. J. Riseman, J. G. Kirkwood, in F. R. Eirich, "Rheology, Theory and Application," Vol. I, Academic Press, New York (1956).
9. M. Kurata, H. Yamakawa, J. Chem. Phys. 29, 311 (1958).
10. H. Kuhn, W. Kuhn, A. Silberberg, J. Polymer Sci. 14, 193 (1954).
11. International Union of Pure and Applied Chemistry, J. Polymer Sci. 10, 129 (1953).

12. A. F. Schick, S. J. Singer, J. Phys. Chem. 54, 1028 (1950).

13. G. V. Schulz, H. Inagaki, R. Kirste, Z. Phys. Chem. N. F. 24, 390 (1960).

14. N. Gralén, Dissertation, Uppsala (1944).

15. H. Luetge, G. Meyerhoff, Makromol. Chem. 68, 180 (1963).

16. K. Nachtigall, G. Meyerhoff, Z. Phys. Chem. (Frankfurt) 30, 35 (1961).

17. G. Meyerhoff, Makromol. Chem. 15, 68 (1955).

18. G. Meyerhoff, Makromol. Chem. 12, 45 (1954).

19. K. Uda, G. Meyerhoff, Makromol. Chem. 67, 168 (1961).

20. R. Signer, J. Liecht, Helv. Chim. Acta 21, 530 (1938).

21. A. Polson, Kolloid-Z. 83, 172 (1938).

22. A. M. Holtzer, H. Benoit, M. Doty, J. Phys. Chem. 58, 624 (1954).

23. J. W. Williams, W. M. Saunders, J. Phys. Chem. 58, 854 (1954).

24. A. G. Ogston, E. F. Woods, Trans. Faraday Soc. 50, 635 (1954).

25. A. G. Ogston, Trans. Faraday Soc. 49, 1481 (1953).

26. F. R. Senti, N. N. Hellman, N. H. Ludwig, G. E. Babcock, R. Tobin, C. A. Glass, B. L. Lamberts, J. Polymer Sci. 17, 527 (1955).

27. L. D. Moore, Jr., G. R. Greear, J. O. Sharp, J. Polymer Sci. 59, 339 (1962).

28. G. V. Schulz, K. Altgelt, H.-J. Cantow, Makromol. Chem. 21, 13 (1956).

29. K. Altgelt, G. V. Schulz, Makromol. Chem. 32, 66 (1959).

30. K. Altgelt, G. V. Schulz, Makromol. Chem. 36, 209 (1960).

31. G. V. Schulz, A. Mula, Proc. Nat. Rubber, Res. Conf., Kuala Lumpur, 1960, pp. 602-610.

32. L. Mandelkern, W. R. Krigbaum, H. A. Scheraga, P. J. Flory, J. Chem. Phys. 20, 1392 (1952).

33. K. H. Schachman, M. A. Lauffer, J. Am. Chem. Soc. 71, 536 (1949).

34. M. A. Lauffer, J. Am. Chem. Soc. 66, 1188 (1944).

35. H. Neurath, Chem. Rev. 30, 357 (1942).

36. M. A. Lauffer, J. Am. Chem. Soc. 66, 1195 (1944).

37. H. K. Schachman, W. J. Kauzman, J. Phys. Chem. 53, 150 (1949).

38. H. W. McCormick, J. Polymer Sci. A 1, 103 (1963).

39. S. J. Singer, J. Chem. Phys. 15, 341 (1947).

40. L. E. Miller, F. A. Hamm, J. Phys. Chem. 57, 110 (1953).

41. H. Ende, G. Meyerhoff, G. V. Schulz, Z. Naturforsch. 13b, 713 (1958).

42. S. Shulman, J. Am. Chem. Soc. 75, 5846 (1953).

43. G. Meyerhoff, Z. Physik. Chem. (Frankfurt) 4, 335 (1955).

44. G. Meyerhoff, Z. Physik. Chem. (Frankfurt) 23, 100 (1960).

45. G. V. Schulz, A. Horbach, Makromol. Chem. 29, 93 (1959).

46. G. V. Schulz, D. Laue, O. Bodmann, Makromol. Chem. 31, 75 (1959).

47. R. Cecil, A. G. Ogston, Biochem. J. 42, 229 (1948).

48. G. V. Schulz, H. A. Ende, Z. Physik. Chem. (Frankfurt) 36, 82 (1963).

49. G. V. Schulz, M. Marx, Makromol. Chem. 14, 52 (1954).

50. M. Cantow, G. Meyerhoff, G. V. Schulz, Makromol. Chem. 49, 1 (1961).

51. O. Bodmann, D. Kranz, G. V. Schulz, Makromol. Chem. 41, 225 (1960).

52. G. V. Schulz, G. Meyerhoff, Z. Elektrochem. 56, 545 (1952).

53. K. J. Ivin, H. A. Ende, G. Meyerhoff, Polymer 3, 129 (1962).

54. K. J. Ivin, J. Polymer Sci. 25, 228 (1957).

55. G. V. Schulz, H. Doll, Z. Elektrochem. 63, 301 (1959).

56. E. D. Kunst, Dissertation, Groningen, 1950.

57. G. V. Schulz, K. V. Guenner, H. Gerrens, Z. Physik. Chem. (Frankfurt) 4, 192 (1955).

58. G. Meyerhoff, Naturwissenschaften 41, 13 (1954).

59. G. Meyerhoff, G. V. Schulz, Makromol. Chem. 7, 294 (1951).

60. G. V. Browning, J. D. Ferry, J. Chem. Phys. 17, 1107 (1949).

61. A. Muenster, J. Polymer Sci. 8, 633 (1952).

62. J. Stamm, J. Am. Chem. Soc. 52, 3047 (1930).

63. J. Stamm, J. Am. Chem. Soc. 52, 3062 (1930).

64. G. Meyerhoff, J. Polymer Sci. 29, 399 (1958).

65. G. Meyerhoff, Makromol. Chem. 32, 249 (1959).

66. J. Hengstenberg, G. V. Schulz, Makromol. Chem. 2, 5 (1948).

67. A. Polson, Kolloid-Z. 87, 149 (1939).

68. J. M. G. Cowie, S. Bywater, J. Polymer Sci. C 30, 85 (1970).

69. T. Svedberg, K. O. Pedersen, The Ultracentrifuge, The Clarendon Press, Oxford (1940).

70. A. G. Polson, Dissertation University of Stellenbosch (1937), South Africa.

71. R. O. Carter, J. Am. Chem. Soc. 63, 1960 (1941).

72. H. G. Tennent, C. F. Vilbrandt, J. Am. Chem. Soc. 65, 424 (1943).

73. K. O. Pedersen, J. J. I. Andersson, unpubl. results, cit. i. Ref. 67.

74. W. R. Krigbaum, A. M. Kotliar, J. Polymer Sci. 32, 323 (1958).

75. A. Oth, Ind. Chim. Belge 20, 423 (1955).

76. T. Svedberg, I.-B. Eriksson-Quensel, J. Am. Chem. Soc. 56, 1700 (1934).

77. L. Krejci, T. Svedberg, J. Am. Chem. Soc. 56, 1706 (1934).

78. T. Svedberg, I.-B. Eriksson, J. Am. Chem. Soc. 55, 2834 (1933).

79. W. F. H. M. Mommaertz, R. G. Parrish, J. Biol. Chem. 188, 545 (1951).

80. P. Johnson, R. Landolt, Nature 165, 430 (1950).

81. G. L. Miller, W. C. Price, Arch. Biochem. Biophys. 10, 467 (1946).

82. W. F. H. M. Mommaertz, Arkiv Kemi Mineral. Geol. 19A, No. 17 (1945).

83. N. W. Pirie, Advan. Enzymol. 5, 1 (1945).

84. M. A. Lauffer, J. Phys. Chem. 44, 1137 (1940).

85. M. A. Lauffer, W. M. Stanley, J. Biol. Chem. 135, 463 (1940).

86. A. G. Ogston, Biochem. J. 37, 78 (1943).

87. A. S. McFarlane, R. A. Kekwick, Biochem. J. 32, 1607 (1938).

88. F. C. Bawden, N. W. Pirie, unpubl. (cit. in Ref. 83).

89. M. M. Huque, D. A. Goring, S. G. Mason, Can. J. Chem. 36, 952 (1958).

90. M. L. Hunt, S. Newman, H. A. Scheraga, P. J. Flory, J. Phys. Chem. 60, 1278 (1956).

91. S. Newman, L. Loeb, C. M. Conrad, J. Polymer Sci. 10, 463 (1953).

92. B. Ingelman, M. S. Halling, Arkiv Kemi 1, 61 (1949).

93. H.-J. Cantow, Z. Naturforsch. 7b, 485 (1952).

94. R. S. J. Manley, Arkiv Kemi 9, 519 (1956).

95. Y. Fujisaki, H. Kobayashi, Chem. High Polymer (Japan) 19, 81 (1962).

96. H.-J. Cantow, Makromol. Chem. 30, 169 (1959).

97. D. Henley, Arkiv Kemi 18, 327 (1961).

98. W. Brown, Arkiv Kemi 18, 227 (1961).

99. H.-J. Cantow, G. V. Schulz, Z. Phys. Chem. (Frankfurt) 2, 117 (1954).

100. D. J. Bell, H. Gutfreund, R. Cecil, A. G. Ogston, Biochem. J. 42, 405 (1945).

101. R. W. G. Wyckoff, J. Biol. Chem. 121, 219 (1937).

102. H. Neurath, A. M. Saum, J. Biol. Chem. 126, 435 (1938).

103. F. C. Bawden, N. W. Pirie, Proc. Roy. Soc. (London), Ser. B 123, 274 (1937).

104. W. M. Stanley, J. Phys. Chem. 42, 55 (1938).

105. H.-J. Cantow, Z. Physik. Chem. (Frankfurt) 7, 58 (1956).

106. W. R. Krigbaum, J. Am. Chem. Soc. 76, 3758 (1954).

107. H. W. McCormick, J. Polymer Sci. 36, 341 (1959).

108. J. W. Breitenbach, E. L. Forster, A. J. Renner, Kolloid-Z. 127, 1 (1952).

109. W. R. Krigbaum, J. E. Kurz, P. Smith, J. Phys. Chem. 65, 1984 (1961).

110. H.-G. Elias, F. Patat, J. Polymer Sci. 29, 141 (1958).

111. A. R. Shultz, J. Am. Chem. Soc. 76, 3422 (1954).

112. H.-G. Elias, F. Patat, Makromol. Chem. 25, 13 (1958).

113. T. Svedberg, I.-B. Eriksson, J. Am. Chem. Soc. 54, 3998 (1933).

114. L. Krejci, T. Svedberg, J. Am. Chem. Soc. 57, 946 (1935).

115. J. Bisschops, J. Polymer Sci. 17, 81 (1955).

116. K. Dialer, R. Kerber, Makromol. Chem. 17, 56 (1955).

117. A. Oth, Bull. Soc. Chim. Belges 64, 484 (1955).

118. A. R. Peacocke, H. R. Schachmann, Biochim. Biophys. Acta 15, 198 (1954).

119. K. Dialer, K. Vogler, F. Patat, Helv. Chim. Acta 35, 869 (1952).

120. J. L. Oncley, G. Scatchard, A. Brown, J. Phys. Chem. 51, 184 (1947).

121. G. V. Schulz, H. Baumann, R. Darskus, J. Phys. Chem. 70, 3647 (1966).

122. R. A. Kekwick, K. O. Pedersen, Biochem. J. 30, 2201 (1936).

123. M. Deidelberger, K. O. Pedersen, J. Gen. Physiol. 19, 95 (1935).

124. G. J. Howard, D. O. Jordan, J. Polymer Sci. 12, 209 (1954).

125. S. Newman, F. Eirich, J. Colloid Sci. 5, 541 (1950).

126. N. Gralen, G. Lagermalm, J. Phys. Chem. 56, 514 (1952).

127. J. A. V. Butler, A. B. Robins, K. V. Shooter, Proc. Roy. Soc. (London) Ser. A 241, 299 (1951).

128. L. Mandelkern, P. J. Flory, J. Chem. Phys. 20, 212 (1952).

129. H. A. Dieu, J. Polymer Sci. 12, 417 (1957).

130. L. Freund, M. Daune, J. Polymer Sci. 29, 161 (1958).

131. D. O. Jordan, A. R. Mathieson, M. R. Porter, J. Polymer Sci. 21, 463 (1956).

132. D. O. Jordan, A. R. Mathieson, M. R. Porter, J. Polymer Sci. 21, 473 (1956).

133. T. A. Orofino, J. W. Mickey, Jr., J. Chem. Phys. 38, 2512 (1963).

134. M. Wales, K. E. Van Holde, J. Polymer Sci. 14, 81 (1954).

135. B. Rosen, P. Kamath, F. Eirich, Discussions Faraday Soc. 11, 135 (1951).

136. D. A. I. Goring, C. Chepeswick, J. Colloid Sci. 10, 440 (1955).

137. L. P. Witnauer, F. R. Senti, M. D. Stern, J. Polymer Sci 16, 1 (1955).

138. R. L. Baldwin, Biochem. J. 55, 644 (1953).

139. H. Malmgren, Acta Chem. Scand. 6, 1 (1952).

140. R. Gehm, Acta Chem. Scand. 5, 270 (1951).

141. J. H. Fessler, A. G. Ogston, Trans. Faraday Soc. 47, 667 (1951).

142. L. W. Nichol, A. B. Roy, Biochemistry 4, 386 (1965).

143. W. Scholtan, Makromol. Chem. 36, 162 (1960).

144. W. Scholtan, Makromol. Chem. 7, 209 (1952).

145. C. G. Holmberg, Arkiv Kemi, Mineral. Geol. 17A, No. 28 (1944).

146. V. L. Koenig, Arch. Biochem. 25, 241 (1950).

147. V. L. Koenig, J. D. Perrings, Arch. Biochem. Biophys. 41, 367 (1952).

148. G. R. Andersson, Arkiv Kemi 20, 513 (1963).

149. J. A. Manson, G. J. Arquette, Makromol. Chem. 37, 187 (1960).

150. H.-G. Elias, Makromol. Chem. 50, 1 (1961).

151. K. Nakanishi, M. Kurata, Bull. Chem. Soc. (Japan) 33, 152 (1960).

152. W. G. Martin, W. H. Cook, C. A. Winkler, Can. J. Chem. 34, 809 (1956).

153. W. Scholtan, Makromol. Chem. 23, 128 (1957).

154. V. L. Koenig, J. D. Perrings, Arch. Biochem. Biophys. 36, 147 (1952).

155. V. L. Koenig, J. D. Perrings, Arch. Biochem. Biophys. 40, 218 (1952).

156. V. L. Koenig, K. O. Pedersen, Arch. Biochem. 25, 97 (1950).

157. I. B. Eriksson-Quensel, Biochem. J. 32, 585 (1938).

158. A. Tiselius, D. Gross, Kolloid-Z. 66, 11 (1934).

159. K. O. Pedersen, Biochem. J. 30, 961 (1936).

160. W. R. Krigbaum, P. J. Flory, J. Am. Chem. Soc. 75, 1775 (1953).

161. M. D. Dayhoff, G. E. Perlmann, D. A. MacInnes, J. Am. Chem. Soc. 74, 2515 (1952).

162. S. N. Chinai, R. J. Samuels, J. Polymer Sci. 19, 463 (1956).

163. S. N. Chinai, J. D. Matlack, A. L. Resnick, R. J. Samuels, J. Polymer Sci. 17, 391 (1955).

165. S. N. Chinai, R. A. Guzzi, J. Polymer Sci. 21 417 (1956).

166. P. Outer, C. I. Carr, B. H. Zimm, J. Chem. Phys. 18, 830 (1950).

167. K. A. Granath, J. Colloid Sci. 13, 308 (1958).

168. O. Lamm, A. Polson, Biochem. J. 30, 528 (1936).

169. J. Oth, V. Desreux, Bull. Soc. Chem. Belges 66, 303 (1957).

170. W. B. Bridgman, J. Am. Chem. Soc. 68, 857 (1946).

171. H. K. Schachman, W. F. Harrington. J. Polymer Sci. 12, 379 (1954).

172. D. McIntyre, A. Wims, L. C. Williams, L. Mandelkern, J. Phys. Chem. 66, 1932 (1962).

173. H. K. Schachman, Protein and Function, Brookhaven Symposia in Biology: Nr. 13, 49 (1960).

174. H. K. Schachman, J. Am. Chem. Soc. 73, 4453 (1951).

175. I.-B. Eriksson-Quensel, T. Svedberg, Biol. Bull. 71, 498 (1936).

176. J. Oth, V. Desreux, Bull. Soc. Chim. Belges 63, 285 (1954).

177. P. J. Rempp, J. Chim. Phys. 54, 432 (1957).

178. K. Dialer, K. Vogler, Makromol. Chem. 6, 191 (1951).

180. C. Rosse, U. Bianchi, V. Mangasco, J. Polymer Sci. 30, 175 (1958).

181. G. Kegeles, S. M. Klainer, W. J. Salem, J. Phys. Chem. 61, 1286 (1957).

182. G. W. Schwert, S. Kaufman, J. Biol. Chem. 179, 655 (1949).

183. J. Feisst, H.-G. Elias, Makromol. Chem. 82, 78 (1965).

184. H.-G. Elias, Z. Physik. Chem. (Frankfurt) 28, 303 (1961).

185. H.-G. Elias, Chem. Ingr.-Tech. 33, 359 (1961).

186. K. Froembling, F. Patat, Makromol. Chem. 25, 41 (1958).

187. H.-G. Elias, Makromol. Chem. 27, 192 (1958).

188. H.-G. Elias, E. Maenner, Makromol. Chem. 40, 207 (1960).

189. H.-G. Elias, Angew. Chem. 73, 209 (1961).

190. F. Patet, H.-G. Elias, Naturwissenschaften 46, 322 (1959).

191. J. Hengstenberg, E. Schuch, Makromol. Chem. 7, 236 (1952).

192. T. A. Ritscher, H.-G. Elias, Makromol. Chem. 30, 48 (1959).

193. C. Sadron, C., P. Rempp, J. Polymer Sci. 29, 127 (1958).

194. A. Muenster, Z. Physik. Chem. (Leipzig) 197, 17 (1951).

195. H. Diener, A. Muenster, Z- Physik. Chem. (Frankfurt) 13, 202 (1957).

196. G. Jacobsson, Acta Chem. Scand. 8, 1843 (1954).

197. W. Burchard, Z. Physik. Chem. (Frankfurt) 42, 293 (1964).

198. W. Burchard, Makromol. Chem. 59, 16 (1963).

199. P. E. Onyon, J. Polymer Sci. 37, 315 (1959).

200. P. E. Onyon, J. Polymer Sci. 22, 13 (1956).

201. H. Kobayashi, J. Polymer Sci. 39, 369 (1959).

202. S. Newman, W. R. Krigbaum, C. Laugier, P. J. Flory, J. Polymer Sci. 14, 451 (1954).

203. G. B. Rathmann, F. A. Bovey, J. Polymer Sci. 15, 544 (1955).

204. I. J. O'Donnell, E. F. Woods, J. Polymer Sci. 21, 397 (1956).

205. S. N. Chinai, R. A. Guzzi, J. Polymer Sci. 41, 475 (1956).

206. H. Kobayashi, J. Polymer Sci. 26, 230 (1957).

207. G. Allen, C. Booth, M. N. Jones, Polymer (London) 5, 195 (1964).

208. J. E. Hansen, M. G. McCarthy, T. J. Dietz, J. Polymer Sci. 7, 77 (1951).

209. S. N. Chinai, A. L. Resnick, H. T. Lee, J. Polymer Sci. 33, 471 (1958).

210. R. Kirste, G. V. Schulz, Z. Physik. Chem. (Frankfurt) 27, 301 (1961).
211. A. F. V. Eriksson, Acta Chem. Scand. 7, 623 (1953).
212. A. F. V. Eriksson, Acta Chem. Scand. 10, 378 (1956).
213. A. Ehrenberg, Acta Chem. Scand. 11, 1257 (1957).
214. R. V. Webber, J. Am. Chem. Soc. 78, 536 (1956).
215. G. V. Schulz, R. Kirste, Z- Physik. Chem. (Frankfurt) 30, 171 (1961).
216. G. V. Schulz, A. Haug, R. Kirste, Z- Physik. Chem. 38, 1 (1963).
217. R. Schumacher, H.-G. Elias, Makromol. Chem. 76, 12 (1964).
218. H.-G. Elias, R. Schumacher, Makromol. Chem. 76, 23 (1964).
219. P. R. Saunders, J. Polymer Sci. 57, 131 (1961).
220. D. W. Carlson, Dissertation, University Delaware, June, 1959.
221. H. C. Beachell, D. W. Carlson, J. Polymer Sci. 40, 543 (1959).
222. R. Signer, P. v. Tavel, Helv. Chim. Acta 21, 535 (1938).
223. H. Mosimann, Helv. Chim. Acta 26, 61 (1943).
224. B. H. Soerbo, Acta Chem. Scand. 7, 1129 (1953).
225. A. Ehrenberg, H. Theorell, Acta Chem. Scand. 9, 1193 (1949).
226. H. Theorell, R. Bonnichsen, Acta Chem. Scand. 5, 1105 (1951).
227. H. G. Boman, Acta Chem. Scand. 5, 1311 (1951).
228. A. Ehrenberg, S. Paléus, Acta Chem. Scand. 9, 538 (1955).
229. V. N. Tsvetkov, V. G. Alsoshin, Zh. Fiz. Khim. 33, 2767 (1959). (Russ. J. Phys. Chem. 33, 619 (1959).
230. R. Chiang, J. Phys. Chem. 69, 1645 (1965).
231. V. Kokle, F. W. Billmeyer, Jr., L. T. Muus, E. J. Newitt, J. Polymer Sci. 62, 251 (1962).
232. J. Bischoff, V. Desreux, Bull. Soc. Chim. Belges 61, 10 (1952).
233. Y. Hachihama, H. Sumitomo, Technol. Rept. Osaka Univ. 3, 385 (1953).
234. H. Vink, Arkiv. Kemi 14, 29 (1957).
235. S. Claesson, W. Bergmann, G. Jayme, Svensk Papperstid 62, 141 (1959).
236. W. H. Stockmayer, L. L. Chan, ACS Meeting, Detroit, 1965, Polymer Preprints 6, 333 (1965).
237. M. J. R. Cantow, R. S. Porter, J. F. Johnson, ACS Meeting, Detroit, 1965, Polymer Preprints 6, 338 (1965).
238. G. M. Guzman, Anales Real Soc. Espan, Fis. Quim. (Madrid) 52B, 377 (1956).
239. J. Schurz, Papier 15, 530 (1961).
240. R. Arnold, S. R. Caplan, Trans. Faraday Soc. 51, 857 (1955).
241. A. Ehrenberg, K. Dalziel, Acta Chem. Scand. 11, 398 (1957).
242. A. Ehrenberg, K. Agner, Acta Chem. Scand. 12, 95 (1958).
243. K. Z. Fattakhov, V. N. Tsetkov, O. V. K. Kallistov, Zh. Eksperim. i. Teor. Fiz. 26, 351 (1954).
244. M. L. Wagner, H. A Scheraga, J. Phys. Chem. 60, 1066 (1956).
245. C. E. H. Bawn, R. F. Freeman, A. R. Kamaliddin, Trans. Faraday Soc. 46, 862 (1950).
246. L. Trossarelli, G. Saini, Atti Accad. Sci. Torino: Classe Sci. Fis. Mat. Nat. 90, 419 (1955-56).
247. G. Saini, L. Trossarelle, Atti Accad. Sci. Torino: Clesse Sci. Fis. Mat. Nat. 90, 431 (1955-56).
248. H. Frind, Faserforsch. Textiltech. 5, 540 (1954).
249. Q. A. Trementozzi, J. Am. Chem. Soc. 76, 5273 (1954).
250. V. E. Eskin, T. I. Volkov, Vysokomolekul. Soedin. 5, 614 (1963).
251. L. G. Longsworth, J. Am. Chem. Soc. 75, 5705 (1953).
252. L. G. Longsworth, J. Phys. Chem. 58, 770 (1954).
253. F. W. Billmeyer, Jr., C. B. DeThan, J. Am. Chem. Soc. 77, 4763 (1955).
254. A. Takahashi, J. Hayashi, I. Kagawa, Kogyo Kagaku Zasshi 60, 1059 (1957).
255. H. H. Weber, Biochim. Biophys. Acta 4, 12 (1950).
256. O. Snellman, T. Erdoes, Biochim. Biophys. Acta 2, 650 (1948).
257. E. J. Cohn, J. L. Oncley, L. E. Strong, W. L. Hughes, Jr., S. H. Armstrong, Jr., J. Clin. Invest. 23, 417 (1944).

259. I. Jullander, Arkiv Kemi 21A, 8 (1945).
260. D. S. Feingold, M. Gehatia, J. Polymer Sci. 22, 783 (1957).
261. G. L. Borgin, J. Am. Chem. Soc. 71, 2247 (1949).
262. J. Y. Chien, L. H. Shih, S. C. Yu, J. Polymer Sci. 29, 117 (1958).
263. J. Y. Chien, L. H. Shih, Z. Physik. Chem. 207, 60 (1957).
264. J. Y. Chien, L. H. Shih, K. T. Shih, Hua Hsueh Hsueh Pao 23, 215 (1957).
265. H.-G. Elias, H. H. Schlubach, Ann. Chem. 627, 126 (1959).
266. W. Scholtan, Makromol. Chem. 14, 169 (1954).
267. G. M. Chetyrkina, V. G. Aldoshin, S. Y. Frenkel, Vysokomolekul. Soedin. 1, 1133 (1959).
268. G. V. Schulz, M. Hoffmann, Makromol. Chem. 23, 220 (1957).
269. H. Mosiman, T. Svedberg, Kolloid-Z. 100, 99 (1942).
270. H. Sumitomo, Y. Hachihama, Chem. High Polymers (Japan) 10, 544 (1953).
271. H. Sumitomo, Y. Yatsuhama, Chem. High Polymers (Japan) 11, 65 (1954).
272. H.-G. Elias, Makromolekuele, Huethig & Wepf, Basel-Heidelberg (1971).
273. E. F. Casassa, W. H. Stockmayer, Polymer 3, 53 (1962).
274. T. G. Fox, J. B. Kinsinger, H. F. Mason, E. M. Schuele, Polymer 3, 71 (1962).
275. E. Cohn-Ginsberg, T. G. Fox, H. F. Mason, Polymer 3, 97 (1962).
276. H. Campbell, P. Johnson, Trans. Faraday Soc. 40, 221 (1944).
277. W. Brown, D. Henley, Makromol. Chem. 79, 68 (1964).
278. W. Brown, D. Henley, J. Oehman, Arkiv Kemi 22, 189 (1964).
279. R. Van Leemput, R. Stein, J. Polymer Sci. A2, 4039 (1964).
280. R. Van Leemput, R. Stein, J. Polymer Sci. A1, 985 (1963).
281. W. Burchard, E. Husemann, Makromol. Chem. 44-46, 358 (1961).
282. T. Matsuo, H. Inagaki, Makromol. Chem. 55, 150 (1962).
283. T. Matsuo, H. Inagaki, Makromol. Chem. 53, 130 (1962).
284. W. Brown, D. Henley, J. Oehman, Makromol. Chem. 62, 164 (1963).
285. W. Brown, D. Henley, J. Oehman, Makromol. Chem. 64, 49 (1963).
286. J. B. Berkowitz, M. Yamin, R. M. Fuoss, J. Polymer Sci. 28, 69 (1958).
287. A. Ciferri, M. Kryszewski, G. Weill, J. Polymer Sci. 27, 167 (1958).
288. R. L. Cleland, J. Polymer Sci. 27, 349 (1958).
289. S. N. Chinai, P. C. Scherer, D. W. Lewi, J. Polymer Sci. 17, 117 (1955).
290. L. Trossarelli, E. Campi, G. Saini, J. Polymer Sci. 35, 205 (1959).
291. L. H. Tung, J. Polymer Sci. A2, 4875 (1964).
292. Q. A. Trementozzi, J. Polymer Sci. 23, 887 (1957).
292a. W. R. Krigbaum, Q. A. Trementozzi, J. Polymer Sci. 28, 295 (1958).
293. Q. A. Trementozzi, J. Polymer Sci. 36, 113 (1959).
294. W. H. Stockmayer, L. D. Moore, Jr., M. Fixman, B. N. Epstein, J. Polymer Sci. 16, 517 (1955).
295. S. N. Chinai, J. Polymer Sci. 25, 413 (1957).
296. L. H. Tung, J. Polymer Sci. 36, 287 (1959).
297. W. Cooper, G. Vaughan, D. E. Eaves, R. W. Madden, J. Polymer Sci. 50, 159 (1961).
298. P. Dreizehn, R. W. Noble, D. F. Waugh, J. Am. Chem. Soc. 84, 4938 (1962).
299. P. Doty, J. H. Bradburg, A. M. Holtzer, J. Am. Chem. Soc. 78, 947 (1956).
300. S. Mao, E. V. Frisman, Vysokomolekul. Soedin. 4, 1839 (1962).
301. H. Boedtker, P. Doty, J. Am. Chem. Soc. 78, 4267 (1956).
302. R. S. Stein, P. Doty, J. Am. Chem. Soc. 68, 159 (1946).
303. U. P. Strauss, P. L. Wineman, J. Am. Chem. Soc. 80, 2366 (1958).

304. H. Boedtker, S. Simmons, J. Am. Chem. Soc. 80, 2550 (1958).

305. A. Holtzer, S. Lowey, J. Am. Chem. Soc. 81, 1370 (1959).

306. R. Bjorklund, S. Katz, J. Am. Chem. Soc. 78, 2123 (1956).

307. V. N. Tsetkov, O. V. Kalistov, Zh. Fiz. Khim. 33, 710 (1959).

308. A. Silberberg, J. Eliassaf, A. K. Katschalsky, J. Polymer Sci. 23, 259 (1957).

309. P. Doty, E. Mishuck, J. Am. Chem. Soc. 69, 1631 (1947).

310. P. Doty, M. Brownstein, W. Schlener, J. Phys. Colloid. Chem. 53, 213 (1949).

311. W. R. Krigbaum. P. J. Flory, J. Am. Chem. Soc. 75, 5254 (1953).

312. M. Morton, T. E. Helminiak. S. D. Gadkary, F. Bueche, J. Polymer Sci. 57, 471 (1962).

313. J. V. Kolseke, S. F. Kurath, J. Polymer Sci. A2, 4123 (1964).

314. G. Sitaramaiah, D. A. I. Goring, J. Polymer Sci. 58, 1107 (1962).

315. R. Chiang, J. Polymer Sci. 28, 235 (1958).

316. W. R. Krigbaum, J. E. Kurz, P. Smith, J. Phys. Chem. 65, 1984 (1961).

317. P. J. Flory, J. Am. Chem. Soc. 65, 372 (1943).

318. J. B. Kinsinger, R. E. Hughes, J. Phys. Chem. 63, 2002 (1959). From R. Chiang, Chapter in "Newer Methods of Polymer Characterization," ed. B. Ke Interscience Publ., New York, 1964, p. 471.

319. F. Danusso, G. Moraglio, Makromol. Chem. 28, 250 (1958).

320. T. A. Orofino, J. J. Flory, J. Phys. Chem. 63, 283 (1959).

321. H. W. McCormick, J. Polymer Sci. 41, 327 (1959).

322. I. Y. Poddubnyi, V. A. Grechanovskii, M. I. Mosevitskii, Vysokomolekul. Soedin. 5, 1042 (1964).

323. I. Y. Poddubnyi, V. A. Grechanovskii, M. I. Mosevitskii, Vysokomolekul. Soedin. 5, 1049 (1964).

324. I. Y. Poddubnyi. V. A. Grechanovskii, A. V. Podalinskii, Vysokomolekul. Soedin. 5, 1588 (1964).

325. W. Burchard, Makromol. Chem. 67, 182 (1963).

326. A. von Haug, G. Meyerhoff, Makromol. Chem. 53, 91 (1962).

327. J. M. G. Cowie, D. J. Worsfold, S. Bywater, Trans. Faraday Soc. 57, 705 (1961).

328. W. R. Krigbaum, D. K. Carpenter, S. Newman, J. Phys. Chem. 62, 1586 (1958).

329. W. R. Krigbaum, D. K. Carpenter, J. Phys. Chem. 59, 1166 (1955).

330. M. Matsumoto, Y. Ohyanagi, J. Polymer Sci. 46, 441 (1960).

331. Y. Ohyanagi, M. Matsumoto, Chem. High Polymer (Japan) 16, 296 (1959).

332. J. Moacanin, J. Appl. Polymer Sci. 1, 272 (1959).

333. G. C. Berry, L. M. Hobbs, V. C. Long, Polymer (London) 5, 31 (1964).

334. W. R. Krigbaum. L. H. Sperling, J. Phys. Chem. 64, 99 (1960).

335. H. J. L. Trap, J. J. Hermans, J. Phys. Chem. 58, 757 (1954).

336. N. S. Schneider, P. Doty, J. Phys. Chem. 58, 762 (1954).

337. S. Bywater, P. Johnson, Trans. Faraday Soc. 47, 195 (1951).

338. E. V. Gouinlock, Jr., P. J. Flory, H. A. Scheraga, J. Polymer Sci. 16, 383 (1955).

339. K. Krishnamurti, T. Svedberg, J. Am. Chem. Soc. 52, 2897 (1930).

340. E. O. Kraemer, J. Phys. Chem. 45, 660 (1941).

341. E. O. Kraemer, J. Phys. Chem. 46, 177 (1942).

342. L. H. Arond, H. P. Frank, J. Phys. Chem. 58, 953 (1954).

343. H. Boedtker, P. Doty, J. Phys. Chem. 58, 968 (1954).

344. T. Homma, H. Fujita, J. Appl. Polymer Sci. 9, 1701 (1965).

345. B. P. Brand, D. A. I. Goring, P. Johnson, Trans. Faraday Soc. 51, 872 (1955).

346. E. A. Kanevskaya, P. I. Zubov, L. V. Ivanova, Yu. S. Liptaov, Vysokomolekul. Soedin. 6, 981 (1964).

347. S. Lowey, C. Cohen, J. Mol. Biol. 4, 293 (1962).

348. M. J. Kronman, M. D. Stern, J. Phys. Chem. 59, 969 (1955).

349. G. Meyerhoff, "Bestimmung des Molekulargewichtes durch Diffusions- und Sedimentationsmessungen (Ultrazentrifuge)." In: Houben-Weyl: "Methoden der organischen Chemie." Vol. III, Part 1, pp. 390-408; Georg Thieme Verlag, Stuttgart, 1955.

350. J. Hengstenberg, "Sedimentation und Diffusion von Makromolekuelen." In: H. A. Stuart: Physik der Hochpolymeren, Vol. II, pp. 411-494; Springer Verlag, Berlin, Goettingen, Heidelberg, 1953.

351. A. L. Geddes, "Determination of Diffusivity," In: A. Weissberger: Technique of Organic Chemistry. Vol. I, Part 1, pp. 551-619; Interscience Publishers, New York, N.Y. 1949.

352. J. B. Nichols, E. D. Bailey, "Determinations with the Ultracentrifuge," In: A. Weissberger: Technique of Organic Chemistry, Vol. I, Part 1, pp. 621-730, Interscience Publishers, New York, N.Y. 1949.

353. P. O. Kinell, B. G. Ranby, Advan. Colloid. Sci. 3, 182 (1950).

354. H. K. Schachman, "Ultracentrifugation in Biochemistry." Academic Press, New York and London; 1959.

355. R. L. Baldwin, K. E. Van Holde, Fortschr. Hochpolymer-Forsch. 1, 451 (1960).

356. H. Fujita, "Mathematical Theory of Sedimentation Analysis," Academic Press, New York 1962.

357. J. W. Williams, "Ultracentrifugal Analysis in Theory and Experiment," Academic Press, New York, London, 1963.

358. M. Gehatia, J. Polymer Sci. 57, 241 (1962).

359. M. Gehatia, Naturwissenschaften 48, 598 (1961).

360. T. Svedberg, H. Rinde, J. Am. Chem. Soc. 46, 2677 (1924).

361. H. Mosimann, R. Signer, Helv. Chim. Acta 27, 1123 (1944).

362. P. W. Bridgman, Proc. Am. Acad. Arts Sci. 61, 57 (1926).

363. J. Oth, V. Desreux, Bull. Soc. Chim. Belges 63, 133 (1954).

364. M. Wales, J. Am. Chem. Soc. 81, 4758 (1959).

365. J. Fujita, J. Chem. Phys. 24, 1084 (1956).

366. H.-G. Elias, Makromol. Chem. 29, 30 (1959).

367. H. Mark, In: Saenger, R., "Der feste Koerper," pp. 65-104; Hirzel Verlag, Zuerich, 1938.

368. Houwink, R. R., J. Prakt. Chem. 157, 15 (1940).

369. I. Jullander, Arkiv Kemi, Mineral, Geol. 19B, No. 4 (1944).

370. I. Jullander, "Ueber die Berechnung von Molekulargewichten aus Messungen von Sedimentationsgeschwindigkeit und Diffusion." In: "The Svedberg," Jubilee Volume, Almquist and Wiksells Boktryckerei AB, Uppsala, 1944.

371. S. J. Singer, J. Polymer Sci. 1, 445 (1946).

372. N. Bauer, "Determination of Density," In: A. Weissberger: Techniques of Organic Chemistry, Vol. I, Part 1; Interscience Publishers, New York, N.Y., pp. 253-296, 1949.

373. H. Kienitz, "Bestimmung der Dichte." In: Houben-Weyl: Methoden der organischen Chemie, Voll. III, Part 1, Georg Thieme Verlag, Stuttgart: pp. 163-217, 1955.

374. W. Geffcken, C. Beckmann, A. Kruis, Z. Physik. Chem. (B)20, 398 (1933).

375. A. B. Lamb, R. E. Lee, J. Am. Chem. Soc. 35, 1666 (1913).

376. D. A. MacInnes, M. O. Dayhoff, B. R. Ray, Rev. Sci. Instr. 22, 642 (1951).

377. D. A. MacInnes, M. O. Dayhoff, J. Am. Chem. Soc. 74, 1017 (1952).

378. J. Stauff, G. Ruemmler, Kolloid-Z. 166, 152 (1959).

379. P. J. Flory, W. R. Krigbaum, J. Chem. Phys. 18, 1086 (1950).

380. P. J. Flory, J. Chem. Phys. 17, 1347 (1949).

381. W. H. Stockmayer, E. F. Casassa, J. Chem. Phys. 20, 1560 (1952).

382. W. R. Krigbaum. P. J. Flory, J. Polymer Sci. 9, 503 (1952).

383. M. v. Smoluchowski, Ann. Physik. 25, 205 (1908).

384. A. Einstein, Ann. Physik. 33, 1275 (1910).

385. P. Debye, J. Appl. Physics 15, 338 (1944).

386. G. V. Schulz, Z. Physik. Chem. 193, 168 (1944).

387. P. J. Flory, "Principles of Polymer Chemistry," Cornell University Press, Ithaca, New York, 1953.

388. L. Mandelkern, P. J. Flory, J. Chem. Phys. $\underline{19}$, 984 (1951).

389. C. Tanford, "Physical Chemistry of Macromolecules," John Wiley and Sons, New York, N.Y., 1961.

390. M. V. Volkenstein, "Confirgurational Statistics of Polymeric Chains," Interscience Publishers, New York-London, 1963.

391. P. J. Flory, W. R. Krigbaum, Ann. Rev. Phys. Chem. $\underline{2}$, 383 (1951).

392. F. T. Wall, L. A. Hiller, Jr., Ann. Rev. Phys. Chem. $\underline{5}$, 267 (1954).

393. J. J. Hermans, Ann. Rev. Phys. Chem. $\underline{8}$, 179 (1957).

394. E. F. Casassa, Ann. Rev. Phys. Chem. $\underline{11}$, 477 (1960).

395. W. H. Stockmayer, Makromol. Chem. $\underline{35}$, 54 (1960).

396. T. B. Grimley, "The Theory of High Polymer Solutions," in "Progress in High Polymers," Vol. I, Eds. J. C. Robb, F. W. Peaker; Academic Press, Inc., New York, N.Y. 1961.

397. R. E. Hughes, C. A. v. Frankenberg, Ann. Rev. Phys. Chem. $\underline{14}$, 291 (1963).

398. P. J. Flory, J. Chem. Phys. $\underline{10}$, 51 (1942).

399. M. L. Huggins, J. Phys. Chem. $\underline{46}$, 151 (1942).

400. J. Feisst, H.-G. Elias, Makromol. Chem. $\underline{82}$, 78 (1965).

401. W. Burchard, M. Nosslir, Makromol. Chem. $\underline{82}$, 109 (1965).

402. H. L. Wagner, K. F. Wissburn, Makromol. Chem. $\underline{81}$, 14 (1965).

403. M. L. Wallach, Polymer Preprints $\underline{6/1}$, 53 (1965).

404. J. Leonard, H. Daoust, J. Phys. Chem. $\underline{69}$, 1174 (1965).

405. P. R. Saunders, J. Polymer Sci. A $\underline{3}$, 1221 (1965).

406. J. Blanchford, R. F. Robertson, J. Polymer Sci. A $\underline{3}$, 1289 (1965).

407. H.-G. Elias, H. Schlumpf, Makromol. Chem. $\underline{85}$, 118 (1965).

408. J. Springer et al. Ber. Bunsenges. Physik. Chem. $\underline{69}$, 494 (1965).

409. M. J. R. Cantow et al. Polymer Preprints $\underline{6/1}$, 338 (1965).

410. M. J. R. Cantow et al. Makromol. Chem. $\underline{87}$, 248 (1965).

411. D. Kranz et al. Biochim. Biophys. Acta $\underline{102}$, 514 (1965).

412. G. Meyerhoff, N. Suetterlin, Makromol. Chem. $\underline{87}$, 258 (1965).

413. M. Kurata, H. Utiyama, K. Kamada, Makromol. Chem. $\underline{88}$, 281 (1965).

414. G. Rehage et al. Makromol. Chem. $\underline{88}$, 232 (1965).

415. R. Kirste, W. Wuderlich, J. Polymer Sci. B $\underline{3}$, 851 (1965).

416. A. Chatterjee, J. Am. Chem. Soc. $\underline{86}$, 3640 (1964).

417. T. Lindahl et al., J. Am. Chem. Soc. $\underline{87}$, 4961 (1965).

418. C. Rossi, E. Bianchi, E. Pedemonte, Makromol. Chem. $\underline{89}$, 95 (1965).

419. H.-G. Elias, O. Etter, Makromol. Chem. $\underline{89}$, 228 (1965).

420. J. B. Kinsinger, L. E. Ballard, J. Polymer Sci. A $\underline{3}$, 3963 (1965).

421. M. L. Wallach, M. A. Kabayama, Polymer Preprints $\underline{7/1}$, 237 (1966).

422. B. R. Jennings, H. G. Jerrard, J. Chem. Phys. $\underline{44}$, 1291 (1966).

423. H.-G. Elias, H. P. Lys, Makromol. Chem. $\underline{92}$, 1 (1966).

424. P. W. Heaps et al., Nature $\underline{209}$, 397 (1966).

425. W. Brown, R. Wirkstroem, European Polymer J. $\underline{1}$, 1 (1965).

426. O. Wetter, Z. Naturforschung $\underline{19b}$, 60 (1964).

427. M. Kalfus, J. Mitus, J. Polymer Sci. A $\underline{1/4}$, 953 (1966).

428. G. Allen, C. Booth, C. Price, Polymer $\underline{7}$, 167 (1966).

429. J. E. Mark, R. A. Wessling, R. E. Hughes, J. Phys. Chem. $\underline{70}$, 1895 (1966).

430. W. H. Stockmayer, L. L. Chan, J. Polymer Sci. A $\underline{4}$, 437 (1966).

431. G. C. Berry, J. Chem. Phys. $\underline{44}$, 4550 (1966).

432. C. Garbuglio et al., Makromol. Chem. $\underline{97}$, 97 (1966).

433. L. J. Fetters, H. Yu, Polymer Preprints $\underline{7/2}$, 443 (1966).

434. H. Fujita et al., Biopolymers $\underline{4}$, 781 (1966).

435. H. Fujita et al., Biopolymers $\underline{4}$, 769 (1966).

436. J. Rosenbloom, E. C. Cox, Biopolymers $\underline{4}$, 747 (1966).

437. H. Elmgren, Arkiv Kemi $\underline{24/3}$, 237 (1965).

438. G. Delmas, D. Patterson, Polymer $\underline{7}$, 513 (1966).

439. M. J. Hunter, J. Phys. Chem. $\underline{70}$, 3285 (1966).

440. J. M. G. Cowie, S. Bywater, J. Macromol. Chem. $\underline{1}$, 851 (1966).

441. M. L. Wallach, M. A. Kabayama, J. Polymer Sci. A-1, $\underline{4}$, 2667 (1966).

442. W. Scholtan, S. Y. Lie, Makromol. Chem. $\underline{98}$, 204 (1966).

443. T. Kataka et al., J. Phys. Chem. $\underline{70}$, 4099 (1966).

444. K. Hanafusa et al., J. Phys. Chem. $\underline{70}$, 4004 (1966).

445. M. Iwama et al., J. Macromol. Chem. $\underline{1}$, 701 (1966).

446. D. K. Carpenter, J. Polymer Sci. $\underline{A-2}$, 923 (1966).

447. C. Tanford et al., J. Am. Chem. Soc. $\underline{89}$, 729 (1967).

448. N. Eliezer, Biopolymers $\underline{5}$, 105 (1967).

449. B. Porsch, M. Kubin, J. Polymer Sci. C $\underline{16}$, 515 (1967).

450. A. Teramoto, T. Yamashita, H. Fujita, J. Chem. Phys. $\underline{46}$, 1919 (1967).

451. M. Bohdanecky et al., J. Polymer Sci. A-2, $\underline{5}$, 343 (1967).

452. M. Nagasawa, Y. Eguchi, J. Phys. Chem. $\underline{71}$, 880 (1967).

453. S. Gundiah, Makromol. Chem. $\underline{104}$, 196 (1967).

454. A. A. Tager, V. M. Andreeva, J. Polymer Sci. C $\underline{16}$, 1145 (1967).

455. M. L. Wallach, J. Polymer Sci. A-2, $\underline{5}$, 653 (1967).

456. J. Llopis, A. Albert, P. Usobiaga, European Polymer J. $\underline{3}$, 259 (1967).

457. J. Naghizadeh, J. Springer, Kolloid-Z. $\underline{215}$, 21 (1967).

458. J. M. Barrales-Rienda, D. C. Pepper, Polymer $\underline{8}$, 337 (1967).

459. T. E. Helminiak, C. L. Brenner, W. E. Gibbs, Polymer Preprints $\underline{8}$, 284 (1967).

460. G. Meyerhoff, J. Polymer Sci. C $\underline{16}$, 1579 (1967).

461. D. H. Juers, H. A. Swenson, S. F. Kurath, J. Polymer Sci. A-2, $\underline{5}$, 361 (1967).

462. R. B. Beevers, Polymer $\underline{8}$, 410 (1967).

463. P. Callot, A. Banderet, J. Chim. Phys. $\underline{64}$, 1260 (1967).

464. S. Lapanje, C. Tanford, J. Am. Chem. Soc. $\underline{89}$, 5030 (1967).

465. G. Meyerhoff, Makromol. Chem. $\underline{107}$, 124 (1967).

466. J. M. Barrales-Rienda, D. C. Pepper, European Polymer J. $\underline{3}$, 535 (1967).

467. N. Kuwahara, T. Okazawa, M. Kaneko, J. Chem. Phys. $\underline{47}$, 3357 (1967).

468. W. J. Closs, J. G. Jerrard, B. R. Jennings, Nature $\underline{215}$, 1196 (1967).

469. W. Wunderlich, Makromol. Chem. $\underline{108}$, 315 (1967).

470. W. Scholtan, S. Y. Lie, Makromol. Chem. $\underline{108}$, 104 (1967).

471. K. Karunakaran, M. Santappa, Makromol. Chem. $\underline{111}$, 20 (1967).

472. J. Noda et al., J. Phys. Chem. $\underline{71}$, 4048 (1967).

473. W. Banks, C. F. Greenwood, D. J. Hourston, Trans. Faraday Soc. $\underline{64}$, 363 (1968). Makromol. Chem. $\underline{111}$, 226 (1968).

474. R. G. Kirste, W. Wunderlich, Z. Physik. Chem. $\underline{58}$, 133 (1968).

475. G. Meyerhoff, U. Moritz, Makromol. Chem. $\underline{109}$, 143 (1967).

476. H.-G. Elias, J. Gerber, Markomol. Chem. $\underline{112}$, 142 (1968).

477. E. Penzel, G. V. Schulz, Makromol. Chem. $\underline{112}$, 260 (1968).

478. S. P. Rao, N. Santappa, J. Polymer Sci. A-1, $\underline{6}$, 95 (1968).

479. T. Kotaka, Y. Murakami, H. Inagaki, J. Phys. Chem. $\underline{72}$, 829 (1968).

480. J. Moakanin et al., J. Macromol. Sci. Chem. $\underline{A1/8}$, 1497 (1967).

481. G. V. Schulz, H. Baumann, Makromol. Chem. $\underline{114}$, 122 (1968).

482. K. Karunakaran, M. Santappa, J. Polymer Sci. A-2, $\underline{6}$, 713 (1968).

483. H. Triebel, Biopolymers $\underline{6}$, 449 (1968).

484. T. Kato, K. Miyaso, M. Nagasawa, J. Phys. Chem. $\underline{72}$, 2161 (1968).

485. G. Cohen, H. Eisenberg, Biopolymers $\underline{6}$, 1077 (1968).

486. T. Kotaka, N. Donkai, J. Polymer Sci. A-2, $\underline{6}$, 1457 (1968).

487. J. Goldfarb, S. Rodriguez, Makromol. Chem. $\underline{116}$, 96 (1968).

488. J. Lilié, J. Springer, K. Ueberreiter, Kolloid-Z. 226, 138 (1968).

489. J. Danon, J. Polymer Sci. C 16, 4071 (1968).

490. J. M. G. Cowie, Polymer 9, 587 (1968).

491. Y. Arnaud, J. Polymer Sci. C 16, 4103 (1968).

492. R. G. Kirste, G. Wild, Makromol. Chem. 121, 174 (1969).

493. T. Norisuye et al., J. Chem. Phys. 49, 4330 (1968).

494. N. Kuwahara, T. Oikawa, M. Kaneko, J. Chem. Phys. 49, 4972 (1968).

495. D. N. Cramod, J. R. Urwin, European Polymer J. 5, 35 and 45 (1969).

496. W. R. Moore, M. Uddin, European Polymer J. 5, 185 (1969).

497. L. A. Utracki, R. Simha, J. Fetters, J. Polymer Sci. A-2, 6, 2051 (1968).

498. M. Heskins, J. E. Guillet, J. Macromol. Sci. A2/8, 1441 (1968).

499. N. Kuwahara et al., J. Polymer Sci. A-2, 7, 285 (1969).

500. K. Matsumura, Makromol. Chem. 124, 204 (1969).

501. N. Kuwahara et al., J. Polymer Sci. C 23, 543 (1968).

502. A. Kotera et al., J. Polymer Sci. C 23, 619 (1968).

503. L. A. Papazian, Polymer 10, 399 (1969).

504. W. Burchard, Polymer 10, 467 (1969).

505. A. H. Rosenberg, W. Studier, Biopolymers 7, 765 (1969).

506. W. Maechtle, H. Fischer, Angew. Makromol. Chem. 7, 147 (1969).

507. J. Danon, R. Derai, European Polymer J. 5, 659 (1969).

508. T. S. Ng, G. V. Schulz, Makromol. Chem. 127, 165 (1969).

509. L. H. Tung, G. W. Knight, J. Polymer Sci. A-2, 7, 1623 (1969).

510. R. Dick, D. Roussel, H. Benoit, J. Chim. Phys. 66, 1612 (1969).

511. M. Kubin, B. Porsch, European Polymer J. 6, 97 (1970).

512. R. Jerome, V. Desreux, European Polymer J. 6, 411 (1970).

513. A. Takahashi et al., J. Phys. Chem. 74, 944 (1970).

514. H. Dautzenberg, Faserforsch. Textiltechn. 21, 117 (1970).

515. W. Maechtle, Angew. Makromol. Chem. 10, 1 (1970).

516. Y. Noguchi et al., J. Chem. Phys. 52, 2651 (1970).

517. G. Tanaka, S. Imai, H. Yamakawa, J. Chem. Phys. 52, 2630 (1070).

518. K. Kamide, K. Sugamiya, C. Nakayama, Makromol. Chem. 132, 75 (1970).

519. G. Adank, H.-G. Elias, Makromol. Chem. 102, 151 (1967).

520. H.-G. Elias, H. Dietschy, Makromol. Chem. 105, 102 (1967).

521. K. D. Goebel, W. G. Miller, Macromolecules 3, 64 (1970).

522. K. Sakato, M. Kurata, Polymer J. 1, 260 (1970).

523. R. L. Arnett, R. Q. Gregg, J. Phys. Chem. 74, 1593 (1970).

524. W. Wunderlich, Angew. Chem. 11, 73 (1970).

525. W. Wunderlich, Angew. Makromol. Chem. 11, 189 (1970).

526. G. Meyerhoff, S. Shimotsuma, Makromol. Chem. 135, 195 (1970).

527. K. Matsumura, Polymer J. 1, 322 (1970).

528. M. Lechner, G. V. Schulz, European Polymer J. 6, 945 (1970).

529. R. L. Cleland, J. L. Wang, Biopolymers 9, 799 (1970).

530. T. C. Laurent, M. Ryan, A. Pietruszkiewicz, Biochim. Biophys. Acta 42, 476 (1960).

531. W. Banks, C. T. Greenwood, J. Sloss, Makromol. Chem. 140, 109 (1970).

532. B. J. Rietveld, J. Polymer Sci. A-2, 8, 1837 (1970).

533. K.-J. Linow, B. Philipp, Faserforsch. Textiltechn. 21, 483 (1970).

534. K. Okita, A. Teramoto, H. Fujita, Polymer J. 1, 582 (1970).

535. R. H. Marchessault, K. Okamura, C. J. Su Macromolecules 3, 735 (1970).

536. H. A. Scheraga, L. Mandelkern, J. Am. Chem. Soc. 75, 179 (1953).

537. J. Cornibert et al., Macromolecules 3, 741 (1970).

538. A. Krasna, J. R. Dawson, J. A. Harpst, Biopolymers 9, 1017 (1970).

539. A. Krasna, Biopolymers 9, 1029 (1970).

540. K. E. Reinert, J. Strassburger, H. Triebel, Biopolymers 10, 285 (1971).

541. K. E. Reinert, Biopolymers 10, 275 (1971).

542. M. Gilbert, F. J. Hybart, J. Polymer Sci. A-1, 9, 227 (1971).

543. V. Petrus, J. Danihel, M. Bohdanecky, European Polymer J. 7, 143 (1971).

544. W. J. Bengough, G. F. Grant, European Polymer J. 7, 203 (1971).

545. W. Banks, C. T. Greenwood, J. Sloss, European Polymer J. 7, 263 (1971).

546. J. R. Urwin, M. Girolamo, Makromol. Chem. 142, 161 (1971).

547. W. Banks, C. T. Greenwood, Makromol. Chem. 144, 135 (1971).

548. G. C. Berry, J. Polymer Sci. A-2, 9, 687 (1971).

549. K. Kamada, H. Sato, Polymer J. 2, 489 (1971).

550. P. F. Mijnlieff, W. J. M. Jaspers, Trans. Faraday Soc. 67/6, 1837 (1971).

551. M. Kozhokaryu, V. S. Skazka, K. G. Berdnikova, Vysokomol. Soedin. 8, 1063 (1966).

552. J. J. Poddubnyi, A. V. Podalinskii, V. A. Grechanovskii, Vysokomolekul. Soedin. 8, 1556 (1966).

553. A. B. Duckova et al., Vysokomolekul. Soedin. 8, 1814 (1966).

554. V. M. Golubev, S. Ya. Frenkel, Vysokomolekul. Soedin. A 9, 1847 (1967).

555. V. N. Tsvetkov et al., Vysokomolekul. Soedin. A 10, 74 (1968).

556. V. N. Tsvetkov et al., Vysokomolekul. Soedin. A 10, 547 (1968).

557. V. M. Golubev, S. Ya. Frenkel, Vysokomolekul. Soedin. A 10, 750 (1968).

558. Yu. Ye. Eizner, A. N. Cherkasov, S. J. Klenin, Vysokomolekul. Soedin. A 10, 1971 (1968).

559. V. N. Tsvetkov et al., Vysokomolekul. Soedin. 7, 1103 (1965).

560. V. N. Tsvetkov et al., Vysokomolekul. Soedin. A 11, 349 (1969).

561. S. R. Rafikov, Vysokomolekul. Soedin. A 11, 1990 (1969).

562. A. A. Tager et al., Vysokomolekul. Soedin. A 12, 1320 (1970).

563. V. N. Tsvetkov et al., Vysokomolekul. Soedin. A 12, 1974 (1970).

564. N. Nakata, Makromol. Chem. 149, 99 (1971).

565. L. J. Fetters, H. Yu, Macromolecules 4, 385 (1971).

566. K. de Groot et al., Kolloid-Z. 246, 578 (1971).

567. K. F. Elgert, E. Seiler, Makromol. Chem. 151, 83 (1972).

568. R. Gabler, N. C. Ford, Jr., F. E. Karasz, Polymer Preprints 12, 776 (1971).

569. K.-J. Linow, B. Philipp, Faserforsch. Textiltechn. 22, 444 (1971).

570. K. Kamada, H. Sato, Polymer J. 2, 593 (1971).

571. O. Kramer, J. E. Frederick, Macromolecules 4, 613 (1971).

572. W. L. Petikolas, Adv. Polymer Sci. 9, 285 (1972).

573. T. Homma, H. Fujita, J. Appl. Polymer Sci. 9, 1701 (1965).

574. M. Abe et al., J. Appl. Polymer Sci. 8, 2549 (1965).

575. L. C. Cerny, R. C. Graham, H. James, Jr., J. Appl. Polymer Sci. 11, 1941 (1967).

576. W. J. Closs, B. R. Jennings, H. G. Jerrard, European Polymer J. 4, 639 (1968).

577. H. Matsuda, H. Aonuma, S. Kuroiwa, J. Appl. Polymer Sci. 14, 335 (1970).

578. M. G. Wirick, H. Waldmann, J. Appl. Polymer Sci. 14, 579 (1970).

579. J. K. Nekrasov et al., Vysokomolekul. Soedin. A 14, 866 (1972).

580. V. N. Tsvetkov et al., Vysokomolekul. Soedin. A 14, 427 (1972).

581. A. Wada, N. C. Ford, F. E. Karasz, Polymer Preprints 13, 191 (1972).

582. K. Lederer, Polymer Preprints 13, 203 (1972).

583. V. N. Tsetkov et al., Vysokomolekul. Soedin. A 14, 369 (1972).

584. D. J. Goldwasser, D. J. Williams, Polymer Preprints 12, 539 (1971).

585. K. Kubo, K. Ogino, Bull. Chem. Soc. (Japan) 44, 997 (1971).

586. T. Lucas, Compt. Rend. C-270, 1377 (1970).

587. M. Gorden, S. Polowinski, British Polymer J. 2, 182 (1970).

588. K. Matsumura, Bull. Chem. Soc. (Japan) 43, 1303 (1970).

589. J. K. Nekrasov, Vysokomolekul. Soedin. A 13, 1707 (1971).

590. M. Abe, M. Iwama, T. Homma, J. Chem. Soc. (Japan) Ind. Chem. Sect. 72, 2313 (1969).

591. V. Bugdahl, Kautschuk, Gummi, Kunststoffe 22, 486 (1969).

592. H. Matsuda, H. Aonuma, S. Kuroiwa, Rept. Progr. Polymer Phys. (Japan) 11, 25 (1968).

593. A. Kotera, H. Matsuda, K. Takemura, Rept. Progr. Polymer Phys. (Japan) 11, 61 (1968).

594. A. Kotera, T. Hamada, Rept. Progr. Polymer Phys. (Japan) 11, 57 (1968).

595. K. Matsumura, Bull. Chem. Soc. (Japan) 42, 1874 (1969).

596. G. Moraglio, F. Zoppi, F. Danusso, Chim. Ind. (Milan) 51, 117 (1969).

597. H. Matsuda, K. Yamano, H. Inagaki, J. Chem. Soc. (Japan) Ind. Chem. Sect. 73, 390 (1970).

598. M. Matsushima, M. Fukatsu, M. Kurata, Bull. Chem. Soc. (Japan) 41, 2570 (1968).

599. A. J. Hyde, A. G. Tanner, J. Colloid Interf. Sci. 28, 179 (1968).

600. A. Červenka et al., Collection Cech. Chem. Commun. 33, 4248 (1968).

601. M. Abe et al., Bull. Chem. Soc. (Japan) 41, 2330 (1968).

602. V. Petrus, Collection Cech. Commun. 33, 119 (1968).

603. K. Kamide et al., Chem. High Polymers (Tokyo) 24, 679 (1967).

604. G. Moraglio, N. Oddo, Chim. Ind. 48, 224 (1966).

605. J. M. G. Cowie, E. L. Cussler, J. Chem. Phys. 46, 4886 (1967).

606. K. F. Elgert, K. Cammann, Z. Anal. Chem. 226, 193 (1967).

607. H. Sato, Bull. Chem. Soc. (Japan) 39, 2335 (1966).

608. H. Sato, Bull. Chem. Soc. (Japan) 39, 2340 (1966).

609. K. Takamizawa, Bull. Chem. Soc. (Japan) 39, 1168 (1966).

610. A. Kotera et al. Bull. Chem. Soc. (Japan) 39, 1192 (1966).

611. C. Trevissoi, R. Francesconi, Chim. Ind. 47, 1184 (1965).

612. G. Moraglio, J. Masini, Chim. Ind. 47, 1050 (1965).

613. C. Trevissoi, R. Francesconi, Chim. Ind. 47, 744 (1965).

614. J. Eigner, P. Doty, J. Mol. Biol. 12, 549 (1965).

615. R. Jaenicke, M. Gehatia, Biochim. Biophys. Acta 45, 217 (1960).

616. A. Soda, T. Fujimoto, M. Nagasawa, J. Phys. Chem. 71, 4274 (1967).

617. M. Pizzoli, G. Ceccorulli, European Polymer J. 8, 769 (1972).

618. M. B. Huglin, "Light Scattering from Polymer Solutions," Academic Press, London, New York (1972).

POLYMOLECULARITY CORRECTION FACTORS

R. E. Bareiss[*]

Editorial office: "Die Makromolekulare Chemie", Mainz

CONTENTS

[*] The author thanks Dr. G. S. Greschner, Institut fuer Physikalische Chemie, Universitaet Mainz, for performing the computer calculations and Professor Dr. B. A. Wolf, Institut fuer Physikalische Chemie, Universitaet Mainz, for submitting to him unpublished calculations.

[**] SF = Schulz-Flory Distribution of Molecular Weight. [***] LN = Logarithmic Normal Distribution of Molecular Weight.

A. LIST OF SYMBOLS USED

a_D: exponent in $D = K_D M^{a_D}$

a_s: exponent in $s = K_s M^{a_s}$

a_η: exponent in $[\eta] = K_\eta M^{a_\eta}$

B, B', B'':   constants

D: diffusion coefficient

D_A: "area average" diffusion coefficient $D_A = (\sum\limits_i W_i D_i^{-0.5})^{-2}$

$<D>_w$:   weight average diffusion coefficient (as index: D_w)

f: frictional coefficient

$[f]$:   intrinsic frictional coefficient $[f] = \dfrac{f}{\eta}$

i: numbers $1, 2 \ldots \ldots i$

k: degree of coupling in Schulz-Flory distributions
                $k = [(<M>_w / <M>_n) - 1]^{-1}$

K_D: constant in $D = K_D M^{a_D}$

K_o: constant for determination of unperturbed dimensions:
                $K_o = \Phi_\Theta \cdot (r_o^2 / M)^{3/2}$

K_s: constant in $s = K_s \cdot M^{a_s}$

K_S: constant in $S^2 = K_S \cdot M^{1+\epsilon}$

K_η: constant in $[\eta] = K_\eta \cdot M^{a_\eta}$

M: molecular weight in g/mol

$<M>_{av}$:   any molecular weight average

$<M>_D$:   molecular weight from D

M_i: molecular weight of component i

M_m: median molecular weight in logarithmic normal distributions

$<M>_n$:   number average molecular weight (as index: M_n)

$<M>_s$:   molecular weight from s

$<M>_{sD}$:   molecular weight from s and D

$<M>_{s\eta}$:   molecular weight from s and η

$<M>_w$:   weight average molecular weight (as index: M_w)

$<M>_\eta$:   viscosity average molecular weight

$<M>_{\eta D}$:   molecular weight from η and D

n: argument $n = w/M$

N_L: Avogadro number

P_o: constant in $[f]_\Theta = P_o \cdot r_o$

q: polymolecularity correction factor

$q_{B,M}$:   q for $<M>_{av}$ in the Baumann equation

$q_{B,S}$:   q for $<S^2>_{av}$ in the Baumann equation

q_{BSF}:   q for the Burchard-Stockmayer-Fixman-equation

q_{CB1}, q_{CB2}:   q's for the Cowie-Bywater equation

q_{Diff}:   q for $D = K_D \cdot M^{a_D}$

q_{KMSH}:   q for $[\eta] = K_\eta \cdot M^{a_\eta}$

q_{Sed}:   q for $s = K_s \cdot M^{a_s}$

r_o: unperturbed end to end distance

s: sedimentation coefficient

s_{max}:   s from maximum of gradient curve

$<s>_w$:   weight average sedimentation coefficient (as index: s_w)

S: radius of gyration

$<S>_{av}$:   any average of the average square radius of gyration

$<S^2>_z$:   z-average square radius of gyration (as index: S_z)

$v_2^\star$:   partial specific volume of the solute (index 2)

w: argument $w = n \cdot M$

W: weight fraction

z: argument $z = w \cdot M$

Γ: gamma function (tabulated e.g. in 32)

ϵ: exponent of the relation between radius of gyration and molecular weight Eq. (9)

η: viscosity (index 1: solvent)

$[\eta]$:   intrinsic viscosity

Θ: theta-temperature; index: theta state

π: pi

ϱ: density of solvent (index 1)

σ: standard deviation of the logarithmic normal distribution

$\sum\limits_i$:   summation over all species i

Φ: Fox-Flory constant $\Phi_\Theta = 2.87 \cdot 10^{23} \, mol^{-1}$ (for end to end distance)

B. INTRODUCTION

Exact conclusions about the correlation of molecular weight and properties dependent on molecular weight for polymers exhibiting a molecular weight distribution (polymolecularity) require the comparison of corresponding averages. E.g. in Staudinger's viscosity law (1), modified by Kuhn (2), Mark (3), Sakurada (4) and Houwink (5) the experimentally determined intrinsic viscosity is a weight average (6-8) but the corresponding molecular weight is the viscosity average molecular weight (9, 10):

$$[\eta] = K_\eta \cdot (\langle M \rangle_\eta)^{a_\eta}$$

where $\qquad [\eta] = \sum_i W_i [\eta]_i \quad \text{and} \quad \langle M \rangle_\eta = \left(\sum_i W_i M_i^{a_\eta} \right)^{1/a_\eta}$ $\qquad\qquad$ (1)

A direct experimental determination of $\langle M \rangle_\eta$ is not possible (exception: the combination of the intrinsic viscosity $[\eta]_\Theta$ and the weight average sedimentation coefficient $\langle s \rangle_{w,\Theta}$ for the Θ-state yields the viscosity average molecular weight $\langle M \rangle_{\eta,\Theta}$ (13) in the Flory-Mandelkern-Scheraga (11, 12) equation). Therefore, for the determination of the constant K_η and the exponent a_η in Eq. (1) usually different molecular weight averages, e.g. $\langle M \rangle_w$, $\langle M \rangle_n$, $\langle M \rangle_{sD}$, $\langle M \rangle_{\eta D}$ or $\langle M \rangle_{s\eta}$ etc.[*] are applied. From the latter $\langle M \rangle_\eta$ can be calculated provided type and width of the molecular weight distribution are known (10). Alternatively a polymolecularity correction factor q can be calculated, which corrects Eq. (1) for the use of $\langle M \rangle_{av}$ instead of $\langle M \rangle_\eta$:

$$[\eta] = K_\eta \cdot (\langle M \rangle_\eta)^{a_\eta} = K_\eta \cdot q_{KMSH} \cdot (\langle M \rangle_{av})^{a_\eta} \qquad\qquad (1a)$$

$$q_{KMSH} = (\langle M \rangle_\eta / \langle M \rangle_{av})^{a_\eta} \qquad\qquad (1b)$$

Correspondingly polymolecularity correction factors can be useful for the sedimentation coefficient - molecular weight relationship:

$$s = K_s \cdot (\langle M \rangle_s)^{a_s} \qquad\qquad (2)$$

and the diffusion coefficient-molecular weight relationship:

$$D = K_D \cdot (\langle M \rangle_D)^{a_D} \qquad\qquad (3)$$

if other than the appropriate molecular weight averages $\langle M \rangle_s$ and $\langle M \rangle_D$ resp. (10) are to be used for the evaluation of the constants K_s and K_D and the exponents a_s and a_D, resp. Analogously to Eq. (1a) the polymolecularity correction factors for Eq. (2) and Eq. (3) are found to be:

$$q_{sed} = (\langle M \rangle_s / \langle M \rangle_{av})^{a_s} \qquad\qquad (2a)$$

$$q_{Diff} = (\langle M \rangle_D / \langle M \rangle_{av})^{a_D} \qquad\qquad (3a)$$

Also for the experimental determination of the unperturbed dimensions of macromolecules, polymolecularity correction factors must be used if the results for polymers with different type and/or width of the molecular weight distribution are to be compared with one another and with theory, resp. (10, 14, 15).

The Burchard-Stockmayer-Fixman procedure (16, 17) for polymolecular polymers and the use of $\langle M \rangle_w$ is given by (10, 14):

$$[\eta]/(\langle M \rangle_w)^{1/2} = \Phi_\Theta \cdot K_0^{1/2} \cdot q_{BSF} + B \cdot (\langle M \rangle_w)^{1/2} \qquad\qquad (4)$$

By plotting $\qquad [\eta]/(q_{BSF} \cdot (\langle M \rangle_w)^{1/2}) \text{ vs. } (\langle M \rangle_w)^{1/2}/q_{BSF}$

of polymer homologues with different width and/or type of molecular weight distribution a straight line is obtained from which by extrapolating $(\langle M \rangle_w)^{1/2}/q_{BSF} \longrightarrow 0$ the unperturbed dimensions can be calculated[*].

---

[*] See this Handbook under "Solution Properties".

The correction factor for Eq. (4) is found to be (14):

$$q_{BSF} = \langle M^{1/2} \rangle_w / (\langle M \rangle_w)^{1/2} \tag{4a}$$

In the equation of Cowie and Bywater (18) two polymolecularity correction factors are necessary if polymolecular polymers are to be investigated. If the averages $\langle M \rangle_w$ and $\langle s \rangle_w$ are to be inserted we obtain (10, 14):

$$((\langle M \rangle_w)^{1/2} / \langle s \rangle_w) \cdot (1 - {}^*v_2 \varrho_1)/(\eta_1 N_L) = P_0 \cdot q_{CB1} + q_{CB2} \cdot B'(\langle M \rangle_w)^{1/2} \tag{5}$$

with

$$q_{CB1} = \langle M^{1-0,5\varepsilon} \rangle_w / ((\langle M \rangle_w)^{1/2} \cdot \langle M^{0,5(1-\varepsilon)} \rangle_w) \tag{5a}$$

$$q_{CB2} = \langle M^{1,5-0,5\varepsilon} \rangle_w / (\langle M \rangle_w \cdot \langle M^{0,5(1-\varepsilon)} \rangle_w) \tag{5b}$$

The unperturbed dimensions are obtained by plotting $(\langle M \rangle_w)^{1/2} / \langle s \rangle_w$ vs. $(\langle M \rangle_w)^{1/2}$ and extrapolating $(\langle M \rangle_w)^{1/2} \to 0$.

Baumann (19) describes a procedure for the evaluation of unperturbed dimensions from the determination of the molecular weight and the radius of gyration for a series of polymer homologues. If the latter exhibit different width and/or type of molecular weight distribution the Baumann equation can be written with two polymolecularity correction factors $q_{B,M}$ (for the molecular weight) and $q_{B,S}$ (for the radius of gyration)(15).

$$(\langle S^2 \rangle_{av} \cdot q_{B,S}/(\langle M \rangle_{av} \cdot q_{B,M}))^{3/2} = K_0^{3/2} + B'' \cdot (\langle M \rangle_{av} \cdot q_{B,M})^{1/2} \tag{6}$$

if $\langle M \rangle_w$ is to be used:

$$q_{B,M_w} = (\langle M^2 \rangle_z)^2 / (\langle M \rangle_w \cdot (\langle M^{3/2} \rangle_z)^2) \tag{6a}$$

and if $\langle S^2 \rangle_z$ is to be applied:

$$q_{B,S_z} = ((\langle M^2 \rangle_z)^3 / (\langle M^{3/2} \rangle_z)^{0/3}) \cdot ((\langle M^{1,5+1,5\varepsilon} \rangle_z)^{2/3} / \langle M^{1+\varepsilon} \rangle_z) \tag{6b}$$

The polymolecularity correction factors (1b), (2a), (3a), (4a), (5a), (5b), (6a) and (6b) are tabulated on the following pages for Schulz-Flory Distributions of molecular weights (20-23)[*]

$$W(M) = [(k/\langle M \rangle_n)^{k+1}/\Gamma(k+1)] \cdot M^k \cdot \exp[-(k/\langle M \rangle_n)M] \tag{7}$$

and for Logarithmic Normal Distributions of molecular weights (24-26)[*]

$$W(M) = [1/(\sigma \sqrt{2\pi})] \cdot (1/M) \cdot \exp[-\ln(M/M_m)^2/(2\sigma^2)] \tag{8}$$

as functions of the width of the distribution expressed as $\langle M \rangle_w / \langle M \rangle_n$ and of the exponent a_η of Eq. (1).

The exponents a_η, a_s, a_D of Eqs. (1), (2), (3) and ε of

$$S^2 = K_s \cdot M^{1+\varepsilon} \tag{9}$$

are interconnected with one another (27-30) by the "Exponents' Rule" (10):

$$a_\eta = 2 - 3a_s = -(1+3a_D) = 1,5\,\varepsilon + 0,5 \tag{10}$$

[*] See this Handbook under L. H. Peebles: "Rates of Polymerization and Depolymerization, Average Molecular Weights and Molecular Weight Distribution of Polymers.

In the following tables the values for $<M>_w/<M>_n$ range from 1.1 to 20.0, those for a_η from 0.4 to 0.9, in order to cover also limiting cases which occur in practical work (e.g. $a_\eta < 0.5$; $\epsilon < 0$. The intervals for both parameters $<M>_w/<M>_n$ and a_η and the number of decimal places of the polymolecularity correction factors are chosen so that for any value of $<M>_w/<M>_n$ and a_η within the above mentioned limits the polymolecularity correction factor can be found by interpolation of the values given in the tables with an accuracy of 2 decimal places. In order to be able to calculate the polymolecularity correction factors for any $<M>_w/<M>_n$ and a_η (or a_s, a_D and ϵ resp.) the exact formulae for all polymolecularity correction factors tabulated are given in the legends of the corresponding tables.

Table 1.1 Polymolecularity Correction Factors for $[\eta] = K_\eta \cdot M^{a_\eta}$ for Schulz-Flory Distributions using Weight Average Molecular Weight $<M>_w$:

$$q_{KMSH,M_w}^{(SF)} = \frac{\Gamma(k+1+a_\eta)}{(k+1)^{a_\eta} \cdot \Gamma(k+1)}$$

$<M>_w/<M>_n$	a_η			
	0.40	0.50	0.70	0.90
1.1	0.9891	0.9887	0.9906	0.9960
1.3	0.9725	0.9716	0.9765	0.9900
1.5	0.9605	0.9594	0.9665	0.9858
1.7	0.9515	0.9502	0.9590	0.9827
2.0	0.9414	0.9400	0.9509	0.9793
2.5	0.9301	0.9287	0.9419	0.9750
3.0	0.9228	0.9213	0.9360	0.9731
4.0	0.9137	0.9123	0.9289	0.9701
5.0	0.9083	0.9069	0.9247	0.9684
10.0	0.8977	0.8964	0.9166	0.9650
20.0	0.8924	0.8913	0.9126	0.9634

Table 1.2 Polymolecularity Correction Factors for $[\eta] = K_\eta \cdot M^{a_\eta}$ for Logarithmic Normal Distributions using Weight Average Molecular Weight $<M>_w$:

$$q_{KMSH,M_w}^{(LN)} = (\langle M \rangle_w/\langle M \rangle_n)^{(a_\eta^2 - a_\eta)/2}$$

$<M>_w/<M>_n$	a_η			
	0.40	0.50	0.70	0.90
1.1	0.9886	0.9882	0.9900	0.9957
1.3	0.9690	0.9677	0.9728	0.9883
1.5	0.9525	0.9506	0.9583	0.9819
1.7	0.9383	0.9358	0.9458	0.9764
2.0	0.9202	0.9170	0.9298	0.9693
2.5	0.8959	0.8918	0.9083	0.9596
3.0	0.8765	0.8717	0.8911	0.9518
4.0	0.8468	0.8409	0.8645	0.9395
5.0	0.8244	0.8178	0.8445	0.9301
6.0	0.8065	0.7993	0.8285	0.9225
8.0	0.7792	0.7711	0.8039	0.9107
10.0	0.7586	0.7499	0.7852	0.9016
20.0	0.6980	0.6877	0.7301	0.8739

Table 1.3 Polymolecularity Correction Factors for $[\eta] = K_\eta \cdot M^{a_\eta}$ for Schulz-Flory Distributions using Number Average Molecular Weight $<M>_n$:

$$q_{KMSH,M_n}^{(SF)} = \frac{(k+1+a_\eta)}{k^{a_\eta} \Gamma(k+1)}$$

$\dfrac{<M>_w}{<M>_n}$	a_η					
	0.40	0.50	0.60	0.70	0.80	0.90
1.1	1.0276	1.0370	1.0474	1.0589	1.0715	1.0852
1.3	1.0802	1.1078	1.1388	1.1733	1.2115	1.2537
1.5	1.1297	1.1750	1.2262	1.2837	1.3480	1.4199
2.0	1.2422	1.3293	1.4296	1.5447	1.6765	1.8274
2.5	1.3419	1.4684	1.6162	1.7888	1.9901	2.2253
3.0	1.4320	1.5958	1.7898	2.0197	2.2922	2.6156
4.0	1.5908	1.8245	2.1078	2.4514	2.8692	3.3782
5.0	1.7290	2.0279	2.3968	2.8529	3.4186	4.1222
6.0	1.8525	2.2129	2.6643	3.2317	3.9469	4.8516
8.0	2.0683	2.5429	3.1523	3.9380	4.9549	6.2762
10.0	2.2548	2.8348	3.5944	4.5937	5.9138	7.6655
15.0	2.6415	3.4586	4.5684	6.0836	8.1624	11.029
20.0	2.9579	3.9860	5.4197	7.4299	10.264	14.280

Table 1.4 Polymolecularity Correction Factors for $[\eta] = K_\eta \cdot M^{a_\eta}$ for Logarithmic Normal Distribution using Number Average Molecular Weight $<M>_n$:

$$q_{KMSH,M_n}^{(LN)} = (\langle M \rangle_w / \langle M \rangle_n)^{(a_\eta^2 + a_\eta)/2}$$

$\dfrac{<M>_w}{<M>_n}$	a_η					
	0.40	0.50	0.60	0.70	0.80	0.90
1.1	1.0270	1.0364	1.0468	1.0583	1.0710	1.0849
1.3	1.0762	1.1034	1.1342	1.1690	1.2079	1.2515
1.5	1.1202	1.1642	1.2149	1.2728	1.3390	1.4144
1.7	1.1602	1.2202	1.2901	1.3713	1.4653	1.5741
2.0	1.2142	1.2968	1.3947	1.5105	1.6472	1.8088
2.5	1.2925	1.4100	1.5524	1.7249	1.9343	2.1890
3.0	1.3602	1.5098	1.6944	1.9226	2.2056	2.5582
4.0	1.4743	1.6818	1.9453	2.2815	2.7132	3.2716
5.0	1.5693	1.8286	2.1652	2.6055	3.1861	3.9593
6.0	1.6515	1.9580	2.3633	2.9040	3.6330	4.6272
8.0	1.7901	2.1810	2.7132	3.4462	4.4691	5.9176
10.0	1.9055	2.3714	3.0200	3.9355	5.2481	7.1614
20.0	2.3136	3.0753	4.2121	5.9445	8.6445	12.9533

Table 1.5 Polymolecularity Correction Factors for $[\eta] = K_\eta \cdot M^{a_\eta}$ for Schulz-Flory Distributions using Molecular Weight from $<s>_w$ and D_A:

$$q^{(SF)}_{KMSH,s_wD_A} = \frac{\Gamma(k+1+a_\eta)}{\Gamma(k+1)} \cdot \left(\frac{\Gamma^3(k+1)}{\Gamma\left(k+(5/3)-(a_\eta/3)\right)\ \Gamma^2\left(k+(7/6)+(a_\eta/6)\right)} \right)^{a_\eta}$$

$<M>_w/<M>_n$	a_η			
	0.40	0.50	0.70	0.90
1.1	1.0001	1.0029	1.0114	1.0236
1.5	1.0007	1.0113	1.0439	1.0907
2.0	1.0012	1.0176	1.0678	1.1404
3.0	1.0018	1.0243	1.0929	1.1926
5.0	1.0024	1.0299	1.1137	1.2363
10.0	1.0029	1.0342	1.1297	1.2700
20.0	1.0032	1.0364	1.1378	1.2871

Table 1.6 Polymolecularity Correction Factors for $[\eta] = K_\eta \cdot M^{a_\eta}$ for Logarithmic Normal Distributions using Molecular Weight from $<s>_w$ and D_A:

$$q^{(LN)}_{KMSH,s_wD_A} = (\langle M \rangle_w/\langle M \rangle_n)^{(8a_\eta^2 - 3a_\eta - a_\eta^3)/12}$$

$<M>_w/<M>_n$	a_η			
	0.40	0.50	0.70	0.90
1.1	1.0001	1.0030	1.0118	1.0245
1.5	1.0005	1.0128	1.0512	1.1086
2.0	1.0009	1.0219	1.0891	1.1927
3.0	1.0015	1.0349	1.1448	1.3222
5.0	1.0021	1.0516	1.2191	1.5056
10.0	1.0031	1.0746	1.3276	1.7958
20.0	1.0040	1.0981	1.4459	2.1418

Table 1.7 Polymolecularity Corrections Factors for $[\eta] = K_\eta \cdot M^{a_\eta}$ for Schulz-Flory Distributions using Molecular Weight from D_A and $[\eta]$:

$$q^{(SF)}_{KMSH,D_A\eta_w} = \frac{[\Gamma(k+1)]^{5a_\eta-1} \cdot [\Gamma(k+1+a_\eta)]^{a_\eta+1}}{[\Gamma\left(k+(7/6)+(a_\eta/6)\right)]^{6a_\eta}}$$

$<M>_w/<M>_n$	a_η			
	0.40	0.50	0.70	0.90
1.1	1.0044	1.0087	1.0231	1.0468
1.5	1.0172	1.0343	1.0909	1.1861
2.0	1.0270	1.0538	1.1426	1.2949
3.0	1.0375	1.0747	1.1983	1.4147
5.0	1.0465	1.0924	1.2458	1.5189
10.0	1.0535	1.1061	1.2831	1.6020
20.0	1.0570	1.1132	1.3022	1.6452

Table 1.8 Polymolecularity Correction Factors for $[\eta] = K_\eta \cdot M^{a_\eta}$ for Logarithmic Normal Distributions using Molecular Weight from D_A and $[\eta]$:

$$q_{KMSH,D_A\eta_w}^{(LN)} = (\langle M \rangle_w / \langle M \rangle_n)^{(5a_\eta^3 + 4a_\eta^2 - a_\eta)/12}$$

$\dfrac{\langle M \rangle_w}{\langle M \rangle_n}$	a_η					
	0.40	0.50	0.60	0.70	0.80	0.90
1.1	1.0045	1.0090	1.0154	1.0239	1.0349	1.0487
1.5	1.0191	1.0387	1.0670	1.1057	1.1572	1.2241
2.0	1.0329	1.0671	1.1173	1.1875	1.2834	1.4130
3.0	1.0526	1.1085	1.1922	1.3131	1.4851	1.7297
5.0	1.0780	1.1629	1.2937	1.4903	1.7850	2.2316
10.0	1.1134	1.2409	1.4454	1.7698	2.2909	3.1532
20.0	1.1500	1.3243	1.6150	2.1016	2.9402	4.4554

Table 1.9 Polymolecularity Correction Factors for $[\eta] = K_\eta \cdot M^{a_\eta}$ for Schulz-Flory Distributions using Molecular Weight from s_{max} and $[\eta]$:

$$q_{KMSH,s_{max}\eta_w}^{(SF)} = \frac{1}{[k+(1/3)+(a_\eta/3)]^{(2a_\eta - a_\eta^2)/2}} \cdot \left(\frac{\Gamma(k+1+a_\eta)}{\Gamma(k+1)}\right)^{1-(a_\eta/2)}$$

$\langle M \rangle_w / \langle M \rangle_n$	a_η			
	0.40	0.50	0.70	0.90
1.1	1.0072	1.0090	1.0122	1.0147
1.5	1.0309	1.0380	1.0500	1.0583
2.0	1.0522	1.0634	1.0815	1.0928
3.0	1.0793	1.0948	1.1187	1.1317
5.0	1.1063	1.1256	1.1535	1.1667
10.0	1.1307	1.1528	1.1833	1.1956
20.0	1.1446	1.1680	1.1995	1.2110

References page IV-130

Table 1.10 Polymolecularity Correction Factors for $[\eta] = K_\eta \cdot M^{a_\eta}$ for Logarithmic Normal Distributions using Molecular Weight from s_{max} and $[\eta]$:

$$q^{(LN)}_{KMSH,s_{max}\eta_w} = (\langle M \rangle_w / \langle M \rangle_n)^{(8a_\eta - 2a_\eta^2 - a_\eta^3)/12}$$

$\langle M \rangle_w / \langle M \rangle_n$	a_η			
	0.40	0.50	0.70	0.90
1.1	1.0226	1.0272	1.0346	1.0393
1.5	1.0998	1.1208	1.1555	1.1781
2.0	1.1766	1.2152	1.2802	1.3234
3.0	1.2941	1.3620	1.4793	1.5591
5.0	1.4589	1.5725	1.7747	1.9167
10.0	1.7166	1.9110	2.2720	2.5366
20.0	2.0198	2.3223	2.9088	3.3569

Table 1.11 Polymolecularity Correction Factors for $[\eta] = K_\eta \cdot M^{a_\eta}$ for Schulz-Flory Distributions using Molecular Weight from $\langle s \rangle_w$ and $[\eta]$:

$$q^{(SF)}_{KMSH,s_w\eta_w} = \frac{1}{[\Gamma(k+(5/3)-(a_\eta/3))]^{1.5a_\eta}} \cdot \frac{[\Gamma(k+1+a_\eta)]^{1-0.5a_\eta}}{[\Gamma(k+1)]^{1-2a_\eta}}$$

$\langle M \rangle_w / \langle M \rangle_n$	a_η			
	0.40	0.50★	0.70	0.90
1.1	0.9980	1.0000	1.0056	1.0122
1.5	0.9925	1.0000	1.0212	1.0459
2.0	0.9885	1.0000	1.0323	1.0702
5.0	0.9811	1.0000	1.0530	1.1153
10.0	0.9786	1.0000	1.0600	1.1307
20.0	0.9773	1.0000	1.0635	1.1385

Table 1.12 Polymolecularity Correction Factors for $[\eta] = K_\eta \cdot M^{a_\eta}$ for Logarithmic Normal Distributions using Molecular Weight from $\langle s \rangle_w$ and $[\eta]$:

$$q^{(LN)}_{KMSH,s_w\eta_w} = (\langle M \rangle_w / \langle M \rangle_n)^{(5a_\eta^2 - 2a_\eta^3 - 2a_\eta)/6}$$

$\langle M \rangle_w / \langle M \rangle_n$	a_η			
	0.40	0.50★	0.70	0.90
1.1	0.9980	1.0000	1.0058	1.0127
1.5	0.9914	1.0000	1.0249	1.0550
2.0	0.9853	1.0000	1.0429	1.0958
5.0	0.9663	1.0000	1.1026	1.2367
10.0	0.9521	1.0000	1.1499	1.3552
20.0	0.9381	1.0000	1.1993	1.4850

★ See Ref. 13.

Table 2.1 Polymolecularity Correction Factors for $s = K_s \cdot M^{a_s}$ for Schulz-Flory Distributions using s_{max} and $<M>_w$:

$$q_{Sed,s_{max}M_w}^{(SF)} = \left(\frac{3k+1+a_\eta}{3(k+1)}\right)^{(2-a_\eta)/3}$$

$\dfrac{<M>_w}{<M>_n}$	a_η and (a_s)			$\dfrac{<M>_w}{<M>_n}$	a_η and (a_s)		
	0.40 (0.53)	0.50 (0.50)	0.90 (0.37)		0.40 (0.53)	0.50 (0.50)	0.90 (0.37)
1.1	0.9738	0.9770	0.9877	5.0	0.7433	0.7746	0.8805
1.5	0.9009	0.9129	0.9533	10.0	0.7056	0.7416	0.8634
2.0	0.8475	0.8660	0.9284	20.0	0.6860	0.7246	0.8547

Table 2.2 Polymolecularity Correction Factors for $s = K_s \cdot M^{a_s}$ for Logarithmic Normal Distributions using s_{max} and $<M>_w$:

$$q_{Sed,s_{max}\,M_w}^{(LN)} = (\langle M \rangle_w / \langle M \rangle_n)^{(11a_\eta - 2a_\eta^2 - 14)/18}$$

$\dfrac{<M>_w}{<M>_n}$	a_η and (a_s)			$\dfrac{<M>_w}{<M>_n}$	a_η and (a_s)		
	0.40 (0.53)	0.50 (0.50)	0.90 (0.37)		0.40 (0.53)	0.50 (0.50)	0.90 (0.37)
1.1	0.9488	0.9535	0.9702	5.0	0.4119	0.4472	0.5996
1.5	0.7998	0.8165	0.8791	6.0	0.3795	0.4000	0.5639
2.0	0.6825	0.7071	0.8023	8.0	0.3179	0.3536	0.5164
3.0	0.5458	0.5774	0.7053	10.0	0.2811	0.3162	0.4811
4.0	0.4658	0.5000	0.6437	20.0	0.1919	0.2236	0.3860

Table 2.3 Polymolecularity Correction Factors for $s = K_s \cdot M^{a_s}$ for Schulz-Flory Distributions using s_{max} and $<M>_n$:

$$q_{Sed,s_{max}\,M_n}^{(SF)} = \left(\frac{3k+1+a_\eta}{3k}\right)^{(2-a_\eta)/3}$$

$\dfrac{<M>_w}{<M>_n}$	a_η and (a_s)			$\dfrac{<M>_w}{<M>_n}$	a_η and (a_s)		
	0.40 (0.53)	0.50 (0.50)	0.90 (0.37)		0.40 (0.53)	0.50 (0.50)	0.90 (0.37)
1.1	1.0246	1.0247	1.0228	5.0	1.7536	1.7321	1.5886
1.5	1.1183	1.1180	1.1061	6.0	1.9005	1.8708	1.6875
2.0	1.2266	1.2247	1.1971	8.0	2.1679	2.1213	1.8600
3.0	1.4213	1.4142	1.3499	10.0	2.4092	2.3452	2.0086
4.0	1.5951	1.5811	1.4776	20.0	3.3902	3.2404	2.5636

References page IV-130

Table 2.4 Polymolecularity Correction Factors for $s = K_s \cdot M^{a_s}$ for Logarithmic Normal Distributions using s_{max} and $<M>_n$:

$$q_{\text{Sed},s_{max} M_n}^{(LN)} = (\langle M \rangle_w / \langle M \rangle_n)^{(5a_\eta - 2a_\eta^2 - 2)/18}$$

$\dfrac{<M>_w}{<M>_n}$	a_η and (a_s)			
	0.40 (0.53)	0.50 (0.50)	0.70 (0.43)	0.90 (0.37)
1.1	0.9983	1.0000	1.0028	1.0047
1.5	0.9928	1.0000	1.0118	1.0200
2.0	0.9878	1.0000	1.0202	1.0345
5.0	0.9718	1.0000	1.0476	1.0819
10.0	0.9599	1.0000	1.0688	1.1192
20.0	0.94814	1.0000	1.0904	1.1577

Table 3.1 Polymolecularity Correction Factors for $D = K_D \cdot M^{a_D}$ for Schulz-Flory Distributions using D_A and $<M>_n$:

$$q_{\text{Diff},D_A M_n}^{(SF)} = \frac{1}{k^{-(a_\eta + 1)/3}} \cdot \left(\frac{\Gamma(k+1)}{\Gamma(k+(7/6)+(a_\eta/6))} \right)^2$$

$\dfrac{<M>_w}{<M>_n}$	a_η and (a_D)			$\dfrac{<M>_w}{<M>_n}$	a_η and (a_D)		
	0.40 (-0.47)	0.50 (-0.50)	0.90 (-0.63)		0.40 (-0.47)	0.50 (-0.50)	0.90 (-0.63)
1.1	0.9723	0.9700	0.9602	5.0	0.5483	0.5230	0.4309
1.5	0.8798	0.8705	0.8324	6.0	0.5069	0.4807	0.3868
2.0	0.7939	0.7790	0.7200	8.0	0.4467	0.4197	0.3253
2.5	0.7292	0.7107	0.6392	10.0	0.4045	0.3773	0.2840
3.0	0.6783	0.6575	0.5781	20.0	0.2955	0.2695	0.1854
4.0	0.6026	0.5790	0.4908				

Table 3.2 Polymolecularity Correction Factors for $D = K_D \cdot M^{a_D}$ for Logarithmic Normal Distributions using D_A and $<M>_n$:

$$q_{\text{Diff},D_A M_n}^{(LN)} = (\langle M \rangle_w / \langle M \rangle_n)^{-(a_\eta^2 + 8a_\eta + 7)/36}$$

$\dfrac{<M>_w}{<M>_n}$	a_η and (a_D)			$\dfrac{<M>_w}{<M>_n}$	a_η and (a_D)		
	0.40 (-0.47)	0.50 (-0.50)	0.90 (-0.63)		0.40 (-0.47)	0.50 (-0.50)	0.90 (-0.63)
1.1	0.9729	0.9707	0.9610	5.0	0.6293	0.6047	0.5112
1.5	0.8899	0.8810	0.8445	10.0	0.5155	0.4870	0.3829
2.0	0.8192	0.8053	0.7490	20.0	0.4223	0.3921	0.2868
3.0	0.7290	0.7094	0.6325				

Table 3.3 Polymolecularity Correction Factors for $D = K_D \cdot M^{a_D}$ for Schulz-Flory Distributions using $<D>_w$ and $<M>_n$:

$$q^{(SF)}_{Diff,D_w\,M_n} = \frac{1}{k^{-(a_\eta+1)/3}} \cdot \frac{\Gamma\left(k+(2/3)-(a_\eta/3)\right)}{\Gamma(k+1)}$$

$\dfrac{<M>_w}{<M>_n}$	a_η and (a_D)			$\dfrac{<M>_w}{<M>_n}$	a_η and (a_D)		
	0.40 (-0.47)	0.50 (-0.50)	0.90 (-0.63)		0.40 (-0.47)	0.50 (-0.50)	0.90 (-0.63)
1.1	0.9877	0.9876	0.9884	7.0	0.6064	0.5959	0.5768
1.5	0.9406	0.9400	0.9429	10.0	0.5285	0.5154	0.4868
2.0	0.8878	0.8862	0.8896	15.0	0.4478	0.4323	0.3945
3.0	0.8017	0.7979	0.7981	20.0	0.3963	0.3796	0.3369
5.0	0.6837	0.6760	0.6659				

Table 3.4 Polymolecularity Correction Factors for $D = K_D \cdot M^{a_D}$ for Logarithmic Normal Distributions using $<D>_w$ and $<M>_n$:

$$q^{(LN)}_{Diff,D_w\,M_n} = (<M>_w/<M>_n)^{(a_\eta^2-a_\eta-2)/18}$$

$\dfrac{<M>_w}{<M>_n}$	a_η and (a_D)			$\dfrac{<M>_w}{<M>_n}$	a_η and (a_D)		
	0.40 (-0.47)	0.50 (-0.50)	0.90 (-0.63)		0.40 (-0.47)	0.50 (-0.50)	0.90 (-0.63)
1.1	0.9882	0.9882	0.9890	7.0	0.7849	0.7841	0.7978
1.5	0.9508	0.9506	0.9540	10.0	0.7509	0.7499	0.7654
2.0	0.9174	0.9170	0.9227	15.0	0.7139	0.7128	0.7302
3.0	0.8722	0.8717	0.8802	20.0	0.6888	0.6877	0.7062
5.0	0.8185	0.8178	0.8296				

Table 4.1 Polymolecularity Correction Factors for the Burchard-Stockmayer-Fixman Procedure (Eq. (4))
for Schulz-Flory Distributions (Eq. (7)) using $<M>_w$:

$$q^{(SF)}_{BSF,M_w} = \frac{1}{(k+1)^{1/2}} \cdot \frac{\Gamma(k+1,5)}{\Gamma(k+1)}$$

and for Logarithmic Normal Distributions (Eq. (8) (31)):

$$q^{(LN)}_{BSF,M_w} = (<M>_w/<M>_n)^{-1/8}$$

$<M>_w/<M>_n$	1.1	1.5	2.0	3.0	5.0	10.0	20.0
$q^{(SF)}_{BSF,\,M_w}$	0.9887	0.9594	0.9400	0.9213	0.9069	0.8964	0.8913
$q^{(LN)}_{BSF,\,M_w}$	0.9882	0.9506	0.9170	0.8717	0.8178	0.7499	0.6877

References page IV-130

Table 4.2.1.1 Polymolecularity Correction Factors for the Cowie-Bywater Procedure (Eq. (5)) for Schulz-Flory Distributions using $<M>_w$ and $<s>_w$:

$$q_{CB1}^{(SF)} = \frac{\Gamma(k+2-0,5\varepsilon)}{(k+1)^{0,5} \cdot \Gamma(k+1,5-0,5\varepsilon)}$$

$\dfrac{<M>_w}{<M>_n}$	a_η and (ε)			$\dfrac{<M>_w}{<M>_n}$	a_η and (ε)		
	0.40 (-0.07)	0.50 (0.00)	0.90 (0.27)		0.40 (-0.07)	0.50 (0.00)	0.90 (0.27)
1.1	1.0129	1.0114	1.0054	5.0	1.1145	1.1026	1.0538
1.5	1.0477	1.0424	1.0209	10.0	1.1287	1.1155	1.0613
2.0	1.0716	1.0638	1.0322	20.0	1.1358	1.1220	1.0650
3.0	1.0955	1.0854	1.0441				

Table 4.2.1.2 Polymolecularity Correction Factors for the Cowie-Bywater Procedure (Eq. (5)) for Schulz-Flory Distributions using $<M>_w$ and $<s>_w$:

$$q_{CB2}^{(SF)} = \frac{k+1,5-0,5\varepsilon}{k+1}$$

$\dfrac{<M>_w}{<M>_n}$	a_η and (ε)			$\dfrac{<M>_w}{<M>_n}$	a_η and (ε)		
	0.40 (-0.07)	0.50 (0.00)	0.90 (0.27)		0.40 (-0.07)	0.50 (0.00)	0.90 (0.27)
1.1	1.0485	1.0455	1.0333	5.0	1.4267	1.4000	1.2933
1.5	1.1778	1.1667	1.1222	10.0	1.4800	1.4500	1.3300
2.0	1.2667	1.2500	1.1833	20.0	1.5067	1.4750	1.3483
3.0	1.3556	1.3333	1.2444				

Table 4.2.2.1 Polymolecularity Correction Factors for the Cowie-Bywater Procedure (Eq. (5)) for Logarithmic Normal Distributions using $<M>_w$ and $<s>_w$ (31):

$$q_{CB1}^{(LN)} = (\langle M\rangle_w / \langle M\rangle_n)^{(1-2\varepsilon)/8}$$

$\dfrac{<M>_w}{<M>_n}$	a_η and (ε)			$\dfrac{<M>_w}{<M>_n}$	a_η and (ε)		
	0.40 (-0.07)	0.50 (0.00)	0.80 (0.27)		0.40 (-0.07)	0.50 (0.00)	0.80 (0.27)
1.1	1.0136	1.0120	1.0056	5.0	1.2561	1.2228	1.0984
1.5	1.0591	1.0520	1.0239	10.0	1.3857	1.3335	1.1438
2.0	1.1032	1.0905	1.0413	15.0	1.4676	1.4029	1.1711
3.0	1.1684	1.1472	1.0662	20.0	1.5287	1.4542	1.1909

<u>Table 4.2.2.2</u> Polymolecularity Correction Factors for the Cowie-Bywater Procedure (Eq. (5)) for Logarithmic Normal Distributions using $<M>_w$ and $<s>_w$ (31):

$$q_{CB2}^{(LN)} = (\langle M \rangle_w / \langle M \rangle_n)^{(1-\varepsilon)/2}$$

$\dfrac{<M>_w}{<M>_n}$	a_η and (ε)			$\dfrac{<M>_w}{<M>_n}$	a_η and (ε)		
	0.40 (-0.07)	0.50 (0.00)	0.90 (0.27)		0.40 (-0.07)	0.50 (0.00)	0.90 (0.27)
1.1	1.0521	1.0488	1.0356	10.0	3.4145	3.1623	2.3263
1.5	1.2414	1.2247	1.1603	12.0	3.7633	3.4641	2.4871
2.0	1.4473	1.4142	1.2894	16.0	4.3873	4.0000	2.7638
3.0	1.7967	1.7321	1.4960	20.0	4.9418	4.4721	2.9995
5.0	2.3593	2.2361	1.8042				

<u>Table 4.3.1.1</u> Polymolecularity Correction Factors for the Baumann Procedure (Eq. (6)) for Schulz-Flory Distributions using $<s^2>_z$:

$$q_{B,S_z}^{(SF)} = (k+1)^3 \cdot (k+2)^2 \cdot (k+3)^2 \cdot \frac{\Gamma^3(k+1)}{\Gamma^{8/3}(k+3,5)} \cdot \frac{\Gamma^{2/3}(k+3,5+1,5\varepsilon)}{\Gamma(k+3+\varepsilon)}$$

$\dfrac{<M>_w}{<M>_n}$	a_η and (ε)			
	0.40 (-0.07)	0.50 (0.00)	0.70 (0.13)	0.90 (0.27)
1.1	1.1016	1.1044	1.1105	1.1174
1.3	1.2287	1.2353	1.2496	1.2654
1.5	1.3050	1.3140	1.3334	1.3549
1.7	1.3539	1.3665	1.3894	1.4147
2.0	1.4069	1.4191	1.4456	1.4747
3.0	1.4884	1.5033	1.5355	1.5709
5.0	1.5427	1.5595	1.5956	1.6352
10.0	1.5784	1.5965	1.6352	1.6775
20.0	1.5950	1.6135	1.6535	1.6971

<u>Table 4.3.1.2</u> Polymolecularity Correction Factors for the Baumann Procedure (Eq. (6)) for Schulz-Flory Distributions using $<M>_w$:

$$q_{B,M_w}^{(SF)} = \frac{\Gamma(k+1) \cdot \Gamma^2(k+4)}{\Gamma(k+2) \cdot \Gamma^2(k+3,5)}$$

$<M>_w / <M>_n$	1.1	1.3	1.5	1.7	2.0
$q_{B,M_w}^{(SF)}$	1.2048	1.5203	1.7520	1.9292	2.1287
$<M>_w / <M>_n$	2.5	3.0	5.0	10.0	20.0
$q_{B,M_w}^{(SF)}$	2.3548	2.5055	2.8071	3.0333	3.1464

References page IV-130

Table 4.3.2.1 Polymolecularity Correction Factors for the Baumann Procedure (Eq. (6)) for Logarithmic Normal Distributions using $<s^2>_z$:

$$q_{B,S_z}^{(LN)} = (\langle M \rangle_w / \langle M \rangle_n)^{0.25\epsilon^2 + 0.5\epsilon + 1.25}$$

$\dfrac{\langle M \rangle_w}{\langle M \rangle_n}$	a_η and (ϵ)			
	0.40 (−0.07)	0.50 (0.00)	0.70 (0.13)	0.90 (0.27)
1.1	1.1231	1.1265	1.1342	1.1429
1.3	1.3764	1.3881	1.4143	1.4443
1.5	1.6385	1.6600	1.7086	1.7649
1.7	1.9083	1.9412	2.0158	2.1032
2.0	2.3259	2.3784	2.4986	2.6410
3.0	3.8109	3.9482	4.2690	4.6612
4.0	5.4097	5.6569	6.2429	6.9751
5.0	7.0989	7.4767	8.3833	9.5353
6.0	8.8637	9.3905	10.6666	12.3105
8.0	12.5824	13.4543	15.5985	18.4217
10.0	16.5112	17.7828	20.9465	25.1832
12.0	20.6159	22.3345	26.6513	32.5127
14.0	24.8731	27.0807	32.6710	40.3509
16.0	29.2651	32.0000	38.9741	48.6526
20.0	38.4030	42.2949	52.3366	66.5101

Table 4.3.2.2 Polymolecularity Correction Factors for the Baumann Procedure (Eq. (6)) for Logarithmic Normal Distributions using $<M>_w$:

$$q_{B,M_w}^{(LN)} = (\langle M \rangle_w / \langle M \rangle_n)^{2.25}$$

$\langle M \rangle_w / \langle M \rangle_n$	1.1	1.3	1.5	1.7	2.0
$q_{B,M_w}^{(LN)}$	1.24	1.80	2.49	3.30	4.76
	2.5	3.0	4.0	5.0	6.0
	7.86	11.84	22.63	37.38	56.34
	7.0	8.0	9.0	10.0	11.0
	79.70	107.63	140.30	177.83	220.36
	12.0	14.0	16.0	18.0	20.0
	268.01	379.13	512.00	667.36	845.90

G. REFERENCES

1. H. Staudinger, W. Heuer, Chem. Ber. 63, 222 (1930).
2. W. Kuhn, Kolloid-Z. 68, 2 (1934).
3. H. Mark, in: "Der feste Koerper" (Vortraege zum 50jaehrigen Bestehen der Physikalischen Gesellschaft Zuerich 1937), ed. by R. Saenger, Hirzel, Leipzig 1938.
4. I. Sakurada, Kasen-Koenshu 5, 33 (1940).
5. R. Houwink, J. Prakt. Chem. 157, 15 (1941).
6. W. Philippoff, "Viskositaet der Kolloide" (Handbuch der Kolloidwissenschaft in Einzeldarstellungen, Vol. IX), Th. Steinkopff, Dresden-Leipzig 1942, p. 333.
7. G. Meyerhoff, Advan. Polymer Sci. 3, 59 (1961).
8. P. Szewczyk, M. Kalfus, J. Macromol. Sci. (Chem.) A7, 737 (1973).
9. P. J. Flory, J. Am. Chem. Soc. 65, 372 (1943).
10. H.-G. Elias, R. Bareiss, J. G. Watterson, Advan. Polymer Sci. 11, 111 (1973).
11. L. Mandelkern, P. J. Flory, J. Chem. Phys. 20, 212 (1952).
12. H. A. Scheraga, L. Mandelkern, J. Am. Chem. Soc. 75, 179 (1953).
13. H.-G. Elias, R. Bareiss, J. Macromol. Sci. (Chem.) A1, 1377 (1967).
14. W. Sutter, A. Kuppel, Makromol. Chem. 149, 271 (1971).
15. R. E. Bareiss, Makromol. Chem. 170, 251 (1973).
16. W. Burchard, Makromol. Chem. 50, 20 (1961).

17. W. H. Stockmayer, M. Fixman, J. Polymer Sci. C 1, 37 (1963).
18. J. M. G. Cowie, S. Bywater, Polymer 6, 197 (1965).
19. H. Baumann, J. Polymer Sci. B 3, 1069 (1965).
20. G. V. Schulz, Z. Physik. Chem. B 30, 379 (1935).
21. P. J. Flory, J. Am. Chem. Soc. 58, 1877 (1936).
22. G. V. Schulz, Z. Physik. Chem. B 43, 25 (1939).
23. B. H. Zimm, J. Chem. Phys. 16, 1099 (1948).
24. W. D. Lansing, E. O. Kraemer, J. Am. Chem. Soc. 57, 1369 (1935).
25. H. Wesslau, Makromol. Chem. 20, 111 (1956).
26. E. P. Honig, J. Phys. Chem. 69, 4418 (1965).
27. N. Yamada, H. Matsuda, H. Nakamura, Rept. Progr. Polymer Phys. Japan 4, 81 (1961).
28. K. G. Berdnikova, G. V. Tarasova, V. S. Skazka, N. A. Nikitin, G. V. Dyuzhev, Vysokomolekul. Soedin. 6, 2057 (1964); Polymer Sci. USSR 6, 2280 (1965).
29. D. M. Crothers, B. H. Zimm, J. Mol. Biol. 12, 525 (1965).
30. H.-G. Elias, Makromol. Chem. 122, 264 (1969).
31. R. E. Bareiss, unpublished calculations.
32. Handbook of Chemistry and Physics, ed. by R. C. Weast, 52nd Ed., The Chemical Rubber Co., Cleveland, Ohio 1971-1972, p. A-191.

POLYMER-SOLVENT INTERACTION PARAMETERS

B. A. Wolf

Universität Mainz

Institut für Physikalische Chemie

Mainz, Germany

The main thermodynamic equations describing binary polymer-solvent systems are of the well known type of Eq. (1), independently derived by Flory and Huggins:

$$\frac{\Delta\mu_1}{RT} = \ln(1 - \phi_2) + (1 - \frac{1}{P_n})\phi_2 + \chi\,\phi_2^2 \tag{1}$$

in which $\Delta\mu_1$ is the chemical potential of the solvent, ϕ_2 the volume fraction of the polymer with a number average degree of polymerization P_n and χ a dimensionless interaction parameter.

Originally the Flory-Huggins constant χ was considered to be independent of ϕ_2 and P_n and to vary inversely with temperature, but application of Eq. (1) to measured $\Delta\mu_1$-values soon demonstrated the oversimplified nature of these assumptions. Nowadays χ is therefore often treated phenomenologically as the reduced residual chemical potential of the solvent and newer theoretical derivations start from the calculation of the residual Gibbs free energy of mixing.

For practical purposes Eq. (1) is normally used as a mere correlation function in order to obtain the system-specific parameter χ from measured $\Delta\mu_1$-values (osmotic pressure, light scattering, vapor pressure ...). Besides Eq. (1) several more theoretical relations containing the Flory-Huggins parameter can serve as the basis for calculating an interaction parameter from some other experimentally determined properties. Among these are the starting equation of the Flory-Huggins theory, i.e. the expression for the Gibbs free energy of mixing and the equations for the chemical potential of the polymer and for the equilibrium swelling of polymer networks; some non-equilibrium properties, such as the stress-strain behavior of swollen gels and the intrinsic viscosity of dissolved polymers are sometimes also included.

Since all these theoretical equations are treated as mere correlation functions, without consideration of the correctness of the theoretical assumptions underlying their derivation, the various interaction parameters thus obtained for the same system from different relations are normally not identical. In the present chapter the symbol χ is therefore reserved for correlation functions based on probably the most frequently measured property, namely the chemical potential $\Delta\mu_1$ of the solvent.

All correlation functions containing volume fractions suffer from the severe limitation that it is impossible to recalculate the measured properties from such tabulated interaction parameters without a knowledge of the partial specific volumes of the components.

For this reason and as a result of theoretical arguments increasing use is made of somewhat modified correlation functions, such as

$$\frac{\Delta\mu_1}{RT} = \ln(1 - \omega_2) + (1 - \frac{1}{P_n})\omega_2 + \chi_{(\omega)}\,\omega_2^2 \tag{2}$$

and

$$\frac{\Delta\mu_1}{RT} = \ln(1 - s_2) + (1 - \frac{1}{P_n})s_2 + \chi_{(seg)}\,s_2^2 \tag{3}$$

in which the temperature- and pressure independent weight fractions (111) ω_2 and segment fractions (41) s_2 of the polymer have replaced the volume fractions. The interaction parameters $\chi_{(\omega)}$ and $\chi_{(seg)}$ defined by Eq. (2) and (3), resp., now contain the entire dependence of $\Delta\mu_1/RT$ on the variables of state, in contrast to χ.

From the foregoing considerations it is obvious that one has to ascertain in any case, which particular correlation function has been used as the basis for the evaluation of a certain interaction parameter and which concentration variables have been chosen.

The concentration dependence of χ (or $\chi_{(\omega)}$ or $\chi_{(seg)}$) can generally be described by a power series with respect to ϕ_2 (or ω_2 or s_2)

$$\chi = \chi_1 + \chi_2\phi_2 + \chi_3\phi_2^2 + \dots \tag{4}$$

For sufficiently dilute solutions the only parameter necessary to characterize a system is χ_1; its value can for instance be calculated from tabulated data for the second osmotic virial coefficient A_2 by means of Eq. (5)

$$\chi_1 = 0.5 - A_2\,\rho_2^2\,\bar{v}_1 \tag{5}$$

in which ρ_2 is the density of the polymer and $\bar{v}_1$ the partial molar volume of the solvent.

In the case of highly concentrated solutions an analogous expansion with respect to ϕ_1 can serve the same purpose.

For probably the best studied system polystyrene/cyclohexane Koningsveld and Kleintjens (70) were able to show that the experimentally determined concentration dependence of χ is well reproduced by Eq. (6), which follows from lattice theory

$$\chi = \alpha + \frac{\beta(1 - \gamma)}{(1 - \gamma\phi_2)^2} \tag{6}$$

where α and γ are constants and β varies linearly with $1/T$ and $1/P_n$ for the T and P_n range under investigation. Similar equations hold for weight fractions and base mol fractions.

Because of the complexity of the material as demonstrated above, the author confines himself to the tabulation of interaction parameters for some representative systems, whereas only references, to be used according to the outlined principles, are given for the rest.

POLYMER-SOLVENT INTERACTION PARAMETERS

Literature References

Cellulose	65,120	Poly(isoprene), 1,4-cis	31
Cellulose acetate	3,57,58,64,85-89,115	--, Hevea rubber	9,10,12,16-20,36,39,45-48,50,
Cellulose nitrate	58,61,85,87,88,115,123		58,59,72,73,81,91,92,114,115,
Collagen	120		130
Ethyl cellulose	85,87,115	--, Gutta Percha, Balata	58
Lignin	22	Poly(isoprene-co-styrene)	105
Poly(acrylonitrile)	83,115	Poly(methyl methacrylate)	18,21,62,67,99,112,113,115
Poly(p-bromostyrene)	28	Poly(methylene)	52,115
Poly(1,3-butadiene)	11,60,114,115	Poly(p-methylstyrene)	28
Poly(butadiene-co-acrylonitrile)	20,114,115,124	Poly(1-octadecene)	122
Poly(butadiene-co-2-methyl-5-vinylpyridine)	72	Poly(oxyethylene)	58,79
Poly(butadiene-co-styrene)	16,20,44,72,114,115,118,131	Poly(oxypropylene)	79
Poly(chloroprene) (Neoprene CN, WRT)	20,114,115,124	Poly(propylene)	66,90,115
Poly(o-chlorostyrene)	77	Poly(styrene)	5,7,8,10-15,21,33,42,43,49,54,
Poly(p-chlorostyrene)	28,77		55,58,64,69,70,76,77,82,84,98-
Poly(chlorotrifluoroethylene)	24,51,63		100,104,108-111, 115-117
Poly(1-decene)	122	Poly(styrene-co-divinylbenzene)	14
Poly(dimethyl siloxane)	25,53,71,78,93 103,115,121	Poly(vinyl acetate)	23,29,64,68,80,97,99,115,120,
Poly(1-dodecene)	122		128,129
Poly(ethylene)	26,27,52,90,94,101,106,115,	Poly(vinyl alcohol)	64,96,107
	119,125,126	Poly(vinyl chloride) *	1,2,21,34,35,58,80,115, 134
Poly(ethylene-co-propylene)	30,56,115	--, chlorinated	58
Poly(1-heptene)	122	Poly(vinyl chloride-co-acrylonitrile)	133
Poly(hexamethylene adipamide)	120,127	Poly(vinyl chloride-co-vinyl acetate)	58
(poly(iminoadipoyliminohexamethylene))		Poly(vinyl isobutyl ether)	95
Poly(hexamethylene sebacamide)	120	Poly(vinylidene cyanide-co-vinyl acetate)	132
(poly(iminosebacoyliminohexamethy-		Serum albumin	120
lene))		Silk	120
Poly(indene), hydrogenated	58	Viscose	120
Poly(isobutene)	4,6,21,32,37,38,40,60,74,	Wool	120
	75,80,82,98,99,102,115		
Poly(isobutene-co-isoprene) (butyl rubber)	16,20,21,72,115		

* see also "physical properties of poly(vinyl chloride)"

Selected Data

Polymer	Solvent	$T °C$	χ_1	$\chi_{1(seg)}$	References
Cellulose acetate	Acetone	25	0.45	--	85
	Pyridine	25	0.28	--	85
	Dioxane	25	0.38	--	85
Cellulose nitrate	Acetone	25	0.27	--	85
	Methyl ethyl ketone	25	0.21	--	85
	Methyl n-amyl ketone	25	0.02	--	85
Ethyl cellulose	Acetone	25	0.46	--	85
	Methyl ethyl ketone	25	0.42	--	85
	Chloroform	25	0.34	--	85
	Carbon tetrachloride	25	0.46	--	85
	Benzene	25	0.48	--	85
Poly(dimethyl siloxane)	Cyclohexane	25	--	0.42	41,78
	Chlorobenzene	25	--	0.47	41,78
Poly(isobutene)	Benzene	27	0.50	--	37
		25	--	0.50	36,75
	Cyclohexane	25	--	0.47	32,36,74
	n-Pentane	25	--	0.49	4,32,36,40
	n-Octane	25	--	0.46	32,40
Poly(isoprene), natural rubber	Benzene	25	0.44	--	58
		25	--	0.42	36,45-47

Selected Data   (Cont'd.)

Polymer	Solvent	T °C	χ_1	$\chi_{1(seg)}$	References
Poly(isoprene), natural rubber (Cont'd.)	Carbon tetrachloride	15-20	0.28	--	58
	Chloroform	15-20	0.37	--	58
	Carbon disulfide	25	0.49	--	58
	Amyl acetate	25	0.49	--	58
Poly(methyl methacrylate)	Chloroform	25	--	0.377	112,113
	Heptanone-4	25	--	0.509	67
	Tetrahydrofuran	25	--	0.447	67
Poly(oxyethylene)	Water	27	0.45	--	58
Poly(styrene)	Toluene	27	0.44	--	58
	Ethyl benzene	25	--	0.40	54,100
	Methyl ethyl ketone	25	--	0.47	5,7,33,42
	Cyclohexane	25	--	0.505	55,76,100,111
Poly(vinyl acetate)	Acetone	25	0.437	--	23
	Methyl ethyl ketone	25	0.429	--	23
	Dioxane	25	0.407	--	23
Poly(vinyl chloride)	Tetrahydrofuran	27	0.14	--	58
	Dioxane	27	0.52	--	58
	Methyl ethyl ketone	26	0.409	--	34
	Cyclohexanone	29.8	0.240	--	34

χ_1:   Partly taken from M.L. Huggins "Physical Chemistry of High Polymers", Wiley, New York, 1958, p. 48 and from C. J. Sheehan and A. L. Bisio, Rubber Chem. Technol. 39, 149 (1966).

$\chi_{1(seg)}$:   Taken from P. J. Flory, Discussions Faraday Soc. 49, 7 (1970).

REFERENCES

1. C. E. Anagnostopoulos, A. Y. Coran, H. R. Gamrath, J. Appl. Polymer Sci. 4, 181 (1960).
2. C. E. Anagnostopoulos, A. Y. Coran, J. Polymer Sci. 57, 1 (1962).
3. W. J. Badgley, H. Mark, J. Phys. Chem. 51, 58 (1947).
4. C. H. Baker, W. B. Brown, G. Gee, J. S. Rowlinson, D. Stubley, R. E. Yeadon, Polymer 3, 215 (1962).
5. C. E. H. Bawn, R. F. J. Freeman, A. R. Kammaliddin, Trans. Faraday Soc. 46, 677 (1950).
6. C. E. H. Bawn, R. D. Patel, Trans. Faraday Soc. 52, 1664 (1956).
7. C. E. H. Bawn, M. A. Wajid, Trans. Faraday Soc. 52, 1658 (1956).
8. J. Biroš, K. Solc, J. Pouchlý, Faserforsch. Textiltechn. 15, 608 (1964).
9. A. F. Blanchard, P. M. Wootten, J. Polymer Sci. 34, 627 (1959).
10. C. Booth, G. Gee, G. R. Williamson, J. Polymer Sci. 23, 3 (1957).
11. C. Booth, G. Gee, M. N. Jones, W. D. Taylor, Polymer 5, 353 (1964).
12. C. Booth, G. Gee, G. Holden, G. R. Williamson, Polymer 5, 343 (1964).
13. R. H. Boundy, R. F. Boyer, "Styrene, Its Polymers, Copolymers, and Derivatives" Reinhold, New York, 1952, p. 345.
14. R. F. Boyer, R. S. Spencer, J. Polymer Sci. 3, 97 (1948).
15. J. W. Breitenbach, H. P. Frank, Monatsh. Chem. 79, 531 (1948).
16. G. M. Bristow, Trans. Faraday Soc. 55, 1246 (1959).
17. G. M. Bristow, J. Polymer Sci. 36, 526 (1959).
18. G. M. Bristow, J. Appl. Polymer Sci. 2, 120 (1959).
19. G. M. Bristow, W. F. Watson, Trans. Faraday Soc. 54, 1567 (1958).
20. G. M. Bristow, W. F. Watson, Trans. Faraday Soc. 54, 1731 (1958).
21. G. M. Bristow, W. F. Watson, Trans. Faraday Soc. 54, 1742 (1958).
22. W. Brown, J. Appl. Polymer Sci. 11, 2381 (1967).
23. G. V. Browning, J. D. Ferry, J. Chem. Phys. 17, 1107 (1949).
24. A. M. Bueche, J. Am. Chem. Soc. 74, 65 (1952).
25. A. M. Bueche, J. Polymer Sci. 15, 97 (1955).
26. E. M. Cernia, C. Mancini, A. Saini, J. Appl. Polymer Sci. 12, 789 (1968).
27. A. Y. Coran, C. E. Anagnostopoulos, J. Polymer Sci. 57, 13 (1962).
28. R. Corneliussen, S. A. Rice, H. Yamakawa, J. Chem. Phys. 38, 1768 (1963).
29. G. R. Cotton, A. F. Sirianni, I. E. Puddington, J. Polymer Sci. 32, 115 (1958).
30. G. Crespi, M. Bruzzone, Chim. Ind. (Milan) 41, 741 (1959).
31. R. E. Cunningham, J. Polymer Sci. 42, 571 (1960).
32. G. Delmas, D. Patterson, T. Somcynsky, J. Polymer Sci. 57, 79 (1962).
33. P. Doty, M. Brownstein, W. Schlener, J. Phys. Chem. 53, 213 (1949).
34. P. Doty, E. Mishuck, J. Am. Chem. Soc. 69, 1631 (1947).
35. P. Doty, H. S. Zable, J. Polymer Sci. 1, 90 (1946).
36. B. E. Eichinger, P. J. Flory, Trans. Faraday Soc. 64, 2035, 2053, 2061, 2066 (1968).
37. P. J. Flory, J. Am. Chem. Soc. 65, 375 (1943).
38. P. J. Flory, H. Daoust, J. Polymer Sci. 25, 429 (1957).
39. P. J. Flory, N. Rabjohn, M. C. Schaffer, J. Polymer Sci. 4, 225 (1949).
40. P. J. Flory, J. L. Ellenson, B. E. Eichinger, Macromolecules 1, 279 (1968).
41. P. J. Flory, Discussions Faraday Soc. 49, 7 (1970).
42. P. J. Flory, H. Hoecker, Trans. Faraday Soc. 67, 2258 (1971).
43. H. P. Frank, H. Mark, J. Polymer Sci. 6, 243 (1951).
44. D. M. French, R. H. Ewart, Anal. Chem. 19, 165 (1947).
45. G. Gee, L. R. G. Treloar, Trans. Faraday Soc. 38, 147 (1942).
46. G. Gee, W. J. C. Orr, Trans. Faraday Soc. 42, 507 (1946).
47. G. Gee, J. B. M. Herbert, R. C. Roberts, Polymer 6, 541 (1965).
48. G. Gee, Chem. Soc. (London) Special Publ. 23, 75 (1968).
49. A. I. Goldberg, W. P. Hohenstein, H. Mark, J. Polymer Sci. 2, 503 (1947).
50. S. L. Gumbrell, R. S. Rivlin, Trans. Faraday Soc. 49, 1945 (1953).
51. H. T. Hall, J. Am. Chem. Soc. 74, 68 (1952).

52. I. Harris, J. Polymer Sci. 8, 353 (1952).
53. R. L. Hauser, C. A. Walker, F. L. Kilbourne, jr., Ind. Eng. Chem. 48, 1202 (1956).
54. H. Hoecker, P. J. Flory, Trans. Faraday Soc. 67, 2270 (1971).
55. H. Hoecker, H. Shih, P. J. Flory, Trans. Faraday Soc. 67, 2275 (1971).
56. E. D. Holly, J. Polymer Sci. A2, 5267 (1964).
57. P. Howard, R. S. Parikh, J. Polymer Sci. C30, 17 (1970).
58. M. L. Huggins, Ann. N. Y. Acad. Sci. 44, 431 (1943).
59. M. L. Huggins, Ind. Eng. Chem. 35, 216 (1943).
60. R. S. Jessup, J. Res. Nat. Bur. Std. 60, 47 (1958).
61. A. L. Jones, Trans. Faraday Soc. 52, 1408 (1956).
62. A. Kagemoto, S. Murakami, R. Fujishiro, Bull. Chem. Soc. (Japan) 40, 11 (1967).
63. H. S. Kaufman, M. S. Muthana, J. Polymer Sci. 6, 251 (1951).
64. T. Kawai, J. Polymer Sci. 32, 425 (1958).
65. T. Kawai, J. Polymer Sci. 37, 181 (1959).
66. J. B. Kinsinger, R. E. Hughes, J. Phys. Chem. 63, 2002 (1959).
67. R. Kirste, G. V. Schulz, Z. Physik. Chem. (Frankfurt) 27, 301 (1961).
68. R. J. Kokes, A. R. Di Pietro, F. A. Long, J. Am. Chem. Soc. 75, 6319 (1953).
69. R. Koningsveld, L. A. Kleintjens, A. R. Shultz, J. Polymer Sci., A2, 8, 1261 (1970).
70. R. Koningsveld, L. A. Kleintjens, Macromolecules, 4, 637 (1971).
71. A. Y. Korolev, K. A. Andrianov, L. S. Utesheva, T. E. Vvedenskaya, Dokl. Akad. Nauk SSSR 89, 65 (1953).
72. G. Kraus, "Rubber World", 135, No. 1, 67 (1956).
73. W. R. Krigbaum, D. K. Carpenter, J. Polymer Sci. 14, 241 (1954).
74. W. R. Krigbaum, P. J. Flory, J. Am. Chem. Soc. 75, 1773 (1953).
75. W. R. Krigbaum, P. J. Flory, J. Am. Chem. Soc. 75, 5254 (1953).
76. W. R. Krigbaum, D. O. Geymer, J. Am. Chem. Soc. 81, 1859 (1959).
77. K. Kubo, K. Ogino, Bull. Chem. Soc. (Japan) 44, 997 (1971).
78. N. Kuwahara, T. Okazawa, M. Kaneko, J. Polymer Sci. C23, 543 (1968).
79. B. N. Malcom, J. S. Rowlinson, Trans. Faraday Soc. 53, 921 (1957)
80. H. Mark, A. V. Tobolsky, "Physical Chemistry of High Polymeric Systems", 2nd ed., Interscience N. Y., 1950, p. 265.
81. S. H. Maron, N. Nakajima, J. Polymer Sci. 40, 59 (1959).
82. S. H. Maron, N. Nakajima, J. Polymer Sci. 42, 327 (1960).
83. C. R. Masson, H. W. Melville, J. Polymer Sci. 4, 337 (1949).
84. E. H. Merz, R. W. Raetz, J. Polymer Sci. 5, 587 (1950).
85. W. R. Moore, J. A. Epstein, A. M. Brown, B. M. Tidswell, J. Polymer Sci. 23, 23 (1957).
86. W. R. Moore, B. M. Tidswell, J. Polymer Sci. 27, 459 (1958).
87. W. R. Moore, B. M. Tidswell, J. Polymer Sci. 29, 37 (1958).
88. W. R. Moore, R. Shuttleworth, J. Polymer Sci. A1, 733 (1963).
89. W. R. Moore, R. Shuttleworth, J. Polymer Sci. A1, 1985 (1963).
90. G. Moraglio, Chim. Ind. (Milan) 41, 984 (1959).
91. L. Mullins, J. Polymer Sci. 19, 225 (1956).
92. L. Mullins, J. Appl. Polymer Sci. 2, 1 (1959).
93. A. Muramoto, Polymer J. (Japan) 1, 450 (1970).
94. M. S. Muthana, H. Mark, J. Polymer Sci. 4, 527 (1949).

95. M. S. Muthana, H. Mark, J. Polymer Sci. 4, 531 (1949).
96. A. Nakajima, K. Furutachi, Kobunshi Kagaku 6, 460 (1949).
97. A. Nakajima, H. Yamakawa, I. Sakurada, J. Polymer Sci. 35, 489 (1959).
98. R. Noel, D. Patterson, T. Somcynsky, J. Polymer Sci. 42, 561 (1960).
99. T. A. Orofino, P. J. Flory, J. Chem. Phys. 26, 1067 (1957).
100. H. J. Palmen, Thesis, Technische Hochschule Aachen, Germany, 1965.
101. D. Patterson, Y. B. Tewari, H. P. Schreiber, J. E. Guillet, Macromolecules 4, 356 (1971).
102. S. Prager, E. Bagley, F. A. Long, J. Am. Chem. Soc. 75, 2742 (1953).
103. F. P. Price, S. G. Martin, J. P. Bianchi, J. Polymer Sci. 22, 49 (1956).
104. G. Rehage, H. J. Palmen, D. Moeller, W. Wefers; Paper presented at IUPAC Symposium on Macromolecules Toronto, Sept. 1968.
105. J. Rehner, jr., R. L. Zapp, W. J. Sparks, J. Polymer Sci. 11, 21 (1953).
106. C. E. Rogers, V. Stannett, M. Szwarc, J. Phys. Chem. 63, 1406 (1959).
107. I. Sakurada, A. Nakajima, H. Fujiwara, J. Polymer Sci. 35, 497 (1959).
108. M. J. Schick, Thesis, Polytechnic Inst. Brooklyn, 1948.
109. M. J. Schick, P. Doty, B. H. Zimm, J. Am. Chem. Soc. 72, 530 (1950).
110. Th. G. Scholte, European Polymer J. 6, 1063 (1970).
111. Th. G. Scholte, J. Polymer Sci., A2, 8, 841 (1970).
112. G. V. Schulz, H. Doll, Ber. Bunsenges. Physik. Chem. 27, 301 (1953).
113. G. V. Schulz, H. Baumann, R. Darskus, J. Phys. Chem. 70, 3647 (1966.).
114. R. L. Scott, M. Magat, J. Polymer Sci. 4, 555 (1949).
115. C. J. Sheehan, A. L. Bisio, Rubber Chem. Technol. 39, 149 (1966).
116. A. R. Shultz, P. J. Flory, J. Am. Chem. Soc. 75, 3888 (1953).
117. A. R. Shultz, P. J. Flory, J. Polymer Sci. 15, 231 (1955).
118. A. G. Shvarts, Kolloid-Z. 19, 376 (1957).
119. I. Sobolev, J. A. Meyer, V. Stannett, M. Swarc, Ind. Eng. Chem. 49, 441 (1957).
120. H. W. Starkweather, jr., J. Appl. Polymer Sci. 2, 129 (1959).
121. L. E. St. Pierre, H. A. Dewhurst, A. M. Bueche, J. Polymer Sci. 36, 105 (1959).
122. P. J. T. Tait, P. J. Livesey, Polymer 11, 359 (1970).
123. H. Takenaka, J. Polymer Sci. 24, 321 (1957).
124. A. V. Tobolsky, I. B. Prettyman, J. H. Dillon, J. Appl. Phys. 15, 380 (1944).
125. Q. A. Trementozzi, J. Polymer Sci. 23, 887 (1957).
126. L. H. Tung, J. Polymer Sci. 24, 333 (1957).
127. L. Valentine, J. Polymer Sci. 23, 297 (1957).
128. V. V. Varadaiah, J. Polymer Sci. 19, 477 (1956).
129. R. H. Wagner, J. Polymer Sci. 2, 21 (1947).
130. G. S. Whitby, A. B. A. Evans, D. S. Pasternack, Trans. Faraday Soc. 38, 269 (1942).
131. J. A. Yanko, J. Polymer Sci. 3, 576 (1948).
132. J. A. Yanko, J. Polymer Sci. 22, 153 (1956).
133. S. G. Zelikman, N. V. Mikhailov, Kolloid-Z. 19, 35 (1957).
134. A. Nakajima, Kobunshi Kagaku 7, 309 (1950).

CONCENTRATION DEPENDENCE OF THE VISCOSITY OF DILUTE POLYMER SOLUTIONS

HUGGINS- AND SCHULZ-BLASCHKE-COEFFICIENTS

N. Sütterlin
Röhm GmbH, Darmstadt, Germany

Contents

A. INTRODUCTION

The concentration dependence of the viscosity of dilute polymer solutions can be expressed in a power series in concentration

$$\eta = \eta_0 (1 + H_1 C + H_2 C^2 + H_3 C^3 + \ldots) \tag{1}$$

wherein η and η_0 denote the viscosity of solution and solvent, respectively. This equation may be rewritten in the more familiar form

$$(\eta - \eta_0)/\eta_0 C = \eta_{sp}/C = [\eta] + k_1 [\eta]^2 C + k_2 [\eta]^3 C^2 + \ldots \tag{2}$$

where $[\eta]$ is the intrinsic viscosity, $k_i (= k_1, k_2 \ldots$ etc.) a dimensionless parameter, and k_1 corresponds to the Huggins coefficient k_H. For low concentrations C we obtain Huggin's (1) well known relationship

$$\eta_{sp}/C = [\eta] (1 + k_H [\eta] C) \tag{3}$$

By an earlier, empirical approach of Schulz and Blaschke (2), Equation (4)

$$\eta_{sp}/C = [\eta] (1 + k_{SB} \eta_{sp}) \tag{4}$$

had been found. This is the only theoretically substantiated equation among the large number of approaches to the mathematical description of the concentration dependence of the viscosity of dilute polymers suggested in the literature (3). From a hydrodynamic point of view, Huggins (1) first arrived at a relationship corresponding to Equation (4) from which he derived his Equation (3) by introducing the approximation

$$\eta_{sp} \approx [\eta] \cdot C \tag{5}$$

which only applies to extremely low polymer concentrations. Despite its approximative character and limited applicability it is widely used. With suitable mathematical approximations it can be proved that all other equations suggested in literature can be derived from Equation (3) and, above all, Equation (4) (4).

An equation of Kraemer (139) is often used for extrapolation because of the small concentration dependence. This relationship is most conveniently written as

$$\frac{\ln \eta_r}{c} = [\eta] - k_K [\eta]^2 C + \ldots$$
$$k_K = 1/2 - k_H$$

Numerous studies revealed that Huggins' coefficient k_H is always higher than Schulz-Blaschke's coefficient k_{SB}, whereas it is the opposite with the viscosity values. Comparing Equations (3) and (4) leads to the quantitative relationship

$$k_H = k_{SB} \frac{[\eta]_{SB} \eta_{sp}}{[\eta]^2_H C} \tag{6}$$

This shows that the coefficient differences largely depend on the investigated specific viscosity region as well as on the concentration. Identical values of k_H and k_{SB} would require an infinitely low specific viscosity as the upper limit of a concentration series. Therefore, the evaluation of viscosimetric measurements by the method of Huggins must remain within constant limits of specific viscosity if reproducible values are to be obtained. Evaluations by the method of Schulz and

Blaschke, however, are not bound to a constant specific viscosity region and also allow a wider range of concentrations than is possible with the method of Huggins. For extrapolation procedures in connection with intrinsic viscosities and Huggins' coefficient k_H see (5).

k_H and k_{SB} not only depend on the intermolecular hydrodynamic interaction but also on many other factors, which certainly include the intermolecular thermodynamic interaction. In many cases experimentally obtained values will not be functions of the polymer - solvent system alone (the situation is even more complicated in the case of copolymers or solvent mixtures due to preferential solvation) but also of the molecular weight, the molecular weight distribution, the velocity gradient and the degree of branching of the chain molecule (6). However, results reported in the literature are often contradictory and but few and timied attempts have been made to mathematically determine such dependencies. What makes any such undertaking particularly difficult is the rather considerable limit of error with such experimentally determined k_H and k_{SB} values.

As regards the polymer-solvent interaction, there seems to be a certain relationship between the magnitude of k_H and the coefficient of expansion α_η (7, 8).

Still less agreement exists on the dependence of the viscosity coefficient on the molecular weight M. k_H and k_{SB} are reported to both increase and decrease with M and also both quantities are observed to be independent of the molecular weight. It seems that, with very low molecular weights, k_H and k_{SB} frequently decrease as M increases (reduction of the segmental density with increasing M), are then largely independent of M within a certain molecular weight range and finally increase along with M (branching ?) (9). However, this assertion is not generally acceptable (6). Attempts at finding a mathematical expression for the dependence of k_H on the molecular weight have been made, for example, by Stern (10), Utracki and Simha (11) as well as by Sotobayashi (12).

Experience, hitherto, has shown that the molecular weight distribution does not significantly influence k_H and k_{SB}. A wide distribution, if at all, normally causes but a slight increase in viscosity coefficient, which is negligible in most cases (6) (cf. (13)).

Available literature reports both an increase (14) and a decrease (15) in the Huggins' coefficient with an increasing shear gradient. Schurz and Pippan (16) suggested a modified Huggins Equation to determine the shear dependence in dilute solutions. On the whole, the influence of the velocity gradient on the intrinsic viscosity is, however, neglected, unless it exceeds a value of $1500 \ s^{-1}$.

Moreover, conformity does not exist in the literature as regards the influence of branching on the viscosity coefficient. Most results indicate that increased values of k_H and k_{SB} are obtained with a higher degree of branching, the condition being, however, a marked change in segmental density, resp. coil volume (17-25).

According to a hydrodynamic theory of Riseman and Ullman (26), the value of k_H would be 3/5 for a coil and 11/15 for rods. $k_H \geq 1$ results for spherical particles, and therefore k_H is often found to be between 0.8 and 1.3 in the case of aggregated chain molecules. The limiting value $k_{H\theta}$ for an undrained coil under θ conditions is according to Sakei (27, 28) $k_{H\theta} = 0.52$ (relatively close to corresponding measurements (86)). In suitable solvents, $k_H < 0.52$ should be obtained with uncharged systems, provided coils of sufficiently high molecular weight are available.

The following tables are a collection of literature references, roughly classified by the intrinsic viscosity $[\eta]$, the dimension of $[\eta]$ being ml/g. Values marked ★ are interpolated mean values taken from the available literature. The above mentioned influences exerted on k_H and k_{SB} are too complex to be tabulated. As regards characteristic features of polymers and methods as well as conditions of measurement not given in the tables, reference is made to the original publications.

B. TABLES

1. HUGGINS COEFFICIENTS

Polymer	Solvent	T[°C]	Remarks	$[\eta]$	k_H	Ref.
1.1 POLY(DIENES) AND POLY(ALKENES)						
Poly(1,3-butadiene), cis	toluene	25				31
Poly(isoprene), cis	toluene	25				31
Natural rubber	benzene	30		354	0.32	29
	n-hexane	30		170	0.35	29
Gutta Percha	toluene	25				31
Poly(chloroprene)	benzene	25		10	0.58★	32
				20	0.47★	
				30-120	0.34★± 0.06	
				130	0.39★	
				150	0.43	
				170	0.46★	
				200	0.52★	
Poly(ethylene)	tetralin	80				33
	p-xylene	100	low pressure	38	0.71	34, 126
				97	0.45	
				327	0.69	
	tetralin	80	high pressure			35
	p-xylene	81	high pressure	35	0.38	34, 126
				54	0.27	
				71	0.27	
				93	0.38	
				141	0.39	
				153	0.65	

Polymer	Solvent	T[°C]	Remarks	[η]	k_H	Ref.
Poly(ethylene) (Cont'd.)	p-xylene	81	high pressure	159	0.89	34, 126
		105	high pressure			36, 126
Poly(isobutene)	benzene	20		24	0.54	8
				36	0.65	
				107	0.87	
		24		128	0.49±0.01	40
		25		27	0.49	8
				70	0.55★	
				100	0.59★	
				120	0.61★	
				183	0.70	
		30		30	0.49★	8
				40	0.50★	
				100	0.57★	
				160	0.64★	
				240	0.70★	
		40		30-320	0.42★±0.02	8
	carbon tetrachloride	30		14- 42	0.36	41
	chlorobenzene	25		155	0.36	8
	cyclohexane	25		19	0.38±0.02	42
				54	0.35±0.04	
				384	0.30±0.01	
	dibutyl ether	25		165	0.33	8
	diisobutylene	25	$M_v = (16-926) \times 10^3$		0.40±0.05	43
	heptane	25		185	0.30	8
	tetrachloroethane	25		276	0.27	8
	tetradecane	25		189	0.33	8
	toluene	25		171	0.36	8
Poly(propylene)	benzene	30	atactic	30	0.43	38
				50	0.43★	
				95	0.42	
				136	0.43	37
				150	0.42	38
				169	0.40	
	carbon tetrachloride	30	atactic	253	0.25	37
	chlorobenzene	30	atactic	165	0.33	37
	chloroform	30	atactic	196	0.31	37
	α-chloronaphthalene	139.2	atactic			39
	cyclohexane	30	atactic	45-330	0.27★±0.02	38
				276	0.24	37
	decalin	30	atactic	277	0.25	37
		135	isotactic			39
	2-ethyl hexyl acetate	30	atactic	120	0.38	37
	n-heptane	30	atactic	205	0.34	37
	isoamyl acetate	34	atactic	26	0.45	38
				38	0.52★	
				65	0.65	
				94	0.82	
				98	0.83	
	isopropyl ether	30	atactic	156	0.36	37
	tetrahydrofuran	30	atactic	176	0.32	37
	tetralin	30	atactic	155	0.36	37
		130	atactic			39
	toluene	30	atactic	30-120	0.36★±0.02	38
				190-220	0.40★±0.03	
				180	0.35	37
	p-xylene	125	atactic			39

1.2 POLY(ACRYLIC ACID) AND POLY(METHACRYLIC ACID) DERIVATIVES

Polymer	Solvent	T[°C]	Remarks	[η]	k_H	Ref.
Poly(acrylic acid)	water (0.1 M NaCl)	25				44
Poly(acrylonitrile)	dimethylformamide	25				51
		30		70-190	0.30	53, 52
				157	0.40	54, 57

Polymer	Solvent	T[°C]	Remarks	$[\eta]$	k_H	Ref.
Poly(acrylonitrile) (Cont' d.)	dimethylformamide	30		106	0.28	56
				111	0.32	
				134-296	0.31	
				170-245	0.31★	55
				309-324	0.33	56
				329-342	0.35	
				239-711	0.33	53
		40				55
		50				55
		60				55, 57
		70				55
Poly(ethyl acrylate)	acetone	30		46-177	0.35	48, 49
Poly(ethyl methacrylate)	benzene	35		145	0.29	61
		44		146	0.29	
		52		148	0.29	
		60		145	0.36	
	n-butanol	45		47	0.70	61
		52		55	0.54	
		60		64	0.45	
		70		72	0.40	
	n-butyl bromide	42		160	0.24	61
		50		149	0.40	
		58		159	0.30	
		65		166	0.25	
	n-butyl chloride	40		142	0.32	61
		47		144	0.30	
		54		146	0.26	
		61		147	0.26	
	ethyl acetate	35		133	0.31	61
				120-970	0.25-0.27	63
		45		134	0.28	61
		55		136	0.33	
		65		136	0.38	
	isoamyl acetate	50		145	0.39	61
		60		149	0.38	
		70		152	0.37	
		80		156	0.32	
	isopropanol	36.9		88	1.06	61
		44		129	0.27	
		51		151	0.39	
		60		183	0.43	
	methyl ethyl ketone	23		157	0.35	61
		25		158	0.34	
		35		157	0.35	
		45		158	0.34	
	methyl n-propyl ketone	40		196	0.20	61
		50		183	0.34	
		60		187	0.32	
		70		189	0.34	
	m-xylene	47		131	0.39	61
		54		134	0.33	
		61		137	0.33	
		70		139	0.29	
Poly(methacrylic acid)	water (0.1 M NaCl)	25				44
		25	isotactic			44
Poly(methacrylonitrile)	dimethylformamide	29.2	anionic	26	0.40	64
				60-143	0.34★±0.03	
Poly(methyl acrylate)	acetone	30		95-298	0.38	45
				333	0.37	30
		40		314	0.38	30
	benzene	25		52-343	0.37	45
		30		383	0.31	30
		35		50-335	0.38	45

References page IV-149

★ dependent on polymerization conditions

Polymer	Solvent	T[°C]	Remarks	$[\eta]$	k_H	Ref.
Poly(methyl acrylate) (Cont'd.)	benzene	40		382	0.38	30
		45		49-330	0.38	45
		55		47-329	0.38	
	chlorobenzene	25		53-305	0.40	45
	chloroform	30		220	0.41	46
		35		215	0.39	
		40		222	0.36	
	ethyl acetate	30		120	0.49	46
		35		118	0.50	
		40		115	0.51	
	methyl ethyl ketone	25		45	0.65	47
				69	0.54	
				85	0.56	
				116	0.39	
		30		108	0.45	46
				302	0.43	30
		35		104	0.44	46
		40		104	0.44	46
	toluene	25		34-147	0.59	45
		30		61	0.59	46
				201	0.33	30
		35		34-160	0.52	45
		40		69	0.49	46
		45		35-169	0.49	45
		55		35-175	0.48	45
Poly(methyl methacrylate)	acetone	25				58
	benzene	20		85	0.25★	59
		25		4	1.65★	60
				5	1.09★	
				9.5	0.69★	
				11	0.56	
				14	0.50	
				15	0.44	
				22	0.35	
				26- 40	0.30★	
				66-331	0.25★±0.01	
				667	0.18	
			90% isotactic	41	0.39	8
				49- 61	0.33	
				73-125	0.30★±0.03	
	butyl acetate	25		7	0.90	60
				10	0.82	
				29	0.67	
				47	0.62	
				64	0.58	
				110	0.47	
			90% isotactic	22	0.67	8
				29	0.53	
				39	0.61	
				52	0.57	
	n-butyl bromide	35		47	0.50	61
		42		56	0.40	
		50		63	0.41	
		58		70	0.35	
	chloroform	25		19	0.32	60
				70	0.28★	
				146	0.25	
				390	0.22★	
				870	0.18	
		30, 60				62
	cyclohexanone	40		138	0.44	61
		50		146	0.45	
		60		152	0.44	

CONCENTRATION DEPENDENCE OF VISCOSITY

Polymer	Solvent	T[$^{\circ}$C]	Remarks	[η]	k_H	Ref.
Poly(methyl methacrylate) (Cont'd.)						
	cyclohexanone	70		163	0.40	61
	dimethylformamide	25		56	0.48	8
	ethyl acetate	25		53	0.41	8
	isoamyl acetate	50		36	0.71	61
		65		44	0.63	
		72.5		46	0.47	
		80		47	0.50	
	methyl ethyl ketone	25		7	0.90	60
				18	0.56	
				27	0.49	
				56	0.40	8
				118	0.35	60
				137	0.33	
				151	0.40	47
				172	0.49	
		30, 60				62
	di-n-propyl ketone	35-50				60
	tetrachloroethane	25		101	0.29	8
	toluene	25		9	0.63	60
				12	0.52	
				24	0.46	
				55	0.41	8
				83	0.33	60
				229	0.25	
				414	0.23	
		30		37	0.43±0.04	42, 62
				80	0.41±0.03	
		60				62
	o-xylene	40		68	0.34	61
		50		72	0.21	
		60		76	0.32	
		70		81	0.27	
	m-xylene	40		55	0.48	61
		50		62	0.67	
		60		71	0.55	
		70		80	0.35	
	p-xylene	40		42	1.31	61
		50		48	1.26	
		60		59	0.93	
		70		64	0.47	
Poly(N-methylol-acrylamide)	water (pH 7)	30		21	0.46	50

1.3 VINYL POLYMERS

Polymer	Solvent	T[$^{\circ}$C]	Remarks	[η]	k_H	Ref.
Poly(4-isopropylstyrene)	toluene	25		44-139	0.34±0.04	93
Poly(styrene)	benzene	25		105	0.37	84
	butyl acetate	25		43	0.38	8
				77	0.36	
				126	0.35	
				150	0.33	
	carbon tetrachloride	25				85
	chloroform	30, 60				62
	cyclohexane	31, 95	trifunctional star molecules	42	0.60	84
		34=θ		16	0.50±0.005	86
				48	0.52±0.05	
				178	0.61	16
		34	branched	29-31	0.86★±0.02	87
		34.3	trifunctional star molecules	43	0.59	84
		34.8=θ	trifunctional star molecules	44	0.59	84
		35				88
		36 ≈ θ		152	0.51	8
		40		174	0.49	8

References page IV-149

Polymer	Solvent	T[$^{\circ}$C]	Remarks	[η]	k_H	Ref.
Poly(styrene) (Cont'd.)	cyclohexane	40, 22	trifunctional star molecules	47	0.53	84
		50		55	0.50±0.005	86
	cis-decalin	20		20	0.41★±0.02	8
				40	0.42★±0.04	
				60	0.43★±0.02	
				80	0.44★±0.02	
				120	0.45★±0.02	
	trans-decalin	20		38	0.60	8
				80	0.71	
		25		50	0.49★±0.01	8
				70	0.50	
				100	0.50★±0.02	
				130	0.51★±0.02	
		30		47	0.40	8
				117	0.42	
	diethyl malonate	34=θ		45	0.47	8
				82	0.50	
		40		45	0.50	8
				83	0.47	
	dimethylformamide	25		19	0.50	8
				28	0.45	
				36	0.44	
				60	0.40	
				76-172	0.39	
		30				52
		95		20-700	0.48	89
	dioctyl phthalate	25		49	0.59	8
				89	0.58	
		35		51	0.57	8
				86	0.50	
		45		45	0.56	8
				54- 87	0.50	
	methyl ethyl ketone	25				85
		30				52, 62
		40				49
		40.3		44-151	0.54±0.09	92
		60				62
	1-methylnaphthalene	25				90
	toluene	21		40-125	0.35±0.03	86
				207	0.33±0.05	
				360	0.30±0.05	
				460	0.25±0.08	
		25				85, 90
		25	anionic	8	0.51	91, 126
				10	0.50	
				14	0.44	
				28	0.39	
				36-272	0.34±0.03	
		25	thermal	41-123	0.34±0.02	91, 126
			branched	56	0.41	87
				71	0.38	
			linear	63- 84	0.31	87
		30		14	0.46±0.02	42, 62, 126
				73-234	0.38±0.015	92
				28	0.41±0.02	42
				31	0.40±0.01	
				73	0.37±0.01	
				78	0.36±0.03	
				98	0.39	84
				203	0.34±0.01	42
		60				62
Poly(vinyl acetate)	acetone	18		68	0.33	77
		25		30	0.46	77, 49, 126

CONCENTRATION DEPENDENCE OF VISCOSITY

Polymer	Solvent	T[°C]	Remarks	[η]	k_H	Ref.
Poly(vinyl acetate) (Cont'd.)	acetone	25		42- 66	0.35★	77, 49, 126
				91	0.37	
				134	0.41	
		30		70	0.35★±0.02	79
				150	0.36★±0.02	
				250	0.37★±0.02	
				350	0.38★±0.02	
				450	0.39★±0.02	
		32		65	0.41	77
		39		64	0.42	77
		46		63	0.44	77
	benzene	30		126	0.17	78
				50-450	0.37±0.03	79
				147	0.25	78
				164	0.28	
				197	0.30	
				230	0.31	
				262	0.45	
				281	0.65	
		30	zero shear plot	124-381	0.34	
			emulsion pol.	77	1.42	79
				133	1.21	
				180	1.10	
				229	1.05	
		35	branched	45	0.28	80
				72-200	0.35★±0.02	
				300	0.37★±0.02	
				400	0.38★±0.02	
				500	0.39★±0.02	
				600	0.40★±0.02	
				700	0.42★±0.02	
		35	linear	137-177	0.33★±0.01	80
				196-250	0.34★±0.01	
				281-570	0.35★±0.01	
		40		204	0.25	78
				243	0.28	
				268	0.36	
				354	0.44	
			zero shear plot	197-413	0.42	
	chlorobenzene	18		71	0.44	77
		25		37- 45	0.35★±0.03	77, 126
				66- 69	0.44★±0.01	
				99-121	0.35★±0.04	
		32		68	0.43	77
		39		67	0.44	77
		46		65	0.40	77
	2-chloroethylene	25				49
	chloroform	18		101	0.34	77
		25		45	0.31	77, 126
				62- 92	0.37★±0.02	
				146	0.28	
		32		99	0.37	77
		39		95	0.39	77
		46		92	0.41	77
	dioxane	18		77	0.27	77
		25		75	0.29	77, 126
		32		74	0.31	77
		39		73	0.32	77
		46		72	0.34	77
	methanol	18		41	0.61	77
		25		30	0.41★±0.04	77, 126
				40	0.49★±0.03	
				50	0.57★±0.03	

References page IV-149

Polymer	Solvent	T[^{o}C]	Remarks	[η]	k_H	Ref.
Poly(vinyl acetate) (Cont'd.)	methanol	30				81
		32		44	0.52	77
		39		45	0.41	77
		46		47	0.39	77
	toluene	18		46	0.57	77
		25		46	0.55	77, 126
		30				81
		32		47	0.56	77
		39		47	0.56	77
		46		48	0.54	77
	water	25	partially alcoholized			82
Poly(vinyl alcohol)	dimethyl sulfoxide	30				65
	water	30		30	0.57★	67, 65
				40	0.59★	
				40	0.90	66
				50	0.60★	67
				60	0.66★	
				70- 80	0.68★	
				77	0.69	66
				90	0.71★	67
				100	0.74★	
				130	0.79★	
				150	0.84★	
				170	0.83★	
		60		32	1.26	66
				59	0.93	
Poly(vinyl chloride)	acetone	20		36	0.42	8
	anisole	25		34	0.46	8
	bromobenzene	25		40	0.40	8
	butyl acetate	25		43	0.38	8
	γ-butyrolactone	25		34	0.43	8
	chlorobenzene	25		39	0.42	8
	chloroform	20		38	0.49	8
	cyclohexanone	20				49, 68
		25		90-130	0.34	70, 69
		30		100-121	0.35	72, 71, 73
	cyclopentanone	20		46	0.33	8
	1, 2-dichloroethane	25		41	0.40	8
	dimethylformamide	25		40	0.30	8
	dioxane	25		33	0.50	8
				40	0.48★	
				50	0.44★	
				60	0.41★	
				70	0.38★	
	ethyl acetate	25		39	0.43	8
	ethyl acetyl acetate	25		34	0.43	8
	methyl ethyl ketone	25		26	0.34	8
				32	0.35	
				47- 83	0.38±0.01	
	morpholine	30		71	0.64	74
				89	0.57	
				108	0.49	
		50		64	0.81	74
				79	0.67	
				98	0.53	
		69		58	0.85	74
				70	0.70	
				87	0.55	
	nitrobenzene	25		41	0.33	8
	tetrachloroethane	25		29	0.42	8
				36- 76	0.37★±0.01	
				93	0.35	
	tetrahydrofuran	20	100% conversion	20-143	0.28★±0.02	75
				49	0.35	8

Polymer	Solvent	T[°C]	Remarks	[η]	k_H	Ref.
Poly(vinyl chloride) (Cont'd.)	tetrahydrofuran	20		171	0.31	75
				178	0.32	
				201-220	0.36★±0.02	
		30		76- 91	0.44★±0.01	74
				107	0.47	
	tetrahydrofuryl alcohol	25		44	0.35	8
	tetramethylene sulfoxide	25		46	0.30	8
Poly(vinyl cyanoethyl ether)	water	30				65
	acetone	30				65
Poly(vinylbiphenyl)	benzene	20		7	0.55	95
				22- 38	0.51★±0.04	
				93-134	0.34★±0.01	
		30		7	0.56	95
				21- 07	0.51★±0.03	
				92-133	0.35★±0.01	
		45		21- 37	0.54★±0.02	95
				91-129	0.36★±0.01	
		65		7	0.51	95
				21- 36	0.51★±0.12	
				91-126	0.36★±0.01	
		75		21- 36	0.43★±0.01	95
				90-122	0.36★±0.01	
Poly(vinyl formate)	acetone	30		50	0.34	76
				113	0.54	
	acetonitrile	30		48	0.30	76
				162	0.51	
	dioxane	30		44	0.30	76
				134	0.55	
	methyl acetate	30		62	0.42	76
				74	0.46	
	methyl formate	30		48	0.30	76
				107	0.46	
Poly(1-vinylnaphthalene)	benzene	20		16	0.46	95
				27	0.45	
				37	0.41★±0.03	
		30		15	0.47	95
				27	0.48	
				37	0.43★±0.02	
		45		15	0.61	95
				26	0.57	
				36	0.43★±0.02	
		65		14	0.75	95
				25	0.46	
				36	0.44★±0.01	
		75		14	0.78	95
				25	0.60	
				35	0.49★±0.01	
Poly(2-vinylnaphthalene)	benzene	20		18	0.50	95
				21	0.47	
				46	0.41	
				67-107	0.35★±0.01	
		30		18	0.50	95
				21	0.49	
				45	0.43	
				66	0.41	
				86	0.38	
				106	0.36	
		45		17	0.58	95
				20	0.57	
				43	0.45	
				63-103	0.39★±0.01	

Polymer	Solvent	T[°C]	Remarks	$[\eta]$	k_H	Ref.
Poly(2-vinylnaphthalene) (Cont'd.)						
	benzene	65		17	0.51	95
				20	0.43	
				43-101	0.38★±0.06	
		75		17	0.48	95
				19	0.44	
				43-100	0.38★±0.06	
Poly(4-vinylpyridine)	ethanol	25				96
Poly(vinyltrimethylsilane)	cyclohexane	25		96-154	0.35±0.03	83
Poly(vinyl 4-trimethylsilyl benzoate)						
	toluene	25		136	0.19	94
Poly(vinyl 3-trimethylsilyl benzoate)						
	toluene	25		214	0.54	94

1.4 POLY(OXIDES)

Polymer	Solvent	T[°C]	Remarks	$[\eta]$	k_H	Ref.
Poly(oxyethylene)	benzene	20		2	≈ 3.7	98
				48	0.4	
	chloroform	20		3	≈ 0.9	98
				82	0.4	
	dimethylformamide	20		3	≈ 0.6	98
				45	0.4	
	toluene	30		3	≈ 2.50	98
				39	0.4	
		35		1.1	8.16	99
				1.9	3.41	
				2.4	2.06	
				5.5	2.3	
	water	20		3	1.1	98
				5	0.4	
		35		1.7	4.95	99
				3.0	2.42	
				5.0	0.93	
				10.8	0.44	
Poly(oxymethylene)	p-chlorophenole	60		135-270	0.30	97
	dimethylformamide	140		69	0.23	97
				102	0.21	
		150				97
Poly(oxypropylene)	benzene	35				100

1.5 POLY(ESTERS)

Polymer	Solvent	T[°C]	Remarks	$[\eta]$	k_H	Ref.
Poly(oxyethyleneoxyadipoyl)	cyclohexanone	25				101
	dimethylformamide	25				101
Poly(oxyethyleneoxysebacoyl)	cyclohexanone	25				101
Poly(oxyethyleneoxyterephthaloyl)	m-cresol/phenol = 1/1	25				102
Poly(oxydecamethyleneoxysebacoyl)						
	cyclohexanone	25				101
Poly(oxytetramethyleneoxyadipoyl)						
	cyclohexanone	25				101
Poly(oxytetramethyleneoxysebacoyl)						
	cyclohexanone	25				101

1.6 POLY(AMIDES)

Polymer	Solvent	T[°C]	Remarks	$[\eta]$	k_H	Ref.
Poly(iminoadipoyliminohexamethylene) (Nylon 66)						
	m-cresol	20				112
	formic acid (90%)	25		83	0.20	113
				100	0.22★±0.01	
				120	0.24★±0.02	
				140	0.27★±0.02	
				160	0.27★±0.02	
				180	0.28★±0.02	
				200	0.29★±0.01	

Polymer	Solvent	T[°C]	Remarks	$[\eta]$	k_H	Ref.
Poly(iminoisophthaloylimino-1,2-phenylene)						
	dimethylformamide (1.18 N LiCl)	25				114
Poly(iminoisophthaloylimino-1,3-phenylene)						
	sulfuric acid	25				114
	dimethylformamide (1.18 N LiCl)	25				114
Poly(iminoisophthaloylimino-1,4-phenylene)						
	sulfuric acid	25				114
	dimethylformamide (1.18 N LiCl)	25				114
Poly(imino-1-oxohexamethylene) (Nylon 6)						
	sulfuric acid (62%)	25		34	1.06	108
				60	0.74	
				195	0.36	
	(60%)	25		52	0.52★±0.1	107
				54	0.44★±0.1	
				57	0.34★±0.06	
	(52.5%)	25		30	1.37	108
				50	0.96	
				135	0.44	
				152	0.39	
	(33%)	25				107
	tricresol (0.019 Vol% H_2O)	25		50	1.1	108, 111
				109	0.49	
				335	0.29	
	formic acid (85%)	25		70	0.32★±0.08	107
				75	0.27★±0.09	
				80	0.22★±0.06	
				85	0.18★±0.02	
				48	0.24	108
				86	0.12	
				460	0.13	
	(68.8%)	25		32	0.38	108
				200	0.32	
		25		210	0.32	110
				240	0.34	
				332	0.36	
				558	0.56	
				764	0.74	
				965	0.89	
				1200	1.26	
				1340	1.31	
				73	0.24	110
				84	0.28	
				101	0.66	
				155	1.36	
				202	1.89	
				245	2.84	
				305	3.36	
	hexafluoroisopropanol (+0.05 M CF_3COONa)	20		200-500	0.23	109
	sulfuric acid (96%)	25		94-104	0.30★±0.07	107
	(93%)	25		63	0.42	108
				125	0.26	
				490	0.23	
				502	0.19	
	(80%)	25		48	0.50	108
				90	0.34	
				325	0.30	

Polymer	Solvent	T[^{o}C]	Remarks	[η]	k_H	Ref.
Poly(iminoterephthaloylimino-1,2-phenylene)						
	sulfuric acid	25				114
	dimethylformamide					
	(1.18 N LiCl)	25				114
Poly(iminoterephthaloylimino-1,3-phenylene)						
	sulfuric acid	25				114
Poly(iminoterephthaloylimino-1,4-phenylene)						
	sulfuric acid	25				114

1.7   OTHERS

Polymer	Solvent	T[^{o}C]	Remarks	[η]	k_H	Ref.
Na-polyphosphate	water (NaBr)	25		50-276	0.29★±0.04	103
Poly(oxydimethylsilylene)	benzene	25		251	0.44	104
	carbon tetrachloride	25		312	0.51	104
	chlorobenzene	25		219	0.61	104
	cyclohexane	25		324	0.37	104
	n-heptane	25		305	0.45	104
	methyl ethyl ketone	25		151	0.80	104
	tetrahydrofuran	25		319	0.39	104
	toluene	25		62-241	0.55★±0.02	104
				146	0.51	105
				577-727	0.46★±0.01	
Poly(thiopropylene)	benzene	20				106

1.8   CELLULOSE AND DERIVATIVES

Polymer	Solvent	T[^{o}C]	Remarks	[η]	k_H	Ref.
Amylose	water (KOH)	25				115
	water (NaOH)	25				115
	water (KOH/NaCl)	25				115
	water (1 N KOH)	25		139	0.19	116
		30		129	0.20	116
		35		121	0.48	116
		40		107	0.88	116
Cellulose	Cd-ethylenediamine	20		339-1395	0.50±0.1	117
	(cadoxen)	25		426-510	0.46★±0.1	118
	Cu-ethylenediamine					
	(cuoxen)	25		440	0.54★±0.02	118
				466	0.59	
				510	0.65	
	Zn-ethylenediamine					
	(zincoxen)	25		248-383	0.47★±0.09	118
Cellulose acetate	acetone	30	acetyl group content 60.44 →50.35			119
	formic acid	30	acetyl group content 60.44 →50.35			119
	pyridine	30	acetyl group content 60.44 →50.35			119
Cellulose triacetate	acetic acid	25		72	0.42	120
				117	0.57	
	chloroform	25		43	0.45	120
				72	0.43	
	tetrachloroethane	25		50	0.34	120
				82	0.38	
Cellulose nitrate	acetone	20		695-1995	0.53±0.08	117
	diethyl adipate	25				121
	isoamyl acetate	25				122
Ethyl cellulose	ethyl acetate	20		41-188	0.65±0.15	117
Hydroxyethyl cellulose	water	0		423	0.57	123
				845	0.74	
				1052	0.80	
				1218	0.87	
		15		402	0.43	123
				794	0.59	
				977	0.68	
				1130	0.75	
		25		381	0.30	123

Polymer	Solvent	T[°C]	Remarks	$[\eta]$	k_H	Ref.
Hydroxyethyl cellulose (Cont'd.)	water	25		754	0.47	123
				895	0.62	
				1062	0.67	
		35		370	0.22	123
				725	0.40	
				875	0.49	
				1010	0.57	
Na-carboxymethyl cellulose	water (0.01 M NaCl)	25	Degr. subst. 0.75			121
	water (0.1 M NaCl)	25	Degr. subst. 0.75			121
	water (0.2 M NaCl)	25	Degr. subst. 0.75			121
	water (1 M NaCl)	25	Degr. subst. $0.5 \rightarrow 2.5$			124
	water (1 N NaOH)	25	Degr. subst. $0.5 \rightarrow 2.5$			124
	water (NaCl)	25				125

2. SCHULZ-BLASCHKE COEFFICIENTS

Polymer	Solvent	T[°C]	Remarks	$[\eta]$	k_{SB}	Ref.
Poly(isoprene)						
Guttapercha	toluene**	25				129
Poly(chloroprene)	benzene	25		19- 208	0.32	127
Poly(methylene)	m-xylene**	120				129
Poly(ethylene)	decalin	135	low pressure	121-1535	0.29	128
Poly(1-butene)	decalin	115	stereoregular	$\eta_{sp}<1$	0.48	130
Poly(isobutene)	diisobutylene**	20				129
Poly(acrylonitrile)	dimethylformamide**	25				129
Poly(methyl methacrylate)	acetone**	25				129
				$\eta_{sp}<1$	0.50	130, 131
		32			$\frac{1}{k_{SB}} = 6.09 \log (\eta_{sp} + 5)-2.05$	132
	benzene**	25				129, 131
	butyl acetate	25				131
	chloroform	20		25-480	0.30	2
	chloroform**	25				129
		27.3				133
		32			$\frac{1}{k_{SB}} = 5.41 \log (\eta_{sp} + 5)-1.15$	132
	methyl ethyl ketone	25		$\eta_{sp}<1$	0.50	130
	toluene**	25				129
				$\eta_{sp}<1$	0.33	130
Poly(ethyl methacrylate)	methyl ethyl ketone	23		$\eta_{sp}<1$	0.33	130
Poly(n-butyl methacrylate)	isopropanol	21.5=θ		$\eta_{sp}<1$	1.00	130
	methyl ethyl ketone	23		$\eta_{sp}<1$	0.36	130
Poly(1-ethylbutyl methacrylate)	isopropanol	27.4=θ		$\eta_{sp}<1$	0.89	130
	methyl ethyl ketone	25		$\eta_{sp}<1$	0.30	130
Poly(n-hexyl methacrylate)	isopropanol	32.6=θ		$\eta_{sp}<1$	0.70	130
	methyl ethyl ketone	23		$\eta_{sp}<1$	0.36	130
Poly(n-octyl methacrylate)	n-butanol	16.8=θ		$\eta_{sp}<1$	0.80	130
	methyl ethyl ketone	23		$\eta_{sp}<1$	0.46	130
Poly(n-lauryl methacrylate)	n-butyl acetate	23		$\eta_{sp}<1$	0.38	130
	isopropyl acetate	13=θ		$\eta_{sp}<1$	0.59	130
Poly(vinyl chloride)	cyclohexanone**	25				129
Poly(vinyl acetate)	benzene	30			$\frac{1}{k_{SB}} = 5.224 \log (\eta_{sp} + 4)-0.65$	132
	toluene**	25				129
Poly(styrene)	benzene**	25				129
	benzene	30			$\frac{1}{k_{SB}} = 10.534 \log (\eta_{sp} + 3)-2.65$	132
	butanone**	25				129

** k_{SB} is roughly calculable from Arrhenius "constant".

Polymer	Solvent	T[°C]	Remarks	$[\eta]$	k_{SB}	Ref.
Poly(styrene) (Cont'd.)	chloroform**	25				129
	cyclohexane	34				129
	ethyl acetate**	25				129
	methyl ethyl ketone	30				133
	toluene**	25				129
		30				133
	m-xylene**	25				129
Poly(oxyethylene)	acetonitrile	25		$M_v > 10^4$	0.22	134
	benzene	10		$M_v > 10^4$	0.21	134
		37		$M_v > 10^4$	0.23	134
	benzyl alcohol	25		$M_v > 10^4$	0.19	134
	carbon tetrachloride	25		$M_v > 10^4$	0.30	134
	dimethylformamide	25		$M_v > 10^4$	0.27	134
	dioxane	25		$M_v > 10^4$	0.27	134
	methanol	25		$M_v > 10^4$	0.26	134
	methyl acetate	25		$M_v > 10^4$	0.27	134
	water	25		$M_v > 10^4$	0.17	134
Poly(oxydimethylsilylene)	butanone**	25				129
	chloroform**	25				129
Poly(iminoadipoyliminohexamethylene) (Nylon 66)						
	formic acid (90%)	25		83	0.20	113
				100	0.22★±0.02	
				120	0.24★±0.02	
				140	0.26★±0.02	
				200	0.28★±0.01	
Cellulose	cadoxen	20		339-1395	0.29±0.08	117
	cuoxam	20		946-1125	0.29	135
	cuoxen	20		190-2800	0.29	136
Cellulose acetate	acetone	25				49
Cellulose nitrate	acetone	20	(13.7% N)	180- 500	0.29	137
				500-1000	0.34	
				1000-1400	0.37	
				1400-2000	0.40	
				695-1995	0.30±0.03	117
		27				49
	ethyl acetate	20	(13.7% N)	200- 600	0.26	137
				600-1200	0.30	
				1200-1800	0.33	
				1800-2400	0.37	
Ethyl cellulose	acetone	20	(Degree of Subst. 2.73)	24- 188	0.35	138
		30	(Degree of Subst. 2.73)	24- 179	0.33	138
		40	(Degree of subst. 2.73)	23- 170	0.32	138
	benzene	20	(Degree of subst. 2.73)	27- 460	0.36	138
		30	(Degree of subst. 2.73)	26- 437	0.35	138
		40	(Degree of subst. 2.73)	362- 418	0.34	138
	isobutanol	20	(Degree of subst. 2.73)	25- 442	0.34	138
		30	(Degree of subst. 2.73)	25- 412	0.33	138
		40	(Degree of subst. 2.73)	25- 388	0.32	138
	n-butyl chloride	20	(Degree of subst. 2.73)	25- 502	0.33	138
		30	(Degree of subst. 2.73)	25- 468	0.33	138
		40	(Degree of subst. 2.73)	24- 433	0.32	138
	ethyl acetate	20		41- 188	0.31±0.05	117
			(Degree of subst. 2.73)	26- 472	0.29	138
		30	(Degree of subst. 2.73)	25- 445	0.28	138
		40	(Degree of subst. 2.73)	25- 424	0.27	138

C. REFERENCES

1. M. L. Huggins, J. Am. Chem. Soc. 64, 2716-18 (1942).

2. G. V. Schulz, F. Blaschke, J. Prakt. Chem. 158, 130-35 (1941); ibid. 159, 146-54 (1941).

3. I. R. Rutgers, Rheolog. Acta (Ergaenzungsh. z. Kolloid-Z. 2, 305-48 (1962).

4. K. J. Linow, B. Philipp, Plaste Kautschuk 18, 721-24 (1971).

5. T. Sakai, J. Polymer Sci. A2 6, 1659-72 (1968).

6. R. Berger, Plaste Kautschuk 14, 11-14 (1967).

7. R. Berger, Makromol. Chem. 102, 24-38 (1967).

** k_{SB} is roughly calculable from Arrhenius "constant".

8. M. Bohdanecky, Collection Czech. Chem. Commun. 35, 1972-90 (1970).

9. See e.g., O. F. Solomon, I. Z. Ciuta, Bul. Inst. Politeh. "Gheorghe Gheorghiu-Dej," Bucuresti 30, 87-95 (1968).

10. M. D. Stern, see 11.

11. L. Utracki, R. Simha, J. Phys. Chem. 67, 1052 (1963).

12. H. Sotobayashi, Makromol. Chem. 73, 235-39 (1964).

13. K. Tompa, Polymer Solutions, p. 268-69, Acad. Press, New York 1956.

14. E. g., L. M. Hobbs, V. C. Long, Polymer (London) 4, 479-91 (1963).

15. E. g., S. L. Kapur, S. Gundiah, J. Polymer Sci. 26, 89-99 (1957).

16. J. Schurz, H. Pippan, Monatsh. Chem. 94, 859-89 (1963).

17. B. Simha, J. Res. Nat. Bur. Std. 42, 409-18 (1949).

18. T. Gillespie, J. Polymer Sci. C 3, 31-37 (1963).

19. F. Eirich, J. Riseman, J. Polymer Sci. 4, 417-34 (1949).

20. L. H. Cragg, C. C. Bigelow, J. Polymer Sci. 16, 177-91 (1955).

21. M. Hoffmann, Makromol. Chem. 24, 245-57 (1957).

22. C. E. H. Bawn, M. B. Huglin, Polymer (London) 3, 615-23 (1962).

23. J. Blackford, R. F. Robertson, J. Polymer Sci. A 3, 1289-1302 and 1303-11 (1965).

24. J. A. Manson, L. H. Cragg, Canad. J. Chem. 30, 482-96 (1952).

25. J. A. Manson, L. H. Cragg, Canad. J. Chem. 36, 858-68 (1958).

26. J. Risemann, R. Ullman, J. Chem. Phys. 19, 578 (1951).

27. T. Sakai, J. Polymer Sci. A-2, 6, 1535 (1968).

28. T. Sakai, Macromolecules 3, 96-97 (1970).

29. S. L. Kapur, S. Gundiah, Makromol. Chem. 26, 119-25 (1958).

30. S. Gundiah, N. V. Viswanathan, S. L. Kapur, J. Sci. Ind. Res. 19B, 191 (1960).

31. G. M. Bristow, J. Polymer Sci. 62, 168-71 (1962).

32. J. Poláček, Collection Czech. Chem. Commun. 25, 2103-14 (1960).

33. L. Nicolas, J. Polymer Sci. 29, 191-217 (1958).

34. Q. A. Trementozzi, J. Polymer Sci. 23, 887-902 (1957).

35. E. Willert, R. Berger, G. Langhammer, Plaste Kautschuk 14(5), 303-08 (1967).

36. H. J. L. Schuurmans, J. Polymer Sci. 57, 557-68 (1962).

37. G. Moraglio, F. Danusso, Ann. Chim. (Rome) 49, 902-10 (1959).

38. F. Danusso, G. Moraglio, Atti Accad. Nazl, Lincei, Rend., Classe Sci. Fis. Mat. Nat. 25, 509-16 (1958).

39. K. Kamide, Kobunshi Kagaku 21, 152-60 (1964).

40. V. P. Budtov, Vysokomolekul. Soedin., Ser. A 9, 765-71 (1967).

41. W. F. Haddon, Jr., R. S. Porter, J. F. Johnson, J. Appl. Polymer Sci. 8, 1371-78 (1964).

42. Lung-Yu Chou, J. L. Zakin, J. Colloid Interface Sci. 25, 547-57 (1967).

43. P. H. Plesch, P. P. Rutherford, Polymer 1, 271-73 (1960).

44. G. Barone, V. Crescenzi, F. Quadrifoglio, V. Vitagliano, Ric. Sci. 36, 477-81 (1966).

45. N. Yokomichi, K. Ogino, T. Nakagawa, Nippon Kagaku Zasshi 87, 233-36 (1966).

46. N. T. Srinivasan, M. Santappa, Makromol. Chem. 27, 61-68 (1958).

47. A. Kotera, T. Saito, Y. Watanabe, M. Ohama, Makromol. Chem. 87, 195-208 (1965).

48. H. Somitomo, Y. Yatsukama, Chem. High Polymer (Japan) 11, 65-70 (1954).

49. M. L. Huggins, "Physical Chemistry of High Polymers," Wiley, New York, p. 82 (1958).

50. H. Kamogawa, T. Sekiya, Kogyo Kagaku Zasshi 63, 1631-35 (1960).

51. N. M. Beder, A. B. Pakshver, Khim. Volokna 3, 21-24 (1961).

52. Y. Shimura, J. Polymer Sci. A-2, 4, 423-35 (1966).

53. T. Shibukawa, K. Nakaguchi, Kobunshi Kagaku 14, 353-58 (1957).

54. S. Uchiyama, Sen-i Gakaishi 22, 373-78 (1966).

55. M. Katayama, T. Ogoshi, Chem. High Polymers (Japan) 13, 114-19 (1956); ibid. 148-52.

56. T. Shibukawa, K. Nakaguchi, Kobunshi Kagaku 14, 203-09 (1957).

57. T. Shibukawa, K. Nakaguchi, Kobunshi Kagaku 14, 291-94 (1957); ibid. 294-97.

58. A. Blumstein, F. W. Billmeyer, Jr., J. Polymer Sci. A-2, 4, 465-74 (1966).

59. G. S. Kolesnikov, Ts èng Han-ming, Vysokomolekul. Soedin. 3, 1210-16 (1961).

60. M. Bohdanecky, Collection Czech. Commun. 31, 4095-107 (1966).

61. P. Vasudevan, M. Santappa, Makromol. Chem. 137, 262-75 (1970).

62. S. Morimoto, A. Machi, Kogyo Gijutsuin Sen I Kogyo Shikensho Kenkyn Hokoku 67, 1-11 (1964).

63. K. Karunakaran, M. Santappa, Current Sci. (India) 33, 551-52 (1964).

64. J. Herz, J. C. Galin, P. Rempp, J. Parrod, Compt. Rend. 261, 1319-21 (1965).

65. M. Negishi, S. Yanagibori, K. Yoshida, M. Shiraishi, Kobunshi Kagaku 14, 239-46 (1957).

66. Y. Sakaguchi, N. Yasuhira, Kobunshi Kagaku 13, 437-40 (1956).

67. M. Matsumoto, K. Imai, Chem. High Polymers (Japan) 12, 402-09 (1955).

68. J. W. Breitenbach, E. L. Forster, A. J. Renner, Kolloid-Z. 127, 1 (1952).

69. F. Danusso, G. Moraglio, S. Gazzeva, Chim. Ind. (Milan) 36, 883 (1954).

70. L. C. Grotz, J. Polymer Sci. B 2, 883-85 (1964).

71. G. M. Guzman, J. M. G. Faton, Anales Real Soc. Espan. Fis. Quim. (Madrid) 54-B, 263 (1958).

72. K. Goto, Y. Ono, S. Furusawa, Chem. High Polymers (Japan) 11, 437-43 (1954).

73. H. Sobue, Y. Tabata, Y. Tajima, Kogyo Kagaku Zasshi 61, 1328-31 (1958).

74. Y. Nakamura, M. Saito, Kobunshi Kagaku 17, 718-21 (1960).

75. F. Krasovec, "J. Stefan," Inst. Repts. (Ljubljana) 3, 203-11 (1956).

76. K. Fujii, S. Imoto, J. Ukida, M. Matsumoto, Kobunshi Kagaku 19, 581-86 (1962).

77. W. E. Moore, M. Murphy, J. Polymer Sci. 56, 519-32 (1962).

78. S. L. Kapur, S. Gundiah, J. Polymer Sci. 26, 89-100 (1957).

79. K. Imai, U. Maeda, M. Matsumoto, Kobunshi Kagaku 14, 419-24 (1957).

80. L. M. Hobbs, V. C. Long, Polymer 4, 479-91 (1963).

81. M. Matsumoto, U. Maida, K. Imai, Kobunshi Kagaku 14, 425-29 (1957).

82. A. Beresniewicz, J. Polymer Sci. 39, 63-79 (1959).

83. A. B. Duchkova, N. S. Nametkin, V. S. Khotimskii, L. E. Gusel' nikov, S. G. Durgar' yan, Vysokomolekul. Soedin. 8, 1814-17 (1966).

84. T. A. Orofino, F. Wenger, J. Phys. Chem. 67, 566-75 (1963).

85. J. R. Urwin, J. M. Stearne, Makromol. Chem. 78, 204-15 (1964).

86. V. P. Budtov, Vysokomolekul. Soedin., Ser. A 9, 765-71 (1967).

87. M. Morton, T. E. Helminiak. S. D. Gadkary, F. Bueche, J. Polymer Sci. 57, 471-82 (1962).

88. H.- G. Elias, O. Etter, Makromol. Chem. 66, 56-72 (1963).

89. V. A. Myagchenkov, E. V. Kuznetsov, V. Ya. Kitkevich, Vysokomolekul. Soedin. 6, 1366-70 (1964).

90. U. Lohmander, R. Stroemberg, Makromol. Chem. 72, 143-58 (1964).

91. H. W. McCormick, J. Colloid Sci. 16, 635-37 (1961).

92. A. I. Goldberg, W. P. Hohenstein, H. Mark, J. Polymer Sci. 2, 503-10 (1947).

93. F. S. Holahan, S. S. Stivala, D. W. Levi, J. Polymer Sci. A 3, 3987-92 (1965).

94. H. Hopff, M. A. Osman, Makromol. Chem. 135, 175-79 (1970).

95. L. A. Utracki, R. Simha, Makromol. Chem. 117, 94-116 (1968).

96. D. O. Jordan, A. R. Mathieson, M. R. Porter, J. Polymer Sci. 21, 473-82 (1956).

97. W. Tummler, Plaste Kautschuk 12, 582-87 (1965).

98. G. Mueh, Kolloid-Z. 196, 140-46 (1964); ibid. 64-65.

99. D. K. Thomas, A. Charlesby, J. Polymer Sci. 42, 195-202 (1960).

100. K. Okazaki, Kogyo Kagaku Zasshi 64, 342-45 (1961).

101. A. Orszagh, F. Fejgin, Polimery 8, 233-36 (1963).

102. H. Sobue, A. Kajiura, Kogyo Kagaku Zasshi 62, 1908-11 (1959).

103. U- P. Strauss, J. Polymer Sci. 33, 291-94 (1958).

104. M. Takeda, A. Yamada, T. Saito, Kobunshi Kagaku 14, 265-69 (1957).

105. M. Takeda, A. Yamada, Kobunshi Kagaku 14, 247-51 (1957).

106. V. E. Eskin, A. E. Nesterov, Vysokomolekul. Soedin. 8, 141-45 (1966).

107. W. Sbrolli, T. Capaccioli, Chim. Ind. (Milan) 42, 243-48 (1960).

108. F. Rybnikář, Faserforsch. u. Textiltech. 9, 500-04 (1958).

109. A. V. Pavlov, S. E. Bresler, S. R. Rafikov, Vysokomolekul. Soedin. 6, 2068-72 (1964).

110. O. Quadrat, M. Bohdanecky, Collection Czech. Chem. Commun. 29, 2469-78 (1964).

111. J. Sebenda, J. Kralicek, Collection Czech. Chem. Commun. 31, 2534-46 (1966).

112. G. Prati, Ann. Chim. (Rome) 47, 51-57 (1957).

113. E. Heim, Faserforsch. u. Textiltech. 11, 513-23 (1960).

114. V. Z. Nikonov, L. B. Sokolov, Vysokomolekul. Soedin. 8, 1529-34 (1966).

115. S. R. Erlander, R. M. Purvinas, Stärke 20, 37-45 (1968).

116. M. K. S. Morsi, C. Sterling, J. Appl. Polymer Sci. 10, 925-28 (1966).

117. L. S. Bolotnikova, T. I. Samsonova, Zh. Prikl. Khim. 38, 2299-303 (1965).

118. M. W. Pandit, R. S. Singh, Indian J. Technol. 6, 172-76 (1968).

119. V. P. Kharitonova, A. B. Pakshver, Kolloid. Zhur. 20, 110-16 (1958).

120. P. Haward, R. S. Parikh, J. Polymer Sci. A 1, 6, 537-46 (1968).

121. U. Lohmander, R. Stroemberg, Makromol. Chem. 72, 143-58 (1964).

122. F. Higashide, J. Nakajima, Kogyo Kagaku Zasshi 65, 258-62 (1962).

123. W. Brown, Arkiv Kemi 18, 227-84 (1961).

124. K. Watanabe, M. Nakamura, Kogyo Kagaku Zasshi 69, 1329-31 (1966).

125. D. A. I. Goring, S. Sitaramaiah, Polymer 4, 7-14 (1963).

126. O. F. Solomon, I. Z. Ciuta, Bul. Inst. Politeh. "Gheorghe Gheorghiu-Dej," Bucuresti 30, 87-95 (1968).

127. J. Poláček, J. Polymer Sci. 39, 469-73 (1959).

128. H. Wesslau, Makromol. Chem. 20, 111-42 (1956).

129. M. Hoffmann, Makromol. Chem. 24, 222-44 (1957); ibid. 245-57.

130. R. J. Valles, M. C. Otzinger, D. W. Levi, J. Appl. Polymer Sci. 4, 92-94 (1960).

131. H. Craubner, Makromol. Chem. 93, 24-32 (1966).

132. V. V. Varadaiah, Ramakrishna Rao, J. Polymer Sci. 19, 379-80 (1956).

133. V. E. Hart, J. Polymer Sci. 17, 215-19 (1955).

134. H.-G. Elias, Makromol. Chem. 99, 291-96 (1966).

135. E. Husemann, G. V. Schulz, J. Makromol. Chem. 1, 197 (1943).

136. M. Marx, Makromol. Chem. 16, 157 (1955); Papier 10, 135 (1956).

137. M. Marx, G. V. Schulz, Makromol. Chem. 31, 140-53 (1959); see also M. Marx-Figini, Makromol. Chem. 36, 220-31 (1960).

138. G. Meyerhoff, N. Suetterlin, unpublished data.

139. E. O. Kraemer, Ind. Eng. Chem. 30, 1200 (1938).

TABLES OF RELATIVE VISCOSITIES CALCULATED FROM

FIKENTSCHER K-VALUES FROM 1 to 100 ★

P. E. Hinkamp
Designed Polymers Research
Dow Chemical Company
Midland, Michigan

The data in the tables are solutions to the Fikentscher equation (1)

$$\log \eta_{rel} = [75k^2/(1 + 1.5kc) + k] c$$

where K = 1.000 k for a concentration, c, of 0.5, 1.0, and 4.0 grams per 100 ml of solution. The computer generated lists give values for η_{rel} as K is raised from 1 to 100 in 0.5 unit steps. They may be applied regardless of the polymer, solvent, or temperature used. To avoid confusion it is important that data on the solvent, temperature, and concentration used for the measurement be included when reporting results.

The concentrations selected are those most commonly found in the literature.

───────────

1. H. Fikentscher, Cellulose Chemie 13, 58 (1932).

CALCULATION OF RELATIVE VISCOSITIES FROM FIKENTSCHER
K-VALUES FOR A GIVEN CONCENTRATION

Concentration = 0.5 g/100 ml.

K	η_{rel}	K	η_{rel}	K	η_{rel}	K	η_{rel}	K	η_{rel}
1.0	1.001	18.5	1.052	36.0	1.162	53.5	1.349		
1.5	1.002	19.0	1.054	36.5	1.167	54.0	1.356		
2.0	1.003	19.5	1.056	37.0	1.171	54.5	1.362		
2.5	1.003	20.0	1.059	37.5	1.175	55.0	1.369		
3.0	1.004	20.5	1.061	38.0	1.179	55.5	1.376		
3.5	1.005	21.0	1.064	38.5	1.184	56.0	1.383		
4.0	1.006	21.5	1.066	39.0	1.188	56.5	1.390		
4.5	1.007	22.0	1.069	39.5	1.193	57.0	1.398		
5.0	1.008	22.5	1.071	40.0	1.197	57.5	1.405		
5.5	1.009	23.0	1.074	40.5	1.202	58.0	1.412		
6.0	1.010	23.5	1.077	41.0	1.207	58.5	1.420		
6.5	1.011	24.0	1.080	41.5	1.212	59.0	1.427		
7.0	1.012	24.5	1.082	42.0	1.217	59.5	1.435		
7.5	1.014	25.0	1.085	42.5	1.222	60.0	1.443		
8.0	1.015	25.5	1.088	43.0	1.227	60.5	1.451		
8.5	1.016	26.0	1.091	43.5	1.232	61.0	1.459		
9.0	1.017	26.5	1.094	44.0	1.237	61.5	1.467		
9.5	1.019	27.0	1.097	44.5	1.242	62.0	1.475		
10.0	1.020	27.5	1.100	45.0	1.247	62.5	1.483		
10.5	1.022	28.0	1.104	45.5	1.253	63.0	1.492		
11.0	1.023	28.5	1.107	46.0	1.258	63.5	1.500		
11.5	1.025	29.0	1.110	46.5	1.264	64.0	1.509		
12.0	1.026	29.5	1.113	47.0	1.269	64.5	1.517		
12.5	1.028	30.0	1.117	47.5	1.275	65.0	1.526		
13.0	1.030	30.5	1.120	48.0	1.281	65.5	1.535		
13.5	1.032	31.0	1.124	48.5	1.286	66.0	1.544		
14.0	1.033	31.5	1.127	49.0	1.292	66.5	1.553		
14.5	1.035	32.0	1.131	49.5	1.298	67.0	1.562		
15.0	1.037	32.5	1.135	50.0	1.304	67.5	1.572		
15.5	1.039	33.0	1.139	50.5	1.310	68.0	1.581		
16.0	1.041	33.5	1.142	51.0	1.317	68.5	1.591		
16.5	1.043	34.0	1.146	51.5	1.323	69.0	1.601		
17.0	1.045	34.5	1.150	52.0	1.329	69.5	1.610		
17.5	1.047	35.0	1.154	52.5	1.336	70.0	1.620		
18.0	1.050	35.5	1.158	53.0	1.342	70.5	1.630		

★ Reprinted from Polymer 8, 381 (1967) with the permission of the publisher.

RELATIVE VISCOSITIES CALCULATED FROM FIKENTSCHER K-VALUES

Concentration = 0.5 g/100 ml.   (Cont'd.)

K	$\eta_{rel.}$	K	$\eta_{rel.}$	K	$\eta_{rel.}$	K	$\eta_{rel.}$
71.0	1.641	78.5	1.809	86.0	2.012	93.5	2.255
71.5	1.651	79.0	1.822	86.5	2.027	94.0	2.273
72.0	1.661	79.5	1.834	87.0	2.042	94.5	2.291
72.5	1.672	80.0	1.847	87.5	2.057	95.0	2.309
73.0	1.683	80.5	1.860	88.0	2.072	95.5	2.328
73.5	1.694	81.0	1.873	88.5	2.088	96.0	2.347
74.0	1.704	81.5	1.886	89.0	2.104	96.5	2.366
74.5	1.716	82.0	1.899	89.5	2.120	97.0	2.385
75.0	1.727	82.5	1.913	90.0	2.136	97.5	2.404
75.5	1.738	83.0	1.926	90.5	2.152	98.0	2.424
76.0	1.750	83.5	1.940	91.0	2.169	98.5	2.444
76.5	1.761	84.0	1.954	91.5	2.186	99.0	2.464
77.0	1.773	84.5	1.968	92.0	2.203	99.5	2.485
77.5	1.785	85.0	1.983	92.5	2.220	100.0	2.506
78.0	1.797	85.5	1.997	93.0	2.237		

Concentration = 1.0 g/100 ml.

K	$\eta_{rel.}$	K	$\eta_{rel.}$	K	$\eta_{rel.}$	K	$\eta_{rel.}$
1.0	1.002	20.5	1.125	40.0	1.423	59.5	2.011
1.5	1.004	21.0	1.130	40.5	1.434	60.0	2.031
2.0	1.005	21.5	1.135	41.0	1.445	60.5	2.052
2.5	1.007	22.0	1.141	41.5	1.456	61.0	2.074
3.0	1.008	22.5	1.146	42.0	1.467	61.5	2.095
3.5	1.010	23.0	1.152	42.5	1.479	62.0	2.118
4.0	1.012	23.5	1.157	43.0	1.490	62.5	2.140
4.5	1.014	24.0	1.163	43.5	1.502	63.0	2.163
5.0	1.016	24.5	1.169	44.0	1.514	63.5	2.186
5.5	1.018	25.0	1.175	44.5	1.527	64.0	2.210
6.0	1.020	25.5	1.182	45.0	1.539	64.5	2.234
6.5	1.022	26.0	1.188	45.5	1.552	65.0	2.258
7.0	1.025	26.5	1.194	46.0	1.565	65.5	2.283
7.5	1.027	27.0	1.201	46.5	1.578	66.0	2.309
8.0	1.030	27.5	1.208	47.0	1.591	66.5	2.334
8.5	1.032	28.0	1.215	47.5	1.605	67.0	2.360
9.0	1.035	28.5	1.222	48.0	1.619	67.5	2.387
9.5	1.038	29.0	1.229	48.5	1.633	68.0	2.414
10.0	1.041	29.5	1.236	49.0	1.647	68.5	2.442
10.5	1.044	30.0	1.243	49.5	1.662	69.0	2.470
11.0	1.047	30.5	1.251	50.0	1.677	69.5	2.498
11.5	1.050	31.0	1.259	50.5	1.692	70.0	2.527
12.0	1.053	31.5	1.266	51.0	1.707	70.5	2.557
12.5	1.057	32.0	1.274	51.5	1.723	71.0	2.587
13.0	1.060	32.5	1.282	52.0	1.738	71.5	2.617
13.5	1.064	33.0	1.291	52.5	1.755	72.0	2.648
14.0	1.068	33.5	1.299	53.0	1.771	72.5	2.680
14.5	1.071	34.0	1.308	53.5	1.788	73.0	2.712
15.0	1.075	34.5	1.316	54.0	1.805	73.5	2.745
15.5	1.079	35.0	1.325	54.5	1.822	74.0	2.778
16.0	1.083	35.5	1.334	55.0	1.839	74.5	2.812
16.5	1.088	36.0	1.344	55.5	1.857	75.0	2.846
17.0	1.092	36.5	1.353	56.0	1.875	75.5	2.881
17.5	1.096	37.0	1.362	56.5	1.893	76.0	2.917
18.0	1.101	37.5	1.372	57.0	1.912	76.5	2.953
18.5	1.105	38.0	1.382	57.5	1.931	77.0	2.990
19.0	1.110	38.5	1.392	58.0	1.951	77.5	3.028
19.5	1.115	39.0	1.402	58.5	1.970	78.0	3.066
20.0	1.120	39.5	1.413	59.0	1.990	78.5	3.105

Concentration = 1.0 g/100 ml. (Cont'd.)

K	$\eta_{rel.}$	K	$\eta_{rel.}$	K	$\eta_{rel.}$	K	$\eta_{rel.}$
79.0	3.145	84.5	3.630	90.0	4.220	95.5	4.942
79.5	3.185	85.0	3.679	90.5	4.280	96.0	5.016
80.0	3.226	85.5	3.729	91.0	4.341	96.5	5.090
80.5	3.268	86.0	3.779	91.5	4.403	97.0	5.166
81.0	3.310	86.5	3.831	92.0	4.466	97.5	5.244
81.5	3.353	87.0	3.884	92.5	4.531	98.0	5.323
82.0	3.398	87.5	3.937	93.0	4.596	98.5	5.403
82.5	3.442	88.0	3.992	93.5	4.663	99.0	5.485
83.0	3.488	88.5	4.048	94.0	4.731	99.5	5.568
83.5	3.534	89.0	4.104	94.5	4.800	100.0	5.653
84.0	3.582	89.5	4.162	95.0	4.871		

Concentration = 4.0 g/100 ml.

K	$\eta_{rel.}$	K	$\eta_{rel.}$	K	$\eta_{rel.}$	K	$\eta_{rel.}$
1.0	1.010	22.0	1.646	43.0	4.102	64.0	13.934
1.5	1.015	22.5	1.674	43.5	4.210	64.5	14.390
2.0	1.021	23.0	1.704	44.0	4.321	65.0	14.862
2.5	1.028	23.5	1.735	44.5	4.436	65.5	15.352
3.0	1.034	24.0	1.766	45.0	4.555	66.0	15.861
3.5	1.041	24.5	1.799	45.5	4.677	66.5	16.388
4.0	1.049	25.0	1.833	46.0	4.804	67.0	16.935
4.5	1.057	25.5	1.867	46.5	4.935	67.5	17.502
5.0	1.065	26.0	1.903	47.0	5.071	68.0	18.091
5.5	1.073	26.5	1.940	47.5	5.211	68.5	18.701
6.0	1.083	27.0	1.978	48.0	5.355	69.0	19.335
6.5	1.092	27.5	2.017	48.5	5.505	69.5	19.993
7.0	1.102	28.0	2.058	49.0	5.660	70.0	20.675
7.5	1.112	28.5	2.100	49.5	5.820	70.5	21.383
8.0	1.123	29.0	2.143	50.0	5.985	71.0	22.119
8.5	1.134	29.5	2.187	50.5	6.156	71.5	22.882
9.0	1.146	30.0	2.233	51.0	6.333	72.0	23.674
9.5	1.158	30.5	2.280	51.5	6.516	72.5	24.497
10.0	1.170	31.0	2.329	52.0	6.706	73.0	25.352
10.5	1.183	31.5	2.379	52.5	6.902	73.5	26.239
11.0	1.197	32.0	2.431	53.0	7.104	74.0	27.160
11.5	1.211	32.5	2.484	53.5	7.314	74.5	28.117
12.0	1.226	33.0	2.540	54.0	7.532	75.0	29.111
12.5	1.241	33.5	2.597	54.5	7.756	75.5	30.143
13.0	1.256	34.0	2.655	55.0	7.989	76.0	31.216
13.5	1.272	34.5	2.716	55.5	8.230	76.5	32.331
14.0	1.289	35.0	2.778	56.0	8.479	77.0	33.488
14.5	1.306	35.5	2.843	56.5	8.737	77.5	34.692
15.0	1.324	36.0	2.910	57.0	9.005	78.0	35.942
15.5	1.343	36.5	2.978	57.5	9.282	78.5	37.242
16.0	1.362	37.0	3.049	58.0	9.568	79.0	38.592
16.5	1.381	37.5	3.122	58.5	9.865	79.5	39.997
17.0	1.402	38.0	3.198	59.0	10.173	80.0	41.456
17.5	1.423	38.5	3.276	59.5	10.492	80.5	42.973
18.0	1.445	39.0	3.356	60.0	10.822	81.0	44.551
18.5	1.467	39.5	3.439	60.5	11.164	81.5	46.192
19.0	1.490	40.0	3.525	61.0	11.518	82.0	47.898
19.5	1.514	40.5	3.614	61.5	11.886	82.5	49.672
20.0	1.539	41.0	3.705	62.0	12.266	83.0	51.517
20.5	1.564	41.5	3.800	62.5	12.661	83.5	53.436
21.0	1.590	42.0	3.898	63.0	13.070	84.0	55.432
21.5	1.618	42.5	3.998	63.5	13.494	84.5	57.509

RELATIVE VISCOSITIES CALCULATED FROM FIKENTSCHER K-VALUES

Concentration = 4.0 g/100 ml. (Cont'd.)

K	$\eta_{rel.}$	K	$\eta_{rel.}$	K	$\eta_{rel.}$	K	$\eta_{rel.}$
85.0	59.669	89.0	80.432	93.0	109.087	97.0	148.817
85.5	61.917	89.5	83.527	93.5	113.369	97.5	154.769
86.0	64.256	90.0	86.750	94.0	117.831	98.0	160.974
86.5	66.690	90.5	90.106	94.5	122.479	98.5	167.442
87.0	69.223	91.0	93.600	95.0	127.321	99.0	174.184
87.5	71.859	91.5	97.239	95.5	132.367	99.5	181.214
88.0	74.603	92.0	101.029	96.0	137.626	100.0	188.543
88.5	77.459	92.5	104.975	96.5	143.105		

THETA-SOLVENTS

H.-G. Elias and H. G. Bührer
Midland Macromolecular Institute
Midland, Mich.

Contents

A. INTRODUCTION

Theta-solvents (θ-solvents) are solvents in which, at a given temperature, a polymer molecule is in the so-called theta-state. In the theta-state, the solution behaves thermodynamically ideal at low concentrations. The theta-temperature may be phenomenologically defined as the critical miscibility temperature at the limit of the infinite molecular weight (44). Since P. J. Flory was the first to show the importance of the theta-state for a better understanding of molecular and technological properties of polymers, theta-temperatures are also called "Flory-temperatures". (The name "van't Hoff-temperature" has been suggested (51), but is not accepted internationally).

The precise definition of a theta-temperature is given thermodynamically. The chemical potential of the solvent, $\Delta\mu_1$, can be split into an ideal and an excess term:

$$\Delta\mu_1 = \Delta\mu_1^{id} + \Delta\mu_1^{exc} \tag{1}$$

where the excess chemical potential is given by the enthalpy of dilution ΔH_1 and the excess entropy of dilution ΔS_1^{exc}:

$$\Delta\mu_1^{exc} = \Delta H_1 - T\Delta S_1^{exc} \tag{2}$$

At the theta-temperature, the excess chemical potential $\Delta\mu_1^{exc}$ and, correspondingly, the excess Free Energy of dilution are zero. This does not imply, however, that both the enthalpy of dilution and the excess entropy of dilution are zero. It means only that both terms on the right-hand side of eq. (2) compensate each other.

The chemical potential itself is not a measurable quantity. However, it may be replaced for instance by the product of the osmotic pressure π and the partial molar volume of the solvent $\bar{v}_1$:

$$-\Delta\mu_1 = \pi v_1 \tag{3}$$

The concentration dependence of the osmotic pressure of solutions of non-electrolytes can be written as a power series

$$\pi = \frac{RT}{M_n} c_2 + A_2 \cdot c_2^2 + A_3 \cdot c_2^3 + \ldots \tag{4}$$

where c_2 = concentration of the solute (polymer), R = gas constant, M_n = number average molecular weight of the polymer, and A_2, A_3 ... are the second, third ... virial coefficients. At low polymer concentrations, the third term of the right hand side of Eq. (4) is negligible. Comparison of equations (1), (3), and (4) then gives

$$A_2 = -\frac{\Delta\mu_1^{exc}}{v_1 c_2^2} \tag{5}$$

Since by definition the excess chemical potential is zero at the theta-temperature, the second virial coefficient is zero, too. With respect to the concentration dependence of the osmotic pressure, the solution will consequently behave as pseudo-ideal at the theta-temperature. To avoid confusion with true ideal solutions (where both enthalpy of dilution and excess entropy of dilution are zero), it was proposed that solutions in the theta-state should be termed pseudo-ideal (99).

A polymer/solvent system may exhibit two theta-temperatures (225). In the case of endothermic solutions, a lowering of temperature will lead to less positive second virial coefficients. The theta-temperature corresponds to an upper critical solution temperature in the limit of infinite molecular weight. In the case of exothermic solutions, the theta-temperature corresponds however to a lower critical solution temperature for infinite molecular weight (see e.g. (113)). Theta-temperatures from exothermic solutions have the symbol θ_+, those from endothermic solutions the symbol θ_-. In the tables, the different theta-temperatures are differentiated by a + or - sign behind the numerical values.

The thermodynamic behavior described above results from the fact that long-range interactions are not present in the theta-state. Long-range interactions are intra-molecular interactions between groups of one and the same polymer molecule separated by many chemical bonds. They correspond to van der Waals forces between different molecules in low molecular chemistry. At the theta-temperature, the polymer molecule thus exhibits its unperturbed dimensions, i. e. dimensions influenced only by short-range interactions between neighboring groups and by skeletal effects (bond distances, valence angles).

The suitability of a solvent as theta-solvent thus depends on the polymer (constitutiton, configuration), on the solvent (constitution) and on temperature, because all these factors influence long-range interactions (interactions with solvent and excluded volume effect). A molecular weight dependence of the theta-temperature is scarcely detectable for single solvents in the high molecular weight range of polymer (see however ref. 79). Not enough measurements are available at present for the low molecular weight range. With suitable mixed solvents at constant temperature, a small variation of the theta composition with molecular weight is measurable (35). This corresponds to a small variation of the theta-temperature with molecular weight at a given solvent composition.

The thermodynamic conditions for the theta-state are fulfilled for single as well as for mixed solvents. In the absence of large variations in temperature, the unperturbed dimensions of a given apolar polymer are approximately the same in different single apolar solvents. However, the dimensions can vary widely in mixed solvents (33), if solvent/non-solvent pairs are used. This behavior may be explained by the non-negligible interactions of non-solvent with polymer groups and interactions between solvent and non-solvent molecules. The observation of only small variations of the unperturbed dimensions with different single solvents at approximately equal theta-temperatures leads to the conclusion that approximately equal interactions must be present in order to get theta conditions at equal temperatures.

Methods to Determine Theta-Solvents

(a)   Phase Equilibria (PE)

The temperature for phase separation, i. e. the critical miscibility temperature, is determined for a number of different concentrations of a polymer of known number-average molecular weight and the maximum critical miscibility temperature T_c is noted. The experiment is repeated for a series of polymers of the same constitution and configuration but of different molecular weights. For large molecular weights, the critical miscibility temperature T_c can be extrapolated to infinite molecular weight according to ref. (44):

$$(1/T_c) = (1/\theta)(1 + (b/\overline{M}_n^{0.5})) \qquad (6)$$

where θ is the theta-temperature, and b a constant for the particular system. If used with mixed solvents, the data must be extrapolated not only to infinite molecular weight but, in addition, to infinite dilution (68). This is not a straightforward procedure, however. The method can be used for separation in two liquid phases only. Care must therefore be taken with crystalline polymers.

(b)   Second Virial Coefficient (A_2)

According to Eq. (5) the second virial coefficient A_2 is zero for a θ-solvent and consequently the slope of a π/c vs c plot is also zero if the solvent is a θ-solvent. The third virial coefficient A_0 may be different from zero however, and measurements should be made at sufficiently low concentrations. All absolute methods for the determination of molecular weights may be used unless they depend on specific solvent behavior, like ebullioscopy and cryoscopy. Methods normally used are osmotic pressure (OP), light scattering (LS), sedimentation equilibrium (SE), and the approach to sedimentation equilibrium [Archibald's method (Arch)].

The concentration dependence of reciprocal apparent molecular weights is determined for a given polymer/solvent pair at different temperatures and the resulting 2nd virial coefficients are plotted against temperature. The dependence of A_2 on temperature is linear only in the neighborhood of the theta-temperature. Care must be taken, therefore, to work close to the theta-temperature.

Alternatively, the second virial coefficient for a given polymer may be measured at a constant temperature using different solvent/non-solvent ratios.

(c)   Concentration Dependence of Hydrodynamic Parameters

The concentration dependence of the sedimentation coefficient (s) and of the diffusion coefficient (D) is not only affected by thermodynamic effects but, in addition, by frictional forces. The effect is more pronounced at higher molecular weights (82) and even at theta-temperatures, therefore, a concentration dependence of s or D will appear. As a result, this method is not very suitable to determine theta-temperatures except for lower molecular weight polymers.

(d)   Cloud Point Titration (CT)

Polymer solutions of different concentrations are titrated with a non-solvent until the first sign of cloudiness. The logarithm of the non-solvent concentration at the cloud point is then plotted against the logarithm of the polymer concentration at the cloud point and extrapolated to 100 % polymer (28). The solvent/non-solvent mixture at this point corresponds to a theta-mixture (29). A knowledge of the molecular weight is not necessary.

(e)   Viscosity-Molecular Weight Relationship (VM)

The method makes use of the fact that the exponent, a, in the Staudinger-Mark-Houwink equation

$$[\eta] = K \cdot M^a \qquad (7)$$

is equal to 0.5 for a random coil in a theta-solvent. A series of polymers of the same type with widely different known molecular weights is used to determine intrinsic viscosities $[\eta]$ at different temperatures. The theta-temperature can be determined either by direct experiment (VM(T-E)), or if it is not in the measurable range by calculation (VM(T-C)) (see Ref. 44, Chapt. XIV).

An alternative method is to change the solvent/non-solvent ratio at a given temperature and to measure intrinsic viscosities of different polymers in different solvent mixtures (6), however, the exponent "a" does not depend linearly on solvent/non-solvent composition.

References page IV-170

The methods suitable for the determination of theta-solvent or theta-compositions are compared below:

| Method | Number of Polymer Samples Required | Conditions for Application of the Method | | Determined Value[+)] |
		Knowledge of Polymer Molecular Weight Required	Type of Solvent	
Phase equilibrium (PE)	$\geqq$ 3	yes	preferably single	θ-temp.
Virial coefficient (A_2)	1	given by the applied method	single and mixed	θ-temp. or θ-comp.
Cloud point titration (CT)	1	no	mixed	θ-comp.
Viscosity-molecular-weight-relationship (VM)	$\geqq$ 3	yes	preferably mixed	θ-comp.

+) temp. = temperature; comp. = composition.

In the following table, theta-solvents for various polymers have been compiled from the literature. They are subdivided into a number of polymer groups. Within each subgroup, the polymers are listed in alphabetical order. A compound like poly(p-bromostyrene) is therefore found in subgroup 1.4. Poly(vinyls) under bromo... and not under -styrene. The structural units of copolymers are also given in an alphabetical order. Poly(butadiene-co-styrene) is placed therefore after poly(butadiene). Making use of this arrangement, the suffixed ortho (o), meta (m), para (p), normal (n), etc., were neglected. However, the suffices di-, tri-, are used for alphabetical arrangement.

Within each group, the atactic polymers are followed by the isotactic and syndiotactic polymers. The microtacticity is expressed in diads.

The theta-solvents are arranged in increasing order of their theta-temperatures. The components of mixed solvents are given in the order: solvent, followed by the non-solvent, or the worse solvent. The composition of mixed solvents is given on a volume/volume basis except where otherwise noted.

Abbreviations used :

A_2 :	Second virial coefficient	S :	Sedimentation coefficient
Arch :	Archibald's Method	SE :	Sedimentation equilibrium
CT :	Cloud point titration	VM :	Viscosity-molecular weight relationship
D :	Diffusion coefficient	VM(T-E) :	θ-Temperature, determined by using VM
LS :	Light scattering	VM(T-C) :	θ-Temperature, calculated using VM
OP :	Osmotic pressure	+ :	θ-Temperatures of exothermic solutions
PE :	Phase equilibrium	- :	θ-Temperatures of endothermic solutions.

B. TABLE OF θ-SOLVENTS FOR POLYMERS

| Polymer | Theta-Solvent | | θ-Temp. (^{o}C) | | Method | References |
	Name	Composition				
1. Main-chain Carbon Polymers						
1.1 Poly(dienes)						
Poly(1,4-butadiene)						
97 % cis	n-Heptane		-1	(-)	PE	200
90 % cis	Hexane/heptane	50 /50	5	(-)	PE, s	91
94.6 % cis	3-Pentanone		10.3	(-)	PE	166
90 % cis			10.6	(-)	PE	1
90 % cis	5-Methyl-2-hexanone		12.6	(-)	PE	1
90 % cis	Hexane/heptane	75 /25	20	(-)	PE	91
90 % cis	Isobutyl acetate		20.5	(-)	VM, A_2(OP)	24
94.6 % cis	5-Methyl-2-hexanone/ 2-pentanone	3 :1	22.3	(-)	PE	166
94.6 % cis	3-Pentanone/2-pentanone	3 :2	30.0	(-)	PE	166
	Butanone		30	(-)	VM	201
94.6 % cis	5-Methyl-2-hexanone/ 2-pentanone	1 :1	32.7	(-)	PE	166
97 % cis	n-Propyl acetate		35.5	(-)	A_2(OP), VM	200
96.4 % cis	5-Methyl-2-hexanone/ 2-pentanone	1 :3	46.2	(-)	PE	166
90 % cis	2-Pentanone		59.7	(-)	PE	1
Poly(butadiene-co-styrene)						
% butadiene 76.1	2-Pentanone		21	(-)	PE	202

Polymer	Theta-Solvent Name	Composition	θ-Temp. ($^{\circ}$C)		Method	References
Poly(butadiene-co-styrene) (Cont'd.)						
% butadiene 75	n-Octane		21	(-)	VM	199
70			21	(-)	PE, VM, s, D	92
76.1	Methyl isobutyl ketone		46	(-)	PE	202
Poly(bromo-1,4-isoprene) cis						
30.2 double bonds subst.	Cyclohexane		~ 20	(-)	A_2(OP, LS)	101
Poly(chloroprene)	Butanone		25	(-)	A_2(LS, OP)	203
	Cyclohexane		45.5	(-)	PE	203
	Cyclopentane		56.3	(-)	PE	203
Poly(1,4-isoprene)						
cis (natural rubber)	2-Pentanone		14.5	(-)	PE	109
	Hexane/isopropanol	1:1	21	(-)	VM	199
	Butanone		25	(-)	VM, s	204
96 % cis	n-Heptane/n-propanol	69.5/30.5 (w/w)	25	(-)	A_2(LS), CT, VM	18
trans (gutta percha)	n-Propyl acetate		60	(-)	PE	109
Poly(1,2-co-3,4-isoprene)						
35 % 1,2, 65 % 3,4	Benzene/isopropanol	55 /45	20	(-)	CT	35
Poly(isoprene-block-co-styrene)	Butanone		44	(-)	VM	226
1.2 Poly(alkenes)						
Poly(acenaphthylene)	Ethylene dichloride		20±2	(-)	PE, VM	127
			30	(-)	A_2	142
			35	(-)	A_2(LS), VM	143
Poly(1-butene)	Toluene		-46	(-)	VM	144
atact.	Isoamyl acetate		23	(-)	VM	144
	Phenetole		61	(-)	VM	144
	Anisole		83	(-)	VM	144
			86.2	(-)	PE	67
isotact.	Cyclohexane/n-propanol	69 /31	35	(-)	VM	129
	Anisole		89.1	(-)	PE	67
Poly(isobutene)	Ethylbenzene		-24.0	(-)	PE	48
	Toluene		-13.0	(-)	PE	48
	Chlorobenzene/n-propanol	79.7/20.3	14.0	(-)	CT	37
	Chloroform/n-propanol	79.5/20.5	14.0	(-)	CT	37
	Ethyl n-caprylate		22	(-)	PE	85
	Benzene		24	(-)	VM	106
			24.0	(-)	PE, VM	48
	Carbon tetrachloride/butanone	66.4/33.6	25.0	(-)	CT	37
	Carbon tetrachloride/dioxane	63.8/36.2	25.0	(-)	CT	37
	Chlorobenzene/n-propanol	76.0/24.0	25.0	(-)	CT	37
	Chloroform/n-propanol	77.1/22.9	25.0	(-)	CT	37
	Cyclohexane/butanone	63.2/36.8	25.0	(-)	CT	37
	Cyclohexane/dioxane	45.1/54.9	25.0	(-)	CT	37
	n-Hexane/n-butanol	76.4/23.6	25.0	(-)	CT	37
	n-Hexane/butanone	63.4/36.6	25.0	(-)	CT	37
	n-Hexane/n-decanol	58.1/41.1	25.0	(-)	CT	37
	n-Hexane/dioxane	51.8/48.2	25.0	(-)	CT	37
	n-Hexane/n-heptanol	62.6/37.4	25.0	(-)	CT	37
	n-Hexane/n-hexanol	68.3/31.7	25.0	(-)	CT	37
	n-Hexane/3-methyl-2-butanone	57.6/42.4	25.0	(-)	CT	37
	n-Hexane/n-octanol	63.7/36.3	25.0	(-)	CT	37
	n-Hexane/n-pentanol	71.7/28.3	25.0	(-)	CT	37
	n-Hexane/n-propanol	80.3/19.7	25.0	(-)	CT	37
	Methyl cyclohexanone/n-butanol	70.8/29.2	25.0	(-)	CT	37
	Methylcyclohexane/n-decanol	52.5/47.5	25.0	(-)	CT	37
	Methylcyclohexane/dioxane	49.0/51.0	25.0	(-)	CT	37
	Methylcyclohexane/n-heptanol	60.5/39.5	25.0	(-)	CT	37
	Methylcyclohexane/n-octanol	56.0/44.0	25.0	(-)	CT	37
	Methylcyclohexane/n-pentanol	65.2/34.8	25.0	(-)	CT	37
	Methylcyclohexane/n-propanol	74.2/25.8	25.0	(-)	CT	37
	Toluene/cyclohexanol	70.7/29.3	25.0	(-)	CT	37

Polymer	Theta-Solvent Name	Composition	θ-Temp. (°C)		Method	References
Poly(isobutene) (Cont' d.)	Ethylbenzene/diphenyl ether	75 /25	26.8	(-)	PE	48
	Ethyl heptanoate		33	(-)	PE	48
	Chlorobenzene/n-propanol	67.5/32.5	49.0	(-)	CT	37
	Chloroform/n-propanol	71.4/28.6	49.0	(-)	CT	37
	Methylcyclohexane/n-butanol	57.9/42.1	49.0	(-)	CT	37
	Ethyl caproate		57	(-)	PE	48
	Ethylbenzene/diphenyl ether	50 /50	76.0	(-)	PE	48
	Phenetole		86.0	(-)	PE, VM	48
	Anisole		105.5	(-)	PE, VM	48
	Diphenyl ether		148	(-)	PE	48
Poly(ethylene)	n-Pentane		~ 85	(+)	PE	145
	Diphenylene oxide		~ 118	(-)	PE	146
	Biphenyl		125	(-)	PE	146
			125	(-)	A_2(LS)	147
			127.5	(-)	PE	148
	n-Hexane		133	(+)	PE	145
	1-Dodecanol		137.3	(-)	PE	148
			138	(-)	PE, VM	146
	Diphenylmethane		142.2	(-)	PE	148
	Bis(2-ethylhexyl) adipate		145	(-)	VM	133
			145	(-)	PE	149
			170	(-)	PE	146
	Bis(2-ethylhexyl) sebacate		150	(-)	PE	146
	n-Decanol		153.3	(-)	PE	148
	Anisole		153.5	(-)	PE	148
	Diphenyl ether		161.4	(-)	PE	112
			163.9	(-)	PE	148
			~ 165	(-)	A_2(LS)	147
	p-Nonyl phenol		162.4	(-)	PE	148
	p-Octyl phenol		174.5	(-)	PE	148
	n-Octanol		180.1	(-)	PE	148
	Benzyl phenyl ether		191.5	(-)	PE	148
	p-tert-Amyl alcohol		199.2	(-)	PE	148
	Nitrobenzene		> 200	(-)	PE	146
	Dibutyl phthalate		> 200	(-)	PE	146
Poly(1-hexene sulfone) 1:1 atact.	Butanone/isopropanol	41.5/58.5	4±5	(-)	A_2(LS), PE	57
	Butanone/n-hexane	29.8/70.2	8±2	(-)	PE	150
	n-Hexylchloride		13±2	(-)	A_2(LS), PE	57
			13±1	(-)	A_2(LS)	150
	Dioxane/n-hexane	40 /60	20.5±1	(-)	VM	151
	Butanone/isopropanol	37 /63	23.5±2	(-)	A_2(LS), PE	57
		37 /63	24±2	(-)	PE	150
Poly(2-methyl-1-pentene sulfone)	Butanone/n-hexane	35.4/64.6	11.5±2	(-)	PE	150
	Butanone/isopropanol	39.5/60.5	22.5±2	(-)	PE	150
Poly(1-octene)	Phenetole		50.4	(-)	PE	152
	n-Pentane		162.5	(+)	PE	153
Poly(1-pentene) atact.	Phenetole		48.3	(-)	A_2(OP)	69
isotact.	Isoamyl acetate		31-32	(-)	VM	154
	Phenetole		55.8	(-)	A_2(OP)	69
	2-Pentanol		62.4	(-)	PE	46
Poly(propylene) atact.	Carbon tetrachloride/n-propanol	74 /26	25.0	(-)	CT	32
	Carbon tetrachloride/n-butanol	67 /33	25.0	(-)	CT	32
	n-Hexane/n-butanol	68 /32	25.0	(-)	CT	32
	n-Hexane/n-propanol	78 /22	25.0	(-)	CT	32
	Methylcyclohexane/n-propanol	69 /31	25.0	(-)	CT	32
	Methylcyclohexane/n-butanol	66 /34	25.0	(-)	CT	32
	i-Amyl acetate		34	(-)	VM	23
	n-Amyl acetate		36.6	(-)	A_2(OP)	155
	n-Butyl acetate		58.5	(-)	A_2(OP)	155
	i-Butyl acetate		65.5	(-)	A_2(OP)	155
	1-Chloronaphthalene		68	(-)	A_2(LS)	103
			74	(-)	PE	60
	n-Propyl acetate		85.5	(-)	A_2(OP)	155

Polymer	Theta-Solvent Name	Composition	θ-Temp. (°C)	Method	References
Poly(propylene) atact. (Cont'd.)	Cyclohexanone		92	(-)	PE	60
	Diphenyl ether		153	(-)	PE, VM	60
			153.3	(-)	PE	61
isotact.	i-Amyl acetate		70	(-)	A_2(Arch)	131
	Biphenyl		125.1	(-)	PE	156
	Diphenyl ether		142.8	(-)	PE	156
			145	(-)	PE, VM	60
			146.2	(-)	PE	61
	Dibenzyl ether		183.2	(-)	PE	156
syndiotact.	i-Amyl acetate		34	(-)	A_2(Arch)	131
Poly(tetrafluoroethylene-co-trifluoronitroso-methane) 1:1 atact.	Trichlorotrifluoroethane (Freon 113)		35	(-)	VM	84

1.3 Poly(acrylics)

Polymer	Theta-Solvent Name	Composition	θ-Temp. (°C)	Method	References
Poly(acrylic acid) atact.	Dioxane		30	(-)	PE, A_2(OP)	85
Poly(acrylonitrile-co-butadiene) 2/3	Butanone/isopropanol	60 /40	20	(-)	s, D	185
	Butanone/acetonitrile	21 /79	19-22	(-)	s, D	185
	Benzene/butanone/cyclohexane	85.5/4.5/10	19-22	(-)	s, D	185
	Ethyl acetate/cyclohexane	67.3/32.7	19-22	(-)	s, D	185
	Butanone/cyclohexane	47.5/52.5	22	(-)	s, D	185
Poly(acrylonitrile-co-styrene) atact. ca. (1/2) = (mol/mol)	Benzene/methanol	66.7/33.3	25	(-)	CT	30
	Dimethylformamide/methanol	44.7/55.3	25	(-)	CT	30
Poly(n-butyl acrylate)	Benzene/methanol	52 /48	25	(-)		186
Poly(n-butyl acrylate-co-vinylidene chloride) atact. 16.7/83.3	Benzyl alcohol		44	(-)	PE	2
Poly(n-butyl methacrylate) atact.	Dimethylformamide		23.6	(-)	PE	187
	Isopropanol		21.5	(-)	A_2(LS), VM	12
			23.7	(-)	PE, A_2(LS)	16
			25	(-)	A_2(LS)	188
	Methyl cellosolve		59.6	(-)	PE	187
	Methyl carbitol		87.9	(-)	PE	187
isotact.	Dimethylformamide		48.1	(-)	PE	187
	Methyl cellosolve		74.9	(-)	PE	187
	Methyl carbitol		115.0	(-)	PE	187
Poly(cyclohexyl methacrylate) atact.	n-Butanol		23.0	(-)	VM	53
Poly(dodecyl methacrylate)	n-Amyl alcohol		29.5	(-)	VM, A_2	123
Poly(ethyl acrylate)	Methanol		20.5	(-)	PE	223
	Ethanol		37.4	(-)	PE	223
	n-Propanol		39.5	(-)	PE	223
	n-Butanol		44.9	(-)	PE	223
Poly(ethyl acrylate-co-vinylidene chloride) 14.9/85.1	Ethyl acetoacetate		49.6	(-)	PE	2
Poly(2-ethylbutyl methacrylate) atact.	Isopropanol		27.4	(-)	PE, A_2(LS)	25
Poly(ethyl methacrylate) atact.	m-Xylene		-3	(-)	VM	189
	Methyl n-propyl ketone		-1	(-)	VM	189
	n-Butyl bromide		0	(-)	VM	189
	Butanone/isopropanol	12.5/87.5	23	(-)	PE, A_2(LS)	14
	Isopropanol		36.9	(-)	PE, A_2(LS)	16
Poly(n-hexyl acrylate-co-vinylidene chloride) 14.5/85.5	Benzyl alcohol		56.8	(-)	PE	2
Poly(n-hexyl methacrylate) atact.	Isopropanol		32.6	(-)	PE, A_2(LS)	10
Poly(N-isopropylacrylamide)	Water		31	(+)	PE	117
Poly(n-lauryl methacrylate) atact.	Isopropyl acetate		13	(-)	PE, A_2(LS)	10a
	n-Pentanol		29.5	(-)	VM	72
			29.5	(-)	PE, A_2(LS)	71
Poly(methacrylic acid) atact.	0.002M HCl in water		30	(-)	VM	58
	0.5N NaCl in water		43	(-)	A_2(LS)	190
	0.05N NaCl in water		68	(-)	A_2(LS)	190
Poly(methacrylonitrile) atact.	Dimethylformamide		29.2	(-)	VM	90
Poly(methyl n-butacrylate)	n-Butanol		13	(-)	A_2(LS)	191
Poly(methyl ethacrylate)	Diisobutyl ketone		11.4	(-)	A_2(LS)	191

Abbreviations page IV-159; References page IV-170

Polymer		Theta-Solvent		θ-Temp. (°C)		Method	References
		Name	Composition				
1.3	Poly(acrylics) (Cont'd.)						
Poly(methyl methacrylate)	atact.	Chloroform		-273±50	(-)	VM(T-C)	47
		Dichloroethane		-233±50	(-)	VM(T-C)	47
		Benzene		-223±50	(-)	VM(T-C)	47
		Methyl methacrylate		-163±50	(-)	VM(T-C)	47
		Butanone		~ -98	(-)	VM(T-C)	47
		Ethyl acetate		~ -98	(-)	VM(T-C)	47
		Toluene		-65±10	(-)	VM(T-C)	47
		Acetone		-55±10	(-)	VM(T-C)	47
		Methyl isobutyrate		-53±10	(-)	VM(T-C)	47
		2-Methyl-4-pentanone		-42	(-)	A_2(LS)	63
		Methyl isovalerate		-37	(-)	A_2(LS)	63
		Butyl acetate		-20	(-)	A_2(LS)	63
		Butanone/isopropanol	58.2/41.8	4	(-)	A_2(LS) (80 % syn)	63
			55 /45	7.0	(-)	A_2(LS) (80 % syn)	102
		2-Heptanone		~ 11	(-)	VM(T-C)	47
		Butanone/isopropanol	55 /45	12.8	(-)	A_2(LS)	63
		Nitrobenzene/isopropanol	51.6/48.4 (w/w)	18	(-)	VM	141
		Benzene/n-hexane	70 /30	20	(-)	CT	52
		Benzene/isopropanol	38 /62	20	(-)	CT	52
		Dioxane/n-hexane	59 /41	20	(-)	CT	35
		3-Methyl-2-butanone/n-hexane	83 /17	20	(-)	CT	36
		Dioxane/cyclohexane	53 /47	20	(-)	A_2, VM	167
		Cyclohexanone/isopropanol	51.6/48.4	21-22	(-)	VM	192
		2-Ethylbutyraldehyde		22	(-)	PE	47
		Butanone/isopropanol	50 /50	22.8	(-)	A_2(LS)	63
		trans-Decalin		23.5	(-)	VM	141
		m-Xylene		24	(-)	VM	141
		Dioxane + 15 % water		25	(-)	A_2(LS)	193
		Acetone/ethanol	47.7/52.3	25	(-)	CT, A_2(LS)	34
		Acetone/methanol	78.1/21.9	25	(-)	CT	34
		Butanone/cyclohexane	59.5/40.5	25	(-)	CT, A_2(LS)	34
		Butanone/n-hexane	70.7/29.3	25	(-)	CT, A_2(LS)	34
		Butanone/isopropanol	50 /50	25	(-)	PE (7 % isot.)	95
			50 /50	25	(-)	VM(T-C)	7
			50 /50	25	(-)	A_2(LS)	11
			58.2/41.8	25	(-)	A_2(LS)	34
		Carbon tetrachloride/n-hexane	99.4/ 0.6	25	(-)	CT	34
		Carbon tetrachloride/methanol	53.3/46.7	25	(-)	CT	34
		Dioxane/cyclohexane	53.4/46.6	25	(-)	CT, A_2(LS)	34
		3-Heptanone		~ 25	(-)	VM(T-C)	47
		Toluene/n-hexane	81.2/18.8	25	(-)	CT	34
		Toluene/methanol	35.7/64.3	26.2	(-)	PE, A_2(LS)	16
		Carbon tetrachloride		~ 27	(-)	VM(T-C)	47
		Butanone/isopropanol	46.8/53.2	28.5	(-)	A_2(LS)	63
		Acetonitrile		30	(-)	VM, A_2(LS)	20
		Butyl chloride		32.6	(-)	A_2(LS)	62
		3-Heptanone		33.7	(-)	PE (7 % isot.)	95
		4-Heptanone		33.8	(-)	PE	47
				34±10	(-)	VM(T-C)	47
		n-Butyl chloride		35.0	(-)	A_2(LS) (80 % syn)	102
		2,2-dimethyl-4-pentanone		~ 35	(-)	PE	47
		n-Butyl chloride		35.4	(-)	A_2(LS)	63
		4-Heptanone		40.4	(-)	A (LS)	63
		Amyl acetate		~ 41	(-)	VM(T-C)	47
		2,4-Dimethyl-3-pentanone		41	(-)	VM(T-C)	7
		Acetonitrile		45	(-)	PE, VM(T-C)	47
		2,4-Dimethyl-3-pentanone		46	(-)	PE	47
		2-Octanone		52	(-)	PE	47
		Isoamyl acetate		57.5	(-)	A_2(LS)	63
		3-Octanone		72	(-)	PE	47
	7 % i	n-Propanol		84.4	(-)	PE	95

Polymer			Theta-Solvent Name	Composition		θ-Temp. (°C)		Method	References
Poly(methyl methacrylate) (Cont'd.)									
7 % i			p-Cymene			159.7	(-)	PE	95
94 % i			Butanone/isopropanol	55	/45	25.0	(-)	A₂(LS)	102
94 % i			n-Butyl chloride			26.5	(-)	A₂(LS)	102
			Acetonitrile			27.6	(-)	PE	64
90 % i			Butanone/isopropanol	50	/50	30.3	(-)	PE	95
90 % i			3-Heptanone			40.0	(-)	PE	95
90 % i			n-Propanol			75.9	(-)	PE	95
90 % i			p-Cumene			152.1	(-)	PE	95
100 % s			Butanone/isopropanol	55	/45	8.0	(-)	A₂(LS)	102
100 % s			n-Butyl chloride			26.5	(-)	A₂(LS)	102
100 % s			n-Propanol			85.2	(-)	PE	95
Poly(methyl methacrylate-block-co- α-methylstyrene)									
			2-Methyl cyclohexanol			62	(-)	PE	137
			3-Heptanone			84	(-)	PE	137
			1-Chloro-n-hexane			93	(-)	PE	137
			2-Octanone			94	(-)	PE	137
Poly(methyl methacrylate-co-styrene)									
atact.	% MMA = 73.9		Benzene/n-hexane	62	/38	20	(-)	CT	36
	= 57.7			59	/41	20	(-)	CT	36
	= 41.9			51	/49	20	(-)	CT	36
	= 23.7			44	/56	20	(-)	CT	36
	= 73.9		Benzene/isopropanol	41	/59	20	(-)	CT	36
	= 57.7			48	/52	20	(-)	CT	36
	= 41.9			51	/49	20	(-)	CT	36
	= 23.7			57	/43	20	(-)	CT	36
	= 73.9		3-Methyl-2-butanone/n-hexane	76	/24	20	(-)	CT	36
	= 57.7			71	/29	20	(-)	CT	36
	= 41.9			66	/34	20	(-)	CT	36
	= 23.7			60	/40	20	(-)	CT	36
random	% sty.	28.5	2-Ethoxyethanol			40.0	(-)	A₂(OP)	194
		29.3				40.0	(-)	A₂(OP), VM	196
		55.2				58.4	(-)	A₂(OP)	194
		56.2				58.4	(-)	A₂(OP), VM	196
		35.9				69.5	(-)	A₂(OP)	195
		69.4				72.8	(-)	A₂(OP)	194
		70.2				72.8	(-)	A₂(OP), VM	196
		48.6				81.0	(-)	A₂(OP)	195
		55.2	Cyclohexanol			61.3	(-)	A₂(OP)	194
		56.2				61.3	(-)	A₂(OP), VM	196
		52				61.6	(-)	A₂, VM	197
		69.4				63.0	(-)	A₂(OP)	194
		70.2				63.0	(-)	A₂(OP), VM	196
		29.3				68.0	(-)	A₂(OP), VM	196
		28.5				68.2	(-)	A₂(OP)	194
		50.0				68.6	(-)	A₂	135
		35.9				80.5	(-)	A₂(OP)	195
		48.6				81.3	(-)	A₂(OP)	195
block	% sty.	35.9	2-Ethoxyethanol			69.5	(-)	A₂(OP)	194
		48.6				81.0	(-)	A₂(OP)	194
		49.6				81.0	(-)	A₂(OP), VM	196
		35.9	Cyclohexanol			80.5	(-)	A₂(OP)	194
		50				81.1	(-)	A₂, VM	197
		48.6				81.3	(-)	A₂(OP)	194
		49.6				81.3	(-)	A₂(OP), VM	196
		50				81.6	(-)	A₂	135
		various				82	(-)	PE	137
		72.4				84	(-)	A₂(OP)	194
		73.2				84	(-)	A₂(OP), VM	196
		84.6				84	(-)	A₂(OP)	194
		85.1				84	(-)	A₂(OP), VM	196
alternat.	% Sty.	50	Cyclohexanol			60.8	(-)	A₂, VM	197

Polymer			Theta-Solvent Name	Composition	θ-Temp. (°C)		Method	References
Poly(n-octyl acrylate-co-vinylidene								
chloride)	atact.	15.5/84.4	Benzyl alcohol		77.9	(-)	PE	2
Poly(n-octyl methacrylate)	atact.		n-Butanol		16.8	(-)	PE, A (LS)	13
Poly(isooctyl methacrylate)	atact.		Acetone/n-heptane	64.1/35.9	25	(-)	CT, OP	198
Poly(isooctyl methacrylate-co-styrene)								
	20/80		Butanone/methanol	99.4/ 0.6	25	(-)	CT, OP	198
Poly(isopropyl acrylate)	atact.		1,2-Butanediol/1,3-butanediol	68.4/31.6 (w/w)	121.0	(-)	PE	75
			n-Decane		166.6	(-)	PE	75
	isotact.		1,2-Butanediol/1,3-Butanediol	68.4/31.6 (w/w)	123.5	(-)	PE	110
			n-Decane		178.0	(-)	PE	110
	syndiot.		n-Decane		168.3	(-)	PE	75

1.4 Vinyl Polymers

Polymer		Theta-Solvent Name	Composition	θ-Temp. (°C)		Method	References
Poly(p-bromostyrene)	atact.	Benzene		20	(-)	VM (a = 0.53)	59
Poly(o-chlorostyrene)		Butanone		25	(-)	PE, VM	134
Poly(p-chlorostyrene)		Isopropyl monochloroacetate		-8.2	(-)	PE	114
		Ethyl benzene		-7.5	(-)	PE	114
		Ethyl monochloroacetate		-1.3	(-)	PE	114
		Carbon tetrachloride/toluene	1:1	13.4	(-)	PE	157
			5:2	32.0	(-)	PE	157
		Carbon tetrachloride		52.3	(-)	PE	114
				58.9	(-)	PE	157
		Cumene		60.9	(-)	PE	114
		Methyl monochloroacetate		65.0	(-)	PE	114
		Isopropyl acetate		65.0	(+)	PE	114
		Ethyl acetate		117	(+)	VM(T-C)	158
		n-Butyl acetate		337	(+)	VM(T-C)	158
Poly(p-chlorostyrene-co-methyl meth-							
acrylate)	51.6/48.4	trans-Decalin/trichloroethylene	78 /22	22.3	(-)	VM	159
Poly(2,5-dichlorostyrene)	atact.	Ethyl acetate/ethanol	93.7/ 6.3 (w/w)	30.5	(-)	A_2(LS)	40
Poly(3,4-dichlorostyrene)	atact.	Butyl acetate/butanol	92.9/ 7.1 (w/w)	32.9	(-)	A_2(LS)	39
Poly(p-methoxystyrene)		Methyl isobutyl ketone		23.4	(-)	CT	132
		t-Butylbenzene		52.2	(-)	PE	132
		i-Amyl acetate		75.0	(-)	PE	132
		Dichlorodecane		92.6	(-)	PE	132
Poly(α-methylstyrene)		trans-Decalin		9.5	(-)	VM	130
		Benzene/methanol	79.4/20.6	30	(-)		86
		Cyclohexane		32.3	(-)	A_2(LS)	160
				33.1	(-)	A_2(LS)	160
				34.3	(-)	A_2(LS)	160
				34.5	(-)	VM	130
				36.8	(-)	A_2(LS)	160
				37.0	(-)	VM	161
				38	(-)	VM, A_2(Arch)	162
				39	(-)	A_2(LS), s	163
Poly(2-methyl-5-vinylpyridine)		Propionitrile		-3.6	(-)	PE	164
		n-Propyl acetate		19.3	(-)	PE, VM	164
		n-Butyl acetate		21.8	(-)	PE, VM	164,165
		Ethyl propionate		25.4	(-)	PE	164
		Isobutyl methyl ketone		37.4	(-)	VM	165
				38.7	(-)	PE	164
		n-Amyl acetate		48.2	(-)	VM	165
				48.4	(-)	PE	164
		Isobutyl acetate		49.0	(-)	PE	164
		Tetrahydronaphthalene		49.5	(-)	PE	164

For Poly(α-methylstyrene):

$x^\star_{i/s} = 0.1, \quad x_{s/s} = 0.9$

$x_{i/s} = 0.19, \quad x_{s/s} = 0.81$

$x_{i/i} = 0.03, \quad x_{i/s} = 0.29,$

$x_{s/s} = 0.68$

$x_{i/i} = 0.11, \quad x_{i/s} = 0.45,$

$x_{s/s} = 0.44$

$x_s = 0.64$

$x_s = 0.64$

★ Explanation : see page II-461

Polymer		Theta-Solvent		θ-Temp. (°C)		Method	References
		Name	Composition				
Poly(2-methyl-5-vinylpyridine) (Cont'd.)		Ethyl-n-butyrate		50.0	(-)	PE	164
		i-Amyl acetate		53.2	(-)	PE	164
		n-Propyl propionate		58.0	(-)	PE	164
Poly(p-isopropylstyrene)	atact.	Dioxane/isopropanol	35 /65	20	(-)	CT	35
Poly(styrene)	atact.	n-Butyl formate		-9	(-)	A_2	169
		1-Chloro-n-decane		+6.6	(-)	A_2(LS)	170
		Hexyl-m-xylene		12.5	(-)	A_2	169
		Cyclohexane/toluene	86.9/13.1	15	(-)	PE	105
		trans-Decalin		18.2	(-)	A_2(LS)	136
		trans-/cis-Decalin	79.6/23.1	19.3	(-)	PE	87
		trans-Decalin		23.8	(-)	A_2(LS)	172
		Benzene/n-hexane	39 /61	20	(-)	CT	52
		Benzene/isopropanol	66 /34	20	(-)	CT	35
		Dioxane/n-hexane	38 /62	20	(-)	CT	52
		Dioxane/isopropanol	55 /45	20	(-)	CT	35
		3-Methyl-2-butanone/n-hexane	52 /48	20	(-)	CT	36
		Butanone/isopropanol	85.7/14.3	23	(-)	A_2(LS, OP)	15
		Benzene/cyclohexanol	38.4/61.6	25	(-)	CT, A_2(LS)	33
		Benzene/n-hexane	34.7/65.3	25	(-)	CT, A_2(LS)	33
		Benzene/methanol	77.8/22.2	25	(-)	CT, A_2(LS)	33
		Benzene/isopropanol	64.2/35.8	25	(-)	CT, A_2(LS)	33
		Butanone/methanol	88.7/11.3	25	(-)	CT, A_2(LS)	33
			89 /11	25	(-)	A_2(OP), VM	89
		Carbon tetrachloride/methanol	81.7/18.3	25	(-)	CT, A_2(LS)	33
		Chlorobenzene/diisopropyl ether	32 /68	25	(-)	CT, A_2(LS)	33
		Chloroform/methanol	75.2/24.8	25	(-)	CT, A_2(LS)	33
			74.7/25.3	25	(-)	A_2(OP), VM	89
		Dioxane/methanol	71.4/28.6	25	(-)	CT, A_2(LS)	33
		Tetrahydrofuran/methanol	71.3/28.7	25	(-)	CT, A_2(LS)	33
		Toluene/methanol	20 /80	25	(-)	A_2(OP), VM	89
		Decalin		29.5	(-)	A_2	169
		Butanone/methanol	88.9/11.1	30	(-)	PE	105
		Toluene/n-heptane	47.6/52.4	30	(-)	PE	105
		Decalin		31	(-)	A_2(LS)	93
		Diethyl malonate		31	(-)	A_2	169
				34.2	(-)	VM	166
				35.9	(-)	A_2	171
		1-Chloro-n-undecane		32.8	(-)	A_2	170,171
		Benzene/methanol	74.0/26.0	34	(-)	VM	5
		Butanone/isopropanol	82.6/17.4	34	(-)	VM	5
		p-Dioxane/methanol	65.1/34.9	34	(-)	VM	5
		Toluene/methanol	75.2/24.8	34	(-)	VM	5
		Cyclohexane		34	(-)	A_2	169
				34	(-)	PE	66
				34.4-		A_2(LS), VM	65
				35.4	(-)		
				34.5	(-)	VM	173
				34.5	(-)	VM	166
				34.8	(-)	A_2	171
				35.0	(-)	A_2(LS)	88
				35	(-)	s, D	9
				35	(-)	D	81
				35	(-)	SE	78
				35	(-)	s	54
		Benzene/n-butanol	58 /42	35	(-)	A_2(LS)	168
		Benzene/heptane	44 /56	35	(-)	A_2(LS)	168
		Benzene/methanol	74.7/25.3	35	(-)	A_2(LS)	168
		Benzene/isopropanol	61 /39	35	(-)	A_2(LS)	168
		Carbon tetrachloride/n-butanol	65 /35	35	(-)	A_2(LS)	168
		Carbon tetrachloride/heptane	53 /47	35	(-)	A_2(LS)	168
		Dioxane/heptane	41.5/58.5	35	(-)	A_2(LS)	168
		Dioxane/methanol	66.5/33.5	35	(-)	A_2(LS)	168
			66.5/33.5	35	(-)	A_2, VM	167

Polymer	Theta-Solvent Name	Composition	θ-Temp. (°C)		Method	References
Poly(styrene) atact. (Cont'd.)						
	Dioxane/isopropanol	51.5/48.5	35	(-)	A_2(LS)	168
	Nitropropane/heptane	42 /58	35	(-)	A_2(LS)	168
	Diethyl malonate/diethyl oxalate	4 : 1	40	(-)	VM	166
		1 : 1	47.4	(-)	VM	166
		1 : 4	52.6	(-)	VM	166
	Cyclohexane/methylcyclohexane	2 : 1	43	(-)	VM	166
		1 : 1	48	(-)	VM	166
		1 : 2	54	(-)	VM	166
	Diethyl oxalate		51.5	(-)	A_2	169
			55.8	(-)	VM	166
	1-Chloro-n-dodecane		58.6	(-)	A_2(LS)	170
	Methylcyclohexane		60	(-)	VM	173
			67.2	(-)	A_2(LS)	174
			68	(-)	A_2	169
			70.5	(-)	VM	166
			70.5	(-)	VM	27
	Butanone/isopropanol	87 /13	67	(-)	A_2(LS)	88
		87 /13	67	(-)	A_2(LS)	50
	Ethylcyclohexane		70	(-)	PE	49
	dl-Terpineol		78.5	(-)	VM	173
	Cyclohexanol		83.5	(-)	A_2	169
			83.5	(-)	VM	173
			83.5	(-)	A_2	135
	3-Methyl-cyclohexanol		98	(-)	VM	173
	dl-Menthol		115	(-)	VM	173
Poly(styrene), comblike branched	Cyclohexane		20.5– 35.5	(-)	A_2(LS), CT	120
Poly(styrene-co-p-isopropylstyrene)						
atact. 66.9 % styrene	Dioxane/isopropanol	48 /52	20	(-)	CT	35
44 % styrene		44 /56	20	(-)	CT	35
12 % styrene		37 /63	20	(-)	CT	35
Poly(vinyl acetate) atact.	Methanol		6	(-)	PE	175
			6	(-)	PE	176
	Ethanol/methanol	80 /20	17	(-)	PE	176
		60 /40	26.5	(-)	PE	176
		50 /50	34	(-)	PE	177
		40 /60	36	(-)	PE	176
	Butanone/isopropanol	73.2/26.8	25	(-)	PE, A_2(LS)	104
	3-Methyl-butanone/n-heptane	73.2/26.8	25	(-)	PE, A_2(LS)	104
	3-Heptanone		29	(-)	PE, A_2(LS)	77
	3-Methyl-butanone/n-heptane	72.7/27.3	30	(-)	PE, A_2(LS)	76
	Acetone/isopropanol	23 /77	30	(-)	PE	177
	6-Methyl-3-heptanone		66	(-)	PE, A_2(LS)	77
Poly(vinyl alcohol)	Water		~ 97	(+)	PE	118
			~ 107	(+)	CT	119
	2 M NaCl in water		101±3	(+)	CT	119
Poly(vinyl alcohol-co-vinyl urethane)	n-Propanol/water					
degree of urethanization (mol %)						
11.5		30 /40	30	(-)	CT	178
8.1		30 /41	30	(+)	CT	178
4.9		30 /40	30	(+)	CT	178
11.5		4 /5	60	(-)	CT	178
8.1		4 /5	60	(+)	CT	178
4.9		4 /5	60	(+)	CT	178
Poly(vinyl benzoate) atact.	Xylene		32.5	(-)	A_2(LS)	96
Poly(vinyl bromide) atact.	Tetrahydrofuran/methanol	83 /17	20	(-)	A_2(LS)	19
Poly(vinyl carbanilate) atact.	Acetone/7.8 % water		20	(-)	A_2(LS)	179
	Butanone/ 56 % water		20	(-)	A_2(LS)	179
	Dioxane/72 % methanol		20	(-)	A_2(LS)	179
	Tetrahydrofuran/71.5 % methanol		20	(-)	A_2(LS)	179
	Diethyl ketone		35	(-)	A_2(LS)	179

Polymer		Theta-Solvent		θ-Temp. (°C)		Method	References
		Name	Composition				
Poly(vinyl chloride)	atact.	Tetrahydrofuran/water	100/11.9	30	(-)	CT	180
			100/ 9.5	30	(-)	CT	180
		Benzyl alcohol		155.4	(-)	PE	97
Poly(vinyl p-chlorobenzoate)	atact.	Butanone/n-butanol	53/47	60	(-)	PE	94
Poly(vinyl cyclohexane)	atact.	Tetrahydrofuran		25	(-)	A_2(LS)	181
Poly(β-vinylnaphthalene)	atact.	Toluene/decalin	43.5/56.5	30.2	(-)	A_2(LS); VM	41
Poly(vinyl pivalate)	atact.	Acetone/methanol	38.2/61.8	20	(-)	CT	26
		Benzene/methanol	33.3/66.7	20	(-)	CT	26
		Butanone/methanol	24.6/75.4	20	(-)	CT	26
Poly(2-vinylpyridine)	atact.	Benzene		15	(-)	VM	128
		Heptane/propanol	59.6/40.4 (w/w)	25.0	(-)	PE	55
Poly(vinylpyrrolidone)		Dioxane		-10	(-)	A_2(LS)	182
		Water/acetone	33.2/66.8	25	(-)	CT, VM	29
Poly(N-vinyl succinamic acid)		0.2N HCl in water		25	(-)	VM	183
Poly(vinylsulfonic acid)	K-salt	0.346M NaBr in water		-0.6	(-)	PE	184
		0.349M KCl in water		5.5	(-)	PE	184
		0.347M KBr in water		5.7	(-)	PE	184
		0.65 M KCl in water		26.0	(-)	PE	184
		1.003M NaCl in water		32.4	(-)	PE	184
		1.008M NaBr in water		40.1	(-)	PE	184
		1.001M KCl in water		44.5	(-)	PE	184
Poly(vinyltriazole)	atact.	Dimethylformamide/dioxane	76.6/23.4	25	(-)	CT	30.73

2. Main-chain Carbon Heteroatom Polymers

2.1 Poly(oxides)

Polymer		Theta-Solvent		θ-Temp. (°C)		Method	References
		Name	Composition				
Poly(oxy-2,6-dimethyl-1,4-phenylene)		Methylene chloride		89.2	(-)	a	205
Poly(oxyethylene)		Acetonitrile/isopropyl ether	45 /55	20	(-)	CT	35
		Chloroform/n-hexane [b]	54 /46	20	(-)	CT	35
			47.4/52.6	20	(-)	CT, A_2(Arch), VM	29
		Nitroethane/isopropyl ether	45 /55	20	(-)	CT	35
		0.45M K_2SO_4 in water		35	(-)	PE	3
				0℃	( )	VM	101
		0.39M $MgSO_4$ in water		45	(-)	PE	3
		Methyl isobutyl ketone		50	(-)	VM	121
		Diethylene glycol diethyl ether		50	(-)	VM	121
Poly(oxyethylethylene)		Isopropanol		29.8	(-)	VM	124
				30	(-)	VM	115
Poly(oxy(2-hydroxytrimethylene)oxy-1,4 phenyleneisopropylidene-1,4- phenylene)		Toluene/cyclohexanone	67.5/32.5	22	(-)	VM	207
		Toluene/tetrahydrofuran	1:2	30	(-)	s	206
Poly(oxypropylene)		Isooctane		50	(-)	A_2(LS)	208
Poly(oxytetramethylene)		Toluene		-28±3	(-)	A_2(OP)	209
		n-Butanol		5	(-)	PE	212
		Butanone		25	(-)	VM	210
		Chlorobenzene		25	(-)	VM	210
		Benzene/acetonitrile	38.5/61.5	25	(-)	CT	31
		Butanone/acetonitrile	61.7/38.3	25	(-)	CT	31
		Carbon tetrachloride/acetonitrile	49.8/50.2	25	(-)	CT	31
		Chlorobenzene/acetonitrile	39.9/60.1	25	(-)	CT	31
		Tetrahydrofuran/acetonitrile	41.3/58.7	25	(-)	CT	31
		Toluene/acetonitrile	39 /61	25	(-)	CT	31
		Heptane/cyclohexane		26	(-)		214
		Ethyl acetate/n-hexane	22.7/77.3	30.4	(-)	PE	211
				∼ 33	(-)	A_2(LS)	213
		Diethyl malonate		33.5	(-)	PE	211
		Isopropanol		44.6	(-)	PE	211

a. Concentration dependence of m.p. in CH_2Cl_2 -   b. Composition depends strongly on quality of n-hexane.

Polymer	Theta-Solvent		θ-Temp. (°C)		Method	References
	Name	Composition				
2.2 Poly(esters)						
Poly(oxycarbonyloxy-1,4-phenyleneiso-propylidene-1,4-phenylene)	Chloroform/ethanol	74.5/25.5	18	(-)	CT	215
	Chloroform		20	(-)	A_2	17
	Dioxane/cyclohexane	63.9/36.1 (w/w)	25	(-)	VM	116
	n-Butyl benzyl ether		170	(-)	A_2(LS)	216
Poly(oxyethyleneoxyadipoyl)	Cyclohexanol		114.5	(-)	PE	111
Poly(oxyethyleneoxysuberoyl)	Cyclohexanol		88	(-)	PE	111
Poly(oxyhexamethyleneoxyadipoyl)	Cyclohexanol		64	(-)	PE	111
Poly(oxytetramethyleneoxyadipoyl)	Cyclohexanol		77	(-)	PE	111
Poly(poly(oxyethylene)oxyadipoyl)						
M_n of PEO = 150	Chloroform/n-hexane	65 /35	20	(-)	CT	35
= 200		63 /37	20	(-)	CT	35
= 300		62 /38	20	(-)	CT	35
= 387		60 /40	20	(-)	CT	35
= 600		59 /41	20	(-)	CT	35
= 2580		57 /43	20	(-)	CT	35
= 7000		56 /44	20	(-)	CT	35
Poly(poly(oxyethylene)oxysebacoyl)						
M_n of PEO = 387	Chloroform/n-hexane	56 /44	20	(-)	CT	35
	Acetonitrile/isopropyl ether	35.8/64.2	20	(-)	CT	35
Poly(poly(oxyethylene)oxysuberoyl)						
M_n of PEO = 387	Chloroform/n-hexane	58 /42	20	(-)	CT	35
	Acetonitrile/isopropyl ether	38.6/61.4	20	(-)	CT	35
Poly(poly(oxyethylene)oxysuccinoyl)						
M_n of PEO = 387	Chloroform/n-hexane	61 /39	20	(-)	CT	35
	Acetonitrile/isopropyl ether	44.6/55.4	20	(-)	CT	35
2.3 Poly(amides)						
Poly(γ-benzyl glutamate)	Dichloroethane/diethylene glycol	80 /20	25	(-)	A_2	140
Poly(iminoadipoyliminohexamethylene) (Nylon 66)	m-Cresol/548 g CCl_4/402 g cyclo-hexane/per liter mixture		20	(-)	s	122
	2.3M KCl in 90 % (vol) formic acid		25	(-)	A_2(LS)	98,38
			25	(-)	VM	38
	Aq. formic acid/KCl +)		25	(-)	A_2(LS)	217
Poly(L-proline)	Water		100	(+)	A_2	138
	Trifluoroethanol		130	(+)	A_2	138
2.4 Poly(ureas) and Poly(urethanes)						
Poly(oxyethyleneoxycarbonylimino-1,4-phenylenemethylene-1,4-phenylene-iminocarbonyl)	Acetone/dimethylformamide	21 /79	25	(-)	VM	218
Poly(ureyleneheptamethylene)	Dichloroacetic acid		46	(-)	VM	43
	90 % (vol) sulfuric acid		46	(-)	VM	43
2.5 Poly(saccharides)						
Amylopectin	Water		25.0	(-)	A_2(LS)	107
Amylose (potato)	Dimethyl sulfoxide/acetone	56.6/43.5	20	(-)	VM	8
	0.33M KCl in water		22.5	(-)	VM	139
	Dimethyl sulfoxide/methanol	49 /51	25.0	(-)	VM	21
	Dimethyl sulfoxide/0.5M KCl	25 /75	25	(-)	VM	21
	0.33M KCl in water		25	(-)	VM	42
	Sodium acetate buffer (0.5N NaOH, 0.5N acetic acid until pH = 5.0)		25	(-)	VM	21
Amylose tricarbanilate	Pyridine/water	86.7/13.3	25	(-)	A_2(LS)	126
Cellulose diacetate	Tetrachloroethane		56.6	(-)	A_2(OP)	224
Cellulose tributyrate	Dodecane/tetralin	75 /25	122	(-)	PE	74
Cellulose tricaproate	Dimethylformamide		41	(-)	A_2(OP)	68
Cellulose tricaprylate	Dimethylformamide		140	(-)	PE	74
	3-Phenylpropanol		48	(-)	PE	74

+) Wt.-% formic acid = 12.1M + 62.2 (M = molarity of KCl).

Polymer	Theta-Solvent Name	Composition	θ-Temp. (°C)		Method	References
2.6 Poly(siloxanes) and Poly(phosphates)						
Poly(metaphosphate)	0.415M aq. NaBr		25	(-)	A_2(LS)	108
Poly(oxydimethylsilylene)						
(Poly(dimethyl siloxane))	Hexane		-173.2	(-)	VM	219
	Heptane		-173.2	(-)	VM	219
	Octane		-143.2	(-)	VM	219
	Nonane		-113.2	(-)	VM	219
	Methylcyclohexane		-113.2	(-)	VM	219
	Methylcyclopentane		-98.2	(-)	VM	219
	Cyclohexane		-81	(-)	A_2(OP)	70
			-68.2	(-)	VM	219
	Xylene		-48.2	(-)	VM	219
			-47	(-)	A_2(OP)	70
	Butyl acetate		-38.2	(-)	VM	219
	Toluene		-33.2	(-)	VM	219
			-30	(-)	A_2(OP)(-)	70
	Propyl acetate		-28.2	(-)	VM	219
	Chlorobenzene		-19.2	(-)	VM	219
	Benzene		-7	(-)	A_2(OP)	70
			-5.2	(-)	VM	219
			-3	(-)	A_2(OP)	70
	Ethyl iodide		2.1	(-)	A_2(LS)	100
	Ethyl acetate		4.8	(-)	VM	219
			18	(-)	A_2(OP)	70
	Butanone		19.8	(-)	VM	219
			20	(-)	PE	45
			20	(-)	A_2(OP)	70
	$C_8F_{18}/C_2Cl_4F_2$	33.17/66.83 (w/w)	22.5	(-)	A_2(LS), PE	22
	Carbon tetrachloride/benzyl alcohol	78.8/21.2	25	(-)	CT	31
	Chloroform/benzyl alcohol	87.2/12.8	25	(-)	CT	31
	Ethylbenzene/benzyl alcohol	85.4/14.6	25	(-)	CT	31
	Methylene chloride/benzyl alcohol	77.3/22.7	25	(-)	CT	31
	Toluene/cyclohexanol	66/34	25	(-)	CT	31
	Toluene/1-nitropropane	64/36	25	(-)	CT, A_2(LS)	31
	Trichloroethylene/benzyl alcohol	78.3/21.7	25	(-)	CT, A_2(LS)	31
	Toluene/benzyl alcohol	83.8/16.2	25	(-)	CT	31
	Toluene/nitroethane	74.1/25.9	25	(-)	CT	31
	Toluene/nitromethane	81.1/18.9	25	(-)	CT	31
	Bromocyclohexane		29.0	(-)	A_2(LS)	100
	Toluene/acetone	5:6	29.5	(-)	A_2(LS)	220
	Bromocyclohexane/phenetole	85.7/14.3	36.3	(-)	A_2(LS)	100
	Bromocyclohexane		36.8	(-)	VM	219
	Chlorobenzene/dimethyl phthalate	88.2/11.8	56.3	(-)	A_2(LS)	100
	Chlorobenzene		68	(-)	A_2(OP)	70
	Bromobenzene		78.3	(-)	VM	219
			78.7	(-)	A_2(LS)	100
	Phenetole		83.0	(-)	PE	45
			83.0	(-)	A_2(OP)	70
			89.3	(-)	VM	219
			89.5	(-)	A_2(LS)	100
Poly(phenyl sesquioxane)	Ethylene dichloride		56	(-)		221
Poly(γ-trifluoropropyl methyl siloxamer)	Cyclohexyl acetate		25	(-)	VM	222
	Methyl hexanoate		72.8	(-)	VM	222

C. REFERENCES

1. M. Abe, H. Fujita, Rep. Prog. Polymer Phys. Japan 7, 42 (1964).
2. M. Asahina, M. Sato, T. Kobayashi, Bull. Chem. Soc. Japan 35, 630 (1962).
3. F. E. Bailey, Jr., R. W. Callard, J. Appl. Polymer Sci. 1, 56 (1959).
4. U. Baumann, H. Schreiber, K. Ressmar, Makromol. Chem. 36, 81 (1960).
5. U. Bianchi, V. Magnasco, C. Rossi, Chim. Ind. (Milan) 40, 263 (1958).
6. U. Bianchi, V. Magnasco, C. Rossi, Ric. Sci. 28, 1412 (1958).
7. M. Bohdanecky, Collection Czech. Chem. Comuun. 29, 876 (1964).
8. W. Burchard, Makromol. Chem. 64, 110 (1963).
9. H. J. Cantow, Makromol. Chem. 30, 169 (1959).

10. S. N. Chinai, J. Polymer Sci. 25, 413 (1957).

10a. S. N. Chinai, J. Polymer Sci. 44, 475 (1959).

11. S. N. Chinai, Ch. W. Bondurant, J. Polymer Sci. 22, 555 (1956).

12. S. N. Chinai, R. A. Guzzi, J. Polymer Sci. 21, 417 (1956).

13. S. N. Chinai, A. L. Resnick, H. T. Lee, J. Polymer Sci. 33, 471 (1958).

14. S. N. Chinai, R. J. Samuels, J. Polymer Sci. 19, 463 (1956).

15. S. N. Chinai, P. C. Scherer, C. W. Bondurant, D. W. Levi, J. Polymer Sci. 22, 527 (1956).

16. S. N. Chinai, R. J. Valles, J. Polymer Sci. 39, 363 (1959).

17. A. de Chirico. Chim. Ind. (Milan) 42, 248 (1960).

18. A. de Chirico. Chim. Ind. (Milan) 46, 53 (1964).

19. A. Ciferri, M. Lauretti, Ann. Chim. (Rome) 48, 198 (1958).

20. E. Cohn-Ginsberg, T. G. Fox, H. Maron, Polymer 3, 97 (1962).

21. J. M. G. Cowie, Makromol. Chem. 59, 189 (1963).

22. V. Cresenci, P. J. Flory, J. Am. Chem. Soc. 86, 141 (1964).

23. F. Danusso, G. Moraglio, Rend. Accad. Naz. Lincei 25, 509 (1958).

24. F. Danusso, G. Moraglio, G. Gianotti, J. Polymer Sci. 51, 475 (1961).

25. F. E. Didot, S. N. Chinai, D. W. Levi, J. Polymer Sci. 43, 557 (1960).

26. J. Dohany, Thesis ETH Zurich, 1963.

27. E. T. Dumitru, L. H. Cragg, cited in ref. (44), p. 615.

28. H.-G. Elias, Makromol. Chem. 33, 140 (1959).

29. H.-G. Elias, Makromol. Chem. 50, 1 (1961).

30. H.-G. Elias, Makromol. Chem. 54, 78 (1962).

31. H.-G. Elias, G. Adank, Makromol. Chem. 69, 241 (1963).

32. Hj. Dietschy, Thesis ETH Zurich, 1966.

33. H.-G. Elias, O. Etter, Makromol. Chem. 66, 56 (1963).

34. H.-G. Elias, O. Etter, Makromol. Chem. 89, 228 (1965).

35. H.-G. Elias, U. Gruber, Makromol. Chem. 78, 72 (1964).

36. H.-G. Elias, U. Gruber, Makromol. Chem. 86, 168 (1965).

37. H.-G. Elias, F. W. Ibrahim, Makromol. Chem. 89, 12 (1965).

38. H.-G. Elias, R. Schumacher, Makromol. Chem. 76, 23 (1964).

39. V. E. Eskin, L. N. Andreeva, Vysokomolekul. Soedin. 3, 435 (1961); ref. J. Polymer Sci. 56, S 39 (1962).

40. V. E. Eskin, K. Z. Gumargalieva, Vysokomolekul. Soedin. 2, 265 (1960).

41. V. E. Eskin, O. Z. Korothina, Vysokomolekul. Soedin. 2, 272 (1960); Polymer Sci. USSR 2, 247 (1961).

42. W. W. Everett, J. F. Foster, J. Am. Chem. Soc. 81, 3464 (1959).

43. J. Feisst, H.-G. Elias, Makromol. Chem. 82, 78 (1965).

44. P. J. Flory, "Principles of Polymer Chemistry", Cornell University Press, Ithaca, N. Y., 1953.

45. P. J. Flory, L. Mandelkern, J. B. Kinsinger, W. B. Shultz, J. Am. Chem. Soc. 74, 3364 (1952).

46. J. E. Mark, P. J. Flory, J. Am. Chem. Soc. 87, 1423 (1965).

47. T. G. Fox, Polymer 3, 111 (1962).

48. T. G. Fox, P. J. Flory, J. Am. Chem. Soc. 73, 1909 (1951).

49. T. G. Fox, P. J. Flory, J. Am. Chem. Soc. 73, 1915 (1951).

50. T. G. Fox, J. B. Kinsinger, H. F. Mason, E. M. Schuele, Polymer 3, 71 (1962).

51. T. B. Grimley, Proc. Roy Soc. (London) A212, 339 (1952).

52. U. Gruber, Thesis, ETH. Zurich, 1964.

53. J. Hakozaki, J. Chem. Soc. Japan, Pure Chem. Sect. (Nippon Kagaku Zasshi) 82, 158 (1961).

54. T. Homma, K. Kawahara, H. Fujita, M. Ueda, Makromol. Chem. 67, 132 (1963).

55. A. J. Hyde, R. B. Taylor, Polymer 4, 1 (1963).

56. H. Inagaki, A. Nakazawa, T. Kotaka, Bull. Inst. Chem. Res., Kyoto University 43, 135 (1965).

57. K. J. Ivin, H. A. Ende, G. Meyerhoff, Polymer 3, 129 (1962).

58. A. Katchalsky, H. Eisenberg, J. Polymer Sci. 6, 145 (1951).

59. W. Kern, D. Braun, Makromol. Chem. 27, 23 (1958).

60. J. B. Kinsinger, R. E. Hughes, J. Phys. Chem. 67, 1922 (1963).

61. J. B. Kinsinger, R. A. Wessling, J. Am. Chem.Soc. 81, 2908 (1959).

62. R. Kirste, G. V. Schulz, Z. Physik. Chem. (Frankfurt) 27, 310 (1960).

63. R. Kirste, G. V. Schulz, Z. Physik. Chem. (Frankfurt) 30, 171 (1961).

64. S. Krause, E. Cohn-Ginsberg, J. Phys. Chem. 67, 1479 (1963).

65. W. R. Krigbaum, D. K. Carpenter, J. Phys. Chem. 59, 1166 (1955).

66. W. R. Krigbaum, P. J. Flory, J. Polymer Sci. 11, 37 (1953).

67. W. R. Krigbaum, J. E. Kurz, P. Smith, J. Phys. Chem. 65, 1984 (1961).

68. W. R. Krigbaum, L. H. Sperling, J. Phys. Chem. 64, 99 (1960).

69. W. R. Krigbaum, J. D. Woods, J. Polymer Sci. A2, 3075 (1964).

70. N. Kuwahara, Y. Miyake, M. Kaneko, J. Furuichi, Rep. Prog. Polymer Phys. (Japan) 5, 1 (1962).

71. H. T. Lee, D. W. Levi, J. Polymer Sci. 47, 449 (1960).

72. D. W. Levi, H. T. Lee, R. J. Valles, J. Polymer Sci. 62, S 163 (1962).

73. M. Lippay, Thesis, ETH Zurich, 1963.

74. L. Mandelkern, P. J. Flory, J. Am. Chem. Soc. 74, 2517 (1952).

75. J. E. Mark, Diss. Abstr. 23, 1205 (1962).

76. M. Matsumoto, Y. Ohyanagi, J. Polymer Sci. 46, 441 (1960).

77. M. Matsumoto, Y. Ohyanagi, J. Polymer Sci. 50, S 1 (1961).

78. H. W. McCormick, J. Polymer Sci. 36, 341 (1959).

79. D. McIntyre, J. H. O'Mara, B. C. Konouck, J. Am. Chem. Soc. 81, 3498 (1959).

80. R. A. Mendelson, J. Polymer Sci. 46, 493 (1960).

81. G. Meyerhoff, Makromol. Chem. 37, 97 (1960).

82. G. Meyerhoff, in J. W. Williams, ed. "Ultracentrifugal Analysis in Theory and Praxis", Academic Press, New York, 1963.

83. J. Moacanin, J. Appl. Polymer Sci. 1, 272 (1959).

84. G. A. Morneau, P. I. Roth, A. R. Shultz, J. Polymer Sci. 55, 609 (1961).

85. S. Newman, W. R. Krigbaum, C. Laugier, P. J. Flory, J. Polymer Sci. 14, 451 (1954).

86. S. Okamura, T. Higashimura, Y. Imanishi, Chem. High Polymers (Japan) 16, 244 (1959).

87. R. Okada, Y. Toyoshima, H. Fujita, Makromol. Chem. 59, 137 (1963).

88. P. Outer, C. I. Carr, B. H. Zimm, J. Chem. Phys. 18, 830 (1950).

89. J. Oth, V. Desreux, Bull. Soc. Chim. Belges 63, 285 (1954).

90. C. G. Overberger, E. M. Pearce, N. Mayes, J. Polymer Sci. 34, 109 (1959).

91. I. Ya. Poddubnyi, V. A. Grechanovskii, M. I. Mosevitskii, Vysokomolekul. Soedin. 5, 1049 (1963); Polymer Sci. USSR 5, 105 (1964).

92. I. Ya. Poddubnyi, V. A. Grechanovskii, M. I. Mosevitskii, A. V. Podalinskii, Vysokomolekul. Soedin. 5, 1042 (1963); Polymer Sci. USSR 5, 97 (1964).

93. C. Reiss, H. Benoit, Compt. Rend 253, 268 (1961).

94. Y. Sakaguchi, J. Nishino, K. Tsugawa, Chem. High Polymers (Japan) 20, 661 (1963).

95. I. Sakurada, A. Nakajima, O. Yoshizaki, K. Nakamae, Kolloid-Z. 186, 41 (1962).

96. I. Sakurada, Y. Sakaguchi, S. Kokuryo, Chem. High Polymers (Kobunshi Kagaku) 17, 227 (1960).

97. M. Sato, Y. Koshiishi, M. Asahina, J. Polymer Sci. B1, 233 (1963).

98. P. R. Saunders, J. Polymer Sci. 57, 131 (1961).

99. G. V. Schulz, H. J. Cantow, Z. Elektrochem. 60, 517 (1956).

100. G. V. Schulz, A. Haug, Z. Physikal. Chem. (Frankfurt) 34, 328 (1962).

101. G. V. Schulz, A. Mula, Makromol. Chem. 44/46, 479 (1961).

102. G. V. Schluz, W. Wunderlich, R. Kirste, Makromol. Chem. 75, 22 (1964).

103. Y. Shiokawa, H. Takeo, K. Takamizawa, T. Oyama, Rep. Prog. Polymer Phys. Japan 7, 49 (1964).

104. A. R. Shultz, J. Am. Chem. Soc. 76, 3422 (1954).

105. A. R. Shultz, Thesis, Cornell University, cited in ref. (44), p. 615.

106. V. S. Skazka, R. A. Zobov, A. Moshpauenko, Vysokomolekul. Soedin. 4, 1257 (1962).

107. C. J. Stacy, J. F. Foster, J. Polymer Sci. 20, 56 (1956).

108. U. P. Strauss, P. L. Wineman, J. Am. Chem. Soc. 80, 2366 (1958).

109. H. L. Wagner, P. J. Flory, J. Am. Chem. Soc. 74, 195 (1952).

110. R. A. Wessling, Diss. Abstr. 23, 1536 (1962).

111. E. A. Zavaglia, F. W. Billmeyer, Jr. Official Digest, 36, 221 (1964).

112. R. Chiang, J. Phys. Chem. 69, 1645 (1965).

113. H.-G. Elias, "Makromolek.", Huethig and Wepf, Basel 1971, p. 188

114. Y. Izumi, S. Otsuka, Y. Miyake, Rep. Progr. Polymer Phys. Japan 12, 13 (1969).

115. C. Booth, R. Orme, Polymer 11, 626 (1970).

116. W. R. Moore, M. Uddin, European Polymer J. 5, 185 (1969).

117. M. Heskins, J. E. Guillet, J. Macromol. Sci. A2, 1441 (1968).

118. H. A. Dieu, J. Polymer Sci. 12, 417 (1954).

119. D. H. Napper, Kolloid-Z. 234, 1149 (1969).

120. D. Decker, Makromol. Chem. 125, 136 (1969).

121. D. R. Beech, C. Booth, J. Polymer Sci. A2, 7, 575 (1969).

122. J. O. Threlkeld, H. A. Ende, J. Polymer Sci. A2, 4, 663 (1966).

123. U. Moritz, G. Meyerhoff, Makromol. Chem. 139, 23 (1970).

124. M. Matsushima, M. Fukatsu, M. Kurata, Bull. Chem. Soc. Japan 41, 2570 (1968).

125. T. E. Helminiak, C. L. Benner, W. E. Gibbs, Polymer Preprints 8, 284 (1967).

126. W. Banks, C. T. Greenwood, J. Sloss, Makromol. Chem. 140, 109 (1970).

127. J. M. Barrales-Rienda, D. C. Pepper, Polymer 8, 337 (1967).

128. A. Dondos, Makromol. Chem. 135, 181 (1970).

129. K. Satyanarayana Sastry, R. D. Patel, European Polymer J. 5, 79 (1969).

130. I. Noda, K. Mizutani, T. Kato, T. Fujimoto, M. Nagasawa, Macromolecules 3, 787 (1970).

131. H. Inagaki, T. Miyamoto, S. Ohta, J. Phys. Chem. 70, 3420 (1966).

132. M. Pizzoli, G. Stea, G. Ceccorulli, G. B. Gechele, European Polymer J. 6, 1219 (1970).

133. H. Voelker, F.-J. Luig, Ang. Makromol. Chem. 12, 43 (1970).

134. K. Matsumura, Polymer J. (Japan) 1, 322 (1970).

135. D. Froelich, H. Benoit, Makromol. Chem. 92, 224 (1966).

136. A. Kotera, H. Matsuda, K. Konishi, K. Takemura, J. Polymer Sci. C23, 619 (1968).

137. T. Kotaka, H. Ohnuma, H. Inagaki, Polymer Preprints 11, 660 (1970).

138. W. L. Mattice, L. Mandelkern, Macromolecules 4, 271 (1971).

139. W. Banks, C. T. Greenwood, Makromol. Chem. 67, 49 (1963).

140. J. Lilie, J. Springer, K. Ueberreiter, Kolloid-Z. 226, 138 (1968).

141. M. Quadrat, M. Bohdanecký, J. Polymer Sci., A2, 5, 1309 (1967).

142. J. Moacanin, A. Rembaum, R. K. Laudenslager, Polymer Preprints 4, (2), 179 (1963).

143. J. Moacanin, A. Rembaum, R. K. Laudenslager, R. Adler, J. Macromol. Sci. [Chem] A1, 1497 (1967).

144. G. Moraglio, G. Gianotti, F. Danusso, European Polymer J. 3, 251 (1967).

145. A. Nakajima, F. Hamada, Rep. Prog. Polymer Phys. Japan 9, 41 (1966).

146. C. J. Stacey, R. L. Arnett, J. Phys. Chem. 69, 3109 (1965).

147. R. Koningsveld, Th. G. Scholte, private communication.

148. A. Nakajima, H. Fujiwara, F. Hamada, J. Polymer Sci., A2, 4, 507 (1966).

149. L. D. Moore, Jr., J. Polymer Sci. 36, 155 (1959).

150. T. W. Bates, J. Biggins, K. J. Ivin, Makromol. Chem. 87, 180 (1965).

151. T. W. Bates, K. J. Ivin, Polymer 8, 263 (1967).

152. J. B. Kinsinger, L. E. Ballard, J. Polymer Sci. A3, 3963 (1965).

153. J. B. Kinsinger, L. E. Ballard, J. Polymer Sci. B2, 879 (1964).

154. G. Moraglio, J. Brzezinski, J. Polymer Sci. B2, 1105 (1964).

155. K. Takamizawa, Y. Kambara, T. Oyama, Rep. Prog. Polymer Phys. Japan 10, 85 (1967).

156. A. Nakajima, A. Saijyo, J. Polymer Sci., A2, 6, 735 (1968).

157. K. Kubo, K. Ogino, T. Nakagawa, Rep. Prog. Polymer Phys. Japan 10, 7 (1967).

158. K. Kubo, K. Ogino, Rep. Prog. Polymer Phys. Japan 10, 111 (1967).

159. R. B. Mohite, S. Gundiah, S. L. Kapur, Makromol. Chem. 108, 52 (1967).

160. J. M. G. Cowie, S.Bywater, J. Polymer Sci., A2, 6, 499 (1968).

161. J. M. G. Cowie, S. Bywater, D. J. Worsfolf, Polymer 8, 105 (1967).

162. B. J. Cottam, J. M. G. Cowie, S. Bywater, Makromol. Chem. 86, 116 (1965).

163. M. Abe, K. Sakato, T. Kageyama, M. Fukatsu, M. Kurata, Bull. Chem. Soc. Japan 41, 2330 (1968).

164. G. B. Gechele, L. Crescentini, J. Polymer Sci. 3, 3599 (1965).

165. C. Garbuglio, L. Crescentini, A. Mula, G. B. Gechele, Makromol. Chem. 97, 97 (1966).

166. M. Abe, H. Fujita, J. Phys. Chem. 69, 3263 (1965).

167. A. Dondos, H. Benoit, J. Polymer Sci. B7, 335 (1969).

168. J. M. G. Cowie, J. Polymer Sci., C, 23, 267 (1968).

169. G. V. Schulz, H. Baumann, Makromol. Chem. 60, 120 (1963).

170. T. A. Orofino, A. Ciferri, J. Phys. Chem. 68, 3136 (1964).

171. T. A. Orofino, J. W. Mickey, Jr., J. Chem. Phys. 38, 2512 (1963).

172. H. Inagaki, H. Suzuki, M. Fujii, T. Matsuo, J. Phys. Chem. 70, 1718 (1966).

173. C. Reiss, H. Benoit, J. Polymer Sci., C, 16, 3079 (1968).

174. A. Kotera, T. Saito, N. Yamaguchi, Annual Meeting Chem. Soc. Japan, 1962; cited in A. Kotera, N. Iso, A. Senuma, T. Hamada, Iupac Symposium Makromol. Prague 1965, Preprint 232.

175. M. Ueda, K. Kajitani, Makromol. Chem. 108, 138 (1967).

176. R. Naito, Chem. High Polymers (Kobunshi Kagaku) 16, 7 (1959).

177. S. Tsuchiya, Y. Sakaguchi, J. Sakurada, Kobunshi Kagaku 18, 346 (1961).

178. I. Sakurada, A. Nakajima, K. Shibatani, J. Polymer Sci. A2, 3545 (1964).

179. W. Burchard, Z. Physik. Chem. New Series 43, 265 (1964).

180. A. Nakajima, K. Kato, Makromol. Chem. 95, 52 (1966).

181. H.-G. Elias, O. Etter, J. Macromol. Chem. 1, 431 (1966).

182. W. Burchard, Habilitationsschrift

183. A. F. Nikolaev, N. V. Daniel, A. M. Toroptseva, I. Varga, N. V. Ivanova, Vysokomolekul. Soedin. 6, 292 (1964); Polymer Sci. USSR 6, 336 (1964).

184. H. Eisenberg, D. Woodside, J. Chem. Phys. 36, 1844 (1962).

185. I. Ya. Poddubnyi, A. V. Podalinskii, V. A. Grechanovskii, Polymer Sci. USSR 8, 1715 (1966).

186. T. Lucas, Bull. Liaison Lab. (Lab, Prof, Peint,) 20, 15 (1967).

187. A. Nakajima, K. Okazaki, Chem. High Polymers (Japan) 22, 791 (1965).

188. R. van Leemput, R. Stein, J. Polymer Sci. A2, 4039 (1964).

189. P. Vasudevan, M. Santappa, Makromol. Chem. 137, 261 (1970).

190. Ye. A. Kanevskaya, P. I. Zubov, L. V. Ivanova, Yu. S. Lipatov, Vysokomolekul. Soedin. 6, 981 (1964); Polymer Sci. USSR 6, 1080 (1964).

191. M. Iwama, H. Utiyama, M. Kurata, J. Macromol. Chem. 1, 701 (1966).

192. O. Quadrat, M. Bohdaneck'y, P. Munk, J. Polymer Sci., C, 16, 95 (1967).

193. Z. Tuzar, P. Kratochvil, Collection Czech. Chem. Commun. 32, 3358 (1967).

194. H. Ohnuma, Y. Murakami, T. Kotaka, Rep. Prog. Polymer Phys. Japan 10, 63 (1967).

195. T. Kotaka, H. Ohnuma, H. Inagaki, Polymer 10, 517 (1969).

196. T. Kotaka, H. Ohnuma, Y. Murakami, J. Phys. Chem. 70, 4099 (1966).

197. T. Kotaka, T. Tanaka, H. Ohnuma, Y. Murakami, H. Inagaki, Polymer J. 1, 245 (1970).

198. M. Kalfus, J. Mitus, J. Polymer Sci., A1, 4, 953 (1966).

199. I. Ya. Poddubnyi, V. A. Grechanovskii, A. V. Podalinskii, J. Polymer Sci., C, 16, 3109 (1968).

200. G. Moraglio, European Polymer J. 1, 103 (1965).

201. D. M. French, A. W. Casey, C. I. Collins, P. Kirchner, Polymer Preprints 7, 447 (1966).

202. T. Homma, K. Kawahara, H. Fujita, J. Appl. Polymer Sci. 8, 2853 (1964).

203. K. Hanafusa, A. Teramoto, H. Fujita, J. Phys. Chem. 70, 4004 (1966).
204. K. H. Elgert, Thesis, Muenchen, 1963.
205. A. R. Shultz, C. R. McCullough, J. Polymer Sci., A2, 7, 1577 (1969).
206. J. Brzezinski, B. Czarnecka, Polimery Tworzywa 14, 117 (1969).
207. S. Krozer, A. Pastusiak, IUPAC Symposium Makromol., Prague 1965, Preprint 609.
208. G. Allen, C. Booth, C. Price, Polymer 7, 167 (1966).
209. S. M. Ali, M. B. Huglin, Makromol. Chem. 84, 117 (1965).
210. N. V. Makletsova, I. V. Epel'baum, B. A. Rozenberg, Ye. B. Lyudvig, Vysokomolekul. Soedin. 7, 70 (1965); Polymer Sci. USSR 7, 73 (1966).
211. J. M. Evans, M. B. Huglin, Makromol. Chem. 127, 141 (1969).
212. J. M. Evans, M. B. Huglin, cited in P. Dreyfuss and M. P. Dreyfuss, Adv. Polymer Sci. 4, 528 (1967).
213. M. Kurata, H. Utiyama, K. Kamada, Makromol. Chem. 88, 281 (1965).
214. R. W. May, R. E. Wetton, cited in P. Dreyfuss, M. P. Dreyfuss, Adv. Polymer Sci. 4, 528 (1967).
215. S. Krozer, Plaste und Kautschuk 15, 175 (1968).
216. G. C. Berry, H. Nomura, K. G. Mayhan, J. Polymer Sci., A2, 5, 1 (1967).
217. P. R. Saunders, J. Polymer Sci. 57, 131 (1962).
218. H. C. Beachell, J. C. Peterson, J. Polymer Sci., A1, 7, 2021 (1969).
219. N. Kuwahara, M. Kaneko, Makromol. Chem. 82, 205 (1965).
220. A. Czuppon, IUPAC Symposium Makromol. Prague 1965, Preprint 459.
221. O. Quadrat, Collection Czech. Chem. Commun. 35, 2564 (1970).
222. R. R. Buch, H. M. Klimisch, O. K. Johannson, J. Polymer Sci., A2, 7, 563 (1969).
223. J. Llopis, A. Albert, P. Usobiaga, European Polymer J. 3, 259 (1967).
224. T. Ikeda, H. Kawaguchi, Rep. Prog. Polymer Phys. Japan 9, 23 (1966).
225. G. Delmas, D. Patterson, Polymer 7, 513 (1966).
226. I. R. Urwin, M. Girolamo, Makromol. Chem. 150, 179 (1971).

FRACTIONATION OF POLYMERS

A. Bello and J. M. Barrales-Rienda
Instituto de Plásticos y Caucho
Patronato "Juan de la Cierva" C.S.I.C.
Madrid, Spain

and

G. M. Guzmán
Universidad Politécnica de Barcelona
Barcelona, Spain

Contents

A. PRINCIPLES OF POLYMER FRACTIONATION

As a general rule, the composition of a polymeric substance is not homogeneous. The differences between the macromolecules of such a substance may be classified according to three main properties: a) molecular weight, b) chemical composition and c) molecular configuration and structure. The fractionation of a polymeric substance means the separation of that substance into its different molecular species, using a suitable experimental technique, in order to obtain homogeneous fractions.

The molecular weight distribution is a general feature for practically all synthetic polymers, as a consequence of the particular nature of the polymerization process by which they are made. Natural polymers usually present molecular weight distribution as a consequence of the degradation processes suffered by the substance during isolation from the living tissues. Additional reasons, such as more or less accidental degradation during processing, improper handling, or routine use, may contribute substantially to an increase in the natural molecular weight distribution already present in the sample.

The existence of molecular weight heterogeneity in macromolecular substances is therefore quite general. It is one of their fundamental properties and is directly responsible for the necessity of using several molecular weight averages for their description. It also exerts a permanent influence on all the properties of the substance, both in solution and in the solid state.

Differences in chemical composition in polymers originate from those reactions which offer several possibilities of substitution along the backbone of the macromolecule, for instance, the synthesis of random, block and graft copolymers and the many partial chemical transformations to which the substances can be subjected.

The third kind of heterogeneity mentioned above refers to differences in the physical configuration of the macromolecules, such as those between linear and branched polymers, and also to differences in the tacticity of the several molecular species present in the mixture, which is usually reflected in varying amounts of amorphous and crystalline materials in the substance.

Most of the experimental techniques developed so far to fractionate polymers refer to fractionation according to molecular weight. Chemical composition and physical structure differences are handled by more or less sophisticated modifications of the solubility method, such as varying the nature of the solvent/non-solvent mixture or the temperature of extraction or by using an appropriate active support (adsorbent).

Those experimental techniques referred to in this table are usually based on the variation of some properties directly related to the molecular size. It is common to classify the fractionation method according to their preparative or analytical character. The latter methods do not isolate fractions; they are mainly intended to explore the molecular weight distribution of the polymer.

The classification of fractionation methods together with the basic idea of each experimental technique is briefly described below.

Reviews of polymer fractionation have been published by Cragg and Hammerschlag (329), Desreux and Oth (366), Schulz (1326), Conrad (299), Hall (581), Channen (277), Guzmán (567) (568), Fuchs and Leugering (472), Käsbauer and Schuch (753), Screaton (1342), Schneider (1310), Cantow (252), Johnson, Porter and Cantow (714), Moll (1020) and Tung (1489). Reviews dealing with some specific aspects of fractionation have been published by Schurz (1337), Samsonov (1295) and Schneider (1311). Reviews concerning theoretical aspects of polymer fractionation have been given in the book of Tompa (1457) and by Voorn (1543), Huggins and Okamoto (667) and Koningsveld (817).

In recent years Gel Permeation Chromatography (GPC) has become a technique of widespread use not only for molecular weight distribution but for preparative fractionation. There exists much information on this technique in recent literature. General reviews on GPC have been published by Cantow, Porter and Johnson (259), Heitz and Kern (613), Altgelt and Moore (31), Johnson and Porter (712) (713), Altgelt (29), Determan (368), Altgelt and Segal (32) and Lambert (860).

B. FRACTIONATION METHODS

I. Fractionation by Solubility

(1) Fractional Precipitation

Addition of non-solvent. Successive precipitation of polymer species from a solution by addition of a miscible non-solvent. The larger molecules precipitate first.

Lowering the temperature. Successive precipitation of polymer species from a solution by controlled cooling. The larger molecules precipitate first.

Solvent volatilization. Successive precipitation of polymer species from a solution of the polymer in a solvent/non-solvent mixture by controlled evaporation of the more volatile solvent. The larger molecules precipitate first.

Pressure variation at lower critical solution temperature. Successive separation is carried out at a lower critical solution temperature (LCST) under isothermal conditions by changing the pressure of the system. Low molecular weight species are precipitated first. The efficiency of the method is poor in the range of low molecular weights.

(2) Turbidimetric Titration

Continuous precipitation of polymer species from a very dilute solution by progressive addition of non-solvent. In the absence of coagulation the amount of polymer precipitated can be measured by the increase in optical density of the solution. The larger molecules precipitate first. This is an analytical method. The method can also be reversed, i.e. the polymer is precipitated first completely and then redissolved by progressive addition of solvent.

(3) Summative Precipitation

Simultaneous precipitation of polymer species from several solutions of the same sample by addition of increasing amounts of non-solvent to the solution. The sum of all the precipitates constitutes a cumulative weight distribution. This is an analytical method.

(4) Cumulative Volume of Precipitate

Successive precipitation of polymer species from a solution by addition of increasing amounts of non-solvent. Fractions are not isolated and the cumulative volume of precipitate is observed and determined after each non-solvent increment. This is an analytical method.

(5) Fractional Solution

Direct extraction. Direct and successive extraction of polymer with a liquid of increasing solvent power; smaller molecules are extracted first.

Film extraction
> Continuous operation. The polymer solution is applied as a thin coating on both sides of a slowly moving belt. On passing through a drying region the solvent evaporates; the thin film on the belt is extracted continuously in a series of tubes containing solvent/non-solvent mixtures of increasing solvent power kept at constant temperature.

> Batch operation. A metal foil is coated thinly with polymer; the foil is cut into strips and extracted successively with solvent/non-solvent mixtures of increasing solvent power. Smaller molecules are extracted first.

Column extraction
The polymer is distributed on an inert support packed in a column. Successive elution then takes place with a liquid of increasing solvent power.

Extraction of a coacervate
Successive extraction of polymer species from a coacervate. Smaller molecules are extracted first.

(6) Distribution Between Immiscible Solvents

Polymer species are distributed according to molecular size between two immiscible liquids of different solvent power. Countercurrent extraction is particularly suitable for this method.

References page IV-221

(7)   Fractional crystallization

Successive separation of species from polymer solutions by crystallization at different temperatures. The crystallization can be carried out under stagnant conditions, but considerable improvement in the efficiency is obtained if the crystallization is induced by fast stirring (stirring crystallization). Under these conditions, the high molecular weight constituent crystallizes first settling on the stirrer as long thin fibrillar crystals. The technique can be coupled with fractional extraction at increasing solution temperatures. The method is poorly reproducible.

II. Fractionation by Chromatography

(1)   Adsorption Chromatography (Chromatography on an active support)

The adsorption of polymer species on an active support depends on the molecular weight.

Frontal analysis. The active support is packed in a column; a solution of polymer passes down the column and is collected when leaving it. The concentration in the effluent changes with the volume collected, and presents successive fronts due to the different adsorption of molecular species on the active support.

Elution analysis.
      Column. A small quantity of polymer is sorbed on the upper portion of the support packed in the column. The elution with a suitable solvent takes place; each component moves down the column with different rate and is completely displaced at some time and collected in the effluent; in the gradient elution method the eluent is a liquid of increasing solvent power.

      Thin layer (TLC). A glass plate is covered with a film of adsorbent, usually silica gel, the film is activated by drying and then a spot of polymer solution is put on the plate, which is then placed vertically in a tank containing a suitable solvent. As solvent moves upwards it carries along each component at different rates. Colorless separated species can be made visible by appropriate methods. When TCL is used combined with a concentration gradient method, the composition of the solvent changes with the development time. This technique allows separation according to chemical composition, molecular weight and stereoregularity.

(2)   "Precipitation" Chromatography (Baker-Williams-Method) (Chromatography on an inactive support)

The support is an inert material packed in a column. On top of the column a small amount of polymer is placed, as in elution chromatography. A temperature gradient is set along the column, the upper part being at higher temperature than the bottom; then the elution takes place with a solvent mixture of increasing solvent power. Polymer species move down the column being in a continuous exchange between a precipitated phase and a saturated solution. This distribution depends on the molecular size.

Precipitation chromatography can also be carried out in the absence of a temperature gradient. It then becomes essentially a continuous fractionation by fractional solution. (See above under Fractional Solution Column Extraction).

(3)   Gel Permeation Chromatography (Chromatography on a porous support)

Gel permeation chromatography is a new separation process based on differences in the depth to which molecules of different chain length are able to diffuse into pores of an expanded and highly crosslinked polymer gel network or expanded silica gel or "gels" of porous glass. As a result of the restrictions imposed by the pore sizes on the larger molecules there is a greater pore volume available to the low molecular species giving them in effect a longer path length and they, therefore, are more strongly retarded during elution. Since polymer solubility and adsorption are not a part of the process the volume required for elution essentially depends only on the chain length and appears to be insensitive to structure.

(4)   Partition Chromatography

Polymer species are distributed between two liquid phases, one of them mobile and the other fixed by absorption in a support. The support consists of strips or sheets of porous paper. The immobile phase is packed in a solumn.

(5)   Ion-exchange Chromatography

The support is an ion-exchange resin, constituting an immobile phase, through which the solution of polymer is passed. This method is appropriate for polymer species bearing electric charges; the molecules are distributed between the liquid phase and the interface, according to their ionic adsorption forces, which depend on the electric charge and the size of the macromolecules.

III. Fractionation by Sedimentation

(1)   Sedimentation Velocity

Sedimentation velocity of polymer species in a high centrifugal field is a function of molecular size. This is usually an analytical method unless a velocity ultracentrifuge of the preparative type is used.

(2)   Sedimentation Equilibrium

At lower rotational speeds of the ultracentrifuge it becomes possible to create an equilibrium situation such that the rate of sedimentation is exactly equal to the rate of back-diffusion of the macromolecules. Larger polymer molecules are then found closer to the bottom of the cell than smaller ones.

(3) Density Gradient Technique

In a preformed or self-generated density gradient the polymer collects around the position of its own density in a band. The width of this band is dependent on the molecular weight of the polymer. A band of polydisperse macromolecules contains more high molecular weight polymer near its maximum than near its "tails". This technique is also capable of separating according to composition.

IV. Fractionation by Diffusion

(1) Thermal Diffusion

A polymer solution is placed between two surfaces, and a high temperature gradient is established between them. The solution is in contact with an upper and a lower reservoir. The temperature gradient gives rise to a thermal circulation of the molecules, producing a separation of polymer species, which migrate towards the lower reservoir. Thermal diffusion is more pronounced for the larger molecules than for the smaller.

(2) Brownian Diffusion

Polymer molecules diffuse at different rates from a solution into a solvent by a boundary, depending on their molecular weights. The translational diffusion constants can be determined by special optical means and related to the inhomogeneity of the sample. This is an analytical method.

V. Fractionation by Ultrafiltration through Porous Membranes

A polymer solution is submitted to ultrafiltration through a series of membranes of different porosity. The rate of diffusion depends on the molecular size and the degree of permeability of the membranes. It is possible to isolate fractions with varying molecular weights at different times.

VI. Fractionation by Zone Melting

A solid solvent is packed in a column. A small amount of polymer is put on top of the solid solvent and dissolved by heating a narrow zone. After solidification, a lower zone is heated, melted, and resolidified. All the column is treated in this way. Polymer species move down the column at different rates depending on their molecular size during the molten stage. At the end, the polymer si distributed throughout the entire column and recovered by cutting the solid into several portions and subliming the solvent.

VII. Electron Microscope Counting Method

It is possible with most glassy amorphous polymers to observe single spherical molecules by means of the electron microscope, if very dilute solutions in a solvent/-non-solvent mixture are sprayed onto a thin substrate. Consequently, the weight and number average molecular weights and molecular weight distributions can be obtained. The method can be recommended and applied only for polymers with molecular weights higher than 500,000.

C. TABLE OF FRACTIONATION SYSTEMS FOR DIFFERENT POLYMERS

Polymer	Method of Fractionation	Solvent or Solvent/Non-solvent Mixture	Remarks	References
1. Main-chain Acyclic Carbon Polymers				
1.1 Poly(dienes)				
Poly(butadiene)				
	Fractional precipitation	Benzene/acetone	28°C Branched polymer	305, 1532
		Benzene/acetone : dioxane followed by acetone-methanol		500
		Benzene/methanol	Low. temp., 1, 4-cis	1
		Benzene/methanol		218, 219, 670, 437
		Benzene/n-butanol		826
		Heptane/acetone		1041
		Pentane/methanol	Carboxy terminated. 30°C	461
		Tetrahydrofuran/water		1430
		Toluene/ethanol		1524
		Toluene/n-butanol		826
		Toluene/methanol		396, 1620

References page IV-221

Polymer	Method of Fractionation	Solvent or Solvent/Non-solvent Mixture	Remarks	References
Poly(butadiene) (Cont'd.)				
	Fractional solution	Chloroform/methanol	Column elution	1343, 1344
		Chloroform, chloroform-ethanol, ethanol, carbon tetrachloride, carbon tetrachloride-chloroform	Column elution, carboxy + hydroxy terminated	878
	Coacervation	Benzene/pentane	(1,4-cis)	591
	Chromatography	Benzene/ethanol	"Precipitation chromat."	307, 308
		Benzene/methanol	"Precipitation chromat." 35-28°C	1219
		Butyl acetate/isopropanol	"Precipitation chromat." 25-55°C	1266
		Chloroform	GPC	878, 1343, 1344, 1345
		o-Dichlorobenzene	GPC	608, 609, 591
		Diisobutene/isooctane	"Precipitation chromat."	671
		Diisobutene/n-propanol	(1,4-cis)	671
		n-Heptane/isooctane	"Precipitation chromat." 40-50°C	618
		Tetrahydrofuran	GPC	179,190,191,192,481, 580, 839, 877, 1238, 1343, 1344, 1345, 1509, 1524
		Toluene	GPC 70°C	943
		Toluene/isopropanol	"Precipitation chromat." 22-44°C	661
		Toluene/n-propanol		671
		1,2,4-Trichlorobenzene	GPC 130°C	662, 943, 944
		Trichloroethylene	GPC 123°C	1447
	Sedimentation velocity	Diethyl ketone	Ultracentrifuge 10,3^{o}C	1516
		n-Heptane-isooctane	Ultracentrifuge (1,4-cis) 20,5^{o}C	1040
		n-Hexane-n-heptane (1:1)	Ultracentrifuge (stereoregular)	307, 1211
		Octane	Ultracentrifuge	219
Poly(chloroprene)				
	Fractional precipitation	Benzene/acetone		601
		Benzene/methanol		585, 1014, 1015, 1016
		2-Butanone/methanol	25°C	339
	Fractional solution	Benzene/isopropanol	30°C	171
		Benzene/methanol	Neoprene, 30°C, cont. film ext.	170, 1322
	Chromatography	Benzene	GPC	920
		Benzene/methanol	"Precipitation chromat." 28-35°C	1219
		Chloroform	GPC	1322
		Tetrahydrofuran	GPC	321
	Sedimentation velocity	2-Butanone	20°C	1322
	Thermal diffusion	Benzene		828, 843
Poly(isoprene)				
	Fractional precipitation	Benzene/acetone	Hevea	601
		Benzene/n-butanol		1222
		Benzene/ethanol	Gutta percha balata, synthetic trans-poly(isoprene)	305
		Benzene/isopropanol	Low temp.	175
		Benzene/methanol	Low temp.	175, 563
		Chloroform/acetone	Pale crepe	248
		Dichloroethane/2-butanone		906
		Toluene/n-butanol	30°C	1221
		Toluene/boiling methanol	Chlorinated natural rubber	26
		Toluene/methanol		125, 193, 1089, 1269, 1547
		Toluene-ethanol (4:1)/ethanol	Rubber	1464
	Fractional solution	Acetone, n-hexane	Guayule, extraction	600, 602, 962
		Acetone	Hevea, extraction	711
		Benzene/methanol	25°C, column extraction, natural rubber	224

Polymer	Method of Fractionation	Solvent or Solvent/Non-solvent Mixture	Remarks	References
Poly(isoprene) (Cont'd.)				
	Turbidimetric titration	Benzene/methanol	20°C	563
		Toluene-ethanol (8 : 2)/ethanol	25°C	1224
	Extraction	Benzene/methanol	Natural rubber	1569
	Chromatography	Benzene/methanol	"Precipitation chromat."	180,1089,1219
		Chloroform	GPC	353
		Cyclohexane	GPC, 58°C	908
		Cyclohexanone	Partition paper	80,151
		o-Dichlorobenzene	GPC, 135°C	352
		Tetrahydrofuran	GPC	1238
		Toluene	GPC	1089
		Trichlorobenzene	GPC	593,662
Poly(1-isopropylidenedicyclopentadiene)				
	Fractional precipitation	Toluene/methanol		272
Poly(1,3-pentadiene)				
	Fractional precipitation	Benzene/2-butanone	(amorph./cryst.) 1,4-cis, isotact.	1098
Poly(perfluorobutadiene)				
	Fractional solution	Hexane, hexafluorobenzene	Extraction	1459,1460
Poly(1,1,2-trichlorobutadiene)				
	Fractional precipitation	Benzene/petroleum ether		1175,1388
1.2 Poly(alkenes)				
Poly(butylethylene)				
	Fractional precipitation	Xylene/triethylene glycol	130°C	1361
Poly(1-butene) see Poly(ethylethylene)				
Poly(cyclohexylethylene)				
	Fractional solution	Benzene	Extraction	1142
Poly(cyclopentylethylene)				
	Fractional solution	Ethyl ether	Extraction	1142
Poly(cyclopropylethylene)				
	Fractional solution	Benzene	Extraction	1142
		Ethyl ether, n-heptane, octane, nonane	Extraction	1143
Poly(ethylethylene)				
	Fractional precipitation	Cyclohexane/acetone	35°C	1297
		Cyclohexanone/cyclohexanol-glycol (3:1)	115°C	1411
		o-Dichlorobenzene/dimethylformamide		1044
		Toluene/methanol	Atactic	1043,1044
		Xylene/triethylene glycol	130°C	1361
	Fractional solution	Benzene/methanol	Extraction	416
	Chromatography	Benzene/2-butanone	"Precipitation chromat." $18-50^\circ$C	576
		Dichlorobenzene	GPC, 125°C	919
		Trichlorobenzene	GPC, 135°C	1268
Poly(ethylene)				
	Fractional precipitation	Amyl acetate	133°C, low.temp.	4
		Benzene/poly(oxyethylene)	75°C	1092
		2-Ethylhexanol (85)/decalin (15)	Solvent volat., low.temp.	767,1566,1567
		Liquid ethylene at 130 atm. and 80°C	Releasing pressure	1513
		Tetralin/benzyl alcohol	Low. temp.	749,794
		Toluene/poly(oxyethylene)	80°C	1107
		Toluene/n-propanol	Low. temp.	1092,1502

Polymer	Method of Fractionation	Solvent or Solvent/Non-solvent Mixture	Remarks	References
Poly(ethylene) (Cont'd.)				
	Fractional precipitation	Toluene/triethylene glycol	107°C	749
		Xylene/poly(oxyethylene)	75°C, 130°C	733,1061,1092,1120, 1121
		Xylene/n-propanol	90°C	52
		Xylene/triethylene glycol		134-137,759,760,1078, 1248, 1361,1486-1488
	Fractional solution	Decalin	150°C	875
		Decalin/2-ethoxyethanol	Column elution	550
		Decalin/ethylene glycol-hexanol	139°C, column elution	809
		Dichlorobenzene/dimethyl phthalate	127°C, glass beads, column elution	1348,1349
		Diisopropyl ether, benzene, xylene	Extraction, waxes	810
		Mesitylene/2-butoxyethanol	Column elution, 126°C	1462
		Petroleum ether	Column elution, low M.W.	512
		Toluene	Column elution	365,367,1092
		Tetralin/2-butoxyethanol	126°C, column elution, analytical and preparative, branched	1462
		Toluene/poly(oxyethylene)	80°C	1105,1106
		Xylene		604,758,759,775
		Xylene/n-butanol	Film extraction	471
		Xylene/2-butoxyethanol	Column elution	285,315,430,606, 1079-1081,1083,1084 1462
		Xylene, chloronaphthalene, perchloro-ethylene	Increasing temp.	1184
		Xylene/diethylene glycol monomethyl ether	Column elution, 126°C	1036,1293,1294,1305
		Xylene/2-ethoxyethanol	Column elution	459,582,620,761,778, 970,1289,1566,1577, 1579
		Xylene/trichloroethylene	Boiling mixt., 46-86°C	775
		Xylene/triethylene glycol	Column elution	136,1486
	Coacervation	Dichlorobenzene/triethylene glycol	138°C	352
		Xylene/poly(oxyethylene)		1120,1341
	Chromatography	Dichlorobenzene	GPC	106,352,355,678,917, 919,1074,1580
		Perchloroethylene	GPC, 110°C	1280
		Tetralin	GPC, 125°C	920
		Toluene	GPC, 25°C, low M.W.	638
		Tetralin/2-butoxyethanol	"Precipitation chromat." 110-160°C	298,558
		Trichlorobenzene	GPC	190,192,296,314,372, 384,385,773,774,971, 1073-1076,1141,1158, 1194,1246,1247,1294, 1316,1491,1571,1573, 1577-1579
	Sedimentation velocity	1-Bromonaphthalene	Ultracentrifuge, 110°C	960
		1-Chloronaphthalene	Ultracentrifuge, 120°C	1026
	Sedimentation equilibrium	Biphenyl	Ultracentrifuge, 123,2°C	1324
	Fractional crystallization	Xylene		757,763,776,818,1185, 1235
		Xylene, perchloroethylene or 1-chloro-naphthalene	Lowering temp. stirred	1184
Poly(1,1-dimethylethylene) (Poly(isobutene))				
	Fractional precipitation	Benzene/acetone		159,161,319,320,338, 449,455,456,465,844, 931,932
		Benzene/methanol	28°C	401,453
		Cyclohexane/butanone	25°C	1162

Polymer	Method of Fractionation	Solvent or Solvent/Non-solvent Mixture	Remarks	References
Poly(1,1-dimethylethylene) (Cont'd.)				
	Fractional precipitation	Heptane/ethanol		181
		Liquid ethylene at 130 atm. releasing pressure	80°C	1513
		Toluene/methanol		1424
		2,4,4-Trimethylpentene/n-butanol		905
	Turbidimetric titration	Butyl acetate, butyl ether, methoxymethanol, methyl propyl ketone, diethyl ketone, ethyl amyl ketone, heptane-n-butanol	Lowering temp.	1022
	Fractional solution	Benzene	Column elution, 24,5°C	950
		Toluene/acetone	Extraction, 30°C	159
		Toluene/methanol	Column elution, charcoal	162,862
	Chromatography	Benzene/acetone	"Precipitation chromat.", glass beads 28-50°C	254,255,1230
		Carbon tetrachloride	GPC	302
		Diisobutene/n-butanol	"Precipitation chromat." 10-60°C	1160
		Hexane/ethanol	"Precipitation chromat." 10-60°C	1160
		Tetrahydrofuran	GPC, 25°C	1109,1110
		Trichlorobenzene	GPC, 150°C	258,1228
		Xylene/n-propanol	"Precipitation chromat." 10-60°C	1160
	Sedimentation velocity	2,2,4-Trimethylpentane		1549
	Thermal diffusion	n-Heptane	40-50°C	850
Poly(1,5-dimethylhexyl)ethylene)				
	Extraction	Boiling (acetone, ethyl acetate, ethyl ether, diisopropyl ether, cyclohexane)		292
	Chromatography	Acetone/ethyl ether, benzene	20-78°C, active support	1203
Poly(3,7-dimethyl-1-octene) see Poly((1,5-dimethylhexyl)ethylene)				
Poly(hexadecylethylene)				
	Fractional precipitation	Xylene/triethylene glycol	130°C	1361
Poly(1-hexene) see Poly(butylethylene)				
Poly(hexylethylene)				
	Fractional precipitation	Cyclohexane/acetone		785
		Xylene/triethylene glycol	130°C	1361
Poly(isobutene) see Poly(1,1-dimethylethylene)				
Poly((2-methylbutyl)ethylene)				
	Extraction	Boiling (acetone, ethyl acetate, ethyl ether, isopropyl ether, and cyclohexane)	stereoisomers	292
Poly((2-ethylbutyl)ethylene)				
	Fractional precipitation	Benzene-nitrobenzene	Low. temp.	913
	Chromatography	Alcohols, aromatic hydrocarbons, ethers	Column elution, active support	1203
Poly(3-methyl-1-pentene) see Poly((1-methylpropyl)ethylene)				
Poly((1-methylpropyl)ethylene				
	Extraction	Boiling (acetone, ethyl acetate and benzene)		1202
	Chromatography	Acetone/ethyl ether, benzene	Column elution, 18-20°C, active support, acetone soluble fraction	1203
Poly(propylene)				
	Fractional precipitation	Benzene/acetone		452
		Benzene/methanol	Atactic	142,786
		Cyclohexane/acetone	Atactic	786,1027
		Cyclohexanone/ethylene glycol or dimethyl phthalate	Isotactic, 130-135°C	786

Polymer	Method of Fractionation	Solvent or Solvent/Non-solvent Mixture	Remarks	References
Poly(propylene) (Cont'd.)				
	Fractional precipitation	Decalin/phenol-decalin/propylene glycol	130°C, N$_2$ atm.	873
	Turbidimetric titration	Chloroform/isopropanol	30°C, atactic	722
		Tetralin/butylcellosolve	110°C	1432, 1434
	Coacervation	Tetralin/poly(oxyethylene)	140°C	1256
	Fractional solution	Decalin/2-ethoxyethanol	Column extraction, 110°C	684
		o-Dichlorobenzene/diethylene glycol mono-butyl ether	Column elution, 166°C	1592, 1593
		o-Dichlorobenzene/dimethyl phthalate	Column elution, 150°C	1348
		o-Dichlorobenzene/diethylene glycol mono-methyl ether	Column elution	347, 641, 969, 1229
		Kerosene/ethylene glycol (10 %) in 2-butoxy-ethanol	Column extr., 156°C, isotactic	336, 429, 1367
		Kerosene/diethylene glycol monomethyl ether	150°C, column extr.	1527, 1576
		Nine solvents of increasing solvent power and b.p.	According to crystallinity	1103
		Tetralin	Column extr., increasing temp.	1378
		Tetralin/diethylene glycol monoethyl ether	Column extr., 176°C, isotactic	1183
		Tetralin/diethylene glycol monoethyl ether-ethylene glycol (75 % - 25 %)	Column extr., 180°C, isotactic	629
		Tetralin/dimethyl phthalate	Column extr., 155-178°C	969
		Tetralin/2-butoxyethanol	Column extr.	1432
		Tetralin/2-ethoxyethanol	Column extr., 170°C	1161
		Xylene/2-ethoxyethanol	165°C	652
		Xylene/methanol	Column extr., 56°C, atactic	1229, 1367
		Xylene/poly(oxyethylene)	Coacervate extraction, 134°C	347
	Extraction	Acetone, ethyl ether, n-pentane, n-hexane, n-heptane, n-octane	Isotactic	903, 1095, 1286
		Boiling (acetone, ethyl ether, n-hexane, n-heptane)	Amorphous and crystalline	484
		Boiling (ethyl ether, n-heptane, methyl-cyclohexane)	Amorphous and crystalline	1526
		n-Heptane	Hot and cold	442
		17 Hydrocarbon fractions with increasing b.p. from 35 to 130°C		1077
	Chromatography	Bayol D/2-butoxyethanol	"Precipitation chromat." 75-170°C	1379
		Dichlorobenzene	GPC, 125°C	919
		Isopropyl ether	High M.W. and crystallinity	1096
		Petroleum ether/methanol	Silica gel, 0-48°C, low M.W.	511
		Tetralin	Column elution, increasing temp. 20 to 150°C, according to tacticity	297
		Tetralin	GPC, 130-145°C	859
		Tetralin/2-butoxyethanol	"Precipitation chromat." 140-180°C	298
		Tetralin/diethylene glycol monobutyl ether	Column elution, 165°C	297
		Trichlorobenzene	GPC, 135°C	296, 336, 1151, 1152
		Trichlorobenzene/dimethyl phthalate	"Precipitation chromat." 55-125°C	68
	Fractional crystallization	n-Heptane	Stirred, opt. act. (amorphous and crystalline)	1144
		Perchloroethylene	75-90°C, isotactic	1184
		Xylene	Stereoregular spn.	731
Poly(propylethylene)				
	Fractional precipitation	Toluene/methanol	30°C, isotactic	345, 1042
Poly(vinylcyclohexane) see Poly(cyclohexylethylene)				

1.3 Poly(acrylic acid) and Derivatives

Polymer	Method of Fractionation	Solvent or Solvent/Non-solvent Mixture	Remarks	References
Poly(acrylamide)				
	Fractional precipitation	Water/methanol	30°C	120, 121, 1334

Polymer	Method of Fractionation	Solvent or Solvent/Non-solvent Mixture	Remarks	References
Poly(acrylamide) (Cont' d.)				
	Fractional precipitation	Water/isopropanol		1536
	Chromatography	Water	GPC, sephadex	1091
	Electron microscopy	Water/n-propanol (dispersant agent)		1242
Poly(acrylic acid)				
	Fractional precipitation	Methanol/water	θ point phase separation, sodium salt	999
		Methanol/ethyl acetate	20°C	1478, 1483
Poly(benzyl acrylate) chromium tricarbonyl				
	Chromatography	Tetrahydrofuran	GPC	1206
Poly(tert-butyl acrylate)				
	Fractional precipitation	Methanol/methanol-water	25°C	708
Poly(1,1-dihydroperfluorobutyl) acrylate				
	Fractional precipitation	Benzotrifluoride/methanol		1252
Poly(ethyl acrylate)				
	Fractional precipitation	Acetone/water-methanol (1:5)	30°C	893-895
		Acetone/methanol		927
Poly(ferrocenylethyl acrylate)				
	Chromatography	Tetrahydrofuran	GPC	1205
Poly(ferrocenylmethyl acrylate)				
	Chromatography	Tetrahydrofuran	GPC	1204, 1207
Poly(isopropyl acrylate)				
	Fractional precipitation	Chlorobenzene/n-hexadecane	$110-115^{o}$C, stereoregular	933
		Benzene/n-hexane	25°C, atactic	933
Poly(methyl acrylate)				
	Fractional precipitation	Acetone/water-methanol		570, 904
		2-Butanone/methanol		833
		Liquid ethylene at 130 atm. and 80°C	Releasing pressure	1513
Poly(propyl acrylate)				
	Fractional precipitation	Acetone/methanol		927

1.4 Poly(methacrylic acid) and Derivatives

Polymer	Method of Fractionation	Solvent or Solvent/Non-solvent Mixture	Remarks	References
Poly(butyl methacrylate)				
	Fractional precipitation	Acetone/methanol	Lowering temp.	287
		Acetone/water		1525
	Chromatography	Benzene	GPC	612
	Thermal diffusion	Benzene		825
Poly(tert-butyl methacrylate)				
	Fractional precipitation	Acetone/methanol		834
	Sedimentation velocity	Butyl acetate	$20-21^{o}$C	141
Poly(carbethoxyphenyl methacrylamide)				
	Sedimentation velocity	Ethyl acetate	20°C	462
Poly(4-chlorophenyl methacrylate)				
	Fractional precipitation	Chloroform/methanol		1465
Poly(cyclohexyl methacrylate)				
	Fractional precipitation	Dioxane/methanol	Lowering temp.	973
Poly(dodecyl methacrylate)				
	Fractional precipitation	Toluene/methanol		1050
	Chromatography	Tetrahydrofuran	GPC, 25°C, branched	1050

Polymer	Method of Fractionation	Solvent or Solvent/Non-solvent Mixture	Remarks	References
Poly(ethylene glycol methacrylate)				
	Fractional precipitation	Ethanol/n-heptane-benzene (1:1)	30°C	187
		Ethanol-methanol (0.7:1)/n-heptane-benzene (1:1)	30°C	186,187,392
Poly(ethyl methacrylate)				
	Fractional precipitation	Acetone/acetone-water (4:1)	Lowering temp.	289,752
		Benzene/n-hexane		1530,1531
Poly(ferrocenylethyl methacrylate)				
	Chromatography	Tetrahydrofuran	GPC	1205
Poly(ferrocenylmethyl methacrylate)				
	Chromatography	Tetrahydrofuran	GPC	1204,1207
Poly(hexyl methacrylate)				
	Fractional precipitation	Acetone/ethanol	Lowering temp.	286
Poly(N-(2-hydroxyethyl)phthalimido methacrylate)				
	Fractional precipitation	Chloroform/petroleum ether	20°C	1622
Poly(isobutyl methacrylate)				
	Fractional precipitation	Acetone/methanol		1518,1521
Poly(isooctyl methacrylate)				
	Sedimentation velocity	Heptane-acetone	θ-Solvent	728
Poly(methacrylic acid)				
	Fractional precipitation	Methanol/ethyl ether	25°C	450,754,984
		Methanol/ethyl acetate-acetic acid		48,466,1100,1101
		Methanol/methyl isobutyl ketone		1574
	Electrophoresis	Water (Na$_2$HPO$_4$)	Stereoregular separation	49
Poly(methyl methacrylate)				
	Fractional precipitation	Acetone/aq. acetone		119
		Aq. acetone solutions	Lowering temp.	119
		Acetone/hexane	Summative method	166,801,941
		Acetone/n-heptane		639
		Acetone/methanol		181,183,295,454,457, 700,704,1395
		Acetone/methanol-water (1:1)		295,457
		Acetone/petroleum ether	20°C	420
		Acetone/skellsolve	Low M.W.	295,457
		Acetone/water		133,288,295, 457,683,1625
		Benzene/cyclohexane		418,420,584,784
		Benzene-chloroform (3:1)/petroleum ether		979
		Benzene/n-hexane	25°C	33,453,584,841,1029
		Benzene/isopropyl alcohol		755
		Benzene/methanol		66,147,572,699,1066, 1169,1170,1272-1275, 1624
		Benzene/petroleum ether	25°C	182,183,331,332, 1240,1431,1485
		Benzene/skellsolve		295,457,842
		2-Butanone-ethanol/cyclohexane		1559
		2-Butanone/n-hexane		262
		2-Butanone/methanol		262,833
		Carbon tetrachloride/skellsolve		295,454,457
		Chloroform/n-heptane	Lowering temp.	1372
		Chloroform/petroleum ether		1466
		Toluene/petroleum ether		317
	Turbidimetric titration	Acetone/water		19,596
		Benzene/cyclohexane		333

Polymer	Method of Fractionation	Solvent or Solvent/Non-solvent Mixture	Remarks	References
Poly(methyl methacrylate) (Cont'd.)				
	Turbidimetric titration	2-Butanone/water-methanol (3:1)		19
		Propanol/butanol	Lowering temp.	1022
	Fractional solution	Acetone/n-hexane	Column extraction, 20°C, glass beads, analytical and preparative	1056
		Acetone/methanol	Column extraction	364,1217,1218,1327
		Benzene/cyclohexane	Column extraction	1327
		Benzene/methanol	Column extraction	98
		2-Butanone/cyclohexane-methanol		1559
		Chloroform/methanol	Column extraction	203
		Toluene/petroleum ether		403
	Chromatography	Acetone	TLC	1010
			Activated carbon + supercel	293
			GPC	976,981
		Acetone/methanol	"Precipitation chromat.", periodic temp. changes	1216,1220
		Benzene/methanol	"Precipitation chromat."	357,572
		2-Butanone/2-butanone-70% ethanol		1559
		Chloroform	GPC, 33°C	77,351
		Chloroform/methanol	TLC	681
		Ethyl acetate	TLC	1010
		Trichloroethylene	GPC, 23°C	1447
		Tetrahydrofuran	GPC	21,65,145,663,664, 977,980,1236,1386
		2,2,2-Trifluoroethanol	GPC, 50°C	1236,1237
	Sedimentation velocity	Acetone	Ultracentrifuge	783
		Ethyl acetate	Ultracentrifuge	419,421
		Ethyl acetate/acetone	Ultracentrifuge	784
		Nitromethane	Ultracentrifuge, θ-temp. 28°C	1275
	Thermal diffusion	Acetone		467
		Benzene	Plate type apparatus	826,827
		Toluene		864,867
Poly(p-methylphenyl methacrylate)				
	Chromatography	Tetrahydrofuran	GPC	1568
Poly(1-naphthyl methacrylate)				
	Sedimentation velocity		$20\text{-}21^{\circ}$C	141

1.5 Other α- and β-Substituted Poly(acrylics) and Poly(methacrylics)

Polymer	Method of Fractionation	Solvent or Solvent/Non-solvent Mixture	Remarks	References
Poly(di-n-butyl itaconate)				
	Chromatography	Benzene/methanol	"Precipitation chromat."	1535
Poly(di-n-decyl itaconate)				
	Chromatography	Benzene/methanol	"Precipitation chromat."	1535
Poly(1,1-(diethoxycarbonyl)ethylene)				
	Fractional precipitation	Tetrahydrofuran/water	50°C	698
	Chromatography	Chloroform	GPC	698
Poly(methyl butacrylate)				
	Fractional solution	Acetone/methanol	Column elution	696
		Benzene/n-hexane	Column elution	696

1.6 Poly(vinyl ethers)

Polymer	Method of Fractionation	Solvent or Solvent/Non-solvent Mixture	Remarks	References
Poly(butoxyethylene)				
	Fractional precipitation	Benzene/methanol		1363
Poly(hexadecyloxyethylene)				
	Fractional precipitation	Heptane/acetone		822

Polymer	Method of Fractionation	Solvent or Solvent/Non-solvent Mixture	Remarks	References
Poly(isobutoxyethylene)				
	Fractional precipitation	Toluene-butanone (1:1)/ethanol	Lowering temp., oppanol C	1304
	Extraction	Acetone, benzene, ethyl ether, methylene chloride		914
Poly(isopropyloxyethylene)				
	Extraction	Acetone, benzene, ethyl ether, methylene chloride		914
Poly(methoxyethylene)				
	Fractional precipiation	Benzene/n-hexane, heptane or decane		929
		Methanol	Lowering temp., 65-3°C	1523
		Water	Raising temp.	929
Poly(vinyl isobutyl ether) see Poly(isobutoxyethylene)				
Poly(vinyl methyl ether) see Poly(methoxyethylene)				

1.7 Poly(vinyl alcohol), Poly(vinyl ketones), Poly(vinyl halides), Poly(vinyl nitriles)

Polymer	Method of Fractionation	Solvent or Solvent/Non-solvent Mixture	Remarks	References
Poly(acrylonitrile)				
	Fractional precipitation	Dimethylformamide/decalin	110 to 90°C	1323
		Dimethylformamide/ethanol	pH 6.0, ageing period	997
		Dimethylformamide/heptane	60°C	129,168,195,291,477, 478,655,762,993
		Dimethylformamide/heptane-ether (1:1)		622,1428
		Dimethylformamide/ligroin	50-60°C	538
		Dimethyl sulfoxide/toluene	35°C	479,480,680, 732,735,800
		Dimethyl sulfoxide-aq. HCl/toluene	35°C	796
		Hydroxyacetonitrile-ethanol/benzene or toluene		795
		60% Nitric acid/butanol	5°C	478
		Dimethylformamide/lauryl alcohol		1408
	Fractional solution	Dimethylformamide/butyl acetate	90°C, film extr.	471
		Dimethylformamide/heptane	Extraction	845
	Thermal diffusion	Dimethylformamide		864,865,867
Poly(chlorotrifluoroethylene)				
	Fractional precipitation	Dichlorobenzotrifluoride/diethyl phthalate (3:1)	Solvent volatization, 150°C, Kel-F	766,1119
Poly(diketene)				
	Extraction	Acetone, dioxane		483,765
		Methanol, ethyl ether, benzene, acetone	Boiling	1600
		Ethyl ether, acetone, benzene, toluene	Boiling	1613
Poly(isopropenyl methyl ketone)				
	Chromatography	Tetrahydrofuran	GPC	918
Poly(phenyl vinyl ketone)				
	Extraction	Acetone	Sol. and insol.	1470
	Chromatography	Tetrahydrofuran	GPC	532
Poly(vinyl alcohol)				
	Fractional precipitation	Water/acetone		374,376,866,1425
		Water/acetone-propanol		1461
		Water/methyl acetate-methanol (3:1)	Film extraction	468
		Water/n-propanol		143,954,1392 1585,1604
	Chromatography	Water	GPC	847
	Sedimentation velocity	Water	Ultracentrifuge	463

Polymer	Method of Fractionation	Solvent or Solvent/Non-solvent Mixture	Remarks	References
Poly(vinyl alcohol) (Cont' d.)				
	Thermal diffusion	Water		467, 864, 866, 868
	Brownian diffusion	Water		463
	Fractional crystallization	Water	Stirring	1602-1604
	Foam method	Water	Amorphous, crystalline	679, 782
Poly(vinyl chloride)				
	Fractional precipitation	Acetone-chlorobenzene-cyclohexane/-methanol		993
		Cyclohexanone/n-butanol		382
		Cyclohexanone/ethylene glycol		498, 968, 1060
		Cyclohexanone/methanol		222, 530, 565
		Cyclohexanone/methanol-water		1619
		Liquid ethylene at 132 bar and 80°C	Releasing pressure	1513
		Tetrahydrofuran/ethanol		631
		Tetrahydrofuran/methanol		139, 530
		Tetrahydrofuran/water		69, 113, 184, 185, 797, 816, 838, 968, 1030, 1085, 1192, 1195, 1196, 1426, 1528, 1537
	Turbidimetric titration	Cyclohexanone/heptane-carbon tetra-chloride (9:1)		1136
	Extraction	Mixtures, ethyl alcohol, methyl ethyl ketone, tetrahydrofuran, chloroform		888
		Acetone, chloroform, tetrahydrofuran	Hot and cold	1123
	Fractional solution	Cyclohexanone/ethylene glycol	Column extraction	497, 498
		Cyclohexanone/methanol	Film extraction	574
		Methyl isobutyl ketone		863
		Tetrahydrofuran/water	44°C	1407
	Chromatography	Benzene-tetrahydrofuran (6:4)	GPC	920
		Cyclohexanone/methanol	"Precipitation chromat.", $25-60^{o}$C	335
		Tetrahydrofuran	GPC, 25°C	275, 436, 528, 529, 566, 886, 1104, 1238, 1271, 1442, 1463, 1467
	Thermal diffusion	Cyclohexanone		575, 864, 867
Poly(vinyl fluoride)				
	Fractional precipitation	Dimethylformamide/water		1614
Poly(vinyl methyl ketone)				
	Extraction	Acetone	Soluble and insoluble	1470

1.8 Poly(vinyl esters)

Poly(vinyl acetate)				
	Fractional precipitation	Acetone/hexane	25°C	453, 1031, 1364
		Acetone/n-heptane	35°C, branched	635, 636
		Acetone/pentane		869
		Acetone/petroleum ether		1003, 1249, 1424
		Acetone/water		157, 470, 1011, 1086, 1128
		Acetone/water-methanol (1:1)		559, 560, 748, 1546
		Benzene/cyclohexane	25°C	1170
		Benzene/isopropanol		407
		Butanone/cyclohexane	25°C	1162
		Chlorobenzene/cyclohexane		1162
		Dioxane/isopropanol		407
		Dioxane/n-heptane	25°C	1162
		Ethyl acetate/cyclohexane	25°C	790, 1162
		Methanol/water	35°C, branched	152, 154, 901

Polymer	Method of Fractionation	Solvent or Solvent/Non-solvent Mixture	Remarks	References
Poly(vinyl acetate) (Cont'd.)				
	Turbidimetric titration	Acetone/water		19, 1048
	Fractional solution	Methyl and ethyl acetate/petroleum ether	Film extraction	143, 469, 471
		Methyl isobutyl ketone		863
	Distribution between two immiscible liquids	Benzene-methanol-water	Countercurrent	1264
	Chromatography	Acetone	Activated carbon + supercel	293
		Benzene/isopropanol	"Precipitation chromat.", 28-60°C	255
		Tetrahydrofuran	GPC	528, 529, 1104, 1299
	Thermal diffusion	Benzene		869
		Chloroform		869
		Dioxane		869
		Ethyl acetate		869
		Isoamyl acetate		869
		Methanol		869
		Toluene		864, 869
	Selective adsorption	Acetone	Carbon	747
Poly(vinyl p-trimethyl silyl benzoate)				
	Fractional precipitation	Benzene/methanol		650
1.9 Poly(styrenes)				
Poly(chlorostyrene)				
	Fractional precipitation	Benzene/methanol	35°C, (ortho, meta and para subst.)	95, 951-953, 848, 1018
		Butanone/methanol	(para subst.)	1429
Poly(p-methoxystyrene)				
	Fractional solution	Butanone/ethanol	Column extr., 25°C, glass beads	266, 1209
Poly(m-methylstyrene)				
	Fractional precipitation	Benzene/methanol	35, 8°C	280
Poly(α-methylstyrene)				
	Fractional precipitation	Benzene/methanol		316, 324, 756
		Toluene/methanol	25°C	1113
	Turbidimetric titration	Methylcyclohexane	Lowering temp.	311, 1234
	Fractional solution	Cyclohexane	Column extraction, increasing temp., 10-22°C	1597
	Coacervation	Toluene/methanol		240
	Chromatography	Benzene/n-hexane	"Precipitation chromat.", 10-58°C	1597
		Tetrahydrofuran	GPC	406, 482, 764, 1588, 1597
	Sedimentation velocity	Cyclohexane	35°C	476, 959, 1597
Poly(styrene)				
	Fractional precipitation	Acetone/methanol-water		455, 1392, 1586
			Anionic monodisperse, 30°C	1611
		Acetone-2-butanone/methanol		603
		Benzene/ethanol		694
		Benzene/butanol	30°C	685
		Benzene/methanol		28, 71, 142, 181, 210, 244, 358, 475, 789-791, 807, 844, 854, 924, 1066, 1087, 1162, 1193, 1241, 1276, 1303, 1396, 1290, 1424, 1468, 1584
			Isotactic	210, 1475
			Star and comb-branched	250, 1624
			Cumulative volume of precipitate	204
		Benzene/n-propanol	Solvent, volatilization, 30°C	683
		Benzene-tetrahydrofuran/methanol	Comb-shaped	1112

Polymer	Method of Fractionation	Solvent or Solvent/Non-solvent Mixture	Remarks	References
Poly(styrene) (Cont'd.)				
	Fractional precipitation	2-Butanone/methanol		6, 426, 455, 569, 579, 705, 769, 884, 974, 1128, 1522, 1564, 1565
			Star and comb-shaped	28, 253, 1276, 1611
		2-Butanone/ethanol		158
		2-Butanone/butanol + 2 % water	Solvent volatilization	294, 531, 542
		Chloroform/methanol		710, 1372
		Carbon tetrachloride/ethanol		377
		Chlorobenzene/cyclohexanol	$8\,^{\circ}$C, isotactic	1514
		o-Dichlorobenzene/poly(oxyethylene) 400	$50\,^{\circ}$C, isotactic	1082
		Ethyl acetate/ethanol	$20\,^{\circ}$C	458
		Methyl acetate	$120\,^{\circ}$C, isothermal at LCST with pressure variation	1069
		Methylcyclohexane	Lowering temp., atactic	734
		Liquid ethylene at 130 atm. and $80\,^{\circ}$C	Releasing pressure	1513
		Toluene/n-decane		1233
		Toluene/methanol		579, 744-746, 1094, 1112, 1131
		Toluene/petroleum ether		587
		Toluene/poly(oxyethylene)	Isotactic	768, 770
	Turbidimetric titration	Benzene/benzene-methanol (1:4)		1287
		Benzene/isopropanol		405
		Benzene/ethanol		1243, 1287
		Benzene/methanol		19, 333, 536, 598, 792, 793, 967, 1243, 1287
		Benzene/5 % water + methanol		1243
		2-Butanone/ethanol	Star-branched	233
		2-Butanone/isopropanol	$30\,^{\circ}$C	124, 1510
		2-Butanone/water-acetone		619
		2-Butanone/water-methanol (3:1)		19
		Cyclohexanol	Lowering temp.	1022
		Chloroform/isopropanol	$30\,^{\circ}$C	722
		Cyclohexane/ethanol		1243
		Cyclohexane-toluene/ethanol		1243
		Dioxane/methanol		1243
		Dioxane/water		1243
		Methylcyclohexane	Lowering temp., anionic	1255, 1396
		Toluene/n-butanol	Lowering temp.	133
		Toluene/methanol		945
	Extraction	Butanone, ethyl ether, tetrahydrofuran, ethyl acetate, benzene and toluene	Cold and hot	1590
		Butyl acetate-n-propanol (7:3)	Increasing temp.	788
	Fractional solution	Acetone/methanol-water		27
		Benzene/ethanol	Column extraction	9
		Benzene/methanol		160, 594, 595
		Butanone/ethanol	Column extraction	1314
		Butanone/methanol	Column extraction, $25\,^{\circ}$C, isotactic	778
		Carbon disulfide/petroleum ether	Film extraction	471
	Coacervation	Benzene/methanol		547
		Butanone/butanol		547
		Chloroform/butanol		547
		Toluene/butanol		547
	Chromatography	Acetone and acetone-tetrahydrofuran (7:3)	TLC, conc. gradient	1139
		Acetone-chloroform (8:2)	TLC, constant conc.	1139
		Benzene	GPC	499, 612, 624, 808, 920
			GPC, $205\,^{\circ}$C, "Supercritical fluid chromat.", 41-62 bar	707
		Benzene, carbon tetrachloride, methanol, petroleum ether		27

Polymer	Method of Fractionation	Solvent or Solvent/Non-solvent Mixture	Remarks	References
Poly(styrene) (Cont' d.)				
	Chromatography	Benzene/ethanol	"Precipitation chromat."	99, 216, 440, 718, 899, 986, 1125, 1126, 1187, 1308, 1331
		Benzene-hexane (34.7 : 65.3)	TLC, gradient conc., θ-temp.=25°C	1138
		Benzene/isopropanol	TLC, gradient conc., θ-temp.=25°C	1137, 1138
		Benzene/methanol		244, 807, 872, 1321
		Benzene, acetone, butanone, ethanol	TLC, gradient conc.	737
		Aromatic and chlorinated solvents	GPC	1025
		2-Butanone/ethanol	"Precipitation chromat."	76, 212-214, 216, 646, 721, 975, 1187, 1322, 1313, 1315
			Charcoal	1610
		2-Butanone/methanol	"Precipitation chromat."	541, 1594
		Carbon tetrachloride	GPC	302, 353
		Chlorobenzene	GPC, 25°C	211
		Chloroform	GPC	926
		Chloroform-cyclohexane (3:1)	GPC	369, 370
		Chloroform-n-hexane (2:8)	GPC	1148
		m-Cresol	GPC, 100°C	176
		Cyclohexane		265, 948
		Cyclohexane-butanone (50:2)	TLC, gradient conc.	681
		o-Dichlorobenzene	GPC, 130°C	106, 165, 352, 608, 609
		Dichloroethane and dichloroethane - isopropanol	GPC, 22°C	1368
		Dioxane-methanol (71.4 : 28.6)	TLC, gradient conc., θ-temp.=25°C	1137, 1140
		Dioxane-propanol (55 : 45)	TLC, gradient conc., θ-temp.=25°C	1140
		Tetrahydrofuran	GPC, 25°C	23-25, 190, 191, 391, 886, 978, 982, 1104, 1283, 1561
			GPC	105, 106, 140, 192, 243, 250, 301, 303, 369, 390, 482, 528, 610, 611, 665, 666, 885, 898, 921, 922, 940, 976, 977, 980, 1129, 1132, 1133, 1149, 1157, 1179, 1325, 1358, 1508
			GPC, 50°C	859, 957, 1386, 1387
		Tetrahydrofuran/acetone		400
		Toluene	GPC, 25°C	234, 370, 871, 892, 1025, 1159, 1606
		Toluene/isopropanol	"Precipitation chromat."	661
		Toluene/methanol		986, 1330
		Trichlorobenzene	GPC	662, 1573
		Trichloroethylene		1447
	Sedimentation velocity	Butanone	Ultracentrifuge	133, 769, 1549, 1552
		Bromoform		1370
		Cyclohexane	Ultracentrifuge, 35°C	227, 228, 508, 829, 916, 958, 1118, 1303, 1325, 1550, 1584, 1587
		Ethyl acetate	Ultracentrifuge	769
		Ethylene bromide		1370
		Toluene	23, 5°C	257
	Sedimentation equilibrium	Cyclohexane	35°C	251, 1324
		Methylcyclohexane	75°C	251
	Sedimentation density gradient	Carbon tetrachloride, cyclohexanol		621
	Thermal diffusion	Butanone		1440
		Dioxane, ethyl benzene, pyridine, styrene, o-xylene, toluene		411
		Toluene, butyl acetate		642
		Toluene		864, 867, 1440

Polymer	Method of Fractionation	Solvent or Solvent/Non-solvent Mixture	Remarks	References
Poly(styrene) (Cont'd.)				
	Thermal diffusion	Toluene	Thermal field flow method	1449
	Brownian diffusion	Benzene, cyclohexane		947
	Membrane diffusion	Benzene, chlorobenzene		988
		Toluene	Maranyl	546
	Zone melting	Benzene		896, 897
		Naphthalene		1180
		Cyclohexane		1285
Poly(styrenesulfonic acid), ammonium salt				
	Fractional solution	Water	Ion exchange resins	597
Poly(styrenesulfonic acid), sodium salt				
	Fractional precipitation	4 N aq. sodium iodide/9 N aq. sodium iodide		934

1.10 Others

Polymer	Method of Fractionation	Solvent or Solvent/Non-solvent Mixture	Remarks	References
Poly(1,2-acenaphthenylene)				
	Fractional precipitation	Benzene/n-hexane	$20^{o}C$, preparative	103
		Benzene/methanol	$20^{o}C$	102, 103, 1013
Asphaltenes				
	Chromatography	Benzene-methanol	GPC	30
Poly((9-acrydinyl)ethylene)				
	Chromatography	0.02 M Hydrochloric acid	Sephadex G-25, GPC	644
Poly(N-allylstearamide) see Poly((stearoyliminomethyl)ethylene)				
Poly(benzylethylene)				
	Extraction	Acetone, ethyl ether, benzene		1458
		Butanone, toluene	Hot	340
Poly(carbazolylethylene)				
	Fractional precipitation	Benzene/methanol		1071, 1377, 1499, 1501, 1503
		Tetrahydrofuran/water		1071, 1503
	Coacervation	Benzene/methanol		852
	Chromatography	Tetrahydrofuran	GPC, $25^{o}C$	626, 1377
Poly(carboxychloromethylethylene)				
	Fractional precipitation	Methanol/water		802
Poly(1-cyano-2-phenylvinylene)				
	Chromatography	Benzene/methanol	Column elution, CaOH : kieselgur (1 : 1)	701
Poly(diphenylethylene)				
	Fractional precipitation	Benzene/methanol		1517
Poly(ferrocenylethylene)				
	Chromatography	Tetrahydrofuran	GPC	1204, 1207
Poly(indolylethylene)				
	Fractional precipitation	Acetone/ethanol	$20^{o}C$	378
Poly(1-iodo-2-phenylvinylene)				
	Fractional precipitation	Benzene/methanol	Film	283
Poly(pyridylethylene)				
	Fractional precipitation	Benzene/n-hexane	(2-subst.), $25^{o}C$	381, 676
		Butanone/ligroin	(2-subst.), $30^{o}C$	51
		Methanol/ethyl ether-benzene (4:1)	(2-methyl-5-subst.)	1479
		Methanol/water	(2-methyl-5-subst.)	1009
		Methanol/toluene	(4-subst.), $15^{o}C$	148
		Nitroethane/benzene	(4-subst.)	443
		tert-Butanol/benzene	(4-subst.)	443

Polymer	Method of Fractionation	Solvent or Solvent/Non-solvent Mixture	Remarks	References
Poly(pyridylethylene) (Cont'd.)				
	Fractional precipitation	Tetrahydrofuran/n-heptane	(2-subst.), atactic and syndiot.	912
	Fractional solution	Benzene/n-heptane	Column extraction	518
		Cyclohexanone/n-heptane	Column extraction, 70° C (2-methyl-5-subst.)	496
	Chromatography	Dimethylformamide	(2-subst.), 50° C	640
	Sedimentation velocity	Water 0.1 N KBr	25° C, (ethyl bromide salt, 4-subst.)	1062
Poly(stearoyliminomethyl)ethylene				
	Fractional solution	Benzene/ethanol	Column extraction	716
Poly(trimethylsilylethylene)				
	Fractional precipitation	Toluene/methanol		389
Poly(vinyl butyral)				
	Fractional precipitation	Isopropanol/water		630
Poly(N-vinylcarbazol) see Poly(carbazolylethylene)				
Poly(vinyl indol) see Poly(indolylethylene)				
Poly(vinyl pyridine) see Poly(pyridylethylene)				
Poly(vinylpyrrolidone)				
	Fractional precipitation	Chloroform/ethyl ether		709
		Ethanol/benzene		709
		Ethanol/petroleum ether		1317
		Methanol/ethyl ether		996
		Water/acetone		249,271,373,492, 709,1317
	Fractional solution	Acetone	Extraction	996
		Water/organic acids	25° C	1347
	Turbidimetric titration	Water/acetone		709
		Water/ammonium sulfate solutions	25° C	709
		Water/sodium sulfate solutions		249,271,619,1318
	Chromatography	Water	GPC, dextran crosslinked with epichlorohydrin	488
	Sedimentation velocity	Water	Ultracentrifuge	996,1319
	Thermal diffusion	Water and ethanol		864,866,868,870
	Ultrafiltration	Water	Amicon XM-4B and UM-1 Diaflo-membranes	75

1.11 Random Copolymers

Polymer	Method of Fractionation	Solvent or Solvent/Non-solvent Mixture	Remarks	References
Poly(acrylamide-co-sodium maleinate)				
	Fractional precipiation	Water/acetone-water		1063,1064
		Water/methanol-water		1063,1064
		Water/acetone		1063
		Water/methanol		1063
Poly(acrylic acid-co-maleic anhydride)				
	Extraction	Acetone	PAA homopolymer	409
Poly(acrylonitrile-co-methyl acrylate)				
	Fractional precipitation	Dimethylformamide/petroleum ether-methylcyclohexane-dioxane (6 : 4 : 1)		1323
Poly(acrylonitrile-co-butadiene)				
	Fractional precipitation	Benzene/methanol		1213,1214
		Butanone/methanol		1213
	Chromatography	Benzene/methanol		1321

Polymer	Method of Fractionation	Solvent or Solvent/Non-solvent Mixture	Remarks	References
Poly(acrylonitrile-co-butadiene)				
	Chromatography	Toluene/methanol	Perbunan, charcoal	862
		Tetrahydrofuran	GPC	1238
	Sedimentation velocity	Acetone-isopropanol (61 : 39)	Ultracentrifuge	1215
Poly(acrylonitrile-co-methyl methacrylate)				
	Fractional precipitation	Dimethylformamide/n-hexane-ethyl ether (2:1)		318
		Dimethylformamide/n-hexane	60°C	1033
		Acetone/methanol		1359
	Fractional solution	Butanone/methanol	Column extraction, cellite	517
	Extraction;precipitation	Chloroform, dimethylformamide/methanol		268
		Chloroform, benzene/methanol		269
Poly(acrylonitrile-co-methylvinylpyridine-co-vinyl acetate)				
	Fractional precipitation	Dimethylformamide/n-hexane-ethyl ether (2:1)		318
Poly(acrylonitrile-co-α-methylstyrene-co-vinyl acetate)				
	Fractional precipitation	Dimethylformamide/n-heptane	70°C	434
	Fractional solution	Dimethylformamide/n-heptane	Column extraction, 70°C	434
Poly(acrylonitrile-co-methyl acrylate)				
	Fractional precipitation	Dimethylformamide/heptane-ethyl ether (1:1)		1427
Poly(acrylonitrile-co-styrene)				
	Fractional precipitation	Butanone/cyclohexane	35°C, fractionation by chemical composition	1445
		Chloroform/methanol		1360,1615
		Tetrahydrofuran/petroleum ether		1340
		Toluene	Lowering temp., 66,0-2,5^{o}C, chemical composition	1446
	Coacervation	Benzene/triethylene glycol	60°C	910,911
		Dichloromethane/triethylene glycol	25°C	910,911
	Fractional solution	Acetone/methanol	Column extraction	910,911
		Dichloromethane/methanol		910,911
		Dichloromethane/triethylene glycol	25°C	911
		Benzene/triethylene glycol	25°C	911
	Chromatography	Butanone/cyclohexane		1321
Poly(acrylonitrile-co-vinyl chloride)				
	Fractional precipitation	Acetone/methanol		1339
Poly(acrylonitrile-co-vinylidene chloride)				
	Fractional precipitation	Tetrahydrofuran/n-heptane		1553
Poly(acrylonitrile-co-acrylic acid-co-butadiene)				
	Chromatography	Acetone/methanol-water (2:3)	"Precipitation chromat.", 42-2°C	876,879
Poly(di-n-alkyl itaconates-co-styrene)				
	Fractional solution	Benzene/methanol	Column extraction	1533
Poly(allylsilane-co-propylene)				
	Fractional precipitation	Xylene	Decreasing temperature (82-0°C)	902
Poly(butadiene-co-ethylene)				
	Extraction	Ethyl acetate, n-hexane and n-pentane	Boiling points	1099
Poly(butadiene-co-isoprene)				
	Fractional precipitation	Benzene/methanol	3,4-Isoprene, 1,2-butadiene	1097
		Benzene/butanone followed by benzene/-methanol	3,4-Isoprene, 1,2-butadiene	1097
	Extraction	Acetone	3,4-Isoprene, 1,2-butadiene	1097

Polymer	Method of Fractionation	Solvent or Solvent/Non-solvent Mixture	Remarks	References
Poly(butadiene-co-propylene)				
	Fractional precipitation	Benzene/butanone		1414
		Benzene/methanol		1414
Poly(butadiene-co-styrene)				
	Fractional precipitation	Benzene/acetone		601
		Benzene/methanol	SBR	414, 647, 1607
			Summative method	533
		Toluene/ethanol	SBR	503
		Toluene/methanol	SBR	194, 362, 711
	Extraction	Hexane/dimethylformamide, dioxane	"Two layer extraction method"	692
	Selective adsorption	Benzene	Carbon black	534
	Chromatography	Benzene/ethanol	"Precipitation chromat."	307, 1506, 1507
		o-Dichlorobenzene	GPC, 80°C	1178
		Tetrahydrofuran	GPC	592, 839, 1178, 1238, 1508
		Toluene/methanol	Charcoal	862
		Toluene/n-butanol, isooctane, heptane	"Precipitation chromat.", 85-25°C	872
	Membrane diffusion	n-Heptane		220
Poly(butadiene-co-styrene)				
	Sedimentation velocity	Methyl n-propyl ketone	SBR	647
		n-Octane	21°C	729, 1215
		n-Propyl ketone	SBR	646
Poly(butadiene-co-vinyl isopropyl ether)				
	Fractional precipitation	Benzene/methanol		1511
Poly(bis-acrylic acid phenylesters-co-styrene)				
	Fractional precipitation	Toluene /methanol	See Ref. for exact constitution of the phenyl ester	742, 743
Poly(1-butene-co-propylene)				
	Chromatography	Tetralin	Column extraction, increasing temp.	1378
Poly(n-butyl acrylate-co-styrene)				
	Chromatography	Butanone/methanol	"Precipitation chromat.", glass beads	928
Poly(butyl methacrylate-co-glycidyl methacrylate-co-methacrylic acid)				
	Fractional precipitation	Dioxane/water	20°C	1391
Poly(cumarone-co-indene)				
	Fractional precipitation	Benzene-ethyl acetate/methanol		1621
Poly(chloroprene-co-dichlorobutadiene)				
	Fractional precipitation	Benzene/methanol	20°C	520
Poly(p-chlorostyrene-co-methyl methacrylate)				
	Fractional precipitation	Benzene/methanol	27°C	1017
Poly(dimethyl itaconate-co-styrene)				
	Chromatography	Benzene/methanol	"Precipitation chromat.", 50-20°C	1534
Poly(divinyl ether-co-maleic anhydride)				
	Fractional precipitation	Acetone/n-hexane	Cyclocopolymer	22
Poly(ethylene-co-1-butene)				
	Chromatography	Dichlorobenzene	GPC	648, 649, 919
Poly(ethylene-co-α-methylstyrene)				
	Fractional precipitation	Benzene/ethanol	Block copolymer	1436
Poly(ethylene-co-propylene)				
	Fractional precipitation	Benzene/butanone		1414
		Butyl ether/butanol	120°C	470
		Isooctane/ethanol	25°C	273
		Xylene/dimethylformamide	85°C	470

Polymer	Method of Fractionation	Solvent or Solvent/Non-solvent Mixture	Remarks	References
Poly(ethylene-co-propylene) (Cont'd.)				
	Turbidimetric titration	Heptane - n-propanol	Decreasing temp., 80-25°C	494
	Fractional solution	Toluene/n-butanol	Column extraction, 58°C	516
		Xylene/cellosolve	Column extraction, 126°C	771
		Xylene/ethylene glycol	Column extraction	516
	Chromatography	o-Dichlorobenzene	GPC	648,649
		1,2,4-Trichlorobenzene	GPC, 130°C	96,273,592,771
Poly(ethylene-co-1-isopropylidene-3a,4,7,7a-tetrahydroindene)				
	Chromatography	1,2,4-Trichlorobenzene	GPC, 130°C	273
Poly(ethylene-co-propylene-co-1-isopropylidene-3a,4,7,7a-tetrahydroindene)				
	Chromatography	1,2,4-Trichlorobenzene	GPC, 130°C	273
Poly(ethylene-co-vinyl acetate)				
	Fractional precipitation	Benzene/isopropanol	25°C	930
Poly(isooctyl methacrylate-co-styrene)				
	Sedimentation velocity	Butanone/methanol	θ solvent	728
Poly(isobutene-co-isoprene)				
	Fractional precipitation	Toluene-acetone (9:1)/acetone	Butyl rubber	1464
Poly(isoprene-co-styrene)				
	Fractional precipitation	Benzene/methanol		1605
Poly(maleic anhydride-co-styrene)				
	Fractional precipitation	Acetone/benzene		438
		Acetone/petroleum ether		415
		Benzene/petroleum ether		1114
Poly(maleic anhydride-co-vinyl ethyl ether)				
	Fractional precipitation	Butanone/methanol-water		682,949
	Chromatography	Butanone	TLC, silica-alumina	682
		Butanone/methanol	"Precipitation chromat."	928
	Sedimentation gradient	Butanone (30%)-1,2-dibromo-1,1-difluoro-ethane (13%)-isopropanol (29.4%)-dichlorooctafluorocyclohexane-1 (27.6%)	Ultracentrifuge	1088
Poly(methyl-12-acrylostearate-co-vinyl chloride)				
	Fractional precipitation	Tetrahydrofuran/methanol		290
Poly(3-methyl butene-1-co-2,4-dimethyl-2,7-octadiene)				
	Extraction	n-Heptane		1201
Poly(methyl methacrylate-co-cetyl methacrylate)				
	Fractional precipitation	Benzene/methanol		144
Poly(methyl methacrylate-co-dimsylsodium modified methyl methacrylate)				
	Fractional precipitation	Chloroform/petroleum ether 25°C		54,55
Poly(methyl methacrylate-co-ethyl acrylate)				
	Fractional precipitation	Acetone/water		522
Poly(methyl methacrylate-co-ethylene dimethacrylate)				
	Fractional solution	Butanone/ethanol	Column extraction, 35°C	730
Poly(methyl methacrylate-co-methacrylic acid)				
	Fractional precipitation	Acetone/petroleum ether-acetone (4:1)		1067
		Acetone-dimethylformamide (19:1)/petroleum ether-acetone		1065
		Dioxane/dioxane-benzene		132,1065
Poly(methyl methacrylate-co-methyl acrylate)				
	Fractional precipitation	Acetone/methanol		833
Poly(methyl methacrylate-co-methyl vinyl ketone)				
	Chromatography	Tetrahydrofuran	GPC	36

Polymer	Method of Fractionation	Solvent or Solvent/Non-solvent Mixture	Remarks	References
Poly(methyl methacrylate-co-2-methyl-5-vinylpyridine)				
	Fractional precipitation	Acetone/n-heptane		803
Poly(methyl methacrylate-co-styrene)				
	Fractional precipitation	Acetone/acetonitrile	25 $^{\circ}$C	1444
		Benzene/methanol	30 $^{\circ}$C	71,380
		Butanone/methanol		832
		Butanone/isopropyl ether		341,830,831,1412
		n-Butyl chloride/cyclohexane	25 $^{\circ}$C	1444
		Toluene/acetonitrile		849,891
		Toluene/hexane-methanol		849,891
	Coacervation	Benzene/triethylene glycol	50 $^{\circ}$C	909
	Fractional solution	Acetone/methanol	Column extraction, 30 $^{\circ}$C	909
		Carbon tetrachloride/methanol	Column extraction, 30 $^{\circ}$C	909
		Dichloromethane/methanol	Column extraction, 25 $^{\circ}$C	909
	Chromatography	Benzene/methanol	"Precipitation chromat.", 15-40 $^{\circ}$C	2
		Benzene/petroleum ether		2
Poly(monobutyl itaconate-co-dibutyl itaconate)				
	Fractional precipitation	Acetone/petroleum ether		131
Poly(1-octadecene-co-1-dodecene)				
	Fractional solution	Benzene/ethanol	Column extraction, glass beads	451
	Chromatography	Benzene/ethanol	"Precipitation chromat.", 23-73 $^{\circ}$C	451
Poly(phenyl vinyl ketone-co-styrene)				
	Chromatography	Tetrahydrofuran	GPC	532
Poly(styrene-co-p-bromostyrene)				
	Fractional precipitation	Chloroform/petroleum ether		623
Poly(styrene-co-p-tert-butylstyrene)				
	Extraction	Toluene/methanol		1147
Poly(styrene-co-divinyl benzene)				
	Fractional precipitation	Acetone-dioxane (1:1)/methanol (or butanone)		1451
	Sedimentation velocity	Cyclohexane	35 $^{\circ}$C	1490
		n-Octane	21 $^{\circ}$C	1212
Poly(styrene-co-2-ethylhexyl acrylate-co-glycidyl acrylate)				
	Fractional precipitation	Butanone/petroleum ether followed by methanol		1254
Poly(styrene-co-p-fluorostyrene)				
	Fractional precipitation	Tetrahydrofuran/methanol		1146
	Extraction	Heptane, butanone, tetrahydrofuran	Boiling	1146
Poly(styrene-co-p-iodostyrene)				
	Fractional solution	Benzene/methanol	Column extraction	208
		Chloroform/methanol		209
Poly(styrene-co-isobutene)				
	Fractional solution	Benzene/isopropanol	25 $^{\circ}$C	343
		Cyclohexane/n-propanol	25 $^{\circ}$C	343
Poly(styrene-co-1-heptene)				
	Fractional solution	Benzene/hexane	Column extraction, 45 $^{\circ}$C	74
Poly(styrene-co-isobutene)				
	Fractional solution	Benzene/isopropanol	25 $^{\circ}$C	343
		Cyclohexane/isopropanol	25 $^{\circ}$C	343
Poly(styrene-co-p-lauroyliminostyrene)				
	Fractional precipitation	Butanone/butanol	Preferential evaporation	1116

Polymer	Method of Fractionation	Solvent or Solvent/Non-solvent Mixture	Remarks	References
Poly(styrene-co-p-methoxystyrene)				
	Fractional solution	Butanone/ethanol	Column extraction, 25°C	1208
Poly(styrene-co-3-methyl-1-butene)				
	Fractional precipitation	Carbon tetrachloride/methanol		1145
Poly(styrene-co-4-methyl-1-pentene)				
	Fractional solution	Benzene/hexane	Column extraction, 45°C	74
	Fractional precipitation	Carbon tetrachloride/methanol		1145
Poly(styrene-co-o-methylstyrene)				
	Extraction	Toluene/methanol		1147
Poly(styrene-co-p-methylstyrene)				
	Fractional precipitation	Benzene/methanol		250
Poly(styrene-co-1-octadecene)				
	Fractional solution	Benzene/hexane	Column extraction, 45°C	74
Poly(styrene co-p-vinylbenzyl chloride)				
	Chromatography	Tetrahydrofuran	GPC	250
Poly(styrene-co-m-vinyltoluene)				
	Fractional precipitation	Butanone/butanol	Solvent volatilization	1019
Poly(styrene-co-p-vinyl-trans-stilbene)				
	Chromatography	Chlorobenzene	GPC, 25°C	211
Poly(tetrafluoroethylene-co-trifluoronitrosomethane)				
	Fractional precipitation	Freon 113/acetone		1051
Poly(vinyl-2-pyridine-co-vinyl-4-pyridine)				
	Fractional precipitation	Tetrahydrofuran-ethanol/n-heptane	20°C	555
Poly(N-vinylpyrrolidone-co-vinylamine)				
	Fractional precipitation	Methanol/acetone-ethyl ether		556
Poly(vinyl acetate-co-maleic anhydride)				
	Fractional precipitation	Acetone/petroleum ether	Butyl ester lactonized copolymer.	1046
Poly(vinyl acetate-co-vinylidene cyanide)				
	Fractional precipitation	Nitromethane/methanol-water (1:1), methanol	50°C	1608
Poly(vinyl acetate-co-vinyl chloride)				
	Fractional precipitation	Tetrahydrofuran/water	25°C	281
	Chromatography	Tetrahydrofuran	GPC	281
Poly(N-vinylcarbazole-co-p-methoxystyrene)				
	Fractional precipitation	Toluene/n-amyl alcohol	25°C, analytical	101

1.12 Blockcopolymers

Polymer	Method of Fractionation	Solvent or Solvent/Non-solvent Mixture	Remarks	References
Poly(acrylamide:methyl methacrylate)				
	Fractional precipitation	Water/methanol followed by water		1381
Poly(isobutene:styrene)				
	Fractional precipitation	Benzene/isopropanol		344
		Cyclohexane/n-propanol		344
	Chromatography	Benzene/isopropanol	"Precipitation chromat.", thermal grad.	344
		Cyclohexane/isopropanol		344
Poly(isoprene:styrene)				
	Turbidimetric titration	Toluene/methanol		1307
	Chromatography	Tetrahydrofuran	GPC, 42°C	330
	Sedimentation analysis	Carbon tetrachloride/cyclohexane		1200

References page IV-221

Polymer	Method of Fractionation	Solvent or Solvent/Non-solvent Mixture	Remarks	References
Poly(hexyl methacrylate :lauryl methacrylate)				
	Fractional precipitation	Benzene/methanol	20°C, anionic	7
Poly(methyl methacrylate :hexyl methacrylate)				
	Fractional precipitation	Benzene/methanol	20°C, anionic	7
Poly(methyl methacrylate :lauryl methacrylate)				
	Fractional precipitation	Benzene/methanol	20°C, anionic	7
		Benzene/methanol-isopropanol (1:1)	20°C, anionic	7
		Benzene/methanol-heptane (1:1)	20°C, anionic	7
Poly(methyl methacrylate :styrene)				
	Fractional precipitation	Benzene-acetone (1:1)/petroleum ether		1512
		Benzene-chlorobenzene (1:1)/methanol		64
		Benzene-chlorobenzene (1:1)/petroleum ether	anionic	464
		Benzene/methanol		683,1021
		Butanone/diisopropyl ether	30°C	840
		Chloroform/methanol		1381
	Turbidimetric titration	Acetone/water		967
		Benzene/methanol	25°C	683
		Butanone/isopropanol		1512
		Butanone/water-methanol (3:1)		1021
	Extraction	Acetone, tetralin, cyclohexane and acetonitrile		1512
		Acetone, tetralin		1021
	Fractional solution	Benzene/methanol	Column extraction	242
	Chromatography	Benzene/methanol	"Precipitation chromat."	1021
		Acetone/water	"Precipitation chromat."	215
		Tetrahydrofuran	GPC	118
		Chloroform-ethyl acetate (25:5) followed (25:11.5)	TLC	736
		Carbon tetrachloride-ethyl acetate (25:5) followed by (25:35)	TLC	736
Poly(methyl methacrylate :octadecyl methacrylate)				
	Fractional precipitation	Benzene/methanol	20°C, anionic	7
Poly(methyl methacrylate :tetrafluoroethylene)				
	Extraction	Butanone/ethyl ether (1:5), butanone		1560
Poly(methyl methacrylate :vinyl acetate)				
	Turbidimetric titration	Acetone/water		967
Poly(α-methylstyrene :isoprene)				
	Chromatography	Toluene	GPC, 70°C, triblock	439
	Sedimentation velocity	Tetrahydrofuran		439
		Cyclohexane		439
Poly(propylene :1-butene)				
	Fractional solution	Tetralin	Column extraction, increasing temp.	309
Poly(propylene : styrene)				
	Chromatography	Acetone/methanol	"Precipitation chromat.", periodic temp. changes, 26-38°C	447
Poly(styrene : butadiene)				
	Fractional precipitation	Benzene/butanone	Anionic, "cross" fractionation	1443
		Cyclohexane/isooctane		1443
	Fractional solution	Benzene	Extraction	1130
	Chromatography	Tetrahydrofuran	GPC, anionic, 25°C	35
			GPC, star branched, anionic	839
			GPC	1178,1284

Polymer	Method of Fractionation	Solvent or Solvent/Non-solvent Mixture	Remarks	References
Poly(styrene : dihydronaphthalene)				
	Extraction	Ethyl acetate, butanone		274
Poly(styrene : dimethyl siloxane)				
	Chromatography	Toluene	GPC, 20°C	349
Poly(styrene : α-methylstyrene)				
	Fractional precipitation	Butanone - benzene (6:1)/methanol + 0.01% $CaCl_2$		64
Poly(styrene : 2-vinylpyridine)				
	Fractional precipitation	Benzene/n-heptane		552, 553
Poly(styrene : 4-vinylpyridine)				
	Fractional precipitation	Benzene/n-heptane		554
Poly(4-vinylbiphenyl : isoprene)				
	Chromatography	Toluene	GPC, 70°C	615
Poly(4-vinylpyridine : 2-vinylpyridine)				
	Fractional precipitation	Tetrahydrofuran -methanol/n-heptane	20°C	555
Poly(2-vinylpyridine : methacrylic acid esters)				
	Turbidimetric titration			1410

1.13 Graftcopolymers

Polymer	Method of Fractionation	Solvent or Solvent/Non-solvent Mixture	Remarks	References
Acrylamide + poly(N-eosin vinylamine hydrochloride)				
	Fractional precipitation	Water/acetone		1381
Acrylamide + poly(vinyl alcohol)				
	Turbidimetric titration	Formic acid/butanone		571
		Water/ethanol		571
Acrylic acid + poly(N-eosin vinylamine hydrochloride)				
	Fractional precipitation	Water/acetone		1381
Acrylic acid + poly(ethylene)				
	Fractional precipitation	Xylene/methanol	130°C	1263
Acrylonitrile + poly(dimethylvinyl siloxane)				
	Extraction	Dimethylformamide	DMF(PAN)	130
		Boiling benzene	C_6H_6(PSiO)	130
Acrylonitrile + poly(N-eosin vinylamine hydrochloride)				
	Fractional precipitation	Water/acetone		1381
Acrylonitrile + poly(styrene-co-maleic anhydride)				
	Fractional solution	Aq. acetone, tetrahydrofuran, dimethyl-formamide	Extraction	1260
Acrylonitrile + poly(pyridylethylene)				
	Extraction	Ethanol	Homopolymer	706
	Fractional precipitation	Ethanol/dimethylformamide		706
Butadiene + poly(ethylene-co-propylene)				
	Chromatography	Toluene	GPC	34
Butyl methacrylate + poly(vinyl acetate)				
	Fractional precipitation	Acetone/methanol - water (1:1)		607
Ethyl methacrylate + chlorinated rubber				
	Fractional precipitation	Butanone/methanol		1250
Ethyl methacrylate + poly(vinyl chloride)				
	Fractional precipitation	Butanone/methanol		1250
Glycidyl methacrylate + poly(vinyl chloride)				
	Extraction; precipitation	Butanone, isopropanol		1253

References page IV-221

Polymer	Method of Fractionation	Solvent or Solvent/Non-solvent Mixture	Remarks	References
Isobutene + poly(chloromethyl styrene)				
	Fractional precipitation	Benzene - butanone (1 : 4)/methanol		805
Isopropylstyrene + poly(methyl methacrylate)				
	Fractional precipitation	Benzene/isopropanol		405
Methyl acrylate + chlorinated rubber				
	Fractional precipitation	Butanone/methanol		1250
Methyl acrylate + poly(vinyl chloride)				
	Fractional precipitation	Butanone/methanol		1250
Methyl acrylate + poly(styrene)				
	Fractional precipitation	Chloroform/methanol		1382
Methyl methacrylate + chlorinated rubber				
	Fractional precipitation	Butanone/methanol		1250
Methyl methacrylate + poly(isobutene)				
	Extraction	Butanone, cyclohexane		201
Methyl methacrylate + poly(p-isopropylstyrene)				
	Fractional precipitation	Benzene/methanol		946
Methyl methacrylate + poly(styrene)				
	Fractional precipitation	Chloroform/methanol		1380
		Benzene - chlorobenzene (1:1)/methanol		284
		Butanone/methanol - water (1:1)		1385
Methyl methacrylate + poly(styrene-co-maleic anhydride)				
	Fractional precipitation	Acetone/methanol		1260
Methyl methacrylate + rubber				
	Fractional precipitation	Acetone/hexane		801
	Extraction	Acetone		79
		Petroleum ether, acetone		47
Methyl methacrylate + poly(vinyl acetate)				
	Fractional precipitation	Acetone/methanol - water (1:1)		607
		Acetone/methanol - water (1:2)		1380
Methyl methacrylate + poly(vinyl alcohol)				
	Fractional precipitation	Benzene/n-butanol		1291
Methyl methacrylate + poly(vinyl benzoate)				
	Fractional precipitation	Acetone/methanol		1383
Methyl methacrylate + poly(vinyl chloride)				
	Fractional precipitation	Butanone/methanol		1250
		Dioxane/methanol		1380
	Extraction	Acetone		279
Styrene + acrylonitrile + poly(vinyl chloride)				
	Extraction	Toluene/dichloroethane		279
Styrene + poly(butadiene)				
	Turbidimetric titration	Benzene/acetone - methanol (9:1)		78
Styrene + poly(butyl methacrylate)				
	Fractional precipitation	Benzene/methanol		607
Styrene + poly(p-chlorostyrene)				
	Fractional precipitation	Benzene/methanol		1008
Styrene + poly(N-eosin vinylamine hydrochloride)				
	Fractional precipitation	Water/acetic acid		1381
Styrene + poly(ethyl acrylate)				
	Extraction	Cyclohexane, ethyl ether, acetonitrile		669
Styrene + poly(ethylene)				
	Fractional precipitation	Toluene/methanol	$90°C$	282

Polymer	Method of Fractionation	Solvent or Solvent/Non-solvent Mixture	Remarks	References
Styrene + poly(isobutene)				
	Fractional precipitation	Cyclohexane/n-propanol		278
	Turbidimetric titration	Toluene/methanol		835
	Extraction	Hexane, cyclohexane		1346
		Butanone, cyclohexane		201
	Chromatography	Benzene - cyclohexane/isopropanol	Fractionation on the basis of M.W., "Precipitation chromat."	342
		Cyclohexane/n-propanol	"Precipitation chromat."	278
Styrene + poly(methyl acrylate)				
	Fractional precipitation	Chloroform/methanol		1384
Styrene + poly(methyl methacrylate)				
	Fractional precipitation	Benzene - acetone (1:1)/methanol (0.1% $CaCl_2$)	Graft mastication	46
		Benzene - chlorobenzene (1:1)/petroleum ether	Graft mastication	46,493
		Benzene/petroleum ether		607
		Benzene/methanol		607
		Chloroform/methanol		1380,1384
	Turbidimetric titration	Benzene - chlorobenzene (1:1)/petroleum ether		1004
	Extraction	Ethyl ether, acetonitrile and benzene		1007
Styrene + poly(propylene)				
	Turbidimetric titration	Benzene/methanol		448,722
	Fractional solution	Benzene/methanol	Column extraction, 15°C	447,448
		Chloroform/isopropanol	Column extraction, 15°C	447,448,723
Styrene + poly(styrene-co-vinyl benzyl)				
	Fractional precipitation	Benzene/methanol		250
Styrene + poly(styrene-co-methyl methacrylate)				
	Fractional precipitation	Benzene/methanol		39
Styrene + poly(vinyl acetate)				
	Fractional precipitation	Benzene/petroleum ether		607
Styrene + poly(vinyl chloride)				
	Fractional precipitation	Benzene/methanol		493,1008
		Tetrahydrofuran/methanol		586
		Tetrahydrofuran/petroleum ether		586
		Tetrahydrofuran/water		586
	Extraction	Benzene (25°C), toluene (25°C), toluene (100°C)		279
		Toluene/dichloroethane	25-95°C	279
Styrene + maleic anhydride + poly(vinyl chloride)				
	Extraction	Butanone, ammonium hydroxide		881
Vinyl acetate + poly(ethyl α-chloroacrylate)				
	Fractional precipitation	Acetone/ethanol		1385
		Butanone/methanol - water (1:1)		1385
Vinyl acetate + poly(butyl methacrylate)				
	Fractional precipitation	Acetone/methanol - water (1:1)		607
Vinyl acetate + poly(methyl methacrylate)				
	Fractional precipitation	Acetone/methanol		1380,1385
		Acetone/methanol - water (1:1)		607
	Fractional solution	Acetone/water	Column extraction	20
	Chromatography	Acetone/acetone - water	"Precipitation chromat.", 15-45°C	589

Polymer	Method of Fractionation	Solvent or Solvent/Non-solvent Mixture	Remarks	References
Vinyl acetate + poly(styrene)				
	Fractional precipitation	Butanone/methanol		1380
Vinyl acetate + poly(vinyl benzoate)				
	Fractional precipitation	Acetone/methanol - water		1383
Vinyl chloride + poly(methyl methacrylate)				
	Fractional precipitation	Dioxane/methanol		1380

1.14 Mixtures of Polymers

Polymer	Method of Fractionation	Solvent or Solvent/Non-solvent Mixture	Remarks	References
Poly(styrene) + poly(butadiene)				
	Chromatography	Tetrahydrofuran	GPC, 23°C	1284
Poly(propylene) + poly(1-butene)				
	Fractional solution	Tetralin	Column elution, increasing temp.	1378
Poly(propylene) + poly(ethylene)				
	Fractional solution	Tetralin	Column elution, increasing temp.	1378
	Fractional crystallization	Perchloroethylene	Stirred solution	1184
Poly(dimethyl siloxane) + poly(styrene)				
	Chromatography	Toluene	GPC, 20°C	349
Poly(p-iodostyrene) + poly(vinyl acetate)				
	Extraction	Benzene		371
Poly(methyl methacrylate) + poly(hexyl methacrylate)				
	Fractional precipitation	Benzene/methanol	Mixture (1:1), 20°C	7
Poly(methyl methacrylate) + poly(lauryl methacrylate)				
	Fractional precipitation	Benzene/methanol	Mixture (1:1), 20°C	7
Poly(methyl methacrylate) + poly(styrene)				
	Fractional precipitation	Benzene/methanol	30°C	71
	Turbidimetric titration	Benzene/methanol	25°C	683
	Chromatography	Trichloroethylene	GPC, 23°C	1447
Poly(styrene) + poly(vinyl acetate)				
	Extraction	Methanol		371
Poly(styrene) + poly(vinyl chloride)				
	Extraction	Cyclohexane		371
Poly(styrene) + paraffins				
	Fractional solution	Chloroform	Column extraction	808
Poly(vinyl acetate) + poly(vinyl chloride)				
	Extraction	Tetrahydrofuran		371

2. Main-chain Carbocyclic Polymers

2.1 Poly(phenylenes)

Polymer	Method of Fractionation	Solvent or Solvent/Non-solvent Mixture	Remarks	References
Poly(benzene-co-anthracene)				
	Fractional solution	Benzene, bromobenzene		150
Poly(bromophenylenes) oligomers				
	Fractional solution	Dioxane/water	Column elution	1336
Poly(p-dichloromethylbenzene-co-benzene)				
	Chromatography	Chloroform	GPC	544
Poly(p-dichloromethylbenzene-co-diphenylmethane)				
	Chromatography	Chloroform	GPC	545
Poly(1,4(1,3)-phenylene-1,3-butadiynylene)				
	Chromatography	o-Dichlorobenzene	GPC, 120°C	1572

Polymer	Method of Fractionation	Solvent or Solvent/Non-solvent Mixture	Remarks	References
2.2 Formaldehyde Resins				
p-Cresol-formaldehyde				
	Fractional precipitation	Benzene/petroleum ether		780
		Toluene/petroleum ether		780
		Tetrahydrofuran/petroleum ether		780
		Benzene/methanol		780
		Tetrahydrofuran - methanol (1:2)/water		780
		Dioxane, dichloroethane, trichloroethylene/ methanol, ethanol, n-propanol, n-butanol, ethyl ether, formamide		1539
Diphenyl ether-formaldehyde				
	Fractional precipitation	Benzene/methanol	30°C	1108
Methylphenol derivatives-formaldehyde				
	Chromatography	Benzene - acetic acid - water (800:240:30)	Paper chromat., 15°C	441
Phenol-formaldehyde				
	Fractional precipitation	Acetone - methanol/petroleum ether		267
		Acetone/petroleum ether		267
		Alkali/acid		267
		Dioxane, dichloroethane, trichloroethylene/- methanol ethanol, n-propanol, n-butanol, ethyl ether, formamide		1539
		Methanol/water		267
	Coacervation	Ethanol or dioxane/salts in water		245
		Methanol, ethanol, n-propanol, Novolac, dioxane/carbon dioxide, water solutions		246
	Chromatography	Tetrahydrofuran	GPC, 25°C	1548
		Benzene - acetic acid - water	Paper chromat., partition	441,1182
Novolac resins				
	Chromatography	Tetrahydrofuran	GPC	1023
Urea-formaldehyde				
	Fractional precipitation	Ethanol - water (1:1)/methanol		1540

3. Main-chain Heteroatom Polymers

3.1 Poly(oxides)

Poly(acetaldehyde) see Poly(oxyethylidene)

Poly(epichlorohydrin) see Poly(oxy(chloromethyl)ethylene)

Poly(ethylene oxide) see Poly(oxyethylene)

Polymer	Method of Fractionation	Solvent or Solvent/Non-solvent Mixture	Remarks	References
Poly(oxy-tert-butylethylene)				
	Fractional crystallization	Benzene	80°C and cooling	12
Poly(oxy(chloromethyl)ethylene)				
	Extraction; precipitation	Acetone (cold), acetone/methanol, methanol/ water		693
Poly(oxy-1,2-cyclohexylene)				
	Fractional precipitation	Benzene/methanol	25°C	966
Poly(oxydecamethylene)				
	Fractional precipitation	Benzene/methanol		1595
Poly(oxy-2,6-dimethyl-1,4-phenylene) (PPO)				
	Fractional precipitation	Chloroform/methanol	20°C, preparative	104,874,1154
		Dichloroethane/nitromethane		1400
		Toluene/methanol	Preparative	100,104,874
	Turbidimetric titration	Chloroform/ethanol	25°C	235
	Fractional solution	Chloroform/methanol	Column extraction	235

Polymer	Method of Fractionation	Solvent or Solvent/Non-solvent Mixture	Remarks	References
Poly(oxy-2,6-dimethyl-1,4-phenylene) (Cont'd.)				
	Chromatography	Benzene	GPC, 50°C	1365
		Dimethyl sulfoxide	GPC	890
		Methylene chloride	GPC, 25°C	720
		Toluene	GPC	8
		Tetrahydrofuran	GPC	528, 1259
	Fractional crystallization	Dioxane	Stirred	8
Poly(oxyethylene)				
	Fractional precipitation	Benzene/ethyl ether	20°C	739, 740
		Benzene/n-heptane	60°C	1267
		Benzene/n-hexane	Lowering temp.	72
		Benzene/isooctane	25°C	13, 127, 128, 174, 199, 1054, 1401-1405
			37°C	1413
			65°C, low M.W.	363
		(7:3)	Lowering temp.	126, 1261
		Chloroform/hexane		70
		Isopropanol/n-heptane	60°C	1267
		Toluene/n-hexane	60°C	359
	Turbidimetric titration	Chloroform/n-hexane		405
		Benzene/n-heptane	35°C	605
	Extraction	Water/hexane	Distribution between two immiscible liquids, countercurrent	263, 264
		Water/chloroform - benzene	Distribution between two immiscible liquids, countercurrent	263
	Chromatography	Benzene	GPC	612
		Butanone- water (1:1)	TLC, Kieselgel	241
		o-Dichlorobenzene	GPC	1557
		o-Dichlorobenzene	GPC, 130°C	608, 609
			Gaschromatography of trimethylsilyl ether derivatives. Low M.W.	444
		Ethylene glycol - methanol (4:1)	TLC, gradient const.	1139
		Methanol followed by methanol/DMF (4:1)	Gradient elution, TLC	1139
		Tetrahydrofuran	GPC	610, 611
		Trichloroethylene	GPC, 23°C	1447
	Fractional crystallization	Benzene/isooctane		199
Poly(oxy(ethylethylene))				
	Fractional precipitation	Benzene/methanol	GPC	200
	Chromatography	Trichlorobenzene	GPC, 90°C	200
Poly(oxyethylidene) (Poly(acetaldehyde))				
	Extraction	(Propion-, butyr, isobutyr- and isovaler-aldehyde) acetone (boiling) Diisopropyl ether, benzene	Amorphous+crystalline	1093
		Methanol, chloroform	Amorphous+crystalline	473
Poly(oxy(2-hydroxytrimethylene)oxy-1,4-phenyleneisopropylidene-1,4-phenylene))				
	Fractional solution	Chloroform/n-hexane	Column extraction	1068
	Chromatography	Tetrahydrofuran	GPC	399
Poly(oxyhexamethylene)				
	Fractional precipitation	Butanone	30°C (phase separation 28°C)	1596
Poly(oxy-2-methyl-1,4-cyclohexylene)				
	Chromatography	Tetrahydrofuran	GPC	820, 821
Poly(oxymethylene) (Poly(formaldehyde))				
	Fractional precipitation	Dimethylformamide	123.5-106°C	373, 643

Polymer	Method of Fractionation	Solvent or Solvent/Non-solvent Mixture	Remarks	References
Poly(oxymethylene) (Cont'd.)				
	Turbidimetric titration	Dimethylformamide	Lowering temp.	1529
	Fractional solution	Phenol/ethylcellosolve	Column extraction, Kisselite	726, 799
		Tetrachloroethane - phenol (3:1)/n-hexanol	120°C	551
		Tetrachloroethane - phenol (3:1)/n-hexanol	Column extraction, 130°C	551
	Extraction	o-Dichlorobenzene	$150\text{-}164^{\circ}$C	923
		Dimethylformamide	Lowering temp., fractional	375
	Fractional crystallization	Phenol	Lowering temp. (diacetate)	1409
		p-Chlorophenol	Stirred, 60°C	798
Poly(oxymethylene(2-hydroxy-5-methyl)-1,3-phenylenemethylene))				
	Fractional precipitation	Benzene/petroleum ether	25°C	741
Poly(oxymethyloxyethylene)				
	Chromatography	Benzene	GPC	772
Poly(oxyphenylethylene)				
	Fractional precipitation	Benzene/isooctane	25°C	12
Poly(oxypropylene)				
	Fractional precipitation	Acetone	Lowering temp.: 0 to -78°C, crystalline + amorphous (d,l-comps.)	485
			0°C, crystalline+ amorphous	487
		Isooctane	60°C, lowering temp.	15
				196
		Isooctane/octamethyl tetrasiloxane	Amorphous, extracted with isooctane at 40°C; crystalline, fractionated at 75°C	16
		Isopropanol/water	$70, 74^{\circ}$C	14, 1519
		Methanol/water	25°C	14
	Extraction	Acetone (0°C), hexane (-78°C)	Optically act.	688-691
		Acetone	0°C (D-L)	1469
	Extraction; precipitation	Acetone (cold); acetone/methanol methanol/water	Amorphous+crystalline	693
	Distribution between two immiscible liquids	Water/hexane	Countercurrent	203, 204
		Water/chloroform - benzene	Countercurrent	263
	Fractional solution	Ethanol/water	Column extraction	306
		Acetone/diisopropyl ether	Column extraction, alumina	1052
	Chromatography	Benzene	Membrane chromat.	988
		Benzene, heptane	Extraction from an active support	486
		Methanol/water	"Precipitation chromat.", $20\text{-}60^{\circ}$C	1320
		Tetrahydrofuran	GPC	837, 977, 985, 1177, 1178, 1320, 1519, 1520
		Toluene	GPC	1320
	Fractional crystallization	Isooctane	Cooling	15, 197, 198, 1232
			40°C	16
Poly(propylene oxide) see Poly(oxy(propylene))				
Poly(oxytetramethylene)				
	Fractional precipitation	Acetone	Lowering temp.	1282
		Benzene/n-hexane	60°C	72
			Lowering temp.	73
		Benzene/methanol		11, 424, 425
		Toluene/methanol		925
	Turbidimetric titration	Ethanol/water		925
	Distribution between immiscible liquids	Cyclohexane - toluene (9:1)/water-methanol		715
	Extraction	Water/acetone		858
		Isopropanol/water	Continuous extraction	170

Polymer	Method of Fractionation	Solvent or Solvent/Non-solvent Mixture	Remarks	References
Poly(oxytetramethylene) (Cont'd.)				
	Fractional solution	Butanone	Column elution, silica gel	856-858
		Isopropanol/water	50°C, low M.W.	171
	Chromatography	Acetone/water		1117
		Tetrahydrofuran	GPC	856, 857
	Sedimentation velocity	Ethyl acetate - n-hexane (22.3:77.7)	30.4°C	1599

3.2 Poly(carbonates)

Polycarbonates (general) Poly(oxycarbonyl oxy .....)

Polymer	Method of Fractionation	Solvent or Solvent/Non-solvent Mixture	Remarks	References
	Fractional precipitation	Methylene chloride/n-heptane		1456
	Turbidimetric titration	Chloroform/methanol		846

Poly(oxycarbonyloxy-1,4-phenylene cyclohexylidene-1,4-phenylene)

Polymer	Method of Fractionation	Solvent or Solvent/Non-solvent Mixture	Remarks	References
	Fractional precipitation	Methylene chloride/hexane		501

Poly(oxycarbonyloxy-1,4-phenyleneisopropylidene-1,4-phenylene)

Polymer	Method of Fractionation	Solvent or Solvent/Non-solvent Mixture	Remarks	References
	Fractional precipitation	Chloroform/methanol	23°C, 25°C	523, 1037, 1038, 1039
		Chloroform/n-octane	30°C	523
		Methylene chloride/n-heptane		155
		Methylene chloride/methanol	0, 20, 25, 30°C	523, 915, 1376
		Methylene chloride/n-octane	25°C, 30°C	523
		Methylene chloride/petroleum ether	25°C	915
		Methylene chloride - cresol/petroleum ether	-20°C	915
		Tetrachloroethane/paraffin oils	25°C	915
	Chromatography	Methylene chloride/n-heptane		1492, 1495, 1496
		Methylene chloride	GPC	633
		Tetrahydrofuran	GPC, Lexan 121-111	393
			GPC	995
	Fractional solution	Methylene chloride/n-hexane		118

Blockcopolycarbonates from
di(4-hydroxyphenyl)isopropylidene and di(4-hydroxy-3,5-dichlorophenyl)isopropylidene

Polymer	Method of Fractionation	Solvent or Solvent/Non-solvent Mixture	Remarks	References
	Turbidimetric titration	Methylene chloride/methanol		814

Blockcopolycarbonates from
di(4-hydroxyphenyl)isopropylidene and di(4-hydroxyphenyl)ethylene

Polymer	Method of Fractionation	Solvent or Solvent/Non-solvent Mixture	Remarks	References
	Turbidimetric titration	Methylene chloride/methanol		814

Blockcopolycarbonates from
di(hydroxyphenyl)isopropylidene and di(4-hydroxyphenyl)sulfone

Polymer	Method of Fractionation	Solvent or Solvent/Non-solvent Mixture	Remarks	References
	Turbidimetric titration	Methylene chloride/methanol		814

3.3 Poly(esters)

Poly(esters) (general)

Polymer	Method of Fractionation	Solvent or Solvent/Non-solvent Mixture	Remarks	References
	Chromatography	Ethanol - cyclohexane (3:7)/cyclohexane	"Precipitation chromat."	1227
		Chloroform/ethanol, ethyl ether/water	Paper chromat., oligomers from aliphatic and aromatic diacids and glycols	50

Poly(adipic acid-co-ethylene glycol) see Poly(oxyethyleneoxyadipoyl)

Poly(adipic acid-co-glycerol)

Polymer	Method of Fractionation	Solvent or Solvent/Non-solvent Mixture	Remarks	References
	Fractional precipitation	Benzene/petroleum ether		267

Poly(adipic acid-co-poly(ethylene glycol))

Polymer	Method of Fractionation	Solvent or Solvent/Non-solvent Mixture	Remarks	References
	Fractional solution	Butanone	Column extraction, silica gel	1612
	Fractional precipitation	Chloroform/n-hexane		405
		Acetonitrile/isopropyl ether		405

Poly(ε-caprolactone) see Poly(oxycarbonylpentamethylene)

Polymer	Method of Fractionation	Solvent or Solvent/Non-solvent Mixture	Remarks	References
Poly(ethylene terephthalate) see Poly(oxyethyleneoxyterephthaloyl)				
Poly(11-hydroxyundecanoic acid) see Poly(oxycarbonyldecamethylene)				
Poly(isophthalic acid-co-maleic anhydride-co-propylene glycol)				
	Chromatography	Acetone/n-heptane	"Precipitation chromat.", 27-50°C	256
Poly(lauric acid-co-glycerol)				
	Fractional precipitation	Toluene/n-heptane	Alkyd resins	1070
Poly(oxyadipoyloxydecamethylene)				
	Fractional precipitation	Benzene/petroleum ether		1244
Poly(oxycarbonyl-1,1-dimethylethylene) (poly(pivalolactone))				
	Chromatography	m-Cresol	GPC, 112°C	202
Poly(oxycarbonylethylene) (poly(β-propiolactone))				
	Fractional precipitation	Chloroform/petroleum ether		334
Poly(oxycarbonyldecamethylene)				
	Fractional precipitation	Benzene/methanol	55°C	900
Poly(oxycarbonylpentamethylene)				
	Fractional precipitation	Benzene/n-heptane		812
		Benzene/isooctane	34°C	1189
		Benzene/petroleum ether	25°C	334
	Chromatography	Benzene/isooctane		1189
Poly(oxyethyleneoxyadipoyl)				
	Fractional precipitation	Benzene/petroleum ether		1244
	Fractional crystallization	Ethanol	Hot	637
Poly(oxyethyleneoxyisophthaloyl)				
	Fractional precipitation	Tetrachloroethane/n-heptane		1591
	Chromatography	Tetrahydrofuran	GPC, 37°C	164
Poly(oxyethyleneoxyphthaloyl) (Poly(ethylene terephthalate))				
	Fractional precipitation	Dichloroethane/petroleum ether		627
Poly(oxyethyleneoxysebacoyl)				
	Fractional crystallization	Ethanol	Hot	637
	Chromatography	Tetrahydrofuran	GPC, 37°C	164
Poly(oxyethyleneoxysuberoyl)				
	Fractional crystallization	Ethanol	Hot	637
	Chromatography	Tetrahydrofuran	GPC, 37°C	164
Poly(oxyethyleneoxyterephthaloyl) (Poly(ethylene terephthalate))				
	Fractional precipitation	o-Chlorophenol/n-heptane		989,990,1034,1035
		m-Cresol/ligroin (b.p. 100°C)	50°C	1500
		Dimethylformamide	Lowering temp.	1500
		Phenol/cyclohexane	70°C	806
		Phenol - tetrachloroethane (1:1)/n-heptane		509,510,1257
		Phenol - tetrachloroethane (1:1)/ligroin	25°C	1544
		Phenol - tetrachloroethane (1:1)/ligroin	65°C	539
	Fractional solution	Phenol/n-heptane	56°C	1155
		Phenol - tetrachloroethane (3:2)/n-nonane	Film extraction, 95°C	471
	Coacervation	Phenol - tetrachloroethane/n-heptane		1494
	Distribution between two immiscible liquids	m-Cresol/petroleum ether		1500
	Chromatography	Trifluoroacetic acid - chloroform (20:80) solvent, trifluoroacetic acid - chloroform (10:90) eluent	Charcoal	232
		o-Chlorophenol	GPC, 50°C	990

Polymer	Method of Fractionation	Solvent or Solvent/Non-solvent Mixture	Remarks	References
Poly(oxyhexamethyleneoxyadipoyl)				
	Chromatography	Tetrahydrofuran	GPC, 37°C	164
Poly(oxyhexamethyleneoxy(α,α'-dibutyl)sebacoyl)				
	Fractional precipitation	Benzene/methanol		115,1581
	Fractional solution	Benzene/methanol		115,1581
Poly(oxyhexamethyleneoxysebacoyl)				
	Turbidimetric titration	Carbon tetrachloride/n-heptane	20°C	824
Poly(oxyisophthaloyloxy-1,4-phenylene-(2-phenyl)phthalimidine-1,4-phenylene)				
	Fractional precipitation	Tetrachloroethane/tetrahydrofuran		386
	Sedimentation velocity	Tetrahydrofuran	20°C	386
Poly(oxyisophthaloyloxy-1,4-phenylenephthalidylidene-1,4-phenylene)				
	Turbidimetric titration	Tetrachloroethane/ethanol		388
	Extraction	Phenol - tetrachloroethane (1:3)/heptane		388
	Chromatography	Tetrachloroethane/ethanol	"Precipitation chromat.", 15-65°C	388
Poly(oxysebacoyloxydecamethylene)				
	Turbidimetric titration	Carbon tetrachloride/n-heptane	20°C	824
Poly(oxysebacoyloxy-1,4-phenylenephthalidylidene-1,4-phenylene)				
	Fractional precipitation			1453
Poly(oxysuccinoyloxyhexamethylene)				
	Chromatography	Butanone - cyclohexane (3:1)/cyclohexane-butanone (1:4)	"Precipitation chromat.", 27-70°C	588
		Butanone/cyclohexane	"Precipitation chromat.", 27-70°C	588
		Butanone/cyclohexane	"Precipitation chromat.", 40-70°C	588
Poly(oxytetramethyleneoxyadipoyl)				
	Chromatography	Tetrahydrofuran	GPC, 37°C	164
Poly(oxytartaroyloxyhexamethylene)				
	Chromatography	Benzene	Urea	781
Poly(oxyterephthaloyloxy-1,4-phenylenefluoren-9-ylidene-1,4-phenylene)				
	Fractional precipitation			1453
Poly(oxyterephthaloyloxy-1,4-phenylenephthalidylidene-1,4-phenylene)				
	Fractional precipitation	Carbon tetrachloride - phenol (3:1)/heptane		387
Poly(phthalic anhydride-co-coconut oil-co-glycerol)				
	Fractional precipitation	Acetone/water	Alkyd resins	446
Poly(phthalic anhydride-co-glycerol)				
	Fractional precipitation	Acetone/water	Alkyds	445
		Butanone - acetone/methanol		177
		Butanone - acetone/methanol - water		177
Poly(phthalic acid-co-pentaerythrytol)				
	Fractional precipitation	Acetone	Lowering temp.	178
Poly(pivalolactone) see Poly(oxycarbonyl-1,1-dimethylethylene)				
Poly(sebacic acid-co-hexanediol-co-hexanetriol)				
	Fractional precipitation	Benzene/methanol		112
Poly(sebacic, adipic, tartaric acid-co-1,6-hexanediol-1,10-decanediol)				
	Fractional precipitation	Benzene/methanol		111,1581
	Fractional solution	Benzene/methanol		111,1581
Poly(succinic-co-pimelic acid-co-hexanediol)				
	Fractional precipitation	Benzene/methanol		112
Poly(succinic acid-co-hexanediol) see Poly(oxysuccinoyloxyhexamethylene)				
Poly(terephthalic acid-co-ethylene glycol) see Poly(oxyethyleneoxyterephthaloyl)				

Polymer	Method of Fractionation	Solvent or Solvent/Non-solvent Mixture	Remarks	References

3.4 Poly(urethanes) and Poly(ureas)

Poly(caprolactone-co-4,4'-diphenylmethanediisocyanate-co-ethylene diamine)

| | Fractional precipitation | Dimethylacetamide/n-heptane - ethyl ether | | 1296 |

Poly(caprolactone-co-trimethylolethane-co-4,4'-diphenylmethanediisocyanate-co-ethylene diamine)

| | Fractional precipitation | Dimethylacetamide/n-heptane - ethyl ether | Branched polymer | 1296 |

Poly(oxycarbonylimino-2,4-toluyleneiminocarbonyl (poly(oxypropylene))

| | Fractional solution | Benzene/isooctane | Column extraction | 686,1251 |

Poly(oxyethyleneoxycarbonylimino-1,4-phenylenemethylene-1,4-phenylene)

| | Fractional solution | Dimethylformamide/acetone | Direct sequential extraction | 123 |
| | Extraction | Dimethylformamide/acetone | $30^{o}C$ | 122 |

Poly(oxytetramethyleneoxycarbonyliminohexamethyleneiminocarbonyl)

| | Fractional precipitation | m-Cresol/n-hexane | | 1111 |
| | Chromatography | Formic acid - water (88%) | Paper partition, $20^{o}C$ | 62 |

Poly(ureylenehexamethyleneureylene-1,4-phenyleneoxycarbonyloxy-1,4-phenylene)

| | Turbidimetric titration | Dimethylformamide | | 813 |

Poly(ureylene-1,5-naphthyleneureylene-1,4-phenyleneoxycarbonyloxy-1,4-phenylene)

| | Turbidimetric titration | Dimethylformamide | | 813 |

Poly(ureylene-1,4-phenylenemethylene-1,4-phenyleneureylene-1,4-phenyleneoxycarbonyloxy-1,4-phenylene)

| | Turbidimetric titration | Dimethylformamide | | 813 |

Poly(ureylene-1,4-phenyleneureylene-1,4-phenyleneoxycarbonyloxy-1,4-phenylene)

| | Turbidimetric titration | Dimethylformamide | | 813 |

Poly(ureylene-1,4-toluylene-ureylene-1,4-phenyleneoxycarbonyloxy-1,4-phenylene)

| | Turbidimetric titration | Dimethylformamide | | 813 |

3.5 Poly(amides) and Poly(imines)

Poly(6-aminocaproic acid-co-hexamethylenediamine-co-adipic acid)

| | Fractional precipitation | Phenol/water | $70-90^{o}C$ | 114 |

Poly((butylimino)carbonyl) (Poly(butylisocyanate))

| | Fractional precipitation | Carbon tetrachloride/methanol | | 1484 |
| | Sedimentation velocity | Carbon tetrachloride | | 1484 |

Poly(ϵ-caprolactam) see Poly((imino(1-oxohexamethylene))

Poly(caprolactam-co-hexamethylenediamine-co-adipic acid-co-azelaic acid)

| | Extraction | Methanol/water | $35^{o}C$ | 1172 |
| | Turbidimetric titration | Methanol/water | | 1172 |

Poly(trans-2,5-dimethyl-1,4-piperazinediylcarbonyl-trans-1,2-cyclohexylene carbonyl)

| | Chromatography | Dichloroethylene - cyclohexane | Column elution, $30^{o}C$ | 1024 |

Poly(ethylimino-trans-fumaroylethylimino-1,4-(2-methyl)phenylenemethylene-1,4-(2-methyl)phenylene))

	Fractional precipitation	Methanol/water		432
		Dimethylformamide/water		432
	Turbidimetric titration	Dimethylformamide/water		431

Poly(fumaric acid-co-N,N'-diethyl-4,4'-diamino-3,3'-dimethyldiphenylmethane) see
Poly(ethylimino-trans-fumaroylethylimino-1,4-(2-methyl)phenylenemethylene-1,4-(2-methyl)phenylene))

Poly(hexamethylene-co-adipic acid) see Poly(iminoadipoyliminohexamethylene)

Poly(hexyliminocarbonyl) (Poly(hexylisocyanate))

| | Coacervate extraction | Toluene/methanol | | 146 |

Poly(iminoadipoyliminodecamethylene) (Nylon 610)

| | Distribution between immiscible liquids | Phenol/water | Continuous, Nylon 610 | 394 |
| | Chromatography | Formic acid - water (88%) | Partition chromatography, $20^{o}C$, Nylon 610 | 62 |

Polymer	Method of Fractionation	Solvent or Solvent/Non-solvent Mixture	Remarks	References
Poly(iminoadipoyliminohexamethylene) (Nylon 66)				
	Fractional precipitation	m-Cresol/cyclohexane		656,719
		Phenol/water	Nylon 66	1439
	Turbidimetric titration	m-Cresol/cyclohexane	Nylon 66	657
	Sedimentation gradient	Carbon tetrachloride, m-cresol, cyclohexane	Ultracentrifuge	1450
	Distribution between immiscible liquids	Phenol/water	Continuous, Nylon 66	394
	Chromatography	Formic acid - water (88%)	Partition chromat., 20°C, Nylon 66	62
Poly(iminoethylene)				
	Curtain electrophoresis	Acetic acid - water		1190
Poly(iminohexamethyleneiminosebacoyl)				
	Chromatography	Formic acid - water (88%)	Paper partition	62
	Mechanical fractionation	Tyler screens	Interfacial condensation	1049
Poly(iminoisophthaloyliminodecamethylene)				
	Turbidimetric titration	p-Cresol/acetone		1389
Poly(imino(1-oxodecamethylene)) (Nylon 10)				
	Distribution between two immiscible liquids	Phenol/water	70°C	754
Poly(imino(1-oxododecamethylene)) (Nylon 12)				
	Chromatography	Tetrahydrofuran	GPC	404
		Hexamethylphosphoramide	GPC, 85°C	1156
Poly(imino(1-oxohexamethylene)) (Nylon 6)				
	Fractional precipitation	m-Cresol/cyclohexane		719
		m-Cresol/ethyl ether	26°C	578
		m-Cresol/ligroin		538
				549,1497
		Phenol - tetrachloroethylene (1:1)/n-heptane	25°C	1245
		Phenol/water	70°C	1582
	Distribution between immiscible liquids	Phenol/water	Continuous, Nylon 6	394
	Turbidimetric titration	HCl/aqueous ammonium sulfate		1223
	Coacervation	Phenol/ethylene glycol/water		1493
	Fractional solution	Formic acid/propyl acetate	Film extraction	471
		Phenol/water	Column extraction, 70°C	505,955
		Water - methanol (81.5:18.5)/phenol	Column extraction, 25°C, linear and branched	956
	Chromatography	m-Cresol	GPC	106,107,300
		Formic acid - water (88%)	Paper partition, 20°C	62
		Hexamethylphosphoramide	GPC, 85 C	1156
	Sedimentation velocity	Hexafluoroisopropanol	Ultracentrifuge	1173
	Sedimentation equilibrium	Formic acid - water (85%) + 2 M of KCl	25°C	1324
	Thermal diffusion	Formic acid - water (90%)		864
	Fractional crystallization	Phenol - tetrachloroethane (1:1)/ethanol - n-heptane	25°C, stirred	163
Poly(imino(1-oxoundecamethylene)) (Nylon 11)				
	Chromatography	Hexamethylphosphoramide	GPC, 85°C, Nylon 11	1156
Poly(iminoterephthaloylimino-1,4-phenylenephthalidylidene-1,4-phenylene)				
	Fractional precipitation	Dimethylformamide/acetone - ethanol (2:1)		1270
		Dimethylformamide/acetone		1270

Polymer	Method of Fractionation	Solvent or Solvent/Non-solvent Mixture	Remarks	References
Poly(iminoterephthaloylimino-1,4-phenylenefluorene-9-ylidene-1,4-phenylene)				
	Fractional precipitation	Dimethylformamide/acetone - ethanol (2:1)		1270
		Dimethylformamide/acetone		1270
Poly(iminotetramethyleneiminoisophthaloyl)				
	Turbidimetric titration	p-Cresol/acetone		1389
Poly(1,4-piperazinediylsebacoyl)				
	Mechanical fractionation	Tyler screens	Interfacial condensation	1049

3.6 Poly(amino acids)

Polymer	Method of Fractionation	Solvent or Solvent/Non-solvent Mixture	Remarks	References
Poly(γ-benzyl-L-glutamate)				
	Fractional precipitation	Dichloroethane/petroleum ether		1480-1482
	Fractional solution	Formic acid	Extraction	1005
		Ethanol (hot)	Extraction	1005
	Chromatography	m-Cresol	GPC, 100°C	515
		m-Cresol - chlorobenzene/benzoic acid	GPC	205
		Methylene chloride/methanol	"Precipitation chromat."	312, 313, 527
	Dialysis	Dioxane		1005
Fibrinogen				
	Sedimentation	Buffer solution	20°C	412
		Phosphate buffer		413
Gelatin				
	Coacervation	Water/ethanol		1231, 1399
Glycogen				
	Sedimentation	Water		223
Poly(hydroxybutyl glutamine)				
	Fractional precipitation	Methanol/ethyl ether		1542
Poly(hydroxybutyl glutamine-co-glycine)				
	Fractional precipitation	Methanol/ethyl ether		37
Poly-N^5-(3-hydroxypropyl)-L-glutamine				
	Chromatography	Water	GPC	1122
Poly(hydroxypropyl glutamine)				
	Fractional precipitation	Methanol/ethyl ether		1542
Poly(hydroxypropyl glutamine-co-L-alanine)				
	Fractional precipitation	Methanol/ethyl ether		1210
Poly(hydroxypropyl glutamine-co-hydroxybutyl glutamine)				
	Fractional precipitation	Methanol/water		1542
Poly(hydroxybutyl glutamine-co-L-serine)				
	Fractional precipitation	Methanol/ethyl ether		668
Poly(γ-methyl-L-glutamate)				
	Chromatography		GPC	67
Poly(L-proline)				
	Chromatography	Water	GPC	495
Poly(sarcosine)				
	Chromatography	Dioxane/water	"Precipitation chromat.", dimethylamide	260
Poly(N-tosyl-L-tyrosine-formal)				
	Fractional precipitation	Acetone/water		1355

3.7 Poly(sulfides) and Poly(sulfones)

Poly(1-butenesulfone) see Poly(sulfonylethylethylene)

Polymer	Method of Fractionation	Solvent or Solvent/Non-solvent Mixture	Remarks	References
Polysulfide rubbers				
	Chromatography	Dioxane	Column elution, silica gel	519
Poly(oxy-1,4-phenylenesulfonyl-1,4-phenyleneoxy-1,4-phenyleneisopropylidene-1,4-phenylene)				
	Fractional precipitation	Dimethyl sulfoxide	80° C, lowering temp.	18
	Chromatography	Tetrahydrofuran, dimethylformamide, chloroform and dimethylacetamide	GPC	1345
Poly(propylenesulfide) see Poly(thiopropylene)				
Poly(sulfonylbutylethylene)				
	Fractional precipitation	Acetone/methanol	Lowering temp.	108,109
Poly(sulfonylethylethylene)				
	Chromatography	Tetrahydrofuran	40° C	229
Poly(sulfonyl-1-methyl-1-propylethylene)				
	Fractional precipitation	Acetone/methanol	Lowering temp.	108
Poly(sulfonylphenylethylene)				
	Fractional precipitation	Tetrahydrofuran/methanol	40° C	417
Poly(tetrathioethylene)				
	Chromatography	Dioxane		514
Poly(thio-1-methyltrimethylene)				
	Chromatography	Benzene	GPC, 45° C	738
Poly(thiopropylene)				
	Fractional precipitation	Benzene/methanol		442

3.8 Poly(silanes) and Poly(siloxanes)

Polymer	Method of Fractionation	Solvent or Solvent/Non-solvent Mixture	Remarks	References
Poly(dimethylsilmethylene)				
	Chromatography	Toluene	GPC	889
Poly(1,1-diphenyl-2,2,3,3,4,4-hexamethylcyclotetrasiloxane)				
	Fractional precipitation	Benzene/methanol		44
Poly(divinyl tetramethyldisiloxane)				
	Fractional precipitation	Butanone/methanol	25° C	564
Poly(oxydi(chlorophenyl)titano-tetra(oxy(chlorophenyl)silylidyne))				
	Fractional precipitation	Carbon tetrachloride/petroleum ether		59
Poly(oxydimethylsilylene)				
	Fractional precipitation	Benzene/methanol	25° C	159,161,337,777, 853,855
		Ethyl acetate/methanol		356,998
	Turbidimetric titration	Butanone/methanol		350
	Fractional solution	Benzene/methanol		43
		Cyclohexane - carbon tetrachloride (3:1)/methanol	25° C, extraction	1191
	Foam fractionation	Benzene		490
	Chromatography	Benzene	GPC, 55° C	490
		Chloroform	GPC	354,998
		Ethanol/methanol, diethyl ether/ethanol, and diethyl ether	Gradient elution, carbon	83
		Toluene	GPC	777,883
		Trichloroethylene	GPC, 25° C	1357,1447
			"Precipitation chromat.", temp. grad., quartz sand, branched polymer	45
	Sedimentation velocity	Heptane	20° C	1191
	Distribution between two immiscible liquids	Methanol/carbon tetrachloride in countercurrent with cyclohexane		93

Polymer	Method of Fractionation	Solvent or Solvent/Non-solvent Mixture	Remarks	References
Poly(oxydiphenylsilylene)				
	Fractional precipitation	Benzene/methanol		1472, 1473
Poly(oxydiphenyltitano-tetra(oxydiphenylsilylene))				
	Fractional precipitation	Carbon tetrachloride/petroleum ether		59
Poly(oxydipropylsilylene)				
	Fractional precipitation	Toluene/methanol		882
Poly(oxyferrooxyhalophenylsilylene)				
	Fractional precipitation	Benzene/petroleum ether		247
Poly(oxyhydroxyalumino-oxydi(chlorophenyl)silylene)				
	Fractional precipitation	Carbon tetrachloride/petroleum ether		61
Poly(oxyhydroxyalumino-oxydiphenylsilylene)				
	Fractional precipitation	Carbon tetrachloride/petroleum ether		61
Poly(oxymethylphenylsilylene)				
	Fractional precipitation	Toluene/isopropanol		238
Poly(oxymethyl-3,3,3-trifluoropropylsilylene)				
	Fractional precipitation	Ethyl acetate/toluene		236, 237
		Benzene/methanol	20°C	1174
Poly(phenylsilsesquioxane)				
	Fractional precipitation	Benzene/methanol		1239
			30°C	616
Polymer from chromium potassium sulfate and phenylsodiumoxydioxysilan				
	Fractional precipitation	Carbon tetrachloride/petroleum ether		60

3.9 Poly(phosphazenes) and Related Polymers

Polymer	Method of Fractionation	Solvent or Solvent/Non-solvent Mixture	Remarks	References
Poly(di(p-chlorophenoxy)phosphazene)				
	Fractional precipitation	Dimethylformamide	Lowering temp.	17
Poly(di(p-phenylphenoxy)phosphazene)				
	Fractional precipitation	Dimethylformamide	Lowering temp.	17
Poly(fluoroalkoxyphosphazene)				
	Fractional solution	1,1,2-Trichloro-1,2,2-trifluoroethane/-acetone	Extraction	577
Poly((2,2,2-trifluoroethoxy)phosphazene)				
	Fractional precipitation	Acetone/cyclohexane	25°C	17
		Methanol	Lowering temp., 40 to 25°C	17
Poly(ureylene-1,4-phenyleneureylene(phenylphosphinylidene))				
	Fractional precipitation	Dimethyl sulfoxide/carbon tetrachloride	25°C	261
Polymer from chromium(II)acetate and potassium diphenylphosphinate				
	Chromatography	Tetrahydrofuran	GPC	361

3.10 Others

Polymer	Method of Fractionation	Solvent or Solvent/Non-solvent Mixture	Remarks	References
Poly(5,5'-dibenzimidazole-2,2'-diyl-1,3-phenylene)				
	Sedimentation equilibrium	Dimethylacetamide	Ultracentrifuge, 40°C	507
	Fractional precipitation	Dimethylacetamide/n-hexane	61°C	617
Poly(2,3-dihydrobenzofuran-2,3-diyl)				
	Fractional precipitation	Benzene/methanol	Optically active	428
Poly(ferrocenediylfurfurylidene)				
	Fractional precipitation	Dioxane/isopropanol - water		1102
Poly(oxadiazolediyl-1,4-phenylene-phthalidylidene-1,4-phenylene)				
	Fractional precipitation	Tetrachloroethane - phenol/n-heptane		1454

Polymer	Method of Fractionation	Solvent or Solvent/Non-solvent Mixture	Remarks	References
Poly(oxybenzoxazole-5,2-diyl-1,4-phenylenephthalidylidene-1,4-phenylenebenzoxazole-2,5-diyl)				
	Fractional precipitation	Tetrachloroethane - phenol/n-heptane		1454
Poly(imides):				
Poly(5,7-dihydro-1,3,5,7-tetraoxobenzo[1,2-c:4,5-c'] dipyrrole-2,6-[1H,3H] -diyl-1,4-phenylenephthalidylidene-1,4-phenylene)				
	Fractional precipitation	Nitrobenzene/dichloroethane	50 and 63°C	823
Poly(oxyphthalimide-5,2-diyl-1,4-phenylenephthalidylidene-1,4-phenylenephthalimide-2,5-diyl)				
	Fractional precipitation	Tetrachloroethane - phenol/n-heptane		1454
Poly(maleimides):				
Poly(1-2,4-dimethylphenyl-2,5-dioxopyrrolidin-3,4-diyl)				
	Fractional precipitation	Ethyl acetate/methanol	21°C	1471
Poly(1-isobutyl-2,5-dioxopyrrolidin-3,4-diyl)				
	Fractional precipitation	Benzene/methanol		1476
Poly(1-tolyl-2,5-dioxopyrrolidin-3,4-diyl)				
	Fractional precipitation	Chloroform/methanol	20°C	1477

3.11 Random Copolymers

Poly(acetaldehyde-co-chloral)				
	Fractional precipitation	Chloroform/methanol		697
Poly(acrolein-co-thiophenylmercaptal)				
	Fractional precipitation	Benzene/methanol		1333
Poly((3,3-bis-(chloromethyl)oxacyclobutane-co-β-propiolactone)				
	Fractional precipitation	Ethylene dichloride/ethanol		1422
Poly(1,4-bis-(dimethylhydroxysilyl)phenylene-co-dimethylsiloxane)				
	Fractional precipitation	Benzene/methanol	20°C	41
Poly(styrene-co-epichlorohydrin)				
	Turbidimetric titration	Benzene/methanol		1000
Poly(tetrahydrofuran-co-diketene)				
	Extraction	Methanol, benzene		1423
Poly(tetrahydrofuran-co-1,2-butene oxide)				
	Chromatography	Tetrahydrofuran	GPC	583
Poly(tetrahydrofuran-co-ε-caprolactone)				
	Fractional precipitation	Benzene/methanol		1598
	Chromatography	Tetrahydrofuran	GPC	1598
Poly(tetrahydrofuran-co-propylene oxide)				
	Chromatography	Tetrahydrofuran	GPC	172,173,583
Poly(tetrahydrofuran-co-n-propyl glycidyl ether)				
	Chromatography	Tetrahydrofuran	GPC	583
Poly(tetrahydrofuran-co-vinyl butyl ether)				
	Extraction	Heptane, acetone		1225

3.12 Blockcopolymers

Epoxy resin : butadiene-styrene rubber				
	Turbidimetric titration	Chloroform/methanol		751
Poly(acetaldehyde : propylene oxide)				
	Fractional solution	Methanol, chloroform	Extraction	474
Poly(ethylene oxide : α-methylstyrene)				
	Chromatography	Tetrahydrofuran	GPC	764
Poly(formaldehyde : acrylonitrile)				
	Extraction	Acetonitrile		491

Polymer	Method of Fractionation	Solvent or Solvent/Non-solvent Mixture	Remarks	References
Poly(formaldehyde : butadiene)				
	Extraction	Toluene, acetone, methanol, water		1115
Poly(formaldehyde : isoprene)				
	Extraction	Toluene, acetone, methanol, water		1115
Poly(formaldehyde : α-methylstyrene)				
	Extraction	Toluene, acetone, methanol, water		1115
Poly(formaldehyde : styrene)				
	Extraction	Toluene, acetone, methanol, water		1115
Poly(formaldehyde : 2-methyl-5-vinylpyridine)				
	Extraction	Toluene, acetone, methanol, water		1115
Poly(α-methylstyrene : propylene sulfide)				
	Chromatography	Tetrahydrofuran	GPC, $45\,^{\circ}C$	1053
Poly(styrene : ethylene oxide)				
	Fractional precipitation	Dimethoxyethane/water	$90\,^{\circ}C$	64
		Chloroform/water		64
Poly(styrene : propylene oxide)				
	Fractional precipitation	Benzene/methanol		1001
Poly(styrene : tetrahydrofuran)				
	Chromatography	Tetrahydrofuran	GPC	1601
Poly(tetrahydrofuran : methyl glycidyl ether)				
	Fractional precipitation	Methanol/water		1226

3.13 Graftcopolymers

Polymer	Method of Fractionation	Solvent or Solvent/Non-solvent Mixture	Remarks	References
Acrylic acid + poly(β-alanine)				
	Turbidimetric titration	Formic acid - water (85:15)/ethanol		57
α-Hydroxyisobutyric acid anhydrosulfite + poly(acrylonitrile)				
	Turbidimetric titration	Dimethylformamide/chloroform		687
Methyl methacrylate + poly(iminoadipoyliminohexamethylene)				
	Fractional precipitation	m-Cresol/methanol		651
		Formic acid/methanol		651
Methyl methacrylate + poly(oxyethylene)				
	Fractional precipitation	Chloroform/ethyl ether		1498
Styrene + poly(oxyfumaroyloxy(methylene)$_n$)			$n = 4, 6, 10$	
	Extraction	Benzene		433
Styrene + poly(oxymaleoyloxyhexamethylene)				
	Fractional precipitation	Benzene/methanol		1114
		Benzene/petroleum ether		1114
Styrene + poly(siloxane)				
	Turbidimetric titration	Benzene/methanol - acetone (3:1)		1002
Poly(epoxide) + poly(organosiloxanes)				
	Turbidimetric titration	Dioxane/water		40

3.14 Mixtures of Polymers

Polymer	Method of Fractionation	Solvent or Solvent/Non-solvent Mixture	Remarks	References
Poly(iminoadipoyliminohexamethylene) + poly(methyl methacrylate)				
	Fractional precipitation	m-Cresol/methanol		651
		Formic acid/methanol		651
		Dichloroacetic acid		651
	Turbidimetric titration	m-Cresol/methanol		651
	Extraction	Chloroform		651
	Sedimentation gradient	Dichloroacetic acid, methanol		651

Polymer	Method of Fractionation	Solvent or Solvent/Non-solvent Mixture	Remarks	References

Poly(iminotetramethyleneiminoisophthaloyl) + poly(iminoisophthaloyliminodecamethylene)

| | Turbidimetric titration | p-Cresol/acetone | | 1389 |

4. Poly(saccharides)

4.1 Poly(saccharides)

Polymer	Method of Fractionation	Solvent or Solvent/Non-solvent Mixture	Remarks	References
Alginates	Fractional precipitation	Water/manganous chloride + calcium chloride	Sodium alginate	961
		Water/sodium chloride	Sodium alginate	599
	Centrifugation	Water	Ethylene diammonium alginates	304
Amylose	Fractional precipitation	Acetone/acetone - water (1:1)	Acetate	673
		Acetone/water	Carbanilate	239
		Dimethylformamide/methanol + 10% $CaCl_2$	Carbanilate	674
		Dimethyl sulfoxide/acetone		506
		Dimethyl sulfoxide/benzene		506
		Dimethyl sulfoxide/n-butanol	$25^{o}C$	81,82,84,87,88
		Dimethyl sulfoxide/ethanol		83,322,506
		Nitromethane/methanol	Acetate	323
		Water/methanol	Glucoamidoethyl	1570
Amylose acetate				
	Fractional precipitation	Dimethyl sulfoxide/butanol	$25^{o}C$	81,82,86,88
		Nitromethane/methanol	$25^{o}C$	323,328
Amylose benzoate				
	Fractional solution	Nitromethane/methanol	$31^{o}C$	1163,1164,1165
Amylose tricarbanilate				
	Fractional precipitation	Acetone/ethanol	$25^{o}C$	85,89,90,91
		Dimethylformamide/methanol + 10% $CaCl_2$	$20^{o}C$	110,674a,674b,1197
	Chromatography	Ethanol/water		1197,1198
Arabinose	Fractional precipitation	Acetone/petroleum ether	Araban acetate	537
	Chromatography	Chloroform/acetone, pyridine, glacial acetic acid	Araban acetate, charcoal	537
Arabinogalactan				
	Chromatography	Water	GPC, Sephadex G-75	1418
Carboxymethyl cellulose (sodium)				
	Fractional precipitation	Water - acetone (7:3) + 0.5% NaCl/acetone	$30^{o}C$,	1166-1168
		Water/acetone		231
		Water/ethanol		540
	Chromatography	0.5% aq. NaCl/acetone	GPC, Sephadex G-200	207,526
Cellulose	Fractional precipitation	Acetone/acetone - water	Preripened regenerated, alkali cellulose	942
		Cadoxen - water (1:1)/n-propanol - water (3:1)		1188
		Cadoxen/water - glycerol (1:1)		188
		Cuene/propanol +)	Varying temp.	1371
		Cuoxam/propanol x)	Varying temp.	1371
		2 N NaOH/methanol	Rayon, summative method	310
		Iron-sodium tartrate/mannitol		703d
		Sulfuric acid (60-65%)/water	$-15^{o}C$	1373
	Fractional solution	Cuene +)	Extraction	1059
		Cadoxen		703a
		Cuoxam x)		1366
		Iron-sodium tartrate solution/NaOH > 5 N	Summative method	703b
	Extraction	Iron-sodium tartrate solution	Regenerated fibers	1176
		Iron-sodium tartrate	Active coal	703c
	Centrifugal sedimentation	Water suspension	Crystals suspension "micel sol"	702

+) Cupriethylene diamine   x) Cuprammonium hydroxide

Polymer	Method of Fractionation	Solvent or Solvent/Non-solvent Mixture	Remarks	References
Cellulose acetate				
	Fractional precipitation	Acetone/95% ethanol	Lowering temp.	63,1199,1390,1421
		Acetone/ethanol	Solvent volatilized	819,1199
		Acetone - ethanol (5:3)/hexane	Triacetate	453,1036
		Acetone/heptane - acetone (1:1)	Diacetate	535
		Acetone/methanol	Column extraction	365
		Acetone/water	25°C	573,1046
		Acetone/water followed by acetone/pentane	"Cross fractionation"	1278
		Acetone - water (3:1)/water		1406
		Aqueous acetone/heptane - acetone (3:1)		1302
		Chloroform - acetone (1:1)/petroleum ether	Triacetate, 25°C	658-660
		Chloroform/ethanol, tetrachoroethylene/ethanol	Diacetate	377
	Turbidimetric titration	Acetone	Adsorption onto activated carbon	994
	Fractional solution	Butanol - ethanol (4:1)/95 % ethanol		167
		Dioxane/hexane mixtures and dioxane/heptane reflux.	Triacetate	327
		Water/water - acetone	Summative method	819
	Chromatography	1,2-Dichloroethane	GPC, Porosil A, triacetate	1417
		Tetrahydrofuran	GPC, several acetates	221,1057,1435
	Sedimentation velocity	Acetone	25°C	535,1549,1551
	Diffusion into porous charcoal	Acetone, dioxane	Countercurrent	1416,1420
	Thermal diffusion	Dichloroethane	Triacetate	864,865,867
Cellulose acetate-butyrate				
	Fractional precipitation	Acetone/water - acetone (1:1) + 2 % salt and then water alone + 2%. salt		1433
		Acetone/isopropyl ether		1047
	Turbidimetric titration	Acetone/ethanol - water (3:1)		1045
Cellulose nitrate				
	Fractional precipitation	Acetone/n-hexane		348
		Acetone - h-hexane (1:1)/n-hexane		590
		Acetone/n-heptane	Lowering temp.	1372
		Acetone/petroleum ether		56,1072
		Acetone/water	"Triangle method"	964
				5,348,963,1554
			0°C	935
			25°C	1452
		Acetone/water followed by acetone/heptane	"Cross fractionation"	3
		Acetone/acetone - water (1:1)		1072,1616-1618
		Acetone/acetone - water (4:1) until pure water	0°C, no light	937-939,1328
		Acetone - 9% water/water		1006,1279,1448
		Ethyl acetate/n-heptane		1301
	Turbidimetric titration	Acetone/methanol - water (9:1)		1134
	Fractional solution	Acetone/methanol - water	Column extraction	365
		Ethyl acetate/95% ethanol		410,625
	Distribution between immiscible liquids	Cellulose triacetate gel in methyl acetate/water, ethanol, butyl acetate		225a
	Chromatography	Acetone	Activated carbon	293
			GPC	976,981
		Acetone/cyclohexane followed by acetone/methanol	Elution chromatography, Starch	225b
		Tetrahydrofuran	GPC	976,977,980,1057, 1058, 1150
	Sedimentation velocity	Acetone	Ultracentrifuge	1135

Polymer	Method of Fractionation	Solvent or Solvent/Non-solvent Mixture	Remarks	References
Cellulose nitrate (Cont'd.)				
	Diffusion into porous charcoal	Acetone and dioxane	Countercurrent	1420
	Membrane diffusion	Acetone		1538
Cellulose trinitrate				
	Fractional precipitation	Acetone/water - acetone		936, 1186, 1329
		Acetone - ethanol (5:3)/hexane	Room temp.	453, 1028, 1032
		Acetone - ethanol (5:3)/Skellsolve	High M.W.	672
		Acetone/hexane	Low. M.W.	672
	Chromatography	Tetrahydrofuran	GPC	10, 395, 983, 1265, 1350- 1354, 1545
Cellulose tributyrate				
	Fractional precipitation	Acetone/water		861
Cellulose tricarbanilate				
	Fractional precipitation	Acetone/acetone - water (1:1)		675
		Acetone/1% aqueous NaCl		222
	Chromatography	Tetrahydrofuran	GPC	222, 1055
Cellulose tripropionate				
	Chromatography	Tetrahydrofuran	GPC	221
Cellulose xanthate				
	Fractional precipitation	10% aq. NaOH/methanol	Viscose fraction	521
Dextran	Fractional precipitation	Water/ethanol	Acid hydrolyzed	1356
		Water/methanol	Low M.W.	53, 513, 1623
	Chromatography	Water	Activated carbib	293
			GPC	189, 543, 885
		Water/ethanol	"Precipitation chromat."	397, 398
		Water/1% diethylene glycol	GPC, porous silica beads	886
	Ultrafiltration	Water	Amicon XM-4B and UM-1 Diaflo- membranes	75
	Sedimentation velocity	Water	Ultracentrifuge, 20°C	397
Diethylaminoethyl amylose hydrochloride				
	Chromatography	0.5 M aq. NaCl	GPC	526
Ethyl cellulose				
	Fractional precipitation	Acetone/water		987
		Ethyl acetate - acetone (1:4)/water		1300
Poly(D-glucose)				
	Fractional precipitation	Water/methanol	Lowering temp.	991
Gum arabic	Fractional precipitation	0.5 aq. NaCl/acetone		1419
Guaran triacetate				
	Fractional precipitation	Chloroform/ethanol	22.5°C	811
	Sedimentation velocity	Acetonitrile	25°C	811
Hydroxyethyl cellulose				
	Fractional precipitation	Water/acetone	25°C	230, 408
Hydroxypropyl cellulose				
	Fractional solution	Ethanol/n-heptane	30°C	1583
	Chromatography	Tetrahydrofuran	GPC, 25°C	1583
Lignins	Chromatography	Dimethyl sulfoxide	GPC	1393
Ligninsulfonic acid, sodium salt				
	Fractional precipitation	Water/ethanol		435
	Fractional solution	Aq. NaCl/95% ethanol	Brownian diffusion	1012

Polymer	Method of Fractionation	Solvent or Solvent/Non-solvent Mixture	Remarks	References
Ligninsulfonic acid, sodium salt (Cont'd.)				
	Extraction	Water/ethanol		435
	Chromatography	Water	GPC, 44°C	189
		0.0001 aq. NaCl	GPC, Sephadex, G-50	561
Methyl cellulose				
	Thermal diffusion	Water		467
Pectins	Fractional precipitation	Dioxane/petroleum ether	Propionate, lowering temp.	1153
	Fractional centrifugation	Ethylenediamine	Amylopectin	1398
	Membrane diffusion	Water		1277
Starch	Fractional precipitation	Thymol/n-butanol		326
	Aqueous leaching	Water	Several temp. and times	92,325,548
	Alkaline leaching	NaOH 0.5 M (water)		92
	Extraction	Aq. NaOH/n-butanol		92,117
	Centrifugation	Dilute alkali		116
Tribenzyl glucans				
	Sedimentation velocity	Benzene	8°C, Benzylated linear α-1,6-glucans	972
Xylan	Fractional precipitation	Dimethyl sulfoxide/ethanol		880

4.2 Graftcopolymers

Polymer	Method of Fractionation	Solvent or Solvent/Non-solvent Mixture	Remarks	References
Acrylamide + cellulose				
	Extraction	Dimethylformamide, water cupraammonium	Cellophane	504
Acrylamide + cellulose acetate				
	Extraction	Acetone		504
Acrylamide + ethyl cellulose				
	Extraction	Methanol	Hot	504
Acrylonitrile + cellulose				
	Extraction	Dimethylformamide, water cupraammonium	Cellophane	504
Acrylonitrile + cellulose acetate				
	Fractional precipitation	Dimethylformamide/chloroform		1262
	Extraction	Dimethylformamide	Acetylated	1262
		Dimethylformamide - dioxane (3.5:2)	Cotton	1262
Acrylonitrile + methyl cellulose				
	Extraction	Water	Hot	270
Acrylonitrile + starch				
	Extraction	Dimethylformamide, γ-butyrolactone, dimethyl sulfoxide		427
	Chromatography	Dimethylformamide	GPC, 75°C	557
Ethyl acrylate + cellulose				
	Extraction	Acetone	55°C	725
Ethylene oxide + cellulose acetate				
	Fractional precipitation	Dimethylformamide/ethyl ether		97
Methyl acrylate + amylose				
	Fractional precipitation	Dimethyl sulfoxide/n-butanol		1171
	Extraction	Acetone, benzene, chloroform, toluene and ethyl acetate		1171
Methyl, ethyl, butyl acrylates + cellulose acetate				
	Extraction; precipitation	Benzene; acetone/water		677
	Turbidimetric titration	Acetone - methanol (6:4)/methanol		677

Polymer	Method of Fractionation	Solvent or Solvent/Non-solvent Mixture	Remarks	References
Methyl methacrylate + cellulose acetate				
	Fractional precipitation	95% Pyridine - 5% acetone/water		169
Methyl methacrylate + ethyl cellulose				
	Fractional precipitation	Acetone/methanol		270
Methyl methacrylate + starch				
	Extraction precipitation	Benzene; aq. dimethylformamide/methanol		270
2-Methyl-5-vinylpyridine + cellulose acetate				
	Extraction	Acetone, methylene chloride, methanol		724
Styrene + benzyl cellulose				
	Fractional precipitation	Benzene/methanol		270
Styrene + cellulose				
	Fractional precipitation	Dioxane/methanol	20°C	1388
Styrene + cellulose acetate				
	Fractional precipitation	Chloroform/n-heptane	Triacetylated viscose	1262
		Pyridine/aq. HCl		1609
	Chromatography	Benzene/methanol	"Precipitation chromat.", 15-60°C, celite	1562
	Extraction	Acetone - water (7:3), benzene		1562, 1563
		Benzene, acetone, 2-ethoxyethanol, water		1609
Styrene + methylated xylan				
	Fractional precipitation	Benzene/petroleum ether		1127
Vinyl acetate + cellulose acetate				
	Extraction; precipitation	Methanol; benzene/methanol		270
2-Vinylpyridine + cellulose				
	Extraction	Benzene		1288
2-Vinylpyridine + cellulose acetate				
	Extraction	Methanol/ acetone - dichloroethane (1:4)		1437
N-Vinylpyrrolidone + cellulose acetate				
	Extraction	Water, acetone		1438

4.3 Mixtures of Polymers

Polymer	Method of Fractionation	Solvent or Solvent/Non-solvent Mixture	Remarks	References
Cellulose acetate + poly(oxyethylene)				
	Fractional precipitation	Dimethylformamide/ethyl ether		97

D. REFERENCES

1. M. Abe, Y. Murakami, H. Fujita, J. Appl. Polymer Sci. 9, 2549 (1965).
2. G.J.K. Acres, F. L. Dalton, J. Polymer Sci., A 1, 2419 (1963).
3. K. Aejmelaeus, Ann. Acad. Sci. Fennicae, Ser. A., 2, 63 (1950).
4. S. L. Aggarwal, L. Marker, M. J. Carrano, J. Appl. Polymer Sci. 3, 78 (1960).
5. J. C. Aggarwal, J. L. McCarthy, J. Indian Chem. Soc. 26, 11 (1949).
6. E. Ahad, Y. Sicotte, J. Polymer Sci., C, 30, 163 (1970).
7. H. Ailhaud, Y. Gallot, A. Skoulios, Makromol. Chem. 140, 179 (1970).
8. P. J. Akers, G. Allen, M. J. Bethell, Polymer 9, 575 (1968).
9. A. Albert, D.C. Pepper, Proc. Roy. Soc. (London) A 263, 75 (1961).
10. W. J. Alexander, Th. E. Muller, J. Polymer Sci., C, 36, 87 (1971).
11. S. M. Ali, M. B. Huglin, Makromol. Chem. 84, 117 (1965).
12. G. Allen, C. Booth, S. J. Hurst, Polymer 8, 385 (1967).
13. G. Allen, C. Booth, S. J. Hurst, M. N. Jones, C. Price, Polymer 8, 391 (1967).
14. G. Allen, C. Booth, M. N. Jones, Polymer 5, 195 (1964).
15. G. Allen, C. Booth, M. N. Jones, Polymer 5, 257 (1964).
16. G. Allen, C. Booth, C. Price, Polymer 8, 397 (1967).
17. G. Allen, C. J. Lewis, S. M. Todd, Polymer 11, 44 (1970).
18. G. Allen, J. McAinsh, C. Strazielle, European Polymer J. 5, 319 (1969).
19. P. E. M. Allen et al., Makromol. Chem. 39, 52 (1960).
20. P. E. M. Allen, G. M. Burnett, J. M. Downer, H. W. Melville, Makromol. Chem. 38, 72 (1960).
21. P. E. M. Allen, R. P. Chaplin, D. O. Jordan, European Polymer J. 8, 271 (1972).
22. V. R. Allen, S. R. Turner, J. Macromol. Sci. Chem. A5, 229 (1971).
23. D. F. Alliet, J. Appl. Polymer Symposia 8, 39 (1969).
24. D. F. Alliet, J. M. Pacco, J. Polymer Sci., C, 21, 199 (1968).
25. D. F. Alliet, J. M. Pacco, Polymer Preprints, 8(2), 1288 (1967).
26. R. Allirot, Compt. Rend. 231, 1065 (1950).
27. T. Altares, D. P. Wyman, V. R. Allen, J. Polymer Sci., A-1, 2, 4533 (1964).

28.   T. Altares, D. P. Wyman, V. R. Allen, K. Meyersen, J. Polymer Sci., A-1, 3, 4131 (1965).

29.   K. H. Altgelt, in "Advances in Chromatography", Editors: J. C. Giddings, R. A. Keller, Vol. 7, Marcel Dekker, New York, 1968.

30.   K. H. Altgelt, J. Appl. Polymer Sci. 9, 3389 (1965).

31.   K. H. Altgelt, J. C. Moore, in Polymer Fractionation, Editor M. J. R. Cantow, Academic Press, London, 1967.

32.   K. H. Altgelt, L. Segal, "Gel Permeation Chromatography", Editors, Marcel Dekker, New York, 1971.

33.   J. M. Alvariño, A. Bello, G. M. Guzmán, Anales Quím. (Madrid), 66, 105 (1970).

34.   A. J. Amass, E. W. Duck, J. R. Hawkins, J. M. Locke, European Polymer J. 8, 781 (1972).

35.   R. J. Ambrose, J. Appl. Polymer Sci. 15, 1297 (1971).

36.   Y. Amerik, J. E. Guillet, Macromolecules 4, 375 (1971).

37.   V. S. Ananthanarayanan, R. H. Andreatta, D. Poland, H. A. Scheraga, Macromolecules 4, 417 (1971).

39.   G. A. Andreyeva, S. P. Mitsengendler, K. I. Sokolova, A. A. Korotkov, Polymer Sci. USSR 8, 2391 (1966); Vysokomolekul. Soedin. 8, 2159 (1966).

40.   K. A. Andrianov et al., Vysokomolekul. Soedin. 3, 1692 (1961).

41.   K. A. Andrianov, V. I. Pakhomov, V. M. Gel' perina, D. N. Mukhina, Polymer Sci. USSR 8, 1787 (1966); Vysokomolekul. Soedin. 8, 1618 (1966).

43.   K. A. Andrianov, I. I. Tverdohlebova, S. A. Pavlova, B. G. Zavin, J. Polymer Sci., C, 22, 741 (1969).

44.   K. A. Andrianov, S. Ye. Yakushkina, L. N. Goniava, Polymer Sci. USSR 8, 2398 (1966); Vysokomolekul. Soedin. 8, 2166 (1966).

45.   K. A. Andrianov, B. G. Zavin, N. V. Pertsova, Polymer Sci. USSR, 10, 49 (1968); Vysokomolekul Soedin. 10A, 46 (1968).

46.   D. J. Angier, R. J. Ceresa, W. F. Watson, J. Polymer Sci. 34, 699 (1959).

47.   D. J. Angier, D. T. Turner, J. Polymer Sci. 28, 265 (1958).

48.   E. V. Anufrieva, T. M. Birshtein, T. N. Nekrasova, O. B. Ptitsyn, T. V. Sheveleva, J. Polymer Sci., C, 16, 3519 (1968).

49.   G. Anzuino, L. Constantino, R. Gallo, V. Vitagliano, J. Polymer Sci., B, 4, 459 (1966).

50.   I. Arendt, H. J. Schenck, Kunststoffe 48, 111 (1958).

51.   S. Arichi, S. Mitsuta, N. Sakamoto, H. Murata, Bull. Chem. Soc. Japan 39, 428 (1966).

52.   R. S. Aries, A. P. Sachs, J. Polymer Sci. 21, 551 (1956).

53.   L. H. Arond, H. P. Frank, J. Phys. Chem. 58, 953 (1954).

54.   F. Arranz, J. C. Galin, Makromol. Chem. 152, 185 (1972).

55.   F. Arranz, M. Galin, J. C. Galin, P. Rempp, IUPAC Boston, Vol. 1, 432 (1971).

56.   H. Asaoka, A. Suzuki, J. Soc. Text. Cell. Ind. (Japan) 11, 32 (1955).

59.   T. P. Avilova, V. T. Bykov, V. P. Marinin, N. P. Shapkin, Polymer Sci. USSR 7, 2377 (1965); Vysokomolekul. Soedin. 7, 2168 (1965).

60.   T. P. Avilova, V. T. Bykov, L. A. Kondratenko, Polymer Sci. USSR 8, 12 (1966); Vysokomolekul. Soedin. 8, 14 (1966).

61.   T. P. Avilova, V. T. Bykov, G. Ya. Zolotar, Polymer Sci. USSR 7, 917 (1965); Vysokomolekul. Soedin. 7, 831 (1965).

62.   C. W. Ayers, Analyst 78, 382 (1953).

63.   W. J. Badgley, H. F. Mark, J. Phys. Chem. 51, 58 (1947).

64.   M. Baer, J. Polymer Sci. A2, 417 (1964).

65.   G. Bagby, R. S. Lehrle, J. C. Robb, Polymer 9, 284 (1968).

66.   G. Bagby, R. S. Lehrle, J. C. Robb, Makromol. Chem. 119, 122 (1968).

68.   M. D. Baijal, R. M. Diller, F. R. Pool, Polymer Preprints 10(2), 1464 (1969).

69.   M. D. Baijal, K. M. Kanppila, Polymer Eng. Sci. 11, 182 (1971).

70.   F. E. Bailey, G. M. Powell, K. L. Smith, Ind. Eng. Chem. 50, 8 (1958).

71.   F. C. Baines, J. C. Bevington, European Polymer J. 3, 593 (1967).

72.   K. Bak, G. Elefante, J. E. Mark, Polymer Preprints 8(2), 998 (1967).

73.   K. Bak, G. Elefante, J. E. Mark, J. Phys. Chem. 71, 4007 (1967).

74.   B. Baker, P. J. T. Tait, Polymer 8, 225 (1967).

75.   R. W. Baker, J. Appl. Polymer Sci. 13, 369 (1969).

76.   C. A. Baker, R. J. P. Williams, J. Chem. Soc. 2352 (1956).

77.   D. G. H. Ballard, J. V. Dawkins, Makromol. Chem. 148, 195 (1971).

78.   G. D. Ballova, V. M. Bulatova, K. A. Vylegzhanina, Ye. I. Yegorova, L. L. Sul' zhenko, G. P. Fratkina, Polymer Sci. USSR 11, 2080 (1969); Vysokomolekul. Soedin. 11A, 1827 (1969).

79.   A. Banderet, W. Kobryner, Compt. Rend. 244, 604 (1957).

80.   T. F. Banigan, Science 117, 249 (1953).

81.   W. Banks, C. T. Greenwood, European Polymer J. 4, 249 (1968).

82.   W. Banks, C. T. Greenwood, European Polymer J. 4, 377 (1968).

83.   W. Banks, C. T. Greenwood, Makromol. Chem. 67, 49 (1963).

84.   W. Banks, C. T. Greenwood, Makromol. Chem. 114, 245 (1968).

85.   W. Banks, C. T. Greenwood, Makromol. Chem. 144, 135 (1971).

86.   W. Banks, C. T. Greenwood, Polymer 10, 257 (1969).

87.   W. Banks, C. T. Greenwood, D. J. Hourston, Makromol. Chem. 111, 226 (1968).

88.   W. Banks, C. T. Greenwood, D. J. Hourston, Trans. Faraday Soc. 64, 363 (1968).

89.   W. Banks, C. T. Greenwood, J. Sloss, European Polymer J. 7, 263 (1971).

90.   W. Banks, C. T. Greenwood, J. Sloss, Makromol. Chem. 140, 109 (1970).

91.   W. Banks, C. T. Greenwood, J. Sloss, Makromol. Chem. 140, 119 (1970).

92.   W. Banks, C. T. Greenwood, J. Thomson, Makromol. Chem. 31, 197 (1959).

93.   D. W. Bannister, C. S. G. Phillips, R. J. P. Williams, Anal. Chem. 26, 1451 (1954).

95.   I. A. Baranovskaya, V. Ye. Eskin, Polymer Sci. USSR 7, 373 (1965); Vysokomolekul. Soedin. 7, 339 (1964).

96.   K. Baranwal, H. Jacobs, J. Appl. Polymer Sci. 13, 797 (1969).

97.   A. Bar-Ilan, A. Zilkha, European Polymer J. 6, 403 (1970).

98.   A. Barlow, R. S. Lehrle, J. C. Robb, D. Sunderland, Polymer 8, 537 (1967).

99.   W. K. R. Barnikol, G. V. Schulz, Z. Physik. Chem. N.F. (Frankfurt) 47, 89 (1965).

100. J. M. Barrales-Rienda, Anales Quim. (Madrid) 66, 767 (1970).

101. J. M. Barrales-Rienda, R. G. Brown, D. C. Pepper, Polymer 10, 327 (1969).

102. J. M. Barrales-Rienda, J. González Ramos, M. V. Dabrio, Anales Chim. (Madrid), 67, 651 (1971).

103. J. M. Barrales-Rienda, D. C. Pepper, Polymer 8, 337 (1967).

104. J. M. Barrales-Rienda, D. C. Pepper, European Polymer J. 3, 535 (1967).

105. C. A. Barson, J. C. Robb, Brit. Polymer J. 3, 53 (1971).

106. G. L. Bata, J. E. Hazell, L. A. Prince, IUPAC Toronto, Preprint A4. 14 (1968).

107. G. L. Bata, J. E. Hazell, L. A. Prince, J. Polymer Sci., C, 30, 157 (1970).

108. T. W. Bates, J. Biggins, K. J. Ivin, Makromol. Chem. 87, 180 (1965).

109. T. W. Bates, K. J. Ivin, Polymer 8, 263 (1967).

110. R. J. E. Cumberbirch, W. G. Harland, J. Textile Inst. 49, T 679 (1958).

111. H. Batzer, Makromol. Chem. 5, 66 (1950).

112. H. Batzer, Makromol. Chem. 12, 145 (1954).

113. H. Batzer, A. Nisch, Makromol. Chem. 22, 131 (1957).

114. H. Batzer, A. Moeschle, Makromol. Chem. 22, 195 (1957).

115. H. Batzer, F. Wiloth, Makromol. Chem. 8, 41 (1952).

116. H. Baum, G. A. Gilbert, J. Colloid Sci. 11, 428 (1956).

117. H. Baum, G. A. Gilbert, H. L. Wood, J. Chem. Soc. 4047 (1955).

118. G. F. Baumann, S. Steingiser, J. Polymer Sci. A1, 3395 (1963).

119. J. H. Baxendale, S. Bywater, M. G. Evans, Trans. Faraday Soc. 42, 675 (1946).

120. B. Baysal, J. Polymer Sci., C, 4, 935 (1963).

121. B. Baysal, G. Adler, D. Ballantine, A. Glines, J. Polymer Sci., B, 1, 257 (1963).
122. H. C. Beachell, J. C. Peterson, Polymer Preprints 8(1), 456 (1967).
123. H. C. Beachell, J. C. Peterson, J. Polymer Sci.,A-1, 7, 2021 (1969).
124. W. H. Beattie, J. Polymer Sci., A-1, 3, 527 (1965).
125. W. H. Beattie, C. Booth, J. Appl. Polymer Sci. 7, 508 (1963).
126. D. R. Beech, C. Booth, J. Polymer Sci., A-2, 7, 575 (1969).
127. D. R. Beech, C. Booth, D. V. Dogson, R. R. Sharpe, J. R. S. Waring, Polymer 13, 73 (1972).
128. D. R. Beech, C. Booth, I. H. Hillier and C. J. Pickles, European Polymer J. 8, 799 (1972).
129. R. B. Beevers, J. Polymer Sci., A-1, 2, 5257 (1964).
130. S. I. Beilin, Y. L. Vollershtein, B. A. Dolgoplosk, V. A. Kargin, M. N. Shvarts, T. V. Fremel, Polymer Sci. USSR 10, 1 (1968); Vysokomolekul. Soedin. 10A, 3 (1968).
131. Ye. A. Bekturov, L. A. Bimendina, S. V. Bereza, Polymer Sci. USSR 12, 2468 (1970); Vysokomolekul. Soedin 12A, 2179 (1970).
132. Ye. A. Bekturov, R. Ye. Legkunets, Polymer Sci. USSR 12, 702 (1970); Vysokomolekul. Soedin. 12A, 626 (1970).
133. I. M. Bel' gouskii, V. M. Gol' berg, I. A. Krasotkina, D. Ya. Toptygin, Polymer Sci. USSR 13, 758 (1971); Vysokomolekul. Soedin. 13A, 666(1971).
134. G. P. Belov, S. K. Kaltochikhina, L. I. Atanasova, V. I. Tsvetkova, Polymer Sci. USSR 8, 2046 (1966); Vysokomolekul. Soedin. 8, 1852 (1966).
135. G. P. Belov, V. I. Kuznetsov, T. I. Solovyeva, N. M. Chirkov, S. S. Ivanchev, Makromol. Chem. 140, 213 (1970).
136. G. P. Belov, A. P. Lisitskaya, N. M. Chirkov, V. I. Tsvetkova, Polymer Sci. USSR 9, 1417 (1967); Vysokomolekul. Soedin. 9, 1269 (1967).
137. G. P. Belov, A. P. Lisitskaya, T. I. Solovyeva, N. M. Chirkov, European Polymer J. 6, 29 (1970).
139. W. I. Bengough, G. F. Grant, European Polymer J. 7, 203 (1971).
140. H. Benoit, Z. Grubisic, P. Rempp, D. Decker, J.-G. Zilliox, J. Chim. Phys. 63, 1507 (1966).
141. K. G. Berdnikova, G. V. Tarasova, V. S. Skazka, N. A. Nikitin, G. V. Dyuzhev, Polymer Sci. USSR 6, 2280 (1964); Vysokomolekul. Soedin. 6, 2057 (1964).
142. D. Berek, D. Lath, V. Durdovic, J. Polymer Sci., C, 16, 659 (1967).
143. A. Beresniewicz, J. Polymer Sci. 35, 321 (1959).
144. S. V. Bereza, Ye. A. Bekturov, S. R. Rafikov, Polymer Sci. USSR 11, 1907 (1969); Vysokomolekul. Soedin. 11A, 1681 (1969).
145. K. C. Berger, G. V. Schulz, Makromol. Chem. 136, 221 (1970).
146. M. N. Berger, B. M. Tidswell, IUPAC Helsinki, Vol. III, Preprint II-63 (1972).
147. H. Berghmans, G. Smets, Makromol. Chem. 115, 187 (1968).
148. J. B. Berkowitz, M. Yamin, R. M. Ruoss, J. Polymer Sci. 28, 69 (1958).
150. A. A. Berlin, V. A. Grigorovskaya, O. G. Sel' skaya, Polymer Sci. USSR 12, 2879 (1970); Vysokomolekul. Soedin. 12A, 2541 (1970).
151. D. E. Bernal, Rev. Gen. Caoutchouc 32, 889 (1955).
152. G. C. Berry, R. G. Craig, Polymer 5, 19 (1964).
153. G. C. Berry, T. G. Fox, J. Macromol. Sci. Chem. A3, 1125 (1969).
154. G. C. Berry, L. M. Hobbs, V. C. Long, Polymer 5, 31 (1964).
155. G. C. Berry, H. Nomura, K. G. Mayhan, J. Polymer Sci., A-2, 5, 1 (1967).
156. G. C. Berry, S. P. Yen, Advan. Chem. Ser. 91, 734 (1969).
157. J. C. Bevington, G. M. Guzmán, H. W. Melville, Proc. Roy. Soc. (London) A 221, 437 (1954).
158. J. P. Bianchi, F. P. Price, B. H. Zimm, J. Polymer Sci. 25, 27 (1957).
159. U. Bianchi, M. Dalpiaz, E. Patrone, Makromol. Chem. 80, 112 (1964).
160. U. Bianchi, V. Magnasco, J. Polymer Sci. 41, 177 (1959).
161. U. Bianchi, E. Patrone, M. Dalpiaz, Makromol. Chem. 84, 230 (1965).
162. U. Bianchi, A. Peterlin, J. Polymer Sci., A-2, 6, 1011 (1968).
163. P. Biernacki, C. Chrzczonowicz, M. Wlodarczyk, European Polymer J. 7, 739 (1971).
164. F. W. Billmeyer, I. Katz, Macromolecules 2, 105 (1969).
165. F. W. Billmeyer, R. N. Kelley, Polymer Preprints 8(2), 1259 (1967).
166. F. W. Billmeyer, W. H. Stockmayer, J. Polymer Sci. 5, 121 (1950).
167. J. Bischoff, V. Desreux, Bull. Soc. Chim. Belges 60, 137 (1951).
168. J. Bisschops, J. Polymer Sci. 17, 81 (1955).
169. W. R. Blackmore, W. Alexander, Can. J. Chem. 39, 1888 (1961).
170. D. E. Blair, Polymer Preprints 11(2), 967 (1970).
171. D. E. Blair, J. Appl. Polymer Sci. 14, 2469 (1970).
172. L. P. Blanchard, M. D. Baijal, Polymer Preprints 7(2), 944 (1966).
173. L. P. Blanchard, M. D. Baijal, J. Polymer Sci., A-1, 5, 2045 (1967).
174. M. J. Blandamer, M. F. Fox, E. Powell, J. W. Stafford, Makromol. Chem. 124, 222 (1969).
175. G. F. Bloomfield, Rubber Chem. Technol. 24, 737 (1951).
176. D. D. Bly, Polymer Preprints 8(2), 1234 (1967).
177. E. G. Bobalek et al., J. Appl. Polymer Sci. 8, 632 (1964).
178. Ye. P. Bogomolova, A. A. Trapeznikova, V. N. Fedotov, Polymer Sci. USSR 12, 1375 (1970); Vysokomolekul. Soedin. 12 A, 1216 (1970).
179. V. Boháčková, E. Fišerová, Z. Gallot-Grubišic, J. Poláček, M. Stolka, J. Chim. Phys. 67, 777 (1970).
180. V. Boháčková, J. Poláček, Z. Grubišic, H. Benôit, J. Chim. Phys. 66, 207 (1969).
181. M. Bohdanecký, Collection Czech. Chem. Commun. 35, 1972 (1970).
182. M. Bohdanecký, Collection Czech. Chem. Commun. 29, 876 (1964).
183. M. Bohdanecký, Collection Czech. Chem. Commun. 31, 4095 (1966).
184. M. Bohdanecký, V. Petrus, P. Kratochvíl, Collection Czech. Chem. Commun. 33, 4089 (1968).
185. M. Bohdanecký, V. Petrus, P. Kratochvíl, Collection Czech. Chem. Commun. 34, 1168 (1969).
186. M. Bohdanecký, Z. Tuzar, Collection Czech. Chem. Commun. 34, 3318 (1969).
187. M. Bohdanecký, Z. Tuzar, M. Štoll, R. Chromeček, Collection Czech. Chem. Commun. 33, 4104 (1968).
188. L. S. Bolotnikova, T. I. Samsonova, Polymer Sci. USSR 6, 590 (1964); Vysokomolekul. Soedin. 6, 533 (1964).
189. K. J. Bombaugh, W. A. Dark, R. N. King, J. Polymer Sci., C, 21, 131 (1968).
190. K. A. Boni, F. A. Sliemers, P. B. Stickney, Polymer Preprints 8(1), 446 (1967).
191. K. A. Boni, F. A. Sliemers, P. B. Stickney, J. Polymer Sci., A-2, 6, 1567 (1968).
192. K. A. Boni, F. A. Sliemers, P. B. Stickney, J. Polymer Sci., A-2, 6, 1579 (1968).
193. C. Booth, Polymer 4, 471 (1963).
194. C. Booth, L. R. Beason, J. Polymer Sci. 42, 99 (1960).
195. C. Booth, L. R. Beason, J. Polymer Sci. 42, 108 (1960).
196. C. Booth, D. V. Dodgson, I. H. Hillier, Polymer 11, 11 (1970).
197. C. Booth, W. C. E. Higginson, E. Powell, Polymer 5, 479 (1964).
198. C. Booth, M. N. Jones, E. Powell, Nature (London) 196, 772 (1962).
199. C. Booth, C. Price, Polymer 7, 85 (1966).
200. C. Booth, R. Orme, Polymer 11, 626 (1970).
201. P. Borrell, G. Riess, A. Banderet, Bull. Soc. Chim. France 2, 354 (1961).
202. C. Borri, S. Brueckner, V. Crescenzi, G. della Fortuna, A. Mariano, P. Scarazzato, European Polymer J. 7, 1515 (1971).
203. E. Borsig, M. Lazár, M. Čapla, Š. Florián, Angew. Makromol. Chem. 9, 89 (1969).
204. R. F. Boyer, J. Polymer Sci. 8, 73, 197 (1952).
205. E. M. Bradbury, C. Crane-Robinson, H. W. E. Rattle, Polymer 11, 277 (1970).
207. D. A. Brant, B. Kwon Min, Macromolecules 2, 1 (1969).
208. D. Braun, D. Chaudhari, Makromol. Chem. 133, 241 (1970).
209. D. Braun, D. Chaudhari, W. Mächtle, Angew. Makromol. Chem. 15, 83 (1971).

210. D. Braun, W. Fischer, Makromol. Chem. 85, 155 (1965).

211. D. Braun, F. -J. Quesada Lucas, W. Neumann, Makromol. Chem. 127, 253 (1969).

212. J. W. Breitenbach, H. G. Burger, Makromol. Chem. 54, 60 (1962).

213. J. W. Breitenbach, H. G. Burger, A. Schindler, Monatsh. Chem. 93, 160 (1962).

214. J. W. Breitenbach, H. Gabler, O. F. Olaj, Makromol. Chem. 81, 32 (1965).

215. J. W. Breitenbach, O. F. Olaj, A. Schindler, Monatsh. Chem. 91, 205 (1960).

216. J. W. Breitenbach, B. A. Wolf, Makromol. Chem. 108, 263 (1967).

218. S. E. Bresler, I. Ya. Poddubnyi, S. Ya. Frenkel, Zh. Tekhn. Fiz. 23, 1521 (1953).

219. S. E. Bresler, I. Y. Poddubnyi, S. Ya. Frenkel, Rubber Chem. Technol. 30, 507 (1957).

220. P. I. Brewer, Polymer 9, 545 (1968).

221. R. J. Brewer, L. J. Tanghe, S. Bailey, J. Polymer Sci., A-1, 7, 1635 (1969).

222. R. J. Brewer, L. J. Tanghe, S. Bailey, J. T. Burr, J. Polymer Sci., A-1, 6, 1697 (1968).

223. W. B. Bridgman, J. Am. Chem. Soc. 64, 2349 (1942).

224. G. M. Bristow, B. Westall, Polymer 8, 609 (1967).

225a. M. C. Brooks, R. M. Badger, J. Am. Chem. Soc. 72, 1705 (1950).

225b. M. C. Brooks, R. M. Badger, J. Am. Chem. Soc. 72, 4384 (1950).

226. L. de Brouckere, E. Bidaine, A. van der Heyden, Bull. Soc. Chim. Belges 58, 418 (1949).

227. F. M. Brower, H. W. McCormick, J. Polymer Sci. 36, 341 (1959).

228. F. M. Brower, H. W. McCormick, J. Polymer Sci. A-1, 1749 (1963).

229. J. R. Brown, J. H. O' Donnell, Macromolecules 5, 109 (1972).

230. W. Brown, Arkiv Kemi 18, 227 (1961).

231. W. Brown, D. Henley, J. Oehman, Arkiv Kemi 22, 189 (1964).

232. S. D. Bruck, J. Polymer Sci. 32, 519 (1958).

233. W. A. J. Bryce, I. G. Meldrum, J. Polymer Sci., C, 16, 3321 (1968).

234. W. A. J. Bryce, G. McGibbon, I. G. Meldrum, Polymer 11, 394 (1970).

235. J. Brzezinski, Z. Czlonkowska-Kohutnicka, O. Kallistov, Sz. Krozer, Plaste Kautschuk 15, 403 (1968).

236. R. R. Buch, H. M. Klimisch, O. K. Johannson, J. Polymer Sci., A-2, 7, 563 (1969).

237. R. R. Buch, H. M. Klimisch, O. K. Johannson, Polymer Preprints 6(2), 1022 (1965).

238. R. R. Buch, H. M. Klimisch, O. K. Johannson, J. Polymer Sci., A-2, 8, 541 (1970).

239. W. Burchard, E. Husemann, Makromol. Chem. 44-46, 358 (1961).

240. D. E. Burge, D. B. Bruss, J. Polymer Sci. A-1, 1927 (1963).

241. K. Burger, Z. Anal. Chem. 196, 259 (1963).

242. G. M. Burnett, P. Meares, C. Paton, Trans. Faraday Soc. 58, 723 (1962).

243. F. W. Burns, B. McCarthy, R. M. O' Connor, D. C. Pepper, IUPAC Helsinki, Vol. I, Preprint 1-30. (1972).

244. H. Burns, D. K. Carpenter, Macromolecules 1, 384 (1968).

245. A. Buzagh, K. Udvarhelyi, F. Horkay, Kolloid Z. 154, 130 (1957).

246. A. Buzagh, K. Udvarhelyi, F. Horkay, Kolloid Z. 157, 53 (1958).

247. V. T. Bykov, T. P. Avilova, N. P. Shapkin, Polymer Sci. USSR 12, 813 (1970); Vysokomolekul. Soedin. 12 A, 724 (1970).

248. S. Bywater, P. Johnson, Trans. Faraday Soc. 47, 195 (1951).

249. H. Campbell, P. O. Kane, I. G. Ottewill, J. Polymer Sci. 12, 611 (1954).

250. F. Candau, P. Rempp, Makromol. Chem. 122, 15 (1969).

251. H. J. Cantow, Makromol. Chem. 30, 81 (1959).

252. M. J. R. Cantow, Editor "Polymer Fractionation", Academic Press, New York (1966).

253. M. J. R. Cantow, R. S. Porter, J. F. Johnson, Makromol. Chem. 49, 1 (1961).

254. M. J. R. Cantow, R. S. Porter, J. F. Johnson, Nature (London) 192, 752 (1962).

255. M. J. R. Cantow, R. S. Porter, J. F. Johnson, J. Polymer Sci., C, 1, 187 (1963).

256. M. J. R. Cantow, R. S. Porter, J. F. Johnson, J. Appl. Polymer Sci. 8, 2963 (1964).

257. M. J. R. Cantow, R. S. Porter, J. F. Johnson, Polymer Preprints 6(1), 338 (1965).

258. M. J. R. Cantow, R. S. Porter, J. F. Johnson, J. Polymer Sci., B, 4, 707 (1966).

259. M. J. R. Cantow, R. S. Porter, J. F. Johnson, J. Macromol. Sci. Rev. Macromol. Chem. 1, 393 (1966).

260. S. R. Caplan, J. Polymer Sci. 35, 409 (1959).

261. C. E. Carraher, T. W. Brandt, Makromol. Chem. 130, 166 (1970).

262. E. F. Casassa, W. H. Stockmayer, Polymer 3, 53 (1962).

263. L. C. Case, J. Phys. Chem. 62, 895 (1958).

264. L. C. Case, Makromol. Chem. 41, 61 (1960).

265. R. H. Casper, G. V. Schulz, J. Polymer Sci., A-2, 8, 893 (1970).

266. G. Ceccorulli, M. Pizzoli, G. Stea, Makromol. Chem. 142, 153 (1971).

267. J. Cepelák, Chem. Prumysl 6, 106 (1956).

268. R. J. Ceresa, Polymer 1, 480 (1960).

269. R. J. Ceresa, Polymer 1, 488 (1960).

270. R. J. Ceresa, Polymer 2, 213 (1961).

271. L. C. Cerny, T. E. Helminiak, J. F. Meier, J. Polymer Sci. 44, 539 (1960).

272. S. Cesca, A. Priola, A. de Chirico, G. Santi, Makromol. Chem. 143, 211 (1971).

273. S. Cesca, A. Roggero, T. Salvatori, A. de Chirico, G. Santi, Makromol. Chem. 133, 161 (1970).

274. G. Champetier et al., J. Polymer Sci. 58, 911 (1962).

275. R. K. S. Chan, Polymer Eng. Sci. 11, 152 (1971).

276. F. S. C. Chang, Polymer Preprints 12(2), 835 (1971).

277. E. W. Channen, Rev. Pure Appl. Chem. 9, 225 (1959).

278. A. Chapiro, P. Cordier, J. Jozefowicz, J. Sebban-Danon, J. Polymer Sci, C, 4, 491 (1963).

279. A. Chapiro, G. Palma, European Polymer J. 3, 151 (1967).

280. A. K. Chaudhuri, D. K. Sarkar, S. R. Palit, Makromol. Chem. 111, 36 (1968).

281. H. R. Chen, L. P. Branchard, IUPAC Leiden, Vol. I, 37 (1970).

282. W. K. W. Chen, H. Z. Friedlander, J. Polymer Sci., C, 4, 1195 (1963).

283. M. I. Cherkashin, P. P. Kisilitsa, O. G.Sel'skaya, A. A. Berlin, Polymer Sci, USSR 10, 233 (1968); Vysokomolekul. Soedin. 10 A, 196 (1968).

284. A. N. Cherkasov, S. I. Klenin, G. A. Andreyeva, Polymer Sci. USSR 12, 1382 (1970); Vysokomolekul. Soedin. 12 A, 1223 (1970).

285. R. Chiang, J. Phys. Chem. 69, 1645 (1965).

286. S. N. Chinai, J. Polymer Sci. 25, 413 (1957).

287. S. N. Chinai, R. A. Guzzi, J. Polymer Sci. 21, 417 (1956).

288. S. N. Chinai, J. D. Matlack, A. L. Resnick, R. L. Samuels, J. Polymer Sci. 17, 391 (1955).

289. S. N. Chinai, R. J. Samuels, J. Polymer Sci. 19, 463 (1956).

290. R. C. L. Chow, C. S. Marvel, J. Polymer Sci., A-1, 5, 2949 (1967).

291. G. Ciampa, H. Schwindt, Chim. Ind. 37, 169 (1955).

292. F. Ciardelli, G. Montagnoli, D. Pini, O. Pieroni, C. Carlini, E. Benedetti, Makromol. Chem. 147, 53 (1971).

293. S. Claesson, Discussions Faraday Soc. 7, 321 (1949).

294. D. Cleverdon, D. Laker, J. Appl. Chem. 1, 6 (1951).

295. E. Cohn-Ginsberg, T. G. Fox, H. F. Mason, Polymer 3, 97 (1962).

296. H. Coll, D. K. Gilding, J. Polymer Sci., A-2, 8, 89 (1970).

297. R. L. Combs, D. F. Slonaker, F. B. Joyner, H. W. Coover, J. Polymer Sci., A-1, 5, 215 (1967).

298. R. L. Combs, D. F. Slonaker, J. T. Summers, Org. Coatings Plastics Preprints 21, 249 (1961).

299. C. M. Conrad, Ind. Eng. Chem. 45, 2511 (1953).

300. W. Conti, E. Sorta, European Polymer J. 8, 475 (1972).

301. A. R. Cooper, A. R. Bruzzone, J. F. Johnson, Makromol. Chem. 138, 279 (1970).

302. A. R. Cooper, J. F. Johnson, Polymer Preprints 12(2), 738 (1971).

303. A. R. Cooper, J. F. Johnson, A. R. Bruzzone, Polymer Preprints 10(2), 1455 (1969).

304. R. E. Cooper, J. Upadhyay, A. Wasserman, J. Polymer Sci. 60, S 46 (1962).

305. W. Cooper, D. E. Eaves, G. Vaughan, J. Polymer Sci. 59, 241 (1962).

306. W. Cooper, G. A. Pope, G. Vaughan, European Polymer J. 4, 207 (1968).

307. W. Cooper, G. Vaughan, D. E. Eaves, R. W. Madden, J. Polymer Sci. 50, 159 (1961).

308. W. Cooper, G. Vaughan, J. Yardles, J. Polymer Sci. 59, S 2 (1962).

309. H. W. Coover, R. L. McConnell, F. B. Joyner, D. F. Slonaker, R. L. Combs, J. Polymer Sci., A-1, 4, 2563 (1966).

310. S. Coppick, O. A. Battista, M. R. Lytton, Ind. Eng. Chem. 42, 2533 (1950).

311. C. F. Cornet, Polymer 9, 7 (1968).

312. A. Cosani, E. Peggion, E. Scoffne, A. S. Verdini, J. Polymer Sci., B, 4, 55 (1965).

313. A. Cosani, E. Peggion, E. Scoffone, A. S. Verdini, Makromol. Chem. 97, 113 (1966).

314. J. A. Cote, M. Shida, J. Polymer Sci., A-2, 9, 421 (1971).

315. B. J. Cottam, J. Appl. Polymer Sci. 9, 1853 (1965).

316. B. J. Cottam, J. M. G. Cowie, S. Bywater, Makromol. Chem. 86, 116 (1965).

317. B. J. Cottam, D. M. Wiles, S. Bywater, Can. J. Chem. 41, 1905 (1963).

318. G. R. Cotton, W. C. Schneider, J. Appl. Polymer Sci. 7, 1243 (1963).

319. F. R. Cottrell, E. W. Merrill, K. A. Smith, J. Polymer Sci., A-2, 7, 1415 (1969).

320. F. R. Cottrell, E. W. Merrill, K. A. Smith, IUPAC Toronto, Preprint A8.1 (1968).

321. J. Čoupek, K. Bouchal, Makromol. Chem. 135, 69 (1970).

322. J. M. G. Cowie, Makromol. Chem. 42, 230 (1960).

323. J. M. G. Cowie, J. Polymer Sci. 49, 455 (1961).

324. J. M. G. Cowie, S. Bywater, J. Polymer Sci., A-2, 6, 499 (1968).

325. J. M. G. Cowie, C. T. Greenwood, J. Chem. Soc. 2862 (1957).

326. J. M. G. Cowie, C. T. Greenwood, J. Chem. Soc. 4640 (1957).

327. J. M. G. Cowie, R. J. Ranson, Makromol. Chem. 143, 105 (1971).

328. J. M. G. Cowie, P. M. Toporowski, Polymer 5, 601 (1964).

329. L. H. Cragg, H. Hammerschlag, Chem. Rev. 39, 79 (1946).

330. D. N. Cramond, J. M. Hammond, J. R. Urwin, European Polymer J. 4, 451 (1968).

331. H. Craubner, Makromol. Chem. 78, 121 (1964).

332. H. Craubner, Makromol. Chem. 89, 105 (1965).

333. H. Craubner, Ber. Bunsenges. Phys. Chem. 74, 1262 (1970).

334. V. Crescenzi, G. Manzini, G. Calzolari, C. Borri, European Polymer J. 8, 449 (1972).

335. M. A. Crook, F. S. Walker, Nature (London) 198, 1163 (1963).

336. P. Crouzet, F. Fine, P. Mangin, J. Appl. Polymer Sci. 13, 205 (1969).

337. C. Cuniberti, J. Polymer Sci., A-2, 8, 2051 (1970).

338. C. Cuniberti, U. Bianchi, Polymer 7, 151 (1966).

339. J. Curchod, T. Ve, J. Appl. Polymer Sci. 9, 3541 (1965).

340. G. F. D'Alelio, A. B. Finestone, L. Taft, T. J. Miranda, J. Polymer Sci. 45, 83 (1960).

341. J. Danon, R. Derai, European Polymer J. 5, 659 (1969).

342. J. Danon, J. Jozefowicz, J. Polymer Sci., C, 16, 4523 (1969).

343. J. Danon, J. Jozefowicz, European Polymer J. 5, 405 (1969).

344. J. Danon, J. Jozefowicz, European Polymer J. 6, 199 (1970).

345. F. Danusso, G. Gianotti, Makromol. Chem. 61, 164 (1963).

347. T. E. Davis, R. L. Tobias, J. Polymer Sci. 50, 227 (1961).

348. W. E. Davis, J. Am. Chem. Soc. 69, 1453 (1947).

349. W. G. Davies, D. P. Jones, Polymer Preprints 11(2), 447 (1970).

350. V. P. Davydova, N. A. Pravikova, T. A. Yakushina, V. I. Yakovleva, Polymer Sci. USSR 8, 479 (1966); Vysokomolekul. Soedin. 8, 436 (1966).

351. J. V. Dawkins, J. Macromol. Sci. B2, 623 (1968).

352. J. V. Dawkins, European Polymer J. 6, 831 (1970).

353. J. V. Dawkins, R. Denyer, J. W. Maddock, Polymer 10, 154 (1969).

354. J. V. Dawkins, M. Hemming, Makromol. Chem. 155, 75 (1972).

355. J. V. Dawkins, J. W. Maddock, European Polymer J. 7, 1537 (1971).

356. J. V. Dawkins, J. W. Maddock, D. Coupe, J. Polymer Sci., A-2, 8, 1803 (1970).

357. J. V. Dawkins, F. W. Peaker, European Polymer J. 6, 209 (1970).

358. P. C. Deb, S. R. Palit, Makromol. Chem. 128, 123 (1969).

359. F. de Candia, Makromol. Chem. 141, 177 (1971).

361. A. D. Delman, J. Kelly, J. Mironov, B. B. Simms, J. Polymer Sci., A-1, 4, 1277 (1966).

362. A. D. Delman, B. B. Simms, A. E. Ruff, J. Polymer Sci. 45, 415 (1960).

363. G. Delmas, J. Appl. Polymer Sci. 12, 839 (1968).

364. V. Desreux, Bull. Soc. Chim. Belges 57, 416 (1948).

365. V. Desreux, Rec. Trav. Chim. 68, 789 (1949).

366. V. Desreux, A. Oth, Chem. Weekbl. 48, 247 (1952).

367. V. Desreux, M. C. Spiegels, Bull. Soc. Chim. Belges 59, 476 (1950).

368. D. Determann, "Gel Chromatography", 2nd. Edition, Springer-Verlag, Berlin (1969).

369. H. Determann, M. Kriewen, Th. Wieland, Makromol. Chem. 114, 256 (1968).

370. H. Determann, G. Lueben, Th. Wieland, Makromol. Chem. 73, 168 (1964).

371. H. Dexheimer, O. Fuchs, Makromol. Chem. 96, 172 (1966).

372. H. Dexheimer, E. Helmes, H. J. Leugering, IUPAC Toronto, Preprint A8. 13 (1968).

373. K. Dialer, K. Vogler, Makromol. Chem. 6, 191 (1951).

374. K. Dialer, K. Vogler, F. Patat, Helv. Chim. Acta 35, 869 (1952).

375. R. Dick, H. Sack, H. Benoit, J. Polymer Sci., C, 16, 4597 (1969).

376. H. A. Dieu, J. Polymer Sci. 12, 417 (1954).

377. A. Dobry, J. Chim. Phys. 35, 387 (1938).

378. Ye. S. Domnina, G. G. Skvortsova, Polymer Sci. USSR 8, 1394 (1966); Vysokomolekul. Soedin. 8, 1268 (1966).

379. A. Dondos, Makromol. Chem. 135, 181 (1970).

380. A. Dondos, H. Benoit, Makromol. Chem. 118, 165 (1968).

381. A. Dondos, H. Benoit, Makromol. Chem. 129, 35 (1969).

382. P. Doty, H. Wagner, S. Singer, J. Phys. Colloid Chem. 51, 32 (1947).

384. E. E. Drott, R. A. Mendelson, Polymer Preprints 12(1), 277 (1971).

385. E. E. Drott, R. A. Mendelson, J. Polymer Sci., A-2, 8, 1373 (1970).

386. L. V. Dubrovina, S. A. Pavlova, V. V. Korshak, Polymer Sci. USSR 8, 2171 (1966); Vysokomolekul. Soedin. 8, 1965 (1966).

387. L. V. Dubrovina, S. A. Pavlova, V. A. Vasnev, S. V. Vinogradova, V. V. Korshak, Polymer Sci. USSR 12, 1484 (1970); Vysokomolekul. Soedin. 12 A, 1308 (1970).

388. L. V. Dubrovina, G. I. Timofeyeva, V. V. Korshak, S. A. Pavlova, Polymer Sci. USSR 6, 2225 (1964); Vysokomolekul. Soedin. 6, 2011 (1964).

389. A. B. Duchkova, N. S. Nametkin, V. S. Khotimskii, L. Ye. Gusel'nikov, S. G. Durgar'yan, Polymer Sci. USSR 8, 2002 (1966); Vysokomolekul. Soedin. 8, 1814 (1966).

390. J. H. Duerksen, A. E. Hamielec, Polymer Preprints 8(2), 1337 (1967).

391. J. H. Duerksen, A. E. Hamielec, J. Polymer Sci., C, 21, 83 (1968).

392. K. Dušek, Collection Czech. Chem. Commun. 34, 3309 (1969).

393. I. Duvdevani, J. A. Biesenberger, M. Tan, J. Polymer Sci., B, 9, 429 (1971).

394. N. Duveau, A. Piguet, J. Polymer Sci. 57, 357 (1962).

395. J. Dyer, L. H. Phifer, J. Polymer Sci., C, 36, 103 (1971).

396. K. C. Eberly, B. L. Johnson, J. Polymer Sci. 3, 283 (1948).

397. K. H. Ebert, M. Brosche, K. F. Elgert, Makromol. Chem. 72, 191 (1964).

398.   K. H. Ebert, G. Schenk, G. Rupprecht, M. Brosche, H. W. Weng,
       D. Heinicke, Makromol. Chem. 96, 206 (1966).

399.   G. D. Edwards, J. Appl. Polymer Sci. 9, 3845 (1965).

400.   B. S. Ehrlich, W. V. Smith, Polymer Preprints 12(2), 847 (1971).

401.   B. E. Eichinger, P. J. Flory, Polymer Preprints 8(2), 973 (1967).

403.   E. J. Elgood et al., J. Appl. Polymer Sci. 8, 882 (1964).

404.   H.-G. Elias, A. Fritz, Makromol. Chem. 114, 31 (1968).

405.   H.-G. Elias, V. Gruber, Makromol. Chem. 78, 72 (1964).

406.   H.-G. Elias, V. S. Kamat, Makromol. Chem. 117, 61 (1968).

407.   H.-G. Elias, F. Patat, Makromol. Chem. 25, 13 (1957).

408.   H. Elmgren, S. Norrby, Makromol. Chem. 123, 265 (1969).

409.   A. A. El' Saied, S. Ya. Mirlina, V. A. Kargin, Polymer Sci. USSR
       11, 314 (1969); Vysokomolekul. Soedin. 11A, 282 (1969).

410.   C. Emery, W. E. Cohen, Aust. J. Appl. Sci. 2, 473 (1951).

411.   A. H. Emery, H. G. Drickamer, J. Chem. Phys. 23, 2252 (1953).

412.   H. A. Ende, J. M. Peterson, G. V. Schulz, Z. Physik. Chem. N. F.
       (Frankfurt) 41, 224 (1964).

413.   H. A. Ende, G. V. Schulz, Z. Physik. Chem. N. F. (Frankfurt)
       33, 143 (1962).

414.   R. Endo, Chem. High Polymers (Japan) 19, 39 (1962).

415.   R. Endo, T. Hinokuma, M. Takeda, J. Polymer Sci., A-2, 6, 665
       (1968).

416.   R. Endo, K. Kimura, M. Takeda, Bull. Chem. Soc. Japan 37, 950
       (1964).

417.   R. Endo, T. Manago, M. Takeda, Bull. Chem. Soc. Japan 39, 733
       (1966).

418.   A. F. V. Eriksson, Acta Chem. Scand. 3, 1 (1949).

419.   A. F. V. Eriksson, Acta Chem. Scand. 7, 623 (1953).

420.   A. F. V. Eriksson, Acta Chem. Scand. 7, 377 (1953).

421.   A. F. V. Eriksson, Acta Chem. Scand. 10, 360, 378 (1956).

422.   V. Ye. Eskin, A. Ye. Nesterov, Polymer Sci. USSR 8, 152 (1966);
       Vysokomolekul. Soedin. 8, 141 (1966).

423.   V. E. Eskin, I. N. Serdjuk, J. Polymer Sci., C, 23, 309 (1968).

424.   J. M. Evans, M. B. Huglin, Makromol. Chem. 127, 141 (1969).

425.   J. M. Evans, M. B. Huglin, R. F. T. Stepto, Makromol. Chem.
       146, 91 (1971).

426.   J. Exner, M. Bohdanecký, Collection Czech. Chem. Commun. 31,
       3985 (1966).

427.   G. F. Fanta, R. C. Burr, C. R. Russel, C. E. Rist, J. Appl. Polymer
       Sci. 10, 929 (1966).

428.   M. Farina, G. Bressan, Makromol. Chem. 61, 79 (1963).

429.   J. G. Fatou, European Polymer J. 7, 1057 (1971).

430.   J. G. Fatou, L. Mandelkern, J. Phys. Chem. 69, 417 (1965).

431.   O. Ya. Fedotova et al., Vysokomolekul. Soedin. 5, 900 (1963).

432.   O. Ya. Fedotova, M. I. Shtil' man, Polymer Sci. USSR 8, 1332
       (1966); Vysokomolekul. Soedin. 8, 1209 (1966).

433.   R. Feinauer, W. Funke, K. Hamann, Makromol. Chem. 82, 123
       (1965).

434.   D. Feldman, C. Uglea, N. Simionescu, J. Polymer Sci., A-1, 7,
       439 (1969).

435.   V. F. Felicetta, A. Ahola, J. L. McCarthy, J. Am. Chem. Soc.
       78, 1899 (1956).

436.   R. E. Felter, E. S. Moyer, L. N. Ray, J. Polymer Sci., B, 7, 529
       (1969).

437.   S. Fernandez Bermudez, J. G. Fatou, J. Royo, Anales Chim. (Madrid)
       66, 779 (1970).

438.   J. D. Ferry et al., J. Colloid Sci. 6, 429 (1951).

439.   L. J. Fetters, M. Morton, Macromolecules 2, 453 (1969).

440.   R. V. Figini, H. Hostalka, K. Hurm, G. Loehr, G. V. Schulz, Z.
       Physik. Chem. N. F. (Frankfurt) 45, 269 (1965).

441.   S. R. Finn, J. W. James, J. Appl. Chem. 6, 466 (1956).

442.   A. P. Firsov, I. V. Yeremina, N. M. Chirkov, Polymer Sci. USSR
       6, 417 (1964); Vysokomolekul. Soedin. 6, 377 (1964).

443.   E. B. Fitzgerald, R. M. Fuoss, Ind. Eng. Chem. 42, 1603 (1950).

444.   J. R. Fletcher, H. E. Persinger, J. Polymer Sci., A-1, 6, 1025
       (1968).

445.   J. R. Fletcher, L. Polgar, D. H. Solomon, J. Appl. Polymer Sci. 8,
       663 (1964).

446.   J. R. Fletcher, L. Polgar, D. H. Solomon, J. Appl. Polymer Sci. 8,
       659 (1964).

447.   Š. Florián, D. Lath, Z. Maňásek, Angew. Makromol. Chem. 13, 43
       (1970).

448.   Š. Florián, D. Lath, V. Ďurďovič, Angew. Makromol. Chem. 10, 189
       (1970).

449.   P. J. Flory, J. Am. Chem. Soc. 65, 372 (1943).

450.   P. J. Flory, J. Am. Chem. Soc. 65, 375 (1943).

451.   D. L. Flowers, W. A. Hewett, R. D. Mullineaux, J. Polymer Sci.,
       A-1, 2, 2305 (1964).

452.   C. M. Fontana, J. Phys. Chem. 63, 1167 (1959).

453.   R. J. Fort, R. J. Hutchinson, W. R. Moore, M. Murphy, Polymer 4,
       35 (1963).

454.   T. G. Fox, Polymer 3, 111 (1962).

455.   T. G. Fox, P. J. Flory, J. Am. Chem. Soc. 70, 2384 (1948).

456.   T. G. Fox, P. J. Flory, J. Phys. Chem. 53, 197 (1949).

457.   T. G. Fox, J. B. Kinsinger, H. F. Mason, C. M. Schuele, Polymer
       3, 71 (1962).

458.   H. P. Frank, J. W. Breitenbach, J. Polymer Sci. 6, 609 (1951).

459.   P. S. Francis, R. C. Cooke, J. H. Elliot, J. Polymer Sci. 31, 453
       (1958).

460.   M. Freeman, P. P. Manning, J. Polymer Sci., A-1, 2, 2017 (1964).

461.   D. M. French, A. W. Casey, C. I. Collins, P. Kirchner, Polymer
       Preprints 7(1), 447 (1966).

462.   S. Ya. Frenkel, S. I. Klenin, Polymer Sci. USSR 6, 1571 (1964);
       Vysokomolekul. Soedin. 6, 1420 (1964).

463.   L. Freund, M. Daune, J. Polymer Sci. 29, 161 (1958).

464.   D. Freyss, P. Rempp, H. Benoit, J. Polymer Sci., B, 2, 217 (1964).

465.   E. V. Frisman, M. A. Sibileva, M. A. Chebishyan, Polymer Sci.
       USSR 9, 1191 (1967).

466.   E. V. Frisman, M. A. Sibileva, N. T. Kim Ngan, T. N. Nekrasova,
       Polymer Sci. USSR 10, 2125 (1968); Vysokomolekul. Soedin. 10 A,
       1834 (1968).

467.   H. Fritzemeier, J. J. Hermans, Bull. Soc. Chim. Belges 57, 136 (1948).

468.   O. Fuchs, Makromol. Chem. 7, 259 (1952).

469.   O. Fuchs, Makromol. Chem. 5, 245 (1951).

470.   O. Fuchs, Makromol. Chem. 58, 65 (1962).

471.   O. Fuchs, Z. Elektrochem. 60, 229 (1956).

472.   O. Fuchs, H. J. Leugering in "Kunststoffe", Vol. 1, Editors R. Nitsche,
       K. A. Wolf, Springer, Berlin, pp. 118 (1962).

473.   H. Fujii, J. Furukawa, T. Saegusa, A. Kawasaki, Makromol. Chem.
       40, 226 (1960).

474.   H. Fujii, T. Fujii, T. Saegusa, J. Furukawa, Makromol. Chem. 63,
       147 (1963).

475.   T. Fujimoto, H. Narukawa, M. Nagasawa, Macromolecules 3, 57
       (1970).

476.   T. Fujimoto, N. Ozaki, M. Nagasawa, J. Polymer Sci., A-1, 3,
       2259 (1965).

477.   Y. Fujisaki, Chem. High Polymers (Japan) 19, 64 (1962).

478.   Y. Fujisaki, H. Kobayashi, Chem. High Polymers (Japan) 18, 305
       (1961).

479.   Y. Fujisaki, H. Kobayashi, Chem. High Polymers (Japan) 19, 49 (1962).

480.   Y. Fujisaki, H. Kobayashi, Chem. High Polymers (Japan) 19, 73 (1962).

481.   B. L. Funt, V. Homef, J. Appl. Polymer Sci. 15, 2439 (1971).

482.   B. L. Funt, D. R. Richardson, J. Polymer Sci., A-1, 8, 1055 (1970).

483.   J. Furukawa et al., Makromol. Chem. 39, 243 (1960).

484.   J. Furukawa et al., Makromol. Chem. 41, 17 (1960).

485.   J. Furukawa, S. Akutsu, T. Saegusa, Makromol. Chem. 81, 100 (1965).

486.   J. Furukawa, S. Akutsu, T. Saegusa, Makromol. Chem. 94, 68 (1966).

487.   J. Furukawa, T. Saegusa, S. Yasui, S. Akutsu, Makromol. Chem. 94,
       74 (1966).

488.   A. Gabert, H. Seide, G. Langhammer, J. Polymer Sci., C, 16,
       3547 (1968).

489.   D. Gaeckle, D. Patterson, Macromolecules 5, 136 (1972).

490. G. L. Gaines, D. G. Le Grand, J. Polymer Sci., B, 6, 625 (1968).

491. J. C. Galin, Makromol. Chem. 124, 118 (1969).

492. A. Gallo, Chim. Ind. (Milan) 35, 487 (1953).

493. J. Gallot, P. Rempp, J. Parrod, J. Polymer Sci., B 1, 329 (1963).

494. L. W. Gamble, W. T. Wipke, T. Lane, J. Appl. Polymer Sci. 9, 1503 (1965).

495. V. Ganser, J. Engel, D. Winklmair, G. Krause, Biopolymers 9, 329 (1970).

496. C. Garbuglio, L. Crescentini, A. Mula, G. B. Gechele, Makromol. Chem. 97, 97 (1966).

497. C. Garbuglio, A. Mula, L. Chinellato, J. Polymer Sci., C, 16, 1529 (1967).

498. C. Garbuglio, G. Pezzin, C. Badoni, Chim. Ind. (Milan), 46, 797 (1964).

499. M. Garcia-Martí, K. H. Reichert, Makromol. Chem. 144, 17 (1971).

500. V. Garten, W. Becker, Makromol. Chem. 3, 78 (1949).

501. T. I. Garmonova, M. G. Vitovskaya, P. N. Laurenke, V. N. Tsvetkov, Ye. V. Korovina, Polymer Sci. USSR 13, 996 (1971); Vysokomolekul. Soedin. 13 A, 884 (1971).

502. G. Gavoret, J. Duclaux, J. Chim. Phys. 42, 41 (1945).

503. G. Gavoret, M. Magat, J. Chim. Phys. 44, 90 (1947).

504. N. Geacintov et al., J. Appl. Polymer Sci. 3, 54 (1960).

505. G. B. Gechele, A. Mattiussi, European Polymer J. 1, 47 (1965).

506. R. Geddes, C. T. Greenwood, A. W. Mac Gregor, A. R. Procter, J. Thompson, Makromol. Chem. 79, 189 (1964).

507. M. Gehatia, D. R. Wiff, IUPAC Boston, Vol. II, 1045 (1971).

508. M. Gehatia, D. R. Wiff, European Polymer J. 8, 585 (1972).

509. K. Gehrke, G. Reinisch, J. Polymer Sci., A-1, 7, 1571 (1969).

510. K. Gehrke, G. Reinisch, IUPAC Praga, Preprint 444 (1965).

511. G. Geiseler, H. P. Baumann, Z. Elektrochem. 62, 209 (1958).

512. G. Geiseler, H. Herold, F. Runge, Erdoel Kohle 7, 357 (1954).

513. K. Gekko, Makromol. Chem. 148, 229 (1971).

514. A. N. Genkin, T. P. Nasonova, I. Y. Poddubnyi, Vysokomolekul. Soedin. 4, 1088 (1962).

515. J. Gerber, H.-G. Elias, Makromol. Chem. 112, 142 (1968).

516. R. German, R. Hank, G. Vaughan, Kautschuk Gummi, Kunststoffe 19, 67 (1966).

517. H. Gerrens, H. Ohlinger, R. Fricker, Makromol. Chem. 87, 209 (1965).

518. J. A. Gervasi, A. B. Gosnell, V. Stannett, Polymer Preprints 8(1), 785 (1967).

519. A. N. Geukin et al., Vysokomolekul. Soedin. 4, 1088 (1962).

520. A. V. Gevorkyan, Ye. S. Yegiyan, N. G. Oganeyan, R. V. Bagdasaryan, Polymer Sci. USSR 12, 1218 (1970); Vysokomolekul. Soedin. 12 A, 1078 (1970).

521. K. K. Ghosh, P. K. Choudhury, Makromol. Chem. 102, 217 (1967).

522. M. Giurgea, C. Ghita, I. Baltog, A. Lupu, J. Polymer Sci., A-2, 4, 529 (1966).

523. G. Gloeckner, Plaste Kautschuk 10, 154 (1963).

525. R. A. Godfrey, G. W. Miller, J. Polymer Sci., A-1, 7, 2387 (1969).

526. K. D. Goebel, D. A. Brant, Macromolecules 3, 634 (1970).

527. K. D. Goebel, W. G. Miller, Macromolecules 3, 64 (1970).

528. D. J. Goedhart, A. Opschoor, IUPAC Leiden, Vol I, 163 (1970).

529. D. J. Goedhart, A. Opschoor, J. Polymer Sci., A-2, 8, 1227 (1970).

530. A. L. Goff, I. P. Yakovlev, V. M. Zhulin, M. G. Gonikberg, Polymer Sci. USSR 11, 1487 (1969); Vysokomolekul. Soedin. 11A, 1309 (1969).

531. A. I. Goldberg, W. P. Hohenstein, H. F. Mark, J. Polymer Sci. 2, 503 (1947).

532. F. J. Golemba, J. E. Guillet, Macromolecules 5, 212 (1972).

533. M. A. Golub, J. Polymer Sci. 11, 281 (1953).

534. M. A. Golub, J. Polymer Sci. 11, 583 (1953).

535. V. M. Golubev, S. Ya. Frenkel, Polymer Sci. USSR 10, 869 (1968); Vysokomolekul. Soedin. 10 A, 750 (1968).

536. G. Gooberman, J. Polymer Sci. 40, 469 (1959).

537. A. E. Googban, H. S. Owens, J. Polymer Sci. 23, 825 (1957).

538. A. Gordienko, Faserforsch. Textiltech. 4, 499 (1953).

540. D. A. I. Goring, G. Sitaramaiah, Polymer 4, 7 (1963).

541. A. B. Gosnell, J. A. Gervasi, A. Schindler, J. Polymer Sci., A-1, 4, 1401 (1966).

542. N. Gralen, G. Lagermalm, J. Phys. Chem. 56, 514 (1952).

543. K. A. Granath, P. Flodin, Makromol. Chem. 48, 160 (1961).

544. N. Grassie, I. G. Meldrum, European Polymer J. 6, 499 (1970).

545. N. Grassie, I. G. Meldrum, European Polymer J. 6, 513 (1970).

546. J. H. S. Green, H. T. Hookway, M. F. Vaughan, Chem. & Ind. 862 (1958).

547. J. H. S. Green, M. F. Vaughan, Chem. & Ind. 29, 829 (1958).

548. C. T. Greenwood, Adv. Carbohydr. Chem. 11, 335 (1956).

549. W. Griehl, H. Lueckert, J. Polymer Sci. 30, 399 (1958).

550. B. M. Grieveson, Makromol. Chem. 84, 93 (1965).

551. H. Grohn, H. Friedrich, J. Polymer Sci., C, 16, 3737 (1968).

552. P. Grosius, Y. Gallot, A. Skoulios, European Polymer J. 6, 355 (1970).

553. P. Grosius, Y. Gallot, A. Skoulios, Makromol. Chem. 127, 94 (1969).

554. P. Grosius, Y. Gallot, A. Skoulios, Makromol. Chem. 132, 35 (1970).

555. P. Grosius, Y. Gallot, A. Skoulios, Makromol. Chem. 136, 191 (1970).

556. R. I. Gruz, T. Yu. Verkhoglyadova, Ye. F. Panarin, S. N. Ushakov, Polymer Sci. USSR 13, 736 (1971); Vysokomolekul. Soedin 13A, 647 (1971).

557. L. A. Gugliemelli, M. O. Weaver, C. R. Russell, C. E. Rist, J. Polymer Sci., B, 9, 151 (1971).

558. J. E. Guillet, R. L. Combs, D. F. Slonaker, H. W. Coover, J. Polymer Sci. 47, 307 (1960).

559. S. Gundiah, Makromol. Chem. 104, 196 (1967).

560. S. Gundiah, N. V. Viswanathan, S. L. Kapur, Indian. J. Chem. 4, 344 (1966).

561. P. R. Gupta, J. L. McCarthy, Macromolecules 1, 236 (1968).

562. P. R. Gupta, J. L. MaCarthy, Macromolecules 1, 495 (1968).

563. A. A. Guryleva, M. P. Dianov, Polymer Sci. USSR 10, 1661 (1968); Vysokomolekul. Soedin. 10 A, 1431 (1968).

564. L. Ye. Gusel' nikov, A. Yu. Koshevnik, I. M. Kosheleva, M. M. Kusakov, E. A. Razumovskaya, Polymer Sci. USSR 7, 950 (1965); Vysokomolekul. Soedin. 7, 860 (1965).

565. A. Guyot, P. Quang. -Tho, J. Chim. Phys. 63, 742 (1966).

566. A. Guyot, P. Rocaniere, J. Appl. Polymer Sci. 13, 2019 (1969).

567. G. M. Guzmán, in "Progress in High Polymers", Editors, J. C. Robb, F. W. Peaker, Heywood, London, 1961.

568. G. M. Guzmán, in "Polymer Handbook", 1st Edition, Editors, J. Brandrup/E.H. Immergut, Interscience, New York, 1966.

569. G. M. Guzmán, J. Polymer Sci. 19, 519 (1956).

570. G. M. Guzmán, Anales Real Soc. Esp. Fis. Quim. 50B, 631 (1954).

571. G. M. Guzmán, F. Arranz, Anales Real Soc. Esp. Fis. Quim. 54B, 445 (1963).

572. G. M. Guzmán, A. Bello, Makromol. Chem. 107, 46 (1967).

573. G. M. Guzmán, J. M. G. Fatou, Anales Real Soc. Esp. Fis. Quim. 53B, 669 (1957).

574. G. M. Guzmán, J. M. G. Fatou, Anales Real Soc. Esp. Fis. Quim. 54B, 263 (1958).

575. G. M. Guzmán, J. M. G. Fatou, Anales Real Soc. Esp. Fis. Quim. 54B, 609 (1958).

576. W. F. Haddon, R. S. Porter, J. F. Johnson, J. Appl. Polymer Sci. 8, 1371 (1964).

577. G. L. Hagnauer, N. S. Schneider, J. Polymer Sci., A-2, 10, 699 (1972).

578. P. Hague, M. B. Huglin, B. L. Johnson, J. Appl. Polymer Sci. 12, 2105 (1968).

579. W. Hahn, W. Muller, R. W. Webber, Makromol. Chem. 21, 131 (1956).

580. A. F. Halasa, H. E. Adams, J. Polymer Sci., C, 30, 169 (1970).

581. R. W. Hall, in "Techniques of Polymer Characterization", Editors, P. W. Allen, Butterworths, London (1959).

582. T. Hama, K. Yamaguchi, T. Suzaki, Makromol. Chem. 155, 283 (1972).

583. J. M. Hammond, J. F. Hooper, W. G. P. Robertson, J. Polymer Sci., A-1, 9, 281 (1971).

584. E. Hamori, L. R. Prusonowski, P. G. Sparks, R. E. Hughes, J. Phys. Chem. 69, 1101 (1965).

585. K. Hanafusa, A. Teramoto, H. Fujita, J. Phys. Chem. 70, 4004 (1966).

586. H. Langner, Makromol. Chem. 119, 37 (1968).

587. J. Hannus, G. Smets, Bull. Soc. Chim. Belges 60, 76 (1951).

588. C. M. Hansen, G. A. Sather, J. Appl. Polymer Sci. 8, 2479 (1964).

589. R. Hardy, P. E. M. Allen, Makromol. Chem. 42, 38 (1960).

590. W. G. Harland, J. Textile Inst. 46, 483 (1955).

591. D. J. Harmon, J. Polymer Sci., C, 8, 243 (1965).

592. D. J. Harmon, J. Appl. Polymer Sci. 11, 1333 (1967).

593. D. J. Harmon, H. L. Jacobs, J. Appl. Polymer Sci. 10, 253 (1966).

594. R. E. Harrington, P. G. Pecoraro, J. Polymer Sci., A-1, 4, 475 (1966).

595. R. E. Harrington, B. H. Zimm, Polymer Preprints 6(1), 346 (1965).

596. I. Harris, R. G. J. Miller, J. Polymer Sci. 7, 377 (1951).

597. N. Hartler, Acta Chem. Scand. 11, 1162 (1957).

598. G. W. Hastings, D. W. Ovenall, F. W. Peaker, Nature (London) 177, 1091 (1956).

599. A. Haugh, Acta Chem. Scand. 13, 601 (1959).

600. E. A. Hauser, D. S. Le Beau, Indian Rubber J. 110, 601 (1946).

601. E. A. Hauser, D. S. Le Beau, J. Phys. Chem. 54, 256 (1950).

602. E. A. Hauser, D. S. Le Beau, Rubber, Chem. Technol. 20, 70 (1947).

603. A. Hauss, Angew. Makromol. Chem. 8, 73 (1969).

604. S. W. Hawkins, H. Smith, J. Polymer Sci. 28, 341 (1958).

605. J. N. Hay, M. Sabir, R. L. T. Steven, Polymer 10, 187 (1969).

606. S. Hayashi, F. Hamada, A. Saijoo, A. Nakajima, Kobunshi Kagaku Japan 24, 769 (1967).

607. P. Hayden, R. Roberts, Int. J. Appl. Radiation Isotopes 5, 269 (1959).

608. J. E. Hazell, L. A. Prince, H. E. Stapelfeldt, Polymer Preprints 8(2), 1303 (1967).

609. J. E. Hazell, L. A. Prince, H. E. Stapelfeldt, J. Polymer Sci., C, 21, 43 (1968).

610. W. Heitz, Makromol. Chem. 145, 141 (1971).

611. W. Heitz, Angew. Makromol. Chem. 10, 115 (1970).

612. W. Heitz, B. Bömer, H. Ullner, Makromol. Chem. 121, 102 (1969).

613. W. Heitz, W. Kern, Angew. Makromol. Chem. 1, 150 (1967).

614. J. Heller, J. Moacanin, J. Polymer Sci., B, 6, 595 (1968).

615. J. Heller, J. F. Schimscheimer, R. A. Pasternak, C. B. Kingsley, J. Moacanin, J. Polymer Sci., A-1, 7, 73 (1969).

616. T. E. Helminiak, C. L. Benner, W. E. Gibbs, Polymer Preprints 8(1), 284 (1967).

617. T. E. Helminiak, C. L. Benner, W. E. Gibbs, Polymer Preprints 11(1), 291 (1970).

618. J. F. Henderson, J. M. Hulme, J. Appl. Polymer Sci. 11, 2349 (1967).

619. J. Hengstenberg, Z. Elektrochem. 60, 236 (1956).

620. P. M. Henry, J. Polymer Sci. 36, 3 (1959).

621. J. J. Hermans, H. A. Ende, J. Polymer Sci., C, 1, 161 (1963).

622. P. Herrent, J. Polymer Sci., 8, 346 (1952).

623. J. Herz, D. Decker-Freyss, P. Rempp, J. Polymer Sci., C, 16, 4035 (1968).

624. G. Heufer, D. Braun, J. Polymer Sci., B, 3, 495 (1965).

625. E. Heuser, Wm. Shockley, R. Kjellgreen, Tappi 33, 92 (1950).

626. T. Higashimura, T. Matsuda, S. Okamura, J. Polymer Sci., A-1, 8, 483 (1970).

627. A. Hilt, K. H. Reichert, K. Hamann, Makromol. Chem. 101, 246 (1967).

628. S. Hirata, H. Hasegawa, A. Kishimoto, J. Appl. Polymer Sci. 14, 2025 (1970).

629. M. Hirooka, H. Kanda, K. Nakaguchi, J. Polymer Sci., B, 1, 701 (1963).

630. S. M. Hirshfield, E. R. Allen, J. Polymer Sci. 39, 554 (1959).

631. M. Hloušek, J. Láníková, Chem. Prumysl 16, 160 (1966).

632. M. Hloušek, J. Láníková, J. Polymer Sci., C, 16, 935 (1967).

633. H. C. Hoare, D. E. Hillmann, Brit. Polymer J. 3, 259 (1971).

634. K. Hoashi, T. Mochizuki, Makromol. Chem. 100, 78 (1967).

635. L. M. Hobbs, S. C. Kothari, V. C. Long, G. C. Sutaria, J. Polymer Sci. 22, 123 (1956).

636. L. M. Hobbs, V. C. Long, Polymer 4, 479 (1963).

637. S. J. Hobbs, F. W. Billmeyer, J. Polymer Sci., A-2, 8, 1387 (1970).

638. H. Hoecker, K. Saeki, Makromol. Chem. 148, 107 (1971).

639. N. Ho.-Duc, H. Daoust, Can. J. Chem. 46, 2456 (1968).

640. N. Ho.-Duc, H. Daoust, A. Gourdenne, Polymer Preprints 12(1), 639 (1971).

641. A. S. Hoffman, B. A. Fries, P. C. Condit, J. Polymer Sci., C, 4, 109 (1963).

642. J. D. Hoffman, B. H. Zimm, J. Polymer Sci. 15, 405 (1955).

643. L. Hoehr, V. Jaacks, H. Cherdron. S. Iwabuchi, W. Kern, Makromol. Chem. 103, 279 (1967).

644. R. B. Homer, M. Shinitzky, Macromolecules 1, 469 (1968).

645. I. Homma et al., Makromol. Chem. 67, 132 (1963).

646. T. Homma, H. Fujita, J. Appl. Polymer Sci. 9, 1701 (1965).

647. T. Homma, K. Kawahara, H. Fujita, J. Appl. Polymer Sci. 8, 2853 (1964).

648. P. J. Holdsworth, A. Keller, I. M. Ward, T. Williams, Makromol. Chem. 125, 70 (1969).

649. P. J. Holdsworth, A. Keller, I. M. Ward, T. Williams, Makromol. Chem. 125, 82 (1969).

650. H. Hopff, M. A. Osman, Makromol. Chem. 143, 289 (1971).

651. H. Hopff, M. De Tomasi, H.-G. Elias, Makromol. Chem. 96, 41 (1966).

652. R. H. Horowitz, Polymer Preprints 3, 167 (1962).

654. K. Hoshino, M. Watanabe, J. Am. Chem. Soc. 73, 4816 (1951).

655. R. C. Houtz, Textile Res. J. 20, 786 (1950).

656. G. J. Howard, J. Polymer Sci. 37, 310 (1959).

657. G. J. Howard, J. Polymer Sci., A, 1, 2667 (1963).

658. P. Howard, R. S. Parikh, IUPAC Prag, Preprint p. 265 (1965).

659. P. Howard, R. S. Parikh, J. Polymer Sci., A-1, 4, 407 (1966).

660. P. Howard, R. S. Parikh, J. Polymer Sci., C, 30, 17 (1970).

661. H. L. Hsieh, J. Polymer Sci., A-1, 3, 191 (1965).

662. H. L. Hsieh, O. F. McKinney, J. Polymer Sci., B, 4, 843 (1966).

663. R. Y. M. Huang, P. Chandramouli, J. Polymer Sci., A-1, 7, 1393 (1969).

664. R. Y. M. Huang, G. Mayer, IUPAC Budapest, Vol. III, Preprint 6/03 (1969).

665. R. Y. M. Huang, J. F. Westlake, J. Polymer Sci., A-1, 8, 49 (1970).

666. R. Y. M. Huang, J. F. Westlake, S. C. Sharma, J. Polymer Sci., A-1, 7, 1729 (1969).

667. M. L. Huggins, H. Okamoto, in "Polymer Fractionation", Editor, M. J. R. Cantow, Academic Press, New York (1966).

668. L. J. Hughes, R. H. Andreatta, H. A. Scheraga, Macromolecules 5, 187 (1972).

669. L. J. Hughes, G. L. Brown, J. Appl. Polymer Sci. 7, 59 (1963).

670. M. B. Huglin, D. H. Whitehurst, D. Sims, J. Appl. Polymer Sci. 12, 1889 (1968).

671. J. M. Hulme, L. A. McLeod, Polymer 3, 153 (1962).

672. M. L. Hunt, H. A. Scheraga, P. J. Flory, J. Phys. Chem. 60, 1278 (1956).

673. E. Husemann, Makromol. Chem. 26, 181, 199 (1958).

674a. E. Husemann, B. Pfannemueller, Makromol. Chem. 49, 214 (1961).

674b. E. Husemann, B. Pfannemueller, Makromol. Chem. 69, 74 (1963).

675. E. Husemann, R. Werner, Makromol. Chem. 59, 43 (1963).

676. A. J. Hyde, R. B. Taylor, Polymer 4, 1 (1963).

677. F. Ide, R. Handa, K. Nakatsuka, Chem. High Polymer (Japan) 21, 57 (1964).

678. K.-H. Illers, Makromol. Chem. 118, 88 (1968).

679. K. Imai, M. Matsumoto, Bull. Chem. Soc. Japan 36, 455 (1963).

680. H. Inagaki, K. Hayashi, T. Matsuo, Makromol. Chem. 84, 80 (1965).

681. H. Inagaki, F. Kamiyama, T. Yagi, Macromolecules 4, 133 (1971).

682. H. Inagaki, H. Matsuda, F. Kamiyama, Macromolecules 1, 520 (1968).

683. H. Inagaki, T. Miyamoto, Makromol. Chem. 87, 166 (1965).

684. H. Inagaki, T. Miyamoto, S. Ohta, J. Phys. Chem. 70, 3420 (1966).

685. H. Inagaki, H. Suzuki, M. Fujii, T. Matsuo, J. Phys. Chem. 70, 1718 (1966).

686. J. D. Ingham, N. S. Rapp, J. Polymer Sci., A-1, 2, 4941 (1964).

687. S. Inoue, K. Tsubaki, T. Yamada, T. Tsuruta, Makromol. Chem. 125, 181 (1969).

688. S. Inoue, I. Tsukuma, M. Kawaguchi, T. Tsuruta, Makromol. Chem. 103, 151 (1967).

689. S. Inoue, T. Tsuruta, N. Yoshida, Makromol. Chem. 79, 34 (1964).

690. S. Inoue, T. Tsuruta, N. Yoshida, Kogyo Kagaku Zassi 67, 1439 (1964).

691. S. Inoue, Y. Yokota, N. Yoshida, T. Tsuruta, Makromol. Chem. 90, 131 (1966).

692. K. Irako, S. Anzai, A. Onishi, Bull. Chem. Soc. Japan 41, 501 (1968).

693. S. Ishida, Bull. Chem. Soc. Japan 33, 727 (1960).

694. S. S. Ivanchev, N. I. Solomko, Polymer Sci., USSR 8, 353 (1966); Vysokomolekul. Soedin. 8, 322 (1966).

695. K. J. Ivin, H. A. Ende, G. Meyerhoff, Polymer 3, 129 (1962).

696. M. Iwama, H. Utiyama, M. Kurata, J. Macromol. Sci. Chem. 1, 701 (1966).

697. T. Iwata, T. Saegusa, H. Fujii, J. Furukawa, Makromol. Chem. 97, 49 (1966).

698. V. Jaacks, G. Franzmann, Makromol. Chem. 143, 283 (1971).

699. J. Jatte, C. Berliner, J. Chim. Phys. 63, 389 (1966).

700. J. Jaffe, C. Berliner, M. Lambert, J. Chim. Phys. 64, 499 (1967).

701. M. Janić, M. J. Beneš, J. Peška, Makromol. Chem. 138, 99 (1970).

702. G. Jayme, H. Knolle, Makromol. Chem. 82, 190 (1965).

703a. G. Jayme, P. Kleppe, Papier 15, 6 (1961); G. Jayme, K. K. Hasvold, Papier 20, 657 (1966).

703b. G. Jayme, P. Kleppe, A. Kunschner, Makromol. Chem. 48, 144 (1961); G. Jayme, K. Kringstad, Angew. Chem. 73, 219 (1961);

703c. G. Jayme, J. Troeften, Naturwissenschaften 52, 496 (1965).

703d. G. Jayme, G. El-Kodsi, Papier 24, 125, 679 (1970); G. Jayme, G. El-Kodsi, Papier 25, 589, 627 (1971).

704. H. H. G. Jellinek, M. D. Luh, Polymer Preprints 8(1), 533 (1967).

705. H. H. G. Jellinek, G. White, J. Polymer Sci. 6, 757 (1951).

706. A. M. Jendrichowska-Bonamour, European Polymer J. 6, 1491 (1970).

707. R. E. Jentoft, T. H. Gouw, J. Polymer Sci., B, 7, 811 (1969).

708. R. Jerome, V. Desreux, European Polymer J. 6 411 (1970).

709. B. Jirgensons, J. Polymer Sci. 8, 519 (1952).

710. R. M. Johnsen, Chem. Scripta 1, 81 (1971).

711. B. L. Johnson, Ind. Eng. Chem. 40, 351 (1948).

712. J. F. Johnson, R. S. Porter, Editors, "Analytical Gel Permeation Chromatography", J. Polymer Sci., C, 21, Interscience, New York (1968).

713. J. F. Johnson, R. S. Porter, "Gel Permeation Chromatography" in "Progress in Polymer Science. Editor A.D. Jenkins, Vol. 2, Pergamon Press, 201 (1970).

714. J. F. Johnson, R. S. Porter, M. J. R. Cantow, "Fractionation" in Encyclopedia Polymer Sci., Technol Editors, H. F. Mark, N. G. Gaylord, N. M. Bikales, Interscience, New York 7, 231 (1967).

715. P. R. Johnston, J. Appl. Polymer Sci. 9, 467 (1965).

716. E. F. Jordan, A. N. Wrigley, J. Polymer Sci., A-1, 2, 3909 (1964).

718. S. Jovanović, J. Romatowski, G. V. Schulz, Makromol. Chem. 85, 187 (1965).

719. J. Juilfs, Kolloid Z. 141, 88 (1955).

720. P. C. Juliano, J. V. Crivello, D. E. Floryan, Polymer Preprints 12(1), 632 (1971).

721. J. L. Jungnickel, F. T. Weiss, J. Polymer Sci. 49, 437 (1961).

722. V. Juraníčová, Š. Florián, D. Berek, European Polymer J. 6, 57 (1970).

723. V. Juraníčová, S. Florián, D. Berek, European Polymer J. 6, 63 (1970).

724. R. I. Kabalyunas, G. D. Shershneva, R. M. Livshits, Z. A. Rogovin, Polymer Sci. USSR 8, 260 (1966); Vysokomolekul. Soedin. 8, 240 (1966).

725. Y. Kakamura, O. Hinojosa, J. C. Arthur, J. Appl. Polymer Sci. 15, 391 (1971).

726. H. Kakiuchi, W. Fukuda, Kogyo Kagaku Zasshi 66, 964 (1963).

728. M. Kalfus, J. Mitus, J. Polymer Sci., A-1, 4, 953 (1966).

729. M. Kalfus, E. Kopytovski, Z. Skupinska, S. Lesniak, Polymer Sci. USSR 7, 1830 (1965); Vysokomolekul. Soedin. 7, 1655 (1965).

730. K. Kamada, H. Sato, Polymer J. (Japan) 2, 593 (1971).

731. P. M. Kamath, L. Wild, Polymer Eng. Sci. 6, 213 (1966).

732. K. Kamide, K. Fujii, H. Kobayashi, Makromol. Chem. 117, 190 (1968).

733. K. Kamide, K. Ohno, T. Kawai, Makromol. Chem. 137, 1 (1970).

734. K. Kamide, K. Sugamiya, T. Ogawa, C. Nakayama, N. Baba, Makromol. Chem. 135, 23 (1970).

735. K. Kamide, T. Terakawa, Makromol. Chem. 155, 25 (1972).

736. F. Kamiyama, H. Matsuda, H. Inagaki, Makromol. Chem. 125, 286 (1969).

737. F. Kamiyama, H. Matsuda, H. Inagaki, Polymer J. 1, 518 (1970).

738. R. Kammereck, L. J. Fetters, M. Morton, Polymer Preprints 11(1), 72 (1970).

739. H. Kämmerer, P. N. Grover, Makromol. Chem. 96, 270 (1966).

740. H. Kämmerer, P. N. Grover, Makromol. Chem. 99, 49 (1966).

741. H. Kämmerer, W. Kern, W. Heuser, J. Polymer Sci. 28, 331 (1958).

742. H. Kämmerer, K. F. Mueck, Makromol. Chem. 136, 161 (1970).

743. H. Kämmerer, K. F. Mueck, Makromol. Chem. 136, 179 (1970).

744. H. Kämmerer, F. Rocaboy, Compt. Rend. 256, 4440 (1963).

745. H. Kämmerer, F. Rocaboy, Makromol. Chem. 72, 76 (1964).

746. H. Kämmerer, K.-G. Steinfort, W. Kern, Makromol. Chem. 70, 173 (1964).

747. P. J. Kangle, E. Pacsu, J. Polymer Sci. 54, 301 (1961).

748. S. L. Kapur, S. Gundiah, J. Polymer Sci. 26, 89 (1957).

749. M. G. Karasev, I. N. Andreyeva, N. M. Domareva, K. I. Kosmatykh, M. G. Karaseva, N. A. Domnicheva, Polymer Sci. USSR 12, 1275 (1970); Vysokomolekul. Soedin. 12 A, 1127 (1970).

751. V. A. Kargin, N. A. Plate, A. S. Dobrynina, Kolloidn. Zh. 20, 332 (1958).

752. K. Karuwakaran, M. Santappa, Makromol. Chem. 111, 20 (1968).

753. F. Käsbauer, E. Schuck in "Kunststoffe", Editor, K. A. Wolf, Vol. I, Springer, Berlin, 1962.

754. A. Katchalski, H. Eisenberg, J. Polymer Sci. 6, 145 (1951).

755. I. Katime, A. Roig, L. M. León, Anales Quim. (Madrid) 67, 811 (1971).

756. T. Kato, K. Miyaso, M. Nagasawa, J. Phys. Chem. 72, 2161 (1968).

757. T. Kawai, Makromol. Chem. 102, 125 (1967).

758. T. Kawai, K. Ehara, H. Sasano, K. Kamide, Makromol. Chem. 111, 271 (1968).

759. T. Kawai, T. Goto, H. Maeda, Kolloid. Z. - Z. Polymere 223, 117 (1968).

760. T. Kawai, T. Hama, K. Ehara, Makromol. Chem. 113, 282 (1968).

761. T. Kawai, M. Hosoi, K. Kamide, Makromol. Chem. 146, 55(1971).

762. T. Kawai, E. Ida, Kolloid Z. - Z. Polymere 194, 40 (1964).

763. T. Kawai, A. Keller, J. Polymer Sci. B 2, 333 (1964).

764. T. Kawai, S. Shiozaki, S. Sonoda, H. Nakagawa, T. Matsumoto, H. Maeda, Makromol. Chem. 128, 252 (1969).

765. A. Kawasaki, J. Furukawa, T. Saegusa, M. Mise, T. Tsuruta, Makromol. Chem. 42, 25 (1960).

766. H. S. Kaufman, E. Solomon, Ind. Eng. Chem. 45, 1779 (1953).

767. H. S. Kaufman, E. K. Walsh, J. Polymer Sci. 26, 124 (1957).

768. K. Kawahara, Chem. High Polymer (Japan) 18, 687 (1961).

769. K. Kawahara, Makromol. Chem. 73, 1 (1964).

770. K. Kawahara, R. Okada, J. Polymer Sci. 56, S 7 (1962).
771. G. I. Keim, D. L. Christman, L. R. Kangas, S. K. Keahey, Macromolecules 5, 217 (1972).
772. T. Kelen, D. Schlotterbeck, V. Jaacks, IUPAC Boston, Vol. II, 649 (1971).
773. A. Keller, Kolloid Z. - Z. Polymere 231, 386 (1969).
774. A. Keller, E. Martuscelli, D. J. Priest, Y. Udagawa, IUPAC Leiden, Vol. II, 843 (1970).
775. A. Keller, A. O' Connor, Polymer 1, 163 (1960).
776. A. Keller, F. M. Willmouth, J. Polymer Sci., A-2, 8, 1443 (1970).
777. H. Kendrick, J. Polymer Sci., A-2, 7, 297 (1969).
778. A. S. Kenyon, I. O. Salyer, J. Polymer Sci. 43, 427 (1960).
779. A. S. Kenyon, I. O. Salyer, J. E. Kurz, D. R. Brown, J. Polymer Sci., C, 8, 205 (1965).
780. W. Kern et al., Makromol. Chem. 6, 206 (1951).
781. W. Kern, H. Schmidt, H. S. von Steinwehr, Makromol. Chem. 16, 74 (1955).
782. K. Kikukawa, S. Nozakura, S. Murahashi, Polymer J. 3, 52 (1972).
783. P. O. Kinell, Acta Chem. Scand. 1, 335 (1947).
784. P. O. Kinell, Acta Chem. Scand. 1, 832 (1947).
785. J. B. Kinsinger, L. E. Ballard, J. Polymer Sci., A-1, 3, 3936 (1965).
786. J. B. Kinsinger, R. E. Hughes, J. Phys. Chem. 63, 2002 (1959).
788. J. Klein, H. Friedel, J. Appl. Polymer Sci. 14, 1927 (1970).
789. J. Klein, E. Killmann, Makromol. Chem. 96, 193 (1966).
790. J. Klein, F. Patat, Makromol. Chem. 97, 189 (1966).
791. J. Klein, G. Weinhold, Angew. Makromol. Chem. 10, 49 (1970).
792. J. Klein, U. Wittenberger, Makromol. Chem. 122, 1 (1969).
793. V. I. Klenin, G. G. Uglanova, Polymer Sci. USSR 11, 2587 (1969); Vysokomolekul. Soedin. 11A, 2273 (1969).
794. A. M. Knebel' man, L. A. Kantor, D. F. Kagan, Polymer Sci. USSR 12, 3117 (1970).
795. H. Kobayashi, J. Polymer Sci. 26, 230 (1957).
796. H. Kobayashi, J. Fujisaki, J. Polymer Sci., B, 1, 15 (1963).
797. T. Kobayashi, Bull. Chem. Soc. Japan 35, 726 (1962).
798. E. Kobayashi, T. Higashimura, S. Okamura, J. Macromol. Sci. Chem. A 1, 1519 (1967).
799. E. Kobayashi, S. Okamura, S. Signer, J. Appl. Polymer Sci. 12, 1661 (1968).
800. H. Kobayashi, K. Sasaguri, Y. Fujisaki, T. Amano, J. Polymer Sci., A-1, 2, 313 (1964).
801. W. Kobryner, A. Bandere, Compt. Rend. 245, 689 (1957).
802. R. Kocher, Ch. Sadron, Makromol. Chem. 10, 172 (1953).
803. S. M. Kochergin, V. P. Barabanov, I. G. Fedorova, Polymer Sci. USSR 8, 1006 (1966); Vysokomolekul. Soedin. 8, 916 (1966).
805. G. Kockelbergh, G. Smets, J. Polymer Sci. 33, 227 (1958).
806. H. M. Koepp, H. Werrer, Makromol. Chem. 32, 309 (1959).
807. A. Kohler, J. G. Zilliox, P. Rempp, J. Polacek, I, Koessler, European Polymer J. 8, 627 (1972).
808. K. Kohlschuetter, K. Unger, K. Vogel, Makromol. Chem. 93, 1 (1966).
809. V. Kokle, F. W. Billmeyer, J. Polymer Sci., C, 8, 217 (1965).
810. H. Koelbel, W. H. E. Mueller, H. Hammer, Makromol. Chem. 70, 1 (1964).
811. J. V. Koleske, S. F. Kurath, J. Polymer Sci., A-1, 2, 4123 (1964).
812. J. V. Koleske, R. D. Lundberg, J. Polymer Sci., A-2, 7, 897 (1969).
813. H. S. Kolesnikov, O. V. Smirnova, V. N. Lamm, A. A. Kudryashov, Polymer Sci. USSR 11, 2518 (1969); Vysokomolekul. Soedin. 11 A, 2211 (1969).
814. H. S. Kolesnikov, O. V. Smirnova, A. K. Mikitayev, V. M. Gradyshev, Polymer Sci. USSR 12, 1617 (1970); Vysokomolekul. Soedin. 12 A, 1424 (1970).
816. M. Kolínsky, V. Jišová, D. Lim, IUPAC Helsinki, Vol. II, Preprint 1/88 (1972).
817. R. Koningsveld, Fortschr. Hochpolymer Forsch. 7, 1 (1970).
818. R. Konigsveld, A. J. Pennings, Rec. Trav. Chim. 83, 552 (1964).
819. H. Konishi, Sen-i-Gakkaishi 17, 1170 (1961).
820. J. Kops, H. Spanggaard, Makromol. Chem. 151, 21 (1972).
821. J. Kops, H. Spanggaard, Polymer Preprints 13(1), 90 (1972).
822. Ye. V. Korneyeva, V. N. Tsvetkov, P. N. Laurenko, Polymer Sci. USSR 12, 1554 (1970); Vysokomolekul. Soedin. 12 A, 1369 (1970).
823. V. V. Korshak, S. A. Paulova, L. V. Boiko, T. M. Babchinitser, S. V. Vinogradova, Ya. S. Vygodskii, N. A. Gulebeva, Polymer Sci. USSR 12, 63 (1970); Vysokomolekul. Soedin. 12 A, 56 (1970).
824. V. V. Korshak, G. I. Tarasova, S. A. Pavlova, Polymer Sci. USSR 13, 1179 (1971); Vysokomolekul. Soedin. 13, 1047 (1971).
825. I. Koessler, J. Krejsa, J. Polymer Sci. 29, 69 (1958).
826. I. Koessler, J. Krejsa, J. Polymer Sci. 57, 509 (1962).
827. I. Koessler, H. Krauserova, J. Polymer Sci., A-1, 4, 1329 (1966).
828. I. Koessler, M. Stolka, J. Polymer Sci. 44, 213 (1960).
829. T. Kotaka, N. Donkai, J. Polymer Sci., A-2, 6, 1457 (1968).
830. T. Kotaka, Y. Murakami, H. Inagaki, J. Phys. Chem. 72, 829 (1968).
831. T. Kotaka, H. Ohnuma, H. Inagaki, Polymer 10, 517 (1969).
832. T. Kotaka, T. Tanaka, H. Ohnuma, Y. Muzakami, H. Inagaki, Polymer J. 1, 245 (1970).
833. A. Kotera, T. Saito, Y. Watanabe, M. Ohama, Makromol. Chem. 87, 195 (1965).
834. M. Kozhokaryu, V. S. Skazka, K. G. Berdnikova, Polymer Sci. USSR 8, 1167 (1966); Vysokomolekul. Soedin. 8, 1063 (1966).
835. P. V. Kozlov et al., Vysokomolekul. Soedin. 2, 1575 (1960)
836. J. J. Krackeler, H. Naidus, J. Polymer Sci., C, 27, 207 (1969).
837. D. Kranz, Kolloid Z. - Z. Polymere 227, 41 (1968).
838. F. Krasovek, J. Stefan Inst. Repts Ljubljana 3, 203 (1956).
839. G. Kraus, C. J. Stacy, J. Polymer Sci., A-2, 10, 657 (1972).
840. S. Krause, J. Phys. Chem. 65, 1618 (1961).
841. S. Krause, E. Cohn-Ginsberg, Polymer 3, 566 (1962).
842. S. Krause, E. Cohn-Ginsberg, J. Polymer Sci., A-1, 2, 1393 (1964).
843. J. Krejsa, Makromol. Chem. 33, 244 (1959).
844. W. R. Krigbaum, P. J. Flory, J. Am. Chem. Soc. 75, 1775 (1953).
845. W. R. Krigbaum, A. M. Kotliar, J. Polymer Sci. 32, 323 (1958).
846. S. Krozer, M. Vainryb, L. Silina, Vysokomolekul. Soedin. 2, 1876 (1960).
847. M. Kubin, P. Špaček, R. Chromáček, Collection Czech. Chem. Commun. 32, 3881 (1967).
848. K. Kubo, K. Ogino, Sci. Papers Coll. Gen. Educ. 16, 193 (1966).
849. L. G. Kudryavtseva, A. D. Litmanovich, A. V. Topchiyev, V. Ya. Shtern, Neftekhimiya 3, 343 (1963).
850. M. M. Kusakov, A. Yu. Koshevnik. D. N. Nekrasov, V. F. Chirkova, L. M. Shol' pina, Polymer Sci. USSR 8, 1141 (1966); Vysokomolekul. Soedin. 8, 1040 (1966).
851. J. E. Kurz, J. Polymer Sci., A-1, 3, 1895 (1965).
852. N. Kuwahara, S. Higashida, M. Nakata, M. Kaneko, J. Polymer Sci., A-2, 7, 285 (1969).
853. N. Kuwahara, M. Kaneko, Makromol. Chem. 82, 205 (1965).
854. N. Kuwahara, M. Kaneko, K. Kubo, Makromol. Chem. 110, 294 (1967).
855. N. Kuwahara, T. Okazawa, M. Kaneko, J, Polymer Sci., C, 23, 543 (1968).
856. A. I. Kuzayev, G. N. Komratov, G. V. Korovina, S. G. Entelis, Polymer Sci. USSR 12, 1167 (1970); Vysokomolekul. Soedin. 12 A, 1033 (1970).
857. A. I. Kuzayev, G. N. Komratov, G. V. Korovina, S. G. Entelis, Polymer Sci. USSR 12, 1124 (1970); Vysokomolekul. Soedin. 12 A, 995 (1970).
858. V. N. Kuznetsov, V. B. Kogan, L. A. Venkstem, S. D. Vogman, T. A. Usatova, V. A. Morozova, Polymer Sci. USSR 12, 3143 (1970); Vysokomolekul. Soedin. 12 A, 2768 (1970).
859. A. Lambert, Polymer 10, 213 (1969).
860. A. Lambert, Brit. Polymer J. 3, 13 (1971).
861. R. F. Landel, J. D. Ferry, J. Phys. Chem. 59, 658 (1955).
862. I. Landler, Compt. Rend. 225, 629 (1947).

863.   W. J. Langford, D. J. Vaughan, Nature (London) 184, 116 (1959).

864.   G. Langhammer, J. Polymer Sci. 29, 505 (1958).

865.   G. Langhammer, Makromol. Chem. 21, 74 (1956).

866.   G. Langhammer, Kolloid Z. 146, 44 (1956).

867.   G. Langhammer, Svensk. Kem. Tidskr. 69, 328 (1957).

868.   G. Langhammer, Naturwiss. 41, 552 (1954).

869.   G. Langhammer, H. Foerster, Z. Physik. Chem. N.F. (Frankfurt) 15, 212 (1958)

870.   G. Langhammer, H. Pfennig, K. Quitzsch. Z. Elektrochem. 62, 458 (1958).

871.   G. Langhammer, K. Quitzsch, Makromol. Chem. 43, 160 (1961).

872.   G. Langhammer, H. Seide, Kolloid. Z.-Z. Polymere 216-217, 264 (1967).

873.   J. Láníková, M. Hloušek, European Polymer J. 6, 25 (1970).

874.   J. Láníková, M. Hloušek, M. Nováková, Pardubice Symposium Preprints 299 (1970).

875.   L. M. Lanovskaya, N. A. Pravikova, Polymer Sci. USSR 11, 2208 (1969); Vysokomolekul Soedin. 11 A, 1938 (1969).

876.   R. D. Law, Polymer Preprints 8(2), 1348 (1967).

877.   R. D. Law, J. Polymer Sci., A-1, 7, 2097 (1969).

878.   R. D. Law, J. Polymer Sci., A-1, 9, 589 (1971).

879.   R. D. Law, J. Polymer Sci., C, 21, 225 (1968).

880.   R. G. Le Bel, D. A. I. Goring, J. Polymer Sci., C, 2, 29 (1963).

881.   P. Lebel, C. Job, J. Polymer Sci., C, 4, 649 (1963).

882.   C. L. Lee, F. A. Emerson, J. Polymer Sci., A-2, 5, 829 (1967).

883.   C. L. Lee, C. L. Frye, O. K. Johannson, Polymer Preprints 10(2), 1361 (1969).

884.   J. Leonard, H. Daoust, J. Phys. Chem. 69, 1174 (1965).

885.   M. Le Page, R. Beau, A. J. De Vries, Polymer Preprints 8(2), 1211 (1967).

886.   M. Le Page, R. Beau, A. J. de Vries, J. Polymer Sci., C, 21, 119 (1968).

887.   J. B. Lesh, K. Schultz, J. D. Porsche, Ind. Eng. Chem. 42, 1376 (1950).

888.   H. Leth-Pedersen, J. Polymer Sci., B, 5, 239 (1967).

889.   G. Levin, J. B. Carmichel, J. Polymer Sci., A-1, 6, 1 (1968).

890.   J. J. Lindberg, P. Sterck, B. Hortling, Suomen Kemistilehti 42 B, 451 (1969).

891.   A. D. Litmanovich, V. Ya. Shtern, Polymer Sci. USSR 7, 1478 (1965); Vysokomolekul. Soedin. 7, 1332 (1965).

892.   J. N. Little, J. L. Waters, K. J. Bombaugh, W. J. Pauplis, Polymer Preprints 10 (1), 326 (1969).

893.   J. Llopis, A. Albert, P. Usobiaga, Anales Real Soc. Esp. Fis. Quim. 62B, 1103 (1966).

894.   J. Llopis, A. Albert, P. Usobiaga, European Polymer J. 3, 259 (1967).

895.   J. Llopis, A. Albert, P. Usobiaga, Anales Real Soc. Esp. Fis. Quim. 63 B, 413 (1967).

896.   J. D. Loconti, J. W. Cahill, J. Polymer Sci. 49, S 2 (1961).

897.   J. D. Loconti, J. W. Cahill, J. Polymer Sci. A-1, 3163 (1963).

898.   G. Loehr, G. V. Schulz, Makromol. Chem. 77, 240 (1964).

899.   G. Loehr, G. V. Schulz, Makromol. Chem. 117, 283 (1968).

900.   F. Lombard, Makromol. Chem. 8, 201 (1952).

901.   V. C. Long, G. C. Bery, L. M. Hobbs, Polymer 5, 517 (1964).

902.   P. Longi, F. Greco, U. Rossi, Makromol. Chem. 116, 113 (1968).

903.   P. Longi, A. Roggero, Ann. Chim. (Rome) 51, 1013 (1961).

904.   B. E. López Madruga, J. M. Barrales-Rienda, G. M. Guzmán, Anales Quím. (Madrid) 67, 153 (1971).

905.   B. V. Losikov, N. I. Kaverina, A. A. Fedyantseva, Khim. Tekhnol. Topliva 3, 51 (1956).

906.   E. G. Lovering, J. Polymer Sci., C, 30, 329 (1970).

908.   E. G. Lovering, W. B. Wright, J. Polymer Sci., A-1, 6, 2221 (1968).

909.   Lj. Lovrić, IUPAC Helsinki, Vol. III, Preprint II-41 (1972).

910.   I. L. Lovric, J. Polymer Sci., A-2, 7, 1357 (1969).

911.   I. L. Lovric, J. Polymer Sci., A-2, 8, 807 (1970).

912.   C. Loucheux, Z. Czlonkowska, J. Polymer Sci., C, 16, 4001 (1968).

913.   P. L. Luisi, F. Pezzana, European Polymer J. 6, 259 (1970).

914.   P. L. Luisi, E. Chiellini, P. F. Franchini, M. Orienti, Makromol. Chem. 112, 197 (1968).

915.   S. Luňák, M. Bohdanecký, Collection Czech. Chem. Commun. 30, 2756 (1965).

916.   H. Luetje, Z. Physik. Chem. N. F. (Frankfurt) 43, 11 (1964).

917.   B. J. Lyons, A. S. Fox, Polymer Preprints 8(2), 1241 (1967).

918.   A. R. Lyons, E. Catterall, Makromol. Chem. 146, 141 (1971).

919.   B. J. Lyons, A. S. Fox, J. Polymer Sci., C, 21, 159 (1968).

920.   D. MacCallum, Makromol. Chem. 100, 117 (1967).

921.   S. Machi, J. Silverman, D. J. Metz, J. Phys. Chem. 76, 930 (1972).

922.   C. Macosko, K. E. Weale, Polymer Preprints 10(2), 562 (1964).

923.   J. Majer, Makromol. Chem. 86, 253 (1965).

924.   A. I. Maklakov, E. I. Nagumanova, Polymer Sci. USSR 7, 2302 (1965); Vysokomolekul. Soedin. 7, 2101 (1965).

925.   N. V. Makletsova, I. V. Epel' baum, B. A. Rozenberg, Ye. B. Lyudvig, Polymer Sci. USSR 7, 73 (1965); Vysokomolekul. Soedin. 7, 70 (1965).

926.   C. P. Malone, H. L. Suchan, W. W. Jau, J. Polymer Sci., B, 7, 781 (1969).

927.   D. Mangaraj, S. к. Patra, Makromol. Chem. 107, 230 (1967).

928.   D. Mangaraj, S. B. Rath, Polymer Preprints 12(1), 573 (1971).

929.   J. A. Manson, G. J. Arquelte, Makromol. Chem. 37, 187 (1960).

930.   K. Mara, T. Imoto, Kolloid. Z. - Z. Polymere 237, 297 (1970).

931.   J. E. Mark, G. B. Thomas, Polymer Preprints 7(2), 438 (1966).

932.   J. E. Mark, G. B. Thomas, J. Phys. Chem. 70, 3588 (1966).

933.   J. E. Mark, R. A. Wessling, R. E. Hughes, J. Phys. Chem. 70, 1895 (1966).

934.   C. A. Marshall, R. A. Mock, J. Polymer Sci. 17, 591 (1955).

935.   M. Marx-Figini, J. Polymer Sci. 30, 119 (1958).

936.   M. Marx-Figini, J. Polymer Sci., C, 28, 57 (1969).

937.   M. Marx-Figini, Makromol. Chem. 32, 233 (1959).

938.   M. Marx-Figini, E. Penzel, Makromol. Chem. 87, 307 (1965).

939.   M. Marx-Figini, G. V. Schulz, Makromol. Chem. 62, 49 (1963).

940.   T. Masuda, K. Kitagawa, T. Inoue, S. Onogi, Macromolecules 3, 116 (1970).

941.   T. Masuda, K. Kitagawa, S. Onogi, Polymer J. 1, 418 (1970).

942.   V. J. Masura, Faserforsch. Textiltech. 13, 517 (1962).

943.   R. D. Mate, H. S. Lundstrom, Polymer Preprints 8(2), 1397 (1967).

944.   R. D. Mate, H. S. Lundstrom, J. Polymer Sci., C, 21, 317 (1968).

945.   A. R. Mathieson, J. Colloid Sci. 15, 387 (1960).

946.   Matlack et al., J. Polymer Sci. 49, 533 (1961).

947.   H. Matsuda, H. Aonuma, S. Kuroiwa, J. Appl. Polymer Sci. 14, 335 (1970).

948.   H. Matsuda, M. Takeshima, S. Shimizu, S. Kuroiwa, Makromol. Chem. 134, 309 (1970).

949.   H. Matsuda, K. Yamano, H. Inagaki, J. Polymer Sci., A-2, 7, 609 (1969).

950.   T. Matsumoto, N. Nishioka, H. Fujita, J. Polymer Sci., A-2, 10, 23 (1972).

951.   K. Matsumura, Bull. Chem. Soc. Japan 42, 1874 (1969).

952.   K. Matsumura, Makromol. Chem. 124, 204 (1969).

953.   K. Matsumura, M. Fukaya, K. Mizuno, Bull. Chem. Soc. Japan, 43, 1881 (1970).

954.   T. Matsuo, H. Inagaki, Makromol. Chem. 55, 150 (1962).

955.   A. Mattiussi, G. B. Gechele, M. Pizzoli, European Polymer J. 2, 383 (1966).

956.   A. Mattiussi, F. Manescalchi, G. B. Gechele, European Polymer J. 5, 105 (1969).

957.   J. A. May, W. B. Smith, J. Phys. Chem. 72, 216, 2993 (1968).

958.   H. W. McCormick, J. Polymer Sci. 36, 341 (1959).

959.   H. W. McCormick, J. Polymer Sci. 41, 327 (1959).

960.   H. W. McCormick, J. Polymer Sci., A-1, 103 (1963).

961. R. H. McDowell, Chem. Ind. 1401 (1958).

962. J. W. Meeks, T. F. Banigan, R. W. Planck, India Rubber World 122, 301 (1950).

963. A. M. Meffroy-Biget, Bull. Soc. Chim. France 458, 465 (1954).

964. A. M. Meffroy-Biget, Compt. Rend. 240, 1707 (1955).

965. B. Meissner, I. Klier, S. Kucharkik, J. Polymer Sci., C, 16, 793 (1967).

966. A. Mele, R. Rufo, L. Santonato, Polymer 10, 233 (1969).

967. H. W. Melville, B. D. Stead, J. Polymer Sci. 16, 505 (1956).

968. Z. Menčík, Chem. Zvesti. 9, 165 (1955).

969. R. A. Mendelson, J. Polymer Sci., A-1, 2361 (1963).

970. R. A. Mendelson, E. E. Drott, J. Polymer Sci., B, 6, 795 (1968).

971. R. A. Mendelson, F. L. Finger, Polymer Preprints 12(1), 370 (1971).

972. J.-P. Merle, A. Sarko, Macromolecules 5, 132 (1972).

973. E. H. Merz, J. Polymer Sci. 3, 790 (1948).

974. E. H. Merz, R. W. Raetz, J. Polymer Sci. 5, 587 (1950).

975. J. C. Meunir, R. Van Leemput, Makromol. Chem. 142, 1 (1971).

976. G. Meyerhoff, Angew. Makromol. Chem. 4/5, 268 (1968).

977. G. Meyerhoff, Ber. Bunsenges. Phys. Chem. 69, 866 (1965).

978. G. Meyerhoff, J. Polymer Sci., C, 21, 31 (1968).

979. G. Meyerhoff, Makromol. Chem. 12, 45 (1954).

980. G. Meyerhoff, Makromol. Chem. 89, 282 (1965).

981. G. Meyerhoff, Makromol. Chem. 134, 129 (1970).

982. G. Meyerhoff, Polymer Preprints 8(2), 1295 (1967).

983. G. Meyerhoff, S. Jovanovic, J. Polymer Sci., B, 5, 495 (1967).

984. G. Meyerhoff, W. Meier, K. Berger, Makromol. Chem. 91, 290 (1966).

985. G. Meyerhoff, U. Moritz, R. G. Kirste, W. Heitz, European Polymer J. 7, 933 (1971).

986. G. Meyerhoff, J. Romatowski, Makromol. Chem. 74, 222 (1964).

987. G. Meyerhoff, N. Suetterlin, Makromol. Chem. 87, 258 (1965).

988. G. Meyerhoff, S. Shimotsuma, Makromol. Chem. 109, 263 (1967).

989. G. Meyerhoff, S. Shimotsuma, Makromol. Chem. 135, 195 (1970).

990. G. Meyerhoff, S.Shimotsuma, IUPAC Boston Vol. II, 1157 (1971).

991. F. Micheel, A. Boeckmann, W. Meckstroth, Makromol. Chem. 48, 15 (1961).

993. N. V. Mikhailov, S, G, Zelikman, Kolloidn. Zh. 18, 717 (1956).

994. B. Miller, E. Pacsu, J. Polymer Sci. 41, 97 (1959).

995. G. W. Miller, S. A. D. Visser, Polymer Preprints 8(1), 641 (1967).

996. L. E. Miller, F. A. Hamm, J. Phys. Chem. 57, 110 (1953).

997. M. L. Miller, J. Polymer Sci. 56, 203 (1962).

998. N. J. Mills, European Polymer J. 5, 675 (1969).

999. A. Minakata, N. Imai, Biopolymers 11, 329 (1972).

1000. Y. Minoura, M. Mitoh, Makromol. Chem. 99, 186 (1966).

1001. Y. Minoura, M. Mitoh, Makromol. Chem. 124, 241 (1969).

1002. Y. Minoura, M. Shundo, J. Enomoto, J. Polymer Sci., A-1, 6, 979 (1968).

1003. G. S. Misra, V. P. Gupta, Makromol. Chem. 71, 110 (1964).

1004. I. Mita, J. Chim. Phys. 59, 530 (1962).

1005. J. C. Mitchell, A. E. Wooddard, P. Doty, J. Am. Chem. Soc. 79, 3955 (1957).

1006. R. L. Mitchell, Ind. Eng. Chem. 45, 2526 (1953).

1007. S. P. Mitsengendler et al., Vysokomolekul. Soedin. 4, 1366 (1962).

1008. S. P. Mitsengendler, K. I. Sokolova, G. A. Andreyeva, A. A. Korotkov, T. Kadyrov, S. I. Klenin, S. Ya. Magarik, Polymer Sci. USSR 9, 1261 (1967); Vysokomolekul. Soedin. 9, 1133 (1967).

1009. M. Miura, Y. Kubota, T. Masuzakawa, Bull. Chem. Soc. Japan 38, 316 (1965).

1010. T. Miyamoto, H. Inagaki, Macromolecules 2, 554 (1969).

1011. K. Mizuhara, K. Hara, T. Imoto, Kolloid. Z. - Z. Polymere 229, 17 (1969).

1012. J. Moacanin et al., J. Am. Chem. Soc. 81, 2054 (1959).

1013. J. Moacanin, A. Rembaum, R. K. Laudenslager, J. Macromol. Sci. Chem., A-1, 1497 (1967).

1014. W. E. Mochel, J. B. Nichols, J. Am. Chem. Soc. 71, 3435 (1949).

1015. W. E. Mochel, J. B. Nichols, Ind. Eng. Chem. 43, 154 (1951).

1016. W. E. Mochel, J. B. Nichols, C. J. Mighton, J. Am. Chem. Soc. 70, 2185 (1948).

1017. R. B. Mohite, S. Gundiah, S. L. Kapur, Makromol. Chem. 108, 52 (1967).

1018. R. B. Mohite, S. Gundiah, S. L. Kapur, Makromol. Chem. 116, 280 (1968).

1019. R. A. Mock et al., J. Polymer Sci. 11, 447 (1953).

1020. W. L. Moll, Kolloid. Z. - Z. Polymere 223, 48 (1968).

1021. P. Molyneux, Makromol. Chem. 43, 31 (1961).

1022. P. Molyneux, Kolloid Z. - Z. Polymere 226, 15 (1968).

1023. P. G. Montague, F. W. Peaker, P. Bosworth, P. Lemon, Brit. Polymer J. 3, 93 (1971).

1024. G. Montaudo, P. Maravigna, P. Finocchiaro, C. G. Overberger, Macromolecules 5, 203 (1972).

1025. J. C. Moore, J. Polymer Sci., A-1, 2, 835 (1964).

1026. L. D. Moore, G. R. Greear, J. O. Sharp, J. Polymer Sci. 59, 339 (1962).

1027. W. R. Moore, G. F. Boden, J. Appl. Polymer Sci. 9, 2019 (1965).

1028. W. R. Moore, J. A. Epstein, J. Appl. Chem. 6, 168 (1956).

1029. W. R. Moore, R. J. Fort, J. Polymer Sci., A-1, 1, 929 (1963).

1030. W. R. Moore, R. J. Hutchinsons, Nature (London) 200, 1095 (1963).

1031. W. R. Moore, M. Murphy, J. Polymer Sci. 56, 519 (1962).

1032. W. R. Moore, G. P. Pearson, Polymer 1, 144 (1960).

1033. W. R. Moore, K.Saito, European Polymer J. 3, 65 (1967).

1034. W. R. Moore, D. Sanderson, Polymer 9, 153 (1968).

1035. W. R. Moore, R. P. Sheldon, J. Textile Inst. 50, T 294 (1959).

1036. W. R. Moore, B. M. Tidswell, J. Appl. Chem. 8, 232 (1958).

1037. W. R. Moore, M. A. Uddin, European Polymer J. 3, 673 (1967).

1038. W. R. Moore, M. A. Uddin, European Polymer J. 6, 185 (1969).

1039. W. R. Moore, M. A. Uddin, European Polymer J. 6, 121 (1970).

1040. G. Moraglio, Chim. Ind. (Milan) 44, 352 (1962).

1041. G. Moraglio, European Polymer J. 1, 103 (1965).

1042. G. Moraglio, G. G. Gianotti, European Polymer J. 5, 781 (1969).

1043. G. Moraglio, G. Gianotti, F. Danusso, European Polymer J. 3, 251 (1967).

1044. G. Moraglio, G. Gianotti, F. Zoppi, U. Bonicelli, European Polymer J. 7, 303 (1971).

1045. D. R. Morey, J. W. Tamblyn, J. Appl. Phys. 16, 419 (1945).

1046. D. R. Morey, J. W. Tamblyn, J. Phys. Chem. 50, 12 (1946).

1047. D. R. Morey, J. W. Tamblyn, J. Phys. Chem. 51, 721 (1947).

1048. D. R. Morey, E. W. Taylor, G. P. Waugh, J. Colloid. Sci. 6, 470 (1951).

1049. P. W. Morgan, S. L. Kwolek, J. Polymer Sci. 62, 33 (1962).

1050. U. Moritz, G. Meyerhoff, Makromol. Chem. 139, 23 (1970).

1051. G. A. Morneau, P. I. Roth, A. R. Shultz, J. Polymer Sci. 55, 609 (1961).

1052. R. J. Morris, H. E. Persinger, J. Polymer Sci. A 1, 1041 (1963).

1053. M. Morton, R. Kamnereck, L. J. Fetters, Macromolecules 4, 11 (1971).

1054. T. M. Moshkina, A. N. Pudovik, Vysokomolekul. Soedin. 5, 1106 (1963).

1055. R. Muggli, H.-G. Elias, K. Muehlethaler, Makromol. Chem. 121, 290 (1960). R. Muggli, Cellulose Chem. Technol. 2, 549 (1968).

1056. A. Mula, L. Chinellato, European Polymer J. 6, 1 (1970).

1057. T. E. Muller, W. J. Alexander, Polymer Preprints 8(2), 1371 (1967).

1058. T. E. Muller, W. J. Alexander, J. Polymer Sci., C, 21, 283 (1968).

1059. W. A. Muller, L. N. Rogers, Ind. Eng. Chem. 45, 2522 (1953).

1060. P. Munk, Collection Czech. Chem. Commun. 32, 787 (1967).

1061. C. Mussa, J. Polymer Sci. 28, 587 (1958).

1062. M. I. Mustafayev, K. V. Aliev, V. A. Kavanov, Polymer Sci. USSR 12, 968 (1970); Vysokomolekul. Soedin. 12 A, 855 (1970).

1063. V. A. Myagchenkov, V. F. Korenkov, L. F. Antonova, A. Ya. Belobokova, Polymer Sci. USSR 12, 1980 (1970); Vysokomolekul. Soedin. 12 A, 1745 (1970).

1064. V. A. Myagchenkov, V. F. Kurenkov, E. V. Kuznetsov, S. Ya. Frenkel, European Polymer J. 6, 63 (1970).

1065. V. A. Myagchenkov, Ye. V. Kuznetsov, W. I. Dominova, Polymer Sci. USSR 6, 1786 (1964); Vysokomolekul. Soedin. 6, 1612 (1964).

1066. V. A. Myagchenkov, Ye. V. Kuznetsov, V. Ya. Kitkevich, Polymer Sci. USSR 6, 1507 (1964); Vysokomolekul. Soedin. 6, 1366 (1964).

1067. V. A. Myagchenkov, Ye. V. Kuznetsov, L. A. Zaltsgendler, Polymer Sci. USSR 7, 2275 (1965); Vysokomolekul. Soedin. 7, 2077 (1965).

1068. G. E. Myers, J. R. Dagon, J. Polymer Sci., A-1, 2, 2631 (1964).

1069. C. D. Myrat, J. S. Rowlinson, Polymer 6, 645 (1965).

1070. T. Nagata, J. Appl. Polymer Sci. 13, 2601 (1969).

1071. J. Naghizadeh, J. Springer, Kolloid. Z. - Z. Polymere 215, 21 (1967).

1072. H. Nakahara, M. Shihanda, J. Soc. Text. Cell. Ind. (Japan) 8, 438 (1952).

1073. N. Nakajima, J. Polymer Sci., A-2, 5, 101 (1966).

1074. N. Nakajima, Polymer Preprints 8(2), 1282 (1967).

1075. N. Nakajima, J. Polymer Sci., C, 21, 153 (1968).

1076. N. Nakajima, Polymer Preprints 12(2), 804 (1971).

1077. A. Nakajima, H. Fujiwara, Bull. Chem. Soc. Japan 37, 909 (1964).

1078. A. Nakajima, F. Hamada, Kolloid. Z. - Z. Polymere 205, 55 (1965).

1079. A. Nakajima, F. Hamada, S. Hayashi, J. Polymer Sci., C, 15, 285 (1966).

1080. A. Nakajima, F. Hamada, S. Hayashi, A. Saijoo, Discussion Meeting of Soc. Polymer Sci., Japan, July 10, 1965, Kobe, Japan.

1081. A. Nakajima, F. Hamada, S. Hayashi, T. Sumida, Kolloid. Z. - Z. Polymere 222, 10 (1968).

1082. A. Nakajima, F. Hamada, T. Shimizu, Makromol. Chem. 90, 229 (1966).

1083. A. Nakajima, S. Hayashi, Kolloid. Z. - Z. Polymere 225, 116 (1968).

1084. A. Nakajima, S. Hayashi, T. Korenaga, T. Sumida, Kolloid. Z. - Z. Polymere 222, 124 (1968).

1085. A. Nakajima, K. Kato, Makromol. Chem. 95, 52 (1966).

1086. A. Nakajima, I. Sakurada, Chem. High Polymers (Japan) 11, 110 (1954).

1087. M. Nakata, Makromol. Chem. 149, 99 (1971).

1088. A. Nakazawa, J. J. Hermans, J. Polymer Sci., A-2, 9, 1871 (1971).

1089. S. Nair, B. C. Sekhar, Chem. Soc. (London) Spec. Publ. 23, 105 (1968).

1091. H. Narita, S. Machida, Makromol. Chem. 97, 209 (1966).

1092. A. Nasini, C. Mussa, Makromol. Chem. 22, 59 (1957).

1093. G. Natta et al., Atti. Accad. Nazl. Lincei 28, 18 (1960).

1094. G. Natta, F. Danusso, G. Moraglio, Makromol. Chem. 20, 37 (1956).

1095. G. Natta, I. Pasquon, A. Zambelli, G. Gatti, Makromol. Chem. 70, 191 (1964).

1096. G. Natta, M. Pegoraro, M. Peraldo, Ricerca Sci. 28, 1473 (1958).

1097. G. Natta, L. Porri, A. Carbonaro, Makromol. Chem. 77, 126 (1964).

1098. G. Natta, L. Porri, A. Carbonaro, G. Stoppa, Makromol. Chem. 77, 114 (1964).

1099. G. Natta, A. Zambelli, I. Pasquon, F. Ciampelli, Makromol. Chem. 79, 161 (1964).

1100. T. N. Nekrasova, E. Churylo, Polymer Sci. USSR 11, 1249 (1969); Vysokomolekul. Soedin. 11 A, 1103 (1969).

1101. T. N. Nekrasova, O. B. Ptitsyn, M. S. Shikanova, Polymer Sci. USSR 10, 1771 (1968); Vysokomolekul. Soedin. 10, 1530 (1968).

1102. E. W. Neuse, K. Koda, E. Carter, Makromol. Chem. 84, 213 (1965).

1103. S. Newman, J. Polymer Sci. 47, 111 (1960).

1104. E. Nichols, Polymer Preprints 12(2), 828 (1971).

1105. L. Nicolás, Compt. Rend. 236, 809 (1953).

1106. L. Nicolás, Compt. Rend. 242, 2720 (1956).

1107. L. Nicolás, Makromol. Chem. 24, 173 (1957).

1108. A. Ninagawa, I. Ijichi, M. Imoto, Makromol. Chem. 107, 196 (1967).

1109. N. Nishida, D. G. Salladay, J. W. White, Polymer Preprints 12(2), 522 (1971).

1110. N. Nishida, D. G. Salladay, J. L. White, J. Appl. Polymer Sci. 15, 1181 (1971).

1111. M. Nishide, M. Sera, Kogyo Kagaku Zasshi 64, 1145 (1961).

1112. I. Noda, T. Horikawa, T. Kato, T. Fujimoto, M. Nagasawa, Macromolecules 3, 795 (1970).

1113. I. Noda, S. Saito, T. Fujimoto, M. Nagasawa, J. Phys. Chem. 71, 4048 (1967).

1114. K. Nöllen, W. Funke, K. Hamann, Makromol. Chem. 94, 248 (1966).

1115. K. Noro, H. Kawazura, T. Moriyama, S Yoshioka, Makromol. Chem. 83, 35 (1965).

1116. N. T. Notley, J. Polymer Sci. A-1, 227 (1963).

1117. E. A. Ofstead, Polymer Preprints 6(2), 674 (1965).

1118. H. Ohnuma, T. Kotaka, H. Inagaki, Polymer 10, 501 (1969).

1119. H. Okamoto, J. Polymer Sci. 37, 173 (1959).

1120. H. Okamoto, J. Polymer Sci. A-2, 3451 (1964).

1121. H. Okamoto, K. Sekikawa, J. Polymer Sci. 55, 597 (1961).

1122. K. Okita, A. Taremoto, H. Fujita, Biopolymers 9, 717 (1970).

1123. N. A. Okladnov, V. I. Zegel' man, V. P. Lebedev, S. V. Svetozar, Ye. N. Zil' berman Skii, Polymer Sci. USSR 12, 349 (1970); Vysokomolekul. Soedin. 12 A, 306 (1970).

1125. O. F. Olaj, J. W. Breitenbach, I. Hofreiter, Makromol. Chem. 91, 264 (1966).

1126. O. F. Olaj, J. W. Breitenbach, B. Wolf, Monatsh. Chem. 95, 1646 (1964).

1127. J. J. O' Malley, R. H. Marchessault, J. Phys. Chem. 70, 3235 (1966).

1128. S. Onogi, S. Kimura, T. Kato, T. Masuda, N. Miyanaga, J. Polymer Sci., C, 15, 381 (1966).

1129. S. Onogi, T. Masuda, K. Kitagawa, Macromolecules 3, 109 (1970).

1130. R. J. Orr, H. L. Williams, J. Am. Chem. Soc. 79, 3137 (1957).

1131. H. Orth, Angew. Makromol. Chem. 23, 83 (1972).

1132. F. T. Osborne, S. Omi, V. Stannett, E. P. Stahel, J. Polymer Sci., A-1, 8, 1657 (1970).

1133. H. W. Osterhoudt, L. N. Ray, Polymer Preprints 8(2), 1220 (1967).

1134. A. Oth, Bull. Soc. Chim. Belges 58, 285 (1949).

1135. J. Oth, V. Desreux, Ricerca Sci. 25, 447 (1955).

1136. J. Oth, V. Desreux, Bull. Soc. Chim. Belges 63, 261 (1954).

1137. E. P. Otocka, Macromolecules 3, 691 (1970).

1138. E. P. Otocka, Polymer Preprints 12(1), 645 (1971).

1139. E. P. Otocka, M. Y. Hellman, Macromolecules 3, 362 (1970).

1140. E. P. Otocka, M. Y. Hellman, P. M. Muglia, Macromolecules 5, 227 (1972).

1141. E. P. Otocka, R. J. Roe. M. Y. Hellman, P. M. Muglia, Macromolecules 4, 507 (1971).

1142. C. G. Overberger, A. E. Borchet, A. Katchman, J. Polymer Sci. 44, 491 (1960).

1143. C. G. Overberger, G. W. Halek, J. Polymer Sci., A-1, 8, 359 (1970).

1144. C. G. Overberger, P. A. Jarovitzky, H. Mukamal, Polymer Preprints 8(1), 401 (1967).

1145. C. G. Overberger, K. Mizamichi, J. Polymer Sci., A 1, 2023 (1963).

1146. C. G. Overberger, S. Nozakura, J. Polymer Sci. A 1, 1445 (1963).

1147. C. G. Overberger, S. Nozakura, J. Polymer Sci. A 1, 1444 (1963).

1148. A. C. Ouano, J. Polymer Sci., A-2, 9, 377 (1971).

1149. A. C. Ouano, J. Polymer Sci., B, 9, 2179 (1971).

1150. A. C. Ouano, A. Broido, E. M. Barrall II, A.C. Javier-Son, Polymer Preprints 12(2), 859 (1971).

1151. A. C. Ouano, P. L. Mercier, Polymer Preprints 8(2), 1389 (1967).

1152. A. C. Ouano, P. L. Mercier, J. Polymer Sci., C, 21, 309 (1968).

1153. H. S. Owens, J. C. Miers, W. D. Maclay, J. Colloid. Sci. 3, 277 (1948).

1154. A. Pacter, K. A. Sharif, J. Polymer Sci., B, 9, 435 (1971).

1155. F. Pailhes, M. Alliot-Lugaz, N. Duveau, J. Kyritsos, J. Polymer Sci., C, 16, 1177 (1967).

1156. R. Panaris, G. Pallas, J. Polymer Sci.,B, 8, 441 (1970).

1157. R. Panaris, A. Peyrouset, IUPAC Helsinki, Vol. III, Preprint II-45 (1972).

1158. R. Panaris, A. Peyrouset, IUPAC Helsinki, Vol. III, Preprint II-46 (1972).

1159. J. Pannell, Polymer 13, 2 (1972).

1160. C. J. Panton, P. H. Plesch, P. P. Rutherford, J. Chem. Soc. 2586 (1964).

1161. P. Parrini, F. Sebastiano, G. Messina, Makromol. Chem. 38, 27 (1960).

1162. F. Patat, J. Klein, Makromol. Chem. 93, 230 (1966).

1163. C. K. Patel, R. D. Patel, Makromol. Chem. 128, 157 (1969).

1164. C. K. Patel, R. D. Patel, Stärke 21, 166 (1969).

1165. C. K. Patel, R. D. Patel, Makromol. Chem. 131, 281 (1970).

1166. J. R. Patel, Makromol. Chem. 134, 263 (1970).

1167. J. R. Patel, C. K. Patel, R. D. Patel, Stärke 19, 330 (1967).

1168. J. R. Patel, R. D. Patel, Polymer 10, 167 (1969).

1169. V. M. Patel, C. K. Patel, K.C. Patel, R. D. Patel, Angew. Makromol. Chem. 18, 39 (1971).

1170. E. Patrone, U. Bianchi, Makromol. Chem. 94, 52 (1966).

1171. S. K. Patra, S. Ghosh, B. K. Patnaik, R. T. Thampy, Chem. Soc. (London) Spec. Publ. 23, 233 (1968).

1172. A. V. Paulov, V. G. Aldoshin, S. Ya. Frenkel, Polymer Sci. USSR 6, 1773 (1964); Vysokomolekul. Soedin. 6, 1600 (1964).

1173. A. V. Paulov, S. Ye. Bresler, S. R. Rafikov, Polymer Sci. USSR 6, 2293 (1964); Vysokomolekul. Soedin. 6, 2068 (1964).

1174. S. A. Paulova, V. I. Pakhomov, I. I. Tverdokhlebova, Polymer Sci. USSR 6, 1408 (1964); Vysokomolekul. Soedin. 6, 1275 (1964).

1175. S. A. Paulova, T. A. Soboleva, A. P. Suprun, Polymer Sci. USSR 6, 144 (1964); Vysokomolekul. Soedin. 6, 122 (1964).

1176. P. Paulusma, D. Vermaas, J. Polymer Sci., C, 2, 488 (1963).

1177. W. A. Pavelich, R. A. Livigni, Polymer Preprints 8(2), 1343 (1967).

1178. W. A. Pavelich, R. A. Livigni, J. Polymer Sci., C, 21, 215 (1968).

1179. F. W. Peaker, J. M. Patel, J. Appl. Polymer Symp. 8, 125 (1969).

1180. F. W. Peaker, J. C. Robb, Nature (London) 182, 1591 (1958).

1181. F. W. Peaker, M. G. Rayner, European Polymer J. 6, 107 (1970).

1182. H. C. Peer, Rec. Trav. Chim. 78, 631 (1959).

1183. M. Pegoraro, Chim. Ind. (Milan) 44, 18 (1962).

1184. A. J. Pennings, J. Polymer Sci., C, 16, 1799 (1967).

1185. A. J. Pennings, A. M. Kiel, Kolloid. Z. - Z. Polymere 205, 160 (1965).

1186. E. Penzel, G. V. Schulz, Makromol. Chem. 113, 64 (1968).

1187. D. C. Pepper, P. P. Rutherford, J. Appl. Polymer Sci. 2, 100 (1959).

1188. J. M. Pereña, E. Riande, G. M. Guzmán, Anales Quím. (Madrid) 68, 441 (1972).

1189. R. Perret, A. Skoulius, Makromol. Chem. 152, 291 (1972).

1190. T. D. Perrine, P. F. Goolsby, J. Polymer Sci., A-1, 3, 3031 (1965).

1191. N. V. Pertsova, K. A. Andrianov, I. I. Tverdokhlebova, S. A. Paulova, Polymer Sci. USSR 12, 1131 (1970); Vysokomolekul. Soedin. 12 A, 1001 (1970).

1192. V. Petrus, Collection Czech. Chem. Commun. 33, 119 (1968).

1193. V. Petrus, I. Danihel, M. Bohdanecký, European Polymer J. 7, 143 (1971).

1194. J. Peyroche, Y. Girard, R. LaPutte, A. Guyot, Makromol. Chem. 129, 215 (1969).

1195. G. Pezzin, G. Sanmartin, F. Zilio-Grandi, J. Appl. Polymer Sci. 11, 1539 (1967).

1196. G. Pezzin, G. Talamini, G. Vidotto, Makromol. Chem. 43, 12 (1961).

1197. B. Pfannemueller, W. Burchard Makromol. Chem. 121, 1 (1969).

1198. B. Pfannemueller, H. Mayerhoefer, R. C. Schulz, Makromol. Chem. 121, 147 (1969).

1199. H. J. Phillipp, C. F. Bjork, J. Polymer Sci. 6, 383, 549 (1951).

1200 J. Phudhomme, S. Bywater, Polymer Preprints 10(2), 518 (1969).

1201. J. Di. Pietro, A. Di. Edwardo, Makromol. Chem. 98, 275 (1966).

1202. P. Pino, F. Ciardelli, G. P. Lorenzi, Makromol. Chem. 70, 182 (1964).

1203. P. Pino, G. Montagnoli, F. Ciardelli, E. Benedetti, Makromol. Chem. 93, 158 (1966).

1204. C. U. Pittman, J. C. Lai, D. P. Vanderpool, M. Good, Macromolecules 3, 746 (1970).

1205. C. U. Pittman, R. L. Voges, W. R. Jones, Macromolecules 4, 291 (1971).

1206. C. U. Pittman, R. L. Voges, J. Eider Macromolecules 4, 302 (1971).

1207. C. U. Pittman, M. Good, J. Lai, D. Vanderpool, in "Polymer Characterization", Editor C. D. Craver, Plenum Press, New York, p. 97 (1971).

1208. M. Pizzoli, G. Ceccorulli, European Polymer J. 8, 769 (1972).

1209. M. Pizzoli, G. Stea, G. Ceccorulli, G. B. Gechele, European Polymer J. 6, 1219 (1970).

1210. K. E. Platzer, V. S. Ananthanarayanan, R. H. Andreatta, H. A. Scheraga, Macromolecules 5, 177 (1972).

1211. I. Ya. Poddubnyi, V. A. Grechanovskii, E. G. Ehrenburg, Makromol. Chem. 94, 268 (1966).

1213. I. Ya. Poddubnyi, A. V. Podalinskii, Polymer Sci. USSR 11, 450 (1969); Vysokomolekul. Soedin. 11 A, 400 (1969).

1214. I. Ya. Poddubnyi, A. V. Podalinskii, V. A. Grechanovskii, Polymer Sci. USSR 8, 1715 (1966); Vysokomolekul. Soedin. 8, 1556 (1966).

1215. I. Ya. Poddubnyi, M. A. Rabinezzon, J. Appl. Polymer Sci. 9, 2527 (1965).

1216. J. Poláček, Collection Czech. Chem. Commun. 28, 1838 (1963).

1217. J. Poláček, Collection Czech. Chem. Commun. 28, 3011 (1963).

1218. J. Poláček, European Polymer J. 6, 81 (1970).

1219. J. Poláček, I. Koessler, J. Vodehnal, J. Polymer Sci., A-1, 3, 2511 (1965).

1220. J. Poláček, L. Sculz, I. Koessler, J. Polymer Sci., C, 16, 1327 (1967).

1221. D. J. Pollock, L. J. Elyash, T. W. De Witt, J. Polymer Sci. 15, 87 (1955).

1222. F. P. Price, J. P. Bianchi, J. Polymer Sci. 15, 355 (1955).

1223 V. M. Polyakova, A. Ye. Fainerman, R. T. Voitsekhouskii, Polymer Sci. USSR 6, 479 (1964); Vysokomolekul. Soedin. 6, 432 (1964).

1224. G. R. Polyakova, N. A. Pravikova, Polymer Sci. USSR 9, 1578 (1967); Vysokomolekul. Soedin. 9, 1405 (1967).

1225. V. A. Ponomarenko, A. M. Khomutov, A. P. Alimov, Polymer Sci. USSR 10, 1203 (1968); Vysokomolekul. Soedin. 10 A, 1038 (1968).

1226. V. A. Ponomarenko, A. M. Khomutov, S. I. Il' chenko, G. N. Goseshkova, V. S. Bogdanov, Polymer Sci. USSR 11, 206 (1969); Vysokomolekul. Soedin. 11 A, 182 (1969).

1227. M. T. Pope, T. J. Weakley, R. J. P. Williams, J. Chem. Soc. 3442 (1959).

1228. R. S. Porter, M. J. R. Cantow, J. F. Johnson, J. Appl. Polymer Sci. 11, 335 (1967).

1229. R. S. Porter, M. J. R. Cantow, J. F. Johnson, Makromol. Chem. 94, 143 (1966).

1230. R. S. Porter, J. F. Johnson, Polymer 5, 201 (1964).

1231. J. Pouradier, A. M. Venet, J. Chim. Phys. 47, 11 (1950).

1232. E. Powell, Polymer 8, 211 (1967).

1233. P. O. Powers, Ind. Eng. Chem. 42, 2558 (1950).

1234. N. A. Pravikova, Y. U. Ryabova, P. Vyrskyi, Vysokomolekul. Soedin. 5, 1165 (1963).

1235. R. B. Prime, B. Wunderlich, J. Polymer Sci., A-2, 7, 2061 (1969).

1236. T. Provder, J. H. Clark, E. E. Drott, Polymer Preprints 12(2), 819 (1971).

1237. T. Provder, J. C. Woodbrey, J. H. Clark, Separ. Sci. 6, 101 (1971).

1238. J. R. Purdon, R. D. Mate, J. Polymer Sci., A-1, 6, 243 (1968).

1239. O. Quadrat, Collection Czech. Chem. Commun. 35, 2564 (1970).

1240. O. Quadrat, Collection Czech. Chem. Commun. 36, 2042 (1971).

1241. O. Quadrat, M. Bohdanecký, Collection Czech. Chem. Commun. 33, 2130 (1968).

1242. D. V. Quayle, Polymer 8, 217 (1967).

1243. W. Rabel, K. Ueberreiter, Kolloid. Z. - Z. Polymere 198, 1 (1964).

1244. S. R. Rafikov, V. V. Korshak, G. N. Chelnokova, Izv. Akad. Nauk SSSR, 642 (1948).

1245. G. Rafler, G. Reinisch, Angew. Makromol. Chem. 20, 57 (1971).

1246. A. Ram, J. Miltz, J. Appl. Polymer Sci. 15, 2639 (1971).

1247. A. Ram, J. Miltz, IUPAC Helsinki, Vol. III, Preprint II-47 (1972).

1248. N. K. Raman, J. J. Hermans, J. Polymer Sci. 35, 71 (1959).

1249. M. R. Rao, V. Kalpagam, J. Sci. Ind. Res. (India) 20, 207 (1961).

1250. S. P. Rao, M. Santappa, J. Polymer Sci., A-1, 6, 95 (1968).

1251. N. S. Rapp, J. D. Ingham, J. Polymer Sci., A-2, 689 (1964).

1252. G. B. Rathmann, F. A. Bovey, J. Polymer Sci. 15, 544 (1955).

1253. A. Ravve, J. T. Khamis, J. Polymer Sci. 61, 185 (1962).

1254. A. Ravve, J. T. Khamis, L. X. Mallavarapu, J. Polymer Sci., A-1, 3, 1775 (1965).

1255. M. G. Rayner, Polymer 10, 827 (1969).

1256. O. Redlich, A. L. Jacobson, W. H. McFadden, J. Polymer Sci., A-1, 393 (1963).

1257. G. Reinisch, G. Rafler, G. I. Timofejewa, Angew. Makromol. Chem. 7, 110 (1969).

1258. P. Rempp, Polymer Preprints 7(1), 141 (1966).

1259. A. Revillon, L. E. St. Pierre, J. Appl. Polymer Sci. 14, 373 (1970).

1260. G. E. J. Reynolds, J. Polymer Sci., C, 16, 3957 (1968).

1261. E. Riande, M. Alonso, J. Mª G. Fatou, Rev. Plasticos (Madrid) 186, 1834 (1971).

1262. G. N. Richards, J. Appl. Polymer Sci. 5, 540 (1961).

1263. J. K. Rieke, G. M. Hart, J. Polymer Sci., C, 1, 117 (1963).

1264. R. Rigamonti, E. Meda, Ricerca. Sci. 25, 457 (1955).

1265. M. Rinaudo, J. P. Merle, European Polymer J. 6, 41 (1970).

1266. W. Ring, H.-J. Cantow, Makromol. Chem. 89, 138 (1965).

1267. W. Ring, H.-J. Cantow, W. Holtrup, European Polymer J. 2, 151 (1966).

1268. W. Ring, W. Holtrup, Makromol. Chem. 103, 83 (1967).

1269. M. Riou, R. Pibarot, Rev. Gen. Caoutchouc 27, 596 (1950).

1270. V. V. Rode, P. N. Gribkova, V. V. Korshak, Polymer Sci. USSR 11, 60 (1969); Vysokomolekul. Soedin. 11 A, 57 (1969).

1271. C. L. Rohn, J. Polymer Sci., A-2, 5, 547 (1967).

1272. A. Roig, J. E. Figueruelo, E. Llano, J. Polymer Sci., B, 3, 171 (1965).

1273. A. Roig, J. E. Figueruelo, E. Llano, J. Polymer Sci.. C, 16, 4141 (1968).

1274. A. Roig, E. Llano, J. E. Figueruelo, Anales Quím. (Madrid) 67, 699 (1971).

1275. A. Roig, E. Llano, J. E. Figueruelo, I. Katime, Anales Quím. (Madrid) 67, 343 (1971).

1276. J. E. L. Roovers, S. Bywater, Polymer Preprints 12(1), 290 (1971).

1277. J. L. Rosenberg, C. O. Beckmann, J. Colloid. Sci. 3, 483 (1948).

1278. A. J. Rosenthal, B. B. White, Ind. Eng. Chem. 44, 2693 (1952).

1279. W. E. Roseveare, L. Poore, Ind. Eng. Chem. 45, 2518 (1953).

1280. J. H. Ross, M. E. Casto, J. Polymer Sci., C, 21, 143 (1968).

1282. B. A. Rozenberg, N. V. Makletsova, I. V. Epel' baum, Ye. B. Lyudvig, S. S. Medvedev, Polymer Sci. USSR 7, 1163 (1965); Vysokomolekul. Soedin. 7, 1051 (1965).

1283. A. Rudin, G. W. Bennett, J. R. Mc Laren, J. Appl. Polymer Sci. 13, 2371 (1969).

1284. J. R. Runyon, D. E. Barnes, J. F. Rudd, L. H. Tung, J. Appl. Polymer Sci. 13, 2359 (1969).

1285. A. M. Ruskin, G. Parravano, J. Appl. Polymer Sci. 8, 565 (1964).

1286. C. A. Russell, J. Appl. Polymer Sci. 4, 219 (1960).

1287. L. G. Ryabova, Z. Ya. Berestneva, N. A. Pravikova, Polymer Sci. USSR 7, 1976 (1965); Vysokomolekul. Soedin. 7, 1796 (1965).

1288. M. U. Sadykov, U. Azizov, Kh. V. Usmanov, T. M. Mirkamilov, Polymer Sci. USSR 10, 376 (1968); Vysokomolekul. Soedin. 10 A, 322 (1968).

1289. S. Saeda, J. Yotsuyanagi, K. Yamaguchi, J. Appl. Polymer Sci. 15, 277 (1971).

1290. N. Sakota, N. Nakamura, K. Nishihara, Makromol. Chem. 129, 56 (1969).

1291. I. Sakurada, S. Matuzawa, Y. Kubota, Makromol. Chem. 68, 115 (1963).

1292. I. Sakurada, A. Nakajima, K. Shibatani, Polymer Sci., A, 2, 3545 (1964).

1293. R. Salovey, M. Y. Hellman, J. Polymer Sci., B, 5, 647 (1967).

1294. R. Salovey, M. Y. Hellman, J. Polymer Sci., A-2, 5, 333 (1967).

1295. G. V. Samsonov, Usp. Khim. 30, 1410 (1961).

1296. H. Sato, Bull. Chem. Soc. Japan 39, 2335 (1966).

1297. K. S. Sastry, R. D. Patel, European Polymer J. 5, 79 (1969).

1298. M. I. Savitskaya, S. Ya. Frenkel, Zh. Fiz. Khim. 32, 1063 (1958).

1299. W. Schabel, E. Schamberg, Makromol. Chem. 104, 9 (1967).

1300. P. C. Scherer, R. D. McNeer, Rayon Synth. Text. 30, 56 (1949).

1301. P. C. Scherer, B. P. Rouse, Rayon Synth. Text. 29, 55, 85 (1948).

1302. P. C. Scherer, R. B. Thompson, Rayon Synth. Text. 31, 51 (1950).

1303. E. Schiedermaier, J. Klein, Angew. Makromol. Chem. 10, 169 (1970).

1304. C. Schildknecht, S. Gross, H. Davidson, Ind. Eng. Chem. 40, 2104 (1948).

1305. A. Schindler, Monatsh. Chem. 95, 868 (1964).

1306. A. Schindler, R. B. Strong, Makromol. Chem. 114, 77 (1968).

1307. S. Schlick, M. Levy, J. Phys. Chem. 64, 883 (1960).

1308. P. J. Schmitt, G. V. Schulz, Makromol. Chem. 121, 184 (1969).

1309. H. Schnecko, W. Dost, W. Kern, Makromol. Chem. 121, 159 (1969).

1310. N. S. Schneider, Anal. Chem. 33, 1829 (1961).

1311. N. S. Schneider, J. Polymer Sci., C, 8, 179 (1965).

1312. N. S. Schneider et al., J. Polymer Sci. 37, 551 (1959).

1313. N. S. Schneider, L. G. Holmes, J. Polymer Sci. 38, 552 (1959).

1314. N. S. Schneider, J. D. Loconti, L. G. Holmes, J. Appl. Polymer Sci. 3, 251 (1960).

1315. N. S. Schneider, J. D. Loconti, L. G. Holmes, J. Appl. Polymer Sci. 5, 354 (1961).

1316. N. S. Schneider, R. J. Traskos, A. S. Hoffman, J. Appl. Polymer Sci. 12, 1567 (1968).

1317. W. Scholtan, Makromol. Chem. 24, 83 (1957).

1318. W. Scholtan, Makromol. Chem. 24, 104 (1957).

1319. W. Scholtan, Makromol. Chem. 36, 162 (1960).

1320. W. Scholtan, D. Kranz, Makromol. Chem. 110, 150 (1967).

1321. W. Scholtan, F. J. Kwoll, Makromol. Chem. 151, 33 (1972).

1322. W. Scholtan, H. Lange, S. Y. Lie, K. Dinges, R. Meyer-Mader, Angew. Makromol. Chem. 14, 43 (1970).

1323. W. Scholtan, H. Marzolph, Makromol. Chem. 55, 52 (1962).

1324. Th. G. Scholte, J. Polymer Sci., A-2, 6, 111 (1968).

1325. Th. G. Scholte, European Polymer J. 6, 51 (1970).

1326. G. V. Schulz, in "Die Physik der Hochpolymeren", Editor H.A. Stuart, Springer, Berlin, Vol. 2 (1953).

1327. G. V. Schulz, K. C. Berger, A. G. R. Scholz, Ber. Bunsenges. Phys. Chem. 68, 856 (1965).

1328. G. V. Schulz, M. Marx-Figini, Makromol. Chem. 14, 52 (1954).

1329. G. V. Schulz, E. Penzel, Makromol. Chem. 112, 260 (1968).

1330. G. V. Schulz, J. Romatowski, Makromol. Chem. 85, 195 (1965).

1331. G. V. Schulz, A. G. R. Scholz, R. V. Figini, Makromol. Chem. 57, 220 (1962).

1332. G. V. Schulz, W. Wunderlich, R. Kirte, Makromol. Chem. 75, 22 (1964).

1333. R. C. Schulz, E. Mueller, W. Kern, Makromol. Chem. 30, 39 (1959).

1334. R. C. Schulz, G. Renner, A. Henglein, W. Kern, Makromol. Chem. 12, 20 (1954).

1335. R. C. Schulz, N. Vollkommer, Makromol. Chem. 116, 288 (1968).

1336. W. W. Schulz, W. C. Purdy, Anal. Chem. 35, 2044 (1963).

1337. J. Schurz, Oesterr. Chemiker Ztg. 56, 311 (1955).

1338. J. Schurz, M. Rebek, H. Spoerk, Angew. Makromol. Chem. 1, 42 (1967).

1339.   J. Schurz, Th. Steiner, H. Streitzig, Makromol. Chem. 23, 141 (1957).

1340.   H. Schuster, M. Hoffmann, K. Dinges, Angew. Makromol. Chem. 9, 35 (1969).

1341.   H. J. L. Schuurmans, J. Polymer Sci. 57, 557 (1962).

1342.   R. M. Screaton, in "Newer Methods of Polymer Characterization", Editor B. Ke, Interscience, New York (1964).

1343.   R. M. Screaton, R. W. Seemann, Polymer Preprints 8(2), 1379 (1967).

1344.   R. M. Screaton, R. W. Seemann, J. Polymer Sci., C, 21, 297 (1968).

1345.   R. M. Screaton, R. W. Seemann, J. Appl. Polymer Symposia 8, 81 (1969).

1346.   J. Sebban-Danon, J. Polymer Sci. 48, 121 (1960).

1347.   B. Sébille, J. Néel, J. Chim. Phys. 60, 475 (1963).

1348.   B. See, T. G. Smith, J. Appl. Polymer Sci. 10, 1625 (1966).

1349.   B. See, T. G. Smith, European Polymer J. 7, 727 (1971).

1350.   L. Segal, J. Polymer Sci., B, 4, 1011 (1966).

1351.   L. Segal, Polymer Preprints 8(2), 1365 (1967).

1352.   L. Segal, J. Polymer Sci., C, 21, 267 (1968).

1353.   L. Segal, J. D. Timpa, J. I. Wadsworth, J. Polymer Sci., A-1, 8, 3577 (1970).

1354.   L. Segal, J. D. Timpa, J. I. Wadsworth, J. Polymer Sci., A-1, 8, 25 (1970).

1355.   E. Selegny, M. Vert, European Polymer J. 7, 1307 (1971).

1356.   F. R. Senti et al., J. Polymer Sci. 17, 527 (1955).

1357.   M. T. Shaw, F. Rodriguez, Polymer Preprints 7(2), 1053 (1966).

1358.   P. J. Sheth, J. F. Johnson, R. S. Porter, Polymer Preprints 12(2), 513 (1971).

1359.   J. Shimura, Bull. Chem. Soc. Japan 40, 273 (1967).

1360.   J. Shimura, I. Mita, H. Kambe, J. Polymer Sci., B, 2, 403 (1964).

1361.   K. Shirayama, T. Matsuda, S. I. Kita, Makromol. Chem. 147, 155 (1971).

1363.   M. F. Shostakovskii, U. Z. Annenkova, A. K. Khaliulli, A. I. Inyotkin, E. A. Gaitseva, V. N. Salaurov, Polymer Sci. USSR 11, 2257 (1969); Vysokomolekul. Soedin. 11 A, 1979 (1969).

1364.   A. R. Shultz, J. Am. Chem. Soc. 76, 3422 (1954).

1365.   A. R. Shultz, A. L. Bridgman, E. M. Hadsell, C. R. McCullough, J. Polymer Sci., A-2, 10, 273 (1972).

1366.   N. V. Shulyatikova, D. I. Mandel' baum, Zh. Prikl. Khim. 24, 264 (1951).

1367.   S. Shyluk, J. Polymer Sci. 62, 317 (1962).

1368.   S. T. Sie, G. W. A. Rijnders, Separ. Sci. 2, 729, 755 (1967).

1370.   R. Signer, H. Gross, Helv. Chim. Acta 17, 726 (1934).

1371.   H. Sihtola, E. Kaila, L. Laamanen, J. Polymer Sci. 23, 809 (1957).

1372.   H. Sihtola, E. Kaila, N. Virkola, Makromol. Chem. 11, 70 (1953).

1373.   C. C. Simionescu, E. Calistru, Faserforsch. Textiltech. 7, 171 (1956).

1376.   G. Sitaramaiah, J. Polymer Sci., A-1, 3, 2743 (1965).

1377.   G. Sitaramaiah, D. Jacobs, Polymer 11, 165 (1970).

1378.   D. F. Slonaker, R. L. Combs, H. W. Coover, J. Macromol. Sci. Chem. A 1, 539 (1967).

1379.   D. F. Slonaker, R. L. Combs, J. E. Guillet, H. W. Coover, J. Polymer Sci., A-2, 4, 523 (1966).

1380.   G. Smets, M. Claesen, J. Polymer Sci. 8, 289 (1952).

1381.   G. Smets, W. de Winter, G. Delzenne, J. Polymer Sci. 55, 767 (1961).

1382.   G. Smets, E. Dysseleer, Makromol. Chem. 91, 160 (1966).

1383.   G. Smets, A. Hertoghe, Makromol. Chem. 17, 189 (1956).

1384.   G. Smets, A. Poot, G. L. Duncan, J. Polymer Sci. 54, 65 (1961).

1385.   G. Smets, J. Roovers, W. van Humbeek, J. Appl. Polymer Sci. 5, 149 (1961).

1386.   W. B. Smith, Polymer Preprints 11(2), 1019 (1970).

1387.   W. B. Smith, J. A. May, Ch. W. Kim, J. Polymer Sci., A-2, 4, 365 (1966).

1388.   T. A. Soboleva, A. P. Suprun, S. A. Paulova, Polymer Sci. USSR 6, 104 (1964); Vysokomolekul. Soedin. 6, 89 (1964).

1389.   L. B. Sokolov, V. Z. Nikonov, G. N. Shilyakova, Polymer Sci. USSR 11, 699 (1969); Vysokomolekul. Soedin. 11 A, 616 (1969).

1390.   A. M. Sookne, M. Harris, Ind. Eng. Chem. 37, 475, 478 (1945).

1391.   M. F. Sorokin, M. M. Babkina, Polymer Sci. USSR 8, 122 (1966); Vysokomolekul. Soedin. 8, 115 (1966).

1392.   H. Sotobayashi, K. Ueberreiter, Z. Elektrochem. 67, 178 (1963).

1393.   T. N. Soundararajan, M. Wayman, IUPAC Toronto, Preprint A 11.3 (1968).

1394.   P. P. Spiegelman, G. Parravano, J. Polymer Sci., A-1, 2, 2245 (1964).

1395.   J. Springer, K. Ueberreiter, E. Moeller, Ber. Bunsenges. Phys. Chem. 69, 494 (1965).

1396.   J. Springer, K. Ueberreiter, W. Weinle, European Polymer J. 6, 87 (1970).

1397.   J. Springer, K. Ueberreiter, R. Wenzel, Makromol. Chem. 96, 122 (1966).

1398.   C. J. Stacy, J. F. Foster, J. Polymer Sci. 25, 39 (1957).

1399.   G. Stainsby, Discussions Faraday Soc. 18, 288 (1954).

1400.   G. D. Staffin, C. C. Price, J. Am. Chem. Soc. 82, 3632 (1960).

1401.   J. W. Stafford, Makromol. Chem. 134, 57 (1970).

1402.   J. W. Stafford, Makromol. Chem. 134, 73 (1970).

1403.   J. W. Stafford, Makromol. Chem. 134, 87 (1970).

1404.   J. W. Stafford, Makromol. Chem. 134, 99 (1970).

1405.   J. W. Stafford, Makromol. Chem. 134, 113 (1970).

1406.   H. Staudinger, T. Eiche, Makromol. Chem. 10, 235 (1953).

1407.   H. Staudinger, M. Haberle, Makromol. Chem. 9, 48 (1952).

1408.   R. Stefani, M. Chevreton, J. Terrier, C. Eyraud, Compt. Rend. 248, 2006 (1959).

1409.   Ya. Steiny, Polymer Sci. USSR 10, 2184 (1968); Vysokomolekul. Soedin. 10 A, 1883 (1968).

1410.   J. K. Stille, M. Kamachi, M. Kurihara, Polymer Preprints 12 (2), 223 (1971).

1411.   S. S. Stivala, R. J. Vallés, D. W. Levi, J. Appl. Polymer Sci. 7, 97 (1963).

1412.   W. H. Stockmayer et al., J. Polymer Sci. 16, 517 (1955).

1413.   C. Strazielle, Makromol. Chem. 119, 50 (1968).

1414.   T. Suminoe, N. Yamazoki, S. Kambara, Chem. High Polymers (Japan) 20, 461 (1963).

1415.   H. Sumitomo, K. Kobayashi, J. Polymer Sci., A-1, 4, 907 (1966).

1416.   H. A. Swenson, Acta Chem. Scand. 9, 572 (1955).

1417.   H. A. Swenson, J. A. Carlson, H. M. Kaustinen, J. Polymer Sci., C, 36, 293 (1971).

1418.   H. A. Swenson, H. M. Kaustinen, J. J. Bachhuber, J. A. Carlson, Macromolecules 2, 142 (1969).

1419.   H. A. Swenson, H. M. Kaustinen, O. A. Kaustinen, N. S. Thompson, J. Polymer Sci., A-2, 6, 1593 (1968).

1420.   H. A. Swenson, A. Rosenberg, Acta Chem. Scand. 10, 1393 (1956).

1421.   D. L. Swanson, J. W. Williams, J. Appl. Phys. 26, 810 (1955).

1422.   K. Tada, T. Saegusa, J. Furukawa, Makromol. Chem. 102, 47 (1967).

1423.   K. Tada, Y. Yamada, T. Saegusa, J. Furukawa, Makromol. Chem. 99, 232 (1966).

1424.   A. A. Tager, A. A. Anikeyeva, V. M. Andreyeva, T. Ya. Gomarova, L. A. Chemoskutova, Polymer Sci. USSR 10, 1926 (1968); Vysokomolekul. Soedin. 10 A, 1661 (1968).

1425.   A. A. Tager, A. A. Anikeyeva, L. V. Adamova, V. M. Andreyeva, J. A. Kuz' mina, M. V. Tsilipotkina, Polymer Sci. USSR 13, 751 (1971); Vysokomolekul. Soedin. 13 A, 659 (1971).

1426.   A. Takahashi, M. Obara, I. Kagawa, Kogyo Kagaku Zasshi 66, 960 (1963).

1427.   M. Takahashi, M. Watanabe, Sen-i-Gakkaishi 17, 111 (1961).

1428.   M. Takahashi, M. Watanabe, Sen-i-Gakkaishi 17, 122 (1961).

1429.   K. Takamizawa, Bull. Chem. Soc. Japan 39, 1186 (1966).

1430. M. Takeda, R. Endo, Y. Matsuura, J. Polymer Sci., C, 23, 487 (1968).

1431. G. Talamini, G. Vidotto, Makromol. Chem. 110, 111 (1967).

1432. S. Tanaka, A. Nakamura, H. Morikawa, Makromol. Chem. 85, 164 (1965).

1433. J. W. Tamblyn, D. R. Morey, R. H. Wagner, Ind. Eng. Chem. 37, 573 (1945).

1434. S. Tanaka, H. Morikawa, J. Polymer Sci., A-1, 3, 3147 (1965).

1435. L. J. Tanghe, W. J. Rebel, R. J. Brewer, J. Polymer Sci., A-1, 8, 2935 (1970).

1436. H. Tanzawa, T. Tanaka, A. Soda, J. Polymer Sci., A-2, 7, 929 (1969).

1437. S. A. Tashmukhamedov, P. P. Larin, R. S. Tillayev, Yu. T. Tash-pulatov, Kh. U. Usmanov, Polymer Sci. USSR 10, 2113 (1968); Vysokomolekul. Soedin. 10 A, 1823 (1968).

1438. S. A. Tashmukhamedov, P. P. Larin, R. S. Tillayev, Yu. T. Tash-pulatov, Kh. U. Usmanov, Polymer Sci. USSR 11, 515 (1969); Vysokomolekul. Soedin. 11 A, 453 (1969).

1439. G. B. Taylor, J. Am. Chem. Soc. 69, 638 (1947).

1440. D. L. Taylor, J. Polymer Sci., A-1, 2, 611 (1964).

1442. M. Tepelekian, P. Quang Tho, A. Guyot, European Polymer J. 5, 795 (1969).

1443. S. Teramachi, Y. Kato, J. Macromol. Sci. Chem. A 4, 1785 (1970).

1444. S. Teramachi, Y. Kato, Macromolecules 4, 54 (1971).

1445. S. Teramachi, M. Nagasawa, J. Macromol. Sci. Chem. A 2, 1169 (1968).

1446. S. Teramachi, H. Tomioka, M. Sotokawa, J. Macromol. Sci. Chem. A 6, 97 (1972).

1447. S. L. Terry, F. Rodriguez, Polymer Preprints 8(2), 1270 (1967).

1448. B. B. Thomas, W. J. Alexander, J. Polymer Sci. 15, 361 (1955).

1449. G. H. Thompson, M. N. Myers, G. Giddings, Anal. Chem. 41, 1219 (1969).

1450. J. O. Threlkeld, H. A. Ende, J. Polymer Sci., A-2, 4, 663 (1966).

1451. C. D. Thurmond, B. H. Zimm, J. Polymer Sci. 8, 477 (1952).

1452. T. E. Timell, Ind. Eng. Chem. 47, 2166 (1955).

1453. G. I. Timofeyeva, L. V. Dubrovina, V. M. Men' shov, Polymer Sci. USSR 12, 1560 (1970); Vysokomolekul. Soedin. 12 A, 1374 (1970).

1454. G. I. Timofeeva, S. A. Pavlova, IUPAC Helsinki, Vol. III, Pre-print II-64 (1972).

1456. M. Tomikawa, Chem. High Polymer (Japan) 20, 11 (1963).

1457. H. Tompa, "Polymer Solutions", Butterworths, London (1956).

1458. A. V. Topchiev, G. I. Chernyi, V. N. Andronov, Dokl. Akad. Nauk SSSR 143, 879 (1962).

1459. M. S. Toy, Polymer Preprints 12 (1), 385 (1971).

1460. M. S. Toy, J. M. Newman, J. Polymer Sci., A-1, 2333 (1969).

1461. A. H. Traaen, J. Appl. Polymer Sci. 7, 581 (1963).

1462. R. T. Traskos, N. S. Schneider, A. S. Hoffman, J. Appl. Polymer Sci. 12, 509 (1968).

1463. W. Trautvetter, Makromol. Chem. 101, 214 (1967).

1464. N. M. Tret' yakova, L. V. Kosmodem' yanskii, R. G. Romanova, E. G. Lazaryants, Polymer Sci. USSR 12, 3127 (1970); Vysokomole-kul. Soedin. 12 A, 2754 (1970).

1465. M. Tricot, V. Desreux, Makromol. Chem. 149, 185 (1971).

1466. E. Trommsdorff, H. Koehle, P. Lagally, Makromol. Chem. 1, 169 (1948).

1467. L. D. Trung, J. Mordini, P. Q. Tho, A. Guyot, European Polymer J. 6, 1187 (1970).

1468. M. V. Tsilipotkina, A. A. Tager, E. B. Makovskaya, U. Partina, Polymer Sci. USSR 12, 1222 (1970); Vysokomolekul. Soedin. 12 A, 1082 (1970)

1469. T. Tsuruta, S. Inoue, N. Yoshida, Y. Yokota, Makromol. Chem. 81, 191 (1965).

1470. T. Tsuruta, R. Fujio, J. Furukawa, Makromol. Chem. 80, 172 (1964).

1471. V. N. Tsvetkov et al., Polymer Sci. USSR 13 A, 705 (1971); Vysokomolekul. Soedin. 13 A, 620 (1971).

1472. V. N. Tsvetkov, K. A. Andrianov, G. I. Okhrimenko, I. N. Shten-nikova, G. A. Fomin, M. G. Vitovskaya, V. I. Pakhomov, A.A. Yarosh, D. N. Andreyev, Polymer Sci. USSR 12, 2146 (1970); Vysokomolekul. Soedin. 12 A, 1892 (1970).

1473. V. N. Tsvetkov, K. A. Andrianov, I. N. Shtennikova, G. I. Okhri-menkom, L. N. Andreyeva, G. A. Fomin, V. I. Pakhomov, Poly-mer Sci. USSR 10, 636 (1968); Vysokomolekul. Soedin. 10 A, 547 (1968).

1474. V. N. Tsvetkov, V. P. Budtov, Polymer Sci. USSR 6, 1332 (1964); Vysokomolekul. Soedin. 6, 1209 (1964).

1475. V. N. Tsvetkov, N. N. Boitsova, Vysokomolekul. Soedin. 5, 1263 (1963).

1476. V. N. Tsvetkov, G. A. Fomin, P. N. Lavrenko, I. N. Shtennikova, T. V. Sheremeteva, L. I. Godunova, Polymer Sci. USSR 10, 1051 (1968); Vysokomolekul. Soedin. 10 A, 93 (1968).

1477. V. N. Tsvetkov, N. N. Kupriyanova, G. V. Tarasova, P. N. Lavrenko, I. I. Migunova, Polymer Sci. USSR 12, 2238 (1970); Vysokomolekul. Soedin. 12 A, 1974 (1970).

1478. V. N. Tsvetkov, S. Ya. Lyubina, T. V. Barskaya, Polymer Sci. USSR 6, 886 (1964); Vysokomolekul. Soedin. 6, 806 (1964).

1479. V. N. Tsvetkov, S. Ya. Lyubina, V. Ye. Bychkovaand, I. A. Strelina, Polymer Sci. USSR 8, 928 (1966); Vysokomolekul. Soedin. 8, 846 (1966).

1480. V. N. Tsvetkov, Yu. V. Mitin, I. N. Shtennikova, V. R. Glushen-kova, G. V. Tarasova, V. S. Skazka, N. A. Nikitin, Polymer Sci. USSR 7, 1216 (1965); Vysokomolekul. Soedin. 7, 1098 (1965).

1481. V. N. Tsvetkov, I. N. Shtennikova, Ye. I. Ryumtsev, V. S. Skaz-ka, Polymer Sci. USSR 7, 1231 (1965); Vysokomolekul. Soedin. 7, 1111 (1965).

1482. V. N. Tsvetkov, I. N. Shtennikova, Ye. I. Ryumsev, G. I. Okhri-menko, Polymer Sci. USSR 7, 1223 (1965); Vysokomolekul. Soedin. 7, 1104 (1965).

1483. V. N. Tsvetkov, V. S. Skazka, G. V. Tarasova, V. M. Yamshchi-kov, S. Ya. Lyubina, Polymer Sci. USSR 10, 81 (1968); Vysokomo-lekul. Soedin. 10 A, 74 (1968).

1484. V. N. Tsvetkov, I. N. Shtennikova, E. I. Ryumsev, Yu. P. Getman-chuk, IUPAC Leiden, Vol. I, 51 (1970).

1485. V. N. Tsvetkov, V. S. Skazka, N. M. Krivoruchko, Vysokomolekul. Soedin. 2, 1045 (1960).

1486. L. H. Tung, J. Polymer Sci. 20, 495 (1956).

1487. L. H. Tung, J. Polymer Sci. 24, 333 (1956).

1488. L. H. Tung, S.P.E. Journal 14, 25 (1958).

1489. L. H. Tung, J. Macromol. Sci. Rev. Macromol. Chem. C 6, 51 (1971).

1490. L. H. Tung, J. Polymer Sci., A-2, 7, 47 (1969).

1491. L. H. Tung, J. C. Moore, G. W. Knight, J. Appl. Polymer Sci. 10, 1261 (1966).

1492. E. Turska, A. Dems, M. Siniarska, Bull. Acad. Polon. Sci. Ser. Sci. Chim. 13, 189 (1965).

1493. E. Turska, M. Laczkowski, J. Polymer Sci. 23, 285 (1957).

1494. E. Turska, T. Skwarsi, Zeszyty Nauk Politechn. Lodz Chem. 5, 21 (1957).

1495. E.Turska, M. Siniarska-Kapuścińska, European Polymer J. 5 (Suppl), 431 (1969).

1496. E. Turska, A. M. Wróbel, Polymer 11, 408 (1970).

1497. Z. Tuzar, P. Kratochvil, M. Bohdanecký, J. Polymer Sci., C, 16, 633 (1967).

1498. M. A. Twaik, M. Tahan, A. Zilkha, J. Polymer Sci., A-1, 7, 2469 (1969).

1499. K. Ueberreiter, W. Bruns, Ber. Bunsenges. Phys. Chem. 68, 541 (1964).

1500. K. Ueberreiter, R. Goetze, Makromol. Chem. 29, 61 (1959).

1501. K. Ueberreiter, J. Melsheimer, J. Springer, Kolloid. Z. - Z. Polymere 234, 989 (1969).

1502. K. Ueberreiter, J. Ortmann, G. Sorge, Makromol. Chem. 8, 21 (1952).

1503. K. Ueberreiter, J. Springer, Z. Phys. Chem. N.F. (Frankfurt) 36, 299 (1963).

1506. C. A. Uraneck, J. E. Burleigh, J. Appl. Polymer Sci. 9, 1273 (1965).

1507. C. A. Uraneck, J. E. Burleigh, Kautschuk, Gummi, Kunststoffe 19, 532 (1966).

1508. C. A. Uraneck, J. E. Burleigh, J. Appl. Polymer Sci. 14, 267 (1970).

1509. C. A. Uraneck, J. N. Short, J. Appl. Polymer Sci. 14, 1421 (1970).

1510. J. R. Urwin et al., Makromol. Chem. 72, 53 (1964).

1511. S. N. Ushakov, S. P. Mitsengendler, V. N. Krasulina, Izv. Akad. Nauk SSSR 3, 366 (1957).

1512. J. R. Urwin, J. M. Stearn, Makromol. Chem. 78, 194 (1964).

1513. U.S. Pat. 2,457,238 (1948).

1514. H. Utiyama, J. Phys. Chem. 69, 4138 (1965).

1515. L. Utracki, R. Simha, Makromol. Chem. 117, 94 (1968).

1516. R. H. Valentine, J. D. Ferry, T. Homma, K. Ninomiya, J. Polymer Sci., A-2, 6, 479 (1968).

1517. D. G. Val'kovskii, S. L. Sosin, V. V. Korshak, S. A. Pavlova, Polymer Sci. USSR 6, 2044 (1964); Vysokomolekul. Soedin. 6, 1848 (1964).

1518. R. J. Valles, Polymer Preprints 6(2), 1041 (1965).

1519. R. J. Valles, Makromol. Chem. 113, 147 (1968).

1520. R. J. Valles, Polymer Preprints 9(1), 752 (1968).

1521. R. J. Valles, E. C. Schramm, J. Polymer Sci., A-1, 3, 3853 (1965).

1522. I. Valyi, A. G. Janssen, H. F. Mark, J. Phys. Chem. 49, 461 (1945).

1523. E. J. Vandenberg, J. Polymer Sci., C, 1, 207 (1963).

1524. D. C. Van Landuyt, C. W. Huskins, J. Polymer Sci., B, 6, 643 (1968).

1525. R. Van Leemput, R. Stein, J. Polymer Sci., A-1, 985 (1963).

1526. J. Van Schooten, H. Van Hoorn, J. Boerma, Polymer 2, 161 (1961).

1527. J. Van Schooten, P. W. O. Wijga, Makromol. Chem. 43, 23 (1961).

1528. I. K. Varma, S. S. Grover, J. Appl. Polymer Sci. 14, 2965 (1970).

1529. N. I. Vasil'ev, V. I. Irzhak, V. I. Kartsovnik, N. S. Yenikolopyan, Polymer Sci. USSR 12, 2276 (1970); Vysokomolekul. Soedin. 12 A, 2006 (1970).

1530. P. Vasudevan, M. Santappa, Makromol. Chem. 137, 261 (1970).

1531. P. Vasudevan, M. Santappa, J. Polymer Sci., A-2, 9, 483 (1971).

1532. G. Vaughan, D. E. Eaves, W. Cooper, Polymer 2, 235 (1961).

1533. J. Veličković, J. Filipović, IUPAC Helsinki, Vol. III, Preprint II-67 (1972).

1534. J. Veličković, D. Jovanović, J. Vukajlović, Makromol. Chem. 129, 203 (1969).

1535. J. Veličković, S. Vasović, Makromol. Chem. 153, 207 (1972).

1536. K. Venkatarao, M. Santappa, J. Polymer Sci., A-1, 8, 1785 (1970).

1537. G. Vidotto, R. Zannetti, L. Cavalli, Makromol. Chem. 146, 159 (1971).

1538. H. Vink, R. Wikstron, Svensk Papperstid. 66, 55 (1963).

1539. R. E. Vogel, Kunststoffe 42, 17 (1952).

1540. R. E. Vogel, Kunststoffe 44, 335 (1954).

1542. P. H. Von Dreele, N. Lotan, V. S. Ananthanarayanan, R. H. Andreatta, D. Poland, H. A. Scheraga, Macromolecules 4, 408 (1971).

1543. M. J. Voorn, Fortschr. Hochpolymer Forsch. 1, 192 (1959).

1544. Yu. P. Vyrskii, O. A. Klapovskaya, N. V. Andrianova, O. F. Alkayeva, Polymer Sci. USSR 10, 1959 (1968); Vysokomolekul. Soedin. 10 A, 1688 (1968).

1545. J. I. Wadsworth, L. Segal, J. D. Timpa, Polymer Preprints 12(2), 854 (1971).

1546. R. H. Wagner, J. Polymer Sci. 2, 21 (1947).

1547. H. L. Wagner, P. J. Flory, J. Am. Chem. Soc. 74, 195 (1952).

1548. E. R. Wagner, R. J. Greff, J. Polymer Sci., A-1, 9, 2193 (1971).

1549. M. Wales, F. T. Adler, K. E. Van Holde, J. Phys. Colloid. Chem. 55, 145 (1951).

1550. M. Wales, S. J. Rehfeld, J. Polymer Sci. 62, 179 (1962).

1551. M. Wales, D. L. Swanson, J. Phys. Colloid. Chem. 55, 203 (1951).

1552. M. Wales, J. W. Williams, J. O. Thompson, R. H. Ewart, J. Phys. Chem. 52, 983 (1948).

1553. M. L. Wallach, Polymer Preprints 10(2), 1248 (1969).

1554. H. A. Wannow, F. Thormann, Kolloid. Z. 112, 94 (1949).

1557. J. L. Waters, Polymer Preprints 6(2), 1061 (1965).

1558. W. F. Watson, J. Polymer Sci. 13, 595 (1954).

1559. T. J. R. Weakley, R. J. P. Williams, J. D. Wilson, J. Chem. Soc. 3963 (1960).

1560. J. K. Weise, Polymer Preprints 12(1), 512 (1971).

1561. P. Weiss, J. Herz, P. Rempp, Z. Gallot, H. Benoit, Makromol. Chem. 145, 105 (1971).

1562. J. D. Wellons, A. Schindler, V. Stannett, Polymer 5, 499 (1964).

1563. J. D. Wellons, V. Stannett, Polymer Preprints 6(1), 368 (1965).

1564. F. Wenger, Makromol. Chem. 64, 151 (1963).

1565. F. Wenger, S. P. S. Yen, Makromol. Chem. 43, 1 (1961).

1566. H. Wesslau, Makromol. Chem. 26, 96, 102 (1958).

1567. H. Wesslau, Makromol. Chem. 20, 111 (1956).

1568. B. Wesslein, G. Gunneby, G. Hellstroem, P. Svedling, IUPAC Helsinki, Vol. I, Preprint 1/46 (1972).

1569. B. Westall, Polymer 9, 243 (1968).

1570. R. L. Whistler, H. J. Roberts, J. Org. Chem. 26, 2458 (1961).

1571. H. Whitaker, G. Hills, J. Appl. Polymer Sci. 13, 1921 (1969).

1572. D. M. White, Polymer Preprints 12(1), 155 (1971).

1573. B. A. Whitehouse, Macromolecules 4, 463 (1971).

1574. N. M. Wiederhorn, A. R. Brown, J. Polymer Sci. 8, 651 (1952).

1576. P. W. O. Wijga, J. Van Schooten, J. Boerma, Makromol. Chem. 36, 115 (1960).

1577. L. Wild, R. Guliaana, J. Polymer Sci., A-2, 5, 1087 (1967).

1578. L. Wild, R. Ranganath, T. Ryle, Polymer Preprints 12(1), 266 (1971).

1579. L. Wild, R. Ranganath, T. Ryle, Polymer Sci., A-2, 9, 2137 (1971).

1580. T. Williams, Y. Udagawa, A. Keller, I. M. Ward, J. Polymer Sci., A-2, 8, 35 (1970).

1581. F. Wiloth, Makromol. Chem. 8, 111 (1952).

1582. F. Wiloth, Makromol. Chem. 14, 156 (1954).

1583. M. Wirick, M. Waldman, J. Appl. Polymer Sci. 14, 579 (1970).

1584. U. Wittenberger, I. Klein, Angew. Makromol. Chem. 8, 133 (1969).

1585. E. Wolfram, M. Nagy, Kolloid. Z. - Z. Polymere 227, 86 (1968).

1586. D. P. Wyman, T. Altares, Makromol. Chem. 72, 68 (1964).

1587. D. P. Wyman, L. J. Elyash, W. J. Frazer, J. Polymer Sci., A-1, 3, 681 (1965).

1588. D. P. Wyman, I. H. Song, Makromol. Chem. 115, 64 (1968).

1589. A. I. Yakubchik, B. I. Tikhumirov, L. N. Mikhailova, Polymer Sci. USSR 7, 1728 (1965); Vysokomolekul. Soedin. 7, 1562 (1965).

1590. A. Yamada, Y. Yamamuro, M. Yanagita, Bull. Chem. Soc. Japan 35, 609 (1962).

1591. R. Yamadera, Ch. Sonoda, J. Polymer Sci., B, 3, 411 (1965).

1592. K. Yamaguchi, Makromol. Chem. 128, 19 (1969).

1593. K. Yamaguchi, Makromol. Chem. 132, 143 (1970).

1594. K. Yamaguchi, S. Saeda, J. Polymer Sci., A-2, 7, 1303 (1969).

1595. K. Yamamoto, H. Fujita, Polymer 8, 517 (1967).

1596. K. Yamamoto, H. Fujita, Polymer 7, 557 (1966).

1597. Y. Yamamoto, I. Noda, M. Nagasawa, Polymer J. 1, 304 (1970).

1598. Y. Yamashita, Polymer Preprints 13(1), 151 (1972).

1599. Y. Yamashita, M. Hirota, H. Matsui, A. Hirao, K. Nobutoki, Polymer J. 2, 43 (1971).

1600. Y. Yamashita, S. Miura, M. Nakamura, Makromol. Chem. 68, 31 (1963).

1601. Y. Yamashita, K. Nobutoki, Y. Nakamura, M. Hirota, Macromolecules 4, 548 (1971).

1602. K. Yamaura, Y. Hoe. S. Matsuzawa, Y. Go, Kolloid. Z. - Z. Polymere 243, 7 (1971).

1603. K. Yamaura, S. Kinugasa, S. Matsuzawa, Kolloid. Z. - Z. Polymere 248, 893 (1971).

1604.   K. Yamaura, S. Matsuzawa, Y. Go, Kolloid Z. - Z. Polymere <u>240</u>, 820 (1970).

1605.   N. Yamazaki et al., Kogyo Kagaku Zasshi <u>64</u>, 1687 (1961).

1606.   W. W. Yau, C. P. Malone, J. Polymer Sci., B, <u>5</u>, 663 (1967).

1607.   J. A. Yanko, J. Polymer Sci. <u>3</u>, 576 (1948).

1608.   J. A. Yanko, J. Polymer Sci. <u>22</u>, 153 (1956).

1609.   H. Yasuda, J. A. Wray, V. Stannett, J. Polymer Sci., C, <u>2</u>, 387 (1963).

1610.   S. J. Yeh, H. L. Frisch, J. Polymer Sci. <u>27</u>, 149 (1958).

1611.   S.-P. S. Yen, Makromol. Chem. <u>81</u>, 152 (1965).

1612.   V. V. Yevreinov, V. I. Gerbich, L. I. Sarynina, S. G. Entelis, Polymer Sci. USSR <u>12</u>, 938 (1970); Vysokomolekul. Soedin. <u>12 A</u>, 829 (1970).

1613.   K. Yoshida, Y. Yamashita, Makromol. Chem. <u>100</u>, 175 (1967).

1614.   A. A. Yulchibayev, Kh. U. Usmanov, A. Kh. Gufarov, A. A. Matya-kuvov, IUPAC Helsinki, Vol. II, Preprint 1/153 (1972).

1615.   Yu. S. Zaitsev, V. D. Yenal' ev, A. I. Yurzhenko, Polymer Sci. USSR <u>12</u>, 2831 (1970); Vysokomolekul. Soedin. <u>12 A</u>, 2500 (1970)

1616.   F. Zapf, Makromol. Chem. <u>3</u>, 164 (1949).

1617.   F. Zapf, Makromol. Chem. <u>10</u>, 35 (1953).

1618.   F. Zapf, Makromol. Chem. <u>10</u>, 61 (1953).

1619.   V. I. Zegel' man, M. N. Shlykova, S. V. Svetozarskii, Ye. N. Zil' berman, Polymer Sci. USSR <u>10</u>, 133 (1968); Vysokomolekul. Soedin. <u>10 A</u>, 114 (1968).

1620.   R. P. Zelinski, C. F. Wofford, J. Polymer Sci., A-1, <u>3</u>, 93 (1965).

1621.   A. C. Zettlemoyer, E. T. Pieski, Ind. Eng. Chem. <u>45</u>, 165 (1953).

1622.   S. Ya. Zhelobayeva, A. V. Troitskaya, T. N. Osipova, S. I. Kle-nin, A. F. Nikolayev, Polymer Sci. USSR <u>13</u>, 951 (1971); Vysoko-molekul. Soedin. <u>13 A</u>, 842 (1971).

1623.   M. Zief, G. Brunner, J. Metzendorff, Ind. Eng. Chem. <u>48</u>, 119 (1956).

1624.   J. G. Zilliox, P. Rempp, J. Parrod, J. Polymer Sci., C, <u>22</u>, 145 (1968).

1625.   A. Živný, J. Pouchlý, K. Solc, Collection Czech. Chem. Commun. <u>32</u>, 2753 (1967).

SOLVENTS AND NON-SOLVENTS FOR POLYMERS

O. Fuchs and H.-H. Suhr [*]

Hoechst AG

Frankfurt (Main)/Germany

Contents

A. INTRODUCTION

The tables contain qualitative data for a selected number of polymers. Since no standard definition for solvent- non-solvent-systems has been used in most of the original sources, the recognition of a certain compound as a solvent or non-solvent is to some extend influenced by personal interpretation. No attempt has been made to edit the original information. Division into only the two classes, solvents and non-solvents, is dictated by the practical point of view. For more quantitative information the user is referred to the tables of θ-solvents and fractionation of polymers in this Handbook.

The arrangement of polymers into classes is based on the chemical structure. Since properties change gradually within a series of homologues polymers were arranged according to increasing complexity regarding the size of the monomer unit and the kind and number of substituents. Only when this principle could not be applied was an alphabetical listing chosen. We believe that a typical solution behavior of similar polymers may be recognized more easily by this arrangement. When formulae are given, they refer to the main structures present in the polymers.

Copolymers have not been included in these tables. In their behavior they resemble more or less the properties of the homopolymer of the dominating monomer, although they generally exhibit higher solubilities than the corresponding homopolymers.

Solubility normally increases with rising temperature, however, negative temperature coefficients are observed.

---

[*] Based on a similar table of K. Meyersen, Hoechst AG, Frankfurt (Main), in the first edition.

Increase in molecular weight reduces solubility.

Increased branching increases the solubility compared to a linear polymer of the same molecular weight.

Properties change gradually within a series of homologous polymers as well as solvents. Solubility or dissolving power may increase, decrease, reach a maximum or minimum.

Certain combinations of two or more solvents may become non-solvents. Conversely mixtures of two or more non-solvents may sometimes become solvents. These possibilities should be considered if new solvent- non-solvent combinations are to be examined.

The classification of a certain compound as a non-solvent does not necessarily imply ability to act as a precipitant since this is influenced also by the nature of the particular solvent of a solvent- non-solvent pair. However, most non-solvents combine both properties.

The list of solvents and non-solvents for each polymer follows as simple arrangement by functional groups.

Homologues and closely related compounds generally have similar properties. When specific solvents or non-solvents are cited it is with the understanding that homologues and compounds with similar structures can be expected to exhibit similar properties. If class names are used, they refer to the most common compounds. Less common compounds, although falling into a class already mentioned, are additionally cited.

Water is a non-solvent for most polymers and is, therefore, only mentioned if similar polymers or derivatives are water soluble.

The data refer to room temperature unless otherwise stated.

The following abbreviations are used:

DMA	N, N-dimethylacetamide	dil.	diluted
DMF	N, N-dimethylformamide	conc.	concentrated
DMSO	dimethyl sulfoxide	aqu.	aqueous
HMTP	hexamethyltrisphosphoramide	elev.	elevated
TMS	tetramethylene sulfone	temp.	temperature
THF	tetrahydrofuran	S. C.	substituent content
sw	swelling	D. S.	degree of substitution
degrad.	degradation	mol. wt.	molecular weight

B. TABLES OF SOLVENTS AND NON-SOLVENTS

Polymer	Solvents	Non-solvents	References
1. Main-chain Acyclic Carbon Polymers			
1.1 Poly(dienes), Poly(acetylenes) (see also 0.1, 0.2)			
Poly(dienes) unsubstituted			
Poly(allene)	benzene, halogenated hydrocarbons.	hexane, methanol.	86
Poly(butadiene)	aliphatic, cycloaliphatic and aromatic hydrocarbons, chlorinated hydrocarbons, THF, higher ketones, higher aliphatic esters.	alcohol, lower ketones and esters, nitromethane, propionitrile, water, dil. acids, dil. alkalies, hypochlorite solutions.	2, 4, 12
Poly(isoprene)	see Poly(butadiene)		
Poly(dienes) substituted			
Poly(2-tert-butyl-1, 3-butadiene)	n-heptane, benzene, chloroform, carbon tetrachloride, diethyl ether, carbon disulfide.	acetone, alcohol.	51, 52
Poly(5, 7-dimethyl-1, 6-octadiene)	see Poly(2-tert-butyl-1, 3-butadiene)		
Poly(1-methoxybutadiene)			
crystalline		heptane, benzene, methanol, dioxane, acetone.	111
Poly(2-chlorobutadiene)	benzene, chlorinated hydrocarbons, chlorobenzene, dioxane, pyridine, cyclohexanone, ethyl acetate, cyclohexane/toluene.	aliphatic hydrocarbons, mineral oils, toluene (sw), alcohols, ketones, water, non-oxidizing conc. acids, incl. hydrogen fluoride.	1, 2, 4, 12
1, 4-cis	hexane, benzene, chloroform, carbon tetrachloride, ether, THF.	methanol, ethanol, acetone.	49
Poly(2-chloromethylbutadiene)	dichloromethane, THF, toluene.	methanol, ethanol.	62
Poly(perfluoro-1, 4-pentadiene)	hexafluorobenzene.		
Poly(acetylenes)			
Poly(acetylene)	isopropylamine, aniline, DMF.	cyclohexane, benzene, toluene, methylene chloride, carbon tetrachloride, methanol, pyridine, acetone, methyl ethyl ketone.	38
Poly(phenylacetylene)			
low mol. wt.	carbon tetrachloride, methanol, acetone.		35, 63
higher mol. wt.	benzene.	methanol.	

Polymer	Solvents	Non-solvents	References
Poly(acetylenes) (Cont'd.)			
Poly(diphenyldiacetylene)	cyclohexane, chloroform, dioxane.	methanol.	64
1.2 Poly(alkenes)			
Poly(methylene)	see Poly(ethylene)		
Poly(ethylene)			
high density	above 80°C: aliphatic, cycloaliphatic and aromatic hydrocarbons, halogenated aliphatic, cycloaliphatic and aromatic hydrocarbons, higher aliphatic esters and ketones, di-n-amyl ether.	all common organic solvents at room temperature, more polar organic solvents even at elevated temperatures, anorganic solvents.	1-4, 12
low density	as above, but temperature 20-30°C lower, depending on degree of branching.	as above.	4
Poly(propylene)			
atactic	hydrocarbons and chlorinated hydrocarbons at room temperature, isoamyl acetate, diethyl ether.	more polar organic solvents with small hydrocarbon group even at elevated temperature.	4
isotactic	see Poly(ethylene)		
Poly(1-butene), isotactic	see Poly(ethylene)		
Poly(isobutene)	aliphatic, cycloaliphatic and aromatic hydrocarbons, chlorinated aliphatic, cycloaliphatic and aromatic hydrocarbons, THF, dioxane, aliphatic ethers, anisole, higher esters, higher alcohols, β,β-dichlorodiethyl ether, carbon disulfide, ethylsulfide.	lower ketones and esters, lower alcohols, lower organic acids, nitromethane, propionitrile.	1, 3, 4, 12, 121
Poly(4-methyl-1-pentene), isotactic	see Poly(ethylene)		
Poly(cyclopentylethylene)	toluene, diethyl ether, chloroform.	methanol.	53
Poly(cyclohexylethylene)			
atactic	aliphatic, cycloaliphatic and aromatic hydrocarbons, chlorinated aliphatic and aromatic hydrocarbons, THF.	alcohols, ethers, dioxane, esters, ketones.	226
stereospecific	at elevated temperature: methyl cyclohexane, decahydronaphthalene, tetrahydronaphthalene, benzene, ethyl benzene, toluene, xylene, chlorobenzene, o-dichlorobenzene, trichlorobenzene.	heptane, methylethylketone, nitrobenzene.	227
Poly(cyclohexylalkene)s	toluene, diethyl ether (partially), chloroform.	methanol.	58
Poly(cyclohexenylethylene)			
atactic	aliphatic hydrocarbons.	ethanol, acetone.	217
isotactic	aromatic hydrocarbons, halogenated hydrocarbons (partially).	aliphatic hydrocarbons, ethanol, acetone, diethylether.	217
Poly(pentenamer) (Poly(cyclopentene))			
⟨CH = CH(CH$_2$)$_3$⟩$_n$	cycloaliphatic hydrocarbons, aromatic hydrocarbons, chlorinated hydrocarbons.	alcohols, ethers, aliphatic ketones.	131
Poly(ethylene), chlorinated, 40 % Cl	at elevated temperature: tetrahydronaphthalene, toluene, xylene, tetrachloroethane, chlorobenzene, cyclohexanone.	methylene chloride, chloroform, methanol, ethanol, butanol, dioxane, THF, acetone, methyl ethyl ketone, methylacetate.	239
Poly(ethylene), chlorinated, 60 % Cl	tetrahydronaphthalene, benzene, toluene, methylene chloride, chloroform, dioxane, THF, cyclohexanone, acetone/carbon disulfide 1 : 1.	aliphatic and cycloaliphatic hydrocarbons, methanol, ethanol, acetone, methyl acetate.	
1.3 Poly(acrylics), Poly(methacrylics)			
1.3.1 Poly(acrylic acids)			
Poly(acrylic acid)			
atactic	methanol, ethanol, ethylene glycol, methoxyethanol, dioxane, formamide, DMF, water, dil. alkali solutions.	aliphatic and aromatic hydrocarbons, esters, ketones, dioxane at higher temperature (sw).	1, 2, 4, 7, 12
isotactic	dioxane/water (80 : 20).	dioxane.	8
Poly(methacrylic acid)	water, aqu. hydrogen chloride (0.002 M, above 30°C), dil. aqu. sodium hydroxide.	hydrocarbons, alcohols, ketones, carboxylic acids, esters.	2, 4
isotactic, syndiotactic	water (partially)		9

Polymer	Solvents	Non-solvents	References
1.3.1 Poly(acrylic acids) (Cont'd.)			
Poly(itaconic acid)	methanol, DMF, water.	benzene, chloroform, carbon tetrachloride, ethanol, THF, dioxane, aniline, acetone, ethyl acetate, carbon disulfide.	115
1.3.2 Poly(acrylates)			
General	aromatic hydrocarbons, chlorinated hydrocarbons, THF, esters, ketones.	aliphatic hydrocarbons, hydrogenated naphthalenes, diethyl ether.	
Poly(methyl acrylate)	see general, and dimethyl tetrahydrofuran, glycolic ester ethers, phosphorus trichloride.	methanol, ethanol, 2-alkoxy ethanols, carbon tetrachloride.	3, 4, 50
Poly(ethyl acrylate)	see general, and methanol, ethanol, butanol, dimethyl tetrahydrofuran, glycol ether.	aliphatic alcohols $C \geq 5$, cyclohexanol, tetrahydrofurfuryl alcohol.	3, 4, 50
Poly(butyl acrylate)	see general, and turpentine, butanol.	methanol, ethanol, cyclohexyl acetate, ethyl acetate.	3
Poly(5-cyano-3-thia-pentyl acrylate)	dioxane, pyridine, acetone, acetonitrile.	solvents of low solubility parameter.	70
1.3.3 Poly(methacrylates)			
General	benzene, toluene, xylene, methylene chloride, chloroform, ethylene chloride, chlorobenzene, isobutanol (hot), cyclohexanol (hot), β-ethoxyethanol, dioxane, methyl ethyl ketone, diisopropyl ketone, cyclohexanone, acetic acid, isobutyric acid, methyl formate, ethyl acetate, cyclohexyl acetate, isobutyl propionate, butyl lactate.	hexane, cyclohexane, gasoline, nujol, castor oil, methanol, ethylene glycol, glycerol, formamide.	
Poly(methyl methacrylate)	see general and ethanol/water, ethanol/carbon tetrachloride, isopropanol/methyl ethyl ketone 1 : 1 above 25 °C, formic acid, nitroethane.	turpentine, linseed oil, hydrogenated naphthalenes, carbon tetrachloride, ethanol (absolute), butylene glycol, diethyl ether, isopropyl ether, higher esters, m-cresol.	1-4, 14, 122
Poly(ethyl methacrylate)	see general and tetraline, turpentine (hot), carbon tetrachloride, ethanol (hot), isopropanol above 37 °C, ethyl ether, formic acid.	linseed oil, isopropyl ether.	4, 122
Poly(n-propyl methacrylate)	see general and cyclohexane (hot), gasoline (hot), turpentine, castor oil (hot), linseed oil (hot), carbon tetrachloride, ethanol, diethyl ether, isopropyl ether, acetone.	formic acid.	122
Poly(n-butyl methacrylate) and	Poly(isobutyl methacrylate) see general and hexane, cyclohexane, gasoline, nujol (hot), turpentine, castor oil (hot), linseed oil (hot), carbon tetrachloride, ethanol (hot), isopropanol above 23.7 °C, diethyl ether, isopropyl ether, acetone.	formic acid.	1, 3, 4, 122
Poly(n-hexyl methacrylate)	isopropanol above 33 °C, methyl ethyl ketone.		4
Poly(2-ethylbutyl methacrylate)	isopropanol above 27 °C, methyl ethyl ketone.		4
Poly(n-octyl methacrylate)	n-butanol, methyl ethyl ketone.		4
Poly(n-lauryl methacrylate)	n-pentanol above 29 °C, isopropyl acetate, n-butyl acetate.		4
Poly(4-tert-butylphenyl methacrylate)	acetone.		4
Poly(bornyl methacrylate)	benzene.	methanol.	78
Poly(β-(N-carbazyl)ethyl methacrylate) crystalline	diphenyl ether (hot), aniline (hot, partially), nitrobenzene (hot).	methyl hexyl ketone.	29
1.3.4 Poly(α, β-disubstituted acrylates)			
Poly(dimethyl itaconate)	benzene, methylene chloride, chloroform, 1, 1-dichloroethane, chlorobenzene, furfurol, THF, dioxane, acetone, methyl ethyl ketone, methyl acetate, acetonitrile, DMF, nitromethane, nitrobenzene.	hexane, isooctane, cyclohexane, toluene, ethylbenzene, carbon tetrachloride, methanol, ethanol, diisopropyl ether, ethyl acetate, amyl acetate, propylene carbonate, water.	98, 105

Polymer	Solvents	Non-solvents	References

1.3.4 Poly(α,β-disubstituted acrylates) (Cont'd.)

Polymer	Solvents	Non-solvents	References
Poly(di-n-butyl itaconate)	hexane, isooctane, cyclohexane, benzene, toluene, ethyl benzene, styrene, 1,1-dichloroethane, methylene chloride, chloroform, carbon tetrachloride, chlorobenzene, ethanol, THF, dioxane, di-isopropyl ether, acetone, methyl ethyl ketone, methyl acetate, ethyl acetate, amyl acetate.	methanol, furfurol, dioxane, propylene carbonate, DMF, acetonitrile, nitromethane, nitrobenzene.	98, 105

1.3.5 Poly(acrylamides), Poly(methacrylamides)

Polymer	Solvents	Non-solvents	References
Poly(acrylamide)	morpholine, water.	hydrocarbons, alcohols, glycols, diethyl ether, THF, esters, DMF, nitrobenzene.	1, 2, 4
Poly(N-isopropylacrylamide)	water (cold).	water (hot).	17, 33
Poly(N,N-dimethylacrylamide)	methanol, water (40°C).		4
Poly(N-(1,1-dimethyl-3-oxobutyl)acrylamide)	toluene, butanol, methyl ethyl ketone.	water (sw).	13
Poly(methacrylamide)	methanol, ethylene glycol, acetone, water.	hydrocarbons, diethyl ether, ester.	1, 2, 4
Poly(N-carbazolylcarbonyl-ethylene) (Poly(9-acryloylcarbazole))			
atactic	benzene (partially).	methanol.	45
tactic	conc. or moderately conc. sulfuric acid, chloroform (partially).		
Poly(morpholinocarbonyl ethylene)	DMF.		4
Poly(piperidinocarbonyl-ethylene) (Poly(acrylopiperidide))	DMF.		4

1.4 Poly(vinyl ethers)

Polymer	Solvents	Non-solvents	References
General	benzene, toluene, methylene chloride, chloroform, carbon tetrachloride, n-butanol, methyl ethyl ketone, cyclohexanone.	heptane.	
Unsubstituted			
Poly(methoxyethylene) (Poly(methyl vinyl ether))			
amorphous	see general and ethanol, isopropanol, acetone, ethylacetate, butyl acetate, water (cold).	mineral oils, ethylene glycol, ethyl ether, water (hot).	1-4, 110
crystalline		methanol, acetone, water.	110, 189
Poly(ethoxyethylene)	see Poly(methoxyethylene)		
Poly(propoxyethylene)			
crystalline		heptane, acetone.	110
Poly(isopropoxyethylene)			
crystalline	see Poly(propoxyethylene)		
Poly(butoxyethylene)	see general and cyclohexane, n-heptane, diethyl ether, bis(β-ethoxy-ethyl) ether.	ethanol, β-ethoxy-ethanol.	110, 123
Poly(isobutoxyethylene)			
amorphous	see general and cyclohexane, n-heptane, ethyl ether, aliphatic alcohols C ≥ 3, methyl ethyl ketone, bis(β-ethoxy-ethyl) ether, isopropyl acetate, carbon disulfide.	methanol, ethanol, β-ethoxy-ethanol.	1-3, 22, 108, 110, 123
crystalline	chloroform above 50°C.	heptane, benzene (sw at 20°C), chloroform (sw at 20°C, isopropanol (hot), n-butanol, bis(β-ethoxy-ethyl) ether, methyl ethyl ketone.	110
Poly(tert-butoxyethylene)			
amorphous	acetone, methyl ethyl ketone.		22, 110
crystalline		heptane, benzene.	
Poly(neopentoxyethylene)			
crystalline		heptane, benzene.	110
Poly(benzyloxyethylene)			
atactic	benzene, toluene, acetone.	methanol, ethanol, water.	36
Substituted			
Poly(carbomethoxymethoxyethylene)	methylene chloride, chloroform.	diethyl ether.	118
Poly(2-methoxyethoxyethylene)			
crystalline	water.	diethyl ether.	110
Poly(2-chloroethoxyethylene)			
crystalline		acetone.	
Poly(2,2,2-trifluoroethoxyethylene)			
crystalline		heptane, benzene, dioxane.	110

Polymer	Solvents	Non-solvents	References

1.5   Poly(vinyl alcohols), Poly(acetals), Poly(vinyl ketones)

Polymer	Solvents	Non-solvents	References
Poly(vinyl alcohol)	glycols (hot), glycerol (hot), piperazine, triethylene-diamine, formamide, DMF, DMSO (hot), water, HMPT.	hydrocarbons, chlorinated hydrocarbons, lower alcohols, THF, dioxane, ethylene glycol formal, ketones, carboxylic acids, esters, ethyl lactate, conc. aqu. salt solutions.	1-4, 12
syndiotactic	water above 160°C, 1,3-propandiol above 160°C.		46
12 % acetyl	water (cold).	hydrocarbons, halogenated hydrocarbons, ketones, carboxylic acids, esters, water (hot).	2
35 % acetyl	water, alcohols, aqu. sol. of tetraalkylammonium-bromide and -iodide.	water.	53
Poly(allyl alcohol)			
lower mol. wt.	methanol, glycerol, cresol, dioxane, THF, pyridine.		1
high mol. wt. (DP > 350)	mixtures of conc. hydrochloric acid and dioxane, methanol, THF.	most organic solvents, water.	
Poly(vinyl formal)	benzene/alcohol (70:30), toluene, methylene chloride, chloroform, carbon tetrachloride/alcohol (70:30), dichloroethylene, dichloroethylene/diacetone alcohol (50:50), 2-chloroethanol, benzyl alcohol, furfurol, THF, dioxane, cyclohexanone, formic acid, acetic acid, DMF.	aliphatic hydrocarbons, aromatic hydrocarbons (sw), methanol, ethanol, dioxane, pyridine (sw), esters, water, dilute acids, acetone (sw).	2, 3, 12
Poly(vinyl acetal)	benzene, benzene/ethanol (1:1), toluene, chloroform, chloroform/methanol (9:1), carbon tetra-chloride, dichloroethylene, 2-chloroethanol, ethanol, butanol, THF, dioxane, ethylene glycol, acetone, cyclohexanone, ethyl acetate, ethyl lac-tate, benzyl acetate.	aliphatic hydrocarbons, methanol (sw), diethyl ether (sw), pyridine (sw), water, dil. acids.	3, 12
Poly(vinyl acetal) (high degree of acetalization)	ethylene dichloride, methanol, ethanol, dioxane, pyridine, acetic acid (glacial), nitromethane.	benzene, acetone.	2
Poly(vinyl butyral)			
acetalization 70 %	alcohols, cyclohexanone, ethyl lactate, ethyl glycol acetate.	aliphatic, cycloaliphatic and aromatic hydrocarbons (sw), methylene chloride, aliphatic ketones, most esters, water.	188
acetalization 77 %	methylene chloride, alcohols, acetone, methyl ethyl ketone, cyclohexanone, lower esters.	aliphatic, cycloaliphatic and aromatic hydrocarbons (sw), methyl isobutyl ketone, higher esters.	
acetalization 83 %	methylene chloride, alcohols, ketones, lower esters.	aliphatic, cycloaliphatic and aromatic hydrocarbons (sw), methanol, higher esters.	
Poly(vinyl cyclohexanone ketal)	chloroform, ethylene chloride, ethanol, butanol, ethyl glycol, THF, cyclohexanone.	aliphatic hydrocarbons, carbon tetrachloride (sw), diethyl ether (sw).	3
Poly(vinyl methyl ketone)	chloroform, pyridine, THF, dioxane, acetone, methyl vinyl ketone, acetic acid, ethyl acetate, DMF.	alcohol, petroleum ether, carbon tetrachloride, diethyl ether, water.	1, 2, 12
Poly(methyl isopropenyl ketone)	dioxane, acetone, esters.	petroleum ether, alcohol, water.	1, 12

1.6   Poly(vinyl halides)

Polymer	Solvents	Non-solvents	References
Poly(vinyl chloride)			
high mol. wt.	THF, acetone/carbon disulfide, methyl ethyl ketone, cyclopentanone, cyclohexanone, DMF, nitrobenzene, DMSO.	aliphatic hydrocarbons, mineral oils, aromatic hydrocarbons (sw), vinyl chloride, alcohols, glycols, aniline (sw), acetone (sw), carboxylic acids, acetic anhydride (sw), esters, nitroparaffins (sw), carbon disulfide, non-oxidizing acids, conc. alkalies.	1-4, 12
lower mol. wt.	toluene, xylene, methylene chloride, ethylene chloride, perchloroethylene/acetone, 1,2-di-chlorobenzene, dioxane, acetone/carbon disulfide, cyclopentanone, cyclohexanone, diisopropyl ketone, mesityl oxide, isophorone, DMF, nitro-benzene, HMPT, tricresyl phosphate.		
chlorinated, 63 % Cl	aromatic hydrocarbons, chloroform, chlorobenzene, THF, dioxane, acetone, cyclohexanone, butyl acetate, nitrobenzene, DMF, DMSO.	aliphatic and cycloaliphatic hydrocarbons, carbon tetrachloride, methyl acetate, nitromethane, organic and inorganic acids.	12

Polymer	Solvents	Non-solvents	References

1.6   Poly(vinyl halides)   (Cont'd.)

Poly(vinylidene chloride) — Solvents: THF (hot), tetrahydronaphthalene (hot), trichloroethane, pentachloroethane, 1,2-dichlorobenzene, trichlorobenzene, tetrahydrofurfuryl alcohol, dioxane, cyclohexanone, DMA, butyl acetate, DMF, N-methylpyrrolidone, benzonitrile. — Non-solvents: hydrocarbons, chloroform, ethyl bromide, vinylidene chloride, alcohols, phenols, THF (cold, sw), dioxane (cold, sw), cyclohexanone (cold, sw), carbon disulfide, conc. and moderately conc. acids and alkalies (except ammonia). — References: 2, 3, 12, 54, 57

Poly(vinyl bromide) — Solvents: THF, cyclohexanone. — Non-solvents: aliphatic and cycloaliphatic hydrocarbons, methanol, ethanol. — References: 4

Poly(vinyl fluoride) — Solvents: cyclohexanone (hot), DMF, DMA (hot), dinitrile, DMSO (hot). — Non-solvents: aliphatic, cycloaliphatic, aromatic hydrocarbons. — References: 59

  chlorinated, 30 % Cl — Solvents: DMF. — Non-solvents: aliphatic, cycloaliphatic, aromatic hydrocarbons.
  chlorinated, 60 % Cl — Solvents: carbon tetrachloride. — Non-solvents: aliphatic, cycloaliphatic, aromatic hydrocarbons.

Poly(vinylidene fluoride) — Solvents: cyclohexanone, γ-butyrolactone, ethylene carbonate, propylene carbonate, DMA, N-methylpyrrolidone, DMSO. — Non-solvents: aliphatic and cycloaliphatic hydrocarbons, alcohols, acetone, methylisobutyl ketone, DMF. — References: 60, 68

Poly(1,2-difluoroethylene) — Solvents: acetone, methylisobutyl ketone, DMF. — Non-solvents: aliphatic and cycloaliphatic hydrocarbons, methanol, ethanol. — References: 60

Poly(chlorotrifluoroethylene) — Solvents: cyclohexane (235°C), benzene (200°C), toluene (142°C), p-xylene (140°C), 1,1,1-trichloroethane (120°C), carbon tetrachloride (114°C), 1,2,3-trifluoropentachloropropane, 1,2,3-trifluoropentachloropentane, 1,1,2,2-tetrafluoro-3,3,4,4-tetrachlorocyclobutane, 1,2-dichlorotrifluorobenzene, 2,5-dinitrotrifluorobenzene (130°C), mesitylene (140°C), HMTP. — Non-solvents: common organic solvents at room temp. — References: 1, 2, 4, 113, 124

Poly(tetrafluoroethylene) — Solvents: perfluorokerosene (350°C), no other solvent known. — References: 1, 2, 79, 218

Poly(3,3,3-trifluoropropylene) — Solvents: hexafluorobenzene, acetone. — Non-solvents: aliphatic and cycloaliphatic hydrocarbons, methanol, ethanol. — References: 89

Poly(hexafluoropropylene) — Solvents: hexafluorobenzene, perfluorodibutyl ether, perfluorodibutylamine. — References: 87

Poly(3,3,4,4,5,5,5-heptafluoro-1-pentene) — Solvents: perfluorohexane. — References: 88

1.7   Poly(vinyl nitriles)

Poly(acrylonitrile) — Solvents: o-, m-, p-phenylene diamine, N-formylhexamethyleneimine, N-nitrosopiperidine, maleic anhydride, chloromaleic anhydride, succinic anhydride, acetic anhydride, citraconic anhydride, γ-butyrolactone, dioxanone, p-dioxanedione, ethylene oxalate, ethylene carbonate, propylene carbonate, 2-oxazolidone, 1-methyl-2-pyridone, 1,5-dimethyl-2-pyrrolidone, ε-caprolactam, DMF, dimethylthioformamide, N-methyl-β-cyanoethylformamide, cyanoacetic acid, α-cyanoacetamide, N-methylacetamide, N,N-diethylacetamide, DMA, dimethylmethoxyacetamide, N,N-dimethyl-α,α,α-trifluoroacetamide, N,N-dimethylpropionamide, N,N,N',N'-tetramethyloxamide, hydroxyacetonitrile, chloroacetonitrile, chloroacetonitrile/water, β-hydroxypropionitrile, malonitrile, fumaronitrile, succinonitrile, adiponitrile, bis(2-cyanoethyl)ether, bis(2-cyanoethyl)sulfide, bis(4-cyanobutyl)sulfone, 1,3,3,5-tetracyanopentane, nitromethane/water (94:6), 1,1,1-trichloro-3-nitro-2-propane, tri(2-cyanoethyl)nitromethane, 3-, 4-nitrophenol, methylene dithiocyanate, trimethylene dithiocyanate, DMSO, tetramethylene sulfoxide, dimethyl sulfone, ethyl methyl sulfone, 2-hydroxyethyl methyl sulfone, ethylene-1,2-bis-(ethyl sulfone), dimethyl phosphite, diethyl phosphite, sulfuric acid, nitric acid, p-phenol sulfonic acid, conc. aqu. lithium chloride, conc. aqu. zinc chloride, conc. aqu. aluminum perchlorate, conc. aqu. sodium thiocyanate, conc. aqu. calcium thiocyanate, molten quaternary ammonium salts and their aqu. solutions. — Non-solvents: hydrocarbons, chlorinated hydrocarbons, alcohols, diethyl ether, ketones, piperazinedione, 1,6-hexanediamine, propyl formate, formamide, methylformamide, n-butylformamide, diethylformamide, dipropylformamide, methoxyacetamide, N-methyldiacetamide, dimethyloxamide, ethylene urea, acetonitrile (sw), acrylonitrile, methoxyacetonitrile, 1-hydroxypropionitrile, 2-methoxypropionitrile, methylmalonitrile, dimethylmalonitrile, 1,1-dimethylsuccinonitrile, suberonitrile, methyl thiocyanate, hexamethylene dithiocyanate, aliphatic nitro-compounds, 1-nitrophenol, diethyl sulfoxide, n-butyl methyl sulfoxide, bis(2-hydroxyethyl) sulfoxide, diethyl sulfone (sw), diallyl sulfone, 3,4-dimethyl sulfolane. — References: 1, 4, 12, 119, 120

	Polymer	Solvents	Non-solvents	References
1.7	Poly(vinyl nitriles) (Cont'd.)			
	Poly(methacrylonitrile)	methylene chloride, pyridine, acetone, cyclo-hexanone, furfural, cyanoacetic acid, acetanhydride, ethylene carbonate, DMF ($> 20^\circ$C), benzonitrile, dinitriles, nitromethane, DMS, HMTP.	aliphatic hydrocarbons, toluene, alcohols, esters, methacrylonitrile.	1, 2, 4
1.8	Poly(vinyl esters)			
	Unsubstituted			
	Poly(vinyl acetate)	benzene, toluene, chloroform, carbon tetrachloride/ethanol, dichloroethylene/ethanol (20:80), chloro-benzene, methanol, ethanol/water, n-butanol/water, allyl alcohol, 2,4-dimethyl-3-pentanol, benzyl alcohol, tetrahydrofurfuryl alcohol, THF, di-methyltetrahydrofuran, dioxane, glycol ethers, glycol ether esters, acetone, methyl ethyl ketone, acetic acid, lower aliphatic esters, vinyl acetate, acetals, acetonitrile, nitromethane, DMF, DMSO.	saturated hydrocarbons, xylene (sw), mesitylene, carbon tetrachloride (sw), ethanol (anhydrous, sw), anhydrous alcohols C > 1, ethylene glycol, cyclo-hexanol, methylcyclohexanol, diethyl ether (anhy-drous, alcohol free), higher esters C > 5, carbon disulfide, water (sw), dil. acids, dil. alkalies.	1, 2, 4
	syndiotactic	chloroform, chlorobenzene.	benzene, acetone.	46
	Poly(vinyl propionate)	see Poly(vinyl acetate)		
	Poly(vinyl n-butyrate) and Poly(vinyl isobutyrate)	see Poly(vinyl acetate) also cyclohexane, amyl alcohol, hexyl acetate.		239
	Poly(vinyl pivalate)	benzene, toluene, acetone, butanone, ethyl acetate.	hexane, methanol, water.	28
	Poly(vinyl caproate)	benzene.		4
	Poly(vinyl caprylate)	aliphatic and aromatic hydrocarbons, acetone.	lower alcohols.	2
	Poly(vinyl laurate)	aliphatic and aromatic hydrocarbons.	lower alcohols, acetone.	2
	Poly(vinyl benzoate)	see Poly(vinyl acetate) also p-xylene.		4, 33
	Substituted			
	Poly(vinyl chloroacetate)	chloroform, chlorobenzene, pyridine, dioxane, cyclohexanone, ethyl acetate.	saturated hydrocarbons, acetone (sw).	2
	Poly(vinyl acetylacetate)	chloroform, acetone, THF, dioxane, ethyl acetate, acetic acid, pyridine, DMF, saturated aqu. sol. of magnesium perchlorate.	benzene, alcohols, diethyl ether, water.	91
1.9	Poly(styrenes)			
	Poly(styrene)	cyclohexane (above 35°C), cyclohexane/acetone, methylcyclohexane/acetone, decahydronaphthalene/diethyl oxalate, benzene, toluene, ethylbenzene, styrene, lower chlorinated aliphatic hydrocarbons, phenol/acetone, THF, dimethyltetrahydrofuran, di-oxane, methyl ethyl ketone, diisopropyl ketone, cyclohexanone, glycol formal, ethyl acetate, butyl acetate, methyl-, ethyl-, n-butyl phthalate, 1-nitropropane, carbon disulfide, tributyl phos-phate, phosphorus trichloride.	saturated hydrocarbons, alcohols, phenol, diols, ethylene chlorohydrin, perfluorobenzene, 1,2,3,4-tetrafluorobenzene (lower than 10°C), diethyl ether, glycol ethers, acetone, acetic acid, isobutyl phthalate, methylhexyl phthalate, tri(chloroethyl) phosphate, tricresyl phosphate.	1-4, 10
	Poly(α-methyl styrene)	see Poly(styrene)		
	Poly(4-chlorostyrene)	methyl ethyl ketone, toluene.		4
	Poly(4-bromostyrene)	benzene		4
	Poly(dichlorostyrene)	toluene		4
	Poly(4-methoxystyrene)	methyl ethyl ketone, toluene.		4
	Poly(2,5-dimethoxystyrene)	benzene, toluene, chloroform.	hexane, methanol.	67
1.10	Others (alphabetically ordered)			
	Poly(acrolein) (redoxpolymerization)	above 60°C: pyridine/water (55:45 to 90:10), above 130°C: nitrobenzene, DMS, sat. stannous chloride solution, γ-butyrolactone (160-170°C), ethylene carbonate (130-135°C), DMF (153°C), DMSO (160-170°C), divinyl sulfone (150-155°C), TMS (160-165°C).	hydrocarbons, chlorinated hydrocarbons, lower al-cohols, diethyl ether, aromatic ketones, esters.	95, 96
	Poly(acrolein) (ionic polymerization)	see 3.1.1		
	Poly(anthrylethylene)	methylene chloride.	methanol.	61

Polymer	Solvents	Non-solvents	References
1.10 **Others** (Cont'd.)			
Poly(N-benztriazolylethylene)	chlorinated hydrocarbons, cyclohexanol, glacial acetic acid, DMF.	hydrocarbons, alcohols, ketones, esters, water.	21
Poly(biphenylethylene)	benzene, toluene, dimethyloxyethane.	methanol.	104
Poly(N-carbazolylethylene) (Poly(N-vinylcarbazole))			
	benzene, toluene, xylene, methylene chloride, chloroform, tetrachloroethane, chlorobenzene, THF, dioxane, cyclohexanone, benzyl acetate, conc. nitric acid, conc. sulfuric acid.	aliphatic hydrocarbons, hydroaromatic hydro-carbons, carbon tetrachloride, trichloroethylene (sw), tetrachloroethylene, 1-chlorotoluene (sw), alcohols, chlorohydrin, diols, glycol monoether, diethyl ether, dimethyltetrahydrofuran, acetals, aliphatic ketones, dil. carboxylic acids, esters, water, dil. alkalies.	1-3, 12, 219
Poly(6-(N-carbazolyl)hexylethylene))	N-methylpyrrolidone (123°C), acetone (partially).	methanol.	32, 99
Poly(5-(N-carbazolyl)pentylethylene))	N-methylpyrrolidone (hot), acetone (partially).	methanol.	32, 99
Poly(diallyldimethylsilane)	benzene.		106
Poly(2,4-dimethyl-6-triazinylethylene)			
	benzene, alcohols, ketones, esters, water (cold), dil. alkali.	aliphatic hydrocarbons, water (ppt. below 73°C).	48
Poly(diphenylphosphinylideneethylene) (Poly(vinyldiphenylphosphine oxide))			
	benzene, toluene, methanol, ethanol.	hexane, water.	56, 76
Poly(diphenylthiophosphinylideneethylene)			
	benzene, chloroform.	methanol.	76
Poly(2-methyl-5-pyridylethylene)	methanol, DMF.		4
Poly(3-morpholinylethylene)	water.		126
Poly(1-nitropropylene)	DMF (partially).		85
Poly(2-pyridylethylene) (Poly(2-vinylpyridine))			
	benzene, chloroform, alcohols/water, ethanol, oc-tanol, THF, dioxane, pyridine, vinylpyridine, acetone, methyl ethyl ketone, nitroethane, glacial acetic acid, aqu. mineral acids.	toluene, carbon tetrachloride, water (sw).	1, 2, 116, 121, 125
crystalline	aromatic hydrocarbons (reflux), chlorinated sol-vents, methanol.	aliphatic hydrocarbons (reflux), diethyl ether, methyl ethyl ketone.	
Poly(4-pyridylethylene)	benzene, methanol, ethanol, ethanol/water (92:8 % wt.), n-propanol, isopropanol/methyl ethyl ketone (86:14 % wt.), tert-butanol, octanol, cyclohexanol, benzyl alcohol, THF, dioxane/water (1:1), pyridine, acetone/water (1:1), nitromethane, aqu. mineral acids.	petroleum ether, diethyl ether, dioxane, acetone, methyl ethyl ketone, ethyl acetate, water.	1, 4, 125
Poly(N-pyrrolidonylethylene)	(solubility depending on small amounts of water) methylene chloride, chloroform, methanol, ethanol, aromatic alcohols, chlorohydrins, pyridine, acetone, glacial acetic acid, chloroacetic acid esters, lactic acid esters, nitromethane, water, dil. acids.	aliphatic and aromatic hydrocarbons, carbon tetra-chloride, trichloroethylene, 1-chlorotoluene, diethyl ether, dipropyl ether, acetone, acetone/water, di-propyl ketone, acetic acid esters, methoxy-butyric acid esters, nitromethane/water.	1-4, 12
Poly(N-thiopyrrolidonylethylene)	DMF.	aliphatic, cycloaliphatic and aromatic hydrocarbons, butanol, methyl ethyl ketone, water.	128
Poly(N-1,2,4-triazolylethylene)	glacial acetic acid, DMF, DMSO, water.	hydrocarbons, chlorinated hydrocarbons, alcohols, ketones, esters.	21
Poly(vinyl sulfate)	aqu. NaCl (0.5M).		4
Poly(vinyl sulfofluoride)	THF, acetone, ethyl acetate, DMF.	diethyl ether.	1
Poly(vinyl sulfonic acid)	water.	hydrocarbons, ketones, esters.	1, 2
sodium salt	water, aqu. sodium chloride (0.5M).	methanol, acetone.	1
2. **Main-chain Carbocyclic Polymers**			
2.1 **Poly(phenylenes)**			
Poly(2,5-dihydroxy-1,4-phenyleneethylene)		common organic solvents, aqu. potassium hydroxide.	114
Poly(2,5-dimethoxy-1,4-phenyleneethylene)			
	bromoform.	common organic solvents.	94, 114
Poly(2,5-dimethyl-1,4-phenyleneethylene)			
low mol. wt.	1,2- and 1,4-dichlorobenzene, chloroform.		16

Polymer	Solvents	Non-solvents	References

2.1   Poly(phenylenes)   (Cont'd.)

Poly(2,5-dimethyl-1,4-phenylenemethylene)

low mol. wt.	1,2- and 1,4-dichlorobenzene (hot), chloroform.		16
Poly(nitrophenylene)		all common organic compounds.	42
Poly(1,4-phenyleneethylene)	biphenyl, chlorinated biphenyls, phenyl ether, benzoyl benzoate, nitrobenzene.		109, 114
Poly(tetramethyl-1,4-phenyleneethylene)		benzene, 1-chloronaphthalene (sw), ethanol, benzyl benzoate (sw).	114

2.2   Others   (alphabetically ordered)

Poly(acenaphthylene)	benzene, toluene, chloroform, carbon tetrachloride, 1,2-dichloroethane (> 30°C).	alcohols, ethers, dimethoxyethane, acetone, carboxylic acids.	2
Poly(1,3-cyclohexadiene)	benzene, xylene.	methanol.	82
Poly(1,5-cyclooctadiene)	aromatic hydrocarbons, chlorinated hydrocarbons, methanol (oligomers), diethyl ether (partially).	methanol.	81

Poly(cyclopentadiene)	benzene, toluene, chloroform, carbon tetrachloride, THF, dioxane.	hexane, petrol. ether, methanol.	103
Poly(endo-dicyclopentadiene)	benzene, chlorobenzene, carbon disulfide.	hexane, methanol, acetone.	185
Poly(1,2-dihydronaphthalene)	bis(phenoxyphenyl) ether (above 290°C).	tetrahydronaphthalene, tetrachloroethane, chloronaphthalene, diphenyl ether, amyl acetate.	186
Poly(dimethylfulvene)	n-hexane, benzene, halogenated hydrocarbons, diethyl ether, acetone.	alcohol.	93
Poly(indene)	aromatic and chlorinated hydrocarbons, pyridine, diethyl ether, dioxane, ketones, drying oils.	aliphatic hydrocarbons (sw), lower alcohols, water, acids, alkalies.	12
Poly(1-isopropylidene-3a,4,7,7a-tetrahydroindene)	benzene, chloroform, THF, carbon disulfide.	hexane, cyclohexane, ethanol, acetone.	187

3.   Main-chain Acyclic Heteroatom Polymers

3.1   Main-chain -C-O-C- Polymers

3.1.1   Poly(oxides)

Unsubstituted

Poly(oxymethylene)	at elevated temp.: benzyl alcohol, phenol, chlorophenols, aniline, formamide, DMF, malodinitrile, γ-butyrolactone, ethylene carbonate, bromobenzene, diphenyl ether, benzyl benzoate.	aliphatic hydrocarbons, lower alcohols, diethyl ether, lower esters.	2, 133, 135
Poly(oxyethylene)	benzene, chloroform, carbon tetrachloride, alcohols, cyclohexanone, esters, DMF, acetonitrile, water (cold), aqu. K_2SO_4 (0.45M above 35°C), aqu. $MgSO_4$ (0.39M above 45°C).	aliphatic hydrocarbons, ethers, dioxane (sw), water (hot).	1, 2, 4
Poly(oxyethylidene) (Poly(acetaldehyde))			
amorphous	aromatic hydrocarbons, chloroform, alcohols, ketones, esters.	aliphatic hydrocarbons.	220
crystalline	chloroform (partially).	aromatic hydrocarbons, chloroform (partially), alcohols, ketones, esters.	
Poly(oxypropylene)			
crystalline	benzene, toluene, carbon tetrachloride, methanol (hot), ethanol, THF, dioxane, acetone, methyl ethyl ketone.	diethyl ether (sw), 2-aminoethanol, ethyl acetate (sw), DMF.	1
Poly(oxypropylidene)	DMF.		117
crystalline	chloroform (partially).	common organic solvents.	
Poly(oxyisopropylidene) (Poly(acetone))			
	chloroform, carbon tetrachloride, acetone.		6
Poly(oxytetramethylene)	benzene, methylene chloride, chloroform, THF, ethanol, acetone/water, thionyl chloride.	petroleum ether, hexane, methanol, water.	27, 134, 136, 137
Poly(oxycyclopentenylene)	benzene, chloroform, carbon disulfide.	methanol.	80
Poly(oxycyclohexenylene)	aliphatic hydrocarbons, paraffin (hot), toluene, methyl acetate, chinese wood oil.	ethanol, 2-ethoxyethanol, acetone.	1
Poly(oxy-1,3-phenylene)	benzene, biphenyl, 3-pentanol, phenyl ether, pyridine, benzophenone, nitrobenzene, DMF, DMSO.	methanol.	101

	Polymer	Solvents	Non-solvents	References
3.1.1	Poly(oxides) (Cont'd.)			
	Substituted			
	Poly(oxy-2,6-dimethyl-1,4-phenylene)			
	amorphous	α-pinene (hot).	α-pinene (cold), methanol, ethanol.	129, 130, 139, 221
	crystalline	benzene, toluene, chloroform, chlorobenzene.	α-pinene (hot), methanol, ethanol, nitromethane.	
	Poly(oxy-2-chloroethylidene)			
	amorphous	chloroform.	methanol.	43
	crystalline		chloroform, methanol.	
	Poly(oxy-2,2,2-trichloroethylidene) (Poly(chloral))			
	crystalline		chloroform, methanol, other common solvents.	117
	Poly(oxy-2,2-bischloromethyl-trimethylene)			
		cyclohexanone.	methanol.	26

Polymeric Structures, not 100 % specific

Poly(dimethylketene)

$\begin{array}{c} \vdash O-C \dashv_n \\ \| \\ C(CH_3)_2 \end{array}$

	Polymer	Solvents	Non-solvents	References
	low mol. wt.	diethyl ether.	methanol.	24, 138
	higher mol. wt.	benzene, chloroform (partially), carbon tetrachloride.		

Poly(acrolein) (ionic polymerization)

$\vdash O - CH(CH=CH_2) \dashv_n$

		benzene, carbon tetrachloride, THF, dioxane, acetone, DMF.	petroleum ether, methanol.	97
	Poly(acrolein) (redox polymerization) see 1.10			
	Poly(α-methylacrolein)	aniline, pyridine, γ-butyrolactone, DMF, nitrobenzene.	hydrocarbons, alcohols.	1
	Poly(1,2-di(epoxyethyl)benzene))			34
		chloroform, THF.		
	Poly(phenylglycidyl ether)	xylene (hot), 1,2-dichlorobenzene (hot), DMF (hot).	common organic solvents at room temperature.	30

Poly(glutardialdehyde)

	low mol. wt.	benzene, chloroform, diethyl ether, THF, dioxane.	petroleum ether, water.	37, 114, 119, 127
	high mol. wt.	benzene (partially), methylene chloride, THF, pyridine.		
	Poly(β-methylglutardialdehyde)			
	low mol. wt.	chloroform, diethyl ether, THF, dioxane.	petroleum ether, water.	37
	Poly(β-phenylglutardialdehyde)			
	low mol. wt.	benzene, choroform, diethyl ether, THF, dioxane.	petroleum ether, water.	37
	Poly(trans-1,2-cyclohexanedicarboxaldehyde)			15
		hexane, benzene, chloroform.	methanol.	
	Poly(2-formyl-Δ^5-dihydropyran)	aromatic hydrocarbons, chloroform, pyridine, THF.		39

	Polymer	Solvents	Non-solvents	References
3.1.2	Poly(carbonates)			
	Poly(oxycarbonyloxyhexamethylene)	benzene, chloroform, acetone.	ethanol, ether.	1
	Poly(oxycarbonyloxy-1,3-phenylene)		common organic solvents.	1
	Poly(oxycarbonyloxy-1,4-phenylene)		common organic solvents.	1
	Poly(oxycarbonyloxy-1,4-phenyleneisopropylidene-1,4-phenylene)			
		methylene chloride, chloroform, m-cresol, THF, dioxane, cyclohexanone, pyridine, DMF.	hydrocarbons, styrene, carbon tetrachloride, acetone, lower esters.	1, 140-142
	chlorinated (8 % Cl and more)	benzene, styrene, methyl methacrylate.		142
	Poly(oxycarbonyloxy-1,4-phenylene-2-pentylidene-1,4-phenylene)			
		benzene, toluene, methylene chloride, chloroform, ethyl acetate, butyl acetate.		1

Polymer	Solvents	Non-solvents	References

3.1.3   Poly(esters) (alphabetically ordered)

Poly(ethylene terephthalate)   see Poly(oxyethyleneoxyterephthaloyl)

Poly(oxycarbonylethylene) (Poly(β-propiolactone))

	chloroform, formic acid.	hydrocarbons.	144, 145

Poly(oxycarbonylpropylene) dichloroethylene, chloroform, trifluoroethanol. 143

Poly(oxycarbonyl-1,4-cyclohexylenecarbonyloxyoctamethylene) - cis and trans

| | chloroform. | | 4 |

Poly(oxycarbonyl-1-vinylethylene) (Poly(β-vinyl-β-propiolactone))

| | chlorinated hydrocarbons, butyrolactone, DMF. | | 146 |

Poly(oxyethyleneoxyterephthaloyl)

| crystalline | chloral hydrate, phenol, phenol/tetrachloro-ethane (1:1 vol.), phenol/2,4,6-trichlorophenol (10:7 vol.), chlorophenol, nitrobenzene, DMSO (hot), halogenated aliphatic carboxylic acids. | hydrocarbons, chlorinated hydrocarbons, aliphatic alcohols, ketones, carboxylic esters, ethers. | 2,4,148 |

| Poly(oxyfumaroyloxyhexamethylene) | chloroform. | | 4 |
| Poly(oxyglutaryloxyhexamethylene) | benzene, toluene, chloroform, chlorobenzene, THF, dioxane. | | 147 |

Poly(oxyhexamethyleneoxy-α,α'-dibutylsebacoyl)

| | benzene. | | 4 |

Poly(oxyhexamethyleneoxysebacoyl) (Poly(hexamethylene sebacate))

| | benzene, toluene, chloroform, chlorobenzene, THF, dioxane. | | 4,147 |

Poly(oxyisophthaloyloxy-4,4'-biphenylene) similar properties as (Poly(oxy-1,4-phenyleneoxyisophthaloyl))

Poly(oxymaleoyloxyhexamethylene)	benzene, chloroform, THF.		4
Poly(oxyoxalyloxyethylene)		benzene, toluene, chloroform, chlorobenzene, THF, dioxane.	147
Poly(oxyoxalyloxyhexamethylene)	chloroform, chlorobenzene, dioxane.	benzene, toluene, THF.	147

Poly(oxy-1,4-phenyleneoxyisophthaloyl)

| | at elevated temp.: m-terphenyl, 2,4,6-trichlorophenol, halogenated biphenyls, halogenated diphenyl oxides, benzo-phenone, halogenated naphthalenes. | common organic solvents. | 132 |

Poly(oxysebacoyloxyhexadecamethylene)

	chloroform.		4
Poly(oxysuccinyloxyhexamethylene)	benzene, toluene, chloroform, chlorobenzene, THF, dioxane.		4,147
Poly(oxyundecanoyl)	benzene, chloroform.		4

3.1.4   Poly(anhydrides)

| Poly(oxyterephthaloyl) | no solvent known. | | 149 |

3.1.5   Poly(urethanes) (alphabetically ordered)

| Poly(urethanes) (general) | phenol, m-cresol, formic acid, sulfuric acid. | saturated hydrocarbons, alcohols, diethyl ether. | 2 |

Poly(oxy-1,4-cyclohexyleneoxycarbonylimino-1,4-phenylenemethylene-1,4-phenyleneiminocarbonyl)

| | THF, dioxane (hot), DMF. | 1,1,2-trichloroethane. | 155 |

Poly(oxyethyleneoxycarbonyliminoethyleneiminocarbonyl)

| | m-cresol. | methylene chloride, chloroform, formic acid. | 152 |

Poly(oxyethyleneoxycarbonyliminohexamethyleneiminocarbonyl)

| | DMF. | diethyl ether. | 151 |

Poly(oxyethyleneoxycarbonylimino-1,3-phenyleneiminocarbonyl)

| | DMSO, m-cresol (hot), sulfuric acid. | xylene, chloroform, chlorobenzene, cyclohexanone (sw), DMF (sw). | 153 |

3.2   Main-chain -O-Heteroatom Polymers

3.2.1   Poly(sulfonates)

Poly(aryl sulfonates)

$$+SO_2-Ar-SO_2-O-Ar-O+_n$$

$Ar = -C_6H_4-; -C_6H_4-C_6H_4-;$
$-C_6H_4-CH_2-C_6H_4-;$
$-C_6H_4-O-C_6H_4-;$
$-C_6H_4-SO_2-C_6H_4-;$

| | DMF | methanol. | 71 |

Polymer	Solvents	Non-solvents	References
3.2.2 Poly(siloxanes) (ordered according to increasing complexity)			
Poly(siloxanes) (general)			
fluids and greases	aromatic and chlorinated hydrocarbons, esters.	lower alcohols, water, moderately conc. acids and alkalies.	12
rubbers		aromatic and chlorinated hydrocarbons (sw), esters (sw).	
Poly(oxydimethylsilylene)	cyclohexane, aliphatic hydrocarbons, cyclohexene, hydrogenated xylene, benzene, toluene, mesitylene, triethylbenzene, methylene chloride, chloroform, carbon tetrachloride, ethyl bromide, trichloroethylene, n-butyl chloride, chlorobenzene, o-fluorotoluene, 1,2-dimethoxyethane, phenetol (above 13°C), 4-chlorotoluene, 1,2-dichlorobenzene, octylamine, methyl ethyl ketone (above 20°C), diethyl ketone, ethyl acetate, isopropyl acetate, amyl acetate, cyclohexyl acetate.	dimethylnaphthalene, 1,2-dichloroethane, 1,4-dibromobutane, lauryl bromide, bromobenzene, dichlorobenzene, 4-bromotoluene, methanol, ethanol, 2-ethoxyethanol, 2-isopropoxyethanol, 2-butoxyethanol, n-decyl alcohol, cyclohexanol, benzyl alcohol, ethylene glycol, chloroethyl ether, dioxane, "Carbitol", diphenyl oxide, aniline, acetone, cyclohexanone, γ-butyrolactone, mesityl oxide, acetophenone, methylacetophenone, ethyl formate, isobutylphenyl acetate, benzyl acetate, ethyl lactate, ethyl benzoate, diethyl phthalate, dibutyl phthalate, acetonitrile, 1-nitropropane, nitrobenzene, 1-nitrotoluene.	2,4
Poly(oxymethylphenylsilylene)	toluene, chloroform, butanol, diethyl ether, acetone (hot), ethyl acetate.	methanol, ethanol, ethylene glycol.	156
Poly(oxydiphenylsilylene)	acetone (hot).	acetone (cold).	156
Poly(oxy-bis(3,4-dichlorophenyl)silylene)			
	benzene, acetone.		157
3.3 Main-chain -C-S-C- and -C-S-N- Polymers			
3.3.1 Poly(sulfides) (alphabetically ordered)			
Poly(dithioethyleneoxyethylene)	1,1,2-trichloroethane (partially), other chlorinated hydrocarbons (partially).	gasoline (sw), kerosine (sw), turpentine (sw), benzene (sw), carbon tetrachloride (sw), alcohols, carbon disulfide, dil. acids, dil. alkalies.	11,12
liquid polymers	benzene, ethylene dichloride.		
Poly(tetrathioethylene)	no solvent known.	gasoline, kerosine, turpentine, benzene (sw), carbon tetrachloride, alcohols, carbon disulfide, dil. acids, dil. alkalies.	11,12
Poly(tetrathioethyleneoxyethylene)	1,1,2-trichloroethane (partially), other chlorinated hydrocarbons (partially).	gasoline (sw), kerosine, turpentine (sw), benzene (sw), carbon tetrachloride (sw), alcohols, carbon disulfide, dil. acids, dil. alkalies.	11,12
liquid polymers	benzene, ethylene dichloride.		
Poly(thio-4,4'-biphenylene)	biphenyl, dimethyl-p-terphenyl, dichlorobiphenyl, hexachlorobiphenyl.		158
Poly(thiodifluoromethylene) $\{ S - CF_2 \}_n$	chloroform/methanol.	conc. nitric acid (reflux), aqu. sodium hydroxide (10 % reflux).	100
Poly(thiophenylene)	biphenyl, dimethyl-p-terphenyl, dichlorobiphenyl, hexachlorobiphenyl.	at reflux temp.: toluene, pyridine, 2,4-lutidine, phenyl oxide, phenyl sulfide.	107,158
3.3.2 Poly(sulfones) (ordered according to increasing complexity)			
Poly(ethylene-co-sulfur dioxide)	no solvent known.	common organic solvents.	1,11
Poly(propylene-co-sulfur dioxide)	conc. nitric acid, conc. sulfuric acid.	common organic solvents.	1,11
Poly(1-butene-co-sulfur dioxide)	acetone, cyclohexanone.	paraffins, cycloparaffins, aromatic hydrocarbons.	1
Poly(butadiene-co-sulfur dioxide)			
Poly(isoprene-co-sulfur dioxide)			
Poly(dimethylbutadiene-co-sulfur dioxide)			
	conc. nitric acid, conc. sulfuric acid.	common organic solvents.	1
Poly(1-pentyne-co-sulfur dioxide)	dioxane.		1
Poly(1-hexyne-co-sulfur dioxide)	dioxane, acetone.		1
Poly(1-heptyne-co-sulfur dioxide)	dioxane, acetone.		1
Poly(oxy-1,4-phenylenesulfonyl-1,4-phenylene)			
	methylene chloride, DMF, DMSO.	aliphatic, cycloaliphatic and aromatic hydrocarbons, halogenated hydrocarbons, alcohols, silicon oil.	159,162

Polymer	Solvents	Non-solvents	References

3.4   Main-chain -C-N-C- Polymers

3.4.1   Poly(amides)

Poly(amides) 1:
Poly(N-butyliminocarbonyl) (Poly(butyl isocyanate))

	benzene, carbon tetrachloride, THF.		25,165
Poly(N-ethyliminocarbonyl)	trifluoroacetic acid, sulfuric acid.		165
Poly(N-m-tolyliminocarbonyl)	DMF.		165
Poly(N-vinyliminocarbonyl)	pyridine, DMF, DMA, DMSO.	chloroform, alcohols, dioxane, carbon disulfide.	31

Poly(amides) 3:
Poly(imino(1-oxo-2,2-dimethyl-3-phenyltrimethylene)) (Poly(3,3-dimethyl-4-phenylazetidine-2-one))

	phenol, trifluoroethanol, sulfuric acid.	chloroform, methanol, HMTP, sol. of $Ca(CNS)_2$ in methanol.	168

Poly(imino(1-oxo-3-methyltrimethylene)) (Poly(4-methylazetidine-2-one))

optical activity < 75 %	phenol, formic acid, sulfuric acid, sol. of $Ca(SCN)_2$ in methanol.	chloroform, methanol, trifluoroethanol, HMTP.	168,169
optical activity > 80 %	chlorobenzene/trichloroacetic acid (1:1), formic acid/trichloroacetic acid (3:7), dichloroacetic acid.	formic acid, sulfuric acid.	169
Poly(imino(1-oxotrimethylene))	glycerol (hot), phenol (hot), formic acid, aqu. chloral hydrate, water > 140°C, nitrophenols, chloroacetic acid, cyanoacetic acid, nitric acid, sulfuric acid, aqu. sol. of $CaBr_2$ or $FeCl_3$.	most organic solvents, dilute bases and acids, phenol (cold), DMF, DMSO.	1,166,167

Poly(amide) 4:
Poly(imino(1-oxotetramethylene)) (Poly(pyrrolidone))

	methanol, benzyl alcohol (hot), m-cresol, o-chlorophenol, α-pyrrolidone (hot), formic acid, acetic acid (hot), dichloroacetic acid, sulfuric acid, aqu. sol. of $Ca(CNS)_2$.	formamide, DMF, water.	170

Poly(amides) 6:
Poly(imino(1-oxohexamethylene))

	m-cresol, chlorophenol, formic acid, acetic acid, trichloroacetic acid, ethylene carbonate, sulfuric acid, phosphoric acid, HMTP.	hydrocarbons, chloroform, alcohols, ethers, ketones, esters.	2,4
chlorinated	o-chlorophenol, DMSO, sulfuric acid.	formic acid.	171

Poly(amide) 6,6:
Poly(iminoadipoyliminohexamethylene)

	room temperature: trifluoroethanol, trichloroethanol, phenol, cresols, chloral hydrate, formic acid, halogenated acetic acids, hydrogen fluoride, hydrogen chloride/methanol, liquid sulfur dioxide, sulfuric acid, phosphoric acid, saturated solutions of alcohol-soluble salts, e.g. calcium chloride, magnesium chloride in methanol. **at 120-180°C:** benzyl alcohol, ethylene chlorohydrin, 1,3-chloropropanol, 2-butene-1,4-diol, diethylene glycol, acetic acid, formamide, N-acetylmorpholin, DMSO.	hydrocarbons, aliphatic alcohols, chloroform, diethyl ether, aliphatic ketones and esters.	1

Poly(amide) 6,10:
Poly(iminoadipoyliminodecamethylene)

	chlorobenzene, diethyl succinate (79°C).		4

Poly(amide) 11:
Poly(imino(1-oxoundecamethylene))

	higher primary alcohols, DMF, DMSO.		1

Aromatic Polyamides:
Poly(iminoethyleneiminoterephthaloyl)

	sulfuric acid	chloroform, ethanol/water (80:20), m-cresol, acetone, formamide, DMF, trifluoroacetic acid.	173

Poly(N-methyliminoethylene-N-methyliminoterephthaloyl)

	chloroform, m-cresol, acetone, formamide, DMF, trifluoroacetic acid, sulfuric acid.	ethanol/water (80:20).	173

Polymer	Solvents	Non-solvents	References

3.4.1 Poly(amides) (Cont'd.)

Aromatic Polyamides :

Poly(N-methyliminotetramethylene-N-methyliminoterephthaloyl)

| | chloroform, ethanol/water (80 : 20) (hot), m-cresol (hot), formamide, sulfuric acid. | acetone, DMF. | 173 |

Poly(iminoisophthaloylimino-4, 4'-biphenylene)

| | conc. sulfuric acid. | methanol, m-cresol (sw), N-methylpyrrolidone (sw), formic acid (85 %) (sw), glacial acetic acid (sw), DMF (sw), DMA (sw), DMSO (sw). | 41 |

Poly(iminotetramethyleneiminoterephthaloyl)

| | trifluoroacetic acid, sulfuric acid. | chloroform, m-cresol, acetone, formamide, DMF. | 173 |

Poly(pyromellitic dianhydride-co-aromatic diamines) (polyamic acids)

| | DMF, DMA, DMSO, tetramethyl urea. | | 172 |

HOOC $\qquad$ COOH

$+$ NH-CO $\qquad$ CO-NH-R $+_n$

R = -C$_6$H$_4$-; -C$_6$H$_4$-C$_6$H$_4$-; -C$_6$H$_4$-O-C$_6$H$_4$-;
    -C$_6$H$_4$-S-C$_6$H$_4$-; -C$_6$H$_4$-SO$_2$-C$_6$H$_4$-;
    -C$_6$H$_4$-CH$_2$-C$_6$H$_4$-; -C$_6$H$_4$-C(CH$_3$)$_2$-C$_6$H$_4$-.

Sulfur and Phosphorus Containing Polyamides

Poly(iminoadipoyliminotrimethylene-1, 4-phenylenephosphinidene-1, 4-phenylenetrimethylene)

| | ethanol, cresol, formic acid, DMF, aqu. hydrochloric acid (5 %). | hexane (sw), chloroform, chlorobenzene, C-7 fluoro-alcohol, acetone, acetonitrile, water. | 84 |

Poly(iminoethyleneiminocarbonyl-1, 4-phenylenesulfonyl-1, 4-phenylene carbonyl)

| | tetrachloroethane/phenol (40:60). | 2-chloroethanol, DMF, m-cresol, 1, 1, 2-trichloro-ethane/formic acid (60 : 40), formic acid. | 163 |

Poly(iminohexamethyleneiminocarbonylethylene-1, 4-phenylene-phosphinylidene-1, 4-phenyleneethylenecarbonyl)

| | ethanol, C-7 fluoroalcohol, cresol, formic acid, DMF. | hexane, chloroform (sw), chlorobenzene, acetone, acetonitrile, water (sw), aqu. hydrochloric acid (5 %, sw). | |

Poly(iminohexamethyleneiminocarbonylethylenethioethylenecarbonyl)

| | cresol, formic acid, DMF. | hexane, chloroform, chlorobenzene, ethanol, C-7 fluoroalcohol (sw), acetone, acetonitrile, water, aqu. hydrochloric acid (5 %). | 84 |

Poly(iminoisophthaloylimino-1, 3-phenylenesulfonyl-1, 3-phenylene)

| | N-methylpyrrolidone, DMA, DMF. | | 164 |

Poly(iminoterephthaloylimino-1, 3-phenylenesulfonyl-1, 3-phenylene)

| | DMA, DMF. | N-methylpyrrolidone. | 164 |

Poly(iminoterephthaloylimino-1, 4-phenylenesulfonyl-1, 4-phenylene)

| | sol. of LiCl in DMF (5 %). | N-methylpyrrolidone, DMF. | 164 |

Poly(iminoterephthaloyliminotrimethylene-1, 4-phenylenephosphinidene-1, 4-phenylenetrimethylene)

| | cresol, formic acid, DMF. | hexane, chloroform, chlorobenzene, ethanol (sw), acetone, acetonitrile, water, hydrochloric acid (5 %, sw). | 84 |

3.4.2 Poly(hydrazides)

Poly(hydrazoadipoylhydrazoisophthaloyl)

| | HMTP, DMSO. | | 72 |

Poly(hydrazoadipoylhydrazosuccinyl)

| | | DMSO (sw). | 72 |

Poly(hydrazoisophthaloylhydrazoterephthaloyl)

| | N-methylpyrrolidone, formic acid (degrad.), di-chloroacetic acid (degrad.), trifluoroacetic acid (degrad.), DMA, DMSO (cold), sulfuric acid (degrad.). | chloroform, trichloroethane (sw), trifluoropropanol, m-cresol, nitrobenzene, water. | 73 |

3.4.3 Poly(ureas)

Poly(ureas)

| | general: phenol, m-cresol, formic acid, sulfuric acid. | alcohols, diethyl ether. | 3 |

Poly(ureylenenonamethylene)

| | sulfuric acid (60 % and 95 %). | m-cresol (sw), glacial acetic acid, sulfuric acid (90 %, sw), aqu. NaOH (45 %, sw). | 175 |

Poly(ureylene-1, 3-phenyleneureylene-2-methyl-5-sulfo-1, 3-phenylene)

| | aqu. NaOH. | water. | 176 |

Polymer	Solvents	Non-solvents	References
3.4.4 Poly(carbodiimides) (arranged according to increasing complexity)			
Poly(diethylcarbodiimide)	formic acid, aqu. mineral acids.		55
Poly(diallylcarbodiimide)	formic acid.		55
Poly(di-n-butylcarbodiimide)	formic acid.		55
Poly(isobutylcarbodiimide)	formic acid.		55
Poly(di-n-hexylcarbodiimide)	formic acid, n-heptane.		55
Poly(diphenylcarbodiimide)	formic acid.		55
Poly(hexamethylenecarbodiimide)	no solvent known.		23
Poly(4,4'-diphenylenemethanecarbodiimide)			
	m-cresol, trichlorophenol/phenol (7:10).		23
Poly(3-methyl-1,4-phenylenecarbodiimide)			
	no solvent known.		23
Poly(1,3-xylylenecarbodiimide)	no solvent known.	xylene, chloroform, trichloroethylene, chloro-	23
Poly(2,2'-dimethyl-biphenylenecarbodiimide)		benzene, ethylene chlorohydrin, cyclohexanone,	
	no solvent known.	formic acid, DMF, DMSO.	23
Poly(2,2'-dimethoxy-biphenylenecarbodiimide)			
	no solvent known.		23
Poly(1,5-naphthylenecarbodiimide)	no solvent known.		23
3.5 Poly(phosphazenes)			
Poly(bis(trifluoroethoxy)phosphazene)	methanol (hot), ethylene glycol dimethyl ether, THF, acetone, cyclohexanone (hot), methyl ethyl ketone, ethyl acetate.	aliphatic hydrocarbons, cyclohexane, aromatic hydrocarbons, ethanol, diethyl ether, dioxane, water.	182-184
Poly(dimethoxyphosphazene)	chloroform, methanol, dimethoxyethane, THF, dioxane, pyridine, DMF, acetonitrile.	benzene, ethanol, diethyl ether, acetone, water.	182
Poly(diphenoxyphosphazene)	benzene (hot), toluene, chlorobenzene, THF, dioxane, DMF.	hexane, ethanol, acetone, DMSO, water.	182
3.6 Poly(silanes), Poly(silazanes)			
Poly(alkyltrisilazane)	cyclohexane, aromatic hydrocarbons, carbon tetrachloride.		179
Poly(dimethylsilylene-1,4-phenylene) (Poly(dimethyl-p-phenylenesilane))			
	aromatic hydrocarbons.	aliphatic hydrocarbons, ethanol.	177
Poly(hexamethylcyclotrisilazane)	aliphatic and aromatic hydrocarbons, carbon tetrachloride.	acetone, DMSO.	180
Poly(iminoethyleneiminodimethylsilylene)			
	xylene.		181
4. Main-chain Heterocyclic Polymers			
4.1 Poly(benzoxazoles), Poly(oxadiazoles), Poly(oxadiazolidines)			
Poly(benzoxazoles)			
Poly(2,5-benzoxazole-diyl)	sulfuric acid.		196
Poly(2,6-benzoxazole-diyl)	sulfuric acid.		196

Poly(dibenzoxazoles)

R = -m-C$_6$H$_4$-	sulfuric acid.	organic solvents.	90
R = -p-C$_6$H$_4$-	sulfuric acid.	organic solvents.	90, 198
R = $+$ CH$_2$ $+_8$	m-cresol, formic acid, sulfuric acid.		198
R = cyclohexylene	sulfuric acid.	formamide, DMF, DMA, DMSO, N-methylpyrrolidone, acetic acid.	195

R = $+$ CH$_2$ $+_4$	tetrachloroethane (hot), chloroform, m-cresol, pyridine (hot), THF (hot), DMF (hot), sulfuric acid (hot).	benzyl alcohol, formic acid.	222

Polymer	Solvents	Non-solvents	References

4.1   Poly(benzoxazoles), Poly(oxadiazoles), Poly(oxadiazolidines)   (Cont'd.)

Poly(dibenzoxazoles)   (Cont'd.)

$R = -m-C_6H_4-$	tetrachloroethane (hot), chloroform, benzyl alcohol (hot), m-cresol (hot), THF, pyridine, sulfuric acid.	DMF (partially), formic acid (partially).	222
$R = -p-C_6H_4-$	chloroform, m-cresol, formic acid (hot), sulfuric acid (hot).	tetrachloroethane, benzyl alcohol, pyridine, THF, DMF.	222

Poly(1,3,4-oxadiazoles)

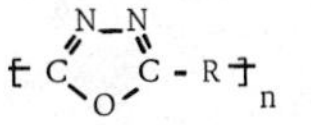

$R = -(CH_2)_8-$	benzene, chloroform, dichloroethylene, m-cresol, N-methylpyrrolidone (hot), DMA (hot), DMSO (hot), DMF, sulfuric acid.		74, 92, 199
$R = -(CH_2)_{4,5,6,7,9,10,12,15,20}-$	N-methylpyrrolidone (hot), DMA (hot), DMSO (hot), sulfuric acid (95 %).		199
R = p-cyclohexylene	sulfuric acid, formic acid (80 %).	DMSO, DMA, N-methylpyrrolidone.	200
R = m-cyclohexylene	sulfuric acid, formic acid (80 %), DMSO, DMA, N-methylpyrrolidone.		200
$R = -o-C_6H_4-$	sulfuric acid.		201
$R = -m-C_6H_4-$	trifluoroacetic acid, sulfuric acid.	chloroform, tetrafluoropropanol, m-cresol, DMF, nitrobenzene.	223-225

Poly(oxadiazolidines)

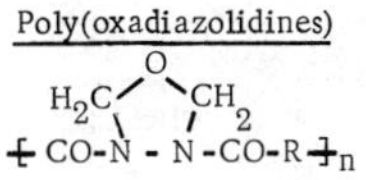

$R = -m-C_6H_4-$	DMSO.	methanol.	75

4.2   Poly(dithiazoles), Poly(benzothiazoles)

Poly(dithiazoles)

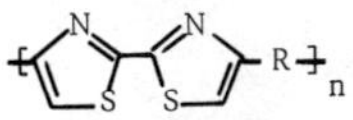

$R = -p-C_6H_4-$	sulfuric acid (cold).	ethanol, diethyl ether, DMF, diphenylmethane, phenyl ether, quinoline, formic acid, DMA, DMSO.	102
$R = -p-C_6H_4-(CH_2)_6-p-C_6H_4-$	sulfuric acid (cold).		102
$M_w = 11.000$	DMF (cold).		
$M_w = 12.000$	DMF (hot).		
$M_w = 27.000$		DMF (hot).	

Poly(benzothiazoles)

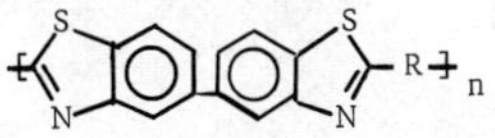

$R = -m-C_6H_4-$	conc. sulfuric acid.		19
R = -3,5-pyridylene-	conc. sulfuric acid.		19
$R = -p-C_6H_4-O-p-C_6H_4-$	conc. sulfuric acid.		19
$R = -p-C_6H_4-CO-p-C_6H_4-$	conc. sulfuric acid.		19
$R = -p-C_6H_4-$	conc. sulfuric acid.		198
R = -1,6-anthrylene-	sulfuric acid (partially).		203
Poly(2,6-benzothiazole-diyl)	sulfuric acid.		196

4.3   Poly(pyromellitimides)

Poly(pyromellitimides)

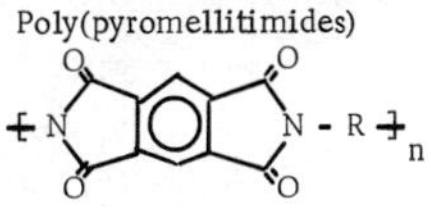

$R = -(CH_2)_9-$	m-cresol	common organic solvents	1, 20

Polymer	Solvents	Non-solvents	References

4.3   Poly(pyromellitimides) (Cont'd.)

R = -m-C_6H_4-; -p-C_6H_4-	conc. sulfuric acid.	common organic solvents.	1, 20
R = -p-C_6H_4-p-C_6H_4-	fuming nitric acid.	common organic solvents.	1, 20
R = -p-C_6H_4-CH_2-p-C_6H_4-; -p-C_6H_4-$C(CH_3)_2$-p-C_6H_4-;			
	conc. sulfuric acid.	common organic solvents.	1, 20
R = -p-C_6H_4-O-p-C_6H_4-	DMA, N-methylcaprolactame, fuming nitric acid.	common organic solvents.	1, 20
R = -p-C_6H_4-S-p-C_6H_4-	fuming nitric acid.	common organic solvents.	1, 20
R = -p-C_6H_4-SO_2-p-C_6H_4-; -m-C_6H_4-SO_2-m-C_6H_4-	conc. sulfuric acid.	common organic solvents.	1, 20

R =
$$\begin{array}{c} -CH_2\diagdown\ \diagup CH_2- \\ \qquad C \\ -CH_2\diagup\ \diagdown CH_2- \end{array}$$
          no solvent known.   190

R = -m-C_6H_3(4-OH)-$C(CH_3)_2$-m-C_6H_3(4-OH)-

          DMF, DMA, DMSO, nitrobenzene, tetramethylene
          sulfone.   191

4.4   Poly(quinoxalines)

Poly(quinoxaline-2,6-diylquinoxaline-6,2-diyl-1,4-phenylene)

          sulfuric acid.   194

Poly(quinoxalines) several different compositions

          1,3-dichloro-1,1,3,3-tetrafluoro-2,2-dihydroxy-   192, 193
          propane, methane sulfonic acid.

4.5   Poly(benzimidazoles)

Poly(benzimidazoles)

R = $-(CH_2)_4$	formic acid, DMSO (partially).	DMF.	18
R = -p-C_6H_4-		DMF, DMSO, formic acid.	18
R = -m-C_6H_4-	formic acid (partially).	DMF, DMSO.	18

Poly(dibenzimidazoles)

R = $-(CH_2)_n$			
n = 4, 7, 8, 11, 20	N-methylpyrrolidone, m-cresol, formic acid (85 %), DMA, DMSO, conc. sulfuric acid.	glacial acetic acid, DMF (sw).	40
R = -cyclohexylene-	m-cresol, N-methylpyrrolidone, DMF, DMSO, acetic acid, formic acid, sulfuric acid.		195
R = -m-C_6H_4-	N-methylpyrrolidone, formic acid, DMF (partially), DMA, DMSO, conc. sulfuric acid.	m-cresol, formic acid (85 %, sw), glacial acetic acid, DMF.	18, 40
R = -p-C_6H_4-	formic acid (2-3 % soluble).	DMF, DMSO.	18, 40
R = -o-biphenylene-	formic acid, DMF, DMSO.		18, 40
R = -p-biphenylene-	formic acid (partially).	DMF, DMSO.	18, 40
R = -3,5-pyridylene-	formic acid (partially), DMF (partially), DMSO.		18, 40
R = -2,5-furylidene-	formic acid (partially), DMF, DMSO.		18, 40
R = -p-C_6H_4-NHCO-RCONH-p-C_6H_4-; -$(CH_2)_8$-; -p-C_6H_4-; -m-C_6H_4-			
	N-methylpyrrolidone, DMF, DMA, DMSO, conc. sulfuric acid.	methanol, m-cresol (sw), formic acid (sw), glacial acetic acid (sw).	18, 40

4.6   Poly(piperazines)

Poly(2,5-diketo-1,4-piperazinediyl-1,4-phenylenesulfonyl-1,4-phenylene)

          formic acid, dichloroacetic acid, trifluoroacetic   DMA, DMF, DMSO.   160
          acid.

Polymer	Solvents	Non-solvents	References

4.6   Poly(piperazines) (Cont'd.)

Poly(2,5-dimethyl-1,4-piperazinediylcarbonyl-1,4-phenylenesulfonyl-1,4-phenylene)

| | sym-tetrachloroethane/phenol (40:60), DMF. | methanol, water. | 1 |

Poly(1,4-piperazinediylcarbonylethylene-1,4-phenylenephosphinylidene-1,4-phenyleneethylenecarbonyl)

| | chloroform, ethanol, C-7 fluoroalcohol, cresol, formic acid, DMF, aqu. hydrochloric acid (5 %). | hexane, chlorobenzene, acetone, acetonitrile (sw). | 84 |

Poly(piperazinediylcarbonyloxy-cis-cyclohexyleneoxycarbonyl)

| | methylene chloride, chloroform, m-cresol, formic acid. | | 152 |

Poly(1,4-piperazinediylcarbonyl-2,5-pyridinediylcarbonyl)

| | phenol, m-cresol, formic acid, sulfuric acid, conc. aqu. HCl. | | 174 |

4.7   Poly(anhydrides)

| Poly(maleic anhydride) | lower aliphatic alcohols, ethers, ketones, nitro-paraffins, dioxane, butanone, acetophenone, acetic anhydride, DMF, acetonitrile, water. | higher paraffins, most chlorinated hydrocarbons, aromatic hydrocarbons. | 83 |

Poly(acrylic anhydride)

$$+ CH_2 +_n$$
(with ring structure containing O, O)

| | DMF, DMSO. | benzene, alcohols, diethyl ether. | 1 |

| Poly(methacrylic anhydride) | DMF, DMSO. | aliphatic hydrocarbons, methanol. | 65 |
| Poly(itaconic acid anhydride) | THF, dioxane, acetone, methyl ethyl ketone, cyclohexanone, acetic acid. | chloroform, diethyl ether. | 238 |

5.   Formaldehyde Resins

Aniline-formaldehyde resins

| low mol. wt. | chlorinated hydrocarbons, methylcyclohexanone. | hydrocarbons, alcohols, aniline (sw). | 12 |
| high mol. wt. | no solvent known. | chlorinated hydrocarbons (sw), aniline. | 12 |

Melamine-formaldehyde resins

very low mol. wt.	alcohol, water.		12
intermediates	pyridine, formalin, formic acids, dil. and conc. acids.		
final resins, high mol. wt.	no solvent known.		

Phenol-formaldehyde resins

Novolaks and low mol. wt.	hydrocarbons, diethyl ether, acetone, esters.		
4-tert-butylphenol and 4-phenylphenol polymers	drying oils.		
final resins	molten phenols (with some decomposition).		
fully cured resins		most common organic compounds, alcohol, water (sw), dil. acids (sw), dil. alkalies (sw).	12

p-Toluenesulfonamide-formaldehyde resins

| | many organic solvents. | aliphatic hydrocarbons, alcohol, water. | 12 |

Urea-formaldehyde resins

very low mol. wt.	alcohol, water.		12
intermediates	pyridine, formalin, formic acid, dil. and conc. acids.		
final resins, high mol. wt.	no solvent known.		

6.   Natural Polymers and Modified Natural Polymers

6.1   Natural Rubber and Derivatives   (see also 1.1)

Natural rubber	hydrocarbons, benzene, toluene, chlorinated hydrocarbons, n-propyl ketone (above 14.5 °C).	alcohols, acetone, carboxylic acids.	2, 4
cyclic	aromatic and chlorinated hydrocarbons.	aliphatic hydrocarbons, oils (sw), alcohol, diethyl ether, acetone, water, acids, alkalies.	12
hydrochlorinated	chlorinated hydrocarbons.	alcohol, water.	12
chlorinated	hydrogenated naphthas, aromatic and chlorinated hydrocarbons, ketones (limited solubility in acetone), esters, glyceryl esters.	aliphatic hydrocarbons, alcohol, diethyl ether (sw), acetone (sw), water, conc. acids, oxidizing acids, moderately conc. alkalies.	12

	Polymer	Solvents	Non-solvents	References
6.2	Gutta Percha			
	Gutta percha	hot petroleum ether, benzene, chloroform.	alcohol, water.	12
6.3	Cellulose and Derivatives			
6.3.1	Cellulose			
	Cellulose	trifluoroacetic acid, N-ethylpyridinium chloride/pyridine (1:1), N-ethylpyridinium chloride/DMF (1:1), aqu. solutions of: cupriethylenediamine, sodium xanthate, tetramethylammonium hydroxide, calcium thiocyanate, alkalies (ice-cold), cupriammonium hydroxide, beryllium perchlorate, zinc chloride (hot), zinc chloride/hydrochloric acid (cold), conc. sulfuric acid, conc. phosphoric acid.	hydrocarbons, mineral oils, water (sw), dil. aqu. alkalies.	2, 4, 12, 204, 205
6.3.2	Cellulose Ethers			
	Methyl cellulose			
	S.C. = 3-10 %	aqu. alkali.	water.	12, 210
	= 22-32 %	aqu. alkali, water (cold), methanol/methylene chloride, N-ethylpyridinium chloride/DMF.	water (hot), methanol, diethyl ether, methylene chloride.	12, 205, 210
	=> 40 %	chloroform, acetone, pyridine, esters, cyclohexanone.	ethanol, diethyl ether, water, aliphatic hydrocarbons.	12, 210
	Ethyl cellulose			
	D.S. = 0.5-o.7	aqu. alkali.	water (cold).	3-5, 12
	= 1.0-1.5	pyridine, formic acid, acetic acid, walter (cold), cuoxam.	ethanol.	210
	= 2	methylene chloride, chloroform, dichloroethylene, chlorohydrins, ethanol, THF.	hydrocarbons, carbon tetrachloride, trichloroethylene, alcohols, diethyl ether, ketones, esters, water.	
	= 2.3	benzene, toluene, alkyl halogenids, alcohols, furan derivatives, ketones, acetic esters, carbon disulfide, nitromethane.	ethylene glycol, acetone (cold).	
	= 3	benzene, toluene, methylene chloride, alcohols, esters.	hydrocarbons, decalin, xylene, carbon tetrachloride, tetrahydrofurfuryl alcohol, diols, n-propyl ether.	
	Isopropyl cellulose			
	D.S. = 0.5	water (cold).	water (hot).	210
	= 2.5	benzene, chloroform, dichloroethane, butanol, THF, dioxane, glycol formal, acetone, cyclohexanone, methyl cyclohexanone, methyl acetate.	hydrocarbons, methanol, ethanol, ethanol/diethyl ether (1:3), propanol.	3
	Butyl cellulose			
	D.S. = 0.8	polar organic solvents.	hydrocarbons, water.	210
	= 1.9	aromatic hydrocarbons, polar organic solvents.	aliphatic hydrocarbons, water.	210
	Benzyl cellulose			
	D.S. = 1.65	methylene chloride, methylglycol acetate, nitrobenzene.	benzene, chlorobenzene, butyl acetate.	210
	= 2	tetralin, benzene/alcohol, benzene/acetone, methylene chloride, dichloroethylene, trichloroethylene, chlorobenzene, chlorohydrins, ethanol/carbon tetrachloride (1:1), tetrahydrofurfuryl alcohol, pyridine, acetone, methyl ethyl ketone, acetates ($< C_5$).	hydrocarbons, carbon tetrachloride, o-chlorotoluene, alcohols ($> C_5$), diethyl ether, dioxane, dipropyl ketone.	3, 12
	= 3	chloroform, chlorobenzene, acetone, methyl ethyl ketone, cyclohexanone, ethyl acetate, butyl acetate.	aliphatic hydrocarbons, aromatic hydrocarbons (sw), carbon tetrachloride, ethanol, diethyl ether, dioxane.	
	Hydroxyethyl cellulose			
	S.C. = < 30 %	alkali.		1, 2, 12
	= 35-55 %	water.		
	=> 55 %	benzene/alcohol, benzene/acetone, chloroform, pyridine, acetone, esters.	benzene, methanol (sw), THF, acetone, methyl acetate.	
	Carboxymethyl cellulose			
	S.C. = 5-10 %	alkali.		12
	= 15-30 %	water (sodium salt).		
	= high	benzene/alcohol, benzene/acetone, chloroform, pyridine, acetone, esters.		

Polymer	Solvents	Non-solvents	References
6.3.2 <u>Cellulose Ethers</u> (Cont'd.)			
Cyanoethyl cellulose			
S.C. = 8-12 %	alkali.		12
= 24-32 %	water.		
= >50 %	chloroform, benzene/alcohol, benzene/acetone, pyridine, acetone, esters.		
6.3.3 <u>Cellulose Esters</u>			
Cellulose triformate	pyridine, formic acid.		3
Cellulose acetate			
D.S. = 0.6-0.8	water.		
= 1.3-1.7	2-methoxyethanol.	acetone, water.	210
= 2 -2.3	methylene chloride/methanol (80:20), chloroform/methanol, benzyl alcohol, phenols, ethylene glycol ethers, dioxane, diethanolamine, pyridine, aniline, acetone, cyclohexanone, formic acid, acetic acid (glacial), methyl acetate, ethyl acetate/nitrobenzene, glycol monoethyl ether acetate, nitromethane.	hydrocarbons, aliphatic ethers, weak mineral acids.	2, 12
Cellulose triacetate	methylene chloride, methylene chloride/ethanol (8:2), chloroform, chloroform / alcohol, trichloroethane, THF, dioxane, acetone, acetone/water (8:2), methyl acetate, ethyl acetate, ethylene glycol ether acetates, ethylene carbonate.	aliphatic hydrocarbons, benzene, dichloroethane, chlorobenzene, o-chlorotoluene, ethanol (absolute), aliphatic ethers, acetone, weak mineral acids.	2-4, 12, 228
Cellulose tripropionate	benzene, dichloroethane, chlorobenzene, acetone, ethyl acetate.	hydrocarbons, ethanol, o-chlorotoluene.	3
Cellulose tributyrate	benzene, xylene (hot), dodecane/tetralin (3:1, above 130 °C), chloroform, carbon tetrachloride, chlorobenzene, ethanol (hot), THF, dioxane, methyl ethyl ketone, cyclohexanone, ethyl acetate, amyl acetate, nitrobenzene.	hexane, cyclohexane, methanol, 2-ethylhexanol, diethyl ether.	210
Cellulose tricaproate	hexane, benzene, xylene, chloroform, carbon tetrachloride, chlorobenzene, ethanol (hot), diethyl ether, THF, dioxane, acetone, cyclohexanone, ethyl acetate, DMF, nitroethane, glacial acetic acid (hot).	methanol, formamid, β-hydroxyethyl acetate, nitromethane.	210
Cellulose tricaprylate	toluene, 3-phenylpropanol (above 48 °C), DMF (above 140 °C).		4
Cellulose stearate			
D.S. = 3	benzene, toluene, tetrachloroethane.	dichloroethane, acetone, ethyl acetate, butyl acetae.	210
Cellulose acetate butyrate			
D.S. (acetate) = 0.5			
D.S. (butyrate) = 2.35	benzene, toluene (hot), chloroform, carbon tetrachloride, tetrachloroethane, methanol (hot), acetone, cyclohexanone, dioxane, aliphatic esters, nitroethane.	aliphatic hydrocarbons, methanol (cold), ethanol, diethyl ether.	210
D.S. (acetate) = 2.1			
D.S. (butyrate) = 0.7	chloroform, dichloroethane, tetrachloroethane, dioxane, acetone, cyclohexanone, methyl acetate, ethyl acetate, nitroethane.	hydrocarbons, benzene/methanol (1:4 and 4:1), carbon tetrachloride, methanol, ethanol, diethyl ether, amyl acetate.	210
Cellulose tricarbanilate	benzene, chloroform, THF, dioxane, acetone, DMF, pyridine, N-ethylpyridinium chloride/pyridine (1:1), N-ethylpyridinium chloride/DMSO (1:1).	methanol, water.	205, 210
6.3.4 <u>Cellulose Nitrate and Sulfate</u>			
Cellulose nitrate			
N 6.8 %	water.		206
N 10.5-12 %	alcohol (lower), alcohol/diethyl ether, acetone, amyl acetate, ethylene glycol ethers, acetic acid (glacial).	higher alcohols, higher carboxylic acids, higher ketones, tricresyl phosphate.	2, 12, 210
N 12.7 %	halogenated hydrocarbons, ethanol/diethyl ether, acetone, methyl amyl ketone, cyclohexanone, methyl acetate, ethyl acetate, ethyl butyrate, ethyl lactate, ethylene glycol ether acetates, ethylene carbonate, furan derivatives, nitrobenzene.	aliphatic hydrocarbons, aromatic hydrocarbons (sw), lower alcohols, higher alcohols (sw), ethylene glycol, diethyl ether, dil. carboxylic acids, water.	2, 4, 12, 210

Polymer	Solvents	Non-solvents	References
6.3.4 <u>Cellulose Nitrate and Sulfate</u> (Cont'd.)			
Cellulose sulfate, sodium salt			
D.S. = 1	water.	ethanol.	210
6.4 <u>Starch and Derivatives</u>			
Amylose	ethylenediamine, nitromethane, N-ethylpyridinium chloride/pyridine (1:1), DMSO, water (hot), aqu. KOH, aqu. KCl (0.33 M, above 25°C).	n-butanol, diethyl ether.	4, 100, 205, 207, 210
Amylose acetate			
D.S. = 2	chloroform, methyl acetate, nitromethane.	water.	4
Amylose tricarbanilate	dioxane, pyridine, acetone.		4
Amylopectin	water, water/butanol.	ethanol, diethyl ether, acetone.	208
Benzyl amylose			
D.S. = 3	benzene, chloroform, THF, dioxane, DMF, DMSO.	methanol, ethanol, water.	209
Starch	N-ethylpyridinium chloride/pyridine (1:1), aqu. sol. of cupriethylenediamine, fused chloral hydrate, water (hot, pressure), liquid ammonia.	diethyl ether, water (sw), alkali (sw).	12, 100, 205, 210
Starch, methyl ether			
D.S. = 0.5-2	water (cold).	water (hot).	210
D.S. = 3	chloroform.		210
6.5 <u>Other Poly(saccharides)</u>			
Alginic acid	water (sparingly), aqu. alkalies, aqu. sodium carbonate.	hydrocarbons, hydrophobic organic compounds, hydrophilic organic compounds (sw), dil. acids, aqu. calcium salt solutions.	12, 100
Alkali and ammonium salts	aqu. alkali, water.	organic hydrophobic compounds.	12
Magnesium and ferrous salts	water.	organic hydrophobic compounds.	
Calcium salts	weak aqu. alkali	water, organic hydrophobic compounds.	
Copper, zinc, aluminum salts	aqu. ammonium hydroxide.	water, organic hydrophobic compounds.	
Beryllium, chromium, ferric salts		aqu. alkali, organic hydrophobic compounds.	
Chitin	formic acid, conc. mineral acids (degrad.)	organic solvents, dil. acids.	100
Glycogen	dil. trichloroacetic acid.	alcohol.	100
Gum arabic	water (warm).	alcohol.	100
Gum tragacanth (Tragacanthin, tragacantic acid)	water.	alcohol, acetone.	100
Heparin	water, dil. alkali solution, alkaline solution of ammonium sulfate.	alcohol, acetone, acetic acid.	100
Pectin	water, dil. acids.	alcohol.	100
6.6 <u>Natural Resins</u>			
Colophony	aromatic hydrocarbons (partially), chloroform, carbon tetrachloride, methanol, ethanol, diethyl ether, acetone, ethyl acetate.	aliphatic hydrocarbons.	3
Copal	hydrocarbons (10-55 %)*, benzene (35-60 %)*, chloroform (30-90 %)*, carbon tetrachloride (15-40 %)*, methanol (20-70 %)*, ethanol (20-100 %)*, diethyl ether (10-90 %)*, acetone (25-90 %)*.		3
Shellac	benzene (10-20 %)*, chloroform (25-40 %)*, carbon tetrachloride (5-15 %)*, methanol (100 %), ethanol (85-98 %)*, diethyl ether (10-25 %)*, acetone (50-80 %)*.	aliphatic hydrocarbons.	3
7. <u>Inorganic Polymers</u>			
Poly(phosphorus nitrile chloride)	cyclohexane, benzene, chloroform, carbon tetrachloride.	aliphatic hydrocarbons.	215
Poly(Na-m-phosphate)	aqu. sol. of LiCl, NaCl or NH_4Cl.	ethanol, acetone, water.	202
Poly(tri-n-butyl tin fluoride)	aliphatic hydrocarbons.		216
Poly(phosphoryl dimethylamide)	carbon tetrachloride (hot), water.	acetone.	197

*) Percentage soluble, depending on source.

C. REFERENCES

1. Houben-Weyl, "Methoden der Organischen Chemie", Band XIV/1 and 2, 4. Auflage, Thieme Verlag, Stuttgart 1961, 1963.
2. H. Dexheimer, O. Fuchs in R. Nitsche and K. A. Wolf, "Struktur und Physikalisches Verhalten der Kunststoffe", Springer Verlag, Berlin-Goettingen-Heidelberg 1961.
3. K. Thinius, "Analytische Chemie der Plaste, "Springer Verlag, Berlin-Goettingen-Heidelberg, 1952.
4. M. Kurata, W. H. Stockmayer, "Advances in Polymer Science", Vol. 3, p. 196, Springer Verlag, Berlin-Goettingen-Heidelberg, 1963.
5. A. Muenster, "Loeslichkeit und Quellung" in H. A. Stuart, "Physik der Hochpolymeren", Vol. II, p. 193, Springer Verlag, Berlin-Goettingen-Heidelberg, 1953.
6. J. Furukawa, T. Saegusa, "Polymerisation of Aldehydes and Oxides", Polymer Reviews, Vol. 3, Interscience, New York, 1963.
7. P. J. Flory, J. E. Osterheld, J. Phys. Chem. 58, 653 (1954).
8. M. L. Miller, K. O'Donnell, J. Skogman, J. Colloid Sci. 17, 649 (1962).
9. E. M. Loebl, J. J. O'Neill, J. Polymer Sci. 45, 538 (1960).
10. R. C. Schulz, R. Wolf, Makromol. Chem. 99, 76 (1966).
11. N. Gaylord, "Polyethers", Part III, "Polyalkylene Sulfides and Other Polythioethers", High Polymers, Vol. XIII, Part III, Interscience, New York-London, 1962.
12. W. J. Roff, "Fibers, Plastics, and Rubbers", Academic Press, New York, 1956.
13. L. E. Coleman, J. F. Bork, D. P. Wyman, ACS Div. Polymer Chem. Polymer Preprints 5, 250 (1964).
14. C. Simionescu, D. Feldman, I. Theil, Plaste Kautschuk 15, 714 (1968).
15. C. G. Overberger, S. Ishida, ACS Div. Polymer Chem., Polymer Preprints 5, 210 (1964).
16. J. E. Moore, ACS Div. Polymer Chem. Polymer Preprints 5, 250 (1964).
17. J. S. Scarpa, D. D. Mueller, I. M. Klotz, J. Am. Chem. Soc. 89, 6024 (1967).
18. H. H. Levine, K. J. Kjeller, and C. G. Delano, ACS Div. Polymer Chemistry, Polymer Preprints 5, 160 (1964).
19. P. Hergenrother, W. Wrasidlo, H. H. Levine, ACS Div. Polymer Chem., Polymer Preprints 5, 153 (1964).
20. C. E. Sroog, S. V. Abramo, C. E. Berr, W. M. Edwards, A. L. Endrey, K. L. Olivier, ACS Div. Polymer Chem., Polymer Preprints 5, 132 (1964).
21. H. Hopff, M. Lippay, Makromol. Chem. 66, 157 (1963).
22. R. J. Kern, J. J. Hawkins, J. D. Calfee, Makromol. Chem. 66, 127 (1963).
23. D. J. Lyman, N. Sadri, Makromol. Chem. 67, 1 (1963).
24. G. F. Pregaglia, M. Binaghi, M. Cambini, Makromol. Chem. 67, 10 (1963).
25. W. Burchard, Makromol. Chem. 67, 182 (1963).
26. I. Penczek, S. Penzcek, Makromol. Chem. 67, 203 (1963).
27. K. Weissermel, E. Noelken, Makromol. Chem. 68, 140 (1963).
28. H. Hopff, J. Dohany, Makromol. Chem. 69, 131 (1963).
29. G. Natta, P. Longi, E. Pellino, Makromol. Chem. 71, 212 (1964).
30. A. Takahashi, S. Kambara, Makromol. Chem. 72, 92 (1964).
31. R. C. Schulz, R. Stenner, Makromol. Chem. 72, 202 (1964).
32. J. Heller, D. J. Lyman, W. A. Hewett, Makromol. Chem. 73, 48 (1964).
33. C. K. Chiklis, J. M. Grasshoff, J. Polymer Sci., A-2, 8, 1617 (1970).
34. C. Aso. Y. Aito, Makromol. Chem. 73, 141 (1964).
35. S. Kambara, H. Noguchi, Makromol. Chem. 73, 244 (1964).
36. H. Yuki. K. Hatada, S. Murahashi, K. Hibino, Polymer J. 1, 271 (1970).
37. K. Meyersen, R. C. Schulz, W. Kern, Makromol. Chem. 58, 204 (1962).
38. Y. Tabata, B. Saito, H. Shibano, H. Sobue, K. Oshima, Makromol. Chem. 76, 89 (1964).
39. H. Ohse, H. Cherdron, F. Korte, Makromol. Chem. 76, 147 (1964).
40. Y. Iwakura, K. Uno, Y. Imai, Makromol. Chem. 77, 33 (1964).
41. Y. Iwakura, K. Uno, Y. Imai, M. Fukui, Makromol. Chem. 77, 41 (1964).
42. H. O. Wirth, R. Mueller, W. Kern, Makromol. Chem. 77, 90 (1964).
43. T. Iwata, G. Wasai, T. Saegusa, J. Furukawa, Makromol. Chem. 77, 229 (1964).
45. J. Heller, C. B. Kingsley, Makromol. Chem. 78, 47 (1964).
46. M. Sumi, K. Matsumura, R. Ohno, S. Nozakura, S. Murahashi, Chem. High Polymers (Tokyo) 24, 606 (1967).
47. J. Pellon, L. H. Schwind, M. J. Guinard, W. M. Thomas, J. Polymer Sci. 55, 161 (1961).
48. A. T. Coscia, R. L. Kugel, J. Pellon, J. Polymer Sci. 55, 303 (1961).
49. C. A. Aufdermarsh, Jr., R. Pariser, J. Polymer Sci. A2, 4727 (1964).
50. G. F. L. Ehlers, J. Polymer Sci. A2, 4989 (1964).
51. W. Marcony, A. Mazzei, S. Cucinella, M. Cesari, J. Polymer Sci., A2, 4261 (1964).
52. J. M. Wilbur, Jr., C. S. Marvel, J. Polymer Sci. A2, 4415 (1964).
53. S. Saito, J. Polymer Sci., A-1, 7, 1789 (1969).
54. R. A. Wessling, J. Appl. Polymer Sci. 14, 1531 (1970).
55. G. C. Robinson, J. Polymer Sci. A2, 3901 (1964).
56. H. R. Allcock, J. Polymer Sci. A2, 4087 (1964).
57. R. A. Wessling, J. Appl. Polymer Sci. 14, 2263 (1970).
58. Y. Iwakura, K. Uno, K. Ichikawa, J. Polymer Sci. A2, 3387 (1964).
59. G. H. Kalb, D. D. Coffman, T. A. Ford, F. L. Johnston, J. Appl. Polymer Sci. 4, 55 (1960).
60. W. S. Durrell, G. Westmoreland, M. G. Moshonas, J. Polymer Sci., A, 3, 2975 (1965).
61. R. H. Michel, J. Polymer Sci. A2, 2533 (1964).
62. J. F. Klebe, J. Polymer Sci. A2, 2673 (1964).
63. I. M. Paushkin, S. A. Nizova, J. Polymer Sci. A2, 2783 (1964).
64. P. Teyssie, A. C. Korn-Girard, J. Polymer Sci. A2, 2849 (1964).
65. J. C. L. Hwa, W. A. Fleming, L. Miller, J. Polymer Sci. A2, 2395 (1964).
66. R. E. Moser, H. Kamogawa, H. Hartmann, H. G. Cassidy, J. Polymer Sci. A2, 2401 (1964).
67. H. Kamogawa, H. G. Cassidy, J. Polymer Sci. A2, 2409 (1964).
68. E. L. Gal'perin et al., Vysokomolekul. Soedin. A 12, 1654 (1970).
70. J. Prager, R. M. McCurdy, G. B. Rathmann, J. Polymer Sci. A2, 1941 (1964).
71. D. W. Thomson, G. F. L. Ehlers, J. Polymer Sci. A2, 1051 (1964).
72. A. H. Frazer, F. T. Wallenberger, J. Polymer Sci. A2, 1137 (1964).
73. A. H. Frazer, F. T. Wallenberger, J. Polymer Sci. A2, 1147 (1964).
74. A. H. Frazer, W. Sweeny, F. T. Wallenberger, J. Polymer Sci. A2, 1157 (1964).
75. A. H. Frazer, F. T. Wallenberger, J. Polymer Sci. A2, 1181 (1964).
76. R. Rabinowitz, R. Marcus, J. Pellon, J. Polymer Sci. A2, 1233 (1964).
78. M. Imoto, T. Otsu, K. Tsuda, T. Ito, J. Polymer Sci. A2, 1407 (1964).
79. R. C. Doban, A. C. Knight, J. H. Peterson, C. A. Sperati, Am. Chem. Soc., Meeting Atlantic City, 1956.
80. R. Bacskai, J. Polymer Sci. A1, 2777 (1963).
81. R. Reichel, C. S. Marvel, R. Z. Greenley, J. Polymer Sci. A1, 2935 (1963).
82. D. A. Frey, M. Hasegawa, C. S. Marvel, J. Polymer Sci. A1, 2057 (1963).
83. J. L. Lang, W. A. Pavelich, H. D. Clarey, J. Polymer Sci. A1, 1123 (1963).
84. J. Pellon, W. G. Carpenter, J. Polymer Sci. A1, 863 (1963).
85. A. V. Topchiev, V. P. Alaniya, J. Polymer Sci. A1, 599 (1963).
86. W. P. Baker, J. Polymer Sci. A1, 655 (1963).

87. R. E. Lowry, D. W. Brown, L. A. Wall, J. Polymer Sci. A-1, 4, 2229 (1966).

88. D. W. Brown, R. E. Lowry, L. A. Wall, J. Polymer Sci., A-1, 8, 2441 (1970).

89. D. McIntyre, D. R. Valentine, Am. Chem. Soc. Meeting, New York, Sept. 1966, Polymer Preprints 7, 1133 (1966).

90. T. Kubota, R. Nakanishi, J. Polymer Sci. B2, 655 (1964).

91. H. Staudinger, M. Haeberle, Makromol. Chem. 9, 52 (1953).

92. M. Hasegawa, T. Unishi, J. Polymer Sci. B2, 237 (1964).

93. H. Mains, J. H. Day, J. Polymer Sci. B1, 347 (1963).

94. L. D. Taylor, H. S. Kolesinski, J. Polymer Sci. B1, 117 (1963).

95. R. C. Schulz, J. Kovacs, W. Kern, Makromol. Chem. 52, 236 (1962).

96. R. Hank, Makromol. Chem. 52, 108 (1962).

97. R. C. Schulz, W. Passmann, Makromol. Chem. 60, 139 (1963).

98. J. Velickovic, J. Vukajlovic-Filipovic, Angew. Makromol. Chem. 13, 79 (1970).

99. Y. L. Gefter, International Series of Monographs on Organic Chemistry, Vol. 6, "Organophosphorus Monomers and Polymers", Pergamon Press, MacMillan Company, New York (1962).

100. B. Jirgensons, "Natural Organic Polymers", Pergamon Press, New York, 1962.

101. G. P. Brown, A. Goldman, ACS Div. of Polymer Chem., Polymer Preprints 4, no. 2, 39 (1963).

102. D. T. Longone, H. H. Un, J. Polymer Sci. A 3, 3117 (1965).

103. S. P. Yen, ACS Div. Polymer Chem., Polymer Preprints 4, 82 (1963).

104. J. Moacanin, A. Rembaum, R. K. Laudenslager, ACS Div. Polymer Chem., Polymer Preprints 4, 179 (1963).

105. J. Velickovic, V. Juranicova, J. Filipovic, IUPAC Intern. Symp. Macromolecules, Leiden, Vol. I, 289 (1970).

106. G. B. Butler, Proc. Conf. High Temperature Polymer and Fluid Research, p. 53, Dayton, Ohio, 1962.

107. H. A. Smith, C. E. Handlovits, Proc. Conf. High Temperature Polymer and Fluid Research, p. 123, Dayton, Ohio 1962.

108. G. Natta, G. Mazzanti, P. Longi, F. Bernardini, J. Polymer Sci. 31, 181 (1958).

109. I. R. Schaefgen, J. Polymer Sci. 41, 133 (1959).

110. E. J. Vandenberg, R. F. Heck, D. S. Breslow, J. Polymer Sci. 41, 519 (1959).

111. R. F. Heck, D. S. Breslow, J. Polymer Sci. 41, 521 (1959).

112. E. C. Winslow, A. Laferriere, J. Polymer Sci. 60, 65 (1962).

113. H. T. Hall, J. Am. Chem. Soc. 74, 68 (1952).

114. C. Aso, Y. Aito, Makromol. Chem. 58, 195 (1962).

115. H. Nakamoto, Y. Ogo, T. Imoto, Makromol. Chem. 111, 104 (1968).

116. G. Natta, G. Mazzanti, G. Dall'Asta, P. Longi, Makromol. Chem. 37, 160 (1960).

117. J. Furukawa, T. Saegusa, H. Fujii, Makromol. Chem. 44/46, 398 (1961).

118. D. D. Coffman, G. H. Kalb, A. B. Ness, J. Org. Chem. 13, 223 (1948).

119. W. W. Moyer, Jr., D. A. Grev, J. Polymer Sci. B1, 29 (1963).

120. G. E. Ham, Ind. Eng. Chem. 46, 390 (1954).

121. S. Arichi, H. Matsuura, Y. Tamimoto, H. Murata, Bull. Chem. Soc. (Japan) 39, 434 (1966).

122. D. E. Strain, R. G. Kennelly, H. R. Dittmar, Ind. Eng. Chem. 31, 382 (1939).

123. C. E. Schildknecht, S. T. Gross, H. R. Davidson, J. M. Lambert, A. O. Zoss, Ind. Eng. Chem. 40, 2104 (1948).

124. DBP 926,163; 928, 793; 929, 931; 929.932 (1955), Farbwerke Hoechst; O. Fuchs et al., Chem. Z. 1956, 1169; 1958, 3756.

125. P. Grosius, Y. Gallot, A. Skoulios, Makromol. Chem. 136, 191 (1970).

126. Brit. P. 849.038 (1960). Dow Chemical, C.A. 55, 12432 (1961).

127. C. G. Overberger, S. Ishida, H. Ringsdorf, J. Polymer Sci. 62, S1 (1962).

128. L. E. Coleman, J. F. Bork, J. Polymer Sci. A-1, 8, 2073 (1970).

129. W. A. Butte, C. C. Price, R. E. Hughes, J. Polymer Sci. 61, 528 (1962).

130. G. F. Endres, A. S. Hay, J. W. Eustance, J. Org. Chem. 28, 1300 (1963).

131. G. Dall'Asta, G. Natta, G. Mazzanti, Chem. Engng. News 42, no. 14, 42 (1964).

132. USP 3.036.990; 3.036.991 (1962), S. W. Kantor, F. F. Holub, General Electric Company.

133. R. G. Alsup. J. O. Punderson, G. F. Leverett, J. Appl. Polymer Sci. 1, 185 (1959).

134. V. N. Kuznetsov, S. D. Vogman et al., Vysokomolekul. Soedin. A 12, 2635 (1970).

135. N. G. Gaylord, "Polyethers", Interscience, New York, 1963, Part I, p. 46.

136. R. C. Schulz, R. Wolf, Makromol. Chem. 99, 76 (1966).

137. T. Saegusa, S. Matsumoto, Y. Hashimoto, Macromolecules 3, 377 (1970).

138. G. Natta, G. Mazzanti, G. F. Pregaglia, M. Dinaghi, M. Cambini, Makromol. Chem. 51, 148 (1962).

139. G. D. Staffin, C. C. Price, J. Am. Chem. Soc. 82, 3632 (1960).

140. W. R. Moore, M. Uddin, European Polymer J. 5, 185 (1969).

141. W. R. Moore, M. Uddin, European Polymer J. 6, 121 (1970).

142. G. S. Kolesnikov et al., Vysokomolekul. Soedin. B 12, 548 (1970).

143. R. H. Marchessault, K. Okamura, C. J. Su, Macromolecules 3, 735 (1970).

144. S. Inoue, Y. Tomoi, T. Tsuruta, J. Furukawa, Makromol. Chem. 48, 229 (1961).

145. H. Cherdron, H. Ohse, F. Korte, Makromol. Chem. 56, 179, 187 (1962).

146. H. Ohse, H. Cherdron, Makromol. Chem. 95, 283 (1966).

147. H. Batzer, H. Lang, Makromol. Chem. 15, 211 (1955).

148. H. Prueckner, Erdoel, Kohle, Erdgas, Petrochemie 16, 188 (1959).

149. F. H. Henglein, H. E. Tarrach, Kunststoffe-Plastics 6, 5 (1959).

151. T. M. Gricenko et al., Khim. Volokna 1970, No. 5, p. 18.

152. E. L. Wittbecker, M. Katz, J. Polymer Sci. 40, 367 (1959).

153. D. J. Lyman, J. Polymer Sci. 45, 49 (1960).

155. D. J. Lyman, J. Polymer Sci. 55, 507 (1961).

156. B. A. Kiselov, I. A. Stepina, Z. P. Ablekova, Soviet Plastics 1970, p. 13.

157. J. B. Ganci, F. A. Bettelheim, J. Polymer Sci. A2, 4011 (1964).

158. US-Patent 3.380.951 (Phillips Petroleum).

159. H. A. Vogel, J. Polymer Sci., A-1, 8, 2035 (1970).

160. Y. Iwakura, S. I. Izawa, F. Hayano, J. Polymer Sci., A-1, 6, 1097 (1968).

162. M. E. A. Cudby, R. G. Feasy, B. E. Jennings, M. E. B. Jones, J. B. Rose, Polymer 6, 589 (1965).

163. B. L. Kwolek, P. W. Morgan, J. Polymer Sci. A2, 2693 (1964).

164. S. A. Suncket, W. A. Murphey, S. B. Spenk, J. Polymer Sci. 40, 389 (1959).

165. V. E. Shashoua, W. Sweeny, R. F. Tietz, J. Am. Chem. Soc. 82, 866 (1960).

166. J. Masamoto, Y. Kaneko, K. Sasaguri, H. Kobayashi, J. Soc. Fiber Sci. Technol. Japan 25, 525 (1969).

167. J. Masamoto, K. Sasaguri, C. Ohizumi, H. Kobayashi, J. Polymer Sci. A-2, 8, 1703 (1970).

168. R. Graf, G. Lohaus, K. Boerner, E. Schmidt, H. Bestian, Angew. Chem. 74, 523 (1962).

169. E. Schmidt, Angew. Makromol. Chem. 14, 185 (1970).

170. H. Sekiguchi, Bull. Soc. Chim. France 1960, 1839.

171. G. Reinisch, K. Dietrich, European Polymer J. 6, 1269 (1970).

172. C. E. Sroog, S. V. Abramo, C. E. Berr, W. M. Edwards, A. L. Endrey, K. L. Olivier, ACS, Div. Polymer. Chem., Polymer Preprints 5/1, 132 (1964); J. Polymer Sci. A3, 1373 (1965).

173. V. E. Shaskoua, W. M. Eareckson, J. Polymer Sci. 40, 343 (1959).

174. V. P. Sarzevskaja et al., Ukr. Khim. Zh. 35, 390 (1969).

175. T. Ikemura, Chem. High Polymers (Tokyo) 26, 845 (1969).

176. H. Rinke, E. Istel in Houben-Weyl, "Methoden der Organischen Chemie", 4. Ed., Vol. 14/2; E. Mueller, Editor, Georg Thieme Verlag, Stuttgart 1963.

177. K. A. Andrianov, V. I. Pachomov, V. M. Gel'perina, Dokl. Akad. Nauk SSSR 162, 79 (1965).

179. L. W. Breed, R. L. Elliott, A. F. Ferris, J. Polymer Sci. A 2, 45 (1964); J. Org. Chem. 27, 1114 (1962).

180. C. R. Krueger, E. G. Rochow, J. Polymer Sci. A2, 3179 (1964).

181. R. Minné, E. G. Rochow, J. Am. Chem. Soc. 82, 5625 (1960).

182. H. R. Allcock, R. L. Kugel, K. J. Valan, Inorg. Chem. 5, 1709 (1966).

183. G. Allen, C. J. Lewis, S. M. Todd, Polymer 11, 31 (1970).

184. G. Allen, C. J. Lewis, S. M. Todd, Polymer 11, 44 (1970).

185. S. Cesca, A. Priola, G. Santi, J. Polymer Sci. B8, 573 (1970).

186. L. A. Wall, L. J. Fetters, S. Straus, J. Polymer Sci. B5, 721 (1967).

187. S. Cesca, A. Roggero, N. Palladino, A. De Chivico, Makromol. Chem. 136, 23 (1970).

188. Brochure "Mowital" of Farbwerke Hoechst A G, Frankfurt/M., 1971.

189. G. F. L. Ehlers, J. D. Ray, J. Polymer Sci. A2, 4989 (1964).

190. J. Heller, J. H. Hodgkin, F. J. Martinelli, J. Polymer Sci. B6, 153 (1968).

191. V. V. Korsak, S. M. Tseitlin, V. I. Azarov, A, I. Parlov, Izv. Akad. Nauk SSSR, Ser. Chim. 1968, No. 1, 226.

192. J. K. Stille, E. L. Mainen, Macromolecules 1, 36 (1968).

193. F. De Schryver, C. S. Marvel, J. Polymer Sci., A-1, 5, 545 (1967).

194. J. K. Stille, J. R. Williamson, J. Polymer Sci. A2, 3867 (1964).

195. S. Inoue, Y. Imai, K. Uno, Y. Iwakura, Makromol. Chem. 95, 236 (1966).

196. Y. Imai, K. Uno, Y. Iwakura, Makromol. Chem. 83, 179 (1965).

197. E. Schwarzmann, I. R. Van Walzer, J. Am. Chem. Soc. 82, 6009 (1960).

198. Y. Imai, I. Kaoka, K. Uno, Y. Iwakura, Makromol. Chem. 83, 167 (1965).

199. Y. Iwakura, K. Uno, S. Hara, Makromol. Chem. 94, 103 (1966).

200. Y. Iwakura, K. Uno, S. Hara, Makromol. Chem. 95, 248 (1966).

201. C. J. Abshire, C. S. Marvel, Makromol. Chem. 44/46, 388 (1961).

202. J. F. McCullough, I. R. Van Walzer, E. J. Griffith, J. Am. Chem. Soc. 78, 4528 (1956).

203. H. Kokelenberg, C. S. Marvel, J. Polymer Sci., A-1, 8, 3235 (1970).

204. H. Valdsar, R. Dunlap, Am. Chem. Soc. Meeting, Atlantic City, N. Y., Sept. 1952.

205. E. Husemann, E. Siefert, Makromol. Chem. 128, 288 (1969).

206. C. E. Schildknecht (Editor), Polymer Processes, Interscience Publ., New Aork 1956, p. 427.

207. C. T. Greenwood, D. J. Hourston, A. R. Proeter, European Polymer J. 6, 293 (1970).

208. K. H. Meyer, Angew. Chem. 63, 155 (1951).

209. G. Keilich, N. Frank, E. Husemann, Makromol. Chem. 143, 275 (1971).

210. E. Ott, H. M. Spurlin (Ed.), "Cellulose and Cellulose Derivatives", 2. Ed., Interscience Publ., New York, 1954/55. In this book more references.

215. R. Knoesel, J. Perrod, H. Benoit, Compt. Rend. 251, 2944 (1960).

216. P. Dunn, D. Oldfield, J. Macromol. Sci. A4, 1157 (1970).

217. W. Marconi, S. Cesca, G. Della Fortuna, J. Polymer Sci. B2, 301 (1964).

218. N. K. I. Symons, J. Polymer Sci. 51, S 21 (1961).

219. G. Sitaramaiah, D. Jacobs, Polymer 11, 165 (1970).

220. J. Furukawa, T. Saegusa, H. Fujii, A. Kawasaki, T. Tatano, Makromol. Chem. 33, 32 (1959).

221. J. M. Barrales–Rienda, D. C. Pepper, European Polymer J. 3, 535 (1967).

222. V. V. Korshak, G. M. Tseitlin, A. I. Pavlow, Vysokomolekul. Soedin. 8, 1599 (1966).

223. A. H. Frazer, W. Sweeny, F. T. Wallenberger, J. Polymer Sci. A2, 1157 (1964).

224. M. Hasegawa, R. Huiski, J. Polymer Sci. B2, 237 (1964).

225. Y. Iwakura, K. Uno, S. Hara, J. Polymer Sci. A3, 45 (1965).

226. H. G. Elias, O. Etter, J. Polymer Sci. A-1, 5, 947 (1967).

227. W. H. McCarty, G. Parravano, J. Polymer Sci. A1, 3, 4029 (1965)

228. P. Howard, R. S. Parikh, J. Polymer Sci., A-1, 6, 537 (1968).

238. S. Nagai, Bull. Chem. Soc. Japan 37, 369 (1964).

239. O. Fuchs, unpublished results.

SPECIFIC REFRACTIVE INDEX INCREMENTS OF POLYMERS IN DILUTE SOLUTION

M. B. Huglin
Department of Chemistry
University of Salford, England

Contents

A. INTRODUCTION

This section presents a compilation of the specific refractive index increments, $\nu = dn/dc$, of polymer solutions.

This is the limiting value at $c = 0$ of the quantity $(n - n_o)/c$, where n and n_o are the refractive indices of the solution and solvent respectively and c (g/ml) denotes the concentration of the polymer in solution. Conditions of constant pressure p, constant temperature T and a fixed wavelength of light in vacuo λ_o are implied.

Specific refractive index increments are needed to calculate molecular weights, shapes and dimensions of polymers in solution using the light scattering equation:

$$\frac{Kc}{R_\theta} = \frac{1}{\langle M \rangle_w P(\theta)} + 2A_2 \cdot c \qquad (1)$$

where

$$K = (2\pi^2/N_A \lambda_o^4)(n_o \cdot \nu)^2 \qquad (2)$$

c is the concentration of the polymer in g/ml, R_θ is the reduced intensity at angle θ, $\langle M \rangle_w$ is the weight-average molecular weight, $P(\theta)$ is the particle scattering factor (see following section), A_2 is the second virial coefficient, N_A is Avogadro's number, and λ_o is the wavelength of the light in vacuo. Since ν appears as a squared term in the light scattering equation, an error of 3 % in ν, for instance, will result in an error of about 6 % in the molecular weight. Therefore this quantity must be determined accurately.

The values of K [$cm^2 g^{-2}$ mol] for six different wavelengths λ_o [nm] are as follows:

λ_o [nm]	K	λ_o [nm]	K
365.0	$18.5 \times 10^{-6} n_o^2 \nu^2$	578.0	$2.94 \times 10^{-6} n_o^2 \nu^2$
435.8	$9.09 \times 10^{-6} n_o^2 \nu^2$	584.3	$2.72 \times 10^{-6} n_o^2 \nu^2$
546.1	$3.68 \times 10^{-6} n_o^2 \nu^2$	643.8	$1.91 \times 10^{-6} n_o^2 \nu^2$

Frequently, the light scattering equation is expressed in the form:

$$\frac{Hc}{\tau} = \frac{1}{\langle M \rangle_w P(90)} + 2A_2 \cdot c$$

where τ is the turbidity of the solution in excess of that of solvent, and $H = (32\pi^3/3N_A \cdot \lambda_o^4)(n_o \cdot \nu)^2$. The value of H is thus $(16\pi/3)K$, examples being $H = 6.18 \times 10^{-5} n_o^2 \nu^2$ and $H = 15.2 \times 10^{-5} n_o^2 \nu^2$ for $\lambda_o = 546.1$ nm and 435.8 nm respectively.

For a copolymer comprising monomers A and B or a terpolymer comprising units A, B and C the specific increment is an additive function of the composition of the polymer expressed in terms of weight fraction w, thus:

$$\nu = w_A \nu_A + w_B \nu_B \qquad (3)$$

$$\nu = w_A \nu_A + w_B \nu_B + w_C \nu_C \qquad (4)$$

Eqs. (3) and (4) have been verified with samples of known composition (5, 6). However, Kambe and Honda (7) have found that for the system styrene - acrylonitrile - methyl methacrylate the value of ν for the terpolymer, from monomer feeds other than the azeotropic composition, changes with conversion, on account of a composition distribution. On the other hand a constant value of ν, independent of conversion, is yielded for the azeotropic case in which feed and polymer have the same composition.

When ν_A and ν_B differ appreciably in a given solvent, Eq. (3) affords a convenient means of estimating composition from the measured value of ν. If this is not the situation, it is possible to introduce changes in the specific increment by using mixed solvents. The appropriate specific increment to be utilised in light scattering is then ν_μ, that is the one at constant chemical potential of low molecular weight solvent. ν_μ is measured after dialysis equilibrium between solution and pure mixed solvent. ν and ν_μ allow one to obtain the coefficient of preferential adsorption γ (ml/g) of the liquid (solvent 1) in the binary solvent mixture, which has the greater affinity for the polymer:

$$\gamma = (\nu_\mu - \nu)/(dn_o/d\phi_1) \qquad (5)$$

In Eq. (5) $dn_o/d\phi_1$, is the variation of the refractive index of the mixed solvent with its composition expressed in terms of the volume fraction ϕ_1 of Liquid 1.

For a copolymer derived from A and B, light scattering always yields the true value of the molecular weight M if $\nu_A = \nu_B$. However, it is not possible to derive the molecular weights M_A and M_B of the constituent portions or the heterogeneity parameters P and Q. If $\nu_A \neq \nu_B$ an apparent value of M is yielded. Tuzar et al. (8) have summarized the criteria for the most sensitive charaterization of a copolymer:

Quantity to be determined	Criterion
M	ν_A and ν_B both large and of same sign ; or $\nu_A = \nu_B$
M_A	$\|\nu_A\|$ large and $\nu_B = 0$
M_B	$\|\nu_B\|$ large and $\nu_A = 0$
Q	$\nu = 0$

Normally, these conditions cannot be satisfied by using single solvents alone, and recourse must be made to the use of mixed solvents and employing ν_μ, $(\nu_A)_\mu$ and $(\nu_B)_\mu$ of the appropriate composition to satisfy them.

The concentration dependence of ν is expressed by the coefficient a_2 in Eq. (6)

$$(n - n_o)/c = \nu + a_2 c \qquad (6)$$

and the following equation for a_2 has been derived by Lorimer (1):

$$a_2 = (3n_o^2 - 2) \nu^2/2n_o(n_o^2 + 2)$$
$$- (n_o^2 - 1)(n_o^2 + 2)(\partial\bar{v}_2/\partial c)/12n_o \qquad (7)$$

Expressions have also been formulated for ν in terms of the partial specific volume of the polymer in solution $\bar{v}_2$, n_o and the refractive index of the polymer in solution n_2. These afford an indirect means of obtaining $\bar{v}_2$ (see refs. 2-5).

Measurement and Calibration

Experimental methods and refinements have been summarized recently elsewhere (4). The main approaches are:

(1) Differential Refractometry: *)

This relies on the image displacement ΔX of beams from the light source, which traverse pure solvent and polymer solution in a divided cell. Adequate temperature control is necessary and it is essential to utilize a stoppered cell to prevent solvent "creep". The method is not an absolute one and necessitates calibration with solutions of accurately known $(n - n_o)$, so that the calibration constant k can be determined for the apparatus:

$$k = (n - n_o)/\Delta X \qquad (8)$$

In Eq. (8), n is the refractive index of the solution of calibrant and n_o that of the solvent. The value of k decreases slightly with increase of λ_o and must therefore be redetermined, if measurements are to be conducted at a different wavelength. A selection of useful calibration data (all involving aqueous solutions) has been gathered from several sources in the literature and is given in Table B. The Brice-Phoenix differential cell is satisfactory for liquids having refractive indices not in excess of about 1.60. At higher values than this the slit image disappears from the field of vision because of total reflection in the glass partition dividing the cell. Kratochvíl and Babka (9) have described a useful modification whereby the partition is replaced with glass of a higher refractive index. This extends the useful range of the differential refractometer so that even such highly refractive liquids as 1-bromonaphthalene can be accommodated.

(2) Interferometry:

In this absolute method light beams of identical geometrical paths traverse two different optical paths and the number (f) of displaced fringes is measured by an optical compensator. If 1 is the length of the troughs, $(n - n_o)$ is given by

$$(n - n_o) = f\lambda_o/1 \qquad (9)$$

When $(n - n_o)$ is very small, for example in extremely dilute solutions, the accuracy can be enhanced by increasing the trough length. A range of troughs of different lengths is normally available with a commercial instrument. The slope of a plot of f versus c, when multiplied by $(1/\lambda_o)$ yields the value of ν. Baltog et al. (10) have developed an improved cell which prevents the formation of a liquid film at the central surface with the trough, thus enabling accurate measurements to be made with volatile liquids.

*) In addition to the commercial differential refractometers in common use, mention should be made of another instrument, the Knauer, which is currently available (Dr. Ing. Herbert Knauer, 637 Oberursel/Ts., Adenauerallee 21, Germany). Three types of model are manufactured and are of the continuous detector type (similar to the Waters).

References page IV-303

Differences in the dispersion of refractive index of the solvent and the solute can lead to discontinuities in the plot of f versus c. This problem can be overcome by measuring f for a large number of solutions of gradually increasing concentration, i.e. by the dropwise addition of concentrated polymer solution to pure solvent (5, 11).

Literature Values

The data in Table are more extensive than those collected by Chiang in the first edition (1966) of the Polymer Handbook and other compilations (4, 12). However, they differ in the respect that values of ν for proteins, copolymers and certain low molecular weight compounds have been omitted on this occasion. In the column devoted to ν, values appropriate to $\lambda_o = 436$ nm and $\lambda_o = 546$ nm appear on the left-hand side and right-hand side respectively, whilst those relating to unspecified wavelengths or different bracketed wavelengths appear in the centre of the column. The following abbreviations are adopted:

Arochlor	(approximately) chlorinated diphenyl	GHCl	guanidine hydrochloride
Cadoxen	triethylenediamine cadmium hydroxide	I	ionic strength
Calgon	sodium hexametaphosphate	MEK	methyl ethyl ketone
Cuoxam	cuprammonium hydroxide	TFP	2, 2', 3, 3'-tetrafluoropropanol
Cuoxen	cupriethylenediamine hydroxide	THF	tetrahydrofuran
DMF	N, N-dimethylformamide	Versene	sodium salt of EDTA
DMSO	dimethyl sulfoxide	Freon E2	$F \dashleftarrow CFCF_2O \dashrightarrow_2 CHFCF_3$ $\overset{|}{CF_3}$
D.N.	degree of neutralization		
D.S.	degree of substitution	λ_o	appreciable red band of the spectrum obtained with a tungsten lamp and a red filter
EDTA	ethylenediamine tetra-acetic acid		
FeTNa	iron - tartaric acid - sodium complex solution	μ	specific increments measured at dialysis equilibrium
Freon 113 $\equiv$ Isceon 113	$Cl_2FC\text{-}CF_2Cl$		

B. REFRACTOMETRIC CALIBRATION DATA

Solute	Concentration (g/100 ml Solution)	Concentration (g/100 g Water)	Temp. ($^{\circ}$C)	$(n - n_o) \times 10^6$ 436 nm	546 nm	578 nm	589 nm
NaCl	0.0938	0.09412	25	173	166	165	165
	0.1034	0.1037	25	190	184	182	182
	0.3362	0.3375	25	614	592	588	587
	0.5602	0.5627	25	1019	982	976	974
	0.6866	0.6900	25	1246	1202	1194	1191
	0.9090	0.9142	25	1645	1586	1576	1572
	1.1240	1.1311	25	2028	1956	1943	1939
	1.6465	1.6595	25	2954	2848	2830	2824
	2.0327	2.0513	25	3634	3504	3481	3474
	3.7307	3.7854	25	6583	6345	6306	6293
	6.7405	6.9092	25	11673	11247	11174	11151
NH_4NO_3	0.07878	0.07906	25	103.5	98.77	-	97.73
	0.1671	0.1677	25	218.9	208.9	-	206.7
	0.3433	0.3450	25	448.5	428.0	-	423.5
	1.0314	1.0408	25	1336	1275	-	1262
	2.8884	2.9479	25	3702	3531	-	3494
	5.8417	6.0732	25	7410	7066	-	6992
Na_2SO_4	0.09566	0.09595	25	153.1	151.0	-	150.5
	0.2198	0.2205	25	349.0	344.1	-	343.0
	0.2686	0.2694	25	425.3	419.3	-	417.9
	0.3939	0.3952	25	620.1	611.5	-	609.4
	0.7508	0.7536	25	1168	1151	-	1148
	1.2640	1.2693	25	1940	1913	-	1906
	1.5417	1.5488	25	2352	2319	-	2311
	2.5322	2.5471	25	3796	3744	-	3731
	5.7875	5.8500	25	8309	8195	-	8168
$SrCl_2$	0.08474	0.08506	25	159.8	155.1	-	154.1
	0.1749	0.1755	25	328.5	318.9	-	316.7
	0.2430	0.2437	25	455.5	442.2	-	439.1
	0.3580	0.3591	25	669.2	649.6	-	645.1

Solute	Concentration (g/100 ml Solution)	(g/100 g Water)	Temp. (^{o}C)	$(n - n_o) \times 10^6$ 436 nm	546 nm	578 nm	589 nm
KCl	0.0696	0.0699	25	100	96	96	96
	0.1067	0.1070	25	153	148	147	146
	0.2799	0.2812	25	399	386	383	382
	0.5964	0.5994	25	845	817	812	810
	1.0794	1.0869	25	1521	1469	1460	1457
	1.4911	1.5037	25	2093	2022	2009	2005
	2.9821	3.0250	25	4135	3994	3969	3962
	3.9969	4.0703	25	5500	5314	5281	5271
	4.4732	4.5647	25	6136	5926	5890	5879
	5.9642	6.1217	25	8105	7828	7781	7766
	6.4680	6.6526	25	8763	8465	8414	8398
Sucrose	0.100	0.100	20	-	-	-	140
			28	-	-	-	140
	0.200	0.200	20	-	-	-	290
			28	-	-	-	290
	0.300	0.301	20	-	-	-	430
			28	-	-	-	430
	0.400	0.402	20	-	-	-	580
			28	-	-	-	570
	0.500	0.502	20	-	-	-	720
			28	-	-	-	720
	0.600	0.604	20	-	-	-	860
			28	-	-	-	860
	0.701	0.705	20	-	-	-	1010
			28	-	-	-	1000
	0.801	0.806	20	-	-	-	1150
			28	-	-	-	1140
	0.902	0.9082	20	-	-	-	1300
			28	-	-	-	1290
	1.002	1.010	20	-	-	-	1440
			28	-	-	-	1430
	1.506	1.523	20	-	-	-	2160
			28	-	-	-	2150
	2.012	2.041	20	-	-	-	2890
			28	-	-	-	2870
	3.030	3.093	20	-	-	-	4340
			28	-	-	-	4300
	3.52	-	25	5100	5000	-	-
	4.056	4.167	20	-	-	-	5810
			28	-	-	-	5760
	5.089	5.263	20	-	-	-	7280
			28	-	-	-	7220
	6.14	-	25	8900	8800	-	-
	7.181	7.527	20	-	-	-	10270
			28	-	-	-	10190
	8.240	8.696	20	-	-	-	11780
			28	-	-	-	11680
	9.306	9.890	20	-	-	-	13300
			28	-	-	-	13190
	10.6	-	25	15400	15200	-	-
	16.3	-	25	23600	23200	-	-

C. TABLE OF SPECIFIC REFRACTIVE INDEX INCREMENTS OF POLYMER SOLUTIONS

Polymer	Solvent	dn/dc [ml/g]			T (°C)	References
		λ_0 = 436 nm	Others	λ_0 = 546 nm		

1.   Main-chain Acyclic Carbon Polymers

1.1   Poly(dienes)

Polymer	Solvent	λ_0 = 436 nm	Others	λ_0 = 546 nm	T (°C)	References
Poly(1,3-butadiene)	Benzene			0.011		57
		0.0117		0.0086	20	58
	Bromoform			-0.115		58
	1-Bromonaphthalene			-0.185		58
	Chloroform			0.083		58
	Cyclohexane	0.144				59, 60
		0.126		0.118		61
pure 1,4		0.121		0.113	25	62
pure 1,2		0.092		0.087	25	62
	Diethyl ketone	0.156			25	63
	n-Heptane	0.151		0.141		61
	THF			0.132		57
Poly(chloroprene)	n-Butyl acetate	0.146			25	64
	Carbon tetrachloride	0.0976			25	64
	trans-Decalin	0.0799			25	64
		0.0820			40	64
		0.0777			10	64
	MEK	0.160			25	64
		0.162		0.156	25	65
			0.152			66
	THF		0.138			66
Poly(cyclopentadiene)	Benzene			0.0394	21	67
	Carbon tetrachloride			0.125	21	67
Poly(isoprene)						
97.8 % cis 1,4	Benzene	0.0143		0.0194	20	58
73-80 % cis 1,4	Bromobenzene			-0.041	20	73
	Chloroform	0.100		0.095	25	68
73-80 % cis 1,4	o-Chlorotoluene			-0.003	20	73
73-80 % cis 1,4	Cyclohexane			0.104	20	73
70-75 % cis 1,4				0.1034	23	34
trans		0.117				59
cyclic				0.131		74, 75
		0.1305		0.1238	25	69
				0.124	27	69
				0.117	7	69
	Cyclohexene	0.0988		0.0943	25	68
		0.0992		0.0947	27	68
		0.0953		0.0909	17	68
		0.0914		0.0869	7	68
(non-extracted)				0.145	20	70
(extracted)				0.0946	20	70
	Decalin	0.0669		0.0605	25	69
73-80 % cis 1,4	n-Hexane			0.168	20	73
		0.1886		0.1802	25	68
73-80 % cis 1,4	Methyl isobutyl ketone			0.139	20	73
cyclic	THF			0.156		73, 74
cis 1,4				0.128	20	70
				0.128	20	70
		0.160 (calc.)			19-21	71
70-75 % cis 1,4	Toluene			0.0308	35	34
73-80 % cis 1,4				0.028	20	73
		0.0308		0.0339	25	68
--, chlorinated	MEK	0.131			35	72
Poly(trichlorobutadiene)	Benzene		0.06		25	76

Polymer	Solvent	dn/dc [ml/g]			T (°C)	References
		λ_0 = 436 nm	Others	λ_0 = 546		

1.2 Poly(alkenes)

Polymer	Solvent	λ_0 = 436 nm	Others	λ_0 = 546	T (°C)	References
Poly(1-butene)	n-Nonane	0.092			35	13
		0.108			80	13
Poly(ethylene)	Biphenyl			-0.174	123.2	14
	1-Chloronaphthalene			-0.195	65-80	15
				-0.199	90	16, 17
				-0.198	90	18
				-0.197	100	19
				-0.188	105	20, 21
				-0.195	110	19
				-0.1957	114	23
				-0.193	120	19
				-0.192	125	24
				-0.195	125	25, 18
				-0.191	125	16
				-0.214	125	26
		-0.245			125	27
				-0.1967	127	23
				-0.1955	128	23
				-0.191	130	19
$\overline{M}_n$ = 12.300				-0.1931	135	22
13.500				-0.1929	135	22
18.300				-0.1932	135	22
25.200				-0.1916	135	22
110.000				-0.1883	135	22
112.000				-0.1879	135	22
				-0.190	135	17, 28
				-0.1956	139	23
				-0.1961	140	23
				-0.191	140	25
				-0.189	140	19
				-0.1967	145	23
				-0.187	150	19
				-0.1943	151	23
	n-Decane			0.0881	65	15
				0.0890	70	15
				0.0901	75	15
				0.0917	80	15
				0.0870	111	23
				0.0881	114	23
				0.095	115	24
				0.0915	120	23
				0.0922	125	23
				0.0937	130	23
				0.0955	138	23
				0.0971	139	23
				0.0985	143	23
				0.0979	146	23
				0.0995	149	23
	o-Dichlorobenzene			0.078	120	29
	Tetralin			-0.0834	50	15
				-0.0811	80	15
		0.104			81.5	30, 20
		-0.0887		-0.075	81.5	25
				-0.0910	95	23
				-0.0918	95	23
				-0.0903	100	23
				-0.0893	104	23
		-0.0805		-0.0691	105	25
				-0.100	105	21 ★
				-0.078	105	29 ★
				-0.072	105	29 ★
				-0.083	105	29 ★

★ Values are erroneously reported as positive in Ref.

Polymer	Solvent	dn/dc [ml/g] $\lambda_o = 436$ nm	Others	dn/dc [ml/g] $\lambda_o = 546$ nm	T (°C)	References
Polyethylene) (Cont'd.)						
	Tetralin			-0.088	110	23
				-0.0866	116	23
				-0.0858	119	23
				-0.075	125	15
				-0.0837	128	23
				-0.0813	138	23
				-0.0801	144	23
$\overline{M}_n = 2.000$	1,2,4-Trichlorobenzene			-0.1085	135	22
18.400				-0.1085	135	22
34.000				-0.1082	135	22
77.000				-0.1073	135	22
110.000				-0.1063	135	22
Poly(isobutene)	Cyclohexane			0.105	20	31
			0.099		25	32
	Decalin	0.043		0.0374	25	33
	n-Decane			0.1036	35	34
	Dibutyl ether			0.124	22.5	35
				0.167	120	35
	n-Heptane			0.135	25	36
		0.140			25	37
		0.1422			25	38
	n-Hexane	0.1008			25	38
	Isoamyl isovalerate	0.1066			15	38
		0.1091			25	38
		0.1129			40	38
	Isooctane	0.140			25.5	39
			0.135		25	32
				0.139	25	40
Poly(1-octene)	Bromobenzene			-0.105	30	41
Poly(propylene)	Benzene			-0.0497	25	42
	Bromobenzene			-0.1125	40	42
				-0.0599	25	43
	n-Butyl chloride			0.1082	25	43
	n-Butyl propionate			0.1101	25	43
	Carbon tetrachloride			0.0112	25	43
	1-Chloronaphthalene	-0.228		-0.227	125	44
				-0.195	140	25
		-0.227			145	27
		-0.231			145	45
isotactic				-0.189	125	46
				-0.188	140	28,47
		-0.2275		-0.2157	145	48
		-0.195	-0.200(365)	-0.184	150	49
	Cyclohexane			0.0570	25	42
	Decalin		0.023		120	50
isotactic	n-Heptane	0.1105			30	51
syndiotactic		0.1077			30	52
	n-Hexane			0.1152	25	42
	Methylcyclohexane			0.0652	25	42
	Toluene			-0.0368	25	42
				-0.0027	25	43
	p-Xylene			-0.0380	25	43
Poly(trifluorochloroethylene)	Mesitylene			0.026	145	53
Poly(vinylcyclohexane)	Cyclohexane			0.101	25	305
	Cyclohexane/dioxane (50.1/49.9 vol.)			0.103	25	305
	Cyclohexane/MEK (64.6/35.4 vol.)			0.138	25	305
	Decalin/dioxane (50.6/49.4 vol.)			0.078	25	305
	Decalin/MEK (59.7/40.3 vol.)			0.075	25	305
	n-Heptane/dioxane (56.8/43.2 vol.)			0.147	25	305
	n-Heptane/MEK (82.6/17.4 vol.)			0.167	25	305
	n-Hexane			0.170	25	305
	THF			0.131	25	305

Polymer	Solvent	dn/dc [ml/g]			T (°C)	References
		λ_o = 436 nm	Others	λ_o = 546 nm		

1.3 Poly(acrylic acid) and Derivatives

Polymer	Solvent	λ_o = 436 nm	Others	λ_o = 546 nm	T (°C)	References
Poly(acrylic acid)						
D.N. = 0.102, I = 0.10	Aqueous NaCl	0.158			30	85
0.335, 0.10		0.179			30	85
0.994, 0.10		0.261			30	85
	Dioxane		0.090		25	86
			0.088		25	86
	HCl 0.1N		0.140 (589)			87
	0.2N		0.146		20-65	77
	Water		0.137			80
Poly(benzyl acrylate)	Benzene	0.062		0.062	20	6
	DMF	0.124	0.132 (365)	0.119	20	6
			0.118 (578)		20	6
			0.117 (589)		20	6
			0.116 (644)		20	6
	MEK	0.1760			20	88
		0.1746		0.1665	20	6
		0.1746	0.1653(578)	0.1665	20	5
			0.1647(589)		20	5
			0.1623(644)		20	5
Poly(n-butyl acrylate)	Acetone	0.112			20	89
	Benzene			-0.0292	30	84
	Chlorobenzene			-0.0525	30	84
	n-Hexane			0.0885	30	84
	THF			0.0651	30	84
	Toluene			-0.0239	30	84
Poly(tert-butyl acrylate)	MEK			0.0818	25	90
Poly(1, 1-dihydroperfluorobutyl acrylate)						
	Benzotrifluoride		0.030		25	91
			0.0311		25	91
Poly(ethyl acrylate)	Acetone			0.107	25	10, 92, 93
		0.109			20	89
	Benzene			-0.042	25	10
	Chloroform			0.0363	25	10
	Dichloroethane			0.023	25	10
	Dioxane			0.036	25	10, 93
	DMF			0.032	25	10
	Ethyl acetate			0.0916	25	10
	MEK	0.0870	0.0854(578)	0.0856	20	5
			0.0853(589)		20	5
			0.0852(644)		20	5
				0.086	25	10, 93
				0.1407	30	94
	n-Methyl pyrrolidone			-0.027	25	10
	Water		0.131		25	95
Poly(isopropyl acrylate)						
isotactic, syndiotactic and atactic :						
	Bromobenzene	-0.103		-0.0870	60	101
isotactic and atactic :	TFP			0.124	25	101
Poly(methyl acrylate)	Acetone	0.113			20	89
	Acetonitrile	0.120		0.118	25	96
		0.121		0.119	32	96
		0.122		0.120	40	96
		0.123		0.122	48	96
		0.127		0.125	56	96
	Benzene	-0.019			30	97
	1, 2-Dibromoethane	-0.0606		-0.0560	25	96
		-0.0598		-0.0528	32	96
		-0.0578		-0.0532	40	96
		-0.0579		-0.0523	48	96

Polymer	Solvent	dn/dc [ml/g]			T (°C)	References
		λ_0 = 436 nm	Others	λ_0 = 546 nm		
Poly(methyl acrylate) (Cont' d.)						
	Dioxane	0.0503		0.0497	25	96
		0.055			30	97
		0.0524		0.0515	32	96
		0.0533		0.0526	40	96
		0.0547		0.0537	48	96
		0.0566		0.0562	56	96
	Ethyl acetate	0.0940		0.0920	19.5	96
		0.0989		0.0967	25	96
		0.0967		0.0964	32	96
		0.102			35	98
		0.102		0.101	40	96
		0.105		0.103	48	96
	Isoamyl acetate	0.077			30	97
	MEK	0.090			room	99
		0.0938	0.0921(578)	0.0923	20	5
			0.0919(589)		20	5
			0.0915(644)		20	5
		0.097			30	97, 100
		0.090			35	72
	Toluene	-0.009			30	97
Poly(n-octadecyl acrylate)	Benzene			-0.0505	30	84
				-0.0269	30	84
				-0.0363	30	84
	n-Butyl acetate			0.0883	30	84
	Carbon tetrachloride			0.0114	30	84
	Chlorobenzene			-0.0682	30	84
				-0.0489	30	84
				-0.0563	30	84
	Cyclohexane			0.0175	30	84
	n-Hexane			0.1096	30	84
				0.1292	30	84
	THF			0.0702	30	84
				0.0898	30	84
				0.0723	30	84
	Toluene			-0.0425	30	84
Poly(n-propyl acrylate)	MEK			0.1102	30	94

1.4   Poly(methacrylic acid) and Derivatives

Polymer	Solvent	dn/dc [ml/g]			T (°C)	References
		λ_0 = 436 nm	Others	λ_0 = 546 nm		
Poly(n-butyl methacrylate)	Acetone	0.1249	0.1257(366)	0.1236	25	105
		0.122			20	89
	Benzene	-0.023		-0.014	25	106
		-0.024		-0.015	25	107
	Bromobenzene	-0.084		-0.073	25	106, 107
	Bromoform	-0.129		-0.118	25	106
	Carbon tetrachloride	0.026		0.027	25	106
	Isopropanol	0.102		0.102	23	108
		0.1059			20	109
		0.1066			23.7	109
		0.1068			25	109
		0.1076			31	109
		0.1097			45	109
	MEK	0.103		0.102	25	106, 107
		0.1059	0.1064(366)	0.1046	25	105
				0.104	23	108
Poly(tert-butyl methacrylate)	n-Butyl acetate		0.09			110
Poly(p-tert-butylphenyl methacrylate)						
	Acetone		0.183			111
Poly(cetyl methacrylate)	Benzene		0.02			112
	n-Butyl acetate	0.095			25	113

SPECIFIC REFRACTIVE INDEX INCREMENTS

Polymer	Solvent		dn/dc [ml/g]			T (°C)	References
		λ_o = 436 nm	Others	λ_o = 546 nm			
Poly(cetyl methacrylate) (Cont'd.)							
	Carbon tetrachloride		-0.015				112
	n-Heptane		0.114			21	112
Poly(4-chlorophenyl methacrylate)							
	Benzene			0.0758		25	114
Poly(cyclohexyl methacrylate)	Cyclohexane			0.0845		25	115
Poly(n-docosyl methacrylate)	o-Dichlorobenzene	0.102		0.090		25	116
	Tetralin	0.094		0.084		25	116
		0.0935				29	116
		0.093		0.082		35	116
		0.091		0.081		50	116
		0.088		0.077		65	116
		0.084		0.070		75	116
		0.076		0.068		85	116
	THF	0.0745		0.0745		25-48	116
Poly(2-ethylbutyl methacrylate)	Isopropanol	0.109		0.105		25	117
	MEK	0.104		0.102		25	117
Poly(ethylene glycol monomethacrylate)							
	Dioxane/water (90/10 vol.)			0.083		25	119
				0.066 (μ)		25	119
	(80/20 vol.)			0.086		25	119
				0.071 (μ)		25	119
	(60/40 vol.)			0.095		25	119
				0.100 (μ)		25	119
	(40/60 vol.)			0.108		25	119
				0.120 (μ)		25	119
	(20/80 vol.)			0.123		25	119
				0.130 (μ)		25	119
	DMF			0.076			120
	Ethanol/water (80/20 vol.)			0.124		25	119
				0.121 (μ)		25	119
	(70/30 vol.)			0.121		25	119
				0.119 (μ)		25	119
	(60/40 vol.)			0.120		25	119
				0.120 (μ)		25	119
	(40/60 vol.)			0.123		25	119
				0.127 (μ)		25	119
	(30/70 vol.)			0.125		25	119
				0.132 (μ)		25	119
	Formamide			0.045			120
	Methanol			0.162			120
	n-Propanol/dioxane (85/15 vol.)			0.105		25	119
				0.108 (μ)		25	119
	(70/30 vol.)			0.100		25	119
				0.103 (μ)		25	119
	(60/40 vol.)			0.096		25	119
				0.097 (μ)		25	119
	(40/60 vol.)			0.091		25	119
				0.090 (μ)		25	119
	(30/70 vol.)			0.087		25	119
				0.086 (μ)		25	119
	n-Propanol/water (95/5 vol.)			0.110		25	121
				0.105 (μ)		25	121
	(90/10 vol.)			0.113		25	121
				0.105 (μ)		25	121
	(80/20 vol.)			0.115		25	121
				0.106 (μ)		25	121
	(75/25 vol.)			0.103 (μ)			120
	(70/30 vol.)			0.119		25	121
				0.109 (μ)		25	121
	(60/40 vol.)			0.122		25	121
				0.110 (μ)		25	121

References page IV-303

Polymer	Solvent		dn/dc [ml/g]		T (°C)	References
		$\lambda_o = 436$ nm	Others	$\lambda_o = 546$ nm		
Poly(ethylene glycol monomethacrylate) (Cont'd.)						
	n-Propanol/water (50/50 vol.)			0.125	25	121
				0.123 (μ)	25	121
	(40/60 vol.)			0.128	25	121
				0.138 (μ)	25	121
	(30/70 vol.)			0.131	25	121
				0.149 (μ)	25	121
	(20/80 vol.)			0.135	25	121
				0.153 (μ)	25	121
	Toluene/methanol (42/58 vol.)			0.101	25	118
				0.063 (μ)	25	118
	Toluene/isopropanol (38/62 vol.)			0.087	25	118
				0.064 (μ)	25	118
Poly(ethyl methacrylate)						
	Ethyl acetate	0.109				122
	Ethyl acetate/ethanol (1/4.5-1/6 vol.)	0.113				122
	MEK			0.104	23	123
	MEK/isopropanol (1/7 vol.)			0.107	23	123
Poly(n-hexyl methacrylate)	MEK	0.105		0.105	23	124
	Isopropanol	0.108		0.106	32.6	124
Poly(isobutyl methacrylate)	Acetone	0.122		0.117	25	103, 104
Poly(isooctyl methacrylate)	n-Heptane		0.109			168
Poly(n-lauryl methacrylate)	n-Amyl alcohol	0.0744		0.0781		125
	n-Butyl acetate	0.092		0.090	23	126
	Isopropyl acetate	0.107		0.104	13	126
Poly(methacrylic acid)	Alcoholic LiBr	0.154			25	127
	Ethanol	0.154			25	127
	HCl 0.1N		0.159 (589)			87
	0.5N			0.137	25	128
	0.045N		0.156 (560)		20	129
			0.162(μ)(560)		20	129
	0.02N		0.158		20-65	77
	0.001N			0.140	25	128
	Methanol			0.134	25	128
	NaBr 0.1M		0.234 (560)		20	129
			0.209(μ)(560)		20	129
	NaCl 0.1M		0.213(μ)(560)		20	129
	NaF 0.1M		0.229 (560)		20	129
			0.219(μ)(560)		20	129
	NaI 0.1M		0.197(μ)(560)		20	129
	Na_2SO_4 0.1M		0.210(μ)(560)		20	129
	$(NH_4)_2Mo_7O_{24}$ 0.01M		0.227		20	129
			0.169(μ)(560)		20	129
	Water			0.142	25	128
			0.153			80
Poly(methyl methacrylate)	Acetone	0.1391			27	130
		0.107			27	131
		0.137		0.134	23±2	132
		0.129				133
			0.131			134
		0.1313	0.134 (366)	0.1293	25	135
		0.136		0.134	25	10, 136
			0.132			92
		0.132			20	89
		0.1305	0.1281(578)	0.1285	20	5
			0.1279(589)		20	5
			0.1276(644)		20	5
				0.132	20	93
				0.135	20	137, 138
		0.1312		0.1292	20	139
				0.1330	25	140, 141

SPECIFIC REFRACTIVE INDEX INCREMENTS

Polymer	Solvent		λ_o = 436 nm	dn/dc [ml/g] Others	λ_o = 546 nm	T (°C)	References
Poly(methyl methacrylate) (Cont'd.)							
isotactic	Acetone		0.136		0.134	20	142
	Acetone/benzene	(88/12 vol.)			0.1170	25	141
		(76/24 vol.)			0.1005	25	141
		(64/36 vol.)			0.0861	25	141
		(52/48 vol.)			0.0734	25	141
		(40/60 vol.)			0.0590	25	141
	Acetone/chloroform	(80/20 vol.)			0.1170	25	140
		(60/40 vol.)			0.1005	25	140
		(52/48 vol.)			0.091	25	140
		(40/60 vol.)			0.0861	25	140
		(20/80 vol.)			0.0734	25	140
	Acetonitrile		0.140		0.137	20	136
	Arochlor		-0.124			20	144
	Benzene		0.0398			27	130
					-0.010		145
			0.002			25	146
			0.0039		0.0110	25	139
					0.004		147
				≈ 0			148
					-0.016	25	10
			0.001		0.007	20	6
	Bromobenzene				-0.058		145
	1-Bromonaphthalene				-0.147		145
					-0.148	25	9
	n-Butyl acetate		0.0987	0.101 (366)	0.0970	25	135
			0.0986		0.0969	25	139
			0.1			25	113
					0.0970	25	149
	n-Butyl chloride		0.0913		0.0898	20	143
			0.0948		0.0934	30	143
			0.0984		0.0970	40	143
atactic			0.0931		0.0914	25	150
			0.1021		0.1001	50	150
syndiotactic			0.1001		0.0986	50	150
isotactic			0.0902		0.0875	25	150
			0.0965		0.0954	50	150
	Carbon tetrachloride				0.023		145
	Chlorobenzene				-0.026		145
			-0.018			25	146
				-0.026			148
	Chloroform		0.0631	0.0635(366)	0.0629	25	135
				0.056			134
			0.0630		0.0628	25	139
				0.050			148
					0.059	25	10, 140
	1,2-Dichloroethane		0.050		0.051		151
			0.050			30	152
					0.050		147
					0.046	25	10
	Dioxane				0.075		154
			0.0720	0.074 (366)	0.0707	25	135, 139
					0.068		145
					0.071	25	93
				0.0703(589)		25	155
					0.07	25	121, 147
				0.070			148
					0.066	25	10
					0.0707	25	156, 157
	Dioxane/water	(95/ 5 vol.)			0.072	25	121
					0.076 (μ)	25	121
		(90/10 vol.)			0.076	25	121
					0.083 (μ)		121

Polymer	Solvent	dn/dc [ml/g]			T (°C)	References
		λ_o = 436 nm	Others	λ_o = 546 nm		
Poly(methyl methacrylate) (Cont'd.)						
	Dioxane/water (85/15 vol.)			0.079	25	121
				0.072	25	118
				0.090 (μ)	25	118, 121
				0.090	30	158
	(82/18 vol.)			0.082	25	121
				0.096 (μ)	25	121
	DMF	0.0600	0.0617(578)	0.0614	20	5, 6
			0.0618(589)		20	5, 6
			0.0619(644)		20	5, 6
			0.055 (365)		20	6
				0.055	25	10
				0.069	30	159
				0.095	90	159
			0.042			148
	Ethyl acetate	0.1200	0.123 (366)	0.1180	25	135, 139
			0.1174(589)		25	155
				0.118	25	10
	Isoamyl acetate	0.0931		0.0911	20	143
		0.0945		0.0926	30	143
		0.0959		0.0942	40	143
		0.0974		0.0957	50	143
	MEK			0.111		154
				0.112	room	160
		0.093				131
			0.111			134
				0.1173	room	163
		0.100			room	99
			0.110			148
		0.1131	0.1110(578)	0.1112	20	5, 6, 88
			0.1107(589)		20	5
			0.1102(644)		20	5
		0.113		0.111	23 ± 2	132
				0.117	25	162
		0.111			25	146
				0.111	25	132, 161
				0.110	25	10, 93
				0.108	25	165
		0.01385			27	130
			0.111 (488)		28	134
		0.114		0.113		151
				0.111		145
		0.113			30	152
			0.059		30	164
			0.131			153
				0.113		147
				0.1147	35	34
		0.113			35	72
	MEK/isopropanol	0.1090		0.1075	10	143
		0.1130		0.1114	20	143
		0.1167		0.1153	30	143
		0.1208		0.1192	40	143
atactic	(55/54 vol.)	0.1169		0.1153	30	150
syndiotactic		0.1165		0.1155	30	150
isotactic		0.1167		0.1152	30	150
	Methyl methacrylate			0.075	30	158
	N-Methylpyrrolidone			0.0176	25	10
	Nitroethane			0.100		154
				0.094	25	161
	TFP			0.141	25	121
	TFP/water (95/ 5 vol.)			0.141	25	121
				0.140 (μ)	25	121
	(90/10 vol.)			0.139	25	121
				0.137 (μ)	25	121

Polymer	Solvent		dn/dc [ml/g]			T (^{o}C)	References
			λ_o = 436 nm	Others	λ_o = 546 nm		
Poly(methyl methacrylate) (Cont'd.)							
	TFP/water	(80/20 vol.)			0.135	25	121
					0.131 (μ)	25	121
		(70/30 vol.)			0.133	25	121
					0.124 (μ)	25	121
		(65/35 vol.)			0.132	25	121
					0.121 (μ)	25	121
	THF		0.0887	0.091 (366)	0.0871	25	135
			0.0886		0.0870	25	139
					0.0902	36	34
			0.09				166
	Toluene		0.004			30	152
			0.010			25	146
			0.0094		0.0157	25	139
				0.004			153
					0.005		147
atactic			0.010			25	146
syndiotactic			0.007			25	146
isotactic			0.014			25	146
	Toluene/methanol	(42/58 vol.)			0.094	25	118
					0.152 (μ)	25	118
					0.096	25	167
					0.154 (μ)	25	167
	Toluene/isopropanol	(38/62 vol.)			0.073	25	118
					0.121 (μ)	25	118
Poly(n-octyl methacrylate)	n-Butanol		0.083		0.080	16.8	169
	MEK		0.107		0.107	23	169
Poly(phenyl methacrylate)	Acetone				0.234		170
	MEK				0.182	25	115

1.5   Other α- and β-Substituted Poly(acrylics) and Poly(methacrylics)

Polymer	Solvent	λ_o = 436 nm	Others	λ_o = 546 nm	T (^{o}C)	References
Poly(acrylamide)	Water		0.149		20-65	77
				0.163		78
			0.195 (589)		30	79
			0.179			80
Poly(N-tert-butylacrylamide)	Methanol	0.249		0.234		81
				0.234	20	82
Poly(di-n-amyl itaconate)	n-Amyl acetate	0.074			20	102
	Benzene	-0.035			20	102
	Carbon tetrachloride	0.018			20	102
	Chloroform	0.036			20	102
	Ethyl acetate	0.102			20	102
	MEK	0.096			20	102
	Toluene	-0.027			20	102
Poly(di-n-butyl itaconate)	Acetone	0.109			20	102
	n-Amyl acetate	0.074			20	102
	Benzene	-0.030			20	102
	Carbon tetrachloride	0.020			20	102
	Chlorobenzene	-0.050			20	102
	Chloroform	0.036			20	102
	Ethyl acetate	0.101			20	102
	Ethylbenzene	-0.021			20	102
	n-Hexane	0.101			20	102
	MEK	0.096			20	102
	Methyl acetate	0.112			20	102
	Styrene	-0.077			20	102
	Toluene	-0.023			20	102
Poly(di-n-decyl itaconate)	n-Amyl acetate	0.079			20	102
	Benzene	-0.041			20	102

Polymer	Solvent	dn/dc [ml/g]			T (°C)	References
		λ_o = 436 nm	Others	λ_o = 546 nm		
Poly(di-n-decyl itaconate) (Cont'd.)						
	Carbon tetrachloride	0.015			20	102
	Toluene	-0.032			20	102
Poly(diethyl itaconate)	n-Amyl acetate	0.074			20	102
	Benzene	-0.018			20	102
	Carbon tetrachloride	0.024			20	102
	Ethyl acetate	0.100			20	102
	MEK	0.096			20	102
	Toluene	-0.014			20	102
Poly(di-n-heptyl itaconate)	Benzene	-0.037			20	102
	Ethyl acetate	0.102			20	102
	MEK	0.095			20	102
	Toluene	-0.029			20	102
Poly(di-n-hexyl itaconate)	n-Amyl acetate	0.075			20	102
	Benzene	-0.036			20	102
	Carbon tetrachloride	0.017			2C	102
	Chloroform	0.035			20	102
	Ethyl acetate	0.102			20	102
	MEK	0.095			20	102
	Toluene	-0.028			20	102
Poly(N, N-dimethylacrylamide)	Methanol		0.194		25	83
	Water		0.150		25	83
Poly(dimethyl itaconate)	Acetone	0.118			20	102
	Benzene	0.000			20	102
	Chlorobenzene	-0.022			20	102
	Chloroform	0.041			20	102
	MEK	0.096			20	102
	Methyl acetate	0.110			20	102
Poly(di-n-nonyl itaconate)	n-Amyl acetate	0.079			20	102
	Benzene	-0.030			20	102
	Toluene	-0.039			20	102
Poly(di-n-octyl itaconate)	n-Amyl acetate	0.078			20	102
	Benzene	-0.039			20	102
	Carbon tetrachloride	0.016			20	102
	Chloroform	0.033			20	102
	Ethyl acetate	0.105			20	102
	MEK	0.096			20	102
	Toluene	-0.030			20	102
Poly(di-n-propyl itaconate)	n-Amyl acetate	0.074			20	102
	Benzene	-0.025			20	102
	Carbon tetrachloride	0.021			20	102
	Chloroform	0.036			20	102
	MEK	0.096			20	102
	Toluene	-0.019			20	102
Poly(di-n-undecyl itaconate)	n-Amyl acetate	0.079			20	102
	Benzene	-0.042			20	102
	Carbon tetrachloride	0.014			20	102
	Chlorobenzene	-0.064			20	102
	Chloroform	0.030			20	102
	Ethyl benzene	-0.030			20	102
	n-Hexane	0.112			20	102
	Styrene	-0.088			20	102
	Toluene	-0.032			20	102
Poly(methacrylamide)	Water		0.209		20-65	77
Poly(methyl n-butacrylate)	n-Butanol	0.0878			13	286
	MEK	0.1192			30	286
Poly(methyl ethacrylate)	Diisobutyl ketone	0.0791			11.4	286
	MEK	0.1357			30	286

Polymer	Solvent	dn/dc [ml/g] λ_o = 436 nm	Others	λ_o = 546 nm	T (°C)	References
Poly(N-octadecylacrylamide)	Benzene			0.0269	30	84
	Chlorobenzene			0.0489	30	84
	Cyclohexane			0.0651	30	84
	n-Hexane			0.1292	30	84
				0.1168	30	84
	THF			0.0898	30	84
	Toluene			0.0154	30	84

1.6   Poly(styrenes)

Polymer	Solvent	dn/dc [ml/g] λ_o = 436 nm	Others	λ_o = 546 nm	T (°C)	References
Poly(p-chlorostyrene)	Benzene	0.095				171
	Chloroform	0.157				171
	Dioxane	0.164				171
	MEK	0.207		0.197	20	172
	Toluene	0.121			35	173
Poly(p-iodostyrene)	Dioxane	0.156	0.166 (365)	0.144	20	174
			0.143 (578)		20	174
			0.142 (589)		20	174
			0.140 (644)		20	174
Poly(p-isopropylstyrene)	Chloroform	0.134		0.124	25	187
	MEK	0.198		0.186	25	187
	THF	0.177		0.167	25	188
Poly(p-isopropylstyrene) hydroperoxide						
	MEK	0.195-0.203		0.184-0.197	25	187
	THF	0.177		0.167	25	189
Poly(p-methoxystyrene)	Benzene/cyclohexanol					
	(61.19/38.81 vol.)			0.112		175
	Benzene/methanol					
	(73.69/26.31 vol.)			0.125		175
	Chloroform/cyclohexane					
	(36.08/63.92 vol.)			0.163		175
	Chloroform/methanol					
	(66.49/33.51 vol.)			0.190		175
	MEK/n-heptane					
	(80.01/19.99 vol.)			0.211		175
	MEK/n-propanol					
	(80.64/19.36 vol.)			0.215		175
	Toluene	0.105		0.101	25	176
	Toluene/cyclohexanol					
	(62.8/37.2 vol.)			0.113		175
Poly(α-methylstyrene)	Benzene	0.1238			25	177
				0.138	25	178
		0.134			30	179
		0.131			30	180
	Carbon tetrachloride	0.178	0.193 (366)	0.168	23	181
	Cyclohexane	0.206			32	183
		0.208			36	183
		0.199			37	182
		0.200			39	179
		0.1967			39	177
		0.210			40	183
		0.212			45	183
		0.214			50	183
		0.204	0.219 (366)	0.192	35	181
	1,2-Dichloroethane	0.1764			25	177
	Toluene			0.1165	23	34
		0.131			25	184
		0.1307			25	177
		0.130	0.135 (366)	0.126	23	181
			0.129		25	185
			0.137			186

Polymer	Solvent	dn/dc [ml/g] λ_0 = 436 nm	Others	λ_0 = 546 nm	T (°C)	References
Poly(styrene)	Isoamyl acetate	0.205			30	97
	n-Amyl acetate	0.205			20	102, 190
	Arochlor			-0.017		144
	Benzene	0.1151				191
		0.111			20	102
		0.1066		0.1034	20	6
		0.113			20	203
		0.1066	0.1030(578)	0.1034	20	5
			0.1028(589)		20	5
			0.1021(644)		20	5
		0.115			25	205
		0.108		0.105	25	106
		0.109		0.105	25	107
			0.109			32
		0.112	0.107 (589)	0.109	25	206
				0.110	25	193
			0.115			194
				0.108		57
				0.1025	25	195
		0.108			25	146, 197
		0.1094		0.1062	25	139
			0.106			148
				0.105	25	10
		0.111			25	198
				0.106	25	145, 196, 199, 201
		0.113			28	192
				0.109	30	204
				0.106		147
		0.101			30	97
		0.115			30	200
				0.1123	30	84
		0.112			35	202
		0.131			70	203
	Benzene/bromoform (95/ 5 vol.)	0.101			25	198
	(90/10 vol.)	0.096			25	198
	(85/15 vol.)	0.099			25	198
	(75/25 vol.)	0.083			25	198
	(50/50 vol.)	0.074			25	198
	(25/75 vol.)	0.036			25	198
	Benzene/(CH$_2$Cl)$_2$CHOH (33/67 vol.)			0.112		207
	Benzene/cyclohexane (82.9/18.1 mol)	0.137			20	208
		0.155			70	208
	(65.3/34.7 mol)	0.155			20	208
		0.172			70	208
	(44.8/55.2 mol)	0.170			20	208
		0.186			70	208
	(23.6/76.4 mol)	0.183			20	208
		0.197			70	208
	(10/90 mol)	0.189			20	208
		0.202			70	208
	(4.8/95.2 mol)	0.191			20	208
	Benzene/cyclohexanol (61.6/38.4 vol.)			0.108	25	209
	(81.8/18.2 mol)	0.121			20	208
		0.134			70	208
	(90.9/9.1 mol)	0.119			20	208
	(99.628/0.372 mol)	0.124			20	208
		0.134			70	208
	Benzene/n-docosane (98.7/1.3 mol)	0.122			20	210
		0.140			70	210

SPECIFIC REFRACTIVE INDEX INCREMENTS

Polymer	Solvent	dn/dc [ml/g]			T ($^{\circ}$C)	References
		λ_o = 436 nm	Others	λ_o = 546 nm		
Poly(styrene) (Cont' d.)	Benzene/n-docosane					
	(95.7/4.3 mol)	0.135			20	210
		0.147			70	210
	(90.3/9.7 mol)	0.147			20	210
		0.150			70	210
	Benzene/n-dodecane					
	(96.2/3.8 mol)	0.123			20	210
		0.139			70	210
	(92.7/7.3 mol)	0.131			20	210
		0.145			70	210
	(86.4/13.6 mol)	0.140			20	210
		0.149			70	210
	Benzene/1-dodecanol					
	(98.1/1.9 mol)	0.116			20	210
	(94.6/5.4 mol)	0.122			20	210
	(89.7/10.3 mol)	o.132			20	210
	Benzene/ethanol					
	(94.3/ 5.7 mol)	0.115			20	210
	(89.1/10.9 mol)	0.119			20	210
		0.131			70	210
	(62.1/37.9 mol)	0.162			20	210
		0.151			70	210
	Benzene/n-heptane (44/56 vol.)	0.130			35	211
	(90/10 vol.)	0.124			35	202
	Benzene/n-hexane					
	(88/12 mol)	0.141			20	210
	(78.6/21.4 mol)	0.160			20	210
		0.174			70	210
	(63.2/36.8 mol)	0.188			20	210
		0.199			70	210
	(55.1/44.9 mol)	0.197			20	210
		0.205			70	210
	Benzene/1-hexanol					
	(92.7/ 7.3 mol)	0.119			20	210
	(84.9/15.1 mol)	0.126			20	210
		0.135			70	210
	(76.6/23.4 mol)	0.136			20	210
	(67.8/32.2 mol)	0.148			20	210
		0.144			70	210
	Benzene/MEK (25.1/74.9 mol)	0.210			20	203
		0.224			70	203
	(50.2/49.8 mol)	0.178			20	203
		0.193			70	203
	(75.1/24.9 mol)	0.145			20	203
		0.162			70	203
	Benzene/methanol					
	(70.9/29.1 mol)	0.154			20	208
	(83.9/16.1 mol)	0.124			20	208
	(92.4/ 7.6 mol)	0.114			20	208
	(22.2/77.8 vol.)			0.145	25	209
	(76.6/23.4 vol.)			0.122	25	193
	(80 /20 vol.)			0.137		207
	(74.7/25.3 vol.)	0.159			35	211
	Benzene/1-octadecanol					
	(99.15/0.85 mol)	0.116			20	210
		0.134			70	210
	Benzene/poly(oxyethylene) 420					
	(95.9/4.1 mol)	0.125			20	210
	Benzene/poly(oxyethylene) 2100					
	(99.41/0.59 mol)	0.120			20	210
	Benzene/isopropanol					
	(35.8/64.2 vol.)			0.159	25	212
	(65 /35 vol.)			0.154		207
	(61 /39 vol.)	0.142			35	211

References page IV-303

Polymer	Solvent	$\lambda_o = 436$ nm	Others	$\lambda_o = 546$ nm	T (°C)	References
Poly(styrene) (Cont'd.)	Bromobenzene	0.0492			18	213
		0.043		0.042	25	106
		0.0495		0.0485		107
			0.0455		25	214
				0.042	25	145, 215
	Bromoform			-0.005		57
		0.015		0.012	25	106
		0.014			30	216
	1-Bromonaphthalene			-0.051		145
				-0.091		57
				-0.0517	25	9
	n-Butyl chloride			0.1997	25	43
	n-Butyl propionate			0.2018	25	43
	Carbon tetrachloride	0.156			20	102
		0.161			25	217
				0.144		154
				0.146		145
		0.159		0.147	25	106
		0.160	0.145(589)	0.150	25	206
				0.1518	30	84
				0.148	30	204
		0.162			35	202
	Carbon tetrachloride/n-butanol (65/35 vol.)	0.176			35	211
	Carbon tetrachloride/n-heptane (53/47 vol.)	0.189			35	211
	Carbon tetrachloride/methanol (18.3/81.7 vol.)			0.162	25	209
	Chlorobenzene			0.079		145
		0.0833			18	213
		0.082			20	102
		0.079			25	146
				0.079	25	215
				0.0848	30	84
			0.079			148
	Chlorobenzene/isopropyl ether (68/32 vol.)			0.192	25	209
	Chloroform			0.158		57
		0.1554			18	213
		0.165			20	102
				0.169	25	26
			0.152			148
		0.161			25	146
	Cyclohexane	0.1795		0.1682	20	219
		0.1798		0.1685	21	219
		0.1801		0.1687	22	219
		0.1804		0.1690	23	219
		0.1810		0.1695	25	139, 219
$\overline{M}_w = 2.18 \times 10^3$		0.168		0.156	25	220
2.92×10^3		0.171		0.160	25	220
4.69×10^3		0.174		0.163	25	220
11.8×10^3		0.177		0.165	25	220
22.2×10^3		0.180		0.167	25	220
167×10^3		0.181		0.167	25	220
		0.1825		0.1709	30	219
		0.181			34.5	221
		0.1840		0.1723	35	219
				0.171	35	209
		0.181			35	202, 222
				0.179	40	218
		0.1855		0.1738	40	219
		0.1870		0.1752	45	219
		0.188			50	223
			0.181			224

SPECIFIC REFRACTIVE INDEX INCREMENTS

Polymer	Solvent	dn/dc [ml/g]			T (°C)	References
		λ_o = 436 nm	Others	λ_o = 546 nm		
Poly(styrene) (Cont'd.)	Cyclohexanone	0.151			30	216
	Cyclohexene	0.165			25	23
	Decalin			0.129	24	217
				0.129	25	215
		0.118			25	23
				0.137	52	23
				0.138	60	23
				0.143	87	23
				0.147	104	23
				0.152	123	23
				0.155	139	23
	cis-Decalin	0.124			12	221
	1,2-Dibromoethane	0.065			30	216
	1,2-Dichloroethane	0.168		0.155		151
				0.161	20	218
		0.168			30	152
				0.168		147
			0.168			152
		0.166			30	200
	Diethyl fumarate/CH_2ClCH_2OH (80/20 vol.)			0.152		207
	Dioxane			0.171		154
				0.176		225
				0.168		145
				0.172		196
		0.185	0.204 (365)	0.173	20	174
			0.171 (578)		20	174
			0.170 (589)		20	174
			0.168 (644)		20	174
		0.1852		0.1731	25	139
			0.1702(589)		25	155
		0.193	0.176 (589)	0.182	25	206
			0.170			148
		0.181			35	97
		0.190			35	202
				0.184		147
	Dioxane/n-heptane (41.5/58.5 vol.)	0.212			35	211
	Dioxane/methanol (28.6/71.4 vol.)			0.191	25	209
	(66.5/33.5 vol.)	0.218			35	211
	Dioxane/isopropanol (51.5/48.5 vol.)	0.212			35	211
	DMF	0.170.			18	213
		0.1739	0.1632(578)	0.1648	20	5, 6
			0.1625(589)		20	5, 6
			0.1604(644)		20	5, 6
			0.145			148
			0.189 (365)		20	6
				0.165	25	215
		0.175			25	146
	Ethyl acetate	0.2310			18	213
		0.234			20	102
				0.2231	25	226
	Ethyl acetate/ethanol (94.81/ 5.19 vol.)			0.2196	25	226
	(91.76/ 8.24 vol.)			0.2210	25	226
	(89.34/10.66 vol.)			0.2238	25	226
	Ethyl benzene	0.111			25	102
	MEK	0.2258		0.2138	7.5	219
		0.2275		0.2155	15	219
		0.2290			18	213
		0.2287		0.2167	20	219
		0.2296		0.2170	20	6

Polymer	Solvent	dn/dc [ml/g] λ_o = 436 nm	Others	λ_o = 546 nm	T (°C)	References
Poly(styrene) (Cont'd.)	MEK			0.214	20	227
		0.243			20	203
			0.232			153
		0.229			20	88, 231
		0.228			20	102, 190
		0.2296	0.2150(578)	0.2170	20	5
			0.2143(589)		20	5
			0.2110(644)		20	5
		0.2298		0.2175	25	139
			0.2146(589)		25	155
				0.2165	25	165
		0.2298		0.2178	25	219
				0.214	25	195
				0.220		145, 154, 196
		0.229				146
		0.226			25	217
		0.231		0.220	25	212
		0.231	0.208 (589)	0.218		151, 232
			0.220 (589)			232
		0.226			25	228
			0.220			148
				0.219	25	215
				0.230		147
		0.227		0.215	25	106
			0.210*			224
		0.229		0.217		107
			0.222 (437)		25	233
		0.232	0.216 (589)	0.223	25	206
		0.238			30	216
		0.232			30	152
		0.223			30	229
		0.229		0.219		230
			0.170		30	164
		0.2309		0.2189	30	219
		0.227			35	97
		0.2321		0.2201	35	219
		0.2333		0.2213	40	219
		0.2345		0.2225	45	219
				0.230	67	218
		0.254			70	203
	MEK/(CH$_2$Cl)$_2$CHOH (46/54 vol.)			0.165		207
	MEK/1-dodecanol (95.4/4.6 mol)	0.238			20	203
		0.253			70	203
	(91.2/8.8 mol)	0.232			20	203
		0.248			70	203
	MEK/1-hexanol (93.3/ 6.7 mol)	0.241			20	203
	(84.8/15.2 mol)	0.238			20	203
		0.251			70	203
	(73.7/26.3 mol)	0.234			20	203
		0.249			70	203
	MEK/methanol (11.3/88.7 vol.)			0.214	25	209
	MEK/isopropanol (11.4/88.6 vol.)			0.211	25	209
	(85 /15 vol.)			0.219		207
	Methylcyclohexane			0.188		26
	Methylcyclohexane/acetone (17 /83 vol.)	0.242			22	234
	(22.2/77.8 vol.)	0.242			22	234
	(28 /72 vol.)	0.239			22	234
	(39 /61 vol.)	0.231			22	234

★ In journal erroneously given 0.110

SPECIFIC REFRACTIVE INDEX INCREMENTS

Polymer	Solvent	dn/dc [ml/g] λ_o = 436 nm	Others	λ_o = 546 nm	T (°C)	References
Poly(styrene) (Cont'd.)	Methylcyclohexane/acetone					
	(50 /50 vol.)	0.228			22	234
	(61 /39 vol.)	0.223			22	234
	(72 /28 vol.)	0.222			22	234
	(83 /17 vol.)	0.217			22	234
	(88.3/11.7 vol.)	0.210			22	234
	Nitropropane	0.208			35	202
	Nitropropane/n-heptane (42/58 vol.)	0.216			35	202
		0.218			35	211
	Phenylcyanide	0.0775			25	146
	Styrene	0.054			20	102
	THF	0.195			30	216
				0.1926	30	84
			0.190			235
				0.198		57
			0.194			166
$\overline{M}_w$ = 2.4 x 10^4		0.205		0.193	25	236
5.31 x 10^4		0.202		0.189	25	236
16.5 x 10^4		0.207		0.194	25	236
54.1 x 10^4		0.208		0.195	25	236
150 x 10^4		0.209		0.196	25	236
570 x 10^4		0.211		0.198	25	236
	THF/methanol (28.7/71.3 vol.)			0.179	25	209
	Toluene	0.111		0.109	room	232
		0.112			12	221
				0.104	20	31, 227
		0.112			20	102
				0.109	20	238
		0.1113	0.1073(578)	0.1079	20	5
			0.1071(589)		20	5
			0.1060(644)		20	5
			0.113			153
		0.112		0.108		230
		0.195			20	231
				0.110	20	240
		0.111			23	237
		0.112			25	239
				0.109		225
		0.1118		0.1079	25	241
		0.1129		0.1091	25	242
				0.1130	25	43
				0.110	25	243
				0.111	25	240
		0.111			25	217, 228
		0.112			25	146
		0.1150		0.1111	25	139
				0.111	28	29
				0.110		147
		0.113			30	152, 216
		0.118			30	229
		0.114			35	97
				0.118	45	243
				0.125	65	243
			0.110 (633)			244
				0.118	67	218
isotactic		0.111			30	245
dispersion	Water		0.250		25	95
			0.241			246
		0.256			25	247
latex		0.257				248
sulfonated	Neutralization equiv. =					
	0.9 meq/g; 0.05M NaCl			0.212	25	247
	1.8 meq/g; 0.05M NaCl			0.207	25	247
	2.0 meq/g; 0.10M NaCl			0.195	25	247
	2.7 meq/g; 0.10M NaCl			0.205	25	247

Polymer	Solvent	dn/dc [ml/g]			T (°C)	References
		λ_o = 436 nm	Others	λ_o = 546 nm		
Poly(styrene-p-sulfonic acid)	Aq. KCl 2.5M			0.0347 (μ)[*]	25	249
--, K-salt, D.N. = 1.0 :	Aq. KCl; I = 0.10	0.197			30	85

1.7 Others

Polymer	Solvent	dn/dc [ml/g]			T (°C)	References
		λ_o = 436 nm	Others	λ_o = 546 nm		
Poly(acenaphthylene)	Benzene	0.225		0.119	25	55
	1, 2-Dichloroethane	0.264		0.238	35	55
	Toluene			0.190	25	56
Poly(acrylonitrile)	γ-Butyrolactone			0.079	25	276
				0.078	30	277
	Dimethylacetamide	0.0767		0.0769	25	278
	DMF	0.0870			18	213
				0.082	20	238
				0.0874	20	281
		0.0877			20	283
		0.0914			20	283
		0.0924			20	283
				0.196	25	284
				0.083	25	215, 276, 278, 280
				0.089		279
		0.080			25	282
				0.084	25	277
				0.087	25	10
	DMSO	0.029		0.031	25	278
				0.043	25	285
				0.092	140	278
				0.0263	25	284
Poly(N-isobutyl maleimide	Chlorobenzene		0.01			577
Poly(2-methyl-5-vinylpyridine)	n-Butyl acetate	0.200		0.188	25	269
	MEK	0.212		0.202	25	269
	Methanol	0.273		0.257	25	270
Poly(vinyl acetate)	Acetone			0.095	20	238
		0.104				287
			0.116		25	32
				0.104	25	289
		0.104			25	288
				0.105	25	201
				0.1021	35	34
	Acetonitrile		0.104			290
	Benzene		0.030			194
				-0.0225	30	84
	n-Butyl acetate			0.0716	30	84
	Chlorobenzene			-0.0426	30	84
			-0.049		25	32
	Dioxane		0.028			290
	Ethyl acetate		0.087		25	32
	Ethyl formate		0.095			290
	MEK		0.080		15	32
				0.080		291
		0.080				252, 287, 292
			0.080		25	32, 293
		0.083			25	205
				0.080	25	289
			0.080		35	32
	Methanol	0.107			25	287
		0.130			25	288
		0.132			25	205, 294
			0.124		25	32
				0.132		289
		0.1319			34-38	295

[*] Expressed in 1/g-segment

SPECIFIC REFRACTIVE INDEX INCREMENTS

Polymer	Solvent	dn/dc [ml/g] λ_o = 436 nm	Others	λ_o = 546 nm	T (°C)	References
Poly(vinyl acetate) (Cont'd.)	Methyl isobutyl ketone		0.068			290
	Methyl isopropyl ketone/n-heptane (3/1 - 6/1 vol.)	0.075		0.075	25	252
	THF			0.0582	30	84
	Trichlorobenzene	0.1030			34-38	296
Poly(vinyl acetate) (partially hydrolyzed)						
Deg. of Hydrolysis = 0.840	Water			0.151		297
0.883				0.155		297
0.964				0.161		297
0.008	Acetone			0.103	25	289
0.029				0.101	25	289
0.049				0.098	25	289
0.119				0.090	25	289
0.008	MEK			0.083	25	289
0.029				0.087	25	289
0.049				0.092	25	289
0.119				0.110	25	289
0.008	Methanol			0.128	25	289
0.029				0.115	25	289
0.049				0.107	25	289
0.119				0.071	25	289
Poly(vinyl alcohol)	Water			0.150	20	298
		0.144			20	299
		0.168	0.164 (578)	0.164	30	287
Poly(vinylbiphenyl)	Benzene	0.182		0.165	24	300
	Carbon tetrachloride	0.220		0.207	24	300
	Dioxane	0.257		0.233	24	300
	Ethylene glycol dimethyl ether	0.301		0.277	24	300
	Ethylene glycol dimethyl ether/ ethylene glycol monomethyl (56/44 vol.)	0.297-0.301		0.271-0.275	30	300
Poly(vinyl bromide)	THF			0.112		251
			C.112 (644)			252
Poly(vinyl n-butyl ether)	MEK	0.0792			25	250
Poly(vinyl n-butyrate)	MEK		0.0684		20	301
Poly(vinyl carbanilate)	Dioxane	0.159		0.155	20	302
Poly(N-vinylcarbazole)	Benzene	0.210		0.190	25	304
		0.219		0.195	25	303
		0.222		0.198	30	303
		0.225		0.201	35	303
		0.228		0.204	40	303
		0.231		0.207	45	303
	Chloroform	0.268		0.232	25	304
	Tetrachloroethane	0.214		0.185	25	304
	THF	0.282		0.262	25	304
Poly(vinyl chloride)	Acetone			0.138	20	238
	Aq. NH$_4$OH; pH 9-10 (latex)			0.200		253
	Cyclohexanone	0.0718	0.0701(578)	0.0705	15.5	259
		0.0723	0.0708(578)	0.0714	20	259
				0.0840	25	255, 256
		0.079	0.076 (589)	0.077	25	206
				0.075	25	10
				0.074	25	254, 257
			0.078			258
		0.0739	0.0716(578)	0.0727	25	259
		0.0758	0.0740(578)	0.0742	25	259
		0.0764	0.0745(578)	0.0746	30	259
		0.0767	0.0753(578)	0.0756	35	259
				0.0905	50	255

Polymer	Solvent	$\lambda_o = 436$ nm	Others	$\lambda_o = 546$ nm	T (°C)	References
Poly(vinyl chloride) (Cont'd.)	Cyclohexanone			0.0840	50	256
				0.0900	75	256
				0.0970	75	255
				0.0940	95	256
	Di-n-butyl phthalate			0.0380	25	256
	Di-2-ethylhexyl phthalate			0.0435	25	256
				0.0465	50	256
				0.0494	80	256
				0.0518	105	256
	Di-2-heptylnonyl phthalate			0.0420	25	256
	Di-n-hexyl phthalate			0.0250	25	256
				0.0260	50	256
				0.0280	70	256
				0.0300	105	256
	Dimethyl phthalate			0.0260	25	256
	Dioxane			0.0900	25	256
				0.0925	25	260
		0.107	0.102 (589)	0.104	25	206
				0.11	25	261
Mol. wt. = 2.04×10^6				0.0850	28.5	262
1.78×10^6				0.0865	28.5	262
	Di-n-propyl phthalate			0.0395	25	256
	DMF			0.069	25	256
				0.078	50	256
				0.0875	75	256
				0.095	95	256
		0.084				258
		0.0816	0.0813(578)	0.0810	25	259
	Tetrachloroethane			0.0432	20	255
				0.0450	25	255
				0.0540	50	255
				0.0630	75	255
	THF	0.1040	0.1010(578)	0.1017	16.5	259
		0.115		0.110	room	265
			0.102			258
				0.106		266
		0.1124			20	263
				0.105	25	261
				0.1123	25	256
		0.119	0.115 (589)	0.116	25	206
				0.1065	25	267
		0.1079	0.1068(578)	0.1052	25	259
		0.1067	0.1046(578)	0.1053	25	259
		0.1093	0.1059(578)	0.1064	30	259
		0.1118	0.1093(578)	0.1087	35	259
				0.1045	35	34
				0.1065		264
	THF/water (91/9 vol.)	0.112 (μ)		0.1085 (μ)	25	267
--, chlorinated 56.4 % wt. Cl :	Tetrachloroethane			0.045	25	255
58.3 % wt. Cl				0.0424	25	255
61.5 % wt. Cl				0.0525	25	255
63.4 % wt. Cl				0.0606	25	255
66.6 % wt. Cl				0.073	25	255
Poly(vinyl ethyl ether)	MEK	0.0736			25	250
Poly(vinyl fluoride)	DMF		≈ 0.02			268
Poly(vinyl isopropyl ether)	MEK	0.0827			25	250
Poly(vinyl methyl ether)	MEK	0.0944			25	250
Poly(α-vinylnaphthalene)	Benzene	0.192		0.175	24	300
	Ethylene dichloride	0.239		0.217	24	300
	Ethylene glycol dimethyl ether	0.268		0.248	24	300
	Toluene	0.162-0.192			24	300

SPECIFIC REFRACTIVE INDEX INCREMENTS

Polymer	Solvent	dn/dc [ml/g] λ_o = 436 nm	Others	λ_o = 546 nm	T (°C)	References
Poly(β-vinylnaphthalene)	Benzene	0.182-0.197		0.167-0.181	24	300
	Carbon tetrachloride	0.215		0.194	24	300
	Ethylene glycol dimethyl ether	0.309		0.284	24	300
	Ethylene glycol dimethyl ether/ ethylene glycol monomethyl ether (56/44 vol.)	0.283		0.258	24	300
Poly(2-vinylpyridine)	Chloroform			0.1372	23	34
				0.1215	35	34
atactic	DMF			0.157		271
isotactic				0.163		271
atactic	Ethanol			0.221		271
isotactic				0.221		271
isotactic	Ethanol + 0.1M sodium acetate			0.222		271
	THF		0.182			272
Poly(4-vinylpyridine)	Chloroform	0.150			25	146
	DMF	0.160			25	146
	Ethanol	0.231				273
	Ethanol/water (95/5 wt.)	0.231				273
	Isopropanol	0.224			25	274
	MEK/isopropanol (85/15 wt.)	0.221			25	274
	(86/14 wt.)	0.224			25	274
	Methanol	0.267			25	274
	Nitromethane		0.180			275
			0.159			275
			0.152			275
	Phenyl cyanide	0.052			25	146
Poly(vinyl pyrrolidone)	Chloroform	0.108		0.108		306
	Methanol			0.183	25	201
				0.176	25	193
	Water			0.135	25	193
		0.185		0.185		306
Poly(vinylsulfonic acid), potassium salt						
	KCl 0.35M			0.01664 (μ)*)	23	308
	0.65M			0.01638 (μ)*)	23	308
Poly(vinyltrimethyl silane)	Carbon tetrachloride			0.048		307
	Cyclohexane			0.083		307

2.   Main-chain Heteroatom Polymers

2.1   Poly(oxides) and Poly(sulfides)

Poly(formaldehyde) see Poly(oxymethylene)

Poly(oxybutylethylene) [Poly(hexamethylene oxide)]

Polymer	Solvent	dn/dc [ml/g] λ_o = 436 nm	Others	λ_o = 546 nm	T (°C)	References
	Diethyl ketone		0.088		25	340
Poly(oxy-tert-butylethylene)	Isooctane			0.114	75	328
Poly(oxy-2,6-dimethyl-1,4-phenylene)						
	Benzene			0.114	25	330
	Chlorobenzene			0.098	25	330
	Chloroform			0.124	25	330
				0.169	25	331
	Toluene			0.1153	25	331
				0.118	25	330
	Xylene			0.1136	25	331
Poly(oxy-2,6-diphenyl-1,4-phenylene)						
	Chloroform			0.226	30	330(a)
	Toluene			0.185	25	331
Poly(oxyethylene) glycol						
Mol. wt. 62	Acetonitrile			0.0964	25	332
100				0.106	25	332

*) Expressed in 1/g-segment

References page IV-303

Polymer	Solvent	dn/dc [ml/g]			T (°C)	References
		λ_o = 436 nm	Others	λ_o = 546 nm		
Poly(oxyethylene) glycol (Cont'd.)						
Mol. wt. 161	Acetonitrile			0.114	25	332
205				0.121	25	332
316				0.123	25	332
407				0.130	25	332
970				0.135	25	332
9,400				0.135	25	332
Mol. wt. 106	Benzene	-0.086				333
194		-0.073				333
282		-0.066				333
810		-0.059				333
3,510				-0.016	25	332
	n-Butanol			0.076		334
	Chloroform/n-hexane (47/53 vol.)			0.091	20	193
Mol. wt. 10,000	Dioxane			0.045		333
	DMF			< 0.05	25	335
Mol. wt. 9,400				0.044	25	332
Mol. wt. 810	MEK			0.092		333
10,000				0.094		333
	Methanol			0.150		334
Mol. wt. 62				0.118	25	332
100				0.127	25	332
161				0.135	25	332
205				0.139	25	332
316				0.141	25	332
445				0.142	25	332
810				0.143		333
1,020				0.144	25	332
3,000				0.149		333
6,000				0.150		333
9,400				0.150	25	332
10,000				0.148		333
31,000				0.150	25	332
				0.143	25	193
				0.152	45	328
Mol. wt. 6,700	Methyl acetate			0.111	25	332
	THF			0.068		334
	Water	0.134		0.132	27	337
				0.139		334
Mol. wt. 62		0.093				333
106		0.108				333
194		0.124				333
300		0.126		0.123		333
600		0.135		0.131		333
600			0.128 (589)		25	338
810		0.136		0.128		333
1,200		0.139		0.134		333
3,000		0.141				333
4,000			0.134 (589)		25	338
6,000			0.134 (589)		25	338
6,000		0.145		0.139		333
9,000			0.134 (589)		25	338
9,400				0.135	25	332
10,000		0.142				333
14,400				0.139	25	336
14,400				0.115	80	336
31,000				0.135	25	332
340,000		0.149			25	339
Poly(oxyethylethylene) [Poly(1-butene oxide)]						
	Benzene			-0.0403	25	326
	Chlorobenzene			-0.0784	25	326
	Ethyl acetate			0.1125	25	326
	n-Hexane			0.1008	25	326

Polymer	Solvent	dn/dc [ml/g] $\lambda_o = 436$ nm	Others	$\lambda_o = 546$ nm	T (°C)	References
Poly(oxyethylethylene) [Poly(1-butene oxide)] (Cont'd.)						
	MEK	0.0863			30	327
	Toluene			-0.0357	25	326
Poly(oxymethylene)	Hexafluoroacetone/water (1/1.7 mol)					
	+ 1 % vol. triethylamine buffer;					
	pH 2.0	0.134		0.128		341
		0.114 (μ)		0.112 (μ)		341
	1H, 1H, 5H-Octafluoropentanol-1			0.156	110	325
Poly(oxymethyleneoxyethylene) [Poly(dioxolane)]						
	Chlorobenzene			0.051		578
	o-Dichlorobenzene			0.075		578
	Dichloroethane			0.020		578
	Dioxane			0.050		578
Poly(oxyoctylethylene)	Chloroform		0.0357		25	329
Poly(oxyphenylethylene) [Poly(styrene oxide)]						
	Benzene			0.085	30	328
Poly(oxypropylene)						
Mol. wt. 67	Acetone	0.085			25	342
125		0.0915			25	342
450		0.096			25	342
1,100		0.099			25	342
1,240		0.099			25	342
2,100		0.100			25	342
3,270		0.100			25	342
3,850		0.101			25	342
	Benzene	-0.0530		-0.0448	25	343
	Chlorobenzene	-0.0658		-0.0638	25	343
	n-Hexane	0.0775		0.0775	25	343
		0.0460		0.0460	40	343
Mol. wt. 9.6×10^5		0.0887		0.0887	46	343
		0.101		0.101	57	343
2.0×10^5		0.0895		0.0895	46	343
		0.104		0.104	57	343
	Isceon 113	0.118		0.115	25	343
	Isooctane	0.0655		0.0655	35	343
	Methanol	0.137		0.135	24±1	300
Mol. wt. 12.2×10^5		0.118		0.118	25	343
12.5×10^5		0.115		0.115	25	343
Poly(oxytetramethylene) [Poly(tetrahydrofuran)]						
	Chlorobenzene	0.070				347
	Ethyl acetate	0.110			25	348
		0.114			30	349
	Ethyl acetate/n-hexane					
	(22.7/77.3 wt.)	0.114			31.8	349
	Isopropanol	0.108			46	351
	Isopropyl acetate	0.098			22.5	352
	MEK	0.102			30	348
				0.091	25	350
				0.095		347
	Methyl acetate			0.101	25	350
	3-Methyl-2-heptanone			0.056	25	350
	2-Pentanone			0.084	25	350
	THF	0.0625		0.0625	25	348
				0.064	25	350
Poly(oxytrimethylene)	MEK		0.0946			353
Poly(thiopropylene)	Benzene			0.100	20	344
				0.0935		345
			0.08095			346

Polymer	Solvent		dn/dc [ml/g]			T (°C)	References
			λ_0 = 436 nm	Others	λ_0 = 546 nm		
2.2	**Poly(sulfones)**						
Poly(1-butene-co-sulfur dioxide)	Chloroform		0.0970	0.0949(578)		20	355
Poly(cyclohexene-co-sulfur dioxide)							
	Chloroform		0.1133	0.1104(578)		20	355
Poly(1-dodecene-co-sulfur dioxide)							
	Chloroform		0.0695	0.0581(578)		20	355
Poly(1-hexene-co-sulfur dioxide)	Chloroform		0.0790	0.0782(578)		20	355
	n-Hexyl chloride		0.0796			7	355
			0.0834			20	355
	MEK/isopropanol (37 /63 vol.)		0.1242			27	355
			0.1261			37	355
	(41.5/58.5 vol.)		0.1208			7	355
			0.1233			20	355
Poly(1-octene-co-sulfur dioxide)	Chloroform		0.0732	0.0718(578)		20	355
Poly(oxy-1,4-phenylenesulfonyl-1,4-phenylene) [Poly(sulfon)]							
	DMF				0.194	25	580
2.3	**Poly(amides), Poly(esters), Poly(urethanes)**						
Poly(N-butyliminocarbonyl) [Poly(butyl isocyanate)]							
	Chloroform				0.054		356
Poly(ethylene terephthalate) see Poly(oxyethyleneoxyterephthaloyl)							
Poly(N-hexyliminocarbonyl) [Poly(hexyl isocyanate)]							
	Chloroform		0.054				358
	THF		0.0978				358
			0.100		0.097		360
Poly(iminoadipoyliminohexamethylene) (Nylon 66)							
	Formic acid	75 %		0.144		25	311
		80 %		0.145		25	311
		85 %	0.141			25	312
		90 %		0.145		25	311, 313
					0.145	25	309
			0.145			25	312
		95 %		0.150			311
		100 %		0.157		25	311
			0.1525				310
	Formic acid + KCl						
		85 % + 2.0M KCl		0.124			313
		90 % + 0.2M KCl		0.143			313
		0.5M KCl		0.140			313
		1.0M KCl		0.136			313
		1.5M KCl		0.131			313
		2.0M KCl		0.126			313
		2.5M KCl		0.122			313
		95 % + 2.0M KCl		0.129			313
	Formic acid + sodium formate						
		75 % + 0.5 M		0.138		25	311
		80 % + 0.5 M		0.136		25	311
		90 % + 0.02M		0.147		25	311
		0.05M		0.146		25	311
		0.10M		0.142		25	311
		0.2 M		0.142		25	311
		0.5 M		0.136		25	311
		0.75M		0.130		25	311
		1.0 M		0.124		25	311
		95 % + 0.5 M		0.136		25	311
		100 % + 0.5 M		0.136		25	311
	Tetrafluoropropanol				0.190	25	314, 315

Polymer	Solvent		dn/dc [ml/g]			T (°C)	References
		λ_o = 436 nm	Others	λ_o = 546 nm			
Poly(iminoadipoyliminohexamethylene) (Nylon 66) (Cont'd.)							
	Tetrafluoropropanol + 0.1N sodium trifluoroacetate buffer	0.190				25	314
Poly(imino(1-oxohexamethylene)) (Nylon 6)							
	Carbon disulfide 95 %			0.0815		25	309
	m-Cresol			-0.016		25	309
	Dichloroacetic acid			0.098		25	309
				0.099		50	309
				0.104		80	309
	Formic acid 90 % + 1.0M KCl			0.136		25	309
	2.0M KCl			0.126		25	309
	Octafluoropentanol			0.192		25	309
	Tetrafluoropropanol/water						
	(95/ 5 vol.)			0.183		25	316
	(90/10 vol.)			0.182		25	316
	(80/20 vol.)			0.180		25	316
	(70/30 vol.)			0.179		25	316
	(90/10 vol.)			0.18		25	316
	Tetrafluoropropanol + 1.27 x 10^{-3} g/ml LiCl (extrapolated to 0 % H_2O)			0.183		25	316
	2.54 x 10^{-3} g/ml LiCl "			0.183		25	316
	4.24 x 10^{-3} g/ml LiCl "			0.183		25	316
	6.36 x 10^{-3} g/ml LiCl "			0.182		25	316
Poly(oxyadipoyloxytetramethylene) [Poly(butylene adipate)]							
	Acetic acid			0.100		50	317
	Diethyl carbonate			0.084		35	317
				0.091		70	317
	Dioxane			0.0518		35	317
	THF			0.0784		35	317
Poly(oxycarbonylethylidene) [Poly(lactic acid)]							
	Bromobenzene			-0.06		85	324
Poly(oxycarbonylimino-1, 4-phenylenemethylene-1, 4-phenyleneiminocarbonyloxyhexamethylene)							
	DMF		0.159				357
	DMF/acetone (90/10 vol.)		0.166				357
			0.163 (μ)				357
	(80/20 vol.)		0.173				357
			0.178 (μ)				357
	(70/30 vol.)		0.181				357
			0.188 (μ)				357
	(60/40 vol.)		0.189				357
			0.200 (μ)				357
	(50/50 vol.)		0.195				357
			0.205 (μ)				357
	(40/60 vol.)		0.203				357
	DMF/toluene (90/10 vol.)		0.154				357
			0.159 (μ)				357
	(80/20 vol.)		0.147				357
			0.160 (μ)				357
	(70/30 vol.)		0.143				357
			0.149 (μ)				357
	(60/40 vol.)		0.139				357
			0.135 (μ)				357
	(50/50 vol.)		0.134				357
			0.125 (μ)				357
	(40/60 vol.)		0.130				357
			0.120 (μ)				357
	(30/70 vol.)		0.125				357
	(25/75 vol.)		0.123				357
Poly(oxycarbonylimino-2, 4-tolyleneiminocarbonyl-oxy(poly(oxy-propylene))							
	Methanol	0.152		0.150		22	361
		0.148		0.145		22	361

Polymer	Solvent	dn/dc [ml/g]			T (°C)	References
		λ_o = 436 nm	Others	λ_o = 546 nm		
Poly(oxycarbonyloxy-1,4-phenyleneisopropylidene-1,4-phenylene)						
[Poly(carbonate) from Bisphenol A]						
	Chloroform	0.164			20	318
				0.162		319
				0.156	30	319a
	Dioxane			0.168	25	320
	Methylene chloride	0.1660	0.1547(553)		7	322
			0.1538(574)		7	322
		0.177			25	321
		0.1773	0.1654(553)		27	322
			0.1643(574)		27	322
	THF	0.1889	0.1765(553)		7	322
			0.1753(574)		7	322
		0.1943	0.1815(553)		27	322
			0.1805(574)		27	322
Poly(oxycarbonylpentamethylene) [Poly(ϵ-caprolactone)]						
Prepared with 1 mol % $(C_2H_5)_3Al$						
	Acetic acid			0.108	50	317
	o-Dichlorobenzene			-0.0595	35	317
	Diethyl carbonate			0.092	35	317
				0.0990	70	317
	Dioxane			0.0489	25	317
				0.0507	35	317
				0.054	50	317
				0.0581	70	317
Prepared with 1 mol % $(C_2H_5)_2AlCl$						
	Acetic acid			0.0940	50	317
	o-Dichlorobenzene			-0.057	35	317
	Diethyl carbonate			0.093	35	317
				0.0939	40	317
				0.101	70	317
	Dioxane			0.0489	25	317
				0.0504	35	317
				0.0581	70	317
	THF			0.0795	35	317
Prepared with 0.61 mol % n-C_4H_9OK						
	Diethyl carbonate			0.0917	35	317
				0.1021	70	317
Poly(oxyethyleneoxyterephthaloyl) [Poly(ethylene terephthalate)]						
	Dichloroacetic acid			0.1021	35	34
	Trifluoroacetic acid			0.278		323
Poly(oxysebacoyloxyoctamethylene)						
	Acetic acid			0.100	50	317
	o-Dichlorobenzene			-0.071	35	317
	Diethyl carbonate			0.090	45	317
				0.100	70	317
	Dioxane			0.050	35	317
				0.054	50	317
				0.054	70	317
	THF			0.076	35	317
Poly(N-tolyliminocarbonyl) [Poly(tolyl isocyanate)]						
	Bromoform		0.017			359
2.4　Poly(silanes) and Poly(silazanes)						
Ludox	Diluted with water	0.0620	0.0606(578)	0.0608	23±2	362
	Diluted with NaCl (0.05M)	0.0592	0.0585(578)	0.0587	23±2	362
Poly(dimethylsilmethylene)	n-Heptane	0.098			25	37
	n-Heptane/n-propanol					
	(75 /25 vol.)	0.115			25	37
	(71 /29 vol.)	0.092			25	37
	(69.5/30.5 vol.)	0.103			25	37

Polymer	Solvent	dn/dc [ml/g] $\lambda_o = 436$ nm	Others	$\lambda_o = 546$ nm	T (°C)	References
Poly(dimethyl siloxane) see Poly(oxydimethylsilylene)						
Poly(dimethylsiltrimethylene)	n-Heptane	0.123			25	37
Poly(diphenylsiltrimethylene)	Toluene	0.112			25	37
	Toluene/cyclohexanol					
	(79.5/20.5 vol.)	0.133			25	37
	(75 /25 vol.)	0.135			25	37
	(70 /30 vol.)	0.145			25	37
	(40 /60 vol.)	0.172			25	37
	(37.6/62.4 vol.)	0.175			25	37
Poly(oxydimethylsilylene)	Benzene			0.098	25	201
	Benzene/triethyl borate					
	(94.44/5.56 vol.)			0.115	25	138
Mol. wt. 1.06×10^6	Bromocyclohexane			0.078	25	363
0.78×10^6				0.079	25	363
0.545×10^6				-0.083	29	364
0.34×10^6				0.081	25	363
0.085×10^6				0.090	25	363
0.033×10^6				0.090	25	363
1892				-0.086	29	364
978				-0.092	29	364
682				-0.098	29	364
150				-0.111	29	364
	Toluene	0.103		0.094		365
		-0.1042		-0.0947	14.6	366
		-0.1037		-0.0941	19.9	366
		-0.1029		-0.0933	25.4	366
		-0.1027		-0.0933	29.9	366
		-0.1023		-0.0927	34.8	366
Mol. wt. 0.545×10^6				-0.084	25	364
0.21×10^6				0.092	25	367
1892				-0.085	25	364
978				-0.094	25	364
682				-0.099	25	364
150				-0.134	25	364
Poly(oxymethyl-3,3,3-trifluoropropylsilylene)						
	Cyclohexyl acetate	0.052			35	368
Poly(oxyphenylsilylene)	Benzene		0.08			369
Poly(silicic acid)	Water	0.0592				362
			0.65			370
Poly(trimethylsiloxanotitanoxanes)						
	Cyclohexane			0.156	25	371

2.5   Others

Ionene Polymers

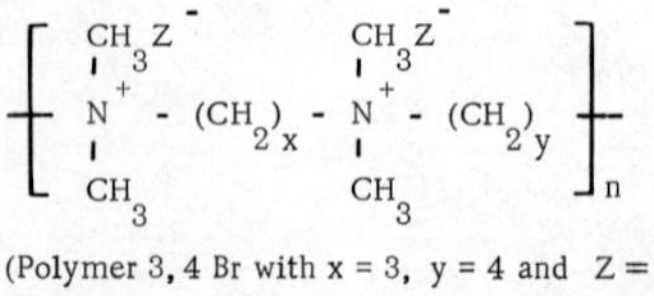

(Polymer 3,4 Br with x = 3, y = 4 and Z = Br)
(Polymer 6,6 Br with x = 6, y = 6 and Z = Br)
(Polymer 6,6 Cl with x = 6, y = 6 and Z = Cl)

	3,4 Br	0.4 M aq. KBr	0.1560-0.1601			572
		0.30M KBr	0.160			572
		0.20M	0.160			572
		0.10M	0.162			572
		0.05M	0.168			572
	6,6 Br	0.4 M KBr	0.1567-0.1609			572
	6,6 Cl	0.4 M KBr	0.1546-0.1599			572
	6,6 Cl	0.4 M KCl	0.1636-0.1668			572

Polymer	Solvent	λ_o = 436 nm	Others	λ_o = 546 nm	T (°C)	References
Poly(5, 7-dihydro-1, 3, 5, 7-tetraoxobenzo[1, 2-c : 4, 5-c'] dipyr-role-2, 6 [1H, 3H] -diyl-1, 4-phenyleneisopropylidene-1, 4-phenylene) (Polyimide from pyromellitic anhydride and 4, 4'-diamino-diphenylpropane)						
	LiBr 0.1N in dimethylacetamide			0.196		574
Poly(fluoroalkoxyphosphazenes)	Freon E2/acetone (10/1 vol.)	0.048			30	579
Poly(iminocarbonyl-3, 4-dicarboxy-1, 5-phenylenecarbonylimino-1, 4-phenyleneisopropylidene-1, 4-phenylene) [Poly(amic acid)]						
	LiBr 0.1N in dimethylacetamide			0.206		574
Poly(iminocarbonyl-3, 4-dicarboxy-1, 5-phenylenecarbonylimino-1, 4-phenyleneoxy-1, 4-phenylene) [Poly(amic acid)]						
	LiBr 0.1N in dimethylacetamide			0.210		573
Sodium polyphosphate	Aq. NaBr 0.1 N		0.109			581
	0.25N		0.105			581
	0.35N		0.102			581
	0.40N		0.101			581

3.   Cellulose and Derivatives

Polymer	Solvent	λ_o = 436 nm	Others	λ_o = 546 nm	T (°C)	References
Carboxymethyl cellulose	NaCl 0.1M			0.147	25	373
Carboxymethyl cellulose, sodium salt						
D.S. = 0.96	Cadoxen 0.237M Cd.	0.1398		0.1896	25	377
		0.1373 (μ)		0.1861 (μ)	25	377
0.94		0.147		0.145	25	404
0.44		0.164		0.162	25	404
0.21		0.175		0.171	25	404
	Aq. NaCl	0.136				128
		0.154				405
		0.158				405
Cellulose	Acetone	0.111			25	374
	Cadoxen	0.186		0.183	25	375, 376
	0.237 M Cd	0.1317		0.1927	25	377
		0.1925 (μ)		0.1890	25	377
	Cuoxam 0.205 M Cu	0.117		0.233	25	378
	Cuoxen 0.0518M Cu	0.1352		0.2574	25	377
	0.0776M Cu			0.2653	25	377
	FeTNa	0.110 (μ)		0.245 (μ)	25	379
Cellulose nitrate						
10.98 % N	Acetone	0.1022		0.0998	25	380
11.89 % N		0.1010		0.0985	25	380
12.55 % N		0.0968		0.0950	25	380
12.9 % N		0.1151			25	382
11.3 % N	Ethyl acetate			0.103	20	381
13.49 % N		0.102			25	383
Cellulose triacetate	Chloroform		0.0406		25	384
			0.0496		55	384
Cellulose tricaproate	1-Chloronaphthalene			0.147	24	385
	Dioxane/water (100/7 vol.)			0.104	63	385
	DMF	0.0442		0.0478	41	385
Cellulose tricarbanilate	Acetone	0.2069		0.1966	7	385
		0.2155		0.2047	25	386
		0.2176		0.2033	27	385
		0.215			30	216
	Benzophenone		0.0148(589)		55	387
	Cyclohexanone	0.1556		0.1666	25	386
		0.137			30	216

Polymer	Solvent		dn/dc [ml/g]			T (°C)	References
			λ_0 = 436 nm	Others	λ_0 = 546 nm		
Cellulose tricarbanilate (Cont'd)							
	Dibutyl ketone				0.210	25	388
	Diethyl ketone		0.1911		0.1797	25	386
	Dioxane		0.1651		0.1558	25	386
			0.168	0.181 (361)	0.156	20	389
	Furfural		0.070			30	216
	MEK		0.188			30	216
	Pyridine		0.089			30	216
	THF		0.182			30	216
Cellulose trinitrate (fully substituted)	Acetone		0.105				253
			0.102		0.101	room	390
			0.1116			25	382
			0.0930		0.0903	25	380
	Ethyl acetate		0.102			30	383
			0.107			25	391
					0.103	20	381
Cellulose xanthate	Aq. NaOH				0.212		392
					0.230		393
					0.190		394
				0.20			395
		1.0M			0.231		378
		2.5M			0.185		396
Cellulose xanthate, sodium salt							
D.S. = 0.25	Aq. NaOH	1.0N	0.205				406
0.48			0.275				406
0.77			0.300				406
0.91			0.425				406
Diethylacetamide cellulose xanthate							
D.S. = 0.40	Aq. DMSO 90%			0.144			397
0.49				0.098			397
0.60				0.144			397
0.80				0.144			397
0.92				0.099			397
1.00				0.096			397
1.22				0.079			397
Ethyl cellulose	Methanol		0.130			25	398
Ethyl hydroxyethyl cellulose	Methanol		0.147				399
Hemicellulose	KOH 0.36M				0.131		400
	NaCl 0.25M/0.5% Calgon		0.127				400
	0.5 M				0.145		400
	0.5 M/0.5% Versene		0.136				400
Hydroxypropyl cellulose	Anhydrous ethanol				0.120	25	401
Methyl cellulose	Water		0.154				402
					0.145		403
Lignin	Aq. NaHCO$_3$; pH 9.65				0.218	25	372
Ligninsulfonic acid, sodium salt	Aq. NaCl 0.1M			0.197			407
				0.200			407
	DMSO			0.117			407
4. Poly(saccharides)							
Amylopectin	DMSO				0.066		408, 409
	DMSO/water (90/10 vol.)				0.074		408, 409
	(80/20 vol.)				0.083		408, 409
	(60/40 vol.)				0.101		408, 409
	(50/50 vol.)				0.110		408, 409
	Ethylenediamine		0.098		0.098		410
	Ethylenediamine hydrate		0.092		0.092		410

Polymer	Solvent		dn/dc [ml/g]			T (°C)	References
			λ_o = 436 nm	Others	λ_o = 546 nm		
Amylopectin (Cont'd.)	Formamide		0.069		0.069		410
	GHCl	4.2M			0.140		408, 409, 411
	KOH	1N	0.142		0.142		410
	NaOH	0.2N	0.142				412
	Water		0.156		0.154		412
			0.155		0.151		413
					0.152		408, 409, 411
Amylopectin acetate	Acetone		0.118				412
	Acetonitrile		0.128				412
	Chloroform		0.051				412
	Dioxane		0.057				412
	Nitromethane		0.088				412
Amylose	DMSO		0.0676		0.0659		414
	GHCl	4.2M	0.118		0.116	25	415
	Aq. KCl			0.146			416
	KOH	1M		0.146			416
	Aq. NaIO$_3$		0.153		0.152		417
Amylose acetate	Nitromethane		0.0875			30	418
			0.0857			room	419
					0.0835	20	420
					0.0835	20	421
Amylose tricarbanilate	Acetone		0.2279		0.2164	27	422
			0.2218		0.2094	7	422
	Dioxane		0.163	0.178 (361)	0.151	20	389
	Pyridine				0.0973	25	423
Araban	Aq. DMF 1 %			0.142			424
	NaCl	2M		0.130			424
Araban acetate	DMF				0.126		425
	MEK				0.0886		425
Chitosan	Water			0.166			426
Dextran	Ba(OH)$_2$	0.0503M	0.1465		0.1441	25	377
			0.1642 (μ)		0.1618 (μ)	25	377
		0.108 M	0.1415		0.1394	25	377
			0.1738 (μ)		0.1708 (μ)	25	377
	Cadoxen	0.237 M Cd	0.1235		0.1219	25	377
			0.1842 (μ)		0.1809 (μ)	25	377
			0.1907 (μ)		0.1875 (μ)	25	377
	Cuoxen	0.0518M Cu	0.1323			25	377
			0.2525 (μ)			25	377
		0.0776M Cu	0.2625 (μ)			25	377
	LiSCN	2.5M			0.129		409
	NaCl	0.1M			0.151		427
		0.5M	0.1476		0.1454	25	377
			0.1470 (μ)		0.1445 (μ)	25	377
		1 M	0.1447		0.1424	25	377
			0.1435 (μ)		0.1413 (μ)	25	377
		2 M	0.1361		0.1342	25	377
			0.1374 (μ)		0.1351 (μ)	25	377
	NaOH	0.25M	0.1456		0.1433	25	377
			0.1584 (μ)		0.1558 (μ)	25	377
		0.75M	0.1437		0.1415	25	377
			0.1646 (μ)		0.1621 (μ)	25	377
	ZnCl$_2$	4 M	0.151		0.147		428
	Water		0.154		0.151		429
					0.151		430
					0.15	25	431
			0.1518	0.1476(578)	0.1481	25	432
			0.154		0.151		428
			0.161			23	237

Polymer	Solvent	λ_o = 436 nm	dn/dc [ml/g] Others	λ_o = 546 nm	T ($^\circ$C)	References
Dextran (Cont'd.)	Water	0.150			20	433
				0.152		409
		0.1504		0.1481		377
		0.1494 (μ)		0.1472 (μ)		377
Dextran tricarbanilate	DMF	0.161		0.151		428
Dialdehyde starch	Water	0.153		0.152		434
Glucomannan triacetate	Nitroethane			0.076	25	435
D-Glucose	Products from irradiated aq. glucose solutions					
$\overline{M}_w$ = 9.8 x 10^3	Water			0.166		436
14 x 10^4				0.200		436
50 x 10^5				0.215		436
25 x 10^5	DMSO/water (98/2 wt.)			-0.135		436
Guaran triacetate	Acetonitrile			0.1200	22.5	437
Gum arabic	Aq. HCl	0.152		0.150	25	438
	NaCl 0.35M; NaOH 0.02N; 0.25% Na salt of EDTA	0.149				439
	Water	0.145				439
Hydroxyethyl starch						
D.S. = 0.8	Aq. NaCl 0.9%	0.136			25	440
		0.160		0.161	25	441
0.06-1.2 (no change, av. value)	Aq. NaCl 0.9%	0.151			20	442
	Water	0.151			20	442
Related compounds:						
Acetyl (D.S. = 0.1)	Water	0.150			20	442
(D.S. = 0.7)		0.147			20	442
(D.S. = 1.0)		0.141			20	442
Hydroxy butyl (D.S. = 0.2)		0.151			20	442
Hydroxy propyl						
(D.S. = 0.5)		0.151			20	442
(D.S. = 1.5)		0.152			20	442
Hydrolysed waxy maize starch		0.152			20	442
α-Keratose	Aq. phosphate buffer; I = 0.2; pH 6.7			0.182		443
	Dichloroacetic acid			0.088		444
	Dichloroacetic acid 0.1M KCl			0.088		444
	Formic acid			0.156		444
	0.1M KCl			0.155		444
	0.5M KCl			0.152		444
Lipopolysaccharide	Water	0.151				445
Mannan	Acetone	0.102		0.102	room	446
	Ethyl acetate	0.101		0.101	room	446
Magnesium alginate	Aq. $MgCl_2$	0.158		0.157	20-25	447
Mucopolysaccharides	Phosphate buffer 0.15M		0.166			448
	Phosphate buffer 0.15M + NaCl 0.2M		0.166			448
	Water		0.181			448
Potassium alginate	Aq.; in presence of KCl	0.160		0.159	20-25	447
Sodium alginate	Aq.; in presence of NaCl	0.165		0.163	20-25	447
Sodium amylose xanthate						
(D.S. = 0.4-0.5)	Aq. NaCl 0.1 M			-0.1460	25	449
	0.25M			-0.1480	25	449
	0.50M			-0.1555	25	449
	1.0 M			-0.1741	25	449

Polymer	Solvent	dn/dc [ml/g]			T (°C)	References
		λ_o = 436 nm	Others	λ_o = 546 nm		
Sodium amylose xanthate (Cont'd.)						
(Ripening time = 0 hr); in air	Aq. NaOH 4 %			0.1700		450
22				0.1547		450
46				0.1471		450
94				0.1522		450
142				0.1938		450
190				0.2564		450
0	under N_2			0.1700		450
45				0.1785		450
93				0.1899		450
140				0.2181		450
188				0.2963		450
Sodium carboxymethyl amylose	Aq. NaCl					
	0.0065 M	0.138			30	450
	0.01 M	0.138			30	450
	0.03 M	0.138			30	450
	0.035 M	0.134			30	450
	1.0 M	0.124			30	450
	2.5 M	0.103			30	450
	0.35 M	0.1325			35	451
Xylan	Cadoxen		0.178		25	452
	DMSO		0.064			453
			0.062			454
	Ethyl acetate	0.097		0.097		455

D. REFERENCES

1.   J. W. Lorimer, Polymer 13, 274 (1972).

2.   M. B. Huglin in "Light Scattering from Polymer Solutions" (Ed. M. B. Huglin) Academic Press, London/New York (1972).

3.   W. Heller, J. Polymer Sci., A-2, 4, 209 (1966).

4.   M. B. Huglin, J. Appl. Polymer Sci. 9, 4003 (1965).

5.   W. Mächtle, H. Fischer, Angew. Makromol. Chem. 7, 147 (1969).

6.   W. Mächtle, Angew. Makromol. Chem. 15, 17 (1971).

7.   Y. Kambe, C. Honda, Angew. Makromol. Chem. 25 163 (1972).

8.   Tuzar, P. Kratochvíl, D. Straková, European Polymer J. 6, 1113 (1970).

9.   P. Kratochvíl, J. Babka, J. Appl. Polymer Sci. 16, 1053 (1972).

10.   I. Baltog, C. Ghita, L. Ghita, European Polymer J. 6, 1299 (1970).

11.   P. Kratochvíl, B. Sedláček, D. Straková, Z. Tuzar, Makromol. Chem. 148, 271 (1971).

12.   M. B. Huglin, J. Appl. Polymer Sci. 9, 3963 (1965).

13.   W. R. Krigbaum, J. E. Kurz, P. Smith, J. Phys. Chem. 65, 1984 (1961).

14.   Th. G. Scholte, J. Polymer Sci., A-2, 6, 91 (1968).

15.   R. Chiang, J. Polymer Sci., C, 8, 295 (1965).

16.   F. W. Billmeyer, Jr., J. Am. Chem. Soc. 75, 6118 (1953).

17.   R. Chiang, J. Polymer Sci. 36, 91 (1959).

18.   L. H. Tung, J. Polymer Sci. 36, 287 (1959).

19.   H. P. Schreiber, M. H. Waldman, J. Polymer Sci. A-2, 1655 (1964).

20.   Q. A. Trementozzi, J. Polymer Sci. 23, 887 (1957).

21.   Q. A. Trementozzi, J. Polymer Sci. 36, 113 (1959).

22.   H. L. Wagner, C. A. J. Hoeve, J. Polymer Sci., A-2, 9, 1763 (1971).

23.   J. Ehl, C. Loucheux, C. Reiss, H. Benoit, Makromol. Chem. 75, 35 (1964).

24.   V. Kokle, F. W. Billmeyer, Jr., L. T. Muus, E. J. Newitt, J. Polymer Sci. 62, 251 (1962).

25.   E. E. Drott, R. A. Mendelson, J. Polymer Sci. B2, 187 (1964).

26.   A. Kotera, N. Iso, A. Senuma, T. Hamada, J. Polymer Sci., A-2, 5, 277 (1967).

27.   A. Kotera, T. Saito, H. Matsuda, A. Wada, Rept. Progr. Polymer Phys. Japan 8, 5 (1965).

28.   R. Chiang, J. Polymer Sci. 28, 235 (1958).

29.   J. E. Kurz, J. Polymer Sci., A, 3, 1895 (1965).

30.   R. A. Mendelson, J. Polymer Sci. 46, 493 (1960).

31.   D. A. Albright, J. W. Williams, J. Phys. Chem. 71, 2780 (1967).

32.   A. A. Tager, A. A. Anikeyeva, V. M. Andreyeva, T. Ya. Gumarova, L. A. Chernoskutova, Polymer Sci. USSR 10, 1926 (1968).

33.   F. R. Cottrell, E. W. Merrill, K. A. Smith, J. Polymer Sci., A-2, 7, 1415 (1969).

34.   D. Coupe, P. A. Curnuck, J. V. Dawkins, ICI Limited, Petrochemicals + Polymer Lab., Runcorn, England - private communication (1969).

35.   G. Delmas, D. Patterson, Polymer 7, 513 (1966).

36.   A. Červenka, M. Marek, K. Šolc, P. Kratochvíl, Collection Czech. Chem. Commun. 33, 4248 (1968).

37.   A. Yu. Koshevnik, M. M. Kusakov, E. A. Razumovskaya, Polymer Sci. USSR 10, 3242 (1968).

38.   T. Matsumoto, N. Nishioka, H. Fujita, J. Polymer Sci., A-2, 10, 23 (1972).

39.   A. Sakanishi, J. Chem. Phys. 48, 3850 (1968).

40.   K. E. van Holde, J. W. Williams, J. Polymer Sci. 11, 243 (1953).

41.   J. B. Kinsinger, L. E. Ballard, J. Polymer Sci., A, 3, 3963 (1965).

42.   H.-G. Elias, H. Dietschy, Makromol. Chem. 105, 102 (1967).

43.   S. Florián, D. Lath, Z. Maňásek, Angew. Makromol. Chem. 13, 43 (1970).

44.   J. B. Kinsinger, R. E. Hughes, J. Phys. Chem. 63, 2002 (1959).

45.   K. Yamaguchi, Makromol. Chem. 128, 19 (1969).

46.   W. E. Weston, F. W. Billmeyer, Jr., J. Phys. Chem. 65, 576 (1961).

47.   L. Westerman, J. Polymer Sci., A-1, 411 (1963).

48.   P. P. Parrini, F. Sebastiano, G. Messina, Makromol. Chem. 38, 27 (1960).

49.   D. K. Carpenter, J. Polymer Sci., A-2, 4, 923 (1966).

50.   H. Janeschitz-Kriegl, U. Daum, Kolloid Z.-Z. Polymer 210, 112 (1966).

51. T. Miyamoto, H. Inagaki, J. Polymer Sci., A-2, 7, 963 (1969).
52. H. Inagaki. T. Miyamoto, S. Ohta, J. Phys. Chem. 70, 3421 (1966).
53. H. T. Hall, J. Polymer Sci. 7, 443 (1951).
54. L. Saunders, L. Spirer, Polymer 6, 635 (1965).
55. J. Moacanin, A. Rembaum, R. K. Laudenslager, R. Adler, J. Macromol. Sci. A1, 1497 (1967).
56. J. M. Barrales-Rienda, D. C. Pepper, Polymer 8, 337 (1967).
57. R. J. Angelo, R. M. Ikeda, M. L. Wallach, Polymer 6, 141 (1965).
58. R. Kuhn, S. B. Liang, H.-J. Cantow, Angew. Makromol. Chem. 18, 101 (1971).
59. W. Cooper, D. E. Eaves, G. Vaughan, J. Polymer Sci. 59, 241 (1962).
60. W. Cooper, G. Vaughan, D. E. Eaves, R. W. Madden, J. Polymer Sci. 50, 159 (1961).
61. Ph. Ribeyrolles, A. Guyot, H. Benoit, J. Chim. Phys. 56, 377 (1959).
62. R. Kratochvíl, D. Straková, P. Schmidt, Angew. Makromol. Chem. 23, 169 (1972).
63. S. Abe, Y. Murakami, H. Fujita, J. Appl. Polymer Sci. 9, 2549 (1965).
64. T. Norisaye, K. Kawahara, A. Teramoto, H. Fujita, J. Chem. Phys. 49, 4330 (1968).
65. K. Hanafusa, A. Teramoto, H. Fujita, J. Phys. Chem. 70, 4004 (1966).
66. J. Curchod, T. Ve, J. Appl. Polymer Sci. 9, 3541 (1965).
67. P. V. French, A. Wassermann, J. Chem. Soc. 1044 (1963).
68. G. V. Schulz, K. Altgelt, H.-J. Cantow, Makromol.Chem. 21, 13 (1956).
69. K. Altgelt, G. V. Schulz, Makromol. Chem. 36, 209 (1959).
70. J. Vavra, J. Polymer Sci., C, 16, 1103 (1967).
71. G. M. Bristow, B. Westall, Polymer 8, 609 (1967).
72. S. P. Rao, M. Santappa, J. Polymer Sci., A-1, 6, 95 (1968).
73. P. J. Reed, J. R. Urwin, Austral. J. Chem. 23, 1743 (1970).
74. V. Boháčkova, J. Polaček, H. Benoit, J. Chim. Phys. 66, 197 (1969).
75. V. Boháčkova, J. Polaček, Grubisic, H. Benoit, J. Chim. Phys. 66, 207 (1969).
76. S. A. Pavlova, T. A. Soboleva, A. P. Suprun, Vysokomolekul. Soedin. 6, 122 (1964).
77. A. Silberberg, J. Eliassaf, A. Katchalsky, J. Polymer Sci. 23, 259 (1957).
78. H.-J. Cantow, Z. Naturforsch. 7B, 485 (1952).
79. T. A. Fadner, H. Morawetz, J. Polymer Sci. 45, 475 (1960).
80. V. A. Myagchenkov, A. K. Vagapova, E. V. Kuznetsov, Vysokomolekul. Soedin. 11, 673 (1969).
81. E. A. S. Cavell, I. T. Gilson, B. R. Jennings, H. G. Jerrard, J. Polymer Sci., A, 2, 3615 (1964).
82. B. R. Jennings, Thesis - Univ. Southampton (1964).
83. L. Trossarelli, M. Meirone, J. Polymer Sci. 57, 445 (1962).
84. E. F. Jordan, Jr., J. Polymer Sci., A-1, 6, 2209 (1968).
85. T. A. Orofino, P. J. Flory, J. Phys. Chem. 63, 283 (1959).
86. S. Newman, W. R. Krigbaum, C. Langier, P. J. Flory, J. Polymer Sci. 14, 451 (1954).
87. Z. Alexandrowicz, J. Polymer Sci. 40, 91 (1959).
88. D. Braun, G. Mott, Angew. Makromol. Chem. 18, 183 (1971).
89. W. Wunderlich, Angew. Makromol. Chem. 11, 189 (1970).
90. R. Jerome, V. Desreux, European Polymer J. 6, 411 (1970).
91. G. B. Rathman, F. A. Bovey, J. Polymer Sci. 15, 544 (1955).
92. M. Giurgea, C. Ghita, I. Baltog, A. Lupu, J. Polymer Sci. 4, 529 (1966).
93. M. Giurgea, L. Pop, I. Baltog, A. Belea, Rev. Roumaine Chim. 11, 467 (1966).
94. D. Mangaraj, S. K. Patra, Makromol. Chem. 107, 230 (1967).
95. H. Wesslau, Makromol. Chem. 69 213 (1963).
96. K. Harvey, M. B. Huglin, University of Salford; to be published.
97. H. Matsuda, K. Yamano, H. Inagaki, J. Polymer Sci., A-2, 7, 609 (1969).
98. K. Karunakaran, M. Santappa, J. Polymer Sci., A-2, 6, 713 (1968).
99. A. Kotera, T. Saito, Y. Watanabe, M. Ohama, Makromol. Chem. 87, 195 (1965).
100. H. Matsuda, H. Inagaki, Bull. Inst. Chem. Res. Kyoto Univ. 46, 48 (1968).
101 R. A. Wessling, J. E. Mark, E. Hamori, R. E. Hughes, J. Phys. Chem. 70, 1903 (1966).
102. J. Veličković, J. Vukajlović-Filipović, Angew. Makromol. Chem. 13, 79 (1970).
103. R. J. Valles, J. Polymer Sci., A, 3, 3853 (1965).
104. R. J. Valles, Makromol. Chem. 100, 167 (1967).
105. R. van Leemput, R. Stein, J. Polymer Sci. A-1, 985 (1963).
106. V. E. Eskin, A. L. Izyumnikov, E. D. Rogozhkina, Yu. P. Vyrskii, Polymer Sci. USSR 7, 1310 (1965).
107. V. E. Eskin, A. L. Izyumnikov, E. D. Rogozhkina, Yu. P. Vyrskii, Polymer Sci. USSR 9, 591 (1967).
108. S. N. Chinai, R. A. Guzzi, J. Polymer Sci. 21 417 (1956).
109. R. van Leemput, R. Stein, J. Polymer Sci. A-2 4039 (1964).
110. M. Kozhokaryu, V. S. Skazka, K. G. Berdnikova, Polymer Sci. USSR 8, 1167 (1966).
111. V. N. Tsvetkov, O. V. Kallistov, Zh. Fiz. Khim. 33, 710 (1959).
112. V. N. Tsvetkov, D.Khardi, I. N. Shtennikova, Ye. V. Korneyeva, G. F. Pirogova, K. Nitrai, Polymer Sci. USSR 11, 392 (1969).
113. S. V. Bereza, S. R. Rafikov, Ye. A. Bekturov, R. Ye. Legkunets, Polymer Sci. USSR 10, 2945 (1968).
114. M. Tricot, V. Desreux, Makromol. Chem. 149, 185 (1971).
115. N. Hadjichristidis, M. Devaleriola, V. Desreux, European Polymer J. 8, 1193 (1972).
116. D. W. Levi, R. J. Valles, E. C. Schramm, J. Polymer Sci. 56, 305 (1962).
117. F. E. Didcot, S. N. Chinai, D. W. Levi, J. Polymer Sci. 43, 557 (1960).
118. Z. Tuzar, P. Kratochvil, Polymer Letters 7, 825 (1969).
119. Z. Tuzar, M. Bohdanecky, Collection Czech. Chem. Commun. 34, 289 (1969).
120. M. Bohdanecky, Z. Tuzar, M. Stoll, R. Chromoček, Collection Czech. Chem. Commun. 33, 4104 (1968).
121. Z. Tuzar, P. Kratochvíl, Collection Czech. Chem. Commun. 32, 3358 (1967).
122. K. Karunakaran, M. Santappa, Makromol. Chem. 111, 20 (1968).
123. S. N. Chinai, R. J. Samuels, J. Polymer Sci. 19, 463 (1956).
124. S. N. Chinai, J. Polymer Sci. 25, 413 (1957).
125. H. T. Lee, D. W. Levi, J. Polymer Sci. 47, 449 (1960).
126. S. N. Chinai, R. A. Guzzi, J. Polymer Sci. 41, 475 (1959).
127. Q. A. Trementozzi, J. Am. Chem. Soc. 76, 5273 (1954).
128. H. J. L. Trap, J. J. Hermans, J. Phys. Chem. 58, 757 (1954).
129. A. Vrij, J. Th. G. Overbeek, J. Colloid Sci. 17, 570 (1962).
130. H. L. Bhatnagar, A. R. Biswas, J. Polymer Sci. 13, 461 (1954).
131. R. Tremblay, Y. Sicotte, M. Rinfret, J. Polymer Sci. 14, 310 (1954).
132. E. S. Cohn, E. M. Schuele, J. Polymer Sci. 14, 309 (1954).
133. J.-Y. Chien, L.-H. Shih, S.-C. Yu, J. Polymer Sci. 29, 117 (1958).
134. J. Bischoff, V. Desreux, Bull. Soc. Chim. Belges 61, 10 (1952).
135. H.-J. Cantow, O. Bodmann, Z. Physik Chem. (Frankfurt) 3, 65 (1955).
136. E. Cohn-Ginsberg, T. G. Fox, H. F. Mason, Polymer 3, 97 (1962).
137. H.-G. Elias, O. Etter, Makromol. Chem. 89, 228 (1965).
138. W. Schnabel, Makromol. Chem. 86, 9 (1965).
139. O. Bodmann, Makromol. Chem. 122, 196 (1969).
140. S. Fujishige, H.-G. Elias, Makromol. Chem. 155, 127 (1972).
141. S. Fujishige, H.-G. Elias, Makromol. Chem. 155, 137 (1972).
142. S. Krause, E. Cohn-Ginsberg, Polymer 3, 565 (1962).
143. R. Kirste, G. V. Schulz, Z. Physik Chem. (Frankfurt) 27, 301 (1961).
144. T. Matsuo, A. Pavan, A. Peterlin, D. T. Turner, J. Colloid Interfac. Sci. 24, 241 (1967).
145. W. Bushuk, H. Benoit, Canad. J. Chem. 36, 1616 (1958).
146. A. Bello, G. M. Guzman, European Polymer J. 2, 85 (1966).
147. T. Kotaka, N. Donkai, H. Ohnuma, H. Inagaki, J. Polymer Sci., A-2, 6, 1803 (1968).

148. J. Lamprecht, C. Strazielle, J. Dayantis, H. Benoit, Makromol. Chem. 148, 285 (1971).

149. L. Mrkvičková-Vaculová, P. Kratochvíl, Collection Czech. Commun. 37, 2015 (1972).

150. G. V. Schulz, W. Wunderlich, R. Kirste, Makromol. Chem. 75, 22 (1964).

151. S. Krause, J. Phys. Chem. 65, 1618 (1961).

152. H. Ohnuma, T. Kotaka, H. Inagaki, Polymer 10, 501 (1969).

153. H. Ohnuma, T. Kotaka, H. Inagaki, Rept. Progr. Polymer Phys. Japan 11, 29 (1968).

154. W. H. Stockmayer, L. D. Moore, Jr., M. Fixman, B. N. Epstein, J. Polymer Sci. 16, 517 (1955).

155. O. Bodmann, Makromol. Chem. 122, 210 (1969).

156. L. Mrkvičková-Vaculová, P. Kratochvíl, Collection Czech. Chem. Commun. 37, 2029 (1972).

157. J. Trekoval, P. Kratochvíl, J. Polymer Sci., A-1, 10, 1391 (1972).

158. A. Červenka, P. Kratochvíl, Collection Czech. Chem. Commun. 34, 2568 (1969).

159. R. Chiang, J. J. Burke, J. O. Threlkeld, T. A. Orofino, J. Phys. Chem. 70, 3591 (1966).

160. F. W. Billmeyer, Jr., C. B. de Than, J. Am. Chem. Soc. 77, 4763 (1955).

161. E. F. Casassa, W. H. Stockmayer, Polymer 3, 53 (1962).

162. S. N. Chinai, J. D. Matlock, A. L. Resnick, R. J. Samuels, J. Polymer Sci. 17, 391 (1955).

163. S. N. Chinai, J. Polymer Sci. 14, 408 (1954).

164. Y. Shimura, H. Kambe, Rept. Progr. Polymer Phys. Japan 9, 103 (1966).

165. U. Gruber, H.-G. Elias, Makromol. Chem. 84, 168 (1965).

166. A. Dondos, Makromol. Chem. 147, 123 (1971).

167. J. T. Guthrie, M. B. Huglin, G. O. Phillips, Makromol. Chem. 149, 309 (1971).

168. M. Kalfus, J. Mitus, J. Polymer Sci., A-1, 4, 953 (1966).

169. S. N. Chinai, A. L. Resnick, H. T. Lee, J. Polymer Sci. 33, 471 (1958).

170. V. N. Tsvetkov, V. G. Aldoshin, Russ. J. Phys. Chem. 33, 619 (1959).

171. R. B. Mohite, S. Gundiah, S. L. Kapur, Makromol. Chem. 116, 280 (1968).

172. G. Greber, J. Tolle, W. Burchard, Makromol. Chem. 71, 47 (1964).

173. K. Matsumura, Makromol. Chem. 124, 204 (1969).

174. D. Braun, W. Chaudhari, W. Mächtle, Angew. Makromol. Chem. 15, 83 (1971).

175. A. Matiussi, E. Conti, G. B. Gechele, European Polymer J. 8, 429 (1972).

176. M. Pizzoli, G. Stea, G. Ceccorulli, G. B. Gechele. European Polymer J. 6, 1219 (1970).

177. K. Okita, A. Teramoto, K. Kawahara, H. Fujita, J. Phys. Chem. 72, 278 (1968).

178. A. M. Kotliar, J. Polymer Sci. 55, 71 (1961).

179. M. Abe, K. Sakato, T. Kageyama, M. Fukatsu, M. Kurata, Bull. Chem. Soc. Japan 41, 2330 (1968).

180. K. Sakato - quoted in ref. 584.

181. P. F. Mijnlieff, D. J. Coumou, J. Colloid Interf. Sci. 27, 553 (1968).

182. J. M. G. Cowie, S. Bywater, D. J. Worsfold, Polymer 8, 105 (1967).

183. T. Kato, K. Miyasu, M. Nagasawa, J. Phys. Chem. 72, 2161 (1968).

184. T. Fujimoto, N. Ozaki, M. Nagasawa, J. Polymer Sci., A, 3, 2259 (1965).

185. D. E. Burge, D. B. Bruss, J. Polymer Sci., A, 1, 1927 (1963).

186. A. F. Sirianni, D. G. Worsfold, S. Bywater, Trans. Faraday Soc. 55, 2124 (1959).

187. J. D. Matlock, S. N. Chinai, R. A. Guzzi, D. W. Levi, J. Polymer Sci. 49, 533 (1961).

188. F. S. Holahan, S. S. Stivala, D. W. Levi, J. Polymer Sci., A, 3, 3987 (1965).

189. F. S. Holahan, S. S. Stivala, D. W. Levi, J. Polymer Sci., A, 3, 3993 (1965).

190. J. Veličkovič, D. Jovanovic, J. Vukajlovič, J. Makromol. Chem. 129, 203 (1969).

191. M. Morton, T. E. Helminiak, S. D. Gadkary, F. Bueche, J. Polymer Sci. 57, 471 (1962).

192. J. C. Spitzbergen, H. C. Beachell, J. Polymer Sci., A-2, 1205 (1964).

193. H.-G. Elias, Makromol. Chem. 50, 1 (1961).

194. V. V. Varadaiah, J. Polymer Sci. 19, 477 (1956).

195. A. J. Hyde, J. H. Ryan, F. T. Wall, T. F. Schatzki, J. Polymer Sci. 33, 129 (1958).

196. B. R. Jennings, H. Plummer, Brit. J. Appl. Phys. 1, 1201 (1968).

197. L. M. Alberino, W. W. Graessley, J. Phys. Chem. 72, 4229 (1968).

198. J. M. G. Cowie, R. Dey, J. T. McCrindle, Polymer J. 2, 88 (1971).

199. R. S. Roche, A. G. Tanner, Angew. Makromol. Chem. 13, 183 (1970).

200. A. Yamamoto, M. Fujii, G. Tanaka, H. Yamakawa, Polymer J. 2, 799 (1971).

201. Y. Ikada, W. Schnabel, Makromol. Chem. 86, 20 (1965).

202. J. M. G. Cowie, S. Bywater, J. Macromol. Chem. 1, 581 (1966).

203. H. Lange, Makromol. Chem. 86, 192 (1965).

204. T. E. Smith, D. K. Carpenter, Macromolecules 1, 204 (1968).

205. W. W. Graessley, R. D. Hartung, W. C. Uy, J. Polymer Sci., A-2, 7, 1919 (1969).

206. G. Wigand, J.-J. Veith, Plaste Kautschuk 16, 671 (1969).

207. R. H. Ewart, C. P. Roe, P. Debye, J. R. McCartney, J. Chem. Phys. 14, 687 (1946).

208. H. Lange, Kolloid Z.-Z. Polymere 201, 123 (1965).

209. H.-G. Elias, O. Etter, Makromol. Chem. 66, 56 (1963).

210. H. Lange, Kolloid Z.-Z. Polymere 199, 128 (1964).

211. J. M. G. Cowie, J. Polymer Sci., C, 23, 267 (1968).

212. B. A. Brice, M. Halwer, R. Speiser, J. Opt Soc. Am. 40, 768 (1950).

213. H. Kambe, I. Mita, Y. Shimura, Rept. No. 381 (Vol. 29, No. 3). Aeronaut. Res. Inst., Univ. of Tokyo (1964).

214. H. Janeschitz-Kriegl, Kolloid-Z. 203, 119 (1965).

215. H. Baumann, H. Lange, Angew. Makromol. Chem. 9, 16 (1969).

216. J. T. Guthrie, M. B. Huglin, G. O. Phillips, Univ. Salford - to be published.

217. J. W. Breitenbach, H. Gabler, Makromol. Chem. 37, 53 (1960).

218. P. Outer, C. J. Carr, B. H. Zimm, J. Chem. Phys. 18, 830 (1950).

219. H.-J. Cantow, Z. Physik. Chem. (Frankfurt) 7, 58 (1956).

220. J. P. Kratochvil in "Characterisation of Macromolecular Structure" - Proc. Conf. at Warrenton, Virginia, USA, Publication No. 1573, Nat. Acad. Sci., Washington D.C., USA (1968).

221. G. C. Berry, J. Chem. Phys. 44, 4550 (1966).

222. J. M. G. Cowie, E. L. Cussler, J. Chem. Phys. 46, 4886 (1967).

223. L. A. Papazian, Polymer 10, 399 (1969).

224. T. Alteres, Jr., D. P. Wyman, V. R. Allen, K. Meyersen, J. Polymer Sci., A, 3, 4131 (1965).

225. Q. A. Trementozzi, R. F. Steiner, P. Doty, J. Am. Chem. Soc. 74, 2070 (1952).

226. R. M. Johnsen, Chem. Scripta 1, 81 (1971).

227. P. M. Doty, B. H. Zimm, H. F. Mark, J. Chem. Phys. 12, 144 (1944).

228. J. W. Breitenbach, H. Gabler, O. F. Olaj, Makromol. Chem. 81, 32 (1965).

229. D. K. Carpenter, W. R. Krigbaum, J. Chem. Phys. 24, 1041 (1956).

230. S. M. Maron, R. L. H. Lou, J. Polymer Sci. 14, 29 (1955).

231. J. Oth, V. Desreux, Bull. Soc. Chim. Belges 63, 285 (1954).

232. H. P. Frank, H. F. Mark, J. Polymer Sci. 17, 1 (1955).

233. P. Doty, R. F. Steiner, J. Chem. Phys. 18, 1211 (1950).

234. J. M. G. Cowie, J. T. McCrindle, European Polymer J. 8, 1185 (1972).

235. P. Grosius, Y. Gallot, A. Skoulios, Makromol. Chem. 127, 94 (1969).

236. G. V. Schulz, H. Baumann, Makromol. Chem. 114, 122 (1968).

237. Y. Tomimatsu, L. Vitello, K. Fong, J. Colloid Interf. Sci. 27, 573 (1968).

238. E. Elbing, A. G. Parts, Makromol. Chem. 82, 270 (1965).

239. J. M. G. Cowie, D. J. Worsfold, S. Bywater, Trans. Faraday Soc. 57, 705 (1961).

240. O. Bodmann, Thesis, Mainz (1955).

241. J. H. O'Mara, D. McIntyre, J. Phys. Chem. 63, 1435 (1959).

242. P. N. Norberg, L.-O. Sundeloef, Makromol. Chem. 77, 77 (1964).

243. T. G. Scholte, J. Polymer Sci., A-2, 10, 519 (1972).

244. B. Chincoli, A. J. Havlik, R. D. Vold - unpublished results quoted in ref. 586.

245. L. Trossarelli, E. Campi, G. Saini, J. Polymer Sci. 35, 205 (1959).

246. T. Alfrey, Jr., E. B. Bradford, J. W. Vanderhoff, J. Opt. Soc. Am. 44, 603 (1954).

247. F. S. Chan, D. A. I. Goring, Canad. J. Chem. 44, 725 (1966).

248. W. B. Dandliker, J. Am. Chem. Soc. 72, 5110 (1950).

249. W. R. Carroll, H. Eisenberg, J. Polymer Sci., A-2, 4, 599 (1966).

250. J. A. Manson, G. J. Arquette, Makromol. Chem. 37, 187 (1960).

251. A. Ciferri, M. Kryszewski, G. Weill, J. Polymer Sci. 27, 167 (1958).

252. A. R. Shultz, J. Am. Chem. Soc. 76, 3422 (1954).

253. H. Benoit, A. M. Holtzer, P. Doty, J. Phys. Chem. 58, 624 (1954).

254. P. Kratochvíl, Collection Czech. Chem. Commun. 29, 2767 (1964).

255. G. Kalz, Plaste Kautschuk 17, 331 (1970).

256. E. Schroeder, Plaste Kautschuk 15, 2 (1968).

257. M. Hlousek, J. Láníková, Chem. Prumysel 16, 160 (1966).

258. C. Wippler - through SOFICA Manual (1962).

259. L. Lapčik, L.-O. Sundeloef, Chem. Scripta 2, 41 (1972).

260. M. Bohdanecký, V. Petrus, P. Kratochvíl, Collection Czech. Chem. Commun. 33, 4089 (1968).

261. G. Kalz - unpublished results quoted in ref. 255.

262. P. Doty, H. Wagner, S. Singer, J. Phys. Chem. 51, 32 (1947).

263. M. Freeman, P. P. Manning, J. Polymer Sci. A-2, 2017 (1964).

264. D. Laker, J. Polymer Sci. 25, 122 (1957).

265. T. Kobayashi, Bull. Chem. Soc. Japan 35, 636 (1962).

266. G. Palma, G. Pezzin, S. Zaramella, J. Polymer Sci., C, 33, 23 (1971).

267. P. Kratochvíl, D. Straková, Makromol. Chem. 154, 325 (1972).

268. M. L. Wallach, M. A. Kabayama, J. Polymer Sci., A-1, 4, 2667 (1966).

269. C. Garbuglio, L. Crescentini, A. Mula, G. B. Gechele, Makromol. Chem. 97, 97 (1966).

270. M. Miura, Y. Kubota, T. Masuzakawa, Bull. Chem. Soc. Japan 38, 316 (1965).

271. C. Loucheux, Z. Czlonkowska, J. Polymer Sci., C, 16, 4001 (1968).

272. G. Muller - quoted in ref. 235.

273. J. B. Berkowitz, M. Yamin, R. M. Fuoss, J. Polymer Sci. 28, 69 (1958).

274. A. G. Boyes, U. P. Strauss, J. Polymer Sci. 22, 463 (1956).

275. G. R. Seely, Macromolecules 2, 302 (1969).

276. R. L. Cleland, W. H. Stockmayer, J. Polymer Sci. 17, 473 (1955).

277. T. Shibukawa, M. Sone, A. Uchida, K. Iwahori, J. Polymer Sci., A-1, 6, 147 (1968).

278. R. Chiang, J. C. Stauffer, J. Polymer Sci., A-2, 5, 101 (1967).

279. W. R. Krigbaum, A. M. Kotliar, J. Polymer Sci. 32, 323 (1958).

280. P. F. Onyon, J. Polymer Sci. 37, 315 (1959).

281. N. Fujisaki, Kobunshi Kagaku 18, 581 (1961).

282. L. H. Peebles, J. Polymer Sci., A, 3, 361 (1965).

283. K. Kamide, H. Kobayashi, Y. Miyazaki, C. Nakayama, Chem. High Polymer (Tokyo) 24, 679 (1967).

284. M. P. Zverev, A. N. Barash, L. P. Nikonorova, L. V. Ivanova, P. I. Zubov, Polymer Sci. USSR 9, 1038 (1967).

285. H. Lange, H. Baumann, Angew. Makromol. Chem. 14, 25 (1970).

286. M. Iwama, H. Utiyama, M. Kurata, J. Macromol. Chem. 1, 701 (1966).

287. M. Matsumoto, Y. Ohyanagi, J. Polymer Sci. 46, 441 (1960).

288. M. Ueda, K. Kajitani, Makromol. Chem. 108, 138 (1967).

289. H.-G. Elias, P. Vogt, Makromol. Chem. 157, 263 (1972).

290. V. Kalpagam, M. Ramakishna, V. S. R. Rao, J. Polymer Sci. A-1, 233 (1963).

291. H.-G. Elias, F. Patat, Makromol. Chem. 25, 13 (1958).

292. L. M. Hobbs, V. C. Long, Polymer 4, 479 (1963).

293. R. U. Howard, Thesis, M. I. T., Cambridge, Mass. USA (1952).

294. W. W. Graessley, H. M. Mittelhauser, J. Polymer Sci., A-2, 5, 431 (1967).

295. F. W. Billmeyer, Jr. - quoted in ref. 582.

296. W. W. Graessley - quoted in ref. 582.

297. A. Beresniewicz, J. Polymer Sci. 39, 63 (1959).

298. H. Inagaki, T. Oyama, Kobunshi Kagaku 1, 20 (1952).

299. H. A. Dieu, J. Polymer Sci. 12, 417 (1954).

300. W. H. Beattie, R. K. Laudenslager, J. Moacanin, N. A. S. A. Tech. Memo No. 33-242, Jet Prop. Lab., California (1966).

301. S. Matsuzawa, H. Tohara, A. Harada, Rept. Progr. Polymer Phys. Japan 12, 29 (1969).

302. W. Burchard, M. Nosseir, Makromol. Chem. 82, 109 (1965).

303. K. Ueberreiter, J. Springer, Z. Physik. Chem. (N. F.) 36, 299 (1963).

304. G. Sitaramaiah, D. Jacobs, Polymer 11, 165 (1970).

305. H.-G. Elias, O. Etter, J. Macromol. Chem. 1, 431 (1966).

306. G. B. Levy, J. Polymer Sci. 17, 247 (1955).

307. A. B. Duchkova, N. S. Nametkin, V. S. Khotimskii, L. E. Gusel'-nikov, S. G. Durgar'yan, Polymer Sci. USSR 8, 2002 (1966).

308. H. Eisenberg, J. Chem. Phys. 44, 137 (1966).

309. H.-G. Elias, R. Schumacher, Makromol. Chem. 76, 23 (1964).

310. A. G. Nasini, C. Ambrosino, L. Trossarelli, Ricerca Sci. 25, 625 (Internat. Symp. Macromol. Chem., Milan-Turin, 1954) (1955).

311. P. R. Saunders, J. Polymer Sci. A-2, 3755 (1964).

312. H. G. Fendler, H. A. Stuart, Makromol. Chem. 25, 159 (1958).

313. P. R. Saunders, J. Polymer Sci. 57, 131 (1962).

314. H. C. Beachell, D. W. Carlson, J. Polymer Sci. 40, 543 (1959).

315. D. W. Carlson, Thesis, Univ. Delaware, USA (1959).

316. Z. Tuzar, P. Kratochvíl, M. Bohdanecký, J. Polymer Sci., C, 16, 633 (1967).

317. M. K. Knecht, H.-G. Elias, Makromol. Chem. 157, 1 (1972).

318. A. De Chirico, Chim. Ind. (Milan) 42, 248 (1960).

319. G. Sitaramaiah, J. Polymer Sci., A, 3, 2743 (1965).

319a. W. R. Moore, M. Uddin, European Polymer J. 5, 185 (1969).

320. A. R. Shultz, P. L. Wineman, M. Kramer, Macromolecules 1, 488 (1968).

321. G. C. Berry, H. Nomura, K. G. Mayhan, J. Polymer Sci., A-2, 5, 1 (1967).

322. G. V. Schulz, A. Horbach, Makromol. Chem. 29, 93 (1959).

323. M. L. Wallach, Makromol. Chem. 103, 19 (1967).

324. A. E. Tonelli, P. J. Flory, Macromolecules 2, 225 (1969).

325. H. L. Wagner, K. F. Kissbrun, Makromol. Chem. 81, 14 (1965).

326. C. Booth, R. Orme, Polymer 11, 626 (1970).

327. M. Matsushima, M. Fakatsu, M. Kurata, Bull. Chem. Soc. Japan 41, 2570 (1968).

328. G. Allen, C. Booth, S. J. Hurst, M. N. Jones, C. Price, Polymer 8, 391 (1967).

329. K. Yamamoto, H. Fujita, Polymer 8, 517 (1967).

330. J. M. Barrales-Rienda, D. C. Pepper, European Polymer J. 3, 535 (1967).

330a. A. R. Shultz, A. L. Bridgman, E. M. Hadsell, C. R. McCullough, J. Polymer Sci., A-2, 10, 273 (1972).

331. P. J. Akers, G. Allen, M. J. Bethell, Polymer 9, 575 (1968).

332. H.-G. Elias, H. Lys, Makromol. Chem. 92, 1 (1966).

333. P. Rempp, J. Chim. Phys. 54, 421 (1957).

334. C. Strazielle, Makromol. Chem. 119, 50 (1968).

335. T. A. Ritscher, H.-G. Elias, Makromol. Chem. 30, 48 (1959).

336. W. Schnabel, U. Borgwadt, Makromol. Chem. 123, 73 (1969).

337. W. Ring, H.-J. Cantow, W. Holtrup, European Polymer J. 2, 151 (1966).

338. W. Heller, T. L. Pugh, J. Polymer Sci. 47, 203 (1960).

339. A. Teramoto, H. Fujita, Makromol. Chem. 85, 261 (1965).

340. K. Yamamoto, H. Fujita, Polymer 7, 557 (1966).

341. W. H. Stockmayer, L.-L. Chan, J. Polymer Sci., A-2, 4, 437 (1966).

342. G. Meyerhoff, U. Moritz, Makromol. Chem. 109, 143 (1967).

343. G. Allen, C. Booth, M. N. Jones, Polymer 5, 195 (1964).

344. V. E. Eskin, A. E. Nesterov, Polymer Sci. USSR 8, 152 (1966).

345. A. Von Raven, H. Heusinger, Paper No. P-53, IUPAC Internat. Conf. on "Chemical Transformations of Polymers", Bratislava, June 22-24 (1971).

346. A. Stokes, European Polymer J. 6, 719 (1970).

347. N. V. Makletsova, I. V. Epel'baum, B. A. Rozenberg, E. B. Lyudvig, Polymer Sci. USSR 7, 73 (1965).

348. D. Sims, Ministry of Defence, E. R. D. E., Waltham Abbey, Essex, England - unpublished results (1966).

349. M. Kurata, H. Utiyama, K. Kamada, Makromol. Chem. 88, 281 (1965).

350. H.-G. Elias, G. Adank, Makromol. Chem. 103, 230 (1967).

351. J. M. Evans, M. B. Huglin, Makromol. Chem. 127, 141 (1969).

352. J. M. Evans, M. B. Huglin, European Polymer J. 6, 1161 (1970).

353. K. Yamamoto, A. Teramoto, H. Fujita, Polymer 7, 267 (1966).

355. K. V. Ivin, H. A. Ende, G. Meyerhoff, Polymer 3, 129 (1962).

356. A. J. Bur, D. E. Roberts, J. Chem. Phys. 51, 406 (1969).

357. Z. Tuzar, H. C. Beachell, Polymer Letters 9, 37 (1971).

358. N. S. Schneider, S. Furusaki, R. W. Lenz, J. Polymer Sci., A-1, 3, 933 (1965).

359. V. N. Tsvetkov, I. N. Shtennikova, E. I. Ryumtsev, L. N. Andreyeva, Yu. P. Getmanchuk, Yu. L. Spirin, R. I. Dryagileva, Polymer Sci. USSR 10, 2482 (1968).

360. H. Plummer, B. R. Jennings, European Polymer J. 6, 171 (1970).

361. J. Moacanin, J. Appl. Polymer Sci. 1, 272 (1959).

362. Gj. Dež</br>elić, J. P. Kratohvil, Kolloid-Z. 173, 38 (1960).

363. G. V. Schulz, A. Haug, Z. Physik. Chem. (Frankfurt) 34, 328 (1962).

364. G. Adank, H.-G. Elias, Makromol. Chem. 102, 151 (1967).

365. H. H. Tokimoto, C. T. Forbes, R. K. Laudenslager, J. Appl. Polymer Sci. 5, 153 (1961).

366. R. Nillson, L.-O. Sundeloef, Makromol. Chem. 66, 11 (1963).

367. W. Schnabel, Makromol. Chem. 103, 103 (1967).

368. R. R. Buch, H. M. Klimisch, O. K. Johannson, J. Polymer Sci., A-2, 7, 563 (1969).

369. V. N. Tsvetkov, K. A. Andrianov, I. N. Shtennikova, G. I. Okhrimenko, L. N. Andreyeva, G. A. Fomin, V. I. Pakhomov, Polymer Sci. USSR 10, 636 (1968).

370. A. Audsley, J. Aveston, J. Chem. Soc. 2320 (1962).

371. D. E. G. Jones, J. W. Lorimer, Polymer 13, 265 (1972).

372. P. R. Gupta, D. A. I. Goring, Canad. J. Chem. 38, 270 (1960).

373. G. Sitaramaiah, D. A. I. Goring, J. Polymer Sci. 58, 1107 (1962).

374. M. Marx-Figini, E. Penzel, Makromol. Chem. 87, 307 (1965).

375. W. Brown, R. Wikstroem, European Polymer J. 1, 1 (1965).

376. D. Henley, Arkiv Kemi 18, 327 (1961).

377. H. Vink, G. Dahlstroem, Makromol. Chem. 109, 249 (1967).

378. N. Gralen, Thesis, Uppsala (1944), quoted in ref. 583.

379. L. Valtasaari, Tappi 48, 627 (1965).

380. R. M Badger, R. H. Blaker, J. Phys. Colloid Chem. 53, 1056 (1949).

381. C. Wippler, J. Chim. Phys. 53, 346 (1956).

382. G. V. Schulz, E. Penzel, Makromol. Chem. 112, 260 (1968).

383. M. L. Hunt, S. Newman, H. A. Scheraga, P. J. Flory, J. Phys. Chem. 60, 1278 (1956).

384. A. Sharples, F. L. Swinton, J. Polymer Sci. 50, 53 (1961).

385. W. R. Krigbaum, L. H. Sperling, J. Phys. Chem. 64, 99 (1960).

386. J. Oehman, Arkiv Kem. 31, 125 (1969).

387. H. Janeschitz-Kriegl, W. Burchard, J. Polymer Sci., A-2, 6, 1953 (1968).

388. J. Oehman, V. P. Shanbhag, Arkiv Kem. 31, 137 (1969).

389. W. Burchard, Makromol. Chem. 88, 11 (1965).

390. S. B. Sellen, M. P. Levi, Polymer 8, 633 (1967).

391. M. M. Huque, D. A. I. Goring, S. G. Mason, Canad. J. Chem. 36, 952 (1958).

392. C. W. Tait, R. J. Vetter, J. M. Swanson, P. Debye, J. Polymer Sci. 7, 261 (1951).

393. S. Claesson, H. H. Bruun, Svensk. Papperstid. 60, 336 (1957).

394. K. K. Ghosh, P. K. Choudhury, Makromol. Chem. 102, 217 (1967).

395. P. F. Onyon, J. Polymer Sci. 37, 295 (1959).

396. K. K. Ghosh, J. Appl. Polymer Sci. 15, 537 (1971).

397. R. H. Cornell, H. A. Swenson, J. Appl. Polymer Sci. 5, 641 (1961).

398. P. C. Scherer, A. Tannenbaum, D. W. Levi, J. Polymer Sci. 43, 531 (1960).

399. R. St. J. Manley, Arkiv Kemi 9, 519 (1956).

400. H. A. Swenson, A. J. Morak, S. Kurath, J. Polymer Sci. 51 231 (1961).

401. M. G. Wirick, M. H. Waldman, J. Appl. Polymer Sci. 14, 579 (1970).

402. W. B. Neely, J. Polymer Sci. A-1, 311 (1963).

403. W. J. Hillend, H. A. Swenson, J. Polymer Sci. A-2, 4921 (1964).

404. W. Brown, D. Henley, J. Oehman, Makromol. Chem. 62, 164 (1963).

405. N. S. Schneider, P. Doty, J. Phys. Chem. 58, 762 (1954).

406. B. Das, P. K. Choudhury, J. Polymer Sci., A-1, 5, 769 (1967).

407. W. G. Yean, D. A. I. Goring, J. Appl. Polymer Sci. 14, 115 (1970).

408. S. R. Erlander, R. Tobin, Makromol. Chem. 111, 194 (1968).

409. S. R. Erlander, R. Tobin, Makromol. Chem. 111, 212 (1968).

410. C. J. Stacy, J. F. Foster, J. Polymer Sci. 20, 57 (1956).

411. S. R. Erlander, H. L. Griffin, Stärke 19, 134 (1967).

412. L. P. Witnauer, F. R. Senti, M. D. Stern, J. Polymer Sci. 16, 1 (1955).

413. P. Debye, J. Phys. Colloid Chem. 51, 18 (1947).

414. W. B. Everett, J. F. Foster, J. Am. Chem. Soc. 81 3459 (1959).

415. H. L. Griffin, S. R. Erlander, F. R. Senti, Stärke 19, 8 (1967).

416. E. F. Paschall, J. F. Foster, J. Polymer Sci. 9, 85 (1952).

417. S. R. Erlander, H. L. Griffin, Stärke 19, 139 (1967).

418. J. M. G. Cowie, J. Polymer Sci. 49, 455 (1961).

419. R. S. Patel, R. D. Patel, Makromol. Chem. 90, 262 (1966).

420. W. Banks, C. T. Greenwood, Stärke 19, 394 (1967).

421. W. Banks, C. T. Greenwood, D. J. Hourston, Trans. Faraday Soc. 64, 363 (1968).

422. W. Burchard, E. Husemann, Makromol. Chem. 44, 358 (1961).

423. W. Banks, C. T. Greenwood, J. Sloss, European Polymer J. 7, 263 (1971).

424. Y. Tomimatsu, K. J. Palmer, A. E. Goodban, W. H. Ward, J. Polymer Sci. 36, 129 (1959).

425. Y. Tomimatsu, K. J. Palmer, J. Polymer Sci. A-1, 1005 (1963).

426. P. J. Van Duin, J. J. Hermans, J. Polymer Sci. 36, 295 (1959).

427. B. Sedlacek, I. Fric, Collection Czech. Chem. Commun. 24, 2834 (1959).

428. W. Burchard, B. Pfannemueller, Makromol. Chem. 121, 18 (1969).

429. F. R. Senti, N. N. Hellman, N. H. Ludwig, G. E. Babcock, R. Tobin, C. A. Glass, B. L. Lamberts, J. Polymer Sci. 17, 527 (1955).

430. L. H. Arond, H. P. Frank, J. Phys. Chem. 58, 953 (1954).

431. K. Frombling, F. Patat, Makromol. Chem. 25, 41 (1958).

432. M. Zebec, Gj. Dež</br>elić, J. Kratohvil, K. F. Schulz, Croat. Chem. Acta 30, 251 (1958).

433. E. Antonini, L. Bellelli, M. R. Bruzzesi, A. Caputo, E. Chiancone, A. Rossi-Fanelli, Biopolymers 2, 27 (1964).

434. S. Levine, H. L. Griffin, F. R. Senti, J. Polymer Sci. 35, 31 (1959).

435. D. H. Juers, H. A. Swenson, S. F. Kurath, J. Polymer Sci., A-2, 5, 361 (1967).

436. J. B. Snell, J. Polymer Sci., A, 3, 2591 (1965).

437. G. V. Koleske, S. F. Kurath, J. Polymer Sci. A-2, 4123 (1964).

438. A. Veis, D. N. Eggenberger, J. Am. Chem. Soc. 76, 1560 (1954).

439. H. A. Swenson, H. M. Kaustinen, O. A. Kaustinen, N. S. Thompson, J. Polymer Sci., A 2, 6, 1593 (1968).

440. L. C. Cerny, R. C. Graham, H. James, Jr., J. Appl. Polymer Sci. 11, 1941 (1967).

441. C. T. Greenwood, D. J. Hourston, Stärke 19 243 (1967).

442. K. A. Granath, R. Stroemberg, A. N. de Belder, Stärke 21, 251 (1969).

443. B. S. Harrap, E. F. Woods, Austral. J. Chem. 11, 581 (1958).

444. B. S. Harrap, E. F. Woods, Austral. J. Chem. 11, 592 (1958).

445. F. C. McIntire, H. W. Sievert, G. H. Barlow, R. A. Finley, A. Y. Lee, Biochem. 6, 2363 (1967).

446. W. Mackie, S. B. Sellen, Polymer 10, 621 (1969).

447. P. Buchner, R. E. Cooper, A. Wasserman, J. Chem. Soc. 3974 (1961).

448. M. B. Mathews. Biochim. Biophys. Acta 35, 9 (1959).

449. A. Pramanik, P. K. Choudhury, J. Polymer Sci., A-1, 6, 1121 (1968).

450. A. Pramanik, P. K. Choudhury, J. Polymer Sci., A-1, 7, 1055 (1969).

451. J. R. Patel, C. K. Patel, R. D. Patel, Stärke 19, 330 (1967).

452. R. Wikstroem, Svensk Papperstid. 71, 399 (1968).

453. D. A. I. Goring, T. E. Timell, J. Phys. Chem. 64, 1426 (1960).

454. R. G. Le Bel, D. A. I. Goring, J. Polymer Sci., C-2, 29 (1963).

455. W. Mackie, D. B. Sellen, Biopolymers 10, 1 (1971).

571. G. Cohen, H. Eisenberg, Biopolymers 4 429 (1966).

572. D. Casson, A. Rembaum, Macromolecules 5, 75 (1972).

573. M. L. Wallach, J. Polymer Sci., A-2, 5, 653 (1967).

574. M. L. Wallach, J. Polymer Sci., A-2, 7, 1995 (1969).

575. L. V. Dubrovina, S. A. Pavlova, V. V. Korshak, Polymer Sci. USSR 8, 827 (1966).

576. S. A. Pavlova, L. V. Dubrovina, G. I. Timoveeva, J. Polymer Sci., C, 16, 2649 (1967).

577. V. N. Tsvetkov, G. A. Fomin, P. N. Lavrenko, I. N. Shtennikova, T. V. Sheremeteva, L. I. Godunova, Polymer Sci. USSR 10, 1051 (1968).

578. N. A. Pravikova, Ye. B. Berman, Ye. B. Lyudvig, A. G. Davtyan, Polymer Sci. USSR 12, 653 (1970).

579. G. L. Hagnauer, N. S. Schneider, J. Polymer Sci., A-2, 10, 699 (1972).

580. G. Allen, J. McAinsh, European Polymer J. 6, 1635 (1970).

581. U. P. Strauss, P. L. Wineman, J. Am. Chem. Soc. 80, 2366 (1958).

582. G. C. Berry, L. M. Hobbs, V. C. Long, Polymer 5, 31 (1964).

583. R. St. John Manley, A. Bengtsson, Svensk. Papperstid. 61, 471 (1958).

584. H. Utiyama, N. Sugi, M. Kurata, M. Tamura, Bull. Inst. Chem. Res., Kyoto Univ. 46, 198 (1968).

585. J. P. Kratohvil, Kolloid-Z.-Z. Polymere 238, 455 (1970).

586. W. Kaye, A. J. Havlik, J. B. McDaniel, Polymer Letters 9, 695 (1971).

DISSYMMETRIES AND PARTICLE SCATTERING FACTORS *

R. Chiang †
Monsanto Company
St. Louis, Missouri

Contents

A. INTRODUCTION

Scattering of light by macromolecules in dilute solutions has been shown to obey the following equation :

$$\frac{Kc}{R_\theta} = \frac{1}{<M>_w P(\theta)} + 2 A_2 c \tag{1}$$

where

$$K = (2\pi^2/N\lambda^4)(n_o \, dn/dc)^2,$$

c is the concentration of the polymer in g/ml, R_θ is the reduced intensity at angle θ, $<M>_w$ is the weight-average molecular weight, $P(\theta)$ is the particle scattering factor, A_2 is the second virial coefficient, N is Avogadro's number, λ is the wavelength of the light in vacuo, and n_o the refractive index of the solvent.

For polymers whose dimensions are 1/20 or less of the wavelength of the incident light, the angular distribution is symmetrical about $\theta = 90^o$; the intensities of the light scattered in the forward direction and in the backward direction are the same. For polymers with dimensions approaching the wavelength of the light, the scattering envelope is not symmetrical but is greater in the forward direction than in the backward direction because of the phase differences of the light scattered from different parts of the molecule (internal interference).

The particle scattering factor, $P(\theta)$, is defined as the ratio of the light intensity, $I(\theta)$, scattered by a polymer in solution at an angle θ to the light intensity, I_θ^o, in the absence of interference; thus,

$$P(\theta) = I_\theta/I_\theta^o = R_\theta/R_\theta^o$$

In practice, the light intensities scattered at an angle θ and its complementary angle $\pi - \theta$ are measured, the ratio of these intensities being referred to as the dissymmetry ratio and denoted by z_θ,

$$z_\theta = I_\theta/I_{\pi-\theta} = P(\theta)/P(\pi-\theta)$$

It has become standard practice to use 45 and 135° for measurements of dissymmetries, although angles other than 45 and 135° are occasionally used.

The Particle Scattering Factor

The particle scattering factor assumes different forms for different geometric shapes of the scattering particles. It may be calculated by the mathematical equations given below (1, 2, 3) :

For monodisperse coils with mean-square end-to-end distance $\overline{r^2}$,

$$P(\theta) = (\frac{2}{x^2}) [e^{-x}-(1-x)]; \quad x = \frac{k^2 v^2 \overline{r^2}}{6} = (2/3)(\frac{\overline{r^2}}{\lambda'^2})[2\pi \sin(\theta/2)]^2 \tag{2a}$$

For rods of length L,

$$P(\theta) = \frac{1}{x} \int_0^{2x} (\frac{\sin w}{w})dw - (\frac{\sin x}{x})^2; \quad x = \frac{kvL}{2} \tag{2b}$$

For discs of diameter D,

$$P(\theta) = \frac{x^o}{1} - \frac{x^2}{6} + \sum_{n=3}^{\infty} \frac{x^{2(n-1)}}{(2n + b_{n-1}/b_{n-2})b_{n-1}} \cos(n+1)\pi; \quad x = \frac{kvD}{2} \tag{2c}$$

b_n being the denominator of nth term; for example, $b_1 = 1$, $b_2 = 6$, etc.

For spheres of diameter D,

$$P(\theta) = [(\frac{3}{x^3})(\sin x - x \cos x)]^2; \quad x = \frac{kvD}{2} \tag{2d}$$

where $k = 2\pi/\lambda'$; $v = 2 \sin(\theta/2)$; and λ' = the wavelength of light in solution (equal to the wavelength in vacuum divided by the refractive index of the solvent).

---

* Reprinted from the first edition

For the corresponding expressions for ellipsoids, cylinders and stiff chains, the literature on this subject should be consulted (4). Tables 2 and 3 give the calculated particle scattering factor for coils, rods and spheres for different values of x (5).

Regardless of the shape of the scattering particle, $P(\theta)$ can be directly determined by using the general equation

$$P(\theta) = 1 - (\overline{S^2}/3)(kv)^2 + \ldots \ldots \tag{3}$$

where $\overline{S^2}$ is the mean-square-radius of gyration, the first term following unity being always equal to 1/3 of $(kv)^2$ multiplied by the mean-square-radius of gyration. This quantity is directly determined by light scattering without any restriction to the shape of the molecule. $\overline{S^2} = \overline{r^2}/6$ for random coils, $\overline{S^2} = L^2/12$ for rods, and $\overline{S^2} = 3D^2/20$ for spheres, etc.

Furthermore it is clear from eq. (3) that $P(\theta)$ approaches unity when θ approaches zero or when the scattering center is small compared to the wavelength of the light (Rayleigh scattering or small particle scattering); otherwise, $P(\theta) < 1$.

Determination of Molecular Weight and Dimension

To obtain the molecular weights and dimensions of molecules from light scattering data, two methods are commonly used: the dissymmetry method and the method of Zimm (6).

Dissymmetry Method. The dissymmetry method consists of the following steps:

1. The intensities of the scattered light at 45, 90, and 135° are measured for polymer solutions at different concentrations.
2. The dissymmetry ratios z_{45} are calculated for each concentration and extrapolated to zero concentration. The dissymmetry at zero concentration is referred to as the intrinsic dissymmetry $[z]$.
3. The correction factor $P(90)$ is taken from Table 4 or 5 from the dissymmetry z_{45} after choosing the proper model.
4. The true molecular weight is obtained by substituting this correction factor $P(90)$ into equation (1).
5. From $[z]$, the values of $\sqrt{\overline{r^2}}/\lambda'$, L/λ', and D/λ' for random coils, rods, and spheres, respectively, are calculated (Tables 4 and 6).

The dissymmetry method is conveniently used when the molecular weight, and hence the dissymmetry, is low; if the dissymmetry is high, extrapolation of the scattering data to zero concentration and zero angle is performed graphically by the method of Zimm.

Zimm Plot. Substituting eq. (3) into eq. (1) and averaging over all molecular weights we have

$$\frac{Kc}{R_\theta} = \frac{1}{\langle M \rangle_w} \left[1 + \frac{\langle \overline{s^2} \rangle_z}{3} \left(\frac{2\pi^2}{\lambda'} \right) \left(2 \sin \frac{\theta}{2} \right)^2 \right] + \ldots \tag{4}$$

Note that the dimensions thus obtained are z-average quantities. In accordance with eq. (4), when Kc/R_θ is plotted against $\sin^2(\theta/2) + k'c$, where k' is an arbitrary constant, familes of straight lines or nearly straight lines are obtained at constant θ and constant c, thus forming a grid-like plot (Zimm plot). With the Zimm plot, it is possible to extrapolate the scattering data to $\theta = 0$ and $c = 0$ simultaneously. The two extrapolated lines intercept the ordinate at the same point $(Kc/R_\theta)_{c=0, \theta=0}$.

The weight-average molecular weight, second virial coefficient, and z-average mean square radius of gyration are calculated from the following equations:

$$\langle M \rangle_w = \frac{1}{\left(\frac{Kc}{R_\theta} \right)_{c=0, \ \theta=0}}$$

$$\langle \overline{s^2} \rangle_z = \frac{3\lambda'^2}{16\pi^2} \left(\frac{\text{slope of the line } c = 0}{\text{intercept}} \right)$$

$$A_2 = \frac{k' \ (\text{slope of the line } \theta = 0)}{2}$$

Effect of Molecular Weight Distribution on $P(\theta)$

The scattering factors given in eq. (2) hold only for monodisperse samples. Significant deviations are encountered for broad molecular weight distributions. For coiled molecules having generalized exponential distribution

$$w(M) = \beta^{(\beta+1)} M^\beta \exp(-\beta M/\langle M \rangle_n)/\langle M \rangle_n^{(\beta+1)} \Gamma(\beta+1)$$

where β is the homogeneity parameter defined by

$$\langle M \rangle_z : \langle M \rangle_w : \langle M \rangle_n = (\beta+2) : (\beta+1) : \beta$$

and $w(M)$ is the weight fraction of a polymer having a molecular weight in the range of M and $M + dM$, the particle scattering factor will be according to Zimm (6)

$$P(\theta) = [2(1+\beta)^{(1+\beta)} - 2(1+\beta -\beta\langle x \rangle)(1+\beta + \langle x \rangle^\beta]/\langle x \rangle^2 \beta (1+\beta + \langle x \rangle)^\beta,$$

where $\langle x \rangle$ is the weight average of x, x being defined in eq. (2a).

The effect of polydispersity on $P(\theta)$ as mentioned above has to be corrected using the correction factors given in Table 5 for different ratios of $\langle M \rangle_w / \langle M \rangle_n$.

B. TABLES OF DISSYMETRIES AND PARTICLE SCATTERING FACTORS

1. Trigonometric Functions Frequently Used in Light Scattering Measurements

θ	$\sin \theta$	$1 + \cos^2 \theta$	$\dfrac{\sin \theta}{1 + \cos^2\theta}$	$\sin (\theta/2)$	$\sin^2 (\theta/2)$
30°	0.500	1.750	0.286	0.259	0.0670
37.5°	0.609	1.628	0.374	0.322	0.104
45°	0.707	1.500	0.471	0.383	0.146
60°	0.866	1.250	0.693	0.500	0.250
75°	0.966	1.067	0.905	0.609	0.371
90°	1.000	1.000	1.000	0.707	0.500
105°	0.966	1.067	0.905	0.793	0.629
120°	0.866	1.250	0.693	0.866	0.750
135°	0.707	1.500	0.471	0.924	0.854
142.5°	0.609	1.628	0.374	0.947	0.897
150°	0.500	1.750	0.286	0.966	0.977

2. Particle Scattering Factor for Spheres and Rods [x is defined in eq. (2)] (5)

x	$P(\theta)$ sphere	rod	x	$P(\theta)$ sphere	rod
0.1	0.998	0.999	2.6	0.219	0.543
0.2	0.998	0.996	2.7	0.191	0.524
0.3	0.978	0.990	2.8	0.165	0.506
0.4	0.968	0.983	2.9	0.141	0.489
0.5	0.949	0.973	3.0	0.120	0.473
0.6	0.929	0.961	3.1	0.0999	0.447
0.7	0.906	0.948	3.2	0.0824	0.443
0.8	0.818	0.933	3.3	0.0671	0.430
0.9	0.848	0.916	3.4	0.0534	0.417
1.0	0.817	0.897	3.5	0.0420	0.406
1.1	0.781	0.878	3.6	0.0320	0.395
1.2	0.745	0.857	3.7	0.0237	0.384
1.3	0.707	0.835	3.8	0.0172	0.375
1.4	0.667	0.813	3.9	0.0117	0.366
1.5	0.627	0.790	4.0	0.00757	0.358
1.6	0.587	0.767	4.1	0.00453	0.350
1.7	0.546	0.745	4.2	0.00231	0.343
1.8	0.506	0.720	4.3	0.000930	0.366
1.9	0.465	0.696	4.4	0.000199	0.329
2.0	0.426	0.673	4.5	0.0000196	0.323
2.1	0.388	0.650	4.6	0.000216	0.317
2.2	0.350	0.627	4.7	0.000740	0.311
2.3	0.316	0.605	4.8	0.00147	0.306
2.4	0.282	0.584	4.9	0.00233	0.300
2.5	0.249	0.563	5.0	0.00336	0.295

3. <u>Particle Scattering Factor for Monodisperse Coils</u> (x is defined in eq. (2)] (5)

$x^{1/2}$	$P(\theta)$	$x^{1/2}$	$P(\theta)$	$x^{1/2}$	$P(\theta)$	$x^{1/2}$	$P(\theta)$
0.1	0.996	1.4	0.573	2.7	0.215	4.0	0.111
0.2	0.987	1.5	0.536	2.8	0.203	4.1	0.106
0.3	0.971	1.6	0.499	2.9	0.192	4.2	0.102
0.4	0.949	1.7	0.466	3.0	0.182	4.3	0.0976
0.5	0.922	1.8	0.434	3.1	0.172	4.4	0.0936
0.6	0.891	1.9	0.404	3.2	0.163	4.5	0.0899
0.7	0.855	2.0	0.377	3.3	0.155	4.6	0.0864
0.8	0.817	2.1	0.352	3.4	0.147	4.7	0.0830
0.9	0.776	2.2	0.328	3.5	0.140	4.8	0.0799
1.0	0.736	2.3	0.307	3.6	0.134	4.9	0.0769
1.1	0.694	2.4	0.287	3.7	0.127	5.0	0.0741
1.2	0.652	2.5	0.269	3.8	0.122		
1.3	0.613	2.6	0.228	3.9	0.116		

4. <u>Dissymetries and Correction Factors for Monodisperse Coils</u> (7)

| D/λ' | Discs | | Spheres | | Rods | | Coils | |
	z_{45}	$\dfrac{1}{P(90)}$	z_{45}	$\dfrac{1}{P(90)}$	z_{45}	$\dfrac{1}{P(90)}$	z_{45}	$\dfrac{1}{P(90)}$
0.04	1.009	1.005	1.009	1.006	1.005	1.004	1.009	1.007
.06	1.017	1.012	1.020	1.014	1.011	1.008	1.022	1.016
.08	1.030	1.021	1.037	1.026	1.020	1.014	1.040	1.028
.10	1.048	1.033	1.058	1.040	1.031	1.022	1.063	1.044
.12	1.069	1.049	1.084	1.059	1.045	1.032	1.090	1.064
.14	1.095	1.067	1.117	1.081	1.061	1.043	1.124	1.088
.16	1.126	1.088	1.156	1.107	1.080	1.057	1.162	1.115
.18	1.163	1.112	1.202	1.138	1.102	1.072	1.206	1.147
.20	1.204	1.141	1.257	1.173	1.126	1.090	1.255	1.183
.22	1.252	1.173	1.320	1.214	1.153	1.109	1.310	1.223
.24	1.307	1.208	1.395	1.260	1.183	1.131	1.370	1.268
.26	1.368	1.249	1.481	1.313	1.216	1.154	1.436	1.317
.28	1.438	1.294	1.582	1.373	1.250	1.180	1.507	1.371
.30	1.516	1.344	1.700	1.440	1.288	1.207	1.583	1.430
.32	1.604	1.400	1.838	1.517	1.328	1.237	1.663	1.495
.34	1.703	1.461	2.000	1.604	1.370	1.269	1.748	1.564
.36	1.813	1.529	2.192	1.702	1.414	1.304	1.838	1.639
.38	1.936	1.605	2.420	1.814	1.460	1.341	1.931	1.720
.40	2.073	1.688	2.692	1.941	1.508	1.379	2.027	1.806
.42	2.224	1.780	3.021	2.086	1.556	1.421	2.126	1.898
.44	2.391	1.881	3.420	2.251	1.606	1.464	2.227	1.996
.46	2.574	1.992	3.912	2.441	1.656	1.510	2.329	2.100
.48	2.775	2.115	4.522	2.660	1.705	1.558	2.433	2.210
.50	2.991	2.249	5.292	2.912	1.754	1.608	2.538	2.325
.52	3.224	2.397	6.276	3.205	1.802	1.660	2.642	2.447
.54	3.472	2.560	7.559	3.548	1.848	1.714	2.747	2.575
.56	3.732	2.738	9.266	3.949	1.892	1.770	2.850	2.709
.58	4.000	2.932	11.59	4.423	1.933	1.827	2.953	2.849
.60	4.273	3.145			1.972	1.886	3.054	2.995
.62	4.544	3.376			2.007	1.946	3.154	3.148
.64	4.808	3.627			2.040	2.006	3.251	3.306
.66	5.059	3.899			2.069	2.068	3.347	3.469
.68	5.291	4.192			2.095	2.130	3.440	3.639
.70	5.498	4.506			2.118	2.193	3.530	3.815
.72	5.678	4.841			2.139	2.256	3.618	3.996
.74	5.829	5.196			2.156	2.318	3.703	4.183
.76	5.951	5.570			2.172	2.380	3.785	4.376

D = major dimension of scattering particle; diameter for discs and spheres; length for rods; root-mean-square end-to-end distance for coils.

D/λ'	Discs z_{45}	$\frac{1}{P(90)}$	Spheres z_{45}	$\frac{1}{P(90)}$	Rods z_{45}	$\frac{1}{P(90)}$	Coils z_{45}	$\frac{1}{P(90)}$
.78	6.045	5.961			2.185	2.442	3.864	4.574
.80	6.114	6.366			2.197	2.503	3.940	4.778
.82	6.162	6.783			2.207	2.563	4.014	4.988
.84	6.195	7.208			2.216	2.622	4.084	5.203
.86	6.215	7.637			2.225	2.681	4.152	5.424
.88	6.227	8.066			2.232	2.738	4.217	5.650
.90	6.235	8.492			2.239	2.795	4.279	5.881
.92	6.243	8.911			2.246	2.850	4.338	6.118
.94	6.252	9.321			2.252	2.905	4.395	6.361
.96	6.265	9.719			2.258	2.958	4.449	6.609
.98	6.281	10.10			2.264	3.012	4.501	6.862
1.00	6.304	10.48			2.270	3.064	4.551	7.121

5. Correction Factors for Polydisperse Coils Having Generalized Exponential Distribution (8)

Monodisperse $\langle M \rangle_w / \langle M \rangle_n = 1$		Deviations for Polydisperse Coils (added to monodisperse)							
z_{45}	$\frac{1000}{P(\theta)}$	$\langle M \rangle_w / \langle M \rangle_n = 1.1$	$\langle M \rangle_w / \langle M \rangle_n = 1.2$	$\langle M \rangle_w / \langle M \rangle_n = 1.5$	$\langle M \rangle_w / \langle M \rangle_n = 2.0$	$\langle M \rangle_w / \langle M \rangle_n = 3.0$	$\langle M \rangle_w / \langle M \rangle_n = 5.0$	$\langle M \rangle_w / \langle M \rangle_n = 10.0$	$\langle M \rangle_w / \langle M \rangle_n = 20.0$
1.02	1014	0	0	0	0	0	0	0	0
1.04	1028	0	0	0	0	0	0	0	1
1.06	1042	0	0	0	0	0	0	0	1
1.08	1056	0	0	0	0	0	0	1	1
1.10	1070	0	0	1	1	1	1	1	2
1.12	1085	0	0	1	1	2	2	2	3
1.14	1099	0	0	1	2	2	3	3	4
1.16	1113	0	1	2	3	3	3	4	4
1.18	1128	1	1	2	3	4	5	5	5
1.20	1142	1	1	3	4	5	6	6	7
1.22	1157	1	2	4	5	6	7	7	7
1.24	1171	1	2	5	6	8	9	9	9
1.26	1186	2	3	5	7	9	10	11	11
1.28	1201	2	3	6	9	11	12	13	13
1.30	1215	2	4	7	10	12	14	14	15
1.32	1230	2	5	8	11	14	15	16	17
1.34	1245	3	5	10	13	16	17	19	19
1.36	1260	3	6	11	14	18	19	21	22
1.38	1275	4	7	12	16	20	22	23	24
1.40	1290	4	7	13	18	22	24	26	27
1.42	1305	4	8	15	20	24	27	28	29
1.44	1320	5	9	16	22	26	29	31	32
1.46	1335	5	10	17	23	28	32	34	35
1.48	1350	6	11	19	26	31	35	37	38
1.50	1366	6	11	21	28	34	38	40	41
1.52	1381	7	12	22	30	36	41	43	45
1.54	1397	8	13	24	32	39	44	47	48
1.56	1412	8	14	26	35	42	47	50	52
1.58	1428	9	15	28	37	45	50	54	56
1.60	1444	9	16	29	40	48	54	58	60

z_{45}	$\dfrac{1000}{P(\theta)}$	$\langle M_w\rangle/\langle M_n\rangle = 1.1$	$\langle M_w\rangle/\langle M_n\rangle = 1.2$	$\langle M_w\rangle/\langle M_n\rangle = 1.5$	$\langle M_w\rangle/\langle M_n\rangle = 2.0$	$\langle M_w\rangle/\langle M_n\rangle = 3.0$	$\langle M_w\rangle/\langle M_n\rangle = 5.0$	$\langle M_w\rangle/\langle M_n\rangle = 10.0$	$\langle M_w\rangle/\langle M_n\rangle = 20.0$
	Monodisperse $\langle M_w\rangle/\langle M_n\rangle = 1$			Deviations for Polydisperse Coils (added to monodisperse)					
1.62	1460	10	18	31	43	51	58	62	64
1.64	1476	11	19	33	45	55	61	66	68
1.66	1492	11	20	35	48	58	65	70	72
1.68	1508	12	21	38	51	62	69	74	76
1.70	1524	13	22	40	54	66	73	79	81
1.72	1540	13	23	42	57	69	78	83	86
1.74	1557	14	25	44	60	73	82	88	91
1.76	1573	15	26	47	63	77	86	93	95
1.78	1590	16	27	49	67	81	91	98	101
1.80	1607	16	29	52	70	85	96	103	106
1.82	1624	17	30	54	74	90	101	108	111
1.84	1641	18	32	57	77	94	106	113	117
1.86	1658	19	33	60	81	99	111	119	123
1.88	1675	20	35	62	85	103	116	125	129
1.90	1693	21	36	65	89	108	122	131	135
1.92	1710	22	38	68	93	113	127	137	141
1.94	1728	23	39	71	97	118	133	143	147
1.96	1746	23	41	74	101	123	139	149	154
1.98	1763	24	43	77	105	129	145	156	161
2.00	1781	25	45	80	109	134	151	162	168
2.02	1799	26	46	83	114	140	157	169	175
2.04	1818	27	48	87	119	145	164	176	182
2.06	1836	28	50	90	123	151	170	183	189
2.08	1855	30	52	94	128	157	177	191	197
2.10	1874	30	53	97	133	163	184	198	205
2.12	1892	32	55	101	138	169	191	206	213
2.14	1911	33	57	104	143	176	198	214	221
2.16	1931	34	59	108	148	182	206	222	230
2.18	1950	35	62	112	153	189	214	230	238
2.20	1969	36	63	115	159	196	221	239	247
2.22	1989	37	66	120	164	203	229	247	256
2.24	2009	38	68	123	170	210	237	256	265
2.26	2029	40	70	128	176	217	246	265	275
2.28	2049	41	72	132	182	225	254	275	284
2.30	2069	42	74	136	188	232	263	284	294
2.32	2090	43	77	140	194	240	272	294	304
2.34	2111	45	79	145	200	248	281	304	315
2.36	2131	46	81	149	206	256	290	314	325
2.38	2152	47	84	154	213	264	300	325	336
2.40	2174	49	86	158	219	272	310	335	347
2.42	2195	50	88	163	226	281	320	346	359
2.44	2217	51	91	168	233	290	330	358	370
2.46	2239	53	94	173	241	299	340	369	382
2.48	2260	54	96	178	248	308	351	381	395
2.50	2283	56	99	183	255	318	362	392	407
2.52	2305	57	102	188	263	327	373	405	420
2.54	2328	59	104	194	270	337	385	417	433
2.56	2351	60	107	199	278	347	396	430	446
2.58	2374	62	110	205	286	358	408	443	460
2.60	2397	64	113	210	294	368	420	457	474
2.62	2420	65	116	216	303	379	433	470	488
2.64	2444	67	119	222	311	390	446	485	503
2.66	2468	69	122	228	320	401	459	499	518

CORRECTION FACTORS FOR POLYDISPERSE COILS

Monodisperse $<M_w>/<M_n> = 1$		Deviations for Polydisperse Coils (added to monodisperse)							
		1.1	1.2	1.5	2.0	3.0	5.0	10.0	20.0
z_{45}	$\dfrac{1000}{P(\theta)}$	$=$ $<M_w>/<M_n>$	$=$ $<M_w>/<M_n>$	$=$ $<M_w>/<M_n>$	$=$ $<M_w>/<M_n>$	$=$ $<M_w>/<M_n>$	$=$ $<M_w>/<M_n>$	$=$ $<M_w>/<M_n>$	$=$ $<M_w>/<M_n>$
2.68	2492	70	125	234	329	412	472	514	533
2.70	2517	72	128	240	338	421	486	528	549
2.72	2542	74	131	246	347	436	499	544	565
2.74	2566	75	135	253	356	448	514	560	581
2.76	2592	77	138	259	366	460	528	576	598
2.78	2617	79	141	266	376	473	543	592	615
2.80	2643	81	145	273	386	486	559	609	633
2.82	2669	83	148	279	396	499	574	626	651
2.84	2695	85	152	287	406	513	590	644	670
2.86	2722	87	155	294	417	527	606	662	689
2.88	2749	89	159	301	428	541	623	680	708
2.90	2776	91	163	309	439	553	640	700	728
2.92	2803	93	166	316	450	570	657	719	748
2.94	2831	95	170	324	461	585	675	739	769
2.96	2859	97	174	332	473	601	694	759	790
2.98	2887	99	178	340	485	616	712	779	811
3.00	2916	102	183	348	498	633	732	801	834
3.02	2945	104	187	357	510	649	751	822	856
3.04	2974	106	191	365	523	666	771	845	880
3.06	3004	108	195	374	536	683	792	868	904
3.08	3034	111	200	383	549	704	812	891	928
3.10	3064	113	204	392	563	719	834	915	954
3.12	3095	115	209	401	577	737	856	939	979
3.14	3126	118	213	411	591	756	878	965	1006
3.16	3157	121	218	421	606	776	901	990	1023
3.18	3189	123	223	430	621	795	925	1016	1060
3.20	3221	126	228	441	636	816	949	1044	1089
3.22	3254	128	233	451	651	836	973	1071	1118
3.24	3287	131	238	461	667	858	999	1099	1147
3.26	3320	134	243	472	684	879	1025	1129	1178
3.28	3354	137	249	483	700	901	1051	1158	1209
3.30	3388	140	254	494	718	924	1079	1189	1241
3.32	3423	143	259	505	735	947	1106	1220	1274
3.34	3458	146	265	517	752	971	1135	1252	1308
3.36	3493	149	271	529	771	995	1164	1285	1343
3.38	3529	152	277	541	789	1020	1194	1319	1378
3.40	3565	155	283	554	808	1046	1225	1353	1415
3.42	3602	159	289	566	827	1072	1256	1388	1452
3.44	3640	162	296	580	848	1099	1289	1425	1491
3.46	3678	165	302	593	868	1126	1321	1462	1530
3.48	3716	169	308	607	889	1154	1356	1500	1570
3.50	3755	172	315	620	910	1183	1390	1539	1611
3.52	3795	176	322	634	932	1213	1426	1580	1654
3.54	3835	180	329	649	954	1243	1462	1621	1698
3.56	3876	183	336	663	977	1274	1500	1663	1742
3.58	3917	187	343	679	1000	1306	1539	1707	1789
3.60	3959	191	350	694	1024	1338	1578	1752	1836
3.62	4001	195	358	710	1049	1372	1619	1798	1885
3.64	4044	199	366	726	1074	1406	1660	1845	1934
3.66	4088	204	374	743	1100	1441	1704	1894	1986
3.68	4132	207	382	760	1126	1477	1747	1943	2038
3.70	4177	212	391	778	1153	1514	1792	1995	2093
3.72	4223	216	399	795	1181	1552	1839	2047	2148
3.74	4269	221	407	814	1210	1591	1886	2101	2205

DISSYMMETRIES AND PARTICLE SCATTERING FACTORS

z_{45}	Monodisperse $\langle M_w \rangle / \langle M_n \rangle = 1$ $\dfrac{1000}{P(\theta)}$	Deviations for Polydisperse Coils (added to monodisperse)							
		$1.1 = \langle M_w \rangle / \langle M_n \rangle$	$1.2 = \langle M_w \rangle / \langle M_n \rangle$	$1.5 = \langle M_w \rangle / \langle M_n \rangle$	$2.0 = \langle M_w \rangle / \langle M_n \rangle$	$3.0 = \langle M_w \rangle / \langle M_n \rangle$	$5.0 = \langle M_w \rangle / \langle M_n \rangle$	$10.0 = \langle M_w \rangle / \langle M_n \rangle$	$20.0 = \langle M_w \rangle / \langle M_n \rangle$
3.76	4317	225	416	832	1238	1631	1935	2157	2264
3.78	4364	231	425	852	1268	1672	1985	2214	2325
3.80	4413	235	435	871	1299	1714	2037	2272	2387
3.82	4462	241	444	891	1331	1758	2089	2333	2452
3.84	4513	245	454	912	1363	1802	2144	2395	2517
3.86	4564	251	464	933	1396	1848	2200	2459	2586
3.88	4616	256	474	954	1430	1894	2258	2525	2655
3.90	4669	262	484	977	1465	1943	2318	2593	2727
3.92	4723	268	494	1000	1500	1993	2378	2662	2801
3.94	4777	273	506	1023	1538	2044	2442	2735	2878
3.96	4833	279	517	1047	1575	2096	2506	2808	2956
3.98	4890	285	528	1071	1614	2150	2572	2885	3037
4.00	4948	291	540	1096	1653	2205	2640	2963	3121

6. <u>Average Dimensions of Polydisperse Coils Having Generalized Exponential Distribution</u> (8)

z_{45}	$(1000/\lambda') \langle \overline{r^2} \rangle_w^{1/2}$								
	$1 = \langle M_w \rangle / \langle M_n \rangle$ (monodisperse)	$1.1 = \langle M_w \rangle / \langle M_n \rangle$	$1.2 = \langle M_w \rangle / \langle M_n \rangle$	$1.5 = \langle M_w \rangle / \langle M_n \rangle$	$2.0 = \langle M_w \rangle / \langle M_n \rangle$	$3.0 = \langle M_w \rangle / \langle M_n \rangle$	$5.0 = \langle M_w \rangle / \langle M_n \rangle$	$10.0 = \langle M_w \rangle / \langle M_n \rangle$	$20.0 = \langle M_w \rangle / \langle M_n \rangle$
1.02	57	54	52	49	46	44	42	41	40
1.04	80	77	71	70	66	62	60	59	58
1.06	98	94	91	85	81	77	74	72	71
1.08	113	108	105	99	93	89	86	84	83
1.10	126	121	117	111	105	100	96	94	93
1.12	138	133	129	121	115	110	106	103	102
1.14	149	143	139	131	124	119	114	112	111
1.16	159	153	149	140	133	127	123	120	118
1.18	168	163	158	149	142	135	131	127	126
1.20	177	171	167	157	150	143	138	135	133
1.22	186	180	175	165	157	150	145	142	140
1.24	194	188	183	173	165	158	152	149	147
1.26	202	195	190	180	172	164	159	155	154
1.28	209	203	197	187	179	171	166	162	160
1.30	216	210	204	194	185	178	172	168	166
1.35	233	227	221	210	201	193	187	183	181
1.40	249	242	237	226	216	208	202	198	196
1.45	264	257	252	241	231	222	216	211	209
1.50	278	271	266	255	245	236	229	225	223
1.55	292	285	279	268	258	249	242	238	236
1.60	304	298	292	281	274	262	255	251	248
1.65	317	310	305	291	284	275	268	263	261

$$(1000/\lambda') \langle \overline{r^2} \rangle_w^{1/2}$$

z_{45}	$\langle M \rangle_w/\langle M \rangle_n = 1$ (monodisperse)	$\langle M \rangle_w/\langle M \rangle_n = 1.1$	$\langle M \rangle_w/\langle M \rangle_n = 1.2$	$\langle M \rangle_w/\langle M \rangle_n = 1.5$	$\langle M \rangle_w/\langle M \rangle_n = 2.0$	$\langle M \rangle_w/\langle M \rangle_n = 3.0$	$\langle M \rangle_w/\langle M \rangle_n = 5.0$	$\langle M \rangle_w/\langle M \rangle_n = 10.0$	$\langle M \rangle_w/\langle M \rangle_n = 20.0$
1.70	329	323	317	307	297	287	280	275	273
1.75	340	335	329	319	309	300	293	288	285
1.80	352	346	341	331	321	312	305	300	297
1.85	363	357	353	343	333	324	317	312	309
1.90	373	368	364	354	345	336	329	324	321
1.95	384	379	375	366	357	348	341	336	333
2.00	395	390	386	377	368	359	353	348	345
2.10	415	411	408	400	391	383	376	372	369
2.20	435	432	429	422	414	406	400	395	393
2.30	454	452	450	444	437	430	424	420	418
2.40	474	472	471	466	460	454	448	444	442
2.50	493	493	492	488	484	478	473	469	467
2.60	512	513	513	511	507	502	498	495	493
2.70	531	533	534	533	531	527	524	521	519
2.80	550	553	555	556	555	553	550	548	547
2.90	570	574	576	579	580	579	578	576	575
3.00	589	595	598	603	606	606	606	605	604
3.10	609	616	620	627	632	634	635	635	635
3.20	629	637	643	652	659	663	665	666	666
3.30	650	659	666	678	687	693	697	699	699
3.40	671	682	690	705	716	725	730	733	734
3.50	693	706	715	732	746	757	765	769	771
3.60	716	730	740	761	778	792	801	807	810
3.70	739	755	767	791	811	828	840	848	851
3.80	764	781	795	823	846	867	881	891	895
3.90	789	809	824	856	883	908	925	937	942
4.00	816	838	855	891	923	951	972	986	993

C. REFERENCES

1. P. Debye, J. Phys. Coll. Chem. 51, 18 (1947).

2. P. Debye, E. W. Anacker, J. Phys. Coll. Chem. 55, 644 (1951).

3. B. H. Zimm, R. S. Stein, P. Doty, Polymer Bulletin 1, 90 (1945).

4. See, for example: (a) E. P. Geiduschek, A. Holtzer: "Application of Light Scattering to Biological Systems: Deoxyribonucleic Acid and the Muscle Proteins," in "Advances in Biological and Medical Physics," ed. by C. A. Tobias, and J. H. Lawrence, Academic Press, New York, 1958, Vol. VI.; (b) A. Peterlin, "Light Scattering by Non-Gaussian Macromolecular Coils," Proceedings of the Interdisciplinary Conference on Electromagnetic Scattering," Pergamon Press, New York, 1963, p. 357.

5. P. Doty, R. F. Steiner, J. Chem. Phys. 18, 1211 (1950).

6. B. H. Zimm, J. Chem. Phys. 16, 1099 (1948).

7. W. H. Beattie, C. Booth, J. Phys. Chem. 64, 696 (1960).

8. W. H. Beattie, C. Booth, J. Polymer Sci. 44, 81 (1960).

DIPOLE MOMENTS OF POLYMERS IN SOLUTION*

W. R. Krigbaum and J. V. Dawkins
Department of Chemistry
Duke University
Durham, N.C.

<u>Contents</u>   Page

A. INTRODUCTION

Each conformation of a polar polymer in solution will have a dipole moment μ, in Debye units D, equal to the vectorial sum of the moments of the N dipolar groups along the chain. The mean-square dipole moment $\overline{\mu^2}$ of a long chain molecule in solution is expressed by (1, 2):

$$\overline{\mu^2} = \varphi N \mu_o^2$$

where μ_o is the dipole moment of the monomeric repeating unit and φ is a constant characterizing the average molecular conformation. Generally N is equal to the number-average degree of polymerization $\overline{P_n}$, but for poly(alkylene oxides):

$$N = \overline{P_n} + 1$$

since the chain molecule has two hydroxyl endgroups. Except in the case of short chains, the dipole moment per monomer unit $(\overline{\mu^2}/N)^{1/2}$ is usually independent of N. If φ is independent of $\overline{P_n}$, solvent, and temperature and, therefore, of the excluded volume effect, its value can be correlated with short range intramolecular interactions. Current theory (3) predicts that when the internal motion of the chain is restricted, φ is less than unity, and also that isotactic, atactic, and syndiotactic forms of the same polymer should have different average moments, although theory cannot furnish a reliable estimate of the nature and magnitude of the difference.

The value of φ is readily determined since $\overline{\mu^2}$ and μ_o^2 can be obtained directly from the experimental measurement of the dielectric constant of the polymer, and of a polar molecule having a structure similar to the monomer unit, in the same solvent at the same temperature (see the article of de Brouckère and Mandel (4) for references to experimental techniques). The interpretation of the results is seriously handicapped because of the lack of an accepted theory of dieletric behaviour. The calculation of the dipole moment from the dielectric constant was pioneered by Debye (5). His theory of dilute solutions can be applied to polar polymers in non-polar solvents. To obtain a correct absolute value for $\overline{\mu^2}$, the internal field must be evaluated by taking into account the interactions of neighboring dipoles along the polymer chain (3). When polar solvents are employed, the internal field problem is more serious and will be influenced by polymer-solvent interactions and intermolecular dipole interactions. Various theories have been proposed for calculating the interaction effects (4) and their application in calculating $\overline{\mu^2}$ is influenced by the extent of solvent polarization.

Since the measured value of the dipole moment may be in considerable error, care should be exercised in the use of $\overline{\mu^2}$ for elucidating polymer structure and conformation. The table contains most of the data in the literature. Wherever possible, experimental conditions have been included and assumptions concerning the constant φ are mentioned. The model compounds used for comparison are given under remarks.

---

* Reprinted from the first Edition.

DIPOLE MOMENTS OF POLYMERS IN SOLUTION

B. DIPOLE MOMENTS OF POLYMERS IN SOLUTION

Polymer	Solvent	Temp. [°C]	$(\overline{\mu^2}/N)^{1/2}$ (D)	φ	Remarks	References
1. Vinyl Polymers						
Poly(styrene)						
isotactic	toluene	38.4	0.44	0.53	μ_o taken as 0.60 D	6
atactic	toluene	38.4	0.36	0.36	μ_o taken as 0.60 D	6
	carbon tetrachloride	25.0	0.26	0.56	μ_o (ethylbenzene) = 0.35 D	7
Poly(o-bromostyrene)				1.21		8
				0.50		8
Poly(o-chlorostyrene)				1.69		8
Poly(p-chlorostyrene)	benzene	30.0 and 50.0	1.45	0.56	μ_o (p-chlorotoluene) = 1.93 D	1
				0.42–0.56		8
isotactic	benzene	30.0	1.22			9
atactic	benzene	30.0	1.22			9
Poly(p-iodostyrene)				0.50		8
Poly(vinyl acetate)	benzene and carbon tetrachloride	25.0	1.61–1.70	0.89–0.94	μ_o (ethyl acetate) = 1.80 D	10
	benzene	20.0	1.70	0.84	μ_o (ethyl acetate) = 1.86 D	11
				0.75–0.80		12
Poly(vinyl bromide)	dioxane	25.0	1.41–1.79	0.53–0.86	μ_o (ethyl bromide) = 1.93 D. $\overline{\mu^2}$ varies non-linearly with N.	13
Poly(vinyl chloride)	dioxane	25.0	1.61–1.68	0.66–0.72	μ_o (ethyl chloride in carbon tetrachloride) = 1.98 D	14
atactic	tetrahydrofuran	20.0	1.31	0.64	μ_o (sec-butyl chloride in tetrahydrofuran) = 1.64 D	15
syndiotactic	tetrahydrofuran	20.0	1.39	0.71	μ_o (sec-butyl chloride in tetrahydrofuran) = 1.64 D	15
atactic	tetrahydrofuran	40.0	1.31	0.62	μ_o (sec-butyl chloride in tetrahydrofuran) = 1.66 D	15
	dioxane	20.0 and 40.0	1.62	0.59	μ_o (sec-butyl chloride in dioxane) = 2.12 D	15
	dioxane	20.0–65.0	1.67–1.75	0.70–0.75	μ_o (sec-butyl chloride) = 2.00 D. φ independent of N and temperature	16
				0.75		1
Poly(vinyl isobutyl ether)						
isotactic	benzene	25.0	1.16	0.90	μ_o = 1.22 D. Isotactic form postulated to be in a helical conformation.	12
atactic	benzene	25.0	1.07	0.77	μ_o = 1.22 D	12
isotactic	benzene	50.0	1.21	0.98	μ_o = 1.22 D. Isotactic form postulated to be in a helical conformation.	12
atactic	benzene	50.0	1.11	0.83	μ_o = 1.22 D	12
isotactic	benzene	30.0	0.99			9
atactic	benzene	30.0	0.98			9
2. Acrylic and Methacrylic Polymers						
Poly(methyl acrylate)	benzene	25.0	1.41–1.44	0.64–0.67	μ_o (methyl propionate in benzene) = 1.76 D	17
Poly(ethyl acrylate)						
atactic	benzene	30.0	1.02			9
syndiotactic	benzene	30.0	1.02			9
Poly(methyl methacrylate)						
isotactic	benzene	30.0	1.43	0.67	μ_o (methyl propionate in benzene) = 1.73 D (reference 20)	18
atactic	benzene	30.0	1.29–1.35	0.56–0.61	μ_o (methyl propionate in benzene) = 1.73 D (reference 20)	18
syndiotactic	benzene	30.0	1.27	0.53	μ_o (methyl propionate in benzene) = 1.73 D (reference 20)	18

Polymer	Solvent	Temp. [°C]	$(\overline{\mu^2}/N)^{1/2}$ (D)	φ	Remarks	References
Poly(methyl methacrylate)	(Cont'd.)					
isotactic	benzene	25.0-65.0	1.40-1.44	0.77-0.81	μ_o (monomer in benzene) = 1.60 D. Isotactic form postulated to be in a helical conformation.	19
atactic	benzene	25.0-65.0	1.33-1.41	0.69-0.78	μ_o (monomer in benzene) = 1.60 D	19
syndiotactic	benzene	25.0-65.0	1.34-1.41	0.70-0.78	μ_o (monomer in benzene) = 1.60 D	19
isotactic	toluene	30.0-90.0	1.29-1.39	0.65-0.75	μ_o (monomer in benzene) = 1.60 D	19
atactic	toluene	30.0-90.0	1.30-1.41	0.66-0.78	μ_o (monomer in benzene) = 1.60 D	19
syndiotactic	toluene	30.0-90.0	1.31-1.42	0.67-0.79	μ_o (monomer in benzene) = 1.60 D	19
atactic	5 solvents	25.0	-	0.62	μ_o (methyl propionate in each solvent) φ independent of solvent	20
	benzene	25.0	1.33-1.52	0.55-0.72	μ_o (methyl isobutyrate in benzene) = 1.80 D	21
	dioxane	23.0	1.50	-		4
				0.53-0.66		8, 12
Poly(ethyl methacrylate)				0.59-0.62		8
Poly(propyl methacrylate)				0.56-0.59		8
Poly(isopropyl methacrylate)				0.61-0.66		8
Poly(butyl methacrylate)				0.55-0.59		8
	4 solvents	23.0	1.39-1.65		$(\overline{\mu^2}/N)^{1/2}$ is temperature dependent	4
Poly(phenyl methacrylate)				0.55		8
Poly(p-chlorophenyl methacrylate)				0.35		8
Poly(dichlorophenyl methacrylate)				0.38		22

Polymer	Solvent	Temp. [°C]	P_n	$(\overline{\mu^2}/N)^{1/2}$ (D)	φ	Remarks	References
3. **Polydienes and Polyoxides**							
Poly(isoprene)							
cis 1,4	benzene	25.0	13,762	0.28	0.70	μ_o (2-methyl-2-butene in benzene) = 0.34 D	23
trans 1,4	benzene	25.0	3,125	0.31	0.82	μ_o (2-methyl-2-butene in benzene) = 0.34 D	23
Poly(chloroprene)	benzene	20.0	280	1.45			11
Poly(oxyethylene)	dioxane	25.0	1-7	1.68-1.29			24
	benzene	20.0	1.0-33.6	1.41-1.09			25
	benzene	20.0	2-227	1.46-1.07			2
	benzene	25.0	4.1-153.0	1.61-1.13			26
	benzene	25.0	4.0-176.2	1.68-1.13			27
endgroup : $-OC_2H_5$	benzene	20.0	2 and 6	1.15 and 1.11			2
	benzene	25.0	1-6	1.14-1.07			28
	benzene	50.0	1-6	1.14-1.09			28
Poly(oxypropylene)	benzene	25.0	6.6-69.0	1.40-1.02			25

C. REFERENCES

1. P. Debye, F. Bueche, J. Chem. Phys. 19, 589 (1951).

2. J. Marchal, H. Benoit, J. Chim. Phys. 52, 818 (1955); J. Polymer Sci. 23, 223 (1957).

3. M. V. Vol'kenstein, "Configurational Statistics of Polymeric Chains", (Translated by S. N. and M. J. Timasheff) p. 331, Interscience: New York, 1963.

4. L. de Brouckère, M. Mandel, "Dielectric Properties of Dilute Polymer Solutions", in Advances in Chemical Physics, ed. by I. Prigogine 1, p. 77, Interscience: New York, 1958.

5. P. Debye, "Polar Molecules", Chemical Catalog Co.: New York, 1929.

6. W. R. Krigbaum, A. Roig, J. Chem. Phys. 31, 544 (1959).

7. C. G. Le Fèvre, R. J. W. Le Fèvre, G. M. Parkins, J. Chem. Soc. (1958) 1468.

8. O. B. Ptitsyn, Soviet Phys.-Usp. (English transl.) 2, 797 (1960).

9. H. A. Pohl, H. H. Zabusky, J. Phys. Chem. 66, 1390 (1962).

10. C. G. Le Fèvre, R. J. W. Le Fèvre, G. M. Parkins, J. Chem. Soc. (1960) 1814.

11. I. Sakurada, S. Lee, Z. Phys. Chem. B43, 245 (1939).

12. M. Takeda, Y. Imamura, S. Okamura, T. Higashimura, J. Chem. Phys. 33, 631 (1960).

13. R. J. W. Le Fèvre, K. M. S. Sundaram, J. Chem. Soc. 1962, 4003.

14. R. J. W. Le Fèvre, K. M. S. Sundaram, J. Chem. Soc. 1962, 1494.

15. A. Kotera, M. Shima, N. Fujisaki, T. Kobayashi, Bull. Chem. Soc. Japan 35, 1117 (1962).

16. Y. Imamura, J. Chem. Soc. Japan 76, 217 (1955).

17. R. J. W. Le Fèvre, K. M. S. Sundaram, J. Chem. Soc. 1963, 3188.

18. R. Bacskai, H. A. Pohl, J. Polymer Sci. 42, 151 (1960).

19. R. Salovey, J. Polymer Sci. 50, S7 (1961).

20. J. Marchal, C. Lapp, J. Polymer Sci. 27, 571 (1958).

21. R. J. W. Le Fèvre, K. M. S. Sundaram, J. Chem. Soc. 1963, 1880.

22. G. P. Mikhailov, J. Polymer Sci. 30, 605 (1958).

23. R. J. W. Le Fèvre, K. M. S. Sundaram, J. Chem. Soc. 1963, 3547.

24. T. Uchida, Y. Kurita, N. Koizumi, M. Kubo, J. Polymer Sci. 21, 313 (1956).

25. G. D. Loveluck, J. Chem. Soc. 1961, 4729.

26. M. Aroney, R. J. W. Le Fèvre, G. M. Parkins, J. Chem. Soc. 1960, 2890.

27. V. Magnasco, G. Dellepiane, C. Rossi, Makromol. Chem. 65, 16 (1963).

28. A. Kutera, K. Suzuki, K. Matsumura, T. Nakano, Y. Oyama, U. Kambayashi, Bull. Chem. Soc. Japan 35, 797 (1962).

HEAT, ENTROPY AND VOLUME CHANGES FOR POLYMER-LIQUID MIXTURES

C. Booth
University of Manchester
Department of Chemistry
Manchester, England

Contents

A. INTRODUCTION

This contribution is intended as a guide to experimental investigations of the thermodynamics of binary mixtures of non-electrolyte homopolymers and liquids. It takes the form of a listing of papers which contain information about heat, entropy and volume changes associated with mixing of polymers and liquids and with dilution of concentrated polymer solutions. Reports of free energy data alone are listed only when it seems useful or interesting so to do. (Extensive compilations of such data already exist (134,156) : see also tables p. IV-131, IV-335, IV-337).

The table is based upon a survey of the appropriate literature (for mixtures of high molecular weight homopolymers and low molecular weight liquids) published since 1962. Coverage outside this main area is less complete. Specifically excluded are papers published prior to 1940 (ref. 4 and 7 discuss early work), and work on oligomer-liquid mixtures, polymer mixtures, ternary systems, polyelectrolyte solutions, and the sorption of water in cellulose and other polymers (for which see ref. 215).

Studies of dilute solutions are excluded (but see p. IV-1, IV-61).
Choice of a boundary to separate dilute solutions from concentrated solutions is arbitrary, but would lie somewhere in the region 2-5 wt-% polymer. Work directed towards the properties of isolated polymer molecules or estimation of the second virial coefficient has been excluded. However we have thought it useful to include reference to measurements of the densities of certain polymer solutions of fairly low concentration.

Methods of investigation are listed (see below) but values of the thermodynamic quantities are not. Generally the data for concentrated solutions do not reduce satisfactorily to simple parameters, nor do current theories (167,216) suggest such a representation. Data for solution processes involving glassy or partially crystalline polymers also defy quantitative tabulation. Selective reviews have been published (127,168,240). This tabulation is indiscriminate.

Methods
General discussions of the thermodynamics of solutions, and of pertinent experimental methods, can be found in standard works (see, for example, ref. 217). Details of applications to polymer solutions can be found in the tabulated references.

Heats of mixing (ΔH_m) of polymers and liquids, and integral heats of dilution (ΔH_d) of polymer solutions by liquids, are measured by calorimetry. Heats of mixing, heats of solution (mixing of polymer with a large amount of solvent) and heats of dilution ($\overline{\Delta H_1}$) are usually quoted. This latter quantity is

$$\overline{\Delta H_1} = \overline{H_1} - H_1^o \qquad (1)$$

where bar denotes a partial molar quantity, i.e.

$$\overline{H_1} = (\delta H/\delta n_1)_{T,P,n_2} \qquad (2)$$

superscript o denotes pure liquid (standard state) and subscript 1 denotes liquid (2 denotes polymer). Provided experimental measurements of ΔH_m or ΔH_d cover a range of concentrations $\overline{\Delta H_1}$ can be calculated. It is necessary that the polymer be liquid (rubbery) in order to derive $\overline{\Delta H_1}$ unequivocally from ΔH_m : for crystalline and glassy polymers ΔH_m contains contributions from processes other than dilution.

Volumes of mixing (ΔV_m) of polymers and liquids are measured directly by dilatometry, or indirectly by determination of densities of pure components and solution (by pyknometry, displacement, floatation etc.). Volumes of mixing (or volumes of solution if appropriate) are usually quoted. Partial molar volumes can be derived from the variation of ΔV_m with concentration (e.g. for liquid or rubbery polymers) or more directly from the concentration dependence of solution density. Estimation of $\overline{v}_2$ from refractive index increments (146) has been used for concentrated solutions (248).

Solvent activities (a_1) can be evaluated by a variety of methods, e.g. osmometry, light or X-ray scattering, sedimentation equilibrium, phase equilibrium, vapour pressure lowering, swelling. The solvent activity is related to the free energy of dilution ($\overline{\Delta F_1} = \overline{F_1} - F_1^o$, or $\mu_1 - \mu_1^o$ using conventional symbolism for chemical potential by

$$\overline{\Delta F_1} = RT \ln a_1 \qquad (3)$$

The variation of a_1 with temperature at constant pressure and composition gives the heat and entropy of dilution

$$\overline{\Delta H}_1 \;=\; R \;(\partial \ln a_1 / \partial (1/T))_{P, n_1, n_2} \tag{4}$$

$$\overline{\Delta S}_1 \;=\; -R[\ln a_1 \;+\; (\partial \ln a_1 / \partial \ln T)_{P, n_1 n_2}] \tag{5}$$

where $\overline{\Delta S}_1 = \overline{S}_1 - S_1^{o}$. The variation of a_1 with pressure at constant temperature and composition gives the volume of dilution ($\overline{\Delta V}_1 = \overline{V}_1 - V_1^{o}$)

$$\overline{\Delta V}_1 \;=\; RT \;(\partial \ln a_1 / \partial P)_{T, n_1, n_2} \tag{6}$$

There are particular difficulties associated with the quantitative interpretation of phase equilibria (both liquid-liquid and crystal-liquid) and generally references to these sources are excluded (but see ref. 195 and 209). Also the derivation of $\overline{\Delta H}_1$ and $\overline{\Delta S}_1$ from the temperature dependence of swelling has often been based upon assumptions concerning the concentration dependence of $\overline{\Delta F}_1$ (e.g. ref. 12, 14, 72). The swelling studies we have listed cover an adequate concentration range, or avoid the problem by subsidiary measurements (e.g. ref. 128). Recent applications of gas-liquid partition chromatography to the determination of solvent activities in polymer solutions (246, 255) fall outside the scope of table. With these reservations we have listed most of the investigations of the temperature dependence of solvent activity, and a number of investigations confined to one temperature. No complete investigations of the pressure dependence of solvent activity meet our criteria for inclusion (but see ref. 105, 224, 256 and 257).

References
References are arranged in rough chronological order, so that the higher numbers refer to the later work.

Abbreviations

CD	- Calorimetry: integral heat of dilution		OS	- Osmometry		
CM	- Calorimetry: heat of mixing (or solution)		LS	- Light scattering	VP	- Vapour pressure lowering (manometry, isopies-
D	- Density (pyknometry, displacement, floa-		XS	- X-ray scattering		tic distillation, vapour pressure osmometry,
	tation etc.)		SD	- Sedimentation-diffusion equilibrium		etc.)
DL	- Dilatometry		PE	- Phase equilibrium (liquid-liquid)	SW	- Swelling.

B. TABLE OF LITERATURE REFERENCES FOR HEAT, ENTROPY AND VOLUME CHANGES FOR POLYMER-LIQUID MIXTURES

Liquid	Method	References	Liquid	Method	References
1. Main-chain Carbon Polymers			Poly(1,3-isoprene), cis (Cont' d.)		
			carbon tetrachloride	SW	11
1.1 Poly(dienes)			chloroform	SW	11
			cyclohexane	SW	11
Poly(1,3-butadiene)			2,4-dimethyl-3-pentanone	SW	11
benzene	CM	22,89	2,2-dimethylpropane	VP	49
	VP	78	ethyl acetate	DL	128
chloroform	VP	127		VP	67,128
dioxane	VP	180		SW	11,128
--, cis			heptane	SW	11
benzene	CD	152	hexane	VP	63
toluene	CD	152	isoprene	VP	137
Poly(chloroprene)			methyl acetate	SW	11,128
benzene	CM	22	methyl alcohol	VP	8
chloroprene	VP	137	2-methylbutane	VP	49
Poly(1,3-isoprene)			2-methylpropane	VP	49
benzene	CM	8,10	octane	VP	192
heptane	CM	8	pentane	VP	49
isoprene	VP	137		SW	11
--, cis			2-pentanone	SW	128
acetone	VP	128	propyl acetate	SW	11
	SW	11,128	toluene	OS,VP	1
benzene	CM	59		SW	11
	DL,OS	6,181	2,2,4-trimethylpentane	VP	192
	VP	6,10,79,141, 181	Poly(1,3-isoprene), trans		
	SW	11,129,141	octane	VP	192
butane	VP	49	toluene	OS	2
2-butanone	DL,VP	128	2,2,4-trimethylpentane	VP	182,192
	SW	11,128			
butyl acetate	SW	11	1.2 Poly(alkenes)		
butyl butyrate	SW	11			
carbon disulfide	SW	11	Poly(ethylene) (see ref. 92 for additional VP data)		
			bromomethane	VP	92

Liquid	Method	References	Liquid	Method	References
Poly(ethylene) (Cont' d.)			Poly(isobutene) (Cont' d.)		
1-chloronaphthalene	CM	164	ethyl ether	CM	100,101
dodecane	VP	107,115	ethyl esters : hexanoate to decanoate	CM	101
heptane	VP	41	ethyl hexadecanoate	CM	101
hexane	VP	92	ethyl tetradecanoate	CM	101
phenyl ether	VP,PE	195	heptane	CM	22,94,100,124,
octacosane	SW	99			172,200
1,2,3,4-tetrahydronaphthalene	CM	164		DL	74,149,175,200
p-xylene	VP	142		OS	242
			hexadecane	CM	100
Poly(ethylene-co-propylene)				DL	175
benzene	VP	130	hexane	CM	100,200
	SW	131		DL	149,175,200
carbon tetrachloride	VP	130		OS	242
chlorobenzene	SW	131	hexyl ether	CM	101
chloromethane	VP	130	methyl alcohol	VP	112
cyclohexane	VP	130	2-methylbutane	CM	100
	SW	131		VP	38
heptane	SW	131	methylcyclohexane	CM	100,124,200
hexane	VP	130		DL	149,200
pentane	VP	130	3-methylpentane	CM	100
tetrachloroethylene	SW	131	2-methylpropane	VP	38
1,1,2-trichloro-1,2,2-trifluoroethane	VP	130	nonane	CM	100
Poly(1-butene)			octane	CM	100
dodecane	VP	115		DL,OS	175
octane	VP	109	pentane	CM	100
Poly(1-decene)				DL	103,175
toluene	VP	212		VP	38,103,171
Poly(1-dodecene)				OS	242
toluene	VP	212	pentyl ether	CM	101
Poly(1-heptene)			propyl ether	CM	101
toluene	VP	212	tetradecane	CM	100
Poly(isobutene)			toluene	CD	104,111
benzene	CD	77,104		CM	22,111,200
	CM	22,94,100,124,		DL	74,149,200
		153,172,200		VP	112
	D	78,87,139,231	tridecane	CM	100
	DL	74,149,174,200	2,2,4-trimethylpentane	CM	22,100,111,172
	OS	65,174		D	231
	VP	60,78,174		DL	74
butane	VP	38		LS	158,173
butyl ether	CM	101		VP	17
butyl propionate	VP	112	undecane	CM	100
carbon tetrachloride	CD	111	Poly(1-octadecene)		
	CM	111,124	toluene	VP	212
	VP	112	Poly(propylene)		
chlorobenzene	CD	104	benzene	VP	108
	CM	94,100,200	carbon tetrachloride	VP	236
	D	87	1-chloronaphthalene	CM	164
	DL	149,200	2,4-dimethyl-3-pentanone	VP	133
	OS	102	dodecane	VP	136,186
cyclohexane	CM	94,100,111,	heptane	VP	108,237
		124,200	hexane	VP	108
	D	139	nonyl alcohol	VP	145
	DL	149,170,200	pentane	VP	237
	OS	65	3-pentanone	VP	125,133
	LS	173	1,2,3,4-tetrahydronaphthalene	CM	164
	VP	29,60,112,170	2,2,4-trimethylpentane	VP	108
decane	CM	100			
	DL	149,175	1.3 Poly(acrylics)		
2,2-dimethylpropane	VP	38			
dodecane	CM	100	Poly(butyl methacrylate)		
ethylbenzene	CM	94	butyl acetate	CM	59

Liquid	Method	References	Liquid	Method	References
1.3　Poly(acrylics)　(Cont' d.)			**Poly(acrylonitrile-co-1,3-butadiene)　(Cont' d.)**		
Poly(methyl methacrylate)			dibutyl sebacate	CM,D	113
acetone	CD	161,176	dicumylmethane	CM,D	113
	D	64,193	ditolylmethane	CM,D	113
	OS	34,193	Poly(vinyl acetate)		
	VP	241	acetone	CM	80,200
benzene	CD	177,202		DL	200
	CM	245		LS	173
	D	64,193		VP	40,80
	OS	34,193	benzene	CD	185
	VP	241		CM	44,245
2-butanone	D	64		VP	40,81,90
butyl acetate	D	64	2-butanone	CD	185
	OS	39		CM,DL	200
carbon tetrachloride	D	64		OS	18
chlorobenzene	CD	177		LS	173
	D	64	butyl acetate	CM	59,200
chloroform	CD	232		DL	200
	CM	210	chlorobenzene	CM	44
	D	64		D	87,53
	DL	230		LS	173
	OS	34	chloroform	CM	44
o-dichlorobenzene	CD	177	1-chloropropane	VP	40
	D	64	3-chloropropene	VP	40
dichloroethane	CM,VP	88	1,2-dichloroethane	CM	44
1,2-dichloroethylene	OS	162	diethyl adipate	CD,D	30
dioxane	CD	161	diethyl malonate	CD,D	30
	D	64	diethyl oxalate	CD,D	30
	OS	34,162	diethyl sebacate	CD,D	30
ethyl acetate	D	64	diethyl succinate	CD,D	30
	OS	39	ethyl acetate	CM	33,200
methyl isobutyrate	CM,VP	88		D	231
3-pentanone	D	64		DL	200
	OS	34		LS	173
tetrahydrofuran	D	64	3-heptanone	CM	150,200
	OS	34		DL	200
toluene	CD	161,232	isopropylamine	VP	40
	D	64	methyl acetate	CM,DL	200
	DL	230	methyl alcohol	CM	44,52
	OS	34		D	231
m-xylene	D	64		LS	173
	OS	34	2-pentanone	CM,DL	200
Poly(methyl methacrylate-co-isobutene)			propyl acetate	CM,DL	200
acetone	CM	213	propyl alcohol	VP	40
benzene	CM	213	propylamine	VP	40
1,2-dichloroethane	CM	213	1,1,2,2-tetrachloroethane	CM	44,52
heptane	CM	213	toluene	CD	68
Poly(methyl methacrylate-co-styrene)				CM	44
acetone	VP	241	1,2,3-trichloropropane	OS	18
benzene	VP	241	vinyl acetate	VP	90
carbon tetrachloride	VP	241	Poly(vinyl acetate-co-vinyl alcohol)		
			acetone	CM,VP	80
1.4　Vinyl Polymers			water	CD	62
			Poly(vinyl alcohol)		
Poly(acrylonitrile)			chloroacetic acid	CM	120
acrylonitrile	VP	137	ethyl alcohol	CM	33
benzene	CM,VP	51	water	CD	61,62
N,N-dimethylformamide	CM	51		CM	33,59
Poly(acrylonitrile-co-methacrylic acid)				LS	239
acrylonitrile	VP	137		VP	91,239
Poly(acrylonitrile-co-1,3-butadiene)			Poly(N-vinyl carbazole) [(poly(9-carbazolyl-ethylene)]		
benzene	CM	22,50			
	VP	50	benzene	CD	176
dibutyl phthalate	CM,D	113		VP	135

Liquid	Method	References	Liquid	Method	References
1.4 Vinyl Polymers (Cont'd.)			carbon tetrachloride (Cont'd.)	LS	122
				VP	16,241
Poly(vinyl chloride)			chlorobenzene	CD	57
cyclohexanone	CD	254		CM	57,159
	CM	114,254		D	28,64,87
	DL	114		OS	138
dichloroethane	CM	59		LS	122
tetrahydrofuran	CD,CM	254		SW	123
Poly(2-vinyl pyridine) [poly(2-pyridyl-ethylene)]			chloroform	CD	58,207,247
				CM	32
ethyl alcohol	LS	119		D	48,64,87
Poly(N-vinyl pyrrolidone) [poly(2-oxo-1-pyrrolidinyl)ethylene]				OS	36
				VP	55
chloroform	CM,D	189		SW	123
water	CM,D	189	cyclohexane	CD	57,75,86,208
	OS	234		CM	32,37,46,57,200,208
1.5 Poly(styrenes)				D	46,48,64,205,231
Poly(p-bromostyrene)				DL	15,74,200,205,229
toluene	VP	110		OS	21,36,76,85,197
Poly(p-bromostyrene-co-p-methylstyrene)				LS	122,155,206,226
toluene	VP	110		SD	205,252
Poly(p-chlorostyrene)				PE	209
ethyl acetate	VP	218		VP	16,56,85,208
toluene	VP	110		SW	123
Poly(α-methylstyrene)			cyclohexene	CD	208
benzene	CM	244		CM	32,208
cumene	VP	148		VP	208
α-methylstyrene	VP	148	decahydronaphthalene	CD	208
Poly(p-methylstyrene)				CM	200,208
toluene	VP	110		D	27
Poly(styrene)				DL	200
acetone	CM	32,46,37		LS	122,158
	D	46,64		VP	220
	VP	55,241	o-dichlorobenzene	D	27,64
benzene	CD	31,75,86,160,207,208,247	1,3-diphenylbutane	CD	121,160
	CM	22,32,35,47,59,96,200,208,243		SW	121,160
	D	27,42,64,139	dioxane	CD	147
	DL	15,74,200		CM	59,200
	OS	15,36,160		D	27,64
	LS	26,122,155,158,173		DL	200
	XS	219		OS	138
	VP	16,31,35,160,208,241		VP	16
	SW	160	ethyl acetate	CD	147
bromobenzene	D	28		CM	32,73
2-butanone	CD	147		D	27,64
	CM	32,73,159		OS	21,36
	D	27,42,48,64		VP	56
	DL	15,227	ethylbenzene	CD	54,57,58,86,121,160,208,251
	OS	15,21,36,227			
	VP	20		CM	33,47,57,83,96,111,200,208,251
butyl acetate	CM	32			
	D	64		D	27,42,121,160,231
	OS	36			
	VP	16		DL	74,200,228
sec-butyl alcohol	SW	123		OS	121,160,228
butylbenzene	CM,DL	200		LS	122
carbon tetrachloride	CD,CM	111			
	D	64			
	OS	36			

Liquid	Method	References	Liquid	Method	References
Poly(styrene) (Cont'd.)			**2. Main-chain Carbon Heteroatom Polymers**		
ethylbenzene (Cont' d.)	VP	54,96,121,160			
	SW	121,160,123	2.1 Poly(oxides) and Poly(esters)		
ethylene glycol ether acetate	CM	32	Epoxy resin : poly(4,4'-bis-glycidylphenyl-		
heptane	SW	123	2,2'-propane)		
hexane	SW	123	chloroform	VP,SW	118
hexyl alcohol	SW	123	dioxane	VP,SW	118
isopentyl alcohol	SW	123	Poly(oxyethylene)		
mesitylene	CM	32,200	benzene	CM	150
	D	64		D	233
	DL	200		VP	190,233
methoxybenzene	CD	147	butyl alcohol	CD	238
nitromethane	VP	16	carbon tetrachloride	CM	140,201
octane	VP	16		DL,VP	201
	SW	123	chloroform	CD	253
pentane	SW	123		CM	201,253
3-pentanone	D	64		DL	201
	VP	16		VP	125,201
propyl acetate	VP	55	chloromethane	CD,CM	253
propyl alcohol	SW	123	dioxane	CM	140
propylbenzene	CM,DL	200	ethyl alcohol	CD	238
propyl ether	VP	16		CM	150
styrene	CM	13,32	methyl alcohol	CD	238
	VP	137		CM	150
1,2,3,4-tetrahydronaphthalene	CM	208	water	CD	163,253
tetrahydrofuran	D	64		CM	66,97,151,211,
toluene	CD	31,57,58,86,147,			253
		199,204,207,208,		D	66,191
		247,251		OS	234
	CM	32,37,46,57,59,		VP	66,191
		96,159,169,200,	Poly(oxycarbonyloxy-1,4-phenyleneiso-		
		208,249,251	propylidene-1,4-phenylene)		
	D	14,27,42,46,	dioxane	CD	176
		48,64,205	Poly(oxyethyleneoxyadipoyl)		
	DL	15,74,200,205,	dioxane	CM	143
		230		VP	143,180
	OS	15,21,36,76,	Poly(oxyethyleneoxysebacoyl)		
		157,222	dioxane	CM,VP	143
	LS	122,206	butyl butyrate	VP	214
	SD	205	Poly(oxyethyleneoxysuccinyl)		
	VP	16,20,31,56,110	dioxane	CM,VP	143
	SW	123	Poly(oxypropylene)		
o-xylene	CM	32	benzene	VP	235
	OS	36	butyl alcohol	CM,DL,SW	126
m-xylene	CM	32,200	sec-butyl alcohol	CM,SW	126
	D	64	tert-butyl alcohol	CM,SW	126
	DL	200	carbon tetrachloride	CM	126,178
	OS	36		DL,VP	178
	VP	16		SW	126
Poly(styrene-co-1,3-butadiene)			chloroform	CM,DL,VP	178
benzene	CM	22,51	cyclohexane	CM,SW	126
	VP	51	ethyl alcohol	CM,DL,SW	126
dibutyl phthalate	CM,D	113	hexane	CM,SW	126
dibutyl sebacate	CM,D	113	isobutyl alcohol	CM,SW	126
dicumylmethane	CM,D	113	isopentyl alcohol	CM,SW	126
ditolylmethane	CM,D	113	isopropyl alcohol	CM,SW	126
			methyl alcohol	CM,D,VP	95
1.6 Others			pentyl alcohol	CM,SW	126
			propyl alcohol	CM,SW	126
Poly(nitro-1,4-phenylene)			water	CM	66,97,179
dioxane	VP	203		VP,D	66
			Poly(oxytrimethyleneoxyadipoyl)		
			dioxane	CM,VP	143

Liquid	Method	References	Liquid	Method	References
2.2 Poly(amides)			**Poly(oxydimethylsilylene)** (Cont' d.)		
			isopropylbenzene	CM,D	184
Poly[L-iminocarbonyl(benzyloxycarbonyl-			mesitylene	CM,D	184
methyl)methylene] [poly(β-benzyl-L-			methyl alcohol	SW	198
aspartate)]			methylcyclohexane	CM,D	184
chloroform	D,VP	144	methyl esters : propionate to decanoate	CM	132
Poly[L-iminocarbonyl(benzyloxycarbonyl-			2-methylpropane	VP	106
ethyl)methylene] [poly(γ-benzyl-L-			nonane	CM	101
glutamate)]			octamethyl cyclotetrasiloxane	CM	43,200
chloroform	CD	194		DL	200
dichloroacetic acid	CD	194		VP	43
	CM	154	octane	CM	101
1,2-dichloroethane	CM	154		VP	223
	D,VP	144	pentane	CM	101
N,N-dimethylformamide	VP	250		VP	106
pyridine	D,VP	144	2-pentanone	SW	198
Poly[imino(1-oxohexamethylene)] (Nylon 6)			poly(1,1-dimethylethylene) oligomer	CM,DL	98
decyl alcohol	VP	214	poly(oxydimethylsilylene) oligomers	CM	101
			propionates : propyl to pentyl	CM	132
2.3 Poly(siloxanes)			propyl acetate	CM	132,200
				DL	200
Poly(oxydimethylsilylene) [poly(dimethyl			propyl alcohol	SW	198
siloxane)]			tetradecane	CM	101
acetates : butyl to decyl	CM	132	tridecane	CM	101,225
acetone	SW	198	toluene	CM	184
benzene	CM	23,124,184		D	184
	D	184		OS	183
	VP	23,223		VP	223
	SW	198		SW	198
bromocyclohexane	CM,DL	200	undecane	CM	101
butane	VP	106	xylenes	SW	198
2-butanone	CM	153,200,225		CM,D	184
	DL	200	**Poly(dimethyl siloxane-co-diphenyl siloxane)**		
	SW	198	**(silicone resin)**		
carbon tetrachloride	CM	124	benzene	VP	223
chlorobenzene	CM	124	ethylbenzene	VP	223
	OS	183	hexane	VP	223
cyclohexane	CM	124,184	octane	VP	223
	D	184	toluene	VP	223
	OS	183			
decane	CM	101	**2.4 Cellulose and Derivatives**		
diethoxymethane	CM	132			
1,2-dimethoxyethane	CM	132	Cellulose acetate (and triacetate) (see also ref. 93)		
			acetone	D	69,165
2,2-dimethylpropane	VP	106		OS	69
dodecane	CM	101		VP	69,116
ethers : propyl to hexyl	CM	132	acetonitrile	D	165
ethyl acetate	CM	132,200,225	aniline	D	165
	DL	200	chloroform	D	165
ethylbenzene	CM,D	184		VP	117
	VP	223	chloromethane	D	165
ethyl esters : propionate to dodecanoate	CM	132		VP	117
ethyl ether	CM	132,225	dioxane	D	165
heptane	CM	101,124,184,		VP	116
		200	methyl acetate	D	165
	D	184		VP	116
	DL	200	phenol	D,OS,VP	69
hexadecane	CM	101	pyridine	D	165
hexamethyl disiloxane	CM	196,200		VP	116
	DL	200	tetrachloroethane	OS	3
hexane	CM	101,200	Cellulose nitrate		
	DL	200	acetone	CD	5,9
	VP	223		CM	24,59
hexyl alcohol	SW	198		D	5,9,139

nβ

nγ

Liquid	Method	References	Liquid	Method	References
Cellulose nitrate (Cont'd.)			Ethyl cellulose (Cont'd.)		
acetone (Cont'd.)	OS	221	benzene	D	187
	VP	5,9,19,45,		VP	70
		71,116	butane	VP	82
acetonitrile	VP	45	2-butanone	D	187
2-butanone	D	139	butyl acetate	D	187
butyl acetate	CM	24,59	carbon tetrachloride	D	187
	D	139	chloroform	D	187
cyclopentanone	VP	45	2,2-dimethylpropane	VP	82
dibutyl phthalate	CM	24	ethyl acetate	D	187
3,3-dimethyl-2-butanone	VP	45	methyl alcohol	VP	70
2,4-dimethyl-3-pentanone	VP	45	2-methylbutane	VP	82
dioxane	VP	45	2-methylpropane	VP	82
ethyl acetate	CM	24,25	pentane	VP	82
	D	139	water	VP	84
ethyl propyl ether	VP	45	Hydroxyethyl cellulose		
2-heptanone	D	139	water	OS	234
methyl acetate	D	139			
	VP	116	2.5 Others		
methyl alcohol	CM	25,59	Dextran		
3-methyl-2-butanone	VP	45	water	OS	234
nitromethane	VP	45	Lignin		
2-pentanone	D	139	N,N-dimethylformamide	VP	166
	VP	45	dioxane	VP	166
propyl acetate	D	139	methyl sulfoxide	VP	166
tritolyl phosphate	CM	24	Starch		
Ethyl cellulose			water	VP	188
acetone	D	187			
	VP	70			

C. REFERENCES

1. K. H. Meyer, E. Wolff, C. G. Boissonas, Helv. Chim. Acta 23, 430 (1940).
2. E. Wolff, Helv. Chim. Acta 23, 439 (1940).
3. O. Hagger, A. J. A. van der Wyk, Helv. Chim. Acta 23, 484 (1940).
4. K. H. Meyer, A. J. A. van der Wyk, Helv. Chim. Acta 23, 488 (1940).
5. E. Calvet, Compt. Rend. 213, 126 (1941); 214, 767 (1942).
6. G. Gee, L. R. G. Treloar, Trans. Faraday Soc. 38, 147 (1942).
7. M. L. Huggins, Ind. Eng. Chem. 35, 216 (1943).
8. J. Ferry, G. Gee, L. R. G. Treloar, Trans. Faraday Soc. 41, 340 (1945).
9. E. Calvet, Bull. Soc. Chim. 12, 553 (1945).
10. G. Gee, W. J. C. Orr, Trans. Faraday Soc. 42, 507 (1946).
11. G. Gee, Trans. Faraday Soc. 42B, 33 (1946).
12. P. Doty, H. S. Zable, J. Polymer Sci. 1, 90 (1946).
13. D. E. Roberts, W. W. Walton, R. S. Jessup, J. Polymer Sci. 2, 420 (1947).
14. R. F. Boyer, R. S. Spencer, J. Polymer Sci. 3, 97 (1948).
15. J. W. Breitenbach, H. P. Frank, Monatsh. 79, 531 (1948).
16. E. C. Baughan, Trans. Faraday Soc. 44, 495 (1948).
17. A. A. Tager, V. A. Kargin, Kolloidn. Zh. 10, 455 (1948).
18. G. V. Browning, J. D. Ferry, J. Chem. Phys. 17, 1107 (1949).
19. H. Campbell, P. Johnson, J. Polymer Sci. 4, 247 (1949).
20. C. E. H. Bawn, R. F. J. Freeman, A. R. Kamaliddin, Trans. Faraday Soc. 46, 677 (1950).
21. M. J. Schick, P. Doty, B. H. Zimm, J. Am. Chem. Soc. 72, 530 (1950).
22. A. A. Tager, V. Sanatina, Kolloidn. Zh. 12, 474 (1950); Rubber Chem. Technol. 24, 773 (1951).
23. M. J. Newing, Trans. Faraday Soc. 46, 613 (1950).
24. S. M. Lipatov, S. I. Meerson, Kolloidn. Zh. 12, 122 (1950).
25. S. M. Lipatov, S. I. Meerson, Kolloidn. Zh. 12, 427 (1950).
26. P. Debye, A. M. Bueche, J. Chem. Phys. 18, 1423 (1950).
27. D. J. Streeter, R. F. Boyer, Ind. Eng. Chem. 43, 1790 (1951).
28. W. Heller, A. C. Thompson, J. Colloid Sci. 6, 57 (1951).
29. L. der Minassian, M. Magat, J. Chim. Phys. 48, 574 (1951).
30. P. Meares, Trans. Faraday Soc. 47, 699 (1951).
31. H. Tompa, J. Polymer Sci. 8, 51 (1952).
32. H. Hellfritz, Makromol. Chem. 7, 191 (1952).
33. A. A. Tager, V. A. Kargin, Kolloidn. Zh. 14, 367 (1952).
34. G. V. Schulz, H. Doll, Z. Elektrochem. 56, 248 (1952).
35. A. A. Tager, Zh. H. Dombek, Kolloidn. Zh. 15, 69 (1953).
36. G. V. Schulz, H. Hellfritz, Z. Elektrochem. 57, 835 (1953).
37. K. von Guenner, G. V. Schulz, Naturwissensch. 40, 164 (1953).
38. S. Prager, E. Bagley, F. A. Long, J. Am. Chem. Soc. 75, 2742 (1953).
39. G. V. Schulz, H. Doll, Z. Elektrochem. 57, 841 (1953).
40. R. J. Kokes, A. R. DiPietro, F. A. Long, J. Am. Chem. Soc. 75, 6319 (1953).
41. J. H. van der Waals, J. J. Hermans, Rec. Trav. Chim. Pays-Bas 69, 971 (1953).
42. M. Griffel, R. S. Jessup, J. A. Cogliano, R. P. Park, J. Res. Nat. Bur. Std. 52, 217 (1954).
43. R. C. Osthoff, W. T. Grubb, J. Am. Chem. Soc. 76, 399 (1954).
44. H. Daoust, M. Rinfret, Can. J. Chem. 32, 492 (1954).
45. E. C. Baughan, A. L. Jones, K. Stewart, Proc. Roy. Soc., A, 225, 478 (1954).
46. G. V. Schulz, K. von Guenner, H. Gerrens, Z. Physik. Chem., NF, 4, 192 (1955).
47. A. A. Tager, R. V. Krivokorytova, P. M. Khodorov, Dokl. Akad. Nauk. SSSR 100, 741 (1955).
48. B. Rosen, J. Polymer Sci. 17, 559 (1955).
49. A. Aitken, R. M. Barrer, Trans. Faraday Soc. 51, 116 (1955).
50. A. A. Tager, L. K. Kosova, Kolloidn. Zh. 17, 391 (1955).
51. A. A. Tager, L. K. Kosova, D. Y. Karlinskaya, I. A. Yurina, Kolloidn. Zh. 17, 315 (1955).

52. M. Parent, M. Rinfret, Can. J. Chem. 33, 971 (1955).
53. A. Horth, M. Rinfret, J. Am. Chem. Soc. 77, 503 (1955).
54. T. V. Gatovskaya, V. A. Kargin, A. A. Tager, Zh. Fiz. Khim. 29, 883 (1955).
55. C. E. H. Bawn, M. A. Wajid, Trans. Faraday Soc. 52, 1658 (1956).
56. K. Schmoll, E. Jenckel, Z. Elektrochem. 60, 756 (1956).
57. E. Jenckel, K. Gorke, Z. Elektrochem. 60, 579 (1956).
58. K. Amaya, R. Fujishiro, Bull. Chem. Soc. (Japan) 29, 270 (1956).
59. S. I. Meerson, S. M. Lipatov, Kolloidn. Zh. 18, 447 (1956).
60. C. E. H. Bawn, R. D. Patel, Trans. Faraday Soc. 52, 1664 (1956).
61. K. Amaya, R. Fujishiro, Bull. Chem. Soc. (Japan) 29, 361 (1956).
62. K. Amaya, R. Fujishiro, Bull. Chem. Soc. (Japan) 29, 830 (1956).
63. V. A. Kargin, T. V. Gatovskaya, Zh. Fiz. Khim. 30, 1852 (1956).
64. G. V. Schulz, M. Hoffmann, Makromol. Chem. 23, 220 (1957).
65. P. J. Flory, H. Daoust, J. Polymer Sci. 25, 429 (1957).
66. G. N. Malcolm, J. S. Rowlinson, Trans. Faraday Soc. 53, 921 (1957).
67. C. Booth, G. Gee, G. R. Williamson, J. Polymer Sci. 23, 3 (1957).
68. H. J. L. Schuurmans, J. J. Hermans, J. Phys. Chem. 61, 1496 (1957).
69. R. Jeffries, Trans. Faraday Soc. 53, 1592 (1957).
70. R. M. Barrer, J. A. Barrie, J. Polymer Sci. 23, 331 (1957).
71. H. Takenaka, J. Polymer Sci. 24, 321 (1957).
72. J. W. Breitenbach, J. Polymer Sci. 23, 949 (1957).
73. A. A. Tager, L. A. Galkina, Nauch. Dokl. Vysshei Skoly, Khim. i Khim. Tekhnol. 1, 357 (1958).
74. A. A. Tager, A. Smirnova, N. Sysueva, Nauch. Dokl. Vysshei Skoly, Khim. i. Khim. Tekhnol. 1, 135 (1958).
75. K. Amaya, R. Fujishiro, Bull. Chem. Soc. (Japan) 31, 19 (1958).
76. G. Rehage, H. Meys, J. Polymer Sci. 30, 271 (1958).
77. M. A. Kabayama, H. Daoust, J. Phys. Chem. 62, 1127 (1958).
78. R. S. Jessup, J. Res. Nat. Bur. Std. 60, 47 (1958).
79. R. M. Barrer, R. R. Ferguson, Trans. Faraday Soc. 54, 989 (1958).
80. A. A. Tager, M. Iovleva, Zh. Fiz. Khim. 32, 1774 (1958).
81. A. F. Sirianni, R. Tremblay, I. E. Puddington, Can. J. Chem. 36, 543 (1958).
82. R. M. Barrer, J. A. Barrie, J. Slater, J. Polymer Sci. 27, 177 (1958).
83. A. A. Tager, M. M. Gur'yanova, Zh. Fiz. Khim. 32, 1958 (1958).
84. R. M. Barrer, J. A. Barrie, J. Polymer Sci. 28, 377 (1958).
85. W. R. Krigbaum, D. O. Geymer, J. Am. Chem. Soc. 81, 1859 (1959).
86. G. V. Schulz, A. Horbach, Z. Physik. Chem., NF, 22, 377 (1959).
87. A. Horth, D. Patterson, M. Rinfret, J. Polymer Sci. 39, 189 (1959).
88. A. A. Tager, M. V. Tsilipotkina, V. K. Doronina, Zh. Fiz. Khim. 33, 335 (1959).
89. R. S. Jessup, J. Res. Nat. Bur. Std. 62, 1 (1959).
90. A. Nakajima, H. Yamakawa, I. Sakurada, J. Polymer Sci. 35, 489 (1959).
91. I. Sakurada, A. Nakajima, H. Fujiwara, J. Polymer Sci. 35, 497 (1959).
92. C. E. Rogers, V. Stanett, M. Szwarc, J. Phys. Chem. 63, 1406 (1959).
93. A. A. Tager, O. Popova, Zh. Fiz. Khim. 33, 593 (1959).
94. C. Watters, H. Daoust, M. Rinfret, Can. J. Chem. 38, 1087 (1960).
95. M. L. Lakhanpal, B. E. Conway, J. Polymer Sci. 46, 75 (1960).
96. S. I. Meerson, S. M. Lipatov, Kolloidn. Zh. 21, 531 (1959).
97. R. G. Cunninghame, G. N. Malcolm, J. Phys. Chem. 65, 1454 (1961).
98. G. Allen, G. Gee, J. P. Nicholson, Polymer 2, 8 (1961).
99. P. J. Flory, A. Ciferri, R. Chiang, J. Am. Chem. Soc. 83, 1023 (1961).
100. G. Delmas, D. Patterson, T. Somcynsky, J. Polymer Sci. 57, 79 (1962).
101. G. Delmas, D. Patterson, D. Bohme, Trans. Faraday Soc. 58, 2116 (1962).
102. J. Leonard, H. Daoust, J. Polymer Sci. 57, 53 (1962).
103. C. H. Baker, W. B. Brown, G. Gee, J. S. Rowlinson, D. Stubley, R. E. Yeadon, Polymer 3, 215 (1962).
104. M. Senez, H. Daoust, Can. J. Chem. 40, 734 (1962).
105. J. S. Ham. M. C. Bolen, J. K. Hughes, J. Polymer Sci. 57, 25 (1962).
106. R. M. Barrer, J. A. Barrie, N. K. Raman, Polymer 3, 595 (1962).
107. T. V. Gatovskaya, G. M. Pavlyuchenko, V. A. Berestnev, V. A. Kargin, Dokl. Akad. Nauk, SSSR 143, 590 (1962); Proc. Acad. Sci. USSR 143, 209 (1962).
108. K. P. Kwei, T. K. Kwei, J. Phys. Chem. 66, 2146 (1962).
109. G. M. Pavlyuchenko, T. V. Gatovskaya, V. A. Kargin, Dokl. Akad. Nauk, SSSR 147, 150 (1962); Proc. Acad. Sci., USSR 147, 790 (1962).
110. R. Corneliussen, S. A. Rice, H. Yamakawa, J. Chem. Phys. 38, 1768 (1963).
111. A. A. Tager, A. I. Podlesnyak, Vysokomolekul. Soedin., A5, 87 (1963); Polymer Sci., USSR 4, 698 (1963).
112. A. A. Tager, M. V. Tsilipotkina, V. Ye. Dreval, O. V. Nechayeva, Vysokomolekul. Soedin. A5, 94 (1963); Polymer Sci., USSR 4, 706 (1963).
113. M. P. Zverev, S. P. Ruchinskii, P. I. Zubov, Dokl. Akad. Nauk, SSSR 149, 128 (1963); Proc. Acad. Sci., USSR 149, 207 (1963).
114. S. I. Meerson, I. M. Zagrayevskaya, Kolloidn. Zh. 25, 202 (1963).
115. T. V. Gatovskaya, G. M. Pavlyuchenko, V. A. Berestnev, V. A. Kargin, Vysokomolekul. Soedin. A5, 960 (1963); Polymer Sci., USSR 5, 9 (1963).
116. W. R. Moore, R. Shuttleworth, J. Polymer Sci., A, 1, 733 (1963).
117. W. R. Moore, R. Shuttleworth, J. Polymer Sci., A, 1, 1985 (1963).
118. T. K. Kwei, J. Polymer Sci., A, 1, 2977 (1963).
119. A. J. Hyde, R. B. Taylor, Makromol. Chem. 62, 204 (1963).
120. S. I. Meerson, I. M. Zagrayevskaya, Kolloidn. Zh. 25, 197 (1963).
121. J. Biros, K. Solc, J. Pouchly, Faserforsch. Textil. 15, 608 (1964).
122. A. A. Tager, V. M. Andreyeva, E. M. Evsina, Vysokomolekul. Soedin. A6, 1901 (1964); Polymer Sci., USSR 6, 2107 (1964).
123. G. Rehage, Kolloid Z. 196, 97 (1964); 199, 1 (1964).
124. G. Delmas, D. Patterson, S. N. Bhattacharyya, J. Phys. Chem. 68, 1468 (1964).
125. G. Allen, C. Booth, G. Gee, M. N. Jones, Polymer 5, 367 (1964).
126. B. E. Conway, J. P. Nicholson, Polymer 5, 387 (1964).
127. C. Booth, G. Gee, M. N. Jones, W. D. Taylor, Polymer 5, 353 (1964).
128. C. Booth, G. Gee, G. Holden, G. R. Williamson, Polymer 5, 343 (1964).
129. G. Butenuth, Rubber, Chem. Technol. 37, 326 (1964).
130. H. K. Frensdorff, J. Polymer Sci., A, 2, 333 (1964).
131. E. D. Holly, J. Polymer Sci., A, 2, 5267 (1964).
132. D. Patterson, J. Polymer Sci., A, 2, 5177 (1964).
133. W. B. Brown, G. Gee, W. D. Taylor, Polymer 5, 362 (1964).
134. J. L. Gardon, Encycl. Polymer Sci. Technol. 3, 833 (1964).
135. K. Ueberreiter, W. Bruns, Ber. Bunsenges. Physik. Chem. 68, 541 (1964).
136. G. M. Pavlyuchenko, T. V. Gatovskaya, V. A. Kargin, Vysokomolekul. Soedin. A6, 1190 (1964); Polymer Sci., USSR 6, 1309 (1964).
137. A. N. Gudkov, N. A. Fermor, N. I. Smirnov, Zh. Prikl. Khim. 37, 2204 (1964); J. Appl. Chem. USSR 37, 2179 (1964).
138. J. Leonard, H. Daoust, J. Phys. Chem. 69, 1174 (1965).
139. W. R. Moore, B. M. Tidswell, Makromol. Chem. 81, 1 (1965).
140. M. L. Lakhanpal, M. Lal, R. K. Sharma, Indian J. Chem. 3, 547 (1965).
141. G. Gee, J. B. M. Herbert, R. C. Roberts, Polymer 6, 541 (1965).
142. A. S. Michaels, R. W. Hausslein, J. Polymer Sci., C, 10, 61 (1965).
143. A. A. Tager, L. Ya. Kasas', Dokl. Akad. Nauk, SSSR 165, 1122 (1965); Proc. Acad. Sci., USSR 165, 892 (1965).
144. P. J. Flory, W. J. Leonard, J. Am. Chem. Soc. 87, 2102 (1965).
145. G. M. Pavlyuchenko, T. V. Gatovskaya, V. A. Kargin, Vysokomolekul. Soedin. 7, 647 (1965); Polymer Sci., USSR 7, 714 (1965).
146. W. Heller, J. Phys. Chem. 69, 1123 (1965).
147. A. Kagemoto, S. Murakami, R. Fujishiro, Bull. Chem. Soc. (Japan) 39, 15 (1966).
148. S. G. Canagaratna, D. Margerison, J. P. Newport, Trans. Faraday Soc. 62, 3058 (1966).
149. C. Cuniberti, U. Bianchi, Polymer 7, 151 (1966).
150. M. L. Lakhanpal, H. L. Taneja, R. K. Sharma, Indian J. Chem. 4, 12 (1966).

151. M. L. Lakhanpal, V. Kapoor, R. K. Sharma, S. C. Sharma, Indian J. Chem. **4**, 59 (1966).

152. A. Kagemoto, S. Murakami, R. Fujishiro, Bull. Chem. Soc. (Japan) **39**, 1814 (1966).

153. U. Bianchi, E. Pedemonte, C. Rossi, Makromol. Chem. **92**, 114 (1966).

154. G. Giacometti, A. Turolla, Z. Physik. Chem., NF, **51**, 108 (1966).

155. H. Benoit, C. Picot, Pure Appl. Chem. **12**, 545 (1966).

156. C. J. Sheehan, A. L. Bisio, Rubber Chem. Technol. **39**, 149 (1966).

157. N. Kuwahara, T. Okazawa, M. Kaneko, J. Chem. Phys. **47**, 3357 (1967).

158. A. A. Tager, V. M. Andreyeva, J. Polymer Sci., C, **16**, 1145 (1967).

159. U. Bianchi, C. Cuniberti, E. Pedemonte, C. Rossi, J. Polymer Sci., A-2, **5**, 743 (1967).

160. J. Pouchly, J. Biros, K. Solc, J. Vondrejsova, J. Polymer Sci., C, **16**, 679 (1967).

161. A. Kagemoto, S. Murakami, R. Fujishiro, Bull. Chem. Soc. (Japan) **40**, 11 (1967).

162. J. Leonard, H. Daoust, Can. J. Chem. **45**, 409 (1967).

163. A. Kagemoto, S. Murakami, R. Fujishiro, Makromol. Chem. **105**, 154 (1967).

164. H. P. Schreiber, M. H. Waldman, J. Polymer Sci., A-2, **5**, 555 (1967).

165. W. R. Moore, J. Polymer Sci., C, **16**, 571 (1967).

166. W. Brown, J. Appl. Polymer Sci. **11**, 2381 (1967).

167. D. Patterson, Rubber Chem. Technol. **40**, 1 (1967).

168. D. Patterson, A. A. Tager, Vysokomolekul. Soedin. A**9**, 1814 (1967); Polymer Sci. USSR **9**, 2051 (1967).

169. G. Allen, R. C. Ayerst, J. R. Cleveland, G. Gee, C. Price, J. Polymer Sci., C, **23**, 127 (1968).

170. B. E. Eichinger, P. J. Flory, Trans. Faraday Soc. **64**, 2061 (1968).

171. B. E. Eichinger, P. J. Flory, Trans. Faraday Soc. **64**, 2066 (1968).

172. A. A. Tager, A. I. Podlesnyak, L. V. Demidova, Vysokomolekul. Soedin. B**10**, 601 (1968).

173. A. A. Tager, A. A. Anikeyeva, V. M. Andreyeva, T. Ya. Gumarova, L. A. Chernoskutova, Vysokomolekul. Soedin. A**10**, 1661 (1968); Polymer Sci. USSR **10**, 1926 (1968).

174. B. E. Eichinger, P. J. Flory, Trans. Faraday Soc. **64**, 2053 (1968).

175. P. J. Flory, J. L. Ellenson, B. E. Eichinger, Macromolecules **1**, 279 (1968).

176. W. Bruns, F. Mehdorn, K. Mirus, K. Ueberreiter, Kolloid Z. **224**, 17 (1968).

177. H. Daoust, A. Hade, Polymer **9**, 47 (1968).

178. R. W. Kershaw, G. N. Malcolm, Trans. Faraday Soc. **64**, 323 (1968).

179. M. L. Lakhanpal, H. G. Singh, H. Singh, S. C. Sharma, Indian J. Chem. **6**, 95 (1968).

180. Yu. S. Lipatov, L. M. Sergeyeva, G. F. Kovalenko, Vysokomolekul. Soedin. B**10**, 205 (1968).

181. B. E. Eichinger, P. J. Flory, Trans. Faraday Soc. **64**, 2035 (1968).

182. T. A. Bogayevskaya, T. V. Gatovskaya, V. A. Kargin, Vysokomolekul. Soedin. B**10**, 376 (1968).

183. N. Kuwahara, T. Okazawa, M. Kaneko, J. Polymer Sci., C, **23**, 543 (1968).

184. S. Morimoto, J. Polymer Sci., A-1, **6**, 1547 (1968).

185. A. Kagemoto, R. Fujishiro, Bull. Chem. Soc. (Japan) **41**, 2201 (1968).

186. T. A. Bogayevskaya, T. V. Gatovskaya, V. A. Kargin, Vysokomolekul. Soedin. A**10**, 1357 (1968); Polymer Sci. USSR **10**, 1574 (1968).

187. W. R. Moore, Contribution to Chem. Soc. Symposium, Solution Properties of Natural Polymers, Edinburgh (1968); Chem. Soc. Special Publication No. 23, p. 185 (1968).

188. M. Masuzawa, C. Stirling, J. Appl. Polymer Sci. **12**, 2023 (1968).

189. J. Goldfarb, S. Rodriguez, Makromol. Chem. **116**, 96 (1968).

190. M. L. Lakhanpal, H. G. Singh, S. C. Sharma, Indian J. Chem. **6**, 436 (1968).

191. M. L. Lakhanpal, K. S. Chhina, S. C. Sharma, Indian J. Chem. **6**, 505 (1968).

192. T. A. Bogayevskaya, T. V. Gatovskaya, Vysokomolekul. Soedin. B**10**, 555 (1968).

193. N. Kuwahara, T. Oikawa, M. Kaneko, J. Chem. Phys. **49**, 4972 (1968).

194. A. Kagemoto, R. Fujishiro, Makromol. Chem. **114**, 139 (1968).

195. R. Koningsveld, A. J. Staverman, J. Polymer Sci., A-2, **6**, 325 (1968).

196. D. Patterson, S. N. Bhattacharyya, P. Picker, Trans. Faraday Soc. **64**, 648 (1968).

197. G. Rehage, H. J. Palmen, D. Moeller, W. Wefers, contribution to IUPAC Symposium, Toronto (1968).

198. R. D. Seeley, Rubber Chem. Technol. **41**, 608 (1968).

199. G. Lewis, A. F. Johnson, J. Chem. Soc. (A), p. 1816 (1969).

200. U. Bianchi, C. Cuniberti, E. Pedemonte, C. Rossi, J. Polymer Sci., A-2, **7**, 855 (1969).

201. G. N. Malcolm, C. E. Baird, G. R. Bruce, K. G. Cheyne, R. W. Kershaw and M. C. Pratt, J. Polymer Sci., A-2, **7**, 1495 (1969).

202. S. Takagi, R. Fujishiro, Rep. Prog. Polymer Phys. (Japan) **12**, 39 (1969).

203. M. Hyodo, K. Kubo, K. Ogino, Rep. Prog. Polymer Phys. (Japan) **12**, 21 (1969).

204. G. Lewis, A. F. Johnson, Polymer **11**, 336 (1970).

205. Th. G. Scholte, J. Polymer Sci., A-2, **8**, 841 (1970).

206. Th. G. Scholte, European Polymer J. **6**, 1063 (1970).

207. S. Morimoto, Nippon Kagaku Zasshi **91**, 31, 117 (1970).

208. A. A. Tager, A. I. Podlesnyak, M. V. Tsilipotkina, L. V. Adamova, A. A. Bakhareva, L. V. Demidova, Vysokomolekul. Soedin. A**12**, 1320 (1970); Polymer Sci. USSR **12**, 1497 (1970).

209. R. Koningsveld, L. A. Kleintjens, A. R. Shultz, J. Polymer Sci., A-2, **8**, 1261 (1970).

210. C. Gerth, F. H. Mueller, Kolloid Z. **241**, 1071 (1970).

211. H. Nakayama, Bull. Chem. Soc. (Japan) **43**, 1683 (1970).

212. P. J. T. Tait, P. J. Livesey, Polymer **11**, 359 (1970).

213. S. A. Tashmukamedov, R. S. Tillayev, Kh. U. Usmanov, Uzb. Khim. Zh. **14**, 44 (1970).

214. T. A. Bogayavskaya, T. V. Gatovskaya, V. A. Kargin, Vysokomolekul. Soedin. A**12**, 243 (1970); Polymer Sci. USSR **12**, 279 (1970).

215. D. Machin, C. E. Rogers, Encycl. Polymer Sci. Technol. **12**, 679 (1970).

216. D. K. Carpenter, Encycl. Polymer Sci. Technol. **12**, 627 (1970).

217. J. H. Hildebrand, J. M. Prausnitz, R. L. Scott, Regular and Related Solutions, Van Nostrand Reinhold Co., New York (1970), Ch. 2

218. K. Kubo, K. Yamatsuta, K. Ogino, Rep. Prog. Polymer Phys. (Japan) **13**, 19 (1970).

219. Y. Taru, K. Yosizaki, E. Wada, K. Kanamaru, Rep. Prog. Polymer Phys. (Japan) **13**, 21 (1970).

220. S. Higashida, N. Kuwahara, M, Kaneko, Rep. Prog. Polymer Phys. (Japan) **13**, 5 (1970).

221. T. Oikawa, N. Kuwahara, M. Kaneko, Rep. Prog. Polymer Phys. (Japan) **13**, 9 (1970).

222. K. Kamada, H. Sato, Rep. Prog. Polymer Phys. (Japan) **13**, 47 (1970).

223. A. Muramoto, Polymer J. **1**, 450 (1970).

224. G. V. Schulz, M. Lechner, J. Polymer Sci., A-2, **8**, 1885 (1970); European Polymer J. **6**, 945 (1970).

225. S. Morimoto, Makromol. Chem. **133**, 197 (1970); Rep. Prog. Polymer Phys. (Japan) **13**, 29 (1970).

226. Th. G. Scholte, J. Polymer Sci., A-2, **9**, 1553 (1971).

227. P. J. Flory, H. Hoecker, Trans. Faraday Soc. **67**, 2258 (1971).

228. H. Hoecker, P. J. Flory, Trans. Faraday Soc. **67**, 2270 (1971).

229. H. Hoecker, H. Shih, P. J. Flory, Trans. Faraday Soc. **67**, 2275 (1971).

230. G. Lewis, A. F. Johnson, J. Chem. Soc. (A), p. 3528 (1971).

231. A. A. Tager, L. V. Adamova, M. V. Tsilipotkina, G. I. Florova, Vysokomolekul. Soedin. A**13**, 654 (1971); Polymer Sci. USSR **13**, 745 (1971).

232. G. Lewis, A. F. Johnson, J. Chem. Soc. (A), p. 3524 (1971).

233. C. Booth, C. J. Devoy, Polymer **12**, 309 (1971).

234. H. Vink, European Polymer J. **7**, 1411 (1971).

235. C. Booth, C. J. Devoy, Polymer **12**, 320 (1971).

236. H. Ochiai, K. Gekko, H. Yamamura, J. Polymer Sci., A-2, **9**, 1629 (1971).

237. A. Takizawa, T. Negishi, K. Ishikawa, Sen-i Gakkaishi 26, 567
      (1971).

238. A. Kagemoto, Y. Itoi, Y. Baba, R. Fujishiro, Makromol. Chem.
      150, 255 (1971).

239. A. A. Tager, A. A. Anikeyeva, L. V. Adamova, V. M. Andreyeva,
      T. A. Kuz'mina, M. V. Tsilipotkina, Vysokomolekul. Soedin. A13,
      659 (1971); Polymer Sci. USSR 13, 751 (1971).

240. A. A. Tager, Vysokomolekul. Soedin. A13, 467 (1971); Polymer Sci.
      USSR 13, 531 (1971).

241. K. Yamatsuta, K. Kubo, K. Ogino, Rep. Prog. Polymer Phys. (Japan)
      14, 29 (1971).

242. T. Okazawa, M. Kaneko, Rep. Prog. Polymer Phys. (Japan) 14, 7
      (1971); Polymer J. 2, 747 (1971).

243. S. Ichihara, A. Komatsu, T. Hata, Polymer J. 2, 644 (1971).

244. S. Ichihara, A. Komatsu, T. Hata, Polymer J. 2, 650 (1971).

245. S. Ichihara, A. Komatsu, T. Hata, Polymer J. 2, 640 (1971.

246. D. Patterson, Y. B. Tewari, H. P. Schreiber, J. E. Guillet, Macro-
      molecules 4, 376 (1971).

247. S. Morimoto, Bull. Chem. Soc. (Japan) 44, 879 (1971).

248. Th. G. Scholte, J. Polymer Sci., A-2, 10, 519 (1972).

249. C. Price, R. C. Williams, R. C. Ayerst, Amorphous Polymers (Wiley-
      Interscience, London, 1972), p. 117.

250. J. H. Rai, W. G. Miller, Macromolecules 5, 45 (1972).

251. S. H. Maron, F. E. Filisko, J. Macromol. Sci., B, 6, 57 (1972).

252. B. J. Reitveld, Th. G. Scholte, J. P. L. Pijpers, Brit. Polymer J. 4,
      109 (1972).

253. S. H. Maron, F. E. Filisko, J. Macromol. Sci., B, 6, 79 (1972).

254. S. H. Maron, F. E. Filisko, J. Macromol. Sci., B, 6, 413 (1972).

255. R. D. Newman, J. M. Prausnitz, J. Phys. Chem. 76, 1492 (1972).

256. L. Zeman, J. Biros, G. Delmas, D. Patterson, J. Phys. Chem. 76,
      1206,1214 (1972), and references therein.

257. B. A. Wolf, J. Polymer Sci., A-2, 10, 847 (1972).

HEATS OF SOLUTION OF SOME COMMON POLYMERS

D. R. Cooper
University of Manchester
Department of Chemistry
Manchester, England

The entropy change accompanying the dissolution of a liquid or rubbery polymer may often be approximated by the entropy of random mixing (1, 2, 3); but the corresponding heat change is less readily estimated. However, a convenient guide to its sign and magnitude is provided by the heat of solution of the polymer in a large excess of the solvent. Experimental heats of solution for a number of polymer/solvent systems are summarised in the following table. All the tabulated values were found by direct calorimetry, and are those appropriate to the formation of infinitely dilute solutions. The physical state of the polymer is specified in each case. Only for liquid or rubbery polymers (jointly denoted L) may the heat of solution be regarded purely as that of mixing the polymer and solvent segments. For an amorphous polymer below its glass transition temperature (G), the glass enthalpy may make a considerable exothermic contribution to the heat of solution. This therefore depends strongly on temperature, and to a lesser extent on the pretreatment of the glass (4, 5). On the other hand, the dissolution of a partially crystalline polymer (C) will tend to be endothermic, to a degree related to the latent heat of fusion of the polymer, and highly dependent on the particular morphology of the sample. The values quoted for such systems can be regarded as typical.

Polymer	M.W.	T[$^{\circ}$C]	State	Solvent	Heat Absorbed, [kJ/kg Polymer]	References
Poly(butadiene)	8×10^4	27	L	Benzene	+ ~ 13	6
Poly(isoprene)	4×10^3	16	L	Benzene	+ 12	7
Poly(ethylene)	3×10^5	80	C	1,2,3,4-Tetrahydronaphthalene	+ 630	8
		130	L		+ 39	
	7×10^5	80	C	1-Chloronaphthalene	+ 320	8
		130	L		+ 36	
Poly(propylene)	-	120	C	1,2,3,4-Tetrahydronaphthalene	+ 320	8
Poly(isobutene)	3×10^4	25	L	Benzene	+ 19	9
				Chlorobenzene	+ 12	
				Cyclohexane	- 0.7	
				Ethyl ether	+ 2.8	
				Pentane	- 3.6	
				Decane	- 0.5	
	2×10^6	-	L	Carbon tetrachloride	+ 4.1	10
Poly(methyl methacrylate)	10^5	25	G	Chloroform	- 84	11
	-	25	C	Chloroform	- 37	
Poly(vinyl acetate)	1.4×10^5	25	G (?)	Benzene	+ 2.3	12
				Chloroform	- 45	
				Methyl alcohol	+ 28	
	4×10^5	30	L (?)	2-Butanone	- 1.8	13
Poly(styrene)	7×10^5	26	G	Cyclohexane	- 8.2	14
				Toluene	- 22	
		26	G	Decahydronaphthalene	- 11	
		160	L	Decahydronaphthalene	+ 5.5	
	10^5	45	G	Chloroform	- 24	15
Poly(oxyethylene), hydroxy-ended	335	27	L	Benzene	+ 16	16
				Methyl alcohol	- 1.2	
	6×10^3	30	C	Chloroform	+ 52	17
				Water	+ 24	
	5×10^3	80	L	Water	- 120	18
--, methoxy-ended	350	6	L	Carbon tetrachloride	- 20	19
				Chloroform	- 180	
Poly(oxypropylene), hydroxy-ended	10^3	27	L	Methyl alcohol	- 7	20
	400	27	L	Water	- 170	21
--, methoxy-ended	2×10^3	6	L	Carbon tetrachloride	- 20	22
				Chloroform	- 100	
Poly(oxydimethylsilylene) (poly(dimethyl siloxane))	10^5	25	L	Benzene	+ 14	23
				Carbon tetrachloride	+ 2.4	
				Cyclohexane	+ 5.1	
				Decane	+ 3.9	24
				Pentane	- 0.9	
	8×10^4	25	L	Ethyl ether	- 1.3	25
	2×10^5	30	L	2-Butanone	+ 15	13

REFERENCES

1. C. Booth, G. Gee, M. N. Jones, W. D. Taylor, Polymer 5, 353 (1964).
2. D. Patterson, Rubber Chem. Technol. 40, 1 (1967).
3. D. K. Carpenter, Encycl. Polymer Sci. Technol. 12, 627 (1970).
4. S. Ichihara, A. Komatsu, T. Hata, Polymer J. 2, 644 (1971).
5. C. Price, R. C. Williams, R. C. Ayerst, "Amorphous Polymers" (Wiley-Interscience, London, 1972), p. 117
6. R. S. Jessup, J. Res. Natl. Bur. Stds. 62, 1 (1959).
7. G. Gee, W. J. C. Orr, Trans. Faraday Soc. 42, 507 (1946).
8. H. P. Schreiber, M. H. Waldman, J. Polymer Sci., A-2, 5, 555 (1967).
9. G. Delmas, D. Patterson, T. Somcynsky, J. Polymer Sci. 57, 79 (1962).
10. A. A. Tager, A. I. Podlesnyak, Polymer Sci. USSR 4, 698 (1963).
11. C. Gerth, F. H. Mueller, Kolloid-Z. 241, 1071 (1970).
12. H. Daoust, M. Rinfret, Can. J. Chem. 32, 492 (1954).
13. U. Bianchi, C. Cuniberti, E. Pedemonte, C. Rossi, J. Polymer Sci., A-2, 7, 855 (1969).
14. A. A. Tager, A. I. Podlesnyak, M. V. Tsilipotkina, L. V. Adamova, A. A. Bakhareva, L. V. Demidova, Polymer Sci. USSR 12, 1497 (1970).

15. S. Morimoto, Bull. Chem. Soc. Japan 44, 879 (1971).
16. M. L. Lakhanpal, H. L. Taneja, R. K. Sharma, Indian J. Chem. 4, 12 (1966).
17. S. H. Maron, F. E. Filisko, J. Macromol. Sci., B, 6, 79 (1972).
18. G. N. Malcolm, J. S. Rowlinson, Trans. Faraday Soc. 53, 921 (1957).
19. G. N. Malcolm, C. E. Baird, G. R. Bruce, K. G. Cheyne, R. W. Kershaw, M. C. Pratt, J. Polymer Sci., A-2, 7, 1495 (1969).
20. M. L. Lakhanpal, B. E. Conway, J. Polymer Sci. 46, 75 (1960).
21. R. G. Cunninghame, G. N. Malcolm, J. Phys. Chem. 65, 1454 (1961).
22. R. W. Kershaw, G. N. Malcolm, Trans. Faraday Soc. 64, 323 (1968).
23. G. Delmas, D. Patterson, S. N. Bhattacharyya, J. Phys. Chem. 68, 1468 (1964).
24. G. Delmas, D. Patterson, D. Bohme, Trans. Faraday Soc. 58, 2116 (1962).
25. D. Patterson, J. Polymer Sci., A, 2, 5177 (1964).

SOLUBILITY PARAMETER VALUES

H. Burrell
Inmont Corporation
Clifton, N. J.

Contents

A. INTRODUCTION

The process of dissolving a polymer in a solvent is governed by the familiar free energy equation

$$\Delta F = \Delta H - T \Delta S$$

where ΔF = the change in Gibb's free energy, ΔH = the heat of mixing, T = the absolute temperature and ΔS = the entropy of mixing. A negative ΔF predicts that a process will occur spontaneously.

Since the dissolution of a polymer is always connected with a large increase in entropy, the magnitude of the heat term ΔH is the deciding factor in determining the sign of the free energy change. Hildebrand and Scott (16a) proposed that

$$\Delta H_M = V_M \left[(\Delta E_1/V_1)^{1/2} - (\Delta E_2/V_2)^{1/2} \right]^2 \phi_1 \phi_2$$

where ΔH_M = overall heat of mixing; V_M = total volume of the mixture; ΔE = energy of vaporization of component 1 or 2; V = molar volume of component 1 or 2; ϕ = volume fraction of component 1 or 2 in the mixture.

The expression $(\Delta E/V)$ is the energy of vaporization per cubic centimeter. This has been variously described as the "internal pressure" or the "cohesive energy density". Bagley (2a) discusses the theory in detail.

If this equation is rearranged as follows

$$\Delta H_M/V_M \phi_1 \phi_2 = \left[(\Delta E_1/V_1)^{1/2} - (\Delta E_2/V_2)^{1/2} \right]^2$$

it may be seen that the heat of mixing per cubic centimeter at a given concentration is equal to the square of the difference between the square roots of the cohesive energy densities of the components. It is therefore convenient to assign to this latter quantity the symbol δ. Expressed mathematically

$$\delta = (\Delta E/V)^{1/2}$$

where ΔE is the energy of vaporization to a gas at zero pressure (i.e., at infinite separation of the molecules). The dimensions of δ are $(\text{cal/cm}^3)^{1/2} = (4.187 \, \text{J}/10^{-6} \, \text{m}^3)^{1/2} = 2.046 \cdot 10^3 \, (\text{J/m}^3)^{1/2}$.

It may thus be seen that the unit heat of mixing of two substances is dependent on $(\delta_1 - \delta_2)^2$. If the heat of mixing is not to be so large as to prevent mixing, then $(\delta_1 - \delta_2)^2$ has to be relatively small. In fact, if $(\delta_1 - \delta_2)^2 = 0$, solution is assured by the entropy factor. As the value approaches zero, $\delta_1 \rightarrow \delta_2$. This is mathematically equivalent to saying that if the δ values of two substances are nearly equal, the substances will be miscible. For this reason, Scott proposed the term δ as the solubility parameter.

No assumptions were made about polarity, solvation or association in deriving the solubility parameter. The solubility parameter governs only the heat of mixing of liquids or amorphous polymers. A non-crystalline polymer will, therefore, according to this theory, dissolve in a solvent of similar δ without the necessity of solvation, chemical similarity, association, or any specially directed intermolecular force. The high entropy change possible with polymer is sufficient reason for solution to occur.

The solubility parameter of a solvent is a readily calculable quantity. The solubility parameter of a polymer (or for that matter of any non-volatile substance) cannot be determined directly because most polymers cannot be vaporized without decomposing. The solubility parameter of a polymer is therefore defined as the same as that of a solvent in which the polymer will mix (a) in all proportions, (b) without heat change, (c) without volume change, and (d) without reaction or any special association (35a).

B.   CALCULATION OF SOLUBILITY PARAMETERS FROM PHYSICAL CONSTANTS

From Heat of Vaporization - It can be shown that

$$\Delta E = \Delta H - RT$$

where ΔH is the latent heat of vaporization at temperature T [K] and R is the gas constant (1.986), so that

$$\Delta E_{25^\circ C} = \Delta H_{25^\circ C} - 592$$

Where ΔH at $25^\circ C$ may be found in the literature, the value may be used to calculate ΔE. This is the most direct way to calculate δ and is also the most accurate. In fact, all other methods are approximations to a degree. However, it should be noted that the most accurate numerical values do not always correspond exactly to observed solubility behavior. Sometimes an empirical shift of a few tenths of a unit seems to be required to describe actual behavior.

For most solvents, direct measurements of the heat of vaporization at the desired temperature have not been made or cannot be found in the literature. In such cases one must resort to one of the various known methods for estimating ΔH as, for example, the Clausius-Clapeyron equation. Perhaps the most convenient of the estimations is Hildebrand's equation

$$\Delta H_{25^\circ C} = 23.7T_b + 0.020T_b^2 - 2950$$

where T_b is the boiling point in K. This is given graphically in "The Solubility of Nonelectrolytes" (16a). Figure 1 is an adaptation of Hildebrand's curve in which RT has already been subtracted from ΔH so that ΔE at $25^\circ C$ may be read directly from the boiling point in $^\circ C$ at 760 millimeters. In each of these cases, the data are reasonably accurate only for liquids which are not hydrogen bonded. They do not yield accurate data for esters, ketones, alcohols, etc.; however, a final correction may be applied to the δ calculated from ΔE read from Figure 1, so that the estimate of the solubility parameter is sufficiently close for practical applications (6). These corrections are as follows: for alcohols, add 1.4 to calculated δ; for esters add 0.6 to calculated δ; for ketones, add 0.5 to calculated δ if the boilint point is under $100^\circ C$. Otherwise nothing is added.

Figure 1, therefore, represents the most convenient source of data for calculating δ.

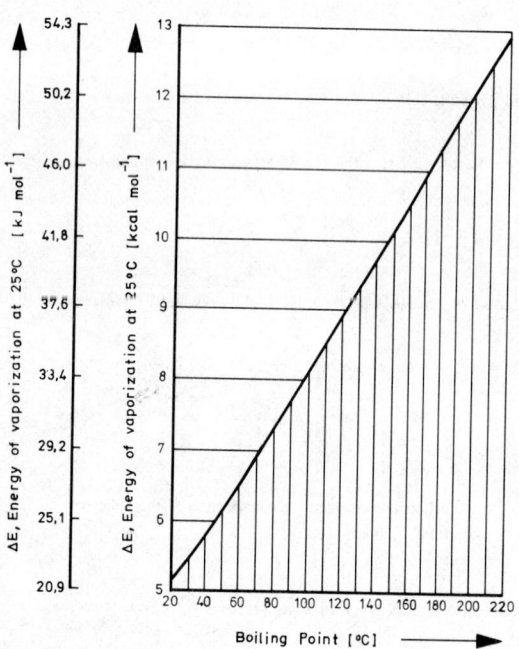

Figure 1 - The solubility parameter δ of a solvent can be calculated from its energy of vaporization. In this curve, ΔE at $25^\circ C$ may be read directly from the boiling point.

From Thermal Coefficients - The solubility parameter may be estimated at T [K] from the coefficient of thermal expansion α and the compressibility β by the equation

$$\delta \simeq (\alpha T/\beta)^{1/2}$$

Since these thermal coefficients are not readily available for most liquids, the method is mostly of theoretical interest.

This equation might provide a means for direct estimation of δ for polymers because α and β should be measurable whereas ΔH, of course, is not.

From Relationship of Pressure to Temperature - δ may also be calculated from $(TdP/dT)^{1/2}$, where dP/dT must be determined at constant volume. The data required here are the same as for the Clausius-Clapeyron equation, and the calculation is somewhat simpler.

From van der Waals' Gas Constant - Tables are available in most handbooks for the van der Waals correction constants to the Gas Law, a and b, where a must be in (liters)^2 (atmospheres). For some liquids these values may be at hand where other data are not available, or it might be desirable to check a δ value obtained from other sources. In such cases, the expression

$$\delta \simeq 1.2a^{1/2}/V$$

may be useful.

From Critical Pressure - The solubility parameter is related to the critical pressure P_c of a substance through the empirical equation

$$\delta \simeq 1.25 P_c^{1/2}$$

where the critical pressure is expressed in atmospheres. This equation gives values of poor accuracy. However, the calculation is quick and simple where a table of critical pressures is at hand.

From Surface Tension - Hildebrand (16a) suggests that

$$\delta = 4.1 (\gamma/V^{1/3})^{0.43}$$

where γ is the surface tension in [erg cm^{-2}]. The correlation with measured values leaves something to be desired and is discussed in detail by Lee (20a).

From Kauri-Butanol Values - Proprietary hydrocarbon solvents are usually mixtures with boiling ranges and unknown molecular weight. Solubility parameter values may be estimated from Kauri-Butanol values (ASTM Method D1133-54T) from the equation

$$\delta = 6.3 + 0.03 (KB)$$

From Structural Formula - Small (35) has published a table of "molar-attraction constants" which allows the estimation of solubility parameter merely from the structural formula of the compound and its density. Small's data are reproduced in Table B1. The molar-attraction constants, G, are additive over the formula

---

and are related to the solubility parameter by the equation

$$\delta = d\,\Sigma G/M$$

where ΣG is the sum for all the atoms and groupings in the molecule, d is the density and M the molecular weight. The table should not be used for alcohols, amines, carboxylic acids or other strongly hydrogen bonded compounds unless such functional groups constitute only a small part of the molecule so that the proportional error is not great. In general, the accuracy is quite good to the first decimal place, which is adequate for practical purposes.

The real value of Small's method lies in its application to polymers. The methods mentioned previously are suitable only for volatile solvents where some physical property, such as the boiling point, can be measured. In this case, however, if the G values are added for the structural configuration of the repeating unit in the polymer chain, the solubility parameter may be calculated.

Small's values in Table B1 will generate solubility parameters consistent with data derived from measurements of heat of evaporation. Data in Tables D1 and D2 are calculated from latent heats or approximated from actual boiling points. Therefore, data derived from Table B1 should agree with data in Tables D1 and D2.

Recently, Hoy (40) reexamined a large number of solvents and calculated their solubility parameters from vapor pressure measurements. He devised a new concept of molar attraction constants consistent with data derived from measurements of vapor pressure. His values differ considerably from Small's values and are reproduced in Table B2. Readers are advised to calculate solubility parameters according to Tables B1 and B2 and to compare the results. Solubility parameters derived from vapor pressure data may be found in Ref. 40.

B1. Group Molar Attraction Constants at 25°C (according to Small - derived from measurement of heat of evaporation)

Group		G	Group		G	Group		G
-CH$_3$		214	Ring,	5-membered	105-115	Br	single	340
-CH$_2$-	single-bonded	133	Ring,	6-membered	95-105	I	single	425
-CH<		28	Conjugation		20-30	CF$_2$	n-fluorocarbons only	150
>C<		-93	H	(variable	80-100	CF$_3$		274
CH$_2$=		190	O	ethers	70	S	sulfides	225
-CH=	double-bonded	111	CO	ketones	275	SH	thiols	315
>C=		19	COO	esters	310	ONO	nitrates	~ 440
CH=C-		285	CN		410	NO$_2$	(aliphatic nitro-compounds)	~ 440
-C=C-		222	Cl	(mean)	260	PO$_4$	(organic phosphates)	~ 500
Phenyl		735	Cl	single	270	Si	(in silicones)	-38
Phenylene	(o,m,p)	658	Cl	twinned as in >CCl$_2$	260			
Naphthyl		1146	Cl	triple as in -CCl$_3$	250			

B2. Group Molar Attraction Constants (according to Hoy - derived from measurement of vapor pressure)

Group		G	Group		G	Group	G
-CH$_3$		147.3	-OH→		225.84	Structure Feature	
-CH$_2$-		131.5	-H	acidic dimer	-50.47	conjugation	23.26
>CH-		85.99	OH	aromatic	170.99	cis	- 7.13
>C<		32.03	NH$_2$-		226.56	trans	-13.50
CH$_2$=		126.54	-NH-		180.03	4-membered ring	77.76
-CH=		121.53	-N-		61.08	5-membered ring	20.99
>C=		84.51	C≡N		354.56	6-membered ring	-23.44
-CH=	aromatic	117.12	NCO		358.66	o-substitution	9.69
-C=	aromatic	98.12	-S-		209.42	m-substitution	6.6
-O-	(ether, acetal)	114.98	Cl$_2$		342.67	p-substitution	40.33
-O-	(epoxide)	176.20	Cl	primary	205.06		22.56
-COO-		326.58	Cl	secondary	208.27		
>C=O		262.96	Cl	aromatic	161.0		62.5
-CHO		292.64	Br		257.88		
(CO)$_2$O		567.29	Br	aromatic	205.60	base value	135.1
			F		41.33		

Reprinted from "New Values of the Solubility Parameters from Vapor Pressure Data", J. Paint Technol. 42, 76 (1970).

C. EXPERIMENTAL DETERMINATIONS OF SOLUBILITY PARAMETERS FOR POLYMERS

As mentioned above, except for Small's method δ cannot be calculated from the physical constants of polymer because of their nonvolatility. A direct estimation is, therefore, the only recourse when the structural formula for the polymer is not accurately known.

<u>From Solubility</u> - A list of solvents of gradually increasing δ can be arranged such that a given polymer is soluble in all of the solvents grouped within a certain range in the list. Gee (13) indicates that it may not be exactly correct to assume that the solubility parameter of the polymer is actually the midpoint of the soluble range. However, for practical purposes this provides an adequate estimation. (see below).

<u>From Swelling Values</u> - Another way of determining δ for polymers is to prepare a sparsely cross-linked form (for example, by copolymerizing about 1 per cent divinyl benzene with styrene) and to immerse samples in a series of liquids of varying solubility parameter. Being cross-linked, the material will not dissolve but will swell to varying degrees. The amount of swelling will be greatest in the liquid which has the same δ as the polymer. By inference, the soluble, uncross-linked polymer also has the same value.

<u>Other Methods</u> - Scatchard (31a) suggested using refractive index measurements, and Small (35) described an osmotic pressure or vapor pressure calculation. Mangaraj (23) used intrinsic viscosity.

<u>Method for Determining Solubility Parameter Ranges of Commercial Polymers</u> - For practical commercial purposes, it has been found expedient to determine experimentally a solubility parameter range for each H-bonded class of solvents. The midpoints of these ranges may be taken as single-valued quantities for some purposes, but such midpoints will not necessarily agree with single-values obtained by other methods (13).

A gram or two of solid polymer is placed in a test tube and an approximate amount of a selected solvent is added such that the final solution would have about the correct solids content for the expected commercial use, e.g. 50 % for alkyds, 20 % for vinyls, etc. The exact amount is often unimportant except for poor solvents; it should be kept in mind that polymers are usually miscible in concentrated solutions although they may form two phases in dilute solution. The mixture may be warmed and stirred to speed up solution, but it should be cooled and observed at room temperature. The resulting mixture should be a single phase, clear and free from gel particles or cloudiness or else the polymer is judged insoluble. The solvents to be used are selected from the Solvent Spectra shown below:

Solvent	δ	Solvent	δ	Solvent	δ
Poorly Hydrogen Bonded		**Moderately Hydrogen Bonded**		**Strongly Hydrogen Bonded**	
n-Pentane	7.0	Diethyl ether	7.4	2-Ethyl hexanol	9.5
n-Heptane	7.4	Diisobutyl ketone	7.8	Methyl isobutyl carbinol	10.0
Methylcyclohexane	7.8	n-Butyl acetate	8.5	2-Ethylbutanol	10.5
Solvesso 150	8.5	Methyl propionate	8.9	n-Pentanol	10.9
Toluene	8.9	Dibutyl phthalate	9.3	n-Butanol	11.4
Tetrahydronaphthalene	9.5	Dioxane	9.9	n-Propanol	11.9
o-Dichlorobenzene	10.0	Dimethyl phthalate	10.7	Ethanol	12.7
1-Bromonaphthalene	10.6	2,3-Butylene carbonate	12.1	Methanol	14.5
Nitroethane	11.1	Propylene carbonate	13.3		
Acetonitrile	11.8	Ethylene carbonate	14.7		
Nitromethane	12.7				

Here a group of solvents has been especially selected so that the δ values increase by reasonably constant steps within each H-bonded class. The object of using these solvent spectra is to establish a solubility parameter range for a polymer rather than a single valued number. This has the advantage of automatically showing the allowable difference which can be tolerated between the absolute values of the polymer and solvent. In carrying out the procedure, it is convenient to select the first trials about 1/3 and 2/3 of the way down any one column; for example, in the poorly H-bonded group toluene and nitroethane would be chosen. If the polymer is soluble in both, there is no need to try intermediate solvents because experience (as well as theory) has shown that the polymer will be soluble in every case; instead, the solvents at the ends of the spectrum should be tried next. If the polymer was soluble in one but not both of the initial trials, the third trial should be about half-way between the two. By successive choices sets of two adjacent solvents will be found, one of which dissolves the polymer and one of which does not. The parameter values of the solvents which do dissolve the polymer mark the ends of the range. The procedure is then repeated for the other two H-bonded classes.

D.   TABLES OF SOLUBILITY PARAMETERS

This section consists of five tables, the first three listing values for solvents and the last two for polymers.

Tables 1 (an alphabetical listing) and 2 (arranged in order of increasing solubility parameter values) contain data on a wide variety of liquids. The values were calculated from heats of vaporization where they were available in the literature, or approximated from Figure 1. Corrections were applied for esters and alcohols as explained above. In some cases, therefore, the values are not exact and may vary a few tenths of $(cal/cm^3)^{1/2}$ from the "true" value. However, the tables have been prepared so as to be consistent and solubility prediction using the values is good. Values for commercial mixed hydrocarbon solvents were estimated from published Kauri-Butanol values.

For practical usage, it has been found necessary to consider a hydrogen bonding parameter as well as the cohesive energy density values (6). Since H-bond energies are difficult to determine and the accuracy is always questionable, it is convenient to arrange solvents qualitatively into three classes:

- Poorly H-bonded (hydrocarbons and their halo-, nitro- and cyano-substitution products);
- Moderately H-bonded (esters, ethers, ketones, glycol monoethers);
- Strongly H-bonded (alcohols, amines, acids, amides, aldehydes).

In Tables 1 and 2, each solvent is classified with the designation p, m or s referring to these categories.

Several schemes have been proposed to quantify the H-bonding parameter. That proposed by Hansen (15a) is scientifically sound and internally consistent. Table 3 is based on values proposed by Hansen which fit the equation

$$\delta_o^2 = \delta_d^2 + \delta_p^2 + \delta_h^2$$

where δ_o = Total solubility parameter; δ_d = component due to dispersion forces; δ_p = component due to polar forces; δ_h = component due to hydrogen bonding.

Table 4 lists solubility parameter ranges for commercial polymers and related materials. They were determined by the method previously discussed, using commercial grades of solvents. Ranges reported are consistent with solvent values given in Tables 1 and 2.

Table 5 lists literature values reported for a variety of polymers. These differ from Table 4 in that they are single values obtained by different methods and under varied conditions. The original literature references should be checked where possible before the selection and use of any particular value.

<u>Table 1 : Solubility Parameters of Solvents</u> (alphabetical list)

Solvent	δ $[J/m^3]^{1/2} \cdot 10^{-3}$	δ $[cal/cm^3]^{1/2}$	H-Bonding Group	Solvent	δ $[J/m^3]^{1/2} \cdot 10^{-3}$	δ $[cal/cm^3]^{1/2}$	H-Bonding Group
Acetaldehyde	21.1	10.3	m	Bromostyrene (ortho)	20.1	9.8	p
Acetic acid	20.7	10.1	s	Butadiene-1,3	14.5	7.1	p
Acetic anhydride	21.1	10.3	s	Butane (normal)	13.9	6.8	p
Acetone	20.3	9.9	m	Butanediol-1,3	23.7	11.6	s
Acetonitrile	24.3	11.9	p	Butanediol-1,4	24.8	12.1	s
Acetophenone	21.7	10.6	m	Butanediol-2,3	22.7	11.1	s
Acetyl chloride	19.4	9.5	m	Butyl acetate (iso)	17.0	8.3	m
Acetylmorpholine (N)	23.7	11.6	m	Butyl acetate (normal)	17.4	8.5	m
Acetylpiperidine (N)	22.9	11.2	s	Butyl acetate (secondary)	16.8	8.2	m
Acetylpyrrolidine (N)	23.3	11.4	s	Butyl acrylate (iso)	17.4	8.5	m
Acrolein	20.1	9.8	s	Butyl acrylate (normal)	18.0	8.8	m
Acrylic acid	24.6	12.0	s	Butyl alcohol (iso)	21.5	10.5	s
Acrylonitrile	21.5	10.5	p	Butyl alcohol (normal)	23.3	11.4	s
Allyl acetate	18.8	9.2	m	Butyl alcohol (secondary)	22.1	10.8	s
Allyl alcohol	24.1	11.8	s	Butyl alcohol (tert.)	21.7	10.6	s
Allyl chloride	18.0	8.8	m	Butylamine (mono, normal)	17.8	8.7	s
Ammonia	33.4	16.3	s	Butyl bromide (normal)	17.8	8.7	m
Amyl acetate (iso)	16.0	7.8	m	Butyl bromide (secondary)	17.2	8.4	m
Amyl acetate (normal)	17.4	8.5	m	Butyl (iso) butyrate (normal)	16.0	7.8	m
Amyl acetate (secondary)	17.0	8.3	m	Butyl (normal) butyrate (normal)	16.6	8.1	m
Amyl alcohol	20.5	10.0	s	Butyl chloride (iso)	16.6	8.1	m
Amyl alcohol (normal)	22.3	10.9	s	Butylene-2,3 carbonate	24.8	12.1	m
Amylamine (normal)	17.8	8.7	s	Butylene (iso)	13.7	6.7	p
Amyl bromide (normal)	15.6	7.6	m	Butyl ether	16.0	7.8	m
Amyl chloride	17.0	8.3	m	Butyl formate (iso)	16.8	8.2	m
Amylene	14.1	6.9	p	Butyl formate (normal)	18.2	8.9	m
Amyl ether (normal, di-)	14.9	7.3	m	Butyl iodide (normal)	17.6	8.6	m
Amyl formate (iso)	16.4	8.0	m	Butyl lactate (normal)	19.2	9.4	m
Amyl formate (normal)	17.4	8.5	m	Butyl methacrylate	16.8	8.2	m
Amyl iodide (normal)	17.2	8.4	m	Butyl stearate	15.3	7.5	m
Anethole (para)	17.2	8.4	m	Butyl propionate	18.0	8.8	m
Aniline	21.1	10.3	s	Butyraldehyde	18.4	9.0	m
Anthracene	20.3	9.9	p	Butyric acid (iso)	21.1	10.3	s
Apco #18 solvent	15.3	7.5	p	Butyric acid (normal)	21.5	10.5	s
Apco #140 solvent	14.9	7.3	p	Butyrolactone	25.8	12.6	m
Apco thinner	16.0	7.8	p	Butyronitrile (iso)	20.1	9.8	p
Aroclor 1248	18.0	8.8	p	Butyronitrile (normal)	21.5	10.5	p
Benzaldehyde	19.2	9.4	m	Caprolactam (epsilon)	26.0	12.7	m
Benzene	18.8	9.2	p	Caprolactone	20.7	10.1	m
Benzonitrile	17.2	8.4	p	Capronitrile	19.2	9.4	p
Benzyl alcohol	24.8	12.1	s	Carbon disulfide	20.5	10.0	p
Bicyclohexyl	17.4	8.5	p	Carbon tetrachloride	17.6	8.6	p
Bromobenzene	20.3	9.9	p	Celanese solvent 601	18.8	9.2	m
Bromonaphthalene	21.7	10.6	p				

SOLUBILITY PARAMETER VALUES

Solvent	δ $[J/m^3]^{1/2} \cdot 10^{-3}$	δ $[cal/cm^3]^{1/2}$	H-Bonding Group	Solvent	δ $[J/m^3]^{1/2} \cdot 10^{-3}$	δ $[cal/cm^3]^{1/2}$	H-Bonding Group
Chloroacetonitrile	25.8	12.6	p	Difluoro-dichloromethane			
Chlorobenzene	19.4	9.5	p	(Freon 12)	11.3	5.5	p
Chloroethyl acetate (beta)	19.8	9.7	m	Difluoro-tetrachloroethane			
Chloroform	19.0	9.3	p	(Freon 112)	16.0	7.8	p
Chlorostyrene (ortho or				Diformylpiperazine (N, N)	31.5	15.4	m
para)	19.4	9.5	p	Dihexyl ether	16.4	8.0	m
Chlorotoluene (para)	18.0	8.8	p	Di-n-hexyl phthalate	18.2	8.9	m
Cresol (meta)	20.9	10.2	s	Diisobutylene	15.8	7.7	p
Cyclobutanedione	22.5	11.0		Diisobutyl ketone	16.0	7.8	m
Cyclohexane	16.8	8.2	p	Diisodecyl phthalate	14.7	7.2	m
Cyclohexanol	23.3	11.4	s	Diisopropyl ether	14.1	6.9	m
Cyclohexanone	20.3	9.9	m	Diisopropyl ketone	16.4	8.0	m
Cyclopentane	17.8	8.7	p	Dimethylacetamide (N, N)	22.1	10.8	m
Cyclopentanone	21.3	10.4	m	Dimethylaniline	19.8	9.7	m
Cymene (para)	16.8	8.2	p	Dimethyl-2, 2-butanediol-1, 2			
				(isobutylene glycol)	22.9	11.2	s
Decahydronaphthalene	18.0	8.8	p	Dimethyl-2, 2-butanediol-1, 3	20.5	10.0	s
Decane (normal)	13.5	6.6	p	Dimethyl carbonate	20.3	9.9	m
Decyl acrylate (iso)	16.8	8.2	m	Dimethyl ether	18.0	8.8	m
Diacetone alcohol	18.8	9.2	m	Dimethylformamide (N, N)	24.8	12.1	m
Diacetone alcohol methyl				Dimethyl malonate	22.5	11.0	m
ether (Pentoxone)	16.8	8.2	m	Dimethylnitroamine (N, N)	26.8	13.1	s
Diacetylpiperazine (N, N)	28.0	13.7	m	Dimethyl oxalate	22.5	11.0	m
Diamyl phthalate	18.6	9.1	m	Dimethyl phosphite	25.6	12.5	m
Dibenzyl ether	19.2	9.4	m	Dimethyl phthalate	21.9	10.7	m
Dibromoethane-1, 2	21.3	10.4	p	Dimethyl siloxanes	10.0-12.1	4.9-5.9	p
Dibromoethylene-1, 2	20.7	10.1	p	Dimethyl sulfide	19.2	9.4	p
Dibutoxyethyl phthalate				Dimethyl sulfone	29.7	14.5	m
(Kronisol)	16.4	8.0	m	Dimethyl sulfoxide	24.6	12.0	m
Dibutylamine	16.6	8.1	s	Dimethyltetramethylene			
Dibutyl furmarate	18.4	9.0	m	sulfone	24.8	12.1	m
Dibutyl maleate	18.4	9.0	m	Dioctyl adipate	17.8	8.7	m
Dibutyl phenyl phosphate	17.8	8.7	m	Dioctyl phthalate	16.2	7.9	m
Dibutyl phthalate	19.0	9.3	m	Dioctyl sebacate	17.6	8.6	m
Dibutyl sebacate	18.8	9.2	m	Dioxane 1, 4	20.5	10.0	m
Dichloroacetic acid	22.5	11.0	s	Dioxolane-1, 3	20.9	10.2	m
Dichlorobenzene (ortho)	20.5	10.0	p	Dipentene	17.4	8.5	p
Dichloroethyl ether	20.1	9.8	m	Diphenyl ether	20.7	10.1	m
Dichloroethylene, cis-1, 2	18.6	9.1	p	Diphenyl 2-ethylhexyl			
Dichloroethylene, trans-1, 2	18.4	9.0	p	phosphate	17.6	8.6	m
Dichloropropane-1, 2	18.4	9.0	p	Dipropylene glycol	20.5	10.0	s
Dichloropropane-2, 2	16.8	8.2	p	Dipropylene glycol mono-			
Diethylacetamide (N, N)	20.3	9.9	m	methyl ether	19.0	9.3	m
Diethylamine	16.4	8.0	s	Dipropyl phthalate	19.8	9.7	m
Diethyl carbonate	18.0	8.8	m	Dipropyl sulfone	23.1	11.3	m
Diethylene glycol	24.8	12.1	s	Dodecane	16.2	7.9	p
Diethylene glycol mono-				Dodecanol-1	20.1	9.8	s
butyl ether (normal)	19.4	9.5	m				
Diethylene glycol mono-				Epichlorohydrin	22.5	11.0	s
ethyl ether	20.9	10.2	m	Ethane	12.3	6.0	p
Diethylene glycol mono-				Ethylacetamide (N)	25.2	12.3	s
ethyl ether acetate	17.4	8.5	m	Ethyl acetate	18.6	9.1	m
Diethylene glycol mono-				Ethyl acrylate	17.6	8.6	m
laurate	17.8	8.7	m	Ethyl alcohol	26.0	12.7	s
Diethyl ether	15.1	7.4	m	Ethylamine	20.5	10.0	s
Diethylformamide (N, N)	21.7	10.6	m	Ethyl amyl ketone	16.8	8.2	m
Diethyl ketone	18.0	8.8	m	Ethylbenzene	18.0	8.8	p
Diethyl maleate	20.3	9.9	m	Ethyl benzoate	16.8	8.2	m
Diethyl oxalate	17.6	8.6	m	Ethyl bromide	19.6	9.6	m
Diethyl phthalate	20.5	10.0	m	Ethyl-2-butanol-1	21.5	10.5	s
Diethyl-2, 2-propanediol-1, 2				Ethyl n-butyrate	17.4	8.5	m
(heptylene glycol)	20.3	9.9	s	Ethyl caprylate	14.9	7.3	m
Diethyl sulfone	25.4	12.4	m	Ethyl chloride	18.8	9.2	m

Solvent	δ $[J/m^3]^{1/2} \cdot 10^{-3}$	δ $[cal/cm^3]^{1/2}$	H-Bonding Group	Solvent	δ $[J/m^3]^{1/2} \cdot 10^{-3}$	δ $[cal/cm^3]^{1/2}$	H-Bonding Group
Ethyl cyanoacetate	22.5	11.0	m	Hydrogen cyanide	24.8	12.1	s
Ethylene bromide	19.8	9.7	p	Iodobenzene	20.7	10.1	p
Ethylene carbonate	30.1	14.7	m	Isophorone	18.6	9.1	m
Ethylene chlorohydrin	25.0	12.2	s	Isoprene	15.1	7.4	p
Ethylene cyanohydrin	31.1	15.2	s				
Ethylenediamine	25.2	12.3	s	Lauryl alcohol	16.6	8.1	s
Ethylene dichloride	20.1	9.8	p	Low Odor Mineral Spirits	14.1	6.9	p
Ethylene glycol	29.9	14.6	s				
Ethylene glycol diacetate	20.5	10.0	m	Maleic anhydride	27.8	13.6	s
Ethylene glycol diethyl ether	17.0	8.3	m	Malononitrile	30.9	15.1	p
Ethylene glycol dimethyl ether	17.6	8.6	m	Mesitylene	18.0	8.8	p
Ethylene glycol methyl ether acetate	18.8	9.2	m	Mesityl oxide	18.4	9.0	m
Ethylene glycol monobenzyl ether	22.3	10.9	m	Methacrylic acid	22.9	11.2	s
				Methane	11.0	5.4	p
Ethylene glycol monobutyl ether	19.4	9.5	m	Methanol	29.7	14.5	s
Ethylene glycol monoethyl ether	21.5	10.5	m	Methoxy (4)-methyl (4)-pentanol-2	17.4	8.5	s
Ethylene glycol monoethyl ether acetate	17.8	8.7	m	Methoxy (4)-methyl (4)-pentanone-2	17.0	8.3	m
Ethylene glycol monomethyl ether	23.3	11.4	m	Methylacetamide	29.9	14.6	s
				Methyl acetate	19.6	9.6	m
Ethylene glycol monophenyl ether	23.5	11.5	m	Methyl acrylate	18.2	8.9	m
Ethylene oxide	22.7	11.1	m	Methylamine	22.9	11.2	s
Ethylformamide (N)	28.4	13.9	s	Methyl amyl acetate	16.4	8.0	m
Ethyl formate	19.2	9.4	m	Methyl amyl ketone	17.4	8.5	m
Ethyl-2-hexanediol-1,3 (octylene glycol)	19.2	9.4	s	Methyl benzoate	21.5	10.5	m
Ethylhexanol	19.4	9.5	s	Methyl bromide	19.6	9.6	m
Ethyl hexyl acrylate	16.0	7.8	m	Methyl n-butyl ketone	17.0	8.3	m
Ethylidene chloride	18.2	8.9	p	Methyl n-butyrate	18.2	8.9	m
Ethyl iodide	19.2	9.4	m	Methyl caprolactone	18.2	8.9	m
Ethyl isobutyl ether	15.3	7.5	m	Methyl chloride	19.8	9.7	m
Ethyl isobutyrate	16.2	7.9	m	Methylcyclohexane	16.0	7.8	p
Ethyl lactate	20.5	10.0	m	Methylcyclohexanone	19.0	9.3	m
Ethyl mercaptan	18.8	9.2	p	Methylene chloride	19.8	9.7	p
Ethyl methacrylate	17.0	8.3	m	Methylene glycolate	25.4	12.4	m
Ethyl morpholine (N)	18.2	8.9	s	Methylene iodide	24.1	11.8	p
Ethyl orthoformate	17.0	8.3	m	Methyl ethyl ketone	19.0	9.3	m
Ethyl propionate	17.2	8.4	m	Methyl ethyl sulfone	27.4	13.4	m
Fluorocarbons, aliphatic	11.3-12.7	5.5-6.2	p	Methylformamide (N)	32.9	16.1	s
Fluorocarbons, aromatic	15.3-16.8	7.5-8.2	p	Methyl formate	20.9	10.2	m
Formamide	39.3	19.2	s	Methyl n-hexyl ketone	17.0	8.3	m
Formic acid	24.8	12.1	s	Methyl iodide	20.9	10.2	m
Formylmorpholine (N)	26.6	13.0	m	Methyl isoamyl ketone	17.2	8.4	m
Formylpiperidine (N)	23.5	11.5	m	Methyl isobutyl carbinol	20.5	10.0	s
Furane	19.2	9.4	m	Methyl isobutyl ketone	17.2	8.4	m
Furfural	22.9	11.2	m	Methyl isobutyrate	17.0	8.3	m
Furfuryl alcohol	25.6	12.5	s	Methyl isopropyl ketone	17.4	8.5	m
				Methyl isovalerate	16.2	7.9	m
Glycerol	33.8	16.5	s	Methyl methacrylate	18.0	8.8	m
				Methyl nonyl ketone	16.0	7.8	m
Heptane (normal)	15.1	7.4	p	Methyl-2-pentanediol-1,3	21.1	10.3	s
Heptyl alcohol (normal)	21.7	10.6	s	Methyl-2-pentanediol-2,4	19.8	9.7	s
Hexamethylphosphoramide	21.5	10.5	s	Methyl-2-pentanediol monoethyl ether (Pentoxol)	17.4	8.5	m
Hexane (normal)	14.9	7.3	p	Methyl propionate	18.2	8.9	m
Hexanediol-2,5	21.1	10.3	s	Methyl propyl ketone	17.8	8.7	m
Hexene-1	15.1	7.4	p	Methyl propyl sulfone	25.6	12.5	m
Hexyl alcohol (normal)	21.9	10.7	s	Methyl pyrrolidone-2 (N)	23.1	11.3	m
Hydrazine	37.3	18.1	s	Methyl salicylate	21.7	10.6	m
Hydrogen	6.9	3.0	p	Methylstyrene (alpha)	17.4	8.5	p
Hydrogenated terphenyl (Monsanto's HB-40)	18.4	9.0	p	Methyltetramethylene sulfone	26.4	12.9	m
				Methyl n-valerate	16.2	7.9	m
				Monofluoro-dichloromethane (Freon 21)	17.0	8.3	p

Solvent	δ $[J/m^3]^{1/2} \cdot 10^{-3}$	δ $[cal/cm^3]^{1/2}$	H-Bonding Group	Solvent	δ $[J/m^3]^{1/2} \cdot 10^{-3}$	δ $[cal/cm^3]^{1/2}$	H-Bonding Group
Monofluoro-trichloromethane				Pyrone (gamma)	27.4	13.4	m
(Freon 11)	15.6	7.6	p	Pyrrolidone (alpha)	30.1	14.7	s
Morpholine	22.1	10.8	s				
				Quinoline	22.1	10.8	s
Naphthalene	20.3	9.9	p				
Neopentane	12.9	6.3	p	Santicizer 8	24.3	11.9	m
Neopentyl glycol	22.5	11.0	s	Shell Sol 72	14.7	7.2	p
Nitrobenzene	20.5	10.0	p	Shell TS28 Solvent	15.1	7.4	p
Nitroethane	22.7	11.1	p	Silicon tetrachloride	15.1	7.4	p
Nitromethane	26.0	12.7	p	Socal Solvent No. 1	16.6	8.1	p
Nitro-n-octane	14.3	7.0	p	Socal Solvent No. 2	16.2	7.9	p
Nitro-1-propane	21.1	10.3	p	Socal Solvent No. 3	15.8	7.7	p
Nitro-2-propane	20.3	9.9	p	Solvesso 100	17.6	8.6	p
Nonyl phenol	19.2	9.4	s	Solvesso 150	17.4	8.5	p
				Styrene	19.0	9.3	p
Octane (normal)	15.6	7.6	p	Styrene oxide	21.5	10.5	m
Octyl alcohol (normal)	21.1	10.3	s	Succinic anhydride	31.5	15.4	s
				Terpene B	17.2	8.4	p
Pentachloroethane	19.2	9.4	p	Tetrachloroethane-1, 1, 2, 2	19.8	9.7	p
Pentane (normal)	14.3	7.0	p	Tetrachloroethylene (perchloro-			
Pentanediol-1, 5	23.5	11.5	s	ethylene)	19.0	9.3	p
Pentanediol-2, 4	22.1	10.8	s	Tetraethylene glycol	20.3	9.9	s
Perchloroethylene	19.0	9.3	p	Tetrahydrofuran	18.6	9.1	m
Perfluoroheptane	11.9	5.8	p	Tetrahydronaphthalene	19.4	9.5	p
Perfluoromethylcyclohexane	12.3	6.0	p	Tetramethylene sulfone	27.4	13.4	m
Phenanthrene	20.1	9.8	p	Tetramethyloxamide	23.3	11.4	m
Phenylhydrazine	25.6	12.5	s	Thiophene	20.1	9.8	m
Pine Oil	17.6	8.6	p	Toluene	18.2	8.9	p
Piperidine	17.8	8.7	s	Tolylenediisocyanate	23.7	11.6	s
Piperidone	27.8	13.6	s	Tributylamine	15.8	7.7	s
Propane	13.1	6.4	p	Trichloroethane-1, 1, 2	19.6	9.6	p
Propiolactone	27.2	13.3	m	Trichloroethylene	18.8	9.2	p
Propionic acid	20.3	9.9	s	Tricresyl phosphate	17.2	8.4	m
Propionic anhydride	20.5	10.0	s	Triethylamine	15.1	7.4	s
Propionitrile	22.1	10.8	p	Triethylene glycol	21.9	10.7	s
Propyl acetate (iso)	17.2	8.4	m	Triethylenetetramine	22.7	11.1	s
Propyl acetate (normal)	18.0	8.8	m	Trifluoro-trichloroethane			
Propyl alcohol	23.5	11.5	s	(Freon 113)	14.9	7.3	p
Propyl alcohol (normal)	24.3	11.9	s	Trimethyl-3, 5, 5-hexanol			
Propylbenzene (normal)	17.6	8.6	p	(nonyl alcohol)	17.2	8.4	s
Propyl bromide	18.2	8.9	m	Triphenyl phosphate	17.6	8.6	m
Propyl butyrate	17.2	8.4	m	Triphenyl phosphite	19.0	9.3	m
Propyl butyrate (iso, iso)	16.2	7.9	m	Tripropylene glycol	18.8	9.2	s
Propyl chloride (iso)	16.6	8.1	m	Tripropylene glycol methyl			
Propyl chloride (normal)	17.4	8.5	m	ether	17.8	8.7	m
Propylene-1, 2 carbonate	27.2	13.3	m	Turpentine	16.6	8.1	p
Propylene glycol	25.8	12.6	s	Valeric acid (normal)	20.1	9.8	s
Propylene glycol methyl ether	20.7	10.1	m	Valeronitrile (normal)	19.6	9.6	p
Propylene oxide	18.8	9.2	m	Varnolene (Varsol #2)	15.6	7.6	p
Propyl ether (di-, normal)	16.0	7.8	m	Vinyl acetate	18.4	9.0	m
Propyl ether (iso)	14.5	7.1	m	Vinyl chloride	16.0	7.8	m
Propyl formate	18.8	9.2	m	Vinyl toluene	18.6	9.1	p
Propyl propionate	17.4	8.5	m	V M & P Naphtha	15.6	7.6	p
Pyridine	21.9	10.7	s	Water	47.9	23.4	s
				Xylene	18.0	8.8	p

Table 2: Solubility Parameters of Solvents (numerical list)

Solvent	δ $[J/m^3]^{1/2} \cdot 10^{-3}$	δ $[cal/cm^3]^{1/2}$	H-Bonding Group	Solvent	δ $[J/m^3]^{1/2} \cdot 10^{-3}$	δ $[cal/cm^3]^{1/2}$	H-Bonding Group
Hydrogen	6.1	3.0	p	Perfluoroheptane	11.9	5.8	p
Dimethyl siloxanes	10.0-12.1	4.9-5.9	p	Ethane	12.3	6.0	p
Methane	11.0	5.4	p	Perfluoromethylcyclohexane	12.3	6.0	p
Difluoro-dichloromethane				Neopentane	12.9	6.3	p
(Freon 12)	11.3	5.5	p	Propane	13.1	6.4	p
Fluorocarbons, aliphatic	11.3-12.7	5.5-6.2	p	Decane (normal)	13.5	6.6	p

Solvent	δ $[J/m^3]^{1/2} \cdot 10^{-3}$	δ $[cal/cm^3]^{1/2}$	H-Bonding Group	Solvent	δ $[J/m^3]^{1/2} \cdot 10^{-3}$	δ $[cal/cm^3]^{1/2}$	H-Bonding Group
Butylene (iso)	13.7	6.7	p	Butyl (normal) butyrate			
Butane (normal)	13.9	6.8	p	(normal)	16.6	8.1	m
Amylene	14.1	6.9	p	Butyl chloride (iso)	16.6	8.1	m
Diisopropyl ether	14.1	6.9	m	Dibutylamine	16.6	8.1	s
Low Odor Mineral Spirits	14.1	6.9	p	Lauryl alcohol	16.6	8.1	s
Nitro-n-octane	14.3	7.0	p	Propyl chloride (iso)	16.6	8.1	m
Pentane (normal)	14.3	7.0	p	Socal Solvent No. 1	16.6	8.1	p
Butadiene-1,3	14.5	7.1	p	Turpentine	16.6	8.1	p
Propyl ether (iso)	14.5	7.1	m	Butyl acetate (secondary)	16.8	8.2	m
Diisodecyl phthalate	14.7	7.2	m	Butyl formate (iso)	16.8	8.2	m
Shell Sol 72	14.7	7.2	p	Cyclohexane	16.8	8.2	p
Amyl ether (normal, di-)	14.9	7.3	m	Cymene (para)	16.8	8.2	p
Apco ≠140 Solvent	14.9	7.3	p	Decyl acrylate (iso)	16.8	8.2	m
Ethyl caprylate	14.9	7.3	m	Diacetone alcohol methyl			
Hexane (normal)	14.9	7.3	p	ether (Pentoxone)	16.8	8.2	m
Trifluoro-trichloroethane				Dichloropropane-2,2	16.8	8.2	p
(Freon 113)	14.9	7.3	p	Ethyl amyl ketone	16.8	8.2	m
Diethyl ether	15.1	7.4	m	Ethyl benzoate	16.8	8.2	m
Heptane (normal)	15.1	7.4	p	Butyl methacrylate	16.8	8.2	m
Hexene-1	15.1	7.4	p	Amyl acetate (secondary)	17.0	8.3	m
Isoprene	15.1	7.4	p	Amyl chloride (normal)	17.0	8.3	m
Shell TS28 Solvent	15.1	7.4	p	Butyl acetate (iso)	17.0	8.3	m
Silicon tetrachloride	15.1	7.4	p	Ethylene glycol diethyl ether	17.0	8.3	m
Triethylamine	15.1	7.4	s	Ethyl methacrylate	17.0	8.3	m
Apco ≠18 Solvent	15.3	7.5	p	Ethyl orthoformate	17.0	8.3	m
Butyl stearate	15.3	7.5	m	Methoxy (4)-methyl			
Ethyl isobutyl ether	15.3	7.5	m	(4)-pentanone-2	17.0	8.3	m
Fluorocarbons, aromatic	15.3-16.8	7.5-8.2	p	Methyl n-butyl ketone	17.0	8.3	m
Amyl bromide (normal)	15.6	7.6	m	Methyl n-hexyl ketone	17.0	8.3	m
Monofluoro-trichloromethane				Methyl isobutyrate	17.0	8.3	m
(Freon 11)	15.6	7.6	p	Monofluoro-dichloromethane			
Octane (normal)	15.6	7.6	p	(Freon 21)	17.0	8.3	p
V M & P Naphtha	15.6	7.6	p	Amyl iodide (normal)	17.2	8.4	m
Varnolene (Varsol ≠2)	15.6	7.6	p	Anethole (para)	17.2	8.4	m
Diisobutylene	15.8	7.7	p	Benzonitrile	17.2	8.4	p
Socal Solvent No. 3	15.8	7.7	p	Butyl bromide (secondary)	17.2	8.4	m
Tributylamine	15.8	7.7	s	Ethyl propionate	17.2	8.4	m
Amyl acetate (iso)	16.0	7.8	m	Methyl isoamyl ketone	17.2	8.4	m
Apco thinner	16.0	7.8	p	Methyl isobutyl ketone	17.2	8.4	m
Butyl (iso) butyrate (normal)	16.0	7.8	m	Propyl acetate (iso)	17.2	8.4	m
Butyl ether	16.0	7.8	m	Propyl butyrate	17.2	8.4	m
Difluoro-tetrachloroethane				Terpene B	17.2	8.4	p
(Freon 112)	16.0	7.8	p	Tricresyl phosphate	17.2	8.4	m
Diisobutyl ketone	16.0	7.8	m	Trimethyl-3,5,5-hexanol			
Ethyl hexyl acrylate	16.0	7.8	m	(nonyl alcohol)	17.2	8.4	s
Methylcyclohexane	16.0	7.8	p	Amyl acetate (normal)	17.4	8.5	m
Methyl nonyl ketone	16.0	7.8	m	Amyl formate (normal)	17.4	8.5	m
Propyl ether (di-, normal)	16.0	7.8	m	Bicyclohexyl	17.4	8.5	p
Vinyl chloride	16.0	7.8	m	Butyl acetate (normal)	17.4	8.5	m
Dioctyl phthalate	16.2	7.9	m	Butyl acrylate (iso)	17.4	8.5	m
Dodecane	16.2	7.9	p	Diethylene glycol monoethyl			
Ethyl isobutyrate	16.2	7.9	m	ether acetate	17.4	8.5	m
Methyl isovalerate	16.2	7.9	m	Dipentene	17.4	8.5	p
Methyl n-valerate	16.2	7.9	m	Ethyl n-butyrate	17.4	8.5	m
Propyl butyrate (iso, iso)	16.2	7.9	m	Methoxy (4)-methyl			
Socal Solvent No. 2	16.2	7.9	p	(4)-pentanol-2	17.4	8.5	s
Amyl formate (iso)	16.4	8.0	m	Methyl amyl ketone	17.4	8.5	m
Dibutoxyethyl phthalate				Methyl isopropyl ketone	17.4	8.5	m
(Kronisol)	16.4	8.0	m	Methyl-2-pentanediol mono-			
Diethylamine	16.4	8.0	s	ethyl ether (Pentoxol)	17.4	8.5	s
Dihexyl ether	16.4	8.0	m	Methylstyrene (alpha)	17.4	8.5	p
Diisopropyl ketone	16.4	8.0	m	Propyl chloride (normal)	17.4	8.5	m
Methyl amyl acetate	16.4	8.0	m	Propyl propionate	17.4	8.5	m

SOLUBILITY PARAMETER VALUES

Solvent	δ $[J/m^3]^{1/2} \cdot 10^{-3}$	δ $[cal/cm^3]^{1/2}$	H-Bonding Group
Solvesso 150	17.4	8.5	p
Butyl iodide (normal)	17.6	8.5	m
Carbon tetrachloride	17.6	8.6	p
Diethyl oxalate	17.6	8.6	m
Dioctyl sebacate	17.6	8.6	m
Diphenyl 2-ethylhexyl phosphate	17.6	8.6	m
Ethyl acrylate	17.6	8.6	m
Ethylene glycol dimethyl ether	17.6	8.6	m
Pine oil	17.6	8.6	p
Propylbenzene (normal)	17.6	8.6	p
Solvesso 100	17.6	8.6	p
Triphenyl phosphate	17.6	8.6	m
Amylamine (normal)	17.8	8.7	s
Butylamine (mono, normal)	17.8	8.7	s
Butyl bromide (normal)	17.8	8.7	m
Cyclopentane	17.8	8.7	p
Dibutyl phenyl phosphate	17.8	8.7	m
Diethylene glycol monolaurate	17.8	8.7	m
Dioctyl adipate	17.8	8.7	m
Ethylene glycol monoethyl ether acetate	17.8	8.7	m
Methyl propyl ketone	17.8	8.7	m
Piperidine	17.8	8.7	s
Tripropylene glycol methyl ether	17.8	8.7	m
Allyl chloride	18.0	8.8	m
Aroclor 1248	18.0	8.8	p
Butyl acrylate (normal)	18.0	8.8	m
Butyl propionate	18.0	8.8	m
Chlorotoluene (para)	18.0	8.8	p
Decahydronaphthalene	18.0	8.8	p
Diethyl carbonate	18.0	8.8	m
Diethyl ketone	18.0	8.8	m
Dimethyl ether	18.0	8.8	m
Ethylbenzene	18.0	8.8	p
Mesitylene	18.0	8.8	p
Methyl methacrylate	18.0	8.8	m
Propyl acetate (normal)	18.0	8.8	m
Xylene	18.0	8.8	p
Butyl formate (normal)	18.2	8.9	m
Di-n-hexyl phthalate	18.2	8.9	m
Ethylidene chloride	18.2	8.9	p
Ethylmorpholine (N)	18.2	8.9	s
Methyl acrylate	18.2	8.9	m
Methyl n-butyrate	18.2	8.9	m
Methyl propionate	18.2	8.9	m
Propyl bromide	18.2	8.9	m
Toluene	18.2	8.9	p
Butyraldehyde	18.4	9.0	m
Dibutyl fumarate	18.4	9.0	m
Dibutyl maleate	18.4	9.0	m
Dichloroethylene, trans-1,2	18.4	9.0	p
Dichloropropane-1,2	18.4	9.0	p
Hydrogenated terphenyl (Monsanto's HB-40)	18.4	9.0	p
Mesityl oxide	18.4	9.0	m
Vinyl acetate	18.4	9.0	m
Diamyl phthalate	18.6	9.1	m
Dichloroethylene, cis-1,2	18.6	9.1	p
Ethyl acetate	18.6	9.1	m
Isophorone	18.6	9.1	m
Tetrahydrofuran	18.6	9.1	m
Vinyltoluene	18.6	9.1	p
Allyl acetate	18.8	9.2	m
Benzene	18.8	9.2	p
Celanese Solvent 601	18.8	9.2	m
Diacetone alcohol	18.8	9.2	m
Dibutyl sebacate	18.8	9.2	m
Ethyl chloride	18.8	9.2	m
Ethylene glycol methyl ether acetate	18.8	9.2	m
Ethyl mercaptan	18.8	9.2	p
Propylene oxide	18.8	9.2	m
Propyl formate	18.8	9.2	m
Trichloroethylene	18.8	9.2	p
Tripropylene glycol	18.8	9.2	s
Chloroform	19.0	9.3	p
Dibutyl phthalate	19.0	9.3	m
Dipropylene glycol mono-methyl ether	19.0	9.3	m
Methylcyclohexanone	19.0	9.3	m
Methyl ethyl ketone	19.0	9.3	m
Perchloroethylene	19.0	9.3	p
Styrene	19.0	9.3	p
Tetrachloroethylene (perchloroethylene)	19.0	9.3	p
Triphenyl phosphite	19.0	9.3	m
Benzaldehyde	19.2	9.4	m
Butyl lactate (normal)	19.2	9.4	m
Capronitrile	19.2	9.4	p
Dibenzyl ether	19.2	9.4	m
Dimethyl sulfide	19.2	9.4	p
Ethyl formate	19.2	9.4	m
Ethyl-2-hexanediol-1,3 (octylene glycol)	19.2	9.4	s
Ethyl iodide	19.2	9.4	m
Furan	19.2	9.4	m
Nonyl phenol	19.2	9.4	s
Pentachloroethane	19.2	9.4	p
Acetyl chloride	19.4	9.5	m
Chlorobenzene	19.4	9.5	p
Chlorostyrene (ortho or para)	19.4	9.5	p
Diethylene glycol monobutyl ether (normal)	19.4	9.5	m
Ethylene glycol monobutyl ether	19.4	9.5	m
Ethylhexanol	19.4	9.5	s
Tetrahydronaphthalene	19.4	9.5	p
Ethyl bromide	19.6	9.6	m
Methyl acetate	19.6	9.6	m
Methyl bromide	19.6	9.6	m
Trichloroethane-1,1,2	19.6	9.6	p
Valeronitrile	19.6	9.6	p
Chloroethyl acetate (beta)	19.8	9.7	m
Dimethylaniline	19.8	9.7	m
Dipropyl phthalate	19.8	9.7	m
Ethylene bromide	19.8	9.7	p
Methyl chloride	19.8	9.7	m
Methylene chloride	19.8	9.7	p
Methyl-2-pentanediol-2,4	19.8	9.7	s
Tetrachloroethane-1,1,2,2	19.8	9.7	p
Acrolein	20.1	9.8	s
Bromostyrene (ortho)	20.1	9.8	p
Butyronitrile (iso)	20.1	9.8	p
Dichloroethyl ether	20.1	9.8	m
Dodecanol-1	20.1	9.8	s

Solvent	δ $[J/m^3]^{1/2} \cdot 10^{-3}$	δ $[cal/cm^3]^{1/2}$	H-Bonding Group	Solvent	δ $[J/m^3]^{1/2} \cdot 10^{-3}$	δ $[cal/cm^3]^{1/2}$	H-Bonding Group
Ethylene dichloride	20.1	9.8	p	Bromonaphthalene	21.7	10.6	p
Phenanthrene	20.1	9.8	p	Butyl alcohol (tert.)	21.7	10.6	s
Thiophene	20.1	9.8	m	Diethylformamide (N, N)	21.7	10.6	m
Valeric acid (normal)	20.1	9.8	s	Heptyl alcohol	21.7	10.6	s
Acetone	20.3	9.9	m	Methyl salicylate	21.7	10.6	m
Anthracene	20.3	9.9	p	Dimethyl phthalate	21.9	10.7	m
Bromobenzene	20.3	9.9	p	Hexyl alcohol (normal)	21.9	10.7	s
Cyclohexanone	20.3	9.9	m	Pyridine	21.9	10.7	m
Diethylacetamide (N, N)	20.3	9.9	m	Triethylene glycol	21.9	10.7	s
Diethyl maleate	20.3	9.9	m	Butyl alcohol (secondary)	22.1	10.8	s
Diethyl-2, 2-propanediol-1, 3 (heptylene glycol)	20.3	9.9	s	Dimethylacetamide (N, N)	22.1	10.8	m
Dimethyl carbonate	20.3	9.9	m	Morpholine	22.1	10.8	s
Methyl caprolactone	20.3	9.9	m	Pentanediol-2, 4	22.1	10.8	s
Naphthalene	20.3	9.9	p	Propionitrile	22.1	10.8	p
Nitro-2-propane	20.3	9.9	p	Quinoline	22.1	10.8	s
Propionic acid	20.3	9.9	s	Amyl alcohol	22.3	10.9	s
Tetraethylene glycol	20.3	9.9	s	Ethylene glycol monobenzyl ether	22.3	10.9	m
Amyl alcohol (iso)	20.5	10.0	s	Cyclobutanedione	22.5	11.0	m
Carbon disulfide	20.5	10.0	p	Dichloroacetic acid	22.5	11.0	s
Dichlorobenzene (ortho)	20.5	10.0	p	Dimethyl malonate	22.5	11.0	m
Diethyl phthalate	20.5	10.0	m	Dimethyl oxalate	22.5	11.0	m
Dimethyl-2, 2-butanediol-1, 3	20.5	10.0	s	Epichlorohydrin	22.5	11.0	s
Dioxane-1, 4	20.5	10.0	m	Ethyl cyanoacetate	22.5	11.0	m
Dipropylene glycol	20.5	10.0	s	Neopentyl glycol	22.5	11.0	s
Ethylamine	20.5	10.0	s	Butanediol-2, 3	22.7	11.1	s
Ethylene glycol diacetate	20.5	10.0	m	Ethylene oxide	22.7	11.1	m
Ethyl lactate	20.5	10.0	m	Nitroethane	22.7	11.1	p
Methyl isobutyl carbinol	20.5	10.0	s	Triethylene tetramine	22.7	11.1	s
Nitrobenzene	20.5	10.0	p	Acetylpiperidine (N)	22.9	11.2	s
Propionic anhydride	20.5	10.0	s	Dimethyl-2, 2-butanediol-1, 2 (isobutylene glycol)	22.9	11.2	s
Acetic acid	20.7	10.1	s	Furfural	22.9	11.2	m
Caprolactone	20.7	10.1	m	Methacrylic acid	22.9	11.2	s
Dibromoethylene-1, 2	20.7	10.1	p	Methylamine	22.9	11.2	s
Diphenyl ether	20.7	10.1	m	Dipropyl sulfone	23.1	11.3	m
Iodobenzene	20.7	10.1	p	Methyl-pyrrolidone-2 (N)	23.1	11.3	m
Propylene glycol methyl ether	20.7	10.1	m	Acetylpyrrolidone (N)	23.3	11.4	s
Cresol (meta)	20.9	10.2	s	Butyl alcohol (normal)	23.3	11.4	s
Diethylene glycol monoethyl ether	20.9	10.2	m	Cyclohexanol	23.3	11.4	s
Dioxolane-1, 3	20.9	10.2	m	Ethylene glycol monomethyl ether	23.3	11.4	m
Methyl formate	20.9	10.2	m	Tetramethyloxamide	23.3	11.4	m
Methyl iodide	20.9	10.2	m	Ethylene glycol monophenyl ether	23.5	11.5	m
Acetaldehyde	21.1	10.3	m	Formylpiperidine (N)	23.5	11.5	m
Acetic anhydride	21.1	10.3	s	Pentanediol-1, 5	23.5	11.5	s
Aniline	21.1	10.3	s	Propyl alcohol (iso)	23.5	11.5	s
Butyric acid (iso)	21.1	10.3	s	Acetylmorpholine (N)	23.7	11.6	m
Hexanediol-2, 5	21.1	10.3	s	Butanediol-1, 3	23.7	11.6	s
Methyl-2-pentanediol-1, 3	21.1	10.3	s	Tolylenediisocyanate	23.7	11.6	s
Nitro-1-propane	21.1	10.3	p	Allyl alcohol	24.1	11.8	s
Octyl alcohol (normal)	21.1	10.3	s	Methylene iodide	24.1	11.8	p
Cyclopentanone	21.3	10.4	m	Acetonitrile	24.3	11.9	p
Dibromoethane-1, 2	21.3	10.4	p	Propyl alcohol (normal)	24.3	11.9	s
Acrylonitrile	21.5	10.5	p	Santicizer 8	24.3	11.9	m
Butyl alcohol (iso)	21.5	10.5	s	Acrylic acid	24.6	12.0	s
Butyric acid (normal)	21.5	10.5	s	Dimethyl sulfoxide	24.6	12.0	m
Butyronitrile (normal)	21.5	10.5	p	Benzyl alcohol	24.8	12.1	s
Ethyl-2-butanol-1	21.5	10.5	s	Butanediol-1, 4	24.8	12.1	s
Ethylene glycol monoethyl ether	21.5	10.5	m	Butylene-2, 3 carbonate	24.8	12.1	m
Hexamethylphosphoramide	21.5	10.5	s	Diethylene glycol	24.8	12.1	s
Methyl benzoate	21.5	10.5	m	Dimethylformamide (N, N)	24.8	12.1	m
Styrene oxide	21.5	10.5	m				
Acetophenone	21.7	10.6	m				

Solvent	δ $[J/m^3]^{1/2} \cdot 10^{-3}$	δ $[cal/cm^3]^{1/2}$	H-Bonding Group	Solvent	δ $[J/m^3]^{1/2} \cdot 10^{-3}$	δ $[cal/cm^3]^{1/2}$	H-Bonding Group
Dimethyltetramethylene sulfone	24.8	12.1	m	Methyl ethyl sulfone	27.4	13.4	m
Formic acid	24.8	12.1	s	Pyrone (gamma)	27.4	13.4	m
Hydrogen cyanide	24.8	12.1	s	Tetramethylene sulfone	27.4	13.4	m
Ethylene chlorohydrin	25.0	12.2	s	Maleic anhydride	27.8	13.6	s
Ethylacetamide (N)	25.2	12.3	s	Piperidone	27.8	13.6	s
Ethylenediamine	25.2	12.3	s	Diacetylpiperazine (N, N)	28.0	13.7	m
Diethyl sulfone	25.4	12.4	m	Ethylformamide (N)	28.4	13.9	s
Methylene glycolate	25.4	12.4	m	Methanol	29.7	14.5	s
Dimethyl phosphite	25.6	12.5	m	Dimethyl sulfone	29.7	14.5	m
Furfuryl alcohol	25.6	12.5	s	Ethylene glycol	29.9	14.6	s
Methyl propyl sulfone	25.6	12.5	m	Methylacetamide	29.9	14.6	s
Phenylhydrazine	25.6	12.5	s	Ethylene carbonate	30.1	14.7	m
Butyrolactone	25.8	12.6	m	Pyrrolidone (alpha)	30.1	14.7	s
Chloroacetonitrile	25.8	12.6	p	Malononitrile	30.9	15.1	p
Propylene glycol	25.8	12.6	s	Ethylene cyanohydrin	31.1	15.2	s
Caprolactam (epsilon)	26.0	12.7	m	Diformylpiperazine (N, N)	31.5	15.4	m
Ethyl alcohol	26.0	12.7	s	Succinic anhydride	31.5	15.4	s
Nitromethane	26.0	12.7	p	Methylformamide (N)	32.9	16.1	s
Methyltetramethylene sulfone	26.4	12.9	m	Ammonia	33.4	16.3	s
Formylmorpholine (N)	26.6	13.0	m	Glycerol	33.8	16.5	s
Dimethylnitroamine (N, N)	26.8	13.1	s	Hydrazine	37.3	18.1	s
Propiolactone	27.2	13.3	m	Formamide	39.3	19.2	s
Propylene-1, 2-carbonate	27.2	13.3	m	Water	47.9	23.4	s

Table 3 : Three Dimensional Solubility Parameters of Solvents $[cal/cm^3]^{1/2}$

Solvent	δ_o	δ_d	δ_p	δ_h	Solvent	δ_o	δ_d	δ_p	δ_h
Acetic acid	10.5	7.10	3.9	6.6	Diethylene glycol monobutyl ether	8.96	7.80	3.4	5.2
Acetic anhydride	10.30	7.50	5.4	4.7	Diethylene glycol monomethyl ether	10.72	7.90	3.8	6.2
Acetone	9.77	7.58	5.1	3.4	Diethyl ether	7.62	7.05	1.4	2.5
Acetonitrile	11.9	7.50	8.8	3.0	Diethylsulfide	8.46	8.25	1.5	1.0
Acetophenone	9.68	8.55	4.2	1.8	Diisobutyl ketone	8.17	7.77	1.8	2.0
Aniline	11.04	9.53	2.5	5.0	Dimethyl-formamide	12.14	8.52	6.7	5.5
Benzaldehyde	10.40	9.15	4.2	2.6	Dimethyl sulfoxide	12.93	9.00	8.0	5.0
Benzene	9.15	8.95	0.5	1.0	Dioxane	10.0	9.30	0.9	3.6
α-Bromonaphthalene	10.25	9.94	1.5	2.0	Dipropyl-amine	7.79	7.50	0.7	2.0
1,3-Butanediol	14.14	8.10	4.9	10.5	Dipropylene glycol	15.52	7.77	9.9	9.0
n-Butanol	11.30	7.81	2.8	7.7	Ethanol	12.92	7.73	4.3	9.5
2-Butoxyethanol	10.25	7.76	3.1	5.9	Ethanolamine	15.48	8.35	7.6	10.4
n-Butyl acetate	8.46	7.67	1.8	3.1	Ethyl acetate	9.10	7.44	2.6	4.5
n-Butyl lactate	9.68	7.65	3.2	5.0	Ethylbenzene	8.80	8.70	0.3	0.7
Butyric acid	9.2 (?)	7.30	2.0	5.2	2-Ethyl-butanol	10.38	7.70	2.1	6.6
γ-Butyrolactone	12.78	9.26	8.1	3.6	Ethylene chloride	9.76	9.20	2.6	2.0
Butyronitrile	9.96	7.50	6.1	2.5	Ethylene glycol	16.30	8.25	5.4	12.7
Carbon disulfide	9.97	9.97	0	0	Ethylene glycol monoethyl ether	11.88	7.85	4.5	7.0
Carbon tetrachloride	8.65	8.65	0	0	Ethylene glycol monoethyl ether acetate	9.60	7.78	2.3	5.2
Chlorobenzene	9.57	9.28	2.1	1.0	Ethylene glycol monomethyl ether	12.06	7.90	4.5	8.0
1-Chlorobutane	8.46	7.95	2.7	1.0	2-Ethyl-hexanol	9.85	7.78	1.6	5.8
Chloroform	9.21	8.65	1.5	2.8	Ethyl lactate	10.5	7.80	3.7	6.1
m-Cresol	11.11	8.82	2.5	6.3	Formamide	17.8	8.4	12.8	9.3
Cyclohexane	8.18	8.18	0	0	Formic acid	12.15	7.0	5.8	8.1
Cyclohexanol	10.95	8.50	2.0	6.6	Furan	9.09	8.70	0.9	2.6
Cyclohexanone	9.88	8.65	4.1	2.5	Glycerol	21.1	8.46	?	?
Cyclohexylamine	9.05	8.45	1.5	3.2	Hexane	7.24	7.23	0	0
Cyclohexylchloride	8.99	8.50	2.7	1.0	Isoamyl acetate	8.32	7.45	1.5	3.4
Diacetone alcohol	10.18	7.65	4.0	5.3	Isobutyl isobutyrate	8.04	7.38	1.4	2.9
o-Dichlorobenzene	9.98	9.35	3.1	1.6	Isophorone	9.71	8.10	4.0	3.6
2, 2-Dichlorodiethyl ether	10.33	9.20	4.4	1.5	Mesityl oxide	9.20	7.97	3.5	3.0
Diethyl amine	7.96	7.30	1.1	3.0	Methanol	14.28	7.42	6.0	10.9
Diethylene glycol	14.60	7.86	7.2	10.0					

Solvent	δ_o	δ_d	δ_p	δ_h	Solvent	δ_o	δ_d	δ_p	δ_h
Methylal	8.52	7.35	0.9	4.2	n-Propanol	11.97	7.75	3.3	8.5
Methylene chloride	9.93	8.91	3.1	3.0	Propylene carbonate	13.30	9.83	8.8	2.0
Methyl ethyl ketone	9.27	7.77	4.4	2.5	Propylene glycol	14.80	8.24	4.6	11.4
Methyl isoamyl ketone	8.55	7.80	2.8	2.0	Pyridine	10.61	9.25	4.3	2.9
Methyl isobutyl carbinol	9.72	7.47	1.6	6.0	Styrene	9.30	9.07	0.5	2.0
Methyl isobutyl ketone	8.57	7.49	3.0	2.0	Tetrahydrofuran	9.52	8.22	2.8	3.9
Morpholine	10.52	9.20	2.4	4.5	Tetralin	9.50	9.35	1.0	1.4
Nitrobenzene	10.62	8.60	6.0	2.0	Toluene	8.91	8.82	0.7	1.0
Nitroethane	11.09	7.80	7.6	2.2	1,1,1-Trichloroethane	8.57	8.25	2.1	1.0
Nitromethane	12.30	7.70	9.2	2.5	Trichloroethylene	9.28	8.78	1.5	2.6
2-Nitropropane	10.02	7.90	5.9	2.0	Water	23.5	6.0	15.3	16.7
Pentanol (1)	10.61	7.81	2.2	6.8	Xylene	8.80	8.65	0.5	1.5

Table 4: Solubility Parameter Ranges of Commercial Polymers [+]
(chemical composition added whereever known)

Polymer	Solubility Parameter Ranges in Solvents which are Hydrogen Bonded					
	$[J/m^3]^{1/2} \cdot 10^{-3}$			$[cal/cm^3]^{1/2}$		
	Poorly	Moderately	Strongly	Poorly	Moderately	Strongly
Poly(acrylics)						
Acryloid B-66 (MMA/BMA copolymer)	17.4-22.7	16.0-24.8	0	8.5-11.1	7.8-12.1	0
Acryloid B-72 (EMA/MA copolymer)	17.4-26.0	18.2-26.0	0	8.5-12.7	8.9-13.3	0
Acryloid B-82 (EA/MMA copolymer)	17.4-22.7	18.2-24.8	0	8.5-11.1	8.9-12.1	0
Poly(butyl acrylate)	14.3-26.0	15.1-24.8	19.4-26.0	7.0-12.7	7.4-12.1	9.5-12.7
Poly(isobutyl methacrylate)	17.4-22.7	17.4-20.3	19.4-23.3	8.5-11.1	8.5- 9.9	9.5-11.4
Poly(n-butyl methacrylate)	15.1-22.7	15.1-20.3	19.4-23.3	7.4-11.1	7.4- 9.9	9.5-11.4
Poly(ethyl methacrylate)	17.4-22.7	16.0-27.2	19.4-23.3	8.5-11.1	7.8-13.3	9.5-11.4
Poly(methacrylic acid)	0	20.3	26.0-29.7	0	9.9	12.7-14.5
Poly(methyl methacrylate)	18.2-26.0	17.4-27.2	0	8.9-12.7	8.5-13.3	0
80 BMA/20 An	17.4-26.0	16.0-25.0	0	8.5-12.7	7.8-12.2	0
75 Isobornyl MA/25 C	15.1-21.7	15.1-17.4	19.4-22.3	7.4-10.6	7.4- 8.5	9.5-10.9
20 MAA/80 - Blown Linseed Oil	17.4	0	0	8.5	0	0
15 MAA/38 EA/47 S	19.4	17.4-18.2	0	9.5	8.5- 8.9	0
15 MAA/27.5 MA/57.5 VAc	0	20.3-30.1	26.0-29.7	0	9.9-14.7	12.7-14.5
15 MAA/17.5 MA/67.8 VAc	0	20.3-30.1	26.0-29.7	0	9.9-14.7	12.7-14.5
58 MAA/42 C	19.4	17.4-25.0	0	9.5	8.5-12.2	0
50 MMA/50 EA	17.4-26.0	16.0-27.0	0	8.5-12.7	7.8-13.2	0
25 MMA/75 EA	0	18.2-22.1	19.4-29.7	0	8.9-10.8	9.5-14.5
40 MMA/40 EA/20 AGE	17.4-26.0	16.0-27.0	21.5	8.5-12.7	7.8-13.2	10.5
45 MMA/45 EA/10 AM	22.7-26.0	18.2-27.0	26.0-29.7	11.1-12.7	8.9-13.2	12.7-14.5
55 MMA/30 EA/15 An	21.7-26.0	17.4-20.3	0	10.6-12.7	8.5- 9.9	0
40 MMA/40 EA/20 An	21.7-26.0	19.0-30.1	0	10.6-12.7	9.3-14.7	0
40 MMA/40 EA/20 tBAMA	17.4-26.0	16.0-27.0	19.4-29.7	8.5-12.7	7.8-13.2	9.5-14.5
40 MMA/40 EA/20 C	17.4-26.0	16.0-27.0	20.5-22.3	8.5-12.7	7.8-13.2	10.0-10.9
40 MMA/40 EA/20 MAA	0	18.2-22.1	19.4-29.7	0	8.9-10.8	9.5-14.5
45 MMA/45 EA/10mAm	17.4-26.0	17.4-27.0	26.0-29.7	8.5-12.7	8.5-13.2	12.7-14.5
40 MMA/40 EA/20 VBE	17.4-26.0	16.0-27.0	26.0	8.5-12.7	7.8-13.2	12.7
Alkyd Resins						
40 % Adipic, Glycerol Phthalate	22.7-26.0	20.3-30.1	26.0-29.7	11.1-12.7	9.9-14.7	12.7-14.5
45 % Linseed, Glycerol Phthalate	14.3-24.3	15.1-22.1	19.4-24.3	7.0-11.9	7.4-10.8	9.5-11.9
28 % Soy, Glycerol Phthalate	17.4-22.7	17.4-27.0	0	8.5-11.1	8.5-13.2	0
30 % Soy, Glycerol Phthalate	17.4-26.0	17.4-30.1	0	8.5-12.7	8.5-14.7	0
45 % Soy, Glycerol Phthalate	14.3-22.7	15.1-22.1	19.4-24.3	7.0-11.1	7.4-10.8	9.5-11.9
38 % Soy-DHC, Glycerol Oil Modified Alkyd	17.4-22.7	16.0-24.8	20.5-23.3	8.5-11.1	7.8-12.1	10.0-11.4
45 % Soy, Pentaerythritol Phthalate	14.3-22.7	15.1-22.1	19.4-24.3	7.0-11.1	7.4-10.8	9.5-11.9
Methacrylated DHC, PE Alkyd	17.4-22.1	17.4-27.0	0	8.5-11.1	8.5-13.2	0
Pelargonic Alkyd	17.4-22.1	16.0-24.8	0	8.5-11.1	7.8-12.1	0

[+] NOTE: See end of Table 4 for monomer identification.

| Polymer | Solubility Parameter Ranges in Solvents which are Hydrogen Bonded | | | | | |
| | $[J/m^3]^{1/2} \cdot 10^{-3}$ | | | $[cal/cm^3]^{1/2}$ | | |
	Poorly	Moderately	Strongly	Poorly	Moderately	Strongly
Alkyd Resins (Cont'd.)						
PE Phthalate Benzoate	19.4-26.0	17.4-30.1	0	9.5-12.7	8.5-14.7	0
PE Tri-(p-tert) Butyl Benzoate Mono-Acid-Phthalate	17.4-26.0	15.1-20.3	19.4-23.3	8.5-12.7	7.4- 9.9	9.5-11.4
PE p-tert-Butyl Benzoate Phthalate	17.4-22.7	16.0-27.2	20.5-22.3	8.5-11.1	7.8-13.3	10.0-10.9
Styrenated Linseed Alkyd	17.4-24.3	15.1-22.1	0	8.5-11.9	7.4-10.8	0
Vinyl Toluene Linseed Alkyd	16.0-21.7	19.0-22.1	0	7.8-10.6	9.3-10.8	0
Amine Resins						
Beckamine P-196 (Urea-Formaldehyde)	18.2-22.7	17.4-22.1	19.4-26.0	8.9-11.1	8.5-10.8	9.5-12.7
Beetle 227-8 (Urea-Formaldehyde)	0	0	18.2-23.3	0	0	8.9-11.4
Cyzak 1006 (Melamine-Formaldehyde)	17.4-22.7	16.0-27.2	21.5-22.3	8.5-11.1	7.8-13.3	10.5-10.9
Cyzak 1007 (Melamine-Formaldehyde)	17.4-22.7	17.4-27.2	0	8.5-11.1	8.5-13.3	0
Cyzak 1013 (Melamine-Formaldehyde)	17.4-22.7	16.0-24.8	0	8.5-11.1	7.8-12.1	0
Cyzak 1026 (Melamine-Formaldehyde)	0	18.2-27.0	0	0	8.9-13.2	0
Hexa-(methoxy methyl)-melamine	17.4-24.1	17.4-30.1	19.4-33.8	8.5-11.8	8.5-14.7	9.5-16.5
Resimene 888 (Melamine-Formaldehyde)	17.4-21.7	15.1-24.8	19.4-26.0	8.5-10.6	7.4-12.1	9.5-12.7
Uformite MX-61 (Benzoguanamine-Formaldehyde)	17.4-22.7	15.1-22.7	19.4-22.7	8.5-11.1	7.4-11.1	9.5-11.1
Cellulose Derivatives						
Alcohol Soluble Butyrate	24.1	17.4-27.0	23.3-29.7	11.8	8.5-13.2	11.4-14.5
Cellulose Acetate, LL-1	22.7-26.0	20.3-30.1	0	11.1-12.7	9.9-14.7	0
Cellulose Acetate-Butyrate	22.7-26.0	17.4-30.1	26.0-29.7	11.1-12.7	8.5-14.7	12.7-14.5
Cellulose Butyrate, 0.5 sec.	22.7-26.0	17.4-30.1	26.0-29.7	11.1-12.7	8.5-14.7	12.7-14.5
Cyanoethyl Cellulose	22.7-26.0	25.0-30.1	0	11.1-12.7	12.2-14.7	0
Ethyl Cellulose, K-200	0	17.4-22.1	19.4-23.3	0	8.5-10.8	9.5-11.4
Ethyl Cellulose, N-22	16.6-22.7	15.1-22.1	19.4-29.7	8.1-11.1	7.4-10.8	9.5-14.5
Ethyl Cellulose, T-10	17.4-19.4	16.0-20.1	19.4-23.3	8.5- 9.5	7.8- 9.8	9.5-11.4
Ethyl Hydroxyethyl Cellulose	16.6-21.3	16.4-21.3	0	8.1-10.4	8.0-10.4	0
Nitrocellulose, RS, 25 cps.	22.7-26.0	16.0-30.1	29.7	11.1-12.7	7.8-14.7	14.5
Nitrocellulose, SS, 0.5 sec.	22.7-26.0	16.0-30.1	26.0-29.7	11.1-12.7	7.8-14.7	12.7-14.5
Epoxy Resins (Epichlorohydrin - Bisphenol A)						
Epon E-72	17.4-21.7	15.1-20.3	19.4-23.3	8.5-10.6	7.4- 9.9	9.5-11.4
Epon 812	18.2-26.0	16.0-30.1	20.5-29.7	8.9-12.7	7.8-14.7	10.0-14.5
Epon 864	19.4-26.0	17.4-30.1	0	9.5-12.7	8.5-14.7	0
Epon 1001	21.7-22.7	17.4-27.2	0	10.6-11.1	8.5-13.3	0
Epon 1004	0	17.4-27.2	0	0	8.5-13.3	0
Epon 1007	0	17.4-27.2	0	0	8.5-13.3	0
Epon 1009	0	17.4-20.3	0	0	8.5- 9.9	0
Epon 1004 Modified Soy Ester	17.4-22.7	16.0-20.3	0	8.5-11.1	7.8- 9.9	0
Epon 1004 DHC Ester	17.4-22.7	16.0-20.3	0	8.5-11.1	7.8- 9.9	0
Hydrocarbon Resins						
Alpex (Cyclized Rubber)	15.1-21.7	16.0	0	7.4-10.6	7.8	0
Gilsonite Brilliant Black	16.0-21.7	16.0-17.4	0	7.8-10.6	7.8- 8.5	0
Gilsonite Selects	16.0-19.4	16.0-17.4	19.4	7.8- 9.5	7.8- 8.5	9.5
Nebony 100	17.4-21.7	16.0-20.3	0	8.5-10.6	7.8- 9.9	0
90 % Nebony 100/10 % Paraformaldehyde	17.4-21.7	15.1-20.3	0	8.5-10.6	7.4- 9.9	0
50 % Nebony 100/50 % Rosin	16.0-22.7	15.1-20.3	19.4-23.3	7.8-11.1	7.4- 9.9	9.5-11.4
Nevchem 100	14.9-21.9	15.1-20.3	0	7.3-10.7	7.4- 9.9	0
Nevchem 140	14.9-22.7	15.1-21.9	0	7.3-11.1	7.4-10.7	0
Nevex 100	13.7-22.7	15.1-21.9	0	6.7-11.1	7.4-10.7	0
Nevillac 10°	13.7-26.0	15.1-31.7	18.6-29.7	6.7-12.7	7.4-15.5	9.1-14.5
Nevillac Hard	15.6-26.0	15.1-31.7	18.6-29.7	7.6-12.7	7.4-15.5	9.1-14.5
Nevillac Soft	15.6-26.0	15.1-31.7	18.6-29.7	7.6-12.7	7.4-15.5	9.1-14.5
Neville LS 685 (Cumarone type)	15.1-21.7	19.0-20.3	0	7.4-10.6	9.3- 9.9	0
Nevindene R-1	16.0-22.7	15.1-21.9	0	7.8-11.1	7.4-10.7	0
Nevindene R-7	15.6-21.7	15.1-21.9	0	7.6-10.6	7.4-10.7	0
Nevinol A	13.7-26.0	15.1-24.3	19.4-22.9	6.7-12.7	7.4-11.9	9.5-11.2
Nevpene	13.7-21.7	15.1-17.8	0	6.7-10.6	7.4- 8.7	0

Polymer	Solubility Parameter Ranges in Solvents which are Hydrogen Bonded $[J/m^3]^{1/2} \cdot 10^{-3}$			$[cal/cm^3]^{1/2}$		
	Poorly	Moderately	Strongly	Poorly	Moderately	Strongly
Hydrocarbon Resins (Cont'd.)						
Panarez 3-210	17.4-21.7	0	0	8.5-10.6	0	0
Petrolatum 125 HMP	17.4-18.2	0	0	8.5- 8.9	0	0
Pliolite NR (Cyclized Rubber)	17.4-21.7	0	0	8.5-10.6	0	0
Pliolite P-1230	19.4-21.7	0	0	9.5-10.6	0	0
Phenolic Resins						
Amberol F-7 (Rosin modified)	17.4-21.7	16.0-20.1	19.4-22.1	8.5-10.6	7.8- 9.8	9.5-10.8
Amberol M-82 (Rosin modified)	15.1-20.7	15.1-20.3	19.4-22.3	7.4-10.1	7.4- 9.9	9.5-10.9
Bakelite BKR-2620	0	17.2-30.1	19.4-29.7	0	8.4-14.7	9.5-14.5
Bakelite CKR-2400	18.2-24.3	16.0-27.2	19.4-29.7	8.9-11.9	7.8-13.3	9.5-14.5
Bakelite CKR-5254	17.4-20.5	16.0-27.2	19.4-22.1	8.5-10.0	7.8-13.3	9.5-10.8
Bakelite CKR-5360	17.4-22.7	16.0-27.2	19.4-23.3	8.5-11.1	7.8-13.3	9.5-11.4
Durez 220 (Terpene modified)	17.4-21.7	16.0-20.1	19.4-23.3	8.5-10.6	7.8- 9.8	9.5-11.4
Durez 550	14.3-24.3	15.1-20.1	19.4-29.7	7.0-11.9	7.4- 9.8	9.5-14.5
Methylon 75202 (Allyl Phenol)	0	18.2-24.8	0	0	8.9-12.1	0
Poly(esters)						
Acid DEG Maleate-Phthalate	22.7-26.0	15.1-30.1	19.4-29.7	11.1-12.7	7.4-14.7	9.5-14.5
Carboxyl Terminated Diethylene Glycol Phthalateisophthalate	22.7-26.0	17.8-30.1	0	11.1-12.7	8.7-14.7	0
Diethylene Glycol Isophthalate	21.7-26.0	25.0-30.1	0	10.6-12.7	12.2-14.7	0
Diethylene Glycol Phthalate	22.7-26.0	20.3-30.1	0	11.1-12.7	9.9-14.7	0
Dipropylene Glycol Phthalate	19.4-26.4	17.4-30.1	0	9.5-12.9	8.5-14.7	0
Hydrogenated Bisphenol A Fumarate-Isophthalate	0	16.0-20.3	0	0	7.8- 9.9	0
Hydrogenated Bisphenol A-PG Fumarate-Isophthalate	18.2-22.7	17.4-20.3	0	8.9-11.1	8.5- 9.9	0
PE Benzoate-Maleate	22.7	17.4-30.1	0	11.1	8.5-14.7	0
Soluble Polyester Adhesives (EG-Terephthalate) 49000	21.7-22.7	21.7-22.7	0	10.6-11.1	10.6-11.1	0
49001	18.2-21.7	19.0-20.3	0	8.9-10.6	9.3- 9.9	0
49002	19.4-21.7	19.0-20.3	0	9.5-10.6	9.3- 9.9	0
TEG-EG Maleate-Terephthalate	21.7-24.1	17.4-29.1	0	10.6-11.8	8.5-14.2	0
Triethylene Glycol Maleate	22.7-26.0	20.3-27.0	0	11.1-12.7	9.9-13.2	0
Poly(amides)						
Nylon, Type 8	0	0	24.3-29.7	0	0	11.9-14.5
Versamid 100	17.4-21.7	17.4-18.2	19.4-23.3	8.5-10.6	8.5- 8.9	9.5-11.4
Versamid 115	17.4-21.7	16.0-20.3	19.4-26.0	8.5-10.6	7.8- 9.9	9.5-12.7
Versamid 900	0	0	0	0	0	0
Versamid 930	0	0	19.4-23.3	0	0	9.5-11.4
Versamid 940	0	0	19.4-23.3	0	0	9.5-11.4
Rosin Derivatives						
Abalyn (Methyl Abietate)	14.3-22.7	15.1-22.1	20.5-26.0	7.0-11.1	7.4-10.8	10.0-12.7
Abitol (Hydroabietyl Alcohol)	14.3-21.9	15.1-22.1	20.5-29.7	7.0-10.7	7.4-10.8	10.0-14.5
Alkydol 160	19.4	17.4-22.1	19.4-26.0	9.5	8.5-10.8	9.5-12.7
Amberol F-7 (Phenol-Formaldehyde modified rosin)	17.4-21.7	16.0-20.1	19.4-22.3	8.5-10.6	7.8- 9.8	9.5-10.9
Amberol 750 (Phenol-Formaldehyde modified rosin)	0	18.2-22.1	19.4-26.0	0	8.9-10.8	9.5-12.7
Amberol 801 (Phenol-Formaldehyde modified rosin)	17.4-22.7	15.1-20.3	0	8.5-11.1	7.4- 9.9	0
Arochem 455 (Bisphenol Epoxy modified)	0	16.0-27.2	19.4-29.7	0	7.8-13.3	9.5-14.5
Arochem 462	19.4	17.4-22.1	19.4-29.7	9.5	8.5-10.8	9.5-14.5
Cellolyn 21	14.3-22.7	15.1-20.3	20.5-24.3	7.0-11.1	7.4- 9.9	10.0-11.9
Cellolyn 95-80T	14.3-22.7	15.1-20.3	20.5-24.3	7.0-11.1	7.4- 9.9	10.0-11.9
Cellolyn 102	16.0-20.5	17.4-22.1	20.5-24.3	7.8-10.0	8.5-10.8	10.0-11.9
Cellolyn 104	16.0-20.5	16.0-22.1	20.5-23.3	7.8-10.0	7.8-10.8	10.0-11.4
Dymerex (Dimerized Rosin)	15.1-21.7	16.0-20.3	19.4-23.3	7.4-10.6	7.8- 9.9	9.5-11.4
Ester Gum (Glycerol Rosin Ester)	14.3-21.7	15.1-22.1	19.4-22.3	7.0-10.6	7.4-10.8	9.5-10.9
17 % Fumarated Rosin	17.4-21.7	15.1-27.2	19.4-29.7	8.5-10.6	7.4-13.3	9.5-14.5
22 % Fumarated Rosin	18.2-21.7	15.1-27.2	19.4-29.7	8.9-10.6	7.4-13.3	9.5-14.5
Hercolyn D (Methyl Hydroabietate)	14.3-24.1	15.1-22.1	20.5-26.0	7.0-11.8	7.4-10.8	10.0-12.7
Lewisol 28 (Maleic modified)	16.0-20.5	15.1-20.3	20.5-21.7	7.8-10.0	7.4- 9.9	10.0-10.6
Neolyn 23	17.4-22.7	17.4-27.2	0	8.5-11.1	8.5-13.3	0

SOLUBILITY PARAMETER VALUES

| Polymer | Solubility Parameter Ranges in Solvents which are Hydrogen Bonded | | | | | |
| | $[\text{J/m}^3]^{1/2} \cdot 10^{-3}$ | | | $[\text{cal/cm}^3]^{1/2}$ | | |
	Poorly	Moderately	Strongly	Poorly	Moderately	Strongly
Rosin Derivatives (Cont'd.)						
Neolyn 40	18.2-26.0	17.4-27.2	20.5-26.0	8.9-12.7	8.5-13.3	10.0-12.7
Pentalyn A (Pentaerythritol Ester of Rosin)	17.4-21.7	15.1-20.3	19.4-23.3	8.5-10.6	7.4- 9.9	9.5-11.4
Pentalyn C (")	15.1-21.9	15.1-20.3	20.5-23.3	7.4-10.7	7.4- 9.9	10.0-11.4
Pentalyn G (")	17.4-21.7	16.0-20.3	19.4-22.3	8.5-10.6	7.8- 9.9	9.5-10.9
Pentalyn H (")	14.3-21.9	15.1-20.3	20.5-23.3	7.0-10.7	7.4- 9.9	10.0-11.4
Pentalyn K (")	17.4-21.7	16.0-20.3	19.4-23.3	8.5-10.6	7.8- 9.9	9.5-11.4
Pentalyn X (")	16.0-20.5	15.1-20.3	20.5	7.8-10.0	7.4- 9.9	10.0
Pentalyn B25 (")	17.4-22.7	15.1-20.3	20.5-24.3	8.5-11.1	7.4- 9.9	10.0-11.9
Pentalyn 255 (")	18.2-20.5	15.1-22.1	20.5-29.7	8.9-10.0	7.4-10.8	10.0-14.5
Pentalyn 802A (")	16.0-20.5	15.1-20.3	0	7.8-10.0	7.4- 9.9	0
Pentalyn 830 (")	17.4-19.4	16.0-22.1	19.4-23.3	8.5- 9.5	7.8-10.8	9.5-11.4
Pentalyn 833 (")	17.4-21.9	15.1-20.3	22.3	8.5-10.7	7.4- 9.9	10.9
Pentalyn 856 (")	17.4-22.7	15.1-22.1	19.4-23.3	8.5-11.1	7.4-10.8	9.5-11.4
Pentalyn 954 (")	16.0-21.9	15.1-20.3	0	7.8-10.7	7.4- 9.9	0
Petrex 7-75T	18.2-26.0	17.4-27.2	23.3-29.7	8.9-12.7	8.5-13.3	11.4-14.5
Petrex 130H	16.0-22.7	15.1-16.2	20.5-24.3	7.8-11.1	7.4- 7.9	10.0-11.9
Poly-pale Ester 1	16.2-21.9	15.1-20.3	20.5-23.3	7.0-10.7	7.4- 9.9	10.0-11.4
Poly-pale Ester 10	14.3-22.7	15.1-20.3	20.5-24.3	7.0-11.1	7.4- 9.9	10.0-11.9
Staybelite Ester 5 (Glycerol Ester of Hydrogenated Rosin)	14.3-21.9	15.1-20.3	20.5-24.3	7.0-10.7	7.4- 9.9	10.0-11.9
Staybelite Ester 10 (Glycerol Ester of Hydrogenated Rosin)	14.3-21.9	15.1-20.3	20.5-24.3	7.0-10.7	7.4- 9.9	10.0-11.9
Vinsol (Oxygenated Rosin)	21.7-29.3	16.0-27.2	19.4-26.0	10.6-11.9	7.8-13.3	9.5-12.7
WW Gum Rosin	17.4-22.7	15.1-22.1	19.4-23.3	8.5-11.1	7.4-10.8	9.5-11.4
Wood Rosin M Grade	15.1-21.7	15.1-22.1	19.4-29.7	7.4-10.6	7.4-10.8	9.5-14.5
Poly(styrene) and Copolymers						
Bakelite's RMD4511 (S/An)	21.7-22.7	19.0	0	10.6-11.1	9.3	0
Buton 100 (S/B)	14.3-21.7	15.1-20.3	0	7.0-10.6	7.4- 9.9	0
Buton 300 (S/B)	17.4-21.7	15.1-20.3	19.4-21.5	8.5-10.6	7.4- 9.9	9.5-10.5
Koppers KTPL-A (Poly(styrene))	16.4-21.7	16.6-20.3	0	8.0-10.6	8.1- 9.9	0
Lytron 810	24.3	20.3-30.1	0	11.9	9.9-14.7	0
Lytron 820	19.4	18.2-30.1	22.3-29.7	9.5	8.9-14.7	10.9-14.5
Marbon 9200 (S/B)	17.4-21.7	19.0-20.3	0	8.5-10.6	9.3- 9.9	0
Parapol S-50	19.4	16.0	0	9.5	7.8	0
Parapol S-60	19.4	16.0	0	9.5	7.8	0
Piccoflex 120	17.4-22.7	16.0-20.3	0	8.5-11.1	7.8- 9.9	0
Shell Polyaldehyde Resin EX39 (Poly(acrolein))	18.2-22.7	19.0-27.2	0	8.9-11.1	9.3-13.3	0
Shell Polyaldehyde Resin EX40 (Poly(acrolein))	18.2-22.7	19.0-27.2	0	8.9-11.1	9.3-13.3	0
85 % Styrene/15 % Acrylamide	0	0	0	0	0	0
85 % Styrene/15 % Acrylonitrile	18.2-22.7	19.0-24.8	0	8.9-11.1	9.3-12.1	0
85 % Styrene/15 % Butenol	17.4-22.7	16.0-20.3	0	8.5-11.1	7.8- 9.9	0
82 % Styrene/18 % Cyclol	19.4	17.4-22.1	0	9.5	8.5-10.8	0
81 % Styrene/11 % 2-Ethyl hexyl acrylate/8 % Acrylic acid	18.2-21.7	16.0-20.3	0	8.9-10.6	7.8- 9.9	0
90 % Styrene/10 % Methacrylic Acid	19.4-21.7	17.4-22.1	0	9.5-10.6	8.5-10.8	0
85 % Styrene/15 % Methyl acrylate	17.4-22.7	16.0-24.8	0	8.5-11.1	7.8-12.1	0
60 % Styrene/40 % Methyl Half Ester of Maleic acid	0	17.4-27.0	21.5-23.3	0	8.5-13.2	10.5-11.4
57 % Styrene/43 % Propyl Half Ester of Maleic acid	18.2-24.1	17.4-27.0	19.4-23.3			
85 % Styrene/15 % Vinyl butyl ether	17.4-22.7	16.0-20.3	0	8.5-11.1	7.8- 9.9	0
Vinyl Polymers						
Acryloid K120N	17.4-26.0	17.4-27.0	0	8.5-12.7	8.5-13.2	0
DQDA 6225	17.4-19.4	0	0	8.5- 9.5	0	0
DQDA 3457	17.4-19.4	0	0	8.5- 9.5	0	0
Elvax 150 (PVAc/E)	16.0-21.7	0	0	7.8-10.6	0	0
Elvax 250 (PVAc/E)	17.4-19.4	0	0	8.5- 9.5	0	0
Exon 470	17.4-22.7	16.0-20.3	0	8.5-11.1	7.8- 9.9	0
Exon 471	17.4-22.7	16.0-24.8	0	8.5-11.1	7.8-12.1	0
Exon 473	17.4-22.7	16.0-20.3	0	8.5-11.1	7.8- 9.9	0
Geon 121 (Poly(vinyl chloride))	21.7-22.7	19.0-20.3	0	10.6-11.1	9.3- 9.9	0
Polycyclol	0	18.2-22.1	19.4-29.7	0	8.9-10.8	9.5-14.5
Poly(vinyl butyl ether)	16.0-21.7	15.1-20.3	19.4-23.3	7.8-10.6	7.4- 9.9	9.5-11.4

Polymer	Solubility Parameter Ranges in Solvents which are Hydrogen Bonded $[J/m^3]^{1/2} \cdot 10^{-3}$			$[cal/cm^3]^{1/2}$		
	Poorly	Moderately	Strongly	Poorly	Moderately	Strongly
Vinyl Polymers (Cont'd.)						
Poly(vinyl ethyl ether)	14.3-22.7	15.1-22.1	19.4-29.7	7.0-11.1	7.4-10.8	9.5-14.5
Poly(vinyl formal) (7/70E)	0	20.3-27.2	0	0	9.9-13.3	0
Poly(vinyl formal) (15/95E)	0	20.3-27.2	0	0	9.9-13.3	0
Poly(vinyl isobutyl ether)	14.3-21.7	15.1-20.3	19.4-23.3	7.0-10.6	7.4- 9.9	9.5-11.4
Saran F-120 (Poly(vinylidene chloride))	19.4-22.7	24.8-30.1	0	9.5-11.1	12.1-14.7	0
Saran F-220 (Poly(vinylidene chloride))	19.4-22.7	22.1-30.1	0	9.5-11.1	10.8-14.7	0
Shawinigan RS3512 (30 % hydrolysed PVAc)	0	0	0	0	0	0
Shawinigan RS3648 (30 % hydrolysed PVAc)	22.7-26.0	22.1-30.1	26.0-29.7	11.1-12.7	10.8-14.7	12.7-14.5
Tedlar (Poly(vinyl fluoride))	0	0	0	0	0	0
63 VAc/33 EHA/4 MAA	0	17.4-20.3	0	0	8.5- 9.9	0
76 VAc/12 EHA/8 C/4 MAA	0	19.0-27.2	0	0	9.3-13.3	0
70 VAc/20 EA/10 Cyclol	18.2-26.0	16.0-30.1	19.4-29.7	8.9-12.7	7.8-14.7	9.5-14.5
46 VBE/27 An/27 mAm	24.1	18.2-22.1	0	11.8	8.9-10.8	0
46 VBE/27 MA/27 mAm	22.9-26.0	18.2-28.4	21.5-29.7	11.2-12.7	8.9-13.9	10.5-14.5
75 Vinylidene Chloride/25 Acrylic acid	0	17.4-25.0	19.4-29.7	0	8.5-12.2	9.5-14.5
Vinylite AYAA (Poly(vinyl acetate))	18.2-26.0	17.4-30.1	29.7	8.9-12.7	8.5-14.7	14.5
Vinylite VAGH (3 % hydrolysed PVAc)	21.7-22.7	16.0-20.3	0	10.6-11.1	7.8- 9.9	0
Vinylite VMCH (VAc/maleic copolymer)	21.7-22.7	16.0-24.8	0	10.6-11.1	7.8-12.1	0
Vinylite VXCC	19.4-22.7	16.0-27.0	0	9.5-11.1	7.8-13.2	0
Vinylite VYHH (Poly(vinyl chloride-co-acetate))	19.0-22.7	16.0-27.1	0	9.3-11.1	7.8-13.3	0
Vinylite VYLF (")	19.4-22.7	16.0-27.0	0	9.5-11.1	7.8-13.2	0
Vinylite XYHL (Poly(vinyl butyral))	0	18.2-22.1	19.4-29.7	0	8.9-10.8	9.5-14.5
Vinylite XYSG (")	0	18.2-22.1	19.4-29.7	0	8.9-10.8	9.5-14.5
Vyset 69	18.2-19.4	17.4-20.3	0	8.9- 9.5	8.5- 9.9	0
Miscellaneous						
Acrylamide Monomer	22.7-26.0	22.1-30.1	19.4-29.7	11.1-12.7	10.8-14.7	9.5-14.5
Bakelite P-47 (Sulfone Resin)	21.7	20.3	0	10.6	9.9	0
Carbowax 4000 (Poly(ethylene oxide))	18.2-26.0	17.4-30.1	19.4-29.7	8.9-12.7	8.5-14.7	9.5-14.5
Chlorinated Rubber	17.4-21.7	16.0-22.1	0	8.5-10.6	7.8-10.8	0
Dammar Gum (dewaxed)	17.4-21.7	16.0-20.3	19.4-22.3	8.5-10.6	7.8- 9.9	9.5-10.9
Hexadecyl Monoester of TMA	0	18.2-27.0	19.4-29.7	0	8.9-13.2	9.5-14.5
Hydrogenated Sperm Oil WX135	18.2-21.7	0	0	8.9-10.6	0	0
Hypalon 20 (Chlorosulfonated poly(ethylene))	16.6-20.1	17.2-18.0	0	8.1- 9.8	8.4- 8.8	0
Hypalon 30 (Chlorosulfonated poly(ethylene))	17.4-21.7	16.0-17.4	0	8.5-10.6	7.8- 8.5	0
Ketone Resin S588 (Poly(butanone-2))	21.7-26.0	17.4-27.0	19.4-29.7	10.6-12.7	8.5-13.2	9.5-14.5
Lexan 100 Polycarbonate Resin	19.4-21.7	19.0-20.3	0	9.5-10.6	9.3- 9.9	0
Lexan 105 Polycarbonate Resin	19.4-21.7	19.0-20.3	0	9.5-10.6	9.3- 9.9	0
Resin #510 (Fossil Coal Resin)	15.1-21.7	15.1-19.0	19.4	7.4-10.6	7.4- 9.3	9.5
Santolite MHP (p-Toluenesulfonamide-Formaldehyde)	21.7-26.0	16.0-30.1	19.4	10.6-12.7	7.8-14.7	9.5
Shellac (Pale-pale)	0	20.3-22.1	19.4-29.7	0	9.9-10.8	9.5-14.5
Silicone DC-23	15.1-17.4	15.1-16.0	19.4-20.5	7.4- 8.5	7.4- 7.8	9.5-10.0
Silicone DC-1107	14.3-19.4	19.0-22.1	19.4-23.3	7.0- 9.5	9.3-10.8	9.5-11.4
Silicone Intermediate Z6018	17.4-22.7	16.2-25.0	20.5-23.3	8.5-11.1	7.9-12.2	10.0-11.4
Soy Oil	14.3-22.7	15.1-22.1	19.4-24.3	7.0-11.1	7.4-10.8	9.5-11.9
Soy Oil, Blown	14.3-22.7	15.1-22.1	19.4-26.0	7.0-11.1	7.4-10.8	9.5-12.7
p-Toluene Sulfonamide	24.1	20.3-30.1	26.0-29.7	11.8	9.9-14.7	12.7-14.5

Note : Identification of Monomer Symbols in Table 4

AGE	Allyl glycidyl ether	DHC	Dehydrated castor oil	MMA	Methyl methacrylate
Am	Acrylamide	EA	Ethyl acrylate	PE	Pentaerythritol
An	Acrylonitrile	E	Ethylene	PG	1,2-Propane glycol
B	Butadiene	EG	Ethylene glycol	PVAc	Poly(vinyl acetate)
tBAMA	tert-Butylamino methacrylate	EHA	Ethylhexyl acrylate	S	Styrene
BMA	Butyl methacrylate	EMA	Ethyl methacrylate	TEG	Triethylene glycol
C	Cyclol (bicyclo-2,2,1-hept-5-ene-2-methanol)	MA	Maleic anhydride	TMA	Trimelletic anhydride
		MAA	Methacrylic acid	VAc	Vinyl acetate
DEG	Diethylene glycol	mAm	Methylol acrylamide	VBE	Vinyl butyl ether

Table 5: Single Value Solubility Parameters of Polymers

Polymer	$\delta[J/m^3]^{1/2} \cdot 10^{-3}$	$\delta[cal/cm^3]^{1/2}$	Method	T °C	References
5.1 Poly(dienes)					
Poly(butadiene)	14.65	7.16	calc.		8
	17.19	8.40			20
	17.09	8.35			34
	17.15	8.38	calc.		35
	17.2-17.6	8.4-8.6	obs.		35
	16.6	8.1			36
	17.6	8.6			21
(emulsion)	17.19	8.40			28
(sodium)	17.60	8.60			28
	16.6	8.1	calc.		17
	17.09	8.35			33
hydrogenated	16.6	8.1	swelling		24
	16.47	8.05			24
	16.6	8.1	av.		24
Poly(butadiene-co-acrylonitrile)					
BUNA N (82/18)	17.90-17.72	8.75-8.66			5
(80/20)	18.4	9.0	calc.		20
	19.4	9.5	obs.		20
BUNA N (72/25)	18.93	9.25	calc.	25	35
	19.19	9.38	obs.		12
	19.4	9.5	obs.		33
	18.2	8.9			36
	19.4	9.5			28
(70/30)	20.26-20.11	9.90-9.83			5
	19.19	9.38			13
(61/39)	21.1	10.3			5
	21.38-21.28	10.45-10.40			5
Poly(butadiene-co-styrene)					
BUNA S (94/4)	16.64-16.45	8.13-8.04			5
(90/10)	17.13	8.37			34
(87.5/12.5)	16.57-16.30	8.10-8.01			5
	16.55	8.09			13
	17.6	8.6			33
	17.31	8.46			35
(85/15)	17.19	8.40			34
	17.35	8.48	calc.		35
	17.4	8.5	obs.		35
	17.39	8.50			28
	17.41	8.51	calc.		20
	17.39	8.50	obs.		20
	17.50	8.55			33
(75/25)	17.29	8.45			34
	17.47	8.54	calc.		35
	16.55	8.09	obs. (lit)		35
	16.49	8.06	obs. (lit)		35
	16.6	8.1			36
	17.50	8.55			28
	17.56	8.58	calc.		20
	17.50	8.55	obs.		20
	17.60	8.60			33
(71.5/28.5)	16.72-16.55	8.17-8.09			5
	17.51	8.56			35
(70/30)	17.35	8.48			34
(60/40)	17.50	8.55			34
	17.70	8.65	calc.		35
	17.74	8.67	obs.		35
	17.74	8.67			28
	17.76	8.68	calc.		20
	17.74	8.67	obs.		20
	17.80	8.70			33

References page IV-359

Polymer	$\delta\,[J/m^3]^{1/2}\cdot 10^{-3}$	$\delta\,[cal/cm^3]^{1/2}$	Method	$T\,^{\circ}C$	References
Poly(butadiene-co-vinylpyridine) (75/25)	19.13	9.35			28
Poly(chloroprene)	18.42	9.00		25	34
	16.59	8.11	calc.		8
	19.19	9.38	calc.		35
	16.74	8.18	obs.		12
	18.93	9.25	obs.		33
	17.6	8.6			36
	16.8	8.2			11
	18.8	9.2			21
	16.76	9.2			28
	17.6	8.19	swelling		13
	17.74-17.54	8.6			5
	15.18	8.67-8.57			5
Poly(isoprene)					
1,4-cis	15.18	7.42	calc.	25	8
	16.64	8.13			12
	16.68	8.15		35	23
	16.57	8.10		35	23
	20.46	10.0	swelling	35	23
	16.57	8.10	av.	35	23
	16.47	8.05	swelling	35	23
	16.57	8.10	swelling	35	23
	16.68	8.15	calc.	35	23
	16.6	8.1	swelling		25
	16.4	8.0			25
	16.47	8.05	av.		25
	16.82	8.22		25	34
	16.68	8.15	calc.	25	35
	16.2	7.9	obs.		35
	16.33	7.98	obs.		35
	17.09	8.35	obs.		35
	16.2	7.9			21
natural rubber	16.2	7.9			13
	16.6	8.1			5
	16.68	8.15			35
	17.09	8.35			33
	17.0	8.3			38
	16.6	8.1			36
	16.4	8.0			5,33
	17.09	8.35			28
	16.33	7.98			13
	16.6	8.1			5
	16.49-16.62	8.06-8.12			5
Gutta Percha	16.6	8.1	calc.		22
chlorinated	19.2	9.4			6
5.2 Poly(alkenes)					
Poly(ethylene)	15.76	7.70			34
	16.6	8.1	calc.		35
	16.0	7.8			16
	16.2	7.9			31
	17.09	8.35			2
	16.4	8.0	calc.		38
	16.2	7.9			36
	16.2	7.9			21
	16.8	8.2	calc.		31
	16.2	7.9	obs.		22
	17.99	8.79	calc.		36a

SOLUBILITY PARAMETER VALUES

Polymer	$\delta\,[\text{J/m}^3]^{1/2} \cdot 10^{-3}$	$\delta\,[\text{cal/cm}^3]^{1/2}$	Method	T °C	References
Poly(isobutene)	14.5	7.1	calc.		8
	16.06	7.85	av.	35	23
	16.0	7.8	swelling		23
	16.47	8.05	swelling		23
	16.25	7.94			5,33
	16.06	7.85		25	34
	15.76	7.70	calc.		35
	16.47	8.05	obs.		35
	16.4	8.0			16
	16.47	8.05			33
	16.6	8.1			6
	16.0	7.8			5
	17.0	8.3	calc.		38
	16.47	8.05			21
	16.47	8.05			28
Poly(isobutene-co-isoprene) butyl rubber	16.06-15.90	7.85-7.77			5
	16.47	8.05			33
	15.76	7.70			35
Poly(methylene)	14.3	7.0	extrap.	20	14
Poly(propylene)	18.8	9.2		25	16
	19.2	9.4	calc.		38

5.3 <u>Poly(acrylics) and Poly(methacrylics)</u>

Polymer	$\delta\,[\text{J/m}^3]^{1/2} \cdot 10^{-3}$	$\delta\,[\text{cal/cm}^3]^{1/2}$	Method	T °C	References
Poly(acrylic acid)					
--, butyl ester	18.0	8.8		35	24
	18.01	8.80	av.		24
	18.52	9.05	swelling		24
	17.4	8.5	calc.		24
	18.6	9.1	swelling		26
	18.52	9.05			26
	19.77	9.66	calc.		36a
--, ethyl ester	19.13	9.35	av.		24
	19.2	9.4	swelling		24
	19.8	9.7	calc.		24
	19.2	9.4	swelling		26
	19.13	9.35			26
	19.19	9.38			14
	18.8	9.2	calc.		17
	20.40	9.97	calc.		36a
--, isobornyl ester	16.8	8.2	calc.		17
--, methyl ester	20.7	10.1	av.		24
	20.77	10.15	swelling		24
	20.1	9.8	calc.		24
	20.77	10.15	swelling		26
	21.3	10.4			26
	20.7	10.1	swelling		26
--, propyl ester	18.52	9.05			24
	18.42	9.00	av.		24
	18.4	9.0	calc.		24
Poly(acrylonitrile)	25.27	12.35	calc.		36a
	25.6	12.5			20
	26.09	12.75	calc.	25	35
	31.5	15.4			36
Poly(α-chloroacrylic acid) methyl ester	20.7	10.1	calc.		35
Poly(methacrylic acid)					
--, butyl ester	17.90	8.75	swelling		26
	17.8	8.7			26
	18.01	8.80	swelling		26
	17.90	8.75			26
	17.0	8.3	calc.		22
--, ethoxyethyl ester	18.4	9.0	swelling		26
	20.3	9.9			26

References page IV-359

Polymer	$\delta\ [J/m^3]^{1/2} \cdot 10^{-3}$	$\delta\ [cal/cm^3]^{1/2}$	Method	$T\ ^{\circ}C$	References
Poly(methacrylic acid) (Cont'd.)					
--, ethyl ester	18.31	8.95	swelling		26
	18.2	8.9			26
	18.6	9.1	calc.		22
--, n-hexyl ester	17.6	8.6	calc.		22
--, isobornyl ester	16.6	8.1			17
--, lauryl ester	16.8	8.2	calc.		22
--, methyl ester	18.58	9.08		25	5
	18.66-18.52	9.12-9.05			5
	19.4	9.5	swelling		26
	19.34	9.45			26
	26.27	12.84			34
	18.93	9.25	calc.		35
	18.4-19.4	9.0-9.5			1
	18.58	9.08			36
	19.50	9.53	calc.		36a
--, octyl ester	17.2	8.4	calc.		22
--, propyl ester	18.0	8.8	calc.		22
--, stearyl ester	16.0	7.8	calc.		22
Poly(methacrylonitrile)	21.9	10.7	calc.		35
	21.0	10.7			36
5.4 **Poly(vinyl halides), Poly(vinyl alcohol), Poly(vinyl ester)**					
Poly(tetrafluoroethylene)	12.7	6.2	calc.		35
	12.7	6.2			36
Poly(vinyl acetate)	19.62	9.59	calc.	25	8
	19.13	9.35		35	27
	20.93	10.23	calc.		36a
	19.2	9.40	Small's Method		27
	18.0	8.80	lit.		27
	22.61	11.05		25	34
	19.2	9.4	calc.		35
	19.2	9.4			36
Poly(vinyl acetate-co-vinyl alcohol)	21.94	10.72	calc.		36a
Poly (vinyl alcohol)	25.78	12.60			34
Poly(vinyl bromide)	19.42	9.49			9
	19.6	9.6	calc.		35
Poly(vinyl chloride)	19.34-19.19	9.45-9.38			5
	19.28	9.42	calc.		8
	20.67	10.10			34
	19.54	9.55	calc.		35
	19.8	9.7	obs.		21
	19.2	9.4			5
	19.8	9.7			6
	22.1	10.8			16
	20.1	9.8			38
	19.50	9.53			36
	19.8	9.7			21
	20.32	9.93	calc.		36a
Poly(vinyl chloride-co-vinyl acetate)					
(87/13)	21.7	10.6	calc.		8
	21.3	10.4			6
(VYHH)	20.42	9.98	calc.		36a
Poly(vinyl chloride-co-vinyl acetate-co-maleic acid)	20.44	9.99	calc.		36a
Poly(vinyl chloride-co-vinyl acetate-co-vinyl alcohol)	20.77	10.15	calc.		36a
Poly(vinylidene chloride)	25.0	12.2			6
Poly(vinylidene cyanide-co-vinyl acetate)	22.67	11.08	calc.		39
Poly(vinyl propionate)	18.01	8.80		35	27
	18.52	9.05	Small's Method		27

Polymer	$\delta[J/m^3]^{1/2} \cdot 10^{-3}$	$\delta[cal/cm^3]^{1/2}$	Method	T °C	References
5.5　Poly(styrenes)					
	17.52	8.56			5
	17.58-17.45	8.59-8.53			5
	20.16	9.85	calc.		8
	17.86-17.92	8.73 or 8.76			15
	17.84, 18.56	8.72, 9.07	visc.		35b
	18.6	9.1			20
	18.72	9.15		35	23
	18.62	9.10	av.		23
	18.62	9.10	swelling		23
	18.66	9.12	calc.		23
	19.09	9.33		25	34
	18.66	9.12	calc.		35
	18.6	9.1	obs. (lit)		35
	17.6-19.8	8.6-9.7	obs. (lit)		35
	17.4	8.5			4
	17.6	8.6			5
	18.4	9.0			16
	19.28	9.42	calc.		36a
	21.1	10.3	calc.		38
	17.52	8.56			36
	17.84	8.72			5, 31
	18.6	9.1			33
	17.6-17.8	8.6-8.7			21
	18.6	9.1			28
Poly(styrene-co-divinylbenzene)	18.6	9.1	obs. (lit)		35
	17.39	8.50			4
5.6　Others					
Alkyd, medium oil length	19.2	9.4			6
Epoxy resin	22.3	10.9			36
Poly(iminohexamethyleneiminoadipoyl)	27.8	13.6			36
Poly(oxydimethylsilylene)	15.04	7.35			34
	14.9	7.3	obs.		6
	15.45	7.55	swelling		25
	15.6	7.6	av.		25
	14.9	7.3			36
	15.59	7.62			14
Poly(oxyethyleneoxyterephthaloyl)	21.9	10.7			36
	21.9	10.7	calc.		35
Poly(thioethylene)	19.23	9.40			34
	19.19	9.38	swelling		13
	18.4	9.0	swelling		13
Poly(thiophenylethylene)	19.0±1.0	9.3±0.5			30
Poly(urethane) (unknown composition)	20.5	10.0	swelling		25
	20.5	10.0	av.		25
5.7　Cellulose and Derivatives					
Benzyl cellulose	25.23	12.33			34
Cellulose	32.02	15.65			34
Cellulose acetate (56 % ac. groups)	27.83	13.60			34
(48 % ac. groups)	27.19	13.29			34
Cellulose diacetate	23.22	11.35	calc.		35
	22.3	10.9			36
	22.3	10.9			21
Cellulose nitrate					
(11.83 % N)	30.39	14.85			34
	21.44	10.48	calc.		35
	21.7	10.6			21
	23.5	11.5			6
(11.4 % N)	21.93	10.72			21
Ethyl cellulose	21.1	10.3			6

E. REFERENCES

1. J. Alfrey, A. O. Goldberg, J. A. Price, J. Colloid Sci. 5, 251 (1950).
2. G. Allen, G. Geen, D. Mangaraj, D. Dims, G. J. Wilson, Polymer 1, 467 (1960).
2a. E. Bagley, et al, J. Paint Tech. 41, 495 (1969); ibid. 43, 35 (1971).
3. R. F. Boyer, R. S. Spencer, J. Polymer Sci. 3, 97 (1948).
4. R. F. Boyer, R. S. Spencer, "High Polymer Physics", Paper 5, Part III, Remoen Press, New York, 1948.
5. G. M. Briston, W. F. Watson, Trans. Faraday Soc. 54, 1731, 1742 (1958).
6. H. Burrell, Official Digest of the Federation of Societies for Paint Technology 27, 726 (1955); 29, 1069, 1159 (1957); 40, 197 (1968).
7. H. Burrell, ACS Preprints, Div. Org. Coatings & Plastics, Sept. 1971.
8. A. T. DiBenedetto, J. Polymer Sci., A1, 3459 (1963).
9. D. Edelson, R. M. Fuoss, J. Am. Chem. Soc. 71, 3548 (1949).
10. H.-G. Elias, O. Etter, Makromol. Chem. 66, 56 (1963).
11. G. Gee, Trans. Faraday Soc. 40, 468 (1944).
12. G. Gee, Trans. Faraday Soc. 38, 418 (1942); see also ref. 13.
13. G. Gee, Trans. Inst. Rubber Ind. 18, 266 (1943).
14. G. Gee, G. Allen, G. Wilson, Polymer (London) 1, 456 (1960).
15. J. H. S. Green, Nature 183, 818 (1959).
15a. C. M. Hansen, J. Paint Tech. 39, 104,505 and 511 (1967).
16. R. A. Hayes, J. Appl. Polymer Sci. 5, 318 (1961).
16a. J. Hildebrand, R. Scott, "The Solubility of Non-electrolytes", 3rd Ed., Reinhold Publishing Corp., N. Y., 1949.
17. L. J. Hughes, private communication.
18. J. B. Kinsinger, private communication.
19. W. R. Krigbaum, private communication.
20. M. Lautout, M. Magat, Z. Physik.Chem. (Frankfurt) 16, 292 (1958).
20a. L.-H. Lee, J. Paint Tech. 42, 365 (1970).
21. M. Magat, J. Chem. Phys. 46, 344 (1949).
22. L. Mandelkern, private communication.
23. D. Mangaraj, S. K. Bhatnagar, S. B. Rath, Makromol.Chem. 67, 75 (1963).
24. D. Mangaraj, S. Patra, S. B. Rath, Makromol. Chem. 67, 84 (1963).
25. D. Mangaraj, Makromol. Chem. 65, 29 (1963).
26. D. Mangaraj, S. Patra, S. Rashid, Makromol. Chem. 65, 39 (1963).
27. D. Mangaraj, S. Patra, P. C. Roy, S. K. Bhatnagar, Makromol. Chem. 84, 225 (1965).
28. H. Mark, A. V. Tobolsky, "Physical Chemistry of High Polymers," Interscience, New York, 1950, p. 263.
29. W. R. Moore, J. Polymer Sci. 5, 91 (1950).
30. A. Noshay, C. C. Price, J. Polymer Sci. 54, 533 (1961).
31. R. B. Richards, Trans. Faraday Soc. 42, 10 (1946).
31a. G. Scatchard, Chem. Revs. 44, 7 (1949).
32. C. Schuerch, J. Am. Chem. Soc. 74, 5061 (1952).
33. R. L. Scott, M. Magat, J. Polymer Sci. 4, 555 (1949).
34. A. G. Shvarts, Kolloid. Zh. 18, 755 (1956); Colloid J. 18, 753 (1956).
35. P. A. Small, J. Appl. Chem. 3, 71 (1953).
35a. H. M. Spurlin, J. Polymer Sci. 3, 714 (1948).
35b. K. W. Suh, D. H. Clarke, J. Polymer Sci., A-1, 5, 1671 (1967).
36. A. V. Tobolsky, "Properties and Structure of Polymers", Wiley, New York, 1960, p. 64, 66.
36a. Union Carbide Corp., private communication.
37. E. E. Walker, J. Appl. Chem. 2, 470 (1952).
38. F. Vocks, J. Polymer Sci. A-2, 5319 (1964).
39. J. A. Yanko, J. Polymer Sci. 22, 153 (1956).
40. K. L. Hoy, J. Paint Technol. 42, 76 (1970).

Note: A complete bibliography will be found in references 6 and 7.

OPTICALLY ACTIVE POLYMERS

M. Goodman and N. Ueyama
University of California, San Diego
La Jolla, Calif.

Table of Content

The optical activity of polymers carrying asymmetric side chains and their low molecular weight analogs is given, by definition:

$$[M]_D = [\alpha]_D \times \frac{\text{Mean Residue Weight}}{100}$$

Abbreviations used:

AcAc — Acetylacetonate
AIBN — Azobisisobutyronitrile
BPO — Benzoyl peroxide
DBP — Dibenzoyl peroxide
DCA — Dichloroacetic acid

DMF — Dimethylformamide
DMSO — Dimethyl sulfoxide
EDC — Ethylene dichloride
M — Monomer
MC — Model compound
NCA — N-Carboxylic anhydride

P — Polymer
R.T. — Room Temperature
TCA — Trichloroacetic acid
TFA — Trifluoroacetic acid
TFEL — Trifluoroethanol
THF — Tetrahydrofuran

Polymer	Polymerization Systems			Polymer (P) Values		Monomer (M) or Model Compound (MC) Values		Optical Activity Measured		Ref.
	Catalyst or Initiator	Solvent	T [°C]	$[\alpha]_D$	$[M]_D$	$[\alpha]_D$	$[M]_D$	Solvent	T [°C]	
1. MAIN-CHAIN ACYCLIC CARBON POLYMERS										
1.1 POLY(ALKENES)										
Poly[(R)(-)3,7-dimethyl-1-octene]										
	Al(i-C4H9)3/TiCl4			-	-20 → -120	-	+14.4 (MC)	toluene	25	4, 7
	Al(i-C4H9)3/TiCl3			-	-35.3 → -144	-	+14.4 (MC)	toluene	25	
	LiAlH4/TiCl4/monomer	benzene	30-40	-78.1	-111	+6.25	+8.88 (MC)	carbon tetrachloride	25	2-6
Poly[(S)(+)5-methyl-1-heptene]										
	Al(i-C4H9)3/TiCl4	isooctane	R.T.	-	+27.4 → +68.1	-	+11.7 (MC)	benzene	25	2-6, 9
Poly[(S)(+)4-methyl-1-hexene]										
	Al(i-C4H9)3/TiCl4	isooctane	R.T.	-	+205 → +288	-	+21.3 (MC)	benzene (MC);toluene (P)	25	2-6, 9
	Al(i-C4H9)3/TiCl3	isooctane	R.T.	-	+149,+279	-	21.3 (MC)	benzene	25	
Poly[(S)4-methyl-1-hexyne]										
	Fe(AcAc)3/Al(i-C4H9)3	isooctane	20-30	-22.9	-	-2.97 (MC)	-	isooctane	25	1
Poly[(S)(+)6-methyl-1-octene]										
	Al(i-C4H9)3/TiCl3	isooctane	R.T.	-	+16.0	-	+13.3 (MC)	benzene	25	4
					+20.4	-	+13.3 (MC)			5
Poly[(S)(+)3-methyl-1-pentene]										
	Al(i-C4H9)3/TiCl4		20-25	-	+29.4,+158	-	-11.4 (MC)	tetralin (P)	20 (MC) 25 (P)	2-6, 8, 9
	Al(i-C4H9)3/TiCl3		20-25	-	+75.8,+157	-	-11.4 (MC)	tetralin (P)	20 (MC) 25 (P)	5
					+161				25 (P)	5
1.2 POLY(ACRYLICS), POLY(METHACRYLICS) and RELATED POLYMERS										
POLY(ACRYLICS)										
Poly(L-bornyl acrylate)										
	AIBN	benzene	65	-	-46.1	-	-56.4 (M)	toluene	25	10
	AIBN	toluene	60	-	-47.2	-	-56.4 (M)	toluene	25	
	AIBN	toluene	35	-	-47.4	-	-56.4 (M)	toluene	25	
	C6H5MgBr	toluene	-70	-	-43.0	-	-56.4 (M)	toluene	25	
	U.V. benzoin	toluene	-78	-	-47.2	-	-56.4 (M)	toluene	25	
Poly(d-sec-butyl-α-bromoacrylate)										
	BPO	dioxane	R.T.	-	+15.3	-	+37.4 (M)	dioxane	30	13
Poly(d-sec-butyl-α-chloroacrylate)										
	BPO		50-100	-	+23.4	-	+55.5 (M)	neat (M) chloroform (P)	22 (M) 21 (P)	13
	BPO	dioxane	R.T.	-	+17.9	-	+42.2 (M)	dioxane	25	14
Poly(L-menthyl acrylate)										
	AIBN		50	-79.0	-	-99.3 (M)	-	benzene	22	11

Polymer	Polymerization Systems			Polymer (P) Values		Monomer (M) or Model Compound (MC) Values		Optical Activity Measured		Ref.
	Catalyst or Initiator	Solvent	T [°C]	$[\alpha]_D$	$[M]_D$	$[\alpha]_D$	$[M]_D$	Solvent	T [°C]	

POLY(METHACRYLICS)

Polymer	Catalyst or Initiator	Solvent	T [°C]	$[\alpha]_D$	$[M]_D$	$[\alpha]_D$	$[M]_D$	Solvent	T [°C]	Ref.
Poly(L-bornyl methacrylate)										
	AIBN	benzene	55	-	-70.8	-	-50.1 (M)	toluene	20	10
Poly(1,2-5,6-diisopropylidene-D-glucofuranosyl-3-methacrylate)										
	BPO	benzene	80	-49.0	-	-32.2 (M)	-	ethanol (M);dichloroethane (P)	25	23,24
	AIBN	benzene	40-50	-48.4	-	-28.8 (M)	-	benzene	20	25-27
Poly(2,3-4,5-diisopropylidene-L-sorbofuranosyl-1-methacrylate)										
	AIBN	benzene		-84.5	-	-8.6 (M)	-	benzene	26	
Poly[(+)1,3-dimethylbutyl methacrylate]										
	BPO	-	82-110	+18.2	+30.9	+23.4 (MC)	+43.5 (MC)	methylene chloride	20	17,18
	AIBN									
Poly(1-menthyl methacrylate)										
	AIBN	benzene	55	-	-193	-	-202 (M)	benzene	25	16,11
	BPO	-	40-85	-	-256.8	-	-202 (M)	chloroform	20	
	C6H5MgBr	toluene	R.T.	-	-158.9	-	-202 (M)	chloroform	20	
	C6H5MgBr	toluene	-75	-	-177.0	-	-202 (M)	benzene	25	
	γ-ray	-	-75	-	-194.0	-	-202 (M)	benzene	25	
Poly(1-α-methylbenzyl methacrylate)										
	U.V. AIBN	-	35	-79.5	-151	-54.4 (M)	-103 (M)	dioxane	25	19-21
	U.V. benzoin	-	-65	-72.5	-138	-54.4 (M)	-103 (M)	dioxane	25	
	n-BuLi	toluene	-60	-100	-190	-54.4 (M)	-103 (M)	dioxane	25	
	AIBN	dioxane	35	-	-147	-	-78.8 (M)	neat (M);dioxane (P)	25	
Poly(2-methylbutyl methacrylate)										
	BPO	-	100	-	+5.93	-	+6.99 (M)	isooctane (M);chloroform (P)	20	16,22
	C6H5MgBr	toluene	R.T.	-	+6.84	-	+6.99 (M)	isooctane (M);chloroform (P)	20	

POLY(ACRYLAMIDES), POLY(METHACRYLAMIDES)

Polymer	Catalyst or Initiator	Solvent	T [°C]	$[\alpha]_D$	$[M]_D$	$[\alpha]_D$	$[M]_D$	Solvent	T [°C]	Ref.
Poly(acryl-L-glutamic acid)										
	AIBN	dioxane	60	-21.0	-	-21.0 (MC)	-	dioxane	25	12
Poly(methacryl-D-alanine)										
	AIBN	dioxane	60	+42.0	-	+48.9 (MC)	-	dioxane	25	28
Poly(methacryl-L-glutamic acid)										
	AIBN	dioxane	60	-23.0	-	-26.0 (MC)	-	dioxane	25	28
Poly[(+)N-methyl-N-α-methylbenzyl-acrylamide]										
	AIBN	-	60		278		504 (M)	benzene	25	29
	n-BuLi	toluene	22		397		504 (M)	benzene	25	
	n-BuLi	toluene	-50		366		504 (M)	benzene	25	
	C6H5MgBr	toluene	-50		337		504 (M)	benzene	25	
Poly[(-)N-propyl-N-α-methylbenzyl-acrylamide]										
	AIBN	toluene	70		-167		-444 (M)	benzene	25	29
	n-BuLi	toluene	22		-318		-444 (M)	benzene	25	

References page IV-374

Polymer	Polymerization Systems			Polymer (P) Values		Monomer (M) or Model Compound (MC) Values		Optical Activity Measured		Ref.
	Catalyst or Initiator	Solvent	T [°C]	$[\alpha]_D$	$[M]_D$	$[\alpha]_D$	$[M]_D$	Solvent	T [°C]	
POLY(ITACONATE)										
Poly[bis-2-methylbutyl itaconate]	-		R.T.	+4.92		+4.97 (M)				15
	BPO	-	40-85	+4.75		+5.57 (M)		heptane	20	16
1.3 POLY(VINYL ETHERS), POLY(VINYL KETONES), POLY(VINYL ESTERS)										
POLY(VINYL ETHERS)										
Poly(1-bornyloxyethylene)	$BF_3 \cdot OEt_2$			-	-142		-122.4 (M)	benzene	25	31
Poly(1-cholesteryloxyethylene)				-	+312		+34.5 (MC)			5
				-	-93		-131.4 (M)	benzene	25	31
Poly(1,2-5,6-diisopropylidene-α-D-glucofuranosyl-3-oxyethylene)	$BF_3 \cdot OEt_2$	n-hexane + CH_2Cl_2	-78	-	+5.7	-	-77.2 (M)	ethanol (M);dichloroethane (P)	25	36,24
Poly(1-menthyloxyethylene)	C_4H_9MgBr	toluene	80	-	-358	-	-121.9 (M)	chloroform (M);benzene (P)	25	31
	$BF_3 \cdot OEt_2$	n-hexane	-78	-	-396	-	-121.9 (M)	chloroform (M);benzene (P)	25	31
	$SnCl_4$	petroleum ether	25	-	-373	-	-121.9 (M)	chloroform (M);benzene (P)	25	
	$Mn\text{-}MoO_3\text{-}H_2SO_4$	n-hexane	-10	-	-357	-	-118.8 (M)	chloroform (M);benzene (P)	25	33
	$Mn\text{-}MoO_3\text{-}H_2SO_4$	n-hexane	-30	-	-353	-	-118.8 (M)	chloroform (M);benzene (P)	25	
Poly(1-α-methylbenzyloxyethylene)	$BF_3 \cdot OEt_2$	propane	-78	-	+68.6	-	-71.9 (M)	benzene	25	35
Poly[(S)2-methylbutyloxyethylene]	$Al(i\text{-}C_4H_9)Cl$	propylene + toluene	-78	-	+4.9	-	+1.1 (MC); +7.6 (M)	neat (M, MC); toluene (P)	25	34
	$Al(i\text{-}C_4H_9)_3/H_2SO_4$	ether	15-20	-	-5.9	-	+1.1 (MC); +7.6 (M)	neat (M, MC); toluene (P)	25	
	$Al(i\text{-}OC_3H_7)_3/H_2SO_4$	ethyl acetate	0-20	-	+6.4	-	+1.1 (MC); +7.6 (M)	neat (M, MC); toluene (P)	25	
Poly[(S)1-methylpropyloxyethylene]	$Al(i\text{-}C_4H_9)Cl$	propylene + toluene	-78	-	+246	-	+13.24 (M)	neat (M);toluene (P)	25	32
	$Al(i\text{-}C_4H_9)_3/H_2SO_4$	diethyl ether	15-20	-	+206	-	+13.67 (M)	neat (M);toluene (P)	25	
POLY(VINYL KETONES)										
Poly[(S)2-methylbutyl vinyl ketone]	spontaneous	-	20-25	-	-10.0	-	+11.5 (MC)	chloroform	25	30
	$LiAlH_4$	toluene	-16-50	-	43.0	-	+11.5 (MC)	chloroform	25	
Poly[(S)3-methylpentyl vinyl ketone]	spontaneous	-	20-25	-	+15.6	-	+15.2 (MC)	chloroform	25	30
	$LiAlH_4$	toluene	0	-	+11.7	-	+15.2 (MC)	chloroform	25	
Poly[(S)1-methylpropyl vinyl ketone]	$LiAlH_4$	toluene	0	-	-118	-	+33.4 (MC)	chloroform	25	30

Polymer	Catalyst or Initiator	Solvent	T [°C]	[α]$_D$	[M]$_D$	[α]$_D$	[M]$_D$	Solvent	T [°C]	Ref.
Poly[(S)1-methylpropyl vinyl ketone] (Cont'd.)										
	spontaneous	-	R.T.	-	-42.5	-	+33.4 (MC)	chloroform	25	
POLY(VINYL ESTERS)										
Poly(d-sec-butyl p-vinyl-benzoate)										
	BPO	-	40	+22.9	-	+24.0 (M)	-	benzene	50	40
Poly(vinyl -1-2-phenylbutyrate)										
	BPO	dioxane	100	-29.1	-	-20.4 (M)	-	dioxane	25	13
1.4 POLY(STYRENES)										
Poly(4-sec-butoxymethylstyrene)										
	BPO	dioxane	55	+10.15	-	+12.07 (M)	-	dioxane	25	41
Poly(2-(2-thio-3-methylpentyl)styrene)										
	AIBN	-	76	+9.57	-	+13.97 (M)	-	neat (M);benzene (P)	24.7	42
1.5 OTHERS										
Poly(N-bornyl maleimide)										
	BPO	THF	50	+8.4	-	-9.6 (M)	-	ethanol (M);THF (P)	25 (P)	43
	AIBN	THF	50	+10.1	-	-9.6 (M)	-	ethanol (M);THF (P)	25 (P)	
	n-BuLi	THF	50	-9.1	-	-9.6 (M)	-	ethanol (M);THF (P)	25 (P)	
Poly[N^3-(1,2-diethoxycarbonylethyl)ureidoethylene]										
	AIBN	benzene	60	+24.3	+63.1	+43.5 (M)	+112.5 (M)	dioxane	25	38
				+31.7	+82.0	+43.5 (M)	+112.5 (M)	benzene	25	
Poly[N^3-(1,3-diethoxycarbonylpropyl)ureidoethylene]										
	AIBN	benzene	60	+7.83	+21.3	+21.6 (M)	+58.9 (M)	benzene	25	38
Poly[(R)3,7-dimethyloctanal divinylacetal] $R = CH_2-CH(CH_3)-(CH_2)_3-CH(CH_3)_2$										
	AIBN	-	80	+3.18	+7.19	+2.48 (MC)	-	chloroform	25	72
	BF$_3$ · OEt$_2$	-	-78	-6.78	-14.32	+2.48 (MC)	-	chloroform	25	
Poly[N^3-(1-ethoxycarbonyl-3-methylbutyl)ureidoethylene]										
	AIBN	benzene	60	+0.5	+1.1	+20.3 (M)	+46.3 (M)	benzene	25	38
Poly[(2-isopropyl-5-methylphenoxycarbonylaminoethylene]										
	AIBN	benzene	95	-60.4	136.0	-79.9 (M)	-179.7 (M)	benzene	25	38
	AIBN	benzene	60	-60.4	136.0	-79.9 (M)	-179.7 (M)	benzene	25	
	γ-ray	benzene	22	-58.5	-131.8	-79.9 (M)	-179.7 (M)	benzene	25	
	t-BuOOH	SO$_2$	-25	-66.2		-79.9 (M)	-179.7 (M)	benzene	25	
Poly[(d)α-phenylethyl isonitrile]										
	DBP-O$_2$	-	60	+365	-	+37.2 (M)	-179.7 (M)	toluene	27	39
Poly[(1)α-phenylethyl isonitrile]										
	H$_2$SO$_4$-O$_2$	-	50	-382	-	-35.8 (M)	-179.7 (M)	toluene	27	39

References page IV-374

2. MAIN-CHAIN ACYCLIC HETEROATOM POLYMERS

2.1 POLY(OXIDES)

Polymer	Catalyst or Initiator	Solvent	T [°C]	$[\alpha]_D$	$[M]_D$	$[\alpha]_D$	$[M]_D$	Solvent	T [°C]	Ref.
Poly[R)(+)citronellal]										
	Al(i-C$_4$H$_9$)$_3$	ether	-78	-82.7~90.1	-127 - -139	-	+11.2 (MC)	chloroform	25	44
	n-BuLi	n-hexane	-78	-91.1	-140	-	+11.2 (MC)	chloroform	25	
	BF$_3$·OEt$_2$	ether	-78	-89.5	-138	-	+11.2 (MC)	chloroform	25	
Poly(D-oxy-2,3-dimethylethylene) $+O-CH-CH-CH_2+_n$ / CH$_3$ CH$_3$										
	Al(i-C$_4$H$_9$)$_3$/H$_2$O	n-heptane	-78	0		+58.8 (M)	-	neat (M);benzene or chloroform (P)	25	47
Poly[R)(+)oxy-5-methoxy-3-methyl1-hexylidene]										
	ZnEt$_2$	n-hexane	-75	+36.7	+52.8	+2.47 (M)	+8.24 (MC)	chloroform	25	45
Poly[(R) oxy-2-methylbutylidene]										
	ZnEt$_2$	n-hexane	-75	+36.7	-	+30.7 (M)	-	chloroform	25	45
Poly[(+)oxy-1-methyl1-2-carbomethoxyethylene] CH$_3$-OCO CH$_3$ / $+O-CH-CH+_n$										
	BF$_3$·OEt$_2$/AlEt$_3$	-	25	-36.9	-	+24.6 (M)	-	neat (M);methylene chloride (P)	25	48
Poly(L-oxypropylene)										
	KOH	-	R.T.	-16±5	-	+15 (M)	-	benzene (P) ether (M)	20 (P) 21 (M)	46
				+25±5		+15 (M)	-	chloroform (P)	20 (P) 21 (M)	
	FeCl$_3$-monomer	ether	R.T.	-20±5	-	+15 (M)	-	benzene (P)	20 (P) 21 (M)	
				+25±5		+15 (M)	-	chloroform (P)	20 (P) 21 (M)	

2.2 POLY(ESTERS)

Polymer	Catalyst or Initiator	Solvent	T [°C]	$[\alpha]_D$	$[M]_D$	$[\alpha]_D$	$[M]_D$	Solvent	T [°C]	Ref.
Poly[L(-) oxycarbonylethylidene]										
	ZnCl$_2$	heptane	175-195	$[\alpha]_{578}$ -153.3	-	$[\alpha]_{578}$ -275.8 (M)	-	chloroform	25	51
Poly(L-oxycarbonylisobutylidene)										
	ZnO	-	180-190	~20	-	~-50 (MC)	-	cyclohexane	25	52
Poly[(R)(+)oxycarbonyl-3-methylhexamethylene]										
	Sb$_2$O$_3$/Zn(CH$_3$COO)$_2$	-	215	+8.55	-	+0.94 (M)	-	TFEL	25	49
Poly[(R)(-)oxycarbonyl-3-methylhexamethylene]										
	Sb$_2$O$_3$/Zn(CH$_3$COO)$_2$	-	215	+3.36	-	-0.96 (M)	-	benzene	25	49
Poly[(R)(-)oxycarbonyl-2-methylpentamethylene] (Poly[(R)(-)-β-methylcaprolactone])										
	Al(i-C$_4$H$_9$)$_3$/H$_2$O	benzene	50	$[\alpha]_{450}$ +32.57	-	$[\alpha]_{350}$ +29.78 (MC)	-	TFEL	25	53
Poly[(R)oxycarbonyl-3-methylpentamethylene]										
	Al(i-C$_4$H$_9$)$_3$/H$_2$O	benzene	50	$[\alpha]_{450}$ +26.95	-	$[\alpha]_D$ +51.77 (M)	-	TFEL (P);chloroform (M)	25	53

Column group headers: Polymerization Systems; Polymer (P) Values; Monomer (M) or Model Compound (MC) Values; Optical Activity Measured

Polymer	Catalyst or Initiator	Polymerization Systems Solvent	Polymer (P) Values T [°C]	[α]_D	[M]_D	Monomer (M) or Model Compound (MC) Values [α]_D	[M]_D	Optical Activity Measured Solvent	T [°C]	Ref.
Poly[(R)oxycarbonyl-4-methylpentamethylene]										
	Al(i-C$_4$H$_9$)$_3$/H$_2$O	benzene	50	[α]$_{450}$ +8.78	-	[α]$_{350}$ -15.91 (MC) [α]$_D$ -36.11 (M)	-	TFEL (P, MC) chloroform (M)	25	53
Poly(D-oxycarbonylpropylene)										
	AlEt$_3$	ether	R.T.	[α]$_{300}$ +19	-	-	-	chloroform		50
	natural polymer		R.T.	[α]$_{300}$ +44	-	-	-	chloroform		50
Poly(oxycarbonyl-1,2,3-trimethoxytetramethylene) (Poly(2,3,4-trimethyl-1-arabonolactone))										
	CH$_3$COCl/HCl	-	R.T.	[α]$_{546}$ +39	-	[α]$_{546}$ +203.5 (M)	-	benzene	21	54

2.3 POLY(URETHANES)

Polymer	Catalyst or Initiator	Solvent	T [°C]	[α]_D	[M]_D	[α]_D	[M]_D	Solvent	T [°C]	Ref.
Poly[oxycarbonylimino-(L)2-isobutylethylene-ureylene-(L)1-isobutylethylene]	Na$_2$CO$_3$	H$_2$O/THF	< 0	-48.2	-	-	-	methanol	20	63
Poly[oxycarbonylimino-(L)2-isobutylethylene-ureylene-(L)1-methylethylene]	Na$_2$CO$_3$	H$_2$O/THF	< 0	-19.2	-	-	-	methanol	20	63
Poly[oxycarbonylimino-(L)2-isobutylethylene-ureylene-(L)1-methylethylene]	Na$_2$CO$_3$	H$_2$O/THF	< 0	-27.2	-	-	-	methanol	20	63
Poly[oxycarbonylimino-(L)2-isobutylethylene-ureylene-(L)1-methylethylene]	Na$_2$CO$_3$	H$_2$O/THF	< 0	-27.2	-	-	-	methanol	20	63
Poly[oxycarbonylimino-(D)2-methylethylene-ureylene-(L)1-isobutylethylene]	Na$_2$CO$_3$	H$_2$O/THF	< 0	-54.7	-	-	-	methanol	20	63
Poly[oxycarbonylimino-(L)2-methylethylene-ureylene-(L)1-isobutylethylene]	Na$_2$CO$_3$	H$_2$O/THF	< 0	-47.9	-	-	-	methanol	20	63
Poly[oxycarbonylimino-(D)2-methylethylene-ureylene-(L)1-isopropylethylene]	Na$_2$CO$_3$	H$_2$O/THF	< 0	-53.2	-	-	-	methanol	20	63
Poly[oxycarbonylimino-(L)2-methylethylene-ureylene-(L)1-isopropylethylene]	Na$_2$CO$_3$	H$_2$O/THF	< 0	-26.0	-	-	-	methanol	20	63
Poly[oxycarbonylimino-(D)2-methylethylene-ureylene-(L)1-methylethylene]	Na$_2$CO$_3$	H$_2$O/THF	< 0	-16	-	-	-	m-cresol	20	63
Poly[oxycarbonylimino-(L)2-methylethylene-ureylene-(L)1-methylethylene]	Na$_2$CO$_3$	H$_2$O/THF	< 0	-50	-	-	-	m-cresol	20	63

2.4 POLY(SULFIDES), POLY(THIOESTERS)

Polymer	Catalyst or Initiator	Solvent	T [°C]	[α]_D	[M]_D	[α]_D	[M]_D	Solvent	T [°C]	Ref.
Poly[(R)(-)thiocarbonyl-3-methylpentamethylene]	n-BuLi	-	R.T.	-5.39	-	-3.21 (M)	-	dioxane	25	57
Poly[(R)(+)thiocarbonyl-2-methylpentamethylene]	n-BuLi	-	190	+15.66	-	+9.03 (M)	-	dioxane	25	57
	n-BuLi	-	190	+18.62	-	-	-	THF	25	57
Poly[(-)thiopropylene]	BF$_3$·OEt$_2$	methylene chloride	-	-62.8	-	-32.6 (M)	-	methylene chloride	22	55
	BF$_3$·OEt$_2$	methylene chloride	-	-126.9	-	-32.6 (M)	-	chloroform	22	55
	KOH	-	-	-159	-	-35 (MC)	-	benzene (P)	18	56
	NaOH	-	15	-189	-	-35 (MC)	-	chloroform (P)	18	56

References page IV-374

2.5 POLY(AMIDES)

Polymer	Polymerization Systems			Polymer (P) Values		Monomer (M) of Model Compound (MC) Values		Optical Activity Measured		Ref.
	Catalyst or Initiator	Solvent	T [°C]	$[\alpha]_D$	$[M]_D$	$[\alpha]_D$	$[M]_D$	Solvent	T [°C]	
Poly[threo-imino-(1-oxo-2,3-dimethyltrimethylene)] $+NHCO-CH-CH-CH+_n$ (CH$_3$ CH$_3$)	K-Pyrrolidone	DMSO	25	-25.0		-50.6 (M)		neat (M);TFA (P)	25	62
Poly[(S)imino(1-oxo-3-ethyltrimethylene)] (Poly(S)β-aminovalerate))	K-Pyrrolidone	DMSO	25	-32.7		-2.4 (M)		neat (M);DCA (P)	25	62
Poly[(-)imino-(1-oxo-3-methylhexamethylene)]	H$_2$O	-	200	+39.7	-	-36.15 (M)		cresol (P);water (M)	25	59,60
	NaH	-	160	+43.07	-	-36.15 (M)		cresol (P);water (M)	25	61
Poly[(+)imino(1-oxo-4-methylhexamethylene)]	NaH		150	+19.24	-			TFEL	25	61
Poly[(-)imino-(1-oxo-5-methylhexamethylene)]	NaH		194	+25.15	-			TFEL	25	61
Poly[(+)imino-(1-oxo-3-methyl-6-isopropylhexamethylene)]	K-C$_6$H$_5$COCl		150	+9.5	-	-53.6 (M)	-	formic acid	19	58
Poly[(S)imino-(1-oxo-3-methyltrimethylene)] (Poly(S)β-aminobutyrate))	K-Pyrrolidone	DMSO	25	-9.3		-8.3 (M)		neat (M);TCA (P)	25	62
				-26.1				DCA (P)	25	62
				-33.5				TCA-chlorobenzene (P)	25	62
Poly[(R)imino-(1-oxo-3-vinyltrimethylene)]	K-Pyrrolidone	DMSO	25	-19.5		+25 (M)		neat (M);DCA (P)	25	62
Poly[(+)iminoterephthaloylimino-2,8-naphthylene-8,2-naphthylene]				$[\alpha]_{578}$ -15		$[\alpha]_{578}$ +66.6 (MC)		THF	21	71
Poly[S(+)N-(2-methylbutyl)iminocarbonyl]	NaCN	DMF	-75	+160	+180.8	+3.08 (M) -6 (MC)	-10.3 (MC)	chloroform	25	64
Poly[N-(phenylethylidene)iminoethylene-N-(phenylethylidene)iminoterephthaloyl]				$[\alpha]_{289}$ -74 $[\alpha]_{289}$ -294					20	70
Poly[N-(phenylethylidene)iminohexamethylene-N-(phenylethylidene)iminodipicolinoyl]				$[\alpha]_{289}$ -139 $[\alpha]_{289}$ -610					20	70
Poly[d-N-(phenylpropyliminocarbonyl)]	NaCN	DMF	R.T.	-468.0	-754.8	+35 (M) +114.4 (MC)	+44.8 (M)	chloroform	25	65
O$_2$N—(NO$_2$)(CH$_3$ CH$_3$)—CO—CONH-R-NH]$_n$										
(-) R = 1,2-phenylene				-4.94				5% LiCl-DMF	20	73
(+) R = 1,3-phenylene				+0.50				5% LiCl-DMF	20	73
(+) R = 1,4-phenylene				+6.53				5% LiCl-DMF	20	73
(+) R = 1,2-pyrazolidine				+5.4				TFEL	20	73
(+) R = 2,5-dimethyl-1,4-piperazinediyl				+28.0				TFEL	20	73

Polymer	Polymerization Systems		Polymer (P) Values			Monomer (M) or Model Compound (MC) Values		Optical Activity Measured		Ref.
	Catalysator or Initiator	Solvent	T [°C]	$[\alpha]_D$	$[M]_D$	$[\alpha]_D$	$[M]_D$	Solvent	T [°C]	

+CO—[benzene ring with Cl, Cl, CH₃, CH₃]—CONH—R—NH+ₙ

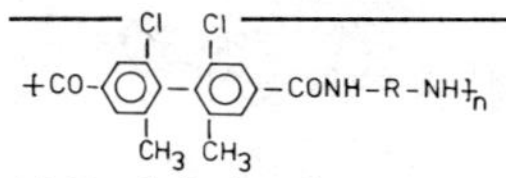										
(-) R = 1,2-phenylene				-19.6	-	-	-	5% LiCl-DMF	20	73
(-) R = 1,3-phenylene				-19.4	-	-	-	5% LiCl-DMF	20	73
(-) R = 1,4-phenylene				-18.2	-	-	-	5% LiCl-DMF	20	73
(-) R = 1,2-piperazinediyl				-6.7	-	-	-	TFEL	20	73
(-) R = 1,4-piperazinediyl				-10.0	-	-	-	TFEL	20	73
(-) R = 2,5-dimethyl-1,4-piperazinediyl				-6.7	-	-	-	TFEL	20	73

2.6 POLY(IMINES)

$\{N-CH-(CH_2)_2\}_n$

Polymer	Catalysator or Initiator	Solvent	T [°C]	$[\alpha]_D$	$[M]_D$	$[\alpha]_D$	$[M]_D$	Solvent	T [°C]	Ref.
Poly[D-conidine]	BF₃·OEt₂	ether	R.T.	-140.8	-	+71.7 (M)	-	chloroform		69
Poly[L-conidine]	BF₃·OEt₂	ether	R.T.	+140.8	-	-71.0 (M)	-	chloroform		69
Poly[(R)imino-2-ethylethylene]	BF₃·OEt₂	-	100	127.4	-	+17.6 (M)	-	neat (M);benzene (P)	23 (P) 18 (M)	67
	AlCl₃	-	100	-108.9	-	+17.6 (M)	-	neat (M);benzene (P)	16 (P) 18 (M)	67
	from (R)-2-amino-1-bromobutane			-109.2	-	-	-	benzene	16	
Poly[(S)iminoisobutylethylene]	BF₃·OEt₂	-	80-100	+66.7	-	-24.6 (M)	-	benzene	20 (P) 32 (M)	68
Poly[L-iminopropylene]	BF₃·OEt₂	-	80	+109.3	-	-12.8 (M)	-		22	66
Poly[D-iminopropylene]	BF₃·OEt₂	-	80	-101.5	-	+12.4 (M)	-		22	66

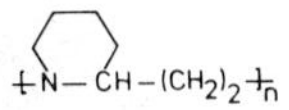

2.7 POLY(AMINO ACIDS)

Polymer	Catalysator or Initiator	Solvent	T [°C]	$[\alpha]_D$	Solvent	T [°C]	Ref.
Poly(O-acetyl-L-hydroxyproline)	NCA	pyridine		+25 (Form I)	90% HCOOH	25	110
				-175 (Form II)	90 % HCOOH after 6 hrs	25	110
Poly(O-acetyl-L-serine)	NCA, triethylamine	nitrobenzene		$[\alpha]_{546}$ +31 ; $[\alpha]_{546}$ +159	DCA ; 10% DCA-90% EDC	25	114
Poly(L-alanine)	NCA, tributylamine	benzene-dioxane		$[m']_D$ +21 ($[Rvac]_D$ +82) ; $[m']_D$ -90	99% CHCl₃-1% DCA		81, 84
Poly(α-L-aspartic acid)	from poly(β-benzyl-L-aspartate) (NCA, 120°C in vac.)		80	+6.2	c = 3 1 N NaOH		87
			80	-11.2	c = 5 1 N NaOH		88
Poly(β-L-aspartic acid)	Active ester method		140	$[\alpha]_{546}$ -14 ; $[\alpha]_{546}$ -23	0.2 M NaCl, pH 2 ; 0.1 M NaCl, pH 4	25 ; 25	89 ; 89

References page IV-374

Polymer	Polymerization Systems			Polymer (P) Values		Monomer (M) or Model Compound (MC) Values		Optical Activity Measured		Ref.
	Catalyst or Initiator	Solvent	T [°C]	$[\alpha]_D$	$[M]_D$	$[\alpha]_D$	$[M]_D$	Solvent	T [°C]	
Poly(D-α-amino-n-butyric acid)										
	NCA, $C_6H_5CH_2NH_2$	nitrobenzene		$[m']_D$ -19 ($[Rvac]_D$ +81)				90% $CHCl_3$ -10% DCA		81,84
				$[m']_D$ +107				TFA		
Poly(L-arginine · HCl)										
	from poly(L-nitroarginine)			-21.3				H_2O	20	85
Poly(β-benzyl-L-aspartate)										
	NCA, NaOMe	chloroform		$[\alpha]_{546}$ -168				chloroform	22	86
				$[\alpha]_{546}$ -18				DCA	22	86
Poly(β-benzyl-D-aspartate)										
	NCA, NaOMe	chloroform		$[\alpha]_{546}$ +174				chloroform	22	86
				$[\alpha]_{546}$ +19				DCA	22	86
Poly(γ-benzyl-L-glutamate)										
	NCA			-30				hydrazine	20	96
				-46				TFA	20	96
	NaOMe	chloroform		$[\alpha]_{546}$ +14				chloroform	22	86,98
	NaOMe	chloroform		$[\alpha]_{546}$ -15				DCA	22	86
	NaOMe	chloroform		$[\alpha]_{546}$ +15				chloroform	25	97
	NaOMe	chloroform		-14				DCA	25	97
Poly(γ-benzyl-D-glutamate)										
	NCA, NaOMe	chloroform		$[\alpha]_{546}$ -18				chloroform	22	86
	NCA, NaOMe	chloroform		$[\alpha]_{546}$ +17				DCA	22	86
Poly(L-benzyl-L-histidine)										
	NCA, triethylamine	dioxane		$[\alpha]_{546}$ +11				chloroform	25	101
	triethylamine			$[\alpha]_{546}$ 208				DCA-$CHCl_3$	25	101
	NCA, triethylamine	dioxane		-21.1				acetic acid	20	102
Poly(S-carbobenzoxymethyl-L-cysteine)										
	NCA, NaOMe	nitrobenzene		$[\alpha]_{546}$ -128				0.2 M NaCl, pH 7	25	90
				$[\alpha]_{546}$ -125				0.2 M NaCl, pH 5	25	
				$[\alpha]_{546}$ -33				0.2 M NaCl, pH 3	25	
Poly(ε-N-carbobenzoxy-L-lysine)										
	NCA, NaOMe	dioxane		$[\alpha]_{546}$ +5				chloroform	25	84,103
				$[\alpha]_{546}$ -25				50% $CHCl_3$ -50% DCA	25	84,103
Poly(3,4-dehydro-L-proline)										
	NCA	pyridine		-500				99% formic acid	25	109
				-1250				TFA	25	109
Poly(L-glutamic acid)										
	from poly(γ-benzyl-L-glutamate) (NCA, aliphatic amine, DMF)									
				+3				dioxane-H_2O, 0.2 M NaCl, pH 4.7	22	84,92
				-77				dioxane-H_2O, 0.2 M NaCl, pH 6.5	22	84,92
				+7				2-chloroethanol-H_2O, 0.1 M RbCl, pH 5		84,93
				-75				2-chloroethanol-H_2O, 0.1 M RbCl, pH 7.5		84,93
	from poly(γ-benzyl-L-glutamate) (NCA, NaOMe)									
				-6				0.2 M NaCl, pH 4.5	20	84,94
				-102				0.2 M NaCl, pH 8.0	20	84,94

Polymer	Polymerization Systems — Catalyst or Initiator, Solvent	Polymer (P) Values — T [°C]	Polymer (P) Values — [α]_D	Polymer (P) Values — [M]_D	Monomer (M) or Model Compound (MC) Values — [α]_D	Monomer (M) or Model Compound (MC) Values — [M]_D	Optical Activity Measured — Solvent	Optical Activity Measured — T [°C]	Ref.
Poly(γ-D-glutamic acid) Active ester method			+23.6				H_2O	23	95
Poly(γ-L-glutamic acid) Active ester method			-23.6				H_2O	23	95
Poly(L-histidine) from poly(1-benzyl-L-histidine) (NCA, NEt₃, dioxane)			$[\alpha]_{546}$ -65				0.2 M NaCl, pH 2.4	25	101
			$[\alpha]_{546}$ -210				0.2 M NaCl, pH 5.85	25	101
			$[\alpha]_{546}$ -67				0.2 M NaCl, pH 6.00	25	101
			$[\alpha]_{546}$ +28				DCA	25	101
			$[\alpha]_{546}$ -20				CH_3COOH	22	102
Poly(δ-hydroxy-O-acetyl-L-α-aminovaleric acid) NCA, triethylamine, nitrobenzene			$[\alpha]_{546}$ +10.1				DMF	25	83
			$[\alpha]_{546}$ -26.5				TFEL	25	
			$[\alpha]_{546}$ -31.4				DCA	25	
Poly(L-hydroxyproline) from poly(O-acetyl hydroxy-L-proline)			-400				H_2O	25	110
Poly(L-leucine) NCA, tributylamine, benzene			$[m']_D$ -12 ($[Rvac]_D$ +96)				benzene		81,84
			$[m']_D$ -111				TFA		81
Poly(L-lysine) from poly(ε-N-carbobenzoxy-L-lysine)			-144				H_2O, pH 7		105
			-35				H_2O, pH 12		105
			-51				H_2O, pH 10.10	21.4	105
			-77				H_2O, pH 9.98	23.2	105
			-108				H_2O, pH 9.72	25.6	105
Poly(L-lysine hydrochloride) from poly(-N-carbobenzoxy-L-lysine)			-78.8				6 N HCl	25	106
Poly(L-methionine) NCA, NaOMe, nitrobenzene			$[\alpha]_{546}$ +24.4				90% CHCl₃-10% DCA	25	107
			$[\alpha]_{546}$ +16.8				DCA	25	107
			$[\alpha]_{546}$ -115.3				TFA	25	107
			$[\alpha]_{546}$ +22				chloroform		108
			$[\alpha]_{546}$ -107				80% TFA-20% DCA		108
Poly(L-methionine-S-carboxymethylthetin) $+NH-CH-CO+_n$ $(CH_2)_2-S^{\oplus}-CH_2-COO^{\ominus}$ / CH_2-COOH $Br^{\ominus}$ from poly(L-methionine)			$[\alpha]_{546}$ -44				HCl, H_2O (pH 1.1~2.0)	25	107
Poly(L-methionine-S-methylsulfonium bromide) $+NH-CH-CO+_n$ $(CH_2)_2-S^{\oplus}<^{CH_3}_{CH_3}$ from poly(L-methionine)			$[\alpha]_{546}$ -74.3				H_2O	24	107
			$[\alpha]_{546}$ -38.6				10 M LiBr	24	107
Poly(N-methyl-L-alanine) NCA, $C_6H_5CH_2CH_2NH_2$, dioxane			-3.04				TFEL	24	82

References page IV-374

OPTICALLY ACTIVE POLYMERS

Polymer	Polymerization Systems		Polymer (P) Values			Monomer (M) or Model Compound (MC) Values		Optical Activity Measured		Ref.
	Catalyst or Initiator	Solvent	T [°C]	$[\alpha]_D$	$[M]_D$	$[\alpha]_D$	$[M]_D$	Solvent	T [°C]	
Poly(γ-methyl-L-glutamate)	NCA			-31.3				DCA	25	99
				-76.1				TFA	25	99
				+10				DMF	25	100
Poly(L-nitroarginine)	NCA	DMF		-16.6				DMF	30	85
Poly(γ-phthalimidomethyl-L-glutamate)	NCA, triethylamine			+12.5				DMF	25	91
Poly(L-pipecolic acid)	NCA, diethylamine [structure: N—CH—CO cyclohexyl]$_n$	benzene or dioxane		-325 (Form I)				acetic acid	25	109
				-50 (Form II)				acetic acid after 2hrs	25	
Poly(L-proline)				+50 (Form I)				acetic acid		111
				-540 (Form II)				acetic acid		111
	NCA, pyridine			+48 (Form I)				10% pyridine-90% H_2O	25	112
				-420 (Form II)				10% pyridine after 50hrs	25	112
				+33 (Form I)				H_2O		113
				-540 (Form II)				0.1 M KCl		113
Poly(L-serine)	from poly(O-acetyl-L-serine)			$[\alpha]_{546}$ -22				H_2O	25	114
				$[\alpha]_{546}$ -7.3				10 M LiBr	25	
				$[\alpha]_{546}$ -25.2				8 M urea	25	
				$[\alpha]_{546}$ -9.3				hydrazine	25	
				$[\alpha]_{546}$ -9.7				DCA	25	
Poly(O-p-tolylsulfonyloxy-L-proline)	NCA	pyridine		0 (Form I)				acetic acid	25	110
				-120 (Form II)						
Poly(ε-N-trifluoroacetyl-L-lysine)	NCA, diethylamine	DMF		$[\alpha]_{546}$ -6.8				methanol	20	104
				$_{546}$ -36.8				DCA	20	
Poly(L-tryptophan)	NCA, diethylamine	dioxane		$[\alpha]_{546}$ +218				DMF	20	115
				$[\alpha]_{546}$ +343				40% DMF-60% DCA	20	
Poly(L-tyrosine)	from poly(O-carbobenzoxy-L-tyrosine) (NCA, 110°C in vac.)			+184				DMF		116
				+73				20% DMF-80% DCA		
				+150				0.15 M NaCl, pH 10.85		
				+28				0.15 M NaCl, pH 12.27		
				+8.7				1 N NaOH	20	117

3. POLY(SACCHARIDES)

Polymer	Polymerization Systems			Polymer (P) Values		Monomer (M) or Model Compound (MC) Values		Optical Activity Measured		Ref.
	Catalyst or Initiator	Solvent	T [°C]	$[\alpha]_D$	$[M]_D$	$[\alpha]_D$	$[M]_D$	Solvent	T [°C]	
Poly[α-1,6-anhydro-2,3,4-tri-O-acetyl-β-D-mannopyranose]	PF_5	methylene chloride	0	+96	-	-	-	DMSO	30	80
Poly[α(1→6')-anhydro-2,3,4-tri-O-benzyl-β-D-galactopyranose]	PF_5	methylene chloride	-78	+105.1	-	-46.1 (M)	-	chloroform	25	75
Poly[α-1,6-anhydro-2,3,4-tri-O-benzyl-β-D-mannopyranose]	PF_5	methylene chloride	-78	+57.5	-	-32 (M)	-	chloroform	30 (P); 25 (M)	80
Poly[α-(1→6')-anhydro-D-galactopyranose] from 2,3,4-tri-O-benzyl derivative	PF_5			+207.1	-	-	-	10% LiOH-0.5 % borate	25	75
(Poly[1,6-anhydro-β-D-glucopyranose])	$CH_2ClCOOH$	-	21-127	+91	-	-	-	H_2O	22	77
Poly[α-1,6-anhydro-D-mannopyranose] from 2,3,4-tri-O-benzyl derivative				+122.8	-	-	-	1% DMSO	30	80
Poly[1,6-anhydro-2,3,4-tri-O-methyl-β-D-glycopyranose]	$BF_3 \cdot OEt_2$	toluene	R.T.	+199	-	-63.7 (M)	-	chloroform	25 (P);20 (M)	78
	$BF_3 \cdot OEt_2/H_2O$	toluene	R.T.	+204	-	-63.7 (M)	-	chloroform	25 (P);20 (M)	
Poly[D-galactose]	HCL	DMSO	40	+79.1	-	-	-	H_2O	20	74
Poly[D-glucose]	HCl	DMSO	40	+90.3	-	-	-	H_2O	20	74
	poly phosphate formamide		50-60	+16	-	-	-	-	-	76
Poly[D-mannose]	HCl	DMSO	40	+102.9	-	-	-	H_2O	20	74
Poly[2,3,6-tri-O-methyl-D-glucose]	$HBr-HCl-P_2O_5$	DMSO	R.T.	+136.6	-	-	-	chloroform	20	79
Poly[D-xylose]	HCl	DMSO	40	+96.1	-	-	-	H_2O	20	74

4. REFERENCES

1. F. Ciardelli, E. Benedetti, O. Pieroni, Makromol. Chem. 103, 1 (1957).

2. P. Pino, G. P. Lorenzi, L. Lardicci, Chim. Ind. (Milan) 42, 712 (1960).

3. P. Pino, G. P. Lorenzi, J. Am. Chem. Soc. 82, 4745 (1960).

4. P. Pino, F. Ciardelli, G. P. Lorenzi, G. Montagnoli, Makromol. Chem. 61, 207 (1963).

5. P. Pino, P. Salvadori, E. Chiellini, P. L. Luisi, Pure Applied Chem. 16, 469 (1968).

6. P. Pino, G. P. Lorenzi, L. Lardicci, F. Ciardelli, Vysokomolekul. Soedin. 3, 1567 (1961).

7. M. Goodman, K. J. Clark, M. A. Stake, A. Abe, Makromol. Chem. 72, 131 (1964).

8. W. J. Bailey, E. T. Yates, J. Org. Chem. 25, 1800 (1960).

9. S. Nozakura, S. Takeuchi, H. Yuki, S. Murahashi, Bull. Chem. Soc. (Japan) 34, 1673 (1961).

10. R. C. Schulz, H. Hilpert, Makromol. Chem. 55, 132 (1962).

11. R. C. Schulz, Z. Naturforsch. 19b, 387 (1964).

12. R. K. Kulkarni, H. Morawetz, J. Polymer Sci. 54, 491 (1961).

13. C. S. Marvel, J. Dec, H. G. Cooke, Jr., J. Am. Chem. Soc. 62, 3499 (1940).

14. J. W. C. Crawford, D. Plant, J. Chem. Soc. 1952, 4492.

15. P. Walden, Z. Phys. Chem. 20, 383 (1896).

16. E. I. Klabunovskii, I. I. Petrov, M. I. Schvartmen, Vysokomolekul. Soedin. 6, 1487 (1964).

17. C. L. Arcus, D. W. West, Chem. Ind. (London) 1958, 230.

18. C. L. Arcus, D. W. West, J. Chem. Soc. 1959, 2699.

19. K. J. Liu, J. S. Lignowski, R. Ullman, Makromol. Chem. 105, 8 (1967).

20. N. Beredjick, C. Schuerch, J. Am. Chem. Soc. 78, 2646 (1956).

21. N. Beredjick, C. Schuerch, J. Am. Chem. Soc. 80, 1933 (1958).

22. E. I. Klabunovskii, M. I. Schvartmen, I. I. Petrov, Vysokomolekul. Soedin. 6, 1579 (1964).

23. T. P. Bird, W. A. P. Black, E. T. Dewar, D. Rutherford, Chem. Ind. (London) 1960, 1331.

24. W. A. P. Black, E. T. Dewar, D. Rutherford, J. Chem. Soc. 1963, 4433.

25. S. Kirmura, M. Imoto, Makromol. Chem. 50, 155 (1961).

26. S. Kimura, K. Hirano, M. Imoto, Kogyo Kagaku Zasshi (Japan) 65, 688 (1962).

27. M. Imoto, S. Kimura, Makromol. Chem. 53, 210 (1962).

28. R. K. Kulkarni, H. Morawetz, J. Polymer Sci. 54, 491 (1961).

29. V. E. Kaiser, R. C. Schulz, Makromol. Chem. 81, 273 (1965).

30. O. Pieroni, F. Ciardelli, C. Botteghi, L. Lardidi, P. Salvadori, P. Pino, J. Polymer Sci. C 22, 993 (1967).

31. G. Anzuino, V. Grescenzi, M. D' alagni, A. M. Liquori, F. Quadrifoglio, F. Ascoli, Communication at the IX. Natl. Meeting of Italian Chem. Soc., Naples, 1962.

32. G. P. Lorenzi, E. Benedetti, E. Chiellini, Chim. Ind. (Milan) 46, 1474 (1964).

33. D. Basagni, A. M. Liquori, B. Pispesa, J. Polymer Sci. B 2, 241 (1964).

34. P. Pino, G. P. Lorenzi, Makromol. Chem. 47, 242 (1961).

35. G. J. Schmitt, C. Schuerch, J. Polymer Sci. 45, 313 (1960).

36. W. A. P. Black, E. T. Dewar, D. Rutherford, Chem. Ind. (London) 1962, 1624.

37. R. C. Schulz, R. H. Jung, Makromol. Chem. 96, 295 (1966).

38. R. C. Schulz, H. Hartman, Makromol. Chem. 65, 106 (1963).

39. F. Millich, G. K. Baker, Macromolecules 2, 122 (1969).

40. C. S. Marvel, C. G. Overberger, J. Am. Chem. Soc. 68, 2106 (1946).

41. C. S. Marvel. C. G. Overberger, J. Am. Chem. Soc. 66, 475 (1944).

42. C. G. Overberger, L. C. Palmer, J. Am. Chem. Soc. 78, 666 (1956).

43. H. Yamaguchi, Y. Minoura, J. Polymer Sci. A-1, 8, 929 (1970).

44. A. Abe, M. Goodman, J. Polymer Sci. A-1, 2193 (1963).

45. A. Abe, M. Goodman, J. Polymer Sci. 59, S 37 (1962).

46. C. C. Price, M. Osgan, R. E. Hughes, C. Shambelan, J. Am. Chem. Soc. 78, 690, 4787 (1956).

47. E. J. Vandenberg, J. Am. Chem. Soc. 83, 3538 (1961).

48. H. Shimazaki, J. Chem. Soc. (Japan) 87, 462 (1966).

49. C. G. Overberger, , S. Ozaki, D. M. Braunstein, Makromol. Chem. 93, 13 (1966).

50. J. R. Shelton, J. B. Lando, D. E. Agostini, Polymer Letters 9, 173 (1971).

51. R. C. Schulz, J. Schwaab, Makromol. Chem. 87, 90 (1965).

52. Y. Iwakura, K. Iwata, S. Matsuo, A. Tohara, Makromol. Chem. 122, 275 (1969).

53. C. G. Overberger, H. Kaye, J. Am. Chem. Soc. 89, 5649 (1967).

54. H. D. K. Drew, W. N. Haworth, J. Chem. Soc. 1927, 775.

55. N. Spassky, P. Sigwalt, Bull. Chim. Soc. France 1967, 4617.

56. T. Tsunetsugu, J. Furukawa, T. Fueno, J. Polymer Sci. A-1, 9, 3541 (1971).

57. C. G. Overberger, J. K. Weise, J. Am. Chem. Soc. 90, 3538 (1968).

58. M. Imoto, H. Sakurai, T. Kono, J. Polymer Sci. 50, 467 (1961).

59. C. G. Overberger, H. Jabloner, J. Polymer Sci. 55, 32 (1961).

60. C. G. Overberger, H. Jabloner, J. Am. Chem. Soc. 85, 3431 (1963).

61. C. G. Overberger, G. M. Parker, J. Polymer Sci. C 22, 387 (1968).

62. E. Schmidt, Angew. Makromol. Chem. 14, 185 (1970).

63. Y. Iwakura, K. Hayashi, K. Inagaki, Makromol. Chem. 110, 84 (1967).

64. M. Goodman, S. Chen, Macromolecules 4, 625 (1971).

65. M. Goodman, S. Chen, Macromolecules 3, 398 (1970).

66. Y. Minoura, M. Takebayashi, C. C. Price, J. Am. Chem. Soc. 81, 4689 (1959).

67. K. Tsuboyama, S. Tsuboyama, M. Yanagita, Bull. Chem. Soc. (Japan) 40, 2954 (1967).

68. S. Tsuboyama, Bull. Chem. Soc. (Japan) 35, 1004 (1962).

69. M. S. Toy, C. C. Price, J. Am. Chem. Soc. 82, 2613 (1960).

70. A. P. Terentev, V. V. Dunina, E. G. Rukhadze, Vysokomolekul Soedin. A-9, 599 (1967).

71. R. C. Schulz, R. H. Jung, Makromol. Chem. 116, 190 (1968).

72. A. Abe, M. Goodman, J. Polymer Sci. A 2, 3491 (1964).

73. C. G. Overberger, T. Yoshimura, A. Ohnishi, A. S. Gomes, J. Polymer Sci. A-1, 8, 2275 (1970).

74. F. Micheel, A. Bockmann, W. Meckstroth, Makromol. Chem. 48, 1 (1961).

75. T. Uryu, H. Libert, J. Zachoval, C. Schuerch, Macromolecules 3, 345 (1970).

76. G. Schramm, H. Grotsch, W. Pollman, Angew. Chem. 74, 53 (1962).

77. J. Das. Carvalho, W. Trins, C. Schuerch, J. Am. Chem. Soc. 81, 4054 (1959).

78. C. C. Tu, C. Schuerch, J. Polymer Sci. B 1, 163 (1963).

79. F. Micheel, A. Bockmann, Makromol. Chem. 51, 97 (1962).

80. J. Frechet, C. Schuerch, J. Am. Chem. Soc. 91, 1161 (1969).

81. A. R. Downie, A. Elliot, W. E. Hanby, B. R. Malcolm, Proc. Roy. Soc. A 242, 325 (1957).

82. M. Goodman, Fu Chen in preparation.

83. M. Goodman, A. M. Felix, Biochemistry 3, 1529 (1964).

84. P. Urnes, P. Doty, Advan. Protein Chem. 16, 401 (1961).

85. T. Hayakawa, Y. Fujiwara, J. Noguchi, Bull. Chem. Soc. (Japan) 40, 1205 (1967).

86. E. R. Blout, R. H. Karlson, J. Am. Chem. Soc. 80, 1259 (1958).

87. M. Frankel, A. Berger, J. Org. Chem. 16, 1513 (1951).

88. A. Berger, E. Katchalski, J. Am. Chem. Soc. 73, 4084 (1951).

89. J. Kovacs, R. Ballina, R. L. Rodin, D. Balasurkamanium, H. Applequist, J. Am. Chem. Soc. 87, 119 (1965).

90. S. Ikeda, Biopolymers 5, 356 (1967).

91. M. Wilchek, A. Fransdorff, M. Sela, Arch. Biochem. Biophys. 113, 742 (1966).

92. W. Moffitt, J. T. Young, Pro. Nat. Acad. Sci. USA 42, 596 (1956).

93. L. Goldstein, E. Katchalski, Bull. Research Council Israel A 9, 138 (1960).

94. A. Wada, Mol. Phys. 3, 409 (1960).

95. J. Kovacs, G. H. Schmit, E. J. Jonson, Can. J. Chem. 47, 3690 (1960).

96. J. T. Yang, P. Doty, J. Am. Chem. Soc. 79, 761 (1957).

97. E. R. Blout, R. H. Karlson, J. Am. Chem. Soc. 78, 941 (1958).

98. W. E. Hanby, S. G. Ealey, J. Eatson, J. Chem. Soc. 1950, 3239.

99. M. Goodman, E. Schmitt, I. Listowoky, F. Boardman, I. G. Rosen, M. Stake, in "Polyamino Acids, Polypeptides and Proteins," (M. A. Stahmann ed.), Univ. Wisconsin Press, Madison, 1962.

100. M. Goodman, E. E. Schmitt, D. Yphantis, J. Am. Chem. Soc. 82, 5507 (1959).

101. K. S. Norland, G. D. Fasman, E. Katchalski, E. R. Blout, Biopolymers 1, 277 (1961).

102. A. Patchornik, A. Berger, E. Katchalski, J. Am. Chem. Soc. 79, 5227 (1957).

103. G. D. Fasman, M. Idelson, E. R. Blout, J. Am. Chem. Soc. 83, 709 (1961).

104. M. Sela, R. Arnon, I. Jacobson, Biopolymers 1, 517 (1963).

105. J. Applequist, P. Doty, in "Polyamino Acids, Polypeptides and Proteins," (M. A. Stahmann ed.), Univ. Wisconsin Press, Madison, 1962.

106. H. Devoe, I. Tinoco, J. Mol. Biol. 4, 518 (1962).

107. G. E. Perlmann, E. Katchalski, J. Am. Chem. Soc. 84, 452 (1962).

108. G. D. Fasman, in "Polyamino Acids, Polypeptides and Proteins", (M. A. Stahmann, ed.), Univ. Wisconsin Press, Madison, 1962.

109. E. Katchalski, A. Berger, J. Kurtz, in "Aspects of Protein Structure," (G. N. Ramachandran ed.) Academic Press, New York, 1963, p. 205.

110. J. Kurtz, G. D. Fasman, A. Berger, E. Katchalski, J. Am. Chem. Soc. 80, 393 (1958).

111. E. Katchalski, I. Z. Steinberg, Ann. Rev. Phys. Chem. 12, 433 (1961).

112. J. Kurtz, A. Berger, E. Katchalski, Nature 178, 1066 (1956).

113. W. F. Harrington, M. Sela, Biochem. Biophys. Acta 27, 24 (1958).

114. G. D. Fasman, E. R. Blout, J. Am. Chem. Soc. 82, 2262 (1960).

115. M. Sela, I. Z. Steinberg, E. Daniel, Biochem. Biophys. Acta 46, 433 (1961).

116. J. D. Coombes, E. Katchalski, P. Doty, Nature 185, 534 (1960).

117. E. Katchalski, M. Sela, J. Am. Chem. Soc. 75, 5284 (1953).

ANISOTROPY OF SEGMENTS AND MONOMER UNITS OF POLYMER MOLECULES

V. N. Tsvetkov and L. N. Andreeva

Institute of High Molecular Weight Compounds
at the Academy of Sciences of the USSR
Leningrad, USSR

Contents   Page

A. INTRODUCTION

Segmental anisotropy $(\alpha_1 - \alpha_2)^x$ of a chain molecule may be determined experimentally from measurements of flow birefringence in the solution of a polymer with a sufficiently high molecular weight so that the conformation of its molecules would correspond to that of a Gaussian coil.

For calculating $(\alpha_1 - \alpha_2)^x$ Kuhn's equation was used (1)

$$\frac{\Delta n}{\Delta \tau} = \left[\frac{\Delta n}{g(\eta - \eta_o)}\right]_{\substack{g \to o \\ c \to o}} \equiv \frac{[n]}{[\eta]} = \frac{4\pi}{45\,kT} \cdot \frac{(n^2 + 2)^2}{n} \cdot (\alpha_1 - \alpha_2)^x \tag{1}$$

where $\Delta \tau$ is the tangential flow stress, Δn is the observed flow birefringence of solution, g is the velocity gradient, c and η are the concentration and the viscosity of the solution, respectively, $[\eta]$ and $[n]$ are intrinsic values of viscosity and flow birefringence of solution, η_o and n are the viscosity and the refractive index of the solvent.

Another method of determining segmental anisotropy (which is used more seldom) is through the measurement of stress birefringence in swollen polymers (74). The stress optical coefficient $e = \Delta n/\Delta p$ (where Δp is the normal stress in the sample) determined experimentally is equal to $\Delta n/2\Delta \tau$ (2), where $\Delta n/\Delta \tau$ is also related to $(\alpha_1 - \alpha_2)^x$ according to Eq. (1). Data obtained by this method are marked in the Table with a symbol sw.p. (swollen polymer).

If the refractive indices of the polymer and the solvent are not equal, the experimental value of $(\alpha_1 - \alpha_2)^x$ includes not only the intrinsic segmental $\alpha_1 - \alpha_2$ but also a part produced by the form effect (3, 4). In this case the value $\alpha_1 - \alpha_2$ may be calculated using $(\alpha_1 - \alpha_2)^x$ and the theoretical equation for macro- and micro-from effects (3, 4).

It is preferable for determining $\alpha_1 - \alpha_2$ to use the "matching" solvent in which the increment of refractive index of the polymer $dn/dc = 0$.

The Table gives the intrinsic segmental anisotropy, $\alpha_1 - \alpha_2$ obtained directly from measurements in the "matching" solvent as well as the $(\alpha_1 - \alpha_2)^x$ values including the form effect. The corresponding figures are marked with a symbol (x).

In some of the reviewed works in which measurements were carried out in "non-matching" solvents, the segmental anisotropy was calculated on the basis of the theory of the form effect (3, 4). In these cases the figures are marked with a symbol (xx).

Intrinsic segmental anisotropy of the polymer chain in various solvents even in the absence of the form effect may differ owing to a "specific" effect of the solvent (see, e.g., poly(vinyl acetate) (5-11)). "Specific" effect may probably include such phenomena as a change in the type of rotation of the side groups (5, 6, 8, 12), a change in the polarizability of its bonds and the short-range ordering of the solvent molecules (11).

Principal values which effect the intrinsic segmental anisotropy of the chain are the anisotropy of the monomer unit and equilibrium rigidity of the chain $<r^2>/L$. Here L is the full length of the extended chain and $<r^2>^{1/2}$ is the mean square end-to-end distance in a θ-solvent.

$a_{\parallel} - a_{\perp}$ is the difference between the polarizabilities of the monomer unit in parallel and perpendicular directions of the chain. Values of $a_{\parallel} - a_{\perp}$ presented in the Table were calculated from equation:

$$a_{\parallel} - a_{\perp} = \lambda L(\alpha_1 - \alpha_2)/<r^2> \tag{2}$$

where λ is the length of the monomer unit in the chain direction.

For cellulose esters $a_{\parallel} - a_{\perp}$ and, correspondingly, $\alpha_1 - \alpha_2$ depend on the degrees of substitution (D.S.) which are also given in the Table.

B.   TABLE OF ANISOTROPY OF SEGMENTS AND MONOMER UNITS OF POLYMER MOLECULES

Polymer	Solvent		$(\alpha_1 - \alpha_2) \times 10^{25}$ cm^3	$(a_\parallel - a_\perp) \times 10^{25}$ cm^3	References
1. Main-chain Acyclic Carbon Polymers					
1.1 Poly(dienes)					
Poly(butadiene)					
1,4-cis	benzene	sw. p.	+ 61.3 to 63	+ 30.8	52-54
	carbon tetrachloride	sw. p.	+ 53.5^x	+ 31.7^x	53
		sw. p.	+ 55.2^x		54
	cyclohexane	sw. p.	+ 57.3	+ 33.9	53
	toluene	sw. p.	+ 72	+ 42.6	53
	p-xylene	sw. p.	+ 86.9	+ 51.4	53
1,4-trans	benzene		+ 71	+ 37.4	52
		sw. p.	+ 70.4		53
	carbon tetrachloride	sw. p.	+ 58.1^x		55
		sw. p.	+ 61.1^x	+ 36.3^x	53
	cyclohexane	sw. p.	+ 57.3^x	+ 33.1^x	53
	toluene	sw. p.	+ 81.6	+ 48.6	53
	p-xylene	sw. p.	+ 101	+ 60.2	53
Poly(chloroprene)	α-bromonaphthalene		+ 110^x		11
	carbon tetrachloride		+ 33		11
	chlorobenzene		+ 64		11
	dichloroethane		+ 39		11
	α-methylnaphthalene		+ 99^x		11
	tetrachloroethylene		+ 46		11
	toluene		+ 67		11
	p-xylene		+ 88		11
Poly(isoprene)					
cis	benzene		+ 48	+ 30.5	52
	squalene	sw. p.	+ 53.1		56
trans	benzene		+ 49	+ 31	52
1.2 Poly(alkenes)					
Poly(1-butene)					
atactic	toluene		+ 33.4		13
isotactic	toluene		+ 25.2		13
Poly(1-decene)					
isotactic	toluene		- 82.5		13
Poly(1-dodecene)					
isotactic	toluene		- 120		13
Poly(ethylene)	tetralin, xylene		+ 60		14
	decalin		+ 30		14
Poly(1-heptene)					
isotactic	toluene		- 24.5		13
Poly(1-hexadecene)					
isotactic	toluene		- 205		13
			- 213		13
Poly(1-hexene)					
atactic	toluene		+ 12.1		13
isotactic	toluene		- 6.5		13
Poly(isobutene)	benzene, chlorobenzene, tetrachloroethylene, m-xylene		+ 45 to + 59		11, 14, 15
	carbon tetrachloride		+ 35		11
	decalin		+ 30		14
	p-xylene		+ 69		11
Poly(1-octadecene)					
isotactic	toluene		- 257		13
Poly(1-octene)					
isotactic	toluene		- 39		13
Poly(1-pentene)					
isotactic	toluene		+ 8.0		13
			+ 9.3		13

Polymer	Solvent		$(\alpha_1 - \alpha_2) \times 10^{25}$ cm^3	$(a_\| - a_\perp) \times 10^{25}$ cm^3	References
1.2 Poly(alkenes) (Cont'd.)					
Poly(diphenylpropylene)	bromoform		+ 80		16
Poly(propylene)					
atactic	benzene, xylene		+ 45		14
	carbon tetrachloride		+ 30	+ 3.5	17
	decalin		+ 30		14
	toluene		+ 55		13
isotactic	carbon tetrachloride		+ 30	+ 3.5	17
Poly(1-tetradecene)					
isotactic	toluene		− 171		13
			− 176		13
1.3 Poly(acrylic acid) and Derivatives					
Poly(acrylic acid)	dioxane		$\approx$ − 0.5[xx]	$\approx$ − 0.1[xx]	18
--, sodium salt	0.0012M NaCl, pH = 7		− 20[xx]		
	water pH = 6.1		− 4[xx] to + 10[xx]		19
Poly(n-butyl acrylate)	benzene		− 11	− 1.5	20
	decalin		− 17.8	− 1.9	21
	toluene		− 10.1	− 1.1	21
			− 6.5	− 0.9	20
Poly(cetyl acrylate)	decalin		− 141	− 6.5	22
	toluene		− 164	− 7.5	21
Poly(cholesteryl acrylate)	benzene		− 360	− 16	23
Poly(decyl acrylate)	decalin		− 74	− 3.7	21
	toluene		− 95	− 4.7	21
Poly(ethyl acrylate)	benzene	sw. p.	+ 3.0	+ 0.36	24
	bromobenzene	sw. p.	+ 10[xx]	+ 1.2[xx]	24
	bromoform	sw. p.	− 37[xx]	− 4.5[xx]	24
	dibromoethane	sw. p.	− 14[xx]	− 1.7[xx]	24
	dimethylformamide	sw. p.	− 11[xx]	− 1.7[xx]	24
Poly(methyl acrylate)	benzene		+ 17	+ 2.5	20
	toluene		+ 16	+ 1.9	21
			+ 26	+ 3.6	20
Poly(octadecyl acrylate)	decalin		− 190	− 6.6	21
	toluene		− 232	− 0.8	21
Poly(octyl acrylate)	decalin		− 57.4	− 4.3	21
	toluene		− 47.9	− 3.6	21
1.4 Poly(methacrylic acid) and Derivatives					
Poly(n-butyl methacrylate)					
atactic	benzene		− 14	− 2.1	25
isotactic	benzene		− 2.0	− 0.3	26
Poly(tert-butyl methacrylate)					
atactic	benzene		+ 2.1	+ 0.3	20
isotactic	benzene		+ 19.3	+ 3.0	20
Poly(p-tert-butylphenyl methacrylate)	bromobenzene		− 90	− 7.5	27, 28
Poly(p-carbethoxy-N-phenylmethacryl-amide	o-toluidine		− 230	− 23	40
Poly(p-carboxyphenyl methacrylate)	dioxane		+ 180[x]	+ 8.0[x]	35
	0.1M NaCl		+ 370[x]		35
Poly(p-(4-cetoxybenzoxy)phenyl meth-acrylate)	bromoform		− 1000[x]	− 40	36
	benzene		− 1600[x]		36
	chloroform		− 1400[x]		36
	carbon tetrachloride		− 3000[x]		37
	tetrahydrofuran		− 890[x]		36
	benzene/heptane, 52/48; 66/34		− 4200[x]		36
Poly(cetyl methacrylate)	benzene		− 160	− 8.9	29
Poly(p-chloro-N-phenylmethacrylamide)	o-toluidine		− 160	− 20	40
Poly(glycol methacrylate)	dimethylformamide		+ 1.0	+ 0.18	30
	ethyl alcohol		− 6.0[xx]	− 1.1[xx]	30
	water		− 6.0[xx]	− 1.1[xx]	30

Polymer	Solvent	$(\alpha_1 - \alpha_2) \times 10^{25}\ cm^3$	$(a_\parallel - a_\perp) \times 10^{25}\ cm^3$	References
1.4 Poly(methacrylic acid) and Derivatives (Cont'd.)				
Poly(hexyl methacrylate)	benzene	− 40	− 4.6	31
	carbon tetrachloride	− 9.7	− 1.1	31
Poly(methacrylic acid)	methanol	+ 50[x]		32
	0.002M HCl	+ 150[x]		32
--, sodium salt	0.012M NaCl pH = 7	+ 150[x]		33
	0.0012M NaCl pH = 7	+ 400[x]		33
	water	+ 56[x] to + 300[x]		34
Poly(p-methylcarboxyphenyl methacrylate)	dibromoethane	− 77	− 7	35
Poly(methyl methacrylate)				
atactic	benzene	+ 2.0	+ 0.3	38
isotactic	benzene	+ 25	+ 3.5	38
Poly(β-naphthyl methacrylate)	tetrabromoethane	− 60	− 8.5	4
Poly(p-(4-nonoxybenzoxy)phenyl methacrylate)	carbon tetrachloride	− 2620[x]		37
Poly(octyl methacrylate)	benzene	− 47	− 5.9	31
	carbon tetrachloride	− 12.5	− 1.6	31
Poly(N-phenylmethacrylamide)	o-toluidine	− 103	− 13	40
Poly(phenyl methacrylate)	bromobenzene	− 10.5	− 1.5	4
Poly(p-(4-propoxybenzoxy)phenyl methacrylate)	dimethylformamide/toluene, 1/1	− 320		36
1.5 Vinyl Polymers				
Poly(acrylonitrile)	dimethylformamide	− 23	− 1.8	4
Poly(p-chlorostyrene)	bromoform	− 230	− 35	41
Poly(2,5-dichlorostyrene)	bromoform	− 265	− 30	41
Poly(3,4-dichlorostyrene)	tetrabromoethane	− 300	− 25	15
Poly(p-methylstyrene)				
atactic	bromoform	− 147	− 20	42
isotactic	bromoform	− 140	− 19	42
Poly(2,5-dimethylstyrene)	bromoform	− 180	− 25	41
Poly(2-methyl-5-vinyl-N-butylpyridinium bromide)	0.01M NaCl	− 900[x]		43
	0.1 M NaCl	− 270[x]		43
Poly(2-methyl-5-vinylpyridine)	bromoform	− 260	− 29	44
Poly(2-methyl-5-vinylpyridinium chloride)	0.1 M HCl	− 300[x]		45
Poly(styrene)				
atactic	bromoform	− 145	− 18	46, 41
isotactic	bromoform	− 224	− 23	46
Poly(vinyl acetate)	acetone	− 20[xx]	− 5.3	7
	sw. p.	− 37[xx]	− 5.3[xx]	10
	benzene	+ 4.0 to + 5.9	+ 0.5 to + 0.8	6, 8, 9, 11
	sw. p.	+ 5.1[xx]	+ 0.75[xx]	10
	bromobenzene sw. p.	+ 9.4[xx]	+ 1.3[xx]	10
	bromoform	− 20[xx]		7
	sw. p.	− 18.7[xx]	− 2.7[xx]	10
	carbon tetrachloride	− 16	− 2.0	9, 11
		− 26		6
	sw. p.	− 25.2[xx]	− 3.6	10
	chlorobenzene	+ 14[xx]	+ 1.75[xx]	11
	chloroform	− 34.9[xx]	− 2.6[xx]	6
		− 24	− 3.0	9, 11
	sw. p.	− 17.9		10
	cyclohexanone	− 23	− 2.9	9, 11
	dichloroethane	− 36	− 4.5	11
		− 39	− 4.9	11
	tetrabromoethane	− 25[xx]	− 3.1[xx]	11
		− 33[xx]		7
	sw. p.	− 46.4[xx]	− 6.6[xx]	10
	toluene	+ 10	+ 1.25	9, 11
		+ 13.5	+ 2.0	8
		+ 19		6
	sw. p.	+ 9.4[xx]	+ 1.3[xx]	10
	o-xylene	+ 2.0	+ 0.25	9, 11
	sw. p.	+ 9.8[xx]	+ 1.4[xx]	10

Polymer	Solvent	$(\alpha_1-\alpha_2) \times 10^{25}$ cm^3	$(a_\parallel - a_\perp) \times 10^{25}$ cm^3	References
1.5 Vinyl Polymers (Cont'd.)				
Poly(vinyl butyral)	chloroform	+ 81		11, 47
	toluene/phenole, 79/21	+ 173		11, 47
Poly(vinyl butyrate)	benzene	− 8.0		11, 47
	carbon tetrachloride	− 36		11, 47
	chloroform	− 48		11, 47
Poly(vinyl chloride)	tetrahydrofuran	+ 40		48
Poly(vinyl cinnamate)	bromoform	− 420	− 35	49
Poly(β-vinylnaphthalene)	tetrabromoethane	− 430	− 30	4
	bromoform	− 440	− 20 to − 30	39
Poly(vinyl propionate)	benzene	− 4.4		11, 47
	carbon tetrachloride	− 31		11, 47
	chloroform	− 40		11, 47
	toluene	+ 1.3		11, 47
Poly(4-vinylpyridine)	bromoform	− 240	− 22	50
Poly(4-vinylpyridinium chloride)	0.1 M HCl	− 260^x		50
	0.05M HCl	− 440^x		50
Poly(vinylpyrrolidone)	benzyl alcohol	− 75	− 10	4
Poly(vinyl stearate)	carbon tetrachloride	− 130	− 4.7	51
1.6 Copolymers, Graft- and Blockcopolymers				
Poly(p-(4-cetoxybenzoxy)phenyl methacrylate- co-cetyl methacrylate)	carbon tetrachloride			
mol.% 81/19		− 2000^x		93
60/40		− 920^x		93
39/61		− 540^x		93
22/78		− 400^x		93
15/85		− 277^x		93
8/92		− 180^x		93
4/96		− 16^x		93
Poly(methyl methacrylate-co-p-tert-butyl- phenyl methacrylate)	chlorobenzene			
mol.% 91/ 9		< + 1.5		94
80/20		− 7.4		94
50/50		− 30.4		94
25/75		− 44		94
Poly(phenylbutyl isocyanate-co-chloral)	carbon tetrachloride			
mol.% 50/50		+ 12^x		60
Poly(styrene-co-p-chlorostyrene)	bromoform			
mol.% 89.7/10.3		− 165		95
83.2/16.8		− 172		95
68.7/31.3		− 198		95
55 /45		− 202		95
39.5/60.5		− 226		95
Poly(styrene-co-N-methylcitraconimide)	bromoform			
mol.% 54/46		− 26		63
48/52		− 34		63
Poly(styrene-co-methyl methacrylate)	bromobenzene			
mol.% 70/30		− 88		96
50/50		− 68		96
30/70	chlorobenzene	− 34		96
Poly(styrene : propylene), atactic	chlorobenzene			
mol.% 64/36		− 8		97
Poly(n-butyl methacrylate + styrene)	bromoform			
mol.% 8/92		+ 1190		98
Poly(tert-butyl methacrylate + styrene)	bromoform			
mol.% 8/92		+ 540		98
Poly(methyl methacrylate + styrene)	bromoform			
mol.% 10/90		+ 700 to + 7000		98, 99
87/13		+ 30		98
(70 to 90)(30 to 10)		+ 100 to + 1100		100
Poly(propylene, atactic + styrene)	chlorobenzene			
mol.% 30/70		+ 22		93

Polymer	Solvent	$(\alpha_1 - \alpha_2) \times 10^{25}$ cm^3	$(a_\parallel - a_\perp) \times 10^{25}$ cm^3	References
1.6 Copolymers, Graft- and Blockcopolymers (Cont'd.)				
Poly(vinyl chloride + styrene)	benzene			
weight % 5.2/94.8		+ 330		43
5.6/94.4		+ 180		43
12.1/87.9		+ 155		43
30.7/69.3		+ 110		
2. Main-chain Carbocyclic Polymers				
Poly(acenaphthylene)				
helix 4_1	bromoform	- 300[xx]	- 13.6[xx]	39
trans	bromoform	- 300[xx]	- 20[xx]	39
Poly(1,2,3-trimethyl-2,3-dihydro-1,6-indendiyl)	bromoform	+ 78		92
Poly(1,2,3-trimethyl-2,3-dihydro-1,6-indenediyl-1,4-phenyleneethylene)	bromoform	+ 142		92
3. Main-chain Heteroatom Polymers				
3.1 Poly(oxides)				
Poly(oxypropylene)	cyclohexanone	+ 18	+ 6	57
3.2 Poly(esters)				
Poly(oxycarbonyloxy-1,4-phenylenecyclo-hexylidene-1,4-phenylene)	bromoform	- 114	- 76	58
Poly(oxyethyleneoxyterephthaloyl)	dichloroethane/phenol, 1/1	+ 70		59
3.3 Poly(amides)				
Poly(n-butyliminocarbonyl) (Poly(butyl isocyanate))	carbon tetrachloride	+ 4100		60
Poly(chlorohexyliminocarbonyl)	carbon tetrachloride	+ 3000		60
Poly(n-tolyliminocarbonyl)	carbon tetrachloride	- 39		60
3.4 Poly(peptides)				
Poly(γ-benzyl-L-glutamate)				
helix	bromoform	+ 13700[x]	+ 11.4[x]	86
helix	dichloroethane	+ 6900	+ 5.8	87
coil	dichloroacetic acid	+ 117[x]	+ 12[x]	88
Poly(S-carbobenzoxymethyl-L-cystein)	dichloroacetic acid	+ 22[x]		86
Poly(ε,N-carbobenzoxy-L-lysine)				
helix	dimethylformamide	+ 3600[x]	+ 12[x]	86
coil	dichloroacetic acid	< +		86
Poly(L-glutamic acid)				
helix	phosphate buffer, 0.1M, pH = 4.2	+ 1900[x]		86
coil	phosphate buffer, 0.1M, pH = 12.5	+ 136[x]		86
3.5 Poly(siloxanes)				
Linear polymers				
Poly(oxydimethylsilylene)	petroleum ether	+ 4.7	+ 0.96	77
Poly(oxymethylphenylsilylene)				
90 : 10	petroleum ether	- 2.3	- 0.47	77
87.5 : 12.5	decalin	- 10	- 2.0	78
75 : 25	benzene	- 13.6	↖ 2.7	79
	decalin	- 21	- 4.0	78
62.5 : 37.5	decalin	- 36[xx]	- 7.2[xx]	78
50 : 50 atactic	benzene	- 25.5	- 5.1	79
	decalin, tetralin	- 52[xx]	- 10[xx]	78
isotactic	decalin, tetralin	- 82[xx]	- 16[xx]	78
Poly(oxyphenylsilylene)	benzene	- 85	- 17	79

References page IV-384

Polymer	Solvent	$(\alpha_1 - \alpha_2) \times 10^{25}$ cm^3	$(a_{\parallel} - a_{\perp}) \times 10^{25}$ cm^3	References
3.5 Poly(siloxanes) (Cont'd.)				
Ladder polymers				
Poly(isobutylphenylsilsesquioxane) (1:1)	benzene	− 840	− 22	80
Poly(isohexylphenylsilsesquioxane) (1:1)	benzene	− 840	− 16	80
Poly(m-chlorophenylsilsesquioxane)	benzene, carbon tetrachloride	− 3600	− 51	80
Poly(3-methyl-1-butenesilsesquioxane)	benzene	− 570[x]	− 6.5[x]	81, 82
	butyl acetate	− 380[x]	− 4.9	81, 82
Poly(phenylsilsesquioxane)	bromoform	− 1060 to − 1800	− 23 to − 31	83-85
4. Main-chain Heterocyclic Polymers				
4.1 Poly(imides)				
Poly(N-2, 4-dimethylphenylmaleimide)	bromoform	− 200	− 12.5	61
Poly(N-isobutylmaleimide)	chlorobenzene	+ 160	+ 10	62
Poly(N-methylcitraconimide)	bromoform	+ 150	+ 16.7	63
Poly(N, N-piperazindiyl-2, 5-diketo-1, 3-pyr-rolidindiylhexamethylene-2, 5-diketo-1, 3-pyrrolidindiyl)	benzyl alcohol	+ 56	+ 31	64
	bromoform/cyclohexanol, 60/40	+ 30	+ 17	64
Poly(p-tolylmaleimide)	bromoform	− 160	− 10	65
4.2 Poly(saccharides)				
Benzyl cellulose	dioxane	+ 294[x]	+ 5[x]	67
Cellulose benzoate	bromobenzene	− 914	− 18	66
	dioxane	− 447[x]		67
Cellulose dimethylphosphonocarbamate				
D.S. = 2.0	0.01 M NaCl	+ 710[x]	+ 16[x]	68
	0.2 M NaCl	+ 640[x]	+ 16[x]	68
Cellulose diphenylacetate	acetophenone	+ 1360	+ 22	69
	dioxane	+ 1030[x]	+ 22[x]	67
Cellulose diphenylphosphonocarbamate	dioxane	+ 626[x]	+ 10[x]	67
Cellulose monophenyl acetate	bromoform	+ 440	+ 8.0	67
Cellulose nitrate				
D.S. = 13.8 %	cyclohexanone	− 820	− 14	70
Cellulose phenylcarbamate	acetophenone	− 1100	− 18	69
	dioxane	− 1530[x]	− 25[x]	67
	benzophenone 55°C	− 742[a]		71
	80°C	− 572[a]		71
Cellulose stearate				
D.S. = 2.0	tetrachloroethane	− 500		72
Cellulose tributyrate	dioxane	+ 61[x]	+ 1.0[x]	67
	tetrachloroethane	− 34	− 1.3	72
Cyanoethylacetyl cellulose	cyclohexanone sw.p.	+ 15	+ 0.4	73, 74
Cyanoethyltrityl cellulose	cyclohexanone sw.p.	+ 220		74
Ethyl cellulose	carbon tetrachloride	+ 430	+ 11	75
	dioxane	+ 396		67
Sulfate cellulose, sodium salt				
D.S. = 0.4	aqueous NaCl 0.2 M	+ 634[x]	+ 16[x]	76
	0.15 M	+ 645[x]	+ 17[x]	76
	0.10 M	+ 680[x]	+ 16[x]	76
	0.01 M	+ 980[x]	+ 16[x]	76
	0.005 M	+ 1300[x]	+ 16[x]	76
	0.001 M	+ 1750[x]	+ 16[x]	76

a) calculated from the experimental data [n] /[η] .

C. REFERENCES

1. W. Kuhn, H. Kuhn, Helv. Chim. Acta 26, 1394 (1943).
2. A. Lodg, Trans. Faraday Soc. 42, 273 (1946).
3. V. N. Tsvetkov, Chapter XIV in Bacon Ke (Ed.) "Newer Methods of Polymer Characterization", Interscience, New York, 1964.
4. V. N. Tsvetkov, V. E. Eskin, S. Ya. Frenkel, "Structure of Macromolecules in Solutions", Nauka, Moskow, 1964.
5. V. N. Tsvetkov, J. Polymer Sci. 57, 727 (1962).
6. E. V. Frisman, V. A. Andreichenko, Vysokomolekul. Soedin. 4, 1559 (1962).
7. E. V. Frisman, An Bao Chzu, Vysokomolekul. Soedin. 4, 1564 (1962).
8. V. N. Tsvetkov, N. N. Boitsova, M. G. Vitovskaja, Vysokomolekul. Soedin. 6, 267 (1964).
9. E. V. Frisman, G. A. Dyuzev, A. K. Dadivanyan, Vysokomolekul. Soedin. 6, 341 (1964).
10. V. N. Tsvetkov, A. E. Grishchenko, L. E. De-Millo, E. N. Rostovskii, Vysokomolekul. Soedin. 6, 384 (1964).
11. E. V. Frisman, A. K. Dadivanyan, J. Polymer Sci. C16, 1001 (1967).
12. V. N. Tsvetkov, Leningrad. Univ. Vestnik Ser., Fiz. Khim. N 22, 39 (1961).
13. W. Philippoff, E. G. M. Tornqvist, J. Polymer Sci., C, N 23, 881 (1968).
14. T. I. Garmonowa, Leningrad. Univ., Vestnik Ser. Fiz. Khim. N 22, 72 (1962).
15. V. N. Tsvetkov, V. E. Bychkova, S. M. Savvon, I. K. Nekrasov, Vysokomolekul. Soedin. 1, 1407 (1959).
16. P. Gramain, J. Leray, H. Benoit, J. Polymer Sci. C16, 3983 (1968).
17. V. N. Tsvetkov, O. V. Kallistov, E. V. Korneeva, I. K. Nekrasov, Vysokomolekul. Soedin. 5, 1538 (1963).
18. V. N. Tsvetkov, S. Ya. Ljubina, T. V. Barskaya, Vysokomolekul. Soedin. 6, 806 (1964).
19. W. Kuhn, H. Oswald, H. Kuhn, Helv. Chim. Acta 36, 1209 (1953).
20. V. N. Tsvetkov, N. N. Boitsova, M. G. Vitovskaja, Vysokomolekul. Soedin. 6, 297 (1964).
21. V. N. Tsvetkov, L. N. Andreeva, E. V. Korneeva, P. N. Lavrenko, Dokl. Akad. Nauk, SSSR 205, 895 (1972).
22. V. N. Tsvetkov, L. N. Andreeva, E. V. Korneeva, P. N. Lavrenko, N. A. Plate, V. P. Shibaev, B. S. Petrukhin, Vysokomolekul. Soedin. 13A, 2226 (1971).
23. V. N. Tsvetkov, E. V. Korneeva, I. N. Shtennikova, P. N. Lavrenko, G. F. Kolbina, D. Hardi, K. Nitrai, Vysokomolekul. Soedin. 14A, 427 (1972).
24. A. E. Grishchenko, E. P. Vorobjeva, Vysokomolekul. Soedin. 15A, 895 (1973).
25. V. N. Tsvetkov, S. Ja. Ljubina, Vysokomolekul. Soedin. 1, 577 (1959).
26. V. N. Tsvetkov, M. G. Vitovskaja, S. Ja. Ljubina, Vysokomolekul. Soedin. 4, 577 (1962).
27. V. N. Tsvetkov, S. Ja. Magarik, Dokl. Akad. Nauk SSSR 115, 911 (1957).
28. V. N. Tsvetkov, I. N. Shtennikova, Zh. Tekhn. Fiz. 29, 885 (1959).
29. V. N. Tsvetkov, D. Hardi, I. N. Shtennikova, E. V. Korneeva, G. F. Pirogova, K. Nitrai, Vysokomolekul. Soedin. 11A, 349 (1969).
30. A. E. Grishchenko, R. I. Ezrielev, Vysokomolekul. Soedin. 14A, 521 (1972).
31. A. E. Grishchenko, M. G. Vitovskaja, V. N. Tsvetkov, L. N. Andreeva, Vysokomolekul. Soedin. 8, 800 (1966).
32. V. N. Tsvetkov, S. Ya. Ljubina, K. L. Bolevskii, Vysokomolekul. Soedin. Sb. Karbotsepnie Soedin. 4, 26 (1963).
33. V. N. Tsvetkov, S. Ya. Ljubina, K. L. Bolevskii, Vysokomolekul. Soedin. Sb. Karbotsepnie Soedin. 4, 33 (1963).
34. W. Kuhn, O. Kuenzle, A. Katchalsky, Helv. Chim. Acta 31, 1994 (1948).
35. E. N. Zakharova, P. N. Lavrenko, G. A. Fomin, I. I. Konstantinov, Vysokomolekul. Soedin. 13A, 1870 (1971).
36. V. N. Tsvetkov, I. N. Shtennikova, E. I. Rjumtsev, G. F. Kolbina, I. I. Konstantinov, Ju. B. Amerik, B. A. Krentsel, Vysokomolekul. Soedin. 11A, 2528 (1969).
37. V. N. Tsvetkov, I. N. Shtennikova, E. I. Rjumtsev, G. F. Kolbina, E. V. Korneeva, I. I. Konstantinov, Yu. B. Amerik, B. A. Krentsel, Vysokomolekul. Soedin. 15A, 2158 (1973).
38. V. N. Tsvetkov, N. N. Boitsova, Vysokomolekul. Soedin. 2, 1176 (1960).
39. V. N. Tsvetkov, M. G. Vitovskaja, P. N. Lavrenko, E. N. Zakharova, I. F. Gavrilenko, N. N. Stefanovskaja, Vysokomolekul. Soedin. 13A, 2532 (1971).
40. V. N. Tsvetkov, V. E. Bychkova, Vysokomolekul. Soedin. 6, 600 (1964).
41. E. V. Frisman, A. M. Martsinovsky, N. A. Domnitcheva, Vysokomolekul. Soedin. 2, 1148 (1960).
42. V. N. Tsvetkov, N. N. Boitsova, Vysokomolekul. Soedin. 5, 1263 (1963).
43. V. N. Tsvetkov, S. Ya. Ljubina, V. E. Bychkova, I. A. Strelina, Vysokomolekul. Soedin. 8, 846 (1966).
44. I. N. Shtennikova, E. V. Korneeva, V. E. Bychkova, G. M. Pavlov, G. S. Sogomonyants, Vysokomolekul. Soedin. 14B, 118 (1972).
45. S. Ja. Ljubina, I. A. Strelina, G. S. Sogomonyants, S. I. Dmitrieva, O. Z. Korotkina, G. V. Tarasova, V. S. Skazka, V. M. Jamshchikov, Vysokomolekul. Soedin. 12A, 1560 (1970).
46. V. N. Tsvetkov, S. Ja. Magarik, Dokl. Akad. Nauk SSSR 127, 840 (1959).
47. E. V. Frisman, A. K. Dadivanyan, G. A. Dyuzhev, Dokl. Acad. Nauk SSSR 153, 1062 (1963).
48. S. P. Micengendler, K. I. Sokolova, G. A. Andreeva, A. A. Korotkov, T. Kadirov, S. I. Klenin, S. Ja. Magarik, Vysokomolekul. Soedin. 9A, 1133 (1967).
49. V. N. Tsvetkov, E. N. Zakharova, G. A. Fomin, P. N. Lavrenko, Vysokomolekul. Soedin. 14A, 1956 (1972).
50. S. Ya. Lyubina, I. A. Strelina, V. S. Skazka, G. V. Tarasova, V. M. Jamshchikov, Vysokomolekul. Soedin. 14A, 1371 (1972).
51. V. S. Skazka, G. A. Fomin, G. V. Tarasova, I. G. Kirillova, V. M. Jamshchikov, A. E. Grishchenko, I. A. Shefer, I. S. Alekseeva, Vysokomolekul. Soedin. 15A, 2561 (1973).
52. I. Ja. Poddubnyi, E. G. Erenburg, M. A. Eryomina, Vysokomolekul. Soedin. 10A, 1381 (1968).
53. M. Fukuda, G. L. Wilkes, R. S. Stein, J. Polymer Sci. A2, 1417 (1971).
54. T. Ishikava, K. Nagai, J. Polymer Sci. A2, 1123 (1969).
55. T. Ishikava, K. Nagai, Polymer J. (Japan) 1, 116 (1970).
56. L. R. G. Treloar, Trans. Faraday Soc. 43, 284 (1947).
57. V. N. Tsvetkov, T. I. Garmonova, R. P. Stankevitch, Vysokomolekul. Soedin. 8, 980 (1966).
58. T. I. Garmonova, M. G. Vitovskaja, P. N. Lavrenko, V. N. Tsvetkov, E. V. Korovina, Vysokomolekul. Soedin. 13, 884 (1971).
59. S. M. Savvon, K. K. Turoverov, Vysokomolekul. Soedin. 6, 205 (1964).
60. V. N. Tsvetkov, I. N. Shtennikova, E. I. Rjumtsev, Ju. P. Getmanchuk, European Polymer J. 1, 767 (1971).
61. V. N. Tsvetkov, G. V. Tarasova, E. L. Vinogradov, N. N. Kurpijanova, V. M. Yamshchikov, V. S. Skazka, V. S. Ivanov, V. K. Smirnova, I. I. Migunova, Vysokomoleku1. Soedin. 13A, 620 (1971).
62. V. N. Tsvetkov, G. A. Fomin, P. N. Lavrenko, I. N. Shtennikova, T. V. Sheremeteva, L. I. Godunova, Vysokomolekul. Soedin. 10A, 903 (1968).
63. M. G. Vitovskaja, V. N. Tsvetkov, L. I. Godunova, T. V. Sheremeteva, Vysokomolekul. Soedin. 9A, 1682 (1967).

64. T. I. Garmonova, M. G. Vitovskaja, S. V. Bushin, T. V. Shereme-
    teva, Vysokomolekul. Soedin. 15A, in press.
65. V. N. Tsvetkov, N. N. Kupriyanova, G. V. Tarasova, P. N. Lavren-
    ko, I. I. Migunova, Vysokomolekul. Soedin. 12A, 1974 (1970).
66. V. N. Tsvetkov, I. N. Shtennikova, Vysokomolekul. Soedin. 6, 1041
    (1964).
67. V. N. Tsvetkov, E. I. Rjumtsev, I. N. Shtennikova, T. V. Peker,
    N. V. Tsvetkova, Dokl. Akad. Nauk SSSR 207, 1173 (1972).
68. E. N. Zakharova, L. I. Kutsenko, V. N. Tsvetkov, V. S. Skazka,
    G. V. Tarasova, V. M. Jamshchikov, Leningrad. Univ. Vestnik
    Ser. Fiz. Khim. N 16, 55 (1970).
69. G. I. Okhrimenko, Dissertation, Leningrad, 1969.
70. V. N. Tsvetkov, I. N. Shtennikova, N. A. Megeritskaya, L. S.
    Bolotnikova, Vysokomolekul. Soedin. 5, 74 (1963).
71. H. Janeschitz-Kriegl, W. Burhard, Adv. Polymer Sci. 6, 170 (1969).
72. V. N. Tsvetkov, S. Ja. Ljubina, I. A. Strelina, S. I. Klenin, V. I.
    Kurljankina, Vysokomolekul. Soedin. 15A, 691 (1973).
73. V. N. Tsvetkov, A. E. Grishchenko, P. A. Slavetskaja, Vysokomo-
    lekul. Soedin. 6, 856 (1964).
74. V. N. Tsvetkov, A. E. Grishchenko, J. Polymer Sci. C16, 3195
    (1968).
75. V. N. Tsvetkov, I. N. Shtennikova, Vysokomolekul. Soedin. 2, 808
    (1960).
76. V. N. Tsvetkov, E. N. Zakharova, M. M. Krunchak, Vysokomolekul.
    Soedin. 10A, 685 (1968).
77. V. N. Tsvetkov, E. V. Frisman, N. N. Boitsova, Vysokomolekul.
    Soedin. 2, 1001 (1960).
78. V. N. Tsvetkov, K. A. Andrianov, E. L. Vinogradov, S. E. Yakush-
    kina, Ts. V. Vardasanidze, Vysokomolekul. Soedin. 9B, 893 (1967).
79. V. N. Tsvetkov, K. A. Andrianov, E. L. Vinogradov, V. I. Pakhomov,
    S. E. Yakushkina, Vysokomolekul. Soedin. 9A, 3 (1967).
80. V. N. Tsvetkov, Makromol. Chem. 160, 1 (1972).
81. V. N. Tsvetkov, K. A. Andrianov, M. G. Vitovskaja, N. N. Maka-
    rova, E. N. Zakharova, S. V. Bushin, P. N. Lavrenko, Vysokomole-
    kul. Soedin. 14A, 369 (1972).
82. V. N. Tsvetkov, K. A. Andrianov, M. G. Vitovskaja, N. N. Maka-
    rova, S. V. Bushin, E. N. Zakharova, P. N. Lavrenko, A. A. Gor-
    bunov, Vysokomolekul. Soedin. 15A, 872 (1973).

83. V. N. Tsvetkov, K. A. Andrianov, E. L. Vinogradov, I. N. Shtenni-
    kova, S. E. Yakushkina, V. I. Pakhomov, J. Polymer Sci. C23, 385
    (1968).
84. V. N. Tsvetkov, K. A. Andrianov, I. N. Shtennikova, G. I. Okhri-
    menko, L. N. Andreeva, G. A. Fomin, V. I. Pakhomov, Vysoko-
    molekul. Soedin. 10A, 547 (1968).
85. V. N. Tsvetkov, K. A. Andrianov, G. I. Okhrimenko, I. N. Shtenni-
    kova, G. A. Fomin, M. G. Vitovskaja, V. I. Pakhomov, A. A.
    Jarosh, D. N. Andrejev, Vysokomolekul. Soedin. 12A, 1892 (1970).
86. V. N. Tsvetkov, I. N. Shtennikova, V. S. Skazka, E. I. Rjumtsev,
    J. Polymer Sci., C 16, 3205 (1968).
87. V. N. Tsvetkov, E. I. Rjumtsev, I. N. Shtennikova, G. I. Okhri-
    menko, Vysokomolekul. Soedin. 8, 1466 (1966).
88. V. N. Tsvetkov, I. N. Shtennikova, E. I. Rjumtsev, G. F. Pirogova,
    Vysokomolekul. Soedin. 9A, 1575 (1967).
92. V. N. Tsvetkov, S. Ja. Magarik, Dokl. Acad. Nauk SSSR 115, 911
    (1957).
93. V. N. Tsvetkov, E. I. Rjumtsev, I. N. Shtennikova, E. V. Korneeva,
    G. I. Okhrimenko, N. A. Mikhailova, A. A. Baturin, Ju. A. Amerik,
    B. A. Krentsel, Vysokomolekul. Soedin. 15A, 2570 (1973).
    European Polymer J. 9, 481 (1973).
94. S. Ja. Magarik, V. N. Tsvetkov, Zh. Fiz. Khim. 33, 835 (1959).
95. T. M. Birshtein, V. P. Budtov, E. V. Frisman, N. K. Janovskaja,
    Vysokomolekul. Soedin. 4, 455 (1962).
96. E. V. Frisman, N. N. Boitsova, Leningrad. Univ. Vestnik Ser. N 4,
    26 (1959).
97. A. Romanov, S. Ja. Magarik, M. Lazar, Vysokomolekul. Soedin.
    9B, 292 (1967).
98. V. N. Tsvetkov, G. A. Andreeva, I. A. Baranovskaja, V. E. Eskin,
    S. I. Klenin, S. Ja. Magarik, J. Polymer Sci. C16, 239 (1967).
99. V. N. Tsvetkov, S. Ja. Magarik, T. Kadirov, G. A. Andreeva,
    Vysokomolekul. Soedin. 10A, 943 (1968).
100. P. Gramain, J. Leray, H. Benoit, J. Polymer Sci. C16, 3983 (1968).

V. PHYSICAL CONSTANTS OF SOME IMPORTANT POLYMERS

PHYSICAL CONSTANTS OF POLY(BUTADIENE)

G. H. Stempel

Oak Ridge Drive,
Cuyahoga Falls, Ohio

Property	Value	Ref.
Activation Energy of Adhesion, $[\text{kJ mol}^{-1}]$		
Poly(butadiene), sodium catalyzed		
$\quad$ to aluminium	21.8	5
$\quad$ to copper	22.2	
Coefficient of Expansion, volume, $[\text{K}^{-1}]$		
Butyl lithium polymer	7.5×10^{-4}	44
$\quad$ (43% 1,4-cis-; 50% 1,4-trans; 7% 1,2-)		
Cohesive Energy Density, $[\text{J cm}^{-3}]^{1/2} \equiv [\text{MJ m}^{-3}]^{1/2}$		
Emulsion	17.2	40
Na catalyzed	17.6	40
1,4-cis	16.85	41
Na catalyzed	17.06	41
Critical Surface Tension of Spreading, γ_c, $[\text{mN m}^{-1}] \equiv [\text{dyn cm}^{-1}]$		4
1,2-	25	
1,4-trans	31	
1,4-cis	32	

Crystallographic Data

14-20, 22

Isomer	Lattice	Monomers per Unit Cell	Cell Dimension [Å]			Cell Angles		
			a	b	c (chain axis)	α	β	γ
1,4-trans-(99-100%)	Pseudohexagonal	1	4.54	4.54	4.9 (Mod. I) *			
			4.88	4.88	4.68 (Mod. II)			
1,4-cis-(98-99%)	Monoclinic	4	4.60	9.50	8.6	90	109	90
1,2-isotactic (99%)	Rhombohedral	18	17.3	17.3	6.5	90	90	120
1,2-syndiotactic (98%)	Orthorhombic	4	10.98	6.60	5.14	90	90	90

* Transition temperature: 75°C.

Property	Value	Ref.
Density, $[\text{g cm}^{-3}] \equiv [\text{Mg m}^{-3}]$	see also Molar Polarizability	
1,4-trans (99-100%) Modification I	0.97	15
$\qquad\qquad\qquad\quad$ Modification II	0.93	15
1,4-cis (98-99%)	1.01	15
1,2-isotactic (99%)	0.96	15
1,2-syndiotactic (98%)	0.96	15
sec-Butyl lithium polymer	0.89	34
$\quad$ 40.5% 1,4-cis; 33.5% 1,4-trans; 26% 1,2-		
n-Butyl lithium polymer	0.895	34
$\quad$ 44% 1,4-cis; 42% 1,4-trans; 14% 1,2		
Dielectric Constant, ϵ, and Loss Factor		42
1,4-trans (99%), 25°C $\qquad \epsilon$	2.51	
$\qquad\qquad\qquad\qquad \tan \delta$	0.002 (50 Hz)	

Property	Value	Ref.
Entropy of Fusion, [J mol^{-1} K^{-1}]		21
1,4-trans (99%) Modification I	26.8	
Modification II	11.3	
1,4-cis (98%)	33.5	
Entropy of Polymerization, [J mol^{-1} K^{-1}]		
at 25°C: butadiene to cis-1,4-polybutadiene (94%)	84.2	6
First Order Transition Temperature, [°C]		
1,4-trans Modification I to Modification II	75	21
Glass Transition Temperature, [°C]		
1,4-cis ("high")	-102	26
1,4-cis (98-99%)	-95	3
1,4-cis	-106	39
1,4-trans	-107	39
1,4-trans (94%)	-83	6
1,2-	-15	39
Heat of Combustion, [kJ (kg CO$_2$)$^{-1}$]		
Emulsion polymer	13876	38

Heat of Combustion and Formation of Three Emulsion Poly(butadienes), [kJ mol^{-1}]

Polymerization Temperature, [°C]	Heat of Combustion	Heat of Formation	
50	-2447	12.91	38
50	-2450	16.20	38
5	-2447	13.81	38

Property	Value	Ref.
Heat of Fusion, [kJ mol^{-1}]		
1,4-cis	2.51	39
1,4-cis	9.2±0.5	21
1,4-trans, Modification I	4.184	37
1,4-trans, Modification I	10.0±4.2	21
1,4-trans, Modification II	4.60	21
1,4-trans, Modification II	4.184	43
Huggins Coefficient,	see Ref.	29
Infrared Absorption Coefficients, [dl cm^{-1} mg^{-1}] x 10^3		12

Isomer	Wave Length [μm]		
	10.35	10.95-10.98	13.5-13.65
1,4-trans	23.3		
1,2-	0.828	26.7	0.231
1,4-cis	0.609	0.107	5.73

Property		Ref.
Infrared Molar Absorptivities, [mol^{-1} cm^{-1}]		25

Isomer	Wave Length [μm]		
	10.3	11.0	12-15.75
1,4-trans	133	2.4	0.86
1,2-	6.7	184	4.7
1,4-cis	4.4	1.9	10.1

Property	Value	Ref.

Melting Temperature, [°C]

1,4-trans (99-100%) Modification I	97	13,15,37
Modification II	145	
1,4-cis (98-99%)	2	
1,2-isotactic (99%)	126	
1,2-syndiotactic	156	

Microstructures of Commercially Available Poly(butadienes)   29

		% 1,4-cis	% 1,4-trans	% 1,2
Cariflex BR 1220	Companie Francaise	~98	< 1	< 1
Ameripol CB 220	Goodrich	~98	< 1	< 1
Taktene	Polymer Corporation, Ltd.	~98	< 1	< 1
JSR BR 01	Japan Synthetic Rubber	> 94	< 5	1
Nipol BR 1220	Japanese Geon	94	4	2
Budene 501	Goodyear	93	4	3
Phillips 1203	Phillips Petroleum	92	5	3
Europrene Cis	A. N. I. C.	92	5	3
Buna CB10	Stereokautschuk-Werke	91	6	3
Buna CB11		91	5	4
Cisdene 100	American Rubber & Chemical	84	10	6
Asadene NF 35 R	Asahi Chemical Industry	38	~49	11
Diene 35 A	Firestone	35	55	10
Solprene 200	Phillips Petroleum	35	55	10
Intene 35 NF	International Synthetic	35	53	11
I. F. P.	Institut Francais du Petrole	26	70	4
M 12 RE	Societe Nationale de Petroles d' Aquitaine	11	71	18
I. F. P.	Institut Francais du Petrole	5	9	86
Trans-4 Phillips	Phillips Petroleum	4	93	3

Molar Polarizability, α $[cm^3]$ x 10^{-25}; **Refractive Index,** n_D^{25}; **and Density,** d^{25} $[g\ cm^{-3}] \equiv [Mg\ m^{-3}]$

$$\alpha = 3M(n^2 - 1)/4\ \pi Nd\ (n^2 + 2)$$

Polymer	Composition Wt. %			n_D^{25}	d^{25}	α x 10^{25}	
	1,4-trans	1,2	1,4-cis				
Emulsion	71	19	10	1.518	0.913	70.6	28
Lithium Catalyzed		Not given		1.516	0.913	70.4	28
1,4-cis		Not given		1.526	0.915	71.4	28
Emulsion -10°C	77.5	15.9	6.6	1.5147	0.8911	71.6[x]	11
Emulsion 25°C	68.3	18.8	12.9	1.5149	0.8920	71.9[x]	11
Emulsion 100°C	54.5	20.1	25.4	1.5160	0.8933	71.9[x]	11

[x] Calculated from data of Ref. 11.

For trans-1,4-poly(butadiene) (91.9 trans-1,4, 5,6 cis-1,4, 2,5 1,2):   33

$$[n]_D^T = 1.5229 - 0.366\ x\ 10^{-3}, \text{ where T is } T°C.$$

Polymerization Data

a) Effect of Polymerization Catalyst on Microstructure of Poly(butadiene)

Catalyst	% 1,4-cis	% 1,4-trans	% 1,2	Ref.
Lithium	35	55	10	32
Sodium	10	25	65	32
Alfin	8	63	29	32
Potassium	15	40	45	32
Cesium	6	35	59	32

Property	Value	Ref.

a) Effect of Polymerization Catalyst on Microstructure of Poly(butadiene) (Cont'd.)

Catalyst	% 1,4-cis	% 1,4-trans	% 1,2	Ref.
n-Butyl lithium/bis (2-methoxyethyl) ether	-	ca. 5.3	94.3	31
$Al(C_2H_5)_3$ - VCl_3	-	99	1	36
$Al(C_2H_5)_3$ - $CoCl_2$	97.9	1.1		

All polymerizations carried out in hydrocarbon solvent.

b) Effect of Polymerization Solvent on Microstructure of Poly(butadiene)

Solvent	% 1,4-cis	% 1,4-trans	% 1,2	Ref.
Hexane	43	50	7	32
Toluene	44	47	9	32
Benzene-$(C_2H_5)_3$N	23	40	37	32
Benzene-THF[x]	13	13	74	32
Heptane-THF[x]	4	4	92	32
THF[x]	0	9	91	32

[x] Tetrahydrofuran; Catalyst: C_2H_5Li

Refractive Index   see Molar Polarizability

Sedimentation Constants for 1,4-cis

Solvent	T [°C]	% cis	s_o	
3-Pentanone	10.3	95	$s_o = 5.30 \times 10^{-3} M^{-0.5}$ svedbergs	2
n-Hexane/n-Heptane, (1:1)	20		$s_o = 3.02 \times 10^{-2} M^{-0.48}$ svedbergs	23

see also Table in this Handbook

Theta Solvents and Temperatures, T [°C]

n-Hexane/n-Heptane, (1:1)	5	23
Isobutyl acetate	20.5	7
5-Methyl-2-hexanone	12.6	1
2-Pentanone	59.7	1
3-Pentanone	10.3	2

Viscosity-Molecular Weight Relationship, $[\eta] = K M^a$, [ml g^{-1}] see also corresponding table in this Handbook

a) Good Solvents

Solvent	T [°C]	Microstructure cis-	trans-	1,2-	Notes	$K \times 10^3$	a	Notes	
Toluene	25	97				30.5	0.725	M_w	7
Toluene	30		"high trans"			29.4	0.753	M_n	8
Cyclohexane	20		79	21		36	0.70	M_w	24
Toluene	25.9	14.8	62	23.2	(emulsion, 50°C)	725	0.45	M_n	10
Toluene	25.9		79.6	19.6	(emulsion, -20°C)	106	0.63	M_n	10
Cyclohexanone	25	10	70	20	(alfin)	12.5	0.77	M_w	27
Toluene	25	35	55	10	(lithium)	14.2	0.80	M_n	30
Toluene	25		ca. 5.3	94.3		90.1	0.81	M_n	31
THF	25		ca. 5.3	94.3		60.1	0.85	M_n	31
THF	25	35	55	10	(lithium)	15.6	0.80	M_n	31

Property	Value	Ref.

b) Theta Conditions (for 1,4-cis)

Solvent	Theta T [°C]	cis-Content, %	$K \times 10^3$	a	Notes	M_w/M_n	
3-Pentanone	10.3	95	152	0.5	M_w	1.43	2
		95	181	0.5	M_n		2
5-Methyl-2-hexanone	12.6	95	150	0.5	M_w		2
			130	0.5	M_w		9
Isobutyl acetate	20.5	98	185	0.5	M_n	1.11	7
n-Hexane/n-Heptane, (1:1)	5	90	138	0.53	M_v		23

REFERENCES

1. M. Abe, H. Fujita, Repts. Progr. Polymer Phys. (Japan) 7, 42 (1964).
2. M. Abe, Y. Murakami, H. Fujita, J. Appl. Polymer Sci. 9, 2549 (1965).
3. M. Baccaredda, E. Butta, Chim. Ind. (Milan) 42, 978 (1960).
4. L. and H. Lee, J. Polymer Sci. A-2, 5, 1103 (1967).
5. S. S. Voyutskii, Yu Y. Markin, G. M. Gorchakova, V. E. Gul', Adhesive Age 8(11), 24 (1965).
6. F. S. Dainton, D. M. Evans, F. E. Hoare, T. P. Melia, Polymer 3, 297 (1962).
7. F. Danusso, G. Moraglio, G. Gianotti, J. Polymer Sci. 51, 475 (1961).
8. R. Endo, Nippon Gomu Kyokaishi 35, 658 (1962).
9. H. Fujita, N. Takaguchi, K. Kawahara, M. Abe, H. Utiyama, K. Kajitani, 12th Polymer Symposium, Nagoya, Japan, November, 1963.
10. B. L. Johnson, R. D. Wolfangel, Ind. Eng. Chem. 41, 1580 (1949).
11. L. Mandelkern, M. Tyron, F. A. Quinn, J. Polymer Sci. 19, 81 (1956).
12. P. Morero, A. Santambrogio, L. Porri, F. Ciampelli, Chim. Ind. (Milan) 41, 758 (1959).
13. G. Natta, Rev. Gen. Caoutchouc 40, 786 (1963).
14. G. Natta, Rubber Plastics Age 38, 495 (1957).
15. G. Natta, Science 147, 269 (1965).
16. G. Natta, P. Corradini, Angew. Chem. 68, 615 (1956).
17. G. Natta, P. Corradini, J. Polymer Sci. 20, 251 (1956).
18. G. Natta, P. Corradini, Nuovo Cimento 15, Suppl. 1, 9 (1960).
19. G. Natta, P. Corradini, Rubber Chem. Technol. 33, 732 (1960).
20. G. Natta, P. Corradini, L. Porri, Atti. Accad. Nazl. Lincei Rend. 20, 728 (1956).
21. G. Natta, G. Moraglio, Rubber Plastics Age 44, 42 (1963); Makromol. Chem. 66, 218 (1963).
22. G. Natta, L. Porri, P. Corradini, D. Morero, Atti. Accad. Nazl. Lincei Rend. 20, 560 (1956).
23. I. Y. Poddubnyi, V. A. Grechanovskii, Vysokomolekul. Soedin. 6, 64 (1964).
24. Ph. Ribeyrolles, A. Guyot, H. Benoit, J. Chim. Phys. 56, 383 (1959).
25. R. S. Silas, J. Yates, V. Thornton, Anal. Chem. 31, 529 (1959).
26. G. S. Trick, J. Appl. Polymer Sci. 3, 253 (1960).
27. R. L. Cleland, J. Polymer Sci. 27, 349 (1958).
28. J. Furukawa, S. Yamashita, T. Kotani, M. Kawashima, J. Appl. Polymer Sci. 13, 2527 (1969).
29. J. Curchod, Rubber Chem. Technol. 43, 1367 (1970).
30. H. E. Adams, Kelly Farhat, B. L. Johnson, Ind. Eng. Chem., Prod. Res. Develop. 5, 126 (1966).
31. J. N. Anderson, M. L. Barzan, H. E. Adams, Rubber Chem. Technol. 45, 1281 (1972).
32. H. E. Adams, R. L. Bebb, L. E. Forman, L. B. Wakefield, Rubber Chem. Technol. 45, 1252 (1972).
33. T. Ishikawa, K. Nagai, Polymer J. 1(1), 117 (1970).
34. G. V. Vinogradov, E. A. Dzyura, A. Ya. Malkin, V. A. Grechanovskii, J. Polymer Sci. A 29, 1153 (1971).
35. G. Natta, L. Porri, P. Corradini, D. Morero, Chim. Ind. (Milan) 40, 366 (1958).
36. Belgian Patent 573,680 to Montecatini (1958).
37. M. Berger, D. J. Buckley, J. Polymer Sci. A 1, 2945 (1963).
38. R. S. Nelson, R. S. Jessup, D. E. Roberts, J. Res. Natl. Bur. Std. 48, 275 (1952).
39. W. S. Bahary, D. I. Sapper, J. H. Lane, Rubber Chem. Technol. 40, 1529 (1967).
40. R. L. Scott, Thesis, Princeton University (1945).
41. V. A. Grigorovskaya, A. V. Shvarts, L. P. Bychova, Vysokomolekul. Soedin. Adgeziya Polimerov. Sb. Stalei, 1963, p. 41.
42. M. Pegoraro, K. Mitoraj, Makromol. Chem. 61, 132 (1963).
43. L. Mandelkern, F. A. Quinn, Jr., J. Polymer Sci. 19, 77 (1956).
44. R. H. Valentine, J. D. Ferry, T. Homma, K. Ninomiya, J. Polymer Sci. A 2(6), 479 (1968).

PHYSICAL CONSTANTS OF DIFFERENT RUBBERS★

Lawrence A. Wood

National Bureau of Standards
Washington, D. C.

CONTENTS

Where a range is given, there are available several observations which differ. In most cases the differences are thought to be real, arising from differences in the rubber rather than from errors of observation. Where a single value is given, it is either because no other observations are available or because there seems to be no significant disagreement among values within the errors of observation. Where values are not given, data have not been found. Where dashes are shown, either the physical measurement is impossible or the constant in question is not adequately defined under the given conditions. 50 phr means "50 parts of carbon black by weight per 100 parts of rubber," corresponding to a volume fraction of about 0.2. The values shown refer to specific vulcanizates cited in the corresponding references. Other vulcanizates may yield a broader range of values.

The metric technical unit "kilogram force" or "kilopond" has been taken as 9.80665 newtons (980,665 dynes).

Where an undefined calorie has been used by the respective authors, it has been taken as the thermochemical calorie equal to 4.1840 joules. Values are given for constants at a temperature of 25°C and a pressure of 1 normal atmosphere = 1.01325×10^5 Pa (where 1 Pa = 1 N/m^2 = 10 dyn/cm^2 = 10^{-5} bar).

1 - POLY(ISOPRENE), NATURAL RUBBER (HEVEA)

Property	Unvulcanized	Ref.	Pure-gum vulcanizate	Ref.	Vulcanizate Containing About 33% Carbon Black (=50 phr) Vol fraction = 0.2 approx.	Ref.	Hard Rubber (Ebonite)	Ref.
Density, [Mg m^{-3}]≡[g cm^{-3}]	0.913 (0.906-0.916)	11 11	0.970 (0.920-1.000)	19, 54 2	1.120 (1.120-1.180)	19 2, 54	1.170 (1.130-1.180)	11, 53 6
Coefficient of Expansion, Volume, ß=(1/V) (∂V/∂T), [K^{-1}]	670×10^{-6}	11	660×10^{-6}	11, 14	530×10^{-6} (450-550×10^{-6})	19 5	190×10^{-6}	11
Thermal								
Glass Transition Temperature [K] (deg C)	210 (-72) (-74 to -69)	13 13	210 (-63) (-72 to -61)	62 62	208 (-65)	8, 27, 43	353 (+80)	53
Heat Capacity, Cp, [kJ kg^{-1} K^{-1}] ∂Cp/∂T, [kJ kg^{-1} K^{-2}]	1.905 3.54×10^{-3}	20, 57 20, 57	1.828	30	1.494	30	1.385	6, 11
Thermal Conductivity, [W m^{-1} K^{-1}]	0.134	11	0.153 (0.14-0.15)	66 3, 54	0.280	51	0.163 0.160-0.180	11 6, 8
Heat of Combustion, [kJ kg^{-1}]	-45,200	11	-44,400	11			-33,000	11
Equilibrium Melting Temperature, [K] (deg C)	301 (28) (30-39)	36, 49 15, 34						
Heat of Fusion of Crystal, [kJ kg^{-1}]	64.0	36, 49						
Optical								
Refractive Index, n_D	1.5191	63	1.5264	11	--		1.6	11
dn_D/dT, [K^{-1}]	-37×10^{-5}	63	-37×10^{-5}	11	--			
Electrical								
Dielectric Constant, (1 kHz)	2.37-2.45	7, 11	2.68 2.5-3.0	7 7, 11			2.82 2.8-2.9	7 6, 7
Dissipation Factor (1 kHz)	0.001-0.003	7	0.002-0.04	7			0.0043-0.009	6, 7
Conductivity (60 sec), [S m^{-1}]	2-57×10^{-15}	7, 11	2-100×10^{-15}	7, 11			2-3000×10^{-15}	6, 7
Mechanical								
Compressibility B = $-\frac{1}{V_o} \frac{∂V}{∂P}$, [$MPa^{-1}$]	515×10^{-6}	61	514×10^{-6}	61	410×10^{-6}	41	240×10^{-6}	53

★This table was originally prepared for the Smithsonian Physical Tables (9th Edition) 1954. It was revised and extended for this Handbook in 1965 and 1973. Contribution of the National Bureau of Standards, not subject to copyright.

Property	Unvulcanized	Ref.	Pure-gum vulcanizate	Ref.	Vulcanizate Containing About 33% Carbon Black (=50phr) Vol fraction = 0.2 approx.	Ref.	Hard Rubber (Ebonite)	Ref.
Compressibility (Cont'd.)								
$\partial B/\partial P$, [MPa^{-2}]	-2.1 x 10^{-6}	61	-2.4 x 10^{-6}	61	-1.8 x 10^{-6}	41	-0.41 x 10^{-6}	53
$\partial B/\partial T$, [MPa^{-1} K^{-1}]	+2.3 x 10^{-6}	61	+2.1 x 10^{-6}	61			+1.1 x 10^{-6}	53
Thermal Pressure Coefficient, $\gamma = \beta/\beta_A$, [MPa K^{-1}]**			1.22	14				
$\partial\gamma/\partial T$, [MPa K^{-2}]***			-0.0052	14				
Bulk Modulus (isothermal), [MPa]*	1940	61	1950	61	2440	41	4170	53
Bulk Modulus (adiabatic), [MPa]*	2270	61	2260	61				
Bulk Wave Velocity v_b, [m s^{-1}] (longitudinal wave)	1580	61	1580	61	1490	23		
			(1500-1580)	23,31,61				
$\partial v_b/\partial T$, [m s^{-1} K^{-1}]	-3	61	-3	61				
Strip (longitudinal wave)								
Velocity v_1 (1 kHz), [m s^{-1}]			45	23	141	23	1540	11,40
			(35-51)	8,11,23				
$\partial v_1/\partial T$, [m s^{-1} K^{-1}]			-0.2	11				
Ultimate Elongation, [%]	--		750-850	1,2	550-650	1,2	6	5
							(3-8)	6
Tensile Strength, [MPa]*	--		17-25	1,2	25-35	1,2	60-80	6
Initial Slope of Stress-Strain Curve,								
Young's Modulus, E (60s), [MPa]*	--		1.3	38,50	3.0-8.0	2	3000	5,40
			(1.0-2.0)	2,38,50,62				
Shear Modulus, G(60s), [MPa]*	--		0.43	50,62	1.3-2.0	9,45	600	40
			(0.3-0.7)	45,62				
Shear Compliance J (60s), [MPa^{-1}]	--		2.3 x 10^{-12}	50,62	0.5-0.7 x 10^{-12}	9,45	1.7 x 10^{-12}	40
			(1.5-3.5 x 10^{-12})	45,62				
Creep Rate, [%/unit log t]	--		2	38,62	8	8		
			(1-3)	21,38,50,56,59,62	(7-12)	8		
Complex Dynamic Shear Modulus G* (1 Hz)	8,27,43,52 65		8,24,25,26, 27,42,43,45,46,52					
Storage Modulus G', [log MPa] (values of log G')	9.61-10 (9.53 to 9.75)-10	65	9.61-10 (9.49 to 9.78)-10	26	0.79 0.27 to 1.12	27	3.05	43
Loss Modulus G'', [log MPa] (values of log G'')	8.46-10 (8.43 to 8.65)-10		7.80-10 (7.72 to 8.48)-10	26	9.83-10 (9.50 to 10.11)-10	27	1.65	43
Loss Tangent G''/G'	0.09 (0.07-0.13)	65	0.016 (0.01-0.05)	26	0.11 (0.10-0.17)	27	0.040	43
Resilience (rebound), [%]	75-77	2,22	75-84	22,25	50 (45-55)	22 2,9,22,25	63-67	5

2 - POLY(BUTADIENE-CO-STYRENE), (SBR, GR-S), (ABOUT 23.5% BOUND STYRENE CONTENT)

Property	Unvulcanized	Ref.	Pure-gum vulcanizate	Ref.	Vulcanizate Containing About 33% Carbon Black (=50 phr) Vol fraction = 0.2 approx.	Ref.
Density, [Mg m^{-3}]≡[g cm^{-3}]	0.933 (0.9325-0.9335)	13 13	0.980 (0.940-1.000)	19 19	1.150	19,54
Coefficient of Expansion, [K^{-1}] volume $\beta=(1/v)(\partial V/\partial T)$	660 x 10^{-6}	13	660 x 10^{-6} (650-700 x 10^{-6})	9,19,54 54	530 x 10^{-6}	19
Thermal						
Glass Transition Temperature, [K] (deg C)	209-214 (-64 to -59)	13	221 (-52)	62	221 (-52)	62
Heat Capacity, Cp, [kJ kg^{-1} K^{-1}]	1.89	48	1.83	30	1.50	30
$\partial Cp/\partial T$, [kJ kg^{-1} K^{-2}]	3.2 x 10^{-3}	48				
Thermal Conductivity, [W m^{-1} K^{-1}]			0.190-0.250	8,51	0.300	51
Heat of Combustion, [kJ (kg CO$_2$)$^{-1}$]	-56.5 x 10^{3}	13				

identical with *[N mm^{-2}]; **[N mm^{-2} K^{-1}]; ***[N mm^{-2} K^{-2}].

Property	Unvulcanized	Ref.	Pure-gum vulcanizate	Ref.	Vulcanizate Containing About 33% Carbon Black (=50 phr) Vol fraction = 0.2 approx.	Ref.
Optical						
Refractive Index n_D	1.5345	13			--	
	(1.534-1.535)	13				
dn_D/dT, [K^{-1}]	-37×10^{-5}	13			--	
Electrical						
Dielectric Constant (1 kHz)	2.5	7	2.66	7		
Dissipation Factor	0.0009	7	0.0009	7		
Mechanical						
Compressibility						
$B=1/V \; \partial V/\partial P_{\emptyset}$, [$MPa^{-1}$]	530×10^{-6}	41	510×10^{-6}	23	400×10^{-6}	41
$\partial B/\partial P$, [MPa^{-1}]	-2.7×10^{-6}	41	-2.0×10^{-6}	41	-1.8×10^{-6}	41
Bulk Modulus (isothermal), [MPa]*	1,890	41	1,960	41	2,500	41
Bulk Wave Velocity, [$m\,s^{-1}$] (longitudinal wave)			1485	23	1510	23
Strip (longitudinal wave)						
Velocity v_1 (1 kHz), [$m\,s^{-1}$]			73	55	161	23
$\partial v_1/\partial T$, [$m\,s^{-1}\,K^{-1}$]			-0.2	55		
Ultimate Elongation, [%]	--		400-600	1,2	400-600	1,2
Tensile Strength, [MPa]*			1.4-3.0	1,2	17-28	1,2
Initial Slope of Stress-Strain Curve Young's Modulus (60s), [MPa]*	--		1.6	38,62	3-6	2
			(1.0-2.0)	38		
Shear Modulus G (60s), [MPa]*	--		0.53	62	2.0	45
			(0.3-0.7)	38	(2.0-2.5)	9,45
Shear Compliance J (60s), [MPa^{-1}]	--		1.9×10^{-12}	62	0.5×10^{-12}	45
Creep Rate (1/J) ($\partial J/\partial \log t$), [%/unit log t]	--		7	38,62	12	8
			(3-10)	38,62		
Complex Dynamic Shear Modulus C* 1 Hz		35,44,65		25,27,31,35,44		25,44,45,54
Storage Modulus G', [log MPa]	9.82-10	65	9.88-10	27	0.94	44
(values of log G')	(9.82 to 9.85)-10		(9.64 to 10.20)-10		(0.39 to 0.94)	
Loss Modulus G'', [log MPa]	8.94-10	65	8.92-10	27	0.28	44
(values of log G'')	(8.56 to 8.94)-10		(8.73 to 8.94)-10			
Loss Tangent G''/G'	0.13	65	0.11	27	0.22	44
	(0.05-0.13)		(0.07-0.18)		(0.14-0.22)	
Resilience (rebound), [%]			65	2,22,25	40	22
					(40-50)	2,25

3 - POLY(ISOBUTENE-CO-ISOPRENE) (BUTYL RUBBER, IIR, GR-I)

Property	Unvulcanized	Ref.	Pure-gum vulcanizate	Ref.	Vulcanizate Containing About 33% Carbon Black (=50 phr) Vol fraction =0.2 approx.	Ref.
Density, [$Mg\,m^{-3}$]$\equiv$[$g\,cm^{-3}$]	0.917	58	0.933	18	1.130	19
			(0.930-0.970)	19		
Coefficient of Expansion, volume $\beta=(1/V)(\partial V/\partial T)$, [$K^{-1}$]			560×10^{-6}	18,47	460×10^{-6}	19
Thermal						
Glass Transition Temperature, [K] (deg C)	202 (-71)	32	210 (-63)	62		
	(-75 to -67)	8,13				
Heat Capacity, Cp, [$kJ\,kg^{-1}\,K^{-1}$]	1.95	28,57,30	1.85	30		
$\partial Cp/\partial T$, [$kJ\,kg^{-1}\,K^{-2}$]	4.4×10^{-3}	28,57				
Thermal Conductivity, [$W\,m^{-1}\,K^{-1}$]			0.130	51	0.230	51
Optical						
Refractive Index n_D	1.5081	13			--	

identical with * [$N\,mm^{-2}$]

Property	Unvulcanized	Ref.	Pure-gum vulcanizate	Ref.	Vulcanizate Containing About 33% Carbon Black (=50 phr) Vol fraction = 0.2 approx.	Ref.
Electrical						
Dielectric Constant (1 kHz)	2.38	7	2.42	7		
Dissipation Factor (1 kHz)	0.003	7	0.0054	7		
Mechanical						
Compressibility,						
$B = -(1/V_0)(\partial V/\partial P)_T$, $[MPa^{-1}]$			508×10^{-6}	47	460×10^{-6}	23
$\partial B/\partial T$, $[MPa^{-1} K^{-1}]$			1.7×10^{-6}	47		
Thermal Pressure Coefficient						
$\gamma = \beta/B$, $[MPa K^{-1}]$ **			1.13	47		
$\partial \gamma/\partial T$, $[MPa K^{-2}]$ ***			-0.0032	47		
Bulk Modulus (isothermal), $[MPa]$ *			1970	47	2170	23
Bulk Wave Velocity, (longitudinal wave)			1485	23	1510	23
Strip (longitudinal wave)						
velocity v_1 (1 kHz), $[m\ s^{-1}]$			100	55	210	23
			(100-111)	8, 23, 55		
$\partial v_1/\partial T$, $[m\ s^{-1} K^{-1}]$			-2.2	55		
Ultimate Elongation, [%]	--		750-950	1, 2	650-850	1, 2
Tensile Strength, $[MPa]$ *	--		18-21	1, 2	18-21	1, 2
Initial Slope of Stress-Strain Curve, Young's						
Modulus (60s), $[MPa]$ *	--		1.0	38, 62	3-4	2
			(0.7-1.5)	2, 38, 62		
Shear Modulus G (60s), $[MPa]$ *	--		0.33	38, 62	1.8	9
			(0.2-0.5)	2, 38, 62		
Shear Compliance J (60s), $[MPa^{-1}]$	--		3.1×10^{-12}	62	0.56×10^{-12}	9
			$(2-4 \times 10^{-12})$	2, 44, 62		
Creep Rate $(1/J)(\partial J/\partial \log t)$, [%/unit log t]	--		4	38, 62		
			2-8	38, 62		
Complex Dynamic Shear Modulus G^* (60 Hz)						
Storage Modulus G', [log MPa] (values of log G')	0.50	52	9.64-10	25	0.56	25
Loss Modulus G'', [log MPa] (values of log G'')	9.98-10	52	9.48-10	25	0.21	25
Loss Tangent G''/G'	0.3	52	0.7	25	0.45	25
Resilience (rebound), [%]			13-16	2, 22	14	2

4 - POLY(CHLOROPRENE) (CR, NEOPRENE)

Property	Unvulcanized	Ref.	Pure-gum vulcanizate	Ref.	Vulcanizate Containing About 33% Carbon Black (=50 phr) Vol fraction = 0.23 approx.	Ref.
Density, $[Mg\ m^{-3}] \equiv [g\ cm^{-3}]$	1.230	4, 13, 58	1.320	19	1.420	19
Coefficient of Expansion,						
volume, $\beta = (1/V)(\partial V/\partial T)$, $[K^{-1}]$	600×10^{-6}	4, 8	$610-720 \times 10^{-6}$	4, 19, 64		
Thermal						
Glass Transition Temperature, [K] (deg C)	228 (-45)	33	228 (-45)	29, 33, 62, 64	230 (-43)	33
Heat Capacity, C_p, $[kJ\ kg^{-1} K^{-1}]$	2.18	4	2.1-2.2	8	1.7-1.8	4, 54
Thermal Conductivity, $[W\ m^{-1} K^{-1}]$	0.192	4	0.192	8, 54	0.210	4
Equilibrium Melting Temperature,						
[K] (deg C)	353 (80)	29, 39				
[K] (deg C)	333 (60)	34				
Heat of Fusion of Crystal, $[kJ\ kg^{-1}]$	94.6	39				
Optical						
Refractive Index n_D	1.558	12			--	
dn_D/dT, $[K^{-1}]$	-36×10^{-5}	12			--	

identical with *$[N\ mm^{-2}]$; **$[N\ mm^{-2} K^{-1}]$; ***$[N\ mm^{-2} K^{-2}]$

Property	Unvulcanized	Ref.	Pure-gum vulcanizate	Ref.	Vulcanizate Containing About 33% Carbon Black (=50 phr) Vol fraction = 0.23 approx.	Ref.
Electrical						
Dielectric Constant (1 kHz)			6.5-8.1	7		
Dissipation Factor (1 kHz)			0.031-0.086	7		
Conductivity, [S m^{-1}]			$3\text{-}1400 \times 10^{-12}$	7		
Mechanical						
Compressibility						
$B=-(1/V_o)(\partial V/\partial P)$, [MPa^{-1}]	480×10^{-6}	23, 41	440×10^{-6}	23, 41	360×10^{-6}	23, 41
$\partial B/\partial P$, [MPa^{-2}]	-2.8×10^{-6}	41	-2.3×10^{-6}	41	-1.7×10^{-6}	23, 41
Bulk Modulus (isothermal), [MPa]*	2080	23	2270	23	2780	23, 41
Bulk Wave Velocity, [m s^{-1}] (longitudinal wave)	1420	23	1520	23		
Strip (longitudinal wave)						
Velocity v_1 (1 kHz), [m s^{-1}]			69	8, 23	196	23
Ultimate Elongation, [%]	--		800-1000	1, 2	500-600	1, 2
Tensile Strength, [MPa]*	--		25-38	1, 2	21-30	1, 2
Initial Slope of Stress-Strain Curve, Young's Modulus E (60s), [MPa]*	--		1.6	38, 62	3-5	2
			(1-3)	2, 38, 62		
Shear Modulus G (60s), [MPa]*	--		0.52	62	1.4	9
			(0.3-1.0)	2, 38, 62		
Shear Compliance J (60s), [MPa^{-1}]	--		2.0×10^{-12}	2, 38, 62	0.7×10^{-12}	9
			$1\text{-}3 \times 10^{-12}$	62		
Creep Rate $(1/J)(\partial J/\partial \log t)$, [%/unit log t]	--		6	38, 62		
			(5-10)	38, 62		
Complex Dynamic Shear Modulus G* (60 Hz)						
Storage Modulus G', [log MPa] (values of log G')			9.81-10	25, 64	0.45	25
Loss Modulus G'', [log MPa] (values of log G'')			9.04-10	25, 64	9.75-10	25
Loss Tangent G''/G'			0.17	25, 64	0.20	25
Resilience (rebound), [%]			60-65	2, 22	48	22
					(40-50)	2, 22

5 - REFERENCES

Summaries and Tabulations of Published Values

1. J. M. Ball, G. C. Maassen, Am. Soc. Testing Materials, Symp. Appl. Synthetic Rubbers, March 2, 1944, p. 27.

2. B. B. S. T. Boonstra, "Properties of Elastomers," Chapter 4 of Vol. III in R. Houwink (Ed.), "Elastomers, Their Chemistry, Physics and Technology," Elsevier Publishing Company, New York, 1948.

3. L. C. Carwile, H. J. Hoge, "Thermal Conductivity of Soft Vulcanized Natural Rubber, Selected Values," in "Advances in Thermophysical Properties at Extreme Temperatures and Pressures," Am. Soc. Mechanical Engineers, New York, 1965. Rubber Chem. Technol. 39, 125 (1966).

4. N. L. Catton, "The Neoprenes," Rubber Chemicals Division, E. I du Pont de Nemours and Company, Wilmington, Delaware, 1953.

5. T. R. Dawson, B. D. Porritt, "Rubber Physical and Chemical Properties," Res. Assoc. Brit. Rubber Manufactures, Croydon, England, 1935.

6. A. R. Kemp, F. S. Malm, "Hard Rubber (Ebonite)," Chapter XVIII in C. C. Davis, J. T. Blake (Ed.) "Chemistry and Technology of Rubber," Reinhold Publishing Corp., New York, 1937.

7. A. T. McPherson, "Electrical Properties of Elastomers and Related Polymers," Rubbers Chem. Technol. (Rubber Reviews), 36, 1230 (1963).

8. A. R. Payne, J. R. Scott, "Engineering Design with Rubber," Interscience Publishers, New York, 1960.

9. I. B. Prettyman, "Physical Properties of Natural and Synthetic Rubber Stocks," Handbook of Chemistry and Physics, Chemical Rubber Pub. Co., Cleveland, Ohio, 44th Ed. 1962, p. 1564.

10. W. J. Roff, J. R. Scott, "Handbook of Common Polymers - Fibers, Films, Plastics and Rubbers," Butterworth, London and CRC Press, Cleveland, Ohio, 1971.

11. L. A. Wood, "Values of the Physical Constants of Rubbers," Proc. Rubber Technology Conference, (Institution of the Rubber Industry, London), 1938, p. 933; Rubber Chem. Technol. 12, 130 (1939).

12. L. A. Wood, "Synthetic Rubbers: A Review of Their Compositions, Properties, and Uses," Natl. Bur. Std. Circ. C427 (1940); Rubber Chem. Technol. 13, 861 (1940); India Rubber World 102, No. 4, 33 (1940).

13. L. A. Wood, "Physical Chemistry of Synthetic Rubbers," Chapter 10 in G. S. Whitby (Ed.) "Synthetic Rubbers," John Wiley, Inc., New York, 1954.

14. G. Allen, V. Bianchi, C. Price, Trans. Faraday Soc. 59, 2493 (1963); Rubber Chem. Technol. 37, 606 (1964).

15. E. H. Andrews, P. J. Owen, A. Singh, Proc. Roy. Soc. (London) A 324, 79 (1971; Rubber Chem. Technol. 45, 1315 (1972).

* identical with [N mm^{-2}]

16. G. W. Becker, H. Oberst, Kolloid-Z. 148, 6 (1956).
17. N. Bekkedahl, Proc. Rubber Technol. Conference (Institution of the Rubber Industry, London) 1938, p. 223, Rubber Chem. Technol. 12, 150 (1939).
18. N. Bekkedahl, J. Res. Nat. Bur. Std. 43, 145 (1949).
19. N. Bekkedahl, F. L. Roth, Natl. Bur. Std. Unpublished observations of density and expansivity, 1948.
20. S. S. Chang, A. B. Bestul, J. Res. Natl. Bur. Std. 75A, 113 (1971).
21. R. Chasset, P. Thirion, Proc. Intern. Conf. Non-crystalline Solids, Delft 1964; J. A. Prins, Ed., p. 345, North Holland Publishing Co., Amsterdam, Interscience Publishers, New York, Rubber Chem. Technol. 39, 870 (1966); Rev. Gen Caoutchouc 44, 1041 (1967).
22. B. B. S. T. Boonstra, Rev. Gen. Caoutchouc 27, 409 (1950); Translated in Rubber Chem. Technol. 24, 199 (1951).
23. W. S. Cramer, I. Silver, NAVORD Report 1778, Feb. 1951, U. S. Naval Ordnance Lab., White Ock, Md.
24. R. A. Dickie, J. D. Ferry, J. Phys. Chem. 70, 2594 (1966).
25. J. H. Dillon, I. B. Prettyman, G. L. Hall, J. Appl. Phys. 15, 309 (1944); Rubber Chem. Technol. 17, 597 (1944).
26. J. D. Ferry, R. G. Mancke, E. Maikawa, Y. Oyanagi, R. A. Dickie, J. Phys. Chem. 68, 3414 (1964).
27. W. P. Fletcher, A. N. Gent, British J. Appl. Phys. 8, 194 (1957).
28. G. T. Furukawa, M. L. Reilly, J. Res. Natl. Bur. Std. 56, 285 (1956) RP 2676.
29. A. N. Gent, J. Polymer Sci. A 3, 3787 (1965).
30. W. H. Hamill, B. A. Mrowca, R. L. Anthony, Ind. Eng. Chem. 38, 106 (1946); Rubber Chem. Technol. 19, 622 (1946).
31. D. G. Ivey, B. A. Mrowca, E. Guth, J. Appl. Phys. 20, 486 (1949); Rubber Chem. Technol. 23, 172 (1950).
32. R. M. Kell, B. Bennett, P. B. Stickney, Rubber Chem. Technol. 31, 499 (1958).
33. R. M. Kell, B. Bennett, P. B. Stickney, J. Appl. Polymer Sci. 2, 8 (1959).
34. W. R. Krigbaum, J. V. Dawkins, G. H. Via, Y. I. Balta, J. Polymer Sci. A 2, 1, 475 (1966).
35. R. G. Mancke, J. D. Ferry, Trans. Soc. Rheology 12, 335 (1968).
36. L. Mandelkern, "Crystallization of Polymers," McGraw-Hill, New York, 1964. also L. Mandelkern, Chem. Rev. 56, 903 (1956).
37. L. Mandelkern, J. Polymer Sci. 47, 494 (1960).
38. G. M. Martin, F. L. Roth, R. D. Stiehler, Trans. Inst. Rubber Ind. 32, 189 (1956); Rubber Chem. Technol. 30, 876 (1957).
39. J. T. Maynard, W. E. Mochel, J. Polymer Sci. 13, 235 (1954).
40. G. Mikhailov, V. Kirilina, Rubber Chem. Technol. 14, 858 (1941).
41. W. H. S. Naunton, "Rubber in Engineering," Ministry of Supply, London, 1945, or Chemical Publishing Co., Brooklyn, N. Y. 1946, p. 30.
42. A. W. Nolle, J. Polymer Sci. 5, 1 (1950).
43. A. R. Payne in P. Mason, N. Wookey, (Ed.) "Rheology of Elastomers," Pergamon Press, London 1958, p. 86, values tabulated in J. D. Ferry, "Viscoelastic Properties of Polymers," John Wiley, New York, 1961, p. 458.

44. A. R. Payne in "The Physical Properties of Polymers," S. C. I. Monograph No. 5, Society of Chemical Industry, London and Macmillan, New York, 1959, p. 273.
45. W. Philipoff, J. Appl. Phys. 24, 685 (1953).
46. D. J. Plazek, J. Polymer Sci. A-2, 4, (5) 745 (1966).
47. C. Price, J. Padget, M. C. Kirkham, G. Allen, Polymer 10, 495 (1969).
48. R. D. Rands, Jr., W. J. Perguson, J. L. Prather, J. Res. Natl. Bur. Std. 33, (1944), RP 1595.
49. D. E. Roberts, L. Mandelkern, J. Am. Chem. Soc. 77, 781 (1955).
50. F. L. Roth, G. W. Bullman, L. A. Wood, J. Res. Natl. Bur. Std. 69A, 347 (1965), Rubber Chem. Technol. 39, 397 (1966).
51. H. Schilling, Kautschuk Gummi 16, 84 (1963).
52. L. Schmieder, K. Wolf, Kolloid-Z. 134, 149 (1953).
53. A. H. Scott, J. Res. Natl. Bur. Std. 14, 99 (1935), RP 760; Rubber Chem. Technol. 8, 401 (1935).
54. A. J. Wildschut, "Technological and Physical Investigations on Natural and Synthetic Rubbers," Elsevier Publishing Company, Inc., New York, 1946.
55. R. S. Witte, B. A. Mrowca, E. Guth, J. Appl. Phys. 20, 481 (1949); Rubber Chem. Technol. 23, 163 (1950).
56. L. A. Wood, J. Rubber Res. Inst. Malaya 22(3), 309 (1969). Rubber Chem. Technol. 43, 1482 (1970).
57. L. A. Wood, N. Bekkedahl, J. Polymer Sci. B 5, 169 (1967).
58. L. A. Wood, N. Bekkedahl, F. L. Roth, J. Res. Natl. Bur. Std. 29, 391 (1942) RP 1507; Ind. Eng. Chem. 34, 1291 (1942); Rubber Chem. Technol. 16, 244 (1943).
59. L. A. Wood, G. W. Bullman, J. Polymer Sci. A-2, 10, 43 (1972).
60. L. A. Wood, G. W. Bullman, G. E. Decker, J. Res. Natl. Bur. Std. 76A, 51 (1972).
61. L. A. Wood, G. M. Martin, J. Res. Natl. Bur. Std. 68A, 259 (1964); Rubber Chem. Technol. 37, 850 (1964).
62. L. A. Wood, F. L. Roth, Proc. 4th Rubber Technol. Conf., London, 1962, p. 328, Institution of the Rubber Industry, London, 1963; Rubber Chem. Technol. 36, 611 (1963).
63. L. A. Wood, L. W. Tilton, Proc. 2nd Rubber Technol. Conf., London, 1948, p. 142, Institution of the Rubber Industry, London; J. Res. Natl. Std. 43, 57 (1949), RP 2004.
64. T. P. Yin, R. Pariser, J. Appl. Polymer Sci. 7, 667 (1963).
65. L. J. Zapas, S. L. Shufler, T. W. de Witt, J. Polymer Sci. 18, 245 (1955); Rubber Chem. Technol. 29, 725 (1956).
66. M. N. Pilsworth, Jr., H. J. Hoge, H. E. Robinson, J. Materials 7 (4), 580 (1972).

PHYSICAL CONSTANTS OF POLY(ETHYLENE)[*]

S. L. Aggarwal
Research and Development Division
The General Tire & Rubber Company
Akron, Ohio

Table I: Property	Value	Ref.
Bond Length (C-C), [Å]	1.53	1, 2, 3
Bond Angle (C $\overset{C}{\diagup}$ C), [deg]	112	1, 2, 3

Chain Branching (short)

Effect on density and refractive index   8, 74

Methyl groups per 1000 C atoms	Density $[Mg\ m^{-3}] \equiv [g\ cm^{-3}]$	Refractive Index n_D^{25}
83	0.91	1.5060
48	0.917	1.5168
46	0.925	1.5152
26	0.929	1.5227
16	0.926	1.5260

Effect on expansion coefficient (mean) and specific volume of crystalline phase   9

Methyl groups per 1000 C atoms	$(1/V_{20}) (\Delta V/\Delta T) \times 10^4$		V_{20}
	-150 to 100°C	0 to 100°C	
0.3	2.47	3.13	1.001
2	2.59	2.95	0.998
17.5	2.61	2.98	1.010
23	2.84	3.42	1.009
37	2.96	3.70	1.017

Effect on long period spacings and crystallinity of completely annealed samples   10

Polyethylene	Methyl groups per 1000 C atoms	Long Period Spacing [Å]	% Crystallinity
Branched (High Pressure) Ref. (7)	60	220	21
	45	200	48
	35	210	50
	28	220	53
	20	230	56
	15	250	59
	10	260	62
Linear (Ziegler Type Catalysts) Ref. (8, 9)	7	320	77
	5	360	79
Linear (Phillips Petroleum Process) Ref. (20)	2	420	88

[*] Data for two basic types of poly(ethylenes): "High Pressure" and "Low Pressure" are tabulated for only those properties which are dependent on short chain branches in the polymer chain. "High Pressure" poly(ethylenes) are more branched (20-30 ethyl and butyl branches per 1000 C atoms), than the "Low Pressure" poly(ethylenes) (< 7 ethyl or butyl branches per 1000 C atoms). Early descriptions of short chain branching in poly(ethylene) were in terms of methyl groups per 1000 C atoms.

Table I: Property	Value	Ref.

Chain Branching (short)   (Cont'd.)

Effect on melting point. (Both the amount and randomness of branching affect the melting point. Experimental conditions used for the following data were not adequate for equilibrium crystallinity and accuracy of melting points).

Methyl Groups per 1000 C atoms	Melting Point [$^{\circ}$C]
87	105
28	113
28	108
8	123
0	132 (?)

Coefficient of Thermal Expansion

(branched polyethylene)

12

Temp. [$^{\circ}$C]	Coefficient of Expansion x 10^5		Specific Volume Ratio ($V_E/V_{25^{\circ}C}$)
	linear	cubical	
-35	10.0	30	0.969
20	13.7	41	0.975
0	18.3	55	0.986
20	23.7	71	0.997
25	24.8	74	1.000
40	29.0	87	1.012
60	33.7	101	1.031
80	40.3	121	1.055
100	46.6	140	1.094
110	51.0	153	1.130
115	25.0	75	1.140
115-150	25.0	75	-
150	25.0	75	1.168

Crystallinity

13, 14

(Depends upon chain branching; the following are some representative values of typical poly(ethylenes)).

Polyethylene Type	Methyl Groups per 1000 C atoms	Melting Point [$^{\circ}$C]	% Crystallinity
Marlex 50 (Phillips, Linear)	0	135	91
Super-Dylan (Ziegler, Linear)	0	130	81
DYNH (Union Carbide/Branched)	25.6	112	52
Marlex 50 (Annealed)			93.8
(Granular)			82.6
Low Density Branched Poly(ethylene) (Annealed)			53
(Granular)			50
(Drawn)			47

60

Crystallization Kinetic Parameters

see table "Rate of Crystallization of Polymers" in this Handbook

Table I: Property   Value   Ref.

Crystallographic Data and Crystallographic Modifications

| Polyethylene sample type | Crystal System | Space Group | Unit Cell Parameters [Å] | | | | Monomers per Unit Cell | Calculated Density [Mg m^{-3}] $\equiv$ [g cm^{-3}] | Ref. |
			a	b	c	Angles [deg]			
"High Pressure"	Orthorhombic	Pnam	7.40	4.93	2.534		2	1.00	1
			7.36	4.92	-		-	1.014	4
	Monoclinic	C2/m	8.09	2.53	4.79	$\beta = 107.9$	2	0.997	5
"Low Pressure" poly(methylene)	Triclinic		4.285	4.820	2.54	$\alpha = 90$ $\beta = 110$ $\gamma = 108$	1	1.00	6

Density, [Mg m^{-3}] $\equiv$ [g cm^{-3}]   (see also Table II and III)

(unless otherwise stated, the values of density are given for 25°C).

Amorphous (from extrapolation of data above the melting point)	0.855	16
Commercial high pressure poly(ethylene)	0.915 - 0.935	17
Experimental high pressure poly(ethylene)	0.940 - 0.970	17
Ziegler process (Ref. 18,19) poly(ethylene)	0.940 - 0.965	17
Phillips process (Ref. 20) poly(ethylene)	0.960 - 0.970	17
Crystal density (theoretical)	see Crystallographic Data and Crystallographic Modifications	

Dimensions of Linear Poly(ethylene) Molecules (unperturbed) in tetralin at 105°C:

$<r_o^2>$ is the weight average mean-square end-to-end distance in [nm]; M_w is the weight average molecular weight; D is the diameter of a spherical segment of the lattice model chain; and r_{max} is the length of the fully extended chain.

M_w x 10^{-5}	$(<r_o^2>/M_w)^{1/2}$	$(<r_o^2>/Dr_{max})$	21
1.25	0.192	0.780	
2.69	0.215	0.986	
4.65	0.190	0.773	

Temperature dependence of $<r_o^2>$
(in long chain paraffinic hydrocarbon solvents) $-d\ln<r_o^2>/dT = 1.2 \times 10^{-3}$ 22

Elastic Compliance   see Table II

Electrical Properties   (see also Table III)

Dielectric constant, at 100 kc at 23°C	2.3	23
Dielectric Loss, tan δ	1×10^{-4} - 1×10^{-3}	24,25

The values of tan δ depend on temperature, and structure of polyethylene (24).

Dielectric Strength, [V cm^{-1}]

-200 to 0°C	7×10^{6}	26
50°C	5.3×10^{6}	
100°C	1.8×10^{6}	

Elongation at Break, [%]   see Table III

Enthalpy, Entropy   see thermodynamic properties

Entropy of Fusion, [J mol^{-1} K^{-1}]

ΔS_u	9.81	28
	9.60	27

Table I: Property	Value	Ref.

Entropy of Fusion, [J mol^{-1} K^{-1}] (Cont'd.)

at constant volume (ΔS_{uv})   7.42   27

7.72   29

Flash Ignition Temperature

ASTM Method E136-58T):   340°C   30

Frictional Properties

Coefficient of Friction μ	Steel Sliding on Polymer		Polymer Sliding on Steel		Polymer Sliding on Polymer	
	Polished	Abraded	Polished	Abraded	Polished	Abraded
Static (M_S)	0.60	0.33	0.60	0.33	0.60	0.33
Kinetic (M_K)	0.60	0.33	0.60	0.33	0.60	0.33
	0.50	0.25				

Glass Transition   (see also Transition and Relaxations Temperatures)

Temperature*, [^{o}C]   -80 to -90° 63

Activation Energy, [kJ mol^{-1}]   46 - 75

*Considerable disagreement exists between different authors on the exact value of transition that can be identified as glass transition temperature. See Ref. (63) for detailed discussion. Ref. (47,48) give glass transition as -125°C. See also table "Glass Transition Temperatures" in this Handbook.

Hardness, Shore D   see Table III

Heat Capacity, specific   see Thermodynamic Properties

Heat of Combustion   32

Density [Mg m^{-3}] = [g cm^{-3}]	Methyl Groups per 1000 C atoms	Heat of Combustion ΔE [kJ kg^{-1}]
0.9391	8.3	-46,412
0.9220	24.7	-46,492
0.9053	46.2	-46,542

Heat of Fusion

From differential thermal analysis data:

Poly(ethylene) Type	Melting Point [^{o}C]	Heat of Fusion [kJ kg^{-1}]	
Marlex 50 (Phillips, Linear)	135	245.3	13
Super-Dylan (Ziegler, Linear)	130	218.6	13
DYNH (Union Carbide, branched)	112	140.6	13
Linear Poly(ethylene) from dilatometric measurements:		280.5	27
from calorimetric measurements:		277.1	33

Infrared Absorption Bands   see Ref.   34, 35, 36, 91, 92

Infrared Functional Groups   see Table II

Intrinsic Viscosity   see Table II

Melt Index   see Table II

Impact Strength, Izod   see Table III

Table I: Property | Value | Ref.

Low Temperature Brittleness	see Table III	
Mechanical Properties	see Table III	
Melting Temperature, [°C]		
Poly(methylene)	136.5 ± 0.5	64
Linear Poly(ethylene)	137.5	27
Linear Poly(ethylene), High Molecular Weight Fraction	138.5	41
From Extrapolation of M.P. of n-paraffins	141 ± 2.4	42

see also Table II and chain branching, effect on melting temperature.

Melt Viscosity	see Table II	
Molecular Properties of Typical Poly(ethylenes)		65, 66, 67

	Low Density	High Density	
	High Pressure Process (17)	Ziegler Process (18, 19)	Phillips Process (20)
Molecular Structure	Branched; 20-30 ethyl and butyl branches/1000 carbon atoms; few long chain branches	Mainly linear; less than >10 ethyl branches per 1000 carbon atoms	Almost linear
Number of double bounds per 1000 carbon atoms	< 0.3	< 0.2	< 0.3
Type of unsaturation, [%]			
Vinylidene	80	30	-
Vinyl	10	45	∼ 100
Trans	10	25	-
Molecular Weight Averages			
$\overline{M}_w$	50.000 - 300.000	50.000 - 300.000	50.000 - 300.000
$\overline{M}_n$	10.000 - 40.000	5.000 - 15.000	5.000 - 15.000
$\overline{M}_w/\overline{M}_n$	2 - 50	4 - 15	4 - 15

Permeability and Diffusion Constants	see corresponding table in this Handbook and Ref.	43, 44, 49, 68, 70, 71
Permeability Factors	for common organic compounds see Ref.	44, 71
	for gases at elevated pressures see Ref.	72
Properties of a Series of Selected Poly(ethylene) Samples	see Table II	
Properties of Typical Poly(ethylenes)	see Table III	
Raman Spectra	see Ref.	73
Refractive Index		
Amorphous, [n_{5461}^{25}]	1.49	45
Crystal, $\alpha \simeq \beta$	1.520	45
γ	1.582	

(α, β and γ are refractive indices along the a, b und c crystallographic directions of the crystal).

Specific Refractivity, r [cm^3 g^{-1}] ≡ [1 kg^{-1}]

$r = v\left(\dfrac{n^2 - 1}{n^2 + 1}\right)$ where v and n are specific volume and refractive index, respectively

Table I: Properties	Value	Ref.

Refractive Index, Specific Refractivity (Cont' d.)

Temp. [^{o}C]	$v[cm^{3}g^{-1}] \equiv [1 kg^{-1}]$	n	r
	Low Density Poly(ethylene) (Alathon-10)		
90	1.159	1.4801	0.3293
100	1.178	1.4693	0.3283
108	1.209	1.4575	0.3297
113	1.239	1.4432	0.3286
118	1.250	1.4392	0.3289
124.4	1.256	1.4368	0.3288
			Av. 0.3290
	High Density Poly(ethylene) (Marlex 50)		
130	1.261	1.4327	0.3273
139.9	1.270	1.4297	0.3297
150.6	1.281	1.4261	0.3283

Dependence of refractive index on chain branching, crystallinity and density   see chain branching

Single Crystal Lamella Thickness

(Long Periods from Low Angle X-Ray Diffraction)   75

Effect of crystallization temperature and solvents

Solvent	Cryst. Temp. [^{o}C]	Conc. [%]	Long Period [Å]	Ref.
	High Density, Poly(ethylene), (Phillips Marlex 50)			
Tetrachloroethylene	50	0.31	112	76
	60	0.25	100	77
p-Xylene	72	0.46	110	77
	78	0.58	140	76
Xylene	50	-	92.5	78, 75
	60	-	102	
	70	-	111.5	
	80		120.5	
	90		150	
	75	0.1	115	79
	80	0.1	125	
	85	0.1	133	
Melt	120	100	190	80
	125	100	223	
	130	100	355	
n-Butyl acetate	105	0.45	147	77
	110	0.57	162	76
Diphenyl ether	120	0.47	202	76
	125	0.37	173	77
	Low Density Poly(ethylene), (Dupont Alathon)			
Butyl Stearate	111	0.25	126, 155	75
	113.5	0.25	166	
Squalene	155.5	0.25	176	
Glycol dipalmitate	155.4	0.25	151	
	118	0.25	164	
	121	0.25	206	
Tripalmitin	121	0.25	188	

Table I: Properties	Value	Ref.

Softening Temperature, Vicat   see Table III

Solution Properties   see Dimensions of linear poly(ethylene) molecules

Solvent-Nonsolvent Systems for Fractionation

Solvent	Nonsolvent	
xylene	triethylene glycol	49, 50
p-xylene	ethylene glycol monoethyl ether	51
tetralin	2-butoxyethanol	52
	poly(ethylene oxide) mol. wt. 200	53, 54

Tensile Modulus   see Table III

Tensile Strength   see Table III

Thermodynamic Properties

    Heat Capacity (0°C), C_p, [kJ kg^{-1} K^{-1}]   59

High density poly(ethylene) (Marlex),	
annealed, 93% crystallinity	1.566
amorphous	2.281
Low density poly(ethylene)	
53% crystallinity	1.989
drawn	1.901

see also Table "Heat Capacity of High Polymers" in this Handbook

high pressure, low density poly(ethylene) from 10-320 K   see Ref.   58

crystalline and amorphous poly(ethylene) from 1-400 K   see Ref.   81

Entropy S, and Enthalpy H of crystalline and amorphous poly(ethylene)   81

Temp. [K]	Crystalline H [J mol^{-1}]	Crystalline S [J mol^{-1} K^{-1}]	Amorphous H [J mol^{-1}]	Amorphous S [J mol^{-1} K^{-1}]
415	6896	32.68	11007	42.62
275	3123	21.85	6347	28.95
255	2734	20.39	5732	26.67
0	0	0	2767	5.31

Entropy S, Enthalpy H and Gibbs Free Energy of high pressure, low density poly(ethylene) from 10-320 K see Ref.   58

Transition and Relaxation Temperatures

There is considerable disagreement in the literature on the phenomena associated with the various transition and relaxation temperatures observed. Transition temperatures and temperatures associated with peaks in dynamic loss are collected together under the above combined heading. The transition and relaxation temperatures associated with amorphous regions of branched and linear poly(ethylenes) are designated as α, β, γ, etc. in the descending temperature order.   82

Designation	Temperature Range [°C]	Approx. Activation Energy [kJ/mol]	
α	60 to 80	>420	
β	-20 to -30 *	160 - 200	83
γ	-80 to -90 *	46 - 75	
	-120 to -130	32	

* These frequently merge depending upon the crystallinity and frequency of the test method

Table I: Properties	Value	Ref.

Transition and Relaxation Temperatures (Cont'd.)

Frequency dependence of relaxation temperatures in dynamic mechanical loss measurements.

84

α [Hz]	T[K]	β [Hz]	T[K]	γ [Hz]	T[K]
High Pressure, Branched Poly(ethylenes)					
0.3	340	0.3	268	1.25	140
1.2	327	4.1	268	8.6	166
39	333	150	253	324	158
150	355	540	265	1.2×10^3	165
200	360	520	280	1.15×10^3	165
600	385	6000	320	1.9×10^4	≤ 200
4×10^4	≥ 360	4×10^4	275	4×10^4	180
1×10^5	≥ 320	1×10^5	283	1×10^5	≤ 190
-	-	1×10^5	285	1×10^5	200
-	-	5×10^5	285	5×10^5	205
2×10^6	360	2×10^6	295	2×10^6	210
-	-	2×10^6	300	2×10^6	210
Low Pressure, Linear Poly(ethylenes)					
0.3	373	-		1.25	153
0.2	368	8	273	10	173
≤ 460	≥ 380	1.1×10^3	295	840 and 1.57×10^3	175
3000	420	-		-	-

Viscosity-Molecular Weight Relationship : see corresponding table in this Handbook

Table II: Properties of a Series of Selected Poly(ethylene) Samples*

Sample No.	Optical Melting Point [°C]	Density[1] [Mg m^{-3}]$\equiv$[g cm^{-3}]	per 100 C Methyl	per 2000 C Vinyl	Trans-unsatuaration	Vinylidene	Carbonyl
PE 1	104.2	0.9142	3.68	0.18	0.15	0.79	0.13
PE 2	112.4	0.9225	2.59	0.32	0.11	0.32	0.02
PE 3	112.2	0.9218	2.48	0.10	0.06	0.29	0.05
PE 4	113.7	0.9232	2.55	0.11	0.06	0.29	n.d.
PE 5	114.0	0.9219	2.46	0.11	0.05	0.30	0.01
PE 6	114.5	0.9228	2.31	0.06	0.05	0.26	n.d.
PE 7	113.5	0.9207	2.59	0.11	0.06	0.33	0.02
PE 8	112.0	0.9188	2.54	0.10	0.06	0.31	n.d.
PE 9	121.5	0.9334	1.40	0.04	0.02	0.11	0.87
PE 10	135.8	0.9549	0.1	1.82	0.04	0.15	< 0.005
PE 11	-	0.9554	0.165	0.86		0.17	

*Data supplied through the courtesy of R. Longworth, Plastics Department, E. I. du Pont de Nemours & Co., Wilmington, Delaware.

Sample No.	Molecular Weight Weight Ave.	Molecular Weight Number Ave.	Intrinsic Viscosity[2] [ml g^{-1}]	Melt Index[3]	Melt Viscosity[4] [Pa s]	Elastic Compliance[5] [MPa^{-1}]
PE 1	510,000[6]	10,700[10]	79.5	1.80	64	63 x 10^{-6}
PE 2	300,000[6]	13,300[10]	75.7	1.95	38	46
PE 3	550,000[6]	19,100[10]	96.1	0.16	620	46
PE 4	225,000[7]	(16,000)[9]	62	19.9	24	22
PE 5	500,000[7]	(18,000)[9]	75	3.30	185	36
PE 6	500,000[7]	(22,000)[9]	82	1.06	73	54
PE 7	300,000[7]	(45,000)[9]	77	2.94	23	31
PE 8	800,000[7]	(55,000)[9]	97	0.21	310	54
PE 9	300,000[7]	(27,000)[9]	73	3.75	153	39 x 10^{-6}
PE 10	144,000[6]	11,500[10]	116	2.92	-	-
PE 11	-	-	-	0.46	-	-

Notes:
1. Samples annealed 1 hour at about 100°C.
2. α-Chloronaphthalene, 125°C.
3. ASTM D-1238-57T.
4. Newtonian melt viscosity at 150°C and 400 Pa.
5. Steady-state elastic compliance from creep recovery at 150°C and 400 Pa.
6. Light scattering after optical clarification by high temperature ultracentrifugation.
7. Preliminary value, subject to revision.
8. Measured by osmometry.
9. "Best guess." Subject to drastic revision.
10. Cryoscopy

Table III: Properties of Typical Poly(ethylenes) (65, 93)

Property	Low Density ASTM Type I[*]	Medium Density ASTM Type II[*]	High Density ASTM Type III[*]
Abrasion Resistance, Taber, [mg/1000 cycles]	10 - 15	6 - 10	2 - 5
Brittlenes, Low Temperature, [°C]	< -118	< -118	< -118 to -73
Coefficient of thermal expansion x 10^{-5} (D696), [K^{-1}]	10	-	13
Density (D792)[**], [Mg m^{-3}] $\equiv$ [g cm^{-3}]	0.910 - 0.925	0.926 - 0.940	0.941 - 0.965
Dielectric Constant at 1 kHz (D150)	2.28	-	2.32
Elongation at Break (D638), [%]	150 - 600	100 - 150	12 - 700
Hardness, Shore D (D1706)	44- 48	45 - 60	55 - 70
Impact strength, Izod (D256), [ft lb/in notch]	> 16	> 16	0.8 - 14
Power Factor at 1 kHz (D150)	< 0.0001	-	< 0.0001
Heat Capacity, [kJ kg^{-1} K^{-1}]	1.916	1.916	1.916
Tensile Modulus (D638), [MPa] $\equiv$ [N mm^{-2}]	55.1 - 172	172 - 379	413 - 1034
Tensile Strength (D638), [MPa] $\equiv$ [N mm^{-2}]	15.2 - 78.6	12.4 - 19.3	17.9 - 33.1
Vicat, Softening Temp. (D1525), [°C]	88 - 100	99 - 124	112 - 132
Volume Resistivity (D257), [cm]	6 x 10^{15}	-	6 x 10^{5}

[*] ASTM designation D1248-72, "Standard Specification for Polyethylene Plastics Molding and Extrusion Materials."

[**] The numbers in parentheses refer to the ASTM Standards, American Society for Testing Materials, 1916 Race Street, Philadelphia, Pa. 19103.

REFERENCES

1. C. W. Bunn, Trans. Faraday Soc. 35, 482 (1939).
2. A. Charlesby, Proc. Phys. Soc. (London) 57, 496 (1945).
3. J. J. Trillat, Compt. Rend. 230, 1522 (1950).
4. E. R. Walter, F. P. Reding, J. Polymer Sci. 21, 561 (1956).
5. K. Tanaka, T. Seto, T. Hara, J. Phys. Soc. (Japan) 17, 873 (1962).
6. A. Turner-Jones, J. Polymer Sci. 62, 53 (1962).
7. W. D. Niegisch, P. R. Swan, J. Appl. Phys. 31, 1906 (1960).
8. M. Baccaredda, G. Schiavinato, J. Polymer Sci. 12, 155 (1954).
9. E. A. Cole, D. R. Holmes, J. Polymer Sci. 46, 245 (1960).
10. C. Sella, Sompt. Rend. 248, 1819 (1959).
11. F. P. Reding, J. Polymer Sci. 32, 487 (1958).
12. F. C. Hahn, M. L. Macht, D. A. Fletcher, Ind. Eng. Chem. 37, 526 (1945).
13. B. Ke, J. Polymer Sci. 42, 15 (1960).
14. S. L. Aggarwal, O. J. Sweeting, Chem. Rev. 57, 665 (1957).
15. S. Buckser, L. H. Tung, J. Phys. Chem. 63, 763 (1958).
16. G. Allen, G. Gee, G. J. Wilson, Polymer 1, 456 (1960).

17. K. W. Doak, A. Schrage, "Polymerization and Copolymerization Processes," in R. A. V. Raff, K. W. Doak, Ed., "Crystalline Olefin Polymers", Part I, Interscience, New York, 1965, p. 301ff.
18. K. Ziegler, Angew. Chem. 64, 323 (1952).
19. K. Ziegler, E. Holzkamp, H. Breil, H. Martin, Angew. Chem. 67, 426, 541 (1955).
20. J. P. Hogan, R. L. Banks, U. S. Patent 2,825,721 assigned to Phillips Petroleum Company, March 4, 1958.
21. Q. A. Trementozzi, J. Polymer Sci. 36, 113 (1959).
22. P. J. Flory, A. Ciferri, R. Chiang, J. Am. Chem. Soc. 83, 1023 (1961).
23. V. L. Lanza, D. B. Herrmann, J. Polymer Sci. 28, 622 (1958).
24. G. P. Mikhailov, S. P. Kabin, T. A. Krylova, Zh. Tekhn. Fiz. 27, 2050 (1957).
25. H. D. Anspon, et al., "Polyethylene" in W. M. Smith, Ed., "Manufacture of Plastics," Reinhold, New York, 1964, p. 150ff.
26. K. H. Stark, C. G. Garton, Nature, 176, 1225 (1955).
27. F. A. Quinn, L. Mandelkern, J. Am. Chem. Soc. 80, 3178 (1958).
28. H. W. Starkweather, Jr., R. H. Boyd, J. Phys. Chem. 64, 410 (1960).
29. L. Mandelkern, "Crystallization of Polymers," McGraw-Hill, New York, 1964, p. 130.
30. G. A. Patten, Mod. Plastics 38, (11), 119 (1961).
31. R. C. Bowers, W. C. Clinton, W. A. Zisman, PB 111185 (Naval Research Laboratory Report 4167), May 19, 1953.
32. Reference 25, p. 160.
33. B. Wunderlich, M. Dole, J. Polymer Sci. 24, 201 (1957).
34. S. Krimm, Fortschr. Hochpolymer Forschung, 2, 51 (1960).
35. J. R. Nielsen, R. F. Holland, J. Mol. Spectr. 4, 488 (1960); 6, 394 (1961).
36. M. P. Groenewege, J. Schuyer, J. Smidt, C. A. F. Tuijnman, "Absorption and Relaxation Spectra of Polyolefins," in R. A. V. Raff, K. W. Doak, Ed., "Crystalline Olefin Polymers," Part I, Interscience, New York, 1965, p. 762.
37. Q. A. Trementozzi, S. Newman, "Dilute Solution Properties," in R. A. V. Raff, K. W. Doak, Ed., Crystalline Olefin Polymers," Part I, Interscience, New York, 1965, p. 406ff.
38. L. H. Tung, J. Polymer Sci. 24, 333 (1957).
39. Q. A. Trementozzi, J. Polymer Sci. 22, 187 (1956).
40. Q. A. Trementozzi, J. Polymer Sci. 23, 887 (1957).
41. R. F. Chiang, P. J. Flory, J. Am. Chem. Soc. 83, 2857 (1961).
42. M. G. Broadhurst, J. Chem. Phys. 36, 2578 (1962).
43. A. S. Michaels, H. J. Bixler, J. Polymer Sci. 50, 413 (1961).
44. V. Stannett, H. Yasuda, "Permeability," in R. A. V. Raff, K. W. Doak, Ed., "Crystalline Olefin Polymers," Part II, Interscience, New York, 1965, p. 131ff.
45. W. M. D. Bryant, J. Polymer Sci. 2, 556 (1947).
46. J. P. Bianchi, W. G. Luetzel, F. P. Price, J. Polymer Sci. 27, 561 (1958).
47. J. A. Faucher, F. P. Reding, "Relationship Between Structure and Fundamental Properties," in R. A. V. Raff, K. W. Doak, Ed., "Crystalline Olefin Polymers," Part I, Interscience, New York, 1965, p. 700ff.
48. F. P. Reding, J. A. Faucher, R. D. Whitman, J. Polymer Sci. 57, 483 (1962).
49. L. H. Tung, J. Polymer Sci. 20, 495 (1956).
50. L. H. Tung, J. Polymer Sci. 24, 333 (1957).
51. P. S. Francis, R. C. Cooke, Jr., J. H. Elliot, J. Polymer Sci. 31, 453 (1958).
52. J. E. Guillet, R. L. Combs, D. F. Slonaker, H. W. Coover, J. Polymer Sci. 47, 307 (1960).
53. I. V. Mussa, J. Polymer Sci. 28, 587 (1958).
54. L. Nicolas, Compt. Rend. 242, 2720 (1956).
55. L. N. Cherkasova, Zh. Fiz. Khim. 33, 1928 (1959).
56. H. C. Raine, R. B. Richards, H. Ryder, Trans. Faraday Soc. 41, 56 (1945).
57. K. Eiermann, Kunststoffe 51, 512 (1961).
58. R. W. Warfield, M. C. Petree, Makromol. Chem. 51, 113 (1962); data used were from different sources: (a) I. V. Sochava, O. N. Trapeznikova, Dokl. Akad. Nauk SSSR, 113, 784 (1957); (b) I. V. Sochava, Dokl. Akad. Nauk SSSR, 130, 126 (1960); (c) M. Dole, W. P. Hettinger, Jr., N. R. Larson, J. A. Wethington, Jr., J. Chem. Phys. 20, 781 (1952).
59. M. Dole, Kolloid-Z. 165, 40 (1959).
60. B. Wunderlich, M. Dole, J. Polymer Sci. 24, 201 (1957).
61. L. Mandelkern, in "Growth and Perfection of Crystals," R. H. Doremus, B. W. Roberts, D. Turnbull, Eds., Wiley, New York, 1958.
62. R. Buchdahl, R. L. Miller, S. Newman, J. Polymer Sci. 36, 215 (1959).
63. R. F. Boyer, Rubber Chem. Technol. (Rubber Reviews) 36, 5 1303-1421 (1963).
64. L. Mandelkern, M. Hellmann, D. W. Brown, D. E. Roberts, F. A. Quinn, Jr., J. Am. Chem. Soc. 75, 4093 (1953).
65. P. D. Ritchie, Ed., "Vinyl and Allied Polymers," Vol. I, Iliffe, London 1968.
66. C. W. Bunn, in "Polythene," A. Renfrew, P. Morgan, Ed., Iliffe, London 1960.
67. D. C. Smith, Ind. Eng. Chem. 48, (7), 1161 (1956).
68. D. V. Brubaker, K. Kammermeyer, Ind. Eng. Chem. 46, 733 (1954).
69. J. A. Meyer, C. Rogers, V. Stannett, M. Swarc, TAPPI, 40, 142 (1957).
70. C. E. Rogers, V. Stannett, M. Szwarc, TAPPI, 44, 715 (1961).
71. M. Salame, S. P. E. Trans. 1, 153 (1961).
72. N. Li, "Permeation of Gases Through Polyethylene Films at Elevated Pressures," Thesis, Stevens Institute of Technology, 1963.
73. P. J. Mendra, Fortschr. Hochpolymer Forsch. 6, 151 (1969).
74. M. Baccaredda, G. Schiavinato, J. Polymer Sci. 12, 155 (1954).
75. P. H. Geil, "Polymer Single Crystals," Interscience, New York, 1963, p. 8611.
76. D. G. Ranby, H. Brumberger, Polymer 1, 399 (1960).
77. B. G. Ranby, F. F. Morehead, N. M. Walter, J. Polymer Sci. 44, 349 (1960).
78. F. P. Price, J. Chem. Phys. 35, 1884 (1961).
79. D. C. Bassett, A. Keller, Phil. Mag. 7, 1553 (1962).
80. L. Mandelkern, A. S. Posner, A. F. Diorio, D. E. Roberts, J. Appl. Phys. 32, 1509 (1961).
81. B. Wunderlich, H. Baur, Fortschr. Hochpolymer Forsch. 7, 2, 151 (1970).
82. R. F. Boyer, "Transitions and Relaxations in Polymers," J. Polymer Sci. C 14, (1966).
83. Reference 63, p. 1409ff.
84. A. E. Woodwad, J. A. Sauer, Fortschr. Hochpolymer Forsch. 1, 136 (1958); data summarized from various sources.
85. L. H. Tung, J. Polymer Sci. 36, 287 (1959).
86. P. Henry, J. Polymer Sci. 36, 3 (1959).
87. R. Chiang, J. Polymer Sci. 36, 91 (1959).
88. E. Duch, L. Kuechler, Z. Electrochem. 60, 218 (1956).
89. J. T. Atkins, L. T. Muns, C. W. Smith, E. T. Pieski, J. Am. Chem. Soc. 79, 5089 (1957).
90. M. O. de La Cuesta, F. W. Billmeyer, J. Polymer Sci. A 1, 1721 (1963).
91. R. G. Snyder, J. Chem. Phys. 47, 1316 (1967).
92. B. E. Read, R. S. Stein, Macromolecules 1, 116 (1968).
93. Trade literature published by Phillips Petroleum Company, Bartlesville, Oklahoma.

PHYSICAL CONSTANTS OF POLY(PROPYLENE)

S. L. Aggarwal
Research and Development Division
The General Tire & Rubber Company
Akron, Ohio

Property	Value	Ref.
Coefficient of Thermal Expansion (D 696), $[K^{-1}]$	6.8×10^{-5}	49
Crystallinity		40, 47

Sample Description and Conditions	Crystalline Weight Fraction
1. Heptane extract of crude polypropylene; "amorphous;" highly atactic	0.14
2. Isotactic, water quenched	0.31
3. Same as 2 above, followed by annealing at $105\,^{\circ}C$ for 1 hr	0.43
4. Same as 2 above, followed by annealing at $160\,^{\circ}C$ for 1/2 hr.	0.65

Property	Value	Ref.
Crystallization Kinetic Parameters	see table "Rate of Crystallization of Polymers" and Ref.	10, 13, 15

Crystallographic Data and Crystallographic Modifications

Stereoisomeric Form	Crystal System	Space Group	a	b	c	α, β or γ (deg.)	Monomers Per Unit	Calculated Density $[g\,cm^{-3}]$	Chain Conformation	
Isotactic I	Monoclinic	C_{2R}^{6}-C2/c	6.65	20.96	6.50	$\beta = 99.3$	12	0.936	3_1 helix	33, 34
			6.67	20.87	6.49	$\beta = 98.2$	12	0.937		31
			6.64	20.88	6.51	$\beta = 98.7$	12	0.940		44
		C_{2h}^{5}-P2i/c	6.69	20.98	6.504	$\beta = 99.5$	12	0.9323		28
		P2i/c	6.60	20.78	6.495	$\beta = 99.62$	12	0.946		43
	Triclinic	C_i^{1}-P1	13.36	6.50	10.99	87,108,99	12	0.934		22
Isotactic II	Hexagonal		12.74	-	6.35	-		0.939		19
		D_3^4-P3$_1$21 or D_3^6-P3$_2$21	6.38	-	6.33	-		0.939		1
Syndiotactic I	Orthorhombic	D_2^5-C222$_1$	14.50	5.60	7.40		8	0.930		6

Property	Value	Ref.
Density at $25\,^{\circ}C$, $[Mg\,m^{-3}] \equiv [g\,cm^{-3}]$	(see also Table II)	30
isotactic, crystalline	0.932 - 0.943	
isotactic, amorphous (from extrapolation of data above melting point)	0.850 - 0.854	
smectic form	0.916	
syndiotactic, crystalline	0.989 - 0.91	
syndiotactic, amorphous (from extrapolation of data above melting point)	0.858	
Dielectric constant	see Table II	

Property	Value	Ref.

Dimensions of Polypropylene Molecules (unperturbed)   see also corresponding table in this Handbook

Radius of gyration, root mean square Z-average,

$<s^2>_Z^{1/2}$, in tetralin at 135°C from light scattering measurements   36

M_w x 10^{-3}	$[\eta]$, [ml/g]	$<s^2>_Z^{1/2}$ [nm]
123	91.	53.4
280	171.	66.6
369	206.	100.8
631	348.	118.1

Electrical Properties   (see Table II)

Elongation at Break, [Percent]   (see Table II)

G-values for Radiation Crosslinking, G(c.l.), and Chain Scission, G(breaks)

Maximum and Minimum Values

	G(breaks)		G(c.l.)	
	Maximum	Minimum	Maximum	Minimum
atactic	0.24	0.10	0.27	0.115
isotactic film	0.21	0.10	0.14	0.069
isotactic flake	0.27	0.10	0.18	0.068

8

at room temperature and in vacuo

	G(c.l.)		G(breaks)
	Solubility Method	Elasticity Method	Solubility or Viscosity Method
atactic	0.12 - 0.27	0.6 - 1.3	0.10 - 0.24
isotactic	0.07 - 0.25		0.10 - 0.24
	0.6		0.9
			5.0

8

G-values in terms of radiolytic gas yields

	Gas	Irradiation T [$^\circ$C]	G(Gas)
atactic	H_2	25	2.34
	CH_4	25	0.095
isotactic	H_2	-196	2.55
	CH_4	-196	0.058
	H_2	25	2.78
	CH_4	25	0.072

Glass Transition Temperature, T_g, [$^\circ$C]   -18   26

Hardness, Shore D   (see Table II)

Heat of Fusion

at T_m on Crystallization

isotactic, Form I, [kJ kg^{-1}]   209   47

[kJ mol^{-1}]   8.79

syndiotactic, [kJ kg^{-1}]   50.2

[kJ mol^{-1}]   2.09

Property	Value	Ref.

Heat Capacity C_p, [kJ kg^{-1} K^{-1}]

isotactic	1.8 - 1.92	
atactic	2.34	46
isotactic $<100^o$ room temperature	4.187 (0.3669 + 0.00242 T), where T is temperature in [^{o}C]	45

Temperature [K]	C_p [J mol^{-1} K^{-1}]
100	26.9
150	38.0
200	47.9
250	58.6
300	(72.6) ★
350	(90.0) ★
400	(113.) ★

★Values in parantheses are either experimentally uncertain or were estimated.

Impact Strenght, Izod   (see Table II)

Infrared Absorption Bands

Assignment in infrared spectrum	see Ref.	14, 16, 37, 21, 24, 42
Used for crystallinity measurements		47

Form	Crystalline Phase Wave Numbers [cm^{-1}]	Amorphous Phase, Wave Numbers [cm^{-1}]
isotactic	809	790
	842	1158
	894	
	997	
syndiotactic	866	1131
	977	1199
		1230

Used for tacticity measurements - ratio of absorption at 997 cm^{-1} to that at 975 cm^{-1}.   17, 23

Low Temperature Brittlenes   (see Table II)

Melt Index (ASTM Method D1238-57T)

Molecular Weight M_w x 10^{-3}	Melt Index (230°C, 2.16 kg)	
142	22.8	
180	7.3	
220	3.5	
292	1.2	
358	0.39	13

Melting Temperature, [^{o}C]

typical highly isotactic	171	12
isotactic crystalline Form I	186	47
syndiotactic	138	47

Property	Value	Ref.
Molecular Properties of Typical Poly(propylenes)		

Molecular Properties of Typical Poly(propylenes)

number of double bonds per 1000 carbon atoms	< 1	
type of unsaturation	vinylidene	
M_w	220.000 - 700.000	
M_n	38.000 - 60.000	
M_w/M_n	5 - 12	

Permeability and Diffusion Coefficients   see corresponding Table in this Handbook and Ref.   18, 41

Properties of Typical Poly(propylenes)   (see Table II)

Refractive Index   1.49   49

Refractive Index Increment, (dn/dc)   36

Temperature [^oC]	Wavelength [nm]	dn/dc [$cm^3 g^{-1}$]
145	436	0.228
145	546	0.216

Sound Velocity, [$m\,s^{-1}$]   2, 25, 38

Unoriented; at 25°C	2.5×10^3	
at 125°C	125×10^3	

Specific Heat   see Heat Capacity

Specific Volume, [$1\,kg^{-1}$] $\equiv$ [$cm^3 g^{-1}$]

Form	$\bar{v}_a$ (amorphous)	$\bar{v}_c$ (crystalline)
isotactic	1.176 - 1.172	1.073 - 1.060
smectic		1.092
syndiotactic	1.165	1.114 - 1.10

Softening Temperature, Vicat   (see Table II)

Tensile Modulus   (see Table II)

Tensile Strength   (see Table II)

Transition and Relaxation Temperatures

The transition and relaxation temperatures associated with peaks in dynamic loss are designated as α, β, γ, etc. in the order of descending temperatures.

Transition	Temperature, [^oC]	Assignment and Remarks	
α	Between 30 - 80	Difficult to resolve	16
β (at 0.2 Hz)	0	Insensitive to changes in crystallinity; long chain motion in amorphous portion	
γ	-80	In atactic polypropylene; hindered movement of $C-CH_3$ units	
δ	< 200	Hindered rotation of CH_3 groups	

Viscosity - Molecular Weight Relationships   see corresponding table in this Handbook

Table II: PROPERTIES OF TYPICAL MAINLY ISOTACTIC POLY(PROPYLENES) (49)[x]

Property	Value
Brittleness (D 746), [$^{\circ}$C]	25
Deflection Temperature (D 648), [$^{\circ}$C]	
at 66 lb/in^2 (4.64 kg/cm^2)	96 - 110
at 264 lb/in^2 (18.6 kg/cm^2)	57 - 63
Density (D 792), [Mg m^{-3}] ≡ [g cm^{-3}]	0.90 - 0.91
Dielectric Constant	
at 1 kHz (D 150)	2.2 - 2.3
Dielectric Strength (D 149), [V mil^{-1}]	610 (430 at 120°C)
[V cm^{-1}] x 10^{-3}	240 (170 at 120°C)
Dissipation Factor	
(60 Hz - 100 MHz) (D 510)	$3 \times 10^{-4} - 1 \times 10^{-3}$
Elongation at Break (D 638), [%]	500 - 900 (30 at -40°C)
Environment Stress Cracking or F_{50} Time (D 1693)	Does not stress crack
Flexural Modulus (D 790), [MPa] ≡ [N mm^{-2}]	1172
Hardness, Shore D (D 1706)	70 - 80
Heat Capacity, [kJ kg^{-1} K^{-1}]	1.926 (see also Table I)
Izod Impact Strength (D 256), [ft lb/in of notch]	0.4 - 6.0 (0.1 - 0.7 at -20°C)
[cm kg/cm notch]	2.2 - 12 (0.55 - 3.9 at -20°C)
Power Factor	
at 1 kHz	< 0.002 - 0.001
Solvent Resistance to Hydrocarbons and Chlorinated Hydrocarbons	Resistant Below about 80°C
Stiffness in Flexure (D 747), [MPa] ≡ [N mm^{-2}]	1172
Tensile Modulus (D 638), [MPa] ≡ [N mm^{-2}]	1032 - 1720
Tensile Strength (D 638), [MPa] ≡ [N mm^{-2}]	29.3 - 38.6
Tensile Yield Elongation (D 638), [%]	11 - 15 (11% at -40°C)
Thermal Conductivity (C 177), [W m^{-1} K^{-1}]	11.7
Vicat Softening Temperature (D 1525), [$^{\circ}$C]	138 - 155
Volume Resistivity (D 257), [Ω cm]	$10^{16} - 10^{17}$

[x] Properties at ambient room temperature, unless denoted otherwise.

The numbers in parentheses refer to the ASTM Standards, American Society for Testing Materials, 1916 Race Street, Philadelphia, Pa., 19103.

REFERENCES

1. E. J. Addink, J. Beintema, Polymer 2, 185 (1961).

2. N. M. Bikales, Editor, "Encyclopedia of Polymer Science and Technology," Interscience, New York, 12, p. 702.

3. F. A. Bovey, "High Resolution NMR of Macromolecules," Academic Press, New York, 1972, pp. 132-42.

4. R. Chiang, J. Polymer Sci. 28, 235 (1958).

5. G. Ciampi, Chim. Ind. (Milan) 38, 298 (1956).

6. P. Vorradini, G. Natta, P. Ganis, P. A. Temussi, J. Polymer Sci. C 16, 2477 (1967).

7. F. Danusso, G. Moragalio, J. Polymer Sci. 24, 161 (1957).

8. M. Dole, "Mechanism of Chemical Effects in Irradiated Polymers" in "Crystalline Olefin Polymers," R. A. V. Raff, K. W. Doak, Editors, Interscience, New York, 1965, p. 907.

9. V. L. Erlich, "Olefin Fibers," in "Encyclopedia of Polymer Science and Technology," Volume 9, pp. 403-40, Interscience Publishers, New York, 1968.

10. B. Falkai, Makromol. Chem. 41, 86 (1960).

11. H. A. Flocke, Kolloid-Z. 180, 118 (1962).

12. L. R. Fortune, G. N. Malcolm, J. Phys. Chem. 64, 934 (1960).

13. H. P. Frank, "Polypropylene," Macdonald Technical and Scientific, London, 1968.

14. G. Gramsberg, Kolloid-Z. 175, 119 (1961).

15. J. H. Griffith, B. G. Ranby, J. Polymer Sci. 38, 107 (1959).

16. M. P. Groenewege, J. Schnuzer, J. Smidt, C. A. F. Tuijnman, in R. A. V. Raff, K. W. Doak, Eds. "Crystalline Olefin Polymers," Part II, p. 798, Interscience, New York, 1965.

17. R. H. Hughes, Am. Chem. Soc., Meeting September 1963, Polymer Division Preprints, 4/2, 697 (1963).

18. D. Jeschke, H. A. Stuart, Z. Naturforsch. 16a, 37 (1961).

19. H. D. Keith, F. J. Padden, Jr., N. M. Walter, H. W. Wyckoff, J. Applied Phys. 30, 1485 (1959).

20. J. B. Kinsinger, R. E. Hughes, J. Phys. Chem. 63, 2002 (1959).

21. C. Y. Liang, M. R. Lytton, C. J. Boone, J. Polymer Sci. 47, 139 (1960); 54, 523 (1961).

22. C. Y. Liang, F. G. Pearson, J. Mol. Spectroscopy 5, 290 (1960).

23. J. P. Luongo, J. Appl. Polymer Sci. 3, 302 (1960).

24. M. P. Macdonald, I. M. Ward, Polymer 2, 341 (1961).

25. T. H. Malim, Iron Age 197, 88 (1966).

26. P. Manaresi, V. Giannella, J. Appl. Polymer Sci. 4, 251 (1960).

27. L. Mandelkern, in R. H. Doremus, B. W. Roberts, D. Turnbull, Eds., "Growth and Perfection of Crystals," John Wiley and Sons, New York, 1958, p. 467; also L. Mandelkern, Chem. Rev. 56, 903 (1956).

28. Z. Mencik, Chem. Prumysl. 10, 377 (1960).

29. R. L. Miller, in R. A. V. Raff, K. W. Doak, Eds., "Crystalline Olefin Polymers," Part I, p. 685, Interscience, New York, 1965.

30. R. L. Miller, "Crystallographic Data for Various Polymers," in this Handbook.

31. R. L. Miller, L. E. Nielsen, J. Polymer Sci. 44, 391 (1960); 55, 643 (1961); also Appendix I in P. H. Geil, "Polymer Single Crystals," Interscience, New York, 1963.

32. G. Natta, Makromol. Chem. 35, 93 (1960).

33. G. Natta, P. Corradini, Nuovo Cimento, Suppl. 15, 40 (1960).

34. G. Natta, P. Corradini, M. Cesari, Atti Accad. Nazl. Lincei, Rend. Classe Sci. Fis, Mat. Nat. 21, 365 (1956).

35. G. Natta, I. Pasquon, P. Corradini, M. Peraldo, M. Pegoraro, A. Zembelli, Atti. Accad. Nazl. Lincei, Rend. Classe Sci. Fis. Mat. Nat. 28, 539 (1960).

36. P. Parrini, F. Sabastiano, G. Messina, Makromol. Chem. 38, 27 (1960).

37. M. Peraldo, M. Farina, Chim. Ind. (Milan) 42, 1349 (1960).

38. H. L. Price, SPE J. 24, 54 (1968).

39. O. Redlich, A. L. Jacobson, M. H. McFadden, J. Polymer Sci. A1, 393 (1963).

40. W. Ruland, Acta Cryst. 14, 1180 (1961).

41. V. Stannett, H. Yasuda, in "Crystalline Olefin Polymers," Part II, pp. 139, 144, 147, K. W. Doak, Editors, Interscience, New York, 1964.

42. M. C. Tobin, J. Phys. Chem. 64, 216 (1960).

43. A. Turner-Jones, J. M. Aizlewood, D. R. Beckett, Macromol. Chem. 75, 134 (1964).

44. Z. W. Wilchinski, J. Appl. Phys. 31, 1969 (1960).

45. R. W. Wilkinson, M. Dole, J. Polymer Sci. 58, 1089 (1962).

46. H. Wilski, T. Grewer, J. Polymer Sci. 66, 33 (1964).

47. B. Wunderlich, "Macromolecular Physics," Academic Press, New York, 1973.

48. B. Wunderlich, H. Baur, Fortschr. Hochpolymer Forsch. 7, 2, 309 (1970).

49. P. D. Ritchie, Editor, "Vinyl and Allied Polymers," Vol. 1, Illiffe, London 1968, p. 143; and from trade literature of various manufacturers of polypropylene in USA.

PHYSICAL CONSTANTS OF POLY(TETRAFLUOROETHYLENE)

C. A. Sperati

Plastics Department
E. I. DuPont de Nemours and Co., Inc.
Parkersburg, West Virginia 26101

Property [**]	Value	Remarks	Ref.
Abrasion Resistance	See Wear Factor		
Bond Angle C-C	See Conformation		
$\quad$ F-C-F [deg]	108		57
Bond Energies, [kJ mol^{-1}] C-F	482		57
$\quad$ C-C	290-335		15
Bond Length, [Å] C-F	1.30-1.32		57
Chemical Resistance	See Solvents		
Coefficient of Friction	$0.224\,W^{0.163}$	W = load in grams recalc from	2
Coefficient of Thermal Expansion			
$\quad$ Linear Expansion, [K^{-1}]	99×10^{-6}	ASTM D 696-44	63
$\quad$ [°F^{-1}]	55×10^{-6}		
$\quad$ Average values for temperatures indicated			31
$\quad\quad$ 25 to -190	86×10^{-6}		
$\quad\quad$ 25 to -150	96		
$\quad\quad$ 25 to -100	112		
$\quad\quad$ 25 to -50	135		
$\quad\quad$ 25 to 0	200		
$\quad\quad$ 25 to 50	124		
$\quad\quad$ 25 to 100	124		
$\quad\quad$ 25 to 150	135		
$\quad\quad$ 25 to 200	151		
$\quad\quad$ 25 to 250	174		
$\quad\quad$ 25 to 300	218		
Cohesive Energy Density	125 (Est) See also Solvent Absorption		65
Compressibility (calculated) [MPa^{-1}]	28.8×10^{-17}		8
$\quad$ [cm^2 dyn^{-1}]	28.8×10^{-12}		
Compressive Strength Properties			
$\quad$ Stress at 1% deformation [MPa]	$47094-231295\,(\%C)[*]+0.040\,(\%C)^2-0.0391\,(\%C)^3$		63
$\quad$ [psi]	$6830.4-335.4\,(\%C)+5.801\,(\%C)^2-0.0391\,(\%C)^3$		
$\quad$ Yield Strength [MPa]	$190.4-0.862\,(K)$	from 4 to 180 K	
$\quad$ [psi]	$27.62-0.125\,(K)$		
Conformation		C-C bond angle is at its minimum energy at 162°	
$\quad$ Energy barrier to rotation			
$\quad$ Gauche-trans energy difference [kJ mol^{-1}]	5.9 ± 1.7		
$\quad$ Trans-trans energy difference [kJ mol^{-1}]	4.6 ± 2.9		
Contact Angle			28
$\quad$ Advancing [deg]	116		
$\quad$ Receding [deg]	92		

See Note 1 [**] $\%C$ = % Crystallinity [*]

Property	Value	Remarks	Ref.
Creep			
Compressive Creep Modulus, [MPa]	186	after 100 h at 0.6895 MPa at 23°C	34, 48
[psi]	27,000	after 100 h at 1000 psi and 23°C	
Flexural Creep Modulus, [MPa]	$2814 - 158.5\ (\%C)^* + 2.919\ (\%C)^2 - 0.1638\ (\%C)^3$	6.895 MPa for 100 h; %C* from 45 to 90	62
[psi]	$408,360 - 22,990\ (\%C)^* + 423.3\ (\%C)^2 + 2375\ (\%C)^3$	1000 psi for 100 h; %C* from 45 to 90	
Tensile Creep Modulus	See Note 2		48
[MPa]	61	after 100 h at 6.865 MPa and 23°C	
[psi]	8900	after 100 h at 1000 psi and 23°C	
Critical Shear Rate for Melt Fracture, $[s^{-1}]$ 10^{-5}		at 380 °C	70
Critical Surface Tension, $[mN\ m^{-1}] \equiv [dyn\ cm^{-1}]$ 18.6			
Crystal Structure Change	See Figure 1 and Transitions		
Crystalline Content, Wt %C	762.25 - (1524.5/Density)	Calc. from	63
Crystallographic Data	See Figure 1		
Phase I		Hexagonal structure similar to phase IV, but little or no lateral congruence	
Separation of chain axes, [Å]	$S = 3.68 \times 10^{-2}\,T + 5688$	T from 40 to 220°C	14
Phase II Triclinic structure			
Repeat distance for 13 CF_2 Groups [Å]	16.9		12, 50
Separation of chain axes, [Å]	5.62		12
Phase III High pressure form, planar zig-zag structure			24
Monoclinic space group B2/m			
	a = 9.50 Å b = 5.06 Å c = 2.62 Å γ = 105.5°	Density $[Mg\ m^{-3}] \equiv [g\ cm^{-3}]$ 2.74 at 12 bar	
Phase IV Hexagonal structure			12, 50
Repeat distance of 15 CF_2 groups [Å]	19.5		
Separation of chain axes, [Å]	5.55		
Deformation Under Load, [%] DEF	$12.93 - 0.3303\ (\%C)^* - 0.002372\ (\%C)^2$	1000 psi for 24 h at 23°C; %C from 50 to 90	62
Density, $[Mg\ m^{-3}] \equiv [g\ cm^{-3}]$, depends on crystalline and void contents			
Completely amorphous (never observed)	2.00 at 23°C		63
Crystalline (from x-ray and IR data)	2.302 at 23°C		55
As polymerized	2.280 - 2.290 at 23°C		63
After melting and recrystallization (void free polymer)	1524.5/(762.25 - %C)	Calc. from	63
Dielectric Constant	See Electrical Properties		
Dielectric Strength	See Electrical Properties		
Dissipation Factor	See Electrical Properties		
Diffusion	See Permeability		
Depolymerization Rate	See Thermal Degradation		

---

*%C = % Crystallinity

Property	Value	Remarks	Ref.
Electrical Properties			18
Dielectric Constant, $[\epsilon]$	(1+0.238 D) / (1-0.119 D) D is Density		22
Dielectric Strength		The exact value depends on the thickness of the test specimen as well as other test variables	
$[MV\ m^{-1}]$	15.7 - 19.7	ASTM D 149-44	63
$[V\ mil^{-1}]$	400 - 500		
Dissipation Factor 60 to 10^9 Hz	See Note 3 2×10^{-4}		
Surface Arc Resistance, [s]	700	ASTM D 495-42	63
Surface Resistivity, 100 %RH [megohms]	3.6×10^6	ASTM D 257-52 T	63
Volume Resistivity, dry, [ohm cm]	10^{19}		63
Equation of State, (P+a) (V-b) = B (T-c)	a = 405.2 bar; b = 0.500 $cm^3\ g^{-1}$;c = 114°C ; B = 0.310 $cm^3\ g^{-1}\ K^{-1}$		38
End Groups	See Note 4		9
Entropy of Fusion	See Melting		
Flammability	VE-O	UL-94	
Flexural Modulus			62
[MPa]	29 (%C) - 1127 at 23°C	%C* from 40 to 90	
[psi]	4210 (%C) - 163,500 at 23°C		
[psi]	1,270 (%C) - 44,500 at 100°C		
Flexural Strength	Does not break in flexure	ASTM D 790-70	
Glass-I Transition	See Transitions alpha		
Glass-II Transition	See Transitions gamma		
Hardness		%C* from 40 to 90	62
Durometer [D Scale]	D = 42 + 0.2 (%C)	ASTM D 676-44 T	
Rockwell [J Scale]	J = 114.6 - 0.433 (%C)	ASTM D 785-47 T	
Scleroscope	S = 133 - 1.3 (%C)		
Heat Capacity, $[kJ\ kg^{-1}\ K^{-1}]$	0.9324 + 1.05 x 10^{-3} T		56, 73
Heat Distortion Temperature, [^{o}C]			
at 0.455 MPa	132	ASTM D 648-56	63
at 66 psi	132		
at 1.820 MPa	60		
at 264 psi	60		
Heat of Crystallization	See Melting-Latent Heat of Fusion		
Heat of Formation, $[kJ\ mol^{-1}]$	813		10, 20
Impact Strength	See Toughness		
Intrinsic Viscosity, $[ml\ g^{-1}]$	300 - 2000	See Solvents	58, 60
Infrared Spectra	See Table 1		40, 45, 74
Lamellar Thickness, $[\mu m]$	0.02 (%C) - 0.9	Calc. from	64, 23, 46
Lattice Energy Per Recurrent Unit (calculated) $[kJ\ mol^{-1}]$	8.5 ± 2.5		8

*%C = % Crystallinity

Property	Value	Remarks	Ref.
Mechanical Loss	See Note 5	ASTM D 2236-70	
Melt Viscosity			
By Capillary Rheometer	See Note 6	ASTM D 2116-66	61
[Pa s]	$1.62 \times 10^{-13} \overline{M}_n^{3.4}$		
[poise]	$1.62 \times 10^{-13} \overline{M}_n^{3.4}$		
By Melt Creep, [Pa s]	4 to 20 $\times 10^{10}$ at 360°C		1, 47
[poise]	4 to 20 $\times 10^{11}$ at 360°C		
Energy of Activation for Melt Flow [kJ mol^{-1}]	63 - 84		1, 60
Melting			
Melting point			
Initial (irreversible - see Note 7) [$^{\circ}$C]	342		41
Second and subsequent (reversible) [$^{\circ}$C]	327		52
Entropy of Melting, [J K^{-1} mol^{-1}]	4.77		66
Corrected to Constant Volume [J K^{-1} mol^{-1}]	3.18		
Volume Expansion during Melting	29%		
Increase of Melting Point with Pressure [K bar^{-1}]	0.152		43
[K atm^{-1}]	0.154		
Latent Heat of Fusion, [kJ kg^{-1}]	57.3		38
Molecular Weight	See Note 8		
$\log \overline{M}_w = 27.5345 - 12.1405$ D			16, 63
$\log \overline{M}_w = 22.324 - 9.967$ Density (ρ)		$\rho = [Mg\ m^{3}] \equiv [g\ cm^{0}]$	
$\log \overline{M}_n = 10.3068 - 5.1318 \log$ H		H = Heat of crystallization in cal/g	67
Particle Size (Average diameter), [μm]			
Dispersion polymer	0.22		37
Granular polymer		ASTM D 1457-69 T	
ASTM Type I	275 - 575		
ASTM Type III Class 1	50		
Permeability to Gases, [m^3 (STP) m s^{-1} m^{-2} Pa^{-1}] $\times 10^{15}$			
Helium	$\log P_{He} = 3.505 - 0.037$ (%C)		62
Carbon Dioxide	$\log P_{CO_2} = 2.605 - 0.037$ (%C)		62
Nitrogen	$\log P_{N_2} = 2.4423 - 0.37$ (%C)		62
Hydrogen	7.35		49
Energy of Activation for Diffusion [kJ mol^{-1}]	27.6		49
Energy of Activation for Permeability [kJ mol^{-1}]	19.7		49

Property	Value	Remarks	Ref.
Phase Diagram	See Figure 1		24, 63, 71

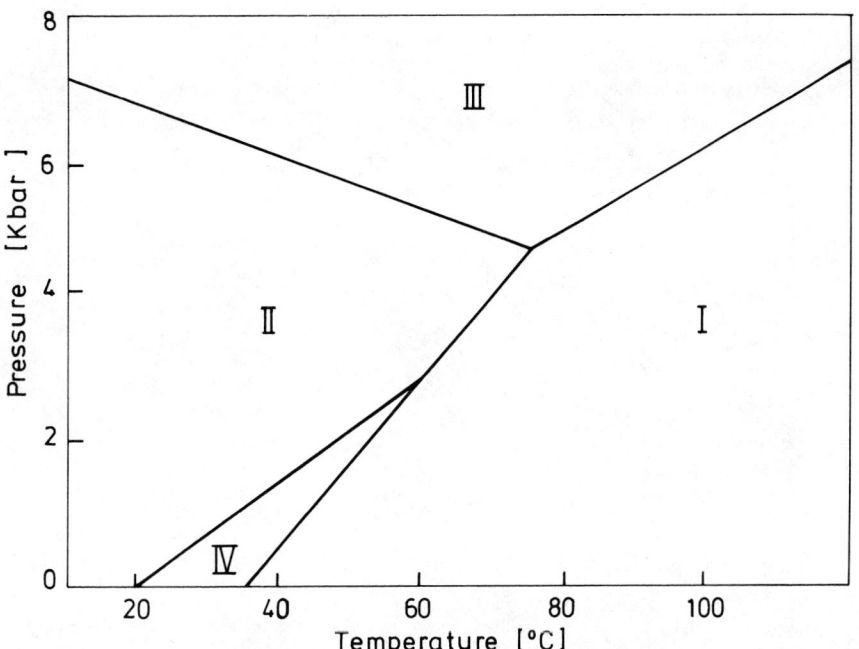

PV Limit, $[\text{Pa m s}^{-1}]$	50 to 250		36
$[\text{psi ft min}^{-1}]$	10,000 to 50,000		

Radiation

    Degradation in Air, [Radicals / 10^6 C Atoms]

$$3.7 \times 10^{-5} D^{0.89}$$

Derived from Ref.   30
D is dose in rads

    Kinetic Chain Length (in vacuo) (CL)   25, 72

$$\log CL = 7.932 - 4640/T - 0.5 \log I$$

T is K, I is megaroentgens per hour

Recovery After Deformation	%Recovery = 111 - 0.87 (%C)	ASTM D-1147	69
	%C from 40 to 90	1/4 in. penetrator, 5 lb preload, 250 lb major load	
Refractive Index, $[n_D^{25}]$	1.376		7
Resistivity	See Electrical Properties		
Service Temperature	See Note 9		19, 68, 69
Solvent Absorption	See Note 10		
	Per Cent Weight Gain = 24.85 - 2.456 HSP	HSP is the Hildebrand solubility parameter	57
Perchloroethylene	[% wt increase] 2.	1000 h in boiling liquid, 60% Cryst.	
Methylene Chloride	[% wt increase] 0.7	1000 h in boiling liquid, 60% Cryst.	
Solvents	See Note 11		
Standard Specific Gravity Relative Molecular Weight			
	See Note 8	ASTM D-1457-69	16

Tensile Strength Properties

Ultimate Strength, [MPa] at 23°C	270 - 0.39 (%C) - 99.3 D	D from 2.170 to 2.280 Mg m^{-3}	
[psi] at 23°C	39,200 - 57 (%C) - 14400 D	%C from 30 to 90	
Ultimate Elongation	200 - 400%		

Property	Value	Remarks	Ref.
Thermal Conductivity, [W m^{-1} K^{-1}]	K = 4.86 x 10^{-4} T + 0.253		56
Thermal Degradation			
Enthalpy of Depolymerization	ΔH = 38,200 - 6.67 (T-360) cal/mol		56
Rate of Depolymerization, [kg s^{-1}]	dM/dt = -3 x 10^{16} M exp (-347 kJ/RT)		59
	R = 8.314 [JK^{-1} mol^{-1}]		
Kinetic Chain Length (CL)	log CL = 8320/T - 9.958	T = K	25, 72
Toughness			
Izod Impact Strength, [N]		ASTM D 256	19
24°C	160		
77°C	greater than 320		
24°C, [ft lb in.$^{-1}$]	3		
77°C	greater than 6		
Tensile Impact Strength		ASTM D 1822-68	61
[kPa m]	TIS = 3379 - 45.1 (%C)		
[ft lb in.$^{-2}$]	TIS = 161 - 21.5 (%C)		
Transitions, [^{o}C]	See Note 12		21
Alpha (Glass I)	126	at 1 Hz	42
Beta I (Crystalline Transition Crystal Disordering Relaxation)			
	19		
Change with Pressure, [K bar^{-1}] 0.0128			
Latent Heat, [kJ kg^{-1}]	13.4		
Volume Expansion, [1 kg^{-1}]	0.0058		
Kinetics of the Transition	See Note 13		
Beta II	30	about one tenth the 19°C transition	46
Entropy Change, Overall (19+30)			
[kJ kg^{-1} K^{-1}]	0.0452		
Due to volume change			
[kJ kg^{-1} K^{-1}]	0.0172		
Corrected to Constant Volume			
[kJ kg^{-1} K^{-1}]	0.0280		
Gamma (Glass II)	-80 at 1 Hz		40
Change of Frequency (F, Hertz) of loss peak with temperature (K)			21
Alpha (Glass I)	log F = 48.3759 - 368 kJ/(2.303 RT)		
Beta (Crystalline Relaxation)	log F = 24.97 - 142 kJ/(2.303 RT)		
Gamma (Glass II)	log F = 22.37 - 75 kJ/(2.303 RT)		
Effect of Temperature on the Transition Pressure			39
Transition between Phases IV and I			
[Pa]	P = (-91 - 57.8 T + 1.99 T^2) x 10^5	T = ^{o}C	
Transition between Phases II and IV			
[Pa]	P = (-820 + 26.9 T + 0.68 T^2) x 10^5	T = ^{o}C	
Triple Point	See Figure 1, (Crystallographic Data and Transitions)		
Unit Cell	See Crystallographic Data		
Water Absorption	per cent increase 0.0	ASTM D 570-59 T	63
Wear Factor	See Note 14		36
[Pa^{-1}]	3000 x 10^{-17}		
[in.^{3} min ft^{-1} lb^{-1} hr^{-1}]	1500 x 10^{-10}		
Weatherability	See Note 15		

NOTES

1. Many of the properties of PTFE are not constant but rather change systematically with crystalline content, molecular weight, void content, temperature, pressure, or some other independent variable. Where available, these interrelationships are shown even though they are usually empirical rather than fundamental.

2. Equations based on time-temperature superposition are given in Ref. 34 for PTFE with a crystalline content approximately 58%. These relationships permit calculation of tensile creep modulus at any stress, time, or temperature in the temperature range of 35-100 and stress levels below the elastic limit. Ref. 33 provides similar data over a wider range of conditions.

3. Precise measurements of dielectric loss (Ref. 58) show peaks that correspond to the gamma transition. The measured values of tan delta, however, were all less than 300×10^{-6} with some as low as 20×10^{-6}.

4. Infrared methods show presence of carboxylic acid monomer and dimer, perfluorovinyl, and acid fluoride end groups in low molecular weight products. The methods available are not sufficiently sensitive to detect end groups in polymers of commercially useful molecular weights. (See Ref. 9).

5. The mechanical loss factor is related to transition behavior. The value depends on temperature and frequency of the dynamic mechanical stress. Values of the logarithmic decrement up to 0.55 have been observed. (See Ref. 42).

6. This relationship is applicable only to PTFE with molecular weight well below that required for plastics applications.

7. The temperature observed for the first (irreversible) melting point decreases with rate of heating. It also differs with the type of polymer. Granular PTFE usually has a higher melting point than the dispersion based products.

8. Standard specific gravity described in ASTM D 1457 and Ref. 16 is a useful inverse measure of relative molecular weight. Estimating actual values of the molecular weight of the highest molecular weight commercial polymers has, to date, involved assumptions and extrapolations. These relatioships, therefore, are at best a first approximation.

9. The recommended maximum service temperature is 260°C (Ref. 19). There are, however, examples of satisfactory use at higher temperatures. Gaskets in totally enclosed systems, for instance, are used repeatedly at 500°C and 15,000 psi (Ref. 17). In compressive strength tests at 4 K, for example, the polymer has been shown to have ductile rather than brittle failure. (See Ref. 68).

10. Weight gain is noted with some chlorinated and fluorinated liquids that contain no hydrogen. The weight increase decreases with increasing crystallinity.

11. PTFE is inert to essentially all common chemicals (see Ref. 52 and 53). It is, however, attacked by molten alkali metals, elemental fluorine, and pure oxygen at elevated temperatures. It can also react under some conditions with alkaline earth oxides and finely divided metals such as aluminum at elevated temperature. Sorption of solvents is mentioned in Note 10. Solutions containing 0.1 to 2% PTFE were obtained in a perfluorokerosene $C_{21}F_{44}$ (FCX-412) in the temperature range 290 - 310. (Ref. 58, 60).

12. Clear evidence for a crystalline transition at about 90°C is presented in a series of papers by Araki (Ref. 3, 4, 5). He also presents evidence for doublets in the transitions (Ref. 4), and points out that the alpha and gamma transitions follow the so-called 2/3 rule when compared with the first order crystalline transitions at 19 and 327°C (Ref. 3).

13. Data on the kinetics of the transitions not amenable to presentation in tabular form are available in Ref. 46.

14. Incorporation of finely divided fillers decreases wear factor as much as 500 to 1000 fold. (See Ref. 36).

15. No change in appearance or properties has been observed after more than fifteen years' exposure outdoors in Florida.

Table I

Infrared Absorption Spectrum of Poly(tetrafluoroethylene)

$[cm^{-1}]$	$[\mu m]$	Intensity	$[cm^{-1}]$	$[\mu m]$	Intensity
3670	2.72	VW	1242	8.05	VS
3570	2.80	VW	1213	8.24	VS
3450	2.90	VW	1152	8.68	VS
3090	3.24	VW	932	10.73	M
2925	3.42	VW	850	11.76	Amorphous
2620	3.82	VW	778	12.85	Amorphous
2590	3.86	VW	738	13.55	Amorphous
2530	3.95	VW	718	13.93	Amorphous
2450	4.08	VW	703	14.22	Amorphous
2367	4.22	M	638	15.67	S
2300	4.35	VW	553	18.08	S
1974	5.07	VW	553	18.08	S
1935	5.17	W	516	19.38	VS
1883	5.31	W	384	26.04	Amorphous
1859	5.38	W			
1792	5.58	M			
1703	5.87	W			
1545	6.47	M			
1451	6.90	M			
1420	7.04	M			

Code: VS - Very Strong; S - Strong; M - Medium; W - Weak; VW - Very Weak

REFERENCES

1. G. Ajroldi, C. Garbuglio, M. Ragazzini, J. Appl. Polymer Sci. 14, 79 (1970).

2. A. J. G. Allan, Lubrication Eng. 14, 211 (1958).

3. Y. Araki, J. Appl. Polymer Sci. 9, 421 (1965).

4. Y. Araki, J. Appl. Polymer Sci. 9, 3575 (1965).

5. Y. Araki, J. Appl. Polymer Sci. 11, 953 (1967).

6. T. W. Bates, W. H. Stockmayer, Macromolecules 1, 17 (1968).

7. F. W. Billmeyer, Jr., J. Appl. Phys. 18, 431 (1947).

8. W. Brandt, J. Chem. Phys. 26, 262 (1957).

9. M. I. Bro, C. A. Sperati, J. Polymer Sci. 38, 289 (1959).

10. W. M. D. Bryant, J. Polymer Sci. 56, 277 (1962).

11. E. S. Clark, L. T. Muus, Z. Krist. 117 (2/3), 108 (1962).

12. E. S. Clark, L. T. Muus, Z. Krist. 117 (2/3), 119 (1962).

13. E. S. Clark, J. Macromol. Sci. B, 1, 795 (1967).

14. E. S. Clark, Previously Unpublished Results.

15. T. L. Cottrell, "The Strengths of Chemical Bonds", 2nd Ed., Butterworths, Washington, D. C. 1958.

16. R. C. Doban, A. C. Knight, J. H. Peterson, C. A. Sperati, Meeting Am. Chem. Soc., Atlantic City, Sept. 1956.

17. R. C. Doban, C. A. Sperati, B. W. Sandt, Soc. Plastics Eng. 11, 17 (1955).

18. DuPont Plastics Department, Electrical/Electronic Design Data for "Teflon" Fluorocarbon Resins Bulletin A-66426 (June 1970).

19. DuPont Plastics Department, "Teflon" Fluorocarbon Resins Mechanical Design Data, Bulletin A-40844.

20. H. C. Duus, Ind. Eng. Chem. 47, 1445 (1955).

21. R. K. Eby, K. M. Sinnott, J. Appl. Phys. 32, 1765 (1961).

22. P. Ehrlich, J. Res. Natl. Bur. Std. 51, 185 (1953).

23. S. Fischer, N. Brown, J. Appl. Phys. 44, 4322 (1973).

24. H. D. Flack, J. Polymer Sci. Part A-2, 10, 1799 (1972).

25. R. E. Florin, M. S. Parker, L. A. Wall, J. Res. Natl. Bur. Std. A, 70 (2), 115 (1966).

26. H. W. Fox, W. A. Zisman, J. Colloid Sci. 5, 514 (1950).

27. R. J. Good, J. A. Kvikstad, W. D. Bailey, J. Colloid Interface Sci. 35, 314 (1971).

28. Imperial Chemical Industries, Ltd., Technical Service Note F-12, Physical Properties of Polytetrafluoroethylene, 2nd Ed. (1968).

29. E. D. Jones, G. P. Koo, J. L. O' Toole, Mod. Plastics 44 (11), 137 (1967).

30. H. S. Judeikis, S. Siegel, IEEE Trans. Nuclear Sci. 14, 237 (1967).

31. R. K. Kirby, J. Res. Natl. Bur. Std. 57, 91 (1956).

32. Y. Kometani, S. Koizumi, S. Fumoto, T. Nakajima, United States Patent 3,462,401 (August 19, 1969).

33. G. P. Koo, "Cold Drawing Behavior of Polytetrafluoroethylene", Thesis from Stevens Institute of Technology, Sc. D. (1969), University Microfilms, Inc., Ann Arbor, Michigan, 69-21, 294.

34. E. P. Koo, E. D. Jones, J. M. Riddell, J. L. O' Toole, Soc. Plastics Eng. J. 21, 1100 (1965).

35. G. P. Koo, M. N. Riddell, J. L. O' Toole, Polymer Eng. Sci. 7, 182 (1967).

36. R. B. Lewis, J. Eng. Ind. 1, 1 (1966).

37. J. F. Lontz, W. B. Happoldt, Jr., Ind. Eng. Chem. 44, 1800 (1952).

38. J. M. Lupton, Meeting Am. Chem. Soc., Chicago, September 1958.

39. G. M. Martin, R. K. Eby, J. Res. Natl. Bur. Std. A 72, 467 (1968).

40. G. Masetti, F. Cabassi, G. Morelli, G. Zerbi, Macomolecules 6, 700 (1973).

41. D. I. McCane, "Encyclopedia of Polymer Science and Technology", Wiley, 13, 623 (1970).

42. N. G. McCrum, B. E. Read, G. Williams", Anelastic and dielectric effects in polymeric solids", John Wiley and Sons, 1967.

43. P. L. McGeer, H. C. Duus, J. Chem. Phys. 20, 1813 (1952).

44. K. Morokuma, J. Chem. Phys. 54, 962 (1971).

45. R. E. Moynihan, J. Am. Chem. Soc. 81, 1045 (1959).

46. R. Natarajan, T. Davidson, J. Polymer Sci. Polym. Phys. Ed. 10, 2209 (1972).

47. A. Nishioka, M. Watanabe, J. Polymer Sci. 24, 298 (1957).

48. J. L. O' Toole, Modern Plastics Encyclopedia 45, 48 (1968).

49. R. A- Pasternak, M. V. Christensen, J. Heller, Macromolecules 3, 366 (1970).

50. R. H. H. Pierce, Jr., E. S. Clark, J. F. Whitney, W. M. D. Bryant, Atlantic City Meeting Am. Chem. Soc., September 18, 1956.

51. C. M. Pooley, D. Tabor, Proc. Roy. Soc. (London), Ser. A 329, 251 (1972).

52. J. C. Reed, E. J. McMahon, J. R. Perkins, Insulation (Libertyville) 10 (5), 35 (1964).

53. M. M. Renfrew, E. E. Lewis, Ind. Eng. Chem. 38, 870 (1946).

54. E. Rydel, Thesis Univ. of Strassbourg, March 22, 1965.

55. A. L. Ryland, J. Chem. Educ. 35, 80 (1958).

56. P. H. Settlage, J. C. Siegle, Phys. Chem. Aerody. Space Flight, Proc. Conf., Philadelphia, 1959, 73 (1961), Pergamm.

57. W. A. Sheppard, C. M. Sharts, Organic Fluorine Chemistry," W. A. Benjamin, Inc., New York, 1969.

58. S. Sheratt, Kirk-Othmer, "Encyclopedia of Chemical Technology", 2nd Ed. 9, 805 (1966).

59. J. C. Siegle, L. T. Muus, Tung Po Lin, H. A. Larsen, J. Polymer Sci. Part A, 2, 391 (1964).

60. C. A. Sperati, Gordon Research Polymers Conference, July 7, 1955.

61. C. A. Sperati, Previously Unpublished Results.

62. C. A. Sperati, J. L. McPherson, Meeting Am. Chem. Soc. Atlantic City, September, 1956.

63. C. A. Sperati, H. W. Starkweather, Jr., Adv. Polymer Sci. 2, 465 (1961).

64. H. W. Starkweather, Jr., Soc. Plastics Eng. Trans. 3, 1 (1963).

65. H. W. Starkweather, Jr., Previously Unpublished Results.

66. H. W. Starkweather, Jr., R. H. Boyd, J. Phys. Chem. 64, 410 (1960).

67. T. Suwa, M. Takehisa, S. Machi, J. Appl. Polymer Sci. 17, 3253 (1973).

68. C. A. Swenson, Rev. Sci. Instr. 25, 834 (1954).

69. P. E. Thomas, J. F. Lontz, C. A. Sperati, J. L. McPherson, Soc. Plastics Engr. J. 12, June (1956).

70. J. P. Tordella, Trans. Soc. Rheo. 7, 231 (1963).

71. C. E. Weir, J. Res. Natl. Bur. Std. 50, 95 (1953).

72. L. A. Wall, Ed., "Fluoropolymers", Wiley-Interscience, 1972.

73. B. Wunderlich, H. Baur, Adv. Polymer Sci. 7, 331 (1970).

74. G. Zerbi, M. Sacci, Macromolecules 6, 692 (1973).

PHYSICAL CONSTANTS OF POLY(ACRYLONITRILE)[*]

W. Fester
Hoechst A.G.
Frankfurt/Main, Germany

Table I: Property	Value		Remarks	Ref.
Birefringence $n_{\parallel} - n_{\perp}$	-0.005		a	1
Coefficient of Expansion				
volume $(1/v) \cdot (dV/dT)_p \times 10^4$, $[K^{-1}]$	3.0 above Tg;	1.4 below Tg		2
	2.8	1.6		3
	3.8	-		4
linear $(1/l) \cdot (dl/dT)_p \times 10^4$, $[K^{-1}]$	1.6	1.0		5
	2±0.4	-		6
Crystallization Temperature, $[^{\circ}C]$				
determined in propylene carbonate	95-100		b	7, 8
Crystallographic Data	see Unit cell dimensions			
Decomposition Temperature, $[^{\circ}C]$	~250		c	9
Density, $[Mg\ m^{-3}] \equiv [g\ cm^{-3}]$				
for flakes at $25^{\circ}C$	1.17-1.18			10, 7
Dielectric Constant (film)				11
10^6 Hz	4.2			
10^3 Hz	5.5			
60 Hz	6.5			
Dissipation Factor (film)				
10^6 Hz	0.033			
10^3 Hz	0.085			
60 Hz	0.113			
Dissolution Temperature, $[^{\circ}C]$				
in propylene carbonate	125-130		b	7, 8
Glass Transition Temperature T_g, $[^{\circ}C]$	87, 103			5
	96.5			3
	87			2
	80			12
	104			13
	85			4, 6
	87			14
Heat of Polymerization, $[kJ\ mol^{-1}]$	-72.4±2.2			15
Infrared Spectrum	see Ref.			29-46
Intrinsic Viscosity, $[ml\ g^{-1}]$			d	16-19

Solvent	T $[^{\circ}C]$	Huggins' coefficient
N, N-Dimethylformamide	25	34
	35	33

[*] Based on a similar table of R. Chiang[†], Monsanto, St. Louis in the first edition.

Table I:Property	Value	Remarks	Ref.
Intrinsic Viscosity, [ml g^{-1}] (Cont'd.)			

	Solvent	$-(d \ln [\eta]/dT)$
	N,N-Dimethylformamide	0.14-0.19
	N,N-Dimethylacetamide	0.27
	Dimethyl sulfoxide	0.08
	60% HNO$_3$	0.05
	γ-Butyrolactone	0.14
		0.13-0.17
	Hydroxyacetonitrile	0.07

Table I:Property	Value	Remarks	Ref.
Melting Point, [$^{\circ}$C]	319	e	13
Nuclear Magnetic Resonance Spectrum	see Ref.		47-49
Properties of Fibers	see Table II		
Refractive Index			
n_d^{25}	1.518	a	3
$n_\perp$	1.510-1.524		11,1
$n_{||}$	1.500-1.520		
$(dn/dT) \times 10^4$, [K^{-1}]	-0.98 below Tg		3
	-1.70 above Tg		
Solvents	dimethylformamide,		20,23
	dimethyl sulfoxide,		
	dimethylacetamide,		
	ethylene carbonate,		
	propylene carbonate,		
	malononitrile,		
	succinonitrile,		
	adiponitrile,		
	γ-butyrolactone,		
	conc. sulfuric and nitric acid		
	conc. salt solutions: LiBr, NaCNS, ZnCl$_2$;		
	see also Table "Solvents-Nonsolvents" in this Handbook		
Stereoregularity	60-50% isotactic	f (see NMR-spectra)	
Unit Cell Dimensions, [$\overset{\circ}{A}$]	orthorhombic a = 10.20 b = 6.1	g	24
	orthorhombic a = 10.55 b = 5.80		23
	orthorhombic a = 21.2 b = 11.6		26
	hexagonal a = 5.99 b = 5.99		27
Viscosity-Average Molecular Weight Relationship	see corresponding table in this Handbook		

Table II: Properties of Acrylic Fibers*
(Ref. 1, 9, 50, 51, 73, 126)

Property	Value	Property	Value	
Breaking Tenacity (dry), [dN tex^{-1}]	1.8-5.4	Elastic Recovery,, [%]		
	(1.8-5.5 p/dtex)	from 2% strain	85-95	
(wet ratio), [%]	75-95	from 5% strain	50-60	
Density, [Mg m^{-3}] = [g cm^{-3}]	1.14-1.19			

The properties of the acrylic fibers are dependent on spinning conditions and the comonomer content in the polymer itself.
Some Trade Marks of Polyacrylonitrile Fibers: Acrilan, Cashmilon, Courtelle, Dolan, Dralon, Euroacryl, Leacryl, Orlon.

Property	Value	Property	Value
Elongation at Break (wet), [%]	15-60	Water Absorption, [%] (21°C 65 rel. humidity)	1-2.5
Initial Modulus, [dN tex^{-1}]	24.5-62 x 10^{-2} 25-62 (p/dtex)	Effects of Acids and Alkalis	good to excellent resistance to mineral acids; fair to good resistance to weak alkali; moderate resistance to strong, cold solutions of alkali.
Knot Strength, from breaking tenacity, [%]	70-80	Effect of Bleaches and Solvents	good resistance to bleaches and common solvents; uneffected by dry cleaning solvents; can be bleached with sodium chlorite
Loop Strength, from breaking tenacity, [%]	45-80		
Sonic Velocity (filaments) depending on orientation, [m s^{-1}] unoriented oriented	2100 3500	Resistance to Mildew, Aging, Sunlight, Abrasion	not attacked by mildew; good resistance to aging, sunlight and abrasion
Sticking Temperature, [°C]	190-330		
Tensile Strength (dry), [MPa]≡[N mm^{-2}]	250-568		

REMARKS

a) $n_{\perp}$ and $n_{\parallel}$ are refractive indices measured with incident light having the vibration vector perpendicular and parallel to the fiber axis, respectively.

b) The dissolution and crystallization temperatures given here are obtained from a free radical poly(acrylonitrile). They are sensitive to chain irregularities in the polymer. Samples of poly(acrylonitrile) obtained from different sources show marked difference in the dissolution and crystallization temperatures, although they have similar IR-spectra, x-ray diffraction patterns, and densities.

c) The thermal decomposition temperature determined by thermogravimetric analysis ranges from 250°C for a poly(acrylonitrile) prepared with an ionic catalyst, to 310°C for a commercial fiber. Pyrolysis of poly(acrylonitrile) carried out in the absence of oxygen at 500-800°C yields HCN and low molecular weight nitriles such as monomer, dimer and methacrylonitrile leaving a residue with a condensed ring structure.

d) The factors which convert the intrinsic viscosity in one solvent (e.g. DMF) into in another (dimethylacetamide, dimethyl sulfone, ethylene carbonate, γ-butyrolactone) can be calculated from data given by Fujisaki and Kobayashi (18).

e) Poly(acrylonitrile) decomposes before melting.

f) Tacticity of PAN was determined by NMR, computing the spectra and by decoupling techniques.

g) The reported unit cell dimensions, especially the c dimension along the chain axis, can only be regarded as estimated because of the diffuse meridian and polar reflections. For more data see Table on "Crystallographic Data for Various Polymers" in this Handbook.

REFERENCES

1.   P. A. Koch, Faserstoff-Tabellen, Ed. 1969, DK 677.494.745.32.
2.   H. J. Kolb, E. F. Izard, J. Appl. Phys. 20, 564 (1949).
3.   R. B. Beevers, J. Polymer Sci. A 2, 5257 (1964).
4.   G. P. Lanzl, quoted in Ref. 1.
5.   W. H. Howard, J. Appl. Polymer Sci. 5, 303 (1961).
6.   C. E. Black, quoted in Ref. 1.
7.   R. Chiang, J. Polymer Sci. A 3, 2019 (1965).
8.   R. Chiang, J. H. Phodes, V. F. Holland, J. Polymer Sci. A 3, 479 (1965).
9.   J. Luenenschloß, E. Hummel, Fasertafel Textil-Praxis 1967.
10. R. Chiang, J. Polymer Sci. A 1, 2765 (1963).
11. M. Harris, "Handbook of Textile Fibers," Harris Research Laboratories 1246, Taylor St., N. W. Washington, D.C. (1954).
12. J. J. Kearvey, E. C. Eberlin, J. Appl. Sci. 3, 47 (1960).
13. W. R. Krigbaum, N. Tokita, J. Polymer Sci. 43, 467 (1960).
14. R. D. Andrews, R. M. Kimmel, J. Polymer Sci. B 3, 167 (1965).
15. L. K. J. Tong, W. C. Kenyon, J. Am. Chem. Soc. 69, 2245 (1947).
16. M. L. Miller, J. Polymer Sci. 56, 203 (1962).

17. H. Kobayashi, J. Polymer Sci. B 1, 299 (1963).
18. Y. Fujisaki and Kobayashi, Chem. High Polymer (Japan), 19, 81 (1962).
19. R. Chiang, J. H. Rhodes, unpublished work (1964).
20. R. C. Houtz, J. Textile Res. 20, 786 (1956).
21. C. H. Bamford, G. C. Eastmond, "Properties of Polymers and Copolymers of Acrylonitrile," in Encyclopedia Polymer Sci. Technol., ed. by H. F. Mark, N. G. Gaylord, N. M. Bikales, Interscience, New York, 1964, Vol. I, p. 407-25.
22. M. L. Miller, J. Polymer Sci. 56, 203 (1962).
23. H. Kobayashi, J. Polymer Sci. B 1, 299 (1963).
24. R. Stefani, M. Chevreton, M. Garnier, M. C. Eyrand, Compt. Rend. 251, 2174 (1960).
25. V. F. Holland, S. B. Mitchell, W. L. Hunter, P. H. Lindenmeyer, J. Polymer Sci. 62, 145 (1962).
26. G. N. Patel, R. D. Patel, J. Polymer Sci. A2, 8, 47-59 (1970).
27. G. Natta, G. Mazzanti, P. Corradini, Atti Accad. Nazl. Lincei, Mem. Classe Sci. Fis., Mat. Nat. 25, 3 (1958).
28. H. Baumann, H. Lange, Angew. Makromol. Chem. 9, 16-34 (1969).

29. H. Tadokoro, S. Murahashi, R. Jamadera, T.-I. Kamei, J. Polymer Sci. A 1, 3029 (1963).

30. R. Yamadera, H. Tadokoro, S. Murahashi, J. Chem. Phys. 41, 1233 (1964).

31. C. Y. Liang, S. Krimm, J. Polymer Sci. 31, 513 (1958).

32. C. Y. Liang, F.G. Pearson, R. H. Marchessault, Spectrochim. Acta. 17, 568 (1961); C. Y. Liang, "IR Spectra: Deuteration and Polarization," Chapter II, in "Newer Methods of Polymer Characterization," ed. by B. Ke, Interscience, New York 1964, p. 47-50.

33. N. Grassie, J. N. Hay, J. Polymer Sci. 56, 189 (1962).

34. Y. Tsuda, Bull. Chem. Soc. (Japan) 34, 1046 (1961).

35. M. Talat-Erben, S. Bywater, Ricerca Sci. 25 A, 11 (1955).

36. N. Grassie, I.C. McNeill, J. Polymer Sci. 27, 207 (1958).

37. N. Grassie, I.C. McNeill, J. Polymer Sci. 30, 37 (1958).

38. N. Grassie, I.C. McNeill, J. Polymer Sci. 33, 171 (1958).

39. W. J. Burlant, C. R. Taylor, J. Phys. Chem. 62, 247 (1958).

40. C. S. H. Chen, N. Colthup, W. Deichert, R. L. Webb, J. Polymer Sci. 45, 247 (1960).

41. A. Bernas, R. Bensasson, I. Rossi, P. Barchewitz, J. Chem. Phys. 59, 1442 (1962).

42. C. A. Levine, G. H. Harris, J. Polymer Sci. 62, 100 (1962).

43. C. L. Stevens, J. C. French, J. Am. Chem. Soc. 76, 4398 (1954).

44. R. Dijkstra, H. J. Backer, Rev. Trav. Chim. 73, 575 (1954).

45. M. L. Miller, ACS Meeting, St. Louis, Polymer Preprint 1, 47 (1961).

46. D. O. Hummel, F. Scholl, "Atlas der Kunststoff-Analyse," Carl Hanser Verlag, Muenchen, Verlag Chemie GmbH, Weinheim/Bergstr. 1968.

47. J. Bargon, K. H. Hellwege, U. Johnson, Kolloid-Z.-Z. Polymere 213, 51-55 (1966).

48. G. Svegliado, G. Talamini, J. Polymer Sci. A 1, 5, 2875 (1967).

49. Yoshino, J. Polymer Sci. B 5, 703 (1967).

50. "1964 Man-Made Fiber Chart," ed. by T. Benton Sevison, Jr., Textile World, 330 W. 42 nd. St. New York, N. Y.

51. R. W. Moncrieff, "Man-Made Fibers," John Wiley, New York (1963).

PHYSICAL CONSTANTS OF POLY(VINYL CHLORIDE)

E. A. Collins and C. A. Daniels
The B. F. Goodrich Chemical Company, Technical Center,
Avon Lake, Ohio

C. E. Wilkes
The B. F. Goodrich Company, Research and Development,
Brecksville, Ohio

Property	Value	Ref.
Birefringence	see Stress Optical Coefficient	

Branching

short chain branching as a function of polymerization temperature

	Ref. 2		Ref. 3		Ref. 4		Ref. 5
P.T. [°C]	$100 \cdot \frac{CH_3}{CH_2}$	P.T. [°C]	$100 \cdot \frac{CH_3}{CH_2}$	P.T. [°C]	$100 \cdot \frac{CH_3}{CH_2}$	P.T. [°C]	$100 \cdot \frac{CH_3}{CH_2}$
50	1.80±0.22[a]	60	0.82	90	0.27	50	0.52
25	1.78±0.22	45	0.56	55-60	0.20	50	0.39
0	1.67±0.23	25	0.47	50	0.20	20	0.40
-20	1.30±0.27	0	0.35	20	0.15	-40	0.23
-30	0.85±0.33	-20	0.46	-15	0.05	-60	0.17
-60	0.65±0.36	-30	0.30	-75	0.0		
		-50	0.33				

[a] confidence limit at 95% probability.

All references use infrared measurement of CH_3 content after $LiAlH_4$ reduction, according to the method of Cotman (1).

long chain branching, [branch/molecule]	1	6
	8-16	7

Coefficient of Thermal Expansion, [K^{-1}]   122

<T_g before annealing	$6.6 - 7.3 \times 10^{-5}$	
after annealing	6.9×10^{-5}	
>T_g	17.0 - 17.5	
PVC-various plastizisers	see Ref.	123

Coefficient of Friction

PVC on steel   25

Plasticizer[xx] [%]	μ (static)	μ (dynamic)
25.9	0.350	0.719
31	0.495	0.787
35.5	0.645	0.857
39.4	0.797	0.925

[xx] Di-2-ethyl hexyl phthalate (DOP), di-iso-decyl phthalate (DIDP), n-octyl n-decyl phthalate (DNODP).

A value of 0.45-60 for unplasticized PVC with steel is reported in the literature   26

Compressibility, [MPa^{-1}] $\times 10^6$ 128

(uniaxially stretched PVC)

percent elongation	$\gamma_{\parallel}$	$\gamma_{\perp}$
85	4.88	8.0
125	4.25	8.45
160	3.57	8.75

Property	Value	Ref.
Creep	see Ref.	50, 144

Crystallinity, [%]   1. from density measurements: (Ref. 5)   2. from X-ray diffraction measurements

T_{polym}	Crystallinity
90	11.3[a]
55-60	11.3
50	13.2
20	15.0
-15	57.3
-75	84.2

T_{polym}	Crystallinity (Ref. 12) [b]	T_{polym}	Crystallinity (Ref. 13) [c]
90°C frac. 1	10.4	50	13[d]
frac. 3	5.4	-20	17
53 frac. 1	10.4[d]	-40	20.5, 20
A	6.5	-60	23, 25
B	8.2		
25	12.2	polym. in propionaldehyde	33[d]
-60	23.7[d]		
polym. in n-butyraldehyde	34.7[d]		

[a] Assumes amorphous density equals 1.385 [g cm^{-3}] and crystalline density equals 1.44 [g cm^{-3}]. (Natta and Corradini (Ref. 10)).

Since the crystalline density for highly crystalline (e.g., single crystal) PVC is considerably greater than 1.44 [g cm^{-3}], as shown in the table on crystallographic data, the above % crystallinity values for the low temperature polymerized PVCs are greatly over-estimated. As pointed out by Kostyuchenko and coworkers (11), if a crystalline density of 1.49 [g cm^{-3}] or greater is used, the calculated percent crystallinity values agree better with x-ray diffraction measurements.

[b] X-ray diffraction method of Hermans and Weidinger (14), double hump amorphous curve.

[c] X-ray diffraction method with Lorentz-polarization and atomic scattering factor corrections.

[d] Using a single hump amorphous x-ray curve and the method of Hermans and Weidinger (14), Lebedev and coworkers (15) (see also Ref. 12) report several commercial PVC crystallinities in the range 20-27%. In general, their method gives higher values than of D D' Amato and Strella (12).

3. from calorimetric measurements (unfractionated polymers)   16

M_n	$T_{polym.}$ [°C]	% Crystallinity
23,200	75	18.4
38,700	65	15.5
53,500	52	15.3
66,700	52	14.4
136,000	25	11.9
155,000	25	11.8

Crystallographic Data	see under Unit Cell

Density, [Mg m^{-3}] $\equiv$ [g cm^{-3}]   4

a. Function of Polymerization Temperature:

$T_{polym.}$ [°C]	M_n	Density [20°]
90	23,750	1.391
55-60	75,000	1.391
50	91,250	1.392
20	172,500	1.393
-15	106,300	1.416
-75	106,300	1.431

b. Dependence on Thermal History: 22

History	Density [20°C]
Original state	1.3743
Quenched from 200°C in ice water	1.3656
Quenched from 130°C to -70°C	1.3716
Quenched from 130°C to -70°C and kept at R.T. for 6.5 hours	1.3743

Property	Value	Ref.

b. Dependence on Thermal History (Cont' d.)

History	Density [20°C]
Annealed at 65°C for 136 hours	1.3745
Annealed at 90°C for 136 hours and slow cooled	1.3834
Annealed at 90°C for 136 hours and quenched to -20°C	1.3817

c. Effect of Temperature on Density:

21

T [°C]	Density
20	1.392
51	1.383
82	1.368
90.6	1.362
97.0	1.357

Dielectric Properties

24

Dielectric Constant [ϵ']

T[°C]	25	40	60	80	90	100	110	120	140
60 Hz	3.50	3.51	3.70	4.25	6.30	10.33	11.89	12.05	11.76
1 kHz	3.39	3.40	3.61	4.09	5.05	7.77	10.21	11.30	11.27
10 kHz	3.29	3.34	3.45	3.89	4.45	5.77	8.50	9.96	10.94

Dielectric Loss Factor [ϵ'']

T[°C]	25	40	60	80	90	100	110	120	140
60 Hz	0.110	0.116	0.125	0.172	0.410	1.20	0.675	0.481	1.65
1 kHz	0.081	0.081	0.080	0.120	0.500	1.415	1.645	0.630	0.319
10 kHz	0.058	0.058	0.050	0.110	0.920	1.37	1.35	1.22	0.490

$T_{polym.}$ = 50°C, $[\eta]$ = 1.17, M_n = 66,700, M_w = 162,000

146

Elongation at Break, [%]

$T_{polym.}$ [°C]	Test Temp. [°C]	$[\eta]^x$	Strain Rate [sec^{-1}]					
			.0026	.020	.20	2.0	20	200
70	25	68	171.5	18.5	13.5	16.5	17.0	13.3
65	25	75	194.5	20.0	14.2	21.0	19.5	16.0
56	25	91.3	196.7	21.0	18.0	21.0	19.5	16.0
50	25	116.9	210.7	23.0	16.5	17.0	18.5	17.0
70	60	68		71.6	20.0	20.0	16.5	24.0
65	60	75		160	23.8	22.5	22.5	27.5
56	60	91.3		207	26.0	25.0	22.0	30.0
50	60	116.9		243	27.0	26.5	22.5	35.0
70	80	68			168	82	60	41
70	100	68				240	168	157

x Intrinsic viscosity [ml/g] in cyclohexanone, 30°C.

Enthalpy, Entropy as function of pressure and temperature see Ref.

109

Property	Value	Ref.
Flory-Huggins Parameter	see Polymer-Solvent Interaction Parameter	
Flow Activation Energy	see Ref.	27, 28
Glass Transition Temperature	see corresponding table in this Handbook	
a. effect of pressure: d Tg/dP, [°C/atm]	0.013	110-113
b. effect of molecular weight and polymerization temperature		117

$T_{polym.}$ [°C]	intrinsic viscosity $[\eta]$ [a]				
	50	75	100	125	150
70	78.5	82	83	83.5 [b]	84.5 [b]
50	80	84	85	86	86.5
40	81	85	87	88.5	89
30	84	87	89	89	90
20	85	88	90	90.5	91
5	86	91	93	94.5	95.5
-15	92	87	100	101.5	102 [b]

[a] in cyclohexanone 30°C. [ml g^{-1}]
[b] Extrapolated from experimental data.

c. effect of polymerization temperature on T_g and T_m (numbers give Ref.)

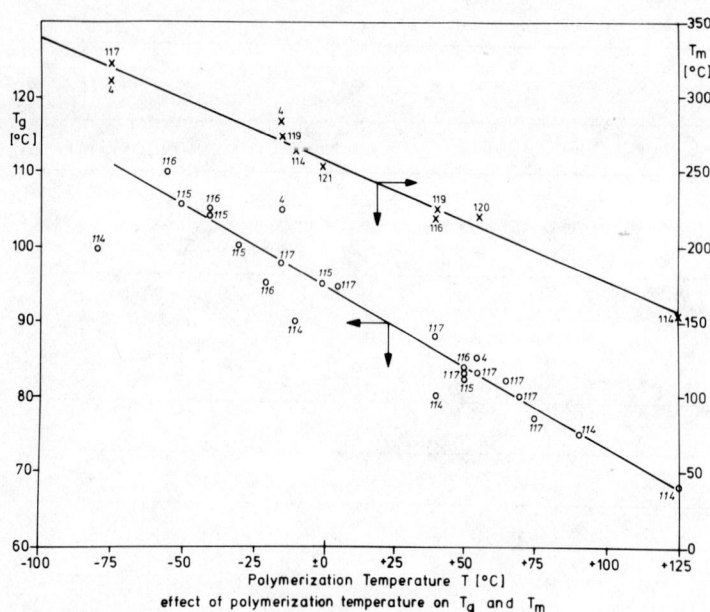

effect of polymerization temperature on T_g and T_m

The lines through the data points represent the best least squares fit of the collective authors' data.

Heat Capacity	see corresponding table in this Handbook and Ref.	124, 126
Heat of Combustion, [kJ kg^{-1}]	-19.900	138

Property	Value	Ref.
Heat of Fusion, [kJ mol^{-1}]	11.3	119
	12.65	134
	2.76	17
	3.28 (most probable value according to Ref. 137)	4
	3.56	135
	3.91	136
Heat of Dilution		
30°C PVC-tetrahydrofuran	see Ref.	16
PVC-cyclohexanone	see Ref.	16
Heat of Polymerization, [kJ mol^{-1}]	-96 to -109	139, 140, 141
Huggins Coefficient	see Ref.	52
Intrinsic Viscosity-Molecular Weight Relationship	see corresponding table in this Handbook and Ref.	53-79
Mark-Houwink Parameters for PVC as a Function of Temperature		80

Solvent	Temp. Range [^{o}C]	a	K [ml g^{-1}]
Cyclohexanone	20-60	0.803	$1.847 \times 10^{-2} - 4.85 \times 10^{-5}$ T
Cyclopentanone	20-60	0.861	$9.086 \times 10^{-5} - 1.55 \times 10^{-5}$ T
THF	20-50	0.851	$1.087 \times 10^{-2} - 1.67 \times 10^{-5}$ T

K values can be calculated at any temperature (T, ^{o}C) for temperatures within the ranges given.

Property	Value	Ref.
Mark-Houwink Parameters	see Intrinsic Viscosity-Molecular Weight Relationship	
Melt Viscosity	see Appendix 1 and Ref. 27, 28	
Permeability	see table "Permeability Coefficients, Diffusion Constants and Solubility Coefficients of Polymers" in this Handbook and Ref.	22, 35-43
Poisson Ratio	0.38	49
Polymer-Solvent-Interaction Parameter	see corresponding table in this Handbook and Ref.	44, 58, 81-93
Refractive Index		50

(PVC (ϱ = 1.384 g/cm^{3}))

Wave Length [nm]	Refractive Index
486.1	1.54806
589.3	1.54151
656.3	1.53843

reciprocal dispersive power V_d = 59.3
critical angle (λ = 589.3 nm) = 56.23
Temperature Coefficient of refractive index = 0.0001142 $^{o}/^{o}$C.

Property	Value	Ref.
Refractive Index Increment	see corresponding table in this Handbook	
Specific Heat	see table "Heat Capacity" in this Handbook and Ref.	124, 126
Specific Refractive Index Increment	see corresponding table in this Handbook	
Second Virial Coefficient	see corresponding table in this Handbook and Ref.	6, 33, 67, 74, 94
Specific Volume, as function of temperature and pressure	see Ref.	109
Spectral Data		
a. Infrared Absorption Bands	see Ref.	95

Property	Value	Ref.
b. Nuclear Magnetic Resonance (high resolution spectra)	see Ref.	96, 97, 98
c. Carbon-13 Magnetic Resonance (chemical shift assigment)	see Ref.	99, 100

Stress Optical Coefficient, [Brewster]

-200 - +50°C	-6.5	142, 143

The value of the S.O.C. becomes positive at the glass trasition but there is some arbitrariness associated with it as above Tg the plots of birefringence change with stress and have an S shape.

Tacticity

(α = fraction syndiotactic dyads)

a. Infrared Analysis

Ref. 101		Ref. 102				Ref. 103		Ref. 4	
$T_{polym.}$ [$^{\circ}$C]	α	$T_{polym.}$ [$^{\circ}$C]	α	α^{a}	α^{b}	$T_{polym.}$ [$^{\circ}$C]	α	$T_{polym.}$ [$^{\circ}$C]	α^{c}
50	0.56	47	0.57	0.67	0.56			90	0.51
25	0.55, 0.56	23	0.59	0.71	0.60	30	0.54	55-60	0.53
0	0.56	-10	0.63	0.76	0.63	0	0.59	50	0.54
-25	0.57	-30	0.66	0.77	0.66	-25	0.69	20	(0.52)
-35	0.58	-63	0.70	0.80	0.73	-35	0.72	-15	0.64
-40	0.59					-70	0.80	-75	0.77
-50	0.59								

[a, c] Method of Germar et al (103).
[b] Modified method of Germar et. al (103).

b. Nuclear Magnetic Resonance Analysis

Ref. 104		Ref. 105		Ref. 106	
$T_{polym.}$ [$^{\circ}$C]	α	$T_{polym.}$ [$^{\circ}$C]	α^{a}	$T_{polym.}$ [$^{\circ}$C]	α
100	0.54	120	0.51	40	0.57
50	0.55	80	0.53	0	0.64
0	0.57	60	0.51	-40	0.65
-78	0.63	5	0.56		
		-20	0.59		
		-50	0.63		
		-78	0.67		

[a] Method of Johnsen (107).

The NMR experiment measures only the soluble portion of the polymer. Highly crystalline, highly syndiotactic PVC's show limited solubility and thus the reported % syndiotacticity values represent lower limits of this parameter (108).

Tensile Modulus, [MPa] $\equiv$ [N mm^{-2}]

-196°C	7584	145[x]
-120°C	5171	145
-75°C	3861	145
20°C	2964	145
30°C	3000	50[xx]
40°C	2930	50
50°C	2427	50
60°C	1551	50
70°C	276	50

[x] Stress-strain measurements at strain rate of 0.00250 sec^{-1}.
[xx] Measured in creep (100 sec., 0.2% strain).

Property	Value	Ref.

Thermal Conductivity λ, [W m^{-1} K^{-1}]

23, 127

T[^{o}C]	λ	T[^{o}C]	λ	T[^{o}C]	λ
-170	0.129	-25	0.155	60	0.164
-150	0.134	0	0.158	70	0.164
-125	0.139	20	0.160	80	0.165
-100	0.144	30	0.161	90	0.165
-75	0.148	40	0.162	100	0.165
-50	0.152	50	0.164		

Thermal conductivity of system PVC/di-2-ethylhexyl phthalate see Ref.

127

Molecular weight, sample polymerization temperature and syndiotactity do not influence the thermal conductivity of polymers appreciably, except in the case where tacticity leads to crystallization.

Unit Cell

PVC Sample Type	Crystal System	Space Group	Unit Cell Parameters [Å]			Monomers per Unit Cell	Calc. Density [gcm^{-3}]	Ref.
			a	b	c			
Commercial, polymerized at 50-60°C	Orthorhombic	Pacm	10.6	5.40	5.10	2	1.44	10
Solution blended high molecular weight, low crystallinity commercial polymer and low molecular weight, high crystallinity polymer	Orthorhombic	Pacm	10.4	5.30	5.10	2	1.48	18
Single crystals, polymerized at -75°C	Orthorhombic		10.32^{a}	5.32^{a}	-		(1.49)^{b}	19
Single crystals, low molecular weight, polymer made in n-butyraldehyde	Orthorhombic	Pacm	10.24	5.24	5.08	2	1.53	20

^{a}Calculated from published d-spacings of major diffraction peaks.
^{b}Calculated density assuming c = 5.10 [Å] and 2 monomers per unit cell.

Unperturbed Dimensions

see corresponding table in this Handbook

Viscosity-Molecular Weight Relationships

see intrinsic viscosity

Appendix I: MELT VISCOSITY OF UNMODIFIED PVC
PREPARED AT VARIOUS POLYMERIZATION TEMPERATURES

Melt viscosity data obtained using a capillary having a 90° entrance angle, length 1.0 in. and diameter .05 in. No corrections applied to the data. All samples contained 2.5 parts dibutyltin dioctylthioglycolate stabilizer per 100 parts resin.

Apparent Viscosity - [Pa · s] x 10^{-4}

T polym. [^{o}C]	L.V.N.★ [ml/g]	Melt Temp. [^{o}C]	Shear Rate [s^{-1}]									
			2.95	7.37	14.7	29.5	73.7	147	295	737	1470	2950
110	30.5	190	--	--	--	1.18	0.72	0.82	0.77	0.61	0.52	0.40
110	30.5	205	--	--	--	0.63	0.39	0.48	0.30	0.22	0.21	0.20
70★★	40.1	140	1128.3	573.1	340.7	202	97.85	54.6	30.0	13.3	6.41	3.08
70	40.1	150	639.83	332.2	200.6	120.8	60.5	35.6	20.44	9.45	4.91	2.47
70	40.1	160	281.4	153.9	97.1	62.4	34.4	21.8	13.41	6.58	3.74	20.4
70	40.1	170	100.9	62.08	42.2	28.7	16.9	11.35	7.65	4.37	2.72	1.62
70	40.1	180	32.12	21.2	15.5	11.42	7.85	5.78	4.38	2.80	1.94	1.27
70	40.1	190	11.95	7.22	5.94	5.15	4.03	3.24	2.65	1.87	1.35	0.95
70	40.1	200	9.03	3.82	2.02	1.86	2.02	1.92	1.4	1.21	.95	.68
70	40.1	210	--	5.31	2.71	1.38	0.89	1.02	0.935	0.775	0.65	0.504
70★★★	49.5	140	2336.4	1209.8	721.6	423.5	191.6	102.1	51.11	20.2	9.29	4.24
70	49.5	150	1433.6	711.03	424.5	223.0	119.92	64.7	34.5	15.2	7.43	3.50
70	49.5	160	839.9	431.9	258.9	155.3	76.6	43.8	24.2	10.72	5.47	2.71

PHYSICAL CONSTANTS OF IMPORTANT POLYMERS

$T_{polym.}$ [°C]	L.V.N.★ [ml/g]	Melt Temp. [°C]	Shear Rate [s^{-1}]									
			2.95	7.37	14.7	29.5	73.7	147	295	737	1470	2950
70	49.5	170	398.4	231.5	142.9	89.6	47.8	29.1	17.2	8.14	4.43	2.32
70	49.5	180	146.6	91.9	58.7	43.2	26.1	17.3	11.2	5.92	3.49	1.98
70	49.5	190	55.34	34.3	25.5	19.4	13.4	9.75	6.58	3.99	2.56	1.52
70	49.5	200	22.1	13.7	11.3	9.5	7.36	5.87	4.48	2.95	2.02	1.33
70	49.5	210	12.7	7.53	5.54	4.54	3.99	3.43	2.82	2.06	1.5	1.04
70	54.8	210	16.23	8.66	8.12	7.31	6.71	5.74	4.71	2.94	20.3	1.31
70	54.8	220	--	--	--	5.26	4.29	3.82	3.24	2.39	1.72	1.18
70	68	160	2464.2	1172.2	705.5	389.9	173.66	89.55	45.76	17.69	8.47	4.01
70	68	170	1196.9	599.14	359.27	218.53	113.21	64.04	34.39	14.44	7.16	3.49
70	68	180	562.8	320.5	205.9	125.6	64.7	37.7	21.8	10.3	5.52	2.92
70	68	190	244.2	144.9	100.8	69.8	40.1	25.5	15.5	7.75	4.40	2.44
70	68	200	94.2	62.1	46.6	35.0	23.7	16.6	11.0	5.94	3.55	2.07
70	68	210	48.1	29.5	23.5	19.6	14.6	11.1	8.02	4.68	2.92	1.73
70	68	220	19.4	15.5	13.2	11.7	9.34	7.51	5.71	3.62	2.51	1.57
65	75.4	180	976.7	522.7	318.4	195.6	103.5	58.7	31.26	13.3	6.87	3.46
65	75.4	190	545.1	254.7	158.9	101	53.6	31.9	19.0	9.25	5.01	2.63
65	75.4	200	271.1	128.5	85.3	59.8	36.3	23.1	14.2	7.2	4.15	2.32
65	75.4	210	95.6	56.7	43.7	33.1	22.1	15.7	10.5	5.53	3.36	2.02
65	75.4	220	52.6	33.2	25.9	21.4	14.9	10.96	7.75	4.49	2.73	1.64
65	86	180	930	448.5	273.3	198	84	45.1	23.22	9.77	4.92	2.46
65	86	190	686.2	381	243.6	148	82	48.1	25.5	11.1	5.7	2.93
65	86	200	229.5	138.7	100.6	67.75	38.75	34.02	13.67	6.77	3.77	2.01
65	86	210	163.9	94.96	71.05	53.3	33.1	21.4	13.5	7.23	4.37	2.47
65	86	220	--	--	--	26.25	17.13	11.8	7.88	4.22	2.56	1.56
56	91.3	180	1898.2	902.1	573.1	292.0	124.2	65.3	35.7	14.86	7.43	3.66
56	91.3	190	1139.9	584.7	373.8	250.4	113.5	60.1	31.5	13.5	6.87	3.53
56	91.3	200	857.7	350	207.1	130	74	43.3	23.4	10.2	5.45	2.85
56	91.3	210	286.9	155.0	112.0	79.2	45.7	28.5	18.7	9.08	4.81	2.54
56	91.3	220	154.9	90.8	55.9	49.5	31.3	18.5	13.1	7.21	4.27	2.37
56	91.3	230	92.7	60.5	48.7	37.7	24.6	17.0	11.1	6.0	3.98	2.48
56	91.3	240	82.3	11.9	54.7	44.6	31.0	24.1	20.1	10.0	6.367	3.611
50	116.9	180	1884	933.9	562.5	304	126.3	67.12	36.11	14.86	7.32	3.39
50	116.9	185	1452.2	730.1	464.3	284.1	126.8	66.1	34.25	14.22	7.0	3.4
50	116.9	190	1002	633.0	398	231.7	116.2	59.96	30.8	13.16	6.58	3.24
50	116.9	200	783.2	410.7	257.4	177.9	87.02	45.1	24.03	10.61	5.57	2.92
50	116.9	205	584.1	318.4	209.6	131.4	78.9	43.5	22.83	9.87	5.25	2.79
50	116.9	210	424.8	226.0	156.0	105.1	59.43	39.3	21.64	9.29	4.93	2.60
50	116.9	220	265.5	147.5	106.12	75.4	44.4	28.02	18.8	8.49	4.62	2.51
50	116.9	230	--	132.9	79.7	58.1	36.0	23.2	14.7	8.34	4.67	2.57
42	147.2	190	1878.6	897.0	537.1	276.7	113.5	58.4	31.8	13.3	6.36	3.04
42	147.2	200	1626.9	703.2	420.8	252.1	108.5	55.9	28.2	12.3	6.15	3.07
42	147.2	210	1352.7	567.6	319.7	206.3	93.3	48.9	25.3	10.7	5.52	2.77
42	147.2	220	968.4	415.3	231.5	142.5	73.5	39.9	21.2	9.19	4.93	2.85
42	147.2	230	--	299	164.5	106.2	60.2	35.2	19.0	8.44	4.55	2.48
42	147.2	240	--	243.6	199.3	157.7	89.7	53.2	33.5	--	--	--
40	122.5	190	2041.3	867.8	511.0	285.6	117.9	58.5	29.2	12.7	6.45	3.14
40	122.5	200	1162	--	323	--	80.8	41.5	21.0	9.09	--	--
40	122.5	205	445.4	244.5	170.5	111.5	64.4	36.6	19.4	8.35	4.37	2.25
40	122.5	210	--	--	--	98.4	53.5	28.7	14.76	6.28	3.28	1.72
40	122.5	215	338.8	185.6	133.4	94.0	53	35.7	20.2	8.73	4.56	2.39
28	208	190	2050.0	916.5	543.5	284.9	117.8	63.7	33.55	14.92	7.25	3.35
28	208	200	1658.7	740.6	452.3	239.6	102.8	52.2	27.4	11.94	6.07	2.87
28	208	210	1104.9	493.4	282.4	167.7	85.5	46.9	25.03	10.44	5.33	2.66
28	208	220	665.6	332.5	214.7	138.4	69.3	39.5	22.1	9.96	5.33	2.73
25	240.5	200	1513.3	647.4	387.4	223.0	94.98	49.08	26.55	11.67	5.84	2.81
25	240.5	210	1503	603	343	200	90.2	46.6	23.5	10.4	5.4	2.7
25	240.5	215	1340.7	562.5	311.5	180	80.44	41.76	21.4	9.49	5.2	2.6
25	240.5	220	1355.7	572.6	332.2	190.9	83.3	43.9	22.5	9.63	5.07	2.57
25	240.5	225	958.4	421.3	250.9	159.3	75.35	40.33	21.03	9.23	4.92	2.60
25	240.5	230	--	409.8	222.0	136.1	67.7	37.5	19.7	8.86	4.62	2.46
25	240.5	240	--	426.4	223.2	132.3	64.2	36.5	20.2	8.53	4.52	2.53

★ Limiting Viscosity Number, Determined in Cyclohexanone 30°C. ★★ Modified with 5 parts Trichloroethylene/100 parts Vinyl Chloride Monomer.

★★★ Modified with 3 parts Trichloroethylene/100 parts Vinyl Chloride Monomer.

REFERENCES

1. J. D. Cotman, Ann. N.Y. Acad. Sci. 57, 417 (1953).
2. G. Boccato, A. Rigo, G. Talamine, F. Zilio-Grandi, Makromol. Chem. 108, 218 (1967).
3. M. Carrega, C. Bonnebat, G. Zednik, Anal. Chem. 42 (14), 1807 (1970).
4. A. Nakajima, H. Hamada, S. Hayashi, Makromol. Chem. 95, 40 (1966).
5. W. Trautvetter, Makromol. Chem. 101, 214 (1967).
6. A. J. de Vries, C. Bonnebat, M. Carrega, Pure Appl. Chem. 25, 209 (1971).
7. S. Krozer, Z. Czlonkowska, Polimery 7, 410 (1962).
8. B. Zimm, W. Stockmayer, J. Chem. Phys. 17, 1301 (1946).
9. B. Zimm, R. W. Kilb, J. Polymer Sci. 37, 19 (1959).
10. G. Natta, P. Corradini, J. Polymer Sci. 20, 251 (1956).
11. A. N. Kostyuchenko, N. A. Okladnov, V. P. Lebedev, J. Polymer Sci. USSR 10, (11), 3025 (1968).
12. R. J. D' Amato, S. Strella, Appl. Polymer Symp. 8, 275 (1969).
13. C. Garbuglio, A. Rodella, G. C. Borsini, E. Gallinella, Chim. Ind. (Milan) 46 (2), 166 (1964).
14. P. H. Hermans, A. Weidinger, Makromol. Chem. 44/46, 24 (1961).
15. V. P. Lebedev, N. A. Okladnov, K. S. Minsker, B. P. Shtarkman, Vysokomolekul. Soedin 7 (4), 655 (1965).
16. S. H. Maron, F. E. Filisko, J. Macromol. Sci. B6, 2, 413 (1972).
17. C. E. Anagnastopoulos, A. Y. Coran, H. R. Gamrath, J. Polymer Sci. 4, 181 (1960).
18. G. Natta, I. W. Bassi, P. Corradini, Rend. Accad. Naz. Lincei 31 (1.2), 1 (1961).
19. A. Nakajima, S. Hayashi, Kolloid-Z. - Z. Polymere 229 (1), 12 (1969).
20. C. E. Wilkes, V. L. Folt, S. Krimm, Macromolecules 6 (2), 235 (1973).
21. K. H. Hellwege, W. Knappe, P. Lehmann, Kolloid-Z. 183 (2), 110 (1962).
22. K. H. Illers, Makromol. Chem. 127, 1 (1969).
23. R. P. Sheldon and Sister K. Lane, Polymer 6, 77 (1965).
24. H. J. Dietrick, M. L. Dannis, B. F. Goodrich Research Center, Brecksville, Ohio, unpublished data.
25. J. B. DeCoste, SPE Antec 15, 232 (1969).
26. A. B. Glanvill, "The Plastics Engineer's Data Book", The Machinery Publishing Co., Ltd. Brighton, Sussex, pp. 181 (1971).
27. E. A. Collins, C. A. Krier, Trans Soc. Rheol. 11 (2), 225 (1967).
28. E. A. Collins, A. P. Metzger, Polymer Eng. Sci. 10 (2), 57 (1970).
29. A. Nakajima, K. Kato, Macromol. Chem. 95, 52 (1966).
30. T. Kobayashi, Bull. Chem. Soc. (Japan) 35 (5), 726 (1962).
31. R. K. S. Chan, Polymer Eng. Sci. 11 (3), 187 (1971).
32. M. D. Baijal, K. M. Kauppila, Polymer Eng. Sci. 11 (3), 182 (1971).
33. C. A. Daniels, E. A. Collins, Polymer Preprints 14 (2), (1973).
34. M. Sato, Y. Koshiishi, M. Asahina, Polymer Letters 1, 233 (1963).
35. A. Lebovis, Mod. Plastics 43, 139 (1966).
36. B. P. Takhomirov, H. B. Hopfenberg, V. Stannett, J. L. Williams, Macromol. Chem. 118, 177 (1968).
37. N. H. Reinkiny, A. E. Barnabeo, W. F. Hale, J. Appl. Polymer Sci. 7, 2135 (1963).
38. W. S. Penn "PVC Technology," Wiley-Interscience (New York, 1971).
39. A. W. Myers, V. Tammela, V. Stannett, M. Szwarc, Mod. Plastics 37 (10), 139 (1960).
40. P. M. Doty, W. H. Aiken, H. Mark, Ind. Eng. Chem. Intern. Edition 38, 788 (1946).
41. L. C. Grotz, J. Appl. Polymer Sci. 9, 207 (1965).
42. R. M. Barrer, R. Mallinder, P. S. L. Wong, Polymer 8 (6), 321 (1967).
43. K. D. Laudenslogel, S. F. Roe, Mod. Parkaging 40, 139 (1967).
44. J. L. Williams, H. B. Hopfenberg, V. Stannett, J. Macromol. Sci. B3 (4), 711 (1969).
45. C. E. Rogers, "Engineering Design for Plastics," E. Bear, ed. Reinhold Publishing Corp., New York, 1964, Chapt. 9.
46. R. W. Roberts, K. Kammermeyer, J. Appl. Polymer Sci. 7, 2183 (1963).
47. Y. Ito, Kobunshi Kagaku 18, 124 (1961).
48. H. Braunisch, H. Lenhart, Kolloid-Z. 177, 24 (1961).
49. R. Raghova, R. M. Caddell, G. S. Y. Yeh, J. Material Sci. 8, 225 (1973).
50. R. M. Ogorkiewicz, "Engineering Properties of Thermoplastics," p. 251, John Wiley and Sons, New York, 1970.
51. M. B. Huglin, ed. "Light Scattering from Polymer Solutions," Academic Press, New York (1972).
52. L. A. Utracki, Polymer J. (Japan) 3 (5) 551 (1972).
53. M. Fournier, Compt. Rend. 222, 1437 (1946).
54. H. Staudinger, J. Schneiders, Ann. 541, 151 (1939).
55. J. Breitenbach, E. Forster, A. Renner, Kolloid-Z. 127, 1 (1952).
56. T. Krasovec, Reports J. Stefan Inst. 3, 203 (1956).
57. Z. Mencik, Chem. Listy 49, 1598 (1955).
58. W. R. Moore, R. J. Hutchinson, Nature 200, 1095 (1963).
59. G. Bier, H. Kramer, Makromol. Chem. 18, 151 (1956).
60. J. Hengstenberg, Angew. Chem. 62, 26 (1950).
61. F. Danusso, G. Moraglio, S. Gazzera, Chem. Ind. (Milan) 36, 883 (1954).
62. A. Oth, Ind. Chim. Belge 20, 423 (1955).
63. D. Harmon, B. F. Goodrich Research Center, Unpublished results.
64. G. M. Guzman, J. M. Fatou, Anales Real. Soc. Espan Fiz. Quim (Madrid) 54 B, 609 (1958).
65. G. Ciapa, H. Schwindt, Makromol. Chem. 21, 169 (1956).
66. G. Pezzin, N. Gligo, J. Appl. Polymer Sci. 10, 1 (1966).
67. M. Bohdanecky, K. Solc, P. Kratochvil, M. Kolinsky, M. Ryska, D. Lim. J. Polymer Sci. A-2, 5, 343 (1967).
68. M. Kolinsky, M. Fyska, M. Bohdanecky, P. Kratochvil, A. Soc, D. Lim, J. Polymer Sci. C 16, 485 (1967).
69. R. M. Fuoss, D. J. Mead, J. Phys. Chem. 47, 59 (1943).
70. J. Vaura, J. Lapcek, J. Sabados, J. Polymer Sci. A-2, 5 (6), 1305 (1967).
71. R. Endo, Kobunshi Kagaku 18, 143 (1961).
72. H. Batzer, A. Nisch, Makromol. Chem. 22, 131 (1957).
73. J. Lyngaae-Jorgensen, J. Polymer Sci. C 33, 23 (1971).
74. T. Kobayashi, Bull. Chem. Soc. (Japan) 35, 726 (1962).
75. A. Takahashi, Kogyo Kagaku Zasshi 66, 960 (1963).
76. M. Freeman, P. Manning, J. Polymer Sci. A-2, 2017 (1964).
77. G. Pezzin, G. Talamini, N. Gligo, Chim. Ind. (Milan) 46, 648 (1964).
78. C. H. Lu, Theses, Virginia Polytech. Inst., (1962).
79. A. Nakazawa, P. Matsuo, H. Inagaki, Bull. Inst. Chem. Res., Koyoto Univ. 44 (4), 354 (1966).
80. S. H. Maron, M. S. Lee, J. Macromol. Sci. B7 (1), 29 (1973); ibid B7 (1), 47 (1973); ibid B7 (1), 61 (1973).
81. A. Nakajima, Kobunshi Kagaku 7, 309 (1950).
82. F. Mencik, Collection Czech. Chem. Commun. 24, 3291 (1959).
83. R. Gautron, C. Wippler, J. Chim. Phys. 58, 754 (1961).
84. E. Schroder, Plaste, Kautschuk 15, 2 (1968).
85. E. J. Mouton, Thesis, Univ. Leiden (1961).
86. J. Breitenbach, E. Forster, A. Renner, Kolloid-Z. 127, 1 (1952).
87. P. Doty, E. Mishuck, J. Am. Chem. Soc. 69, 1631 (1947).

88. P. Doty, H. Zable, J. Polymer Sci. 1, 90 (1946).

89. W. R. Moore, R. J. Hutchinson, J. Appl. Polymer Sci. 8, 2619 (1964).

90. R. L. Adelman, J. M. Klein, J. Polymer Sci. 31, 77 (1958).

91. A. Oth, Compt. Tend., 27 Congres Intern. Chem. Ind., Bruxelles (Sept. 1954).

92. J. B. Haehn, B. F. Goodrich Chemical Company, Technical Center, Avon Lake, Ohio, Unpublished results.

93. P. Kratochvil, Collection Czech. Chem. Commun. 29, 2767 (1964).

94. A. Rudin, J. Beuschop, J. Appl. Polymer Sci. 45, 2881 (1971).

95. S. Krimm, V. L. Folt, J. J. Shipman, A. R. Berens, J. Polymer Sci. A-1, 2621 (1963).

96. F. Heatley, F. A. Bovey, Macromolecules 2 (3), 241 (1969).

97. T. Yoshino, J. Komiyama, J. Polymer Sci. B3, 311 (1965).

98. U. Johnsen, K. Kolbe, Kolloid-Z. 221, 64 (1967).

99. C. J. Carman, A. R. Tarpley, Jr., J. H. Goldstein, J. Am. Chem. Soc. 93, 2864 (1971).

100. C. J. Carman, Macromolecules 6 (5), 725 (1973).

101. J. Stokr, B. Schneider, M. Kolinsky, M. Ryska, D. Lim. J. Polymer Sci. A-1, 5, 2013 (1967).

102. H. U. Pohl, D. O. Hummel, Makromol. Chem. 113, 203 (1968).

103. V. H. Germar, K. H. Hellwege, U. Johnsen, Makromol. Chem. 60, 106 (1963).

104. F. A. Bovey, F. P. Hood, E. W. Anderson, R. L. Kornegay, J. Phys. Chem. 71 (2), 312 (1967).

105. G. Talamini, G. Vidotto, Makromol. Chem. 100, 48 (1967).

106. L. Cavalli, G. C. Borsini, G. Carraro, G. Confalonieri, J. Polymer Sci. A-1, 8, 801 (1970).

107. U. Johnsen, J. Polymer Sci. 54, S 6 (1961).

108. C. E. Wilkes, Macromolecules 4, 443 (1971).

109. R. G. Griskey, C. A. Gellner, M. W. Din. Mod. Plastics 43, 119 (1966).

110. L. A. Chandler, E. A. Collins, B. F. Goodrich Chemical Company, Technical Center, Avon Lake, Ohio, Unpublished data.

111. P. Heydemann, H. D. Guicking, Kolloid-Z. 193, 16 (1963).

112. H. Bianchi, A. Turturro, G. Basile, J. Phys. Chem. 71, 3555 (1967).

113. K. H. Hellwege, W. Knappe, P. Lehmann, Kolloid-Z. 183 (2), 110 (1962).

114. F. P. Reding, E. R. Walter, F. J. Welch, J. Polymer Sci. 56, 225 (1962).

115. G. Pezzin, Plastics and Polymers 37 (130), 295 (1969).

116. G. A. Garbuglio, G. C. Borsivi, E. Gallinella, Chem. Ind. (Milan) 46, 166 (1964).

117. E. A. Collins, L. A. Chandler, P. R. Kumler, B. F. Goodrich Chemical Company, Technical Center, Avon Lake, Ohio, Unpublished data (DTA 1°C./min. T_g taken at inflection point, polymers unmodified).

118. J. E. Clark, Polymer Eng. Sci. 7, 137 (1967).

119. D. Kockott, Kolloid-Z., Z. Polymere 198, 17 (1964).

120. D. E. Witenhafer, J. Macromol. Sci. B4 (4), 915 (1970).

121. A. Michel, A. Guyot, J. Polymer Sci. C 33, 75 (1971).

122. M. L. Dannis, Mod. Plastics 31 (7), 120 (1954).

123. L. Dunlap, J. Polymer Sci. C 30, 561 (1970).

124. R. Hoffman, W. Knappe, Kolloid-Z. - Z. Polymere 240, 784 (1970).

125. K. H. Hellwege, J. Hennig, W. Knappe, Kolloid-Z. 188, 121 (1963).

126. B. Wunderlich, W. Baur, Adv. Polymer Sci. 7, 151 (1970).

127. K. Eiermann, Kunststoffe 51, 512 (1961).

128. J. Hennig, Kolloid-Z. 202 (2), 127 (1965).

129. J. Hennig, Kunststoffe 57 (5), 385 (1967).

130. M. Hattori, Kolloid-Z.-Z. Polymere 202 (1), (1965).

131. R. H. Shoulberg, J. Appl. Polymer Sci. 7, 1597 (1963).

132. R. Berlot, Plast. Mod. Elastomeres 18 (4), 231, 233, 235 (1966).

133. R. C. Steere, J. Appl. Polymer Sci. 10, 1673 (1966).

134. L. Nielson, "Mechanical Properties of Polymers," Reinhold Publishing Co., New York, 1962.

135. J. A. Juijn, Thesis, Technische Hochschule, Delft (1972).

136. R. S. Colborne, J. Applied Polymer Sci. 14, 127 (1970).

137. H. Wilski, Kolloid-Z.-Z. Polymere 238, 426 (1970).

138. E. A. Collins, Unpublished data.

139. G. C. Sinke & D. R. Stull, J. Phys. Chem. 62, 397 (1958).

140. R. M. Joshi, Indian J. Chem. 2, 125 (1964).

141. R. M. Joshi & B. J. Zwolinski, J. Polymer Sci. B 3, 779 (1965).

142. R. D. Andrews, V. Chatre, J. Appl. Phys. 40, 4266 (1969).

143. R. D. Andrews, Y. Kazama, J. Appl. Phys. 39, 4891 (1968).

144. G. Turner, Brit. Plastics 37, 682 Dec. (1964).

145. J. Dyment, H. Ziebland, J. Appl. Chem. 8, 203 (1958).

146. H. H. Bowerman, Proceedings of Fourth Annual Plastics Conferencs, Eastern Michigan University, Sept. 24, 1969.

       H. H. Bowerman, Unpublished data.

PHYSICAL CONSTANTS OF POLY(VINYL ACETATE)

Martin K. Lindemann
C. S. Tanner Co.
Greenville, S. C.

Property	Value	Remarks	Ref.
Absorption of Water, [%]	approx. 3 - 6	at 20°C for 24 to 144 hours	26
Coefficient of Thermal Expansion, [K^{-1}]			
cubic	6.7×10^{-4}		15, 25
linear, below T_g	7×10^{-5}		17, 27
above T_g	22×10^{-5}		
Cohesive Energy Density, [$MJ\ m^{-3}]^{1/2}$	18.6 - 19.09		2, 3, 4, 5
Compressibility, [cm^3/g atm]	18×10^{-6}	in the glassy state	1
Decomposition Temperature, [$^{\circ}$C]	150		15
Density, [$Mg\ m^{-3}] = [g\ cm^{-3}$]			
at 20°C	1.191		16, 24, 26
at 25°C	1.19		
at 50°C	1.17		
at 120°C	1.11		
at 200°C	1.05		
Dielectric Constant	3.5	at 50°C and 2×10^3 kHz	34
	8.3	at 150°C and 2×10^3 kHz	
Dielectric Dissipation Factor, [tan δ]	1.5×10^2	at 50°C and 2×10^3 kHz	34
	26×10^2	at 120°C and 2×10^3 kHz	
Dielectric Strength, [$V\ cm^{-1}$] $\times 10^{-3}$	394	at 30°C	16
	307	at 60°C	
Dipole Moment, [eSU (per monomer unit)]	at 20°C 2.3×10^{-18}		22, 23
	at 150°C 1.77×10^{-18}		
Dynamic Mechanical Loss Peak, [$^{\circ}$C]	70	at 100 Hz	33
Elongation at Break, [%]	10 - 20	at 20°C, 0% RH	27
Glass Transition Temperature, [$^{\circ}$C]	28 - 31		25
pressure dependence, [$K\ bar^{-1}$]	0.022		18
Hardness, [Shore units]	L 80 - 85	at 20°C	27
Heat Capacity, [$kJ\ kg^{-1}$]	1.465	at 30°C	21
Heat Distortion Point, [$^{\circ}$C]	50		16
Heat of Polymerization, [$kJ\ mol^{-1}$]	87.5		31, 32
Index of Refraction, [n_D]			28
at 20.7°C	1.4669		
at 30.8°C	1.4657		
at 52.1°C	1.4600		
at 80°C	1.4480		
at 142°C	1.4317		
Infrared Spectrum	see Ref.		39
Interfacial Tension, [$mN\ m^{-1}$]	14.5	at 20°C, with poly(ethylene)	9
	8.4	at 20°C, with poly(dimethylsiloxane)	11
	9.9	at 20°C, with poly(isobutene)	9
	4.2	at 20°C, with poly(styrene)	11
Internal Pressure, [$J\ m^{-3}$]			
at 0°C	2.554×10^8		6, 29, 30

Property	Value	Remarks	Ref.
Internal Pressure, $[J\ m^{-3}]$ (Cont'd.)			
at $28^{\circ}C$	3.978×10^{8}		6, 29, 30
at $60^{\circ}C$	4.187×10^{8}		
at $20^{\circ}C$	2.847×10^{8}		
at $40^{\circ}C$	4.313×10^{8}		
Intrinsic Viscosity-Molecular Weight Relationship			
	see corresponding Table in this Handbook and Ref.		40-51
Modulus of Elasticity, $[MPa] \equiv [N\ mm^{-2}]$	$1274.9 - 2255.6$	ASTM-D-256	27
Notched Impact Strength, $[cm\ kg/cm^{2}]$	> 100		27
Nuclear Magnetic Resonance Spectra	see Ref.		36
Raman Spectrum	see Ref.		37
Softening Temperature, $[^{\circ}C]$	$35 - 50$		7
Solubility Parameter	see Cohesive Energy Density		
Solution Properties	see Ref.		35
Specific Heat	see Heat Capacity		
Specific Volume, $[1\ kg^{-1}]$	$0.823 + 6.4 \times 10^{-4} \times t$	$t = 100\text{-}200^{\circ}C$	9, 13
	0.84	at $T_g = 28^{\circ}C$	19
Surface Resistance, $[\Omega\ cm^{-1}]$	5×10^{-11}		15
Surface Tension, $[mN\ m^{-1}]$	at $20^{\circ}C$ 36.5		9, 10, 11
	at $140^{\circ}C$ 28.6		
	at $180^{\circ}C$ 25.9		
Tensile Strength, $[MPa] \equiv [N\ mm^{-2}]$	$29.4 - 49.0$	at $20^{\circ}C$	16
Thermal Conductivity, $[W\ m^{-1}\ K^{-1}]$	0.159		16
Ultraviolet Spectra	see Ref.		38
Young's Modulus, $[MPa] \equiv [N\ mm^{-2}]$	600	at $25^{\circ}C$ and 50% RH	8

REFERENCES

1. M. M. Martynyuk, V. K. Semenchenko, Kolloid Zh. 26, 83 (1964).
2. H. Daoust, M. Rinfret, J. Colloid Sci. 7, 11 (1952).
3. D. Managaraj, S. Patra, P. C. Roy, S. K. Bhatnagar, Makromol. Chem. 84, 225 (1965).
4. P. A. Small, J. Appl. Chem. 3, 71 (1953).
5. G. Allen, Polymer 2, 375 (1961).
6. D. R. Paul, A. T. Dibenedetto, J. Polymer Sci. C 16, 1269 (1967).
7. K. Schmieder, K. Wolf, Kolloid Z. 134, 149 (1953).
8. J. Galperin, W. Arnheim, J. Appl. Polymer Sci. 11, 1259 (1967).
9. S. Wu, J. Colloid Interface Sci. 31, 153 (1969).
10. S. Wu, J. Phys. Chem. 74, 632 (1970).
11. S. Wu, J. Polymer Sci. C 34, 19 (1971).
12. S. Wu, J. Phys. Chem. 72, 3332 (1968).
13. J. R. Collier, Ind. Eng. Chem. 61, (9), 50 (1969).
14. R. F. Boyer, J. Macromol. Sci. Phys. B 7, (3), 487 (1973).
15. Mowilith, Polyvinylacetat, Farbwerke Hoechst AG, Frankfurt, 1969, pp. 214-215.
16. T. P. G. Shaw, Encyclopedia of Chemical Technology, Interscience Publishers, New York, 1955, Vol. 14, p. 692.
17. R. F. Clash, L. M. Rynkiewicz, Ind. Eng. Chem. 36, 279 (1944).
18. J. M. O'Reilly, J. Polymer Sci. 57, 429 (1962).
19. A. A. Miller, Macromolecules 4, 757 (1971).
20. G. Adam, Kolloid Z. 195, 1 (1964).
21. H. Wilski, Kolloid Z. 210, 37 (1966).
22. D. J. Meed, R. M. Fuoss, J. Am. Chem. Soc. 63, 2839 (1941).
23. O. Broens, F. H. Mueller, Kolloid Z. 141, 20 (1955).
24. M. K. Lindemann, Encyclopedia of Polymer Science and Technology, Interscience Publishers, New York, 1971, Vol. 15, p. 618.
25. S. Saito, Kolloid Z.-Z. Polymere 189, 116 (1963).
26. C. E. Schildknecht, "Vinyl and Related Polymers", Wiley, New York, 1952, p. 336.
27. H. R. Hacobi, Ullmans Encyklopaedie der Technischen Chemie 3rd Ed., Urban & Schwarzenberg, Munich, 1960, Vol. 11, pp. 34-35.
28. O. Broens, F. H. Mueller, Kolloid Z. 140, 121 (1955).
29. S. Fujimoto, Mem. Defense Acad. Math. Phys. Chem. Eng. (Yokosuka, Japan) 7, 1247 (1967).
30. G. Allen, D. Sims, G. J. Wilson, Polymer 2, 375 (1961).
31. R. M. Joshi, Makromol. Chem. 66, 114 (1963).
32. R. M. Joshi, J. Polymer Sci. 56, 313 (1962).
33. H. Gramberg, Adhesion 10, (3), 97 (1966).
34. H. Thurn, K. Wolf, Kolloid Z. 148, 16 (1956).
35. Ref. 24, pp. 633-636.
36. Ref. 24, p. 551.
37. Ref. 24, p. 548.

38. Ref. 24, p. 550.

39. Ref. 24, p. 547.

40. P. C. Scherer, A. Tannenbaum, D. W. Levi, J. Polymer Sci.
     43, 531 (1960).

41. W. R. Moore, M. Murphy, J. Polymer Sci. 56, 519 (1962).

42. H.-G. Elias, Kunststoffe Plastics 8 441 (1961).

43. S. N. Chinai, P. C. Scherer, D. W. Levi, J. Polymer Sci.
     17, 117 (1955).

44. M. Matsumoto, Y. Ohyanagi, J. Polymer Sci. 46, 441 (1960).

45. R. H. Wagner, J. Polymer Sci. 2, 21 (1947).

46. A. Nakajima, Kobunshi Kagaku 11, 142 (1954).

47. M. R. Rao, V. Kalpagam, J. Polymer Sci. 49, S-14 (1961).

48. H.-G. Elias, F. Patat, Makromol. Chem. 25, 13 (1957).

49. E. Patrone, U. Bianchi, Makromol. Chem. 94, 52 (1966).

50. S. Gundiah, N. V. Viswanathan, S. L. Kapur, Indian J. Chem.
     4, 344 (1966).

51. D. Goedhart, A. Opschoor, J. Polymer Sci. A-2, 8, 1227 (1970).

PHYSICAL CONSTANTS OF POLY(METHYL METHACRYLATE)★

W. Wunderlich

Röhm GmbH,

Darmstadt, Germany

Property	Value	Remarks	Ref.
Birefringence, $(\alpha_1 - \alpha_2)$ [cm^3]			
segmental anisotropy	2×10^{-25}	benzene	1
	20×10^{-25}	benzene, isotactic	
Coefficient of Thermal Expansion, [K^{-1}]			
linear	7×10^{-5}	$0 - 50\,^{\circ}C$	2
volume	2.60×10^{-4}	$< T_g$	3
	2.55×10^{-4}	$< T_g$	4
	2.72×10^{-4}	$< T_g$	5
	2.25×10^{-4}	$< T_g$	6
	5.80×10^{-4}	$> T_g$	3
	5.60×10^{-4}	$> T_g$	4
	5.80×10^{-4}	$> T_g$	5
	5.75×10^{-4}	$> T_g$	6
Compressibility, [MPa^{-1}]			7
	245×10^{-6}	$T = 20\,^{\circ}C$	
	290×10^{-6}	$60\,^{\circ}C$	
	355×10^{-6}	$100\,^{\circ}C$	
	390×10^{-6}	$109.3\,^{\circ}C$	
	500×10^{-6}	$119.8\,^{\circ}C$	
	530×10^{-6}	$129.7\,^{\circ}C$	
	585×10^{-6}	$139.3\,^{\circ}C$	
Density, [$Mg\ m^{-3}$] $\equiv$ [$g\ cm^{-3}$]	1.195	$0\,^{\circ}C$	8
	1.190	$20\,^{\circ}C$	9
	1.188	$25\,^{\circ}C$	10
	1.150	T_g	11
Diffusion O_2 in PMMA, [$cm^2\ s^{-1}$]	0.5×10^{-8}	0.21 bar, $25\,^{\circ}C$	12
	1.0×10^{-8}	0.4 bar, $25\,^{\circ}C$	12
	1.4×10^{-8}	1 bar, $25\,^{\circ}C$	12
H_2O in PMMA, [$g\ cm/cm^2\ h$]	5.2×10^{-8}	permeability	13
Dielectric constant	3.6	50 Hz, $25\,^{\circ}C$	14
	3.0	1 KHz, $25\,^{\circ}C$	14
	2.6	1 MHz, $25\,^{\circ}C$	14
	2.57	30 GHz, $25\,^{\circ}C$	15,16
	2.59	138 GHz, $25\,^{\circ}C$	16
Dissipation Factor, [$\tan \delta$]	0.062	50 Hz, $25\,^{\circ}C$	14
	0.055	1 KHz, $25\,^{\circ}C$	14
	0.014	1 MHz, $25\,^{\circ}C$	14
	0.007	30 GHz, $25\,^{\circ}C$	15,16
	0.010	138 GHz, $25\,^{\circ}C$	16
Entropy of Polymerization, [$kJ\ mol^{-1}\ K^{-1}$]★★	-117	$25\,^{\circ}C$ to solid polymer	17
	-40.3	$-63\,^{\circ}C$	17,18
Extinction Modulus, [mm^{-1}]	5×10^{-5}	633 nm	2
	1×10^{-4}	400 nm	2

★Figures are valid for normal, radical polymerized PMMA if no tacticities are given. ★★mol: monomer unit

Property	Value	Remarks	Ref.
Glass Transition Temperature T_g, [K]		dilatometric	
	378		6, 11
	377		5

T [°C]	Tacticity (triad analysis)			Ref.
	iso	hetero	syn.	
41.5	0.95		0.05	5
54.3	0.73	0.16	0.11	19
61.6	0.62	0.20	0.18	19
104.0	0.06	0.37	0.56	5
114.2	0.10	0.31	0.59	19
120.0	0.10	0.20	0.70	5
125.6	0.09	0.36	0.64	19

Property	Value	Remarks	Ref.
Heat Capacity, [kJ kg^{-1} K^{-1}]	0.585	-173 °C	24
	0.878	-100 °C	24
	1.255	0 °C	24
	1.42 (1.47)	25 °C	24, 23
	1.72 (2.02)	100 °C	24, 23
	2.05 (2.76)	120 °C	24, 25
	2.38 (3.35)	180 °C	24, 25
	4.69 (2.35)	240 °C	25, 36
	10.55 (2.50)	300 °C	25, 36
Heat of Combustion, [kJ kg^{-1}]	-26,200		21
Heat of Polymerization, [kJ mol^{-1}]★	-57.8	to solid polymer	17
Modulus of Elasticity, [MPa]≡[N mm^{-2}]	3300	25 °C	2
Modulus of Shear, [MPa]≡[N mm^{-2}]		dynamic	
	1700	25 °C, 10 Hz	2
Mechanical Loss Factor, [tan δ]	0.08	25 °C, 10Hz	20
Partial Specific Volume, [1 kg^{-1}]≡[cm^3 g^{-1}]	0.812	20 °C in monomer	9
	0.833	70 °C in monomer	9
	0.819	25 °C in monomer	22
Refractive Index			
n_D	1.492	λ = 589 nm	2
n_e	1.494	λ = 546 nm	2, 9
n_g	1.502	λ = 436 nm	2, 9
$-dn_D/dT$	1.1	$< T_g$	2
	2.1	$> T_g$	2
Abbé's number, [r_D]	58.0		2
Resistivity, [Ohm cm]	$>10^{15}$		2
Solvents		see tables on "Solvents - Nonsolvents" in this Handbook	
Thermal Conductivity, [W m^{-1} K^{-1}]	0.193	0 - 50 °C	26, 27
Theta Solvents and Temperatures			
acetonitrile	θ = 30 °C		28
n-butyl chloride	θ = 35.0 °C		29
heptanone-4	θ = 40.4 °C		29
isoamyl acetate	θ = 57.4 °C		29

★ mol: monomer unit

Property	Value	Remarks	Ref.
Theta Solvents and Temperatures (Cont'd.)			
n-butyl chloride	$\theta = 35.0\,^\circ C$	syndiotactic	30
	$\theta = 26.5\,^\circ C$	isotactic	30
heptanone-3	$\theta = 30.0\,^\circ C$	syndiotactic	31
	$\theta = 40.0\,^\circ C$	isotactic	31
n-propanol	$\theta = 75.9\,^\circ C$	syndiotacitc	31
	$\theta = 84.4\,^\circ C$	isotactic	31
Unit Cell		see corresponding table in this Handbook	
Unperturbed Dimensions			
radius of gyration		see corresponding table in this Handbook	32
Velocity of Sound, $[m\ s^{-1}]$	2700	2 MHz	33
Viscosity-Molecular Weight Relationship, $[\eta] = KM^a\ [cm^3\ g^{-1}]$		see also table "Viscosity-Molecular Weight Relationship" in this Handbook	

Solvent	T[$^\circ$C]	K x 10^3	a	
acetone	20	10	0.68	32
benzene	20	5.5	0.76	34
n-butyl acetate	25	25	0.58	32
n-butyl chloride	35 = θ	50	0.50	32
chloroform	20	5.5	0.79	35

Molecular weights were determined by means of light scattering and sedimentation.

REFERENCES

1. V. N. Tsetkov, N. N. Boitzkova, Vysokomolekul. Soedin 3, 1176 (1960).
2. G. Schreyer, "Konstruieren mit Kunststoffen," Carl Hanser, Muenchen 1972.
3. P. Heydemann, D. H. Guicking, Kolloid-Z. 193, 16 (1963).
4. J. Hennig, Diplomarbeit TH Darmstadt 1961.
5. J. C. Wittmann, A. J. Kovacs, J. Polymer Sci. C 16, 4443 (1969).
6. S. Loshaek, J. Polymer Sci. 15, 391 (1955).
7. K. H. Hellwege, W. Knappe, P. Lehmann, Kolloid-Z. 183, 110 (1962).
8. H. J. Kolb, E. F. Izard, J. Appl. Phys. 20, 564 (1949).
9. D. Panke, W. Wunderlich, Roehm GmbH, Darmstadt, Unpublished results.
10. W. G. Gall, N. G. McCrum, J. Polymer Sci. 50, 489 (1961).
11. S. S. Rogers, L. Mandelkern, J. Phys. Chem. 61, 945 (1957).
12. R. E. Barker, Jr., J. Polymer Sci. 58, 553 (1962).
13. F. Bueche, J. Polymer Sci. 14, 414 (1954).
14. G. Schreyer, Kunststoffe 55, 771 (1965).
15. E. Amrhein, Kolloid-Z. 216-7, 38 (1967).
16. W. Zeil, E. Sistig, W. Frank, V. Hoffmann, Ber. Bunsenges. Physik. Chem. 74, 883 (1970).
17. H. Sawada, J. Macromol. Sci. C 3, 313 (1969).
18. R. W. Worfield, M. C. Petree, J. Polymer Sci. A 1, 1701 (1963).
19. E. V. Thompson, J. Polymer Sci. A-2, 4, 199 (1966).
20. J. Heiboer, Kolloid-Z. 148, 36 (1956).
21. K. Krekeler, P. M. Klinke, Kunststoffe 55, 758 (1965).
22. G. V. Schulz, M. Hoffmann, Makromol. Chem. 22, 220 (1957).
23. E. C. Bernhard, "Processing of Thermoplastic Materials," Reinhold, New York, 1959.
24. R. Hoffmann, W. Knappe, Kolloid-Z. 247, 763 (1971).
25. R. G. Griskey, J. O. Hubbel, J. Appl. Polymer Sci. 12, 853 (1968).
26. E. Calvet, J. P. Bros, H. Pinelle, Compt. Rend. 260, 1164 (1965).
27. K. Eiermann, Kolloid-Z. 198, 5 (1964).
28. E. Cohn-Ginsberg, T. G. Fox, H. Maron, Polymer 3, 97 (1962).
29. R. G. Kirste, G. V. Schulz, Z. Physik. Chem. (Frankfurt) 27, 301 (1961).
30. G. V. Schulz, W. Wunderlich, R. G. Kirste, Makromol. Chem. 75, 22 (1964).
31. J. Sakurada, A. Nakajima, O. Yoshizaki, K. Nakamae, Kolloid-Z. 186, 41 (1962).
32. G. V. Schulz, R. G. Kirste, Z- Physik. Chem. (Frankfurt) 30, 171 (1961).
33. F. Pluemer, Kolloid-Z. 176, 176 (1961).
34. H.-J. Cantow, G. V. Schulz, Z. Physik. Chem. (Frankfurt) 1, 365 (1954).
35. H.-J. Cantow, G. Meyerhoff, Z. Elektrochem. 56, 904 (1952).
36. V. Bares, B. Wunderlich, J. Polymer Sci. A2, 11, 861 (1973).

PHYSICAL CONSTANTS OF POLY(STYRENE)

J. F. Rudd
DIG Physical Research,
The Dow Chemical Company,
Midland, Michigan

Property	Value	Remarks	Ref.
Birefringence Dispersion		A = 0.8905	21
$\dfrac{\Delta n(\lambda)}{\Delta n(546\ nm)}$	$A + \dfrac{B}{\lambda^2} + \dfrac{C}{\lambda^4}$	B = 0.275 x 10^{-9} cm^2 B = 0.275 x 10^{-9} cm^2 C = 0.153 x 10^{-18} cm^4	
Coefficient of Thermal Expansion, [K^{-1}]			
linear	6-8 x 10^{-5}	< T_g (unoriented)	3
volume	1.7-2.1 x 10^{-4}	< T_g	7a
	5.1-6.0 x 10^{-4}	> T_g	7a
Coefficient of Friction	0.515	20-80 $^\circ$C	7b, 7c
	0.744	100 $^\circ$C	7b
	> 2	120 $^\circ$C	7b
	0.25	100 $^\circ$C	7c
	2.65	110-120 $^\circ$C	7c
Compressibility, [MPa^{-1}]	220 x 10^{-6}		3
Compressive Modulus, [MPa] $\equiv$ [N mm^{-2}]	3000	(unoriented)	18a
Crystallographic Data	see under unit cell		
Density, [Mg m^{-3}] $\equiv$ [g cm^{-3}]			
amorphous	1.04-1.065		3, 4
crystalline	1.111		2
	1.12		1, 5
dρ/dT [g cm^{-3} K^{-1}]	-2.65 x 10^{-4}	< T_g	6
dρ/dT [g cm^{-3} K^{-1}]	-6.05 x 10^{-4}	> T_g	6
Dielectric Constant, [e]			
amorphous	2.49-2.55	at 1 kHz (curve flat to 1 GHz)	3
crystalline	2.61	at 1 kHz (curve flat to 1 GHz)	16a
Dissipation Factor			
amorphous	15 x 10^{-4}	at 1 kHz (curve flat to 1 GHz)	3
crystalline	3 x 10^{-4}	at 1 kHz (curve flat to 1 GHz)	16a
Dynamic Mechanical Loss Peaks			
(1Hz)	T_β = 300 K		16b
	T_γ = 138 K		16b
(7 kHz)	T_δ = 48 K		16c
Glass-Transition Temperature, T_g, [$^\circ$C]	80		6
	90	T_g(K) = 373 - (1.0 x 10^5/M_v)	8
	(100)		9
Heat Capacity, C_p [kJ kg^{-1} K^{-1}]	1.185 (1.139)	0 $^\circ$C	3, 12
	1.256 (1.394)	50 $^\circ$C	3, 12
	1.838 (1.821)	100 $^\circ$C	3, 12
dC_p/dT [kJ kg^{-1} K^{-2}]	4.04 x 10^{-3}	50 $^\circ$C	3

Property	Value	Remarks	Ref.
Heat of Combustion, $[kJ\ mol^{-1}]$★	-4.33×10^3		13a, 14
Heat of Fusion, $[kJ\ mol^{-1}]$★	8.37 ± 0.08		10
crystalline	9.00		2
Heat of Polymerization, $[kJ\ mol^{-1}]$★	-67.4	to solid polymer	15
	-69.9	to solid polymer	13a
Heat of Solution, $[kJ\ mol^{-1}]$★	-3.60	in monomer	13a
Infrared Spectrum		see Ref.	13b
Melting Point, T_m, $[^{\circ}C]$	240		10
	(250)		11
Melt Viscosity-Molecular Weight Relationship	$\log \eta_T = 3.4 \log M - k$		

	T $[^{\circ}C]$	M range	k	
atactic	217	$\geq$ 38,000	13.04	31
isotactic	281	100,000-600,000	14.42	32a

Nuclear Magnetic Resonance Line Width, [gauss]   32b

T $[^{\circ}C]$	Value
0	6.7
50	6.5
100	5.0
150	1.0

Property	Value	Remarks	Ref.
Nuclear Magnetic Resonance	peaks [ppm] at 3.0, 3.5 (aromatic), 8.4 (CH_2) (referred to tetramethylsilane as + 10.0)		32c
Optical Dispersion, $\eta_F - \eta_C$	1.92×10^{-2}	λ = 486.1 nm	3
		λ = 656.3 nm	3
Poisson's Ratio	0.325		3
	0.33		20
Refractive Index, n_D	1.59-1.60	λ = 589.3 nm	3
$dn_D/dT [K^{-1}]$	-1.42×10^{-4}		3
Resistivity, [Ohm cm]	$10^{20} - 10^{22}$		17a
Sonic Velocity, $[m\ s^{-1}]$	2100		3
Stress-Optical Coefficient, [brewsters]	10.1	monofilament	22
	9.5	extruded sheet	22
	8.3, 8.7	compression moulded	22
Tensile Modulus, $E\ [MPa] \equiv [N\ mm^{-2}]$	3200	unoriented	19
	3400		18a
	4200	oriented monofilament	18b, 18c
$dE/dT\ [N\ mm^{-2}\ K^{-1}] \equiv [MPa\ K^{-1}]$	-4.48		19
Thermal Conductivity, $[W\ m^{-1}\ K^{-1}]$	0.105	$0\ ^{\circ}C$	3
	0.116	$50\ ^{\circ}C$	3
	0.128	$100\ ^{\circ}C$	3

★
mol: monomer unit

Property	Value	Remarks	Ref.
Thermogravimetric Analysis			17b
(in vacuo) (10^oC/min.)			

Wt. Loss [%]	T[^{o}C]	Wt. Loss [%]	T[^{o}C]
0	275	40	390
5	330	60	408
10	343	80	422
20	360	100	450

Property	Value	Remarks	Ref.
Unit Cell, [$\overset{o}{A}$]		rhombohedral system; 18 monomer units in cell	
a_o	21.9		1
	22.08		2
b_o	21.9		1
	22.08		2
c_o	6.63		1
	6.626		2
Ultraviolet Spectrum			
(in vacuo)	Absorption bands at 260, 194 and 80 nm.		33

Viscosity-Molecular Weight Relationships for Poly(styrene)[a]   see also corresponding table in this Handbook

	[η]	M				
Solvent	T[^{o}C]	Solvent	Method	K x 10^3	a	
			atactic			
benzene	30	benzene	osmotic	9.7	0.74	23
benzene	25	toluene	osmotic	11.3	0.73	24
toluene	25	butanone	osmotic	13.4	0.71	
toluene	25	butanone	light scattering	17.	0.69	25
toluene	30	butanone-isopropyl alcohol	light scattering	9.23	0.72	26
benzene	30	toluene	osmotic	10.6	0.735	27
toluene	30	toluene	osmotic	11.0	0.725	
toluene	25	cyclohexanone	sedimentation velocity	9.77	0.73	28
			isotactic			
benzene	30	toluene	osmotic	10.6	0.735	27
toluene	30	toluene	osmotic	11.0	0.725	
benzene	room temp.	toluene	osmotic	9.5	0.77	29
benzene	25	toluene	osmotic	9.79	0.744	30
toluene	25	butanone	light scattering	17.	0.69	25

[a] Constants determined by method and in solvent given in these columns; [η] = KM^a, in [ml/g].

REFERENCES

1. G. Natta, Makromol. Chem. 35, 94 (1960).
2. R. L. Miller, L. E. Nielsen, J. Polymer Sci. 55, 643 (1961).
3. R. H. Boundy, R. F. Boyer, eds. "Styrene, Its Polymers, Co-polymers and Derivatives," Reinhold Publishing Corp., New York, 1952.
4. G. Natta, J. Polymer Sci. 16, 143 (1955).
5. G. Natta, P. Pino, P. Corradini, F. Danusso, E. Mantica, G. Mazzanti, G. Moraglio, J. Am. Chem. Soc. 77, 1708 (1955).
6. W. Patnode, W. J. Scheiber, J. Am. Chem. Soc. 61, 3449 (1939).
7a. R. S. Spencer, G. D. Gilmore, J. Appl. Phys. 20, 502 (1949).

7b. H. J. Karam, W. C. Sager, K. S. Hyun, The Dow Chemical Company, unpublished data.
7c. M. Lund, The Dow Chemical Company, unpublished data.
8. F. P. Reding, J. A. Faucher, R. D. Whitman, J. Polymer Sci. 57, 483 (1962).
9. T. G. Fox, P. J. Flory, J. Appl. Phys. 21, 581 (1950). The effect of residual monomer, low-molecular-weight polymer, ethylbenzene, and plasticizers being to depress T_g, the value given by these authors represents the limiting glass-transition temperature at high molecular weights.
10. R. Dedeurwaerder, J. F. M. Oth, J. Chim. Phys. 56, 940 (1959).

11. T. W. Campbell, A. C. Haven, Jr., J. App. Polymer Sci. 1, 73 (1959).

12. K. Ueberreiter, E. Otto-Laupenmuhlen, Z. Naturforsch. 8a, 664 (1953).

13a. D. E. Roberts, W. W. Walton, R. S. Jessup, J. Polymer Sci. 2, 420 (1947).

13b. C. Y. Liang, S. Krimm, J. Polymer Sci. 27, 241 (1958).

14. J. W. Breitenbach, J. Derkosch, Monatsh. 81, 698 (1950).

15. L. K. J. Tong, W. O. Kenyon, J. Am. Chem. Soc., 69, 1402 (1947).

16a. F. L. Saunders, R. C. Mildner, The Dow Chemical Company, unpublished data.

16b. K.-H. Illers, Z. Electrochem. 65, 679 (1961).

16c. J. M. Crissman, A. E. Woodward, J. A. Sauer, J. Polymer Sci. A-3, 3, 2693 (1965).

17a. P. Woodland, The Dow Chemical Company, unpublished data.

17b. J. G. Cobler, The Dow Chemical Company, unpublished data.

18a. "Strength and Stiffness", in Plastics Design Data, Dow Technical Chemical Publication, 1965 revision.

18b. L. Kin, The Dow Chemical Company, unpublished data.

18c. R. D. Andrews, J. F. Rudd, J. Appl. Phys. 28, 1091 (1957).

19. J. F. Rudd, E. F. Gurnee, J. Appl. Phys. 28, 1096 (1957).

20. L. E. Nielsen, "Mechanical Properties of Polymers," Reinhold Publishing Corp., New York, 1962.

21. E. F. Gurnee, J. Polymer Sci. A-2, 5, 817 (1967).

22. J. F. Rudd, R. D. Andrews, J. Appl. Phys. 31, 818 (1960).

23. R. H. Ewart, H. C. Tingey, Abstracts 111th Am. Chem. Soc. Meet., April 1947; quoted in Fox and Flory, J. Am. Chem. Soc. 73, 1915 (1951).

24. C. E. H. Bawn, R. F. J. Freeman, A. R. Kamaliddin, Trans. Faraday Soc. 46, 1107 (1950).

25. P. Outer, C. I. Carr, B. H. Zimm, J. Chem. Phys. 18, 830 (1950).

26. S. N. Chinai, P. C. Scherer, C. W. Bondurant, D. W. Levi, J. Polymer Sci. 22, 527 (1956).

27. F. Danusso, G. Moraglio, J. Polymer Sci. 24, 161 (1957).

28. H. W. McCormick, J. Polymer Sci. 36, 341 (1959).

29. F. Ang, J. Polymer Sci. 25, 126 (1957).

30. W. R. Krigbaum, D. K. Carpenter, S. Newman, J. Phys. Chem. 62, 1586 (1958).

31. T. G. Fox, P. J. Flory, J. Polymer Sci. 14, 315 (1954).

32a. J. Boon, G. Challa, P. H. Hermans, Makromol. Chem. 74, 129 (1964).

32b. R. Kosfeld, Kolloid-Z. 172, 182 (1960).

32c. F. A. Bovey, G. V. D. Tiers, G. Filipovich, J. Polymer Sci. 38, 73 (1959).

33. R. H. Partridge, J. Chem. Phys. 47, 4223 (1967).

PHYSICAL CONSTANTS OF POLY(OXYMETHYLENE)

K. H. Burg and G. Sextro

Hoechst A.G.
Frankfurt/M., Germany

Property	Value	Ref.
Absorption		
homopolymer (density 1.418-1.419 g cm^{-3})		
a) Equilibrium Concentration, 60°C		
aniline, [wt.-%]	9.60	1
m-cresol, [wt.-%]	16.48	1
dimethylformamide, [wt.-%]	5.73	1
water (DIN 53 472, after 96 hours), [mg]	20-30	3
[wt.-%]	ca. 0.14-0.21	
b) Gain in Volume, 60°C, [vol.-%]		
aniline	13.7	1
m-cresol	23.0	1
dimethylformamide	8.6	1
c) Average Sorption Coefficient, 25°C		
water	see Ref.	2
Coefficient of Linear Thermal Expansion (VDE 0304), [K^{-1}]		
20-100°C	1.0-1.4 x 10^{-4}	3
amorphous > Tg	7.2 x 10^{-4}	6
Cohesive Energy, 25°C, [kJ mol^{-1}]	9.88	4
Cohesive Energy Density, [J m^{-3}]$^{1/2} \cdot 10^{-3}$		
25°C	22.8	1
liquid, 25°C, theoretical value	20.9	4
Crystallinity, [%]		
a) homopolymer		
20°C from density-data	64-69	5
25°C from x-ray-data	77	6
135°C from x-ray-data	75	6
157°C from x-ray-data	67	6
Delrin 150, 20°C from density data	64.0	5
Delrin 550, 20°C from density-data	68.7	5
b) copolymer		
20°C	56-59	5
Hostaform C 2520, M$_w$ 80000, 20°C from density data	56.6	5
Hostaform C 9020, M$_w$ 58000, 20°C from density data	58.7	5
Density, [Mg m^{-3}] $\equiv$ [g cm^{-3}]		
a) homopolymer		
Delrin 500	1.420±0.005	7
after annealing	1.434±0.004	7
Delrin 150, 20°C	1.427	5
Delrin 550, 20°C	1.435	5

Property	Value	Ref.

Density, $[Mg\ m^{-3}] \equiv [g\ cm^{-3}]$ (Cont' d.)

   b) copolymer, $20^{o}C$

      Hostaform C 2520, M_w 80000   1.412   5

      Hostaform C 9020, M_w 58000   1.416   5

Depolymerization, Energy of Activation, $[kJ\ mol^{-1}]$

   a) poly(oxymethylene)diol

thermal depolymerization 100-120°C	113	9
130-150°C	82.9±12.6	8
135-190°C	83.8	11
170-285°C	109	10
cationic depolymerization -30 to +33°C	80.0±9.6	12
130-150°C	87.5±12.6	8
anionic depolymerization 130-150°C	27.2±6.3	8
135-170°C	62.0±7.5	13

   b) poly(oxymethylene)diacetate

thermal depolymerization 178-190°C	234	11
240-340°C	134	10

Depolymerization Rate, $[min^{-1}]$

   poly(oxymethylene)diol

$$V_{depol.} = k_{therm} [POM] + k_{bas} [POM] [Amine]$$ 13

$T\ [^{o}C]$	$k_{therm} \times 10^{3}$	$k_{bas} \times 10^{3}$
135	1.6±0.2	4.1±0.2
150	3.5±0.1	7.7±0.5
160	5.0±0.5	11.6±1.0
170	7.4±1.0	17.0±2.0

Dielectric Constant (10^{3} Hz)

   (DIN 53 483, ASTM D 150-59 T)   3.6-4.0   3

Dielectric Loss, tan δ (10^{3} Hz) $(15 - 50) 10^{-4}$ 3

   (DIN 53 493, ASTM D 150-59T)   see also Ref.   7, 14, 15, 16

Dielectric Strength, $[KV\ mm^{-1}]$

   (DIN 53 481) (50 Hz, 0.5 KV/s, film 0.2 mm)   50-70   3

Diffusion (films)

   a) water

      activation energy, apparent, $[kJ\ mol^{-1}]$

copolymer 0-80°C	41.5	18
copolymer, branched, 50°C	44.4	2

      diffusion coefficient, average, 25°C, $[cm^{2}\ s^{-1}]$

copolymer	2.7×10^{-8}	17
copolymer, branched	7.0×10^{-8}	17
copolymer, linear and branched	see Ref.	2

   b) carbon dioxide

      activation energy, $[kJ\ mol^{-1}]$

copolymer	48.9	17
copolymer, branched	38.5	17

      diffusion coefficient, $[cm^{2}\ s^{-1}]$

copolymer	1.4×10^{-9}	17
copolymer, branched	4.4×10^{-9}	17

Property	Value	Ref.
Elongation at Break, [%]		
(DIN 53 455, ASTM D 638)	20-70	3
Enthalpy, [kJ kg^{-1}]		
homopolymer 0 K	0	19
50	6.909	19
100	29.69	19
200	102.6	19
300	222.5	19
copolymer 0 K	0	19
50	6.948	19
100	29.40	19
200	100.9	19
300	211.4	19
Entropy, [kJ kg^{-1} K^{-1}]		
homopolymer 0 K	0	19
50	0.2097	19
100	0.5167	19
200	1.007	19
300	1.487	19
copolymer 0 K	0	19
50	0.2123	19
100	0.5153	19
200	0.9974	19
300	1.439	19
Entropy of Fusion, [kJ mol^{-1} K^{-1}]		
of homopolymer at constant pressure	8.21 x 10^{-3}	20
of homopolymer at constant volume	4.98 x 10^{-3}	20
Entropy of Polymerization, [kJ mol^{-1} K^{-1}]		
formaldehyde (gas) $\longrightarrow$ poly(oxymethylene) (solid) (from heat-capacity-measurements)	-0.175±0.0008	19
(from equilibrium pressure measurements of gaseous formaldehyde)	-0.129	19
	-0.183	21
trioxane (gas) $\longrightarrow$ poly(oxymethylene) (solid)	-0.155	22
Flexural Stress at Conventional Deflection, [MPa]$\equiv$[N mm^{-2}]		
(DIN 53452)	105-120	3
Glass Transition Temperature	see second order transition points	
Hardness Ball Indentation, [MPa]$\equiv$[N mm^{-2}]		
(DIN 53456 E/ISO 2039) after 30 sec.	130-160	3
Heat Capacity, [kJ kg^{-1} K^{-1}]		
homopolymer 0 K	0	19
50	0.3310	19
100	0.5558	19
200	0.9598	19
300	1.425	19
copolymer 0 K	0	19
50	0.3320	19
100	0.5461	19
200	0.8846	19
300	1.369	19

Property	Value	Ref.
Heat Deflection Temperature under Load, [$^{\circ}$C]		
Martens (DIN 53 458)	65-70	3
Heat of Fusion, [kJ kg^{-1}]		
poly(trioxymethylene)	234	16
poly(tetraoxymethylene)	222	16
poly(pentaoxymethylene)	203	16
poly(oxymethylene), theoretical value for 100% crystallinity	316-335	15
copolymer (trioxane/2% ethylene oxide)	181-192	15
Heat of Polymerization, [kJ mol^{-1}]		
formaldehyde (gas 1 bar, 25°C) $\longrightarrow$ poly(oxymethylene)		
(values obtained from equilibrium pressure data)	51.1	25
	59.9	26
	68.2	27
	72.0	21
Heat of Solution, [kJ mol^{-1}]		
of water in branched copolymer, 50°C	19.3	2
Heat of Sublimation	see Lattice Energy	
Huggins Coefficient	see Viscosity-Molecular Weight-Relationship	
Impact Strength, [Nmm/mm^2]		
(DIN 53 453)	no failure	3
Impact Strength (notched), [Nmm/mm^2]		
(DIN 53 453/ASTM D 256)	6-10	3
Infra-Red Spectra	see Ref.	28
Lattice Energy, [kJ mol^{-1}]		
(heat of sublimation of poly(oxymethylene) at 0 K; theoretical value)	15.9±1.7	4
Melt Index (M.I.) - Molecular Weight) (M_w) - Relationship, [dg min^{-1}]		
M = 6 200-129 000; 190°C	M.I. = 1.30 x 10^{18} x $M_w^{-3.55}$	29
Melting Point, [$^{\circ}$C]		
a) homopolymer		
Delrin 150	181	5
extended chain fibers	182.5±0.5	30
folded chain spherulites	174	30
as a function of crystallization temperature	see Ref.	31
from solution polymerization of trioxane, ionically initiated		
fibrous product	186-187	30
powdery product	178-180	30
from vacuum sublimation of trioxane	195-200	30
b) poly(oxymethylene) dimethyl ethers		
monomer	-105.0	4
dimer	-69.7	4
trimer	-42.5	4
tetramer	-9.8	4
pentamer	18.3	4

Property	Value	Ref.
Melting Point, [^{o}C] (Cont'd.)		
c) copolymer		
Hostaform C 2520, M_w = 80000	167	5
Hostaform C 9020, M_w = 58000	170	5
various kinds and contents of comonomers	see Ref.	32
copolymers with 0/2/4/6/7/ mol% dioxolane	see Ref.	32
Modulus, Tensile (1 mm/min), [MPa]$\equiv$[N mm^{-2}]		
(DIN 53 457)	2600-3400	3
NMR-Sectra	see Ref.	33, 34
Oligomers R-[OCH$_2$]$_n$-R,		
a) boiling points R = -OCH$_3$, [^{o}C]		
n = 2	105.0	4
n = 3	155.9	4
n = 4	201.8	4
n = 5	242.3	4
b) density R = -OOC-CH$_3$, 25°C, [g cm^{-3}]		
n = 1	1.1283	35
n = 2	1.1610	35
n = 3	1.1823	35
poly(oxymethylene), liquid, 25°C, theoretical value	1.328	4
c) heat of vaporization	see Ref.	4
d) infrared-data R = -OOCCH$_3$		35

wave number [cm^{-1}]	absorption [cm^{-1} x mol^{-1}]
1760	395
1755	330
1755	330

(n = 1, n = 2, n = 3 correspond to the three rows above)

Property	Value	Ref.
e) melting points		
oligo (oxymethylene)dimethyl ether n = 2-5	see Ref.	4
f) vapor pressure		
oligo (oxymethylene)dimethyl ether n = 2-5	see Ref.	4
Permeability	see corresponding table in this Handbook	
Permeability, apparent activation energy, [kJ mol^{-1}]		
water		
homopolymer, melt index 2.3; 50°C	23.0	2
copolymer, linear, melt index 2.5; 50°C	29.7	2
copolymer, branched, melt index 3.9; 50°C	25.1	2
copolymer, 25°C	13.0	17
copolymer, branched, 25°C	6.3	17
carbon dioxide		
copolymer, 25°C	31.4	17
copolymer, branched	25.1	17
Refractive Index, [n_D]		
extended chain crystals (fibers), different samples		
$n_{\parallel}$	1.545-1.553	30
$n_{\perp}$	1.489	30

Property	Value	Ref.

Resistance to Tracking

   (DIN 53 480/VDE 0303)

KA 3 b

KA > 600   3

Schulz-Blaschke-Coefficient   see Viscosity-Molecular Weight-Relationship

Second Order Transition Point

  glass transition temperature

      homopolymer

	198	6
	188	cit. in 23
	190	cit. in 23
	199	24

α-Relaxation, activation energy, [kJ mol^{-1}]
(for POM single crystal)

$T < 40\,^{\circ}C$	155	36
$40\,^{\circ}C < T < 90\,^{\circ}C$	105	36
$T > 90\,^{\circ}C$	159	36

Solvents

Solvent	dissolving Temp. [$^{\circ}C$]	gel Temp. [$^{\circ}C$]	
phenol	109	58	1
o-chlorophenol	-	70	1
aniline	130	102	1
cyclopentanone	-	112	1
ethylene carbonate	145	117	1
cyclohexanone	-	119	1
benzyl alcohol	132	119	1
acetic anhydride	-	120	1
formamide	150	135	1

Specific Volumes

  homopolymer

      crystalline phase (varying temperatures [t, $^{\circ}C$])
      amorphous phase (varying temperatures [t, $^{\circ}C$])

$V_K = 0.669 + 7.85 \times 10^{-5} t + 5.93 \times 10^{-7} t^2 \, [cm^3/g]$ cit. in 38

$V_a = 0.792 + 3.07 \times 10^{-4} t \, [cm^3/g]$ cit. in 38

$V_a = 0.7553 + 5.78 \times 10^{-4} (t-20) \, [cm^3/g]$ 5

  copolymer

      varying amount (X) of comonomer (wt.-% dioxolane, 20$^{\circ}C$)
      varying temperatures (X = 2.9 wt.-%; Hostaform C)

$V_a^{20} = 0.7553 + 0.08674\, X \, [cm^3/g]$ 5

$V_a = 0.7578 + 5.78 \times 10^{-4} (t-20) \, [cm^3/g]$ 5

Stress-Strain Curves   see Ref.   39

Surface Resistance, [ohm]

   (film 0.2 mm) (DIN 53 482) $> 10^{13}$ 3

Tensile Strength (50 mm/min), [MPa]$\equiv$[N mm^{-2}]

   (DIN 53 455/ASTM D 638)   65-72   3

Tensile Strength at Break, [%]

   (DIN 53 455/ASTM D 638)   20-70   3

Thermal Conductivity at 20$^{\circ}C$, [W m^{-1} K^{-1}]

   (VDE 0304)   0.292   3

Property	Value	Ref.

Trioxane

Boiling point, [$^{\circ}$C]	115	40
Density, 65°C, [g cm^{-3}]	1.17	40
Dielectric constant, 20°C, 1.6 MHz	3.2-3.4	40
Entropy of gas, [kJ kg^{-1} K^{-1}]		
0 K	0	22
50	0.244	
100	0.573	
200	1.064	
273.16	1.373	
298.16	1.477	
Enthalpy of gas, [kJ kg^{-1}]		
0 K	0	22
50	7.97	
100	32.29	
200	104.9	
273.16	178.2	
298.16	208.0	
Melting point, [$^{\circ}$C]	62-64	38
Molecular weight	90.08	40
Solvents	water, alkohols and ethers	
Specific Heat, [kJ kg^{-1} K^{-1}]		22
0 K	0	
50	0.372	
100	0.5767	
200	0.8783	
273.16	1.138	
298.16	1.237	
Unit Cell	see corresponding table in this Handbook	

Vicat Softening Point, [$^{\circ}$C]	160-170	3

(DIN 53 460/ASTM D 1525-65 T)

Viscosity-Molecular Weight Relationships, $[\eta] = K \times M^a$, [ml g^{-1}]   see also corresponding table in this Handbook

Solvent	T[$^{\circ}$C]	K x 10^3	a	No of Samples FR[*]	WP[**]	Molecular Weight range x 10^{-3}	Method	
p-chlorophenol, 2% α-pinen	130	54.3	0.66	?	?	?	OS	cit. in 41
	60	41.3[***]	0.724	-	4	62-129	LS	29
1H,1H,5H- octafluoropentanol-1	110	13.3	0.81	-	4	62-129	LS	29
phenol-tetrachloroethane (25/75 wt.%)/								
2% α-pinen	90	12.16	0.64	8	-	1.1-92	EG	42
phenol	90	11.3	0.76	?	-	?	OS	cit. in 43

[*] fractionated   [**] polymer unfractionated   [***] $[\eta]_{inh}$

Huggins Coefficient k_H, $\eta_{sp/c} = \eta + k_H [\eta]^2 c$

hexafluoroacetone hydrate 25°C	0.41	44
hexafluoroacetone-water-mixtures	see Ref.	44
hexafluoroacetone-sesquihydrate + triethylamine, 25°C	0.31	45
hexafluoroacetone-water 1:1.7 mol + triethylamine, 25°C	0.30±0.02	46
p-chlorphenol/2% α-pinene, 60°C	0.40	47

Schulz-Blaschke-Coefficient K_{SB}, $\eta_{sp/c} = K_{SB} [\eta] \eta_{sp} + [\eta]$

hexafluoroacetone-hydrate, 25°C	0.29	44
hexafluoroacetone-hydrate/water-mixtures, 25°C	see Ref.	44
phenol-tetrachloroethane 1:3 (per weight), 90°C	0.23	43
	0.21	48

Property	Value	Ref.
Volume Resistance, [ohm cm]		
(film 0.2 mm) (DIN 53 482/ASTM D 257-61)	$>10^{15}$	3

REFERENCES

1. R. G. Alsup, J. O. Punderson, G. F. Leverett, J. Appl. Polymer Sci. 1, 185 (1959).
2. G. F. Hardy, J. Polymer Sci. A-2, 5, 671 (1967).
3. VDI, Richtlinien 2477, "Polyacetale fuer die Feinwerktechnik," VDI Verlag Duesseldorf, Grundruck Oktober 1973.
4. R. H. Boyd, J. Polymer Sci. 50, 133 (1961).
5. H. Wilski, Makromol. Chem. 150, 209 (1971).
6. Y. Aoki, A. Nobuta, A. Chiba, M. Kaneko, Polymer J. 2, 502 (1971).
7. B. E. Read, G. Williams, Polymer 2, 239 (1961).
8. M. Raetzsch, G. Eckhardt, Plaste Kautschuk 20, 424 (1973).
9. Y. Iwasa, T. Imoto, Nippon Kagaku Zasshi 84, 31 (1963).
10. L. A. Dudina, N. S. Enikolopyan, Polymer Sci. USSR, 5, 36 (1964).
11. N. Grassie, R. S. Roche, unpublished
12. J. Mejzlik, Makromol. Chem. 59, 184-88 (1963).
13. H. Pennewiss, V. Jaacks, W. Kern, Makromol. Chem. 103, 285-88 (1967).
14. Y. Ishida, Kolloid-Z. 171, 149 (1960).
15. H. Wilski, Kolloid-Z. 248, 867 (1971).
16. A. Tanaka, S. Uemura, Y. Ishida, J. Polymer Sci. (Phys.) 10, 2093 (1972).
17. J. L. Williams, V. Stannett, J. Appl. Polymer Sci. 14, 1949 (1970).
18. M. Braden, J. Polymer Sci. A-1, 6, 1227 (1968).
19. F. S. Dainton, D. M. Evans, F. E. Hoare, T. P. Melia, Polymer 3, 263 (1962).
20. H. W. Starkweather, Jr., R. H. Boyd, J. Phys. Chem. 64, 410 (1960).
21. J. B. Thompson (see J. F. Walker, "Formaldehyde," 3rd Ed. p. 180, Reinhold, New York 1964).
22. G. A. Clegg, T. P. Melia, A. Tyson, Polymer 9, 75 (1968).
23. F. S. Stehling, L. Mandelkern, J. Polymer Sci. B-7, 255 (1969).
24. K. Miki, S. Yamane, M. Kaneko, "Proc. Fifth Intern. Congress Rheology," Vol. 3, University of Tokyo Press, p. 335 (1970).

25. F. S. Dainton, K. J. Ivin, D. A. G. Walmsley, Trans. Faraday Soc. 55, 61 (1959).
26. J. F. Walker, "Formaldehyde," 3rd Ed., p. 180 Reinhold, New York, 1964.
27. Y. Iwasa, T. Imoto, J. Chem. Soc. Japan, Pure Chem. Sect. 84, 29 (1963).
28. A. Novak, E. Whalley, Trans. Faraday Soc. 55, 1484 (1959).
29. H. L. Wagner, K. F. Wissbrun, Makromol. Chem. 81, 14 (1965).
30. M. Jaffe, B. Wunderlich, Kolloid-Z. 216/217, 203 (1967).
31. K. F. Wissbrun, J. Polymer Sci. A-2, 4, 827 (1966).
32. M. Inoue, J. Appl. Polymer Sci. 8, 2225 (1964).
33. D. Fleischer, R. C. Schulz, Makromol. Chem. 166, 103 (1972).
34. Y. Yamashita, T. Asakura, M. Okada, K. Ito, Makromol. Chem. 129, 1 (1969).
35. J. Majer, Makromol. Chem. 82, 169 (1965).
36. K. Miki, Polymer J. 1, 432 (1970).
37. F. J. Boerio, D. D. Cornell, J. Polymer Sci. (Phys.) 11, 391 (1973).
38. J. Majer, O. Hainová, Kolloid-Z. 201, 23 (1965).
39. W. H. Linton, H. H. Goodman, J. Appl. Polymer Sci. 1, 179 (1959).
40. J. F. Walker, "Formaldehyde," 3rd Ed., p. 192, Reinhold, New York, 1964.
41. A. Tanaka, S. Uemura, Y. Ishida, J. Polymer Sci. A-2, 8, 1585 (1970).
42. K. Doerffel, H. Friedrich, H. Grohn, D. Wimmers, Plaste-Kautschuk 12/9, 524 (1965).
40. W. Thuemmler, Plaste-Kautschuk 12, 382 (1965).
44. N. Grassie, R. S. Roche, J. Polymer Sci. C, 16, 4207 (1968).
45. L. Hoehr, V. Jaacks, H. Cherdron, S. Iwabuchi, W. Kern, Makromol. Chem. 103, 279 (1967).
46. W. H. Stockmayer, Lock Lim Chan, J. Polymer Sci. A-2, 4, 437 (1966).
47. E. Kobayashi, S. Okamura, R. Signer, J. Appl. Polymer Sci. 12, 1661 (1968).
48. H. Grohn, H. Friedrich, J. Polymer Sci. C, 16, 3737 (1968).

PHYSICAL CONSTANTS OF POLY(OXYETHYLENEOXYTEREPHTHALOYL) (POLY(ETHYLENE TEREPHTHALATE))★

E. L. Ringwald

E. L. Lawton

Monsanto Triangle Park Development Center, Inc.
Research Triangle Park, North Carolina

Table I: Property	Value	Ref.
Birefringence of Filaments, [Sodium Light] ★★		1
Draw Ratio		
2.0	0.040	
3.0	0.092	
4.0	0.167	
5.0	0.193	
Bursting Strength, [g cm^{-2}]		
1 mil film, 23°C	46×10^{2} (66 lb in^{-2})	2, 3
Coefficient of Friction (film),		
kinetic, film to film	0.45	2, 3
Coefficient of Thermal Conductivity (film), [W m^{-1} K^{-1}]		
100°C	37.5×10^{-3}	
Coefficient of Thermal Expansion (film), [K^{-1}]		2, 3
20 - 50°C	1.7×10^{-5}	
Coefficient of Volume Expansion, [K^{-1}]		9
crystalline, below T_g	1.71×10^{-4}	
crystalline, above T_g	3.94×10^{-4}	
Conductivity for Direct Current		

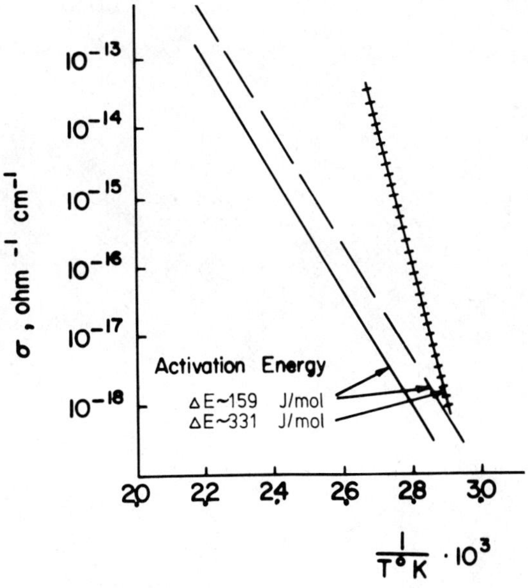
σ, ohm $^{-1}$ cm^{-1}
10^{-13}
10^{-14}
10^{-15}
10^{-16}
10^{-17}
10^{-18}
Activation Energy
ΔE~159 J/mol
ΔE~331 J/mol
2.0
2.2
2.4
2.6
2.8
3.0
$\frac{1}{T°K} \cdot 10^3$

Figure I. Direct Current Conductivity at Various Temperatures and Degrees of Orientation and Crystallinity (44)

——— Amorphous
— — — Crystalline
++++++ Oriented & Crystalline

★ Properties of films are given for Mylar Type A (DuPont) biaxially oriented and crystalline films. Properties of filaments are given for various products.

★★ Value sensitive to semi-crystalline morphology.

Table I: Property	Value	Ref.

Crystallographic Data see Unit Cell

Density, $[Mg\ m^{-3}] \equiv [g\ cm^{-3}]$

Form	Sample Density	Estimate of % Crystallinity by		Ref.
		Density	Infrared	
Amorphous, non-oriented	1.335	0	2	5
Partly cryst., non-oriented	1.385	42	48	5
Highly cryst., oriented	1.390	46	65	5
	1.389	45	76	5
	1.381	38	80	
Calculated, crystal	1.455			6

Dielectric Strength (film), $[V\ cm^{-1}]$ — 2, 3

23°C, 60 Hz 2.95×10^6 (7500 V mil^{-1})
150°C, 60 Hz 2.75×10^6 (5000 V mil^{-1})

Dielectric Constant (film), — 2, 3

23°C, 60 Hz 3.30
23°C, 1 kHz 3.25
23°C, 1 MHz 3.0
23°C, 1 GHz 2.8
150°C, 60 Hz 3.7

Dissipation Factor (film) — 2, 3

23°C, 60 Hz 0.0025
23°C, 1 kHz 0.0050
23°C, 1 MHz 0.016
23°C, 1 GHz 0.003
150°C, 60 Hz 0.00040

Elastic Constants (filaments), $[MPa^{-1}] \cdot 10^{-4}$ — 7

Oriented

S_{11} transverse 16
S_{33} longitudinal 0.71
S_{44} torsional 14
S_{12} -5.8
S_{13} -0.31

Unoriented

S_{11} or S_{33} extensional 4.4
S_{44} torsional 11

Entropy of Fusion, $[J\ mol^{-1}\ K^{-1}]$ 42.7 (10.2 [cal mol K^{-1}]) — 2, 3

Folding Endurance (film), [cycles] — 2, 3

23°C 300000

Glass Transition Temperature★★, [°C]

amorphous 67 — 9, 10
crystalline 81 — 10
crystalline and oriented 125 — 11

Heat Capacity see Specific Heat

Heat of Combustion, $[kJ\ kg^{-1}]$ -2.16×10^4 (-9300 BTU lb^{-1}) — 12

★★
Value sensitive to semi-crystalline morphology.

Table I: Property	Value	Ref.
Heat of Fusion, [kJ kg^{-1}]	121	13
	113	13
	129	14
	133	8
	144	15
Heat of Sorption (film), [kJ kg^{-1} mol^{-1}]		16
carbon dioxide	-3.1 x 10^4	
methane	-2.3 x 10^4	
Hygroscopic Coefficient of Expansion (film), [cm cm^{-1} %R. H.$^{-1}$]		2, 3
20 - 92% RH	1.1 x 10^{-5}	
Impact Strength (film), [kg cm m^{-1}]		2, 3
23°C	2.4 x 10^5 (6.0 kg cm mil^{-1})	
Infrared Spectra	see Ref.	46, 47, 48
Insulation Resistance (film), [Mohm mfds]		17, 18, 19
100°C	5000	
130°C	400	
150°C	100	
Melting Point, [$^\circ$C]		
commercial PET	250 - 265	2, 3
effect of diethylene glycol content	see Figure II	21

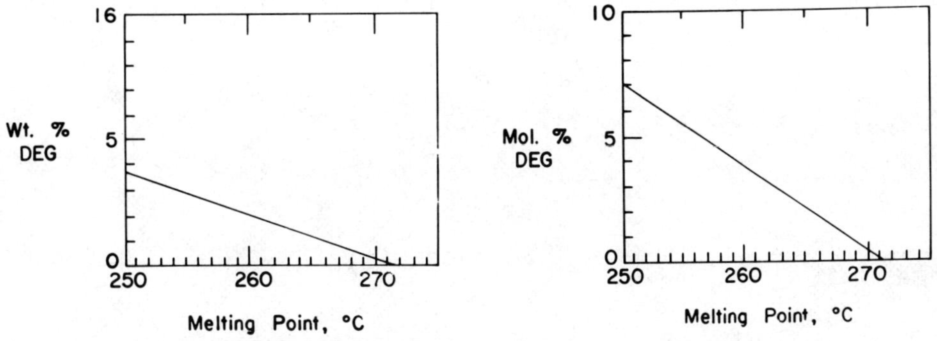

Figure II.   Effect of Diethylene Glycol (DEG) Content on Poly(ethylene terephthalate) (PET)
Melting Point (21, 2) - Melting Point [$^\circ$C] = 271 - 3 $\cdot$ (Mol % DEG)

Melt Viscosity versus Intrinsic Viscosity

Melt Viscosity [Pa s], 280°C★	Intrinsic Viscosity in s-tetrachloroethane/phenol (40/60), [ml g^{-1}], 30°C
0.45	10
4.5	20
25.0	30
95.0	40
115.0	50
145.0	60
800.0	70
1180.0	80
2000.0	90

★ Taken from Figure 14 of Ref. 2

Table I: Property	Value	Ref.

Moisture Absorption, [%]   18, 19

immersion in water at 25°C for 1 week   0.8

Oligomers - Acyclic

$A = -OC- \bigcirc -CO-$

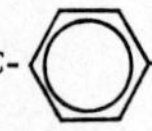

$G = -OCH_2 CH_2 O -$

Structure	Molecular Weight	Melting Point [°C]	
H[GA]$_1$ OH	210.2	178	22
H[GA]$_2$ OH	402.4	200 - 205	22
H[GA]$_3$ OH	594.6	219 - 223	22
		179 - 183	23
		186	24
H[GA]$_1$ -G-H	254.2	109 - 110	22
H[GA]$_2$ -G-H	446.4	173 - 174	22
H[GA]$_3$ -G-H	638.6	200 - 205	22
H[GA]$_4$ -G-H	830.8	213 - 216	22
H[GA]$_5$ -G-H	1023.0	218 - 220	22
HO-A-[GA]$_1$ OH	358.3	>360	22
HO-A [GA]$_2$ OH	550.5	280 - 281	22
HO-A-[GA]$_3$ OH	742.7	268 - 270	22
HO-A-[GA]$_4$ OH	934.9	252 - 255	22
HO-A-[GA]$_5$ OH	1127.1	233 - 236	22

Oligomers - Cyclic

Cyclic Dimer

    melting point, [°C]   175, 224   23
                           229   25

    unit cell [Å]   a = 8.58   25
                 b = 12.75
                 c = 8.01
                 β = 90.7°

Cyclic Trimer (B-Type Crystal)

    melting point, [°C]   319   23
                            317 - 320   26
                            321   27

    crystalline transition, [°C]   199   27
    (A-type ⟶ B-type)   195   28

Cyclic Tetramer

    melting point, [°C]   326   23

Cyclic Pentamer

    melting point, [°C]   256   23

Permeability (film)   see corresponding table in this Handbook

Poisson's Ratio (oriented filament)   7

    extensional   0.44
    transverse   0.37

Raman Spectra   29, 30

    Frequency, [cm^{-1}]   1730
                            1618
                            1096
                            857
                            632
                            278

Table I: Property	Value	Ref.

Refractive Index (film), [Na Light]

amorphous, 25°C	1.5760	2
crystalline and biaxially oriented, 23°C	1.64	

Resistivity   2, 3

Surface, [ohm cm^{-2}]

23°C, 30% R.H.	2×10^{16}
23°C, 80% R.H.	2×10^{12}

Volume, [ohm cm]

23°C	10^{18}
150°C	10^{14}

Service Temperature, [°C]   -60 - +150   2, 3

Shrinkage (film), [%]   3

150°C, 30 min.   2 - 3

Solubility Constants (film), [cm^3 STP cm^{-3} Pa^{-1}] x 10^7 16

3 mil, 25°C nitrogen	4.3 (0.044 cm^3 STP cm^{-3} atm^{-1})
oxygen	7.5 (0.076 cm^3 STP cm^{-3} atm^{-1})
methane	19.7 (0.200 cm^3 STP cm^{-3} atm^{-1})
argon	0.8 (0.0077 cm^3 STP cm^{-3} atm^{-1})

Solvents   see table "Solvents-Nonsolvents" in this Handbook

Solvent-Nonsolvent for Fractionation

Solvent	Nonsolvent	
o-chlorophenol	hexane	31
phenol/tetrachloroethane (1:1)	gasoline	32
phenol/chlorobenzene (1:1)	gasoline	33
phenol/dichloroethane (2:3)	benzene	34
phenol	cyclohexane	35
phenol	ethanol	36

Sonic Velocity (filaments), [$m\ s^{-1}$]   37

10 K cycles s^{-1} unoriented	1400
highly oriented	5900

Specific Heat, [$kJ\ kg^{-1}\ K^{-1}$]   8

$C_p = 4.184\ (A + Bx\ T°C)$

Condition of Polymer	A	B x 10^4	Effective Temperature, T[°C]
Molten polymer	0.3243	5.65	270 to 290
Flake	0.2502	9.40	-20 to 60
Yarn (undrawn)	0.2469	10.07	-5 to 60
Yarn (drawn)	0.2482	9.89	-10 to 55
Yarn (drawn + annealed)	0.2431	9.23	-10 to 80
	0.2502	9.31	100 to 200

Stick Point Temperature, [°C]   230 to 240   4

Stress-Strain Curves for Filaments   see Ref. 38   38

Table I: Property	Value	Ref.
Surface Tension, $[mN\ m^{-1}] \equiv [dyn\ cm^{-1}]$		
solid/liquids, 25°C	39.5	39
	42.1	40
molten, 290°C	27±3	41
Tear Strength (film), $[g\ m^{-1}]$		2, 3
Initial, 23°C	23.6×10^{6} (600 g mil^{-1})	
Propagating, 23°C	0.59×10^{6} (15 g mil^{-1})	
Tensile Strength (film), $[MPa] \equiv [N\ mm^{-2}]$		2, 3
23°C	1.73×10^{-4} (25000 lb in^{-2})	
Thermal Conductivity (film), $[W\ m^{-1}\ K^{-1}]$		42
33°C	1.47×10^{-5}	
Thermal Diffusivity (film), $[cm^{2}\ s^{-1}]$		42
33°C	9.29×10^{-4}	
Torsional Modulus, $[MPa] \equiv [N\ mm^{-2}]$		7
oriented filament	720 (0.072 dyn cm^{-2} x 10^{11})	
Transition Temperatures, $[°C]$		
from second moment (NMR) measurements on fibers		43, 7
unoriented	120 ± 5	43, 7
oriented	140 ± 10	43, 7
Unit Cell, $[\mathring{A}]$		
X-ray Diffraction (291 K) a	4.56	6
b	5.94	
c	10.75	
α	98.5°	
β	118°	
γ	112°	
Electron Diffraction, a	4.52	45
b	5.98	
c	10.77	
α	101°	
β	118°	
γ	111°	
Viscosity-Molecular Weight Relationship	see table II, III, Figures 3 and 4	
Young's Modulus, $[MPa] \equiv [N\ mm^{-2}]$		7
Oriented Filament extensional	1.41×10^{-4}	
transverse	0.063×10^{-4}	
Zero Strength Temperature (film), $[°C]$	248	4

Table II Intrinsic Viscosity - Molecular Weight Relationships (Unfractionated)

$[\eta] = K \times M^a$

Line Number	Solvent	T[°C]	K × 10⁴	a	Molecular Weight Range × 10⁻³	Method	Ref.
1	trifluoroacetic acid	30	4.33	0.68	26 - 118	Light Scattering	49
2	tetrachloroethane/phenol (5:3)	30	2.29	0.73	26 - 118	Light Scattering	49
3	o-chlorophenol	25	6.56	0.73	12 - 25	Osmometry	50
4	o-chlorophenol	25	3.0	0.77	13 - 28	End Group	51
5	tetrachloroethane/phenol (1:1)	25	2.1	0.82	5 - 25	End Group	52
-	tetrachloroethane/phenol (1:1)	20	1.27	0.86	5 - 21	End Group	53
6	tetrachloroethane/phenol (1:1)	20	7.55	0.685	3 - 30	End Group	35
-	phenol/2,4,6-trichlorophenol (10:7)	29.8	2.10	0.80	1 - 8	End Group	54

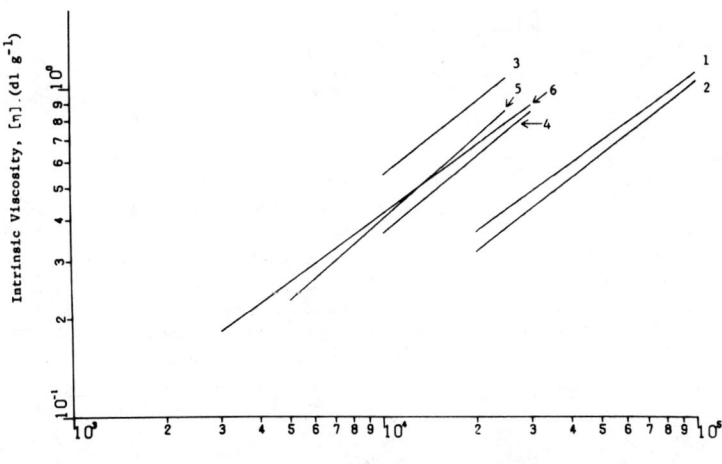

Molecular Weight, M

Table III Intrinsic Viscosity - Molecular Weight Relationships (Fractionated)

$[\eta] = K \times M^a$

Line Number	Solvent	T[°C]	K × 10⁴	a	Molecular Weight Range × 10⁻³	Method	Ref.
1	o-chlorophenol	25	1.9	0.81	15 - 38	End Group	55
2	dichloroacetic acid	25	67	0.47	15 - 38	End Group	55
3	trifluoroacetic acid	25	14	0.64	15 - 38	End Group	55
3	phenol/tetrachloroethane (2:3)	25	14	0.64	15 - 38	End Group	55
4	o-chlorophenol	25	4.25	0.69	20 - 100	Sedimentation & Diffusion	56
5	tetrachloroethane/phenol (1:1)	20	0.9	0.87	5 - 21	End Group	53
-	phenol/dichloroethane (2:3)	20	0.92	0.85	8 - 30	End Group	34

Table III Intrinsic Viscosity - Molecular Weight Relationships (Fractionated)

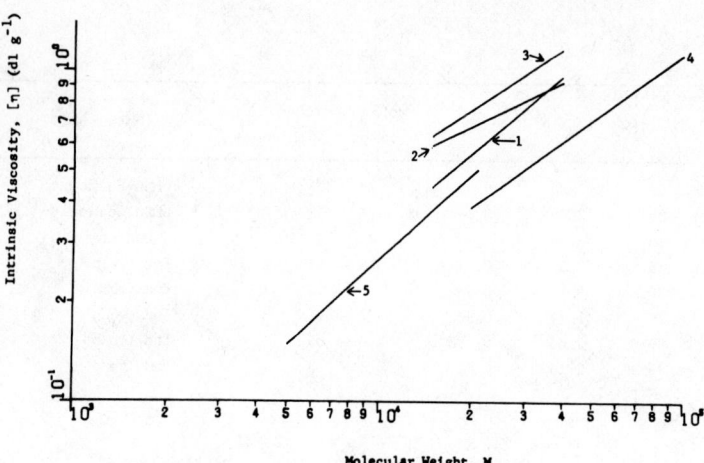

REFERENCES

1. G. Farrow, J. Bagley, Textile Res. J. 32, 587 (1962).
2. C. J. Heffelfinger, K. L. Knox, "Polyester Films," in "The Science and Technology of Polymer Films," Vol. II, edit. O. J. Sweeting, Wiley-Interscience, New York, 1971, p. 587.
3. J. M. Hawthorne, C. J. Heffelfinger, K. L. Knox, "Polyester Films," "Encyclopedia of Polymer Science," Vol. 11. edit. H. F. Mark, N. G. Gaylord, Interscience Publishers, New York, 1969, p. 42.
4. B. V. Petukhov, "The Technology of Polyester Fibers," Macmillan Co., New York, N. Y. 1963, p. 388.
5. A. B. Thompson, D. W. Woods, Nature 176, 78 (1955).
6. R. de P. Daubeny, C. W. Bunn, C. J. Brown, Proc. Roy. Soc. (London) A 226, 531 (1954).
7. I. M. Ward, J. Macromol. Sci. B 1, 667 (1967).
8. C. W. Smith, M. Dole, J. Polymer Sci. 20, 37 (1956).
9. H. J. Kolb, E. F. Izard, J. Appl. Phys. 20, 564 (1949).
10. O. B. Edgar, R. Hill, J. Polymer Sci. 8, 1 (1952).
11. D. W. Woods, Nature 174, 753 (1954).
12. R. B. LeBlanc, J. Am. Assoc. Textile Chem. Color 2, 123 (1970).
13. R. C. Roberts, Polymer 10, 113 (1969).
14. I. Kirshenbaum, J. Polymer Sci. A 3, 1869 (1965).
15. P. E. Slade, T. A. Orofino, in "Analytical Calorimetry," eds. R. S. Porter, J. F. Johnson, Plenum Press, New York, 1968, p. 63.
16. W. R. Vieth, H. H. Alcalay, A. J. Frabetti, J. Appl. Polymer Sci. 8, 2125 (1964).
17. DuPont Technical Bulletin No. 1-2-53.
18. DuPont Technical Bulletin No. M-1A.
19. D. D. Lanning, Prod. Eng., July, 187 (1956).
20. G. W. Taylor, Polymer 3, 543 (1962).
21. R. Jannssen, H. Ruysschaert, R. Vroom, Makromol. Chem. 77, 153 (1964).
22. H. Zahn, G. B. Gleitsmann, Angew. Chem. 75, 772 (1963).
23. L. H. Peebles, Jr., M. W. Huffman, C. T. Ablett, J. Polymer Sci. A-1, 7, 479 (1969).

24. H. Binder, U. S. Pat. 2,855,432.
25. H. Repin, E. Papanikolau, J. Polymer Sci. A-1, 7, 3426 (1969).
26. F. L. Hamb, L. C. Trent, J. Polymer Sci. B 5, 1057 (1967).
27. E. Ito, S. Okajima, J. Polymer Sci. B 7, 483 (1969).
28. G. L. Binns, et al, Polymer 7, 583 (1966).
29. A. J. Melveger, J. Polymer Sci. A-2, 10, 317 (1972).
30. G. E. McGraw, Polymer Preprints 11, 1122 (1970).
31. W. R. Moore, R. P. Sheldon, J. Textile Ins. 50, T294 (1959).
32. A. Gordienko, Faserforsch. Textiltechn. 4, 199 (1953).
33. A. A. Geller, A. A. Konkin, V. A. Myagkov, Khim. Volokna 3, 10 (1960).
34. Ye. V. Kuznetsova, A. O. Vizel, I. M. Zhermergorn, S. S. Tyulenev, Vysokomolekul. Soedin. 2, 205 (1960).
35. M. M. Koepp, H. Werner, Makromol. Chem. 32, 79 (1959).
36. E. Turska-Kusmierz, T. Skwarski, Prace Inst. Wlokiennictwa 2, 49 (1953).
37. W. W. Mosaley, Jr., J. Appl. Polymer Sci. 3, 226 (1960).
38. A. B. Thompson, J. Polymer Sci. 34, 741 (1959).
39. D. H. Kaelble, J. Adhesion 2, 66 (1970).
40. S. Wu, Polymer Preprints 11, 1291 (1970).
41. H. T. Patterson, K. H. Hu, T. H. Grindstaff, Polymer Preprints 11, 1299 (1970).
42. R. C. Steere, J. Appl. Phys. 37, 3338 (1966).
43. S. Nohara, Chem. High Polymer (Japan) 14, 318 (1957).
44. L. E. Amborski, J. Polymer Sci. 62, 331 (1962).
45. Y. Y. Tomashpol' skii, G. S. Morkova, Polymer Sci. USSR 6, 316 (1964).
46. S. Krimm, Fortschr. Hochpolymer Forsch. 2, 51 (1960).
47. J. L. Koenig, M. J. Hannon, J. Macromol. Sci. B 1, 119 (1967).
48. T. R. Manley, D. A. Williams, J. Polymer Sci. C 22, 1009 (1969).
49. M. L. Wallach, Makromol. Chem. 103, 19 (1967).
50. J. Marshall, A. Todd, Trans. Faraday Soc. 49, 67 (1953).
51. I. M. Ward, Nature 180, 141 (1957).
52. A. Conix, Makromol. Chem. 26, 226 (1958).
53. W. Griehl, S. Neve, Faserforsch. Textiltechn. 5, 423 (1954).
54. N. G. Gaylord, S. Rosenbaum, J. Polymer Sci. 39, 545 (1959).
55. W. R. Moore, D. Sanderson, Polymer 9, 153 (1968).
56. Von G. Meyerhoff, S. Shimotsuma, Makromol. Chem. 135, 195 (1970).

PHYSICAL CONSTANTS OF
POLY(IMINO(1-OXOHEXAMETHYLENE) (POLYAMIDE 6)
AND
POLY(IMINOHEXAMETHYLENEIMINOADIPOYL) (POLYAMIDE 66)

R. Pflüger

BASF AG
Anwendungstechnische Abteilung Kunststoffe
Ludwigshafen, Germany

	Value		
Property	Polyamide 6	Polyamide 66	Ref.
Abrasion Resistance	see References		30, 80
Absorption, [%]			
water, mouldings 20-90°C	9.5±0.5	8.5±0.5	8, 37, 28, 63, 91, 104
ethanol, mouldings 20°C	9-17	9-12	8
butanol, mouldings 20°C	4-16	2-9	8
glycol, mouldings 20°C	6-13	2-10	8
methanol, mouldings 20°C	10-19	10-14	8
propanol, mouldings 20°C	2-18	2-12	8
Adhesive Bond Strength, tensile, [MPa]$\equiv$[N mm^{-2}]			
PA-aluminium		68	77
PA-steel		70	77
PA-copper		76	77
Birefringence, [Δ_n]			
$n_{\parallel}$	1.580	1.582	103
$n_{\perp}$	1.530	1.519	103
Bulk Modulus	see Compressibility		
Coefficient of Friction (dry)			
a) average depth of roughness R_v in [μm] [*]			
0.1	0.35	0.5	30
0.5	0.33	0.43	
1	0.32	0.38	
2	0.32	0.35	
4	0.36	0.38	
6	0.43	0.46	
b) average surface pressure in [MPa]$\equiv$[N mm^{-2}] (R_v optimal)			
0.02	0.4	0.4	30
0.05	0.36	0.36	
0.1	0.35	0.35	
0.15	0.40	0.40	
1.0	0.43	0.44	
5.0	0.48	0.50	
15.0	0.48	0.51	
($R_v < 0.3$ μm) <0.1		0.52	
>1.0		stick-slip-motion	30
Coefficient of Thermal Expansion, [K^{-1}]			
crystalline 20°C	7-10 x 10^{-5}	7-10 x 10^{-5}	8
100°C	10-14 x 10^{-5}	10-14 x 10^{-5}	

[*] average surface pressure 0.1 MPa; surface temperature $<40°$C.

Property	Value		Ref.
	Polyamide 6	Polyamide 66	

Coefficient of Thermal Expansion, $[K^{-1}]$ (Cont'd.)

monoclinic α a	6.2×10^{-5}		107
α c	23.4×10^{-5}		
α b	-4.5×10^{-5}		
triclinic α a		2.1×10^{-5}	107
α c		22.0×10^{-5}	

Cohesive Energy   see solubility parameter

Compliance   see modulus, resp. compressibility

Compressibility $[MPa^{-1}]$

a. dependence on temperature and pressure (mouldings annealed 150 h, 125°C)

20 T [°C] 50 [MPa]		62×10^{-6}	101
100		58	101
150		54	101
300		50	101
120 50		125	101
100		115	101
150		95	101
300		75	101
200 >100		>300	36

b. dependence on water content

<1.5%		137×10^{-6}	99
> ~2%		226	99
melt	335	280	79, 42

Compressive Strength, $[MPa] \equiv [N \; mm^{-2}]$

temperature 20°C,	strain: 1%	14	6
(moulded; 2.5% H_2O)	strain: 2%	28	6
	strain: 4%	56	6
	strain: 6%	70	6

Crystallographic Data   see unit cell dimension

Density, $[Mg \; m^{-3}] \equiv [g \; cm^{-3}]$

crystalline α, monoclinic	1.24		108
α, monoclinic	1.23		43
α, monoclinic	1.21		75, 86
γ, hexagonal	1.13		105
γ, monoclinic	1.17		5
γ, pseudohexagonal	1.155		83
α, triclinic		1.220	97
α, triclinic		1.24	
β, triclinic		1.248	
amorphous γ	1.09		48
α	1.11		48
mouldings, amorphous	1.10	1.09	72
crystalline	1.12-1.14	1.13-1.145	8

Property	Value		Ref.
	Polyamide 6	Polyamide 66	

Dielectric Constant e, Dielectric Loss tan δ   11

$e_r/10^4$ x tan δ Polyamide 6 Frequency in [Hz]

T[°C]	10^2	10^3	10^4	10^5	10^6	10^7	10^8	10^9
-30	3.1/ 140	3.1/ 100	3.0/ 100	3.0/ 160	3.0/ 200	3.0/ 180	3.0/ 120	3.0/ 55
0	3.2/ 120	3.2/ 130	3.2/ 140	3.2/ 170	3.1/ 220	3.1/ 230	3.1/ 160	3.0/ 100
30	3.5/ 65	3.5/ 100	3.4/ 140	3.4/ 190	3.3/ 240	3.2/ 300	3.2/ 300	3.1/ 200
60	5.2/ 840	4.6/ 600	4.2/ 520	3.8/ 540	3.6/ 550	3.4/ 550	3.3/ 480	3.2/ 260
90	15.4/2900	11.5/1650	8.8/1500	7.1/1900	5.7/2100	4.4/1900	3.7/1250	3.4/ 550
25(65 RH)	13.0/2100	9.7/2200	7.2/2250	5.6/2000	4.5/1600	3.9/1000	3.5/ 580	3.3/ 320

Polyamide 66 Frequency in [Hz]

T[°C]	10^2	10^3	10^4	10^5	10^6	10^7	10^8	10^9
-30	3.1/ 120	3.1/ 105	3.1/ 105	3.0/ 130	3.0/ 165	3.0/ 160	3.0/ 100	3.0/ 49
0	3.3/ 110	3.3/ 120	3.2/ 135	3.2/ 160	3.1/ 200	3.0/ 200	3.0/ 160	3.0/ 81
30	3.6/ 85	3.5/ 125	3.4/ 180	3.4/ 215	3.2/ 250	3.1/ 255	3.1/ 220	3.0/ 135
60	5.0/ 810	4.6/ 590	4.3/ 460	4.0/ 390	3.7/ 370	3.5/ 360	3.3/ 320	3.1/ 240
90	10 /2000	8.9/1450	7.6/1300	6.2/1450	5.0/1600	4.0/1300	3.4/ 810	3.2/ 440
25(65 RH)	11 /1650	8.6/1550	6.6/1480	4.9/1270	4.0/ 950	3.4/ 580	3.2/ 300	3.0/ 235

Dielectric Strength, [V cm^{-1}] x 10^{-4}

(VDE 0303, part 2)

	Polyamide 6	Polyamide 66	Ref.
dry	150	150	9
dry 100°C	40	40	31
humid ($\sim$2% H_2O)	80	90	9

Diffusion   see table "Permeability Constants" in this Handbook and Ref. 8, 12a, 25, 27, 28, 79, 90

Elongation, (ISO-R 527, DIN 53 455), ultimate, [%]

		Polyamide 6	Polyamide 66	Ref.
moulded (s = 3 mm), dry		100	40	8
moulded (s = 3 mm), humid ($\sim$2% H_2O)		150-200	150	8
moulded at yield,		10	6	104
dry film, NK 23/50, s = 25 μm	longitudinal	300		109
	transverse	330		109
biaxially oriented	longitudinal	70		109
	transverse	50		109
monomer cast dry		3-70		7
monofilaments dry		20-25	20-25	
monofilaments humid		20-27	20-24	24

Enthalpy, [kJ kg^{-1}]

temperature T [°C]
(ref. to 20°C)

	Polyamide 6	Polyamide 66	Ref.
60	60	85	8, 79
100	135	170	8, 79
200	380	420	8, 79
250	585	590	8, 79
300	750	850	8, 79

pressure [MPa]$\equiv$[N mm^{-2}]
  T = 25°C

	Polyamide 6	Polyamide 66	Ref.
100		70	36
200		140	36

  T =150°C

	Polyamide 6	Polyamide 66	Ref.
0		250	36
100		340	36
200		420	36

Property	Value Polyamide 6	Polyamide 66	Ref.
Entropy of Fusion, $[J\ mol^{-1}\ K^{-1}]$			
crystalline	44-47.5	83-86	103
constant volume		67	2
Equation of State★			
$(p + \pi)(v - \omega) = \dfrac{RT}{M_n}$ π [bar]	2040	3500	58, 42★★
ω [cm^3 g^{-1}]	0.817	0.817	58, 42★★
M_n	105	105	58, 42★★
Glass Transition Temperature	see corresponding table in this Handbook and Ref. 7, 41, 47, 55, 56, 88		
Hardness			
Rockwell hardness (ASTM-D-785)			
dry, 23°C	M 81-90	M 90	8
		M 79	28
		R 118	28
humid (~2,5% H$_2$O), 23°C	M 37	M 53-60	8
		M 59	28
		R 108	28
Ball indentation hardness Hc (DIN 53 456), [MPa]$\equiv$[N mm^{-2}]			
dry, 23°C	150	170	7
dry, nucleated, 23°C	160		7
humid (~2% H$_2$O), 23°C	70	100	7
Heat Capacity	see corresponding table in this Handbook and Ref. 47, 66, 82, 87, 94, 103		
Heat of Combustion, [kJ kg^{-1}]	-31.900	-31.900	7
Heat of Crystallization, [kJ kg^{-1}]	-46.5	-54	53
Heat Distortion Temperature, (ISO-R 75, DIN 53 461), [°C]			
moulded, dry, 0.46 [N mm^{-2}]$\equiv$[MPa]	170-190	>200	51, 7
moulded, dry, 18.5 [N mm^{-2}]	65-75	90-110	51, 7
moulded, annealed, 0.46 [N mm^{-2}]		243	28
moulded, annealed, 18,5 [N mm^{-2}]		105	28
Heat of Formation	see Enthalpy		
Heat of Fusion, [kJ kg^{-1}]			
crystalline from ΔH_m	188		26
crystalline from ΔH_m	191	196	53
crystalline from ΔH_m for α-structure	230-278		48
crystalline calculated from group contribution	193	193	103
amorphous, annealed 8 h at 50°C ($\varrho = 1.111$)	45		48
Heat of Sorption, [J mol^{-1}]			
H$_2$O < 150°C	-45.6 x 10^3		91
melt	-76.8 x 10^3	-58.5 x 10^3	33, 73
Impact Strength, Izod (ASTM D 256), [ft lb in^{-1}]			
dry, 23°C	1.0-2.5	0.9-1.3	8, 28
dry, -40°C	-	0.6	8, 28
humid (2,5% H$_2$O), 23°C	5-22	2.0-3.5	8, 28

★ valid for $[\eta] = 1$ to 3,1; R = 82.06 [cm^3 bar mol^{-1} K^{-1}]; v = specific volume [1 kg^{-1}].   ★★ Basis for calculation.

Property	Value Polyamide 6	Polyamide 66	Ref.
Infrared Spectrum		see Ref. 46	
Mechanical Loss, tan δ			
(γ = 0.1-10 Hz) dry 20°	0.016	0.016	41
dry 40°	0.048	0.026	41
dry 60°	0.175	0.083	41
dry 80°	0.127	0.127	41
dry 100°	0.064	0.064	41
($<10^4$ Hz) dry 23°		< 0.02	68
(10^3 Hz) 0.72% H_2O		0.12	68
(10^3 Hz) 120°		0.2	68
Melting Point (ASTM D 789), [°C]	200-220	250-260	28
equilibrium melting point	231		64
α-crystalline	260	< 270	48
pressure dependence		see Ref. 42, 58	
Melt Viscosity		see Ref. 12, 78, 79, 85	
Modulus, [MPa]≡[N mm^{-2}]			
a. Bulk modulus		see compressibility	
b. Loss modulus G''		see mechanical loss	
c. Modulus of elasticity (Young) (ISO-R 527, DIN 53455)			
23°, dry	3000	3300	7
23°, humid (2% H_2O)	1600	2000	7
100°, dry		600	28
d. Modulus of shear G, (DIN 53445)			
dry, 23°C	1100	1500	8
dry, nucleated	1500	1700	8
dry, 150°C	180	300	8
dry, 200°C	80	150	8
e. Tensile modulus, apparent (DIN 53444, ISO-DR 748) (1000 h)			
dry, 100°	125	170	8
humid, 23°	230	470	
Mold Shrinkage, [%]	0.7-1.5	0.8-1.5	8, 4, 24, 42, 52
Moisture Absorption		see Absorption and Ref. 7, 8, 69, 79, 96, 104	
Nuclear Magnetic Resonance Spectrum, wide line		see Ref. 67, 95	
Oligomers		see Ref. 23, 38, 70, 79, 84	
Oxygen Index (ASTM-D-2863)			
dry	27%	28-29%	7, 28
humid (50% RH)		29-31.3%	28
Permeability		see table "Permeability Coefficients" in this Handbook and Ref. 8, 12a, 13, 14, 22, 28, 34, 39, 54, 60, 61, 93, 106	
Poisson Ratio			
extruded rod; room temp.			
a. compressive axial strain 0.1		0.1-0.25	62
0.2		0.3-0.36	62
0.4		0.40-0.44	62
0.6		0.43-0.45	62
-		0.38	103

Property	Value		Ref.
	Polyamide 6	Polyamide 66	

Poisson Ratio (Cont'd.)

b. vs. temperature 100°C	~0.47	~0.47	
melt	0.50	0.50	103

Refractive Index n_D

single crystals

α, calc.		1.475	20
β, calc.		1.565	20
γ, obs.		1.58	20
mouldings	1.53	1.53	7, 28, 104

Resistance to Tracking (DIN 53 480, IEC-Publ. 112-1971)

KA-method

dry	KA 3c	KA 3c	8
humid	KA 3b	KA 3b	8
KB-method, dry and humid	> 600	> 600	8

Shrinkage   see Mold shrinkage

Solubility parameter, $[J\,m^{-3}]^{1/2} \cdot 10^{-3}$

amorphous, 25°C	27.8	27.8	16, 19, 103

Solvents   see table "Solvents - Nonsolvents" in this Handbook
and Ref. 3, 7, 8, 21, 28, 79, 111

Sonic Velocity, 20°C, $[m\,s^{-1}]$

	1400-2300 (fibers)	2770 (mouldings)	57, 79a

Surface Tension, critical, $[mN\,m^{-1}]$

	46	46	76

Tensile Strength (ISO R 527, DIN 53 455), $[MPa] \equiv [N\,mm^{-2}]$ see also Ref. 32, 98

dry, 23°C	80, 90★	90	8
dry, 150°C	15	25	
wet 23°C, 2% water	45	70	
wet 23°C, 8% water	22	35	
monofilaments 23°C, 50% R.H.			
draw ratio 4.3:1	475-550	575-600	24
5.0:1	675-775	650-850	

Thermal Conductivity, $[W\,m^{-1}\,K^{-1}]$

crystalline (wet) 30°C	0.43	0.43	40
amorphous (wet) 30°C	0.36	0.36	40
melt 250°C	0.21		65

Unit Cell Dimensions

PA	Crystal system	Space Group	Unit Cell parameters [Å]			Angles			Units in cell	Density $[Mg\,m^{-3}] \equiv [g\,cm^{-3}]$	
			a	b	c	α	β	γ			
6	α-monoclinic	C_2^2	9.56	8.01	17.24			67.5	8	1.23	43
	α-monoclinic	C_2^2	4.81	7.61	17.1			79.5	4	1.21	75
	α-monoclinic		9.65	8.11	17.2			66.3			86
	β-monoclinic		4.8	3.6-4.1							81
	γ-mesomorphous (pseudohexagonal)		4.8	4.8	16.6			60			59
	γ-rhombic		4.79	4.79	16.70			60	2	1.133	105
	γ-monoclinic		9.33	4.78	16.88			121	4	1.164	4
	γ-monoclinic		9.14	4.84	16.68			121	4	1.188	50

★nucleated.

Property	Value		Ref.
	Polyamide 6	Polyamide 66	

Unit Cell Dimensions (Cont'd.)

PA	Crystal system	Space Group	Unit cell parameters [Å]			Angles			Units in Cell	Density [Mg m^{-3}]≡[g cm^{-3}]	
			a	b	c	α	β	γ			
66	α-triclinic	C_1^1	4.9	5.4	17.2	48.5	77	63.5	1	1.24	20
	β-triclinic	C_1^1	4.9	8.0	17.2	90	77	67.5	2	1.248	20
	pseudohexagonal							60	(>160°C)		15, 92

Viscosity-Molecular-Weight-Relationships

see corresponding table in this Handbook
and Ref. 18, 21, 29, 35, 44, 45, 74, 89, 102

Volume, Specific, [1 kg^{-1}]

	Polyamide 6	Polyamide 66	Ref.
20°C, amorphous	0.91	0.916	72
270°C, melt	1.032	1.018	8, 42
20°C, crystalline, α-monoclinic	0.813		43
α-triclinic		0.82	97
Change on Melting	0.02-0.03	0.02-0.03	7, 42
		0.11	2

Volume, Molar, [cm^3 mol^{-1}]

	Polyamide 6	Polyamide 66	Ref.
20°C, amorphous	103	207.5	72
20°C, amorphous, calc. from group contribution	104.2	208.3	103

Volume Resistivity, [Ohm cm]

	Polyamide 6	Polyamide 66	Ref.
relative humidity 0%	10^{15}	10^{15}	51
20°C (saturated) 100%	4×10^8	10^9	51
temperature dry, 60°C	8×10^{11}	6×10^{11}	51
dry, 100°C	3×10^9	3×10^9	51
dry, 150°C		10^7	51
saturated at 60% R.H., 20°C		2×10^{11}	51
at 50% R.H., 100°C		4×10^7	51

REFERENCES

1. N. Adams, J. Textile Inst. 47, T 530 (1956).
2. G. Allen, J. Appl. Chem. 14, 1 (1964).
3. G. J. van Amerongan, J. Polymer Sci. 6, 471 (1951).
4. B. Giertz, Kunststoff-Rundschau 20, 8 (1973).
5. H. Arimoto, M. Ishibashi, M. Hirai, Y. Chatani, J. Polymer Sci. A-3, 317 (1965).
6. E. Baer, J. R. Knox, T. J. Linton, R. E. Maier, SPE 16, 398 (1960).
7. "BASF Kunststoffe" 3rd Ed. 1970 and other literature on Ultramid, BASF AG, Ludwigshafen/Rh.
8. BASF: Ultramid, Technical Data 1970.
9. BASF-Table "Kunststoffe in der Elektrotechnik," 1971.
10. R. Bennewitz, Faserforsch.-Textiltechn. 5, 155 (1954).
11. K. Bergmann, in G. Schreyer "Konstruieren mit Kunststoffen," C. Hanser, Muenchen 1972.
12. E. C. Bernhardt, Ed. "Processing of Thermoplastic Materials," Reinh. Publ. Corp. New York 1959.
12a. H. J. Bixler, O. J. Sweeting in "The Science and Technology of Polymer Films," Vol. II, S. 121, Wiley Interscience Publ. New York 1971.
13. H. Braunisch, B. Hoffmann, Kaeltetechnik 13, 59 (1961).

14. H. Braunisch, H. Lenhart, Kolloid-Z. 177, 24 (1961).
15. R. Brill, J. Prakt. Chem. 161, 49 (1943); Makromol. Chem. 18/19, 294 (1956).
16. G. M. Bristow, et al, Trans Faraday Soc. 54, 1742 (1958).
17. Brit. Nylon Spinners, Technical Manual.
18. R. Bruessau, N. Goetz, BASF, Ludwigshafen/Rh. unpublished experiments.
19. C. W. Bunn, J. Polymer Sci. 16, 323 (1955).
20. C. W. Bunn, E. V. Garner, Proc. Roy. Soc. (London) A 189, 39 (1947).
21. J. J. Burke, T. A. Orofino, J. Polymer Sci. A-2, 7, 1 (1969).
22. B. Carlowitz, "Kunststofftabellen," Fritz Schiffmann OHG, Bensberg-Frankenhorst, 1963.
23. P. Čefelin, J. Stehlíček, O. Wichterle, Collection Czech. Chem. Commun. 24, 516 (1959).
24. G. Conzelman, J. Zarate, BASF AG, Ludwigshafen/Rh.; unpublished experiments.
25. J.-L. Dauba, Plastiques Mod. Elastom. 17, 133 (1965).
26. M. Dole, B. Wunderlich, Makromol. Chem. 34, 29 (1959).
27. B. Doležel, Chem. Prumysl 31, 218, 483 (1956); 32, 380 (1956).
28. DuPont: Zytel Nylon Resins, Design Handbook 1972.

29. H.-G. Elias, R. Schumacher, Makromol. Chem. 76, 23 (1964).
30. G. Erhard, E. Strickle, Kunststoffe 62, 2, 232, 282 (1972).
31. K. Feser, M. Glueck, H. J. Mair, Kunststoffe 60, 155 (1970).
32. F. Fischer in "Konstruieren mit Kunststoffen," Part 2 (Ed. G. Schreyer) C. Hanser, Muenchen 1972.
33. O. Fukumoto, J. Polymer Sci. 22, 263 (1956).
34. R. Gabler, Kunststoffe Plastics 1, 5 (1956).
35. W. Griehl, Faserforsch. Textiltechn. 5, 155 (1954).
36. R. G. Griskey, J. K. P. Shou, Modern Plastics 45, June, 148 (1968).
37. H. Hannes, Kolloid-Z. - Z. Polymere 250, 765 (1972).
38. D. Heikens, P. H. Hermans, H. A. Veldhofen, Makromol. Chem. 30, 154 (1959).
39. W. Heilman, Ind. Eng. Chem. 48, 821 (1956).
40. K.-H. Hellwege, R. Hoffmann, W. Knappe, Kolloid-Z. - Polymere 226, 109 (1968).
41. H. Hendus, K. Schmieder, G. Schnell, K. A. Wolf, Festschrift Carl Wurster, BASF 1960.
42. E. Heyman, 27. ANTEC, SPE-J 15, 373 (1969).
43. D. R. Holmes, C. W. Bunn, C. J. Smith, J. Polymer Sci. 17, 159 (1955).
44. K. Hoshino, W. Watanabe, J. Chem. Soc. (Japan), Pure Chem. Sect. 70, 24 (1949).
45. G. J. Howard, J. Polymer Sci. 37, 310 (1959).
46. D. Hummel, "Atlas der Kunststoffanalyse," C. Hanser, Muenchen, Verlag Chemie GmbH, Weinheim 1968.
47. K.-H. Illers, BASF, Ludwigshafen/Rh. unpubl. work 1966-72.
48. K.-H. Illers, H. Haberkorn, Makromol. Chem. 142, 31 (1971).
49. K.-H. Illers, H. Haberkorn, Makromol. Chem. 146, 267 (1971).
50. K.-H. Illers, H. Haberkorn, P. Simák, Makromol. Chem. 158, 285 (1972).
51. Imperial Chemical Industries Ltd., Plastics Division, Welwyn Garden City, Great Britain, Technical Literature Maranyl (1970).
52. K.-H. Inderfurth, "Nylon Technology," McGraw-Hill Book Co. Inc., New York 1953.
53. M. Inoue, J. Polymer Sci. A 1, 2697 (1963).
54. Y. Ito, Kobunshi Kagaku 18, 124 (1961).
55. H. Jacobs, Thesis TH Aachen 1960.
56. J. Janáček, J. IUPAC-Symposium Prag 1965; Preprint P 54.
57. M. Jambrich, I. Diačik, I. Mitterpach, Faserforsch. Textiltechn. 23, 1, 28 (1972).
58. Y. Katayama, K. Yoneda, Rev. Elec. Commun. Lab. 20, 921 (1972).
59. Y. Kinoshita, Makromol. Chem. 33, 1 (1959).
60. H. Klein, H.-P. Weiss, BASF, Ludwigshafen/Rh: unpublished experiments 1972.
61. H. Klein, BASF, Ludwigshafen/Rh; unpublished measurements 1972.
62. I. Krause, A. J. Segreto, H. Przirembel, L. Mach, Mater. Sci. Eng. 1, 239 (1966).
63. H. Kunze, Thesis, TH Aachen 1958.
64. F. N. Liberti, B. Wunderlich, J. Polymer Sci. A-2, 6, 833 (1968).
65. P. Lohe, Kolloid-Z. 203, 115 (1965).
66. F. C. Magne, H. J. Portas, H. Wakeman, J. Am. Chem. Soc. 69, 1896 (1947).
67. D. W. McCall, E. W. Anderson, Polymer 4, 93 (1963).
68. N. G. McCrum, B. E. Read, G. Williams, "Unelastic and Dielectric Effects in Polymer Solids," J. Wiley, London 1967.
69. E. Meier, Thesis, TH Aachen 1968.
70. V. I. Meškov, A. A. Temkina, L. P. Matovea, Soviet. Beitr. Faserforsch. Textiltechn. 3, 285 (1966).
71. R. W. Moncrieff, "Artificial Fibers," Wiley Inc. New York 1950.
72. A. Mueller, R. Pflueger, Kunststoffe 50, 203 (1960).
73. N. Ogata, Makromol. Chem. 42, 52 (1960).
74. N. Ogata, Bull. Chem. Soc. (Japan) 33, 1584 (1960).
75. A. Okada, Chem. High Polymers (Tokyo) 7, 122 (1950).
76. D. K. Owens, R. C. Wendt, J. Appl. Polymer Sci. 13, 1741 (1969).
77. J. Pellon, W. G. Carpenter, J. Polymer Sci. A 1, 863 (1963).
78. G. Pezzin, G. B. Gechele, J. Appl. Polymer Sci. 8, 2195 (1964).
79. R. Pflueger, in Kunststoff-Handbuch, Vol. 6, Polyamide, C. Hanser, Muenchen, 1966.
79a. F. Pluemer, Kolloid-Z. 176, 108 (1961).
80. H. H. Racké in G. Schreyer "Konstruieren mit Kunststoffen," Hanser, Muenchen 1972.
81. A. Reichle, A. Prietzschk, Angew. Chem. 74, 562 (1962).
82. P. N. Richardson, SPE-J. 16, 1324 (1960).
83. L. G. Roldan, H. S. Kaufman, J. Polymer Sci. B 1, 603 (1963).
84. M. Rothe, F.-W. Kunitz, Ann. Chem. 609, 88 (1957).
85. K. Ruppmich, E. Zahn, Kunststofftechnik 12, 5 (1973).
86. C. Ruscher, H. J. Schroeder, Faserforsch. Textiltechn. 11, 165 (1960).
87. S. Satoh, J. Sci. Res. Inst. (Tokyo) 43, 79 (1948).
88. J. A. Sauer et al, J. Polymer Sci. 44, 23 (1960) and J. Colloid Sci. 12, 363 (1957).
89. J. R. Schaefgen, P. J. Flory, J. Am. Chem. Soc. 70, 2709 (1948).
90. E.-O. Schmalz, H. Grundke, Faserforsch. Textiltechn. 20, 377 (1969).
91. E.-O. Schmalz, Faserforsch. Textiltech. 20, 533 (1969).
92. G. F. Schmidt, H. A. Stuart, Z. Naturforsch. 13a, 222 (1958).
93. K. Schneider in Vieweg-Mueller "Kunststoff-Handbuch," Vol. 6, "Polyamide", Hanser, Muenchen 1966.
94. R. J. Shaw, Chem. Eng. Data 14, 461 (1969).
95. W. P. Slichter, J. Appl. Phys. 26, 1099 (1955).
96. H. W. Starkweather, G. E. Moore, J. E. Hansen, T. M. Roder, R. E. Brooks, J. Polymer Sci. 21, 189 (1956).
97. H. W. Starkweather, R. E. Moynihan, J. Polymer Sci. 22, 363 (1956).
98. H. W. Starkweather, R. Brooks, J. Appl. Polymer Sci. 1, 236 (1959).
99. H. W. Starkweather, J. Appl. Polymer Sci. 2, 129 (1959).
100. H. W. Starkweather, J. F. Whitney, D. R. Johnson, J. Polymer Sci. A 1, 715 (1963).
101. H. Tautz, L. Strobel, Kolloid-Z. Z. Polymere 202, 33 (1965).
102. G. B. Taylor, J. Am. Chem. Soc. 69, 635 (1947).
103. D. W. van Krevelen, P. J. Hoftyzer: "Properties of Polymers-Correlation with Chemical Structure," Elsevier Publ. Comp., Amsterdam 1972.
104. VDI/VDE-Werkstoffblatt 2479, Part 1, VDI-Verlag, Duesseldorf 1974.
105. C. D. Vogelsong, J. Polymer Sci. A 1, 1055 (1963).
106. R. Waack, Ind. Eng. Chem. 47, 2524 (1955).
107. J. H. Wakelin, A. Sutherland, L. R. Beck, J. Polymer Sci. 42, 278 (1960).
108. L. G. Wallner, Makromol. Chem. 79, 279 (1948).
109. C.-D. Weiske, Kunststoffe 61, 518 (1971).
110. R. C. Wilhoit, M. Dole, J. Phys. Chem. 57, 14 (1953).
111. German Patent application DAS 1,694,984.

PROPERTIES OF CELLULOSE MATERIALS*

E. Treiber

Svenska Träforskningsinstitutet
Stockholm, Sweden

CONTENTS

		Page				Page
A.	Unit Cell Dimensions	V- 87	L.	Concentration Dependence of Viscosity		V-109
B.	State of Order, Crystallinity	V- 90	M.	Ratio of Intrinsic Viscosities in Different Solvents		V-111
C.	X-Ray Orientation	V- 94	N.	Geometric Dimensions in Solution		V-111
D.	Density	V- 94	O.	Second Virial Coefficients		V-112
E.	Thermal Properties	V- 94	P.	Diffusion and Sedimentation Data		V-112
F.	Refractive Indices	V- 98	Q.	First Order Rate Constants and Energy of Activation of Homogeneous Hydrolysis		V-112
G.	Infrared Spectrum	V- 98	R.	Electrical Properties		V-112
H.	Nonglucose Carbohydrates in Cellulosic Materials	V-100	S.	Physical and Mechanical Properties		V-115
I.	Solvents for Cellulose	V-101	T.	References		V-116
K.	Viscosity-Molecular Weight Relationships	V-103				

V-87

A. UNIT CELL DIMENSIONS [a]

Cellulose Modification	Crystal System	Remarks	Dimension of Unit Cell in [Å] a	b (fiber axis)	c	β	Space Group	Notes on Origin and Preparation of Sample	Ref.
Cellulose I	monoclinic	Meyer-Misch cell	8,35	10,3	7,9	84°	P 2$_1$		277, 459
			8,23	10,28	7,84	84°			5
			8,203	10,295	7,836	84°23'			411
			8,22	10,30	7,81	85,1°	P 2$_1$(C$_2^2$)	mean value of measurements of different samples [b] ramie	234
			8,167	10,306	7,844	84°5'20" [c]		ramie	204
			8,174±0,007 [c]	not measured	7,889±0,005 [c]	83°28' ±0' [c]		cotton	454
			8,171±0,032	not measured	7,846±0,019	83°37' ±8'		ramie	454
			8,181±0,039	not measured	7,873±0,025	82°46' ±24'		linen	454
		Relationship between	8,205±0,035	not measured	7,908±0,011	81°51' ±11'		bacterial cellulose	454
		the Meyer-Misch unit	8,212	not measured	7,882	83°21'		valonia	454
		cell and the unit	8,283		7,963	83°33'		cladophora prolifera	454
		cell suggested by	8,17	10,34	7,85	83,6°		ramie	191
		Ellis and Warwicker	8,20	10,34	7,84	82°		valonia	92
		for cellulose	8,20	10,30	7,90	83,3°		surgical cotton	85
		mean value:	8,198±0,034 [c]	10,317±0,021 [c]	7,873±0,042 [c]	83,34±0,91 [c]			

*Based on a table in the first edition by T. Lukanoff and B. Philipp, Deutsche Akademie der Wissenschaften, Institut fuer Faserstoff-Forschung, Teltow-Seehof, DDR.

Cellulose Modification	Crystal System	Remarks	a	b (fiber axis)	c	β	Space Group	Notes on Origin and Preparation of Sample	Ref.
Cellulose I (Cont'd.)		comparison of two well-crystallized samples	8.26	10.30	7.83	85.2°		ramie	234
			8.18	10.30	7.79	85.0°		ramie	356
		Sponsler-Dore-cell;Sauter cell	10.8	10.4	11.8	85°	P 1	bacterial cellulose	
	triclinic (monoclinic)	modif. Sponsler-Dore cell (Ellis-Warwicker (154))	2 x 8.20	10.34	2 x 7.84	82°	P 1	valonia (electron diffraction)	92
			16.43	10.33	15.70	96°58'	P 1	Chaetomorpha melagonium	315
Cellulose II	monoclinic	Andress-cell	8.14	10.3	9.14	62°	$P2_1$	dry	6,354,278
			8.02	10.3	9.03	62.8°		surgical cotton, mercerized, dry	85
			8.015	10.298	9.126	62°37.5'		dry	204
			7.965		9.242	62°9'		dry	235
			7.963	10.28	9.268	62.14°		dry	220
		corrected value!	7.94		8.82	64.0°		bone dry	133,411
			7.641	10.280	9.200	63.51°		bone dry	198
			7.979	10.292	9.242	62°8'		bone dry	411
			7.917		9.150	62°45'		mean value of different samples;dry	411
			7.970±0.055		9.219±0.088	62°14'±10'		ramie, mercerized, dry	454
			8.059		9.382	61°45'		linen, mercerized, dry	454
			8.014		9.149	62°26'		mercerized bacterial cellulose, dry	454
			7.831-7.965		9.134-9.291	62°36'-63°13'		different viscose fibers, dry	454
			7.955		9.167	63°3'		cuprammonium rayon, dry	454
			7.917		9.083	62°42'		Fortisan, dry	454
			7.936		9.077	63°		high tenacity viscose rayon, dry	454
			8.001		9.061	63°23'		cellopentaose)	454
			8.164	10.302	9.162	62°10'		mercerized ramie and Fortisan, swollen in water	204
			8.185		9.222	61°53'		ramie, mercerized, swollen in water	235
		corrected value!	8.35		9.10	61°18'		swollen in water	133
			8.121-8.573		9.277-9.550	61.41-58.63°		mercerized cellulose I and regenerated cellulose, swollen in water	220
			8.033-8.238	10.3	9.303-9.414	61.22-59.91°		different treated samples, swollen in water	198
			10.03	10.4	9.98	52°		"water cellulose" or cellulose hydrate II	354,128
Cellulose III	tetragonal (?)		15.5		15.5	90°		average values	146
	monoclinic	III α	7.8	10.3	10.0	58°		extreme limits of measurements with different samples obtained via NH_3-cellulose from native cellulose	88
			7.22-7.82	10.28	9.93-10.0	57°7'-58°6'			237
			7.87	10.3	10.1	58°		decomposition of ethylamine cellulose from cotton	25
			8.05	10.4	10.27	57.5°			372
	monoclinic	III β	7.63-7.72	10.28	9.80-9.89	58°8'-60°6'		extreme limits of measurements with different samples obtained via NH_3-cellulose from mercerized cellulose	237

Cellulose Modification	Crystal System	Remarks	Dimension of Unit Cell in [Å]				Space Group	Notes on Origin and Preparation of Sample	Ref.
			a	b (fiber axis)	c	β			
Cellulose IV (high temperature cellulose)	tetragonal		7.95	10.2	7.95	90°			166
	tetragonal (?)		8.068		7.946	90°			454
	orthorhombic		8.11	10.3	7.9	90°			227
			8.12	10.3	7.99	90°			86
			7.94-8.19	10.28	7.92-8.02	90°		extreme limits from measurements of different samples.	237,236
Cellulose X	monoclinic		8.10	10.3	8.16	78° 18'		obtained by treatment of cellulose I or II with 40% HCl	88,85,83
	orthorhombic		8.12	10.3	7.99	90°			88,86

a) For critical comments on the crystal structure (unit cell), proposed for cellulose I and II, see Ref. 88, 154, 205 and 360. b) Additional data may be found in Ref. 360 and 210. c) standard deviation.

B. STATE OF ORDER, CRYSTALLINITY

Remarks:

Within the rather broad spectrum of results from various physical and chemical methods for determination of average "degree of order" (so called "crystallinity") of cellulose samples, only X-ray diffraction data are presented in the table B 3 in detail.

Definitions:

For comparing quantitatively samples of different origin with regard to state of order, the following terms are used:

Degree of Crystallinity after Hermanns and Weidinger (135-138, 141, 142): $D C = \dfrac{I_{cr}}{I_{cr} + I_{am}}$

I_{cr} = scattering intensity of ordered regions; I_{am} = scattering intensity of disordered regions

Crystallinity Index after ANT-WUORINEN (7): $CR\ I = \dfrac{Am\ W}{Cr\ H}$

Am W = half angular width in radians of the overlapping 101 and 002 reflections, measured in radians; Cr H = ratio of crystalline height to total distance between zero-line and total blackness

Crystallinity Ratio (index of order) after ANT-WUORINEN and VISAPÄÄ (11, 436): $Cr\ R\ (I\ O) = 1 - \dfrac{Am\ H}{Cr\ H} = 1 - \dfrac{Am\ H}{Tot.\ H - Am\ H}$

Am H = height of interference minimum at $2\Theta = 18\text{-}19^{\circ}$ for cellulose I and $2\Theta = 13\text{-}15^{\circ}$ for cellulose II
Tot. H = height of interference maximum at $2\Theta = 22\text{-}23^{\circ}$ for cellulose I and $2\Theta = 19\text{-}22^{\circ}$ for cellulose II

Table B 1 Average Ordered Fraction Measured By Various Techniques (451)

Technique	Cotton	Mercerized cotton	Wood pulps	Regenerated cellulose
X-ray diffraction	0.73	0.51	0.60	0.35
Density	0.64	0.36	0.50	0.35
Deuteration or moisture regain	0.58	0.41	0.45	0.25
Acid hydrolysis	0.90	0.80	0.85	0.70
Periodate oxidation	0.92	0.90	0.92	0.80
Iodine sorption	0.87	0.68	0.85	0.60
Formylation	0.79	0.65	0.75	0.35

Table B 2 Average Ordered Fraction of Cotton and Linters Measured By Various Techniques (451)

Technique	Average crystallinity value for cotton and linters (%)	Number of publications involved in the average crystyllinity' value	Approximate average crystallinity value for wood pulps	Approximate average crystallinity value for regenerated celluloses
Physical:				
X-ray diffraction	73	25	60	35
Density	64	18	50	35
Absorption and "Swelling Chemical":				
Deuteration	58	5	45	25
Moisture regain	58	-	50	25
Hailwood-Horrobin	67	3	55	35
Non-freezing water	c. 85	1	-	50
Acid hydrolysis	90	26	85	70
Alcoholysis	c. 90	4	-	-
Periodate oxidation	92	4	92	80
Nitrogen tetroxide oxidation	c. 70	1	-	50
Formylation	79	7	-	35
Iodine sorption	87	9	85	60
"Non-swelling" Chemical:				
Chromic acid oxidation	99.7	1	-	-
Thallation	99.6	2	-	-

Most of these values are general averages of several published values: the number of publications involved in each average is given in the third column. The term "crystallinity value" does not imply that any particular method measures crystallinity or order in any strictly defined sense, or that the various methods measure precisely the same type or level of order/disorder. With the chemical methods for instance, the value given is merely the fraction of the material that is not readily accessible to, and thus not able to react with the particular reagent.

Table B 3 Degree of Crystallinity (D C), Crystallinity Index (Cr I) and Crystallinity Ratio (Cr R (I O))

Type of Cellulose	D C	Ref.	Cr I	Ref.	Cr R (I O)	Ref.	% crystallinity (other methods)	Ref.
Ramie	0.60 - 0.71	138, 141, 142, 187			0.62 - 0.69	187	44 - 47	437
Ramie, mercerized	0.54 - 0.74	342 187			0.747 a)	187	45	437
Jute, Corchorus olitorius	0.49	140						
Jute, Corchorus capsularis	0.58 - 0.66	343					65 ± 3	456
	0.47 - 0.67	343						
Flax, bleached	average 0.67; range 0.64-0.69	7	average 0.74; range 0.70-0.77	7			70	444
Cotton, different origins	0.69 - 0.71	141			0.727 a)	187	79	373
	0.708	222					47	437
Cotton, mercerized	average 0.70; range 0.66-0.74	7	average 0.70; range 0.66-0.74	7			51	444
							42	437
Cotton, linters	0.70	141	average 0.72-0.73; range 0.69-0.76	7, 8, 439	0.82 - 0.86	11, 436	57	87
	0.727	222						
Cotton linters, mercerized	average 0.48; range 0.41-0.55	7	average 0.43; range 0.32-0.50	7, 9	0.56-0.59	436, 9, 12		
Cellulose from valonia ventricosa	average 0.68; range 0.65-0.70	138						
Cellulose from Douglas fir:								
normalwood					0.543	232		
compression wood					0.464			
Cellulose from Cotton wood:								
normal wood					0.545	232		
tension wood					0.666			
Cellulose from Pinus radiata (5, 10 and 15 annual ring)	0.73 - 0.55	339						
Cellulose from Quercus prinus:								
Xylem-cellulose					0.5107	33		
Phloem-cellulose					0.4337			
Rhytidome-cellulose					0.4309			
Bacterial cellulose (Acetobacter xylinum)	0.40	138						
Different wood pulps	0.62 - 0.70	138, 141, 7 187	0.61 - 0.68	7, 8, 9, 439	0.58 - 0.72 0.619 a)	11, 436, 12, 187 187	56 - 68	63
Linen							66 ± 3	456
Hemp							60 ± 3	
Manila hemp							42 ± 4	456
Sisal							38 ± 2	
Dissolving pulp	0.638	222						
Acetate grade wood pulp	0.49	140	0.35 - 0.43	9	0.45 - 0.56	436, 12	52	87
Different wood pulps, mercerized								
Acetate grade wood pulp, mercerized							49	87

Type of Cellulose	D C	Ref.	Cr I	Ref.	Cr R (I O)	Ref.	% crystallinity (other methods)	Ref.
Different cellulose samples								
Viscose rayon, normal grade	0.38 - 0.40	138, 141, 142, 252, 418 7	0.27 - 0.79	7, 439, 9, 418			36 - 44 35	351 87
average 0.27; range 0.13-0.36								
Viscose staple fibers, normal grade	0.30 - 0.39 0.400-0.408	138, 141, 418, 386 222	0.30 - 0.79	7, 418	0.34 - 0.38	8, 9, 20	33 - 36	374
average 0.27; range 0.30-0.52								
Viscose fibers, mercerized	0.40 - 0.52	142, 140, 252	0.36 - 0.43	439, 9	0.45 - 0.56	9, 20		
High tenacity viscose rayon	0.23 - 0.56	138, 141, 142, 252, 418, 26	0.28 - 0.78	439, 418, 218				
High wet modulus (HWM) viscose fibers (polynosics)	0.27 - 0.44	138, 141, 252, 418, 386	0.30 - 0.76	439, 418, 218			38	374
Polynosic-fiber							35.6	437
Cellophane	0.30 - 0.40	138, 141, 386					48	374
Regenerated cellulose precipitated from viscose as flake	0.45 - 0.53 0.43	139 412					26.9-27.1	316
Precipitated β-cellulose	0.50 - 0.62	412						
Cuprammonium rayon	0.40 - 0.44 0.406	137, 138, 142 222						
Cuprammonium rayon, mercerized	0.49	140						
Saponified acetate fibers	average 0.38	141						
Fortisan	0.39 - 0.54	141, 418, 386 139	0.33 - 0.86	439, 418, 218				
"amorphous" powder, obtained by dry-grinding viscose rayon	0.08	130, 131						
"amorphous" powder, obtained by dry-grinding viscose rayon, recrystallized	0.35 - 0.40	131						
Cellulose particles, obtained by acid hydrolysis ("cellulosic micelles")					0.85[a]	187		
Linters, treated with 80% ethylenediamine					0.46[a]	188		

[a] Crystallinity index according to Jayme & Knolle (187).

References page V-116

Table B 4 Crystallite Sizes

Type of cellulose	diameter [Å]	wide and thick [Å]	range of length [Å]	Ref.
Cotton	50 - 60			100
	50		500	74
different samples	58 ± 3			379
	60			165
	64			398
	50 - 100			352
	100			334
	146			148
	150			152
		62 x 49		315
		64 x 50		296
		93 x 32		221
		110 x 25		31
		110 x 69 - 72		437
		130 x 25		302
			400 - 2550	451
Ramie	16 - 26			318
	50 - 100			122
	59			398, 315
	68			148
	70			219
	< 400			155
		61 x 167		437
			600 - 2500	451
Hemp				148
	44			
	49 ± 3		189	456
Flax	51.5			148
Jute	55			148
	49 ± 4		174	456
Bamboo	< 70			100
Valonia		148 x 53		21
Chaetomorpha melagonium		169 x 114		315
Acetobacter xylinum		84 x 70		315
Tunicin		84 x 34		315
Linen	52 ± 3		195	456
Manila hemp	39 ± 2		140	456
Sisal	38 ± 3		130	456
Normal viscose fibers	50 - 70			374
HWM-viscose fibers	76			374
Polynosic fibers	93			374
isotropic "Hermans"-viscose fibers	50 - 200			410

C. X-RAY ORIENTATION

For evaluation of x-ray orientation of cellulose (157a, 219a) the "orientation factor" f_x (40a, 126a, 131) is used in the following table. As existing data are largely a function of the stretching conditions in the spinning process, only ranges of f_x are presented for regenerated cellulose fibers.

Nature of Material	Data	References
Ramie	0.97	131
Ramie, mercerized without tension	0.90	131
Ramie, mercerized and reorientated	0.98	131
Viscose rayon, normal grade	0.31 - 0.91	418, 218, 131, 439
Viscose staple fibers, normal grade	0.30 - 0.67	418, 386
Lilienfeld rayon	0.70 - 0.94	418, 131, 439
High-tenacity viscose cord rayon	0.39 - 0.998	142, 418, 172, 439
High wet modulus (HWM) viscose fibers	0.45 - 0.993	418, 172, 218, 439, 386
Isotropic viscose model fibers, without stretching	0.06 - 0.15	134
Isotropic viscose model fibers, after intense stretching	up to 0.90	134
Cuprammonium rayon	0.86	131
Fortisan	0.7 - 0.99	418, 172, 218, 386, 439

D. DENSITY

Type of Cellulose	$d_t^4 [g\ ml^{-1}] \equiv [Mg\ m^{-3}]$	References
Cellulose I	1.582 - 1.630[a]	198, 411, 233, 50, 208, 129, 253
Cellulose II	1.583 - 1.62[a]	411, 50, 199, 129
Cellulose IV	1.61[a]	411
Cotton	1.545 - 1.585	389, 445, 293, 70, 4, 127
Ramie	ca. 1.55	129, 338
Flax	1.541	377
Hemp	1.541	377
Jute	1.532	377
Wood pulps	1.535 - 1.547	120, 203, 157, 208, 201, 80, 4
Cuprammonium fibers	1.519 - 1.531	293, 70, 127
Polynosics	1.489 - 1.528	206, 54
Viscose fibers (and films) (rayon and staple)	1.508 - 1.548	293, 70, 231, 290, 289, 192
High tenacity viscose fibers	1.498 - 1.524	290, 174

[a] Calculated from x-ray data.

E. THERMAL PROPERTIES

Property	Type of Cellulose	value	Remarks	Ref.
Decomposition point, [°C]	cellulose	200 - 270		242
	cotton	150		299
	viscose rayon	180		299
Start of thermal degradation, [°C]	linters	225		362
	bleached sulphite pulp	225		
	kraft pulp	240		
	filter paper	220	under nitrogen	40
Fast endothermal degradation	linters	≈ 300		362
	bleached sulphite pulp	≈ 330		
Heat capacity, [kJ kg^{-1} K^{-1}]	cellulose	1.34		170
	cotton	1.22		254
		1.15-1.18	0 - 34°C	274
		1.32	0 - 100°C	
		1.327 - 1.257		387
		1.273 - 1.357		103
	mercerized cotton	1.235		254

References page V-116

Property	Type of Cellulose	Value	Remarks	Ref.
Heat capacity, $[kJ\ kg^{-1}\ K^{-1}]$ (Cont'd.)	ramie	1.775		291
	flax	1.344 - 1.348		103
	hemp	1.327 - 1.352		103
	jute	1.357		103
	viscose rayon	1.357		103
		1.415 - 1.595		291
	cuprammonium fibers	1.357		103
	paper	1.17 - 1.32		274
Heat of crystallization, $[kJ\ kg^{-1}]$	cellulose, extrapolated to 100% crystallinity	105.9 ± 5.0		51
	cellulose I	121.8		401
	cellulose II	134.8		401
Heat of recrystallization, $[kJ\ kg^{-1}]$	amorphous cellulose → cellulose II	41.9	by wetting	130
Heat of transition, $[kJ\ kg^{-1}]$	cellulose I → cellulose II	38.1		229
Heat of combustion, $[kJ\ kg^{-1}]$	linters	-17438.9	30°C, constant pressure	190
	wood pulp	-17472.4		
Heat of formation, $[kJ\ kg^{-1}]$	cellulose	5949.7		190
Heat of solution, $[kJ\ kg^{-1}]$ of dry material	cotton in Cuene [e)]	108.0		51
	cotton in Et_3PhNOH	142.5		248
	rayon in Cuene	95.5		51
	rayon in Et_3PhNOH	138.3 - 153		288
	ball milled cellulose in Et_3PhNOH	243		248
	cellulose II in Et_3PhNOH	182.7		248
Differential heat of sorption of liquid water[a)], $[kJ\ kg^{-1}]$	cotton	1235;1194-1327	at zero regain	110,309,345
		1090	at zero regain, from sorption isotherm	402
		1332	at zero regain, 27°C	107
		972-1040;964-1006	at zero regain	16,129
		1047	at zero regain, 25°C	14
	Egyptian cotton	1214	at zero regain, 26°C	384
	cotton, dried by solvent exchange	1047	at zero regain, 26°C	384
	cotton mercerized	1173	at zero regain	345
	linters	1143	at zero regain	22
	linters mercerized	1327	at zero regain	22
	flax	1248, 1206	at zero regain	110,16
	wood pulps	1173-1256;1098	at zero regain	22,298
	Fortisan	1235;1169	at zero regain	345,16
	cuprammonium fibers	1214;1115	at zero regain	
	normal viscose fibers	1173-1215;1050;984; 917-1022;1277	at zero regain	
	viscose fiber	1549	at zero regain, 26°C	384
	viscose fiber, dried by solvent exchange	1172	at zero regain, 26°C	384
	cotton	293	at 60% R.H. [d)]	110
		201-281	at 65% R.H.	299
	cotton, mercerized	230	at 60% R.H.	345
	viscose rayon	318	at 60% R.H.	110
		268-297	at 65% R.H.	110
Heat evolved by 1 g of fibers going from 40% R.H. to 70% R.H. in Joule [d)]	cotton	84		299
	viscose rayon	168		299
Integral heat of wetting [b)] of dry material $[kJ\ kg^{-1}]$ Wetting in water at zero regain	cotton	33.9 - 56.2	at 20-25°C	110,345,129, 14,22,57,76, 230
		40.6	at 20°C	258
	average value	46.5	at 20-25°C	451
	bleached	40.0;39.0		258,350
	mercerized	62.0-78.8;62.6	at 20-25°C	345,22,230, 297,258
	mercerized, average value	75.2	at 20-25°C	451
	ball milled	90.5	at 20-25°C	441

Property	Type of Cellulose	Value	Remarks	Ref.
Integral heat of wetting (Cont'd.)				
Wetting in water at zero regain				
	linters	44.8 - 57.8	at 20-25°C	51, 22, 441, 310
	ramie	34.4 - 48.6	at 20-25°C	129, 230, 76
	flax	39.4 - 54.4	at 20-25°C	110, 76
	wood pulps	49.4 - 71.2	at 20-25°C	22, 57, 230, 76
	regenerated cellulose	83.8 - 106.4	at 20-25°C	22
	Fortisan	76.2	at 20-25°C	345
	cuprammonium rayon	93.8	at 20-25°C	110
	viscose fibers	68.2 - 105.5	at 20-25°C	51, 110, 345, 22, 230, 76, 442, 322
		68.2 - 82.5	at 20-25°C	350
	polynosic fibers	83.1 - 87.8	at 20°C	55
Wetting in methanol	cotton	7.3	at 20-25°C	451
Wetting in bases, acids and salt solutions, $\Delta H_{salt} - \Delta H_{H_2O}$ in [kJ kg^{-1}]				
	cotton	10.2	5% NaOH	228
		62.2	17.5% NaOH	
		62.0	18% NaOH	24
		83.8	18% NaOH	196
	cotton, mercerized	56.4	18% NaOH	228
	ramie	67	20% NaOH	319
	wood pulp	11.2	5% NaOH	228
		78.3	17.5% NaOH	228
	wood pulp, different degree of refining	67.0 - 81.0	20% NaOH, 25°C	369
	dissolving pulp	$\approx$ 68	18% NaOH, 20°C	440
	different viscose fibers	25 - 47	5% NaOH	350
		31.8 - 36.4	5% NaOH	175
		62.8 - 67.8	17.5% NaOH	175
	cotton	4.6	0.1 M H$_2$SO$_4$	258
		3.5	3 M NaCl	197
		3.1	3 M KCl	
		5.4	3 M MgCl$_2$	
		3.8	3 M CaCl$_2$	
Flammability of fibers (ignition temperature; ^{o}C)				
	cotton	390, 400		77, 102
	viscose rayon	420		77
limiting oxygen index	cotton	0.184		102
	viscose rayon	0.197		102
Thermal conductivity $\lambda^{c)}$, [W m^{-1} K^{-1}]	cotton	0.071	density 0.5 g/ml	463
	rayon	0.054 - 0.07		102
	sulphite pulp, wet	0.8		111
	sulphite pulp, dry	0.067		
	laminated kraft paper	0.13		404
	different papers	0.029-0.17	compare Fig. 1	405
	alkali cellulose	0.0581	density 0.2 g/ml	420
		0.0465		90
		0.0674		28

Property	Type of Cellulose	Value	Remarks	Ref.

Thermal conductivity (Cont'd.)

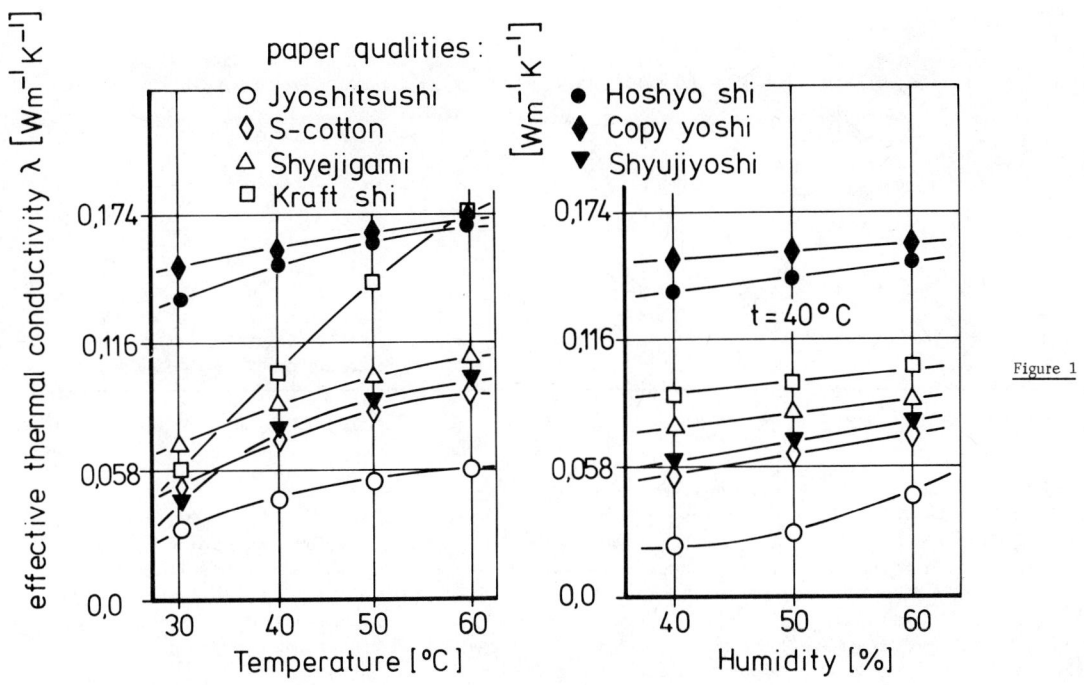

Figure 1

Thermal expansion coefficient

Linear expansion α: [K^{-1}]	different papers	$2 - 7.5 \times 10^{-6}$	machine direction	226
		$7.9 - 16.2 \times 10^{-6}$	cross machine direction	
	cotton, rayon	80×10^{-6}		458
Cubic expansion, [$ml \cdot g^{-1} \cdot K^{-1}$]	various celluloses	$5.1 - 6.0 \times 10^{-4}$		341
	linters	4×10^{-4}	$-30°C$ to $+25°C$	421

Transition points

Glass transition [$°C$]	cellulose	230		18
		220 - 245		105
Secondary transitions	cellulose	19 - 33		341
		25		443
	paper	$-20; +15, \approx 35, +60$		225, 226, 18
	cellulose	≈ 200		18

a) Differential heat of sorption (= heat of absorption) is the heat evolved when one gram of (liquid) water is absorbed by an infinite mass of material at a given moisture regain and is expressed in [$kJ\ kg^{-1}$] of water absorbed. Values for the absorption from the vapour can be obtained by adding the latent heat of vaporization of water. The values for cellulosic material average between 2500 and 3000 [$kJ\ kg^{-1}$] in the middle range of humidities.

b) Integral heat of sorption (= heat of wetting) is the heat evolved when one gram of a cellulosic material is completely wetted at a given moisture regain [$kJ\ kg^{-1}$].
The heat of wetting decreases very rapidly with increasing moisture content of the sample and at a moisture content $>20\%$ there is practically no heat of wetting.

c) The thermal conductivity depends largely on relative humidity and packing density of the material.

d) R.H. = relative humidity.

e) Cuene = Cupriethylendiamine.

PROPERTIES OF CELLULOSE MATERIALS

F. REFRACTIVE INDICES

Type of Cellulose	n	n	n - n	References[*]
Ideally oriented native cellulose fiber (cellulose I)	1.618	1.544	0.074	129
Ramie	1.595 - 1.601	1.525 - 1.534	0.061 - 0.071	129, 99, 194, 337, 17, 98
Flax	1.571 - 1.595	1.517 - 1.532	0.054 - 0.067	337, 98, 144
Hemp	1.585 - 1.591	1.526 - 1.530	0.055 - 0.065	144, 97
Cotton	1.576 - 1.595	1.527 - 1.534	0.045 - 0.062	99, 337, 98, 145, 174, 276
Jute	1.577	1.536	0.041	98
Ideally oriented regenerated cellulose fiber (cellulose II)	1.578	1.523	0.055[a] / 0.043[b]	129, (32) / 129
Fortisan			0.045	149
High tenacity rayon	1.542 - 1.553	1.509 - 1.513	0.029 - 0.043	174, (337, 417, 29)
Polynosics	1.556 - 1.570	1.518 - 1.531	0.036 - 0.043	174, 206, (32)
Normal viscose fibers	1.529 - 1.547	1.512 - 1.520	0.013 - 0.034	348, (337, 174, 417, 29, 206, 23)
Cuprammonium fibers	1.548 - 1.571	1.519 - 1.534	0.018 - 0.037	129, 337, 145, 149, (337, 98)

[*] References in parenthesis give additional data.   [a] bone dry.   [b] conditioned.

G. INFRARED SPECTRUM[*]

The following table gives selected data on position and - as far as possible - assignment of IR-absorption bands of cellulose I and cellulose II according to decreasing wave number. Data on cellulose III and cellulose IV were omitted, as the few references (260, 262, 151, 361, 321a) available barely include assigments, and as it is generally agreed that the IR-spectra of cellulose III and cellulose IV resp. closely resemble those of cellulose II and cellulose I resp. Even between cellulose I and cellulose II the differences are rather small, the bands being generally mor diffuse with cellulose II. Differences in intensity of absorption with conversion of cellulose I to cellulose II or between "Valonia type" and "ramie type" of cellulose I are marked in the following table by underlining (increase) or bracketing (decrease) of the wave number concerned.

Position of Absorption Bands in $[\text{cm}^{-1}]$				Assignment	References
Valonia, Bacterial Cellulose	Cellulose I	Ramie, Cotton	Cellulose II		
		6770	6770	OH stretching, overtone	27
	5190		5190	absorbed H_2O, overtone	27
	4780		4780	OH and CH deformation + OH stretching	27
			4560	observed with cellophane only	27
	4365		4365	C-O stretching + OH stretching or CH_2 bending + CH_2 stretching	27
	4235		4235	OH and CH deformation + CH and CH_2 stretching	27
	3990		3970	C-O stretching + CH and CH_2 stretching	27
	3125-3660			free OH and bonded OH stretching	161
			3484-3490 ⎫	OH stretching of crystalline part of	260, 262, 361
			3444-3450 ⎭	cellulose II after elimination of the amorphous part	240, 259
	3200-3400		3200-3400	OH stretching, bands of H bonds	260, 151
	3401-3405⊥			OH stretching	260, 422, 261
			3374-3394	OH stretching, maximum of absorption depending on origin of cellulose II	38
	3375			OH stretching	240, 353
			3350-3360	broad band of OH stretching, obviously not resolved completely	262, 255
3350∥		(3350)∥		OH stretching	240, 422, 353
	3340-3345		(3340)	OH stretching	361, 261, 162
			3322	OH stretching	259
	3305⊥-3309⊥			OH stretching	260, 240, 261, 353
		3300∥		OH stretching	422
	3275			OH stretching	260, 240, 353
3243-3245				OH stretching	260, 261, 353
			3163	OH stretching	259

[*] These spectroscopical data were kept in this edition because of the large amount or information which apparently cannot be found elsewhere.

Valonia, Bacterial Cellulose	Cellulose I	Ramie, Cotton	Cellulose II	Assignment	References
	2970 ‖			CH stretching	240
		2967 ‖		antisym. CH_2 stretching	422
	2960			antisym. CH_2 stretching	396, 397
	2945			antisym. CH_2 stretching	151, 240
2914⊥		2910 ⊥		CH stretching	151, 240, 422, 396
	2900		2900	broad band of CH stretching	361
2897⊥				CH stretching	151, 240
		2880-2890		CH stretching	396, 397
	2870 ⊥			CH stretching	151, 240
	2853 ‖			sym. CH_2 stretching	240, 422, 257
	1760			C=O band from ester groups	161
	1730-1740			C=O stretching of carboxyl or lacton groups	161, 10, 13, 321
	1635-1670			absorbed H_2O	151, 161, 422, 162, 397, 10, 241
	1550-1650			COO' stretching	397
	1590			COO' stretching	161
	1560			COO' stretching	150
	1530			COO' stretching	10
	1455 ⊥			OH in plane bending	241
	1446 ⊥			OH in plane bending, "crystallinity band" (321)	422
	1440			OH in plane bending	151
	1426-1430 ‖		(1426-1430 ‖)	CH_2 bending	151, 361, 422, 353, 396, 257, 241, 93, 271, 438
	1380		1380	CH bending, same intensity with cellulose I and II	361, 241
	1350-1450		1350-1450	COO' stretching	397
	1370		1370	COO' stretching	397, 150, 93
	1374 ‖			CH bending	241
	1370			CH bending	151, 161
	1365			CH bending	422
	1350-1355		1355		361, 93
	1340		(1340)	OH in plane bending	396, 93, 438
1336				OH in plane bending	241
	1330-1335			OH in plane bending	151, 161, 422
		1328		OH in plane bending	241
	1320		(1320)	OH in plane bending or CH bending	161, 396, 397, 438, 461
1317 ⊥				CH_2 wagging	241
	1315			CH_2 wagging	151
	1310			OH bending	422
	1290		1290		438
1282		(1282 ‖)	1280	CH bending	151, 241
		1275 ⊥			241
	1250				241
	1240		1240		361, 397
	1230-1235		1235	OH in plane bending?	151, 241, 438
	1210		1210		361
	1205 ⊥		(1205)	OH in plane bending	151, 241, 438
	1170		(1170)		438
	1164		(1164)	stretching of C-O in ring or bending of C-OH	161, 162
	1162 ‖			antisym. bridge oxygen stretching	151, 241
	1125 ‖		(1125)		241
	1120			antisym. in phase ring wagging	151
	1119			C-O-C stretching	161
	1115		1115	"association band"	361
	1110		(1110)	antisym. in phase ring stretching	241, 271
	1060		~1065	OH bending	161, 93
	1058 ‖			C-O stretching	151
1040					93
	1035 ‖			C-O stretching	241
	1025			C-O stretching	151
	1015 ‖			C-O stretching	241
	(1005)		1005		438
	1000 ‖			C-O stretching, or C-C stretching	161, 241

Position of Absorption Bands in [cm^{-1}]

Position of Absorption Bands in $[cm^{-1}]$			Assignment	References
Valonia, Bacterial Cellulose	Cellulose I Ramie, Cotton	Cellulose II		
	(985-990)	990	C-O stretching	151, 271
		970		396
	(900-910)	900	CH bending or CH_2 stretching, "amorphous band"	161, 396, 397, 321, 438
	(893-895∥)	895	antisym. out-of-phase stretching	151, 241, 93
~ 800			ring breathing	241
~ 740 ⊥			CH_2 rocking	241
~ 700 ⊥			OH out-of-plane bending	241
663 ⊥			OH out-of-plane bending	257, 241
~ 650			OH out-of-plane bending	161, 422
	620	620		438
	560			361, 438
	520	525		361, 397
	500			397
	450	460		361, 397
	430	425		361, 397

H. NONGLUCOSE CARBOHYDRATES IN CELLULOSIC MATERIALS

Data were obtained by estimation of the "pentosan content" by destillative dehydratation and subsequent determination of the furfural obtained by volumetry, gravimetry, or colorimetry.

Because of the vaste of literature and the impossibility to discriminate between xylose and arabinose units by this method, data are given as a short survey only, omitting references to the original literature.

Nature of Material	Pentosan Content, [%]
Cotton linters, raw	1.5- 2
Cotton linters, purified	0.1- 0.5
Jute	9 -11
Hemp	3 -10
Flax	3 - 8
Wood pulps	
Coniferous sulfate, paper grade	5 -10
Coniferous prehydrolysis sulfate, rayon grade	2 - 4
Coniferous prehydrolysis sulfate, cord grade	1.5- 2.5
Coniferous prehydrolysis sulfate, acetate grade	0.2- 0.5
Coniferous sulfite, paper grade	4 - 7
Coniferous sulfite, rayon grade	2.5- 5
Coniferous sulfite-soda, rayon grade	2 - 4
Coniferous sulfite, alkali refined	1 - 2
Deciduous sulfate, paper grade	18 -28
Deciduous prehydrolysis sulfate, rayon grade	3 - 6
Deciduous (gumwood) prehydrolysis **sulfate, cord grade**	1.5- 2.5
Deciduous sulfite, paper grade	8 -20
Deciduous sulfite, rayon grade	3.5- 7
Annual plants, sulfate, paper grade	18 -31
Annual plants, prehydrolysis sulfate, rayon grade	3 - 8
Annual plants, sulfite, rayon grade	3 - 6

I. SOLVENTS FOR CELLULOSE

Dissolution of cellulose is always preceded by swelling and reaction with the hydroxyl groups of the carbohydrate chain. The following table somewhat arbitrarily covers "one-step solution processes" in solvents, where chemical reaction (formation of addition or complex compounds) and dissolution occur simultaneously. "Two-step solution processes" via well defined cellulose derivatives (esters like cellulose nitrate or acetate, ethers like methyl-cellulose or carboxymethyl-cellulose) are not included.

Solvent	Optimal Conditions for Dissolution	Remarks	Analytical and/or Technical Application	Ref.
Mineral acids				
HCl, HBr	35.5-42% HCl	fast hydrolytic degradation		455a
	~66% HBr	fast hydrolytic degradation		81
H_2SO_4	~71.2-72% H_2SO_4 (10.5 M) (0°C)			81,82
H_3PO_4	~83% H_3PO_4 (14.1 M)	less hydrolytic degradation than with HCl and H_2SO_4, incomplete dissolution of high-DP native cellulose	approximate estimation of chain length distribution by fractional dissolution	81,82
Trifluoroacetic acid	100%	small (?) degradation		425
Inorganic salts in aqueous solution [a]				
$LiCl$, $(ZnCl_2)$		strong swelling, dissolution accompanied by degradation	swelling in $ZnCl_2$-solution for manufacture of vulcanized fiber material ("vegetable parchment")	403,238
$Ca(SCN)_2$	62-68% $Ca(SCN)_2$ ~125°C	dissolution accompanied by degradation, high-DP native cellulose is barely dissolved	analytical separation of cotton and rayon	453,91
LiSCN, NaSCN, LiI, NaI, KI $K_2[HgI_4]$	~10% of salt, 100°C	low DP cellulose (β-cellulose) to about 50% dissolved		200,451 91
Strong bases in aqueous solution				
LiOH	7% LiOH -5°C	degraded cotton cellulose nearly completely dissolved		71
NaOH	9-10% NaOH -5°C	cellulose of low and medium DP and high accessibility only is dissolved	characterization of pulp and rayon by alkali solubility	71,347
NaOH + ZnO [d]	10% NaOH + 4% ZnO; 0-10°C 12% NaOH + 5% ZnO; ~10°C	nearly complete dissolution of rayon and hydrolytic degraded pulp, native wood pulp dissolved to about 50%	incomplete fractionation by dissolution	71,413,207
NaOH + BeO [d]	10-12% NaOH + 2.5% BeO; ≈20°C	≈20°C extraction of low DP cellulose from wood pulp	characterization of dissolving pulp	71,413
Tetraalkyl bases having a molecular weight of approximately >150				
Tetraethylammoniumhydroxide and some higher homologs	2.3 N TEAH; ambient temperature	complete dissolution of regenerated cellulose and of wood pulp up to DP 1500	viscosimetric estimation of DP	73,243,244
Trimethylbenzylammonium-hydroxide ("Triton B"), Dimethyldibenzylammonium-hydroxide ("Triton F")	Triton B: 3.5 M; 20°C ≈2 M; 0°C Triton F: 2 M; 20°C >1.8 M; 0°C	complete dissolution of native and regenerated cellulose	viscosimetric estimation of DP; production of filaments on an experimental scale	47,243
Tetraalkylphosphonium hydroxides				242
Tetraalkylarsonium hydroxides				242
Trialkylsulfonium hydroxides				380
Trialkylselenonium hydroxides				238

Metal complex solutions

Solvent	Optimal Conditions for Dissolution	Remarks	Analytical and/or Technical Application	Ref.
$[Cu(NH_3)_4](OH)_2$ "Cuoxam", "Cuam"	for viscosimetric estimation of DP usually 15 g/l Cu, 200 g/l NH_3; 20°C	complete and fast dissolution of even high DP native cellulose; considerable degradation by oxygen	viscosimetric estimation of DP, production of filaments on a technical scale; most widly used solvent for cellulose	371, 53
modif. Cuoxam		addn. of NaOH and CuCl		308
$[Cu(en)_2](OH)_2$ "Cupriethylenediamine", "Cuene", "CED"	for viscosimetric estimation of DP usually 0.5 mol/l Cu; 1 mol/l ethylenediamine 20°C	complete and fast dissolution of even high DP native cellulose; solution less sensitive to oxydative degradation than cellulose in cuoxam	viscosimetric determination of DP, investigation of physico-chemical properties of cellulose solutions	409, 147
Cu: Biuret: Alkali	0.12-0.5 mol Cu + equivalent amount of biuret, in 1.6 N KOH	7-8% cellulose of DP 800 are dissolved; solutions are sensitive against oxidative degradation	regeneration in form of films and fibers is possible	182
$[Co(en)_3](OH)_2$ "Cooxene"	6.8% (by weight) Co, 26.6% ethylenediamine, 1.2% glycerol	6-8% cellulose of DP 650 are dissolved by the solvent specified		176, 177
$[Ni(NH_3)_6](OH)_2$ "Nioxam"	3.76% Ni; 30.4% NH_3	4% cellulose of DP 620 are dissolved by the solvent specified		177, 178
$[Ni(en)_3](OH)_2$ "Nioxene"	~8% Ni in 30-40% aqu. soln. of ethylenediamine	2% cellulose of DP 620 are dissolved		177, 178
$[Zn(en)_3](OH)_2$ "Zincoxene"	4.5% Zn; 35.8% ethylenediamine 0°C	2.9% cellulose of DP 550 are dissolved		177, 179
$[Cd(en)_3](OH)_2$ "Cadoxene"	~4.5% Cd in 30% aqu. soln. of ethylenediamine	3% cellulose of DP 6?0 are dissolved rapidly to a colorless, clear solution	viscosimetric estimation of DP, investigation of physico-chemical properties of cellulose solutions [b]	177, 180
	4.5-5.2% Cd in 30% aqu. soln. of ethylenediamine + 0.2-0.5 M NaOH	cotton cellulose of DP 1200 is completely dissolved	characterization of rayon pulps by turbidimetry and particle counting	126
complex solution of the Na-salt of Fe-tartaric acid in diluted NaOH solution "FeTNa[c]) or in tartrate + NaOH. (FeTNa and FeTNa mod.)	390 g/l complex (Fe:tartaric acid: NaOH = 1:3:6) + 2.0 N NaOH + 30 g/l Na-tartrate	even native cellulose of high DP is completely dissolved; oxydative degradation is very small	estimation of DP; investigation of physico-chemical properties of cellulose solutions, characterization of pulps, morphological investigation of swelling and dissolution of cellulose	177, 181

Non-aqueous systems, nitrogen compounds containing

Solvent	Optimal Conditions for Dissolution	Remarks	Analytical and/or Technical Application	Ref.
Liquid NH_3, containing NH_4 SCN, NaSCN, $NaNO_3$ or NaI				358
Pyridine, containing Chloral				279
Acetonitrile/N_2O_4	10-30% N_2O_4; 70-90% N_2O_4			94
DMSO/N_2O_4	4 parts DMSO + 1-1.7 parts N_2O_4		spinnable solution	424
Dimethylformamide/N_2O_4		degradation negligible		370
Dimethylacetamide/N_2O_4		degradation negligible		370
SO_2/amines				114
SO_2/DMSO				114
SO_2/Dimethylformamide			high viscose, transparent and spinnable solutions	114
SO_2/Diethylamine/Acetonitrile	25% SO_2, 12% DEA, 63% Acetonitrile	dissolves only Cell I with a DP <1000		329
DMSO/Methylamine	16.5% CH_3NH_2, 5°C	dissolves cellulose with DP <500		214
N-Ethylpyridinium chloride + pyridine	75°C	very slight degradation, dissolves cellulose with DP <6500		164

a) dissolution effect: Chlorides: Li, Sn, Sb; Bromides: Li, Na, Ca; Zn, Sn; Iodides: Li, Na, Ca; Nitrates: Li; Thiocyanates: Li, Sr, Mg, Mn; Mercuri-iodides: Li, Na, K, Mn, Al; Perchlorates: Be, [b] Cellulose solutions in cadoxen may be diluted, e.g. 1:1 (Henley) 126). [c] EWNN in the German literature. [d] hydroxonionic compounds.

References page V-116

K. VISCOSITY-MOLECULAR WEIGHT RELATIONSHIPS

The constants K_m and a of the Mark-Houwink Equation

$$[\eta] = K_m \cdot M^a$$

are given in Table K 1.1, K 1.2 and K 2. as well as the corresponding constant K'_m relating intrinsic viscosity and degree of polymerization P

$$[\eta] = K'_m \, P^a \qquad (K'_m = K_m \cdot M_o^a; \ M_o = \text{molecular weight of the monomeric unit})$$

Dimension:

$$[\eta]_{ml/g} = 100 \, [\eta]_{dl/g} = 1000 \, Z_{\eta \, 1/g}$$

Abbreviations:

F	= Fractionated
U	= Unfractionated
OS	= Osmotic measurements
S	= Measurement by Sedimentation
SD	= Measurement by Sedimentation-Diffusion
LS	= Measurement by Light Scattering
K	= from kinetic data
comp.	= viscosimetric comparison
NC	= cellulose nitrate
CN	= Cellulose nitrate used for calibration
FeTNa	= Na-salt of tartratoferric acid in complex solution (EWNN)
Cuam	= Cuprammonium hydroxide
Cuen	= Cupriethylenediamine
Cadoxene	= Cadmiumethylenediamine

Symbols:

c	concentration
η	viscosity of the solution
η_o	viscosity of the solvent
$\eta_r = \eta/\eta_o$	relative viscosity
$\eta_{sp} = (\eta - \eta_o)/\eta_o$ or $\eta_r - 1$	specific viscosity
η_{sp}/c	reduced viscosity
$[\eta] = \lim\limits_{\substack{c=0 \\ G=0}} \eta_{sp}/c$	intrinsic viscosity
G	rate of shear (velocity gradient)

Remarks:

In general, the relative viscosity (η/η_o) of a polymer solution is strongly dependent on the polymer concentration c. In addition, it depends on the solvent power of the solvent, degree of substitution (of cellulose derivatives), shear rate in the viscosimeter and temperature. Measurements of viscosity can be disturbed by gel particles and, in the case of non-stable solutions, by degradation, for example by air oxydation.

The determination of the molecular weight (M_η) or degree of polymerization (P_η) by means of viscosity measurements is not an absolute determination method. Calibration with an absolute method is required. With monodisperse calibration samples all calibration methods would be reliable. However, as even fractionated samples of cellulose derivatives are polydisperse to some degree a precise correlation between M_η and $M_{\text{number average}}$ cannot be obtained. Therefore, viscosity average molecular weights (or P_η) may differ appreciably from number-average values obtained by common osmotic measurements. However, they are always rather close to the weight-average molecular weights (M_w) calculated from light scattering or ultracentrifugal measurements. Another reliable method to determine M_η is the measurement of random degradation.

Methods of calculating M_w from viscosity measurements of non-dilute solutions are entirely empirical. An interconversion nomograph is given in Fig. 2 (419). Otherwise the measurements are made in dilute solutions and the intrinsic viscosity $[\eta]$ is obtained by graphical or numerical extrapolation to zero concentration. The concentration dependence can be approximated by many formulas, but the following are perhaps more frequently used:

Martin's Equation:
$$\log\frac{\eta_{sp}}{c} = \log [\eta] + k' \, [\eta] \, c \qquad (\eta_{sp} = \frac{\eta}{\eta_o} - 1)$$

Huggins' Equation:
$$\frac{\eta_{sp}}{c} = [\eta] + k'' \, [\eta]^2 \cdot c$$

Schulz-Blaschke-Equation:
$$\frac{\eta_{sp}}{c} = [\eta] + k''' \cdot \eta_{sp} \cdot [\eta]$$

The last equation in the range η_{sp} 0.3 - 0.5 is very suitable for an evaluation of $[\eta]$ by measurements at a single concentration. Different values for the three constants are listed in Table L.

Solutions of chain polymers show a non-Newtonian behaviour which makes the viscosity a function of the rate of shear, G, at which it is measured. The dependence may be expressed by the relation

$$[\eta]_G = [\eta]_{G=0} (1 - x \cdot G \cdot [\eta]^\epsilon) \qquad \text{(for cellulose nitrate } x = 3.6 \cdot 10^{-6}. \quad \epsilon = 1.4 \text{ (250)).}$$

In practice it is customary to extrapolate either the apparent viscosity to zero rate of apparent shear (435) or the reduced viscosity to zero rate of shear. To eliminate the tedious procedure of determining the intrinsic viscosity at zero rate of shear, it has even been suggested that a conventional viscosity number be determined at a fixed rate of shear $[\eta]_{c=0}^{G=conv.}$

Figure 2

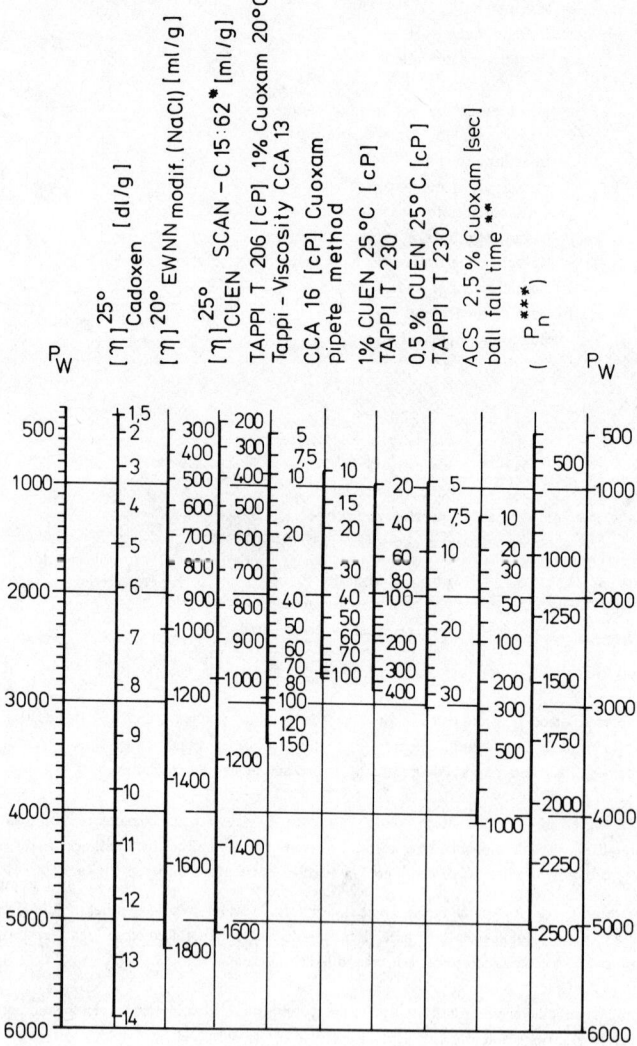

*very close to Zellcheming Merkblatt IV/36/51.   **seconds x 21.9 = cp$_{(ACS)}$.   ***conversion into P_n is general undefined due to polydispersity dependence.

Table K 1.1 K_m and K'_m derived from osmotic measurements on fractionated samples. The K_m (K'_m)-values are depending on the polydispersity of the samples! (see explanation in the preface)

Substance	Solvent	K'_m [ml g⁻¹]	$K_m \cdot 10^2$ [ml g⁻¹]	a	viscosity range [η]·10⁻² [ml g⁻¹]	$\bar{G}$ [s⁻¹]	T [°C]	Nitrogen Content [%]	Nature of Sample	Method of calibration	Ref.
Cellulose nitrate	acetone	1.02	0.377	1.0	6.5 – 14.6	1200	20	~ 12.5	F	OS	163
		1.29	0.508	0.98	0.77 – 8.0	-	20	~ 13.4	F	OS	449
			0.28	1.0	-	-	20	-	F	OS	365
		1.06	0.374	1.0	-	-	-	~ 13.4	F	OS	193
		1.37		0.98	1.7 – 9.0	-	-	-	F	OS	35
		1.02±4%	0.356	1.0	1.82 – 3.85	1200	20	~ 12.5	F	OS	450
		3.09	2.82	0.83	0.8 – 7.5	-	20	13.3–13.9	F	OS	303
		4.46	4.37	0.82	1.0 – 7.0	-	20	~ 13.5	F	OS	304
		1.20	-	1.0	-	-	-	~ 14.15	F	OS	245
		1.93	1.1	0.91	1.12 – 30	G → 0	25	13.9	F, U	OS	167,168
		1.14	0.42	1.0	1.12 – 6.8	G → 0	25	13.9	F, U	OS	167,168
		0.91	0.336	1.0	6.8 – 30	G → 0	25	13.9	F, U	OS	167,168
		1.1	0.378	1.0	6.5 – 24.5	-	25	13.8	F	OS	153,381
		1.42	0.713	0.933	1.0 – 15	-	25	14.1	F	OS	68
		1.13	-	1.0	-	-	-	13.7	F	OS	217
	mean:	1.36	-	0.948	1.0 – 18.0	-	20	13.8	F	OS	414
		2.10	2.53	0.795	-	-	25	12.1	F	OS	295
		1.56	0.84	0.92	1.35 – 5.59	-	25	13.8	F	OS	324
		1.33	1.43	0.81	0.94 – 5.68	-	25	13.8	F	OS	324
	ethyl acetate	2.13	1.22	0.92	1 – 50	G → 0	25	13.9	F, U	OS	167,168
		1.33	0.51	0.98	1.53 – 6.87	-	25	13.8	F	OS	324
		1.80	1.83	0.82	1.25 – 7.16	-	25	12.5	F	OS	324
		1.32	0.38	1.03	1 – 50	G → 0	25	13.9	F, U	OS	201
		0.276	0.063	1.37	1 – 15	2500	30	13.5	F	OS	158
	butyl acetate	1.41	0.58	0.969	<7	-	25	14.1	F	OS	68
		2.01	0.96	0.94	1.75 – 7.53	-	25	13.8	F	OS	324
		2.49	3.0	0.79	1.47 – 7.88	-	25	12.5	F	OS	324
		1.41	0.92	0.905	-	-	25	12.1	F	OS	295
	ethyl lactate	1.52	0.525	1.0	2.12 – 19	200 – 500	25	13.7	F	OS	368
		1.2	0.38	1.03	1.19 – 16.0	G → 0	25	13.9	F, U	OS	167,168
	n-amyl methyl ketone	2.74	3.61	0.780	-	-	25	12.1	F	OS	295
		1.92	1.03	0.92	1.79 – 9.90	-	25	13.8	F	OS	324
		1.85	1.95	0.82	1.34 – 7.76	-	25	12.5	F	OS	324
	cyclohexanone	2.00	2.24	0.81	-	-	25	12.1	F	OS	295
		3.26	3.06	0.82	1.77 – 6.02	-	25	13.8	F	OS	324
		2.61	3.96	0.75	1.25 – 6.12	-	25	12.5	F	OS	324
Cellulose	cuoxam	0.5	0.308	1.0	0.9 – 9	-	20	-	-	OS (acet.)	392
		3.19	11.3	0.657	0.2 – 4	-	25	-	F	comp.	68
		2.91	10.1	0.661	0.2 – 4	-	25	-	F	comp.	68
	cuene	1.33	-	0.905	1 – 21.4	G → 0	25	-	F, U	comp.	167,168

Substance	Solvent	K'_m [ml g^{-1}]	K_m · 10^2 [ml g^{-1}]	a	viscosity range [η]·10^{-2} [ml g^{-1}]	G [s^{-1}]	T [°C]	Nitrogen Content [%]	Nature of Sample	Method of Calibration	Ref.
Cellulose (Cont'd.)	cuene (Cont'd.)	0.807	0.498	1.0	1 - 2.4	G → 0	25	-	FU	comp.	167,168
		0.64	0.395	1.0	2.4 - 21.4	G → 0	25	-	FU	comp.	167,168
		0.54	0.334	1.0	<10	-	-	-	F	QS (CN)	265
	cadoxene	0.668	-	1.0	<10	-	25	-	-	comp.	335
		0.435	-	1.0	0.5 - 7.5	-	-	-	-	comp.	385
		mean: 0.7585	-	0.96	1.9 - 6.5	-	20	-	FU	comp.	307,381,184
	FeTNa (= EWNN)										
	EWNN	0.913	-	1.0	-	-	20	-	U	comp.	185
	EWNN SNE	0.907	-	1.0	-	-	20	-	U	comp.	185
	EWNN mod.	1.1	-	0.9	2.5 - 22	-	20	-	FU	comp.	calculated from 186
	EWNN mod. (NaCl)	2.145	-	0.805	2.5 - 22	-	20	-	FU	comp.	calculated from 186

Table K 1.2 K_m and K'_m derived from absolute measurements, giving the weight average (M_w, P_w)

Substance	Solvent	K'_m [ml g^{-1}]	K_m · 10^2 [ml g^{-1}]	a	viscosity range [η]·10^{-2} [ml g^{-1}]	G [s^{-1}]	T [°C]	Nitrogen Content [%]	Nature of Sample	Method of Calibration	Ref.
Cellulose nitrate	acetone	1.43	0.735	0.93	-	-	-	13,4	F	SD	193
		1.5	0.863	0.91	3.9 - 50	-	20	13,8	F	SD	281
		3.85	4.32	0.79	2.37 - 45.5	1300	20	13,8	F, U	SD	366
		1.8	1.085	0.9	2.37 - 45.5	1300	20	13,8	F, U	SD	366
		1.5	0.904	0.9	2.37 - 45.5	1300; G → 0	20	13,8	F, U	SD	265,283
		0.82	0.250	1.0	3 - 45	-	20	13,8	F, U	SD	283
		4.46	5.94	0.76	>12	-	-	13,8	F, U	SD	270
		0.82	0.282	1.0	<12	-	-	13,8	U	SD	270
		0.592	-	1.0	3.9 - 6	-	-	≈ 12,0	F, U	S	215
		1.43	-	0.98	0.45 - 22	-	-	-	-	LS	20
		0.495 (0.574)	0.173	1.0 (0.98)	1.23 - 30.3	G → 0	25	13,8	F	LS	153
		0.83	0.377	0.95	0.9 - 18	-	25	13,8	F	LS	286
		2.19	1.66	0.86	1.7 - 5.2	-	-	13,8	F	LS	160
		0.95	-	1.0	-	G → 0	-	-	F	LS	217
		-	0.98	0.91	0.3 - 68	-	-	-	-	LS	169
		-	0.28	1.0	-	-	20	-	-	SD	282
		a)	-	-	-	1300	20	-	-	LS, SD	333
	ethyl acetate	1.25 / 1.25	0.482 / 0.437	0.99 / 1.0	4.3 - 64	500	-	13,5	U	SD	312
		0.685	0.235	1.0	1.23 - 36.3	G → 0	-	13,8	F	LS	153
		0.755	0.25	1.01	1.07 - 15	2500	30	13,5	F	LS	158
	butyl acetate	0.81	0.28	1.0	2.12 - 19	200 - 500	25	13,7	F	LS	368
		-	21	0.71	-	-	-	-	-	-	300
		0.78	0.283	0.99	0.9 - 18	-	25	13,8	F	LS	286
		27.8	-	0.572	30 - 53	<1000	30	13,8	FU	LS	407
	tetrahydrofuran	-	-	0.89	-	-	-	calc. 14,16	-	-	158

Substance	Solvent	K'_m [ml g^{-1}]	$K_m \cdot 10^2$ [ml g^{-1}]	a	viscosity range $[\eta]\cdot10^{-2}$ [ml g^{-1}]	$\bar{G}$ [s^{-1}]	T [°C]	Nitrogen Content [%]	Nature of Sample	Method of Calibration	Ref.
Cellulose	cuoxam	8.08	56.4	0.523	4.7 – 16.8	-	20	0	U	SD	108
		0.524	0.85	0.81	4.7 – 16.8	-	20	0	U	SD	19
		0.85	1.70	0.77	4.7 – 16.8	-	20	0	U	SD	131
		0.435	0.268	1.0	1.75 – 19.7	500	-	0	U	SD (comp.)	313
		0.68	0.70	0.9	4.8 – 28	-	-	0	U	SD (CN)	264
		0.53	0.543	0.9	1 – 28	-	-	0	U	SD (CN)	265
		1.14	0.33	1.15	4.7 – 16.8	-	-	0	U	SD	15
		0.16	0.099	1.0	-	-	-	0	F, U	SD, LS (CN)	284
		1.58	3.3	0.76	>4.25	-	-	0	F, U	SD (CN)	270
		0.29	0.18	1.0	<4.25	-	-	0	F, U	SD (CN)	270
		0.384	0.237	1.0	1.2 – 5.05	-	-	0	U	S	215
		0.80	1.30	0.81	0.3 – 6	-	25	0	F	LS (CN)	286
	cuen			0.80		-	-	0	F	LS	61
		0.59	0.364	1.0	2.3 – 27.2	500	-	0	U	SD (CN)	312
		0.98	1.01	0.9	1.9 – 28	G → 0	-	0	F, U	SD (CN)	264
		0.82	0.85	0.9	1.9 – 28	G → 0	-	0	F, U	SD (CN)	265
		0.58	0.358	1.0	6 – 18	-	-	0	U	SD (CN)	60
		0.248	0.153	1.0	-	-	20	0	U	SD (CN)	284
		2.44	5.10	0.76	>600	-	20	0	F, U	SD (CN)	270
		0.448	0.277	1.0	<600	-	-	0	F, U	SD (CN)	270
		0.891	0.936	0.936	1.9 – 21.3	-	-	0	U	SD, LS	346
		1.7		0.8	1.7 – 3.9	-	20	0	-	K (OS)	430
	cadoxen	2.5	5.47	0.75	<18	500	25	0	U	SD	125
		0.51	0.315	1.01	2 – 11	-	20	0	U	SD	184
		0.38	0.235	1.0	2 – 11	-	20	0	U	SD	183
		0.201	0.124	1.0	>5.3	-	-	0	F, U	SD, LS	284
		1.98	4.13	0.76	<5.3	-	-	0	F, U	SD (CN)	270
		0.365	0.225	1.0	-	-	-	0	F, U	SD (CN)	270
		0.525	0.427	0.94	<10	-	20	0	F	SD (CN)	270
		0.712	0.593	0.94	>10	-	20	0	F	SD (CN)	270
		3.64	9.33	0.72	<18	-	20	0	F	SD (CN)	125
	FeTNa (= EWNN)	1.8 (1.75)	3.56 (3.38)	0.77	<18	500	25	0	U	LS	125
		1.84	3.85	0.76	0.4 – 1.2	500	25	0	F	SD, LS	41
		0.84	0.519	1.0	3 – 16	G → 0	20	0	U	SD	30
		0.925	0.571	1.0	3 – 16	G → 0	20	0	U	SD	30
		0.814	0.502	1.0	3 – 16	G → 0	20	0	U	SD	30
		0.66	0.407	1.01	6.6 – 20	G → 0	20	0	U	SD	60, 30
		0.575	0.355	1.0	<13	-	20	0	U	SD	183
		0.49	0.303	1.0	<16	-	20	0	U	SD	183
		0.325	0.201	1.0	<8	-	20	0	U	SD	183
		0.64 – 0.75★		1.0	-	-	-	0	U	comp.	2
		2.0		0.84	2.4 – 21.2	-	-	0	-	comp.	346
		2.74	5.31	0.78★★	4.9 – 31.5	-	-	0	-	LS	427

★ FeTNa of different composition.　　★★ FeTNa mod.

a) $$\log \frac{P_w}{100} = 0.0814 + 0.9240 \cdot \log \frac{[\eta]_{1300}}{100} + 0.1170 \left(\log \frac{[\eta]_{1300}}{100} \right)^2$$

Table K. 2   Solvent systems and cellulose derivatives of minor interest for the determination of P or molecular weight of cellulose

Substance	Solvent	K' [ml g⁻¹]	$K_m \cdot 10^2$ [ml g⁻¹]	a	molecular weight range (P)	$\bar{G}$ [s⁻¹]	T [°C]	Nature of Sample	Method of Calibration	Ref.
Cellulose	zincoxene	0,767	0,585	0.936	-	-	20	U	OS	357
	tetraethylammonium hydroxide (2.3 M)	0,42	0,26	1.0	-	-	20	U	OS (via CN)	391
					265 - 1300	-	20		OS; comp.	216
	LiOH-solution	≈ 0,55	≈ 0,34	1.0	-	-	20	U	OS	391
	Ca(SCN)$_2$-solution	0,8	≈ 0,5	1.0	-	-	20	U	OS	391
	H$_3$PO$_4$-solution	1,8 - 2,1	≈ 1,3	1.0	-	-	20	U	OS	391
	H$_2$SO$_4$ (50%)-solution	2,03	1,71	0.94	>150	(49,72)	25	U	K	433
		1,77		0.92	>150	-	20	U	K	434
	10% NaOH-solution	0,7	0,43	1.0	-	-	20	U	OS	391
Cellulose sulfates (Na-Cell · S; S = 7,2%)	1% NaCl-solution	1,7	21	0.64	60 - 185	-	-	F	SD	326
Cellulose xanthate (viscose)	NaOH-solution	1,7		0.8	< 600	-	20	U	comp.	416
	9% NaOH	0,3		0.81	< 450			U	comp.	3
	1 M NaOH	a)		1.0				U	comp.	393
Carboxymethyl cellulose	NaCl-solution at infinite ionic strength		0,19	0.60	≈ 600 - 4000	G → 0	0	U	comp.	89
Hydroxyethyl cellulose (MS = 1,67; DS = 0,88)	water	1,1		0.87	330 - 2600	G → 0	25	F	LS, SD	43
	cadoxen	1,3		0.79	330 - 2600	G → 0	25	F	LS	44
Hydroxypropyl cellulose	ethanol	0,121	2,68	1.17		G → 0	25	F	LS	44
Sulfoethyl cellulose	2% NaOH		66	0.95					OS	455
	6% NaOH		28	0.56					OS	332
Methyl cellulose (28% CH$_3$O-)	water	7,6	5,23	0.63	100 - 1600	550	20	F	SD	332
(22-33% CH$_3$O-)	chloroform	1,06 - 1,1	1,51	1.0	75 - 200			F	SD	423
(36-45,6% CH$_3$O-)	m-cresol	1,1	0,45	1.0	200 - 450			F	comp.	382
(24-45,6% CH$_3$O-)	methanol	1,25 - 1,4		1.0	20 - 500			F	comp.	395
Ethyl cellulose	acetone		1,6	0.65			25	F	LS	395
Cellulose acetate (sec.) (DS = 2,73 - 2,77)	acetone	2,44	0,58	1.05	100 - 560		20	F	SD	359
	acetone			0.9			30		DS	285
(DS = 2,25 - 2,38)	tetrachloroethane			0.760				F	comp.	331
(DS = 2,86)	methylene chloride/ethanol 80/20	1,56		0.82	100 - 560		25		comp.	69
	m-cresol	0,63		0.90			25	F	comp.	104
	chloroform	0,53		0.834				F	comp.	331
	methylene chloride/acetone		0,17	1.0	100 - 150			F	OS	69
	methylene chloride		0,22	1.0	48 - 450		25	F	OS	394
		1,45		0.95	150 - 560			F	LS	394
				0.83				F	LS	156
Cellulose tributyrate	methyl ethyl ketone		1,82	0.8	370 - 850		30	F	LS	78
Cellulose tricaproate	dioxane		12,5	0.57	90 - 2850		35	F	LS	460
Cellulose tricarbanilate	dioxane		0,0813	0.97	600 - 4180	G → 0	25	F	LS	159

Substance	Solvent	K'_m [ml g⁻¹]	$K_m \cdot 10^2$ [ml g⁻¹]	a	molecular weight range (P)	$\bar{G}$ [s⁻¹]	T [°C]	Nature of Sample	Method of Calibration	Ref.
Cellulose tricarbanilate (Cont'd.)	acetone		0,143	0,91	600 - 4180	G → 0	25	F	LS	376
			0,466	0,84	≥1000	G → 0	20	F	LS	49
2,4-Dinitrobenzoyl cellulose	cyclohexanone		0,191	0,86	600 - 4180	G → 0	25	F	LS	376
	nitrobenzene		25,9	0,5		200	25	F	OS	383

a)
$$[\eta] = 0{,}0167 \cdot P_w^{0,82} + (0{,}0062 \cdot DS_{2,3} - 0{,}002 \cdot DS_6) \cdot P_w^{0,94}$$

($DS_{2,3,6}$ = degree of substitution on carbon atom $C_{2,3}$ or C_6 respectively).

L. CONCENTRATION DEPENDENCE OF VISCOSITY

For explanation see under Remarks this Table section K

Substance	Solvent	k' Martin Coefficient	Ref.	k" Huggins Coefficient	Ref.	k''' Schulz-Blaschke Coefficient	Ref.
Cellulose	cuoxam	0,1303	263	0,37	414	0,29	269
		0,12	72			0,1552	364
		0,127 - 0,107^g)	65			0,239 - 0,335	269
	cuen	0,132; 0,11^d)	414			average: 0,287	
		0,157 (0,172)^h)	46				
		0,13 - 0,15	323			0,33	430
		0,127^b)	67			0,278 - 0,296	269
		0,18	72			average: 0,290	
		0,138	301				
		0,14	317				
		0,143	414				
		0,145	314				
		0,161 ± 0,013	247				
	cadoxen			0,26 - 0,39	41	0,280	307
				0,50	37	0,29	37
	FeTNa (EWNN)	0,169 - 0,247	346	0,372 - 0,553	346	0,339 - 0,41^i)	189
		0,177 - 0,179	426			0,355	2
						0,338 - 0,339	426
						average: 0,338	
from theoretical considerations:		2,3 · k' ≤ k"		0,38		≤ 0,38	
Cellulose nitrate	acetone	0,247	263	mean value 0,606	167	0,315 ± 12%	163
		0,20 - 0,22^a)	75	0,66^b)	414	0,307 ± 22,5%	303
		0,15 - 0,20^a)	72	0,541^b)	406	0,30 ± 0,01	327
		0,20	1	0,36 - 0,73	328	0,27 - 0,30	328
						0,30 - 0,49^c)	268
						0,25 - 0,52	269
						average: 0,371	

Substance	Solvent	Martin Coefficient k	Ref.	Huggins Coefficient k''	Ref.	Schulz-Blaschke Coefficient k'''	Ref.
Cellulose nitrate (Cont'd.)	acetone (Cont'd.)					0.247	364
						0.33 d); 0.35; 0.40	266
						0.32 ($P_w < 2000$)	333
						0.40 ($P_w > 2000$)	333
						0.36 - 0.52	202
						0.22 - 0.30	328
						0.27	246
	butyl acetate	0.180	263, 75, 1, 364	0.50	414		
		0.244	414	0.3 - (1.1)	328		
		0.297 - 0.314 e) / (≈0.4) f)	349				
		0.164 - 0.205 e)	112				
	ethyl acetate			0.337	168		
				0.35 b)	292,3		
				0.295 b)	406		
				0.40	292		
	ethyl lactate	0.18	1				
	cellosolve	0.20	263				
Cellulose acetate	acetone	0.19	72	0.70	26	0.33 ± 0.06	34
	chloroform, methyl cellosolve	0.178	263				
	sym. tetrachloroethane	0.178	263				
	chloroform	0.15	263				
Cellulose triacetate				0.427 - 0.450	156		
Cellulose tricarbanilate				0.45 - 0.60	376		
Methyl cellulose	toluene/ethanol 80/20	0.231	375	0.398	171		
Ethyl cellulose		0.111	263				
Cellulose xanthate	2.2 N NaOH; 0°C	0.12 - 0.16	320				
		0.14 ± 0.02	89				
Hydroxyethyl cellulose	water	0.231	375			0.40	432
	1 M NaOH					0.37	432
	cadoxen			0.42	42		
Carboxymethyl cellulose	1% NaCl	0.16	1				
	1 M NaCl					0.35	432
	1 M NaOH					0.43	432
	1.5 N NaOH	0.148	336				
	1 M HCl					0.47	432

a) different fractions.   b) for $G = 0$.   c) the higher values are valid for mercerized and oxydized cellulose samples.   d) regenerated alkali cellulose.   e) cellulose nitrate from viscose rayon.

f) degraded samples.   g) depending on P.   h) cuoxam with 25.4 g Cu/l and 176 g NH_3/l.   i) FeTNa-solutions of different compositions.

M. RATIO OF INTRINSIC VISCOSITY IN DIFFERENT SOLVENTS

(only references are given for less important ratios)

$\dfrac{[\eta] \text{ NC*-Acetone}}{[\eta] \text{ Cell-Cuen}}$

Data	Ref.
1.46	168
1.42	167
1.18-1.90	415
1.83	265
1.55	84
1.83	66
1.9	330

$\dfrac{[\eta] \text{ NC-Acetone}}{[\eta] \text{ Cell-Cuam}}$

Data	Ref.
2.2	392
2.6	366
2.09-2.63	84
2.17	415
2.83	270
2.9	330

$\dfrac{[\eta] \text{ NC-Acetone}}{[\eta] \text{ Cell-FeTNa}}$

Data	Ref.
1.27	60
1.6	330

$\dfrac{[\eta] \text{ NC-Acetone}}{[\eta] \text{ Cell-Cadoxene}}$

Data	Ref.
2.25	270

$\dfrac{[\eta] \text{ NC-Acetone}}{[\eta] \text{ Ethyl Acetate}}$ (167,168,415, 324,448,267, 294)

$\dfrac{[\eta] \text{ Cell-Cuen}}{[\eta] \text{ Cell-FeTNa}}$

Data	Ref.
0.88	426
0.91	330

$\dfrac{[\eta] \text{ Cell-Cuam}}{[\text{NC-Ethyl Acetate}]}$

Data	Ref.
0.35	312

$\dfrac{[\eta] \text{ Cell-Cuam}}{[\eta] \text{ Cell-Cadoxene}}$

Data	Ref.
0.89	284
0.80	306

$\dfrac{[\eta] \text{ NC-Acetone}}{[\eta] \text{ NC-Butyl Acetate}}$ (415,324,294, 314)

$\dfrac{[\eta] \text{ Cell-Cuen}}{[\eta] \text{ Cell-Cadoxene}}$

Data	Ref.
0.81	124

$\dfrac{[\eta] \text{ Cell-Cuam}}{[\eta] \text{ Cell-FeTNa}}$

Data	Ref.
0.45	108
0.59	330

$\dfrac{[\eta] \text{ Cell-Cuam}}{[\eta] \text{ Cell-10\% NaOH}}$

Data	Ref.
0.72	462

$\dfrac{[\eta] \text{ Cell-Cadoxene}}{[\eta] \text{ Cell-FeTNa}}$ (284,184)

* NC = Cellulose nitrate   Cell = Cellulose   Cuen, Cuam, FeTNa see under Solvents for Cellulose

N. GEOMETRIC DIMENSIONS IN SOLUTION*

Data are given on KUHN-KUHN equivalent chain length A_m, the radius of gyration $<s^2>^{1/2}$, and the end-to-end distance $<r^2>^{1/2}$. Molecular weight and in some cases values of A_m, $<s^2>^{1/2}$ and $<r^2>^{1/2}$ are designated according to the method used for the measurement.

<u>Abbreviations:</u> OS = osmotic measurement LS = light scattering SD = sedimentation-diffusion UC = equilibrium measurement in ultracentrifuge

Solvent System	Remarks on Sample	Temperature of Measurement [°C]	$\overline{M} \times 10^{-3}$	A_m [nm]	$<s^2>^{1/2}$ [nm]	$<r^2>^{1/2}$ [nm]	Ref.**
Cellulose-Cuam			195.6	5			325,(400)
Cellulose-Cadoxene				7			59,(48)
	linters	25	- (OS)		89.5 (LS)	219.0 (LS)	125
			1415 (UC)				
			945 (LS)				
			745 (SD)				
			152 (OS)		48.5 (LS)	119.0 (LS)	125
			430 (UC)				
			290 (LS)				
			210 (SD)				
Cellulose nitrate-acetone			13 (SD)			20***	284
			23.5 (SD)			28***	
			46 (SD)			55***	
			90 (SD)			75***	
			178 (SD)			115***	
			262 (SD)			130***	
			456 (SD)			185***	
			630 (SD)			205***	
			2490 (SD)			530***	

* See also table on "Unperturbed Dimensions" in this Handbook.   ** References in parenthesis give additional data.   *** Different methods of calculation.

O.   SECOND VIRIAL COEFFICIENTS

see corresponding table in this Handbook and References 125, 303

P.   DIFFUSION AND SEDIMENTATION DATA

see corresponding table in this Handbook and References 60, 123, 125, 429a.

Q.   FIRST ORDER RATE CONSTANTS AND ENERGY OF ACTIVATION OF HOMOGENEOUS HYDROLYSIS ★

Hydrolyzing Acid	Medium Concentration [%]	Temperature [$^{\circ}$C]	Concentration of Cellulose [%]	Rate Constant x 10^6 [min^{-1}]	Energy of Activation [$kJ\ mol^{-1}$]	Origin and Preparation of Cellulose Sample	Ref.
H_2SO_4	50	20.5	1	123		viscose rayon	213
		30	1	580			
		40	1	2850			
	51	18	0.5	30.5		cotton	95
		30	0.6	234			
	52	18	0.57	40	117	native ramie	96
		30	0.57	272			
		40	1.5	1850		cotton linters	212
	65	0.12	5.65	14		cotton	80
		20	1.27	400			
H_3PO_4	78	12	not given	0.329	127	purified cotton, partially degraded, reprecipitated after dissolution in H_3PO_4	363
		20		1.27			
		29		6.10			
		40		39.1			
	81.25	20	not given	0.85	146	acetate grade sulfite pulp	256
		30		7.1			
		40		37			
		20	not given	1.0	142	prehydrolyzed sulfate pulp	256
		30		8.9			
		40		38			
	86	0.12	0.07	0.083		purified cotton	80
		20	0.07	average 2.8 range 2.2 - 4			
	92.3	25	0.10	7.18		wood pulp, unbeaten	390

★   Additional data may be found in Ref. 317a, 363a, 390.

R.   ELECTRICAL PROPERTIES

Definitions:

specific resistance ρ [ohm cm] (= cm/mhO) is defined as the resistivity between opposite faces of a 1-cm cube.

mass specific resistance R_s [ohm g cm^{-2}] is defined as the resistance in ohms between the ends of a specimen 1 cm long and of mass 1 g

$$R_s = \rho \cdot d \text{ (d = density of the material in g/cm}^3) \qquad (299, 249)$$

The electrical resistance depends on moisture content M, impurities, temperature t, polarization effects and the average field strength.   For most hygroscopic textile fibers the equation $R \cdot M^n = K$ fits the experimental results (M = moisture content, %; n and K are constants).   The change of R_s with temperature t and moisture content can be described by the equation:

$$\log R_s = \log (R_s)_{t,\ M\ =\ 0} - (a - b \cdot M) \cdot t + \frac{c \cdot t^2}{2} \text{ (t = temperature in }^{\circ}\text{C; a, b and c are constants)}$$

dielectric constant ϵ is the ratio between the capacity of a condenser in the medium C_p and the capacity of this condenser in vacuum C_o ($\epsilon = C_p / C_o$).

dielectric loss factor is defined as the tangent of the loss angle δ.   Both quantities are depending on frequency f, moisture content M, temperature t, impurities, fiber orientation and packing density.

At optical frequencies, the following equations exist:

$$\epsilon = n^2(1 - k^2) \text{ and } \tan \delta = 2 k/(1 - k^2) \text{ (n = refractive index and k = absorption index)}$$

Zeta potential (electrokinetic petential or electrophoretic mobility) [mV].   (Negative) surface charge of cellulose fibers, dispersed and moving in water, expressed in [mV].   The ζ-potential depends on the kind of the fiber material (cooking and bleaching conditions as well as crystallinity, carboxyl and lignin content), the mechanical treatment [beating], drying conditions and ions content in the system and pH.

Dielectric strength [kV/mm]. Limiting electrical field strength, at which a breakdown occurs; depends on the thickness and imperfections of the sample, the shape of the voltage curve, the kind and arrangement of the electrodes, the testing time and the other testing conditions.

Remarks:

As existing data are largely depending on the method of measurement as well as on moisture, purity etc. of the material investigated, only some representative data are presented in this table. For further details see the surveys of HEARLE (299, 115) and CLAUSSNITZER (64) as well as the references in parenthesis.

Property	Material	Value	Remarks	Ref.
Specific resistance ρ [ohm cm]	insulating paper	$\approx 2 \times 10^{14}$	air dry	58
		5×10^{17}	M < 1%	428
		10^{12}	M = 12%	428
		10^{14}	20°C, air dry	239
		10^{11}	100°C, air dry	239
	insulating kraft paper	4×10^{17}	20°C	447
		1.2×10^{17}	100°C	447
	pure cellulose	10^{18}		211,305
	cotton	$\approx 2 \times 10^{7}$	52% R.H.[a)], 30°C, single fiber	143
		$\approx 1 \times 10^{7}$	62% R.H., 30°C	
		$\approx 7 \times 10^{5}$	75% R.H., 30°C	
		$\approx 3 \times 10^{5}$	85% R.H., 30°C	
	vicose rayon	$\approx 3 \times 10^{6}$	75% R.H., 30°C	
		$\approx 1 \times 10^{6}$	85% R.H., 30°C	
	--, yarn	$1.3 - 3.5 \times 10^{6}$	75% R.H., 30°C, nominal size 150/75	
	regenerated cellulose	10^{17}		211
	Cellophane, Cuprophane	$1 \times 10^{10} - 3 \times 10^{12}$	65% R.H., 20°C	64
Mass specific resistance R_s [ohm g cm^{-2}]	raw cotton	8×10^{5}	75% R.H., 25°C	446,(116)
		5×10^{5}		
	--, washed	$12 - 37 \times 10^{6}$		
	cotton	7×10^{6}	65% R.H., 20°C	117
	--, purified	9.3×10^{5}	M = 10%, 20°C	116
	ramie	$6 \times 10^{4} - 1.8 \times 10^{6}$	M = 10%, 20°C	116
	--, purified	5.7×10^{6}	M = 10%, 20°C	
	hemp	1×10^{7}	M = 10%, 20°C	
	jute, purified	2.5×10^{8}	M = 10%, 20°C	
	flax, purified	6.6×10^{6}	M = 10%, 20°C	
	viscose rayon	$10^{8} - 10^{9}$	M = 10%, 20°C	
	cuprammonium rayon	2.6×10^{8}	M = 10%, 20°C	
	Fortisan	3.5×10^{7}	M = 10%, 20°C	
	cotton, mercerized	6.6×10^{6}	M = 10%, 20°C	
Dielectric constant ϵ	pure cellulose	5.5 - 8.1	10^{6} kHz	64
		7.5	bone dry, 0.1 kHz, 20°C	408
		7.2	bone dry, 1 kHz, 20°C	
		7.0	bone dry, 10 kHz, 20°C	
		5.8	bone dry, 10^{4} kHz, 20°C	
		5.6	bone dry, 10^{5} kHz, 20°C	
	cotton linters	6.1	1 kHz, 25°C	251
		2.76	unbeaten, 1 kHz, density 1 g/ml	429
		2.85 - 3.05	beaten, 1 kHz, density 1 g/ml	
		3.2	0% R.H., 1 kHz	118,(299,79)
		7.1	45% R.H., 1 kHz	
		18.0	65% R.H., 1 kHz	
		3.0	0% R.H. 100 kHz	
		4.4	45% R.H., 100 kHz	
		6.0	65% R.H., 100 kHz	
		2.50	$\parallel$ fiber-axis at optical frequencies	299
		2.34	$\perp$ fiber-axis at optical frequencies	
	wood pulp fiber	5.5		239
		5.7		280
	alpha wood pulp	6.15	vacuum dried at 30°C, 1 kHz, 30°C	195
	bleached sulfite pulp	2.90	unbeaten, 1 kHz, density 1 g/ml	429
		3.03	beaten, 1 kHz, density 1 g/ml	

Property	Material	Value	Remarks	Ref.
Dielectric constant ϵ (Cont'd.)	paper	2.2 - 2.7		239
		1.2 - 4	1 kHz	211
	regenerated cellulose	6.7 - 7.5	films, regenerated from viscose, 1 kHz, 20-70°C	52, (399)
		7.5	1 kHz	211
	cellophane film	7.7	dry, 0.06 kHz, 25°C	399,(378)
		7.3	10 kHz, 25°C	
		6.7	10^3 kHz, 25°C	
		4.04	3×10^6 kHz, M = 0%	378
	viscose rayon	3.5	0% R.H., 100 kHz	118,(299,79)
		4.7	45% R.H., 100 kHz	
		5.3	65% R.H., 100 kHz	
		3.6	0% R.H., 1 kHz	
		5.4	45% R.H., 1 kHz	
		8.4	65% R.H., 1 kHz	
		2.37	∥ fiber-axis at optical frequencies	299
		2.31	⊥ fiber-axis at optical frequencies	
	regenerated fibers	2.0	1 kHz, density 0.5 g/ml	429
Dielectric loss factor ($\approx$ power factor) tan δ	cellulose	0.015	20°C, 0.1 kHz	408
		0.02	20°C, 1 kHz	
		0.03	20°C, 10 kHz	
		0.045	20°C, 10^2 kHz	
		0.065	20°C, 10^3 kHz	
		0.08	20°C, 10^4 kHz	
		0.07	20°C, 10^5 kHz	
	alpha-cellulose	0.0015	99% α-cellulose, 20°C	203
	unbleached sulphate pulp	0.00125	acid washed, 80°C	
	wood pulp	0.001	60°C, 94-95% α-cellulose, 5-6% pentosans, Na < 22 ppm	287
		0.002	120°C	
	kraft paper	0.0026 - 0.0028	30°C, 0.006-0.18% H_2O	62
	insulating paper	0.0025	50 Hz	239
	regenerated cellulose	0.010		211
	cellophane	0.02 - 0.3		64
		0.015	0°C, 0.1 kHz	388
		0.06	0°C, 10^3 kHz	
		0.009	M = 0%, 60 Hz	344
		0.016	M = 0%, 10 kHz	
		0.062	M = 0%, 10^3 kHz	
	viscose staple fibers	0.007	packing density 50%, 0% R.H., 1 kHz	118,(299,79)
		0.08	packing density 50%, 45% R.H., 1 kHz	
		0.40	packing density 50%, 65% R.H., 1 kHz	
		0.015	packing density 50%, 0% R.H., 100 kHz	
		0.02	packing density 50%, 45% R.H., 100 kHz	
		0.03	packing density 50%, 65% R.H., 10^2-5 x 10^3 kHz	
Dielectric strength (break-down voltage) [kV/mm]	native cellulose fiber	50	dry	280
	insulating paper	7.7 - 9.2	20°C, 50 Hz, D = 0.11 - 0.12 mm[b], 0.71 - 0.83 g/ml, R.H. = 65%	119
	condenser paper	4.7 - 12.1	oven dry, DC	273
	cellophane	30 - 50	D = 1 mm, 50 Hz	64
Isoelectric point [pH]	wood pulp	2.75		355
	cotton	2.5	0.02 M HCl-KCl-buffer	113
Zeta-potential [- mV]	cellulose	21.0	fines from filter paper, Whatman No. 1	272,(275)
		30		340
	cotton	21.1	fines	272,(275)
	Cross-Bevan cellulose	22.9	fines	
	chlorite holocellulose	32.0	fines from Eucalyptus reagnans	
	chemical pulps, bleached	17 - 33	fines	
	--, unbleached	23.5 - 33.7	fines	
	mechanical and semimechanical pulps	17.2 - 46.1	fines	
	beech kraft pulp	19.8	17° SR[c]	173

References page V-116

55 1881 a2 b4 c5 15 65

Property	Material	Value	Remarks	Ref.
Zeta-potential (Cont' d.)	beech kraft pulp (Cont' d.)	15.9	51° SR, demineralized	173
		20.6	oven dried at 130°C	
	unbleached wood sulfite pulp	9	distilled water, 20°C	224
	unbleached spruce sulfite pulp	23	distilled water	120
	unbleached sulfate pulp	18	distilled water	120
	bleached birch sulfate pulp	21	distilled water	120
	viscose rayon	26	$NaBH_4$-reduced sample, distilled water	121

Energy of activation for electrical conduction

$[\text{kJ mol}^{-1}]$

	high alpha cellulose	44.4		305

$R_s \cdot M^n = K$

		n	log K		
	cotton	11.4	16.6	between 30 and 90% R.H.	116,(299)
	purified cotton	10.7	16.7	between 30 and 90% R.H.	
	ramie	12.3	18.6	between 30 and 90% R.H.	
	viscose rayon	11.6	19.6	between 30 and 90% R.H.	
	cotton	1.9	16.0	for M < 3.5%	
	viscose rayon	0.87	14.0	for M < 6%	

$\log R_s = (\log R)_{t, M=0} - (a - b \cdot M) \cdot t + \dfrac{c \cdot t^2}{2}$

		a	b	c	
	cotton	0.0863	0.00535	0.00035	299
	viscose rayon	0.0707	0.00186	0.00037	

$k = \dfrac{\epsilon - 1}{\epsilon + 2} \cdot \dfrac{1}{d}$

		k		
	unbeaten bleached sulfite pulp	0.4785	2 kHz, d = 0.4848 g/ml	429
	regenerated cellulose fiber	0.5162	2 kHz, d = 0.488 g/ml	

a) R.H. = relative humidity. b) D = sample thickness. c) $^\circ$SR = freeness (degree Schopper-Riegler).

S. PHYSICAL AND MECHANICAL PROPERTIES

1. Breaking strength is the maximum load required to break a fiber. Breaking length is the length of the freely hanging fiber, which will cause the fiber to break under its own weight.

   Strength in N/mm^2 = strength in g/den x 9 x 9.807 x sp. gr. of the fiber
   Strength in $lb/inch^2$ = strength in g/den x sp. gr. x 12.861
   Strength in g/tex = breaking strength in kilometers = strength in g/den x 9

2. Modulus of elasticity is the ratio of unit stress to unit strain (conversion factor to SI-units see under breaking strength).

3. Moisture regain is the percentage of water in the fiber at standard conditions (20°C, 65% R.H.) calculated for the constant oven dry weight of the fiber.

4. Water imbibition is the weight of water in per cent retained by the wet fiber after centrifuging at standard conditions.

All data of dry fibers are given for standard conditions (65% R.H. and 20°C). Data are taken from references (418, 106, 457, 452).

	Cotton	Flax	Jute	Viscose rayon		Viscose Staple High wet modulus (HWM) =		Fibers
				Normal	High Tenacity	Normal	Polynosics	Cross-linked
Breaking strength (g/den) dry	3.0- 4.9	4.5- 9.0	~3.4	1.6- 2.5	4.0- 6.1	1.8- 3.0	3.0- 6.0	2.0- 2.3
Breaking length (km) dry	27 - 44	40.5- 81.0	~30.6	14.5-22.5	36 - 54	16 - 27	27 - 54	18.0-20.7
wet/dry (%)	100 -110	102 -106	99 -104	15 - 60	65 - 80	45 - 65	70 - 80	65 - 75
Breaking elongation dry (%)	8 - 10	2.5- 3	1.7- 1.9	25 - 30	9 - 20	14 - 30	7 - 15	12 - 16
Breaking elongation wet (%)	11	2.5- 3	1.7- 1.9	20 - 40	14 - 26	22 - 30	8 - 20	14 - 19
Modulus of elasticity (g/den)	42 - 82	200	195	33 - 77	69 -100	48 - 68	30 - 90	35 - 42
Moisture regain (%)	7.1- 8.5	10.0	12	12	14	13	12.1- 12.7	13
Water imbibition (%)	48		85 -120	85 -120	45 - 70	95 -100	55 - 70	40 - 65

T. REFERENCES

1. F. A. Abadie-Maumert, Ø. Ellefsen, Norsk. Skogind. <u>11</u>, 540 (1957).
2. W. B. Achwal, T. V. Narayan, U. M. Purao, Tappi <u>50</u>, [6,] 90A (1967).
3. W. J. Alexander, O. Goldschmid, R. L. Mitchell, Ind. Eng. Chem. <u>49</u>, 1303 (1957).
4. B. Alinče, L. Kuniak, Papir Celulosa <u>19</u>, 67 (1964).
5. K. R. Andress, Z. Phys. Chem. (B) <u>2</u>, 380 (1929).
6. K. R. Andress, Z. Phys. Chem. (B) <u>4</u>, 190 (1929).
7. O. Ant-Wuorinen, Paperi Puu <u>37</u>, 335 (1955).
8. O. Ant-Wuorinen, A. Visapää, Paperi Puu <u>38</u>, 523 (1956).
9. O. Ant-Wuorinen, A. Visapää, Paperi Puu <u>40</u>, 313 (1958).
10. O. Ant-Wuorinen, A. Visapää, Paperi Puu <u>42</u>, 367 (1960).
11. O. Ant-Wuorinen, A. Visapää, Paperi Puu <u>43</u>, 105, 207, 289, 343 (1961).
12. O. Ant-Wuorinen, A. Visapää, Paperi Puu <u>44</u>, 337 (1962).
13. O. Ant-Wuorinen, A. Visapää, Paperi Puu <u>45</u>, 35 (1963).
14. G. H. Argue, O. Maass, Can. J. Res. (B) <u>12</u>, 564 (1935).
15. M. I. Arkhipov, Izv. Vysshykh Uchebn. Zavedenii, Khim. i Khim. Tekhnol. <u>3</u>, 1109 (1960).
16. D. K. Ashpole, Nature <u>169</u>, 37 (1952).
17. K. Atsuki, S. Okajima. J. Soc. Chem. Ind. (Japan) <u>40</u>, 360B (1937).
18. E. L. Back, E. I. E. Didriksson, Svensk Papperstidn. <u>72</u>, 687 (1969).
19. W. Badgley, B. I. Frilette, H. Mark, Ind. Eng. Chem. <u>37</u>, 227 (1945).
20. R. M. Badger, R. H. Blaker, J. Phys. Chem. <u>53</u>, 1056 (1949).
21. V. Balashov, R. D. Preston, Nature <u>176</u>, 64 (1955); R. D. Preston, A. B. Wardop, E. Nicolai, Nature <u>162</u>, 957 (1948).
22. E. Balcerzyk, K. Hempel, Przegl. Papiern. <u>20</u>, 309 (1964).
23. N. Barakat, A. M. Hindeleh, Textile Res. J. <u>34</u>, 581 (1964).
24. T. Barratt, J. W. Lewis, J. Textile Inst. <u>13</u>, T 113 (1922).
25. A. J. Barry, F. C. Peterson, A. King, J. Am. Chem. Soc. <u>58</u>, 333 (1936).
26. A. Bartovics, H. Mark, J. Am. Chem. Soc. <u>65</u>, 1901 (1943); H. J. Philipp, C. F. Bjoerk, J. Polymer Sci. <u>6</u>, 383, 549 (1951).
27. K. H. Bassett, C. Y. Liang, R. H. Marchessault, J. Polymer Sci. A <u>1</u>, 1687 (1963).
28. W. J. Beek, G. Marrucci, S. H. Davis, Chem. Eng. Sci. <u>23</u>, 1347 (1968).
29. V. A. Berestnev, N. I. Grechushkina, M. B. Lytkina, V. A. Kargin, Khim. Volokna <u>1962</u>, (3), 45.
30. W. Bergmann, Thesis, TH Darmstadt (1958).
31. S. M. Betrabet, E. Daruwalla, H. T. Lokhande, Textile Res. J. <u>36</u>, 684 (1966).
32. B. E. Bingham, Makromol. Chem. <u>77</u>, 139 (1964).
33. A. P. Binotto, W. K. Murphey, B. E. Cutter, Wood & Fiber <u>3</u>, 179 (1971).
34. K. H. Bischoff, B. Philipp, Faserforsch. Textiltechn. <u>17</u>, 115 (1966).
35. R. H. Blaker, R. M. Badger, R. M. Noyes, J. Phys. Chem. <u>51</u>, 574 (1947).
36. L. S. Bolotnikova, T. I. Samsonova, Vysokomolekul. Soedin <u>6</u>, 533 (1964).
37. L. S. Bolotnikova, T. I. Samsonova, Zh. Prikl. Khim. <u>38</u>, 2299 (1965).
38. P. Bouriot, Bull. Inst. Textile France <u>1962</u>, 1197.
39. F. C. Brenner, V. Frilette, H. Mark, J. Am. Chem. Soc. <u>70</u>, 877 (1948).
40. A. Broido, Pyrodynamics <u>4</u>, 243 (1966).
40a. J. de Booys, P. H. Hermans, Kolloid-Z. <u>97</u>, 229 (1941).
41. W. Brown, R. Wikstroem, European Polymer J. <u>1</u>, 1 (1965).

42. W. Brown, D. Henley, J. Oehman, Makromol. Chem. <u>62</u>, 164 (1963).
43. W. Brown, D. Henley, J. Oehman, Arkiv Kemi <u>22</u>, 189 (1964).
44. W. Brown, D. Henley, J. Oehman, Makromol. Chem. <u>64</u>, 49 (1963).
45. B. L. Browning, L. O. Sell, Tappi <u>39</u>, 489 (1956).
46. B. L. Browning, L. O. Sell, W. M. Abel, Tappi <u>37</u>, 273 (1954).
47. T. Brownsett, D. A. Clibbens, J. Textile Inst. <u>32</u>, T 57 (1941).
47a. N. Buchner, Papier <u>14</u>, 126 (1960).
48. W. Burchard, Z. Phys. Chem. (Frankfurt) <u>42</u>, 293 (1964).
49. W. Burchard, E. Husemann, Makromol. Chem. <u>44/46</u>, 358 (1961).
50. A. Burgeni, O. Kratky, Z. Phys. Chem. (Leipzig) B <u>4</u>, 401 (1929).
51. E. Calvet, P. H. Hermans, J. Polymer Sci. <u>6</u>, 33 (1951).
52. A. Campbell, Proc. Roy. Soc. (London) A <u>78</u>, 196 (1907).
53. E. K. Carver, E. C. Bingham, H. Bradshaw, C. S. Venables, Ind. Eng. Chem. Analyt. Edit. <u>1</u>, 49 (1929). D. A. Clibbens, A. H. Little, J. Textile Inst. <u>27</u>, T 285 (1936). A. F. Martin, Ind. Eng. Chem. <u>45</u>, 2497 (1953).
54. Centre Recherches Ind. Textile de Rouen, Bull. Inst. Textile France <u>1962</u>, 945.
55. J. Chabert, Bull. Inst. Textile France <u>20</u>, 335 (1966).
56. M. Chang, Tappi <u>55</u>, 1253 (1972).
57. M. M. Chilikin, Zh. Prikl. Khim <u>3</u>, 221 (1930); W. B. Chambpell, Ind. Eng. Chem. <u>26</u>, 218 (1934); W. G. MacMillan, R. R. Mukherjee, M. K. Sen, J. Textile Inst. <u>37</u>, T 13 (1946); S. M. Lipatov, D. V. Zharkovskii, J. M. Zagraevskaya, Kolloidn. Zh. <u>21</u>, 526 (1959); K. P. Mishchenko, S. L. Talmud, V. I. Yakimova, Vysokomolekul. Soedin <u>1</u>, 662 (1959); D. V. Zharkovskii, Dokl. Akad. Nauk. Belorussk. SSR <u>3</u>, 492 (1959), through Chem. Zentr. <u>1961</u>, 17966; Kh. U. Usmanov, I. Kh. Khakimov, Uzbeksk. Khim. Zh. <u>1959</u> (5), 30; Kh. U. Usmanov, A. A. Yul' chibaev, Sh. Nadzhimutdinov, Vysokomolekul. Soedin <u>3</u>, 1217 (1961).
58. H. F. Church, J. Soc. Chem. Ind. (London) <u>66</u>, 221 (1947).
59. S. Claesson, Polymer <u>3</u>, 471 (1962).
60. S. Claesson, W. Bergmann, G. Jayme, Svensk Papperstidn. <u>62</u>, 141 (1959).
61. S. Claesson, 11. Nordiska Kemistmoete 20-25.8.62 Åbo, Finland.
62. F. M. Clark, Ind. Eng. Chem. <u>44</u>, 887 (1952).
63. G. L. Clark, H. C. Terford, Anal. Chem. <u>27</u>, 888 (1955).
64. W. Claussnitzer, in Landolt-Boernstein, "Zahlenwerte und Funktionen," 6. Ed. Vol. IV (Technik), part 3, Springer-Verlag, Berlin 1957.
65. C. M. Conrad, V. W. Tripp, T. Mares, J. Phys. Colloid Sci. <u>55</u>, 1474 (1951).
66. C. M. Conrad, V. W. Tripp, T. Mares, J. Phys. Colloid Chem. <u>55</u>, 238 (1956).
67. K. H. Cram, J. C. Whitwell, Textile Res. J. <u>28</u>, 849 (1958).
68. R. J. E. Cumberbirch, W. G. Harland, J. Textile Inst. <u>49</u>, T 679 (1958); W. G. Harland, J. Textile Inst. <u>49</u>, T 478 (1958).
69. R. J. E. Cumberbirch, W. G. Harland, Shirley Inst. Mem. <u>31</u>, 199 (1958).
70. G. F. Davidson, J. Textile Inst. <u>18</u>, T 175 (1927).
71. G. F. Davidson, J. Textile Inst. <u>25</u>, T 174 (1934); <u>27</u>, T 112 (1936); <u>28</u>, T 27 (1937).
72. N. E. Davis, J. H. Elliott, J. Colloid Sci. <u>4</u>, 313 (1949).
73. A. Dehnert, W. Koenig, Cellulosechem. <u>6</u>, 1 (1925).
74. H. and H. Dolmetsch, Melliand Textilber. <u>48</u>, 1449 (1967).

75. G. J. Doyle, G. Harbottle, R. Badger, R. M. Noyes, J. Phys. Colloid Chem. 51, 567 (1947).

76. A. V. Dumanskii, E. F. Nekryach, Kolloidn. Zh. 17, 171 (1955).

77. E. I. DuPont de Nemours "The Flammability of Textile Fibers," Bull. X-45 Wilmington (1955).

78. N. P. Dymarchuk, K. P. Mishchenko, T. V. Fomina, Zh. Prikl. Khim. 37, 2263 (1964).

79. I. S. Eifer, E. I. Berner, Khim. Volokna 1963, [4], 45.

80. A. af Ekenstam, Chem. Ber. 69, 553 (1936).

81. A. af Ekenstam, Chem. Ber. 69, 549, 553 (1936); R. Haller, Textilrundschau 2, 39 (1947).

82. A. af Ekenstam, Svensk. Papperstidn. 45, 81 (1942).

83. Ø. Ellefsen, Norelco Repr. 7, 104 (1960).

84. Ø. Ellefsen, F. A. Abadie-Maumert, Norsk. Skogind. 10, 238 (1956).

85. Ø. Ellefsen, J. Gjønnes, N. Norman, Acta Chem. Scand. 13, 853 (1959).

86. Ø. Ellefsen, N. Norman, J. Polymer Sci. 58, 769 (1962).

87. Ø. Ellefsen, K. Kringstad, B. A. Tønnesen, Norsk. Skogind. 18, 419 (1964).

88. Ø. Ellefsen, B. A. Tønnesen as well as D. W. Jones, in N. M. Bikales, L. Segal, "Cellulose and Cellulose Derivatives," Part IV, Wiley-Interscience, New York 1971.

89. H. Elmgren, Arkiv Kemi 24, 237 (1965).

90. F. Endress, Chem. Ing. Techn. 23, 265 (1959).

91. H. Erbring, H. Geinitz, Kolloid-Z. 84, 25 (1938).

92. D. G. Fisher, J. Mann, J. Polymer Sci. 42, 189 (1960).

93. F. H. Fórziati, J. W. Rowen, J. Res. Natl. Bur. Std. 46, 38 (1951).

94. W. F. Fowler, C. C. Unruh, P. A. McGee, W. G. Kenyon, J. Am. Chem. Soc. 69, 1636 (1947).

95. K. Freudenberg, G. Blomquist, Chem. Ber. 68, 2070 (1935).

96. K. Freudenberg, W. Kuhn, W. Duerr, F. Bolz, G. Steinbrunn, Chem. Ber. 63, 1510 (1930).

97. A. Frey-Wyssling, Jahrb. Wiss. Botanik 65, 201 (1926).

98. A. Frey-Wyssling, Kolloidchem. Beihefte 23, 40 (1927).

99. A. Frey-Wyssling, Helv. Chim. Acta 19, 911, 981 (1936).

100. A. Frey-Wyssling, Protoplasma 27, 372 (1937).

101. W. E. Gloor, E. D. Klug, in E. Ott, M. W. Grafflin, H. M. Spurlin, Eds. "Cellulose and Cellulose Derivatives," Part III, Interscience, New York 1955, p. 1440.

102. H. Goerlach, Chemiefasern 22, 611 (1972).

103. W. Goetze, F. Winkler, Faserforsch. Textiltechn. 18, 119 (1967).

104. V. M. Golubev, S. Ya. Frenkel, Vysokomolekul. Soedin A 9, 1847 (1967).

105. D. A. I. Goring, Pulp Paper Mag. Can. 64, T 517 (1963); V. A. Kargin, P. V. Kozlov, Wang Nai-Chang, Dokl. Akad. Nauk. SSSR 130, 356 (1960).

106. R. S. Govil, Man-Made Textiles 41 (9), 45 (1964).

107. N. Gralén, Thesis, Uppsala 1944.

108. N. Gralén, J. Linderot, Svensk Papperstidn. 59, 14 (1956).

109. J. Greyson, A. A. Levi, J. Polymer Sci. A 1, 3333 (1963).

110. J. C. Guthrie, J. Textile Inst. 40, T 489 (1949).

111. S. T. Han, T. Ulmanen, Tappi 41, 185 (1958).

112. W. G. Harland, J. Textile Inst. 46, T 464 (1955).

113. M. Harris, A. M. Sookne, in Matthews-Mauersberger "Textile Fibers," 5. Ed. J. Wiley, New York 1947.

114. K. Hata, K. Yokota, J. Sci. Fiber Sci. Techn. Japan 24, 420 (1968).

115. J. W. S. Hearle, R. H. Peters, "Moisture in Textiles," Butterworths, London 1960; J. S. W. Hearle, J. Textile Inst. 43 P 194 (1952); 44 T 129 (1953).

116. J. W. S. Hearle, J. Textile Inst. 44, T 117 (1953).

117. J. W. S. Hearle, J. Textile Inst. 48, P 40 (1957).

118. J. W. S. Hearle, Textile Res. J. 24, 307 (1954).

119. H. Heering, in E. v. Rzika, "Starkstromtechnik", Taschenbuch fuer Elektrotechniker, part I, Berlin 1951.

120. C. Heinegård, priv. communication.

121. C. Heinegård, E. Treiber, priv. communication.

122. J. Hengstenberg, H. Mark, Z. Krist. 69, 271 (1928).

123. J. Hengstenberg, in H. A. Stuart, Ed. "Die Physik der Hochpolymeren," Vol. II, Springer, Berlin 1953, pp. 464-77.

124. D. Henley, Svensk Papperstidn. 63, 143 (1960).

125. D. Henley, Arkiv Kemi 18, 327 (1961).

126. D. Henley, Svensk Papperstidn. 63, 143 (1960); A. Donetzhuber, Svensk Papperstidn. 63, 447 (1960); H. Reimers, Papier 16, 566 (1962).

126a. J. J. Hermans, P. H. Hermans, D. Vermaas, A. Weidinger, Rec. Trav. Chim. 65, 427 (1946).

127. P. H. Hermans, J. J. Hermans, D. Vermaas, Kolloid-Z. 109, 9 (1944); J. Polymer Sci. 1, 149, 156, 162 (1946).

128. P. H. Hermans, A. Weidinger, J. Colloid Sci. 1, 185 (1946); H. Sobue, H. Kiessig, K. Hess, Z. Phys. Chem. (B) 43, 309 (1939).

129. P. H. Hermans, "Contribution to the Physics of Cellulose Fibers," Elesevier, New York 1946.

130. P. H. Hermans, A. Weidinger, J. Am. Chem. Soc. 68, 2547 (1946).

131. P. H. Hermans, "Physics and Chemistry of Cellulose Fibers," Elesevier, New York 1949.

132. P. H. Hermans, A. Weidinger, J. Am. Chem. Soc. 68, 2547 (1946).

133. P. H. Hermans, A. Weidinger, loc. cit. 198.

134. P. H. Hermans, J. J. Hermans, D. Vermaas, A. Weidinger, J. Polymer Sci. 1, 393 (1946); 2, 632 (1947).

135. P. H. Hermans, A. Weidinger, J. Appl. Phys. 19, 491 (1948).

136. P. H. Hermans, A. Weidinger, Bull. Soc. Chim. Belges 57, 123 (1948).

137. P. H. Hermans, A. Weidinger, Kolloid-Z. 115, 103 (1949); 120, 3 (1951).

138. P. H. Hermans, A. Weidinger, J. Polymer Sci. 4, 135 (1949).

139. P. H. Hermans, A. Weidinger, J. Polymer Sci. 5, 565 (1950).

140. P. H. Hermans, A. Weidinger, J. Polymer Sci. 6, 533 (1951).

141. P. H. Hermans, Makromol. Chem. 6, 25 (1951).

142. P. H. Hermans, A. Weidinger, Textile Res. J. 31, 558, 571 (1961).

143. S. P. Hersh, D. J. Montgomery, Textile Res. J. 22, 805 (1953).

144. A. Herzog, Textile Forsch. 4, 58 (1922).

145. A. Herzog, "Die Unterscheidung von natuerlichen und kuenstlichen Seiden," Th. Steinkopff, Dresden, 1920, p. 59; A. Herzog, "Die mikroskopische Untersuchung der Seide und Kunstseide," Springer, Berlin 1924, pp. 66-69.

146. K. Hess, J. Gundermann, Chem. Ber. 70 B, 1788 (1937).

147. K. Hess, W. Weltzien, E. Messmer, Ann. Chem. 435, 1 (1924); F. L. Straus, R. M. Levy, Paper Trade J. 114, 23 (1942).

148. A. N. J. Heyn, Textile Res. J. 19, 163 (1949); J. Am. Chem. Soc. 70, 3138 (1948).

149. A. N. J. Heyn, Textile Res. J. 22, 513 (1952).

150. H. G. Higgins, J. Polymer Sci. 28, 645 (1958).

151. H. G. Higgins, C. M. Stewart, K. H. Harrington, J. Polymer Sci. 51, 59 (1961).

152. C. W. Hock, J. Polymer Sci. 8, 425 (1952).

153. A. M. Holtzer, H. Benoit, P. Doty, J. Phys. Chem. 58, 635 (1954).

154. G. Honjo, M. Watanabe, Nature 181, 326 (1958); K. C. Ellis, J. O. Warwicker, J. Polymer Sci. 56, 339 (1962); D. W. Jones, J. Polymer Sci. 32, 371 (1958); 42, 173 (1960).

155. R. Hosemann, Z. Physik 114, 133 (1939).

156. P. Howard, R. S. Parikh, J. Polymer Sci. A-1, 6, 537 (1968).

157. J. A. Howsmon, Textile Res. J. 19, 152 (1949).

157a. J. A. Howsmon, W. A. Sisson, in E. Ott, H. M. Spurlin,
       M. W. Grafflin, "Cellulose and Cellulose Derivatives,"
       2nd Ed. Interscience, New York 1954, pp. 291-316.
158.   M. L. Hunt, S. Newman, H. A. Scheraga, P. J. Flory, J.
       Phys. Chem. 60, 1278 (1956).
159.   L. Huppenthal, S. Claesson, Roczn. Chemii 39, 1867 (1965).
160.   M. M. Huque, D. A. J. Goring, S. G. Mason, Can. J.
       Chem. 36, 952 (1958).
161.   F. G. Hurtubise, Can. Textile J. 76, 53 (1959).
162.   F. G. Hurtubise, H. Kraessig, Analyt. Chem. 32, 177 (1960).
163.   E. Husemann, G. V. Schulz, Z. Phys. Chem. (B) 52, 1
       (1942).
164.   E. Husemann, E. Siefert, Makromol. Chem. 128, 288 (1969).
165.   E. Husemann, A. Carnap, J. Makromol. Chem. 1, 16 (1943);
       J. Gjønnes, N. Norman, Acta Chem. Scand. 12, 2028 (1958).
166.   K. Hutino, I. Sakurada, Naturwissenschaften 28, 577 (1940).
167.   E. H. Immergut, J. Schurz, H. Mark, Monatsh. Chem. 84,
       219 (1953).
168.   E. H. Immergut, B. G. Rånby, H. F. Mark, Ind. Eng.
       Chem. 45, 2483 (1953).
169.   E. H. Immergut, F. R. Eirich, Ind. Eng. Chem. 45, 2500
       (1953).
170.   "International Critical Tables", Vol. II, p. 237.
171.   N. Iso, D. Yamamoto, Bull. Chem. Soc. (Japan) 41, 1064
       (1968).
172.   D. S. Jackson, A. Sandig, Textile Res. J. 31, 421 (1961).
173.   G. Jacquelin, H. Bourlas, Tech. Rech. Papet 3, 49 (1964).
174.   J. Jacquemart, Bull. Inst. Textile France 1962, 963.
175.   V. Jancarik, L. Kuniak, Faserforsch. Textiltech. 20, 491
       (1969).
176.   G. Jayme, Papier 5, 244 (1951).
177.   G. Jayme, in N. M. Bikales, L. Segal, "Cellulose and
       Cellulose Derivatives", Wiley Interscience, New York 1971.
178.   G. Jayme, K. Neuschäffer, Papier 9, 563 (1955).
179.   G. Jayme, K. Neuschäffer, Papier 11, 47 (1957).
180.   G. Jayme, K. Neuschäffer, Makromol. Chem. 23, 71 (1957);
       Naturwissenschaften 44, 62 (1957).
181.   G. Jayme, V. Verburg, Reyon, Zellwolle, Chemiefasern 32,
       193, 275 (1954); L. Valtasaari, Paperi Puu 39, 243, 250, 252
       (1957); G. Jayme, G. El-Kodsi, Papier 22, 120 (1968);
       G. Jayme, Zellcheming, Mitt. Fachaussch. 19, 17 (1971).
182.   G. Jayme, F. Lang, Kolloid-Z. 150, 5 (1957).
183.   G. Jayme, P. Kleppe, Papier 15, 492 (1961).
184.   G. Jayme, P. Kleppe, Papier 15, 6 (1961).
185.   G. Jayme, J. Troeften, Melliand Textilber. 47, 1432 (1966).
186.   G. Jayme, G. El-Kodsi, Papier 24, 410, 501 (1970).
187.   G. Jayme, H. Knolle, Papier 18, 249 (1964).
188.   G. Jayme, E. Roffael, E. Oltus, Papier 23, 129 (1969).
189.   G. Jayme, G. El-Kodsi, Papier 24, 679 (1970).
190.   R. S. Jessup, E. I. Proser, J. Res. Natl. Bur. Std. 44, 385
       (1950).
191.   D. W. Jones, J. Polymer Sci. 32, 371 (1958); 42, 173 (1960).
192.   J. Juilfs, Forschungsber. Wirtsch. Verkehrsminist. Nordrhein-
       Westfahlen, Nr. 381, 38, 44 (1957).
193.   I. Jullander, Arkiv Kemi 21, 8 (1945).
194.   K. Kanamaru, Helv. Chim. Acta 17, 1037, 1425 (1934).
195.   D. E. Kane, J. Polymer Sci. 18, 405 (1955).
196.   N. E. Karasev, N. P. Dymarchuk, K. P. Mishchenko,
       Zh. Prikl. Khim. 39, 2301 (1966).
197.   N. E. Karasev, N. P. Dymarchuk, K. P. Mishchenko, J. Appl.
       Chem. USSR 40, [5], 1032 (1967).
198.   W. Kast, R. Schwarz, Z. Elektrochem. 56, 228 (1952).
199.   W. Kast, Z. Elektrochem. 57, 525 (1953).
200.   I. R. Katz, I. Seiberlich, Paper Trade J. 110, [7], 37 (1940).
201.   T. Kawai, K. Kamide, J. Polymer Sci. 54, 343 (1961).
202.   A. Keil, R. Jacob, G. Heinrich, Abh. DAW Berlin 1963, 99.

203.   E. Kelk, I. O. Wilson, Proc. IEE 112, 602 (1965).
204.   H. Kiessig, Z. Elektrochem. 54, 320 (1950).
205.   H. Kiessig, Z. Phys. Chem. (B) 43, 79 (1939); A. Frey-
       Wyssling, Biochim. Biophys. Acta 18, 166 (1955).
206.   E. Klein, K. Bosarge, J. Polymer Sci. C 2, 515 (1963).
207.   T. N. Kleinert, V. Moessmer, W. Wincor, Monatsh. Chem.
       87, 82 (1956); T. N. Kleinert, O. Dioszegi, R. M. Dlouhy,
       Svensk Papperstidn. 62, 834 (1959).
208.   N. V. Klenkova, O. M. Kulakova, L. A. Volkova, Zh.
       Prikl. Khim. 36, 166 (1963).
209.   W. T. Koch, W. B. Kunz, J. T. Massengale, Tappi 46, 569
       (1963).
210.   P. A. Koch, in Landolt-Boernstein, "Zahlenwerte und Funktio-
       nen, " Vol. IV, part 1, Springer, Berlin 1955.
211.   G. T. Kohmann, Ind. Eng. Chem. 31, (1939) 807.
212.   A. A. Konkin, R. A. Krylova, Z. A. Rogovin, Kolloid-Z.
       15, 246 (1953).
213.   A. A. Konkin, L. I. Novikova, Nauchn. Issled. Tr. Vses.
       Nauchn. Issled. Inst. Iskusstvennogo Volokna, 1957, (3), 3.
214.   A. Koura, H. Schleicher, B. Philipp, Faserforsch. Textil-
       tech. 23, 128 (1972).
215.   E. O. Kraemer, Ind. Eng. Chem. 30, 1200 (1938).
216.   H. Kraessig, E. Siefert, Makromol. Chem. 14, 1 (1954).
217.   H. Kraessig, Makromol. Chem. 26, 1 (1958).
218.   H. Kraessig, W. Kitchen, J. Polymer Sci. 51, 123 (1961).
219.   O. Kratky, Kolloid-Z. 120, 24 (1951).
219a. O. Kratky, in H. A. Stuart, "Die Physik der Hochpolymeren",
       Springer, Berlin 1955, Vol. III, pp. 288-305.
220.   O. Kratky, E. Treiber, Z. Elektrochem. 55, 716 (1951).
221.   O. Kratky, H. Sembach, Angew. Chem. 67, 603 (1955).
222.   O. Kratky, A. Krausz, Kolloid-Z. 151, 14 (1957).
223.   W. R. Krigbaum, L. H. Sperling, J. Phys. Chem. 64, 99 (1960).
224.   T. E. Khripunova, V. I. Yurev, S. S. Pozin, Probl. Fiz.
       Khim. Mekh. Volok. 1967, 473.
225.   J. Kubát, C. Pattyranie, Nature 215, 390 (1967).
226.   J. Kubát, S. Martin-Loef, Ch. Soeremark, Svensk. Papperstidn.
       72, 763 (1969).
227.   T. Kubo, Z. Phys. Chem. (A) 187, 297 (1940).
228.   L. Kuniak, B. Alinče, Papier 21, 248 (1967).
229.   K. Lauer, Kolloid-Z. 121, 139 (1951).
230.   K. Lauer, R. Doederlein, C. Jäckel, O. Wilde, J. Makromol.
       Chem. 1, 76 (1943).
231.   K. Lauer, U. Westerman, Kolloid-Z. 107, 89 (1944).
232.   C. L. Lee, Forest Prod. J. 1961, 108.
233.   Ch. Legrand, Thesis, Paris 1953.
234.   Ch. Legrand, Compt. Rend. 226, 1983 (1948); Acta Cryst. 5,
       800 (1952).
235.   Ch. Legrand, Compt. Rend. 227, 529 (1948).
236.   Ch. Legrand, Compt. Rend. 233, 407 (1951).
237.   Ch. Legrand, J. Chem. Phys. 48, 33 (1951); J. Polymer Sci.
       7, 333 (1951); Thesis, Paris 1953.
238.   K. Letters, Kolloid-Z. 58, 229 (1932); N. M. Bikales, L.
       Segal, "Cellulose and Cellulose Derivatives", Vol. V, part IV,
       Wiley Interscience, New York 1971.
239.   H. Liander, E. Bjoerkman, Svensk. Papperstidn. 55, 597 (1952).
240.   C. Y. Liang, R. H. Marchessault, J. Polymer Sci. 35, 529
       (1959); 37, 385 (1959).
241.   C. Y. Liang, R. H. Marchessault, J. Polymer Sci. 39, 269
       (1959).
242.   Th. Lieser, "Kurzes Lehrbuch der Cellulosechemie," Borntraeger,
       Berlin 1953.
243.   Th. Lieser, E. Leckzyck, Ann. Chem. 522, 56 (1936).
244.   Th. Lieser, Ann. Chem. 528, 276 (1937).
245.   C. H. Lindsley, M. B. Frank, Ind. Eng. Chem. 45, 2491 (1953).
246.   K. J. Linow, Diplomarbeit TU Dresden 1966.
247.   K. J. Linow, B. Philipp, Faserforsch. Textiltech. 18, 492 (1967).

248. S. M. Lipatov, D. V. Zharkovskii, J. M. Zagraevskaya, Kolloidn. Zh. 21, 526 (1959).

249. O. Lochmueller, Faserforsch. Textiltech. 16, 7 (1965).

250. U. Lohmander, Å. Svensson, Makromol. Chem. 65, 202 (1963).

251. H. A. de Luca, W. B. Campbell, O. Maas, Can. J. Res. B 16, 273 (1938).

252. A. Lude, Ann. Sci. Textiles Belges 1961, (2), 36.

253. W. J. Lyons, J. Chem. Phys. 9, 377 (1941).

254. F. C. Magne, H. J. Portas, H. Wakeham, J. Am. Chem. Soc. 69, 1896 (1947).

255. J. Mann, H. J. Marrinan, Trans. Faraday Soc. 52, 481, 487, 492 (1956).

256. R. H. Marchessault, G. B. Ranby, Svensk. Papperstidn. 62, 230 (1959).

257. R. H. Marchessault, Pure Appl. Chem. 5, 107 (1962).

258. V. Marčokova, M. Pašteka, Faserforsch. Textiltech. 19, 164 (1968).

259. H. J. Marrinan, J. Mann, J. Appl. Chem. (London) 4, 201 (1954).

260. H. J. Marrinan, J. Mann, J. Polymer Sci. 21, 301 (1956).

261. H. J. Marrinan, J. Mann, J. Polymer Sci. 27, 595 (1958).

262. H. J. Marrinan, J. Mann, J. Polymer Sci. 32, 357 (1958).

263. A. F. Martin, Am. Chem. Soc. Meeting, Memphis, April 20-24, 1942.

264. M. Marx, Makromol. Chem. 16, 157 (1955).

265. M. Marx, Papier 10, 135 (1956).

266. M. Marx-Figini, Makromol. Chem. 50, 196 (1961).

267. M. Marx-Figini, G. V. Schulz, Makromol. Chem. 54, 102 (1962).

268. M. Marx-Figini, Papier 13, 572 (1959).

269. M. Marx, G. V. Schulz, Makromol. Chem. 31, 140 (1959).

270. M. Marx-Figini, Papier 16, 551 (1962).

271. A. W. McKenzie, H. G. Higgins, Svensk Papperstidn. 61, 893 (1958).

272. A. W. McKenzie, APPITA 22, 82 (1968).

273. K. Meier, "Elektrische Eigenschaften von Starkstromkondensatoren," SEV-Bull. 49 (1958).

274. K. Meltzer, Thesis, TH Darmstadt 1928.

275. J. Melzer, Papier 26, 305 (1972).

276. R. Meredith, J. Textile Inst. 37, T 205 (1946).

277. K. H. Meyer, L. Misch, Chem. Ber. 70 B, 266 (1937); Helv. Chim. Acta 20, 232 (1937).

278. K. H. Meyer, N. P. Badenhuizen, Nature 140, 281 (1937).

279. K. H. Meyer, M. Studer, A. J. A. van der Wyk, Monatsh. Chem. 81, 151 (1950).

280. K. Meyer, H. Mark, "Makromolekulare Chemie," Akad. Verlag, Leipzig 1950.

281. G. Meyerhoff, Naturwissenschaften 41, 13 (1954).

282. G. Meyerhoff, J. Polymer Sci. 29, 399 (1958).

283. G. Meyerhoff, Makromol. Chem. 32, 249 (1959).

284. G. Meyerhoff, Fortsch. Hochpolymer Forsch. 3, 59 (1961).

285. G. Meyerhoff, N. Suetterlin, Makromol. Chem. 87, 258 (1965).

286. R. J. C. Michie, Polymer 2, 446 (1961).

287. B. G. Milov, S. Kh. Kitaeva, Z. I. Ukrainskaya, V. T. Renne, M. N. Morozova, L. K. Borodulina, Bumazh. Prom. 4, 3 (1965).

288. N. V. Mikhailov, E. Z. Fainberg, Dokl. Akad. Nauk. SSSR 109, 1160 (1956); J. Polymer Sci. 30, 259 (1958).

289. N. V. Mikhailov, M. N. Zav'yalova, V. O. Gorbacheva, Khim. Volokna 1960 (1), 19.

290. N. V. Mikhailov, E. Z. Fainberg, M. Kozler, Vysokomolekul. Soedin 2, 1044 (1960).

291. N. V. Mikhailov, E. Z. Fainberg, Vysokomolekul. Soedin 4, 230 (1962).

292. R. L. Mitchell, Ind. Eng. Chem. 45, 2526 (1953).

293. W. Moll, Z. Ver. Deut. Chem. Beih. Nr. 47, 105 (1943).

294. W. R. Moore, J. A. Epstein, A. M. Brown, B. M. Tidswell, J. Polymer Sci. 23, 23 (1957).

295. W. R. Moore, G. D. Edge, J. Polymer Sci. 47, 469 (1960).

296. F. F. Morehead, Textile Res. J. 20, 549 (1950).

297. J. L. Morrison, W. B. Campbell, O. Maass, Can. J. Res. B 18, 168 (1940).

298. W. Morrow, Tappi 42, 167 (1959).

299. W. E. Morton, J. W. S. Hearle, "Physical Properties of Textile Fibers", Butterworths, London 1962.

300. H. Mosimann, Helv. Chim. Acta 26, 369 (1943).

301. W. A. Mueller, L. N. Rogers, Ind. Eng. Chem. 45, 2522 (1953).

302. S. M. Mukherjee, J. Sikorski, H. J. Woods, J. Textile Inst. 43, T 196 (1952).

303. A. Muenster, Z. Phys. Chem. (Leipzig) 197, 17 (1951).

304. A. Muenster, J. Polymer Sci. 8, 633 (1952).

305. E. J. Murphy, Can. J. Phys. 41, 1022 (1963).

306. H. Nadziakiewicz, H. Jedlinska, Polimery 7, 85 (1962).

307. H. Nadziakiewicz, H. Jedlinska, Polimery 7, 131 (1962).

308. E. Naujoks, Melliand Textilber. 46, 731 (1965).

309. S. M. Neale, W. A. Stringfellow, Trans Faraday Soc. 37, 525 (1941).

310. E. F. Nekryach, F. Semchenko, Ukr. Khim. Zh. 26, 700 (1960).

311. S. B. Newman, in G. M. Kline, Ed. "Analytical Chemistry of Polymers," Part III, John Wiley, New York 1962, pp. 286-287.

312. S. Newman, L. Loeb, C. M. Conrad, J. Polymer Sci. 10, 463 (1953).

313. S. Newman, L. Loeb, C. M. Conrad, J. Polymer Sci. 54, 343 (1961).

314. L. Nicolas, Bull. Assoc. Techn. Ind. Papetiére 5, 427 (1951).

315. I. Nieduszynski, R. D. Preston, Nature 225, 273 (1970).

316. S. Nomura, S. Kawabata, H. Kawai, Y. Yamaguchi, A. Fukushima, H. Takahara, J. Polymer Sci. A-2, 7, 325 (1969).

317. NORME FRANCAISE, NF. T 12-005 (1953).

317a. L. I. Novikova, A. A. Konkin, Zh. Prikl. Khim. 32, 1081 (1959); A. A. Konkin, A. G. Yashunskaya, E. M. Bychkova, Nauchn. Issled. Tr., Vses, Nauch.-Issled. Inst. Iskusstvennogo Volokna 1955 (2), 3; A. A. Konkin, N. I. Kaplan, Z. A. Rogovin, Zh. Prikl. Khim. 28, 729 (1955); R. A. Martin, E. Pacsu, Textile Res. J. 26, 192 (1956); L. Joergensen, Thesis, Trondheim 1950.

318. M. Oberlin, J. Mering, Compt. Rend. (Paris) 238, 1046 (1954).

319. I. Okamura, Naturwissenschaften 21, 393 (1933).

320. L. Oldsberg, O. Samuelson, Svensk Papperstidn. 60, 745 (1957).

321. R. T. O'Connor, E. F. Du Prè, O. Mitcham, Textile Res. J. 28, 382 (1958).

321a. W. Otting, "Spektrale Zuordnungstafel der Infrarot-Absorptionsbanden," Springer Heidelberg 1963.

322. Ya. V. Pak, Kh. U. Usmanov, Kolloidn. Zh. 18, 233 (1956).

323. A. Parisot, J. Cyrot, J. Textile Inst. 42, P 783 (1951).

324. G. P. Pearson, W. R. Moore, Polymer 1, 144 (1960).

325. A. Peterlin, in H. A. Stuart, Ed. "Die Physik der Hochpolymeren," Vol. II, Springer Berlin 1953, p. 556.

326. G. A. Petropavlovskii, M. M. Krunchak, J. Appl. Chem. USSR 39, 2201 (1966).

327. B. Philipp, K. J. Linow, Zellstoff Papier 14, 321 (1965).

328. B. Philipp, G. Gloeckner, K. J. Linow, Faserforsch. Textiltech. 18, 311 (1967).

329. B. Philipp, H. Schleicher, I. Laskowski, Faserforsch. Textiltech. 23, 60 (1972).

330. B. Philipp, K. J. Linow, to be published.

331. H. J. Philipps, C. F. Bjork, J. Polymer Sci. 6, 549 (1951).

332. E. A. Plisko, L. A. Nudga, Vysokomolekul. Soedin 9B, 825 (1967).

333. S. Poller, Faserforsch. Textiltech. 20, 71 (1969).

334. G. Porod, loc. cit. E. Treiber, "Die Chemie der Pflanzen-zellwand," Springer Berlin 1957.

335. G. Prati, Tinctoria 59, 233, 279, 329 (1962).

336. G. Prati, A. Seves, Ric. Doc. Tess. 3, 112 (1966).

337. J. M. Preston, Trans. Faraday Soc. 29, 65 (1933).

338. J. M. Preston, M. V. Nimkar, J. Textile Inst. 41, T 446 (1950).

339. R. D. Preston, P. H. Hermans, A. Weidinger, J. Exptl. Botany 1, 344 (1950).

340. G. Rabinov, E. Heymann, J. Phys. Chem. 47, 655 (1943).

341. M. V. Ramiah, D. A. Goring, J. Polymer Sci. C 11, 27 (1965).

342. P. K. Ray, Textile Res. J. 37, 434 (1967).

343. P. K. Ray, J. Appl. Polymer Sci. 13, 2593 (1969).

344. W. Reddish, Trans. Faraday. Soc. 46, 459 (1950).

345. W. H. Rees, J. Textile Inst. 39, T 351 (1948).

346. M. Rinaudo, Papetrie 90, 479 (1968).

347. U. Roessner, Deut. Textiltech. 16, 304 (1966).

348. L. Rose, J. D. Griffiths, J. Textile Inst. 39, P 265 (1948).

349. W. E. Roseveare, L. Poore, Ind. Eng. Chem. 45, 2518 (1953).

350. H. Ruck, Papier 23, 872 (1969).

351. H. Ruck, H. Kraessig, Norelco Reptr. 7, 71 (1960).

352. B. G. Ranby, Makromol. Chem. 13, 40 (1954); B. G. Ranby, E. Ribi, Experientia 6, 12 (1950).

353. B. D. Saksena, K. C. Aggarwal, G. S. Jauhrit, J. Polymer Sci. 62, 347 (1962).

354. I. Sakurada, K. Hutino, Kolloid-Z. 77, 346 (1936).

355. G. Sarret, Thesis, University of Grenoble, France 1967.

356. E. Sauter, Z. Phys. Chem. (B) 35, 83 (1937); 43, 294 (1939).

357. V. P. Saxena, H. L. Bhatnagar, A. B. Biswas, J. Appl. Polymer Sci. 7, 181 (1963).

358. P. C. Scherer, J. Am. Chem. Soc. 53, 4009 (1931).

359. P. C. Scherer, A. Tanenbaum, D. W. Levi, J. Polymer Sci. 33, 531 (1960).

360. E. Schiebold, Kolloid-Z. 108 248 (1944); 120, 55 (1951).

361. B. Schneider, J. Vodnanski, Collection Czech. Chem. Commun. 28, 2080 (1963).

362. R. Schneider, O. Toeppel, Papier 25, 849 (1971).

363. G. V. Schulz, H. J. Loehmann, J. Prakt. Chem. 157, 238 (1941).

363a. G. V. Schulz, E. Husemann, Z. Phys. Chem. B 52, 23 (1942).

364. G. V. Schulz, F. Blaschke, J. Prakt. Chem. 158, 130 (1941).

365. G. V. Schulz, Z. Phys. Chem. 193 168 (1944).

366. G. V. Schulz, M. Marx, Makromol. Chem. 14, 52 (1954).

368. J. Schurz, Papier 15, 10 (1961).

369. K. Schwabe, U. Trobisch, Zellstoff Papier 5, 55 (1956).

370. R. G. Schweiger, Chem. Ind. (London) 1969, [10], 296

371. E. Schweizer, J. Prakt. Chem. 72, 109 (1857).

372. L. Segal, L. Loeb, J. J. Creely, J. Polymer Sci. 13, 193 (1954).

373. L. Segal, J. J. Creely, A. E. Martin, C. M. Conrad, Textile Res. J. 29, 786 (1959).

374. R. Segula, Melliand Textilber. 51, 743 (1970).

375. A. Seves, G. Prati, G. Vecchio, Ric. Doc. Tess. 4 13 (1967).

376. V. P. Shanbhag, Arkiv Kemi 29, 1 (1968).

377. V. I. Sharkov, V. P. Levanova, Vysokomolekul. Soedin 1, 1027 (1959).

378. T. M. Shaw, J. J. Windle, J. Appl. Phys. 21, 956 (1950); H. W. Verseput, Tappi 34, 572 (1951); D. E. Kane, J. Polymer Sci. 18, 405 (1955). W. Reddish, Trans. Faraday Soc. 46, 459 (1950).

379. S. G. Shenouda, A. Viswanathan, J. Appl. Polymer Sci. 16, 395 (1972).

380. R. S. Shutt, US-Patent 2,371,359.

381. H. Sihtola, B. Kyrklund, L. Laamanen, I. Palenius, Paperi Puu 45, 225, 229, (1963).

382. R. Signer, P. v. Tavel, Helv. Chim. Acta 21, 535 (1938).

383. Cr. I. Simionescu, C. V. Uglea, Chem. Technol. Cellul. 3, 649 (1969).

384. S. R. Sivaraja Iyer, N. T. Baddi, Cell. Chem. Technol. 3, 561 (1969).

385. D. K. Smith, R. F. Bampton, W. J. Alexander, Ind. Eng. Chem. Process Design Div. 2, 57 (1963).

386. J. K. Smith, W. J. Kitchen, D. B. Mutton, J. Polymer Sci. C 2, 499 (1963).

387. H. Sommer, F. Winkler, in H. Sommer "Handbuch der Werk-stoffpruefung," Vol. V, Springer Berlin 1960.

388. H. Staeger, B. Frischmuth, F. Held, Schweiz. Verband Materialpruefung Technik. Report No. 41, Zuerich 1946.

389. A. J. Stamm, L. A. Hansen, J. Phys. Chem. 41, 1007 (1937).

390. A. J. Stamm, W. E. Cohen, J. Phys. Chem. 42, 921 (1938).

391. H. Staudinger, G. Daumiller, Chem. Ber. 70, 2508 (1937).

392. H. Staudinger, G. Daumiller, Ann. Chem. 529, 219 (1937).

393. H. Staudinger, F. Zapf, J. Prakt. Chem. 156, 261 (1940).

394. H. Staudinger, K. Eder, J. Prakt. Chem. 159, 39 (1941).

395. H. Staudinger, F. Reinecke, Ann. Chem. 535, 47 (1938).

396. B. I. Stepanov, R. G. Zhbankov, R. Marupov, Vysokomole-kul. Soedin 3, 1633 (1961).

397. B. I. Stepanov, R. G. Zhbankov, Zavodsk. Lab. 29, 696 (1963).

398. C. Sterling, B. J. Spit, Am. J. Botany 44, 851 (1957).

399. W. N. Stoops, J. Am. Chem. Soc. 56, 1480 (1934).

400. H. A. Stuart, in H. A. Stuart, Ed. "Die Physik der Hochpoly-meren," Vol. II, Springer Berlin 1953, p. 654-64.

401. H. A. Stuart, "Die Physik der Hochpolymeren," Vol. III, Springer Berlin 1955.

402. J. B. Tailor, J. Textile Inst. 43, T 489 (1952).

403. Th. Taylor, Brit. Patent 787 (1859).

404. T. Terada, N. Ito, Y. Goto, Kami Pa Gikyoshi 23, 191 (1969).

405. K. Terasaki, K. Matsuura, Japan Tappi 26, 511 (1972).

406. T. E. Timell, Svensk Papperstidn. 57, 913 (1954).

407. T. E. Timell, Svensk Papperstidn. 60, 836 (1957).

408. W. Trapp, L. Pungs, Holzforsch. 10, 65 (1956).

409. W. Traube, Chem. Ber. 44, 3319 (1911); ibid. 54, 3220 (1921); ibid. 55, 1899 (1922).

410. E. Treiber, Protoplasma 40, 166 (1951).

411. E. Treiber, "Die Chemie der Pflanzenzellwand," Springer Berlin 1957.

412. E. Treiber, W. Berndt, M. Ruck, H. Toplak, Chem. Ing. Technik 26, 687 (1954).

413. E. Treiber, B. Abrahamson, Papier 13, 253 (1959).

414. E. Treiber, B. Abrahamson, Holzforsch. 13, 161 (1959).

415. E. Treiber, B. Abrahamson, V. Holta, Svensk Papperstidn. 62, 459 (1959).

416. E. Treiber, J. Rehnstroem, Svensk Papperstidn. 62, 1 (1959).

417. E. Treiber, Lecture at 2nd Intern. Man-Made Fiber Con-ference, Dornbirn, July 15, 1963.

418. E. Treiber, Chemiefasern 14, 25 (1964).

419. E. Treiber, Mittlg. Fachausschuesse 18, 18 (1970).

420. E. Treiber, Mittlg. Fachausschuesse 20, 15 (1972).

421. M. V. Tsilipotkina, A. A. Tager, B. S. Petrov, G. Pusto-baeva, Vysokomolekul. Soedin 4, 1844 (1962).

422. M. Tsuboi, J. Polymer Sci. 25, 159 (1957).

423. K. Uda, G. Meyerhoff, Makromol. Chem. 47, 168 (1961).

424. US-Patent 3,236,669

425. H. Valdsaar, R. Dunlap, Am. Chem. Soc. Meeting Atlantic City N.J. 14-19 Sept. 1952.

426. L. Valtasaari, Paperi Puu $\underline{39}$, 243, 250, 252 (1957).

427. L. Valtasaari, Tappi $\underline{48}$, 627 (1965).

428. H. Veith, Frequenz $\underline{3}$, 165, 216, (1949).

429. A. Venkateswaran, J. Appl. Polymer Sci. $\underline{9}$, 1127 (1965).

429a. H. Vink, Arkiv Kemi $\underline{11}$, 29 (1957).

430. H. Vink, Arkiv Kemi $\underline{14}$, 195 (1959).

431. H. Vink, Arkiv Kemi $\underline{20}$, 513 (1963).

432. H. Vink, Makromol. Chem. $\underline{67}$, 105 (1963).

433. H. Vink, Svensk. Papperstidn. $\underline{70}$, 734 (1967).

434. H. Vink, Makromol. Chem. $\underline{94}$, 1 (1966).

435. H. Vink, in N. M. Bikales, L. Segal "Cellulose and Cellulose Derivatives," Vol. V, part IV, Wiley Interscience, New York, 1971.

436. A. Visapää, Faserforsch. Textiltech. $\underline{15}$, 579 (1964).

437. A. Viswanathan, V. Venkatakrishnan, J. Appl. Polymer Sci. $\underline{13}$, 571, 785 (1969).

438. J. Vodnanski, M. Slabina, B. Schneider, Collection Czech. Chem. Commun. $\underline{28}$, 3245 (1963).

439. H. Vosters, Svensk Papperstidn. $\underline{65}$, 65 (1962).

440. L. Waengberg, E. Treiber, loc. cit. I. Czajlik, E. Treiber, Holzforsch. $\underline{23}$, 133 (1969).

441. M. Wahba, J. Phys. Colloid. Chem. $\underline{52}$, 1197 (1948).

442. M. Wahba, J. Phys. Colloid. Chem. $\underline{54}$, 1148 (1950).

443. M. Wahba, S. Nashed, K. Aziz, J. Textile Inst. $\underline{49}$, T 519 (1958).

444. J. H. Wakelin, H. S. Virgin, E. Crystal, J. Appl. Phys. $\underline{30}$, 1654 (1959).

445. H. Wakeham. Textile Res. J. $\underline{19}$, 595 (1949).

446. A. C. Walker, M. H. Quell, J. Textile Inst. $\underline{24}$, T 123 (1933).

447. A. Wallraff, Elektrotech. Z. $\underline{63}$, 539 (1942).

448. M. Wandel, Thesis, Stuttgart 1957.

449. H. A. Wannow, Kolloid-Z. $\underline{104}$, 103 (1944).

450. H. A. Wannow, F. Thormann, Kolloid-Z. $\underline{112}$, 94 (1949).

451. J. O. Warwicker, R. Jeffries, R. L. Colbran, R. N. Robinson, "A Review of the Literature on the Effect of Caustic Soda and Other Swelling Agents on the Fine Structure of Cotton," Shirley Inst. Pamphlet No. 93, Altrincham 1966.

452. W. Wegener, in H. Sommer, Ed. "Handbuch der Werkstoffpruefung," Vol. V, Springer Berlin 1960, pp. 394, 428.

453. P. P. von Weimarn, Kolloid-Z. $\underline{11}$, 41 (1912).

454. H. J. Wellard, J. Polymer Sci. $\underline{13}$, 471 (1954).

455. M. G. Wirick, M. Waldman, Am. Chem. Soc. Meeting, Sept. 9-12, 1968.

455a. R. Willstaetter, L. Zechmeister, Chem. Ber. $\underline{46}$, 2401 (1913). A. af Ekenstam, "Ueber die Celluloseloesungen in Mineralsäuren," Lund 1936. L. P. Vyrodova, V. I. Sharkov, Zh. Prikl. Khim. $\underline{39}$, [3], 682 (1966).

456. A. Wlochowicz, Przegl. Wlok. $\underline{22}$, 471 (1968).

457. N. S. Wooding, in I. W. S. Hearle, Ed. "Fibre Structure," Butterworth, London, 1963, pp. 460, 466.

458. E. Wuenneberg, W. Fischer, A. Sapper, Z. Phys. Chem. A $\underline{151}$, 1 (1930).

459. A. v. d. Wyk, K. H. Meyer, J. Polymer Sci. $\underline{2}$, 583 (1946).

460. N. P. Zakurdaeva, S. A. Minyailo, E. K. Podgorodetskii, Vysokomolekul. Soedin $\underline{10}$ B, 538 (1968).

461. R. G. Zhbankov, V. I. Nepochatych, R. Marupov, Z. A. Rogovin, Vysokomolekul. Soedin $\underline{4}$, 1696 (1962).

462. W. Zimmermann, Melliand Textilber. $\underline{23}$, 73 (1942).

463. E. Talpay, B. Talpay, "Der Baumwolltest", 2 ed. Bachmann, Bremen.

VI. PHYSICAL DATA OF OLIGOMERS

PHYSICAL DATA OF OLIGOMERS

M. Rothe
Universität Mainz
Organisch-Chemisches Institut
Mainz, Germany

CONTENTS[*]

[*] References follow after each subgroup.

PHYSICAL DATA OF OLIGOMERS

Oligomers are defined as the low members of the polymeric-homologous series, with molecular weights up to about 1000 - 2000. They are easily accessible by stepwise synthesis or by separation from the polymers. Being low molecular compounds of defined molecular weights oligomers represent ideal model substances for the polymers. Deep insight is gained into the relation between chain length and physical properties by physical studies of whole series of oligomers. In this way certain physical data of the polymers could be explained for the first time. In chemical respect oligomers on principle must have the same properties as the related polymers; they are, however, much easier accessible to all studies owing to their strictly defined and comparatively simpler structure.

Homologous oligomers due to their low molecular weights differ sufficiently in their physical properties so that they can be separated into chemical individuals. Therefore they can be used for elucidating the structure of polymers and - in close relation - for the investigation of the mechanism of polymerization. Oligomers are intermediates in polycondensation and polyaddition reactions and are present in more or less significant amounts in the polymers due to the equilibrium between different chains and between chains and rings. Therefore important conclusions on the structure of the related high polymers can be drawn from isolation and structure of oligomers along with a comparison of chemical properties of polymers and oligomers. In this way exact evidence can be gained on the type of linkage between the monomer units in the polymer and on the structure of unknown endgroups. Finally, unequivocal evidence on the mechanism of polymer formation is to be expected from the behaviour of oligomers under the conditions of polycondensation.

The following tables give the physical properties of the most important linear, cyclic, and branched oligomers which are significant for polymer chemistry. Derivatives and co-oligomers (such as co-oligopeptides, -saccharides and -nucleotides) as well as organosilicon and inorganic oligomers are not included. Only those literature references are included which indicate the best methods of synthesis and the most important properties of the oligomers concerned.

1. OLIGOMERS CONTAINING MAIN CHAIN ACYCLIC CARBON ONLY

1.1 OLIGO(OLEFINS)

1.1.1 OLIGO(METHYLENES) AND OLIGO(ETHYLENES)

1.1.1.1 n-Alkanes $H[CH_2]_n H$

n	Mol. Wt.	m.p. [$^\circ$C]	b.p. [$^\circ$C/mm]	[$d_4/^\circ$C]	References
1	16.0	-182.6	-161.6	0.4240/-164	21, 22, 75, 84
2	30.1	-183.3	-88.5	0.5462/-89	21, 75, 84
3	44.1	-187.1	-42.2	0.5824/-45	21, 22, 75, 84
4	58.1	-138.4	-0.5	0.6011/0	21, 75, 84
5	72.2	-129.7	36.1	0.6263/20	5, 7, 13, 21, 75, 84
6	86.2	-94.0	68.7	0.6594/20	3, 5, 7, 21, 75, 84
7	100.2	-90.5	98.4	0.6838/20	5, 7, 13, 21, 22, 75, 84
8	114.2	-56.8	125.7	0.7026/20	3, 5, 7, 13, 21, 84, 87
9	128.3	-53.5	150.8	0.7177/20	5, 7, 13, 21, 84, 87
10	142.3	-29.7	174.1	0.7301/20	3, 5, 7, 18, 21, 84, 87
11	156.3	-25.6	195.9	0.7402/20	3, 5, 7, 13, 21, 77, 83, 84, 87
12	170.3	-9.7	216.3	0.7487/20	5, 7, 17, 21, 84, 87
13	184.4	-5.2	235.5	0.7563/20	5, 7, 13, 21, 77, 83, 84, 87
14	198.4	5.5	253.6	0.7627/20	5, 7, 17, 21, 84, 87
15	212.4	10.1	270.7	0.7684/20	3, 5, 7, 21, 77, 83, 84, 87
16	226.4	18.1	287.1	0.7733/20	3, 5, 7, 15, 16, 19, 21, 78, 84, 87
17	240.5	22.1	302.6	0.7767/22	3, 5, 7, 13, 14, 21, 75, 77, 83, 84
18	254.5	28.2	317.4	0.7768/28	3, 5, 7, 14, 15, 17, 19, 21, 75, 84
19	268.5	32.0	331.6	0.7776/32	1, 3-5, 7, 14, 15, 19, 21, 77, 83, 84
20	282.6	36.6	345.1	0.7550/70	1, 3-5, 7, 14, 15, 17-19, 21, 82, 88
21	296.6	40.4	215/15	0.7583/70	1-4, 14, 15, 19, 21, 22, 77, 83
22	310.6	44.4	224-225/15	0.7631/70	1-4, 7, 14, 21, 78, 83
23	324.6	47.7	234/15	0.7641/70	1-4, 14, 21, 77, 83
24	338.7	51.1	243/15	0.7657/70	1-4, 14, 17, 21, 83
25	352.7	53.7	259/15	0.7693/70	1-4, 5, 14, 21, 77, 83
26	366.7	57.0	262/15	0.7704/70	1-4, 7, 14, 21, 83
27	380.7	59.0	270/15	0.7732/70	1-4, 8, 14, 21, 77, 83
28	394.8	61.4	279-281/15	0.7750/70	1, 2, 4, 13-15, 17, 19, 21, 83
29	408.8	63.6	286/15	0.7755/70	1-4, 14, 21, 77, 83
30	422.8	66.0	304/15	0.7795/70	1-4, 7, 14, 18, 21, 82, 83
31	436.9	67.9	302/15	0.7678/100	1-4, 8, 77, 83, 85
32	450.9	69.5	310/15	0.7645/100	1, 2, 4, 15, 16, 19, 20, 83
33	464.9	71.3			1, 2, 4, 5, 77, 83
34	478.9	72.9			1-4, 7, 77, 83, 85
35	493.0	74.7	331/15	0.7814/74	1, 2, 4, 7, 8, 10, 15, 19, 77, 83, 85

References page VI-5

n	Mol. Wt.	m.p. [°C]	b.p. [°C/mm]	$[d_4/°C]$	References
36	507.0	76.1	298.4/3	0.7783/90	1, 2, 4, 7, 10, 13-15, 19, 77, 78, 83
37	531.0	77.7			75
38	535.0	79.2			75, 77, 83
39	549.1	80.5			12, 75, 77, 83
40	563.1	81.5	$150/10^{-4}$		6, 7, 10, 17, 18, 21, 75, 77, 82, 83
41	577.1	81.7			75
42	591.2	82.9			76
43	605.2	85.5	332/3	0.7812/90	10, 22, 77, 83, 86
44	619.2	86.6			3, 11, 13, 15, 19, 77, 80, 89
46	647.2	88.2			77, 80, 91
50	703.4	92.3	420-422/15		6, 7, 11, 15, 21, 77, 80, 82, 90
52	731.4	94.2			77, 80, 90
54	759.5	95.2			2, 7, 77, 80, 90
60	843.6	99.4	$250/10^{-4}$		6, 7, 11, 21, 77, 80, 82, 90
62	871.7	100.7			2, 7, 11, 77, 80, 90
64	899.7	102.3			2, 11, 22, 77, 80, 90
66	927.8	103.8			9, 77, 80, 90
67	941.8	104.3			9, 77, 80, 90
70	983.9	105.5	$300/10^{-4}$		6, 7, 11, 21, 77, 80, 82, 90
82	1152.2	110.5			77, 80, 81
94	1320.6	114.1-114.5			15, 19, 20, 77, 79, 80
100	1404.7	115.4			21, 77, 80-82
120	1685.3	119.0			82
140	1965.8	121.0			82

1.1.1.2 Cycloalkanes $\left[CH_2\right]_n$

n	Mol. Wt.	m.p. [°C]	b.p. [°C/mm]	$[d_4/°C]$	References
3	42.1	-127.5	-32.8	0.7352/-80	23, 24
4	56.1	< -80	13.1/774	0.698/0	25
5	70.1	-93.9	49.3	0.745/20	26
6	84.2	6.6	80.7	0.7784/20	27-29
7	98.2	-8.0	118.1	0.8098/20	29-31
8	112.2	13.5	149/749	0.8349/20	32-34
9	126.2	9.7	69/14	0.8534/15.2	29, 34
10	140.3	10.8	201	0.8577/20	29, 35, 36
11	154.3	-7.3	91/12	0.8591/20	29, 37
12	168.3	62-63		0.861	29, 38, 39
13	182.4	23.5	128/20	0.861	29, 34, 38
14	196.4	55-55.5	131/11	0.8259/79	29, 38, 40, 41
15	210.4	65-66	147/12	0.8240/78	29, 38, 40
16	224.4	62-63	170-171/20	0.819/79	38, 40-42
17	238.5	66-67		0.8200/77	40, 43, 44
18	252.5	74-75		0.8201/76	40-42, 45
19	266.5	81-82			40
20	280.5	61-62			18, 41
21	294.6	63-64			41
22	308.6	52-53	177/0.4	0.8174/75	38, 43
23	322.6	49-50	177/0.1	0.8259/64	46
24	336.6	50-51	222-228/0.6		38, 40, 41, 45
25	350.7	53-54			47
26	364.7	44-46	218-219/0.5	0.8120/78	38, 40, 43
27	378.7	47-48			41
28	392.8	48	213-214/0.25	0.8243/58	38, 41, 48
29	406.8	47	215/1.1	0.8232/64	38
30	420.8	57-58	230/0.2	0.8233/69	18, 41, 42, 45, 48
32	448.9	59-60		0.8261/70	41, 48
34	476.9	66-67	230-240/0.3	0.8229/76	48, 92
36	505.0	70-71			41, 45
40	561.1	76-77			18, 41

n	Mol. Wt.	m.p. [$^{\circ}$C]	b.p. [$^{\circ}$C/mm]	[$d_4/^{\circ}$C]	References
42	589.2	75-76			45
45	631.2	80-81			41
50	701.4	87-88			18
54	757.5	90-91			41

1.1.2 OLIGO(PERFLUOROMETHYLENES) AND OLIGO(PERFLUOROETHYLENES)

1.1.2.1 Perfluoro-n-alkanes $F[CF_2]_n F$

n	Mol. Wt.	m.p. [$^{\circ}$C]	b.p. [$^{\circ}$C/mm]	[$d_4/^{\circ}$C]	References
1	88.0	-183.7	-128/754	1.619/-129	49
2	138.0	-94	-78.3	1.590/-78	49, 50
3	188.0	-183	-38		51, 52
4	238.0		-2.0		52-54
5	288.1	-125.4	29.3	1.620	52, 54-56
6	338.1	-86.3	57.2	1.6995	52, 57, 58
7	388.1	-78	82.2	1.7333	52, 56, 59
8	438.1	-65	103.3	1.776/25	52, 59
9	488.1	-16	125.3	1.799/25	52, 59
10	538.1	36	144.2	1.770/45	52, 59
11	588.1	57	161	1.745/70	52, 59
12	638.1	75	178	1.670/113.5	52, 59
16	838.2	125	232		54, 59

1.1.2.2 Perfluoro-cycloalkanes $[CF_2]_n$

n	Mol. Wt.	m.p. [$^{\circ}$C]	b.p. [$^{\circ}$C]	References
3	150.0	-80	-31.5	52, 60, 61
4	200.0	-38.7	-5	61-63
5	250.0	9.9-10.2	23.5	51, 61, 64
6	300.0	58.2		51, 56, 65
7	350.0		80.0	51

1.1.3 OLIGO(ISOBUTYLENES)
(References 66-74)

	Compound	Mol. Wt.	m.p. [$^{\circ}$C]	b.p. [$^{\circ}$C/mm]	d_4^{20}
Monomer:	isobutylene $CH_2=C(CH_3)_2$	56.1		-6	
Dimers:	2.4.4-trimethyl-1-pentene $CH_2=C-CH_2-C(CH_3)_3$ \ CH_3	112.2	-93.6	101.4	0.7150
	2.4.4-trimethyl-2-pentene $CH_3-C=CH-C(CH_3)_3$ \ CH_3	112.2	-106	104.9	0.7211
	3.4.4-trimethyl-2-pentene $CH_3-CH=C-C(CH_3)_3$ \ CH_3	112.2		112.3	0.7392
	2.3.4-trimethyl-2-pentene $(CH_3)_2CH-C=C(CH_3)_2$ \ CH_3	112.2	-113.4	116.3	0.7434
	2.3.4-trimethyl-1-pentene $CH_2=C-CH-CH(CH_3)_2$ \ CH_3 CH_3	112.2		108	0.729
	2.3.3-trimethyl-1-pentene $CH_2=C-C(CH_3)_2-CH_2-CH_3$ \ CH_3	112.2	-69	108.4	0.7352
Trimers:	1.1-dineopentylethylene $CH_2=C[CH_2C(CH_3)_3]_2$	168.3		85-86/40	0.7599

REFERENCES

1. F. Krafft, Chem. Ber. 15, 1687, 1711 (1882); 19, 2218 (1886); 40, 4479 (1907).

2. A. Gascard, Ann. Chim. (Paris) (9) 15, 332 (1921).

3. A. Mueller, Proc. Roy. Soc. (London) Ser. A 120, 437 (1928); 127, 417 (1930); 138, 514 (1932).

4. J. H. Hildebrand, A. Wachter, J. Am. Chem. Soc. 51, 2487 (1929).

5. G. S. Parks, H. M. Hoffmann, S. B. Thomas, J. Am. Chem. Soc. 52, 1032 (1930).

6. W. H. Carothers, J. W. Hill, J. E. Kirby, R. A. Jacobsen, J. Am. Chem. Soc. 52, 5279 (1930).

7. W. E. Garner, K. van Bitter, A. M. King, J. Chem. Soc. 1931, 1533.

8. H. Staudinger, F. Staiger, Chem. Ber. 68, 707 (1935).

9. F. Francis, A. M. King, J. A. V. Willis, J. Chem. Soc. 1937, 999.

10. H. J. Bacher, J. Strating, Rec. Trav. Chim. Pays-Bas 59, 933 (1940).

11. W. F. Syer, R. F. Patterson, J. L. Keays, J. Am. Chem. Soc. 66, 179 (1944).

12. E. Stenhagen, B. Tägtstroem, J. Am. Chem. Soc. 66, 846 (1944).

13. A. K. Doolittle, R. H. Peterson, J. Am. Chem. Soc. 73, 2145 (1951).

14. A. A. Schaerer, C. J. Busso, A. E. Smith, L. B. Shinner, J. Am. Chem. Soc. 77, 2017 (1955).

15. P. R. Templin, Ind. Eng. Chem. 48, 154 (1956).

16. F. Sondheimer, Y. Amiel, J. Am. Chem. Soc. 79, 5817 (1957).

17. F. Sondheimer, Y. Amiel, R. Wolovsky, J. Am. Chem. Soc. 79, 6263 (1957).

18. F. Sondheimer, R. Wolovsky, D. A. Ben-Efraim, J. Am. Chem. Soc. 83, 1686 (1961).

19. A. Odajima, J. A. Sauer, A. E. Woodward, J. Phys. Chem. 66, 718 (1962).

20. D. W. MaCCall, D. C. Douglass, E. W. Anderson, Ber. Bunsenges. Physik. Chem. 67, 336 (1963).

21. J. D. Downer, K. J. Beynon, in Rodd's "Chemistry of Carbon Compounds", Vol. I, Part A, 2nd Ed., Elsévier Publ. Co., Amsterdam-London-New York, 1964, p. 367.

22. R. Simha, A. J. Havilik, J. Am. Chem. Soc. 86, 197 (1964).

23. M. Trautz, K. Winkler, J. Prakt. Chem. 104, 37 (1922).

24. R. A. Ruehrwein, T. M. Powell, J. Am. Chem. Soc. 68, 1063 (1946).

25. G. B. Heisig, J. Am. Chem. Soc. 63, 1698 (1941).

26. J. G. Aston, H. L. Fink, S. C. Schumann, J. Am. Chem. Soc. 65, 341 (1943).

27. J. G. Aston, G. J. Szasz, H. L. Fink, J. Am. Chem. Soc. 65, 1135 (1943).

28. R. A. Raphael, in E. R. Rodd, Ed., "Chemistry of Carbon Compounds", Vol. II, Part A, 1st Ed., Elsevier, Amsterdam-London-New York, 1953, p. 124.

29. L. Ruzicka, P. A. Plattner, H. Wild, Helv. Chim. Acta 29, 1611 (1946).

30. E. P. Kohler, M. Tishler, H. Potter, H. T. Thompson, J. Am. Chem. Soc. 61, 1057 (1939).

31. R. Willstätter, T. Kametaka, Chem. Ber. 41, 1480 (1908).

32. N. D. Zelinsky, M. G. Freimann, Chem. Ber. 63, 1485 (1930).

33. L. Ruzicka, H. A. Boekenoogen, Helv. Chim. Acta 14, 1319 (1931).

34. L. Ruzicka, P. A. Plattner, H. Wild, Helv. Chim. Acta 28, 395 (1945).

35. P. A. Plattner, J. Huelstkamp, Helv. Chim. Acta 27, 220 (1944).

36. W. Hueckel, A. Gercke, A. Gross, Chem. Ber. 66, 563 (1933).

37. P. A. Plattner, Helv. Chim. Acta 27, 801 (1944).

38. L. Ruzicka, M. Stoll, H. W. Huyser, H. A. Boekenoogen, Helv. Chim. Acta 13, 1152 (1930).

39. R. Wolovsky, F. Sondheimer, J. Am. Chem. Soc. 84, 2844 (1962).

40. J. Dale, A. J. Hubert, G. S. D. King, J. Chem. Soc. 1963, 73.

41. F. Sondheimer, Y. Amiel, R. Wolovsky, J. Am. Chem. Soc. 81, 4600 (1959).

42. L. Ruzicka, W. Bruegger, C. F. Seidel, H. Schinz, Helv. Chim. Acta 11, 496 (1928).

43. L. Ruzicka, G. Giacomello, Helv. Chim. Acta 20, 548 (1937).

44. L. Ruzicka, W. Bruegger, M. Pfeiffer, H. Schinz, M. Stoll, Helv. Chim. Acta 9, 499 (1926).

45. F. Sondheimer, R. Wolovsky, J. Am. Chem. Soc. 84, 260 (1962).

46. R. Stoll, Helv. Chim. Acta 16, 493 (1933).

47. A. J. Hubert, J. Dale, J. Chem. Soc. 1963, 86.

48. L. Ruzicka, M. Huerbin, M. Furter, Helv. Chim. Acta 17, 78 (1934).

49. O. Ruff, Angew. Chem. 46, 739 (1939).

50. E. L. Pace, J. G. Aston, J. Am. Chem. Soc. 70, 566 (1948).

51. J. H. Simons, L. P. Block, J. Am. Chem. Soc. 61, 2962 (1939).

52. R. N. Haszeldine, J. Chem. Soc. 1953, 3761.

53. J. A. Brown, W. H. Mears, J. Phys. Chem. 62, 961 (1958).

54. W. B. Burford, R. D. Fowler, J. M. Hamilton, Jr., H. C. Anderson, C. E. Weber, R. G. Sweet, Ind. Eng. Chem. 39, 319 (1947).

55. L. L. Burger, G. H. Cady, J. Am. Chem. Soc. 73, 4245 (1951).

56. F. L. Mohler, E. G. Bloom, J. H. Lengel, C. E. Wise, J. Am. Chem. Soc. 71, 337 (1949).

57. V. E. Stiles, G. H. Cady, J. Am. Chem. Soc. 74, 3771 (1952).

58. R. D. Dunlap, C. J. Murphy, R. G. Bedford, J. Am. Chem. Soc. 80, 83 (1958).

59. R. N. Haszeldine, F. Smith, J. Chem. Soc. 1950, 3617; 1951, 603.

60. B. Atkinson, J. Chem. Soc. 1952, 2684.

61. J. D. Park, A. F. Benning, F. Downing, J. F. Laucius, R. C. McHarness, Ind. Eng. Chem. 39, 354 (1947).

62. J. R. Lacher, G. W. Tompkin, J. D. Park, J. Am. Chem. Soc. 74, 1693 (1952).

63. G. Bier, R. Schäff, K. H. Kahrs, Angew. Chem. 66, 285 (1954).

64. N. Fukuhara, L. A. Bigelow, J. Am. Chem. Soc. 63, 2792 (1941).

65. H. J. Christoffers, E. C. Lingafelter, G. H. Cady, J. Am. Chem. Soc. 69, 2502 (1947).

66. F. C. Whitmore, S. N. Wren, J. Am. Chem. Soc. 53, 3136 (1931).

67. C. O. Tongberg, J. D. Pickens, M. R. Fenske, F. C. Whitmore, J. Am. Chem. Soc. 54, 3706, 3710 (1932).

68. F. C. Whitmore, K. C. Laughlin, J. Am. Chem. Soc. 54, 4011 (1932).

69. F. C. Whitmore, K. C. Laughlin, J. F. Matuszeski, J. D. Surmatis, J. Am. Chem. Soc. 63, 757 (1941).

70. J. A. Dixon, N. C. Cook, F. C. Whitmore, J. Am. Chem. Soc. 70, 3363 (1948).

71. F. C. Whitmore, C. D. Wilson, J. V. Capinjola, C. O. Tongberg, G. H. Fleming, R. V. McGrew, J. N. Cosby, J. Am. Chem. Soc. 63, 2035 (1941).

72. F. L. Howard, T. W. Mears, A. Fookson, P. Pomerantz, D. B. Broocks, J. Res. Natl. Bur. Std. 38, 365 (1947).

73. P. D. Bartlett, G. L. Fraser, R. B. Woodward, J. Am. Chem. Soc. 63, 498 (1941).

74. M. J. Batujew, A. P. Meschtscherjakow, A. D. Matwejewa, Dokl. Akad. Nauk, SSSR, _1958_, 75; through C. 1960, 1870.

75. F. D. Rossini, K. S. Pitzer, R. L. Arnett, R. M. Braun, G. C. Pimental, "Selected Values of Physical and Thermodynamic Properties of Hydrocarbons and Related Compounds", Carnegie Press, Pittsburgh, Pa., 1953.

76. G. L. Clark, H. A. Smith, Ind. Eng. Chem. _23_, 700 (1931).

77. P. J. Flory, A. Vrij, J. Am. Chem. Soc. _85_, 3548 (1963).

78. R. A. Orwell, P. J. Flory, J. Am. Chem. Soc. _89_, 6814 (1967).

79. R. R. Reinhard, J. A. Dixon, J. Org. Chem. _30_, 1450 (1965).

80. M. G. Broadhurst, J. Chem. Phys. _36_, 2578 (1962).

81. G. Ställberg, S. Ställberg-Stenhagen, E. Stenhagen, Acta. Chem. Scand. _6_, 313 (1952).

82. W. Heitz, Th. Wirth, R. Peters, G. Strobl, E. W. Fischer, Makromol. Chem. _162_, 63 (1972).

83. M. G. Broadhurst, J. Res. Natl. Bur. Std. _66A_, 241 (1962).

84. M. Kurata, S. Isida, J. Chem. Phys. _23_, 1126 (1955).

85. S. H. Piper, A. C. Chibnall, S. J. Hopkins, A. Pollard, J. A. B. Smith, E. F. Williams, Biochem. J. _25_, 2072 (1931).

86. W. M. Mazee, Rec. Trav. Chim. Pays-Bas _67_, 197 (1948).

87. H. L. Finke, M. E. Gross, G. Waddington, H. M. Huffmann, J. Am. Chem. Soc. _76_, 333 (1954).

88. A. A. Schaerer, G. G. Baylé, W. M. Mazee, Rec. Trav. Chim. Pays-Bas _75_, 513 (1956).

89. B. G. Ranby, F. F. Morehead, N. M. Walter, J. Polymer Sci. _44_, 349 (1960).

90. G. Egloff, "Physical Constants of Hydrocarbons", Vol. 1 and 5, A. C. S. Monograph Series, Reinhold Publ. Corp., New York, 1953.

91. M. Brini, Bull. Soc. Chim. France _1955_, 996.

92. T. Tasumi, T. Shimanouchi, R. F. Schaufele, Polymer J. _6_, 740 (1971).

1.2 OLIGO(DIENES)

1.2.1 OLIGOMERS OF BUTADIENE

No.	Oligomers	Mol. Wt.	m.p. [°C]	b.p. [°C/mm.]	n_D^{20}	Ref.
1	4-Vinyl-1-cyclohexene CH=CH₂	108.2		129-130		1-7
2	cis-1,5-Cyclooctadiene	108.2	-70.1	150.8/755	1.4936	3-5
	trans-1,5-Cyclooctadiene		-62			7-10
3	1,5,9-Cyclododecatriene	162.3				6,10-12
	all-trans-		34	237.5	1.5005	
	trans,trans,cis-		-16.8	241.5	1.5078	
	cis,cis,trans		-9 to -8	244	1.5129	
4	1-Vinyl-3,7-cyclodecadien	162.3		100-110/14		10
5	trans-3-Methyl-1,4,6-heptatriene CH₃CH–CH=CHCH=CH₂ CH=CH₂	108.2		117-118	1.4670	13,14
6	1,3,6-Octatriene CH₂=CH-CH=CH-CH₂-CH=CH-CH₃	108.2		129	1.4743	33
7	1,3,7-Octatriene CH₂=CH-CH=CH-CH₂-CH₂-CH=CH₂	108.2		124-125	1.4682	31,32
8	2,4,6-Octatriene CH₃-CH=CH-CH=CH-CH=CH-CH₃	108.2		65-66/40	1.5131/27	34

1.2.2 OLIGOMERS OF DERIVATIVES OF BUTADIENE

No.	Monomer	Oligomers	Mol. Wt.	m.p. [°C]	b.p. [°C/mm]	n_D/°C	Ref.
1	2-Chlorobutadiene-(1,3) CH₂=CCl-CH=CH₂	1,5-(or 1,6-)dichlorocyclooctadiene-1,5	177.1		92-94/3.8		9
2	Isoprene CH₂=C-CH=CH₂ CH₃	dipentene (2 isomers) CH₂ C-CH₃ CH₃	136.3		163-167 167-170		15-17
		diprene CH₂ C-CH₃ CH₃	136.3		173-174		15,16

No.	Monomer	Oligomers	Mol. Wt.	m.p. [$^{\circ}$C]	b.p. [$^{\circ}$C/mm]	$n_D/^{\circ}$C	Ref.
2	Isoprene (Cont'd.) $CH_2=C-CH=CH_2$ with CH_3	2,4-dimethyl-4-ethenyl-1-cyclohexene	136.3		42-44/10	1.4630/20	11, 29, 30
		2,7-dimethyl-1,3,6-octatriene (trans) $CH_3-C=CH-CH_2-CH=CH-C=CH_2$	136.3		54/10	1.4786/20	11, 29, 30
3	1,3-Diphenyl-1,3-buta-diene $CH=CH-C=CH_2$ with C_6H_5 C_6H_5	1,3,4-triphenyl-4-trans-styrylcyclohexene	411.6	137-138			18, 19
4	Hexafluoro-1,3-butadiene $CF_2=CF-CF=CF_2$	cyclic dimer	324.1	40	80		20, 21

No.	Oligomers	Mol. Wt.	m.p. [$^{\circ}$C]	b.p. [$^{\circ}$C/mm]	$n_D/^{\circ}$C	Ref.
	1.2.3 OLIGOMERS OF CYCLOPENTADIENE					
1	Dicyclopentadiene	132.2	33	70/15		22
2	Tricyclopentadiene	198.3	68	130-145/15		22, 23
	1.2.4 OLIGOMERS OF ALLENE					
1	1,2-Dimethylene-cyclobutane	80.1		73-74	1.4721/20	24-27
2	2,7-Dimethylene-$\Delta^{9,10}$-octalin	160.3		107-108/13	1.5248/25	24, 28

REFERENCES

1. W. E. Vaughan, J. Am. Chem. Soc. 54, 3863 (1952).
2. N. E. Duncan, G. Janz, J. Chem. Phys. 20, 1644 (1952).
3. J. C. Hillyer, J. V. Smith, Jr., Ind. Eng. Chem. 45, 1133 (1955).
4. K. Ziegler, H. Wilms, Ann. Chem. 567, 1 (1950); K. Ziegler, H. Sauer, L. Bruns, H. Froitzheim-Kuehlborn, J. Schneider, Ann. Chem. 589, 122 (1954).
5. G. Wilke, Angew. Chem. 69, 397 (1957); G. Wilke, E. W. Mueller, M. Kroener, ibid. 73, 33 (1961).
6. G. Wilke, Angew. Chem. 75, 10 (1963).
7. G. Pajarr, D. Fiumani, M. Morr, Gazz. Chim. Ital. 92, 1452 (1962); through C. 1965, No. 2, 990.
8. E. Vogel, Angew. Chem. 68, 413 (1956).
9. A. E. Cope, W. J. Bailey, J. Am. Chem. Soc. 70, 2305 (1948).
10. H. W. B. Reed, J. Chem. Soc. 1954, 1931.
11. G. Wilke, J. Polymer Sci. 38, 45 (1959).
12. H. Breil, P. Heimbach, M. Kroener, H. Mueller, G. Wilke, Makromol. Chem. 69, 18 (1963).
13. S. Tanaka, K. Mabuchi, N. Shimazaki, J. Org. Chem. 29, 1626 (1964).
14. S. Otsuka, T. Kikuchi, T. Taketomi, J. Am. Chem. Soc. 85, 3709 (1963).
15. T. Wagner-Jauregg, Ann. Chem. 488, 176 (1931).
16. O. Aschan, Ann. Chem. 439, 221 (1924).
17. C. Walling, J. Peisach, J. Am. Chem. Soc. 80, 5819 (1958).
18. W. Herz, E. Lewis, J. Org. Chem. 23, 1646 (1958).
19. T. L. Jacobs, M. H. Goodrow, J. Org. Chem. 23, 1653 (1958).
20. I. L. Karle, J. Karle, T. B. Owen, R. W. Broge, A. H. Fox, J. L. Hoard, J. Am. Chem. Soc. 86, 2523 (1964).
21. M. Prober, W. T. Miller, J. Am. Chem. Soc. 71, 598 (1949).
22. K. Alder, G. Stein, Ann. Chem. 496, 204 (1932); 504, 216 (1933).
23. H. Staudinger, H. A. Bruson, Ann. Chem. 447, 109 (1926).
24. S. V. Lebedev, Zh. Fiz. Khim. Obshch. 45, 1357 (1913).
25. R. N. Meinert, C. D. Hurd, J. Am. Chem. Soc. 52, 4540 (1930).
26. K. Alder, O. Ackermann, Chem. Ber. 87, 1567 (1954).
27. A. T. Blomquist, J. A. Verdol, J. Am. Chem. Soc. 78, 109 (1956).
28. B. Weinstein, A. H. Fenselau, Tetrahedron Letters 1963, 1463.
29. J. Itakura, H. Tanaka, Makromol. Chem. 123, 274 (1969).
30. L. I. Zakharkin, Dokl. Akad. Nauk SSSR, 131, 1069 (1960).
31. E. J. Senutny, J. Am. Chem. Soc. 89, 6793 (1967).
32. S. Takahashi, T. Shibano, N. Hagihara, Tetrahedron Letters 1967, 2451.
33. H. Takahasi, S. Tai, M. Yamaguchi, J. Org. Chem. 30, 1661 (1965).
34. T. Alderson, E. L. Jenner, R. V. Lindsey, Jr., J. Am. Chem. Soc. 87, 5638 (1965).

1.3 OLIGOMERS WITH ALIPHATIC SIDE CHAINS WHICH IN ADDITION CONTAIN HETEROATOMS

1.3.1 OLIGOMERIC ACRYLIC DERIVATIVES

1.3.1.1 OLIGO(ACRYLIC ACIDS)

1.3.1.1.1 $H[-CH_2CH(COOH)-]_n CH_3$

1.3.1.1.2 $(CH_3)_2C-[-CH_2CH-]_n-C-(CH_3)_2$
               $\overset{|}{COOH}\;\;\;\;\overset{|}{COOH}\;\;\;\overset{|}{COOH}$

n	Mol. Wt.	Steric structure	m.p. [°C]	b.p. [°C/mm]	Ref.
1	88.1		-46	153/760	
2	160.2	isotactic	128		1-3
		syndiotactic	141		1-3
3	232.2	isotactic	162-165		3
		heterotactic	143-147		3
		syndiotactic	142-146		3

n	Mol. Wt.	m.p. [°C]	Ref.
1	246.3	30-35	4
2	318.3	70-80 ★	4
3	390.4	181-183	4
4	462.5	80-82	5

★ p-Nitrobenzylester

1.3.1.2 OLIGO(ACRYLATES)

1.3.1.2.1 $H[-CH_2CH(COOCH_3)-]_n CH_3$

n	Mol. Wt.	Steric structure	b.p. [°C/mm]	$n_D/°C$	Ref.
1	102.1		91.5/742	1.3838/20	6
2	188.2	isotactic	64/0.7	1.4258	3, 6
		syndiotactic			
3	274.3	isotactic	150-160/2	1.4405	3, 6
		heterotactic	150-160/2	1.4416	3, 6
		syndiotactic	150-160/2	1.4433	3, 6

1.3.1.2.2 $RO[-CH_2CH(COOR)-]_n H$

1.3.1.2.3 $CH_2=C-[-CH_2CH-]_n H$
            $\overset{|}{COOCH_3}\;\;\;\overset{|}{COOCH_3}$

R	n	Mol. Wt.	b.p. [°C/mm]	n_D^{25}	Ref.
CH_3	1	118.1	55-56/20		7
CH_3	2	204.2	90/1	1.433	7
C_2H_5	1	146.2	75-77/20		7
C_2H_5	2	246.3	82/0.11	1.446	7

n	Mol. Wt.	b.p. /Torr	Ref.
1	172.1	81/1	7,8
2	258.2	120/0.7	8
3	344.3	166-168/0.3	8
4	430.4	180/0.02	8

1.3.1.3 OLIGO(ACRYLONITRILES)

1.3.1.3.1 $H[-CH_2CH(CN)-]_n H$

n	Mol. Wt.	b.p. [°C/mm]	$n_D/°C$	Ref.
1	55.1	97.1	1.3689/15	
2	108.1	135/12	1.4312/25	9
3	161.2	195-196/2.5	1.4609/20	9,10

1.3.1.3.2 $H[-CH_2CH(CN)-]_n CH_3$

n	Mol. Wt.	m.p. [°C]	b.p. [°C/mm]	$d_4/°C$	$n_D/°C$	Ref.
1	69.1	-71.5	107-108	0.773/60	1.3720/20	
2	122.2	8.8 [a]	106/2	0.9051/60	1.4191/60	10, 11
		50 [b]	94/2	0.8940/60	1.4155/60	10, 11
3	175.2 [c]	37-39				10
		45-46				10
		80-81				10
4	228.3 [d]	130-135				10
		158				10

a) isotactic.   b) syndiotactic.   c) 3 isomers.   d) 2 isomers.

1.3.1.3.3 $H[-CH(CN)CH_2-]_n CH_2 CN$

n	Mol. Wt.	b.p. [°C/mm]	n_D/°C	Ref.
1	94.1	287.4	1.4347/20	
2	147.2		1.4644/25	9, 12
3	200.2	95/0.01		9, 12

1.3.1.3.4 1,2-Dicyanocyclobutanes $[-CH_2CH(CN)-CH(CN)CH_2-]_n$

n	Mol. Wt.	m.p. [°C]	b.p.[°C/mm]	n_D^{25}	Ref.
1 cis	106.1	0	108-115/3-4	1.4628	13, 14
trans	106.1	62	140-145/3-4		13, 14

1.3.1.3.5 Unsaturated Oligomers

No.	Oligomers	Mol. Wt.	m.p. [°C]	b.p.[°C/mm]	n_D^{25}	Ref.
1	2,3-Dihydromuconitrile NCCH₂CH=CHCH₂CN (trans)	106.1	76			14, 15
2	1,1,4,4-Tetracyanoethyl-1,4-dicyano-trans-2-butene (NCCH₂CH₂)₂C(CN)-CH ‖ (NCCH₂CH₂)₂C(CN)-CH	318.4	240			14
3	α-Methylene-ϐ-methyladiponitrile CH₃CH(CN)CH₂CH₂C(CN)=CH₂	134.2		148/25	1.4502	16

1.3.2 OLIGOMERIC METHACRYLIC DERIVATIVES

1.3.2.1 OLIGO(METHACRYLIC ACIDS)

1.3.2.1.1 $H\left[-CH_2-C(CH_3)-\atop COOH\right]_n C(CH_3)_2 COOH$

n	Mol. Wt.	m.p. [°C]	Ref.
2	260.3	Oil	19
3	346.4	45-48	18, 19
4	432.5	56-59	19

1.3.2.1.2 Anionic Oligo(methacrylic acids) $CH_3O\left[-CH_2C(CH_3)-\atop COOH\right]_n H$

n	Mol. Wt.	m.p. [°C]	Ref.
2	204.2	82	17
3	290.3	160	17

1.3.2.2 OLIGO(METHACRYLATES)

1.3.2.2.1 Anionic Oligo(methacrylates) $CH_3O\left[-CH_2C(CH_3)-\atop COOCH_3\right]_n H$

n	Mol. Wt.	b.p. [°C/mm]	d^{20}	Ref.
1	132.2	147	0.9749	19
2	232.3	241	1.0540	19
3	332.4	116/0.1	1.1045	19
4	432.5	190/0.1	1.12	19
6	632.8		1.565	19

1.3.2.2.2 Unsaturated Oligo(methacrylates) $CH_3OOC-C-\atop CH_2 \left[-CH_2CH_2CH(COOCH_3)-\right]_n CH_3$

n	Mol. Wt.	b.p. [°C/mm]	n_D^{25}	Ref.
1	188.2	107/7	1.4445	16, 20
2	276.3	173/3	1.4588	16, 20

1.3.2.3 OLIGO(METHACRYLONITRILES)

1.3.2.3.1 Anionic Oligo(methacrylonitriles) $CH_3O\left[-CH_2C(CH_3)-\atop CN\right]_n H$

n	Mol. Wt.	b.p. [°C/mm]	n_D	Ref.
1	99.1			
2	166.2	92/0.5	1.438	21, 34
3	233.3	165/0.5	1.464	21, 34
4	300.4	235/0.5	1.478	21, 34
5	367.5	300-305/0.5	1.488	21, 34

1.3.2.3.2 1,2-Dicyanocyclo(1,2)dimethylbutanes $[-CH_2C(CH_3)(CN)-C(CH_3)(CN)CH_2-]_n$

n	Mol. Wt.	m.p. [°C]	b.p. [°C/mm]	Ref.
1 cis	134.2	107-108	170/25	16
trans	134.2	90.3	120/25	16

No.	Monomer	Oligomers	Mol. Wt.	m.p. [°C]	b.p. [°C/mm]	Ref.

1.3.3 OLIGO(VINYL ALDEHYDES) AND OLIGO(VINYL KETONES)

No.	Monomer	Oligomers	Mol. Wt.	m.p. [°C]	b.p. [°C/mm]	Ref.
1	Acrolein $CH_2=CH-CHO$	2-formyl-2,3-dihydropyran	112.1		145-148	22-25
		3-formyl-5,6-dihydropyran	112.1		77-78/12	26,27
2	α-Methylacrolein $CH_2=C-CHO$ $\overset{|}{CH_3}$	2-formyl-2,5-dimethyl-2,3-dihydropyran	140.2	-75	166/750	24,28
3	α-Ethylacrolein $CH_2=C-CHO$ $\overset{|}{C_2H_5}$	2-formyl-2,5-diethyl-2,3-dihydropyran	168.2		195	24
4	Methyl vinyl ketone $CH_2=CH-CO-CH_3$	6-methyl-2-acetyl-2,3-dihydropyran	128.2		68/13	29
5	2-Methyl-1-butene-(3)-one $CH_2=C-COCH_3$ $\overset{|}{CH_3}$	2,6-dimethyl-3,7-dioxo-1-octene $CH_3COCH(CH_3)CH_2CH_2COC=CH_2$ $\overset{\quad\quad\quad\quad\quad\quad\quad\quad\quad|}{\quad\quad\quad\quad\quad\quad\quad\quad\quad CH_3}$	168.2		83-85/17	30

1.3.4 OLIGOMERIC N-VINYL DERIVATIVES

No.	Monomer	Oligomers	Mol. Wt.	m.p. [°C]	b.p. [°C/mm]	Ref.
1	N-Vinylpyrrolidone $CH_2=CH$	1,3-bis-[N-pyrrolidone-(2)-yl]-1-butene	212.3	75		31
2	N-Vinylpyridine $CH_2=CH$	sym-tri(4-pyridyl)cyclohexane	315.4	228.5		32,33

REFERENCES

1. K. Auwers, J. Thorpe, Ann. Chem. 285, 335 (1895).
2. E. Moeller, Chem. Ber. 43, 3250 (1910).
3. H. G. Clark, Makromol. Chem. 86, 107 (1965).
4. H. Kämmerer, A. Jung, Makromol. Chem. 101, 284 (1967).
5. H. Kämmerer, N. Oender, Makromol. Chem. 111, 67 (1968).
6. D. Lim, O. Wichterle, J. Polymer Sci. 29, 579 (1958).
7. B. A. Feit, European Polymer J. 3, 523 (1967).
8. B. A. Feit, European Polymer J. 8, 321 (1972).
9. H. Zahn, P. Schaefer, Chem. Ber. 92, 736 (1959); Makromol. Chem. 30, 225 (1959).
10. T. Takata, M. Taniyama, Chem. High Polymers (Japan) 16, 693 (1959); T. Takata, H. Ishii, Y. Nishiyama, M. Taniyama, Chem. High Polymers (Japan) 18, 235 (1961); through C. 1964, 729-731.
11. H. G. Clark, Makromol. Chem. 63, 69 (1963).
12. R. C. Houtz, Textile Res. J. 20, 786 (1950).
13. E. C. Coyner, W. S. Hillmann, J. Am. Chem. Soc. 71, 324 (1949).
14. N. Takashima, C. C. Price, J. Am. Chem. Soc. 84, 489 (1962).
15. E. Reppe, Ann. Chem. 596, 133 (1956).
16. C. J. Albisetti, D. C. England, M. J. Hogsed, R. M. Joyce, J. Am. Chem. Soc. 78, 472 (1956).
17. Th. Voelker, A. Neumann, V. Baumann, Makromol. Chem. 63, 182 (1963); G. Schreyer, Th. Voelker, Makromol. Chem. 63, 202 (1963).
18. H. Kämmerer, J. S. Shukla, G. Scheuermann, Makromol. Chem. 116, 72 (1968).
19. H. Kämmerer, J. S. Shukla, Makromol. Chem. 116, 62 (1968).
20. E. Trommsdorf, Angew. Chem. 68, 355 (1956).
21. B. A. Feit, J. Wallach, A. Zilkha, J. Polymer Sci. A 2, 4743 (1964).
22. K. Alder, A. Rueden, Chem. Ber. 74, 920 (1941).
23. S. Potnis, K. Shorara, R. C. Schulz, W. Kern, Makromol. Chem. 63, 78 (1963).
24. H. Schulz, H. Wagner, Angew. Chem. 62, 105 (1950).
25. S. M. Scherlin, A. J. Berlin, T. A. Sserebrennikowa, F. E. Rabinowitsch, Zh. Obshch. Khim. 8, 22 (1938); through C. 1939, I, 1971.
26. R. Hall, Chem. Ind. 1955, 1772.
27. G. Dumas, P. Rumpff, Compt. Rend. 242, 2574 (1956).
28. G. G. Stoner, J. S. McNulty, J. Am. Chem. Soc. 72, 1531 (1950).
29. K. Alder, H. Offermanns, E. Rueden, Chem. Ber. 74, 905 (1941).
30. H. Staudinger, B. Ritzenthaler, Chem. Ber. 67, 1773 (1934).
31. J. W. Breitenbach, F. Galinovsky, H. Nesvadba, E. Wolf, Naturwissenschaften 42, 155, 440 (1955).
32. F. Longo, J. W. Bassi, F. Greco, M. Cambini, Tetrahedron Letters 1964, 995.
33. A. Segre, Tetrahedron Letters 1964, 1001.
34. B. A. Feit, E. Heller, A. Zilkha, J. Polymer Sci. A-1, 4, 1499 (1966).

1.4 OLIGO(STYRENES)

No.	Monomer	Oligomers	Mol. Wt.	m.p. [$^\circ$C]	b.p. [$^\circ$C/mm]	n_D^{20}	Ref.
1	Styrene	3-methyl-1-phenylindane (2 stereoisomers)	208.3 208.3	9.5 25.5	168-169/16 157/12	1.5810 1.5809	1-5
		1,3-diphenyl-1-butene	208.3 208.3		181-182/20 134-135/1	 1.5930	1-3, 6
		1,4-diphenyl-1-butene	208.3		124		6
		1,2-diphenylcyclobutane	208.3			1.5913	7
2	α-Methylstyrene	1,3,3-trimethyl-1-phenylindane	236.4	53	158-160/10		8-10
		4-methyl-2,4-diphenyl-2-pentene	236.4	52	166-167/15	1.5728	8, 11, 12
3	α-Methyl-p-methoxy-styrene		296.4		237-240/18	1.5703	9, 13
4	α-Methyl-p-aminostyrene	1,3,3-trimethyl-4',6-diamino-phenylindane	266.4	93-94			9
			266.4	173			9, 14
5	α-Methyl-p-carboxy-styrene	1,3,3-trimethyl-1-phenylindane-4',6-dicarboxylic acid	348.4	297			9, 14
6	α-p-Dimethylstyrene	1,3,3,4',6-pentamethyl-1-phenylindane	264.4	40	142-144/0.8		9, 15, 16

No.	Monomer	Oligomers	Mol. Wt.	m.p. [°C]	b.p. [°C/mm]	n_D^{20}	Ref.
7	α-Ethylstyrene $CH_2=C-C_2H_5$	1,3-diethyl-3-methyl-1-phenylindane	264.4		104-106/0.3	1.5642	17
		3,5-diphenyl-5-methyl-2-heptene $CH_3-C(C_2H_5)CH_2C=CHCH_3$	264.4		133-135/1.2	1.5434/25	17
8	α-m-Dimethylstyrene $CH_2=C-CH_3$	1,3,3,3',7-pentamethyl-1-phenylindane	264.4	57			9
9	p-Bromstyrene $CH_2=CH-$⟨⟩$-Br$	cis-1,3-di(p-bromophenyl)-1-butene $Br-$⟨⟩$-CH=CHCH(CH_3)$⟨⟩$-Br$	366.1	67-68			18, 19
10	Stilben ⟨⟩$-CH=CH-$⟨⟩	1,2,3,4-tetraphenylcyclobutane	360.5	164-165			20-22
		1,2-diphenyl-1-p-tolylethane $C_6H_5CHCH_2C_6H_5$	272.4	42-43			20

REFERENCES

1. P. E. Spoerri, M. J. Rosen, J. Am. Chem. Soc. 72, 4918 (1950).

2. M. J. Rosen, J. Org. Chem. 18, 1701 (1953).

3. B. B. Corson, J. Dorsky, J. E. Nickels, W. M. Kutz, H. J. Thayer, J. Org. Chem. 19, 17 (1954).

4. M. J. Rosen, Org. Syn. 35, 83 (1955).

5. R. Stoemer, H. Kootz, Chem. Ber. 61, 2330 (1928).

6. R. Fittig, E. Erdmann, Ann. Chem. 216, 179 (1883); H. Stobbe, G. Posujak, Ann. Chem. 371, 287 (1909).

7. I. S. Bengelsdorf, J. Org. Chem. 25, 1468 (1960).

8. E. Bergmann, H. Taubadel, H. Weiss, Chem. Ber. 64, 1493 (1931).

9. J. C. Petropoulos, J. J. Fisher, J. Am. Chem. Soc. 80, 1938 (1958).

10. L. M. Adams, R. J. Lee, F. T. Wadsworth, J. Org. Chem. 24, 1186 (1959).

11. F. S. Dainton, R. H. Tomlinson, J. Chem. Soc. 1953, 151.

12. J. M. Van der Zanden, Th. R. Rix, Rec. Trav. Chim. 75, 1343 (1956).

13. J. M. Van der Zanden, Th. R. Rix, Rec. Trav. Chim. 75, 1166 (1956).

14. J. v. Braun, E. Anton, W. Haensel, G. Werner, Ann. Chem. 472, 1 (1929).

15. M. Tiffenear, Ann. Chim. Phys. 10, 197 (1907).

16. V. N. Ipatieff, H. Pines, R. C. Oldberg, J. Am. Chem. Soc. 70, 2123 (1948).

17. C. G. Weiberger, E. M. Pearce, D. Tanner, J. Am. Chem. Soc. 80, 1761 (1958).

18. G. L. Goerner, J. W. Pearce, J. Chem. Soc. 73, 2304 (1951).

19. J. Hukki, Acta. Chem. Scand. 3, 279 (1949).

20. D. S. Brakmann, P. H. Plesch, J. Chem. Soc. 1958, 3563; Chem. Ind. 1955, 255.

21. M. Pailer, O. Mueller, Monatsh. Chem. 79, 615 (1948).

22. Y. D. Fulton, J. D. Dunitz, Nature (London) 160, 161 (1947).

2. OLIGOMERS CONTAINING HETEROATOMS IN THE MAIN CHAIN

2.1 OLIGOMERS CONTAINING O IN THE MAIN CHAIN

2.1.1 OLIGO(ETHERS) AND OLIGO(ACETALS)

2.1.1.1 Oligo(formaldehyde) Derivatives

2.1.1.1.1 Oligo(formaldehyde) Dihydrates $H[OCH_2]_n OH$

n	Mol. Wt.	m.p. [$^{\circ}$C]	Ref.
4	138.1	95-105 (d)	1, 2
8	258.1	115-120 (d)	1, 2

2.1.1.1.2 Oligo(formaldehyde) Diacetates $CH_3CO-[OCH_2]_n-OCOCH_3$

n	Mol. Wt.	m.p. [$^{\circ}$C]	b.p. [$^{\circ}$C]	d/$^{\circ}$C	n_D/$^{\circ}$C	Ref.
1	132.1	-23	39-40	1.128/24	1.4025/24	2, 3, 4
2	162.1	-13	60-62	1.158/24	1.4124/24	3, 4
3	192.2	-3	84	1.179/24	1.4185/24	3
4	222.2	7	102-104	1.195/24	1.4233/24	2, 3
5	252.2	17	124-126	1.204/24	1.4258/24	3
8	342.3	32-34		1.216/36	1.4297/36	2, 3
9	372.3	40-43		1.353/15		3
10	402.4	52-53.5				3
11	432.4	65.5-67				3
12	462.4	73-75				2, 3
14	522.5	84-86				3
15	552.5	90.5-92		1.364/25		3
16	582.5	93-95				2, 3
17	612.6	98.5-99.5		1.370/15		3
19	672.6	107-109		1.390/15		3
20	702.6	111-112				2, 3
22	762.7	116-118		1.465		3

2.1.1.1.3 Oligo(formaldehyde) Dimethyl Ethers $CH_3-[OCH_2]_n-OCH_3$

n	Mol. Wt.	m.p. [$^{\circ}$C]	d/$^{\circ}$C	Ref.
1	76.1	-105	0.8538/20	5, 12, 15
2	106.1	-69.7	0.9597/25	5, 13, 14, 15
3	136.2	-42.5	1.0242/25	5
4	166.2	-9.8	1.0671/25	5
5	196.2	18.3	1.1003/25	5

2.1.1.1.4 Oligo(formaldehyde) Dipropyl Ethers $C_3H_7-[OCH_2]_n-OC_3H_7$

C_3H_7	n	Mol. Wt.	m.p. [$^{\circ}$C]	b.p. [$^{\circ}$C/mm]	d/$^{\circ}$C	n_D/$^{\circ}$C	Ref.
n-	1	132.2		137.2-137.6	0.83325/25	1.3913/25	6, 7, 8
	2	162.2		67/11	0.89725/25	1.4004/25	6, 9
	3	192.2		97/11	0.94325/25	1.4086/23	6, 9
	4	222.3	-15 to -13		0.99025/25	1.4137/26	6
	5	252.3	8 - 8.5		1.01424/25	1.4181/26	6
i-	1	132.2		117-119	0.8242/20	1.3864/20	10
	2	162.2		39.5-41/23	0.8897/20	1.3971/20	10
	3	192.2		68.2-68.5/3	0.9348/20	1.4035/20	10
	4	222.3		93.5-94.5/3	0.9751/20	1.4117/20	10
	5	252.3		120-123/3	1.0275/20	1.4235/20	10
	6	282.3	23.4-24.3	159-163/3.7	1.101/26	1.4467/26	10

2.1.1.1.5　Oligo(formaldehyde) Diallyl Ethers　$CH_2=CHCH_2-[OCH_2]_n-OCH_2CH=CH_2$

n	Mol. Wt.	m.p. [$^\circ$C]	b.p. [$^\circ$C/mm]	d_{25}^{25}	$n_D/^\circ$C	Ref.
1	128.2		138-139		1.4226/21	6, 11
2	158.2		75-76.5/15	0.946	1.4280/25	6
3	188.2		58.5-64/0.3	0.992	1.4320/25	6
4	218.3	-4.3	82-87/0.3	1.027	1.4350/25	6
5	248.3	15.5	105-107/0.3	1.059	1.4377/25	6
6	278.3	22.5	144-155/0.4	1.079	1.4411/25	6

2.1.1.1.6　Cyclic Oligo(formaldehydes)　$[OCH_2]_n$

n	Mol. Wt.	m.p. [$^\circ$C]	b.p. [$^\circ$C]	Ref.
3	90.1	67-68	114.5	16, 17
4	120.1	114		17, 18, 19, 47, 48
5	150.2	61		31, 45, 46
6	180.2	72.5		47
15	450.4	68-70		31

2.1.1.2　Oligomers of Higher Aldehydes

2.1.1.2.1　Cyclic Oligo(aldehydes)　$[OCHR]_n$

R	n	Mol. Wt.	m.p. [$^\circ$C]	b.p. [$^\circ$C]	Ref.
CH_3	3	132.2	12.6	125	20
CH_3	4	176.2	246.2	112-115 (subl.)	20, 21
C_6H_5	3	318.4	175-176	75	35

2.1.1.3　Oligo(ethylene Oxides)

2.1.1.3.1　Linear Oligo(ethylene Oxides)　$H[OCH_2CH_2]_nOH$

n	Mol. Wt.	m.p. [$^\circ$C]	b.p. [$^\circ$C/mm]	d_{20}^{20}	n_D^{20}	Ref.
1	62.1	-12.6	197.8	1.113	1.4324	23, 24, 25
2	106.1	-10.1	245	1.120	1.4477	23, 24, 25, 26
3	150.2	-9.4	122-123/0.1	1.1274/15	1.4568	23, 25, 26
4	194.2	-9.4	144-145.5/0.1	1.127	1.4604	23, 24, 25, 26
5	238.3	-8.7	174-176/0.14		1.4629	25, 26, 27
6	282.3	2.1	185-186/0.015	1.127	1.4647	23, 24, 25, 26, 27, 28
7	326.4	7.7	241-244/0.6		1.4663	25, 26
8	370.4		206-209/0.015			27
9	414.5	29.8-30.3				37
10	458.6		220-223/0.01			27
15	678.8	38.6-38.8				37
18	811.0	32.0-32.4				37
27	1207.5	44.6-44.8				37
36	1604.0	43.0-43.4				37
45	2000.4	49.9-50.1				37

2.1.1.3.2　Cyclic Oligo(ethylene Oxides)　$[OCH_2CH_2]_n$

n	Mol. Wt.	m.p. [$^\circ$C]	b.p. [$^\circ$C]	$d_4/^\circ$C	Ref.	n	Mol. Wt.	m.p. [$^\circ$C]	Ref.
1	44.1	-111.3	10.7	1.8922/6	29	6	264.3	39-40	39, 41
2	88.1	11.7	101.5	1.0336/20	30	7	308.4	Oil	39
4	176.2	16			40, 31	8	352.4	19	39
5	220.3	Oil			31, 39				

2.1.1.3.3 Oligo(styrene oxides)

Monomer	Oligomers	Mol. Wt.	m.p. [°C]	b.p. [°C/mm]	Ref.
Styrene oxide	trans-2,5-diphenyl-1,4-dioxane	240.3	177		42-44
$CH_2\text{-}CH\text{-}C_6H_5$ (O)	$[OCH_2CH(C_6H_5)]_2$				
	cis-2,5-diphenyl-1,4-dioxane	240.3	121-122		42,43
	2-benzyl-4-phenyl-1,3-dioxolane	240.3	37-40	150-151/1	43

$C_6H_5\text{-}CH\text{-}O$ / $CH_2\text{-}O$ CH-CH_2-C_6H_5

2.1.1.4 Oligomeric Polymethylene Oxides

2.1.1.4.1 $H[O(CH_2)_x]_n OH$

n	x	Mol. Wt.	m.p. [°C]	b.p. [°C/mm]	n_D^{60}	Ref.
3	4	234.3	33	182-184/2	1.4484	33
	5	276.4	29	188-190/1	1.4495	33
	6	318.5	56	209-210/0.9	1.4538	33
	10	486.8	80	290-300/1	1.4518	33

2.1.1.4.2 $H(CH_2)_x[O(CH_2)_x]_n H$

n	x	Mol. Wt.	m.p. [°C]	b.p. [°C/mm]	d_4^{20}	n_D^{20}	Ref.
2	3	160.3		180	0.836	1.4090	33
3	3	218.3		240-250			33
2	4	202.3		234-236	0.843	1.4226	24,32,33
3	4	274.4		137/1.4	0.878	1.4357	32,33
4	4	346.6		190/1	0.895	1.4393	32,33
2	5	244.4		120/0.2	0.843	1.4318	33
3	5	330.6		160/0.3	0.870	1.4416	33
4	5	416.7		200/0.2	0.890	1.4471	33
2	6	286.5		130/0.4	0.842	1.4379	33
3	6	386.7		184/0.6	0.867	1.445	33
4	6	486.8	20.5	260/1.5	0.879	1.451	33
2	10	454.8	40	233/0.3	0.819/60°	1.4370/60°	33
3	10	611.1	50	310-330/2	0.834/60°	1.4416/60°	33
4	10	767.4	59		0.850/60°	1.4457/60°	33

2.1.1.4.3 Cyclic Oligomers

	Mol. Wt.	m.p. [°C]	b.p. [°C]	Ref.		Mol. Wt.	m.p. [°C]	b.p. [°C/mm]	Ref.
$[O(CH_2)_3]_4$	232.3	70		38	$[OCH_2C(CH_2Cl)_2CH_2]_3$	465.1	122		34
$[OCH_2C(CH_3)_2CH_2]_4$	344.5	157		38	$[OCH(CH_3)CH_2]_4$	232.3		105/5	49

2.1.1.5 Higher Cyclic Oligo(acetals)

2.1.1.5.1 Cyclic Oligo(alkylene formals) $[OCH_2O(CH_2)_x]_n$

x	n	Mol. Wt.	m.p. [°C]	b.p. [°C/mm]	Ref.	x	n	Mol. Wt.	m.p. [°C]	Ref.
5	1	116.2		40-44/11	22	9	2	344.5	68-69	22
	2	232.3	55-56		22	10	2	372.6	93-94	22
6	2	260.4	71-72		22	14	2	484.8	103.5-104	22

PHYSICAL DATA OF OLIGOMERS

2.1.1.5.2 Cyclic Oligomer of o-Formyl(phenylacetaldehyde)

	n	Mol. Wt.	m.p. [°C]	Ref.
	3	444.5	284-285	36

REFERENCES

1. H. Staudinger, W. Kern, in "Die hochmolekularen organischen Verbindungen", Springer, Berlin-Goettingen-Heidelberg, 1960, p. 248.
2. W. Kern, Angew. Chem. 73, 177 (1961).
3. H. Staudinger, R. Signer, H. Johner, M. Luethy, W. Kern, D. Russidis, O. Schweitzer, Ann. Chem. 474, 145 (1929).
4. M. Descudé, Ann. Chim. (Paris) 29, 502 (1903).
5. R. H. Boyd, J. Polymer Sci. 50, 133 (1961).
6. R. F. Webb, A. J. Derke, L. S. A. Smith, J. Chem. Soc. 1962, 4307.
7. M. Ghysels, Bull. Soc. Chim. Belges 33, 57 (1924).
8. A. J. Vogel, J. Chem. Soc. 1948, 623.
9. A. Rieche, H. Groß, Chem. Ber. 93, 259 (1960).
10. E. Klein, J. K. Smith, R. J. Eckert, Jr., J. Appl. Polymer Sci. 7, 383 (1963).
11. J. A. Trillat, R. Cambier, Bull. Soc. Chim. France 11, 757 (1894).
12. E. Fischer, G. Giebe, Chem. Ber. 30, 3054 (1897).
13. M. Descudé, Compt. Rend. 138, 1705 (1904).
14. J. Loebering, R. Fleischmann, Chem. Ber. 70, 1680 (1937).
15. T. Uchida, Y. Kurita, M. Kubo, J. Polymer Sci. 19, 365 (1956).
16. V. Jaacks, W. Kern, Makromol. Chem. 52, 37 (1962).
17. J. F. Walker, "Formaldehyde," 3rd Ed., Reinhold Publ. Corp. New York, 1964.
18. H. Staudinger, M. Luethy, Helv. Chim. Acta 8, 66 (1925).
19. K. Hayashi, H. Ochi, M. Nishii, Y. Miyake, S. Okamura, J. Polymer Sci. A 1, 427 (1963).
20. A. W. Johnson, et al., in "Chemistry of Carbon Compounds", Ed. E. H. Rodd, Vol. I, Part A, Elsevier, New York, 1951, p. 476.
21. T. S. Patterson, G. M. Holmes, J. Chem. Soc. 1935, 904.
22. J. W. Hill, W. H. Carothers, J. Am. Chem. Soc. 57, 925 (1935).
23. K. J. Rauterkus, H. G. Schimmel, W. Kern, Makromol. Chem. 50, 166 (1961); K. J. Rauterkus, W. Kern, Chimia 16, 114 (1962).
24. P. Rempp, Bull. Soc. Chim. France 1957, 844.
25. Y. Kuroda, M. Kubo, J. Polymer Sci. 26, 323 (1957); T. Uchida, Y. Kurita, N. Koizumi, M. Kubo, J. Polymer Sci. 21, 313 (1956).

26. A. F. Gallaugher, H. Hibbert, J. Am. Chem. Soc. 58, 813 (1936).
27. S. Perry, H. Hibbert, Can. J. Res. A 14, 77 (1936); B 14, 82 (1936).
28. R. Fordyce, E. L. Lovell, H. Hibbert, J. Am. Chem. Soc. 61, 1905, 1912, 1916 (1939).
29. A. W. Johnson, A. G. Long, C. E. Dalgliesh, in "Chemistry of Carbon Compounds", Ed. E. H. Rodd, Vol. I, Part A, Elsevier, New York, 1951, p. 670.
30. G. R. Ramage, E. H. Rodd, J. K. Landquist, in "Chemistry of Carbon Compounds", Ed. E. H. Rodd, Vol. IV, Part C, Elsevier, New York, 1951, p. 1529.
31. K. H. Burg, H. D. Hermann, H. Rehling, Makromol. Chem. 111, 181 (1968).
32. T. P. Hobin, Polymer 9, 65 (1968).
33. T. P. Hobin, Polymer 6, 403 (1965).
34. Y. Arimatsu, J. Polymer Sci. A-1, 4, 728 (1966).
35. G. B. Stampa, Macromolecules 2, 203 (1969).
36. S. Tagami, T. Kagiyama, C. Aso, Polymer J. 2, 101 (1971).
37. B. Boemer, H. Heitz, W. Kern, J. Chromatog. 53, 51 (1970).
38. J. B. Rose, J. Chem. Soc. 1956, 542.
39. J. Dale, P. O. Kristiansen, Chem. Commun. 1971, 670; Acta Chem. Scand. 26, 1471 (1972).
40. D. G. Stewart, D. Y. Waddan, E. T. Borrows, B.P. 785 229 (1957).
41. C. J. Pedersen, J. Am. Chem. Soc. 89, 7017 (1967).
42. L. A. Bryan, W. M. Smedley, R. K. Summerbell, J. Am. Chem. Soc. 72, 2206 (1950).
43. R. K. Summerbell, M. J. Kland-English, J. Am. Chem. Soc. 77, 5095 (1955).
44. S. Kondo, L. P. Blanchard, J. Polymer Sci. B 7, 621 (1969).
45. M. Nishii, K. Hayashi, S. Okamura, J. Polymer Sci. B 7, 891 (1969).
46. Y. Chatani, K. Kitahama, H. Tadokoro, T. Yamauchi, Y. Miyake, J. Macromol. Sci. Phys. B 4, 61 (1970).
47. Y. Chatani, T. Ohno, T. Yamauchi, Y. Miyake, J. Polymer Sci. A 2, 11, 369 (1973).
48. Y. Chatani, T. Uchida, H. Tadokoro, K. Hayashi, M. Nishii, S. Okamura, J. Macromol. Sci. Phys. B 2, 567 (1968).
49. J. L. Down, J. Lewis, B. Moore, G. Wilkinson, J. Chem. Soc. 1959, 3767.

2.1.2 OLIGO(CARBONATES)

2.1.2.1 Cyclic Oligo(diol carbonates) $[O(CH_2)_x O-CO]_n$

x	n	Mol. Wt.	m.p. [°C]	Ref.	x	n	Mol. Wt.	m.p. [°C]	Ref.
4	2	232.2	175-176	1	9	2	372.5	95-95.5	2
5	2	260.3	117-118	2	10	1	200.3	10-11	2
6	2	288.3	128-129	2	10	2	400.6	105-106	2
7	2	316.4	97-98	2	11	1	214.3	40-41	2
8	1	172.2	21.5-23	2	11	2	428.6	97-97.5	2
8	2	344.5	116-117	2	12	1	228.3	11-12	2
9	1	186.3	34-35	2	12	2	456.7	93-95	2

2.1.2.2 Cyclic Oligo(diphenol carbonates) $[-O-\bigcirc-A-\bigcirc-O-CO]_4$

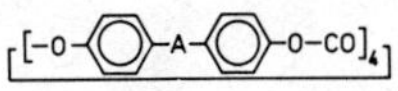

A	Mol. Wt.	m.p. [°C]	Ref.
C(CH$_3$)$_2$	1017.2	375	3
-S-	977.1	320-322	3

REFERENCES

1. W. H. Carothers, F. J. van Natta, J. Am. Chem. Soc. __52__, 314 (1930).
2. J. W. Hill, W. H. Carothers, J. Am. Chem. Soc. __55__, 5031 (1933).
3. H. Schnell, L. Bottenbruch, Makromol. Chem. __57__, 1 (1962).

2.1.3 OLIGO(ESTERS)

2.1.3.1 Oligoesters of Hydroxy Acids

2.1.3.1.1 Oligoesters of L-lactic acid

2.1.3.1.1.1 Oligo(lactic acid) methyl esters
H[OCH(CH$_3$)CO]$_n$OCH$_3$

2.1.3.1.1.2 Acetyl oligo(lactic acid) methyl esters
CH$_3$CO-[OCH(CH$_3$)CO]$_n$OCH$_3$

n	Mol. Wt.	b.p. [°C/mm]	Ref.
1	104.1	58-59/29	
2	176.2	79-80/2	53

n	Mol. Wt.	b.p. [°C/mm]	$[\alpha]_D^{25}$ in CH$_2$Cl$_2$	(conc.)	Ref.
1	146.1	67-68/13	-47.3	(1.41)	53
2	218.2	87-88/0.5	-65.6	(1.98)	53
3	290.3	144-146/1	-81.8	(1.53)	53

2.1.3.1.1.3 Cyclic Oligoesters $[OCH(CH_3)CO]_n$

n	Config.	Mol. Wt.	m.p. [°C]	b.p. [°C/mm]	$[\alpha]_D$ in C$_6$H$_6$	Ref.
2	D	144.1	98.7	150/25	+281.6 (c=0.82, 16°C)	3
2	L	144.1	98.7	150/25	-280 (c=0.58, 18°C)	3,4
2	DL	144.1	128	256		1-7

2.1.3.1.2 Oligoesters of L-Isovaleric Acid

2.1.3.1.2.1 Oligo(isovaleric acid) methyl esters

H-[O-CH-CO]$_n$-OCH$_3$
          |
      CH(CH$_3$)$_2$

2.1.3.1.2.2 Acetyl oligo(isovaleric acid) methyl esters

CH$_3$CO-[O-CH-CO]$_n$-OCH$_3$
              |
          CH(CH$_3$)$_2$

n	Mol. Wt.	b.p. [°C/mm]	Ref.
1	132.2	81-82/45	51
2	232.3	106.5-107/1	51
3	332.4	151-153/1	51
4	432.5	197-199.5/1	51
5	532.6	207-212	51

n	Mol. Wt.	b.p. [°C/mm]	$[\alpha]_D^{25}$ in CHCl$_3$	(conc.)	Ref.
1	174.2	89-89.5	-30.9	(1.78)	51
2	274.3	101-101.5	-43.8	(1.14)	51
3	374.4	153-154	-49.0	(1.21)	51
4	474.5	186-188	-53.6	(1.14)	51
5	574.6	230-232	-60.8	(0.86)	51

2.1.3.1.3 Oligoesters of Pivalic Acid

2.1.3.1.3.1 Oligo(pivalic acids)   H[OCH$_2$C(CH$_3$)$_2$CO]$_n$OH

2.1.3.1.3.2 Oligo(pivalic acid) isobutyl esters
H[OCH$_2$C(CH$_3$)$_2$CO]$_n$-OCH$_2$CH(CH$_3$)$_2$

n	Mol. Wt.	m.p. [°C]	Ref.
1	118.1	123-124	50
2	218.2	71.5-72.5	50
3	318.4	67-68	50
4	418.5	111.5-112.5	50
5	518.6	119.5-120.5	50
6	618.7	138-139	50

n	Mol. Wt.	m.p. [°C]	b.p. [°C/mm]	Ref.
2	274.4		115-117/0.5	49
3	374.5		175-177/1	49
4	474.6	75.5-76		49
5	574.7	97-98		49
6	674.8	120.5-121.5		49
7	774.9	134-135		49

2.1.3.1.4 Cyclic Oligoesters of ω-Hydroxy Acids $[O(CH_2)_x CO]_n$

x	n	Mol. Wt.	m.p. [°C]	b.p. [°C/mm]	Ref.
1	2	116.1	86-87		1
5	1	114.1	5	104-106/10	8-10
	2	228.3	112-113	130/0.13	8, 9
	3	342.4		250/0.2	9
6	1	128.2		80-82/11	10
	2	256.3	41	135/0.25	9
	3	384.5		202/0.2	9
7	1	142.2		72-73/11	10
	2	284.4	93	152/0.13	9
8	1	156.2	31-31.5	86-87/10	9-11
	2	312.5	57-58	158/0.03	9, 10
	3	468.7	20	240/0.1	9
9	1	170.3	6.4	100/10	9-12
	2	340.5	97	192/0.3	9, 12-14
	3	510.8	29	270/0.3	9
10	1	184.3	3	116/10	9-12
	2	368.6	74		9, 12
11	1	198.3	2	130/10	10, 11, 15
	2	396.6	101		9
12	1	212.3	27.5	143/10	9-12, 16
	2	424.7	84		9, 14
13	1	226.4	33-33.7	165/15	9, 11, 16
	2	452.7	107		9, 12
14	1	240.4	37-37.5	169/10	9-11, 16
	2	480.7	88		9, 15
15	1	254.4	35.5-36.5	188/15	9, 11, 16
	2	508.8	108		9
16	1	268.4	42-43	194/15	9, 11, 16
	2	536.9	97		9
17	1	282.5	37	143/0.25	9
	2	564.9	114		9
22	1	348.6	36	175/0.2	9
	2	697.2	105		9

2.1.3.1.5 Oligoesters of Aromatic Hydroxy Acids

2.1.3.1.5.1 Linear Oligoesters of Salicylic Acids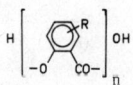

2.1.3.1.5.2 Cyclic Oligoesters of Salicyclic Acid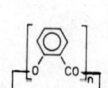

R	n	Mol. Wt.	m.p. [°C]	Ref.
H	2	258.2	148-149	17-19
m-CH₃	2	286.3	162	20
p-CH₃	2	286.3	128-129	20

n	Mol. Wt.	m.p. [°C]	Ref.
2	240.2	234	17, 19
3	360.3	200	17, 21
4	480.4	298-300	17
6	720.7	375 (d)	17

2.1.3.1.5.3 Cyclic Oligoesters of Cresotic Acid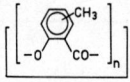

-CH₃	n	Mol. Wt.	m.p. [°C]	Ref.	-CH₃	n	Mol. Wt.	m.p. [°C]	Ref.
o-	2	268.3	240 (d)	20, 22, 23	m-	4	536.5	305 (d)	22
	3	402.4	264-265	22	p-	2	268.3	235-235.5	22, 23
	4	536.5	299-300	22		3	402.4	244.5-245	22
m-	2	268.3	255 (d)	22, 23		4	536.5	347 (d)	22
	3	402.4	207-207.5	22					

2.1.3.1.5.4 Cyclic Oligoesters of Thymotic Acid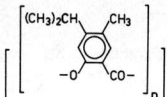

2.1.3.1.5.5 Cyclic Oligoesters of Thiosalicylic Acid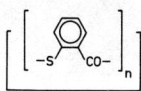

n	Mol. Wt.	m.p. [°C]	Ref.
2	352.4	207	24
3	528.7	217	24

n	Mol. Wt.	m.p. [°C]	Ref.
2	272.3	176-177	25, 26
3	408.5	257-258	25
4	544.7	288-290	25

2.1.3.1.5.6 Linear Oligoesters of p-(β-Hydroxy)-ethoxybenzoic Acid $H[OCH_2CH_2O\bigcirc CO]_n OR$

R	n	Mol. Wt.	m.p. [°C]	Ref.	R	n	Mol. Wt.	m.p. [°C]	Ref.
H	1	182.2	177	27	$-CH_2CH_2OH$	1	226.2	77	28
	2	346.3	165	27		2	390.4	114	28
	3	510.5	183	27		3	554.6	136	28
	4	674.7	192	27		4	718.7	156	28

2.1.3.2 Oligoesters of Diols and Dicarboxylic Acids

2.1.3.2.1 Cyclic Oligo(alkylene succinates) $[O(CH_2)_x O-CO(CH_2)_2 CO]_n$

2.1.3.2.2 Cyclic Oligo(neopentyl adipate) $[OCH_2C(CH_3)_2 CH_2 O-CO(CH_2)_4 CO]_n$

x	n	Mol. Wt.	m.p. [°C]	b.p. [°C/mm]	Ref.
2	2	288.3	131		29, 30
3	1	158.2	81	94-100/2	29
	2	316.3	138		29
4	1	172.2	42	95-96/2	29
	2	344.4	121		29
5	1	186.2	19	88-89/1	29
	2	372.4	87		29
6	1	200.2	-15	108-110/2	29
	2	400.5	110		29
7	1	214.3	49	116-118/1-2	29
	2	428.5	86		29
8	1	228.3	71		29
	2	456.6	109		29
10	1	256.3	58	135-140/2	29
	2	512.7	109		15, 29

n	Mol. Wt.	m.p. [°C]	Ref.
3	428.5	127-128	47

2.1.3.2.3 Cyclic Oligo(alkylene sebacates) $[O(CH_2)_x O-CO(CH_2)_8 CO]_n$

x	n	Mol. Wt.	m.p. [°C]	Ref.
2	1	228.3	40-41	15, 29
	2	456.6	80-81	15
3	1	242.3	14	29, 31
	2	484.6	113-113.5	15, 31

2.1.3.2.4 Cyclic Oligo(ethylene dicarboxylates) $[O(CH_2)_2 O-CO(CH_2)_x CO]_n$

x	n	Mol. Wt.	m.p. [°C]	Ref.	x	n	Mol. Wt.	m.p. [°C]	Ref.
7	1	214.3	52	29	10	1	256.3	18	29
	2	428.5	145	29		2	512.7	95-96	15
8	1	228.3	40-41	15, 29	11	1	270.4	-8	29
	2	456.6	80-81	15		2	540.7	145-146	15
9	1	242.3	35	15	12	2	568.8	102-103	15
	2	484.6	143	15					

2.1.3.2.5 Cyclic Oligo(ethylene isophthalates) $[OCH_2CH_2O-CO\bigcirc CO]_n$

n	Mol. Wt.	m.p. [°C]	Ref.
2	384.4	325-327	34

2.1.3.2.6 Oligo(ethylene terephthalates)

2.1.3.2.6.1 Oligomeric Hydroxy Acids $H[OCH_2CH_2O-CO\bigcirc CO]_nOH$

n	Mol. Wt.	m.p. [$^{\circ}$C]	Ref.
1	210.2	178	35
2	402.4	200-205	36
3	594.6	219-223	36

2.1.3.2.6.2 Oligomeric Diols $H[OCH_2CH_2O-CO\bigcirc CO]_n-OCH_2CH_2OH$

n	Mol. Wt.	m.p. [$^{\circ}$C]	Ref.
1	254.2	109-110	35, 36
2	446.4	173-174	35, 36
3	638.6	200-205	35, 36
4	830.8	213-216	35, 36
5	1023.0	218-220	36

2.1.3.2.6.3 Oligomeric Dicarboxylic Acids $HOOC\bigcirc CO-[OCH_2CH_2O-CO\bigcirc CO]_nOH$

n	Mol. Wt.	m.p. [$^{\circ}$C]	Ref.
1	358.3	>360	36, 37, 38
2	550.5	284-286	33, 36, 37, 38
3	742.7	274-276	33, 37, 38
4	934.9	252-255	38
5	1127.1	233-236	38

2.1.3.2.6.4 Oligomeric Dicarboxylates

$CH_3OOC\bigcirc CO-[OCH_2CH_2O-CO\bigcirc CO]_nOCH_3$

n	Mol. Wt.	m.p. [$^{\circ}$C]	Ref.
1	386.3	168-170	35, 52
2	578.5	194-198	35, 52
3	770.7	215-217	52
4	962.8	231-232	52
5	1155.0	242-243	52
6	1347.1	248-249	52
7	1539.3	250-252	52
8	1731.5	251-253	52
9	1923.7	253-254	52
10	2115.8	254-256	52

2.1.3.2.6.5 Cyclic Oligoesters $[OCH_2CH_2O-CO\bigcirc CO]_n$

n	Mol. Wt.	m.p. [$^{\circ}$C]	Ref.
2	384.0		32, 33, 46
3	576.5	319	15, 33, 39, 40, 43, 44, 46
4	768.7	328-331	33, 40, 43, 46
5	960.9	264	33, 40, 46
6	1152.7	306	33, 46
7	1344.8	238-240	33

2.1.3.2.7 Cyclic Oligoesters of Diethylene Glycol $[OC\bigcirc COO-X-OCO\bigcirc CO-OCH_2CH_2OCH_2CH_2O]_n$

X	n	Mol. Wt.	m.p. [$^{\circ}$C]	Ref.
$-(CH_2)_2O(CH_2)_2-$	1	472.4	221-223	45
$-CH_2CH_2-$	1	428.4	173-179	40, 43

2.1.3.2.8 Oligo(tetramethylene terephthalates)

2.1.3.2.8.1 Oligomeric Diols

$H[O(CH_2)_4O-CO\bigcirc CO]_n-O(CH_2)_4OH$

n	Mol. Wt.	m.p. [$^{\circ}$C]	Ref.
2	530.5	138-140	43

2.1.3.2.8.2 Oligomeric Dicarboxylic Acids

$HOOC\bigcirc CO[O(CH_2)_4O-CO\bigcirc CO]_nOH$

n	Mol. Wt.	m.p. [$^{\circ}$C]	Ref.
1	385.3	259	43

2.1.3.2.8.3 Cyclic Oligoesters

$[O(CH_2)_4OCO\bigcirc CO]_n$

n	Mol. Wt.	m.p. [$^{\circ}$C]	Ref.
4	660.4	247-249	43

2.1.3.2.9 Oligo(1,4-cyclohexylenedimethylene terephthalates)

2.1.3.2.9.1 Oligomeric Hydroxy Acids

$H[OCH_2\langle H\rangle CH_2O-CO\bigcirc CO]_nOH$

n	Config.	Mol. Wt.	m.p. [$^{\circ}$C]	Ref.
1	trans	292.3	175-178	41

2.1.3.2.9.2 Oligomeric Dicarboxylic Acids

$HOOC\bigcirc CO-[OCH_2\langle H\rangle CH_2O-CO\bigcirc CO]_nOH$

n	Config.	Mol. Wt.	m.p. [$^{\circ}$C]	Ref.
1	trans	440.4	>310	41
1	cis	440.4	>300	41

2.1.3.2.9.3 Cyclic Oligoesters $[OCH_2\langle H\rangle CH_2O-CO\bigcirc CO]_n$

n	Mol. Wt.	m.p.	Ref.
3	823.0	288-298	42

REFERENCES

1. C. A. Bischoff, P. Walden, Ann. Chem. 279, 45 (1893); Chem. Ber. 26, 262 (1893).
2. C. A. Bischoff, P. Walden, Ann. Chem. 279, 71 (1893); Chem. Ber. 27, 2949 (1894).
3. E. Jungfleisch, M. Godchot, Compt. Rend. 141, 111 (1905); 142, 632 (1906); 144, 425 (1908).
4. J. Kleine, K.-H. Kleine, Makromol. Chem. 30, 23 (1959).
5. R. Dietzel, R. Krug, Chem. Ber. 58, 1307 (1925).
6. J. Wislicenus, Ann. Chem. 167, 318 (1873).
7. R. Eder, F. Kutter, Helv. Chim. Acta 9, 557 (1926).
8. F. J. van Natta, J. W. Hill, W. H. Carothers, J. Am. Chem. Soc. 56, 455 (1934).
9. M. Stoll, A. Rouvé, Helv. Chim. Acta 18, 1087 (1935).
10. R. Huisgen, H. Ott, Tetrahedron 6, 253 (1959).
11. H. Hunsdiecker, H. Erlbach, Chem. Ber. 80, 129 (1947).
12. E. W. Spanagel, W. H. Carothers, J. Am. Chem. Soc. 58, 654 (1936).
13. W. H. Lycan, R. Adams, J. Am. Chem. Soc. 51, 3450 (1929).
14. W. H. Carothers, F. J. van Natta, J. Am. Chem. Soc. 55, 4714 (1933).
15. F. L. Hamb, L. C. Trent, J. Polymer Sci. B 5, 1057 (1967).
16. L. Ruzicka, M. Stoll, Helv. Chim. Acta 11, 1159 (1928).
17. W. Baker, W. D. Ollis, T. S. Zealley, J. Chem. Soc. 1951, 201.
18. A. Einhorn, Chem. Ber. 44, 437 (1911).
19. G. Schroeter, Chem. Ber. 52, 2224 (1919).
20. R. Anschuetz, Ann. Chem. 439, 8 (1924).
21. A. Einhorn, H. Pfeiffer, Chem. Ber. 34, 2952 (1901); A. Einhorn, C. Mettler, Chem. Ber. 35, 3644 (1902).
22. W. Baker, B. Gilbert, W. D. Ollis, T. S. Zealley, J. Chem. Soc. 1951, 209.
23. L. Anschuetz, G. Gross, Chem. Ber. 77, 644 (1944).
24. W. Baker, B. Gilbert, W. D. Ollis, J. Chem. Soc. 1952, 1443.
25. W. Baker, A. S El-Nawary, W. D. Ollis, J. Chem. Soc. 1952, 3163.
26. R. Anschuetz, E. Rhodius, Chem. Ber. 47, 2733 (1914).
27. M. Ishibashi, M. Hirai, Chem. High Polymers (Tokyo) 21, 231 (1964).
28. M. Ishibashi, M. Hirai, Chem. High Polymers (Tokyo) 21, 235 (1964).
29. E. W. Spanagel, W. H. Carothers, J. Am. Chem. Soc. 57, 929 (1935).
30. W. H. Carothers, G. L. Dorough, J. Am. Chem. Soc. 52, 711 (1930).
31. M. Stoll, A. Rouvé, Helv. Chim. Acta 19, 253 (1936).
32. H. Repin, E. Papanikolau, J. Polymer Sci. A-1, 7, 3426 (1969).
33. H. Zahn, H. Repin, Chem. Ber. 103, 3041 (1970).
34. C. E. Berr, J. Polymer Sci. 40, 591 (1955).
35. H. Zahn, R. Krzikalla, Makromol. Chem. 23, 31 (1957).
36. H. Zahn, C. Borstlap, G. Valk, Makromol. Chem. 64, 18 (1963).
37. H. Zahn, B. Seidel, Makromol. Chem. 29, 70 (1959).
38. H. Zahn, B. Gleitsmann, Angew. Chem. 75, 772 (1963).
39. S. D. Ross, E. R. Coburn, W. A. Leach, W. B. Robinson, J. Polymer Sci. 13, 406 (1954).
40. J. Goodman, B. F. Nesbitt, Polymer 1, 384 (1960); J. Polymer Sci. 48, 423 (1960).
41. H. Zahn, G. Valk, Makromol. Chem. 64, 37 (1963).
42. H. Zahn, G. Valk, Polymer Letters 1, 105 (1963).
43. E. Meraskentis, H. Zahn, J. Polymer Sci. A-1, 4, 1890 (1966); Chem. Ber. 103, 3034 (1970).
44. G. L. Binns, J. S. Frost, F. S. Smith, E. C. Yeadon, Polymer 7, 583 (1966).
45. T. Morimoto, Y. Fujita, Takeda, J. Polymer Sci. A-1, 6, 1044 (1968).
46. L. H. Peebles, Jr., M. W. Huffman, C. T. Ablett, J. Polymer Sci. A-1, 7, 479 (1969).
47. I. S. Megna, A. Koroscil, J. Polymer Sci. A-1, 7, 1371 (1969).
48. M. Goodman, M. D' Alagni, J. Polymer Sci. B 5, 515 (1967).
49. Y. Iwakura, K. Hayashi, K. Iwata, S. Matsuo, Makromol. Chem. 108, 300 (1967); 110, 90 (1967).
50. Y. Iwakura, K. Hayashi, K. Iwata, S. Matsuo, Makromol. Chem. 123, 245 (1969).
51. Y. Iwakura, K. Iwata, S. Matsuo, A. Tohara, Makromol. Chem. 146, 33 (1971).
52. R. Penisson, H. Zahn, Makromol. Chem. 133, 25 (1970).
53. S. Matsuo, Y. Iwakura, Makromol. Chem. 152, 203 (1972).

2.1.4 OLIGO(URETHANES)

2.1.4.1 Oligo(alkylene urethanes)

2.1.4.1.1 Oligo(ethylene urethanes)

2.1.4.1.1.1 Acetoxy-oligo(ethylene urethanes)

$CH_3CO-[OCH_2CH_2NHCO]_nOCH_3$

n	Mol. Wt.	m.p. [°C]	Ref.
2	248.2	69-70	15
3	335.3	98-99	15
4	422.4	138.5-139.5	15
5	509.5	164-165	15
6	596.6	174-176	15
7	683.6	185-186	15

2.1.4.1.1.2 Hydroxy-oligo(ethylene urethanes)

$H[OCH_2CH_2NHCO]_nOCH_3$

n	Mol. Wt.	m.p. [°C]	b.p. [°C/mm]	Ref.
1	119.1	~0	116-117/2	15
2	206.2	62-63		15
3	293.3	101.5-102.5		15
4	380.4	129-130		15
5	467.4	152-153		15
6	554.5	168.5-169.5		15

2.1.4.1.2 Oligo[(3-methyl)ethylene urethanes]

2.1.4.1.2.1 Acetoxy-oligo[(3-methyl)ethylene urethanes] $CH_3CO[OCH_2CH(CH_3)NHCO]_nOCH_3$

n	Mol. Wt.	m.p. [°C]	b.p. [°C/mm]	$[\alpha]_D$ in EtOH	(conc.)	Ref.
1	175.2		109-110/3	-23.2	(1.39)	11
2	276.3	93-98		-30.2	(0.32)	11
3	377.4	113-117		-31.8	(0.77)	11
4	478.5	143-147		-30.9	(1.14)	11
5	579.6	153-154		-30.6	(0.42)	11
6	680.7	167				11
7	781.8	173				11

2.1.4.1.2.2 Hydroxy-oligo[(3-methyl)ethylene urethanes] $H[OCH_2CH(CH_3)NHCO]_nOCH_3$

n	Mol. Wt.	m.p. [°C]	b.p. [°C/mm]	$[\alpha]_D$ in EtOH	(conc.)	Ref.
1	133.2		109-110/2	-15.8	(2.8)	11, 16
2	234.3	60-64				11
3	335.4	86-87				11
4	436.5	120-122				11
5	537.6	135-137				11
6	638.7	152-153				11

2.1.4.1.3 Oligo[(3-isopropyl)ethylene urethanes]

2.1.4.1.3.1 Acetoxy-oligo[(3-isopropyl)ethylene urethanes] $CH_3CO\left[-OCH_2\underset{CH(CH_3)_2}{CHNHCO-}\right]_nOCH_3$

n	Mol. Wt.	m.p. [°C]	b.p. [°C/mm]	$[\alpha]_D$ in EtOH	(conc.)	Ref.
1	203.2		117.5-118/1	-22.2	(0.79)	10
2	332.4	111.5-112.5		-33.5	(2.0)	10
3	461.6	142-143		-28.4	(2.1)	10
4	590.7	156.5-158		-4.2	(1.9)	10

2.1.4.1.4 Oligo[(3-isobutyl)ethylene urethanes]

2.1.4.1.4.1 Acetoxy-oligo[(3-isobutyl)ethylene urethanes] $CH_3CO\left[-OCH_2\underset{CH_2CH(CH_3)_2}{CHNHCO-}\right]_nOCH_3$

n	Mol. Wt.	m.p. [°C]	b.p. [°C/mm]	$[\alpha]_D^{25}$ in EtOH	(conc.)	Ref.
1	217.3		128-129/1			9
2	360.5	73-74		-41.7	(2.0)	9, 14
3	503.6	97-98		-42.9	(1.0)	9, 14
4	646.8	136-138		-41.5	(1.4)	9, 14
5	790.0	149-150		-38.5	(2.0)	9, 13
6	933.2	162-164		-35.9	(1.0)	9, 13
7	1076.4	168-169		-30.5	(2.2)	9, 13
8	1219.5	180-181				12
9	1362.7	185.5-186.5				12

2.1.4.1.4.2 Hydroxy-oligo[(3-isobutyl)ethylene urethanes] $H\left[-OCH_2\underset{CH_2CH(CH_3)_2}{CHNHCO-}\right]_nOCH_3$

n	Mol. Wt.	b.p. [°C/mm]	$[\alpha]_D$ in EtOH	(conc.)	Ref.
1	175.3	119-121/1	-34.1	(2.6)	14
2	318.5	oil	-47.2	(2.3)	14
3	461.6	oil	-45.0	(0.99)	14

2.1.4.1.5   Oligo[(β-benzyl)ethylene urethanes]

2.1.4.1.5.1   Acetoxy-oligo[(β-benzyl)ethylene urethanes] $CH_3CO\left[-OCH_2CHNHCO-\atop CH_2C_6H_5\right]_n OCH_3$

n	Mol. Wt.	m.p. [$^{\circ}$C/mm]	$[\alpha]_D$ in EtOH	(conc.)	Ref.
2	428.5	104.5-106.5	-17.9	(1.2)	11, 14
3	605.7	138-139	-18.1	(0.71)	11, 14
4	782.9	149-150			11

2.1.4.1.5.2   Hydroxy-oligo[(β-benzyl)ethylene urethanes] $H\left[-OCH_2CHNHCO-\atop CH_2C_6H_5\right]_n OCH_3$

n	Mol. Wt.	b.p. [$^{\circ}$C/mm]	$[\alpha]_D$ in EtOH	(conc.)	Ref.
1	209.2	51.5-53	-25.9	(0.72)	11, 14
2	386.4	130-131	-27.8	(0.77)	11, 14
3	563.6	153-154			11

2.1.4.1.6   Oligo(trimethylene urethanes)

2.1.4.1.6.1   Acetoxy-oligo(trimethylene urethanes)
$CH_3CO-[O(CH_2)_3NHCO]_n OCH_3$

n	Mol. Wt.	m.p. [$^{\circ}$C]	Ref.
2	276.3	57.5-58.5	18
3	377.4	95-97	18
4	478.5	118.5-120	18
5	579.6	133.5-135	18

2.1.4.1.6.2   Hydroxy-oligo(trimethylene urethanes)
$H[O(CH_2)_3NHCO]_n OCH_3$

n	Mol. Wt.	m.p. [$^{\circ}$C]	Ref.
1	133.2	oil	18
2	234.3	oil	18
3	335.4	76.5-79	18
4	436.5	106.5-108.5	18
5	537.6	126-128	18

2.1.4.1.7   Oligo(pentamethylene urethanes)

2.1.4.1.7.1   Acetoxy-oligo(pentamethylene urethanes)
$CH_3CO-[O(CH_2)_5NHCO]_n OCH_3$

n	Mol. Wt.	m.p. [$^{\circ}$C]	Ref.
2	332.4	62-63.5	1
3	461.6	97-98.5	1
4	590.7	107-109	1
5	719.9	123-124.5	1
6	849.0	129.5-131	1
7	978.2	136-138	1
8	1107.4	140-143	1

2.1.4.1.7.2   Hydroxy-oligo(pentamethylene urethanes)
$H[O(CH_2)_5NHCO]_n OCH_3$

n	Mol. Wt.	m.p. [$^{\circ}$C]	b.p. [$^{\circ}$C/mm]	Ref.
1	161.2		131-134/0.2	1
2	290.4	74-75		1
3	419.5	99-101		1
4	548.7	112		1
5	677.8	116		1
6	807.0	122-125		1
7	936.2	131-134.5		1
8	1065.3	139-140.5		1

2.1.4.2   Oligo(urethanes) of Diisocyanates and Glycols

2.1.4.2.1   Oligo(urethanes) of hexamethylene diisocyanate and diglycols

2.1.4.2.1.1   Diol oligo(urethanes) $HO[(CH_2)_2X(CH_2)_2O-CO-NH(CH_2)_6NH-CO-O]_n-(CH_2)_2X(CH_2)_2OH$

X	n	Mol. Wt.	m.p. [$^{\circ}$C]	Ref.	X	n	Mol. Wt.	m.p. [$^{\circ}$C]	Ref.
O	1	380.4	68-69	2,3	O	7	2026.3	122-124	2,3
	2	654.8	103-104	2		15	4221.0	119-123	2,3
	3	929.1	123-125	2,3	S	1	412.6	105-106	2
	4	1203.4	120-123	2		3	993.4	132-134	2
	5	1478.7	123-124	2,3		7	2155.5	133-135	2

2.1.4.2.1.2 Cyclic Oligo(urethanes) $[(CH_2)_2X(CH_2)_2O-CO-NH(CH_2)_6NH-CO-O]_n$

X	n	Mol. Wt.	m.p. [°C]	Ref.	X	n	Mol. Wt.	m.p. [°C]	Ref.
O	1	274.3	138	4, 17	O	5	1371.5	136	17
	2	548.6	170	17		6	1645.8	145-148	17
	3	722.9	131-133	17		7	1920.1	134	17
	4	1097.2	151-153	17	S	1	290.4	128	4

2.1.4.2.2 Oligo(urethanes) of Diisocyanates and 1,4-Butanediol

2.1.4.2.2.1 Diol Oligo(urethanes) $HO[(CH_2)_4O-CO-NH-X-NH-CO-O]_n-(CH_2)_4OH$

X	n	Mol. Wt.	m.p. [°C]	Ref.	X	n	Mol. Wt.	m.p. [°C]	Ref.
$-(CH_2)_6-$	1	348.4	103-105	2, 3, 5, 6	$-(CH_2)_6-$	9	2414.9	177-179	2, 3
	2	606.8	146	2, 3, 5	(biphenyl-OCH₃, OCH₃)	1	476.5	133.5	2, 3
	3	865.1	162-163	2, 3, 5		2	862.9	127-128	2, 3
	4	1123.4	169-170	2, 3		3	1249.4	164-166	2, 3
	5	1381.7	171-173	2, 3		7	2795.0	190-195	2, 3
	6	1640.0	173-174	2, 3		15	5886.3	210-215	2, 3
	7	1898.3	175-176	2, 3					

2.1.4.2.2.2 Diamine Oligo(urethanes) $H[NH(CH_2)_6NH-CO-O-(CH_2)_4O-CO]_n-NH(CH_2)_6NH_2 \cdot 2\,HBr$

n	Mol. Wt.	m.p. [°C]	Ref.
1	536.4	228	5
2	794.7	226	5
3	1053.0	218	5
4	1311.4	212	5

2.1.4.2.2.3 Cyclic Oligo(urethanes) $[NH(CH_2)_6NH-CO-O(CH_2)_4O-CO]_n$

n	Mol. Wt.	m.p. [°C]	Ref.
1	258.3	164	4, 6, 7
2	516.6	198	6, 8

REFERENCES

1. Y. Iwakura, K. Hayashi, K. Iwata, Makromol. Chem. 89, 214 (1965).

2. W. Kern, H. Kalsch, K. J. Rauterkus, H. Sutter, Makromol. Chem. 44-46, 78 (1961).

3. W. Kern, Angew. Chem. 71, 585 (1959).

4. W. Kern, K. J. Rauterkus, W. Weber, Makromol. Chem. 43, 98 (1961).

5. H. Zahn, M. Dominik, Makromol. Chem. 44-46, 290 (1961).

6. H. Zahn, M. Dominik, Chem. Ber. 94, 125 (1961).

7. O. Bayer, Ann. Chem. 549, 286 (1941); Angew. Chem. 59A, 257 (1947).

8. W. Kern, K. J. Rauterkus, W. Weber, W. Heitz, Makromol. Chem. 57, 241 (1962).

9. K. Iwata, Y. Iwakura, K. Hayashi, Makromol. Chem. 112, 242 (1968).

10. K. Iwata, Y. Iwakura, K. Hayashi, Makromol. Chem. 116, 250 (1968).

11. K. Iwata, Y. Iwakura, Makromol. Chem. 135, 165 (1970).

12. K. Iwata, Y. Iwakura, Makromol. Chem. 134, 321 (1970).

13. Y. Iwakura, K. Hayashi, K. Iwata, Makromol. Chem. 108, 296 (1967).

14. Y. Iwakura, K. Hayashi, K. Iwata, Makromol. Chem. 93, 274 (1966).

15. Y. Iwakura, K. Hayashi, K. Iwata, Makromol. Chem. 95, 217 (1966).

16. Y. Iwakura, K. Hayashi, K. Iwata, Makromol. Chem. 104, 46 (1967).

17. W. Heitz, H. Hoecker, W. Kern, H. Ullner, Makromol. Chem. 150, 73 (1971).

2.2 OLIGOMERS CONTAINING S AND Se IN THE MAIN CHAIN

2.2.1 OLIGO(SULFIDES) AND OLIGO(SELENIDES)

2.2.1.1 Cyclic Oligo(sulfides) and Oligo(selenides) $[XCH_2]_n$

X	n	Mol. Wt.	m.p. [°C]	Ref.
S	3	138.3	215-218	1-7
	4	184.4	49-50	1, 5, 6, 8
	5	230.5	120-121	6, 9
Se	3	279.0	226-228	1
	4	372.0	80-81	1

2.2.1.2 Cyclic Oligo(sulfides) containing ether bridges $[OCH_2-(SCH_2)_n]$

n	Mol. Wt.	m.p. [°C]	Ref.
4	214.4	106-108	6, 10
5	260.5	163-165	6, 10

REFERENCES

1. M. Russo, L. Mortillaro, L. Credali, C. DeChecchi, J. Polymer Sci. B, 4, 248 (1966).
2. A. W. v. Hofmann, Ann. Chem. 145, 360 (1868).
3. E. Baumann, Chem. Ber. 23, 60 (1890).
4. R. W. Bonst, E. W. Constable, Org. Syn. 16, 81 (1936).
5. M. Schmidt, K. Blaettner, Angew. Chem. 71, 4078 (1959).
6. L. Credali, M. Russo, Polymer 8, 469 (1967).

7. E. Gipstein, E. Wellisch, O. J. Sweeting, J. Polymer Sci. B, 1, 239 (1963).
8. M. Russo, L. Mortillaro, C. DeChecchi, G. Kalle, M. Mammi, J. Polymer Sci. B, 3, 501 (1965).
9. M. Russo, L. Mortillaro, L. Credali, C. DeChecchi, J. Polymer Sci. B, 3, 455 (1965).
10. M. Russo, L. Mortillaro, L. Credali, C. DeChecchi, J. Chem. Soc. C, 1966, 428.

2.3 OLIGOMERS CONTAINING N IN THE MAIN CHAIN

2.3.1 OLIGO(AMIDES)

2.3.1.1 Oligoamides of β-Alanine (Nylon 3)

2.3.1.1.1 Linear Oligoamides $H[NH(CH_2)_2CO]_nOH$

n	Mol. Wt.	m.p. (d) [$^{\circ}$C]	Ref.
1	89.1	206	1
2	160.2	212	2, 4
3	231.2	>255	3, 4
4	302.3	>260	4
5	373.4	>310	5
6	444.5	>320	5

2.3.1.1.2 Cyclic Oligoamides $[NH(CH_2)_2CO]_n$

n	Mol. Wt.	m.p. [$^{\circ}$C]	Ref.
1	71.1	74-74.5	6, 7
2	142.2	298-299	4, 8, 9
3	213.2	>350	4, 10
4	284.3	>350	4, 10
5	355.4	>350	5
6	426.5	>350	5

2.3.1.2 Oligoamides of γ-Aminobutyric Acid (Nylon 4)

2.3.1.2.1 Linear Oligoamides $H[NH(CH_2)_3CO]_nOH$

n	Mol. Wt.	m.p. [$^{\circ}$C]	Ref.
1	103.1	204-205	11
2	188.2	186	4, 12
3	273.4	198-199	4
4	358.4	202-203	4

2.3.1.2.2 Cyclic Oligoamides $[NH(CH_2)_3CO]_n$

n	Mol. Wt.	m.p. [$^{\circ}$C]	Ref.
1	85.1	24	13, 14
2	170.2	283	4, 70, 71
3	255.3	242-243	4, 10
4	340.4	255	4, 10
6	510.6	295	10

2.3.1.2.3 Anionic Oligoamides $C_6H_5CO-[-NH-(CH_2)_3-CO-]_n-N\!\!\underset{O=C}{\overset{|}{\big|}}(CH_2)_3$

n	Mol. Wt.	m.p. [$^{\circ}$C]	Ref.
0	189.2	92.5	15
1	274.3	121	16

2.3.1.3 Oligoamides of δ-Aminovaleric Acid (Nylon 5)

2.3.1.3.1 Linear Oligoamides $H[NH(CH_2)_4CO]_nOH$

n	Mol. Wt.	m.p. [$^{\circ}$C]	Ref.
1	117.2	160-162	17
2	216.3	178	18
3	315.4	184-185.5	18
4	414.5	196.5-198.5	18

2.3.1.3.2 Cyclic Oligoamides $[NH(CH_2)_4CO]_n$

n	Mol. Wt.	m.p. [$^{\circ}$C]	Ref.
1	98.1	39-40	14, 17, 19
2	198.3	295-296	8, 18
3	297.4	329-331	18
4	396.5	266-267	18

PHYSICAL DATA OF OLIGOMERS

2.3.1.4 Oligoamides of ε-Aminocaproic Acid (Nylon 6)

2.3.1.4.1 Linear Oligoamides
H[NH(CH$_2$)$_5$CO]$_n$OH

n	Mol. Wt.	m.p. [$^{\circ}$C]	Ref.
1	131.2	206-208	20, 21
2	244.3	198-199	22-24
3	357.5	203-204	23-25
4	470.7	206-207	23, 24
5	583.8	207-208	23, 69
6	697.0	209-210	23, 24
7	810.1	209-210	23, 69
8	923.3	210-211	23
9	1036.5	208-211	26, 69
10	1149.6	212-213	26
11	1262.8	209-212	26, 69
12	1375.9	211-213	26
13	1489.1	210-211	69
15	1715.3	207-208	69
17	1941.5	209-211	69
21	2394.5	212-213	69
25	2847.0	212-213.5	69

2.3.1.4.2 End-group protected Oligoamides
C$_2$H$_5$CO-[NH(CH$_2$)$_5$CO]$_n$-NHC$_3$H$_7$

n	Mol. Wt.	m.p. [$^{\circ}$C]	Ref.
1	228.3	106	31
2	341.5	149	31
3	454.7	172	31
4	567.8	181	31
5	681.0	187	31
6	794.2	191	31
7	907.3	197	31
8	1020.5	200	31
9	1133.6	201	31
10	1246.8	201	31
11	1360.0	204	31
12	1473.1	202	31
13	1586.3	200	31
14	1699.3	205	31
15	1812.6	206	31
16	1925.8	206	31

2.3.1.4.3 Cyclic Oligoamides [NH(CH$_2$)$_5$CO]$_n$

n	Mol. Wt.	m.p. [$^{\circ}$C]	TCW[*] R$_F$	EAW[**]	Ref.
1	113.2	69.5	0.56	0.88	14, 17, 27
2	226.3	348	0.48	0.81	4, 24, 28, 29
3	339.5	244	0.34	0.77	4, 24, 28, 29
4	452.6	256-257	0.25	0.69	4, 24, 28, 29
5	565.8	254	0.16	0.59	4, 24, 29
6	678.9	260	0.10	0.49	4, 24, 29, 60
7	792.1	243	0.08	0.40	29, 30, 60
8	905.3	254	0.04		29, 30, 60
9	1018.4	240	0.02		29, 30, 68

[*] Solvent:Tetrahydrofuran/cyclohexane/water (186:14:10)
[**] Solvent:Ethyl acetate/acetone/water (10:10:2)

2.3.1.4.4 Anionic Oligoamides
CH$_3$CO-[NH-(CH$_2$)$_5$-CO-]$_n$-N(CH$_2$)$_5$ with O=C

n	Mol. Wt.	b.p. [$^{\circ}$C/mm]	R$_F$[*]	Ref.
0	135.1	133-134/16		32
1	268.3		0.76	33
2	381.5		0.67	33
3	494.7		0.58	33

[*] Solvent:Tetrahydrofuran/cyclohexane/water (186:14:10)

2.3.1.5 Oligoamides of Higher ω-Amino Acids (Nylon 7 - 12)

2.3.1.5.1 Linear Oligoamides H[NH(CH$_2$)$_x$CO]$_n$OH

x	n	Mol. Wt.	m.p. [$^{\circ}$C]	Ref.
6	1	145.2	195	34, 35
	2	272.4	205-208	36
7	1	159.2	188	4, 37
	2	300.4	191-192	4, 38
8	1	173.3	184	4, 34
	2	328.5	184-186	4
10	1	201.3	186-187	34, 38
	2	384.6	187-188	4, 40
	3	567.9	191	5, 40
	4	751.2	177-179	40, 63
	5	934.5	177-178	63
	10	1850.8	181-182	63, 67
11	1	215.3	186-187	41
	2	412.6	192-193	31

2.3.1.5.2 Cyclic Oligoamides [NH(CH$_2$)$_x$CO]$_n$

x	n	Mol. Wt.	m.p. [$^{\circ}$C]	Ref.
6	1	127.2	29-30	14, 17, 42
	2	254.4	236-237	4, 10
7	1	141.2	72-73	42, 13, 14, 34
	2	282.4	277	4, 10
8	1	155.3	138-139	14, 42
	2	310.5	201	4, 10
9	1	169.3	162	14, 42
	2	338.4	230	10
10	1	183.3	155	42, 43
	2	366.6	188-189	4, 44
	3	549.9	184	44, 60
11	1	197.3	155	13, 14, 42, 45
	2	394.6	212	60, 61
	3	531.9	175	31

References page VI-28

2.3.1.6 Oligoamides of Anthranilic Acid

2.3.1.6.1 Linear Oligoamides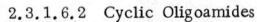

2.3.1.6.2 Cyclic Oligoamides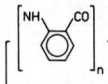

n	Mol. Wt.	m.p. [°C]	Ref.
1	137.1	146.1	46-48
2	256.3	203-204	46
3	375.4	228	46
4	494.5	d	46

n	Mol. Wt.	m.p. [°C]	Ref.
2	238.3	330	47

2.3.1.7 Cyclic Diamides of Aliphatic Dicarboxylic Acids and Diamines $[NH(CH_2)_x NH\text{-}CO(CH_2)_x CO]_n$

x	n	Mol. Wt.	m.p. [°C]	Ref.	x	n	Mol. Wt.	m.p. [°C]	Ref.
3	1	170.2	268	49,50	7	1	282.4	256	49
4	1	198.3	295	49,51	8	1	310.5	205	8,49
5	1	226.3	303	49	9	1	338.5	224	49
6	1	254.3	234	49,51	10	1	366.6	191	8,49

2.3.1.8 Oligoamides of Adipic Acid and Hexamethylene Diamine (Nylon 6.6)

2.3.1.8.1 Linear Oligoamino Acids $H[NH(CH_2)_6 NH\text{-}CO(CH_2)_4 CO]_n OH$

n	Mol. Wt.	m.p. [°C]	Ref.	n	Mol. Wt.	m.p. [°C]	Ref.
1	244.3	193	52,53	4	923.3	245-248	31,63,65
2	470.7	221-222	52,53,63	5	1149.5	247-251	31,63,65
3	696.9	246-248	52,53,63				

2.3.1.8.2 End-group Protected Oligoamides

2.3.1.8.2.1 Oligomeric Dicarboxylic Acid Dipropylamides
$C_3H_7 NH\text{-}CO(CH_2)_4 CO\text{-}[NH(CH_2)_6 NH\text{-}CO(CH_2)_4 CO]_n \text{-}NHC_3H_7$

n	Mol. Wt.	m.p. [°C]	Ref.
1	454.7	229-230	64
2	681.0	239-242	64
3	907.3	248-250	64
4	1133.6	257-260	64
5	1359.9	258-261	64

2.3.1.8.2.2 Oligomeric Dipropionyl Diamines
$C_2H_5 CO\text{-}[NH(CH_2)_6 NH\text{-}CO(CH_2)_4 CO]_n \text{-}NH(CH_2)_6 NH\text{-}COC_2H_5$

n	Mol. Wt.	m.p. [°C]	Ref.
1	454.7	202-203	64
2	681.0	226-228	64
3	907.3	229-231	64

2.3.1.8.3 Oligomeric Diamines
$H[NH(CH_2)_6 NH\text{-}CO(CH_2)_4 CO]_n \text{-}NH(CH_2)_6 NH_2 \cdot 2 HCl$

n	Mol. Wt.	m.p. [°C]	Ref.
1	415.4	248-250	65
2	641.7	249-251	65
3	868.0	252-254	65
4	1094.4	251-254	65
5	1320.7	252-255	65

2.3.1.8.4 Oligomeric Dicarboxylic Acids
$HOOC(CH_2)_4 CO\text{-}[NH(CH_2)_6 NH\text{-}CO(CH_2)_4 CO]_n OH$

n	Mol. Wt.	m.p. [°C]	Ref.
1	372.4	188-189	56
2	598.7	200-205	56,57
3	825.0	210-211	56
4	1051.3	214-219	31

2.3.1.8.5 Cyclic Oligoamides $\left[NH(CH_2)_6NH\text{-}CO(CH_2)_4CO\right]_n$

n	Mol. Wt.	m.p. [°C]	R_F TCW *	EAW **	Ref.	n	Mol. Wt.	m.p. [°C]	R_F TCW *	EAW **	Ref.
1	226.3	254	0.47	0.51	51, 53-55, 62	3	678.9	235	0.13	0.23	53, 55, 62
2	452.6	243-244	0.28	0.38	53-55, 62	4	905.3	273	0.06	0.16	53, 60, 62

* Solvent:Tetrahydrofuran/cyclohexane/water (186:14:10)   ** Solvent:Ethyl acetate/acetone/water (10:10:2)

2.3.1.9 Oligoamides of Sebacic Acid and Hexamethylenediamine (Nylon 6.10)

2.3.1.9.1 Linear Oligoamino Acids
$H\left[NH(CH_2)_6NH\text{-}CO(CH_2)_8CO\right]_n OH$

n	Mol. Wt.	m.p. [°C]	Ref.
1	300.4	187-188	58
2	582.8	196-199	63, 67
3	865.3	208-210	63, 67

2.3.1.9.2 Cyclic Oligoamides
$\left[NH(CH_2)_6NH\text{-}CO(CH_2)_8CO\right]_n$

n	Mol. Wt.	m.p. [°C]	Ref.
1	282.4	227-228	51
2	564.8	225	58, 60

2.3.1.9.3 Oligomeric Diamines
$H\left[NH(CH_2)_6NH\text{-}CO(CH_2)_8CO\right]_n\text{-}NH(CH_2)_6NH_2$

n	Mol. Wt.	m.p. [°C]	Ref.
1	398.6	129-131	57, 58
2	681.1	177-179	58
3	963.5	203-210	58

2.3.1.9.4 Oligomeric Dicarboxylic Acids
$HOOC(CH_2)_8CO\text{-}\left[NH(CH_2)_6NH\text{-}CO(CH_2)_8CO\right]_n OH$

n	Mol. Wt.	m.p. [°C]	Ref.
1	494.6	156-159	57, 58
2	767.1	182-184	57, 58
3	1049.5	181-185	58

2.3.1.10 Oligoamides of Oxalic Acid and Hexamethylenediamine
$\left[NH(CH_2)_6NH\text{-}COCO\right]_n$

n	Mol. Wt.	m.p. [°C]	Ref.
1	170.2	232	66
2	340.4	303	66
3	510.6	345	66

2.3.1.11 Oligoamides of Terephthalic Acid and Pentamethylenediamine
$\left[NH(CH_2)_5NH\text{-}CO\text{-}\bigcirc\text{-}CO\right]_n$

n	Mol. Wt.	m.p. [°C]	Ref.
2	464.3	407	59

REFERENCES

1. J. H. Ford, Org. Syn., Coll. Vol. III, 34; W. J. Taylor, J. Chem. Soc. 1928, 1898.
2. H. Th. Hanson, E. L. Smith, J. Biol. Chem. 175, 883 (1948).
3. E. Adams, N. C. Davis, E. L. Smith, J. Biol. Chem. 199, 845 (1952).
4. M. Rothe, Habilitationsschrift, Universitaet Halle, Germany, 1960.
5. M. Rothe, unpublished data.
6. R. W. Holley, A. D. Holley, J. Am. Chem. Soc. 71, 2129 (1949).
7. S. Searles, Jr., R. E. Wann, Chem. Ind. 1964, 2097.
8. M. Rothe, R. Timler, Chem. Ber. 95, 783 (1958).
9. H. K. Hall, Jr., J. Am. Chem. Soc. 80, 6404 (1962).
10. M. Rothe, Angew. Chem. 74, 725 (1962).
11. C. C. DeWitt, Org. Syn. Coll. Vol. II, 25 (1943).
12. R. L. Evans, F. Irreverre, J. Org. Chem. 24, 863 (1959); S. Murahashi, H. Sekiguchi, H. Yuki, Compt. Rend. 248, 1521 (1959).
13. K. Dachs, E. Schwarz, Angew. Chem. 74, 540 (1962).
14. R. Huisgen, H. Brade, H. Walz, I. Glogger, Chem. Ber. 90, 1437 (1957).
15. S. J. Kanewskaja, Chem. Ber. 69, 266 (1936).
16. H. Sekiguchi, Bull. Soc. Chim. (France) 1960, 1835.
17. O. Wallach, Ann. Chem. 312, 171 (1900); 324, 281 (1902).
18. M. Rothe, R. Hoßbach, Makromol. Chem. 70, 140 (1964).
19. N. Yoda, A. Miyake, J. Polymer Sci. 43, 117 (1960).
20. S. Gabriel, A. Maass, Chem. Ber. 32, 1266 (1899).
21. J. C. Eck, Org. Syn. Coll. Vol. II, 28 (1943); C. Y. Myers, L. E. Miller, Org. Syn. 32, 13 (1952).
22. G. M. van der Want, A. Staverman, Rec. Trav. Chim. 71, 379 (1952).
23. H. Zahn, D. Hildebrand, Chem. Ber. 90, 320 (1957).
24. M. Rothe, F.-W. Kunitz, Ann. Chem. 609, 88 (1957).
25. G. M. van der Want, A. Staverman, P. Inklaar, Rec. Trav. Chim. 71, 1252 (1952).
26. H. Zahn, D. Hildebrand, Chem. Ber. 92, 1963 (1959).
27. C. S. Marvel, Org. Syn. Coll. Vol. II, 371 (1943).
28. P. H. Hermans, Rec. Trav. Chim. 72, 798 (1953).

29. I. Rothe, M. Rothe, Chem. Ber. 88, 284 (1955); M. Rothe, J. Polymer Sci. 30, 227 (1958).

30. H. Zahn, J. Kunde, Ann. Chem. 70, 189 (1958).

31. H. Zahn, G. B. Gleitsmann, Angew. Chem. 75, 772 (1963).

32. H. A. Offe, Z. Naturforschung 2b, 183 (1947).

33. K. Gehrke, Faserforsch. Textiltechn. 13, 557 (1962).

34. D. D. Coffman, N. L. Cox, E. L. Martin, W. E. Mochel, F. J. van Natta, J. Polymer Sci. 3, 85 (1948).

35. C. F. Horn, B. T. Freure, H. Vineyard, H. J. Decker, Angew. Chem. 74, 531 (1962).

36. M. Rothe, R. Hoßbach, unpublished data.

37. A. G. Goldsobel, Chem. Ber. 27, 3121 (1894).

38. T. Gäumann, H. H. Guenthard, Helv. Chim. Acta 35, 53 (1952).

39. M. Genas, Angew. Chem. 74, 535 (1962).

40. H. Zahn, H. Roedel, J. Kunde, J. Polymer Sci. 36, 539 (1959).

41. A. Neuberger, Proc. Roy. Soc. A, 158, 84 (1937).

42. L. Ruzicka, M. Kobelt, O. Haeflinger, V. Prelog, Helv. Chim. Acta 32, 544 (1949).

43. W. Ziegenbein, W. Lang, Angew. Chem. 74, 943 (1962).

44. H. Zahn, J. Kunde, Chem. Ber. 94, 2470 (1961).

45. G. Wilke, Angew. Chem. 75, 10 (1963).

46. H. Meyer, Ann. Chem. 351, 267 (1907).

47. G. Schroeter, O. Eisleb, Ann. Chem. 367, 101 (1909); Chem. Ber. 52, 2224 (1919).

48. E. Mohr, F. Koehler, H. Ulrich, J. Prakt. Chem. 79, 281 (1909); 80, 1 (1909).

49. J. Dale, R. Coulon, J. Chem. Soc. 1964, 182.

50. G. J. Glover, H. Rapoport, J. Am. Chem. Soc. 86, 3397 (1964).

51. H. Stetter, J. Marx, Ann. Chem. 607, 59 (1957).

52. H. Zahn, F. Schmidt, Makromol. Chem. 36, 1 (1959).

53. I. Rothe, M. Rothe, Makromol. Chem. 68, 206 (1963).

54. H. Zahn, F. Schmidt, Chem. Ber. 92, 1381 (1959); H. Zahn, P. Miro, F. Schmidt, Chem. Ber. 90, 1411 (1957).

55. M. Rothe, I. Rothe, H. Bruenig, K.-D. Schwenke, Angew. Chem. 71, 700 (1959).

56. H. Zahn, W. Lauer, Makromol. Chem. 23, 85 (1957).

57. C. D. Cowell, Chem. Ind. 1954, 577.

58. H. Zahn, G. B. Gleitsmann, Makromol. Chem. 60, 45 (1963).

59. H. K. Livingston, R. L. Gregory, Polymer 13, 297 (1972).

60. P. Kusch, H. Zahn, Angew. Chem. 77, 720 (1965).

61. H. Zahn, H.-D. Stolper, G. Heidemann, Chem. Ber. 98, 3251 (1965).

62. H. Zahn, P. Kusch, Chem. Ber. 98, 2588 (1965).

63. P. Kusch, Kolloid-Z. 208, 138 (1966); H. Klostermeyer, J. Halstrøm, P. Kusch, J. Foehles, W. Lunkenheimer, Peptides 1967, 113; North-Holland Publ. Co., Amsterdam.

64. H. Zahn, O. P. Garg, Kolloid-Z. 208, 132 (1966).

65. H. Zahn, P. Kusch, J. Shah, Kolloid-Z. 216/7, 298 (1967); H. Zahn, Z. Ges. Textil-Ind. 66, 928 (1964).

66. O. Vogl, A. C. Knight, Macromolecules 1, 311 (1968).

67. H. Zahn, P. Kusch, Z. Ges. Textil-Ind. 69, 880 (1967).

68. M. Rothe, U. Kress, unpublished data.

69. M. Rothe, W. Dunkel, J. Polymer Sci., Polymer Letters 5, 589 (1967).

70. M. Rothe, I. Rothe, Makromol. Chem 85, 307 (1965).

71. G. I. Glover, R. B. Smith, H. Rapoport, J. Am. Chem. Soc. 87, 2003 (1965).

2.3.2 OLIGO(PEPTIDES)

2.3.2.1 Oligopeptides of Glycine

2.3.2.1.1 Linear Oligopeptides $H[NHCH_2CO]_n OH$

n	Mol. Wt.	m.p. [°C]	Ref.
1	75.1	233-236	1
2	132.1	210-215	2
3	189.2	235	3, 4
4	246.2	240	4, 5
5	303.3	270	4, 5
6	360.3	d	6

2.3.2.1.2 Cyclic Oligopeptides $[NHCH_2CO]_n$

n	Mol. Wt.	m.p. [°C]	Ref.
2	114.1	309	2, 7
4	228.2	>330	8, 9
5	285.3	>330	9
6	342.3	>355	8, 10, 11

2.3.2.2 Oligopeptides of Sarcosine

2.3.2.2.1 Linear Oligopeptides $H[N(CH_3)CH_2CO]_n OH$

n	Mol. Wt.	m.p. [°C]	Ref.
1	89.1	211-213	1
2	160.2	190-191	13, 17, 21

2.3.2.2.2 Cyclic Oligopeptides $[N(CH_3)CH_2CO]_n$

n	Mol. Wt.	m.p. [°C]	Ref.
2	142.2	147	13, 27
3	213.3	221	51
4	284.4	>350	51
5	355.5	∼265	51
6	426.5	295	17, 51
8	568.7	328	51
9	639.8	>320	17
12	853.0	>320	17

PHYSICAL DATA OF OLIGOMERS

2.3.2.3 Oligopeptides of L-Alanine

2.3.2.3.1 Linear Oligopeptides $H[NHCH(CH_3)CO]_n OH$

n	Mol. Wt.	m.p. [^{o}C]	Specific Rotation				
			$[\alpha]_D$	[^{o}C]	Conc.	Solvent	Ref.
1	89.1		+14.5	25	10	6 N HCl	1
2	160.2	298	-37.3	24	2	0.5 N HCl	6,16
3	231.3	257-263	-85.4	23	2	0.5 N HCl	14,15
4	302.3	269-272	-131.0	25	2	0.5 N HCl	16
5	373.4		-149.7	23	2	0.5 N HCl	16
6	444.5		-156.6	23	0.9	0.5 N HCl	16
7	515.6	320					15

2.3.2.3.2 Cyclic Oligopeptides $\left[NHCH(CH_3)CO\right]_n$

n	Mol. Wt.	m.p. [^{o}C]	Specific Rotation				
			$[\alpha]_D$	[^{o}C]	Conc.	Solvent	Ref.
2	142.2	297	-28.8	20	2	Water	6

2.3.2.4 Oligopeptides of L-Valine, L-Leucine, and L-Isoleucine

2.3.2.4.1 Linear Oligopeptides $H[NHCHRCO]_n OH$

R	n	Mol. Wt.	Specific Rotation				
			$[\alpha]_D$	[^{o}C]	Conc.	Solvent	Ref.
-CH(CH$_3$)$_2$	1	117.2	+28.8	20	3.4	6 N HCl	1
	2	216.3	+10.8	25	2	H$_2$O	18,19
	3	315.4	-41.8	21	2.7	1 N HCl	19
-CH$_2$CH(CH$_3$)$_2$	1	131.2	+13.9	25	9.1	4.5 N HCl	1
	2	244.3	-13.4	23	1	1 N NaOH	18,22-24
	3	357.5	-51.4	20	3.1	1 N NaOH	26
	4	470.7	-90.0	20	7.6	1 N NaOH	26
-CH(CH$_3$)CH$_2$CH$_3$	1	131.2	+40.6	20		6 N HCl	1
	2	244.3	+17.1	25	1	H$_2$O + 1 eq HCl	20

2.3.2.4.2 Endgroup Protected Oligopeptides $(CH_3)_3COCO-\left[NHCH(CHCH_2CH_3)CO\atop \quad CH_3\right]_n-OCH_3$

n	Mol. Wt.	m.p. [^{o}C]	Specific Rotation		
			$[\alpha]_D^{24}$ in TFE*	$[\alpha]_D^{24}$ in HFA*	Ref.
2	358.4	158-159	-41.8	-37.6	56-58
3	471.6	190-191	-56.2	-62.5	56-58
4	584.8	240	-69.0	-70.8	56,57
5	698.0	240	-74.4	-95.8	56,57
6	811.2	240	-68.4	-98.0	56,57
7	924.4	240	+13.9	-103.9	56,57
8	1037.6	240	+24.6	-107.8	56,57

*TFE = Trifluoroethanol, HFA = Hexafluoroacetone sesquihydrate

2.3.2.4.3 Cyclic Oligopeptides $\left[NHCHRCO\right]_n$

R	n	Mol. Wt.	m.p. [^{o}C]	Specific Rotation				
				$[\alpha]_D$	[^{o}C]	Conc.	Solvent	Ref.
-CH$_2$CH(CH$_3$)$_2$	2	226.3	277	+48.7	20	8.7	CH$_3$COOH	25

2.3.2.5 Oligopeptides of L-Hydroxy and L-Mercapto Amino Acids

2.3.2.5.1 Linear Oligopeptides $H[NHCHRCO]_nOH$

R	n	Mol. Wt.	m.p. [°C]	$[\alpha]_D$	[°C]	Conc.	Solvent	Ref.
$-CH_2OH$	1	105.1		+14.5	25	9.3	N HCl	1
	2	192.2		+14.2	25	7	N HCl	28
$-CH_2C_6H_4OH$	1	181.2		-7.3	25	4	6.1 N HCl	1
	2	344.4		+30.1	19	4	H_2O + 1 eq HCl	29,30
	3	507.6	181-182					31
$-CH_2SH$	1	121.1		+7.6	26	12	N HCl	1
· 1.5 HCl	2	224.2		+35	22	1	0.2 N HCl	37
$-CH_2CH_2SCH_3$	1	149.2		+23.4	20	5	3 N HCl	1
	2	280.4		+27.0	24	2	H_2O	32-34
	3	411.6		-70.0	16	1	H_2O	33

2.3.2.5.2 Cyclic Oligopeptides $[NHCHRCO]_n$

R	n	Mol. Wt.	m.p. [°C]	$[\alpha]_D$	[°C]	Conc.	Solvent	Ref.
$-CH_2OH$	2	174.2	247	-67.5	25	2.2	H_2O	35
$-CH_2C_6H_4OH$	2	326.4	277-280	-223.8	20	2.4	NaOH	36

2.3.2.5.3 Oligopeptides of Cystine $CyS = \begin{bmatrix} -NHCHCO- \\ CH_2S \end{bmatrix}$

	Mol. Wt.	$[\alpha]_D$	[°C]	Conc.	Solvent	Ref.
H-CyS-CyS-OH	222.3	-29	25	1	N HCl	38
H-CyS-CyS-OH H-CyS-CyS-OH	444.6	-58.7	26	1	N HCl	37, 38

2.3.2.6 Oligopeptides of Basic L-Amino Acids $H[NHCHRCO]_nOH$

R	n	Mol. Wt.	R_F*	$[\alpha]_D$	[°C]	Conc.	Solvent	Ref.
$-(CH_2)_4NH_2$	1	146.2	0.24	+25.7	25	1.6	6 N HCl	1
	2	274.4	0.17	+5.6	25	2	6 N HCl	39, 40
	3	402.6	0.12	-2.2	24	2	0.5 N HCl	40, 41
	4	530.7	0.09					40
	5	658.9	0.06					40

*Solvent: n-Butanol/acetic acid/water/pyridine (30:6:24:20)

2.3.2.7 Oligopeptides of Acidic L-Amino Acids

2.3.2.7.1 α- and γ-Oligopeptides of Glutamic Acid $H[NH-X-CO]_nOH$

X	n	Mol. Wt.	$[\alpha]_D$	[°C]	Conc.	Solvent	Ref.
CH_2CH_2COOH -CH-	1	147.1	+31.2	22.4	1	6 N HCl	1
	2	276.3	+18.2	24	1-2	0.5 N HCl	42-44
	3	405.4	-7.2	19	1.4	H_2O	45

X	n	Mol. Wt.	$[\alpha]_D$	[°C]	Specific Rotation Conc.	Solvent	Ref.
COOH	2	276.3	+3.8	24	1-2	0.5 N HCl	43, 44, 46, 47
-CHCH$_2$CH$_2$-	3	405.4	-7.2	24	2	0.5 N HCl	48

2.3.2.7.2 Oligopeptides of Benzyloxycarbonyl-Aspartic Acid and Glutamic Acid C$_6$H$_5$CH$_2$OCO-[NHCHRCO]$_{n-1}$-NHCHR'COOC$_2$H$_5$

R	R'	n	Mol. Wt.	m.p. [°C]	$[\alpha]_D$	Specific Rotation Conc.	Ref.
-CH$_2$COOCH$_3$	-CH$_2$COOC$_2$H$_5$	2	452.5	80-81	+17.9	3.1	49
		3	581.6	127-128	-1.01	3.1	49
		4	710.7	143-144	-13.1	1.1	49
		5	839.8	161-163	-19.1	0.75	49
		6	968.9	175-178	-26.6	0.7	49
		8	1227.2	207 (d)	-35.1	0.5	49
		11	1614.5	224 (d)	-42.9	0.45	49
		14	2001.9	233 (d)	-46.0	0.3	49
-CH$_2$CH$_2$COOCH$_3$	-CH$_2$CH$_2$COOC$_2$H$_5$	2	480.5	86	-12.4	2	50
		3	623.6	124	-18.0	2	50
		4	766.8	139	-21.3	2	50
		5	910.0	200	-22.7	2	50
		6	1053.1	250	-26.7	2	50
		7	1196.3	259	-28.7	2	50
		9	1482.5	d	-32.6	2	50
		11	1768.8	d	-35.6	2	50

2.3.2.7.3 Oligopeptides of Benzyloxycarbonyl-γ-Glutamic Acid-α-Methyl Ester C$_6$H$_5$CH$_2$OCO-$\begin{bmatrix}$ NHCHCH$_2$CH$_2$CO $\\$ COOCH$_3$ $\end{bmatrix}_n$OCH$_3$

n	Mol. Wt.	m.p. [°C]	$[\alpha]_D^{20}$ (c=5, DMF)	Ref.	n	Mol. Wt.	m.p. [°C]	$[\alpha]_D^{20}$ (c=5, DMF)	Ref.
2	452.4	97-98	-25.2	59	5	881.7	175-179	-30.1	59
3	595.5	135-137	-26.1	59	6	1024.8	195-198	-31.5	59
4	738.6	135-139	-28.2	59	7	1153.9	206-213	-32.8	59

2.3.2.7.4 Oligopeptides of Benzyloxycarbonyl-γ-Glutamic Acid-α-tert.-Butyl Ester C$_6$H$_5$CH$_2$OCO-$\begin{bmatrix}$ NHCHCH$_2$CH$_2$CO $\\$ COOC(CH$_3$)$_3$ $\end{bmatrix}_n$-OC(CH$_3$)$_3$

n	Mol. Wt.	m.p. [°C]	$[\alpha]_D$	Specific Rotation [°C] (c=1, CH$_3$OH)	Ref.
2	561.7	83-84	-25.4	22	71
3	751.9	79	-30.2	21	71
4	937.1	87-89	-33.2	22	71
5	1122.4	102-103.5	-36.8	23	71
6	1307.6	123-125	-39.9	22.5	71
7	1504.8	130-132	-43.7	21	71

2.3.2.8 Oligopeptides of L-Proline

2.3.2.8.1 Linear Oligopeptides

$$H\left[N-CHCO \right]_n OH$$

n	Mol. Wt.	m.p. [°C]	E_F-Werte[*]					$[\alpha]^{22}_D$ (c=1, H_2O)	Ref.
1	115.2	220-222	1.00					-86.5	1
2	212.3	144-145	0.84					-171	52, 62, 69
3	309.4	122	0.72					-220	62, 69, 53
4	406.5	170	0.60					-291	62, 69
5	503.6	189	0.52	1.00				-338	62, 69
6	600.7	209		0.91				-374	62, 69
7	697.8	228		0.77				-394	62, 69
8	794.9	260		0.72				-412	62, 69
9	892.0	260		0.66				-435	62, 69
10	989.1	280		0.60	1.00			-456	62, 69
11	1086.2	280			0.92			-464	62, 69
12	1183.3	280			0.83			-472	62, 69
13	1280.4	300			0.75	1.00		-480	69
14	1377.5	300			0.70	0.94		-491	69
15	1474.6	300			0.64	0.90		-496	62, 69
16	1571.7	300				0.86		-499	69
17	1668.8	300				0.81	1.00	-505	69
18	1765.9	300					0.95	-509	69
19	1863.0	300					0.90	-513	69
20	1960.1	300					0.84	-520	69
22	2154.5							-523	70
25	2445.7							-526	70
28	2737.0							-528	70
31	3028.5							-532	70
34	3320.1							-530	70
37	3611.4							-528	70
40	3902.7							-532	70

[*] Paper electrophoresis; buffer: $HCOOH/CH_3COOH/H_2O$ (1:1:3), 1000 - 5000 V, 40-115mA, 40-240 min.

2.3.2.8.2 Endgroup Protected Oligopeptides

2.3.2.8.2.1 tert.-Butyloxycarbonyl Oligoprolines

$$(CH_3)_3COCO-\left[N-CHCO \right]_n -OH$$

n	Mol. Wt.	m.p. [°C]	R_F[*]	Specific Rotation $[\alpha]^{22}_D$, c=1 (in $CHCl_3$)	Ref.
1	215.3	137-138	0.84	-105	62, 72
2	312.4	191-193	0.68	-138	62
3	409.5	219-220	0.60	-173	62
4	506.6	178	0.53	-186	62
5	603.7	224	0.47	-208	62
6	700.8	247	0.41	-232	62
7	797.9	267	0.37	-242	62
8	894.0	>280	0.33	-256	62
9	991.1	>280	0.29	-263	62
10	1088.2	>280	0.25	-267	62
11	1185.3	>300	0.23	-269	62
12	1282.4	>300	0.22	-272	62
15	1573.7	>300	0.22	-280	62

[*] Solvent: n-Butanol/acetone/acetic acid/ammonia, 1:4/water (9:3:2:2:4)

2.3.2.8.2.2 tert.-Amyloxycarbonyl Oligoprolines

$$C_5H_{11}OCO-\left[\overset{\square}{N}-CHCO\right]_n-OH$$

n	Mol. Wt.	m.p. [$^{\circ}$C]	$[\alpha]_D$	[$^{\circ}$C]	Specific Rotation Conc.	Solvent	Ref.
1	229.3	94-95.5	-47.2	21	1.7		63-68
2	326.4	157-159	-113	25	2.0	EtOH	64-68
3	423.5	200-201.5	-170	25	2.2	EtOH	64-68
4	520.6	207.5-208.5	-227	24	2.1	EtOH	64-68
5	617.7	224.5-226	-248	24	2.3	EtOH	64-68
6	714.8	240-242	-291	24	2.1	EtOH	64-68
8	908.0	280	-318	22	0.9	EtOH	64-68

REFERENCES

1. J. P. Greenstein, M. Winitz, "Chemistry of the Amino Acids", Vol. III, J. Wiley & Sons, New York, 1961.
2. E. Fischer, E. Fourneau, Chem. Ber. 34, 2868 (1901).
3. E. Fischer, Chem. Ber. 36, 2982 (1903).
4. M. Rothe, H. Bruenig, G. Eppert, J. Prakt. Chem. 8, 323 (1959).
5. E. Fischer, Chem. Ber. 37, 2486 (1904).
6. E. Fischer, Chem. Ber. 39, 453 (1906).
7. H. F. Schott, J. B. Larkin, L. B. Rockland, M. S. Dunn, J. Org. Chem. 12, 490 (1947).
8. R. Schwyzer, B. Iselin, W. Rittel, P. Sieber, Helv. Chim. Acta 39, 872 (1956).
9. M. Rothe, G. Luedke, unpublished data.
10. J. C. Sheehan, W. L. Richardson, J. Am. Chem. Soc. 76, 6329 (1954); J. C. Sheehan, M. Goodman, W. L. Richardson, J. Am. Chem. Soc. 77, 6391 (1955).
11. D. G. Ballard, C. H. Bamford, F. J. Weymouth, Proc. Roy. Soc. 227 A, 155 (1955).
12. B. F. Erlanger, E. Brand, J. Am. Chem. Soc. 73, 3508 (1951).
13. F. Sigmund, F. Liedl, Hoppe-Seyler's Z. Physiol. Chem. 202, 268, 275 (1931).
14. E. Brand, B. F. Erlanger, H. Sachs, J. Am. Chem. Soc. 73, 3510 (1951).
15. H. Zahn, A. Meißner, Ann. Chem. 636, 132 (1960).
16. E. Brand, B. F. Erlanger, H. Sachs, J. Am. Chem. Soc. 74, 1849 (1952).
17. M. Rothe, R. Theysohn, unpublished data.
18. M. A. Nyman, R. M. Herbst, J. Org. Chem. 15, 108 (1950).
19. S. Shankman, Y. Schvo, J. Am. Chem. Soc. 80, 1164 (1958).
20. T. Sugimura, W. K. Paik, unpublished data.
21. S. P. Datta, R. Lebermann, B. R. Ratin, Trans. Faraday Soc. 55, 1982 (1959).
22. E. Fischer, Chem. Ber. 39, 2893 (1906).
23. F. H. Carpenter, D. T. Gish, J. Am. Chem. Soc. 74, 3818 (1952).
24. F. C. McKay, N. F. Albertson, J. Am. Chem. Soc. 79, 4686 (1957).
25. E. Fischer, A. H. Koelker, Ann. Chem. 354, 39 (1907).
26. E. Abderhalden, R. Fleischmann, Fermentforsch. 9, 524 (1928).
27. Y. Shibata, T. Asahina, Bull. Chem. Soc. Japan 1, 71 (1927).
28. J. S. Fruton, J. Biol. Chem. 146, 463 (1942).
29. M. Bergmann, L. Zervas, L. Salzmann, H. Schleich, Hoppe-Seyler's Z. Physiol. Chem. 224, 17 (1934).
30. J. L. Bailey, J. Chem. Soc. 1950, 3461.
31. A. E. Barkdoll. W. F. Ross, J. Am. Chem. Soc. 66, 951 (1944).
32. C. A. Dekker, S. P. Taylor, J. S. Fruton, J. Biol. Chem. 180, 155 (1949).
33. M. Brenner, R. W. Pfister, Helv. Chim. Acta 34, 2085 (1951).
34. H. B. Milne, C.-H. Peng, J. Am. Chem. Soc. 79, 645 (1957).
35. E. Fischer, Chem. Ber. 40, 1501 (1907); E. Fischer, W. A. Jacobs, Chem. Ber. 39, 2942 (1906).
36. E. Fischer, W. Schrauth, Ann. Chem. 354, 21 (1907).
37. J. P. Greenstein, J. Biol. Chem. 121, 9 (1937).
38. N. Izumiya, J. P. Greenstein, Arch. Biochem. Biophys. 52, 203 (1954); R. Wade, M. Winitz, J. P. Greenstein, J. Am. Chem. Soc. 78, 373 (1956).
39. B. F. Erlanger, E. Brand, J. Am. Chem. Soc. 73, 4025 (1951); 72, 3314 (1950).
40. S. G. Waley, J. Watson, J. Chem. Soc. 1953, 475.
41. E. Brand, B. F. Erlanger, J. Polatnik, H. Sachs, D. Kirschenbaum, J. Am. Chem. Soc. 73, 4027 (1957).
42. M. Bergmann, L. Zervas, Chem. Ber. 65, 1192 (1932).
43. S. Goldschmidt, C. Jutz, Chem. Ber. 89, 518 (1956).
44. H. Sachs, E. Brand, J. Am. Chem. Soc. 75, 4608 (1953).
45. D. A. Rowlands, G. T. Young, Biochem. J. 65, 516 (1957).
46. V. Bruckner, M. Szekerke, J. Kovacs, Hoppe-Seyler's Z. Physiol. Chem. 309, 25 (1957).
47. W. L. LeQuesne, G. T. Young, J. Chem. Soc. 1950, 1959.
48. H. Sachs, E. Brand, J. Am. Chem. Soc. 76, 1811 (1954).
49. M. Goodman, F. Boardman, J. Listowsky, J. Am. Chem. Soc. 85, 2483, 2491 (1963).
50. M. Goodman, E. E. Schmitt, D. A. Yphantis, J. Am. Chem. Soc. 84, 1283, 1288 (1962).
51. J. Dale, K. Titlestad, Chem. Commun. 1969, 656.
52. N. C. Davis, E. L. Smith, J. Biol. Chem. 200, 373 (1953).
53. M. Rothe, K.-D. Steffen, I. Rothe, Angew. Chem. 77, 347 (1965).
54. J. Kapfhammer, A. Matthes, Hoppe-Seyler's Z. Physiol. Chem. 223, 43 (1934).
55. E. Abderhalden, H. Nienburg, Fermentforsch. 13, 573 (1933).
56. C. Toniolo, Biopolymers 10, 1707 (1971).
57. U. Widmer, G. P. Lorenzi, Chimia 25, 236 (1971).
58. J. E. Shields, S. T. McDowell, J. Pavlos, G. R. Gray, J. Am. Chem. Soc. 90, 3549 (1968).
59. M. Kajtár, M. Hollósi, Acta Chim. Acad. Sci. Hung. 65, 403 (1970).
60. M. Kajtár, M. Hollósi, G. Snatzke, Tetrahedron 27, 5659 (1971).
61. S. M. McElvain, P. M. Laughton, J. Am. Chem. Soc. 73, 448 (1951).
62. M. Rothe, R. Theysohn, K.-D. Steffen, Tetrahedron Letters 1970, 4063.
63. S. Sakakibara, M. Shin, M. Fujimo, Y. Shimonishi, S. Inoue, N. Inukai, Bull. Chem. Soc. Japan 38, 1522 (1965).
64. M. Miyoshi, T. Kimura, S. Sakakibara, Bull. Chem. Soc. Japan 43, 2941 (1970).

65. T. Isemura, H. Okabayashi, S. Sakakibara, Biopolymers 6, 307
     (1968).

66. H. Okabayashi, T. Isemura, S. Sakakibara, Biopolymers 6, 323
     (1968).

67. H. Okabayashi, T. Isemura, Bull. Chem. Soc. Japan 43, 20
     (1970).

68. H. Okabayashi, T. Isemura, Bull. Chem. Soc. Japan 43, 359
     (1970).

69. M. Rothe, J. Mazának, Tetrahedron Letters 1972, 3795

70. M. Rothe, J. Mazának, unpublished data.

71. J. Meienhofer, P. M. Jacobs, H. A. Godwin, I. H. Rosenberg,
     J. Org. Chem. 35, 4137 (1970).

72. E. Schnabel, Ann. Chem. 702, 188 (1967).

2.3.3 OLIGO(IMINES)

2.3.3.1 Linear Oligo(imines)

2.3.3.1.1 Linear Oligo(ethylene imines) $H[NHCH_2CH_2]_n NH_2$

n	Mol. Wt.	m.p. [$^\circ$C]	b.p. [$^\circ$C/mm]	d_{20}^{20}	n_D^{25}	Ref.
1	60.1	8.5	116.5	0.8994	1.4536	1,2
2	103.1		207.1	0.9586	1.4810	1-3
3	146.2	12.0	277.9	0.9839	1.4951	1-4
4	189.3		333	0.9994	1.5015	1-3
7	318.5		109-110/8.5		1.5132	2,5
9	404.7		199-200/1		1.5161	2,5

2.3.3.1.2 Linear Oligo(alkylene imines) $R-[NH(CH_2)_6 NH(CH_2)_{10}]_n -R'$

R	R'	n	Mol. Wt.	m.p. [$^\circ$C]	Ref.
H	OH	1	272.5	64-67	6
		2	526.9	70-72	6
		3	781.4	85-86	6
$(CH_2)_{10}OH$	OH	1	428.8	99-100	6
		2	683.2	90-93	6
		3	937.7	80-84	6
H	$NH(CH_2)_6 NH_2$	1	370.7	70-72	6
		2	625.1	86-88	6
		3	879.6	81-85	6

2.3.3.2 Cyclic Oligo(imines) $[NH(CH_2)_x]_n$

x	n	Mol. Wt.	m.p. [$^\circ$C]	b.p. [$^\circ$C/mm]	$d_4/^\circ$C	Ref.
2	1	43.1		55-56	0.8321/24	2,7,8
	2	86.1	104	145-146		8,9
	4	172.3	35	110/10^{-4}		10
3	1	57.1		63	0.8436/20	11
	2	114.2	14-15	186-188		8,11-13
4	1	71.1		88	0.8520/22.5	14
	2	142.2		95/12	0.9020/18	8,13
5	1	85.2	-9	106	0.8606/20	14
	2	170.3		108-110/12	0.9195/13	8
6	1	99.2		138	0.8864/21	8,15,16
	2	198.3	72			13,17,18
	3	297.5	42			17
	4	396.7	59-60			17
	5	495.9	45			17
	6	595.1	67-68			17
7	1	113.1	-33	162	0.8895/20	16,19
	2	226.2	26		0.9012/30	19
8	1	127.2		90/24	0.9021/21	16
	2	254.5	55			20
9	1	141.3			0.8982/21	16
	2	282.5	38			19
13	1	197.4	50-51			16
	2	394.7	52			21

REFERENCES

1. A. L. Wilson, Ind. Eng. Chem. 27, 870 (1935).
2. G. D. Jones, A. Langsjoen, M. M. Ch. Neumann, J. Zomlefer, J. Org. Chem. 9, 125 (1944).
3. H. B. Jonassen, T. B. Crumpler, T. D. O. Brien, J. Am. Chem. Soc. 67, 1709 (1945).
4. F. G. Mann, J. Chem. Soc. 1934, 461.
5. G. S. Whitby, N. Wellman, V. W. Floutz, L. H. Stephens, Ind. Eng. Chem. 42, 445 (1950).
6. H. Zahn, G. B. Gleitsmann, Makromol. Chem. 63, 129 (1963).
7. C. F. H. Allen, F. W. Spangler, E. R. Webster, Org. Syn. 30, 38 (1950).
8. J. v. Braun, G. Blessing, F. Zobel, Chem. Ber. 57, 185 (1924).
9. A. Ladenburg, J. Abel, Chem. Ber. 21, 758 (1888).
10. H. Stetter, K.-H. Mayer, Chem. Ber. 94, 1410 (1961).
11. C. C. Howard, W. Marckwald, Chem. Ber. 32, 2031, 2038 (1899).

12. E. L. Buehle, J. Am. Chem. Soc. 65, 29 (1943).
13. H. Stetter, J. Marx, Ann. Chem. 607, 59 (1958); H. Stetter, H. Spangenberger, Chem. Ber. 91, 1982 (1958).
14. T. S. Stevens, in E. R. Rodd, Ed., "Chemistry of Carbon Compounds", Vol. IV, Part A, 1st ed., Elsevier, Amsterdam-London-New York, 1953, p. 61; N. Campbell, ibid. p. 570.
15. K. Ziegler, Ph. Orth, Chem. Ber. 66, 1867 (1933).
16. L. Ruzicka, M. Kobelt, O. Haefliger, V. Prelog, Helv. Chim. Acta 32, 544 (1949).
17. H. Zahn, H. Spoor, Chem. Ber. 89, 1296 (1956); 92, 1375 (1959).
18. M. A. Th. Rogers, Nature (London) 177, 128 (1956).
19. A. Mueller, E. Srepel, E. Funder-Fritzsche, F. Dicker, Monatsh. Chem. 83, 386 (1952).
20. A. Mueller, L. Kindlmann, Chem. Ber. 74, 416 (1941).
21. A. Mueller, Chem. Ber. 67, 295 (1934).

2.3.4　OLIGO(UREAS)

2.3.4.1　Oligo(methylene ureas)　$H[NHCONHCH_2]_n NHCONH_2$

n	Mol. Wt.	m.p. [$^{\circ}$C]	Ref.
1	132.1	218 (d)	1,2
2	204.2	227	1
3	276.3	231-233	1,2
5	420.5	236 (d)	2

2.3.4.2　Oligo(methylene thioureas)　$H[NHCSNHCH_2]_n NHCSNH_2$

n	Mol. Wt.	m.p. [$^{\circ}$C]	Ref.
1	164.3	198 (d)	3,4
2	252.4	210 (d)	3
3	340.5	215 (d)	3

2.3.4.3　Oligo(methylol thioureas)　$R[NHCSNHCH_2]_n OH$

R	n	Mol. Wt.	m.p. [$^{\circ}$C]	Ref.
H	1	106.2	104-105	3,4
	2	194.3	190-192 (d)	3

R	n	Mol. Wt.	m.p. [$^{\circ}$C]	Ref.
-CH$_2$OH	1	136.2	92	3
	2	224.3	132	3,4

REFERENCES

1. A. A. Wanscheidt, S. K. Naumova, J. P. Melnikowa, Zh. Obshch. Khim. 10, 1968 (1940); through C. 1941 II 184.
2. H. Kadowaki, Bull. Chem. Soc. Japan 11, 248 (1936).

3. H. J. Becher, F. Griffel, Chem. Ber. 91, 691 (1958).
4. H. Staudinger, K. Wagner, Makromol. Chem. 12, 168 (1954).

3.　CARBON CHAIN OLIGOMERS CONTAINING MAIN CHAIN CYCLIC UNITS

3.1　Oligo(cyclopentylenes)

n	Mol. Wt.	m.p. [$^{\circ}$C]	b.p. [$^{\circ}$C/mm]	d_4^{20}	Ref.
1	70.1	-93.9	42.3	0.7510	
2	138.3		190	0.8646	1,2
3	206.3		293-294	0.9177	1,2
4	274.5		369-370	0.9564	1,2
6	410.7	143-146	235/0.1		1

3.2 Oligo(spiranes)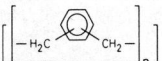

R	n	Mol. Wt.	m.p. [°C]	b.p. [°C/mm]	Ref.
-CH$_2$-	5	216.3		78/0.3	3
-(CH$_2$)$_2$-	3	244.4		84/0.05	3
	5	324.5	61		3
	7	406.6	107-108		3
-(CH$_2$)$_3$-	3	272.5	52.5	120-122/0.05	3
	5	352.6	79.5		3
	7	432.7	103-104		3
	9	512.8	138-140		3

3.3 Oligo(xylylenes)

3.3.1 Linear Oligo(xylylenes)

	n	Mol. Wt.	m.p. [°C]	b.p. [°C/mm]	Ref.
o-	1	106.2	-25	144	
	2	210.3	66.5	177-178/20	5-7
m-	1	106.2	-47.4	139	
	2	210.3		296	7,8
p-	1	106.2	13-14	138	
	2	210.3	82	178/18	7,9-12
	3	314.5	141-142		10,13

3.3.2 Cyclic Oligo(xylylenes)

	n	Mol. Wt.	m.p. [°C]	b.p. [°C]	Ref.
o-	2	208.3	112-112.5		5,6,14
	3	312.5	184.5		5,6
	4	416.6	205		15
m-	2	208.3	132-133	290	10
p-	2	208.3	285-287		8,11-13,19
	3	312.5	166-167		10,13
	4	416.6	179-182		13

3.4 Oligo(stilbenes)

n	Mol. Wt.	m.p. [°C]	Ref.
1	180.2	124	17
2	282.4	265	16-18
3	384.5	350-352	17
4	486.6	360-374	16,17
5	588.7	395-400	16,17
6	690.9	415-420	16,17
7	793.0	425	16,17

3.5 Oligo(benzyls)

n	Mol. Wt.	m.p. [°C]	Ref.
1	168.2	25.1	
2	258.3	82	20,23
3	348.4	90	21,23
4	438.6	120	23
5	528.7	153-156	22,23

3.6 Oligo-(2,5-dimethyl-benzyls)

n	Mol. Wt.	m.p. [°C]	Ref.
1	224.3	60-61	22,24
2	342.5	153-154	22
4	578.8	234-236	22
6	815.1	262-266	22

3.7 Oligo(2,3,5,6-tetramethyl-benzyls)

n	Mol. Wt.	m.p. [°C]	Ref.
1	280.5	155-156	22,25
2	426.7	263-264	22,26
3	572.9	305-307	22
4	719.1	335-337	22

3.8 Oligo(p-phenylene oxides)

n	Mol. Wt.	m.p. [°C]	Ref.
1	170.2	28	
2	262.3	73	23,27
3	354.4	105	23,27
4	446.5	142	23,27

3.9 Oligo(p-phenylene sulfides)

n	Mol. Wt.	m.p. [°C]	Ref.
1	186.4	-21.5	
2	294.5	80	23,28
3	402.6	110	23
4	510.7	150	23

3.10 Oligo(p-phenoxyphenylmethanes) H—[—⬡—CH₂—⬡—O—]ₙ—⬡—CH₂—⬡   3.11 Oligo(diphenylmethanes) H—[—⬡—CH₂—⬡—]ₙ—H

n	Mol. Wt.	m.p. [$^{\circ}$C]	Ref.
1	350.4	59	23
2	532.7	101	23

n	Mol. Wt.	m.p. [$^{\circ}$C]	Ref.
1	168.2	25.1	
2	334.5	115	4,29
3	500.7	180	4
4	666.9	240	4

REFERENCES

1. J. v. Braun, J. Reitz-Kopp, Chem. Ber. 74, 1105 (1941).
2. G. E. Goheen, J. Am. Chem. Soc. 63, 744 (1941).
3. E. Buchta, K. Geibel, Ann. Chem. 678, 53 (1964).
4. M. Busch, W. Weber, J. Prakt. Chem. 146, 1 (1936).
5. W. Baker, R. Banks, D. R. Lyon, F. G. Mann, J. Chem. Soc. 1945, 27.
6. A. C. Cope, S. W. Fenton, J. Am. Chem. Soc. 73, 1668 (1951).
7. G. H. Coleman, W. H. Holst, R. D. Maxwell, J. Am. Chem. Soc. 58, 2310 (1936).
8. M. Swarcz, J. Chem. Phys. 16, 128 (1948).
9. T. Reichstein, R. Oppenauer, Helv. Chim. Acta 16, 1373 (1933).
10. W. Baker, J. F. W. McOmie, J. M. Norman, J. Chem. Soc. 1951, 1114.
11. C. J. Brown, A. C. Farthing, Nature (London) 164, 915 (1949); J. Chem. Soc. 1953, 3261, 3265, 3270.
12. D. J. Cram, H. Steinberg, J. Am. Chem. Soc. 73, 5691 (1951).
13. L. A. Errede, R. S. Gregorian, J. M. Hoyt, J. Am. Chem. Soc. 82, 5218 (1960).
14. L. A. Errede, J. Am. Chem. Soc. 83, 949 (1961).
15. E. D. Bergmann, Z. Pelchowicz, J. Am. Chem. Soc. 75, 4281 (1953).
16. G. Drefahl, G. Ploetner, Chem. Ber. 94, 907 (1961); 91, 1274 (1958).

17. G. Drefahl, R. Kuehmstedt, H. Oswald, H.-H. Hoerhold, Makromol. Chem. 131, 89 (1970).
18. T. W. Campbell, R. N. McDonald, J. Org. Chem. 24, 1246 (1959).
19. H. E. Winberg, F. S. Fawcett, W. E. Mochel, C. W. Theobald, J. Am. Chem. Soc. 82, 1428 (1960).
20. A. B. Galun, A. Kaluszymer, E. D. Bergmann, J. Org. Chem. 27, 1426 (1961).
21. E. Connerade, Bull. Soc. Chim. Belges 44, 411 (1935).
22. G. Montaudo, F. Bottino, S. Caccamese, P. Finocchiaro, G. Bruno, J. Polymer Sci. A-1, 8, 2453 (1970).
23. G. Montaudo, G. Bruno, P. Maravigna, P. Finocchiaro, G. Centineo, J. Polymer Sci., in press.
24. R. C. Huston, D. T. Ewing, J. Am. Chem. Soc. 37, 2394 (1915).
25. C. M. Welch, H. A. Smith, J. Am. Chem. Soc. 73, 4391 (1951).
26. H. Kaemmerer, M. Harris, Makromol. Chem. 66, 215 (1963).
27. H. Staudinger, F. Staiger, Ann. Chem. 517, 67 (1935).
28. E. Bourgeois, A. Fonassin, Bull. Soc. Chim. France 9, 941 (1911).
29. N. Wolf, Chem. Ber. 14, 2031 (1881).

3.12 PHENOL-FORMALDEHYDE AND RELATED OLIGOMERS

3.12.1 Linear Phenol-Formaldehyde Oligomers

3.12.1.1 Phenol-formaldehyde Oligomers

n	Mol. Wt.	m.p. [$^{\circ}$C]	Ref.
1	200.2	119-120	1-4
2	306.4	158-159	2-6
3	412.5	161-162	3,4

n	Mol. Wt.	m.p. [$^{\circ}$C]	Ref.
4	518.6	148-150	4,7
5	624.7	203-204	4
6	730.8	213-214	4

3.12.1.2 p-Cresol-formaldehyde Oligomers

3.12.1.2.1

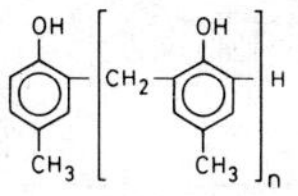

n	Mol. Wt.	m.p. [°C]	Ref.		n	Mol. Wt.	m.p. [°C]	Ref.
1	228.3	126	8-10, 56-58		6	829.1	225-230	12
2	348.4	214-215	8-11, 56-58		7	949.2	167-170	12
3	468.6	173	8, 9, 56, 57		8	1069.4	205-210	12
4	588.7	130	12, 56, 57		9	1189.5	175-180	12
5	708.9	215-217	12		11	1429.8	245	12

3.12.1.2.2

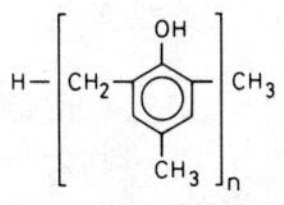

n	Mol. Wt.	m.p. [°C]	Ref.		n	Mol. Wt.	m.p. [°C]	Ref.
1	136.1	70-72	61		4	496.7	204-205	11
2	256.3	148	11, 13-16		5	616.9	152-154	61
3	376.5	183-184	11					

3.12.1.3 p-tert. Butylphenol-formaldehyde Oligomers

3.12.1.3.1

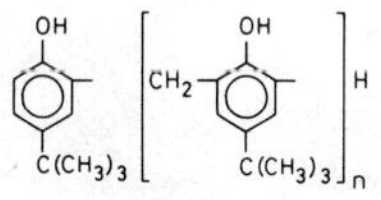

n	Mol. Wt.	m.p. [°C]	Ref.		n	Mol. Wt.	m.p. [°C]	Ref.
1	312.5	156	17, 18, 62		6	1123.6	249-250	17, 18
2	474.7	218-220	17, 18, 62		7	1285.9	253-256	18
3	636.9	211	17, 18, 62		8	1448.1	224-226	17
4	799.2	216-217	17, 18, 62		10	1772.6	140	17
5	961.4	250	17, 18, 62					

3.12.1.3.2

n	Mol. Wt.	m.p. [°C]	Ref.		n	Mol. Wt.	m.p. [°C]	Ref.
0	220.4	70-71	61		2	460.7	82-84	61
1	340.5	125-130	61		4	701.0	188-190	61

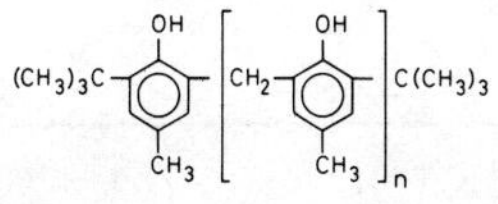

3.12.1.4 2,4,6-Mesitol-formaldehyde Oligomers

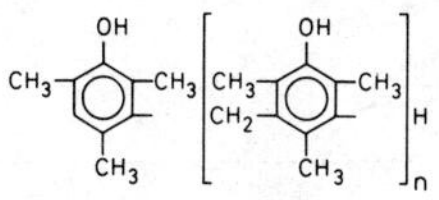

n	Mol. Wt.	m.p. [°C]	Ref.
1	284.4	188	19
2	432.6	257	19

3.12.1.5 4-Carboxy-phenol-formaldehyde Oligomers

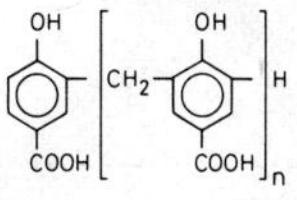

n	Mol. Wt.	m.p. [°C]	Ref.
1	288.3	305-307	59
2	438.4	310	59

3.12.1.6 4-Bromo-phenol-formaldehyde Oligomers

n	Mol. Wt.	m.p. [°C]	Ref.
1	358.1	183-184	1, 60
2	543.1	237-238	60

3.12.1.7 p-Cresol-acetaldehyde Oligomers

R	n	Mol. Wt.	m.p. [°C]	Ref.
H	1	241.3	141	20, 21
H	2	373.5	214-215	21
CH$_3$	1	270.4	135-135.5	22
CH$_3$	2	404.6	154-155	63
CH$_3$	3	538.7	204-205	63

3.12.1.8 Oligomers with Carbonyl and Sulfonyl Bridges

3.12.1.8.1

n	Mol. Wt.	m.p. [°C]	Ref.
1	242.3	106-107	23, 24, 56-58
2	376.4	122-123	24, 56, 57

3.12.1.8.2

n	Mol. Wt.	m.p. [°C]	Ref.
1	347.2	179	25
2	517.4	238-240	25
3	687.6	246-252	25

3.12.2 Oligomeric Phenol Alcohols

3.12.2.1 Phenol Monoalcohols

n	Mol. Wt.	m.p. [°C]	Ref.
1	124.1	86-87	3, 26
2	230.3	121.5-123	3, 27, 28

3.12.2.2 Phenol Dialcohols

n	Mol. Wt.	m.p. [°C]	Ref.
1	154.2	101	29-32
2	260.3	126-127	28

3.12.2.3 p-Cresol Monoalcohols

n	Mol. Wt.	m.p. [°C]	Ref.
1	138.2	107	8, 33, 34
2	258.3	148	8, 9, 35
4	498.6		36

3.12.2.4 p-Cresol Dialcohols

n	Mol. Wt.	m.p. [°C]	Ref.
1	168.2	133-134	13, 37
2	288.3	151.5	9, 35, 38, 39
3	408.5	203	9, 39, 40

3.12.2.5 p-tert. Butylphenol Dialcohols

n	Mol. Wt.	m.p. [°C]	Ref.
1	210.3	74-75	41
2	372.5	117-118	18

3.12.2.6 o-Hydroxydibenzyl Ethers

n	Mol. Wt.	m.p. [°C]	Ref.
2	318.4	85	42

References page VI-42

3.12.3 Cyclic Phenol-formaldehyde Oligomers

3.12.3.1

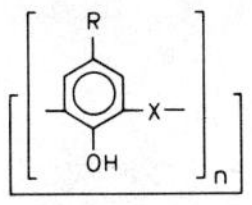

X	R	n	Mol. Wt.	m.p. [^oC]	Ref.
$-CH_2-$	$-CH_3$	4	480.6	300	36, 43
	$-C(CH_3)_3$	4	648.9	300	18, 44
	$-C_6H_5$	4	728.9	300-360 (d)	18
	$-C_6H_{11}$	4	753.1	330	18
	$-CH_2C_6H_5$	4	785.0	330	18
	$-C(CH_3)_2CH_2CH_3$	4	705.0	280	18
	$-C(CH_3)_2CH_2C(CH_3)_2$	4	813.2	333	18
$-CH_2OCH_2-$	$-CH_3$	4	600.7	264-266	42
	$-C(CH_3)_3$	3	576.8	245	45

3.12.3.2

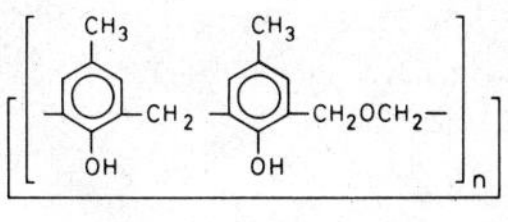

n	Mol. Wt.	m.p. [^oC]	Ref.
2	540.7	325-330	46

3.12.3.3

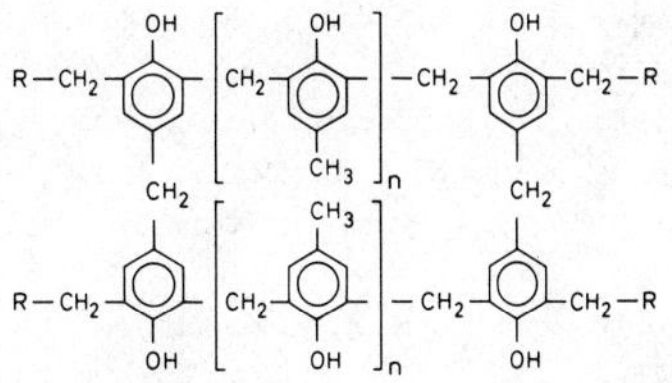

R	n	Mol. Wt.	Ref.
H	2	916.2	47
	1	1201.5	43, 47

3.12.4 Branched Phenol-formaldehyde Oligomers

3.12.4.1

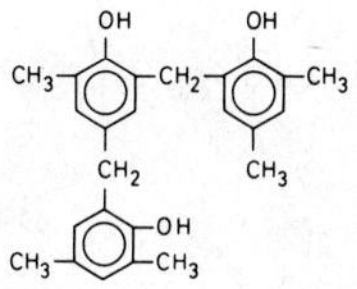

Mol. Wt.	m.p. [^oC]	Ref.
376.5	158	48

3.12.4.2

R_1	R_2	Mol. Wt.	m.p. [^oC]	Ref.
H	H	411.5		49
H	CH_3	453.6	184-187	30, 50
CH_3	CH_3	495.6	185-186	51

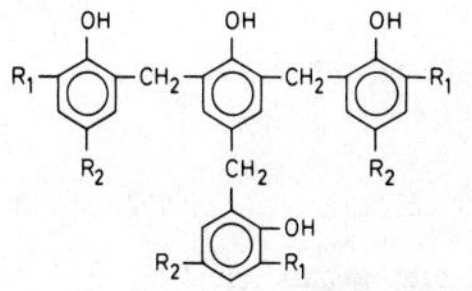

3.12.4.3

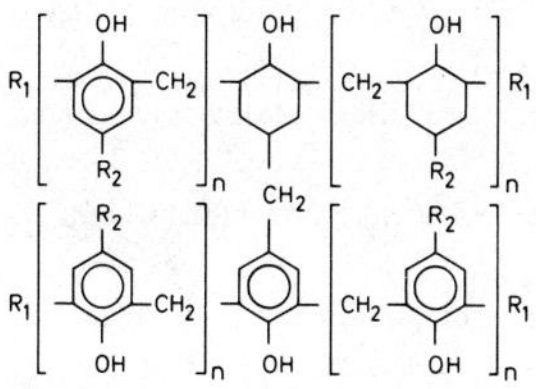

R_1	R_2	n	Mol. Wt.	m.p. [^oC]	Ref.
H	H	1	624.7	185-187	52, 53
CH_3	CH_3	1	737.0	207-208	6, 53, 54
CH_3	CH_3	2	1217.5	190-191	54

PHYSICAL DATA OF OLIGOMERS

3.12.5 Hydroquinone Oligomers

3.12.5.1

3.12.5.2

n	Mol. Wt.	m.p. [$^{\circ}$C]	Ref.
1	300.4	97-98	64
2	462.5	146-147	64

R	n	Mol. Wt.	m.p. [$^{\circ}$C]	Ref.
H	0	300.4	97-98	64
	1	462.5	192-193	64
CH$_3$	0	328.4	162-163	64
	1	490.6	242-243	64
	2	652.8	283-284	64
	3	815.0	306-307	64
	4	977.2	326-328	64

REFERENCES

1. C. A. Buehler, D. E. Copper, E. O. Scrudder, J. Org. Chem. 8, 316 (1943).
2. S. R. Finn, J. W. James, C. J. S. Standen, J. Appl. Chem. 4, 497 (1954); S. R. Finn, G. Lewis, N. J. L. Megson, J. Soc. Chem. Ind. 69, 551 (1950); S. R. Finn, G. Lewis, J. Appl. Chem. 1, 524 (1954).
3. D. A. Fraser, R. W. Hall, P. A. Jenkins, A. L. Raum, J. Appl. Chem. 7, 689 (1957).
4. H. Kaemmerer, H. Lenz, Makromol. Chem. 27, 162 (1958).
5. H. L. Bender, A. G. Franham, J. W. Guyer, F. N. Apel, T. B. Gibb, Jr., Ind. Eng. Chem. 44, 1619 (1952).
6. A. C. Davis, B. T. Hayes, R. F. Hunter, J. Appl. Chem. 3, 312 (1953).
7. S. Seto, H. Horiuchi, A. Takahashi, J. Chem. Soc. Japan, Ind. Chem. Sect. 58, 378 (1955).
8. H. Kaemmerer, W. Rausch, Makromol. Chem. 24, 152 (1957).
9. M. Koebner, Z. Angew. Chem. 46, 251 (1953).
10. N. J. L. Megson, A. A. Drummond, J. Soc. Chem. Ind. 49, 251 T (1930).
11. E. Ziegler, I. Hontschitz, L. Milowitz, Monatsh. Chem. 78, 334 (1948).
12. H. Kaemmerer, W. Rausch, H. Schweikert, Makromol. Chem. 56, 123 (1962).
13. K. v. Auwers, Chem. Ber. 40, 2524 (1907).
14. K. Fries, K. Kann, Ann. Chem. 353, 335 (1907).
15. K. Hultzsch, Chem. Ber. 74, 898 (1941).
16. A. Zinke, E. Ziegler, Chem. Ber. 74, 541 (1941).
17. H. Kaemmerer, K. Haberer, Monatsh. Chem. 95, 1589 (1964).
18. A. Zinke, R. Kretz, E. Leggewie, K. Hoessinger, Monatsh. Chem. 83, 1213 (1952).
19. S. R. Finn, J. W. G. Musty, J. Soc. Chem. Ind. 69, Suppl. No. 1, S 3 (1950).
20. E. Adler, H. v. Euler, G. J. Gie, Arkiv Kemi 16A, No. 12 (1943).
21. L. M. Debing, Trans. Electrochem. Soc. 90, 277 (1946).
22. H. v. Euler, E. Adler, J. O. Cedwall, Arkiv Kemi 15A, No. 19 (1942).
23. A. v. Bayer, V. Drewsen, Ann. Chem. 212, 344 (1882).
24. H. Kaemmerer, G. Buesing, H.-G. Haub, Makromol. Chem. 66, 82 (1963).
25. H. Kaemmerer, M. Harris, Makromol. Chem. 62, 18 (1963); J. Polymer Sci. A, 2, 4003 (1964).
26. S. Lederer, J. Prakt. Chem. 50, 223 (1894).
27. S. R. Finn, J. W. James, C. J. S. Standen, Chem. Ind. 1954, 188; J. Appl. Chem. 4, 296 (1954).
28. A. T. Carpenter, R. F. Hunter, J. Appl. Chem. 3, 486 (1953).
29. S. R. Finn, J. W. G. Musty, J. Appl. Chem. 2, 88 (1952).
30. J. H. Freeman, J. Am. Chem. Soc. 74, 6257 (1952).
31. H. Kaemmerer, M. Grossmann, Chem. Ber. 86, 1492 (1953).
32. J. Reese, Angew. Chem. 64, 399 (1952).
33. H. v. Euler, E. Adler, G. Eklund, O. Toerngren, Arkiv Kemi 15B, No. 9 (1942).
34. O. Manasse, Chem. Ber. 27, 2409 (1894).
35. P. Maitland, D. C. Pepper, J. Soc. Chem. Ind. 61, 66 (1942).
36. B. T. Hayes, R. F. Hunter, J. Appl. Chem. 8, 743 (1958).
37. F. Ullmann, K. Brittner, Chem. Ber. 42, 2539 (1909).
38. E. Adler, Arkiv Kemi 14B, No. 23 (1941).
39. S. R. Finn, G. J. Lewis, J. Soc. Chem. Ind. 69, 132 (1950).
40. S. Kyrning, Arkiv Kemi 15A, No. 2 (1941).
41. F. Hanus, E. Fuchs, J. Prakt. Chem. 153, 327 (1939).
42. H. Kaemmerer, M. Dahm, Kunststoffe-Plastics 6, 1 (1959).
43. B. T. Hayes, R. F. Hunter, Chem. Ind. 1956, 193.
44. A. Zinke, E. Ziegler, Chem. Ber. 77, 264 (1944).
45. K. Hultzsch, Kunststoffe 52, 19 (1962).
46. H. v. Euler, E. Adler, B. Bergstroem, Arkiv Kemi 14B, No. 30 (1941).
47. R. F. Hunter, C. Turner, Chem. Ind. 1957, 72.
48. E. Ziegler, Monatsh. Chem. 79, 142 (1948).
49. E. Ziegler, Oester. Chem. Ztg. 49, 92 (1948).
50. A. T. Carpenter, R. F. Hunter, J. Appl. Chem. 1, 217 (1951).
51. A. T. Carpenter, R. F. Hunter, J. Chem. Soc. 1954, 2731.
52. H. Kaemmerer, H. Lenz, Kunststoffe 51, 26 (1961).
53. A. C. Davis, B. T. Hayes, R. F. Hunter, J. Appl. Chem. 7, 521 (1957).
54. R. F. Hunter, C. Turner, J. Appl. Chem. 7, 528 (1957).
55. H. v. Euler, E. Adler, S. v. Kispeczy, A. M. Fagerlund, Arkiv Kemi 14A, No. 10 (1940).
56. H.-J. Eichhoff, H. Kaemmerer, D. Weller, Makromol. Chem. 129, 109 (1969).
57. H.-J. Eichhoff, H. Kaemmerer, D. Weller, Makromol. Chem. 132, 163 (1970).
58. H. Kaemmerer, G. Gros, H. Schweikert, Makromol. Chem. 143, 135 (1971).
59. H. Kaemmerer, W. Lotz, Makromol. Chem. 145, 1 (1971).
60. H. Kaemmerer, G. Gros, Makromol. Chem. 149, 85 (1971).
61. M. B. Huglin, G. J. Knight, W. W. Wright, Makromol. Chem. 152, 67, 83 (1972).
62. T. Cairns, G. Eglinton, Nature (London) 196, 535 (1962).
63. H. Kaemmerer, A. Kiegel, unpublished results.
64. G. Manecke, D. Zeipner, Makromol. Chem. 129, 183 (1969).

3.13 OLIGO(PHENYLENES)

3.13.1 o-Oligo(phenylenes)

3.13.1.1 Linear o-Oligo(phenylenes)

n	Mol. Wt.	m.p. [°C]	b.p. [°C]	Ref.
3	230.3	59	332	1,2
4	306.4	119	420	1,3-6
6	458.6	217		3,5
8	610.8	320		5

3.13.1.2 Cyclic o-Oligo(phenylenes)

n	Mol. Wt.	m.p. [°C]	Ref.
2	152.2	111	4,7-9
3	228.3	196.5	10,11
4	304.4	233	4,5
6	456.6	335	5
8	608.8	425	5

3.13.2 m-Oligo(phenylenes)

3.13.2.1 Linear m-Oligo(phenylenes)

n	Mol. Wt.	m.p. [°C]	b.p. [°C]	Ref.
3	230.3	89	365	12-14
4	306.4	86.5-87.5	419	3,6,15-17
5	382.5	117-117.5		12,15

n	Mol. Wt.	m.p. [°C]	b.p. [°C]	Ref.
6	458.6	148		3,15,17,18
8	610.8	129-131		15,17
9	689.9	195-200		18

3.13.2.2 Cyclic m-Oligo(phenylenes)

n	Mol. Wt.	m.p. [°C]	Ref.
6	456.6	509.5-511	19

3.13.2.3 Oligo(3-methyl-m-phenylenes)

n	Mol. Wt.	m.p. [°C]	b.p. [°C]	Ref.
2	182.3	9-9.5	280	20-23
3	272.4	65		12

3.13.3 p-Oligo(phenylenes)

3.13.3.1 p-Oligo(phenylenes)

n	Mol. Wt.	m.p. [°C]	b.p. [°C/mm]	Solubility (g/l toluene)	Ref.
1	78.1	5.5	80.1	∞	
2	154.2	71	156	430	24
3	230.3	215	376	7.4	3,12,14,25,26
4	306.4	322	428/10	0.12	3,12,16,21,25,26
5	382.5	395		< 0.005	12,16,21,25,27,28
6	458.6	465			3,29-31

3.13.3.2 Oligo(3-methyl-p-phenylenes)

n	Mol. Wt.	m.p. [°C]	b.p. [°C]	Ref.
1	92.1	-95	110.6	
2	182.3		273-274	22,23,32
3	272.4	43		33

3.13.3.3 Oligo(2,5-dimethyl-p-phenylenes)

n	Mol. Wt.	m.p. [°C]	b.p. [°C]	Solubility (g/l toluene)	Ref.
1	106.2	13.3	138.4	∞	
2	210.3	53-54		700	20,34
3	314.5	182-183		28	34
4	418.6	264-266		1.1	34
5	522.8	307-309		0.24	34

3.13.3.4 Oligo(2,3,5,6-tetramethyl-p-phenylenes)

n	Mol. Wt.	m.p. [°C]	b.p. [°C]	Solubility (g/l toluene)	Ref.
1	134.2	79.2	196-198	∞	
2	266.4	136-137		365	34, 35
3	398.6	270-272		24	34
4	530.8	270-272		3.4	34

3.13.3.5 Oligo(2,2'-dimethyl-biphenylenes)

n	Mol. Wt.	m.p. [°C]	Solubility (g/l toluene)	Ref.
1	182.3	24		20, 23, 36
2	362.5	96	490	37

3.13.3.6 Oligo(3,3'-dimethyl-biphenylenes)

n	Mol. Wt.	m.p. [°C]	b.p. [°C]	Solubility (g/l toluene)	Ref.
1	182.3	9-9.5	280	∞	20-23, 38
2	362.5	76.5		512	27, 37
3	542.8	142		87	39

n	Mol. Wt.	m.p. [°C]	Solubility (g/l toluene)	Ref.
4	723.0	273	< 6.5	39, 40
5	903.3	285	~2	40
6	1083.5	298	< 0.8	40

3.13.3.7 Oligo(3,3''-dimethyl-p-terphenylenes)

n	Mol. Wt.	m.p. [°C]	Solubility (g/l toluene)	Ref.
1	258.4	140-141	4.9	27, 37
2	514.7	258	0.4	37

3.13.3.8 Oligo(2,5-dimethoxy-p-phenylenes)

n	Mol. Wt.	m.p. [°C]	b.p. [°C]	Ref.
1	138.2	56	212.6	
2	274.3	105		20, 41
3	410.5	189		41
4	546.6	246		41

3.13.3.9 Oligo(3,3'-dimethoxy-biphenylenes)

n	Mol. Wt.	m.p. [°C]	b.p. [°C]	Solubility (g/l toluene)	Ref.
1	214.3	36	328		42
2	426.5	158		22	38, 43

3.13.3.10 Oligo(2',3''-dimethoxy-p-quaterphenylenes)

n	Mol. Wt.	m.p. [°C]	Solubility (g/l toluene)	Ref.
1	366.5	183-184	13	43
2	730.9	276-277	~ 0.2	43

3.13.4 Oligo(p-quinones)

n	Mol. Wt.	m.p. [°C]	Ref.
1	108.1	116.5	
2	214.2	194	41

n	Mol. Wt.	m.p. [°C]	Ref.
3	320.3	> 230	41
4	426.3	230-270	41

REFERENCES

1. W. Bachmann, H. T. Clarke, J. Am. Chem. Soc. 49, 2089 (1927).
2. C. F. H. Allen, F. D. Pingert, J. Am. Chem. Soc. 64, 1365 (1942).
3. J. A. Cade, A. Pilbeam, Tetrahedron 20, 519 (1964); J. Chem. Soc. 1964, 114.
4. W. S. Rapson, R. G. Shuttleworth, J. N. van Niekerk, J. Chem. Soc. 1943, 326.

5. G. Wittig, G. Lehmann, Chem. Ber. 90, 875 (1957).
6. S. T. Bowden, J. Chem. Soc. 1931, 1111.
7. W. Baker, M. P. V. Boarland, J. F. W. McOmie, J. Chem. Soc. 1954, 1476.
8. W. C. Lothrop, J. Am. Chem. Soc. 63, 1187 (1941).
9. G. Wittig, W. Herwig, Chem. Ber. 87, 1511 (1954).
10. C. Mannich, Chem. Ber. 40, 159 (1907).
11. D. G. Copeland, K. E. Dean, D. McNeil, J. Chem. Soc. 1960, 1689.
12. M. Busch, W. Weber, J. Prakt. Chem. 146, 1 (1936).
13. A. Gillam, D. H. Hey, J. Chem. Soc. 1939, 1170; H. France, J. M. Heilbron, D. H. Hey, J. Chem. Soc. 1939, 1288; 1938, 1364.
14. G. F. Woods, J. W. Tucker, J. Am. Chem. Soc. 70, 2174 (1948).
15. R. L. Alexander, Jr., J. Org. Chem. 21, 1464 (1956).
16. G. F. Woods, F. T. Reed, J. Am. Chem. Soc. 71, 1348 (1949).
17. M. Bennett, N. B. Sunshine, G. F. Woods, J. Org. Chem. 28, 2514 (1963); W. Davey, D. H. Maass, J. Chem. Soc. 1963, 4386.
18. L. Silverman, W. Houk, Anal. Chem. 27, 1956 (1955).
19. H. A. Staab, F. Binnig, Tetrahedron Letters 1964, 319.
20. F. Ullmann, G. M. Meyer, O. Loewenthal, E. Gilli, Ann. Chem. 332, 38 (1904).
21. E. Mueller, T. Toepel, Chem. Ber. 72, 273 (1939).
22. G. F. Woods, A. L. van Artsdale, F. T. Reed, J. Am. Chem. Soc. 72, 3221 (1950).
23. E. A. Johnson, J. Chem. Soc. 1957, 4155.
24. E. Clar, "Polycyclic Hydrocarbons," Academic Press, New York, Springer, Berlin, 1964.
25. O. Gerngross, M. Dunkel, Chem. Ber. 57, 739 (1924); O. Gerngross, C. Schachnow, R. Jonas, Chem. Ber. 57, 747 (1924).

26. H. O. Wirth, K. H. Goenner, W. Kern, Makromol. Chem. 63, 53 (1963).
27. H. O. Wirth, K. H. Goenner, R. Stueck, W. Kern, Makromol. Chem. 63, 30 (1963).
28. T. W. Campbell, R. N. McDonald, J. Org. Chem. 24, 730 (1959).
29. P. Kovacić, R. M. Lange, J. Org. Chem. 29, 2416 (1964).
30. T. Nozaki, M. Tamura, Y. Harada, K. Saito, Bull. Chem. Soc. Japan 33, 1329 (1960).
31. R. Pummerer, K. Bittner, Chem. Ber. 57, 84 (1924); 64, 2477 (1931); R. Pummerer, L. Seligsberger, Chem. Ber. 64, 2477 (1931).
32. F. Mayer, K. Freitag, Chem. Ber. 54, 347 (1921).
33. H. O. Wirth, H. Hefner, W. Kern, unpublished results.
34. H. O. Wirth, F. U. Herrmann, W. Kern, Makromol. Chem. 80, 120 (1964).
35. E. Marcus, W. M. Lauer, R. T. Arnold, J. Am. Chem. Soc. 80, 3742 (1958).
36. D. M. Hall, M. S. Leslie, E. E. Turner, J. Chem. Soc. 1950, 711.
37. W. Kern, M. Seibel, H. O. Wirth, Makromol. Chem. 29, 164 (1959).
38. W. Schlenk, M. Brauns, Chem. Ber. 48, 661 (1914).
39. W. Kern, W. Gruber, H. O. Wirth, Makromol. Chem. 37, 198 (1960).
40. W. Heitz, R. Ullrich, W. Kern, Makromol. Chem. 98, 29 (1966).
41. H. Erdtmann, M. Granath, G. Schultz, Acta Chem. Scand. 8, 1442 (1954).
42. N. Kornblum, Org. Syn. Coll. Vol. III, 3rd ed., 1962, 295.
43. W. Kern, H. W. Ebersbach, I. Ziegler, Makromol. Chem. 31, 154 (1959).

4. OLIGOMERS CONTAINING HETEROCYCLIC RINGS IN THE MAIN CHAIN

4.1 HETEROCYCLIC OLIGOMERS

4.1.1 Oligo(2, 5-thienylenes)

n	Mol. Wt.	m.p. [°C]	b.p. [°C]	Ref.
1	84.1	-38.4	84.1	
2	166.3	32-33.5	260	1,2
3	248.4	94-95.5		1,3
4	330.5	215-216		1,3
5	412.5	256-257		1,3
6	494.8	304		3
7	576.9	326-328		3

4.1.2 Oligo(2, 6-pyridylenes)

n	Mol. Wt.	m.p. [°C]	b.p. [°C]	Ref.
1	79.1	-42	115.5	
2	156.2	70.1	273-275	4,5
3	233.3	88-89	370	4-6
4	310.4	219-220		5
5	387.5	265		5
6	464.5	350		5

4.1.3 Oligo(3, 5-pyridylenes)

n	Mol. Wt.	m.p. [°C]	b.p. [°C]	Ref.
2	156.2	68	291-292/736	9
3	233.3	249-251		9

4.1.4 Oligo(2, 6-quinolylenes)

n	Mol. Wt.	m.p. [°C]	b.p. [°C]	Ref.
1	129.2	-15.6	237.1	
2	256.3	144		7,8
3	383.5	267-269		8
4	510.6	348-350		8

REFERENCES

1. J. W. Sease, L. Zechmeister, J. Am. Chem. Soc. 69, 270 (1947).
2. W. Borsche, B. G. B. Scholten, Chem. Ber. 50, 596 (1917).
3. W. Steinkopff, R. Leitsmann, K. H. Hoffmann, Ann. Chem. 546, 180 (1941).
4. G. T. Morgan, F. H. Burstall, J. Chem. Soc. 1932, 20.
5. F. H. Burstall, J. Chem. Soc. 1938, 1662.
6. F. H. Case, W. H. Butte, J. Org. Chem. 26, 4415 (1961).
7. H. Weidel, Monatsh. Chem. 8, 120 (1887).
8. S. G. Waley, J. Chem. Soc. 1938, 2008.
9. M. Busch, W. Weber, J. Prakt. Chem. 146, 1 (1936).

4.2 OLIGO(SACCHARIDES)

4.2.1 Oligomeric Pentoses

4.2.1.1 Oligo(β-xylopyranoses) (Ref. 1-3)

n	Mol. Wt.	m.p. [°C]	$[\alpha]_D^{25}$ in H_2O
1	150.1	153	+19.2
2	282.2	186-187	-25.6
3	414.4	215-216	-48.1
4	546.5	224-226	-61.9
5	678.6	240-242	-72.9
6	810.7	237-242	-78.5
7	942.8	240-242	-74.0

4.2.1.2 Oligo(β-xylopyranose acetates) (Ref. 1-3)

n	Mol. Wt.	m.p. [°C]	Specific Rotation $[\alpha]_D^{25}$	in $CHCl_3$
2	534.5	155.5-156	-74.5	0.9
3	750.7	109-110	-84.3	0.6
4	966.9	201-202	-93.7	0.8
5	1183.1	249-250	-97.5	1.1
6	1399.3	260-261	-102.0	1.5

4.2.2 Oligomeric Hexoses

4.2.2.1 Malto-oligooses (Ref. 4-6)

n	Mol. Wt.	m.p. [°C]	$[\alpha]_D^{15}$ in H_2O
1	180.2	146	+52.6
2	342.3	160-165	+136.0
3	504.4		+160.0
4	666.6		+177.0
5	828.7		+180.3
6	990.9		+184.7
7	1153.0		+186.4

4.2.2.2 Cyclodextrins (Ref. 4-6)

n	Mol. Wt.	$[\alpha]_D$ (c=1, H_2O)
6	972.8	+149.0
7	1135.0	+158.8
8	1297.1	+170.0

4.2.2.3 Cello-oligooses (Ref. 7-14)

n	Mol. Wt.	m.p. [°C]	$[\alpha]_D$	Specific Rotation [°C]	c in H_2O
1	180.2	150	+52.5	20	4
2	342.3	225 (d)	+34.6	20	8
3	504.4	238 (d)	+21.6	26	4
4	666.6	253 (d)	+16.5	23	3.4
5	828.7	267 (d)	+11.0	30	4.1
6	990.9	278 (d)	+10.0	30	1.2
7	1153.0	286 (d)	+7.3	30	0.1

4.2.2.4 Cello-oligoose acetates (Ref. 7-14)

n	Mol. Wt.	m.p. [°C]	$[\alpha]_D^{20-25}$ (c=5, in $CHCl_3$)
1	390.4	113	+101.6
2	678.6	229.5	+41.0
3	966.9	223-224	+22.6
4	1255.1	230-234	+13.4
5	1543.4	240-241	+4.2
6	1831.6	252-255	-0.2
7	2119.8	263-266	-4.4

4.2.2.5   Isomalto-oligooses (Ref. 15)

n	Mol. Wt.	m.p. [°C]	$[\alpha]_D$ in H_2O
1	180.2	146	+52.6
2	342.3	225 (d)	
3	504.4		
4	666.6		+153
5	828.7		+160
6	990.9		+163

4.2.2.6   Gentio-oligooses (Ref. 15)

n	Mol. Wt.	m.p. [°C]	$[\alpha]_D$ in H_2O
1	180.2	150	+52.5
2	342.3	190-195	+9.6
3	504.4		-10.5
4	666.6		-19.5

4.2.2.7   Galakto-oligooses (Ref. 15)

n	Mol. Wt.	m.p. [°C]	$[\alpha]_D$ in H_2O
1	180.2	167	+52.5
2	342.3	210-211	+173
3	504.4	115-120	+58
4	666.6		+53

4.2.2.8   Manno-oligooses (Ref. 16,17)

n	Mol. Wt.	m.p. [°C]	Specific Rotation $[\alpha]_D$	[°C]	c, H_2O
1	180.2	132			
2	342.3	193-194	-7.7	25	0.9
3	504.4	137-137.5	-23.3		1.3
4	666.6	232-234	-31		1.6

4.2.3   Oligomeric Amino Sugars

4.2.3.1   N-Acetyl Chito-oligooses (Ref. 18,19)

n	Mol. Wt.	m.p. [°C]	Specific Rotation $[\alpha]_D$	c in H_2O
1	221.2			
2	424.4	260-262 (d)	+17.2	0.5
3	627.6	290-311 (d)	+2.2	0.9
4	830.8	290-300 (d)	-4.1	1.0
5	1034.0	285-295 (d)	-9.1	1.0
6	1237.2		-11.4	0.8
7	1440.4		-12.6	0.3

REFERENCES

1. R. L. Whistler, C.-C. Tu, J. Am. Chem. Soc. 74, 4334 (1952).

2. R. H. Marchessault, T. E. Timell, J. Polymer Sci. C, 2, 49 (1963).

3. C. T. Bishop, Canad. J. Chem. 33, 1073 (1955).

4. W. J. Whelan, J. M. Bailey, P. J. P. Roberts, J. Chem. Soc. 1953, 1293.

5. J. M. Bailey, W. J. Whelan, S. Peart, J. Chem. Soc. 1950, 3692.

6. K. Freudenberg, F. Cramer, Chem. Ber. 83, 296 (1950).

7. E. E. Dickey, M. L. Wolfrom, J. Am. Chem. Soc. 71, 825 (1949).

8. M. L. Wolfram, J. C. Dacons, J. Am. Chem. Soc. 74, 5331 (1952).

9. L. Zechmeister, G. Tóth, Chem. Ber. 64, 854 (1931).

10. K. Hess, K. Dziengel, Chem. Ber. 68, 1594 (1935).

11. C. S. Hudson, J. M. Johnson, J. Am. Chem. Soc. 37, 1276 (1915).

12. R. Willstätter, L. Zechmeister, Chem. Ber. 46, 2401 (1913); 62, 722 (1929).

13. K. Freudenberg, G. Blomquist, Chem. Ber. 68, 2070 (1935).

14. H. Staudinger, E. V. Leopold, Chem. Ber. 67, 479 (1934).

15. W. Walter, in H. M. Rauen, Ed., "Biochemisches Taschenbuch", Part 1, 2nd ed., Springer, Berlin, 1964, p. 98.

16. R. L. Whistler, C. G. Smith, J. Am. Chem. Soc. 74, 3795 (1952).

17. R. L. Whistler, J. Z. Stein, J. Am. Chem. Soc. 73, 4187 (1951).

18. S. A. Barker, A. B. Forster, M. Stacey, J. M. Webber, J. Chem. Soc. 1958, 2218.

19. H. P. Lenk, M. Wenzel, E. Schuette, Hoppe-Seyler's Z. Physiol. Chem. 326, 116 (1961).

VII. PHYSICAL PROPERTIES OF MONOMERS AND SOLVENTS

PHYSICAL PROPERTIES OF MONOMERS

D. Fleischer*

Institut für Makromolekulare Chemie
Technische Hochschule Darmstadt
Darmstadt, Germany

This table contains some of the principal physical properties of monomers with carbon-carbon double bonds. The monomers are arranged alphabetically many derivatives being grouped together under the same parent compound by placing the substituting atoms or groups after the parent name. Compounds are listed under their most commonly used name in polymer chemistry. Therefore, the names used are not always in agreement with the rules of the International Union of Chemistry. No special annotations have been added to distinguish between international union and common names.

Molecular Weights: are computed according to the International Atomic Weight values.
Densities: are relative to water (specific gravity), otherwise it has the dimension grams per milliliter at the indicated temperature. A superscript indicates the temperature of the liquid.
Melting and Boiling Points: are given in degrees centigrade ($^{\circ}$C). The boiling points are stated at atmospheric pressure unless otherwise indicated by a superscript which shows the pressure (in mbar) under which the compound boils at the given temperature.
Refractive Index: is determined at the stated temperature and is reported for the D line of the spectrum of sodium (n_D).
Vapor pressure: is given in mbar at the temperature indicated by superscript.

The physical data in this table were collected from the open literature, technical information sheets and patents. Various sources sometimes gave different values which have been added in parentheses.

Abbreviations:

aa	acetic acid	Chl	chloroform	eth	ether	i insoluble
Ac	acetone	calc	calculated	exp	explodes	m meta
Alc	alcohol	cond	concentrated	frz	freezes	o ortho
Bz	benzene	d	decomposes	h	hot	os organic solvents

p para · polym polymerizes · ss slightly soluble · s soluble · subl sublimes · unst unstable · unsym unsymmetrical · vs very soluble · w water · < below · ∞ soluble in all proportions

Name	Synonyms and Formula	Mol. Wt.	Density	Boiling Point[$^{\circ}$C]	Melting Point[$^{\circ}$C]	Refractive Index	Vapor pressure (mbar)	Solubility w alc eth	
Acetylene	ethyne HC:CH	26.04	0.6181[-82]	-83.6;stbl	-81.8	1.0005[0]	-	ss ss -	s:Chl
--,chloro-	ethynyl chloride HC:CCl	60.48	0.002 (760mm)	-32	-126	-	-	d ss -	
--,dichloro-	ClC:CCl	94.93	-	exp[19;760]	-66	-	-	- s s	
--,diphenyl-	C6H5C:CC6H5	178.24	0.9657[100]	170;300	63.5	-	-	i ss vs	
--,phenyl-	C6H5C:CH	102.14	0.9299[20/20]	142-3	-	1.548[20]	-	i s s	
Acrolein	propenal;acrylaldehyde;acrylic aldehyde; CH2:CHCHO	56.06	0.8406[20/20];0.8410[20]	52.7;52.5-3.5	-86.9;-87.7	1.3998[20] 1.4013[20]	285.0[20]	vs s s	
--,1-chloro-	2-chloropropenal;CH2:CClCHO	90.51	1.199[20]	40[30]	-	1.463[20]	-	- - vs	vs:CCl4
--,1-methyl-	methacrolein;CH2:C(CH3)CHO;2-methyl-propenal	70.09	0.837[20];0.8474[20/20]	68.4	-81.5	1.4169[20] 1.4191[20]	158.5[20]	∞ ∞ ∞	
--,2-methyl-	2-butenal;crotonaldehyde;CH3CH:CHCHO	70.09	0.8575[15];0.8495[25]	104-5	-74	1.4388[17.3] 1.4373[20]	-	s vs vs	∞:Bz
--,-,1-chloro-	2-chloro-2-butenal;CH3CH:CClCHO	104.54	1.1404[23]	147-8	-	1.478[23]	-	ss s s	s:CCl4
Acrylaldehyde	see Acrolein								
Acrylamide	propenamide;CH2:CHCONH2	71.08	1.122[30]	125[25];116.5[10]	84.8	-	0.19[40]	vs s s	vs:Chl
--,1-methyl-	methacrylamide;CH2:C(CH3)CONH2	85.11	-	-	-	-	-	- s ss	

*Present adress: Hoechst A.G. Frankfurt/M, Germany

Name	Synonyms and Formula	Mol.Wt.	Density	Boiling Point [°C]	Melting Point [°C]	Refractive Index	Vapor pressure (mbar)	w	alc	eth	
Acrylamide (Cont' d.)											
—,2-methyl-	2-butenamide;crotonamide;$CH_3CH{:}CHCONH_2$	85,11	-	sub at;140^{13}	158;161,5cor	1.4420^{165}	-	ss	s	ss	s:Bz
Acrylates	see Acrylic acid, esters										
Acrylic acid	propenoic acid;$CH_2{:}CHCO_2H$;ethylene-carboxylic acid	72,06	$1.0472^{20/20}$ 1.0511^{20}	$48,5^{15};39^{10};141,3$	12,3;13,0	1.4224^{20}	$4,13^{20}$	∞	∞	∞	s:Ac
—,allyl ester	$CH_2{:}CHCO_2CH_2CH{:}CH_2$	112,13	$1.0452^{20};0.9441^{20}$	$122;72^{27}$	-	$1.4390^{9};1.4320^{20}$	-	ss	s	s	
—,benzyl ester	$CH_2{:}CHCO_2CH_2C_6H_5$	162,19	$1.068^{20};1.0573^{20}$	$113{-}4^{19};94^{6};228$	-	$1.513^{24};1.5143^{20}$	-	l	s	s	
—,butyl ester	$CH_2{:}CHCO_2C_4H_9$	128,17	$0.8986^{20};0.894$	$39^{10};69^{50};$	-64.6	$1.4187^{25};$	$4,4^{20}$	l	s	s	
—,2-cyanoethyl ester	$CH_2{:}CHCO_2CH_2CH_2CN$	125,13	$1.0690^{20/20}$	146-8(polym);110(polym)	-16.9	$1.4156^{25};1.4433^{20}$	$0,04^{20}$				
—,cyclohexyl ester	$CH_2{:}CHCO_2C_6H_{11}$	154,21	1.0275^{20}	103	-	1.4673^{20}	-	l	∞	∞	
—,ethyl ester	$CH_2{:}CHCO_2C_2H_5$	100,12	$0.924^{20};0.9230^{20/20}$	$182{-}4;88^{20}$	-75;-71.2	$1.4054^{20};1.4068^{25}$	$39,3^{20}$	ss	∞	∞	s:MeOH
—,2-ethyl hexyl ester	$CH_2{:}CHCO_2CH_2CH(C_2H_5)C_4H_9$	184,28	0.8869^{20}	99,6-8	-90	$1.4350^{20};1.4330^{25}$	-		∞	∞	
—,isobutyl ester	$CH_2{:}CHCO_2CH_2CH(CH_3)_2$	128,17	0.8896^{20}	$90^{50};215(polym)$		1.4150^{20}	-	ss	s	s	
—,methyl ester	$CH_2{:}CHCO_2CH_3$	86,09	$0.9558^{20};0.9535^{20}$	$62^{70};132$	<-75	$1.3984^{20};1.4040^{20}$	$90,9^{20}$	ss	s	s	vs:Chl
—,chloride	propenyl chloride;$CH_2{:}CHCOCl$;acrylic chloride	90,51	$1.14;1.1136^{20}$	75-6	-	1.4343	-	d	d	.	vs:Chl
—,1-chloro-	2-chloropropenoic acid;$CH_2{:}CClCO_2H$	106,51	-	176-81(d)	65	-	-	s	s	s	
—,—,ethyl ester	$CH_2{:}CClCO_2C_2H_5$	134,56	1.1404^{20}	$51{-}3^{3};55$	-	1.4384^{20}	-	.	vs	vs	
—,—,methyl ester	$CH_2{:}CClCO_2CH_3$	120,54	1.189^{20}	$52^{8};57{-}9$	-	1.4442^{20}	-	.	vs	vs	
—,—,phenyl ester	$CH_2{:}CClCO_2C_6H_5$	182,61	-	$91{-}3^{3}$	-	1.5808^{20}	-				
—,2-chloro-cis	3-chloropropenoic acid;$ClCH{:}CHCO_2H$	106,51	-	$107^{17.5}$	63-4	-	-	.	s	s	
—,2-chloro-trans	3-chloropropenoic acid;$ClCH{:}CHCO_2H$	106,51	-	94^{18}	86;92-4	-	-	.	.	.	
—,1,2-dichloro-	$ClCH{:}CClCO_2H$	140,96	-	-	85-6	-	-	vs	vs	vs	ssBz
—,2,2-dichloro-	$Cl_2C{:}CHCO_2H$	140,96	-	subl.	76-7	1.3869^{20}	-	ss	.	s	vs:Chl
—,2,2-difluoroethyl ester	$F_2C{:}CHCO_2C_2H_5$	136,09	-	90,5-91.7	-	-	-	.	vs	vs	
—,1-fluoro-	$CH_2{:}CFCO_2H$	90,05	-	-	51,5-2	-	-				
—,—,methyl ester	$CH_2{:}CFCO_2CH_3$	104,08	-	92,5-3.5	-	1.3870^{25}	-				
—,—,2-difluoro-	$F_2C{:}CFCO_2H$	126,03	-	-	35,5-6,5	-	-				
—,1-methyl-	see Methacrylic acid										
—,2-methyl-cis	isocrotonic acid;2-butenoic acid;	86,09	$1.0312^{15};1.0267^{20}$	$74^{15};171,9(d)$	14-5	$1.4457^{20};1.4483^{14};$	-	vs	vs	s	
—,2-methyl-trans	crotonic acid;$CH_3CH{:}CHCO_2H$	86,09	$1.018^{15};0.9604^{77}$	189;185	72	$1.4228^{79,7};1.4249^{77}$	-	vs	vs	s	
Acrylic aldehyde	see Acrolein										
—,1-trifluoro-methyl-	$CH_2{:}C(CF_3)CO_2H$	140,07	-	146-8	50,2	-	-				
—,—,methyl ester	$CH_2{:}C(CF_3)CO_2CH_3$	154,09	-	103,8-5,0	-	1.3370^{20}	-				
Acrylonitrile	propenenitrile;$CH_2{:}CHCN$; vinyl cyanide	53,06	$0.8060^{20};0.8375^{20/20}$	77,3;77,5-9	-83,8;-82	$1.3911^{20};1.3888^{25}$	$110,6^{20}$	s	∞	∞	∞:Ac
—,1-chloro-2-difluoro-	$F_2C{:}CClCN$	123,48	-	63	-	1.3793^{24}	-				
—,1-fluoro-	$CH_2{:}CFCN$	71,04	-	17,7-18	-	1.3162^{30}	-				
—,1-methyl-	methacrylonitrile $CH_2{:}C(CH_3)CN$	67,09	$0.805^{25};0.7998^{20}$	90,3	-35,8;-40	$1.4013^{20};1.4007^{40}$	$164^{40};21,3^{16}$	l	∞	∞	
—,2-methyl-	2-butenenitrile;$CH_3CH{:}CHCN$;propenyl-cyanide;butenoicnitrile	67,09	$0.8239^{20};0.826^{23}$	118-9	-51,5	$1.4242^{20};1.4225^{16}$	$64,4^{20};20,3^{23}$	l	vs	s	

Name	Synonyms and Formula	Mol.Wt.	Density	Boiling Point[°C]	Melting Point[°C]	Refractive Index	Vapor pressure (mbar)	Solubility w	alc	eth
Acryl chloride	see Acrylic acid, chloride									
Allyl acetate	$CH_3CO_2CH_2CH:CH_2$	100.12	0.9276^{20}	103.5	-	1.4049^{20}	-	ss	∞	∞
Allylacetic acid	4-pentenoic acid;$CH_2:CHCH_2CH_2CO_2H$	100.12	0.9843^{18};0.9804^{20}	188-9;93^{20}	-22.5	$1.4341^{7.5}$; 1.4281^{20}	-	ss	vs	vs	s:Chl
Allylacetonitrile	allylmethylcyanide;4-pentenenitrile; $CH_2:CHCH_2CH_2CN$	81.12	1.1803^{13};0.8239^{24}	140;6C-1^{40}	-	1.4213^{14}	-	i	∞	∞
Allyl acrylate	see Acrylic acid, allyl ester									
Allyl alcohol	2-propen-1-ol;$CH_2:CHCH_2OH$	58.08	0.8540^{20};$0.8533^{20/20}$	97	-129	1.4135^{20}	22.9^{25}	∞	∞	∞	s:Chl
--,2-bromo-	$CH_2:CBrCH_2OH$	136.98	1.6;1.621^{20}	152;153-4^{755};62^{21}	-	1.500^{18}	-	d	d	s
--,2-chloro-	$CH_2:CClCH_2OH$	92.53	1.1618^{20}	136-40;756	-	1.4588^{20}	-	s	s	.
--,3-chloro-	$CHCl:CHCH_2OH$	92.53	1.1769^{20}	153	-	1.4638^{20}	-	ss	.	∞
--,3-methyl-	2-buten-1-ol;crotyl alcohol propenyl-carbinol;$CH_3CH:CHCH_2OH$	72.12	0.8726^{0};0.854^{20}; 0.8521^{20}	118;117-121	<-30	1.4240^{20}	-	vs	vs	∞
Allylaldehyde	see Acrolein									
Allylamine	2-propenylamine;3-aminopropene; $CH_2:CHCH_2NH_2$	57.10	0.7627^{20};0.7621^{20}	58;53.2	-88.2	1.41943^{22}; 1.4205^{20}	-	∞	∞	∞	s:Chl
Allyl bromide	3-bromopropene;$CH_2:CHCH_2Br$	120.98	1.398^{20}	71.3;70^{753}	-119.4	1.4654^{20}; 1.4697^{25}	-	i	.	.	s:Chl; s:CCl$_4$
--,1-bromo-	2,3-dibromopropene;$CH_2:CBrCH_2Br$	199.88	1.934^{2};2.0346^{25}	140;37^{11};141-1.5	-	1.5416^{25}	-	i	s	.	s:Chl
Allylcarbylamine	see Allyl isocyanide									
Allyl chloride	3-chloropropene;$CH_2:CHCH_2Cl$;1-chloro-2-propene	76.53	0.9376^{20};$0.9392^{20/20}$	45	-134.5	1.4160^{20}	392.3^{20}	1	∞	∞	∞:Ac ∞:Bz
--,1-chloro-	2,3-dichloropropene;$CH_2:CClCH_2Cl$	110.97	1.211^{20};1.236^{0}	94	-	1.4603^{20}	-	1	s	s	s:Chl
Allyl cyanide	3-butenenitrile;$CH_2:CHCH_2CN$;vinylaceto-nitrile	67.09	0.8318^{20};$0.8359^{20/20}$	118.9	-8.4	1.40602^{20}	-	ss	s	s
Allyl ether	diallyl ether;$CH_2:CHCH_2OCH_2CH:CH_2$	98.15	0.8260^{20}	94^{761}	-	1.4163^{20}	-	1	∞	∞
Allyl ethyl ether	3-ethoxypropene;$CH_2:CHCH_2OC_2H_5$	86.14	0.7651^{20}	66^{761}	-	1.3881^{20}	-	1	∞	∞
Allyl formate	$CH_2:CHCH_2OOCH$	86.09	0.9460^{18};0.948^{18}	83.6	-	-	-	ss	vs	vs
Allyl fluoride	3-fluoropropene;$CH_2:CHCH_2F$	60.07	-	-10;-3	-	1.5540^{21}	-	ss	vs	vs
Allyl iodide	3-iodopropene;$CH_2:CHCH_2I$	167.98	1.8494^{22};1.8454^{22}	102;103.1	-99.3	1.5530^{20};	-	i	s	s	s:Chl
Allyl isocyanide	3-isocyanopropene;$CH_2:CHCH_2NC$;allyl carbylamine	67.09	0.794^{17}	106;98	-	-	-	ss	∞	∞
Allyl isopropyl ether	$CH_2:CHCH_2OC_3H_7$	100.16	0.7764^{20}	83-4	-	1.3946^{20}	-	.	vs	.
Allyl methyl cyanide	see Allylacetonitrile									
Allyl methyl ether	3-methoxypropene;$CH_2:CHCH_2OCH_3$	72.11	$0.77^{11/11}$	46;42.5-3^{757}	-	1.3803^{20};1.3778^{20}	-	1	∞	∞
Allyl phenyl ether	$CH_2:CHCH_2OC_6H_5$	134.18	0.9832^{15};0.9811^{20}	191.7^{12};760	-	1.5218^{20};1.5223^{20}	-	1	s	s	s:Bz
--,4-chloro-	$ClC_6H_4OCH_2CH:CH_2$	168.63	1.131^{15}	106-7^{12};d	-	1.5348^{25}	-	1	s	s
--,2,4,6-tribromo-	$Br_3C_6H_4OCH_2CH:CH_2$	370.88	-	-	33-4	-	-	1	1	1
Allyl propyl ether	$CH_2:CHCH_2OC_3H_7$	100.16	0.7670^{15};0.7764^{20}	90-2^{12}	-	1.3919^{20}	-	1	∞	∞
Allyl-2-tolyl ether	$CH_2:CHCH_2OC_6H_4CH_3$	148.21	0.9698^{15}	205-8;85^{13}	-	1.5188^{15}	-	1	s	s
Allyl-3-tolyl ether	$CH_2:CHCH_2OC_6H_4CH_3$	148.21	0.965^{15};0.9564^{20}	211-4;93.5^{13}	-	1.5179^{20}	-	1	.	.	s:Ac
Allyl-4-tolyl ether	$CH_2:CHCH_2OC_6H_4CH_3$	148.21	$0.9728^{15/15}$	214.5;97-8^{16}	-	1.5157^{24}	-	1	.	.	s:Ac

PHYSICAL PROPERTIES OF MONOMERS

Name	Synonyms and Formula	Mol. Wt.	Density	Boiling Point [°C]	Melting Point [°C]	Refractive Index	Vapor pressure (mbar)	w	alc	eth	(solvents)
Biisopropenyl	see 1,3-Butadiene-,2,3-dimethyl-										
Bromoethene	see Vinyl bromide										
Bromoprene	see 1,3-Butadiene-,2-bromo-										
1,2-Butadiene	methylallene:CH_2:C:$CHCH_3$	54.09	0.676^0;0.652^{20}	10.85	-136.2	$1.4205^{1,3}$; 1.4205^{20}	-	i	∞	∞	s:Ac
-,4-bromo-	CH_2:C:$CHCH_2Br$	133.00	1.4255^{20}	109-111		1.5248^{20}	-	i	·	·	vs:OS
-,4-chloro-	CH_2:C:$CHCH_2Cl$	88.54	0.9891^{20}	88		1.4775^{20}	-	ss	s	s	
-,4-iodo-	CH_2:C:$CHCH_2I$	179.99	1.7129^{20}	130		1.5709^{20}	-	·	·	s	∞:CCl_4
-,3-methyl-	CH_2:C:$C(CH_3)_2$	68.12	0.6833^{20};0.6804^{-6}	40.5-41.5	-120	1.4166^{20}	-	i	∞	∞	
1,3-Butadiene	vinylethylene;erythrene;biethylene;bivinyl; CH_2:CHCH:CH_2	54.09	0.6211^{20};0.650^{-2}	-4.4	-108.9	1.4292^{-25}	-	i	s	s	
-,2-bromo-	bromoprene;CH_2:CBrCH:CH_2	133.00	1.397^{20}	$42\text{-}3^{165}$		1.4988^{20}	-	i	s	s	vs:Chl
-,1-chloro-	CH_2:CHCH:CHCl	88.54	0.9606^{20}	68		1.4712^{20}	-	·	s h	s	s:Ac
-,2-methyl-	CH_2:$C(CH_3)$:CHCl	102.57	0.9710^{20}	$107;45.4^{100}$		1.4792^{20}	-	·	vs	·	s:Chl
-,3-methyl-	CH_2:$C(CH_3)$CH:CHCl	102.57	0.9543^{20}	$99\text{-}100^{100}$		1.4719^{20}	-	i	s	s	
-,2-chloro-	CH_2:CHCCl:CH_2;chloroprene	88.54	0.9583^{20}	$59.4;6.4^{100}$		1.4583^{20}	-	ss	·	∞	s:OS
-,3-methyl-	CH_2:c(CH_3)CCl:CH_2	102.57	0.9593^{20}	93		1.4686^{15}	-	i	s	vs	vs:Chl
-,1,2-dichloro-	CH_2:CHCCl:CHCl	122.98	1.1991^{20}	$60\text{-}5^{105};35^{40}$		1.5078^{20};1.4960^{20}	-	i	s	s	vs:CCl_4
-,2,3-dichloro-	CH_2:CClCCl:CH_2	122.98	1.1829^{20}	98		1.4890^{20}	-	i	s	s	vs:Chl
-,2,3-dimethyl-	biisopropenyl;CH_2:$C(CH_3)C(CH_3)$:CH_2	82.15	0.7267^{20}	75.9;68-78	-76	1.4394^{20}	-		s		
-,2-fluoro-	fluoroprene;CH_2:CHCF:CH_2	72.08	0.843^{-4}	12		1.400^4	-				
-,hexachloro-	CCl_2:CClCCl:CCl_2	260.76	1.6820^{20}	$215;101^{20}$	-21	1.5542^{20}	-	i	s	s	
-,hexafluoro-	CF_2:CFCF:CF_2	162.04	1.553^{-20}	6	-132	1.378^{-20}	-				
-,2-iodo-	iodoprene;CH_2:CICH:CH_2	179.99	1.7278^{20}	111-3		1.5616^{20}	-				
-,2-methyl-	see Isoprene										
2,3-Butadien-1-ol	CH_2:C:$CHCH_2OH$	70.09	0.9164^{20}	126-8		1.4759^{20}	-	s	vs	vs	vs:OS
2-Butenal	see Acrolein, 2-methyl-										
2-Butenamide	see Acrylamide, 2-methyl-					37.7					
1-Butene	1-butylene;ethylethylene;CH_3CH_2CH:CH_2	56.11	0.668^0;0.5951^{20}	-6.3^{758}	-185.4	1.3962^{20}	84	i	vs	vs	s:Bz
-,4-bromo-	$BrCH_2CH_2CH$:CH_2	135.01	1.3230^{20}	98.5	-157.3	1.4622^{20}	-	i	s	s	s:Bz
-,2,3-dimethyl-	$(CH_3)_2CHC(CH_3)$:CH_2	84.16	0.6803	55.67	-115.2	1.3995^{20}	-	i	s	s	s:Ac
-,3,3-dimethyl-	$(CH_3)_3CCH$:CH_2	84.16	0.6529^{20}	41.2	-137.6	1.3763^{20}	-	i	s	s	s:Chl
-,2-methyl-	$CH_3CH_2C(CH_3)$:CH_2	70.14	0.6623^{20};0.6504^{20}	31.2	-168.5	1.3778^{20};1.3874^{20}	-	i	∞	∞	s:Bz
-,3-methyl-	$(CH_3)_2CHCH$:CH_2	70.14	0.6272^{20}	20.0	-138.9	1.3643^{20}	-	i	vs	vs	s:Bz
2-Butene, cis	2-butylene;pseudobutylene;CH_3CH:$CHCH_3$	56.11	0.6213	3.7	-105.6	1.3931^{-25}	-	i	s	s	s:Bz
2-Butene, trans	CH_3CH:$CHCH_3$	56.11	0.6042^{20}	0.9		1.3848^{-25}	-	i	s	s	s:Bz
-,2,3-dimethyl-	tetramethylethylene;$(CH_3)_2C$:$C(CH_3)_2$	84.16	0.7080^{20};0.712	73.2	-74.3	1.4122^{20}	-	i	s	s	s:Chl
-,2-methyl-	$(CH_3)_2C$:$CHCH_3$	70.14	0.6623^{20}	38.6^{16}	-133.8	1.3874^{20}	-	i	s	s	s:Bz
-,1,4-diol-	$HOCH_2CH$:$CHCH_2OH$	88.11	1.0698	$235;132^{16}$	4	1.4782^{20};1.477^{25}	-	s	vs	·	
2-Butenenitrile	see Acrylonitrile, 2-methyl-										
3-Butenenitrile	see Allyl cyanide										
3-Butenoic acid	see Vinyl acetic acid										
2-Buten-1-ol	see Allyl alcohol, 3-methyl-										
Butylene	see Butene										
Butyric acid											
-,ethenyl ester	see Vinyl butyrate										

Name	Synonyms and Formula	Mol.Wt.	Density	Boiling Point[°C]	Melting Point[°C]	Refractive Index	Vapor pressure (mbar)	Solubility w	alc	eth	
Chloroethene	see Vinyl chloride										
Chloroprene	see 1,3-Butadiene, 2-chloro-										
Citraconic acid	see Maleic acid, 2-methyl-										
Crotonaldehyde	see Acrolein, 2-methyl-										
Crotonamide	see Acrylamide, 2-methyl-										
Crotonic acid	see Acrylic acid, 2-methyl-										
Crotononitrile	see Acrylonitrile, 2-methyl-										
Diallylamine	$(CH_2:CHCH_2)_2NH$	97.16	0.7874^{20}	111.0		1.4387^{20}	-	s	s		
Dibromopropene	see Allyl bromide, 1-bromo-										
Dimethylallene	see 1,2-Butadiene, 3-methyl-										
Erythrene	see 1,3-Butadiene										
Ethylene	Ethene; $CH_2:CH_2$	28.05	$0.566^{-102};0.384^{-10}$	-103.7	-169.2frz-181	1.363^{-100}	-	s	ss	s	ss:Bz;ss:Ac
--, bromo-	see Vinyl bromide										
--, chloro-	see Vinyl chloride										
--,1-chloro-1-fluoro-	$CH_2:CClF$	80.48	-	-24	-169	-	$102.6^{25};233.3^{50}$				
--, chlorotrifluoro-	$CF_2:CFCl$	116.47	-	-29	-157.5	-	$56.0^{-};666.5^{100}$				
--,1,1-dibromo-	vinylidene bromide; $CH_2:CBr_2$	185.86	2.178^{21}	92		-	-	i	vs	vs	s:Chl
--,1,2-dibromo, cis	acetylene dibromide; CHBr:CHBr	185.86	2.2464^{20}	112.5	-53	1.5428^{20}	-	i	vs	vs	s:Chl
--,1,2-dibromo, trans	acetylene dibromide; CHBr:CHBr	185.86	2.2308^{20}	108	-6.5	1.5505^{18}	-	i	s	vs	vs:Chl
--,1,1-dichloro-	vinylidene chloride; $CH_2:CCl_2$	96.94	1.2129^{20}	31.7	-122.1	1.4249^{20}	$286.6^{0};^{25}$	i	vs	vs	vs:Chl
							798.5^{25}				
--,1,2-dichloride, cis	acetylene dichloride; CHCl:CHCl	96.94	1.2837^{20}	60.3	-80.5	1.4490^{20}	-	ss	8	8	vs:Chl
--,1,2-dichloro, trans	acetylene dichloride; CHCl:CHCl	96.94	1.2565^{20}	47.5	-50	1.4454^{20}	-	ss	8	8	vs:Chl
--,1,1-dimethyl-	see Propene, 2-methyl-										
--,1,1-diphenyl-	$(C_6H_5)_2C:CH_2$	180.25	$1.028^{16};1.0206^{22}$	$277;94-5^{11}$	$9;8.2$	1.6100^{20}	-	i	s		s:Chl
--, fluoro-	see Vinyl fluoride										
--, iodo-	see Vinyl iodide										
--, phenyl-	see Styrene										
--, tetrabromo-	perbromoethylene; $Br_2C:CBr_2$	343.66	-	$226-7;100^{15}$	56.5	-	-	i	s	s	vs:Chl
--, tetrachloro-	perchloroethylene; $Cl_2C:CCl_2$	165.83	1.6227^{20}	$121;14^{10}$	-19;frz-22	$1.5044^{20};1.5053^{20}$	-	i	8	8	vs:Chl
--, tetrafluoro-	perfluoroethylene; $F_2C:CF_2$	100.02	1.519^{20}	-76.3	142.5	-	-	i	i	i	
--, tetraiodo-	periodoethylene; $I_2C:CI_2$	531.64	2.983	subl.	192	-	-	i	ss	ss	vs:CS_2
--, tetramethyl-	see 2-Butene; 2,3-dimethyl-										
--, tribromo-	ethynil tribromide; CHBr:CBr	264.76	$2.708^{20.5}$	$163-4;75^{15}$		1.6045^{16}	-	ss	vs	s	s:Chl
--, trichloro-	ethynil trichloride; CHCl:CCl_2	131.39	1.4642^{20}	87	$-73;$frz$-86.4;-88$	1.4773^{20}	-	ss	8	8	s:Chl
Ethylenecarboxylic acid	see Acrylic acid										
Ethylene sulfonic acid	vinyl sulfonic acid; $CH_2:CHSO_3H$	108.10	$1.3921^{25};1.4003^{20/20}$	$100.5^{1};125^{1}$	-	1.4493^{20}	-	s	s	.	ss:Chl
--, n-amyl ester	$CH_2:CHSO_3(CH_2)_4CH_3$	178.23	1.087^{20}	$131^{17};15$	-	1.4412^{20}	-				
--, n-butyl ester	$CH_2:CHSO_3(CH_2)_3CH_3$	164.21	1.122^{20}	117^{15}	-	1.4416^{20}	-				
--, ethyl ester	$CH_2:CHSO_3C_2H_5$	136.16	1.1831^{25}	76^{15}	-	1.4316^{25}	-				
--, n-hexyl ester	$CH_2:CHSO_3C_6H_{13}$	192.26	1.050^{20}	146^{15}	-	1.4430^{20}	-				
--, isoamyl ester	$CH_2:CHSO_3CH_2CH_2CH(CH_3)_2$	178.23	1.082^{20}	124^{15}	-	1.4415^{20}	-				

Name	Synonyms and Formula	Mol.Wt.	Density	Boiling Point[°C]	Melting Point[°C]	Refractive Index	Vapor pressure (mbar)	Solubility w	alc	eth
Ethylene sulfonic acid (Cont' d.)										
--, isobutyl ester	CH_2:$CHSO_3CH_2CH(CH_3)_2$	164.21	1.1898^{25}	$78^{5.5}$	-	1.4258^{25}	-			
--, isopropyl ester	CH_2:$CHSO_3CH(CH_3)_2$	152.20	1.132^{20}	70^{4}	-	1.4321^{20}	-			
--, methyl ester	CH_2:$CHSO_3CH_3$	122.13	1.248^{20}	91^{15}	-	1.4316^{20}	-			
--, phenyl ester	CH_2:$CHSO_3C_6H_5$	184.20	1.1657^{25}	46^{2}	-	1.4258^{25}	-			
--, n-propyl ester	CH_2:$CHSO_3C_3H_7$	152.20	1.156^{20}	110^{18}	-	1.4368	-			
Ethylethylene	see 1-Butene									
Fluoroprene	see 1,3-Butadiene, 2-fluoro-									
Fumaric acid	butenedioic acid, trans;HO_2CCH:$CHCO_2H$	116.07	1.635^{20}	$165^{1.7}$ subl.;290	286-7	-	-	ss	s	ss; ss:Ac; ss:CCl$_4$
--, di-n-amyl ester	$C_5H_{11}O_2CCH$:$CHCO_2C_5H_{11}$	256.33	0.9681^{20}		-	1.4496^{20}				
--, di-n-butyl ester	$C_4H_9O_2CCH$:$CHCO_2C_4H_9$	228.29	0.9869^{20}	162^{7}	-	1.4469^{20}				
--, diethyl ester	$C_2H_5O_2CCH$:$CHCO_2C_2H_5$	177.18	1.0521^{20}	138^{8}	-	1.4408^{20}		i	s	s; s:Ac
--, diisoamyl ester	$C_5H_{11}O_2CCH$:$CHCO_2C_5H_{11}$	256.33	0.9655^{20}	214^{11}	-	1.4479^{20}		.	s	s; s:Ac
--, diisobutyl ester	$C_4H_9O_2CCH$:$CHCO_2C_4H_9$	228.29	0.9760^{20}	166;160^{5}	-	1.4432				
--, diisopropyl ester	$C_3H_7O_2CCH$:$CHCO_2C_3H_7$	200.24	-	170^{5};122^{5}	-	-		s	s	s; s:Ac;s:Bz
--, dimethyl ester	CH_3O_2CCH:$CHCO_2CH_3$	144.13	-	225-6	102	-				
--, dinitrile ester	trans-1,2-Dicyanoethylene;Fumaronitrile; NCCH:CHCN	78.07	0.9416^{111}	192	96.8	1.4349^{111}		s	s	s; s:Ac;s:Bz
--, diphenyl ester	$C_6H_5O_2CCH$:$CHCO_2C_6H_5$	268.27	-	219^{14}	161-2	-		i	ss	. ; ss:Bz
--, dipropyl ester	$C_3H_7O_2CCH$:$CHCO_2C_3H_7$	200.24	1.0129^{20}	110^{5}	-	1.4439^{20}		s	s	s
--, bromo-	HO_2CCH:$CBrCO_2H$	194.98	-	200	185-6	-				
--, chloro-	HO_2CCH:$CClCO_2H$	150.52	-	subl.^{d}	192-3	1.4571^{20}		vs	vs	vs; vs:Bz
--, -, diethyl ester	$C_2H_5O_2CCH$:$CClCO_2C_2H_5$	206.63	1.1880^{25}	250;136^{19}	-			i	vs	vs
--, -, dimethyl ester	CH_3O_2CCH:$CClCO_2CH_3$	178.57	1.2899^{25}	224;108^{15}	-			.	vs	s
--, dimethyl-	$HO_2C(CH_3)C$:$C(CH_3)CO_2H$	144.13	-	-^{d}	241			ss;h	ss	. ; l:Bz
--, methyl-	mesaconic acid;$HO_2CC(CH_3)$:$CHCO_2H$	130.10	1.466^{20}	205	204.5			vs	vs	vs; ss:Bz; ss:Chl
--, -, diethyl ester	$C_2H_5O_2CC(CH_3)$:$CHCO_2C_2H_5$	186.21	$1.0453^{20/20}$	229;$93\text{-}5^{10}$	-	$1.4488^{20/20}$		.	s	s; s:Bz
--, -, dimethyl ester	$CH_3O_2CC(CH_3)$:$CHCO_2CH_3$	158.16	1.1153;1.0914^{20}	203.5;100^{16}	-	1.4512^{20}		ss	s	s; s:Bz
Hemiterpene	see Isoprene									
Iodoprene	see 1,3-Butadiene, 2-iodo-									
Isobutenyl chloride	see Propene, 3-chloro-2-methyl-									
Isobutylene	see Propene, 2-methyl-									
Isocrotonic acid	see Acrylic acid, 2-methyl-cis									
Isocrotyl chloride	see Propene, 1-chloro-2-methyl-									
Isoprene	2-methyl-1,3-butadiene; hemiterpene; CH_2:$CHC(CH_3)$:CH_2	68.12	0.6810^{20}	34	-146	1.4220^{20}	-	i	∞	∞; ∞:Bz; ∞:Ac
Itaconic acid	Succinic acid, methylene; CH_2:$C(CO_2H)CH_2CO_2H$	130.10	1.632	d	175	-	-	s	s	ss; s:Chl
Maleic acid	butenedioic acid-cis;HO_2CCH:$CHCO_2H$	116.07	1.609^{20}	-	139.4;130.5	-	-	vs	vs	s; vs:Ac; ss:Bz

Name	Synonyms and Formula	Mol.Wt.	Density	Boiling Point [°C]	Melting Point [°C]	Refractive Index	Vapor pressure (mbar)	Sol. w	alc	eth	other
Maleic acid (Cont'd.)											
--, di-n-amyl ester	$C_5H_{11}O_2CCH{:}CHCO_2C_5H_{11}$	256.33	0.9741^{20}	161^{10}	-	1.4475^{20}	-	-	-	-	
--, di-n-butyl ester	$C_5H_{10}O_2CCH{:}CHCO_2C_4H_9$	228.29	0.9938^{20}	147.5^{12}	-85	1.4454^{20}	-	i	s	s	
--, diethyl ester	$C_4H_9O_2CCH{:}CHCO_2C_2H_5$	172.18	$1.0662^{20};1.064^{25}$	$223;219;105\text{-}6^{14}$	-11.2;-8.8	$1.4416^{20};1.4401^{20}$	-			s	
--, diisoamyl ester	$C_5H_{11}O_2CCH{:}CHCO_2C_5H_{11}$	256.33	0.9714^{20}	157^{13}	-	1.4459^{20}	-				
--, diisobutyl ester	$C_4H_9O_2CCH{:}CHCO_2C_2H_5$	228.29	0.9820^{20}	125^{5}	-	1.4418^{20}	-	i	s		
--, dimethyl ester	$CH_3O_2CCH{:}CHCO_2CH_3$	144.13	1.1502^{20}	$109^{15};202^{17}$	-19	$1.4423^{20};1.4416^{20}$	-	i	vs	vs	vs:AC;
--, diphenyl ester	$C_6H_5O_2CCH{:}CHCO_2C_6H_5$	268.27	-	226	73	-	-	i	vs	vs	vs:Bz
--, di-n-propyl ester	$C_3H_7O_2CCH{:}CHCO_2C_3H_7$	200.24	1.0245^{20}	126^{12}	-	1.4433^{20}	-				
--, bromo-	$HO_2CCH{:}CBrCO_2H$	194.98	-	d	138-41	-	-	vs/h	vs	vs	ss:Bz;
--, chloro-	$HO_2CCH{:}CClCO_2H$	150.52	-	-	108;114	-	-	vs	vs	vs	ss:Chl
--,--, diethyl ester	$C_2H_5O_2CCCl{:}CHCO_2C_2H_5$	206.63	1.1741^{20}	$235;125^{19}$	-		-	.	vs	vs	vs:AC;
--,--, dimethyl ester	$CH_3O_2CCCl{:}CHCO_2CH_3$	178.57	1.2775^{25}	106.5^{18}	-		-	.	vs	s	vs:AC;
--, dichloro-	$HO_2CCCl{:}CClCO_2H$	184.97	-	119-20	-		-	ss	s	s	
--, dihydroxy-	$HO_2CCOH{:}COHCO_2H$	148.07	-	155	-		-	ss	ss	ss	
--, methyl-	citraconic acid;$HO_2CCH{:}C(CH_3)CO_2H$	130.10	1.617^{25}	-	93-3.8		-	vs	.	ss	ss:Bz;
--,--, diethyl ester	$C_2H_5O_2CCH{:}C(CH_3)CO_2C_2H_5$	186.21	1.0491^{20}	$230.3;120^{20}$	-	$1.4468^{20};1.4442^{20}$	-	ss	s	s	ss:Chl
--,--, dimethyl ester	$CH_3O_2CCH{:}C(CH_3)CO_2CH_3$	158.16	$1.1153^{20};1.1491^{20}$	$210.5;94\text{-}5^{11}$	-	$1.4473^{20};1.4486^{20}$	-	ss	s	s	
Mesaconic acid	see Fumaric acid, methyl-										
Methacrolein	see Acrolein, 1-methyl-										
Methacrylamide	see Acrylamide, 1-methyl-										
Methacrylates	see Methacrylic acid, ester										
Methacrylic acid	2-methyl-propenoic acid;$CH_2{:}C(CH_3)CO_2H$	86.09	1.0153^{20}	$162\text{-}3^{757};60^{12}$	16	$1.4314^{20};1.4288^{25}$	s	s	∞	∞	
--, allyl ester	$CH_2{:}C(CH_3)CO_2CH_2CH{:}CH_2$	126	0.935^{20}	$42\text{-}5^{15}$	<-76	1.4365^{20}	-	i	∞	∞	
--, butyl ester	$CH_2{:}C(CH_3)CO_2C_4H_9$	142.20	0.8936^{20}	$165\text{-}8;52;160^{11}$	-	$1.4240^{20};1.4215^{25}$	-	ss	∞	∞	
--, ethyl ester	$CH_2{:}C(CH_3)CO_2C_2H_5$	114.15	$0.9135^{25};0.911^{20}$	$117.5\text{-}9.5;30^{18}$	-	$1.4147^{20};1.4115^{25}$	-	i	∞	∞	
--, hexyl ester	$CH_2{:}C(CH_3)CO_2C_6H_{13}$	170.24	0.885^{25}	198-240	-	1.429^{20}	-		∞	∞	
--, isobutyl ester	$CH_2{:}C(CH_3)CO_2CH_2CH(CH_3)_2$	142.20	0.8858^{20}	$155;45^{11}$	-	$1.4199^{20};1.418^{24}$	-	i	∞	∞	
--, isopropyl ester	$CH_2{:}C(CH_3)CO_2CH(CH_3)_2$	128.17	0.8847^{20}	$126\text{-}7^{32}$	-	$1.4122^{20};1.412^{24}$	-	i	∞	∞	
--, methyl ester	methyl methacrylate;$CH_2{:}C(CH_3)CO_2CH_3$	100.12	$0.9440^{20};0.940^{25}$	100-101;24	-48	$1.4142^{25};1.4118^{20}$	42.7^{20}	ss	∞	∞	
--,--, propyl ester	$CH_2{:}C(CH_3)CO_2C_3H_7$	128.17	0.9022^{20}	141	-	1.4191^{20}	-	i	∞	∞	
Methacrylonitrile	see Acrylonitrile, 3-methyl-										
Methoxyethene	see Vinyl methyl ether										
Methoxystyrene	see Vinyl methyl ether,1-phenyl-										
Methyl isopropenyl ketone	1-methylvinyl methyl ketone; $CH_3COC(CH_3){:}CH_2$	84.12	$0.8527^{20};0.8550^{20/20}$	$98;38^{85}$	-54	1.4220^{20}	-	ss	∞	∞	
Methyloxirene	see Propene, 1,2-epoxy-										
4-Pentenenitrile	see Allylacetonitrile										
4-Pentenoic acid	see Allylacetic acid										

PHYSICAL PROPERTIES OF MONOMERS

Name	Synonyms and Formula	Mol.Wt.	Density	Boiling Point[°C]	Melting Point[°C]	Refractive Index	Vapor pressure (mbar)	w	alc	eth	Solubility
Perbromoethylene	see Ethylene, tetrabromo-										
Perfluoropropene	see Propene, hexafluoro-										
Propenal	see Acrolein										
Propenamide	see Acrylamide										
Propene	propylene;methylethylene;CH_2:$CHCH_3$	42.07	0.5139^{-20};0.5193^{20}	-47.8	-185.2	1.3567^{-70}	301.8^{38}	vs	vs	.	vs:aa
--,3-amino-	see Allylamine										
--,1-bromo-	propenyl bromide;CHBr:$CHCH_3$	120.98	1.4291^{20};1.4133^{20}	57.8;60.2	-113;-116.6	1.4560^{20};1.4519^{16}	-	i	.	.	s:Chl;s:Ac
--,2-bromo-	isopropenyl bromide;CH_2:$CBrCH_3$	120.98	1.362^{20}	48.4^{748}	-124.8	1.4440^{20};1.4467	-	i	.	s	s:Chl;s:Ac
--,3-bromo-	see Allyl bromide										
--,1-chloro, cis	propenyl chloride;1-chloropropylene	76.53	0.9347^{20}	32.8	-134.8	1.4055^{20}	-	i	.	.	s:Chl;s:Ac
--,1-chloro, trans	CHCl:$CHCH_3$	76.53	0.9350^{20}	37.4^{760};738	-99	1.4054^{20};1.404^6	-	i	.	.	s:Chl;s:Ac
--,2-chloro-	isopropenyl chloride;CH_2:$CClCH_3$	76.53	0.9017^{20};0.918^9	22.6^{760};23^{738}	-137.4;138.6	1.404^6;1.3973^{20}	-	i	.	s	s:Chl;s:Ac
--,3-chloro-	see Allyl chloride										
--,1-chloro-2-methyl-	isocrotyl chloride;CHCl:$C(CH_3)_2$	90.55	0.925^{16};0.9186^{20}	68^{754}	-	1.4221^{20}	-	∞	∞	∞	vs:Chl
--,3-chloro-2-methyl-	isobutenyl chloride;CH_2:$C(CH_3)CH_2Cl$	90.55	0.9165^{20};0.925	71.5-2.5	-	1.4291^{20};1.427	-	∞	∞	∞	vs:Chl
--,2,3-dibromo-	see Allyl bromide, 1-bromo-										
--,1,1-dichloro-	Cl_2C:$CHCH_3$	110.97	1.1864^{25};$1.1764^{19.5/0}$	76-7;78^{757}	-	1.4430^{25}	-	i	.	s	s:Chl;s:Ac
--,1,2-dichloro-	CHCl:$CClCH_3$	110.97	1.1818^{20}	77.0	-	1.4471^{20}	-	i	vs	vs	vs:CCl_4
--,2,3-dichloro-	see Allyl chloride,1-chloro-										
--,1,2-epoxy-	allylene oxide;methyloxirene;CH_3C:CH (O)	56.06	-	63	-	-	-	ss	∞	∞	
--,3-ethoxy-	see Allyl ethyl ether										
--,3-fluoro-	see Allyl fluoride										
--,hexafluoro-	perfluoropropene;CF_2:$CFCF_3$	150.02	1.583^{-40}	-29.4	-156.2	-	-				
--,3-iodo-	see Allyl iodide										
--,3-methoxy-	see Allyl methyl ether										
--,2-methyl-	isobutylene;CH_2:$C(CH_3)_2$	56.11	$0.6266^{-6.6}$;0.5942^{20} liq	-6.9	-141	1.3926^{-25};1.3814^{-25}	13.3^{-82};933.1^{-9}	i	vs	vs	
--,-,tetramer	tetraisobutylene;$(C_4H_8)_4$	224.44	0.7944^{20}	243-6;109.5^{15}	-98	1.4482^{20}	-				
--,-,trimer	triisobutylene;$(C_4H_8)_3$	168.33	0.7590^{20}	179-81;56^{10}	-76	1.4314	-				
--,2-phenyl-	see Styrene,1-methyl-										
--,1,1,2-trichloro-	CCl_2:$CClCH_3$	145.42	$1.387^{14/14}$;1.382^{20}	118;41^{52}	-	1.4827^{20}	-	i	s	s	s:Chl
--,1,2,3-trichloro-	CHCl:$CClCH_2Cl$	145.42	$1.414^{20/20}$	142;$32-3^{14}$	-	1.5020^{20}	-	i	vs	vs	s:Chl
--,3,3,3-trichloro-	CH_2:$CHCCl_3$	145.42	$1.369^{20/20}$	114-5;57^{103}	-30	1.4827^{20}	-	i	s	s	s:Chl
Propenenitrile	see Acrylonitrile										
Propenoic acid	see Acrylic acid										
2-Propen-1-ol	see Allyl alcohol										
Propenyl carbinol	see Allyl alcohol, 3-methyl-										
Propionic acid; --, ethenyl ester	see Vinyl propionate										
Propenyl cyanide	see Acrylonitrile, 2-methyl-										
Propylene	see Propene										
Propylene aldehyde	see Acrolein, 2-methyl-										

Name	Synonyms and Formula	Mol. Wt.	Density	Boiling Point[°C]	Melting Point[°C]	Refractive Index	Vapor pressure (mbar)	Solubility w	alc	eth	
Styrene	ethenylbenzene;vinylbenzene;phenylethylene $C_6H_5CH:CH_2$	104.15	$0.9060^{20};0.9075^{20/20}$	$145.2;33.6^{10}$	-30.6	1.5468^{20}	6.7^{20}	i	s	s	s:CS$_2$; ∞:Bz
-,o-amino-	$C_6H_5NH_2CH:CH_2$	149.15	-	$104^{12};2.5$	-	1.608^{15}					
-,p-amino-	$NH_2C_6H_4CH:CH_2$	149.15	-	81.5^{75}	-	$1.6070^{25};19.5$					
-,1-bromo-	$C_6H_5CBr:CH_2$	183.05	1.4025^{23}	$160;86-7^{14}$	-43.5	$1.5881^{20};22$		i	∞	∞	
-,2-bromo-	$C_6H_5CH:CHBr$	183.05	$1.4269^{16};1.0984^{25}$	$107^{23};71^{3};219$sl.d.	-7	$1.6096^{20};1.5990^{22}$					vs:Chl
-,m-bromo-	$C_6H_4BrCH:CH_2$	183.05	1.4160^{20}	$75;90-4^{3}$	-	$1.5933^{20};1.5903^{20}$					
-,o-bromo-	$C_6H_4BrCH:CH_2$	183.05	1.4059^{20}	$206;12-65^{3}$	frz.-52.8	$1.5927^{20};1.5914^{20}$		i	.	.	vs:Chl
-,p-bromo-	$BrC_6H_4CH:CH_2$	183.05	$1.3984^{18};1.0984^{25}$	$88;89^{17}$	4.5	$1.5947^{20};1.5950^{20}$					
-,1-chloro-	$C_6H_5CCl:CH_2$	138.60	1.1016^{15}	$199;76^{18}$	-23	$1.5612^{20};1.523^{17}$		i	s	s	
-,2-chloro-	$C_6H_5CH:CHCl$	138.60	1.1095^{20}	$199;90^{18}$	-	$1.5648^{20};1.5736^{25}$		i	s	s	
-,m-chloro-	$C_6H_4ClCH:CH_2$	138.60	$1.1168^{20};1.090^{20}$	$62-3;51;57^{3}$	-	$1.5625^{20};1.5619^{20}$		i		s	s:Ac
-,o-chloro-	$C_6H_4ClCH:CH_2$	138.60	1.100^{20}	$188.7;60-1^{4}$	-63.15	1.5649^{20}		.	s	s	s:Ac
-,p-chloro-	$ClC_6H_4CH:CH_2$	138.60	$1.0868^{20};1.1554^{20}$	$192.0;53-4;74^{12};0.15$	-15.90	1.5660^{20}		i	s	s	∞:Bz;Ac
-,o-cyano-	$C_6H_4(CN)CH:CH_2$	129.16	-	$53^{3.5}$	-	1.5756^{20}					
-,m-cyano-	$C_6H_4(CN)CH:CH_2$	129.16	-	83.5	-	1.5630^{20}					s:Bz
-,p-cyano-	$NCC_6H_4CH:CH_2$	129.16	-	$90^{1.5}$	-15	1.575^{25}					s:Bz
-,1,2-difluoro-	$C_6H_5CF:CHF$	141.08	$1.0177^{20};1.025^{20}$	$86.2-90.2^{60}$	-	1.5061^{20}					
-,m-fluoro-	$C_6H_4FCH:CH_2$	122.14	1.0282^{20}	$30-1;32^{32}$	-	1.5170^{20}		i	s	s	s:Bz
-,o-fluoro-	$C_6H_4FCH:CH_2$	122.14	1.024^{20}	$32-4;46^{50}$	-	1.5200^{20}		i	s	s	s:Bz
-,p-fluoro-	$FC_6H_4CH:CH_2$	122.14	$1.0468^{35};1.0353^{31}$	$296;67.4;29-30^{4}$	-34.5	1.5158^{31}					
-,m-hydroxy-	vinylphenol;$C_6H_4OHCH:CH_2$	120.15	$1.0468^{18};1.0468^{36}$	$114-6^{15};15^{14}$	29	1.5804^{35}		s	vs	vs	s:CS$_2$
-,o-hydroxy-	vinylphenol;$C_6H_4OHCH:CH_2$	120.15	1.0609^{20}	$108;77^{15};101^{14}$	73.5	$1.577^{35};1.5783^{27}$					
-,p-hydroxy-	vinylphenol;$C_6H_4OHCH:CH_2$	120.15	-	-	-	-					
-,m-iodo-	$C_6H_4ICH:CH_2$	233.05	-	73^{3}	-	1.6390^{20}					
-,m-methoxy-	m-vinylanisole;$CH_2:CHC_6H_4OCH_3$	134.18	0.9999^{17}	$114-6^{17};89^{14}$	-	$1.5540^{20};1.5586^{23}$		i	s	s	s:Bz
-,o-methoxy-	o-vinylanisole;$CH_2:CHC_6H_4OCH_3$	134.18	1.0049^{13}	$195-9;84^{16}$	29	1.5388^{13}		i	s	s	s:Ac;s:Bz
-,p-methoxy-	p-vinylanisole;$CH_2:CHC_6H_4OCH_3$	134.18	1.0001^{20}	$204-5;96^{16}$	-	1.4642^{20}		i	s	s	s:Bz
-,1-methyl-	2-phenyl-1-propene;$C_6H_5C(CH_3):CH_2$	118.18	$0.9082^{20};0.9165^{10}$	$72;163.4^{30}$	-23.2	$1.5386^{20};1.5358^{25}$	2.5^{20}	s	s	s	s:MeOH;s:CS$_2$
-,m-methyl-	$C_6H_4(CH_3)CH:CH_2$	118.18	$0.9164^{15.5/15.5}$	$164;171-2.5^{3}$	-	1.5411^{20}					
-,o-methyl-	$C_6H_4(CH_3)CH:CH_2$	118.18	$0.9165^{15.5/15.5}$	$170;52^{9}$	-	1.5444^{20}					
-,p-methyl-	$C_6H_4(CH_3)CH:CH_2$	118.18	$0.9261^{15.5/15.5}$	$173;66^{18}$	58	$1.5420^{20};1.5402^{25}$					
-,2-nitro-	1,2-nitrovinylbenzene;$C_6H_5CH:CHNO_2$	149.15	-	$250;150^{14}$	-	-					
-,m-nitro-	$NO_2C_6H_4CH:CH_2$	149.15	-	100^{3}	-5	1.5818^{20}		i$_h$		vs	vs:Chl;
-,o-nitro-	$C_6H_4NO_2CH:CH_2$	149.15	-	-	13.5	-		ss			vs:CS$_2$
-,p-nitro-	$NO_2C_6H_4CH:CH_2$	149.15	-	d^{17}	29	-			h	vs	s:Chl
-,m-trifluoromethyl-	$C_6H_4(CF_3)CH:CH_2$	172.14	-	55	-	1.4655^{20}			vs		
Tetraisobutylene	see Propene, 2-methyl-tetramer..										
Tetramethylethylene	see 2-Butene, 2,3-dimethyl-										
Triallylamine	$(CH_2:CHCH_2)_3N$	137.22	$0.809^{20};0.800^{20}$	$155-6;149.5$	<-70	1.4502^{20}		h		s	s:Bz;s:Ac
Triisobutylene	see Propene, -methyl-, trimer										
Vinyl acetate	acetic acid;ethenyl ester;$CH_3CO_2CH:CH_2$	86.09	$0.9317^{20};0.9338^{20/20}$	72.5	-93.2	1.3959^{20}	118.2^{20}	s$_h$	∞	s	s:Bz;s:Ac

PHYSICAL PROPERTIES OF MONOMERS

Name	Synonyms and Formula	Mol.Wt.	Density	Boiling Point[°C]	Melting Point[°C]	Refractive Index	Vapor pressure (mbar)	Solubility w	alc	eth	
Vinyl acetic acid	3-butenoic acid;CH_2:$CHCH_2CO_2H$	86.09	1.0091^{20}	169^{764};163;69-70^{12}	-35	1.4239^{20};1.4252^{20}	-	s			s:Ac:s:Chl
Vinylacetonitrile	see Allyl cyanide										
Vinylanisole	see Styrene, methoxy-										
Vinylbenzene	see Styrene										
Vinyl bromide	bromoethene;CH_2:CHBr	106.96	1.4933^{20}(liq);1.5167^{14}	15.8	-139.54	1.4410^{20}	-	i	s	s	∞:OS
Vinyl butyl ether	butoxyethene;CH_2:$CHOC_4H_9$	100.16	0.7742^{25};$C.7888^{20}$	93.8	-112.7;-92	1.3997^{25};1.4026^{20}	56.0^{20}	i	vs	∞	vs:Ac
Vinyl butyl sulfide	CH_2:$CHSC_4H_9$	116.21	-	142	-	1.474^{25}	-				
Vinyl butyrate	butyric acid,ethenyl ester;CH_2:$CHO_2CC_3H_7$	114.14	$0.9022^{20/20}$;$0.8994^{20/20}$	116.5	-	1.411^{15}	19.3^{20}	ss	s	vs	
Vinyl chloride	chloroethene;CH_2:CHCl	62.50	0.99176^{20};0.9106^{20}	-13.37	-153.79	1.399^{15};1.3700^{20} calc;	133.3;-55.8	ss	s	vs	
Vinyl 2-chloro-ethyl ether	CH_2:$CHOCH_2CH_2Cl$	106.55	1.0475^{20};0.0493^{20}	109.1	-69.7	1.4378^{20}	24.0^{20}	.	vs	vs	
Vinyl crotonate	2-butenoic acid,ethenyl ester;	112.12	$0.9439^{20/20}$;$0.941^{20/20}$	133.7;23-4^7	-	1.450^{20}	0.7	s	s		
Vinyl cyanide	see Acrylonitrile CH_3CH:$CHCO_2CH_2$										
Vinyl ether	divinyl ether;CH_2:$CHOCH$:CH_2;ethenyl oxethene	70.09	0.773^{20};$0.774^{20/20}$	39;28.3	-101	1.3989^{20}	-	i		∞	
-,hexachloro-	Cl_2C:$CClOCCl$:CCl_2	276.76	1.654^{21}	210	-	-					
Vinylethylene	see 1,3-Butadiene										
Vinyl ethyl ether	C_2H_5OCH:CH_2	72.11	0.7589^{20};$0.7541^{20/20}$	35-6	-115.8	1.3767^{20};1.3739^{25}	567.9^{20}	ss	s		
-,1-chloro-	C_2H_5OCCl:CH_2	106.55	1.02^{22}	122-3;760		1.4378^{20}		.	vs		
-,2-chloro-	$ClCH$:CH OCH:CH_2	106.55	1.0475^{25}	108;760	-	1.4558^{17}		s	vs		
-,1,2-dichloro-	C_2H_5OCCl:CHCl	141.00	1.1972^{25}	128.2	-	1.4018^{20}		s			
-,1-ethyl-	$C_2H_5OC(C_2H_5)$:CH_2	100.16	-	87	-	1.3927^{20}					
-,1-methyl-	$C_2H_5OC(CH_3)$:CH_2	86.14	-	62	-						
-,1-phenyl-	$C_2H_5OC(C_6H_5)$:CH_2	148.21	-	211.96;12.5	-						
Vinyl ethyl ketone	CH_2:$CHCOC_2H_5$	84.12	0.8468	102;740;60	-	1.4192^{25}		s			
Vinyl ethyl sulfide	C_2H_5SCH:CH_2	88.16	0.8468	91.5	-	1.4631^{25}					
Vinyl 2-ethyl hexoate	$C_4H_9CH(C_2H_5)CO_2CH$:CH_2	170.24	$0.8751^{20/20}$	185.5	-						
Vinyl 2-ethylhexyl ether	$C_4H_9CH(C_2H_5)CH_2OCH$:CH_2	156.27	$0.8102^{20/20}$	174	-	1.4247^{25}	0.4^{20}				s:Ac
Vinyl fluoride	fluoroethene;CH_2:CHF	46.04	$0.9651^{20/20}$	-72.2;-88	-100			i	s		
Vinyl formate	HCO_2CH:CH	72.06	-	46.6;745	-161	1.4757^{20}	360.0^{20}	i	s		
Vinyl heptafluorobutyrate	$C_3F_7CO_2CH$:CH_2	300.09	-	78-9;745	-	1.3086^{20}					
Vinylidene bromide	see Ethylene,1,1-dibromo-										
Vinylidene chloride	see Ethylene,1,1-dichloro-										
Vinyl iodide	iodoethene;CH_2:CHI	153.95	2.08;2.037^{20}	56	-	1.5385^{20}		i	∞		
Vinyl isoamyl ether	$(CH_3)_2CHCH_2CH_2OCH$:CH_2	114.19	0.7826^{20}	112-3;760	-	1.4072^{25}		i	s	vs	
Vinyl isobutyl ether	$(CH_3)_2CHCH_2OCH$:CH_2	100.16	0.7645^{20};$0.7692^{20/20}$	83	-112;-132.3	1.3966^{20};1.3998^{25}	76.0^{20}	ss	vs	∞	vs:Ac;vs:Bz
Vinyl isopropyl ether	$(CH_3)_2CHOCH$:CH_2	86.14	$0.7534^{20/20}$	55.6	-140	1.3840^{25};1.3830^{25}		ss	vs	vs	
Vinyl 2-methoxyethyl ether	$CH_3OCH_2CH_2OCH$:CH_2	86.29	0.8967	109	-83	1.4072^{25}					
Vinyl methyl ether	methoxyethene;CH_3OCH:CH_2	58.08	0.7725^{20};$0.7500^{20/20}$	1C;8	-122	1.3730^0;1.3947^{25}	-	ss	vs	vs	vs:Ac;vs:Bz
-,1-amyl-	$CH_3OC(C_5H_{11})$:CH_2	128.23	-	144.5	-	1.4284^{25}					

Name	Synonyms and Formula	Mol.Wt.	Density	Boiling Point [°C]	Melting Point [°C]	Refractive Index	Vapor pres- sure (mbar)	Sol. w	Sol. alc	Sol. eth
Vinyl methyl ether			-	38	-	1.3816^{20}	-			
--,1-methyl-	$CH_3OC(C_5H_{11}):CH_2$	72.11	-	-	-	1.5422^{20}	-			
--,1-phenyl-	1-methoxystyrene; $CH_3OC(C_6H_5):CH_2$	134.18	-	196	-	$1.4084^{20};1.4086^{20}$	-			
Vinyl methyl ketone	$CH_2:CHCOCH_3$	69.08	$0.8636^{20};0.8393^{25}$	$32;81.4$	-	1.4835^{20}	-			
Vinyl methyl sulfide	$CH_2:CHSCH_3$	74.13	-	66.8^{745}	-	1.3095^{20}	-			
Vinyl pentafluoro- propionate	$C_2F_5CO_2CH:CH_2$	194.07	-	58	-	-	-			
Vinyl phenyl ether	$C_6H_5OCH:CH_2$	120.15	0.9770^{20}	$155\text{-}7$	-	1.5226^{20}	-	i	.	vs
--,1-bromo-	$C_6H_5OCBr:CH_2$	199.06	-	121^{13}	-	1.5700^{20}	-			
--,1-chloro-	$C_6H_5OCCl:CH_2$	154.60	-	106^{20}	-	1.5511^{20}	-			
--,1-methyl-	$C_6H_5OC(CH_3):CH_2$	134.18	-	169^{14}	-	1.5172^{23}	-			
--,1-phenyl-	$C_6H_5OC(C_6H_5):CH_2$	196.25	-	151^{14}	-	1.1073^{20}	-			
Vinyl phenyl sulfide	$C_6H_5SCH:CH_2$	136.20	$0.9173^{20/20}$	$201;78^{4}$	-	1.5883^{20}	46.7^{20}			
Vinyl propionate	$C_2H_5CO_2CH:CH_2$			95	-	-	-			
Vinyl propyl ether	$C_3H_7OCH:CH_2$	86.13	0.9985^{20}	$65.5\ d$	-	1.3908^{20}	-			
2-Vinylpyridine	$CH_2:CHC_5H_4N$	105.14	0.9985^{20}	$158\text{-}9;50\text{-}5^{4};79\text{-}82^{29}$	-	1.5494^{20}	-	ss	vs	vs ; vs:Chl;vs:Ac
3-Vinylpyridine	$CH_2:CHC_5H_4N$	105.14		68^{18}	-	1.5530^{20}	-	h	h	
4-Vinylpyridine	$CH_2:CHC_5H_4N$	105.14	0.9800^{20}	$59^{12};65^{15}$	-	$1.5499^{25};1.5449^{20}$	-	s	s	ss
1-Vinyl-2-pyrrolidone	$CH_2CH_2CH_2CONCH:CH_2$	111	1.04^{25}	$94\text{-}6^{13\text{-}4}$	-	1.5120^{25}	-	∞	∞	
Vinyl sulfide	ethenylthioethene;divinyl sulfide; $CH_2:CHSCH:CH_2$	86.15	$0.9174^{15};0.9125^{20}$	$86;84^{759}$	-	0.9174^{15}	-	ss		s:Ac
Vinylsulfonic acid	see Ethylene sulfonic acid									
Vinyl thioethers	see Vinyl corresponding sulfides									
Vinyl trifluoroacetate	$CF_3CO_2CH:CH_2$	140.06	1.2031^{25}	$39.5^{757};40.5$	-	$1.3106^{25};1.3151^{25}$	-			

ISOREFRACTIVE AND ISOPYCNIC SOLVENT PAIRS

H.-G. Elias

Midland Macromolecular Institute

Midland, Mich.

Isorefractive solvents are solvents having the same refractive index, and isopycnic solvents, those which have the same density.

The determination of molecular weights in mixed solvents by non-colligative methods, such as light scattering and ultracentrifugation will lead only to <u>apparent</u> molecular weights, if non-isorefractive solvent components are used. The observed apparent increase or decrease of molecular weights depends in sign and magnitude on preferential solvation as well as on the refractive index increment of solvent component 2 in solvent component 1 at a fixed polymer concentration. The effect will disappear, if isorefractive solvent pairs are used.

A similar effect may be observed in ultracentrifugal experiments with nonisopycnic solvent pairs. To suppress these effects, isopycnic/isorefractive solvent pairs should be used. A table of isorefractive and isopycnic solvent pairs was prepared (for $25\,^{\circ}$C), starting with 392 commonly used solvents (1). A solvent pair was classified as isorefractive and isopycnic if the deviations between the respective components were not higher as $\pm$ 0.002 in refractive index and $\pm$ 0.015 g/ml in density. No check on compatibility of the components was made.

Solvent 1	Solvent 2	Refractive Index		Density	
		1	2	1	2
Acetone	ethanol	1.357	1.359	0.788	0.786
Ethyl formate	methyl acetate	1.358	1.360	0.916	0.935
Ethanol	propionitrile	1.359	1.363	0.786	0.777
2,2-Dimethylbutane	2-methylpentane	1.366	1.369	0.644	0.649
2-Methylpentane	n-hexane	1.369	1.372	0.649	0.655
	2,3-dimethylbutane	1.369	1.372	0.649	0.657
	3-methylpentane	1.369	1.374	0.649	0.660
2,3-Dimethylbutane	n-hexane	1.372	1.372	0.657	0.655
	3-methylpentane	1.372	1.374	0.657	0.660
n-Hexane	3-methylpentane	1.372	1.374	0.655	0.660
Isopropyl acetate	2-chloropropane	1.375	1.376	0.868	0.865
2-Butanone	butyraldehyde	1.377	1.378	0.801	0.799
Butyraldehyde	butyronitrile	1.378	1.382	0.799	0.786
Propyl ether	butyl ethyl ether	1.379	1.380	0.753	0.746
2,4-Dimethylpentane	2-methylhexane	1.379	1.382	0.799	0.786
Acetaldehyde-diethyl acetal	butyl ethyl ether	1.379	1.380	0.753	0.746
Propyl acetate	ethyl propionate	1.382	1.382	0.883	0.888
	isobutyl formate	1.382	1.383	0.883	0.881
	1-chloropropane	1.382	1.386	0.883	0.890
Butyronitrile	tert-butanol	1.382	1.385	0.786	0.781
Ethyl propionate	isobutyl formate	1.382	1.383	0.888	0.881
	1-chloropropane	1.382	1.386	0.888	0.890
2-Methylhexane	n-heptane	1.382	1.385	0.674	0.680
	3-methylhexane	1.382	1.386	0.674	0.683
Propanol	3-methyl-2-butanone	1.383	1.386	0.806	0.807
	2-pentanone	1.383	1.387	0.806	0.804
Isobutyl formate	1-chloropropane	1.383	1.386	0.881	0.890
	sec-butyl acetate	1.383	1.387	0.881	0.868
	butyl formate	1.383	1.387	0.881	0.888
Diethylamine	n-propylamine	1.384	1.386	0.702	0.713
n-Heptane	3-methylhexane	1.385	1.386	0.680	0.683
	2,3,3-trimethylbutane	1.385	1.387	0.680	0.686
	2,2,4-trimethylpentane	1.385	1.389	0.680	0.687
	2,3-dimethylpentane	1.385	1.389	0.680	0.691
3-Methylhexane	2,3,3-trimethylbutane	1.386	1.387	0.683	0.686
	2,2,4-trimethylpentane	1.386	1.389	0.683	0.687
	2,3-dimethylpentane	1.386	1.389	0.683	0.691
1-Chloropropane	butyl formate	1.386	1.387	0.890	0.888
3-Methyl-2-butanone	2-pentanone	1.386	1.390	0.807	0.802
	3-pentanone	1.386	1.390	0.807	0.810
n-Propylamine	diisopropylamine	1.386	1.390	0.713	0.712
	sec-butylamine	1.386	1.390	0.713	0.720
2,3,3-Trimethylbutane	2,2,4-trimethylpentane	1.387	1.389	0.686	0.683
	2,3-dimethylpentane	1.387	1.389	0.686	0.691

1) H.-G. Elias, F. W. Ibrahim, Makromol. Chem. 65, 127 (1963).

ISOREFRACTIVE AND ISOPYCNIC SOLVENT PAIRS

Solvent 1	Solvent 2	Refractive Index		Density	
		1	2	1	2
sec-Butyl acetate	methyl butyrate	1.387	1.391	0.868	0.875
Butyl formate	n-dodecane	1.387	1.391	0.888	0.775
Isobutyl acetate	methyl butyrate	1.388	1.391	0.871	0.875
	butyl acetate	1.388	1.392	0.871	0.877
2, 2, 4-Trimethylpentane	2, 3-dimethylpentane	1.389	1.389	0.687	0.691
Diisopropylamine	sec-butylamine	1.390	1.390	0.712	0.720
2-Pentanone	3-pentanone	1.390	1.390	0.802	0.810
	4-methyl-2-pentanone	1.390	1.394	0.802	0.797
	2-methyl-1-propanol	1.390	1.394	0.802	0.798
3-Pentanone	4-methyl-2-pentanone	1.390	1.394	0.810	0.797
	2-methyl-1-propanol	1.390	1.394	0.810	0.798
Methyl butyrate	butyl acetate	1.391	1.392	0.875	0.877
	2-chlorobutane	1.391	1.395	0.875	0.868
2-Chloro-2-methylpropane	2-chlorobutane	1.392	1.395	0.872	0.868
Butyl acetate	2-chlorobutane	1.392	1.395	0.877	0.868
4-Methyl-2-pentanone	2-methyl-1-propanol	1.394	1.394	0.797	0.798
	valeronitrile	1.394	1.395	0.797	0.795
	2-butanol	1.394	1.395	0.797	0.803
	2-hexanone	1.394	1.395	0.797	0.810
	1-butanol	1.394	1.397	0.797	0.812
	methacrylonitrile	1.394	1.398	0.797	0.795
	3-methyl-2-pentanone	1.394	1.398	0.797	0.808
2-Methyl-1-propanol	valeronitrile	1.394	1.395	0.798	0.795
	2-butanol	1.394	1.395	0.798	0.803
	2-hexanone	1.394	1.395	0.798	0.810
	butanol	1.394	1.397	0.798	0.812
	methacrylonitrile	1.394	1.398	0.798	0.795
	3-methyl-2-pentanone	1.394	1.398	0.798	0.808
Octane	2, 2, 5-trimethylhexane	1.395	1.397	0.698	0.703
2-Butanol	butanol	1.395	1.397	0.803	0.812
	methacrylonitrile	1.395	1.398	0.803	0.795
	3-methyl-2-pentanone	1.395	1.398	0.803	0.808
	2, 4-dimethyl-3-pentanone	1.395	1.399	0.803	0.805
2-Hexanone	butanol	1.395	1.397	0.810	0.812
	methacrylonitrile	1.395	1.398	0.810	0.795
	3-methyl-2-pentanone	1.395	1.398	0.810	0.808
	2, 4-dimethyl-3-pentanone	1.395	1.399	0.810	0.805
Valeronitrile	methacrylonitrile	1.395	1.398	0.795	0.795
	3-methyl-2-pentanone	1.395	1.398	0.795	0.808
	2, 4-dimethyl-3-pentanone	1.395	1.399	0.795	0.805
2-Hexanone	3-methyl-2-pentanone	1.395	1.398	0.810	0.808
Isobutylamine	triethylamine	1.395	1.399	0.729	0.723
	n-butylamine	1.395	1.399	0.729	0.736
2-Chlorobutane	isobutyl n-butyrate	1.395	1.399	0.868	0.860
Butyric acid	2-methoxy-ethanol	1.396	1.400	0.955	0.960
Butanol	3-methyl-2-pentanone	1.397	1.398	0.812	0.808
	2, 4-dimethyl-3-pentanone	1.397	1.399	0.812	0.805
1-Chloro-2-methylpropane	isobutyl n-butyrate	1.397	1.399	0.872	0.860
	amyl acetate	1.397	1.400	0.872	0.871
	1-chlorobutane	1.397	1.400	0.872	0.881
2, 5, 5-Trimethylhexane	2, 2, 3-trimethylpentane	1.397	1.401	0.703	0.712
Methyl methacrylate	3-methyl-2-pentanone	1.398	1.398	0.795	0.808
Methacrylonitrile	2, 4-dimethyl-3-pentanone	1.398	1.399	0.795	0.805
	2-methyl-2-butanol	1.398	1.404	0.795	0.805
3-Methyl-2-pentanone	2, 4-dimethyl-3-pentanone	1.398	1.399	0.808	0.805
Triethylamine	n-butylamine	1.399	1.399	0.723	0.736
	2, 2, 3-trimethylpentane	1.399	1.401	0.723	0.712
	n-nonane	1.399	1.401	0.723	0.714
Triethylamine	dipropylamine	1.399	1.401	0.723	0.736
n-Butylamine	n-dodecane	1.399	1.400	0.736	0.746
Isobutyl n-butyrate	amyl acetate	1.399	1.400	0.860	0.871
	isoamyl acetate	1.399	1.403	0.860	0.868
	1-chlorobutane	1.399	1.401	0.860	0.875
1-Nitropropane	propionic anhydride	1.399	1.400	0.995	1.007

Solvent 1	Solvent 2	Refractive Index		Density	
		1	2	1	2
Amyl acetate	1-chlorobutane	1.400	1.400	0.871	0.881
	tetrahydrofuran	1.400	1.404	0.871	0.885
n-Dodecane	dipropylamine	1.400	1.403	0.746	0.736
	cyclopentane	1.400	1.404	0.746	0.740
1-Chlorobutane	tetrahydrofuran	1.400	1.404	0.871	0.885
2,2,3-Trimethylpentane	n-nonane	1.401	1.403	0.712	0.714
Isovaleric acid	2-ethoxyethanol	1.402	1.405	0.923	0.926
	valeric acid	1.402	1.406	0.923	0.936
Dipropylamine	cyclopentane	1.403	1.404	0.736	0.740
	methylcyclopentane	1.403	1.407	0.736	0.744
n-Nonane	2,2,4-trimethyl-1-pentene	1.403	1.407	0.714	0.712
Isoamyl acetate	tributyl borate	1.403	1.407	0.868	0.854
2-Pentanol	2-methyl-2-butanol	1.404	1.404	0.804	0.805
	3-methyl-1-butanol	1.404	1.404	0.804	0.805
	4-heptanone	1.404	1.405	0.804	0.813
	2-heptanone	1.404	1.406	0.804	0.811
2-Methylbutanol	3-methyl-1-butanol	1.404	1.404	0.805	0.805
	capronitrile	1.404	1.405	0.805	0.801
	4-heptanone	1.404	1.405	0.805	0.813
	2-heptanone	1.404	1.406	0.805	0.811
	pentanol	1.404	1.408	0.805	0.810
	3-methyl-2-butanol	1.404	1.408	0.805	0.815
3-Methyl-1-butanol	capronitrile	1.404	1.405	0.805	0.801
	4-heptanone	1.404	1.405	0.805	0.813
	2-heptanone	1.404	1.406	0.805	0.811
	pentanol	1.404	1.408	0.805	0.810
	3-methyl-2-butanol	1.404	1.408	0.805	0.815
Cyclopentane	methylcyclopentane	1.404	1.407	0.740	0.744
Capronitrile	4-heptanone	1.405	1.405	0.801	0.813
	2-heptanone	1.405	1.406	0.801	0.811
	2-pentanol	1.405	1.407	0.801	0.804
	pentanol	1.405	1.408	0.801	0.810
	3-methyl-2-butanol	1.405	1.408	0.801	0.815
	4-methyl-2-pentanol	1.405	1.409	0.801	0.802
	3-isopropyl-2-pentanone	1.405	1.409	0.801	0.808
	2-methyl-1-butanol	1.405	1.409	0.801	0.815
4-Heptanone	2-heptanone	1.405	1.406	0.813	0.811
	pentanol	1.405	1.408	0.813	0.810
	3-methyl-2-butanol	1.405	1.408	0.813	0.815
	4-methyl-2-pentanol	1.405	1.409	0.813	0.802
	3-isopropyl-2-pentanone	1.405	1.409	0.813	0.808
	2-methyl-1-butanol	1.405	1.409	0.813	0.815
2-Ethoxy-ethanol	valeric acid	1.405	1.406	0.926	0.936
2-Heptanone	pentanol	1.406	1.408	0.811	0.810
	3-methyl-2-butanol	1.406	1.408	0.811	0.815
	4-methyl-2-pentanol	1.406	1.409	0.811	0.802
	3-isopropyl-2-pentanone	1.406	1.409	0.811	0.808
	2-ethoxy-ethanol	1.406	1.409	0.811	0.815
	2-methyl-1-butanol	1.406	1.409	0.811	0.815
	amyl ether	1.406	1.410	0.811	0.799
2-Pentanol	pentanol	1.407	1.408	0.804	0.810
	3-methyl-2-butanol	1.407	1.408	0.804	0.815
	4-methyl-2-pentanol	1.407	1.409	0.804	0.802
	3-isopropyl-2-pentanone	1.407	1.409	0.804	0.808
	2-methyl-1-butanol	1.407	1.409	0.804	0.815
	amyl ether	1.407	1.410	0.804	0.799
2,2,4-Trimethyl-1-pentene	n-decane	1.407	1.409	0.712	0.726
Tributyl borate	isoamyl isovalerate	1.407	1.410	0.854	0.853
	allyl alcohol	1.407	1.411	0.854	0.847
Pentanol	3-methyl-2-butanol	1.408	1.408	0.810	0.815
	4-methyl-2-pentanol	1.408	1.409	0.810	0.802
	3-isopropyl-2-pentanone	1.408	1.409	0.810	0.808
	2-methyl-1-butanol	1.408	1.409	0.810	0.815
	amyl ether	1.408	1.410	0.810	0.799

ISOREFRACTIVE AND ISOPYCNIC SOLVENT PAIRS

Solvent 1	Solvent 2	Refractive Index		Density	
		1	2	1	2
3-Methyl-2-butanol	4-methyl-2-pentanol	1.408	1.409	0.815	0.802
	3-isopropyl-2-pentanone	1.408	1.409	0.815	0.808
	2-methyl-1-butanol	1.408	1.409	0.815	0.815
	amyl ether	1.408	1.410	0.815	0.799
4-Methyl-2-pentanol	3-isopropyl-2-pentanone	1.409	1.409	0.802	0.808
	2-methyl-1-butanol	1.409	1.409	0.802	0.815
	amyl ether	1.409	1.410	0.802	0.799
3-Isopropyl-2-pentanone	2-methyl-1-butanol	1.409	1.409	0.808	0.815
	amyl ether	1.409	1.410	0.808	0.799
2-Methyl-1-butanol	amyl ether	1.409	1.410	0.815	0.799
Isoamyl isovalerate	allyl alcohol	1.410	1.411	0.853	0.847
Amyl ether	2-octanone	1.410	1.414	0.799	0.814
2,4-Dimethyldioxane	allyl chloride	1.412	1.413	0.935	0.932
	caproic acid	1.412	1.415	0.935	0.923
Diethyl malonate	ethyl cyanoacetate	1.412	1.415	1.051	1.056
Allyl chloride	capric acid	1.413	1.415	0.932	0.923
2-Octanone	3-methyl-2-heptanone	1.414	1.415	0.814	0.818
	hexanol	1.414	1.416	0.814	0.814
	2-pentanol	1.414	1.416	0.814	0.826
	caprylonitrile	1.414	1.418	0.814	0.810
	2-heptanol	1.414	1.418	0.814	0.818
3-Octanone	3-methyl-2-heptanone	1.414	1.415	0.830	0.818
	2-pentanol	1.414	1.416	0.830	0.826
3-Methyl-2-heptanone	hexanol	1.415	1.416	0.818	0.814
	2-pentanol	1.415	1.416	0.818	0.826
	caprylonitrile	1.415	1.418	0.818	0.810
	2-heptanol	1.415	1.418	0.818	0.818
Hexanol	2-pentanol	1.416	1.416	0.814	0.826
	caprylonitrile	1.416	1.418	0.814	0.810
	2-heptanol	1.416	1.418	0.814	0.818
	3-methyl-2-pentanol	1.416	1.420	0.814	0.823
	2-ethyl-1-butanol	1.416	1.420	0.814	0.829
2-Pentanol	2-heptanol	1.416	1.418	0.826	0.818
	3-methyl-2-pentanol	1.416	1.420	0.826	0.823
	2-ethyl-1-butanol	1.416	1.420	0.826	0.829
Dibutylamine	allylamine	1.416	1.419	0.756	0.758
Caprylonitrile	2-heptanol	1.418	1.418	0.810	0.818
	3-methyl-2-pentanol	1.418	1.420	0.810	0.823
	heptanol	1.418	1.422	0.810	0.818
2-Heptanol	3-methyl-2-pentanol	1.418	1.420	0.818	0.823
	2-ethyl-1-butanol	1.418	1.420	0.818	0.829
	heptanol	1.418	1.422	0.818	0.818
	3-isopropyl-2-heptanone	1.418	1.423	0.818	0.815
Allylamine	methylcyclohexane	1.419	1.421	0.758	0.765
3-Methyl-2-pentanol	2-ethyl-1-butanol	1.420	1.420	0.823	0.829
	heptanol	1.420	1.422	0.823	0.818
	3-isopropyl-2-heptanone	1.420	1.423	0.815	0.815
2-Ethyl-1-butanol	heptanol	1.420	1.422	0.829	0.818
	3-isopropyl-2-heptanone	1.420	1.423	0.829	0.815
Methylcyclohexane	cyclohexane	1.421	1.424	0.765	0.774
Heptanol	3-isopropyl-2-heptanone	1.422	1.423	0.818	0.815
3-Isopropyl-2-heptanone	octanol	1.423	1.427	0.815	0.821
	3-methyl-2-pentanol	1.423	1.427	0.815	0.824
3-Chloro-2-methyl-1-propene	caprylic acid	1.425	1.426	0.917	0.905
Caprylic acid	N-methyl-alaninenitrile	1.426	1.429	0.905	0.895
Octanol	3-methyl-2-pentanol	1.427	1.427	0.821	0.824
1-Chlorooctane	1-chloro-2-ethylhexane	1.428	1.430	0.867	0.872
2-Methyl-7-ethylnonane	2-methyl-7-ethyl-4-undecanone	1.433	1.435	0.830	0.832
Butyrolactone	chloro-tert-butanol	1.434	1.436	1.051	1.059
	1,3-propanediol	1.434	1.438	1.051	1.049
	diethyl maleate	1.434	1.438	1.051	1.064
4-n-Propyl-5-ethyldioxane	N-methyl-morpholine	1.435	1.436	0.927	0.924
2-Methyl-7-ethyl-4-undecanone	2-methyl-7-ethyl-4-nonanol	1.435	1.438	0.832	0.829

Solvent 1	Solvent 2	Refractive Index		Density	
		1	2	1	2
2-Methyl-7-ethyl-4-undecanone	6-ethyl-2-nonanol	1.435	1.438	0.832	0.836
6-Ethyl-3-octanol	5-ethyl-2-nonanol	1.435	1.438	0.832	0.830
Chloro-tert-butanol	1,3-propanediol	1.436	1.438	1.059	1.049
	diethyl maleate	1.436	1.438	1.059	1.064
N-Methyl-morpholine	dibutyl sebacate	1.436	1.440	0.924	0.932
2-Methyl-7-ethyl-4-nonanol	5-ethyl-2-nonanol	1.438	1.438	0.829	0.830
	6-ethyl-3-octanol	1.438	1.438	0.829	0.836
	butanethiol	1.438	1.440	0.829	0.837
	2-methyl-7-ethyl-4-undecanol	1.438	1.442	0.829	0.829
	ethyl sulfide	1.438	1.442	0.829	0.831
	6-ethyl-3-decanol	1.438	1.441	0.829	0.838
5-Ethyl-2-nonanol	6-ethyl-3-octanol	1.438	1.438	0.830	0.836
	butanethiol	1.438	1.440	0.830	0.837
	2-methyl-7-ethyl-4-undecanol	1.438	1.442	0.830	0.829
	ethyl sulfide	1.438	1.442	0.830	0.831
	6-ethyl-3-decanol	1.438	1.441	0.830	0.838
1,3-Propanediol	diethyl maleate	1.438	1.438	1.049	1.064
Methyl salicylate	2-methyl-7-ethyl-1-undecanol	1.438	1.442	0.836	0.829
	ethyl sulfide	1.438	1.442	0.836	0.831
	butanethiol	1.438	1.442	0.836	0.837
6-Ethyl-3-octanol	6-ethyl-3-decanol	1.438	1.441	0.836	0.838
Butanethiol	6-ethyl-3-decanol	1.440	1.441	0.837	0.838
	2-methyl-7-ethyl-4-undecanol	1.440	1.442	0.837	0.829
	ethyl sulfide	1.440	1.442	0.837	0.831
	mesityl oxide	1.440	1.442	0.837	0.850
6-Ethyl-3-decanol	2-methyl-7-ethyl-4-undecanol	1.441	1.442	0.838	0.829
	ethyl sulfide	1.441	1.442	0.838	0.831
	mesityl oxide	1.441	1.442	0.838	0.850
1-Chlorododecane (techn.)	mesityl oxide	1.441	1.442	0.862	0.850
	butyl steareate	1.441	1.442	0.862	0.854
	1-chlorotetradecane (techn.)	1.441	1.445	0.862	0.858
2-Methyl-7-ethyl-4-undecanol	ethyl sulfide	1.442	1.442	0.829	0.831
	2-butyloctyl-3-aminopropyl ether	1.442	1.446	0.829	0.842
Mesityl oxide	butyl stearate	1.442	1.442	0.850	0.854
	1-chlorotetradecane (techn.)	1.442	1.445	0.850	0.858
	2-butyloctyl-3-aminopropyl ether	1.442	1.446	0.850	0.842
Butyl stearate	1-chlorotetradecane (techn.)	1.442	1.445	0.850	0.858
	2-butyloctyl-3-aminopropyl ether	1.442	1.446	0.854	0.842
Ethyl sulfide	2-butyloctyl-3-aminopropyl ether	1.442	1.446	0.831	0.842
1,3-Butanediol sulfite	1,2-dichloroethane	1.444	1.444	1.231	1.245
	trans-1,2-dichloroethylene	1.444	1.444	1.231	1.257
1,2-Dichloroethane	trans-1,2-dichloroethylene	1.444	1.444	1.231	1.257
1-Chlorotetradecane (techn.)	2-butyloctyl-3-aminopropyl ether	1.445	1.446	0.857	0.842
	1-chlorohexadecane (techn.)	1.445	1.448	0.857	0.859
Diethylene glycol	formamide	1.445	1.446	1.128	1.129
	ethylene glycol diglycidyl ether	1.445	1.447	1.128	1.134
2-Butyloctyl-3-aminopropyl ether	3-lauroxy-1-propylamine	1.446	1.447	0.842	0.840
Formamide	ethylene glycol diglycidyl ether	1.446	1.447	1.129	1.134
2-Methylmorpholine	cyclohexanone	1.446	1.448	0.951	0.943
	1-amino-2-propanol	1.446	1.448	0.951	0.961
Dipropyleneglycol-monoethyl ether	tetrahydrofurfuryl alcohol	1.446	1.450	1.043	1.050
1-Amino-2-methyl-2-pentanol	2-butylcyclohexanone	1.449	1.453	0.904	0.901
Tetrahydrofurfurylalcohol	dipropyleneglycol-monoethyl ether	1.450	1.452	1.050	1.047
3-Methyl-5-ethyl-2,4-heptanediol	2-propylcyclohexanone	1.452	1.452	0.922	0.923
	4-methylcyclohexanone	1.452	1.454	0.922	0.908
	3-methylcyclohexanol	1.452	1.455	0.922	0.913
	2,2'-dimethyl-2,2'-dipropyldiethanolamine	1.452	1.456	0.922	0.922
2-Propylcyclohexanone	4-methylcyclohexanol	1.452	1.454	0.923	0.908
	3-methylcyclohexanol	1.452	1.455	0.923	0.913
	2,2'-dimethyl-2,2'-dipropyldiethanolamine	1.452	1.456	0.923	0.922
	1,8-cineole	1.452	1.456	0.923	0.921
4-Methylcyclohexanol	2,2'-dimethyl-2,2'-dipropyldiethanolamine	1.454	1.456	0.908	0.922
3-Methylcyclohexanol	2,2'-dimethyl-2,2'-dipropyldiethanolamine	1.455	1.456	0.913	0.922
Cyclohexylamine	1-chloroeicosane (techn.)	1.456	1.459	0.862	0.872

Solvent 1	Solvent 2	Refractive Index		Density	
		1	2	1	2
1-Chloroeicosane (techn.)	oleic acid	1.459	1.459	0.872	0.887
Oleic acid	2-(β-ethyl)butylcyclohexanone	1.459	1.461	0.887	0.892
	2-butylcyclohexanol	1.459	1.462	0.887	0.898
	2-(β-ethyl)-hexylcyclohexanone	1.459	1.463	0.887	0.892
(1,1',2,2'-Tetramethyl)-diethanolamine	1-aminopropanol	1.459	1.459	0.973	0.965
	N-(n-butyl)-diethanolamine	1.459	1.461	0.973	0.965
Carbon tetrachloride	4,5-dichloro-1,3-dioxolane-2-one	1.459	1.461	1.584	1.591
2-(β-Ethyl)-butylcyclohexanone	2,4-bis-(α-phenylethyl)-phenylmethyl ether	1.461	1.462	0.892	0.898
	2-(β-ethyl)-hexylcyclohexanone	1.461	1.463	0.892	0.892
N-(n-Butyl)-diethanolamine	cyclohexanol	1.461	1.465	0.965	0.968
2-Butylcyclohexanol	2-(β-ethyl)-hexylcyclohexanone	1.462	1.463	0.898	0.892
	2-ethylcyclohexanol	1.462	1.463	0.898	0.908
N-β-Oxypropyl-morpholine	fluorobenzene	1.462	1.463	1.013	1.020
Fluorobenzene	N(2-hydroxyethyl)-2-hydroxybutylamine	1.463	1.467	1.020	1.027
d-α-Pinene	1-α-pinene	1.464	1.465	0.855	0.855
	trans-decahydronaphthalene	1.464	1.468	0.855	0.867
m-Fluorotoluene	p-fluorotoluene	1.465	1.467	0.994	0.994
	o-fluorotoluene	1.465	1.468	0.994	0.995
1-α-Pinene	trans-decahydronaphthalene	1.465	1.468	0.855	0.867
p-Fluorotoluene	o-fluorotoluene	1.467	1.468	0.994	0.995
N-(2-hydroxyethyl)-2-hydroxybutylamine	N-(2-hydroxyethyl)-2-hydroxypropylamine	1.467	1.468	1.027	1.042
	3-allyloxy-2-hydroxypropylamine	1.467	1.469	1.027	1.017
	di-(2-hydroxybutyl)-ethanolamine	1.467	1.469	1.027	1.018
	di-(2-hydroxypropyl)-ethanolamine	1.467	1.469	1.027	1.042
	di-(2-hydroxypropyl)-ethanolamine	1.468	1.469	1.042	1.042
3-Allyloxy-2-hydroxypropylamine	di-(2-hydroxybutyl)-ethanolamine	1.469	1.469	1.017	1.018
cis-Decahydronaphthalene	1-methoxy-1-butene-3-yn	1.479	1.480	0.893	0.902
	n-dodecyl-4-tert-butylphenyl ether	1.479	1.482	0.893	0.881
	n-dodecylphenyl ether	1.479	1.482	0.893	0.891
	4-dodecyl-4-methylphenyl ether	1.479	1.483	0.893	0.889
1-Methoxy-1-butene-3-yn	n-dodecylphenyl ether	1.480	1.482	0.902	0.891
	n-dodecyl-4-methylphenyl ether	1.480	1.483	0.902	0.889
n-Dodecyl-4-tert-butylphenyl ether	n-dodecylphenyl ether	1.482	1.482	0.881	0.891
	n-dodecyl-4-methylphenyl ether	1.482	1.483	0.881	0.889
n-Dodecylphenyl ether	n-dodecyl-4-methylphenyl ether	1.482	1.483	0.891	0.889
Butylbenzene	dioctylbenzene (90 % p.; 10 % m.)	1.487	1.487	0.856	0.856
	p-cymene	1.487	1.488	0.856	0.853
	isopropylbenzene	1.487	1.489	0.856	0.857
	tert-butylcumene (80 % p.; 15 % m.)	1.487	1.490	0.856	0.856
	n-propylbenzene	1.487	1.490	0.856	0.856
	sec-butylbenzene	1.487	1.490	0.856	0.856
	hexyl-m-xylene	1.487	1.490	0.856	0.860
	tert-butylbenzene	1.487	1.490	0.856	0.862
	isopropylethylbenzene (35 % p.; 60 % m.)	1.487	1.491	0.856	0.856
p-Cymene	isopropylbenzene	1.488	1.489	0.853	0.857
	tert-butylcumene (80 % p.; 15 % m.)	1.488	1.490	0.853	0.856
	n-propylbenzene	1.488	1.490	0.853	0.858
	sec-butylbenzene	1.488	1.490	0.853	0.858
	hexyl-m-xylene (mainly 1,3,5-)	1.488	1.490	0.853	0.860
	tert-butylbenzene	1.488	1.490	0.853	0.862
	isopropylethylbenzene (35 % p.; 60 % m.)	1.488	1.491	0.853	0.856
	tert-butyltoluene (85 % p.; 10 % m.)	1.488	1.491	0.853	0.858
	hexylcumene (90 % p.; 5 % m.)	1.488	1.492	0.853	0.863
	octyltoluene (96 % p.; 2 % m.)	1.488	1.492	0.853	0.866
Isopropylbenzene	tert-butylcumene (80 % p.; 15 % m.)	1.489	1.490	0.857	0.856
	n-propylbenzene	1.489	1.490	0.857	0.858
	sec-butylbenzene	1.489	1.490	0.857	0.858
	hexyl-m-xylene (mainly 1,3,5-)	1.489	1.490	0.857	0.860
	tert-butylbenzene	1.489	1.490	0.857	0.862
	isopropylethylbenzene (35 % p.; 60 % m.)	1.489	1.491	0.857	0.856
	tert-butyltoluene (80 % p.; 15 % m.)	1.489	1.491	0.857	0.858
tert-Butylcumene (80 % p.; 15 % m.)	n-propylbenzene	1.490	1.490	0.856	0.858
	sec-butylbenzene	1.490	1.490	0.856	0.858
	hexyl-m-xylene (mainly 1,3,5-)	1.490	1.490	0.856	0.860

Solvent 1	Solvent 2	Refractive Index		Density	
		1	2	1	2
tert-Butylcumene (80 % p.; 15 % m.)	tert-butylbenzene	1.490	1.490	0.856	0.862
	isopropylethylbenzene (35 % p.; 60 % m.)	1.490	1.491	0.856	0.856
	tert-butyltoluene (80 % p.; 15 % m.)	1.490	1.491	0.856	0.858
	hexylcumene (90 % p.; 5 % m.)	1.490	1.492	0.856	0.863
	octyltoluene (96 % p.; 2 % m.)	1.490	1.492	0.856	0.866
	octylcumene (90 % p.; 4 % m.)	1.490	1.492	0.856	0.869
	dihexylbenzene	1.490	1.492	0.856	0.870
	p-xylene	1.490	1.493	0.856	0.857
	1,3-diethylbenzene	1.490	1.493	0.856	0.860
	tert-butyl-m-xylene (mainly 1,3,5-)	1.490	1.493	0.856	0.862
	ethylbenzene	1.490	1.493	0.856	0.863
	octylethylbenzene (80-90 % p.; 10 % m.)	1.490	1.493	0.856	0.866
	isopropyl-m-xylene (mainly 1,3,5-)	1.490	1.494	0.856	0.860
	toluene	1.490	1.494	0.856	0.862
n-Propylbenzene	sec-butylbenzene	1.490	1.490	0.858	0.858
	hexyl-m-xylene (mainly 1,3,5-)	1.490	1.490	0.858	0.860
	tert-butylbenzene	1.490	1.490	0.858	0.862
	isopropylethylbenzene (35 % p.; 60 % m.)	1.490	1.491	0.858	0.856
	tert-butyltoluene (80 % p.; 15 % m.)	1.490	1.491	0.858	0.858
	hexylcumene (90 % p.; 5 % m.)	1.490	1.492	0.858	0.863
	octyltoluene (96 % p.; 2 % m.)	1.490	1.492	0.858	0.866
	octylcumene (90 % p.; 4 % m.)	1.490	1.492	0.858	0.869
	dihexylbenzene (85 % p.; 10 % m.)	1.490	1.492	0.858	0.870
	p-xylene	1.490	1.493	0.858	0.857
	1,3-diethylbenzene	1.490	1.493	0.858	0.860
	tert-butyl-m-xylene (mainly 1,3,5-)	1.490	1.493	0.858	0.862
	ethylbenzene	1.490	1.493	0.858	0.862
	octylethylbenzene (80-90 % p.; 10 % m.)	1.490	1.493	0.858	0.866
	isopropyl-m-xylene (mainly 1,3,5-)	1.490	1.494	0.858	0.860
	toluene	1.490	1.494	0.858	0.860
sec-Butylbenzene	hexyl-m-xylene (mainly 1,3,5-)	1.490	1.490	0.858	0.860
	tert-butylbenzene	1.490	1.490	0.858	0.862
	isopropylethylbenzene (35 % p.; 60 % m.)	1.490	1.491	0.858	0.856
	tert-butyltoluene (80 % p.; 15 % m.)	1.490	1.491	0.858	0.858
	hexylcumene (90 % p.; 5 % m.)	1.490	1.492	0.858	0.863
	octyltoluene (96 % p.; 2 " m.)	1.490	1.492	0.858	0.866
	octylcumene (90 % p.; 4 % m.)	1.490	1.492	0.858	0.869
	dihexylbenzene (85 % p.; 10 % m.)	1.490	1.492	0.858	0.870
	p-xylene	1.490	1.493	0.858	0.857
	1,3-diethylbenzene	1.490	1.493	0.858	0.860
	tert-butyl-m-xylene (mainly 1,3,5-)	1.490	1.493	0.858	0.862
	ethylbenzene	1.490	1.493	0.858	0.863
	octylethylbenzene (80-90 % p.; 10 % m.)	1.490	1.493	0.858	0.866
	isopropyl-m-xylene (mainly 1,3,5-)	1.490	1.494	0.858	0.860
	toluene	1.490	1.494	0.858	0.862
Hexyl-m-xylene (mainly 1,3,5-)	tert-butylbenzene	1.490	1.490	0.860	0.862
	isopropylethylbenzene (35 % p.; 60 % m.)	1.490	1.491	0.860	0.856
	tert-butyltoluene (85 % p.; 10 % m.)	1.490	1.491	0.860	0.858
	hexylcumene (90 % p.; 5 % m.)	1.490	1.492	0.860	0.863
	octyltoluene (96 % p.; 2 % m.)	1.490	1.492	0.860	0.866
	octylcumene (90 % p.; 4 % m.)	1.490	1.492	0.860	0.869
	dihexylbenzene (85 % p.; 10 % m.)	1.490	1.492	0.860	0.870
	p-xylene	1.490	1.493	0.860	0.857
	1,3-diethylbenzene	1.490	1.493	0.860	0.860
	tert-butyl-m-xylene (mainly 1,3,5-)	1.490	1.493	0.860	0.862
	ethylbenzene	1.490	1.493	0.860	0.863
	octylethylbenzene (80-90 % p.; 10 % m.)	1.490	1.493	0.860	0.866
	isopropyl-m-xylene (mainly 1,3,5-)	1.490	1.494	0.860	0.860
	toluene	1.490	1.494	0.860	0.862
tert-Butylbenzene	isopropylethylbenzene (35 % p.; 60 % m.)	1.490	1.491	0.862	0.856
	tert-butyltoluene (80 % p.; 15 % m.)	1.490	1.491	0.862	0.858
	hexylcumene (90 % p.; 5 % m.)	1.490	1.492	0.862	0.863
	octyltoluene (96 % p.; 2 % m.)	1.490	1.492	0.862	0.866
	octylcumene (90 % p.; 4 % m.)	1.490	1.492	0.862	0.869

Solvent 1	Solvent 2	Refractive Index		Density	
		1	2	1	2
tert-Butylbenzene	dihexylbenzene (85 % p.; 10 % m.)	1.490	1.492	0.862	0.870
	p-xylene	1.490	1.493	0.862	0.857
	1,3-diethylbenzene	1.490	1.493	0.862	0.860
	tert-butyl-m-xylene (mainly 1,3,5-)	1.490	1.493	0.862	0.862
	ethylbenzene	1.490	1.493	0.862	0.863
	octylethylbenzene (80-90 % p.; 10 % m.)	1.490	1.493	0.862	0.866
	isopropyl-m-xylene (mainly 1,3,5-)	1.490	1.494	0.862	0.860
	toluene	1.490	1.494	0.862	0.862
Isopropylethylbenzene (35 % p.; 60 % m.)	tert-butyltoluene (80 % p.; 15 % m.)	1.491	1.491	0.856	0.858
	hexylcumene (90 % p.; 5 % m.)	1.491	1.492	0.856	0.863
	octyltoluene (96 % p.; 2 % m.)	1.491	1.492	0.856	0.866
	octylcumene (90 % p.; 4 % m.)	1.491	1.492	0.856	0.869
	dihexylbenzene (85 % p.; 10 % m.)	1.491	1.492	0.856	0.870
	p-xylene	1.491	1.493	0.856	0.857
	1,3-diethylbenzene	1.491	1.493	0.856	0.860
	tert-butyl-m-xylene (mainly 1,3,5-)	1.491	1.493	0.856	0.862
	ethylbenzene	1.491	1.493	0.856	0.863
	octylethylbenzene (80-90 % p.; 10 % m.)	1.491	1.493	0.856	0.866
	isopropyl-m-xylene (mainly 1,3,5-)	1.491	1.494	0.856	0.860
	toluene	1.491	1.494	0.856	0.862
	tert-butyl ethylbenzene (70 % p.; 25 % m.)	1.491	1.495	0.856	0.854
	m-xylene	1.491	1.495	0.856	0.860
	hexylethylbenzene (70 % p.; 25 % m.)	1.491	1.495	0.856	0.868
tert-Butyltoluene (85 % p.; 10 % m.)	hexylcumene (90 % p.; 5 % m.)	1.491	1.492	0.858	0.863
	octyltoluene (96 % p.; 2 % m.)	1.491	1.492	0.858	0.866
	octylcumene (90 % p.; 4 % m.)	1.491	1.492	0.858	0.869
	dihexylbenzene (85 % p.; 10 % m.)	1.491	1.492	0.858	0.870
	p-xylene	1.491	1.493	0.858	0.857
	1,3-diethylbenzene	1.491	1.493	0.858	0.860
	tert-butyl-m-xylene (mainly 1,3,5-)	1.491	1.493	0.858	0.862
	ethylbenzene	1.491	1.493	0.858	0.863
	octylethylbenzene (80-90 % p.; 10 % m.)	1.491	1.493	0.858	0.866
	isopropyl-m-xylene (mainly 1,3,5-)	1.491	1.494	0.858	0.860
	toluene	1.491	1.494	0.858	0.862
	tert-butylethylbenzene (70 % p.; 25 % m.)	1.491	1.495	0.858	0.854
	m-xylene	1.491	1.495	0.858	0.860
	hexylethylbenzene (70 % p.; 25 % m.)	1.491	1.495	0.858	0.868
1-Phenyl-1-hydroxyphenyl ether	1,3-dimorpholyl-2-propanol	1.491	1.493	1.081	1.094
Hexylcumene (90 % p.; 5 % m.)	octyltoluene (96 % p.; 2 % m.)	1.492	1.492	0.863	0.866
	octylcumene (90 % p.; 4 % m.)	1.492	1.492	0.863	0.869
	dihexylbenzene (85 % p.; 10 % m.)	1.492	1.492	0.863	0.870
	p-xylene	1.492	1.493	0.863	0.857
	1,3-diethylbenzene	1.492	1.493	0.863	0.860
	tert-butyl-m-xylene (mainly 1,3,5-)	1.492	1.493	0.863	0.862
	ethylbenzene	1.492	1.493	0.863	0.863
	octylethylbenzene (80-90 % p.; 10 % m.)	1.492	1.493	0.863	0.866
	isopropyl-m-xylene (mainly 1,3,5-)	1.492	1.494	0.863	0.860
	toluene	1.492	1.494	0.863	0.862
	tert-butylethylbenzene (70 % p.; 25 % m.)	1.492	1.495	0.863	0.854
	m-xylene	1.492	1.495	0.863	0.860
	hexylethylbenzene (70 % p.; 25 % m.)	1.492	1.495	0.863	0.868
	1,4-diethylbenzene	1.492	1.496	0.863	0.858
	isopropylbenzene	1.492	1.498	o.863	0.857
Dihexylbenzene (85 % p.; 10 % m.)	octylcumene (90 % p.; 4 % m.)	1.492	1.492	0.870	0.869
	p-xylene	1.492	1.493	0.870	0.857
	1,3-diethylbenzene	1.492	1.493	0.870	0.860
	tert-butyl-m-xylene (mainly 1,3,5-)	1.492	1.493	0.870	0.862
	ethylbenzene	1.492	1.493	0.870	0.863
	octylethylbenzene (80-90 % p.; 10 % m.)	1.492	1.493	0.860	0.866
	isopropyl-m-xylene (mainly 1,3,5-)	1.492	1.494	0.870	0.860
	toluene	1.492	1.494	0.870	0.862
	m-xylene	1.492	1.495	0.870	0.860
	hexylethylbenzene (70 % p.; 25 % m.)	1.492	1.495	0.870	0.868

Solvent 1	Solvent 2	Refractive Index		Density	
		1	2	1	2
Dihexylbenzene (85 % p.; 10 % m.)	1,4-diethylbenzene	1.492	1.496	0.870	0.858
	isopropylbenzene	1.492	1.498	0.870	0.857
Octyltoluene (96 % p.; 2 % m.)	octylcumene (90 % p.; 4 % m.)	1.492	1.492	0.866	0.869
	dihexylbenzene (85 % p.; 10 % m.)	1.492	1.492	0.866	0.870
	p-xylene	1.492	1.493	0.866	0.857
	tert-butyl-m-xylene (mainly 1,3,5-)	1.492	1.493	0.866	0.862
	ethylbenzene	1.492	1.493	0.866	0.863
	octylethylbenzene (80-90 % p.; 10 % m.)	1.492	1.493	0.866	0.866
	isopropyl-m-xylene (mainly 1,3,5-)	1.492	1.494	0.866	0.860
	toluene	1.492	1.494	0.866	0.862
	tert-butylethylbenzene (70 % p.; 25 % m.)	1.492	1.495	0.866	0.854
	m-xylene	1.492	1.495	0.866	0.860
	hexylethylbenzene (70 % p.; 25 % m.)	1.492	1.495	0.866	0.868
	1,4-diethylbenzene	1.492	1.496	0.866	0.858
	isopropylbenzene	1.492	1.498	0.866	0.857
Octylcumene (90 % p.; 4 % m.)	p-xylene	1.492	1.493	0.866	0.857
	1,3-diethylbenzene	1.492	1.493	0.866	0.860
	tert-butyl-m-xylene (mainly 1,3,5-)	1.492	1.493	0.866	0.862
	ethylbenzene	1.492	1.493	0.866	0.863
	octylethylbenzene (80-90 % p.; 10 % m.)	1.492	1.493	0.866	0.866
	isopropyl-m-xylene (mainly 1,3,5-)	1.492	1.494	0.866	0.860
	toluene	1.492	1.494	0.866	0.862
	tert-butylethylbenzene (70 % p.; 25 % m.)	1.492	1.495	0.866	0.854
	m-xylene	1.492	1.495	0.866	0.860
	hexylethylbenzene (70 % p.; 25 % m.)	1.492	1.495	0.866	0.868
	1,4-diethylbenzene	1.492	1.496	0.866	0.858
	isopropylbenzene	1.492	1.498	0.866	0.857
p-Xylene	1,3-diethylbenzene	1.493	1.493	0.857	0.860
	tert-butyl-m-xylene (mainly 1,3,5-)	1.493	1.493	0.857	0.862
	ethylbenzene	1.493	1.493	0.857	0.863
	octylethylbenzene (80-90 % p.; 10 % m.)	1.493	1.493	0.857	0.866
	isopropyl-m-xylene (mainly 1,3,5-)	1.493	1.494	0.857	0.860
	toluene	1.493	1.494	0.857	0.854
	tert-butylethylbenzene (70 % p.; 25 % m.)	1.493	1.495	0.857	0.854
	m-xylene	1.493	1.495	0.857	0.860
	hexylethylbenzene (70 % p.; 25 % m.)	1.493	1.495	0.857	0.868
	1,4-diethylbenzene	1.493	1.496	0.857	0.858
	mesitylene	1.493	1.497	0.857	0.861
	hexyltoluene (70 % p.; 25 % m.)	1.493	1.497	0.857	0.870
	isopropylbenzene	1.493	1.498	0.857	0.857
1,3-Diethylbenzene	tert-butyl-m-xylene (mainly 1,3,5-)	1.493	1.493	0.860	0.862
	ethylbenzene	1.493	1.493	0.860	0.863
	octyltoluene (96 % p.; 2 % m.)	1.493	1.493	0.860	0.866
	octylethylbenzene (80-90 % p.; 10 % m.)	1.493	1.493	0.860	0.866
	isopropyl-m-xylene (mainly 1,3,5-)	1.493	1.494	0.860	0.860
	toluene	1.493	1.494	0.860	0.860
	tert-butyl-ethylbenzene (70 % p.; 15 % m.)	1.493	1.495	0.860	0.854
	m-xylene	1.493	1.495	0.860	0.860
	1,4-diethylbenzene	1.493	1.496	0.860	0.858
	mesitylene	1.493	1.497	0.860	0.861
	hexyltoluene (70 % p.; 25 % m.)	1.493	1.497	0.860	0.870
	isopropylbenzene	1.493	1.498	0.860	0.857
tert-Butyl-m-xylene (mainly 1,3,5-)	ethylbenzene	1.493	1.493	0.862	0.863
	octylethylbenzene (80-90 % p.; 10 % m.)	1.493	1.493	0.862	0.866
	isopropyl-m-xylene (mainly 1,3,5-)	1.493	1.494	0.862	0.860
	toluene	1.493	1.494	0.862	0.862
	tert-butyl-ethylbenzene (70 % p.; 25 % m.)	1.493	1.495	0.862	0.854
	m-xylene	1.493	1.495	0.861	0.860
	hexylethylbenzene (70 % p.; 25 % m.)	1.493	1.495	0.862	0.868
	1,4-diethylbenzene	1.493	1.496	0.862	0.858
	mesitylene	1.493	1.497	0.862	0.861
	hexyltoluene (70 % p.; 25 % m.)	1.493	1.497	0.862	0.870
	isopropylbenzene	1.493	1.498	0.862	0.857

Solvent 1	Solvent 2	Refractive Index		Density	
		1	2	1	2
Ethylbenzene	octylethylbenzene (80-90 % p.; 10 % m.)	1.493	1.493	0.863	0.866
	isopropyl-m-xylene (mainly 1,3,5-)	1.493	1.494	0.863	0.860
	toluene	1.493	1.494	0.863	0.862
	tert-butylethylbenzene (70 % p.; 25 % m.)	1.493	1.495	0.863	0.854
	m-xylene	1.493	1.495	0.863	0.860
	hexylethylbenzene (70 % p.; 25 % m.)	1.493	1.495	0.863	0.868
	1,4-diethylbenzene	1.493	1.496	0.863	0.858
	mesitylene	1.493	1.497	0.863	0.861
	hexyltoluene (70 % p.; 25 % m.)	1.493	1.497	0.863	0.870
	isopropylbenzene	1.493	1.498	0.863	0.857
Octylethylbenzene (80-90 % p.; 10 % m.)	isopropyl-m-xylene (mainly 1,3,5-)	1.493	1.494	0.866	0.860
	toluene	1.493	1.494	0.866	0.860
	tert-butyl-ethylbenzene (70 % p.; 25 % m.)	1.493	1.495	0.866	0.854
	m-xylene	1.493	1.495	0.866	0.860
	hexylethylbenzene (70 % p.; 25 % m.)	1.493	1.495	0.866	0.868
	1,4-diethylbenzene	1.493	1.496	0.866	0.858
	mesitylene	1.493	1.497	0.866	0.861
	hexyltoluene (70 % p.; 25 % m.)	1.493	1.497	0.866	0.870
	isopropylbenzene	1.493	1.498	0.866	0.857
Isopropyl-m-xylene (mainly 1,3,5-)	toluene	1.494	1.494	0.860	0.862
	tert-butyl-ethylbenzene (70 % p.; 25 % m.)	1.494	1.495	0.860	0.854
	m-xylene	1.494	1.495	0.860	0.860
	hexylethylbenzene (70 % p.; 25 % m.)	1.494	1.495	0.860	0.868
	1,4-diethylbenzene	1.494	1.496	0.860	0.858
	mesitylene	1.494	1.497	0.860	0.861
	hexyltoluene (70 % p.; 25 % m.)	1.494	1.497	0.860	0.870
	benzene	1.494	1.498	0.860	0.874
Toluene	tert-butylethylbenzene (70 % p.; 25 % m.)	1.494	1.495	0.862	0.854
	m-xylene	1.494	1.495	0.862	0.854
	hexylethylbenzene (70 % p.; 25 % m.)	1.494	1.495	0.862	0.868
	1,4-diethylbenzene	1.494	1.496	0.862	0.858
	mesitylene	1.494	1.497	0.862	0.861
	hexyltoluene (70 % p.; 25 % m.)	1.494	1.497	0.862	0.870
	benzene	1.494	1.498	0.862	0.874
tert-Butylethylbenzene (70 % p.; 25 % m.)	m-xylene	1.495	1.495	0.854	0.860
	1,4-diethylbenzene	1.495	1.496	0.854	0.858
	mesitylene	1.495	1.497	0.854	0.861
m-Xylene	hexylethylbenzene (70 % p.; 25 % m.)	1.495	1.495	0.860	0.868
	1,4-diethylbenzene	1.495	1.496	0.860	0.858
	mesitylene	1.495	1.497	0.860	0.861
	benzene	1.495	1.498	0.860	0.874
Hexylethylbenzene (70 % p.; 25 % m.)	1,4-diethylbenzene	1.495	1.496	0.868	0.858
	mesitylene	1.495	1.497	0.868	0.861
	hexyltoluene (70 % p.; 25 % m.)	1.495	1.497	0.868	0.870
	benzene	1.495	1.497	0.868	0.874
1,4-Diethylbenzene	mesitylene	1.496	1.497	0.858	0.861
	hexyltoluene (70 % p.; 25 % m.)	1.496	1.498	0.858	0.870
Mesitylene	ethylbenzene	1.497	1.497	0.861	0.870
Hexyltoluene (70 % p.; 25 % m.)	benzene	1.497	1.498	0.870	0.874
	1,2-diethylbenzene	1.497	1.501	0.870	0.876
Benzene	mesitylene	1.498	1.498	0.874	0.874
	1,2-diethylbenzene	1.498	1.501	0.874	0.876
Mesitylene	1,2-diethylbenzene	1.498	1.501	0.874	0.876
1,2-Diethylbenzene	o-xylene	1.501	1.503	0.867	0.876
β-Picoline	phenetole	1.504	1.505	0.953	0.961
Phenetole	pyridine	1.505	1.507	0.961	0.978
Cyclohexylcumene (50 % p.; 20 % m.)	cyclohexylethylbenzene (60 % p.; 20 % m.)	1.516	1.520	0.917	0.923
Benzyl acetate	chloro-tert-butylbenzene	1.518	1.521	1.051	1.039
Cyclohexylethylbenzene (60 % p.; 20 % m.)	cyclohexyltoluene	1.520	1.523	0.923	0.923
2-Furfurol	thiophene	1.524	1.526	1.057	1.059
Benzyl alcohol	m-cresol	1.538	1.542	1.041	1.037
m-Cresol	benzaldehyde	1.542	1.544	1.037	1.041
m-Toluidine	o-toluidine	1.566	1.570	0.985	0.994

REFRACTIVE INDICES OF COMMON SOLVENTS

H.-G. Elias
Midland Macromolecular Institute
Midland, Mich.

Measurements which depend upon the difference in refractive index between the polymer and the solvent will, in general, give greater accuracy as the refractive index increment between polymer and solvent is increased. The magnitude of the increment may be either positive or negative. Systems involving refractive index increments are those of light scattering and ultracentrifugation when either schlieren or interference optics are used. A table of commonly used solvents, arranged according to increasing refractive index, will be useful in practical work with many different polymers.

Solvents Arranged According to Increasing Refractive Index

Solvent	n	Solvent	n	Solvent	n
Trifluoroacetic acid	1.283	sec-Butyl acetate	1.387	Tetrahydrofuran	1.404
Trifluoroethanol	1.290	Butyl formate	1.387	Capronitrile	1.405
Octafluoro-1-pentanol	1.316	Isobutyl acetate	1.388	4-Heptanone	1.405
Dodecafluoro-1-heptanol	1.316	2,2,4-Trimethylpentane	1.389	2-Ethoxyethanol	1.405
Methanol	1.326	2,3-Dimethylpentane	1.389	2-Heptanone	1.406
Acetonitrile	1.342	Acetic anhydride	1.389	Valeric acid	1.406
Ethyl ether	1.352	Diisopropylamine	1.390	Diisobutylene	1.407
Acetone	1.357	2-Butylamine	1.390	Methylcyclopentane	1.407
Ethyl formate	1.358	2-Pentanone	1.390	Isoamyl ether	1.407
Ethanol	1.359	3-Pentanone	1.390	2-Pentanol	1.407
Methyl acetate	1.360	Nitroethane	1.390	Tributyl borate	1.407
Propionitrile	1.363	Methyl n-butyrate	1.391	Pentanol	1.408
2,2-Dimethylbutane	1.366	Butyl acetate	1.392	3-Methyl-2-butanol	1.408
Isopropyl ether	1.367	2-Nitropropane	1.392	Diethyl oxalate	1.408
2-Methylpentane	1.369	4-Methyl-2-pentanone	1.394	Decane	1.409
Formic acid	1.369	2-Methyl-1-propanol	1.394	4-Methyl-2-pentanol	1.409
Ethyl acetate	1.370	Octane	1.395	3-Isopropyl-2-pentanone	1.409
Acetic acid	1.370	Isobutylamine	1.395	2-Methyl-1-butanol	1.409
Propionaldehyde	1.371	Valeronitrile	1.395	Butyric anhydride	1.409
n-Hexane	1.372	2-Butanol	1.395	Amyl ether	1.410
2,3-Dimethylbutane	1.372	2-Hexanone	1.395	Isoamyl isovalerate	1.410
3-Methylpentane	1.374	2-Chlorobutane	1.395	1-Chloropentane	1.410
2-Propanol	1.375	Butyric acid	1.396	Allyl alcohol	1.411
Isopropyl acetate	1.375	2,2,5-Trimethylhexane	1.397	2,4-Dimethyldioxane	1.412
Propyl formate	1.375	Dibutyl ether	1.397	Ethyl lactate	1.412
2-Chloropropane	1.376	Butanol	1.397	Diethyl malonate	1.412
2-Butanone	1.377	Acrolein	1.397	3-Chloropropene	1.413
Butyraldehyde	1.378	1-Chloro-2-methylpropane	1.397	Ethyleneglycol diacetate	1.413
2,4-Dimethylpentane	1.379	Methacrylonitrile	1.398	2-Octanone	1.414
Propyl ether	1.379	3-Methyl-2-pentanone	1.398	3-Octanone	1.414
Acetaldehyde diethyl acetal	1.379	Triethylamine	1.399	3-Methyl-2-heptanone	1.415
Butyl ethyl ether	1.380	Butylamine	1.399	Caproic acid	1.415
Nitromethane	1.380	2,4-Dimethyl-3-pentanone	1.399	4-Methyldioxane	1.415
Trifluoropropanol	1.381	Isobutyl n-butyrate	1.399	1,2-Propyleneglycol-1-monobutyl ether	1.415
2-Methylhexane	1.382	1-Nitropropane	1.399	Ethyl cyanoacetate	1.415
Butyronitrile	1.382	Dodecane	1.400	Dibutylamine	1.416
Propyl acetate	1.382	Amyl acetate	1.400	Hexanol	1.416
Ethyl propionate	1.382	1-Chlorobutane	1.400	2-Pentanol	1.416
2-Methyl-2-propanol	1.383	2-Methoxyethanol	1.400	1,1-Dichloroethane	1.416
Propanol	1.383	Propionic anhydride	1.400	Heptachlorodiethyl ether	1.416
Isobutyl formate	1.383	2,2,3-Trimethylpentane	1.401	3-Methoxy-propylamine	1.417
Diethyl carbonate	1.383	1-Chlorobutane	1.401	Caprylonitrile	1.418
Heptane	1.385	β-Methoxypropionitrile	1.401	2-Heptanol	1.418
tert-Butanol	1.385	Isovaleric acid	1.402	Allylamine	1.419
Propionic acid	1.385	Nonane	1.403	1,2-Propyleneglycol carbonate	1.419
3-Methylhexane	1.386	Dipropylamine	1.403	2-Heptanol	1.420
Propylamine	1.386	Isoamyl acetate	1.403	3-Methyl-2-pentanol	1.420
3-Methyl-2-butanone	1.386	Cyclopentane	1.404	2-Ethyl-1-butanol	1.420
1-Chloropropane	1.386	2-Methyl-2-butanol	1.404	1-Chloro-2-methyl-1-propene	1.420
2,2,3-Trimethylbutane	1.387	3-Methyl-1-butanol	1.404	p-Dioxane	1.420

Methylcyclohexane	1.421		Diethyleneglycol mono-β-hydroxyiso-			Morpholyl N-(ethylhydroxy) ethylamine	1.485
4-Hydroxy-4-methyl-2-pentanone	1.421		propylether	1.448		2-Ethylidenecyclohexanone	1.486
Heptanol	1.422		1-Amino-2-methyl-2-pentanol	1.449		Butylbenzene	1.487
3-Isopropyl-2-heptanone	1.423		Tetrahydrofurfuryl alcohol	1.450		Dioctylbenzene (90 % p.; 10 % m.)	1.487
Cyclohexane	1.424		2-Propylcyclohexanone	1.452		p-Cymene	1.488
2-Bromopropane	1.424		2-Aminoethanol	1.452		Isopropylbenzene	1.489
3-Chloro-2-methyl-1-propene	1.425		1,4-Butanediol glycidyl ether	1.452		Furfurylalcohol	1.489
Caprylic acid	1.426		4-Chloro-1,3-dioxolane-2-one	1.452		tert-Butylcumene (80 % p.; 15 % m.)	1.490
Ethylene carbonate	1.426		1-Chlorooctadecane (techn.)	1.453		n-Propylbenzene	1.490
Octanol	1.427		2-Butylcyclohexanone	1.453		sec-Butylbenzene	1.490
3-Methyl-2-heptanol	1.427		Ethylenediamine	1.454		Hexyl-m-xylene (mainly 1,3,5-)	1.490
N,N-Dimethylformamide	1.427		2-(β-Methyl) propylcyclohexanone	1.454		tert-Butylbenzene	1.490
Sulfuric acid	1.427		4-Methylcyclohexanol	1.454		Dibutyl phthalate	1.490
1-Chlorooctane	1.428		3-Methylcyclohexanol	1.455		Isopropylethylbenzene (35 % p.; 60 % m.)	1.491
Triisobutylene	1.429		Bis(2-chloroethyl) ether	1.455		tert-Butyltoluene (85 % p.; 10 % m.)	1.491
N-Methyl-alaninnitrile	1.429		Cyclohexyl	1.456		1-Phenyl-1-hydroxyphenylethane	1.491
1,2-Ethanediol	1.429		1,8-Cineole	1.456		Hexylcumene (90 % p.; 5 % m.)	1.492
1-Chloro-2-ethylhexane	1.430		2,2'-Dimethyl-2,2'-dipropyl diethanol-			Octyltoluene (96 % p.; 2 % m.)	1.492
Ethylcyclohexane	1.431		amine	1.456		Octylcumene (90 % p.; 4 % m.)	1.492
1,2-Propanediol	1.431		1,3-Butanediol glycidyl ether	1.456		Dihexylbenzene (85 % p.; 10 % m.)	1.492
1-Bromopropane	1.431		1-Chloroeicosane (techn.)	1.459		p-Xylene	1.493
2-Methyl-7-ethyl-4-nonanone	1.433		Oleic acid	1.459		1,3-Diethylbenzene	1.493
Ethyleneglycol monoallyl ether	1.434		(1,1',2,2'-Tetramethyl)-diethanolamine	1.459		tert-Butyl-m-xylene (mainly 1,3,5-)	1.493
Butyrolactone	1.434		3-Aminopropanol	1.459		Ethylbenzene	1.493
2-Methyl-7-ethyl-undecanone	1.435		Carbon tetrachloride	1.459		Octylethylbenzene (80-90 % p.; 10 % m.)	1.493
4-n-Propyl-5-ethyldioxane	1.435		3-Methyl-5-ethyl-2,4-heptanediol	1.459		1,3-Dimorpholyl-2-propanol	1.493
1,2-Dichloroisobutane	1.435		2-(β-Ethyl)-butylcyclohexanone	1.461		1,1,2,2-Tetrachloroethane	1.493
1,2-Propyleneglycol sulfite	1.435		2-Methylcyclohexanol	1.461		Isopropyl-m-xylene (mainly 1,3,5-)	1.494
N-Methyl-morpholine	1.436		N-(n-Butyl)-diethanolamine	1.461		Toluene	1.494
Chloro-tert-butanol	1.436		4,5-Dichloro-1,3-dioxolane-2-one	1.462		Benzyl ethyl ether	1.494
Epichlorohydrin	1.436		2-Butylcyclohexanol	1.462		tert-Butylethylbenzene (70 % p.; 25 % m.)	1.495
Triethyleneglycol monobutyl ether	1.437		N-β-Hydroxypropyl-morpholine	1.462		m-Xylene	1.495
2-Methyl-7-ethyl-4-nonanol	1.438		2-(β-Ethyl)-hexylcyclohexanone	1.463		Hexylethylbenzene (70 % p.; 25 % m.)	1.495
5-Ethyl-2-nonanol	1.438		2-Ethylcyclohexanol	1.463		1,4-Diethylbenzene	1.496
6-Ethyl-3-octanol	1.438		Fluorobenzene	1.463		2,3-Dichlorodioxane	1.496
1,3-Propanediol	1.438		d-α-Pinene	1.464		Mesitylene	1.497
Diethyl maleate	1.438		1-α'-Pinene	1.465		Hexyltoluene (90 % p.; 5 % m.)	1.497
Butanethiol	1.440		Cyclohexanol	1.465		2-Iodopropane	1.497
Dibutyl sebacate	1.440		m-Fluorotoluene	1.465		Benzene	1.498
2-Chloroethanol	1.440		p-Fluorotoluene	1.467		Propyl benzoate	1.498
6-Ethyl-3-decanol	1.441		N-(2-Hydroxyethyl)-2-hydroxybutylamine	1.467		α-Picoline	1.499
1-Chlorododecane (techn.)	1.441		4-Chloromethyl-1,3-dioxolane-2-one	1.467		1,2-Diethylbenzene	1.501
3-Methyl-2,4-pentanediol	1.441		trans-Decahydronaphthalene	1.468		Pentachloroethane	1.501
Dimethyl maleate	1.441		o-Fluorotoluene	1.468		1-Iodopropane	1.502
2-Methyl-7-ethyl-4-undecanol	1.442		N-(2-Hydroxyethyl)2-hydroxypropyl-			o-Xylene	1.503
Ethyl sulfide	1.442		amine	1.468		Ethyl benzoate	1.503
Mesityl oxide	1.442		3-Allyloxy-2-hydroxypropylamine	1.469		β-Picoline	1.504
Butyl stearate	1.442		Di-(2-hydroxybutyl)-ethanolamine	1.469		Tetrachloroethylene	1.504
Cyclohexene	1.443		Di-(2-hydroxypropyl)-ethanolamine	1.469		Phenetole	1.505
Lauryl glycidyl ether	1.443		D-Limonene	1.471		Pyridine	1.507
Dibutyl maleate	1.444		2-(α-Hydroxybutyl)-cyclohexanol	1.473		Iodoethane	1.512
1,3-Butyleneglycol sulfite	1.444		1,2,3-Trichloroisobutane	1.473		Phenyl methallyl ether	1.514
1,2-Dichloroethane	1.444		Decahydronaphthalene	1.474		Anisole	1.515
Glycol sulfite	1.444		1,2,3-Propanetriol	1.474		Methyl benzoate	1.515
Chloroform	1.444		Trichloroethylene	1.475		Cyclohexylcumene (60 % p.; 25 % m.)	1.516
1-Chlorotetradecane (techn.)	1.445		N(β-Hydroxyethyl)-morpholine	1.476		Diallyl phthalate	1.517
Diethylene glycol	1.445		Dimethyl sulfoxide	1.476		Benzyl acetate	1.518
cis-1,2-Dichloroethylene	1.445		cis-Decahydronaphthalene	1.479		Cyclohexylethylbenzene (60 % p.; 20 % m.)	1.520
2-Butyloctyl-3-aminopropyl ether	1.446		2-(α-Hydroxyethyl)-cyclohexanol	1.479		2-Methyl-4-tert-butyl-phenol	1.521
2-Methyl-morpholine	1.446		1-Methoxy-1-butene-3-yn	1.480		Phenyl acetonitrile	1.521
Formamide	1.446		2-Butylidenecyclohexanone	1.481		(Chloro-tert-butyl)benzene	1.521
3-Lauryl-1-hydroxypropylamine	1.447		N(β-Chloroallyl)-morpholine	1.481		Methyl salicylate	1.522
Ethylene glycol diglycidyl ether	1.447		n-Dodecyl-4-tert-butyl phenylether	1.482		Cyclohexyltoluene (50 % p.; 20 % m.)	1.523
1-Chlorohexadecane (techn.)	1.448		n-Dodecyl phenyl ether	1.482		Chlorobenzene	1.523
Cyclohexanone	1.448		n-Dodecyl 4-methylphenyl ether	1.483		Furfural	1.524
1-Amino-2-propanol	1.448		N-Hydroxyethyl-1,3-propanediamine	1.483		Octachlorodiethyl ether	1.525

Benzonitrile	1.526	m-Dichlorobenzene	1.543	o-Chloroaniline	1.586
Thiophene	1.526	Benzaldehyde	1.544	Bromoform	1.587
Nonachlorodiethyl ether	1.529	Styrene	1.545	Thiophenol	1.588
Iodomethane	1.530	Nitrobenzene	1.550	2,4-Bis(α-phenylethyl) phenylmethyl	
4-Phenyldioxane	1.530	o-Dichlorobenzene	1.551	ether	1.590
3-Phenyl-1-propanol	1.532	Bromobenzene	1.557	Carbon disulfide	1.628
Acetophenone	1.532	o-Nitroanisole	1.560	1,1,2,2-Tetrabromoethane	1.633
Benzyl alcohol	1.538	m-Toluidine	1.566	Methylene iodide	1.749
1,2-Dibromoethane	1.538	Benzyl benzoate	1.568		
1,2,3,4-Tetrahydronaphthalene	1.539	o-Toluidine	1.570		
m-Cresol	1.542	1-Methoxyphenyl-1-phenylethane	1.571		
β,β-Di(butylxanthogenic acid) diethylester	1.543	Aniline	1.583		

PHYSICAL CONSTANTS OF THE MOST COMMON SOLVENTS

Solvent	b.p. [°C]	m.p. [°C]	Specific Gravity* [°C]	Viscosity at Various Temp., centipoise [°C]			n_D at Various Temp. [°C]	
			20	20	25	30	20	25
Acetic acid	117.9	16.60	1.0492	1.21		1.040	1.37160	1.36995
Acetone	56.24	-95.35	0.7899	0.324	0.316	0.2954	1.35880	1.35609
Benzene	80.10	5.5	0.87865	0.52	0.6028	0.564	1.50110	1.49790
Benzyl alcohol	205.3	-15.3	1.0419	5.8		4.650	1.5396	1.5371
Butyl acetate	126.5	-77.9	0.8825	0.732	0.688	0.637	1.39406	
n-Butyl alcohol	117.25	-89.53	0.8098	2.948		2.271	1.39931	1.3970
Carbon tetrachloride	76.54	-22.99	1.5940	0.969	0.8876	0.843	1.46010	1.45759
m-Cresol	202.2	11.5	1.0336	20.8		9.807	1.5438	
Chloroform	61.7	-63.55	1.4832	0.568	0.542	0.514	1.4459	
Cyclohexane	80.74	6.55	0.77855	0.979	0.898	0.825	1.42623	1.42354
Cyclohexanone	155.65	-16.40	0.9478	2.453 (15°)		1.803	1.4507	
Dimethylformamide	153.0	-60.48	0.9487				1.4305	1.4269
1,4-Dioxane	101.32	11.80	1.0337	1.26		1.087	1.42241	1.42025
Ethyl acetate	77.06	-83.58	0.9003	0.455	0.426	0.400	1.37239	1.36979
Ethyl alcohol	78.5	-114.5	0.7893	1.200	1.078	1.003	1.3611	1.35941
Ethyl ether	34.48	-116.3	0.71378	0.233	0.222		1.35272	
Formic acid	100.70	8.4	1.220	1.804	1.966	1.465	1.37140	1.36938
n-Heptane	98.43	-90.61	0.68376	0.409	0.386	0.364	1.38777	1.38512
n-Hexane	68.95	-95.3	0.6603	0.326	0.294	0.278	1.37506	1.37226
Methanol	64.96	-93.9	0.7914	0.597	0.547	0.510	1.3288	1.32663
Methyl ethyl ketone	79.60	-86.35	0.8054	0.423 (15°)		0.365	1.3788	1.37612
n-Octane	125.66	-56.8	0.7025	0.5458	0.5136	0.472	1.39743	1.39505
n-Propyl alcohol	97.4	-126.5	0.8035	2.256		1.722	1.3850	1.3835
Tetrahydrofuran	64-65	-65.0	0.8892	0.486		0.438	1.4050	1.4040
Tetralin	207.57	-35.80	0.9702	2.202	2.003		1.54135	1.53919
Toluene	110.62	-95.0	0.8669	0.590	0.5516	0.526	1.4961	1.49413
Water	100.0	0	0.99823	1.0050	0.8937	0.8007	1.33299	
p-Xylene	138.35	13.26	0.8611	0.648	0.605	0.568	1.49581	1.49325

* Specific gravity is referred to water at 4°C.

VIII. CONTEMPORARY THERMOPLASTIC MATERIALS

CONTEMPORARY THERMOPLASTIC MATERIAL - PROPERTY CHART

V. A. Matonis

Monsanto Company, Indian Orchard, Ma.

Contents

This table is intended to give a survey of mechanical, electrical and optical properties and to show which magnitudes may be expected. Average data for a number of commercial polymers were collected and arranged according to either increasing or decreasing values. The reader is adviced to use these data with caution. They depend to a large extent on sample history and exact composition. Catalysts and methods of preparation may influence the data in the case of thermosetting resins.

1. Tensile Modulus, ASTM D-638 $[MPa] \cdot 10^{-3} \equiv [N\ mm^{-2}] \cdot 10^{-3}$
To convert to $[psi]$, multiply by 1.45×10^5

Material	High	Low	Material	High	Low
Phenol-formaldehyde resin	6.9	5.2	ABS, high impact	2.4	1.4
Poly(ester), rigid cast resin	4.4	2.1	Poly(aryl ether)	2.2	
Poly(styrene), general purpose	4.1	2.8	Poly(chlorotrifluoroethylene)	2.1	1.0
Poly(styrene), high heat	4.1	2.8	Ethyl cellulose	2.1	.7
Poly(vinyl chloride)	4.1	2.4	Poly(methyl methacrylate)/Poly(vinyl chloride), alloy	1.9	
Poly(vinyl formal)	4.1	2.4	Poly(propylene)	1.6	1.1
Poly(oxymethylene)	3.6	3.0	Cellulose nitrate	1.5	1.3
Epoxy cast resins	3.5	2.4	Cellulose propionate	1.5	.4
Poly(oxymethylene), copolymer	3.4	2.6	Poly(methyl-1-pentene)	1.4	1.1
Poly(styrene-co-acrylonitrile)	3.4	2.8	Cellulose acetate butyrate	1.4	.3
Poly(styrene), high impact	3.4	1.4	Nylon 11	1.3	
Poly(phenylene sulfide)	3.3		Nylon 12	1.2	
Poly(vinyl chloride), chlorinated	3.3	2.5	Poly(ethylene), high density	1.2	.4
Poly(imide)	3.1		Poly(vinylidene fluoride)	.83	
Poly(methyl methacrylate), molding grade	3.1	2.6	Poly(vinylidene chloride)	.55	.34
Poly(vinyl chloride-co-propylene)	3.1	2.4	Poly(tetrafluoroethylene)	.45	.26
Poly(α-methylstyrene-co-methyl methacrylate)	3.0		Ionomers	.41	.14
ABS, medium impact	2.9	2.1	Poly(ethylene), medium density	.38	.17
Nylon 66	2.9	1.2	Poly(styrene-co-butadiene)	.35	.01
Cellulose acetate	2.8	.4	Poly(tetrafluoroethylene-co-hexafluoropropylene)	.34	
Poly(vinyl chloride), ABS modified	2.7	2.1	Poly(ethylene), low density	.26	.10
Poly(phenylene oxide)	2.6	2.4	Poly(1-butene)	.18	
Poly(carbonate) - ABS alloy	2.6	2.1	Poly(ethylene-co-vinyl acetate)	.08	.01
Poly(sulfone)	2.5		Poly(ethylene-co-ethyl acrylate)	.05	.03
Poly(carbonate)	2.4	2.1			

2. Compressive Modulus, ASTM D-695 $[MPa] \cdot 10^{-3} \equiv [N\ mm^{-2}] \cdot 10^{-3}$

To convert to $[psi]$, multiply by 1.45×10^{5}

Material	High	Low	Material	High	Low
Poly(oxymethylene)	4.6		Poly(propylene)	2.1	1.0
Poly(vinyl chloride), chlorinated	4.1	2.3	Poly(methyl methacrylate)/Poly(vinyl chloride) alloy	2.1	
Poly(imide)	3.7		Nylon 6	1.7	
Poly(styrene-co-acrylonitrile)	3.7		ABS, medium impact	1.7	1.4
Poly(methyl methacrylate), molding grade	3.2	2.6	ABS, high impact	1.4	1.0
Poly(oxymethylene), copolymer	3.1		Nylon 11	1.2	
Poly(phenylene oxide)	2.6		Poly(vinylidene fluoride)	.83	
Poly(sulfone)	2.6	2.3	Poly(styrene-co-butadiene)	.83	.02
Poly(α-methylstyrene-co-methyl methacrylate)	2.6	1.7	Poly(1-butene)	.2	
Poly(carbonate)	2.4				

3. Percent Elongation at Fail, ASTM D-638

Material	High	Low	Material	High	Low
Poly(styrene-co-butadiene)	1000	300	Poly(aryl ether)	90	25
Poly(ethylene), high density	1000	20	Cellulose acetate butyrate	88	40
Poly(ethylene-co-vinyl acetate)	900	750	Poly(styrene), high impact	80	2
Poly(ethylene), low density	800	90	Poly(oxymethylene), copolymer	75	30
Poly(ethylene-co-ethyl acrylate)	750	700	Poly(oxymethylene)	60	20
Poly(propylene)	700	200	Cellulose acetate	70	6
Poly(ethylene), medium density	600	50	Poly(vinyl chloride), chlorinated	65	5
Poly(ethylene-co-propylene) (Poly-allomer)	500	400	ABS, high impact	60	5
Ionomers	450	350	Cellulose nitrate	45	40
Poly(tetrafluoroethylene)	450	250	Ethyl cellulose	40	5
Poly(vinyl butyral)	450	150	Poly(vinyl chloride)	40	2
Poly(1-butene)	380	300	ABS, medium impact	25	5
Poly(tetrafluoroethylene-co-hexafluoropropylene)	330	250	Poly(methyl-1-pentene)	22	13
Nylon 6	300	200	Poly(vinyl formal)	20	5
Poly(vinylidene fluoride)	300	100	Poly(vinyl chloride), ABS modified	20	5
Nylon 66	300	60	Cycloaliphatic epoxy	10	2
Nylon 11	300		Poly(methyl methacrylate), molding grade	10	2
Nylon 12	300		Poly(imide)	8	5
Poly(vinylidene chloride)	250		Epoxy cast resins	6	3
Poly(chlorotrifluoroethylene)	250	80	Poly(ester), rigid cast resin	< 5	
Poly(methyl methacrylate)/(Poly(vinyl chloride)) alloy	150		Poly(styrene-co-acrylonitrile)	3.7	1.5
Poly(carbonate) - ABS alloy	150	10	Poly(phenylene sulfide)	3.0	
Poly(vinyl chloride-co-propylene)	140	100	Poly(α-methylstyrene-co-methyl methacrylate)	3.0	
Poly(carbonate)	130	100	Poly(styrene), high heat	2.5	1.4
Poly(phenylene oxide)	100	50	Poly(styrene), general purpose	2.5	1.0
Poly(sulfone)	100	50	Phenol-formaldehyde resin	1.5	1.0
Cellulose propionate	100	30	Melamine-phenol resin	.8	.4

4. Tensile Strength, ASTM D-638 $[MPa] \equiv [N\ mm^{-2}]$

To convert to $[psi]$, multiply by 145

Material	High	Low	Material	High	Low
Poly(ester), rigid cast	90	41	Poly(methyl methacrylate), molding grade	76	48
Epoxy cast resins	90	28	Poly(phenylene sulfide)	74	
Nylon 66	83	77	Poly(imide)	72	
Poly(vinyl formal)	83	69	Poly(phenylene oxide)	72	
Poly(styrene-co-acrylonitrile)	83	62	Poly(sulfone)	70	
Cycloaliphatic epoxy resin	83	55	Poly(oxymethylene)	69	
Poly(styrene), high heat	83	45	Poly(α-methylstyrene-co-methyl methacrylate)	69	
Poly(styrene), general purpose	83	34	Poly(carbonate)	65	55
Nylon 6	81	69	Nylon 12	64	55

4. <u>Tensile Strength, ASTM D-638 [MPa]</u>   (Cont'd.)

Material	High	Low	Material	High	Low
Poly(vinyl chloride), chlorinated	62	52	Poly(tetrafluoroethylene)	45	17
Poly(vinyl chloride)	62	34	ABS, high impact	43	24
Cellulose acetate	62	13	Poly(chlorotrifluoroethylene)	41	31
Poly(oxymethylene), copolymer	61		Poly(methyl methacrylate)/Poly(vinyl chloride), alloy	38	
Poly(carbonate) - ABS alloy	57	52	Poly(propylene)	38	30
Nylon 11	55		Poly(ethylene), high density	38	21
Cellulose nitrate	55	48	Ionomers	34	24
Phenol-formaldehyde resin	55	48	Poly(vinylidene chloride)	34	21
Melamine-phenol resin	55	41	Poly(1-butene)	30	26
Poly(vinyl chloride-co-propylene)	55	34	Poly(methyl-1-pentene)	28	24
Ethyl cellulose	55	14	Poly(ethylene-co-propylene) (Poly-allomer)	26	21
Cellulose propionate	54	14	Poly(ethylene), medium density	24	8
Poly(aryl ether)	52		Poly(tetrafluoroethylene-co-hexafluoropropylene)	21	19
Poly(vinylidene fluoride)	51	38	Poly(styrene-co-butadiene)	21	4
ABS, medium impact	48	38	Poly(vinyl butyral)	21	3
Poly(vinyl chloride), ABS modified	48	37	Poly(ethylene-co-vinyl acetate)	19	10
Cellulose acetate butyrate	48	18	Poly(ethylene), low density	16	4
Poly(styrene), high impact	48	10	Poly(ethylene-co-ethyl acrylate)	14	11

5. <u>Compressive Strength, ASTM D-695 [MPa]</u> $\equiv$ [N mm^{-2}]
   To convert to [psi], multiply by 145

Material	High	Low	Material	High	Low
Melamine-formaldehyde resin	310	276	Poly(styrene), general purpose	110	79
Cellulose acetate	248	14	Poly(styrene), high heat	110	79
Cellulose nitrate	241	152	Poly(α-methylstyrene-co-methyl methacrylate)	103	76
Ethyl cellulose	241	69	Poly(sulfone)	99	
Melamine-phenol resin	207	179	Poly(vinyl chloride)	90	55
Poly(ester), rigid cast	207	90	Nylon 6	90	
Phenol-formaldehyde resin	207	69	ABS, medium impact	86	54
Epoxy cast resins	172	103	Poly(vinyl chloride-co-propylene)	81	53
Poly(imide)	>165		Poly(carbonate) - ABS alloy	76	59
Poly(vinyl chloride), chlorinated	152	62	Poly(carbonate)	74	71
Cellulose propionate	152	17	Poly(styrene), high impact	62	26
Cellulose acetate butyrate	151	14	Poly(vinylidene fluoride)	60	
Poly(urethane), thermoplastic elastomer	138		Poly(propylene)	55	38
Cycloaliphatic epoxy resin	138	103	ABS, high impact	55	31
Poly(oxymethylene)	124		Poly(chlorotrifluoroethylene)	51	32
Poly(methyl methacrylate), molding grade	124	83	Poly(methyl methacrylate)/Poly(vinyl chloride) alloy	43	
Poly(styrene-co-acrylonitrile)	117	96	Poly(ethylene), high density	25	19
Poly(phenylene oxide)	113	110	Poly(vinylidene chloride)	19	14
Poly(oxymethylene), copolymer	110		Poly(tetrafluoroethylene)	12	5

6. <u>Flexural (Yield) Strength, ASTM D-790 [MPa]</u> $\equiv$ [N mm^{-2}]
   To convert to [psi], multiply by 145

Material	High	Low	Material	High	Low
Poly(ester), rigid cast resin	159	59	Poly(vinyl chloride)	110	69
Epoxy cast resins	145	92	Cellulose acetate	110	14
Poly(phenylene sulfide)	138		Poly(sulfone)	106	
Poly(α-methylstyrene-co-methyl methacrylate)	131	110	Poly(vinyl chloride-co-propylene)	106	69
Poly(methyl methacrylate), molding grade	131	90	Poly(imide)	103	
Poly(vinyl formal)	124	117	Poly(phenylene oxide)	103	
Poly(vinyl chloride), chlorinated	117	100	Phenol-formaldehyde resin	103	83
Poly(styrene-co-acrylonitrile)	117	96	Poly(carbonate) - ABS alloy	100	72
Poly(styrene), high heat	117	69	Poly(oxymethylene)	97	

6. Flexural (Yield) Strength, ASTM D-790 [MPa]   (Cont'd.)

Material	High	Low	Material	High	Low
Melamine-formaldehyde resin	97	76	Melamine-phenol resin	69	55
Poly(styrene), general purpose	96	60	ABS, high impact	69	41
Poly(oxymethylene), copolymer	90		Poly(chlorotrifluoroethylene)	64	51
Cycloaliphatic epoxy resin	90	69	Cellulose acetate butyrate	64	12
ABS, medium impact	90	65	Poly(methyl methacrylate)/Poly(vinyl chloride), alloy	60	
Poly(carbonate)	88	84	Poly(propylene)	55	41
Poly(styrene), high impact	83	34	Poly(vinyl chloride), ABS modified	52	
Ethyl cellulose	83	28	Poly(ethylene), medium density	48	33
Cellulose propionate	79	20	Poly(vinylidene chloride)	43	29
Poly(aryl ether)	76		Poly(ethylene-co-ethyl acrylate)	25	21
Cellulose nitrate	76	62			

7. Hardness, ASTM D-785 & ASTM D-1706

This table is ordered according to decreasing maximum hardness within each sub-group:
Rockwell E, M, L and R, and Shore Durometer D and A

Material	High	Low	Material	High	Low
Melamine-phenol resin	E 100	E 95	Nylon 6	R 115	
Poly(imide)	E 58	E 45	Poly(aryl ether)	R 115	
Phenol-formaldehyde resin	M115		ABS, medium impact	R 115	R107
Poly(ester), rigid cast resin	M115	M 70	Cellulose Nitrate	R 115	R 95
Epoxy cast resins	M110	M 80	Ethyl cellulose	R 115	R 50
Poly(methyl methacrylate), molding grade	M105	M 85	Cellulose acetate butyrate	R 115	R 31
Poly(oxymethylene)	M 94		Poly(vinyl chloride), ABS modified	R 113	R102
Poly(styrene-co-acrylonitrile)	M 90	M 80	Poly(propylene)	R 110	R 80
Poly(styrene), high heat	M 90	M 65	Nylon 11	R 108	
Poly(vinyl formal)	M 85		Nylon 12	R 106	
Poly(oxymethylene), copolymer	M 85	M 78	ABS, high impact	R 105	R 75
Poly(styrene), general purpose	M 80	M 65	Poly(methyl methacrylate)/Poly(vinyl chloride), alloy	R 104	
Poly(carbonate)	M 78	M 73	Poly(chlorotrifluoroethylene)	R 95	R 75
Poly(α-methylstyrene-co-methyl methacrylate)	M 75		Poly(ethylene-co-propylene) (Poly(allomer))	R 85	R 80
Poly(sulfone)	M 69		Poly(tetrafluoroethylene-co-hexafluoropropylene)	R 25	
Poly(vinylidene chloride)	M 65	M 50	Poly(vinyl chloride)	D 85	D 65
Poly(styrene), high impact	M 60	M 20	Poly(vinylidene fluoride)	D 80	
Poly(vinyl chloride)	M 55	M 18	Poly(ethylene), high density	D 70	D 60
Poly(methyl-1-pentene)	L 74	L 67	Poly(1-butene)	D 65	
Cellulose acetate	R 115	R 34	Ionomers	D 65	D 50
Poly(phenylene sulfide)	R 115		Poly(ethylene), medium density	D 60	D 50
Poly(vinyl chloride), chlorinated	R 115		Poly(ethylene), low density	D 46	D 41
Cellulose propionate	R 115	R 10	Poly(ethylene-co-vinyl acetate)	D 38	D 17
Nylon 66	R 115		Poly(ethylene-co-ethyl acrylate)	D 36	D 27
Poly(phenylene oxide)	R 115		Poly(vinyl butyral)	A100	A 10
Poly(carbonate) - ABS alloy	R 115	R 106	Poly(styrene-co-butadiene)	A 90	A 40

8. Deflection Temperature under Load, ASTM D-648

$^{\circ}$C at 1,820 [N mm^{-2}] (264 psi) ; $^{\circ}$F = 9/5 ($^{\circ}$C) + 32

Material	High	Low	Material	High	Low
Poly(imide)	360		Melamine-formaldehyde resin	148	145
Epoxy cast resins	260	127	Poly(carbonate)	140	129
Cycloaliphatic epoxy resin	>250		Poly(phenylene sulfide)	137	
Poly(ester), rigid cast	204	60	Phenol-formaldehyde resin	127	116
Poly(phenylene oxide)	191		Poly(carbonate) - ABS alloy	127	104
Poly(sulfone)	174		Poly(oxymethylene)	124	
Melamine-phenol resin	154	141	Poly(methyl methacrylate), molding grade	121	74
Poly(aryl ether)	149		Poly(vinyl chloride), chlorinated	112	94

8. Deflection Temperature under Load, ASTM D-648 (Cont'd.)

Material	High	Low	Material	High	Low
Poly(styrene), high heat	111	93	Poly(styrene), general purpose	71	60
Poly(oxymethylene), copolymer	110		Poly(tetrafluoroethylene-co-hexafluoropropylene)	70	
Poly(α-methylstyrene-co-methyl methacrylate)	105		Nylon 6	68	60
Poly(styrene-co-acrylonitrile)	104	88	Poly(propylene)	66	54
Nylon 66	104	66	Poly(vinylidene chloride)	66	54
ABS, high impact	96	88	Poly(1-butene)	60	54
ABS, medium impact	96	88	Poly(ethylene), high density	60	43
Poly(styrene), high impact	93	85	Poly(ethylene-co-propylene) (Poly(allomer))	56	51
Cellulose acetate butyrate	91	48	Nylon 11	55	
Poly(vinylidene fluoride)	91		Nylon 12	55	49
Ethyl cellulose	88	46	Poly(ethylene), medium density	49	41
Cellulose acetate	87	47	Poly(ethylene), low density	41	32
Poly(vinyl chloride), ABS modified	82	68	Poly(ethylene-co-vinyl acetate)	34	
Cellulose propionate	78	54	Poly(vinyl formal)	33	16
Poly(methyl methacrylate)/Poly(vinyl chloride), alloy	77		Poly(vinyl butyral)	17	
Poly(vinyl chloride-co-propylene)	77	71	Poly(chlorotrifluoroethylene)★	126	
Poly(vinyl chloride)	77	60	Ionomers★	49	38
Cellulose nitrate	71	60			

★ at 0.455 [N mm^{-2}] (66 psi)

9. Mold Shrinkage (Linear) [%]

Material	Low	High	Material	Low	High
Silicone, cast resin	0	0.6	Poly(carbonate) - ABS alloy	0.5	0.9
Poly(phenylene sulfide)	0.1		Ethyl cellulose	0.5	0.9
Poly(vinyl chloride)	0.1	0.4	Nylon 6	0.6	1.4
Epoxy cast resins	0.1	0.4	Poly(phenylene oxide)	0.7	
Poly(styrene-co-butadiene)	0.1	0.5	Poly(sulfone)	0.7	
Poly(methyl methacrylate), molding grade	0.1	0.8	Poly(aryl ether)	0.7	
Poly(styrene), high heat	0.1	0.8	Poly(ethylene-co-vinyl acetate)	0.7	1.1
Poly(vinyl formal)	0.15	0.4	Poly(methyl methacrylate)/Poly(vinyl chloride), alloy	0.8	
Poly(vinyl chloride-co-propylene)	0.2	0.5	Poly(vinylidene chloride)	0.8	1.2
Poly(styrene), general purpose	0.2	0.6	Phenyl-formaldehyde resin	1.0	1.2
Poly(styrene), high impact	0.2	0.6	Poly(chlorotrifluoroethylene)	1.0	1.5
Poly(styrene-co-acrylonitrile)	0.2	0.7	Poly(ethylene-co-propylene) (Poly(allomer))	1.0	2.0
Poly(α-methylstyrene-co-methyl methacrylate)	0.2	0.8	Poly(propylene)	1.0	2.5
Cellulose propionate	0.3	0.6	Melamine-formaldehyde resin	1.1	1.2
Cellulose acetate butyrate	0.3	0.6	Nylon 11	1.2	
Poly(vinyl chloride), chlorinated	0.3	0.7	Nylon 66	1.5	
Cellulose acetate	0.3	0.8	Poly(ethylene-co-ethyl acrylate)	1.5	3.5
Nylon 12	0.3	1.5	Poly(ethylene), low density	1.5	5.0
Ionomers	0.3	2.0	Poly(ethylene), medium density	1.5	5.0
Poly(1-butene)	0.3	2.6	Poly(oxymethylene)	2.0	2.5
Melamine-phenol resin	0.4	1.0	Poly(ethylene), high density	2.0	5.0
Poly(vinyl chloride), ABS modified	0.4	1.5	Poly(oxymethylene), copolymer	2.5	
Poly(carbonate)	0.5	0.7	Poly(vinylidene fluoride)	3.0	
ABS, high impact	0.5	0.8	Poly(tetrafluoroethylene-co-hexafluoropropylene)	3.0	6.0
ABS, medium impact	0.5	0.8			

10. Water Absorption, ASTM D-570 [%]
After 24 hours

Material	Low	High	Material	Low	High
Poly(chlorotrifluoroethylene)	.00		Poly(ethylene), high density	<.01	
Poly(tetrafluoroethylene)	.00		Poly(propylene)	<.01	
Poly(ethylene), medium density	<.01		Poly(ethylene-co-propylene) (Poly(allomer))	<.01	

10. Water Absorption, ASTM D-570 [%]   (Cont'd.)

Material	Low	High	Material	Low	High
Poly(1-butene)	< 0.1		Poly(styrene-co-acrylonitrile)	.20	.30
Poly(tetrafluoroethylene-co-hexafluoropropylene)	.01		Poly(carbonate) - ABS alloy	.20	.35
Poly(ethylene), low density	< .015		Poly(sulfone)	.22	
Poly(vinyl chloride), chlorinated	.02	.15	Poly(oxymethylene), copolymer	.22	
Poly(styrene), general purpose	.03	.10	Poly(aryl ether)	.25	
Poly(vinyl chloride)	.03	.40	Nylon 12	.25	
Poly(ethylene-co-ethyl acrylate)	.04		Poly(oxymethylene)	.25	
Poly(vinylidene fluoride)	.04		ABS, medium impact	.30	
Poly(ethylene-co-vinyl acetate)	.05	.13	Poly(methyl methacrylate), molding grade	.30	.40
Poly(styrene), high impact	.05	.20	Melamine-formaldehyde resin	.30	.50
Poly(styrene), high heat	.05	.40	Melamine-phenol resin	.30	.65
Poly(phenylene oxide)	.06		Poly(imide)	.32	
Poly(vinyl chloride-co-propylene)	.07	.40	Nylon 11	< .40	
Epoxy cast resins	.08	.15	Ethyl cellulose	.8	1.8
Poly(vinylidene chloride)	.10		Cellulose acetate butyrate	.9	2.0
Phenol-formaldehyde resin	.10	.20	Poly(vinyl formal)	1.0	1.3
Poly(methyl methacrylate)/Poly(vinyl chloride), alloy	.13		Cellulose nitrate	1.0	2.0
Poly(carbonate)	.15	.18	Poly(vinyl butyral)	1.0	2.0
Poly(ester), rigid cast resin	.15	.60	Nylon 6	1.3	1.9
Poly(styrene-co-butadiene)	.19	.39	Cellulose propionate	1.3	2.0
Poly(vinyl chloride), ABS modified	.20		Ionomers	1.4	
ABS, high impact	.20		Nylon 66	1.5	
Poly(α-methylstyrene-co-methyl methacrylate)	.20		Cellulose acetate	1.7	4.0

11. Volume Resistivity ASTM D-257 [Ohm · cm]

Material	High	Low	Material	High	Low
Poly(tetrafluoroethylene)	1.0×10^{19}		Poly(vinyl chloride-co-propylene)	$> 1.0 \times 10^{15}$	
Poly(ethylene)	5.0×10^{18}	1.0×10^{17}	Poly(ethylene-co-propylene) (Poly(allomer))	$> 1.0 \times 10^{15}$	
Poly(tetrafluoroethylene-co-hexyfluoropropylene)	$> 2.0 \times 10^{18}$		Poly(ester), rigid cast	1.0×10^{15}	
Poly(chlorotrifluoroethylene)	1.2×10^{18}		Poly(ethylene-co-vinyl acetate)	1.0×10^{15}	
Poly(phenylene oxide)	1.0×10^{17}		Poly(oxymethylene)	1.0×10^{15}	
Poly(imide)	$> 1.0 \times 10^{16}$★		Poly(α-methylstyrene-co-methyl methacrylate)	1.0×10^{15}	
Poly(propylene)	$> 1.0 \times 10^{16}$★		Poly(vinyl chloride), chlorinated	1.0×10^{15}	
Ionomers	$> 1.0 \times 10^{16}$★		Nylon 66	1.0×10^{15}	1.0×10^{14}
Poly(styrene), general purpose	$> 1.0 \times 10^{16}$★		Silicone, cast resin	1.0×10^{15}	1.0×10^{14}
Poly(styrene), high impact	$> 1.0 \times 10^{16}$★		Poly(styrene), high heat	1.0×10^{15}	1.0×10^{14}
Poly(styrene-co-acrylonitrile)	$> 1.0 \times 10^{16}$★		Nylon 6	1.0×10^{15}	1.0×10^{13}
Poly(vinyl chloride)	$> 1.0 \times 10^{16}$		Epoxy cast resins	1.0×10^{15}	1.0×10^{12}
Poly(sulfone)	5.0×10^{16}		Cellulose acetate butyrate	1.0×10^{15}	1.0×10^{10}
ABS, high impact	4.8×10^{16}	1.0×10^{16}	Poly(methyl methacrylate), molding grade	$> 1.0 \times 10^{14}$	
Poly(carbonate) - ABS alloy	4.0×10^{16}	2.2×10^{16}	Poly(vinylidene fluoride)	2.0×10^{14}	
ABS, medium impact	2.7×10^{16}		Poly(oxymethylene), copolymer	1.0×10^{14}	
Poly(styrene-co-butadiene)	2.5×10^{16}	5.0×10^{13}	Ethyl cellulose	1.0×10^{14}	1.0×10^{12}
Poly(carbonate)	2.1×10^{16}		Cellulose acetate	1.0×10^{14}	1.0×10^{10}
Poly(aryl ether)	1.5×10^{16}		Nylon 11	1.0×10^{13}	
Cellulose nitrate	1.5×10^{16}		Nylon 12	1.0×10^{13}	
Cellulose propionate	1.0×10^{16}	1.0×10^{12}	Cycloaliphatic epoxy resin	1.0×10^{13}	
Poly(ethylene-co-ethyl acrylate)	1.0×10^{16}		Poly(urethane), thermoplastic elastomer	1.1×10^{12}	2.0×10^{10}
Poly(methyl-1-pentene)	1.0×10^{16}		Phenol-formaldehyde resin	1.0×10^{12}	1.0×10^{10}
Poly(vinylidene chloride)	1.0×10^{16}	1.0×10^{14}	Poly(vinyl butyral)	0.5×10^{12}	

★ above the sensitivity of the instrument used.

12. Dielectric Strength, Short Time, ASTM D-149 [V cm^{-1}] · 10^{-3}
To convert to volts/mil., multiply by 2.54

Material	High-Low	Material	High-Low
Poly(vinyl chloride), chlorinated	590-480	Poly(styrene-co-acrylonitrile)	200-160
Poly(vinyl chloride)	590-160	Poly(methyl methacrylate), molding grade	200-160
Ionomers	430-350	Epoxy cast resins	200-160
Poly(ethylene), low density	390-180	Poly(tetrafluoroethylene)	200-160
Poly(ethylene), medium density	390-180	Poly(ester), rigid cast	200-150
Poly(ethylene-co-propylene) (Poly(allomer))	370-320	ABS, high impact	200-140
Poly(ethylene-co-vinyl acetate)	310-240	Ethyl cellulose	200-140
Poly(styrene), general purpose	280-200	Poly(carbonate) - ABS alloy	200-140
Poly(propylene)	260-200	ABS, medium impact	200-140
Poly(vinyl chloride), ABS modified	240	Poly(vinyl formal)	200
Poly(phenylene sulfide)	240	Poly(α-methylstyrene-co-methyl methacrylate)	180
Nylon 66	240	Nylon 12	180
Poly(chlorotrifluoroethylene)	240-200	Cellulose propionate	180-120
Poly(tetrafluoroethylene-co-hexafluoroethylene)	240-200	Poly(aryl ether)	170
Poly(styrene), high heat	240-160	Nylon 11	170
Poly(vinylidene chloride)	240-160	Poly(sulfone)	170
Poly(styrene), high impact	240-120	Poly(methyl methacrylate)/Poly(vinyl chloride), alloy	160
Cellulose nitrate	240-120	Nylon 6	160
Cellulose acetate	240-100	Poly(carbonate)	160
Poly(imide)	220	Phenol-formaldehyde resin	160-120
Silicone, cast resin	220	Cellulose acetate butyrate	160-100
Poly(ethylene-co-ethyl acrylate)	220-180	Poly(oxymethylene)	150
Poly(phenylene oxide)	220-160	Poly(vinyl butyral)	140
Poly(styrene-co-butadiene)	200-170	Melamine-phenol resin	130- 90
Poly(oxymethylene), copolymer	200	Poly(vinylidene fluoride)	100
Poly(ethylene), high density	200-190		

13. Dielectric Constant, ASTM D-150
At 60 Hz

Material	Low	High	Material	Low	High
Poly(tetrafluoroethylene)	2.0		Poly(sulfone)	3.14	
Poly(tetrafluoropropylene-co-hexafluoropropylene)	2.1		Cycloaliphatic epoxy resin	3.2	
Poly(methyl-1-pentene)	2.12		Poly(vinyl chloride)	3.2	3.6
Poly(propylene)	2.2		Poly(vinyl chloride), chlorinated	3.3	3.8
Poly(chlorotrifluoroethylene)	2.24		Poly(methyl methacrylate), molding grade	3.3	3.9
Poly(ethylene), low density	2.25	2.35	Poly(imide)	3.4	
Poly(ethylene), medium density	2.25	2.35	Poly(α-methylstyrene-co-methyl methacrylate)	3.4	
Poly(ethylene-co-propylene) (Poly(allomer))	2.3		Epoxy cast resins	3.5	5.0
Poly(ethylene), high density	2.30	2.35	Cellulose acetate butyrate	3.5	6.4
Ionomers	2.4	2.5	Cellulose acetate	3.5	7.5
Poly(carbonate) - ABS alloy	2.4	4.5	Poly(oxymethylene)	3.7	
ABS, high impact	2.4	5.0	Poly(oxymethylene), copolymer	3.7	
ABS, medium impact	2.4	5.0	Nylon 11	3.7	
Poly(styrene), general purpose	2.45	3.1	Poly(vinyl formal)	3.7	
Poly(styrene), high heat	2.45	3.4	Cellulose propionate	3.7	4.3
Poly(styrene), high impact	2.45	4.75	Nylon 6	3.8	
Poly(ethylene-co-vinyl acetate)	2.5	3.16	Poly(methyl methacrylate)/Poly(vinyl chloride), alloy	4.0	
Poly(styrene-co-butadiene)	2.5	3.4	Nylon 66	4.0	
Poly(phenylene oxide)	2.58		Nylon 12	4.2	
Poly(styrene-co-acrylonitrile)	2.6	3.4	Poly(vinylidene chloride)	4.5	6.0
Poly(ethylene-co-ethyl acrylate)	2.7	2.9	Phenol-formaldehyde resin	5.0	6.5
Silicone, cast resin	2.75	4.2	Poly(vinyl butyral)	5.60	
Poly(carbonate)	2.97	3.17	Cellulose nitrate	7.0	7.5
Ethyl cellulose	3.0	4.2	Melamine-phenol resin	7.0	7.7
Poly(ester), rigid cast resin	3.00	4.36	Melamine-formaldehyde resin	7.9	11.0
Poly(vinyl chloride-co-propylene)	3.1	3.7	Poly(vinylidene fluoride)	8.4	
Poly(aryl ether)	3.14				

14. Dissipation Factor, ASTM D-150
At 60 Hz

Material	Low	High	Material	Low	High
Poly(methyl-1-pentene)	.00007		Cycloaliphatic epoxy resin	.005	
Poly(styrene), general purpose	.0001	.0006	Ethyl cellulose	.005	.020
Poly(tetrafluoroethylene)	.0002		Poly(α-methylstyrene-co-methyl methacrylate)	.006	
Poly(tetrafluoroethylene-co-hexafluoropropylene)	<.0003★		Poly(aryl ether)	.006	
Poly(phenylene oxide)	.00035		Poly(styrene-co-acrylonitrile)	.006	.008
Poly(styrene), high impact	.0004	.002	Poly(vinyl chloride)	.007	.020
Poly(propylene)	<.0005★		Poly(vinyl chloride-co-propylene)	.008	.010
Poly(ethylene), low density	<.0005★		Nylon 66	.01	
Poly(ethylene), medium density	<.0005★		Poly(ethylene-co-ethyl acrylate)	.01	.02
Poly(ethylene), high density	<.0005★		Cellulose propionate	.01	.04
Poly(ethylene-co-propylene) (Poly(allomer))	<.0005★		Cellulose acetate butyrate	.01	.04
Poly(styrene), high heat	.0005	.003	Cellulose acetate	.01	.06
Poly(sulfone)	.0008		Poly(vinyl formal)	.013	
Poly(carbonate)	.0009		Poly(urethane), thermoplastic elastomer	.015	.048
Ionomers	.001	.003	Poly(vinyl chloride), chlorinated	.019	.021
Silicone, cast resin	.001	.025	Nylon 11	.03	
Poly(chlorotrifluoroethylene)	.0012		Poly(vinylidene chloride)	.03	.045
Poly(styrene-co-butadiene)	.002	.003	Nylon 6	.03	.07
Epoxy cast resins	.002	.010	Poly(methyl methacrylate)/Poly(vinyl chloride),alloy	.04	
Poly(ester), rigid cast resin	.003	.040	Nylon 12	.04	
Poly(carbonate) - ABS alloy	.003	.007	Poly(methyl methacrylate), molding grade	.04	.06
ABS, high impact	.003	.008	Melamine-formaldehyde resin	.048	.162
ABS, medium impact	.003	.008	Poly(vinylidene fluoride)	.049	
Poly(ethylene-co-vinyl acetate)	.003	.02	Phenol-formaldehyde resin	.06	.10
Poly(oxymethylene)	.0048		Cellulose nitrate	.09	.12
Poly(oxymethylene), copolymer	.005		Poly(vinyl butyral)	.115	

★ below the sensitivity of the instrument used

15. Arc Resistance, ASTM D-495 [s]

Material	High	Low	Material	High	Low
Poly(chlorotrifluoroethylene)	>360		Poly(ester), rigid cast resin	135	115
Cellulose acetate	310	50	Poly(styrene), high heat	135	60
Poly(oxymethylene), copolymer	240		Silicone, cast resin	130	115
Poly(ethylene), medium density	235	200	Poly(oxymethylene)	129	
Poly(imide)	230		Poly(sulfone)	122	75
Poly(tetrafluoroethylene)	>200		Poly(carbonate) - ABS alloy	120	70
Cellulose propionate	190	175	Epoxy cast resins	120	45
Poly(propylene)	185	136	Poly(carbonate)	120	10
Poly(aryl ether)	>180		Nylon 12	110	
Melamine-phenol resin	180	130	Poly(styrene-co-butadiene)	95	
Poly(tetrafluoroethylene-co-hexafluoropropylene)	>165		Ionomers	<90	
Poly(ethylene), low density	160	135	ABS, high impact	85	50
Poly(α-methyl styrene-co-methyl methacrylate)	150		ABS, medium impact	85	50
Poly(styrene-co-acrylonitrile)	150	100	Ethyl cellulose	80	60
Melamine-formaldehyde resin	145	100	Poly(vinyl chloride)	80	60
Nylon 66	140	130	Poly(phenylene oxide)	75	
Poly(styrene), general purpose	140	60	Poly(vinylidene fluoride)	70	50
Poly(styrene), high impact	140	20	Poly(methyl methacrylate)/Poly(vinyl chloride), alloy	25	

16. Transmittance, ASTM D-1003 [%]

Material	High	Low	Material	High	Low
Cellulose acetate butyrate	95	75	Ethyl cellulose	88	
Poly(methyl methacrylate), molding grade	> 92		Poly(styrene-co-acrylonitrile)	88	76
Cellulose nitrate	92	89	Ionomers	85	75
Poly(styrene), general purpose	92	87	Poly(ethylene), medium density	80	10
Cellulose propionate	92	80	Poly(ethylene-co-vinyl acetate)	80	0
Poly(carbonate)	91	82	Poly(ethylene), low density	57	35
Poly(methyl-1-pentene)	> 90		Poly(styrene), high impact	75	0
Poly(α-methylstyrene-co-methyl methacrylate)	90		Poly(ethylene), high density	40	0
Poly(styrene), high heat	90	88	ABS, medium impact	33	
Poly(propylene)	90	55	ABS, high impact	28	

17. Haze, ASTM D-1003 [%]

Material	Low	High	Material	Low	High
Poly(styrene), general purpose	< .1	3.0	Poly(methyl methacrylate), molding grade	< 3	
Poly(styrene-co-acrylonitrile)	.1	.4	Ionomers	3	17
Poly(carbonate)	.5		Poly(ethylene), low density	4	50
Ethyl cellulose	< 1.0		Poly(methyl-1-pentene)	< 5	
Poly(propylene)	1.0	3.5	Poly(sulfone)	5	
Cellulose nitrate	2	4	Poly(ethylene), high density	10	50
Cellulose acetate butyrate	2	5	Poly(styrene), high impact	> 77	
Cellulose propionate	2	5	ABS, medium impact	100	
Poly(ethylene), medium density	2	40	ABS, high impact	100	
Poly(ethylene-co-vinyl acetate)	2	40			

IX. SUBJECT INDEX

This index only contains physical constants of polymers, while the names of polymers may be found in the table of the corresponding constant. Only polymers of Section V are individually listed in this index. The page numbers given for them do not represent the only place of reference. Additional data for these polymers may be found in the appropriate table section of the particular constant concerned.

A

B

C

D

E

F

G

H

I

K

L

M

N

O

P

S

T

U

V

W

X

Y

Z